Advances in Experimental Medicine and Biology

Volume 1318

Advances in Experimental Medicine and Biology provides a platform for scientific contributions in the main disciplines of the biomedicine and the life sciences. This series publishes thematic volumes on contemporary research in the areas of microbiology, immunology, neurosciences, biochemistry, biomedical engineering, genetics, physiology, and cancer research. Covering emerging topics and techniques in basic and clinical science, it brings together clinicians and researchers from various fields.

Advances in Experimental Medicine and Biology has been publishing exceptional works in the field for over 40 years, and is indexed in SCOPUS, Medline (PubMed), Journal Citation Reports/Science Edition, Science Citation Index Expanded (SciSearch, Web of Science), EMBASE, BIOSIS, Reaxys, EMBiology, the Chemical Abstracts Service (CAS), and Pathway Studio.

2019 Impact Factor: 2.450 5 Year Impact Factor: 2.324

More information about this series at http://www.springer.com/series/5584

Nima Rezaei
Editor

Coronavirus Disease - COVID-19

Springer

Editor
Nima Rezaei MD, PhD
Research Center for Immunodeficiencies,
Children's Medical Center
Tehran University of Medical Sciences
Tehran, Iran

Network of Immunity in Infection,
Malignancy and Autoimmunity (NIIMA)
Universal Scientific Education and Research
Network (USERN)
Tehran, Iran

ISSN 0065-2598 ISSN 2214-8019 (electronic)
Advances in Experimental Medicine and Biology
ISBN 978-3-030-63760-6 ISBN 978-3-030-63761-3 (eBook)
https://doi.org/10.1007/978-3-030-63761-3

This Springer imprint is published by the registered company Springer Nature Switzerland AG
The registered company address is: Gewerbestrasse 11, 6330 Cham, Switzerland

This book would not have been possible without the continuous encouragement of my family.

I wish to dedicate it to my daughters, Ariana and Arnika, with the hope that we learn enough from today to make a brighter future for the next generation.

This book is also dedicated to honoring the memory of our brave fallen doctors and nurses who fought against COVID-19.

Preface

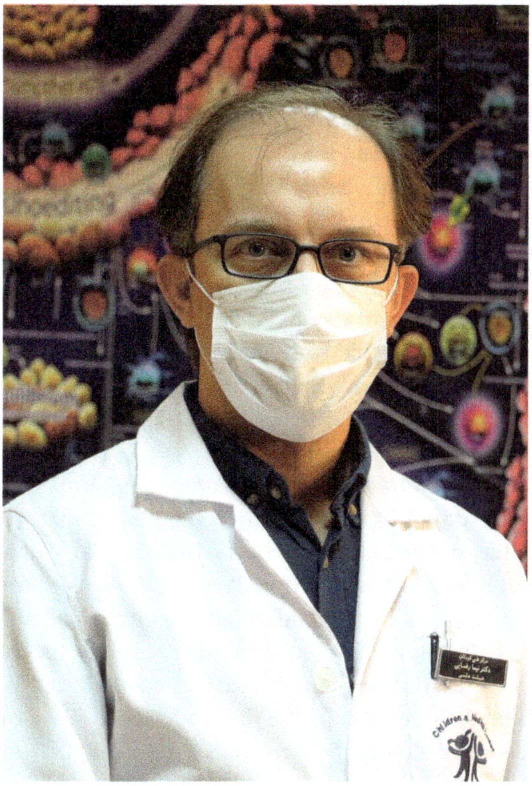

On March 2, 2020, I had an hour-long meeting with three of my colleagues, including the head of the hospital, head of infant intensive care unit (ICU), and the consultant of the hospital, at the head office of the Children's Medical Center, the Pediatrics Center of Excellence (Tehran, Iran), to discuss the novel coronavirus disease 2019 (COVID-19) outbreak. A few days later, I realized that all of them were positive for severe acute respiratory syndrome coronavirus 2 (SARS-CoV-2), confirmed by real-time RT PCR (rRT-PCR) assays. Two of them presented with only low-grade fever and malaise, while the third one suffered from anosmia in addition to cough and myalgia. So, it raised several questions in my mind, while I also had some concerns about being affected by SARS-CoV-2, not only because of that meeting with three physicians who were positive for the virus, but also for meeting and talking closely with two other faculty members of our hospital, who were both recently infected by COVID-19. So, am I positive for COVID-19? Fortunately, in the 2 weeks after that meeting and after 8 months of working in the pediatric referral hospital during the pandemic, I had not presented any symptom of the disease, until October 2020, when I faced headache, backache, chills, and malaise, which made me to qurantine myself for 18 days. As I was quite sure about my symptoms, I did not test for SARS-CoV-2 during my quarantine. However, SARS-CoV-2 IgG had been incresead slightly, when I got back to the hospital. So, I am not among more than a hundred million of reported cases, which might show that the number of COVID-19 is understimated!

Meanwhile, the global challenges and concerns still remain, while many questions are yet to be answered! Is being asymptomatic not equal to being affected? How does the virus spread? Does the outbreak situation change by weather in different seasons? Who are at a higher risk of infection and mortality? Are healthy individuals without any underlying disease protected!? How can we protect ourselves? And the immune system: Friend or Foe?

We, the world outside of China, heard the news of a Chinese city affected by pneumonia of unknown origin in December 2019. We also saw this city, in response to the increasing number of patients presented with this unusual pneumonia, construct a temporary hospital in less than 2 weeks, but did not realize that this disease could spread far beyond its boundaries, strike near us, and provide an experience far worse than what happened in the city when affected the first time. Wuhan was that city, and with the discovery of a coronavirus as the pathogen behind that, COVID-19 was the name assigned to that unusual pneumonia.

This novel SARS-CoV-2 proved to be unique in terms of transmissibility and mortality. A proof of its being highly contagious is that while we were all obsessed with the movement of the SARS-CoV-2 from Wuhan to all territories worldwide, it has been very difficult for most of us to track the chronological order of its global spread after affecting Wuhan. However, its consequences are hitting us; 5 months have passed since the the World Health Organization (WHO) declared COVID-19 a pandemic on March 11, 2020, while many countries are fighting the first and second wave of the disease. It should be mentioned that there were only about 100,000 diagnosed cases within a 3-month period since the beginning of the outbreak; in the month of April, about 3 million new cases were reported; now, there are more than 120 million officially reported cases and more than 2.5 million deaths from all over the world (mid-March 2021); and it's still unclear when the pandemic will end! The pandemic has profoundly affected not only human health but also human behavior and thought. No curative therapy so far! One of my aforementioned colleagues, who was positive for SARS-CoV-2 in March 2020 and cured, faced reinfection 2 months later with positive PCR again! There are some other reports, especially from healthcare workers who are exhausted from the continuous long-term fight against COVID-19, which shows the potential risk of reinfection after decreasing the SARS-CoV-2 antibody level. So, how effective will the vaccines be?

It's about a century since the Spanish influenza pandemic of 1918. How much does this situation differ from that time? And how much have we remembered and learned from the pitfalls we had faced? It should also be mentioned that this novel coronavirus (SARS-CoV-2) is the third one in the twenty-first century that has brought us outbreaks after severe acute respiratory syndrome (SARS) and the Middle East respiratory syndrome (MERS) in 2003 and 2012, respectively, but how much have we prepared ourselves for this third one? Can we make sure that we will never face such a situation in the future?

Certain countries, which were at the top of the list of the number of infected individuals, implemented the policy of closing their borders to restrict travel from outside. However, they simply forgot the key point that this virus does not know any border; thus, a borderless solution is needed when the entire world is affected by such a global challenge! We are all living on the same Earth, and the world's complex problem should be considered a human problem in general; therefore, the solution can't be country-based. Indeed, it's not limited to a specific field of science. Such a complex problem involves not only medical scientists but also other scientists from formal science to physical and chemical science to biological and social science. Thus, there is a need to work together to solve complex problems, like COVID-19.

It is now more than a year that the SARS-CoV-2 is an inhabitant of the human heart, lungs, intestines, and brain in an unsatisfactory manner, asking its origins, evolution, and pathogenesis has become habituated to human beings, in the hope to hear and learn something for enhancing their preparedness for the next wave of the pandemic and the fourth outbreak of the century. Inevitably, there have been massive amounts of data published on this curious subject. In addition to the unexpected rapid flow of publications made available within only a few months, I have a proof of such curiosity. When I was preparing the proposal for this book in March when the outbreak just reached us, I could only collect about ten evidence-based chapter titles, but since then, as time passed, there was a lot I had to include in the book. And now that the book is about to be published, it has more than 50 chapters.

After a rapid introduction to COVID-19 as a global challenge (Chap. 1), the book provides general discussions over characteristics, ecology, and evolution of coronaviruses (Chaps. 2 and 3). Then, it goes into the details about epidemiological (Chaps. 4 and 5), genetic (Chaps. 6 and 7), immunological (Chaps. 8 and 9), oxidative stress (Chap. 10), and diagnostic and prognostic (Chaps. 22–24) aspects of COVID-19. Chap. 11 takes a general view of clinical manifestations of COVID-19, while Chap. 20 and Chap. 21 link to the involvement of individual systems. During the COVID-19 pandemic, pediatrics and geriatrics shaped a sharp contrast in terms of disease outcomes, so Chaps. 12 and 13 separately discuss these specific populations. Pregnant women and neonates are other populations treated specially under the pandemic condition (Chap. 14). The COVID-19 problem has shown its worst scenarios in the case of pre-existing conditions, in particular, cardiovascular diseases (Chap. 15), hypertension (Chap. 16), and cancer (Chaps. 17 and 18). Furthermore, it is expected to be complicated if it occurs concurrently with tropical infections (Chap. 19). The book contains several chapters concerning the treatment of COVID-19, ranging from supportive ventilator support and nutrition therapy to the development of potential virus- and host-based therapies, immune-based therapies, photobiomodulation, and antiviral photodynamic therapy (Chaps. 25–30). The relations of COVID-19 to dentistry, hematology, ophthalmology, and pharmacy lie in Chaps. 33–36, while its general implications to the healthcare setting take place in Chap. 32. As the virus is transmitted, misinformation and rumors spread quickly, resulting in the infodemic (Chap. 37). The book also discusses social issues posed by COVID-19 , for example, social isolation, quarantine, lockdown (Chap. 38), prejudice, and discrimination (Chap. 39), and the resulting consequences on mental health (Chaps. 40–42), education (Chaps. 43 and 44), tourism industry (Chap. 45), and economy (Chap. 46). It also makes attempts to explain the bioinformatic approaches (Chap. 47), innovation and ingenuity (Chap. 48), globalization (Chaps. 49 and 50), social and scientific networking (Chap. 51), interdisciplinary approaches (Chap. 52), and art integration (Chap. 53) as solutions to the problems of COVID-19. Of course, many difficulties remain to be dissolved, merely as an example of the challenges for vaccine development (Chap. 31).

The *COVID-19* Book is the result of the valuable contribution of more than 200 scientists and clinicians from more than 100 well-known universities/institutes worldwide. I would like to hereby acknowledge the expertise of all contributors for generously devoting their time and considerable effort in preparing their respective chapters. I would also like to express my gratitude to Springer Nature for providing me the opportunity to publish the book.

Finally, I hope that this timely book will be comprehensible, cogent, and of special value for researchers and clinicians who wish to extend their knowledge on COVID-19.

Tehran, Iran Nima Rezaei

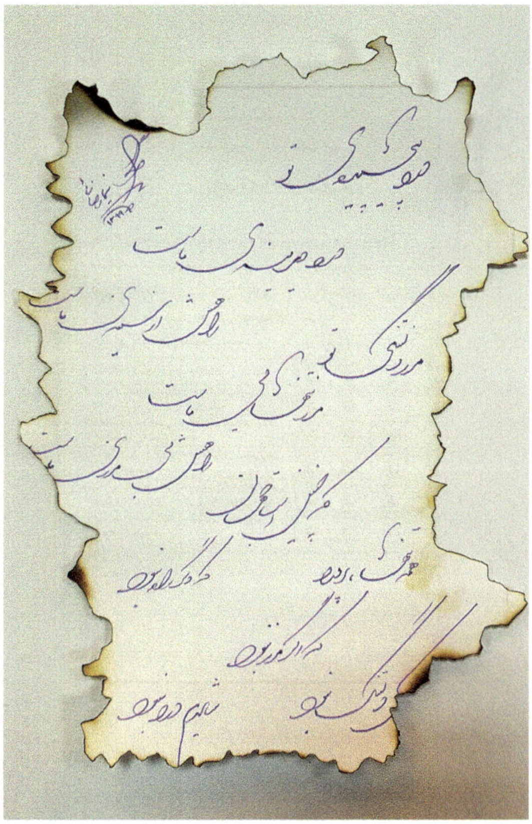

Your complex pain
is our old pain
Its care could be in our brain

Your sadness
is our loneliness
Its solution should be borderless
This is our world without rain
all alone, full of pain
Uncertain, no gain

If there is no border
the world might have a new order
No complain, no pain…

Acknowledgment

I would like to express my gratitude to the editorial assistant of this book, Dr. Amene Saghazadeh. Without doubt, the book would not have been completed without her contribution.

Nima Rezaei

Contents

Introduction on Coronavirus Disease (COVID-19) Pandemic: The Global Challenge

1

Nima Rezaei, Saboura Ashkevarian, Mahsa Keshavarz Fathi, Sara Hanaei, Zahra Kolahchi, Seyedeh-Sanam Ladi Seyedian, Elham Rayzan, Mojdeh Sarzaeim, Aida Vahed, Kawthar Mohamed, Sarah Momtazmanesh, Negar Moradian, Zahra Rahimi Pirkoohi, Noosha Sameeifar, Mahsa Yousefpour, Sepideh Sargoli, Saina Adiban, Aida Vahed, Niloufar Yazdanpanah, Heliya Ziaei, and Amene Saghazadeh

Abstract

By driving the ongoing pandemic of coronavirus disease 2019 (COVID-19), coronaviruses have become a significant change in twenty-first-century medicine, healthcare systems, education, and the global economy. This chapter rapidly reviews the origin, immunopathogenesis, epidemiology, diagnosis, clinical manifestations, and potential

N. Rezaei (✉) · M. K. Fathi · S. Hanaei · E. Rayzan · M. Sarzaeim · H. Ziaei · A. Saghazadeh
Universal Scientific Education and Research Network (USERN), Tehran, Iran

Research Center for Immunodeficiencies, Children's Medical Center, Tehran University of Medical Sciences, Tehran, Iran
e-mail: rezaei_nima@tums.ac.ir

S. Ashkevarian
Universal Scientific Education and Research Network (USERN), Tehran, Iran

Research Institute of Applied Sciences, Academic Center of Education, Culture and Research (ACECR), Shahid Beheshti University, Tehran, Iran

Z. Kolahchi · K. Mohamed · S. Momtazmanesh · N. Moradian · N. Yazdanpanah
Universal Scientific Education and Research Network (USERN), Tehran, Iran

School of Medicine, Tehran University of Medical Sciences, Tehran, Iran

S.-S. Ladi Seyedian
Universal Scientific Education and Research Network (USERN), Tehran, Iran

USC Institute of Urology, USC/Norris Comprehensive Cancer Center, University of Southern California, Los Angeles, CA, USA

A. Vahed · M. Yousefpour · S. Sargoli · S. Adiban · A. Vahed
Universal Scientific Education and Research Network (USERN), Tehran, Iran

Z. R. Pirkoohi
Universal Scientific Education and Research Network (USERN), Tehran, Iran

Shahid Beheshti University, Tehran, Iran

N. Sameeifar
Universal Scientific Education and Research Network (USERN), Tehran, Iran

School of Medicine, Shahid Beheshti University of Medical Sciences, Tehran, Iran

N. Rezaei (ed.), *Coronavirus Disease - COVID-19*, Advances in Experimental Medicine and Biology 1318, https://doi.org/10.1007/978-3-030-63761-3_1

therapeutics of COVID-19. It would also explore the effects of the introduction of a single virus, the so-called severe acute respiratory syndrome coronavirus 2 (SARS-CoV-2), on the public health preparedness planning.

Keywords

Borderless · Coronavirus · COVID-19 · Global · Immune system · Pandemic · SARS-CoV-2

1.1 Introduction

Coronaviruses are enveloped viruses composed of non-segmented, single-stranded, and positive-sense RNA, which is the largest genome among RNA viruses. They belong to the order *Nidovirales* having the common characteristics: gene expression through a 3′ nested set of multiple subgenomic mRNAs (sg mRNAs), a ribosomal frameshifting mechanism for the expression, a unique set of enzymatic activities, a virion envelope formation, and a multi-span membrane protein in the virion. According to antigenic relationships, 28 species of coronaviruses are included in three groups. Species of group 3 rely on the bird host, while species of the other groups occur in mammals, including mouse, rat, bat, cow, dog, pig, puffin, horse, and human (Masters 2006).

Human coronaviruses (hCoVs) date back to the 1960s with the identification of hCoV-229E and hCoV-OC43 from the respiratory tract and nasopharyngeal wash of patients with the common cold (Su et al. 2016). They, however, appeared to be highly pathogenic to humans in the early 2000s, when a virus, SARS-CoV, with the characteristic features of hCoVs caused an outbreak of severe acute respiratory syndrome (SARS) and resulted in 776 deaths with a case fatality rate (CFR) of 9.6%. Bioinformatics information indicates that SARS-CoV does possess nothing new compared to any of the previous three groups of coronaviruses, though owning features in relation to all the three groups of coronaviruses. SARS-CoV is therefore supposed to be produced by multiple genetic recombination events. Subsequently, hCoV-NL63 and hCoV-HKU1 first occurred in 2004 and 2005, and both correlated with a form of mild respiratory disease. hCoVs highly pathogenic to humans continued by the outbreak of the Middle East respiratory syndrome (MERS) with 584 deaths and CFR of 36% in 2012. By driving the ongoing pandemic of coronavirus disease 2019 (COVID-19), which so far has been the cause of death of more than 560,000 people worldwide, they have become a significant change in the twenty-first-century medicine, healthcare system, education, and economy.

1.2 The Source of the Infection

Phylogenic analyses split coronaviruses into four subgroups: alpha, beta, gamma, and delta. Before the novel coronavirus 2019 appeared, six hCoVs were reported relating to either alpha-(HCoV-229E and HCoV-NL63) or betacoronaviruses (HCoV-OC43, HCoV-HKU1, MERS-CoV, and SARS-CoV) (Shereen et al. 2020). The new HCoV that causes COVID-19 shares significant similarities with the SARS-CoV from the structural and pathological points of view and is, therefore, classified as a betacoronavirus and called SARS-CoV-2. HCoVs leading to outbreaks are apparently of zoonotic origin circulated in animals before transmission to humans. SARS-CoV-2 first occurred in late December 2019, in Wuhan, China. Notably, evidence shows the existence of SARS-CoV in bats, palm civets, and raccoon dogs. Also, MERS-CoV occurs in the primary host of camels, while bats might transmit it. Human epidemiological data connects a relatively high proportion of first- and second-generation human cases of COVID-19 to the Huanan Seafood Wholesale Market in Wuhan (Huang et al. 2020a). Following whole-genome sequencing, which demonstrated 96.2% similarity, SARS-CoV-2 was proven to be associated with Bat-CoV-RatG13 (SARS-like bat coronavirus) (Zhou et al. 2020b; Paraskevis et al. 2020). Although the primary host had been indicated, it was imperative to determine the intermediate hosts between bats and humans to control the pandemic. Sequence and structural alignment of angiotensin-converting enzyme 2 (ACE2) among human, nonhuman primates, domestic animals, wild animals, and rodents indicate that SARS-CoV-2 may not infect chicken, while nonhuman primates may be an intermediate host for transmission (Luan et al. 2020). Analysis with the binding model of S protein, receptor-binding domain region (RBD region), and ACE2 showed that snakes, pangolins, and turtles could serve as possible intermediate hosts (Liu et al. 2020b). However, recent studies have demonstrated that it is unlikely for snakes and turtles to be intermediate hosts, and researchers suggested pangolins as potential targets (Liu et al. 2020a). On a recent study that isolated coronavirus, they detected Malayan pangolins to exhibit a high sequence identity (100% in the E gene, 98.2% in the M gene, 90.4% in the S gene, and 96.7% in the N gene); therefore, it was suggested that SARS-CoV-2 could perhaps arise as a result of recombination between Pangolin-CoV-like virus and Bat-CoV-RatG13-like virus (Xiao et al. 2020).

1.3 Immunopathogenesis of COVID-19

1.3.1 Virus Entry and Spread

1.3.1.1 The Virus Binding to Its Cell Receptors

SARS-CoV and SARS-CoV-2 do the action of cell entry with a high level of equivalence, arising from the same cleavage junctions, the highly similar same sequence (96%) of their main protease, a high degree (76%) of similarity in the amino acid sequence of their S protein, similar S2′ cleavage site, and similar residues essential for binding ACE2 (Saghazadeh and Rezaei 2020b). Compared to that of SARS-CoV, the S protein of SARS-CoV-2 has acquired a polybasic cleavage site, O-linked glycans, and an S_B subunit with higher affinity for binding ACE2. These differences may, in part, explain the higher transmission of SARS-CoV-2, which as of July 11, 2020, has infected 12 million people worldwide, compared to SARS-CoV with less than 9000 confirmed cases during the entire outbreak.

The viral infection occurs when the virus acts on host cells. In the case of SARS-CoV-2, the RBD of the spike (S) protein on the viral envelope seems to orchestrate such a mutual action with the cell-surface receptor, e.g., ACE2 (Jahanshahlu and Rezaei 2020b). It requires host proteases for the cleavage of the S protein into two subunits, S1 and S2, that undertake the attachment of the virus to the membrane and then the fusion of cellular and viral membranes. Endocytosis is another mechanism that contributes to viral internalization, as described for the SARS-CoV (Li et al. 2020b). Whether or not involving clathrin, endocytosis is accompanied by the fusion of membrane with vesicles that transport viral particles and genome followed by the release of the virus into the cell.

1.3.1.2 Antigen Presentation

Self-/nonself-recognition is mastered by the major histocompatibility complex (MHC). Particularly speaking, the presentation of endogenous antigens to cytotoxic CD8+ T cells is mediated by MHC class I, while the presentation of exogenous antigens to helper CD4+ T cells is carried out by MHC class II. Studies have associated MHC polymorphisms with a spectrum of immune-mediated conditions, including aging, atopic diseases, autoimmune diseases, and neurological diseases. Of interest to here is the association of these polymorphisms with infectious diseases, such as human immunodeficiency virus (HIV), hepatitis B virus (HBV), hepatitis C virus (HCV), and tuberculosis, which is a potential source of variation in response to vaccines and, therefore, can help the development of vaccinomics (Saghazadeh and Rezaei 2019).

Upon the cell entry of the virus, the presentation of viral antigens by antigen-presenting cells

(APCs) leads to the activation of immune responses. Less is understood about the mechanism of presentation of SARS-CoV-2, and all our current knowledge is regarding the presentation of SARS-CoV. The presentation of SARS-CoV involves MHC class I and, to a lesser extent, MHC class II. As reviewed in (Li et al. 2020b), MHC polymorphisms confer susceptibility to (HLA-B*4601, HLA-B*0703, HLA-DR B1*1202, and HLA-Cw*0801) and protection against (HLA-DR0301, HLA-Cw 1502, and HLA-A*0201) SARS-CoV. It would strengthen the importance of MHC molecules in the immunopathogenesis of SARS-CoV and possibly in that of SARS-CoV-2.

1.3.1.3 Pathological Findings

Acute respiratory distress syndrome (ARDS) and multiorgan dysfunction syndrome (MODS) are the most common cause of death in patients with COVID-19.

Lung

Lungs present evidence of acute or organizing phase of diffuse alveolar damage (DAD), such as desquamation of pneumocytes, hyaline membrane formation, vascular congestion, pulmonary edema, mononuclear cell infiltrates, and multinucleated syncytial cells (Xu et al. 2020). Consolidation characterized by abundant neutrophilic infiltration may appear as well and indicates bacterial pneumonia superimposed on SARS-CoV-2 pneumonia (Tian et al. 2020). A variety of immune cells expressing CD3, CD4, CD8, CD20, CD79a, CD5, CD38, and CD68 may exist as proven by immunohistochemistry (Luo et al. 2020). SARS-CoV-2 inclusions in the nucleus and cytoplasm are not visible.

Kidney

Kidney examination reveals diffuse proximal tubule injury as manifested by brush border loss, non-isometric vacuolar degeneration, and necrosis (Su et al. 2020). Also, it displays hemosiderin granules, pigmented casts, and erythrocyte aggregations that may obstruct the flow in the capillaries but lacks any signs of vasculitis, interstitial inflammation, and hemorrhage. Of note, tubular cells and podocytes contain clusters of SARS-CoV-2-like particles, as demonstrated by electron microscopy.

Liver

The liver shows moderate microvesicular steatosis, mild sinusoidal dilation, and mild portal and lobular inflammatory activity (Xu et al. 2020; Tian et al. 2020). These changes do not distinguish SARS-CoV-2 infection from drug-induced liver injury.

Heart

Interstitial infiltrates of mononuclear cells, mild focal edema, interstitial fibrosis, and myocardial hypertrophy might occur in the heart tissue (Xu et al. 2020; Tian et al. 2020).

1.3.2 Role of the Immune System During COVID-19

The respiratory effects of COVID-19 are well-known, but COVID-19 can affect other systems, mainly the gastrointestinal system (Gu et al. 2020), central nervous system (Saleki et al. 2020; Jahanshahlu and Rezaei 2020a), and cardiovascular system (Zheng et al. 2020b; Hessami et al. 2020). Its ability might lie partly in multiple routes, e.g., fecal-oral transmission, airborne transmission, and contact, through which it can spread, and also, partly it can make in through pulmonary and systemic immune responses. Research has revealed that an overactivated immune system occurs during COVID-19 and plays a role in determining the outcome of the disease (Bahrami et al. 2020; Basiri et al. 2020b; Sahu et al. 2020; Yazdanpanah et al. 2020a; Saghazadeh and Rezaei 2020a; Nasab et al. 2020), while inborn errors of immunity are not particularly vulnerable to COVID-19 and what is likely to happen to high-risk population, i.e., older adults and people with other comorbid conditions (Babaha and Rezaei 2020). Therefore, it seems that the immune system starts attacking the body in response to the trigger of COVID-19.

1.3.2.1 Antiviral Immune Dysfunction

When a pathogen invades the body, innate immune receptors initiate a cascade of immune responses through recognition of pathogen-associated molecular patterns (PAMPs). Type I interferon (IFN) responses are an essential element of antiviral immunity and success in the control of viral replication (Saghazadeh and Rezaei 2017). Both SARS-CoV and MERS-CoV possess structural and nonstructural proteins that prevent these responses from being generated by either directly or indirectly interfering with IFN signaling pathways (Prompetchara et al. 2020). Inhibition of type I IFN responses in the early phase of infection leads to progression to severe infection, and this may be true for SARS-CoV-2. Also, antiviral immunity depends on the effector functions of cytotoxic lymphocytes, e.g., cytotoxic T lymphocytes (CTLs) and natural killer (NK) cells. In people with COVID-19, CTLs and NK cells not only are reduced in number but also are functionally exhausted as indicated by the increased expression of exhaustion markers NKG2A on NK cells and PD-1 and Tim-3 on T cells (Diao et al. 2020; Zheng et al. 2020a; Fathi and Rezaei 2020).

1.3.2.2 Cytokine Storm

Cytokine storm syndromes (CSS) refer to conditions associated with the maladaptive release of cytokines (Ye et al. 2020), including malignancy, ARDS, cytokine release syndrome, familial hemophagocytic lymphohistiocytosis, systemic juvenile idiopathic arthritis, and macrophage activation syndrome. In particular, the cytokine storm might occur in infections with the influenza virus and Epstein-Barr virus. Disease progression in patients with COVID-19 is accompanied by immune dysregulation, and this is not restricted to the antiviral immune dysfunction mentioned above but extended to pro-inflammatory cytokine upregulation (Rokni et al. 2020). The latter, along with multiorgan involvement, are the main characteristics of cytokine storm syndrome, which are seen in severe COVID-19. For this, severe COVID-19 is supposed to be a cytokine storm syndrome. The supposition is supported by the biomarkers of CSS that hitherto have been associated with the severity of COVID-19, including inflammatory markers (CRP and procalcitonin), tissue injury markers (LDH, AST, ALT), cytokines and chemokines (IL6, I1B, IFNY, IL2Ra), fibrin degradation products (D-dimer, fibrinogen), and macrophage activation markers (ferritin) (Henderson et al. 2020).

1.4 Epidemiology

1.4.1 A Historical Review of SARS Pandemic

In late 2002, an emerging infectious disease-causing pneumonia was introduced to the world. In March 2003, a CoV was identified as the causative agent, and a month later (April 16, 2003), the WHO verified that SARS-CoV was the definitive origin of the disease. SARS is defined as a respiratory infection with a history of high fever, cough, and dyspnea and may lead to respiratory distress syndrome. The WHO reported the outbreak on February 11, 2003, when the confirmed cases exceeded 300, and declared the disease as a global alert on March 12, 2003. The first case was reported from China in November 2002, and then in February 2003, the virus spread to Hong Kong, Singapore, Toronto, and Hanoi. It was found that on February 21, 2003, an infected physician transported the infection to Hong Kong. In the hotel he was staying in, transmission to other guests occurred and subsequently contributed to other countries' involvement (Cherry 2004; Skowronski et al. 2005). It seems that SARS-CoV, a beta genera virus, had an animal reservoir, an Asian civet cat. The transmission was mainly via human to human contact with respiratory droplets, and most happened in markets. The Chinese government's nontransparency resulted in a global epidemic.

SARS diagnosis is based on clinical manifestations and laboratory findings. Dynamic changes in chest X-rays are compatible with the infection stages. Contact with confirmed cases and travel to endemic areas are clues in diagnosis. The treatment was mainly supportive, but some drugs showed clinical improvements like lopinavir, ritonavir, and remdesivir (Peeri et al. 2020). However, no specific drug or vaccine was developed. The WHO measures were limited to traditional epidemic controlling tools.

Patients were mostly from healthcare providers, and most transmissions happened in the hospital setting (Skowronski et al. 2005). Many medical procedures, like intubation, which exposed them to respiratory discharge, put them at risk of infection (Peeri et al. 2020). The disease also had a tremendous psychological impact on the healthcare staff. Confronting with a disease with unknown origin made immense distress (Sim and Chua 2004). During the pandemic, governments dealt with significant economic challenges. Quarantine limited the trading to essential everyday needs, and the tourism industry was among the most affected industries (Fan 2003).

The pandemic peak was the last week of May 2003. Finally, the epidemic led to more than 8000 infected cases with about 10% mortality rate, and 29 countries were affected. Presumably, the last patients were infected in July 2003, and the disease was controlled after 7 months (Cherry 2004). Up to now, no other cases of SARS infection have been reported (Peeri et al. 2020).

1.4.2 The Current State of COVID-19

Undoubtedly, the eyes can easily judge how powerful the global influence of the COVID-19 outbreak is, starting from December 2019 in Wuhan, China, spreading across countries worldwide in a few months, and turning into a pandemic (Hanaei and Rezaei 2020; Jabbari et al. 2020). However, the measurement of such an influence is necessary for controlling the present pandemic and also would help in being prepared for its next waves. When defining the epidemiology of an epidemic, different factors matter, e.g., the number of infected people, the transmissibility of the infection, and the spectrum of clinical severity (Lipsitch et al. 2020).

1.4.2.1 Epidemiological Characteristics

By July 12, 2020, COVID-19 has affected around 13 million people worldwide. A systematic review (Park et al. 2020) summarizes the main epidemiological characteristics of COVID-19 as follows: the basic reproduction number, 1.9–6.5; the incubation period, 4–6 days; and the case fatality rate outside of China, 0.3–1.4%.

1.4.2.2 Routes of Transmission

The transmission of SARS-CoV-2 is mainly dependent on respiratory droplets, followed by aerosols. Other routes that possibly mediate the virus transmission include the eye-nose and fecal-oral transmission routes.

1.4.2.3 The Spectrum of Infection

The spectrum of clinical severity covers both asymptomatic and symptomatic individuals. A symptomatic infection might cause either mild to moderate symptoms of infection, severe infection that requires hospital admission, a critical infection that requires intensive care unit (ICU) admission, or even death. Among symptomatic individuals, about 20% develop severe to critical infection, and the remainders are diagnosed as mild to moderate infection. Very concerning is that current data, though not sufficient for calculation of the role of asymptomatic individuals in the transmission of infection, clearly reveals that the challenge of containing the infection largely lies not in detecting asymptomatic cases but who can carry high concentrations of the live SARS-CoV-2 in the absence of symptoms. Reports provide different estimations of the ratio of asymptomatic individuals, ranging from 30% to more than 50% (Nishiura et al. 2020; Gandhi et al. 2020).

1.4.2.4 High-Risk Population

Adults aged 70 and above, male sex, and preexisting medical conditions, e.g., cancer, diabetes, cardiovascular disease, chronic respiratory disease, and hypertension, are the factors known to be associated with death from COVID-19 (Ahmadi et al. 2020; Hessami et al. 2020; Basiri et al. 2020b; Shamshirian and Rezaei 2020). Moreover, reports have indicated that the mortality of COVID-19 is not equal in different ethnicities. Accordingly, in the USA, most of the deceased patients were Blacks or indigenous Americans. Even after age adjustment, the mortality rates of Blacks and indigenous Americans were 3.8 and 3.2 times as high as Whites, respectively. On the other hand, the mortality of other races came after these two (https://www.apmresearchlab.org/covid/deaths-by-race).

1.5 Diagnosis and Clinical Manifestations of COVID-19

1.5.1 Diagnosis

Diagnosis of COVID-19 is mainly based on epidemiological history, clinical manifestations, and some auxiliary examinations, such as nucleic acid detection, computed tomography (CT) scan, immune identification technology, and blood culture (Li et al. 2020b). Recent outbreaks with RNA viruses have presented great difficulties with diagnosis (Basiri et al. 2020a). It was particularly the case during the 2015–2016 pandemic of Zika virus (ZIKV) due to low viral titers and transient infection (Faye et al. 2008).

1.5.1.1 Nucleic Acid Test

The two commonly used nucleic acid detection technologies for SARS-CoV-2 are real-time quantitative polymerase chain reaction (RT-qPCR) and high-throughput sequencing. The reliable identification method for SARS-CoV-2 is virus blood culture and high-throughput sequencing of the whole genome (Li et al. 2020b). Diagnosis by screening and confirmation steps by the molecular method have been designed and validated with the use of synthetic nucleic acid technology. It was followed by the development of numerous nucleic acid assays (Corman et al. 2020). Diagnosis of suspected cases is confirmed by RNA assays with real-time PCR, using respiratory samples. RNA-based molecular tests require to set up facilities and instruments, with appropriate biosafety measures, and skilled laboratory technicians at a high cost.

1.5.1.2 Serologic Diagnosis

Serologic tests are comparatively easier to perform, requiring less technical expertise and equipment compared to nucleic acid detection (Rashid et al. 2020). Samples are the blood that is collected in tubes, which poses a less potential risk to the staff handling the samples. The use of serologic methods requires appropriate interpretations of the results and understanding of the strengths and limitations of such tests. There is much to determine the value of serological testing in COVID-19 diagnosis and monitoring. More comprehensive evaluations of the performance of serologic tests are rapidly underway. Considerations for the use of serologic methods for COVID-19 require the correct and appropriate interpretations of the results and understanding the strengths and limitations of such tests (Rashid et al. 2020).

1.5.1.3 Imaging

Some patients with positive chest CT findings may present with negative results of RT-PCR for SARS-CoV-2 (Xie et al. 2020a). It has been suggested that CT could play a role in the COVID-19 case ascertainment. CT abnormalities might predate RT-PCR positivity in symptomatic patients and those without symptoms, which subsequently tested positive by RT-PCR (Hessami et al. 2020; Corman et al. 2020).

1.5.2 Clinical Manifestations

Even though coronaviruses usually have mild symptoms in human beings, betacoronaviruses such as SARS- and MERS-CoV can cause severe forms associated with high mortality rates (Peeri et al. 2020; Sabino-Silva et al. 2020). SARS-CoV-2, which is responsible for the recent outbreak, is an emerging betacoronavirus that has clinical manifestations with various severity in five stages, from asymptomatic infection to critical form (Bulut and Kato 2020; Verity et al. 2020; Lotfi and Rezaei 2020). Although the global concern about this viral infection has been increased obviously due to the high transmissibility of the disease, it has a relatively low mortality rate compared to previous coronaviruses (Petrosillo et al. 2020). Most cases are categorized in the mild to the moderate classification of the disease, while a low number of patients develop a severe or critical form of the disease. The disease may be asymptomatic up to 14 days and even more up to 24 days in some cases (Ather et al. 2020; Peng et al. 2020); but the mean incubation period reported for this viral infection is 5.2 days, which is relatively similar to SARS and MERS (Bulut and Kato 2020). Therefore, patients who have an asymptomatic form of the infection and also patients during the asymptomatic incubation period can be carriers and transmit the disease without awareness. The most common symptoms of this viral infection are fever, followed by dry cough, fatigue, and shortness of breath (Hu et al. 2020). Among upper airway manifestations, pharyngodynia, nasal congestion, and rhinorrhea are more common (Lovato and de Filippis 2020). In more severe cases, viral pneumonia can cause ARDS, and also these patients can have major complications such as acute cardiac injury (ACI), acute kidney injury (AKI), and shock (Hu et al. 2020). In a study of 1590 patients with COVID-19, these comorbid conditions resulted in poorer clinical outcomes, especially in the presence of more than one comorbidity (Guan et al. 2020a). Epidemiological studies reveal that a higher mortality rate occurs in geriatric patients and patients who have preexisting comorbid conditions and that the median survival time might be reduced to 5 days in old patients (Bulut and Kato 2020).

Other less prevalent clinical manifestations include headache, diarrhea, nausea, vomiting, myalgia, and arthralgia (Guan et al. 2020b; Huang et al. 2020b). In the mild form of the disease, alterations in the sense of smell and taste are commonly reported by the patients, and the clinicians should consider these symptoms as initial apparent symptoms of the disease (Spinato et al. 2020). Most of the patients have alterations in their chest CT. Bilateral ground-glass opacities (GGO), especially in the periphery of the lungs, are a common finding among most of the patients due to pneumonia (Han et al. 2020).

Potential neurological manifestations have been reported in COVID-19 patients, such as dizziness, acute cerebrovascular diseases, unstable walking, malaise, and impaired consciousness (Wang et al. 2020b). Keratoconjunctivitis is another infrequent clinical symptom that might appear as an initial clinical symptom in COVID-19 patients (Cheema et al. 2020).

1.6 Potential Therapeutics for COVID-19

1.6.1 Interferon-Alpha and Interferon-Beta

Both interferon-alpha (IFN-α) and interferon-beta (IFN-β) have been shown to affect coronaviruses, especially MERS-CoV. Although most in vitro studies and animal studies reported controversial effects of IFN-α and IFN-β against the SARS-CoV-2, some publications have shown promising results of IFN-α and IFN-β in combination with ribavirin and/or lopinavir/ritonavir (Sanders et al. 2020). According to the study by Ivan Fan-Ngai Hung et al., in patients with mild to moderate COVID-19 infection, the early triple antiviral therapy (ribavirin and lopinavir/ritonavir) with IFN-β-1b was safer than using the lopinavir/ritonavir alone and had better results in palliating the symptoms, reducing the duration of hospitalization, and SARS-CoV-2 viral shedding (Hung et al. 2020). Therefore, IFN-β might be a good choice in combination with other antivirals, but further research is needed to confirm its actual effects.

1.6.2 Antivirals

In combat with a newly emerged disease that does not have any specific known treatment, the tendency is to use the most probably potent available drugs on patients due to the time-consuming process of de novo drug discovery. This process is named drug repurposing or drug repositioning that is defined as using an approved drug for a new disease that is not on its indication list. Considering the matter of time in the current pandemic condition, the best way to find potential candidates for repurposing method might be computational or in silico approach. After a drug is proposed by computational methods, the next step is to investigate the efficacy of that candidate by in vitro studies. If the candidate got approval in this level, it could enter the observational studies or clinical trials, in which the latter is the most valid and reliable method to confirm a drug candidate.

Table 1.1 provides a summary of the drugs most frequently entered into clinical trials. Chloroquine and hydroxychloroquine are of the very first proposed candidates that entered the trials. However, the computational methods have not reported any specific target for them. Lopinavir and ritonavir gain a low score in computational methods in addition to the fact that in observational studies and clinical trials, no specific benefit more than standard care was reported. Ribavirin is another candidate that is acceptable by computational methods by targeting two components of COVID-19, but due to its toxicity and high dose requirement, it might not be the right choice. Although remdesivir was a promising candidate in both computational and in vitro studies, the overall number of cases is not enough to statistically confirm its effectiveness. Favipiravir has the same condition as remdesivir, but the reports of its effectiveness are on moderate cases of the disease, and there is not enough evidence for its effect on more severe cases. The other potential candidate to mention is atazanavir, which has successfully targeted six components of COVID-19 in computational methods. Furthermore, besides its positive results in in vitro studies in decreasing viral replication, it is reported that atazanavir has some anti-inflammatory effects because it diminishes the IL-6 and TNF-α secretion from infected monocytes. However, there is no ongoing clinical trial on this potential candidate.

1.6.3 Convalescent Plasma and IVIG

In urgent conditions such as outbreaks when there is no approved medication, the last hope of increasing the survival of the patients is to use convalescent plasma or immunoglobulins (Pourahmad et al. 2020). It is usually performed by taking the plasma of a survived patient, which contains several types of immunoglobulins, and injecting it into patients that represent severe forms of the disease. This method might be useful if it was used within 48 h of the ICU admission of the patient. Many positive changes might occur, such as reducing the need for mechanical ventilation, reducing the length of hospital stay, and recovery promotion (Chen et al. 2020; Xie et al. 2020b).

1.6.4 Monoclonal Antibodies

Monoclonal antibodies are produced by one specific clone of B cells. Therefore, they have a monovalent affinity and only identify one epitope of an antigen. While most of the literature regarding the treatment of the COVID-19 infection with monoclonal antibodies is related to the previous knowledge of SARS and MERS, some studies express the effectiveness of anti-IL-6 antibodies against the COVID-19 infection. Tocilizumab is an IL-6 receptor antagonist that lowered the use of oxygen supplement in 75% of patients in a study conducted in China. Another relevant monoclonal antibody is sarilumab, in which its efficacy is not approved yet, and it is in phase II/III clinical trial (Lu et al. 2020).

1.6.5 Corticosteroids

When an enhanced immune response arises, it causes several injuries to the body, and the use of corticosteroids lies in reducing these injuries. However, the corticosteroids might cause some side effects like lowering the pace of the viral

Table 1.1 Antivirals repurposed by computational methods

Antivirals	Computational methods	In vitro studies	Observational/clinical studies
Chloroquine/ hydroxychloroquine	- No specific target, the probability of targeting 3CLpro, PLP, Nsp3b, Nsp3c, and E-channel - The effectiveness of the drugs that have no specific targets is controversial, and they are not recommended	Inhibiting the glycosylation of host receptors, proteolytic processing, and endosomal acidification, therefore, the viral entry prevention (Sanders et al. 2020)	No clear benefit was present, and some side effects like ventricular arrhythmias are reported (Mehra et al. 2020)
Lopinavir/ritonavir	- Low docking scores - The failure of the drug-target interaction according to molecular dynamics simulation	3CLpro inhibition in other coronavirus strains (Sanders et al. 2020)	No benefit was present apart from standard care (Cao et al. 2020).
Ribavirin	Targets PLP and RdRp	Inhibiting viral replication in high concentrations only (Sanders et al. 2020)	Due to its toxicity and high dose requirement, this drug has less chance in the treatment of COVID-19 (Sanders et al. 2020)
Remdesivir	Targeting 3CLpro and RdRp	In vitro activity against several *Coronaviridae*, including SARS-CoV-2 with an EC50 value of 0.77 Mm (Sanders et al. 2020)	Reported good effectiveness against the COVID-19 infection, but more studies are needed to statistically confirm its effectiveness (Wang et al. 2020)
Favipiravir	Targeting RdRp	Having an EC50 of 61.88 μM/L against SARS-CoV-2 in Vero E6 cells (Sanders et al. 2020)	- Useful in the clinical recovery of moderate cases at day 7, more RCTs are needed to confirm further the results (Sanders et al. 2020) - Median viral clearance time (4 days) and chest image improvement rate (91.43%) (Lu et al. 2020)
Atazanavir	Targeting 3CLpro, RdRp, helicase, 3′to5′exonuclease, 2′-O-ribose methyltransferase, and EndoRNAse	- Diminished viral replication in the human epithelial pulmonary cells infected with COVID-19 (Fintelman-Rodrigues et al. 2020) - Diminished secretion of IL-6 and TNF-alpha in SARS-CoV-2-infected human monocytes (Fintelman-Rodrigues et al. 2020)	No trial was found

Mohamed et al. (2020b)

clearance and, consequently, lengthening the disease period. The use of the corticosteroids in the treatment of COVID-19 is controversial, and almost all studies do not recommend it (Sanders et al. 2020).

1.6.6 Vaccine Development

Vaccine development (for both therapeutic and preventive purposes) is a long time-consuming process that needs to become faster during epidemics. The COVID-19 vaccine candidates that have entered the clinical trials are DNA-based, RNA-based, live attenuated virus, and inactivated virus vaccine platforms as of writing this manuscript (Craven 2020, 28 May). RNA-based and DNA-based vaccine development platforms and recombinant-subunit developing platforms have a potential faster process than other platforms. The potential fast speed of RNA and DNA vaccines is because the development process of these vaccines does not require fermentation or culture. Furthermore, next-generation sequencing and reverse genetics are considered as tools that can accelerate the vaccine development process in the time of epidemics. However, there are more challenges in developing a vaccine for COVID-19 itself. For example, whether to use a full-length spike protein or RBD as the target has remained an unsolved issue (Sharifkashani et al. 2020). Furthermore, regarding the recent vaccine candidates' preclinical studies, it has been observed that the candidates can intensify the pulmonary damage both directly and as a consequence of the surge in the level of antibodies (Lurie et al. 2020). By considering these obstacles in this way, the idea of vaccine repurposing might help to speed up this process. As an illustration, the Bacillus Calmette-Guerin (BCG) live-attenuated vaccine is one of the first candidates proposed to be effective against COVID-19, and some clinical trials are underway to investigate its effectiveness (medicine 2020a, b).

1.6.7 Nutrition

As new trends in the diet have emerged, so have the new trends in noncommunicable diseases. Nutrition therapy and its immunomodulatory effects have been a topic of crucial importance in the context of chronic diseases that lack a specific treatment, including cardiovascular diseases, cancer, and metabolic disorders. The importance of nutrition therapy extends to acute conditions that increase nutritional requirements, for example, infections and organ failure, as well as to other conditions that increase the risk of undernutrition like aging (Hirbod-Mobarakeh et al. 2014). Besides, micronutrients, e.g., trace elements and vitamins, crucially contribute to the regulation of immune responses, interestingly antiviral immune responses (Mahmoudi and Rezaei 2019). Nutritional therapy is, therefore, of high priority during the pandemic of COVID-19, causing an acute respiratory infection associated with fever and multiorgan failure. In particular, it might be beneficial to older people with preexisting chronic conditions, who are considered as the most susceptible vulnerable to COVID-19 and adverse outcomes.

Other efforts to combat COVID-19 occur in multiple disciplines of immunotherapy (Fathi and Rezaei 2020; Lotfi et al. 2020), regenerative medicine (Basiri et al. 2020b), medical biotechnology (Rezaei 2020b), picotechnology (Rabiee et al. 2020), and telemedicine (Moazzami et al. 2020). However, the quality of clinical studies does not meet the requirements (Rzymski et al. 2020). The observation of the same phenotype of disease in family members and individuals with specific genetic defects might establish a genetic basis of disease that would deserve the attention of clinical pharmacists (Yousefzadegan and Rezaei 2020; Darbeheshti and Rezaei 2020).

1.7 COVID19 Pandemic: Global Challenges, Prevention, and Preparedness for the Next Pandemic

The primary prevention steps included isolation of confirmed cases, quarantine of suspected individuals, and traveler screening. Scientific collaborations facilitated the identification of the virus, and the genome was sequenced in mid-April 2003 (Skowronski et al. 2005). The genome sequencing gave on to the development of sophisticated diagnosis tests and targets for medication and vaccine (Chow et al. 2003).

1.7.1 Global Challenges of Coronavirus Pandemic (Lessons Learned from the 2003 SARS Pandemic)

1.7.1.1 Animal Source Control

There are some similarities and differences between SARS-CoV and SARS-CoV-2. Some animal sources, including *Rhinolophus sinicus* (bat) and civet cats, have been suggested as SARS-CoV sources. They had coronaviruses genetically homologous to SARS-CoV (Yuan et al. 2010; Wang et al. 2005). The probable sources of SARS-CoV include *Rhinolophus affinis*, snakes, and pangolin (Gralinski and Menachery 2020; Zhou et al. 2020a; Ahmed et al. 2020; Liu et al. 2020a). Prohibition of wild animal trading and slaughtering is the critical benefit of virus source identification (Peeri et al. 2020; Zhong and Zeng 2008).

1.7.1.2 Early Detection, Isolation, and Tracing Close Contacts

Early detection of COVID-19 cases makes a basis for its containment. Isolation of the confirmed cases can hamper contacts and, ultimately, the transmission of the disease (Lau et al. 2004). The median incubation period of COVID-19 is 4.8 days; however, the median time from symptom onset to confirmed diagnosis of COVID-19 in severe cases is 8 days. Therefore, early detection plays a substantial role in the reduction of COVID-19 burden (Yang et al. 2020; Bi et al. 2020). Improvement of diagnostic methods and decreasing the time of detection are necessary. In addition to symptoms, travel and contact history should be taken into consideration for early case detection (Kim et al. 2020). Early case identification must be followed by tracing the close contacts, which requires self-quarantine and follow-up for manifestations (Abdullah et al. 2003).

1.7.1.3 Rigorous Infection Control

COVID-19 could be transmitted from both symptomatic cases and those in the incubation period who do not have clinical manifestations (Li et al. 2020a; Rothe et al. 2020). In order to control the infection, early isolation and treatment of confirmed cases and strategies for the prevention of virus transmission from asymptomatic patients must be taken into consideration (Helmy et al. 2020). Quarantine, social distancing, and closing borders are among the control and prevention strategies (Cowling and Leung 2020). The general population must receive practical guidelines on personal hygiene, including cough or squeeze etiquette, frequent handwashing, wearing a face mask, using a disinfectant, and staying home. Wearing personal protection equipment (PPE), including mask, gown, and shield, is also necessary for healthcare workers (Moradian et al. 2020; Yan et al. 2020).

1.7.1.4 Vaccine Development

Vaccination is considered as the most effective approach for the reduction of incidence and mortality of many infectious diseases (Clem 2011). Although the process of development, evaluation, and approval of vaccines usually requires around 1 year or more (Verch et al. 2018; Wang et al. 2020a), many investigators are trying to develop vaccines for COVID-19 prevention. They have accelerated testing various vaccine strategies in clinical trials (Thanh Le et al. 2020).

Dissemination of Information

During an outbreak, dissemination of correct information could be beneficial for the prevention of public panic, keeping clinicians updated on the last practical guidelines, and providing data for epidemiological analysis (Song and Karako 2020). Aside from the role of international agencies such as the World Health Organization (WHO), online registries and timely publication of cases and experimental studies are essential for the dissemination of scientific information (Zhong and Zeng 2008).

Recurrence

There were five viral infection outbreaks before SARS-CoV-2, in the twenty-first century, including SARS-CoV in 2003 and MERS-CoV in 2011. SARS-CoV-2 pandemic is the sixth outbreak of the century. Control of virus transmission through preventive measures for general populations and healthcare workers, as well as early case detection, isolation, treatment, and tracing close contacts, could hopefully prevent a second wave (Hiscott et al. 2020). However, it has been hypothesized that termination of social distancing before the end of the pandemic leads to the initiation of a second wave with higher mortality than the first wave (Goscé et al. 2020; Podile and Basu 2020; Leung et al. 2020). SARS-CoV-2 has shown the risk of reinfection (Jabbari and Rezaei 2020) and might have recurring seasonal cycles after the first wave (Podile and Basu 2020).

Psychosocial Support
COVID-19 has induced public panic, and COVID-19-related lockdown and quarantine have considerably restricted economic activities and social relationships. All of these factors can affect the mental health of the general population (Jeong et al. 2016; Brooks et al. 2020; Rubin and Wessely 2020). Healthcare workers are also at higher risk of stress and burnout (Robertson et al. 2004; Lai et al. 2020; Yazdanpanah et al. 2020b; Rezaei 2020a). Older people, children, pregnant women, and patients with prior chronic diseases, especially psychiatric diseases, are the other specific populations at high risk of mental stress, and they need special attention (Doraiswamy et al. 2020; Ghosh et al. 2020; Yao et al. 2020; Wang et al. 2020c; Mirbeyk and Rezaei 2020). Media, healthcare organizations, and authorities all should address the psychosocial effects of COVID-19 by providing appropriate content and supportive regulations (Dubey et al. 2020).

1.7.2 Preparing for the Next COVID-19 Pandemic

Another global wave of the COVID-19 is predicted by many scientists to occur mainly in the fall and winter. The next assault might be much more detrimental than the first wave (Sun 2020). Several respiratory viruses and coronaviruses similar to COVID-19 are found to behave seasonally (Monto et al. 2020; Dowell and Ho 2004). Whether the novel coronavirus 2019 will follow the same pattern and recur harder in fall and winter is disputable. The significant potential danger of another deadly COVID-19 outbreak mandates authorities and individuals to get ready for it.

Governmental and health authorities can play a critical role in enforcing our preparedness. First and foremost, from the first wave of the COVID-19 pandemic, we learned that the most critical tool in managing COVID-19 is rapid and mass testing in addition to contact tracing (Sanchez 2020). When it comes to COVID-19, even a small number of infected patients can be indicative of clusters, which need to be detected and isolated. Moreover, constant testing and contact tracing can help health authorities to know the number of infected and exposed individuals, which can be of immense help in anticipating the pattern of the pandemic and getting prepared adequately. Second, when the COVID-19 struck, we faced a global shortage of personal protective equipment (PPE), ventilators, intensive care unit (ICU) beds, testing kits, and other resources (Todd 2020). Large quantities of PPE, testing kits, and ventilators should be on hand to be used whenever necessary. Third, protecting the vulnerable population who are more likely to be affected by COVID-19, including the elderly, people with a disability or long-term health condition, and those in disadvantaged communities, should be one of the main priorities of governments (Banerjee 2020). Planning for maintaining their physical and mental health, in addition to reducing adverse socioeconomic effects of another potential pandemic, is crucial in our preparedness for another COVID-19 pandemic. Designating financial resources to minimize health inequalities (Anyane-Yeboa et al. 2020), ensuring their food security, and providing them with hygiene products, together with the formation of support networks, can facilitate helping these groups.

Individuals, in addition to governmental and health authorities, should also gear up for another strike of the COVID-19. Initially, they should maintain social distancing and regular hand washing, which might become the new healthy lifestyle until we found a solution for this yet evolving challenge to prevent the spread of the virus. Additionally, by targeting other diseases for which we have a solution, we can be more prepared for facing another wave of COVID-19 pandemic. In the cold seasons, when there is a high chance for COVID-19 to recur, every year, the healthcare system should also treat patients infected with seasonal flu. Individuals must receive their flu vaccine to lower the incidence of other respiratory infections in these seasons.

Most importantly, people should be mentally prepared for new waves of COVID-19. They should try to maintain their mental health and consider the possibility of occurrence of another similar or even deadlier wave. This possibility should be taken into account when they are making their personal and professional plans (Miller 2020).

1.8 The Economy of COVID-19

As the world becomes more interconnected, and the global village comes to reality, not only the occurrence of pandemic becomes more probable but also the economic impact of diseases increases (Fernandes 2020). A pandemic can affect the economy in various ways. The most apparent channel is increased mortality and morbidity that leads to decreased human capital in the future or raised health expenditure for disabled individuals. However, in the COVID-19 pandemic situation, the correlation between mortality rate and economic impacts is not reliable; this pandemic mostly affects the economy through other ways (Fernandes 2020). The new integrated world has interconnected networks of supply chain routes. It means that a demand or supply shock in one place of the world will affect the whole world. China is responsible for 16% of global GDP. By locking down the manufactures of China, many supply chains were disrupted. By spreading the outbreak to other parts of the world, especially countries with high impacts on the economy such as the USA, Germany, and France, the balance between demand and supply of the market was disrupted and created an economic shock (Richard Baldwin 06 March 2020). Many manufactures and companies shut down around the world; some of them will not reopen ever. It leads to an increased unemployment rate and poverty (McKee and Stuckler 2020). Car companies have been closed, and even manufacturers of luxury goods are shutting. These are happening due to two main factors, reduced demands and shortages of supplies (Fernandes 2020). Many companies also had to fire their employees to prevent bankruptcy. By closing the manufactures and reducing travel rate, the demands of oil dropped significantly. On the other hand, the two major oil producers, Saudi and Russia, argued and started to produce high amounts of oil. Thus, oil prices decreased dramatically (Nicola et al. 2020).

Tourism was also influenced. Traveling between and across the countries decreased due to quarantine and social distancing. It has been predicted that nearly 50 million jobs are at risk in tourism (Nicola et al. 2020). Also, disruption of the airline industry will lead to losing 44% of its revenue (Açikgöz and Günay 2020).

The United Nations Conference on Trade and Development (UNCTAD) announced that only in 2020, the COVID-19 pandemic could potentially cost about $2 trillion around the world (Development 2 March 2020). Also, GDP may reduce by 7% in the first 6 months of 2020, predicted by World Economic Prospects. By returning economic activities to normal and in the absence of any new outbreaks, the GDP will stop decreasing in the second half of the year (Prospects April 2020).

Despite the adverse effects of the COVID-19 pandemic on the global economy, it has some benefits too. By disrupting the old supply chain routes, new chain mechanisms will appear and make new economic opportunities. Online shopping and online educating will improve that will lead to significant progress in digital technology and Internet network. Even reduced oil prices will result in increased economic growth in the long-term for some countries that are oil importers (Açikgöz and Günay 2020).

1.9 Conclusion

Like other hCoVs leading to outbreaks, the SARS-CoV-2 is apparently of zoonotic origin circulated in animals before transmission to humans. The respiratory effects of COVID-19 are well-known, but COVID-19 can affect other systems, mainly the gastrointestinal system, central nervous system, and cardiovascular system. Its ability might lie partly in multiple routes, e.g., fecal-oral transmission, airborne transmission, and contact, through which it can spread, and also, partly it can make in through pulmonary and systemic immune responses. Research has revealed that immune abnormalities occur during COVID-19 and play a role in determining the outcome of the disease. Recent viral outbreaks have presented great difficulties with diagnosis, including COVID-19. Diagnosis of COVID-19 is mainly based on epidemiological history, clinical manifestations, and some auxiliary examinations, such as nucleic acid detection, CT scan, immune identification technology, and blood culture. The spectrum of clinical severity covers both asymptomatic and symptomatic individuals. A symptomatic infection might cause either mild to moderate symptoms of infection, severe infection that requires hospital admission, a critical infection that requires intensive care unit (ICU) admission, or even death. Among symptomatic individuals, about 20% develop severe to critical infection, and the remainders are diagnosed as mild to moderate infection. The most common symptoms of this viral infection are fever, followed by dry cough, fatigue, and shortness of breath. Other less prevalent clinical manifestations include headache, diarrhea, nausea, vomiting, myalgia, and arthralgia. In the mild form of the disease, alterations in the sense of smell and taste are commonly reported by the patients, and the clinicians should consider these symptoms as initial apparent symptoms of the disease. Current therapeutics for COVID-19 include interferons, antivirals, convalescent plasma and IVIG, monoclonal antibodies, corticosteroids, and nutrition therapy (Saghazadeh and Rezaei 2020b). Also, the immune system and the cell-virus interaction seem promising targets for the effective treatment of COVID-19 (Sharifkashani et al. 2020; Mansourabadi et al. 2020; Pashaei and Rezaei 2020).

Fig. 1.1 The borderless solution to the world's pain [*Adapted with permission from the Association of Science and Art (ASA), USERN; Contributors: Nima Rezaei (lyrics) and Saina Adiban (graphic design)*]

Fig. 1.2 The endless art and the borderless solution to the world' pain [*Adapted with permission from the Association of Science and Art (ASA), USERN; Contributors: Nima Rezaei (lyrics), Mahsa Yousefpour (calligraphy), Sepideh Sargoli (painting ideas), and Siavosh Mahraeen (graphic design)*]

In addition to the lessons learned from the previous coronavirus outbreaks, public health preparedness planning is necessary to focus on animal source control, early detection, isolation and tracing close contacts, rigorous infection control, vaccine development, dissemination of accurate information, recurrence management, and psychosocial support. The COVID-19 pandemic has deeply influenced the global economy, mainly involving supply chain routes, oil prices, and the tourism sector.

In this manner, the COVID-19 pandemic has induced panic at the international level. As of writing this, the world's panic is spreading globally, posing a complex problem with its pain (Fig. 1.1), asking for humanity to reorder the connections and, therefore, the world (Fig. 1.2) to resolve any complaint derived from loneliness to the well-being and being prepared by knowing that borderless communication would bring us about confidence for uncertain times (Mohamed et al. 2020a; Momtazmanesh et al. 2020; Moradian et al. 2020).

References

Abdullah AS, Tomlinson B, Cockram CS, Thomas GN (2003) Lessons from the severe acute respiratory syndrome outbreak in Hong Kong. Emerg Infect Dis 9(9):1042

Açikgöz Ö, Günay A (2020) The early impact of the Covid-19 pandemic on the global and Turkish economy. Turk J Med Sci 50(SI-1):520–526. https://doi.org/10.3906/sag-2004-6

Ahmadi M, Saffarzadeh N, Habibi MA, Hajiesmaeili F, Rezaei N (2020, Jann 1) Colon cancer and SARS-CoV-2: impact of ACE2 expression in susceptibility to COVID-19. bioRxiv

Ahmed W, Angel N, Edson J, Bibby K, Bivins A, O'Brien JW, Choi PM, Kitajima M, Simpson SL, Li J (2020) First confirmed detection of SARS-CoV-2 in untreated wastewater in Australia: a proof of concept for the wastewater surveillance of COVID-19 in the community. Sci Total Environ 728:138764

Anyane-Yeboa A, Sato T, Sakuraba A (2020) Racial disparities in COVID-19 deaths reveal harsh truths about structural inequality in america. J Intern Med 288:479. https://doi.org/10.1111/joim.13117

Ather A, Patel B, Ruparel NB, Diogenes A, Hargreaves KM (2020) Coronavirus disease 19 (COVID-19): implications for clinical dental care. J Endod 46(5): 584–595. https://doi.org/10.1016/j.joen.2020.03.008

Babaha F, Rezaei N (2020) Primary immunodeficiency diseases in COVID-19 pandemic: a predisposing or protective factor? Am J Med Sci. https://doi.org/10.1016/j.amjms.2020.07.027

Bahrami A, Vafapour M, Moazzami B, Rezaei N (2020, Jul 8) Hyperinflammatory shock related to COVID-19 in a patient presenting with multisystem inflammatory syndrome in children: first case from Iran. J Paediatr Child Health

Banerjee D (2020) 'Age and ageism in COVID-19': elderly mental health-care vulnerabilities and needs. Asian J Psychiatr 51:102154. https://doi.org/10.1016/j.ajp.2020.102154

Basiri A, Heidari A, Nadi MF, Fallahy MTP, Nezamabadi SS, Sedighi M, Saghazadeh A, Rezaei N (2020a) Microfluidic devices for detection of RNA viruses. Rev Med Virol n/a (n/a):e2154. https://doi.org/10.1002/rmv.2154

Basiri A, Pazhouhnia Z, Beheshtizadeh N, Hoseinpour M, Saghazadeh A, Rezaei N (2020b, Jun 20) Regenerative medicine in COVID-19 treatment: real opportunities and range of promises. Stem Cell Rev Rep 20:1–13

Bi Q, Wu Y, Mei S, Ye C, Zou X, Zhang Z, Liu X, Wei L, Truelove SA, Zhang T, Gao W, Cheng C, Tang X, Wu X, Wu Y, Sun B, Huang S, Sun Y, Zhang J, Ma

T, Lessler J, Feng T (2020) Epidemiology and transmission of COVID-19 in 391 cases and 1286 of their close contacts in Shenzhen, China: a retrospective cohort study. Lancet Infect Dis 20:911. https://doi.org/10.1016/S1473-3099(20)30287-5

Brooks SK, Webster RK, Smith LE, Woodland L, Wessely S, Greenberg N, Rubin GJ (2020) The psychological impact of quarantine and how to reduce it: rapid review of the evidence. Lancet 395:912

Bulut C, Kato Y (2020) Epidemiology of COVID-19. Turk J Med Sci 50(SI-1):563–570. https://doi.org/10.3906/sag-2004-172

Cao B, Wang Y, Wen D, Liu W, Wang J, Fan G, Ruan L, Song B, Cai Y, Wei M, Li X, Xia J, Chen N, Xiang J, Yu T, Bai T, Xie X, Zhang L, Li C, Yuan Y, Chen H, Li H, Huang H, Tu S, Gong F, Liu Y, Wei Y, Dong C, Zhou F, Gu X, Xu J, Liu Z, Zhang Y, Li H, Shang L, Wang K, Li K, Zhou X, Dong X, Qu Z, Lu S, Hu X, Ruan S, Luo S, Wu J, Peng L, Cheng F, Pan L, Zou J, Jia C, Wang J, Liu X, Wang S, Wu X, Ge Q, He J, Zhan H, Qiu F, Guo L, Huang C, Jaki T, Hayden FG, Horby PW, Zhang D, Wang C (2020) A trial of lopinavir–ritonavir in adults hospitalized with severe Covid-19. N Engl J Med 382(19):1787–1799. https://doi.org/10.1056/NEJMoa2001282

Cheema M, Aghazadeh H, Nazarali S, Ting A, Hodges J, McFarlane A, Kanji JN, Zelyas N, Damji KF, Solarte C (2020) Keratoconjunctivitis as the initial medical presentation of the novel coronavirus disease 2019 (COVID-19). Can J Ophthalmol 55:e125. https://doi.org/10.1016/j.jcjo.2020.03.003

Chen L, Xiong J, Bao L, Shi Y (2020) Convalescent plasma as a potential therapy for COVID-19. Lancet Infect Dis 20(4):398–400. https://doi.org/10.1016/S1473-3099(20)30141-9

Cherry JD (2004) The chronology of the 2002–2003 SARS mini pandemic. Paediatr Respir Rev 5(4):262–269

Chow KYC, Hon CC, Hui RKH, Wong RTY, Yip CW, Zeng F, Leung FCC (2003) Molecular advances in severe acute respiratory syndrome-associated coronavirus (SARS-CoV). Genomics Proteomics Bioinformatics 1(4):247–262

Clem AS (2011) Fundamentals of vaccine immunology. J Global Infect Dis 3(1):73–78. https://doi.org/10.4103/0974-777X.77299

Corman VM, Landt O, Kaiser M, Molenkamp R, Meijer A, Chu DKW, Bleicker T, Brünink S, Schneider J, Schmidt ML (2020) Detection of 2019 novel coronavirus (2019-nCoV) by real-time RT-PCR. Eurosurveillance 25(3):2000045

Cowling BJ, Leung GM (2020) Epidemiological research priorities for public health control of the ongoing global novel coronavirus (2019-nCoV) outbreak. Eurosurveillance 25(6):2000110

Craven J (2020, May 28) COVID-19 vaccine tracker. https://www.raps.org/news-and-articles/news-articles/2020/3/covid-19-vaccine-tracker

Darbeheshti F, Rezaei N (2020) Genetic predisposition models to COVID-19 infection. Med Hypotheses 142:109818

Diao B, Wang C, Tan Y, Chen X, Liu Y, Ning L, Chen L, Li M, Liu Y, Wang G (2020) Reduction and functional exhaustion of T cells in patients with coronavirus disease 2019 (COVID-19). Front Immunol 11:827

Doraiswamy S, Cheema S, Mamtani R (2020) Older people and epidemics: a call for empathy. Age Ageing 49(3):493–493

Dowell SF, Ho MS (2004) Seasonality of infectious diseases and severe acute respiratory syndrome-what we don't know can hurt us. Lancet Infect Dis 4(11):704–708. https://doi.org/10.1016/S1473-3099(04)01177-6

Dubey S, Biswas P, Ghosh R, Chatterjee S, Dubey MJ, Chatterjee S, Lahiri D, Lavie CJ (2020) Psychosocial impact of COVID-19. Diabetes Metab Syndr 14:779. https://doi.org/10.1016/j.dsx.2020.05.035

Fan EX (2003) SARS: economic impacts and implications. Asian Development Bank, Manila

Fathi N, Rezaei N (2020) Lymphopenia in COVID-19: therapeutic opportunities. Cell Biol Int 44:1792

Faye O, Faye O, Dupressoir A, Weidmann M, Ndiaye M, Sall AA (2008) One-step RT-PCR for detection of Zika virus. J Clin Virol 43(1):96–101

Fernandes N (2020) Economic effects of coronavirus outbreak (COVID-19) on the World Economy SSRN: https://ssrn.com/abstract=3557504 or https://doi.org/10.2139/ssrn.3557504

Fintelman-Rodrigues N, Sacramento CQ, Lima CR, da Silva FS, Ferreira A, Mattos M, de Freitas CS, Soares VC, Dias SD, Temerozo JR, Miranda M (2020, Jan 1) Atazanavir inhibits SARS-CoV-2 replication and pro-inflammatory cytokine production. bioRxiv

Gandhi M, Yokoe DS, Havlir DV (2020) Asymptomatic transmission, the Achilles' heel of current strategies to control Covid-19. N Engl J Med 382(22):2158–2160. https://doi.org/10.1056/NEJMe2009758

Ghosh R, Dubey MJ, Chatterjee S, Dubey S (2020) Impact of COVID-19 on children: special focus on psychosocial aspect. Education 31:34

Goscé L, Phillips PA, Spinola P, Gupta DRK, Abubakar PI (2020) Modelling SARS-COV2 spread in London: approaches to lift the lockdown. J Infect 81:260. https://doi.org/10.1016/j.jinf.2020.05.037

Gralinski LE, Menachery VD (2020) Return of the coronavirus: 2019-nCoV. Viruses 12(2):135. https://doi.org/10.3390/v12020135

Gu J, Han B, Wang J (2020) COVID-19: gastrointestinal manifestations and potential fecal–oral transmission. Gastroenterology 158(6):1518–1519

Guan W-j, Liang W-h, Zhao Y, Liang H-r, Chen Z-s, Li Y-m, Liu X-q, Chen R-c, Tang C-l, Wang T, Ou C-q, Li L, Chen P-y, Sang L, Wang W, Li J-f, Li C-c, Ou L-m, Cheng B, Xiong S, Ni Z-y, Xiang J, Hu Y, Liu L, Shan H, Lei C-l, Peng Y-x, Wei L, Liu Y, Hu Y-h, Peng P, Wang J-m, Liu J-y, Chen Z, Li G, Zheng Z-j, Qiu S-q, Luo J, Ye C-j, Zhu S-y, Cheng L-l, Ye F, Li S-y, Zheng J-p, Zhang N-f, Zhong N-s, He J-x (2020a) Comorbidity and its impact on 1590 patients with COVID-19 in China: a nationwide analysis. Eur Respir J 55(5):2000547. https://doi.org/10.1183/13993003.00547-2020

Guan W-j, Ni Z-y, Hu Y, Liang W-h, Ou C-q, He J-x, Liu L, Shan H, Lei C-l, Hui DSC, Du B, Li L-j, Zeng G, Yuen K-Y, Chen R-c, Tang C-l, Wang T, Chen P-y, Xiang J, Li S-y, Wang J-l, Liang Z-j, Peng Y-x, Wei L, Liu Y, Hu Y-h, Peng P, Wang J-m, Liu J-y, Chen Z, Li G, Zheng Z-j, Qiu S-q, Luo J, Ye C-j, Zhu S-y, Zhong N-s (2020b) Clinical characteristics of coronavirus disease 2019 in China. N Engl J Med 382(18):1708–1720. https://doi.org/10.1056/NEJMoa2002032

Han R, Huang L, Jiang H, Dong J, Peng H, Zhang D (2020) Early clinical and CT manifestations of coronavirus disease 2019 (COVID-19) pneumonia. Am J Roentgenol 215:1–6. https://doi.org/10.2214/AJR.20.22961

Hanaei S, Rezaei N (2020) COVID-19: developing from an outbreak to a pandemic. Arch Med Res 51:582

Helmy YA, Fawzy M, Elaswad A, Sobieh A, Kenney SP, Shehata AA (2020) The COVID-19 pandemic: a comprehensive review of taxonomy, genetics, epidemiology, diagnosis, treatment, and control. J Clin Med 9(4):1225. https://doi.org/10.3390/jcm9041225

Henderson LA, Canna SW, Schulert GS, Volpi S, Lee PY, Kernan KF, Caricchio R, Mahmud S, Hazen MM, Halyabar O (2020) On the alert for cytokine storm: immunopathology in COVID-19. Arthritis Rheum 72:1059

Hessami A, Shamshirian A, Heydari K, Pourali F, Alizadeh-Navaei R, Moosazadeh M, Abrotan S, Shojaei L, Sedighi S, Shamshirian D, Rezaei N (2020, Oct 16) Cardiovascular diseases burden in COVID-19: Systematic review and meta-analysis. Am J Emerg Med

Hirbod-Mobarakeh A, Mahmoudi M, Rezaei N (2014) Nutrition, immunity, and aging. In: Immunology of aging. Springer, Berlin, pp 263–284

Hiscott J, Alexandridi M, Muscolini M, Tassone E, Palermo E, Soultsioti M, Zevini A (2020) The global impact of the coronavirus pandemic. Cytokine Growth Factor Rev 53:1. https://doi.org/10.1016/j.cytogfr.2020.05.010

Hu Y, Sun J, Dai Z, Deng H, Li X, Huang Q, Wu Y, Sun L, Xu Y (2020) Prevalence and severity of corona virus disease 2019 (COVID-19): a systematic review and meta-analysis. J Clin Virol 127:104371

Huang C, Wang Y, Li X, Ren L, Zhao J, Hu Y, Zhang L, Fan G, Xu J, Gu X, Cheng Z, Yu T, Xia J, Wei Y, Wu W, Xie X, Yin W, Li H, Liu M, Xiao Y, Gao H, Guo L, Xie J, Wang G, Jiang R, Gao Z, Jin Q, Wang J, Cao B (2020a) Clinical features of patients infected with 2019 novel coronavirus in Wuhan, China. Lancet (London) 395(10223):497–506. https://doi.org/10.1016/s0140-6736(20)30183-5

Huang X, Wei F, Hu L, Wen L, Chen K (2020b) Epidemiology and clinical characteristics of COVID-19. Arch Iran Med March 23(4):268–271. https://doi.org/10.34172/aim.2020.09

Hung IF-N, Lung K-C, Tso EY-K, Liu R, Chung TW-H, Chu M-Y, Ng Y-Y, Lo J, Chan J, Tam AR, Shum H-P, Chan V, Wu AK-L, Sin K-M, Leung W-S, Law W-L, Lung DC, Sin S, Yeung P, Yip CC-Y, Zhang RR, Fung AY-F, Yan EY-W, Leung K-H, Ip JD, Chu AW-H, Chan W-M, Ng AC-K, Lee R, Fung K, Yeung A, Wu T-C, Chan JW-M, Yan W-W, Chan W-M, Chan JF-W, Lie AK-W, Tsang OT-Y, Cheng VC-C, Que T-L, Lau C-S, Chan K-H, To KK-W, Yuen K-Y (2020) Triple combination of interferon beta-1b, lopinavir-ritonavir, and ribavirin in the treatment of patients admitted to hospital with COVID-19: an open-label, randomised, phase 2 trial. Lancet 395(10238):1695–1704. https://doi.org/10.1016/S0140-6736(20)31042-4

Jabbari P, Rezaei N (2020) With risk of reinfection, is COVID-19 here to stay? Disaster Med Public Health Prep:1. https://doi.org/10.1017/dmp.2020.274

Jabbari P, Jabbari F, Ebrahimi S, Rezaei N (2020, Jun) COVID-19: a chimera of two pandemics. Disaster Med Public Health Prep 14(3):e38–9

Jahanshahlu L, Rezaei N (2020a) Central nervous system involvement in COVID-19. Arch Med Res 51:721

Jahanshahlu L, Rezaei N (2020b) Monoclonal antibody as a potential anti-COVID-19. Biomed Pharmacother 129:110337

Jeong H, Yim HW, Song Y-J, Ki M, Min J-A, Cho J, Chae J-H (2016) Mental health status of people isolated due to middle east respiratory syndrome. Epidemiol Health 38:e2016048

Kim JY, Choe PG, Oh Y, Oh KJ, Kim J, Park SJ, Park JH, Na HK, Oh M-d (2020) The first case of 2019 novel coronavirus pneumonia imported into Korea from Wuhan, China: implication for infection prevention and control measures. J Korean Med Sci 35(5):e61

Lai J, Ma S, Wang Y, Cai Z, Hu J, Wei N, Wu J, Du H, Chen T, Li R (2020) Factors associated with mental health outcomes among health care workers exposed to coronavirus disease 2019. JAMA Netw Open 3(3):e203976–e203976

Lau JT, Tsui H, Lau M, Yang X (2004) SARS transmission, risk factors, and prevention in Hong Kong. Emerg Infect Dis 10(4):587

Leung K, Wu JT, Liu D, Leung GM (2020) First-wave COVID-19 transmissibility and severity in China outside Hubei after control measures, and second-wave scenario planning: a modelling impact assessment. Lancet 395(10233):1382–1393. https://doi.org/10.1016/S0140-6736(20)30746-7

Li J-Y, You Z, Wang Q, Zhou Z-J, Qiu Y, Luo R, Ge X-Y (2020a) The epidemic of 2019-novel-coronavirus (2019-nCoV) pneumonia and insights for emerging infectious diseases in the future. Microbes Infect 22(2):80–85. https://doi.org/10.1016/j.micinf.2020.02.002

Li X, Geng M, Peng Y, Meng L, Lu S (2020b) Molecular immune pathogenesis and diagnosis of COVID-19. J Pharm Anal 10(2):102–108. https://doi.org/10.1016/j.jpha.2020.03.001

Lipsitch M, Swerdlow DL, Finelli L (2020) Defining the epidemiology of Covid-19—studies needed. N Engl J Med 382(13):1194–1196

Liu P, Jiang J-Z, Wan X-F, Hua Y, Li L, Zhou J, Wang X, Hou F, Chen J, Zou J, Chen J (2020a) Are pangolins the intermediate host of the 2019 novel coronavirus (SARS-CoV-2)? PLoS Pathog 16(5):e1008421. https://doi.org/10.1371/journal.ppat.1008421

Liu Z, Xiao X, Wei X, Li J, Yang J, Tan H, Zhu J, Zhang Q, Wu J, Liu L (2020b) Composition and divergence of coronavirus spike proteins and host ACE2 receptors predict potential intermediate hosts of SARS-CoV-2. J Med Virol 92:595. https://doi.org/10.1002/jmv.25726

Lotfi M, Rezaei N (2020) SARS-CoV-2: a comprehensive review from pathogenicity of the virus to clinical consequences. J Med Virol 92:1864

Lotfi M, Hamblin MR, Rezaei N (2020) COVID-19: transmission, prevention, and potential therapeutic opportunities. Clin Chim Acta 508:254

Lovato A, de Filippis C (2020) Clinical presentation of COVID-19: a systematic review focusing on upper airway symptoms. Ear Nose Throat J 99:569.:0145561320920762. https://doi.org/10.1177/0145561320920762

Lu C-C, Chen M-Y, Chang Y-L (2020) Potential therapeutic agents against COVID-19: what we know so far. J Chin Med Assoc Latest Articles 83:534

Luan J, Lu Y, Jin X, Zhang L (2020) Spike protein recognition of mammalian ACE2 predicts the host range and an optimized ACE2 for SARS-CoV-2 infection. Biochem Biophys Res Commun 526(1):165–169. https://doi.org/10.1016/j.bbrc.2020.03.047

Luo W, Yu H, Gou J, Li X, Sun Y, Li J, Liu L (2020) Clinical pathology of critical patient with novel coronavirus pneumonia (COVID-19). Preprints 2020:2020020407

Lurie N, Saville M, Hatchett R, Halton J (2020) Developing Covid-19 vaccines at pandemic speed. N Engl J Med 382(21):1969–1973. https://doi.org/10.1056/NEJMp2005630

Mahmoudi M, Rezaei N (2019) Nutrition and immunity. Springer, Cham

Mansourabadi AH, Sadeghalvad M, Mohammadi-Motlagh HR, Rezaei N (2020) The immune system as a target for therapy of SARS-CoV-2: a systematic review of the current immunotherapies for COVID-19. Life Sci 258:118185. https://doi.org/10.1016/j.lfs.2020.118185

Masters PS (2006) The molecular biology of coronaviruses. Adv Virus Res 66:193–292

McKee M, Stuckler D (2020) If the world fails to protect the economy, COVID-19 will damage health not just now but also in the future. Nat Med 26(5):640–642. https://doi.org/10.1038/s41591-020-0863-y

medicine USnlo (2020a) BCG vaccination to protect healthcare workers against COVID-19 (BRACE). https://clinicaltrials.gov/ct2/show/NCT04327206

medicine USnlo (2020b) Reducing health care workers absenteeism in Covid-19 pandemic through BCG vaccine (BCG-CORONA). https://clinicaltrials.gov/ct2/show/NCT04328441

Mehra MR, Desai SS, Ruschitzka F, Patel AN (2020) Hydroxychloroquine or chloroquine with or without a macrolide for treatment of COVID-19: a multinational registry analysis. Lancet. https://doi.org/10.1016/S0140-6736(20)31180-6

Miller AM (2020) How you should prepare for a potential 'second wave' of the coronavirus, according to experts. Business Insider France. https://www.businessinsider.fr/graphique-du-jour-titres-les-plus-ecoutes-sur-spotify-2017. Accessed 25 May 2020

Mirbeyk M, Rezaei N (2020) The impact of COVID-19 on pregnancy and neonatal health: a systematic review

Moazzami B, Razavi-Khorasani N, Moghadam AD, Farokhi E, Rezaei N (2020) COVID-19 and telemedicine: immediate action required for maintaining healthcare providers Well-being. J Clin Virol 126:104345

Mohamed K, Rodríguez-Román E, Rahmani F, Zhang H, Ivanovska M, Makka SA, Joya M, Makuku R, Islam MS, Radwan N (2020a) Borderless collaboration is needed for COVID-19—a disease that knows no borders. Infect Control Hosp Epidemiol 41:1–2

Mohamed K, Yazdanpanah N, Saghazadeh A, Rezaei N (2020b) Computational drug discovery and repurposing for the treatment of COVID-19: a systematic review. Bioinorg Chem 104490. https://doi.org/10.1016/j.bioorg.2020.104490

Momtazmanesh S, Ochs HD, Uddin LQ, Perc M, Routes JM, Vieira DN, Al-Herz W, Baris S, Prando C, Rosivall L (2020) All together to fight COVID-19. Am J Trop Med Hyg 102(6):1181–1183

Monto AS, DeJonge PM, Callear AP, Bazzi LA, Capriola SB, Malosh RE, Martin ET, Petrie JG (2020) Coronavirus occurrence and transmission over 8 years in the HIVE cohort of households in Michigan. J Infect Dis 222:9. https://doi.org/10.1093/infdis/jiaa161

Moradian N, Ochs HD, Sedikies C, Hamblin MR, Camargo CA, Martinez JA, Biamonte JD, Abdollahi M, Torres PJ, Nieto JJ (2020) The urgent need for integrated science to fight COVID-19 pandemic and beyond. J Transl Med 18(1):1–7

Nasab MG, Saghazadeh A, Rezaei N (2020) SARS-CoV-2-a tough opponent for the immune system. Arch Med Res 51:589. https://doi.org/10.1016/j.arcmed.2020.05.020

Nicola M, Alsafi Z, Sohrabi C, Kerwan A, Al-Jabir A, Iosifidis C, Agha M, Agha R (2020) The socio-economic implications of the coronavirus and COVID-19 pandemic: a review. Int J Surg 78:185.:S1743-9191(1720)30316-30312. https://doi.org/10.1016/j.ijsu.2020.04.018

Nishiura H, Kobayashi T, Miyama T, Suzuki A, Jung S-M, Hayashi K, Kinoshita R, Yang Y, Yuan B, Akhmetzhanov AR, Linton NM (2020) Estimation of the asymptomatic ratio of novel coronavirus infections (COVID-19). Int J Infect Dis 94:154–155. https://doi.org/10.1016/j.ijid.2020.03.020

Paraskevis D, Kostaki EG, Magiorkinis G, Panayiotakopoulos G, Sourvinos G, Tsiodras S (2020) Full-genome evolutionary analysis of the novel corona virus (2019-nCoV) rejects the hypothesis of emergence as a result of a recent recombination event. Infect Genet Evol: J Mol Epidemiol Evol Genet Infect Dis 79:104212. https://doi.org/10.1016/j.meegid.2020.104212

Park M, Cook AR, Lim JT, Sun Y, Dickens BL (2020) A systematic review of COVID-19 epidemiology based on current evidence. J Clin Med 9(4):967

Pashaei M, Rezaei N (2020) Immunotherapy for SARS-CoV-2: potential opportunities. Expert Opin Biol Ther 20:1–5. https://doi.org/10.1080/14712598.2020.1807933

Peeri NC, Shrestha N, Rahman MS, Zaki R, Tan Z, Bibi S, Baghbanzadeh M, Aghamohammadi N, Zhang W, Haque U (2020) The SARS, MERS and novel coronavirus (COVID-19) epidemics, the newest and biggest global health threats: what lessons have we learned? Int J Epidemiol 49:717

Peng X, Xu X, Li Y, Cheng L, Zhou X, Ren B (2020) Transmission routes of 2019-nCoV and controls in dental practice. Int J Oral Sci 12(1):1–6

Petrosillo N, Viceconte G, Ergonul O, Ippolito G, Petersen E (2020) COVID-19, SARS and MERS: are they closely related? Clin Microbiol Infect 26:729

Podile AR, Basu A (2020) Opportunities, challenges and directions in science and technology for tackling COVID-19. Trans Indian Natl Acad Eng 5:1–5. https://doi.org/10.1007/s41403-020-00111-z

Pourahmad R, Moazzami B, Rezaei N (2020) Efficacy of Plasmapheresis and Immunoglobulin Replacement Therapy (IVIG) on Patients with COVID-19. SN Compr Clin Med 2:1407–1411. https://doi.org/10.1007/s42399-020-00438-2

Prompetchara E, Ketloy C, Palaga T (2020) Immune responses in COVID-19 and potential vaccines: lessons learned from SARS and MERS epidemic. Asian Pac J Allergy Immunol 38(1):1–9

Rabiee N, Rabiee M, Bagherzadeh M, Rezaei N (2020) COVID-19 and picotechnology: potential opportunities. Med Hypotheses 144:109917

Rashid ZZ, Othman SN, Samat MNA, Ali UK, Wong KK (2020) Diagnostic performance of COVID-19 serology assays. Malays J Pathol 42(1):13–21

Rezaei N (2020a) COVID-19 affects healthy pediatricians more than pediatric patients. Infect Control Hosp Epidemiol 41:1. https://doi.org/10.1017/ice.2020.139

Rezaei N (2020b) COVID-19 and medical biotechnology. Avicenna J Med Biotechnol 12(3):139

Richard Baldwin BWDM (2020, March 06) Economics in the time of COVID-19. A VoxEU.org Book, CEPR Press

Robertson E, Hershenfield K, Grace SL, Stewart DE (2004) The psychosocial effects of being quarantined following exposure to SARS: a qualitative study of Toronto health care workers. Can J Psychiatr 49(6):403–407

Rokni M, Hamblin MR, Rezaei N (2020) Cytokines and COVID-19: friends or foes? Hum Vaccin Immunother 16:1–3

Rothe C, Schunk M, Sothmann P, Bretzel G, Froeschl G, Wallrauch C, Zimmer T, Thiel V, Janke C, Guggemos W (2020) Transmission of 2019-nCoV infection from an asymptomatic contact in Germany. N Engl J Med 382(10):970–971

Rubin GJ, Wessely S (2020) The psychological effects of quarantining a city. BMJ 368:m313

Rzymski P, Nowicki M, Mullin GE, Abraham A, Rodríguez-Román E, Petzold MB, Bendau A, Sahu KK, Ather A, Naviaux A-F (2020) Quantity does not equal quality: scientific principles cannot be sacrificed. Int Immunopharmacol 86:106711

Sabino-Silva R, Jardim ACG, Siqueira WL (2020) Coronavirus COVID-19 impacts to dentistry and potential salivary diagnosis. Clin Oral Investig 24(4):1619–1621

Saghazadeh A, Rezaei N (2017) Implications of toll-like receptors in Ebola infection. Expert Opin Ther Targets 21(4):415–425

Saghazadeh A, Rezaei N (2019) In: Rezaei N (ed) Introductory chapter: immunogenetics, immunogenetics. IntechOpen. https://doi.org/10.5772/intechopen.85505. Available from: https://www.intechopen.com/books/immunogenetics/introductory-chapter-immunogenetics

Saghazadeh A, Rezaei N (2020a) Immune-epidemiological parameters of the novel coronavirus – a perspective. Expert Rev Clin Immunol 16(5):465–470. https://doi.org/10.1080/1744666x.2020.1750954

Saghazadeh A, Rezaei N (2020b) Towards treatment planning of COVID-19: rationale and hypothesis for the use of multiple immunosuppressive agents: anti-antibodies, immunoglobulins, and corticosteroids. Int Immunopharmacol 84:106560. https://doi.org/10.1016/j.intimp.2020.106560

Sahu KK, Siddiqui AD, Rezaei N, Cerny J (2020) Challenges for management of immune thrombocytopenia during COVID-19 pandemic. J Med Virol 92:2277

Saleki K, Banazadeh M, Saghazadeh A, Rezaei N (2020) The involvement of the central nervous system in patients with COVID-19. Reviews in the Neurosciences 1, no. ahead-of-print

Sanchez E (2020) COVID-19 science: why testing is so important. American Heart Association. https://www.heart.org/en/news/2020/04/02/covid-19-science-why-testing-is-so-important. Accessed 25 May 2020

Sanders JM, Monogue ML, Jodlowski TZ, Cutrell JB (2020) Pharmacologic treatments for coronavirus disease 2019 (COVID-19): a review. JAMA 323(18):1824–1836. https://doi.org/10.1001/jama.2020.6019

Shamshirian D, Rezaei N (2020) Cardiovascular diseases burden in COVID-19: systematic review and meta-analysis

Sharifkashani S, Bafrani MA, Khaboushan AS, Pirzadeh M, Kheirandish A, Yavarpour Bali H, Hessami A, Saghazadeh A, Rezaei N (2020) Angiotensin-converting enzyme 2 (ACE2) receptor and SARS-CoV-2: potential therapeutic targeting. Eur J Pharmacol 884:173455. https://doi.org/10.1016/j.ejphar.2020.173455

Shereen MA, Khan S, Kazmi A, Bashir N, Siddique R (2020) COVID-19 infection: origin, transmission, and characteristics of human coronaviruses. J Adv Res 24:91–98. https://doi.org/10.1016/j.jare.2020.03.005

Sim K, Chua HC (2004) The psychological impact of SARS: a matter of heart and mind. CMAJ 170(5):811–812

Skowronski DM, Astell C, Brunham RC, Low DE, Petric M, Roper RL, Talbot PJ, Tam T, Babiuk L (2005) Severe acute respiratory syndrome (SARS): a year in review. Annu Rev Med 56:357–381

Song P, Karako T (2020) COVID-19: real-time dissemination of scientific information to fight a public health emergency of international concern. Biosci Trends 14:1

Spinato G, Fabbris C, Polesel J, Cazzador D, Borsetto D, Hopkins C, Boscolo-Rizzo P (2020) Alterations in smell or taste in mildly symptomatic outpatients with SARS-CoV-2 infection. JAMA 323(20):2089–2090. https://doi.org/10.1001/jama.2020.6771

Su S, Wong G, Shi W, Liu J, Lai ACK, Zhou J, Liu W, Bi Y, Gao GF (2016) Epidemiology, genetic recombination, and pathogenesis of coronaviruses. Trends Microbiol 24(6):490–502. https://doi.org/10.1016/j.tim.2016.03.003

Su H, Yang M, Wan C, Yi L-X, Tang F, Zhu H-Y, Yi F, Yang H-C, Fogo AB, Nie X, Zhang C (2020) Renal histopathological analysis of 26 postmortem findings of patients with COVID-19 in China. Kidney Int 98(1):219–227. https://doi.org/10.1016/j.kint.2020.04.003

Sun LH (2020) CDC director warns second wave of coronavirus is likely to be even more devastating. The Washington Post. https://www.washingtonpost.com/health/2020/04/21/coronavirus-secondwave-cdcdirector/. Accessed 25 May 2020

Thanh Le T, Andreadakis Z, Kumar A, Gomez Roman R, Tollefsen S, Saville M, Mayhew S (2020) The COVID-19 vaccine development landscape. Nat Rev Drug Discov 19(5):305–306. https://doi.org/10.1038/d41573-020-00073-5

Tian S, Xiong Y, Liu H, Niu L, Guo J, Liao M, Xiao S-Y (2020) Pathological study of the 2019 novel coronavirus disease (COVID-19) through postmortem core biopsies. Mod Pathol 33:1–8

Todd B (2020) Where are the masks? Am J Nurs 120(6):18–19. https://doi.org/10.1097/01.NAJ.0000668700.23009.1e

Verch T, Trausch JJ, Shank-Retzlaff M (2018) Principles of vaccine potency assays. Bioanalysis 10(3):163–180. https://doi.org/10.4155/bio-2017-0176

Verity R, Okell LC, Dorigatti I, Winskill P, Whittaker C, Imai N, Cuomo-Dannenburg G, Thompson H, Walker PGT, Fu H (2020) Estimates of the severity of coronavirus disease 2019: a model-based analysis. Lancet Infect Dis 20:669

Wang M, Yan M, Xu H, Liang W, Kan B, Zheng B, Chen H, Zheng H, Xu Y, Zhang E (2005) SARS-CoV infection in a restaurant from palm civet. Emerg Infect Dis 11(12):1860

Wang Y, Zhang D, Du G, Du R, Zhao J, Jin Y, Fu S, Gao L, Cheng Z, Lu Q, Hu Y, Luo G, Wang K, Lu Y, Li H, Wang S, Ruan S, Yang C, Mei C, Wang Y, Ding D, Wu F, Tang X, Ye X, Ye Y, Liu B, Yang J, Yin W, Wang A, Fan G, Zhou F, Liu Z, Gu X, Xu J, Shang L, Zhang Y, Cao L, Guo T, Wan Y, Qin H, Jiang Y, Jaki T, Hayden FG, Horby PW, Cao B, Wang C (2020) Remdesivir in adults with severe COVID-19: a randomised, double-blind, placebo-controlled, multicentre trial. Lancet 395(10236):1569–1578. https://doi.org/10.1016/S0140-6736(20)31022-9

Wang F, Kream RM, Stefano GB (2020a) An evidence based perspective on mRNA-SARS-CoV-2 vaccine development. Med Sci Monit 26:e924700. https://doi.org/10.12659/msm.924700

Wang H-Y, Li X-L, Yan Z-R, Sun X-P, Han J, Zhang B-W (2020b) Potential neurological symptoms of COVID-19. Ther Adv Neurol Disord 13:1756286420917830. https://doi.org/10.1177/1756286420917830

Wang S, Wen X, Liu B, Dong Y, Cui M (2020c) Psychological influence of Coronovirus disease 2019 (COVID-19) pandemic on the general public, medical workers and patients with mental disorders and its countermeasures. Psychosomatics. https://doi.org/10.1016/j.psym.2020.05.005

Xiao K, Zhai J, Feng Y, Zhou N, Zhang X, Zou J-J, Li N, Guo Y, Li X, Shen X, Zhang Z, Shu F, Huang W, Li Y, Zhang Z, Chen R-A, Wu Y-J, Peng S-M, Huang M, Xie W-J, Cai Q-H, Hou F-H, Liu Y, Chen W, Xiao L, Shen Y (2020) Isolation and characterization of 2019-nCoV-like coronavirus from Malayan Pangolins. bioRxiv:2020.2002.2017.951335. https://doi.org/10.1101/2020.02.17.951335

Xie X, Zhong Z, Zhao W, Zheng C, Wang F, Liu J (2020a) Chest CT for typical 2019-nCoV pneumonia: relationship to negative RT-PCR testing. Radiology 296:E41.200343

Xie Y, Cao S, Dong H, Li Q, Chen E, Zhang W, Yang L, Fu S, Wang R (2020b) Effect of regular intravenous immunoglobulin therapy on prognosis of severe pneumonia in patients with COVID-19. J Infect:S0163–4453(0120)30172–30179. https://doi.org/10.1016/j.jinf.2020.03.044

Xu Z, Shi L, Wang Y, Zhang J, Huang L, Zhang C, Liu S, Zhao P, Liu H, Zhu L (2020) Pathological findings of COVID-19 associated with acute respiratory distress syndrome. Lancet Respir Med 8(4):420–422

Yan Y, Shin WI, Pang YX, Meng Y, Lai J, You C, Zhao H, Lester E, Wu T, Pang CH (2020) The first 75 days of novel coronavirus (SARS-CoV-2) outbreak: recent advances, prevention, and treatment. Int J Environ Res Public Health 17(7):2323. https://doi.org/10.3390/ijerph17072323

Yang Y, Lu Q, Liu M, Wang Y, Zhang A, Jalali N, Dean N, Longini I, Halloran ME, Xu B, Zhang X (2020, Jan 1) Epidemiological and clinical features of the 2019 novel coronavirus outbreak in China. MedRxiv

Yao H, Chen J-H, Xu Y-F (2020) Patients with mental health disorders in the COVID-19 epidemic. Lancet Psychiatry 7(4):e21

Yazdanpanah F, Hamblin MR, Rezaei N (2020a) The immune system and COVID-19: friend or foe? Life Sci 256:117900. https://doi.org/10.1016/j.lfs.2020.117900

Yazdanpanah N, Saghazadeh A, Rezaei N (2020b) Anosmia: a missing link in the neuroimmunology of coronavirus disease 2019 (COVID-19). Rev Neurosci 31:691. https://doi.org/10.1515/revneuro-2020-0039

Ye Q, Wang B, Mao J (2020) The pathogenesis and treatment of theCytokine Storm'in COVID-19. J Infect 80(6):607–613

Yousefzadegan S, Rezaei N (2020) Case report: death due to COVID-19 in three brothers. Am J Trop Med Hyg 102(6):1203–1204

Yuan J, Hon C-C, Li Y, Wang D, Xu G, Zhang H, Zhou P, Poon LL, Lam TT-Y, Leung FC-C (2010) Intraspecies diversity of SARS-like coronaviruses in Rhinolophus sinicus and its implications for the origin of SARS coronaviruses in humans. J Gen Virol 91(4):1058–1062

Zheng M, Gao Y, Wang G, Song G, Liu S, Sun D, Xu Y, Tian Z (2020a) Functional exhaustion of antiviral lymphocytes in COVID-19 patients. Cell Mol Immunol 17(5):533–535

Zheng Y-Y, Ma Y-T, Zhang J-Y, Xie X (2020b) COVID-19 and the cardiovascular system. Nat Rev Cardiol 17(5):259–260

Zhong NS, Zeng GQ (2008) Pandemic planning in China: applying lessons from severe acute respiratory syndrome. Respirology 13:S33–S35. https://doi.org/10.1111/j.1440-1843.2008.01255.x

Zhou P, Yang XL, Wang XG, Hu B, Zhang L, Zhang W, Si HR, Zhu Y, Li B, Huang CL, Chen HD (2020a) Discovery of a novel coronavirus associated with the recent pneumonia outbreak in humans and its potential bat origin. bioRxiv. Cold Spring Harb Lab 2020:22–914952

Zhou P, Yang XL, Wang XG, Hu B, Zhang L, Zhang W, Si HR, Zhu Y, Li B, Huang CL, Chen HD, Chen J, Luo Y, Guo H, Jiang RD, Liu MQ, Chen Y, Shen XR, Wang X, Zheng XS, Zhao K, Chen QJ, Deng F, Liu LL, Yan B, Zhan FX, Wang YY, Xiao GF, Shi ZL (2020b) A pneumonia outbreak associated with a new coronavirus of probable bat origin. Nature 579(7798):270–273. https://doi.org/10.1038/s41586-020-2012-7

Coronaviruses: What Should We Know About the Characteristics of Viruses?

2

Wei Ji

Abstract

The ongoing coronavirus disease (COVID-19) caused by severe acute respiratory syndrome coronavirus-2 (SARS-CoV-2) is highly contagious and fatal, posing a direct threat to human health and the global economy. Most strategies to prevent, control, and eradicate COVID-19 are established based on the specific characteristics of the pathogen. The quest for interruption and eradication of COVID-19 has moved research forward in understanding fundamental aspects of the virus genome, proteome, replication mechanisms, and virus-host interactions, which pave the way for the development of effective antiviral drugs and vaccines. This chapter provides an overview of recent progress in human coronavirus taxonomy, molecular features of the SARS-CoV-2 genome and proteome, and virus life cycle.

Keywords

COVID-19 · Genome · Human coronavirus · Life cycle · SARS-CoV-2 · Severe acute respiratory syndrome · Structural and nonstructural proteins · Taxonomy

W. Ji (✉)
Medical school, Liaocheng University, Liaocheng, China

Network of Immunity in Infection, Malignancy and Autoimmunity (NIIMA), Universal Scientific Education and Research Network (USERN), Liaocheng, China

2.1 Introduction

A novel pathogen caused the current outbreak of coronavirus disease 2019 (COVID-19), severe acute respiratory syndrome coronavirus 2 (SARS-CoV-2), previously known as 2019 novel coronavirus (2019-nCoV), which is similar yet unlike to SARS-CoV. SARS-CoV caused 774 deaths in 8096 infected individuals around the world between November 2002 and July 2003 (Shi and Hu 2008; Organization 2004; Peiris et al. 2004). SARS-CoV-2 infected approximately 5.9 million people and killed more than 367,000 people worldwide between December 2019 and May 2020. Although the two viruses have similar genome organization, SARS-CoV-2 has higher morbidity and manifest as acute respiratory tract infection with extrapulmonary involvement, such as cardiovascular complications and multi-organ failure (Chan et al. 2015; Wang et al. 2020b, c; Yang et al. 2020; Assiri et al. 2013; Rota et al. 2003; Zheng et al. 2020).

SARS-CoV-2 is the seventh member of human coronaviruses (HCoV). The taxonomy of these viruses will be discussed later (see Classification section). HCoVs share similar genome architectures, comprised of a basic set of genes 5′-replicase-structure proteins-3′. SARS-CoV-2

N. Rezaei (ed.), *Coronavirus Disease - COVID-19*, Advances in Experimental Medicine and Biology 1318, https://doi.org/10.1007/978-3-030-63761-3_2

genome consists of 5'UTR/nsp1-16/S/3a/E/M/6/7a/7b/8/N/10/3'UTR organization. Untranslated regions (UTRs) 5'UTR and 3'UTR are crucial for RNA replication and transcription. Nonstructural proteins nsp1-16 mediate RNA synthesis. Structural proteins, including the S (spike), E (envelope), M (membrane), and nucleoprotein (N), are involved in virus assembly and cell entrance. Accessory proteins 3a, 6, 7a, 7b, 8, and 10 can interact with nonstructural proteins and structural proteins, with undefined functions (Li et al. 2020). Altogether, these components fill different roles in the virus life cycle and can be considered as drug targets.

2.2 Classification

Coronaviruses are a group of related enveloped RNA viruses sorted into four groups based on phylogenetic clustering, which belong to the subfamily *Orthocoronavirinae*, under the family *Coronaviridae*, order *Nidovirales* (Chan et al. 2015; Burrell et al. 2017). According to the proposal of the International Committee on Taxonomy of Viruses (ICTV), these groups – alpha-, beta-, gamma- and delta-coronaviruses – have been conferred the taxonomic position of genera.

Coronaviruses can infect a variety of birds and mammals (Fig. 2.1). Alphacoronaviruses have been isolated from humans, bats, murine, mink, alpaca, porcine, and camel; betacoronaviruses from humans, mink, tiger, dog, civet, murine, bat, hedgehog, rabbit, rat, camel, horse, pig, and bovine; gammacoronaviruses from turkey, pheasant, duck, Canada goose, dolphin, and whale; and delta coronaviruses from night heron, wigeon, common moorhen, sparrow, quail, munia, magpie robin, Chinese bulbul, thrush, white-eye, penguin, *Calidris*, falcon, houbara, pigeon, and porcine.

Notably, seven of these viruses are related to human disease, called human coronavirus (HCoV) (Fig. 2.1). Four HCoVs (alphacoronaviruses HCoV-229E and HCoV-NL63 and betacoronaviruses HCoV-HKU1 and HCoV-OC43) typically cause mild, self-limiting respiratory tract infection with high morbidity rates (Chan et al. 2015; Gaunt et al. 2010; Pene et al. 2003; Vabret et al. 2003; Woo et al. 2005; Abdul-Rasool and Fielding 2010). The remaining three categorically harmful HCoVs are SARS-CoV, Middle East respiratory syndrome coronavirus (MERS-CoV), and SARS-CoV-2. The SARS-CoV is no longer circulating in humans (WHO 2020b). MERS-CoV is lingering in humans in the Middle East (WHO 2020a), while SARS-CoV-2 is still raging in humans worldwide and has caused a corrupting impact on our health and society.

2.3 Virion Structure and Genome Organization

Like other coronaviruses, SARS-CoV-2 virion particles are roughly spherical or moderately pleiomorphic, and the diameters range from 80 to 160 nm (Liu et al. 2020a). The virus particle is lipid-enveloped and carries external spike protein on the membrane surface, shaped like a crown (crown = corona) under electron microscopy (Fig. 2.2a) (Pyrc et al. 2007; Liu et al. 2020a). Closed within the lipid bilayer is the nucleocapsid (viral core) formed by nucleic acid and nucleocapsid protein (Liu et al. 2020a).

Like other hCoVs, SARS-CoV-2 is a single-stranded, non-segmented, positive-sense RNA virus. The genome of SARS-CoV-2 is approximately 30 kb, and the genome of hCoVs is the largest among all RNA viruses (Kim et al. 2020a). The genome of hCoVs is a conserved organization, consisting of a basic set of genes 5′-replicase-structure proteins-3′ (Fig. 2.3), and can be used to generate a genomic RNA (gRNA) and different sub-genomic RNAs (sgRNAs). These mRNAs all have a 5′-methyl-cap structure, a 3′poly A tail, and one to several open reading frames (ORFs) between them. 5′ Two-thirds of the genome contains ORF 1a and 1b, which share a small region of overlap. ORF1a and ORF1b are both derived from gRNA, coding for polyprotein 1a (PP1a) and Polyprotein 1b (PP1b). PP1a and PP1b undergo posttranslational cleavage by virus-encoded proteases – papain-like proteinase and

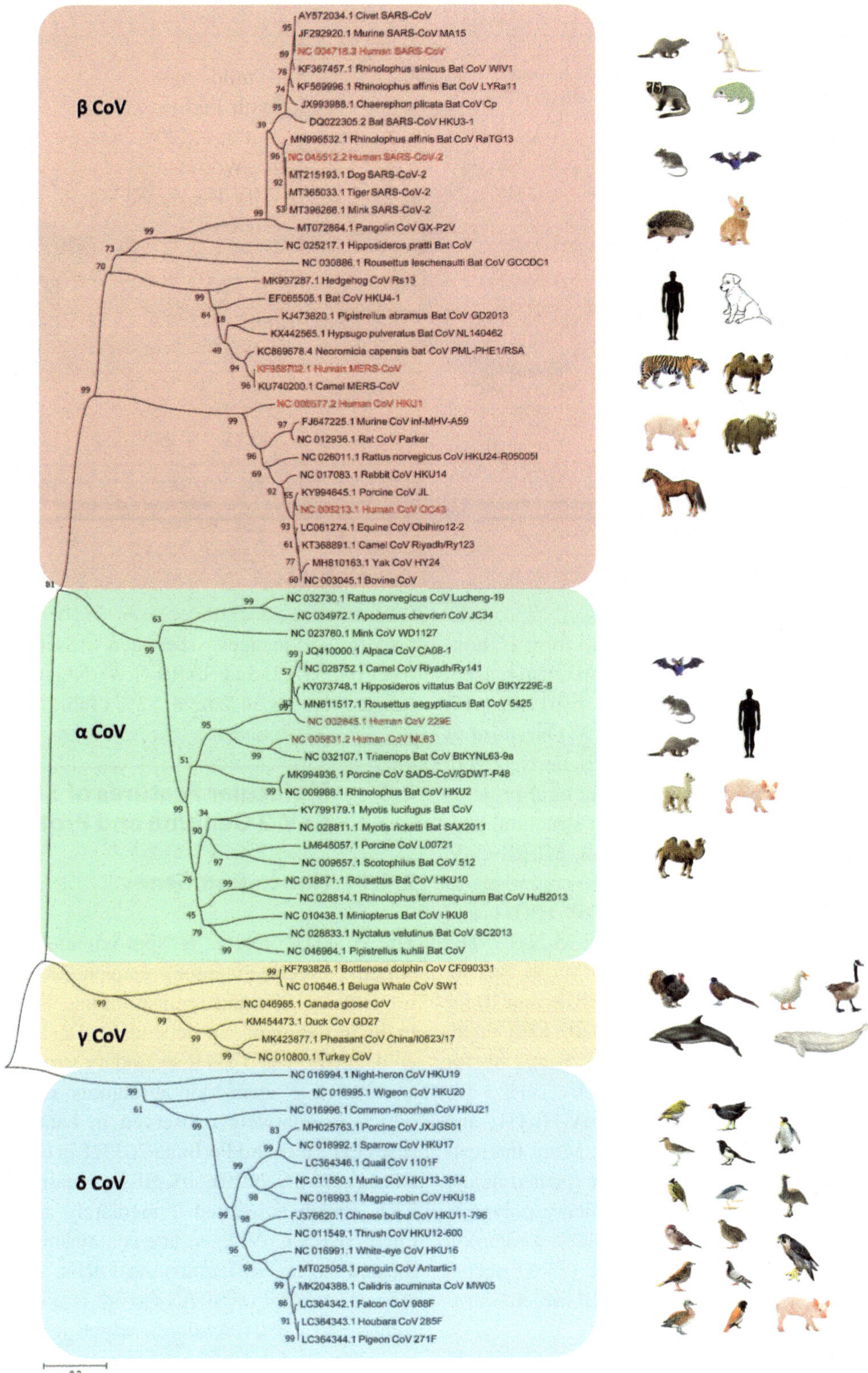

Fig. 2.1 Taxonomy of coronaviruses. The phylogenetic tree was inferred from complete RdRp sequences of the 76 coronaviruses, which were grouped into four genera (α, β, γ, δ). Coronaviruses originating from different hosts are listed on the right side. The tree was constructed by the neighbor-joining method with 1000 replicates bootstrapping using MEGA7. Human coronaviruses (SARS-CoV, SARS-CoV-2, MERS-CoV, HKU1, OC43, 229E, and NL63) are indicated in red

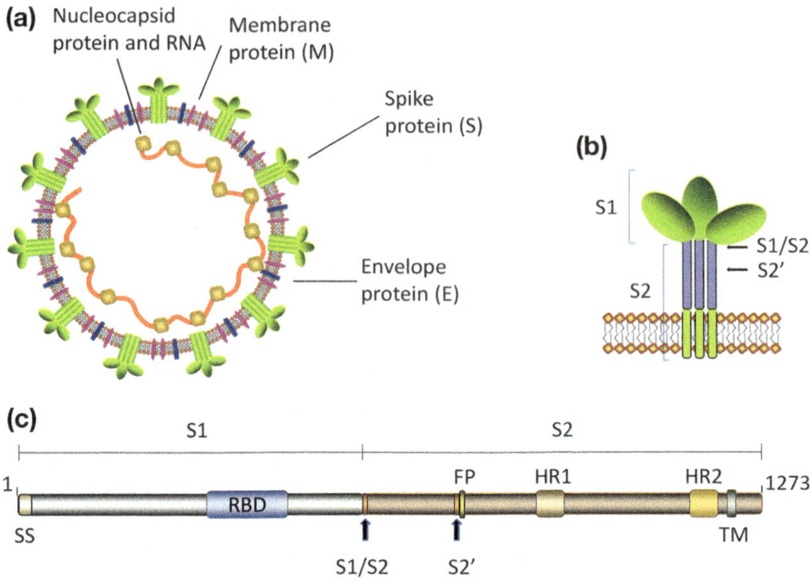

Fig. 2.2 SARS-CoV-2 structure and spike protein. (**a**) Diagram showing RNA genome and the four major structural proteins: spike protein (S), an envelope protein (E), membrane protein (M), and nucleocapsid protein (N). (**b**) Schematic of trimer spike protein structure. (**c**) Linear representation of monomeric spike protein

main protease (Mpro,3CLpro) – to form 16 non-structural proteins: nsp1, nsp2, nsp3, nsp4, nsp5, nsp6, nsp7, nsp8, nsp9, nsp10, nsp11, nsp12, nsp13, nsp14, nsp15, and nsp16. 3' One-third of the genome contains distinct ORFs, derived from different sgRNAs, coding for structural proteins and accessory proteins. The main structural proteins of SARS-CoV, SARS-CoV-2, MERS-CoV, HCoV-229E, and HCoV-NL63 comprise of 5'-S-E-M-N-3′, and that of HCoV-HKU1 and HCoV-OC43 contain 5'-HE-S-E-M-N-3′. The accessory proteins are 3a, 3b, 6, 7a, 7b, 8a, 8b, and 9b for SARS-CoV; 3a, 6, 7a, 7b, 8, and 10 for SARS-CoV-2 (Taiaroa et al. 2020; Kim et al. 2020a; Li et al. 2020); 3, 4a, 4b, 5, and 8b for MERS-CoV; 4a and 4b for HCoV-229E; 3 for HCoV-NL63; 4 and N2 for HCoV-HKU1; and ns2 and ns12.9 for HCoV-OC43. More interestingly, these accessory proteins are located in any positions downstream of the replicase polyprotein, except between E and M. Besides, a common transcription regulatory sequence (TRS) occurs upstream of the 5′ end of each ORF, which is crucial for sgRNA constitution.

SARS-CoV-2 is 80% identical to SARS-CoV, at the whole-genome level, which has been relatively systematically studied and may serve as a mirror for SARS-CoV-2 investigations (Zhou et al. 2020). Besides, the similarity of different protein sequences between SARS-CoV-2 (NC_045512.2) and SARS-CoV (NC_004718.3) ranges from 68.34% to 99.83% (Table 2.1).

2.4 Molecular Features of SARS-CoV-2 Genome and Proteins

2.4.1 Noncoding Genes

The 5′ UTR region is approximately 265-nt, which contains cis-acting sequences necessary for transcription and replication and is involved in the initiation of cap-dependent translation (Liu et al. 2007). The 5'UTR secondary structure consists of five stem-loop structures (SL1-SL5), which is relatively conserved in betacoronaviruses (Masters and Perlman 2013; Liu et al. 2007; Rangan et al. 2020a, b). SL3 contains a short TRS, which is located immediately adjacent to ORF. SARS-CoV-2 genome contains nine TRSs, located before different nine ORFs, separately (Fig. 2.3). TRS of SRAS-CoV-2 is a conserved sequence (5'-ACGAAC-3′), which is consistent with that of SARS-CoV (Masters and Perlman 2013). SL5 contains the AUG start codon of ORF1ab (Rangan et al. 2020b).

The 3' UTR region is approximately 229-nt, which possesses a polyadenylate tail and is

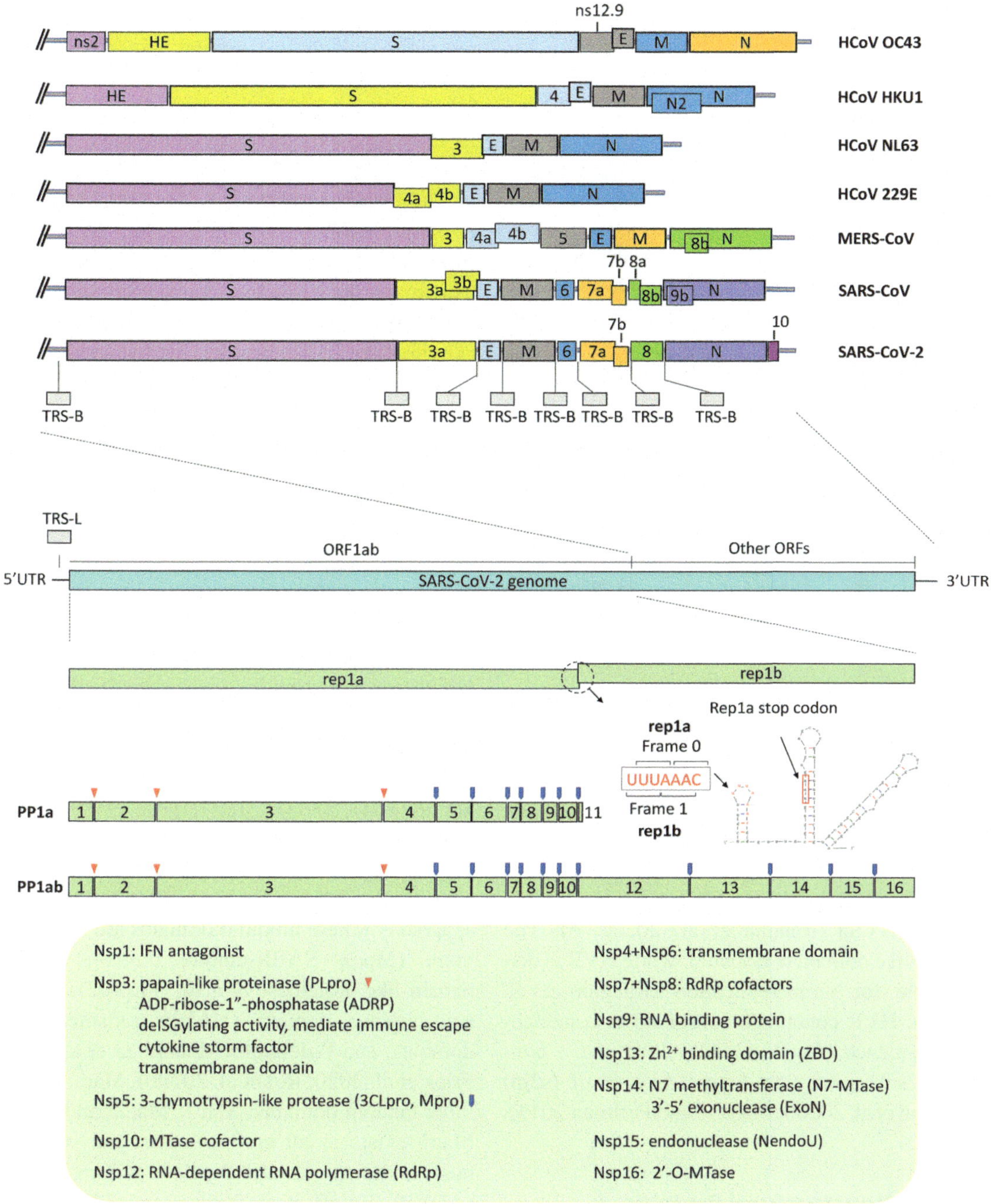

Fig. 2.3 Genome organization and protein products of SARS-CoV-2. A schematic of the complete genome of SARS-CoV-2 (NC_045512.2) is shown in the middle. Replicase genes and products are present at the bottom. Other genes from the seven human coronaviruses are placed on the top panel, which are OC43 (NC_006213.1), HKU1 (NC_006577.2), NL63 (NC_005831.2), 229E (NC_002645.1), MERS-CoV (NC_019843.3), SARS-CoV (NC_004718.3), and SARS-CoV-2 (NC_045512.2). The RNA structure of the frameshift motif was calculated using the UNAFold web server

Table 2.1 SARS-CoV-2 (NC_045512.2) protein features

SARS-CoV-2	Residues (amino acid)	Molecular size (KDa, without modifications)	Similarity with SARS-CoV in protein sequence level
nsp1	180	19.78	84.44%
nsp2	638	70.52	68.34%
nsp3	1945	217.28	75.97%
nsp4	500	56.19	80.00%
nsp5	306	33.8	96.08%
nsp6	290	33.04	88.15%
nsp7	83	9.24	98.80%
nsp8	198	21.89	97.47%
nsp9	113	12.38	97.35%
nsp10	139	14.79	97.12%
nsp11	13	1.33	84.62%
nsp12	932	106.67	96.35%
nsp13	601	66.86	99.83%
nsp14	527	59.82	95.07%
nsp15	346	38.82	88.73%
nsp16	298	33.33	93.29%
S	1273	141.2	75.96%
E	75	8.37	94.74%
M	222	25.15	90.54%
N	419	45.64	90.52%

considered important for virus replication and potentially translation (Rangan et al. 2020b; Taiaroa et al. 2020). The 3′ UTR includes switch-like domain (mutually exclusive formation of a pseudoknot and stem-loop) and hypervariable region (HVR) (Rangan et al. 2020a, b). The switch-like domain is essential, while HVR is dispensable for viral replication (Rangan et al. 2020b). HVR contains a conserved octa-nucleotide sequence (5'-GGAAGAGC-3′) and a conserved subregion stem-loop II-like motif (s2m) (Rangan et al. 2020b; Masters and Perlman 2013).

2.4.2 Nonstructural Proteins

The Nsp1 gene encodes a 19.78 kDa protein, which consists of approximately 180 amino acids. Through binding to 40S and 80S ribosomes and blocking mRNA entry channels, nsp1 can shutdown capped mRNA translation, including mRNAs coding for antiviral defense factors (Thoms et al. 2020). A C-terminal of nsp1 is crucial for ribosome binding and translation suppression, which is similar to SARS-CoV (Kamitani et al. 2009). Mutations (K164A/H165A) in this motif abrogate its binding capacity to the ribosome subunit (Thoms et al. 2020). It indicates that the K/H residues in this motif are crucial for ribosome binding and translation suppression, which is similar to SARS-CoV (Kamitani et al. 2009). This mechanism confers nsp1 to facilitate immune evasion through blocking type I interferon innate immune response (Thoms et al. 2020). Besides, nsp1 shares 88.44% identity with that of SARS-CoV at the protein level.

Nsp2 gene encodes a 638-amino-acid protein. It shares 68.34% identity with that of SARS-CoV at the protein level. The function of nsp2 has not been investigated in SARS-CoV-2 yet but reported in SARS-CoV. The deletion of the nsp2 gene in the SARS-CoV genome attenuates viral growth and RNA synthesis (Graham et al. 2005), which indicates that nsp2 is dispensable for viral replication and provides a new scenery of attenuated live vaccine design.

Nsp3 gene codes a 217.28 kDa multi-transmembrane protein that consists of 1945 amino acids and is the largest of replicase-transcriptase complex (RTC) proteins (Angeletti et al. 2020; Wu et al. 2020a). It contains multiple tandem domains, performing different functions, separately. These modular domains are macrodomain (Mac), SARS-unique domain (SUD), papain-like proteinase domain (PLpro2), nucleic acid-binding domain (NAB), transmembrane domains, and Y domain (Alhammad et al. 2020; Frick et al. 2020; Rut et al. 2020b). Mac contains three tandem domains: Mac1, Mac2, and Mac3. Mac1 exists in all coronaviruses and is previously known as ADP-ribose-1″-phosphatase (ADRP) (Egloff et al. 2006; Putics et al. 2005; Saikatendu et al. 2005). Mac1 functions as a highly efficient mono-adenosine diphosphate (ADP) ribosylhydrolase enzyme, which possesses the capacity of binding ADP-ribose and hydrolyzing single mono-ADP-ribose unit (post-translational modification) from protein substrate (Alhammad et al. 2020; Frick et al. 2020). Besides, a bioinformatic analysis suggested that

Mac may remove mono-ADP-ribosylation from the STAT1, which may be related to cytokine storm observed of COVID-19 (Claverie 2020). PLpro2 acts as a cysteine protease, which recognizes the LXGG motif in polyproteins 1a and 1ab between SARS-CoV-2 proteins nsp1 and nsp2, nsp2, and nsp3 and nsp3 and nsp4 (nsp1/nsp2, nsp2/nsp3, nsp3/nsp4) (Rut et al. 2020a). PLpro2 carries out a precise cleavage straight after this motif to separate nsp1, nsp2, and nsp3 from polyproteins (Rut et al. 2020a). Also, PLpro2 harbors the deISGylating activity to regulate immune evasion, similar to SARS-CoV and MERS-CoV (Clasman et al. 2020; Ratia et al. 2014; Mielech et al. 2014). The ISGylation is defined as conjugation of ISG15 to target proteins and mediate antiviral response, which can be reversed using deISGylation. This mechanism facilitates PLpro2 to block interferon-responsive factor 3 (IRF3) nuclear translocation and further attenuate type I interferon responses, which may be considered as an antiviral target (Shin et al. 2020). SARS-CoV PLpro2 has another scheme of hydrolyzing K48-linked Ub chains to mediate immune escape, in which SARS-CoV-2 PLpro2 does not possess (Rut et al. 2020a; Shin et al. 2020). The functions of other domains remain unclear and need further investigation.

Nsp4 is a 56.19 kDa protein that consists of 500 amino acids. This protein contains multiple transmembrane helices, anchoring in the intracellular membrane (Wu et al. 2020a). SARS-CoV-2 nsp4 is 80% identical with SARS-CoV nsp4 at the amino acid level. The function of nsp4 has not been investigated in SARS-CoV-2 yet but reported in SARS-CoV. SARS-CoV nsp4 binding with nsp3 is essential but not sufficient for membrane rearrangement, which is crucial for RNA replication (Sakai et al. 2017).

Nsp5 gene encodes a 33.8 kDa protein consisting of 306 amino acids. This protein is a conserved 3-chymotrypsin-like protease (3CLpro) and shares a 96.08% identity between SARS-CoV and SARS-CoV-2 at the amino acid level (Tahir ul Qamar et al. 2020; Kneller et al. 2020). 3CLpro is the main protease (Mpro) in cleaving polyprotein1a and polyprotein 1ab at 11 distinct sites to produce 13 nonstructural proteins: nsp4–nsp16 (Kneller et al. 2020; Zhang et al. 2020). Mpro consists of three domains: I, II, III (Kneller et al. 2020; Zhang et al. 2020; Jin et al. 2020a). Domains I/II (residues 8–184) are the catalytic domains, and domain III (residues 201–303) mediates the dimerization of nsp5, which is the necessary prerequisite to the catalytic activity (Zhang et al. 2020; Kneller et al. 2020). Based on its importance to viral replication, nsp5 can be designed as a drug target (Jin et al. 2020a, b; Zhang et al. 2020; Tahir ul Qamar et al. 2020; Elfiky et al. 2020).

Nsp6 is a 33.04 kDa protein that consists of 290 amino acids. This protein contains multiple transmembrane domains, together with nsp3 and nsp4, positioning the RTC in the intracellular membrane (Benvenuto et al. 2020; Wu et al. 2020a). SARS-CoV-2 nsp6 is 88.15% identical with SARS-CoV nsp6 at the amino acid level, and the function of nsp6 needs further lucubration.

Nsp7 is a 9.24 kDa protein that consists of 83 amino acids. Nsp8 is a 21.89 kDa protein that consists of 198 amino acids. Nsp7 and nsp8 mediate the formation of accessory subunits of RNA-dependent RNA polymerase (RdRp) (Peng et al. 2020; Hillen et al. 2020; Romano et al. 2020). Polymerase assay in vitro showed that the absence of either accessory subunit makes replication impossible (Peng et al. 2020). Besides, nsp7 and nsp8 are conserved between SARS-CoV-2 and SARS-CoV, with 98.8% and 97.47% identity at the amino acid level, respectively.

Nsp9 is a 12.38 kDa protein that consists of 113 amino acids. Nsp9 takes part in viral genome reproduction via binding single-stranded RNA (Littler et al. 2020). Even more remarkably, it interacts with NKRF (NF-KB repressor) and facilitates both IL-8 and IL-6 induction, which is involved in cytokine storm syndromes and elevated mortality in COVID-19 patients (Mehta et al. 2020; Li et al. 2020; Wang et al. 2020a). Besides, nsp9 shares 97.35% protein sequence identity and similar crystal structure between SARS-CoV-2 and SARS-CoV (Littler et al. 2020).

Nsp10 is a 14.79 kDa protein that consists of 139 amino acids. It is a zinc finger protein that can bind nonspecific RNA and interact with nsp16 as its cofactor (Rosas-Lemus et al. 2020; Li et al. 2020; Krafcikova et al. 2020). Nsp10 also interacts with nsp14 as its cofactor. More impressive is that nsp10 elevates both IL-8 and IL-6 production in lung epithelial A549 cells, reflecting that it is a potential virulence factor of hyper inflammation similar to nsp9 (Li et al. 2020). Besides, nsp10 is conserved between SARS-CoV-2 and SARS-CoV, with 97.12% identity at the protein sequence level. Following behind nsp10, nsp11 is an accessory remnant from polyprotein 1a cleavage. It is a 1.33 kDa peptide that consists of 13 amino acids and shares 84.62% identity with that of SARS-CoV at the protein sequence level.

Nsp12 is a 106.67 kDa protein that consists of 932 amino acids. It is a core catalytic subunit of the polymerase complex, which also needs the involvement of an nsp7-nsp8 heterodimer and an additional nsp8 subunit (Peng et al. 2020; Hillen et al. 2020). Nsp12, nsp7, and nsp8 are remarkably and equally crucial for polymerase activity because the absence of either will inactivate the enzyme capacity (Peng et al. 2020). Nsp12 contains an N-terminal nucleotidyltransferase (NiRAN) domain and a C-terminal RdRp domain, with an interface domain between them (Peng et al. 2020). The RdRp activity can be terminated in vitro by tenofovir and emtricitabine (two FDA-approved HIV drugs for pre-exposure prophylaxis, PrEP), which may be designated as a potential PrEP therapy against COVID-19 (Copertino Jr. et al. 2020; Jockusch et al. 2020). Besides, nsp12 is 96.35% identical to that of SARS-CoV at the protein sequence level.

Nsp13 is a 66.86 kDa protein that consists of 601 amino acids. Nsp13 is the most conserved protein between SARS-CoV-2 and SARS-CoV, with 99.83% identity at the amino acid level. Nsp13 of both viruses is a multifunctional protein, comprising an N-terminal Zn^{2+} binding domain (ZBD) and a C-terminal helicase domain (Mirza and Froeyen 2020; Romano et al. 2020). SARS-CoV nsp13 can unwind duplex RNA or DNA with a 5′ to 3′ directionality in an NTP-dependent manner, and this catalytic efficiency can be enhanced (twofold) by nsp12 (Adedeji et al. 2012a). It also contains 5′-triphosphatase activity, required for the first step of 5'cap synthesis (Adedeji et al. 2012b; Ivanov et al. 2004; Tanner et al. 2003). These functions of SARS-CoV-2 nsp13 need further investigation and confirmation.

Nsp14 is a 59.82 kDa protein that consists of 527 amino acids. Nsp14 is a conserved protein with a 95.07% identity between SARS-CoV-2 and SARS-CoV at the amino acid level. Nsp14 of each virus is a bifunctional enzyme with an N-terminal 3′-5′ exonuclease (ExoN) domain and a C-terminal guanine-N7 methyltransferase (N7-MTase) domain (Minskaia et al. 2006; Romano et al. 2020). ExoN can proofread the elongating RNA and excise the mismatched base (Ogando et al. 2019; Eckerle et al. 2010). N7-MTase is S-adenosylmethionine (SAM)-dependent methyltransferases responsible for gRNA and sgRNA m7GpppA cap synthesis (Krafcikova et al. 2020). SARS-CoV study showed that the nsp10 cofactor enhances the ExoN activity of nsp14 (>35-fold) without influence upon its N7-MTase activity (Bouvet et al. 2012). Yeast two-hybrid analysis provides evidence for SARS-CoV-2 nsp14-nsp10 interaction. However, the function behind the complex needs further investigation (Li et al. 2020).

Nsp15 is a 38.82 kDa protein that consists of 346 amino acids. It is an endonuclease (NendoU) composed of an N-terminal oligomerization domain, a middle domain, and a C-terminal NendoU catalytic domain (Kim et al. 2020b). NendoU activity is $Mn2^{+}$-dependent, which cuts single-stranded RNA substrates with specificity cleavage downstream of uridylate residues and releases 2′-3′ cyclic phosphates (Kim et al. 2020b; Bhardwaj et al. 2006). Also, the nsp15 protein sequence is relatively conserved, with 88.73% identity between SARS-CoV-2 and SARS-CoV.

Nsp16 is a 33.33 kDa protein that consists of 298 amino acids. It is an m7GpppA-specific, SAM-dependent 2'-O-MTase, only active with the help of cofactor nsp10 (Decroly et al. 2011; Krafcikova et al. 2020). Nsp16 protein sequence is conserved, with 93.29% identity between

SARS-CoV-2 and SARS-CoV. SARS-CoV nsp10/nsp16 MTase activity requires the involvement of Mg^{2+} (Bouvet et al. 2010), and the divalent cation is present in SARS-CoV-2 nsp16, outside the active site, with an undefined function (Decroly et al. 2011).

2.5 Structural Proteins and Accessory Proteins

The spike gene encodes a 141.20 kDa surface structural protein that consists of 1273 amino acids. Spike monomer protein will go through a process from oligomerization to trimerization, along with the glycosylation, which will enlarge the monomeric molecular size to approximately 250 kDa (Shang et al. 2020a). Mass spectrometry studies have shown that there are 22 N-linked glycan sites in the monomer, with each trimer displaying 66 N-linked glycosylation sites (Watanabe et al. 2020). The mature spike protein is a trimeric class I viral fusion protein containing two functional subunits (Watanabe et al. 2020) (Fig. 2.2b). Subunit 1 (S1) is responsible for receptor binding, while subunit 2 (S2) drives the membrane fusion between virion and host cell (Hoffmann et al. 2020; Watanabe et al. 2020). Removal of S1 crown is necessary for the S2 conformational changes, leading to membrane fusion (Walls et al. 2017).

Spike monomer is a transmembrane protein consisting of a large N-terminal ectodomain, transmembrane domain (TM), and a tiny C-terminal ectodomain (Fig. 2.2c). The N-terminal ectodomain includes different subdomains: SS (signal sequence), receptor-binding domain (RBD), S1/S2 cleavage site, S2'locus, fusion peptide (FP), heptad repeat 1 (HR1), and heptad repeat 2 (HR2) (Wrapp et al. 2020). The signal sequence mediates the protein inserted into the endoplasmic reticulum (ER) (Masters and Perlman 2013). RBD specifically binds to hACE2 as a virus receptor and determines its extensive potential host tropism (Wan et al. 2020; Guo et al. 2020; Liu et al. 2020b). The receptor-binding motif (RBM) is the cassette containing 14 residues that directly interact with hACE2 (Graham and Baric 2010). Changes in RBM can lead to efficient cross-species transmission (Graham and Baric 2010; Song et al. 2005; Li et al. 2005). hACE2 binding affinity of SARS-CoV-2 RBD is higher than that of SARS-CoV, while the entire SARS-CoV-2 spike has a comparable or lower binding affinity to hACE2 than SARS-CoV spike (Shang et al. 2020a). This inconsistent state is caused by RBD switchable heteromorphism between a standing-up position and a lying-down position. RBD of SARS-CoV is mostly in the standing-up state, which is more accessible to ACE2 receptor (Gui et al. 2017; Yuan et al. 2017); while RBD of SARS-CoV-2 spike is mostly in the lying-down state, which is less accessible and benefits immune evasion through epitope masking (Walls et al. 2020; Wrapp et al. 2020; Shang et al. 2020a). Although the veiled RBD may exhibit poor recognition for hACE2 and insufficient entrance into host cells, it evolved to possess two characteristics: an RBD with higher hACE2 binding affinity and a furin motif (S1/S2 cleavage site), allowing its spike to be preactivated (Shang et al. 2020a). S1/S2 cleavage site (Q_{677}TNSPRRAR↓SV_{687}) is recognized by the furin-like host like protease, which cuts S protein into similar-sized S1 and S2. Mutation of the S1/S2clevage site from Q677TNSPRRAR↓SV687 to Q677TILR↓SV683 can abrogate the furin-like protease cleavage. S2' is another cleavage site located in half of the S2 position, which is recognized and cleaved by cell surface transmembrane serine protease called TMPRSS2. Abrogating furin cleavage at S1/S2 sites, SARS-CoV-2 remains infectious by TMPRSS2 cleavage in S2' locus, similar to SARS-CoV (Walls et al. 2020). After the cleavage, hydrophobic FP is exposed and initiates viral and cellular membrane fusion (Tang et al. 2020; Ling et al. 2020). Heptad repeat region1 (HR1) and heptad repeat region 2 (HR2) then facilitate the termination procedure of membrane fusion (Tang et al. 2020; Xia et al. 2020; Ling et al. 2020). SARS-CoV-2 shows higher membrane fusion capacity than SARS-CoV (Xia et al. 2020). Also, viral-neutralizing epitopes exist within the spike protein, especially in RBD motif, which is sufficient to raise a protective antibody response in infected human patients (Wu et al. 2020b, c; Ku et al. 2020). This phenomenon provides potential opportunity to cure COVID-19.

The envelope gene (E) codes for an 8.37 kDa one transmembrane protein that consists of 75 amino acids (Bianchi et al. 2020; Thomas 2020). It is the smallest virion structural protein, which shares 94.74% identity with that of SARS-CoV at the protein sequence level. The computational analysis demonstrated that envelope protein of both viruses might function as a gated proton channel (Wilson et al. 2004; Sarkar and Saha 2020) and contains a C-terminal Bcl-2 homology 3 (BH3)-like motif, which is responsible for interaction with antiapoptotic host protein Bcl-xL and inducing apoptosis (Navratil et al. 2020).

The membrane (M) protein is a 25.15 kDa transmembrane protein that consists of 222 amino acids. Bioinformatic analyses showed that it contains a single three-transmembrane domain (Thomas 2020), and it shares a 90.54% identity with that of SARS-CoV at the protein sequence level. SARS-CoV-2 genome-wide yeast two-hybrid screens and co-immunoprecipitations (co-IP) experiments showed that M protein could interact with S and N proteins, which needs further investigations (Li et al. 2020).

Nucleocapsid phosphoprotein (N) is a 45.64 kDa protein that consists of 419 amino acids. It shares a 90.52% identity with that of SARS-CoV at the protein sequence level. The N protein of both viruses contains an N-terminal RNA binding domain and a C-terminal domain (Dinesh et al. 2020; Kang et al. 2020). The RNA binding domain of both viruses captures the RNA genome to form the ribonucleoprotein complex, and the C-terminal domain of SARS-CoV binding M protein then serves as a bridge for linking RNA genome to the viral membrane (Dinesh et al. 2020). The SARS-CoV-2 proteomic analysis demonstrated that N protein could interact with proteins M and E, and the mechanism of which needs intensive investigations (Li et al. 2020).

SARS-CoV-2 has six accessory proteins (3a, 6, 7a, 7b, 8, and 10), and SARS-CoV contains eight accessory proteins (3a, 3b, 6, 7a, 7b, 8a, 8b, and 9b). Previously study of SARS-CoV showed that accessory proteins are dispensable for RNA replication, while some of them are related to virus pathogenesis through modulation of interferon signaling pathways and production of pro-inflammatory cytokines (Liu et al. 2014). Although functions of accessory proteins of SRAS-CoV-2 are still indistinct, the proteomic survey demonstrated that some of these proteins could interact with host proteins, which provides insights into their purposes in the life cycle (Li et al. 2020).

2.6 SARS-CoV-2 Life Cycle

2.6.1 Virus Entry and Uncoating

Recently, significant progress has been made in understanding the critical steps of the SARS-CoV-2 life cycle, although many steps remain enigmatic (Fig. 2.4a). Once SARS-CoV-2 reaches the target cell, the infection cycle begins with the attachment and entry process, which is mediated by the interaction of spike glycoproteins with hACE2 receptor and the following membrane fusion (Shang et al. 2020a). Binding of the SARS-CoV S1 subunit to the hACE2 primes endocytosis of virus particle to form a double-membrane vesicle. Then, cleavage of S2' locus is mediated by lysosomal protease cathepsin L and the host cell serine protease TMPRSS2, co-localizing with and binding to hACE2 (Shulla et al. 2011; Walls et al. 2020). This cleavage process disassociates S1 from S2 and then facilitates S2 conformational rearrangement with exposure of hydrophobic fusion peptide within S2 trimer. This fusion peptide penetrates the cell membrane, which initiates the membrane fusion event between SARS-CoV-2 and acidified endosome. Meanwhile, S2 conformational change primes the interaction between HR1 and HR2 in each monomer, which forms an antiparallel helix-coiled bundle (Xia et al. 2020). The formation of this trimer of hairpins brings the juxtaposition of viral and cellular membranes closer, resulting in ultimate fusion and subsequent release of the viral core into the cytosol (Ling et al. 2020; Tang et al. 2020; Xia et al. 2020).

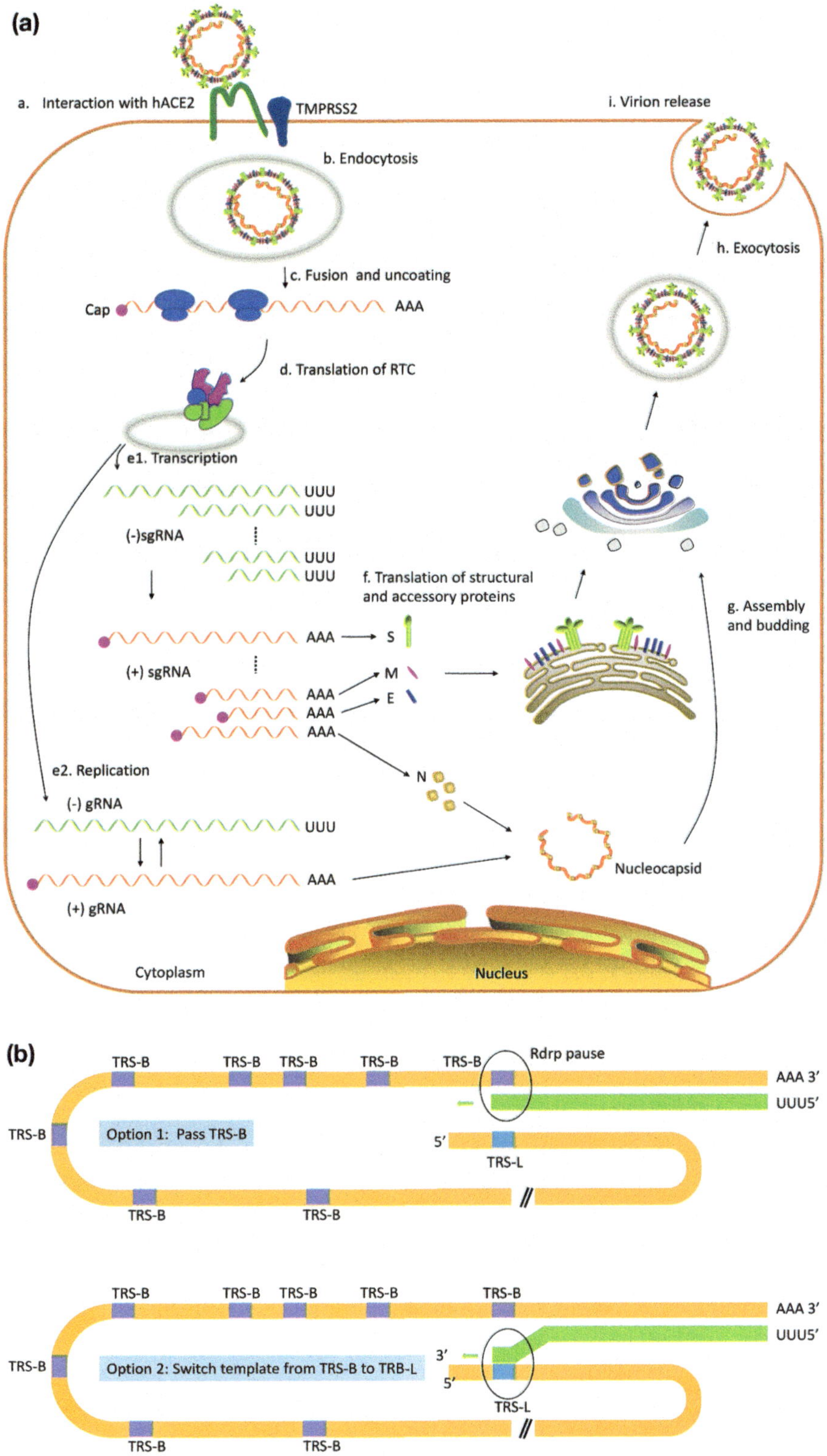

Fig. 2.4 Overview of SARS-CoV-2 life cycle (**a**) and transcription (**b**)

2.6.2 Viral Replicase Expression and RNA Synthesis

After the delivery of viral core into the cytosol, cap-dependent translation is initiated and gives rise to two polyproteins (pp1a and pp1ab) from the viral genome. Translation begins at 5'UTR and continues until the ribosome encounters the heptanucleotide slippery sequence 5'-UUUAAAC-3′ (Fig. 2.3) (Kelly and Dinman 2020). If the ribosome passes the heptamer sequence without frameshift, the translation will terminate at the UAA stop codon, which is 12-nt downstream of the slippery motif. If ribosome pauses in the slippery motif with codon frameshifting, the translation will terminate at another UAA stop codon, which is 8084-nt downstream of the frameshifting signals. The first strategy generates a truncated PP1a, and the latter produces PP1ab. PP1a was automatically hydrolyzed into 11 nonstructural proteins – nsp1–nsp11 – and PP1ab produces 15 nonstructural proteins: nsp1–nsp10 and nsp12–nsp16. The proteolysis was mediated by nsp3 (PLpro) and nsp5 (Mpro) (Rut et al. 2020a; Kneller et al. 2020; Tahir ul Qamar et al. 2020). PLpro performs separation of nsp1, nsp2, and nsp3, while Mpro executes the other protein isolation. The mature nsps assemble into RTC to carry out RNA synthesis.

RNA synthesis begins with the replication and transcription forming distinct negative-strand RNAs, which are used as templates for further synthesis of the capped gRNA and sgRNAs (Nomburg et al. 2020; Davidson et al. 2020; Kim et al. 2020a). The gRNA acts as a genome and the mRNA for coding RTC proteins, and sgRNAs function as mRNAs for encoding structural and accessory proteins. Each sgRNA contains a 5′end leader sequence (~70-nt) and a body RNA, which is identical to the genome segment 3′ (Kim et al. 2020a).

The negative-strand RNA synthesis begins with RTC at the 3′end of gRNA (Masters and Perlman 2013). When RdRp crosses TRS in the body (TRS-B), the elongation process pauses with two following options (Fig. 2.4b) (Kim et al. 2020a): RdRp may continue to elongate the negative-sense RNA intermediates. Alternatively, it may switch the template from TRS-B to TRS in the leader (TRS-L). This mechanism results in discontinuous negative-strand RNA synthesis and fusion of leader RNA to body RNA. Additionally, the negative-strand gRNA replication requires RdPR to pass through all TRS-B sites in gRNA. Each negative-strand RNA intermediate is equipped with a 5′ oligouridylate tracts and 3′ anti-leader, which then mediates the synthesis of one gRNA and eight sgRNAs (ORF S, ORF 3a/3b, ORF E, ORF M, ORF6, ORF 7a/7b, ORF8, ORF N) (Davidson et al. 2020; Nomburg et al. 2020; Kim et al. 2020a).

Beyond this canonical TRS-mediated template-switching mechanism, there exist other noncanonical mechanisms, similar to other coronaviruses (Kim et al. 2020a; Irigoyen et al. 2016; Viehweger et al. 2019). Sequencing analyses showed that minor noncanonical transcripts owe junction sites different from canonical transcripts, such as ORF7b and ORF10 without the TRS motif (Kim et al. 2020a; Nomburg et al. 2020; Davidson et al. 2020). The function of these noncanonical transcripts is still unclear, which needs further investigation.

2.6.3 Virion Assembly and Release

After transcription, sgRNAs immediately give rise to accessory and the structural proteins that build progeny viruses. The membrane-associated proteins S, E, and M are initially inserted into the ER and then transported into Golgi. Meanwhile, N protein integrates the viral genome to form nucleocapsid, which can coalesce with envelope components via interaction with M and E proteins (Liang et al. 2020). The assembled virions are then transported via vesicles and released out of the cells. In this process, spike trimer of mature virions may undergo proteolytical preactivation at S1/S2 cleavage site by furin for more accessible cell entrance (Shang et al. 2020b).

2.7 Conclusion

The novel SARS-CoV-2 is still raging worldwide, with a corrupting impact on human health. Specific antiviral drugs and efficient commercial vaccines for controlling and eradicating SARS-CoV-2 are at the preliminary stage of development. These developments require knowledge of essential viral characteristics. In the context of this chapter, we reviewed the classification of coronaviruses, compared the genome architecture of human coronaviruses, discussed the molecular features of the SARS-CoV-2 genome and proteins, and depicted SARS-CoV-2 life cycles. These are crucial for further understanding of the evolutionary and pathogenic mechanisms of SARS-CoV-2 and developing better strategies for diagnosis, prevention, and treatment of COVID-19.

References

Abdul-Rasool S, Fielding BC (2010) Understanding human coronavirus HCoV-NL63. Open Virol J 4:76–84. https://doi.org/10.2174/1874357901004010076

Adedeji AO, Marchand B, Te Velthuis AJW, Snijder EJ, Weiss S, Eoff RL, Singh K, Sarafianos SG (2012a) Mechanism of nucleic acid unwinding by SARS-CoV helicase. PLoS One 7(5):e36521–e36521. https://doi.org/10.1371/journal.pone.0036521

Adedeji AO, Singh K, Sarafianos SG (2012b) Structural and biochemical basis for the difference in the helicase activity of two different constructs of SARS-CoV helicase. Cell Mol Biol (Noisy-le-Grand) 58(1):114–121

Alhammad YMO, Kashipathy MM, Roy A, Johnson DK, McDonald P, Battaile KP, Gao P, Lovell S, Fehr AR (2020) The SARS-CoV-2 conserved macrodomain is a highly efficient ADP-ribosylhydrolase enzyme. bioRxiv:2020.2005.2011.089375. https://doi.org/10.1101/2020.05.11.089375

Angeletti S, Benvenuto D, Bianchi M, Giovanetti M, Pascarella S, Ciccozzi M (2020) COVID-2019: the role of the nsp2 and nsp3 in its pathogenesis. J Med Virol 92(6):584–588. https://doi.org/10.1002/jmv.25719

Assiri A, Al-Tawfiq JA, Al-Rabeeah AA, Al-Rabiah FA, Al-Hajjar S, Al-Barrak A, Flemban H, Al-Nassir WN, Balkhy HH, Al-Hakeem RF, Makhdoom HQ, Zumla AI, Memish ZA (2013) Epidemiological, demographic, and clinical characteristics of 47 cases of Middle East respiratory syndrome coronavirus disease from Saudi Arabia: a descriptive study. Lancet Infect Dis 13(9):752–761. https://doi.org/10.1016/S1473-3099(13)70204-4

Benvenuto D, Angeletti S, Giovanetti M, Bianchi M, Pascarella S, Cauda R, Ciccozzi M, Cassone A (2020) Evolutionary analysis of SARS-CoV-2: how mutation of non-structural protein 6 (NSP6) could affect viral autophagy. J Infect. https://doi.org/10.1016/j.jinf.2020.03.058

Bhardwaj K, Sun J, Holzenburg A, Guarino LA, Kao CC (2006) RNA recognition and cleavage by the SARS coronavirus endoribonuclease. J Mol Biol 361(2):243–256. https://doi.org/10.1016/j.jmb.2006.06.021

Bianchi M, Benvenuto D, Giovanetti M, Angeletti S, Ciccozzi M, Pascarella S (2020) Sars-CoV-2 envelope and membrane proteins: differences from closely related proteins linked to cross-species transmission? Preprints

Bouvet M, Debarnot C, Imbert I, Selisko B, Snijder EJ, Canard B, Decroly E (2010) In vitro reconstitution of SARS-coronavirus mRNA cap methylation. PLoS Pathog 6(4):e1000863

Bouvet M, Imbert I, Subissi L, Gluais L, Canard B, Decroly E (2012) RNA 3′-end mismatch excision by the severe acute respiratory syndrome coronavirus nonstructural protein nsp10/nsp14 exoribonuclease complex. Proc Natl Acad Sci 109(24):9372–9377. https://doi.org/10.1073/pnas.1201130109

Burrell CJ, Howard CR, Murphy FA (2017) Chapter 31 – coronaviruses. In: Burrell CJ, Howard CR, Murphy FA (eds) Fenner and White's medical virology, 5th edn. Academic, London, pp 437–446. https://doi.org/10.1016/B978-0-12-375156-0.00031-X

Chan JF, Lau SK, To KK, Cheng VC, Woo PC, Yuen K-Y (2015) Middle East respiratory syndrome coronavirus: another zoonotic betacoronavirus causing SARS-like disease. Clin Microbiol Rev 28(2):465–522

Clasman JR, Everett RK, Srinivasan K, Mesecar AD (2020) Decoupling deISGylating and deubiquitinating activities of the MERS virus papain-like protease. Antivir Res 174:104661

Claverie J-M (2020) The likely role of de-mono-ADP-ribosylation of STAT1 by the SARS-CoV-2 nsp3 protein in the cytokine storm syndrome of COVID-19

Copertino Jr DC, Lima B, Duarte R, Wilkin T, Gulick R, de Mulder Rougvie M, Nixon D (2020) Antiretroviral drug activity and potential for pre-exposure prophylaxis against COVID-19 and HIV infection

Davidson AD, Williamson MK, Lewis S, Shoemark D, Carroll MW, Heesom K, Zambon M, Ellis J, Lewis PA, Hiscox JA, Matthews DA (2020) Characterisation of the transcriptome and proteome of SARS-CoV-2 using direct RNA sequencing and tandem mass spectrometry reveals evidence for a cell passage induced in-frame deletion in the spike glycoprotein that removes the furin-like cleavage site. bioRxiv:2020.2003.2022.002204. https://doi.org/10.1101/2020.03.22.002204

Decroly E, Debarnot C, Ferron F, Bouvet M, Coutard B, Imbert I, Gluais L, Papageorgiou N, Sharff A, Bricogne G, Ortiz-Lombardia M, Lescar J, Canard B (2011) Crystal structure and functional analysis of the SARS-coronavirus RNA cap 2'-O-methyltransferase nsp10/nsp16 complex. PLoS Pathog 7(5):e1002059–e1002059. https://doi.org/10.1371/journal.ppat.1002059

Dinesh DC, Chalupska D, Silhan J, Veverka V, Boura E (2020) Structural basis of RNA recognition by the SARS-CoV-2 nucleocapsid phosphoprotein. bioRxiv:2020.2004.2002.022194. https://doi.org/10.1101/2020.04.02.022194

Eckerle LD, Becker MM, Halpin RA, Li K, Venter E, Lu X, Scherbakova S, Graham RL, Baric RS, Stockwell TB, Spiro DJ, Denison MR (2010) Infidelity of SARS-CoV Nsp14-exonuclease mutant virus replication is revealed by complete genome sequencing. PLoS Pathog 6(5):e1000896. https://doi.org/10.1371/journal.ppat.1000896

Egloff M-P, Malet H, Putics A, Heinonen M, Dutartre H, Frangeul A, Gruez A, Campanacci V, Cambillau C, Ziebuhr J (2006) Structural and functional basis for ADP-ribose and poly (ADP-ribose) binding by viral macro domains. J Virol 80(17):8493–8502

Elfiky A, Ibrahim N, Elshemey W (2020) Drug repurposing against MERS CoV and SARS-COV-2 PLpro in silico

Frick DN, Virdi RS, Vuksanovic N, Dahal N, Silvaggi NR (2020) Molecular basis for ADP-ribose binding to the macro-X domain of SARS-CoV-2 Nsp3. bioRxiv:2020.2003.2031.014639. https://doi.org/10.1101/2020.03.31.014639

Gaunt ER, Hardie A, Claas ECJ, Simmonds P, Templeton KE (2010) Epidemiology and clinical presentations of the four human coronaviruses 229E, HKU1, NL63, and OC43 detected over 3 years using a novel multiplex real-time PCR method. J Clin Microbiol 48(8):2940–2947. https://doi.org/10.1128/jcm.00636-10

Graham RL, Baric RS (2010) Recombination, reservoirs, and the modular spike: mechanisms of coronavirus cross-species transmission. J Virol 84(7):3134–3146. https://doi.org/10.1128/jvi.01394-09

Graham RL, Sims AC, Brockway SM, Baric RS, Denison MR (2005) The nsp2 replicase proteins of murine hepatitis virus and severe acute respiratory syndrome coronavirus are dispensable for viral replication. J Virol 79(21):13399–13411. https://doi.org/10.1128/jvi.79.21.13399-13411.2005

Gui M, Song W, Zhou H, Xu J, Chen S, Xiang Y, Wang X (2017) Cryo-electron microscopy structures of the SARS-CoV spike glycoprotein reveal a prerequisite conformational state for receptor binding. Cell Res 27(1):119–129

Guo H, Hu B-J, Yang X-L, Zeng L-P, Li B, Ouyang S-Y, Shi Z-L (2020) Evolutionary arms race between virus and host drives genetic diversity in bat SARS related coronavirus spike genes. bioRxiv:2020.2005.2013.093658. https://doi.org/10.1101/2020.05.13.093658

Hillen HS, Kokic G, Farnung L, Dienemann C, Tegunov D, Cramer P (2020) Structure of replicating SARS-CoV-2 polymerase. bioRxiv:2020.2004.2027.063180. https://doi.org/10.1101/2020.04.27.063180

Hoffmann M, Kleine-Weber H, Schroeder S, Krüger N, Herrler T, Erichsen S, Schiergens TS, Herrler G, Wu N-H, Nitsche A, Müller MA, Drosten C, Pöhlmann S (2020) SARS-CoV-2 cell entry depends on ACE2 and TMPRSS2 and is blocked by a clinically proven protease inhibitor. Cell 181(2):271–280.e278. https://doi.org/10.1016/j.cell.2020.02.052

Irigoyen N, Firth AE, Jones JD, Chung BY-W, Siddell SG, Brierley I (2016) High-resolution analysis of coronavirus gene expression by RNA sequencing and ribosome profiling. PLoS Pathog 12(2)

Ivanov KA, Thiel V, Dobbe JC, van der Meer Y, Snijder EJ, Ziebuhr J (2004) Multiple enzymatic activities associated with severe acute respiratory syndrome coronavirus helicase. J Virol 78(11):5619–5632

Jin Z, Du X, Xu Y, Deng Y, Liu M, Zhao Y, Zhang B, Li X, Zhang L, Peng C (2020a) Structure of M pro from SARS-CoV-2 and discovery of its inhibitors. Nature:1–5

Jin Z, Zhao Y, Sun Y, Zhang B, Wang H, Wu Y, Zhu Y, Zhu C, Hu T, Du X (2020b) Structural basis for the inhibition of SARS-CoV-2 main protease by antineoplastic drug Carmofur. Nat Struct Mol Biol:1–4

Jockusch S, Tao C, Li X, Anderson TK, Chien M, Kumar S, Russo JJ, Kirchdoerfer RN, Ju J (2020) Triphosphates of the two components in DESCOVY and TRUVADA are inhibitors of the SARS-CoV-2 polymerase. bioRxiv:2020.2004.2003.022939. https://doi.org/10.1101/2020.04.03.022939

Kamitani W, Huang C, Narayanan K, Lokugamage KG, Makino S (2009) A two-pronged strategy to suppress host protein synthesis by SARS coronavirus Nsp1 protein. Nat Struct Mol Biol 16(11):1134

Kang S, Yang M, Hong Z, Zhang L, Huang Z, Chen X, He S, Zhou Z, Zhou Z, Chen Q, Yan Y, Zhang C, Shan H, Chen S (2020) Crystal structure of SARS-CoV-2 nucleocapsid protein RNA binding domain reveals potential unique drug targeting sites. Acta Pharm Sin B. https://doi.org/10.1016/j.apsb.2020.04.009

Kelly JA, Dinman JD (2020) Structural and functional conservation of the programmed −1 ribosomal frameshift signal of SARS-CoV-2. bioRxiv:2020.2003.2013.991083. https://doi.org/10.1101/2020.03.13.991083

Kim D, Lee J-Y, Yang J-S, Kim JW, Kim VN, Chang H (2020a) The architecture of SARS-CoV-2 transcriptome. Cell 181(4):914–921.e910. https://doi.org/10.1016/j.cell.2020.04.011

Kim Y, Jedrzejczak R, Maltseva NI, Wilamowski M, Endres M, Godzik A, Michalska K, Joachimiak A (2020b) Crystal structure of Nsp15 endoribonuclease NendoU from SARS-CoV-2. Protein Sci n/a (n/a). https://doi.org/10.1002/pro.3873

Kneller DW, Phillips G, O'Neill HM, Jedrzejczak R, Stols L, Langan P, Joachimiak A, Coates L, Kovalevsky A (2020) Structural plasticity of the SARS-CoV-2 3CL M pro active site cavity revealed by room temperature X-ray crystallography

Krafcikova P, Silhan J, Nencka R, Boura E (2020) Structural analysis of the SARS-CoV-2 methyltransferase complex involved in coronaviral RNA cap creation. bioRxiv:2020.2005.2015.097980. https://doi.org/10.1101/2020.05.15.097980

Ku Z, Ye X, Toa Salazar G, Zhang N, An Z (2020) Antibody therapies for the treatment of COVID-19. Antibody Ther. https://doi.org/10.1093/abt/tbaa007

Li W, Zhang C, Sui J, Kuhn JH, Moore MJ, Luo S, Wong SK, Huang IC, Xu K, Vasilieva N (2005) Receptor and viral determinants of SARS-coronavirus adaptation to human ACE2. EMBO J 24(8):1634–1643

Li J, Guo M, Tian X, Liu C, Wang X, Yang X, Wu P, Xiao Z, Qu Y, Yin Y, Fu J, Zhu Z, Liu Z, Peng C, Zhu T, Liang Q (2020) Virus-host interactome and proteomic survey of PMBCs from COVID-19 patients reveal potential virulence factors influencing SARS-CoV-2 pathogenesis. bioRxiv:2020.2003.2031.019216. https://doi.org/10.1101/2020.03.31.019216

Liang Q, Li J, Guo M, Tian X, Liu C, Wang X, Yang X, Wu P, Xiao Z, Qu Y (2020) Virus-host interactome and proteomic survey of PMBCs from COVID-19 patients reveal potential virulence factors influencing SARS-CoV-2 pathogenesis. bioRxiv

Ling R, Dai Y, Huang B, Huang W, Yu J, Lu X, Jiang Y (2020) In silico design of antiviral peptides targeting the spike protein of SARS-CoV-2. Peptides 130:170328. https://doi.org/10.1016/j.peptides.2020.170328

Littler DR, Gully BS, Colson RN, Rossjohn J (2020) Crystal structure of the SARS-CoV-2 non-structural protein 9, Nsp9. bioRxiv:2020.2003.2028.013920. https://doi.org/10.1101/2020.03.28.013920

Liu P, Li L, Millership JJ, Kang H, Leibowitz JL, Giedroc DP (2007) A U-turn motif-containing stem-loop in the coronavirus 5′ untranslated region plays a functional role in replication. RNA 13(5):763–780. https://doi.org/10.1261/rna.261807

Liu DX, Fung TS, Chong KK, Shukla A, Hilgenfeld R (2014) Accessory proteins of SARS-CoV and other coronaviruses. Antivir Res 109:97–109. https://doi.org/10.1016/j.antiviral.2014.06.013

Liu C, Yang Y, Gao Y, Shen C, Ju B, Liu C, Tang X, Wei J, Ma X, Liu W (2020a) Viral architecture of SARS-CoV-2 with post-fusion spike revealed by Cryo-EM. bioRxiv

Liu Y, Hu G, Wang Y, Zhao X, Ji F, Ren W, Gong M, Ju X, Li C, Hong J, Zhu Y, Cai X, Wu J, Lan X, Xie Y, Wang X, Yuan Z, Zhang R, Ding Q (2020b) Functional and genetic analysis of viral receptor ACE2 orthologs reveals a broad potential host range of SARS-CoV-2. bioRxiv:2020.2004.2022.046565. https://doi.org/10.1101/2020.04.22.046565

Masters P, Perlman S (2013) Coronaviridae, pp 825–858. Fields virology 1

Mehta P, McAuley DF, Brown M, Sanchez E, Tattersall RS, Manson JJ (2020) COVID-19: consider cytokine storm syndromes and immunosuppression. Lancet 395(10229):1033–1034

Mielech AM, Kilianski A, Baez-Santos YM, Mesecar AD, Baker SC (2014) MERS-CoV papain-like protease has deISGylating and deubiquitinating activities. Virology 450–451:64–70. https://doi.org/10.1016/j.virol.2013.11.040

Minskaia E, Hertzig T, Gorbalenya AE, Campanacci V, Cambillau C, Canard B, Ziebuhr J (2006) Discovery of an RNA virus 3′→5′ exoribonuclease that is critically involved in coronavirus RNA synthesis. Proc Natl Acad Sci U S A 103(13):5108–5113. https://doi.org/10.1073/pnas.0508200103

Mirza MU, Froeyen M (2020) Structural elucidation of SARS-CoV-2 vital proteins: computational methods reveal potential drug candidates against main protease, Nsp12 polymerase and Nsp13 helicase. J Pharm Anal. https://doi.org/10.1016/j.jpha.2020.04.008

Navratil V, Lionnard L, Longhi S, Hardwick JM, Combet C, Aouacheria A (2020) The severe acute respiratory syndrome coronavirus 2 (SARS-CoV-2) envelope (E) protein harbors a conserved BH3-like sequence. bioRxiv:2020.2004.2009.033522. https://doi.org/10.1101/2020.04.09.033522

Nomburg J, Meyerson M, DeCaprio JA (2020) Noncanonical junctions in subgenomic RNAs of SARS-CoV-2 lead to variant open reading frames. bioRxiv:2020.2004.2028.066951. https://doi.org/10.1101/2020.04.28.066951

Ogando NS, Ferron F, Decroly E, Canard B, Posthuma CC, Snijder EJ (2019) The curious case of the Nidovirus exoribonuclease: its role in RNA synthesis and replication fidelity. Front Microbiol 10(1813). https://doi.org/10.3389/fmicb.2019.01813

Organization WH (2004) Summary of probable SARS cases with onset of illness from 1 November 2002 to 31 July 2003 based on data as of the 31 December 2003. Computer software] http://www.who.int/csr/sars/country/table2004_04_21/en/index.html

Peiris J, Guan Y, Yuen K (2004) Severe acute respiratory syndrome. Nat Med 10(12):S88–S97

Pene F, Merlat A, Vabret A, Rozenberg F, Buzyn A, Dreyfus F, Cariou A, Freymuth F, Lebon P (2003) Coronavirus 229E-related pneumonia in immunocompromised patients. Clin Infect Dis 37(7):929–932. https://doi.org/10.1086/377612

Peng Q, Peng R, Yuan B, Zhao J, Wang M, Wang X, Wang Q, Sun Y, Fan Z, Qi J, Gao GF, Shi Y (2020) Structural and biochemical characterization of nsp12-nsp7-nsp8 core polymerase complex from COVID-19 virus. bioRxiv:2020.2004.2023.057265. https://doi.org/10.1101/2020.04.23.057265

Putics A, Filipowicz W, Hall J, Gorbalenya AE, Ziebuhr J (2005) ADP-ribose-1″-monophosphatase: a conserved coronavirus enzyme that is dispensable for viral replication in tissue culture. J Virol 79(20):12721–12731

Pyrc K, Berkhout B, van der Hoek L (2007) The novel human coronaviruses NL63 and HKU1. J Virol 81(7):3051–3057

Rangan R, Watkins AM, Kladwang W, Das R (2020a) De novo 3D models of SARS-CoV-2 RNA elements and small-molecule-binding RNAs to guide drug discovery. bioRxiv:2020.2004.2014.041962. https://doi.org/10.1101/2020.04.14.041962

Rangan R, Zheludev IN, Das R (2020b) RNA genome conservation and secondary structure in SARS-CoV-2 and SARS-related viruses. bioRxiv:2020.2003.2027.012906. https://doi.org/10.1101/2020.03.27.012906

Ratia K, Kilianski A, Baez-Santos YM, Baker SC, Mesecar A (2014) Structural basis for the ubiquitin-linkage specificity and deISGylating activity of SARS-CoV papain-like protease. PLoS Pathog 10(5):e1004113

Romano M, Ruggiero A, Squeglia F, Maga G, Berisio R (2020) A structural view at SARS-CoV-2 RNA replication machinery: RNA synthesis, proofreading and final capping

Rosas-Lemus M, Minasov G, Shuvalova L, Inniss NL, Kiryukhina O, Wiersum G, Kim Y, Jedrzejczak R, Maltseva NI, Endres M, Jaroszewski L, Godzik A, Joachimiak A, Satchell KJF (2020) The crystal structure of nsp10-nsp16 heterodimer from SARS-CoV-2 in complex with S-adenosylmethionine. bioRxiv:2020.2004.2017.047498. https://doi.org/10.1101/2020.04.17.047498

Rota PA, Oberste MS, Monroe SS, Nix WA, Campagnoli R, Icenogle JP, Peñaranda S, Bankamp B, Maher K, Chen M-h, Tong S, Tamin A, Lowe L, Frace M, DeRisi JL, Chen Q, Wang D, Erdman DD, Peret TCT, Burns C, Ksiazek TG, Rollin PE, Sanchez A, Liffick S, Holloway B, Limor J, McCaustland K, Olsen-Rasmussen M, Fouchier R, Günther S, Osterhaus ADME, Drosten C, Pallansch MA, Anderson LJ, Bellini WJ (2003) Characterization of a novel coronavirus associated with severe acute respiratory syndrome. Science 300(5624):1394–1399. https://doi.org/10.1126/science.1085952

Rut W, Lv Z, Zmudzinski M, Patchett S, Nayak D, Snipas SJ, El Oualid F, Huang TT, Bekes M, Drag M, Olsen SK (2020a) Activity profiling and structures of inhibitor-bound SARS-CoV-2-PLpro protease provides a framework for anti-COVID-19 drug design. bioRxiv:2020.2004.2029.068890. https://doi.org/10.1101/2020.04.29.068890

Rut W, Zmudzinski M, Lyu Z, Nayak D, Snipas SJ, Bekes M, Huang TT, Olsen SK, Drag M (2020b) Activity profiling of SARS-CoV-2-PLpro protease provides structural framework for anti-COVID-19 drug design. bioRxiv:2020.2004.2029.068890. https://doi.org/10.1101/2020.04.29.068890

Saikatendu KS, Joseph JS, Subramanian V, Clayton T, Griffith M, Moy K, Velasquez J, Neuman BW, Buchmeier MJ, Stevens RC (2005) Structural basis of severe acute respiratory syndrome coronavirus ADP-ribose-1 ″-phosphate dephosphorylation by a conserved domain of nsP3. Structure 13(11):1665–1675

Sakai Y, Kawachi K, Terada Y, Omori H, Matsuura Y, Kamitani W (2017) Two-amino acids change in the nsp4 of SARS coronavirus abolishes viral replication. Virology 510:165–174. https://doi.org/10.1016/j.virol.2017.07.019

Sarkar M, Saha S (2020) Structural insight into the putative role of novel SARS CoV-2 E protein in viral infection: a potential target for LAV development and therapeutic strategies. bioRxiv:2020.2005.2011.088781. https://doi.org/10.1101/2020.05.11.088781

Shang J, Wan Y, Luo C, Ye G, Geng Q, Auerbach A, Li F (2020a) Cell entry mechanisms of SARS-CoV-2. Proc Natl Acad Sci:202003138. https://doi.org/10.1073/pnas.2003138117

Shang J, Wan Y, Luo C, Ye G, Geng Q, Auerbach A, Li F (2020b) Cell entry mechanisms of SARS-CoV-2. Proc Natl Acad Sci 117(21):11727–11734

Shi Z, Hu Z (2008) A review of studies on animal reservoirs of the SARS coronavirus. Virus Res 133(1):74–87. https://doi.org/10.1016/j.virusres.2007.03.012

Shin D, Mukherjee R, Grewe D, Bojkova D, Baek K, Bhattacharya A, Schulz L, Widera M, Mehdipour A, Tascher G (2020) Inhibition of papain-like protease PLpro blocks SARS-CoV-2 spread and promotes antiviral immunity

Shulla A, Heald-Sargent T, Subramanya G, Zhao J, Perlman S, Gallagher T (2011) A transmembrane serine protease is linked to the severe acute respiratory syndrome coronavirus receptor and activates virus entry. J Virol 85(2):873–882

Song H-D, Tu C-C, Zhang G-W, Wang S-Y, Zheng K, Lei L-C, Chen Q-X, Gao Y-W, Zhou H-Q, Xiang H (2005) Cross-host evolution of severe acute respiratory syndrome coronavirus in palm civet and human. Proc Natl Acad Sci 102(7):2430–2435

Tahir Ul Qamar M, Alqahtani SM, Alamri MA, Chen L-L (2020) Structural basis of SARS-CoV-2 3CLpro and anti-COVID-19 drug discovery from medicinal plants. J Pharm Anal. https://doi.org/10.1016/j.jpha.2020.03.009

Taiaroa G, Rawlinson D, Featherstone L, Pitt M, Caly L, Druce J, Purcell D, Harty L, Tran T, Roberts J (2020) Direct RNA sequencing and early evolution of SARS-CoV-2. bioRxiv

Tang T, Bidon M, Jaimes JA, Whittaker GR, Daniel S (2020) Coronavirus membrane fusion mechanism offers a potential target for antiviral development. Antivir Res 178:104792–104792. https://doi.org/10.1016/j.antiviral.2020.104792

Tanner JA, Watt RM, Chai Y-B, Lu L-Y, Lin MC, Peiris JM, Poon LL, Kung H-F, Huang J-D (2003) The severe acute respiratory syndrome (SARS) coronavirus NTPase/helicase belongs to a distinct class of 5′ to 3′ viral helicases. J Biol Chem 278(41):39578–39582

Thomas S (2020) The structure of the membrane protein of SARS-CoV-2 resembles the sugar transporter semiSWEET

Thoms M, Buschauer R, Ameismeier M, Koepke L, Denk T, Hirschenberger M, Kratzat H, Hayn M, Mackens-Kiani T, Cheng J, Stürzel CM, Fröhlich T, Berninghausen O, Becker T, Kirchhoff F, Sparrer KMJ, Beckmann R (2020) Structural basis for translational shutdown and immune evasion by the Nsp1 protein of SARS-CoV-2. bioRxiv:2020.2005.2018.102467. https://doi.org/10.1101/2020.05.18.102467

Vabret A, Mourez T, Gouarin S, Petitjean J, Freymuth F (2003) An outbreak of coronavirus OC43 respiratory infection in Normandy, France. Clin Infect Dis 36(8):985–989. https://doi.org/10.1086/374222

Viehweger A, Krautwurst S, Lamkiewicz K, Madhugiri R, Ziebuhr J, Hölzer M, Marz M (2019) Direct RNA nanopore sequencing of full-length coronavirus genomes provides novel insights into structural variants and enables modification analysis. Genome Res 29(9):1545–1554

Walls AC, Tortorici MA, Snijder J, Xiong X, Bosch B-J, Rey FA, Veesler D (2017) Tectonic conformational changes of a coronavirus spike glycoprotein promote membrane fusion. Proc Natl Acad Sci 114(42):11157–11162

Walls AC, Park Y-J, Tortorici MA, Wall A, McGuire AT, Veesler D (2020) Structure, function, and antigenicity of the SARS-CoV-2 spike glycoprotein. Cell

Wan Y, Shang J, Graham R, Baric RS, Li F (2020) Receptor recognition by the novel coronavirus from Wuhan: an analysis based on decade-long structural studies of SARS coronavirus. J Virol 94(7). https://doi.org/10.1128/JVI.00127-20

Wang S, Chen S, Lu H, Zhang X, Tan Y, Ling Y, Lu G, Liu F, Yi Z, Jia X (2020a) Viral and host factors related to the clinic outcome of the SARS-CoV-2 infection

Wang T, Du Z, Zhu F, Cao Z, An Y, Gao Y, Jiang B (2020b) Comorbidities and multi-organ injuries in the treatment of COVID-19. Lancet 395(10228):e52. https://doi.org/10.1016/S0140-6736(20)30558-4

Wang W, Xu Y, Gao R, Lu R, Han K, Wu G, Tan W (2020c) Detection of SARS-CoV-2 in different types of clinical specimens. JAMA. https://doi.org/10.1001/jama.2020.3786

Watanabe Y, Allen JD, Wrapp D, McLellan JS, Crispin M (2020) Site-specific analysis of the SARS-CoV-2 glycan shield. bioRxiv:2020.2003.2026.010322. https://doi.org/10.1101/2020.03.26.010322

WHO (2020a) MERS-CoV case report. https://www.who.int/emergencies/mers-cov/en/

WHO (2020b) SARS-CoV case report. https://www.who.int/csr/don/archive/disease/severe_acute_respiratory_syndrome/en/

Wilson L, Mckinlay C, Gage P, Ewart G (2004) SARS coronavirus E protein forms cation-selective ion channels. Virology 330(1):322–331

Woo PC, Lau SK, Tsoi HW, Huang Y, Poon RW, Chu CM, Lee RA, Luk WK, Wong GK, Wong BH, Cheng VC, Tang BS, Wu AK, Yung RW, Chen H, Guan Y, Chan KH, Yuen KY (2005) Clinical and molecular epidemiological features of coronavirus HKU1-associated community-acquired pneumonia. J Infect Dis 192(11):1898–1907. https://doi.org/10.1086/497151

Wrapp D, Wang N, Corbett KS, Goldsmith JA, Hsieh CL, Abiona O, Graham BS, McLellan JS (2020) Cryo-EM structure of the 2019-nCoV spike in the prefusion conformation. Science 367(6483):1260–1263. https://doi.org/10.1126/science.abb2507

Wu C, Liu Y, Yang Y, Zhang P, Zhong W, Wang Y, Wang Q, Xu Y, Li M, Li X, Zheng M, Chen L, Li H (2020a) Analysis of therapeutic targets for SARS-CoV-2 and discovery of potential drugs by computational methods. Acta Pharm Sin B. https://doi.org/10.1016/j.apsb.2020.02.008

Wu Y, Li C, Xia S, Tian X, Wang Z, Kong Y, Gu C, Zhang R, Tu C, Xie Y (2020b) Fully human single-domain antibodies against SARS-CoV-2. bioRxiv

Wu Y, Wang F, Shen C, Peng W, Li D, Zhao C, Li Z, Li S, Bi Y, Yang Y, Gong Y, Xiao H, Fan Z, Tan S, Wu G, Tan W, Lu X, Fan C, Wang Q, Liu Y, Zhang C, Qi J, Gao GF, Gao F, Liu L (2020c) A noncompeting pair of human neutralizing antibodies block COVID-19 virus binding to its receptor ACE2. Science:eabc2241. https://doi.org/10.1126/science.abc2241

Xia S, Liu M, Wang C, Xu W, Lan Q, Feng S, Qi F, Bao L, Du L, Liu S, Qin C, Sun F, Shi Z, Zhu Y, Jiang S, Lu L (2020) Inhibition of SARS-CoV-2 (previously 2019-nCoV) infection by a highly potent pan-coronavirus fusion inhibitor targeting its spike protein that harbors a high capacity to mediate membrane fusion. Cell Res 30(4):343–355. https://doi.org/10.1038/s41422-020-0305-x

Yang X, Yu Y, Xu J, Shu H, Ja X, Liu H, Wu Y, Zhang L, Yu Z, Fang M, Yu T, Wang Y, Pan S, Zou X, Yuan S, Shang Y (2020) Clinical course and outcomes of critically ill patients with SARS-CoV-2 pneumonia in Wuhan, China: a single-centered, retrospective, observational study. Lancet Respir Med 8(5):475–481. https://doi.org/10.1016/S2213-2600(20)30079-5

Yuan Y, Cao D, Zhang Y, Ma J, Qi J, Wang Q, Lu G, Wu Y, Yan J, Shi Y (2017) Cryo-EM structures of MERS-CoV and SARS-CoV spike glycoproteins reveal the dynamic receptor binding domains. Nat Commun 8:15092

Zhang L, Lin D, Sun X, Curth U, Drosten C, Sauerhering L, Becker S, Rox K, Hilgenfeld R (2020) Crystal structure of SARS-CoV-2 main protease provides a basis for design of improved α-ketoamide inhibitors. Science 368(6489):409–412. https://doi.org/10.1126/science.abb3405

Zheng Y-Y, Ma Y-T, Zhang J-Y, Xie X (2020) COVID-19 and the cardiovascular system. Nat Rev Cardiol 17(5):259–260

Zhou P, Yang XL, Wang XG, Hu B, Zhang L, Zhang W, Si HR, Zhu Y, Li B, Huang CL, Chen HD, Chen J, Luo Y, Guo H, Jiang RD, Liu MQ, Chen Y, Shen XR, Wang X, Zheng XS, Zhao K, Chen QJ, Deng F, Liu LL, Yan B, Zhan FX, Wang YY, Xiao GF, Shi ZL (2020) A pneumonia outbreak associated with a new coronavirus of probable bat origin. Nature 579(7798):270–273. https://doi.org/10.1038/s41586-020-2012-7

Ecology and Evolution of Betacoronaviruses

3

Eduardo Rodríguez-Román and Adrian J. Gibbs

Abstract

The crown-like outline of the virions of coronaviruses will long endure as the iconic image of 2020 – the year of the COVID-19 pandemic. This major human health emergency has been caused by a betacoronavirus, as have others in the past. In this chapter, we outline the taxonomy of betacoronaviruses and their properties, both genetic and biological. We discuss their recombinational and mutational histories separately to show that the sequence of the RaTG13 bat virus isolate is the closest currently known full-length genetic homolog of that of the severe acute respiratory syndrome-related coronavirus 2 (SARS-CoV-2). However, the RaTG13 bat virus and SARS-CoV-2 have probably diverged over 20 years. We discuss the ecology of their pangolin and bat hosts and conclude that, like other recent viral pandemics, the underlying cause of the SARS-CoV-2 emergence is probably the relentless growth of the world's human population and the overexploitation and disturbance of the environment.

E. Rodríguez-Román (✉)
Center for Microbiology and Cell Biology, Instituto Venezolano de Investigaciones Científicas, Caracas, Venezuela

A. J. Gibbs
Emeritus Faculty, Australian National University, Canberra, Australia

Keywords

Betacoronaviruses · Coronaviruses · Ecology · Emergence · Evolution · Phylogeny · SARS-CoV-2

3.1 Introduction

Viruses are the most abundant biological entities around the globe (Suttle 2007). A recently emerged virus called severe acute respiratory syndrome coronavirus 2 (SARS-CoV-2) causes the human coronavirus disease 2019 (COVID-19). SARS-CoV-2 belongs to the species *Severe acute respiratory syndrome-related coronavirus* in the subgenus *Sarbecovirus*, genus *Betacoronavirus* of the family *Coronaviridae* (Gorbalenya et al. 2020a).

As described by the International Committee on Taxonomy of Viruses (ICTV), the family *Coronaviridae* is divided into 2 subfamilies, 5 genera, 26 subgenera, and 46 species (ICTV online). However, only members of the genera *Alphacoronavirus* and *Betacoronavirus* have been reported to infect humans, as shown in Fig. 3.1. There are three subgenera of betacoronaviruses, and these contain 14 species officially recognized by the ICTV.

N. Rezaei (ed.), *Coronavirus Disease - COVID-19*, Advances in Experimental Medicine and Biology 1318, https://doi.org/10.1007/978-3-030-63761-3_3

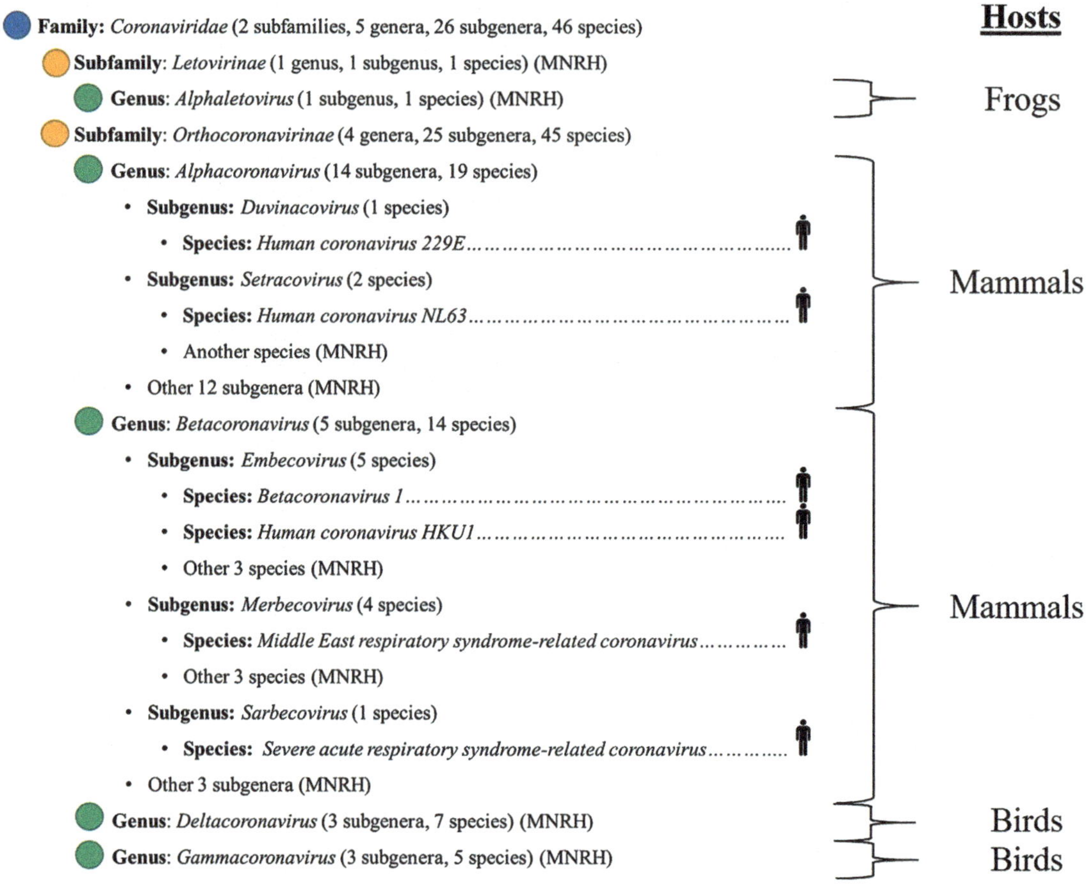

Fig. 3.1 Current taxonomy of the family *Coronaviridae*. Dots represent the taxon family (blue), subfamily (yellow), and genus (green). Coronavirus species reported in humans (person symbol). MNRH stands for taxon or taxa, where members have not been reported in humans. (Prepared with data from ICTV)

Coronaviruses have single-stranded positive-sense RNA genomes that are the largest known for RNA viruses (Anthony et al. 2017).

3.1.1 The Emergence of an Infectious Disease

The emergence of an infectious disease depends on the pathogen, the host, and their environment (Agrios 2005). In recent work (Frutos et al. 2020), the three required conditions for the emergence of human infectious diseases were defined as follows: (i) the pathogen must be able to infect and reproduce in human beings; (ii) there must be interactions between human beings and the pathogen reservoir or the intermediary host; and (iii) there must be the possibility of the human-human infection. The present chapter provides an overview of the discovery of human coronaviruses, their origins, natural reservoirs, and the ability of coronaviruses to cause spillovers and host switching.

3.1.2 A Brief History of the Discovery of Human Coronaviruses

The first coronavirus was discovered in 1937 (Beaudette 1937) and called avian infectious bronchitis virus, a member of the then genus *Coronavirus*, as seen in the first report of the ICTV in 1971 (Fenner 1975). It was associated with avian bronchitis disease and caused death in chick embryos (Beaudette 1937; Fabricant 1998). This virus is now classified as the type of the species *Avian coronavirus* (genus *Gammacoronavirus*) (ICTV online). The first two coronaviruses of humans were discovered in 1965 during an electron microscopy study of the causes of common respiratory diseases of humans. The virions of two viruses were morphologically similar to those of the coronavirus previously reported in chickens, avian infectious bronchitis virus (Almeida and Tyrrell 1967). The coronaviruses OC43 (species *Betacoronavirus 1*) and 229E (species *Human coronavirus 229E*) were thus the first coronaviruses (hCoVs) to be shown to cause disease in humans (Almeida and Tyrrell 1967). June Almeida, the codiscoverer of these first hCoVs, suggested the name "coronavirus" as each virion appeared to have a crown of knob-shaped projections from its surface when viewed under an electron microscope, and thus the name of this group of viruses was established (Henry 2020).

About 40 years passed before another coronavirus that causes disease in humans was discovered. It was SARS-CoV (species *Severe acute respiratory syndrome-related coronavirus*), the etiological agent of severe acute respiratory syndrome (SARS), which emerged in 2002 and caused severe pneumonia and death in some individuals (Drosten et al. 2003). SARS-CoV was the first hCoV to cause an international health emergency, as declared by the World Health Organization (WHO). It was transmitted at a moderate, but significant, rate from human to human and associated with a case fatality rate (CFR) of 9.6% as reported in July 2003 (Mahase 2020). Fortunately, the SARS-CoV was controlled by quarantine and rigorous tracing of contacts. After the SARS epidemic, two other hCoVs were identified almost at the same time, NL63 (species *Human coronavirus NL63*) in 2004, causing croup and bronchiolitis (Fouchier et al. 2004; van der Hoek et al. 2004), and HKU1 (species *Human coronavirus HKU1*) in 2005, causing pneumonia (Woo et al. 2005). The sixth human coronavirus was MERS-CoV (species *Middle East respiratory syndrome-related coronavirus*), which emerged in 2012 (Zaki et al. 2012), the etiological agent of MERS and the second hCoV to cause a health emergency. MERS-CoV caused 858 deaths between 2012 and 2019, with a CFR of 34% (Mahase 2020).

Finally, in December of 2019, the emergence of a new coronavirus causing pneumonia in Wuhan, China, was announced (Wu et al. 2020). Although more infectious than SARS-CoV, its case fatality rate was much lower. Initially, the virus was called 2019-nCoV, but it is now known as SARS-CoV-2 (Gorbalenya et al. 2020a), the etiologic agent of the disease COVID-19, and is the seventh hCoV to be reported. COVID-19 was declared as a pandemic by the WHO on March 11, 2020. As of May 31, 2020, at least 6.2 million people have been infected, and more than 372,000 of these have died (WHO 2020; Worldometer coronavirus website 2020).

3.2 Origin and Evolution of SARS-CoV-2 and Other Betacoronaviruses

All viruses with RNA genomes, riboviruses, are currently placed in the viral realm *Riboviria*, a taxon defined by their RNA-dependent RNA polymerase gene (Gorbalenya et al. 2020b). A distinctive characteristic of most riboviruses is their ability to generate genetic diversity quickly and, therefore, to evolve rapidly (Domingo 2019). They can form large populations and have short generation times (Duffy et al. 2008; González-Candelas et al. 2018).

Organisms, including viruses, evolve by the mutation of individual nucleotides in their genes and also by genetic recombination, in which genome regions of coinfecting viruses are exchanged. The most evident clues about the origins of any virus come from phylogenetic comparisons of its genome sequence with those of other related viruses. Point mutations are of primary value for phylogenetic analysis, whereas recombination often confuses the interpretation of such analyses.

In mid-May 2020, we downloaded from the GenBank and GISAID databases all the full-length coronavirus genomes closest to that of SARS-CoV-2, together with a representative selection of related coronaviruses. It was performed by BLAST searches of the databases using the SARS-CoV-2 Wuhan-Hu-1 sequence (NC_045512) as a query. Over 100 full-length matching sequences were identified and aligned using MAFFT with its L option. A neighbor-joining (NJ) phylogeny of these sequences identified eight genomic sequences closest to NC_045512 in the SARS-CoV-2 lineage. Eleven others, more distant, included the SARS-CoV reference sequence (NC_004718), and an outgroup of ten other coronavirus genomes included that of MERS-CoV. These 30 representative betacoronavirus sequences (Table 3.1) were aligned and checked for recombination using the Recombination Detection Program (RDP) 4.95 (Martin et al. 2015). Recombination was detected in, and between, all betacoronaviruses, but not between them and the outgroup sequences. The 10 outgroup sequences were therefore removed, and the remaining 20 sequences were realigned using MAFFT-L and BioEdit (Hall 1999) to create an alignment of concatenated open reading frames (ORFs) (concats).

The phylogeny of the complete sarbecovirus concats (Fig. 3.2a) shows that they form into two lineages, SARS-1 and SARS-2, with the individual tips at a range of distances from the midpoint root, a sign of recombination and topology that is often not supported statistically. Homologous recombination is widely accepted as playing an essential role in the evolution, emergence, and

Table 3.1 Coronavirus sequences compared phylogenetically

Accession code	Host	Isolate	Country
EF065505	Bat	HKU41	China
EPI_ISL_410721	Pangolin	Guangdong/1/2019	China
EPI_ISL_412977	Bat	RmYN02	China
GQ153540	Bat	HKU3-5	HK
GQ153543	Bat	HKU3-8	HK
KF636752	Bat	Zhejiang/2013	China
KU762337	Bat-*Rous*	GCCDC1 346	China
KX442565	Bat-*Hypsugo*	HKU25/NL140462	China
KY417145	Bat	Rf4092	China
KY417148	Bat	Rs4247	China
KY417152	Bat	Rs9401	China
KY674941	Human	HKU1/N091663B	USA
KY770859	Bat	Anlong112	China
KY770860	Bat	Jiyan84	China
MG772933	Bat	ZC45	China
MG772934	Bat	ZXC21	China
MK211374	Bat	SC2018	China
MK211376	Bat	YN2018B	China
MK211377	Bat	YN2018C	China
MN481964	Human	Riyadh_58_2014	Saudi
MN996528	Human	WIV04 (SARS-CoV-2)	China
MN996532	Bat	RaTG13	China
MT040333	Pangolin	GX-P4L	China
MT040336	Pangolin	GXP5E	China
NC_004718	Human	Tor2 (SARS-CoV)	Canada
NC_009021	Bat	HKU91	China
NC_030886	Bat-*Rous*	GCCDC1 356	China
NC_034440	Bat	PDF2180	Uganda
NC_039207	Bat	Erinaceus/VMC/DEU/2012	Germany
NC_045512	Human	Wuhan-Hu-1 (SARS-CoV-2)	China

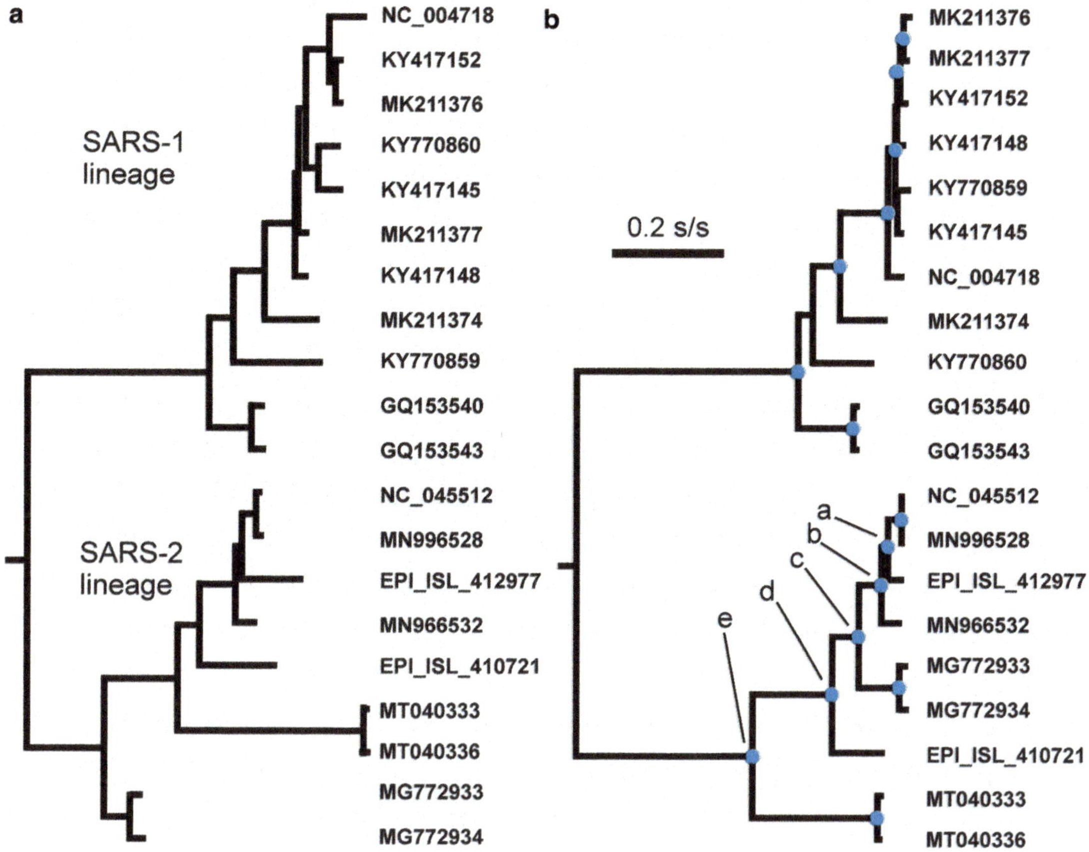

Fig. 3.2 ML phylogenies of 20 sarbecovirus concat sequences using (**a**) the complete concat sequences and (**b**) only nucleotides 1–11,502. The nodes marked "a"–"e" are discussed, and nodes with blue disk have >0.99 SH support

epidemiology of viruses, especially in those with positive RNA genomes, and can lead to rapid changes in genetic diversity (Holmes 2009; Knipe and Howley 2013; Simon-Loriere and Holmes 2011). It has been shown to contribute to changes and expansion of viral host ranges, increases of virulence, acquisition of new hosts by evading host immunity, and resistance to antivirals (Simon-Loriere and Holmes 2011). The alignment of SARS-1 and SARS-2 lineage concats was therefore checked for recombinants using RDP 4.95.

The recombination map (Fig. 3.3) shows that the two SARS-CoV-2 concats (from NC_045512 and MN996528) together with those of RmYN02 and RaTG13 (EPI_ISL_412977 and MN996532) share the same two recombinant regions ("a" and "b"), both of them related to SARS-1 lineage bat viruses. Significantly, however, the concat of RmYN02 has additional recombinant regions most closely related to regions from two other coronaviruses. The positions and statistical support for these recombinant regions are shown in Table 3.2. Further recombinant regions were found in the other SARS-CoV-2 lineage concats, but, significantly, none were in the 5′ terminal third (1-11502 nts) of the sequences. Thus, importantly, this region was available to obtain a phylogeny of the SARS-CoV-2 lineage based on point mutations alone and not confounded by recombination.

Figure 3.2b shows the maximum likelihood (ML) phylogeny calculated from nts 1-11502 of the 20 SARS-1 and SARS-2 lineage concats. It can be seen that, compared with Fig. 3.2a phylogeny, its tips are now at similar distances from the root; the SARS-CoV-2s, RmYN02, and RaTG13 concats now form the sister group to

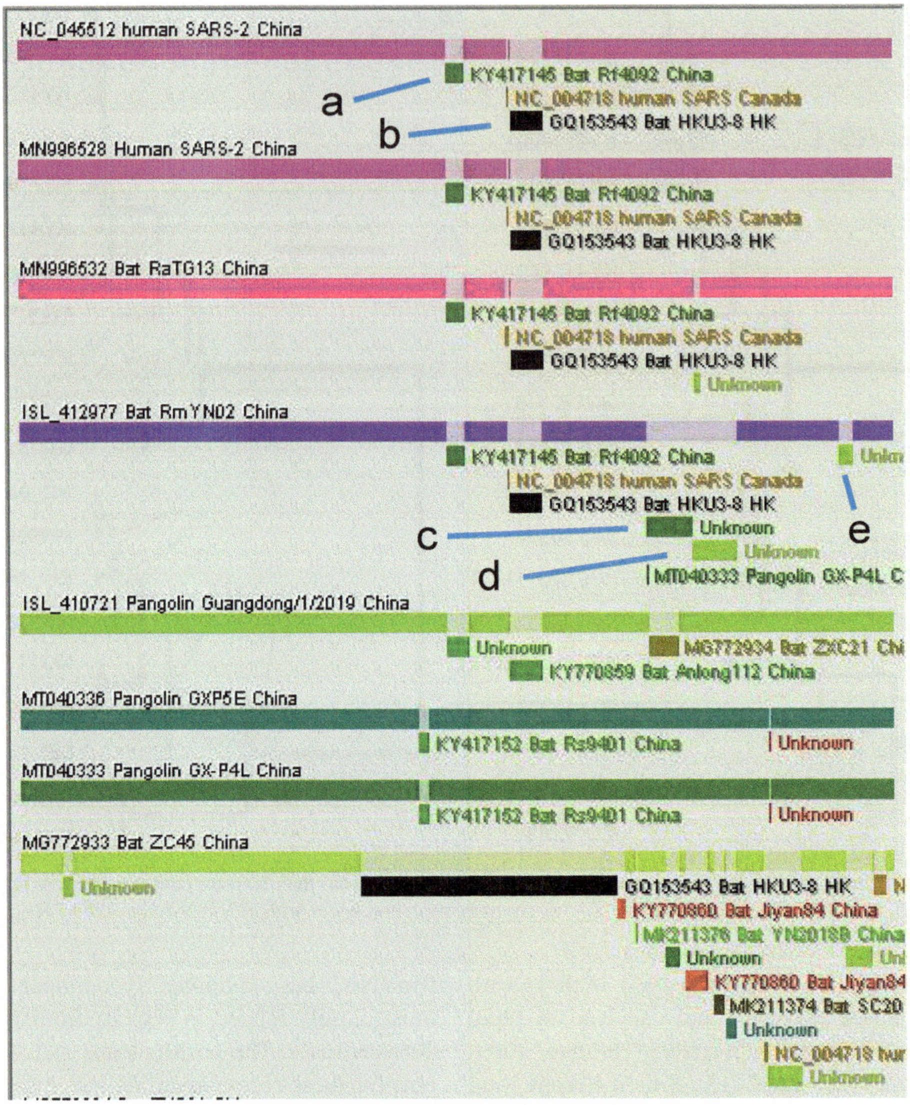

Fig. 3.3 Screenshot of part of the recombination map of the RDP analysis of 20 concats from the same SARS-1 and SARS-2 lineage genomes shown in Fig. 3.2a. The details of the recombinant regions in four of the concats closest to, and including, that of SARS-CoV-2 are shown in Table 3.2

those of the two other bat sequences, MG772933 and MG772934 (ZC45 and ZXC21), and concats of the three isolates from pangolins (EPI_ISL_410721, MT040333, and MT040336) form an outgroup. The two recombinant regions found in the four concats, SARS-CoV-2s, EPI-ISL_412977, and MN996532, were not found in the MG772933, MG772934, or more distant sequences and so must have been acquired by the SARS-CoV-2 progenitor between nodes "b" and "c" (Fig. 3.2b). The additional recombinant regions found in the RmYN02 concat were not present in the other SARS-CoV-2 concats and so must have been acquired after the three diverged at node "a" (Fig. 3.2b).

It is noteworthy that recombinants "a" and "b" had smaller Consensus Recombinant Score (CRS) values (i.e., support) than recombinants "c," "d," and "e"; as predicted by the phylogeny, the first two were acquired before the others. Also

Table 3.2 Positions and support for the recombinant regions of the SARS-CoV-2, RaTG13, and RmYN02 sequences shown in Fig. 3.3

	Position of the recombinant region		Consensus Recombinant Scores (CRS)				
				Major parent		Minor parent	
Region (Fig. 3.3)	Start (nts)	End (nts)	Recombinant CRS	Sequence	CRS	Sequence	CRS
a	14405	15024	0.353	MT040333	0.329	KY417145	0.317
b	16549	17613	0.393	MT040336	0.373	GQ153543	0.234
c	21040	22598	0.745	MN996528	0.135	MT040333	0.120
d	22644	24078	0.810	MN996528	0.034	ISL_410721	0.156
e	27550	28016	0.697	MN996528	0.162	ISL_410721	0.141

interesting is the fact that minor "parents" (i.e., the nearest sequence tested) of recombinant "a" and "b" were coronaviruses of bats from the SARS-1 lineage, whereas those of "c," "d," and "e" were all coronaviruses from pangolins of the SARS-2 lineage. However, it is essential to realize that, although those were the hosts from which they were isolated, the viruses may have infected other hosts en route.

3.2.1 Genome Evolution

Mutation is the fundamental evolutionary mechanism driving diversification in biological entities. Although RNA virus populations generally have high mutation rates, not all mutations in a virus population have the same fate. Some mutations are lost over time by natural selection or genetic drift, while others increase in frequency and eventually become fixed in the virus population (Pagán 2018). Generally, the rate of fixation determines the long-term rate of evolution (Pagán 2018; Vandamme 2009) and results from the combined effects of mutation rates, generation times, effective population size, and fitness in viruses (Duffy et al. 2008), and it can be calculated using Bayesian phylogenetic inference or regression methods (Bouckaert et al. 2019).

The mutation rates of RNA viruses are in the range from 1×10^{-3} to 1×10^{-6} mutations/nucleotide/replication round (m/n/r) (González-Candelas et al. 2018), and generally, this results in a rate of between 1×10^{-2} and 1×10^{-5} nucleotide substitutions/site/year (s/s/y) (Duffy et al. 2008; González-Candelas et al. 2018; Hanada et al. 2004; Jenkins et al. 2002). These rates result from low-fidelity copying by their RdRps, which lack a proofreading mechanism (Drake and Holland 1999). Although coronaviruses possess the largest genomes (26–32 kb) among riboviruses (Anthony et al. 2017; Gorbalenya et al. 2006), they have lower mutation rates than others as they have proofreading activity provided by the 3′ exoribonuclease (ExoN) (Minskaia et al. 2006; Zhang and Holmes 2020). Currently, the coronaviruses are the only group of RNA viruses known to have polymerases with this type of exonuclease activity (Sanjuán and Domingo-Calap 2019).

The genetic inactivation of ExoN activity in engineered SARS-CoV genomes results in viable mutants that have a 15- to 20-fold increases in mutation rates, up to 18 times greater than those tolerated for fidelity mutants of other RNA viruses, and thus this protein is essential for replication fidelity and coronavirus genetic diversity (Denison et al. 2011; Gribble et al. 2020). ExoN has been determined as the first (non-RdRp) viral protein involved in the sensitivity of RNA viruses to mutagens and could be considered as a good target for therapeutics in coronaviruses (Smith et al. 2013) by, for example, inducing lethal mutagenesis in coronavirus populations (Perales et al. 2011).

The effect of ExoN is to produce a mutation rate for SARS-CoV genomes of around 9.06×10^{-7} substitutions/site/replication round (s/n/r), i.e., three orders of magnitude less than the average of other RNA viruses (1×10^{-4}) (Cuevas et al. 2009; Drake 1991; Eckerle et al. 2010), and more like that of some ssDNA viruses such as the bacterial virus ΦX174 (Eckerle et al. 2010; Sanjuán et al. 2010). However, when ExoN is inactivated, the mutation rate for SARS-CoV increases to around 1.26×10^{-5} s/n/r (Eckerle et al. 2010), which is closer to that of other RNA viruses and near the limit as the virus population cannot fix/accumulate more, so it suffers an error catastrophe (Belshaw et al. 2011; Duffy et al. 2008; Sanjuán and Domingo-Calap 2019).

Knowledge of the rates of virus evolution allows the times of coalescence (or divergence) of taxa or lineages, and the date of their most recent common ancestor (tMRCA), to be inferred. Note that although all mutations contribute to the "molecular clock" of a population, most of the most recent mutations are quickly lost (Duchêne et al. 2014), so that the older a population is, the more diverse it will be, and its average evolutionary rate will be less.

Short-term rates of evolution of the current populations have been estimated for the five human betacoronaviruses (Table 3.3). It has been estimated that the tMRCA for the hCoVs OC43

Table 3.3 Evolutionary rate estimates of human betacoronaviruses

Virus	Evolution rate (x10^{-3} s/s/y)	References
SARS-CoV-2	1.04 (0.71–1.40)	Virological website
SARS-CoV	0.80–2.38	Zhao et al. (2004)
MERS-CoV	0.63 (0.14–1.1)	Cotten et al. (2013)
	1.12 (0.88–1.37)	Cotten et al. (2014)
	0.96 (0.83–1.09)	Dudas et al. (2018)
hCoV-OC43	0.43 (0.27–0.60)	Vijgen et al. (2005)
hCoV-HKU1	0.62 (0.42–0.78)[a]	Al-Khannaq et al. (2016)

Prepared with data from http://virological.org/

[a]Only the S gene

and HKU1 is around 120 and 70 years before present (ybp), respectively (Al-Khannaq et al. 2016; Forni et al. 2017; Vijgen et al. 2005), and the tMRCA for hCoVs SARS-CoV and MERS-CoV has been estimated to be around 1985–1998 and 2006, respectively (Forni et al. 2017). Similarly, the tMRCA date for the divergence of SARS-CoV-2 and RaTG13 or RmYN02 was around 40–70 ypb and 37.02 ybp (18.19–55.85), respectively (Boni et al. 2020; Nielsen et al. 2020). Likewise, if all viruses of the SARS-2 lineage (Fig. 3.2b) are evolving at the same rate as that of the complete SARS-CoV-2 genome (1.04×10^{-3} s/s/y) (Table 3.3), one can estimate the dates of their divergences. The nonrecombinant 1-11502 nts region of the SARS-CoV-2 genome is evolving 9.2% faster than the complete genome. Thus, it has an evolutionary rate of 1.135×10^{-3} s/s/y. Therefore, using the pairwise evolutionary (i.e., patristic) distances between the SARS-2 lineage viruses, the nodes marked "a" to "e" in Fig. 3.2b are dated as 1996.8 CE (23.2 years before present; ybp), 1987.3 CE (32.7 ybp), 1945.1 CE (74.9 ybp), 1919.3 CE (100.7 ybp), and 1800 CE (220 ybp), respectively. The standard deviation of the branch length estimates was less than 2%, and the most recent estimates are probably the most accurate because, as explained above, all mutations contribute to the "molecular clock," but most are quickly lost (Duchêne et al. 2014), and therefore times to the older dates are overestimated. However, likely, the two recombinant regions characteristic of the SARS-CoV-2s, EPI_ISL_412977, and MN996532 sequences were acquired by their shared progenitor more than 20 years ago!

It is noteworthy that the tMRCA for all members of the subfamily *Orthocoronavirinae* is estimated to be around 300 million years ago, which is believed to coincide with the separation of the classes Mammalia and Aves (Forni et al. 2017). It means that the evolutionary rate of the coronaviruses since their origins is many orders of magnitude smaller than the evolutionary rate of their extant populations.

As already described, coronaviruses can undergo homologous recombination between viral genomes of the same species (intraspecific recombination) (Gorbalenya et al. 2020a; Lai 1996; Luk et al. 2019; Tao et al. 2017). Recombination in genomes from members grouped in different coronavirus species (interspecific recombination) has been suggested to play an important role in the diversification of this family (Tao et al. 2017); note that the "minor parent" of recombinants "a" and "b" in Table 3.2 are in the SARS-1 lineage, but the more recent ones are all from the SARS-2 lineage. The frequency of recombination in betacoronaviruses has been determined to be 25% or more in the entire genome. It is the highest determining recombination frequency for non-segmented single-stranded positive RNA viruses (Baric et al. 1990). Recombination is not just an experimental artifact and is frequent under natural conditions (Decaro et al. 2009; Herrewegh et al. 1998; Hon et al. 2008; Kahn and McIntosh 2005; Lau et al. 2010, 2011), and as a result, the recombination of the spike protein gene may be the crucial event that changes the host range of coronaviruses (El-Duah et al. 2019).

Coronaviruses with closely similar sequences to human isolates have been found in wild animals and may be a source of human isolates (Azhar et al. 2014; Corman et al. 2016; Guan et al. 2003; Tao et al. 2017). The coronavirus that is genetically most similar to SARS-CoV-2 is RaTG13 (96.1% nt identity) (Zhou et al. 2020), which was isolated from a bat. Recently, the RmYN02 coronavirus was also isolated from a bat, but even though it shares the highest nt identity in the nonstructural region (ORF 1ab), 97.2%, its RBD region has only 61.3% of nt identity with the SARS-CoV-2 region (Zhou et al. 2020), as it is recombinant in that region (Fig. 3.3).

The sequence of coronavirus Guangdong/1/2019, isolated from a Malayan pangolin (*Manis javanica*) has an nt identity of 91.02% with the whole genome, but interestingly its RBD region shares 97.4% of aa identity with SARS-CoV-2, including six key residues in the site probably involved in the interaction between the subunit S1 of the spike protein and the cell receptor in humans (ACE2) (Andersen et al. 2020; Lam et al. 2020). The 12 nts insertion in the SARS-CoV-2 genome encoding -PRRA-, between the S1 and S2 subunits, makes it a site for cleavage by furin and other proteases, and it is not found in other sarbecoviruses (Andersen et al. 2020). This distinctive feature has contributed to conspiracy theories about the origin of SARS-CoV-2. The recombination history we report above for SARS-CoV-2 and its nearest relatives resolves those suggested by others (Lau et al. 2020; Xiao et al. 2020) and indicates that the most reliable comparison of full-length homologs is between the SARS-CoV-2 and RaTG13 sequences, and the slightly closer comparison between the SARS-CoV-2 and RmYN02 sequences is only valid for nts 1-11502.

We, therefore, compared them directly (Fig. 3.4). It can be seen that there are more synonymous (S) than non-synonymous (NS) changes (van Dorp et al. 2020), but both are distributed genome-wide. As expected, most of the differences, especially NS ones, are in the spike protein gene, especially its RBD region and an adjacent "-PRRA- " insertion (Andersen et al.

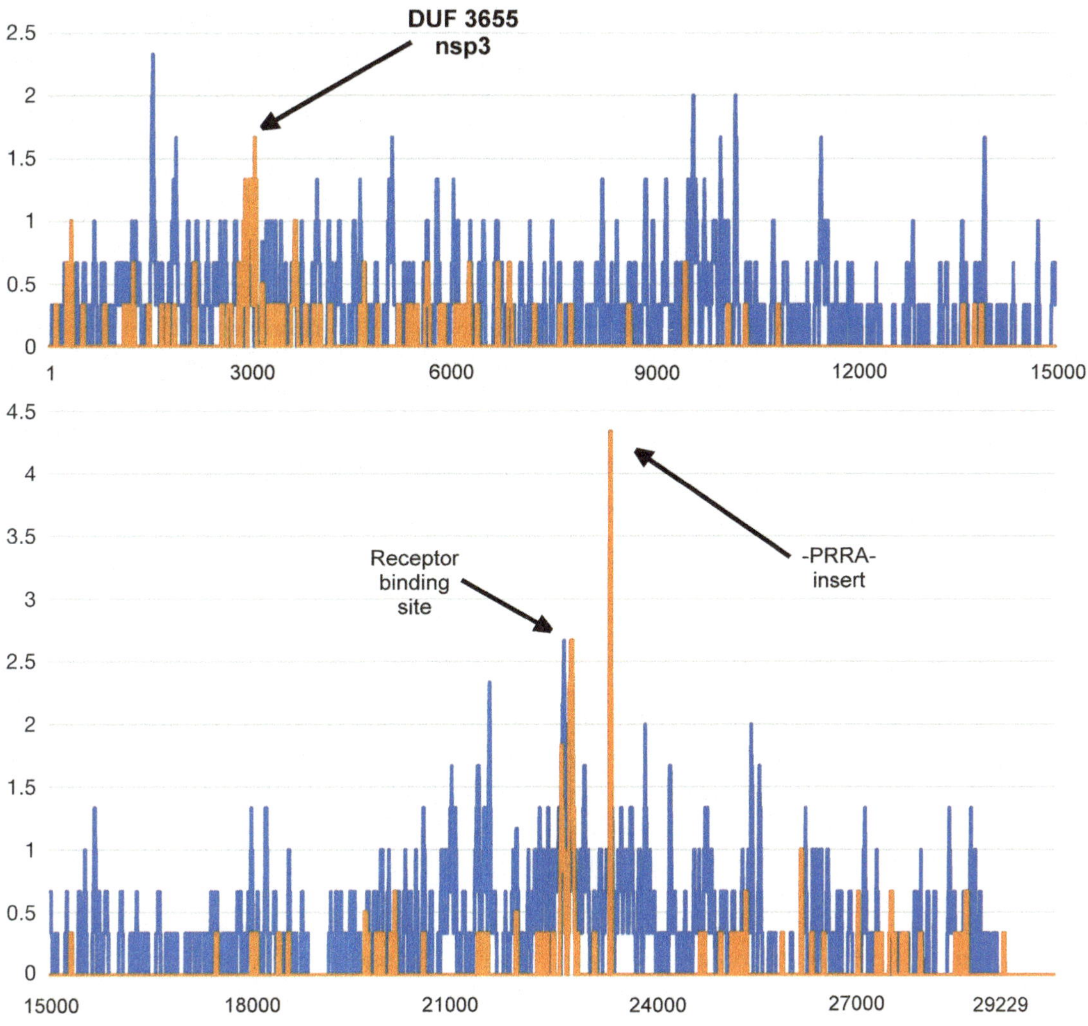

Fig. 3.4 A histogram that scans the S/NS differences between the SARS-CoV-2 and RaTG13 concat sequences using the DnDscan method (Gibbs et al. 2007). The sequences were scanned codon by codon, and their S/NS differences determined one nucleotide position at a time, before calculating sliding running sums for ten codons at each codon position. S and NS differences are in blue and orange, respectively, with their lengths indicating the score (NB the "PRRA" insert is of four NS differences)

2020), but there is also a region of NS differences around codon 1000, which seems to have evaded scrutiny so far. It is in the nsp3 region and named "DUF (domain of unknown function) 3655"; it is a "disordered binding region" (Prates et al. 2020) and is N′ terminally adjacent to the ADP-ribose phosphatase. Most of the amino acid differences in that region are conservative, but notably, three involve proline and would therefore change the shape of the protein!

3.3 Ecology of Betacoronaviruses

3.3.1 Natural Hosts of Coronaviruses

In a previous section, we briefly reviewed the discovery of human coronaviruses. Only seven coronaviruses are currently known to infect and cause diseases in humans. However, 46 coronavirus species are recognized by the ICTV. So, what are the other coronavirus species? Where are they? What are their natural hosts?

Coronaviruses infect a wide range of animals, mainly birds, and mammals, including humans, pets, and livestock. Only members in the genus *Alphaletovirus* (subfamily *Letovirinae*) have been reported to infect animals other than birds and mammals (Bukhari et al. 2018), and alphacoronaviruses and betacoronaviruses only infect mammals (Cui et al. 2019; Shaw et al. 2019), and infections usually cause respiratory illness in humans and gastroenteritis in other mammals (Cui et al. 2019). No biological vectors have been reported for coronaviruses (Weiss and Navas-Martin 2005). Among mammals, bats are thought to be the natural hosts for both alpha- and betacoronaviruses (Calisher et al. 2006; Li et al. 2005; Olival et al. 2017; Woo et al. 2012), and from them, spillover events can occur once or several times to intermediate hosts that finally aid the transmission from animals to humans, as has been documented for human betacoronaviruses such as SARS-CoV and MERS-CoV (Azhar et al. 2014; Cui et al. 2019; Han et al. 2016; Kan et al. 2005; Tu et al. 2004). However, for hCoVs -OC43 and -HKU1, the natural hosts are rodents (Cui et al. 2019). In particular, in the case of betacoronaviruses, all intermediate hosts have been mammals (Cui et al. 2019) because the members grouped in this viral genus infect only mammals. Switching hosts is rare for any virus and particularly rare if the potential host species are from different higher taxa. For SARS-CoV and MERS-CoV, it was concluded that palm civets and camelids were the intermediate hosts, respectively, enabling a switch from bats to humans. Bovines were thought to be the intermediate hosts of hCoV-OC43, and it is still unknown what was the intermediate host to HKU1 (Cui et al. 2019).

Zoonoses are thought to be responsible for about 60–75% of emerging infectious diseases (EIDs), mainly originating directly from wildlife species (Valitutto et al. 2020), and particularly viruses constitute approximately 25% of EIDs (Jones et al. 2008). Recent studies found that 11% of mammals (class Mammalia) are hosts of viruses with "zoonotic potential" and 58% of those mammals host at least one zoonotic virus (Johnson et al. 2020). In addition, domestic mammals, represented by 12 species such as cats, dogs, pigs, and horses, host 50% of zoonotic virus diversity. Meanwhile, bats (order Chiroptera) host 30% of them (Johnson et al. 2020).

Bats are the natural hosts of SARS-like coronaviruses (Li et al. 2005), but probably do not transmit coronaviruses to humans directly (El-Duah et al. 2019). They are, however, probably the natural hosts for all presently known coronavirus lineages and host of the common ancestor for coronaviruses in other species (Anthony et al. 2017; Vijaykrishna et al. 2007).

The large diversity of coronavirus metagenomes has been reported mostly from bats (Anthony et al. 2017). To date, bats are diversified and grouped in 1411 species (~22% extant species mammals) distributed in all around the world excepting Antarctica (ASM Mammal Diversity Database 2020). However, although all orthocoronaviruses are believed to have diverged when mammals and birds diverged around 300 million years ago (Forni et al. 2017), the earliest bat fossils are only around 50 million years old (Gunnell and Simmons 2005). Metagenomic surveys reveal the possibility of earlier lineages of coronaviruses, and coronavirus metagenomes that are phylogenetically basal have been isolated from a Chinese water skink (*Tropidophorus sinicus*) and a frog (*Microhyla fissipes*) (Shi et al. 2018).

Interestingly, the diversity of zoonotic viruses found in mammals is correlated with the diversity of mammals grouped in different orders (Johnson et al. 2020). Previous work found the same correlation between bats and coronaviruses, and specifically the coronavirus' α- and β-diversity is higher when the bat's α- and β-diversity is greater (Anthony et al. 2017). It has been determined that, on average, there are 2.67 coronaviruses per host species (Anthony et al. 2017). If we extrapolate this average to the number of current bat species, a total potential richness of coronaviruses in bats could be determined to be 3767 species, with a minimum of 1411 bat coronavirus species, assuming one coronavirus species per bat species. However, many of them are still undescribed (Anthony et al. 2017).

New coronaviruses are continually evolving and being described. For example, when SARS-CoV-2 emerged, the last ICTV dataset of coronaviruses reported 2505 coronaviruses (Gorbalenya et al. 2020a), and the list continues to grow (Valitutto et al. 2020).

3.3.2 The Environment

The loss of natural habitats in wild animals and the high human population density correlate with the emergence of infectious diseases (Afelt et al. 2018; Frutos et al. 2020; Rocklöv and Sjödin 2020). When natural habitats are reduced by natural or anthropic events, wild animals tend to migrate to anthropogenic areas.

The emergence or reemergence of SARS-related coronaviruses (SARSr-CoVs) from bats in Asia was suggested to have resulted from deforestation and other environmental pressures (Afelt et al. 2018). However, it is possible to be more specific. The diet of pangolins is mostly termites and ants and also can include the eggs, larvae, and pupae of some insects (Lim 2008; Ashokkumar et al. 2017; Swart et al. 1999). Furthermore, it has been reported that bats in the genus *Rhinolophus* from Asia eat ants and, in rainy seasons, termites (Eckrich and Neuweiler 1988). Pangolins and insectivorous bats are both nocturnal. It can create a link between these species from their shared nocturnal habits and diet that change between seasons. Interestingly, insect larvae and the DNA from plant-insect pests have been found in the feces of bats in the genus *Rhinolophus* (Baroja et al. 2019), which may suggest a link between the use of agricultural land, the nocturnal habit of insectivorous bats and pangolins, and the emergence of SARS-CoV-2.

Is it possible to surmise more specifically why two coronaviruses have recently emerged from bats to infect humans? It is most likely that the burgeoning human population is involved as it directly affects the ecology of all organisms on Earth and affects the interface between bat and human habitats and populations (e.g., land-use changes and the spread of factory farming) and indirectly through pollution generated by humans and the climate changes it causes. Several bats in the genus *Rhinolophus*, including the species *R. affinis*, have been recorded/found in both forest and rubber (*Hevea brasiliensis*) plantations in the southeast of Asia (Phommexay et al. 2011). Recently, it has been suggested that the origin of SARS-CoV-2 was not in Wuhan (Zhang et al. 2020).

Rubber plantations in Southeast Asia now cover 8% of the region that includes all of Cambodia and Laos, northwest Vietnam, northeast Thailand, Shan State in Myanmar, and Xishuangbanna Prefecture in southern Yunnan, China (Fox et al. 2018). In 2014, 29% of the Xishuangbanna Prefecture alone was a monoculture of rubber (Fox et al. 2014). Interestingly, pangolins (*M. javanica*) inhabit secondary forests, monocultures, or urban areas without significant preferences (Lim 2008). A recent study found that the highest density of termitaria and termite-infested decaying logs was rubber plantations and forest (Karawita et al. 2020), and virome studies demonstrate the presence of coronaviruses in bat stool samples (Ge et al. 2012). Furthermore, rubber plantations with high termite activity were identified as the preferred foraging habitat of Indian pangolins (*M. crassicaudata*) (Karawita et al. 2018; Perera et al. 2017).

In laboratory studies of the use of guano from insectivorous bats as an enhancer of crop growth, a significant quantity of ants and other larvae were found in humus and guano (Sridhar et al. 2006). The omnivorous behavior of ants is well known, and many of them feed on carrion or mammal feces in search of insect eggs and larvae (Heo et al. 2009; Clark and Blom 1991).

In the early 1950s, Chinese researchers managed to develop rubber varieties that were more suitable for production in the highlands of Southeast Asia. These varieties were introduced by the Chinese government in Xishuangbanna Prefecture on a large scale at the collective/community level (Fox et al. 2018).

Thirty years later, the Chinese government dismantled these agricultural communes and returned the land to individual farmers. Rubber seedlings and subsidies were also granted to each farmer, while these became economically productive (Fox and Castella 2013). Agricultural land was then part of individual farms, while forest land continued to be controlled by the state. In 1983, forest and contracted land policy were implemented in Yunnan province in order to stabilize forest land and wooded fields by granting land titles and demarcation of land, seeking to change the forest management that belonged then to the state, and individuals were now responsible for forest regeneration in the province (Fox and Castella 2013). Both the plots of each farmer and the forests that belonged to the community were leased or contracted to individual households, who were granted long-term use rights (Jianchu et al. 2006).

With the latter came a rubber price protection plan, the introduction of the Household Responsibility System and the introduction of new technologies, and technical training to plant rubber even at elevations below 700 m and on sloping land. All this encouraged small farmers to plant rubber as a cash crop between the 1980s and the mid-1990s. Then, in 2002, a program called "Grain for Green" encouraged farmers to continue planting even on degraded slopes in order to increase forest cover and protect the environment. Particularly in Xishuangbanna, rubber trees were included as forest cover. In addition to the increase in rubber prices and the desire of individual producers to obtain money through this route, it led to a significant increase in the rubber plantation in the pending areas not yet covered, in steep areas below 700 m, and even in the village forests. The rubber monoculture became ubiquitous in the area (Fox and Castella 2013; Fox et al. 2018).

All the conditions described above, including the overexploitation of lands, the monocultures of rubber, the replacement of secondary forests by rubber plantation, the shared habitat preferences of pangolins and bats as well as their nocturnal activity and diet, even the presence of coronaviruses in bat feces, and the presence of food larvae in bat feces, create all the conditions for new pathogens to emerge and to spread eventually into humans.

3.4 Conclusion

The probability of coincidence of bats, pangolins, and humans resulting from the disturbance of their environment, generated by the promotion of rubber production, may have been the triggers for the SARS-CoV-2 virus to emerge and produce the COVID-19 pandemic. Just as significant changes, in the past, of pig, duck, chicken, and even horse, populations probably generated viral epidemics (Morse et al. 2012). The importance of rapid genome sequencing and public data sharing can provide smarter and better epidemiological surveillance (Eden et al. 2020). It requires strong international collaboration and transparency in publishing data (Mohamed et al. 2020). It has been reported that the western and southwest provinces in China, including Yunnan, were less affected by COVID-19, even when few or no measures were used to control the spread of the disease (Leung et al. 2020). Serological assays of the population of Yunnan, particularly those living in Xishuangbanna Prefecture, may indicate whether a SARS-CoV-2-like virus was present before December 2019. Eventually, an international research effort is necessary to develop faster "gene-sequence-to-candidate-vaccine" methods.

Acknowledgments The authors would like to thank Paul Scotti for constructive comments.

References

Afelt A, Frutos R, Devaux C (2018) Bats, coronaviruses, and deforestation: toward the emergence of novel infectious diseases? Front Microbiol 9:702. https://doi.org/10.3389/fmicb.2018.00702

Agrios GN (2005) Plant pathology, 5th edn. Elsevier Academic Press, Burlington

Al-Khannaq MN, Ng KT, Oong XY, Pang YK, Takebe Y, Chook JB, Hanafi NS, Kamarulzaman A, Tee KK (2016) Molecular epidemiology and evolutionary histories of human coronavirus OC43 and HKU1 among patients with upper respiratory tract infections in Kuala Lumpur, Malaysia. Virol J 13:33. https://doi.org/10.1186/s12985-016-0488-4

Almeida JD, Tyrrell DA (1967) The morphology of three previously uncharacterized human respiratory viruses that grow in organ culture. J Gen Virol 1(2):175–178. https://doi.org/10.1099/0022-1317-1-2-175

Andersen KG, Rambaut A, Lipkin WI, Holmes EC, Garry RF (2020) The proximal origin of SARS-CoV-2. Nat Med 26(4):450–452. https://doi.org/10.1038/s41591-020-0820-9

Anthony SJ, Johnson CK, Greig DJ, Kramer S, Che X, Wells H, Hicks AL, Joly DO, Wolfe ND, Daszak P, Karesh W, Lipkin WI, Morse SS, Mazet JAK, Goldstein T, Consortium P (2017) Global patterns in coronavirus diversity. Virus Evol 3(1):vex012. https://doi.org/10.1093/ve/vex012

Ashokkumar M, Valsarajan D, Suresh MA, Kaimal AR, Chandy G (2017) Stomach contents of the Indian Pangolin Manis crassicaudata (Mammalia: Pholidota: Manidae) in tropical forests of southern India. J Threat Taxa 9(5):10246–10248

Azhar EI, El-Kafrawy SA, Farraj SA, Hassan AM, Al-Saeed MS, Hashem AM, Madani TA (2014) Evidence for camel-to-human transmission of MERS coronavirus. N Engl J Med 370(26):2499–2505

Baric RS, Fu K, Schaad MC, Stohlman SA (1990) Establishing a genetic recombination map for murine coronavirus strain A59 complementation groups. Virology 177(2):646–656. https://doi.org/10.1016/0042-6822(90)90530-5

Baroja U, Garin I, Aihartza J, Arrizabalaga-Escudero A, Vallejo N, Aldasoro M, Goiti U (2019) Pest consumption in a vineyard system by the lesser horseshoe bat (Rhinolophus hipposideros). PLoS One 14(7):e0219265. https://doi.org/10.1371/journal.pone.0219265

Beaudette F (1937) Cultivation of the virus of infectious bronchitis. J Am Vet Med Assoc 90:51–60

Belshaw R, Sanjuán R, Pybus OG (2011) Viral mutation and substitution: units and levels. Curr Opin Virol 1(5):430–435. https://doi.org/10.1016/j.coviro.2011.08.004

Boni MF, Lemey P, Jiang X, Lam TT-Y, Perry B, Castoe T, Rambaut A, Robertson DL (2020) Evolutionary origins of the SARS-CoV-2 sarbecovirus lineage responsible for the COVID-19 pandemic. bioRxiv

Bouckaert R, Vaughan TG, Barido-Sottani J, Duchêne S, Fourment M, Gavryushkina A, Heled J, Jones G, Kühnert D, De Maio N, Matschiner M, Mendes FK, Müller NF, Ogilvie HA, du Plessis L, Popinga A, Rambaut A, Rasmussen D, Siveroni I, Suchard MA, Wu CH, Xie D, Zhang C, Stadler T, Drummond AJ (2019) BEAST 2.5: an advanced software platform for Bayesian evolutionary analysis. PLoS Comput Biol 15(4):e1006650. https://doi.org/10.1371/journal.pcbi.1006650

Bukhari K, Mulley G, Gulyaeva AA, Zhao L, Shu G, Jiang J, Neuman BW (2018) Description and initial characterization of metatranscriptomic nidovirus-like genomes from the proposed new family Abyssoviridae, and from a sister group to the Coronavirinae, the proposed genus Alphaletovirus. Virology 524:160–171. https://doi.org/10.1016/j.virol.2018.08.010

Calisher CH, Childs JE, Field HE, Holmes KV, Schountz T (2006) Bats: important reservoir hosts of emerging viruses. Clin Microbiol Rev 19(3):531–545. https://doi.org/10.1128/CMR.00017-06

Clark WH, Blom PE (1991) Observations of ants (Hymenoptera: Formicidae: Myrmicinae, Formicinae, Dolichoderinae) utilizing carrion. Southwest Nat 36(1):140–142

Corman VM, Eckerle I, Memish ZA, Liljander AM, Dijkman R, Jonsdottir H, Ngeiywa KJJ, Kamau E, Younan M, Al Masri M (2016) Link of a ubiquitous human coronavirus to dromedary camels. Proc Natl Acad Sci 113(35):9864–9869

Cotten M, Watson SJ, Kellam P, Al-Rabeeah AA, Makhdoom HQ, Assiri A, Al-Tawfiq JA, Alhakeem RF, Madani H, AlRabiah FA, Al Hajjar S, Al-nassir WN, Albarrak A, Flemban H, Balkhy HH, Alsubaie S, Palser AL, Gall A, Bashford-Rogers R, Rambaut A, Zumla AI, Memish ZA (2013) Transmission and evolution of the Middle East respiratory syndrome coronavirus in Saudi Arabia: a descriptive genomic study. Lancet 382(9909):1993–2002. https://doi.org/10.1016/S0140-6736(13)61887-5

Cotten M, Watson SJ, Zumla AI, Makhdoom HQ, Palser AL, Ong SH, Al Rabeeah AA, Alhakeem RF, Assiri A, Al-Tawfiq JA (2014) Spread, circulation, and evolution of the Middle East respiratory syndrome coronavirus. MBio 5(1):e01062–01013

Cuevas JM, Duffy S, Sanjuán R (2009) Point mutation rate of bacteriophage ΦX174. Genetics 183(2):747–749

Cui J, Li F, Shi Z-L (2019) Origin and evolution of pathogenic coronaviruses. Nat Rev Microbiol 17(3):181–192

Decaro N, Mari V, Campolo M, Lorusso A, Camero M, Elia G, Martella V, Cordioli P, Enjuanes L, Buonavoglia C (2009) Recombinant canine coronaviruses related to transmissible gastroenteritis virus of Swine are circulating in dogs. J Virol 83(3):1532–1537. https://doi.org/10.1128/JVI.01937-08

Denison MR, Graham RL, Donaldson EF, Eckerle LD, Baric RS (2011) Coronaviruses: an RNA proofreading machine regulates replication fidelity and diversity. RNA Biol 8(2):270–279. https://doi.org/10.4161/rna.8.2.15013

Domingo E (2019) Virus as populations: composition, complexity, quasi species, dynamics, and biological implications. Academic Press, New York

Drake JW (1991) A constant rate of spontaneous mutation in DNA-based microbes. Proc Natl Acad Sci U S A 88(16):7160–7164. https://doi.org/10.1073/pnas.88.16.7160

Drake JW, Holland JJ (1999) Mutation rates among RNA viruses. Proc Natl Acad Sci 96(24):13910–13913

Drosten C, Günther S, Preiser W, Van Der Werf S, Brodt H-R, Becker S, Rabenau H, Panning M, Kolesnikova L, Fouchier RA (2003) Identification of a novel coronavirus in patients with severe acute respiratory syndrome. N Engl J Med 348(20):1967–1976

Duchêne S, Holmes EC, Ho SY (2014) Analyses of evolutionary dynamics in viruses are hindered by a time-dependent bias in rate estimates. Proc Biol Sci 281(1786):20140732. https://doi.org/10.1098/rspb.2014.0732

Dudas G, Carvalho LM, Rambaut A, Bedford T (2018) MERS-CoV spillover at the camel-human interface. Elife 7. https://doi.org/10.7554/eLife.31257

Duffy S, Shackelton LA, Holmes EC (2008) Rates of evolutionary change in viruses: patterns and determinants. Nat Rev Genet 9(4):267–276. https://doi.org/10.1038/nrg2323

Eckerle LD, Becker MM, Halpin RA, Li K, Venter E, Lu X, Scherbakova S, Graham RL, Baric RS, Stockwell TB, Spiro DJ, Denison MR (2010) Infidelity of SARS-CoV Nsp14-exonuclease mutant virus replication is revealed by complete genome sequencing. PLoS Pathog 6(5):e1000896. https://doi.org/10.1371/journal.ppat.1000896

Eckrich M, Neuweiler G (1988) Food habits of the sympatric insectivorous bats Rhinolophus rouxi and Hipposideros lankadiva from Sri Lanka. J Zool 215(4):729–737

Eden JS, Rockett R, Carter I, Rahman H, de Ligt J, Hadfield J, Storey M, Ren X, Tulloch R, Basile K, Wells J, Byun R, Gilroy N, O'Sullivan MV, Sintchenko V, Chen SC, Maddocks S, Sorrell TC, Holmes EC, Dwyer DE, Kok J, Group -nS (2020) An emergent clade of SARS-CoV-2 linked to returned travellers from Iran. Virus Evol 6(1):veaa027. https://doi.org/10.1093/ve/veaa027

El-Duah P, Meyer B, Sylverken A, Owusu M, Gottula LT, Yeboah R, Lamptey J, Frimpong YO, Burimuah V, Folitse R, Agbenyega O, Oppong S, Adu-Sarkodie Y, Drosten C (2019) Development of a whole-virus ELISA for serological evaluation of domestic livestock as possible hosts of human coronavirus NL63. Viruses 11(1):43. https://doi.org/10.3390/v11010043

Fabricant J (1998) The early history of infectious bronchitis. Avian Dis 42(4):648–650

Fenner F (1975) The classification and nomenclature of viruses. Summary of results of meetings of the International Committee on Taxonomy of Viruses in Madrid, September 1975. Intervirology 6(1):1–12. https://doi.org/10.1159/000149448

Forni D, Cagliani R, Clerici M, Sironi M (2017) Molecular evolution of human coronavirus genomes. Trends Microbiol 25(1):35–48. https://doi.org/10.1016/j.tim.2016.09.001

Fouchier RA, Hartwig NG, Bestebroer TM, Niemeyer B, de Jong JC, Simon JH, Osterhaus AD (2004) A previously undescribed coronavirus associated with respiratory disease in humans. Proc Natl Acad Sci U S A 101(16):6212–6216. https://doi.org/10.1073/pnas.0400762101

Fox J, Castella J-C (2013) Expansion of rubber (Hevea brasiliensis) in Mainland Southeast Asia: what are the prospects for smallholders? J Peasant Stud 40(1):155–170

Fox J, Castella J-C, Ziegler AD, Westley SB (2014) Rubber plantations expand in mountainous Southeast Asia: what are the consequences for the environment? East-West Center, Honolulu

Fox J, Castella JC, Hurni K (2018) As rubber plantations expand in the highlands of Southeast Asia, what happens to local farmers? East-West Center, Honolulu

Frutos R, Lopez Roig M, Serra-Cobo J, Devaux CA (2020) COVID-19: the conjunction of events leading to the coronavirus pandemic and lessons to learn for future threats. Front Med 7:223

Ge X, Li Y, Yang X, Zhang H, Zhou P, Zhang Y, Shi Z (2012) Metagenomic analysis of viruses from bat fecal samples reveals many novel viruses in insectivorous bats in China. J Virol 86(8):4620–4630. https://doi.org/10.1128/JVI.06671-11

Gibbs MJ, Wayper P, Fourment ML, Wood JT, Ohshima K, Armstrong JS, Gibbs AJ (2007) The variable codons of H3 influenza A virus haemagglutinin genes. Arch Virol 152(1):11–24. https://doi.org/10.1007/s00705-006-0834-8

González-Candelas F, Patiño-Galindo JÁ, Valiente-Mullor C (2018) Population genomics of human viruses. In: Population genomics: microorganisms. Springer, Cham, pp 267–296

Gorbalenya AE, Enjuanes L, Ziebuhr J, Snijder EJ (2006) Nidovirales: evolving the largest RNA virus genome. Virus Res 117(1):17–37. https://doi.org/10.1016/j.virusres.2006.01.017

Gorbalenya AE, Baker SC, Baric RS, de Groot RJ, Drosten C, Gulyaeva AA, Haagmans BL, Lauber C, Leontovich AM, Neuman BW, Penzar D, Perlman

S, Poon LLM, Samborskiy DV, Sidorov IA, Sola I, Ziebuhr J, Coronaviridae Study Group of the International Committee on Taxonomy of V (2020a) The species severe acute respiratory syndrome-related coronavirus: classifying 2019-nCoV and naming it SARS-CoV-2. Nat Microbiol 5(4):536–544. https://doi.org/10.1038/s41564-020-0695-z

Gorbalenya AE, Krupovic M, Mushegian AR, Kropinski AM, Siddell SG, Varsani A, Adams MJ, Davison AJ, Dutilh BE, Harrach B (2020b) The new scope of virus taxonomy: partitioning the virosphere into 15 hierarchical ranks. Nat Microbiol 5(5):668–674

Gribble J, Pruijssers AJ, Agostini ML, Anderson-Daniels J, Chappell JD, Lu X, Stevens LJ, Routh AL, Denison MR (2020) The coronavirus proofreading exoribonuclease mediates extensive viral recombination. BioRxiv

Guan Y, Zheng BJ, He YQ, Liu XL, Zhuang ZX, Cheung CL, Luo SW, Li PH, Zhang LJ, Guan YJ, Butt KM, Wong KL, Chan KW, Lim W, Shortridge KF, Yuen KY, Peiris JS, Poon LL (2003) Isolation and characterization of viruses related to the SARS coronavirus from animals in southern China. Science 302(5643):276–278. https://doi.org/10.1126/science.1087139

Gunnell GF, Simmons NB (2005) Fossil evidence and the origin of bats. J Mamm Evol 12(1–2):209–246

Hall TA (1999) BioEdit: a user-friendly biological sequence alignment editor and analysis program for Windows 95/98/NT. In: Nucleic acids symposium series, vol 41. London: Information Retrieval Ltd., c1979-c2000, pp 95–98

Han HJ, Yu H, Yu XJ (2016) Evidence for zoonotic origins of Middle East respiratory syndrome coronavirus. J Gen Virol 97(2):274–280. https://doi.org/10.1099/jgv.0.000342

Hanada K, Suzuki Y, Gojobori T (2004) A large variation in the rates of synonymous substitution for RNA viruses and its relationship to a diversity of viral infection and transmission modes. Mol Biol Evol 21(6):1074–1080. https://doi.org/10.1093/molbev/msh109

Henry R (2020) Etymologia: coronavirus. Emerg Infect Dis 26(5):1027

Heo CC, Mohamad AR, Rosli H, Nurul Ashikin A, Chen CD, John J, Hiromu K, Baharudin O (2009) Ants (Hymenoptera: Formicidae) associated with pig carcasses in Malaysia. Trop Biomed 26(1):106–109

Herrewegh AA, Smeenk I, Horzinek MC, Rottier PJ, de Groot RJ (1998) Feline coronavirus type II strains 79-1683 and 79-1146 originate from a double recombination between feline coronavirus type I and canine coronavirus. J Virol 72(5):4508–4514

Holmes EC (2009) The evolution and emergence of RNA viruses. Oxford University Press, New York

Hon CC, Lam TY, Shi ZL, Drummond AJ, Yip CW, Zeng F, Lam PY, Leung FC (2008) Evidence of the recombinant origin of a bat severe acute respiratory syndrome (SARS)-like coronavirus and its implications on the direct ancestor of SARS coronavirus. J Virol 82(4):1819–1826. https://doi.org/10.1128/JVI.01926-07

Jenkins GM, Rambaut A, Pybus OG, Holmes EC (2002) Rates of molecular evolution in RNA viruses: a quantitative phylogenetic analysis. J Mol Evol 54(2):156–165. https://doi.org/10.1007/s00239-001-0064-3

Jianchu X, Fox J, Melick D, Fujita Y, Jintrawet A, Jie Q, Thomas D, Weyerhaeuser H (2006) Land use transition, livelihoods, and environmental services in Montane Mainland Southeast Asia. Mt Res Dev 26(3):278–284

Johnson CK, Hitchens PL, Pandit PS, Rushmore J, Evans TS, Young CCW, Doyle MM (2020) Global shifts in mammalian population trends reveal key predictors of virus spillover risk. Proc Biol Sci 287(1924):20192736. https://doi.org/10.1098/rspb.2019.2736

Jones KE, Patel NG, Levy MA, Storeygard A, Balk D, Gittleman JL, Daszak P (2008) Global trends in emerging infectious diseases. Nature 451(7181):990–993. https://doi.org/10.1038/nature06536

Kahn JS, McIntosh K (2005) History and recent advances in coronavirus discovery. Pediatr Infect Dis J 24 (11 Suppl):S223–227, discussion S226. https://doi.org/10.1097/01.inf.0000188166.17324.60

Kan B, Wang M, Jing H, Xu H, Jiang X, Yan M, Liang W, Zheng H, Wan K, Liu Q, Cui B, Xu Y, Zhang E, Wang H, Ye J, Li G, Li M, Cui Z, Qi X, Chen K, Du L, Gao K, Zhao YT, Zou XZ, Feng YJ, Gao YF, Hai R, Yu D, Guan Y, Xu J (2005) Molecular evolution analysis and geographic investigation of severe acute respiratory syndrome coronavirus-like virus in palm civets at an animal market and on farms. J Virol 79(18):11892–11900. https://doi.org/10.1128/JVI.79.18.11892-11900.2005

Karawita H, Perera P, Gunawardane P, Dayawansa N (2018) Habitat preference and den characterization of Indian Pangolin (Manis crassicaudata) in a tropical lowland forested landscape of Southwest Sri Lanka. PLoS One 13(11):e0206082. https://doi.org/10.1371/journal.pone.0206082

Karawita H, Perera P, Dayawansa N, Dias S (2020) Dietary composition and foraging habitats of the Indian Pangolin (Manis crassicaudata) in a tropical lowland forest-associated landscape in Southwest Sri Lanka. Glob Ecol Conserv 21:e00880

Knipe DM, Howley P (2013) Fields virology. Lippincott Williams & Wilkins, Philadelphia

Lai MM (1996) Recombination in large RNA viruses: coronaviruses. In: seminars in virology, vol 6. Academic, pp 381–388

Lam TT, Shum MH, Zhu HC, Tong YG, Ni XB, Liao YS, Wei W, Cheung WY, Li WJ, Li LF, Leung GM, Holmes EC, Hu YL, Guan Y (2020) Identifying SARS-CoV-2 related coronaviruses in Malayan pangolins. Nature 583:282. https://doi.org/10.1038/s41586-020-2169-0

Lau SK, Li KS, Huang Y, Shek CT, Tse H, Wang M, Choi GK, Xu H, Lam CS, Guo R, Chan KH, Zheng BJ, Woo PC, Yuen KY (2010) Ecoepidemiology and complete genome comparison of different strains of severe acute respiratory syndrome-related Rhinolophus bat coronavirus in China reveal bats as a reservoir for acute, self-limiting infection that allows recombination events.

J Virol 84(6):2808–2819. https://doi.org/10.1128/JVI.02219-09

Lau SK, Lee P, Tsang AK, Yip CC, Tse H, Lee RA, So LY, Lau YL, Chan KH, Woo PC, Yuen KY (2011) Molecular epidemiology of human coronavirus OC43 reveals evolution of different genotypes over time and recent emergence of a novel genotype due to natural recombination. J Virol 85(21):11325–11337. https://doi.org/10.1128/JVI.05512-11

Lau SK, Luk HK, Wong AC, Li KS, Zhu L, He Z, Fung J, Chan TT, Fung KS, Woo PC (2020) Early release-possible bat origin of severe acute respiratory syndrome coronavirus 2

Leung K, Wu JT, Liu D, Leung GM (2020) First-wave COVID-19 transmissibility and severity in China outside Hubei after control measures, and second-wave scenario planning: a modelling impact assessment. Lancet 395(10233):1382–1393. https://doi.org/10.1016/S0140-6736(20)30746-7

Li W, Shi Z, Yu M, Ren W, Smith C, Epstein JH, Wang H, Crameri G, Hu Z, Zhang H, Zhang J, McEachern J, Field H, Daszak P, Eaton BT, Zhang S, Wang LF (2005) Bats are natural reservoirs of SARS-like coronaviruses. Science 310(5748):676–679. https://doi.org/10.1126/science.1118391

Lim N (2008) Ecological research and conservation of Sunda pangolin Manis javanica in Singapore. In: Proceedings of the workshop on trade and conservation of Pangolins Native to South and Southeast Asia. TRAFFIC Southeast Asia, Singapore Zoo, Singapore, pp 90–93

Luk HKH, Li X, Fung J, Lau SKP, Woo PCY (2019) Molecular epidemiology, evolution and phylogeny of SARS coronavirus. Infect Genet Evol 71:21–30. https://doi.org/10.1016/j.meegid.2019.03.001

Mahase E (2020) Coronavirus covid-19 has killed more people than SARS and MERS combined, despite lower case fatality rate. BMJ 368:m641. https://doi.org/10.1136/bmj.m641

Martin DP, Murrell B, Golden M, Khoosal A, Muhire B (2015) RDP4: detection and analysis of recombination patterns in virus genomes. Virus Evol 1(1):vev003. https://doi.org/10.1093/ve/vev003

Minskaia E, Hertzig T, Gorbalenya AE, Campanacci V, Cambillau C, Canard B, Ziebuhr J (2006) Discovery of an RNA virus 3′->5′ exoribonuclease that is critically involved in coronavirus RNA synthesis. Proc Natl Acad Sci U S A 103(13):5108–5113. https://doi.org/10.1073/pnas.0508200103

Mohamed K, Rodríguez-Román E, Rahmani F, Zhang H, Ivanovska M, Makka SA, Joya M, Makuku R, Islam MS, Radwan N, Rahmah L, Goda R, Abarikwu SO, Shaw M, Zoghi S, Irtsyan S, Ling I, Cseprekal O, Faten AB, Hazar Sayar E, Soloukey C, Grancini G, Rezaei N (2020) Borderless collaboration is needed for COVID-19-a disease that knows no borders. Infect Control Hosp Epidemiol 41:1–2. https://doi.org/10.1017/ice.2020.162

Morse SS, Mazet JA, Woolhouse M, Parrish CR, Carroll D, Karesh WB, Zambrana-Torrelio C, Lipkin WI, Daszak P (2012) Prediction and prevention of the next pandemic zoonosis. Lancet 380(9857):1956–1965. https://doi.org/10.1016/S0140-6736(12)61684-5

Nielsen R, Wang H, Pipes L (2020) Synonymous mutations and the molecular evolution of SARS-Cov-2 origins. bioRxiv

Olival KJ, Hosseini PR, Zambrana-Torrelio C, Ross N, Bogich TL, Daszak P (2017) Host and viral traits predict zoonotic spillover from mammals. Nature 546(7660):646–650. https://doi.org/10.1038/nature22975

Pagán I (2018) The diversity, evolution and epidemiology of plant viruses: a phylogenetic view. Infect Genet Evol 65:187–199. https://doi.org/10.1016/j.meegid.2018.07.033

Perales C, Martín V, Domingo E (2011) Lethal mutagenesis of viruses. Curr Opin Virol 1(5):419–422. https://doi.org/10.1016/j.coviro.2011.09.001

Perera P, Karawita K, Pabasara M (2017) Pangolins (Manis crassicaudata) in Sri Lanka: a review of current knowledge, threats and research priorities. J Trop For Environ 7(1):1–14

Phommexay P, Satasook C, Bates P, Pearch M, Bumrungsri S (2011) The impact of rubber plantations on the diversity and activity of understorey insectivorous bats in southern Thailand. Biodivers Conserv 20(7):1441–1456

Prates ET, Garvin MR, Pavicic M, Jones P, Shah M, Alvarez C, Kainer D, Demerdash O, Amos BK, Geiger A (2020) Functional immune deficiency syndrome via intestinal infection in COVID-19. bioRxiv

Rocklöv J, Sjödin H (2020) High population densities catalyse the spread of COVID-19. J Travel Med 27(3). https://doi.org/10.1093/jtm/taaa038

Sanjuán R, Domingo-Calap P (2019) Genetic diversity and evolution of viral populations. Reference Module in Life Sciences

Sanjuán R, Nebot MR, Chirico N, Mansky LM, Belshaw R (2010) Viral mutation rates. J Virol 84(19):9733–9748. https://doi.org/10.1128/JVI.00694-10

Shaw L, Wang A, Dylus D, Meier M, Pogacnik G, Dessimoz C (2019) The phylogenetic range of bacterial and viral pathogens of vertebrates. Mol Ecol

Shi M, Lin XD, Chen X, Tian JH, Chen LJ, Li K, Wang W, Eden JS, Shen JJ, Liu L, Holmes EC, Zhang YZ (2018) The evolutionary history of vertebrate RNA viruses. Nature 556(7700):197–202. https://doi.org/10.1038/s41586-018-0012-7

Simon-Loriere E, Holmes EC (2011) Why do RNA viruses recombine? Nat Rev Microbiol 9(8):617–626. https://doi.org/10.1038/nrmicro2614

Smith EC, Blanc H, Surdel MC, Vignuzzi M, Denison MR (2013) Coronaviruses lacking exoribonuclease activity are susceptible to lethal mutagenesis: evidence for proofreading and potential therapeutics. PLoS Pathog 9(8):e1003565. https://doi.org/10.1371/journal.ppat.1003565

Sridhar K, Ashwini K, Seena S, Sreepada K (2006) Manure qualities of guano of insectivorous cave bat Hipposideros speoris. Trop Subtrop Agroecosyst 6(2):103–110

Suttle CA (2007) Marine viruses – major players in the global ecosystem. Nat Rev Microbiol 5(10):801–812. https://doi.org/10.1038/nrmicro1750

Swart J, Richardson P, Ferguson J (1999) Ecological factors affecting the feeding behaviour of pangolins (Manis temminckii). J Zool 247(3):281–292

Tao Y, Shi M, Chommanard C, Queen K, Zhang J, Markotter W, Kuzmin IV, Holmes EC, Tong S (2017) Surveillance of bat coronaviruses in Kenya identifies relatives of human coronaviruses NL63 and 229E and their recombination history. J Virol 91(5):e01953. https://doi.org/10.1128/JVI.01953-16

Tu C, Crameri G, Kong X, Chen J, Sun Y, Yu M, Xiang H, Xia X, Liu S, Ren T, Yu Y, Eaton BT, Xuan H, Wang LF (2004) Antibodies to SARS coronavirus in civets. Emerg Infect Dis 10(12):2244–2248. https://doi.org/10.3201/eid1012.040520

Valitutto MT, Aung O, Tun KYN, Vodzak ME, Zimmerman D, Yu JH, Win YT, Maw MT, Thein WZ, Win HH, Dhanota J, Ontiveros V, Smith B, Tremeau-Brevard A, Goldstein T, Johnson CK, Murray S, Mazet J (2020) Detection of novel coronaviruses in bats in Myanmar. PLoS One 15(4):e0230802. https://doi.org/10.1371/journal.pone.0230802

van der Hoek L, Pyrc K, Jebbink MF, Vermeulen-Oost W, Berkhout RJ, Wolthers KC, Wertheim-van Dillen PM, Kaandorp J, Spaargaren J, Berkhout B (2004) Identification of a new human coronavirus. Nat Med 10(4):368–373. https://doi.org/10.1038/nm1024

van Dorp L, Acman M, Richard D, Shaw LP, Ford CE, Ormond L, Owen CJ, Pang J, Tan CCS, Boshier FAT, Ortiz AT, Balloux F (2020) Emergence of genomic diversity and recurrent mutations in SARS-CoV-2. Infect Genet Evol 83:104351. https://doi.org/10.1016/j.meegid.2020.104351

Vandamme A-M (2009) Basic concepts of molecular evolution. In: The phylogenetic handbook, vol 2. Cambridge University Press, Cambridge, pp 3–32

Vijaykrishna D, Smith GJ, Zhang JX, Peiris JS, Chen H, Guan Y (2007) Evolutionary insights into the ecology of coronaviruses. J Virol 81(8):4012–4020. https://doi.org/10.1128/JVI.02605-06

Vijgen L, Keyaerts E, Moës E, Thoelen I, Wollants E, Lemey P, Vandamme AM, Van Ranst M (2005) Complete genomic sequence of human coronavirus OC43: molecular clock analysis suggests a relatively recent zoonotic coronavirus transmission event. J Virol 79(3):1595–1604. https://doi.org/10.1128/JVI.79.3.1595-1604.2005

Weiss SR, Navas-Martin S (2005) Coronavirus pathogenesis and the emerging pathogen severe acute respiratory syndrome coronavirus. Microbiol Mol Biol Rev 69(4):635–664. https://doi.org/10.1128/MMBR.69.4.635-664.2005

Woo PC, Lau SK, Chu CM, Chan KH, Tsoi HW, Huang Y, Wong BH, Poon RW, Cai JJ, Luk WK, Poon LL, Wong SS, Guan Y, Peiris JS, Yuen KY (2005) Characterization and complete genome sequence of a novel coronavirus, coronavirus HKU1, from patients with pneumonia. J Virol 79(2):884–895. https://doi.org/10.1128/JVI.79.2.884-895.2005

Woo PC, Lau SK, Lam CS, Lau CC, Tsang AK, Lau JH, Bai R, Teng JL, Tsang CC, Wang M, Zheng BJ, Chan KH, Yuen KY (2012) Discovery of seven novel Mammalian and avian coronaviruses in the genus deltacoronavirus supports bat coronaviruses as the gene source of alphacoronavirus and betacoronavirus and avian coronaviruses as the gene source of gammacoronavirus and deltacoronavirus. J Virol 86(7):3995–4008. https://doi.org/10.1128/JVI.06540-11

Wu F, Zhao S, Yu B, Chen YM, Wang W, Song ZG, Hu Y, Tao ZW, Tian JH, Pei YY, Yuan ML, Zhang YL, Dai FH, Liu Y, Wang QM, Zheng JJ, Xu L, Holmes EC, Zhang YZ (2020) A new coronavirus associated with human respiratory disease in China. Nature 579(7798):265–269. https://doi.org/10.1038/s41586-020-2008-3

Xiao K, Zhai J, Feng Y, Zhou N, Zhang X, Zou JJ, Li N, Guo Y, Li X, Shen X, Zhang Z, Shu F, Huang W, Li Y, Chen RA, Wu YJ, Peng SM, Huang M, Xie WJ, Cai QH, Hou FH, Chen W, Xiao L, Shen Y (2020) Isolation of SARS-CoV-2-related coronavirus from Malayan pangolins. Nature 583:286. https://doi.org/10.1038/s41586-020-2313-x

Zaki AM, van Boheemen S, Bestebroer TM, Osterhaus AD, Fouchier RA (2012) Isolation of a novel coronavirus from a man with pneumonia in Saudi Arabia. N Engl J Med 367(19):1814–1820. https://doi.org/10.1056/NEJMoa1211721

Zhang YZ, Holmes EC (2020) A genomic perspective on the origin and emergence of SARS-CoV-2. Cell 181(2):223–227. https://doi.org/10.1016/j.cell.2020.03.035

Zhang X, Tan Y, Ling Y, Lu G, Liu F, Yi Z, Jia X, Wu M, Shi B, Xu S, Chen J, Wang W, Chen B, Jiang L, Yu S, Lu J, Wang J, Xu M, Yuan Z, Zhang Q, Zhao G, Wang S, Chen S, Lu H (2020) Viral and host factors related to the clinical outcome of COVID-19. Nature 583:437. https://doi.org/10.1038/s41586-020-2355-0

Zhao Z, Li H, Wu X, Zhong Y, Zhang K, Zhang YP, Boerwinkle E, Fu YX (2004) Moderate mutation rate in the SARS coronavirus genome and its implications. BMC Evol Biol 4:21. https://doi.org/10.1186/1471-2148-4-21

Zhou P, Yang XL, Wang XG, Hu B, Zhang L, Zhang W, Si HR, Zhu Y, Li B, Huang CL, Chen HD, Chen J, Luo Y, Guo H, Jiang RD, Liu MQ, Chen Y, Shen XR, Wang X, Zheng XS, Zhao K, Chen QJ, Deng

F, Liu LL, Yan B, Zhan FX, Wang YY, Xiao GF, Shi ZL (2020) A pneumonia outbreak associated with a new coronavirus of probable bat origin. Nature 579(7798):270–273. https://doi.org/10.1038/s41586-020-2012-7

Websites

ASM Mammal Diversity Database, American Society of Mammalogist; Chiroptera: https://mammaldiversity.org/#Y2hpcm9wdGVyYSZnbG9iYWxfc2VhcmNoPXRydWUmbG9vc2U9dHJ1ZQ (consulted: May 21, 2020)

ICTV online: https://talk.ictvonline.org/. Consulted on May 21, 2020

Virological.: http://virological.org/. Consulted on May 21, 2020

WHO: WHO Director-General's opening remarks at the media briefing on COVID-19—11 March 2020. World Health Organization website. https://www.who.int/dg/speeches/detail/who-director-general-s-opening-remarks-atthe-media-briefing-on-covid-19—11-march-2020. Published March 11, 2020. Accessed April 12, 2020

Worldometer.: https://www.worldometers.info/coronavirus/. Consulted on May 31, 2020

The Epidemiologic Aspects of COVID-19 Outbreak: Spreading Beyond Expectations

4

Sara Hanaei, Farnam Mohebi, Maziar Moradi-Lakeh, Parnian Jabbari, Surinder Kumar Mehta, Liudmyla S. Kryvenko, Livio Luongo, Loďc Dupré, and Nima Rezaei

Abstract

The coronavirus disease 2019 (COVID-19) outbreak started in late 2019 in Wuhan, Hubei Province of China, and quickly spread to the surrounding regions and neighboring countries. A novel coronavirus, the so-called severe acute respiratory syndrome coronavirus 2 (SARS-CoV-2), was found to be responsible for this outbreak potentially originating from pangolins. In China, the outbreak lasted for 1 month until it seemed to be controlled after affecting over 81,000 individuals and causing deaths in over 4200 patients. Subsequently, and after affecting over 118,000 individuals and causing over 4200 deaths, the condition was officially announced as a pandemic by the World Health Organization (WHO). In the meantime, the epidemic curve took a downtrend in China, the original epicenter of the pandemic, but started to rise in other countries with a steep slope. Among over 215 affected countries, the USA, European countries (Italy, Germany, Spain, France, the UK), Iran, and South Korea had the highest frequencies in the matters of infected patients and deaths.

S. Hanaei (✉) · P. Jabbari
Research Center for Immunodeficiencies, Tehran University of Medical Sciences, Tehran, Iran

Network of Immunity in Infection, Malignancy and Autoimmunity (NIIMA), Universal Scientific Education and Research Network (USERN), Tehran, Iran
e-mail: sara.hanaei@gmail.com

F. Mohebi
Research Associate, Non-Communicable Diseases Research Center, Endocrinology and Metabolism Population Sciences Institute, Tehran University of Medical Sciences, Tehran, Iran

M. Moradi-Lakeh
Preventive Medicine and Public Health Research Center, Iran University of Medical Sciences, Tehran, Iran

S. K. Mehta
Department of Chemistry & Centre of Advanced Studies in Chemistry, Panjab University, Chandigarh, India

L. S. Kryvenko
Kharkiv National Medical University, Kharkiv, Ukraine

L. Luongo
Department of Experimental Medicine, University of Campania "L. Vanvitelli", Naples, Italy

L. Dupré
Ludwig Boltzmann Institute, Rare and Undiagnosed Diseases, Vienna, Austria

N. Rezaei (✉)
Research Center for Immunodeficiencies, Children's Medical Center, Tehran University of Medical Sciences, Tehran, Iran

Network of Immunity in Infection, Malignancy and Autoimmunity (NIIMA), Universal Scientific Education and Research Network (USERN), Tehran, Iran

Department of Immunology, School of Medicine, Tehran University of Medical Sciences, Tehran, Iran
e-mail: rezaei_nima@tums.ac.ir

N. Rezaei (ed.), *Coronavirus Disease - COVID-19*, Advances in Experimental Medicine and Biology 1318, https://doi.org/10.1007/978-3-030-63761-3_4

Importantly, different countries took different policies when encountered with an outbreak, especially in the matter of accuracy of the report and timing of the action. A part of the delays in reporting was expected, including the lag in the chain of reporting, the shortcomings of tests, missed patients, and inadequate testing facilities. However, there were also political and nontechnical reasons that caused the reporting to be inaccurate. Surveillance seems to be less of a reason for the observed in poor management, and it mostly originated from human decision-making failures and political issues. Besides, the culture of populations and their trust in their governments played an important role on how they reacted to the COVID-19 pandemic and acquired policies. Finally, the characteristics of the world today indicate the danger of probable upcoming outbreaks, and policymakers should utilize the existing opportunities, particularly the advancements in technology and media, to prevent or adequately manage them.

Keywords

COVID-19 · Epidemiology · Outbreak · Pandemic · Policymakers · Prevalence

4.1 Introduction

As defined by the World Health Organization (WHO), a disease outbreak is the occurrence of disease cases over average expectancy (Kafieh et al. 2020). When the outbreak develops to be less localized and more likely to cause public panics, it would be called an epidemic. Furthermore, after developing to spread to several countries and even continents and affecting a significant proportion of the population, it would be called a pandemic (US Department of Health Human Services. Principles of Epidemiology in Public Health Practice Third Edition An Introduction to Applied Epidemiology and Biostatistics 2013). As investigated so far, the world has encountered about 20 pandemics, 9 of which occurred in the past two centuries (twentieth and twenty-first). These recent outbreaks and pandemics were different types of flu (Spanish, Asian, Hong Kong, and swine), human immunodeficiency virus infection and acquired immune deficiency syndrome (HIV/AIDS), severe acute respiratory syndrome (SARS), Ebola, and the Middle East respiratory syndrome (MERS) (LePan 2020).

Most recently, a novel coronavirus, severe acute respiratory syndrome coronavirus 2 (SARS-CoV-2), was discovered as the cause of a critical respiratory disease called coronavirus disease 2019 (COVID-19). Though similarly related to a potentially fatal respiratory disorder, the SARS-CoV-2 has shown considerable genetic dissimilarity with causes of SARS and MERS (Ahanchian et al. 2020). The novel coronavirus, in particular, seems more talented in the manipulation of the host immune system that can inhibit antiviral pathways and induce hyper-inflammation (Bahrami et al. 2020; Yazdanpanah et al. 2020a; Fathi and Rezaei 2020; Nasab et al. 2020). It has been shown to affect other vital organs/systems, including but not limited to the blood coagulation system and central nervous system, resulting in a variety of complications (Sahu et al. 2020; Fathi and Rezaei 2020; Lotfi and Rezaei 2020; Saleki et al. 2020) that have made the treatment of disease more challenging (Jahanshahlu and Rezaei 2020; Saghazadeh and Rezaei 2020; Lotfi et al. 2020; Mohamed et al. 2020b).

COVID-19 has caused a propagated pandemic (Hanaei and Rezaei 2020), mainly transmitted from person to person. It has affected almost all countries around the world, from children to adults and from community people to health professionals (Mohamed et al. 2020b). Although COVID-19 seems not being more severe in neonates and children, pregnant women (Mirbeyk and Rezaei 2020), and immunodeficient patients (Ahanchian et al. 2020) than the general population, its adverse effects particularly relate to older people and people with comorbidities, and that might depend on genetic background (Yousefzadegan and Rezaei 2020; Darbeheshti and Rezaei 2020; Ahmadi et al. 2020). Given the trends in the aging population and noncommunicable diseases, COVID-19 has introduced a global challenge that needs all people (Momtazmanesh et al. 2020), disciplines (Moradian et al. 2020; Rezaei 2020; Rabiee et al. 2020; Moazzami et al. 2020; Basiri et al. 2020b), and organizations working together to test their knowledge from the previous pandemics (Jabbari et al. 2020), collaborate with its solving process (Mohamed et al. 2020a), and brand high-standard lessons (Rzymski et al. 2020) for the global immunity in the future (Kafieh et al. 2020).

This chapter provides a review of the characteristics and epidemiological measures of COVID-19 in selected countries, followed by reviewing the adopted policies, shortcomings in confronting the pandemic and the probable underlying reasons, and the susceptibility of the communities to future outbreaks.

4.2 Coronavirus-Caused Outbreaks: SARS, MERS, and COVID-19

In November 2002, the SARS outbreak started in Guangdong province of China. It was not until later in 2003 that the SARS coronavirus (SARS-CoV) was confirmed as the cause of this outbreak. SARS-CoV originated from bats and then transmitted to other animals such as civet or cats and human. The disease manifested with flu-like symptoms, which were all nonspecific for SARS-CoV. After affecting 8437 patients and causing 813 deaths in 26 countries, the outbreak was controlled in July 2003 (Moazzami et al. 2020). The epidemic curves of the SARS outbreak (2002–2003) are illustrated in Figs. 4.1 and 4.2.

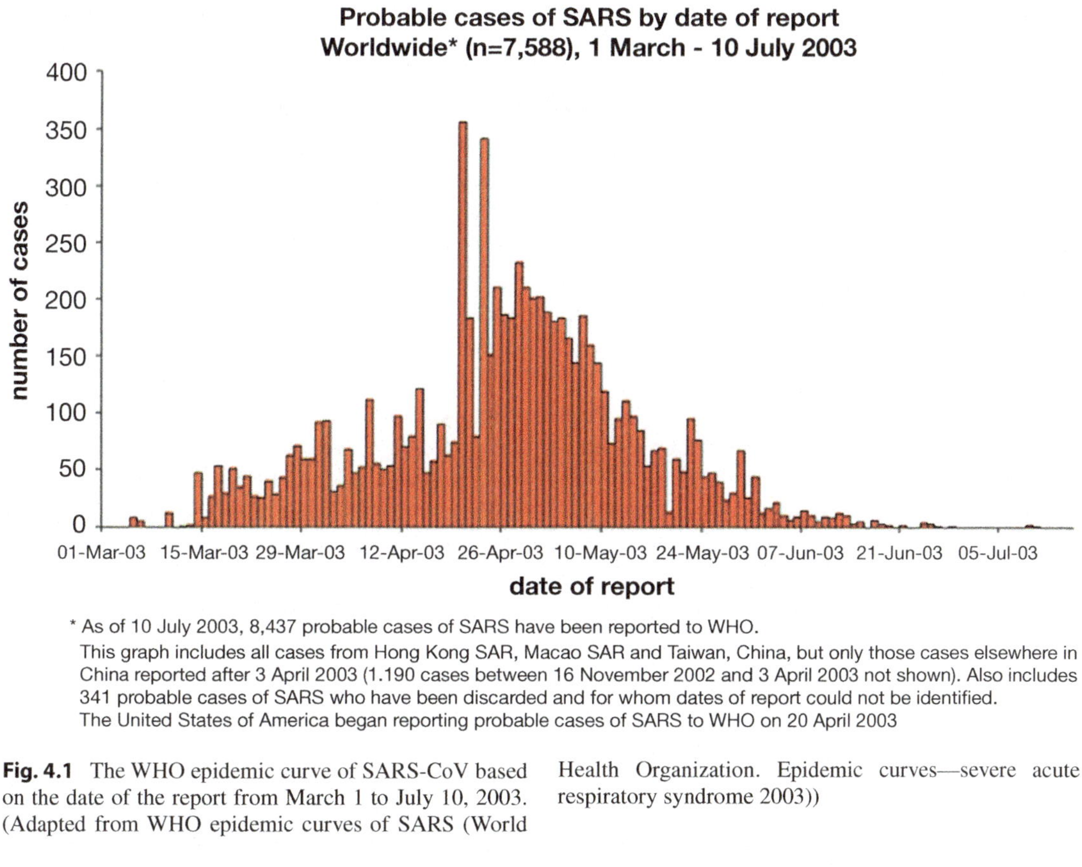

Fig. 4.1 The WHO epidemic curve of SARS-CoV based on the date of the report from March 1 to July 10, 2003. (Adapted from WHO epidemic curves of SARS (World Health Organization. Epidemic curves—severe acute respiratory syndrome 2003))

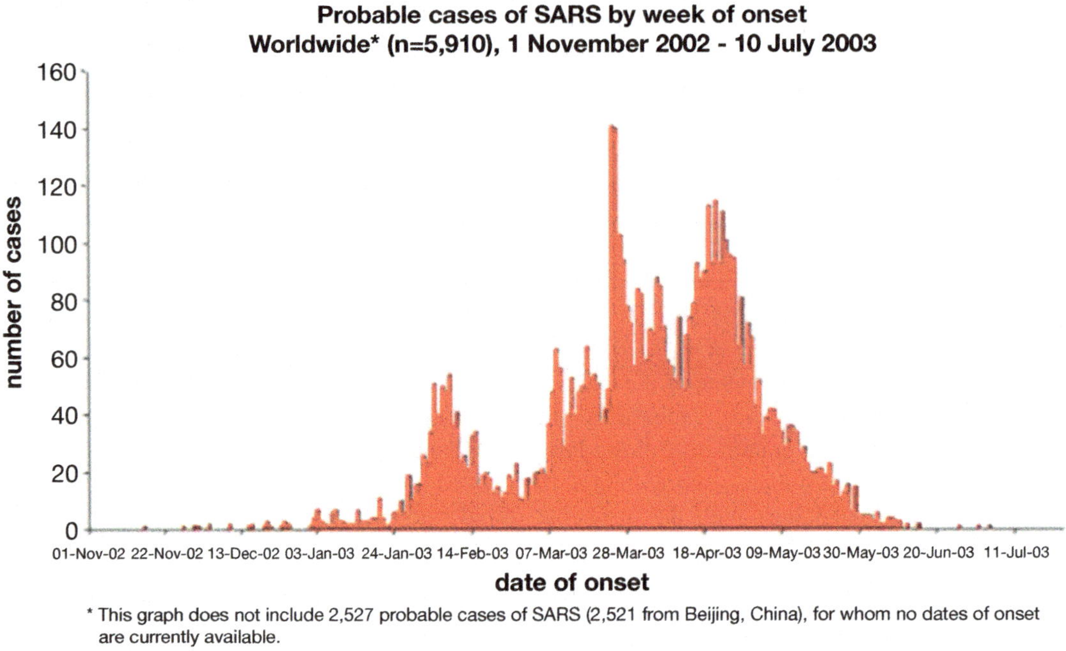

Fig. 4.2 The WHO epidemic curve of SARS-CoV based on the date of onset from November 1, 2002, to July 10, 2003. (Adapted from WHO epidemic curves of SARS (World Health Organization. Epidemic curves—severe acute respiratory syndrome 2003))

In 2012, after almost a decade of peace, a respiratory outbreak, similar to SARS, started in Saudi Arabia. It correlated mainly with the symptoms of fever, cough, and shortness of breath. It was later considered as the MERS outbreak to be caused by a new type of coronavirus (MERS-CoV), probably originating from bats and then transmitted to animals such as camels and then to humans with a zoonotic nature. The MERS outbreak has 2519 affected patients and 866 deaths in 27 countries since 2012 and continues (World Health Organization. Middle East respiratory syndrome coronavirus (MERS-CoV), Fact Sheets. 2019). The epidemic curves of the MERS outbreak from 2012 to 2019 are shown in Fig. 4.3.

While still struggling with MERS, in late December of 2019, the first reports of a new type of pneumonia from Wuhan, Hubei Province of China, appeared. Although the etiology of this type of pneumonia was not clear at the beginning, it was confirmed on January 7, 2020, that the novel coronavirus (2019-nCoV or SARS-CoV-2) was responsible for COVID-19. The WHO officially called this outbreak "a public health emergency of international concern" on January 30, 2020 (Pourahmad et al. 2020). It was first suspected that the virus originated from a common source of seafood. However, later on, less than 15% of new confirmed cases by January 23 had exposures with that source, and thus the hypothesis was rejected (Mansourabadi et al. 2020). The bats were the next suspect. However, it was later proposed that pangolin might be the intermediate host between bats and humans because pangolin-CoV has considerable genome similarities with both BatCoV RaTG13 and 2019-nCoV (Zhang et al. 2020). Despite the initial animal source of the disease, it was then reported that human-to-human transmission would play a major role in the spread of COVID-19 (Mansourabadi et al. 2020). The incubation period of COVID-19 was at first estimated between 2 and 10 days (WHO) and then updated to be 2.4 and 15.5 days (Yazdanpanah et al. 2020b; Backer et al. 2020). During this period, the infected asymptomatic patient might act as a silent carrier of the disease

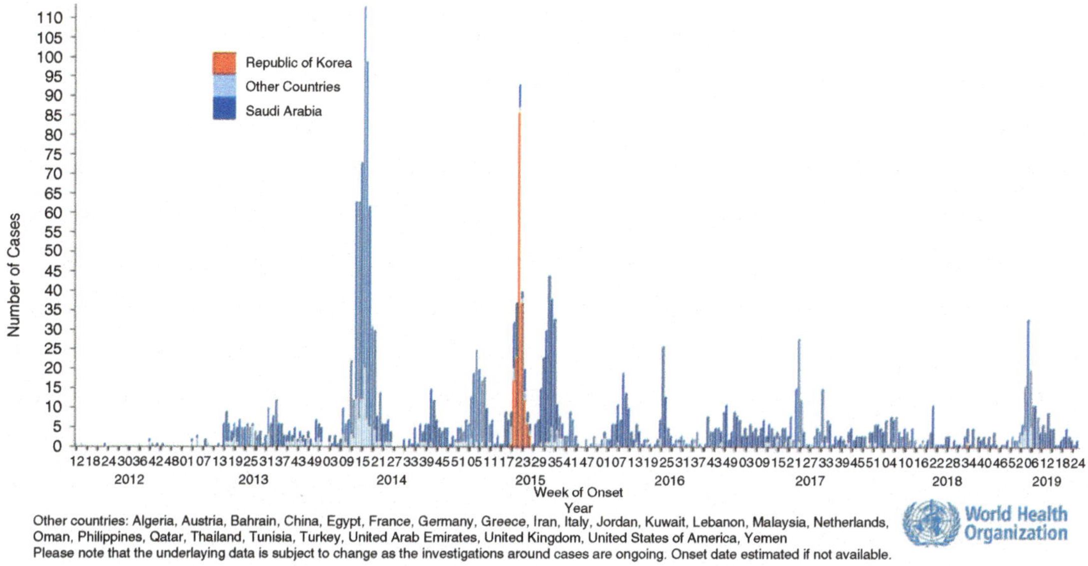

Fig. 4.3 The WHO epidemic curve of confirmed MERS-CoV. (Adapted from WHO MERS maps and epicurves (World Health Organization. Emergencies, Middle East respiratory syndrome coronavirus (MERS-CoV), MERS-CoV maps and epicurves))

and transmit the virus to other individuals, albeit the lower relative probability compared to symptomatic patients.

While the outbreak was quickly developing in China, other countries, including Thailand, Japan, and South Korea, were the firsts to be affected (Yazdanpanah et al. 2020b). The number of affected countries increased rapidly with thousands of confirmed patients and deaths until March 11, 2020, when the WHO officially characterized the COVID-19 outbreak a pandemic. By that time, over 118,000 patients were infected, and over 4200 deaths were caused by the disease in more than 100 countries (Rabiee et al. 2020). The characteristics of these three outbreaks caused by coronaviruses are summarized in Table 4.1.

4.3 The Characteristics of COVID-19

Below are the epidemiologic measurements calculated considering the world population of 7,797,000,000 (Figs. 4.4, 4.5, and 4.6).

4.3.1 Total Affected Cases of COVID-19

By January 5, 2021, a total number of 83,326,479 patients were confirmed for COVID-19. Disease distribution in countries based on the total number of reported cases is presented in Fig. 4.7.

4.3.2 Prevalence Rate

According to the total number of patients and the world population, the prevalence rate of COVID-19 is 10,687 per million population in the world (by January 05, 2021). However, the adjusted prevalence based on the country might be biased, as the accuracy of prevalence in some countries would be a matter of suspect due to the sources of bias (see Sect. 4.6).

4.3.3 Mortality Rate

By January 05, 2021, the COVID-19 caused 1,831,703 deaths among all the confirmed

Table 4.1 Comparison of outbreaks caused by coronaviruses: SARS, MERS, and COVID-19 Update date: 01/05/2021

	SARS	MERS	COVID-19
Time period	2002–2003	2012–present	2019–present
Agent	SARS-CoV	MERS-CoV	SARS-CoV-2
Source	Bats, cats, civet	Bats, camels	Unknown (bats, pangolins)
Reported cases	8437	2494	83,326,479
Reported deaths	813	858	1,831,703
Number of involved countries/regions	26	27	> 215
Incubation period	1–14 days (Organization 2003)	2–14 days (Organization 2018)	2.4–15.5 days

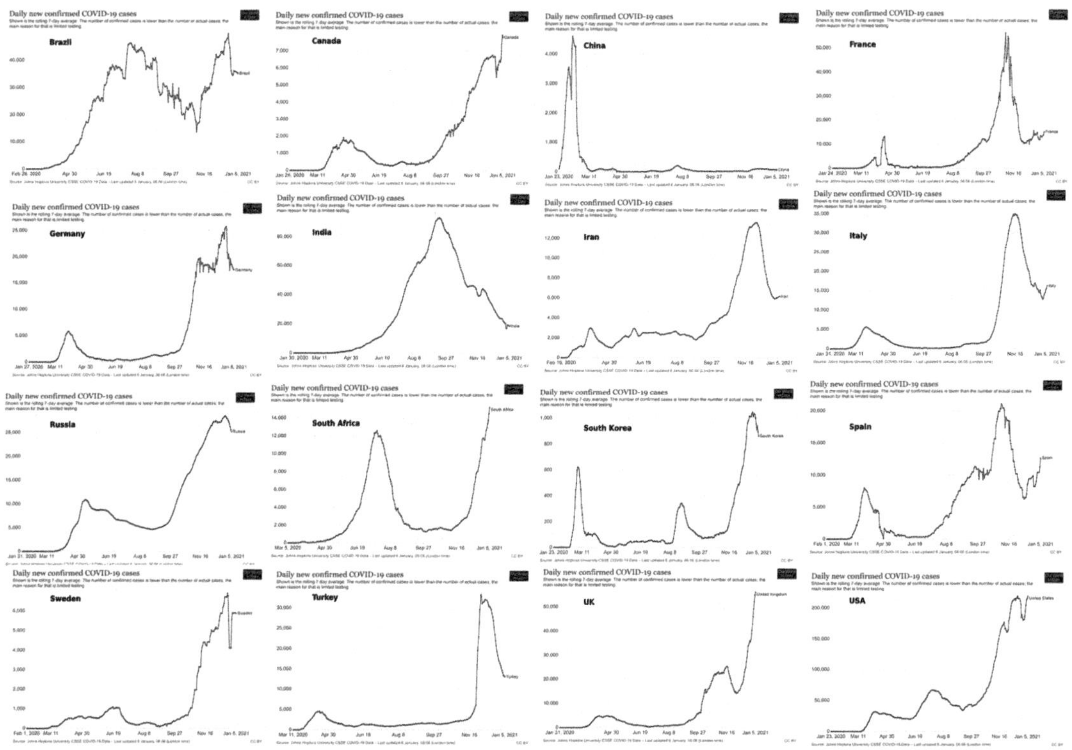

Fig. 4.4 Epidemic curves of daily new cases in 16 selected countries, from January 21, 2020, to January 05, 2021. (Adapted from Our World in Data (Niestadt 2020))

cases, which indicates a mortality rate of 4.45% by this date. Interestingly, the primary estimation of the mortality rate has been different. As reported by the general director of the WHO, Dr. Tedros Adhanom Ghebreyesus, on March 3, COVID-19 had resulted in death in 3.4% of all confirmed cases of this disease by the time (Hanaei and Rezaei 2020). The distribution of total COVID-19 deaths in the world is illustrated in Fig. 4.8. However, and similar to the prevalence rate, the mortality rate adjusted for the population size of countries would be a matter of suspicion due to potential sources of bias (see Sect. 4.6).

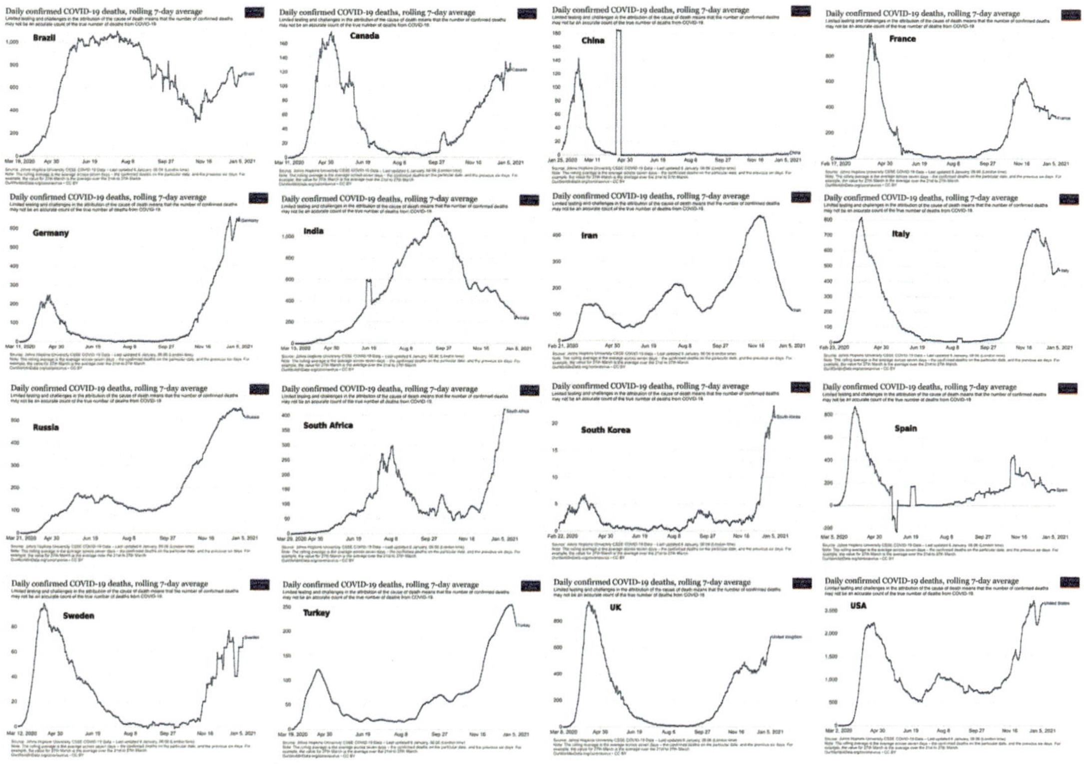

Fig. 4.5 Epidemic curves of daily new deaths in 16 selected countries, from January 21, 2020, to January 05, 2021. (Adapted from Our World in Data (Niestadt 2020))

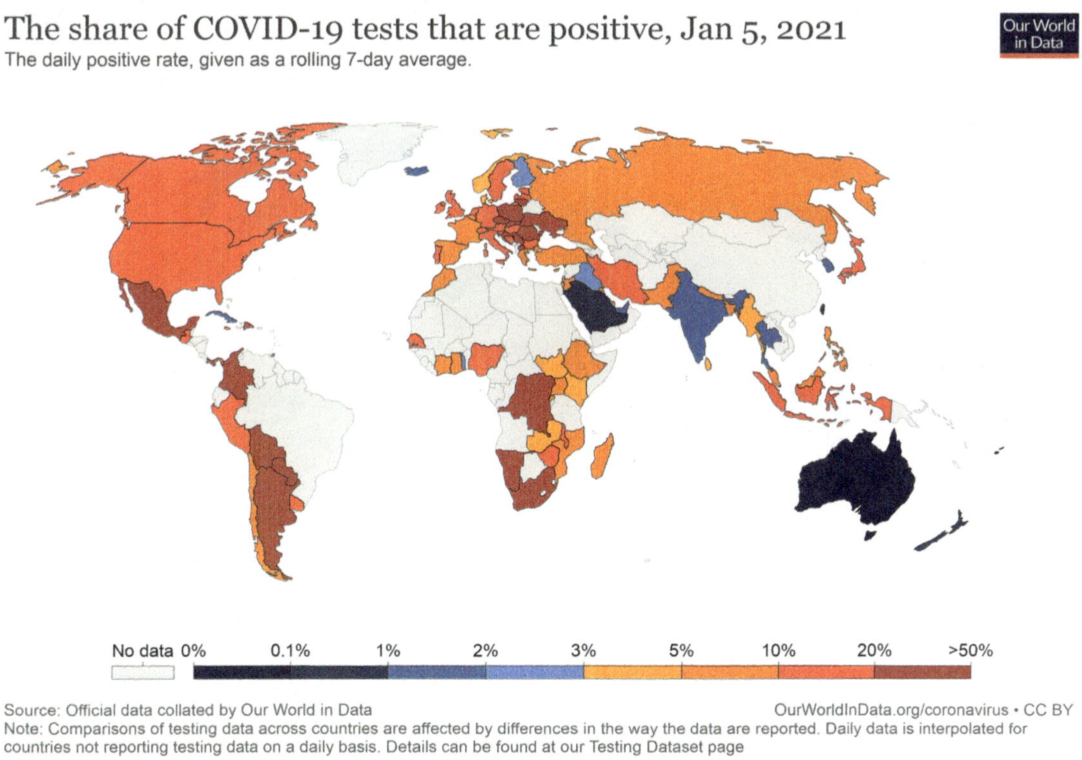

Fig. 4.6 Daily COVID-19 tests performed worldwide. (Adapted from Our World in Data (Niestadt 2020))

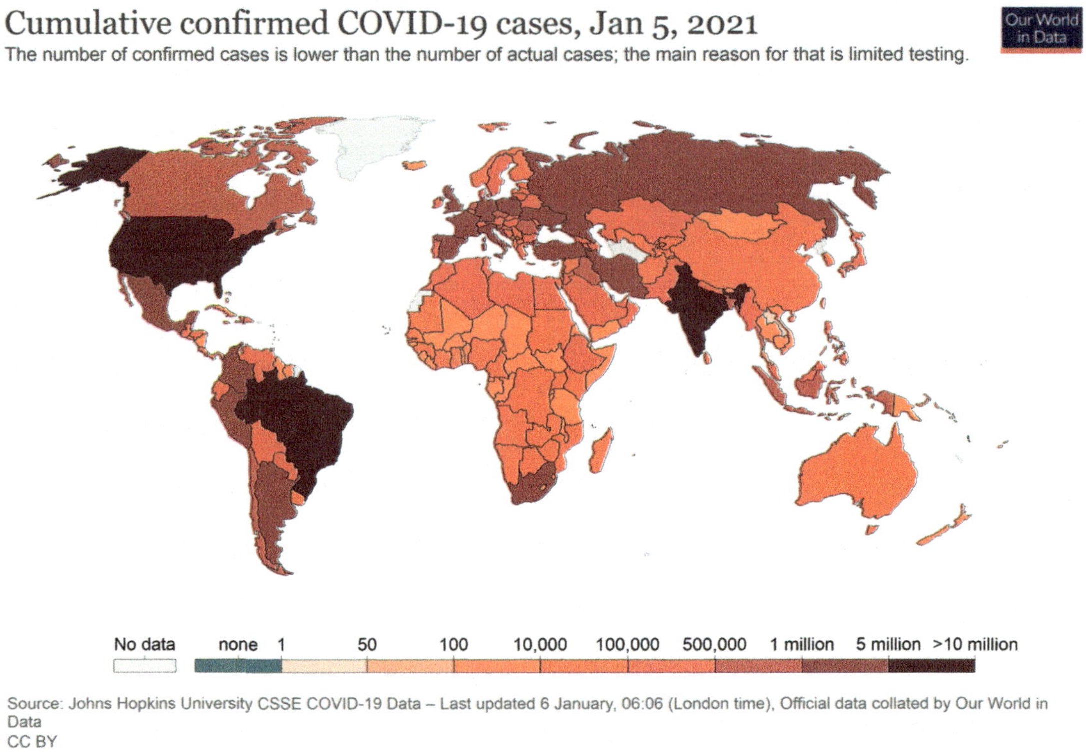

Fig. 4.7 Total confirmed cases of COVID-19 in countries. (Adapted from Our World in Data (Niestadt 2020))

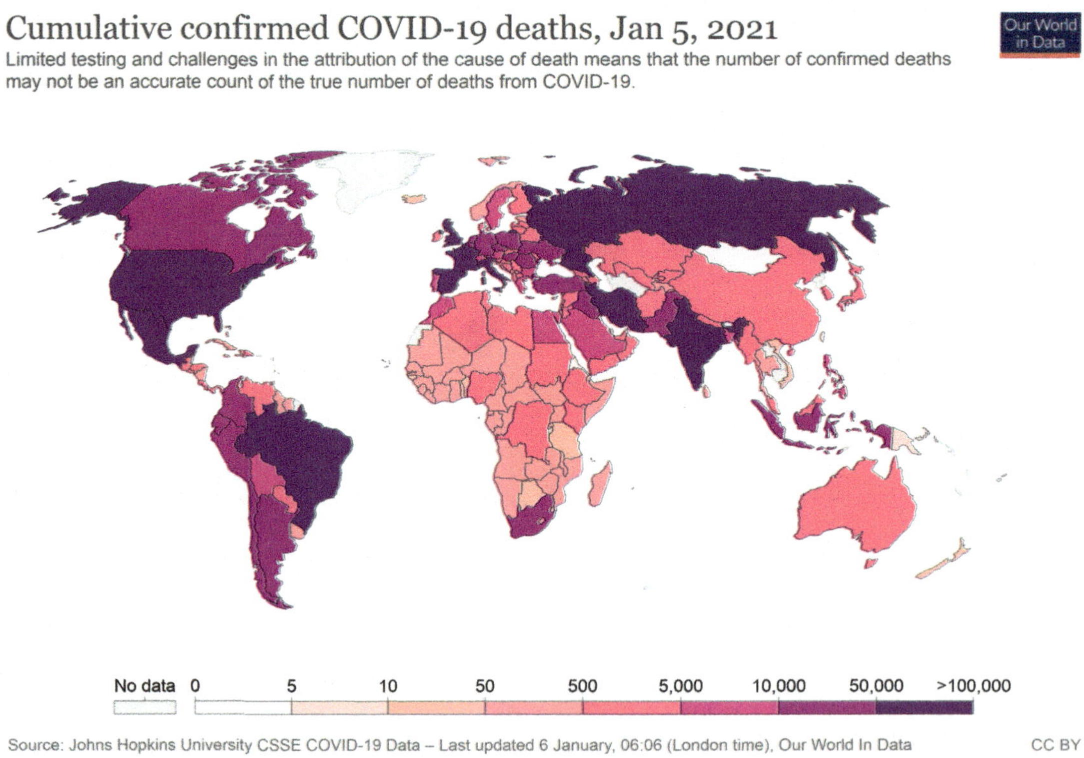

Fig. 4.8 Total deaths caused by COVID-19 in countries. (Adapted from Our World in Data (Niestadt 2020))

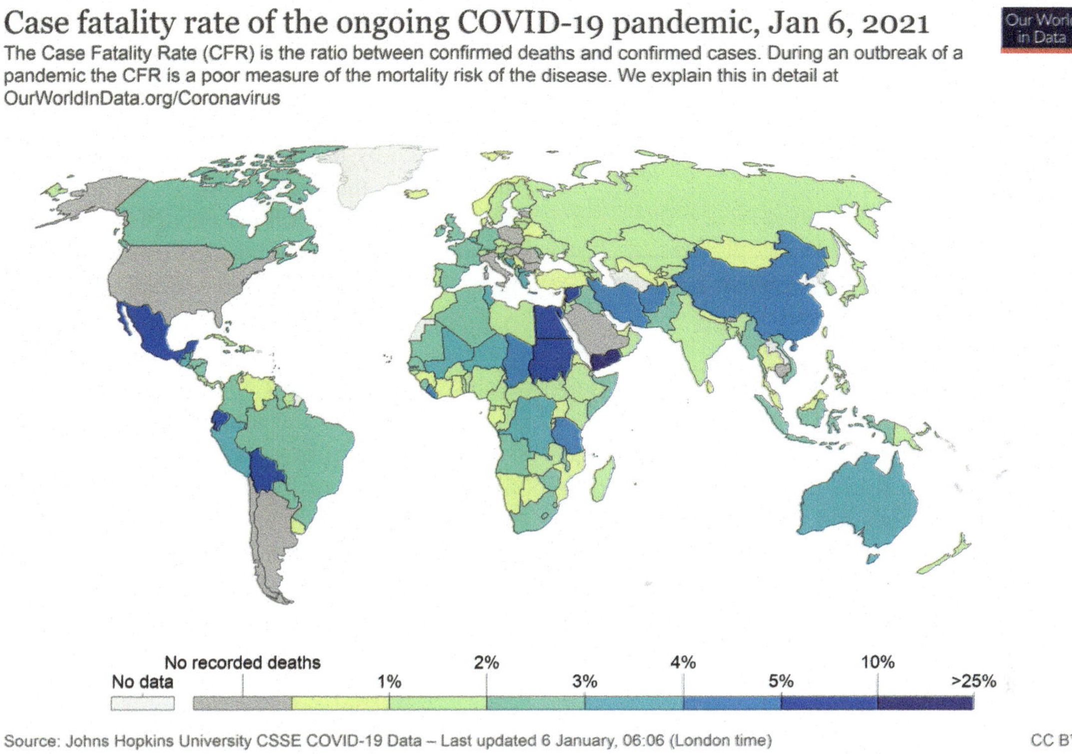

Fig. 4.9 Case-fatality rate in countries with more than 1000 affected patients. (Adapted from Our World in Data (Niestadt 2020))

4.3.4 Case-Fatality Rate

Out of 91,744,365 confirmed cases, 65,612,144 patients have recovered (until the time of writing this document), and 1,963,594 of them died. This indicates the case-fatality rate of 2.14% worldwide. The case-fatality rate was estimated to be 2.3% in the reports from China (Novel 2020) and also in the first report of the WHO on January 29, 2020 (Saghazadeh and Rezaei 2020); however, and as mentioned by the WHO as well, the rate would be more sensible than the initial reports when the disease spread to other countries. The distribution of the worldwide case-fatality rate is illustrated in Fig. 4.9.

4.3.5 COVID-19 Virulence

By January 10, 2021, out of 90,739,150 total confirmed cases of COVID-19, 1,943,214 were deceased of the disease, and 108,525 were in critical condition. Therefore, the virulence of COVID-19 would be 2.26% by this date.

4.3.6 Serial Interval

The average time it takes for a disease to be transmitted from a person to the next person is called the serial interval. COVID-19 is estimated to have a serial interval of 4.0–5.0 days (Li et al. 2020).

4.3.7 R_0 (The Reproduction Ratio or the Reproduction Number)

R_0 represents the number of people that an infected person passes the disease to in its next generation, in which its length is primarily determined by the serial interval measure of the disease. During the COVID-19 outbreak, the initial estimates of R_0 fall between 2.0 and 2.5, as reported by the WHO (Organization and Organization 2020). However, the Imperial College London reported it to be 3.0 (Walker et al. 2020).

4.4 Relevant Reporting Sources and Predictions

Different sources have attempted to collect daily records, as reported by countries' officials. It would be an essential effort in the documentation and keeping the world updated about the condition of one another and also evaluating the efficacy of the implemented strategies. Accordingly, one of the most frequently used sources was the daily situation reports of the WHO (Basiri et al. 2020a). A considerable number of public people, as well as official stakeholders of countries, may rely on any of these sources to stay updated not only about the status of their own country but also about the other countries. Therefore, the consistency between different sources shall be taken into consideration. The epidemic curves provided by different sources are considerably similar to each other, with insignificant differences that might be explained as different timing of taking original reports from countries or delays in updating the websites.

Other data sources are worth attention: (i) the Johns Hopkins report, which follows the general trends of the WHO with slightly higher numbers that is probably due to considering "presumptive positive cases" in its counting as well (CSSEGISandData/COVID-19 2020), and (ii) the European Centre for Disease Prevention and Control (ECDC), which is another source of data based on different resources including the individual level metrics as well and is very similar to the WHO and provides data not only about Europe but all over the world.

Besides the reporting, sources have also provided estimations and forecasts about the future of COVID-19. Although all predictions seem to be mathematically robust in methods, the numbers are very different. For example, the CDC estimated a 200,000 death (Fink 2020) in the USA versus the Imperial College London that predicted it to be approximately 2.2 million (Ferguson et al. 2020). The existing gap is probably caused by the inputs of the models and the uncertainties that come with it. As mentioned previously, there is still not enough certainty and census on the numbers for COVID-19 that determine the number of deaths, consisting of the at-risk population, infection rate, and fatality rate. Besides, there are basic dissimilarities in the reporting methods and statistics of COVID-19 that every entry of the simple equation to calculate death number is wobbly and, thus, broad in the range. Much of the mentioned factors differ in unalike contexts and situations. As a result, there could be no one single number to fit all countries, populations, genders, and age groups. It is not possible, at least yet, to make a collective predicted number of deaths.

4.5 COVID-19 Pandemic Details in China and Selected Countries

As described in detail, the statistics of COVID-19 vary by reporting source, and it is not yet clear how close to reality they are. Nevertheless, reviewing what is available would at least support the claim that the COVID-19 pandemic is one of the biggest disasters of history. Besides, comparisons should be made cautiously as completeness and, in some cases, the validity of the reports are not apparent and most likely are away from perfect. It might lie in different policies and contexts that have resulted in different denominators and, of course, numerators. Iceland, for instance, tests a large population of asymptomatic patients, and on-demand testing is available, while there are countries where the number of testing facilities is far less than the daily infected people such as some European countries with centralized, socialized healthcare systems. Finally, many people are already supposed to experience similar symptoms this time of year due to common cold or flu and are now considered a probable case of mild COVID-19 (Rokni et al. 2020). Although they could not be precisely distinguished from COVID-19 cases when reviewing the reports, it should be considered in interpreting the numbers. The epidemic curves of some highly infected countries are illustrated in Fig. 4.4 (cumulative cases) and Fig. 4.5 (cumulative deaths).

4.6 Bias in Estimating the Extent of COVID-19 Pandemic

The reported numbers of COVID-19 are incomplete, if not invalid. Healthcare systems and authorities are dealing with a disease accompanied by an unknown extent, and they probably just know the tip of the iceberg. Additionally, they are dealing with scarce and limited financial and medical resources to prevent the spread and diagnose and manage the affected patients. All these set creating reliable and accurate reports and datasets at a low priority for authorities, though being critical in decision-making and evaluating the situation and effectiveness of interventions. However, the trade-off exists, and it is not glass clear on what is the most efficient approach to save more people: to invest in reports or to allocate its funding to the medical and economic management of the situation, where it is closer to the battlefield. Besides, the accuracy and availability of tests are increasing alongside with the prevalence and incidence of the disease. Thus, the measures of case detections rates could be different by even 20-fold due to multiple reasons, and either a decline or an increase is a specific indication of progress/deterioration in the situation. Noteworthy, the reporting and estimating biases are nonrandom and, therefore, could not be neglected when comparing the countries. In general, the biases could be categorized as selective and nonselective biases.

4.6.1 Selective Bias or Political Bias

Considering some political reasons or seeking financial benefits, the official reports might be either delayed or underestimated in some countries. Although this may primarily be in line with the benefit, it will have a much destructive effect when the outbreak spread with a significant number of patients and deaths. The reasons are probably the same as the reasons for delays observed in implementing interventions to deal with COVID-19, described in Sect. 4.7. Country-level policies also play a critical role in this context; some countries may have proceeded with an extensive screening protocol for seeking suspicious patients and then coming up with confirming the diagnosis and treatment of patients. It resulted in significantly higher reports from developed countries such as the USA or European countries, which changed the maps of disease spread in a short time (Figs. 4.4, 4.5, 4.7, and 4.8).

4.6.2 Nonselective Bias or Technical Bias

Lack of adequate facilities for diagnosis, weak clinical information and surveillance systems, and underdeveloped death registration systems end up with undercounting of cases and deaths. Therefore, the incidence and disease spread could not be accurately calculated in countries with such systems. Lack of adequate resources for screening and diagnosing all the patients, especially in low-middle-income countries, may result in overestimation of disease severity and mortality, as specific testing for COVID-19 would be limited to hospitalized patients in these countries. They were mainly those with severe diseases or comorbidities (de Sousa et al. 2014). A similar condition might be dominated in some other countries with a low number of reports, especially the low-middle-income ones.

Another source of nonselective underreporting is the diagnostic suspicion bias. The diagnosis of COVID-19 can be missed, as the clinical manifestations of the disease are similar to other types of pneumonia. It was especially the case at the beginning of the COVID-19 pandemic. Therefore, having the background of community-acquired pneumonia (CAP) and common cold in mind, especially in the fall and winter, the physicians could have misdiagnosed some of the first cases of COVID-19 with the aforementioned conditions without specific testing for this disease (de Sousa et al. 2014).

There are also difficulties coming as might be expected with daily reports. For example, there is a lag time between the time of getting the disease and being presented as a confirmed case in the reports. It is due to the expected delay in the chain of reporting and also the time it takes to develop the symptoms and refer to a healthcare center. On top of all, there are errors in testing as a part of their nature. It includes not only false-positive cases, which are less of a concern when the prevalence is high and more concerning in low prevalence of the disease that it could exceed the actual numbers, but also false-negative cases that considerably decrease the accuracy of case detection rates.

4.7 Confronting the COVID-19 Outbreak and the Impact on the Expansion of COVID-19

Countries have adopted different interventions, strategies, and policies to deal with the COVID-19 pandemic. These strategies focus on either preventing and controlling the disease itself or managing the side effects of the former strategies such as economic and social impacts of lockdown. This chapter mainly focuses on the strategies to directly deal with COVID-19. The interventions could be generally categorized as health information (surveillance and intervention information), controlling the transmission (either prevention or mitigation), treating patients, protecting healthcare workers, and coordination.

Some factors significantly impact the national and global response to outbreaks and pandemics and cause delays, including disease severity, number of affected countries, at-risk population size, disease novelty, transmission way, perception by the public, time of the year, and to what extent it is affecting the USA (Hoffman and Silverberg 2018). Even though some delays seem to be necessary or inevitable before taking action, more could be categorized as a failure to perform the best. Of note, recent outbreaks had resulted in efforts to build surveillance and response infrastructure and systems. Thus, the delays seem to be more caused by delays in action mobilization, in comparison with detecting the outbreak, and multiple reasons could cause that. It is not a clear border between whether governments, authorities, or even people were denying the crisis at a personal level, or they decided not to do what is best to control the spread of the disease, though they understood how severe the condition is. These shortcomings could not merely get limited to poor governance, but they are also originating from the natural human failures in making decisions and dealing with fast-changing circumstances and crises.

4.7.1 Policymaker Shortcomings

The shortcuts, as mentioned earlier in decision-making, are mainly based on past experiences and recognized patterns, particularly in conditions that humans find similar to what they are dealing with in the time. Thus, in the best-case scenario, countries and governments could have implemented strategies based on recognized pandemics and outbreaks, including SARS, MERS, and even Spanish flu. However, not only were they different, but also the world today is fundamentally dissimilar to when these pandemics affected (see Sect. 4.8). Besides, their experience of flu, which was mainly compared with COVID-19 at the beginning of its spread to justify no drastic early actions, seems to be less lethal. Therefore, the superficial comparisons of elsewhere and other time comparisons seemed not to help the countries to deal perfectly with COVID-19. Moreover, it blinded some to take the condition less seriously. On top of all, many countries considered China to be significantly unalike them, politically, socially, and even genetically, that they did not consider the possibility of experiencing what China was going through in their home. Of course, authorities' rational decision-making might be clouded by the tendency of being famous and only a means of good news, that they avoided implying restrictive and not-very-welcomed-by-public policies even in their confrontations with a pandemic. The popularity and confirmation seeking (or even populism) were also presented as finding a victim who caused the outbreak by authorities. Among all inclinations, there could also be an optimism bias that comes for authorities with an illusion of control, mentioned by Donald Low, professor of practice in public policy at Hong Kong University of Science and Technology, which probably affected successful and wealthy government more.

4.7.2 General Population Shortcomings

Government and high-level policymakers are not the only confronters with pandemics; people play a substantial role as well. Importantly, culture and history come to the spotlight whenever people are making decisions. The differences in culture in dealing with COVID-19 could be simply observed in different countries' mask-wearing behavior. While it is quite acceptable to not wear masks in public places in the UK and the USA, a bare face could rarely be seen in most Asian countries. All these happen despite almost consistent advice from the WHO about who should wear a mask. Mask-wearing looks to be a cultural norm when digging more in-depth in the culture of these countries. It might be due to being impolite to sneeze or cough openly, avoiding air pollution, having recent and painful histories, and experiences of pandemics and outbreaks. It, results in a higher public awareness or more cautious people, being altruistic in culture and caring not to make others sick, and whether people are considered social pariahs or are stigmatized if wearing a mask or not attending social gatherings or even fashion trends. Of course, some cultures are more cautious and anxious toward any threat. The traces of culture were also presented in the panic-buying behavior, empty store shelves, and "the toilet paper shortage." Among the most critical factors, different cultures come with different hygienic and preventive behavior and self-care, including in the time of a pandemic. The culture also leaves its traces in how people react to mandatory lockdown; as western is historically considered a land of liberty, people would likely be less welcoming in dealing with the obligatory lockdown. Besides, it is less of an option for the authorities as well. Finally, the public trust in governments and social and financial security is also a powerful driver on how people react to COVID-19 reports and acquired policies by authorities as well, presented in the percentage of people who could not follow the social distancing and quarantine recommendations because they needed to work daily to provide for their family. Similarly, the media and its coverage also play a critical role in determining the behavior of people.

Understanding and recognizing the causes of delay and shortage in responses to COVID-19 not only provides a great opportunity for improvement for further preparedness but also is a great driver on why some countries suffered less from this pandemic and some others could not handle it.

4.8 The Psychological Explanation on Determining the Extent of an Outbreak

By causing a significant number of affected patients and deaths, an outbreak could potentially resemble a significant loss for a country or even the whole world. Therefore, the political, healthcare stockholders and the public population may react in five stages of grief (Axelrod 2006).

4.8.1 Denial and Isolation

The COVID-19 outbreak started with one or a few cases in each country and then rapidly developed to affect a considerable proportion of the country. However, surprisingly, it is commonly observed that no serious action or official announcement has been made until the beginning of the steep slope. This could be explained that potential political concerns might put one in the denial stage in the interval between confirmation of the first case and the official announcement of the outbreak in one country. It is also considered that in this interval, the regional reports on the COVID-19 situation would be underestimated.

4.8.2 Anger

Encountering the growing number of affected patients and deaths, the effect of denial would be masked sooner or later. The next reaction might be projecting the anger to find the victim who caused the outbreak. The anger stage of grief, in the context of the COVID-19 outbreak, has shown itself in blaming the first affected countries, neighboring countries, and airlines for starting the outbreak in one country. This shall be carefully differentiated from the scientific seeking of the source of the outbreak, which is a critical step in controlling it. While the causal source of the outbreak shall be quickly found to be eliminated, this shall not be sidelined with blaming one another for causing the outbreak.

4.8.3 Bargaining

In the time of the COVID-19 outbreak, there were numerous bargaining reactions in almost every country. Some examples could be as follows: if the government has started the isolation earlier…, if the borders were closed sooner…, if the people were united…, if we did not take the travel…, and many other buts and ifs. This stage is more commonly taken by the public population rather than the political or healthcare stockholders.

4.8.4 Depression

The fourth stage would be experienced by the countries' officials and the public population. It has shown itself in the COVID-19 outbreak as finding no other solution for overcoming the problem and surrendering the natural selection rule.

4.8.5 Acceptance

It is the most critical and crucial stage for controlling an outbreak. While the other four were the common reactions, which would not result in overcoming the COVID-19 outbreak, the fifth stage would manifest itself in accepting the occurrence of the outbreak and taking a rational strategy to prevent, treat, and control the outbreak.

The most crucial point in encountering these five stages is losing the valuable gold time between stages 1 and 4 to reach stage 5. In other words, the COVID-19 outbreak spread quite fast in the interval between stages 1 and 4, which resulted in unexpected infections and deaths in many countries and underestimation of true prevalence in those countries. The sooner one can overcome the four stages, the sooner the outbreak could be controlled, and the less loss would be experienced. Figure 4.10 illustrates the five stages of grief in an outbreak.

In other words, the COVID-19 outbreak spread quite fast in the interval between stages 1–4, which resulted in unexpected infections and deaths in many countries, and underestimation of true prevalence in those countries. The sooner one can overcome the 4 stages, the sooner the outbreak could be controlled, and the less loss would be experienced.

4.9 Social Impacts of the COVID-19 Pandemic

Other than devastative economic consequences for countries (Chap. 52), COVID-19 left mental and social impacts as well. Although the effects of COVID-19 pandemic might last for decades to an unknown extent, in here, a brief review of social and mental possible effects is presented.

4.9.1 Social Impacts

The social impacts of COVID-19 could be generally categorized as educational, religious, and political impacts of the pandemic. As of the polit-

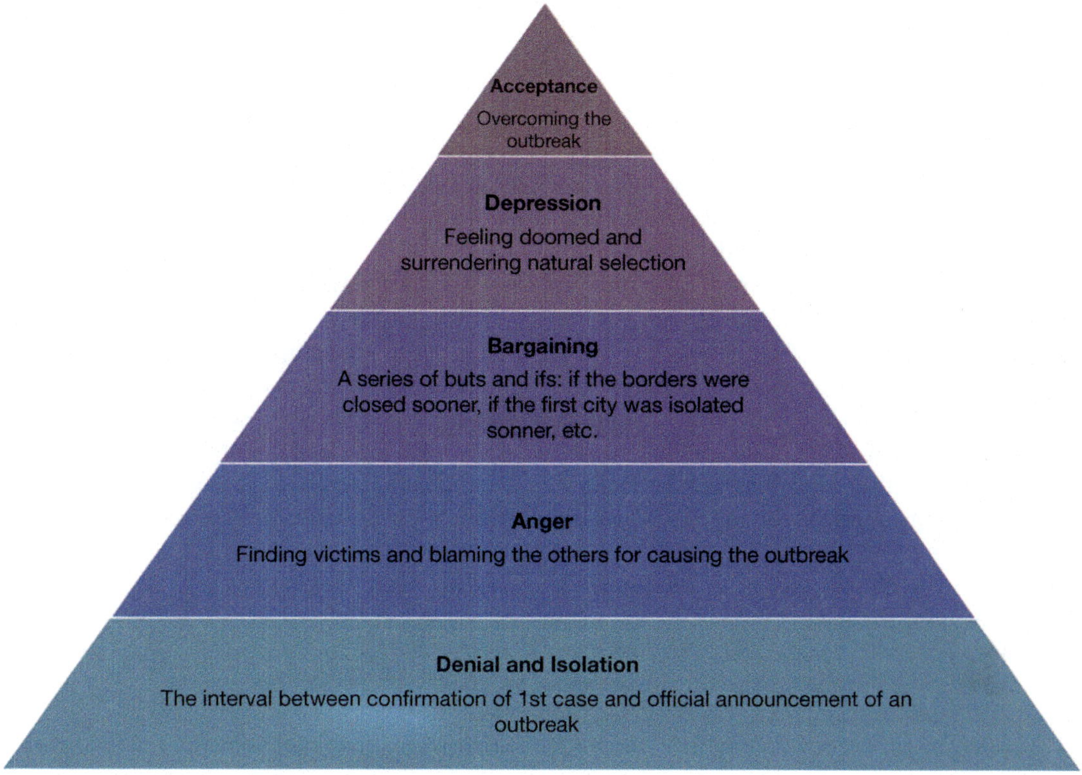

Fig. 4.10 Five stages of grief in encountering an outbreak

ical impacts, some might argue that civil rights and democracy are damaged due to adopted policies of mitigation, suppression, and the quarantine efforts. Besides, the false claims, wrong responses, and enfeebled nations are likely to have an impact on the political power of governments. On the other hand, people project tendencies to appreciate some authorities and service provides, even to the extent of heroism, including healthcare workers and specific authorities. The impact of these could continue as increased demand for the education and paths leading to these social roles. The pandemic has caused schools to close, and over 1.5 billion students worldwide (Buhalis 2020) were banned from studies, with the alternative of home-schooling with an unknown implementation rate and effectiveness. On top of all, the educational impact of pandemics seems to last for decades (Almond 2006). It has become a prerequisite that offline classrooms should transform into an online mode. Although the shift may take a longer time in developing countries than developed countries, however, it will boost the education system efficiently. Additionally, the COVID-19 pandemic could also have religious effects as social and religious rituals, and gestures are banned. Other than the direct explained consequences, these could result in unexpected, or even unknown, outcomes. However, predicting the future social effects is out the scope of this chapter.

4.9.2 Mental Impacts

Nations have a lot on their hands with limited resources in dealing with the physical impacts of pandemics that they mostly and deliberately neglect the mental health impacts of them. Thus, COVID-19, like any other catastrophic event, impairs the mental health of people. The mental problems that are failed to care for would probably be difficult to manage later as well as a result of accumulating and aggregating. The management of mental health problems gets further complicated because the COVID-19 pandemic, like most other social problems, would, in all likelihood, hurt vulnerable populations the most. The vulnerable populations consist of, but are not limited to, i, elders, with limited social encounters, who are now more restricted and would feel more alone than ever; ii, Asian people, as they are minorities in some countries and they are insulted and acted on because of the source of the pandemic; iii, people who are dealing directly with the COVID-19, such as recovered patients, as they could be stigmatized or avoided by people; iv, individuals with a history of psychiatric disorders including generalized anxiety disorder, obsessive-compulsive behaviors, and agoraphobia, as they are quite probable to deteriorate or present harmful behaviors during the crisis; and v, healthcare workers and authorities, who are already under relatively higher social pressure and feel responsible for what happens. Moreover, inequalities seem to widen as the economic shocks always hit people with the lowest income the most. Besides, they are more likely to have uncontrolled chronic diseases that increase the risk of severe infection and poor prognosis if affected by COVID-19.

4.10 The Future Outbreaks

Outbreaks are either human-made or natural; in either case, humans should get exposed to the underlying cause, get infected, and transmit it to others, and the pathogen expands to cause a major epidemic (Shamshirian and Rezaei 2020). There exist barriers between each level of pathogen emergence; however, the world today is an excellent place for infectious diseases to make an outbreak.

4.10.1 Global Trades, Travels, and Transactions

Almost all places in the world are at most 1 day apart with a flight. It provides merely a Utopia for infectious diseases where they could infect communities without any immunity or previous exposure to them as they could rapidly and easily spread among all. Besides, healthcare professionals are also unfamiliar with over-expected infections, and it would take them a while to respond properly to the disease.

4.10.2 Climate Change

The world is changing, and the extent of change and its impacts are not yet totally understood and predicted. As the weather personality changes, we would experience sharper fluctuations that theoretically provide a great opportunity and breeding situation for both infectious diseases in a hot and humid environment and for those that survive in cold weather.

4.10.3 Urbanization

Almost all countries are having positive rates of urbanization. Most of these cities have limited facilities of what we typically think of when we consider cities; they are densely populated locations with unhygienic and unplanned structures, and many people live in slums. Besides, the land use required to expand the cities increases the chance of getting exposed to zoonosis diseases as well.

4.10.4 Poverty

Although extreme poverty has declined, poverty seems to persist. Poverty comes with lower hygiene, lower health, and lower access to the healthcare system, and all these make a great combination as a target for infectious diseases to spread and not to be controlled.

4.10.5 Demographic Shifts

Life expectancy has increased both in developed countries and developing countries with different extents. Older people come with higher susceptibility to infectious diseases. Besides, a significant percentage of them have comorbidities that increase their vulnerability to infectious diseases even more.

4.10.6 Behavior Shifts

There are new trends in human behavior that has never been present in history. Thus, the outcomes are unpredictable. Besides, the previous existing behaviors, like eating behaviors, could also result in different outcomes and, as a result, outbreaks of infectious diseases.

Nevertheless, healthcare systems, surveillance methods and structures, medicine, and health technology are all advancing as well with an unexperienced speed. It is not still crystal clear which side would win the opportunity, but authorities could utilize the opportunities that come with the advancements, particularly in technology and social media, to tackle rising outbreaks more effectively and efficiently than ever.

4.11 Conclusion

The epidemiology of COVID-19 showed how the world is unprepared in facing an outbreak. Even the wealthiest and most developed countries, which ranked highest based on the Global Health Security Index (Mirbeyk and Rezaei 2020) and had the most well-equipped health systems, were damaged tremendously by COVID-19. However, the coronavirus 2019 pandemic can give countries a psychological and, hopefully, structural head start and teach nations important lessons to be utilized in facing future epidemics more effectively and efficiently.

In general, countries should stay committed to the necessary steps of preparation, prevention, and protection in facing any future epidemics. Countries are encouraged and called to prepare themselves for any probable outbreaks based on the predictions of researchers and scientists; besides, they should also plan on how to prevent the spread and how to protect their population if an outbreak comes up. It could range from basic backup equipment and testing facilities, intersectoral coordination of efforts to minimize waste of disorganization in time of emergency, or having the power of becoming independent on essential requirements in the time of closed borders. Among all, the importance of truth-telling mainly came to spotlight the most during the COVID-19 pandemic. It directly affected the responses of authorities and people and weakened the strength and timeliness of reactions and people's trust in officials where it was neglected. Although sharing bad news may cause fear and anxiety, the nations should stay focused on reacting adequately to the situation, and not getting panic attacks, and accepting where they are, even if they are in bad condition. Interventions are effective when they are designed for the real situation, not the wished situation. On top of all, the suppressed information would usually come to light. Thus, the lack of truth-telling might cause the nations to lose their trust in the powers. In the protection phase, unfortunately, marginalized communities who were among the most vulnerable population are usually neglected, but they are a common and important reason why the actions fail. Of course, nations should not forget about the late-onset effects of the pandemics, including economic, social, and mental health impacts, and should plan to address them ahead of the time they get complicated or unable to be controlled.

References

Ahanchian H, Moazzen N, Faroughi MSD, Khalighi N, Khoshkhui M, Aelami MH, Haghi NSM, Rezaei N (2020) COVID-19 in a child with primary specific antibody deficiency

Ahmadi M, Saffarzadeh N, Habibi MA, Hajiesmaeili F, Rezaei N (2020) Colon cancer and SARS-CoV-2: impact of ACE2 expression in susceptibility to COVID-19. bioRxiv

Almond D (2006) Is the 1918 influenza pandemic over? Long-term effects of in utero influenza exposure in the post-1940 US population. J Polit Econ 114(4):672–712

Axelrod J (2006) The 5 stages of loss and grief. Psych Central

Backer JA, Klinkenberg D, Wallinga J (2020) Incubation period of 2019 novel coronavirus (2019-nCoV) infections among travellers from Wuhan, China, 20–28 January 2020. Eur Secur 25(5)

Bahrami A, Vafapour M, Moazzami B, Rezaei N (2020) Hyperinflammatory shock related to COVID-19 in a patient presenting with multisystem inflammatory syndrome in children: first case from Iran. J Paediatr Child Health. https://doi.org/10.1111/jpc.15048

Basiri A, Heidari A, Nadi MF, Fallahy MTP, Nezamabadi SS, Sedighi M, Saghazadeh A, Rezaei N (2020a) Microfluidic devices for detection of RNA viruses. Rev Med Virol:e2154. https://doi.org/10.1002/rmv.2154

Basiri A, Pazhouhnia Z, Beheshtizadeh N, Hoseinpour M, Saghazadeh A, Rezaei N (2020b) Regenerative medicine in COVID-19 treatment: real opportunities and range of promises. Stem Cell Rev Rep:1–13

Buhalis D (2020) BRACE the future and time to reignite, refocus, redesign and reengineer our global tourism industry: the new 4Rs of tourism!

CSSEGISandData/COVID-19 (2020). https://github.com/CSSEGISandData/COVID-19/issues

Darbeheshti F, Rezaei N (2020) Genetic predisposition models to COVID-19 infection. Med Hypotheses

de Sousa R, Reusken C, Koopmans M (2014) MERS coronavirus: data gaps for laboratory preparedness. J Clin Virol 59(1):4–11

Fathi N, Rezaei N (2020) Lymphopenia in COVID-19: therapeutic opportunities. Cell Biol Int

Ferguson N, Laydon D, Nedjati Gilani G, Imai N, Ainslie K, Baguelin M, Bhatia S, Boonyasiri A, Cucunuba Perez Z, Cuomo-Dannenburg G (2020) Report 9: impact of non-pharmaceutical interventions (NPIs) to reduce COVID19 mortality and healthcare demand

Fink S (2020) Worst-case estimates for U.S. Coronavirus deaths. The New York Times

Hanaei S, Rezaei N (2020) COVID-19: developing from an outbreak to a pandemic. Arch Med Res. https://doi.org/10.1016/j.arcmed.2020.04.021

Hoffman SJ, Silverberg SL (2018) Delays in global disease outbreak responses: lessons from H1N1, Ebola, and Zika. Am J Public Health 108(3):329–333

Jabbari P, Jabbari F, Ebrahimi S, Rezaei N (2020) COVID-19: a chimera of two pandemics. Disaster Med Public Health Prep:1–2. https://doi.org/10.1017/dmp.2020.223

Jahanshahlu L, Rezaei N (2020) Monoclonal antibody as a potential anti-COVID-19. Biomed Pharmacother:110337

Kafieh R, Arian R, Saeedizadeh N, Minaee S, Yadav SK, Vaezi A, Rezaei N, Javanmard SH (2020) COVID-19 in Iran: a deeper look into the future. medRxiv

LePan N (2020) A visual history of pandemics. World Economic Forum. https://www.weforum.org/agenda/2020/03/a-visual-history-of-pandemics/. Accessed 31 Mar 2020

Li Q, Guan X, Wu P, Wang X, Zhou L, Tong Y, Ren R, Leung KS, Lau EH, Wong JY (2020) Early transmission dynamics in Wuhan, China, of novel coronavirus–infected pneumonia. N Engl J Med

Lotfi M, Rezaei N (2020) SARS-CoV-2: a comprehensive review from pathogenicity of the virus to clinical consequences. J Med Virol

Lotfi M, Hamblin MR, Rezaei N (2020) COVID-19: transmission, prevention, and potential therapeutic opportunities. Clin Chim Acta

Mansourabadi AH, Sadeghalvad M, Mohammadi-Motlagh HR, Rezaei N (2020) The immune system as a target for therapy of SARS-CoV-2: a systematic review of the current immunotherapies for COVID-19. Life Sci 258:118185. https://doi.org/10.1016/j.lfs.2020.118185

Mirbeyk M, Rezaei N (2020) The impact of COVID-19 on pregnancy and neonatal health: a systematic review

Moazzami B, Razavi-Khorasani N, Dooghaie Moghadam A, Farokhi E, Rezaei N (2020) COVID-19 and telemedicine: immediate action required for maintaining healthcare providers well-being. J Clin Virol 126:104345. https://doi.org/10.1016/j.jcv.2020.104345

Mohamed K, Rodríguez-Román E, Rahmani F, Zhang H, Ivanovska M, Makka SA, Joya M, Makuku R, Islam MS, Radwan N (2020a) Borderless collaboration is needed for COVID-19—a disease that knows no borders. Infect Control Hosp Epidemiol:1–2

Mohamed K, Yazdanpanah N, Saghazadeh A, Rezaei N (2020b) Computational drug discovery and repurposing for the treatment of Covid-19: a systematic review. Available at SSRN 3583748

Momtazmanesh S, Ochs HD, Uddin LQ, Perc M, Routes JM, Vieira DN, Al-Herz W, Baris S, Prando C, Rosivall L (2020) All together to fight COVID-19. Am J Trop Med Hyg 102(6):1181–1183

Moradian N, Ochs HD, Sedikies C, Hamblin MR, Camargo CA, Martinez JA, Biamonte JD, Abdollahi M, Torres PJ, Nieto JJ (2020) The urgent need for integrated science to fight COVID-19 pandemic and beyond. J Transl Med 18(1):1–7

Nasab MG, Saghazadeh A, Rezaei N (2020) SARS-CoV--2–a tough opponent for the immune system. Arch Med Res

Niestadt M (2020) COVID-19 and the tourism sector

Novel CPERE (2020) The epidemiological characteristics of an outbreak of 2019 novel coronavirus diseases (COVID-19) in China. Zhonghua liu xing bing xue za zhi 41(2):145

Organization WH, Organization WH (2020) Report of the who-China joint mission on coronavirus disease 2019 (covid-19)

Pourahmad R, Moazzami B, Rezaei N (2020) Efficacy of plasmapheresis and immunoglobulin replacement therapy (IVIG) on patients with COVID-19. SN Compr Clin Med:1–5

Rabiee N, Rabiee M, Bagherzadeh M, Rezaei N (2020) COVID-19 and picotechnology: potential opportunities. Med Hypotheses 144:109917. https://doi.org/10.1016/j.mehy.2020.109917

Rezaei N (2020) COVID-19 and medical biotechnology. Avicenna J Med Biotechnol 12(3):139–139

Rokni M, Hamblin MR, Rezaei N (2020) Cytokines and COVID-19: friends or foes? Hum Vaccin Immunother:1–3

Rzymski P, Nowicki M, Mullin GE, Abraham A, Rodríguez-Román E, Petzold MB, Bendau A, Sahu KK, Ather A, Naviaux A-F (2020) Quantity does not equal quality: scientific principles cannot be sacrificed. Int Immunopharmacol 86:106711

Saghazadeh A, Rezaei N (2020) Towards treatment planning of COVID-19: rationale and hypothesis for the use of multiple immunosuppressive agents: anti-antibodies, immunoglobulins, and corticosteroids. Int Immunopharmacol 84:106560. https://doi.org/10.1016/j.intimp.2020.106560

Sahu KK, Siddiqui AD, Rezaei N, Cerny J (2020) Challenges for management of immune thrombocytopenia during COVID-19 pandemic. J Med Virol. https://doi.org/10.1002/jmv.26251

Saleki K, Banazadeh M, Saghazadeh A, Rezaei N (2020) The involvement of the central nervous system in patients with COVID-19. Rev Neurosci 1 (ahead-of-print)

Shamshirian D, Rezaei N (2020) Cardiovascular diseases burden in COVID-19: systematic review and meta-analysis

US Department of Health Human Services (2013) Principles of epidemiology in public health practice third edition an introduction to applied epidemiology and biostatistics. Atlanta, Georgia, USA, pp 8. Available on the website: http://www.cdc.gov/ophss/csels/dsepd/SS197

Walker PG, Whittaker C, Watson O, Baguelin M, Ainslie K, Bhatia S, Bhatt S, Boonyasiri A, Boyd O, Cattarino L (2020) The global impact of covid-19 and strategies for mitigation and suppression. WHO Collaborating Centre for Infectious Disease Modelling, MRC Centre for Global Infectious Disease Analysis, Abdul Latif Jameel Institute for Disease and Emergency Analytics, Imperial College London

World Health Organization (2019) Middle East respiratory syndrome coronavirus (MERS-CoV), Fact Sheets. https://www.who.int/en/news-room/fact-sheets/detail/middle-east-respiratory-syndrome-coronavirus-(mers-cov)

Worst-Case Estimates for U.S. Coronavirus Deaths (2020) The New York Times. https://www.nytimes.com/2020/03/13/us/coronavirus-deaths-estimate.html. Accessed 18 Mar 2020

Yazdanpanah F, Hamblin MR, Rezaei N (2020a) The immune system and COVID-19: friend or foe? Life Sci:117900

Yazdanpanah N, Saghazadeh A, Rezaei N (2020b) Anosmia: a missing link in the neuroimmunology of coronavirus disease 2019 (COVID-19). Rev Neurosci. https://doi.org/10.1515/revneuro-2020-0039

Yousefzadegan S, Rezaei N (2020) Case report: death due to COVID-19 in three brothers. Am J Trop Med Hyg 102(6):1203–1204

Zhang T, Wu Q, Zhang Z (2020) Probable pangolin origin of SARS-CoV-2 associated with the COVID-19 outbreak. Curr Biol

5 The Incubation Period of COVID-19: Current Understanding and Modeling Technique

Char Leung

Abstract

This chapter aims to answer the following questions regarding the incubation period of COVID-19. Why is understanding the incubation period of COVID-19 important? How long is the incubation time, and what are the associating factors? How should the incubation period be modeled given the current pandemic situation? Where should we go from here? As a critical epidemiological metric, the incubation period is of public health and clinical importance. While the incubation time of COVID-19 is generally similar to that of SARS and MERS, recent studies identifying factors that impact the incubation period of COVID-19, travel history, for example, only tell part of the story. Therefore, in addition to reviewing current findings, this chapter also explores the modeling technique and future research directions of the incubation period of COVID-19.

C. Leung (✉)
Deakin University, Burwood, VIC, Australia

Network of Immunity in Infection, Malignancy and Autoimmunity (NIIMA), Universal Scientific Education and Research Network (USERN), Burwood, VIC, Australia
e-mail: char.leung@deakin.edu.au

Keywords

COVID-19 · Epidemiology · Incubation period · MERS · Modeling · SARS

5.1 Introduction

The pandemic of coronavirus disease 2019 (COVID-19) originated in Wuhan, China, as pneumonia with unknown etiology. It was declared on the 30th of January 2020 by the World Health Organization (WHO) as the sixth Public Health Emergency of International Concern. The causative agent was later discovered and named as SARS-CoV-2 (previously known as 2019-nCoV). It is the seventh coronavirus pathogenic to humans, after SARS-CoV and MERS-CoV, as well as strains 229E, NL43, HKU1, and OC43 that cause the common cold.

While the pathogenesis, etiology, and epidemiology of COVID-19 remain an active area of research, SARS-CoV-2 is molecularly related to SARS-like coronaviruses detected in bats from Yunnan province (96.3% similarity) (Paraskevis et al. 2020) and Zhoushan (88% identity) (Lu et al. 2020). Consequently, some of the virologic and epidemiological features of the severe acute respiratory syndrome (SARS) also hold for COVID-19. For instance, both viruses are believed to be transmitted through respiratory droplets and direct contact, as well as, most

N. Rezaei (ed.), *Coronavirus Disease - COVID-19*, Advances in Experimental Medicine and Biology 1318, https://doi.org/10.1007/978-3-030-63761-3_5

likely, by the fecal-oral route (Gu et al. 2020). Both viruses use receptor angiotensin-converting enzyme 2 (ACE2) for host cell entry and serine protease TMPRSS2 for spike protein priming (Hoffmann et al. 2020). In contrast, the basic reproduction number of COVID-19 is higher, compared to SARS (Liu et al. 2020a). The differences in the incubation period between COVID-19, SARS, and MERS (Middle East respiratory syndrome) are not fully understood. It remains a research area of public health and clinical importance, despite limited investigation. This chapter focuses on the incubation period of COVID-19 by drawing on current studies as well as the modeling technique.

5.2 The Importance of Understanding the Incubation Period

The term incubation period, also known as the incubation period of a disease, is defined as the time interval between exposure to the pathogen and symptom onset. A less commonly used term, the incubation period of infection, refers to the time from exposure to infection if the infection does not occur directly after exposure (Rothman et al. 2012).

Several reasons call for an urgent understanding of the incubation period of COVID-19. In addition to the determination of the length of quarantine, it would help to (Armenian and Lilienfeld 1983):

- Judge the dynamics of an epidemic: while there is no direct association between the incubation period and the latent period (the time interval between exposure to the pathogen and the host becoming infectious), the incubation period can be used to understand the dynamics of COVID-19, along with the serial interval. It has been suggested that the infectious profile of COVID-19 more closely resembles that of influenza than of SARS (He et al. 2020).
- Estimate date of common exposure: at the beginning of the outbreak of COVID-19 in Wuhan, the time interval between exposure to the Huanan Seafood Wholesale Market and symptom onset was used to speculate the source of infection (Li et al. 2020). It is important because the late identification of the common source may cause a higher number of infections (Huang et al. 2020).
- Assess the risk of developing the disease: the incubation period can also be used as part of the case definition for assessing the risk of developing the disease in those who have close contacts with COVID-19 patients. It has been suggested that the incubation period can help the differential diagnosis between MERS and influenza (Nishiura et al. 2012). It should not be overlooked because COVID-19 could be mistaken for influenza, and patients were not aware of the need for hospitalization (Leung 2020a).
- Study unusual cases: despite being a respiratory disease, some COVID-19 cases have been reported to be associated with Guillain-Barré syndrome (GBS) (El Otmani et al. 2020). While GBS is commonly believed to be a postinfectious disease, a case has been reported in China that the symptom onset of GBS occurred 7 days before that of COVID-19, suggestive of a para-infectious profile (Zhao et al. 2020). The incubation period of COVID-19 should be useful in understanding the relationship between GBS and COVID-19 from an epidemiological perspective.

To date, research on the incubation period of COVID-19 mainly focuses on the following: i, the median and/or mean of the incubation period; ii, the statistical distribution of the incubation period; and iii, factors associated with the incubation period. The incubation period of almost all diseases is right-skewed; thus, the median also serves a suitable summary statistic to describe the data. The variation of the incubation period can be represented by the interquartile range (IQR), confidence interval, or credible interval. The statistical distribution of the incubation provides further analysis from a probabilistic perspective. While the relationship between viral replication and symptom onset is not entirely known, the incubation period of many diseases follows a log-normal distribution, as proposed by Sartwell (Sartwell 1950).

5.3 Existing Findings of the Incubation Period of COVID-19

The first observation of the incubation period of COVID-19 was reported by the National Health Commission of China, suggesting an incubation time between 1 and 14 days (Leung 2020b). To date, most studies concerning the incubation period of COVID-19 focus on patients in China; hence clinical data accessible to researchers outside of China remain very limited. Table 5.1 summarizes the findings reported in different studies.

The size of the sample considered in these studies varies between 6 and 2907. Single-center studies can recruit only a small number of patients, while studies with data gathered from public sources consider a more significant number of patients, although biases may arise from the search strategy. Consequently, the reported mean and median incubation time varies greatly. Also, comparative studies on cohorts with different characteristics report very different incubation times. Thirteen mean incubation periods were reported varying between 1.8 and 7.2 days, whereas 16 median incubation periods were reported varying between 2 and 12 days. Different forms of interval estimates have also been reported, including credible intervals, confidence intervals, ranges, and IQRs. Against highly diverse estimates, it is not surprising that the incubation time of COVID-19 is believed to be similar to that of SARS and MERS (Jiang et al. 2020). The mean and median incubation time of SARS were reported to be 3.8 to 6.9 days (Cai et al. 2006; Cowling et al. 2007; Kuk and Ma 2005; Donnelly et al. 2003; Lau et al. 2010; Lessler et al. 2009) and 4 days (Lessler et al. 2009), respectively, whereas the mean and median incubation time of MERS were reported to be 5 to 6.9 days (Virlogeux et al. 2016a; Cauchemez et al. 2014) and 5.2 days (Assiri et al. 2013), respectively.

The differences in the reported incubation time of COVID-19 may be explained by a variety of factors such as variation of host immune responses and virus replication. In the remainder of this section, we review existing observations regarding the determinants of the incubation period of COVID-19. While the clinical data evidence many of these findings, the pathogenesis of many of these observations requires further research. For instance, it was found that travelers to Hubei had significantly shorter incubation periods than non-travelers (Leung 2020b). Another study found that patients who regularly lived in Wuhan had more extended incubation periods compared to those infected outside Wuhan (Xiao et al. 2020). Because Hubei residents and travelers to Hubei might have been exposed to the pathogen multiple times during their stay in Hubei (Yu et al. 2020b), incubation time may be associated with the dose of exposure, as in the case of typhoid (Fine 2003).

Similar observations concerning the relationship between viral load and incubation period have also been proposed. The incubation period of tertiary patients was significantly more extended than that of secondary patients, while the viral load was undetectable in tertiary patients on day 14 after the exposure (Xu et al. 2020a). However, the authors also noted that the median age of the tertiary patients was significantly higher than that of the secondary patients, implying that age may confound the relation between tertiary transmission and incubation time. Moreover, the authors also reported that tertiary patients had more comorbidities, suggestive of comorbidities being another confounder. Notwithstanding this finding, another study suggested that there is no relationship between an incubation time and comorbidities such as chronic obstructive pulmonary disease, diabetes, hypertension, cardiovascular disease, and cerebrovascular disease (Guan et al. 2020a).

The role viral load plays in the determination of the incubation period remains uncertain, although some studies may help clarify. A study suggested high viral load in the respiratory secre-

Table 5.1 Studies concerning the incubation time of COVID-19

Study	Sample size	Location	The incubation period (days)	Interval estimates (days)
Backer et al. (2020)	88	China	6.4 (mean)	5.6–7.7 (95% CI)
Chen et al. (2020)	12 children and adolescents	Chongqing, China	8 (mean)	1–13 (range)
Guan et al. (2020b)	291	China	4 (median)	2–7 (IQR)
Guan et al. (2020a)	1590	China	3.6 (mean)	4.2 (SD)
Han et al. (2020b)	32	Shaanxi, China	5 and 4 (median) for children and adults	3–12, 2–12 (range) for children and adults
Huang et al. (2020)	6	Anhui, China	2 (median)	1–4 (range)
Ki and 2019-nCoV (2020))	10	South Korea	3.9 (mean), 3 (median)	0–15 (range)
Lauer et al. (2020)	181	25 countries	5.1 (median)	4.5–5.8 (95% CI)
Leung (2020b)	175	China	1.8 and 7.2 (mean) for patients with and without travel history to Hubei	ND
Li et al. (2020)	425	China	5.2 (mean)	4.1–7.0 (95% CI)
Linton et al. (2020)	158	China	5 and 5.6 (mean) excluding/ including Wuhan residents	4.2–6 and 5–6.3 (95% CI) excluding/ including Wuhan residents
Nie et al. (2020)	2907	China	5 (median)	2–8 (IQR)
Qian et al. (2020)	88	Zhejiang, China	6 (median)	3–8 (IQR)
Qiu et al. (2020)	71	Hunan, China	6 (median)	1–32 (range)
Sanche et al. (2020)	24	China	4.2 (mean)	3.5–5.1 (95% CI)
Shen et al. (2020)	6 children	Hunan, China	7.5 (median)	1–6 (range)
Song et al. (2020)	21	Beijing, China	8 (median)	5–12 (IQR)
Sun et al. (2020)	55	Beijing, China	7.5 (median)	5–11.8 (IQR)
Tian et al. (2020)	262	Beijing, China	6.7 (median)	ND
Wang et al. (2020)	483	Henan, China	7 (median), 7.4 (mean)	4.1–9.4 (IQR)
Xiao et al. (2020)	9	Chongqing, China	7 (mean)	2.59 (SD)
Xiao et al. (2020)	2555	China	8.98 (mean)	7.98–9.9 (95% CI)
Xu et al. (2020a)	51	Jiangsu, China	8, 8, and 12 (median) for imported, secondary, and tertiary infection	4–10, 4–11, and 9–14 (IQR) for imported, secondary, and tertiary infection
Xu et al. (2020b)	56	Zhejiang, China	4 (median)	3–5 (IQR)

(continued)

Table 5.1 (continued)

Study	Sample size	Location	The incubation period (days)	Interval estimates (days)
Yu et al. (2020b)	132	Shanghai, China	7.2 (median)	6.4–7.9 (95% CI)
Zhang et al. (2020a)	49	China, excluding Hubei	5.2 (mean)	1.8–12.4 (95% CI)
Zhang et al. (2020b)	8	Beijing, China	2.1 (mean)	ND

CI confidence interval, *IQR* interquartile range, *ND* not determined, *SD* standard deviation

tions of patients during incubation (Chan et al. 2020), as explained by the finding that SARS-CoV-2 produced higher amounts of infectious particles than SARS-CoV in human lung tissues (Chu et al. 2020). Nevertheless, most studies concluded that patients are infectious during the incubation period (Qian et al. 2020; Yu et al. 2020a; Zhang et al. 2020a), including the observation that the serial interval of COVID-19 is shorter than the incubation period (Park et al. 2020). In contrast, it has been observed that patients with SARS are not infectious during the incubation period (Li et al. 2003; Zeng et al. 2009).

While findings of the association between viral load and disease severity in COVID-19 remain conflicting (He et al. 2020; Liu et al. 2020b), at least three studies have demonstrated that disease severity is not correlated to the incubation period (Tian et al. 2020; Guan et al. 2020b; Sun et al. 2020), as opposed to the case of SARS (Virlogeux et al. 2015) and MERS (Virlogeux et al. 2016b). It is important to note that disease severity may be defined differently among studies, varying from clinical features such as dyspnea to the use of diagnosis guideline that includes lab findings (Metlay et al. 2019).

Other features concerning the incubation period of COVID-19 have also been observed. For example, there is no significant difference in the incubation time between females and males (Wang et al. 2020; Nie et al. 2020; Xiao et al. 2020). Some other proposed features represent preliminary results because of the small sample size involved. For instance, there is not yet evidence to show any significant difference in the incubation period between children (n = 7) and adults (n = 25) (Han et al. 2020b), although children present relatively mild symptoms (L'Huillier and Asner 2020; Castagnoli et al. 2020) as well as lower prevalence of elevated inflammatory markers and lymphocytopenia (Ludvigsson 2020). In contrast, a study on a large sample found that incubation periods of elderly patients were significantly more extended (Xiao et al. 2020). Nevertheless, circumstantial evidence suggests that the immune response is associated with the incubation time. A case series proposed that the long-term use of glucocorticoids might cause more extended incubation periods (Han et al. 2020a).

Interestingly, an immunocompromised patient with MERS was found to have longer incubation periods (Kim et al. 2017). It is suspected that multiple courses of chemotherapy and autologous peripheral blood stem cell transplantation prolonged the incubation period. Whether these are coincident observations or the impact of immunosuppression requires further research in immunology.

5.4 Modeling the Distribution of the Incubation Period

While summary statistics such as the median and the mean provide useful information about the incubation period, the statistical distribution reflects variations in infectious dose, in replication times of the pathogen, and levels of susceptibility among members of the host population (Fine 2003).

Just like the case of other diseases, modeling the distribution of the incubation period of COVID-19 is usually subject to incomplete information. Although the date of first symptom onset is usually known, the date of exposure to the infectious source is usually not precisely known. For example, the moment of infection by a virus via a mosquito bite is not directly observable, whereas the onset of symptoms is always known for symptomatic patients (Fourié et al. 2018). In the case of COVID-19, a patient living outside Hubei yet infected during her temporary stay in Hubei can only report a time interval within which the exact time of exposure falls. Even worse, a patient who always lives in Hubei may not be able to recall any possible exposure given human-to-human transmission of SARS-CoV-2. Unless the exact time of exposure or symptom onset is known, incubation period data are usually interval-censored.

While the exclusion of incomplete observations is an option (Zhang et al. 2020a; Ki and 2019-nCoV TFf 2020), maximum likelihood estimation (MLE) can be applied to estimate the incubation period distribution based on interval-censored data. Let E and S be the exact date of exposure and symptom onset, and then the incubation period, T, is defined as $T = S - E$. The following likelihood function, $L_{i,1}$, can be applied to an individual i whose time of exposure and symptom onset is precisely known:

$$L_{i,1} = f(T;\theta)$$

where f is the probability density function of T and θ is the set of distribution parameters.

In the case of interval-censored data, if the time of exposure of individual i is only known to fall between E_L and E_R, where $E_R > E_L$, the likelihood function is given by:

$$L_{i,2} = F(T_R;\theta) - F(T_L;\theta)$$

where F is the cumulative distribution function of T, $T_R = S - E_L$ and $T_L = S - E_R$.

Given a dataset consisting of n observations of the above types, the sum of the log-likelihood, l, for estimating the incubation period is given by:

$$l = \sum_{n}^{i=1} \ln\left[(1-\gamma_i)L_{i,1} + \gamma_i L_{i,2}\right]$$

where $\gamma_i = 1$ if observation i is interval-censored.

The literature on the distribution of the incubation period of COVID-19 remains very scant. While approaches used vary greatly among studies, the lognormal distribution is a popular choice of F (Wang et al. 2020; Lauer et al. 2020; Li et al. 2020; Linton et al. 2020; Zhang et al. 2020a). Weibull (Backer et al. 2020; Leung 2020b; Xiao et al. 2020), beta (Zhu and Chen 2020), and gamma (Yu et al. 2020b) distribution have also been assumed. In contrast, the incubation period of SARS and MERS mainly follows the lognormal and gamma distribution (Donnelly et al. 2003; Cowling et al. 2007; Lau et al. 2010; Lessler et al. 2009; Assiri et al. 2013; Virlogeux et al. 2016a). Nevertheless, the choice of F depends on the data. In many studies, models with different underlying distributions are estimated, and the model selection relies on predetermined criteria (Leung 2020b; Backer et al. 2020).

The median of the incubation period can be obtained from the quantile function of the estimated distribution. That is, given the set of estimated, $\hat{\theta}$, the median is given by $F^{-1}(0.5;\hat{\theta})$ where F^{-1} is the inverse function of F, i.e., the quantile function. The 95% confidence interval of the median requires bootstrapping a nonparametric technique. It can be done by creating K bootstrap samples. These bootstrap samples are then used to estimate K medians of incubation periods, following the abovementioned estimation procedure. The 95% confidence interval is then the 2.5% and 97.5% percentile of these medians.

If F is lognormal as in Sartwell's model, the following parametric method might be used (Gilbert 1987). The median of a lognormal distribution is $e^{\hat{\mu}}$ where $\hat{\mu}$ is the estimated log-transformed mean. The 95% confidence interval of the median is then $\exp\left\{\hat{\mu} \pm t_{0.975,n-1} se(\hat{\mu})\right\}$, where t is the quantile of Student t-distribution and se is the standard error of $\hat{\mu}$.

This approach can further be generalized to study the factors associated with the incubation period. For example, to see any difference in the incubation period distribution between travelers to Hubei and non-travelers, the likelihood functions were parameterized as follows (Leung 2020b):

$$L_{i,1} = f\left(T;,\theta_1 + I\delta_1;,\theta_2 + I\delta_2\right)$$

$$L_{i,2} = F\left(T_R;,\theta_1 + I\delta_1;,\theta_2 + I\delta_2\right) - F\left(T_L;,\theta_1 + I\delta_1;,\theta_2 + I\delta_2\right)$$

where $\theta_1 + I\delta_1$ and $\theta_2 + I\delta_2$ are the shape and scale parameter of Weibull distribution, respectively, with I being the binary variable with value 1 or 0 for travelers to Hubei and non-travelers and δ_1 and δ_2 are parameters indicating the impact of the binary variable.

Other MLE approaches for estimating incubation periods have also been proposed. For instance, the likelihood function can further be modified to accommodate doubly interval-censored data where the time of both exposure and symptom onset is not precisely known (Reich et al. 2009). The incubation period of postinfectious diseases such as Guillain-Barre syndrome can also be estimated by modifying this approach (Leung 2019).

5.5 Future Research Directions

While the comparison between COVID-19, SARS, and MERS remains an exciting theme, other areas of research regarding the incubation period of COVID-19 should be on the agenda. As discussed in the previous section, more research is needed to verify whether the incubation period is related to the infectious dose. This also holds for the relationship between the viral load and the incubation time, as suggested by contradicting findings.

The incubation time may also be determined by the age of the patient, as in the case of campylobacteriosis (Awofisayo-Okuyelu et al. 2017). Because of weaker immune systems, pediatric and geriatric patients are expected to experience more extended incubation periods. While longer incubation periods have been observed in the elderly, existing literature shows an insignificant difference in the incubation period between children and adults, perhaps due to small sample sizes. Nonetheless, more research is needed to explore the relationship between the immune system and incubation time.

Finally, most of the studies have concerned the incubation period of COVID-19 patients in China. Similar to SARS (Lau et al. 2010), variations in incubation time between regions may exist. With COVID-19 raging in more than 200 countries, comparative epidemiologic analyses serve as another important research direction.

5.6 Conclusion

This chapter reviewed existing literature concerning the incubation period of COVID-19, one of the critical epidemiological metrics with clinical significance. Studies report very different incubation times of COVID-19, partly owing to the characteristics of the cohorts as well as the method of sampling. As a result, the incubation period of COVID-19 is not significantly different from that of SARS and MERS. While the travel history and the level of transmission have been found to be associated with the incubation time, the mechanism behind these observations requires further extensive research. Moreover, many other factors, such as viral load, comorbidities, age, and immunosuppression, may pose as factors determining the length of the incubation period. Because of confounding relationships, existing studies remain inconclusive and more evidence is necessary for clarification. Modeling techniques have also been discussed in this chapter. With the presence of incomplete information, maximum likelihood estimation serves as a useful technique to model incubation periods by recognizing the interval-censored data structure.

References

Armenian HK, Lilienfeld AM (1983) Incubation period of disease. Epidemiol Rev 5:1–15. https://doi.org/10.1093/oxfordjournals.epirev.a036254

Assiri A, McGeer A, Perl TM, Price CS, Al Rabeeah AA, Cummings DA, Alabdullatif ZN, Assad M, Almulhim A, Makhdoom H, Madani H, Alhakeem R, Al-Tawfiq JA, Cotten M, Watson SJ, Kellam P, Zumla AI, Memish ZA, Team KM-CI (2013) Hospital outbreak of Middle East respiratory syndrome coronavirus. N Engl J Med 369(5):407–416. https://doi.org/10.1056/NEJMoa1306742

Awofisayo-Okuyelu A, Hall I, Adak G, Hawker JI, Abbott S, McCARTHY N (2017) A systematic review and meta-analysis on the incubation period of Campylobacteriosis. Epidemiol Infect 145(11):2241–2253. https://doi.org/10.1017/S0950268817001303

Backer JA, Klinkenberg D, Wallinga J (2020) Incubation period of 2019 novel coronavirus (2019-nCoV) infections among travellers from Wuhan, China, 20–28 January 2020. Euro Surveill 25(5). https://doi.org/10.2807/1560-7917.ES.2020.25.5.2000062

Cai QC, Xu QF, Xu JM, Guo Q, Cheng X, Zhao GM, Sun QW, Lu J, Jiang QW (2006) Refined estimate of the incubation period of severe acute respiratory syndrome and related influencing factors. Am J Epidemiol 163(3):211–216. https://doi.org/10.1093/aje/kwj034

Castagnoli R, Votto M, Licari A, Brambilla I, Bruno R, Perlini S, Rovida F, Baldanti F, Marseglia GL (2020) Severe acute respiratory syndrome coronavirus 2 (SARS-CoV-2) infection in children and adolescents: a systematic review. JAMA Pediatr. https://doi.org/10.1001/jamapediatrics.2020.1467

Cauchemez S, Fraser C, Van Kerkhove MD, Donnelly CA, Riley S, Rambaut A, Enouf V, van der Werf S, Ferguson NM (2014) Middle East respiratory syndrome coronavirus: quantification of the extent of the epidemic, surveillance biases, and transmissibility. Lancet Infect Dis 14(1):50–56. https://doi.org/10.1016/S1473-3099(13)70304-9

Chan JF, Yuan S, Kok KH, To KK, Chu H, Yang J, Xing F, Liu J, Yip CC, Poon RW, Tsoi HW, Lo SK, Chan KH, Poon VK, Chan WM, Ip JD, Cai JP, Cheng VC, Chen H, Hui CK, Yuen KY (2020) A familial cluster of pneumonia associated with the 2019 novel coronavirus indicating person-to-person transmission: a study of a family cluster. Lancet 395(10223):514–523. https://doi.org/10.1016/S0140-6736(20)30154-9

Chu H, Chan JF, Wang Y, Yuen TT, Chai Y, Hou Y, Shuai H, Yang D, Hu B, Huang X, Zhang X, Cai JP, Zhou J, Yuan S, Kok KH, To KK, Chan IH, Zhang AJ, Sit KY, Au WK, Yuen KY (2020) Comparative replication and immune activation profiles of SARS-CoV-2 and SARS-CoV in human lungs: an ex vivo study with implications for the pathogenesis of COVID-19. Clin Infect Dis. https://doi.org/10.1093/cid/ciaa410

Cowling BJ, Muller MP, Wong IO, Ho LM, Louie M, McGeer A, Leung GM (2007) Alternative methods of estimating an incubation distribution: examples from severe acute respiratory syndrome. Epidemiology 18(2):253–259. https://doi.org/10.1097/01.ede.0000254660.07942.fb

Donnelly CA, Ghani AC, Leung GM, Hedley AJ, Fraser C, Riley S, Abu-Raddad LJ, Ho LM, Thach TQ, Chau P, Chan KP, Lam TH, Tse LY, Tsang T, Liu SH, Kong JH, Lau EM, Ferguson NM, Anderson RM (2003) Epidemiological determinants of spread of causal agent of severe acute respiratory syndrome in Hong Kong. Lancet 361(9371):1761–1766. https://doi.org/10.1016/S0140-6736(03)13410-1

El Otmani H, El Moutawakil B, Rafai MA, El Benna N, El Kettani C, Soussi M, El Mdaghri N, Barrou H, Afif H (2020) Covid-19 and Guillain-Barré syndrome: More than a coincidence! Rev Neurol (Paris). https://doi.org/10.1016/j.neurol.2020.04.007

Fine PE (2003) The interval between successive cases of an infectious disease. Am J Epidemiol 158(11):1039–1047. https://doi.org/10.1093/aje/kwg251

Fourié T, Grard G, Leparc-Goffart I, Briolant S, Fontaine A (2018) Variability of Zika virus incubation period in humans. Open Forum Infect Dis 5(11):ofy261

Gilbert R (1987) Statistical methods for environmental pollution monitoring. Van Nostrand Reinhold Company, New York

Gu J, Han B, Wang J (2020) COVID-19: gastrointestinal manifestations and potential fecal-Oral transmission. Gastroenterology. https://doi.org/10.1053/j.gastro.2020.02.054

Guan WJ, Liang WH, Zhao Y, Liang HR, Chen ZS, Li YM, Liu XQ, Chen RC, Tang CL, Wang T, Ou CQ, Li L, Chen PY, Sang L, Wang W, Li JF, Li CC, Ou LM, Cheng B, Xiong S, Ni ZY, Xiang J, Hu Y, Liu L, Shan H, Lei CL, Peng YX, Wei L, Liu Y, Hu YH, Peng P, Wang JM, Liu JY, Chen Z, Li G, Zheng ZJ, Qiu SQ, Luo J, Ye CJ, Zhu SY, Cheng LL, Ye F, Li SY, Zheng JP, Zhang NF, Zhong NS, He JX, Covid-19 CMTEGf (2020a) Comorbidity and its impact on 1590 patients with Covid-19 in China: a nationwide analysis. Eur Respir J. https://doi.org/10.1183/13993003.00547-2020

Guan WJ, Ni ZY, Hu Y, Liang WH, Ou CQ, He JX, Liu L, Shan H, Lei CL, Hui DSC, Du B, Li LJ, Zeng G, Yuen KY, Chen RC, Tang CL, Wang T, Chen PY, Xiang J, Li SY, Wang JL, Liang ZJ, Peng YX, Wei L, Liu Y, Hu YH, Peng P, Wang JM, Liu JY, Chen Z, Li G, Zheng ZJ, Qiu SQ, Luo J, Ye CJ, Zhu SY, Zhong NS (2020b) Clinical characteristics of coronavirus disease 2019 in China. N Engl J Med. https://doi.org/10.1056/NEJMoa2002032

Han Y, Jiang M, Xia D, He L, Lv X, Liao X, Meng J (2020a) COVID-19 in a patient with long-term use of glucocorticoids: a study of a familial cluster. Clin Immunol 214:108413. https://doi.org/10.1016/j.clim.2020.108413

Han YN, Feng ZW, Sun LN, Ren XX, Wang H, Xue YM, Wang Y, Fang Y (2020b) A comparative-descriptive analysis of clinical characteristics in 2019-coronavirus-infected children and adults. J Med Virol. https://doi.org/10.1002/jmv.25835

He X, Lau EHY, Wu P, Deng X, Wang J, Hao X, Lau YC, Wong JY, Guan Y, Tan X, Mo X, Chen Y, Liao B, Chen W, Hu F, Zhang Q, Zhong M, Wu Y, Zhao L, Zhang F, Cowling BJ, Li F, Leung GM (2020) Temporal dynamics in viral shedding and transmissibility of COVID-19. Nat Med. https://doi.org/10.1038/s41591-020-0869-5

Hoffmann M, Kleine-Weber H, Schroeder S, Krüger N, Herrler T, Erichsen S, Schiergens TS, Herrler G, Wu NH, Nitsche A, Müller MA, Drosten C, Pöhlmann S (2020) SARS-CoV-2 cell entry depends on ACE2 and TMPRSS2 and is blocked by a clinically proven protease inhibitor. Cell 181(2):271–280.e278. https://doi.org/10.1016/j.cell.2020.02.052

Huang L, Zhang X, Wei Z, Zhang L, Xu J, Liang P, Xu Y, Zhang C, Xu A (2020) Rapid asymptomatic transmission of COVID-19 during the incubation period demonstrating strong infectivity in a cluster of youngsters aged 16–23 years outside Wuhan and characteristics of young patients with COVID-19: a prospective contact-tracing study. J Infect. https://doi.org/10.1016/j.jinf.2020.03.006

Jiang X, Rayner S, Luo MH (2020) Does SARS-CoV-2 has a longer incubation period than SARS and MERS? J Med Virol 92(5):476–478. https://doi.org/10.1002/jmv.25708

Ki M, 2019-nCoV TFf (2020) Epidemiologic characteristics of early cases with 2019 novel coronavirus (2019-nCoV) disease in Korea. Epidemiol Health 42:e2020007. https://doi.org/10.4178/epih.e2020007

Kim SH, Ko JH, Park GE, Cho SY, Ha YE, Kang JM, Kim YJ, Huh HJ, Ki CS, Jeong BH, Park J, Jang JH, Kim WS, Kang CI, Chung DR, Song JH, Peck KR (2017) Atypical presentations of MERS-CoV infection in immunocompromised hosts. J Infect Chemother 23(11):769–773. https://doi.org/10.1016/j.jiac.2017.04.004

Kuk AY, Ma S (2005) The estimation of SARS incubation distribution from serial interval data using a convolution likelihood. Stat Med 24(16):2525–2537. https://doi.org/10.1002/sim.2123

L'Huillier AG, Asner SA (2020) Pediatric impact of COVID-19. Rev Med Suisse 16(N°691–2):839–841

Lau EH, Hsiung CA, Cowling BJ, Chen CH, Ho LM, Tsang T, Chang CW, Donnelly CA, Leung GM (2010) A comparative epidemiologic analysis of SARS in Hong Kong, Beijing and Taiwan. BMC Infect Dis 10:50. https://doi.org/10.1186/1471-2334-10-50

Lauer SA, Grantz KH, Bi Q, Jones FK, Zheng Q, Meredith HR, Azman AS, Reich NG, Lessler J (2020) The incubation period of coronavirus disease 2019 (COVID-19) from publicly reported confirmed cases: estimation and application. Ann Intern Med. https://doi.org/10.7326/M20-0504

Lessler J, Reich NG, Brookmeyer R, Perl TM, Nelson KE, Cummings DA (2009) Incubation periods of acute respiratory viral infections: a systematic review. Lancet Infect Dis 9(5):291–300. https://doi.org/10.1016/S1473-3099(09)70069-6

Leung C (2019) Estimated incubation period for mosquito-borne disease-related Guillain-Barre syndrome. Clin Epidemiol Glob Health. https://doi.org/10.1016/j.cegh.2019.08.007

Leung C (2020a) Clinical features of deaths in the novel coronavirus epidemic in China. Rev Med Virol:e2103. https://doi.org/10.1002/rmv.2103

Leung C (2020b) The difference in the incubation period of 2019 novel coronavirus (SARS-CoV-2) infection between travelers to Hubei and nontravelers: the need for a longer quarantine period. Infect Control Hosp Epidemiol:1–3. https://doi.org/10.1017/ice.2020.81

Li Q, Zeng G, Ou JM, Guo GP (2003) Epidemiological study of the transmission chain of a severe acute respiratory syndrome outbreak. Zhonghua Yi Xue Za Zhi 83(11):906–909

Li Q, Guan X, Wu P, Wang X, Zhou L, Tong Y, Ren R, Leung KSM, Lau EHY, Wong JY, Xing X, Xiang N, Wu Y, Li C, Chen Q, Li D, Liu T, Zhao J, Liu M, Tu W, Chen C, Jin L, Yang R, Wang Q, Zhou S, Wang R, Liu H, Luo Y, Liu Y, Shao G, Li H, Tao Z, Yang Y, Deng Z, Liu B, Ma Z, Zhang Y, Shi G, Lam TTY, Wu JT, Gao GF, Cowling BJ, Yang B, Leung GM, Feng Z (2020) Early transmission dynamics in Wuhan, China, of novel coronavirus-infected pneumonia. N Engl J Med 382(13):1199–1207. https://doi.org/10.1056/NEJMoa2001316

Linton NM, Kobayashi T, Yang Y, Hayashi K, Akhmetzhanov AR, Jung SM, Yuan B, Kinoshita R, Nishiura H (2020) Incubation period and other epidemiological characteristics of 2019 novel coronavirus infections with right truncation: a statistical analysis of publicly available case data. J Clin Med 9(2). https://doi.org/10.3390/jcm9020538

Liu Y, Gayle AA, Wilder-Smith A, Rocklöv J (2020a) The reproductive number of COVID-19 is higher compared to SARS coronavirus. J Travel Med 27(2). https://doi.org/10.1093/jtm/taaa021

Liu Y, Yan LM, Wan L, Xiang TX, Le A, Liu JM, Peiris M, Poon LLM, Zhang W (2020b) Viral dynamics in mild and severe cases of COVID-19. Lancet Infect Dis. https://doi.org/10.1016/S1473-3099(20)30232-2

Lu R, Zhao X, Li J, Niu P, Yang B, Wu H, Wang W, Song H, Huang B, Zhu N, Bi Y, Ma X, Zhan F, Wang L, Hu T, Zhou H, Hu Z, Zhou W, Zhao L, Chen J, Meng Y, Wang J, Lin Y, Yuan J, Xie Z, Ma J, Liu WJ, Wang D, Xu W, Holmes EC, Gao GF, Wu G, Chen W, Shi W, Tan W (2020) Genomic characterisation and epidemiology of 2019 novel coronavirus: implications for virus origins and receptor binding. Lancet 395(10224):565–574. https://doi.org/10.1016/S0140-6736(20)30251-8

Ludvigsson JF (2020) Systematic review of COVID-19 in children shows milder cases and a better prognosis than adults. Acta Paediatr. https://doi.org/10.1111/apa.15270

Metlay JP, Waterer GW, Long AC, Anzueto A, Brozek J, Crothers K, Cooley LA, Dean NC, Fine MJ, Flanders SA, Griffin MR, Metersky ML, Musher DM, Restrepo MI, Whitney CG (2019) Diagnosis and treatment of adults with community-acquired pneumonia. An official clinical practice guideline of the American thoracic society and infectious diseases society of America. Am J Respir Crit Care Med 200(7):e45 – e67. https://doi.org/10.1164/rccm.201908-1581ST

Nie X, Fan L, Mu G, Tan Q, Wang M, Xie Y, Cao L, Zhou M, Zhang Z, Chen W (2020) Epidemiological characteristics and incubation period of 7,015 confirmed cases with coronavirus disease 2019 outside Hubei Province in China. J Infect Dis. https://doi.org/10.1093/infdis/jiaa211

Nishiura H, Mizumoto K, Ejima K, Zhong Y, Cowling B, Omori R (2012) Incubation period as part of the case definition of severe respiratory illness caused by a novel coronavirus. Euro Surveill 17(42)

Paraskevis D, Kostaki EG, Magiorkinis G, Panayiotakopoulos G, Sourvinos G, Tsiodras S (2020) Full-genome evolutionary analysis of the novel corona virus (2019-nCoV) rejects the hypothesis of emergence as a result of a recent recombination event. Infect Genet Evol 79:104212. https://doi.org/10.1016/j.meegid.2020.104212

Park M, Cook AR, Lim JT, Sun Y, Dickens BL (2020) A systematic review of COVID-19 epidemiology based on current evidence. J Clin Med 9(4). https://doi.org/10.3390/jcm9040967

Qian GQ, Yang NB, Ding F, Ma AHY, Wang ZY, Shen YF, Shi CW, Lian X, Chu JG, Chen L, Ren DW, Li GX, Chen XQ, Shen HJ, Chen XM (2020) Epidemiologic and clinical characteristics of 91 hospitalized patients with COVID-19 in Zhejiang, China: a retrospective, multi-Centre case series. QJM. https://doi.org/10.1093/qjmed/hcaa089

Qiu C, Deng Z, Xiao Q, Shu Y, Deng Y, Wang H, Liao X, Liu H, Zhou D, Zhao X, Zhou J, Wang J, Shi Z, Long D (2020) Transmission and clinical characteristics of coronavirus disease 2019 in 104 outside-Wuhan patients, China J Med Virol. https://doi.org/10.1002/jmv.25975

Reich NG, Lessler J, Cummings DA, Brookmeyer R (2009) Estimating incubation period distributions with coarse data. Stat Med 28(22):2769–2784. https://doi.org/10.1002/sim.3659

Rothman K, Lash T, Greenland S (2012) Mod Epidemiol. third edn. Wolters Kluwer,

Sanche S, Lin YT, Xu C, Romero-Severson E, Hengartner N, Ke R (2020) High contagiousness and rapid spread of severe acute respiratory syndrome Coronavirus 2. Emerg Infect Dis 26(7). https://doi.org/10.3201/eid2607.200282

Sartwell PE (1950) The distribution of incubation periods of infectious disease. Am J Hyg 51(3):310–318. https://doi.org/10.1093/oxfordjournals.aje.a119397

Shen Q, Guo W, Guo T, Li J, He W, Ni S, Ouyang X, Liu J, Xie Y, Tan X, Zhou Z, Peng H (2020) Novel coronavirus infection in children outside of Wuhan, China. Pediatr Pulmonol. https://doi.org/10.1002/ppul.24762

Sun L, Shen L, Fan J, Gu F, Hu M, An Y, Zhou Q, Fan H, Bi J (2020) Clinical features of patients with coronavirus disease 2019 (COVID-19) from a designated Hospital in Beijing, China. J Med Virol. https://doi.org/10.1002/jmv.25966

Tian S, Hu N, Lou J, Chen K, Kang X, Xiang Z, Chen H, Wang D, Liu N, Liu D, Chen G, Zhang Y, Li D, Li J, Lian H, Niu S, Zhang L, Zhang J (2020) Characteristics of COVID-19 infection in Beijing. J Infect 80(4):401–406. https://doi.org/10.1016/j.jinf.2020.02.018

Virlogeux V, Fang VJ, Wu JT, Ho LM, Peiris JS, Leung GM, Cowling BJ (2015) Brief report: incubation period duration and severity of clinical disease following severe acute respiratory syndrome coronavirus infection. Epidemiology 26(5):666–669. https://doi.org/10.1097/EDE.0000000000000339

Virlogeux V, Fang VJ, Park M, Wu JT, Cowling BJ (2016a) Comparison of incubation period distribution of human infections with MERS-CoV in South Korea and Saudi Arabia. Sci Rep 6:35839. https://doi.org/10.1038/srep35839

Virlogeux V, Park M, Wu JT, Cowling BJ (2016b) Association between severity of MERS-CoV infection and incubation period. Emerg Infect Dis 22(3):526–528. https://doi.org/10.3201/eid2203.151437

Wang P, Lu JA, Jin Y, Zhu M, Wang L, Chen S (2020) Statistical and network analysis of 1212 COVID-19 patients in Henan. China Int J Infect Dis. https://doi.org/10.1016/j.ijid.2020.04.051

Xiao Z, Xie X, Guo W, Luo Z, Liao J, Wen F, Zhou Q, Han L, Zheng T (2020) Examining the incubation period distributions of COVID-19 on Chinese patients with different travel histories. J Infect Dev Ctries 14(4):323–327. https://doi.org/10.3855/jidc.12718

Xu T, Chen C, Zhu Z, Cui M, Dai H, Xue Y (2020a) Clinical features and dynamics of viral load in imported and non-imported patients with COVID-19. Int J Infect Dis 94:68–71. https://doi.org/10.1016/j.ijid.2020.03.022

Xu XW, Wu XX, Jiang XG, Xu KJ, Ying LJ, Ma CL, Li SB, Wang HY, Zhang S, Gao HN, Sheng JF, Cai HL, Qiu YQ, Li LJ (2020b) Clinical findings in a group of patients infected with the 2019 novel coronavirus (SARS-Cov-2) outside of Wuhan, China: retrospective case series. BMJ 368:m606. https://doi.org/10.1136/bmj.m606

Yu P, Zhu J, Zhang Z, Han Y, Huang L (2020a) A familial cluster of infection associated with the 2019 novel coronavirus indicating potential person-to-person transmission during the incubation period. J Infect Dis. https://doi.org/10.1093/infdis/jiaa077

Yu X, Sun X, Cui P, Pan H, Lin S, Han R, Jiang C, Fang Q, Kong D, Zhu Y, Zheng Y, Gong X, Xiao W, Mao S, Jin B, Wu H, Fu C (2020b) Epidemiological and clinical characteristics of 333 confirmed cases with coronavirus disease 2019 in Shanghai, China. Transbound Emerg Dis. https://doi.org/10.1111/tbed.13604

Zeng G, Xie SY, Li Q, Ou JM (2009) Infectivity of severe acute respiratory syndrome during its incubation period. Biomed Environ Sci 22(6):502–510. https://doi.org/10.1016/S0895-3988(10)60008-6

Zhang J, Litvinova M, Wang W, Wang Y, Deng X, Chen X, Li M, Zheng W, Yi L, Wu Q, Liang Y, Wang X, Yang J, Sun K, Longini IM Jr, Halloran ME, Wu P, Cowling BJ, Merler S, Viboud C, Vespignani A, Ajelli M, Yu H (2020a) Evolving epidemiology and transmission dynamics of coronavirus disease 2019 outside Hubei province, China: a descriptive and modelling study. Lancet Infect Dis. https://doi.org/10.1016/s1473-3099(20)30230-9

Zhang MQ, Wang XH, Chen YL, Zhao KL, Cai YQ, An CL, Lin MG, Mu XD (2020b) Clinical features of 2019 novel coronavirus pneumonia in the early stage from a fever clinic in Beijing. Zhonghua Jie He He Hu Xi Za Zhi 43(0):E013. https://doi.org/10.3760/cma.j.issn.1001-0939.2020.0013

Zhao H, Shen D, Zhou H, Liu J, Chen S (2020) Guillain-Barré syndrome associated with SARS-CoV-2 infection: causality or coincidence? Lancet Neurol 19(5):383–384. https://doi.org/10.1016/S1474-4422(20)30109-5

Zhu Y, Chen YQ (2020) On a statistical transmission model in analysis of the early phase of COVID-19 outbreak. Stat Biosci:1–17. https://doi.org/10.1007/s12561-020-09277-0

Coronavirus: Pure Infectious Disease or Genetic Predisposition

6

Farzaneh Darbeheshti, Hassan Abolhassani, Mohammad Bashashati, Saeid Ghavami, Sepideh Shahkarami, Samaneh Zoghi, Sudhir Gupta, Jordan S. Orange, Hans D. Ochs, and Nima Rezaei

Abstract

Severe acute respiratory syndrome coronavirus 2 (SARS-CoV-2), which causes novel coronavirus disease (COVID-19), is the seventh pathogenic coronavirus recently discovered in December 2019 in Wuhan, China. To date, our knowledge about its effect on the human host remains limited. It is well known that host genetic factors account for the individual differences in the susceptibility to infectious diseases. The genetic susceptibility factors to COVID-19 and its severity are associated with several unanswered questions. However, the experience gained from an earlier strain of coronavirus, SARS-CoV-1, which shows 78% genetic similarity to SARS-CoV-2 and uses the same receptor to bind to host cells, could provide some clues. It, therefore, seems possible to assemble new evidence in order to solve a potential genetic predisposition puzzle for COVID-19. In this chapter, the puzzle pieces, including virus entry receptors, immune response, and inflammation-related genes, as well as the probable genetic predisposition models to COVID-19, are discussed.

Keywords

COVID-19 · Genetic predisposition · Genetic susceptibility · SARS-CoV-1 · SARS-CoV-2 · Polymorphism

F. Darbeheshti
Department of Genetics, School of Medicine, Tehran University of Medical Sciences, Tehran, Iran

Medical Genetics Network (MeGeNe), Universal Scientific Education and Research Network (USERN), Tehran, Iran

H. Abolhassani
Division of Clinical Immunology, Department of Laboratory Medicine, Karolinska Institute at Karolinska University Hospital Huddinge, Stockholm, Sweden

Research Center for Immunodeficiencies, Children's Medical Center, Tehran University of Medical Sciences, Tehran, Iran

Primary Immunodeficiency Diseases Network (PIDNet), Universal Scientific Education and Research Network (USERN), Stockholm, Sweden

M. Bashashati
Division of Gastroenterology, Department of Internal Medicine, Texas Tech University Health Sciences Center, El Paso, TX, USA

Network of Immunity in Infection, Malignancy and Autoimmunity (NIIMA), Universal Scientific Education and Research Network (USERN), El Paso, TX, USA

S. Ghavami
Department of Human Anatomy and Cell Science, Max Rady College of Medicine, Rady Faculty of Health Sciences, University of Manitoba, Winnipeg, MB, Canada

Faculty of Medicine, Katowice School of Technology, Katowice, Poland

N. Rezaei (ed.), *Coronavirus Disease - COVID-19*, Advances in Experimental Medicine and Biology 1318, https://doi.org/10.1007/978-3-030-63761-3_6

6.1 Introduction

Infectious diseases exert one of the most substantial selective pressures on human genomes by eliminating individuals with high susceptibility from the population (Hill 2012). Reports presume that virus-human interactions are involved in 30% of human genome evolution (Kenney et al. 2017).

The pathogen-centered approaches have conventionally managed infectious diseases. However, the attention given to the roles of the host genetic background in the burden of infectious diseases has become more remarkable over the past decade. Scientists are looking for substantial inter-individual variations in susceptibility and vulnerability to infectious diseases (Burgner et al. 2006). The advent of high-throughput sequencing technologies and the expansion of genome-wide association (GWA) studies have uncovered genetic predisposition to infections.

The hypothesis-free approach in GWA studies is capable of mapping millions of single nucleotide polymorphisms (SNPs) across different human genomes. However, GWA studies have a main methodological limitation in detecting rare variants with high penetrance, and the available GWA data can explain only a small proportion of the genetic variance of infectious diseases (Schork et al. 2009). Host genetic variants have been demonstrated to be influential on viral disease susceptibilities, such as the CCR5Δ32 mutation in resistance against HIV and mutations in the *FUT2* gene as a protective factor against norovirus infection.

Several single gene defects in immunity-related genes have been identified to affect individual predisposition to severe viral infection with Mendelian inheritance, such as STAT1/2, IRF7/9 (severe influenza disease), IFNAR1/2 (a severe disease caused by yellow fever), and FCGR3A (severe herpes viral infections) with autosomal recessive (AR) inheritance, and POLR3A/C/F (a severe disease caused by yellow fever) with autosomal dominant (AD) inheritance (Tangye et al. 2020). Notably, there are a few genetic polymorphisms that result in susceptibility to limited infections, for instance, SNP rs2070874 of the *IL4* gene for RSV infection and rs2430561 of IFNG for tuberculosis (Shen et al. 2013; Zhang et al. 2016).

Accurate identification of host genetic variability in the susceptibility to exogenous pathogens needs merging insights from other

S. Shahkarami
Medical Genetics Network (MeGeNe), Universal Scientific Education and Research Network (USERN), Tehran, Iran

Gene center, Ludwig-Maximilians-University of Munich, Munich, Germany

S. Zoghi
Research Center for Immunodeficiencies, Children's Medical Center, Tehran University of Medical Sciences, Tehran, Iran

Ludwig Boltzmann Institute for Rare and Undiagnosed Diseases, Vienna, Austria

S. Gupta
Department of Medicine, Division of Basic and Clinical Immunology, University of California, Irvine, CA, USA

J. S. Orange
Immunology, Allergy, and Rheumatology, Baylor College of Medicine and the Texas Children's Hospital, Houston, TX, USA

H. D. Ochs
School of Medicine, Department of Pediatrics, University of Washington, Seattle, WA, USA

Universal Scientific Education and Research Network (USERN), Seattle, WA, USA

N. Rezaei (✉)
Research Center for Immunodeficiencies, Children's Medical Center, Tehran University of Medical Sciences, Tehran, Iran

Department of Immunology, School of Medicine, Tehran University of Medical Sciences, Tehran, Iran

Network of Immunity in Infection, Malignancy and Autoimmunity (NIIMA), Universal Scientific Education and Research Network (USERN), Tehran, Iran

Children's Medical Center Hospital, Tehran, Iran
e-mail: rezaei_nima@tums.ac.ir

fields, including immunology and epidemiology, in order to achieve a systematic understanding of the concept. This chapter discusses the potential genetic variants involved in the earlier strain of coronavirus, namely, severe acute respiratory syndrome-related coronavirus (SARS-CoV)-1, since SARS-CoV-2 has 78% genetic identity to SARS-CoV-1 (Lu et al. 2020; Ren et al. 2020), and therefore it may serve as possible genetic predisposition models for SARS-CoV-2 infection.

The recently emerged SARS-CoV-2 is an enveloped single-stranded RNA betacoronavirus, which recently emerged in December 2019 in Wuhan but rapidly turned into a pandemic affecting people worldwide (Hanaei and Rezaei 2020). SARS-CoV-2 infection shows a wide range of disease severities among different patients (Lotfi and Rezaei 2020). This respiratory disease presents mainly with mild to moderate symptoms. In China, 41% and 26% of affected patients needed hospital admissions and intensive care unit (ICU) care, respectively; mortality was reported in 4.3% of patients (Wang et al. 2020a). However, individuals who have underlying medical conditions (e.g., chronic lung/kidney/liver diseases, severe heart conditions, metabolic syndrome, and secondary immunosuppression) and older people have a higher risk of developing severe symptoms, including respiratory failure. Both these high-risk populations, i.e., older people and people with comorbidities, have the functionally impaired immune system, compared to the general population. Therefore, there has been an interest in understanding the immunopathogenesis of COVID-19 (Nasab et al. 2020; Saghazadeh and Rezaei 2020a; Yazdanpanah et al. 2020), and consequently, hyperinflammation including a cytokine storm appeared to play a role (Bahrami et al. 2020).

Influenza, respiratory syncytial virus (RSV), and pneumonia represent the leading causes of respiratory infection-related death worldwide (Ascough et al. 2018). Concerning these three respiratory infectious diseases, a small percentage of patients experience severe symptoms and need hospital admission. Genetic studies have helped to uncover variants involved in the pathogenesis of respiratory infections.

A challenging issue about COVID-19 is the development of severe symptoms in adults and even adolescents without any underlying disease. This issue has attracted geneticists' attention worldwide to pinpoint the involving genetic factors that make individuals more vulnerable to COVID-19. It is postulated that available knowledge concerning the genetic predisposition factors to SARS-CoV-1 may also be right for SARS-CoV-2, but this needs careful DNA analysis and the use of bioinformatics to find variants in genes that result in facilitating viral entry and abnormal immune responses or allow situations of immune dysregulation that causes an overreaction of innate or adaptive immunity, as in cytokine storm. Below, we will discuss the molecular targets and genetic variants that might be involved in the development and severity of COVID-19 infection.

6.2 Genetic Variants Relating to Virus Binding and Entry Receptors

6.2.1 Angiotensin-Converting Enzyme (ACE)

Angiotensin-converting enzyme 1 (*ACE1*) and 2 (*ACE2*) genes occur on chromosomes 17q and Xp, respectively. These homologous genes encode the essential enzymes that are part of the renin-angiotensin system with a 42% identical protein sequence. ACE1 regulates blood pressure and the balance of fluids and salts by targeting angiotensin I. ACE2, located on the outer surface of cells, modulates the levels of angiotensin II. Also, this protein has been established as a functional receptor for the spike glycoprotein of SARS-CoV-1 and SARS-CoV-2 (Li et al. 2003).

6.2.1.1 ACE1

ACE1 is highly expressed on microvascular endothelium and is probably involved in angiopathy of the lung. It has been assumed that *ACE1* variants could be associated with the progression of SARS-CoV-1 (Itoyama et al. 2004). This gene has an insertion (I) or a deletion (D) polymorphism in the 287-bp Alu repeat sequence in intron 16 (rs1799752). The D allele demonstrates an enhanced ACE1 expression and activity (Rigat et al. 1990).

The association of this polymorphism with pathological states such as myocardial infarction has been documented (Baudin 2002). Concerning SARS-CoV-1 infection, the D allele was statistically associated with hypoxemic status in Vietnamese patients (Itoyama et al. 2004). In Chinese patients, individuals with the DD genotype had higher mortality of acute lung injury (Lu et al. 2011). Moreover, a meta-analysis of ten related studies shows evidence supporting a causal association between D allele and pneumonia vulnerability, and that individuals with the I/I genotype have a high chance to survive from acute respiratory distress syndrome (Wang et al. 2015). However, based on another study, there is no significant association between the ACE1 I/D polymorphism and poor outcomes after SARS-coronavirus infection (Chan et al. 2005).

6.2.1.2 ACE2

The large-scale structural analyses of an atomic-level investigation have revealed a strong interaction between the receptor-binding domain (RBD) of both SARS-CoV-1 and SARS-CoV-2 spike protein with ACE2 as their host receptor, which is responsible for human-to-human transmission (Letko and Munster 2020). A high ACE2 expression has been detected in different cell types, such as epithelial cells of the tongue, alveolar cells, enterocytes, a subpopulation of macrophages, and small intestine (Xu et al. 2020). The spike protein binds to ACE2, and then a transmembrane protease (TMP) cleaves the S1 and S2 subunits of the spike protein that is critical for the virus entry into target cells. Therefore, genetic variation in ACE2 and/or TMP that could affect its expression, protein conformation, and stability are potential genetic predisposing factors to COVID-19. The recent systematic analysis of coding-region variants in the *ACE2* gene, as well as using the GTEx database, has revealed that some SNPs are associated with a higher RNA expression of *ACE2*, which might cause different susceptibility to COVID-19 among patients (Cao et al. 2020).

Studies have investigated the significance of ACE2 variants in susceptibility and vulnerability of individuals to SARS-CoV-1. Generally, 30% of SARS-CoV-1-infected patients required intensive care admission, and interestingly males infected were more likely to manifest severe symptoms (Booth et al. 2003). Moreover, the mortality rates have been estimated at 21.9% and 13.2% of male and female patients, respectively (Karlberg et al. 2004). This evidence points to the intriguing coincidence that the *ACE2* gene is located on the X chromosome.

The assessment of variants in the protein-coding region of *ACE2* in a large human cohort revealed some polymorphisms that either likely protect or predispose individuals to the virus (Stawiski et al. 2020). However, the importance of variants in the human *ACE2* gene for susceptibility to both SARS-CoV-1 and SARS-CoV-2 has not been comprehensively examined. Chiu et al. reported no associations between ACE2 common genetic variants and SARS-CoV-1 susceptibility or outcome (Chiu et al. 2004).

6.2.2 CD147

A transmembrane glycoprotein referred to as CD147/Basigin, which belongs to the immunoglobulin superfamily, is involved in virus infection and tumor development. This protein is highly expressed in infected and inflamed tissues. CD147 has a functional role in facilitating the SARS-CoV-1 invasion of the host cells. Interestingly, it has been presumed that SARS nuclear (N) protein interacts with CD147 located on the membrane of the endoplasmic reticulum (ER), thereby facilitating the virus particles in attaching to and budding into the ER (Chen et al. 2005). Recently, it was documented that the SARS-CoV-2 spike protein also binds to CD147 and facilitates the viral invasion (Wang et al. 2020b). Therefore, both SARS-CoV-1 and

SARS-CoV-2 are expected to invade host cells via CD147.

There is no specific evidence for the role of CD147 polymorphisms in predisposition to coronavirus infection. However, a functional polymorphism of the *CD147* gene has been reported in acute coronary syndrome (ACS) in several independent studies. The rs8259 AA genotype shows a significant association not only with the high expression of CD147 mRNA and protein but also with ACS patients (Yan et al. 2015). It provided evidence for the possible role of rs8259 for the individual difference in the susceptibility to viral infection and the severity of the disease.

Although ACE2 has been widely known as a critical receptor involved in virus invasion, anti-ACE2 antibodies may lead to severe adverse events, since ACE2 is widely expressed in a variety of tissues (Tipnis et al. 2000). On the other hand, Meplazumab, a humanized anti-CD147 antibody, has shown safety in preclinical research (Wang et al. 2020b). Therefore, it has been suggested that blocking CD147 might effectively inhibit the viruses from invading host cells.

6.2.3 Dipeptidyl Peptidase 4 (DPP4)

DPP4, also known as CD26, is a type-II transmembrane protein that is expressed in various tissues, such as the intestine, respiratory tract, liver, kidney, and fibroblasts of the lung. Moreover, DPP4 is expressed on different immune cells, including dendritic cells, macrophages, NK cells, activated T cells, and B cells (Shao et al. 2020). It is involved in the immune response and autoinflammatory reaction by modulating the production of chemokines and cytokines (Shao et al. 2020). The significant association of the genetic variants in the *DPP4* gene with some human diseases has been reported, such as diabetes (Ahmed et al. 2016) and myocardial infarction (Aghili et al. 2012).

The MERS-CoV, which causes Middle East respiratory syndrome (MERS), recognizes DPP4 as the primary receptor (Raj et al. 2013). Interestingly, four DPP4 polymorphisms (rs1229319529, rs780673235, rs764879525, rs763246423) have been documented to reduce the binding efficiency of MERS-CoV spike protein to DPP4 (Kleine-Weber et al. 2020). It has been suggested that the DPP4 polymorphisms affect the course of MERS-CoV infection.

Recently, a high-affinity binding between SARS-CoV-2 spike protein and DPP4 has been reported, using bioinformatics approaches and protein docking based on crystal structures (Li et al. 2020). However, the affinity of the SARS-CoV-2 spike protein with human DPP4 (−34.8 kcal/mol) is lower than its affinity with ACE2 (−39.2 kcal/mol). It is well known that the coronavirus spike protein can be associated with a wide range of molecules on cell surfaces. These molecules could play the role of co-receptors or attachment factors, which play a predominant role in viral entry (Cui et al. 2019). It is conceivable that DPP4 acts as a co-receptor for SARS-CoV-2, though ACE2 is considered the primary receptor for SARS-CoV-2. Hence, the host susceptibility to SARS-CoV-2 might be influenced by protein interactions between the spike protein and cell-surface co-receptors. Moreover, DPP4 inhibitors have been recently proposed as a potential therapeutic strategy for COVID-19 infection (Strollo and Pozzilli 2020).

6.2.4 Transmembrane Serine Protease 2 (TMPRSS2)

As mentioned above, the coronavirus spike glycoprotein, which is embedded in the viral envelope, facilitates membrane fusion through mediating receptor recognition. The spike protein consists of two subunits, S1 and S2, which are involved in binding to the receptor and fusion with the cellular membrane, respectively. After membrane fusion, the coronavirus genome is delivered into the cellular cytoplasm to translation and replication.

TMPRSS2 proteolytically primes the fusion proteins of the human influenza virus (Böttcher et al. 2006), metapneumovirus (Shirogane et al. 2008), and coronavirus (Glowacka et al. 2011). The cleavage of the coronavirus spike glycoprotein by TMPRSS2 produces the fusion-catalyzing forms, which can boost or can be critical for viral infectivity (Bergeron et al. 2005; Iwata-Yoshikawa et al. 2019). Interestingly, a recent in vitro study indicated that Camostat Mesylate, a clinically proven TMPRSS2 inhibitor, partially blocks SARS-CoV-2 entry into cells (Hoffmann et al. 2020).

Some variants in the 3′ UTR region of the *TMPRSS2* gene show a significant impact on its expression in the lung. Different inter-population frequencies of these variants are described. The haplotype with higher expression of TMPRSS2 is characterized by three SNPs (rs2070788, rs9974589, rs7364083), whose minor allele frequencies (MAFs) are significantly increased in Italians compared with East Asians (Asselta et al. 2020). These observations suggest a role of TMPRSS2 polymorphisms in modulating COVID-19 severity, which, however, need additional experimental investigations.

6.3 Genes with Immune Response-Related Functions

6.3.1 Mannose-Binding Lectin (MBL)

Mannose-binding lectin (MBL) is a serum protein that plays a pivotal role in the first-line host defense and the innate immune response (Ezekowitz 2003). MBL binds to microbial surfaces through its carbohydrate recognition domains (CRDs) and induces at least two protective responses, including the activation of the complement cascade and opsonophagocytosis. Genetic deficiency of MBL increases susceptibility to both bacterial and viral infectious diseases such as HIV, Neisseria meningitidis, and influenza A (Garred et al. 1997; Hartshorn et al. 1993; Hibberd et al. 1999). The different vulnerability of individuals to the coronavirus infections might lie in the critical role of MBL in the innate immune response, especially during the time before producing specific antibodies, since the latter are not detected until about 10 days after the onset of symptoms (Peiris et al. 2003).

Some functional SNPs in the promoter and coding regions of the *MBL* gene may affect its serum level (Madsen et al. 1995). Two polymorphisms have been identified in exon 1 (rs1800450) and promoter (rs7096206) of the *MBL* gene, which has a remarkable association with SARS-CoV-1 susceptibility (Ip et al. 2005; Zhang et al. 2005). The higher frequency of haplotypes associated with low or deficient serum levels of MBL in patients than in healthy controls suggests MBL as a first-line host defense against SARS-CoV-1 infection. Moreover, the SNPs associated with low serum levels of MBL have been reported to be increasingly present in several respiratory infectious diseases, such as a higher risk of invasive pneumococcal disease (Roy et al. 2002) and more frequent hospital admissions in patients with chronic obstructive pulmonary disease (Yang et al. 2003). Of note, MBL deficiency is considered by some as one the most common primary immunodeficiencies, in which the extreme deficiency can be detected in more than 10% of the healthy population (Engler et al. 2010).

6.3.2 Type C Lectin

Two closely related transmembrane proteins named DC-SIGN (CD209) and L-SIGN (CLEC4M) are dendritic cell-specific intracellular adhesion molecules that directly recognize a wide range of microorganisms (HIV, Ebola, hepatitis C, and SARS-CoV-1 coronavirus). Both genes have an extended neck region encoded by tandem repeats in the extracellular domain region, which play a crucial role in pathogen recognition of these receptors. While *DC-SIGN* has a relatively constant repeat, the *L-SIGN* gene shows remarkable polymorphisms among different individuals that could affect ligand-binding affinity. The polymorphism (rs4804803), identified in the DC-SIGN promoter, could alter its expression (Chan et al. 2010a). A demographic investigation of the variation in tandem repeat lengths performed in 52 different populations indicated that this region is one of the targets of the selective pressure exerted by pathogens (Barreiro et al. 2005).

Several functional and GWA studies have evaluated the effect of polymorphisms on the pathogen-binding ability and genetic susceptibility to infectious diseases. A genetic-risk association study revealed that homozygosity in L-SIGN tandem repeat (rs71179137) is a significant protective factor for SARS-CoV-1 infection (Chan et al. 2006). However, these findings have not been supported by a case-controlled study in northern China (Zhi et al. 2007). Investigations related to DC-SIGN polymorphisms show a significant correlation between the G allele of the rs4804803 polymorphism and lower levels of promoter activity (Chan et al. 2010a). This variant shows association with several important infectious diseases, such as *Mycobacterium tuberculosis* and HIV-1. Moreover, the G allele could protect the lung from injury during the progression of SARS-CoV-1 infection (Chan et al. 2010b), and patients with the AA genotype had a 60% chance for developing severe symptoms (Chan et al. 2010a). In contrast, another study reported that no genetic predisposition allele in the lectin gene cluster is involved in SARS-CoV-1 infection (Li et al. 2008).

6.3.3 Human Leukocyte Antigen (HLA)

The human major histocompatibility complex, also known as human leukocyte antigen (HLA), consists of the most polymorphic genes. The studies concerning the association between HLA and human diseases, including infectious diseases, have been carried out for nearly 40 years. The products of both HLA class I and II encoding genes share the responsibility of presenting viral antigens on the cell surface for recognition by T cells. Hence, they might be critical components of the adaptive immune system against SARS-CoV-2.

By studying the *HLA* type *A, B, DR*, and *DQ* alleles, a strong association between *HLA-B*0703* and *DRB1*0301* alleles and SARS-CoV-1 infection has been reported among 90 Chinese patients (Ng et al. 2004). On the other hand, in the Vietnamese population, a case-controlled study comprised 44 SARS-CoV-1-infected patients, 103 staff members who had contact with SARS-CoV-1-infected patients but had not developed the infection, and 50 healthy individuals without contact history indicated that *DRB1*12* has a significant positive association with developing SARS-CoV-1 (Keicho et al. 2009). The *DRB1*12* allele tendency to develop infection was also demonstrated in a large study from southern China (Xiong et al. 2008). Furthermore, in a Taiwanese association study, the *B*4601* allele was significantly associated with the severity of SARS-CoV-1 (Lin et al. 2003).

6.4 Inflammation-Related Genes

SARS-CoV-1 and other acute respiratory viruses commonly trigger an inflammatory cascade in local tissues. The upregulation of inflammatory chemokines recruits immune cells into the infected tissue. It has been suggested that variants in inflammation-related genes could be involved in the quality and quantity of immune responses against coronavirus infections (see the following section for more details).

6.4.1 C-C Motif Chemokine Ligand 2 (CCL2)

Chemokines have prominent roles in cell trafficking during immune responses. The CCL2 is involved in both innate and adaptive immunity through chemoattraction of monocytes and T helper cell polarization, respectively. CCL2 is one of the chemokines contributing to the SARS-CoV-1 infection. It is upregulated during the early stage of inflammation in both the epithelial cells of the lung and in monocytes (Law et al. 2005). Moreover, CCL2 overexpression in plasma of patients with SARS-CoV-1 is significantly associated with more severe symptoms (Wong et al. 2004).

Interestingly, the role of a functional SNP (rs1024611 G/A) in the transcriptional activity of CCL2 promoter has been well characterized. The G allele induces higher CCL2 mRNA and protein expression as well as more infiltration of leukocytes into tissues compared with the A allele (McDermott et al. 2005). It has been documented that the G allele could contribute to inter-individual differences in both susceptibility and vulnerability to SARS-CoV-1 infection (Tu et al. 2015).

6.4.2 Interleukins

IL-12 is an inflammatory cytokine that is released by activated dendritic cells. It has a crucial role in inducing cell-mediated immunity through stimulating T-cell and NK-cell proliferation. IL-12 receptor (IL-12R) consists of B1 and B2 subunits, which form the IL-12R complex on activated T and NK cells. A case-controlled study including 115 SARS-CoV-1-infected patients and 2 control groups (141 healthy individuals with close contacts with SARS-CoV-1-infected patients and 155 individuals without contact history) uncovered a significant association between the risk of susceptibility to SARS-CoV-1 infection and c.1664 CT/TT genotypes on *IL12RB1* gene (rs1331950321) in Chinese population (Tang et al. 2008). This variant has been confirmed to cause the missense SNP (p. P534L) in the extracellular coding sequence of the IL12RB1 protein.

IL-6 gene polymorphisms are other potential factors associated with disease progression. The two polymorphisms, c.-174G/C (rs1800795) and c.-572C/G (rs1800797), have been shown to affect both the transcription and secretion levels of IL-6 (Tong et al. 2010). Interestingly, rs1800795 was significantly associated with the severity of pneumonia (Ulhaq and Soraya 2020). It is shown that carriers of IL-6 c.-174 C allele have higher IL-6 levels and 2.42-fold higher risk for septic shock induced by pneumonia (Feng et al. 2015; Jerrard-Dunne et al. 2003). Moreover, the haplotype spanning from −1363 to +4835 from the transcription start site of the *IL-6* gene has the association with susceptibility to acute lung injury (Flores et al. 2008) (Mao et al. 2017). It is suggested that *IL-6* polymorphisms could be involved in COVID-19 progression. The suppression of the IL-6 signaling cascade might be a promising therapeutic strategy against severe SARS-CoV-2 infection (Ulhaq and Soraya 2020).

6.4.3 Interferon-Gamma (IFN-γ)

The interferon-mediated immunopathological events play a pivotal role in immune responses against SARS-CoV-1 infection (Cameron et al. 2007). IFN-γ is another inflammatory cytokine that is produced by T and NK cells to drive Th1 responses. Also, IFN- γ induces the production of free radicals and pro-inflammatory cytokines like tumor necrosis factor (TNF)-α through activating monocytes and macrophages.

The IFN-γ c.874A allele (rs2430561) was significantly overrepresented in 476 SARS-CoV-1-infected patients when compared with the 449 healthy controls. Notably, a dose-dependent manner indicated that the IFN-γ c.874A allele is associated with susceptibility to SARS-CoV-1 (Lim et al. 2006). Interestingly, this allele variant has been previously reported to be associated with several infectious diseases such as hepatitis B virus, tuberculosis, and parvovirus infection (Ben-Ari et al. 2003; Tso et al. 2005). The T allele of IFN-γ c.874A>T provides a binding site for the transcription factor nuclear factor κ-enhancer of activated B cells (NF-κB), which could regulate the expression of IFN-γ (Pravica et al. 2000). It is suggested that the downregulation of IFN-γ by the c.874A allele impairs the antiviral responses.

In the context of the different severities of viral infections, the inflammatory response could get out of control due to the genetic predisposition to immune dysregulation. Thanks to new insight into the molecular events that precipitate an IFN-γ-related cytokine storm, several molecular targets have been linked to the uncontrolled inflammation in H5N1 influenza and SARS-CoV-1 infections (Vaninov 2020; Huang et al. 2005). Studies have suggested several germline mutations in inflammasome (NLRP3 and IL1B), genes (TNF and IFNB1), and cytokine receptor (TNFRSF1B and IL4R) genes that might lead to hemophagocytic lymphohistiocytosis and macrophage activation syndrome, which, in turn, correlate with a higher chance of developing the clinical complication of cytokine storm after viral infections (Canna and Behrens 2012).

6.4.4 IFN-Inducible Genes

Type I IFNs have been commonly applied as an antiviral drug, notably for treating hepatitis C virus (HCV) infection. The role of host genetic factors in response to IFN treatment in HCV-infected patients has been documented. A promoter variant in the IFN-inducible *MX1* gene (encoding IFN-induced GTP-binding protein MXA) is involved in the patient's response to IFN treatment in the Japanese (Suzuki et al. 2004) and Caucasian populations (Knapp et al. 2003). It has also been observed that SNPs in IFN-inducible genes (*OAS-1*, *MX1*, and *PKR*) affect not only the efficiency of IFN treatment but also the natural course of HCV infection. It is presumed that these SNPs could be involved in susceptibility to SARS-CoV-1 infection. An investigation among Vietnamese SARS-CoV-1-infected patients revealed that the G allele of non-synonymous A/G SNP in exon 3 of the *OAS-1* gene (rs3741981, p.S162G) is associated with SARS-CoV-1 infection (Hamano et al. 2005). In addition, the G allele of c.-88 G/T polymorphism (rs2071430) in the promoter of the *MX1* gene is found more frequently in a hypoxemic group than in the non-hypoxemic group of SARS-CoV-1. Interestingly, the G allele shows lower promoter activity compared with T allele (Fernández-Arcás et al. 2004).

6.4.5 RANTES

RANTES chemokine, also known as CCL5, plays a critical role in the recruitment and migration of T cells at the site of inflammation during acute infections. The CCR5 (RANTES receptor) is targeted to block HIV entry to the host cell. Regarding the SARS-CoV-1 epidemic in China, it has been uncovered that the G allele of RANTES c.-28 C/G polymorphism (rs2107538) is associated with more severe symptoms, as defined by admission to ICU or deaths (Ng et al. 2007). The SNP is located at the NF-κB binding site and could be involved in the regulation of RANTES expression (Moriuchi et al. 1997). It should be noted that many chemokines not only participate in antiviral immune response but are also in cell damage and organ failure. Higher levels of RANTES induce lung inflammation and result in lymphopenia and increased risk of secondary infection among SARS-CoV-1-positive patients (Jiang et al. 2005).

Taken together, the considerable evidence for susceptible variants that affect SARS-CoV-1 infection provides some clues and meaningful predictions to solve a potential genetic predisposition puzzle related to COVID-19 (Fig. 6.1). However, these studies have provided some conflicting results, possibly due to low study power, population stratification, and different study designs. Future large-scale genomic studies will hopefully provide more precise insight into the role of the genetic background of the host as to the susceptibility and vulnerability to COVID-19. These investigations are fundamental steps to develop personalized genomic approaches to the prevention and treatment of infectious diseases.

6.5 Autophagy-Related Genes

Autophagy is the cellular response to stress (Brun et al. 2020; Eghtedardoost et al. 2020) and is responsible for removing damaged organelles and misfolded proteins through lysosomal digestion. Autophagy includes three major forms: macroautophagy (hereafter autophagy), microautophagy, and chaperone-mediated autophagy (CME) (Emami et al. 2019). This natural process is a conserved mechanism among species (Shojaei et al. 2020).

A recent investigation has shown that autophagy is involved in the regulation of allergic asthma through several mechanisms, including regulation of fibrosis (McAlinden et al. 2020; Zeki et al. 2016). The genetic polymorphisms in autophagy-related 5 (*ATG5*) and

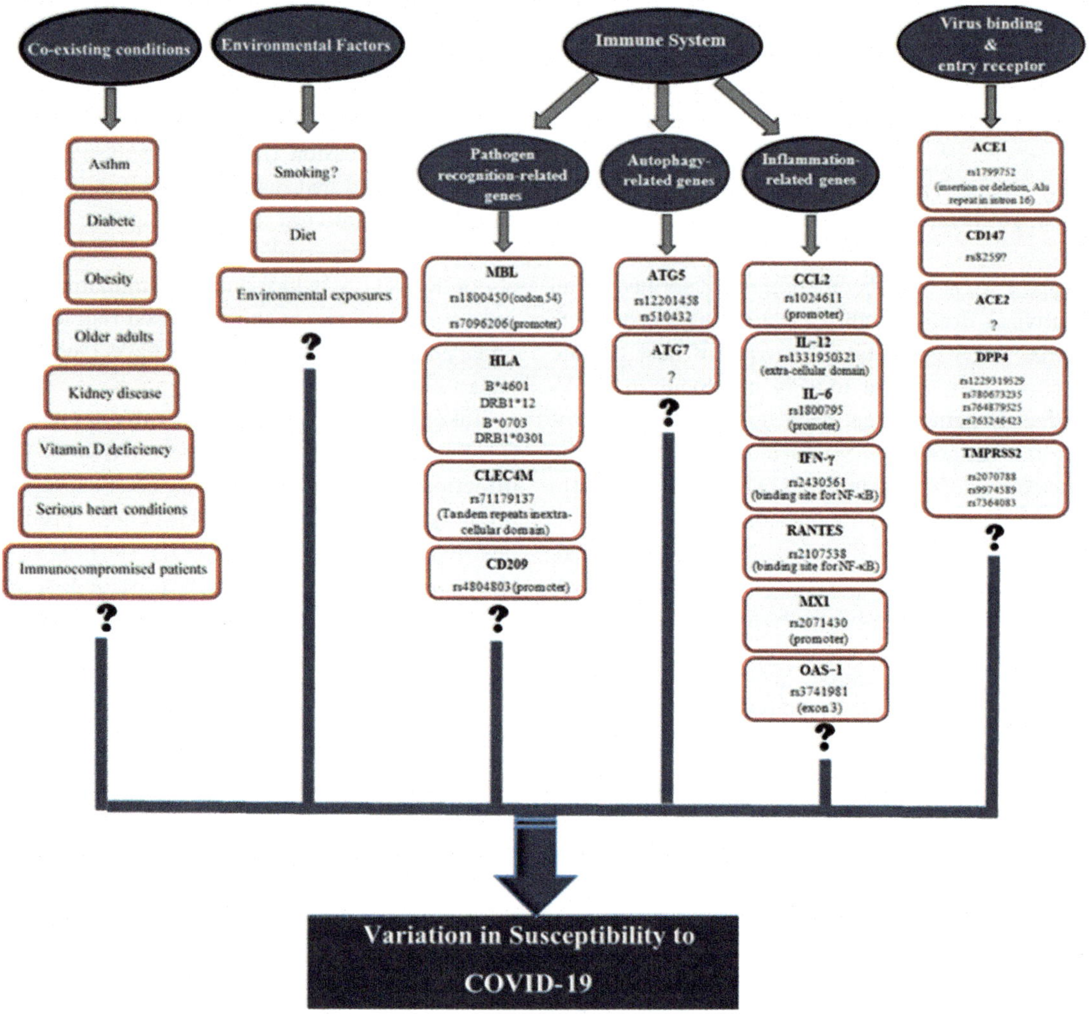

Fig. 6.1 The potential predisposition puzzle for COVID-19. Genetic variants are mostly based on SARS-CoV-1 evidence

autophagy-related 7 (*ATG7*) genes are associated with neutrophilic airway inflammation contributing to the pathogenesis of adult asthma (Pham et al. 2016). The functional single nucleotide polymorphisms in the *ATG5* gene, rs12201458 and rs510432, were found to be associated with asthma severity and pathogenesis (Martin et al. 2012).

Several lines of evidence showed that asthma increases the symptoms and severity of disease in patients who are infected with SARS-CoV-2 (Pennington 2020; Johnston 2020; Abrams et al. 2020; Abrams and Szefler 2020). As an example, asthma was significantly associated with longer intubation time in COVID-19 patients in ICU. It mainly increases the intubation time in younger COVID-19 patients (Mahdavinia et al. 2020). Therefore, it is suggested that there is probably an indirect link between genetic polymorphisms of the autophagy-related genes and COVID-19. The variants that influence the risk of asthma through autophagy genes could potentially increase the severity of COVID-19 pathogenesis.

6.6 Genetic Predisposition Models for Coronavirus Infection

Most variants discussed in this chapter are relatively common; however, their impact on disease susceptibility is overall weak. On the other hand, their additive contribution may increase their impact on susceptibility, severity, or prognosis of the disease. Moreover, it should be noted that other rare variants which cannot be detected by conventional association studies may also affect disease outcome.

Interestingly, there are published reports strongly suggesting potential genetic susceptibility for a fatal form of SARS-CoV-2 infection in multiple members of a family. Death in three 54- to 66-year-old brothers from COVID-19 without any underlying risk-associated conditions, who lived in separate locations in Tehran, Iran, provides strong evidence for the existence of highly penetrant genetic factors (Yousefzadegan and Rezaei 2020). Moreover, in a report from China, four family members died of COVID-19 (Hui et al. 2020). These anecdotal observations suggest that there could be rare variants with high power of penetrance that might explain a dominant inheritance in a subset of patients developing severe manifestations without a history of underlying diseases (Darbeheshti and Rezaei 2020).

The X-linked ACE2 variants that had been linked to the severity of COVID-19 infection may explain the higher prevalence and severity of COVID-19 in men than in women (Cai 2020), since X-chromosome-linked traits usually have less severe manifestations in females because of X inactivation.

SARS-CoV-2 infection tends to be less severe in children. Accordingly, some theories have been proposed, such as low expression of ACE2 in children or having healthier lungs. However, an epidemiological study of SARS-CoV-2-infected patients in China has reported that approximately 6% of cases in children were severe (Dong et al. 2020). Whether severe presentation in some children is the product of genetic susceptibility or underlying acquired factors needs further investigation.

The genetic predisposition to infectious diseases could also result from gene-environment interactions, as a multi-factorial model. In the case of COVID-19, smoking might be a potential environmental risk factor that could exhibit the additive effect in combination with susceptible genetic variants. Some medical conditions, mainly with underlying genetic factors, such as diabetes, obesity, asthma, chronic obstructive pulmonary disease, and vitamin D deficiency, have consequential effects on the course of SARS-CoV-2 infection. Moreover, age is an independent risk factor to suffer more severe symptoms due to natural physiological changes associated with aging.

6.7 Conclusion

COVID-19 is a multisystem infectious disease (Jahanshahlu and Rezaei 2020a; Saleki et al. 2020) that has transmitted between people worldwide and triggered reactions and awareness among a variety of professionals oriented toward containing the disease spread along with its management and treatment (Rezaei 2020a, b; Lotfi et al. 2020; Moazzami et al. 2020). It causes mild to moderate symptoms in nearly 80% of people, even in specific populations who are usually treated special, e.g., children, pregnant women, and immunodeficient patients (Ahanchian et al. 2020; Mirbeyk and Rezaei 2020). However, it may have severe symptoms in people with underlying diseases, like cancer (Ahmadi et al. 2020), and surprisingly, in people without any pre-existing problems (Yousefzadegan and Rezaei 2020; Sahu et al. 2020), and this would further worsen the situation that no formal treatment exists in practice (Fathi and Rezaei 2020; Jahanshahlu and Rezaei 2020b; Saghazadeh and Rezaei 2020b; Mohamed et al. 2020b; Rabiee et al. 2020). In this condition, the genomic data would be a great tool in the tailoring of personalized medicine for the disease, for example, through stem cells (Basiri et al. 2020). For this purpose, well-controlled investigations of rare variants with complete penetrance involving host vulnerability and response to SARS-CoV-2 are essential for predicting disease sensitivity, progression, and therapeutic strategies. Powerful bioinformatics techniques and high-throughput sequencing, now broadly available, have allowed affordable and rapid exploration of genetic predisposition to infectious diseases, including COVID-19, and are expected to contribute to the genetic basis of disease susceptibility continuously. Then, the translation of large-scale genomic results into practice is necessary, which would not be possible until a leap allows international and interdisciplinary collaboration (Momtazmanesh et al. 2020; Mohamed et al. 2020a; Moradian et al. 2020). The advantages of such collaboration are not only deemed worthy in the pandemic condition but are useful to drive a highly qualified personalized medicine into the future of infectious diseases (Rzymski et al. 2020; Kafieh et al. 2020), though inevitably rooting in the past pandemic situations (Jabbari et al. 2020).

References

Abrams EM, Szefler SJ (2020) Managing Asthma during COVID-19: an example for other chronic conditions in children and adolescents. J Pediatr 222:221

Abrams EM, Geert W'J, Yang CL (2020) Asthma and COVID-19. CMAJ 192:E551

Aghili N, Devaney JM, Alderman LO, Zukowska Z, Epstein SE, Burnett MS (2012) Polymorphisms in dipeptidyl peptidase IV gene are associated with the risk of myocardial infarction in patients with atherosclerosis. Neuropeptides 46:367–371

Ahanchian H, Moazzen N, Faroughi MSD, Khalighi N, Khoshkhui M, Aelami MH, Haghi NSM, Rezaei N (2020) COVID-19 in a child with primary specific antibody deficiency. https://doi.org/10.21203/rs.3.rs-28155/v1

Ahmadi M, Saffarzadeh N, Habibi MA, Hajiesmaeili F, Rezaei N (2020) Colon cancer and SARS-CoV-2: impact of ACE2 expression in susceptibility to COVID-19. bioRxiv. https://doi.org/10.1101/2020.06.11.146878

Ahmed RH, Huri HZ, Al-Hamodi Z, Salem SD, Al-absi B, Muniandy S (2016) Association of DPP4 gene polymorphisms with type 2 diabetes mellitus in Malaysian subjects. PLoS One 11:e0154369

Ascough S, Paterson S, Chiu C (2018) Induction and subversion of human protective immunity: contrasting influenza and respiratory syncytial virus. Front Immunol 9:323

Asselta R, Paraboschi EM, Mantovani A, Duga S (2020) ACE2 and TMPRSS2 variants and expression as candidates to sex and country differences in COVID-19 severity in Italy. Aging 12:10087

Bahrami A, Vafapour M, Moazzami B, Rezaei N (2020) Hyperinflammatory shock related to COVID-19 in a patient presenting with multisystem inflammatory syndrome in children: first case from Iran. J Paediatr Child Health. https://doi.org/10.1111/jpc.15048

Barreiro LB, Patin E, Neyrolles O, Cann HM, Gicquel B, Quintana-Murci L (2005) The heritage of pathogen pressures and ancient demography in the human innate-immunity CD209/CD209L region. Am J Hum Genet 77:869–886

Basiri A, Pazhouhnia Z, Beheshtizadeh N, Hoseinpour M, Saghazadeh A, Rezaei N (2020) Regenerative medicine in COVID-19 treatment: real opportunities and range of promises. Stem Cell Rev Rep 16:1–13

Baudin B (2002) New aspects on angiotensin-converting enzyme: from gene to disease. Clin Chem Lab Med 40:256–265

Ben-Ari Z, Mor E, Papo O, Kfir B, Sulkes J, Tambur AR, Tur-Kaspa R, Klein T (2003) Cytokine gene polymorphisms in patients infected with hepatitis B virus. Am J Gastroenterol 98:144–150

Bergeron E, Vincent MJ, Wickham L, Hamelin J, Basak A, Nichol ST, Chrétien M, Seidah NG (2005) Implication of proprotein convertases in the processing and spread of severe acute respiratory syndrome coronavirus. Biochem Biophys Res Commun 326:554–563

Booth CM, Matukas LM, Tomlinson GA, Rachlis AR, Rose DB, Dwosh HA, Walmsley SL, Mazzulli T, Avendano M, Derkach P (2003) Clinical features and short-term outcomes of 144 patients with SARS in the greater Toronto area. JAMA 289:2801–2809

Böttcher E, Matrosovich T, Beyerle M, Klenk H-D, Garten W, Matrosovich M (2006) Proteolytic activation of influenza viruses by serine proteases TMPRSS2 and HAT from human airway epithelium. J Virol 80:9896–9898

Brun P, Tarricone E, Di Stefano A, Mattiuzzo E, Mehrbod P, Ghavami S, Leonardi A (2020) The regulatory activity of autophagy in conjunctival fibroblasts and its possible role in vernal keratoconjunctivitis. J Allergy Clin Immunol 146:1210-1213.e9

Burgner D, Jamieson SE, Blackwell JM (2006) Genetic susceptibility to infectious diseases: big is beautiful, but will bigger be even better? Lancet Infect Dis 6:653–663

Cai H (2020) Sex difference and smoking predisposition in patients with COVID-19. Lancet Respir Med 8:e20

Cameron MJ, Ran L, Xu L, Danesh A, Bermejo-Martin JF, Cameron CM, Muller MP, Gold WL, Richardson SE, Poutanen SM (2007) Interferon-mediated immunopathological events are associated with atypical innate and adaptive immune responses in patients with severe acute respiratory syndrome. J Virol 81:8692–8706

Canna SW, Behrens EM (2012) Making sense of the cytokine storm: a conceptual framework for understanding, diagnosing, and treating hemophagocytic syndromes. Pediatr Clin 59:329–344

Cao Y, Li L, Feng Z, Wan S, Huang P, Sun X, Wen F, Huang X, Ning G, Wang W (2020) Comparative genetic analysis of the novel coronavirus (2019-nCoV/ SARS-CoV-2) receptor ACE2 in different populations. Cell Discov 6:1–4

Chan KCA, Tang NLS, Hui DSC, Chung GTY, Wu AKL, Chim SSC, Chiu RWK, Lee N, Choi KW, Sung YM (2005) Absence of association between angiotensin converting enzyme polymorphism and development of adult respiratory distress syndrome in patients with severe acute respiratory syndrome: a case control study. BMC Infect Dis 5:26

Chan VSF, Chan KYK, Chen Y, Poon LLM, Cheung ANY, Zheng B, Chan K-H, Mak W, Ngan HYS, Xu X (2006) Homozygous L-SIGN (CLEC4M) plays a protective role in SARS coronavirus infection. Nat Genet 38:38–46

Chan KYK, Xu M-S, Ching JCY, So TMK, Lai S-T, Chu C-M, Yam LYC, Wong ATY, Chung PH, Chan VSF (2010a) CD209 (DC-SIGN)− 336A> G promoter polymorphism and severe acute respiratory syndrome in Hong Kong Chinese. Hum Immunol 71:702–707

Chan KY, Xu MS, Ching JC, Chan VS, Ip YC, Yam L, Chu CM, Lai ST, So KM, Wong TY (2010b) Association of a single nucleotide polymorphism in the CD209 (DC-SIGN) promoter with SARS severity. Hong Kong Med J = Xianggang yi xue za zhi 16:37

Chen Z, Mi L, Xu J, Yu J, Wang X, Jiang J, Xing J, Shang P, Qian A, Li Y (2005) Function of HAb18G/CD147 in invasion of host cells by severe acute respiratory syndrome coronavirus. J Infect Dis 191:755–760

Chiu RWK, Tang NLS, Hui DSC, Chung GTY, Chim SSC, Chan KCA, Sung Y-m, Chan LYS, Tong Y-k, Lee W-s (2004) ACE2 gene polymorphisms do not affect outcome of severe acute respiratory syndrome. Clin Chem 50:1683–1686

Cui J, Li F, Shi Z-L (2019) Origin and evolution of pathogenic coronaviruses. Nat Rev Microbiol 17:181–192

Darbeheshti F, Rezaei N (2020) Genetic predisposition models to COVID-19 infection. Med Hypotheses 142:109818

Dong Y, Mo X, Hu Y, Qi X, Jiang F, Jiang Z, Tong S (2020) Epidemiological characteristics of 2143 pediatric patients with 2019 coronavirus disease in China. Pediatrics 145:e20200702

Eghtedardoost M, Ghazanfari T, Sadeghipour A, Hassan ZM, Ghanei M, Ghavami S (2020) Delayed effects of sulfur mustard on autophagy suppression in chemically-injured lung tissue. Int Immunopharmacol 80:105896

Emami A, Shojaei S, da Silva Rosa SC, Aghaei M, Samiei E, Vosoughi AR, Kalantari F, Kawalec P, Thliveris J, Sharma P, Zeki AA, Akbari M, Gordon JW, Ghavami S (2019) Mechanisms of simvastatin myotoxicity: the role of autophagy flux inhibition. Eur J Pharmacol 862:172616

Engler RJM, Klote MM, Spooner CE, Reardon GT, Halsey JF, Nelson MR (2010) Mannan-binding lectin (MBL) serum levels in a healthy adult population: gender differences in prevalence of MBL deficiency. J Allergy Clin Immunol 125:AB10

Ezekowitz RA (2003) Role of the mannose-binding lectin in innate immunity. J Infect Dis 187:S335–SS39

Fathi N, Rezaei N (2020) Lymphopenia in COVID-19: therapeutic opportunities. Cell Biol Int 44:1792

Feng B, Mao Z-r, Pang K, Zhang S-l, Li L (2015) Association of tumor necrosis factor α− 308G/A and interleukin-6− 174G/C gene polymorphism with pneumonia-induced sepsis. J Crit Care 30:920–923

Fernández-Arcás N, Blanco A, Jesus Gaitan M, Nyqvist M, Alonso A, Reyes-Engel A (2004) Differential transcriptional expresion of the polymorphic myxovirus resistance protein A in response to interferon-alpha treatment. Pharmacogenet Genomics 14:189–193

Flores C, Ma S-F, Maresso K, Wade MS, Villar J, Garcia JGN (2008) IL6 gene-wide haplotype is associated with susceptibility to acute lung injury. Transl Res 152:11–17

Garred P, Madsen HO, Balslev U, Hofmann BO, Pedersen C, Gerstoft J, Svejgaard A (1997) Susceptibility to HIV infection and progression of AIDS in relation to variant alleles of mannose-binding lectin. Lancet 349:236–240

Glowacka I, Bertram S, Müller MA, Allen P, Soilleux E, Pfefferle S, Steffen I, Tsegaye TS, He Y, Gnirss K (2011) Evidence that TMPRSS2 activates the severe acute respiratory syndrome coronavirus spike protein for membrane fusion and reduces viral control by the humoral immune response. J Virol 85:4122–4134

Hamano E, Hijikata M, Itoyama S, Quy T, Phi NC, Long HT, Van Ban V, Matsushita I, Yanai H, Kirikae F (2005) Polymorphisms of interferon-inducible genes OAS-1 and MxA associated with SARS in the Vietnamese population. Biochem Biophys Res Commun 329:1234–1239

Hanaei S, Rezaei N (2020) COVID-19: developing from an outbreak to a pandemic. Arch Med Res 51:582

Hartshorn KL, Sastry K, White MR, Anders EM, Super M, Ezekowitz RA, Tauber AI (1993) Human mannose-binding protein functions as an opsonin for influenza A viruses. J Clin Invest 91:1414–1420

Hibberd ML, Sumiya M, Summerfield JA, Booy R, Levin M, Meningococcal Research Group (1999) Association of variants of the gene for mannose-binding lectin with susceptibility to meningococcal disease. Lancet 353:1049–1053

Hill AVS (2012) Evolution, revolution and heresy in the genetics of infectious disease susceptibility. Philos Trans R Soc B: Biol Sci 367:840–849

Hoffmann M, Kleine-Weber H, Schroeder S, Krüger N, Herrler T, Erichsen S, Schiergens TS, Herrler G, Wu N-H, Nitsche A (2020) SARS-CoV-2 cell entry depends on ACE2 and TMPRSS2 and is blocked by a clinically proven protease inhibitor. Cell 181:271

Huang K-J, Ih-Jen S, Theron M, Wu Y-C, Lai S-K, Liu C-C, Lei H-Y (2005) An interferon-γ-related cytokine storm in SARS patients. J Med Virol 75:185–194

Hui XU, Zhiming B, Walsh M (2020) Four deaths in one family show danger of Wuhan home quarantine policy. Available at: https://www.caixinglobal.com

Ip WKE, Chan KH, Law HKW, Tso GHW, Kong EKP, Wong WHS, To YF, Yung RWH, Chow EY, Au KL (2005) Mannose-binding lectin in severe acute respiratory syndrome coronavirus infection. J Infect Dis 191:1697–1704

Itoyama S, Keicho N, Quy T, Phi NC, Long HT, Van Ban V, Ohashi J, Hijikata M, Matsushita I, Kawana A (2004) ACE1 polymorphism and progression of SARS. Biochem Biophys Res Commun 323:1124–1129

Iwata-Yoshikawa N, Okamura T, Shimizu Y, Hasegawa H, Takeda M, Nagata N (2019) TMPRSS2 contributes to virus spread and immunopathology in the airways of murine models after coronavirus infection. J Virol 93:e01815–e01818

Jabbari P, Jabbari F, Ebrahimi S, Rezaei N (2020) COVID-19: a chimera of two pandemics. Disaster Med Public Health Prep 14:1–3

Jahanshahlu L, Rezaei N (2020a) Central nervous system involvement in COVID-19. Arch Med Res 51:721

Jahanshahlu L, Rezaei N (2020b) Monoclonal antibody as a potential anti-COVID-19. Biomed Pharmacother 129:110337

Jerrard-Dunne P, Sitzer M, Risley P, Steckel DA, Buehler A, von Kegler S, Markus HS (2003) Interleukin-6 promoter polymorphism modulates the effects of heavy alcohol consumption on early carotid artery atherosclerosis: the Carotid Atherosclerosis Progression Study (CAPS). Stroke 34:402–407

Jiang Y, Xu J, Zhou C, Wu Z, Zhong S, Liu J, Luo W, Chen T, Qin Q, Deng P (2005) Characterization of cytokine/chemokine profiles of severe acute respiratory syndrome. Am J Respir Crit Care Med 171:850–857

Johnston SL (2020) Asthma and COVID-19: is asthma a risk factor for severe outcomes? Allergy 75:1543

Kafieh R, Arian R, Saeedizadeh N, Minaee S, Yadav SK, Vaezi A, Rezaei N, Javanmard SH (2020) COVID-19 in Iran: a deeper look into the future. medRxiv. https://doi.org/10.1101/2020.04.24.20078477

Karlberg J, Chong DSY, Lai WYY (2004) Do men have a higher case fatality rate of severe acute respiratory syndrome than women do? Am J Epidemiol 159:229–231

Keicho N, Itoyama S, Kashiwase K, Phi NC, Long HT, Van Ban V, Hoa BK, Le Hang NT, Hijikata M, Sakurada S (2009) Association of human leukocyte antigen class II alleles with severe acute respiratory syndrome in the Vietnamese population. Hum Immunol 70:527–531

Kenney AD, Dowdle JA, Bozzacco L, McMichael TM, St. Gelais C, Panfil AR, Sun Y, Schlesinger LS, Anderson MZ, Green PL (2017) Human genetic determinants of viral diseases. Annu Rev Genet 51:241–263

Kleine-Weber H, Schroeder S, Krüger N, Prokscha A, Naim HY, Müller MA, Drosten C, Pöhlmann S, Hoffmann M (2020) Polymorphisms in dipeptidyl peptidase 4 reduce host cell entry of Middle East respiratory syndrome coronavirus. Emerg Microbes Infect 9:155–168

Knapp S, Yee LJ, Frodsham AJ, Hennig BJW, Hellier S, Zhang L, Wright M, Chiaramonte M, Graves M, Thomas HC (2003) Polymorphisms in interferon-induced genes and the outcome of hepatitis C virus infection: roles of MxA, OAS-1 and PKR. Genes Immun 4:411–419

Law HKW, Cheung CY, Ng HY, Sia SF, Chan YO, Luk W, Nicholls JM, Peiris JSM, Lau YL (2005) Chemokine up-regulation in sars-coronavirus–infected, monocyte-derived human dendritic cells. Blood 106:2366–2374

Letko MC, Munster V (2020) Functional assessment of cell entry and receptor usage for SARS-CoV-2 and other lineage B betacoronaviruses. Nature microbiology 5:562–9.

Li W, Moore MJ, Vasilieva N, Sui J, Wong SK, Berne MA, Somasundaran M, Sullivan JL, Luzuriaga K, Greenough TC (2003) Angiotensin-converting enzyme 2 is a functional receptor for the SARS coronavirus. Nature 426:450–454

Li H, Tang NLS, Chan PKS, Wang CY, Hui DSC, Luk C, Kwok R, Huang W, Sung JJY, Kong QP (2008) Polymorphisms in the C-type lectin genes cluster in chromosome 19 and predisposition to severe acute respiratory syndrome coronavirus (SARS-CoV) infection. J Med Genet 45:752–758

Li Y, Zhang Z, Yang L, Lian X, Xie Y, Li S, Xin S, Cao P, Lu J (2020) The MERS-CoV receptor DPP4 as a candidate binding target of the SARS-CoV-2 spike. iScience 23:101160

Lim W, Peiris JSM, Lau YL, Chong WP, Ip WKE, Chow EY, Au KL, Tso GHW, Yung RWH, Ng MW (2006) The interferon gamma gene polymorphism+ 874 A/T is associated with severe acute respiratory syndrome. BMC Infect Dis 6:82

Lin M, Tseng H-K, Trejaut JA, Lee H-L, Loo J-H, Chu C-C, Chen P-J, Su Y-W, Lim KH, Tsai Z-U (2003) Association of HLA class I with severe acute respiratory syndrome coronavirus infection. BMC Med Genet 4:9

Lotfi M, Rezaei N (2020) SARS-CoV-2: a comprehensive review from pathogenicity of the virus to clinical consequences. J Med Virol 92:1864

Lotfi M, Hamblin MR, Rezaei N (2020) COVID-19: transmission, prevention, and potential therapeutic opportunities. Clin Chim Acta 508:254

Lu X-M, Chen G-J, Yang Y, Qiu H-B (2011) Angiotensin-converting enzyme polymorphism affects outcome of local Chinese with acute lung injury. Respir Med 105:1485–1490

Lu R, Zhao X, Li J, Niu P, Yang B, Wu H, Wang W, Song H, Huang B, Zhu N (2020) Genomic characterisation and epidemiology of 2019 novel coronavirus: implications for virus origins and receptor binding. Lancet 395:565–574

Madsen HO, Garred P, Thiel S, Kurtzhals JA, Lamm LU, Ryder LP, Svejgaard A (1995) Interplay between promoter and structural gene variants control basal

serum level of mannan-binding protein. J Immunol 155:3013–3020
Mahdavinia M, Foster KJ, Jauregui E, Moore D, Adnan D, Andy-Nweye AAB, Khan S, Bishehsari F (2020) Asthma prolongs intubation in COVID-19. J Allergy Clin Immunol Pract 8:2388
Mao Z-R, Zhang S-L, Feng B (2017) Association of IL-10 (-819T/C,-592A/C and-1082A/G) and IL-6-174G/C gene polymorphism and the risk of pneumonia-induced sepsis. Biomarkers 22:106–112
Martin LJ, Gupta J, Jyothula SS, Butsch Kovacic M, Biagini Myers JM, Patterson TL, Ericksen MB, He H, Gibson AM, Baye TM, Amirisetty S, Tsoras AM, Sha Y, Eissa NT, Hershey GK (2012) Functional variant in the autophagy-related 5 gene promotor is associated with childhood asthma. PLoS One 7:e33454
McAlinden KD, Kota A, Haghi M, Ghavami S, Sharma P (2020) Pharmacologic inhibition of vacuolar H(+) ATPase attenuates features of severe Asthma in mice. Am J Respir Cell Mol Biol 62:117–120
McDermott DH, Yang Q, Kathiresan S, Adrienne Cupples L, Massaro JM, Keaney JF Jr, Larson MG, Vasan RS, Hirschhorn JN, O'Donnell CJ (2005) CCL2 polymorphisms are associated with serum monocyte chemoattractant protein-1 levels and myocardial infarction in the Framingham Heart Study. Circulation 112:1113–1120
Mirbeyk M, Rezaei N (2020) The impact of COVID-19 on pregnancy and neonatal health: a systematic review. https://doi.org/10.21203/rs.3.rs-25861/v1
Moazzami B, Razavi-Khorasani N, Moghadam AD, Farokhi E, Rezaei N (2020) COVID-19 and telemedicine: immediate action required for maintaining healthcare providers well-being. J Clin Virol 126:104345
Mohamed K, Rodríguez-Román E, Rahmani F, Zhang H, Ivanovska M, Makka SA, Joya M, Makuku R, Islam MS, Radwan N (2020a) Borderless collaboration is needed for COVID-19—A disease that knows no borders. Infect Control Hosp Epidemiol 41:1–2
Mohamed K, Yazdanpanah N, Saghazadeh A, Rezaei N (2020b) Computational drug discovery and repurposing for the treatment of Covid-19: a systematic review. Available at SSRN 3583748
Momtazmanesh S, Ochs HD, Uddin LQ, Perc M, Routes JM, Vieira DN, Al-Herz W, Baris S, Prando C, Rosivall L (2020) All together to Fight COVID-19. Am J Trop Med Hyg 102:1181–1183
Moradian N, Ochs HD, Sedikies C, Hamblin MR, Camargo CA, Alfredo Martinez J, Biamonte JD, Abdollahi M, Torres PJ, Nieto JJ (2020) The urgent need for integrated science to fight COVID-19 pandemic and beyond. J Transl Med 18:1–7
Moriuchi H, Moriuchi M, Fauci AS (1997) Nuclear factor-kappa B potently up-regulates the promoter activity of RANTES, a chemokine that blocks HIV infection. J Immunol 158:3483–3491
Nasab MG, Saghazadeh A, Rezaei N (2020) SARS-CoV--2–a tough opponent for the immune system. Arch Med Res 51:589
Ng MHL, Lau K-M, Li L, Cheng S-H, Chan WY, Hui PK, Zee B, Leung C-B, Sung JJY (2004) Association of human-leukocyte-antigen class I (B* 0703) and class II (DRB1* 0301) genotypes with susceptibility and resistance to the development of severe acute respiratory syndrome. J Infect Dis 190:515–518
Ng MW, Zhou G, Chong WP, Lee LWY, Law HKW, Zhang H, Wong WHS, Fok SFS, Zhai Y, Yung RWH (2007) The association of RANTES polymorphism with severe acute respiratory syndrome in Hong Kong and Beijing Chinese. BMC Infect Dis 7:50
Peiris JSM, Lai ST, Poon LLM, Guan Y, Yam LYC, Lim W, Nicholls J, Yee WKS, Yan WW, Cheung MT (2003) Coronavirus as a possible cause of severe acute respiratory syndrome. Lancet 361:1319–1325
Pennington E (2020) Asthma increases risk of severity of COVID-19. Cleve Clin J Med. https://doi.org/10.3949/ccjm.87a.ccc002
Pham DL, Kim SH, Losol P, Yang EM, Shin YS, Ye YM, Park HS (2016) Association of autophagy related gene polymorphisms with neutrophilic airway inflammation in adult asthma. Korean J Intern Med 31:375–385
Pravica V, Perrey C, Stevens A, Lee J-H, Hutchinson IV (2000) A single nucleotide polymorphism in the first intron of the human IFN-γ gene:: absolute correlation with a polymorphic CA microsatellite marker of high IFN-γ production. Hum Immunol 61:863–866
Rabiee N, Rabiee M, Bagherzadeh M, Rezaei N (2020) COVID-19 and picotechnology: potential opportunities. Med Hypotheses 144:109917
Raj VS, Mou H, Smits SL, Dekkers DHW, Müller MA, Dijkman R, Muth D, Demmers JAA, Zaki A, Fouchier RAM (2013) Dipeptidyl peptidase 4 is a functional receptor for the emerging human coronavirus-EMC. Nature 495:251–254
Ren L-L, Wang Y-M, Wu Z-Q, Xiang Z-C, Guo L, Xu T, Jiang Y-Z, Xiong Y, Li Y-J, Li X-W (2020) Identification of a novel coronavirus causing severe pneumonia in human: a descriptive study. Chin Med J 133:1015
Rezaei N (2020a) COVID-19 affects healthy pediatricians more than pediatric patients. Infect Control Hosp Epidemiol 41:1–1
Rezaei N (2020b) COVID-19 and medical biotechnology. Avicenna J Med Biotechnol 12:139
Rigat B, Hubert C, Alhenc-Gelas F, Cambien F, Corvol P, Soubrier F (1990) An insertion/deletion polymorphism in the angiotensin I-converting enzyme gene accounting for half the variance of serum enzyme levels. J Clin Invest 86:1343–1346
Roy S, Knox K, Segal S, Griffiths D, Moore CE, Welsh KI, Smarason A, Day NP, McPheat WL, Crook DW (2002) MBL genotype and risk of invasive pneumococcal disease: a case-control study. Lancet 359:1569–1573
Rzymski P, Nowicki M, Mullin GE, Abraham A, Rodríguez-Román E, Petzold MB, Bendau A, Sahu KK, Ather A, Naviaux A-F (2020) Quantity does not equal quality: Scientific principles cannot be sacrificed. Int Immunopharmacol 86:106711

Saghazadeh A, Rezaei N (2020a) Immune-epidemiological parameters of the novel coronavirus–a perspective. Expert Rev Clin Immunol 16:1–6

Saghazadeh A, Rezaei N (2020b) Towards treatment planning of COVID-19: rationale and hypothesis for the use of multiple immunosuppressive agents: anti-antibodies, immunoglobulins, and corticosteroids. Int Immunopharmacol 84:106560

Sahu KK, Siddiqui AD, Rezaei N, Cerny J (2020) Challenges for management of immune thrombocytopenia during COVID-19 pandemic. J Med Virol 92:2277

Saleki K, Banazadeh M, Saghazadeh A, Rezaei N (2020) The involvement of the central nervous system in patients with COVID-19. Rev Neurosci 31:453

Schork NJ, Murray SS, Frazer KA, Topol EJ (2009) Common vs. rare allele hypotheses for complex diseases. Curr Opin Genet Dev 19:212–219

Shao S, Xu QQ, Yu X, Pan R, Chen Y (2020) Dipeptidyl peptidase 4 inhibitors and their potential immune modulatory functions. Pharmacol Ther 209:107503

Shen C, Jiao W-W, Feng W-X, Wu X-R, Xiao J, Miao Q, Sun L, Wang B-B, Wang J, Liu F (2013) IFNG polymorphisms are associated with tuberculosis in Han Chinese pediatric female population. Mol Biol Rep 40:5477–5482

Shirogane Y, Takeda M, Iwasaki M, Ishiguro N, Takeuchi H, Nakatsu Y, Tahara M, Kikuta H, Yanagi Y (2008) Efficient multiplication of human metapneumovirus in Vero cells expressing the transmembrane serine protease TMPRSS2. J Virol 82:8942–8946

Shojaei S, Koleini N, Samiei E, Aghaei M, Cole LK, Alizadeh J, Islam MI, Vosoughi AR, Albokashy M, Butterfield Y, Marzban H, Xu F, Thliveris J, Kardami E, Hatch GM, Eftekharpour E, Akbari M, Hombach-Klonisch S, Klonisch T, Ghavami S (2020) Simvastatin increases temozolomide-induced cell death by targeting the fusion of autophagosomes and lysosomes. FEBS J 287:1005–1034

Stawiski EW, Diwanji D, Suryamohan K, Gupta R, Fellouse FA, Sathirapongsasuti F, Liu J, Jiang Y-P, Ratan A, Mis M (2020) Human ACE2 receptor polymorphisms predict SARS-CoV-2 susceptibility. bioRxiv. https://doi.org/10.1101/2020.04.07.024752

Strollo R, Pozzilli P (2020) DPP4 inhibition: preventing SARS-CoV-2 infection and/or progression of COVID-19? Diabetes Metab Res Rev 36:e3330

Suzuki F, Arase Y, Suzuki Y, Tsubota A, Akuta N, Hosaka T, Someya T, Kobayashi M, Saitoh S, Ikeda K (2004) Single nucleotide polymorphism of the MxA gene promoter influences the response to interferon monotherapy in patients with hepatitis C viral infection. J Viral Hepat 11:271–276

Tang F, Liu W, Zhang F, Xin Z-T, Wei M-T, Zhang P-H, Yang H, Ly H, Cao W-C (2008) IL-12 RB1 genetic variants contribute to human susceptibility to severe acute respiratory syndrome infection among Chinese. PLoS One 3:e2183

Tangye SG, Al-Herz W, Bousfiha A, Chatila T, Cunningham-Rundles C, Etzioni A, Franco JL, Holland SM, Klein C, Morio T (2020) Human inborn errors of immunity: 2019 update on the classification from the international union of immunological societies expert committee. J Clin Immunol 40:24–64

Tipnis SR, Hooper NM, Hyde R, Karran E, Christie G, Turner AJ (2000) A human homolog of angiotensin-converting enzyme cloning and functional expression as a captopril-insensitive carboxypeptidase. J Biol Chem 275:33238–33243

Tong Y, Wang Z, Geng Y, Liu J, Zhang R, Lin Q, Li X, Huang D, Gao S, Hu D (2010) The association of functional polymorphisms of IL-6 gene promoter with ischemic stroke: analysis in two Chinese populations. Biochem Biophys Res Commun 391:481–485

Tso HW, Ip WK, Chong WP, Tam CM, Chiang AKS, Lau YL (2005) Association of interferon gamma and interleukin 10 genes with tuberculosis in Hong Kong Chinese. Genes Immun 6:358–363

Tu X, Chong WP, Zhai Y, Zhang H, Zhang F, Wang S, Liu W, Wei M, Siu NHO, Yang H (2015) Functional polymorphisms of the CCL2 and MBL genes cumulatively increase susceptibility to severe acute respiratory syndrome coronavirus infection. J Infect 71:101–109

Ulhaq ZS, Soraya GV (2020) Anti-IL-6 receptor antibody treatment for severe COVID-19 and the potential implication of IL-6 gene polymorphisms in novel coronavirus pneumonia. Available at SSRN 3592878

Vaninov N (2020) In the eye of the COVID-19 cytokine storm. In: Nature Publishing Group

Wang H, Zhang K, Qin H, Yang L, Zhang L, Cao Y (2015) Genetic association between CD143 rs4340 polymorphism and pneumonia risk: a meta analysis. Medicine 94:e883

Wang D, Hu B, Hu C, Zhu F, Liu X, Zhang J, Wang B, Xiang H, Cheng Z, Xiong Y (2020a) Clinical characteristics of 138 hospitalized patients with 2019 novel coronavirus–infected pneumonia in Wuhan, China. JAMA 323:1061–1069

Wang K, Chen W, Zhou Y-S, Lian J-Q, Zhang Z, Du P, Gong L, Zhang Y, Cui H-Y, Geng J-J (2020b) SARS-CoV-2 invades host cells via a novel route: CD147-spike protein. bioRxiv

Wong CK, Lam CWK, Wu AKL, Ip WK, Lee NLS, Chan IHS, Lit LCW, Hui DSC, Chan MHM, Chung SSC (2004) Plasma inflammatory cytokines and chemokines in severe acute respiratory syndrome. Clin Exp Immunol 136:95–103

Xiong P, Zeng X, Song MS, Jia SW, Zhong MH, Xiao LL, Lan W, Cai C, Wu XW, Gong FL (2008) Lack of association between HLA-A,-B and-DRB1 alleles and the development of SARS: a cohort of 95 SARS-recovered individuals in a population of Guangdong, southern China. Int J Immunogenet 35:69–74

Xu H, Zhong L, Deng J, Peng J, Dan H, Zeng X, Li T, Chen Q (2020) High expression of ACE2 receptor of 2019-nCoV on the epithelial cells of oral mucosa. Int J Oral Sci 12:1–5

Yan J, Mao Y, Wang C, Wang Z (2015) Association study between an SNP in CD147 and its expression with acute coronary syndrome in a Jiangsu Chinese population. Medicine 94:e1537

Yang IA, Seeney SL, Wolter JM, Anders EM, McCormack JG, Tunnicliffe AM, Rabnott GC, Shaw JG, Dent AG, Kim ST (2003) Mannose-binding lectin gene polymorphism predicts hospital admissions for COPD infections. Genes Immun 4:269–274

Yazdanpanah F, Hamblin MR, Rezaei N (2020) The immune system and COVID-19: friend or foe? Life Sci 256:117900

Yousefzadegan S, Rezaei N (2020) Case report: death due to COVID-19 in three brothers. Am J Trop Med Hyg 102:1203–1204

Zeki AA, Yeganeh B, Kenyon NJ, Post M, Ghavami S (2016) Autophagy in airway diseases: a new frontier in human asthma? Allergy 71:5–14

Zhang H, Zhou G, Zhi L, Yang H, Zhai Y, Dong X, Zhang X, Gao X, Zhu Y, He F (2005) Association between mannose-binding lectin gene polymorphisms and susceptibility to severe acute respiratory syndrome coronavirus infection. J Infect Dis 192:1355–1361

Zhang M, Lu Y, Zhang X, Lu A, Wang L, Chen C (2016) I nterleukin-4 polymorphism is associated with severity of respiratory syncytial virus infection. J Paediatr Child Health 52:25–29

Zhi L, Zhou G, Zhang H, Zhai Y, Yang H, Zhang F, Wang S, Wei M, Cao W, He F (2007) Lack of support for an association between CLEC4M homozygosity and protection against SARS coronavirus infection. Nat Genet 39:692–693

Genetic Polymorphisms in the Host and COVID-19 Infection

7

Joris R. Delanghe, Marc L. De Buyzere, and Marijn M. Speeckaert

Abstract

The outbreak of the COVID-19 pandemic shows a marked geographical variation in its prevalence and mortality. The question arises if the host genetic variation may (partly) affect the prevalence and mortality of COVID-19. We postulated that the geographical variation of human polymorphisms might partly explain the variable prevalence of the infection. We investigated some candidate genes that have the potential to play a role in the immune defense against COVID-19: complement component 3 (C3), galactoside 2-alpha-L-fucosyltransferase 2 (FUT2), haptoglobin (Hp), vitamin D binding protein (DBP), human homeostatic iron regulator protein (HFE), cystic fibrosis transmembrane conductance regulator (CFTR), and angiotensin-converting enzyme 1 (ACE1). In a univariate approach, ACE1 D/I, C3, CFTR, and HFE polymorphisms correlated significantly with COVID-19 prevalence/mortality, whereas Hp and FUT2 polymorphism did not show any significant correlations. In a multivariate analysis, only ACE1 D/I and C3 polymorphisms were determinants for COVID-19 prevalence/mortality. The other polymorphisms (CFTR, DBP, FUT2, HFE, and Hp) did not correlate with COVID-19 prevalence/mortality. Whereas ACE1 D/I polymorphism shows functional links with ACE2 (which is the receptor for the virus) in COVID-19, C3 can act as a critical step in the virus-induced inflammation. Our findings plead against a bystander role of the polymorphisms as a marker for historical migrations, which comigrate with causal genes involved in COVID-19 infection. Further studies are required to assess the clinical outcome of COVID-19 in C3S and ACE1 D allele carriers and to study the role of C3 and ACE1 D/I polymorphisms in COVID-19 and their potential effects on treatment response.

Keywords

Complement component 3 · COVID-19 · Cystic fibrosis · Haptoglobin · Human

J. R. Delanghe (✉)
Department of Diagnostic Sciences, Ghent University, Ghent, Belgium

Network of Immunity in Infection, Malignancy and Autoimmunity (NIIMA), Universal Scientific Education and Research Network (USERN), Ghent, Belgium
e-mail: joris.delanghe@ugent.be

M. L. De Buyzere
Department of Internal Medicine, Ghent University, Ghent, Belgium

M. M. Speeckaert
Department of Internal Medicine, Ghent University, Ghent, Belgium

Research Foundation-Flanders (FWO), Ghent, Belgium

N. Rezaei (ed.), *Coronavirus Disease - COVID-19*, Advances in Experimental Medicine and Biology 1318, https://doi.org/10.1007/978-3-030-63761-3_7

homeostatic iron regulator · Secretor gene · Vitamin D binding protein

7.1 Introduction

The COVID-19 pandemic in 2019–2020 shows a marked geographical variation in its prevalence and mortality. In particular, some Western European countries were severely affected by the virus, whereas the number of COVID-19 casualties was much lower among Eastern European countries. As so far, no viral mutations have been associated with geographical variations of the pandemic; the question arises if the host genetic variation may (partly) affect the outcome and the mortality of COVID-19. We, therefore, postulated that the marked geographical variation of some immune-related human polymorphisms might partly explain the variable prevalence of COVID-19.

Genetic polymorphisms are the consequence of human evolution, which have been modified by a natural selection process (Langlois and Delanghe 1996; Speeckaert et al. 2006; Delanghe et al. 2015) and usually function to retain a variety of forms in a population living in a varied environment. Polymorphisms are seen in many plasma proteins and are related to biodiversity, genetic variation, and adaptation. Some polymorphic candidate genes (Table 7.1), which may play a role in immune defense against COVID-19, can be put forward.

Complement component 3 (C3) is a central component of the innate immune system, which

Table 7.1 The investigated human genetic polymorphisms

Polymorphism	Genotypes	Remark
C3	C3 FF, C3 FS, C3 SS	Supports complement activation C3a is an anaphylatoxin Stimulates mast cell degranulation Plays a role in chemotaxis A precursor of the stimulating acylation protein Involved in C5 cleavage Assembly of membrane attack complexes and deposition of activated C3 molecules on the microbe surface, which is necessary for phagocytosis Enhances immune response The representative of the first principal component
DBP	Gc 1-1, Gc 2-1, Gc 2-2	Regulator of vitamin D status Involved in fatty acid transport, actin scavenging, macrophage activation, and chemotaxis,
HFE	Wild type, C282Y heterozygous, C282Y homozygous	Demonstrates immune activities that bridge innate and adaptive immunity Relates to MHC I antigen presentation
FUT2	Se/Se, Se/se, se/se	The secretor status has consequences for the immune defense: nonsecretors have reduced susceptibility to the most common strains of norovirus Expression of the Lewis blood group antigens is affected by secretor status
Hp	Hp 1-1, Hp 2-1, Hp 2-2	Involved in iron conservation and immune response (e.g., interaction between B and T lymphocytes)
CFTR	Wt, heterozygous delta F508 and homozygous mutation delta F508	CFTR mutation is associated with increased resistance to secreting diarrheas
ACE	DD, DI, II	Structurally similar to ACE2, which is the SARS-CoV-2 receptor

ACE angiotensin-converting enzyme, *C3 (F, S)* complement component 3 (fast, slow), *CFTR* cystic fibrosis transmembrane conductance regulator, *DBP* vitamin D binding protein, *FUT2* galactoside 2-alpha-L-fucosyltransferase 2, *Gc* group-specific component, *HFE* human homeostatic iron regulator protein, *MHC* major histocompatibility complex forms in association with other complement proteins, a significant host mechanism for the detection and clearance of potential pathogens (Delanghe et al. 2014). This 185 kDa protein interacts with at least 25 different soluble and membrane-bound proteins and supports the activation of all the three pathways of complement activation: the classical, alternative, and lectin pathway. C3a is an anaphylatoxin that stimulates mast cell degranulation and plays an essential role in chemotaxis. Through its interaction with carboxypeptidase B, it serves as a precursor of the acylation stimulating protein (ASP). Binding between C3b and factor B or the plasma glycoprotein properdin, in the presence of factor D, is responsible for the amplification of C3a convertase activity. This process can result in C5 cleavage, assembly of membrane attack complexes (MAC), and deposition of thousands of activated C3 (C3b) molecules on the microbe surface, which is necessary for the phagocytosis of foreign particles, to enhance humoral immune responses to antigens and probably to eliminate self-reactive B cells (Carroll 1998). As C3 fulfills a vital role in the immune defense, it can be postulated that C3 polymorphism has significant clinical consequences. A single base substitution in C3 defines two alleles – S (slow) and F (fast) – based on their differential electrophoretic mobility (Alper and Propp 1968; Teisberg 1970). Those two alleles give rise to three phenotypes: C3SS, C3FS, and C3FF. The single-nucleotide polymorphism (SNP) rs2230199, responsible for the molecular difference between those two variants, is found in nucleotide position 394 (C-G), resulting in the amino acid substitution R-G (Poznansky et al. 1989). The allelic frequency of C3F varies markedly among races: ± 1% among Asians, ± 5% among Africans, and ± 20% among Caucasians (Delanghe et al. 2014). Therefore, C3 genotyping is more critical in white populations.

Angiotensin-converting enzyme 1 (ACE1) is characterized by a polymorphism in its intron 16, which largely determines ACE expression in tissues and plasma. ACE1 shows significant structural homology with ACE2, which is the receptor for the SARS-CoV-2 virus to gain entry into the cells (Delanghe et al. 2020a). Recently, an association between the D-allele of ACE1 and COVID-19 prevalence and mortality has been described (Delanghe et al. 2020a, b).

The secretor status refers to the presence or absence of ABO blood group antigens in body fluids. People who secrete these antigens in their body fluids are referred to as secretors, while people who do not are called nonsecretors. The secretor status is controlled by the *FUT2* gene (also called the *Se* gene). Approximately 80% of Caucasian people are secretors, whereas ± 20% are nonsecretors. The secretor status has consequences for the immune defense: e.g., nonsecretors have reduced susceptibility to the most common strains of norovirus (Rydell et al. 2011). Expression of the Lewis blood group antigens is also affected by secretor status (Henry et al. 1995).

Haptoglobin (Hp) is a hemoglobin-binding plasma protein. The *Hp* gene is highly polymorphic in humans with strong evidence of functionally distinct biochemical phenotypes. In all human populations, three major Hp phenotypes (Hp 1-1, Hp 2-1, and Hp 2-2) are present (Langlois and Delanghe 1996). The Hp polymorphism has an important biological and clinical significance and plays a role in parasitic, bacterial, and viral infections (Delanghe et al. 1998; Kasvosve et al. 2010).

Vitamin D binding protein (DBP), originally known as group-specific component-globulin (Gc-globulin), is a plasma protein that is encoded by the *GC* gene. It is the major plasma carrier protein of vitamin D. Apart from its specific sterol-binding capacity and a determinant of vitamin D status, DBP fulfills several other biological functions, e.g., fatty acid transport, actin scavenging, macrophage activation, and chemotaxis. DBP polymorphism plays a role in many diseases, including infections (Speeckaert et al. 2006; Delanghe et al. 2015).

Viruses depend on iron in order to replicate within living host cells efficiently. Some viruses selectively infect iron-acquiring cells or influence cellular iron metabolism via human hemochromatosis protein (HFE) or hepcidin (Schmidt 2020). The hemochromatosis molecule HFE is a major histocompatibility complex (MHC)-like molecule that plays a major role in iron metabolism. The C282Y polymorphism is associated with hereditary hemochromatosis and consists of a change of cysteine to tyrosine at position 282. Evidence is growing that the HFE protein plays a role in immunity (Barton et al. 2015; Reuben et al. 2017).

The common delta F508 cystic fibrosis transmembrane conductance regulator (CFTR) mutation in European populations might be a consequence of the impact of selective pressures generated by the transmission of pathogens from domesticated animals to man. A mutation that conferred increased resistance to the diseases caused by pathogens transmitted by cattle would have constituted a selective advantage. This selective advantage would be determined by increased resistance to Cl(−)-secreting diarrheas of delta F508 CFTR heterozygotes (Alfonso-Sánchez et al. 2010; Borzan et al. 2014).

In the present study, prevalence and mortality of COVID-19 were compared with epidemiological data on some plasma protein polymorphisms in 49 countries.

7.2 Methods

Prevalence and mortality data (per 1,000,000 inhabitants) of the COVID-19 from a large number of European, Mediterranean, and Middle East countries were included in the present study: Algeria, Austria, Belarus, Belgium, Bosnia and Herzegovina, Bulgaria, Cyprus, Croatia, the Czech Republic, Denmark, Djibouti, Egypt, Estonia, Ethiopia, Finland, France, Germany, Greece, Hungary, Iceland, Iran, Israel, Italy, Jordan, Latvia, Lithuania, Luxemburg, Malta, Moldova, Morocco, the Netherlands, Norway, Oman, Poland, Portugal, Romania, Russia, Saudi Arabia, Serbia, Slovakia, Slovenia, Spain, Sweden, Switzerland, Tunisia, Turkey, and the UK. Data reported on April 14, 2020, by Johns Hopkins were analyzed (www.worldometers.info/coronavirus/countries). Furthermore, the time interval since the start of COVID-19 in each country was recorded in order to synchronize the data.

Data on the geographical variation of some immune system-related human plasma protein polymorphisms on the deletion/insertion (D/I) polymorphism of the *ACE1* gene (Saab et al. 2007), *HFE* gene (Lucotte and Dieterlen 2003), *C3* gene (Cavalli-Sforza et al. 1994; Delanghe et al. 2014), *Hp* gene (Cavalli-Sforza et al. 1994; Langlois and Delanghe 1996), *FUT2* gene (Cavalli-Sforza et al. 1994), *GC* gene (Cavalli-Sforza et al. 1994; Speeckaert et al. 2006; Delanghe et al. 2015), and the *CFTR* mutation (Farrell 2008) were collected from the literature. For the *CFTR* and *HFE* genes (typical European polymorphisms), only genetic data from European countries could be retrieved.

The various genetic polymorphisms were compared with both COVID-19 prevalence and mortality in a univariate and multivariate regression model. Statistical analysis was carried out using a Medcalc® package (Medcalc, Mariakerke, Belgium).

7.3 Results

7.3.1 Univariate Regression Analysis

In a univariate approach, correlation coefficients between COVID-19 prevalence and the investigated polymorphisms showed a large variety. Whereas the Hp and FUT2 polymorphisms did not show any significant correlations, the ACE D/I, C3, CFTR, and HFE polymorphisms showed highly significant results with COVID-19 prevalence (Table 7.2). As illustrated in Fig. 7.1, the log-transformed prevalence of COVID-19 in 40 countries (on April 13, 2020) negatively correlated with the C3S allele frequency: log (prevalence; no. of cases/10^6 inhabitants) = 8.897–0.077 (C3S allele frequency, %), $r^2 = 0.174$; $p = 0.003$. The model improved when the onset of the epidemic in each country was taken into account: log (prevalence; no. of cases/10^6 inhabitants) = 9.586–0.077 (C3S allele frequency, %) – 0.013 (days since January 1, 2020), $r^2 = 0.273$; $p = 0.001$. The correlations obtained between the various genetic polymorphisms and COVID-19

Table 7.2 Univariate models in predicting COVID-19 prevalence

Polymorphism (allele)	(Y = COVID-19 prevalence on April 13, 2020, n/10[6] inhabitants)	N (countries)	R^2	p-value
ACE D/I (D)	5.410–0.047 × (D allele frequency, %)	44	0.230	0.001
C3 (S)	8.897–0.077 × (S allele frequency, %)	48	0.174	0.003
HFE (C282Y)	2.487–0.011 × (C282Y allele frequency, %)	33	0.282	0.001
Hp (Hp1)	2.930–0.007 × (Hp1 allele frequency, %)	49	0.004	n.s.
FUT2 (Se)	3.085–0.008 × (Se allele frequency, %)	49	0.029	n.s.
DBP (DBP1)	8.180–0.075 × (DBP1 allele frequency, %)	49	0.244	<0.001
CFTR (delta F508)	2.737 + 0.0003 × (delta F508 frequency, %)	31	0.195	0.013

ACE angiotensin-converting enzyme, *C3 (S)* complement component 3 (slow), *CFTR* cystic fibrosis transmembrane conductance regulator, *DBP* vitamin D binding protein, *FUT2* galactoside 2-alpha-L-fucosyltransferase 2, *HFE* human homeostatic iron regulator protein, *n.s.* not significant

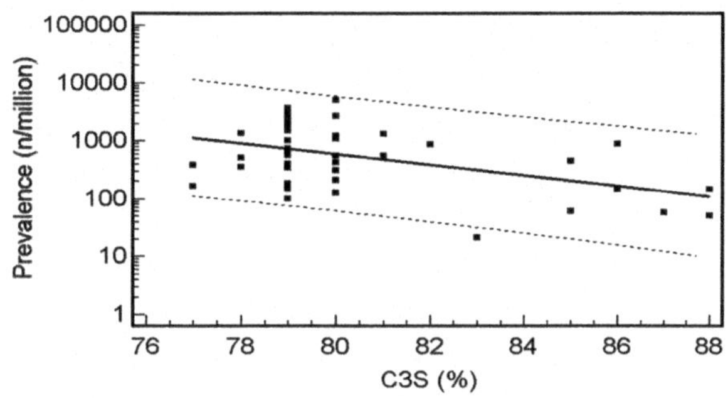

Fig. 7.1 Prevalence of COVID-19 infection in 48 countries (on April 13, 2020) vs. the C3S allele frequency (%). Log (prevalence, no. of cases/106 inhabitants) = 8.897–0.077 (C3S allele frequency, %), $r^2 = 0.1737$; $p = 0.003$

Table 7.3 Univariate models in predicting COVID-19 mortality (*n*/10[6] inhabitants)

Polymorphism (allele)	(Y = COVID-19 mortality on April 13, 2020, *n*/10[6] inhabitants)	*N* (countries)	R^2	*p*-value
ACE D/I (D)	5.448–0.057 × (D allele frequency, %)	44	0.227	0.001
C3 (S)	9.635–0.094 × (S allele frequency, %)	48	0.139	0.009
HFE (C282Y)	1.889 + 0.012 × (C282Y allele frequency, %)	36	0.235	0.003
Hp (Hp1)	1.054 + 0.014 × (Hp1 allele frequency, %)	49	0.013	n.s.
FUT2 (Se)	2.634–0.01 × (Se allele frequency, %)	49	0.004	n.s.
DBP (DBP1)	7.424–0.073 × (DBP1 allele frequency, %)	49	0.158	0.005
CFTR (delta F508)	5.447–0.057 × (delta F508 frequency, %)	31	0.227	0.001

ACE angiotensin-converting enzyme, *C3 (S)* complement component 3 (slow), *CFTR* cystic fibrosis transmembrane conductance regulator, *DBP* vitamin D binding protein, *FUT2* galactoside 2-alpha-L-fucosyltransferase 2, *HFE* human homeostatic iron regulator protein, *n.s.* not significant

mortality were quite similar to the ones obtained for COVID-19 prevalence (Table 7.3). Highly significant correlations were observed for ACE D/I C3, CFTR, and HFE polymorphisms. COVID-19-associated mortality correlated with the C3S allele frequency as follows: log (mortality; no. of cases/10[6] inhabitants) = 9.635–0.094 (C3S allele frequency, %), $r^2 = 0.139$; $p = 0.009$. Also, this correlation improved when the onset of the epidemic was taken into account: 4.955–0.055 (C3S allele frequency, %) – 0.026 (days since January 1, 2020), $r^2 = 0.457$; $p < 0.0001$. Neither FUT2 nor Hp1 polymorphism showed a significant correlation with COVID-19 mortality.

7.3.2 Clustering of Genetic Polymorphisms

The C3S allele frequency showed a strong correlation with the GC1 allele frequency ($r = 0.543$, $p = 0.0001$) and with the FUT2 Se allele frequency ($r = -0.3423$, $p = 0.02$). No significant correlation was found between the C3S allele and the ACE1 D allele frequency ($r = 0.157$), the Hp1 allele frequency ($r = 0.175$), the CFTR frequency ($r = -0.075$), and the *HFE* gene ($r = -0.263$) mutation. On the other hand, in the subgroup of European countries, the *CFTR* mutation frequency showed a strong correlation with the HFE allele frequency: $r = 0.831$; $p < 0.001$.

7.3.3 Multivariate Regression Analysis

In a multivariate regression analysis (backward procedure), only C3S and ACE D/I provided to be determinant for the COVID-19 prevalence. A determination coefficient of 0.453 was calculated for the model by considering the onset of the COVID-19 pandemic in each country. Table 7.4 summarizes the data of the multivariate analysis. The other investigated genetic polymorphisms [C3 (F and S alleles), HFE (*C282Y* mutation), Hp (Hp1 and Hp2 alleles), and DBP (Gc1 and Gc 2 alleles)] did not show a significant correlation with COVID-19 prevalence. Similarly, only the C3 and ACE D/I polymorphism were significant determinants for COVID-19 mortality. The model calculated a determination coefficient of 0.457 when taking into account the onset of the pandemic in each country (Table 7.5). Here again, the other investigated polymorphisms [C3 (F and S alleles), HFE (*C282Y* mutation), Hp (Hp1 and Hp2 alleles), and vitamin D binding protein (Gc 1 and Gc 2 alleles)] did not show a significant correlation with COVID-19 mortality.

Table 7.4 Multivariate model for predicting COVID-19 prevalence (n/10^6 inhabitants) in 42 countries ($r^2 = 0.453$, $p < 0.001$)

Parameter	Coefficient	*t*	*p*-value
Constant	11.655		
C3S allele frequency	−0.085	−3.588	0.001
ACE D allele frequency	−0.026	−2.508	0.016
Start of pandemic (days since January 1, 2020)	−0.011	−2.159	0.037

ACE angiotensin-converting enzyme, *C3 (S)* complement component 3 (slow)

Table 7.5 Multivariate model for predicting COVID-19 mortality (n/10^6 inhabitants) in 42 countries ($r^2 = 0.457$, $p < 0.001$)

Parameter	Coefficient	*t*	*p*-value
Constant	12.878		
C3S allele frequency	−0.095	−2.879	0.007
ACE D allele frequency	−0.031	−2.136	0.016
Start of pandemic (days since January 1, 2020)	−0.011	−2.159	0.037

ACE angiotensin-converting enzyme, *C3 (S)* complement component 3 (slow)

7.4 Discussion

The present study shows that C3 polymorphism may be regarded as an independent confounder in the spread of COVID-19 and associated outcomes of disease. Multiple associations between genetic polymorphisms of C3 and disease have been described (Delanghe et al. 2014). As an example, clinical outcome following renal (Damman et al. 2012; Pallet and Thervet 2012) and liver transplantation (Dhillon et al. 2010) is determined by the C3 polymorphism. Although complement is a key player of protective immunity against pathogens, its excessive or deregulated activation may result in collateral tissue injury (Risitano et al. 2020). A mouse model of severe acute respiratory syndrome (SARS)-CoV-2 demonstrated that dysregulated complement activation contributed to immunopathology, as C3 knockout (C3−/−) mice showed less lung damage and systemic inflammation with similar viral loads as compared to control mice (Gralinski et al. 2018). Moreover, increased serum and lung C5a and C5b-9 concentrations were measured in a murine model of MERS-CoV infection. Administration of a murine C5a antibody resulted in a decreased cytokine response with improved spleen and lung damage (Jiang et al. 2018). Preliminary data of a small case series of four COVID-19 patients with severe pneumonia or ARDS showed promising results from a combination treatment, including anticomplement C5 therapy (eculizumab). However, these findings should be confirmed in more extensive trials, e.g., the ongoing SOLID-C19 trial (Diurno et al. 2020). C3 inhibitors may also have the potential to ameliorate lung injury by simultaneously blocking C3a and C5a generation, as well as intrapulmonary C3 activation and IL-6 release from alveolar macrophages or other cells that express C3a receptors (C3aRs) and/or C5a receptors (C5aRs) (Risitano et al. 2020). Also, the compstatin C3 inhibitor AMY-101 can interfere with IL-6 release, as demonstrated in ex vivo whole-blood infection models with the compstatin C3 inhibitor AMY-101 (Mastellos et al. 2019). So, C3 inhibition in combination with anti-IL-6 regimens could be a successful combination in the search for the treatment of inflammatory lung complications of SARS-CoV-2 infection, as well as of the systemic inflammation affecting the microvascular beds of the kidney, brain, and other vital organs. However, at this moment, there are only limited clinical data on the role of complement activation in the development of SARS-CoV-2-associated ARDS, and there are a lot of unanswered questions concerning the therapeutic use of complement inhibitors for COVID-19 (Risitano et al. 2020). On the other hand, the C3 polymorphism is a pure representative of the first principal component of European gene frequencies (Sokal et al. 1991; Cavalli-Sforza et al. 1994). However, also the *FUT2* gene (Se/se) is a genetic polymorphism that shows a high correlation coefficient with the same first principal component (which explains 28.1% of total genetic variance in Europe) (Cavalli-Sforza et al. 1994). This polymorphism was subsequently eliminated as a determining factor in the multiple regression analysis. The sharp contrast observed between FUT2 (showing no correlation with COVID-19 infection) and C3 (showing a good correlation) proves that the correlation between C3 polymorphism and COVID-19 prevalence/mortality cannot be attributed to a passive migration of this polymorphism with the group of genes belonging to the first principal component group. The C3 polymorphism can be regarded as a key step in the virus-induced inflammatory process.

Our findings confirm the described association between ACE D/I polymorphism and COVID-19 (Delanghe et al. 2020a, b). This polymorphism shows functional links with ACE2 (Kuba et al. 2006; Yan et al. 2020), which plays a key role in COVID-19 infection as the SARS-CoV-2 virus uses this receptor to gain entry into the cells. ACE2-expressing cells are susceptible to COVID-19 infection (Zou et al. 2020). As carriers of the D allele have reduced expression of ACE2, they may have a lower COVID-19 infection risk (Delanghe et al. 2020a, b).

The DBP polymorphism showed a correlation with the COVID-19 prevalence and mortality in a univariate model, which, however, was lost in the multivariate regression model. The presence of microangiopathy and microthrombi in different vascular beds has been observed in COVID-19 patients with multi-organ injury and failure (Liu et al. 2020). The presence of viral elements within endothelial cells (COVID-19-endothelitis) and an accumulation of inflammatory cells, with evidence of endothelial and inflammatory cell death, have recently been demonstrated (Varga et al. 2020). Endothelial injury is associated with the release of toxic intracellular molecules into the circulation, including F-actin. As F-actin is angiopathic, an effective actin scavenger system consisting of gelsolin and DBP is activated to clear F-actin from the circulation rapidly. More specifically, plasma gelsolin depolymerizes F-actin into G-actin, which binds to the highly abundant circulating protein, DBP. The actin-VDBP complexes are removed from the circulation by the reticuloendothelial system, primarily in the liver (Lind et al. 1986; Lee and Galbraith 1992; Meier et al. 2006). As demonstrated in a murine model, a lack of free DBP and increased concentrations and/or prolonged exposure to DBP-actin complexes might induce endothelial cell injury and death, particularly in the lung microvasculature (Ge et al. 2014). Although the major DBP phenotypes have an equal actin-binding affinity, the plasma DBP concentrations are relatively highest in Gc1–1 carriers, which could lead to a more rapidly sequestering of free actin and a better outcome (Speeckaert et al. 2006).

In the subgroup of European countries, remarkably, there was a relationship between COVID-19 prevalence/mortality and the *CFTR* mutation. Like the *C282Y* mutation, this mutation is regarded as a Celtic mutation, showing a marked east-west gradient in Europe (Lucotte and Dieterlen 2003; Farrell 2008). Viral respiratory tract infections are more severe in cystic fibrosis patients than in the general population. As demonstrated during the 2009 influenza A/H1N1 pandemic, this vulnerable group had an increased risk of complications with a negative impact on lung function (Colombo et al. 2011; Bucher et al. 2016). However, at this moment, most patients with cystic fibrosis are doing an exceptional job (social isolation) avoiding SARS-CoV-2 infection, as only a limited number of patients have been infected so far (Colombo et al. 2020). Both HFE and CFTR polymorphisms were eliminated in the multiple regression models for COVID-19 prevalence and mortality. Identification of potentially protective factors, e.g., long-term antibiotic treatment with azithromycin, is not yet possible. Also, the Hp polymorphism did not show a significant relationship. As this polymorphism is strongly linked to iron metabolism and iron conservation, it is unlikely that iron-related mechanisms play a role in effective immune mechanisms against COVID-19.

Our findings are remarkable as all plasma protein polymorphisms investigated show a certain east-west gradient in Europe (Sokal 1991; Sokal et al. 1991; Cavalli-Sforza et al. 1994; Delanghe et al. 2012), which is pleading against the bystander role of the polymorphism as a biomarker for historical migrations in the remote past, which passively comigrates with causal human genetic factors involved in COVID-19 infection. Genetic polymorphisms with a substantial geographical variation (a low frequency in one part of the world, and a high one somewhere else) have a better chance to be selected during evolution (Bradbury 2004).

7.5 Conclusion

In conclusion, we have demonstrated an association between various genetic polymorphisms of the host and COVID-19 infection/mortality. These associations can be explained by variable host-virus interactions and a variable immune response in the host. The strength of the association between the investigated polymorphisms and COVID-19 infection/mortality is further highlighted by the variability in national policies in the studied group of countries during the early phase of the COVID-19 outbreak and also the variable onset of the infection in these countries. Further studies are required to assess the clinical outcome of COVID-19 infection in C3S carriers and to study the exact role of the C3 and the ACE D/I polymorphisms in COVID-19 infection and its therapy response. Eventually, this could lead to drug development and more personalized medicine in COVID-19 infection.

References

Alfonso-Sánchez MA, Pérez-Miranda AM, García-Obregón S, Peña JA (2010) An evolutionary approach to the high frequency of the Delta F508 CFTR mutation in European populations. Med Hypotheses 74:989–992

Alper CA, Propp RP (1968) Genetic polymorphism of the third component of human complement (C'3). J Clin Invest 47:2181–2191

Barton JC, Edwards CQ, Acton RT (2015) HFE gene: structure, function, mutations, and associated iron abnormalities. Gene 574:179–192

Borzan V, Tomašević B, Kurbel S (2014) Hypothesis: possible respiratory advantages for heterozygote carriers of cystic fibrosis linked mutations during dusty climate of last glaciation. J Theor Biol 363:164–168

Bradbury J (2004) Ancient footsteps in our genes: evolution and human disease. Gene variants selected during evolution may underlie many common diseases. Lancet 363:952–953

Bucher J, Boelle PY, Hubert D, Lebourgeois M, Stremler N, Durieu I, Bremont F, Deneuville E, Delaisi B, Corvol H, Bassinet L, Grenet D, Remus N, Vodoff MV, Boussaud V, Troussier F, Leruez-Ville M, Treluyer JM, Launay O, Sermet-Gaudelus I (2016) Lessons from a French collaborative case-control study in cystic fibrosis patients during the 2009 a/H1N1 influenza pandemy. BMC Infect Dis 16:55

Carroll MC (1998) The role of complement and complement receptors in induction and regulation of immunity. Annu Rev Immunol 16:545–568

Cavalli-Sforza LL, Menozzi P, Piazza A (1994) The history and geography of human genes. Princeton University Press, Princeton, pp 1–1088

Colombo C, Battezzati PM, Lucidi V, Magazzù G, Motta V, Alicandro G, Taccetti G, Repetto T (2011) Influenza a/H1N1 in patients with cystic fibrosis in Italy: a multicentre cohort study. Thorax 66:260–261

Colombo C, Burgel PR, Gartner S, van Koningsbruggen-Rietschel S, Naehrlich L, Sermet-Gaudelus I, Southern KW (2020) Impact of COVID-19 on people with cystic fibrosis. Lancet Respir Med 8:e35–e36

Damman J, Daha MR, Leuvenink HG, van Goor H, Hillebrands JL, Dijk MC, Hepkema BG, Snieder H, Born J, de Borst MH, Bakker SJ, Navis GJ, Ploeg RJ, Seelen MA (2012) Association of complement C3 gene variants with renal transplant outcome of deceased cardiac dead donor kidneys. Am J Transplant 12:660–668

Delanghe JR, Langlois MR, Boelaert JR, Van Acker J, Van Wanzeele F, van der Groen G, Hemmer R, Verhofstede C, De Buyzere M, De Bacquer D, Arendt V, Plum J (1998) Haptoglobin polymorphism, iron metabolism and mortality in HIV infection. AIDS 12:1027–1032

Delanghe J, Speeckaert M, De Buyzere ML, Langlois M, Torck M (2012) Human plasma protein polymorphisms and the persistence of cultural diversity. Proc Natl Acad Sci U S A 109:E2914

Delanghe JR, Speeckaert R, Speeckaert MM (2014) Complement C3 and its polymorphism: biological and clinical consequences. Pathology 46:1–10

Delanghe JR, Speeckaert R, Speeckaert MM (2015) Behind the scenes of vitamin D binding protein: more than vitamin D binding. Best Pract Res Clin Endocrinol Metab 29:773–786

Delanghe JR, Speeckaert MM, De Buyzere ML (2020a) COVID-19 infections are also affected by human ACE1 D/I polymorphism. Clin Chem Lab Med 58:1125–1126

Delanghe JR, Speeckaert MM, De Buyzere ML (2020b) The host's angiotensin-converting enzyme polymorphism may explain epidemiological findings in COVID-19 infections. Clin Chim Acta 505:192–193

Dhillon N, Walsh L, Krüger B, Mehrotra A, Ward SC, Godbold J, Radwan M, Schiano T, Murphy B, Schröppel B (2010) Complement component C3 allotypes and outcomes in liver transplantation. Liver Transpl 16:198–203

Diurno F, Numis FG, Porta G, Cirillo F, Maddaluno S, Ragozzino A, De Negri P, Di Gennaro C, Pagano A, Allegorico E, Bressy L, Bosso G, Ferrara A, Serra C, Montisci A, D'Amico M, Schiano Lo Morello S, Di Costanzo G, Tucci AG, Marchetti P, Di Vincenzo U, Sorrentino I, Casciotta A, Fusco M, Buonerba C, Berretta M, Ceccarelli M, Nunnari G, Diessa Y, Cicala S, Facchini G (2020) Eculizumab treatment in patients with COVID-19: preliminary results from real life ASL Napoli 2 Nord experience. Eur Rev Med Pharmacol Sci 24:4040–4047

Farrell PM (2008) The prevalence of cystic fibrosis in the European Union. J Cyst Fibros 7:450–453

Ge L, Trujillo G, Miller EJ, Kew RR (2014) Circulating complexes of the vitamin D binding protein with G-actin induce lung inflammation by targeting endothelial cells. Immunobiology 219:198–207

Gralinski LE, Sheahan TP, Morrison TE, Menachery VD, Jensen K, Leist SR, Whitmore A, Heise MT, Baric RS (2018) Complement activation contributes to severe acute respiratory syndrome coronavirus pathogenesis. MBio 9

Henry S, Oriol R, Samuelsson B (1995) Lewis histo-blood group system and associated secretory phenotypes. Vox Sang 69:166–182

Jiang Y, Zhao G, Song N, Li P, Chen Y, Guo Y, Li J, Du L, Jiang S, Guo R, Sun S, Zhou Y (2018) Blockade of the C5a-C5aR axis alleviates lung damage in hDPP4-transgenic mice infected with MERS-CoV. Emerg Microbes Infect 7(77)

Kasvosve I, Speeckaert MM, Speeckaert R, Masukume G, Delanghe JR (2010) Haptoglobin polymorphism and infection. Adv Clin Chem 50:23–46

Kuba K, Imai Y, Penninger JM (2006) Angiotensin-converting enzyme 2 in lung diseases. Curr Opin Pharmacol 6:271–276

Langlois MR, Delanghe JR (1996) Biological and clinical significance of haptoglobin polymorphism in humans. Clin Chem 42:1589–1600

Lee WM, Galbraith RM (1992) The extracellular actin-scavenger system and actin toxicity. N Engl J Med 326:1335–1341

Lind SE, Smith DB, Janmey PA, Stossel TP (1986) Role of plasma gelsolin and the vitamin D-binding protein in clearing actin from the circulation. J Clin Invest 78:736–742

Liu PP, Blet A, Smyth D, Li H (2020) The science underlying COVID-19: implications for the cardiovascular system. Circulation 142(1):68–78

Lucotte G, Dieterlen F (2003) A European allele map of the C282Y mutation of hemochromatosis: Celtic versus Viking origin of the mutation? Blood Cells Mol Dis 31:262–267

Mastellos DC, Ricklin D, Lambris JD (2019) Clinical promise of next-generation complement therapeutics. Nat Rev Drug Discov 18:707–729

Meier U, Gressner O, Lammert F, Gressner AM (2006) Gc-globulin: roles in response to injury. Clin Chem 52:1247–1253

Pallet N, Thervet E (2012) The genetics of kidney transplantation. Hum Genet 131:317–323

Poznansky MC, Clissold PM, Lachmann PJ (1989) The difference between human C3F and C3S results from a single amino acid change from an asparagine to an aspartate residue at position 1216 on the alpha chain of the complement component, C3. J Immunol 143:3860–3862

Reuben A, Chung JW, Lapointe R, Santos MM (2017) The hemochromatosis protein HFE 20 years later: an emerging role in antigen presentation and in the immune system. Immun Inflammation Dis 5:218–232

Risitano AM, Mastellos DC, Huber-Lang M, Yancopoulou D, Garlanda C, Ciceri F, Lambris JD (2020) Complement as a target in COVID-19? Nat Rev Immunol 20:343–344

Rydell GE, Kindberg E, Larson G, Svensson L (2011) Susceptibility to winter vomiting disease: a sweet matter. Rev Med Virol 21:370–382

Saab YB, Gard PR, Overall AD (2007) The geographic distribution of the ACE II genotype: a novel finding. Genet Res 89:259–267

Schmidt SM (2020) The role of iron in viral infections. Front Biosci (Landmark Ed) 25:893–911

Sokal (1991) The continental population structure of Europe. Ann Rev Anthropol 20:119–140

Sokal RR, Oden NL, Wilson C (1991) Genetic evidence for the spread of agriculture in Europe by demic diffusion. Nature 351:143–145

Speeckaert M, Huang G, Delanghe JR, Taes YE (2006) Biological and clinical aspects of the vitamin D binding protein (Gc-globulin) and its polymorphism. Clin Chim Acta 372:33–42

Teisberg P (1970) High voltage agarose gel electrophoresis in the study of C 3 polymorphism. Vox Sang 19:47–56

Varga Z, Flammer AJ, Steiger P, Haberecker M, Andermatt R, Zinkernagel AS, Mehra MR, Schuepbach RA, Ruschitzka F, Moch H (2020) Endothelial cell infection and endotheliitis in COVID-19. Lancet 395:1417–1418

www.worldometers.info/coronavirus/countries

Yan R, Zhang Y, Li Y, Xia L, Guo Y, Zhou Q (2020) Structural basis for the recognition of SARS-CoV-2 by full-length human ACE2. Science 367:1444–1448

Zou X, Chen K, Zou J, Han P, Hao J, Han Z (2020) Single-cell RNA-seq data analysis on the receptor ACE2 expression reveals the potential risk of different human organs vulnerable to 2019-nCoV infection. Front Med 14:185–192

8 How COVID-19 Has Globalized: Unknown Origin, Rapid Transmission, and the Immune System Nourishment

Amene Saghazadeh and Nima Rezaei

Abstract

The novel coronavirus disease (COVID-19) profoundly influences T-cell immunity. The counts of total T cells and T-cell subsets, especially CD4+ and CD8+ T cells, are decreased in patients with COVID-19. Also, the function of these cells becomes less effective as the expression of immune inhibitory receptors, such as Tim3 and PD-1, increases over time during the disease. Kinetic analyses show that the T-cell profile changes dynamically, so does the COVID-19 stages. As COVID-19 continues to deteriorate and progresses to severe/critical condition, the lymphocyte count steadily decreases. Therefore, the ability of COVID-19 to escape the immune system might lie in its power to profoundly diminish T-cell effective function, which is necessary for the establishment of a robust antiviral immunity. Also, COVID-19 is associated with increased numbers of monocytes and macrophages, and as the disease progresses from a mild form to a severe/critical condition, the macrophage population becomes denser. Monitoring the expression of cytokines associated with macrophage activation, mainly interleukin (IL)-6 and IL-10, indicates that the course of COVID-19 consists of two stages and the transition between disease stages occurs by the end of the first week after onset of symptoms. At the initial stage, the immune military recognizes the severe acute respiratory syndrome coronavirus 2 (SARS-CoV-2) as nonself and thus fires macrophages at the lungs against the virus. The first flame can control disease progression effectively. However, a trained immunocompetent system would maintain the fire of macrophages over an extended time. It lies in its immune memory in tissue-resident macrophages, especially alveolar macrophages, making a professionally trained immune system more likely to be feared by COVID-19 than an untrained immune system. In this manner, the trained immunocompetent system commits such a failure that causes the lungs to come down

A. Saghazadeh
Research Center for Immunodeficiencies, Children's Medical Center Hospital, Tehran University of Medical Sciences, Tehran, Iran

Systematic Review and Meta-analysis Expert Group (SRMEG), Universal Scientific Education and Research Network (USERN), Tehran, Iran

N. Rezaei (✉)
Research Center for Immunodeficiencies, Children's Medical Center Hospital, Tehran University of Medical Sciences, Tehran, Iran

Department of Immunology, School of Medicine, Tehran University of Medical Sciences, Tehran, Iran

Network of Immunity in Infection, Malignancy and Autoimmunity (NIIMA), Universal Scientific Education and Research Network (USERN), Tehran, Iran
e-mail: rezaei_nima@tums.ac.ir

N. Rezaei (ed.), *Coronavirus Disease - COVID-19*, Advances in Experimental Medicine and Biology 1318, https://doi.org/10.1007/978-3-030-63761-3_8

rapidly. The fact that younger age groups, including neonates and children, are less susceptible to COVID-19 than older age groups reflects that the natural affinities of the immune system that has not been trained thoroughly would be standard in combatting against COVID-19 whereas the higher affinities of the trained immune system for rapid activation of immune responses might raise faults – the lungs come down.

Keywords

COVID-19 · Cytokine storm · Interleukin-6 · Lung · Macrophage · Monocyte

8.1 Introduction

In 2019, after the lapse of 8 years, a modified strain of coronaviruses was born in the Chinese city of Wuhan, hence giving it the name as WH-Human 1 coronavirus. When compared with the two other most recent outbreaks caused by virulent strains of betacoronaviruses, the novel strain, also known as 2019 novel coronavirus (2019-nCoV), severe acute respiratory syndrome coronavirus 2 (SARS-CoV-2), or human coronavirus 2019 (HCoV-19), is more serious.

The novel coronavirus causes a multicolor disease, the so-called coronavirus disease 2019 (COVID-19), linked to a spectrum of clinical severity from asymptomatic infection to severe and critical infection (Lotfi and Rezaei 2020). COVID-19 can involve multiple organs/systems (Jahanshahlu and Rezaei 2020a; Saleki et al. 2020; Shamshirian and Rezaei 2020). The quality and quantity of inflammatory responses appear to mainly, if not wholly, explain the disease phenotype that varies from individual to individual (Yazdanpanah et al. 2020a; Saghazadeh and Rezaei 2020a; Yazdanpanah et al. 2020b; Bahrami et al. 2020; Nasab et al. 2020). The evidence behind this derives partly from studies showing a similar picture of the disease in family members who share the immunogenetic background (Yousefzadegan and Rezaei 2020; Darbeheshti and Rezaei 2020) and partly from studies reporting the worst-case scenarios in people with comorbidities known to cause chronic inflammation, e.g., cardiovascular diseases and cancer (Shamshirian and Rezaei 2020; Ahmadi et al. 2020). Interestingly, despite initial thoughts, there have been only sporadic reports of COVID-19 in patients with primary immunodeficiency disorders (Ahanchian et al. 2020), and so these patients appear to be less susceptible to COVID-19 than the general population (Babaha and Rezaei 2020). These observations are why rethinking the role of the immune system in the pathogenesis of COVID-19 is critical to define specific treatments (Fathi and Rezaei 2020; Jahanshahlu and Rezaei 2020b; Mansourabadi et al. 2020; Mohamed et al. 2020b; Pashaei and Rezaei 2020; Pourahmad et al. 2020; Saghazadeh and Rezaei 2020b; Basiri et al. 2020b).

COVID-19 has affected people worldwide in less than 3 months, with no exception noted (Mirbeyk and Rezaei 2020; Rezaei 2020a). As a result, it has been a challenge in healthcare systems, making its prevention, diagnosis, and treatment as a priority area for research in various fields (Basiri et al. 2020a; Rabiee et al. 2020; Rezaei 2020b; Mohamed et al. 2020b; Sharifkashani et al. 2020; Lotfi et al. 2020; Moazzami et al. 2020; Sahu et al. 2020). Not only has it urged seniors to remember the previous pandemics and share their valuable lessons, but also COVID-19 made novices to think differently about globalization and how it can overcome limits and contribute to the success of humanity in fighting against the current and with a risk of reinfection, possibly future outbreaks (Moradian et al. 2020; Hanaei and Rezaei 2020; Jabbari et al. 2020; Kafieh et al. 2020; Mohamed et al. 2020a; Momtazmanesh et al. 2020; Rzymski et al. 2020; Jabbari and Rezaei 2020).

First, the present chapter reviews the when, what, and where of COVID-19. Then, it provides a detailed discussion of the current literature on the immune system abnormalities associated with COVID-19, making doubt on the role of the immune system in COVID-19. Evidence of the hyper inflammation orchestrated by cytokines and macrophages in patients with severe

COVID-19 would further strengthen the contribution of the immune system to COVID-19. Finally, this chapter gives an overview of the current condition in different countries struggling COVID-19, leading to the conclusion that the world is globalized, the globalized world embraces the globalized problems, and the globalized problems call for a globalized solution.

8.2 When, What, and Where of COVID-19

8.2.1 The Family *Coronaviridae*

The subfamily *Orthocoronavirinae* is within the family *Coronaviridae*, along with the subfamily *Coronavirinae*. The genera of the subfamily *Orthocoronavirinae* are alphacoronavirus, betacoronavirus, gammacoronavirus, and deltacoronavirus. *Betacoronavirus* is a genus with subgenera *Sarbecovirus* and *Merbecovirus* to which SARS-CoV and MERS-CoV species belong (Millán-Oñate et al. 2020). The first caused the outbreak of severe acute respiratory syndrome (SARS) in 2002–2003, and the other one succeeded in opening the outbreak of the Middle East respiratory syndrome (MERS) in 2012–2013. The SARS and MERS outbreak correlated with a death rate of 10% and 34%, respectively, indicating their high potential of pathogenicity for humans. They caused together about 1600 deaths. However, the recent outbreak of pneumonia of unknown etiology, which seems to be the result of a novel coronavirus, the so-called SARS-CoV-2, has to date caused over 3 million cases and claimed more than 200,000 deaths in only 4 months (Fig. 8.1).

8.2.2 What Have Coronaviruses Brought for Human Beings?

Since the mid-1960, seven coronaviruses have occurred in humans by consideration of this new one. Four coronaviruses, including 229E, NL63, OC43, and HKU1, are endemic in humans and correlate commonly with mild symptoms affecting the respiratory system, gastrointestinal system, and the nervous system. Three most recent human coronaviruses, SARS-CoV, MERS-CoV, and SARS-CoV-2, appear to be originated by zoonotic transmission and can cause potentially fatal condition characterized by respiratory failure.

Human coronaviruses are positive-sense single-stranded RNA viruses that possess two main groups of proteins: structural and nonstructural proteins. Structural proteins include the spike (S), nucleocapsid (N), matrix (M), and envelope (E). Nonstructural proteins (NSP) include proteases, such as nsp3 and nsp5, and RNA-dependent RNA polymerase (RdRP) such as nsp12. The spike protein is the key to binding the virus to its cell surface receptor. Sequence analyses reveal that the spike glycoprotein of the SARS-CoV-2 has a high sequence identity of greater than 70% to that of the 2002 SARS-CoV (Fig. 8.2).

8.2.3 SARS-CoV-2: Where Does It Come from, and How Does It Go?

8.2.3.1 In 2015, the application of reverse genetics of SARS-CoV led to the development of a recombinant virus that through binding the same surface receptor caused infection in human lung parenchymal cells but was resistant to SARS-CoV vaccine and monoclonal antibodies targeting SARS-CoV spike protein

In 2015, a group of American, European, and Chinese researchers obtained a SARS-like coronavirus WIV1 (SL-CoV-WIV1) from Chinese horseshoe bats, isolated the spike protein of SL-CoV-WIV1, the RsSHC014-CoV sequence, and developed a hybrid virus utilizing the RsSHC014-CoV sequence and the SARS-CoV (mouse-adapted) backbone (Menachery et al. 2015).

Despite very similarities, compared to human SARS-CoV, SL-CoV-WIV1 expresses 14 different residues involved in binding the ACE2 receptor (Menachery et al. 2015). Also, the lentivirus

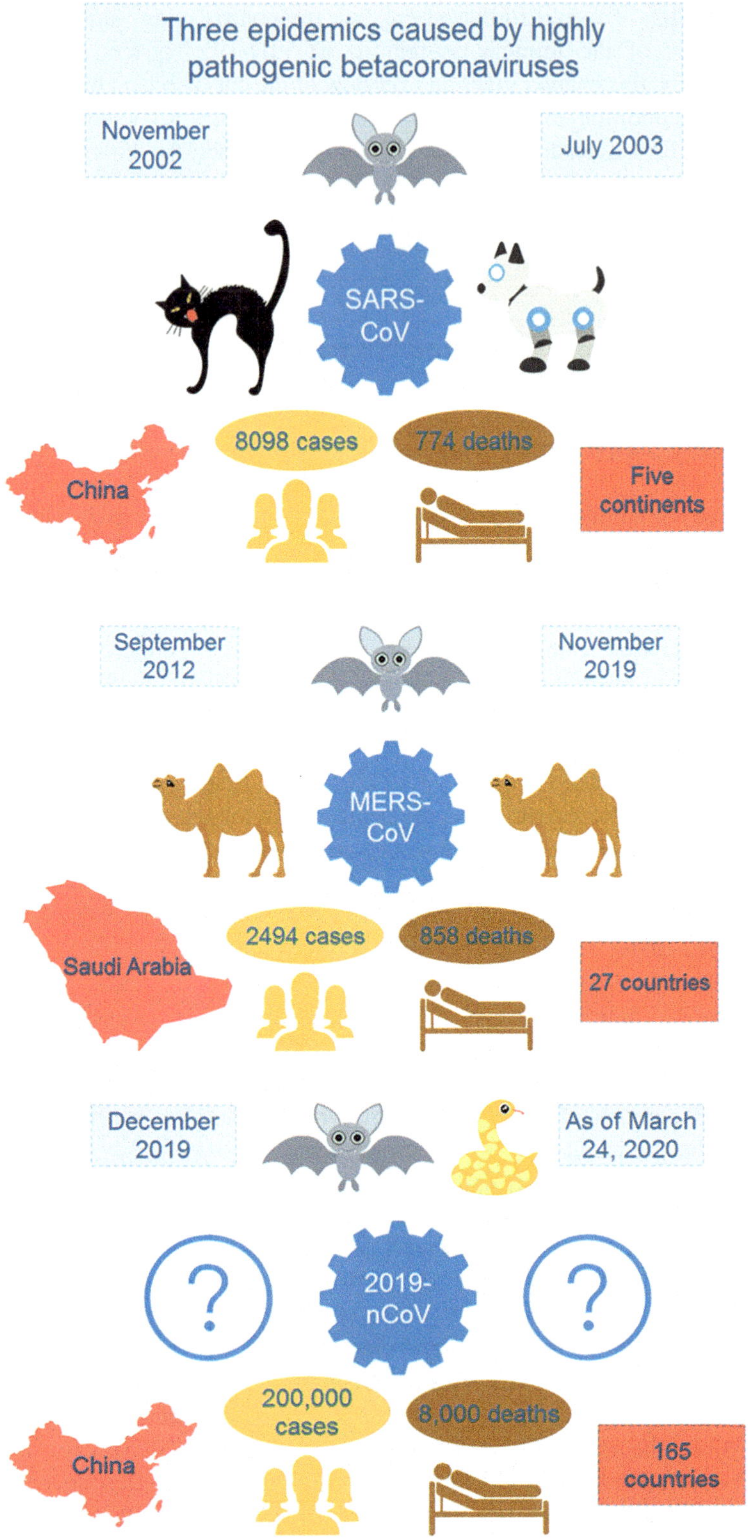

Fig. 8.1 Three epidemics caused by virulent human coronaviruses: SARS-CoV, MERS-CoV, and 2019-nCoV

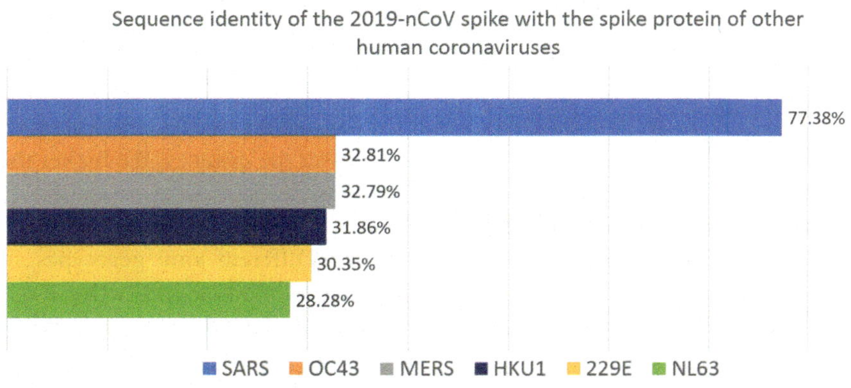

Fig. 8.2 The sequence identity of the 2019-nCoV spike with the spike protein of other human coronaviruses

expressing the spike protein of SL-CoV-WIV1 could not bind human ACE2, while human SARS-CoV can bind the ACE2 receptor. Consequently, the authors used the SARS-CoV backbone to create a chimeric virus composed of the spike protein of SL-CoV-WIV1. They named this novel virus the chimeric CoV or SHC014-MA15. SHC014-MA15 required the ACE2 receptor for the cell entry; could use its orthologs in humans, civet, and bats; and replicated in the human epithelial airway cell line Calu-3 2B4 and primary human airway epithelial (HAE) cultures enough so that a severe infection occurred.

Ten-week-old mice faced weight loss and death upon infection with SHC014-MA15 (Menachery et al. 2015). Weight loss, but no death, occurred in mice infected with SARS-CoV Urbani (SARS-MA15). Lungs from both mice infected with SHC014-MA15 and mice infected with SARS-MA15 revealed the same viral titers. However, pathological findings were different; SHC014-MA15 mainly involved the parenchyma, while SARS-MA15 affected both airways and the parenchyma. For 12-month-old mice, both SARS-MA15 and SHC014-MA15 infection caused weight loss and death. However, death occurred with SHC014-MA15 to a lesser extent than with SARS-MA15 (Menachery et al. 2015). ACE2-deficient mice developed no sign of infection upon exposure to SARS-MA15, indicating the critical role of the ACE2 receptor in the disease progression (Menachery et al. 2015).

None of the four human monoclonal antibodies targeting SARS-CoV spike protein showed significant potential as SHC014-MA15 therapeutics (Menachery et al. 2015). Not only double-inactivated whole SARS-CoV vaccine (DIV) did not offer protection against SHC014-MA15 but also was pathogenic to the aged animals (Menachery et al. 2015). Aged mice vaccinated with DIV developed eosinophilia traffics in the lungs. Also, their serum could not neutralize SHC014-MA15. A live attenuated SHC014-MA15 vaccine could bring both young and aged mice cross-protection against SARS-CoV (Menachery et al. 2015). However, it required a secondary antigen boost at 28 days postinoculation (dpi), and its minimum protective dose caused death in the aged mice.

8.2.3.2 In 2020, genome analyses underscore the characteristic features of the SARS-CoV-2 – they add the view that SARS-CoV-2 has been produced by natural selection

Recently, an American-European research group (Andersen et al. 2020) proposed that SARS-CoV-2 is the result of a natural selection that might occur either in animals before transmission to humans or in humans after transmission from animals. The proposition was based on specific characteristics of the virus.

8.2.3.3 Phylogenetic analyses estimated the origin of SARS-CoV-2 from the same isolate to be about 2 years ago

A study of 276 coronavirus genomes found five SARS-CoV-2 genome sequences with very high bootstrap support (Ji et al. 2020). Well-supported clades A and B (100%) contained coronavirus strains isolated from bats in China, Kenya, and Bulgaria. The clade C included 267 coronavirus genomes and was on a phylogenetic tree with 67% bootstrap support. By consideration of 0.000094 substitutions per site and the SARS-CoV-2 length of 29,865 bp, the evolutionary rate was measured as 0.0038 substitutions per site per year. The time of the most recent common ancestor (TMRCA) approximates the origin of current SARS-CoV-2 sequences from the same kind about 2 years ago.

8.2.3.4. Homologous recombination: a possible mechanism that may increase the cross-species transmission of SARS-CoV-2 spike glycoprotein

There are a variety of viral infections supposed to result from homologous recombination, including the human immunodeficiency virus (Ji et al. 2020). The high similarity in the viral genome between the SARS-CoV-2 and bat SARS-like CoVs posed the potential of recombination events between coronaviruses from bats and another species before being transmitted to humans. Analysis of codon usage patterns indicated a biased occurrence of synonymous codon among the SARS-CoV-2, bat SARS-like CoV ZC45, and Chinese snakes. As shown in Fig. 8.3, the squared Euclidean distance is useful for the measurement of dissimilarity between the SARS-CoV-2 and its potential wildlife reservoirs, where the lesser the distance, the higher the strength of convergent forces.

8.2.3.5 Different viral genotypes shed light on the evolution of the SARS-CoV-2 – it has not been perfected yet and might become converted into a more pathogenic species

Analysis of 27 genomes of SARS-CoV-2 isolates from different patients has associated the virus to at least six major genotypes (Zhang et al. 2020b). Using the higher number of genome samples ($N = 97$), there were 95 variable genome sites, containing only up to 3 mutations at coordinates 8750, 28,112, and 29,063 in most cases (Zhang et al. 2020c). Genome three coordinate-based data could define two major types to which SARS-CoV-2 strains belong (Fig. 8.4), and SARS-CoV-2 strains are mostly of type II strains. Of note, the coordinates of interest retain the same codons for type I SARS-CoV-2 strains and the BatCoV RaTG13. Such a relation is weak for type II SARS-CoV-2 strains. As evidenced by functional analysis, mutations were either non-synonymous (28112) or synonymous at (8750 and 29,063), and those synonymous mutations in type II SARS-CoV-2 strains appear to have higher translational efficiencies relative to those

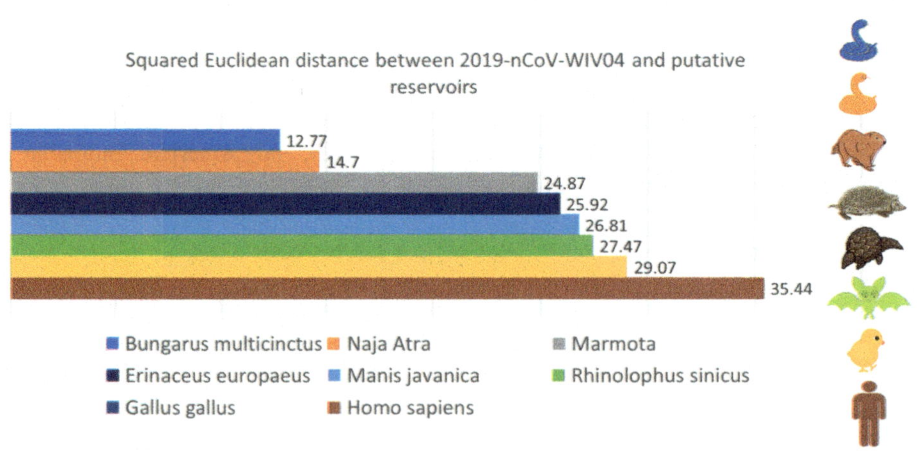

Fig. 8.3 The squared Euclidean distance between 2019-nCoV-WIV04 and putative reservoirs

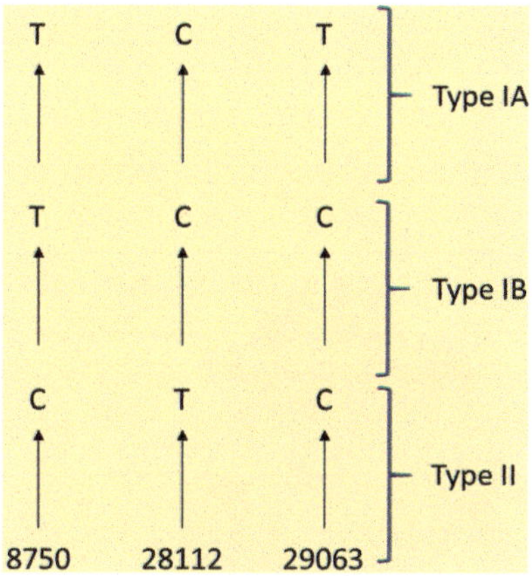

Fig. 8.4 2019-nCoV genotypes

in type I SARS-CoV-2 strains. Higher translational efficiencies might increase the production rate of type II SARS-CoV-2 particles, and consequently, we might expect that type II SARS-CoV-2 strains tend to become in a corresponding degree more contagious than type I SARS-CoV-2 strains.

8.3 Neither the Window of Viral Detection nor Viral Loads Reflect COVID-19 Severity

A study of 173 respiratory specimens (including posterior oropharyngeal saliva and endotracheal aspirate samples) showed that patients with COVID-19 have a median viral load of 105.2 copies per milliliter at presentation with the highest levels being detected within the first week and then decreasing slowly to become undetectable mostly within the next 2 weeks (To et al.). Thus, the usual window of detection of viral RNA might not be more than 3 weeks after onset. However, viral RNA lasts for longer than 20 days in one-third of patients, and this prolonged detection of viral RNA shows no correlation with disease severity. Also, there was no significant difference between initial and peak viral load, and neither initial viral load nor peak viral load was related to disease severity. Patients with severe COVID-19 display a median viral load of 106.17 copies per milliliter, which increases to a peak of 106.91 copies per milliliter. The corresponding values were 105.11 and 106.29 copies per milliliter for patients with mild disease.

8.4 The Older the Age Is, the Higher the Viral Load of SARS-CoV-2 May Increase

In general, the lack of correlation between disease severity and viral loads reveals to us that the SARS-CoV-2 would replicate a similar degree in patients with severe and non-severe disease. Thus, what divides patients into severe and non-severe groups is the individual response to the infection. Interestingly, there is a direct association between age and peak viral loads (To et al.). In particular, the presence of this correlation and the statistics that represent the older people as the most vulnerable group corroborate the determining role of age in individual response to SARS-CoV-2.

8.5 Immune Profile of Peripheral Blood in Patients with COVID-19

COVID-19 occurs in three forms (Wu and McGoogan 2020). In most cases, fighting against infection would be successful, and the disease results in mild symptoms that either does not progress to pneumonia or would progress to mild pneumonia. In 14% of cases, the virus brings a severe threat to the respiratory function and causes serious complications ranging from dyspnea, breathing rate higher than 30 breaths per minute, and peripheral oxygen saturation of 93% or less to the ratio of PaO2 to FiO2 less than 300 and lung infiltrates higher than 50% within 24 to 48 h. The battle would be prolonged and critical to 5% of cases who develop respiratory failure, septic shock, and multiple organ dysfunction or failure.

Regardless of whether or not they have a diagnosis of severe or critical COVID-19, the peripheral blood specimens from patients with COVID-19 present an extremely dysregulated

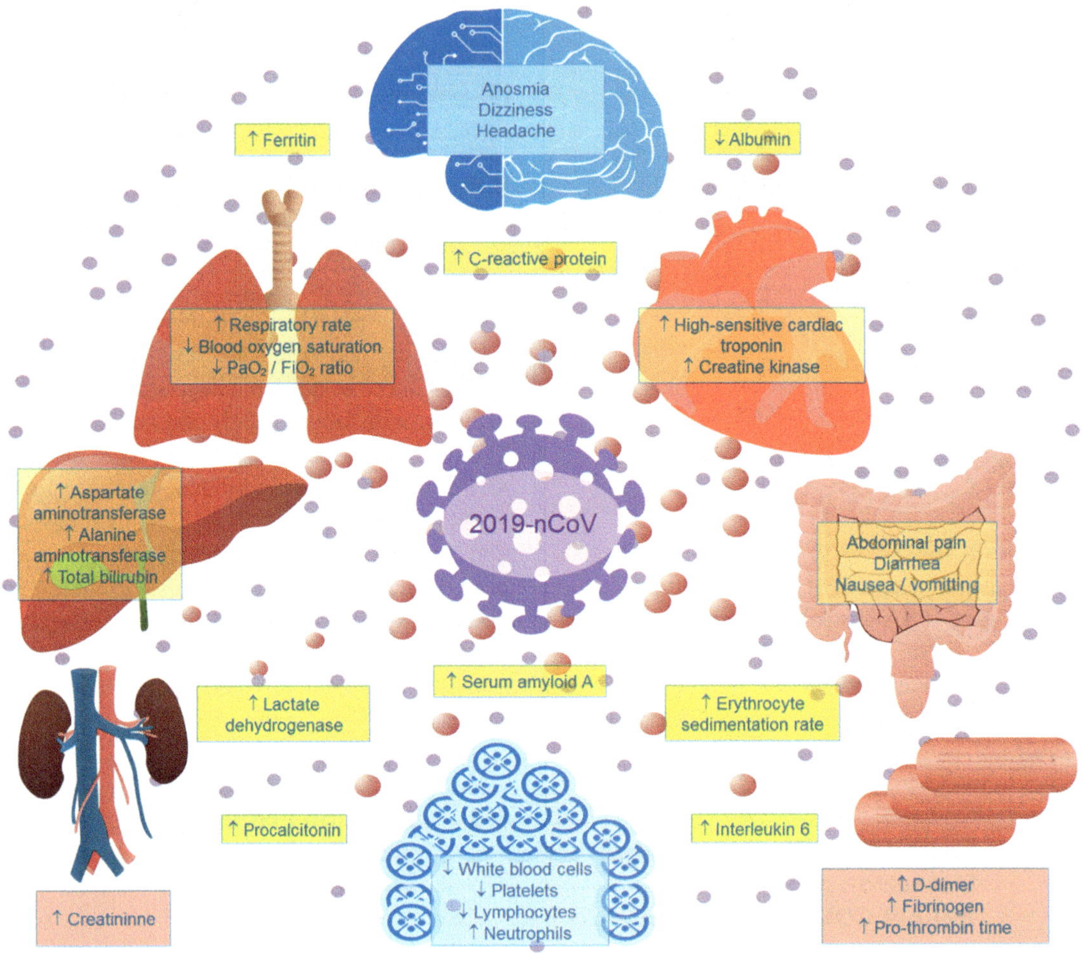

Fig. 8.5 The body burden of COVID-19

milieu, marked by the production of various inflammatory mediators, tissue injury-related enzymes, and biomarkers of hypercoagulation (Wang et al. 2020b). Figure 8.5 gives an overview of the body burden of and biomarker changes well-associated with COVID-19. Below, we will provide a rapid review of COVID-19-related changes in the immune system.

8.5.1 White Blood Cell

The majority of cases with COVID-19 show normal white blood cell counts (WBC) (Guan et al. 2020; Wang et al. 2020a). Among patients with abnormal WBC counts, low counts are about six times more prevalent than high counts (Guan et al. 2020). More than 60% of patients with severe COVID-19 have low WBC counts, while less than 30% of patients with mild and moderate disease show low WBC counts (Guan et al. 2020). The median count of WBC is higher in patients who require admission to the intensive care unit (ICU) than patients who do not need the ICU (Huang et al. 2020; Wang et al. 2020a). However, another study found no such difference between ICU and not-ICU patients (Zhou et al. 2020).

8.5.1.1 Lymphocytes

Lymphopenia is a common laboratory finding (Fathi and Rezaei 2020) present early on since the symptoms appear, even in patients with mild COVID-19 (Guan et al. 2020) and even in those who have normal WBC count (Cai et al. 2020). On admission, patients whose conditions progressed to the critically ill, patients who required admission to ICU, and patients who died from disease had a median lymphocyte count lower than that of patients whose condition was mild or severe, patients who do not require the ICU, and patients who survived (Huang et al. 2020; Wang et al. 2020a; Yang et al. 2020a; Chen et al. 2020b; Zhou et al. 2020). Up to about 96% of the population with severe and critical COVID-19 has lymphopenia (Guan et al. 2020). A lymphocyte count of less than 1100 per mm^3 can predict elevation in alanine aminotransferase (ALT) and, therefore, COVID-19-related hepatic injury (Li et al. 2020b).

T Cells

T-Cell Pattern in Relation to the Deterioration of COVID-19

As the disease continues to deteriorate, the lymphocyte count steadily decreases (Fig. 8.6). The study (Jiang et al. 2020) placed patients with COVID-19 into three categories – remission (N = 74), deterioration (N = 26), and death (N = 12) – and provided a peripheral T-cell profile of patients in three general stages: early, advanced, and late stage. Throughout the stages, total lymphocyte count (TLC) in remitting patients was steadily higher than that in patients of deterioration and death groups. TLC in patients of the death group was lower than patients of the other two groups and steadily decreased during the disease progression. The initial normal TLC decreased in the advanced stage in patients in the deterioration group. For all three subsets of T cells, e.g., CD3+ T cells, CD4+ T cells, and CD8+ T cells, in the advanced stage of COVID-19, patients who obtained the remission had greater counts compared with patients who deteriorated, and these patients, in turn, had greater counts than those who died (Fig. 8.7). The three groups of patients did not differ in B-cell counts. During a follow-up of 1 month, counts of CD4+ T and CD8+ T cells averaged above normal in most of the remission patients. On the contrary, the CD4+ T- and CD8+ T-cell values were below normal in most of the patients in the deterioration and death groups.

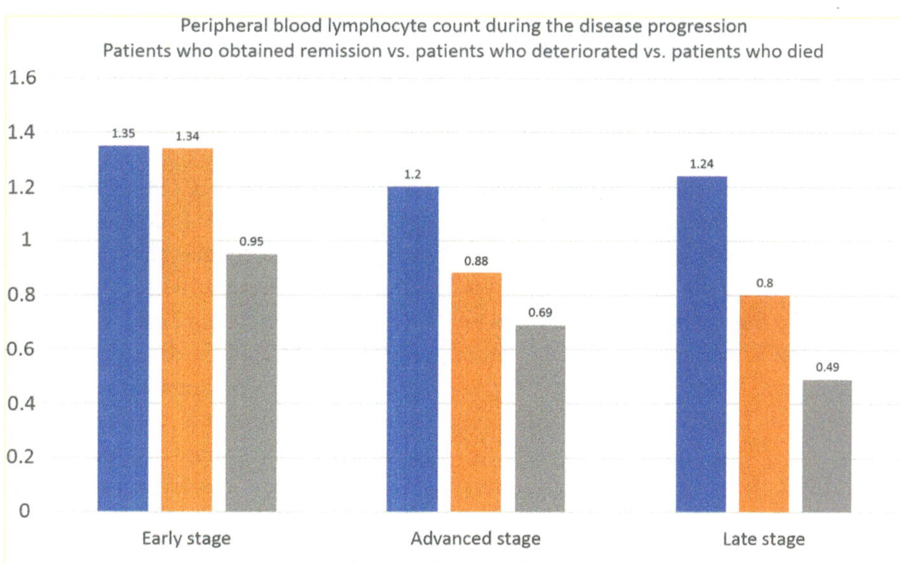

Fig. 8.6 Peripheral blood lymphocyte count during the progression of COVID-19

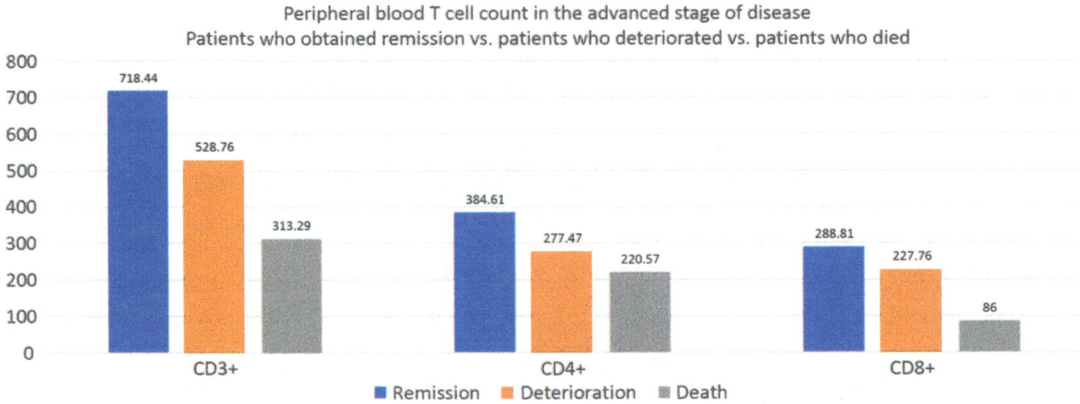

Fig. 8.7 Peripheral blood T-cell count in the advanced stage of COVID-19

T-Cell Pattern in Relation to the Severity of COVID-19

A study of 40 patients with COVID-19 monitored the kinetics of lymphocyte counts within 16 days after the onset of disease and compared that between severe and mild disease (Liu et al. 2020). On the first 6 days after the onset of disease, patients with severe COVID-19 had lower counts of total lymphocyte, CD3+ T cells, CD4+ T cells, and CD8+ T cells, pointing to the lowest levels on the fourth to the sixth day. Since the seventh day, the counts in the severe group began to increase so that on the sixteenth day, there was no difference between severe and mild COVID-19 patients in counts of total lymphocytes.

T-Cell Pattern in Relation to the Criticality of COVID-19

A retrospective study of 522 patients with confirmed COVID-19 reveals that more than 70% of patients had reduced counts of total T cells, CD4+ T cells, and CD8+ T cells (Diao et al. 2020). Patients admitted to the ICU had lower counts of these cells than patients not admitted to ICU. Among non-ICU patients, those who were in a severe or critical state of disease produced lower counts of total T cells, CD4+ T cells, and CD8+ T cells compared to their counterparts. Interestingly, an analysis of total T-cell counts and T-cell subsets in different age groups reflected that the count of these cells changes in an age-dependent manner such that the highest counts of total T cells, CD4+ T cells, and CD8+ T cell are seen in patients with less than 20 years of age while the lowest counts of these cells occur in patients of 60 years and above.

T Cells Become Reduced, Exhausted, and Overwhelmed in COVID-19

The change in T cells is profound. Not only the counts of total T cells and T-cell subsets decrease, but the function of these cells becomes less effective as the expression of immune inhibitory receptors increases over time during the disease. Individuals with COVID-19 have a higher proportion of T cells expressing programmed cell death protein 1 (PD-1), an immune checkpoint that contributes to CD8+ T-cell exhaustion during chronic infections and cancers through affecting T-cell receptor signaling, that controls without COVID-19. Among COVID-19 patients, those who require ICU would, in turn, have a higher proportion of PD-1+ CD8+ T cells compared to non-ICU patients. A follow-up of a few cases provided evidence that the counts of CD8+ T cells start increasing, while PD-1 moves in the direction of decreasing expression throughout the prodromal phase of the disease (Diao et al. 2020).

A study of 33 patients with severe COVID-19 showed that compared to patients who do not require ICU, the counts of total T cells, CD4+ T cells, and CD8+ T cells were lower in patients admitted to ICU, who, in turn, had lower counts than healthy controls (Zhou et al. 2020). Despite lower counts, T cells become activated in patients with severe COVID-19, as indicated by an increased expression of CD69, CD38, and CD44 on both CD4+ T and CD8+ T cells, OX40 (CD134) on CD4+ T cells, and 41BB (CD137) on CD8+ T cells. In parallel, they will become exhausted as represented in the increased expression of Tim3 and PD-1 on both CD4+ T and CD8+ T cells. Intracellular staining highlighted the higher percentage of CD4+ T cells expressing IL-6 and GM-CSF, but not TNFα, in patients with severe COVID-19 than healthy controls and patients admitted to ICU than patients not requiring ICU. Also, CD8+ T cells expressing GM-CSF, but not IL-6 and TNFα, were increased in ICU than non-ICU patients. CD4+ T cells that express both IFNγ and GM-CSF existed merely in patients with severe COVID-19 who are admitted to ICU.

Besides lower counts of lymphocyte subsets (CD3+ T cells, CD4+ T cells, and CD8+ T cells), there is an increased CD4/CD8 ratio in the majority of critically ill patients with COVID-19 admitted to ICU (Wang et al. 2020c). Generally, an increased CD4/CD8 ratio relates to autoimmune diseases, while a decreased ratio commonly occurs in viral infections. These lines of evidence support the idea that COVID-19 tends to develop an autoimmune-like condition rather than merely resemble a viral infection.

B Cells and Natural Killer (NK) Cells

At all the time points within the clinical course of the disease, patients with severe COVID-19 did not differ in the number of B cells and natural killer (NK) cells from patients with mild COVID-19 (Liu et al. 2020) (Fig. 8.8). Among patients with severe COVID-19, ICU and non-ICU patients did not differ in counts of B and NK cells (Zhou et al. 2020). However, when the condition progressed to critically ill, the number of NK cells decreased below normal levels (Wang et al. 2020c).

8.5.1.2 Monocytes

Patients with severe COVID-19, either requiring or not requiring ICU, had lower counts of monocytes than healthy controls (Zhou et al. 2020). However, the proportion of monocytes expressing both CD14 and CD16 was higher in patients with severe COVID-19, especially those who require the ICU (Zhou et al. 2020). Generally, these monocytes, known as intermediate cytokines, exhibit a pro-inflammatory behavior (Zhou et al. 2020). In particular, peripheral blood monocytes from ICU patients with COVID-19 express high levels of IL-6 and GM-CSF (Zhou et al. 2020).

8.5.1.3 Neutrophils

Studies indicate that on admission, the median count of neutrophils is higher in i, patients who were in a severe condition compared to patients who were in a mild condition; ii, patients whose condition was or progressed to critically ill compared to patients whose condition was mild or severe or not progressed to critically; and iii, patients admitted to the ICU compared to patients not admitted to the ICU (Huang et al. 2020; Wang et al. 2020a; Chen et al. 2020b; Liu et al. 2020). Kinetics of neutrophil counts during the clinical course of 16 days following SARS-CoV-2 infection corroborates sustained higher counts in patients with severe disease compared to patients with mild disease (Liu et al. 2020). The study (Liu et al. 2020) identified neutrophil to CD8+ T-cell ratio, neutrophil to lymphocyte ratio, WBC counts, and neutrophil counts as the best predictors of COVID-19 severity.

8.5.2 Immunoglobulins

At the baseline, patients with severe and mild forms of COVID-19 showed no difference in levels of immunoglobulins IgA, IgG, and IgM (Liu et al. 2020). The detection of both IgM and IgG tends to increase by time (Zhang et al. 2020d). For IgM, the detection rates were estimated using ELISA to be about 50% and 80% on day 0 and 5. The corresponding rates for IgG were 80% and 100%. More precisely, a kinetic study of 49 patients with COVID-19 estimated that IgM and IgG simultaneously happen, and their detection rates rise at the median of 10 days after the onset of infection (Li et al. 2020a). In particular, S-IgM and N-IgM exist in about 39% and 51% of patients at the median of 10 (range: 3–20 days)

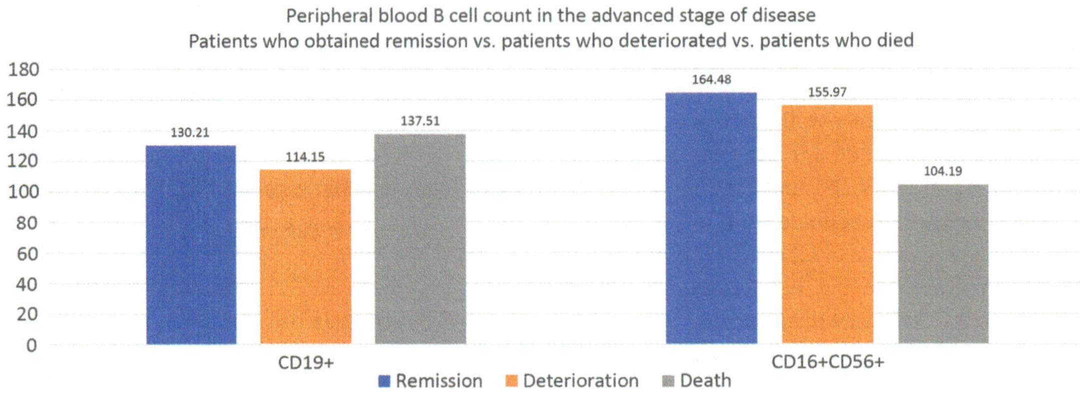

Fig. 8.8 Peripheral blood B-cell count in the advanced stage of COVID-19

and 9 (range: 2–16 days) days after the onset of infection (Li et al. 2020a). S-IgG and S-IgM appeared in about 63% and 41% of patients at a median of 9 days (range: 2–22 days) (Li et al. 2020a). Only about 10% of patients must produce high levels of antibodies, and in these patients, S-IgG is the only antibody that remained high over 30 days after infection (Li et al. 2020a).

8.5.3 Complement Factors

Baseline levels of complements C3 and C4 were not different between patients with severe and mild forms of COVID-19 (Liu et al. 2020).

8.6 Cytokine Storm: A Defensive Strategy Protecting the Home Against SARS-CoV-2 or a Self-Medicated Disaster Hitting the Home by SARS-CoV-2

8.6.1 Cytokines

8.6.1.1 Baseline Levels of Cytokines in Relation to the Severity of COVID-19

Patients with COVID-19 show dysregulation of peripheral levels of cytokines and chemokines in the acute phase of infection (Huang et al. 2020). There were higher plasma levels of IFNγ and TNFα, IL-4, IL-10, IL-1β, IL-1RA, IL-7, IL-8, IL-9, IL-10, basic FGF, GCSF, GM-CSF, IP-10, MCP1, MIP1A, MIP1B, PDGF, and VEGF (Huang et al. 2020). Patients who need ICU admission produce higher levels of IL-2, IL-7, IL-10, GCSF, IP-10, MCP1, MIP1A, and TNFα compared with patients who do not require critical care (Huang et al. 2020).

Of 48 cytokines investigated (Yang et al. 2020b), on admission, patients with mild and severe COVID-19 display increased concentrations of different cytokines, including IFNγ, IL-1RA, IL-2RA, IL-6, IL-10, IL-18, HGF, MCP3, MIG, M-CSF, G-CSF, MIG-1A, CTACK, and IP-10. When compared to patients having a mild form of the disease, patients with severe COVID-19 had higher levels of IP-10, MCP-3, IL-1RA, and MIG at different time points during the disease (Yang et al. 2020b). All the four cytokines, except for MIG, showed association with the PaO_2 to FaO_2 ratio, and their combination yielded good AUCs for the prediction of disease progression (Yang et al. 2020b). However, IP10 was the most consistently elevated cytokine in patients with severe COVID-19 and the only cytokine associated with viral load, as determined by the cycle threshold (CT) value of qRT-PCR (Yang et al. 2020b).

A retrospective study of patients with COVID-19 ($n = 522$) confirmed higher levels of cytokines IL-6, IL-10, and TNFα, but not IFNγ, IL-2, and IL-4 (Diao et al. 2020). Patients admitted to ICU also had higher levels of the three cytokines than non-ICU patients (Diao et al. 2020).

8.6.1.2 Tracing Kinetics of Changes in Cytokine Levels in COVID-19

Cytokines Add Their Character to the Progression of COVID-19
A negative association between cytokine levels and T-cell counts (total and CD4+ and CD8+ subsets), as well as the reduction in cytokine levels during treatment, poses the accumulation of cytokines as a critical inhibitor of T-cell survival and proliferation (Diao et al. 2020).

The Changes of IL-6 and IL-10 Are Most Remarkable During COVID-19
The study (Liu et al. 2020) monitored serum levels of cytokines, e.g., IL-2, IL-4, IL-6, IL-10, IFNγ, and TNFα, during the clinical course of disease in patients with severe and mild COVID-19. For the mild group, a straight line described no fluctuations in concentrations of all cytokines. The severe group showed a very different profile of cytokines. For the cytokines IL-2, IL-4, IFNγ, and TNFα, there was one peak concentration on the fourth to the sixth day. The most potent fluctuations occurred in the concentrations of IL-6 and IL-10.

Another kinetic study of cytokines (IL-2, IL-4, IL-6, IL-10, IL-17, IFNγ, TNFα, and IL-12P70) highlighted IL-6 as the only cytokine that increased in more than 50% of patients with COVID-19 ($N = 49$), with the median detection time of 7 days (range: 2–26 days) (Li et al. 2020a). Analysis of individual patient data indicated that for the most COVID-19-associated cytokines, e.g., IL-6, IL-10, and IFNγ, the peak concentration occurs at 8 days, which begins to decrease within the subsequent 2–3 days (Li et al. 2020a). Higher levels of IL-6 occurred in patients with severe than those with mild disease. Higher levels of IL-10 happened in patients with an underlying condition.

8.6.1.3 IL-6: The Well-Documented Cytokine in COVID-19

IL-6 Levels in Relation to the Severity of COVID-19
Overall, more than 50% of patients display increased levels of IL-6. Of note, patients with higher COVID-19 severity, as defined by chest computed tomography (CT) scan scores, exhibit significantly higher concentrations of IL-6 compared to cases with lower disease severity (Wang et al. 2020b).

IL-6 Levels in Relation to the Criticality of COVID-19
Among the investigated cytokines, IL-6 seems to be of high value because its increase was observed in all critically ill patients admitted to ICU (Wang et al. 2020c), while the increase of other cytokines, e.g., IL-10, IL-4, and IFNγ, existed in less than half of those patients. Monitoring patient status along with measurement of IL-6 levels showed that the increase of IL-6 could predict that respiratory condition may worsen shortly and pulmonary inflammation may progress as represented in chest CT (Wang et al. 2020c).

IL-6 Levels in Relation to the Viral Load of COVID-19
The study of patients with throat-swab specimen that tests positive showed that the SARS-CoV-2 nucleic acid could be detected in critically ill patients, but not in patients having a mild or severe form of the disease (Chen et al. 2020b). Critically ill patients with COVID-19 show tenfold higher levels of IL-6 compared to patients with mild and severe COVID-19, and SARS-CoV-2 RNAaemia relates directly with IL-6 levels (Chen et al. 2020b). In this manner, IL-6 levels of 100 pg/ml or more are considered as a biomarker of mortality and RNAaemia in patients with COVID-19 (Chen et al. 2020b).

IL-6 Levels in Relation to the Prognosis of COVID-19
There is a direct linear correlation between serum IL-6 and CT imaging scores and the reduction of IL-6 levels during treatment (Wang et al. 2020b).

8.6.2 Inflammatory Markers

Inflammatory markers, mainly C-reactive protein (CRP), erythrocyte sedimentation rate (ESR), lactate dehydrogenase (LDH), procalcitonin (PCT), serum amyloid A (SAA), and ferritin, rise in patients with COVID-19.

8.6.2.1 Inflammatory Markers in Relation to the Severity of COVID-19

Patients with a severe form of the disease and patients admitted to ICU show higher baseline levels of inflammatory markers, particularly CRP, compared to patients with mild COVID-19 and patients not admitted to ICU (Liu et al. 2020; Zhou et al. 2020). Compared to patients with lower chest CT scan scores, patients with higher chest CT scan scores have higher levels of CRP, ESR, and serum amyloid A (Wang et al. 2020b). However, CRP seems to be the most important one because of its meaningful correlation with CT scan score in terms of the higher the CRP levels are, the higher the CT scan score, and therefore the more the COVID-19 is severe.

8.6.2.2 Inflammatory Markers in Relation to the Complications of COVID-19

Cardiac complications, including tachycardia, electrocardiography abnormities, diastolic dysfunction, elevated myocardial enzymes, and acute myocardial injury, are prevalent in patients with COVID-19. High CRP levels can predict those who will develop cardiac complications (Xu et al. 2020a).

8.6.2.3 Inflammatory Markers in Relation to the Prognosis of COVID-19

Treatment reduces CRP and ESR levels, and therefore, CRP and ESR offer an excellent prognostic biomarker. In particular, elevated CRP levels are predictive of liver injury, as indicated in high ALT levels (Li et al. 2020b).

8.6.3 Differentially Expressed Genes in Bronchoalveolar Lavage Fluid and Peripheral Blood from Patients with COVID-19

The near-total absence of viral reads and ACE2 expression in PBMC provides evidence that SARS-CoV-2 does not directly infect PBMC, despite the presence of a high number of viral reads in BALF samples (Xiong et al. 2020). Quantitative transcriptome analysis selected about 1000 and 500 differentially expressed genes (DEGs) in BALF and PBMC among patients with COVID-19 compared to controls. The upregulated genes in BALF include those involved in viral invasion strategies such as cotranslational protein targeting to membrane, protein targeting to the endoplasmic reticulum, and viral transcription. While the downregulated genes can act mainly on immune cell activation, the upregulated genes in PBMC play a role in activating the immune responses, especially complement activation, humoral immune response mediated by circulating immunoglobulin, and B cell-mediated immunity, and in inducing inflammatory processes and regulation of acute inflammatory response. Also, genes differentially downregulated in PBMC from patients with COVID-19 are recognized for their participation in processes related to the maturation of immune cells, such as chemotaxis and migration.

Pathways enriched by upregulated genes in BALF include ribosome, protein processing in the endoplasmic reticulum, phagosome, pentose phosphate pathway, carbon metabolism, and lysosome (Xiong et al. 2020). Upregulated genes in PBMC contribute to cell cycle and P53 signaling pathways. In contrast, downregulated genes take part in viral protein interaction with cytokine and cytokine receptor, NF-kappa B (NF-kB) signaling pathway, Toll-like receptor (TLR) signaling pathway, and IL-17 signaling pathway. P53 signaling plays a role in cell cycle control and apoptosis, and its upregulation and mediated apoptosis may explain reduced counts of PBMC, e.g., lymphocytes in patients with COVID-19.

As for their gene expression, patients with COVID-19 show increased expression of cytokines IL-10, CCL2/MCP-1, CXCL10/IP-10, CCL3/MIP-1A, and CCL4/MIP1B in BALF (Xiong et al. 2020). The change in the expression of IL-10, IL-36RN, IL-36G, TNFSF15, CCL5, TNFSF10, CXCL1, and IL-33 depends on disease severity.

8.7 The Scenario of How the Lungs Come Down in Few Days by Macrophage Missiles in COVID-19

8.7.1 Human Studies Reveal the Kinetics of Two-Phase Lung Disease in Patients Who Died of SARS-CoV

Pathological findings of the lungs from patients who died from SARS-CoV include diffuse alveolar damage (DAD), hyperplasia of pulmonary epithelial cells, infiltration of giant cells, alveolar and interstitial macrophages, squamous metaplasia of bronchial mucosa, septal inflammation, SARS-CoV particles, and RNA (Nicholls et al. 2003; Franks et al. 2003).

Lung histopathology changes between the two phases during SARS-CoV infection (Franks et al. 2003). The more the primary injury is damaging, the fewer cells remain to aid the lungs to repair during the second phase.

8.7.1.1 Phase 1 (0–10 Days)

It is an acute phase of DAD accompanied by airspace edema and bronchiolar injury.

The acute phase of DAD (mild to marked) causes between 50 and 75% involvement of the lung parenchyma, interstitial and intra-alveolar edema, interstitial inflammatory cell infiltrate, and pulmonary vascular congestion. Airspace edema (moderate to marked) causes between 50 and 100% involvement of the lung parenchyma and correlates with the accumulation of eosinophilic material within alveolar spaces. Bronchiolar injury arising from fibrin collection within lumens correlates with a loss of cilia, the denudation of the bronchial epithelium, and fibrinoid deposits along the basement membrane.

8.7.1.2 Phase 2 (After 10 Days)

It is the organizing phase of DAD associated with type II pneumocyte hyperplasia, squamous metaplasia, multinucleated giant cells (e.g., type II pneumocytes and macrophages), and acute bronchopneumonia.

The organizing phase of DAD (mild to marked), including interstitial and airspace fibroblast proliferation, orchestrates the repair by hyperplasia of type II pneumocytes and squamous metaplasia. It may show the involvement of the lung parenchyma with a range of 1% to 100%. Type II pneumocytes showing hyperplasia are indicated by cytomegaly, nucleomegaly, chromatin clearing, and prominent nucleoli. Multinucleated cells, CD68+ macrophages, pancytokeratin-positive/TTF-1-positive pneumocytes, and multinucleated cells are commonly present in alveolar spaces. Acute bronchopneumonia (mild to marked) causes between 50 and 75% involvement of the lung and is attributed to the trafficking of neutrophils and macrophages in alveoli and airway passages.

8.7.2 Immunological Cells Invade the Lungs and Occupy its Nation Within 3–10 Days in Patients with COVID-19

A rapid review of recorded COVID-19 cases of death without any preexisting condition reveals an astonishingly rapid progression of lung lesions leading to respiratory failure and death. The intervals between symptom onset and assisted mechanical ventilation range from 3 to 10 days (Chen et al. 2020a).

A 66-year-old male with severe COVID-19 and marked brightness of the lungs had acute respiratory distress syndrome (ARDS), finally leading to lethal respiratory failure (Luo et al. 2020). Immunohistological findings of the case include interstitial inflammatory cell infiltrates consisting of lymphocytes, plasma cells, and mononuclear cells and focal accumulation of immune cells expressing CD3, CD4, CD8, CD20, CD79a, CD5, and CD38.

A confirmed case of COVID-19 was a 50-year-old male who complained of fever, chills, cough, and shortness of breath (Xu et al. 2020b). His breathing became more difficult, as reflected in increased interstitial shadows in both lungs, and ultimately, he died. As confirmed by the tissue examination, mononuclear cell infiltration composed dominantly of lymphocytes is the immune feature of both lungs. Consistently, peripheral blood demonstrated that CD4+ and CD8+ T cells, though generally decreased in numbers, become hyperactive. The other immunological characteristics of the disease include induction of T helper (Th) 17 cell development (represented with increased levels of inflammatory C-C chemokine receptor 6 (CCR6)) and enhancement of effectivity of CD8+ T cells (indicated with increased frequency of T cells expressing cytotoxic granules).

There is also a pathological report of two cases with lung cancer who died of 2019-nCoV infection (Tian et al. 2020). We have to overlook this report because the impact the preexisting lung disease can have on pathological findings is pervasive.

8.7.3 Macrophage Population Properties in COVID-19

8.7.3.1 Blood Monocytes Are Larger and More Complex in Patients with COVID-19

Using flow cytometry, the study (Zhang et al. 2020a) remarked a foreword scatter (FSC)-high population of atypical, large, vacuolated monocytes in the peripheral blood from patients with COVID-19 ($n = 28$). These cells included an expansion of monocytes, M1 macrophages, and M2 macrophages, because of the expression of surface markers related to monocytes, CD14 and CD16, and also a variety of markers including CD11b, CD68, CD80, CD163, and CD206 that define M1 and M2 macrophages. Besides, they could produce cytokines that are expressed by M1 (IL-6 and TNFα) and M2 macrophages (IL-10). The COVID-19-related monocyte population contained a similar number of classical monocytes but did include higher numbers of intermediate and nonclassical monocytes. Neither controls without COVID-19 nor patients with other infectious diseases, including HIV, tuberculosis, malaria, H1N1, and HTNV, exhibited such an FSC-high population of monocytes with multiple phenotypes.

8.7.3.2 Blood Monocytes Correlate with the Severity and Criticality of COVID-19

Patients who required admission to ICU possessed a higher proportion of FSC-high monocytes than of FSC-low monocytes (Zhang et al. 2020a). Besides, the increased percentage of the FSC-low population of monocytes corresponded to discharge from the hospital.

8.7.3.3 Blood Monocytes Express the Lower Amount of ACE2 in Patients with COVID-19

ACE2 is present on different human monocyte cell lines (THP-1 and U939) and murine macrophage cell (RAW264.7) lines (Zhang et al. 2020a). Patients with COVID-19 displayed lower levels of ACE2 on monocytes/macrophages than controls without COVID-19.

8.7.3.4 BALF Macrophages Are Different Among Patients with Severe and Mild COVID-19

The study (Liao et al. 2020) compared the BALF-derived specimens of eight people who had never been diagnosed with COVID-19 with that of those who had bilateral pneumonia and suffered from mild ($N = 3$ and age range 35–37 years) or severe COVID-19 ($N = 3$ and age range 62–66 years). Generally, it showed that different immune cells form 36 different cell clusters in the BALF. Precisely, the proportions of T cells and natural killer (NK) cells in patients with mild COVID-19 exceed those of patients with severe COVID-19, which in turn are higher than those of control subjects (Figs. 8.9 and 8.10). Also, the BALF in patients with severe COVID-19 would harbor a macrophage population that is widely different from that in patients with mild COVID-19 and control subjects in both the amount and cellular phenotype. The severe COVID-19 BALF includes a higher number of macrophages characterized by two predominant cell types: monocyte-derived macrophages expressing Ficolin 1 (FCN1) and pro-fibrotic macrophages representing secreted phosphoprotein 1 (SPP1) (Fig. 8.11).

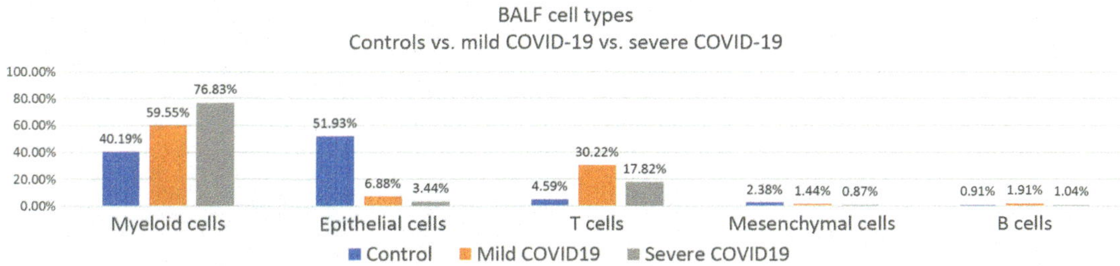

Fig. 8.9 The distribution of discrete cell types in the bronchoalveolar lavage fluid (BALF): controls without COVID-19, patients with mild COVID-19, and patients with severe COVID-19

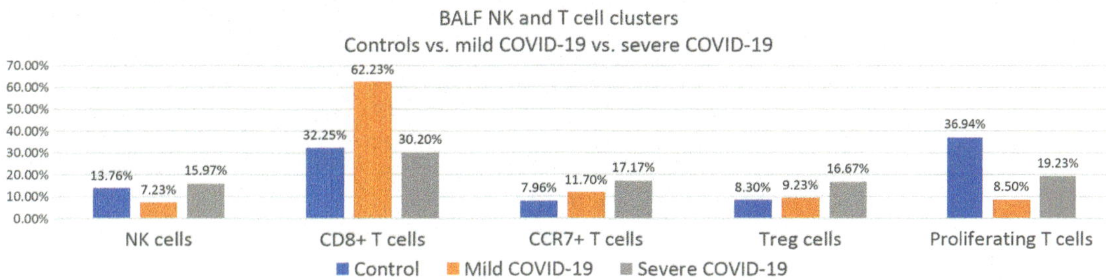

Fig. 8.10 The distribution of natural killer (NK) and T-cell clusters in the bronchoalveolar lavage fluid: controls without COVID-19, patients with mild COVID-19, and patients with severe COVID-19

8.7.3.5 SARS-CoV-2-Positive Macrophages Are Present in the Lung and Other Organs

The study (Munster et al. 2020) of eight adult rhesus macaques has shown that inoculation with SARS-CoV-2 could cause irregular breathing patterns, weight loss, and fever. Animals displayed pulmonary infiltrates in the lower lobes on one dpi, which progressed to the caudal lobes by three dpi. Also, they showed increased levels of cytokines, including IL-6, IL-10, IL-1RA, IL-15, MCP1, and MIP1B and high viral loads in the BALF. Necropsy confirmed interstitial pneumonia characterized by gross lung lesions, thickening of the alveolar septa, edema fluid, fibrin, hyaline membrane formation, type II pneumocyte hyperplasia, alveoli with macrophages and neutrophils, and perivascular infiltration of lymphocytes. There were alveolar macrophages and type II pneumocyte positive for the SARS-CoV-2 antigen. Interestingly, antigen-positive macrophages were present in tissues other than in those the respiratory system, including the mediastinal lymph nodes and the intestinal tract.

8.7.4 Anti-COVID-19 Macrophages: Anti-inflammatory Macrophages that Induce Tolerance to Coronaviruses Resembling the Novel One in Bats

SARS-CoV is a bat-loving microparasite; bats serve as a natural reservoir for SARS-CoV (Li et al. 2005). Bat macrophages do not differ from mice macrophages in the amount of production of pro-inflammatory cytokines, such as IL-1β, TNF-α, and IFNβ. However, bat macrophages act differently in terms of higher levels of the anti-inflammatory cytokine, IL-10, they can produce (Kacprzyk et al. 2017). The production of IL-10 is an effort to balance the heavyweight of pro-inflammatory cytokines. The maintenance of this balance might have determined

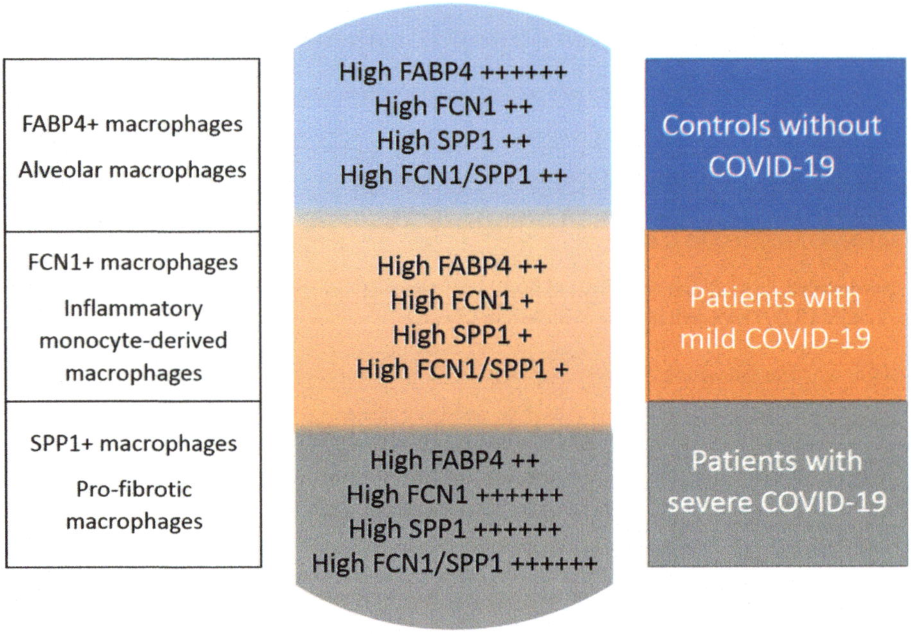

Fig. 8.11 The distribution of different cell clusters in the macrophage population of the bronchoalveolar lavage fluid (BALF): controls without COVID-19, patients with mild COVID-19, and patients with severe COVID-19

that the tolerance to some deadly viral infections is present in bats while it is absent in other mammals.

8.7.5 Pro-COVID-19 Macrophages: ACE2-Hyperactivated Macrophages

As real-time world statistics disseminate it, the COVID-19 infectivity largely depends on the initial host condition with a case fatality rate of 0.9% in patients without comorbid conditions compared with a range of 7.6% to 13.2% in patients with a comorbid condition. Among patients with different comorbid conditions, patients with preexisting cardiovascular disease and hypertension have the highest mortality rate. COVID-19 infectivity depends on its receptor, i.e., ACE2 receptor. Monocyte-derived macrophages in patients with hypertension and prehypertension exhibit higher ACE2 activity compared with individuals with normal blood pressure (Keidar et al. 2007).

8.8 Macrophage-Mediate Cell Pyroptosis and Inflammation

8.8.1 Cell Pyroptosis: A Pro-inflammatory Programmed Cell Death

Pattern recognition receptors (PRRs) are the innate immune receptors that can detect both invading pathogens and endogenous dangers through the recognition of pathogen-associated molecular patterns (PAMPs) and damage-associated molecular patterns (DAMPs). PRRs are of two main categories: transmembrane and cytosolic. Toll-like receptors and C-type lectin receptors (CLRs) take place at the cell membrane (transmembrane), while retinoic acid-inducible gene (RIG)-like receptors (RLRs) and NOD-like receptors (NLRs) are located inside the cell (cytosolic). Upon recognition of PAMPs and DAMPs, PRRs mostly tend to trigger signal transduction pathways linked to the induction of inflammatory responses (Walle and Lamkanfi 2016). However, some NLRs, NLRP3, NLRP1B, and NLRC4 signal the formation of multiprotein complexes, called inflammasomes (Walle and Lamkanfi 2016). In particular, the assembly of the NLRP3 inflammasome consisting of carboxy-terminal leucine-rich repeats (LRRs), amino-terminal caspase recruitment (CARD) or pyrin domain (PYD), adaptor protein apoptosis-associated speck-like protein containing a CARD (ASC), and pro-caspase 1 occurs in response to bacteria, fungi, and viruses and results in the recruitment of caspase 1 (Walle and Lamkanfi 2016). Upon activation, caspase 1 induces cleavage of precursors of pro-inflammatory cytokines, mainly pro-IL-1β and pro-IL-18, and of pore-forming effector proteins of Gasdermin (GSDM) family members, such as GSDMD. In this manner, caspase 1 causes inflammation and the formation of pores in the cells scheduled for pyroptosis. It makes a stressed cellular environment so that neither is favorable for the pathogen replication nor favorable to maintain the cell viability. Such cells release DAMPs and, in this manner, would undergo a programmed pro-inflammatory cell death called pyroptosis.

8.8.2 Cell Pyroptosis: A Double-Edged Sword

Whereas host-mediated suicide seems to benefit the host by its work that is the degradation of intracellular pathogens (Brodsky and Medzhitov 2011), it may become a severe struggle for life when the body gets confused due to chronic inflammation overload. For example, human immunodeficiency virus (HIV) infects CD4+ T cells and induces caspase 1-mediated pyroptosis of these cells. HIV, therefore, correlates with CD4+ T-cell depletion and chronic inflammation, which both of them are not good for the host (Doitsh et al. 2014).

8.8.3 Cell Pyroptosis in Macrophages

Pyroptosis can occur in various cells, CD4+ T cells, keratinocytes, epithelial cells, endothelial cells, and neurons, though the primary cells known to undergo pyroptosis are a professional type of myeloid lineage, including macrophages, neutrophils, and dendritic cells. Infection of macrophages and activation of NLRP3 inflammasome occurs with various RNA viruses, including EV71, H1N1, H7N9 influenza A virus, and Zika. In particular, for both SARS-CoV and MERS-CoV, there is evidence of the use of cell pyroptosis in human macrophages.

8.8.3.1 Cell Pyroptosis in Macrophages During SARS-CoV Infection

As summarized in Fig. 8.12, SARS-CoV possesses several accessory proteins that affect the immune system in different ways to prevent viral clearance. Of which, ORF8b can induce pyroptosis in macrophages through its direct interaction with the NLRP3 inflammasome (Shi et al. 2019).

8.8.3.2 Cell Pyroptosis in Macrophages During MERS-CoV Infection

MERS-CoV can infect both THP-1 macrophages and monocytes, as indicated by the presence of MERS nucleocapsid protein (Jiang et al. 2019). Upon MERS-CoV infection, the expression of pro-IL-1β and pro-caspase 1 increases in THP-1 monocytes and to a higher degree in THP-1 macrophages. In parallel, MERS-CoV infection in these cells led to the increased expression of complement factors, i.e., C3, C3aR, and C5aR1, which play a crucial role in TLR-mediated IL-1β transcription. In hDPP4 transgenic mice infected with MERS-CoV, there is evidence in favor of pyroptosis, such as the increase in the expression of pro-IL1β in the lung, IL-1β in serum, and caspase 1 in the spleen (but not lung). Also, MERS-CoV-infected mice revealed an inflammatory response, characterized by higher serum levels of pro-inflammatory cytokines, e.g., IL-6, IFNγ, and TNFα, and increased expression of CD68 and IFNγR in the lung and spleen. Treatment with an antibody targeting C5aR1 could limit pyroptosis and inflammatory response – through decreasing the expression of IL-1β in serum, caspase 1 in the spleen, IFNγ in serum, and CD68 and IFNγR in the lung and spleen.

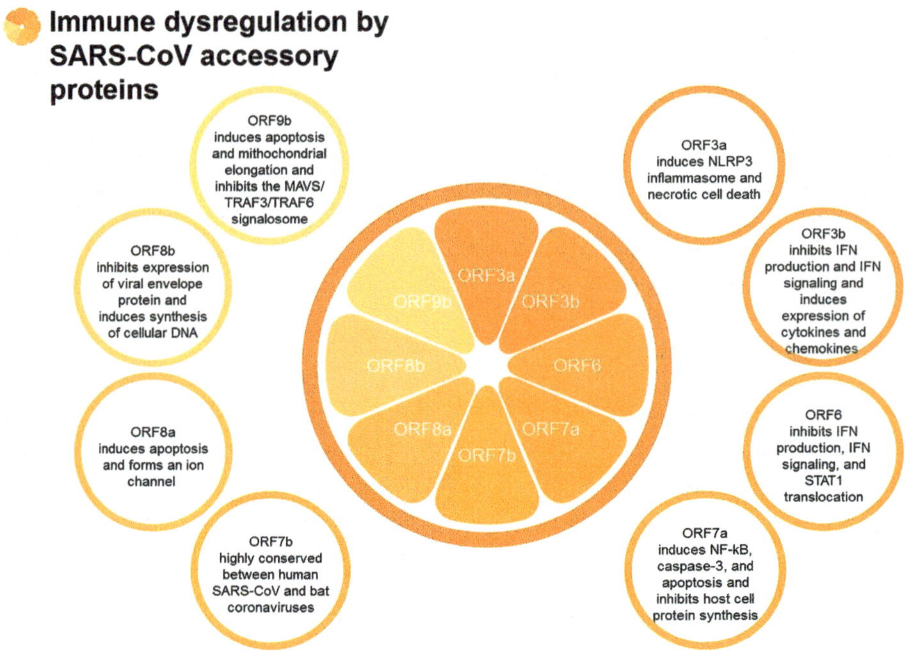

Fig. 8.12 Immune dysregulation by SARS-CoV accessory proteins

8.9 Ravaged by Macrophage Missiles Under the Cytokine Storm: A Lethal Disease Model for COVID-19

The discussion, as mentioned above, reveals that all adult patients with mild, severe, and critical COVID-19 have developed, from one prototype, inflammation and that it is possible to attribute the severity and criticality of COVID-19 to the wildness of this inflammation.

Cytokine release syndrome (CRS) (Shimabukuro-Vornhagen et al. 2018) refers to inflammatory conditions where target cell lysis or T-cell activation is accompanied by the release of cytokines, especially IFNγ and TNFα, that immediately involve immune cells such as endothelial cells, macrophages, and dendritic cells to commence the production of IL-6. IL-6, in turn, can induce immune cells to produce cytokines. In this manner, IL-6 would develop a cytokine storm through a positive feedback loop of cytokine production. In clinical practice, there are different grading systems for CRS is according to the presence and/or severity of fever, constitutional symptoms, hypotension, organ toxicities, shock, hypoxia, and the need for mechanical ventilation.

COVID-19 seems to present a disease that is ideally similar in immune and clinical features with that of CRS. The course of COVID-19 consists of two stages. The first stage causes T-cell reduction, putting the responsibility on the remaining T cells that combat viral infection, making them overwhelmed and exhausted. Also, a variety of cytokines and inflammatory markers rise in response to infection in this stage. The second stage occurs in two main phenotypes that determine the fate of the patient. The favored phenotype of remission/survival leads to an increase in T-cell numbers and a decrease in concentrations of cytokines and inflammatory markers. The unfavored phenotype of deterioration/ death corresponds to the continuing decrease in T-cell numbers and an increase in concentrations of cytokines and inflammatory markers, pro-inflammatory monocytes, and neutrophils. Eventually, COVID-19 can lead to organ dysfunction or failure and coagulopathy, even in patients who reported no history of immune disorders. For both phenotypes, immunoglobulin levels begin to rise in the second stage.

COVID-19 is associated with increased numbers of monocytes and macrophages, and as the disease progresses from a mild form to a severe/critical condition, the macrophage population becomes denser. Monitoring the expression of cytokines associated with macrophage activation indicates that the transition between disease stages occurs by the end of the first week after onset of symptoms. An adult immunocompetent system fires macrophage missile in two phases at the initial and transition states. At the initial stage, the immune military recognizes the SARS-CoV-2 as nonself and thus fires macrophages at the lungs against the virus. The first flame can control disease progression effectively. However, a trained immunocompetent system would maintain the fire of macrophages over an extended time. It lies in its immune memory in tissue-resident macrophages, especially alveolar macrophages, making a professionally trained immune system more likely to be feared by COVID-19 than an untrained immune system (Xing et al. 2020). In this manner, the trained immunocompetent system commits such a failure that causes the lungs to come down rapidly. The fact that younger age groups, including neonates and children, are less susceptible to COVID-19 than older age groups reflects that the natural affinities of the immune system that has not been trained thoroughly would be standard in combatting against COVID-19 whereas the higher affinities of the trained immune system for rapid activation of immune responses might raise faults – the lungs come down.

8.10 Epidemiological Eye of SARS-CoV-2 Has No Exception to Any Region Worldwide

A quick look at the CDC report of COVID-19 cases reveals that the virus has rapidly spread around the world but at varying rates in different countries from 1 to 1,005,808 cases as of writing this. The exceptions and low rates do not necessarily reflect no cases because test kits are not available as demanded by some countries. Besides, asymptomatic patients and patients who tested negative for the coronavirus worsen the situation by increasing the proportion of concealed carriers who handle viral shedding. Four common causes of negative nucleic acid amplification tests (NAAT) such as RT-PCR include low specimen quality, very early or late collection of the specimen, and technical issues (World Health 2020). Virus gene mutations are of technical issues causing the negative result.

8.11 SARS-CoV-2: An International Problem Starting from Wuhan on December 11, 2019, Spreading to 199 Countries and Territories Around the World and Two International Conveyances as of March 28, 2020

The first case occurred in Wuhan on December 11, 2019. The Guardian reports on the first cases of Thailand, Japan, China, South Korea, Taiwan, and the USA on January 22, 2020; Australia and Canada on January 26, 2020; Italy, Sweden, Russia, and the UK on January 31, 2020; Iran on February 19, 2020; Pakistan, Brazil, Georgia, Greece, North Macedonia, Norway, and Romania on February 26, 2020; Argentina, Chile, Jordan, and Ukraine on March 3, 2020; Albania, Cyprus, and Brunei on March 09, 2020; Kazakhstan, Ethiopia, Sudan, Guinea, Kenya and Antigua, and Barb on March 13, 2020; and Dominica, Grenada, Mozambique, and Syria on March 22, 2020 (Amin et al. 1997).

No doubt, COVID-19 is a challenge for low-income countries and countries with the highest population. However, it has raised concerns worldwide. For example, more than 1 month after the beginning of the outbreak, Italy was the first European country affected by COVID-19. On March 27, 2020, the number of total cases in Italy was similar to that of China (81,285 vs. 80,589). However, the number of total deaths was more than twofold for Italian patients compared to Chinese patients (8215 vs. 3287). By considering the country's population, which is estimated to be about 60.48 million in Italy and 1.386 billion in China, the COVID-19-related deaths correspond to 0.01% of the Italian population and 0.0002% of the Chinese population (>57 times greater in Italy than China). It might be due to the age distribution of the population in Italy and China because people aged 60 or older are the main target of COVID-19 and about 23% of the Italian population is 65 years and older, with a corresponding rate of 12% in China.

8.12 Facilitators of SARS-CoV-2 Transmission

8.12.1 Stability on Surfaces

As summarized in Table 8.1, aerosol and surface stability are not different between the SARS-CoV-1 and SARS-CoV-2 concerning the duration of viability, titers of viable virus, and the half-life of viable virus. However, the transmissibility of SARS-CoV-2 is higher than that of SARS-CoV-1. Some attribute this to the higher tropism of SARS-CoV-2 on the upper respiratory tract and the higher viral shedding in asymptomatic individuals (van Doremalen et al. 2020).

Table 8.1 The stability and viability of SARS-CoV-2 and SARS-CoV-1 in the air and on different surfaces

Parameter	Medium	SARS-CoV-2	SARS-CoV-1
Duration of viability, hours	Aerosols	3	3
	Plastic	72	72
	Stainless steel	48	48
	Copper	4	8
	Cardboard	24	8
Titers of viable virus, $TCID_{50}$ per liter of air or milliliter of medium	Aerosols	$10^{3.5}$–$10^{2.7}$	$10^{4.3}$–$10^{3.5}$
	Plastic	$10^{3.7}$–$10^{0.6}$	$10^{3.4}$–$10^{0.7}$
	Stainless steel	$10^{3.7}$–$10^{0.6}$	$10^{3.6}$–$10^{0.6}$
	Copper	$10^{3.2}$–$10^{1.8}$	$10^{3.4}$–$10^{2.0}$
	Cardboard	$10^{2.8}$–$10^{0.8}$	$10^{2.7}$–$10^{0.8}$
The half-life of viable virus, median (2.5%–97.5% quantile range) hours	Aerosols	1.09 (0.64–2.64)	1.18 (0.78–2.43)
	Plastic	6.81 (5.62–8.17)	7.55 (6.29–9.04)
	Stainless steel	5.63 (4.59–6.86)	4.16 (3.30–5.22)
	Copper	0.77 (0.43–1.19)	1.50 (0.93–2.66)
	Cardboard	3.46 (2.34–5.00)	0.59 (0.32–1.21)

8.12.2 Multiple Shedding Routes

A study of 15 patients with COVID-19 detected the SARS-CoV-2 on day 0 in oral swab ($n = 8$), anal swab ($n = 4$), blood ($n = 6$), and serum ($n = 3$) (Zhang et al. 2020d). All serum positive patients showed SARS-CoV-2 in the blood, and all blood positive patients had no positive swabs. Two of eight oral swab-positive patients showed the SARS-CoV-2 in the anal swab. There were three cases with a severe form of the disease, of whom two were serum positive and one was oral swab positive. On day 5, the rate of oral swab positive was reduced by 50% ($n = 4$), and the rate of anal swab positive increased by 25% ($n = 6$). These statistics imply three points about the SARS-CoV-2: a. it involves the respiratory system as well as the gastrointestinal system, b. it tends primarily to enter the upper respiratory system, and c. it, in severe cases, spread via the bloodstream.

8.12.3 The Near-Presentation Detection of Viral RNA Gives Us No Opportunity to Recognize Asymptomatic People

Compared to asymptomatic patients, patients display high viral loads in the throat and to a higher level in the nose immediately after the symptoms appear (Zou et al. 2020). Besides, compared to patients with a mild-to-moderate form of the disease, patients with a severe form of the disease have lower cycle threshold (Ct) values of Orf1b on RT-PCR, which, in turn, correspond to higher viral RNA copy number. Precisely, samples with a Ct value of 40 tested negative and the Ct values of 30.76, 27.67, 24.56, and 21.48 assigned to the viral RNA copies per milliliter of 1.5 × 10^4, 1.5 × 10^5, 1.5 × 10^6, and 1.5 × 10^7, respectively.

8.12.4 Children Contribute Significantly to the Asymptomatic Population of COVID-19

A retrospective study of 36 children with COVID-19 showed that about half of children had a moderate form of disease commonly presented with fever and dry cough and the other half were asymptomatic (Qiu et al.). The proportion of asymptomatic patients is a determining factor affecting human-to-human transmission and, therefore, the efficiency level of interventions. Such a high asymptomatic proportion of children infected with COVID-19 make children as a facilitator of transmission (Kelvin and Halperin).

8.12.5 The High Prevalence of SARS-CoV and MERS-CoV in Animals

A meta-analysis calculated the pooled prevalence of MERS-CoV and SARS-CoV detected by RT-PCR to be higher than 10% and 14% among camels and bats, respectively (Bonilla-Aldana et al. 2020). Having such a large-scale natural reservoir paves the way for coronaviruses to sustain their survival, and therefore, new strains continue to emerge.

8.13 Mathematical Models from Dynamics of the Primary Coming of COVID-19 to Preparedness for the Next Coming of COVID-19

Mathematical models can estimate current and future prevalence rates and allow them to evaluate the efficiency of different interventions on the projected prevalence. For example, Wuhan adopted the strategy of travel restrictions for the control of the spread of SARS-CoV-2 on January 23. The mathematical model calculated that whereas one individual with SARS-CoV-2 infection was able to produce the average number of 2.35 new infections per day at 1 week before the date of implementation of the strategy adopted, the infectivity rate dropped to 1.05 at 1 week after that date (Kucharski et al. 2020).

Using simulation (Prem et al.), Prem et al. assessed the effect of physical distancing plans aiming to limit social mixing on the prevention of secondary outbreak of SARS-CoV-2 in 2020. The extension of school closures and working holidays could induce a significant decrease in the cumulative infections by end-2020 and the peak incidence and a delay in the peak of the outbreak. When returning to work set out to at the beginning of April, we could witness the best effect of physical distancing correlated with greater than 90% and 20% reduction in the median incidence by mid-2020 and end-2020, and the biggest effects occurred among school-aged children and older people. People are permitted to return to work on March 10. So, there is a concern about the magnitude and peak of the second outbreak of COVID-19.

Koo and colleagues (Koo et al.) presented a simulation modeling of SARS-CoV-2 human-to-human transmission in Singapore. They investigated the effect of four interventions (I1, quarantine; I2, quarantine plus school closure; I3, quarantine plus workplace distancing; I4, the combined intervention of quarantine, school closure, and workplace distancing) versus no intervention on the simulated transmission of SARS-CoV-2 under different rates of infectivity and asymptomatic infections. With the assumption of the asymptomatic ratio of 7.5% and R_0 of 1.5–2.5, the median cumulative number of infected individuals at day 80 was estimated to fall within the ranges of 279,000–1,207,000 for no intervention, 15000–520,000 for I1, 10,000–466,000 for I2, 4000–320,000 for I3, and 1800–258,000 for I4. Though the first local transmission could be successfully aflected, this model leading to the recommendation of the combined strategy to control the early spread of SARS-CoV-2 in Singapore would help with preparedness for the second coming of SARS-CoV-2.

8.14 National Healthcare Services Close to Kneeling During the COVID-19 Pandemic: The Need for an International Solution Rather than a Merely National Solution

With the emergence of COVID-19, national healthcare systems have come face to face with their weaknesses, calling for a variety of remedies.

Continuing from Italy – where the healthcare system fears it is going to collapse in regions worst impacted by COVID-19 – there is a substantial need for timely intervention, national coordination, flexible financing, public-private partnerships, and human resource management (Armocida et al.).

Spain, with the first cases reported on February 15, 2020, was the second European country affected by COVID-19. On March 27, 2020, 4365 people had died out of 57,786 cases, meaning the death rate is greater than 7%. A recent crisis of cases with about 7000 cases per day for the last 4 days has made the very challenging times for the Spanish healthcare system concerning different aspects, including the governance, finance, service delivery, medicines and equipment, health workers, and media (Legido-Quigley et al. 2020).

The top of the CDC report has been assigned to the USA, with more than 10,000 cases being detected per day since March 23, 2020. Projections by the IHME (IHME) provided information that the 2019-nCoV pandemic would reach its peak in the second week of April and that the crisis of daily deaths would be on April 14, 2020 (team). For the preparedness for the peak of deaths, the USA's hospital system planned to reduce the demand for elective care and devoted the system to the care of patients with COVID-19.

The UK was an example of a currently low-incidence country with a total of about 17,000 cases and 1000 death as of March 28, 2020. When the number of cases per day was increasing day-by-day, NHS workers raised their concerns about the imminent crisis of deaths due to lack of proper testing of symptomatic cases, contact tracing, and quarantine (Horton 2020). The UK has now placed sixth in the CDC report, with more than 150,000 confirmed cases and 20,000 deaths as of April 28, 2020. The UK has published a plan of a national program of community health workers that aims to provide supportive care for people who are especially vulnerable to COVID-19, e.g., older people and people with preexisting conditions (Haines et al.). It is featured by its large scale and offers in-home care or virtual services to address people's needs. The program will be held by young community health workers (CHW), who can gain immunity after being exposed to the virus, and if they get sick, their immunity will hopefully guide them to be healthy again. CHWs have the responsibility to find those who need advanced care. They will have in-home visits to get vital signs such as temperature, blood pressure, and oxygen saturation and train people on hygiene issues.

The above examples lead us to the conclusion that distinct nations share similar needs while requiring different means corresponding to different demands. The development and implementation of strategies for dealing with the 2019-nCoV pandemic rely on financial incentives as well as nonfinancial incentives. Now, all feel much stress on their knees, from politicians to healthcare workers; international collaboration can enforce nonfinancial incentives more than ever it did. It might be that the eye of COVID-19 could not expect from humanities.

8.15 Conclusion

The Wuhan-born novel coronavirus is only 4 months old, but it has affected more than 3 million people worldwide. With a case fatality of 6.9%, the novel coronavirus disease, COVID-19, has claimed more than 200,000 lives, whereas the previous outbreaks of SARS and MERS by coronaviruses caused a total of fewer than 2000 deaths. Besides, coronaviruses have introduced diseases to humans for 60 years ago, but none of them could spread such rapidly.

Older age groups have appeared especially vulnerable during the outbreak of COVID-19. Interestingly, there is a direct association between age and peak viral loads. By contrast, no correlation exists between disease severity and viral loads. It reveals to us that the novel coronavirus would replicate to a similar degree in patients with severe and non-severe forms of disease and that what divides patients into severe and non-severe groups is the individual response to the infection. Moreover, age has a determining role in individual response to COVID-19.

The present chapter provided evidence that the COVID-19 can escape the immune system due to its power to profoundly diminish T-cell effective function, which is necessary for the establishment of a robust antiviral immunity. Also, it characterized a lethal model of disease for COVID-19 through macrophage missiles and cytokine storm by the trained immunocompetent system (Rokni et al. 2020). The fact that younger age groups, including neonates and children, are less susceptible to COVID-19 than older age groups reflects that the natural affinities of the immune system that has not been trained thoroughly would be standard in combatting against COVID-19 whereas the higher affinities of the trained immune system for rapid activation of immune responses might raise faults – the lungs come down.

Early or late, COVID-19 has affected more than 210 countries as of writing this. Simulations suggest that the world should be prepared for the next COVID-19 outbreak, while many countries are combatting and national healthcare services close to kneeling during the first COVID-19 outbreak. No one can entertain doubt in the necessity of national solutions. However, the causative agent of the COVID-19 outbreak is a doubtful species that has captured the globe without exception. No national solution can change the fact that humans have a limited tolerance to physical distancing strategies and travel restrictions. A universal solution that aims to solve the problem of COVID-19 worldwide would be helpful to all affected nations (Mohamed et al. 2020a; Momtazmanesh et al. 2020; Moradian et al. 2020).

References

Ahanchian H, Moazzen N, Faroughi MSD, Khalighi N, Khoshkhui M, Aelami MH, Haghi NSM, Rezaei N (2020) COVID-19 in a child with primary specific antibody deficiency

Ahmadi M, Saffarzadeh N, Habibi MA, Hajiesmaeili F, Rezaei N (2020) Colon cancer and SARS-CoV-2: impact of ACE2 expression in susceptibility to COVID-19. bioRxiv

Amin RH, Frank RN, Kennedy A, Eliott D, Puklin JE, Abrams GW (1997) Vascular endothelial growth factor is present in glial cells of the retina and optic nerve of human subjects with nonproliferative diabetic retinopathy. Invest Ophthalmol Vis Sci 38(1):36–47

Andersen KG, Rambaut A, Lipkin WI, Holmes EC, Garry RF (2020) The proximal origin of SARS-CoV-2. ARTIC Network 17

Babaha F, Rezaei N (2020) Primary immunodeficiency diseases in COVID-19 pandemic: a predisposing or protective factor? Am J Med Sci. https://doi.org/10.1016/j.amjms.2020.07.027

Bahrami A, Vafapour M, Moazzami B, Rezaei N (2020) Hyperinflammatory shock related to COVID-19 in a patient presenting with multisystem inflammatory syndrome in children: first case from Iran. J Paediatr Child Health. https://doi.org/10.1111/jpc.15048

Basiri A, Heidari A, Nadi MF, Fallahy MTP, Nezamabadi SS, Sedighi M, Saghazadeh A, Rezaei N (2020a) Microfluidic devices for detection of RNA viruses. Reviews in Medical Virology n/a (n/a):e2154. https://doi.org/10.1002/rmv.2154

Basiri A, Pazhouhnia Z, Beheshtizadeh N, Hoseinpour M, Saghazadeh A, Rezaei N (2020b) Regenerative medicine in COVID-19 treatment: real opportunities and range of promises. Stem Cell Rev Rep:1–13. https://doi.org/10.1007/s12015-020-09994-5

Bonilla-Aldana DK, Cardona-Trujillo MC, García-Barco A, Holguin-Rivera Y, Cortes-Bonilla I, Bedoya-Arias HA, Patiño-Cadavid LJ, Paniz-Mondolfi A, Zambrano LI, Dhama K (2020) MERS-CoV and SARS-CoV infections in animals: a systematic review and meta-analysis of prevalence studies

Brodsky IE, Medzhitov R (2011) Pyroptosis: macrophage suicide exposes hidden invaders. Curr Biol 21(2):R72–R75

Cai Y, Hao Z, Gao Y, Ping W, Wang Q, Peng S, Zhao B, Sun W, Zhu M, Li K (2020) Clinical characteristics of seven cases infected with SARS-CoV-2 in the perioperative period of lung resection: a retrospective study from a single thoracic Department in Wuhan, China

Chen N, Zhou M, Dong X, Qu J, Gong F, Han Y, Qiu Y, Wang J, Liu Y, Wei Y (2020a) Epidemiological and clinical characteristics of 99 cases of 2019 novel coronavirus pneumonia in Wuhan, China: a descriptive study. Lancet 395(10223):507–513

Chen X, Zhao B, Qu Y, Chen Y, Xiong J, Feng Y, Men D, Huang Q, Liu Y, Yang B (2020b) Detectable serum SARS-CoV-2 viral load (RNAaemia) is closely associated with drastically elevated interleukin 6 (IL-6) level in critically ill COVID-19 patients. medRxiv

Darbeheshti F, Rezaei N (2020) Genetic predisposition models to COVID-19 infection. Med Hypotheses 142:109818. https://doi.org/10.1016/j.mehy.2020.109818

Diao B, Wang C, Tan Y, Chen X, Liu Y, Ning L, Chen L, Li M, Liu Y, Wang G (2020) Reduction and Functional Exhaustion of T Cells in Patients with Coronavirus Disease 2019 (COVID-19). medRxiv

Doitsh G, Galloway NLK, Geng X, Yang Z, Monroe KM, Zepeda O, Hunt PW, Hatano H, Sowinski S, Muñoz-Arias I (2014) Cell death by pyroptosis drives CD4 T-cell depletion in HIV-1 infection. Nature 505(7484):509–514

Fathi N, Rezaei N (2020) Lymphopenia in COVID-19: therapeutic opportunities. Cell Biol Int 44(9):1792–1797. https://doi.org/10.1002/cbin.11403

Franks TJ, Chong PY, Chui P, Galvin JR, Lourens RM, Reid AH, Selbs E, McEvoy CPL, Hayden CDL, Fukuoka J (2003) Lung pathology of severe acute respiratory syndrome (SARS): a study of 8 autopsy cases from Singapore. Hum Pathol 34(8):743–748

Guan W-j, Ni Z-y, Hu Y, Liang W-h, Ou C-q, He J-x, Liu L, Shan H, Lei C-I, Hui DSC (2020) Clinical Characteristics of Coronavirus Disease 2019 in China. N Engl J Med

Hanaei S, Rezaei N (2020) COVID-19: developing from an outbreak to a pandemic. Arch Med Res. https://doi.org/10.1016/j.arcmed.2020.04.021

Horton R (2020) Offline: COVID-19 and the NHS-"a national scandal". Lancet 395(10229):1022. https://doi.org/10.1016/S0140-6736(20)30727-3

Huang C, Wang Y, Li X, Ren L, Zhao J, Hu Y, Zhang L, Fan G, Xu J, Gu X, Cheng Z, Yu T, Xia J, Wei Y, Wu W, Xie X, Yin W, Li H, Liu M, Xiao Y, Gao H, Guo L, Xie J, Wang G, Jiang R, Gao Z, Jin Q, Wang J, Cao B (2020) Clinical features of patients infected with 2019 novel coronavirus in Wuhan, China. Lancet 395(10223):497–506. https://doi.org/10.1016/S0140-6736(20)30183-5

Jabbari P, Rezaei N (2020) With risk of reinfection, is COVID-19 here to stay? Disaster Med Public Health Prep:1. https://doi.org/10.1017/dmp.2020.274

Jabbari P, Jabbari F, Ebrahimi S, Rezaei N (2020) COVID-19: a chimera of two pandemics. Disaster Med Public Health Prep:1–2. https://doi.org/10.1017/dmp.2020.223

Jahanshahlu L, Rezaei N (2020a) Central nervous system involvement in COVID-19. Arch Med Res. https://doi.org/10.1016/j.arcmed.2020.05.016

Jahanshahlu L, Rezaei N (2020b) Monoclonal antibody as a potential anti-COVID-19. Biomed Pharmacother 129:110337. https://doi.org/10.1016/j.biopha.2020.110337

Ji W, Wang W, Zhao X, Zai J, Li X (2020) Homologous recombination within the spike glycoprotein of the newly identified coronavirus may boost cross-species transmission from snake to human. J Med Virol

Jiang Y, Li J, Teng Y, Sun H, Tian G, He L, Li P, Chen Y, Guo Y, Li J (2019) Complement receptor C5aR1 inhibition reduces pyroptosis in hDPP4-transgenic mice infected with MERS-CoV. Viruses 11(1):39

Jiang W, Zeng Z, Fan D, Zhang Y, Chen Z, Jiang W, Xie S, Xia H, Jiang Y-A, Liu X-A (2020) Decreased counts of T lymphocyte subsets predict prognosis in SARS-CoV-2-infected pneumonia in Wuhan, China: A Retrospective Study (February 20, 2020)

Kacprzyk J, Hughes GM, Palsson-McDermott EM, Quinn SR, Puechmaille SJ, O'Neill LAJ, Teeling EC (2017) A potent anti-inflammatory response in bat macrophages may be linked to extended longevity and viral tolerance. Acta Chiropterologica 19(2):219–228

Kafieh R, Arian R, Saeedizadeh N, Minaee S, Yadav SK, Vaezi A, Rezaei N, Javanmard SH (2020) COVID-19 in Iran: a deeper look into the future. medRxiv

Keidar S, Strizevsky A, Raz A, Gamliel-Lazarovich A (2007) ACE2 activity is increased in monocyte-derived macrophages from prehypertensive subjects. Nephrol Dial Transplant 22(2):597–601

Kucharski AJ, Russell TW, Diamond C, Liu Y, Edmunds J, Funk S, Eggo RM, Sun F, Jit M, Munday JD (2020) Early dynamics of transmission and control of COVID-19: a mathematical modelling study. Lancet Infect Dis

Legido-Quigley H, Mateos-García JT, Campos VR, Gea-Sánchez M, Muntaner C, McKee M (2020) The resilience of the Spanish health system against the COVID-19 pandemic. The Lancet Public Health

Li W, Shi Z, Yu M, Ren W, Smith C, Epstein JH, Wang H, Crameri G, Hu Z, Zhang H (2005) Bats are natural reservoirs of SARS-like coronaviruses. Science 310(5748):676–679

Li B, Feng F, Yang G, Liu A, Yang N, Jiang Q, Zhang H, Wang T, Li P, Mao Y (2020a) Immunoglobulin G/M and cytokines detections in continuous sera from patients with novel coronaviruses (2019-nCoV) infection. Available at SSRN 3543609

Li L, Li S, Xu M, Zheng S, Duan Z, Chen Y, Li J (2020b) The level of plasma C-reactive protein is closely related to the liver injury in patients with COVID-19. medRxiv

Liao M, Liu Y, Yuan J, Wen Y, Xu G, Zhao J, Chen L, Li J, Wang X, Wang F (2020) The landscape of lung bronchoalveolar immune cells in COVID-19 revealed by single-cell RNA sequencing. medRxiv

Liu J, Li S, Liu J, Liang B, Wang X, Li W, Wang H, Tong Q, Yi J, Zhao L (2020) Longitudinal characteristics of lymphocyte responses and cytokine profiles in the peripheral blood of SARS-CoV-2 infected patients

Lotfi M, Rezaei N (2020) SARS-CoV-2: a comprehensive review from pathogenicity of the virus to clinical consequences. J Med Virol. https://doi.org/10.1002/jmv.26123

Lotfi M, Hamblin MR, Rezaei N (2020) COVID-19: transmission, prevention, and potential therapeutic opportunities. Clinica Chimica Acta 508:254–266. https://doi.org/10.1016/j.cca.2020.05.044

Luo W, Yu H, Gou J, Li X, Sun Y, Li J, Liu L (2020) Clinical pathology of critical patient with novel coronavirus pneumonia (COVID-19)

Mansourabadi AH, Sadeghalvad M, Mohammadi-Motlagh HR, Rezaei N (2020) The immune system as a target for therapy of SARS-CoV-2: a systematic review of the current immunotherapies for COVID-19. Life Sci 258:118185. https://doi.org/10.1016/j.lfs.2020.118185

Menachery VD, Yount BL Jr, Debbink K, Agnihothram S, Gralinski LE, Plante JA, Graham RL, Scobey T, Ge X-Y, Donaldson EF (2015) A SARS-like cluster of circulating bat coronaviruses shows potential for human emergence. Nat Med 21(12):1508

Millán-Oñate J, Rodriguez-Morales AJ, Camacho-Moreno G, Mendoza-Ramírez H, Rodríguez-Sabogal IA, Álvarez-Moreno C (2020) A new emerging zoonotic virus of concern: the 2019 novel coronavirus (COVID-19). Infection 24(3)

Mirbeyk M, Rezaei N (2020) The impact of COVID-19 on pregnancy and neonatal health: a systematic review

Moazzami B, Razavi-Khorasani N, Dooghaie Moghadam A, Farokhi E, Rezaei N (2020) COVID-19 and telemedicine: immediate action required for maintaining healthcare providers well-being. J Clin Virol 126:104345. https://doi.org/10.1016/j.jcv.2020.104345

Mohamed K, Rodríguez-Román E, Rahmani F, Zhang H, Ivanovska M, Makka SA, Joya M, Makuku R, Islam MS, Radwan N, Rahmah L, Goda R, Abarikwu SO, Shaw M, Zoghi S, Irtsyan S, Ling I, Cseprekal O, Faten AB, Hazar Sayar E, Soloukey C, Grancini G, Rezaei N (2020a) Borderless collaboration is needed for COVID-19-A disease that knows no borders. Infect Control Hosp Epidemiol:1–2. https://doi.org/10.1017/ice.2020.162

Mohamed K, Yazdanpanah N, Saghazadeh A, Rezaei N (2020b) Computational drug discovery and repurposing for the treatment of Covid-19: a systematic review. Available at SSRN 3583748

Momtazmanesh S, Ochs HD, Uddin LQ, Perc M, Routes JM, Vieira DN, Al-Herz W, Baris S, Prando C, Rosivall L, Abdul Latiff AH, Ulrichs T, Roudenok V, Aldave Becerra JC, Salunke DB, Goudouris E, Condino-Neto A, Stashchak A, Kryvenko O, Stashchak M, Bondarenko A, Rezaei N (2020) All together to fight COVID-19. Am J Trop Med Hyg 102(6):1181–1183. https://doi.org/10.4269/ajtmh.20-0281

Moradian N, Ochs HD, Sedikies C, Hamblin MR, Camargo CA Jr, Martinez JA, Biamonte JD, Abdollahi M, Torres PJ, Nieto JJ, Ogino S, Seymour JF, Abraham A, Cauda V, Gupta S, Ramakrishna S, Sellke FW, Sorooshian A, Wallace Hayes A, Martinez-Urbistondo M, Gupta M, Azadbakht L, Esmaillzadeh A, Kelishadi R, Esteghamati A, Emam-Djomeh Z, Majdzadeh R, Palit P, Badali H, Rao I, Saboury AA, Jagan Mohan Rao L, Ahmadieh H, Montazeri A, Fadini GP, Pauly D, Thomas S, Moosavi-Movahed AA, Aghamohammadi A, Behmanesh M, Rahimi-Movaghar V, Ghavami S, Mehran R, Uddin LQ, Von Herrath M, Mobasher B, Rezaei N (2020) The urgent need for integrated science to fight COVID-19 pandemic and beyond. J Transl Med 18(1):205. https://doi.org/10.1186/s12967-020-02364-2

Munster V, Feldmann F, Williamson B, Van Doremalen N, Perez-Perez L, Schultz J, Meade-White K, Okumura A, Callison J, Brumbaugh B (2020) Respiratory disease and virus shedding in rhesus macaques inoculated with SARS-CoV-2. BioRxiv

Nasab MG, Saghazadeh A, Rezaei N (2020) SARS-CoV-2-A tough opponent for the immune system. Arch Med Res. https://doi.org/10.1016/j.arcmed.2020.05.020

Nicholls JM, Poon LLM, Lee KC, Ng WF, Lai ST, Leung CY, Chu CM, Hui PK, Mak KL, Lim W (2003) Lung pathology of fatal severe acute respiratory syndrome. Lancet 361(9371):1773–1778

Pashaei M, Rezaei N (2020) Immunotherapy for SARS-CoV-2: potential opportunities. Expert Opin Biol Ther:1–5. https://doi.org/10.1080/14712598.2020.1807933

Pourahmad R, Moazzami B, Rezaei N (2020) Efficacy of plasmapheresis and immunoglobulin replacement therapy (IVIG) on patients with COVID-19. SN Comprehensive Clin Med:1–5

Rabiee N, Rabiee M, Bagherzadeh M, Rezaei N (2020) COVID-19 and picotechnology: potential opportunities. Med Hypotheses 144:109917. https://doi.org/10.1016/j.mehy.2020.109917

Rezaei N (2020a) COVID-19 affects healthy pediatricians more than pediatric patients. Infect Control Hosp Epidemiol 1. https://doi.org/10.1017/ice.2020.139

Rezaei N (2020b) COVID-19 and medical biotechnology. Avicenna J Med Biotechnol 12(3):139

Rokni M, Hamblin MR, Rezaei N (2020) Cytokines and COVID-19: friends or foes? Human Vaccines & Immunotherapeutics:1–3

Rzymski P, Nowicki M, Mullin GE, Abraham A, Rodríguez-Román E, Petzold MB, Bendau A, Sahu KK, Ather A, Naviaux A-F (2020) Quantity does not equal quality: scientific principles cannot be sacrificed. Int Immunopharmacol 86:106711

Saghazadeh A, Rezaei N (2020a) Immune-epidemiological parameters of the novel coronavirus - a perspective. Expert Rev Clin Immunol 16(5):465–470. https://doi.org/10.1080/1744666x.2020.1750954

Saghazadeh A, Rezaei N (2020b) Towards treatment planning of COVID-19: rationale and hypothesis for the use of multiple immunosuppressive agents: anti-antibodies, immunoglobulins, and corticosteroids. Int Immunopharmacol 84:106560. https://doi.org/10.1016/j.intimp.2020.106560

Sahu KK, Siddiqui AD, Rezaei N, Cerny J (2020) Challenges for management of immune thrombocytopenia during COVID-19 pandemic. J Med Virol. https://doi.org/10.1002/jmv.26251

Saleki K, Banazadeh M, Saghazadeh A, Rezaei N (2020) The involvement of the central nervous system in patients with COVID-19. Rev Neurosci 31(4):453–456. https://doi.org/10.1515/revneuro-2020-0026

Shamshirian D, Rezaei N (2020) Cardiovascular diseases burden in COVID-19: systematic review and meta-analysis

Sharifkashani S, Bafrani MA, Khaboushan AS, Pirzadeh M, Kheirandish A, Yavarpour Bali H, Hessami A, Saghazadeh A, Rezaei N (2020) Angiotensin-converting enzyme 2 (ACE2) receptor and SARS-CoV-2: potential therapeutic targeting. Eur J Pharmacol 884:173455. https://doi.org/10.1016/j.ejphar.2020.173455

Shi C-S, Nabar NR, Huang N-N, Kehrl JH (2019) SARS-coronavirus open reading frame-8b triggers intracellular stress pathways and activates NLRP3 inflammasomes. Cell Death Dis 5(1):1–12

Shimabukuro-Vornhagen A, Gödel P, Subklewe M, Stemmler HJ, Schlößer HA, Schlaak M, Kochanek M, Böll B, von Bergwelt-Baildon MS (2018) Cytokine release syndrome. J Immunother Cancer 6(1):56–56. https://doi.org/10.1186/s40425-018-0343-9

Tian S, Hu W, Niu L, Liu H, Xu H, Xiao S-Y (2020) Pulmonary pathology of early phase 2019 novel coronavirus (COVID-19) pneumonia in two patients with lung cancer. J Thorac Oncol

van Doremalen N, Bushmaker T, Morris DH, Holbrook MG, Gamble A, Williamson BN, Tamin A, Harcourt JL, Thornburg NJ, Gerber SI, Lloyd-Smith JO, de Wit E, Munster VJ (2020) Aerosol and surface stability of SARS-CoV-2 as compared with SARS-CoV-1. N Engl J Med. https://doi.org/10.1056/NEJMc2004973

Walle LV, Lamkanfi M (2016) Pyroptosis. Curr Biol 26(13):R568–R572

Wang D, Hu B, Hu C, Zhu F, Liu X, Zhang J, Wang B, Xiang H, Cheng Z, Xiong Y (2020a) Clinical characteristics of 138 hospitalized patients with 2019 novel coronavirus–infected pneumonia in Wuhan, China. Jama

Wang H, Luo S, Shen Y, Li M, Zhang Z, Dong Y, Zhou H, Lin L, Guo W, Kang Z (2020b) Multiple enzyme release, inflammation storm and hypercoagulability are prominent indicators for disease progression in COVID-19: a multi-centered, correlation study with CT imaging score

Wang W, He J, Wu S (2020c) The definition and risks of cytokine release syndrome-like in 11 COVID-19-infected pneumonia critically ill patients: disease characteristics and retrospective analysis. medRxiv

World Health Organization (2020) Laboratory testing for coronavirus disease 2019 (COVID-19) in suspected human cases: interim guidance, March 2 2020. World Health Organization.

Wu Z, McGoogan JM (2020) Characteristics of and important lessons from the coronavirus disease 2019 (COVID-19) outbreak in China: summary of a report of 72 314 cases from the Chinese Center for Disease Control and Prevention. Jama

Xing Z, Afkhami S, Bavananthasivam J, Fritz DK, D'Agostino MR, Vaseghi-Shanjani M, Yao Y, Jeyanathan M (2020) Innate immune memory of tissue-resident macrophages and trained innate immunity: re-vamping vaccine concept and strategies. J Leukoc Biol

Xiong Y, Liu Y, Cao L, Wang D, Guo M, Guo D, Hu W, Yang J, Tang Z, Zhang Q (2020) Transcriptomic characteristics of bronchoalveolar lavage fluid and peripheral blood mononuclear cells in COVID-19 patients. Available at SSRN 3549993

Xu H, Hou K, Xu H, Li Z, Chen H, Zhang N, Xu R, Fu H, Sun R, Wen L (2020a) Acute myocardial injury of patients with coronavirus disease 2019. medRxiv

Xu Z, Shi L, Wang Y, Zhang J, Huang L, Zhang C, Liu S, Zhao P, Liu H, Zhu L (2020b) Pathological findings of COVID-19 associated with acute respiratory distress syndrome. Lancet Respir Med

Yang X, Yu Y, Xu J, Shu H, Liu H, Wu Y, Zhang L, Yu Z, Fang M, Yu T (2020a) Clinical course and outcomes of critically ill patients with SARS-CoV-2 pneumonia in Wuhan, China: a single-centered, retrospective, observational study. Lancet Respir Med

Yang Y, Shen C, Li J, Yuan J, Yang M, Wang F, Li G, Li Y, Xing L, Peng L (2020b) Exuberant elevation of IP-10, MCP-3 and IL-1ra during SARS-CoV-2 infection is associated with disease severity and fatal outcome. medRxiv

Yazdanpanah F, Hamblin MR, Rezaei N (2020a) The immune system and COVID-19: friend or foe? Life Sci 256:117900. https://doi.org/10.1016/j.lfs.2020.117900

Yazdanpanah N, Saghazadeh A, Rezaei N (2020b) Anosmia: a missing link in the neuroimmunology of coronavirus disease 2019 (COVID-19). Rev Neurosci. https://doi.org/10.1515/revneuro-2020-0039

Yousefzadegan S, Rezaei N (2020) Case Report: death due to COVID-19 in Three Brothers. Am J Trop Med Hyg 102(6):1203–1204. https://doi.org/10.4269/ajtmh.20-0240

Zhang D, Guo R, Lei L, Liu H, Wang Y, Wang Y, Dai T, Zhang T, Lai Y, Wang J (2020a) COVID-19 infection induces readily detectable morphological and inflammation-related phenotypic changes in peripheral blood monocytes, the severity of which correlate with patient outcome. medRxiv

Zhang L, Shen FM, Chen F, Lin Z (2020b) Origin and evolution of the 2019 novel coronavirus. Clin Infect Dis

Zhang L, Yang J-R, Zhang Z, Lin Z (2020c) Genomic variations of COVID-19 suggest multiple outbreak sources of transmission. medRxiv

Zhang W, Du R-H, Li B, Zheng X-S, Yang X-L, Hu B, Wang Y-Y, Xiao G-F, Yan B, Shi Z-L (2020d) Molecular and serological investigation of 2019-nCoV infected patients: implication of multiple shedding routes. Emerg Microbes Infect 9(1):386–389

Zhou Y, Fu B, Zheng X, Wang D, Zhao C, Sun R, Tian Z, Xu X, Wei H (2020) Pathogenic T cells and inflammatory monocytes incite inflammatory storm in severe COVID-19 patients. Natl Sci Rev

Zou L, Ruan F, Huang M, Liang L, Huang H, Hong Z, Yu J, Kang M, Song Y, Xia J (2020) SARS-CoV-2 viral load in upper respiratory specimens of infected patients. N Engl J Med 382(12):1177–1179

Potential Antiviral Immune Response Against COVID-19: Lessons Learned from SARS-CoV

9

Mahzad Akbarpour, Laleh Sharifi, Amir Reza Safdarian, Pooya Farhangnia, Mahdis Borjkhani, and Nima Rezaei

Abstract

Virus and host innate immune system interaction plays a significant role in forming the outcome of viral diseases. Host innate immunity initially recognizes the viral invasion and induces a rapid inflammatory response, and this recognition activates signaling cascades that trigger the release of antiviral mediators. This chapter aims to explore the mechanisms by which newly emerged coronavirus called severe acute respiratory syndrome coronavirus 2 (SARS-CoV-2) activates the host immune system. Since SARS-CoV-2 shares similarities with SARS-CoV that caused the epidemic of SARS in 2003, the pathogenesis of both viruses could be at least very similar. For this, this chapter provides a synthesis of literature concerning antiviral immunity in SARS-CoV and SARS-CoV-2. It includes the presentation of epitopes linked to SARS-CoV-2 as well as the ability of SARS-CoV-2 to cause proteolytic activation and interact with angiotensin-converting enzyme 2 (ACE2) via molecular mimicry. This chapter characterizes various mechanisms that this virus may engage in escaping the host immunity, ended by a discussion of humoral immune responses against SARS-CoV-2.

Laleh Sharifi, Amir Reza Safdarian and Pooya Farhangnia contributed equally with all other contributors. M. Akbarpour and N. Rezaei share senior authorship.

M. Akbarpour (✉)
Department of Surgery, Northwestern University Feinberg School of Medicine, Chicago, IL, USA

Network of Immunity in Infection, Malignancy and Autoimmunity (NIIMA), Universal Scientific Education and Research Network (USERN), Tehran, Iran

Immunology Board for Transplantation and Cell-Based Therapeutics (ImmunoTACT), Universal Scientific Education and Research Network (USERN), Tehran, Iran
e-mail: mahzad.akbarpour@northwestern.edu

L. Sharifi
Network of Immunity in Infection, Malignancy and Autoimmunity (NIIMA), Universal Scientific Education and Research Network (USERN), Tehran, Iran

Uro-Oncology Research Center, Tehran University of Medical Sciences, Tehran, Iran

A. R. Safdarian
Immunology Board for Transplantation and Cell-Based Therapeutics (ImmunoTACT), Universal Scientific Education and Research Network (USERN), Tehran, Iran

N. Rezaei (ed.), *Coronavirus Disease - COVID-19*, Advances in Experimental Medicine and Biology 1318, https://doi.org/10.1007/978-3-030-63761-3_9

Keywords

COVID-19 · Epitope spreading · Humoral response · Innate immunity · SARS-CoV- 2

9.1 Introduction

Coronaviruses (CoV) have been the cause of three twenty-first-century infectious epidemics: severe acute respiratory syndrome (SARS), Middle East respiratory syndrome (MERS), and coronavirus disease 2019 (COVID-19). SARS-CoV-2, which is known to cause COVID-19, is similar to the SARS-CoV, the causative agent of SARS, in many aspects. Notably, both these viruses bind to the same cell surface receptor angiotensin-converting enzyme 2 (ACE2). For this, this chapter provides a synthesis of literature concerning antiviral immunity in SARS-CoV and SARS-CoV-2. In particular, we review the current evidence of potential mechanisms by which SARS-CoV-2 activates the host immune response, with emphasis on viral antigen presentation, immunodominant epitopes, and possible molecular mimicry characteristics. It will finally summarize antibody-mediated immune responses against SARS-CoV-2 and their diagnostic and clinical value.

9.2 Innate Immune Recognition of Coronaviruses

Interactions between the virus and the host innate immune system play a significant role in determining the outcome of viral diseases. During the early phases of viral invasion, innate immunity can limit the virus replication and control its spread within the host by the production of type I IFNs (Saghazadeh and Rezaei 2017), complement, and other mediators (Katze et al. 2002). The host defense and clearance of viral infections profoundly depend on IFN expression (Ben Addi et al. 2008). Expression of IFNs and subsequent downstream signals result in reprogramming cells into an antiviral state, infection control, and viral clearance (Lazear et al. 2019). In this phase, overactivation of antiviral responses by innate immunity can lead to tissue damage (Garcia-Sastre and Biron 2006). Also, initial innate immune responses play a role in determining adaptive immune responses. Therefore, comprehensive knowledge of how the host innate immune system counteracts the viruses is critical for understanding the signaling cascades regulating pathogenesis, virulence, and disease outcomes.

Invasion of viruses is initially recognized by innate immunity that induces a rapid inflammatory response. This recognition activates signaling cascades that, in turn, trigger the release of antiviral mediators (Takeuchi and Akira 2007).

Department of Pathology, School of Medicine, Alborz University of Medical Sciences, Alborz, Iran

Department of Immunology and Laboratory Medicine, Ghaem Hospital, Alborz, Iran

Department of Immunology, Cancer Biomedical Center, Tehran, Iran

P. Farhangnia
Network of Immunity in Infection, Malignancy and Autoimmunity (NIIMA), Universal Scientific Education and Research Network (USERN), Tehran, Iran

Immunology Board for Transplantation and Cell-Based Therapeutics (ImmunoTACT), Universal Scientific Education and Research Network (USERN), Tehran, Iran

Department of Immunology, School of Medicine, Iran University of Medical Sciences, Tehran, Iran

M. Borjkhani
Immunology Board for Transplantation and Cell-Based Therapeutics (ImmunoTACT), Universal Scientific Education and Research Network (USERN), Tehran, Iran

National Institute of Genetic Engineering and Biotechnology (NIGEB), Tehran, Iran

N. Rezaei
Network of Immunity in Infection, Malignancy and Autoimmunity (NIIMA), Universal Scientific Education and Research Network (USERN), Tehran, Iran

Department of Immunology, School of Medicine, Tehran University of Medical Sciences, Tehran, Iran

Research Center for Immunodeficiencies, Children's Medical Center, Tehran University of Medical Sciences, Tehran, Iran

The host innate immune system detects invading viruses using pattern recognition receptors (PRRs) that can recognize pathogen-associated molecular patterns (PAMPs). At present, the known PRRs mainly include toll-like receptors (TLR), RIG-I-like receptors (RLR), NOD-like receptors (NLR), C-type lectin-like receptors (CLmin), and free-molecule receptors in the cytoplasm, such as stimulator of interferon (IFN) genes (STING), gamma-interferon-inducible protein (IFI16), Cyclic GMP-AMP synthase (cGAS), and DNA-dependent activator of interferon regulatory factors (DAI) (Takaoka et al. 2007; Li et al. 2020a, b). RNA viruses, such as SARS-CoV, SARS-CoV-2, and MERS-CoV, are detected by endosomal RNA PRRs, including TLR3 and TLR7 and/or cytoplasmic RNA sensors, namely, RIG-I and melanoma differentiation-associated protein 5 (MDA5) (Felsenstein et al. 2020).

9.2.1 TLRs

They are famous members of PRRs that are expressed on/in immune and nonimmune cells, specialized to detect PAMPs or damage-associated molecular patterns (DAMPs) (Kawai and Akira 2010). At present, ten TLRs have been detected in humans. TLRs 1, 2, 4, 5, 6, and 10 occur on the cell surface, while TLRs 3, 7, 8, and 9 are present on endosomes. Triggering these receptors by specific ligands results in double binding of TLRs, interactions of their cytoplasmic domains known as the Toll-interleukin 1 receptor (TIR), and initiating signal transduction by recruiting adaptor molecules (Martin and Wesche 2002).

Different TLRs can promote diverse biological functions using varied adapter proteins, such as myeloid differentiation primary response gene (MyD88), TIR domain-containing adaptor protein (TIRAP), TIR domain-containing adaptor protein-inducing interferon β (TRIF), and TRIF-related adaptor molecule (TRAM). These adapter proteins have a common Toll/Interleukin-1 receptor (TIR) domain (Kawai and Akira 2010). MyD88 is recognized as the most famous TIR family member, which is utilized by almost all TLRs except TLR3. Upon PAMP recognition, MyD88 mediates the recruitment of interleukin receptor-associated kinases (IRAKs) that are intracellular signaling molecules essential for the transduction of TLR signals. IRAK4 is crucial to activating nuclear factor-kappa B (NF-κB) and mitogen-activated protein kinase (MAPK) downstream of MyD88. IRAK interacts with tumor necrosis factor receptor (TNFR)-associated factor 6 (TRAF6), which facilitates NEMO (NF-κB essential modifier)/IKKγ (IκB kinase-gamma) ubiquitination to activate NF-kB (Akira et al. 2006; Hacker and Karin 2006; Hacker et al. 2006). TRIF is another adapter protein that recruits to TLR3 and TLR4 and triggers the transcription factors IFN regulatory factor 3 (IRF3) and NF-κB to stimulate the expression of type I IFNs and inflammatory cytokines (Ermolaeva et al. 2008). TRAM and TIRAP function to recruit MyD88 adaptor protein to TLR2 and TLR4 as well as TRIF to the TLR4. Therefore, the TLR signaling pathways are classified as the MyD88-dependent and MyD88-independent (TRIF-dependent) signaling pathways, which activate the gene expression of pro-inflammatory cytokines and IFNs (Akira et al. 2006).

Dendritic cells are the most important antigen-presenting cells (APCs) in the body that express different TLRs and cytoplasmic PRRs, making these cells the best sensors of PAMP and DAMP between all cell types. The plasmacytoid dendritic cells (pDCs) are a subset of dendritic cells with similar morphology to antibody-producing plasma cells that are the primary source of the antiviral cytokines and type I IFNs produced in response to viral infections. This capacity of pDCs is because of the presence of the numerous endosomal TLRs (TLRs 3, 7, 8, and 9), which recognize internalized nucleic acids of viruses (Abbas et al. 2017).

An animal model study showed that TLR3-/-, TLR4-/-, and TRAM-/- mice are more sensitive to coronavirus infection than wild-type mice, but interestingly, they only experience a transient weight loss without mortality. In contrast, TRIF-/- mice show a highly increased susceptibility to SARS-CoV infection with weight loss; increased viral titers; extra infiltration of inflammatory cell and neutrophils; an abnormal pro-inflammatory cytokine, chemokine, and interferon-stimulated gene (ISG) signaling cascade; and morbidity (Totura et al. 2015).

However, MyD88-/- mice in response to SARS-CoV show similar viral titers, weight loss, and mortality, but the outcomes in downstream cellular signaling are very different (Sheahan et al. 2008). After infection with SARS-CoV, both MyD88- and TRIF-deficient mice present a lack of cytokine and chemokine signaling compared to wild-type mice, but on day 4, TRIF-deficient mice present an elevated pro-inflammatory cytokine and ISG signaling. This alteration can be explained by a compensatory mechanism of innate immune signaling resulting from the absence of innate immune response and higher viral titers of SARS-CoV (Totura et al. 2015). Deletion of both arms of TLR signaling results in lethal SARS-CoV disease in mice. Therefore, counterbalanced TLR signaling using both MyD88-dependent and MyD88-independent pathways is crucial to establishing an innate antiviral response against SARS-CoV.

RNA viruses, such as SARS-CoV, SARS-CoV-2, and MERS-CoV, are detected by endosomal TLR3 and TLR7 that direct translocation of the transcription factors NF-κB and IRF3 and increase the production of pro-inflammatory cytokines such as interleukin (IL)-1, IL-6, and tumor necrosis factor-alpha (TNFα) as well as type I IFNs (de Wit et al. 2016; Prompetchara et al. 2020). Interestingly, it has been found that in vitro stimulation of TLR3 by its synthetic ligand polyinosinic-polycytidylic acid (poly I:C) can lead to the secretion of IFNβ and protection against murine coronavirus infection (Mazaleuskaya et al. 2012). In addition to sensing RNA of coronavirus, viral coat proteins can serve as a PAMP that is usually recognized by TLR2 and TLR4 on the cell surface (Kumar et al. 2011). Interaction of coronavirus and cells leads to the production of large quantities of chemokines and cytokines in response to CoV infection by sensing viral RNA and S and M proteins via TLRs (Dosch et al. 2009; Wang and Liu 2016).

9.2.2 RLRs

RLRs consist of the cytosolic PRRs such as RIG-I, MDA5, and laboratory of genetics and physiology 2 (LGP2). These receptors share the DExD/H-box helicase domain and a C-terminal domain of the repressor domain (RD). RIG-I and MDA5 also share common signaling structures, including N-terminal tandem caspase activation and recruitment domains (CARDs). Interaction between their DARDs and CARDs of the essential signaling adapter protein of the RLRs (also known as IPS-1, VISA, or Cardif) initiates the signaling pathway activating TBK1, and IKKε protein kinases result in IRF3 and NF-κB activation (Kawai et al. 2005; Meylan et al. 2005; Xu et al. 2005). Activation of IRF3 and NF-κB would induce transcription of innate immune response genes such as IFNs and direct antiviral mediators as well as pro-inflammatory genes regulating the innate immune response to viral infections. IFNs then induce hundreds of ISGs that their products have antiviral and immunomodulatory functions creating an antiviral state.

9.3 Antigen Presentation Pathways

Presentation of viral antigens is mediated by major histocompatibility complex (MHC; or human leukocyte antigen (HLA) in humans) class I molecules. Viral linear peptides are loaded in the groove of MHC-I molecules. Finally, the peptide-MHC complex is displayed on the surface of cells, especially APCs. This complex is recognized by cytotoxic T lymphocytes (CTLs). Subsequently, a specific antiviral immune response is initiated. There are various components involved in the antigen presentation pathway with the MHC-I molecule, including *proteasome*, tapasin, TAP, *Calreticulin*, and ERP57. The broad spectrum of viruses escapes from antigen presentation pathways by targeting key components of the MHC-I pathway and inhibiting them (Hansen and Bouvier 2009). There is no clear information about how SARS-CoV-2 escapes from antigen presentation pathways. However, there is evidence that MERS-CoV reduces the expression of genes associated with antigen presentation pathways (Menachery et al. 2018). Considering this, SARS-CoV-2 may also use a similar mechanism to reduce the presentation of viral antigens.

Studies show that the N protein of SARS-CoV interacts with the HLA-A*2402 molecule groove (Liu et al. 2010). SARS-CoV-2 antigens are, however, more likely to be presented with MHC-I molecules.

Previous studies have shown the association between HLA alleles and susceptibility to COVID-19. The presence of HLA-B*46:01 allele is associated with vulnerability and severe infection of COVID-19, whereas HLA-B*15:03 allele may provide antiviral immunity due to its high ability to presenting SARS-CoV-2 peptides. Also, HLA-A*02:02 and HLA-C*12:03 alleles function as presenter of SARS-CoV-2 conserved peptides. In contrast, the HLA-A*25: 01 and HLA-C*01: 02 alleles are less likely to present SARS-CoV-2 epitopes (Nguyen et al. 2020). Research has shown the association between HLA-B*46:01 alleles and the severity of SARS infection (Tseng et al. 2003). The presence of HLA-CW*1502 and HLA-DR*0301 alleles in people infected with SARS infection is associated with resistance to infection (Wang et al. 2011). HLA-B*0703 and HLA-B60 alleles may play a role in vulnerability to SARS infection (Ng et al. 2004). These findings may be true in the case of COVID-19, too.

Identifying potential T-cell epitopes will help scientists for suitable vaccine design. In order to identify the COVID-19 T-cell epitopes, a study using computational tools and machine learning has predicted presentable viral peptides for MHC-I and MHC-II. In this study, 405 viral peptides were identified. One of the findings was that HLA-C alleles present the majority of COVID-19 antigens. However, antigens are presented with HLA-A, HLA-B, and HLA-DR as well. When a T-cell epitope with more than one-third of the common HLA alleles among the Chinese population is present, it can be considered a potential epitope as well as a suitable candidate for vaccine design. S protein, in terms of presentation and interaction with MHC, is ranked among top potential candidates. Open reading frame (ORF) 1ab, M, N, and E proteins work in the next grade, respectively (Fast et al. 2020). The information about the association between HLA allele and COVID-19 can lead to the production of diagnostic kits and effective vaccination among the human population (Shi et al. 2020).

9.4 Epitope Spreading and Molecular Immune Mimicry

The concept of epitope spreading refers to the release of new antigens as a result of tissue damage or alteration in self-proteins. New antigens activate lymphocytes and antibody responses with different specificities. Repetition of this cycle causes more tissue damage and the targeting of new epitopes by the autoreactive lymphocytes (Rosenblum et al. 2015). Epitope spreading is one of the mechanisms of infection-induced autoimmunity (Vanderlugt and Miller 2002).

The relationship between epitope spreading and COVID-19 is not obvious. The role of viruses in autoimmune diseases has been well demonstrated. For example, coronaviruses have been associated with the incidence or recurrence of multiple sclerosis (MS) (Boucher et al. 2007). The proposed mechanism is epitope spreading (Vanderlugt and Miller 2002).

Briefly, in epitope spreading, a persistent source of viral infection activates the virus-specific T helper 1 (TH1) cells. The release of inflammatory cytokines by these cells triggers macrophage recruitment and tissue damage, which in turn leads to the release of self-antigens. The self-antigens are processed by APCs and primed to self-reactive TH1 cells. Continuous damage and further release of self-peptides cause the expansion of self-reactive immune responses (Vanderlugt and Miller 2002).

An accurate understanding of the mechanisms of COVID-19 immunopathology may reveal the association between autoimmunity as a result of epitope spreading and COVID-19. The lung is one of the organs most affected by COVID-19.

Infection with COVID-19 is associated with hyper-inflammation, infiltration of neutrophils and macrophages, diffuse alveolar damage, hyaline membrane formation, and inflexibility of the alveolar wall of the lung. Finally, these mechanisms cause respiratory failure (Cao 2020). Furthermore, one study reported that in patients with COVID-19, T lymphocytes and natural killer (NK) cells become exhausted as represented in the increased expression of the NK group 2 member A (NKG2A) marker on the surface of these cells. In contrast, the expression of CD107a, IL-2, IFNγ, Granzyme B, and TNFα is decreased (Cao 2020; Zheng et al. 2020). Taken together, regarding the link between epitope spreading and COVID-19, it is suggested that in the acute phase of the disease, virus-specific TH1 cells are activated in the lungs, resulting in tissue damage that, in turn, induces the release of self-antigens and epitope spreading. Subsequently, the autoreactive immune responses are induced. Meanwhile, in the chronic phase of the disease, exhausted T and NK cells would cause immune tolerance associated with anti-COVID-19 responses. Overall, it is possible to suggest an association between epitope spreading and induced autoimmune disease in SARS-CoV-2 infection.

The antigenic similarity between the parasite and the host is the concept of the molecular mimicry introduced by Damian for the first time. Structural similarities between the microorganisms and host antigens cause the failure of immune tolerance and the activation of the immune system against the host antigens. Therefore, molecular mimicry is considered to be a factor involved in the initiation of autoimmunity (Maoz-Segal and Andrade 2015). The most obvious example of molecular mimicry is the similarity of structure between the M protein of *Streptococcus pyogenes* and the myosin of the human heart and the resultant rheumatic heart disease (Ellis et al. 2005). The similarities of sequence and structure between SARS-CoV-2 antigens and human proteins are essential in understanding the pathogenesis and mechanisms of virus escape from the immune system and thereby finding novel therapeutic and preventive approaches with minimum side effects in humans.

A study has shown molecular mimicry between SARS-CoV spike (S) proteins and human proteins. The immunogenic and pathogenic regions of the virus spike proteins were evaluated using in silico evaluation. Four pathogenic regions, e.g., region 1 (residues 294–259), region 2 (residues 658–715), region 3 (residues 893–941), and region 4 (residues 1127–1184), had sequence homology with hydroxy acid oxidase, human Golgi autoantigen, angrgm-52, and pallidin, respectively. Among these areas, region 3 had the highest homology with the human protein. Also, residues 490–502 (GYQPYRVVVLSFEE) related to S protein had sequence homology with bradykinin. Anti-synthetic peptide D07 and D08 antibodies cross-reacted with A549 cellular lysate. Interestingly, the anti-synthetic peptide D10 antibody showed cross-reactivity with bradykinin protein. The D07/D08 and D10 peptides were related to the pathogenic region 3 and the S protein, respectively (Hwa et al. 2007). The homology of the genomic sequence between SARS-CoV-2 and SARS-CoV is 79.6% (Zhou et al. 2020a, b). Such molecular mimicry might, therefore, occur with SARS-CoV-2.

Another study reported that SARS-CoV produces autoantibodies against adrenocorticotropic hormone (ACTH) via molecular mimicry. Inhibition of ACTH impaired adrenocortical responses and instead increased levels of inflammatory cytokines (Wheatland 2004).

The molecular mimicry between human coronavirus 229E and myelin basic protein may cause MS. Probably, molecular mimicry causes FcγR activity in coronaviruses S proteins. Mimicry of IgG-specific Fc receptor accounts for binding virus nonspecific IgG to the virus. The steric effect resulting from this binding may help the virus to escape from the antibody-dependent cellular cytotoxicity (ADCC) and neutralization by complements (Chew et al. 2003).

A study based on bioinformatics analysis examined 37 SARS-CoV-2 proteins and their homology with human proteins. Eight of the 37 proteins lacked immunogenic peptides. Twelve of the 37 proteins that were related to spike (N = 6) and nonstructural proteins (nsp) (N = 6) had the highest number of immunogenic peptides. All of the proteins had at least one peptide similar to that of human proteins, except the nucleocapsid (N) phosphoprotein. Human peptides similar to viral peptides are present in the brain, lung, kidney, eye, gastrointestinal tract, lymphocyte B and plasma cell, spleen, liver, placenta, testis, pituitary gland, thyroid, heart, and skeletal muscle (Lyons-Weiler 2020). Considering that immunogenic proteins of SARS-CoV-2 have peptides similar to human proteins, there is a possibility of autoimmunity as a result of molecular mimicry. According to these findings, SARS-CoV-2 infection and vaccination may induce organ-specific autoimmune diseases.

According to Cappello's hypothesis, diabetes and high blood pressure induce chronic stress on endothelial cells. As a result of stress, the abnormal expression of some proteins, such as heat shock proteins (HSPs), increases in these cells. In this condition, the cells and tissues undergo molecular mimicry. In this way, the anti-SARS-CoV-2 antibody may cross-react with the abnormally expressed proteins (Cappello 2020). This hypothesis can be explained by the fact that in a bacterial infection, T lymphocyte responses and antibodies against bacterial HSPs occur. Interestingly, these responses may cross-react with self-HSPs and cause autoimmunity via molecular mimicry (Rajaiah and Moudgil 2009). Viruses that lack HSP, including the SARS-CoV-2, may mediate molecular mimicry-induced autoimmune diseases in individuals with a history of a bacterial infection.

A growing body of evidence suggests that the number of regulatory T cells (Tregs) in patients with COVID-19 is decreased. Cytokine storm, autoimmune diseases, and unregulated inflammatory responses result from reduced regulatory T-cell number in patients with COVID-19 (Qin et al. 2020). Recently, a study reported molecular mimicry between human nervous system proteins and SARS-CoV-2 proteins. The study claims that respiratory failure in COVID-19 is due to molecular mimicry between viral proteins and the pre-Botzinger complex (pre-BotC). Pre-BotC-related proteins include Disabled homolog1 (DAB1), Apoptosis-inducing factor mitochondrial1 (AIFM1), and Surfeit locus protein1 (SURF1), which structurally are similar to SARS-CoV-2 Hexamer peptides. The QSQASS, LNEVAK, and SAAEAS virus peptides are similar to DAB-1, AIFM-1, and SURF-1 proteins, respectively (Lucchese and Flöel 2020).

There is another hypothesis related to the molecular mimicry in SARS-CoV-2, which suggests that the molecular mimicry between RRARSVAS peptides at the S1/S2 cleavage site of the virus and human peptide related to epithelial sodium channel α-subunit (ENaC-α) may be a way for virus activation and entry to host cells. ACE2 receptor is an entry receptor for SARS-CoV-2. Proteolytic cleavage of S1/S2 by proteases is important for activating and facilitating virus entry into the cell. The S1 subunit interacts with the ACE2 receptor, while the cleavage of S2 by the host proteases is necessary for the virus entry into cell and fusion of the virus-host cell membranes. Furin protease is involved in the activation of ENaC-α via cutting the peptide bond between the residues of the arginine and serine amino acids. The presence of the RRARSVAS sequence in SARS-CoV-2 has been identified by bioinformatics analysis. However, this sequence does not exist in SARS-CoV. S1/S2 site is activated proteolytically like ENaC-α. Interestingly, ENaC-α expression occurs in cells such as nasal epithelial cells, type 2 lung alveolar cells, and colonic enterocytes. These cells are involved in the pathophysiology of COVID-19 (Anand et al. 2020). Therefore, SARS-CoV-2 might exploit the human protease network for proteolytic activation and interaction with the ACE2 receptor via molecular mimicry. Therefore, the use of protease inhibitors may be considered as a therapeutic approach in COVID-19.

9.5 Antigen Epitopes and Immunodominant Epitopes

The genetic analysis ranks SARS-CoV-2 with the highest closeness to two bat-derived SARS-like coronaviruses called BAT-SL-COVZC 45 and Bat-SL-COVZXC 21 (about 88% identity). The genetic identity of SARS-CoV-2 with SARS-CoV and MERS-CoV was about 79% and 50%, respectively. Phylogenetic analysis has displayed that SARS-CoV-2 belongs to the subgenus *Sarbecovirus* of the genus *Betacoronavirus*.

According to homology models, SARS-CoV-2 and SARS-CoV have a similar receptor-binding domain (RBD) to bind the same receptor. Some differences exist yet (Lu et al. 2020). The RBD of SARS-CoV on S protein is crucial for interaction with the ACE2 receptor and virus entry into host cells (Li et al. 2003, 2005). Interestingly, ACE2 receptor tissue distribution is consistent with possible routes of entry for the SARS-CoV (Hamming et al. 2004). Convergent evolution between SARS-CoV and SARS-CoV-2 RBDs has been proven through structural analysis. The majority of residues in RBDs are highly conserved and have similar side chain properties (Shang et al. 2020). There are 380 amino acid substitutions between protein sequences of SARS-CoV-2 and corresponding sequences of SARS-like CoVs. In comparison with SARS-CoV-2, SARS-CoV has the protein 8a. Furthermore, the protein 8b is 121 amino acids in length in SARS-CoV-2 but is shorter in SARS-CoV, with 84 amino acids. The protein 3b contains 22 amino acids in SARS-CoV-2, whereas it is 154 amino acids in SARS-CoV (Wu et al. 2020). Dipeptidyl peptidase 4 (DPP4) or cluster of differentiation 26 (CD26) mediates MERS-COV entry into cells host (Lu et al. 2013). For this reason, a hypothesis has been proposed that CD26 inhibition may prevent the progression of SARS-CoV-2 infection (Chen et al. 2020; Strollo and Pozzilli 2020; Li et al. 2020a, b). Recently, an in silico study has shown that glutamic acid insertion in the position of 484 of SARS-CoV-2 S protein is essential for the interaction between CD26 and S protein (Li et al. 2020a, b).

It has been reported that the structural proteins of SARS-CoV-2 and SARS-CoV are genetically similar. The amount of similarity for E, M, N, and S proteins is estimated to be about 94.7%, 90.1%, 90.6%, and 76.0%, respectively. Given the lack of mutations, SARS-CoV-related lymphocyte B and T epitopes are similar to those of SARS-CoV-2, about 23% and 16%, respectively (Ahmed et al. 2020).

One study reported that CR3022, a specific human SARS-CoV antibody, could bind to SARS-CoV-2 RBD. Probably a high similarity between RBD epitopes of SARS-CoV and SARS-CoV-2 has caused cross-reaction (Tian et al. 2020). Another study shows that RBDs of SARS-CoV-2 and SARS-CoV have a highly conserved cryptic epitope, which causes cross-reaction via the CR3022 (Yuan et al. 2020). Other SARS-CoV neutralizing antibodies, such as M396 and CR3014, failed to bind to the SARS-CoV-2 spike protein (Tian et al. 2020). Because the recognized RBD epitope by the CR3022 and the ACE2 receptor does not overlap, the CR3022 antibody did not compete with the ACE2 receptor for binding to the SARS-CoV-2 RBD epitope (Yuan et al. 2020). However, SARS-CoV-2 could not be neutralized by this antibody (Vabret et al. 2020; Yuan et al. 2020). Other SARS-CoV neutralizing antibodies compete with the ACE2 receptor (Berry et al. 2004; Sui et al. 2004; van den Brink et al. 2005; Prabakaran et al. 2006; ter Meulen et al. 2006). The m396, S230.15, S109.8, S227.14, S230.15, 80R, CR3022, CR3014, 33G4, 35B5, and 30F9 monoclonal antibodies block the interaction between RBD and cellular ACE2 receptor by recognizing epitopes on SARS-CoV RBD (Jiang et al. 2020). Considering the similarity of RBD antigen between SARS-CoV-2 and SARS-CoV, SARS-CoV neutralizing antibodies can be considered as a treatment and prevention of SARS-CoV-2.

Peptide sequences called mimotope, which mimic carbohydrate, lipid, or protein antigens, induce both humoral and cellular immune responses (Hung and Yu 2019). Recent evidence reveals that IgM mimotope heptapeptides are identical to peptides of SARS-CoV-2. AQTGIAV and TKGPHEF epitopes are conserved among

SARS-CoV and SARS-CoV-2 (Shivarov et al. 2020). SARS-CoV-correlated TTSTALG epitope is part of CD4⁺ T-cell epitope, which is presented in the context of HLA-DRB1*04:01 (Yang et al. 2009). VKGDDVR epitope is part of the immunodominant sites of the SARS-CoV S protein (He et al. 2004). Spike-related TTLDSKT mimotope is specific for SARS-CoV-2, and it may be a putative immunodominant peptide inducing an immune response (Shivarov et al. 2020). It has now been demonstrated that of ten dominant SARS-CoV B-cell epitope regions, six regions had a high percent identity ($\geq$ 90%) with those of SARS-CoV-2. 888FGAGAALQIPFAMQMAYRFNGIG909 B-cell immunodominant epitope, which is related to surface glycoprotein, had a 100% identity between SARS-CoV and SARS-CoV-2. Besides, of 45 dominant SARS-CoV T-cell epitope regions, 14 regions had a 100% identity between SARS-CoV and SARS-CoV-2. These epitopes are associated with surface glycoprotein, membrane protein, and nucleocapsid phosphoprotein (Grifoni et al. 2020). It has been demonstrated that the S protein-specific $CD4^+$ T cells are present in 83% of patients with COVID-19 (Braun et al. 2020). The S protein of SARS-CoV-2 also induces S protein-specific $CD4^+$ and $CD8^+$ T cells, and it is an immunodominant epitope candidate for the vaccine (Weiskopf et al. 2020). Unlike the main protease, there were antibody responses against the N protein and spike RBD. Furthermore, spike RBD induces SARS-CoV-2-specific T-cell responses (Dong et al. 2020). Unlike memory B-cell responses, memory T-cell responses against SARS-CoV S protein last 6 years after infection in convalescent SARS-CoV patients (Tang et al. 2011). Recently, it has been shown that COVID-19 convalescent individuals have RBD-specific antibodies with potent antiviral activity (Robbiani et al. 2020).

Kazuma Kiyotani et al. used bioinformatics analysis for screening potential T-cell epitopes for the SARS-CoV-2. 2013 and 1399 T-cell epitopes were identified for HLA Class I and II, respectively. These results highlighted that 781 HLA Class I and 418 HLA Class II epitopes were common between SARS-CoV and SARS-CoV-2. Additionally, 36 HLA-Class I and 10 HLA-Class II epitopes were common between MERS-CoV and SARS-CoV-2 (Kiyotani et al. 2020). It has been shown that epitope sequences for bat RaTG13 coronavirus and SARS-CoV have a 100% sequence identity with SARS-CoV-2 epitopes (Tilocca et al. 2020). Similarly, bat RaTG13 coronavirus and SARS-CoV-2 use the human ACE2 receptor for cell entry (Shang et al. 2020).

Another study found that 29 out of 34 SARS-CoV-2 epitopes are unique, and 5 epitopes are shared between SARS-CoV-2 and SARS-CoV. RBD of SARS-CoV-2 contained seven epitopes, and notably, one of them was homologous with SARS-CoV. In this study, according to epitope score and surface accessibility parameters, there were 11 antibody epitopes ranked with high scores for SARS-CoV-2. One epitope (KYFKNHTSP) belonged to a conserved region but is located outside RBD. In contrast, AWNRKR and PKKS sequences occurred inside RBD. Finally, this study suggested that most of the antibodies against the SARS-CoV spike protein do not become capable of neutralizing SARS-CoV-2 because of novel B-cell epitopes on the surface of SARS-CoV-2 (Zheng and Song 2020).

The identity and similarity of the surface glycoprotein of SARS-CoV-2 and SARS-CoV are estimated to be about 76.3% and 87.3%, respectively. SARS-CoV-2-correlated VVNQNAQAL CTL epitope has a 100% identity with SARS-CoV (Baruah and Bose 2020). There is a highly similar peptide sequence in RBD of spike protein, which mediates binding to the human ACE2 receptor between SARS-CoV-2 and SARS-CoV (Qiu et al. 2020). A study has indexed 933 SARS-CoV-2 epitopes, which are absent in the *Homo sapiens* proteome. It explains that these are foreign epitopes that cause the immune system to respond and can be considered as potential candidates for vaccine design against the virus (Lucchese 2020). A growing body of evidence reveals that the SARS-CoV-2 furin cleavage site is different from other coronaviruses. Sequence alignment has shown the insertion of four new amino acid residues (PRRA) into SARS-CoV-2 S protein at 681–684 positions. This sequence did not exist in the prior coronaviruses (Coutard et al. 2020; Hoffmann et al. 2020a, b; Walls et al. 2020; Wang et al. 2020; Zhou et al. 2020a, b; Kim et al. 2020a, b). PRRA-based vaccine design may be considered as an avenue for SARS-CoV-2 treatment.

According to central dogma theory, DNA is transcribed to mRNA, and mRNA is translated to protein (Li and Xie 2011). A recent study has delineated the SARS-CoV-2 transcriptome and epitranscriptome map. Results indicate that there are 41 potential RNA modification sites with an AAGAA motif in SARS-CoV-2 RNA. RNA modification may contribute to viral survival and escape from host immune responses (Kim et al. 2020a, b). RNA modification also suppresses innate immune responses (Karikó et al. 2005). It seems that SARS-CoV-2 has changed its epitopes and thereby promoted evolution via RNA modification and, consequently, virus evasion from the immune system.

9.6 Potential Immune Evasion Mechanisms

Viruses have developed various mechanisms to escape host immunity. These mechanisms include the following: i, interfering with recognition by $CD8^+$ T cells by limiting the antigen processing and presentation through blocking transporter associated with antigen processing (TAP) and deletion of MHC class I molecules from the endoplasmic reticulum (ER) in APCs; ii, synthesizing homologs of cytokine receptors; iii, production of immunosuppressive cytokines; and iv, killing or functional impairment of immune cells (Abbas et al. 2017). RNA viruses have evolved mechanisms for evading immune recognition and damping immune responses. Current observations point out that coronaviruses are specialized to escape immune detection at the early stage of infection as they are apt to have an extended incubation period (2–11 days) compared to influenza virus (1–4 days) (Lessler et al. 2009).

Our current knowledge about the evasion mechanisms of the novel SARS-CoV-2 is limited. However, as a member of the *Betacoronavirus* genus, these mechanisms might resemble those of SARS-CoV and MERS-CoV. Most evasion mechanisms depend on the blockade of innate immunity and, in particular, preventing induction and signaling of type I IFNs as well as the ways to increase IFN resistance (Kindler et al. 2016). All coronaviruses induce low levels of IFNs in most cell types (Menachery et al. 2014). Loading of viral double-stranded RNA (dsRNA) inside binary membrane vesicles is the most important mechanism that prevents viral antigen from being recognized by innate endosomal receptors (Knoops et al. 2008; Van Hemert et al. 2008). Similar to large RNA viruses, coronaviruses have sufficient genetic space for encoding several proteins to interfere with innate immune signaling pathways to escape the host responses. Coronaviruses encode a protease accomplished to block the ubiquitination of key signaling molecules to stop the induction of IFNs. SARS and MERS coronaviruses encode a papain-like protease (PLpro), which is capable of preventing IRF3 activation by deubiquitination of TRAF, TBK1, RIG-I, as well as STING (Sun et al. 2012; Bailey-Elkin et al. 2014; Chen et al. 2014; Mielech et al. 2014). Viral proteins, such as N protein, sequester IFN inducing RNA PAMPs (Kopecky-Bromberg et al. 2007; Lu et al. 2011). M protein obstructs the establishment of the TRAF3/TBK1 complex (Siu et al. 2009). Besides, M protein inhibits activation of the interferon-stimulated response element (ISRE) after its stimulation with IFNs (Yang et al. 2013). Nsp1 interacts with host mRNA translation and degradation, as well as the reduction in STAT1 phosphorylation (Lokugamage et al. 2015). The ADP-ribose-1″-monophosphatase domain of nsp3 with an unknown mechanism results in IFN resistance (Kuri et al. 2011). ORF4a interacts with the RLR cofactor PACT (Yang et al. 2013; Siu et al. 2014) and binds TBK1 and IKKε and thereby can inhibit ISRE activation after stimulation with IFNs (Yang et al. 2013, 2015; Matthews et al. 2014). ORF6 inhibits STAT1 nuclear introduction by sequestering karyopherinα2 to intracellular membranes (Frieman et al. 2007; Kopecky-Bromberg et al. 2007). ORF9b causes the proteasomal degradation of TRAF6, TRAF3, and inhibits activation of mitochondrial antiviral-signaling protein (MAVS), which is crucial for triggering nuclear translocation of IRF3 upon the induction of cytoplasmic RNA sensors (Shi et al. 2014). Similar evasion mechanisms are expected for SARS-CoV-2, SARS-CoV, and MERS-CoV, but some differences are present, for instance. ORF6 and ORF3b proteins of SARS-CoV-2 are shortened and probably lost their anti-interferon roles. This difference could be a reason for the significant sensitivity of novel coronavirus to IFNα in vitro (Lokugamage et al. 2020).

It has been found in an animal model study that delays in IFN induction contributes to the activation of monocytes/macrophages and pro-inflammatory cytokines in the lungs leading to vascular leakage and dampened adaptive immune responses (Channappanavar et al. 2016).

It seems that the host immune system develops mechanisms to compensate for the effect of immune escape of viruses partially. Interestingly, in the condition that most cells become IFN-deficient after coronavirus infection, pDCs produce high levels of IFNs in response to SARS-CoV and MERS-CoV (Cervantes-Barragan et al. 2012; Channappanavar et al. 2016). It is found by a mouse coronavirus model study that TLR7 plays an important role to induce IFN production by pDCs in coronavirus infections (Cervantes-Barragan et al. 2007). However, at the next phase, virus-infected cells undergo death, and PRRs sense virus particles released in the intracellular milieu. After the development of innate immune responses and secretion of pro-inflammatory responses, adaptive immune responses initiate defense against viral infection by $CD4^+$ T cell-derived cytokines, $CD8^+$ T cell-mediated cytotoxicity, and antibodies produced by B cells. In addition to depletion of lymphocytes during cytokine storm due to the recruitment of innate immune cells to the lung and causing hyper-inflammation (Cron and Chatham 2020), SARS-CoV-2 can also evade adaptive immune responses by induction of T-cell apoptosis (Yi et al. 2020).

9.7 Humoral Immune Responses

Humoral immune responses against COVID-19 are the result of B cells' cooperation with T follicular helper cells (Tay et al. 2020). The plasma containing neutralizing antibodies from recovered individuals has been used to treat patients with severe COVID-19. With the help of a high-scale sequential RNA sequence B-cell sequencing method in patients recovering from COVID-19, neutralizing antibodies against SARS-CoV-2 were identified. These antibodies are against SARS-CoV-2 spike glycoprotein. Measuring these antibodies, which are of the IgM and IgG classes, is used in conjunction with the RT-PCR test to diagnose the disease. Patients with a severe form of the disease show a high titer of IgG (Long et al. 2020). It raises the antibody-dependent enhancement (ADE) phenomenon in SARS-CoV-2 infection. The ADE phenomenon occurs in viral infections and enhances virus entry and causes severe inflammatory responses. Unfortunately, it has previously been shown that monoclonal antibodies neutralizing MERS spike protein (which binds to receptors at the human cell surface) enhance virus entry. Therefore, this important pathological effect of anti-SARS-CoV-2 antibodies should be considered in vaccine production and antibody therapy (Cao 2020). Indirect enzyme-linked immunosorbent assay (ELISA) is used to assess the profile of IgM, IgG, and IgA in SARS-CoV-2 infection (Guo et al. 2020). These antibodies appear 4 to 7 days after the onset of symptoms. IgM is present until 12 weeks, while IgG is present for a more extended time, so IgG likely plays a protective role (Li et al. 2020a, b). The SARS-CoV on the fourth day after the onset of symptoms causes seroconversion and, in most patients, lasts up to 14 days. There is a specific, long-term neutralizing IgG response for up to 2 years after infection. For MERS infection, seroconversion occurs in the second or third week after the onset of symptoms. In both types of coronavirus infection, a weak and delayed antibody response is associated with severe complications and consequences.

Concerning SARS-CoV-2, studies investigating serological testing are limited. In a preliminary study, peak production of specific IgM occurred on the ninth day after the onset of the disease and then switched to IgG in the second week. Interestingly, anti-SARS-CoV-2 antibodies in sera from patients with confirmed SARS-CoV-2 infection cross-reacted with SARS-CoV. Also, in vitro plaque assay showed that all patients' sera were able to neutralize SARS-CoV-2. Therefore, a humoral immune response could be a successful response. Of course, the relevance of the kinetics or titer of specific antibodies to the severity of the disease must be examined, and the production of more and stronger neutralizing antibodies requires a stronger response of T cells (Prompetchara et al. 2020). IgM, IgA, and IgG might be detected on the first day of symptom onset. However, in general, anti-SARS-CoV-2 antibodies can be evaluated in the acute phase of the disease on the fifth day (Guo et al. 2020).

IgG production and dynamics are different between male and female patients because IgG production is more robust in women in the early stages and at the onset of the disease. Moreover, in the severe stage of the disease, the anti-SARS-CoV-2 IgG titer is higher in women. Therefore, the complications and outcomes of the disease are different between men and women (Zeng et al. 2020). The exact mechanism is not fully understood yet (Fig. 9.1).

The monitoring of anti-SARS-CoV-2 antibodies offers a good way to predict disease progression. In general, anti-SARS-CoV-2 IgG might be detectable on day 4 after the onset of the disease, and the highest level appears in the fourth week, which is similar to the dynamics anti-SARS-CoV IgG follows. Previous studies have shown that in severe SARS, the humoral and serological immune response is more robust, and the level of IgG is higher compared to mild SARS. Therefore, the strong and robust immune response against SARS-CoV-2 might reflect the severity of the disease. Anti-SARS-CoV-2 IgG is involved in both the defense and clearance of the virus and severe tissue damage (Zhang et al. 2020).

Studies on SARS and MERS show that specific antiviral antibodies can be evaluated in 80 to 100% of patients within 2 weeks of the onset of the disease. Currently, little is known about the anti-SARS-CoV-2 antibody responses, and more research is needed on the clinical application of serological tests. Serum samples of patients with COVID-19 showed no cross-reaction with the S1 subunit of the SARS-CoV S protein. However, there was some cross-reactivity of serum samples of patients with COVID-19 to SARS-CoV N antigens. All patients tested positive for virus-specific IgG approximately 17–19 days after the onset of symptoms, and more than 90% of patients showed virus-specific IgM approximately 20–20 days after the onset of symptoms. Studies show that a subset of patients may not develop long-term antibodies to SARS-CoV-2. It remains to be addressed whether these patients are prone to reinfection. The titer of neutralizing antibodies against SARS-CoV-2 in discharged and convalescent patients is closely related to the number of N protein-specific T cells. The presence of specific antibodies against RBD of S protein of SARS-CoV-2 is an important factor for neutralizing the SARS-CoV-2 in convalescent patients. Anti-S-RBD IgG can be useful in the study of the ability of serum neutralization in COVID-19 patients (Ni et al. 2020).

Generally, anti-SARS-CoV-2 humoral immunity has completely dual and exciting effects, which both increase antiviral defense and increase immunity and cause more virus entry and pathogenicity, so this dual effect should be considered in plasma therapy and vaccine discussion. The most important question about the humoral immune response to COVID-19 is precisely how long the antibodies are present and how long they can protect people and why in some individuals, the antibody titer decreases despite the previous infection and a new infection occurs again. Also, if or not plasma therapy transmits the virus remains an open question (Table 9.1).

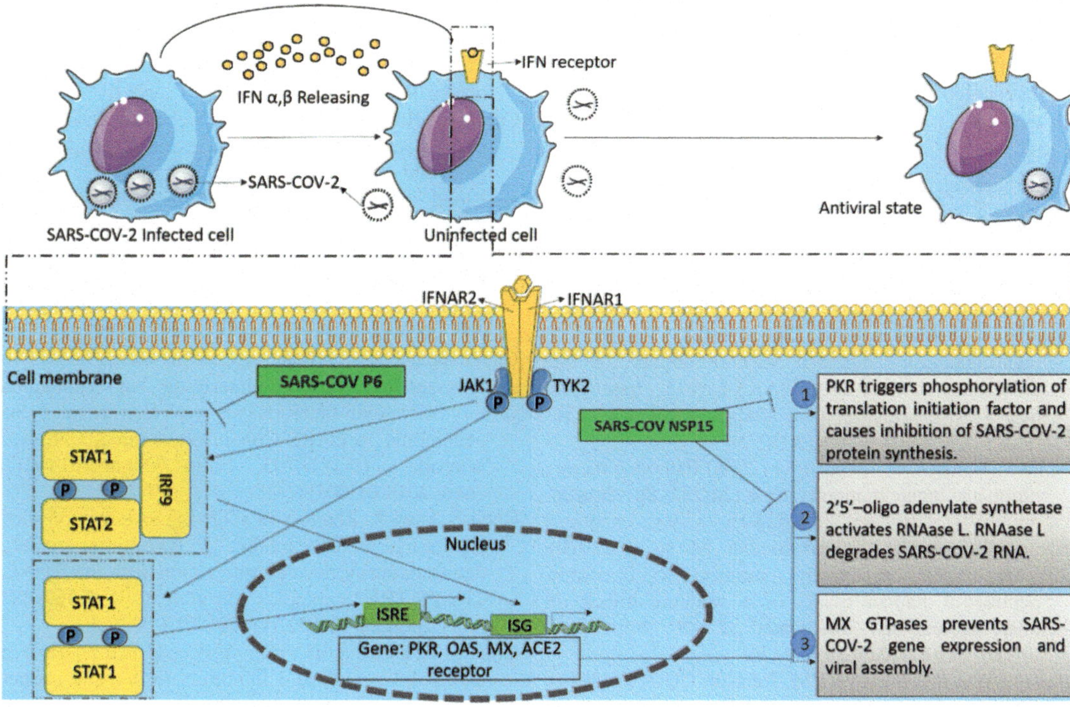

Fig. 9.1 Proposed mechanisms of type I IFN (IFN-α, IFN-β) antiviral defense against SARS-CoV-2 and immune evasion. This effect is via the JAK-STAT signaling pathway. This pathway causes not only an antiviral response but also upregulation of SARS-CoV-2 receptor ACE2 (not shown). The transcription of target genes has been shown IFN, interferon; IFNAR, IFN-α/β receptor; JAK1, Janus kinase 1; TYK2, tyrosine kinase 2; IRF9, interferon regulatory factor 9; ISRE, interferon-stimulated response element; ISG, interferon-stimulated gene; OAS, 2–5-oligoadenylate synthetase; PKR, protein kinase RNA-activated; STAT1, signal transducer and activator of transcription 1; ACE2, angiotensin-converting enzyme 2

Table 9.1 Diagnostic criteria for COVID-19

	Window period	Early phase	Active phase	End phase	Early phase	Previous infection	Recovery
PCR	+	+	+	+	–	–	–
IgM	–	+	+	–	+	–	+
IgG	–	–	+	+	–	+	+

9.8 Conclusion

The COVID-19 pandemic is an ongoing issue that affects the lives of most people around the world with considerable impact on the economy, transport, interpersonal and social relationships, education, and every aspect of the individual's daily life. Therefore, it is of the utmost importance to understand the immunopathology of disease. Although the immune system plays an essential role in fighting COVID-19, paradoxically, it could also be harmful. Most critically ill patients with intensive care needs that develop ARDS have high levels of inflammatory cytokines in their circulation and develop cytokine release syndrome. Although scientists are making an effort to develop proper preventive and therapeutic intervention strategies to overcome the SARS-CoV-2 and more than 200 clinical trials of COVID-19 have been recorded in https://clinicaltrials.gov/, it might take a long time to examine their efficacy in vitro and in vivo. Altogether, better understanding the interaction of the SARS-CoV-2 and the host immunity will bring in-depth knowledge for effective vaccine design, therapeutic protocols, and preventive strategies.

References

Abbas AK, Lichtman AH, Shiv P (2017) Cellular and molecular immunology. Elsevier, Philadelphia

Ahmed S, Quadeer AA, McKay M (2020) Preliminary identification of potential vaccine targets for the COVID-19 coronavirus (SARS-CoV-2) based on SARS-CoV immunological studies. Viruses 12:254. https://doi.org/10.3390/v12030254

Akira S, Uematsu S, Takeuchi O (2006) Pathogen recognition and innate immunity. Cell 124:783–801. https://doi.org/10.1016/j.cell.2006.02.015

Anand P, Puranik A, Aravamudan M et al (2020) SARS-CoV-2 selectively mimics a cleavable peptide of human ENaC in a strategic hijack of host proteolytic. Machinery 2:4–8. https://doi.org/10.1101/2020.04.29.069476

Bailey-Elkin BA, Knaap RCM, Johnson GG et al (2014) Crystal structure of the Middle East respiratory syndrome coronavirus (MERS-CoV) papain-like protease bound to ubiquitin facilitates targeted disruption of deubiquitinating activity to demonstrate its role in innate immune suppression. J Biol Chem 289:34667–34682. https://doi.org/10.1074/jbc.M114.609644

Baruah V, Bose S (2020) Immunoinformatics-aided identification of T cell and B cell epitopes in the surface glycoprotein of 2019-nCoV. J Med Virol 92:495–500. https://doi.org/10.1002/jmv.25698

Ben Addi A, Lefort A, Hua X et al (2008) Modulation of murine dendritic cell function by adenine nucleotides and adenosine: involvement of the A(2B) receptor. Eur J Immunol 38:1610–1620. https://doi.org/10.1002/eji.200737781

Berry JD, Jones S, Drebot MA et al (2004) Development and characterisation of neutralising monoclonal antibody to the SARS-coronavirus. J Virol Methods 120:87–96. https://doi.org/10.1016/j.jviromet.2004.04.009

Boucher A, Desforges M, Duquette P, Talbot PJ (2007) Long-term human coronavirus-myelin cross-reactive T-cell clones derived from multiple sclerosis patients. Clin Immunol 123:258–267. https://doi.org/10.1016/j.clim.2007.02.002

Braun J, Loyal L, Frentsch M et al (2020) Presence of SARS-CoV-2 reactive T cells in COVID-19 patients and healthy donors. In: medRxiv 2020.04.17.20061440. https://doi.org/10.1101/2020.04.17.20061440

Cao X (2020) COVID-19: immunopathology and its implications for therapy. Nat Rev Immunol 20:269–270. https://doi.org/10.1038/s41577-020-0308-3

Cappello F (2020) Is COVID-19 a proteiform disease inducing also molecular mimicry phenomena? Cell Stress Chaperones 25(3):381–382. https://doi.org/10.1007/s12192-020-01112-1

Cervantes-Barragan L, Züst R, Weber F et al (2007) Control of coronavirus infection through plasmacytoid dendritic-cell-derived type I interferon. Blood 109(3):1131–1137. https://doi.org/10.1182/blood-2006-05-023770

Cervantes-Barragan L, Lewis KL, Firner S et al (2012) Plasmacytoid dendritic cells control T-cell response to chronic viral infection. Proc Natl Acad Sci 109:3012–3017. https://doi.org/10.1073/pnas.1117359109

Channappanavar R, Fehr AR, Vijay R et al (2016) Dysregulated type I interferon and inflammatory monocyte-macrophage responses cause lethal pneumonia in SARS-CoV-infected mice. Cell Host Microbe 19:181–193. https://doi.org/10.1016/j.chom.2016.01.007

Chen X, Yang X, Zheng Y et al (2014) SARS coronavirus papain-like protease inhibits the type I interferon signaling pathway through interaction with the STING-TRAF3-TBK1 complex. Protein Cell 5:369–381. https://doi.org/10.1007/s13238-014-0026-3

Chen C-F, Chien C-H, Yang Y-P et al (2020) Role of dipeptidyl peptidase 4 inhibitors in diabetic patients with coronavirus-19 infection. J Chin Med Assoc. https://doi.org/10.1097/JCMA.0000000000000338. Publish Ahead of Print

Chew FT, Ong SY, Hew CL (2003) Severe acute respiratory syndrome coronavirus and viral mimicry. Lancet 361:2081. https://doi.org/10.1016/S0140-6736(03)13652-5

Coutard B, Valle C, de Lamballerie X et al (2020) The spike glycoprotein of the new coronavirus 2019-nCoV contains a furin-like cleavage site absent in CoV of the same clade. Antivir Res 176:104742. https://doi.org/10.1016/j.antiviral.2020.104742

Cron RQ, Chatham WW (2020) The rheumatologist's role in Covid-19. J Rheumatol 200334:639. https://doi.org/10.3899/jrheum.200334

de Wit E, van Doremalen N, Falzarano D, Munster VJ (2016) SARS and MERS: recent insights into emerging coronaviruses. Nat Rev Microbiol 14:523–534. https://doi.org/10.1038/nrmicro.2016.81

Dong C, Ni L, Ye F et al (2020) Characterization of antiviral immunity in recovered individuals infected by SARS-CoV-2. In: medRxiv 2020.03.17.20036640. https://doi.org/10.1101/2020.03.17.20036640

Dosch SF, Mahajan SD, Collins AR (2009) SARS coronavirus spike protein-induced innate immune response occurs via activation of the NF-kappaB pathway in human monocyte macrophages in vitro. Virus Res 142:19–27. https://doi.org/10.1016/j.virusres.2009.01.005

Ellis N, Li Y, Hildebrand W et al (2005) T cell mimicry and epitope specificity of cross-reactive T cell clones from rheumatic heart disease. J Immunol 175:5448–5456. https://doi.org/10.4049/jimmunol.175.8.5448

Ermolaeva MA, Michallet MC, Papadopoulou N et al (2008) Function of TRADD in tumor necrosis factor receptor 1 signaling and in TRIF-dependent inflammatory responses. Nat Immunol 9:1037–1046. https://doi.org/10.1038/ni.1638

Fast E, Altman RB, Chen B (2020) Potential T-cell and B-cell epitopes of 2019-nCoV. In: bioRxiv 2020.02.19.955484. https://doi.org/10.1101/2020.02.19.955484

Felsenstein S, Herbert JA, McNamara PS, Hedrich CM (2020) COVID-19: immunology and treatment options. Clin Immunol 215:108448. https://doi.org/10.1016/j.clim.2020.108448

Frieman M, Yount B, Heise M et al (2007) Severe acute respiratory syndrome coronavirus ORF6 antagonizes STAT1 function by sequestering nuclear import factors on the rough endoplasmic reticulum/Golgi membrane. J Virol 81:9812–9824. https://doi.org/10.1128/JVI.01012-07

Garcia-Sastre A, Biron CA (2006) Type 1 interferons and the virus-host relationship: a lesson in detente. Science 312(5775):879–882. https://doi.org/10.1126/science.1125676

Grifoni A, Sidney J, Zhang Y et al (2020) A sequence homology and bioinformatic approach can predict candidate targets for immune responses to SARS-CoV-2. Cell Host Microbe 27:671–680. https://doi.org/10.1016/j.chom.2020.03.002

Guo L, Ren L, Yang S et al (2020) Profiling early humoral response to diagnose novel coronavirus disease (COVID-19). Clin Infect Dis 71:778. https://doi.org/10.1093/cid/ciaa310

Hacker H, Karin M (2006) Regulation and function of IKK and IKK-related kinases. Sci STKE 2006:re13. https://doi.org/10.1126/stke.3572006re13

Hacker H, Redecke V, Blagoev B et al (2006) Specificity in toll-like receptor signalling through distinct effector functions of TRAF3 and TRAF6. Nature 439:204–207. https://doi.org/10.1038/nature04369

Hamming I, Timens W, Bulthuis MLC et al (2004) Tissue distribution of ACE2 protein, the functional receptor for SARS coronavirus. A first step in understanding SARS pathogenesis. J Pathol 203:631–637. https://doi.org/10.1002/path.1570

Hansen TH, Bouvier M (2009) MHC class I antigen presentation : learning from viral evasion strategies. Nat Rev Immunol 9:503–513. https://doi.org/10.1038/nri2575

He Y, Zhou Y, Wu H et al (2004) Identification of Immunodominant sites on the spike protein of severe acute respiratory syndrome (SARS) coronavirus: implication for developing SARS diagnostics and vaccines. J Immunol 173:4050–4057. https://doi.org/10.4049/jimmunol.173.6.4050

Hoffmann M, Kleine-Weber H, Pöhlmann S (2020a) A multibasic cleavage site in the spike protein of SARS-CoV-2 is essential for infection of human lung cells. Mol Cell 78:779–784.e5. https://doi.org/10.1016/j.molcel.2020.04.022

Hoffmann M, Kleine-Weber H, Schroeder S et al (2020b) SARS-CoV-2 cell entry depends on ACE2 and TMPRSS2 and is blocked by a clinically proven protease inhibitor. Cell 181:271–280.e8. https://doi.org/10.1016/j.cell.2020.02.052

Hung J-T, Yu AL (2019) Chapter 4 - GD2-targeted immunotherapy of Neuroblastoma. In: Ray SKBT-N (ed) Neuroblastoma. Academic Press, pp 63–78. https://doi.org/10.1016/B978-0-12-812005-7.00004-7

Hwa KY, Lin WM, Hou YI, Yeh TM (2007) Molecular mimicry between SARS coronavirus spike protein and human protein. Proc Front Converg Biosci Inf Technol FBIT 2007:294–298. https://doi.org/10.1109/FBIT.2007.108

Jiang S, Hillyer C, Du L (2020) Neutralizing antibodies against SARS-CoV-2 and other human coronaviruses. Trends Immunol 41:545. https://doi.org/10.1016/j.it.2020.04.008

Karikó K, Buckstein M, Ni H, Weissman D (2005) Suppression of RNA recognition by toll-like receptors: the impact of nucleoside modification and the evolutionary origin of RNA. Immunity 23:165–175. https://doi.org/10.1016/j.immuni.2005.06.008

Katze MG, He Y, Gale M Jr (2002) Viruses and interferon: a fight for supremacy. Nat Rev Immunol 2:675–687. https://doi.org/10.1038/nri888

Kawai T, Akira S (2010) The role of pattern-recognition receptors in innate immunity: update on toll-like receptors. Nat Immunol 11:373–384. https://doi.org/10.1038/ni.1863

Kawai T, Takahashi K, Sato S et al (2005) IPS-1, an adaptor triggering RIG-I- and Mda5-mediated type I interferon induction. Nat Immunol 6:981–988. https://doi.org/10.1038/ni1243

Kim SY, Jin W, Sood A et al (2020a) Glycosaminoglycan binding motif at S1/S2 proteolytic cleavage site on spike glycoprotein may facilitate novel coronavirus (SARS-CoV-2) host cell entry. In: bioRxiv 2020.04.14.041459. https://doi.org/10.1101/2020.04.14.041459

Kim D, Lee J-Y, Yang J-S et al (2020b) The architecture of SARS-CoV-2 transcriptome. Cell 181:914–921.e10. https://doi.org/10.1016/j.cell.2020.04.011

Kindler E, Thiel V, Weber F (2016) Interaction of SARS and MERS coronaviruses with the anti-viral interferon response. In: Advances in virus research. Elsevier, pp 219–243. https://doi.org/10.1016/bs.aivir.2016.08.006

Kiyotani K, Toyoshima Y, Nemoto K, Nakamura Y (2020) Bioinformatic prediction of potential T cell epitopes for SARS-CoV-2. J Hum Genet 65:569. https://doi.org/10.1038/s10038-020-0771-5

Knoops K, Kikkert M, van den Worm SHE et al (2008) SARS-coronavirus replication is supported by a reticulovesicular network of modified endoplasmic reticulum. PLoS Biol 6(9):e226. https://doi.org/10.1371/journal.pbio.0060226

Kopecky-Bromberg SA, Martínez-Sobrido L, Frieman M et al (2007) Severe acute respiratory syndrome coronavirus open reading frame (ORF) 3b, ORF 6, and nucleocapsid proteins function as interferon antagonists. J Virol 81(2):548–557. https://doi.org/10.1128/JVI.01782-06

Kumar H, Kawai T, Akira S (2011) Pathogen recognition by the innate immune system. Int Rev Immunol 30:16–34. https://doi.org/10.3109/08830185.2010.529976

Kuri T, Eriksson KK, Putics A et al (2011) The ADP-ribose-1 ″-monophosphatase domains of severe acute respiratory syndrome coronavirus and human coronavirus 229E mediate resistance to anti-viral interferon responses. J Gen Virol 92:1899–1905. https://doi.org/10.1099/vir.0.031856-0

Lazear HM, Schoggins JW, Diamond MS (2019) Shared and distinct functions of type I and type III Interferons. Immunity 50:907–923. https://doi.org/10.1016/j.immuni.2019.03.025

Lessler J, Reich NG, Brookmeyer R et al (2009) Incubation periods of acute respiratory viral infections: a systematic review. Lancet Infect Dis 9:291–300. https://doi.org/10.1016/S1473-3099(09)70069-6

Li G-W, Xie XS (2011) Central dogma at the single-molecule level in living cells. Nature 475:308–315. https://doi.org/10.1038/nature10315

Li W, Moore MJ, Vasilieva N et al (2003) Angiotensin-converting enzyme 2 is a functional receptor for the SARS coronavirus. Nature 426:450–454. https://doi.org/10.1038/nature02145

Li F, Li W, Farzan M, Harrison SC (2005) Structure of SARS coronavirus spike receptor-binding domain complexed with receptor. Science 309:1864–1868. https://doi.org/10.1126/science.1116480

Li G, Fan Y, Lai Y et al (2020a) Coronavirus infections and immune responses. J Med Virol 92:424–432. https://doi.org/10.1002/jmv.25685

Li X, Geng M, Peng Y et al (2020b) Molecular immune pathogenesis and diagnosis of COVID-19. J Pharm Anal 10:102–108. https://doi.org/10.1016/j.jpha.2020.03.001

Liu J, Wu P, Gao F et al (2010) Novel Immunodominant peptide presentation strategy : a featured HLA-A * 2402-restricted cytotoxic T-lymphocyte epitope stabilized by Intrachain hydrogen bonds from severe acute respiratory syndrome coronavirus nucleocapsid protein. J Virol 84:11849–11857. https://doi.org/10.1128/JVI.01464-10

Lokugamage KG, Narayanan K, Nakagawa K et al (2015) Middle East respiratory syndrome coronavirus nsp1 inhibits host gene expression by selectively targeting mRNAs transcribed in the nucleus while sparing mRNAs of cytoplasmic origin. J Virol 89:10970–10981. https://doi.org/10.1128/JVI.01352-15

Lokugamage KG, Hage A, Schindewolf C et al (2020) SARS-CoV-2 sensitive to type I interferon pretreatment. BioRxiv. https://doi.org/10.1101/2020.03.07.982264

Long Q-X, Liu B-Z, Deng H-J et al (2020) Antibody responses to SARS-CoV-2 in patients with COVID-19. Nat Med 26:845. https://doi.org/10.1038/s41591-020-0897-1

Lu X, Pan J, Tao J, Guo D (2011) SARS-CoV nucleocapsid protein antagonizes IFN-β response by targeting initial step of IFN-β induction pathway, and its C-terminal region is critical for the antagonism. Virus Genes 42(1):37–45. https://doi.org/10.1007/s11262-010-0544-x

Lu G, Hu Y, Wang Q et al (2013) Molecular basis of binding between novel human coronavirus MERS-CoV and its receptor CD26. Nature 500:227–231. https://doi.org/10.1038/nature12328

Lu R, Zhao X, Li J et al (2020) Genomic characterisation and epidemiology of 2019 novel coronavirus: implications for virus origins and receptor binding. Lancet 395:565–574. https://doi.org/10.1016/S0140-6736(20)30251-8

Lucchese G (2020) Epitopes for a 2019-nCoV vaccine. Cell Mol Immunol 17:539–540. https://doi.org/10.1038/s41423-020-0377-z

Lucchese G, Flöel A (2020) Molecular mimicry between SARS-CoV-2 and respiratory pacemaker neurons. Autoimmun Rev 102556:102556. https://doi.org/10.1016/j.autrev.2020.102556

Lyons-Weiler J (2020) Pathogenic priming likely contributes to serious and critical illness and mortality in COVID-19 via autoimmunity. J Transl Autoimmun 3:100051. https://doi.org/10.1016/j.jtauto.2020.100051

Maoz-Segal R, Andrade P (2015) Chapter 3 - molecular mimicry and autoimmunity. In: Shoenfeld Y, Agmon-Levin N, Rose AA (eds) Infection and autoimmunity, 2nd edn. Academic Press, Amsterdam, pp 27–44. https://doi.org/10.1016/B978-0-444-63269-2.00054-4

Martin MU, Wesche H (2002) Summary and comparison of the signaling mechanisms of the toll/interleukin-1 receptor family. Biochim Biophys Acta (BBA) - Molecular Cell Res 1592:265–280. https://doi.org/10.1016/S0167-4889(02)00320-8

Matthews KL, Coleman CM, van der Meer Y et al (2014) The ORF4b-encoded accessory proteins of Middle East respiratory syndrome coronavirus and two related bat coronaviruses localize to the nucleus and inhibit innate immune signalling. J Gen Virol 95:874. https://doi.org/10.1099/vir.0.062059-0

Mazaleuskaya L, Veltrop R, Ikpeze N et al (2012) Protective role of toll-like receptor 3-induced type I interferon in murine coronavirus infection of macrophages. Viruses 4:901–923. https://doi.org/10.3390/v4050901

Menachery VD, Eisfeld AJ, Schafer A et al (2014) Pathogenic influenza viruses and coronaviruses utilize similar and contrasting approaches to control interferon-stimulated gene responses. MBio 5:e01174–e01114. https://doi.org/10.1128/mBio.01174-14

Menachery VD, Schäfer A, Burnum-johnson KE et al (2018) MERS-CoV and H5N1 influenza virus antagonize antigen presentation by altering the epigenetic landscape. Proc Natl Acad Sci 115:E1012–E1021. https://doi.org/10.1073/pnas.1706928115

Meylan E, Curran J, Hofmann K et al (2005) Cardif is an adaptor protein in the RIG-I anti-viral pathway and is targeted by hepatitis C virus. Nature 437:1167–1172. https://doi.org/10.1038/nature04193

Mielech AM, Kilianski A, Baez-Santos YM et al (2014) MERS-CoV papain-like protease has deISGylating and deubiquitinating activities. Virology 450:64–70. https://doi.org/10.1016/j.virol.2013.11.040

Ng MHL, Lau K-M, Li L et al (2004) Association of human-leukocyte-antigen class I (B*0703) and class II (DRB1*0301) genotypes with susceptibility and resistance to the development of severe acute respiratory syndrome. J Infect Dis 190:515–518. https://doi.org/10.1086/421523

Nguyen A, David JK, Maden SK et al (2020) Human leukocyte antigen susceptibility map for SARS-CoV-2. J Virol 94. https://doi.org/10.1128/JVI.00510-20

Ni L, Ye F, Cheng M-L et al (2020) Antibody responses to SARS-CoV-2 in patients with COVID-19. Immunity 52:971–977.e3. https://doi.org/10.1016/j.immuni.2020.04.023

Prabakaran P, Gan J, Feng Y et al (2006) Structure of severe acute respiratory syndrome coronavirus receptor-binding domain Complexed with neutralizing antibody. J Biol Chem 281:15829–15836. https://doi.org/10.1074/jbc.M600697200

Prompetchara E, Ketloy C, Palaga T (2020) Immune responses in COVID-19 and potential vaccines: lessons learned from SARS and MERS epidemic. Asian Pacific J Allergy Immunol 38:1–9. https://doi.org/10.12932/ap-200220-0772

Qin C, Zhou L, Hu Z et al (2020) Dysregulation of immune response in patients with coronavirus 2019 (COVID-19) in Wuhan. China Clin Infect Dis 71:762. https://doi.org/10.1093/cid/ciaa248

Qiu T, Mao T, Wang Y et al (2020) Identification of potential cross-protective epitope between a new type of coronavirus (2019-nCoV) and severe acute respiratory syndrome virus. J Genet Genomics 47:115–117. https://doi.org/10.1016/j.jgg.2020.01.003

Rajaiah R, Moudgil KD (2009) Heat-shock proteins can promote as well as regulate autoimmunity. Autoimmun Rev 8:388–393. https://doi.org/10.1016/j.autrev.2008.12.004

Robbiani DF, Gaebler C, Muecksch F et al (2020) Convergent antibody responses to SARS-CoV-2 infection in convalescent individuals. In: bioRxiv 2020.05.13.092619. https://doi.org/10.1101/2020.05.13.092619

Rosenblum MD, Remedios KA, Abbas AK (2015) Mechanisms of human autoimmunity. J Clin Invest 125:2228–2233. https://doi.org/10.1172/JCI78088

Saghazadeh A, Rezaei N (2017) Implications of toll-like receptors in Ebola infection. Expert Opin Ther Targets 4:415–425. https://doi.org/10.1080/14728222.2017.1299128

Shang J, Ye G, Shi K et al (2020) Structural basis of receptor recognition by SARS-CoV-2. Nature 581:221–224. https://doi.org/10.1038/s41586-020-2179-y

Sheahan T, Morrison TE, Funkhouser W et al (2008) MyD88 is required for protection from lethal infection with a mouse-adapted SARS-CoV. PLoS Pathog 4:e1000240. https://doi.org/10.1371/journal.ppat.1000240

Shi C-S, Qi H-Y, Boularan C et al (2014) SARS-coronavirus open reading frame-9b suppresses innate immunity by targeting mitochondria and the MAVS/TRAF3/TRAF6 signalosome. J Immunol 193(6):3080–3089. https://doi.org/10.4049/jimmunol.1303196

Shi Y, Wang Y, Shao C et al (2020) COVID-19 infection: the perspectives on immune responses. Cell Death Differ 27:1451–1454. https://doi.org/10.1038/s41418-020-0530-3

Shivarov V, Petrov PK, Pashov AD (2020) Potential SARS-CoV-2 preimmune IgM epitopes. Front Immunol 11:932. https://doi.org/10.3389/fimmu.2020.00932

Siu K-L, Kok K-H, M-HJ N et al (2009) Severe acute respiratory syndrome coronavirus M protein inhibits type I interferon production by impeding the formation of TRAF3· TANK· TBK1/IKKε complex. J Biol Chem 284(24):16202–16209. https://doi.org/10.1074/jbc.M109.008227

Siu K-L, Yeung ML, Kok K-H et al (2014) Middle east respiratory syndrome coronavirus 4a protein is a double-stranded RNA-binding protein that suppresses PACT-induced activation of RIG-I and MDA5 in the innate anti-viral response. J Virol 88(9):4866–4876. https://doi.org/10.1128/JVI.03649-13

Strollo R, Pozzilli P (2020) DPP4 inhibition: preventing SARS-CoV-2 infection and/or progression of COVID-19? Diabetes Metab Res Rev 36. https://doi.org/10.1002/dmrr.3330

Sui J, Li W, Murakami A et al (2004) Potent neutralization of severe acute respiratory syndrome (SARS) coronavirus by a human mAb to S1 protein that blocks receptor association. Proc Natl Acad Sci 101:2536–2541. https://doi.org/10.1073/pnas.0307140101

Sun L, Xing Y, Chen X et al (2012) Coronavirus papain-like proteases negatively regulate anti-viral innate immune response through disruption of STING-mediated signaling. PLoS One 7(2):e30802. https://doi.org/10.1371/journal.pone.0030802

Takaoka A, Wang Z, Choi MK et al (2007) DAI (DLM-1/ZBP1) is a cytosolic DNA sensor and an activator of innate immune response. Nature 448:501–505. https://doi.org/10.1038/nature06013

Takeuchi O, Akira S (2007) Signaling pathways activated by microorganisms. Curr Opin Cell Biol 19:185–191. https://doi.org/10.1016/j.ceb.2007.02.006

Tang F, Quan Y, Xin Z-T et al (2011) Lack of peripheral memory B cell responses in recovered patients with severe acute respiratory syndrome: a six-year follow-up study. J Immunol 186:7264 LP–7267268. https://doi.org/10.4049/jimmunol.0903490

Tay MZ, Poh CM, Rénia L et al (2020) The trinity of COVID-19: immunity, inflammation and intervention. Nat Rev Immunol 20:363–374. https://doi.org/10.1038/s41577-020-0311-8

ter Meulen J, van den Brink EN, Poon LLM et al (2006) Human monoclonal antibody combination against

SARS coronavirus: synergy and coverage of escape mutants. PLoS Med 3:e237. https://doi.org/10.1371/journal.pmed.0030237

Tian X, Li C, Huang A et al (2020) Potent binding of 2019 novel coronavirus spike protein by a SARS coronavirus-specific human monoclonal antibody. Emerg Microbes Infect 9:382–385. https://doi.org/10.1080/22221751.2020.1729069

Tilocca B, Soggiu A, Sanguinetti M et al (2020) Comparative computational analysis of SARS-CoV-2 nucleocapsid protein epitopes in taxonomically related coronaviruses. Microbes Infect 22:188. https://doi.org/10.1016/j.micinf.2020.04.002

Totura AL, Whitmore A, Agnihothram S et al (2015) Toll-like receptor 3 signaling via TRIF contributes to a protective innate immune response to severe acute respiratory syndrome coronavirus infection. MBio 6:e00638–e00615. https://doi.org/10.1128/mBio.00638-15

Tseng HK, Trejaut JA, Lee H et al (2003) Association of HLA class I with severe acute respiratory syndrome coronavirus infection. BMC Med Genet 4:9. https://doi.org/10.1186/1471-2350-4-9

Vabret N, Britton GJ, Gruber C et al (2020) Immunology of COVID-19: current state of the science. Immunity 52:910. https://doi.org/10.1016/j.immuni.2020.05.002

van den Brink EN, Ter Meulen J, Cox F et al (2005) Molecular and biological characterization of human monoclonal antibodies binding to the spike and nucleocapsid proteins of severe acute respiratory syndrome coronavirus. J Virol 79:1635–1644. https://doi.org/10.1128/JVI.79.3.1635-1644.2005

Van Hemert MJ, van den Worm SHE, Knoops K et al (2008) SARS-coronavirus replication/transcription complexes are membrane-protected and need a host factor for activity in vitro. PLoS Pathog 4(5):e1000054. https://doi.org/10.1371/journal.ppat.1000054

Vanderlugt CL, Miller SD (2002) Epitope spreading in immune-mediated diseases: implications for immunotherapy. Nat Rev Immunol 2:85–95. https://doi.org/10.1038/nri724

Walls AC, Park Y-J, Tortorici MA et al (2020) Structure, function, and antigenicity of the SARS-CoV-2 spike glycoprotein. Cell 181:281–292.e6. https://doi.org/10.1016/j.cell.2020.02.058

Wang Y, Liu L (2016) The membrane protein of severe acute respiratory syndrome coronavirus functions as a novel cytosolic pathogen-associated molecular pattern to promote Beta interferon induction via a toll-like-receptor-related TRAF3-independent mechanism. MBio 7:e01872–e01815. https://doi.org/10.1128/mBio.01872-15

Wang S-F, Chen K-H, Chen M et al (2011) Human-leukocyte antigen class I Cw 1502 and class II DR 0301 genotypes are associated with resistance to severe acute respiratory syndrome (SARS) infection. Viral Immunol 24:421–426. https://doi.org/10.1089/vim.2011.0024

Wang Q, Qiu Y, Li J-Y et al (2020) A unique protease cleavage site predicted in the spike protein of the novel pneumonia coronavirus (2019-nCoV) potentially related to viral transmissibility. Virol Sin 35:337. https://doi.org/10.1007/s12250-020-00212-7

Weiskopf D, Schmitz KS, Raadsen MP et al (2020) Phenotype of SARS-CoV-2-specific T-cells in COVID-19 patients with acute respiratory distress syndrome. In: medRxiv 2020.04.11.20062349, vol 5. https://doi.org/10.1101/2020.04.11.20062349

Wheatland R (2004) Molecular mimicry of ACTH in SARS - implications for corticosteroid treatment and prophylaxis. Med Hypotheses 63:855–862. https://doi.org/10.1016/j.mehy.2004.04.009

Wu A, Peng Y, Huang B et al (2020) Genome composition and divergence of the novel coronavirus (2019-nCoV) originating in China. Cell Host Microbe 27:325–328. https://doi.org/10.1016/j.chom.2020.02.001

Xu LG, Wang YY, Han KJ et al (2005) VISA is an adapter protein required for virus-triggered IFN-beta signaling. Mol Cell 19:727–740. https://doi.org/10.1016/j.molcel.2005.08.014

Yang J, James E, Roti M et al (2009) Searching immunodominant epitopes prior to epidemic: HLA class II-restricted SARS-CoV spike protein epitopes in unexposed individuals. Int Immunol 21:63–71. https://doi.org/10.1093/intimm/dxn124

Yang Y, Zhang L, Geng H et al (2013) The structural and accessory proteins M, ORF 4a, ORF 4b, and ORF 5 of Middle East respiratory syndrome coronavirus (MERS-CoV) are potent interferon antagonists. Protein Cell 4:951–961. https://doi.org/10.1007/s13238-013-3096-8

Yang Y, Ye F, Zhu N et al (2015) Middle East respiratory syndrome coronavirus ORF4b protein inhibits type I interferon production through both cytoplasmic and nuclear targets. Sci Rep 5:17554. https://doi.org/10.1038/srep17554

Yi Y, Lagniton PNP, Ye S et al (2020) COVID-19: what has been learned and to be learned about the novel coronavirus disease. Int J Biol Sci 16(10):1753–1766. https://doi.org/10.7150/ijbs.45134

Yuan M, Wu NC, Zhu X et al (2020) A highly conserved cryptic epitope in the receptor binding domains of SARS-CoV-2 and SARS-CoV. Science 368(6491):630–633. https://doi.org/10.1126/science.abb7269

Zeng F, Dai C, Cai P et al (2020) A comparison study of SARS-CoV-2 IgG antibody between male and female COVID-19 patients: a possible reason underlying different outcome between gender. J Med Virol 92:2050. https://doi.org/10.1002/jmv.25989

Zhang B, Zhou X, Zhu C et al (2020) Immune phenotyping based on neutrophil-to-lymphocyte ratio and IgG predicts disease severity and outcome for patients with COVID-19. In: medRxiv 2020.03.12.20035048. https://doi.org/10.1101/2020.03.12.20035048

Zheng M, Song L (2020) Novel antibody epitopes dominate the antigenicity of spike glycoprotein in SARS-CoV-2 compared to SARS-CoV. Cell Mol Immunol 17:536–538. https://doi.org/10.1038/s41423-020-0385-z

Zheng M, Gao Y, Wang G et al (2020) Functional exhaustion of anti-viral lymphocytes in COVID-19 patients. Cell Mol Immunol 17:533–535. https://doi.org/10.1038/s41423-020-0402-2

Zhou H, Chen X, Hu T et al (2020a) A novel bat coronavirus closely related to SARS-CoV-2 contains natural insertions at the S1/S2 cleavage site of the spike protein. Curr Biol 30:2196–2203.e3. https://doi.org/10.1016/j.cub.2020.05.023

Zhou P, Yang X-L, Wang X-G et al (2020b) A pneumonia outbreak associated with a new coronavirus of probable bat origin. Nature 579:270–273. https://doi.org/10.1038/s41586-020-2012-7

COVID-19 and Cell Stress

10

Abdo A. Elfiky, Ibrahim M. Ibrahim, Fatma G. Amin, Alaa M. Ismail, and Wael M. Elshemey

Abstract

The present century will undoubtedly be marked with the COVID-19 global health crisis. It is not time yet to talk about the total number of deaths and hospitalizations, as they are enormously growing daily. Understanding the nature of COVID-19-induced pneumonia is vital in order to deal with the associated health complications. Cell stress is an established mechanism known to be associated with infection and cancer. Different proteins crucial for cellular response to stress are reported to be a possible target to stop the infection and to reduce the chemo-resistance in cancer. Heat shock protein (HSP) families of chaperones play an essential role in cells both in normal state and under stress. The upregulation of HSP5A, also termed GRP78 or Bip, is reported in different viral infections. This chapter introduces the current knowledge about severe acute respiratory syndrome coronavirus 2 (SARS-CoV-2), which has caused the COVID-19 pandemic, and cell stress aimed at defining possible strategies to combat the COVID-19 pandemic.

Keywords

Cell stress · Chaperone · COVID-19 · GRP78 · SARS-CoV-2 · Spike

A. A. Elfiky (✉) · I. M. Ibrahim · A. M. Ismail
Biophysics Department, Faculty of Science, Cairo University, Giza, Egypt
e-mail: abdo@sci.cu.edu.eg

F. G. Amin
Physics Department, Faculty of Science, Alexandria University, Alexandria, Egypt

W. M. Elshemey
Biophysics Department, Faculty of Science, Cairo University, Giza, Egypt

Physics Department, Faculty of Science, Islamic University in Madinah, Medina, Saudi Arabia

10.1 Introduction

A novel human coronavirus, which broke up in the city of Wuhan in China last year, is the reason for a significant outbreak of the coronavirus disease 2019 (COVID-19), which is currently widely spread worldwide (Elfiky 2020a; Yang 2020). Coronaviruses are common in humans and several other vertebrates. They are known to cause respiratory, enteric, hepatic, and neurologic diseases (Elfiky et al. 2017). Human epidemics of coronaviruses have emerged several times during the past 20 years, such as the severe acute respiratory syndrome coronavirus (SARS-CoV) and Middle East respiratory syndrome coronavirus (MERS-CoV), each one leaving about 800 deaths. HKU1, OC43, 229E, and NL63 are milder strains of human coronaviruses

N. Rezaei (ed.), *Coronavirus Disease - COVID-19*, Advances in Experimental Medicine and Biology 1318, https://doi.org/10.1007/978-3-030-63761-3_10

associated with flu-like symptoms (van den Brand et al. 2015; Hilgenfeld and Peiris 2013). Former human coronaviruses, in comparison with the current COVID-19 pandemic, show several significant differences and similarities. Both MERS-CoV and SARS-CoV have much higher case death rates (36% and 10%, respectively). Although the current SARS-CoV-2 shares 79% of its genome with SARS-CoV, it appears to be much more transmissible (de Wit et al. 2016). Similar to other emerged human coronaviruses, the bat is still a prospective species of origin for SARS-CoV-2 (Li et al. 2005b; Zhou et al. 2020). However, former human coronaviruses, such as SARS-CoV and MERS-CoV, used intermediate hosts, such as civet cat and dromedary camel, respectively (Cui et al. 2019). This fact suggests that SARS-CoV-2 is probably transmitted to humans through an intermediate host (Fig. 10.1)

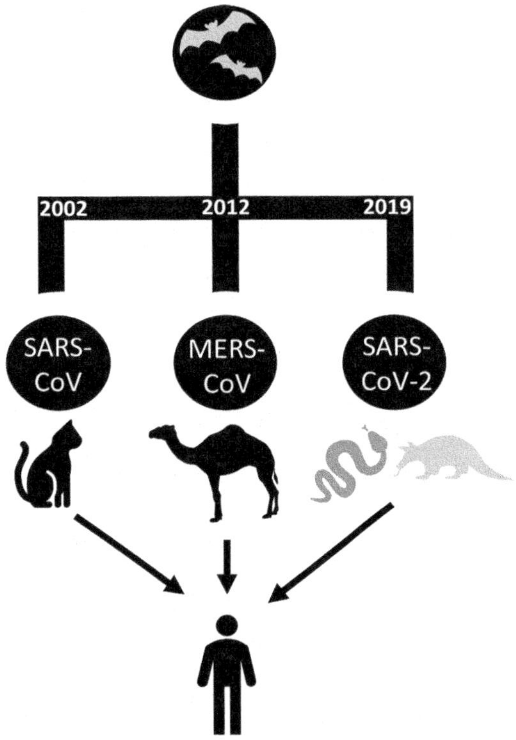

Fig. 10.1 The common reservoir (bat) and the intermediate animal hosts (civet cat, dromedary camel, and unknown animal (snake and pangolin are suspected)) identified for the three different human coronaviruses SARS-CoV, MERS-CoV, and SARS-CoV-2

(Elfiky 2020a). Researchers are still competing to explore SARS-CoV-2 secrets to relive its associated COVID-19 pneumonia (Song et al. 2018).

Mainly, SARS-CoV-2 encodes at least four major structural proteins: spike protein (S), membrane protein (M), an envelope protein (E), and nucleocapsid protein (N) (Wu et al. 2020). Spike protein, which is a type I glycoprotein, protrudes from the surface of the virion and mediates the attachment and virus entry to the host cell. Spike protein has attracted considerable attention because of its function in receptor recognition (Huang et al. 2015; Belouzard et al. 2012).

10.2 SARS-CoV-2 Cell Entry

The first step in coronavirus infection is the binding of the viral homo-trimeric spike protein to the human receptors found at the membrane of the host cell (Ismail and Elfiky 2020; Yan et al. 2020a). Angiotensin-converting enzyme 2 (ACE2) is the primary host cell receptor for some human coronavirus strains, including SARS-CoV-2 (South et al. 2020; Fang et al. 2020). Because the binding of SARS-CoV-2 spike glycoprotein and ACE2 receptor is a critical step for virus entry, virus-receptor binding affinity is under intensive study through different approaches.

It was reported that the angiotensin-converting enzyme 2 (ACE2) binds to the receptor-binding motif in the receptor-binding domain (RBD) of the SARS-CoV spike and functions as a receptor for SARS-CoV (Li et al. 2003, 2005a). ACE2 is widely distributed in the heart, liver, testis, kidney, intestine, and other tissues. It has the physiological functions of controlling heart and kidney function and regulating blood pressure (Anguiano et al. 2017). The structure of human ACE2 in complex with a membrane protein, sodium-dependent neutral amino acid transporter (B0AT1), shows that ACE2 is a dimer (Yan et al. 2020b).

Recently, it has been found that human ACE2 corroborated the entry of SARS-CoV-2 into the cells (Zhou et al. 2020; Letko et al. 2020). Moreover, the crystal structure of the SARS-CoV-2 spike receptor-binding domain bound with ACE2 (PDB ID: 6M0J) indicates how the receptor-binding domain of SARS-CoV-2 interacts with ACE2 and suggests that it is probable that two trimeric spike proteins bind to an ACE2 dimer (Lan et al. 2020). The SARS-CoV-2-ACE2 complex includes a higher number of links, a more extensive interface area, and reduced interface residue fluctuations relative to SARS-CoV. These findings reveal an exceptional evolutionary recognition exerted by coronaviruses toward host recognition. Consequently, the versatility of cell receptor binding strategies has immediate implications for therapeutic approaches.

The glycoprotein spike on the external surface of coronaviruses is responsible for the attachment and entry of the virus to host cells. The receptor-binding domain (RBD) is loosely attached among the virus particle; therefore, it may infect multiple hosts (Raj et al. 2013; Perlman and Netland 2009). Other coronaviruses mostly recognize aminopeptidases or carbohydrates as a critical receptor for entry to human cells (Wang et al. 2013). The entry mechanism of a coronavirus depends upon cellular proteases, which include human airway trypsin-like protease (HAT), cathepsins, and transmembrane protease serine 2 (TMPRSS2) that split the spike protein and establish further penetration changes (Glowacka et al. 2011; Bertram et al. 2011).

Both SARS-CoV and SARS-CoV-2 enter the host cell mainly through the ACE2 receptor (Wan et al. 2020). The SARS-CoV-2 primarily infects lower airways and binds to ACE2 on alveolar epithelial cells. Both viruses are vigorous inducers of inflammatory cytokines (Vaninov 2020). The "cytokine storm" or "cytokine cascade" is the postulated mechanism for organ damage. The virus activates immune cells and boosts the secretion of inflammatory cytokines and chemokines into pulmonary vascular endothelial cells (Andersen et al. 2020; Vaninov 2020).

10.3 Cell Stress

It is a stressful time! COVID-19 creates a plethora of stresses from the sub-cellular level up to the whole globe. Cells can be exposed to a wide range of environmental challenges that impact different sub-cellular levels. Various stressors like xenobiotics, ionizing radiation, hypoxia, chemical toxins, heat stress, oxidative stress, or infectious agents, mainly viruses, lead to perturbation in normal cellular functions and homeostasis (Galluzzi et al. 2018). Cells execute various adaptive mechanisms to cope with stress and re-establish homeostasis, which include heat shock response, unfolded protein response, and autophagy.

Cells mount to appropriate defensive response based on stress duration, cell type, and macromolecular damage. If the stress exceeds the repair capacity and becomes uncontrollable, the cell will turn to apoptosis or programmed cell death (Mehrbod et al. 2019; Samali et al. 2010). The coordination between all regulatory mechanisms is a decision-making for cell fate whether to repair the damage and restore homeostasis or directed to cell demise.

Heat shock response, in general, is a protective response to handle cell stimuli (e.g., heat shock, heavy metals, and oxidative stress) where cells activate the transient expression of chaperones or heat shock proteins (HSP). The chaperons alleviate the heat shock damaged consequences by refolding the unfolded or aggregated proteins. It confers the cells a thermotolerance to gain more resistance for various lethal factors (Samali et al. 2010; Richter et al. 2010).

One member of the heat shock protein 70 (HSP70) is the HSPA5, also termed glucose-regulated protein 78 (GRP78) or binding immunoglobulin protein (Bip). It can be found in the lumen of the endoplasmic reticulum (ER) in all eukaryotic cells (Ibrahim et al. 2019; Lee 2014). HSPA5 is a 654 amino acid protein that corrects the folding, assembles, and prevents the transport of misfolded proteins (Hendershot et al. 1994; Haas 1991; Gething and Sambrook 1992). HSPA5 is a water-soluble protein with small groups of hydrophobic amino acid patches. These patches are responsible for their role in recognition of unfolded proteins (Ting and Lee 1988). HSPA5 has two domains, substrate-binding domain (SBD) at the C-terminal and ATP, or nucleotide, binding domain (ABD or NTP) at the N-terminal (Lindquist and Craig 1988). In the case of cell stress, such as cancer or viral infection, the expression of HSPA5 is increased (Ibrahim et al. 2019).

Autophagy is a multistep process required for the capture and turnover of unfolded proteins or their aggregates, organelles (e.g., endoplasmic reticulum, mitochondria, peroxisome), and invasive pathogens (e.g., bacteria, viruses) through delivering them to the lysosome to be degraded and recycled. The clearance of unwanted materials is governed through three main routes: macroautophagy (Feng et al. 2014; Choi et al. 2018), microautophagy (Mijaljica et al. 2011), and chaperone-mediated autophagy (Kaushik and Cuervo 2018).

Microautophagy is the destruction of cytoplasmic materials through a direct engulfment by the lysosome. During the chaperone-mediated autophagy, the unfolded proteins which have a KFERQ motif are recognized by the cytosolic chaperone (HSPA8). Chaperone translocates the substrates to the lysosome via lysosomal-associated membrane protein 2A (LAMP2A). On the other hand, macroautophagy is the primary route. It begins with the sequestration of cargoes of undesirable cellular constituents by double-membrane vesicle of autophagosome followed by a subsequent fusion with the lysosome to be degraded. The cargoes are degraded into its main components, e.g., sugars, nucleosides, amino acids, and fatty acids, and released into the cytosol to enable their recycling (Feng et al. 2014; Choi et al. 2018). P62 or sequestosome-1 is a critical adaptor molecule that delivers cargoes to autophagosomes; it is a modulator of autophagosome biogenesis. Self-polymerization of p62 is essential for its binding to the cargoes. The binding of its ZZ domain stimulates the polymerization of p62 to the N-terminal arginine of arginylated substrates. For example, in the case of ER-phagy (autophagic degradation of the endoplasmic reticulum), the heat shock protein HSPA5 is N terminally arginylated and bound to the ZZ domain of p62 allowing its polymerization. The complex of polymerized p62 along with the ER transmembrane E3 ligase tripartite motif-containing 13 (TRIM13) drives the ER compartment to autophagosome leading to its lysosomal degradation (Cha-Molstad et al. 2017; Ji et al. 2019). Thus, HSPA5 not only reliefs the ER stress by initiating the UPR but also plays a critical role in ER-phagy.

10.3.1 HSPA5 Functions in Healthy Versus Stressed Cells

Typically, the function of HSPA5 is to bind to misfolded and unfolded proteins and start ER-associated degradation (ERAD), which is responsible for unfolded protein response (UPR) (Ibrahim et al. 2019; Little et al. 1994; Pfaffenbach and Lee 2011). It binds to unfolded proteins through its SBD and prevents their aggregation. This process requires energy, which can be obtained through the hydrolysis of ATP in the NBD domain (Luo et al. 2006). Under normal cell conditions (no stress), HSPA5 is attached to three proteins, which are protein kinase RNA-like endoplasmic reticulum kinase (PERK), inositol-requiring enzyme 1 alpha (IRE1α), and activating transcription factor 6 (ATF6). These three proteins are UPR transmembrane sensors for stress. Under stress conditions such as the accumulation of unfolded proteins in ER, HSPA5 is released from these proteins, activating them, which leads to a reduction in protein translation and enhancement of correct folding (Ibrahim et al. 2019; Pfaffenbach and Lee 2011; Sepulveda et al. 2018).

A more detailed description of what happens is as follows. Activation of PERK requires its dimerization to gain its autophosphorylation activity. After that, it phosphorylates the alpha subunit of eukaryotic translation initiation factor 2 (eIF2α). That, in turn, inhibits the initiation of protein translation and prevents protein synthesis, leading to a reduction in the entry of newly formed proteins into ER. In the case of the activated form of ATF6, it migrates to the Golgi apparatus where its cleavage occurs. The cleaved part, then, moves to the nucleus and acts as an active transcription factor leading to the upregulation of protein transcription that boosts the folding capacity of the ER, such as HSPA5 (Ibrahim et al. 2019; Wang et al. 2009).

On the other hand, the activated form of IRE1α has an endoribonuclease activity. It breaks an intron of length 26-base from the mRNA of X-box binding protein 1 (XBP-1). This protein is a transcriptional factor, which targets genes involved in ERAD, such as protein disulfide isomerase (PDI), ER degradation-enhancing α-mannosidase-like protein (EDEM), endoplasmic reticulum-localized DnaJ 4 (ERdj4), p58, and DnaJ (Wang et al. 2009).

10.3.2 HSPA5 over the Cell Membrane

Under stress, HSPA5 can be found on the surface of the cells where it can interact with a bunch of ligands or other proteins. Cell-surface HSPA5 (CS-HSPA5) plays an essential role in migration, invasion, apoptosis, signaling, and immunity. Some of the ligands that bind to CS-HSPA5 are α2 macroglobulin, Isthmin, Par-4, and plasminogen Kringle 5 (Ibrahim et al. 2019). The proposed mechanism of GRP78 translocation from the ER to the cell surface was described by Tsai et al. (2018). There is a tetra-peptide, KDEL, in the C-terminal of HSPA5 that maintains it inside the ER (Munro and Pelham 1987). Under ER stress conditions, ER chaperones' surface expression is rapidly increased, while the intracellular levels do not change. Thus, it cannot be described by the oversaturation of KDEL receptors' retrieval mechanisms (Tsai et al. 2018). SRC is a non-receptor protein-tyrosine kinase that belongs to SRC family kinases and has many roles in cancer cells (Yeatman 2004) and also has a function in the relocalization of ER chaperones to the cell surface (Tsai et al. 2018). IRE1α can activate SRC through phosphorylation of Y419 amino acid (Tsai et al. 2018). Upon SRC activation, it triggers a signaling cascade reducing, partially, the retrograde chaperones' trafficking from Golgi to ER, allowing escaping of a subfraction of ER chaperones to the cell surface (Tsai et al. 2018).

10.3.3 HSPA5 in Cancer and Viral Infection

HSPA5 has many roles in different cancer types. In breast cancer, the overexpression of HSPA5 reduces the sensitivity to gemcitabine by inhibiting apoptosis (Xie et al. 2016). Besides, HSPA5 overexpression in ovarian carcinoma protects paclitaxel and cisplatin treatments (Zhang et al. 2015; Chen and Xu 2017). On the other hand, overexpression of HSPA5 in pancreatic ductal adenocarcinoma (PDAC) leads to a decrease in patient survivability (Niu et al. 2015).

In addition to the previous roles of CS-HSPA5, it can also act as a receptor for some viruses (Ni et al. 2011; Ibrahim et al. 2019).

Zika virus (ZIKV), which is a flavivirus related to microcephaly, uses its envelope protein for cellular attachment. One of the receptors that help its entry is HSPA5 (Smit et al. 2011; Ojha et al. 2018; Elfiky and Ibrahim 2020). Another virus that can handle the CS-HSPA5 as its receptor is the dengue virus (DENV) (Jindadamrongwech et al. 2004). Moreover, the Japanese encephalitis virus (JEV), which is a neurotropic virus that causes encephalitis in humans with high mortality reaching 30%, can use CS-HSPA5 as its receptor in addition to its two receptors, heparan sulfate proteoglycans (HSPGs) and glycosaminoglycans (GAGs). CS-HSPA5 can bind to JEV envelope domain III (Nain et al. 2017; Chien et al. 2008; Chiou et al. 2005). The MERS-CoV is a human coronavirus that causes severe lower respiratory tract infection, with mortality reaching 36%. MERS-CoV can utilize CS-HSPA5, which promotes its entry in the presence of its functional receptor dipeptidyl peptidase 4 (DPP4) (Chu et al. 2018; Chan et al. 2015). Ebola virus (EBOV) belongs to the filovirus family and has very high mortality reaching 90%. When the cell is under stress, the HSPA5 is overexpressed and escapes ER retention reaching the cell surface (Elfiky 2020b). Two EBOV glycoproteins (GP1 and GP2) may accumulate in the ER and cause cell stress (Bhattacharyya and Hope 2011). GP1 is the viral protein required for virus entry. Overexpression of HSPA5 is reported to increase EBOV infectivity. Recently, an *in silico* study provided a prediction of the binding site between GRP78 and GP1. This would pave the route for the design of drugs to prevent this binding (Elfiky 2020b).

HSPA5 also has a role related to the viruses inside the cell. Human papillomavirus (HPV) is reported as the leading cause of cervical, head, anal, and throat cancers. It has more than 150 different strains with HPV16 as the most common cause of cervical cancer worldwide (Elfiky 2020d). One of the nonstructural proteins in HPV16 is E6, which is reported as the element responsible for cancer cell proliferation (Kaliamurthi et al. 2018). One of the host cell proteins that bind to E6 protein is HSPA5. This binding increases the E6 lifetime *in vivo* because HSPA5 protects it from degradation through the proteasome degradation pathway (Ajiro and Zheng 2015). E6 destabilization is a strategy to prevent cancer proliferation. An *in silico* study provided a prediction of the binding site between the HSPA5 and E6 protein aiming to show the possible targeting binding site (Elfiky 2020d).

10.3.4 CS-HSPA5 Targeting

Severe side effects are accompanying the treatment for cancer; therefore, searching for other ways for cancer treatment is essential (Fennelly and Schneider 1995; Alexandrescu et al. 2005). One of the methods is to use specific ligands that can bind to cancer cells only (Landon and Deutscher 2003). These ligands can be used as a vehicle to transport chemotherapeutics specifically to cancer cells. Three different types of vehicles are used to target CS-HSPA5: peptides, monoclonal antibodies, and natural compounds Elfiky et al. 2020a). Pep42 is a cyclic peptide of 13 amino acids (CTVALPGGYVRVC). This peptide binds specifically to CS-HSPA5. Besides, it is mainly hydrophobic, supporting the primary role of HSPA5, which can bind to unfolded hydrophobic amino acid patches (Kim et al. 2006).

Since CS-HSPA5 acts as a receptor for many viruses, as indicated above, it may be an excellent strategy to try to decrease or inhibit the CS-HSPA5 expression over the cell membrane to fight the infection. Many natural compounds can reduce the expression of CS-HSPA5. Iridoid glucoside catapol, which is extracted from *Radix rehmannia*, is reported to inhibit the expression of HSPA5 in human aortic endothelial cells (Hu et al. 2019). An antitumor flavonoid extracted from *Sophora flavescens* Aiton named kurarinone decreases the expression of HSPA5 in a non-small cell lung cancer cell line (Yang et al. 2018). One of the most exciting HSPA5 downregulators is triptolide. It is the bioactive compound in *Tripterygium wilfordii* Hook F, which can decrease the expression of HSPA5 in leukemia cells (Li et al. 2016). This reduction of HSPA5 by triptolide causes cell death by either apoptosis or autophagy (Chan et al. 2017). Polyphenol epigallocatechin-3-gallate (EGCG) (found in green tea) decreases the CS-HSPA5; however, it increases the HSPA5 expression in the ER (Martinotti et al. 2018).

10.4 COVID-19 Inhibition by Targeting CS-HSPA5 and/or Spike Protein

As mentioned before, the spike protein is a vital viral element that aids in the attachment and virus internalization to the host cell (Ha et al. 2020). A significant amount of host cell receptors are targets for viruses, including the cell-surface HSPA5. Inhibiting the interaction that happens between the COVID-19 spike protein and the host cell receptor HSPA5 would probably decrease the rate of viral infection. Besides, a vaccine against the COVID-19 spike protein would inhibit viral infection. HSPA5 against SARS-CoV-2 spike protein docking has been performed using the Haddock software in four different regions of the spike (Ibrahim et al. 2020). Each region of the spike (predicted to be the binding site due to its hydrophobic and cyclic nature resembling the Pep42) has been utilized as the binding site to HSPA5, using its active residues that have hydrophobic character.

Additionally, protein-protein docking has been performed to test the four regions of the spike that fit tightly in the HSPA5 substrate-binding domain β (SBDβ). The docking pose demonstrates the involvement of the SBDβ of HSPA5 and the receptor-binding domain of the coronavirus spike in recognition of the host cell receptor (C480-C488 region) (Ibrahim et al. 2020). So targeting both HSPA5 and SARS-CoV-2 spike is essential in fighting COVID-19. Additionally, the SARS-CoV-2 spike binding site to HSPA5 is conserved in four different strains of human coronaviruses and hence cross-vaccination is suggested (Elfiky 2020e, Elfiky et al. 2020b). Another work following the same method investigated 13 natural compounds for their binding affinity to the CS-HSPA5 SBDβ domain, including daidzein, genistein, formononetin, biochanin A, palmitic acid, linolenic acid, chlorogenic acid, hydroxytyrosol, caffeic acid, caffeic acid phenethyl ester, p-coumaric acid, cinnamaldehyde, and thymoquinone. The results reveal that the phytoestrogens (daidzein, genistein, formononetin, and biochanin A) have a binding affinity ranging from moderate to high compared with Pep42, which is the reference HSPA5 substrate selected by the author (Elfiky 2020c).

10.5 Conclusion

Although the COVID-19 pandemic is not globally characterized by stress (due to the lockdown and quarantine), it is associated with cellular stress. Human cells respond to stress through some key regulators, such as chaperone protein HSPA5. Cell response to stress may be beneficial to the virulence of the SARS-CoV-2, and hence an external supplement of chaperone inhibitors may help to fight COVID-19. In the current chapter, we summarized the current knowledge of the possible link between cell stress and COVID-19 infectivity. The use of HSPA5 inhibitors, in addition to SARS-CoV-2 spike protein targets, can alleviate SARS-CoV-2 infectivity.

References

Ajiro M, Zheng Z-MJM (2015) E6^ E7, a novel splice isoform protein of human papillomavirus 16, stabilizes viral E6 and E7 oncoproteins via HSP90 and GRP78. mBio 6(1):e02068–e02014

Alexandrescu D, Dutcher J, Wiernik PJ (2005) Metastatic melanoma: is biochemotherapy the future? Med Oncol 22(2):101–111

Andersen KG, Rambaut A, Lipkin WI, Holmes EC, Garry RF (2020) The proximal origin of SARS-CoV-2. Nat Med 26(4):450–452. https://doi.org/10.1038/s41591-020-0820-9

Anguiano L, Riera M, Pascual J, Soler MJ (2017) Circulating ACE2 in cardiovascular and kidney diseases. Curr Med Chem 24(30):3231–3241. https://doi.org/10.2174/0929867324666170414162841

Belouzard S, Millet JK, Licitra BN, Whittaker GR (2012) Mechanisms of coronavirus cell entry mediated by the viral spike protein. Viruses 4(6):1011–1033. https://doi.org/10.3390/v4061011

Bertram S, Glowacka I, Muller MA, Lavender H, Gnirss K, Nehlmeier I, Niemeyer D, He Y, Simmons G, Drosten C, Soilleux EJ, Jahn O, Steffen I, Pohlmann S (2011) Cleavage and activation of the severe acute respiratory syndrome coronavirus spike protein by human airway trypsin-like protease. J Virol 85(24):13363–13372. https://doi.org/10.1128/JVI.05300-11

Bhattacharyya S, Hope T (2011) Full-length Ebola glycoprotein accumulates in the endoplasmic reticulum. Virol J 8(1):11

Cha-Molstad H, Yu JE, Feng Z, Lee SH, Kim JG, Yang P, Han B, Sung KW, Yoo YD, Hwang J, McGuire T, Shim SM, Song HD, Ganipisetti S, Wang N, Jang JM, Lee MJ, Kim SJ, Lee KH, Hong JT, Ciechanover A, Mook-Jung I, Kim KP, Xie XQ, Kwon YT, Kim BY (2017) p62/SQSTM1/Sequestosome-1 is an N-recognin of the N-end rule pathway which modulates autophagosome biogenesis. Nat Commun 8(1):102. https://doi.org/10.1038/s41467-017-00085-7

Chan JF, Lau SK, To KK, Cheng VC, Woo PC, Yuen K-Y (2015) Middle East respiratory syndrome coronavirus: another zoonotic betacoronavirus causing SARS-like disease. Clin Microbiol Rev 28(2):465–522

Chan SF, Chen YY, Lin JJ, Liao CL, Ko YC, Tang NY, Kuo CL, Liu KC, Chung JG (2017) Triptolide induced cell death through apoptosis and autophagy in murine leukemia WEHI-3 cells in vitro and promoting immune responses in WEHI-3 generated leukemia mice in vivo. Environ Toxicol 32(2):550–568

Chen T, Xu S (2017) Chronic exposure of cisplatin induces GRP78 expression in ovarian cancer. In: Proceedings of the 2017 4th international conference on biomedical and bioinformatics engineering. ACM, pp 35–38

Chien Y-J, Chen W-J, Hsu W-L, Chiou S-S (2008) Bovine lactoferrin inhibits Japanese encephalitis virus by binding to heparan sulfate and receptor for low density lipoprotein. Virology 379(1):143–151

Chiou SS, Liu H, Chuang CK, Lin CC, Chen WJ (2005) Fitness of Japanese encephalitis virus to Neuro-2a cells is determined by interactions of the viral envelope protein with highly sulfated glycosaminoglycans on the cell surface. J Med Virol 76(4):583–592

Choi Y, Bowman JW, Jung JU (2018) Autophagy during viral infection – a double-edged sword. Nat Rev Microbiol 16(6):341–354. https://doi.org/10.1038/s41579-018-0003-6

Chu H, Chan C-M, Zhang X, Wang Y, Yuan S, Zhou J, Au-Yeung RK-H, Sze K-H, Yang D, Shuai H (2018) Middle East respiratory syndrome coronavirus and bat coronavirus HKU9 both can utilize GRP78 for attachment onto host cells. J Biol Chem 293(30):11709–11726

Cui J, Li F, Shi ZL (2019) Origin and evolution of pathogenic coronaviruses. Nat Rev Microbiol 17(3):181–192. https://doi.org/10.1038/s41579-018-0118-9

de Wit E, van Doremalen N, Falzarano D, Munster VJ (2016) SARS and MERS: recent insights into emerging coronaviruses. Nat Rev Microbiol 14(8):523–534. https://doi.org/10.1038/nrmicro.2016.81

Elfiky AA (2020a) Anti-HCV, nucleotide inhibitors, repurposing against COVID-19. Life Sci 248:117477. https://doi.org/10.1016/j.lfs.2020.117477

Elfiky AA (2020b) Ebola virus glycoprotein GP1-host cell-surface HSPA5 binding site prediction. Cell Stress Chaperones 25(3):541–548. https://doi.org/10.1007/s12192-020-01106-z

Elfiky AA (2020c) Natural products may interfere with SARS-CoV-2 attachment to the host cell. (just-accepted):1–16

Elfiky AA (2020d) Human papillomavirus E6: host cell receptor, GRP78, binding site prediction 92(12):3759-3765. https://doi.org/10.1002/jmv.25737

Elfiky AA (2020e) SARS-CoV-2 Spike-Heat Shock Protein A5 (GRP78) recognition may be related to the immersed human coronaviruses. Frontiers in Pharmacology. (In press). https://doi.org/10.3389/fphar.2020.577467

Elfiky AA, Ibrahim IM (2020) Zika virus envelope – heat shock protein A5 (GRP78) binding site prediction. J Biomol Struct Dyn:1–13. https://doi.org/10.1080/07391102.2020.1784794

Elfiky AA, Mahdy SM, Elshemey WM (2017) Quantitative structure-activity relationship and molecular docking revealed a potency of anti-hepatitis C virus drugs against human corona viruses. J Med Virol 89(6):1040–1047. https://doi.org/10.1002/jmv.24736

Elfiky AA, Ibrahim IM, Ismail AM, Elshemey WM (2020a) A possible role for GRP78 in cross vaccination against COVID-19. Journal of Infection. https://doi.org/10.1016/j.jinf.2020.09.004

Elfiky AA, Baghdady AM, Ali SA, Ahmed MI (2020b) GRP78 targeting: Hitting two birds with a stone. Life Sciences 260:118317. https://doi.org/10.1016/j.lfs.2020.118317

Fang L, Karakiulakis G, Roth M (2020) Are patients with hypertension and diabetes mellitus at increased risk for COVID-19 infection? Lancet Respir Med 8(4):e21. https://doi.org/10.1016/S2213-2600(20)30116-8

Feng Y, He D, Yao Z, Klionsky DJ (2014) The machinery of macroautophagy. Cell Res 24(1):24–41. https://doi.org/10.1038/cr.2013.168

Fennelly D, Schneider J (1995) Role of chemotherapy dose intensification in the treatment of advanced ovarian cancer. Oncology 9(10):922

Galluzzi L, Yamazaki T, Kroemer G (2018) Linking cellular stress responses to systemic homeostasis. Nat Rev Mol Cell Biol 19(11):731–745. https://doi.org/10.1038/s41580-018-0068-0

Gething M-J, Sambrook J (1992) Protein folding in the cell 355(6355):33–45

Glowacka I, Bertram S, Muller MA, Allen P, Soilleux E, Pfefferle S, Steffen I, Tsegaye TS, He Y, Gnirss K, Niemeyer D, Schneider H, Drosten C, Pohlmann S (2011) Evidence that TMPRSS2 activates the severe acute respiratory syndrome coronavirus spike protein for membrane fusion and reduces viral control by the humoral immune response. J Virol 85(9):4122–4134. https://doi.org/10.1128/JVI.02232-10

Ha DP, Van Krieken R, Carlos AJ, Lee AS (2020) The stress-inducible molecular chaperone GRP78 as potential therapeutic target for coronavirus infection. J Infect:S0163-4453(0120)30398-30394. https://doi.org/10.1016/j.jinf.2020.06.017

Haas I (1991) BiP – a heat shock protein involved in immunoglobulin chain assembly. Curr Topics Microbiol Immunol 167:71–82

Hendershot LM, Valentine VA, Lee AS, Morris SW, Shapiro DN (1994) Localization of the gene encoding human BiP/GRP78, the endoplasmic reticulum cognate of the HSP70 family, to chromosome 9q34. Genomics 20(2):281–284. https://doi.org/10.1006/geno.1994.1166

Hilgenfeld R, Peiris M (2013) From SARS to MERS: 10 years of research on highly pathogenic human coronaviruses. Antivir Res 100(1):286–295. https://doi.org/10.1016/j.antiviral.2013.08.015

Hu H, Wang C, Jin Y, Meng Q, Liu Q, Liu Z, Liu K, Liu X, Sun H (2019) Catalpol inhibits homocysteine-induced oxidation and inflammation via inhibiting Nox4/NF-κB and GRP78/PERK pathways in human aorta endothelial cells. Inflammation 42(1):64–80

Huang X, Dong W, Milewska A, Golda A, Qi Y, Zhu QK, Marasco WA, Baric RS, Sims AC, Pyrc K, Li W, Sui J (2015) Human coronavirus HKU1 spike protein uses O-acetylated sialic acid as an attachment receptor determinant and employs hemagglutinin-esterase protein as a receptor-destroying enzyme. J Virol 89(14):7202–7213. https://doi.org/10.1128/JVI.00854-15

Ibrahim IM, Abdelmalek DH, Elfiky AA (2019) GRP78: a cell's response to stress. Life Sci 226:156–163. https://doi.org/10.1016/j.lfs.2019.04.022

Ibrahim IM, Abdelmalek DH, Elshahat ME, Elfiky AA (2020) COVID-19 spike-host cell receptor GRP78 binding site prediction. J Infect 80(5):554–562. https://doi.org/10.1016/j.jinf.2020.02.026

Ismail AM, Elfiky AA (2020) SARS-CoV-2 spike behavior in situ: a Cryo-EM images for a better understanding of the COVID-19 pandemic. Signal Transduction and Targeted Therapy 5(252)

Ji CH, Kim HY, Heo AJ, Lee SH, Lee MJ, Kim SB, Srinivasrao G, Mun SR, Cha-Molstad H, Ciechanover A, Choi CY, Lee HG, Kim BY, Kwon YT (2019) The N-Degron pathway mediates ER-phagy. Mol Cell 75(5):1058–1072. e1059. https://doi.org/10.1016/j.molcel.2019.06.028

Jindadamrongwech S, Thepparit C, Smith D (2004) Identification of GRP 78 (BiP) as a liver cell expressed receptor element for dengue virus serotype 2. Arch Virol 149(5):915–927

Kaliamurthi S, Selvaraj G, Kaushik AC, Gu K-R, Wei D-Q (2018) Designing of CD8+ and CD8+-overlapped CD4+ epitope vaccine by targeting late and early proteins of human papillomavirus. Biologics 12:107

Kaushik S, Cuervo AM (2018) The coming of age of chaperone-mediated autophagy. Nat Rev Mol Cell Biol 19(6):365–381. https://doi.org/10.1038/s41580-018-0001-6

Kim Y, Lillo AM, Steiniger SC, Liu Y, Ballatore C, Anichini A, Mortarini R, Kaufmann GF, Zhou B, Felding-Habermann B (2006) Targeting heat shock proteins on cancer cells: selection, characterization, and cell-penetrating properties of a peptidic GRP78 ligand. Biochemistry 45(31):9434–9444

Lan J, Ge J, Yu J, Shan S, Zhou H, Fan S, Zhang Q, Shi X, Wang Q, Zhang L, Wang X (2020) Structure of the SARS-CoV-2 spike receptor-binding domain bound to the ACE2 receptor. Nature 581(7807):215–220. https://doi.org/10.1038/s41586-020-2180-5

Landon LA, Deutscher SL (2003) Combinatorial discovery of tumor targeting peptides using phage display. J Cell Biochem 90(3):509–517

Lee AS (2014) Glucose-regulated proteins in cancer: molecular mechanisms and therapeutic potential. Nat Rev Cancer 14(4):263–276. https://doi.org/10.1038/nrc3701

Letko M, Marzi A, Munster V (2020) Functional assessment of cell entry and receptor usage for SARS-CoV-2 and other lineage B betacoronaviruses. Nat Microbiol 5(4):562–569. https://doi.org/10.1038/s41564-020-0688-y

Li W, Moore MJ, Vasilieva N, Sui J, Wong SK, Berne MA, Somasundaran M, Sullivan JL, Luzuriaga K, Greenough TC, Choe H, Farzan M (2003) Angiotensin-converting enzyme 2 is a functional receptor for the SARS coronavirus. Nature 426(6965):450–454. https://doi.org/10.1038/nature02145

Li F, Li W, Farzan M, Harrison SC (2005a) Structure of SARS coronavirus spike receptor-binding domain complexed with receptor. Science (New York, NY) 309(5742):1864–1868. https://doi.org/10.1126/science.1116480

Li W, Shi Z, Yu M, Ren W, Smith C, Epstein JH, Wang H, Crameri G, Hu Z, Zhang H, Zhang J, McEachern J, Field H, Daszak P, Eaton BT, Zhang S, Wang LF (2005b) Bats are natural reservoirs of SARS-like coronaviruses. Science (New York, NY) 310(5748):676–679. https://doi.org/10.1126/science.1118391

Li C, Zhang B, Lv W, Lai C, Chen Z, Wang R, Long X, Feng X (2016) Triptolide inhibits cell growth and GRP78 protein expression but induces cell apoptosis in original and radioresistant NPC cells. Oncotarget 7(31):49588

Lindquist S, Craig E (1988) The heat-shock proteins 22(1):631–677

Little E, Ramakrishnan M, Roy B, Gazit G, Lee AS (1994) The glucose-regulated proteins (GRP78 and GRP94): functions, gene regulation, and applications. Crit Rev Eukaryotic Gene Exp 4(1):1–18

Luo S, Mao C, Lee B, Lee AS (2006) GRP78/BiP is required for cell proliferation and protecting the inner cell mass from apoptosis during early mouse embryonic development. Mol Cell Biol 26(15):5688–5697

Martinotti S, Ranzato E, Burlando B (2018) (−)-Epigallocatechin-3-gallate induces GRP78 accumulation in the ER and shifts mesothelioma constitutive UPR into proapoptotic ER stress. J Cell Physiol 233(10):7082–7090

Mehrbod P, Ande SR, Alizadeh J, Rahimizadeh S, Shariati A, Malek H, Hashemi M, Glover KKM, Sher AA, Coombs KM, Ghavami S (2019) The roles of apoptosis, autophagy and unfolded protein response in arbovirus, influenza virus, and HIV infections. Virulence 10(1):376–413. https://doi.org/10.1080/21505594.2019.1605803

Mijaljica D, Prescott M, Devenish RJ (2011) Microautophagy in mammalian cells: revisiting a 40-year-old conundrum. Autophagy 7(7):673–682. https://doi.org/10.4161/auto.7.7.14733

Munro S, Pelham HR (1987) A C-terminal signal prevents secretion of luminal ER proteins. Cell 48(5):899–907

Nain M, Mukherjee S, Karmakar SP, Paton AW, Paton JC, Abdin M, Basu A, Kalia M, Vrati S (2017) GRP78 is an important host factor for Japanese encephalitis virus entry and replication in mammalian cells. J Virol 91(6):e02274–e02216

Ni M, Zhang Y, Lee AS (2011) Beyond the endoplasmic reticulum: atypical GRP78 in cell viability, signalling and therapeutic targeting. Biochem J 434(2):181–188

Niu Z, Wang M, Zhou L, Yao L, Liao Q, Zhao Y (2015) Elevated GRP78 expression is associated with poor prognosis in patients with pancreatic cancer. Sci Rep 5(1):1–12

Ojha CR, Rodriguez M, Lapierre J, Karuppan M, Kumar M, Branscome H, Kashanchi F, El-Hage N (2018) Complementary mechanisms potentially involved in the pathology of Zika virus. Front Immunol 9:2340

Perlman S, Netland J (2009) Coronaviruses post-SARS: update on replication and pathogenesis. Nat Rev Microbiol 7(6):439–450. https://doi.org/10.1038/nrmicro2147

Pfaffenbach KT, Lee AS (2011) The critical role of GRP78 in physiologic and pathologic stress. Curr Opin Cell Biol 23(2):150–156

Raj VS, Mou H, Smits SL, Dekkers DH, Muller MA, Dijkman R, Muth D, Demmers JA, Zaki A, Fouchier RA, Thiel V, Drosten C, Rottier PJ, Osterhaus AD, Bosch BJ, Haagmans BL (2013) Dipeptidyl peptidase 4 is a functional receptor for the emerging human coronavirus-EMC. Nature 495(7440):251–254. https://doi.org/10.1038/nature12005

Richter K, Haslbeck M, Buchner J (2010) The heat shock response: life on the verge of death. Mol Cell 40(2):253–266. https://doi.org/10.1016/j.molcel.2010.10.006

Samali A, Fulda S, Gorman AM, Hori O, Srinivasula SM (2010) Cell stress and cell death. Int J Cell Biol 2010:245803. https://doi.org/10.1155/2010/245803

Sepulveda D, Rojas-Rivera D, Rodriguez DA, Groenendyk J, Köhler A, Lebeaupin C, Ito S, Urra H, Carreras-Sureda A, Hazari Y (2018) Interactome screening identifies the ER luminal chaperone Hsp47 as a regulator of the unfolded protein response transducer IRE1α. Mol Cell 69(2):238–252. e237

Smit JM, Moesker B, Rodenhuis-Zybert I, Wilschut J (2011) Flavivirus cell entry and membrane fusion. Virus 3(2):160–171

Song W, Gui M, Wang X, Xiang Y (2018) Cryo-EM structure of the SARS coronavirus spike glycoprotein in complex with its host cell receptor ACE2. PLoS Pathog 14(8):e1007236. https://doi.org/10.1371/journal.ppat.1007236

South AM, Diz DI, Chappell MC (2020) COVID-19, ACE2, and the cardiovascular consequences. Am J Physiol Heart Circ Physiol 318(5):H1084–H1090. https://doi.org/10.1152/ajpheart.00217.2020

Ting J, Lee AS (1988) Human gene encoding the 78,000-dalton glucose-regulated protein and its pseudogene: structure, conservation, and regulation. DNA 7(4):275–286

Tsai Y-L, Ha DP, Zhao H, Carlos AJ, Wei S, Pun TK, Wu K, Zandi E, Kelly K, Lee AS (2018) Endoplasmic reticulum stress activates SRC, relocating chaperones to the cell surface where GRP78/CD109 blocks TGF-β signaling. Proc Natl Acad Sci U S A 115(18):E4245–E4254

van den Brand JM, Smits SL, Haagmans BL (2015) Pathogenesis of Middle East respiratory syndrome coronavirus. J Pathol 235(2):175–184. https://doi.org/10.1002/path.4458

Vaninov N (2020) In the eye of the COVID-19 cytokine storm. Nat Rev Immunol 20(5):277. https://doi.org/10.1038/s41577-020-0305-6

Wan Y, Shang J, Graham R, Baric RS, Li F (2020) Receptor recognition by the novel coronavirus from Wuhan: an analysis based on decade-long structural studies of SARS coronavirus. J Virol 94(7):e00127–e00120. https://doi.org/10.1128/JVI.00127-20

Wang M, Wey S, Zhang Y, Ye R, Lee AS (2009) Role of the unfolded protein response regulator GRP78/BiP in development, cancer, and neurological disorders. Antioxidants Redox Signal 11(9):2307–2316

Wang N, Shi X, Jiang L, Zhang S, Wang D, Tong P, Guo D, Fu L, Cui Y, Liu X, Arledge KC, Chen YH, Zhang L, Wang X (2013) Structure of MERS-CoV spike receptor-binding domain complexed with human receptor DPP4. Cell Res 23(8):986–993. https://doi.org/10.1038/cr.2013.92

Wu A, Peng Y, Huang B, Ding X, Wang X, Niu P, Meng J, Zhu Z, Zhang Z, Wang J, Sheng J, Quan L, Xia Z, Tan W, Cheng G, Jiang T (2020) Genome composition and divergence of the novel coronavirus (2019-nCoV) originating in China. Cell Host Microbe 27(3):325–328. https://doi.org/10.1016/j.chom.2020.02.001

Xie J, Tao Z-H, Zhao J, Li T, Wu Z-H, Zhang J-F, Zhang J, Hu X-C (2016) Glucose regulated protein 78 (GRP78) inhibits apoptosis and attentinutes chemosensitivity of gemcitabine in breast cancer cell via AKT/mitochondrial apoptotic pathway. Biochem Biophys Res Commun 474(3):612–619

Yan R, Zhang Y, Li Y, Xia L, Guo Y, Zhou Q (2020a) Structural basis for the recognition of SARS-CoV-2 by full-length human ACE2. Science (New York, NY) 367(6485):1444–1448. https://doi.org/10.1126/science.abb2762

Yan R, Zhang Y, Li Y, Xia L, Zhou Q (2020b) Structure of dimeric full-length human ACE2 in complex with B_0T1. bioRxiv:2020.2002.2017.951848. https://doi.org/10.1101/2020.02.17.951848

Yang L (2020) China confirms human-to-human transmission of coronavirus

Yang J, Chen H, Wang Q, Deng S, Huang M, Ma X, Song P, Du J, Huang Y, Wen Y (2018) Inhibitory effect of kurarinone on growth of human non-small cell lung cancer: an experimental study both in vitro and in vivo studies. Front Pharmacol 9:252

Yeatman TJ (2004) A renaissance for SRC. Nat Rev Cancer 4(6):470–480

Zhang L-Y, Li P-L, Xu A, Zhang X-C (2015) Involvement of GRP78 in the resistance of ovarian carcinoma cells to paclitaxel. Asian Pac J Cancer Prev 16(8):3517–3522

Zhou P, Yang XL, Wang XG, Hu B, Zhang L, Zhang W, Si HR, Zhu Y, Li B, Huang CL, Chen HD, Chen J, Luo Y, Guo H, Jiang RD, Liu MQ, Chen Y, Shen XR, Wang X, Zheng XS, Zhao K, Chen QJ, Deng F, Liu LL, Yan B, Zhan FX, Wang YY, Xiao GF, Shi ZL (2020) A pneumonia outbreak associated with a new coronavirus of probable bat origin. Nature 579(7798):270–273. https://doi.org/10.1038/s41586-020-2012-7

Clinical Manifestations of COVID-19

11

Mahsa Eskian and Nima Rezaei

Abstract

Severe acute respiratory syndrome coronavirus 2 (SARS-CoV-2) is the third coronavirus causing an outbreak in the twenty-first century. It is related to a contagious coronavirus disease (COVID-19), which its high pace of spreading allowed it to lie to the whole world and be turned into a pandemic only a few months after the identification of the first case. Currently, the reverse transcription-polymerase chain reaction (RT-PCR) test of throat swap is the gold standard of diagnosis; however, several studies have reported false-negative results with non-ideal sensitivity. Because this pandemic constitutes a significant burden on global healthcare systems and due to the high transmission rate of the virus, an accurate diagnosis algorithm is needed to reduce the missing case number. A comprehensive clinical examination and taking a history of all systems (not just limited to the respiratory system) combined with hematologic laboratory tests and chest imaging can lead to a sensitive diagnosis, severity assessment, and RT-PCT test interpretation. This chapter focuses on clinical characteristics, hematologic laboratory, and chest imaging features in COVID-19.

M. Eskian (✉)
Network of Immunity in Infection, Malignancy and Autoimmunity (NIIMA), Universal Scientific Education and Research Network (USERN), Tehran, Iran

Department of Radiology, Center for Evidence-Based Imaging, Brigham and Women's Hospital, Harvard Medical School, Boston, MA, USA

Research Center for Immunodeficiencies, Children's Medical Center, Tehran University of Medical Sciences, Tehran, Iran
e-mail: meskian@bwh.harvard.edu

N. Rezaei (✉)
Network of Immunity in Infection, Malignancy and Autoimmunity (NIIMA), Universal Scientific Education and Research Network (USERN), Tehran, Iran

Research Center for Immunodeficiencies, Children's Medical Center, Tehran University of Medical Sciences, Tehran, Iran

Department of Immunology, School of Medicine, Tehran University of Medical Sciences, Tehran, Iran
e-mail: rezaei_nima@tums.ac.ir

Keywords

Clinical characteristics · COVID-19 · Imaging · Laboratory · SARS-CoV-2 · Symptoms

11.1 Introduction

Coronaviruses are enveloped, positive-sense, single-stranded RNA viruses, capable of rapid mutation and recombination and widely existing

N. Rezaei (ed.), *Coronavirus Disease - COVID-19*, Advances in Experimental Medicine and Biology 1318, https://doi.org/10.1007/978-3-030-63761-3_11

in nature. They are classified into alphacoronaviruses and betacoronaviruses, which both have their gene source from bats and are mainly present in mammals such as bats, rodents, civets, and humans, and gammacoronaviruses and deltacoronaviruses, which both have their gene source from birds and mainly occur in birds (Woo et al. 2005; Lau et al. 2015). Betacoronaviruses have proven to be highly pathogenic and contagious pathogens, leading to respiratory, digestive, hepatic, and nervous system disorders (Weiss and Leibowitz 2011; Chen et al. 2020d).

Two of coronaviruses can cause acute respiratory distress syndrome (ARDS), acute respiratory syndrome coronavirus (SARS-CoV) and Middle East respiratory syndrome coronavirus (MERS-CoV), which caused two outbreaks in 2002 in China and 2012 in the Middle East, respectively. SARS-CoV-2 is the seventh beta-coronavirus known to infect humans and cause coronavirus disease in 2019 (COVID-19). At the end of the year 2019, it was discovered in Wuhan, China, with the potential of causing acute respiratory distress syndrome (ARDS) and transmission between humans (Cleary et al. 2020; Phan et al. 2020; Lotfi et al. 2020).

COVID-19 pandemic is the third outbreak of coronavirus in the twenty-first century (Organization 2020; Jabbari et al. 2020).

In December 2019, the first case of COVID-19 pneumonia was reported in Wuhan, China. Infection rapidly spread worldwide in almost 3 months, and World Health Organization announced COVID-19 outbreak as a pandemic and a public health emergency of international concern with a total of 63,965,092 confirmed cases and 1,488,120 total death worldwide as of Dec 3, 2020 (Hick and Biddinger 2020; Lotfi and Rezaei 2020; Hanaei and Rezaei 2020; Organization 2020).

Currently, the source and pathogenesis of the COVID-19 remain unclear, and there are no uniform diagnostic and treatment standards. Unfortunately, in certain patients, the disease progresses rapidly, and respiratory failure can occur within a short time, even leading to death (Liu et al. 2020). Thus, early identification of patients is necessary to prevent and control this outbreak (Zhu et al. 2020d).

The reverse transcription-polymerase chain reaction (RT-PCR) for viral nucleic acid is the current gold standard method to detect and diagnose COVID-19. However, several studies have reported false-negative results and reduced sensitivity for this test (Huang et al. 2020a; Suzuki et al. 2020; Basiri et al. 2020a). For instance, in Wu et al. study with 80 confirmed COVID-19 cases, more than 10% passed 3 tests (with at least 1-day sampling time interval) before they got positive results. Therefore, if the suspected cases are excluded based on even two consecutively negative respiratory pathogenic nucleic acid test results, about 10% of the infected patients will be missed (Wu et al. 2020a). Xu et al. also reported that the detection of the nucleic acid of the SARS-CoV-2 was still negative in some patients on the sixth to the eighth day after the onset of disease (Xu et al. 2020b). In another study, despite negative nucleic acid test results, all 56 patients showed high specific IgG concentrations, suggesting SARS-CoV-2 infection (Dong et al. 2020). It has also been reported that the SARS-CoV-2 was detected by RT-PCR in the upper respiratory tract, even after full resolution of symptoms (Lescure et al. 2020).

Studies reported the sensitivity of the RT-PCR method on throat swab samples ranging from 30% to 60% (Ai et al. 2020). Moreover, it has been reported that the number of days between symptom onset and positive RT-PCR ranged from 1 to 19 days (median 14 days) (Himoto et al. 2020), which can be due to the testing errors or sample collection method of throat swap (Dong et al. 2020). COVID-19 targets human angiotensin-converting enzyme 2 (ACE2) (Sharifkashani et al. 2020; Ahmadi et al. 2020) and infects intrapulmonary epithelial cells more than cells of the upper airways, which may be the reason for the false-negative results of throat swab specimens for RT-PCR assays (Veeramachaneni et al. 2020).

Therefore, non-ideal sensitivity and specificity of RT-PCR testing of the nasopharyngeal sample, as well as the limited number of RT-PCR kits, call for the necessity of diagnostic algorithm

Table 11.1 Systems involved by COVID-19 and associated clinical findings

System or test	Common findings
General	Fatigue, fever, chills, headache, and muscle ache
Respiratory	Dry cough, dyspnea, and chest distress
CNS and PNS	Agnosia, anosmia, altered mental status, acute cerebrovascular disease, epilepsy, central respiratory failure, enhanced tendon reflexes, and ankle clonus
GI	Anorexia, nausea, vomiting, diarrhea, and abdominal pain
Cardiac	Acute myocardial injury, arrhythmia, and cardiac shock
Cutaneous	Petechiae, erythema, vesicles, pustules, papulovesicular exanthem, urticarial, maculopapular eruptions, livedo, and necrosis
Hematologic laboratory tests	Decreased: lymphocyte count and thrombocyte count Increased: CRP, ESR, PCT, D-dimer, prothrombin time, LDH, serum ferritin, AST, ALT, cardiac troponin I, CK-MB, and NT-proBNP
Chest CT scan	Multifocal, peripheral, bilateral ground-glass opacities and consolidation, air bronchogram sign, crazy-paving pattern, and pleural thickening

for patients with suspected COVID-19 pneumonia. Effective triage and early detection of COVID-19 are also essential for disease control, isolation, psychological reassurance for patients, and effective treatment. The combining of clinical, laboratory, and imaging findings is necessary to assist clinicians anywhere in the globe in suspecting the possibility of COVID-19, pretest probability assessment, and accurate interpretation of diagnostic testing. We have tried to address this algorithm in this chapter (Table 11.1).

11.2 Respiratory Manifestations and Constitutional Symptoms

As reported, patients with COVID-19 develop clinical symptoms after an incubation period ranging from 1 to 14 days, mostly 4 to 7 days which is longer in milder or early asymptomatic cases and shorter in severe or rapidly progressive cases (Guan et al. 2020; Tian et al. 2020).

Since the diameter of the SARS-CoV-2 is very small, about 60 to 140 nanometers, it can penetrate the lung terminal structures, including the interlobular and alveolar septum. In healthy lung tissue, ACE2 is mainly expressed by club cells of distal bronchioles and type I and type II alveolar epithelial cells, which are involved in preventing ARDS. Cub cells secrete proteins protective against airway inflammation and oxidative stress, and type II pneumocytes are the defender of the alveolus by synthesizing and recycling all components of the surfactant (Bombardini and Picano 2020). ACE2 serves as a receptor for SARS-CoV-2, which plays a crucial role in lung injury. The binding of COVID-19 spike protein to ACE2, activates the enzyme, thus activating the renin-angiotensin system to cause lung injury, which in turn may contribute to ARDS by triggering vasoconstriction, edema, lymphocyte infiltration, apoptosis, and fibrosis in the lung interstitium (Kuba et al. 2005; Rezaei 2020b; Saghazadeh and Rezaei 2020b).

Therefore, the common symptoms of COVID-19 patients are consistent with the manifestation of lower respiratory tract infections. By contrast, upper respiratory tract symptoms are less common in these patients. General symptoms of viral infection and pneumonia happen in the vast majority of COVID-19-infected patients, including fever, fatigue, cough (mostly without sputum), also chills, dyspnea, nausea, headache, anorexia, and muscle ache. On the other hand, upper respiratory tract signs and symptoms, including rhinorrhea, sneezing, or sore throat, have been rarely reported in COVID-19 cases (Zheng et al. 2020; Zhang et al. 2020d; Tian et al. 2020; Zhang et al. 2020b). In a meta-analysis, including 50,466 patients with COVID-19, fever incidence was 89.1%, 72.2% had a cough, and the incidence of muscle soreness or fatigue was 42.5% (Sun et al. 2020). In another meta-analysis of Overall, 31 articles and 46,959 patients, including 10 English articles and 21 Chinese articles, the most common clinical manifestations were fever (87.3%), cough (58.1%), dyspnea

(38.3%), muscle soreness or fatigue (35.5%), and chest distress (31.2%) (Cao et al. 2020). In other metal analysis studies of COVID-19 cases, the same symptoms have been reported as the most common symptoms of COVID-19 pneumonia (Zhu et al. 2020b; Li et al. 2020b; Rodriguez-Morales et al. 2020; Zhu et al. 2020a).

Although the initiative for COVID-19 screening started from fever clinics since fever, cough, and shortness of breath are the most emphasized symptoms, it increases the risk of omitting those patients with other symptoms and average body temperature, mainly middle-aged and elderly patients, or cases at the first days of the infection or immunosuppressed patients (Jazieh et al. 2020; Young et al. 2020). In a study of 1,099 laboratory-confirmed patients, fever presentation increased from 43.8% on admission to 88.7% during hospitalization (Guan et al. 2020). In another study of 202 hospitalized patients with COVID-19, more than 20% of patients were afebrile on admission (Huang et al. 2020b). It is noteworthy to mention that not all patients with COVID-19 had high temperature during their first visit, although they have experienced a feverish feeling which may be due to the use of over-the-counter antipyretic drugs (Zhu et al. 2020f). Afebrile cases have also been reported in SARS-CoV and MERS-CoV cases (Zumla et al. 2015). Thus, since afebrile patients may be missed if the surveillance case definition focuses on fever detection, it should be noted that fever is not a reliable symptom for the suspicion or screening out the COVID-19.

11.3 Central Nervous System Manifestations

Nervous system involvement has been reported in SARS-CoV and MERS-CoV cases. The SARS-CoV-2 may also enter the central nervous system (CNS) through the hematogenous or retrograde neuronal route, causing neurological symptoms including headache, altered mental status, anosmia, acute cerebrovascular disease, epilepsy, and central respiratory failure, which have been reported in studies of COVID-19 patients (Mao et al. 2020; Yazdanpanah et al. 2020b; Jahanshahlu and Rezaei 2020a; Saleki et al. 2020). Although the most common brain imaging abnormalities reported in COVID-19 cases were leptomeningeal enhancement and bilateral frontotemporal hypoperfusion, which can be due to the hypoxia (Helms et al. 2020; Kandemirli et al. 2020; Ebrille et al. 2020), autopsy results of patients with neurological symptoms confirmed COVID-19 nucleic acid existence in the brain tissue and the brain tissue damage including hyperemesis, edema, and degeneration (Mao et al. 2020).

The neurotropism of SARS-CoV-2 explains the potential mechanism behind the neural invasion of COVID-19 (Hamming et al. 2004). It might lie in the expression of ACE2 on nervous system cells, cytokine storm syndrome (Mehta et al. 2020; Rokni et al. 2020), microvascular thrombosis, and hypoxemia. Neurological symptoms reported in a high percentage of COVID-19 patients ranging from 20% to as high as 80% in severe cases admitted to the intensive care unit (ICU) (Kandemirli et al. 2020; Helms et al. 2020; Mao et al. 2020). Diffuse corticospinal tract signs, including enhanced tendon reflexes and ankle clonus, have also been reported in COVID-19 patients, which may be due to the existence of the SARS-CoV-2 in cerebrospinal fluid (Arabi et al. 2017; Helms et al. 2020).

Severe cases of COVID-19 are more prone to develop neurological symptoms (Mao et al. 2020). Acute ischemic stroke has been reported in young COVID-19 patients (Beyrouti et al. 2020; Oxley et al. 2020), which may be due to the coagulation disorder since a higher level of D-dimer and large vessel occlusion also have been reported in such patients (Thachil et al. 2020). Moreover, severe COVID-19 cases are more prone to develop intracranial cytokine storms, thus acute necrotizing encephalopathy, which is known as a complication of COVID-19 (Li et al. 2020c). Therefore, due to the high mortality rate of neurological injury during COVID-19, patients should be examined for neurological symptoms during the early phase of symptoms, and patients with neurological involvement should be closely monitored and also receive anticoagulant therapies in the case of coagulopathy.

Olfactory and taste disorders have been reported in COVID-19 patients, even in some patients, as the only presenting symptoms (Hjelmesaeth and Skaare 2020). ACE2 is also expressing on cells in nasal and oral tissue (Xu et al. 2020a; Hamming et al. 2004; Fini 2020), and seems to be the potential mechanism of this finding. Although in magnetic resonance imaging studies of patients with anosmia, the olfactory bulb was found to be normal (Galougahi et al. 2020), but the first cranial nerve branches dysfunction can't be rolled out as the potential cause of anosmia. Since necrotic changes was present in olfactory bulb and cranial nerve tracts of COVID-19 patients (Han et al. 2020).

FDG-PET scan also showed redused metabolic activity in the orbitofrontal cortex of COVID-19 patients (Karimi-Galougahi et al. 2020). Moreover, smell loss in these patisnts happens in the absence of mechanical nasal obstruction and lasts longer than common viral rhinitis (Akerlund et al. 1995).

It seems that the virus reaches the brain through peripherally located olfactory dendrites within receptor cells, thus inducing central anosmia (Brann and Firestein 2014). In transgenic mice, SARS-CoV entered the brain through the olfactory bulb since couple hours following infection, viral antigen was detected and was most abundant in the regions of the brain that are connected with the olfactory bulb, including piriform and infralimbic cortices, basal ganglia, and midbrain (Baig et al. 2020; Brann et al. 2020). These findings suggest that impaired neural function due to direct neurotropism of COVID-19 is the main cause of anosmia and probably also agnosia (Hjelmesaeth and Skaare 2020; Karimi-Galougahi et al. 2020).

Anosmia or agnosia has been reported mostly in the mild form of COVID-19. In a study of a total of 417 mild-to-moderate COVID-19 patients, 85.6% and 88.0% of patients reported olfactory and gustatory dysfunctions, respectively (Lechien et al. 2020; Yan et al. 2020). Even it has been reported that using the presence of smell or taste change with fever and muscle ache in outpatient mild-to-moderate cases can give a high sensitivity up to 90% for COVID-19 diagnoses (Menni et al. 2020; Roland et al. 2020). However, it should be noted that smell loss related to COVID-19 is mostly presented suddenly and in the absence of nasal obstruction and rhinitis symptoms (Vaira et al. 2020).

11.4 Gastrointestinal System Manifestations

Some patients with COVID-19 are presented with gastrointestinal (GI) symptoms, including nausea, vomiting, and diarrhea as the initial symptom (Wang et al. 2020a; Zhou et al. 2020; Zhang et al. 2020b). Even abdominal pain in some patients was the only symptom of COVID-19, which may be due to pneumonia affecting lower lobes and/or pleural effusion (Dane et al. 2020). Also in other Coronavirus infections inclusding MERS-CoV and SARS-CoV, GI symptoms have been reported in up to 30% of patients (Assiri et al. 2013; Zhou et al. 2017). GI symptoms in COVID-19 can be explained by the fact that ACE2, as the receptor of the SARS-CoV-2, is also present in the GI tract, including epithelial cells of the tongue, oral cavity, esophagus, and enterocytes in the ileum and colon epithelium (Wan et al. 2020; Xu et al. 2020a).

In a study of 1,141 confirmed COVID-19 cases, 183 (16%) presented with just GI symptoms (Luo et al. 2020). In a meta-analysis of 4,243 COVID-19 patients from 6 countries, the pooled prevalence of all GI symptoms was 17.6%; anorexia was the most common GI symptom (26.8%), followed by diarrhea (12.5%), nausea/vomiting (10.2%), and abdominal pain/discomfort (9.2%) (Cheung et al. 2020). Overall GI symptom presentation varies from 3% to 40% of COVID-19 patients (Zhu et al. 2020a; Li et al. 2020b). However, recall bias may be the reason behind the relatively low percentage of GI symptom reporting in some studies. Moreover, most studies reported hospitalized COVID-19 patients, and GI presentation may have a higher prevalence in milder outpatient cases, since it has been reported that GI symptoms in COVID-19 was associated with longer illness duration but lower ICU admission rate and lower mortality rate (Nobel et al. 2020).

Viral replication was seen in stool samples of patients with MERS-CoV infection (Zhou et al. 2017). COVID-19 virus also could be detected in the stool samples of patients during the first week of symptoms (Young et al. 2020; Zhang et al. 2020a). In a meta-analysis of 60 studies including 4,243 patients, up to 50% of COVID-19 patients tested of viral RNA in stool samples reported positive for viral RNA (Cheung et al. 2020) regardless of GI symptoms or even when the throat swap testing was negative (Qian et al. 2020; Zhang et al. 2020a). In Ling et al. study, the viral RNA could be detected in the stool of 81.8% (54/66) cases with the negative results for throat swabs (Magrone et al. 2020). Therefore, routine RT-PCR testing of stool in patients with COVID-19, especially those presenting with digestive symptoms, is recommended since it also seems to be a safer sample for medical staff compared to a throat swap sample.

Moreover, it should be noted that viral presence in stool samples could raise a serious concern on the isolation policy for the COVID-19 patients, particularly during the recovery phase, since patients with GI symptoms also presented higher rates of familial clustering, which may be due to the aerosols generated from the toilet flushing of the shared toilets (Jin et al. 2020; Zhou et al. 2020). Therefore, toilet isolation should be implemented in households with positive cases of COVID-19, and fecal-oral route transmission should be considered for prevention strategies.

Patients with GI symptoms are more likely to present elevated liver tests, such an aspartate aminotransferase (AST) and alanine aminotransferase (ALT), thus liver injury (Jin et al. 2020; Zhou et al. 2020). However, it seems to be a result of systemic inflammatory response rather than a direct viral invasion to liver cells since ACE2 is highly expressed in bile duct cells but not liver cells (Chai et al. 2020). Moreover, COVID-19 patients with liver damage had higher inflammatory indexes, such as elevated C-reactive protein (CRP) and procalcitonin (PCT), and more likely to have a fever and severe pneumonia, which may be related to the severe immune response (Wang et al. 2020a; Huang et al. 2020a; Bahrami et al. 2020; Yazdanpanah et al. 2020a; Saghazadeh and Rezaei 2020a; Nasab et al. 2020). Liver enzymes need to be monitored in COVID-19 patients regardless of the GI symptoms since liver injury was reported to be significantly higher in COVID-19 patients compared to other respiratory virus infections (Jin et al. 2020; Zhao et al. 2020). Moreover, liver damage was reported to be associated with a significantly higher risk of ICU admission and death in COVID-19 patients (Hajifathalian et al. 2020).

These findings suggest that viral shedding from the intestinal tract can be a potential route for infection transmission. Thus, toilet isolation and washing hands after anal hygiene should be considered to reduce fecal-oral route viral transmission, especially in patients with GI symptoms. Moreover, a stool sample can be another way of diagnosis of COVID-19, especially when the throat swab results are negative despite highly suspicious of infection or when the medical stuff protection is not adequate for the throat swap sample. It should be noticed that an unexplained abdominal pain or GI symptoms could be criteria for COVID-19 testing for reducing missed cases.

11.5 Cardiovascular System Manifestations

Cardiac involvement has been reported in viral infections such as influenza, parvovirus B-19 (Fung et al. 2016), and COVID-19 as well (Shamshirian and Rezaei 2020). In COVID-19, up to 12% of patients were reported to have cardiac injury presenting as abnormal cardiac enzyme level, acute myocardial injury (AMI), and arrhythmia-induced cardiac shock (Ebrille et al. 2020; Kir et al. 2020). In a meta-analysis, including 50,466 patients with COVID-19, 49.4% of the patients presented with myocardial enzyme spectrum abnormalities, which manifested as an increase in cardiac enzymes or lactate dehydrogenase (LDH) levels (Zhu et al. 2020b).

These findings can be explained by the fact that ACE2 is highly expressed in the heart cells (Sodhi et al. 2018) and ACE2 in circulation and local tissues has a significant protective effect on the cardiovascular system (Cole-Jeffrey et al. 2015). It has been shown that ACE2 is expressing in sinoatrial nodal cells (Ferreira et al. 2011). Moreover, conduction disturbances such as ventricular fibrillation have been noted with overexpression of ACE2 receptor in experimental mice models (Danilczyk and Penninger 2006). AMI reported in COVID-19 patients can be a result of the accumulation of LDH, hypoxia caused by respiratory failure, or blood hypercoagulability. Inflammatory markers such as CRP and PCT were also reported to be higher in COVID-19 patients with cardiac injury, suggesting cytokine storm as another cause of myocardial injury (Shi et al. 2020c; Shi et al. 2020b).

Cardiac injury has been reported as an independent risk factor for infection severity and mortality in COVID-19 (Shi et al. 2020c). Thus, monitoring cardiac enzymes, including cardiac troponin I, creatine kinase myocardial band (CK-MB), and N-terminal prohormone of brain natriuretic peptide (NT-proBNP), is highly recommended in severe cases of COVID-19, especially in those who have coagulopathy (Shi et al. 2020c; Shi et al. 2020b).

11.6 Other Systems

Among patients with COVID-19, ocular involvement has been reported mostly as conjunctivitis and, in severe cases (Wu et al. 2020b), even has been reported as the only symptom of patients (Scalinci and Trovato Battagliola 2020). Chen et al. reported that in 535 COVID-19 patients, 27 patients (5%) presented with conjunctival congestion (Chen et al. 2020c).

Reported skin manifestations of COVID-19 include erythematous lesions, dengue-like petechiae (Joob and Wiwanitkit 2020), urticaria, and varicella-like papulovesicular exanthem (Marzano et al. 2020) and chilblain-like lesions (de Masson et al. 2020). In a study of 375 COVID-19 patients with cutaneous manifestations, the most common ones were areas of erythema with vesicles or pustules (pseudo-chilblain) (19%), other vesicular eruptions (9%), urticarial lesions (19%), maculopapular eruptions (47%), and livedo or necrosis (6%) (Galvan Casas et al. 2020).

Some studies have reported that laboratory or imaging confirmed COVID-19 cases who were clinically asymptomatic (Dong et al. 2020; Cao et al. 2020; Shi et al. 2020a). The potential explanation for clinically asymptomatic patients can be the false-positive laboratory tests, especially in cases who had negative imaging as well (Li et al. 2020a), or the SARS-CoV-2 did not sufficiently stimulate the immune response to develop symptoms, or the clinical symptoms were not significant enough for the patient to recall or seek medical care for them. Other explanations can be that the virus just replicated in the upper respiratory system and the immune response was fast and efficient enough to control the infection before the virus reaches the other systems or lower respiratory system; thus, no clinically significant symptoms developed, but the throat swap test was positive.

Since asymptomatic cases have the ability of disease transmission, it raises concerns regarding the medical staff primarily who work with COVID-19 non-infected cases, especially immunocompromised patients. Therefore, we recommend a detailed clinical symptom screening, especially for atypical symptoms, which are more common in the mild form of COVID-19, including anosmia, agnosia, or general fatigue and even using imaging for screening out asymptomatic infected medical staff, especially who are working with high-risk non-infected patients.

11.7 Laboratory Findings

In addition to clinical symptoms, laboratory results also give us some hints in COVID-19 diagnosis. Since SARS-CoV-2 binds the ACE2 receptor, organs that express this receptor, including the myocardium, vascular endothelium, and liver, show involvement by the virus and thus impaired biochemical results. Abnormal blood enzymes, including counts of lymphocytes and neutrophils, CRP, ESR, D-dimer, fibrinogen, LDH, serum ferritin, ALT, and AST, have been reported in COVID-19.

Lymphopenia is the most common reported laboratory abnormality finding in COVID-19 patients especially in severe cases, and the severity of lymphocytopenia is associated with the severity of infection (Rodriguez-Morales et al. 2020; Li et al. 2020b). In a meta-analysis, including 50,466 patients with COVID-19, most patients showed normal WBC counts (64.7%) with lymphopenia (47.6%). In another meta-analysis of 38 studies involving 3,062 COVID-19 patients, normal WBC count (69.7%), lymphopenia (56.5%), elevated CRP (73.6%), and elevated ESR (65.6%) were the most common laboratory findings (Zhu et al. 2020a).

Noticeably, the percentages of lymphocytes but not absolute counts of lymphocytes were lower in severe patients when compared to non-severe COVID-19 patients (Fathi and Rezaei 2020). SARS-CoV-2 has been shown to infect lymphocytes and destruct them, especially in severe cases (Zhao et al. 2020; Zhang et al. 2020b). Therefore, even with normal or increased lymphocyte numbers, especially in the early stage of infection, the number of lymphocytes decreases as the disease progresses. Critically ill patients with MERS infection also presented with lymphocytopenia (He et al. 2005). Rising neutrophil count related to cytokine storm induced by viral invasion and a falling lymphocyte count causes an increase in neutrophil-to-lymphocyte ratio in patients with COVID-19, especially in the severe cases (Zheng et al. 2020). Moreover, lower lymphocytes and higher neutrophil count were significantly associated with ICU admission and mortality rate in COVID-19 (Chen et al. 2020b). Therefore neutrophil-to-lymphocyte ratio, which is also a well-known marker of systemic inflammation in other infections, can be used in diagnosis and prognosis prediction in COVID-19.

Routine hemostasis tests, including raised D-dimer, prothrombin time, and thrombocyte count, can be additional useful tools for improving early diagnosis and disease severity assessment, since these factors were significantly associated with higher ICU admission and mortality rate in COVID-19 patients (Chen et al. 2020a). It is well known that inflammation is closely associated with thrombosis. Since proinflammatory cytokines are modulators of coagulation (Vazquez-Garza et al. 2017) and coagulation is a line of defense against severe infections, hypercoagulable state happens in several infections (Zhang et al. 2020b; Wang et al. 2020a). Moreover, ACE2 is expressed in arterial endothelial cells; SARS-CoV-2 can directly attack vascular endothelial cells and activate the coagulation system after vascular endothelial injury (Bellosta et al. 2020). Elevated von Willebrand factor levels reported in COVID-19 patients is another evidence of endothelial injury in these patients (Panigada et al. 2020). Moreover, hypoxia in severe COVID-19 cases can stimulate thrombosis through increasing blood viscosity and hypoxia-inducible transcription factors (HIFS) (Gupta et al. 2019). It should be noted that hypercoagulable state in COVID-19 leads to excessive consumption of coagulation factors and thrombocytopenia, thus disseminated intravascular coagulation (DIC) (Levi and Ten Cate 1999). It poses COVID-19 as a threat to patients vulnerable to thrombocytopenia (Sahu et al. 2020).

Secondary to the coagulopathy, microthrombosis and microcirculation disorder happen, which lead to deep venous thrombosis, limb ischemia, intermuscular vein thrombosis, pulmonary embolism, myocardial injury, and cerebrovascular events, which have been reported in COVID-19 patients with coagulopathies (Ji et al. 2020; Goyal et al. 2020). Autopsy results also showed that SARS-CoV-2 could cause multiple organ thrombosis events (Wichmann et al. 2020). Therefore, routine monitoring of hemostasis tests can be a useful tool for early establishing anticoagulant treatments, thus preventing thrombotic events and death.

11.8 Imaging Findings

Besides the clinical signs and symptoms and laboratory findings, imaging is another way of diagnosis and severity assessment of COVID-19. Based on studies reporting chest imaging in COVID-19 patients, the density of the lung lesions in viral types of pneumonia, including COVID-19 infection, is not specific or predominant enough to be diagnosed by conventional radiographs. Therefore, chest computed tomography(CT) examination is recommended for diagnosis and follow-up of the COVID-19 patients, especially in screening outpatients with a highly suspected disease with typical clinical manifestations but initial negative RT-PCR test or in patients who may not develop the typical symptoms such as immunosuppressed or oncologic patients (Zhang et al. 2020c; Cao et al. 2020). The sensitivity of chest CT scan in COVID-19 suspected cases has been reported to increase over time after symptom onset up to 97% (Wang et al. 2020c; Ai et al. 2020).

The chest CT scan findings of COVID-19 pneumonia reflect a typical lung injury of viral pneumonia, which is mostly characterized by multifocal, peripheral, bilateral, lower lung ground-glass opacities (GGO) and consolidation (Salehi et al. 2020; Wang et al. 2020a; Zhang et al. 2020b). In a meta-analysis of 4,121 patients with COVID-19, the most common CT scan findings were ground-glass opacities (68.1%), air bronchogram sign (44.7%), crazy-paving pattern (35.6%), consolidation (32.0%), and pleural thickening (27.1%) (Zhu et al. 2020c). Lymphadenopathy, pleural effusion, pericardial effusion, cavitation, halo sign, and pneumothorax are some of the uncommon but possible findings associated with poor prognosis in COVID-19 patients (Zhu et al. 2020c; Qin et al. 2020). Peripheral lung involvement is a common finding in COVID-19 chest imaging, which can be due to the SARS-CoV-2 tendency to affect the terminal bronchioles and lung parenchyma around them (Nie et al. 2020). However, it should be noted that coronavirus pneumonia cannot be excluded according to single-lobe involvement on CT, but multilobe involvement is more common (Zhu et al. 2020e).

As COVID-19 infection progresses, the chest CT manifestations change as well, which helps the differential diagnosis and evaluation of the progression. The median time interval between symptom onset and imaging finding is 7 days (Yang et al. 2020; Li et al. 2020a), which is not significantly different between the elderly group and the young group (Zhu et al. 2020e). The halo sign appears in early stage and then rapidly changed into GGO within a few days due to the invasion of the pulmonary interstitium; thus, during the first week, GGO is the main radiological demonstration. It should be noted that false-positive imaging may happen at this time. During the second week, as the disease progresses, ground-glass opacities transform into multifocal consolidation and increase in size and number (peaked around days 6 to 11). Therefore, fewer GGO lesions and more consolidation involvement are likely an indicator of disease progression. Crazy-paving patterns, septal thickening, and residual parenchymal bands are other findings of the second week. After 14 days in mild-to-moderate recovered patients, the consolidation is gradually absorbed, and no crazy-paving pattern is present anymore; thus in this stage, extensive GGO could be observed again as the demonstration of the consolidation absorption. This absorption stage can be continued beyond 26 days (Pan et al. 2020; Salehi et al. 2020).

A normal chest CT scan cannot exclude the diagnosis of COVID-19. In Yang et al. study of 149 confirmed COVID-19 patients, 11.4% had normal chest CT on admission, and 8% remained negative 10 days later (Yang et al. 2020). In another study of 877 COVID-19 patients, no radiographic or CT abnormality was found in 157 (17.9%) with non-severe cases and in 5 (2.9%) of severe patients (Guan et al. 2020).

Therefore, scattered ground-glass opacities and consolidations, especially in the periphery of the lungs, are the main chest CT scan feature of COVID-19, which can be used for diagnosis and progression follow-up of COVID-19. However, a CT scan is unlikely a reliable stand-alone tool to rule out COVID-19; thus in screening patients, clinical manifestations, laboratory examination, exposure history, RT-PCR test, and chest imaging should be combined for comprehensive analysis.

11.9 Disease Course and Severity Spectrum

COVID-19 shows a biphasic pattern in non-severe cases. The first phase is characterized by fever, cough, fatigue, and other systemic and respiratory symptoms. During this phase, which usually lasts for 7 days since starting symptoms, the disease progresses due to the uncontrolled viral replication; thus, the majority of the patients show positive results for upper respiratory specimens. The second phase is usually during the second week when, in mild cases, symptoms, including fever, mostly begin to relieve and imaging features start to resolve (Chen et al. 2020b). However, severe or hospitalized cases show persistent fever and disease progression with worsening radiologic imaging during the second phase (Chen et al. 2020b; Lescure et al. 2020). Therefore, whether the COVID-19 is going to progress to a critical situation or not is an issue of the first few days of symptoms; thus, we suggest more attention to the disease progression at an early stage, rather than neglecting the mild illness.

Prognosis of COVID-19 infeciton can be determined considering degree of fever (mostly > 38°C) as an indicator of severe inflammatory cytokine secretion, respiration rate, chest pain as the result of inflammation of pleura, progressive hypoxemia, dyspnea as the result of severe damage to alveoli lung invasion and abdominal pain as an indicator of viral replication in the gut; which all are associated with more severe or critical COVID-19 infection (Henry et al. 2020).

Laboratory abnormalities can serve as prognostic factors for COVID-19. In particular, decreased lymphocyte count (Wang et al. 2020b; Zheng et al. 2020), increased white blood cell (WBC) count (Deng et al. 2020), increased inflammatory markers (CRP, erythrocyte sedimentation rate (ESR), PCT, LDH), and abnormal blood coagulation function are associated with severe COVID-19 (Zhu et al. 2020b; Liu et al. 2020). Increased LDH levels also have been reported as an independent risk factor for developing severe or critical COVID-19 (Jin et al. 2020). LDH is also known as the independent risk factor for developing ARDS (Yu et al. 2020). Moreover, organ failure is one of the common and serious complications of severe COVID-19, and thus high ALS, AST, and creatinine indicate organ damage and poor prognosis (Zhang et al. 2020d). Thus, assessment of clinical symptoms and laboratory abnormalities can be used to early identification of high risk patients who might become critically ill later and need close observation, hospitalization, or more aggressive treatments.

11.10 Conclusion

COVID-19 pandemic is the most lethal outbreak of coronavirus in the twenty-first century, and its rapid spread has surpassed both MERS and SARS in the number of confirmed cases and mortality. Based on the high chance of person-to-person transmission, early and accurate diagnosis of patients can lead to more efficient isolation; thus, it is necessary to prevent and control the spread of the pandemic. Moreover, an efficient diagnosis and screening tool leads to a reliable screening out, thus decreasing the medical and psychological burden of a high number of susceptible cases, including the older adults, health-care providers (Rezaei 2020a; Moazzami et al. 2020), pregnant women (Mirbeyk and Rezaei 2020), and individuals having specific genetic profiles (Yousefzadegan and Rezaei 2020; Ahanchian et al. 2020; Darbeheshti and Rezaei 2020; Babaha and Rezaei 2020).

Based on this chapter, the diagnosis of COVID-19 cannot be made by relying on only one tool or criteria. A combination of exposure history, clinical features, throat or stool sample RT-PCR testing, laboratory, and imaging findings is necessary to improve the accuracy of the diagnosis of infection and re-infection with

COVID-19 (Jabbari and Rezaei 2020). Moreover, the clinical examination should not be restricted to the respiratory system, and attention should also be paid to other systems including GI, cardiac, and peripheral and central nervous system to prevent missing cases with atypical symptoms. Clinical, laboratory, and imaging prognostic factors are also helpful in the identification of poor prognosis patients before they develop a critical situation. It offers an opportunity for the establishment of timely treatment, which is important, especially when no specific treatment exists. All the scientific community is, however, endeavoring to justify the use of existing drugs and other therapeutic options (Mohamed et al. 2020b; Pourahmad et al. 2020; Saghazadeh and Rezaei 2020b) and find new therapeutic platforms and targets (Mansourabadi et al. 2020; Fathi and Rezaei 2020; Jahanshahlu and Rezaei 2020b; Basiri et al. 2020b; Pashaei and Rezaei 2020; Rabiee et al. 2020) in parallel. If distributed globally, this parallelism would generate the science promising to prepare for future pandemics by providing high-quality research (Rzymski et al. 2020; Momtazmanesh et al. 2020; Mohamed et al. 2020a; Kafieh et al. 2020; Moradian et al. 2020).

References

Ahanchian H, Moazzen N, Faroughi MSD, Khalighi N, Khoshkhui M, Aelami MH, Haghi NSM, Rezaei N (2020) COVID-19 in a child with primary specific antibody deficiency

Ahmadi M, Saffarzadeh N, Habibi MA, Hajiesmaeili F, Rezaei N (2020) Colon cancer and SARS-CoV-2: impact of ACE2 expression in susceptibility to COVID-19. bioRxiv

Ai T, Yang Z, Hou H, Zhan C, Chen C, Lv W, Tao Q, Sun Z, Xia L (2020) Correlation of chest CT and RT-PCR testing in coronavirus disease 2019 (COVID-19) in China: a report of 1014 cases. Radiology 200642. https://doi.org/10.1148/radiol.2020200642

Akerlund A, Bende M, Murphy C (1995) Olfactory threshold and nasal mucosal changes in experimentally induced common cold. Acta Otolaryngol 115(1):88–92. https://doi.org/10.3109/00016489509133353

Arabi YM, Balkhy HH, Hayden FG, Bouchama A, Luke T, Baillie JK, Al-Omari A, Hajeer AH, Senga M, Denison MR, Nguyen-Van-Tam JS, Shindo N, Bermingham A, Chappell JD, Van Kerkhove MD, Fowler RA (2017) Middle East respiratory syndrome. N Engl J Med 376(6):584–594. https://doi.org/10.1056/NEJMsr1408795

Assiri A, Al-Tawfiq JA, Al-Rabeeah AA, Al-Rabiah FA, Al-Hajjar S, Al-Barrak A, Flemban H, Al-Nassir WN, Balkhy HH, Al-Hakeem RF, Makhdoom HQ, Zumla AI, Memish ZA (2013) Epidemiological, demographic, and clinical characteristics of 47 cases of Middle East respiratory syndrome coronavirus disease from Saudi Arabia: a descriptive study. Lancet Infect Dis 13(9):752–761. https://doi.org/10.1016/s1473-3099(13)70204-4

Babaha F, Rezaei N (2020) Primary immunodeficiency diseases in COVID-19 pandemic: a predisposing or protective factor? Am J Med Sci. https://doi.org/10.1016/j.amjms.2020.07.027

Bahrami A, Vafapour M, Moazzami B, Rezaei N (2020) Hyperinflammatory shock related to COVID-19 in a patient presenting with multisystem inflammatory syndrome in children: first case from Iran. J Paediatr Child Health. https://doi.org/10.1111/jpc.15048

Baig AM, Khaleeq A, Ali U, Syeda H (2020) Evidence of the COVID-19 virus targeting the CNS: tissue distribution, host-virus interaction, and proposed neurotropic mechanisms. ACS Chem Neurosci 11(7):995–998. https://doi.org/10.1021/acschemneuro.0c00122

Basiri A, Heidari A, Nadi MF, Fallahy MTP, Nezamabadi SS, Sedighi M, Saghazadeh A, Rezaei N (2020a) Microfluidic devices for detection of RNA viruses. Rev Med Virol:e2154. doi:https://doi.org/10.1002/rmv.2154

Basiri A, Pazhouhnia Z, Beheshtizadeh N, Hoseinpour M, Saghazadeh A, Rezaei N (2020b) Regenerative medicine in COVID-19 treatment: real opportunities and range of promises. Stem Cell Rev Rep:1–13. https://doi.org/10.1007/s12015-020-09994-5

Bellosta R, Luzzani L, Natalini G, Pegorer MA, Attisani L, Cossu LG, Ferrandina C, Fossati A, Conti E, Bush RL, Piffaretti G (2020) Acute limb ischemia in patients with COVID-19 pneumonia. J Vasc Surg. https://doi.org/10.1016/j.jvs.2020.04.483

Beyrouti R, Adams ME, Benjamin L, Cohen H, Farmer SF, Goh YY, Humphries F, Jager HR, Losseff NA, Perry RJ, Shah S, Simister RJ, Turner D, Chandratheva A, Werring DJ (2020) Characteristics of ischaemic stroke associated with COVID-19. J Neurol Neurosurg Psychiatry. https://doi.org/10.1136/jnnp-2020-323586

Bombardini T, Picano E (2020) Angiotensin-converting enzyme 2 as the molecular bridge between epidemiologic and clinical features of COVID-19. Can J Cardiol 36(5):784.e781–784.e782. https://doi.org/10.1016/j.cjca.2020.03.026

Brann D, Tsukahara T, Weinreb C, Logan DW, Datta SR (2020) Non-neural expression of SARS-CoV-2 entry genes in the olfactory epithelium suggests mechanisms underlying anosmia in COVID-19 patients. bioRxiv:2020.2003.2025.009084. https://doi.org/10.1101/2020.03.25.009084

Brann JH, Firestein SJ (2014) A lifetime of neurogenesis in the olfactory system. Front Neurosci 8:182. https://doi.org/10.3389/fnins.2014.00182

Cao Y, Liu X, Xiong L, Cai K (2020) Imaging and clinical features of patients with 2019 novel coronavirus SARS-CoV-2: a systematic review and meta-analysis. J Med Virol. https://doi.org/10.1002/jmv.25822

Chai X, Hu L, Zhang Y, Han W, Lu Z, Ke A, Zhou J, Shi G, Fang N, Fan J, Cai J, Fan J, Lan F (2020) Specific ACE2 expression in cholangiocytes may cause liver damage after 2019-nCoV infection. bioRxiv:2020.2002.2003.931766. https://doi.org/10.1101/2020.02.03.931766

Chen G, Wu D, Guo W, Cao Y, Huang D, Wang H, Wang T, Zhang X, Chen H, Yu H, Zhang X, Zhang M, Wu S, Song J, Chen T, Han M, Li S, Luo X, Zhao J, Ning Q (2020a) Clinical and immunological features of severe and moderate coronavirus disease 2019. J Clin Invest 130(5):2620–2629. https://doi.org/10.1172/jci137244

Chen J, Qi T, Liu L, Ling Y, Qian Z, Li T, Li F, Xu Q, Zhang Y, Xu S, Song Z, Zeng Y, Shen Y, Shi Y, Zhu T, Lu H (2020b) Clinical progression of patients with COVID-19 in Shanghai, China. J Infect 80(5):e1–e6. https://doi.org/10.1016/j.jinf.2020.03.004

Chen L, Deng C, Chen X, Zhang X, Chen B, Yu H, Qin Y, Xiao K, Zhang H, Sun X (2020c) Ocular manifestations and clinical characteristics of 535 cases of COVID-19 in Wuhan, China: a cross-sectional study. Acta Ophthalmol. https://doi.org/10.1111/aos.14472

Chen Y, Liu Q, Guo D (2020d) Emerging coronaviruses: genome structure, replication, and pathogenesis. J Med Virol 92(4):418–423. https://doi.org/10.1002/jmv.25681

Cheung KS, Hung IF, Chan PP, Lung KC, Tso E, Liu R, Ng YY, Chu MY, Chung TW, Tam AR, Yip CC, Leung KH, Yim-Fong Fung A, Zhang RR, Lin Y, Cheng HM, Zhang AJ, To KK, Chan KH, Yuen KY, Leung WK (2020) Gastrointestinal manifestations of SARS-CoV-2 infection and virus load in fecal samples from the Hong Kong cohort and systematic review and meta-analysis. Gastroenterology. https://doi.org/10.1053/j.gastro.2020.03.065

Cleary SJ, Pitchford SC, Amison RT, Carrington R, Robaina Cabrera CL, Magnen M, Looney MR, Gray E, Page CP (2020) Animal models of mechanisms of SARS-CoV-2 infection and COVID-19 pathology. Br J Pharmacol. https://doi.org/10.1111/bph.15143

Cole-Jeffrey CT, Liu M, Katovich MJ, Raizada MK, Shenoy V (2015) ACE2 and microbiota: emerging targets for cardiopulmonary disease therapy. J Cardiovasc Pharmacol 66(6):540–550. https://doi.org/10.1097/fjc.0000000000000307

Dane B, Brusca-Augello G, Kim D, Katz DS (2020) Unexpected findings of coronavirus disease (COVID-19) at the lung bases on abdominopelvic CT. AJR Am J Roentgenol:1–4. https://doi.org/10.2214/ajr.20.23240

Danilczyk U, Penninger JM (2006) Angiotensin-converting enzyme II in the heart and the kidney. Circ Res 98(4):463–471. https://doi.org/10.1161/01.RES.0000205761.22353.5f

Darbeheshti F, Rezaei N (2020) Genetic predisposition models to COVID-19 infection. Med Hypotheses 142:109818. https://doi.org/10.1016/j.mehy.2020.109818

de Masson A, Bouaziz JD, Sulimovic L, Cassius C, Jachiet M, Ionescu MA, Rybojad M, Bagot M, Duong TA (2020) Chilblains are a common cutaneous finding during the COVID-19 pandemic: a retrospective nationwide study from France. J Am Acad Dermatol. https://doi.org/10.1016/j.jaad.2020.04.161

Deng Y, Liu W, Liu K, Fang YY, Shang J, Zhou L, Wang K, Leng F, Wei S, Chen L, Liu HG (2020) Clinical characteristics of fatal and recovered cases of coronavirus disease 2019 (COVID-19) in Wuhan, China: a retrospective study. Chin Med J (Engl). https://doi.org/10.1097/cm9.0000000000000824

Dong X, Cao YY, Lu XX, Zhang JJ, Du H, Yan YQ, Akdis CA, Gao YD (2020) Eleven faces of coronavirus disease 2019. Allergy. https://doi.org/10.1111/all.14289

Ebrille E, Lucciola MT, Amellone C, Ballocca F, Orlando F, Giammaria M (2020) Syncope as the presenting symptom of COVID-19 infection. HeartRhythm Case Rep. https://doi.org/10.1016/j.hrcr.2020.04.015

Fathi N, Rezaei N (2020) Lymphopenia in COVID-19: therapeutic opportunities. Cell Biol Int 44(9):1792–1797. https://doi.org/10.1002/cbin.11403

Ferreira AJ, Moraes PL, Foureaux G, Andrade AB, Santos RA, Almeida AP (2011) The angiotensin-(1-7)/Mas receptor axis is expressed in sinoatrial node cells of rats. J Histochem Cytochem 59(8):761–768. https://doi.org/10.1369/0022155411411712

Fini MB (2020) Oral saliva and CVID-19. Oral Oncol:104821

Fung G, Luo H, Qiu Y, Yang D, McManus B (2016) Myocarditis. Circ Res 118(3):496–514. https://doi.org/10.1161/CIRCRESAHA.115.306573

Galougahi MK, Ghorbani J, Bakhshayeshkaram M, Naeini AS, Haseli S (2020) Olfactory bulb magnetic resonance imaging in SARS-CoV-2-induced anosmia: the first report. Acad Radiol. https://doi.org/10.1016/j.acra.2020.04.002

Galvan Casas C, Catala A, Carretero Hernandez G, Rodriguez-Jimenez P, Fernandez Nieto D, Rodriguez-Villa Lario A, Navarro Fernandez I, Ruiz-Villaverde R, Falkenhain D, Llamas Velasco M, Garcia-Gavin J, Baniandres O, Gonzalez-Cruz C, Morillas-Lahuerta V, Cubiro X, Figueras Nart I, Selda-Enriquez G, Romani J, Fusta-Novell X, Melian-Olivera A, Roncero Riesco M, Burgos-Blasco P, Sola Ortigosa J, Feito Rodriguez M, Garcia-Doval I (2020) Classification of the cutaneous manifestations of COVID-19: a rapid prospective nationwide consensus study in Spain with 375 cases. Br J Dermatol. https://doi.org/10.1111/bjd.19163

Goyal P, Choi JJ, Pinheiro LC, Schenck EJ, Chen R, Jabri A, Satlin MJ, Campion TR Jr, Nahid M, Ringel JB, Hoffman KL, Alshak MN, Li HA, Wehmeyer GT, Rajan M, Reshetnyak E, Hupert N, Horn EM, Martinez FJ, Gulick RM, Safford MM (2020) Clinical characteristics of Covid-19 in New York City. N Engl J Med. https://doi.org/10.1056/NEJMc2010419

Guan WJ, Ni ZY, Hu Y, Liang WH, Ou CQ, He JX, Liu L, Shan H, Lei CL, Hui DSC, Du B, Li LJ, Zeng G, Yuen KY, Chen RC, Tang CL, Wang T, Chen PY, Xiang J, Li SY, Wang JL, Liang ZJ, Peng YX, Wei L, Liu Y, Hu YH, Peng P, Wang JM, Liu JY, Chen Z, Li G, Zheng ZJ, Qiu SQ, Luo J, Ye CJ, Zhu SY, Zhong NS (2020) Clinical characteristics of coronavirus disease 2019 in China. N Engl J Med 382(18):1708–1720. https://doi.org/10.1056/NEJMoa2002032

Gupta N, Zhao YY, Evans CE (2019) The stimulation of thrombosis by hypoxia. Thromb Res 181:77–83. https://doi.org/10.1016/j.thromres.2019.07.013

Hajifathalian K, Krisko T, Mehta A, Kumar S, Schwartz R, Fortune B, Sharaiha R (2020) Gastrointestinal and hepatic manifestations of 2019 novel coronavirus disease in a large cohort of infected patients from New York: clinical implications. Gastroenterology. https://doi.org/10.1053/j.gastro.2020.05.010

Hamming I, Timens W, Bulthuis ML, Lely AT, Navis G, van Goor H (2004) Tissue distribution of ACE2 protein, the functional receptor for SARS coronavirus. A first step in understanding SARS pathogenesis. J Pathol 203(2):631–637. https://doi.org/10.1002/path.1570

Hanaei S, Rezaei N (2020) COVID-19: developing from an outbreak to a pandemic. Arch Med Res. https://doi.org/10.1016/j.arcmed.2020.04.021

He Z, Zhao C, Dong Q, Zhuang H, Song S, Peng G, Dwyer DE (2005) Effects of severe acute respiratory syndrome (SARS) coronavirus infection on peripheral blood lymphocytes and their subsets. Int J Infect Dis 9(6):323–330. https://doi.org/10.1016/j.ijid.2004.07.014

Helms J, Kremer S, Merdji H, Clere-Jehl R, Schenck M, Kummerlen C, Collange O, Boulay C, Fafi-Kremer S, Ohana M, Anheim M, Meziani F (2020) Neurologic features in severe SARS-CoV-2 infection. N Engl J Med. https://doi.org/10.1056/NEJMc2008597

Henry BM, de Oliveira MHS, Benoit J, Lippi G (2020) Gastrointestinal symptoms associated with severity of coronavirus disease 2019 (COVID-19): a pooled analysis. Intern Emerg Med. https://doi.org/10.1007/s11739-020-02329-9

Hick JL, Biddinger PD (2020) Novel coronavirus and old lessons – preparing the health system for the pandemic. N Engl J Med 382(20):e55. https://doi.org/10.1056/NEJMp2005118

Himoto Y, Sakata A, Kirita M, Hiroi T, Kobayashi KI, Kubo K, Kim H, Nishimoto A, Maeda C, Kawamura A, Komiya N, Umeoka S (2020) Diagnostic performance of chest CT to differentiate COVID-19 pneumonia in non-high-epidemic area in Japan. Jpn J Radiol 38(5):400–406. https://doi.org/10.1007/s11604-020-00958-w

Hjelmesaeth J, Skaare D (2020) Loss of smell or taste as the only symptom of COVID-19. Tidsskr Nor Laegeforen 140(7). https://doi.org/10.4045/tidsskr.20.0287

Huang C, Wang Y, Li X, Ren L, Zhao J, Hu Y, Zhang L, Fan G, Xu J, Gu X, Cheng Z, Yu T, Xia J, Wei Y, Wu W, Xie X, Yin W, Li H, Liu M, Xiao Y, Gao H, Guo L, Xie J, Wang G, Jiang R, Gao Z, Jin Q, Wang J, Cao B (2020a) Clinical features of patients infected with 2019 novel coronavirus in Wuhan, China. Lancet 395(10223):497–506. https://doi.org/10.1016/s0140-6736(20)30183-5

Huang R, Zhu L, Xue L, Liu L, Yan X, Wang J, Zhang B, Xu T, Ji F, Zhao Y, Cheng J, Wang Y, Shao H, Hong S, Cao Q, Li C, Zhao XA, Zou L, Sang D, Zhao H, Guan X, Chen X, Shan C, Xia J, Chen Y, Yan X, Wei J, Zhu C, Wu C (2020b) Clinical findings of patients with coronavirus disease 2019 in Jiangsu province, China: a retrospective, multi-center study. PLoS Negl Trop Dis 14(5):e0008280. https://doi.org/10.1371/journal.pntd.0008280

Jabbari P, Jabbari F, Ebrahimi S, Rezaei N (2020) COVID-19: a chimera of two pandemics. Disaster Med Public Health Prep:1–2. https://doi.org/10.1017/dmp.2020.223

Jabbari P, Rezaei N (2020) With risk of reinfection, is COVID-19 here to stay? Disaster Med Public Health Prep:1. https://doi.org/10.1017/dmp.2020.274

Jahanshahlu L, Rezaei N (2020a) Central nervous system involvement in COVID-19. Arch Med Res. https://doi.org/10.1016/j.arcmed.2020.05.016

Jahanshahlu L, Rezaei N (2020b) Monoclonal antibody as a potential anti-COVID-19. Biomed Pharmacother = Biomedecine & pharmacotherapie 129:110337. https://doi.org/10.1016/j.biopha.2020.110337

Jazieh AR, Alenazi TH, Alhejazi A, Al Safi F, Al Olayan A (2020) Outcome of oncology patients infected with coronavirus. JCO Glob Oncol (6):471–475. https://doi.org/10.1200/go.20.00064

Ji M, Yuan L, Shen W, Lv J, Li Y, Li M, Lu X, Hu L, Dong W (2020) Characteristics of disease progress in patients with coronavirus disease 2019 in Wuhan, China. Epidemiol Infect 148:e94. https://doi.org/10.1017/s0950268820000977

Jin X, Lian JS, Hu JH, Gao J, Zheng L, Zhang YM, Hao SR, Jia HY, Cai H, Zhang XL, Yu GD, Xu KJ, Wang XY, Gu JQ, Zhang SY, Ye CY, Jin CL, Lu YF, Yu X, Yu XP, Huang JR, Xu KL, Ni Q, Yu CB, Zhu B, Li YT, Liu J, Zhao H, Zhang X, Yu L, Guo YZ, Su JW, Tao JJ, Lang GJ, Wu XX, Wu WR, Qv TT, Xiang DR, Yi P, Shi D, Chen Y, Ren Y, Qiu YQ, Li LJ, Sheng J, Yang Y (2020) Epidemiological, clinical and virological characteristics of 74 cases of coronavirus-infected disease 2019 (COVID-19) with gastrointestinal symptoms. Gut 69(6):1002–1009. https://doi.org/10.1136/gutjnl-2020-320926

Joob B, Wiwanitkit V (2020) COVID-19 can present with a rash and be mistaken for dengue. J Am Acad Dermatol 82(5):e177. https://doi.org/10.1016/j.jaad.2020.03.036

Kafieh R, Arian R, Saeedizadeh N, Minaee S, Yadav SK, Vaezi A, Rezaei N, Javanmard SH (2020) COVID-19 in Iran: a deeper look into the future. medRxiv

Kandemirli SG, Dogan L, Sarikaya ZT, Kara S, Akinci C, Kaya D, Kaya Y, Yildirim D, Tuzuner F, Yildirim MS, Ozluk E, Gucyetmez B, Karaarslan E, Koyluoglu I, Demirel Kaya HS, Mammadov O, Kisa Ozdemir I, Afsar N, Citci Yalcinkaya B, Rasimoglu S, Guduk DE, Kedir Jima A, Ilksoz A, Ersoz V, Yonca Eren M, Celtik N, Arslan S, Korkmazer B, Dincer SS, Gulek E, Dikmen I, Yazici M, Unsal S, Ljama T, Demirel I, Ayyildiz A, Kesimci I, Bolsoy Deveci S, Tutuncu M, Kizilkilic O, Telci L, Zengin R, Dincer A, Akinci IO, Kocer N (2020) Brain MRI findings in patients in the intensive care unit with COVID-19 infection. Radiology:201697. https://doi.org/10.1148/radiol.2020201697

Karimi-Galougahi M, Yousefi-Koma A, Bakhshayeshkaram M, Raad N, Haseli S (2020) (18) FDG PET/CT scan reveals hypoactive orbitofrontal cortex in anosmia of COVID-19. Acad Radiol. https://doi.org/10.1016/j.acra.2020.04.030

Kir D, Mohan C, Sancassani R (2020) HEART BRAKE – an unusual cardiac manifestation of coronavirus disease 2019 (COVID-19). JACC Case Rep. https://doi.org/10.1016/j.jaccas.2020.04.026

Kuba K, Imai Y, Rao S, Gao H, Guo F, Guan B, Huan Y, Yang P, Zhang Y, Deng W, Bao L, Zhang B, Liu G, Wang Z, Chappell M, Liu Y, Zheng D, Leibbrandt A, Wada T, Slutsky AS, Liu D, Qin C, Jiang C, Penninger JM (2005) A crucial role of angiotensin converting enzyme 2 (ACE2) in SARS coronavirus-induced lung injury. Nat Med 11(8):875–879. https://doi.org/10.1038/nm1267

Lau SK, Woo PC, Li KS, Tsang AK, Fan RY, Luk HK, Cai JP, Chan KH, Zheng BJ, Wang M, Yuen KY (2015) Discovery of a novel coronavirus, China Rattus coronavirus HKU24, from Norway rats supports the murine origin of Betacoronavirus 1 and has implications for the ancestor of Betacoronavirus lineage A. J Virol 89(6):3076–3092. https://doi.org/10.1128/jvi.02420-14

Lechien JR, Chiesa-Estomba CM, De Siati DR, Horoi M, Le Bon SD, Rodriguez A, Dequanter D, Blecic S, El Afia F, Distinguin L, Chekkoury-Idrissi Y, Hans S, Delgado IL, Calvo-Henriquez C, Lavigne P, Falanga C, Barillari MR, Cammaroto G, Khalife M, Leich P, Souchay C, Rossi C, Journe F, Hsieh J, Edjlali M, Carlier R, Ris L, Lovato A, De Filippis C, Coppee F, Fakhry N, Ayad T, Saussez S (2020) Olfactory and gustatory dysfunctions as a clinical presentation of mild-to-moderate forms of the coronavirus disease (COVID-19): a multicenter European study. Eur Arch Otorhinolaryngol. https://doi.org/10.1007/s00405-020-05965-1

Lescure FX, Bouadma L, Nguyen D, Parisey M, Wicky PH, Behillil S, Gaymard A, Bouscambert-Duchamp M, Donati F, Le Hingrat Q, Enouf V, Houhou-Fidouh N, Valette M, Mailles A, Lucet JC, Mentre F, Duval X, Descamps D, Malvy D, Timsit JF, Lina B, van-der-Werf S, Yazdanpanah Y (2020) Clinical and virological data of the first cases of COVID-19 in Europe: a case series. Lancet Infect Dis. https://doi.org/10.1016/s1473-3099(20)30200-0

Levi M, Ten Cate H (1999) Disseminated intravascular coagulation. N Engl J Med 341(8):586–592. https://doi.org/10.1056/nejm199908193410807

Li K, Fang Y, Li W, Pan C, Qin P, Zhong Y, Liu X, Huang M, Liao Y, Li S (2020a) CT image visual quantitative evaluation and clinical classification of coronavirus disease (COVID-19). Eur Radiol. https://doi.org/10.1007/s00330-020-06817-6

Li LQ, Huang T, Wang YQ, Wang ZP, Liang Y, Huang TB, Zhang HY, Sun W, Wang Y (2020b) COVID-19 patients' clinical characteristics, discharge rate, and fatality rate of meta-analysis. J Med Virol. https://doi.org/10.1002/jmv.25757

Li YC, Bai WZ, Hashikawa T (2020c) The neuroinvasive potential of SARS-CoV2 may play a role in the respiratory failure of COVID-19 patients. J Med Virol. https://doi.org/10.1002/jmv.25728

Liu W, Tao ZW, Wang L, Yuan ML, Liu K, Zhou L, Wei S, Deng Y, Liu J, Liu HG, Yang M, Hu Y (2020) Analysis of factors associated with disease outcomes in hospitalized patients with 2019 novel coronavirus disease. Chin Med J 133(9):1032–1038. https://doi.org/10.1097/cm9.0000000000000775

Lotfi M, Hamblin MR, Rezaei N (2020) COVID-19: transmission, prevention, and potential therapeutic opportunities. Clinica Chimica Acta 508:254–266. https://doi.org/10.1016/j.cca.2020.05.044

Lotfi M, Rezaei N (2020) SARS-CoV-2: a comprehensive review from pathogenicity of the virus to clinical consequences. J Med Virol. https://doi.org/10.1002/jmv.26123

Luo S, Zhang X, Xu H (2020) Don't overlook digestive symptoms in patients with 2019 novel coronavirus disease (COVID-19). Clin Gastroenterol Hepatol 18(7):1636–1637. https://doi.org/10.1016/j.cgh.2020.03.043

Magrone T, Magrone M, Jirillo E (2020) Focus on receptors for coronaviruses with special reference to angiotensin-converting enzyme 2 as a potential drug target – a perspective. Endocr Metab Immune Disord Drug Targets. https://doi.org/10.2174/1871530320666200427112902

Mansourabadi AH, Sadeghalvad M, Mohammadi-Motlagh HR, Rezaei N (2020) The immune system as a target for therapy of SARS-CoV-2: a systematic review of the current immunotherapies for COVID-19. Life Sci 258:118185. https://doi.org/10.1016/j.lfs.2020.118185

Mao L, Jin H, Wang M, Hu Y, Chen S, He Q, Chang J, Hong C, Zhou Y, Wang D, Miao X, Li Y, Hu B (2020) Neurologic manifestations of hospitalized patients with coronavirus disease 2019 in Wuhan, China. JAMA Neurol. https://doi.org/10.1001/jamaneurol.2020.1127

Marzano AV, Genovese G, Fabbrocini G, Pigatto P, Monfrecola G, Piraccini BM, Veraldi S, Rubegni P, Cusini M, Caputo V, Rongioletti F, Berti E, Calzavara-Pinton P (2020) Varicella-like exanthem as a specific

COVID-19-associated skin manifestation: multicenter case series of 22 patients. J Am Acad Dermatol. https://doi.org/10.1016/j.jaad.2020.04.044

Mehta P, McAuley DF, Brown M, Sanchez E, Tattersall RS, Manson JJ (2020) COVID-19: consider cytokine storm syndromes and immunosuppression. Lancet 395(10229):1033–1034. https://doi.org/10.1016/s0140-6736(20)30628-0

Menni C, Valdes A, Freydin MB, Ganesh S, El-Sayed Moustafa J, Visconti A, Hysi P, Bowyer RCE, Mangino M, Falchi M, Wolf J, Steves C, Spector T (2020) Loss of smell and taste in combination with other symptoms is a strong predictor of COVID-19 infection. medRxiv:2020.2004.2005.20048421. https://doi.org/10.1101/2020.04.05.20048421

Mirbeyk M, Rezaei N (2020) The impact of COVID-19 on pregnancy and neonatal health: a systematic review.

Moazzami B, Razavi-Khorasani N, Dooghaie Moghadam A, Farokhi E, Rezaei N (2020) COVID-19 and telemedicine: Immediate action required for maintaining healthcare providers well-being. J Clin Virol 126:104345. https://doi.org/10.1016/j.jcv.2020.104345

Mohamed K, Rodríguez-Román E, Rahmani F, Zhang H, Ivanovska M, Makka SA, Joya M, Makuku R, Islam MS, Radwan N, Rahmah L, Goda R, Abarikwu SO, Shaw M, Zoghi S, Irtsyan S, Ling I, Cseprekal O, Faten AB, Hazar Sayar E, Soloukey C, Grancini G, Rezaei N (2020a) Borderless collaboration is needed for COVID-19 – a disease that knows no borders. Infect Control Hosp Epidemiol:1–2. https://doi.org/10.1017/ice.2020.162

Mohamed K, Yazdanpanah N, Saghazadeh A, Rezaei N (2020b) Computational drug discovery and repurposing for the treatment of Covid-19: a systematic review. Available at SSRN 3583748

Momtazmanesh S, Ochs HD, Uddin LQ, Perc M, Routes JM, Vieira DN, Al-Herz W, Baris S, Prando C, Rosivall L, Abdul Latiff AH, Ulrichs T, Roudenok V, Aldave Becerra JC, Salunke DB, Goudouris E, Condino-Neto A, Stashchak A, Kryvenko O, Stashchak M, Bondarenko A, Rezaei N (2020) All together to fight COVID-19. Am J Trop Med Hyg 102(6):1181–1183. https://doi.org/10.4269/ajtmh.20-0281

Moradian N, Ochs HD, Sedikies C, Hamblin MR, Camargo CA Jr, Martinez JA, Biamonte JD, Abdollahi M, Torres PJ, Nieto JJ, Ogino S, Seymour JF, Abraham A, Cauda V, Gupta S, Ramakrishna S, Sellke FW, Sorooshian A, Wallace Hayes A, Martinez-Urbistondo M, Gupta M, Azadbakht L, Esmaillzadeh A, Kelishadi R, Esteghamati A, Emam-Djomeh Z, Majdzadeh R, Palit P, Badali H, Rao I, Saboury AA, Jagan Mohan Rao L, Ahmadieh H, Montazeri A, Fadini GP, Pauly D, Thomas S, Moosavi-Movahed AA, Aghamohammadi A, Behmanesh M, Rahimi-Movaghar V, Ghavami S, Mehran R, Uddin LQ, Von Herrath M, Mobasher B, Rezaei N (2020) The urgent need for integrated science to fight COVID-19 pandemic and beyond. J Transl Med 18(1):205. https://doi.org/10.1186/s12967-020-02364-2

Nasab MG, Saghazadeh A, Rezaei N (2020) SARS-CoV-2-A tough opponent for the immune system. Arch Med Res. https://doi.org/10.1016/j.arcmed.2020.05.020

Nie S, Han S, Ouyang H, Zhang Z (2020) Coronavirus disease 2019-related dyspnea cases difficult to interpret using chest computed tomography. Respir Med 167:105951. https://doi.org/10.1016/j.rmed.2020.105951

Nobel YR, Phipps M, Zucker J, Lebwohl B, Wang TC, Sobieszczyk ME, Freedberg DE (2020) Gastrointestinal symptoms and COVID-19: case-control study from the United States. Gastroenterology. https://doi.org/10.1053/j.gastro.2020.04.017

Organization WH (2020) Coronavirus disease 2019 (COVID-19) Situation Report-130

Oxley TJ, Mocco J, Majidi S, Kellner CP, Shoirah H, Singh IP, De Leacy RA, Shigematsu T, Ladner TR, Yaeger KA, Skliut M, Weinberger J, Dangayach NS, Bederson JB, Tuhrim S, Fifi JT (2020) Large-vessel stroke as a presenting feature of Covid-19 in the young. N Engl J Med 382(20):e60. https://doi.org/10.1056/NEJMc2009787

Pan F, Ye T, Sun P, Gui S, Liang B, Li L, Zheng D, Wang J, Hesketh RL, Yang L, Zheng C (2020) Time course of lung changes at chest CT during recovery from coronavirus disease 2019 (COVID-19). Radiology 295(3):715–721. https://doi.org/10.1148/radiol.2020200370

Panigada M, Bottino N, Tagliabue P, Grasselli G, Novembrino C, Chantarangkul V, Pesenti A, Peyvandi F, Tripodi A (2020) Hypercoagulability of COVID-19 patients in intensive care unit. A report of thromboelastography findings and other parameters of hemostasis. J Thromb Haemost. https://doi.org/10.1111/jth.14850

Pashaei M, Rezaei N (2020) Immunotherapy for SARS-CoV-2: potential opportunities. Expert Opin Biol Ther:1–5. https://doi.org/10.1080/14712598.2020.1807933

Phan LT, Nguyen TV, Huynh LKT, Dao MH, Vo TAN, Vu NHP, Pham HTT, Nguyen HT, Nguyen TT, Le HQ, Nguyen TV, Nguyen QH, Huynh TP, Nguyen SN, Nguyen AH, Nguyen NT, Nguyen TNT, Nguyen LT, Luong QC, Cao TM, Pham QD (2020) Clinical features, isolation, and complete genome sequence of severe acute respiratory syndrome coronavirus 2 from the first two patients in Vietnam. J Med Virol. https://doi.org/10.1002/jmv.26075

Pourahmad R, Moazzami B, Rezaei N (2020) Efficacy of plasmapheresis and immunoglobulin replacement therapy (IVIG) on patients with COVID-19. SN Comprehensive Clin Med:1–5

Qian GQ, Yang NB, Ding F, Ma AHY, Wang ZY, Shen YF, Shi CW, Lian X, Chu JG, Chen L, Wang ZY, Ren DW, Li GX, Chen XQ, Shen HJ, Chen XM (2020) Epidemiologic and clinical characteristics of 91 hospitalized patients with COVID-19 in Zhejiang, China: a retrospective, multi-centre case series. QJM. https://doi.org/10.1093/qjmed/hcaa089

Qin C, Liu F, Yen TC, Lan X (2020) 18F-FDG PET/CT findings of COVID-19: a series of four highly suspected cases. Eur J Nucl Med Mol Imaging 47(5):1281–1286. https://doi.org/10.1007/s00259-020-04734-w

Rabiee N, Rabiee M, Bagherzadeh M, Rezaei N (2020) COVID-19 and picotechnology: potential opportunities. Med Hypotheses 144:109917. https://doi.org/10.1016/j.mehy.2020.109917

Rezaei N (2020a) COVID-19 affects healthy pediatricians more than pediatric patients. Infect Control Hosp Epidemiol 1. https://doi.org/10.1017/ice.2020.139

Rezaei N (2020b) COVID-19 and medical biotechnology. Avicenna J Med Biotechnol 12(3):139

Rodriguez-Morales AJ, Cardona-Ospina JA, Gutierrez-Ocampo E, Villamizar-Pena R, Holguin-Rivera Y, Escalera-Antezana JP, Alvarado-Arnez LE, Bonilla-Aldana DK, Franco-Paredes C, Henao-Martinez AF, Paniz-Mondolfi A, Lagos-Grisales GJ, Ramirez-Vallejo E, Suarez JA, Zambrano LI, Villamil-Gomez WE, Balbin-Ramon GJ, Rabaan AA, Harapan H, Dhama K, Nishiura H, Kataoka H, Ahmad T, Sah R (2020) Clinical, laboratory and imaging features of COVID-19: a systematic review and meta-analysis. Travel Med Infect Dis 34:101623. https://doi.org/10.1016/j.tmaid.2020.101623

Rokni M, Hamblin MR, Rezaei N (2020) Cytokines and COVID-19: friends or foes? Human Vaccines Immunother:1–3

Roland LT, Gurrola JG II, Loftus PA, Cheung SW, Chang JL (2020) Smell and taste symptom-based predictive model for COVID-19 diagnosis. Int Forum Allergy Rhinol. https://doi.org/10.1002/alr.22602

Rzymski P, Nowicki M, Mullin GE, Abraham A, Rodríguez-Román E, Petzold MB, Bendau A, Sahu KK, Ather A, Naviaux A-F (2020) Quantity does not equal quality: scientific principles cannot be sacrificed. Int Immunopharmacol 86:106711

Saghazadeh A, Rezaei N (2020a) Immune-epidemiological parameters of the novel coronavirus – a perspective. Expert Rev Clin Immunol 16(5):465–470. https://doi.org/10.1080/1744666x.2020.1750954

Saghazadeh A, Rezaei N (2020b) Towards treatment planning of COVID-19: rationale and hypothesis for the use of multiple immunosuppressive agents: anti-antibodies, immunoglobulins, and corticosteroids. Int Immunopharmacol 84:106560. https://doi.org/10.1016/j.intimp.2020.106560

Sahu KK, Siddiqui AD, Rezaei N, Cerny J (2020) Challenges for management of immune thrombocytopenia during COVID-19 pandemic. J Med Virol. https://doi.org/10.1002/jmv.26251

Salehi S, Abedi A, Balakrishnan S, Gholamrezanezhad A (2020) Coronavirus disease 2019 (COVID-19): a systematic review of imaging findings in 919 patients. AJR Am J Roentgenol:1–7. https://doi.org/10.2214/ajr.20.23034

Saleki K, Banazadeh M, Saghazadeh A, Rezaei N (2020) The involvement of the central nervous system in patients with COVID-19. Rev Neurosci 31(4):453–456. https://doi.org/10.1515/revneuro-2020-0026

Scalinci SZ, Trovato Battagliola E (2020) Conjunctivitis can be the only presenting sign and symptom of COVID-19. IDCases 20:e00774. https://doi.org/10.1016/j.idcr.2020.e00774

Shamshirian D, Rezaei N (2020) Cardiovascular diseases burden in COVID-19: systematic review and meta-analysis

Sharifkashani S, Bafrani MA, Khaboushan AS, Pirzadeh M, Kheirandish A, Yavarpour Bali H, Hessami A, Saghazadeh A, Rezaei N (2020) Angiotensin-converting enzyme 2 (ACE2) receptor and SARS-CoV-2: potential therapeutic targeting. Eur J Pharmacol 884:173455. https://doi.org/10.1016/j.ejphar.2020.173455

Shi H, Han X, Jiang N, Cao Y, Alwalid O, Gu J, Fan Y, Zheng C (2020a) Radiological findings from 81 patients with COVID-19 pneumonia in Wuhan, China: a descriptive study. Lancet Infect Dis 20(4):425–434. https://doi.org/10.1016/s1473-3099(20)30086-4

Shi S, Qin M, Cai Y, Liu T, Shen B, Yang F, Cao S, Liu X, Xiang Y, Zhao Q, Huang H, Yang B, Huang C (2020b) Characteristics and clinical significance of myocardial injury in patients with severe coronavirus disease 2019. Eur Heart J. https://doi.org/10.1093/eurheartj/ehaa408

Shi S, Qin M, Shen B, Cai Y, Liu T, Yang F, Gong W, Liu X, Liang J, Zhao Q, Huang H, Yang B, Huang C (2020c) Association of cardiac injury with mortality in hospitalized patients with COVID-19 in Wuhan, China. JAMA Cardiol. https://doi.org/10.1001/jamacardio.2020.0950

Sodhi CP, Wohlford-Lenane C, Yamaguchi Y, Prindle T, Fulton WB, Wang S, McCray PB Jr, Chappell M, Hackam DJ, Jia H (2018) Attenuation of pulmonary ACE2 activity impairs inactivation of des-Arg(9) bradykinin/BKB1R axis and facilitates LPS-induced neutrophil infiltration. Am J Physiol Lung Cell Mol Physiol 314(1):L17–l31. https://doi.org/10.1152/ajplung.00498.2016

Sun P, Qie S, Liu Z, Ren J, Li K, Xi J (2020) Clinical characteristics of hospitalized patients with SARS-CoV-2 infection: a single arm meta-analysis. J Med Virol. https://doi.org/10.1002/jmv.25735

Suzuki T, Kutsuna S, Nakamura K, Ide S, Moriyama Y, Saito S, Morioka S, Ishikane M, Kinoshita N, Hayakawa K, Ohmagari N (2020) Difficulty of down-scaling the precautions for coronavirus disease-19 based on negative throat polymerase chain results in the early phase of infection. J Infect Chemother. https://doi.org/10.1016/j.jiac.2020.05.002

Thachil J, Tang N, Gando S, Falanga A, Cattaneo M, Levi M, Clark C, Iba T (2020) ISTH interim guidance on recognition and management of coagulopathy in COVID-19. J Thromb Haemost 18(5):1023–1026. https://doi.org/10.1111/jth.14810

Tian S, Hu N, Lou J, Chen K, Kang X, Xiang Z, Chen H, Wang D, Liu N, Liu D, Chen G, Zhang Y, Li D, Li J, Lian H, Niu S, Zhang L, Zhang J (2020) Characteristics of COVID-19 infection in Beijing. J Infect 80(4):401–406. https://doi.org/10.1016/j.jinf.2020.02.018

Vaira LA, Deiana G, Fois AG, Pirina P, Madeddu G, De Vito A, Babudieri S, Petrocelli M, Serra A, Bussu F, Ligas E, Salzano G, De Riu G (2020) Objective evaluation of anosmia and ageusia in COVID-19 patients: single-center experience on 72 cases. Head Neck 42(6):1252–1258. https://doi.org/10.1002/hed.26204

Vazquez-Garza E, Jerjes-Sanchez C, Navarrete A, Joya-Harrison J, Rodriguez D (2017) Venous thromboembolism: thrombosis, inflammation, and immunothrombosis for clinicians. J Thromb Thrombolysis 44(3):377–385. https://doi.org/10.1007/s11239-017-1528-7

Veeramachaneni GK, Thunuguntla V, Bobbillapati J, Bondili JS (2020) Structural and simulation analysis of hot spot residues interactions of SARS-CoV 2 with human ACE2 receptor. J Biomol Struct Dyn:1–16. https://doi.org/10.1080/07391102.2020.1773318

Wan Y, Shang J, Graham R, Baric RS, Li F (2020) Receptor recognition by the novel coronavirus from wuhan: an analysis based on decade-long structural studies of SARS coronavirus. J Virol 94(7). https://doi.org/10.1128/jvi.00127-20

Wang D, Hu B, Hu C, Zhu F, Liu X, Zhang J, Wang B, Xiang H, Cheng Z, Xiong Y, Zhao Y, Li Y, Wang X, Peng Z (2020a) Clinical characteristics of 138 hospitalized patients with 2019 novel coronavirus-infected pneumonia in Wuhan, China. JAMA. https://doi.org/10.1001/jama.2020.1585

Wang L, He W, Yu X, Hu D, Bao M, Liu H, Zhou J, Jiang H (2020b) Coronavirus disease 2019 in elderly patients: characteristics and prognostic factors based on 4-week follow-up. J Infect 80(6):639–645. https://doi.org/10.1016/j.jinf.2020.03.019

Wang Y, Dong C, Hu Y, Li C, Ren Q, Zhang X, Shi H, Zhou M (2020c) Temporal changes of CT findings in 90 patients with COVID-19 pneumonia: a longitudinal study. Radiology 296:E55–E64. https://doi.org/10.1148/radiol.2020200843

Weiss SR, Leibowitz JL (2011) Coronavirus pathogenesis. Adv Virus Res 81:85–164. https://doi.org/10.1016/b978-0-12-385885-6.00009-2

Wichmann D, Sperhake JP, Lutgehetmann M, Steurer S, Edler C, Heinemann A, Heinrich F, Mushumba H, Kniep I, Schroder AS, Burdelski C, de Heer G, Nierhaus A, Frings D, Pfefferle S, Becker H, Bredereke-Wiedling H, de Weerth A, Paschen HR, Sheikhzadeh-Eggers S, Stang A, Schmiedel S, Bokemeyer C, Addo MM, Aepfelbacher M, Puschel K, Kluge S (2020) Autopsy findings and venous thromboembolism in patients with COVID-19. Ann Intern Med doi:https://doi.org/10.7326/m20-2003

Woo PC, Lau SK, Chu CM, Chan KH, Tsoi HW, Huang Y, Wong BH, Poon RW, Cai JJ, Luk WK, Poon LL, Wong SS, Guan Y, Peiris JS, Yuen KY (2005) Characterization and complete genome sequence of a novel coronavirus, coronavirus HKU1, from patients with pneumonia. J Virol 79(2):884–895. https://doi.org/10.1128/jvi.79.2.884-895.2005

Wu J, Liu J, Zhao X, Liu C, Wang W, Wang D, Xu W, Zhang C, Yu J, Jiang B, Cao H, Li L (2020a) Clinical characteristics of imported cases of COVID-19 in Jiangsu Province: a multicenter descriptive study. Clin Infect Dis. https://doi.org/10.1093/cid/ciaa199

Wu P, Duan F, Luo C, Liu Q, Qu X, Liang L, Wu K (2020b) Characteristics of ocular findings of patients with coronavirus disease 2019 (COVID-19) in Hubei Province, China. JAMA Ophthalmol. https://doi.org/10.1001/jamaophthalmol.2020.1291

Xu H, Zhong L, Deng J, Peng J, Dan H, Zeng X, Li T, Chen Q (2020a) High expression of ACE2 receptor of 2019-nCoV on the epithelial cells of oral mucosa. Int J Oral Sci 12(1):8. https://doi.org/10.1038/s41368-020-0074-x

Xu J, Wu R, Huang H, Zheng W, Ren X, Wu N, Ji B, Lv Y, Liu Y, Mi R (2020b) Computed tomographic imaging of 3 patients with coronavirus disease 2019 pneumonia with negative virus real-time reverse-transcription polymerase chain reaction test. Clin Infect Dis. https://doi.org/10.1093/cid/ciaa207

Yan CH, Faraji F, Prajapati DP, Boone CE, DeConde AS (2020) Association of chemosensory dysfunction and Covid-19 in patients presenting with influenza-like symptoms. Int Forum Allergy Rhinol. https://doi.org/10.1002/alr.22579

Yang W, Cao Q, Qin L, Wang X, Cheng Z, Pan A, Dai J, Sun Q, Zhao F, Qu J, Yan F (2020) Clinical characteristics and imaging manifestations of the 2019 novel coronavirus disease (COVID-19): a multi-center study in Wenzhou city, Zhejiang, China. J Infect 80(4):388–393. https://doi.org/10.1016/j.jinf.2020.02.016

Yazdanpanah F, Hamblin MR, Rezaei N (2020a) The immune system and COVID-19: friend or foe? Life Sci 256:117900. https://doi.org/10.1016/j.lfs.2020.117900

Yazdanpanah N, Saghazadeh A, Rezaei N (2020b) Anosmia: a missing link in the neuroimmunology of coronavirus disease 2019 (COVID-19). Rev Neurosci. https://doi.org/10.1515/revneuro-2020-0039

Young BE, Ong SWX, Kalimuddin S, Low JG, Tan SY, Loh J, Ng OT, Marimuthu K, Ang LW, Mak TM, Lau SK, Anderson DE, Chan KS, Tan TY, Ng TY, Cui L, Said Z, Kurupatham L, Chen MI, Chan M, Vasoo S, Wang LF, Tan BH, Lin RTP, Lee VJM, Leo YS, Lye DC (2020) Epidemiologic features and clinical course of patients infected with SARS-CoV-2 in Singapore. JAMA. https://doi.org/10.1001/jama.2020.3204

Yousefzadegan S, Rezaei N (2020) Case report: death due to COVID-19 in three brothers. Am J Trop Med Hyg 102(6):1203–1204. https://doi.org/10.4269/ajtmh.20-0240

Yu T, Cai S, Zheng Z, Cai X, Liu Y, Yin S, Peng J, Xu X (2020) Association between clinical manifestations and prognosis in patients with COVID-19. Clin Ther. https://doi.org/10.1016/j.clinthera.2020.04.009

Zhang J, Wang S, Xue Y (2020a) Fecal specimen diagnosis 2019 novel coronavirus-infected pneumonia. J Med Virol. https://doi.org/10.1002/jmv.25742

Zhang JJ, Dong X, Cao YY, Yuan YD, Yang YB, Yan YQ, Akdis CA, Gao YD (2020b) Clinical characteristics of 140 patients infected with SARS-CoV-2 in Wuhan, China. Allergy. https://doi.org/10.1111/all.14238

Zhang L, Zhu F, Xie L, Wang C, Wang J, Chen R, Jia P, Guan HQ, Peng L, Chen Y, Peng P, Zhang P, Chu Q, Shen Q, Wang Y, Xu SY, Zhao JP, Zhou M (2020c) Clinical characteristics of COVID-19-infected cancer patients: a retrospective case study in three hospitals within Wuhan, China. Ann Oncol. https://doi.org/10.1016/j.annonc.2020.03.296

Zhang X, Cai H, Hu J, Lian J, Gu J, Zhang S, Ye C, Lu Y, Jin C, Yu G, Jia H, Zhang Y, Sheng J, Li L, Yang Y (2020d) Epidemiological, clinical characteristics of cases of SARS-CoV-2 infection with abnormal imaging findings. Int J Infect Dis 94:81–87. https://doi.org/10.1016/j.ijid.2020.03.040

Zhao D, Yao F, Wang L, Zheng L, Gao Y, Ye J, Guo F, Zhao H, Gao R (2020) A comparative study on the clinical features of COVID-19 pneumonia to other pneumonias. Clin Infect Dis. https://doi.org/10.1093/cid/ciaa247

Zheng F, Tang W, Li H, Huang YX, Xie YL, Zhou ZG (2020) Clinical characteristics of 161 cases of corona virus disease 2019 (COVID-19) in Changsha. Eur Rev Med Pharmacol Sci 24(6):3404–3410. https://doi.org/10.26355/eurrev_202003_20711

Zhou J, Li C, Zhao G, Chu H, Wang D, Yan HH, Poon VK, Wen L, Wong BH, Zhao X, Chiu MC, Yang D, Wang Y, Au-Yeung RKH, Chan IH, Sun S, Chan JF, To KK, Memish ZA, Corman VM, Drosten C, Hung IF, Zhou Y, Leung SY, Yuen KY (2017) Human intestinal tract serves as an alternative infection route for Middle East respiratory syndrome coronavirus. Sci Adv 3(11):eaao4966. https://doi.org/10.1126/sciadv.aao4966

Zhou Z, Zhao N, Shu Y, Han S, Chen B, Shu X (2020) Effect of gastrointestinal symptoms on patients infected with coronavirus disease 2019. Gastroenterology. https://doi.org/10.1053/j.gastro.2020.03.020

Zhu J, Ji P, Pang J, Zhong Z, Li H, He C, Zhang J, Zhao C (2020a) Clinical characteristics of 3,062 COVID-19 patients: a meta-analysis. J Med Virol. https://doi.org/10.1002/jmv.25884

Zhu J, Zhong Z, Ji P, Li H, Li B, Pang J, Zhang J, Zhao C (2020b) Clinicopathological characteristics of 8697 patients with COVID-19 in China: a meta-analysis. Fam Med Community Health 8(2). https://doi.org/10.1136/fmch-2020-000406

Zhu J, Zhong Z, Li H, Ji P, Pang J, Li B, Zhang J (2020c) CT imaging features of 4121 patients with COVID-19: a meta-analysis. J Med Virol. https://doi.org/10.1002/jmv.25910

Zhu N, Zhang D, Wang W, Li X, Yang B, Song J, Zhao X, Huang B, Shi W, Lu R, Niu P, Zhan F, Ma X, Wang D, Xu W, Wu G, Gao GF, Tan W (2020d) A novel coronavirus from patients with pneumonia in China, 2019. N Engl J Med 382(8):727–733. https://doi.org/10.1056/NEJMoa2001017

Zhu T, Wang Y, Zhou S, Zhang N, Xia L (2020e) A comparative study of chest computed tomography features in young and older adults with corona virus disease (COVID-19). J Thorac Imaging. https://doi.org/10.1097/rti.0000000000000513

Zhu W, Xie K, Lu H, Xu L, Zhou S, Fang S (2020f) Initial clinical features of suspected coronavirus disease 2019 in two emergency departments outside of Hubei, China. J Med Virol. https://doi.org/10.1002/jmv.25763

Zumla A, Hui DS, Perlman S (2015) Middle East respiratory syndrome. Lancet 386(9997):995–1007. https://doi.org/10.1016/s0140-6736(15)60454-8

12 Pediatrics and COVID-19

Tuna Toptan, Sandra Ciesek, and Sebastian Hoehl

Abstract

Viral respiratory tract infections are prevalent in children. They have substantial effects on childhood morbidity throughout the world, especially in developing countries. In this chapter, we describe the preliminary characteristics of pediatric COVID-19 and discover that severe and critical disease in children is rare. Many children remain asymptomatic. The reason why severity increases with progressing age and largely spares children is not yet known. In the search for possible explanations, we explore key differences between the pediatric and adult immune responses to new pathogens, and in host factors, such as ACE2 abundance.

Keywords

ACE2 · Adolescents · Children · COVID-19 · Immunity · Pediatrics · SARS-CoV-2

T. Toptan · S. Ciesek · S. Hoehl (✉)
Institute of Medical Virology, University Hospital Frankfurt, Goethe University, Frankfurt, Germany
e-mail: Sebastian.hoehl@kgu.de

12.1 Introduction

Severe acute respiratory syndrome coronavirus 2 (SARS-CoV-2), the virus that causes coronavirus disease 2019 (COVID-19), is a single-stranded, enveloped RNA virus primarily affecting the respiratory tract. Older adults with underlying medical conditions are disproportionately prone to severe disease. Children have a low risk of severe disease.

In children under 5 years of age, however, other respiratory infections are among the leading causes of death (World Health Organization). Among them, viruses that are transmitted through droplets are important pathogens. Many of them are enveloped RNA viruses, as is SARS-CoV-2. Examples include human respiratory syncytial virus (RSV), influenza viruses, and measles *morbillivirus*. RSV is considered the most critical respiratory tract pathogen of early childhood and the primary cause of bronchiolitis in children younger than 1 year (Kliegman and Geme 2019). From 1997 to 2006, 24% of all lower respiratory tract infection hospitalizations in children younger than 5 years of age was attributed to RSV in the United States. Infants younger than 3 months of age have the highest rate of hospitalization (Stockman et al. 2012). Children are also disproportionately affected by the influenza-associated disease. Worldwide, between 28,000 and 111,500 deaths in children younger than 5 years of age are estimated to be attributable to

N. Rezaei (ed.), *Coronavirus Disease - COVID-19*, Advances in Experimental Medicine and Biology 1318, https://doi.org/10.1007/978-3-030-63761-3_12

influenza, and young children below the age of 2 are particularly affected. Low-to-middle-income countries bear a disproportionate amount of the global burden (Nair et al. 2011). Children with chronic conditions such as pulmonary or cardiac disease, as well as neurologic, renal, and neuromuscular disorders, and other disorders, are at an increased risk for severe disease (Kondrich and Rosenthal 2017). In measles, children below 5 years of age are also at a markedly increased risk for complications, including pneumonia, the most common cause of death in measles (Perry and Halsey 2004).

Children and adolescents are also vulnerable to disease by the seven human coronaviruses: OC43, 229E, NL63, HKU1, MERS, SARS, and SARS-CoV-2. The seasonal coronaviruses OC43, HKU1 (both betacoronavirus), 229E, and NL63 (both alphacoronavirus) usually cause asymptomatic or mild cold-like illness in adults and account for approximately 15% of all common colds. They may also cause lower respiratory tract infections, such as bronchiolitis and pneumonia, and may cause exacerbations of asthma and chronic obstructive pulmonary disease (COPD) in adults. NL63 is a common cause of croup in young children (Sung et al. 2010). In a study from Norway, seasonal coronaviruses, most commonly NL63, were present in 9.1% of all respiratory tract infections of hospitalized children, varying with the season. More than two-thirds were detected with other viruses. A control group also had a similar rate of 10.2% (Heimdal et al. 2019), demonstrating that coronavirus infection is prevalent in children. The pandemic caused by the betacoronavirus severe acute respiratory distress virus (SARS virus) in 2002 and 2003 also affected children, but the severe disease was rare. Younger children commonly had a mild, nonspecific illness and more commonly had coryza or gastrointestinal symptoms than adults. No deaths or cases of acute respiratory distress syndrome occurred in children with SARS younger than 12 years of age. Respiratory distress and hypoxemia were observed in 10–20% of all patients, the case fatality rate in adults was 10–17%, and it was highest in those older than 65 years (>50%) (Kliegman and Geme 2019; Stockman et al. 2007). In 2012, the first cases of Middle East respiratory syndrome (MERS) emerged on the Arab Peninsula. Most cases are located within Saudi Arabia and the United Arab Emirates. Camels were identified as the intermediate host, and cases were often related to exposure to these animals. Pediatric MERS cases are rare but are reported in household contacts. In a small case series of 11 children, most were discovered during contact investigations and were asymptomatic. Only two were symptomatic; both had preexisting medical conditions (Memish et al. 2014). In another study of 280 household contacts of 26 index patients with MERS, 3 children younger than 18 years of age were among the 12 probable secondary infections. All of them were asymptomatic (Drosten et al. 2014).

In this chapter, we will characterize the novel coronavirus disease COVID-19 in children and adolescents and explore the hypotheses of why children are less susceptible to severe disease.

12.2 COVID-19 in Children and Adolescents

SARS and MERS in children commonly present as a mild or asymptomatic disease. COVID-19 also appears to be a mild, self-limiting disease in most children, but children of all age groups are also vulnerable to severe disease. It should be noted that the total number of children with laboratory-confirmed SARS-CoV-2 infection might be underreported. It is reasonable to assume that, due to the lower frequency of typical symptoms, they may not meet the local criteria for testing (McCarthy 2020).

12.2.1 Pediatric COVID-19 Is Often Asymptomatic, and Severe Disease Is Rare

The literature on pediatric COVID-19 is only beginning to evolve, and many aspects are still poorly understood. The data that is available today stems mainly from outbreaks in regions with advanced medical systems in China, the

United States, and Europe. The outcome and burden of COVID-19 in children in the developing world without access to advanced medical care must also be studied and considered. The characterization of pediatric COVID-19 is far from being complete.

However, from the reports available, two main conclusions can already be drawn: (i) most children are asymptomatic or have mild symptoms, and even though children of all age groups are susceptible, (ii) severe disease is rare (Castagnoli et al. 2020).

A summary of 72,314 cases from the Chinese Center for Disease Control and Prevention was published in late February 2020. Most cases included in that study were symptomatic patients with suspected or confirmed COVID-19. Children under the age of 19 made up only 2% of total cases (Wu and McGoogan 2020).

In a large study of more than 2135 laboratory-confirmed (34.1%) or suspected (65.9%) cases of COVID-19 in children from China, the largest study on COVID-19 in children to date in May of 20, over 90% had an asymptomatic, mild, or moderate disease. Severe disease, defined by the occurrence of hypoxia, and critical disease, defined by the occurrence of acute respiratory distress syndrome (ARDS) or multi-organ failure, developed in 5.8% of these children. Children below the age of 1 were more likely to develop the critical or severe disease (8.8% and 1.9%, respectively) than other age groups (in comparison: >15 years of age, 2.7% and 0.3%, respectively) (Dong et al. 2020). A brief report on 171 children with confirmed SARS-CoV-2 infection from Wuhan Children's Hospital, Wuhan, China, was published in March 2020 (Lu et al. 2020). Common signs and symptoms in these children included fever (41.5%), cough, and pharyngeal erythema. Diarrhea, vomiting, and rhinorrhea were also recorded but were less frequent. In total, 15.8% of children were classified as asymptomatic infection, 19.3% as upper respiratory tract infection, and 64.9% as pneumonia in this hospitalized cohort (Lu et al. 2020). First reports from the United States in early April of 2020 also confirmed a milder course for children with COVID-19. The hospitalization rate was also lower in children than adults (5.7% and 10%, respectively). Interestingly, the percentage of patients with confirmed SARS-CoV-2 infection who exhibited symptoms was lower in the pediatric group compared to the adult group for fever, cough, shortness of breath, myalgia, sore throat, headache, nausea/vomiting, abdominal pain, and diarrhea. Only in the case of runny nose, similar relative numbers were reported.

Pediatric deaths from COVID-19 are rare. In China, of 1023 COVID-19-related deaths, none was under the age of 10 years, and just 1 case was reported in the age group 10 to 19 years with a case fatality rate of 0.2% in this group (Surveillances 2020). Out of 1625 COVID-19-related mortalities reported from Italy, none were under 30 years of age (Onder et al. 2020). Similarly, no deaths were reported in those younger than 9 years of age among the 72,314 cases from China (Wu and McGoogan 2020). In a case series of 48 children admitted to North American pediatric intensive care units (PICUs), the ICU mortality was less than 5% at the end of the follow-up period (Shekerdemian et al. 2020).

Due to the low numbers of children with severe or critical COVID-19, factors associated with an unfavorable outcome are not well studied. In the case series from North American PICUs, 80% of patients had significant long-term underlying medical conditions, but they were commonly medically complex and associated with developmental delay and/or genetic abnormalities. Other symptoms warranting PICU admission included a vaso-occlusive crisis in the setting of sickle cell anemia, diabetic ketoacidosis, seizures, and circulatory collapse. In the study of 171 hospitalized children from Wuhan, China, 3 children with preexisting conditions (hydronephrosis, leukemia on maintenance chemotherapy, and intussusception) required mechanical ventilation, of whom 1 child, a 10-month-old with intussusception and multi-organ failure, died (Lu et al. 2020). For the four seasonal coronaviruses, OC43, NL63, HKU1, and 229E, younger age, underlying pulmonary pathology, and immunocompromised conditions are factors associated with severe lower respiratory tract infection (Ogimi et al. 2019).

12.2.2 What Role Do Children Play in the Spread of SARS-CoV-2?

Data on the susceptibility of children to contract an infection with SARS-CoV-2 is contradictory. In current studies (May of 2020), where children are rarely going to school during the pandemic, most children acquire infection through close contact with their parents or other family members (Lu et al. 2020). In a study by Bi et al., data from 391 SARS-CoV-2 cases and 1286 close contacts from Shenzhen, China, were examined (Bi et al. 2020). Children were as likely to be infected with SARS-CoV-2 as adults (infection rate of 7.4% in children under 10 years of age vs. 6.6% in the population average) (Bi et al. 2020). In an analysis of contact surveys from Wuhan and Shanghai, however, children aged 14 or younger were less susceptible to SARS-CoV-2 infection than adults between 15 and 64 years of age (odds ratio 0.34 95%CI 0.24–0.49) (Zhang et al. 2020b). Other studies also indicate that infection may occur less frequently in children when compared to adults. In a population-based study from Iceland, targeted testing by PCR using oropharyngeal or nasopharyngeal swabs were positive in 6.7% (38 of 564) of children under the 10 years of age and in 13.7% (1183 of 8635) of individuals who were 10 years of age or older (Gudbjartsson et al. 2020).

As children are often asymptomatic or paucisymptomatic, it could be hypothesized that they are also less contagious. However, in adults, 44% of transmissions are believed to occur before the occurrence of symptoms (He et al. 2020). A comparative study between children and adults demonstrated no significant difference in the viral loads from upper respiratory samples (Terry et al. 2020). As the viral load is associated with infectivity in vitro (Wolfel et al. 2020), transmission rates may, therefore, be similar in children and adults.

It is of great epidemiological interest to determine whether children are as susceptible to be infected with SARS-CoV-2 as are adults and whether they are as infectious when shedding the virus. Public health strategies commonly include school closures as part of the strategy. These have long-term repercussions on the well-being of the students, who themselves have a low, albeit not a non-existing, risk of severe disease or death. A higher degree of depressive symptoms in children in a primary school in China (Xie et al. 2020) is attributed to reductions in both outdoor activities and social interactions. The long-lasting effects of the measures implemented to curb the pandemic on children have yet to be evaluated. The effectiveness of school closures on the pandemic can be challenging to assess, as they are usually applied as a package of several measures, also including social distancing. The data that is available at this time is often of low quality (Viner et al. 2020). The study of contact surveys from Wuhan and Shanghai that discovered that young children were less susceptible to infection with SARS-CoV-2 also determined that while school closures by themselves cannot interrupt transmission, they can reduce peak incidence by 40–60% (Zhang et al. 2020b).

12.2.3 Why Is the Case Fatality Rate Low in Children?

As SARS-CoV-2 has only been causing disease in humans for a few months at the time of writing, the distinct presentation of and host response to COVID-19 in the elderly, adults, adolescents, children, toddlers, and newborns are not yet well studied. Our endeavor to highlight and explain these differences is merely cautious speculation.

It is reasonable to assume that the underlying pathophysiology leading to a lower fatality rate among children with COVID-19 is multifactorial. Age-specific host factors and fewer concurrent comorbidities, such as cardiovascular disease, diabetes, and hypertension that are commonly associated with severe disease in adults, may reduce the risk of severe disease in children. Accumulated long-term environmental factors, such as the inhalation of cigarette smoke and fine particles, in addition to the general aging process of the lungs, may increase the risk of severe pulmonary disease in older adults. In these patients, reduced muscle function and impaired mucociliary clearance restrict the host immune cells to eliminate pathogens (Lowery et al. 2013). In addition to basal levels of inflammation due to increased inflammasome activation and neutrophil infiltration, older individuals also exhibit impaired toll-like receptor (TLR) function, interferon (IFN) production, and antiviral resistance, which overall increase the susceptibility to respiratory infections (Shaw et al. 2013; Iwasaki et al. 2017).

The age-dependent severity of COVID-19 may be attributed to the adaptation and development of the immune response throughout a lifetime. Immune ontogeny is a dynamic and complicated process, which involves age-dependent maturation of innate and adaptive immune systems and may be modified by genetic and environmental factors. The microbiome content and immune priming via frequent exposure to other pathogens and regular immunizations may indirectly facilitate the containment of this infection. In other words, immune preparedness and response of children to SARS-CoV2 infection differ from adults. Understanding these differences is also crucial for developing safe and effective interventions.

12.2.4 The Immune Response to Viral Infections Differs Between Age Groups

The immune system is not fully functional at birth, and especially adaptive responses to antigenic challenges are relatively immature during the early stages of life. As the protective maternal antibodies wane, young children become more susceptible to infections. Nevertheless, the innate immune response of children to new antigens is highly adaptable, which may, in part, be explained by innate immune memory.

Innate immune memory or trained immunity is proposed as a primitive form of host defense mediated by epigenetic changes, including chromatin structure rearrangements, DNA methylation, and non-coding RNAs (Netea et al. 2019), and it differs from adaptive immune memory which requires additional gene recombination events. Memory B cells (MBCs) of young children are less mutated than MBCs in adults and produce natural antibodies (mostly IgM isotype) with broad reactivity and variable affinity independent of previous antigen encounters (Carsetti et al. 2020). CD27-negative IgM B cells, which are capable of rapid activation during infections, were found to more abundant in children (Grimsholm et al. 2020). Another source of natural antibodies is B1 cells. These low-affinity IgM-producing cells play a critical role in preventing viral dissemination, especially during infancy and early childhood (Simon et al. 2015). Trained immunity enables innate immune cells to respond to the same or new pathogenic stimulus by ameliorated transcription of the genes relevant for host defense and the production of pro-inflammatory cytokines (Netea et al. 2016). The initial challenge, either by vaccines such as Bacille Calmette-Guérin (BCG) (Benn et al. 2013), oral polio, and measles (Goodridge et al. 2016) or an infection, may reprogram the metabolic and the epigenetic landscape of the innate cells and allow transcription factors to access the promoter and enhancer regions of pro-inflammatory genes upon re-stimulation (Netea et al. 2020). Therefore, children challenged by regular vaccinations and frequent infections may acquire a potent trained immunity and limit the virus infection by activation of local innate immune responses during the early phases of the infection.

12.2.5 What Are the Immune Components That Play a Role in Disease Severity During SARS-CoV-2 Infection? Moreover, How Does This Differ in Children?

In SARS-CoV-2 infection, the virus binds angiotensin-converting enzyme 2 (ACE2) receptor for entry (Hoffmann et al. 2020; Walls et al. 2020), which is primed by the cleavage of viral spike (S) protein with TMPRSS2 and cathepsin proteases (Hoffmann et al. 2020). It leads to the fusion of viral and cellular membranes and the release of the viral genome into the cytoplasm. During this period, viral genomic RNA intermediates may act as pathogen-associated molecular patterns (PAMPs) that can be recognized by the cytoplasmic and endosomal pattern recognition receptors (PRRs) (Kato et al. 2006). Activation of PRRs initiates a signaling cascade that leads to the production of other inflammatory cytokines and type I interferons (IFN). The latter induces some key IFN-stimulated genes that limit viral replication through a variety of mechanisms (Schneider et al. 2014). However, like many other viruses, coronaviruses have evolved multiple immune evasion mechanisms to limit the early induction of type I IFN and thus promote viral replication and persistence (Sun et al. 2012; Zust et al. 2011; Hackbart et al. 2020).

Interestingly, the production of type I and III IFNs was found to be dampened in adult COVID-19 patients, unlike infections with other respiratory viruses influenza and RSV (Blanco-Melo et al. 2020). As a result, delayed-type I IFN production and uncontrolled viral replication lead to pyroptosis, which is a highly inflammatory form of programmed cell death and commonly seen with cytopathic viruses (Fink and Cookson 2005). Release of PAMPs and damage-associated molecular patterns (DAMPs) from infected/dying cells to extracellular milieu activates tissue-resident macrophages and leads to the expression of pro-inflammatory chemokines and cytokines, notably IL-6, IL-10, TNF-α, and IL-1β (Wen et al. 2020; Zhang et al. 2020a; Hadjadj et al. 2020; Merad and Martin 2020). These molecules recruit monocytes, macrophages, neutrophils, NK cells, and cytotoxic T cells to the lungs causing overproduction of inflammatory cytokines and tissue damage (Channappanavar and Perlman 2017). Under normal circumstances, pro-inflammatory cytokines prime adaptive T and B cell immune responses that are capable of resolving the infection. However, in some patients with COVID-19, dysfunctional adaptive immune responses as a result of T cell exhaustion, lymphopenia (Tay et al. 2020), and the influx of innate immune cells in the lungs, heart, spleen, lymph nodes, and kidney (Diao et al. 2020; Xu et al. 2020) contribute to severe disease.

In contrast, lymphopenia and increased inflammatory markers are rare in children (Ludvigsson 2020). The cytokine profile in children during SARS-CoV-2 infection showed upregulation of IL-1β but not TNF-α and IL-6 (Ng et al. 2004), which are markedly elevated in severe COVID-19 in adults (Chen et al. 2020a). On the contrary, the majority of SARS-CoV-2-infected children also do not present with immune dysregulation and pro-inflammatory response signatures (Lu et al. 2020; Liu et al. 2020). However, critical host immune factors that underlie the development of severe inflammatory responses are poorly defined.

In regions with a high incidence of COVID-19, cases with diverse clinical presentations that are characterized by excessive inflammation, referred to as the multisystem inflammation syndrome, have recently emerged in children and adolescents (Jones et al. 2020; Verdoni et al. 2020). The World Health Organization (WHO) recently (May 15th, 2020) issued a clear brief that offers a case definition for this syndrome (World Health 2020) that includes fever of at least 3 days, elevated markers of inflammation, no other apparent microbial cause, and evidence of COVID-19, as well as other signs. Criteria for Kawasaki disease, an acute vasculitis of the medium caliber vessels, may be fulfilled. Features of toxic shock syndrome and macrophage activation syndrome have also been observed, as has been abdominal (Dallan et al. 2020) and cardiac manifestations. Share clinical features with patients with severe COVID-19, such as leucopenia with marked lymphocytopenia, thrombocytopenia, and increased ferritin, are observed (Verdoni et al. 2020). Excessive inflammation in the context of COVID-19, therefore, can also be observed in children and adolescents in the context of COVID-19, but while pulmonary symptoms are paramount in older adults, these may be secondary or even absent in children and adolescents.

12.2.6 Immaturity of the Resident Immune System of the Lung

It is relatively difficult to obtain samples from the lower airways of healthy children, and most of the studies on the resident immune system of the lung have been carried out in animal models. Thus, the number of studies describing the immune system components of the respiratory system is limited. In this section, we will focus on the immune components involved in the pathogenesis of COVID-19 described above.

Respiratory immunity comprises epithelial cells, neutrophils, alveolar macrophages, dendritic cells, innate lymphocytes, and adaptive components, including different subsets of T and B cells (Iwasaki et al. 2017). The composition of these cells was determined through bronchoalveolar lavage (BAL) sampling before elective surgery. The profile of leukocytes and alveolar macrophages in the lower airway of children does not change significantly after 3 years of age (Grigg and Riedler 2000). The proportion of B cells, T cells, and natural killer cells in BAL fluid of healthy children seem to be within the normal range for adults. However, due to a higher absolute number of CD8+ cells, CD4/CD8 ratios seem to be lower in children when compared to adults (Ratjen et al. 1995). Since the lung airways are separated mainly from blood vessels and thus have minimal blood cell contaminants, it is reasonable to postulate that the BAL CD8+ T cells are mostly resident effector and memory T cells. Resident CD8+ cells are capable of mediating an enhanced recall response to secondary infection with the same or related pathogen (Hikono et al. 2006). Thus, higher levels of resident CD8+ T cells may provide superior immune protection against SARS-CoV-2 in children. On the other hand, the exact distribution of T cell subpopulations is not defined.

Alveolar macrophages produce low levels of inflammatory cytokines, maintain high phagocytic activity toward infectious agents, and generally suppress inflammation and adaptive immunity (Grigg and Riedler 2000). Despite similar absolute cell numbers among different age groups, the ability of alveolar macrophages to suppress inflammation and T cell activation changes with age. Cultured alveolar macrophages obtained by BAL from infants under 2 years of age expressed less HLA-DR antigen, showed limited phagocytic function, and produced less interleukin IL-1 and TNF secretion following TLR4 stimulation compared to children aged between 2 and 17 (Grigg et al. 1999). However, there are no published data on the suppressive receptor phenotype of alveolar macrophages in healthy children.

12.2.7 ACE-2 Maturity and Abundance

The abundance and maturity of viral receptors and host factors priming the entry may have an impact on the age-related difference in COVID-19 incidence and potentially also on disease severity. ACE2 is a membrane-associated aminopeptidase expressed in many tissues, including lung, vascular endothelia, renal and hearth, and small intestine and testes (Harmer et al. 2002; Hamming et al. 2004), as well as monocytes, neutrophils, and lymphocytes. ACE2 is involved in the conversion of angiotensin II, and it is a crucial modulator of the renin-angiotensin system (RAS) (Tikellis and Thomas 2012). Other coronaviruses, including SARS-CoV (Li et al. 2003) and human CoV-NL63 (Hofmann et al. 2005), use the ACE2 receptor for entry and subsequently downregulate its protein expression similar to SARS-CoV2 infection in Caco2 cells (Bojkova et al. 2020). In mice models, recombinant ACE2 was shown to exhibit a protective role in the lungs, and thus, its reduced expression was implicated to play a role in inflammation and severe lung injury (Kuba et al. 2005). However, a comparative study of biomarkers of host defense in bronchoalveolar lavage fluid showed no age-dependent differences in the RAS between neonates, children, adults, and older adults with acute respiratory distress syndrome (ARDS) (Schouten et al. 2019). ACE2 levels decrease with aging and in patients with chronic diseases, as reported by a recent study (Chen et al. 2020b). It suggests that relatively higher expression of ACE2 receptors in lung pneumocytes in children may have a protective effect on severe clinical manifestations due to SARS-CoV-2 infection. On the other hand, receptor levels and their role in virus entry may be distinct between the lung and nasal environments. Regarding the latter, a recent study showed an age-dependent expression of ACE2 in the nasal epithelium in a cohort of 305 individuals aged between 4 and 60 years (Bunyavanich et al. 2020). ACE2 gene expression was found to be lowest in children. However, this study did not include individuals older than 60 years of age as well as cofactors involved in the entry process, such as TMPRSS2 protease expression. If any of these age-dependent differences of ACE2 expression have an actual impact on the susceptibility of children to be infected with SARS-CoV-2 is yet unknown.

12.3 Conclusion

In the pandemic of SARS-CoV-2, a small proportion of children have severe disease, and mortality is low. Even during the current pandemic, children are reported to be at a higher risk of death from influenza than from SARS-CoV2 infection in the United States (Shekerdemian et al. 2020). Children often remain asymptomatic, and their role in the transmission of the virus is inconclusive. Despite the plethora of research conducted since January 2020, the global burden of COVID-19 itself and the burden on the health of children that results from imposed interventions to combat the pandemic has not been comprehensively evaluated. Underreporting of COVID-19 in children due to its commonly asymptomatic or paucisymptomatic presentation can be expected, especially in low-income settings, where testing capabilities are reduced.

We do not yet know why the severe disease is rare in children, and the underlying pathophysiology may be multifactorial. Changes in the immune response throughout a lifetime may contribute to the higher risk of severe disease in older adults. Mainly young children rely on an innate response to the common encounter of new antigens and respond to viral infections with MBCs that produce antibodies with broad reactivity and variable affinity independent of previous antigen encounter. Immune dysregulation frequently occurs in adults, with a marked increase in TNF-α and IL-6 levels. However, recently we have also seen cases of immune dysregulation occurring in children. The multisystem inflammatory syndrome may be associated with COVID-19 in children, but causation has not been proven yet in May of 2020. Age-dependent differences in lung-resident immune cells, as well as in ACE2 abundance, have been proposed. Whether these differences have any impact on the susceptibility or clinical course of SARS-CoV-2 infection in children remained to be addressed. Further research on SARS-CoV-2 and COVID-19 is required to determine the molecular mechanisms underlying age-dependent differences in disease severity and develop improved care for at-risk patients.

References

Benn CS, Netea MG, Selin LK, Aaby P (2013) A small jab – a big effect: nonspecific immunomodulation by vaccines. Trends Immunol 34(9):431–439. https://doi.org/10.1016/j.it.2013.04.004

Bi Q, Wu Y, Mei S, Ye C, Zou X, Zhang Z, Liu X, Wei L, Truelove SA, Zhang T, Gao W, Cheng C, Tang X, Wu X, Sun B, Huang S, Sun Y, Zhang J, Ma T, Lessler J, Feng T (2020) Epidemiology and transmission of COVID-19 in 391 cases and 1286 of their close contacts in Shenzhen, China: a retrospective cohort study. Lancet Infect Dis 20:911. https://doi.org/10.1016/s1473-3099(20)30287-5

Blanco-Melo D, Nilsson-Payant BE, Liu W-C, Møller R, Panis M, Sachs D, Albrecht RA, tenOever BR (2020) SARS-CoV-2 launches a unique transcriptional signature from in vitro, ex vivo, and in vivo systems. bioRxiv:2020.2003.2024.004655. https://doi.org/10.1101/2020.03.24.004655

Bojkova D, Klann K, Koch B, Widera M, Krause D, Ciesek S, Cinatl J, Munch C (2020) Proteomics of SARS-CoV-2-infected host cells reveals therapy targets. Nature 583:469. https://doi.org/10.1038/s41586-020-2332-7

Bunyavanich S, Do A, Vicencio A (2020) Nasal gene expression of angiotensin-converting enzyme 2 in children and adults. JAMA 323:2427. https://doi.org/10.1001/jama.2020.8707

Carsetti R, Quintarelli C, Quinti I, Mortari EP, Zumla A, Ippolito G, Locatelli F (2020) The immune system of children: the key to understanding SARS-CoV-2 susceptibility? Lancet Child Adolesc Health

Castagnoli R, Votto M, Licari A, Brambilla I, Bruno R, Perlini S, Rovida F, Baldanti F, Marseglia GL (2020) Severe acute respiratory syndrome coronavirus 2 (SARS-CoV-2) infection in children and adolescents: a systematic review. JAMA Pediatr 174:882. https://doi.org/10.1001/jamapediatrics.2020.1467

Channappanavar R, Perlman S (2017) Pathogenic human coronavirus infections: causes and consequences of cytokine storm and immunopathology. Semin Immunopathol 39(5):529–539. https://doi.org/10.1007/s00281-017-0629-x

Chen G, Wu D, Guo W, Cao Y, Huang D, Wang H, Wang T, Zhang X, Chen H, Yu H, Zhang M, Wu S, Song J, Chen T, Han M, Li S, Luo X, Zhao J, Ning Q (2020a) Clinical and immunological features of severe and moderate coronavirus disease 2019. J Clin Invest 130(5):2620–2629. https://doi.org/10.1172/jci137244

Chen JJ, Xia X, Liu K, Yu Z, Tao W, Gong W, Han JJ (2020b) Individual variation of the SARS-CoV2 receptor ACE2 gene expression and regulation. Preprints (2020030191)

Dallan C, Romano F, Siebert J, Politi S, Lacroix L, Sahyoun C (2020) Septic shock presentation in adolescents with COVID-19. Lancet Child Adolesc Health 4:e21. https://doi.org/10.1016/s2352-4642(20)30164-4

Diao B, Wang C, Tan Y, Chen X, Liu Y, Ning L, Chen L, Li M, Liu Y, Wang G, Yuan Z, Feng Z, Wu Y, Chen Y (2020) Reduction and functional exhaustion of T cells in patients with coronavirus disease 2019 (COVID-19). medRxiv:2020.2002.2018.20024364. https://doi.org/10.1101/2020.02.18.20024364

Dong Y, Mo X, Hu Y, Qi X, Jiang F, Jiang Z, Tong S (2020) Epidemiological characteristics of 2143 pediatric patients with 2019 coronavirus disease in China. Pediatrics 145:e20200702. https://doi.org/10.1542/peds.2020-0702

Drosten C, Meyer B, Müller MA, Corman VM, Al-Masri M, Hossain R, Madani H, Sieberg A, Bosch BJ, Lattwein E, Alhakeem RF, Assiri AM, Hajomar W, Albarrak AM, Al-Tawfiq JA, Zumla AI, Memish ZA (2014) Transmission of MERS-coronavirus in household contacts. N Engl J Med 371(9):828–835. https://doi.org/10.1056/NEJMoa1405858

Fink SL, Cookson BT (2005) Apoptosis, pyroptosis, and necrosis: mechanistic description of dead and dying eukaryotic cells. Infect Immun 73(4):1907–1916. https://doi.org/10.1128/IAI.73.4.1907-1916.2005

Goodridge HS, Ahmed SS, Curtis N, Kollmann TR, Levy O, Netea MG, Pollard AJ, van Crevel R, Wilson CB (2016) Harnessing the beneficial heterologous effects of vaccination. Nat Rev Immunol 16(6):392–400. https://doi.org/10.1038/nri.2016.43

Grigg J, Riedler J (2000) Developmental airway cell biology. The "normal" young child. Am J Respir Crit Care Med 162(2 Pt 2):S52–S55. https://doi.org/10.1164/ajrccm.162.supplement_1.maic-14

Grigg J, Riedler J, Robertson CF, Boyle W, Uren S (1999) Alveolar macrophage immaturity in infants and young children. Eur Respir J 14(5):1198–1205. https://doi.org/10.1183/09031936.99.14511989

Grimsholm O, Piano Mortari E, Davydov AN, Shugay M, Obraztsova AS, Bocci C, Marasco E, Marcellini V, Aranburu A, Farroni C, Silvestris DA, Cristofoletti C, Giorda E, Scarsella M, Cascioli S, Barresi S, Lougaris V, Plebani A, Cancrini C, Finocchi A, Moschese V, Valentini D, Vallone C, Signore F, de Vincentiis G, Zaffina S, Russo G, Gallo A, Locatelli F, Tozzi AE, Tartaglia M, Chudakov DM, Carsetti R (2020) The interplay between CD27(dull) and CD27(bright) B cells ensures the flexibility, stability, and resilience of human B cell memory. Cell Rep 30 (9):2963–2977. e2966. doi:https://doi.org/10.1016/j.celrep.2020.02.022

Gudbjartsson DF, Helgason A, Jonsson H, Magnusson OT, Melsted P, Norddahl GL, Saemundsdottir J, Sigurdsson A, Sulem P, Agustsdottir AB, Eiriksdottir B, Fridriksdottir R, Gardarsdottir EE, Georgsson G, Gretarsdottir OS, Gudmundsson KR, Gunnarsdottir TR, Gylfason A, Holm H, Jensson BO, Jonasdottir A, Jonsson F, Josefsdottir KS, Kristjansson T, Magnusdottir DN, le Roux L, Sigmundsdottir G, Sveinbjornsson G, Sveinsdottir KE, Sveinsdottir M, Thorarensen EA, Thorbjornsson B, Löve A, Masson G, Jonsdottir I, Möller AD, Gudnason T, Kristinsson KG, Thorsteinsdottir U, Stefansson K (2020) Spread of SARS-CoV-2 in the Icelandic population. N Engl J Med 382:2302. https://doi.org/10.1056/NEJMoa2006100

Hackbart M, Deng X, Baker SC (2020) Coronavirus endoribonuclease targets viral polyuridine sequences to evade activating host sensors. Proc Natl Acad Sci U S A 117(14):8094–8103. https://doi.org/10.1073/pnas.1921485117

Hadjadj J, Yatim N, Barnabei L, Corneau A, Boussier J, Pere H, Charbit B, Bondet V, Chenevier-Gobeaux C, Breillat P, Carlier N, Gauzit R, Morbieu C, Pene F, Marin N, Roche N, Szwebel T-A, Smith N, Merkling S, Treluyer J-M, Veyer D, Mouthon L, Blanc C, Tharaux P-L, Rozenberg F, Fischer A, Duffy D, Rieux-Laucat F, Kerneis S, Terrier B (2020) Impaired type I interferon activity and exacerbated inflammatory responses in severe Covid-19 patients. medRxiv:2020.2004.2019.20068015. https://doi.org/10.1101/2020.04.19.20068015

Hamming I, Timens W, Bulthuis MLC, Lely AT, Navis GJ, van Goor H (2004) Tissue distribution of ACE2 protein, the functional receptor for SARS coronavirus. A first step in understanding SARS pathogenesis. J Pathol 203(2):631–637. https://doi.org/10.1002/path.1570

Harmer D, Gilbert M, Borman R, Clark KL (2002) Quantitative mRNA expression profiling of ACE 2, a novel homologue of angiotensin converting enzyme. FEBS Lett 532(1–2):107–110. https://doi.org/10.1016/s0014-5793(02)03640-2

He X, Lau EHY, Wu P, Deng X, Wang J, Hao X, Lau YC, Wong JY, Guan Y, Tan X, Mo X, Chen Y, Liao B, Chen W, Hu F, Zhang Q, Zhong M, Wu Y, Zhao L, Zhang

F, Cowling BJ, Li F, Leung GM (2020) Temporal dynamics in viral shedding and transmissibility of COVID-19. Nat Med 26:672. https://doi.org/10.1038/s41591-020-0869-5

Heimdal I, Moe N, Krokstad S, Christensen A, Skanke LH, Nordbø SA, Døllner H (2019) Human coronavirus in hospitalized children with respiratory tract infections: a 9-year population-based study from Norway. J Infect Dis 219(8):1198–1206. https://doi.org/10.1093/infdis/jiy646

Hikono H, Kohlmeier JE, Ely KH, Scott I, Roberts AD, Blackman MA, Woodland DL (2006) T-cell memory and recall responses to respiratory virus infections. Immunol Rev 211:119–132. https://doi.org/10.1111/j.0105-2896.2006.00385.x

Hoffmann M, Kleine-Weber H, Schroeder S, Kruger N, Herrler T, Erichsen S, Schiergens TS, Herrler G, Wu NH, Nitsche A, Muller MA, Drosten C, Pohlmann S (2020) SARS-CoV-2 cell entry depends on ACE2 and TMPRSS2 and is blocked by a clinically proven protease inhibitor. Cell 181(2):271–280. e278. https://doi.org/10.1016/j.cell.2020.02.052

Hofmann H, Pyrc K, van der Hoek L, Geier M, Berkhout B, Pohlmann S (2005) Human coronavirus NL63 employs the severe acute respiratory syndrome coronavirus receptor for cellular entry. Proc Natl Acad Sci U S A 102(22):7988–7993. https://doi.org/10.1073/pnas.0409465102

Iwasaki A, Foxman EF, Molony RD (2017) Early local immune defences in the respiratory tract. Nat Rev Immunol 17(1):7–20. https://doi.org/10.1038/nri.2016.117

Jones VG, Mills M, Suarez D, Hogan CA, Yeh D, Bradley Segal J, Nguyen EL, Barsh GR, Maskatia S, Mathew R (2020) COVID-19 and Kawasaki disease: novel virus and novel case. Hosp Pediatr 10:537. https://doi.org/10.1542/hpeds.2020-0123

Kato H, Takeuchi O, Sato S, Yoneyama M, Yamamoto M, Matsui K, Uematsu S, Jung A, Kawai T, Ishii KJ, Yamaguchi O, Otsu K, Tsujimura T, Koh CS, Reis e Sousa C, Matsuura Y, Fujita T, Akira S (2006) Differential roles of MDA5 and RIG-I helicases in the recognition of RNA viruses. Nature 441(7089):101–105. https://doi.org/10.1038/nature04734

Kliegman R, Geme JWS (2019) Nelson textbook of pediatrics. Elsevier, Philadelphia

Kondrich J, Rosenthal M (2017) Influenza in children. Curr Opin Pediatr 29(3):297–302. https://doi.org/10.1097/mop.0000000000000495

Kuba K, Imai Y, Rao S, Gao H, Guo F, Guan B, Huan Y, Yang P, Zhang Y, Deng W, Bao L, Zhang B, Liu G, Wang Z, Chappell M, Liu Y, Zheng D, Leibbrandt A, Wada T, Slutsky AS, Liu D, Qin C, Jiang C, Penninger JM (2005) A crucial role of angiotensin converting enzyme 2 (ACE2) in SARS coronavirus–induced lung injury. Nat Med 11(8):875–879. https://doi.org/10.1038/nm1267

Li W, Moore MJ, Vasilieva N, Sui J, Wong SK, Berne MA, Somasundaran M, Sullivan JL, Luzuriaga K, Greenough TC, Choe H, Farzan M (2003) Angiotensin-converting enzyme 2 is a functional receptor for the SARS coronavirus. Nature 426(6965):450–454. https://doi.org/10.1038/nature02145

Liu W, Zhang Q, Chen J, Xiang R, Song H, Shu S, Chen L, Liang L, Zhou J, You L, Wu P, Zhang B, Lu Y, Xia L, Huang L, Yang Y, Liu F, Semple MG, Cowling BJ, Lan K, Sun Z, Yu H, Liu Y (2020) Detection of Covid-19 in children in early January 2020 in Wuhan, China. N Engl J Med 382(14):1370–1371. https://doi.org/10.1056/NEJMc2003717

Lowery EM, Brubaker AL, Kuhlmann E, Kovacs EJ (2013) The aging lung. Clin Interv Aging 8:1489–1496. https://doi.org/10.2147/CIA.S51152

Lu X, Zhang L, Du H, Zhang J, Li YY, Qu J, Zhang W, Wang Y, Bao S, Li Y, Wu C, Liu H, Liu D, Shao J, Peng X, Yang Y, Liu Z, Xiang Y, Zhang F, Silva RM, Pinkerton KE, Shen K, Xiao H, Xu S, Wong GWK (2020) SARS-CoV-2 infection in children. N Engl J Med 382:1663. https://doi.org/10.1056/NEJMc2005073

Ludvigsson JF (2020) Systematic review of COVID-19 in children shows milder cases and a better prognosis than adults. Acta Paediatr 109:1088. https://doi.org/10.1111/apa.15270

McCarthy A (2020) COVID-19 and children. Understanding the science of SARS-CoV-2 infection in infants, children, teens. Harvard Medical School. https://hms.harvard.edu/news/covid-19-children

Memish ZA, Al-Tawfiq JA, Assiri A, AlRabiah FA, Al Hajjar S, Albarrak A, Flemban H, Alhakeem RF, Makhdoom HQ, Alsubaie S, Al-Rabeeah AA (2014) Middle East respiratory syndrome coronavirus disease in children. Pediatr Infect Dis J 33(9):904–906. https://doi.org/10.1097/inf.0000000000000325

Merad M, Martin JC (2020) Pathological inflammation in patients with COVID-19: a key role for monocytes and macrophages. Nat Rev Immunol 20:355. https://doi.org/10.1038/s41577-020-0331-4

Nair H, Brooks WA, Katz M, Roca A, Berkley JA, Madhi SA, Simmerman JM, Gordon A, Sato M, Howie S (2011) Global burden of respiratory infections due to seasonal influenza in young children: a systematic review and meta-analysis. Lancet 378(9807):1917–1930

Netea MG, Giamarellos-Bourboulis EJ, Domínguez-Andrés J, Curtis N, van Crevel R, van de Veerdonk FL, Bonten M (2020) Trained immunity: a tool for reducing susceptibility and severity of SARS-CoV-2 infection. Cell 181:969. https://doi.org/10.1016/j.cell.2020.04.042

Netea MG, Joosten LA, van der Meer JW (2016) Adaptation and memory in innate immunity. Semin Immunol 28(4):317–318. https://doi.org/10.1016/j.smim.2016.07.002

Netea MG, Schlitzer A, Placek K, Joosten LAB, Schultze JL (2019) Innate and adaptive immune memory: an evolutionary continuum in the Host's response to pathogens. Cell Host Microbe 25(1):13–26. https://doi.org/10.1016/j.chom.2018.12.006

Ng PC, Lam CW, Li AM, Wong CK, Cheng FW, Leung TF, Hon EK, Chan IH, Li CK, Fung KS, Fok TF (2004) Inflammatory cytokine profile in children with severe acute respiratory syndrome. Pediatrics 113(1 Pt 1):e7–e14. https://doi.org/10.1542/peds.113.1.e7

Ogimi C, Englund JA, Bradford MC, Qin X, Boeckh M, Waghmare A (2019) Characteristics and outcomes of coronavirus infection in children: the role of viral factors and an immunocompromised state. J Pediatric Infect Dis Soc 8(1):21–28. https://doi.org/10.1093/jpids/pix093

Onder G, Rezza G, Brusaferro S (2020) Case-fatality rate and characteristics of patients dying in relation to COVID-19 in Italy. JAMA. https://doi.org/10.1001/jama.2020.4683

WHO Global Health Observatory (GHO) data – Causes of childhood mortality. https://www.who.int/gho/child_health/mortality/causes/en/

Perry RT, Halsey NA (2004) The clinical significance of measles: a review. J Infect Dis 189(Suppl 1):S4–S16. https://doi.org/10.1086/377712

Ratjen F, Bredendiek M, Zheng L, Brendel M, Costabel U (1995) Lymphocyte subsets in bronchoalveolar lavage fluid of children without bronchopulmonary disease. Am J Respir Crit Care Med 152(1):174–178. https://doi.org/10.1164/ajrccm.152.1.7599820

Schneider WM, Chevillotte MD, Rice CM (2014) Interferon-stimulated genes: a complex web of host defenses. Annu Rev Immunol 32:513–545. https://doi.org/10.1146/annurev-immunol-032713-120231

Schouten LR, van Kaam AH, Kohse F, Veltkamp F, Bos LD, de Beer FM, van Hooijdonk RT, Horn J, Straat M, Witteveen E, Glas GJ, Wieske L, van Vught LA, Wiewel MA, Ingelse SA, Cortjens B, van Woensel JB, Bos AP, Walther T, Schultz MJ, Wösten-van Asperen RM (2019) Age-dependent differences in pulmonary host responses in ARDS: a prospective observational cohort study. Ann Intensive Care 9(1):55. https://doi.org/10.1186/s13613-019-0529-4

Shaw AC, Goldstein DR, Montgomery RR (2013) Age-dependent dysregulation of innate immunity. Nat Rev Immunol 13(12):875–887. https://doi.org/10.1038/nri3547

Shekerdemian LS, Mahmood NR, Wolfe KK, Riggs BJ, Ross CE, McKiernan CA, Heidemann SM, Kleinman LC, Sen AI, Hall MW, Priestley MA, McGuire JK, Boukas K, Sharron MP, Burns JP (2020) Characteristics and outcomes of children with coronavirus disease 2019 (COVID-19) infection admitted to US and Canadian Pediatric Intensive Care Units. JAMA Pediatr 174:868. https://doi.org/10.1001/jamapediatrics.2020.1948

Simon AK, Hollander GA, McMichael A (2015) Evolution of the immune system in humans from infancy to old age. Proc Biol Sci 282(1821):20143085. https://doi.org/10.1098/rspb.2014.3085

Stockman LJ, Curns AT, Anderson LJ, Fischer-Langley G (2012) Respiratory syncytial virus-associated hospitalizations among infants and young children in the United States, 1997-2006. Pediatr Infect Dis J 31(1):5–9. https://doi.org/10.1097/INF.0b013e31822e68e6

Stockman LJ, Massoudi MS, Helfand R, Erdman D, Siwek AM, Anderson LJ, Parashar UD (2007) Severe acute respiratory syndrome in children. Pediatr Infect Dis J 26(1):68–74. https://doi.org/10.1097/01.inf.0000247136.28950.41

Sun L, Xing Y, Chen X, Zheng Y, Yang Y, Nichols DB, Clementz MA, Banach BS, Li K, Baker SC, Chen Z (2012) Coronavirus papain-like proteases negatively regulate antiviral innate immune response through disruption of STING-mediated signaling. PLoS One 7(2):e30802. https://doi.org/10.1371/journal.pone.0030802

Sung JY, Lee HJ, Eun BW, Kim SH, Lee SY, Lee JY, Park KU, Choi EH (2010) Role of human coronavirus NL63 in hospitalized children with croup. Pediatr Infect Dis J 29(9):822–826. https://doi.org/10.1097/INF.0b013e3181e7c18d

Surveillances V (2020) The epidemiological characteristics of an outbreak of 2019 novel coronavirus diseases (COVID-19)—China, 2020. China CDC Wkly 2(8):113–122

Tay MZ, Poh CM, Rénia L, MacAry PA, Ng LFP (2020) The trinity of COVID-19: immunity, inflammation and intervention. Nat Rev Immunol 20:363. https://doi.org/10.1038/s41577-020-0311-8

Terry C. Jones BM, Veith T, Zuchowski M, Hofmann J, Stein A, Edelmann A, Corman VM, Drosten C (2020) An analysis of SARS-CoV-2 viral load by patient age. https://virologie-ccm.charite.de/fileadmin/user_upload/microsites/m_cc05/virologie-ccm/dateien_upload/Weitere_Dateien/analysis-of-SARS-CoV-2-viral-load-by-patient-age.pdf

Tikellis C, Thomas MC (2012) Angiotensin-converting enzyme 2 (ACE2) is a key modulator of the renin angiotensin system in health and disease. Int J Pept 2012:256294–256294. https://doi.org/10.1155/2012/256294

Verdoni L, Mazza A, Gervasoni A, Martelli L, Ruggeri M, Ciuffreda M, Bonanomi E, D'Antiga L (2020) An outbreak of severe Kawasaki-like disease at the Italian epicentre of the SARS-CoV-2 epidemic: an observational cohort study. Lancet 395:1771. https://doi.org/10.1016/s0140-6736(20)31103-x

Viner RM, Russell SJ, Croker H, Packer J, Ward J, Stansfield C, Mytton O, Bonell C, Booy R (2020) School closure and management practices during coronavirus outbreaks including COVID-19: a rapid systematic review. Lancet Child Adolesc Health 4(5):397–404. https://doi.org/10.1016/s2352-4642(20)30095-x

Walls AC, Park Y-J, Tortorici MA, Wall A, McGuire AT, Veesler D (2020) Structure, function, and antigenicity of the SARS-CoV-2 spike glycoprotein. Cell 181(2):281–292.e286. https://doi.org/10.1016/j.cell.2020.02.058

Wen W, Su W, Tang H, Le W, Zhang X, Zheng Y, Liu X, Xie L, Li J, Ye J, Dong L, Cui X, Miao Y, Wang D, Dong J, Xiao C, Chen W, Wang H (2020) Immune cell profiling of COVID-19 patients in the recovery stage by single-cell sequencing. Cell Discov 6(1):31. https://doi.org/10.1038/s41421-020-0168-9

Wolfel R, Corman VM, Guggemos W, Seilmaier M, Zange S, Muller MA, Niemeyer D, Jones TC, Vollmar P, Rothe C, Hoelscher M, Bleicker T, Brunink S, Schneider J, Ehmann R, Zwirglmaier K, Drosten C, Wendtner C (2020) Virological assessment of hospitalized patients with COVID-2019. Nature 581:465. https://doi.org/10.1038/s41586-020-2196-x

World Health Organization (2020) Multisystem inflammatory syndrome in children and adolescents with COVID-19: scientific brief, 15 May 2020. World Health Organization, Geneva

Wu Z, McGoogan JM (2020) Characteristics of and important lessons from the coronavirus disease 2019 (COVID-19) outbreak in China: summary of a report of 72 314 cases from the Chinese Center for Disease Control and Prevention. JAMA 323:1239. https://doi.org/10.1001/jama.2020.2648

Xie X, Xue Q, Zhou Y, Zhu K, Liu Q, Zhang J, Song R (2020) Mental health status among children in home confinement during the coronavirus disease 2019 outbreak in Hubei Province, China. JAMA Pediatr 174:898. https://doi.org/10.1001/jamapediatrics.2020.1619

Xu Z, Shi L, Wang Y, Zhang J, Huang L, Zhang C, Liu S, Zhao P, Liu H, Zhu L, Tai Y, Bai C, Gao T, Song J, Xia P, Dong J, Zhao J, Wang FS (2020) Pathological findings of COVID-19 associated with acute respiratory distress syndrome. Lancet Respir Med 8(4):420–422. https://doi.org/10.1016/S2213-2600(20)30076-X

Zhang D, Guo R, Lei L, Liu H, Wang Y, Wang Y, Dai T, Zhang T, Lai Y, Wang J, Liu Z, He A, O'Dwyer M, Hu J (2020a) COVID-19 infection induces readily detectable morphological and inflammation-related phenotypic changes in peripheral blood monocytes, the severity of which correlate with patient outcome. medRxiv:2020.2003.2024.20042655. https://doi.org/10.1101/2020.03.24.20042655

Zhang J, Litvinova M, Liang Y, Wang Y, Wang W, Zhao S, Wu Q, Merler S, Viboud C, Vespignani A, Ajelli M, Yu H (2020b) Changes in contact patterns shape the dynamics of the COVID-19 outbreak in China. Science 368:1481. https://doi.org/10.1126/science.abb8001

Zust R, Cervantes-Barragan L, Habjan M, Maier R, Neuman BW, Ziebuhr J, Szretter KJ, Baker SC, Barchet W, Diamond MS, Siddell SG, Ludewig B, Thiel V (2011) Ribose 2′-O-methylation provides a molecular signature for the distinction of self and non-self mRNA dependent on the RNA sensor Mda5. Nat Immunol 12(2):137–143. https://doi.org/10.1038/ni.1979

Geriatrics and COVID-19 13

Mona Mirbeyk, Amene Saghazadeh, and Nima Rezaei

Abstract

Since December 2019, a novel coronavirus called severe acute respiratory syndrome coronavirus 2 (SARS-CoV-2) has begun to infect people. The virus first occurred in Wuhan, China, but the whole world is now struggling with the pandemic. Over 13 million confirmed cases and 571,000 deaths have been reported so far, and this number is growing. Older people, who constitute a notable proportion of the world population, are at an increased risk of infection because of altered immunity and chronic comorbidities. Thus, appropriate health care is necessary to control fatalities and spread of the disease in this specific population. The chapter provides an overview of diagnostic methods, laboratory and imaging findings, clinical features, and management of COVID-19 in aged people. Possible mechanisms behind the behavior of SARS-CoV-2 in the elderly include immunosenescence and related impaired antiviral immunity, mature immunity and related hyper-inflammatory responses, comorbidities and their effects on the functioning of critical organs/systems, and the altered expression of angiotensin-converting enzyme 2 (ACE2) that acts as an entry receptor for SARS-CoV-2. This evidence defines the herding behavior of COVID-19 in relation to ACE2 under the influence of immune dysregulation. Then, identifying the immunogenetic factors that affect the disease susceptibility and severity and as well as key inflammatory pathways that have the potential to serve as therapeutic targets needs to remain an active area of research.

Keywords

ACE2 · COVID-19 · Elderly · Geriatrics · Herding behavior · Immunity

M. Mirbeyk · A. Saghazadeh
Research Center for Immunodeficiencies, Children's Medical Center, Tehran University of Medical Sciences, Tehran, Iran

Systematic Review and Meta-analysis Expert Group (SRMEG), Universal Scientific Education and Research Network (USERN), Tehran, Iran

N. Rezaei (✉)
Research Center for Immunodeficiencies, Children's Medical Center, Tehran University of Medical Sciences, Tehran, Iran

Department of Immunology, School of Medicine, Tehran University of Medical Sciences, Tehran, Iran

Network of Immunity in Infection, Malignancy and Autoimmunity (NIIMA), Universal Scientific Education and Research Network (USERN), Tehran, Iran
e-mail: rezaei_nima@tums.ac.ir

N. Rezaei (ed.), *Coronavirus Disease - COVID-19*, Advances in Experimental Medicine and Biology 1318, https://doi.org/10.1007/978-3-030-63761-3_13

13.1 Introduction

While humanity was ever proud of his superheroic facilities, a little virus, distinguished first in Wuhan, China, could have fatal effects on the global economy, diplomacy, and especially social life. In December 2019, Chinese health authorities reported pneumonia of unknown origin associated with an exposure to a seafood market in some cases (Rabi et al. 2020). The causative pathogen was unknown, so researchers started to investigate its origin, reservoir, and suspected intermediate hosts. Investigations revealed that the viral genome sequences share about 80% similarity to that of severe acute respiratory syndrome coronavirus (SARS-CoV). The virus was then called SARS-CoV-2, and the disease caused by this virus was named coronavirus disease 2019 (COVID-19) (Zhou et al. 2020; Gorbalenya et al. 2020; Rabi et al. 2020). Over the last 6 months, this novel coronavirus has continued to infect people with a great contagion over the boundaries, with more than 13 million confirmed cases as of July 13, 2020.

SARS-CoV and SARS-CoV-2 belong to the *Coronaviridae* family (*Coronavirinae* subfamily), possessing spikes on the superficial part and positive-stranded RNA in the central part (Coronaviridae—Positive Sense RNA Viruses—Positive Sense RNA Viruses. 2011). *Coronavirinae* subfamily contains four main genera: *Alphacoronavirus*, *Betacoronavirus*, *Gammacoronavirus*, and *Deltacoronavirus*. Gammacoronaviruses and deltacoronaviruses occur in pigs and birds, while alphacoronaviruses and betacoronaviruses are detected in mammals (Coronaviridae—Positive Sense RNA Viruses—Positive Sense RNA Viruses. 2011; Velavan and Meyer 2020). The history of coronaviruses dates back to 54 years ago when the virus was isolated from nasal washing specimens of patients with the common cold (Tyrrell and Bynoe 1966). Until this day, seven types of coronaviruses have been discovered with the ability to infect humans, among which the natural reservoir of five types are probably bats (Cui et al. 2019). From these seven types, four types, including human coronavirus (HCoV)-NL63, HCoV-229E, HCoV-OC43, and HCoV-HKU1, mainly cause the common cold, and three others, including the Middle East respiratory syndrome coronavirus (MERS-CoV), SARS-CoV, and SARS-CoV-2, can represent with fatal pneumonia (Yang et al. 2020; Su et al. 2016; Zhu et al. 2020). SARS-CoV-2 siblings, e.g., MERS-CoV and SARS-CoV, mainly involve the respiratory system and could cause horrifying outbreaks with massive mortality during the past two decades (Hanaei and Rezaei 2020; Jabbari et al. 2020).

Detection of the origin, hosts, and transmission routes of the virus can help to stop the virus from spreading and infecting more people. Further investigations found that the SARS-CoV-2 genome shares a 96% similarity with bat-CoV (Zhou et al. 2020) and put forward pangolins as the intermediate host of SARS-CoV-2 because of more than 90% concordance of pangolin-CoV genome with both bat-CoV and SARS-CoV-2 (Zhang et al. 2020). SARS-CoV-2 transmission mainly occurs through human-to-human contact by inhaling droplets and aerosols (Lotfi et al. 2020). Current evidence suggests the angiotensin-converting enzyme 2 (ACE2) as a cell surface receptor for the virus (Wilder-Smith and Freedman 2020; Wilson and Chen 2020; Sharifkashani et al. 2020; Rezaei 2020b) and the immune responses as the source of variability in the vulnerability of people to infection (Saghazadeh and Rezaei 2020a; Yazdanpanah et al. 2020a; Bahrami et al. 2020; Lotfi and Rezaei 2020; Nasab et al. 2020). Older adults are a high-risk group to this infection as they have an immune system with a decreased ability to fight with infectious diseases (Huang et al. 2020). It is important to notice the clinical features of the disease in aged patients to reduce total mortality.

13.2 Elderlies: Too Frail for COVID-19

Geriatrics is a medical subspecialty aimed at providing health care for old patients. The elderly can be defined in different ways, though it mainly applies to people aged 65 years or over (Orimo et al. 2006; Singh and Bajorek 2014; (WHO)

2002), and that some divide old ages into early elderly (65–74 years old) and late elderly (75 years and older). The World Health Organization (WHO) claims that the world's population is going to be old very soon and that the percentage of the people with the age of 60 or more would surpass the 22% of the whole population by 2050 ((WHO) Updated May 2017).

Children are different from adults in the risk of developing and dying from COVID-19. Studies show that only 2% of cases with confirmed COVID19 are in the age group 0–19 years (Wu and McGoogan 2020). About 15% of patients over age 80 die. Also, patients with severe COVID-19 are older than patients with non-severe COVID-19 (Wang et al. 2020a). Therefore, children and the elderly have high contrast in the behavior of COVID19 (Figs. 13.1 and 13.2). Below are about possible mechanisms behind this behavior of COVID-19.

13.2.1 The Aged Immune System

The human body is a multiphysiological system, and all of its systems would be affected by the aging process, including the immune system, which has an essential role in fighting with infection. The impact of aging on the immune system, which is referred to as immunosenescence, involves both innate and adaptive immunity (Nikolich-Zugich 2018; Fuentes et al. 2017; Wong and Goldstein 2013) and extends over the skin and mucosal immune system (Lichterfeld-Kottner et al. 2020).

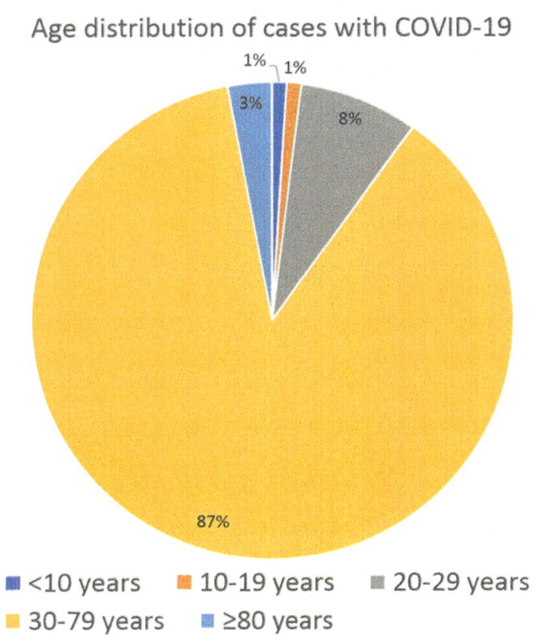

Fig. 13.1 Age distribution of patients with COVID-19

Aging-associated changes in in the immune system include, but are not limited to, the acquisition of old cells in the immune system and other organs/systems, reduction in naïve T cells releasing from the thymus, mRNA processing changes, and dysregulation of immune receptors (Pietilä et al. 2015; Walford 1964; Fuentes et al. 2017). Natural killer cells (NK Cells) are an essential member of the innate immune system that play a significant role not only in eliminating malignant cells but also removing cells infected with viruses. They are affected by aging, as indicated by decreased expression of activating receptors and CD56 marker on their surface (Baume et al. 1992; Carson et al. 1997; Saito et al. 1993). Similarly, as we get older, T cells undergo reductions in their number, function, and surface receptors. Inevitably, the ability of these cells to recognize infectious or tumoral cells would also be impaired (Larbi et al. 2006, 2014; Vallejo 2005).

On the other hand, COVID-19 itself would influence the immune system. It has been shown to decrease T cell numbers and that this might be related to disease severity and adverse outcomes of disease (Qin et al. 2020; Fathi and Rezaei 2020b). NK cells are also reduced in the count and cytotoxic function during this infection (Yaqinuddin and Kashir 2020). Given that hyper-inflammation in COVID-19 is a strong predictor of poor outcomes, inflamm-aging, known as a baseline subclinical mild chronic inflammation produced by aging, can also worsen the outcome of COVID-19 (Bonafè et al. 2020). Altogether, it is not surprising that COVID-19 can cause worst-case scenarios in the context of an aged immune system.

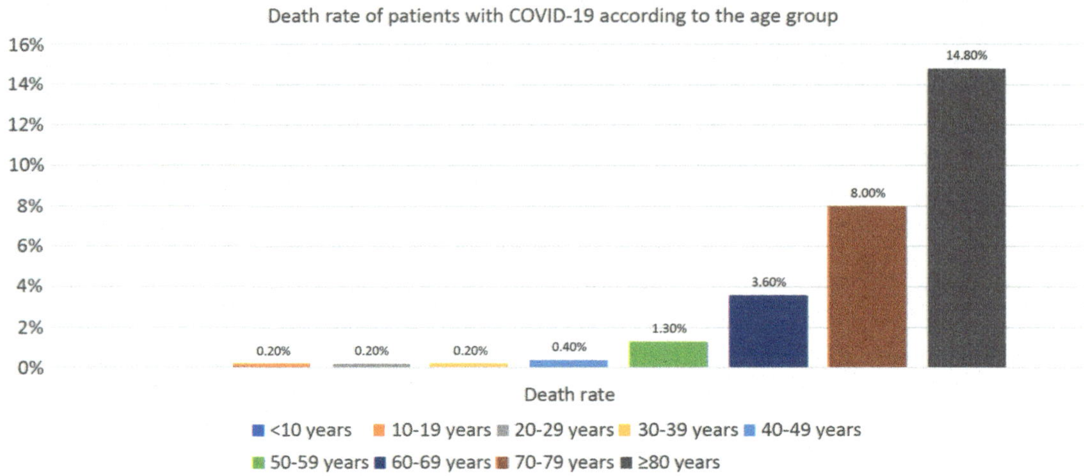

Fig. 13.2 The death rate of patients with COVID-19 according to the age group

13.2.2 The Increased Risk of Chronic Diseases

The frailty of old patients is not only due to the old immune system. Aging is a significant risk factor for chronic diseases, including hypertension, cardiac diseases, cerebrovascular diseases, neurodegeneration, and cancer (Pawelec et al. 2014). Recent investigations have shown that SARS-CoV-2 could cause higher mortality or poorer prognosis in patients with comorbidities, in particular, hypertension, diabetes, and cardiovascular diseases, compared with patients without comorbidities (Guan et al. 2020; Chen et al. 2020a, 2020b; Singhal 2020; Grasselli et al. 2020; Du et al. 2020). Of note, hypertension is associated with about 2.5-fold increased risk of mortality related to COVID-19 (Lippi et al. 2020). Moreover, many elderlies deal with loneliness, which can lead to poor hygiene and neglect of disease symptoms at early stages and, therefore, presenting to hospitals with progressed manifestations. Also, living in care facilities would significantly affect person-to-person transmission.

13.2.3 The Mature Immunity

Generally speaking, T helper (Th) 2 cell and T regulatory (Treg) cell responses and Th17 cytokines decrease with age, whereas Th1 cell immune responses and pattern recognition receptor (PRR)-mediated signaling increase with age. In this manner, a sophisticated immune system, through the production of pro-inflammatory cytokines (IFNγ, TNFα, IL-2, IL-1β), leads to the recruitment of hyper-inflammatory immune responses under pathologic conditions. By contrast, an immature immune system induces hypo--inflammatory responses that are maintained by anti-inflammatory and immune-regulatory cytokines (IL-4, IL-5, IL-10, IL-13, IL-35, and TGFβ) and diminished PRR-mediated signaling. Due to decreased communication between the innate and adaptive immune system in the immature immune system, all responses generated from the adaptive immune system are declined, including T cell proliferation, cytotoxic T cell responses, memory T cells, and CD8+ T and CD4+ T cell activation (Maddux and Douglas 2015). Soluble factors known to maintain immune responses skewed to Th2 cell type come from both the maternal origin, e.g., TGFB, progesterone, and prostaglandin, and the child's own origin, such as adenosine.

Dendritic cells (DCs) are professional myeloid cells that can detect pathogen-associated molecular patterns (PAMPs) and damage-associated molecular patterns (DAMPs) released by invading pathogens or damaged cells and mediate their death through apoptosis, necrosis, or pyroptosis.

Notably, DCs perform apoptosis of their own kind as being programmed. A decreased function of DCs in the clearance of apoptotic/necrotic DCs results in the accumulation of cells that should be dead but are alive, and this would cause critical conditions such as autoimmunity and tumorigenesis. Like mature DCs, immature DCs display the ability of up taking apoptotic/necrotic cells (Kushwah et al. 2010). The ability of immature DCs is, however, unique – whereas apoptotic cells are required inflammatory responses for being detectable by mature DCs, no evident inflammatory response is necessary for their recognition by immature DCs. Uptake of apoptotic DCs would be appropriate for immature DCs to be educated of a tolerogenic character, as indicated by increased expression of TGFβ that is a critical driver of Treg cell induction by naïve CD4+ T cells. More interestingly, cord DCs can, upon exposure to PAMPs, elicit immune responses compatible with those of adult DCs (Langrish et al. 2002). While due to the below--required expression of cytokines, IL12P70 and IFNY, inducing differentiation of naïve CD4+ T cells into Th2 cell dichotomy (Langrish et al. 2002), cord DCs are in an immaturity--enforcement capacity against the pro--inflammatory Th1 cell polarization.

A systematic review of cytokine expression in children has confirmed that the expression of pro--inflammatory cytokines, including IFNY, IL6, IL10, and TNFA, increases with age (Decker et al. 2017). A closer look at individual studies included in the review reveals to us an age--associated increase most being pronounced for the percentage of CD4+ T and CD8+ T cells that express IFNY and TNFA.

13.2.4 The Aberrant Expression of ACE2

13.2.4.1 ACE2: A Receptor for SARS-CoV2 (2019-nCoV) Cell Entry That Is Present on Lungs, Heart, Kidney, and Testis

An insufficient response to classical angiotensin--converting enzyme (ACE) inhibitors among patients with hypertension and cardiovascular diseases posed the possibility that the target may be wrong, i.e., ACE. Efforts led to the recognition of ACE2 (angiotensin-converting enzyme 2) that acts as a human carboxypeptidase that utilizes zinc. The lack of or deficiency of ACE2 correlates with diabetic nephropathy, heart dysfunction, and reduced cardiac contractility and accumulation of angiotensin II and hypoxia-induced genes (Guy et al. 2003).

ACE2 is primarily present on specific tissue like renal, cardiac, and testicular tissue. Besides, its expression exists in other tissue like the liver, lung, pancreas, ovary, colon, small intestine, placenta, and retina (Bindom and Lazartigues 2009). Thereby, ACE2 can ubiquitously exert anti--fibrotic and anti-inflammatory effects through the degradation of its primary substrate, angiotensin II, and also by inhibition of TGFB and MIF (Bindom and Lazartigues 2009). However, ACE2 can mediate the degradation of different substrates, as shown in Fig. 13.3.

Antiserum that contains antibodies against human ACE2 could hinder the entry of pseudotyped SARS-2-S. It supports the notion that SARS-2-S utilizes ACE2 as an entry receptor (Hoffmann et al. 2020).

13.2.4.2 Dual-Function ACE2: Zinc Metalloproteinase and Microbial Collagenase

ACE2 and ACE share 42% sequence identity and 61% sequence conservation around the active site (Guy et al. 2003). Sequence alignment has confirmed that ACE2 possesses an active site with high similarity to that of testicular ACE (tACE), and therefore like other ACE family members, ACE2 can function as a zinc metalloproteinase (Guy et al. 2003). However, it dissociated tACE and ACE2 on a zinc ligand; for tACE, there is a third zinc ligand that occurs within the Glu-(Xaa)3-Asp motif, and for ACE2, the motif lacks the aspartate residue and has obtained a glutamate residue (Guy et al. 2003). This motif, Glu-(Xaa)3-Glu, allows ACE2 to function as a microbial collagenase (Guy et al. 2003).

Having a basic structure like ACE and on the other side having a zinc-binding motif (HEXXH)

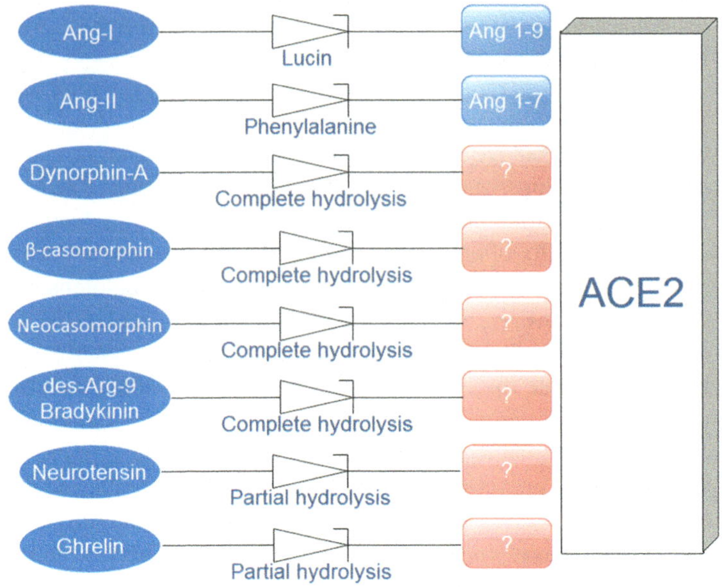

Fig. 13.3 ACE2 substrates and products

resembling the M32 family of peptidases have ranked ACE2 as the first mammalian carboxypeptidase structurally related to ACE, neurolysin, and PfuCP (Guy et al. 2003).

13.2.4.3 ACE2 Is Not Sensitive to ACEI

ACE2 functions as a carboxypeptidase that removes one single amino acid from the C terminus of its substrates, e.g., angiotensin I, angiotensin II, and des-Arg9-bradykinin. ACE is a peptidyl dipeptidase that removes a dipeptide from the C terminus of its substrates, angiotensin I and bradykinin, to reduce respective vasodilatory effects. Also, ACE2 and ACE differ in their ligand-binding pockets that mainly occur at the S2' site (Guy et al. 2003).

13.2.4.4 ACE2 Exists in Membrane-Bound and Soluble Forms

ACE2 is a type I transmembrane protein having a long extracellular N-terminal domain and a short intracellular C-terminal tail. Its soluble form can interfere with ACE2 binding to the SARS-CoV spike protein. ADAM metallopeptidase domain 17, ADAM17, is a TNFA convertase that mediates ACE2 shedding in the soluble form (Lambert et al. 2005). However, ADAM17-mediated ACE2 cleavage does not influence SARS-CoV cell entry as TMPRSS2-mediated ACE2 cleavage did it significantly (Heurich et al. 2014).

13.2.4.5 ACE2 Expression Decreases with Age

The study of rats has confirmed the expression of ACE2 in the lungs, kidneys, heart, and the gastrointestinal system. For the lungs, the ACE2 expression occurs in both the bronchus and pulmonary parenchyma, alveolar type I epithelium, alveolar type II epithelium, bronchiolar epithelium, endothelium, and smooth muscle cells of the pulmonary vascular structure. The expression of ACE2 in the rat lungs significantly decreases with age, and that this decrease is relatively higher in males than females (Xudong et al. 2006). Both the elderly and the male population are at higher risk of dying from COVID-19 (Saghazadeh and Rezaei 2020a).

13.2.4.6 The Herding Behavior of COVID19 in Relation to ACE2 Under the Cytokine Storm

When a large number of people gather in the same event, a real concern arises regarding the risk of injury and death. For example, the event

"2015 Mina stampede" occurred in Mina when people were performing the rites of Hajj. What that caused people to stampede is not apparent. Whatever the primary cause was, the movement of a large number of people led to a disaster associated with 769 deaths. Such behavior of people arising from a sudden fear is referred to as panic behavior. Statistical physicists developed several models for the prediction of human behavior in real crowds and led to the conclusion that individual-centered behavior, like collective behavior, would influence the optimization of the escape from the crowds (Helbing et al. 2000). Suppose a two-exit room filled with fire. A traditional fluid-flow model believes that people will be able to use both doors correctly and succeeded in exit safely. The individual-centered model predicts that a large number of people may leave the room using only one exit, resulting in the blocking of that exit, which is known as the herding behavior (Low 2000).

The SARS-CoV-2 enters the body. Upon its recognition, the immune system would elicit the cytokine storm in the peripheral blood. In a cytokine-filled body, the only optimal strategy is cell surface receptors through which the virus can enter the cell and escape the immune responses. Thus, the virus will find organs, e.g., exit doors, that express the highest density of ACE2 expression. The virus reaches the vital organs, lungs, heart, and kidney, and bind to transmembrane ACE2 as rapidly as possible as much as possible—reduction of functional ACE2, in turn, causes cardiac dysfunction, inflammation, and fibrosis.

The herding behavior of SARS-CoV-2 can explain the most prominent epidemic characteristics of COVID-19. Higher death rates among adults and older adults and patients with underlying conditions, e.g., cancer, diabetes, cardiovascular diseases, and hypertension, linked to limited or aberrant ACE2 expression, can be predicted by occupying ACE2 and blocking its function to the life-threatening extent. Children and young adults who have none of those, as mentioned earlier, conditions have a good repertoire of ACE2. Therefore, when the virus is occupying ACE2, there is a subset of ACE2 that remained functional. It would allow the immune responses to clear the viral infection effectively.

In this manner, the herding behavior of COVID-19 depends on the individual-centered factors, e.g., age and underlying condition, and collective behavior, e.g., the cytokine storm. Three main points of intervention are as follows: 1. hindering the virus binding to ACE2 throughout the disease course, 2. downregulation of cytokine release within the mild stage of the disease and before a severe stage of disease appears, and 3. upregulation of ACE2 expression during a severe stage of the disease to extend the deadline for immunity to act effectively.

13.3 COVID-19 in Aged Patients

History taking and physical examination are the keys in diagnosis, followed by laboratory testing and further paraclinical diagnostics. Understanding disease features is mandatory to diagnose disease in its earliest stages, thereby providing optimal health care and disease management. COVID-19 can manifest in a variety of forms, from a common cold to a severe form of disease like respiratory distress syndrome. Old ages (mostly aged 65 and over) and the presence of comorbidities crucially predict poor prognosis and an increased risk of mortality. The role of immunogenetic background is still under investigation in this regard (Babaha and Rezaei 2020; Darbeheshti and Rezaei 2020; Yousefzadegan and Rezaei 2020).

13.3.1 Clinical Features

The incubation period means the interval time from exposure to a virus source or infected people until the initiation of disease symptoms. It varies from about 1 day to 2 weeks, and the median incubation time is about 4 or 5 days (Xu et al. 2020a; Lauer et al. 2020; Nie et al. 2020). Overall, the most common symptoms of the disease are fever and cough (mostly dry cough).

Fatigue, myalgia, shortness of breath, and dyspnea are also common among patients with COVID-19. Other symptoms can include conjunctival and nasal congestion, sore throat, chills, skin rashes, rhinorrhea, nausea or vomiting, sputum production, headache, diarrhea and hemoptysis, loss of smell sensation, loss of taste sensation, and altered mental status. Moreover, it must be noticed that asymptomatic patients have also been reported (Wang et al. 2020d; Lai et al. 2020; Breslin et al. 2020). Asking a history of exposure to symptomatic or suspicious patients or exposure to the Huanan seafood market would be useful.

In elderly patients, the disease does not include very different clinical manifestations from what is seen commonly. However, probably because of aging and its related comorbidities, these manifestations are more severe compared with younger adults. Besides, old patients may develop more atypical manifestations (Olde Rikkert et al. 2020); for example, in some cases, changes in mental status and orientation have been represented as initial symptoms of COVID-19 in aged patients (Wang et al. 2020c; Ward et al. 2020). Dyspnea, tachypnea, tachycardia, and chest discomfort may occur in elderlies (Nanda et al. 2020).

13.3.2 Laboratory Findings

Laboratory findings can be useful for predicting the prognosis and outcome of the disease. Although a significant number of patients can represent disease with normal laboratory data such as normal complete blood count, the following changes in laboratory data of patients might have been seen: abnormal white cell count, lymphocytopenia, low platelet count, hyponatremia, and increases in c-reactive protein (CRP), aminotransferases, erythrocyte sedimentation rate (ESR), lactate dehydrogenase (LDH), D-dimer, procalcitonin, creatine kinase, and IL-6. In addition to dyspnea and lymphocytopenia, high levels of CRP, ESR, LDH, and IL-6 might be predictive of developing severe or critical conditions.

13.3.3 Imaging Findings

Chest CT scan findings of patients may include ground-glass opacities (GGO), patchy infiltrations, consolidation, pleural effusion, pleural thickening, crazy-paving pattern, and interlobular septal thickening. It seems extensive pulmonary involvement occurs in elderlies more than younger patients.

13.3.4 Diagnosis

We can consider a person suspected to have COVID-19 when that person has had recent travel to Wuhan, China, or at least one of the people whom he is in close contact with, recently recognized as a confirmed case of COVID-19, begins to develop fever or common respiratory symptoms of COVID-19 or chest radiogram of pneumonia (Kim et al. 2020).

Confirmation of COVID-19 in a patient occurs when the pharyngeal swab specimen turns positive for real-time reverse transcription--polymerase chain reaction (RT-PCR) for the SARS-CoV-2 genome (Kim et al. 2020). However, because of probable false negatives for RT-PCR, it is better to repeat the test (Long et al. 2020). Moreover, a chest CT scan is a sensitive diagnostic method for early diagnosis (Wang et al. 2020b; Xu et al. 2020b). Elderlies with mild or uncommon symptoms of the disease can undergo diagnostic assessments for the virus (Nikolich-Zugich et al. 2020). Although immunoglobulins are not used for diagnosis of the disease, positive IgG antibody levels for SARS-CoV-2 can be highly suggestive for recent infection (Gosch et al. 2020). An active area of research is devoted to finding diagnostic methods for more rapid, reliable detection of SARS--CoV-2. Microfluidic devices offer promising platforms in this context (Basiri et al. 2020a).

13.3.5 Management

Overall, there is not a specific treatment or drug for COVID-19, and most of the treatments and care are supportive (Nikolich-Zugich et al. 2020). Considering complications related to altered mental status, such as falling, and underlying diseases is an essential factor to help manage the disease in elderlies. Conventional therapies are high-flow nasal oxygen, non-invasive positive-pressure ventilation, taking care of underlying illness and comorbidities, and treatment of co-infections. Routine use of corticosteroids is controversial (Saghazadeh and Rezaei 2020b); however, in severe cases, it may lead to a better outcome (Veronese et al. 2020). In patients with hypertension, the use of angiotensin-converting enzyme inhibitors (ACEIs) or angiotensin-2 receptor blockers (ARBs) can lead to a better outcome (Meng et al. 2020). Patients with coagulopathy who undergo low molecular weight heparin may have better outcomes (Tang et al. 2020). The use of other drugs, such as antiretrovirals, is uncertain and may need further investigations (Ford et al. 2020). Considering respiratory rehabilitation may also be useful in the post-acute phase (Liu et al. 2020).

13.3.6 The Immune System, Inflammation, and Therapeutic Opportunities

Research shows that while the immune system is expected to elicit antiviral immune responses, the SARS-CoV-2 causes the immune system to induce pro-inflammatory signaling to the extent that the functioning of vital organs/systems, especially the central nervous system and the cardiovascular system, is impaired in the shocking production of these inflammatory cytokines (Bahrami et al. 2020; Jahanshahlu and Rezaei 2020a; Saleki et al. 2020; Yazdanpanah et al. 2020b). Moreover, immune cells having the responsibility of viral clearance show markers of exhaustion during the disease course. Therefore, there has been an increasing interest in identifying critical inflammatory targets and evaluating the adoptive transfer of immune cells as potential therapeutic opportunities (Fathi and Rezaei 2020a; Basiri et al. 2020b; Jahanshahlu and Rezaei 2020b; Mansourabadi et al. 2020; Pashaei and Rezaei 2020). However, the heavy burden of the pandemic (Rezaei 2020a) and the concern that it tends to stay (Jabbari and Rezaei 2020) have attracted much attention from other disciplines around the possible ways and means for prevention and management of the disease, too (Sahu et al. 2020; Moazzami et al. 2020; Rabiee et al. 2020; Mohamed et al. 2020; Momtazmanesh et al. 2020; Moradian et al. 2020).

13.4 Conclusion

Aging affects the immune system function, therefore increasing the risk of infection and adverse outcomes. Moreover, aging is a significant risk factor for chronic comorbidities that can put an aged person in a fragile state. Old patients with COVID-19 tend to develop severe illness and have a higher risk of death. They also are more likely to have abnormal laboratory findings and extensive pulmonary lesions. Old age and pre-existing chronic diseases such as hypertension, cardiovascular diseases, COPD, and diabetes can increase the risk of mortality. It is important to consider mild or atypical symptoms in the setup of an early diagnostic assessment of COVID-19 in an old patient. Monitoring and controlling underlying illness might help to have a better outcome in old patients with COVID-19.

References

(WHO) WHO (2002) Proposed working definition of an older person in Africa for the MDS Project. https://www.who.int/healthinfo/survey/ageingdefnolder/en/. Accessed May 2 2020

(WHO) WHO (Updated May 2017) 10 facts on ageing and health. https://www.who.int/features/factfiles/ageing/en/. Accessed May 2 2020.

Babaha F, Rezaei N (2020) Primary immunodeficiency diseases in COVID-19 pandemic: a predisposing or protective factor? Am J Med Sci. https://doi.org/10.1016/j.amjms.2020.07.027

Bahrami A, Vafapour M, Moazzami B, Rezaei N (2020) Hyperinflammatory shock related to COVID-19 in a patient presenting with multisystem inflammatory syndrome in children: first case from Iran. J Paediatr Child Health. https://doi.org/10.1111/jpc.15048

Basiri A, Heidari A, Nadi MF, Fallahy MTP, Nezamabadi SS, Sedighi M, Saghazadeh A, Rezaei N (2020a) Microfluidic devices for detection of RNA viruses. Rev Med Virol n/a (n/a):e2154. https://doi.org/10.1002/rmv.2154

Basiri A, Pazhouhnia Z, Beheshtizadeh N, Hoseinpour M, Saghazadeh A, Rezaei N (2020b) Regenerative medicine in COVID-19 treatment: real opportunities and range of promises. Stem Cell Rev Rep:1–13. https://doi.org/10.1007/s12015-020-09994-5

Baume DM, Robertson MJ, Levine H, Manley TJ, Schow PW, Ritz J (1992) Differential responses to interleukin 2 define functionally distinct subsets of human natural killer cells. Eur J Immunol 22(1):1–6. https://doi.org/10.1002/eji.1830220102

Bindom SM, Lazartigues E (2009) The sweeter side of ACE2: physiological evidence for a role in diabetes. Mol Cell Endocrinol 302(2):193–202

Bonafè M, Prattichizzo F, Giuliani A, Storci G, Sabbatinelli J, Olivieri F (2020) Inflamm-aging: why older men are the most susceptible to SARS-CoV-2 complicated outcomes. Cytokine Growth Factor Rev. https://doi.org/10.1016/j.cytogfr.2020.04.005

Breslin N, Baptiste C, Gyamfi-Bannerman C, Miller R, Martinez R, Bernstein K, Ring L, Landau R, Purisch S, Friedman AM, Fuchs K, Sutton D, Andrikopoulou M, Rupley D, Sheen J-J, Aubey J, Zork N, Moroz L, Mourad M, Wapner R, Simpson LL, D'Alton ME, Goffman D (2020) Coronavirus disease 2019 infection among asymptomatic and symptomatic pregnant women: two weeks of confirmed presentations to an affiliated pair of new York City hospitals. Am J Obstet Gynecol MFM 2(Suppl):100118. https://doi.org/10.1016/j.ajogmf.2020.100118

Carson WE, Fehniger TA, Caligiuri MA (1997) CD56bright natural killer cell subsets: characterization of distinct functional responses to interleukin-2 and the c-kit ligand. Eur J Immunol 27(2):354–360. https://doi.org/10.1002/eji.1830270203

Chen N, Zhou M, Dong X, Qu J, Gong F, Han Y, Qiu Y, Wang J, Liu Y, Wei Y, Xia J, Yu T, Zhang X, Zhang L (2020a) Epidemiological and clinical characteristics of 99 cases of 2019 novel coronavirus pneumonia in Wuhan, China: a descriptive study. Lancet 395(10223):507–513. https://doi.org/10.1016/s0140-6736(20)30211-7

Chen Y, Li T, Ye Y, Chen Y, Pan J (2020b) Impact of fundamental diseases on patients with COVID-19. Disaster Med Public Health Prep:1–6. https://doi.org/10.1017/dmp.2020.139

Coronaviridae—Positive Sense RNA Viruses—Positive Sense RNA Viruses. (2011). https://talk.ictvonline.org/ictv-reports/ictv_9th_report/positive-sense-rna-viruses-2011/w/posrna_viruses/222/coronaviridae. Accessed 26 Apr 2020

Cui J, Li F, Shi ZL (2019) Origin and evolution of pathogenic coronaviruses. Nat Rev Microbiol 17(3):181–192. https://doi.org/10.1038/s41579-018-0118-9

Darbeheshti F, Rezaei N (2020) Genetic predisposition models to COVID-19 infection. Med Hypotheses 142:109818. https://doi.org/10.1016/j.mehy.2020.109818

Decker M-L, Grobusch MP, Ritz N (2017) Influence of age and other factors on cytokine expression profiles in healthy children—a systematic review. Front Pediatr 5:255

Du Y, Tu L, Zhu P, Mu M, Wang R, Yang P, Wang X, Hu C, Ping R, Hu P, Li T, Cao F, Chang C, Hu Q, Jin Y, Xu G (2020) Clinical features of 85 fatal cases of COVID-19 from Wuhan: a retrospective observational study. Am J Respir Crit Care Med. https://doi.org/10.1164/rccm.202003-0543OC

Fathi N, Rezaei N (2020a) Lymphopenia in COVID-19: therapeutic opportunities. Cell Biol Int 44(9):1792–1797. https://doi.org/10.1002/cbin.11403

Fathi N, Rezaei N (2020b) Lymphopenia in COVID-19: therapeutic opportunities. Cell Biol Int

Ford N, Vitoria M, Rangaraj A, Norris SL, Calmy A, Doherty M (2020) Systematic review of the efficacy and safety of antiretroviral drugs against SARS, MERS or COVID-19: initial assessment. J Int AIDS Soc 23(4):e25489. https://doi.org/10.1002/jia2.25489

Fuentes E, Fuentes M, Alarcon M, Palomo I (2017) Immune system dysfunction in the elderly. An Acad Bras Cienc 89(1):285–299. https://doi.org/10.1590/0001-3765201720160487

Gorbalenya AE, Baker SC, Baric RS, de Groot RJ, Drosten C, Gulyaeva AA, Haagmans BL, Lauber C, Leontovich AM, Neuman BW, Penzar D, Perlman S, Poon LLM, Samborskiy D, Sidorov IA, Sola I, Ziebuhr J (2020) Severe acute respiratory syndrome-related coronavirus: The species and its viruses – a statement of the Coronavirus Study Group. bioRxiv:2020.2002.2007.937862. doi:https://doi.org/10.1101/2020.02.07.937862

Gosch M, Singler K, Kwetkat A, Heppner HJ (2020) Geriatrics in times of corona. Z Gerontol Geriatr 53(3):228–232. https://doi.org/10.1007/s00391-020-01725-2

Grasselli G, Zangrillo A, Zanella A, Antonelli M, Cabrini L, Castelli A, Cereda D, Coluccello A, Foti G, Fumagalli R, Iotti G, Latronico N, Lorini L, Merler S, Natalini G, Piatti A, Ranieri MV, Scandroglio AM, Storti E, Cecconi M, Pesenti A, Nailescu A, Corona A, Zangrillo A, Protti A, Albertin A, Forastieri Molinari A, Lombardo A, Pezzi A, Benini A, Scandroglio AM, Malara A, Castelli A, Coluccello A, Micucci A, Pesenti A, Sala A, Alborghetti A, Antonini B, Capra C, Troiano C, Roscitano C, Radrizzani D, Chiumello D, Coppini D, Guzzon D, Costantini E, Malpetti E, Zoia E, Catena E, Agosteo E, Barbara E, Beretta E, Boselli E, Storti E, Harizay F, Della Mura F, Lorini FL, Donato Sigurtà F, Marino F, Mojoli F, Rasulo F, Grasselli G, Casella G, De Filippi G, Castelli G, Aldegheri G, Gallioli G, Lotti G, Albano G, Landoni G, Marino G, Vitale G, Battista Perego G, Evasi G, Citerio G, Foti G, Natalini G, Merli G, Sforzini I, Bianciardi L, Carnevale L, Grazioli L, Cabrini L, Guatteri L, Salvi L, Dei Poli M, Galletti M, Gemma M, Ranucci M, Riccio M, Borelli M, Zambon M, Subert M, Cecconi M, Mazzoni MG,

Raimondi M, Panigada M, Belliato M, Bronzini N, Latronico N, Petrucci N, Belgiorno N, Tagliabue P, Cortellazzi P, Gnesin P, Grosso P, Gritti P, Perazzo P, Severgnini P, Ruggeri P, Sebastiano P, Covello RD, Fernandez-Olmos R, Fumagalli R, Keim R, Rona R, Valsecchi R, Cattaneo S, Colombo S, Cirri S, Bonazzi S, Greco S, Muttini S, Langer T, Alaimo V, Viola U (2020) Baseline characteristics and outcomes of 1591 patients infected with SARS-CoV-2 admitted to ICUs of the Lombardy region, Italy. JAMA 323 (16):1574–1581. doi:https://doi.org/10.1001/jama.2020.5394

Guan WJ, Liang WH, Zhao Y, Liang HR, Chen ZS, Li YM, Liu XQ, Chen RC, Tang CL, Wang T, Ou CQ, Li L, Chen PY, Sang L, Wang W, Li JF, Li CC, Ou LM, Cheng B, Xiong S, Ni ZY, Xiang J, Hu Y, Liu L, Shan H, Lei CL, Peng YX, Wei L, Liu Y, Hu YH, Peng P, Wang JM, Liu JY, Chen Z, Li G, Zheng ZJ, Qiu SQ, Luo J, Ye CJ, Zhu SY, Cheng LL, Ye F, Li SY, Zheng JP, Zhang NF, Zhong NS, He JX (2020) Comorbidity and its impact on 1590 patients with COVID-19 in China: a nationwide analysis. Eur Respir J 55(5). https://doi.org/10.1183/13993003.00547-2020

Guy JL, Jackson RM, Acharya KR, Sturrock ED, Hooper NM, Turner AJ (2003) Angiotensin-converting enzyme-2 (ACE2): comparative modeling of the active site, specificity requirements, and chloride dependence. Biochemistry 42(45):13185–13192

Hanaei S, Rezaei N (2020) COVID-19: developing from an outbreak to a pandemic. Arch Med Res. https://doi.org/10.1016/j.arcmed.2020.04.021

Helbing D, Farkas I, Vicsek T (2000) Simulating dynamical features of escape panic. Nature 407(6803):487–490

Heurich A, Hofmann-Winkler H, Gierer S, Liepold T, Jahn O, Pöhlmann S (2014) TMPRSS2 and ADAM17 cleave ACE2 differentially and only proteolysis by TMPRSS2 augments entry driven by the severe acute respiratory syndrome coronavirus spike protein. J Virol 88(2):1293. https://doi.org/10.1128/JVI.02202-13

Hoffmann M, Kleine-Weber H, Schroeder S, Krüger N, Herrler T, Erichsen S, Schiergens TS, Herrler G, Wu N-H, Nitsche A (2020) SARS-CoV-2 cell entry depends on ACE2 and TMPRSS2 and is blocked by a clinically proven protease inhibitor. Cell

Huang C, Wang Y, Li X, Ren L, Zhao J, Hu Y, Zhang L, Fan G, Xu J, Gu X, Cheng Z, Yu T, Xia J, Wei Y, Wu W, Xie X, Yin W, Li H, Liu M, Xiao Y, Gao H, Guo L, Xie J, Wang G, Jiang R, Gao Z, Jin Q, Wang J, Cao B (2020) Clinical features of patients infected with 2019 novel coronavirus in Wuhan, China. Lancet 395(10223):497–506. https://doi.org/10.1016/s0140-6736(20)30183-5

Jabbari P, Rezaei N (2020) With risk of reinfection, is COVID-19 Here to stay? Disaster Med Public Health Prep:1. https://doi.org/10.1017/dmp.2020.274

Jabbari P, Jabbari F, Ebrahimi S, Rezaei N (2020) COVID-19: a chimera of two pandemics. Disaster Med Public Health Prep:1–2. https://doi.org/10.1017/dmp.2020.223

Jahanshahlu L, Rezaei N (2020a) Central Nervous System Involvement in COVID-19. Arch Med Res. https://doi.org/10.1016/j.arcmed.2020.05.016

Jahanshahlu L, Rezaei N (2020b) Monoclonal antibody as a potential anti-COVID-19. Biomed Pharmacother 129:110337. https://doi.org/10.1016/j.biopha.2020.110337

Kim ES, Chin BS, Kang CK, Kim NJ, Kang YM, Choi JP, Oh DH, Kim JH, Koh B, Kim SE, Yun NR, Lee JH, Kim JY, Kim Y, Bang JH, Song KH, Kim HB, Chung KH, Oh MD (2020) Clinical course and outcomes of patients with severe acute respiratory syndrome coronavirus 2 infection: a preliminary report of the first 28 patients from the Korean cohort study on COVID-19. J Korean Med Sci 35(13):e142. https://doi.org/10.3346/jkms.2020.35.e142

Kushwah R, Wu J, Oliver JR, Jiang G, Zhang J, Siminovitch KA, Hu J (2010) Uptake of apoptotic DC converts immature DC into tolerogenic DC that induce differentiation of Foxp3+ Treg. Eur J Immunol 40(4):1022–1035

Lai CC, Liu YH, Wang CY, Wang YH, Hsueh SC, Yen MY, Ko WC, Hsueh PR (2020) Asymptomatic carrier state, acute respiratory disease, and pneumonia due to severe acute respiratory syndrome coronavirus 2 (SARS-CoV-2): facts and myths. J Microbiol Immunol Infect. https://doi.org/10.1016/j.jmii.2020.02.012

Lambert DW, Yarski M, Warner FJ, Thornhill P, Parkin ET, Smith AI, Hooper NM, Turner AJ (2005) Tumor necrosis factor-α convertase (ADAM17) mediates regulated ectodomain shedding of the severe-acute respiratory syndrome-coronavirus (SARS-CoV) receptor, angiotensin-converting enzyme-2 (ACE2). J Biol Chem 280(34):30113–30119

Langrish CL, Buddle JC, Thrasher AJ, Goldblatt D (2002) Neonatal dendritic cells are intrinsically biased against Th-1 immune responses. Clin Exp Immunol 128(1):118–123

Larbi A, Dupuis G, Khalil A, Douziech N, Fortin C, Fulop T Jr (2006) Differential role of lipid rafts in the functions of CD4+ and CD8+ human T lymphocytes with aging. Cell Signal 18(7):1017–1030. https://doi.org/10.1016/j.cellsig.2005.08.016

Larbi A, Fortin C, Dupuis G, Berrougui H, Khalil A, Fulop T (2014) Immunomodulatory role of high-density lipoproteins: impact on immunosenescence. Age (Dordr) 36(5):9712. https://doi.org/10.1007/s11357-014-9712-6

Lauer SA, Grantz KH, Bi Q, Jones FK, Zheng Q, Meredith HR, Azman AS, Reich NG, Lessler J (2020) The incubation period of coronavirus disease 2019 (COVID-19) from publicly reported confirmed cases: estimation and application. Ann Intern Med 172(9):577–582. https://doi.org/10.7326/m20-0504

Lichterfeld-Kottner A, El Genedy M, Lahmann N, Blume-Peytavi U, Buscher A, Kottner J (2020) Maintaining skin integrity in the aged: a systematic review. Int J Nurs Stud 103:103509. https://doi.org/10.1016/j.ijnurstu.2019.103509

Lippi G, Wong J, Henry BM (2020) Hypertension in patients with coronavirus disease 2019 (COVID-19): a pooled analysis. Pol Arch Intern Med 130(4):304–309. https://doi.org/10.20452/pamw.15272

Liu K, Zhang W, Yang Y, Zhang J, Li Y, Chen Y (2020) Respiratory rehabilitation in elderly patients with COVID-19: a randomized controlled study. Complement Ther Clin Pract 39:101166. https://doi.org/10.1016/j.ctcp.2020.101166

Long C, Xu H, Shen Q, Zhang X, Fan B, Wang C, Zeng B, Li Z, Li X, Li H (2020) Diagnosis of the coronavirus disease (COVID-19): rRT-PCR or CT? Eur J Radiol 126:108961. https://doi.org/10.1016/j.ejrad.2020.108961

Lotfi M, Rezaei N (2020) SARS-CoV-2: a comprehensive review from pathogenicity of the virus to clinical consequences. J Med Virol. https://doi.org/10.1002/jmv.26123

Lotfi M, Hamblin MR, Rezaei N (2020) COVID-19: transmission, prevention, and potential therapeutic opportunities. Clinica chimica acta; Int J Clin Chem 508:254–266. https://doi.org/10.1016/j.cca.2020.05.044

Low DJ (2000) Following the crowd. Nature 407(6803):465–466

Maddux AB, Douglas IS (2015) Is the developmentally immature immune response in paediatric sepsis a recapitulation of immune tolerance? Immunology 145(1):1–10

Mansourabadi AH, Sadeghalvad M, Mohammadi-Motlagh HR, Rezaei N (2020) The immune system as a target for therapy of SARS-CoV-2: a systematic review of the current immunotherapies for COVID-19. Life Sci 258:118185. https://doi.org/10.1016/j.lfs.2020.118185

Meng J, Xiao G, Zhang J, He X, Ou M, Bi J, Yang R, Di W, Wang Z, Li Z, Gao H, Liu L, Zhang G (2020) Renin-angiotensin system inhibitors improve the clinical outcomes of COVID-19 patients with hypertension. Emerg Microbes Infect 9(1):757–760. https://doi.org/10.1080/22221751.2020.1746200

Moazzami B, Razavi-Khorasani N, Dooghaie Moghadam A, Farokhi E, Rezaei N (2020) COVID-19 and telemedicine: immediate action required for maintaining healthcare providers Well-being. J Clin Virol: The Official Publication of the Pan American Society for Clinical Virology 126:104345. https://doi.org/10.1016/j.jcv.2020.104345

Mohamed K, Rodríguez-Román E, Rahmani F, Zhang H, Ivanovska M, Makka SA, Joya M, Makuku R, Islam MS, Radwan N, Rahmah L, Goda R, Abarikwu SO, Shaw M, Zoghi S, Irtsyan S, Ling I, Cseprekal O, Faten AB, Hazar Sayar E, Soloukey C, Grancini G, Rezaei N (2020) Borderless collaboration is needed for COVID-19-a disease that knows no borders. Infect Control Hosp Epidemiol:1–2. https://doi.org/10.1017/ice.2020.162

Momtazmanesh S, Ochs HD, Uddin LQ, Perc M, Routes JM, Vieira DN, Al-Herz W, Baris S, Prando C, Rosivall L, Abdul Latiff AH, Ulrichs T, Roudenok V, Aldave Becerra JC, Salunke DB, Goudouris E, Condino-Neto A, Stashchak A, Kryvenko O, Stashchak M, Bondarenko A, Rezaei N (2020) All together to fight COVID-19. Am J Trop Med Hyg 102(6):1181–1183. https://doi.org/10.4269/ajtmh.20-0281

Moradian N, Ochs HD, Sedikies C, Hamblin MR, Camargo CA Jr, Martinez JA, Biamonte JD, Abdollahi M, Torres PJ, Nieto JJ, Ogino S, Seymour JF, Abraham A, Cauda V, Gupta S, Ramakrishna S, Sellke FW, Sorooshian A, Wallace Hayes A, Martinez-Urbistondo M, Gupta M, Azadbakht L, Esmaillzadeh A, Kelishadi R, Esteghamati A, Emam-Djomeh Z, Majdzadeh R, Palit P, Badali H, Rao I, Saboury AA, Jagan Mohan Rao L, Ahmadieh H, Montazeri A, Fadini GP, Pauly D, Thomas S, Moosavi-Movahed AA, Aghamohammadi A, Behmanesh M, Rahimi-Movaghar V, Ghavami S, Mehran R, Uddin LQ, Von Herrath M, Mobasher B, Rezaei N (2020) The urgent need for integrated science to fight COVID-19 pandemic and beyond. J Transl Med 18(1):205. https://doi.org/10.1186/s12967-020-02364-2

Nanda A, Vura N, Gravenstein S (2020) COVID-19 in older adults. Aging Clin Exp Res:1–4. https://doi.org/10.1007/s40520-020-01581-5

Nasab MG, Saghazadeh A, Rezaei N (2020) SARS-CoV-2-a tough opponent for the immune system. Arch Med Res. https://doi.org/10.1016/j.arcmed.2020.05.020

Nie X, Fan L, Mu G, Tan Q, Wang M, Xie Y, Cao L, Zhou M, Zhang Z, Chen W (2020) Epidemiological characteristics and incubation period of 7,015 confirmed cases with coronavirus disease 2019 outside Hubei Province in China. J Infect Dis. https://doi.org/10.1093/infdis/jiaa211

Nikolich-Zugich J (2018) The twilight of immunity: emerging concepts in aging of the immune system. Nat Immunol 19(1):10–19. https://doi.org/10.1038/s41590-017-0006-x

Nikolich-Zugich J, Knox KS, Rios CT, Natt B, Bhattacharya D, Fain MJ (2020) SARS-CoV-2 and COVID-19 in older adults: what we may expect regarding pathogenesis, immune responses, and outcomes. Geroscience 42(2):505–514. https://doi.org/10.1007/s11357-020-00186-0

Olde Rikkert MGM, Vingerhoets RW, van Geldorp N, de Jong E, Maas H (2020) Atypical clinical picture of COVID-19 in older patients. Ned Tijdschr Geneeskd 164

Orimo H, Ito H, Suzuki T, Araki A, Hosoi T, Sawabe M (2006) Reviewing the definition of "elderly". Geriatr Gerontol Int 6. https://doi.org/10.1111/j.1447-0594.2006.00341.x

Pashaei M, Rezaei N (2020) Immunotherapy for SARS-CoV-2: potential opportunities. Expert Opin Biol Ther:1–5. https://doi.org/10.1080/14712598.2020.1807933

Pawelec G, Goldeck D, Derhovanessian E (2014) Inflammation, ageing and chronic disease. Curr Opin Immunol 29:23–28. https://doi.org/10.1016/j.coi.2014.03.007

Pietilä M, Neuvonen M, Borodulin K, Korpela K, Sievänen T, Tyrväinen L (2015) Relationships between exposure to urban green spaces, physical activity and self-rated health. J Outdoor Recreat Tour 10:44–54. https://doi.org/10.1016/j.jort.2015.06.006

Qin C, Zhou L, Hu Z, Zhang S, Yang S, Tao Y, Xie C, Ma K, Shang K, Wang W, Tian DS (2020) Dysregulation of immune response in patients with COVID-19 in Wuhan. China Clin Infect Dis. https://doi.org/10.1093/cid/ciaa248

Rabi FA, Al Zoubi MS, Kasasbeh GA, Salameh DM, Al-Nasser AD (2020) SARS-CoV-2 and coronavirus disease 2019: what we know so far. Pathogens 9(3). https://doi.org/10.3390/pathogens9030231

Rabiee N, Rabiee M, Bagherzadeh M, Rezaei N (2020) COVID-19 and picotechnology: potential opportunities. Med Hypotheses 144:109917. https://doi.org/10.1016/j.mehy.2020.109917

Rezaei N (2020a) COVID-19 affects healthy pediatricians more than pediatric patients. Infect Control Hosp Epidemiol 1. https://doi.org/10.1017/ice.2020.139

Rezaei N (2020b) COVID-19 and medical biotechnology. Avicenna J Med Biotechnol 12(3):139

Saghazadeh A, Rezaei N (2020a) Immune-epidemiological parameters of the novel coronavirus–a perspective. Expert Rev Clin Immunol:1–6

Saghazadeh A, Rezaei N (2020b) Towards treatment planning of COVID-19: rationale and hypothesis for the use of multiple immunosuppressive agents: anti-antibodies, immunoglobulins, and corticosteroids. Int Immunopharmacol:106560

Sahu KK, Siddiqui AD, Rezaei N, Cerny J (2020) Challenges for management of immune thrombocytopenia during COVID-19 pandemic. J Med Virol. https://doi.org/10.1002/jmv.26251

Saito S, Morii T, Enomoto M, Sakakura S, Nishikawa K, Narita N, Ichijo M (1993) The effect of interleukin 2 and transforming growth factor-beta 2 (TGF-beta 2) on the proliferation and natural killer activity of decidual CD16- CD56bright natural killer cells. Cell Immunol 152(2):605–613. https://doi.org/10.1006/cimm.1993.1316

Saleki K, Banazadeh M, Saghazadeh A, Rezaei N (2020) The involvement of the central nervous system in patients with COVID-19. Rev Neurosci 31(4):453–456. https://doi.org/10.1515/revneuro-2020-0026

Sharifkashani S, Bafrani MA, Khaboushan AS, Pirzadeh M, Kheirandish A, Yavarpour Bali H, Hessami A, Saghazadeh A, Rezaei N (2020) Angiotensin-converting enzyme 2 (ACE2) receptor and SARS-CoV-2: potential therapeutic targeting. Eur J Pharmacol 884:173455. https://doi.org/10.1016/j.ejphar.2020.173455

Singh S, Bajorek B (2014) Defining 'elderly' in clinical practice guidelines for pharmacotherapy. Pharm Pract (Granada) 12(4):489. https://doi.org/10.4321/s1886-36552014000400007

Singhal T (2020) A review of coronavirus Disease-2019 (COVID-19). Indian J Pediatr 87(4):281–286. https://doi.org/10.1007/s12098-020-03263-6

Su S, Wong G, Shi W, Liu J, Lai ACK, Zhou J, Liu W, Bi Y, Gao GF (2016) Epidemiology, genetic recombination, and pathogenesis of coronaviruses. Trends Microbiol 24(6):490–502. https://doi.org/10.1016/j.tim.2016.03.003

Tang N, Bai H, Chen X, Gong J, Li D, Sun Z (2020) Anticoagulant treatment is associated with decreased mortality in severe coronavirus disease 2019 patients with coagulopathy. J Thromb Haemost 18(5):1094–1099. https://doi.org/10.1111/jth.14817

Tyrrell DAJ, Bynoe ML (1966) Cultivation of viruses from a high proportion of patients with colds. Lancet 287(7428):76–77. https://doi.org/10.1016/S0140-6736(66)92364-6

Vallejo AN (2005) CD28 extinction in human T cells: altered functions and the program of T-cell senescence. Immunol Rev 205:158–169. https://doi.org/10.1111/j.0105-2896.2005.00256.x

Velavan TP, Meyer CG (2020) The COVID-19 epidemic. Tropical Med Int Health 25(3):278–280. https://doi.org/10.1111/tmi.13383

Veronese N, Demurtas J, Yang L, Tonelli R, Barbagallo M, Lopalco P, Lagolio E, Celotto S, Pizzol D, Zou L, Tully MA, Ilie PC, Trott M, López-Sánchez GF, Smith L (2020) Use of corticosteroids in coronavirus disease 2019 pneumonia: a systematic review of the literature. Front Med (Lausanne) 7:170. https://doi.org/10.3389/fmed.2020.00170

Walford RL (1964) The immunologic theory of aging. The Gerontologist 4(4):195–197. https://doi.org/10.1093/geront/4.4.195

Wang H, Luo S, Shen Y, Li M, Zhang Z, Dong Y, Zhou H, Lin L, Guo W, Kang Z (2020a) Multiple enzyme release, inflammation storm and hypercoagulability are prominent indicators for disease progression in COVID-19: a multi-centered, Correlation Study with CT Imaging Score

Wang K, Kang S, Tian R, Zhang X, Zhang X, Wang Y (2020b) Imaging manifestations and diagnostic value of chest CT of coronavirus disease 2019 (COVID-19) in the Xiaogan area. Clin Radiol 75(5):341–347. https://doi.org/10.1016/j.crad.2020.03.004

Wang L, He W, Yu X, Hu D, Bao M, Liu H, Zhou J, Jiang H (2020c) Coronavirus disease 2019 in elderly patients: characteristics and prognostic factors based on 4-week follow-up. J Infect 80(6):639–645. https://doi.org/10.1016/j.jinf.2020.03.019

Wang Y, Liu Y, Liu L, Wang X, Luo N, Li L (2020d) Clinical outcomes in 55 patients with severe acute respiratory syndrome coronavirus 2 who were asymptomatic at hospital admission in Shenzhen, China. J Infect Dis 221(11):1770–1774. https://doi.org/10.1093/infdis/jiaa119

Ward CF, Figiel GS, McDonald WM (2020) Altered mental status as a novel initial clinical presentation for COVID-19 infection in the elderly. Am J Geriatr Psychiatry. https://doi.org/10.1016/j.jagp.2020.05.013

Wilder-Smith A, Freedman DO (2020) Isolation, quarantine, social distancing and community containment: pivotal role for old-style public health measures in the novel coronavirus (2019-nCoV) outbreak. J Travel Med 27(2). https://doi.org/10.1093/jtm/taaa020

Wilson ME, Chen LH (2020) Travellers give wings to novel coronavirus (2019-nCoV). J Travel Med 27(2). https://doi.org/10.1093/jtm/taaa015

Wong C, Goldstein DR (2013) Impact of aging on antigen presentation cell function of dendritic cells. Curr Opin Immunol 25(4):535–541. https://doi.org/10.1016/j.coi.2013.05.016

Wu Z, McGoogan JM (2020) Characteristics of and important lessons from the coronavirus disease 2019 (COVID-19) outbreak in China: summary of a report of 72 314 cases from the Chinese Center for Disease Control and Prevention. Jama

Xu XW, Wu XX, Jiang XG, Xu KJ, Ying LJ, Ma CL, Li SB, Wang HY, Zhang S, Gao HN, Sheng JF, Cai HL, Qiu YQ, Li LJ (2020a) Clinical findings in a group of patients infected with the 2019 novel coronavirus (SARS-Cov-2) outside of Wuhan, China: retrospective case series. BMJ 368:m606. https://doi.org/10.1136/bmj.m606

Xu YH, Dong JH, An WM, Lv XY, Yin XP, Zhang JZ, Dong L, Ma X, Zhang HJ, Gao BL (2020b) Clinical and computed tomographic imaging features of novel coronavirus pneumonia caused by SARS-CoV-2. J Infect 80(4):394–400. https://doi.org/10.1016/j.jinf.2020.02.017

Xudong X, Junzhu C, Xingxiang W, Furong Z, Yanrong L (2006) Age-and gender-related difference of ACE2 expression in rat lung. Life Sci 78(19):2166–2171

Yang Y, Peng F, Wang R, Guan K, Jiang T, Xu G, Sun J, Chang C (2020) The deadly coronaviruses: the 2003 SARS pandemic and the 2020 novel coronavirus epidemic in China. J Autoimmun 109:102434. https://doi.org/10.1016/j.jaut.2020.102434

Yaqinuddin A, Kashir J (2020) Innate immunity in COVID-19 patients mediated by NKG2A receptors, and potential treatment using Monalizumab, Cholroquine, and antiviral agents. Med Hypotheses 140:109777. https://doi.org/10.1016/j.mehy.2020.109777

Yazdanpanah F, Hamblin MR, Rezaei N (2020a) The immune system and COVID-19: friend or foe? Life Sci:117900

Yazdanpanah N, Saghazadeh A, Rezaei N (2020b) Anosmia: a missing link in the neuroimmunology of coronavirus disease 2019 (COVID-19). Rev Neurosci. https://doi.org/10.1515/revneuro-2020-0039

Yousefzadegan S, Rezaei N (2020) Case report: death due to COVID-19 in three brothers. Am J Trop Med Hyg 102(6):1203–1204. https://doi.org/10.4269/ajtmh.20-0240

Zhang T, Wu Q, Zhang Z (2020) Probable pangolin origin of SARS-CoV-2 associated with the COVID-19 outbreak. Curr Biol 30(7):1346–1351.e1342. https://doi.org/10.1016/j.cub.2020.03.022

Zhou P, Yang X-L, Wang X-G, Hu B, Zhang L, Zhang W, Si H-R, Zhu Y, Li B, Huang C-L, Chen H-D, Chen J, Luo Y, Guo H, Jiang R-D, Liu M-Q, Chen Y, Shen X-R, Wang X, Zheng X-S, Zhao K, Chen Q-J, Deng F, Liu L-L, Yan B, Zhan F-X, Wang Y-Y, Xiao G-F, Shi Z-L (2020) A pneumonia outbreak associated with a new coronavirus of probable bat origin. Nature 579(7798):270–273. https://doi.org/10.1038/s41586-020-2012-7

Zhu N, Zhang D, Wang W, Li X, Yang B, Song J, Zhao X, Huang B, Shi W, Lu R, Niu P, Zhan F, Ma X, Wang D, Xu W, Wu G, Gao GF, Tan W (2020) A novel coronavirus from patients with pneumonia in China, 2019. N Engl J Med 382(8):727–733. https://doi.org/10.1056/NEJMoa2001017

Coronavirus Diseases in Pregnant Women, the Placenta, Fetus, and Neonate

14

David A. Schwartz and Amareen Dhaliwal

Abstract

Severe acute respiratory syndrome coronavirus-2 (SARS-CoV-2), the etiological agent of coronavirus disease 2019 (COVID-19), is similar to two other coronaviruses, severe acute respiratory syndrome coronavirus (SARS-CoV) and Middle East respiratory syndrome coronavirus (MERS-CoV), in causing life-threatening respiratory infections and systemic complications in both children and adults. As the COVID-19 pandemic has continued to spread globally, increasing numbers of pregnant women have become infected, raising concern not only for their health but also for the health of their infants. This chapter discusses the effects of coronavirus infections, e.g., MERS, SARS, and COVID 19, on pregnancy and describes the evolving knowledge of COVID 19 among pregnant women. The physiological changes that occur in pregnancy, especially changes in the immune system, are reviewed in terms of their effect on susceptibility to infectious diseases. The effects of COVID-19 on the placenta, fetus, and neonate are also reviewed, including potential clinical outcomes and issues relating to testing and diagnosis. The potential mechanisms of vertical transmission of the virus between pregnant women and their infants are analyzed, including intrauterine, intrapartum, and postpartum infections. Several recent studies have reported the detection of SARS-CoV-2 in tissues from the fetal side of the placenta, permitting the diagnosis of transplacental infection of the fetus by SARS-CoV-2. Placentas from infected mothers in which intrauterine transplacental transmission of SARS-CoV-2 has occurred demonstrate an unusual combination of pathology findings which may represent risk factors for placental as well as fetal infection.

Keywords

COVID-19 · Fetus · Maternal coronavirus infection · Maternal death · Maternal health · MERS · Neonatal infection · Perinatal infection · Pregnancy · SARS · SARS-CoV-2 · Vertical transmission

14.1 Introduction

No one was expecting the coronavirus disease 2019 (COVID-19) pandemic to occur, although the possibility of a cataclysmic global event due

D. A. Schwartz (✉)
Department of Pathology, Medical College of Georgia, Augusta University, Augusta, GA, USA

A. Dhaliwal
Boston University School of Medicine, Boston, MA, USA

N. Rezaei (ed.), *Coronavirus Disease - COVID-19*, Advances in Experimental Medicine and Biology 1318, https://doi.org/10.1007/978-3-030-63761-3_14

to an emerging infectious agent had been discussed in many professional circles. As the Ebola virus disease epidemic was coming to a close in West Africa in 2015, Margaret Chan, MD, the former director-general of the World Health Organization (WHO), presciently stated, "Everyone needs to remember that Ebola was not a worst-case scenario. Preparedness for the future means preparedness for a very severe disease that spreads via the airborne route or can be transmitted during the incubation period before an infected person shows telltale signs of illness." By the close of the West African Ebola epidemic, a minimum of 28,616 persons developed the infection, and 11,323 died, and these deaths included pregnant and lactating mothers, as well as their fetuses, neonates, and infants (Schwartz et al. 2019). Pregnant women and their infants continued to die from the Ebola virus during successive outbreaks in DR Congo (Schwartz 2018, 2019a, b, 2020b).

Pregnant women are significantly affected by emerging infections and constitute a particular group of persons at high risk during outbreaks of novel disease agents (Schwartz 2019a, 2020b). They often bear the burden of the most severe effects of infectious disease not only to themselves, including morbidity and death, but also the risk and consequences of congenital fetal and neonatal infection. Heightened risk for adverse obstetrical outcomes due to emerging infections results from various factors – an altered immune state; changes in the maternal cardiovascular and respiratory systems including increased oxygen consumption, heart rate, and stroke volume and decreased lung capacity; and other physiological changes in pregnancy all increase the likelihood for the severe maternal disease. Moreover, pregnant women often comprise a significant subset of the reproductive age female population in some areas where emerging infections strike. The past 50 years are filled with numerous examples of the vulnerability of pregnant women and their infants to emerging infections. Acquired immunodeficiency syndrome (AIDS) was a devastating event that resulted in both maternal morbidity and mortality but also perinatal infections and deaths. Multiple outbreaks and epidemics of Ebola virus disease in Africa have caused high levels of maternal death as well as the demise of almost every infected fetus and newborn. The Zika virus did not result in maternal deaths or, in many cases, even symptoms; instead, it stealthily crossed the placenta to cause fatal infections, malformations, neurological injuries, and death. It is no wonder that the emergence of the severe acute respiratory syndrome coronavirus-2 (SARS-CoV-2) in Wuhan, China, in December 2019 and the subsequent COVID-19 pandemic has triggered worldwide anxiety and fear among pregnant women and their families. As a newly identified zoonotic virus, the SARS-CoV-2 represents an unknown quantity in terms of its consequences to pregnant women, and the effects of emerging infections may be unknown and difficult to predict.

The recognition of COVID-19 and its rapid global spread has led to widespread concern for its effects on pregnant women and the fetus. Experiences from two previous coronavirus diseases, e.g., Middle Eastern respiratory syndrome (MERS) caused by the Middle East respiratory syndrome coronavirus (MERS-CoV) and severe acute respiratory syndrome (SARS) caused by the severe acute respiratory syndrome coronavirus (SARS-CoV), demonstrated that these highly pathogenic coronaviruses could result in severe maternal illness and death, as well as pregnancy loss. Fortunately, there were no confirmed cases of intrauterine transmission of either the SARS or MERS coronaviruses to the fetus, although the numbers studied were relatively small (Schwartz and Graham 2020).

There have been rapid advances in understanding the medical and epidemiological features of COVID-19 in pregnant women and their infants, but a great deal remains to be understood. This chapter examines the current state of knowledge of the three coronaviruses affecting pregnancy – SARS-CoV, MERS-CoV, and SARS-CoV-2 – and compares and contrasts their effects on outcomes of the mother, fetus, and neonate.

14.2 Pregnancy, the Immune System, and Infectious Disease

The maternal immune system during pregnancy plays a vital role in the potential increased susceptibility of pregnant women to certain infectious diseases. Immunologic changes occur during pregnancy, including alterations in the maternal immune system to tolerate fetal antigens by suppressing cell-mediated immunity while retaining healthy humoral immunity. Thus, while the pregnant woman is not technically immunosuppressed, these changes may induce a state of increased susceptibility to specific intracellular pathogens. These include hepatitis E, influenza, herpes simplex, malaria, and toxoplasmosis (Kourtis et al. 2014; Jamieson et al. 2006).

Pregnancy profoundly affects maternal immunological cells and pathways, which can affect the course of infectious diseases in women. It has been proposed that the immunological changes that occur as a result of pregnancy consist of a finely tuned series of adaptations that have been termed the "immunological clock of pregnancy" by Aghaeepour and colleagues (Aghaeepour et al. 2017). The total number of T cells, including CD3+, CD4+, and CD8+ lymphocytes, is diminished in the peripheral blood. It likely results from an increase in estrogen and progesterone that develops at the close of the first trimester and results in reversible thymic involution. There are also diminished levels of Th1 and Th2 cytokines that contribute to maintaining and balancing the proinflammatory response to infectious agents. This results in cells that have reduced activity following being stimulated without any apparent bias toward a Th2 phenotype. Peripheral NK cells have cytotoxic function and trigger responses to an infectious agent by secreting cytokines such as IFNγ, thereby enabling the adaptive response. These cells are decreased in number and have reduced activity after 20 weeks of gestation (Raj et al. 2014; Pazos et al. 2012). The functions of B cells are also influenced by pregnancy, and it has been found that estrogen reduces both the number and activity of committed B-cell precursors in the bone marrow and the numbers of CD19+ cells in the peripheral blood (Pazos et al. 2012; Medina et al. 1993).

Together with the decrease in T-cell and NK cell function and numbers, there is an increase in some elements of innate immunity. Increased numbers of phagocytic immune cells that include neutrophils, monocytes, and dendritic cells occur in the maternal blood from as early as 20 weeks' gestation (Kraus et al. 2012). A concomitant increase in phagocytic activity has been described. There are also increased levels of defensins, polypeptides that inhibit microbes in a variety of ways in the maternal serum throughout pregnancy (Pazos et al. 2012; Kraus et al. 2012).

Being infected with SARS-CoV-2 might affect typical physiological alterations that occur in the immune response during pregnancy. Severe cases of COVID-19 infection have been found to be associated with a life-threatening cytokine storm, which is referred to as an uncontrolled and excessive release of proinflammatory cytokines. These include increased levels of the interleukins IL-2, IL-7, IL-10, granulocyte colony-stimulating factor (G-CSF), interferon-γ-inducible protein 10 (IP-10), monocyte chemoattractant protein 1 (MCP-1), macrophage inflammatory protein 1 alpha (MIP-1α), and tumor necrosis factor (TNF). Because pregnant women in the first and third trimester are in a proinflammatory state, there is the possibility that a cytokine storm induced by SARS-CoV-2 may result in a more severe inflammatory state during pregnancy (Liu et al. 2020a).

14.3 Severe Acute Respiratory Syndrome (SARS)

Severe acute respiratory syndrome coronavirus (SARS) emerged in late November 2002 from bats and civet cats, leading to human transmission and suspected outbreaks from animal markets in Guangzhou, China. SARS presented with flu and pneumonia-like symptoms and was initially believed to be pneumonia or an influenza outbreak. In February 2003, a traveling physician unknowingly initiated the SARS epidemic in Hong Kong. Soon after, the virus was isolated, recognized as a global threat, and quarantine and travel restrictions were placed. By July 11, 2003, the WHO announced the end of reported human-to-human transmission (Tong 2006). During these few months, the virus reached 29 countries and led to 8427 suspected SARS cases, with 916 deaths and a case fatality rate of 9.6% (Bell et al. 2004). Additionally, healthcare workers accounted for approximately one-third of these cases (Schwartz and Graham 2020).

SARS has been considered to be less infective than the common cold, having a basic reproduction number (R0) of 2–3. SARS has an incubation period of 6.4 days, after which clinical disease develops in three phases: initial flu-like symptoms, pneumonia-like symptoms, and, lastly, acute respiratory distress syndrome. Children, especially those under 12, are less likely to display significant symptoms or progress to the third phase of SARS (Li and Ng 2005). Common symptoms of SARS include fever (100%), chills or rigor (50%), headache (40%), cough (60%), and coryza (50%). Some cases also reported dyspnea, dizziness, sputum production, sore throat, anorexia, nausea, vomiting, diarrhea, abdominal pain, febrile seizure, and rash (Srikantiah et al. 2005; Li and Ng 2005). Mortality was notably higher in elderly patients with comorbid illnesses (Cao et al. 2003). Despite the high case fatality rate of SARS, less than 10% of cases were children, and an even smaller proportion was under 12 years of age. Fortunately, there were no SARS-related deaths reported to occur in children.

The clinical obstetrical outcomes among pregnant women infected with SARS-CoV in Hong Kong were worse than the outcomes among infected women who were not pregnant (Maxwell et al. 2017). Pregnant women with SARS had higher rates of miscarriage, preterm birth, intrauterine growth retardation, and gastrointestinal symptoms (Li and Ng 2005) (Table 14.1). There has been no evidence of vertical transmission of SARS between an infected mother and her fetus or neonate (Schwartz and Graham 2020). The cases of adverse perinatal outcomes have been considered to result from maternal disease, including pneumonia and its systemic effects. Maternal pneumonia from other causes and infections has been shown to be associated with increased risk of preterm labor, intrauterine growth restriction, perinatal death, and neonatal death (Schwartz and Graham 2020).

A case-control study was performed at the Princess Margaret Hospital in Hong Kong in order to investigate the effects of SARS on pregnancy. It compared the clinical obstetrical outcomes of 10 pregnant women with those of 40 non-pregnant women who had the infection (Lam et al. 2004). In the cohort of pregnant women with SARS, there were three maternal deaths (maternal mortality rate of 30%), but no deaths occurred in the cohort of non-pregnant infected women ($P = 0.006$). Other complications, including renal failure ($P = 0.006$) and disseminated intravascular coagulopathy ($P = 0.006$), occurred more frequently in pregnant women with SARS as compared with non-pregnant women. There were six pregnant women with SARS who required admission to the intensive care unit (ICU) (60%), and four needed endotracheal intubations (40%), compared to an intubation rate of 12.5% and an ICU admission rate of 17.5% ($P = 0.012$) in the non-pregnant cohort of women. Not surprisingly, pregnant women with SARS needed a longer time in the hospital than did women in the non-pregnant group. Adverse pregnancy outcomes were common. Four out of five women presenting during the first trimester had sustained miscarriage, and four women presented after the second trimester had to undergo preterm delivery. Five surviving neonates were not infected with SARS, but three had complications due to preterm delivery.

In another study from Hong Kong, Wong et al. (2004) reviewed the outcomes of 12 pregnant patients with SARS. This cohort had complications that included maternal death (3 of 12; case fatality rate of 25%), first trimester spontaneous miscarriage (four of seven), preterm delivery after 24 weeks (four of five), and two cases of intrauterine growth restriction. Specimens of cord blood, placental tissue, and neonatal serology all tested negative for SARS. There was no evidence for vertical transmission of the virus (Wong et al. 2004).

Table 14.1 Reports of some cases of SARS during pregnancy

Study	Number of cases	Maternal outcomes	Neonatal outcomes
Lam et al. (2004)	10	Maternal death (30%), ICU admission (60%), endotracheal intubation (40%), fever, lymphopenia, renal failure, and DIC Four mothers presenting after the second trimester had preterm deliveries	4/5 mothers in the first trimester had a miscarriage 3/5 surviving neonates with complications of prematurity Neonates uninfected
Wong et al. (2004)	12	Maternal death (3) Miscarriage (1) Preterm delivery (4),	Intrauterine growth restriction (2) All neonates, cord blood, and placental tissues tested were uninfected
Zhang et al. (2003)	5	Fever (5), chills (3), cough (4), secondary infection (3), high transaminases (3), thrombocytopenia (2), hypoalbuminemia (4), abnormal chest X-ray (5), and ICU admission (1)	Neonates asymptomatic and uninfected (5)
Yudin et al. (2005)	1	Fever, cough, patchy infiltrate on chest X-ray requiring hospital admission	Neonate was asymptomatic and uninfected

A cohort of five primigravida women in their second and third trimester of pregnancy who were infected with SARS in Guangzhou, China, were noted to have fever (100%), chills or rigor (60%), cough (80%), secondary community-acquired infection (60%), elevated transaminases (60%), thrombocytopenia (40%), hypoalbuminemia (80%), and abnormal chest X-rays (100%). One patient required intensive care. In this group, there were no maternal deaths reported, and all neonates showed no evidence of SARS infection (Zhang et al. 2003).

In Canada, there was a case report of a 33-year-old pregnant woman with SARS who presented at 31 weeks' gestation with fever, cough, and patchy infiltrates. She required a 21-day hospital course without respiratory support, and later she delivered a healthy, uninfected female neonate (Yudin et al. 2005).

The 2002–2003 epidemic of SARS led to poor obstetrical outcomes, including maternal deaths, miscarriages, preterm deliveries, and respiratory distress. Fortunately, there was no evidence of vertical transmission of the virus to the fetus or neonate (Schwartz and Graham 2020). Instead, it is suggested that the effects of viral pneumonia and hypoxia on the neonate and mother were the leading causes of adverse outcomes.

14.4 Middle East Respiratory Syndrome (MERS)

It was first reported in September 2012 in Saudi Arabia from a male patient who had died several months previously with severe pneumonia and multiorgan failure (Milne-Price et al. 2014). There have now been 2519 confirmed cases of MERS, resulting in 866 deaths globally (WHO 2020). While 27 countries have reported confirmed cases of MERS, approximately 80% of these have originated in Saudi Arabia (Assiri et al. 2016). All identified persons who have developed MERS are linked to travel or residence in countries along the Arabian Peninsula.

Isolated as a sporadically zoonotic coronavirus which resides in dromedary camels, the virus was believed to spread from camels to humans and had the potential for intrafamilial human-to-human transmission (Al Awaidy and Khamis 2019; Memish et al. 2013). Similar to other coronaviruses, bats are believed to be the natural reservoir of MERS-CoV, and dromedary camels are assumed to be as possible intermediary hosts and a source of infection to humans (Schwartz and Graham 2020). As of January 2020, MERS accounts for 2298 confirmed cases and 811 deaths. MERS-CoV presents similarly to SARS-CoV but with a drastically higher mortality rate –

greater than 34% – and cases continue to occur (WHO 2020; Assiri et al. 2016).

Nosocomial outbreaks of MERS in healthcare settings have resulted from poor infection control and prevention and are characteristic of MERS transmission (Hui 2017). The risk factors for MERS infections include direct exposure to dromedary camels, exposure to seropositive household contact(s), diabetes mellitus, heart disease, and smoking. Asymptomatic cases are likely to occur as a study of 10,000 participants in Saudi Arabia was found to have a MERS-CoV seroprevalence of 0.15%. Clinically, MERS presents with fever, chills, cough, arthralgia, and gastrointestinal symptoms, which can rapidly progress to pneumonia-like symptoms and respiratory distress. MERS is more likely to lead to respiratory illness in people over 55, immunocompromised status, with other comorbidities or with hypoalbuminemia (Hui 2017; Hajjar et al. 2013).

There are limited data available on the prevalence and clinical features of MERS occurring in pregnant women and their infants. Similar to many other viral infections, it appears probable that the immunological and physiological changes that accompany pregnancy may alter susceptibility to the MERS-CoV and the severity of clinical illness (Schwartz and Graham 2020). As previously discussed, pregnant women infected with the related coronavirus SARS-CoV have increased morbidity and mortality when compared to non-pregnant women, suggesting that pregnant women with MERS are also at risk for developing severe clinical outcomes. However, only 11 pregnancy-associated cases have been described, but significantly 91% have had adverse maternal and infant outcomes.

From the period between November 2012 and February 2016, there were 1308 cases of MERS reported to occur by the Saudi Arabia Ministry of Health. Among them were five pregnant women with ages ranging from 27 to 34 years, all of whom had adverse obstetrical outcomes (Assiri et al. 2016). All five pregnant women had exposure during the second or third trimesters, and all required intensive care. Two mothers died, and there were two cases of perinatal death: one stillborn infant and one neonatal death occurring just after delivery by emergency cesarean section. Two of the five women were healthcare workers, consistent with the high risk for acquiring MERS infections in healthcare settings (Table 14.2).

Alfaraj et al. (2019) reported two pregnant women infected with MERS-CoV as confirmed by nasopharyngeal swab testing by RT-PCR. One mother was a 29-year-old woman who was at 6 weeks' gestation with no underlying medical conditions. The second case was a 39-year-old mother who was at 24 weeks' gestation and had several comorbidities that included hypertension and end-stage renal disease requiring hemodialysis. She presented to the hospital after contact with a person infected with MERS-CoV during an active outbreak. Both women fully recovered from the infection and then tested negative for MERS-CoV. The first patient delivered a healthy, full-term infant, while there is no information on the status of the second patient's delivery and her fetus (Alfaraj et al. 2019).

Malik et al. (2016) reported a 32-year-old pregnant woman with MERS at 34 weeks' gestation who had recent contact with farm animals. She presented with fever, back pain, and dyspnea. She developed respiratory decline resulting in ICU-level care with mechanical ventilation. An emergency C-section delivery resulted in a healthy uninfected newborn. After delivery, the patient's symptoms mildly improved but then later declined and resulted in maternal death (Malik et al. 2016).

The one confirmed case of pregnancy-associated MERS outside of the Middle East occurred in South Korea. Jeong et al. (2017) and Park et al. (2016)) reported that in 2015 a 39-year-old pregnant woman in the third trimester was exposed and infected following contact with a patient having MERS. Although she developed an episode of abrupt vaginal bleeding and rupture of membranes, the mother fully recovered. Her baby was born healthy at 37 weeks, 5 days' gestation. Testing of the infant's blood failed to reveal IgG, IgM, or IgA antibodies to MERS-CoV (Jeong et al. 2017; Park et al. 2016).

While conducting an epidemiologic investigation during the 2012 MERS outbreak in Zarqa,

Table 14.2 Reports of some cases of MERS during pregnancy

Study	Number of cases	Maternal outcomes	Fetal outcomes
Assiri et al. (2016)	5	Patients reporting shortness of breath and pneumonia-like symptoms later complicated by respiratory distress, acute renal failure, ICU-level care, intubation, mechanical ventilation, and maternal and neonatal deaths	Stillbirth (1) Neonatal death (1)
Alfaraj et al. (2019)	2	One asymptomatic mother infected in the first trimester One mother infected during the second trimester with end-stage renal disease, hypertension, and dialysis	Term, healthy neonates (2)
Malik et al. (2016)	1	Third trimester mother with MERS and hypoxemia requiring emergency C-section and later resulting in maternal death	Healthy neonate
Payne et al. (2014)	1	Second trimester with exposure to MERS, fever, cough, and other symptoms; mother had MERS-CoV antibodies	Stillbirth
Jeong et al. (2017) and Park et al. (2016)	1	Third trimester infection with MERS developed vaginal bleeding and rupture of membranes	Healthy neonate
Alserehi et al. (2016)	1	Respiratory failure at 32 weeks' gestation leading to acute respiratory distress syndrome and mechanical ventilation, MERS infection	Healthy neonate

Jordan, Payne et al. (2014) found that at 5-months gestation, a stillbirth had occurred from maternal exposure to MERS-CoV. The pregnant woman had developed fever, headache, fatigue, and cough together with vaginal bleeding and abdominal pain. On the seventh day of symptoms, there was an intrauterine fetal demise. The mother reported that she had unprotected contact with family members who later tested positive for the virus. Later it was confirmed that she had antibodies to MERS-CoV, and this became the first case of stillbirth related to maternal infection with MERS-CoV (Payne et al. 2014).

Alserehi et al. (2016) described a 33-year-old woman who was 32 weeks' pregnant and presented cough, fever, and dyspnea, which led to admission for pneumonia. Due to respiratory failure, an emergency cesarean section was performed, which resulted in the delivery of a healthy newborn. The mother developed acute respiratory distress syndrome requiring mechanical ventilation and tested positive for MERS-CoV by tracheal aspirate. Her symptoms resolved after 4 days of treatment. The neonate was kept in isolation precaution during delivery and tested negative for MERS-CoV (Alserehi et al. 2016).

Among these 11 cases of confirmed cases of MERS during pregnancy, the mean maternal age was 33.2 years, and the mean gestational age was 26.3 weeks. In two of the cases, the source of infection was believed to result from contact with family members who had tested positive for MERS-CoV. The source was unknown in three cases, and in six of the pregnant women, the cause of infection was healthcare-associated (two of these patients were healthcare workers). In one case, the source was thought to be due to animal exposure. Six of the pregnant women received intensive care. Of the three mothers who died, two were exposed to MERS-CoV during the third trimester, and one was exposed in the second trimester. The infant death rate among the 11 pregnancies was 27%. There did not appear to be a correlation of fetal survival with the timing of maternal MERS-CoV infection and gestational age; however, more data are needed to conclude this relationship. Alfaraj et al. (2019) stated that the case fatality rate for the 11 women with MERS, which was 27%, was not statistically different from the overall case fatality rate from MERS in the general population (35%). There was only one case of both maternal and fetal death (Schwartz and Graham 2020).

14.5 COVID-19 in Pregnant Women

Since the identification of SARS-CoV-2 in Wuhan, China, and the resulting global pandemic that was declared on March 11, 2020, there has been an increasing attention to the effect of COVID-19 on pregnant women and their infants. It has been estimated that 116 million babies will be born during the 9 months following the initial recognition of SARS-CoV-2 (UNICEF 2020). Those countries with the highest expected numbers of births since the start of the COVID-19 pandemic include India with 20.1 million expected births, China with 13.5 million, Nigeria with 6.4 million, Pakistan with 5 million, and Indonesia with 4 million. Unfortunately, some of these countries had high maternal or neonatal mortality rates that predated the COVID-19 pandemic. Also, these metrics will probably increase either directly or indirectly as a result of both infections of pregnant women and conditions existing that result from the newly identified coronavirus infection. In the United States, the sixth highest country for expected births, there will be approximately 3.3 million babies born between March 11 and December 16, 2020. COVID-19 has caused significant changes in the way in which prenatal and delivery care for pregnant women are being handled in the United States (Davis-Floyd et al. 2020a, b). These changes, and the potential direct threats from COVID-19 to pregnant women, the fetus, and newborns, have created tremendous concern among women of reproductive age, their families, and healthcare providers.

There is much remaining to understand how COVID-19 affects pregnant women, their pregnancies, the fetus, and neonate. There is currently no evidence that pregnant women are more likely to become infected by SARS-CoV-2 than are non-pregnant women. However, recent data from the CDC indicates that pregnancy is associated with increased risk for an intensive care unit admission and need for mechanical ventilation (Ellington et al. 2020).

14.5.1 Studies from China

Data on the effects of COVID-19 on pregnant women and their infants mainly come from investigations from China, the source of the pandemic. Following the initial descriptions of pneumonia caused by COVID-19 from Wuhan, reports became available of individual, small groups, and cohorts of pregnant women with COVID-19 infection in China (Chen et al. 2020; Li et al. 2020; Schwartz 2020a; Zhang et al. 2020). Based upon these early studies, the spectrum of maternal disease during the third trimester of pregnancy was variable, with some pregnant women having COVID-19 being asymptomatic and others only mild symptoms which were similar to those of non-pregnant infected adults – fever, nonproductive dry cough, and occasionally malaise, dyspnea, headache, and myalgias. Pregnant women with COVID-19 pneumonia and abnormal chest radiographs were reported but did not require intensive care. Comorbid conditions, many of them pregnancy-related such as gestational diabetes, preeclampsia, pregnancy-induced hypertension, uterine scarring, and uterine atony, were noted but appeared to have no significant impact on the effect of COVID-19 on the maternal outcome. Perinatal complications were frequent – these included preterm labor, premature rupture of membranes, intrauterine growth restriction, low birth weight, fetal distress, feeding intolerance, asphyxia, respiratory distress, pneumonia, and, in one infant, neonatal death that did not appear to be related to COVID-19. There were no maternal deaths or severe disease attributable to COVID-19. Importantly, these early studies did not demonstrate any intrauterine fetal infections with SARS-CoV-2 (Yang et al. 2020a; Schwartz 2020a, c).

In addition to individual case reports and small numbers of patients, analyses of medium-sized cohorts of pregnant women with confirmed COVID-19 and their infants from China began to appear in the medical literature. These included reports by Schwartz et al. (2020a) of 38 pregnant women in China infected with SARS-CoV-2 during the third trimester. Liu et al. (2020b) described 13 hospitalized pregnant women in multiple hospitals and cities throughout China. Zeng et al. (2020) reported 33 mother-infant dyads from Wuhan, and Zhang et al. (2020) described an analysis of 16 women from Wuhan and Qianjing. In many cases, testing was performed from the placenta, umbilical cord blood, amniotic fluid, breast milk, maternal blood, and vaginal secretions, and all were negative for SARS-CoV-2 infection, No evidence for intrauterine infection of the fetus was identified in any of these studies or others (Yu and Chen 2020). In a more recent retrospective study of 116 infected pregnant mothers from 25 hospitals conducted in China between January 20 and March 24, 2020, the most frequent symptoms included fever in 59/116 (50.9%) and cough in 22/116 (28.4%). No symptoms were present in 27/116 (23.3%) of the mothers. The median age at presentation was 38 weeks. Abnormal radiologic findings were identified in 104/108 (96.3%) of cases. Eight mothers (6.9%) developed severe pneumonia. One of the eight women presenting in the first trimester and early second trimester had a missed spontaneous abortion. Among 99 pregnant women, there were 21 (21.2%) preterm deliveries, including 6 having preterm premature rupture of membranes (PPROM). Spontaneous preterm birth before 37 weeks of pregnancy occurred in 6/99 (6.1%) pregnancies. There was a single occurrence of severe neonatal asphyxia that resulted in neonatal death. There were no maternal deaths and no cases of vertical transmission (Yan et al. 2020).

Like other adults, COVID-19 pneumonia among pregnant women has been an important cause of morbidity. Kasraeian et al. (2020) performed a meta-analysis of 87 pregnant women with COVID-19 and pneumonia from China. Approximately 65% of the pregnant women in this study were exposed to an infected person. The most common presenting symptom was fever (86%), followed by cough (68%). Seventy percent of women had lymphopenia. Mild or moderate symptoms of COVID-19 were present in 78% of women, and one-third had underlying comorbid conditions. Obstetric complications related to pregnancy were common and included prematurity (61%) and fetal distress (31%) as well as premature rupture of membranes (14%). There were no maternal deaths (Kasraeian et al. 2020).

Another review of pregnant women with COVID-19 from China examined the features of 114 pregnant women from 17 case reports/series and one case-control study (Yang et al. 2020b). This analysis of published cases revealed that the most frequent symptoms were fever (87.5%) and cough (53.8%). Symptoms that were less frequent includedfatigue (22.5%), myalgia (16.3%), dyspnea (11.3%), diarrhea (8.8%), and sore throat (7.5%). Two pregnant women were asymptomatic upon admission, developing initial symptoms a few days later. The large majority of pregnant women (96.5%) had either mild symptoms or described as having a mild or regular type of COVID-19 upon admission, based on the Novel Coronavirus Pneumonia Diagnosis and Treatment Protocols created by the Chinese government. Six patients (5.3%) developed a severe or critical type of COVID-19, including one case of multiple organ failure who required extracorporeal membrane oxygenation (ECMO). There were two cases of induced abortion performed because of patient decisions. Cesarean sections were performed in the majority of patients (91%) for a variety of reasons that included preeclampsia, fetal distress, history of prior operative delivery, and unknown risk of intrapartum mother-to-child transmission by vaginal delivery. The only case-control study examined revealed that there were no significant differences in preeclampsia, gestational diabetes mellitus, and premature rupture of the membrane between pregnant women with and without COVID-19 infection (Yang et al. 2020b).

14.5.2 Studies from North America and Europe

As COVID-19 spread from Asia to Europe and North America, additional information derived from analyses of large cohorts of pregnant women with COVID-19 were published, and the clinical spectrum of the infection in pregnant women and infants became more defined.

The cardiopulmonary disease is the most potentially life-threatening complication of COVID-19 in pregnant women. In order to evaluate the severe manifestations of COVID-19 in pregnancy in the United States, a cohort study examining severe infection among 64 hospitalized pregnant women with severe COVID-19 at 12 different institutions in four states was performed (Pierce-Williams et al. 2020). All 64 women having critical COVID-19 disease had been administered either prophylactic or therapeutic anticoagulation during their admission. Forty-four (69%) women had severe disease, and 20 (31%) had a critical disease. There were multiple pre-existing comorbidities present among women in this cohort, including 25% with a pulmonary condition and 17% with cardiac disease, and a mean BMI was 34 kg/m^2. The mean gestational age at the onset of symptoms was 29 ± 6 weeks, and the women had a mean gestational age at the time of hospital admission of 30 ± 6 weeks. Eighty-one percent of women were treated with hydroxychloroquine, and 9% of women with severe disease and 65% of women with the critical disease received remdesivir. In cases of respiratory compromise, intubation usually occurred around hospital day 9, and the peak of respiratory support for patients having severe disease occurred on hospital day 8. In patients having critical disease, prone positioning was performed in 20% of cases. The rate of acute respiratory distress syndrome (ARDS) was 70%, with 20% of women needing re-intubation. There was a single case of maternal cardiac arrest. No women developed cardiomyopathy, and there were no maternal deaths. Thirty-two (50%) women delivered during their hospitalization for COVID-19 – 34% of severe and 85% of critical patients. Among the pregnant women with critical COVID-19, 88% (15/17) delivered preterm during their disease course, with 94% of them via cesarean. Among all critically ill pregnant women in this cohort, 75% (15/20) delivered preterm. There were no stillbirths or neonatal deaths or cases of vertical transmission.

In New York City, Breslin et al. (2020) described a cohort of 43 pregnant women with COVID-19 presenting to two hospitals during an approximately 2-week period. There were 14 women (32.6%) who were asymptomatic at the time of the initial evaluation. Among these women, ten (71.4%) developed COVID-19 symptoms either during their admission and delivery or in the postpartum period. In addition to this group, there were 29 pregnant women (67.4%) who initially presented with symptoms of COVID-19 infection. Three of these women eventually needed antenatal admission for viral symptoms, and one patient re-presented with worsening respiratory disease and required oxygen supplementation for 6 days postpartum following a successful induction of labor. Utilizing the criteria of Wu and McGoogan (2020), 37 women (86%) had mild disease, 4 (9.3%) had severe disease, and there were 2 (4.7%) pregnant women with the critical disease.

In a large prospective national study of a cohort of 427 pregnant women with COVID-19 from the United Kingdom, it was found that upon hospital admission, the majority of women had potential risk factors for poor outcomes (Knight et al. 2020). These included 233 (56%) pregnant women who were from black or other ethnic minority groups, 281 (69%) who were obese or overweight, 175 (41%) who were 35 years of age or older, and 145 (34%) who had pre-existing comorbidities. Among the 427 pregnant women, 266 (62%) delivered or had a pregnancy loss at the time of the study, and 196 (73%) gave birth at term. There were 41 (10%) women who were admitted to the hospital that required respiratory support, and five (1%) women died. Among the 265 infants tested for SARS-CoV-2, there were 12 (5%) whose infants tested positive for viral RNA, with 6 of them testing positive in the first 12 h after birth (Knight et al. 2020).

Coagulopathy has emerged as a potentially significant complication of COVID-19 in pregnant women. Vlachodimitropoulou Koumoutsea et al. (2020) reported two women with COVID-19 in their third trimester of pregnancy. The patients developed coagulopathy before delivery, as indicated by elevated D-dimers, progressive thrombocytopenia, declining fibrinogen, and rising activated partial thromboplastin time (APTT), and postpartum hemorrhage in one of the mothers. Postpartum coagulopathy has also been described with COVID-19. Two pregnant women with COVID-19 had developed severe COVID-19 pneumonia that required intubation and emergency cesarean sections; the initial evidence of having coagulopathy did not occur until 3 days following delivery when they developed signs of pulmonary microthrombi (Tutiya et al. 2020).

Cardiomyopathy has also been reported as a complication of COVID-19 in pregnant women. In a report of seven pregnant women with SARS-CoV-2 infection, two developed findings of cardiomyopathy that included moderately reduced left ventricular ejection fractions of 40%–45% and hypokinesis. One of the women required CPR and mechanical ventilation (Juusela et al. 2020).

14.5.3 Pregnancy as a Heightened Risk Factor for Women of Reproductive Age with COVID-19

During the initial months of the COVID-19 pandemic, published reports of individual cases, clusters, and small cohorts of pregnant women infected with SARS-CoV-2 originated mostly from China. These reports did not demonstrate that pregnant women were at a heightened risk for having a poor outcome due to their infection status when compared with non-pregnant individuals. However, as the pandemic spread globally, the severe and life-threatening maternal disease was reported with increasing frequency (Schwartz 2020c). On June 26, 2000, the US Centers for Disease Control and Prevention published their analysis of a national cohort of women of reproductive age with and without COVID-19 (Ellington et al. 2020). From January 22 to June 7, 2020, the CDC received surveillance reports of 91,412 women of reproductive age (15–44 years) with COVID-19 for whom data on pregnancy status were available, including 8207 (9.0%) women who were pregnant. Both symptomatic pregnant and non-pregnant women with COVID-19 had similar frequencies of cough (>50%) and shortness of breath (30%); however, the pregnant women reported less fever, headache, diarrhea, muscle aches, and chills. Several chronic conditions not associated with pregnancy were more frequently reported among pregnant women than among non-pregnant women – these included chronic lung disease, diabetes mellitus, and cardiovascular disease. Approximately one-third (31.5%) of pregnant women with COVID-19 had been hospitalized compared with 5.8% of non-pregnant women with COVID-19. Following statistical adjustment for the presence of underlying medical conditions, age, and race/ethnicity, pregnant women with COVID-19 were more likely to require ICU admission and mechanical ventilation than were non-pregnant women with COVID-19.

At the end of June 2020, the Centers for Disease Control and Prevention modified their list of heightened risk groups for COVID-19 to include pregnant women based partly on the finding that approximately 9% of laboratory-confirmed COVID-19 cases occurred in women of childbearing age in the US population in which approximately 5% of women of childbearing age are pregnant at any given time (CDC 2020).

14.5.4 Maternal Mortality

At the start of the pandemic, it was initially believed that COVID-19 could not produce life-threatening diseases in pregnant women. However, this was dispelled when seven maternal deaths resulting from COVID-19 were first reported in April 2020 in Iran. In a cohort of nine pregnant women diagnosed during the late second or third trimester with COVID-19, seven women died, one remained severely ill and ventilator-dependent at the time of publication, and one eventually recovered following a prolonged hospitalization (Hantoushzadeh et al. 2020). None of these women had pre-existing comorbidities that were higher than baseline population risk, and none had renal disease, asthma, or chronic hypertension. At the time of their hospital admissions, all of these patients had normal blood pressures, excluding a comorbid diagnosis of preeclampsia.

Maternal deaths occurred in 5 among 427 (1%) pregnant women with COVID-19 in a national cohort of hospitalized patients reported from the United Kingdom (Knight et al. 2020).

14.5.5 Vertical Transmission of Viruses

As is the case with other newly identified emerging infections, a significant concern during the COVID-19 pandemic is whether an infected pregnant woman can transmit the virus to her infant. This process of mother-to-infant infection is termed vertical transmission and, depending on the agent, can occur by three different mechanisms, e.g., intrauterine, intrapartum, and postpartum routes. Maternal-fetal viral transmission can occur through two significant mechanisms – the hematogenous route and an ascending route. The hematogenous route is characteristic of most viral agents where mother-to-fetus transmission occurs. In this mechanism, virus that is circulating in the maternal bloodstream during pregnancy enters the placenta through maternal blood perfusing the placenta through the uterine arterioles. The virus can then transit the maternal-placental interface to reach the fetal vessels in the chorionic villus tree, and then to be transmitted through the umbilical blood vessels to the fetus. Hematogenous vertical transmission occurs with viruses such as rubella, human immunodeficiency virus, cytomegalovirus, Ebola virus, parvovirus, and Zika virus and requires that the virus be present in the maternal blood – viremia. The ascending route of intrauterine fetal infection develops when microbial agents present in the lower genital tract ascend the cervicovaginal tract to reach the pregnant uterine cavity. They can then infect the uterine lining and placental membranes, infect the amniotic fluid, and cause fetal infection. This mechanism almost always works in the presence of a bacterial infection (Schwartz 2020d).

Intrapartum transmission can occur with some infections around the time of labor and delivery when the fetus passes through an infected birth canal. This type of vertical transmission can occur with herpes simplex virus. Intrapartum transmission causes 85% of vertical HSV infections and is the basis for performing cesarean section delivery in infected mothers. Human papillomavirus and human immunodeficiency virus can also be transmitted to the infant through exposure in an infected birth canal during labor and delivery.

Postpartum vertical transmission of viruses develops following delivery. It can be caused by the contaminative transmission of a virus to the neonate from an infected mother through respiratory secretions and fomites, skin-to-skin-contact, and breast milk. Respiratory viruses can be transmitted by this mechanism, as well as those viral agents present in breast milk, including HIV, cytomegalovirus (CMV), Ebola virus, and others.

During the early phases of the pandemic, the vast majority of neonates who were tested for SARS-CoV-2 following delivery to a mother with COVID-19 were negative for the virus (Simoes and Leal 2020; Schwartz 2020a). A few infants were noted to have positive test results following delivery, but due to the time-lapse between delivery and testing, the source of infection could not be determined (Schwartz 2020e; Zeng et al. 2020). More recently, neonates have been described with positive nasopharyngeal swab results for SARS-CoV-2 using RT-PCR within the first 24 h of life. As COVID-19 has continued to spread throughout the world, there have been increasing numbers of neonates found to have positive tests for the virus, and it has become important to identify the source of their infections.

In the setting of a new emerging infection affecting pregnant women and their infants, it can be challenging to determine whether neonatal infections have occurred through intrauterine transmission including the transplacental or ascending routes, by intrapartum infection during delivery, or by postpartum infection from the mother, other infected persons, or the environment.

In order to guide in determining whether a neonate testing positive for COVID-19 has acquired the infection by transplacental transmission, Schwartz et al. (2020b) have recommended that placentas from infected newborns be examined for direct evidence of infection. Techniques including immunohistochemistry using antibodies to viral antigens and nucleic acid analysis such as in situ hybridization and RNAscope that detect target RNA molecules of the virus within intact cells have the advantage of identifying virus within specific cell types in defined anatomic compartments of the placenta. As a result, these techniques can definitively localize a virus to such fetal cells such as the syncytiotrophoblast, Hofbauer cells (fetal-derived villous stromal macrophages), extravillous trophoblast, and chorionic villous endothelial cells. Using these meothds of placental evaluation, there have been at least several neonates that appear to have acquired SARS-CoV-2 infection in utero from transplacental transmission of the coronavirus from an infected mother (Schwartz and Thomas 2020; Schwartz and Moriotti 2020).

In situ RNA hybridization was recently used to identify and localize SARS-CoV-2 in the placentas from two infected neonates testing positive for COVID-19 in Italy (Patane et al. 2020). Both placentas showed chronic intervillositis and the presence of CD-68-positive macrophages in the intervillous and the villous space. The placental tissues were evaluated using in situ hybridization with RNAscope technology, a method that enables the detection of the SARS-CoV-2 spike protein mRNA by using the V-nCoV2019-S probe. Both placentas showed the coronavirus in the syncytiotrophoblast, confirming the presence of SARS-CoV-2 on the fetal side of the placenta. It provided evidence that that intrauterine infection of fetal cells within the placenta had occurred in the intrauterine environment and before delivery. This study was the first to demonstrate SARS-CoV-2 infection in chorionic villus tissue of placentas from infected neonates, thereby confirming intrauterine fetal exposure and infection with the coronavirus (Schwartz et al. 2020b).

Schwartz and colleagues have proposed that the identification of the virus in chorionic villous tissue of the placenta, using either in situ nucleic acid hybridization methods as was performed by Patane et al. (2020) or by immunohistochemistry using virus-specific antibodies to detect viral antigen, establishes confirmatory evidence of fetal infection (Schwartz et al. 2020b; Schwartz and Thomas 2020). In the setting of an infected maternal-infant dyad, this finding should represent a definitive criterion for confirming that transplacental maternal-fetal infection has occurred.

14.5.6 The Placenta

The placenta is the largest of fetal organs, and its examination has been of great importance in previous outbreaks of emerging infections in evaluating the existence and potential mechanisms of intrauterine vertical transmission of agents such as the Zika virus and Ebola virus (Schwartz 2017; Muehlenbachs et al. 2017). The information currently available regarding placental pathology findings in mothers with COVID-19, the effects of SARS-CoV-2 on the placenta, and the role of the placenta in intrauterine infection of the fetus is rapidly increasing, but is still preliminary (Schwartz and Thomas 2020; Schwartz and Morotti 2020).

Baergen and Heller (2020) analyzed 20 placentas from neonates whose mother had tested positive for SARS-CoV-2. Ten of the 20 cases (50%) demonstrated pathological findings that included some evidence of fetal vascular malperfusion or fetal vascular thrombosis. Fetal vascular malperfusion occurred in nine cases (45%) and was the most common abnormal process. These findings included intramural fibrin deposition in three placentas, foci of villous stromal-vascular karyorrhexis in two placentas, and multiple lesions in four placentas. A few placentas contained recent, intramural, nonocclusive thrombi. The fetal vascular malperfusion was low grade in all of the placentas where it was present. Additional findings included meconium within macrophages, microscopic findings of maternal vascular malperfusion in five placentas, and focal increase in perivillous fibrin deposition. In one placenta from a neonate whose mother had pneumonia and acute hypoxia, there was evidence of an ascending infection with acute chorioamnionitis and funisitis. Four placentas demonstrated chronic villitis of unknown etiology. The villitis was high grade in two placentas and was associated with the occurrence of obliterative vasculopathy in one case. All infants tested negative for COVID-19 (Baergen and Heller 2020). Thus, it would be difficult to associate any of these findings with the maternal-fetal transmission of the virus or with fetal infection.

In another study of placentas, Shanes et al. (2020) examined 16 placentas from deliveries to women with COVID-19, including 15 placentas from third trimester live-born neonates and one from a second trimester (16 weeks) intrauterine fetal demise. In comparison to a cohort of control placentas, the third trimester placentas were more likely to demonstrate at least one feature of maternal vascular malperfusion (12 of 15 placentas), especially abnormal or injured maternal vessels, and intervillous thrombi. The placenta from the case of intrauterine fetal demise had villous edema and a retroplacental hematoma. All infants were found to be uninfected with SARS-CoV-2 by nasopharyngeal and throat swab (Shanes et al. 2020).

Schwartz and Morotti (2020) have recently identified a pattern of placental pathology findings occurring in placentas from neonates who have acquired SARS-Co-2 via the intrautrerine tranplacental route. Placentas from these mother-infant dyads have demonstrated an unusual pathology finding termed chronic histiocytic intervillositis in which histiocytes accumulate in the intervillous space, often accompanied bu other types of inflammatory cells. Also occurring in these placentas is trophoblast necrosis, as well as infection of the syncytiotrophoblast with SARS-CoV-2 as demonstrated by immunohistochemistry and/or RNA in situ hybridization. This group of findings appears to represent pathology risk factors for coronavirus infection of the placenta and transplacental transmission of SARS-CoV-2 from the pregnant mother to the fetus.

14.5.7 Fetal and Neonatal Infections

There has been intense interest in the infection status of infants delivered to pregnant women with COVID-19. There have been approximately 217 neonates delivered to pregnant women with COVID-19 reported in the literature (Schwartz et al. 2020b). Thus far, the large majority of infants born to pregnant women infected with SARS-CoV-2 have been uninfected throughout the world. However, there remains much to be learned about this issue. This apparent low frequency of perinatal infection is not surprising given the information available on vertical transmission with other RNA respiratory viruses, including the coronaviruses SARS-CoV and MERS-CoV, as well as respiratory syncytial virus, influenza, parainfluenza, and human metapneumovirus (Schwartz and Graham 2020; Schwartz and Dhaliwal 2020). The inhibition of respiratory RNA viruses to undergo intrauterine maternal-fetal transmission is likely the result of both viral and host factors and involves the ability of the virus to penetrate the maternal-fetal interface, including the placenta, avoiding the innate immune system, and tropism of the virus to host cells.

Among the neonates that have been reported with positive testing for SARS-CoV-2, some have been asymptomatic (Alzamora et al. 2020; Schwartz et al. 2020a; Wang et al. 2020; Yu et al. 2020; Zeng et al. 2020), and other neonates have had symptomatic illness. The symptomatic infants with positive testing for SARS-CoV-2 have demonstrated variations in the onset of illness, ranging from within 1 or 2 h after delivery to within the first 24–72 h of life and up to many days to weeks following birth (Alzamora et al. 2020; Gordon et al. 2020; Schwartz et al. 2020a; Zaigham and Andersson 2020). Some of these cases have been considered to be suspicious for vertical transmission. However, the mechanisms and timing of acquiring the infection remained unknown.

A recent analysis of 67 neonates delivered to pregnant women with COVID-19 demonstrated neonatal complications that included respiratory distress or pneumonia in 18%, disseminated intravascular coagulation in 3%, asphyxia in 2%, and two cases of perinatal death (Zimmermann and Curtis 2020).

Recently there have been 19 neonates testing positive for COVID-19 reported by Schwartz et al. (2020a) from ten hospitals in multiple provinces throughout Iran, among whom there were infants with early and late positive testing for the virus. These 19 neonates demonstrated significant differences in the epidemiology, clinical characteristics, results of testing, and clinical outcomes. Analysis of possible sources of neonatal infection demonstrated that in 11 of the 19 neonates, at least one family member tested positive for COVID-19. It included eight neonates whose mothers were the only parent with confirmed COVID-19. One mother/father pair had confirmed COVID-19, and in one case, only the grandmother had COVID-19. In one neonate, both parents were tested, but only the father had confirmed infection with the mother testing negative. In five cases, the mothers were tested negative for COVID-19. In 4 of the 15 neonates (27%) who had at least one family member tested, there was no identifiable family source of infection. Preterm delivery frequently occurred among these neonates; 12 of 19 infants (63%) were delivered before 37 weeks' gestation, including two very preterm deliveries at 28 weeks and a set of triplets at 29 weeks' gestation. Two preterm infants in this cohort tested positive for SARS-CoV-2 infection just 1 and 2 h following cesarean section delivery. Respiratory distress was the most common presenting medical problem, occurring in 12 of the 19 (63%) neonates. Twenty-five percent of the neonates with respiratory distress (3 of 12) had already been discharged home and required readmission with the initial onset of respiratory disease. Bradycardia, icterus, lung hemorrhage, thrombocytopenia, and abdominal distension with poor feeding were also reported. Three of 19 (16%) of neonates did not have symptoms or findings that were referable to their positive COVID-19 status. A significant observation in this study was that 7 of 19 (37%) neonates who were discharged from the hospital as healthy subsequently required readmission as they developed signs and symptoms of COVID-19. Among these seven neonates, the average approximate time interval between discharge and readmission for symptoms of COVID-19 was 12 days. Another significant finding in this cohort of infected neonates was the frequent occurrence of disparities between initial testing of neonates for COVID-19 and the results of subsequent tests. If repeat testing for COVID-19 has not been performed among infants in this cohort, infections in 4 of 19 infants (21%) would have been undiagnosed. The neonatal mortality rate from COVID-19 among these 19 infants was 10.5%, though not included the death of 2 triplets whose cause of death could not be confirmed.

14.6 Conclusion

All of the three major pathogenic coronavirus infections occurring in humans – SARS-CoV, MERS-CoV, and SARS-CoV-2 – have now been demonstrated to cause adverse obstetrical outcomes when they occur in pregnant women. Although it was initially believed at the start of the COVID-19 pandemic in China that pregnant women might be spared the life-threatening effects of the virus, as the infection spread, this has been shown to not be the situation. Pregnant women with COVID-19 can, in the minority of cases, develop critical illness including severe pneumonia requiring intensive care and mechanical ventilation, cardiovascular and thrombotic disease, multiple organ dysfunction, and obstetrical near-miss events as well as maternal death. Preterm birth is a frequent complication, and neonates delivered to infected mothers can develop complications including pneumonia and, rarely, death. Early onset of neonatal COVID-19 may be the result of vertical infection in some cases, and placental infection with SARS-CoV-2 has been described in a few cases supporting the occurrence of intrauterine maternal-fetal transmission.

Many questions remain about the effects of COVID-19 on pregnant women and their infants including risk factors for adverse maternal outcomes, frequency and pathological mechanisms of transmission of the virus between mother and infant, and factors that determine whether infected mothers will transmit the virus to the fetus and neonate. It will be of importance to address these questions and others as the pandemic proceeds.

References

Aghaeepour N, Ganio EA, McIlwain D, Tsai AS, Tingle M, Van Gassen S, Gaudilliere DK, Baca Q, McNeil L, Okada R, Ghaemi MS, Furman D, Wong RJ, Winn VD, Druzin ML, El-Sayed YY, Quaintance C, Gibbs R, Darmstadt GL, Shaw GM, Stevenson DK, Tibshirani R, Nolan GP, Lewis DB, Angst MS, Gaudilliere B (2017) An immune clock of human pregnancy. Sci Immunol 2(15). https://doi.org/10.1126/sciimmunol.aan2946

Al Awaidy ST, Khamis F (2019) Middle East respiratory syndrome coronavirus (MERS-CoV) in Oman: current situation and going forward. Oman Med J 34(3):181–183. https://doi.org/10.5001/omj.2019.36

Alfaraj SH, Al-Tawfiq JA, Memish ZA (2019) Middle East respiratory syndrome coronavirus (MERS-CoV) infection during pregnancy: report of two cases & review of the literature. J Microbiol Immunol Infect 52(3):501–503. https://doi.org/10.1016/j.jmii.2018.04.005

Alserehi H, Wali G, Alshukairi A, Alraddadi B (2016) Impact of Middle East respiratory syndrome coronavirus (MERS-CoV) on pregnancy and perinatal outcome. BMC Infect Dis 16:105. https://doi.org/10.1186/s12879-016-1437-y

Alzamora MC, Paredes T, Caceres D, Webb CM, Valdez LM, La Rosa M (2020) Severe COVID-19 during pregnancy and possible vertical transmission. Am J Perinatol 37(8):861–865. https://doi.org/10.1055/s-0040-1710050

Assiri A, Abedi GR, Al Masri M, Bin Saeed A, Gerber SI, Watson JT (2016) Middle East respiratory syndrome coronavirus infection during pregnancy: a report of 5 cases from Saudi Arabia. Clin Infect Dis 63(7):951–953. https://doi.org/10.1093/cid/ciw412

Baergen RN, Heller DS (2020) Placental pathology in Covid-19 positive mothers: preliminary findings. Pediatr Dev Pathol 23(3):177–180. https://doi.org/10.1177/1093526620925569

Bell DM, World Health Organization Working Group on I, Community Transmission of S (2004) Public health interventions and SARS spread, 2003. Emerg Infect Dis 10(11):1900–1906. https://doi.org/10.3201/eid1011.040729

Breslin N, Baptiste C, Gyamfi-Bannerman C, Miller R, Martinez R, Bernstein K, Ring L, Landau R, Purisch S, Friedman AM, Fuchs K, Sutton D, Andrikopoulou M, Rupley D, Sheen JJ, Aubey J, Zork N, Moroz L, Mourad M, Wapner R, Simpson LL, D'Alton ME, Goffman D (2020) COVID-19 infection among asymptomatic and symptomatic pregnant women: Two weeks of confirmed presentations to an affiliated pair of New York City hospitals. Am J Obstet Gynecol MFM:100118. doi:https://doi.org/10.1016/j.ajogmf.2020.100118

Cao B, Liu ZY, Wang MZ, Cai BQ, Xu ZJ, Bai Y, Zhu YJ, Xu WB, Lu WX, Li LY, Li TS, Wang Z (2003) Clinical diagnosis, treatment and prognosis of elderly SARS patients. Zhongguo Yi Xue Ke Xue Yuan Xue Bao 25(5):547–549

CDC (2020) Coronavirus disease 2019 (COVID-19). People of any age with underlying medical conditions. https://www.cdc.gov/coronavirus/2019-ncov/need-extra-precautions/people-with-medical-conditions.html. Accessed June 25 2020

Chen H, Guo J, Wang C, Luo F, Yu X, Zhang W, Li J, Zhao D, Xu D, Gong Q, Liao J, Yang H, Hou W, Zhang Y (2020) Clinical characteristics and intrauterine vertical transmission potential of COVID-19 infection in nine pregnant women: a retrospective review of medical records. Lancet 395(10226):809–815. https://doi.org/10.1016/S0140-6736(20)30360-3

Davis-Floyd R, Gutschow K, Schwartz DA (2020a) The impacts of COVID-19 on birth practices in the United States. Pract Anthropol. https://practicinganthropology.sfaa.net/2020/05/27/the-impacts-of-covid-19-on-birth-practices-in-the-united-states-by-robbie-davis-floyd-kim-gutschow-and-david-a-schwartz/

Davis-Floyd R, Gutschow K, Schwartz DA (2020b) Pregnancy, birth and the COVID-19 pandemic in the United States. Med Anthropol 39(5):413–427. https://doi.org/10.1080/01459740.2020.1761804

Ellington S, Strid P, Tong VT, Woodworth K, Galang RR, Zambrano LD, Nahabedian J, Anderson K, Gilboa SM (2020) Characteristics of women of reproductive age with laboratory-confirmed SARS-CoV-2 infection by pregnancy status – United States, January 22-June 7, 2020. MMWR Morb Mortal Wkly Rep 69(25):769–775. https://doi.org/10.15585/mmwr.mm6925a1

Gordon M, Kagalwala T, Rezk K, Rawlingson C, Ahmed MI, Guleri A (2020) Rapid systematic review of neonatal COVID-19 including a case of presumed vertical transmission. BMJ Paediatr Open 4(1):e000718. https://doi.org/10.1136/bmjpo-2020-000718

Hajjar SA, Memish ZA, McIntosh K (2013) Middle East respiratory syndrome coronavirus (MERS-CoV): a perpetual challenge. Ann Saudi Med 33(5):427–436. https://doi.org/10.5144/0256-4947.2013.427

Hantoushzadeh S, Shamshirsaz AA, Aleyasin A, Seferovic MD, Aski SK, Arian SE, Pooransari P, Ghotbizadeh F, Aalipour S, Soleimani Z, Naemi M,

Molaei B, Ahangari R, Salehi M, Oskoei AD, Pirozan P, Darkhaneh RF, Laki MG, Farani AK, Atrak S, Miri MM, Kouchek M, Shojaei S, Hadavand F, Keikha F, Hosseini MS, Borna S, Ariana S, Shariat M, Fatemi A, Nouri B, Nekooghadam SM, Aagaard K (2020) Maternal death due to COVID-19. Am J Obstet Gynecol 223(1):109 e101–109 e116. https://doi.org/10.1016/j.ajog.2020.04.030

Hui DS (2017) Epidemic and emerging coronaviruses (severe acute respiratory syndrome and Middle East respiratory syndrome). Clin Chest Med 38(1):71–86. https://doi.org/10.1016/j.ccm.2016.11.007

Jamieson DJ, Theiler RN, Rasmussen SA (2006) Emerging infections and pregnancy. Emerg Infect Dis 12(11):1638–1643. https://doi.org/10.3201/eid1211.060152

Jeong SY, Sung SI, Sung JH, Ahn SY, Kang ES, Chang YS, Park WS, Kim JH (2017) MERS-CoV infection in a pregnant woman in Korea. J Korean Med Sci 32(10):1717–1720. https://doi.org/10.3346/jkms.2017.32.10.1717

Juusela A, Nazir M, Gimovsky M (2020) Two cases of coronavirus 2019-related cardiomyopathy in pregnancy. Am J Obstet Gynecol MFM:100113. doi:https://doi.org/10.1016/j.ajogmf.2020.100113

Kasraeian M, Zare M, Vafaei H, Asadi N, Faraji A, Bazrafshan K, Roozmeh S (2020) COVID-19 pneumonia and pregnancy; a systematic review and meta-analysis. J Matern Fetal Neonatal Med:1–8. https://doi.org/10.1080/14767058.2020.1763952

Knight M, Bunch K, Vousden N, Morris E, Simpson N, Gale C, O'Brien P, Quigley M, Brocklehurst P, Kurinczuk JJ, Group UKOSSS-C-IiPC (2020) Characteristics and outcomes of pregnant women admitted to hospital with confirmed SARS-CoV-2 infection in UK: national population based cohort study. BMJ 369:m2107. https://doi.org/10.1136/bmj.m2107

Kourtis AP, Read JS, Jamieson DJ (2014) Pregnancy and infection. N Engl J Med 370(23):2211–2218. https://doi.org/10.1056/NEJMra1213566

Kraus TA, Engel SM, Sperling RS, Kellerman L, Lo Y, Wallenstein S, Escribese MM, Garrido JL, Singh T, Loubeau M, Moran TM (2012) Characterizing the pregnancy immune phenotype: results of the viral immunity and pregnancy (VIP) study. J Clin Immunol 32(2):300–311. https://doi.org/10.1007/s10875-011-9627-2

Lam CM, Wong SF, Leung TN, Chow KM, Yu WC, Wong TY, Lai ST, Ho LC (2004) A case-controlled study comparing clinical course and outcomes of pregnant and non-pregnant women with severe acute respiratory syndrome. BJOG 111(8):771–774. https://doi.org/10.1111/j.1471-0528.2004.00199.x

Li AM, Ng PC (2005) Severe acute respiratory syndrome (SARS) in neonates and children. Arch Dis Child Fetal Neonatal Ed 90(6):F461–F465. https://doi.org/10.1136/adc.2005.075309

Li Y, Zhao R, Zheng S, Chen X, Wang J, Sheng X, Zhou J, Cai H, Fang Q, Yu F, Fan J, Xu K, Chen Y, Sheng J (2020) Lack of vertical transmission of severe acute respiratory syndrome coronavirus 2, China. Emerg Infect Dis 26(6):1335–1336. https://doi.org/10.3201/eid2606.200287

Liu H, Wang LL, Zhao SJ, Kwak-Kim J, Mor G, Liao AH (2020a) Why are pregnant women susceptible to COVID-19? An immunological viewpoint. J Reprod Immunol 139:103122. https://doi.org/10.1016/j.jri.2020.103122

Liu W, Wang J, Li W, Zhou Z, Liu S, Rong Z (2020b) Clinical characteristics of 19 neonates born to mothers with COVID-19. Front Med 14(2):193–198. https://doi.org/10.1007/s11684-020-0772-y

Malik A, El Masry KM, Ravi M, Sayed F (2016) Middle East respiratory syndrome coronavirus during pregnancy, Abu Dhabi, United Arab Emirates, 2013. Emerg Infect Dis 22(3):515–517. https://doi.org/10.3201/eid2203.151049

Maxwell C, McGeer A, Tai KFY, Sermer M (2017) No. 225-management guidelines for obstetric patients and neonates born to mothers with suspected or probable severe acute respiratory syndrome (SARS). J Obstet Gynaecol Can 39(8):e130–e137. https://doi.org/10.1016/j.jogc.2017.04.024

Medina KL, Smithson G, Kincade PW (1993) Suppression of B lymphopoiesis during normal pregnancy. J Exp Med 178(5):1507–1515. https://doi.org/10.1084/jem.178.5.1507

Memish ZA, Zumla AI, Al-Hakeem RF, Al-Rabeeah AA, Stephens GM (2013) Family cluster of Middle East respiratory syndrome coronavirus infections. N Engl J Med 368(26):2487–2494. https://doi.org/10.1056/NEJMoa1303729

Milne-Price S, Miazgowicz KL, Munster VJ (2014) The emergence of the Middle East respiratory syndrome coronavirus. Pathog Dis 71(2):121–136. https://doi.org/10.1111/2049-632X.12166

Muehlenbachs A, de la Rosa Vazquez O, Bausch DG, Schafer IJ, Paddock CD, Nyakio JP, Lame P, Bergeron E, McCollum AM, Goldsmith CS, Bollweg BC, Prieto MA, Lushima RS, Ilunga BK, Nichol ST, Shieh WJ, Stroher U, Rollin PE, Zaki SR (2017) Ebola virus disease in pregnancy: clinical, histopathologic, and Immunohistochemical findings. J Infect Dis 215(1):64–69. https://doi.org/10.1093/infdis/jiw206

Park MH, Kim HR, Choi DH, Sung JH, Kim JH (2016) Emergency cesarean section in an epidemic of the middle east respiratory syndrome: a case report. Korean J Anesthesiol 69(3):287–291. https://doi.org/10.4097/kjae.2016.69.3.287

Patane L, Morotti D, Giunta MR, Sigismondi C, Piccoli MG, Frigerio L, Mangili G, Arosio M, Cornolti G (2020) Vertical transmission of COVID-19: SARS-CoV-2 RNA on the fetal side of the placenta in pregnancies with COVID-19 positive mothers and neonates at birth. Am J Obstet Gynecol MFM:100145. https://doi.org/10.1016/j.ajogmf.2020.100145

Payne DC, Iblan I, Alqasrawi S, Al Nsour M, Rha B, Tohme RA, Abedi GR, Farag NH, Haddadin A, Al Sanhouri T, Jarour N, Swerdlow DL, Jamieson DJ,

Pallansch MA, Haynes LM, Gerber SI, Al Abdallat MM, Jordan M-CIT (2014) Stillbirth during infection with Middle East respiratory syndrome coronavirus. J Infect Dis 209(12):1870–1872. https://doi.org/10.1093/infdis/jiu068

Pazos M, Sperling RS, Moran TM, Kraus TA (2012) The influence of pregnancy on systemic immunity. Immunol Res 54(1–3):254–261. https://doi.org/10.1007/s12026-012-8303-9

Pierce-Williams RAM, Burd J, Felder L, Khoury R, Bernstein PS, Avila K, Penfield CA, Roman AS, DeBolt CA, Stone JL, Bianco A, Kern-Goldberger AR, Hirshberg A, Srinivas SK, Jayakumaran JS, Brandt JS, Anastasio H, Birsner M, O'Brien DS, Sedev HM, Dolin CD, Schnettler WT, Suhag A, Ahluwalia S, Navathe RS, Khalifeh A, Anderson K, Berghella V (2020) Clinical course of severe and critical COVID-19 in hospitalized pregnancies: a US cohort study. Am J Obstet Gynecol MFM:100134. https://doi.org/10.1016/j.ajogmf.2020.100134

Raj RS, Bonney EA, Phillippe M (2014) Influenza, immune system, and pregnancy. Reprod Sci 21(12):1434–1451. https://doi.org/10.1177/1933719114537720

Schwartz DA (2017) Viral infection, proliferation, and hyperplasia of Hofbauer cells and absence of inflammation characterize the placental pathology of fetuses with congenital Zika virus infection. Arch Gynecol Obstet 295(6):1361–1368. https://doi.org/10.1007/s00404-017-4361-5

Schwartz DA (2018) Clinical trials and administration of Zika virus vaccine in pregnant women: Lessons (that should have been) learned from excluding immunization with the Ebola vaccine during pregnancy and lactation. Vaccines (Basel) 6(4):81. https://doi.org/10.3390/vaccines6040081

Schwartz DA (2019a) Maternal and infant death and the rVSV-ZEBOV vaccine through three recent Ebola virus epidemics-West Africa, DRC Équateur and DRC Kivu: 4 years of excluding pregnant and lactating women and their infants from immunization. Current Trop Med Rep 6(4):213–222. https://doi.org/10.1007/s40475-019-00195-w

Schwartz DA (2019b) Pregnant women, vaccine development for emerging and epidemic viral infections and maternal immunization: human rights and the global survival of mothers and infants. Curr Trop Med Rep 6(4):179–185. https://doi.org/10.1007/s40475-019-00196-9

Schwartz DA (2020a) An analysis of 38 pregnant women with COVID-19, their newborn infants, and maternal-fetal transmission of SARS-CoV-2: maternal coronavirus infections and pregnancy outcomes. Arch Pathol Lab Med 144(7):799–805. https://doi.org/10.5858/arpa.2020-0901-SA

Schwartz DA (2020b) Being pregnant during the Kivu Ebola virus outbreak in DR Congo: the rVSV-ZEBOV vaccine and its accessibility by mothers and infants during humanitarian crises and in conflict areas. Vaccines (Basel) 8(1):38. https://doi.org/10.3390/vaccines8010038

Schwartz DA (2020c) The effects of pregnancy on women with COVID-19: maternal and infant outcomes. Clin Infect Dis 71(16):2042–2044. https://doi.org/10.1093/cid/ciaa559

Schwartz DA (2020d) The pathology of pregnancy. In: Strayer DS, Saffitz JE (eds) Rubin's Pathology. Clinicopathologic Foundations of Medicine, 8th edn. Wolters Kluwer, Philadelphia, pp 555–581

Schwartz DA (2020e) Vertical transmission of severe acute respiratory syndrome coronavirus 2 from the mother to the infant. JAMA Pediatr. 174(10):104–1005. https://doi.org/10.1001/jamapediatrics.2020.2135

Schwartz DA, Dhaliwal A (2020) Infections in pregnancy with COVID-19 and other respiratory RNA virus diseases are rarely, if ever, transmitted to the fetus: experiences with coronaviruses, parainfluenza, metapneumovirus, respiratory syncytial virus, and influenza. Arch Pathol Lab Med 144(8):920–928. https://doi.org/10.5858/arpa.2020-0211-SA

Schwartz DA, Graham AL (2020) Potential maternal and infant outcomes from coronavirus 2019-nCoV (SARS-CoV-2) infecting pregnant women: lessons from SARS, MERS, and other human coronavirus infections. Viruses 12(2):194. https://doi.org/10.3390/v12020194

Schwartz DA, Morotti D (2020) Placental pathology of COVID-19 with and without fetal and neonatal infection: trophoblast necrosis and chronic histiocytic intervillositis as risk factors for transplacental transmission of SARS-CoV-2. Viruses 12(11):1308. https://doi.org/10.3390/v12111308

Schwartz DA, Thomas KM (2020) Characterizing COVID-19 maternal-fetal transmission and placental infection using comprehensive molecular pathology. EBioMedicine 60:102983. https://doi.org/10.1016/j.ebiom.2020.102983

Schwartz DA, Anoko JN, Abramowitz S (2019) Pregnant in the Time of Ebola: women and Their Children in the 2013–2015 West African Epidemic. Springer, New York/Berlin

Schwartz DA, Mohagheghi P, Beigi B, Zafaranloo N, Moshfegh F, Yazdani A (2020a) Spectrum of neonatal COVID-19 in Iran: 19 infants with SARS-CoV-2 perinatal infections with varying test results, clinical findings and outcomes. J Matern Fetal Neonatal Med Aug 12:1–10. https://doi.org/10.1080/14767058.2020.1797672

Schwartz DA, Morotti D, Beigi B, Moshfegh F, Zafaranloo N, Patanè L (2020b) Confirming vertical fetal infection with COVID-19: Neonatal and pathology criteria for early onset and transplacental transmission of SARS-CoV-2 from infected pregnant mothers. Arch Pathol Lab Med. 144(12):1451-1456. https://doi.org/10.5858/arpa.2020-0442-SA

Shanes ED, Mithal LB, Otero S, Azad HA, Miller ES, Goldstein JA (2020) Placental pathology in COVID-19. Am J Clin Pathol 154(1):23–32. https://doi.org/10.1093/ajcp/aqaa089

Simoes ESAC, Leal CRV (2020) Is SARS-CoV-2 vertically transmitted? Front Pediatr 8:276. https://doi.org/10.3389/fped.2020.00276

Srikantiah P, Charles MD, Reagan S, Clark TA, Pletz MW, Patel PR, Hoekstra RM, Lingappa J, Jernigan JA, Fischer M (2005) SARS clinical features, United States, 2003. Emerg Infect Dis 11(1):135–138. https://doi.org/10.3201/eid1101.040585

Tong TR (2006) Severe acute respiratory syndrome coronavirus (SARS-CoV). In: Tabor E (ed) Perspectives in medical virology, vol 16. Elsevier, pp 43–95. https://doi.org/10.1016/S0168-7069(06)16004-8

Tutiya CT, Siaulys MM, Kondo MM, Miglioli-Galvao L, E CAG, Pinheiro CC, Torloni MR, de Mello FB (2020) Possible formation of pulmonary microthrombi in the early puerperium of pregnant women critically ill with COVID-19: two case reports. Case Rep Womens Health 27:e00237. https://doi.org/10.1016/j.crwh.2020.e00237

UNICEF (2020) Pregnant mothers and babies born during COVID-19 pandemic threatened by strained health systems and disruptions in services. https://www.unicef.org/press-releases/pregnant-mothers-and-babies-born-during-covid-19-pandemic-threatened-strained-health

Vlachodimitropoulou Koumoutsea E, Vivanti AJ, Shehata N, Benachi A, Le Gouez A, Desconclois C, Whittle W, Snelgrove J, Malinowski AK (2020) COVID-19 and acute coagulopathy in pregnancy. J Thromb Haemost 18(7):1648–1652. https://doi.org/10.1111/jth.14856

Wang S, Guo L, Chen L, Liu W, Cao Y, Zhang J, Feng L (2020) A case report of neonatal COVID-19 infection in China. Clin Infect Dis 71(15):853–857. https://doi.org/10.1093/cid/ciaa225

WHO (2020) MERS Situation Update

Wong SF, Chow KM, Leung TN, Ng WF, Ng TK, Shek CC, Ng PC, Lam PW, Ho LC, To WW, Lai ST, Yan WW, Tan PY (2004) Pregnancy and perinatal outcomes of women with severe acute respiratory syndrome. Am J Obstet Gynecol 191(1):292–297. https://doi.org/10.1016/j.ajog.2003.11.019

Wu Z, McGoogan JM (2020) Characteristics of and important lessons from the coronavirus disease 2019 (COVID-19) outbreak in China: Summary of a report of 72314 cases from the Chinese Center for Disease Control and Prevention. JAMA 323(13):1239–1242. https://doi.org/10.1001/jama.2020.2648

Yan J, Guo J, Fan C, Juan J, Yu X, Li J, Feng L, Li C, Chen H, Qiao Y, Lei D, Wang C, Xiong G, Xiao F, He W, Pang Q, Hu X, Wang S, Chen D, Zhang Y, Poon LC, Yang H (2020) Coronavirus disease 2019 in pregnant women: a report based on 116 cases. Am J Obstet Gynecol 223(1):111 e111–111 e114. https://doi.org/10.1016/j.ajog.2020.04.014

Yang H, Wang C, Poon LC (2020a) Novel coronavirus infection and pregnancy. Ultrasound Obstet Gynecol 55(4):435–437. https://doi.org/10.1002/uog.22006

Yang Z, Wang M, Zhu Z, Liu Y (2020b) Coronavirus disease 2019 (COVID-19) and pregnancy: a systematic review. J Matern Fetal Neonatal Med:1–4. https://doi.org/10.1080/14767058.2020.1759541

Yu Y, Chen P (2020) Coronavirus disease 2019 (COVID-19) in neonates and children from China: a review. Front Pediatr 8:287. https://doi.org/10.3389/fped.2020.00287

Yu N, Li W, Kang Q, Xiong Z, Wang S, Lin X, Liu Y, Xiao J, Liu H, Deng D, Chen S, Zeng W, Feng L, Wu J (2020) Clinical features and obstetric and neonatal outcomes of pregnant patients with COVID-19 in Wuhan, China: a retrospective, single-Centre, descriptive study. Lancet Infect Dis 20(5):559–564. https://doi.org/10.1016/S1473-3099(20)30176-6

Yudin MH, Steele DM, Sgro MD, Read SE, Kopplin P, Gough KA (2005) Severe acute respiratory syndrome in pregnancy. Obstet Gynecol 105(1):124–127. https://doi.org/10.1097/01.AOG.0000151598.49129.de

Zaigham M, Andersson O (2020) Maternal and perinatal outcomes with COVID-19: a systematic review of 108 pregnancies. Acta Obstet Gynecol Scand 99(7):823–829. https://doi.org/10.1111/aogs.13867

Zeng L, Xia S, Yuan W, Yan K, Xiao F, Shao J, Zhou W (2020) Neonatal early-onset infection with SARS-CoV-2 in 33 neonates born to mothers with COVID-19 in Wuhan, China. JAMA Pediatr 174(7):722–725. https://doi.org/10.1001/jamapediatrics.2020.0878

Zhang JP, Wang YH, Chen LN, Zhang R, Xie YF (2003) Clinical analysis of pregnancy in second and third trimesters complicated severe acute respiratory syndrome. Zhonghua Fu Chan Ke Za Zhi 38(8):516–520

Zhang L, Jiang Y, Wei M, Cheng BH, Zhou XC, Li J, Tian JH, Dong L, Hu RH (2020) Analysis of the pregnancy outcomes in pregnant women with COVID-19 in Hubei Province. Zhonghua Fu Chan Ke Za Zhi 55(3):166–171. https://doi.org/10.3760/cma.j.cn112141-20200218-00111

Zimmermann P, Curtis N (2020) COVID-19 in children, pregnancy and neonates: a review of epidemiologic and clinical features. Pediatr Infect Dis J 39(6):469–477. https://doi.org/10.1097/INF.0000000000002700

15 COVID-19 in Patients with Hypertension

Thiago Quinaglia, Mahsima Shabani, and Nima Rezaei

Abstract

Hypertension has been listed in several case series and retrospective cohorts as a potential risk factor for the incidence and severity of the new coronavirus (SARS-CoV-2)-associated disease (COVID-19). The debate is noteworthy because almost one billion people around the globe are estimated to have hypertensive diseases, according to the Global Burden of Disease study. Considering the SARS-CoV-2's high infectivity rates, a possible interaction between COVID-19 and hypertension is worrisome. Additionally, antihypertensive drugs, especially the renin-angiotensin-aldosterone system (RAAS) inhibitors, could also influence the natural course of COVID-19 infection. Not only can these associations hold from an epidemiologic standpoint, a mechanistic scenario possibly exists. Hypertension and antihypertensive drugs can increase the expression of transmembrane angiotensin-converting enzyme (ACE)-2 receptors, the entry target of the viruses, thus facilitating infectivity. On the other hand, an increase in ACE-2 could be protective considering the anti-inflammatory, antithrombotic effects of the ACE-2-angiotensin 1–7/Mas pathway. So far, little is known about the whole picture. Observational studies appear to indicate at least a twofold increased risk of mortality for hypertensive patients with COVID-19; however, the previous and continued use of RAAS inhibitors may be protective in this subgroup of patients. The scarcity of randomized clinical trials precludes evidence-based decision-making. At least one randomized study in a non-specified sub-analysis demonstrated no relationship between an angiotensin-converting enzyme inhibitor and incidence or severity of the disease. It is reflected mainly by observational studies and, therefore, by international cardiology societies' guidelines, which state that antihypertensive drugs, par-

T. Quinaglia
Department of Medicine, Division of Cardiology, Johns Hopkins University, Baltimore, MD, USA

M. Shabani (✉)
Department of Medicine, Division of Cardiology, Johns Hopkins University, Baltimore, MD, USA

Network of Immunity in Infection, Malignancy and Autoimmunity (NIIMA), Universal Scientific Education and Research Network (USERN), Baltimore, MD, USA
e-mail: mshabani@jhmi.edu

N. Rezaei
Network of Immunity in Infection, Malignancy and Autoimmunity (NIIMA), Universal Scientific Education and Research Network (USERN), Baltimore, MD, USA

Research Center for Immunodeficiencies, Children's Medical Center, Tehran University of Medical Sciences, Tehran, Iran

Department of Immunology, School of Medicine, Tehran University of Medical Sciences, Tehran, Iran

N. Rezaei (ed.), *Coronavirus Disease - COVID-19*, Advances in Experimental Medicine and Biology 1318, https://doi.org/10.1007/978-3-030-63761-3_15

ticularly RAAS inhibitors, should not be discontinued unless necessary on a case-by-case basis.

Keywords

Coronavirus disease · COVID-19 · Hypertension · Renin-angiotensin inhibitors · SARS-CoV-2

15.1 Introduction

The new coronavirus (SARS-CoV-2)-derived respiratory infection, known as COVID-19, has placed the whole world in a pandemic crisis emerging in December 2019. Although an incalculable effort is being devoted ubiquitously to tackle its spread, there is no clear solution to this health calamity in the foreseeable future. Effective preventive measures, such as vaccines, may only be available in no less than 12–18 months, considering it usually takes many years for such therapies to be developed and called useful and beneficial. In parallel, drug treatments with general applicability to COVID-19 patients are still unavailable, and randomized clinical trials (RCTs) are still ongoing. Thus, it is of paramount importance to understand the natural course of the disease and its pathophysiology, the effective non-pharmacological measures, and groups with the highest risk. This knowledge allows resources to be adequately allocated and directed to the most vulnerable populations.

At this point, observational evidence indicates that older age and baseline burden of comorbidities are associated with a more severe disease progression. Together with diabetes, lung disease, and coronary artery disease (CAD), hypertension predicted higher rates of acute respiratory distress syndrome (ARDS), acute kidney injury (AKI), cardiac injury, admissions to the intensive care unit (ICU), and death. In multivariate analyses, the predictive power of these parameters fades in some studies except for age in some studies (Wu et al. 2020; Zhou et al. 2020; Ruan et al. 2020; Cheng et al. 2020). However, a study designed to address the specific question of whether hypertension predicts clinical outcomes demonstrated an unambiguous association. In this large retrospective cohort ($n = 2877$), hypertension status is associated with a twofold increase in all-cause mortality (HR 2.12, 95% CI: 1,17–3.82) adjusted for age, sex, diabetes history, CAD, renal failure, heart failure, asthma, chronic obstructive pulmonary disease, and stroke (Gao et al. 2020). Moreover, untreated hypertensive patients ($n = 140$) had a significantly higher mortality rate than treated ones ($n = 730$) (HR 2.17, 95% CI 1.03–4.57, $P = 0.041$). The type of antihypertensive treatment—RAAS inhibitor vs. non-RAAS inhibitor—did not seem to change the clinical outcome in this study, but in a meta-analysis of four studies presented in the same publication ($n = 2242$), RAAS inhibitors tended to decrease mortality (relative risk 0.65, 95% CI 0.45–0.94, $p = 0.02$).

Currently, almost 1 billion people worldwide have hypertension (Forouzanfar et al. 2017), and considering the high infectivity rates of COVID-19, this population is possibly under significant vulnerability. If hypertension leads to more severe disease, it is essential to understand how this risk can be mitigated. Hypertension treatment, whether with RAAS inhibitors or not, could be one way of addressing this issue, but RCTs are needed to confirm. In this chapter, we aim to investigate the relation between COVID-19 and hypertension through a review of available literature by mid-2020.

15.1.1 Epidemiology

As well as a higher risk of COVID-19 infection within the hypertensive population, COVID-19 progression to severe illness and case fatality rates (CFRs) are reported to be higher in patients with hypertension. Studies indicate a high yet heterogeneous prevalence of hypertension among COVID-19 patients, from 56.6% in New York, USA (Richardson et al. 2020), to 15%–30% in individual Chinese studies (Zhou et al. 2020; Rodriguez-Morales et al. 2020) and 15.6% in a

21-study meta-analysis in Chinese population (Guan et al. 2020). Compared with all infected patients or non-severe patients, higher rates of hypertension were observed in patients with severe disease (23.7% in one study; 36.4 vs. 15.1% in another) (Zhou et al. 2020; Mao et al. 2020), patients with ICU admission (49% in one study; 58.3 vs. 21.6% another study) (Grasselli et al. 2020; Zhou et al. 2020), patients with ARDS (risk difference, 13.7%) (Wu et al. 2020), or in COVID-19 non-survivors (48 vs. 23%) (Zhou et al. 2020). Two meta-analyses studying 46,248 patients in Wuhan, China, and 1527 patients from several centers in China reported the most common comorbidity in infected patients to be hypertension (17% and 17.1%) and estimated pooled odds ratios of 2.36 (1.46–3.83) and 2.03 (1.54–2.68) for hypertension in patients with severe disease compared to non-severe disease (Yang et al. 2020; Li et al. 2020). Odds of inhospital mortality was also increased to 3.05-fold (1.75–5.92) in patients with hypertension vs. non-hypertensive subjects (Zhou et al. 2020).

Even though available data directs toward a higher incidence of COVID-19 among hypertensive individuals, most studies are descriptive and not adjusted for other confounding factors, on top of all older age. The Italian survey on 1591 COVID-19 patients pointed to this debate as showing that patients with hypertension were more aged than normotensive cases (Grasselli et al. 2020). A hypertension-focused study on COVID-19 patients showed that hypertensive patients are also older and more likely to have diabetes or cerebrovascular disease and higher levels of acute-phase reactants. Accordingly, multivariate analysis with age and gender as covariates did not report any association between hypertension and COVID-19 progression or mortality (Huang et al. 2020b). Despite a 52.9% prevalence of hypertension within COVID-19 patients in two hospitals in London, age- and gender-adjusted models did not show a significant time-to-event association between hypertension and death or critical care admission individually (Galloway et al. 2020). Likewise, hypertension was not included in a lethality risk score designed by a Mexican study, and higher weights were attributed to age, obesity, and diabetes (Bello-Chavolla et al. 2020). In the CORONADO multicenter study in France, in fully adjusted models for death, neither hypertension nor antihypertensive drugs were associated with primary outcomes or death in COVID-19 patients (Cariou et al. 2020). However, one study with the specific design to understand the association of hypertension and all-cause mortality showed a twofold increase in the ratio, even in the fully adjusted model. This study also showed that untreated hypertensive patients have worse outcomes and, in an accompanying meta-analysis, that RAAS inhibitors may decrease mortality (Gao et al. 2020). The data is observational and should not drive clinical decisions, but currently, it is the best evidence available.

15.1.2 Pathogenesis

Similar to SARS-CoV, part of SARS-CoV-2 infectivity is initiated by the binding of this virus to its functional receptor, transmembrane angiotensin-converting enzyme (ACE)-2 receptor, which is expressed on the lung, intestine, kidney, myocytes, and vascular endothelium (Zou et al. 2020). Viral surface spike glycoprotein (S-glycoprotein) binding to ACE-2 triggers the internalization of virion mediated by host cell protease TMPRSS2 (Hoffmann et al. 2020). Replication of the virus within the host cell leads to decreased expression of ACE-2 on the surface of cells and consequently suppressed angiotensin-2 degradation. As observed in SARS-CoV infection, angiotensin-2 accumulation results in acute lung injury using the type 1 receptor (ATR-1) (Kai and Kai 2020). The hypokalemia also supported the overactivity of angiotensin-2/ATR-1 in COVID-19 patients. A single-center study in China reported a high prevalence of hypokalemia among COVID-19 infected patients, including 39 severe hypokalemia and 69 hypokalemic patients among 175 admitted patients. They have also suggested that sustained correction of hypokalemia reflected by diminished urine K+ loss may be a favorable prognostic factor for COVID-19 outcome (Chen et al. 2020).

Another possible point of interaction between the SARS-CoV-2 mechanism of infection and comorbidities, including hypertension, is the endothelium. Reports of cardiac involvement, hypoxemia with apparently healthy lungs, and severe thrombotic complications (Fox et al. 2020) may be a consequence of diffuse endotheliitis expressed as localized or disseminated vasoconstriction, thrombotic macro- and microangiopathy, and a pro-inflammatory systemic response. The endothelium mainly expresses ACE-2, and there has been evidence of viral inclusions, inflammation, and apoptosis of endothelial cells in several organs and tissues, including the lungs, heart, kidney, and intestines (Varga et al. 2020; Inciardi et al. 2020). Moreover, endothelial cells operate as sentinels of the innate immune system and have a somewhat similar function to that of macrophages, e.g., antigen presentation and cytokine secretion (Shao et al. 2020). Thus, the advent of an endotheliitis could culminate with the commonly cited cytokine storm of COVID-19. Most of the comorbidities related to COVID-19 were previously described to be associated with endothelial dysfunction (Flammer et al. 2012). From this theoretical viewpoint, endothelial function impairment would increase vulnerability to viral effects and severity of the disease once installed but not necessarily increase the incidence of the disease.

15.2 Lessons Learned from Other Types of Pneumonia

15.2.1 Hypertension

Even though the OR of mortality due to influenza in hypertensive patients vs. normotensive patients (OR 3.53) is lower than that of COVID-19 (OR 7.39) (Kang et al. 2020), comorbidities like hypertension are also considered to be a significant risk and prognostic factor for acute viral illnesses like influenza. A systematic review on the risk factors of severe outcomes in patients with seasonal or pandemic influenza reported that although the association between hypertension and mortality from seasonal flu is not significant, the risk of death due to pandemic flu was increased (OR 1.34; 95% CI, 1-10-2.10) in patients with hypertension (Mertz et al. 2013). Also, a meta-analysis on 637 patients with MERS-CoV showed a pooled remarkable 48% (95% CI, 31–65%) prevalence of hypertension (Badawi and Ryoo 2016). On the other hand, there are studies on patients with community-acquired pneumonia that show when adjusted for other comorbidities like age, the association between hypertension and pneumonia fades away (Koivula et al. 1994). Finally, when studying the role of hypertension in illnesses, the predictive value of hypotension for the outcome of pneumonia, especially the early drop in blood pressure during the disease, should also be taken into account (Schulte-Hubbert et al. 2020). Controlling hypertension in these patients can be a double-edged sword.

15.2.2 Renin-Angiotensin-Aldosterone System Inhibitors

In addition to hypertension as a risk factor for the development or progression of acute viral respiratory illnesses, antihypertensive drugs are frequently reported to affect host response to these infections. The immunomodulatory effect of angiotensin receptor blocker (ARB) in host defense against acute critical viral infections is an expanding area of research, frequently pointed during previous pandemics of influenza, Ebola, and other pneumonia outbreaks. Promising results have been observed in the treatment of patients with acute respiratory distress syndrome (ARDS), sepsis, influenza, and Ebola with ARBs, mainly when used with statins (Fedson 2016). An exciting retrospective study in a Texas tertiary care center a year before the emergence of SARS-CoV-2 investigated the effect of angiotensin-converting enzyme inhibitors (ACEIs)/ARB use before/during admission of at least 24 h in all comers >18 years of age and a PCR-based viral respiratory infection. The study showed no association with pneumonia diagnosis, as assessed by an infiltrate in the chest radiograph. Among the 1055 patient admitted with viral pneumonia, the results showed that although baseline history of ACEI use was associated with higher mortality or intubation, continued ACEI use during admission lowered the risk of intubation or death (OR = 0.25; 95% CI, 0.09–0.64) and shorter length of stay (Henry et al. 2018). However, it is not clear whether the possible discontinuation of ACEI during admission, due to shock and organ failure, could have contributed to the increased mortality/intubation and lengths of stay.

In specific populations, these results can vary. In a retrospective study on 50,119 patients over 65 years with nonspecific pneumonia (both viral and bacterial), prior use and inpatient use of ACEI (OR 0.88 and 0.58) or ARB (OR 0.73 and 0.47) were associated with decreased 30-day mortality rates (Mortensen et al. 2012). Also, both prior and inpatient ACEI/ARB use were associated with shorter hospitalization periods. However, in a coronary artery disease population, with a previous history of revascularization, a case-control study (1666 patients with pneumonia and 33,315 time-matched control subjects) demonstrated no association between hospital admissions for community-acquired pneumonia and the use of ACEI (rate ratio 0.98, 95% CI, 0.69–1.40). Likewise, a lack of association was seen for ARB (RR 1.02, 95% CI, 0.70–1.49).

15.2.3 Meta-analysis and Clinical Trials on ACEI/ARB Use and Pneumonia Outcomes

A comprehensive meta-analysis was performed with clinical trials, cohorts, and case-control studies on the effect of ACEI/ARB on pneumonia outcome (incidence and pneumonia-related death). A lower pooled odds of pneumonia incidence was reported in patients treated with ACEI (OR in trials, 0.69; cohorts, 0.58; case-controls, 0.66). For ARBs, the decrease in odds of pneumonia was less significant (OR in trials, 0.90; cohorts, 0.95) (Caldeira et al. 2012). Previous stroke and use of ACEI were associated with a 54% reduction in the risk of pneumonia compared with controls (see below). As for heart failure, ACEI inhibitors were associated with a significant 37% reduction in the risk of pneumonia (0.63, CI, 0.47 to 0.84). For patients with chronic kidney disease, ACEI did not significantly influence the incidence of pneumonia compared with controls. In none of these patients' categories (stroke, heart failure, or chronic kidney disease), ARBs reduced the risk for pneumonia, according to the meta-analysis. The use of ACEI yielded a reduction in mortality related to pneumonia (OR, 0.73; 95% CI, 0.58 to 0.92), although this benefit was less robust than for the incidence reduction. Most of the benefits, however, seemed to exist in Asian ethnicity populations.

A couple of RCTs reported the effects of ACEI in poststroke populations and the incidence of pneumonia. Investigators of the Perindopril Protection Against Recurrent Stroke Study (PROGRESS), a large-scale randomized trial (n = 6105), performed a secondary analysis to evaluate the risk of incident pneumonia in patients with a previous history of cerebrovascular disease. During a median follow-up of 3.9 years, 261 patients developed pneumonia (viral or bacterial). Perindopril use was not associated with a lower risk of pneumonia ($p = 0.09$). The authors note, however, that among participants of Asian ethnicity, there was a relative risk reduction of pneumonia (47%, 14–67%; $p = 0.01$) as compared with non-Asian participants (5%, 27 to 29%; $p = 0.7$) (Ohkubo et al. 2004). In a randomized placebo-controlled trial, an ACEI was tested as a preventive measure for pneumonia in elderly patients on tube-feeding due to cerebrovascular-related dysphagia (n = 71). All patients had a history of hospitalization within 3 months of enrollment. Results showed that a low dose of lisinopril resulted in increased mortality (adjusted OR, 7.79; $p = 0.018$) and did not reduce the incidence rate of pneumonia (Lee et al. 2015). These data are controversial but show that ACEI does not prevent the incidence of pneumonia in the overall cerebrovascular-diseased population, except perhaps in an Asian ethnicity, and can potentially be harmful in a very fragile and pneumonia-susceptible population such as those with dysphagia.

15.3 Hypertension Per Se and COVID-19

Aside from hypertension, the top common comorbidities observed in COVID-19 patients are diabetes, coronary heart disease, and cerebrovascular diseases. A common denominator of these comorbidities is the use of ACEI/ARB for various indications. It is still not explained whether uncontrolled hypertension is individually a risk factor for COVID-19 progression or the drugs used in controlled hypertension are involved in the disease susceptibility/progression. Therefore, current data do not provide strong evidence on the individual association of hypertension with COVID-19 severity. Herein, we discuss possible theories underneath this hypothesis.

Like many other multifactorial diseases, immune dysregulation is involved in the pathogenesis of hypertension. The pro-inflammatory state in chronic diseases can be the common logic underlying the association of COVID-19 and many chronic diseases. In terms of hypertension, both endogenous stress and antigens derived from exogenous risk factors, including unhealthy diets, trigger immune responses toward a pro-inflammatory status. Innate immune pathways hyperactivate the inflammatory response through TLR-mediated inflammasome activation; IL-6, IL-1, and TNF-α production; and alternative stimulation of the complement pathway. Additionally, humoral immunity through autoantibody production, the cellular immune response by CD4+ Th17 and CD8+ T cells, and suppressed regulatory T-cell activity influence the homeostasis in blood pressure control by their direct effect on the heart and blood vessels or indirectly through the kidney and autonomic pathway (Drummond et al. 2019). The ongoing hemodynamic stress itself stimulates the immune response, and thus, this vicious cycle brings a sustained inflammatory environment in the circulation of hypertensive patients. Also, polymorphisms in immunomodulation-related molecules, e.g., SH2BP3 observed in genome-wide association studies, further highlight the role of immune dysregulation in hypertension development (Huan et al. 2015).

This immunity-hypertension interaction could partly explain why patients with hypertension are more susceptible to developing severe forms of COVID-19 infection. Host inflammatory cytokine storm is a single prognostic factor leading to the rapid progression of the disease. The level of many chemokines was shown to be elevated in patients with COVID-19, among which, specific cytokines that are also elevated in hypertensive individuals had higher levels in patients with the severe form of the disease (Huang et al. 2020a; Wu et al. 2020).

Another possible explanation of how hypertension and COVID-19 infection severity are linked is that genetic polymorphisms in ACE-2 pathway-related genes can be the common denominator of both disorders (Fang et al. 2020). Some variants observed within the entry complex of SARS-CoV-2, especially ACE-2, were suggested to work as boosters of spike protein binding to cells, especially in men, including many rare variants not frequently seen in the general population (Darbani 2020). Variants in ACE-2 were previously reported to be associated with hypertension and cardiac function (Patel et al. 2012).

A causal association between hypertension and COVID-19 is not clearly understood. COVID-19 could impair homeostasis of blood pressure allowing deflagration or aggravation of previous hypertension. The infection-derived vascular and cardiac injuries resulting in the extreme form of cardiogenic or septic shock are believed to lower the systolic blood pressure (Kang et al. 2020). However, a considerable proportion of patients with COVID-19 develop acute kidney injury (AKI) (36.6%) (Hirsch et al. 2020), which may cause secondary hypertension. Moreover, overactivity of the angiotensin-2/ATR1 pathway increases the downstream signaling and can potentially increase blood pressure through salt and water retention, as well as arterial stiffening. Another possible interaction would be the stimulation of the vicious cycle of inflammation-hypertension (previously explained) by the host response to SARS-CoV-2.

One consequence of COVID-19, or any other major pandemic, regards the metabolic outcomes of confinement and the consequent sedentary lifestyle. Confinement status can affect diet, physical activity, medication compliance, and access to regular medical provider visits and thereby could cause uncontrolled hypertension (Martinez-Ferran et al. 2020).

15.4 Antihypertensive Drugs and COVID-19

The early finding of a high prevalence of COVID-19 among hypertensive patients prompted researchers to investigate if the antihypertensive medications are mediating this association. A comprehensive retrospective study on 12,594 patients tested for COVID-19 in New York showed that among tested patients, there was no significant difference between patients under treatment with any of the five types of antihypertensive agents and their matched non-treated patients for hypertension and other possible confounders, in being diagnosed with COVID-19 (Reynolds et al. 2020). The five groups of medications studied were ACEI, ARB, beta-blockers, thiazide diuretics, and calcium channel blockers. Also, the risk of developing a severe form of the disease was not different between patients with or without treatment. Similarly, a study on hypertensive patients reported that although hypertensive COVID-19 patients are more likely to need critical care or mechanical ventilation compared to normotensive patients, there is no difference in risk between those on ACEI/ARB treatment and those without (Hu et al. 2020).

Doubt lies over the importance of hypertension itself as well as drugs used to treat hypertension, i.e., ACEIs and ARBs, on the incidence and severity of COVID-19. ACEI and ARBs upregulate ACE-2 receptors in animal models (Ishiyama et al. 2004; Ferrario et al. 2005), but it is not clear if this is also true in humans. Some cross-sectional studies in different clinical populations have demonstrated the absence of difference in plasmatic ACE-2 activity or urinary ACE-2 levels related to the use of these drugs (Epelman et al. 2009; Walters et al. 2017; Ramchand et al. 2018). Also, soluble ACE-2 detected in plasma and urine may not reflect precisely transmembrane ACE-2 at the tissue level. Thus, an increase in the plasma level of ACE-2 may not correlate with an augmentation of ACE-2 receptor activity, the actual point of virus action. Furthermore, to date, there seems to be no data regarding the effects of ACEI or ARB on the expression of ACE-2 receptors by type 2 pneumocytes, the primary target of the virus. It should also be noted that any significant change in the level of these molecules and activation of pathways (either of ACE-2 and angiotensin 1–7 or ACE and angiotensin-2) requires chronic use of this group of drugs (Luque et al. 1996).

On the other hand, increased expression of the ACE homolog, ACE-2, which presents a higher affinity to angiotensin-2, could be beneficial. Angiotensin-2 triggers the release of the soluble domain of ACE-2 into plasma. On its turn, soluble ACE-2 could sequester SARS-CoV-2 within the circulation and impede viral attachment and entry to cells. Circulating ACE-2 also mediates the cleavage of angiotensin-2 to angiotensin 1–7 and alamandine. These autacoids, angiotensin 1–7 and alamandine, have anti-inflammatory effects on the liver, lungs, and cardiovascular system through their interaction with the Mas receptor and Mas-related G-protein receptor pathway. In addition to anti-inflammatory effects, angiotensin 1–7 has protective effects against thrombosis and fibrosis, two other complications of COVID-19 infection (Simões e Silva et al. 2013; Passos-Silva et al. 2015). The promoted increase of ACE-2 by ACEI or ARB could have a protective effect considering that ACE-2 cleaves angiotensin-2, a molecule known to cause vasoconstriction, inflammation, and fibrosis and to worsen lung injury leading to ARDS, as demonstrated with other coronaviruses (Kuba et al. 2005). As opposed to angiotensin-2, angiotensin 1–7/mas receptor signaling blocks all alveolar epithelial cell apoptotic signals and thereby promotes epithelium survival through upregulation of MKP-2 phosphatase (Gopallawa and Uhal 2016).

ARBs are known to antagonize the activity of angiopoietin-2 and could be used to treat host response, although with no direct effect on viral entry. The balance of angiopoietin-1 and angiopoietin-2 is theoretically known to contribute to the integrity of the endothelium layer. Angiopoietin-1/Tie2 downstream signaling suppresses angiopoietin-2 expression. However, pro-inflammatory cytokines accumulated in acute illness trigger the release of angiopoietin-2 from Weibel-Palade bodies, inhibiting Tie2 signaling and further stimulating expression of angiopoietin-2 (Fedson 2016). Increased level of angiopoietin-2 is associated with higher mortality in sepsis patients and accompanied by severe acute lung injury (Calfee et al. 2012). Like in other viral outbreaks, RAAS inhibitors, especially ARBs, act in favor of maintaining the endothelial barrier from acute damage caused by respiratory viruses (Saavedra 2020).

Due to the lack of substantial evidence, the current clinical decision has relied on information from case series and observational studies. A population-based case-control study performed on 6272 patients from Lombardy (Northern Italy) revealed no increased relationship between the use of RAAS blockers and susceptibility of COVID-19 or any other antihypertensive drug (Mancia et al. 2020). Similar results were reported from a cohort of New York patients who underwent COVID-19 testing (n = 12,594) (Reynolds et al. 2020). In another recent retrospective cohort of 18,472 participants tested for COVID-19 carried out in Florida and Ohio (United States), 12.4% of the total population was taking either ACEI or ARBs. The use of these drugs was not associated with the likelihood of having a positive test for COVID-19. However, in secondary and exploratory analysis, results suggested a possible association with worse outcomes for those who tested positive, namely, hospital admissions, ICU admissions, or mechanical ventilation necessity. When weighted by propensity score for the presence of underlying diseases, the association remained significant for hospital admissions and the use of either ACEIs or ARBs.

Moreover, a higher likelihood of ICU admissions was observed for those patients taking ACEI. The authors highlighted that due to the study's observational nature, general conclusions could not be drawn (Mehta et al. 2020). One case series from China corroborates with those findings. The authors demonstrated an unadjusted association of ACEI/ARB use and increased troponin, a marker of severity of the disease, but no association with mortality (Guo et al. 2020).

The controversy increases as new studies are published. A recent multicenter and retrospective survey pointed in the opposite direction. In this cohort of 1128 hypertensive and hospitalized COVID-19 patients (Hubei, China), a group of 188 taking ACEI/ARB had reduced all-cause mortality compared to the group not taking ACEI/ARB in a median follow-up of 28 days. The unadjusted results were similar to the propensity score-matched analysis (in a 1:2 matching ratio) (adjusted HR, 0.37; 95% CI, 0.15–0.89; $p = 0.03$). In a further subgroup propensity score-matched analysis, ACEI/ARB use is associated with a reduced risk of mortality compared with the use of other antihypertensive drugs (n = 745) (adjusted HR, 0.29; 95%CI, 0.12–0.69; $p = 0.01$). The use of ACEI (OR, 0.33; 95% CI, 0.20 to 0.54), but not ARB, was associated with reduced inhospital mortality in an international registry (n = 8910) of 169 hospitals and three continents (Asia, Europe, and North America). These analyses were adjusted for age and sex, repeated for the subgroup of hypertensive patients, and adjusted for statin use (which also associated with reduced mortality), but results remained consistent. While these data cannot confirm that these classes of drugs reduce mortality during COVID-19 infection, they at least suggest that their use does not increase mortality. Altogether, these results likely reflect the fact that there is a greater proportion of patients with underlying comorbidities and cardiovascular disease among the COVID-19 cases, which is a consistent finding throughout studies.

The first RCTs on COVID-19 are beginning to be published. A non-pre-specified interim analysis of the RASTAVI trial assessed the effect of ramipril on the risk incidence or severity of

COVID-19. One-hundred and two patients undergoing transcatheter aortic valve replacement were allocated to ramipril (n = 50) or control groups (n = 52). The results showed no difference in any of the outcomes (incidence or severity). Eleven patients (10.8%) were diagnosed with COVID-19 (six in the control group and five in the ramipril group, HR = 1.150 [95% CI, 0.351–3.768]) (Amat-Santos et al. 2020). Many other clinical trials are on the pipeline. They involve the evaluation of the effects of not only ACEI and ARB in the treatment of COVID-19 but also spironolactone, another RAAS inhibitor, and other antihypertensive treatments, including calcium channel blockers and alpha-1 receptor antagonists (Table 15.1).

So far, data have shown that ACEI/ARB use does not lead to a higher mortality rate, although the relationship with hospital and ICU admissions is debatable. Also, discontinuation or change in antihypertensive drug should not be recommended due to possible consequent harm. Any escalation in the level of medications in patients under chronic treatment can cause an imbalance in blood pressure control culminating with rebound hypertension or hypotensive states due to unadjusted dose of alternative drugs, which could aggravate the hemodynamic changes brought by infection (Vaduganathan et al. 2020; Zhang et al.).

15.5 Guideline Recommendations

Although there is still debate in the role of history of ACEI/ARB use in COVID-19 infection risk or progression, current guidelines recommend continuation but not initiation (unless otherwise implicated) of this group of drugs in patients with COVID-19.

15.5.1 European Society of Cardiology: Council on Hypertension

As of March 13, 2020, the ESC council on hypertension stated that given the lack of strong evidence supporting the harm or benefit of ACEI/ARB use in COVID-19 infection or progression, patients who are on these medications should continue their antihypertensive therapy. Abruption in ACEI/ARB use for the prevention of disease or initiation of the drug as a preventive means of inflammatory injuries in newly diagnosed COVID-19 patients is not recommended by this society (De Simone 2020).

15.5.2 European Society of Hypertension

On April 15, 2020, the ESC COVID-19 task force released a statement on the conditional use of ACEI/ARB blockers in COVID-19 patients. According to ESH, "There is no clear evidence that hypertension *per se* is associated with an increased risk of infection by COVID-19 and the available evidence did not support a deleterious effect of RAS blockers in COVID-19 infections," and thus same predictive strategies should be applied in hypertensive patients based on age and other comorbidities. Moreover, in stable infected patients, discontinuation of medication is not recommended. However, in patients with hemodynamic instability, a case-based decision on the adjustment of blood pressure-lowering agents is necessary.

15.5.3 The American Heart Association, American College of Cardiology, and Heart Failure Society of America

A joint statement released on March 17, 2020, HFSA/AHA/ACC prioritized the use of the highest standard of care for patients with cardiovascular diseases. Therefore, previously administered RAAS antagonists are recommended to be continued in patients with underlying hypertension. However, although the addition or discontinuation of RAAS antagonists is not advised in patients with COVID-19, they have highlighted individualized therapeutic decision-making based on hemodynamic status and clinical pre-

Table 15.1 Ongoing clinical trials as of mid-2020 according to ClinicalTrials.gov

Title	Interventions	Sponsor/collaborators	Phases	Enrollment	Study type	Locations	URL
Hydroxychloroquine or Diltiazem-Niclosamide for the Treatment of COVID-19	Other: standard of care (SOC)|drug, hydroxychloroquine|drug, association of diltiazem and niclosamide	University Hospital, Lille|I-site University Lille North Europe	Phase 3	480	Interventional		NCT04372082
Ramipril for the Treatment of COVID-19	Drug: ramipril 2.5 MG Oral capsule|drug: placebo oral capsule	University of California, San Diego|Pfizer	Phase 2	560	Interventional		NCT04366050
Prazosin to Prevent COVID-19 (PREVENT-COVID Trial)	Drug: prazosin|other, standard of care	Johns Hopkins University|Fast Grants	Phase 2	220	Interventional	Johns Hopkins Hospital, Baltimore, Maryland, United States	NCT04365257
Pilot Clinical Trial of the Safety and Efficacy of Telmisartan for the Mitigation of Pulmonary and Cardiac Complications in COVID-19 Patients	Drug: telmisartan 40 mg|drug, placebo	University of Hawaii	Phase 2	40	Interventional	University of Hawaii -Manoa, John A Burns School of Medicine UH Clinics at Kakaako, Honolulu, Hawaii, United States	NCT04360551
Efficacy of Hydroxychloroquine, Telmisartan, and Azithromycin on the Survival of Hospitalized Elderly Patients with COVID-19	Drug: hydroxychloroquine|drug, azithromycin|drug, telmisartan	University Hospital, Strasbourg, France	Phase 3	1600	Interventional	CHU de Strasbourg, Strasbourg, France	NCT04359953
Prognosis of Coronavirus Disease 2019 (COVID-19) Patients Receiving Antihypertensives	Drug: angiotensin-converting enzyme inhibitors (ACE-I) and angiotensin II receptor blockers (ARBs)	Hakeam Abdulaziz Hakeam|Buraidah Central Hospital|King Khalid University Hospital|King Faisal Specialist Hospital & Research Center		226	Observational	King Faisal Specialist Hospital & Research Centre, Riyadh, Saudi Arabia	NCT04357535

Telmisartan for Treatment of COVID-19 Patients	Drug: telmisartan arm will receive 80 mg Telmisartan twice daily plus standard care	Laboratorio Elea Phoenix S.A.	Phase 2	400	Interventional	Hospital de Cl√≠nicas 'Jos√© de San Mart√≠n', Facultad de Medicina, Universidad de Buenos Aires, Buenos Aires, Ciudad De Buenos Aires, Argentina	NCT04355936
Efficacy of Captopril in Covid-19 Patients with Severe Acute Respiratory Syndrome (SARS) CoV-2 Pneumonia (CAPTOCOVID)	Drug: captopril 25 mg	Assistance Publique - Hopitaux de Paris	Phase 2	230	Interventional	CH Victor Dupuy- Argenteuil, Argenteuil, France|H√¥pital Avicenne, Bobigny, France|H√¥pital Avicenne, Bobigny, France|H√¥pital Avicenne, Bobigny, France|H√¥pital Antoine B√©cl√®re, Clamart, France|CH de Compi√®gne-Noyon, Compi√®gne, France|Groupe hospitalier Sud Ile de France, Melun, France|H√¥pital de la Piti√©- Salp√™tri√®re, Paris, France|H√¥pital Tenon, Paris, France|CHRU de Tours, H√¥pital Bretonneau, Tours, France|H√¥pital de Tours, Tours, France	NCT04355429
Amiodarone or Verapamil in COVID-19 Hospitalized Patients with Symptoms	Drug: amiodarone|drug, verapamil	Nicolaus Copernicus University	Phase 2|phase 3	804	Interventional	Nicolaus Copernicus University, Bydgoszcz, Poland	NCT04351763

(continued)

Table 15.1 (continued)

<table>
<tr><th>Title</th><th>Interventions</th><th>Sponsor/collaborators</th><th>Phases</th><th>Enrollment</th><th>Study type</th><th>Locations</th><th>URL</th></tr>
<tr><td>Austrian Corona Virus Adaptive Clinical Trial (COVID-19)</td><td>Drug: chloroquine or hydroxychloroquine|drug, lopinavir/ritonavir|other, best standard of care|drug, rivaroxaban|drug, thromboprophylaxis|drug, candesartan|drug, non-RAS blocking antihypertensives|drug, clazakizumab|drug, placebo for clazakizumab</td><td>Medical University of Vienna|Kaiser Franz Josef Hospital|SMZ-Ost Donauspital|Otto Wagner Hospital|Hospital Hietzing|Wilhelminenspital Vienna|Medical University Innsbruck</td><td>Phase 2|phase 3</td><td>500</td><td>Interventional</td><td>Medical University of Innsbruck, Innsbruck, Tirol, Austria|Medical University of Vienna, Vienna, Austria|Wilhelminenspital, Vienna, Austria|SMZ S√°d Kaiser Franz Josef Spital, Vienna, Austria|KH Hietzing, Vienna, Austria|SMZ Baumgartner H√∂he Otto Wagner Spital, Vienna, Austria|SMZ Ost Donauspital, Vienna, Austria</td><td>NCT04351724</td></tr>
<tr><td>Spironolactone in COVID-19 Induced ARDS</td><td>Drug: spironolactone 100 mg|drug, placebo oral tablet</td><td>Istanbul University-Cerrahpasa</td><td>Phase 4</td><td>60</td><td>Interventional</td><td>Istanbul University-Cerrahpa≈üa, Istanbul, Turkey</td><td>NCT04345887</td></tr>
<tr><td>Angiotensin-Converting Enzyme Inhibitors in Treatment of COVID-19</td><td>Drug: ACEIs|drug, conventional treatment</td><td>Tanta University</td><td>Phase 3</td><td>60</td><td>Interventional</td><td>Tanta University, Tanta, Egypt</td><td>NCT04345406</td></tr>
<tr><td>Do Angiotensin Receptor Blockers Mitigate Progression to Acute Respiratory Distress Syndrome with SARS-CoV-2 Infection</td><td>Drug: losartan</td><td>Sharp HealthCare</td><td>Phase 4</td><td>200</td><td>Interventional</td><td>Sharp Grossmont Hospital, La Mesa, California, United States|Sharp Chula Vista Medical Center, San Diego, California, United States|Sharp Coronado Hospital, San Diego, California, United States|Sharp Memorial Hospital, San Diego, California, United States</td><td>NCT04340557</td></tr>
</table>

Valsartan for Prevention of Acute Respiratory Distress Syndrome in Hospitalized Patients with SARS-COV-2 (COVID-19) Infection Disease	Drug: valsartan (Diovan)|drug, placebo oral tablet	Radboud University	Phase 4	651	Interventional	Jeroen Bosch Ziekenhuis, ‘s-Hertogenbosch, Netherlands|Rijnstate, Arnhem, Netherlands	NCT04335786
Study of Open-Label Losartan in COVID-19	Drug: losartan	University of Kansas Medical Center	Phase 1	50	Interventional	University of Kansas Medical Center, Kansas City, Kansas, United States	NCT04335123
Renin-Angiotensin System Inhibitors and COVID-19		Societa Italiana dell'Ipertensione Arteriosa		2000	Observational	Spedali Civili di Brescia, Brescia, Italy	NCT04331574
Coronavirus (COVID-19) ACEI/ ARB Investigation	Drug: thiazide or thiazide-like diuretics|drug, calcium channel blockers|drug, ACE inhibitor|drug, angiotensin receptor blocker	National University of Ireland, Galway, Ireland	Phase 4	2414	Interventional	University Hospital Galway, Galway, Ireland	NCT04330300
Losartan for Patients with COVID-19 Requiring Hospitalization	Drug: losartan|other, placebo	University of Minnesota|Bill and Melinda Gates Foundation	Phase 2	200	Interventional	Hennepin County Medical Center, Minneapolis, Minnesota, United States|M Health Fairview University of Minnesota Medical Center, Minneapolis, Minnesota, United States|University of Minnesota, Minneapolis, Minnesota, United States	NCT04312009

(continued)

Table 15.1 (continued)

Title	Interventions	Sponsor/collaborators	Phases	Enrollment	Study type	Locations	URL
Losartan for Patients with COVID-19 NOT Requiring Hospitalization	Drug: losartan|other, placebo	University of Minnesota|Bill and Melinda Gates Foundation	Phase 2	580	Interventional	Hennepin County Medical Center, Minneapolis, Minnesota, United States|M Health Fairview University of Minnesota Medical Center, Minneapolis, Minnesota, United States|University of Minnesota, Minneapolis, Minnesota, United States	NCT04311177

sentations of each admitted patient (Bozkurt et al. 2020). As of April 24, the latest updates in cardiology released by ACC further emphasized the necessity to continue ACEI/ARB in patients with COVID-19 and underlying hypertension, referring to a more recent retrospective study that showed a reduced risk of all-cause mortality in patients admitted for COVID-19 (Zhang et al. 2020).

15.6 Conclusion

The relationship between hypertension and the new coronavirus disease seems far more complicated than could be understood by the first observational publications. Thus far, evidence indicates that hypertension does not increase the incidence of COVID-19 but might be related to increased mortality due to the disease. The paucity of RCTs on antihypertensive use also precludes definite conclusions on evidence-driven clinical decisions. Based on the best data available, the continuation of antihypertensive drugs should be encouraged in patients under chronic therapy. Change of dosage or drug suspension should be done on case-based need or due to hemodynamic instability. There is still controversy on whether the benefits of RAAS inhibitors outweigh their harm. The initiation of RAAS inhibitors is not currently recommended to attenuate the acute course of COVID-19. However, a plethora of clinical trials is on the way to address the use of antihypertensives in general and RAAS inhibitors specifically. Until then, clinical decisions should be based on the best evidence available and a case-by-case setting.

References

Amat-Santos IJ, Santos-Martinez S, Lopez-Otero D, Nombela-Franco L, Gutierrez-Ibanes E, Del Valle R, Munoz-Garcia E, Jimenez-Diaz VA, Regueiro A, Gonzalez-Ferreiro R, Benito T, Sanmartin-Pena XC, Catala P, Rodriguez-Gabella T, Delgado-Arana JR, Carrasco-Moraleja M, Ibanez B, San Roman JA (2020) Ramipril in High Risk Patients with COVID-19. J Am Coll Cardiol. doi:https://doi.org/10.1016/j.jacc.2020.05.040

Badawi A, Ryoo SG (2016) Prevalence of comorbidities in the Middle East respiratory syndrome coronavirus (MERS-CoV): a systematic review and meta-analysis. Int J Infect Dis 49:129–133. https://doi.org/10.1016/j.ijid.2016.06.015

Bello-Chavolla OY, Bahena-López JP, Antonio-Villa NE, Vargas-Vázquez A, González-Díaz A, Márquez-Salinas A, Fermín-Martínez CA, Naveja JJ, Aguilar-Salinas CA (2020) Predicting mortality due to SARS-CoV-2: A mechanistic score relating obesity and diabetes to COVID-19 outcomes in Mexico. J Clin Endocrinol Metab. https://doi.org/10.1210/clinem/dgaa346

Bozkurt B, Kovacs R, Harrington B (2020) HFSA/ACC/AHA statement addresses concerns Re: using RAAS antagonists in COVID-19. American College of Cardiology (ACC)

Caldeira D, Alarcão J, Vaz-Carneiro A, Costa J (2012) Risk of pneumonia associated with use of angiotensin converting enzyme inhibitors and angiotensin receptor blockers: systematic review and meta-analysis. BMJ 345:e4260. https://doi.org/10.1136/bmj.e4260

Calfee CS, Gallagher D, Abbott J, Thompson BT, Matthay MA (2012) Plasma angiopoietin-2 in clinical acute lung injury: prognostic and pathogenetic significance. Crit Care Med 40(6):1731–1737. https://doi.org/10.1097/CCM.0b013e3182451c87

Cariou B, Hadjadj S, Wargny M, Pichelin M, Al-Salameh A, Allix I, Amadou C, Arnault G, Baudoux F, Bauduceau B, Borot S, Bourgeon-Ghittori M, Bourron O, Boutoille D, Cazenave-Roblot F, Chaumeil C, Cosson E, Coudol S, Darmon P, Disse E, Ducet-Boiffard A, Gaborit B, Joubert M, Kerlan V, Laviolle B, Marchand L, Meyer L, Potier L, Prevost G, Riveline JP, Robert R, Saulnier PJ, Sultan A, Thébaut JF, Thivolet C, Tramunt B, Vatier C, Roussel R, Gautier JF, Gourdy P (2020) Phenotypic characteristics and prognosis of inpatients with COVID-19 and diabetes: the CORONADO study. Diabetologia:1–16. https://doi.org/10.1007/s00125-020-05180-x

Chen D, Li X, Song Q, Hu C, Su F, Dai J (2020) Hypokalemia and clinical implications in patients with coronavirus disease 2019 (COVID-19). med Rxiv:2020.2002.2027.20028530. doi:https://doi.org/10.1101/2020.02.27.20028530

Cheng Y, Luo R, Wang K, Zhang M, Wang Z, Dong L, Li J, Yao Y, Ge S, Xu G (2020) Kidney disease is associated with in-hospital death of patients with COVID-19. Kidney Int 97(5):829–838. https://doi.org/10.1016/j.kint.2020.03.005

Darbani B (2020) The expression and polymorphism of entry machinery for COVID-19 in human: juxtaposing population groups, gender, and different tissues. Int J Environ Res Public Health 17(10). https://doi.org/10.3390/ijerph17103433

De Simone G (2020) Position statement of the ESC council on hypertension on ACE-inhibitors and angiotensin receptor blockers. European society of cardiology, Council on hypertension,

Drummond GR, Vinh A, Guzik TJ, Sobey CG (2019) Immune mechanisms of hypertension. Nature Rev Immunol 19(8):517–532. https://doi.org/10.1038/s41577-019-0160-5

Epelman S, Shrestha K, Troughton RW, Francis GS, Sen S, Klein AL, Tang WH (2009) Soluble angiotensin-converting enzyme 2 in human heart failure: relation with myocardial function and clinical outcomes. J Card Fail 15(7):565–571. https://doi.org/10.1016/j.cardfail.2009.01.014

Fang L, Karakiulakis G, Roth M (2020) Are patients with hypertension and diabetes mellitus at increased risk for COVID-19 infection? Lancet Respir Med 8(4):e21. https://doi.org/10.1016/s2213-2600(20)30116-8

Fedson DS (2016) Treating the host response to emerging virus diseases: lessons learned from sepsis, pneumonia, influenza and Ebola. Ann Transl Med 4(21):421–421. https://doi.org/10.21037/atm.2016.11.03

Ferrario CM, Jessup J, Chappell MC, Averill DB, Brosnihan KB, Tallant EA, Diz DI, Gallagher PE (2005) Effect of angiotensin-converting enzyme inhibition and angiotensin II receptor blockers on cardiac angiotensin-converting enzyme 2. Circulation 111(20):2605–2610. https://doi.org/10.1161/CIRCULATIONAHA.104.510461

Flammer AJ, Anderson T, Celermajer DS, Creager MA, Deanfield J, Ganz P, Hamburg NM, Luscher TF, Shechter M, Taddei S, Vita JA, Lerman A (2012) The assessment of endothelial function: from research into clinical practice. Circulation 126(6):753–767. https://doi.org/10.1161/CIRCULATIONAHA.112.093245

Forouzanfar MH, Liu P, Roth GA, Ng M, Biryukov S, Marczak L, Alexander L, Estep K, Hassen Abate K, Akinyemiju TF, Ali R, Alvis-Guzman N, Azzopardi P, Banerjee A, Barnighausen T, Basu A, Bekele T, Bennett DA, Biadgilign S, Catala-Lopez F, Feigin VL, Fernandes JC, Fischer F, Gebru AA, Gona P, Gupta R, Hankey GJ, Jonas JB, Judd SE, Khang YH, Khosravi A, Kim YJ, Kimokoti RW, Kokubo Y, Kolte D, Lopez A, Lotufo PA, Malekzadeh R, Melaku YA, Mensah GA, Misganaw A, Mokdad AH, Moran AE, Nawaz H, Neal B, Ngalesoni FN, Ohkubo T, Pourmalek F, Rafay A, Rai RK, Rojas-Rueda D, Sampson UK, Santos IS, Sawhney M, Schutte AE, Sepanlou SG, Shifa GT, Shiue I, Tedla BA, Thrift AG, Tonelli M, Truelsen T, Tsilimparis N, Ukwaja KN, Uthman OA, Vasankari T, Venketasubramanian N, Vlassov VV, Vos T, Westerman R, Yan LL, Yano Y, Yonemoto N, Zaki ME, Murray CJ (2017) Global burden of hypertension and systolic blood pressure of at least 110 to 115 mm Hg, 1990–2015. JAMA 317(2):165–182. https://doi.org/10.1001/jama.2016.19043

Fox SE, Akmatbekov A, Harbert JL, Li G, Quincy Brown J, Vander Heide RS (2020) Pulmonary and cardiac pathology in African American patients with COVID-19: an autopsy series from New Orleans. Lancet Respir Med. https://doi.org/10.1016/S2213-2600(20)30243-5

Galloway JB, Norton S, Barker RD, Brookes A, Carey I, Clarke BD, Jina R, Reid C, Russell MD, Sneep R, Sugarman L, Williams S, Yates M, Teo J, Shah AM, Cantle F (2020) A clinical risk score to identify patients with COVID-19 at high risk of critical care admission or death: an observational cohort study. J Infect. https://doi.org/10.1016/j.jinf.2020.05.064

Gao C, Cai Y, Zhang K, Zhou L, Zhang Y, Zhang X, Li Q, Li W, Yang S, Zhao X, Zhao Y, Wang H, Liu Y, Yin Z, Zhang R, Wang R, Yang M, Hui C, Wijns W, McEvoy JW, Soliman O, Onuma Y, Serruys PW, Tao L, Li F (2020) Association of hypertension and antihypertensive treatment with COVID-19 mortality: a retrospective observational study. Eur Heart J. https://doi.org/10.1093/eurheartj/ehaa433

Gopallawa I, Uhal BD (2016) Angiotensin-(1–7)/mas inhibits apoptosis in alveolar epithelial cells through upregulation of MAP kinase phosphatase-2. Am J Physiol Lung Cell Mol Physiol 310(3):L240–L248. https://doi.org/10.1152/ajplung.00187.2015

Grasselli G, Zangrillo A, Zanella A, Antonelli M, Cabrini L, Castelli A, Cereda D, Coluccello A, Foti G, Fumagalli R, Iotti G, Latronico N, Lorini L, Merler S, Natalini G, Piatti A, Ranieri MV, Scandroglio AM, Storti E, Cecconi M, Pesenti A (2020) Baseline characteristics and outcomes of 1591 patients infected with SARS-CoV-2 admitted to ICUs of the Lombardy Region, Italy. JAMA. https://doi.org/10.1001/jama.2020.5394

Guan WJ, Ni ZY, Hu Y, Liang WH, Ou CQ, He JX, Liu L, Shan H, Lei CL, Hui DSC, Du B, Li LJ, Zeng G, Yuen KY, Chen RC, Tang CL, Wang T, Chen PY, Xiang J, Li SY, Wang JL, Liang ZJ, Peng YX, Wei L, Liu Y, Hu YH, Peng P, Wang JM, Liu JY, Chen Z, Li G, Zheng ZJ, Qiu SQ, Luo J, Ye CJ, Zhu SY, Zhong NS (2020) Clinical characteristics of coronavirus disease 2019 in China. N Engl J Med. https://doi.org/10.1056/NEJMoa2002032

Guo T, Fan Y, Chen M, Wu X, Zhang L, He T, Wang H, Wan J, Wang X, Lu Z (2020) Cardiovascular implications of fatal outcomes of patients with coronavirus disease 2019 (COVID-19). JAMA Cardiol. https://doi.org/10.1001/jamacardio.2020.1017

Henry C, Zaizafoun M, Stock E, Ghamande S, Arroliga AC, White HD (2018) Impact of angiotensin-converting enzyme inhibitors and statins on viral pneumonia. Proceedings (Baylor University Medical Center) 31(4):419–423. https://doi.org/10.1080/08998280.2018.1499293

Hirsch JS, Ng JH, Ross DW, Sharma P, Shah HH, Barnett RL, Hazzan AD, Fishbane S, Jhaveri KD (2020) Acute kidney injury in patients hospitalized with COVID-19. Kidney Int. https://doi.org/10.1016/j.kint.2020.05.006

Hoffmann M, Kleine-Weber H, Krüger N, Müller M, Drosten C, Pöhlmann S (2020) The novel coronavirus 2019 (2019-nCoV) uses the SARS-coronavirus receptor ACE2 and the cellular protease TMPRSS2 for entry into target cells. bioRxiv:2020.2001.2031.929042. https://doi.org/10.1101/2020.01.31.929042

Hu J, Zhang X, Zhang X, Zhao H, Lian J, Hao S, Jia H, Yang M, Lu Y, Xiang D, Cai H, Zhang S, Gu J, Ye C, Yu G, Jin C, Zheng L, Yang Y, Sheng J (2020) COVID-19 patients with hypertension have more severity condition, and ACEI/ARB treatment have no influence on the clinical severity and outcome. J Infect. https://doi.org/10.1016/j.jinf.2020.05.056

Huan T, Esko T, Peters MJ, Pilling LC, Schramm K, Schurmann C, Chen BH, Liu C, Joehanes R, Johnson AD, Yao C, Ying SX, Courchesne P, Milani L, Raghavachari N, Wang R, Liu P, Reinmaa E, Dehghan A, Hofman A, Uitterlinden AG, Hernandez DG, Bandinelli S, Singleton A, Melzer D, Metspalu A, Carstensen M, Grallert H, Herder C, Meitinger T, Peters A, Roden M, Waldenberger M, Dörr M, Felix SB, Zeller T, Vasan R, O'Donnell CJ, Munson PJ, Yang X, Prokisch H, Völker U, van Meurs JB, Ferrucci L, Levy D (2015) A meta-analysis of gene expression signatures of blood pressure and hypertension. PLoS Genet 11(3):e1005035. https://doi.org/10.1371/journal.pgen.1005035

Huang C, Wang Y, Li X, Ren L, Zhao J, Hu Y, Zhang L, Fan G, Xu J, Gu X, Cheng Z, Yu T, Xia J, Wei Y, Wu W, Xie X, Yin W, Li H, Liu M, Xiao Y, Gao H, Guo L, Xie J, Wang G, Jiang R, Gao Z, Jin Q, Wang J, Cao B (2020a) Clinical features of patients infected with 2019 novel coronavirus in Wuhan, China. Lancet (London, England) 395(10223):497–506. https://doi.org/10.1016/S0140-6736(20)30183-5

Huang S, Wang J, Liu F, Liu J, Cao G, Yang C, Liu W, Tu C, Zhu M, Xiong B (2020b) COVID-19 patients with hypertension have more severe disease: a multi-center retrospective observational study. Hypertens Res. https://doi.org/10.1038/s41440-020-0485-2

Inciardi RM, Lupi L, Zaccone G, Italia L, Raffo M, Tomasoni D, Cani DS, Cerini M, Farina D, Gavazzi E, Maroldi R, Adamo M, Ammirati E, Sinagra G, Lombardi CM, Metra M (2020) Cardiac Involvement in a patient with coronavirus disease 2019 (COVID-19). JAMA Cardiol. https://doi.org/10.1001/jamacardio.2020.1096

Ishiyama Y, Gallagher PE, Averill DB, Tallant EA, Brosnihan KB, Ferrario CM (2004) Upregulation of angiotensin-converting enzyme 2 after myocardial infarction by blockade of angiotensin II receptors. Hypertension 43(5):970–976. https://doi.org/10.1161/01.HYP.0000124667.34652.1a

Kai H, Kai M (2020) Interactions of coronaviruses with ACE2, angiotensin II, and RAS inhibitors-lessons from available evidence and insights into COVID-19. Hypertens Res:1–7. https://doi.org/10.1038/s41440-020-0455-8

Kang Y, Chen T, Mui D, Ferrari V, Jagasia D, Scherrer-Crosbie M, Chen Y, Han Y (2020) Cardiovascular manifestations and treatment considerations in covid-19. Heart. https://doi.org/10.1136/heartjnl-2020-317,056

Koivula I, Sten M, Mäkelä PH (1994) Risk factors for pneumonia in the elderly. Am J Med 96(4):313–320. https://doi.org/10.1016/0002-9343(94)90060-4

Kuba K, Imai Y, Rao S, Gao H, Guo F, Guan B, Huan Y, Yang P, Zhang Y, Deng W, Bao L, Zhang B, Liu G, Wang Z, Chappell M, Liu Y, Zheng D, Leibbrandt A, Wada T, Slutsky AS, Liu D, Qin C, Jiang C, Penninger JM (2005) A crucial role of angiotensin converting enzyme 2 (ACE2) in SARS coronavirus-induced lung injury. Nat Med 11(8):875–879. https://doi.org/10.1038/nm1267

Lee JS, Chui PY, Ma HM, Auyeung TW, Kng C, Law T, Ng LK, Tam KF, Tang WH, Chan BY, Tong MC, Wong KT, Yuen YH, Yuk KL, Kwok T (2015) Does low dose angiotensin converting enzyme inhibitor prevent pneumonia in older people with neurologic dysphagia–a randomized placebo-controlled trial. J Am Med Dir Assoc 16(8):702–707. https://doi.org/10.1016/j.jamda.2015.05.009

Li B, Yang J, Zhao F, Zhi L, Wang X, Liu L, Bi Z, Zhao Y (2020) Prevalence and impact of cardiovascular metabolic diseases on COVID-19 in China. Clin Res Cardiol 109(5):531–538. https://doi.org/10.1007/s00392-020-01626-9

Luque M, Martin P, Martell N, Fernandez C, Brosnihan KB, Ferrario CM (1996) Effects of captopril related to increased levels of prostacyclin and angiotensin-(1–7) in essential hypertension. J Hypertens 14(6):799–805. https://doi.org/10.1097/00004872-199,606,000-00017

Mancia G, Rea F, Ludergnani M, Apolone G, Corrao G (2020) Renin-Angiotensin-Aldosterone System Blockers and the Risk of Covid-19. N Engl J Med. https://doi.org/10.1056/NEJMoa2006923

Mao L, Jin H, Wang M, Hu Y, Chen S, He Q, Chang J, Hong C, Zhou Y, Wang D, Miao X, Li Y, Hu B (2020) Neurologic manifestations of hospitalized patients with coronavirus disease 2019 in Wuhan, China. JAMA Neurol. https://doi.org/10.1001/jamaneurol.2020.1127

Martinez-Ferran M, de la Guía-Galipienso F, Sanchis-Gomar F, Pareja-Galeano H (2020) Metabolic impacts of confinement during the COVID-19 pandemic due to modified diet and physical activity habits. Nutrients 12 (6). doi:https://doi.org/10.3390/nu12061549

Mehta N, Kalra A, Nowacki AS, Anjewierden S, Han Z, Bhat P, Carmona-Rubio AE, Jacob M, Procop GW, Harrington S, Milinovich A, Svensson LG, Jehi L, Young JB, Chung MK (2020) Association of use of angiotensin-converting enzyme inhibitors and angiotensin II receptor blockers with testing positive for coronavirus disease 2019 (COVID-19). JAMA Cardiol. https://doi.org/10.1001/jamacardio.2020.1855

Mertz D, Kim TH, Johnstone J, Lam P-P, Science M, Kuster SP, Fadel SA, Tran D, Fernandez E, Bhatnagar N, Loeb M (2013) Populations at risk for severe or complicated influenza illness: systematic review and meta-analysis. BMJ 347:f5061–f5061. https://doi.org/10.1136/bmj.f5061

Mortensen EM, Nakashima B, Cornell J, Copeland LA, Pugh MJ, Anzueto A, Good C, Restrepo MI, Downs JR, Frei CR, Fine MJ (2012) Population-based study of statins, angiotensin II receptor blockers, and angiotensin-converting enzyme inhibitors on pneumonia-related outcomes. Clin Infect Dis 55(11):1466–1473. https://doi.org/10.1093/cid/cis733

Ohkubo T, Chapman N, Neal B, Woodward M, Omae T, Chalmers J, Perindopril Protection Against Recurrent Stroke Study Collaborative G (2004) Effects of an angiotensin-converting enzyme inhibitor-based regimen on pneumonia risk. Am J Respir Crit Care Med 169(9):1041–1045. https://doi.org/10.1164/rccm.200309-1219OC

Passos-Silva DG, Brandan E, Santos RA (2015) Angiotensins as therapeutic targets beyond heart disease. Trends Pharmacol Sci 36(5):310–320. https://doi.org/10.1016/j.tips.2015.03.001

Patel SK, Wai B, Ord M, MacIsaac RJ, Grant S, Velkoska E, Panagiotopoulos S, Jerums G, Srivastava PM, Burrell LM (2012) Association of ACE2 genetic variants with blood pressure, left ventricular mass, and cardiac function in Caucasians with type 2 diabetes. Am J Hypertens 25(2):216–222. https://doi.org/10.1038/ajh.2011.188

Ramchand J, Patel SK, Srivastava PM, Farouque O, Burrell LM (2018) Elevated plasma angiotensin converting enzyme 2 activity is an independent predictor of major adverse cardiac events in patients with obstructive coronary artery disease. PLoS One 13(6):e0198144. https://doi.org/10.1371/journal.pone.0198144

Reynolds HR, Adhikari S, Pulgarin C, Troxel AB, Iturrate E, Johnson SB, Hausvater A, Newman JD, Berger JS, Bangalore S, Katz SD, Fishman GI, Kunichoff D, Chen Y, Ogedegbe G, Hochman JS (2020) Renin-angiotensin-aldosterone system inhibitors and risk of Covid-19. N Engl J Med. https://doi.org/10.1056/NEJMoa2008975

Richardson S, Hirsch JS, Narasimhan M, Crawford JM, McGinn T, Davidson KW, Barnaby DP, Becker LB, Chelico JD, Cohen SL, Cookingham J, Coppa K, Diefenbach MA, Dominello AJ, Duer-Hefele J, Falzon L, Gitlin J, Hajizadeh N, Harvin TG, Hirschwerk DA, Kim EJ, Kozel ZM, Marrast LM, Mogavero JN, Osorio GA, Qiu M, Zanos TP (2020) Presenting characteristics, comorbidities, and outcomes among 5700 patients hospitalized With COVID-19 in the New York City Area. JAMA. https://doi.org/10.1001/jama.2020.6775

Rodriguez-Morales AJ, Cardona-Ospina JA, Gutierrez-Ocampo E, Villamizar-Pena R, Holguin-Rivera Y, Escalera-Antezana JP, Alvarado-Arnez LE, Bonilla-Aldana DK, Franco-Paredes C, Henao-Martinez AF, Paniz-Mondolfi A, Lagos-Grisales GJ, Ramirez-Vallejo E, Suarez JA, Zambrano LI, Villamil-Gomez WE, Balbin-Ramon GJ, Rabaan AA, Harapan H, Dhama K, Nishiura H, Kataoka H, Ahmad T, Sah R (2020) Clinical, laboratory and imaging features of COVID-19: a systematic review and meta-analysis. Travel Med Infect Dis:101623. https://doi.org/10.1016/j.tmaid.2020.101623

Ruan Q, Yang K, Wang W, Jiang L, Song J (2020) Clinical predictors of mortality due to COVID-19 based on an analysis of data of 150 patients from Wuhan, China. Intensive Care Med 46(5):846–848. https://doi.org/10.1007/s00134-020-05991-x

Saavedra JM (2020) Angiotensin receptor blockers and COVID-19. Pharmacol Res 156:104832. https://doi.org/10.1016/j.phrs.2020.104832

Schulte-Hubbert B, Meiswinkel N, Kutschan U, Kolditz M (2020) Prognostic value of blood pressure drops during the first 24 h after hospital admission for risk stratification of community-acquired pneumonia: a retrospective cohort study. Infection 48(2):267–274. https://doi.org/10.1007/s15010-020-01391-x

Shao Y, Saredy J, Yang WY, Sun Y, Lu Y, Saaoud F, Drummer C, Johnson C, Xu K, Jiang X, Wang H, Yang X (2020) Vascular endothelial cells and innate immunity. Arterioscler Thromb Vasc Biol 40(6):e138–e152. https://doi.org/10.1161/ATVBAHA.120.314330

Simões e Silva A, Silveira K, Ferreira A, Teixeira M (2013) ACE2, angiotensin-(1–7) and Mas receptor axis in inflammation and fibrosis. Br J Pharmacol 169(3):477–492. https://doi.org/10.1111/bph.12159

Vaduganathan M, Vardeny O, Michel T, McMurray JJV, Pfeffer MA, Solomon SD (2020) Renin–angiotensin–aldosterone system inhibitors in patients with Covid-19. New England Journal of Medicine 382(17):1653–1659. https://doi.org/10.1056/NEJMsr2005760

Varga Z, Flammer AJ, Steiger P, Haberecker M, Andermatt R, Zinkernagel AS, Mehra MR, Schuepbach RA, Ruschitzka F, Moch H (2020) Endothelial cell infection and endotheliitis in COVID-19. Lancet 395(10234):1417–1418. https://doi.org/10.1016/S0140-6736(20)30937-5

Walters TE, Kalman JM, Patel SK, Mearns M, Velkoska E, Burrell LM (2017) Angiotensin converting enzyme 2 activity and human atrial fibrillation: increased plasma angiotensin converting enzyme 2 activity is associated with atrial fibrillation and more advanced left atrial structural remodelling. Europace 19(8):1280–1287. https://doi.org/10.1093/europace/euw246

Wu C, Chen X, Cai Y, Xia J, Zhou X, Xu S, Huang H, Zhang L, Zhou X, Du C, Zhang Y, Song J, Wang S, Chao Y, Yang Z, Xu J, Zhou X, Chen D, Xiong W, Xu L, Zhou F, Jiang J, Bai C, Zheng J, Song Y (2020) Risk factors associated with acute respiratory distress syndrome and death in patients with coronavirus disease 2019 pneumonia in Wuhan, China. JAMA Intern Med. https://doi.org/10.1001/jamainternmed.2020.0994

Yang J, Zheng Y, Gou X, Pu K, Chen Z, Guo Q, Ji R, Wang H, Wang Y, Zhou Y (2020) Prevalence of comorbidities in the novel Wuhan coronavirus (COVID-19) infection: a systematic review and meta-analysis. Int J Infect Dis. https://doi.org/10.1016/j.ijid.2020.03.017

Zhang P, Zhu L, Cai J, Lei F, Qin J-J, Xie J, Liu Y-M, Zhao Y-C, Huang X, Lin L, Xia M, Chen M-M, Cheng X, Zhang X, Guo D, Peng Y, Ji Y-X, Chen J, She Z-G, Wang Y, Xu Q, Tan R, Wang H, Lin J, Luo P, Fu S, Cai H, Ye P, Xiao B, Mao W, Liu L, Yan Y, Liu M, Chen M, Zhang X-J, Wang X, Touyz RM, Xia J, Zhang B-H, Huang X, Yuan Y, Rohit L, Liu PP, Li H (2020) Association of inpatient use of angiotensin converting enzyme inhibitors and angiotensin ii receptor blockers with mortality among patients with hypertension hospitalized with COVID-19. Circ Res 126:1671–1681. doi:https://doi.org/10.1161/CIRCRESAHA.120.317134

Zhou F, Yu T, Du R, Fan G, Liu Y, Liu Z, Xiang J, Wang Y, Song B, Gu X, Guan L, Wei Y, Li H, Wu X, Xu J, Tu S, Zhang Y, Chen H, Cao B (2020) Clinical course and risk factors for mortality of adult inpatients with COVID-19 in Wuhan, China: a retrospective cohort study. Lancet 395(10229):1054–1062. https://doi.org/10.1016/s0140-6736(20)30566-3

Zou X, Chen K, Zou J, Han P, Hao J, Han Z (2020) Single-cell RNA-seq data analysis on the receptor ACE2 expression reveals the potential risk of different human organs vulnerable to 2019-nCoV infection. Front Med 14(2):185–192. https://doi.org/10.1007/s11684-020-0754-0

16 COVID-19 and Cardiovascular Diseases

Babak Geraiely, Niloufar Samiei, Parham Sadeghipour, Azita H. Talasaz, Seyedeh Hamideh Mortazavi, and Roya Sattarzadeh Badkoubeh

Abstract

We herein seek to expound on up-to-the-minute information regarding cardiovascular disease in the era of coronavirus disease 2019 (COVID-19) by highlighting acute myocardial injury caused by COVID-19 and probing into its pathophysiology, clinical signs, diagnostic tests, and treatment modalities. We aim to share the latest research findings vis-à-vis cardiovascular disease patients with confirmed or suspected COVID-19 on the association between hypertension and this infectious disease along with the relevant recommendations; describe the mechanism of coronary artery disease in such patients together with the necessary measures in the setting of non-ST-segment elevation acute coronary syndrome, ST-segment elevation myocardial infarction, and chronic coronary syndrome; discuss tachy- and bradyarrhythmias in the COVID-19 setting alongside their treatments; elucidate coagulopathies, venous thromboembolism, and its prophylactic measures in the context of this infection; set out the cardiopulmonary resuscitation protocol as well as the pertinent safety concerns during the current pandemic; and, finally, explicate drug-drug interactions between COVID-19 and cardiovascular medication in hypertension, acute coronary syndrome, heart failure, venous thromboembolism, and arrhythmias.

Keywords

Cardiovascular disease · COVID-19 · Outbreak · SARS-CoV-2

B. Geraiely · R. Sattarzadeh Badkoubeh (✉)
Department of Cardiology, Imam Khomeini Hospital Complex, Tehran University of Medical Sciences, Tehran, Iran
e-mail: satarzad@tums.ac.ir

N. Samiei
Heart Valve Disease Research Center, Rajaie Cardiovascular, Medical, and Research Center, Iran University of Medical Sciences, Tehran, Iran

P. Sadeghipour
Cardiovascular Intervention Research Center, Rajaie Cardiovascular, Medical, and Research Center, Iran University of Medical Sciences, Tehran, Iran

A. H. Talasaz · S. H. Mortazavi
Tehran Heart Center, Tehran University of Medical Sciences, Tehran, Iran

16.1 Introduction

Since the first report on coronavirus disease 2019 (COVID-19) in the Chinese city of Wuhan, a considerable number of articles on the cardiovascular manifestations of this infection and its impact on patients with cardiovascular disease have been published.

This chapter aims to review these articles with a focus on their applicable and practical points

N. Rezaei (ed.), *Coronavirus Disease - COVID-19*, Advances in Experimental Medicine and Biology 1318, https://doi.org/10.1007/978-3-030-63761-3_16

vis-à-vis the management of such patients. In addition, this chapter discusses the precautionary measures and changes proposed by internationally renowned cardiovascular societies with respect to the concerns about the intra-hospital spread of this infection and the resultant alterations in the treatment methods of non-COVID-19 patients with cardiovascular disease. This chapter also highlights key points concerning drug-drug interactions between COVID-19 and cardiovascular medications and thereafter introduces the cardiovascular effects of the pharmaceutical therapies applied thus far in the current epidemic.

Needless to say, given the novelty of the SARS-CoV-2 virus and, thus, the hitherto unknown aspects of it, a sizeable portion of the recommendations in the new guidelines and reported in this chapter are based on expert opinion and are liable to modifications and updates over time.

16.2 Acute Myocardial Injury

16.2.1 Introduction

Acute myocardial injury as myocarditis has been previously reported in studies on Middle East respiratory syndrome (MERS) resulting in acute heart failure (Alhogbani 2016).

The genomic proximity between Middle East respiratory syndrome coronavirus (MERS-CoV) and severe acute respiratory syndrome coronavirus 2 (SARS-CoV-2), which is the strain of the coronavirus that causes coronavirus disease 2019 (COVID-19), leads to the hypothesis that acute myocardial injury occurs through similar mechanisms in COVID-19, too. The role of angiotensin-converting enzyme 2 (ACE2) receptors as the main receptors in the pathophysiology of COVID-19 also supports the theory that the coronavirus can harm the myocardium since the latter abounds with ACE2 receptors (Zheng and Ma 2020).

Acute myocardial injury can lead to inadequate tissue perfusion and, accompanied by hypoxia secondary to pulmonary involvement, exacerbate the disease and further complicate hemodynamic control (2020 ESC Guidance for the diagnosis and management of CV Disease in the context of the COVID-19 Pandemic. Available at https://www.escardio.org/Education/COVID-19-and-Cardiology/ESC-COVID-19-Guidance 2020).

16.2.2 Definition

Myocardial injury is defined as at least a single measurement of cardiac troponin above the 99th percentile of the upper-reference limit (URL) (Thygesen et al. 2018). In some studies, this definition is propounded clearly irrespective of any change in electrocardiography (ECG) or evidence of myocardial injury in echocardiography (Shi et al. 2020).

16.2.3 Incidence

One of the earliest investigations to report the manifestation of myocarditis as acute myocardial injury during the COVID-19 outbreak was a study of 41 patients hospitalized in the Chinese city of Wuhan. The results showed elevated levels of high-sensitivity troponin I (hs-TnI) in five patients, of whom four presented with a more severe form of the infection and needed hospitalization in the intensive care unit (ICU) (Huang et al. 2020).

Another study in Wuhan on 416 patients reported elevated levels of hs-TnI in 82 (19.7%). These patients were older, had higher values of inflammatory markers, exhibited more severe lung involvement, had a higher need for mechanical respiratory ventilators, presented with more acute respiratory distress syndrome (ARDS) symptoms, and were more frequently complicated by renal complications, electrolyte disorders, and coagulopathy. Mortality was reported in 51% of these patients, whereas only 4.5% of those without this marker of myocardial injury expired (Shi et al. 2020). The incidence of myocardial injury has also been reported to range from 7% to 28% in some other studies (Lippi et al. 2020a; Bhatraju et al. 2020). The key point in the interpretation of these findings is that most of these investigations were conducted on hospitalized patients.

16.2.4 Pathophysiology

Various mechanisms have been posited for the occurrence of acute myocardial injury in COVID-19; they include the direct invasion of myocardial cells by the virus, stress-induced cardiomyopathy (Takotsubo cardiomyopathy), ischemic injury caused by atherosclerotic plaque rupture-induced Type 1 myocardial infarction (MI), cytokine storm-induced inflammatory response, and hypoxia created by severe lung involvement resulting in Type 2 MI (Fig. 16.1) (Tersalvi et al. 2020). Other salient postulated mechanisms are acute kidney injury, hypervolemia, and tachyarrhythmias (2020 ESC Guidance for the diagnosis and management of CV Disease in the context of the COVID-19 Pandemic. Available at https://www.escardio.org/Education/COVID-19-and-Cardiology/ESC-COVID-19-Guidance 2020).

The myocarditis hypothesis is reinforced by a few case reports. The presence of ACE2 receptors, which are the direct receptors of the coronavirus, in the heart further supports the theory of myocarditis as the cause of acute myocardial injury. Based on this postulation, the invasion of myocardial cells by the virus triggers an inflammatory response with the secretion of T cells, which in turn aggravates this vicious circle of cell injury (2020 ESC Guidance for the diagnosis and management of CV Disease in the context of the COVID-19 Pandemic. Available at https://www.escardio.org/Education/COVID-19-and-Cardiology/ESC-COVID-19-Guidance 2020).

A case report described a patient who presented with fulminant myocarditis and was treated with corticosteroids and human immunoglobulin. For this patient, who had cardiomegaly and bilateral pleural effusion in computed tomography (CT) scan, ST-segment elevation in ECG, an elevated troponin T level, and an ejection frac-

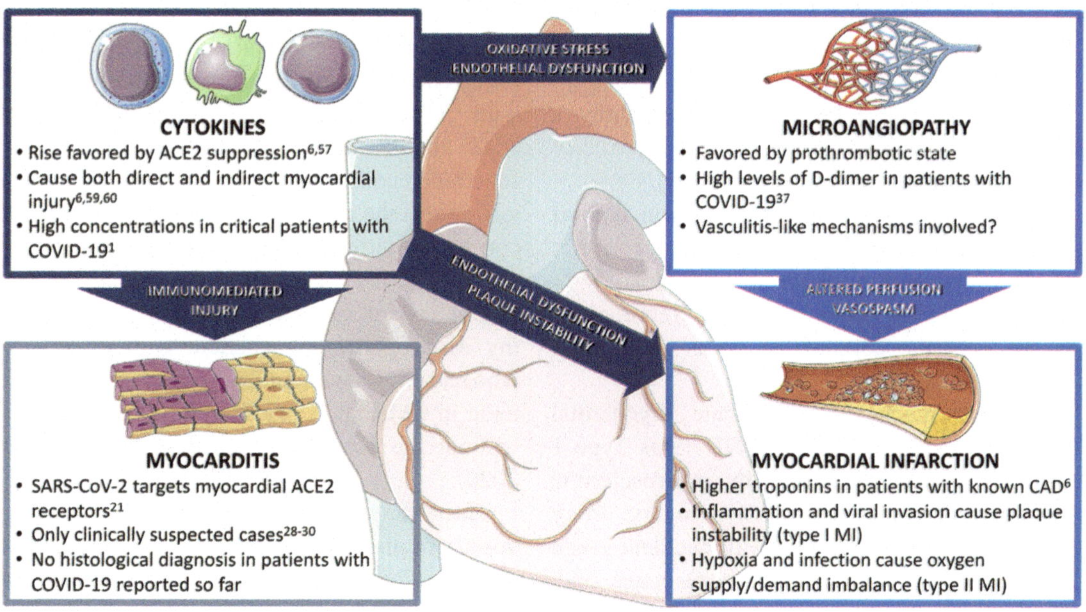

Fig. 16.1 Possible mechanisms explaining troponin elevation in patients with COVID-19
ACE2 Angiotensin-converting enzyme 2; *CAD* Coronary artery disease, *COVID-19* coronavirus disease 2019, *MI* Myocardial infarction; *SARS-CoV-2* Severe acute respiratory syndrome coronavirus 2
Adapted with permission (Tersalvi et al. 2020; https://www.elsevier.com/authors/permission-request-form)

tion of 27% in echocardiography, a sputum sample was checked for the genomes of 13 diseases including viral diseases suspected of causing myocarditis as well as mycoplasma pneumonia and chlamydia, and the results were positive for only the COVID-19 genome. The patient fully recovered from cardiac complications after aggressive anti-inflammatory treatment (Hu et al. 2020).

A case report from Italy described a patient who suffered cardiogenic shock immediately after the COVID-19 symptom onset and needed extracorporeal membrane oxygenation (ECMO). Endomyocardial biopsy and ultrastructural studies revealed viral particles and typical morphology of the virus in the cardiac interstitium, but there was no virus inside the myocardial cells themselves (Tavazzi et al. 2020).

Another case report from Italy described a patient presenting with extensive ST elevation in ECG and severe left ventricular hypertrophy and hypokinesia. The patient underwent cardiovascular magnetic resonance imaging (CMR), which illustrated edema and gadolinium enhancement, in favor of myocarditis. Supportive treatment for COVID-19 and heart failure proved effective (Inciardi et al. 2020).

There have also been studies indicating a higher likelihood of myocarditis in the presence of a higher viral load (Liu et al. 2020b).

Despite the abovementioned cases in support of the myocarditis hypothesis in the occurrence of acute myocardial injury, no autopsy study or myocardial biopsy study has hitherto reported the presence of the coronavirus or its genome in myocardial cells.

Another mechanism for acute myocardial injury in the presence of COVID-19 is Type 1 MI, which can be caused by atherosclerotic plaque rupture and epicardial coronary artery obstruction. Also involved in the incident could be cytokines, acute inflammatory response, and, of course, the theory of the direct viral damage at the plaque site (Thygesen et al. 2018; Tersalvi et al. 2020). Type 2 MI is another mechanism, which occurs following reduced oxygen supply due to hypoxia, on the one hand, and increased oxygen demand due to fever and tachycardia, on the other (Tersalvi et al. 2020; Thygesen et al. 2018).

A study on 18 patients with COVID-19 and ST-segment elevation reported MI in 8 subjects, of whom 6 were confirmed via coronary angiography. Additionally, ten patients had non-coronary myocardial injury. In-hospital death was reported in four patients with MI and nine patients with non-coronary myocardial injury, which was indicative of poor prognosis in such patients (Bangalore et al. 2020).

Another hypothesis regarding the occurrence of acute myocardial injury in the presence of COVID-19 is microangiopathy. Cytokine storms, oxidative stressors, and processes related to ACE2 receptors can expedite the course of coagulopathy (Tersalvi et al. 2020; Yin et al. 2020). Almost half of the patients with COVID-19 have an elevated level of D-dimer, which is a product of the coagulation cascade (Guan et al. 2020b). Moreover, the phenomenon of vasculitis, which has been postulated in the association between COVID-19 and Kawasaki disease, can induce microangiopathy as the cause of cardiac and renal involvement (Tersalvi et al. 2020; Giray et al. 2016).

Stress-induced cardiomyopathy (Takotsubo cardiomyopathy) could also be a culprit in the occurrence of acute myocardial injury in patients with COVID-19 insofar as while it creates new ST-T abnormalities and transient cardiac dysfunction incompatible with the coronary artery territory in the absence of stenotic lesions in the coronary arteries, it is accompanied by elevated troponin levels. This disorder is usually reversible. A case of reverse Takotsubo cardiomyopathy was reported in a patient with COVID-19 who had edema without late gadolinium enhancement in CMR and some lymphocyte infiltrations without the coronavirus in biopsy (Sala et al. 2020).

Heart failure, whether diagnosed or undiagnosed, in the presence of COVID-19 can be accompanied by aggravated symptoms and acute myocardial injury. This issue was confirmed by a study on four patients hospitalized in New York (Fried et al. 2020).

The presence of underlying heart failure or cardiovascular risk factors such as hypertension, diabetes, and a history of coronary artery disease (CAD) was reported to be more frequent in patients requiring hospitalization than in those who did not, which suggests that it could create higher susceptibility to acute myocardial injury (Zhou et al. 2020).

An investigation in China on 187 patients with COVID-19, of whom 144 recovered and 43 expired, reported underlying cardiovascular diseases such as cardiomyopathy, hypertension, and CAD in 35.5% and elevated troponin T (denoting myocardial injury) in 27.8% of the study population. The mortality rate was significantly different between the patients with and without these two parameters (Fig. 16.2) (Guo et al. 2020).

16.2.5 Clinical Signs

Most patients with COVID-19 experiencing acute myocardial injury often present with the typical signs of the infection such as fever, cough, and dyspnea. Few are the patients who present with primary cardiac signs and symptoms such as chest pain and palpitation (Liu et al. 2020a).

16.2.6 Diagnostic Tests

16.2.6.1 Blood Tests

Routine diagnostic tests include complete blood counts; white blood cell differential counts; levels of serum electrolytes, urea, creatinine, and glucose; and liver function tests.

16.2.6.2 Chest X-Ray and CT Scanning

In this group of patients, cardiomegaly or bilateral pleural effusion, often not among the typical manifestations of COVID-19, may be visualized (2020 ESC Guidance for the diagnosis and management of CV Disease in the context of the COVID-19 Pandemic. Available at https://www.escardio.org/Education/COVID-19-and-Cardiology/ESC-COVID-19-Guidance 2020).

16.2.6.3 ECG

All patients with COVID-19 who need special care should undergo admission ECG. New ST-T abnormalities, QRS morphological changes, and

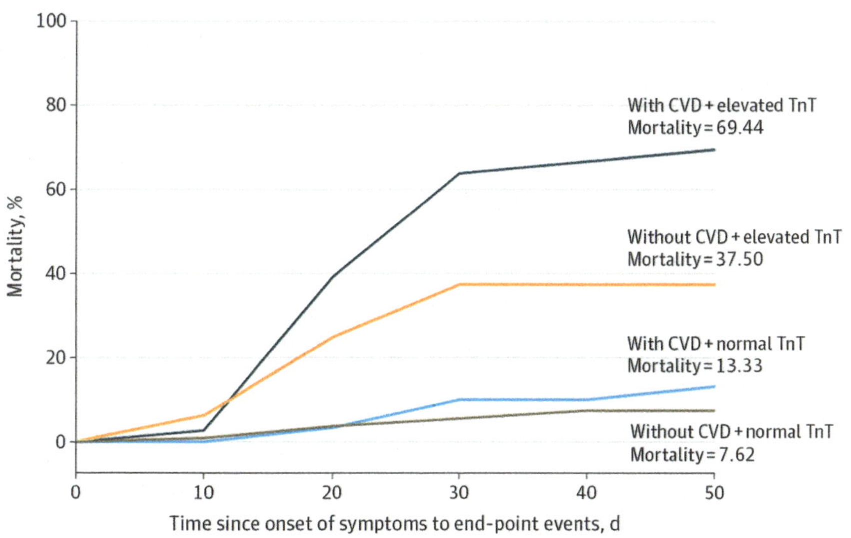

No. at risk					
Without CVD + normal TnT (n = 105)	102	86	41	10	0
Without CVD + elevated TnT (n = 16)	15	12	7	1	0
With CVD + normal TnT (n = 30)	29	25	10	4	0
With CVD + elevated TnT (n = 36)	34	20	8	2	0

Fig. 16.2 Mortality of patients with coronavirus 2019 (COVID-19) with/without cardiovascular disease (CVD) and with/without elevated troponin T (TnT) levels. Adapted with permission (Guo et al. 2020; https://jamanetwork.com/pages/cc-by-license-permissions)

cardiac block may indicate myocardial injury (Caforio et al. 2013). The measurement of the QT-segment in admission ECG can assist in the monitoring of pharmaceutical side effects. ECG changes in patients with COVID-19 and myocardial injury include ST-segment elevation or depression, T-wave inversion, and Q-wave formation (Shi et al. 2020; Inciardi et al. 2020; Bangalore et al. 2020). No specific ECG abnormality has been defined in patients with COVID-19 (2020 ESC Guidance for the diagnosis and management of CV Disease in the context of the COVID-19 Pandemic. Available at https://www.escardio.org/Education/COVID-19-and-Cardiology/ESC-COVID-19-Guidance 2020).

16.2.6.4 Troponin

The troponin test in patients with COVID-19 is valuable in the determination of myocardial injury and the prognosis of the disease. A rise in the troponin level is considered drastic or progressive when it exceeds the 99th percentile URL (Zhou et al. 2020).

16.2.6.5 Natriuretic Peptides

Elevations in the level of brain natriuretic peptide (BNP) or N-terminal prohormone brain natriuretic peptide (NT-proBNP) in patients with suspected heart failure can aid the diagnosis. The measurement of NT-proBNP to determine prognosis in patients with COVID-19 in a previous study showed that its elevated level had an incremental value to hs-TnI (Shi et al. 2020).

Myocardial injury is usually associated with elevated levels of troponin T and I; nonetheless, the levels of BNP and NT-proBNP are allied to the severity of the created hemodynamic stress (2020 ESC Guidance for the diagnosis and management of CV Disease in the context of the COVID-19 Pandemic. Available at https://www.escardio.org/Education/COVID-19-and-Cardiology/ESC-COVID-19-Guidance 2020).

16.2.6.6 Imaging

Echocardiography, CMR, and angiography are drawn upon not only to assess regional and global cardiac motion but also to diagnose other etiologies of cardiac involvement such as valvular and congenital diseases.

Echocardiography

In order to perform echocardiography based on clinical and paraclinical issues suggestive of heart involvement and to consider both patient and operator benefits, the Iranian Society of Echocardiography (ISE) (Sattarzadeh Badkoubeh et al. 2020) recommended the use of this feasible imaging modality in the presence of the following indications:

- Shock state
- New arrhythmias (except for isolated premature ventricular contractions [PVCs] or premature atrial contractions [PACs]) or block in ECG
- Cardiomegaly in CT scan
- More-than-mild pericardial effusion in CT scan
- Elevated cardiac biomarkers defined as:
  - NT-pro-BNP more than 450 pg/dL in patients aged below 50 years
  - NT-pro-BNP more than 900 pg/dL in patients aged between 50 and 75 years
  - NT-pro-BNP more than 1800 pg/dL in patients aged over 75 years
  - An otherwise unexplained elevated hs-TnI level of more than the 99th percentile
- New ischemic changes in ECG
- Generalized edema
- Unexplained deterioration in the clinical status of patients with previous cardiovascular disease

Also according to the recommendations of the ISE, it is also reasonable to perform echocardiography at the discretion of the consultant cardiologist while considering the appropriateness criteria (Sattarzadeh Badkoubeh et al. 2020).

16.2.6.7 Endomyocardial Biopsy

This diagnostic test may be useful in carefully selected patients.

16.2.7 Diagnosis

Acute myocardial injury may be indicated by the patient's history and signs such as acute chest pain, dyspnea, palpitation, and syncope. The diagnosis can be further bolstered by cardiac examinations, ECG, and the measurement of troponin, BNP, and NT-proBNP. The presence of a single clinical criterion together with one of the following items in patients without a history of CAD hints at the presence of non-coronary myocardial injury (Caforio et al. 2013):

- Elevated levels of natriuretic peptides
- Imaging evidence of abnormalities in regional or global cardiac motion
- Significant ECG changes (e.g., significant ST-segment elevation)
- Appearance of edema or late gadolinium enhancement in CMR

Nevertheless, in the era of the COVID-19 pandemic, the assessment of a patient with suspected myocardial injury differs slightly with the standard diagnostic assessment, particularly given the possibility of the spread of the infection in the hospital (Driggin et al. 2020). For instance, it should be determined whether further evaluations aimed at the diagnosis of acute myocardial injury can exert a significant therapeutic impact on the course of the disease (e.g., Type 1 MI). Furthermore, in the allocation of personnel and resources, society's health should be prioritized.

In patients with no suspicion of acute coronary syndrome (ACS) and with hemodynamic stability, further diagnostic tests and multiple imaging are unnecessary because a specific treatment for non-coronary myocardial injury has yet to be formulated. Patients with slightly elevated troponin levels can be followed just clinically. In contrast, in patients with heart failure signs, it is advisable that the threshold for the use of imaging methods be lowered, specifically for transthoracic echocardiography, which is portable and readily available.

Endomyocardial biopsy may prove useful in younger patients or in those experiencing cardiogenic shock or life-threatening arrhythmias without coronary involvement (Caforio et al. 2013).

16.2.8 Differential Diagnosis

The coronary causes of myocardial injury constitute the most significant differential diagnoses. Chronic troponin elevations have been reported in patients with renal disease and in patients at high risk of cardiovascular complications. The level of troponin should be defined for any specific population under study on the basis of sex (Jaffe et al. 2011; Ford et al. 2016).

Cardiogenic shock can concur with other kinds of shocks such as ARDS and sepsis (2020 ESC Guidance for the diagnosis and management of CV Disease in the context of the COVID-19 Pandemic. Available at https://www.escardio.org/Education/COVID-19-and-Cardiology/ESC-COVID-19-Guidance 2020). Such cases require the measurement of central venous oxygen saturation or even invasive monitoring, in addition to routine imaging procedures, for better differentiation between the etiologies of shock state (2020 ESC Guidance for the diagnosis and management of CV Disease in the context of the COVID-19 Pandemic. Available at https://www.escardio.org/Education/COVID-19-and-Cardiology/ESC-COVID-19-Guidance 2020).

16.2.9 Treatment

A proven definitive treatment has yet to emerge for COVID-19; nonetheless, numerous treatment modalities have been suggested and tried to diminish morbidity and mortality in patients with this infection. Some of these therapies have antiviral properties, but evidence regarding their efficacy is conflicting (Sanders et al. 2020). In some trials, immunomodulatory drugs have been used in the hope of ameliorating the severe inflammatory response among critically ill patients.

Glucocorticoids have been used not only for COVID-19-induced pneumonia (Veronese et al. 2020) but also for suspected myocarditis (Hu et al. 2020; Zeng et al. 2020). Still, the recent treatment reviews and guidelines, in the main, do not recommend glucocorticoids merely because a patient is COVID-19-positive (Sanders et al. 2020; COVID-19 Treatment Guidelines Panel. Coronavirus Disease 2019 (COVID-19) Treatment Guidelines. National Institutes of Health. Available at https://covid19treatment-guidelines.nih.gov/.; Massachusetts General Hospital COVID-19 Treatment Guidance. The General Hospital Corporation 2020 Version 3.1: 1-16. Available at https://advances.massgeneral.org/research-and-innovation/article.aspx?id=1184).

Intravenous immunoglobulin has been administered in the treatment of both severe COVID-19-induced pneumonia (Xie et al. 2020) and COVID-19-induced myocarditis (Hu et al. 2020; Zeng et al. 2020). Nevertheless, the very latest treatment guidance and guidelines do not recommend their use (Massachusetts General Hospital COVID-19 Treatment Guidance. The General Hospital Corporation. 2020 Version 3.1: 1-16. Available at https://advances.massgeneral.org/research-and-innovation/article.aspx?id=1184; COVID-19 Treatment Guidelines Panel. Coronavirus Disease 2019 (COVID-19) Treatment Guidelines. National Institutes of Health. Available at https://covid19treatment-guidelines.nih.gov/.).

In immunomodulatory strategies vis-à-vis cytokine storms, particularly against interleukin-1 (IL-1) and IL-6, drugs are used (Conti et al. 2020; Zhang et al. 2020a).

It is vital that further observations on myocardial enzyme curves and imaging studies be undertaken in patients having received treatment with these drugs with a view to correlating immunomodulation with myocardial protection in COVID-19 (Tersalvi et al. 2020).

There is as yet no appropriate treatment for myocardial injury in tandem with COVID-19. Treatment for patients suffering from COVID-19 with myocardial injury, including those with suspected myocarditis, should, therefore, be supportive and seek to treat heart failure and arrhythmias and also to prevent the use of drugs with cardiotoxic effects (Caforio Accessed on May 14, 2020).

With respect to angiotensin-converting enzyme inhibitors (ACEIs) and angiotensin II receptor blockers (ARBs), the available data are conflicting. An early hypothesis postulated that treatment with ACEIs/ARBs could increase the risk of severe COVID-19 (Diaz 2020) and even place patients with diabetes and hypertension at the risk of death (Fang et al. 2020). On the other hand, in support of the protective effects of ARBs on the pulmonary endothelium in the wake of acute injury caused by the virus, a review article cited the therapeutic efficacy of ARBs in the treatment of patients with sepsis, pneumonia, and influenza and reported that ARBs in combination with statins were effective in lowering mortality in the Ebola pandemic (Fedson 2016). It has also been posited that the protective effects of ACE2 are bolstered by ARBs (Saavedra 2020).

Losartan is currently under study for its ability to reduce lung injury in hospitalized and outpatient individuals with COVID-19 (ClinicalTrials.gov. Randomized Controlled Trial of Losartan for Patients With COVID- 19 Not Requiring Hospitalization. Identifier: NCT04311177. March 17, 2020. https://clinicaltrials.gov/ct2/show/NCT04311177; ClinicalTrials.gov. Randomized Controlled Trial of Losartan for Patients With COVID- 19 Requiring Hospitalization. Identifier: NCT04312009. March 17, 2020. https://clinicaltrials.gov/ct2/show/NCT04312009).

A collaborative registry on 8910 patients found no evidence concerning an increased risk of in-hospital mortality following the use of ACEIs/ARBs (Mehra and Desai 2020).

Overall, in patients with heart failure, there is no need to discontinue ACEIs/ARBs unless there are contraindications. In patients with cardiovascular disease who contract COVID-19, any alterations in the therapeutic regimen should be in keeping with the latest reported evidence and at the discretion of the treating physician and team (Statement from the American Heart Association, Heart Failure Society of America, American College of Cardiology. Patients taking ACE-i and ARBs who contract COVID-19 should continue treatment, unless otherwise advised by their physician 2020).

The initiation of ACEIs/ARBs in patients with COVID-19 has not been recommended yet (Massachusetts General Hospital COVID-19

Treatment Guidance. The General Hospital Corporation 2020 Version 3.1: 1–16. Available at https://advances.massgeneral.org/research-and-innovation/article.aspx?id=1184).

Of course, the existing literature also contains recommendations against the use of ACEIs or even ARBs in the case of pneumonitis and ARDS due to the possibility of vasoplegia and renal failure (Mehra and Ruschitzka 2020).

In regard to patients on left ventricular assist devices that contract COVID-19, a formidable challenge arises if they develop vasodilation, followed by diarrhea and dehydration and, consequently, the low-flow state. On the other hand, excess fluid volumes in these patients cause right-sided heart failure (Caforio Accessed on May 14, 2020).

Apropos heart transplantation recipients, in a retrospective study conducted in a center in China on 87 such patients, the follow-up indicated that appropriate preventive measures could decrease the likelihood of infection and COVID-19 (Ren et al. 2020).

The main question is, however, whether or not the severity of COVID-19 and the duration of the disease differ between patients with implanted hearts and those not on immunosuppressants.

The salient discussion point is the advisability of heart transplantation during the COVID-19 pandemic. On the one hand, there is the risk of infection among patients during the pre- and post-implantation hospital stay, and on the other hand, there is the risk of mortality among patients with heart failure on the waiting list. The latest recommendation is to proceed with implantation and continue with the existing therapeutic regimen in patients with negative COVID-19 tests that have had no symptoms or direct contact with known or suspected cases of COVID-19 in the preceding 2–4 weeks. Regarding heart donors, the COVID-19 test should be negative, and if it is positive, a minimum of 2 weeks of negative COVID-19 status prior to organ harvesting is required. What also poses a challenge is a scenario whereby donors are asymptomatic or have had no chance to undergo a COVID-19 test. The treatment for heart transplantation recipients infected with mild COVID-19 is supportive and includes the maintenance of immunosuppression, the reduction of the dose of antimetabolites (mycophenolate or azathioprine), and the continuation of the therapeutic course based on the severity of the disease and the availability of drugs (Clerkin et al. 2020; Guidance for Cardiothoracic Transplant and Mechanical Circulatory Support Centers regarding SARS CoV-2 infection and COVID-19 2020; American Society of Transplantation. COVID-19 (Coronavirus): FAQs for Organ Transplantation. Updated February 29, 2020. Available from: https://www.myast.org/sites/default/files/COVID19%20FAQ%20Tx%20Centers%20030220-1.pdf).

Protease inhibitors are used in the treatment of COVID-19, but they can raise the level of calcineurin inhibitors (Ren et al. 2020).

The golden rule is that patients with chronic heart failure should avail themselves of protective measures so as to avoid infection (2020 ESC Guidance for the diagnosis and management of CV Disease in the context of the COVID-19 Pandemic. Available at https://www.escardio.org/Education/COVID-19-and-Cardiology/ESC-COVID-19-Guidance 2020). Moreover, irrespective of their COVID-19 status, patients suffering from chronic heart failure should strictly adhere to guideline-directed medical therapy. The overuse of intravenous fluids and the prescription of drugs that can cause electrolyte imbalances (e.g., nonsteroidal anti-inflammatory drugs [NSAIDs]) should be avoided in old patients with COVID-19 (Mehra and Ruschitzka 2020). Heart failure specialists need a well-structured approach to patients with COVID-19 and heart failure, an instance of which is presented in Table 16.1 (Mehra and Ruschitzka 2020). It would also be beneficial to consider telemedicine, if feasible, to provide medical advice and to follow stable patients with cardiovascular disease (2020 ESC Guidance for the diagnosis and management of CV Disease in the context of the COVID-19 Pandemic. Available at https://www.escardio.org/Education/COVID-19-and-Cardiology/ESC-COVID-19-Guidance 2020).

Table 16.1 Clinical cardiovascular concerns in COVID-19 illness

COVID-19 infection	Concern	Interpretation
Asymptomatic or early mild disease with constitutional symptoms (fever, dry cough, diarrhea, and headache)	Should background cardiovascular medications be modified?	There is no clear evidence that ACEI or ARBs should be discontinued NSAIDs should be used with caution or, ideally, avoided
Moderate disease with pulmonary complications and shortness of breath (including hypoxia)	Is there a cardiovascular contribution to the lung complications?	Check troponin (evidence of myocardial injury and prognosis) Check natriuretic peptides Consider cardiac echocardiography to evaluate for evidence of underlying structural heart disease, high filling pressures Avoid overuse of intravenous fluids, which may worsen underlying pulmonary edema
Advanced-stage disease with hypoxia, vasoplegia, and shock	Is there evidence of cardiogenic contribution to shock, and what therapy may be potentially curative?	Check for evidence of hyperinflammation or a cytokine release storm (elevated troponin, natriuretic peptides, CRP, and serum ferritin of >1000 ng/ml) (measure IL-6 levels if available) If the cardiac function is reduced (LVEF <50%), consider supportive care with inotropic therapy but move to consider anti-cytokine therapy with drugs such as tocilizumab and corticosteroids

Note that therapy in COVID-19 remains experimental

ACEi Angiotensin-converting enzyme inhibitors, *ARB* Angiotensin receptor blockers, *CRP* C-reactive protein, *IL* interleukin, *LVEF* left ventricular ejection fraction

Adapted with permission and modified (Mehra and Ruschitzka 2020; https://www.elsevier.com/authors/permission-request-form)

16.3 Hypertension and COVID-19

A meta-analysis of four studies on COVID-19 showed that this infection in individuals with hypertension and those with chronic kidney disease manifested itself with greater severity (Guan et al. 2020a). These findings were confirmed by a subsequent cohort study in China (Henry and Lippi 2020). Nonetheless, further scrutiny revealed that the main reason for such conclusions was the absence of age adjustment in the data analyses. Indeed, individuals aged over 60, of whom 50% suffer from hypertension, are prone to more severe forms of this infection; therefore, a higher reported incidence of hypertension among this group of patients with SARS-CoV-2 cannot be deemed unexpected.

Overall, there is currently no evidence indicating that hypertension per se is associated with the severity of COVID-19 or an increase in mortality (2020 ESC Guidance for the diagnosis and management of CV Disease in the context of the COVID-19 Pandemic. Available at https://www.escardio.org/Education/COVID-19-and-Cardiology/ESC-COVID-19-Guidance 2020).

Since the early days of the COVID-19 pandemic, hypotheses have been proposed as to the negative impacts of ACE inhibitors and ARBs on the treatment course of this infection (Fang et al. 2020). Nonetheless, robust evidence has yet to emerge in the support of such hypotheses. In some animal studies, ACE inhibitors and ARBs even played protective roles against severe lung damage. Thus, patients with hypertension who already consume such drugs should continue the medication in keeping with the latest guidelines (2020 ESC Guidance for the diagnosis and management of CV Disease in the context of the COVID-19 Pandemic. Available at https://www.escardio.org/Education/COVID-19-and-Cardiology/ESC-COVID-19-Guidance 2020; Vaduganathan et al. 2020; Li et al. 2020), as endorsed by the latest statements of the International Society of Hypertension (ISH) (Kuster et al. 2020).

During the COVID-19 pandemic, it is recommended that patients with hypertension who are in quarantine monitor their blood pressure at home and refrain from referring to hospitals as much as possible.

With regard to patients with hypertension who are hospitalized for COVID-19, the following measures are recommended:

- Antihypertensive agents may have to be temporarily discontinued in severely ill patients due to hypotension or acute renal failure.
- Intravenous antihypertensive agents in the case of patients requiring invasive ventilation are recommended only if there is persistently severe hypertension. Blood pressure in this group of patients should be merely reduced to below 160/100 mm Hg and not the ideal value.
- Hospitalized patients with hypertension may be at risk of arrhythmias secondary to left ventricular hypertrophy or other cardiac diseases, especially in hypoxic conditions. Hence, the measurement and adjustment of the serum potassium level in this group of patients are vitally important since hypokalemia, which is associated with advanced COVID-19, can increase the risk of arrhythmias (2020 ESC Guidance for the diagnosis and management of CV Disease in the context of the COVID-19 Pandemic. Available at https://www.escardio.org/Education/COVID-19-and-Cardiology/ESC-COVID-19-Guidance 2020).

16.4 COVID-19 and CAD

CAD has various incidence rates in different populations, which is probably why the incidence of CAD among patients with COVID-19 reportedly varies between 4.2% and 25% based on the population under study (Wang et al. 2020a). The incidence of CAD is higher in patients in the ICU than in other persons (Chen et al. 2020).

Myocardial injury, accompanied by elevations in troponin or ECG changes, based on various investigations, has an incidence rate of 7.2% to 20%. Nonetheless, only a few patients with COVID-19 with elevated troponin levels exhibit the signs and symptoms of ACS (Shi et al. 2020). It has been reported that elevated troponin levels in combination with a history of cardiovascular disease increase mortality among patients with COVID-19 to the highest possible level (Guo et al. 2020).

MI may occur in patients with COVID-19 through such different mechanisms as hypoxia, microvascular disorders, hypercoagulability-induced thrombosis, and cytokine storms, which may lead to plaque instability in coronary arteries (Libby et al. 2018).

Patients suffering from fever or hypoxia need an increase in their cardiac output; consequently, those with coronary artery stenosis may develop Type 2 myocardial ischemia. A large body of evidence indicates a significant correlation between a history of CAD and the risk and severity of COVID-19 (Guan et al. 2020b; Zhou et al. 2020; Madjid et al. 2020; Wu and McGoogan 2020).

CAD management during the COVID-19 pandemic requires special considerations.

16.4.1 Non-ST-Segment Elevation Acute Coronary Syndrome (NSTE-ACS)

Upon hospital entrance, all patients with NSTE-ACS need to undergo the SARS-CoV-2 test in order that the treatment team can implement comprehensive protective measures accordingly. If feasible, those with a positive test should be transferred to a COVID-19-designated center.

According to the guidelines of the European Society of Cardiology (ESC), very-high-risk patients with NSTE-ACS that exhibit the following features should be managed via an immediate invasive strategy:

- Unstable hemodynamics or cardiogenic shock
- Chest pain refractory to medical therapy
- Malignant arrhythmias or cardiac arrest
- Mechanical complications of MI
- Acute heart failure
- Recurrent intermittent ST elevation

Otherwise, particularly, in instances where the healthcare system is overwhelmed or catheterization laboratories or their operators are in short supply, many patients with NSTE-ACS can be managed with conservative therapy and be discharged from the hospital early. Nonetheless, very-high-risk patients, similar to patients with ST-segment elevation myocardial infarction (STEMI), need immediate invasive treatment (2020 ESC Guidance for the diagnosis and management of CV Disease in the context of the COVID-19 Pandemic. Available at https://www.escardio.org/Education/COVID-19-and-Cardiology/ESC-COVID-19-Guidance 2020).

16.4.2 STEMI

The COVID-19 pandemic must not delay the timely reperfusion of patients with STEMI. According to the current guidelines, patients with less than 12 h after the onset of symptoms presenting with ST elevations in at least two contiguous leads need reperfusion. The safety of the treatment team should be accorded priority throughout the process (2020 ESC Guidance for the diagnosis and management of CV Disease in the context of the COVID-19 Pandemic. Available at https://www.escardio.org/Education/COVID-19-and-Cardiology/ESC-COVID-19-Guidance 2020).

Upon hospital entrance, all patients with STEMI not having undergone the SARS-CoV-2 test should be considered to have a positive COVID-19 status and, as such, be given the test as expeditiously as possible. Needless to say, safety protocols should be adhered to while awaiting the test result so as to prevent the spread of the infection (2020 ESC Guidance for the diagnosis and management of CV Disease in the context of the COVID-19 Pandemic. Available at https://www.escardio.org/Education/COVID-19-and-Cardiology/ESC-COVID-19-Guidance 2020).

Provided that the safety of the medical team and other patients is guaranteed, in COVID-19-designated centers, the preferred method of reperfusion is primary percutaneous coronary intervention (PCI) (2020 ESC Guidance for the diagnosis and management of CV Disease in the context of the COVID-19 Pandemic. Available at https://www.escardio.org/Education/COVID-19-and-Cardiology/ESC-COVID-19-Guidance 2020). Nevertheless, published experiences during the current pandemic show that the implementation of safety guidelines has delayed primary PCI procedures (Tam et al. 2020; Abdi et al. 2020), which is why fibrinolysis, if not contraindicated, may be preferred when a timely primary PCI procedure is not feasible (2020 ESC Guidance for the diagnosis and management of CV Disease in the context of the COVID-19 Pandemic. Available at https://www.escardio.org/Education/COVID-19-and-Cardiology/ESC-COVID-19-Guidance 2020).

In the primary PCI scenario, it is advisable that, in the presence of subtotal non-culprit lesions and unstable lesions, complete revascularization be done as much as possible so that staged procedures can be avoided and the length of hospital stay be curtailed. Still, the treatment of non-culprit lesions without the mentioned features should be postponed to after the pandemic termination (2020 ESC Guidance for the diagnosis and management of CV Disease in the context of the COVID-19 Pandemic. Available at https://www.escardio.org/Education/COVID-19-and-Cardiology/ESC-COVID-19-Guidance 2020).

16.4.3 Chronic Coronary Syndrome (CCS)

Patients with CCS are often at low risk of cardiovascular incidents; their diagnostic and interventional procedures should, therefore, be postponed as much as possible unless they become clinically unstable. It is also recommended that the clinical follow-up of these patients be done telemedically (2020 ESC Guidance for the diagnosis and management of CV Disease in the context of the COVID-19 Pandemic. Available at https://

www.escardio.org/Education/COVID-19-and-Cardiology/ESC-COVID-19-Guidance 2020).

Evidence indicates that NSAIDs may worsen the treatment course of patients with COVID-19 (Basille et al. 2017). However, the use of low-dose ASA in patients with CCS should not be discontinued because it does not appear to have considerable anti-inflammatory effects. Statin use should be continued unless there is rhabdomyolysis or a drastic increase in liver enzymes.

A précis of the recommendations of the ESC for the treatment of CCS during the COVID-19 pandemic is presented below (2020 ESC Guidance for the diagnosis and management of CV Disease in the context of the COVID-19 Pandemic. Available at https://www.escardio.org/Education/COVID-19-and-Cardiology/ESC-COVID-19-Guidance 2020):

- Medications should be continued in patients with CCS during the pandemic.
- Follow-up with tele-health is preferred in patients with CCS.
- Revascularization in low- or intermediate-risk patients should be postponed.
- Noninvasive testing in these patients should be postponed.
- CT angiography is preferred to functional testing.
- Screening for SARS-COV-2 should be considered in patients undergoing cardiac surgery.
- PCI may be preferred to coronary artery bypass graft surgery, if feasible, during the COVID-19 pandemic.

16.5 Arrhythmias and COVID-19

Patients with COVID-19 are prone to various kinds of tachy- and bradyarrhythmias. These arrhythmias can be triggered by numerous factors, the most significant of which are electrolyte disturbances and hypoxia secondary to the severe forms of COVID-19. The risk of arrhythmias in the milder forms of this infection and in recovering patients has yet to be determined (Lakkireddy et al. 2020). Arrhythmias in patients with COVID-19 can also be due to medication and exacerbated ischemia secondary to an increased oxygen demand in the myocardial tissue.

A study conducted in a hospital in the Chinese city of Wuhan, without clarifying the types of arrhythmias, reported cardiac rhythm alterations in 17% of the entire patients and in 44% of the patients in the ICU (Wang et al. 2020a).

During the COVID-19 pandemic, the management of arrhythmias requires special considerations.

16.5.1 Tachyarrhythmias

16.5.1.1 Supraventricular Arrhythmias

No special report has thus far been published on the incidence rate of the various types of paroxysmal supraventricular tachycardias (PSVTs) during the COVID-19 pandemic; nonetheless, this infection could theoretically lead to the occurrence or exacerbation of these tachycardias.

The pharmacological therapy for PSVTs, based on the existing guidelines, is the use of adenosine for acute termination or the use of beta-blockers and calcium channel blockers for maintenance therapy. Needless to say, bradycardia, which can prolong QT, should be prevented. The option of catheter ablation should be postponed to the post-pandemic era (2020 ESC Guidance for the diagnosis and management of CV Disease in the context of the COVID-19 Pandemic. Available at https://www.escardio.org/Education/COVID-19-and-Cardiology/ESC-COVID-19-Guidance 2020).

16.5.1.2 Atrial Fibrillation (AF) and Flutter

AF can be caused or worsened by COVID-19 in that the infection creates fever, inflammation, hypoxia, increased adrenergic tone, and electrolyte disturbances such as hypokalemia and hypomagnesemia. Research shows that the occurrence of AF in patients with ARDS and sepsis is associated with an increased risk of stroke, heart failure, and death (Boriani et al. 2019; Walkey et al. 2014).

In the event of AF occurrence in patients with COVID-19, the following points should be considered:

- In patients who are hemodynamically unstable due to new-onset AF, electrical cardioversion is recommended. Of course, cardioversion can be efficacious provided that other underlying factors such as hypokalemia, hypomagnesemia, acidosis, increased sympathetic tone, volume overload, and accompanying bacterial infections are effectively treated. Amiodarone is the drug of choice for the maintenance of the sinus rhythm, but its concurrent administration with hydroxychloroquine and azithromycin should be avoided as much as possible.
- In patients who are hemodynamically stable and need antiviral drugs, it is advisable that antiarrhythmic agents be discontinued and the method of rate control be drawn upon.
- Anticoagulants should be administered in accordance with the CHA_2DS_2-VASc Score (2020 ESC Guidance for the diagnosis and management of CV Disease in the context of the COVID-19 Pandemic. Available at https://www.escardio.org/Education/COVID-19-and-Cardiology/ESC-COVID-19-Guidance 2020).

16.5.1.3 Ventricular Arrhythmias

In a recent study conducted in a hospital in the Chinese city of Wuhan, 5.9% of the patients hospitalized for COVID-19 developed malignant ventricular arrhythmias (ventricular tachycardias [VTs] and ventricular fibrillations [VFs]). Additionally, the incidence of such arrhythmias was higher among patients with elevated troponin levels (Guo et al. 2020). These findings show that ventricular arrhythmias are indicative of myocardial injury, necessitating more aggressive antiviral and immunosuppressive treatment (2020 ESC Guidance for the diagnosis and management of CV Disease in the context of the COVID-19 Pandemic. Available at https://www.escardio.org/Education/COVID-19-and-Cardiology/ESC-COVID-19-Guidance 2020).

Torsades de pointes (TdP), a fatal ventricular arrhythmia, can occur in those patients with COVID-19 that are on QT-prolonging medications. The risk of the occurrence of this arrhythmia is higher in the presence of hypokalemia, hypomagnesemia, renal failure, bradycardia, left ventricular hypertrophy, and heart failure, as well as in the female sex. The prevention of these ventricular arrhythmias in the context of SARS-CoV-2 assumes even a greater significance (2020 ESC Guidance for the diagnosis and management of CV Disease in the context of the COVID-19 Pandemic. Available at https://www.escardio.org/Education/COVID-19-and-Cardiology/ESC-COVID-19-Guidance 2020).

Treatment of Ventricular Arrhythmias

In ventricular arrhythmia cases in which the patient with COVID-19 is unresponsive and breathless, the cardiopulmonary resuscitation (CPR) guideline in this chapter should be implemented.

The occurrence of TdP requires the implementation of the following measures:

- Discontinuation of all QT-prolonging medications
- Normalization of the serum potassium level (> 4.5 meq/L)
- Intravenous injection of magnesium
- Augmentation of the heart rate by discontinuing bradycardia-inducing drugs
- Use of isoproterenol or the implantation of temporary pacemakers, if need be

In patients with sustained monomorphic VTs, the following measures are recommended:

- In patients receiving QT-prolonging medications, electrical cardioversion is advisable; and if such patients require antiarrhythmic agents, lidocaine should be administered.
- In patients with structural heart disease, intravenous amiodarone is a suitable option with the proviso that it not be administered concur-

rently with hydroxychloroquine and azithromycin.
- In critically ill patients with COVID-19 that sustain repeated VTs and VFs, amiodarone is the drug of choice (2020 ESC Guidance for the diagnosis and management of CV Disease in the context of the COVID-19 Pandemic. Available at https://www.escardio.org/Education/COVID-19-and-Cardiology/ESC-COVID-19-Guidance 2020).

16.5.2 Bradyarrhythmias

As yet, there has been no specific report on the occurrence of bradyarrhythmias in patients with COVID-19. Nevertheless, theoretically and in particular, if there is myocardial involvement, such patients could experience exacerbated conduction system abnormalities or sinus node dysfunction or even develop new-onset high-degree atrioventricular (AV) block. In critically ill patients in the ICU, transient bradycardia and asystole might occur during the intubation, tracheal suction, or rotation of the patient for breathing in the prone position possibly due to increased vagal tone (Boriani et al. 2019). It is, however, vital that hypoxemia be ruled out (2020 ESC Guidance for the diagnosis and management of CV Disease in the context of the COVID-19 Pandemic. Available at https://www.escardio.org/Education/COVID-19-and-Cardiology/ESC-COVID-19-Guidance 2020).

The risk of AV block and bundle branch block may be increased by some COVID-19 medications such as hydroxychloroquine (more likely with chloroquine) and fingolimod. Some of these effects might manifest themselves after a few weeks. Accordingly, patients who have recovered from COVID-19 should be warned that in the event of presyncope or syncope signs and symptoms, they need to seek medical help (2020 ESC Guidance for the diagnosis and management of CV Disease in the context of the COVID-19 Pandemic. Available at https://www.escardio.org/Education/COVID-19-and-Cardiology/ESC-COVID-19-Guidance 2020).

In the case of refractory symptomatic bradycardia due to AV block or sinus node dysfunction, the following points are recommended:

- All medications leading to bradycardia should be discontinued.
- Isoprenaline and atropine should be administered.
- Temporary pacemaker implantation should be considered.
- After recovery from COVID-19, the patient should undergo evaluation concerning the need for permanent pacemaker implantation since bradycardia might be transient and pacemaker implantation during an infection can result in the superimposed bacterial infection of the device (2020 ESC Guidance for the diagnosis and management of CV Disease in the context of the COVID-19 Pandemic. Available at https://www.escardio.org/Education/COVID-19-and-Cardiology/ESC-COVID-19-Guidance 2020).

16.6 Coagulopathy in the COVID-19 Outbreak

Coagulopathy is believed to predict poor outcomes in patients with COVID-19 (Guan et al. 2020b). Early reports on comparisons between COVID-19 survivors and non-survivors indicated meaningfully higher levels of fibrin degradation products (e.g., D-dimer) and longer prothrombin and activated partial thromboplastin times in the latter (Tang et al. 2020b). For instance, Tang et al. reported that disseminated intravascular coagulation occurred in 71.4% of the non-survivors and 0.6% of the survivors during hospitalization in their study (Tang et al. 2020b). Interestingly, none of the patients developing thrombotic complications in the ICU in a study by Klok et al. suffered disseminated intravascular coagulation (Klok et al. 2020). What, thus, remains unclear is whether the observed coagulopathy is merely a marker of the severity of the disease (e.g., in septic shock) or whether the virus exerts a direct impact on the coagulation cascade (Tang et al. 2020b; Iba et al. 2019). Several investigations have implicated coagulopathy indices in the survival prediction of patients with COVID-19 and postulated that the D-dimer level (Tang et al. 2020b; Guan et al. 2020b), the prothrombin time (Tang et al. 2020b), and thrombocytopenia (Lippi et al. 2020b) may be potential prognostic factors. Contrariwise, bleeding complications appear to be atypical in patients infected with COVID-19 (Thachil et al. 2020).

16.6.1 Venous Thromboembolism (VTE)

Thrombotic events in patients with COVID-19 are more likely to occur in those recommended to take a prolonged bed rest while consuming the pertinent therapeutic regimen. A few reports are available on the prevalence of different thrombotic complications in patients suffering from COVID-19. In three academic/teaching hospitals in the Netherlands, Klok et al. observed a 31% (95% CI: 20 to 41%) incidence rate of thrombotic complications, the majority of which were VTE (27% [95% CI: 17 to 37%]), compared with arterial thrombotic events in 3.7% (95% CI: 0 to 8.2%) (Klok et al. 2020). A similar high incidence rate of pulmonary emboli was also observed by Poissy et al., who reported a rate of 20.6%, which was twice the rate during the same time interval in the previous year (Poissy et al. 2020). The diagnosis of VTE may become further complicated in patients hospitalized for COVID-19 by a variety of reasons ranging from the inapplicability of the D-dimer test to the constraints in the transfer of patients to imaging wards or even positioning them for imaging studies (Bikdeli et al. 2020). Fortunately, VTE diagnosis can be facilitated, albeit partially, by such diagnostic measures as right ventricular enlargement/dysfunction in echocardiography or deep VTE in lower limbs on ultrasound (Bikdeli et al. 2020). A consensus has yet to emerge vis-à-vis the treatment of patients without a definite diagnosis, but intermediate- to full-dose anticoagulation has been suggested by various experts (Bikdeli et al. 2020).

A meticulously devised prophylaxis strategy is needed in view of not only the symptom overlap between pulmonary embolism and acute respiratory disease in patients with COVID-19 (Danzi et al. 2020) but also the abovementioned challenges in the diagnosis and treatment of pulmonary emboli. Apropos of VTE prophylaxis, Tang et al. performed a retrospective analysis on 449 patients hospitalized for severe COVID-19 with the objective of assessing the validity of the sepsis-induced coagulopathy score and the D-dimer level in the risk stratification of such patients (Tang et al. 2020a). Their results showed no 28-day mortality benefit in heparin (low-molecular-weight or unfractionated) users by comparison with nonusers. The authors, however, observed a meaningful improvement in 28-day mortality in patients with a minimum sepsis-induced coagulopathy score of 4 or a D-dimer level of greater than 3.0 μg/mL due to heparin prophylaxis and, therefore, recommended prophylaxis based on risk stratification.

Other risk stratification tools (e.g., Caprini and IMPROVE) have also been recommended for use (Bikdeli et al. 2020). Nevertheless, the recommendations of the International Society on Thrombosis and Haemostasis (ISTH) appear to be liberal insofar as they advocate the administration of low-molecular-weight heparin in all patients hospitalized for COVID-19 (including those that are not critically ill) that do not have contraindications (platelet count $\leq 25 \times 109$/L or active bleeding) (Thachil et al. 2020). Despite the current paucity of data to suggest this routine VTE prophylaxis at this juncture, the case series by Klok et al. confirmed the high incidence of VTE (27%) in patients hospitalized for COVID-19, which might be judged as justification for the liberal approach. Of note, mechanical prophylaxis has been suggested for patients with contraindications for pharmacological prophylaxis (Bikdeli et al. 2020).

VTE prophylaxis merits more attention, although even the earliest reports on the COVID-19 pandemic underscored its significance. In their short report on 1026 patients hospitalized for COVID-19, Wang et al. showed that while more than 40% of their study population had a minimum Padua prediction score of 4 (i.e., high risk for (Wang et al. 2020b) VTE), only 7% received appropriate treatment (Wang et al. 2020b).

16.7 CPR

CPR, which is deemed the terminal procedure of every critical medical process, has the primary objective of restoring life with the fewest sequelae and disabilities (Morrison et al. 2015). Nonetheless, the prominent medical aspect of CPR notwithstanding, under no circumstances can its ethical and legal facets be disregarded (Morrison et al. 2015).

The considerable health burden begotten by COVID-19 has created a stressful situation for both medical staff and patients (Castelletti 2020). Several procedures in CPR known as aerosol-generating procedures with a high risk of transmission and infection aggravate the stress and, as a result, inadvertently undermine CPR efficacy and increase the risk of staff contamination (Resuscitation Council of United Kingdom 2020; Edelson et al. 2020). A well-thought-out series of prognostications and preparations is, therefore, vitally important in the treatment of patients in need of CPR while striving to confer protection against COVID-19 not only to them and their families but also to the hospital staff and other patients (Edelson et al. 2020; Resuscitation Council of United Kingdom 2020). This is probably why CPR recommendations were among the earliest guidelines formulated by international societies in the immediate wake of the outbreak.

Indubitably, evidence concerning COVID19 is still scant, with the major recommendations based on expert consensus and previous viral outbreaks. In writing the following recommendations, we have drawn upon two major sources: available guidelines from eminent international scientific societies (Edelson et al. 2020; Resuscitation Council of United Kingdom 2020) and experts' personal experiences during the current pandemic.

16.7.1 CPR Protocol

1. The basic principles of CPR protocols have certainly not been altered and should be adhered to as strictly as before (Kleinman et al. 2015).
2. Prone CPR is advocated if administered by experienced CPR personnel in order that the risk of aerosol transmission is minimized. In brief, chest compression is applied at the mid-thoracic vertebrae (T7 level) between the two scapulae with the patient in the prone position. Placing a stiff board between the patient's chest and the mattress can augment the compression. If necessary, defibrillation can be administered by placing one of the pads at the patient's left midaxillary line and the other one on the right scapula (Cave et al. 2010).
3. Given the risk of infection spread during the use of masks and bag-valve masks, CPR should be performed, as much as feasible, via chest compression and defibrillation, if need be (chest compression-only CPR). During chest compression, the mask should be fixed on the patient's face so as to limit aerosol spread. If the AutoPulse (ZOLL, CA) or LUCAS (LUCASTM2 Chest Compression System, JOLIFE AB Inc., Lund, Sweden) is available, the use of either one for chest compression is recommended.
4. Airway management requires the observation of the following points (Orser 2020):
   (a) Patients should be intubated only by highly experienced CPR team members and, if available, with the aid of a video laryngoscope. If the patient needs respiratory support before the arrival of the airway team's supervisor, a mask or a bag-valve can be used but only for the purposes of passive fixation. Put differently, without ventilation and bagging, the patient receives oxygen through the mask.
   (b) Intubation should be avoided as long as the patient is conscious. Because of the potential virus aerosolization by atomized local anesthetics, the GlideScope or similar devices should be utilized in their stead, if feasible.

(c) In pre-arrest cases, anesthesia should be induced via rapid-sequence induction (RSI) by an experienced physician. RSI should be tailored to the patient's condition (e.g., presence of severe hypoxia, intolerance of apnea for 30 s, and contraindications to the use of neuromuscular paralyzers). A low tidal volume is prudent in manual ventilation.
(d) It is advisable that 5 min of preoxygenation with 100% oxygen precede RSI in order to avoid manual ventilation as much as possible. Aerosol spread is a potential consequence of manual ventilation.
(e) The placement of a high-efficiency hydrophobic filter between the face mask and the respiratory tract or the respiratory mask or the bag-valve mask is highly recommended (e.g., the Laerdal Resuscitation Bag).

16.7.2 Safety Concerns

1. The suggested personal protective equipment (PPE) by this document is presumed to be essential for any center involved in CPR; still, the complexity of the situation may preclude some scenarios. The aforementioned centers should seek to address shortages and modify the recommendations accordingly while ensuring maximal possible protection.
2. CPR is required by all healthcare workers whether directly (e.g., physicians and nurses) or indirectly (e.g., nurse aides, technicians, infection control committees, hospital housekeepers, and security staff); hence the significance of the presentation and education of the present guidelines, in conjunction with other recommendations and directives, should be endorsed by the authorities of treatment centers to their personnel.
3. All patients presenting with cardiorespiratory arrest to the emergency department should be considered suspicious COVID-19 cases and be administered CPR in accordance with the present guidelines.
4. CPR administration should be done by a minimum number of staff members.
5. CPR administration should not involve less experienced personnel such as students as much as is achievable.
6. Complete PPE packages should be placed at specified spots in each ward. PPE should be donned prior to CPR initiation. The package should contain as a minimum the following items: one long surgical gown or one waterproof one-piece gown, two pairs of gloves, one N95 or one FFP2 or FFP3 mask, one hat, one pair of goggles or one face shield, and one pair of shoe covers. Under no circumstances should CPR be attempted before the team members have donned these items.
7. During and after CPR, contaminated pieces of equipment, not least those containing respiratory secretions (e.g., laryngoscopes and patient masks), should be set down on a labeled tray or in a marked receptacle rather than on the patient's bedside.
8. Upon CPR termination, all pieces of reusable equipment such as laryngoscopes and stethoscopes should be sterilized in keeping with the infection control protocol of the hospital. Additionally, all pieces of nonreusable equipment such as chest leads and endotracheal tubes should also be considered infectious waste and disposed of in designated bins.

Upon CPR termination, all PPE should be doffed in compliance with the pertinent protocol and be discarded in the nearest bin for infectious waste (yellow). While wearing PPE, the person should not be permitted in areas outside the CPR designated zone (Barati et al. 2020).

16.8 Drug-Drug Interactions Between COVID-19 and Cardiovascular Medications

Numerous studies are underway on the management of patients with COVID-19 (Gabutti et al. 2020). Potential therapeutic pathways, which are the main focus of treatment strategies, consist of viral entry blockage (chloroquine, hydroxychloroquine, umifenovir, and interferon), viral RNA synthesis inhibition (remdesivir, favipiravir, and ribavirin), virus replication blockage (protease inhibitors), and passive immunotherapy (mainly tocilizumab as an interleukine-6 inhibitor) (Gasmi et al. 2020). The National Institutes of Health (NIH) states that there are insufficient data to recommend chloroquine, hydroxychloroquine, remdesivir, or tocilizumab for the treatment of patients with COVID-19. However, the NIH recommends against treatment with interferon and protease inhibitors including lopinavir/ritonavir and atazanavir (https://covid19treatmentguidelines.nih.gov). Therapeutic options are not without adverse events and could interact with other medications. However, remdesivir though metabolization by cytochrome 2C8 (CYP2C8), CYP3A4, and CYP2D6 does not show significant interactions due to rapid clearance.

Patients with cardiovascular disease, particularly the severe forms of the disease, are more vulnerable to COVID-19 and are at greater risk of mortality (2020 ESC Guidance for the diagnosis and management of CV Disease in the context of the COVID-19 Pandemic. Available at https://www.escardio.org/Education/COVID-19-and-Cardiology/ESC-COVID-19-Guidance 2020). Therefore, interactions between COVID-19 medications and cardiovascular drugs could be of vital importance in the management of such patients. These interactions are assessed in the following sections, comprising hypertension, ACS, heart failure, VTE, and arrhythmias.

16.8.1 Hypertension

16.8.1.1 ACEIs/ARBs

ACEIs and ARBs are the medications that attract the most attention in this era. The earliest reports on COVID-19 considered patients with a history of ACEI/ARB medication to be at a higher risk of COVID-19, leading to the formulation of many hypotheses as to the reason behind this susceptibility. SARS-CoV-2 attaches to ACE2 receptors and is then internalized to induce infection. ACE2 also degrades angiotensin II (Ang II) and increases the concentration of Ang 1–7 (South et al. 2020; Ferrario et al. 2005; Zhang et al. 2020b). The latter has some beneficial anti-fibrotic effects. Thus, it appears that ACEIs/ARBs by increasing the concentration of ACE2 could have more positive than negative outcomes (South et al. 2020). This advantage is more pronounced in the use of ARBs in animal models, where losartan is reported to increase ACE2 in cardiomyocytes (Ferrario et al. 2005). Consequently, cardiovascular organizations have issued recommendations to continue ACEIs/ARBs in patients with definite indications of hypertension, heart failure, diabetes, or chronic kidney disease (Zhang et al. 2020b). Still, due to the metabolic interactions of ARBs, patients' blood pressure should be monitored (Dixon et al. 2020). Losartan is metabolized via 2C9, and its effects could be influenced by protease inhibitors. Ritonavir could increase the formation of losartan metabolites. The concentration of valsartan could, however, be increased by protease inhibitors due to hepatic transporter interactions. ACEIs, candesartan, eprosartan, and telmisartan do not have significant interactions with protease inhibitors (Giguere et al. 2019).

16.8.1.2 Diuretics

The most imperative issue to consider vis-à-vis the use of diuretics in patients with COVID-19 is the potential electrolyte disturbances. Chloroquine and hydroxychloroquine are known to prolong QTc and could place the patient at the risk of TdP. Electrolyte disturbances, mainly hypokalemia and hypomagnesemia, could increase the risk of TdP. Hence, monitoring the levels of electrolytes is mandated in these patients to prevent QTc prolongation. Furthermore, indapamide has extra interactions with protease inhibitors due to the similar metabolic pathway of cytochrome P (CYP) 3A4 and 2C19, necessitating caution in its titration and dosing in patients with COVID-19 (Giguere et al. 2019).

16.8.1.3 Alpha-Adrenoreceptor Antagonists

Interactions have been reported on metabolism and transporter levels, which could affect the serum concentration and hypotensive effects of alpha-blockers. As a result, this class of medications should be initiated at a low dose and gradually titrated, in tandem with careful blood pressure monitoring (Giguere et al. 2019).

16.8.1.4 Dihydropyridine Calcium Channel Blockers

Amlodipine exposure is reported to be increased by protease inhibitors. Thus, vigilant initiation and titration are highly recommended in patients receiving lopinavir/ritonavir or atazanavir as COVID-19 management (Giguere et al. 2019).

16.8.2 ACS

16.8.2.1 P2Y12 Inhibitors

According to the guidelines, clopidogrel remains the agent of choice in patients receiving fibrinolytics. Nonetheless, in those receiving protease inhibitors, the conversion of clopidogrel into active metabolites could be impaired. For patients undergoing PCI, more potent P2Y12 inhibitors are recommended. Still, the use of ticagrelor is not recommended with protease inhibitors due to the metabolic interaction and the increased risk of bleeding. Thus, prasugrel is the preferred option provided that the increased tendency of bleeding in these patients is taken into account (Dixon et al. 2020).

16.8.2.2 Beta-Blockers

The use of beta-blockers in patients with COVID-19 is associated with two important concerns. Firstly, these agents have metabolic interactions via CYP2D6 and 2C9 pathways, requiring the monitoring of the heart rate and blood pressure (Dixon et al. 2020; Giguere et al. 2019). Secondly, as bradycardia is a risk factor for the development of TdP, it is suggested to maintain the heart rate above 70 beats per minute to reduce the likelihood of QTc prolongation.

16.8.2.3 Statins

High-intensity statins constitute the cornerstone of the management of ACS. Nevertheless, statins have metabolic interactions with medications administered to patients with COVID-19. For patients on lopinavir/ritonavir, atorvastatin (20 mg) and rosuvastatin (10 mg) are recommended, respectively. Tocilizumab also has interactions with statins, which are dependent on cytochromes for their metabolisms. For indications other than ACS, the administration of simvastatin and lovastatin should be prohibited due to the increased risk of toxicity in patients receiving lopinavir/ritonavir (Dixon et al. 2020).

16.8.3 Heart Failure

ACEIs/ARBs, beta-blockers, and mineralocorticoid-receptor antagonists (MRAs) constitute the mainstay of heart failure management. For MRAs, it should be noted that eplerenone is contraindicated owing to metabolic interactions with protease inhibitors, which could increase the risk of toxicity, mainly hyperkalemia. Accordingly, spironolactone is the MRA of choice in patients with heart failure. In patients receiving ivabradine and protease inhibitors, the risk of hypotension and bradycardia could be increased significantly (Dixon et al. 2020); consequently, the concomitant administration is not recommended. Sacubitril valsartan should also be initiated at a low dose and titrated slowly in patients on protease inhibitors as the effects of both drugs might be increased (Giguere et al. 2019).

16.8.4 VTE

16.8.4.1 Anticoagulants

Based on the guidelines of the ESC, direct oral anticoagulants (DOACs) should be considered the oral agents of choice except for those with mechanical valves or antiphospholipid syndrome (2020 ESC Guidance for the diagnosis and management of CV Disease in the context of the COVID-19 Pandemic. Available at https://www.escardio.org/Education/COVID-19-and-Cardiology/ESC-COVID-19-Guidance 2020). Meanwhile, for patients receiving lopinavir/ritonavir, DOACs are not recommended. While there is not enough evidence for edoxaban and dabigatran, an elevation in the serum concentration of apixaban and rivaroxaban has been reported. From a pharmacokinetic point of view, between 20% and 25% of apixaban is metabolized by CYP3A4 compared with rivaroxaban, of which 50% is metabolized by CYP3A4; therefore, a dose reduction of 50% is suggested as an alternative for those receiving apixaban. Needless to say, in patients on 2.5 mg twice daily of apixaban, its use is not recommended. For dabigatran, a dose reduction of 50% is suggested as ritonavir is considered to be a P-glycoprotein inhibitor. However, pharmacokinetic studies have indicated that a 2-h separation in administration obviates the need to reduce the dose of dabigatran. Edoxaban requires no dose adjustment for AF and is deemed an option for VTE at a dose of 30 mg once daily; still, its concomitant administration with antiretroviral agents is not recommended due to a lack of evidence. Warfarin may be considered in patients with COVID-19 with the proviso that the international normalized ratio (INR) be carefully monitored (Health and Services 2013). In hemodynamically unstable patients, low-molecular-weight heparin is indicated, and unfractionated heparin is an agent of choice for those at risk of bleeding or kidney failure. Therefore, DOACs should be considered for stable patients at discharge in need of anticoagulation while taking heed of drug interactions (Bikdeli et al. 2020).

16.8.5 Arrhythmias

16.8.5.1 Calcium Channel Blockers

The interaction between protease inhibitors and diltiazem necessitates the close monitoring of patients for the heart rate and blood pressure (Giguere et al. 2019). Owing to the known cardiac adverse effects of chloroquine and hydroxychloroquine, ECG monitoring is recommended to monitor patients for QTc prolongation, TdP, and AV block. A protocol has been advocated to monitor QT at baseline before the initiation of therapy even among those managed as outpatients. The Tisdale risk score can be employed to estimate the risk of QTc prolongation in patients receiving therapy (Simpson et al. 2020). Dose reduction or treatment discontinuation should be performed when QTc is more than 500 ms or when there is an increase in QTc of greater than 60 ms. Other medications with known QT-prolonging effects should be substituted with alternatives. Furthermore, caution is advised for the treatment of patients receiving antiarrhythmics for QTc prolongation effects. However, QT monitoring alone cannot prevent fatal tachyarrhythmias (Huang 2017) because chloroquine and hydroxychloroquine could also place the patient in danger of disturbances in conduction velocity and calcium homeostasis in cardiomyocytes (Jeevaratnam 2020). QTc prolongation is also a problem associated with lopinavir/ritonavir, while AV block is more common with atazanavir; consequently, baseline ECG and its monitoring during treatment have been recommended. It is also advisable that patients on COVID-19 treatment be monitored closely for their digoxin levels (Dixon et al. 2020).

Tables 16.2, 16.3 and 16.4 present cardiovascular medication interactions with antiviral agents used for COVID-19.

Table 16.2 Cardiovascular medication interactions with antiviral agents used for COVID-19

Lopinavir/Ritonavir			
Drug Class	**Drug**	**Interaction**	**Recommendation**
Statins	Simvastatin/Lovastatin	↑risk of adverse effects (hepatotoxicity and myopathy)	Contraindicated
	Rosuvastatin	Rosuvastatin AUC ↑ 2.1-fold and Cmax ↑ 4.7 fold	Not to exceed 10 mg daily
	Atorvastatin	Atorvastatin AUC ↑ 5.9-fold and Cmax ↑ 4.7 fold	Not to exceed 20 mg daily
Oral Anticoagulants (OACs)	Rivaroxaban	↑risk of bleeding	Contraindicated
	Apixaban	↑risk of bleeding	Use is not recommended / Reduce dose by 50%
	Dabigatran	↑risk of bleeding	May be suggested as an option with 2 hour interval
	Edoxaban	↑risk of bleeding	Use is not recommended
	Warfarin	↓	Monitor INR more frequently
Antiplatelets (P2Y12 inhibitors)*	Clopidogrel	Clopidogrel active metabolite AUC ↓ with impaired platelet inhibition	P2Y12 inhibitor of choice with fibrinolytics, platelet function test suggested
	Ticagrelor	↑risk of bleeding	Contraindicated
	Prasugrel	↑risk of bleeding	Not recommended in high risk of bleeding
Antiarrhythmics	Amiodarone	↑	If possible use alternative; if coadministered, reduce dose and monitor BP, LFT, TFT, and ECG periodically
	Lidocaine	↑	Use with caution; monitor ECG and adjust dose if necessary
	Flecainide	↑	Use is not recommended
	Propafenone	↑	Use is not recommended
	Quinidine	↑	Use with caution; monitor ECG and adjust dose if necessary
Beta-blockers	Metoprolol/Atenolol/Carvedilol/Bisoprolol/Propranolol	↑risk of bradycardia and hypotension	Monitor HR/BP; if necessary consider dose reduction
Calcium Channel Blockers	Amlodipine/Verapamil	↑risk of bradycardia/hypotension)	Monitor BP (HR for verapamil)
	Diltiazem		Dose adjustment based on HR/BP
Angiotensin receptor antagonists	Losartan	↓	Monitor BP
	Valsartan	↑	Monitor BP, SCr, K
	Sacubitril/Valsartan	↑	
Alpha-blockers	Terazosin/Prazosin	↑ risk of hypotension	Monitor BP; dose adjustment may be necessary
Diuretics	Indapamide	↑	Monitor BP and serum electrolytes
	Eplerenone	↑risk of adverse effects	Contraindicated
Phosphodiesterase 5 inhibitors	Sildenafil	↑ risk of hypotension, priapism	Use is not recommended
	Tadalafil	↑ risk of hypotension, priapism	Stop tadalafil ≥ 24 hours before Lopinavir/ritonavir initiation and restart 7 days after Lopinavir/ritronavir initiation with 20 mg daily
Cardiac glycosides	Digoxin	↑risk of toxicity	Monitor digoxin concentration if signs of toxicity develops
Miscellaneous	Ranolazine	Risk of QT prolongation	Contraindicated
	Bosentan	Bosentan AUC ↑ 422% and Cmax ↑ 512%	Stop bosentan ≥36 hours before Lopinavir/ritonavir initiation and restart 10 days after Lopinavir/ritonavir initiation with 62.5 mg daily or every other day
	Ivabradine	↑risk of bradycardia	Contraindicated
	Colchicine	↑risk of toxicity	Use half dose; do not co-administer if renal or hepatic impairment

Red, orange, and yellow cells indicate contraindicated use, use with caution/consider dose adjustment, and use under monitoring, respectively

AUC Area under the curve, *BP* Blood pressure, *ECG* Electrocardiogram, *HR* Heart rate, *INR* International normalized ratio, *LFT* Liver function test, *OAC* Oral anticoagulant, *TFT* Thyroid function test

Table 16.3 Cardiovascular medication interactions with antiviral agents used for COVID-19

Atazanavir/Ritonavir			
Drug Class	**Drug**	**Interaction**	**Recommendation**
Statins	Simvastatin/Lovastatin	↑risk of adverse effects (hepatotoxicity and myopathy)	Contraindicated
	Rosuvastatin	Rosuvastatin AUC ↑ 3-fold and Cmax ↑ 7 fold	Monitor for toxicities Max : 10 mg daily
	Atorvastatin	↑	Monitor for toxicities Max : 20 mg daily
Oral Anticoagulants (OACs)	Rivaroxaban	↑risk of bleeding	Contraindicated
	Apixaban		Use is not recommended Use half dose
	Dabigatran		Use is not recommended Use half dose
	Edoxaban	↑	30 mg Daily
	Warfarin	↓	OAC of choice, Monitor INR more frequently
Antiplatelets (P2Y12 inhibitors)	Clopidogrel	Clopidogrel active metabolite AUC↓ with impaired platelet inhibition	P2Y12 inhibitor of choice/ due to increased risk of bleeding by the others
	Ticagrelor	↑risk of bleeding	Contraindicated
Antiarrhythmics	Amiodarone	↑	If possible use alternative; if coadministered, reduce dose and monitor BP, LFT, TFT, and ECG periodically
	Lidocaine	↑	Use with caution; monitor ECG and adjust dose if necessary
	Flecainide	↑	Use is not recommended
	Propafenone	↑	Use is not recommended
	Quinidine	↑	Use with caution; monitor ECG and adjust dose if necessary
Beta-blockers	Metoprolol/Atenolol/ Carvedilol/Bisoprolol/ Propranolol	↑ risk of bradycardia and hypotension	Monitor HR/BP; if necessary consider dose reduction
Calcium Channel Blockers	Amlodipine/Verapamil	↑risk of bradycardia/ hypotension	Monitor BP (HR for verapamil)
	Diltiazem		Use half dose; monitor BP and HR
Alpha-blockers	Prazosin/Terazosin	↑ risk of hypotension	Monitor BP; dose adjustment may be necessary
Diuretics	Eplerenone	↑risk of adverse effects	Contraindicated
Phosphodiesterase 5 inhibitors	Sildenafil	↑risk of hypotension, priapism	Use is not recommended
	Tadalafil		Stop tadalafil ≥ 24 hours before Lopinavir/ritonavir initiation and restart 7 days after Lopinavir/ritonavir initiation with 20 mg daily
Cardiac glycosides	Digoxin	↑risk of toxicity	Monitor digoxin concentration if signs of toxicity develops
Miscellaneous	Ranolazine	Risk of QT prolongation	Contraindicated
	Bosentan	↑	Stop bosentan ≥ 36 hours before ART initiation and restart 10 days after ART initiation with 62.5 mg daily or every other day
	Ivabradine	↑risk of bradycardia	Contraindicated
	Colchicine	↑risk of toxicity	Use half dose; do not co-administer if renal or hepatic impairment

Red, orange, and yellow cells indicate contraindicated use, use with caution/consider dose adjustment, and use under monitoring, respectively

ART antiretroviral, *AUC* Area under the curve, *BP* Blood pressure, *ECG* Electrocardiogram, *HR* Heart rate, *INR* International normalized ratio, *LFT* Liver function test, *OAC* oral anticoagulant, *TFT* thyroid function test

Table 16.4 Cardiovascular medication interactions with antiviral agents used for COVID-19

Chloroquine Phosphate/ Hydroxychloroquine			
Drug Class	**Drug**	**Interaction**	**Recommendation**
Beta-blockers	Metoprolol/Atenolol/ Carvedilol/ Bisoprolol/ Propranolol	↑risk of bradycardia and hypotension	Monitor HR/BP; if necessary consider dose reduction
Cardiac glycosides	Digoxin	↑risk of toxicity	Monitor digoxin concentration if signs of toxicity develop
Oral Anticoagulants (OACs)	Rivaroxaban Apixaban	↑risk of bleeding	Use with caution
	Dabigatran Edoxaban	↑risk of bleeding	Use is not suggested
	Warfarin	↓	Monitor INR more frequently
Antiarrhythmics	Quinidine/ Disopyramide/ Procainamide/ Flecainide/ Propafenone/Amiodarone/ Sotalol	↑risk of QT prolongation and torse de point	Maintain serum Mg ≥ 2 and K ≥ 4 Monitor QTc before initiation
Other QT-prolonging agents	Fluoroquinolones (ciprofloxacin, levofloxacin) Antipsychotics (haloperidol, quetiapine, thioridazine) Antidepressants (citalopram, escitalopram) Antiemetics (metoclopramide, domperidone, ondansetron) Ranolazine Methadone	↑ risk of QT prolongation and torse de point	Use alternative agents

Red, orange, and yellow cells indicate contraindicated use, use with caution/consider dose adjustment, and use under monitoring, respectively

BP Blood pressure, *HR* Heart rate, *INR* International normalized ratio, *OAC* oral anticoagulant

16.9 Conclusion

This chapter discussed various aspects of COVID-19 in the setting of CVD as well as COVID-19-induced cardiac complications and presented the latest pertinent evidence-based findings and guideline recommendations. In light of the most recent evidence on the serious interactions between COVID-19 and CVD as well as the notable drug-drug interactions between COVID-19 and cardiovascular medication, it is advisable that sufficient heed be paid to the points set out in this chapter. Needless to say, given the novelty of the COVID-19 virus and its many hitherto unknown features, all the aforementioned findings and their respective recommendations are subject to alteration as future multinational trials with large sample volumes further contribute to the current pool of knowledge.

References

Abdi S, Salarifar M, Mortazavi SH, Sadeghipour P, Geraiely B (2020) COVID-19 sends STEMI to quarantine!? Clin Res Cardiol:1–2. https://doi.org/10.1007/s00392-020-01664-3

Alhogbani T (2016) Acute myocarditis associated with novel middle east respiratory syndrome coronavirus. Ann Saudi Med 36(1):78–80. https://doi.org/10.5144/0256-4947.2016.78

American Society of Transplantation. COVID-19 (Coronavirus): FAQs for organ transplantation. Updated February 29, 2020. Available from: https://www.myast.org/sites/default/files/COVID19%20FAQ%20Tx%20Centers%20030220-1.pdf

Bangalore S, Sharma A, Slotwiner A, Yatskar L, Harari R, Shah B, Ibrahim H, Friedman GH, Thompson C, Alviar CL, Chadow HL, Fishman GI, Reynolds HR, Keller N, Hochman JS (2020) ST-segment elevation in patients with Covid-19 – a case series. N Engl J Med. https://doi.org/10.1056/NEJMc2009020

Barati S, Garjani K, Payandemehr P, Totonchi Z, Zanganehfar ME, Sadeghipour P, Amin A, Akhondi V, Javadi AM, Sadeghipour P (2020) Iranian heart association task force on cardiopulmonary resuscitation guidelines on the COVID-19 outbreak. Res Cardiovasc Med 9(1):3–5. https://doi.org/10.4103/rcm.rcm_10_20

Basille D, Plouvier N, Trouve C, Duhaut P, Andrejak C, Jounieaux V (2017) Non-steroidal anti-inflammatory drugs may worsen the course of community-acquired pneumonia: a cohort study. Lung 195(2):201–208. https://doi.org/10.1007/s00408-016-9973-1

Bhatraju PK, Ghassemieh BJ, Nichols M, Kim R, Jerome KR, Nalla AK, Greninger AL, Pipavath S, Wurfel MM, Evans L, Kritek PA, West TE, Luks A, Gerbino A, Dale CR, Goldman JD, O'Mahony S, Mikacenic C (2020) Covid-19 in critically ill patients in the Seattle region – case series. N Engl J Med. https://doi.org/10.1056/NEJMoa2004500

Bikdeli B, Madhavan MV, Jimenez D, Chuich T, Dreyfus I, Driggin E, Nigoghossian C, Ageno W, Madjid M, Guo Y, Tang LV, Hu Y, Giri J, Cushman M, Quere I, Dimakakos EP, Gibson CM, Lippi G, Favaloro EJ, Fareed J, Caprini JA, Tafur AJ, Burton JR, Francese DP, Wang EY, Falanga A, McLintock C, Hunt BJ, Spyropoulos AC, Barnes GD, Eikelboom JW, Weinberg I, Schulman S, Carrier M, Piazza G, Beckman JA, Steg PG, Stone GW, Rosenkranz S, Goldhaber SZ, Parikh SA, Monreal M, Krumholz HM, Konstantinides SV, Weitz JI, Lip GYH (2020) COVID-19 and thrombotic or thromboembolic disease: implications for prevention, antithrombotic therapy, and follow-up. J Am Coll Cardiol. https://doi.org/10.1016/j.jacc.2020.04.031

Boriani G, Fauchier L, Aguinaga L, Beattie JM, Blomstrom Lundqvist C, Cohen A, Dan GA, Genovesi S, Israel C, Joung B, Kalarus Z, Lampert R, Malavasi VL, Mansourati J, Mont L, Potpara T, Thornton A, Lip GYH (2019) European Heart Rhythm Association (EHRA) consensus document on management of arrhythmias and cardiac electronic devices in the critically ill and post-surgery patient, endorsed by Heart Rhythm Society (HRS), Asia Pacific Heart Rhythm Society (APHRS), Cardiac Arrhythmia Society of Southern Africa (CASSA), and Latin American Heart Rhythm Society (LAHRS). Europace 21(1):7–8. https://doi.org/10.1093/europace/euy110

Caforio LA Coronavirus disease 2019 (COVID-19): myocardial injury. In UpToDate, Mancini D (ed) UpToDate, Walthman, MA. Retrieved from http://www.uptodate.com/home. Accessed on 14 May 2020

Caforio AL, Pankuweit S, Arbustini E, Basso C, Gimeno-Blanes J, Felix SB, Fu M, Helio T, Heymans S, Jahns R, Klingel K, Linhart A, Maisch B, McKenna W, Mogensen J, Pinto YM, Ristic A, Schultheiss HP, Seggewiss H, Tavazzi L, Thiene G, Yilmaz A, Charron P, Elliott PM (2013) Current state of knowledge on aetiology, diagnosis, management, and therapy of myocarditis: a position statement of the European Society of Cardiology Working Group on Myocardial and Pericardial Diseases. Eur Heart J 34(33):2636–2648, 2648a–2648d. https://doi.org/10.1093/eurheartj/eht210

Castelletti S (2020) A shift on the front line. N Engl J Med. https://doi.org/10.1056/NEJMp2007028

Cave DM, Gazmuri RJ, Otto CW, Nadkarni VM, Cheng A, Brooks SC, Daya M, Sutton RM, Branson R, Hazinski MF (2010) Part 7: CPR techniques and devices: 2010 American Heart Association Guidelines for Cardiopulmonary Resuscitation and Emergency Cardiovascular Care. Circulation 122(18 Suppl 3):S720–S728. https://doi.org/10.1161/circulationaha.110.970970

Chen T, Wu DCH, Yan W, Yang D, Chen G, Ma K, Xu D, Yu H, Wang H, Wang T (2020) Clinical characteristics of 113 deceased patients with coronavirus disease 2019: retrospective study. BMJ (Clin Res Ed) 368:m1295. https://doi.org/10.1136/bmj.m1295

Clerkin KJ, Fried JA, Raikhelkar J, Sayer G, Griffin JM, Masoumi A, Jain SS, Burkhoff D, Kumaraiah D, Rabbani L, Schwartz A, Uriel N (2020) Coronavirus disease 2019 (COVID-19) and cardiovascular disease. Circulation. https://doi.org/10.1161/circulationaha.120.046941

ClinicalTrials.gov. Randomized Controlled Trial of Losartan for Patients With COVID- 19 Requiring Hospitalization. Identifier: NCT04312009. March 17, 2020. https://clinicaltrials.gov/ct2/show/NCT04312009

Conti P, Gallenga CE, Tete G, Caraffa A, Ronconi G, Younes A, Toniato E, Ross R, Kritas SK (2020) How to reduce the likelihood of coronavirus-19 (CoV-19 or SARS-CoV-2) infection and lung inflammation mediated by IL-1. J Biol Regul Homeost Agents 34(2). https://doi.org/10.23812/Editorial-Conti-2

COVID-19 Treatment Guidelines Panel. Coronavirus Disease (2019) (COVID-19) Treatment Guidelines. National Institutes of Health. Available at https://covid19treatmentguidelines.nih.gov/

Danzi GB, Loffi M, Galeazzi G, Gherbesi E (2020) Acute pulmonary embolism and COVID-19 pneumonia: a random association? Eur Heart J. https://doi.org/10.1093/eurheartj/ehaa254

Diaz JH (2020) Hypothesis: angiotensin-converting enzyme inhibitors and angiotensin receptor blockers may increase the risk of severe COVID-19. J Travel Med. https://doi.org/10.1093/jtm/taaa041

Dixon DL, Van Tassell BW, Vecchie A, Bonaventura A, Talasaz AH, Kakavand H, D'Ascenzo F, Perciaccante A, Castagno D, Ammirati E, Biondi-Zoccai G, Stevens MP, Abbate A (2020) Cardiovascular considerations in treating patients with coronavirus disease 2019 (COVID-19). J Cardiovasc Pharmacol 75(5):359–367. https://doi.org/10.1097/fjc.0000000000000836

Driggin E, Madhavan MV, Bikdeli B, Chuich T, Laracy J, Biondi-Zoccai G, Brown TS, Der Nigoghossian C, Zidar DA, Haythe J, Brodie D, Beckman JA, Kirtane AJ, Stone GW, Krumholz HM, Parikh SA (2020) Cardiovascular considerations for patients, health care workers, and health systems during the COVID-19 pandemic. J Am Coll Cardiol 75(18):2352–2371. https://doi.org/10.1016/j.jacc.2020.03.031

Edelson DP, Sasson C, Chan PS, Atkins DL, Aziz K, Becker LB, Berg RA, Bradley SM, Brooks SC, Cheng A, Escobedo M, Flores GE, Girotra S, Hsu A, Kamath-Rayne BD, Lee HC, Lehotzky RE, Mancini ME, Merchant RM, Nadkarni VM, Panchal AR, Peberdy MAR, Raymond TT, Walsh B, Wang DS, Zelop CM, Topjian A (2020) Interim Guidance for Basic and Advanced Life Support in Adults, Children, and Neonates With Suspected or Confirmed COVID-19: From the Emergency Cardiovascular Care Committee and Get With the Guidelines((R))-Resuscitation Adult and Pediatric Task Forces of the American Heart Association in Collaboration with the American Academy of Pediatrics, American Association for Respiratory Care, American College of Emergency Physicians, The Society of Critical Care Anesthesiologists, and American Society of Anesthesiologists: Supporting Organizations: American Association of Critical Care Nurses and National EMS Physicians. Circulation. https://doi.org/10.1161/circulationaha.120.047463

ESC Guidance for the diagnosis and management of CV Disease in the context of the COVID-19 Pandemic (2020). Available at https://www.escardio.org/Education/COVID-19-and-Cardiology/ESC-COVID-19-Guidance

Fang L, Karakiulakis G, Roth M (2020) Are patients with hypertension and diabetes mellitus at increased risk for COVID-19 infection? Lancet Respir Med 8(4):e21. https://doi.org/10.1016/s2213-2600(20)30116-8

Fedson DS (2016) Treating the host response to emerging virus diseases: lessons learned from sepsis, pneumonia, influenza and Ebola. Ann Transl Med 4(21):421. https://doi.org/10.21037/atm.2016.11.03

Ferrario CM, Jessup J, Chappell MC, Averill DB, Brosnihan KB, Tallant EA, Diz DI, Gallagher PE (2005) Effect of angiotensin-converting enzyme inhibition and angiotensin II receptor blockers on cardiac angiotensin-converting enzyme 2. Circulation 111(20):2605–2610. https://doi.org/10.1161/circulationaha.104.510461

Ford I, Shah AS, Zhang R, McAllister DA, Strachan FE, Caslake M, Newby DE, Packard CJ, Mills NL (2016) High-sensitivity cardiac troponin, statin therapy, and risk of coronary heart disease. J Am Coll Cardiol 68(25):2719–2728. https://doi.org/10.1016/j.jacc.2016.10.020

Fried JA, Ramasubbu K, Bhatt R, Topkara VK, Clerkin KJ, Horn E, Rabbani L, Brodie D, Jain SS, Kirtane A, Masoumi A, Takeda K, Kumaraiah D, Burkhoff D, Leon M, Schwartz A, Uriel N, Sayer G (2020) The variety of cardiovascular presentations of COVID-19. Circulation. https://doi.org/10.1161/circulationaha.120.047164

Gabutti G, d'Anchera E, Sandri F, Savio M, Stefanati A (2020) Coronavirus: Update Related to the Current Outbreak of COVID-19. Infect Dis Ther:1–13. https://doi.org/10.1007/s40121-020-00295-5

Gasmi A, Noor S, Tippairote T, Dadar M, Menzel A, Bjorklund G (2020) Individual risk management strategy and potential therapeutic options for the COVID-19 pandemic. Clin Immunol (Orlando, Fla) 215:108409. https://doi.org/10.1016/j.clim.2020.108409

Giguere P, Nhean S, Tseng AL, Hughes CA, Angel JB (2019) Getting to the heart of the matter: a review of drug interactions between HIV Antiretrovirals and cardiology medications. Can J Cardiol 35(3):326–340. https://doi.org/10.1016/j.cjca.2018.12.020

Giray T, Bicer S, Kucuk O, Col D, Yalvac Z, Gurol Y, Yilmaz G, Sac A, Mogol Y (2016) Four cases with Kawasaki disease and viral infection: aetiology or association. Infez Med 24(4):340–344

Guan WJ, Liang WH, Zhao Y, Liang HR, Chen ZS, Li YM, Liu XQ, Chen RC, Tang CL, Wang T, Ou CQ, Li L, Chen PY, Sang L, Wang W, Li JF, Li CC, Ou LM, Cheng B, Xiong S, Ni ZY, Xiang J, Hu Y, Liu L, Shan H, Lei CL, Peng YX, Wei L, Liu Y, Hu YH, Peng P, Wang JM, Liu JY, Chen Z, Li G, Zheng ZJ, Qiu SQ, Luo J, Ye CJ, Zhu SY, Cheng LL, Ye F, Li SY, Zheng JP, Zhang NF, Zhong NS, He JX (2020a) Comorbidity and its impact on 1590 patients with COVID-19 in China: a nationwide analysis. Eur Respir J 55(5). https://doi.org/10.1183/13993003.00547-2020

Guan WJ, Ni ZY, Hu Y, Liang WH, Ou CQ, He JX, Liu L, Shan H, Lei CL, Hui DSC, Du B, Li LJ, Zeng G, Yuen KY, Chen RC, Tang CL, Wang T, Chen PY, Xiang J, Li SY, Wang JL, Liang ZJ, Peng YX, Wei L, Liu Y, Hu YH, Peng P, Wang JM, Liu JY, Chen Z, Li G, Zheng ZJ, Qiu SQ, Luo J, Ye CJ, Zhu SY, Zhong NS (2020b) Clinical characteristics of coronavirus disease 2019 in China. N Engl J Med 382(18):1708–1720. https://doi.org/10.1056/NEJMoa2002032

Guidance for Cardiothoracic Transplant and Mechanical Circulatory Support Centers regarding SARS CoV-2 infection and COVID-19 (2020)

Guo T, Fan Y, Chen M, Wu X, Zhang L, He T, Wang H, Wan J, Wang X, Lu Z (2020) Cardiovascular implications of fatal outcomes of patients with coronavirus disease 2019 (COVID-19). JAMA Cardiol. https://doi.org/10.1001/jamacardio.2020.1017

Health UD, Services H (2013) Panel on antiretroviral guidelines for adults and adolescents. Guidelines for the use of antiretroviral agents in HIV-1-infected adults and adolescents. US, Washington, DC

Henry BM, Lippi G (2020) Chronic kidney disease is associated with severe coronavirus disease 2019 (COVID-19) infection. Int Urol Nephrol. https://doi.org/10.1007/s11255-020-02451-9

https://covid19treatmentguidelines.nih.gov

https://jamanetwork.com/pages/cc-by-license-permissions

https://www.elsevier.com/authors/permission-request-form

Hu H, Ma F, Wei X, Fang Y (2020) Coronavirus fulminant myocarditis treated with glucocorticoid and human immunoglobulin. Eur Heart J. https://doi.org/10.1093/eurheartj/ehaa190

Huang CL (2017) Murine electrophysiological models of cardiac Arrhythmogenesis. Physiol Rev 97(1):283–409. https://doi.org/10.1152/physrev.00007.2016

Huang C, Wang Y, Li X, Ren L, Zhao J, Hu Y, Zhang L, Fan G, Xu J, Gu X, Cheng Z, Yu T, Xia J, Wei Y, Wu W, Xie X, Yin W, Li H, Liu M, Xiao Y, Gao H, Guo L, Xie J, Wang G, Jiang R, Gao Z, Jin Q, Wang J, Cao B (2020) Clinical features of patients infected with 2019 novel coronavirus in Wuhan, China. Lancet (London, England) 395(10223):497–506. https://doi.org/10.1016/s0140-6736(20)30183-5

Iba T, Levy JH, Warkentin TE, Thachil J, van der Poll T, Levi M (2019) Diagnosis and management of sepsis-induced coagulopathy and disseminated intravascular coagulation. J Thromb Haemost 17(11):1989–1994. https://doi.org/10.1111/jth.14578

Inciardi RM, Lupi L, Zaccone G, Italia L, Raffo M, Tomasoni D, Cani DS, Cerini M, Farina D, Gavazzi E, Maroldi R, Adamo M, Ammirati E, Sinagra G, Lombardi CM, Metra M (2020) Cardiac involvement in a patient with coronavirus disease 2019 (COVID-19). JAMA Cardiol. https://doi.org/10.1001/jamacardio.2020.1096

Jaffe AS, Vasile VC, Milone M, Saenger AK, Olson KN, Apple FS (2011) Diseased skeletal muscle: a noncardiac source of increased circulating concentrations of cardiac troponin T. J Am Coll Cardiol 58(17):1819–1824. https://doi.org/10.1016/j.jacc.2011.08.026

Jeevaratnam K (2020) Chloroquine and hydroxychloroquine for COVID-19: implications for cardiac safety. Eur Heart J Cardiovasc Pharmacother. https://doi.org/10.1093/ehjcvp/pvaa041

Kleinman ME, Brennan EE, Goldberger ZD, Swor RA, Terry M, Bobrow BJ, Gazmuri RJ, Travers AH, Rea T (2015) Part 5: adult basic life support and cardiopulmonary resuscitation quality: 2015 American Heart Association Guidelines Update for Cardiopulmonary Resuscitation and Emergency Cardiovascular Care. Circulation 132(18 Suppl 2):S414–S435. https://doi.org/10.1161/cir.0000000000000259

Klok FA, Kruip M, van der Meer NJM, Arbous MS, Gommers D, Kant KM, Kaptein FHJ, van Paassen J, Stals MAM, Huisman MV, Endeman H (2020) Incidence of thrombotic complications in critically ill ICU patients with COVID-19. Thromb Res. https://doi.org/10.1016/j.thromres.2020.04.013

Kuster GM, Pfister O, Burkard T, Zhou Q, Twerenbold R, Haaf P, Widmer AF, Osswald S (2020) SARS-CoV2: should inhibitors of the renin-angiotensin system be withdrawn in patients with COVID-19? Eur Heart J 41(19):1801–1803. https://doi.org/10.1093/eurheartj/ehaa235

Lakkireddy DR, Chung MK, Gopinathannair R, Patton KK, Gluckman TJ, Turagam M, Cheung J, Patel P, Sotomonte J, Lampert R, Han JK, Rajagopalan B, Eckhardt L, Joglar J, Sandau K, Olshansky B, Wan E, Noseworthy PA, Leal M, Kaufman E, Gutierrez A, Marine JM, Wang PJ, Russo AM (2020) Guidance for cardiac electrophysiology during the coronavirus (COVID-19) Pandemic from the Heart Rhythm Society COVID-19 task force; electrophysiology section of the American College of Cardiology; and the electrocardiography and arrhythmias Committee of the Council on clinical cardiology. Am Heart Assoc Heart Rhythm. https://doi.org/10.1016/j.hrthm.2020.03.028

Li J, Wang X, Chen J, Zhang H, Deng A (2020) Association of Renin-Angiotensin System Inhibitors with Severity or risk of death in patients with hypertension hospitalized for coronavirus disease 2019 (COVID-19) infection in Wuhan, China. JAMA Cardiol. https://doi.org/10.1001/jamacardio.2020.1624

Libby P, Loscalzo J, Ridker PM, Farkouh ME, Hsue PY, Fuster V, Hasan AA, Amar S (2018) Inflammation, immunity, and infection in Atherothrombosis: JACC review topic of the week. J Am Coll Cardiol 72(17):2071–2081. https://doi.org/10.1016/j.jacc.2018.08.1043

Lippi G, Lavie CJ, Sanchis-Gomar F (2020a) Cardiac troponin I in patients with coronavirus disease 2019 (COVID-19): evidence from a meta-analysis. Prog Cardiovasc Dis. https://doi.org/10.1016/j.pcad.2020.03.001

Lippi G, Plebani M, Henry BM (2020b) Thrombocytopenia is associated with severe coronavirus disease 2019 (COVID-19) infections: a meta-analysis. Clin Chim Acta 506:145–148. https://doi.org/10.1016/j.cca.2020.03.022

Liu K, Fang YY, Deng Y, Liu W, Wang MF, Ma JP, Xiao W, Wang YN, Zhong MH, Li CH, Li GC, Liu HG (2020a) Clinical characteristics of novel coronavirus cases in tertiary hospitals in Hubei Province. Chin Med J 133(9):1025–1031. https://doi.org/10.1097/cm9.0000000000000744

Liu Y, Yang Y, Zhang C, Huang F, Wang F, Yuan J, Wang Z, Li J, Li J, Feng C, Zhang Z, Wang L, Peng L, Chen L, Qin Y, Zhao D, Tan S, Yin L, Xu J, Zhou C, Jiang C, Liu L (2020b) Clinical and biochemical indexes from 2019-nCoV infected patients linked to viral loads and lung injury. Sci China Life Sci 63(3):364–374. https://doi.org/10.1007/s11427-020-1643-8

Madjid M, Safavi-Naeini P, Solomon SD, Vardeny O (2020) Potential effects of coronaviruses on the cardiovascular system: a review. JAMA Cardiol. https://doi.org/10.1001/jamacardio.2020.1286

Massachusetts General Hospital COVID-19 Treatment Guidance. The General Hospital Corporation. 2020 Version 3.1: 1–16. Available at https://advances.massgeneral.org/research-and-innovation/article.aspx?id=1184

Mehra MR, Desai SS (2020) Cardiovascular disease, drug therapy, and mortality in Covid-19. N Engl J Med. https://doi.org/10.1056/NEJMoa2007621

Mehra MR, Ruschitzka F (2020) COVID-19 illness and heart failure: a missing link? JACC Heart Fail. https://doi.org/10.1016/j.jchf.2020.03.004

Morrison LJ, Gent LM, Lang E, Nunnally ME, Parker MJ, Callaway CW, Nadkarni VM, Fernandez AR, Billi JE, Egan JR, Griffin RE, Shuster M, Hazinski MF (2015) Part 2: evidence evaluation and management of conflicts of interest: 2015 American Heart Association Guidelines Update for Cardiopulmonary Resuscitation and Emergency Cardiovascular Care.

Circulation 132(18 Suppl 2):S368–S382. https://doi.org/10.1161/cir.0000000000000253

Orser BA (2020) Recommendations for endotracheal intubation of COVID-19 patients. Anesth Analg 130(5):1109–1110. https://doi.org/10.1213/ane.0000000000004803

Poissy J, Goutay J, Caplan M, Parmentier E, Duburcq T, Lassalle F, Jeanpierre E, Rauch A, Labreuche J, Susen S (2020) Pulmonary embolism in COVID-19 patients: awareness of an increased prevalence. Circulation. https://doi.org/10.1161/circulationaha.120.047430

ClinicalTrials.gov. Randomized Controlled Trial of Losartan for Patients With COVID- 19 Not Requiring Hospitalization. Identifier: NCT04311177. March 17, 2020. https://clinicaltrials.gov/ct2/show/NCT04311177

Ren ZL, Hu R, Wang ZW, Zhang M, Ruan YL, Wu ZY, Wu HB, Hu XP, Hu ZP, Ren W, Li LC, Dai FF, Liu H, Cai X (2020) Epidemiologic and clinical characteristics of heart transplant recipients during the 2019 coronavirus outbreak in Wuhan, China: a descriptive survey report. J Heart Lung Transplant 39(5):412–417. https://doi.org/10.1016/j.healun.2020.03.008

Resuscitation Council of United Kingdom (2020, March) Resuscitation Council UK Statement on COVID-19 in relation to CPR and resuscitation in first aid and community settings. https://www.resus.org.uk/media/statements/resuscitation-council-uk-statements-on-covid-19-coronavirus-cpr-and-resuscitation/covid-community/

Saavedra JM (2020) Angiotensin receptor blockers and COVID-19. Pharmacol Res 156:104832. https://doi.org/10.1016/j.phrs.2020.104832

Sala S, Peretto G, Gramegna M, Palmisano A, Villatore A, Vignale D, De Cobelli F, Tresoldi M, Cappelletti AM, Basso C, Godino C, Esposito A (2020) Acute myocarditis presenting as a reverse Tako-Tsubo syndrome in a patient with SARS-CoV-2 respiratory infection. Eur Heart J 41(19):1861–1862. https://doi.org/10.1093/eurheartj/ehaa286

Sanders JM, Monogue ML, Jodlowski TZ, Cutrell JB (2020) Pharmacologic treatments for coronavirus disease 2019 (COVID-19): a review. JAMA. https://doi.org/10.1001/jama.2020.6019

Sattarzadeh Badkoubeh R, Almassi N, Farahani MM, Movassaghi N, Khosravi A, Mirdamadi A, Kamali M, Manshouri S, Tabrizi MT, Farrashi M, Samiei N, committee ISEs (2020) The Iranian Society of Echocardiography (ISE) statement on performing echocardiography during the COVID-19 pandemic. Curr Probl Cardiol:100620. https://doi.org/10.1016/j.cpcardiol.2020.100620

Shi S, Qin M, Shen B, Cai Y, Liu T, Yang F, Gong W, Liu X, Liang J, Zhao Q, Huang H, Yang B, Huang C (2020) Association of Cardiac Injury with Mortality in hospitalized patients with COVID-19 in Wuhan. China JAMA Cardiol. https://doi.org/10.1001/jamacardio.2020.0950

Simpson T, Kovacs R, Stecker E (2020, May 29) Ventricular arrhythmia risk due to hydroxychloroquine-azithromycin treatment for COVID-19. Cardiol Mag. https://www.acc.org/latest-in-cardiology/articles/2020/03/27/14/00/ventricular-arrhythmia-risk-due-to-hydroxychloroquine-azithromycin-treatment-for-covid-19

South AM, Diz DI, Chappell MC (2020) COVID-19, ACE2, and the cardiovascular consequences. Am J Physiol Heart Circ Physiol 318(5):H1084–h1090. https://doi.org/10.1152/ajpheart.00217.2020

Statement from the American Heart Association, Heart Failure Society of America, American College of Cardiology. Patients taking ACE-i and ARBs who contract COVID-19 should continue treatment, unless otherwise advised by their physician. (2020)

Tam CF, Cheung KS, Lam S, Wong A, Yung A, Sze M, Lam YM, Chan C, Tsang TC, Tsui M, Tse HF, Siu CW (2020) Impact of coronavirus disease 2019 (COVID-19) outbreak on ST-segment-elevation myocardial infarction Care in Hong Kong, China. Circ Cardiovasc Qual Outcomes 13(4):e006631. https://doi.org/10.1161/circoutcomes.120.006631

Tang N, Bai H, Chen X, Gong J, Li D, Sun Z (2020a) Anticoagulant treatment is associated with decreased mortality in severe coronavirus disease 2019 patients with coagulopathy. J Thromb Haemost. https://doi.org/10.1111/jth.14817

Tang N, Li D, Wang X, Sun Z (2020b) Abnormal coagulation parameters are associated with poor prognosis in patients with novel coronavirus pneumonia. J Thromb Haemost 18(4):844–847. https://doi.org/10.1111/jth.14768

Tavazzi G, Pellegrini C, Maurelli M, Belliato M, Sciutti F, Bottazzi A, Sepe PA, Resasco T, Camporotondo R, Bruno R, Baldanti F, Paolucci S, Pelenghi S, Iotti GA, Mojoli F, Arbustini E (2020) Myocardial localization of coronavirus in COVID-19 cardiogenic shock. Eur J Heart Fail. https://doi.org/10.1002/ejhf.1828

Tersalvi G, Vicenzi M, Calabretta D, Biasco L, Pedrazzini G, Winterton D (2020) Elevated troponin in patients with coronavirus disease 2019: possible mechanisms. J Card Fail. https://doi.org/10.1016/j.cardfail.2020.04.009

Thachil J, Tang N, Gando S, Falanga A, Cattaneo M, Levi M, Clark C, Iba T (2020) ISTH interim guidance on recognition and management of coagulopathy in COVID-19. J Thromb Haemost. https://doi.org/10.1111/jth.14810

Thygesen K, Alpert JS, Jaffe AS, Chaitman BR, Bax JJ, Morrow DA, White HD (2018) Fourth universal definition of myocardial infarction (2018). J Am Coll Cardiol 72(18):2231–2264. https://doi.org/10.1016/j.jacc.2018.08.1038

Vaduganathan M, Vardeny O, Michel T, McMurray JJV (2020) Renin-angiotensin-aldosterone system inhibitors in patients with Covid-19. N Engl J Med 382(17):1653–1659. https://doi.org/10.1056/NEJMsr2005760

Veronese N, Demurtas J, Yang L, Tonelli R, Barbagallo M, Lopalco P, Lagolio E, Celotto S, Pizzol D, Zou L, Tully MA, Ilie PC, Trott M, Lopez-Sanchez GF,

Smith L (2020) Use of corticosteroids in coronavirus disease 2019 pneumonia: a systematic review of the literature. Front Med 7:170. https://doi.org/10.3389/fmed.2020.00170

Walkey AJ, Hammill BG, Curtis LH, Benjamin EJ (2014) Long-term outcomes following development of new-onset atrial fibrillation during sepsis. Chest 146(5):1187–1195. https://doi.org/10.1378/chest.14-0003

Wang D, Hu B, Hu C, Zhu F, Liu X, Zhang J, Wang B, Xiang H, Cheng Z, Xiong Y, Zhao Y, Li Y, Wang X, Peng Z (2020a) Clinical Characteristics of 138 Hospitalized Patients With 2019 Novel Coronavirus-Infected Pneumonia in Wuhan, China. JAMA. https://doi.org/10.1001/jama.2020.1585

Wang T, Chen R, Liu C, Liang W, Guan W, Tang R, Tang C, Zhang N, Zhong N, Li S (2020b) Attention should be paid to venous thromboembolism prophylaxis in the management of COVID-19. Lancet Haematol. https://doi.org/10.1016/s2352-3026(20)30109-5

Wu Z, McGoogan JM (2020) Characteristics of and important lessons from the coronavirus disease 2019 (COVID-19) outbreak in China: summary of a report of 72 314 cases from the Chinese Center for Disease Control and Prevention. JAMA 323(13):1239–1242. https://doi.org/10.1001/jama.2020.2648

Xie Y, Cao S, Dong H, Li Q, Chen E, Zhang W, Yang L, Fu S, Wang R (2020) Effect of regular intravenous immunoglobulin therapy on prognosis of severe pneumonia in patients with COVID-19. J Infect. https://doi.org/10.1016/j.jinf.2020.03.044

Yin S, Huang M, Li D, Tang N (2020) Difference of coagulation features between severe pneumonia induced by SARS-CoV2 and non-SARS-CoV2. J Thromb Thrombolysis. https://doi.org/10.1007/s11239-020-02105-8

Zeng J-H, Liu Y-X, Yuan J, Wang F-X, Wu W-B, Li J-X, Wang L-F, Gao H, Wang Y, Dong C-F (2020) First case of COVID-19 complicated with fulminant myocarditis: a case report and insights. Infection:1. https://doi.org/10.1007/s15010-020-01424-5

Zhang C, Wu Z, Li JW, Zhao H, Wang GQ (2020a) The cytokine release syndrome (CRS) of severe COVID-19 and Interleukin-6 receptor (IL-6R) antagonist tocilizumab may be the key to reduce the mortality. Int J Antimicrob Agents:105954. https://doi.org/10.1016/j.ijantimicag.2020.105954

Zhang P, Zhu L, Cai J, Lei F, Qin JJ, Xie J, Liu YM, Zhao YC, Huang X, Lin L, Xia M, Chen MM, Cheng X, Zhang X, Guo D, Peng Y, Ji YX, Chen J, She ZG, Wang Y, Xu Q, Tan R, Wang H, Lin J, Luo P, Fu S, Cai H, Ye P, Xiao B, Mao W, Liu L, Yan Y, Liu M, Chen M, Zhang XJ, Wang X, Touyz RM, Xia J, Zhang BH, Huang X, Yuan Y, Rohit L, Liu PP, Li H (2020b) Association of Inpatient use of angiotensin converting enzyme inhibitors and angiotensin II receptor blockers with mortality among patients with hypertension hospitalized with COVID-19. Circ Res. https://doi.org/10.1161/circresaha.120.317134

Zheng YY, Ma YT (2020) COVID-19 and the cardiovascular system. Nat Rev Cardiol 17(5):259–260. https://doi.org/10.1038/s41569-020-0360-5

Zhou F, Yu T, Du R, Fan G, Liu Y, Liu Z, Xiang J, Wang Y, Song B, Gu X, Guan L, Wei Y, Li H, Wu X, Xu J, Tu S, Zhang Y, Chen H, Cao B (2020) Clinical course and risk factors for mortality of adult inpatients with COVID-19 in Wuhan, China: a retrospective cohort study. Lancet (London, England) 395(10229):1054–1062. https://doi.org/10.1016/s0140-6736(20)30566-3

How Prevalent Is Cancer in Confirmed Cases with Coronaviruses and Severe Acute Respiratory Syndromes?

17

Maryam Fotouhi, Elham Samami, Sahar Mohseni, Amir Nasrollahizadeh, Mohammad Haddadi, Mona Mirbeyk, Amene Saghazadeh, and Nima Rezaei

Abstract

Novel coronavirus disease 2019 (COVID-19) has posed a crucial hazard to global health. The new species share similarities with the two previously emerged entities: severe acute respiratory syndrome (SARS) and the Middle East respiratory syndrome (MERS) that have caused outbreaks in 2002 and 2012, respectively. Interestingly, all of these coronaviruses can cause potentially fatal respiratory syndromes, though behave differently in patients with cancer compared to patients without cancer. Accordingly, the present chapter aims to, through a systematic investigation, estimate the prevalence of cancer among COVID-19, SARS, and MERS confirmed cases. Our analysis based on data from 78 studies with SARS, MERS, and COVID-19 confirmed cases showed that the prevalence of cancer (4.94%) stands at

M. Fotouhi · A. Nasrollahizadeh · M. Haddadi · M. Mirbeyk
Systematic Review and Meta-analysis Expert Group (SRMEG), Universal Scientific Education and Research Network (USERN), Tehran, Iran

School of Medicine, Tehran University of Medical Sciences, Tehran, Iran

E. Samami
Systematic Review and Meta-analysis Expert Group (SRMEG), Universal Scientific Education and Research Network (USERN), Tehran, Iran

School of Medicine, Iran University of Medical Sciences, Tehran, Iran

S. Mohseni
Systematic Review and Meta-analysis Expert Group (SRMEG), Universal Scientific Education and Research Network (USERN), Tehran, Iran

School of Nursing and Midwifery, Tehran University of Medical Sciences, Tehran, Iran

A. Saghazadeh
Systematic Review and Meta-analysis Expert Group (SRMEG), Universal Scientific Education and Research Network (USERN), Tehran, Iran

Research Center for Immunodeficiencies, Children's Medical Center, Tehran University of Medical Sciences, Tehran, Iran

N. Rezaei (✉)
Research Center for Immunodeficiencies, Children's Medical Center, Tehran University of Medical Sciences, Tehran, Iran

Department of Immunology, School of Medicine, Tehran University of Medical Sciences, Tehran, Iran

Network of Immunity in Infection, Malignancy and Autoimmunity (NIIMA), Universal Scientific Education and Research Network (USERN), Tehran University of Medical Sciences, Tehran, Iran
e-mail: rezaei_nima@tums.ac.ir

N. Rezaei (ed.), *Coronavirus Disease - COVID-19*, Advances in Experimental Medicine and Biology 1318, https://doi.org/10.1007/978-3-030-63761-3_17

fourth place after hypertension (20.8%), diabetes (11.39%), and cardiovascular diseases (7.46%). According to the findings of the present study, comorbidities are significantly more common in patients with MERS compared to patients with COVID-19 and SARS, and this was the cancer case as well. Further studies need to address whether or not patients with coronaviruses and cancer are different from patients with coronaviruses without cancer in terms of clinical manifestations, laboratory findings, outcomes, and men to women ratio.

Keywords

Cancer · Comorbidities · COVID-19 · MERS · SARS · Systematic review

17.1 Introduction

Novel coronavirus disease 2019 (COVID-19) has posed a crucial hazard to global health as declared by the World Health Organization (WHO) as a pandemic on March 11, 2020. Though the disease can involve multiple systems, it would predominantly affect the respiratory system. Patients with COVID-19 frequently report at least a preexisting condition, such as cardiovascular disease, diabetes, hypertension, and pulmonary disease (Wang et al. 2004a; Shi et al. 2020c, d). Moreover, patients with severe COVID-19 are more likely to report comorbidities than patients with non-severe COVID-19. The findings propose that people with comorbidities are at higher risk of infection and adverse outcomes for COVID-19 (Wang et al. 2020a). In particular, studies show that COVID-19 in cancer patients is different from the general population in terms of higher risk for poor outcomes (Zhang et al. 2020d) such as intensive care unit (ICU) admission, invasive ventilation, and death (Liang et al. 2020). Consistently, a recent investigation has estimated a considerable case fatality rate of COVID-19 to be about 5.6% in cancer patients compared to 2.8% in the general population (Wu and McGoogan 2020a).

Severe acute respiratory syndrome (SARS) and the Middle East respiratory syndrome (MERS) are the other two human coronaviruses that have caused outbreaks in 2002 and 2012, respectively. The new species share similarities with the two previously emerged entities. Interestingly, all of these coronaviruses behave differently in patients with cancer compared to patients without cancer. Accordingly, the present chapter aimed to, through a systematic investigation, estimate the prevalence of cancer among COVID-19, SARS, and MERS confirmed cases.

17.2 Methods

A systematic review of the literature was conducted according to the Preferred Reporting Items for Systematic Review and Meta-Analyses (PRISMA) statement (Moher et al. 2009).

17.2.1 Literature Search

A systematic search was conducted in PubMed, Google Scholar, Scopus, Web of Science, and Cochrane Library to include studies published by April 17, 2020. The combinations of the following keywords were used: 2019 novel coronavirus, novel coronavirus, novel coronavirus 2019, 2019-nCoV, COVID-19, SARS-CoV-2, SARS-CoV, severe acute respiratory syndrome, MERS-CoV, and cancer.

17.2.2 Study Selection

The results of the initial search strategy were first screened by title and abstract. The full texts of potentially relevant articles were examined for inclusion and exclusion criteria. Original observational studies, i.e., case-control, cohort, case-series, and cross-sectional studies, that investigated the prevalence and/or clinical features of COVID-19, SARS, and MERS in adult cancer patients (18 years of age or older) were included in the systematic review. Other types of articles, e.g., case reports, nonhuman studies, editorial, letter-to-editor, and also studies that provided no adequate information for meta-analysis, were excluded.

17.2.3 Data Extraction

Two groups of authors independently extracted the following data from each included publication: author, location, year of publication, study groups (COVID-19, SARS, and MERS), number of participants, age, clinical manifestations (e.g., fever, cough, sputum production, chill, palpitation, abdominal pain, diarrhea, lack of appetite, sore throat, chest pain, dizziness, shortness of breath, respiratory failure, nasal congestion, headache, hemoptysis, myalgia, fatigue, nausea, vomiting), number of comorbidities (cancer, hypertension, diabetes, cardiovascular disease, liver diseases, pulmonary diseases, cerebrovascular diseases, thyroid diseases, chronic kidney diseases, immunodeficiency, endocrine diseases, and cardio-cerebrovascular diseases), and laboratory findings. In case of disagreements, the two groups of authors discussed the study, and if the issue remained, the senior author was consulted.

17.2.4 Quality Assessment

The Newcastle-Ottawa scale (NOS) was used for assessment of the quality of cohort (Wells et al. 2014), case-control (Wells et al. 2016), and cross-sectional studies (Margulis et al. 2014). Also, a quality assessment tool was applied for case series (Murad et al. 2018). Two authors independently performed a quality assessment, and controversies were resolved by discussion between the authors and expert opinions.

17.2.5 Statistical Analysis and Publication Bias

All analyses were performed in STATA MP version 15. The random-effect model of analysis was used to estimate the pooled prevalence of cancer among confirmed cases of COVID-19, SARS, and MERS. The statistics test I2% was calculated for the measurement of heterogeneity across studies. I2 values of 50% and above were considered as significant heterogeneity. Testing for funnel plot asymmetry using Egger's regression test was carried out for the assessment of publication bias.

17.3 Results

17.3.1 Study Selection and Characteristics

After duplicate removal, a total of 1852 discrete records remained for title/abstract screening. Two groups of reviewers independently screened the literature. After screening based on abstract and title, 354 articles were selected for detailed review; of these, 265 were excluded with reasons for exclusion outlined in Fig. 17.1. Finally, 78 studies were included in the meta-analysis (Ai et al. 2020; Ajlan et al. 2014; Al-Tawfiq et al. 2014; Arabi et al. 2014; Assiri et al. 2013; Booth et al. 2003; Cai et al. 2020; Cao et al. 2020; Chan et al. 2020; Chen et al. 2020a, b, c, d, e; Cui et al. 2020; Diao et al. 2020; Du et al. 2020a, b; Gritti et al. 2020; Guan et al. 2020a, b ; Guo et al. 2020; Ho et al. 2003; Hu et al. ; Huang et al. 2020a; Huang et al. 2020b; Hui et al. 2020; Jazieh et al. 2020; Jiang et al. 2020; Kenneth W. Tsang and Article 2003; Kim et al. 2020; Kim et al. 2015; Kui et al. 2020; Lei et al. 2020; Lescure et al. 2020; Li et al. 2020a, b; Liu et al. 2020a, b; Nelson Lee et al. 2003; Omrani et al. 2014; Omrani-Nava et al. 2020 ; Peiris et al. 2003a; Prevention 2015; Prof 2003b; Qi et al. 2020; Qin et al. 2020; Shi et al. 2020a, b, c ; Su et al. 2020; Cho et al. 2016; Tan et al. 2020; Tsui et al. 2003;

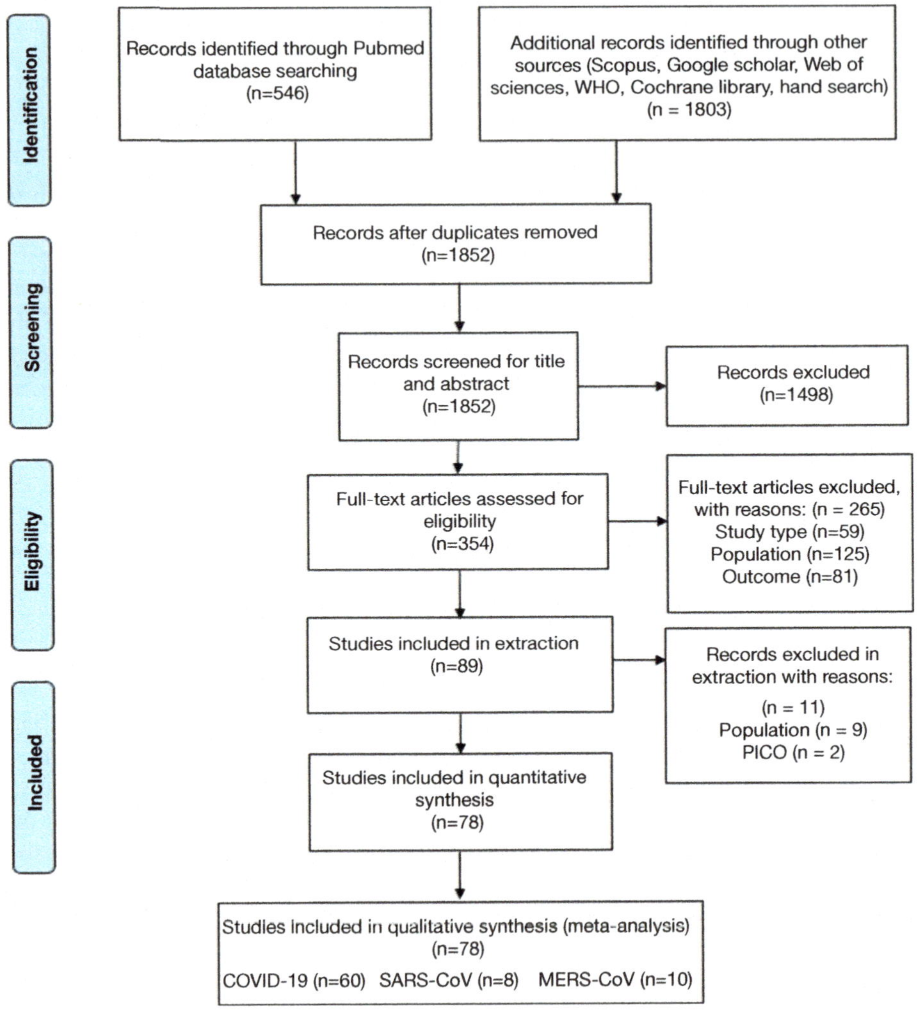

Fig. 17.1 PRISMA flow diagram of study selection

Wang et al. 2020b, c, d, e, f, 2004b, 2020g; Wu et al. 2020a, b, c; Xu et al. 2020a, b; Yang et al. 2020a, b; Zhang et al. 2020a, b, c, d ; Zhang et al. 2020e, f, g; Zhao et al. 2020; Zhou et al. 2020; 房晓伟 et al. 2020). The main characteristics of the included studies and the results of quality assessment are summarized in Tables 17.1, 17.2, 17.3, 17.4 and 17.5.

17.3.2 Demographical Characteristics and Clinical Manifestations

Seventy-eight studies with a total of 11043 patients with confirmed SARS (n = 856), MERS (n = 635), and COVID-19 (n = 9552) were included in the present systematic review and meta-analysis. The mean age of participants across studies was 50.83 years. It was 42.27, 55.38, and 51.29 years in SARS, MERS, and COVID-19 subgroups. Male to female ratio was more than one in total and in COVID-19 and MERS subgroups, while it was less than one in patients infected with SARS (Table 17.6).

Overall, fever was the most prevalent manifestation detected in 84.58% of patients. In particular, it was more common in SARS (99.76%) than in COVID-19 (84.16%) and MERS (75.61%) subgroups. After the fever, the most common clinical manifestations included cough

Table 17.1 Characteristics of the included studies on cancer in cases confirmed as SARS, MERS, and COVID-19 by real-time RT-PCR

First author	Year	Virus	N	Sex (M/F)	Age (mean, SD)	Number of cancer patients
Kenneth W. Tsang et al.	2003	SARS	10	5/5	52.2 ±11	1
Nelson Lee et al.	2003	SARS	138	66/72	39.3 ±16.8	2
J S M Peiris, CM Chu, et al.	2003	SARS	75	36/39	39.54 ±12.3363	1
J S M Peiris, ST Lai, et al.	2003	SARS	50	22/28	45.25±1136	1
Ping Tim Tsui et al.	2003	SARS	323	127/196	41 ±14	1
Wang Y et al.	2004	SARS	44	15/29	45.8545 ±18.5932	7
James C. Ho et al.	2003	SARS	72	30/42	44.75±12.39	2
Christopher M. Booth et al.	2003	SARS	144	56/88	45.33±17.22	9
Kim, K. H. et al.	2015	MERS	185	111/74	55±12	51
Kyung Min Kim et al.	2015	MERS	36	20/16	51.25±13	9
Amr M. Ajlan et al.	2014	MERS	7	5/2	49.85 ±21.22	2
Jaffar A. Al-Tawfiq et al.	2014	MERS	17	11/6	56.25±20.32	1
Yaseen M. Arabi et al.	2014	MERS	12	8/4	59.25±14.37	1
Abdullah Assiri et al.	2013	MERS	47	36/11	55±18.02	1
Omrani, Ali S, et al.	2014	MERS	44	32/12	65.5 ±18.2	2
Korea Centers for Disease Control and Prevention*	2015	MERS	186	111/75	54.33±17.93	43
Sun Young Cho et al.	2016	MERS	82	53/29	54.11±14.88	12
Abdul-Rahman Jazieh	2020	MERS	19	12/7	59±19.5	12
Jin-Wei Ai et al.	2020	COVID-19	102	50/52	50.38 ±16.86	1
Qingxian Cai et al.	2020	COVID-19	298	149/149	47±4.88	4
Xiaoping Chen et al.	2020	COVID-19	25	11/14	51.4 ±11.6	1
Xiaohua Chen et al.	2020	COVID-19	48	37/11	64.8 ±18.1	6
Xu Chen et al.	2020	COVID-19	291	145/146	46.3±18.62	2
Pengfei Cui et al.	2020	COVID-19	35	0/35	61.5 ±11.2	2
Bo Diao et al.	2020	COVID-19	85	48/37	28.25±14.54	1
Ying Huang et al.	2020	COVID-19	36	25/11	69.22 ±6.94	1
Wanli Jiang et al.	2020	COVID-19	112	52/60	53.00 ±17.87	14
Yu Shi et al.	2020	COVID-19	487	259/228	46 ±19	5
Wang Wenjun1	2020	COVID-19	11	10/1	59.66±19.5	1
Hui Hui et al.	2020	COVID-19	41	19/22	48.66±22.27	1
Jing Li et al.	2019	COVID-19	47	28/19	61±14.52	7

(continued)

Table 17.1 (continued)

First author	Year	Virus	N	Sex (M/F)	Age (mean, SD)	Number of cancer patients
Xun Li et al.	2020	COVID-19	25	10/15	71.48 ±12.42	2
Lei Liu et al.	2020	COVID-19	153	93/20	52.33±18.71	12
Ru Liu et al.	2020	COVID-19	41	17/24	39.1 ±9.2	2
Xiaolong Qi et al.	2020	COVID-19	52	17/35	37±16.01	2
Heshui Shi et al.	2020	COVID-19	81	42/39	49.5 ±11	4
L Wang et al.	2020	COVID-19	18	10/8	41±20.91	1
J Wu et al.	2020	COVID-19	80	39/41	46.10 ±15.42	1
Y Xu et al.	2020	COVID-19	45	29/16	56.7 ±15.4	3
W Yang et al.	2020	COVID-19	149	81/68	45.11 ±13.35	2
X Yang et al.	2020	COVID-19	52	35/17	59.7 ±13.3	2
B Zhang et al.	2020	COVID-19	82	54/28	72.5±11.31	6
G Zhang et al.	2020	COVID-19	221	108/113	53.16±19.77	9
J Zhang et al.	2020	COVID-19	140	71/69	56.5±11.85	6
JC Zhang et al.	2020	COVID-19	80	34/46	54.5±12.4	7
Z Zhao et al.	2020	COVID-19	75	42/33	45.33±15.87	1
FANG Xiao-wei et al.	2020	COVID-19	79	45/34	45.1 ±16.6	1
JFW Chan et al.	2020	COVID-19	7	4/3	46.16 ±20.52	1
Nanshan Chen et al.	2020	COVID-19	99	67/32	55.5 ±13.1	1
Wei-Jie Guan et al.	2020	COVID-19	1099	640/459	46.66±17.07	10
Chaolin Huang et al.	2020	COVID-19	41	30/11	49.33±13.06	1
Liu, Kui, et al.	2019	COVID-19	137	61/76	54.25±12.07	2
Dawei Wang et al.	2020	COVID-19	138	75/63	55.33±19.47	10
Fei Zhou et al.	2020	COVID-19	191	119/72	56.33±15.68	2
B. Cao et al.	2020	COVID-19	199	120/79	58.33±14.18	6
Jun Chen et al.	2020	COVID-19	249	126/123	50.33±20.87	1
Rong-Hii Du et al.	2020	COVID-19	179	97/82	57.6±13.7	4
Du Y et al.	2020	COVID-19	85	62/23	65.8±14.2	6
Giuseppe Gritti et al.	2020	COVID-19	21	18/3	62.75±7.14	1
Wei-jie Guan et al.	2020	COVID-19	1590	904/674	48.9±16.3	130
Tao Guo et al.	2020	COVID-19	187	91/96	58.5±14.66	13
Bo Hu et al.	2020	COVID-19	50	34/16	60.62±4.34	2

Eu Suk Kim et al.	2020	COVID-19	28	15/13	42.6±13.4	1
Shaoqing Lei et al.	2020	COVID-19	34	14/20	54±4.78	9
Francois-Xavier Lescure et al.	2020	COVID-19	5	3/2	47±18.08	1
Versa Omrani-Nava et al.	2020	COVID-19	93	51/42	56.3±15.2	4
Un Liu et al.	2020	COVID-19	88	45/43	54.7512.84	2
Jian Wu et al.	2020	COVID-19	80	39/41	46.10±15.42	1
Zhongliang Wang et al.	2020	COVID-19	69	32/37	46.33±20.44	4
Yan Zhang et al.	2020	COVID-19	258	138/120	63.33±10.43	12
Luwen Wang et al.	2020	COVID-19	116	67/49	55±23.27	12
Lang Wang et al.	2020	COVID-19	339	166/173	71±8	15
Rui Zhang et al.	2020	COVID-19	120	43/77	45.4±15.6	7
Xiaoli Zhang et al.	2020	COVID-19	645	328/317	45.3384±4.3381	6
Xi Xu et al.	2020	COVID-19	90	39/51	51±13.81	2
Meizhu Chen et al.	2020	COVID-19	97	42/55	47.5±13.06	6
Hua Su et al.	2020	COVID-19	26	19/7	68.92±13.03	6
Chaomin Wu et al.	2020	COVID-19	201	128/73	51.33±12.69	1

Table 17.2 Quality assessment of cross-sectional studies

First author, year	Selection	Comparability	Exposure/outcome	Total score
Jin-Wei, 2020	****	**	**	********
Qingxian Cai,2020	****	**	***	*********
Xiaoping Chen,2020	****	**	***	*********
Xiaohua Chen, 2020	****	**	***	*********
Xu Chen, 2020	****	**	***	*********
Pengfei Cui, 2020	****	**	**	********
Bo Diao, 2020	****	**	***	*********
Ying Huang, 2020	****	**	**	********
Wanli Jiang, 2020	****	**	***	*********
Kim, K. H., 2015	****	*	**	*******
Shi y, 2020	****	**	***	*********
Wang Wenjun, 2020	****	**	**	********
Wang Y, 2004	****	**	***	*********
Hui Hui, 2020	****	**	**	********
Kyung Min Kim, 2015	****	**	***	*********
Jing Li, 2019	****	**	***	*********
Xun Li,2020	****	**	**	********
Lei Liu, 2020	****	**	***	*********
Ru Liu,2020	****	**	***	*********
Xiaolong Qi, 2020	****	**	**	********
J Wu, 2020	****	**	**	********
Y Xu, 2020	****	**	**	********
X Yang, 2020	****	**	**	********
B Zhang, 2020	****	**	**	********
J Zhang, 2020	****	**	***	*********
JC Zhang, 2020	****	**	***	*********
Abdullah Assiri, 2013	****	**	**	********
Nanshan Chen,2020	****	**	**	********
Wei-Jie Guan, 2020	****	**	***	*********
Chaolin Huang,2020	****	**	***	*********
Liu, Kui, 2019	****	**	**	********
J S M Peiris, 2003	****	**	**	********
Korea Centers for Disease Control and Prevention, 2015	****	**	**	********
J S M Peiris, 2003	****	**	**	********
Sun Young Cho, 2016	****	**	**	********
Jun Chen, 2020	****	**	***	*********
Du Y, 2020	****	**	**	********
Giuseppe Gritti, 2020	****	**	**	********
Wei-jie Guan, 2020	****	**	***	*********
Abdul-Rahman Jazieh, 2020	****	**	**	********
Eu Suk Kim, 2020	****	**	**	********
Versa Omrani-Nava, 2020	****	**	***	*********
Jian Wu, 2020	****	**	**	********
Luwen Wang, 2020	****	**	***	*********
Lang Wang, 2020	****	**	***	*********
Rui Zhang,2020	****	**	***	*********
Xiaoli Zhang, 2020	****	**	***	*********
Xi Xu,2020	****	**	**	********
Meizhu Chen, 2020	****	**	***	*********
Hua Su, 2020	****	**	**	********

Table 17.3 Quality assessment of cohort studies

First author, year	Selection	Comparability	Exposure/outcome	Total score
James C. Ho, 2003	****	**	***	*********
Heshui Shi, 2020	****	**	***	*********
W Yang, 2020	****	**	***	*********
房晓伟, 2020	****	**	***	*********
Amr M. Ajlan, 2014	****	**	***	*********
Nelson Lee, 2003	****	**	***	*********
Omrani Ali S, 2014	****	**	***	*********
Fei Zhou, 2020	****	**	***	*********
B. Cao, 2020	****	**	***	*********
Rong-Hii Du, 2020	****	**	***	*********
Shaoqing Lei, 2020	****	**	***	*********
Yan Zhang, 2020	****	**	***	*********
Chaomin Wu, 2020	****	**	***	*********

Table 17.4 Quality assessment of case-control studies

First author, year	Selection	Comparability	Exposure	Total score
Jaffar A. Al-Tawfiq, 2014	****	**	***	*********

Table 17.5 Quality assessment of case-series studies

First author, year	Domain	Selection	Ascertainment	Causality	Reporting	Total score
L Wang, 2020	*	*	**	*	*	******
G Zhang, 2020	*	*	**	*	*	******
Z Zhao, 2020	*	*	**	*	*	******
Yaseen M. Arabi, 2014	*	*	**	*	*	******
Christopher M. Booth, 2003	*	*	**	*	*	******
Jasper Fuk-Woo Chan, 2020	*	*	**	*	*	******
Kenneth W. Tsang, 2003	*	*	**	*	*	******
Ping Tim Tsui, 2003	*	*	**	*	*	******
Dawei Wang, 2020	*	*	**	*	*	******
Tao Guo, 2020	*	*	**	*	*	******
Bo Hu, 2020	*	*	**	*	*	******
Francois-Xavier Lescure, 2020	*	*	**	*	*	******
Un Liu, 2020	*	*	**	*	*	******
Zhongliang Wang, 2020	*	*	**	*	*	******

Table 17.6 Pooled prevalence of comorbidities and clinical manifestations

Variable	Group	Number of studies	Number of patients	Prevalence (%)	95%CI	I²%	P
Age	COVID-19	60	9552	51.29*	51.15–51.44	98.70	0.00
	SARS	8	856	42.27*	42.09–42.44	77.63	0.00
	MERS	10	635	55.38*	55.13–55.62	56.07	0.015
	Total	78	11043	50.83*	50.69–50.96	98.48	0.00
Male	COVID-19	60	5179	54.21	54.06–54.37	73.32	0.00
	SARS	8	357	41.70	41.42–41.98	0.00	0.54
	MERS	10	399	62.83	62.40–63.26	7.22	0.38
	Total	78	5935	53.74	53.58–53.89	75.79	0.00
Cancer patients	COVID-19	60	388	4.06	3.98–4.13	76.81	0.00
	SARS	8	24	2.80	2.55–3.05	65.15	0.01
	MERS	10	134	21.10	20.22–21.98	90.03	0.00
	Total	78	546	4.94	4.83–5.05	83.03	0.00
Any comorbidity	COVID-19	46	2788	36.47	36.10–36.83	96.13	0.00
	SARS	6	93	19.01	18.27–19.76	72.31	0.00
	MERS	7	214	55.01	52.99–57.03	97.06	0.00
	Total	59	3095	36.31	35.95–36.68	96.76	0.00
Other comorbidities							
Hypertension	COVID-19	52	1829	21.34	21.12–21.56	90.09	0.00
	SARS	3	18	4.69	4.56–4.83	0	–
	MERS	5	77	28.51	27.21–29.82	71.91	0.01
	Total	60	1924	20.86	20.63–21.08	91.88	0.00
Diabetes	COVID-19	54	971	10.89	10.77–11.00	74.34	0.00
	SARS	8	40	4.67	4.42–4.92	55.11	0.03
	MERS	9	165	29.83	27.76–31.91	97.43	0.00
	Total	71	1176	11.39	11.21–11.56	88.75	0.00
Cardiovascular disease	COVID-19	46	568	6.98	6.83–7.12	88.18	0.00
	SARS	7	30	3.84	3.52–4.15	69.90	0.00
	MERS	10	115	18.11	16.71–19.50	92.31	0.00
	Total	63	713	7.46	7.29–7.63	88.51	0.00
Hepatic diseases	COVID-19	32	198	2.68	2.64–2.72	54.75	0.00
	SARS	3	4	1.52	1.40–1.63	0	–
	MERS	3	3	0.96	0.92–1.00	0	–
	Total	38	205	2.58	2.54–2.61	49.74	0.00
Pulmonary diseases	COVID-19	43	444	5.76	5.59–5.93	90.98	0.00
	SARS	4	20	3.08	2.72–3.44	68.86	0.02
	MERS	8	117	19.21	18.28–20.13	89.37	0.00
	Total	55	581	6.48	6.30–6.66	91.36	0.00
Cerebrovascular diseases	COVID-19	27	212	3.90	4.59–7.46	73.87	0.00
	MERS	3	7	2.90	1.08–25	–	–
	Total	30	219	3.86	4.55–7.32	72.66	0.00
Thyroid diseases	COVID-19	5	45	5.49	5.37–5.61	46.58	0.11
	MERS	1	1	0.54	–	–	–
	Total	6	46	4.58	4.42–4.73	82.15	0.00
Chronic kidney diseases	COVID-19	23	95	2.14	2.08–2.20	50.18	0.00
	SARS	2	6	1.30	1.29–1.31	–	–
	MERS	6	53	13.02	11.43–14.60	92.29	0.00
	Total	31	154	2.90	2.75–3.06	74.80	0.00
Immunodeficiency	COVID-19	3	3	0.24	0.22–0.26	–	–
Endocrine diseases	COVID-19	5	45	6.25	6.12–6.91	87.02	0.00
	MERS	1	8	9.75	–	–	–
	Total	6	53	6.86	6.50–7.22	85.82	0.00

(continued)

Table 17.6 (continued)

Variable	Group	Number of studies	Number of patients	Prevalence (%)	95%CI	I²%	P
Cardio--cerebrovascular diseases	COVID-19	5	149	23.53	23.17–23.90	43.20	0.13
Symptoms							
Fever	COVID-19	48	6960	84.16	83.96–84.37	93.91	0.00
	SARS	5	414	99.76	99.72–99.77	–	–
	MERS	7	245	75.61	73.63–77.59	94.03	0.00
	Total	60	7621	84.58	84.36–84.79	94.62	0.00
Cough	COVID-19	46	4958	62.18	61.87–62.49	95.85	0.00
	SARS	5	247	59.23	57.70–60.75	9238	0.00
	MERS	7	127	39.19	36.15–42.24	96.68	0.00
	Total	58	5332	61.18	60.86–61.51	96.05	0.00
Sputum production	COVID-19	23	1193	29.64	29.41–29.87	88.50	0.00
	SARS	2	41	27.70	26.93–28.47	–	–
	MERS	5	52	18.05	16.31–19.79	88.72	0.00
	Total	30	1286	28.83	28.58–29.08	90.17	0.00
Chill	COVID-19	8	373	11.28	11.20–11.37	49.86	0.05
	SARS	4	227	55.77	53.73–57.80	96.64	0.00
	Total	12	600	16.16	15.66–16.67	97.73	0.00
Palpitation	COVID-19	5	41	6.72	6.49–6.95	65.58	0.02
Stomachache	COVID-19	11	27	2.00	1.91–2.09	44.10	0.06
	SARS	1	5	3.47	–	–	–
	MERS	3	33	13.75	13.54–13.95	–	–
	Total	15	65	3.75	3.55–3.96	74.72	0.00
Diarrhea	COVID-19	43	532	6.90	6.79–7.01	81.18	0.00
	SARS	4	67	16.46	15.65–17.26	94.46	0.00
	MERS	6	50	15.77	15.19–16.35	26.42	0.24
	Total	53	649	7.69	7.57–7.82	83.28	0.00
Inappetence	COVID-19	14	357	18.75	18.05–19.44	96.69	0.00
	SARS	1	10	20	–	–	–
	MERS	1	24	12.90	–	–	–
	Total	16	391	18.27	17.65–18.88	96.30	0.00
Sore throat	COVID-19	24	453	10.43	10.25–10.60	90.82	0.00
	SARS	5	68	16.30	15.74–16.87	63.55	0.04
	MERS	4	20	7.63	6.85–8.41	61.34	0.05
	Total	33	541	10.77	10.60–10.94	89.51	0.00
Chest pain	COVID-19	17	221	12.25	11.68–12.82	91.78	0.00
	SARS	1	15	10.41	–	–	–
	MERS	2	8	12.5	11.51–13.48	–	–
	Total	20	244	12.13	11.61–12.64	90.61	0.00
Dizziness	COVID-19	9	74	5.68	5.45–5.90	63.66	0.00
	SARS	4	74	18.18	16.45–19.91	96.12	0.00
	Total	13	148	8.66	8.14–9.17	88.76	0.00
Shortness of breath	COVID-19	39	1830	24.84	24.49–25.18	97.40	0.00
	SARS	4	79	28.31	26.28–30.34	95.99	0.00
	MERS	5	63	52.94	47.72–58.15	95.77	0.00
	Total	48	1972	25.39	25.04–25.74	97.28	0.00
Respiratory failure	COVID-19	13	424	19.46	18.62–20.30	97.95	0.00
	SARS	1	6	8.33	–	–	–
	MERS	1	2	11.76	–	–	–
	Total	15	432	19.05	18.24–19.86	97.57	0.00

(continued)

Table 17.6 (continued)

Variable	Group	Number of studies	Number of patients	Prevalence (%)	95%CI	I^2%	P
Nasal congestion	COVID-19	8	176	4.45	4.41–4.49	86.92	0.00
Headache	COVID-19	25	660	10.80	10.69–10.90	89.58	0.00
	SARS	2	58	37.66	36.31–39.01	–	–
	MERS	3	9	11.84	11.05–12.63	–	–
	Total	30	727	11.46	11.32–11.61	90.29	0.00
Hemoptysis	COVID-19	18	129	3.15	3.04–3.27	79.31	0.00
	SARS	3	98	37.26	34.89–39.63	–	–
	MERS	4	27	10.03	9.63–10.43	0	0.48
	Total	25	254	5.50	5.21–5.78	91.74	0.00
Myalgia	COVID-19	35	994	15.32	15.11–15.54	91.77	0.00
	SARS	5	238	57.07	56.39–57.74	54.53	0.07
	MERS	6	84	27.54	26.52–28.56	74.91	0.00
	Total	46	1316	18.27	17.95–18.56	94.43	0.00
Fatigue	COVID-19	31	2337	34.80	34.43–35.16	98.34	0.00
	SARS	1	25	50	–	–	–
	MERS	1	1	14.28	–	–	–
	Total	33	2363	34.89	34.53–35.25	98.24	0.00
Nausea	COVID-19	14	248	5.67	5.58–5.77	81.19	0.00
	MERS	2	11	18.64	17.30–19.98	–	–
	Total	16	259	5.85	5.74–5.95	80.53	0.00
Vomiting	COVID-19	13	62	4.05	3.93–4.17	22.09	0.23
	MERS	3	12	14.45	12.76–16.14	–	–
	Total	16	74	4.59	4.41–4.77	39.96	0.06

*The mean age of participants (years)

(61.18%), fatigue (34.89%), shortness of breath (25.39%), myalgia (18.27%), headache (11.46%), sore throat (10.77%), chest pain (10.41%), diarrhea (7.69%), nausea (5.85%), hemoptysis (5.50%), vomiting (4.59%), and nasal congestion (4.45%).

17.3.3 The Prevalence of Cancer in COVD-19, SARS, and MERS Confirmed Cases

Cancer was detected in 546 patients, including 24 SARS, 134 MERS, and 388 COVID-19 confirmed cases. The pooled prevalence of cancer was 4.94% in total (95% CI 4.83 to 5.05). In detail, it was estimated to be about 21.10% (95% CI 20.22 to 21.98) in MERS, 4.06% (95% CI 3.98 to 4.13) in COVID-19, and 2.80% (95% CI 2.55 to 3.05) in SARS patients (Fig. 17.2). Across the studies included, the pooled prevalence of any comorbidities was 36.31% (95% CI 35.95 to 36.68). It was 55.01% (95% CI 52.99 to 57.03) in MERS, 36.47% (95% CI 36.10 to 36.83) in COVID-19, and 19.01% (95% CI 18.27 to 19.76) in SARS subgroups. In patients with COVID-19, the most common comorbidities were hypertension, diabetes, and cardiovascular disease with the pooled prevalence of 21.34% (95% CI 21.12 to 21.56), 10.89% (95% CI 10.77 to 11.00), and 6.98% (95% CI 6.83 to 7.12), respectively. In the case of SARS, the three most common conditions were the same but with a significantly lower prevalence of 4.69% (95% CI 4.56 to 4.83), 4.67% (95% CI 4.42 to 4.92), and 3.84% (95% CI 3.52 to 4.15). At the specific subgroup of MERS, diabetes, hypertension, and pulmonary diseases were the first three common conditions with the estimated prevalence of 29.83% (95% CI 27.76 to 31.91), 28.51% (95% CI 27.21 to 29.82), and 19.21% (95% CI 18.28 to 20.13).

As summarized in Table 17.6, an I2 value of 83.3% indicated a significant level of heterogeneity across studies, and thus, the random-effect model was used for pooled analyses. No evidence of publication bias was detected when visualiz-

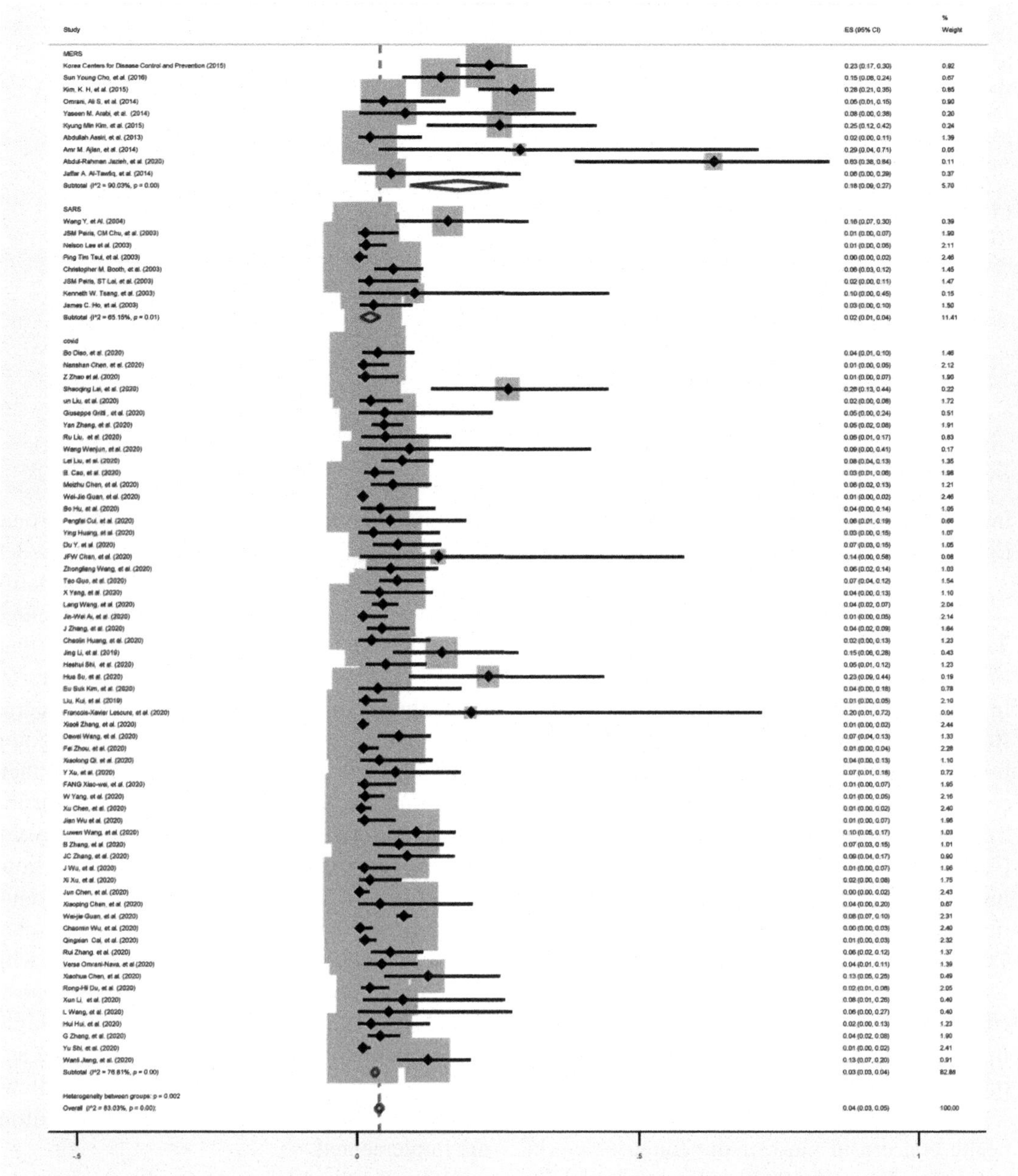

Fig. 17.2 Pooled prevalence of cancer among cases confirmed as SARS, MERS, and COVID-19 by real-time RT-PCR

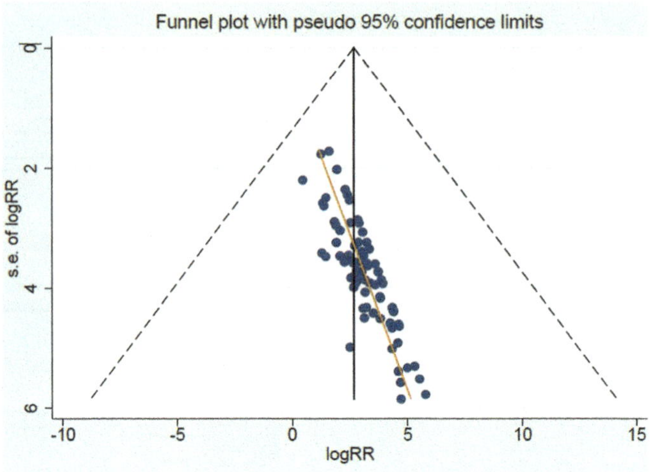

Fig. 17.3 Funnel plot for the assessment of publication bias across studies included in the analysis of the pooled prevalence of cancer among cases confirmed as SARS, MERS, and COVID-19 by real-time RT-PCR

ing the funnel plot (Figs. 17.3) and also using Egger's test ($P = 0.996$).

17.4 Discussion

In summary, our analysis based on data from 78 studies with SARS, MERS, and COVID-19 confirmed cases showed that the prevalence of cancer (4.94%) stands at fourth place after hypertension (20.86 %), cardiovascular diseases (7.46%), and diabetes (11.39%). We observed just barely men more than women. This observation occurred in each of SARS, MERS, and COVID-19 subgroups as well. It could be attributed to a more robust immune response in women compared to men (Belmonte et al. 2020) or arise from the increased exposure of men to Chinese trade markets than women.

The outbreak of COVID-19 is now a public health concern in most of the countries worldwide. By June 10, 2020, this pandemic has affected 213 countries and territories around the world, with a total of 7,145,539 confirmed cases and a death toll of 408,025 deaths (WHO 10 June 2020). SARS epidemic that took place in November 2002 in Guangdong province of southern China was very close to the ongoing pandemic of COVID-19. Additionally, the MERS outbreak first happened to Saudi Arabia in September 2012 and was caused by a coronavirus (Zhong et al. 2003; Mackay and Arden 2015). COVID-19 is associated with a mortality rate lower than that of MERS and SARS, though it has claimed the lives of people more than MERS and SARS integrated. Altogether, all these three coronaviruses have raised concerns for patients with cancer who are undergoing active therapy about their safety due to their weak immune system. Cancer treatment during this pandemic has been discussed in several studies because clinicians should make decisions to continue or postpone it. Some authors do not recommend patients to delay or hold their chemotherapy (Desai et al. 2020). Others are with the opinion that active treatment can be postponed for patients with no severe sign or high-degree tumors until new protocols for safe and effective methods like tele-treatment and follow-ups without unnecessary in-person consulting are implemented.

With all lessons learned from the previous epidemics triggered by coronaviruses, i.e., SARS and MERS, the world has been globally affected by the third outbreak of the coronavirus origin, called COVID-19, since December 2019 (Jabbari et al. 2020). The outbreak rapidly turned into a pandemic state associated with more than 400,000 deaths worldwide as of writing this (Hanaei and Rezaei 2020).

Like SARS and MERS, COVID-19 can cause potentially fatal respiratory syndromes. Recent research confirms that respiratory manifestations are common in patients with COVID-19, and what initially appeared to be the most clinically relevant presentation of COVID-19 has turned out to be one of the many manifestations seen in patients with COVID-19. Such a multiple system involvement (Yazdanpanah et al. 2020c; Shamshirian and Rezaei 2020; Jahanshahlu and Rezaei 2020a; Saleki et al. 2020) seems to depend on the amplification of transduction pathways involved in the production of pro-inflammatory cytokines (Basiri et al. 2020b; Yazdanpanah et al. 2020b; Bahrami et al. 2020b; Rokni et al. 2020) that might have a genetic basis (Darbeheshti and Rezaei 2020; Yousefzadegan and Rezaei 2020).

Studies suggest a dual role for the immune system in the pathogenesis of COVID-19, when antiviral immunity is present on the one hand and, on the other hand, patients having elevated inflammatory markers are most seriously affected by COVID-19 (Rokni et al. 2020; Bahrami et al. 2020a; Yazdanpanah et al. 2020a; Saghazadeh and Rezaei 2020a; Lotfi and Rezaei 2020; Nasab et al. 2020). The latter, i.e., the negative role of the immune system in COVID-19, is further strengthened by the evidence that not only patients with primary immunodeficiency disorders appear to be prone to COVID-19, not more than the general population, but they seem to be less likely to develop the severe disease (Babaha and Rezaei 2020). As a result, recent research says the immune system is the bane of COVID-19 pathogenesis and remarks on the potential benefits derived from its targeting (Sharifkashani et al. 2020; Pourahmad et al. 2020; Mansourabadi et al. 2020; Pashaei and Rezaei 2020; Fathi and Rezaei 2020; Jahanshahlu and Rezaei 2020b; Basiri et al. 2020b; Saghazadeh and Rezaei 2020b).

Cancer is closely linked to chronic inflammation, and this might be why patients with cancer are more vulnerable to COVID-19 infection compared to the general population (Liang et al. 2020). Zhang and colleagues reported that 53.6% of the cancer patients developed severe events, along with a fatality rate of 28.6% (Zhang et al. 2020d), while about 5% of patients with COVID-19 in the general population develop the critical condition with a fatality rate of 2.3% (Wu and McGoogan 2020b). Also, patients with cancer display the altered expression of angiotensin--converting enzyme 2 (ACE2), which serves as a cell surface receptor for the coronavirus causing COVID-19, and this might also play a role in predisposing these patients to severe COVID-19 (Sharifkashani et al. 2020; Ahmadi et al. 2020).

According to the findings of the present study, patients with cancer are at a greater risk of SARS, MERS, and COVID-19, and comorbidities are significantly more common in patients with MERS compared to patients with COVID-19 and SARS, and this is the cancer case as well. Further studies are necessary to address if patients with coronaviruses and cancer are different from patients with coronaviruses without cancer in terms of clinical manifestations, laboratory findings, outcomes, and men to women ratio.

17.5 Conclusion

We are reminded by the pandemic of COVID-19, which could affect people globally and make them practice social distancing which we hope is for a temporary time (Jabbari and Rezaei 2020), that our research needs to be ideal in terms of its quality (Rzymski et al. 2020), especially in the short term of the pandemic when we search for a reliable diagnostic method and specific treatment and prevention (Mohamed et al. 2020b; Rezaei 2020; Rabiee et al. 2020; Kafieh et al. 2020; Lotfi et al. 2020; Basiri et al. 2020a, b), and the backbone of global science would be a chance to outperform (Momtazmanesh et al. 2020; Mohamed et al. 2020a; Moradian et al. 2020).

References

Ahmadi M, Saffarzadeh N, Habibi MA, Hajiesmaeili F, Rezaei N (2020) Colon cancer and SARS-CoV-2: impact of ACE2 expression in susceptibility to COVID-19. bioRxiv

Ai J, Chen J, Wang Y, Liu X, Fan W, Qu G, Zhang M, et al. (2020) The cross-sectional study of hospitalized coronavirus disease 2019 patients in Xiangyang, Hubei province. medRxiv

Ajlan AM, Ahyad RA, Jamjoom LG, Alharthy A, Madani TA (2014) Middle East respiratory syndrome coronavirus (MERS-CoV) infection: chest CT findings. Am J Roentgenol 203(4):782–787

Al-Tawfiq JA, Hinedi K, Ghandour J, Khairalla H, Musleh S, Ujayli A, Memish ZA (2014) Middle East respiratory syndrome coronavirus: a case-control study of hospitalized patients. Clin Infect Dis 59(2):160–165

Arabi YM, Arifi AA, Balkhy HH, Najm H, Aldawood AS, Ghabashi A, Hawa H, Alothman A, Khaldi A, Al Raiy B (2014) Clinical course and outcomes of critically ill patients with Middle East respiratory syndrome coronavirus infection. Ann Int Med 160(6):389–397

Assiri A, Al-Tawfiq JA, Al-Rabeeah AA, Al-Rabiah FA, Al-Hajjar S, Al-Barrak A, Flemban H, Al-Nassir WN, Balkhy HH, Al-Hakeem RF (2013) Epidemiological, demographic, and clinical characteristics of 47 cases of Middle East respiratory syndrome coronavirus disease from Saudi Arabia: a descriptive study. Lancet Infect Dis 13(9):752–761

Babaha F, Rezaei N (2020) Primary immunodeficiency diseases in COVID-19 pandemic: a predisposing or protective factor? Am J Med Sci. https://doi.org/10.1016/j.amjms.2020.07.027

Bahrami A, Vafapour M, Moazzami B, Rezaei N (2020a) Hyperinflammatory shock related to COVID-19 in a patient presenting with multisystem inflammatory syndrome in children: first case from Iran. J Paediatrics Child Health. https://doi.org/10.1111/jpc.15048

Bahrami A, Vafapour M, Moazzami B, Rezaei N (2020b) Hyperinflammatory shock related to COVID-19 in a patient presenting with multisystem inflammatory syndrome in children: first case from Iran. J Paediatrics Child Health

Basiri A, Heidari A, Nadi MF, Fallahy MTP, Nezamabadi SS, Sedighi M, Saghazadeh A, Rezaei N (2020a) Microfluidic devices for detection of RNA viruses. Reviews Med Virol n/a (n/a):e2154. https://doi.org/10.1002/rmv.2154

Basiri A, Pazhouhnia Z, Beheshtizadeh N, Hoseinpour M, Saghazadeh A, Rezaei N (2020b) Regenerative medicine in COVID-19 treatment: real opportunities and range of promises. Stem Cell Rev Rep:1–13. https://doi.org/10.1007/s12015-020-09994-5

Belmonte RL, Corbally M-K, Duneau DF, Regan JC (2020) Sexual Dimorphisms in Innate Immunity and Responses to Infection in Drosophila melanogaster. Front Immunol 10:3075

Booth CM, Matukas LM, Tomlinson GA, Rachlis AR, Rose DB, Dwosh HA, Walmsley SL, Mazzulli T, Avendano M, Derkach P (2003) Clinical features and short-term outcomes of 144 patients with SARS in the greater Toronto area. JAMA 289(21):2801–2809

Cai Q, Huang D, Ou P, Yu H, Zhu Z, Xia Z, Su Y, Ma Z, et al. (2020) COVID-19 in a designated infectious diseases hospital outside Hubei Province, China. medRxiv

Cao B, Wang Y, Wen D, Liu W, Wang J, et al. (2020) A trial of lopinavir–ritonavir in adults hospitalized with severe Covid-19. N Engl J Med 382:1787

Chan JF-W, Yuan S, Kok K-H, To KK-W, Chu H, Yang J, Xing F, Liu J, Yip CC-Y, Poon RW-S, Tsoi H-W, Lo SK-F, Chan K-H, Poon VK-M, Chan W-M, Ip JD, Cai J-P, Cheng VC-C, Chen H, Hui CK-M, Yuen K-Y (2020) A familial cluster of pneumonia associated with the 2019 novel coronavirus indicating person-to-person transmission: a study of a family cluster. Lancet 395(10223):514–523. https://doi.org/10.1016/S0140-6736(20)30154-9

Chen J, Qi T, Liu L, Ling Y, Qian Z, Li T, Li F, Xu Q, et al. (2020a) Clinical progression of patients with COVID-19 in Shanghai, China. J Inf Secur

Chen N, Zhou M, Dong X, Qu J, Gong F, Han Y, Qiu Y, Wang J, Liu Y, Wei Y (2020b) Epidemiological and clinical characteristics of 99 cases of 2019 novel coronavirus pneumonia in Wuhan, China: a descriptive study. Lancet 395(10223):507–513

Chen X, Ling J, Mo P, Zhang Y, Jiang Q, Ma Z, Cao Q, et al. (2020c) Restoration of leukomonocyte counts is associated with viral clearance in COVID-19 hospitalized patients. medRxiv

Chen X, Zhao B, Qu Y, Chen Y, Xiong J, Feng Y, Men D, et al. (2020d) Detectable serum SARS-CoV-2 viral load (RNAaemia) is closely associated with drastically elevated interleukin 6 (IL-6) level in critically ill COVID-19 patients. medRxiv

Chen X, Zheng F, Qing Y, Ding S, Yang D, Lei C, Yin Z, et al. (2020e) Epidemiological and clinical features of 291 cases with coronavirus disease 2019 in areas adjacent to Hubei, China: a double-center observational study. medRxiv

Cho SY, Kang JM, Ha YE, Park GE, Lee JY, Ko JH, Lee JY, Kim JM, Kang CI, Jo IJ, Ryu JG (2016, Sep 3) MERS-CoV outbreak following a single patient exposure in an emergency room in South Korea: an epidemiological outbreak study. Lancet 388(10048):994–1001

Cui P, Chen Z, Wang T, Dai J, Zhang J, Ding T, Jiang J, et al. (2020) Clinical features and sexual transmission potential of SARS-CoV-2 infected female patients: a descriptive study in Wuhan, China. medRxiv

Darbeheshti F, Rezaei N (2020) Genetic predisposition models to COVID-19 infection. Med Hypotheses

Desai A, Sachdeva S, Parekh T, Desai R (2020) COVID-19 and cancer: lessons from a pooled meta-analysis. JCO Glob Oncol 6:557–559. https://doi.org/10.1200/go.20.00097

Diao B, Feng Z, Wang C, Wang H, Liu L, Wang C, et al. (2020) Human kidney is a target for novel severe acute respiratory syndrome coronavirus 2 (SARS-CoV-2) infection. medRxiv

Du R, Liang L, Yang C, Wang W, (2020a) Predictors of mortality for patients with COVID-19 pneumonia caused by SARS-CoV-2: a prospective cohort study. Am **J Respir** Crit Care Med

Du Y, Tu L, Zhu P, Mu M, Wang R, Yang P, ... (2020b) Clinical features of 85 fatal cases of COVID-19 from Wuhan: a retrospective observational study. Am J Respir Crit Care Med 201:1372

Fathi N, Rezaei N (2020) Lymphopenia in COVID-19: therapeutic opportunities. Cell Biol Int 44(9):1792–1797. https://doi.org/10.1002/cbin.11403

Gritti G, Raimondi F, Ripamonti D, Riva I, Landi F, ... (2020) Use of siltuximab in patients with COVID-19 pneumonia requiring ventilatory support. medRxiv

Guan W-j, Ni Z-y, Hu Y Liang W-h, Ou C-q, He J-x, Liu L, Shan H, Lei C-l, Hui DS (2020a) Clinical characteristics of 2019 novel coronavirus infection in China. MedRxiv

Guan W, Liang W, Zhao Y, Liang H, ... (2020b) Comorbidity and its impact on 1590 patients with Covid-19 in China: a nationwide analysis. Eur Respir J

Guo T, Fan Y, Chen M, Wu X, Zhang L, He T, ... (2020) Cardiovascular implications of fatal outcomes of patients with coronavirus disease 2019 (COVID-19). JAMA

Hanaei S, Rezaei N (2020) COVID-19: developing from an outbreak to a pandemic. Arch Med Res. https://doi.org/10.1016/j.arcmed.2020.04.021

Ho JC, Ooi GC, Mok TY, Chan JW, Hung I, Lam B, Wong PC, Li PC, Ho PL, Lam WK, Ng CK, Ip MS, Lai KN, Chan-Yeung M, Tsang KW (2003) High-dose pulse versus nonpulse corticosteroid regimens in severe acute respiratory syndrome. Am J Respir Crit Care Med 168(12):1449–1456. https://doi.org/10.1164/rccm.200306-766OC

Huang C, Wang Y, Li X, Ren L, Zhao J, Hu Y, Zhang L, Fan G, Xu J, Gu X, Cheng Z, Yu T, Xia J, Wei Y, Wu W, Xie X, Yin W, Li H, Liu M, Xiao Y, Gao H, Guo L, Xie J, Wang G, Jiang R, Gao Z, Jin Q, Wang J, Cao B (2020a) Clinical features of patients infected with 2019 novel coronavirus in Wuhan, China. Lancet 395 (10223):497-506. https://doi.org/10.1016/S0140-6736(20)30183-5

Huang Y, Zhou H, Yang R, Xu Y, Feng X, Gong P (2020b) Clinical characteristics of 36 non-survivors with COVID-19 in Wuhan, China. medRxiv

Hui H, Zhang Y, Yang X, Wang X, He B, Li L, Li H, ... (2020) Clinical and radiographic features of cardiac injury in patients with 2019 novel coronavirus pneumonia. medRxiv

Jabbari P, Rezaei N (2020) With risk of reinfection, is COVID-19 here to stay? Disaster Medicine and Public Health Preparedness:1. https://doi.org/10.1017/dmp.2020.274

Jabbari P, Jabbari F, Ebrahimi S, Rezaei N (2020) COVID-19: a chimera of two pandemics. Disaster Med Public Health Prep:1–2. https://doi.org/10.1017/dmp.2020.223

Jahanshahlu L, Rezaei N (2020a) Central Nervous System Involvement in COVID-19. Arch Med Res. https://doi.org/10.1016/j.arcmed.2020.05.016

Jahanshahlu L, Rezaei N (2020b) Monoclonal antibody as a potential anti-COVID-19. Biomed Pharmacother 129:110337. https://doi.org/10.1016/j.biopha.2020.110337

Jazieh AR, Alenazi TH, Alhejazi A, Al Safi F, Al Olayan A (2020) Outcome of oncology patients infected with coronavirus. JCO Glob Oncol

Jiang W, Zeng Z, Fan D, Zhang Y, Chen Z, ... (2020) Decreased counts of T lymphocyte subsets predict prognosis in SARS-CoV-2-infected pneumonia in Wuhan, China: a retrospective study. ... A Retrospective Study ...

Kafieh R, Arian R, Saeedizadeh N, Minaee S, Yadav SK, Vaezi A, Rezaei N, Javanmard SH (2020) COVID-19 in Iran: a deeper look into the future. medRxiv

Kim KM, Ki M, Cho S-I, Sung M, Hong JK, Cheong H-K, Kim J-H, Lee S-E, Lee C, Lee K-J, Park Y-S, Kim SW, Choi BY (2015) Epidemiologic features of the first MERS outbreak in Korea: focus on Pyeongtaek St. Mary's Hospital. Epidemiol Health 37:e2015041–e2015041. https://doi.org/10.4178/epih/e2015041

Kim E, Chin B, Kang C, Kim N, et al. (2020) ... with severe acute respiratory syndrome coronavirus 2 infection: a preliminary report of the first 28 patients from the korean cohort study on COVID-19. J Korean .... ncbi.nlm.nih.gov

Kui L, Fang Y-Y, Deng Y, Liu W, Wang M-F, Ma J-P, Xiao W, Wang Y-N, Zhong M-H, Li C-H (2020) Clinical characteristics of novel coronavirus cases in tertiary hospitals in Hubei Province. Chin Med J

Lee N, Hui D, Wu A, Chan P, Cameron P, Joynt GM, Ahuja A, Yung MY, Leung CB, K.F. To, Lui SF, Szeto CC et al (2003) A major outbreak of severe acute respiratory syndrome in Hong Kong. N Engl J Med. https://doi.org/10.1056/NEJMoa030685

Lei S, Jiang F, Su W, Chen C, Chen J, Mei W, Zhan L, et al. (2020) Clinical characteristics and outcomes of patients undergoing surgeries during the incubation period of COVID-19 infection. EClinicalMedicine. Elsevier

Lescure F, Bouadma L, Nguyen D, Parisey M, et al. (2020) Clinical and virological data of the first cases of COVID-19 in Europe: a case series. Lancet Infect Dis

Li J, Zhang Y, Wang F, Liu B, Li H, Tang G, Chang Z, et al. (2020a) Sex differences in clinical findings among patients with coronavirus disease 2019 (COVID-19) and severe condition. medRxiv

Li X, Wang L, Yan S, Yang F, Xiang L, Zhu J, Shen B, (2020b) Clinical characteristics of 25 death cases infected with COVID-19 pneumonia: a retrospective review of medical records in a single medical center, Wuhan, China. medRxiv

Liang W, Guan W, Chen R, Wang W, Li J, Xu K, Li C, Ai Q, Lu W, Liang H (2020) Cancer patients in SARS-

CoV-2 infection: a nationwide analysis in China. Lancet Oncol 21(3):335–337

Liu L, Liu W, Wang S, Zheng S (2020a) A preliminary study on serological assay for severe acute respiratory syndrome coronavirus 2 (SARS-CoV-2) in 238 admitted hospital patients. medRxiv

Liu R, Ming X, Zhu H, Song L, Gao Z, Gao L, Jia S, et al. (2020b) Association of Cardiovascular Manifestations with In-hospital Outcomes in Patients with COVID-19: a hospital staff data. medRxiv

Lotfi M, Rezaei N (2020) SARS-CoV-2: A comprehensive review from pathogenicity of the virus to clinical consequences. J Med Virol. https://doi.org/10.1002/jmv.26123

Lotfi M, Hamblin MR, Rezaei N (2020) COVID-19: Transmission, prevention, and potential therapeutic opportunities. Clinica chimica acta; Int J Clin Chem 508:254–266. https://doi.org/10.1016/j.cca.2020.05.044

Mackay IM, Arden KE (2015) MERS coronavirus: diagnostics, epidemiology and transmission. Virol J 12(1):222

Mansourabadi AH, Sadeghalvad M, Mohammadi-Motlagh HR, Rezaei N (2020) The immune system as a target for therapy of SARS-CoV-2: A systematic review of the current immunotherapies for COVID-19. Life Sci 258:118185. https://doi.org/10.1016/j.lfs.2020.118185

Margulis AV, Pladevall M, Riera-Guardia N, Varas-Lorenzo C, Hazell L, Berkman ND, Viswanathan M, Perez-Gutthann S (2014) Quality assessment of observational studies in a drug-safety systematic review, comparison of two tools: the Newcastle–Ottawa scale and the RTI item bank. Clinical Epidemiol 6:359

Mohamed K, Rodríguez-Román E, Rahmani F, Zhang H, Ivanovska M, Makka SA, Joya M, Makuku R, Islam MS, Radwan N, Rahmah L, Goda R, Abarikwu SO, Shaw M, Zoghi S, Irtsyan S, Ling I, Cseprekal O, Faten AB, Hazar Sayar E, Soloukey C, Grancini G, Rezaei N (2020a) Borderless collaboration is needed for COVID-19-A disease that knows no borders. Infection control and hospital epidemiology:1-2. https://doi.org/10.1017/ice.2020.162

Mohamed K, Yazdanpanah N, Saghazadeh A, Rezaei N (2020b) Computational drug discovery and repurposing for the treatment of Covid-19: A systematic review. Available at SSRN 3583748

Moher D, Liberati A, Tetzlaff J, Altman DG (2009) Preferred reporting items for systematic reviews and meta-analyses: the PRISMA statement. Ann Int Med 151(4):264–269

Momtazmanesh S, Ochs HD, Uddin LQ, Perc M, Routes JM, Vieira DN, Al-Herz W, Baris S, Prando C, Rosivall L, Abdul Latiff AH, Ulrichs T, Roudenok V, Aldave Becerra JC, Salunke DB, Goudouris E, Condino-Neto A, Stashchak A, Kryvenko O, Stashchak M, Bondarenko A, Rezaei N (2020) All together to Fight COVID-19. Am J Trop Med Hyg 102(6):1181–1183. https://doi.org/10.4269/ajtmh.20-0281

Moradian N, Ochs HD, Sedikies C, Hamblin MR, Camargo CA Jr, Martinez JA, Biamonte JD, Abdollahi M, Torres PJ, Nieto JJ, Ogino S, Seymour JF, Abraham A, Cauda V, Gupta S, Ramakrishna S, Sellke FW, Sorooshian A, Wallace Hayes A, Martinez-Urbistondo M, Gupta M, Azadbakht L, Esmaillzadeh A, Kelishadi R, Esteghamati A, Emam-Djomeh Z, Majdzadeh R, Palit P, Badali H, Rao I, Saboury AA, Jagan Mohan Rao L, Ahmadieh H, Montazeri A, Fadini GP, Pauly D, Thomas S, Moosavi-Movahed AA, Aghamohammadi A, Behmanesh M, Rahimi-Movaghar V, Ghavami S, Mehran R, Uddin LQ, Von Herrath M, Mobasher B, Rezaei N (2020) The urgent need for integrated science to fight COVID-19 pandemic and beyond. J Transl Med 18(1):205. https://doi.org/10.1186/s12967-020-02364-2

Murad MH, Sultan S, Haffar S, Bazerbachi F (2018) Methodological quality and synthesis of case series and case reports. BMJ Evid Based Med 23(2):60–63

Nasab MG, Saghazadeh A, Rezaei N (2020) SARS-CoV-2-A tough opponent for the immune system. Arch Med Res. https://doi.org/10.1016/j.arcmed.2020.05.020

Omrani AS, Saad MM, Baig K, Bahloul A, Abdul-Matin M, Alaidaroos AY, Almakhlafi GA, Albarrak MM, Memish ZA, Albarrak AM (2014) Ribavirin and interferon alfa-2a for severe Middle East respiratory syndrome coronavirus infection: a retrospective cohort study. Lancet Infect Dis 14(11):1090–1095

Omrani-Nava V, Maleki I, Ahmadi A, Moosazadeh M (2020) Evaluation of hepatic enzymes changes and association with prognosis in COVID-19 patients. Hepatitis Monthly. hepatmon.com

Pashaei M, Rezaei N (2020) Immunotherapy for SARS-CoV-2: potential opportunities. Expert Opin Biol Ther:1–5. https://doi.org/10.1080/14712598.2020.1807933

Peiris JSM, Chu C-M, Cheng VC-C, Chan K, Hung I, Poon LL, Law K-I, Tang B, Hon T, Chan C (2003a) Clinical progression and viral load in a community outbreak of coronavirus-associated SARS pneumonia: a prospective study. Lancet 361(9371):1767–1772

Peiris JSM, Lai ST, Poon LLM, Guan Y, Yam LYC, Lim W, Nicholls J, Yee WKS, Yan WW, Cheung MT, Cheng VCC, Chan KH, Tsang DNC, Yung RWH, Ng TK, Yuen KY, SARS study group (2003b) Coronavirus as a possible cause of severe acute respiratory syndrome. Lancet. https://doi.org/10.1016/S0140-6736(03)13077-2

Pourahmad R, Moazzami B, Rezaei N (2020) Efficacy of Plasmapheresis and Immunoglobulin Replacement Therapy (IVIG) on Patients with COVID-19. SN Comprehensive Clin Med:1–5

Prevention KCfDCa (2015) Middle East Respiratory Syndrome Coronavirus Outbreak in the Republic of Korea, 2015. Osong Public Health Res Perspect. https://doi.org/10.1016/j.phrp.2015.08.006

Qi X, Jiang Z, Yu Q, Shao C, Zhang H, Yue H, Ma B, et al. (2020) Machine learning-based CT radiomics model for predicting hospital stay in patients with pneumonia associated with SARS-CoV-2 infection: a multicenter study. medRxiv

Qin X, Qiu S, Yuan Y, Zong Y, Tuo Z, Li J, et al. (2020) Clinical Characteristics and Treatment of Patients Infected with COVID-19 in Shishou, China. China (February 18 ...

Rabiee N, Rabiee M, Bagherzadeh M, Rezaei N (2020) COVID-19 and picotechnology: Potential opportunities. Med Hypotheses 144:109917. https://doi.org/10.1016/j.mehy.2020.109917

Rezaei N (2020) COVID-19 and Medical Biotechnology. Avicenna J Med Biotechnol 12(3):139

Rokni M, Hamblin MR, Rezaei N (2020) Cytokines and COVID-19: friends or foes? Hum Vaccin Immunother:1–3

Rzymski P, Nowicki M, Mullin GE, Abraham A, Rodríguez-Román E, Petzold MB, Bendau A, Sahu KK, Ather A, Naviaux A-F (2020) Quantity does not equal quality: scientific principles cannot be sacrificed. Int Immunopharmacol 86:106711

Saghazadeh A, Rezaei N (2020a) Immune-epidemiological parameters of the novel coronavirus - a perspective. Expert Rev Clin Immunol 16(5):465–470. https://doi.org/10.1080/1744666x.2020.1750954

Saghazadeh A, Rezaei N (2020b) Towards treatment planning of COVID-19: rationale and hypothesis for the use of multiple immunosuppressive agents: anti-antibodies, immunoglobulins, and corticosteroids. Int Immunopharmacol 84:106560. https://doi.org/10.1016/j.intimp.2020.106560

Saleki K, Banazadeh M, Saghazadeh A, Rezaei N (2020) The involvement of the central nervous system in patients with COVID-19. Rev Neurosci 31(4):453–456. https://doi.org/10.1515/revneuro-2020-0026

Shamshirian D, Rezaei N (2020) Cardiovascular diseases burden in COVID-19: systematic review and meta-analysis.

Sharifkashani S, Bafrani MA, Khaboushan AS, Pirzadeh M, Kheirandish A, Yavarpour Bali H, Hessami A, Saghazadeh A, Rezaei N (2020) Angiotensin-converting enzyme 2 (ACE2) receptor and SARS-CoV-2: Potential therapeutic targeting. Eur J Pharmacol 884:173455. https://doi.org/10.1016/j.ejphar.2020.173455

Shi H, Han X, Jiang N, Cao Y, Alwalid O, Gu J, et al. (2020a) Radiological findings from 81 patients with COVID-19 pneumonia in Wuhan, China: a descriptive study. Lancet Infect Dis 20:425

Shi Q, Zhao K, Yu J, Feng J, Zhao K, Zhang X, Chen X, et al. (2020b) Clinical characteristics of 101 non-surviving hospitalized patients with COVID-19: a single center, retrospective study. medRxiv

Shi Q, Zhao K, Yu J, Feng J, Zhao K, Zhang X, Chen X, Hu P, Hong Y, Li M (2020c) Clinical characteristics of 101 non-surviving hospitalized patients with COVID-19: A single center, retrospective study. medRxiv

Shi Y, Yu X, Zhao H, Wang H, Zhao R, Cai Q, Sun S, Fang H, Wang Y, Hu Z (2020d) Host susceptibility to severe COVID-19: a retrospective analysis of a host risk score: findings of 487 cases outside Wuhan. Crit Care 24:108

Su H, Yang M, Wan C, Yi L, Tang F, Zhu H, Yi F, et al. (2020) Renal histopathological analysis of 26 post-mortem findings of patients with COVID-19 in China. Kidney International. Elsevier

Tan C, Zheng X, Huang Y, Liu J (2020) Key to successful treatment of COVID-19: accurate identification of severe risks and early intervention of disease progression. medRxiv

Tsang KW, Ho PL, Ooi GC, Yee WK, Wang T, Chan-Yeung M, Lam WK, Seto WH, Yam LY, Cheung TM, Wong PC, Lam B et al (2003) A cluster of cases of severe acute respiratory syndrome in Hong Kong. N Engl J Med. https://doi.org/10.1056/NEJMoa030666

Tsui PT, Kwok ML, Yuen H, Lai ST (2003) Severe acute respiratory syndrome: clinical outcome and prognostic correlates. Emerg Infect Dis 9(9):1064–1069. https://doi.org/10.3201/eid0909.030362

Wang Y-H, Lin A-S, Chao T-Y, Lu S-N, Liu J-W, Chen S-S, Lin M-C (2004a) A cluster of patients with severe acute respiratory syndrome in a chest ward in southern Taiwan. Intensive Care Med 30(6):1228–1231

Wang YH, Lin AS, Chao TY, Lu SN, Liu JW, Chen SS, Lin MC (2004b) A cluster of patients with severe acute respiratory syndrome in a chest ward in southern Taiwan. Intensive Care Med 30(6):1228–1231. https://doi.org/10.1007/s00134-004-2311-8

Wang D, Hu B, Hu C, Zhu F, Liu X, Zhang J, Wang B, Xiang H, Cheng Z, Xiong Y (2020a) Clinical characteristics of 138 hospitalized patients with 2019 novel coronavirus–infected pneumonia in Wuhan, China. JAMA 323(11):1061–1069

Wang D, Hu B, Hu C, Zhu F, Liu X, Zhang J, Wang B, Xiang H, Cheng Z, Xiong Y, Zhao Y, Li Y, Wang X, Peng Z (2020b) Clinical characteristics of 138 hospitalized patients with 2019 novel coronavirus–infected pneumonia in Wuhan, China. JAMA 323(11):1061–1069. https://doi.org/10.1001/jama.2020.1585

Wang L, Gao Y, Zhang G (2020c) The clinical dynamics of 18 cases of COVID-19 outside of Wuhan, China. Eur Respir J

Wang L, He W, Yu X, Hu D, Bao M, Liu H, Zhou J, et al. (2020d) Coronavirus Disease 2019 in elderly patients: characteristics and prognostic factors based on 4-week follow-up. J Infect

Wang L, Li X, Chen H, Yan S, Li D, Li Y et al (2020e) Coronavirus disease 19 infection does not result in acute kidney injury: an analysis of 116 hospitalized patients from Wuhan, China. Am J Nephrol 51(5):343–348

Wang W, He J, Wu S (2020f) The definition and risks of cytokine release syndrome-like in 11 COVID-19-infected pneumonia critically ill patients: disease characteristics and retrospective …. medRxiv

Wang Z, Yang B, Li Q, Wen L, Zhang R (2020g) Clinical Features of 69 Cases with Coronavirus Disease 2019 in Wuhan, China. Clin Infect Dis. https://doi.org/10.1093/cid/ciaa272

Wells G, Shea B, O'Connell D, Peterson J, Welch V, Losos M, Tugwell P (2014) Newcastle-Ottawa quality assessment scale cohort studies. Ontario: University of Ottawa

Wells GA, Shea B, O'connell D, Peterson J, Welch V, Losos M, Tugwell P (2016) The Newcastle-Ottawa Scale (NOS) for assessing the quality of nonrandomised studies in meta-analyses. 2013.

WHO (2020) Coronavirus disease (COVID-19), Situation report – 142. https://www.who.int/docs/default--source/coronaviruse/situation-reports/20200610-covid-19-sitrep-142.pdf?sfvrsn=180898cd_2

Wu Z, McGoogan JM (2020a) Characteristics of and important lessons from the coronavirus disease 2019 (COVID-19) outbreak in China: summary of a report of 72 314 cases from the Chinese Center for Disease Control and Prevention. JAMA

Wu Z, McGoogan JM (2020b) Characteristics of and important lessons from the coronavirus disease 2019 (COVID-19) outbreak in China: summary of a report of 72 314 cases from the Chinese Center for Disease Control and Prevention. JAMA 323(13):1239–1242

Wu C, Chen X, Cai Y, Zhou X, Xu S, et al. (2020a) Risk factors associated with acute respiratory distress syndrome and death in patients with coronavirus disease 2019 pneumonia in Wuhan, China. JAMA Intern Med 180:934

Wu J, Liu J, Zhao X, Liu C, Wang W, et al. (2020b) Clinical characteristics of imported cases of coronavirus disease 2019 (COVID-19) in Jiangsu Province: a multicenter descriptive study. Clin Infect Dis

Wu J, Liu J, Zhao X, Liu C, Wang W, et al. (2020c) Clinical characteristics of imported cases of COVID-19 in Jiangsu Province: a multicenter descriptive study. Clin Infect Dis

Xu X, Yu C, Qu J, Zhang L, Jiang S, Huang D, et al. (2020a) Imaging and clinical features of patients with 2019 novel coronavirus SARS-CoV-2. Eur J Med 47:1–6

Xu Y, Xu Z, Liu X, Cai L, Zheng H, Huang Y, Zhou L, et al. (2020b) Clinical findings in critical ill patients infected with SARS-Cov-2 in Guangdong Province, China: a multi-center, retrospective, observational study. medRxiv

Yang W, Cao Q, Qin L, Wang X, Cheng Z, Pan A, Dai J, et al. (2020a) Clinical characteristics and imaging manifestations of the 2019 novel coronavirus disease (COVID-19): a multi-center study in Wenzhou city, Zhejiang, China. J Infect 80(4):388-393

Yang X, Yu Y, Xu J, Shu H, Liu H, Wu Y, Zhang L, et al. (2020b) Clinical course and outcomes of critically ill patients with SARS-CoV-2 pneumonia in Wuhan, China: a single-centered, retrospective, observational study. Lancet Respir

Yazdanpanah F, Hamblin MR, Rezaei N (2020a) The immune system and COVID-19: friend or foe? Life Sci 256:117900. https://doi.org/10.1016/j.lfs.2020.117900

Yazdanpanah F, Hamblin MR, Rezaei N (2020b) The immune system and COVID-19: friend or foe? Life Sci: 117900

Yazdanpanah N, Saghazadeh A, Rezaei N (2020c) Anosmia: a missing link in the neuroimmunology of coronavirus disease 2019 (COVID-19). Rev Neurosci. https://doi.org/10.1515/revneuro-2020-0039

Yousefzadegan S, Rezaei N (2020) Case Report: Death due to COVID-19 in Three Brothers. Am J Trop Med Hyg 102(6):1203–1204

Zhang B, Zhou X, Qiu Y, Feng F, Feng J, Jia Y, Zhu H, et al. (2020a) Clinical characteristics of 82 death cases with COVID-19. medRxiv

Zhang G, Hu C, Luo L, Fang F, Chen Y, Li J, Peng Z, et al. (2020b) Clinical features and outcomes of 221 patients with COVID-19 in Wuhan, China. medRxiv

Zhang J, Dong X, Cao Y, Yuan Y, Yang Y, Yan Y, et al. (2020c) Clinical characteristics of 140 patients infected by SARS-CoV-2 in Wuhan, China. Allergy

Zhang L, Zhu F, Xie L, Wang C, Wang J, Chen R, Jia P, Guan H, Peng L, Chen Y (2020d) Clinical characteristics of COVID-19-infected cancer patients: a retrospective case study in three hospitals within Wuhan, China, Ann Oncol

Zhang R, Ouyang H, Fu L, Wang S, Han J, Huang K, Jia M, Song Q, Fu Z (2020e) CT features of SARS-CoV-2 pneumonia according to clinical presentation: a retrospective analysis of 120 consecutive patients from Wuhan city. Eur Radiol. https://doi.org/10.1007/s00330-020-06854-1

Zhang X, Cai H, Hu J, Lian J, Gu J, Zhang S et al (2020f) Epidemiological, clinical characteristics of cases of SARS-CoV-2 infection with abnormal imaging findings. Int J Infect Dis 94:81–87

Zhang Y, Cui Y, Shen M, Zhang J, Liu B, Dai M, Chen L, et al. (2020g) Comorbid diabetes mellitus was associated with poorer prognosis in patients with COVID-19: a retrospective cohort study. medRxiv

Zhao Z, Xie J, Yin M, Yang Y, He H, Jin T, Li W, Zhu X, et al. (2020) Clinical and laboratory profiles of 75 hospitalized patients with novel coronavirus disease 2019 in Hefei, China. medRxiv

Zhong N, Zheng B, Li Y, Poon L, Xie Z, Chan K, Li P, Tan S, Chang Q, Xie J (2003) Epidemiology and cause of severe acute respiratory syndrome (SARS) in Guangdong, People's Republic of China, in February, 2003. Lancet 362(9393):1353–1358

Zhou F, Yu T, Du R, Fan G, Liu Y, Liu Z, Xiang J, Wang Y, Song B, Gu X (2020) Clinical course and risk factors for mortality of adult inpatients with COVID-19 in Wuhan, China: a retrospective cohort study. Lancet

房晓伟, 梅清, 杨田军, 张蕾, 杨云, 王胤中, 童飞, et al. (2020) 2019 新型冠状病毒感染的肺炎 79 例临床特征及治疗分析. html.rhhz.net

18 COVID-19 in Patients with Cancer

Ali Nowroozi, Sepideh Razi, Kamal Kant Sahu, Fabio Grizzi, Jann Arends, Mahsa Keshavarz-Fathi, and Nima Rezaei

Abstract

With more than 5 million cases and 333,212 deaths, COVID-19 (or SARS-CoV-2) continues to spread. General symptoms of this disease are similar to that of many other viral respiratory diseases, including fever, cough, dyspnea, and fatigue, with a chance of progression to more severe complications. However, the virus does not affect all people equally, and cases with comorbidities such as malignancies, cardiovascular diseases, respiratory diseases, and kidney diseases are at higher risk of developing severe events, including requiring intensive ventilation, intensive care unit (ICU) admission, and death. Patients with cancer are more likely to be infected with COVID-19, which is possibly due to their immunological dysfunction or frequent clinic visits. Also, there is a higher chance that these patients experience severe events because of the medication they receive. In this chapter, we will review the main clinical manifestations of COVID-19 in patients

A. Nowroozi · M. Keshavarz-Fathi
Cancer Immunology Project (CIP), Universal Scientific Education and Research Network (USERN), Tehran, Iran

School of Medicine, Tehran University of Medical Sciences, Tehran, Iran

S. Razi (✉)
Cancer Immunology Project (CIP), Universal Scientific Education and Research Network (USERN), Tehran, Iran

Student Research Committee, School of Medicine, Tehran, Iran

K. K. Sahu
Department of Medicine, Saint Vincent Hospital, Worcester, MA, USA

Cancer Immunology Project (CIP), Universal Scientific Education and Research Network (USERN), Worcester, MA, USA

F. Grizzi
Department of Immunology and Inflammation, Humanitas Clinical and Research Center, Milan, Italy

Cancer Immunology Project (CIP), Universal Scientific Education and Research Network (USERN), Milan, Italy

J. Arends
Department of Medicine I, Medical Center – University of Freiburg, Faculty of Medicine, University of Freiburg, Freiburg, Germany

Cancer Immunology Project (CIP), Universal Scientific Education and Research Network (USERN), Freiburg, Germany

N. Rezaei (✉)
Research Center for Immunodeficiencies, Children's Medical Center, Tehran University of Medical Sciences, Tehran, Iran

Department of Immunology, School of Medicine, Tehran University of Medical Sciences, Tehran, Iran

Cancer Immunology Project (CIP), Universal Scientific Education and Research Network (USERN), Stockholm, Sweden
e-mail: rezaei_nima@tums.ac.ir

N. Rezaei (ed.), *Coronavirus Disease - COVID-19*, Advances in Experimental Medicine and Biology 1306, https://doi.org/10.1007/978-3-030-63761-3_18

with cancer. Recommendations and challenges for managing resources, organizing cancer centers, treatment of COVID-19-infected cancer patients, and performing cancer research during this pandemic will also be discussed.

Keywords

Cancer · COVID-19 · Malignancy · Neoplasm · Pandemic · SARS-CoV-2

18.1 Introduction

In late December 2019, novel coronavirus disease (COVID-19) caused by the severe acute respiratory syndrome coronavirus 2 (SARS-CoV-2) was first presented in the Hubei province of China as pneumonia of unknown cause. As of May 22, 2020, COVID-19 has more than 5,118,416 confirmed cases, and 333,212 deaths have been reported worldwide (Johns Hopkins University 2020), and the situation has been declared as a pandemic (World Health Organization 2020; Hanaei and Rezaei 2020). SARS-CoV-2 is a member of the *Coronaviridae* family, which includes severe acute respiratory syndrome (SARS) and Middle East respiratory syndrome (MERS) (first started to spread in 2002 and 2012, respectively), two other coronaviruses that have caused epidemics (Jabbari et al. 2020). The most common symptoms of the disease include fever, cough, dyspnea, and fatigue (Cao et al. 2020). Also, lymphopenia occurs in 57% of the cases, which results in the immune system dysregulation in the patients (Cao et al. 2020; Fathi and Rezaei 2020; Saghazadeh and Rezaei 2020a; Nasab et al. 2020). The most frequent severe complications of COVID-19 are acute respiratory distress syndrome (ARDS) and acute cardiac injury, which happen in 29% and 14% of cases, respectively. Moreover, 29% of cases will eventually require intensive care, and the case fatality rate is 7% among infected patients (Jordan et al. 2020; Zhou et al. 2020; Wu and McGoogan 2020). Studies reported that some risk factors are associated with severe outcomes of COVID-19, including high age and preexisting comorbidities such as cardiovascular and respiratory diseases and cancer (Cao et al. 2020; Shamshirian and Rezaei 2020). It has been reported that patients with cancer are more prone to infections and are considered as a high-risk group due to their immunosuppressed state and altered expression of angiotensin-converting enzyme 2 (ACE2), known as a cell surface receptor for the SARS-CoV-2 (Arjeyni et al. 2017; Sharifkashani et al. 2020; Ahmadi et al. 2020; Rezaei 2020b; Jahanshahlu and Rezaei 2020b). Also, they are more likely to be infected with COVID-19, which is possibly a result of their frequent visits to medical centers. It has been shown that there is a high chance that patients with cancer develop severe events such as requiring intensive ventilation, intensive care unit (ICU) admission, or death (Liang et al. 2020). It might be a result of the clinical therapies they receive. Therefore, it is important to know the degree of risk that patients with cancer are facing during the pandemic and come up with solutions for managing cancer care in this period. In this chapter, we will review the main clinical manifestations of COVID-19 in cancer patients. Recommendations and challenges for managing resources, organizing cancer centers, treatment of COVID-19-infected cancer patients, and performing cancer research during this pandemic will also be discussed.

18.2 Respiratory Infections in Patients with Cancer

Like all other chronic conditions, cancer makes patients more susceptible to severe complications from an acute disease such as a respiratory infection. Oncology patients frequently require chemotherapy or radiation therapy, which both suppress the immune system. Some patients receive immunosuppressants following transplants to prevent rejection. Additionally, hematologic malignancies directly affect the bone marrow and the immune system, causing defects in the body's defense mechanisms. Studies provide evidence that although children with cancer are less likely to develop the severe illness than adults, infants and younger children are more likely to develop severe clinical manifestations than older children because of the immaturity of the immune system. Children, as well as adult patients with cancer, are at higher risk of developing severe complications from an infection due to both the treatment they receive, the malignancy itself, the social distancing, and the number of people visiting oncology departments. The use of telehealth for noncritical outpatient visits, such as for children in follow-up or survivorship clinics, has been aspectable to be implemented to protect children who require hospital visits (Yang et al. 2020; Zhang et al. 2020; Extance 2020; Dong et al. 2020; Moazzami et al. 2020). Previous data showed that influenza has a much more strong effect on patients with cancer than the general population and hospitalization and mortality rates among cancer patients are four times and ten times higher, respectively (Bitterman et al. 2018). Viral respiratory tract infections are usually self-limiting in healthy individuals, and treatment is generally supportive (Kuchar et al. 2015). Although common symptoms of these diseases are generally mild (e.g., sneezing, coughing, and nasal congestion) (Kuchar et al. 2015; Kennedy et al. 2012; Wat 2004; Eccles 2005) and resolve within few days, immunosuppressed patients might progress to more adverse conditions such as lower respiratory tract infection (LRTI) and pneumonia (Kim et al. 2014). LRTI occurs in 30–50% of adults with cancer who are immunocompromised (Hijano et al. 2018; Kim et al. 2007; Hakim et al. 2016; Fisher et al. 2018), and the overall mortality rate among cases with malignancies who develop pneumonia is approximately between 11% and 15% but might reach as high as 75% (Chemaly et al. 2006; Kim et al. 2019; Hijano et al. 2018). Also, studies showed that hematopoietic stem cell transplantation and myelosuppression cause the highest morbidity and mortality rates among these patients (Hijano et al. 2018).

18.3 COVID-19 in Patients with Cancer

18.3.1 Risk Factors for Severe Complications

Liang et al. conducted a study (Liang et al. 2020) on 1590 COVID-19-positive patients, of which 18 had a history of cancer. The results showed a higher prevalence of severe events (ICU admission requiring invasive ventilation or death) among patients with cancer history (39%) compared to the general population (8%, $p < 0.003$). The significant difference was still present when severe complications were characterized by both the abovementioned events and the physician's evaluation (50% vs. 16% in patients with a history of cancer and the general population of COVID-19-infected cases, respectively, $p < 0.008$). Other comorbid conditions were also analyzed as risk factors for severe events, such as hypertension and chronic obstructive pulmonary disease (COPD). However, none of the odds were as high as that of cancer. Moreover, none of the other risk factors were significantly related to severe outcomes in cases with a cancer history. This study also showed that patients who had undergone surgery or chemotherapy within 1 month before the study presented a significantly higher rate of severe events compared to those with no history of surgery or chemotherapy in the past month. According to most studies, lung cancer appears to be the most prevalent malignancy among patients with cancer and COVID-19 infection (Zhang et al. 2020; Yu et al. 2020; Liang et al. 2020). However, in one study, colorectal cancer was the most prevalent, followed by lung cancer. A study in Zhongnan Hospital of Wuhan University (Yu et al. 2020) estimated that out of 1524 patients with cancer, only 0.79% (95% CI = 0.3%–1.2%) were infected with COVID-19. It was higher than the general infection rate in Wuhan over the same time (0.37%) (OR = 2.31, 95% CI = 1.89–3.02), and out of 12 patients with cancer infected with COVID-19, 1 patient (8%) required ICU admission, and 3 patients (25%) died, rather unusual results which are probably due to study limitations such as the small population of the study.

Another study was conducted in Wuhan University's Renmin Hospital, which involved 37 patients with cancer infected with COVID-19. The results demonstrated that 54% of the patients had a severe/critical status, which is much higher than that of general COVID-19-infected cases, and the mortality rate of 13.5% was observed in these cancer patients, which is also higher than the general population. The results of this study also reported that the history of receiving anticancer therapy did not affect the severity of COVID-19 in patients with cancer, which is in contrast to what Liang et al. have found in their study. In another study of 28 cancer patients infected with COVID-19 (out of a total of 1276 COVID-19 cases), 54% developed the severe condition, 21% eventually required ICU care, 36% encountered life-threatening complications, and 29% of the patients died (median time from admission to death = 16 days). The results also reported that severe events were more frequent in patients with stage IV cancer (70%) compared to patients with less advanced disease (44%). Even though this difference was not proved to be significant by statistical analysis, Zhang et al. believe that the cancer stage might affect the clinical outcome of the disease. Patchy consolidations on first computed tomography (CT) after admission were also related to a higher risk of developing severe complications after adjustment. Also, the outcomes of this study demonstrated that the rate of severe complications was significantly higher in patients who had received antitumor treatment within 14 days of their COVID-19 diagnosis, which is compatible with the results of the Liang et al. study. There is a possibility that genes affecting the immune system function contribute to the development of a severe form of COVID-19 (Yousefzadegan and Rezaei 2020; Darbeheshti and Rezaei 2020; Ahanchian et al. 2020; Babaha and Rezaei 2020). Further research is, however, necessary to investigate the issue in cancer patients.

To summarize, current data on epidemiologic statistics of COVID-19-infected cancer patients are inconclusive, and further research is needed.

18.3.2 Clinical Manifestations

Although studies regarding clinical features of COVID-19-infected patients with cancer are few, they give us a reasonable insight into what we should expect and what should be done. Early symptoms of COVID-19 in patients with cancer are almost identical to those in the general population, including fever, dry cough, fatigue, dyspnea, lymphopenia, and elevated high-sensitivity C-reactive protein (hs-CRP) (Rothan and Byrareddy 2020; Zhang et al. 2020). However, it has been reported that in patients with lung cancer, dyspnea presents early in the course of the disease (approximately 1 day after fever), which is much earlier than in the general population (5–8 days after fever) (Zhang et al. 2020; Wang et al. 2020a). Also, Zhang et al. reported that lymphopenia was more prevalent in cancer patients than in the general COVID-19-infected population (82% vs. 57%, respectively) (Cao et al. 2020; Zhang et al. 2020). Anemia and hypoproteinemia might also be present due to the nutritional deterioration of cancer patients (Zhang et al. 2020). According to Liang et al.'s study, patients with cancer seem to be more likely to experience polypnea and present a more severe baseline CT scan. Also, cancer patients face severe events much earlier than the general population (13 days vs. 43 days, $p < 0.001$, age-adjusted). However, more research is needed to prove these results because, in this study, patients with cancer had a higher median age and were more likely to have a history of smoking. Furthermore, the study population was small, and 12 out of 18 cancer cases did not have an active malignancy and were in post-resection routine follow-up (Liang et al. 2020).

Patients with cancer have an immunosuppressed status and are therefore less likely to suffer from severe lung injuries associated with inflammatory cytokines. Studies report that ARDS presents in 29% of patients with cancer, which is comparable to the general population (29%). Even though ARDS in cancer patients is as common as in general COVID-19-infected patients, mortality rates appear to be much higher. One possible explanation for a lower than expected ARDS prevalence in cancer patients is that the immune system and inflammatory cytokines play an important role in ARDS. Hence, immunosuppressed patients are less likely to develop the condition. Still, data on clinical features and outcomes of COVID-19 in cancer patients is limited, and further studies are needed (Xu et al. 2020; Stebbing et al. 2020; Conti et al. 2020; Cao et al. 2020; Zhang et al. 2020). On the other hand, due to impaired immunity, relatively high age, and other concurrent diseases of patients with cancer, the direct effects of the virus will probably appear more severely, which is yet to be proven.

18.4 Medical Care of Patients with Cancer During COVID-19 Pandemic: Challenges and Recommendations

COVID-19 is a new disease, and guidelines of medical care for patients infected with the virus are being updated rapidly. Furthermore, literature regarding cancer care during pandemic infections is limited (Al-Shamsi et al. 2020; Battershill 2006). Therefore, it is necessary to be aware of the most recent findings on the matter. It is also necessary that every structure created in health systems because of the pandemic be flexible enough so that it can be updated according to the most recent guidelines and protocols.

18.4.1 Managing Resources

One of the main points of focus during a pandemic is how to manage limited and rapidly depleting resources (Ueda et al. 2020).

Naturally, healthcare providers face a huge workload because of the spreading disease, and human resources are of utmost importance. Human resources must be distributed carefully so that both the pandemic and other health matters could be managed at the same time. Health institutions and organizations must not allow other departments and tasks to be overshadowed by the pandemic, known as the "distraction effect" (Cortiula et al. 2020; Schrag et al. 2020).

National University Cancer Institute, Singapore (NCIS), has taken its experience during the SARS epidemic into action and developed a comprehensive guideline for managing healthcare workers and hospitals (Ngoi et al. 2020). The main aim of the plan is to keep workforce loss as low as possible. All staff was divided into two groups so that in case of quarantine, there would be a backup team. Subgroups of each physician were distributed geographically to reduce interaction between individuals, and each outpatient division had its own set of departments (e.g., counter, triage, and consultation rooms). To further increase the workforce at the hospital, all community services of NCIS, such as home chemotherapy and nursing, were canceled. Also, cancer surveillance meetings were postponed, and nonresident referrals were ceased to lessen the patient load. In the case of the inpatient setting, all surgeries other than cancer-related ones were deferred by 3 months. Negative pressure isolation rooms were redesigned for COVID-19 suspected or confirmed cases. Blood was also reserved for special situations, as social distancing measures had diminished blood stocks.

A similar study (Jazieh et al. 2020) described actions taken by a King Abdulaziz Medical City's oncology department in Saudi Arabia during the 2015 MERS pandemic. A leadership committee was created, and all outpatient schedules were canceled for 4 days so that the situation could be analyzed more thoroughly. Most of the meetings of the staff were via the Internet. Patients of the outpatients' clinic were divided into three groups, based on the possibility of postponing their appointments: urgent cases which their visits could not be deferred, intermediate cases which could have had their appointments delayed for a few weeks, and follow-up cases which were postponed to latest time possible. All clinic visitors were screened when entering the clinic, and suspected cases were isolated and triaged. To be able to handle the delayed chemotherapies, the working hours of the infusion department were increased. The cancer center decided not to admit any new patients, and all cases that required admission were transferred to another hospital. Moreover, stable cancer patients already admitted were discharged. Training courses for all hospital staff were directed to update their knowledge regarding the ongoing outbreak and reducing their stress. Nonmedical staff were put off-work for a few days and were advised to minimize contact with the medical team upon return. It was told if any of the staff showed symptoms of the disease, do not show up for work. After the control of the outbreak, recovery was initiated in three phases: i. immediate phase, completing all services that were not available at the alternate hospital; ii. intermediate phase, restoring all previous activities; and iii. long-term phase, implementing the lessons learned from the outbreak in the hospital workflow and structure.

Generally, it is recommended that only vital medical services be provided and unnecessary appointments such as elective surgeries, nonurgent matters, and routine follow-ups, especially in low-risk patients, be canceled or postponed (Al-Shamsi et al. 2020; The Cancer Letter 2020; Mukherjee et al. 2003; Wang et al. 2020b; Shankar et al. 2020; Hanna et al. 2020; Cortiula et al. 2020; Ueda et al. 2020; Wei et al. 2020; Bartlett et al. 2020; Schrag et al. 2020; Ganatra et al. 2020). This action not only helps to allocate human resources more efficiently and reduces the chance of cross-infection, but it also enables healthcare personnel to be able to provide drugs and treatment to those who are truly in need. However, health centers must be prepared for the heavy load of patients that has been shifted from the pandemic period to a later time (Cortiula et al. 2020). Using over the phone assessments or online videoconferences could also reduce incoming patient flow to the clinic and the burden it carries (Hanna et al. 2020).

Drug shortage is likely during a pandemic, and medical centers must be prepared for it. The US Food and Drug Administration (FDA) released a statement on February 27, 2020, indicating the shortage of only one drug (US Food and Drug Administration 2020a). On March 21, 2020, FDA's drug shortage list (US Food and Drug Administration) included 26 cancer medications (Al-Shamsi et al. 2020), and as of April 9, 2020, a total of 103 drugs are indicated "currently in shortage." It must be mentioned that some cancer drugs have no alternatives, and the shortage of their supplies might cause disastrous effects on patients (Alpert and Jacobson 2019).

18.4.2 Cancer Centers During the Pandemic

Cancer centers must execute a plan specifically designed for the time of the pandemic to continue cancer treatment, while maximizing safety and minimizing the risk of infection and disease spread. Specific personnel must be designated to supervise the execution, thus ensuring compliance with the protocols (Al-Shamsi et al. 2020).

General screening procedures are necessary. Temperatures and symptoms of COVID-19 must be checked routinely both in staff and throughout the complex. Additionally, past contact and travel history of every individual have to be acquired (Ngoi et al. 2020; Wang et al. 2020b; Hanna et al. 2020; Bitar et al. 2020; Ueda et al. 2020; Pino et al. 2020; Chen et al. 2020). Suspected cases should be separated and transferred to an isolated area for further evaluation by the pandemic team (Ngoi et al. 2020; Bitar et al. 2020; You et al. 2020; Ueda et al. 2020; Chen et al. 2020). If a patient reports previous contact or travel history but is asymptomatic, it is recommended to be asked for 14-day self-isolation and appointment reschedule. If not feasible, the patient has to be isolated, and extra protective measures must be carried out (Bitar et al. 2020). Suspected staff should be put off duty until disease resolution. Restricting the number of companions allowed with a patient to a maximum of one companion will prevent casual gatherings and reduce the risk of cross-infection (Al-Shamsi et al. 2020; Lambertini et al. 2020). Furthermore, some cancer centers have prohibited guest visits to the hematology department (Cortiula et al. 2020). Utilizing Internet-based services such as appointment scheduling systems is advantageous as they reduce the need for on-site appearance. Also, on-phone screening assists in preventing the spread of the disease in ambulatory clinics. Patients could be asked about symptoms or recent contacts, a day before they visit the clinic (Al-Shamsi et al. 2020; You et al. 2020; Cinar et al. 2020; Rivera et al. 2020). Routine checking for COVID-19 symptoms, blood tests, and performing CT imaging before admission are also recommended to better identify patients at risk of carrying the virus (Wang et al. 2020b).

For better organization of the medical team, the facility could be divided into different sections, based on the degree of virus contamination (Chen et al. 2020). Different protocols could then be prepared according to each section's situation.

18.4.3 Cancer Care During the Pandemic

Based on a case-by-case evaluation and if it is medically safe, some professionals recommend that all cancer treatments be delayed to time as late as possible (Lambertini et al. 2020; Ganatra et al. 2020). Although cancer medication may result in worsened outcome in patients infected with COVID-19 (Hanna et al. 2020), some suggest that cancer centers must carry on with providing treatment for all patients as before, but with intense supervision and monitoring for COVID-19 and reduced dosage of immunosuppressing medication (Cortiula et al. 2020; Zhang et al. 2020; Salako et al. 2020). However, due to the increased burden on healthcare systems, this seems to be practically impossible (Lambertini et al. 2020). Limited resources oblige cancer treatment facilities to adopt new strategies in treating patients with cancer. Many healthcare centers have decided not to continue chemotherapy or immunotherapy for patients with advanced malignancy in the maintenance phase or to receive long-time treatments (Wang et al. 2020b; Schrag et al. 2020; Hanna et al. 2020), while some professionals suggest otherwise and recommend continuing treatment for patients with the advanced disease if no COVID-19 symptoms are present (Cortiula et al. 2020). Unlike advanced malignancies, which are a matter of debate, treating patients with a high chance of cure is vastly recommended (Al-Shamsi et al. 2020; The Cancer Letter 2020; Hanna et al. 2020; Bitar et al. 2020; You et al. 2020; Ueda et al. 2020; Pino et al. 2020; Simcock et al. 2020; Schrag et al. 2020). For example, a resectable tumor might grow and become inoperable if neglected (Cortiula et al. 2020). In non-curable cases, patients with lower age and higher life expectancy should be prioritized (You et al. 2020).

Exposure to the medical environment increases the chance of infection (Hanna et al. 2020; Wang et al. 2020a; Chen et al. 2020). During cancer treatment, many procedures involve direct contact with surfaces and other individuals (e.g., visits to the clinic, surgeries, and chemotherapy sessions) and consequently put patients and physicians at risk of infection (Kutikov et al. 2020). As previously stated, the infection rate among patients with cancer appears to be higher than the general population, and it is speculated that this is a consequence of frequent hospital visits of these patients, which increases their chance of encountering the virus and acquiring the disease (Yu et al. 2020). Outpatient visits can especially be challenging due to lower protective measures applied and a higher number of patients. Therefore, the general target is to minimize patients' contact with the hospital and clinic environment, while providing all the necessary treatment (You et al. 2020). Telemedicine (over the phone or online medical consultation) is a practical and beneficial approach to minimize direct contact (Ngoi et al. 2020; Kutikov et al. 2020; Wang et al. 2020b; Hanna et al. 2020; Bitar et al. 2020; You et al. 2020; Ueda et al. 2020; Pino et al. 2020; Simcock et al. 2020; Spicer et al. 2020; Cinar et al. 2020), especially in follow-up patients who are disease-free and where in-person visits are not essential (Lambertini et al. 2020).

It is recommended that patients scheduled for anticancer treatment be under observation and isolated from other patients 7 days before receiving medication (Zhang et al. 2020). If possible, intravenous treatment could be substituted with oral medication to lower the risk of nosocomial infection (Wang et al. 2020b; Al-Shamsi et al. 2020; Hanna et al. 2020; Bitar et al. 2020; You et al. 2020; Pino et al. 2020; Cinar et al. 2020). Patients could also receive chemotherapy at home, if practicable (Al-Shamsi et al. 2020; Shereen and Salman 2019; American Society of Clinical Oncology 2020; You et al. 2020; Cinar et al. 2020). Otherwise, sessions should be rearranged so that it requires fewer hospital visits (Cinar et al. 2020). It is best if patients could obtain medicine with minimum person-to-person contacts, such as drive-through pharmacies or delivery services (Al-Shamsi et al. 2020; Ngoi et al. 2020; Spicer et al. 2020). If blood work is required during the treatment process, the number of draws must be reduced to a minimum and, if suitable, at home (Al-Shamsi et al. 2020; American Society of Clinical Oncology 2020).

Some cancer patients require a transplant for treatment. The pandemic has made organizing a transplant difficult due to several reasons. The donor is required to be present at the hospital for evaluation and preparation. Both visiting a medical center and the possible need for traveling increase the risk of infection for the donor. Additionally, if the donor is infected with COVID-19, the transplant cannot be made until the pathogen is completely cleared from the body. Therefore, some institutions are assigning backup donors for patients for situations where the original donor is not available (Burki 2020).

Managing patients undergoing radiotherapy is a much more complicated task than chemotherapy patients. Patients receiving radiation must visit their health center daily to continue treatment with adjusted doses (Wei et al. 2020), and in many cases, the sessions cannot be halted midway through. Hypofractionated radiation or short-course radiotherapy is recommended to reduce the risk of virus transmission (Ngoi et al. 2020; Hanna et al. 2020; You et al. 2020; Spicer et al. 2020; Cinar et al. 2020). Before radiation delivery, all patients must be screened for COVID-19 and divided into different groups, based on their screening results and status. If the patients have been tested negative in screening and are in the early stages of the disease or need emergency radiation treatment, the best choice is to be treated at the nearest radiotherapy center to avoid unnecessary travel and overpopulating major healthcare centers. However, if the tumor is locally advanced, it is best if radiotherapy is deferred and, instead, neoadjuvant chemotherapy is practiced. Treatment centers must clean and disinfect facilities in-between patients, and protective equipment must be provided both for staff and patients, according to the latest guidelines and protocols, and strict hand hygiene must be maintained by the medical team (Chen et al. 2020; Wei et al. 2020).

Educating individuals is invaluable during infection outbreaks. Approximately 18–33% of infected individuals are estimated to be asymptomatic (Mizumoto et al. 2020; Nishiura et al. 2020). Therefore, patients must be taught about general precautionary methods such as social distancing, proper hand hygiene, control measures and devices, and COVID-19 disease itself (i.e., ways of transmission, symptoms, and disease course), based on the most recent information and protocols (Shankar et al. 2020; Oh 2020; Bitar et al. 2020; Ueda et al. 2020; Lotfi et al. 2020).

Cancer affects patients in emotional and psychological aspects. Amid reduced in-person contact and possible quarantines in a pandemic, it is important to maintain patients' mental health and provide emotional support so that they understand they are not alone (Hanna et al. 2020). Moreover, patients with cancer are more worried about acquiring the infection and its complications than healthy individuals, even young patients who have a relatively low risk of severe complications from the disease (Casanova et al. 2020; Verity et al. 2020). It has been reported that some patients might cancel all treatment plans in fear of acquiring an infection (Chen et al. 2004). Thus, healthcare professionals must make sure that patients receive all necessary medication and should help them overcome these fears. Patients also understand that their relatives are worried about them and feel this burden on themselves (Casanova et al. 2020). Psychology departments could help to lower these effects on individuals by taking extra measures and running extra supportive programs and psychological assessments in this period (Al-Shamsi et al. 2020). The use of telemedicine could also support patients during this period (Lambertini et al. 2020; Cinar et al. 2020).

18.4.4 Managing COVID-19-Infected Cancer Patients

Diagnosing viral infections in patients with cancer is challenging. Some symptoms of the disease, such as fatigue and dyspnea, are common in patients with cancer, especially when they receive medication (Spicer et al. 2020; Cinar et al. 2020). Additionally, radiologic findings of infection could resemble the progression of the tumor or drug-mediated toxicity related to cancer treatment (Lambertini et al. 2020). Hence, screening and diagnostic protocols differ from the general population (Ngoi et al. 2020).

Patients confirmed for COVID-19 must be isolated immediately and consulted with the COVID-19 pandemic team (You et al. 2020; Spicer et al. 2020). Cancer treatment might be significantly associated with the severe clinical condition in COVID-19-infected patients (Zhang et al. 2020; Simcock et al. 2020). However, recent research showed that some groups of medication, including steroids and cytotoxic drugs, are not associated with severe outcomes (Russell et al. 2020) and even might help to accelerate recovery in COVID-19 patients (Saghazadeh and Rezaei 2020b). Oncologists mostly recommend to cease further anticancer therapy and prioritize COVID-19 treatment based on a case-by-case evaluation. They agree to resume anticancer therapy only when no medical signs of the pathogen are present (Al-Shamsi et al. 2020; Yang et al. 2020; Bitar et al. 2020; You et al. 2020; Chen et al. 2020; Cinar et al. 2020) (i.e., the resolution of all symptoms and two consecutive negative molecular tests on nasopharynx swabs) (Centers for Disease Control and Prevention 2020). If testing is not available, at least 3 days and at least 7 days must pass after symptoms relief and initial symptoms appearance, respectively (Cinar et al. 2020). Some expand the criteria for stopping cancer treatment even more and suggest that patients in close contact with a COVID-19-infected person should also stop receiving cancer medication (Shankar et al. 2020). However, in some cases where it is life-saving, cancer treatment could continue on physicians' decisions (Cinar et al. 2020).

COVID-19 management has to be rapid and intense in cancer patients, especially when other comorbidities among older adults, pregnant women, and children are present (Liang et al. 2020; Mirbeyk and Rezaei 2020). The level of support provided for patients must be determined based on cancer-related life expectancy and other comorbid conditions. Even with the highest levels of ventilatory assistance, patients with advanced malignancies are unlikely to survive the infection. Therefore, treatment priority has to be discussed with the patient in the early stages of the disease (Spicer et al. 2020).

Malnutrition must especially be kept in mind when providing supportive care for patients. While malnutrition could increase the risk of acquiring an infection (Farhadi and Ovchinnikov 2018), infections could result in malnutrition via various mechanisms such as gastrointestinal disorders (e.g., loss of appetite, diarrhea) (Li et al. 2020) and metabolism imbalance (Handu et al. 2020). Additionally, malnutrition is common in patients with cancer and is associated with poor prognosis and reduced quality of life (Lee et al. 2016; Jager-Wittenaar et al. 2011; Van Cutsem and Arends 2005; Hill et al. 2011; Norman et al. 2010; Capuano et al. 2010). Therefore, early detection and treatment of any nutritional impairment are crucial in COVID-19-infected cancer patients.

18.4.5 Managing Hematological Malignancies During COVID-19

Among the cancers, hematological malignancies belong to a special subset for their underlying differences in the prevalence and biology when compared to solid tumors. Unlike solid tumors, blood cancers like leukemia, myelodysplastic syndromes, and multiple myeloma primarily involve the bone marrow. Such patients may have neutropenia, pancytopenia, or thrombocytopenia of variable degrees to start with, and treatment with chemotherapy, radiotherapy, or hematopoietic stem cell transplantation could worsen the severity and/or prolong the duration of bone marrow suppression (Sahu et al. 2020b). Though having hematological malignancies does not increase the risk to acquire the SARS-CoV-2 but once acquired, these patients tend to decompensate more rapidly than others (Gavillet et al. 2020). Hong et al. from China noted a similar observation with a rapid spread and transmission of the SARS-CoV-2 among the patients at their hematology ward (Hong M). They noted extremely high mortality (>50%) for hospitalized patients with combined COVID-19 and hematologic malignancy. It is believed that perturbation in both innate and adaptive immunity could lead to higher mortality in patients of hematologic malignancies undergoing chemotherapy on acquiring SARS-CoV-2 infection.

There is no fixed protocol available on how to manage hematological patients during the COVID-19 pandemic. However, international societies like the American Society of Hematology, the European Hematology Association, and other tertiary cancer center experts have laid down interim guidelines to guide hematologists and transplant specialists (https://www.hematology.org/covid-19). In general, patients with stable cancer are suggested to watch and wait for strategy. Patients with chronic myeloid leukemia and chronic lymphoid leukemia are classic examples. On the other side, patients who need urgent treatment like acute leukemias can be offered oral chemotherapies or targeted agents that can be easily taken at home, thereby minimizing hospital visits. These agents can serve as an excellent bridging therapy till the pandemic curve declines when health resources could be redirected back to patients with hematological cancers for blood transfusions and in hospital admissions due to febrile neutropenia, etc. (Sahu et al. 2020a). A classic example of the less intense but effective oral chemotherapeutic agent is oral Venetoclax for patients more than 70 years with acute myeloid leukemia (Pollyea et al. 2019). Data sharing, staying accurate and up to date with the latest developments, and following the local guidelines would be the key factors to help hematologists navigate through this health crisis.

18.5 Cancer Research During the Pandemic

COVID-19 pandemic will affect almost all research areas globally. With the increasing need for human resources, researchers are less likely to participate in studies, and most of the research community will put their attention on COVID-19-related matters. Furthermore, universities are closed in many cities in response to the outbreak, and a large proportion of students are not available for helping in researches. Ongoing trials will have to be prioritized, and inevitably some will be canceled or postponed (The Lancet 2020). Many researchers have already stopped enrolling new participants in trials (Spicer et al. 2020), and many others will be forced to do so (Burki 2020). FDA (US Food and Drug Administration 2020b) and the European Medicines Agency (EMA) (European Medicines Agency 2020) have released guidelines and proposed recommendations for ongoing trials during the COVID-19 pandemic.

With no definitive cure for many types of cancers, especially in advanced stages, cancer research plays an important role in the development of oncology treatment, and with the ongoing pandemic, new trials are less likely to start or will be delayed for a while (Lambertini et al. 2020; Burki 2020). Also, follow-up appointments are a necessity in most trials, which may not be possible during the pandemic, both due to health risks and workload of the medical team. Another challenge during a pandemic is providing investigational drugs to research teams. Delays in drug deliveries might cause irreversible damages to the course of the trial. Maintaining patients' safety and accuracy of the trial at the same time require preparation of new instructions, conducted by the joint decision of the medical team and sponsors of the trial (Kutikov et al. 2020; Lambertini et al. 2020).

Some centers allow therapeutic trials to continue as prior (Schrag et al. 2020), albeit with restricted access to high-risk areas such as isolation wards and the ICU (Ngoi et al. 2020). Recruiting for trials should only be considered when the benefit from the treatment is likely (Ueda et al. 2020). Travel limitations and quarantines are likely to interfere with research procedures such as tests and follow-ups. If possible, tests could be carried out in local clinics of the patients, and telemedicine could be used for follow-ups and instructions (Ngoi et al. 2020; Lambertini et al. 2020). If a center cannot handle the extra workload of researches, patients can be distributed among other less crowded health centers for treatment and follow-up continuation (Lambertini et al. 2020).

Even though the pandemic has hindered research, professionals worldwide are reaching each other and cooperating more than ever. Oncologists from different countries share their opinions with their colleagues to help in producing efficient and beneficial protocols in the path to better managing the global emergency (Schrag et al. 2020).

18.6 Conclusion

COVID-19 (or SARS-CoV-2) has spread all over the globes and has infected millions of people, including healthcare providers (Rezaei 2020a). Patients with comorbidities such as cardiovascular diseases, respiratory diseases, and malignancies are at higher risk of severe events from this disease. Patients with cancer are at increased risk of complications for any respiratory infection, and COVID-19 is not different.

Cancer patients are more likely to be infected with COVID-19, which is probably because of their frequent visits to medical centers. Also, there is a higher chance that these patients develop severe events due to the medications they receive. It is, however, a matter of dispute, and further studies are needed to prove the correlation between anticancer therapy and worsened COVID-19 outcomes. Early symptoms are almost identical in patients with and without cancer, though immune dysregulation seems to be more severe in cancer patients, as lymphopenia is more common compared to the general population. Additionally, symptoms are likely to emerge earlier and with more intensity in these patients than expected. Even though ARDS in cancer patients is as common as in general COVID-19-infected patients, mortality rates appear to be much higher. One possible explanation for low ARDS prevalence in cancer patients is that the immune system and inflammatory cytokines play an important role in ARDS. Hence, immunosuppressed patients are less likely to develop the condition. However, data on clinical features and outcomes of COVID-19 in cancer patients is still limited, and further studies are necessary.

Cancer treatment centers face a demanding challenge during the pandemic. With limited resources, healthcare facilities must manage to handle the ongoing COVID-19 situation, as well as other health services, while minimizing the risk of infection and complications. Guests' visits to the hospital could be limited during the pandemic. Unnecessary services such as elective surgeries and follow-ups of low-risk patients should be postponed. Some activities, such as appointment scheduling, could be done over the phone or the Internet instead of in-person. Healthcare centers must educate both the staff and the patients and provide them with necessary personal protective equipment such as masks and gloves, according to the most recent guidelines. For further prevention, facilities must perform screenings at entrance points, check patients and staff routinely for symptoms, and refer all suspicious cases to the pandemic team.

Patients' visits to hospitals must be as few as possible. Whether to cancel some treatments or not is debated. In both cases, prioritization seems to be the answer. Cancer treatment with the intent of cure and emergency treatments must be prioritized over treatment for advanced diseases and long-term medications. Telemedicine can be beneficial during this period. The use of oral medicine requires fewer hospital visits compared to chemotherapy, and so it is preferred during the pandemic period. Additionally, home delivery of medications and at-home medical services such as chemotherapy infusion and blood tests are beneficial, if feasible. Patients requiring radiotherapy could receive it at the nearest radiation center, and physicians could switch to hypofractionated doses.

Cancer treatment teams must make sure of their patient's psychological health and provide support and consultations if necessary, preferably via telemedicine (Moazzami et al. 2020).

As a result of its influence on the immune system (Rokni et al. 2020; Bahrami et al. 2020; Yazdanpanah et al. 2020a), COVID-19 could affect different organs (Yazdanpanah et al. 2020b; Jahanshahlu and Rezaei 2020a; Saleki et al. 2020) and therefore has been associated with a spectrum of clinical signs and symptoms (Lotfi and Rezaei 2020), making its diagnosis challenging (Basiri et al. 2020a). It is especially true in the cases of reinfection (Jabbari and Rezaei 2020) as well as in patients with cancer because symptoms and findings of the disease are common in cancer patients. If COVID-19 is confirmed in a cancer patient, oncology medication is recommended to be stopped, unless it is lifesaving. Treatment could resume when the disease is cleared.

Research is vital in oncology, and unfortunately, cancer research will probably face difficulties during the pandemic. It is, however, promising that the universal impact of COVID-19 has pushed researchers to become more cooperative than before (Momtazmanesh et al. 2020; Mohamed et al. 2020a; Rzymski et al. 2020; Moradian et al. 2020; Kafieh et al. 2020), providing promising opportunities for COVID-19 treatment (Sharifkashani et al. 2020; Rezaei 2020b; Rabiee et al. 2020; Pourahmad et al. 2020; Mansourabadi et al. 2020; Pashaei and Rezaei 2020; Jahanshahlu and Rezaei 2020b; Basiri et al. 2020b; Mohamed et al. 2020b). Also, as stated before, literature regarding COVID-19-infected cancer patients is limited, and the cancer research community could put its effort on this matter to increase oncology and COVID-19 knowledge simultaneously.

References

Ahanchian H, Moazzen N, Faroughi MSD, Khalighi N, Khoshkhui M, Aelami MH, Haghi NSM, Rezaei N (2020) COVID-19 in a Child with Primary Specific Antibody Deficiency

Ahmadi M, Saffarzadeh N, Habibi MA, Hajiesmaeili F, Rezaei N (2020) Colon cancer and SARS-CoV-2: impact of ACE2 expression in susceptibility to COVID-19. bioRxiv

Al-Shamsi HO, Alhazzani W, Alhuraiji A, Coomes EA, Chemaly RF, Almuhanna M, Wolff R, Nuhad IK, Chua MLK, Hotte SJ, Meyers BM, Elfiki T, Curigliano G, Eng C, Grothey A, Xie C (2020) A practical approach to the Management of Cancer Patients during the novel coronavirus disease 2019 (COVID-19) pandemic: an international collaborative group. Oncologist 25. https://doi.org/10.1634/theoncologist.2020-0213

Alpert A, Jacobson M (2019) Impact of oncology drug shortages on chemotherapy treatment. Clin Pharmacol Ther 106(2):415–421. https://doi.org/10.1002/cpt.1390

American Society of Clinical Oncology (2020) ASCO Coronavirus Resources. https://www.asco.org/asco-coronavirus-information. Accessed 15 Apr 2020

Arjeyni Y, Goudarzi H, Eslami G, Faghihloo E (2017) Viral respiratory infections in patients with Cancer. Int J Cancer Manag 10(2):e8084. https://doi.org/10.5812/ijcp.8084

Babaha F, Rezaei N (2020) Primary immunodeficiency diseases in COVID-19 pandemic: a predisposing or protective factor? Am J Med Sci. https://doi.org/10.1016/j.amjms.2020.07.027

Bahrami A, Vafapour M, Moazzami B, Rezaei N (2020) Hyperinflammatory shock related to COVID-19 in a patient presenting with multisystem inflammatory syndrome in children: first case from Iran. J Paediatr Child Health. https://doi.org/10.1111/jpc.15048

Bartlett DL, Howe JR, Chang G, Crago A, Hogg M, Karakousis G, Levine E, Maker A, Mamounas E, McGuire K, Merchant N, Shibata D, Sohn V, Solorzano C, Turaga K, White R, Yang A, Yoon S, Society of Surgical O (2020) Management of cancer surgery cases during the COVID-19 pandemic: considerations. Ann Surg Oncol. https://doi.org/10.1245/s10434-020-08461-2

Basiri A, Heidari A, Nadi MF, Fallahy MTP, Nezamabadi SS, Sedighi M, Saghazadeh A, Rezaei N (2020a) Microfluidic devices for detection of RNA viruses. Rev Med Virol n/a (n/a):e2154. https://doi.org/10.1002/rmv.2154

Basiri A, Pazhouhnia Z, Beheshtizadeh N, Hoseinpour M, Saghazadeh A, Rezaei N (2020b) Regenerative medicine in COVID-19 treatment: real opportunities and range of promises. Stem Cell Rev Rep:1–13. https://doi.org/10.1007/s12015-020-09994-5

Battershill PM (2006) Influenza pandemic planning for cancer patients. Curr Oncol 13(4):119–120

Bitar N, Kattan J, Kourie HR, Mukherji D, Saghir NE (2020) The Lebanese Society of Medical Oncology (LSMO) statement on the care of patients with cancer during the COVID-19 pandemic. Future Oncol 16:615. https://doi.org/10.2217/fon-2020-0252

Bitterman R, Eliakim-Raz N, Vinograd I, Zalmanovici Trestioreanu A, Leibovici L, Paul M (2018) Influenza vaccines in immunosuppressed adults with cancer. Cochrane Database Syst Rev 2:CD008983. https://doi.org/10.1002/14651858.CD008983.pub3

Burki TK (2020) Cancer care in the time of COVID-19. Lancet Oncol 21:628. https://doi.org/10.1016/S1470-2045(20)30201-1

Cao Y, Liu X, Xiong L, Cai K (2020) Imaging and clinical features of patients with 2019 novel coronavirus SARS-CoV-2: a systematic review and meta-analysis. J Med Virol 92:1449. https://doi.org/10.1002/jmv.25822

Capuano G, Gentile PC, Bianciardi F, Tosti M, Palladino A, Di Palma M (2010) Prevalence and influence of malnutrition on quality of life and performance status in patients with locally advanced head and neck cancer before treatment. Support Care Cancer 18(4):433–437. https://doi.org/10.1007/s00520-009-0681-8

Casanova M, Pagani Bagliacca E, Silva M, Patriarca C, Veneroni L, Clerici CA, Spreafico F, Luksch R, Terenziani M, Meazza C, Podda M, Biassoni V, Schiavello E, Chiaravalli S, Puma N, Bergamaschi L, Gattuso G, Sironi G, Massimino M, Ferrari A (2020) How young patients with cancer perceive the Covid-19 (coronavirus) epidemic in Milan, Italy: is there room for other fears? Pediatr Blood Cancer e28318. https://doi.org/10.1002/pbc.28318

Centers for Disease Control and Prevention (2020) Ending Home Isolation for Immunocompromised Persons with COVID-19. https://www.cdc.gov/coronavirus/2019-ncov/hcp/ending-isolation.html. Accessed 17 Apr 2020

Chemaly RF, Ghosh S, Bodey GP, Rohatgi N, Safdar A, Keating MJ, Champlin RE, Aguilera EA, Tarrand JJ, Raad II (2006) Respiratory viral infections in adults with hematologic malignancies and human stem cell transplantation recipients: a retrospective study at a major cancer center. Medicine (Baltimore) 85(5):278–287. https://doi.org/10.1097/01.md.0000232560.22098.4e

Chen W, Su XY, Wang VJ, Wang EC, Xu R, Zhong S, Sun G, Xia J (2020) Novel coronavirus international public health emergency: guidance on radiation oncology facility operation. Adv Radiat Oncol 5:560. https://doi.org/10.1016/j.adro.2020.03.008

Chen YM, Perng RP, Chu H, Tsai CM, Whang-Peng J (2004) Impact of severe acute respiratory syndrome on the status of lung cancer chemotherapy patients and a correlation of the signs and symptoms. Lung Cancer 45(1):39–43. https://doi.org/10.1016/j.lungcan.2004.01.002

Cinar P, Kubal T, Freifeld A, Mishra A, Shulman L, Bachman J, Fonseca R, Uronis H, Klemanski D, Slusser K, Lunning M, Liu C (2020) Safety at the time of the COVID-19 pandemic: how to keep our oncology patients and healthcare workers safe. J Natl Compr Cancer Netw 18:1–6. https://doi.org/10.6004/jnccn.2020.7572

Conti P, Ronconi G, Caraffa A, Gallenga CE, Ross R, Frydas I, Kritas SK (2020) Induction of pro-inflammatory cytokines (IL-1 and IL-6) and lung inflammation by Coronavirus-19 (COVI-19 or SARS-CoV-2): anti-inflammatory strategies. J Biol Regul Homeost Agents 34(2). https://doi.org/10.23812/CONTI-E

Cortiula F, Pettke A, Bartoletti M, Puglisi F, Helleday T (2020) Managing COVID-19 in the oncology clinic and avoiding the distraction effect. Ann Oncol 31:553. https://doi.org/10.1016/j.annonc.2020.03.286

Darbeheshti F, Rezaei N (2020) Genetic predisposition models to COVID-19 infection. Med Hypotheses 142:109818. https://doi.org/10.1016/j.mehy.2020.109818

Dong Y, Mo X, Hu Y, Qi X, Jiang F, Jiang Z, Tong S (2020) Epidemiological characteristics of 2143 pediatric patients with 2019 coronavirus disease in China. Pediatrics:e20200702. https://doi.org/10.1542/peds.2020-0702

Eccles R (2005) Understanding the symptoms of the common cold and influenza. Lancet Infect Dis 5(11):718–725. https://doi.org/10.1016/S1473-3099(05)70270-X

European Medicines Agency (2020) Guidance on the management of clinical trials during the COVID-19 (Coronavirus) pandemic. https://ec.europa.eu/health/sites/health/files/files/eudralex/vol-10/guidanceclinicaltrials_covid19_en.pdf. Accessed 18 Apr 2020

Extance A (2020) Covid-19 and long term conditions: what if you have cancer, diabetes, or chronic kidney disease? BMJ 368:m1174. https://doi.org/10.1136/bmj.m1174

Farhadi S, Ovchinnikov R (2018) The relationship between nutrition and infectious diseases: a review. Biomed Biotechnol Res J (BBRJ) 2(3):168–172. https://doi.org/10.4103/bbrj.bbrj_69_18

Fathi N, Rezaei N (2020) Lymphopenia in COVID-19: therapeutic opportunities. Cell Biol Int 44(9):1792–1797. https://doi.org/10.1002/cbin.11403

Fisher BT, Danziger-Isakov L, Sweet LR, Munoz FM, Maron G, Tuomanen E, Murray A, Englund JA, Dulek D, Halasa N, Green M, Michaels MG, Madan RP, Herold BC, Steinbach WJ (2018) A Multicenter consortium to define the epidemiology and outcomes of inpatient respiratory viral infections in pediatric hematopoietic stem cell transplant recipients. J Pediatric Infect Dis Soc 7(4):275–282. https://doi.org/10.1093/jpids/pix051

Ganatra S, Hammond SP, Nohria A (2020) The novel coronavirus disease (COVID-19) threat for patients with cardiovascular disease and cancer. JACC CardioOncol 2:350. https://doi.org/10.1016/j.jaccao.2020.03.001

Gavillet M, Carr Klappert J, Spertini O, Blum S (2020) Acute leukemia in the time of COVID-19. Leuk Res 92:106353. https://doi.org/10.1016/j.leukres.2020.106353

Hakim H, Dallas R, Zhou Y, Pei D, Cheng C, Flynn PM, Pui CH, Jeha S (2016) Acute respiratory infections in children and adolescents with acute lymphoblastic leukemia. Cancer 122(5):798–805. https://doi.org/10.1002/cncr.29833

Hanaei S, Rezaei N (2020) COVID-19: developing from an outbreak to a pandemic. Arch Med Res 51:582. https://doi.org/10.1016/j.arcmed.2020.04.021

Handu D, Moloney L, Rozga M, Cheng F (2020) Malnutrition care during the COVID-19 pandemic: considerations for registered dietitian nutritionists evidence analysis center. J Acad Nutr Diet. https://doi.org/10.1016/j.jand.2020.05.012

Hanna TP, Evans GA, Booth CM (2020) Cancer, COVID-19 and the precautionary principle: prioritizing treatment during a global pandemic. Nat Rev Clin Oncol 17:268. https://doi.org/10.1038/s41571-020-0362-6

Hijano DR, Maron G, Hayden RT (2018) Respiratory viral infections in patients with cancer or undergoing hematopoietic cell transplant. Front Microbiol 9:3097. https://doi.org/10.3389/fmicb.2018.03097

Hill A, Kiss N, Hodgson B, Crowe TC, Walsh AD (2011) Associations between nutritional status, weight loss, radiotherapy treatment toxicity and treatment outcomes in gastrointestinal cancer patients. Clin Nutr 30(1):92–98. https://doi.org/10.1016/j.clnu.2010.07.015

Jabbari P, Jabbari F, Ebrahimi S, Rezaei N (2020) COVID-19: a chimera of two pandemics. Disaster Med Public Health Prep:1–2. https://doi.org/10.1017/dmp.2020.223

Jabbari P, Rezaei N (2020) With risk of reinfection, is COVID-19 Here to stay? Disaster Med Public Health Prep:1. https://doi.org/10.1017/dmp.2020.274

Jager-Wittenaar H, Dijkstra PU, Vissink A, van der Laan BF, van Oort RP, Roodenburg JL (2011) Malnutrition and quality of life in patients treated for oral or oropharyngeal cancer. Head Neck 33(4):490–496. https://doi.org/10.1002/hed.21473

Jahanshahlu L, Rezaei N (2020a) Central nervous system involvement in COVID-19. Arch Med Res 51:721. https://doi.org/10.1016/j.arcmed.2020.05.016

Jahanshahlu L, Rezaei N (2020b) Monoclonal antibody as a potential anti-COVID-19. Biomed Pharmacother 129:110337. https://doi.org/10.1016/j.biopha.2020.110337

Jazieh AR, Al Hadab A, Al Olayan A, AlHejazi A, Al Safi F, Al Qarni A, Farooqui F, Al Mutairi N, Alenazi TH (2020) Managing oncology services during a major coronavirus outbreak: lessons from the Saudi Arabia experience. JCO Glob Oncol (6):518–524. https://doi.org/10.1200/GO.20.00063

Johns Hopkins University (2020) Coronavirus COVID-19 Global Cases by the Center for Systems Science and Engineering (CSSE) at Johns Hopkins University (JHU). https://coronavirus.jhu.edu/map.html. Accessed 7 Apr 2020

Jordan RE, Adab P, Cheng KK (2020) Covid-19: risk factors for severe disease and death. BMJ 368:m1198. https://doi.org/10.1136/bmj.m1198

Kafieh R, Arian R, Saeedizadeh N, Minaee S, Yadav SK, Vaezi A, Rezaei N, Javanmard SH (2020) COVID-19 in Iran: A Deeper Look Into The Future. medRxiv

Kennedy JL, Turner RB, Braciale T, Heymann PW, Borish L (2012) Pathogenesis of rhinovirus infection. Curr Opin Virol 2(3):287–293. https://doi.org/10.1016/j.coviro.2012.03.008

Kim YJ, Boeckh M, Englund JA (2007) Community respiratory virus infections in immunocompromised patients: hematopoietic stem cell and solid organ transplant recipients, and individuals with human immunodeficiency virus infection. Semin Respir Crit Care Med 28(2):222–242. https://doi.org/10.1055/s-2007-976494

Kim YJ, Guthrie KA, Waghmare A, Walsh EE, Falsey AR, Kuypers J, Cent A, Englund JA, Boeckh M (2014) Respiratory syncytial virus in hematopoietic cell transplant recipients: factors determining progression to lower respiratory tract disease. J Infect Dis 209(8):1195–1204. https://doi.org/10.1093/infdis/jit832

Kim YJ, Lee ES, Lee YS (2019) High mortality from viral pneumonia in patients with cancer. Infect Dis (Lond) 51(7):502–509. https://doi.org/10.1080/23744235.2019.1592217

Kuchar E, Miskiewicz K, Nitsch-Osuch A, Szenborn L (2015) Pathophysiology of clinical symptoms in acute viral respiratory tract infections. Adv Exp Med Biol 857:25–38. https://doi.org/10.1007/5584_2015_110

Kutikov A, Weinberg DS, Edelman MJ, Horwitz EM, Uzzo RG, Fisher RI (2020) A war on two fronts: Cancer Care in the Time of COVID-19. Ann Intern Med 172:756. https://doi.org/10.7326/M20-1133

Lambertini M, Toss A, Passaro A, Criscitiello C, Cremolini C, Cardone C, Loupakis F, Viscardi G, Meattini I, Dieci MV, Ferrara R, Giusti R, Maio MD (2020) Cancer care during the spread of coronavirus disease 2019 (COVID-19) in Italy: young oncologists' perspective. ESMO Open 5(2). https://doi.org/10.1136/esmoopen-2020-000759

Lee JLC, Leong LP, Lim SL (2016) Nutrition intervention approaches to reduce malnutrition in oncology patients: a systematic review. Support Care Cancer 24(1):469–480. https://doi.org/10.1007/s00520-015-2958-4

Li T, Zhang Y, Gong C, Wang J, Liu B, Shi L, Duan J (2020) Prevalence of malnutrition and analysis of related factors in elderly patients with COVID-19 in Wuhan, China. Eur J Clin Nutr 74:871. https://doi.org/10.1038/s41430-020-0642-3

Liang W, Guan W, Chen R, Wang W, Li J, Xu K, Li C, Ai Q, Lu W, Liang H, Li S, He J (2020) Cancer patients in SARS-CoV-2 infection: a nationwide analysis in China. Lancet Oncol 21(3):335–337. https://doi.org/10.1016/S1470-2045(20)30096-6

Lotfi M, Hamblin MR, Rezaei N (2020) COVID-19: transmission, prevention, and potential therapeutic opportunities. Clinica chimica acta 508:254–266. https://doi.org/10.1016/j.cca.2020.05.044

Lotfi M, Rezaei N (2020) SARS-CoV-2: a comprehensive review from pathogenicity of the virus to clinical consequences. J Med Virol 92:1864. https://doi.org/10.1002/jmv.26123

Mansourabadi AH, Sadeghalvad M, Mohammadi-Motlagh HR, Rezaei N (2020) The immune system as a target for therapy of SARS-CoV-2: a systematic review of the current immunotherapies for COVID-19. Life Sci 258:118185. https://doi.org/10.1016/j.lfs.2020.118185

Mirbeyk M, Rezaei N (2020) The impact of COVID-19 on pregnancy and neonatal health: a systematic review

Mizumoto K, Kagaya K, Zarebski A, Chowell G (2020) Estimating the asymptomatic proportion of coronavirus disease 2019 (COVID-19) cases on board the diamond princess cruise ship, Yokohama, Japan, 2020. Euro Surveill 25(10). https://doi.org/10.2807/1560-7917.ES.2020.25.10.2000180

Moazzami B, Razavi-Khorasani N, Dooghaie Moghadam A, Farokhi E, Rezaei N (2020) COVID-19 and telemedicine: immediate action required for maintaining healthcare providers well-being. J Clin Virol 126:104345. https://doi.org/10.1016/j.jcv.2020.104345

Mohamed K, Rodríguez-Román E, Rahmani F, Zhang H, Ivanovska M, Makka SA, Joya M, Makuku R, Islam MS, Radwan N, Rahmah L, Goda R, Abarikwu SO, Shaw M, Zoghi S, Irtsyan S, Ling I, Cseprekal O, Faten AB, Hazar Sayar E, Soloukey C, Grancini G, Rezaei N (2020a) Borderless collaboration is needed for COVID-19-a disease that knows no borders. Infect Control Hosp Epidemiol 41:1–2. https://doi.org/10.1017/ice.2020.162

Mohamed K, Yazdanpanah N, Saghazadeh A, Rezaei N (2020b) Computational drug discovery and repurposing for the treatment of Covid-19: a systematic review. Available at SSRN 3583748

Momtazmanesh S, Ochs HD, Uddin LQ, Perc M, Routes JM, Vieira DN, Al-Herz W, Baris S, Prando C, Rosivall L, Abdul Latiff AH, Ulrichs T, Roudenok V, Aldave Becerra JC, Salunke DB, Goudouris E, Condino-Neto A, Stashchak A, Kryvenko O, Stashchak M, Bondarenko A, Rezaei N (2020) All together to fight COVID-19. Am J Trop Med Hyg 102(6):1181–1183. https://doi.org/10.4269/ajtmh.20-0281

Moradian N, Ochs HD, Sedikies C, Hamblin MR, Camargo CA Jr, Martinez JA, Biamonte JD, Abdollahi M, Torres PJ, Nieto JJ, Ogino S, Seymour JF, Abraham A, Cauda V, Gupta S, Ramakrishna S, Sellke FW, Sorooshian A, Wallace Hayes A, Martinez-Urbistondo M, Gupta M, Azadbakht L, Esmaillzadeh A, Kelishadi R, Esteghamati A, Emam-Djomeh Z, Majdzadeh R, Palit P, Badali H, Rao I, Saboury AA, Jagan Mohan Rao L, Ahmadieh H, Montazeri A, Fadini GP, Pauly D, Thomas S, Moosavi-Movahed AA, Aghamohammadi A, Behmanesh M, Rahimi-Movaghar V, Ghavami S, Mehran R, Uddin LQ, Von Herrath M, Mobasher B, Rezaei N (2020) The urgent need for integrated science to fight COVID-19 pandemic and beyond. J Transl Med 18(1):205. https://doi.org/10.1186/s12967-020-02364-2

Mukherjee RK, Back MF, Lu JJ, Shakespeare TP, Wynne CJ (2003) Hiding in the bunker: challenges for a radiation oncology department operating in the severe acute respiratory syndrome outbreak. Australas Radiol 47(2):143–145. https://doi.org/10.1046/j.0004-8461.2003.01165.x

Nasab MG, Saghazadeh A, Rezaei N (2020) SARS-CoV-2-a tough opponent for the immune system. Arch Med Res 51:589. https://doi.org/10.1016/j.arcmed.2020.05.020

Ngoi N, Lim J, Ow S, Jen WY, Lee M, Teo W, Ho J, Sundar R, Tung ML, Lee YM, Ngo E, Lim SW, Ong J, Lim F, Bonney G, Vellayappan B, Ho F, Tey J, Chan N, de Mel S, Poon M, Lee SY, Koh LP, Liu TC, Tan LK, Wong A, Wong A, Lim SE, Yeoh A, Wong JE, Tan D, Goh BC, Chng WJ, Soo R, Chee CE, Chee YL, Lee SC, Jeyasekharan AD, National University Cancer Institute S (2020) A segregated-team model to maintain cancer care during the COVID-19 outbreak at an academic center in Singapore. Ann Oncol doi:https://doi.org/10.1016/j.annonc.2020.03.306, 31, 840

Nishiura H, Kobayashi T, Suzuki A, Jung SM, Hayashi K, Kinoshita R, Yang Y, Yuan B, Akhmetzhanov AR, Linton NM, Miyama T (2020) Estimation of the asymptomatic ratio of novel coronavirus infections (COVID-19). Int J Infect Dis 94:154. https://doi.org/10.1016/j.ijid.2020.03.020

Norman K, Stobaus N, Smoliner C, Zocher D, Scheufele R, Valentini L, Lochs H, Pirlich M (2010) Determinants of hand grip strength, knee extension strength and functional status in cancer patients. Clin Nutr 29(5):586–591. https://doi.org/10.1016/j.clnu.2010.02.007

Oh WK (2020) COVID-19 infection in Cancer patients: early observations and unanswered questions. Ann Oncol 31:838. https://doi.org/10.1016/j.annonc.2020.03.297

Pashaei M, Rezaei N (2020) Immunotherapy for SARS-CoV-2: potential opportunities. Expert Opin Biol Ther 20:1–5. https://doi.org/10.1080/14712598.2020.1807933

Pino L, Perez C, Cardona A, Triana I (2020) Cancer Center recommendations to mitigate COVID-19 impact in patients with Cancer: low-resource settings version. JCO Glob Oncol (6):569–570. https://doi.org/10.1200/GO.20.00093

Pollyea DA, Amaya M, Strati P, Konopleva MY (2019) Venetoclax for AML: changing the treatment paradigm. Blood Adv 3(24):4326–4335. https://doi.org/10.1182/bloodadvances.2019000937

Pourahmad R, Moazzami B, Rezaei N (2020) Efficacy of Plasmapheresis and immunoglobulin replacement therapy (IVIG) on patients with COVID-19. SN Comprehensive Clin Med:1–5

Rabiee N, Rabiee M, Bagherzadeh M, Rezaei N (2020) COVID-19 and picotechnology: potential opportunities. Med Hypotheses 144:109917. https://doi.org/10.1016/j.mehy.2020.109917

Rezaei N (2020a) COVID-19 affects healthy pediatricians more than pediatric patients. Infect Control Hosp Epidemiol 1:1106. https://doi.org/10.1017/ice.2020.139

Rezaei N (2020b) COVID-19 and medical biotechnology. Avicenna J Med Biotechnol 12(3):139

Rivera A, Ohri N, Thomas E, Miller R, Knoll MA (2020) The impact of COVID-19 on radiation oncology clinics and cancer patients in the U.S. Adv Radiat Oncol. doi:https://doi.org/10.1016/j.adro.2020.03.006

Rokni M, Hamblin MR, Rezaei N (2020) Cytokines and COVID-19: friends or foes? Hum Vaccin Immunother:1–3

Rothan HA, Byrareddy SN (2020) The epidemiology and pathogenesis of coronavirus disease (COVID-19) outbreak. J Autoimmun:102433. doi:https://doi.org/10.1016/j.jaut.2020.102433

Russell B, Moss C, George G, Santaolalla A, Cope A, Papa S, Van Hemelrijck M (2020) Associations between immune-suppressive and stimulating drugs and novel COVID-19-a systematic review of current evidence. Ecancermedicalscience 14:1022. https://doi.org/10.3332/ecancer.2020.1022

Rzymski P, Nowicki M, Mullin GE, Abraham A, Rodríguez-Román E, Petzold MB, Bendau A, Sahu KK, Ather A, Naviaux A-F (2020) Quantity does not equal quality: scientific principles cannot be sacrificed. Int Immunopharmacol 86:106711

Saghazadeh A, Rezaei N (2020a) Immune-epidemiological parameters of the novel coronavirus - a perspective. Expert Rev Clin Immunol 16(5):465–470. https://doi.org/10.1080/1744666x.2020.1750954

Saghazadeh A, Rezaei N (2020b) Towards treatment planning of COVID-19: rationale and hypothesis for the use of multiple immunosuppressive agents: anti-antibodies, immunoglobulins, and corticosteroids. Int Immunopharmacol 84:106560. https://doi.org/10.1016/j.intimp.2020.106560

Sahu KK, Jindal V, Siddiqui AD (2020a) Managing COVID-19 in patients with Cancer: a double blow for oncologists. JCO Oncol Pract 16(5):223–225. https://doi.org/10.1200/op.20.00167

Sahu KK, Siddiqui AD, Rezaei N, Cerny J (2020b) Challenges for management of immune thrombocytopenia during COVID-19 pandemic. J Med Virol 92:2277. https://doi.org/10.1002/jmv.26251

Salako O, Okunade K, Allsop M, Habeebu M, Toye M, Oluyede G, Fagbenro G, Salako B (2020) Upheaval in cancer care during the COVID-19 outbreak. Ecancermedicalscience 14:ed97. https://doi.org/10.3332/ecancer.2020.ed97

Saleki K, Banazadeh M, Saghazadeh A, Rezaei N (2020) The involvement of the central nervous system in patients with COVID-19. Rev Neurosci 31(4):453–456. https://doi.org/10.1515/revneuro-2020-0026

Schrag D, Hershman DL, Basch E (2020) Oncology practice during the COVID-19 pandemic. JAMA 323:2005. https://doi.org/10.1001/jama.2020.6236

Shamshirian D, Rezaei N (2020) Cardiovascular diseases burden in COVID-19: Systematic review and meta-analysis

Shankar A, Saini D, Roy S, Mosavi Jarrahi A, Chakraborty A, Bharti SJ, Taghizadeh-Hesary F (2020) Cancer care delivery challenges amidst coronavirus disease - 19 (COVID-19) outbreak: specific precautions for Cancer patients and Cancer care providers to prevent spread. Asian Pac J Cancer Prev 21(3):569–573. https://doi.org/10.31557/APJCP.2020.21.3.569

Sharifkashani S, Bafrani MA, Khaboushan AS, Pirzadeh M, Kheirandish A, Yavarpour Bali H, Hessami A, Saghazadeh A, Rezaei N (2020) Angiotensin-converting enzyme 2 (ACE2) receptor and SARS-CoV-2: potential therapeutic targeting. Eur J Pharmacol 884:173455. https://doi.org/10.1016/j.ejphar.2020.173455

Shereen NG, Salman D (2019) Delivering chemotherapy at home: how much do we know? Br J Community Nurs 24(10):482–484. https://doi.org/10.12968/bjcn.2019.24.10.482

Simcock R, Thomas TV, Estes C, Filippi AR, Katz MA, Pereira IJ, Saeed H (2020) COVID-19: global radiation oncology's targeted response for pandemic preparedness. Clin Transl Radiat Oncol 22:55–68. https://doi.org/10.1016/j.ctro.2020.03.009

Spicer J, Chamberlain C, Papa S (2020) Provision of cancer care during the COVID-19 pandemic. Nat Rev Clin Oncol 17:329. https://doi.org/10.1038/s41571-020-0370-6

Stebbing J, Phelan A, Griffin I, Tucker C, Oechsle O, Smith D, Richardson P (2020) COVID-19: combining antiviral and anti-inflammatory treatments. Lancet Infect Dis 20(4):400–402. https://doi.org/10.1016/S1473-3099(20)30132-8

The Cancer Letter (2020) What to expect: Oncology's response to coronavirus in Italy "It's like being in a war". https://cancerletter.com/articles/20200311_1/. Accessed 9 Apr 2020

The Lancet O (2020) COVID-19: global consequences for oncology. Lancet Oncol 21(4):467. https://doi.org/10.1016/S1470-2045(20)30175-3

U.S. Food & Drug Administration (2020a) Coronavirus (COVID-19) Supply Chain Update. https://www.fda.gov/news-events/press-announcements/coronavirus-covid-19-supply-chain-update?utm_campaign=022720_statement_coronavirus%20%28covid-19%29%20supply%20chain%20update&utm_medium=email&utm_source=eloqua. Accessed 9 Apr 2020

U.S. Food & Drug Administration (2020b) FDA Guidance on Conduct of Clinical Trials of Medical Products during COVID-19 Public Health Emergency. https://www.fda.gov/media/136238/download. Accessed 18 Apr 2020

Ueda M, Martins R, Hendrie PC, McDonnell T, Crews JR, Wong TL, McCreery B, Jagels B, Crane A, Byrd DR, Pergam SA, Davidson NE, Liu C, Stewart FM (2020) Managing Cancer care during the COVID-19 pandemic: agility and collaboration toward a common goal. J Natl Compr Cancer Netw 18:1–4. https://doi.org/10.6004/jnccn.2020.7560

Van Cutsem E, Arends J (2005) The causes and consequences of cancer-associated malnutrition. Eur J Oncol Nurs 9(Suppl 2):S51–S63. https://doi.org/10.1016/j.ejon.2005.09.007

Verity R, Okell LC, Dorigatti I, Winskill P, Whittaker C, Imai N, Cuomo-Dannenburg G, Thompson H, Walker PGT, Fu H, Dighe A, Griffin JT, Baguelin M, Bhatia S, Boonyasiri A, Cori A, Cucunuba Z, FitzJohn R, Gaythorpe K, Green W, Hamlet A, Hinsley W, Laydon D, Nedjati-Gilani G, Riley S, van Elsland S, Volz E, Wang H, Wang Y, Xi X, Donnelly CA, Ghani AC, Ferguson NM (2020) Estimates of the severity of coronavirus disease 2019: a model-based analysis. Lancet Infect Dis doi:https://doi.org/10.1016/S1473--3099(20)30243-7, 20, 669

Wang D, Hu B, Hu C, Zhu F, Liu X, Zhang J, Wang B, Xiang H, Cheng Z, Xiong Y, Zhao Y, Li Y, Wang X, Peng Z (2020a) Clinical characteristics of 138 hospitalized patients with 2019 novel coronavirus-infected pneumonia in Wuhan, China. JAMA 323:1061. https://doi.org/10.1001/jama.2020.1585

Wang Z, Wang J, He J (2020b) Active and effective measures for the Care of Patients with Cancer during the COVID-19 spread in China. JAMA Oncol 6:631. https://doi.org/10.1001/jamaoncol.2020.1198

Wat D (2004) The common cold: a review of the literature. Eur J Intern Med 15(2):79–88. https://doi.org/10.1016/j.ejim.2004.01.006

Wei W, Jiang H, Chen W, Zhou Y, Guo S, Zhong G, Zhou H, Wang S, Zhang H, Jiang W, Liu ZG (2020) How should we implement radiotherapy for cancer patients in China during the endemic period of COVID-19? Radiother Oncol 147:100–102. https://doi.org/10.1016/j.radonc.2020.03.016

World Health Organization (2020) Rolling updates on coronavirus disease (COVID-19). https://www.who.int/emergencies/diseases/novel-coronavirus-2019/events-as-they-happen. Accessed 7 Apr 2020

Wu Z, McGoogan JM (2020) Characteristics of and important lessons from the coronavirus disease 2019 (COVID-19) outbreak in China: summary of a report of 72314 cases from the Chinese Center for Disease Control and Prevention. JAMA 323:1239. https://doi.org/10.1001/jama.2020.2648

Xu Z, Shi L, Wang Y, Zhang J, Huang L, Zhang C, Liu S, Zhao P, Liu H, Zhu L, Tai Y, Bai C, Gao T, Song J, Xia P, Dong J, Zhao J, Wang FS (2020) Pathological findings of COVID-19 associated with acute respiratory distress syndrome. Lancet Respir Med 8(4):420–422. https://doi.org/10.1016/S2213-2600(20)30076-X

Yang G, Zhang H, Yang Y (2020) Challenges and countermeasures of integrative Cancer therapy in the epidemic of COVID-19. Integr Cancer Ther 19:1534735420912811. https://doi.org/10.1177/1534735420912811

Yazdanpanah F, Hamblin MR, Rezaei N (2020a) The immune system and COVID-19: friend or foe? Life Sci 256:117900. https://doi.org/10.1016/j.lfs.2020.117900

Yazdanpanah N, Saghazadeh A, Rezaei N (2020b) Anosmia: a missing link in the neuroimmunology of coronavirus disease 2019 (COVID-19). Rev Neurosci 31:691. https://doi.org/10.1515/revneuro-2020-0039

You B, Ravaud A, Canivet A, Ganem G, Giraud P, Guimbaud R, Kaluzinski L, Krakowski I, Mayeur D, Grellety T, Lotz JP (2020) The official French guidelines to protect patients with cancer against SARS-CoV-2 infection. Lancet Oncol 21:619. https://doi.org/10.1016/S1470-2045(20)30204-7

Yousefzadegan S, Rezaei N (2020) Case report: death due to COVID-19 in three brothers. Am J Trop Med Hyg 102(6):1203–1204. https://doi.org/10.4269/ajtmh.20-0240

Yu J, Ouyang W, Chua MLK, Xie C (2020) SARS-CoV-2 transmission in patients with Cancer at a tertiary Care Hospital in Wuhan, China. JAMA Oncol 6:1108. https://doi.org/10.1001/jamaoncol.2020.0980

Zhang L, Zhu F, Xie L, Wang C, Wang J, Chen R, Jia P, Guan HQ, Peng L, Chen Y, Peng P, Zhang P, Chu Q, Shen Q, Wang Y, Xu SY, Zhao JP, Zhou M (2020) Clinical characteristics of COVID-19-infected cancer patients: a retrospective case study in three hospitals within Wuhan. China Ann Oncol 31:894. https://doi.org/10.1016/j.annonc.2020.03.296

Zhou F, Yu T, Du R, Fan G, Liu Y, Liu Z, Xiang J, Wang Y, Song B, Gu X, Guan L, Wei Y, Li H, Wu X, Xu J, Tu S, Zhang Y, Chen H, Cao B (2020) Clinical course and risk factors for mortality of adult inpatients with COVID-19 in Wuhan, China: a retrospective cohort study. Lancet 395(10229):1054–1062. https://doi.org/10.1016/S0140-6736(20)30566-3

COVID-19 and Tropical Infection: Complexity and Concurrence

19

Pathum Sookaromdee and Viroj Wiwanitkit

Abstract

COVID-19 is a newly emerging pandemic caused by a novel coronavirus. After its first report in China in December 2019, the disease already spread and affected more than 200 countries worldwide. It correlates with different phenotypes ranging from an acute febrile illness to severe respiratory problems. Often, patients with COVID-19 suffer from metabolic disorders, and this can result in a more severe clinical course. COVID-19 might also co-occur with other common diseases in different settings. In tropical countries, COVID-19 has already affected thousands of local populations. Tropical diseases such as dengue and tuberculosis can modify the clinical presentation of COVID-19 and result in difficulty in the diagnosis and treatment of the patients. The complexity of concurrence between COVID-19 and tropical diseases is, thus, a matter of concern in tropical medicine. This chapter is devoted to discussing problems surrounding the management of COVID-19 in tropical countries. To exemplify the effects of COVID-19 on tropical countries, the authors would show how COVID-19 has affected Indochina, a large tropical area.

Keywords

COVID-19 · Dengue · Indochina · Infection · Tropical · Tuberculosis

19.1 Introduction

COVID-19 is a newly emerging viral infection that has spread worldwide and caused a pandemic. It is caused by a novel coronavirus, SARS-CoV-2, which seems to be transmitted from wild animals to humans. COVID-19 was firstly reported in patients with pneumonia of unknown origin in Wuhan, China, in December 2019. After its first report in China in December 2019, the disease spread rapidly to many areas around the world (Hsia 2020). Moreover, it was declared by the World Health Organization (WHO) a public health emergency (Cucinotta 2020). COVID-19 has already affected countries all over the world, with more than six million confirmed cases by the end of May. Such a worldwide spread of disease calls for urgent international collaboration to manage the present crisis.

P. Sookaromdee
Private Academic Practice, Pune, India

V. Wiwanitkit (✉)
Department of Community Medicine, Dr. DY Patil University, Pune, India

Systematic Review and Meta-analysis Expert Group (SRMEG), Universal Scientific Education and Research Network (USERN), Pune, India

N. Rezaei (ed.), *Coronavirus Disease - COVID-19*, Advances in Experimental Medicine and Biology 1318, https://doi.org/10.1007/978-3-030-63761-3_19

Table 19.1 Some atypical clinical presentations in COVID-19

Atypical presentations	Examples
Uncommon clinical symptom	Diarrhea
	Skin rash
	Conjunctivitis
	Anosmia
	Muscle weakness
	Unconsciousness
	Sudden death
Uncommon laboratory presentation	Eosinophilia
	Anemia
	Impaired liver function
	Impaired renal function
Non-severe presentation	Asymptomatic presentation
	Mild presentation

Table 19.2 Interrelationship between COVID-19 and some common tropical infections

Tropical infections	Details
Dengue	Mimicry skin lesion, petechiae
	Similar clinical presentation
Tuberculosis	Hidden clinical problem
	Increased mortality
Hepatitis B infection	Increased mortality

COVID-19 often involves the respiratory system. It can cause acute febrile illness and severe respiratory problems. The patients might develop illness after the exposure in a few days to weeks (Rothan 2020). As a highly contagious respiratory disease, human-to-human COVID-19 transmission quickly occurs (Guo 2020). Patients with underlying diseases such as diabetes mellitus and hypertension are more likely to experience a more severe clinical course (Guo 2020). In addition to classical presentations, some COVID-19 patients might manifest atypical clinical features (Table 19.1 and 19.2) that are difficult-to-diagnose cases (Tin 2020). Sometimes, the infection might be mild or asymptomatic, making it spread quickly in the community. Closed contact is the leading way of COVID-19 spread (Singhal 2020), and therefore, social distancing is recommended as an approach to disease prevention (Nicola 2020).

COVID-19 commonly co-occurs with other medical diseases. Considering the rapid spread of disease, we can conclude that this high prevalence of COVID-19 co-occurring with other medical diseases is simply due to having a large population affected by COVID-19 (Rodriguez-Morales 2020). In particular, COVID-19 has affected thousands of local populations in tropical countries. So, it is statistically possible that COVID-19 might also co-occur with tropical diseases, and this concurrence can lead to atypical and severe clinical features of COVID-19, making it difficult to diagnose and treat. Therefore, concurrence between COVID-19 and tropical diseases such as dengue and tuberculosis is a complex issue in clinical medicine.

The management of COVID-19 in tropical countries is of high importance. Governmental public health agencies usually try several attempts for disease control (Wu 2020). According to the public health structure, primary medical care centers serve as frontlines in the management of outbreaks like COVID-19. However, several reasons limit the performance of these centers, particularly in developing countries (Sohrabi 2020), for example, limited knowledge of primary healthcare workers, lack of adequate personnel protective equipment (PPE), and lack of well-skilled primary care center administrators. This chapter is devoted to discussing problems surrounding the management of COVID-19 in tropical countries. To exemplify the effects of COVID-19 on tropical countries, the authors would show how COVID-19 has affected Indochina, a large tropical area.

19.2 COVID-19 and Tropical Infection: Complexity and Concurrence

The nationwide outbreak of infections is quite possible in developing countries because of poverty, low education, and limited infrastructures, while the COVID-19 pandemic has affected the globe, including several tropical countries. There are tropical diseases endemic to the local population. The existence of COVID-19 in tropical countries is a paramount public health concern because COVID-19 might concurrently affect patients with tropical diseases.

19.2.1 COVID-19 and Tropical Infection

19.2.1.1 Dengue

Dengue is a mosquito-borne infection. The mosquito vector is *Aedes* species. This disease is characterized by acute febrile illness and hemorrhagic complications. Laboratory findings include hemoconcentration, thrombocytopenia, and atypical lymphocytosis. The clinical presentation of dengue is similar to that of COVID-19, and this can result in misdiagnosis. The misdiagnosis of COVID-19 as dengue can result in delayed diagnosis of COVID-19 and, therefore, the poor clinical outcome of COVID-19. Saavedra-Velasco (2020) and Yan (2020) reported that COVID-19 could cause false positives in dengue screening tests. Thrombocytopenia is a common laboratory finding in both dengue and COVID-19 and causes difficulty in the differential diagnosis. Patients who have thrombocytopenia present with skin rash and petechiae. As Joob and Wiwanitkit (Joob 2020) noted, petechiae in COVID-19 might be misinterpreted as skin rash due to dengue (Fig. 19.1). So, we need a diagnostic approach for COVID-19 in endemic zones (Saavedra-Velasco 2020). The coinfection between COVID-19 and dengue might become a dangerous combination in tropical countries (Lorenz et al. 2020), requiring a public health plan for the management of co-epidemic of dengue and COVID-19 in tropical countries (Navarro 2020).

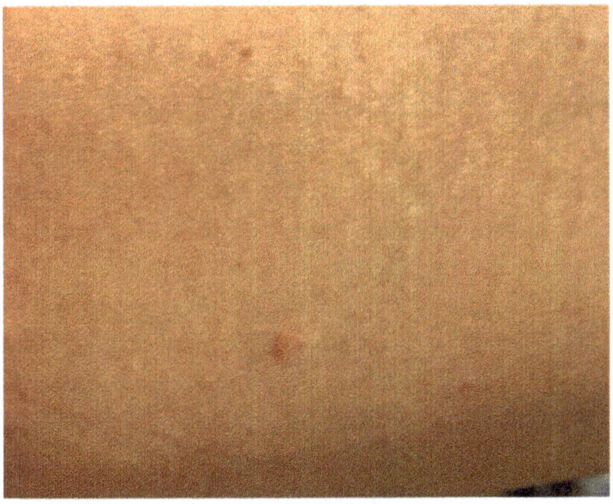

Fig. 19.1 Skin rash in a patient that is difficult to differentiate between dengue and COVID-19

19.2.1.2 Chikungunya

Chikungunya is a mosquito-borne infection characterized by acute febrile illness. Since arthralgia commonly occurs in both COVID-19 and chikungunya, it would be necessary to consider COVID-19 as a differential diagnosis in patients with fever and arthralgia.

19.2.1.3 Zika Virus Infection

It is a mosquito-borne infection. The mosquito vector for the Zika virus infection is *Aedes* species. It is characterized by acute febrile illness and hemorrhagic complications. The patient might also have a low platelet count. Zika virus infection is easily missed as dengue and COVID-19. From infodemiology, the pattern of Google Trends in the early outbreak stage is the same as that previously observed in the Zika virus infection (Mavragani 2020). Nevertheless, there is still no official report on concurrent COVID-19 and Zika virus infection.

19.2.1.4 Yellow Fever

Yellow fever is a mosquito-borne infection characterized by acute febrile illness and jaundice. In the yellow fever endemic area, there is a chance for co-occurrence between COVID-19 and yellow fever (Chauhan et al. 2020). Nevertheless, there is still no official report on concurrent COVID-19 and yellow fever. It might be because yellow fever is presently under strict control by the yellow fever vaccination problem.

19.2.1.5 Malaria

Malaria is a mosquito-borne infection. The mosquito vector is *Anopheles* species. Malaria is an important global public health threat. Millions of people living in tropical countries are at risk for malaria. Malaria is characterized by acute febrile illness. The misinterpretation of malaria as

COVID-19 is possible if blood test examination is not available for malaria. In the malaria-endemic area, there is a chance for co-occurrence between COVID-19 and malaria. Wang et al. (Wang 2020) noted that preparedness is essential for malaria-endemic regions during the COVID-19 pandemic since the combination of two diseases might occur and could result in severe clinical problems.

Nevertheless, there is still no official report on concurrent COVID-19 and malaria. Another critical issue regarding the relationship between malaria and COVID-19 is the use of the antimalarial drug for COVID-19 treatment. Many reports are showing that chloroquine is a possible useful drug for the management of COVID-19 (Luna 2020). Smit et al. concluded that chloroquine could be used safely for COVID-19 and suggested that the drug should be used under monitoring the corrected QT (Smit 2020). However, the margin of safety of the drug is narrow, particularly in children (Smit 2020).

19.2.1.6 Cholera

Cholera is a gastrointestinal infection caused by *Vibrio cholerae*. This bacterial infection is still an important public health problem in many developing countries. Watery diarrhea is a common problem in cholera. The cholera patient might die from severe dehydration if excessive diarrhea occurs. Since COVID-19 might initially present with diarrhea, the misdiagnosis of COVID-19 as cholera is possible. However, cholera does not present with acute febrile illness, making it easier to differentiate between cholera and COVID-19. The sanitation control model for cholera can be adapted for COVID-19 (Signorelli 2020). There is no report on the co-occurrence of COVID-19 and cholera.

19.2.1.7 Salmonellosis

Salmonellosis is a gastrointestinal bacterial infection. Similar to cholera, this bacterial infection is a significant public health problem in many developing countries. Patients with salmonellosis present with fever and diarrhea. Leukopenia is also a common laboratory finding. The misdiagnosis of COVID-19 as salmonellosis is, therefore, possible. There is no report on the co-occurrence of COVID-19 and salmonellosis. However, Wilson (Wilson 2020) and Meyer et al. (Meyer 2020) noted that salmonellosis could be imported in the COVID-19 disaster.

19.2.1.8 Leptospirosis

Leptospirosis is a spirochetal infection. A history of exposure to contaminated floodwater is important for diagnosis. Patients with leptospirosis usually have fever and myalgia. Renal dysfunction is common among patients with leptospirosis. Since fever and myalgia are common in patients with COVID-19, the misdiagnosis of COVID-19 as leptospirosis is possible. No report on the concurrence of COVID-19 and leptospirosis exists.

19.2.1.9 Tuberculosis

Tuberculosis is an important mycobacterial infection caused by *Mycobacterium tuberculosis*. It usually affects the lung and can cause a chronic condition. Although tuberculosis is observed worldwide in both tropical and nontropical areas, the high prevalence of tuberculosis mainly occurs in tropical countries. Therefore, it would be quite possible to see the concurrence between COVID-19 and tuberculosis in tropical countries. There is a report on the poor prognosis of COVID-19 in a Chinese patient with tuberculosis (Chen 2020). Adepoju (Adepoju 2020) noted that a worst-case scenario might be a patient with tuberculosis, HIV infection, and COVID-19.

19.2.1.10 Hepatitis B Infection

Hepatitis B viral (HBV) infection is common in several tropical countries. HBV can cause acute febrile illness or chronic disease. A large portion of the population in tropical countries is seropositive for HBV, and therefore the concurrence between COVID-19 and HBV is possible. Zhong et al. (Zhong 2020) observed the fatal outcome of COVID-19 in a case previously infected with HBV. Also, COVID-19 might present with hepatitis (Sookaromdee 2020) and, therefore, is prone to be misdiagnosed as HBV.

19.2.2 COVID-19 and Cancer in Tropical Countries

Evidence is mounting that the behavior of COVID-19 is much more aggressive among patients with malignancy than otherwise healthy people.

19.2.2.1 HBV-Related Hepatocellular Carcinoma

As already mentioned, HBV can cause chronic infection. Chronic HBV carriers are at risk for developing hepatocellular carcinoma. Therefore, the concurrence between COVID-19 and hepatocellular carcinoma might also be possible. There are reports of death from COVID-19 in cases with HBV-related hepatocellular carcinoma (Zhong 2020).

19.2.2.2 Cholangiocarcinoma

Cholangiocarcinoma is a malignant disease of the biliary tract characterized by severe jaundice at presentation. There is an extremely high prevalence of cholangiocarcinoma in Southeast Asia's tropical areas, and the important underlying etiologies are infection with liver flukes, poor dietary habits, and ingesting raw fish. There is no report on the concurrence of COVID-19 and cholangiocarcinoma.

19.2.3 COVID-19 and Hereditary Disorders in Tropical Countries

Sporadic genetic diseases are observed in clinical practice. In tropical countries, there are some common genetic diseases. The co-occurrence between COVID-19 and tropical genetic disease is interesting.

19.2.3.1 Tropical Hemoglobin Disorder

Inherited hemoglobin disorders are the most common group of genetic diseases in tropical countries. In particular, sickle cell disease is highly prevalent in tropical Africa, while hemoglobin E in Southeast Asia. Thus, it seems possible that COVID-19 can co-occur alongside tropical hemoglobin disorders, and this co-occurrence might change the severity of each condition, as suggested in Beerkens (2020).

There is a report of a patient with sickle cell anemia who developed thrombohemostatic disorder and vaso-occlusion following COVID-19 infection (Nur 2020).

19.2.3.2 Tropical Red Blood Cell Enzymatic Disorder

Glucose-6-phosphate dehydrogenase (G6PD) deficiency is a common red blood cell enzymatic disorder in tropical countries. G6PD deficiency causes red blood cells sensitive to some circumstances, including infections. Hemolysis can quickly occur if the patient is subjected to stress. There are some reports on the co-occurrence between COVID-19 and G6PD deficiency. Aydemir (2020) proposed that G6PD deficiency might increase the severity of COVID-19. Since the antimalarial drug is an alternative drug that is used for the treatment of COVID-19, hemolytic crisis due to using the antimalarial drug in COVID-19 therapy is already reported (Beauverd Y 2020). It is suggested that if the antimalarial drug is selected as a therapeutic alternative, screening for G6PD deficiency is necessary.

19.3 COVID-19 and Role of Primary Care Center: View on Common Pitfalls

The public health system in developing tropical countries is usually weaker than developed Western countries. The frontline for combating with COVID-19 is the primary healthcare worker. In tropical countries, primary healthcare workers have to work hard for the management of the influx of patients (Palacios-Cruz 2020). Several problems for primary care management of COVID-19 in tropical countries are observed. The common problems are limited knowledge of primary care workers, insufficient protective devices, and lack of well-skilled primary care center administrators.

19.3.1 Knowledge of Primary Healthcare Worker

The data on the newly emerging disease is usually limited. Primary healthcare workers in developing countries usually work and live far from the center of the country. In this condition, there is a low chance for continuous education and updating of knowledge. The only data that the local government wants is usually fed to the local primary care workers in remote primary care centers. Knowledge of the local primary care worker is usually not up-to-date. May workers have missed information and out-of-date knowledge (Apaijitt 2020). Hence, we might expect poor knowledge generalization to local people in the rural area of developing countries.

19.3.2 The Vision of Local Primary Healthcare Center Administrator

During the crisis, the admin vision is very important (Khanna 2020). In developing countries, poor vision is common. Since administrators of primary healthcare centers have poor vision, they usually do according to their desire. For example, in the primary health center, an administrator might order that worker work while giving primary care to the patient carefully. This concept is against the standard principle of social distancing. In this manner, the local administrator might focus on the patient's satisfaction score, which is a criterion for accreditation of primary care center, rather than keeping the standard of infection control. How to seek an excellent primary care administrator is a common problem in tropical developing countries where corruption is a rooted problem.

19.3.3 Lack of Infrastructures

As already mentioned, infrastructures are usually limited and not available in primary healthcare units in poor tropical developing countries. Often, the workplace is small such that it does not allow distancing during the COVID-19 outbreak. Additionally, protective devices for practitioners are usually insufficient. It corresponds with an increased risk of infecting a primary care worker in occupational settings.

Since the primary health work during the COVID-19 outbreak in developing tropical countries is usually under stress and limitations, the success might not be easily achieved (Palacios-Cruz 2020). The pitfalls might occur and have to be managed. Common pitfalls that should be mentioned include the following.

19.3.4 Underdiagnosis of COVID-19

As already mentioned, the lack of knowledge among medical personnel is common in developing countries. Also, some COVID-19 patients might be asymptomatic or mildly symptomatic. The underdiagnosis of COVID-19 is a common pitfall in primary care practice in developing countries. The use of criteria for the selection of patients for further surveillance might be useful, but the main problem is the lack of update of the criteria. Commonly, the criteria are out-of-date, resulting in underdiagnosis of COVID-19.

19.3.5 Overdiagnosis of COVID-19

Similarly, overdiagnosis of COVID-19 is also possible. Some private medical centers in developing countries might try to make money by exaggeration of information that causes fearfulness among local people. In this condition, patients are presumptively diagnosed without confirmation, and this is associated with incorrect infection statistics.

19.3.6 Poor Reporting System

The reporting system is essential for surveillance and planning for coping with an outbreak situation. In the remote area, the communication system is usually weak, and the reporting system is usually delayed.

19.3.7 Poor Disease Control Activity

Local social contexts might also increase the difficulty in disease control. For example, in the remote area of Indochina, the local people usually believe in animism. They usually believe in ghosts and do not follow modern infection control precautions. The ritual object of chasing a virus is an excellent example of a strange belief that might result in poor disease control. How to adjust the new standards to the local belief is a critical issue to be considered for further primary healthcare planning.

19.4 COVID-19 in Indochina

Indochina is the tropical zone between China and India. There are many developing tropical countries in this area. Due to the short distance from China, COVID-19 firstly affected this area after its first occurrence in China. Thailand is located in the Indochina peninsula and is the second country in the world affected by COVID-19. At present, all countries in Indochina are already affected by COVID-19. The epidemiological data of those tropical countries in Indochina is interesting.

19.4.1 Thailand

Thailand is the second country in the world affected by COVID-19 since January 2020. The first case of COVID-19 in Thailand, which is also the first case of COVID-19 outside China, was a Chinese tourist. In the first phase, Chinese tourists were the leading group responsible for the movement of infection in Thailand. After unsuccessful disease control in Thailand, the local spread of diseases occurred, and now there are many clusters of COVID-19 related to different activities. An interesting cluster is the Thai-boxing-related COVID-19 cluster. It is the leading cause of widespread disease in Thailand. Finally, another important report is the first possible transmission of disease from a corpse to humans (Sriwijitalai 2020). This case report calls for better management of dead bodies in infectious outbreaks.

19.4.2 Lao

Lao is a poor developing country in Indochina. COVID-19 is a prominent local public health challenge to Lao, leading to the early lockdown of Lao.

19.4.3 Myanmar

Similar to Lao, Myanmar is a poor developing country in Indochina. At first, Myanmar was free from COVID-19, but it is now affected by the disease due to not controlling for migration.

19.4.4 Cambodia

Cambodia is a country in Indochina. This country is very famous for its ancient world wonder, Angkor Wat, and for this, it is a top-rated destination for many Western tourists (Chakraborty 2020). The current situation in Cambodia clearly shows that Western tourists, not Chinese tourists, have imported COVID-19 in this country.

19.4.5 Vietnam

Vietnam is another country in Indochina. This country shares an international border with China. An interesting observation is that the first case of COVID-19 in Vietnam was an infant (Hai 2020).

Lessons learned from the Indochina situation should be mentioned. As already noted, Indochina is the first foci of the outbreak outside China. Therefore, a long history of the outbreak is derived. An apparent reason for disease occurrence in Indochina is the unsuccessful control of immigration; hence, the importation of disease by foreigners occurs. It is a similar problem to any area worldwide. The poor local infrastructure in this area is noted, but the mortality rate of local people is not as high as many developed countries in Europe. It might be hypothesized that natural resistance and cross-protective immunity might exist among the local population living in Indochina.

19.5 Conclusion

COVID-19 is a global public health threat. Tropical countries are already affected by the new disease. In tropical medicine, the co-occurrence of COVID-19 and the endemic tropical disease is possible, and it is an interesting issue for further study in tropical medicine. Some tropical diseases are already reported on their concurrence with COVID-19. The clinical inter-relationship is usually demonstrable. Regarding tropical public health, the challenge is the proper management of the outbreak in the situation of many limitations. Lessons learned from tropical countries might be useful and adaptable elsewhere. International collaboration might be a clue for solving the present global crisis.

References

Adepoju (2020) Tuberculosis and HIV responses threatened by COVID-19. Lancet HIV 7(5):e319–e320

Apaijitt (2020) Knowledge of coronavirus disease 2019 (COVID-19) by medical personnel in a rural area of Thailand: an observation. Infect Control Hosp Epidemiol

Aydemir (2020) Is glucose-6-phosphate dehydrogenase enzyme deficiency a factor in Coronavirus-19 (COVID-19) infections and deaths? Pathog Glob Health 114(3):109–110

Beauverd YAY, Assouline B, Samii K (2020) COVID-19 infection and treatment with hydroxychloroquine cause severe haemolysis crisis in a patient with glucose-6-phosphate dehydrogenase deficiency. Eur J Haematol 105(3):357–359

Beerkens (2020) COVID-19 pneumonia as a cause of acute chest syndrome in an adult sickle cell patient. Am J Hematol 95(7):E154–E156

Chakraborty (2020) OVID-19 outbreak: migration, effects on society, global environment and prevention. Sci Total Environ

Chauhan RP, Dessie ZG, Noreddin A, El Zowalaty ME (2020, Apr) Systematic review of important viral diseases in Africa in light of the 'One Health' concept. Pathogens 9(4):301

Chen (2020) Epidemiological and clinical characteristics of 99 cases of 2019 novel coronavirus pneumonia in Wuhan, China: a descriptive study. Lancet 395(10223):507–513

Cucinotta (2020) WHO declares COVID-19 a pandemic. Acta Biomed 91(1):157–160

Guo (2020) The origin, transmission and clinical therapies on coronavirus disease 2019 (COVID-19) outbreak – an update on the status. Mil Med Res

Hai (2020) The first infant case of COVID-19 acquired from a secondary transmission in Vietnam. Lancet Child Adolesc Health 4(5):405–406

Hsia (2020) Emerging new coronavirus infection in Wuhan, China: situation in early 2020. Case Study Case Rep

Joob (2020) COVID-19 can present with a rash and be mistaken for dengue. J Am Acad Dermatol 82(5):e177

Khanna (2020) COVID-19 pandemic: lessons learned and future directions. Indian J Ophthalmol 68(5):703–710

Lorenz C, Azevedo TS, Chiaravalloti-Neto F (2020) COVID-19 and dengue fever: a dangerous combination for the health system in Brazil. Travel Med Infect Dis 101659–101659

Luna D (2020) Rapid and severe Covid-19 pneumonia with severe acute chest syndrome in a sickle cell patient successfully treated with Tocilizumab. Am J Hematol 95(7):876–878

Mavragani (2020) JMIR Public Health Surveill

Meyer (2020) Imported infections in rural Mid-West United States – a report from a tertiary care center. Mo Med 117(2):89–94

Navarro (2020) COVID-19 and dengue, co-epidemics in Ecuador and other countries in Latin America: pushing strained health care systems over the edge. Travel Med Infect Dis 37:101656

Nicola (2020) Evidence based management guideline for the COVID-19 pandemic – review article. Int J Surg 77:206–216

Nur (2020) Vaso-occlusive crisis and acute chest syndrome in sickle cell disease due to 2019 novel coronavirus disease (COVID-19). Am J Hematol 95(6):725–726

Palacios-Cruz (2020) COVID-19, a worldwide public health emergency. Rev Clin Esp

Rodriguez-Morales (2020) Clinical, laboratory and imaging features of COVID-19: a systematic review and meta-analysis. Travel Med Infect Dis

Rothan (2020) The epidemiology and pathogenesis of coronavirus disease (COVID-19) outbreak. J Autoimmun

Saavedra-Velasco (2020) Coinfection between dengue and covid-19: need for approach in endemic zones. In: Rev Fac Cien Med Univ Nac Cordoba

Signorelli (2020) COVID-19: hygiene and public health to the front. In: Acta Biomed, vol 91, pp 7–8

Singhal (2020) Review of coronavirus Disease-2019 (COVID-19). Indian J Pediatr 87(4):281–286

Smit (2020) Chloroquine for SARS-CoV-2: implications of its unique pharmacokinetic and safety properties. Clin Pharmacokinet 59(6):659–669

Sohrabi (2020) World health organization declares global emergency: a review of the 2019 novel coronavirus (COVID-19). Int J Surg 76:71–76

Sookaromdee (2020) Hepatitis and novel Wuhan coronavirus infection: observation. Med J DY Patil Vidyapeeth 13(2):184–185

Sriwijitalai (2020) OVID-19 in forensic medicine unit personnel: observation from Thailand. J Forensic Leg Med

Tin (2020) Atypical presentations of COVID-19. J Health Sci Med Res 10(03):136–142

Wang (2020) Preparedness is essential for malaria-endemic regions during the COVID-19 pandemic. Lancet 395(10230):1094–1096

Wilson (2020) What goes on board aircraft? Passengers include Aedes, Anopheles, 2019-nCoV, dengue, Salmonella, Zika, et al. Travel Med Infect Dis

Wu (2020) The SARS-CoV-2 outbreak: what we know. Int J Infect Dis 94:44–48

Yan (2020) Covert COVID-19 and false-positive dengue serology in Singapore. Lancet Infect Dis

Zhong (2020) Clinical characteristics and immunosuppressants management of coronavirus disease 2019 in solid organ transplant recipients. Am J Transplant 20(7):1916–1921

Neurologic Manifestations of COVID-19

20

Farnaz Delavari, Farnaz Najmi Varzaneh, and Nima Rezaei

Abstract

Neurological manifestations of novel coronavirus disease (COVID-19) are reported to occur in as much as 37% of the affected patients. These manifestations range from headache and dizziness to altered mental status and consciousness, anosmia, ageusia, sensory disturbances, and stroke. The mechanisms by which the neurological symptoms arise are not yet determined but may either proceed as an indirect consequence of systemic hyperinflammation or result from the direct invasion of the virus to neural and glial cells. The neural invasion can explain both the retrograde pathway of encephalitis and the early manifestation of anosmia by invading the olfactory bulb. Moreover, in the case of attacking the brain stem, it may take part in the early apnea manifestation reported by patients. Additionally, neurotropism of the virus could be the cause of acute hemorrhagic encephalitis. Hyperinflammation can have acute and prolonged effects in the nervous system, such as acute demyelination and predisposition to multiple sclerosis. Moreover, the pro-inflammatory state contributes to hypercoagulation, which in turn could result in cerebrovascular injuries in COVID-19 patients. This chapter would discuss that the neurologic manifestations of the COVID-19 are to be looked at as a multifactorial entangled phenomenon.

Keywords

Betacoronavirus · Blood-brain barrier · Brain stem · Central nervous system · Coronavirus infections · Middle East respiratory syndrome · Olfactory bulb · Pandemics · Severe acute

F. Delavari (✉)
University of Geneva, Geneva, Switzerland

Interactive Research Education and Training Association (IRETA), Universal Scientific Education and Research Network (USERN), Tehran, Iran

F. N. Varzaneh
The Russell H. Morgan Department of Radiology and Radiological Sciences, Johns Hopkins University, Baltimore, MD, USA

Network of Immunity in Infection, Malignancy and Autoimmunity (NIIMA), Universal Scientific Education and Research Network (USERN), Baltimore, MD, USA

N. Rezaei (✉)
Research Center for Immunodeficiencies, Children's Medical Center, Tehran University of Medical Sciences, Tehran, Iran

Department of Immunology, School of Medicine, Tehran University of Medical Sciences, Tehran, Iran

Network of Immunity in Infection, Malignancy and Autoimmunity (NIIMA), Universal Scientific Education and Research Network (USERN), Tehran, Iran
e-mail: rezaei_nima@tums.ac.ir

N. Rezaei (ed.), *Coronavirus Disease - COVID-19*, Advances in Experimental Medicine and Biology 1318, https://doi.org/10.1007/978-3-030-63761-3_20

respiratory syndrome · Viral diseases · Virus internalization · Viral tropism

20.1 Introduction

Since the beginning of novel coronavirus disease (COVID-19) pandemic, studies have been trying to unravel the multifaceted manifestations of the disease. Caused by the virus that is known as severe acute respiratory syndrome coronavirus 2 (SARS-CoV-2), the main manifestation of COVID-19 has been documented as respiratory symptoms. However, as high as 37% of patients with COVID-19 have reported what here is debated as neurological symptoms (Berger 2020; Bohmwald et al. 2018; Bolay et al. 2020; Burke et al. 2020). These symptoms range from more general, such as headaches and dizziness, to more specific such as altered mental states and disturbed consciousness and even more focally such as sensory disturbances and stroke-like symptoms (Carod-Artal 2020; Federico 2020; Gklinos 2020). These neurological manifestations have been reported in more severely affected patients; however, some specific alterations such as anosmia have been reported to occur before the emergence of any other symptoms. Moreover, the postmortem study of a report conducted in China (the first hotbed for the virus) has indicated brain tissue inflammation as well as lesions caused by neural degeneration. Additionally, in March 2020, 4 months after the official emergence of the disease, the first case for SARS-CoV-2-induced viral encephalitis was reported (Bohmwald et al. 2018; Hovanec and Flanagan 1983; Kalinke et al. 2011; Vilensky 2014; Li et al. 2016). Therefore, we suggest that a comprehensive understanding of COVID-19 must include awareness of neural manifestations of the disease (Liu et al. 2020; Mamo et al. 2020).

The concept of a viral infection partaking neurological manifestations is not new to clinicians (Hankins and Rosekrans 2004; Hovanec and Flanagan 1983; Kalinke et al. 2011; Prevention 2020; Whitley and Gnann 2002). Many well-understood viral infections have been accused of structural and functional neural damage. Given the fact that SARS-CoV-2 is a newly emerged virus, prior knowledge on how other pathogens descended from the same family behave has been contributing significantly to our understanding of the disease. Namely, both SARS-CoV and Middle East respiratory syndrome (MERS)-CoV have been associated with polyneuropathy, meningitis, and cerebral inflammation. Furthermore, the reports extend to vascular damage in the central nervous system (CNS) as well as demyelination (Aurelian 2005; Hovanec and Flanagan 1983; Kalinke et al. 2011; Li et al. 2016). More interestingly, some of these neurological manifestations are not temporally accompanying the main respiratory symptoms but have the chorology of their own. The mechanism by which the novel coronavirus causes neurological manifestations is still under debate. The two branches of the dispute revolved around the question of whether the virus attacks the nervous system directly or the damage to the nervous system is merely the consequence of the body's immune reaction to the disease. The importance of this significant question would be implicated in management strategies regarding these symptoms as well as percussions in patients with a definite diagnosis (Whittaker et al. 2020; Wu et al. 2020; Yin et al. 2020). Moreover, the implication of general care may be significantly influenced (Ng Kee Kwong et al. 2020; Bajunaid et al. 2020; Mohile et al. 2020; Sheraton et al. 2020; Siddiqui and Khan 2020; Sweiss et al. 2020). Here we discuss the possible routes through which the SARS-CoV-2 makes way to the nervous system. We then discuss how, given the current and prior knowledge, the neural damage might be caused. We further go into discussing the possible long-term neurologic outcomes of the disease.

20.2 Highway to the Brain

The interaction between known viral infections and the CNS has suggested two main routes of transmission for the pathogen: hematogenous spread of the virus, which happens when a substantial amount of virus is present in the bloodstream, and retrograde viral spread with means of the peripheral nervous system (PNS). In the case of the novel coronavirus, both routes seem imaginable (Algahtani et al. 2016; Dickey et al. 2016; Plaisted et al. 2016; Li et al. 2017; Grist et al. 2018).

20.2.1 The Hematogenous Route

The hematogenous route requires the invasion of endothelial cells of the blood-brain-barrier (BBB) or infection of leukocytes that then, in part, can act as carriers of the virus. Given the significant evidence pointing toward endothelial adherence of the virus, the former would not be unexpected, while the latter is reinforced by the evidence pointing toward immune cell invasion by SARS-CoV-2. However, both of these routes would require the patients to be in advanced stages of the disease with high viral loads (Grist et al. 2018; Agarwal et al. 2020; Barzegar et al. 2020).

20.2.2 The Neural Pathway and Anosmia

Since the beginning of the COVID-19 outbreak, the primary route of transmission has been the upper respiratory tract. Correspondingly, the first possible suggested suspect for neural transmitting routes has been the olfactory bulb, which is situated right above the nasal cavity (Chan et al. 2020). The olfactory bulb has been suggested as a pathway to the brain for many viral infections before (Algahtani et al. 2016). Namely, the well-known respiratory virus influenza has been known to invade the CNS through the olfactory bulb, both by the cell-to-cell synaptic transmission of infected cells and channeling through olfactory unsheathing cells (Kalinke et al. 2011; van Riel et al. 2015; Vilensky 2014; Williams et al. 2004). One of the main symptoms suggesting this route as the main transmitting route has been the frequently reported symptom of anosmia. The frequency of reporting a loss of sense of smell and taste has been prominent to the extent that it was one of the first neurologic symptoms added to the list of Centers for Disease Control and Prevention (CDC) (Montalvan et al. 2020; Robinson and Busl 2020; Nath 2020; Li et al. 2020b).

Moreover, anosmia in patients with COVID-19 happens early in the course of the disease, especially when the disease has not progressed into a constitutional state. Given the well-established knowledge that the COVID-19 colonizes in the nasal cavity, the convergent evidence once again reinforces the olfactory bulb as a major route of neural transmission of the disease. In regard to previous knowledge in rhinoviral infections, the less reported anosmia in patients with influenza, adenovirus, or rhinovirus has been mostly associated with rhino symptoms. Interestingly, the reported anosmia in patients with the COVID-19 seems to happen in the absence of significant nasal congestion or rhinorrhea. It, in turn, has fortified the researchers to look at the novel coronavirus as a neurotropic viral infection (Giacomelli et al. 2020; Prevention 2020; Xydakis et al. 2020).

20.3 Neural Invasion

The angiotensin-converting enzyme 2 (ACE2) receptor, known to function as a cell surface receptor for the SARS-CoV-2, is present in a variety of human tissues, including glial cells and neurons in the brain. Therefore, SARS-CoV-2 can migrate from the systemic circulation into the cerebral circulation. The relatively slow-flow movement of blood in the brain would make the SARS-CoV-2 more interactive with brain capillary endothelial cells, leading to endothelial damage (Baig et al. 2020). The proposed neurovirulent theory of COVID-19 may have raised from the alerting anosmia but have not stayed limited to local nerves. The fact that the novel coronavirus might infiltrate the CNS opens various possible scenarios, in which the virus infects cells (both neurons and glial cells). The so-called neurotropism makes possible the initiation of many neurological manifestations, which may progress to neurological diseases. Prior knowledge provides support for the neurovirulent properties of viruses. Depending on how much profound the virus neurotropism is, the consequences could range from minor neurological symptoms to dire encephalitis, which can be fatal (Aurelian 2005; Hankins and Rosekrans 2004; Whitley and Gnann 2002). In regard to respiratory infections, one particular example could be the family of the human coronavirus (HCoV) itself. Studies have previously demonstrated the presence of HCoV RNA in the brain tissue of infected patients.

Moreover, the persistent presence of the virus in the CNS has been established by studying the infected mice models. The incubating neuroinvasion of this family has opened new insights to look into the pathogenesis of neural disorders in patients with COVID-19 (Leslie-Mazwi et al. 2020; Waldman et al. 2020; Heman-Ackah et al. 2020; Baig et al. 2020; Zayet et al. 2020; Deliwala et al. 2020). It may result in prolonged immune-mediated damage that can cause delayed neurological manifestations (see sect. 5).The neural cell damage caused by the viral invasion would be resulting from both heightened host immune response and the viral replication itself. As a consequence of viral replication, the loss of CNS gray matter might happen, and this, most likely, would result in what was priorly known as typical encephalitis. Given the novelty of COVID-19, at the time of this manuscript, the evidence of encephalitis presentations in COVID-19 remains sparse. However, to the extent of our knowledge, it remains among rare manifestations of the disease with a self-retaining nature. Patients with SARS-CoV-2 encephalitis have shown to deteriorate in consciousness and show signs of meningeal irritations, though CSF specimen tested negative for virus particles (Ye et al. 2020).

20.4 The Brain-Lung Axis

Not to ignore the role of hypoxia and the damage it induces, some recent publications point toward a neural-centered mechanism for hypoxia occurring in COVID-19. Aside from a simple presentation of viral encephalitis, some studies suggest that the neurological involvement of COVID-19 is more of a tangled web and, more interestingly, attribute some respiratory symptoms of the disease to the involvement of the CNS (Li et al. 2020a). This theory may have originated from early translational mice studies, showing the presence of the early SARS-CoV or MERS-CoV in the brain stem of the transnasally infected mice. The notion that the virus invades the brain before the lung in these mice models further fed the theory that the manifested respiratory distress might have a more central origin (Li et al. 2016; McCray et al. 2007; Netland et al. 2008; Saleki et al. 2020). To this date, there is no original investigation confirming the role of medullary invasion in respiratory distress. However, the presence of shortness of breath as a common symptom, even in mild cases of the disease, may indicate a central mechanism for COVID-19 (Li et al. 2020d; Li et al. 2020c). A concurrent mechanism is imaginable in the human population where infection of regions such as the nucleus of the solitary tract and nucleus ambiguous could result in a malfunction in receipt of sensory information from the mechanic receptors in the respiratory tracts and, consequently, aberrant innervation to the airway smooth muscles.

20.5 Neuroinflammatory Manifestations

While suppressing antiviral pathways, the novel coronavirus induces an exacerbated state of inflammation in many patients, and this state of hyperinflammation may involve a storm of cytokines and give rise to cases of unexpected mortalities (Bahrami et al. 2020; Nasab et al. 2020; Saghazadeh and Rezaei 2020a; Yazdanpanah et al. 2020; Fathi and Rezaei 2020; Mehta et al. 2020; Lotfi and Rezaei 2020). Therefore, when evaluating the neurological manifestations of the disease, a state of neuroinflammation, although not specific to the nervous system, must not be overlooked. The presence of headaches has highlighted this mechanism of action as it is one of the most prominent symptoms of COVID-19. Intuitively, the invasion of the CNS with neurotropic signaling causes the glial cells to call for an inflammatory state. Accordingly, in vitro studies have shown that primary glial cells that are disposed to earlier coronavirus strains (CoV) are predisposed to produce inflammatory factors such as IL-6 and TNF-α. The immune cell signaling in the brain will cause collateral brain damage (Moriguchi et al. 2020). It may provide an educated explanation for otherwise unexplained neural lesions. Interestingly, acute demyelination in the white matter without embolic or large vascular originations has been reported (Brun et al. 2020). In a series of 11 critically ill patients with COVID-19 who underwent brain MRI for persistently depressed mental status, 10 patients had diffuse leukoencephalopathy, of which, 4 patients had an additional finding, intracranial microhemorrhage. Therefore, diffuse leukoencephalopathy and microhemorrhages were suggested as late complications of critically ill COVID-19 patients and are likely related to hypoxemia (Radmanesh et al. 2020a). It is worth to note that the same group of authors published a study earlier in March and reviewed brain imaging of 242 patients with COVID-19. Patients with COVID-19 displayed scatted white matter changes, but not diffuse leukoencephalopathy (Radmanesh et al. 2020b). These findings suggest the correlation of diffuse leukoencephalopathy with end-stage and critically ill COVID-19 conditions. Thus, the neural impact of the COVID-19 relates merely to neither the onset nor the course of the disease. The presence of early CoV has been established in chronic neurological disease with an implication of neuroinflammation, including Parkinson's disease (PD) and multiple sclerosis (MS). These examples are two of the widely accepted neurological disease for which prior exposure to an infectious pathogen has been discussed as triggers. Namely, MS has been nominated as a possible consequence of several neurotropic viruses. Albeit, there has not yet been enough evidence to attribute a specific virus to these neurologic disorders. However, a significant association with MS exacerbation and HCoV-229E infection has been implied. Moreover, it has been shown that the pattern for seasonal HCoV infection fits the macro pattern observed for MS exacerbations (Hovanec and Flanagan 1983; Arbour et al. 2000). Overall, these findings indicate the possibility of a delayed neurologic manifestation in COVID-19, which, if not looked out for, may remain obscure (Soma et al. 2018; Hascup and Hascup 2020; Papa et al. 2020; Plaze et al. 2020; Schirinzi et al. 2020).

20.6 Neurovascular Manifestations

One of the less expected characteristics of COVID-19 has been the hematologic effect of the virus (Levi et al. 2020). The hypercoagulopathic state observed in these patients follows pro-inflammation and is a phenotypic disseminated intravascular coagulation (DIC) with thrombocytopenia, increased D-dimer, and prolonged prothrombin time (Asadi-Pooya 2020; Bostanciklioglu 2020; Byrnes et al. 2020; Chen et al. 2020; Kuroda 2020; Liguori et al. 2020; Wilson and Jack 2020). This hypercoagulative state is proposed to be a cause for thrombotic microangiopathy. Notably, neurovascular thrombotic accidents have been reported in patients with COVID-19. Since the outbreak, the overload on the health system has caused the incidence of cases reported as a stroke to fall. However, COVID-19 is suggested to be a cause for stroke itself. To this date, most stroke cases reported in COVID-19 are reported to be ischemic, but rare cases of hemorrhagic stroke and cerebral venous sinus thrombosis have also been reported (Hughes et al. 2020). Intracranial punctuate microhemorrhage is, however, not that rare. Its pathophysiology lies in the development of antiphospholipid antibodies, thrombotic vasculopathy in dermal blood vessels, and, eventually, cerebral vasculitis (Shoskes et al. 2020). The potential mechanism for the stroke phenomena roots back to the increased D-dimer levels of the patients. Complementary theories point toward possible cardiac injury, stating that the cardiac injury predisposes the patients to cardiac embolism and a consequent stroke (Mehta et al. 2020; Brun et al. 2020; Hovanec and Flanagan 1983). The neurovirulent theory also contributes to the stroke manifestations of the disease by causing acute hemorrhagic necrotizing encephalopathy

(Markus and Brainin 2020). Moreover, the overall state of hypoxia can contribute to ischemic brain damage, reported in patients with COVID-19. The multifactorial nature of cerebrovascular incidents makes it challenging to pinpoint the direct pathway the virus can cause stroke in patients with COVID-19. Future pathological assessments, as well as epidemiologic evaluations, can shed light on this matter.

20.7 Conclusion

COVID-19 emerged in late 2019. It began spreading in Wuhan, China, and, shortly after that, reached almost all countries and people of all ages, professionals, and conditions and consequently named a pandemic by the World Health Organization (WHO) (Hanaei and Rezaei 2020; Lotfi et al. 2020; Rezaei 2020a; Mirbeyk and Rezaei 2020). Compared to the other hCoVs causing outbreaks, this new one is more easily transmitted among people so that despite the implementation of a variety of control measures at the national level, the disease has continued to spread out with more than 13 million people affected worldwide (Lotfi et al. 2020; Jabbari et al. 2020). This highly transmissible disease, when no effective drugs and vaccines are present (Saghazadeh and Rezaei 2020b), has been a killer of more than 500,000 people. Such a condition has been a motive to work in collaborative teams to facilitate knowledge integration between different disciplines effectively. It is not only in the hope of offering a cure (Basiri et al. 2020; Fathi and Rezaei 2020; Jahanshahlu and Rezaei 2020b; Moazzami et al. 2020; Mohamed et al. 2020a; Mohamed et al. 2020b; Momtazmanesh et al. 2020; Moradian et al. 2020; Rabiee et al. 2020; Rezaei 2020b; Kafieh et al. 2020) but also to provide a more in-depth look of means and ways to manage patients and healthcare providers' well-being (Moazzami et al. 2020). The proof of this motivation is that scholar metrics reveal an exponential growth of literature on COVID-19. Though not 100% qualified for treating the condition (Rzymski et al. 2020), it has crucially contributed to the understanding of the disease as a multifactorial disease where the genetic background (Ahanchian et al. 2020; Yousefzadegan and Rezaei 2020) and environmental factors play a role in individual susceptibility to infection and adverse outcomes and more interestingly as a multisystem infectious disorder that lies in its capability of causing a combination of hyperinflammation and hypercoagulation making it quite competent to actually all organs/systems where its cell surface receptor(s) is present (Ahmadi et al. 2020; Sahu et al. 2020).

In this chapter, we reviewed the neurological manifestations of COVID-19 and the possible pathophysiologic pathways they occur through. The knowledge on COVID-19 is relatively sparse, and its neurological manifestations have been less visited. However, the importance of clinical consideration for the neurological manifestations must be highlighted. Given that one of the earliest manifestations of SARS-CoV-2 could be neurologic, a better understanding of these phenomena could open opportunities for early detection and early intervention. Moreover, given the incubated resistance possibility for the virus in neural cells, the delayed neurological manifestations must be anticipated (Ogier et al. 2020; Baig et al. 2020). On the other hand, we discuss the possibility that some neurological manifestations of COVID-19 might include delayed but chronic neurodegenerative diseases such as multiple sclerosis. Thus, the moral considerations do not stop with recovery for people who are confronted with the virus. The long-term consequence of SARS-CoV-2 infection could be dire neurologically. This knowledge may alter the view on epidemiologic concepts such as herd immunity (Li et al. 2020b; Stoessl et al. 2020; Kim et al. 2020; Gandhi et al. 2020; Salari et al. 2020). Finally, we would like to emphasize that knowledge is critical by pointing to the fact that a vast body of literature on the neurologic manifestations of COVID-19 is built upon prior extensive studies on SARS and MERS. Similarly, a more critical view on the more recent COVID-19 might result in a more prompt and appropriate response to a later emerging threat (Jahanshahlu and Rezaei 2020a).

References

Agarwal S, Sabadia S, Abou-Fayssal N, Kurzweil A, Balcer LJ, Galetta SL (2020) Training in neurology: flexibility and adaptability of a neurology training program at the epicenter of COVID-19. Neurology. https://doi.org/10.1212/wnl.0000000000009675

Ahanchian H, Moazzen N, Faroughi MSD, Khalighi N, Khoshkhui M, Aelami MH, Haghi NSM, Rezaei N (2020) COVID-19 in a child with primary specific antibody deficiency

Ahmadi M, Saffarzadeh N, Habibi MA, Hajiesmaeili F, Rezaei N (2020) Colon Cancer and SARS-CoV-2: impact of ACE2 expression in susceptibility to COVID-19. bioRxiv

Algahtani H, Subahi A, Shirah B (2016) Neurological complications of Middle East respiratory syndrome coronavirus: a report of two cases and review of the literature. Case Rep Neurol Med 2016:3502683. https://doi.org/10.1155/2016/3502683

Arbour N, Day R, Newcombe J, Talbot PJ (2000) Neuroinvasion by human respiratory coronaviruses. J Virol 74(19):8913–8921. https://doi.org/10.1128/jvi.74.19.8913-8921.2000

Asadi-Pooya AA (2020) Seizures associated with coronavirus infections. Seizure 79:49–52. https://doi.org/10.1016/j.seizure.2020.05.005

Aurelian L (2005) HSV-induced apoptosis in herpes encephalitis. Curr Top Microbiol Immunol 289:79–111. https://doi.org/10.1007/3-540-27320-4_4

Bahrami A, Vafapour M, Moazzami B, Rezaei N (2020) Hyperinflammatory shock related to COVID-19 in a patient presenting with multisystem inflammatory syndrome in children: first case from Iran. J Paediatr Child Health. https://doi.org/10.1111/jpc.15048

Baig AM, Khaleeq A, Ali U, Syeda H (2020) Evidence of the COVID-19 virus targeting the CNS: tissue distribution, host-virus interaction, and proposed neurotropic mechanisms. ACS Chem Neurosci 11(7):995–998. https://doi.org/10.1021/acschemneuro.0c00122

Bajunaid K, Sabbagh AJ, Ajlan A, Al-Jehani H, Alnaami I, Khormi YH, Bin Mahfoodh M, Barnawi A, Bin Salamah A, Alobaid A, Bafaquh MS, Alturki AY, Alkhani A, Al-Habib H, Baeesa S, Al-Habib A (2020) Consensus statement of the Saudi Association of Neurological Surgery (SANS) on triage of neurosurgery patients during COVID-19 pandemic in Saudi Arabia. Neurosciences (Riyadh, Saudi Arabia) 25(2):148–151. https://doi.org/10.17712/nsj.2020.2.20200054

Barzegar M, Mirmosayyeb O, Nehzat N, Sarrafi R, Khorvash F, Maghzi AH, Shaygannejad V (2020) COVID-19 infection in a patient with multiple sclerosis treated with fingolimod. Neurology(R) neuroimmunology & neuroinflammation 7 (4). https://doi.org/10.1212/nxi.0000000000000753

Basiri A, Pazhouhnia Z, Beheshtizadeh N, Hoseinpour M, Saghazadeh A, Rezaei N (2020) Regenerative medicine in COVID-19 treatment: real opportunities and range of promises. Stem Cell Rev Rep:1–13

Berger JR (2020) COVID-19 and the nervous system. J Neurovirol:1–6. https://doi.org/10.1007/s13365-020-00840-5

Bohmwald K, Gálvez NMS, Ríos M, Kalergis AM (2018) Neurologic alterations due to respiratory virus infections. Front Cell Neurosci 12:386. https://doi.org/10.3389/fncel.2018.00386

Bolay H, Gül A, Baykan B (2020) COVID-19 is a real headache! Headache. https://doi.org/10.1111/head.13856

Bostancıklioglu M (2020) Severe acute respiratory syndrome coronavirus 2 is penetrating to dementia research. Curr Neurovasc Res. https://doi.org/10.2174/1567202617666200522220509

Brun G, Hak J-F, Coze S, Kaphan E, Carvelli J, Girard N, Stellmann J-P (2020) COVID-19—white matter and globus pallidum lesions. Neurol Neuroimmunol Neuroinflammation 7(4):e777. https://doi.org/10.1212/NXI.0000000000000777

Burke JF, Chan AK, Mummaneni V, Chou D, Lobo EP, Berger MS, Theodosopoulos PV, Mummaneni PV (2020) Letter: the coronavirus disease 2019 global pandemic: a neurosurgical treatment algorithm. Neurosurgery. https://doi.org/10.1093/neuros/nyaa116

Byrnes S, Bisen M, Syed B, Huda S, Siddique Z, Sampat P, Russo R, Oueida Z, Johri G, Dargon I (2020) COVID-19 encephalopathy masquerading as substance withdrawal. J Med Virol. https://doi.org/10.1002/jmv.26065

Carod-Artal FJ (2020) Neurological complications of coronavirus and COVID-19. Revista de neurologia 70 (9):311-322. doi:https://doi.org/10.33588/rn.7009.2020179

Chan JF-W, Yuan S, Kok K-H, To KK-W, Chu H, Yang J, Xing F, Liu J, Yip CC-Y, Poon RW-S (2020) A familial cluster of pneumonia associated with the 2019 novel coronavirus indicating person-to-person transmission: a study of a family cluster. Lancet 395(10223):514–523

Chen S, Lu H, Liu Z, Yuan W (2020) Comment on "central nervous system involvement by severe acute respiratory syndrome coronavirus –2 (SARS-CoV-2)". J Med Virol. https://doi.org/10.1002/jmv.25991

Deliwala S, Abdulhamid S, Abusalih MF, Al-Qasmi MM, Bachuwa G (2020) Encephalopathy as the sentinel sign of a cortical stroke in a patient infected with coronavirus Disease-19 (COVID-19). Cureus 12(5):e8121. https://doi.org/10.7759/cureus.8121

Dickey LL, Worne CL, Glover JL, Lane TE, O'Connell RM (2016) MicroRNA-155 enhances T cell trafficking and antiviral effector function in a model of coronavirus-induced neurologic disease. J Neuroinflammation 13(1):240. https://doi.org/10.1186/s12974-016-0699-z

Fathi N, Rezaei N (2020) Lymphopenia in COVID-19: therapeutic opportunities. Cell Biol Int 44:1792–1797

Federico A (2020) Brain awareness week, CoVID-19 infection and neurological sciences. Neurol Sci 41(4):747–748. https://doi.org/10.1007/s10072-020-04338-0

Gandhi S, Srivastava AK, Ray U, Tripathi PP (2020) Is the collapse of the respiratory Center in the Brain Responsible for respiratory breakdown in COVID-19 patients? ACS Chem Neurosci 11(10):1379–1381. https://doi.org/10.1021/acschemneuro.0c00217

Giacomelli A, Pezzati L, Conti F, Bernacchia D, Siano M, Oreni L, Rusconi S, Gervasoni C, Ridolfo AL, Rizzardini G, Antinori S, Galli M (2020) Self-reported olfactory and taste disorders in patients with severe acute respiratory coronavirus 2 infection: a cross-sectional study. Clin Infect Dis. https://doi.org/10.1093/cid/ciaa330

Gklinos P (2020) Neurological manifestations of COVID-19: a review of what we know so far. J Neurol. https://doi.org/10.1007/s00415-020-09939-5

Grist JJ, Marro B, Lane TE (2018) Neutrophils and viral-induced neurologic disease. Clinical immunology (Orlando, Fla) 189:52-56. https://doi.org/10.1016/j.clim.2016.05.009

Hanaei S, Rezaei N (2020) COVID-19: developing from an outbreak to a pandemic. Arch Med Res

Hankins DG, Rosekrans JA (2004) Overview, prevention, and treatment of rabies. Mayo Clin Proc 79(5):671–676. https://doi.org/10.4065/79.5.671

Hascup ER, Hascup KN (2020) Does SARS-CoV-2 infection cause chronic neurological complications? GeroScience. https://doi.org/10.1007/s11357-020-00207-y

Heman-Ackah SM, Su YS, Spadola M, Petrov D, Chen HI, Schuster J, Lucas T (2020) Neurologically devastating Intraparenchymal hemorrhage in COVID-19 patients on extracorporeal membrane oxygenation: a case series. Neurosurgery. https://doi.org/10.1093/neuros/nyaa198

Hovanec DL, Flanagan TD (1983) Detection of antibodies to human coronaviruses 229E and OC43 in the sera of multiple sclerosis patients and normal subjects. Infect Immun 41(1):426–429

Hughes C, Nichols T, Pike M, Subbe C, Elghenzai S (2020) Cerebral venous sinus thrombosis as a presentation of COVID-19. Eur J Case Rep Intern Med 7(5):001691

Jabbari P, Jabbari F, Ebrahimi S, Rezaei N (2020) COVID-19: A Chimera of Two Pandemics. Disaster Med Public Health Prep:1–3. https://doi.org/10.1017/dmp.2020.223

Jahanshahlu L, Rezaei N (2020a) Central nervous system involvement in COVID-19. Arch Med Res. https://doi.org/10.1016/j.arcmed.2020.05.016

Jahanshahlu L, Rezaei N (2020b) Monoclonal antibody as a potential anti-COVID-19. Biomed Pharmacother 129:110337

Kafieh R, Arian R, Saeedizadeh N, Minaee S, Yadav SK, Vaezi A, Rezaei N, Javanmard SH (2020) COVID-19 in Iran: a deeper Look into the future. medRxiv

Kalinke U, Bechmann I, Detje CN (2011) Host strategies against virus entry via the olfactory system. Virulence 2(4):367–370. https://doi.org/10.4161/viru.2.4.16138

Kim BJ, Kim ES, Shin MJ, Kim HB, Lee HY, Hong KS, Park HK, Lee J, Sohn SI, Hwang YH, Ko SB, Park JM, Rha JH, Kwon SU, Kim JS, Heo JH, Lee BC, Yoon BW, Bae HJ (2020) Management of Acute Stroke Patients amid the coronavirus disease 2019 pandemic: scientific statement of the Korean stroke society. J Stroke. https://doi.org/10.5853/jos.2020.01291

Kuroda N (2020) Epilepsy and COVID-19: associations and important considerations. Epilepsy Behav E&B 108:107122. https://doi.org/10.1016/j.yebeh.2020.107122

Leslie-Mazwi TM, Fargen KM, Levitt M, Derdeyn CP, Feske SK, Patel AB, Hirsch JA (2020) Preserving access: a review of stroke Thrombectomy during the COVID-19 pandemic. AJNR Am J Neuroradiol. https://doi.org/10.3174/ajnr.A6606

Levi M, Thachil J, Iba T, Levy JH (2020) Coagulation abnormalities and thrombosis in patients with COVID-19. Lancet Haematol 7(6):e438–e440. https://doi.org/10.1016/S2352-3026(20)30145-9

Li K, Wohlford-Lenane C, Perlman S, Zhao J, Jewell AK, Reznikov LR, Gibson-Corley KN, Meyerholz DK, McCray PB Jr (2016) Middle East respiratory syndrome coronavirus causes multiple organ damage and lethal disease in mice transgenic for human dipeptidyl peptidase 4. J Infect Dis 213(5):712–722. https://doi.org/10.1093/infdis/jiv499

Li Z, Zhao K, Lan Y, Lv X, Hu S, Guan J, Lu H, Zhang J, Shi J, Yang Y, Song D, Gao F, He W (2017) Porcine Hemagglutinating encephalomyelitis virus enters neuro-2a cells via Clathrin-mediated endocytosis in a Rab5-, cholesterol-, and pH-dependent manner. J Virol 91(23). https://doi.org/10.1128/jvi.01083-17

Li Y-C, Bai W-Z, Hashikawa T (2020a) The neuroinvasive potential of SARS-CoV2 may play a role in the respiratory failure of COVID-19 patients. J Med Virol 92(6):552–555. https://doi.org/10.1002/jmv.25728

Li YC, Bai WZ, Hashikawa T (2020b) The neuroinvasive potential of SARS-CoV2 may play a role in the respiratory failure of COVID-19 patients. J Med Virol. https://doi.org/10.1002/jmv.25728

Li YC, Bai WZ, Hashikawa T (2020c) The neuroinvasive potential of SARS-CoV2 may play a role in the respiratory failure of COVID-19 patients. J Med Virol 92(6):552–555

Li Z, Huang Y, Guo X (2020d) The brain, another potential target organ, needs early protection from SARS-CoV-2 neuroinvasion. Sci China Life Sci 63(5):771–773

Liguori C, Pierantozzi M, Spanetta M, Sarmati L, Cesta N, Iannetta M, Ora J, Genga Mina G, Puxeddu E, Balbi O, Pezzuto G, Magrini A, Rogliani P, Andreoni M, Biagio Mercuri N (2020) Subjective neurological symptoms frequently occur in patients with SARS-CoV2 infection. Brain Behav Immun. https://doi.org/10.1016/j.bbi.2020.05.037

Liu K, Pan M, Xiao Z, Xu X (2020) Neurological manifestations of the coronavirus (SARS-CoV-2) pandemic 2019-2020. J Neurol Neurosurg Psychiatry 91(6):669–670. https://doi.org/10.1136/jnnp-2020-323177

Lotfi M, Rezaei N (2020) SARS-CoV-2: a comprehensive review from pathogenicity of the virus to clinical consequences. J Med Virol

Lotfi M, Hamblin MR, Rezaei N (2020) COVID-19: transmission, prevention, and potential therapeutic opportunities. Clin Chim Acta

Mamo J, Feroz B, Mahmood S (2020) Covid-19: protecting patients in hospital for neurorehabilitation and their therapists. BMJ (Clinical Research ed) 369:m1630. https://doi.org/10.1136/bmj.m1630

Markus HS, Brainin M (2020) COVID-19 and stroke—a global world stroke organization perspective. Int J Stroke 15(4):361–364. https://doi.org/10.1177/1747493020923472

McCray PB Jr, Pewe L, Wohlford-Lenane C, Hickey M, Manzel L, Shi L, Netland J, Jia HP, Halabi C, Sigmund CD, Meyerholz DK, Kirby P, Look DC, Perlman S (2007) Lethal infection of K18-hACE2 mice infected with severe acute respiratory syndrome coronavirus. J Virol 81(2):813–821. https://doi.org/10.1128/jvi.02012-06

Mehta P, McAuley DF, Brown M, Sanchez E, Tattersall RS, Manson JJ (2020) COVID-19: consider cytokine storm syndromes and immunosuppression. Lancet 395(10229):1033–1034. https://doi.org/10.1016/S0140-6736(20)30628-0

Mirbeyk M, Rezaei N (2020) The impact of COVID-19 on pregnancy and neonatal health: a systematic review

Moazzami B, Razavi-Khorasani N, Moghadam AD, Farokhi E, Rezaei N (2020) COVID-19 and telemedicine: immediate action required for maintaining healthcare providers Well-being. J Clin Virol 126:104345

Mohamed K, Rodríguez-Román E, Rahmani F, Zhang H, Ivanovska M, Makka SA, Joya M, Makuku R, Islam MS, Radwan N (2020a) Borderless collaboration is needed for COVID-19—A disease that knows no borders. Infect Control Hosp Epidemiol:1–2

Mohamed K, Yazdanpanah N, Saghazadeh A, Rezaei N (2020b) Computational drug discovery and repurposing for the treatment of Covid-19: a systematic review. Available at SSRN 3583748

Mohile NA, Blakeley JO, Gatson NTN, Hottinger AF, Lassman AB, Ney DE, Olar A, Schiff D, Shih HA, Strowd R, van den Bent MJ, Ziu M (2020) Urgent considerations for the neuro-oncologic treatment of patients with gliomas during the COVID-19 pandemic. Neuro-Oncology. https://doi.org/10.1093/neuonc/noaa090

Momtazmanesh S, Ochs HD, Uddin LQ, Perc M, Routes JM, Vieira DN, Al-Herz W, Baris S, Prando C, Rosivall L (2020) All together to fight COVID-19. Am J Trop Med Hyg 102(6):1181–1183

Montalvan V, Lee J, Bueso T, De Toledo J, Rivas K (2020) Neurological manifestations of COVID-19 and other coronavirus infections: a systematic review. Clin Neurol Neurosurg 194:105921. https://doi.org/10.1016/j.clineuro.2020.105921

Moradian N, Ochs HD, Sedikies C, Hamblin MR, Camargo CA, Martinez JA, Biamonte JD, Abdollahi M, Torres PJ, Nieto JJ (2020) The urgent need for integrated science to fight COVID-19 pandemic and beyond. J Transl Med 18(1):1–7

Moriguchi T, Harii N, Goto J, Harada D, Sugawara H, Takamino J, Ueno M, Sakata H, Kondo K, Myose N, Nakao A, Takeda M, Haro H, Inoue O, Suzuki-Inoue K, Kubokawa K, Ogihara S, Sasaki T, Kinouchi H, Kojin H, Ito M, Onishi H, Shimizu T, Sasaki Y, Enomoto N, Ishihara H, Furuya S, Yamamoto T, Shimada S (2020) A first case of meningitis/encephalitis associated with SARS-Coronavirus-2. Int J Infect Dis 94:55–58. https://doi.org/10.1016/j.ijid.2020.03.062

Nasab MG, Saghazadeh A, Rezaei N (2020) SARS-CoV-2–a tough opponent for the immune system. Arch Med Res

Nath A (2020) Neurologic complications of coronavirus infections. Neurology 94(19):809–810. https://doi.org/10.1212/wnl.0000000000009455

Netland J, Meyerholz DK, Moore S, Cassell M, Perlman S (2008) Severe acute respiratory syndrome coronavirus infection causes neuronal death in the absence of encephalitis in mice transgenic for human ACE2. J Virol 82(15):7264–7275. https://doi.org/10.1128/jvi.00737-08

Ng Kee Kwong KC, Mehta PR, Shukla G, Mehta AR (2020) COVID-19, SARS and MERS: a neurological perspective. J Clin Neurosci. https://doi.org/10.1016/j.jocn.2020.04.124

Ogier M, Andéol G, Sagui E, Bo GD (2020) How to detect and track chronic neurologic sequelae of COVID-19? Use of auditory brainstem responses and neuroimaging for long-term patient follow-up. Brain Behav Immunity Health:100081. https://doi.org/10.1016/j.bbih.2020.100081

Papa SM, Brundin P, Fung VSC, Kang UJ, Burn DJ, Colosimo C, Chiang HL, Alcalay RN, Trenkwalder C (2020) Impact of the COVID-19 pandemic on Parkinson's disease and movement disorders. Mov Disord Clin Pract 7(4):357–360. https://doi.org/10.1002/mdc3.12953

Plaisted WC, Zavala A, Hingco E, Tran H, Coleman R, Lane TE, Loring JF, Walsh CM (2016) Remyelination is correlated with regulatory T cell induction following human Embryoid body-derived neural precursor cell transplantation in a viral model of multiple sclerosis. PLoS One 11(6):e0157620. https://doi.org/10.1371/journal.pone.0157620

Plaze M, Attali D, Petit AC, Blatzer M, Simon-Loriere E, Vinckier F, Cachia A, Chrétien F, Gaillard R (2020) Repurposing of chlorpromazine in COVID-19 treatment: the reCoVery study. L'Encephale. https://doi.org/10.1016/j.encep.2020.04.010

Prevention CfDCa (2020) Symptoms of Coronavirus. https://www.cdc.gov/coronavirus/2019-ncov/symptoms-testing/symptoms.html

Rabiee N, Rabiee M, Bagherzadeh M, Rezaei N (2020) COVID-19 and Picotechnology: potential opportunities. Med Hypotheses 144:109917

Radmanesh A, Derman A, Lui YW, Raz E, Loh JP, Hagiwara M, Borja MJ, Zan E, Fatterpekar GM (2020a) COVID-19-associated diffuse leukoencephalopathy and microhemorrhages. Radiology 297:202040–202040

Radmanesh A, Raz E, Zan E, Derman A, Kaminetzky M (2020b) Brain imaging use and findings in COVID-19: a single academic center experience in the epicenter of disease in the United States. Am J Neuroradiol

Rezaei N (2020a) COVID-19 affects healthy pediatricians more than pediatric patients. Infect Control Hosp Epidemiol.:1–1

Rezaei N (2020b) COVID-19 and medical biotechnology. Avicenna J Med Biotechnol 12(3):139–139

Robinson CP, Busl KM (2020) Neurologic manifestations of severe respiratory viral contagions. Crit Care Explor 2(4):e0107. https://doi.org/10.1097/cce.0000000000000107

Rzymski P, Nowicki M, Mullin GE, Abraham A, Rodríguez-Román E, Petzold MB, Bendau A, Sahu KK, Ather A, Naviaux A-F (2020) Quantity does not equal quality: scientific principles cannot be sacrificed. Int Immunopharmacol 86:106711

Saghazadeh A, Rezaei N (2020a) Immune-epidemiological parameters of the novel coronavirus–a perspective. Expert Rev Clin Immunol:1–6

Saghazadeh A, Rezaei N (2020b) Towards treatment planning of COVID-19: rationale and hypothesis for the use of multiple immunosuppressive agents: anti-antibodies, immunoglobulins, and corticosteroids. Int Immunopharmacol:106560

Sahu KK, Siddiqui AD, Rezaei N, Cerny J (2020) Challenges for management of immune thrombocytopenia during COVID-19 pandemic. J Med Virol. https://doi.org/10.1002/jmv.26251

Salari M, Zali A, Ashrafi F, Etemadifar M, Sharma S, Hajizadeh N, Ashourizadeh H (2020) Incidence of anxiety in Parkinson's disease during the coronavirus disease (COVID-19) pandemic. Move Disord. https://doi.org/10.1002/mds.28116

Saleki K, Banazadeh M, Saghazadeh A, Rezaei N (2020) The involvement of the central nervous system in patients with COVID-19. Rev Neurosci 31(4):453–456. https://doi.org/10.1515/revneuro-2020-0026

Schirinzi T, Cerroni R, Di Lazzaro G, Liguori C, Scalise S, Bovenzi R, Conti M, Garasto E, Mercuri NB, Pierantozzi M, Pisani A, Stefani A (2020) Self-reported needs of patients with Parkinson's disease during COVID-19 emergency in Italy. Neurol Sci:1–3. https://doi.org/10.1007/s10072-020-04442-1

Sheraton M, Deo N, Kashyap R, Surani S (2020) A review of neurological complications of COVID-19. Cureus 12(5):e8192. https://doi.org/10.7759/cureus.8192

Shoskes A, Migdady I, Fernandez A, Ruggieri P, Rae-Grant A (2020) Cerebral microhemorrhage and purpuric rash in COVID-19: the case for a secondary Microangiopathy. J Stroke Cerebrovasc Dis 29(10):105111

Siddiqui R, Khan NA (2020) Proposed intranasal route for drug Administration in the Management of central nervous system manifestations of COVID-19. ACS Chem Neurosci. https://doi.org/10.1021/acschemneuro.0c00288

Soma T, Saito N, Kawaguchi M, Sasai K (2018) Feline coronavirus antibody titer in cerebrospinal fluid from cats with neurological signs. J Vet Med Sci 80(1):59–62. https://doi.org/10.1292/jvms.17-0399

Stoessl AJ, Bhatia KP, Merello M (2020) Movement disorders in the world of COVID-19. Mov Disord 35(5):709–710. https://doi.org/10.1002/mds.28069

Sweiss NJ, Korsten P, Syed H, Syed A, Baughman RP, Yee AMF, Culver DA, Sosenko T, Azuma A, Bonella F, Costabel U, Drake WP, Drent M, Lower EE, Israel-Biet D, Mostard RLM, Nunes H, Rottoli P, Spagnolo P, Wells AU, Wuyts WA, Judson MA (2020) When the game changes: guidance to adjust sarcoidosis management during the COVID-19 pandemic. Chest. https://doi.org/10.1016/j.chest.2020.04.033

van Riel D, Verdijk R, Kuiken T (2015) The olfactory nerve: a shortcut for influenza and other viral diseases into the central nervous system. J Pathol 235(2):277–287. https://doi.org/10.1002/path.4461

Vilensky JA (2014) The neglected cranial nerve: nervus terminalis (cranial nerve N). Clin Anat (New York, NY) 27(1):46–53. https://doi.org/10.1002/ca.22130

Waldman G, Mayeux R, Claassen J, Agarwal S, Willey J, Anderson E, Punzalan P, Lichtcsien R, Bell M, Przedborski S, Ulane C, Roberts K, Williams O, Lassman AB, Lennihan L, Thakur KT (2020) Preparing a neurology department for SARS-CoV-2 (COVID-19): early experiences at Columbia University Irving medical center and the New York Presbyterian Hospital in new York City. Neurology 94(20):886–891. https://doi.org/10.1212/wnl.0000000000009519

Whitley RJ, Gnann JW (2002) Viral encephalitis: familiar infections and emerging pathogens. Lancet (London, England) 359(9305):507–513. https://doi.org/10.1016/s0140-6736(02)07681-x

Whittaker A, Anson M, Harky A (2020) Neurological manifestations of COVID-19: a review. Acta Neurol Scand. https://doi.org/10.1111/ane.13266

Williams SK, Franklin RJM, Barnett SC (2004) Response of olfactory ensheathing cells to the degeneration and regeneration of the peripheral olfactory system and the involvement of the neuregulins. J Comp Neurol 470(1):50–62. https://doi.org/10.1002/cne.11045

Wilson MP, Jack AS (2020) Coronavirus disease 2019 (COVID-19) in neurology and neurosurgery: a scoping review of the early literature. Clin Neurol Neurosurg 193:105866. https://doi.org/10.1016/j.clineuro.2020.105866

Wu Y, Xu X, Chen Z, Duan J, Hashimoto K, Yang L, Liu C, Yang C (2020) Nervous system involvement after infection with COVID-19 and other coronaviruses. Brain Behav Immun. https://doi.org/10.1016/j.bbi.2020.03.031

Xydakis MS, Dehgani-Mobaraki P, Holbrook EH, Geisthoff UW, Bauer C, Hautefort C, Herman P, Manley GT, Lyon DM, Hopkins C (2020) Smell and taste dysfunction in patients with COVID-19. Lancet Infect Dis. https://doi.org/10.1016/S1473-3099(20)30293-0

Yazdanpanah F, Hamblin MR, Rezaei N (2020) The immune system and COVID-19: friend or foe? Life Sci 256:117900

Ye M, Ren Y, Lv T (2020) Encephalitis as a clinical manifestation of COVID-19. Brain Behav Immun. https://doi.org/10.1016/j.bbi.2020.04.017

Yin R, Feng W, Wang T, Chen G, Wu T, Chen D, Lv T, Xiang D (2020) Concomitant neurological symptoms observed in a patient diagnosed with coronavirus disease 2019. J Med Virol. https://doi.org/10.1002/jmv.25888

Yousefzadegan S, Rezaei N (2020) Case report: death due to COVID-19 in three brothers. Am J Trop Med Hyg 102(6):1203–1204

Zayet S, Ben Abdallah Y, Royer PY, Toko-Tchiundzie L, Gendrin V, Klopfenstein T (2020) Encephalopathy in patients with COVID-19: 'Causality or coincidence?'. J Med Virol. https://doi.org/10.1002/jmv.26027

21

Autoimmune Processes Involved in Organ System Failure Following Infection with SARS-CoV-2

Steven E. Kornguth and Robert J. Hawley

Abstract

During the COVID-19 pandemic associated with high incidence, transmissibility, and mortality, this chapter focuses on three phases of the disease: initial exposure, initiation of the immune response to the agent, and finally, an inflammatory/autoimmune-like presentation with pulmonary, neurological, and renal failure and disseminated intravascular coagulation which occurs in a small proportion of the patients. The elegant demonstration of the site of interaction between the spike (S) protein of the severe acute respiratory syndrome coronavirus 2 (SARS-CoV-2), which is the causative agent of COVID-19, and the ACE (angiotensin-converting enzyme) 2 receptor of cells distributed throughout the body has enabled research efforts to develop pharmacological and immune countermeasures to the viral phase of the disease. This chapter rapidly reviews the molecular and structural organization of SARS-CoV-2 and its interaction with ACE2. It is followed by a discussion over the role of the major histocompatibility complex (MHC) in recognition of the virus. The importance of rapid compartmentation of the viral genome into the target cells as opposed to the binding constant of the virus for the ACE receptor is discussed. Host factors affecting the immune response to the virus are examined, and the subsequent inflammatory dysregulation enabling the cytokine storm leading to system organ failure is described. Finally, the similarities of the clinical effects of the murine hepatitis virus-JHM (a coronavirus) on multi-organ systems (liver, brain, clotting cascade) as described by Perlman and colleagues permit insights regarding the role of the interaction between the host and the virus in developing the clinical presentation of the inflammatory/autoimmune disorders that occur in multiple sclerosis, neuromyelitis optica, and more interestingly, during the third phase of COVID-19.

Keywords

Autoimmune · COVID-19 · Immune dysregulation · Immune response · SARS-CoV-2

S. E. Kornguth (✉)
Department of Neurology, Dell Medical School, University of Texas at Austin, Austin, TX, USA

Biological Safety and Security, Frederick, MD, USA
e-mail: Steve_kornguth@utexas.edu

R. J. Hawley
Biological Safety and Security, Frederick, MD, USA

Department of Kinesiology and Health Education, University of Texas at Austin, Austin, TX, USA

N. Rezaei (ed.), *Coronavirus Disease - COVID-19*, Advances in Experimental Medicine and Biology 1318, https://doi.org/10.1007/978-3-030-63761-3_21

21.1 Introduction

A new human coronavirus (hCoV), known as severe acute respiratory syndrome coronavirus 2 (SARS-CoV-2), is supposed to be a responsible pathogen of the pandemic of coronavirus disease 2019 (COVID-19). The pandemic has caused worldwide severe morbidity and mortality with ensuing economic concerns resulting in extreme hardship and devastating conditions. We examine the stages of disease exposure, replication of the virus, and initiation of the immune response with eventual dysregulation of the immune response and potential treatment(s). The chapter moves from the molecular and structural organization of SARS-CoV-2 to its interaction with the host. It includes the cellular receptors binding to the virus and the role of the major histocompatibility complex (MHC) of the host target organs interacting with the virus. The factors in the host immune response to the virus are examined, and the subsequent inflammatory dysregulation enabling the cytokine storm leading to systemic inflammation, multi-organ dysfunction, and critical illness is described. Populations that are most vulnerable to or resilient to COVID-19 are considered from a population perspective. Various therapeutics are also briefly discussed.

21.2 Ecology and History of SARS-CoV-2

A recently emerged virus is now a reality. It is challenging even robust healthcare systems of many areas worldwide (approximately 217 countries and territories with ten million cases, with over 499,000 deaths, as of June 28, 2020) and in the United States (over 2,500,000 cases with 127,000 deaths, as of June 28, 2020) since being recognized during an outbreak in Wuhan, Hubei province, China, in December 2019. The definitive origin of the virus remains unknown at present. Comparative genomic analysis data shows that its origin is unlikely a result of laboratory manipulation. Reasonable origin(s) of the virus may include "natural selection in an animal host before zoonotic transfer; and natural selection in humans following zoonotic transfer" (Andersen et al. 2020). This virus is a novel human coronavirus named severe acute respiratory syndrome coronavirus 2 (SARS-CoV-2). It is one of the seven known human coronaviruses and causes coronavirus disease 2019 (COVID-19). Coronaviruses are members of the subfamily *Orthocoronavirinae*, in the family *Coronaviridae*, order *Nidovirales*, and realm *Riboviria* (Gorbalenya et al. 2020). Virus particles are enveloped, have a positive-sense single-stranded ribonucleic acid (RNA) genome, and a nucleocapsid of helical symmetry. Coronaviruses are the largest among known RNA viruses having a genome size ranging from about 27 to 34 kilobases (Woo et al. 2010).

Coronaviruses may cause illness in mammals, birds, or humans. There are four genera: *Alpha-*, *Beta-*, *Gamma-*, and *Deltacoronavirus*. Alpha- and betacoronaviruses infect several mammalian species (as well as domestic cats), including humans, while gamma- and deltacoronaviruses infect birds (Corman et al. 2018). Human coronaviruses can cause respiratory infections with a broad spectrum of severity from the common cold to more severe respiratory diseases as referred to as severe acute respiratory syndrome (SARS) and Middle East respiratory syndrome (MERS). Both the SARS and MERS viruses belonged to the genus *Betacoronavirus* and were responsible for outbreaks involving high case fatality rates.

SARS was first identified in November 2002 in the Guangdong province of southern China. However, it could spread to 26 countries in North America, South America, Europe, and Asia, infected 8098 people, and claimed 774 lives during the outbreak. The virus is thought to have spread from bats to civets (carnivores of the family Viverridae – small mammals that resemble a cat in appearance) before the first human patient was infected. Infections were primarily through person-to-person transmission. No known cases of SARS have been reported since 2004, as declared by the Centers for Disease Control and Prevention.

In 2012, Saudi Arabia reported the first case infected with MERS-CoV, a cousin to SARS

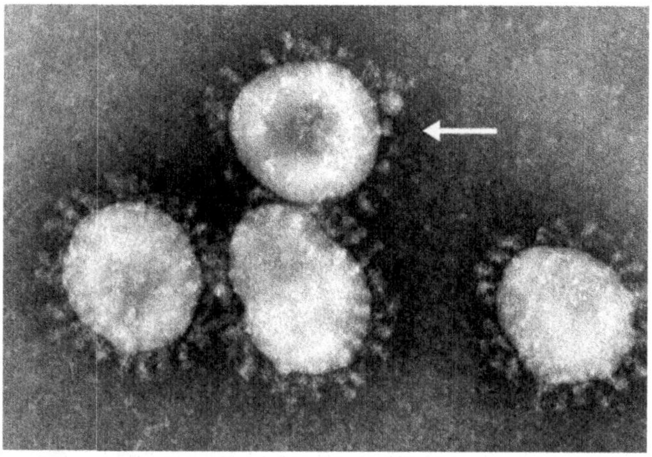

Fig. 21.1 Electron microscopy image of SARS-CoV. The arrow points at a single virion (virus particle). (Adapted from the Centers for Disease Control and Prevention's (CDC) Public Health Image Library (PHIL), identification number 4814 (https://phil.cdc.gov/Details.aspx?pid=15523))

derived from bats. MERS-CoV could cause a respiratory illness that spread to 27 countries in Europe, Africa, Asia, and North America. The MERS virus can be transmitted between animals and humans, thus known as zoonoses. Although the origin(s) of the virus is not fully understood, analyzing different virus genomes has suggested MERS-CoV be originated in bats and then transmitted to dromedary camels. Since 2012, there have been 2494 reported cases of MERS and 858 deaths from the virus. Infections occurred primarily from close human-to-human contact, according to the World Health Organization (WHO).

SARS-CoV-2 is the seventh coronavirus causing disease, i.e., COVID-19, in humans. The virus is classified as a betacoronavirus of group 2B by the WHO (Hui et al. 2020). SARS-CoV-2 is the seventh coronavirus that can infect humans. SARS-CoV, MERS-CoV, and SARS-CoV-2 can cause severe disease, while HKU1, NL63, OC43, and 229E are associated with mild symptoms (Corman et al. 2018). SARS CoV-2, like the MERS and SARS coronaviruses, likely evolved from a virus previously found in animals.

21.3 Structure of SARS-CoV-2

Looking under transmission electron microscopy (TEM), virus particles are surrounded by a crown resembling the sun's corona. The crown or corona-like appearance resulted in the virus being classified as coronavirus from the Latin *corona* (meaning "crown" or "halo"). The crown appearance is due to the surface covering in club-shaped protein spikes (Figs. 21.1, 21.2, and 21.3).

The spike protein (S) is the largest structure on the virion surface, is heavily glycosylated, and uses an N-terminal signal sequence to gain access to the endoplasmic reticulum (ER) and mediate attachment to host cell receptors. The host cell furin-like protease cleaves the S protein into two functional subunits. The S1 subunit is responsible for binding to the host cell receptor, while the S2 subunit is responsible for the fusion of the viral and cell membrane (Yan et al. 2020). The article by He et al. (He et al. 2020) discusses the spike 1 protein attachment via one of three C-terminal domains with the ACE2-receptor binding domain located on CTD-1 moiety. The RNA is the genome of the virus with about 30 kilobases. The viral genome contains 15 genes, among which the S gene codes for a protein on the virion envelope. The genome of SARS-CoV-2 shares about 80% identity with that of SARS-CoV (causative agent of SARS) (Fehr and Perlman 2015). The nucleocapsid protein (N), helically symmetrical configuration, binds to RNA in vitro and is heavily phosphorylated. The N protein binds the viral genome in a beads-on-a-string-type arrangement. These protein interactions likely help bind the viral genome to a replicase-transcriptase complex (RTC) and subsequently package the encapsulated genome into viral particles (Fehr and Perlman 2015). The envelope protein (E) of about 8–12 kilodaltons (kDa) is found in small quanti-

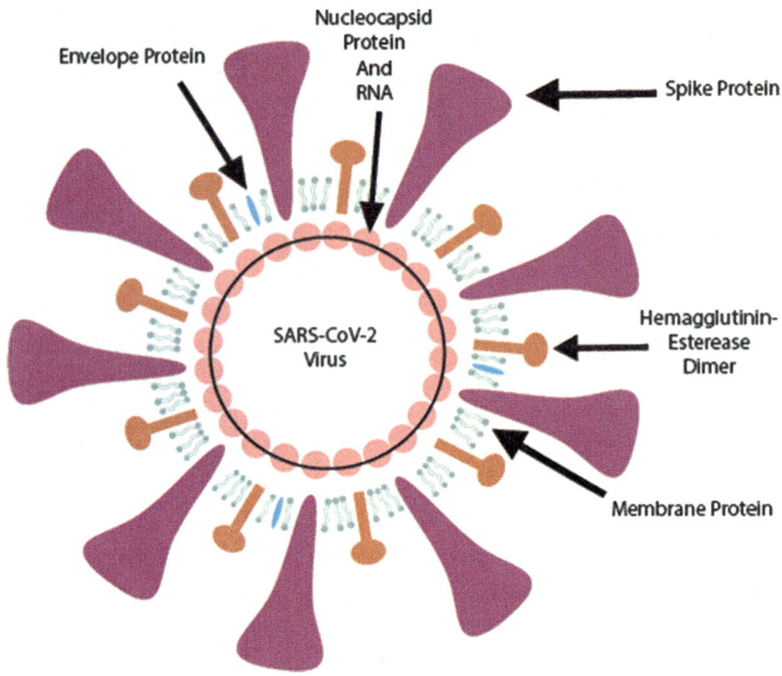

Fig. 21.2 SARS-CoV-2 structure. SARS-CoV-2 consists of a spike protein (S), hemagglutinin-esterase dimer (HE), a membrane glycoprotein (M), an envelope protein (E), a nucleocapsid protein (N), and RNA. (Adapted from https://www.prosci-inc.com/covid-19/)

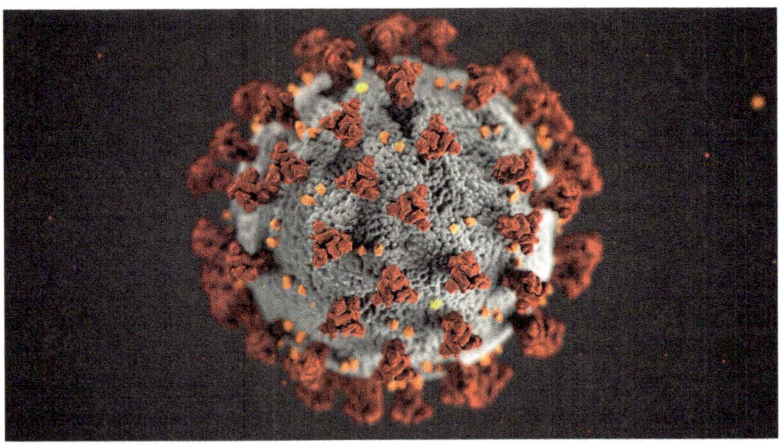

Fig. 21.3 SARS-CoV-2 ultrastructure illustration. (Adapted from the Centers for Disease Control and Prevention's (CDC) Public Health Image Library (PHIL), identification number 2871 (https://phil.cdc.gov/Details.aspx?pid=2871))

ties within the virus. It is most likely a transmembrane protein having ion channel activity. The E protein is necessary for viral replication, facilitating assembly, and release of the virus, but not for pathogenesis (Nieto-Torres et al. 2014).

The membrane protein (M) of about 20–30 kDa is the most abundant structural protein in the virus and is thought to give shape to the virus. Most M proteins do not contain signal sequences, and studies suggest it exists as a dimer in the virion. M proteins may have two different arrangements to enable them to promote membrane curvature as well as bind to the nucleocapsid (Neuman et al. 2011). The hemagglutinin-esterase (HE) dimer protein is present in a subset of betacoronaviruses, including this one. The HE protein has acetyl-esterase activity, binds to sialic acids on surface glycoproteins, and acts as a hemagglutinin (Klausegger et al. 1999). These protein activities are thought to enhance S protein-mediated cell entry and the virus spread through the mucosa (Cornelissen et al. 1997).

21.4 Infectivity and Virulence Factors of SARS-CoV-2

A cluster of infections was initially reported on December 31, 2019, when the WHO China Country Office was informed. Chinese authorities isolated a new coronavirus, initially called novel coronavirus 2019, nCoV-2019, on January 7, 2020. The WHO received detailed information from the National Health Commission of China on January 11 and 12, 2020, regarding the outbreak associated with exposures in one seafood market in Wuhan. The clinical spectrum described at that time included mainly fever and, to a lesser extent, difficult breathing. Chest radiographs showed diffuse pneumonic infiltrates. This novel human coronavirus was subsequently named SARS-CoV-2. The disease, COVID-19, caused by this agent, is characterized by a range of symptoms that include fever, cough, dyspnea (difficult or labored breathing, shortness of breath), and myalgia (pain in a muscle or group of muscles) in most cases. The dynamics of this newly emerged and rapidly growing infectious disease outbreak is challenging because of the limited amount of clinical and epidemiological data available. Early estimates of the exponential growth rate of the outbreak were 0.1–0.14/day with a doubling time of 6–7 days. The reproductive number (R_0), the average number of secondary cases due to infection by an index case, ranged from 2.2 to 2.7 (Li et al. 2020).

Using mathematical model approaches to understand the outbreak dynamics in Wuhan by examining domestic travel and infection data, results showed that the doubling time early in the epidemic was 2.3–3.3 days. The authors assumed a serial interval of 6–9 days and calculated a median R_0 value of 5.7 (95% CI 3.8–8.9). To halt transmission of the virus, Sanche and colleagues indicated that active surveillance, contact tracing, quarantine, and initial robust social (physical) distancing efforts are needed (Sanche et al. 2020). SARS CoV-2 can cause a severe respiratory illness like SARS and MERS based on evidence from clinics in Wuhan, and data suggest that the virus, in general, is less pathogenic than SARS-CoV, and much less than MERS-CoV (Chen 2020). Since coronaviruses have error-prone RNA-dependent RNA polymerases (RdRP), mutations and recombination events frequently occur. It results in a quasispecies assortment that is closely associated with adaptive evolution and the capacity to cause disease (Cui et al. 2019). Higher pathogenicity is often associated with lower transmissibility, which may also apply to a specific virus of different subtypes and strains – the influenza virus is one such example.

The H1N1 virus (the outbreak of 2009) binds to receptors in the upper respiratory tract causing relatively mild disease (case fatality rate of 0.03%) and became endemic in the population. The H7N9 virus (the outbreak of 2013) binds to receptors in the lower respiratory tract and had a fatality rate of approximately 40%. The H7N9 has so far resulted in only a few small clusters of human-to-human transmission. Also, both measles virus and rhinovirus have strong transmissibility but a low mortality rate. Interestingly, SARS-CoV-2 uses the same receptor (ACE2) as CoV-NL63 (a coronavirus causing common colds in human adults and children) but causes disease of very different severity. With SARS-CoV-2, there has been some evidence suggesting that an individual with highly severe COVID-19 will only cause a few infections. By contrast, while the molecular mechanism is not understood, individuals with a moderate disease or latent infection can occasionally cause many infections. A potential consequence of this is that viral mutations that are of less importance at the individual level may eventually confer a high risk for the population. Thus, the current SARS-CoV-2 virus seems to have relatively low pathogenicity and moderate transmissibility. To further refine the risk assessment and subsequent risk management procedures to benefit SARS-CoV-2 mitigation and prevention efforts, further information on the biological and epidemiological characteristics of the virus are hastily needed. Anti-coronaviral drugs and vaccines are vigorously being pursued

(Colson et al. 2020; Grein et al. 2020; Amanat and Krammer 2020; Sanders et al. 2020).

21.5 Coronavirus Diagnostics and Medical Countermeasures

Molecular assay technologies help detect the presence of a virus by identifying a small section of the virus genome, then amplifying that portion until there is an adequate number of nucleotides for detection. These assays are available for the diagnosis of acute infection. This process can decrease testing wait time from days, if not hours, to as little as 5 min for positive results and 13 min for negative results. As previously mentioned, SARS-CoV-2 consists of four major proteins, including a protruding glycoprotein (spike), membrane, envelope, and a nucleocapsid. Given its multi-protein structure, scientists are utilizing reverse transcription-polymerase chain (PC) reaction (RT-PCR) assays and real-time quantitative PCR assays (qRT-PCR). The CDC provides test kits for public health laboratories in order to perform real-time (RT) PCR (rRT-PCR) detection for SARS-CoV-2. The protocol for CDC 2019-novel coronavirus (2019-nCoV) testing is intended for the possible qualitative detection of nucleic acid from the SARS-CoV-2. The traditional PCR test serves as a workhorse for diagnostics.

An antibody test helps identify a past infection. Antibody testing for the coronavirus allows for the determination of the scope of the disease it caused in various geographic areas. An enzyme-linked immunosorbent assay was recently developed using recombinant antigens derived from a portion of the spike protein of SARS-CoV-2. This assay is sensitive and specific and permits for the screening and identification of individuals showing COVID-19 seroconversion (the period during which a specific antibody response develops and antibodies become detectable in the blood) using human plasma/serum as early as 2 days following onset of symptoms (Amanat et al. 2020). The availability of therapeutic neutralizing antibodies against SARS-CoV-2 offers benefits for the treatment of patients during the current pandemic and the possible future re-emergence of the virus. However, the neutralization capability of serum from post-SARS-CoV-2-infected individuals is uncertain. The mechanism of action of the neutralizing antibodies is not fully understood, and there is a concern about the consequences of potential molecular interactions and pulmonary injury (Zhou and Zhao 2020).

Gao and associates from China recently reported the development of an inactivated SARS-CoV-2 virus vaccine candidate (PiCoVacc), shown to induce SARS-CoV-2-specific neutralizing antibodies in mice, rats, and nonhuman primates. The SARS-CoV-2 coronavirus was isolated from a COVID-19-infected patient, cultured in Vero cells, inactivated with β-propiolactone, and purified by Ion-exchange chromatography and size exclusion chromatography. Purified viruses were mixed with Al(OH)3 adjuvant to serve as the SARS-CoV-2 vaccine candidate (Zhang et al. 2020).

21.6 SARS-CoV-2 and the Human Host

Several discrete events are required for viral replication in a susceptible host following inhalation of aerosol particles released by an infected individual into the local environment of a second host. The first event following inhalation (primary route of infection) is the binding of the inhaled virus to specific cell receptors located on the external surface of cells in the alveoli, oral, and nasal mucosa. Additional organ involvement later in the disease process may include the stomach, small intestines, arterial and venous endothelial cells, brain, and kidneys. Upon binding of the SARS-CoV-2 virus to the specific receptors (angiotensin-converting enzyme 2, ACE2), a protease is required to act upon the spike protein, thereby facilitating the fusion of the virus coat membrane with the host cell membrane and uptake of the viral genome into the host cell. The RNA genome of the virus is then localized into the endoplasmic reticulum of the host cell with subsequent replication of the virus in the endoplasmic reticulum and release of mature enveloped virus from infected cells into the interstitial and vascular compartments of the host. This sequence of events is the first phase of the infectivity of a host with SARS-CoV-2.

Data now being accrued and analyzed indicate that the majority of individuals infected with SARS-CoV-2 (60–70%) recovers with minimal clinical symptoms, including fever, perception of breathing difficulties, and GI problems. The remaining group of infected individuals is comprised of those who have significant clinical illnesses and a subset of those who succumb to pulmonary, cardiovascular, and renal failure. This last cohort (those who die) exhibits multi-organ system failure associated with cytokine storms and are felled by dysregulated immune/inflammatory disease seen with toxic shock syndrome. A cytokine (small peptides that are important in cell signaling) storm is a hyperimmune response whereby too many cytokines are released into the blood very rapidly; this may be severe or life-threatening and lead to multiple organ failure.

We, therefore, will consider the problem from the perspective that there are potentially three stages of pathophysiology following exposure of human hosts to the SARS-CoV-2 agent. The first stage is the initiation of infection of a host by the agent, which includes the event of infection from a zoonotic source (bat or pangolin) or human vector. The second stage is the production of new SARS-CoV-2 virus particles in host organ systems, including the lungs, brain, and kidneys, and initiation of an immune response to the viral agent and control of the infectious process in the host. The third stage is the major inflammatory and hyperimmune toxic shock associated with a cytokine storm that can lead to death from multiple organ failure processes in a small percentage of the infected population. A fundamental concept that the authors wish to emphasize is the principle that clinical disease results from an interaction between a stressor (virus in the SARS-CoV-2 pandemic) and host vulnerability/susceptibility factors (e.g., MHC complex, inflammatory response factors in the current pandemic). The first phase of the initiation of infectivity, as described in the above paragraph, is a function primarily of the nature of the SARS-CoV-2 virus with some contribution from host genomic and phenotypic factors. The second and third phases primarily rely on host factors, including the molecular/structural properties of the serine protease, the MHC class 1 and 2 types of the immune defense system of the host, as well as of inflammatory response modifiers on activated macrophage and CD4+ and CD8+ T cells of the host.

21.7 Interaction Between the SARS-CoV-2 and the ACE2 Receptor-Binding Domain (RBD)

The next section will describe the first event of the infectious process, namely, the binding of the SARS-CoV-2 to the ACE2 surface marker on multiple organ systems, including the alveoli of the lung. Multiple lines of evidence demonstrate that the initial interaction of the SARS-CoV-2 with host alveolar epithelium occurs with the ACE2 RBD. The evidence consists of the demonstration that the amino acid sequence of the SARS-CoV-2 spike protein (S), between positions 331 to 524, is the region that binds to the ACE2 receptor-binding region (Tai et al. 2020; Wan et al. 2020). The primary region on human ACE2 that binds the spike protein is between 31 and 38. The primary region of the spike protein that binds ACE2 RBD is between 442 and 487 (Wan et al. 2020; Muth et al. 2018). The critical sites of interactions between the ACE2 RBD and the SARS-CoV-2 include electrostatic forces (e.g., E 484 on the spike with K31 on ACE; K417 with D 30) and pi or hydrogen bonding (e.g., Y 453 with H34; F486 with Y83) (Wang et al. 2020). The single-letter codes for the amino acids are as follows: E is glutamate; D is aspartate; K is lysine; Y is tyrosine; H is histidine; F is phenylalanine. Also, complexes consist of several aromatic or hydrophobic amino acids on the SARS-CoV-2 C-terminal domain (e.g., Y 489 and F 486) interacting with human ACE2 (e.g., F 28, L 79, M82, Y83) (Wang et al. 2020).

There is a 29 nucleotide (nt) region (ORF 8) involved in binding of the SARS-CoV-2 to the ACE2, and modification or deletion of this ORF 8 can result in altered binding and subsequent infection rates by the virus for humans (Muth et al. 2018). Significant insight into the interaction of SARS-CoV-2 with ACE2 was the demonstration of the binding interface between the viral protein and the ACE2 obtained from high-resolution examination of cryo-electron microscope images revealing the binding of SARS-CoV-2 to ACE2 (Wrapp et al. 2020; Yan et al. 2020).

The question then arose whether binding of SARS-CoV-2 to ACE2 is necessary and sufficient for infection of the cell by the agent. It was, therefore, necessary to explore the dissociation constant (Kd) of SARS-Co-V-2 to ACE2 and compare that with that of angiotensin to ACE2. The Kd of SARS-CoV-2 for the RBD of ACE2 is 15 mM (Jackson et al. 1986; Hoffmann et al. 2020), while the Kd for the ACE2 interaction with angiotensin ranges from 57 to 84 picomolar. Therefore, ACE RBD is many magnitudes more favorable to binding angiotensin than for SARS-CoV-2. It implies that there are factors other than the binding of SARS-CoV-2 to ACE2 involved in the initial stage of infection. Additional factors include a serine protease (TMPRSS2) that cleaves the S protein and thereby facilitates entry of SARS-CoV-2 RNA into the host cell (Hoffmann et al. 2020), the MHC organ transplant marker, and a glucose regulatory protein 78 (Ibrahim et al. 2020).

One of these additional factors is a serine protease in the vicinity of the RBD of cells to be infected. Upon binding of the SARS-CoV-2 to the CE2 RBD, a serine protease (TMPRSS2) cleaves the S protein of the viral spike into two fragments S1 and S2. The cleavage site where the protease hydrolyzes the S spike protein to the S1 and S2 fragments contains two adjacent arginine groups (Walls et al. 2020). The S2' site allows fusion of the viral membrane with the cellular membrane of the infected cell, while the SARS-CoV-2 S protein engages the ACE2 (Hoffmann et al. 2020; Li et al. 2005). The short time interval between the association of the SARS-CoV-2 S protein and subsequent cleavage of S to S1 and S2 and resultant membrane fusion provides an insight into the advantage of the SARS-CoV-2 for entry into the host cells. Even the Kd of ACE2 for angiotensin appears to lower the probability of binding of ACE by the virus. The rapid action of the protease with the immediate fusion of viral and cell membrane changes the problem from one of equilibrium on-off kinetics to one of compartmentation. Further evidence that the serine protease component of the host membrane is critical in the infectivity mechanism is the demonstration that serine protease inhibitors (camostat mesylate combined with E-64d) fully inhibit the entry of SARS-CoV-2 into cells. These compounds inhibited the uptake of SARS-CoV-2 into lung cells, further supporting the critical role of the serine protease in infectivity (Hoffmann et al. 2020).

The evidence that the initial step is the interaction between SARS-CoV-2 and ACE2 RBD of pulmonary epithelium comes from the cryo-electron microscopic ultrastructure revealing binding domains. Also, it is supported by the critical role of a serine protease in cleaving the viral spike protein, which results in the fusion of the viral membrane with the cell membrane and uptake of the viral genome into the endoplasmic reticulum of the cell where replication of the virus and subsequent release of membrane-bound viral particles into the interstitial space surrounding the infected cell occurs.

Additional factors contributing to the stabilization of the SARS-CoV-2 complex with ACE2 are the high surface area involved in the interaction between the virus and RBD of ACE2 and the significant number of electrostatic interactions (e.g., lysine or arginine with glutamate or aspartate) in the binding domains (Yan et al. 2020).

21.8 The Potential Roles that MHC Molecules Play in Infection

The prior section describes the first stage of infection of a host resulting from the inhalation of aerosol-containing viral particles from an infected carrier. That was the first stage of disease transmission. The second stage involves virus replication in the host and activation of immune responses to mediate viral clearance and end the viremic stage. The activation of the immune response involves the activation of B and T cells of the immune system. The immune vulnerability of the host to SARS-CoV-2 was proposed (Zhu et al. 2020) to be a result of genetic variation in the mutational rates of viral genomic regions of SARS-CoV, MERS-CoV, and SARS-CoV-2. They demonstrated the relative mutational rates of the viral genomic regions enriched for abundant CD4+ T cell epitopes, CD8+ T cell epitopes, and linear antibody epitopes in the three viruses. Of interest is that the mutational rate of CD8 epitope-enriched regions of 2019-nCoV is higher than those of CD4 epitope regions in SARS and MERS. These authors suggested that this observation results from the selection pressure "inflicted" by CD8+ T cytotoxic cells related to a function of MHC type. Their study compared the susceptibility of a European American (EA) population and a primarily Hong Kong (HK) Chinese population with MHC characteristics. The number of immunogenic binding epitopes across binned genomes (500 bp) was examined for the genomic position (nt), MHC 1 or 2 binder intensity, and MHC mutation rate. Differences in these parameters between the EA and HK populations, at least in part, explain different group susceptibility patterns. This study provides evidence that the MHC structure of the CD4+ T cells and CD8+ T cells appears to regulate the response of the immune system to clear the viral burden from the infected patients eventually. There is an extensive literature that describes the role of MHC type in determining the susceptibility of a host to viruses (HIV, hepatitis B and C, dengue fever, and shock), bacteria (*M. tuberculosis*), and a prion (variant Creutzfeldt-Jakob) (Chapman and Hill 2012).

Clearance of the infectious SARS-CoV-2 is dependent upon the activated CD4+ and the CD8+ T cells and the B cells as described above but also on circulating neutralizing antibodies in the serum (Casadevall and Pirofski 2020) (Zhang et al. 2005). Convalescent sera obtained from patients who have recovered from the viremic stage of the disease have been injected into patients during the first 3 days following infection with SARS-CoV-2. The timing of the injection of the neutralizing antisera appears to be valuable insofar as treatment during advanced clinical disease with COVID-19 is less efficacious. That may be because the titer of the neutralizing antibodies may not be sufficient to result in clearance or that access of the injected antibodies to the enormous viral load is limited (Casadevall and Pirofski 2020).

21.9 Cytokine Storm: Result of the Interaction of SARS-CoV-2 with ACE2

The literature is replete with information implicating a strong role that pathogenic T cells and inflammatory monocytes play in inciting a postinfectious cytokine storm in patients with COVID-19 and also in mice initially infected with various coronaviruses (Zhou et al. 2020; Perlman and Dandekar 2005). The clinical manifestations of the cytokine storm can include severe respiratory distress, including weak oxygen exchange in the pulmonary alveoli; gastrointestinal, renal, hepatic, and central nervous system disease (demyelination); disseminated intravascular coagulation (DIC); and multi-organ failure leading to death (Perlman and Dandekar 2005). It is referred to as a dysregulated immune response by Perlman.

The phenomena mentioned above occur during or after viral clearance, and immune cells (macrophages, CD4+ T cells, and CD8+ T cells) have infiltrated into the organs. The disease stage is a function of the host response to the viral insult rather than to the properties of the virus alone. The severe phase of disease occurs in 1–8% of the population infected with COVID-19. In patients with immune dysregulation resulting from COVID-19, the CD4+ T cells are activated to function as pathogenic T helper cells (TH1) that generate granulocyte-macrophage colony stimulating factor (GM-CSF). These cells produce cytokines and induce proinflammatory CD14 and CD16 monocytes that have a high expression of interleukin-6 (IL-6) (Zhou et al.

2020). The TH1 and inflammatory monocytes accumulate in the lung and cause immune injury (Zhou et al. 2020). Evidence that it is the postinfectious immune dysregulation rather than the primary viral infection that causes organ failure is the observation that monoclonal anti-TH1 or anti-IL-6 blocks the inflammatory response.

The sequence of events involved in the severe cytokine storm following viral infection is as follows: i, the introduction of double-stranded RNA into a vulnerable host where the double-stranded RNA is perceived as a pathogen by the innate immune system (Perlman and Dandekar 2005); ii, initiation of viral clearance by the host T cells that kill virally infected cells and B cells that produce soluble immunoglobulins that bind the viral particles; iii, generation of proinflammatory cytokines that attract neutrophils and macrophages to the sites of viral infection; and iv, release of cellular proteases (e.g., metalloproteinase) that enhance attack on infected cells but also noninfected cells in the vicinity of the inflammation. As Perlman and Dandekar (Perlman and Dandekar 2005) point out, the intense inflammation may compromise the normal function of the organs affected by the intense inflammation. Alternatively, autoimmune responses become dominant in and generate long-term disease of the central nervous system or other organ systems such as the liver and kidneys. The activated T cells may contribute to normal tissue injury nonspecifically (bystander activation). Normal tissue proteins that have epitopes similar to those of the virus could evoke immune responses to these epitopes in normal cells resulting in autoimmune disease that progresses over time and involves multiple organ systems. An example of a coronavirus infection that causes a demyelinating disease similar to multiple sclerosis is seen when murine hepatitis virus (MHC JHM 2.2 V-1) is injected into C57 BL/6 mice. The virus has tropism for oligodendrocytes (Perlman and Dandekar 2005). Demyelination and hind limb paralysis, similar to experimental allergic encephalomyelitis (EAE), are seen at 7 days postinfection. The virus is cleared by day 12 with clearing affected by the CD8+ and the CD4+ T cells. Gamma interferon is required for the demyelination and virus clearing by CD8+ T cells but not the CD4+ T cells. Lymphocyte and macrophage infiltration are seen in the demyelinated region. Activated macrophages and microglia are observed in the perivascular spaces in the region of demyelination (Perlman and Dandekar 2005). The demyelination has shown to result from dysregulation of the immune response rather than primarily from the viremia. The authors surmise that the processes described here are consistent with an autoimmune disease that results from the progression of the disease from a primary viral infectious disease to autoimmune processes in the cytokine storm phase.

21.10 Neural Pathology of Coronaviruses: From Experimental Studies to Human Studies

The single-stranded RNA coronavirus MHV-JHM 2.2 A1 has been shown to cause a multiple sclerosis-like CNS demyelination. It is of particular interest to us because of the demonstration that the renin-angiotensin system (RAS), including ACE2, is involved in the inflammatory disease myelin oligodendrocyte glycoprotein-induced allergic encephalomyelitis (MOG, a model or multiple sclerosis) (Stegbauer et al. 2009; Guo et al. 2017) and because of research done in one of our laboratories (SK) in the early 1970s as well. Stegbauer et al. observed that the RAS components, including renin, angiotensinogen, ACE2, angiotensin II type 1 receptor, and then type 2 receptor, are all present in the central nervous system (Stegbauer et al. 2009). The MOG EAE exhibits changes in Th1 responses, infiltration of macrophages, and antigen-presenting cells (Stegbauer et al. 2009). In this experimental model for MS, the RAS activity was increased, as was ATR 1 R and chemokine expression levels. Losartan, an inhibitor of the ATR1 R, blocked this inflammatory response. The presence of the ACE2 and related components of the RAS system in neurons occurred in some brain regions, for example, the motor cortex and raphe nucleus (Xia and Lazartigues 2008; Doobay et al. 2007; Abiodun and Ola 2020). These observations recalled our investigation into the potential role of viruses and the inflammatory response in the initiation of multiple sclerosis (MS) in humans (Johnson et al. 1976).

The Kornguth laboratory was funded by a contract from the National Institutes of Health to obtain the brains of patients who died with MS. The brains recovered postmortem were portioned into demyelinated regions and other non-demyelinated regions. Portions of each were processed for examinations in an electron microscope, while others were processed for separation in the cesium chloride density gradient. The spinal fluid from the patients obtained postmortem was cleared by centrifugation, and the pellets were examined. Of particular interest was the presence of crystals, paracrystalline arrays, and rigid tubules in the brain and spinal fluid pellets. In the brain sections examined by electron microscopy, 38- to 41-nm-wide tubules in the paracrystalline array were observed. In the pellets of the cerebrospinal fluid, these long tubules in the rigid array had internal diameters 41 to 45 nm. The proteins present in the tubules from the spinal fluid samples at a density of 1.24–1.26 and had the following discrete molecular weights: 40 kDa, 45 kDa, 88–130 kDa, and 300 kDa as determined by gradient gel electrophoresis (Johnson et al. 1976). Other investigators reported observing 15–26 nm paracrystalline arrays in the nuclei and cytoplasm of mononuclear cells or macrophages in perivenous demyelinating lesions of MS brain and attributed these to a paramyxovirus origin (Prineas 1972). A confirmatory report of the presence of 50–60 nm diameter hollow tubes in MS brain appeared in 1979 (Kirk 1979). Kirk (Kirk 1979) concluded that the tubular arrays were not of paramyxovirus origin but rather were proteinaceous aggregates that developed from host cell sources.

MS lesions frequently contain infiltrates of CD4+ and CD8+ T cells and increased levels of interferon-gamma, a cytokine that mainly plays a role in innate and adaptive immunity against infections, in particular, viral infections. Additionally, the MHC subtype affects the susceptibility of a host to MS. These factors are all seen in the development of severe SARS-CoV-2 disease, the third stage of the dysregulated immune initiated disease. In this chapter, the authors considered clinical disease manifestation to result from an interaction between stressor (virus in the case of SARS-CoV-2) and host vulnerability (including MHC levels, CD4+ and CD8+ T cells, increased permeability of the blood-brain barrier, and other comorbid stressors (concurrent existence of MS, traumatic brain injury (TBI), multi-organ disease)). From this perspective, the highly variable responses of persons to the SARS-CoV-2 virus ranging from no clinical disease in the majority of patients to death in 1–3 percent of the infected population may be more readily understood.

The presence of ACE2, other components of RAS, and appropriate receptors of ATr1R, as well as infiltration of monocytes and macrophages into lesions of the CNS in MOG and MS, suggests to the authors that common components of RAS confer susceptibility to the cytokine storm phase 3 of COVID-19 and to MS and related inflammatory responses in autoimmune diseases and TBI. These three related neuropathologies have been of significant interest to the authors (Demock and Kornguth 2019; Kornguth et al. 2017).

21.11 Conclusion

While fear plays a role in this economic and social situation, until we have an effective vaccine, human behavior will be vital to blunting the spread and impact of the disease. If the transmission of the virus from person-to-person could be substantially and dependently interrupted ($R_0 < 1$), then this outbreak could be controlled, but this achievement requires the efforts of all members of society. People should be tested for the presence of the coronavirus when they meet specific criteria: those admitted to the hospital and healthcare workers who have COVID-19 symptoms, people who have symptoms and who live in long-term care facilities, who are 65 years or older, first responders, or those more likely to get seriously ill because of other health conditions, such as heart, lung, or kidney disease, high blood pressure, diabetes, and cancer, or are on immunosuppression therapy. Currently, the duration and nature of immunity generated in response to SARS-CoV-2 infection are unknown. Serological assays will allow us to study the immune response(s) to SARS-CoV-2 qualitatively and quantitatively. Through the measurement of antibody responses and the determination of seroconversion, we should be able to establish if antibody titers correlate with protection.

The hope is that everyone continues to subscribe to and apply the scientific risk-based published guidelines for mitigation (https://www.cdc.gov/coronavirus/2019-ncov/prevent-getting-sick/index.html; https://www.cdc.gov/coronavirus/2019-ncov/prevent-getting-sick/prevention-H.pdf). If not, the resurgence of infections is likely. Any adjustment of strategies for mitigation and containment to control this pandemic and any potential future outbreaks must be scientifically and medically risk-based.

Acknowledgments This manuscript is dedicated to Margaret Kornguth and Evelyn Marie Hawley. The authors thank them for their lifelong support of the scientific and research endeavors of their respective spouses. EMH succumbed to coronavirus disease 2019 (COVID-19) on April 17, 2020. We want to graciously acknowledge Doctors David R. Franz and John T. Pinto for their critical and valuable review of the manuscript.

References

Abiodun OA, Ola MS (2020) Role of brain renin angiotensin system in neurodegeneration: an update. Saudi J Biol Sci 27(3):905–912. https://doi.org/10.1016/j.sjbs.2020.01.026

Amanat F, Krammer F (2020) SARS-CoV-2 vaccines: status report. Immunity 52(4):583–589. https://doi.org/10.1016/j.immuni.2020.03.007

Amanat F, Duehr J, Huang C, Paessler S, Tan GS, Krammer F (2020) Monoclonal antibodies with neutralizing activity and Fc-effector functions against the Machupo virus glycoprotein. J Virol 94(5). https://doi.org/10.1128/JVI.01741-19

Andersen KG, Rambaut A, Lipkin WI, Holmes EC, Garry RF (2020) The proximal origin of SARS-CoV-2. Nat Med 26(4):450–452. https://doi.org/10.1038/s41591-020-0820-9

Casadevall A, Pirofski LA (2020) The convalescent sera option for containing COVID-19. J Clin Invest 130(4):1545–1548. https://doi.org/10.1172/JCI138003

Chapman SJ, Hill AV (2012) Human genetic susceptibility to infectious disease. Nat Rev Genet 13(3):175–188. https://doi.org/10.1038/nrg3114

Chen J (2020) Pathogenicity and transmissibility of 2019-nCoV-A quick overview and comparison with other emerging viruses. Microbes Infect 22(2):69–71. https://doi.org/10.1016/j.micinf.2020.01.004

Colson P, Rolain JM, Raoult D (2020) Chloroquine for the 2019 novel coronavirus SARS-CoV-2. Int J Antimicrob Agents 55(3):105923. https://doi.org/10.1016/j.ijantimicag.2020.105923

Corman VM, Muth D, Niemeyer D, Drosten C (2018) Hosts and sources of endemic human coronaviruses. Adv Virus Res 100:163–188. https://doi.org/10.1016/bs.aivir.2018.01.001

Cornelissen LA, Wierda CM, van der Meer FJ, Herrewegh AA, Horzinek MC, Egberink HF, de Groot RJ (1997) Hemagglutinin-esterase, a novel structural protein of torovirus. J Virol 71(7):5277–5286

Cui J, Li F, Shi ZL (2019) Origin and evolution of pathogenic coronaviruses. Nat Rev Microbiol 17(3):181–192. https://doi.org/10.1038/s41579-018-0118-9

Demock M, Kornguth S (2019) A mechanism for the development of chronic traumatic encephalopathy from persistent traumatic brain injury. J Exp Neurosci 13:1179069519849935. https://doi.org/10.1177/1179069519849935

Doobay MF, Talman LS, Obr TD, Tian X, Davisson RL, Lazartigues E (2007) Differential expression of neuronal ACE2 in transgenic mice with overexpression of the brain renin-angiotensin system. Am J Physiol Regul Integr Comp Physiol 292(1):R373–R381. https://doi.org/10.1152/ajpregu.00292.2006

Fehr AR, Perlman S (2015) Coronaviruses: an overview of their replication and pathogenesis. Methods Mol Biol 1282:1–23. https://doi.org/10.1007/978-1-4939-2438-7_1

Gorbalenya AE, Baker SC, Baric RS, de Groot RJ, Drosten C, Gulyaeva AA, Haagmans BL, Lauber C, Leontovich AM, Neuman BW, Penzar D, Perlman S, Poon LLM, Samborskiy DV, Sidorov IA, Sola I, Ziebuhr J, Grp CS (2020) The species severe acute respiratory syndrome-related coronavirus: classifying 2019-nCoV and naming it SARS-CoV-2. Nat Microbiol 5(4):536–544. https://doi.org/10.1038/s41564-020-0695-z

Grein J, Ohmagari N, Shin D, Diaz G, Asperges E, Castagna A, Feldt T, Green G, Green ML, Lescure FX, Nicastri E, Oda R, Yo K, Quiros-Roldan E, Studemeister A, Redinski J, Ahmed S, Bernett J, Chelliah D, Chen D, Chihara S, Cohen SH, Cunningham J, D'Arminio Monforte A, Ismail S, Kato H, Lapadula G, L'Her E, Maeno T, Majumder S, Massari M, Mora-Rillo M, Mutoh Y, Nguyen D, Verweij E, Zoufaly A, Osinusi AO, DeZure A, Zhao Y, Zhong L, Chokkalingam A,

Elboudwarej E, Telep L, Timbs L, Henne I, Sellers S, Cao H, Tan SK, Winterbourne L, Desai P, Mera R, Gaggar A, Myers RP, Brainard DM, Childs R, Flanigan T (2020) Compassionate use of Remdesivir for patients with severe Covid-19. N Engl J Med 382:2327. https://doi.org/10.1056/NEJMoa2007016

Guo X, Namekata K, Kimura A, Harada C, Harada T (2017) The renin-angiotensin system regulates neurodegeneration in a mouse model of optic neuritis. Am J Pathol 187(12):2876–2885. https://doi.org/10.1016/j.ajpath.2017.08.012

He J, Tao H, Yan Y, Huang SY, Xiao Y (2020) Molecular mechanism of evolution and human infection with SARS-CoV-2. Viruses 12(4). https://doi.org/10.3390/v12040428

Hoffmann M, Kleine-Weber H, Schroeder S, Kruger N, Herrler T, Erichsen S, Schiergens TS, Herrler G, Wu NH, Nitsche A, Muller MA, Drosten C, Pohlmann S (2020) SARS-CoV-2 cell entry depends on ACE2 and TMPRSS2 and is blocked by a clinically proven protease inhibitor. Cell 181:271. https://doi.org/10.1016/j.cell.2020.02.052

Hui DS, E IA, Madani TA, Ntoumi F, Kock R, Dar O, Ippolito G, McHugh TD, Memish ZA, Drosten C, Zumla A, Petersen E (2020) The continuing 2019-nCoV epidemic threat of novel coronaviruses to global health - the latest 2019 novel coronavirus outbreak in Wuhan, China. Int J Infect Dis 91:264–266. https://doi.org/10.1016/j.ijid.2020.01.009

Ibrahim IM, Abdelmalek DH, Elshahat ME, Elfiky AA (2020) COVID-19 spike-host cell receptor GRP78 binding site prediction. J Infect 80:554. https://doi.org/10.1016/j.jinf.2020.02.026

Jackson B, McGrath BP, Matthews PG, Wong C, Johnston CI (1986) Differential renal function during angiotensin converting enzyme inhibition in renovascular hypertension. Hypertension 8(8):650–654. https://doi.org/10.1161/01.hyp.8.8.650

Johnson T, Knobeloch L, Sunderland E, Juhl U, Steinhart C, Scott G, Varkais J, Fuccillo D, Sever J, Kornguth S (1976) Crystals, paracrystals, and rigid tubules in multiple sclerotic brain and spinal fluid. Lab Investig 35(3):264–271

Kirk J (1979) Pseudoviral hollow-cored vesicles in multiple sclerosis brain. Acta Neuropathol 48(1):63–66. https://doi.org/10.1007/bf00691793

Klausegger A, Strobl B, Regl G, Kaser A, Luytjes W, Vlasak R (1999) Identification of a coronavirus hemagglutinin-esterase with a substrate specificity different from those of influenza C virus and bovine coronavirus. J Virol 73(5):3737–3743

Kornguth S, Rutledge N, Perlaza G, Bray J, Hardin A (2017) A proposed mechanism for development of CTE following concussive events: head impact, water hammer injury, Neurofilament release, and autoimmune processes. Brain Sci 7(12). https://doi.org/10.3390/brainsci7120164

Li F, Li W, Farzan M, Harrison SC (2005) Structure of SARS coronavirus spike receptor-binding domain complexed with receptor. Science 309(5742):1864–1868. https://doi.org/10.1126/science.1116480

Li Q, Guan X, Wu P, Wang X, Zhou L, Tong Y, Ren R, Leung KSM, Lau EHY, Wong JY, Xing X, Xiang N, Wu Y, Li C, Chen Q, Li D, Liu T, Zhao J, Liu M, Tu W, Chen C, Jin L, Yang R, Wang Q, Zhou S, Wang R, Liu H, Luo Y, Liu Y, Shao G, Li H, Tao Z, Yang Y, Deng Z, Liu B, Ma Z, Zhang Y, Shi G, Lam TTY, Wu JT, Gao GF, Cowling BJ, Yang B, Leung GM, Feng Z (2020) Early transmission dynamics in Wuhan, China, of novel coronavirus-infected pneumonia. N Engl J Med 382(13):1199–1207. https://doi.org/10.1056/NEJMoa2001316

Muth D, Corman VM, Roth H, Binger T, Dijkman R, Gottula LT, Gloza-Rausch F, Balboni A, Battilani M, Rihtaric D, Toplak I, Ameneiros RS, Pfeifer A, Thiel V, Drexler JF, Muller MA, Drosten C (2018) Attenuation of replication by a 29 nucleotide deletion in SARS-coronavirus acquired during the early stages of human-to-human transmission. Sci Rep 8(1):15177. https://doi.org/10.1038/s41598-018-33487-8

Neuman BW, Kiss G, Kunding AH, Bhella D, Baksh MF, Connelly S, Droese B, Klaus JP, Makino S, Sawicki SG, Siddell SG, Stamou DG, Wilson IA, Kuhn P, Buchmeier MJ (2011) A structural analysis of M protein in coronavirus assembly and morphology. J Struct Biol 174(1):11–22. https://doi.org/10.1016/j.jsb.2010.11.021

Nieto-Torres JL, DeDiego ML, Verdia-Baguena C, Jimenez-Guardeno JM, Regla-Nava JA, Fernandez-Delgado R, Castano-Rodriguez C, Alcaraz A, Torres J, Aguilella VM, Enjuanes L (2014) Severe acute respiratory syndrome coronavirus envelope protein ion channel activity promotes virus fitness and pathogenesis. PLoS Pathog 10(5):e1004077. https://doi.org/10.1371/journal.ppat.1004077

Perlman S, Dandekar AA (2005) Immunopathogenesis of coronavirus infections: implications for SARS. Nat Rev Immunol 5(12):917–927. https://doi.org/10.1038/nri1732

Prineas J (1972) Paramyxovirus-like particles associated with acute demyelination in chronic relapsing multiple sclerosis. Science 178(4062):760–763. https://doi.org/10.1126/science.178.4062.760

Sanche S, Lin YT, Xu C, Romero-Severson E, Hengartner N, Ke R (2020) High contagiousness and rapid spread of severe acute respiratory syndrome coronavirus 2. Emerg Infect Dis 26(7):1470. https://doi.org/10.3201/eid2607.200282

Sanders JM, Monogue ML, Jodlowski TZ, Cutrell JB (2020) Pharmacologic treatments for coronavirus disease 2019 (COVID-19): a review. JAMA. https://doi.org/10.1001/jama.2020.6019

Stegbauer J, Lee DH, Seubert S, Ellrichmann G, Manzel A, Kvakan H, Muller DN, Gaupp S, Rump LC, Gold R, Linker RA (2009) Role of the renin-angiotensin system in autoimmune inflammation of the central nervous system. Proc Natl Acad Sci U S

A 106(35):14942–14947. https://doi.org/10.1073/pnas.0903602106
Tai W, He L, Zhang X, Pu J, Voronin D, Jiang S, Zhou Y, Du L (2020) Characterization of the receptor-binding domain (RBD) of 2019 novel coronavirus: implication for development of RBD protein as a viral attachment inhibitor and vaccine. Cell Mol Immunol 17:613. https://doi.org/10.1038/s41423-020-0400-4
Walls AC, Park YJ, Tortorici MA, Wall A, McGuire AT, Veesler D (2020) Structure, function, and antigenicity of the SARS-CoV-2 spike glycoprotein. Cell 181(2):281–292. e286. https://doi.org/10.1016/j.cell.2020.02.058
Wan Y, Shang J, Graham R, Baric RS, Li F (2020) Receptor recognition by the novel coronavirus from Wuhan: an analysis based on decade-long structural studies of SARS coronavirus. J Virol 94(7). https://doi.org/10.1128/JVI.00127-20
Wang Q, Zhang Y, Wu L, Niu S, Song C, Zhang Z, Lu G, Qiao C, Hu Y, Yuen KY, Wang Q, Zhou H, Yan J, Qi J (2020) Structural and functional basis of SARS-CoV-2 entry by using human ACE2. Cell 181:894. https://doi.org/10.1016/j.cell.2020.03.045
Woo PC, Huang Y, Lau SK, Yuen KY (2010) Coronavirus genomics and bioinformatics analysis. Viruses 2(8):1804–1820. https://doi.org/10.3390/v2081803
Wrapp D, Wang N, Corbett KS, Goldsmith JA, Hsieh CL, Abiona O, Graham BS, McLellan JS (2020) Cryo-EM structure of the 2019-nCoV spike in the prefusion conformation. Science 367(6483):1260–1263. https://doi.org/10.1126/science.abb2507
Xia H, Lazartigues E (2008) Angiotensin-converting enzyme 2 in the brain: properties and future directions. J Neurochem 107(6):1482–1494. https://doi.org/10.1111/j.1471-4159.2008.05723.x
Yan R, Zhang Y, Li Y, Xia L, Guo Y, Zhou Q (2020) Structural basis for the recognition of SARS-CoV-2 by full-length human ACE2. Science 367(6485):1444–1448. https://doi.org/10.1126/science.abb2762
Zhang JS, Chen JT, Liu YX, Zhang ZS, Gao H, Liu Y, Wang X, Ning Y, Liu YF, Gao Q, Xu JG, Qin C, Dong XP, Yin WD (2005) A serological survey on neutralizing antibody titer of SARS convalescent sera. J Med Virol 77(2):147–150. https://doi.org/10.1002/jmv.20431
Zhang J, Zeng H, Gu J, Li H, Zheng L, Zou Q (2020) Progress and prospects on vaccine development against SARS-CoV-2. Vaccines (Basel) 8(2). https://doi.org/10.3390/vaccines8020153
Zhou G, Zhao Q (2020) Perspectives on therapeutic neutralizing antibodies against the novel coronavirus SARS-CoV-2. Int J Biol Sci 16(10):1718–1723. https://doi.org/10.7150/ijbs.45123
Zhou YF, Fu B, Zheng X, Wang D et al (2020) Pathogenic T cells and Inflammatory monocytesincite inflammatory storm in severe COVID-19 patients. Nat Sci Rev 7(6):998–1002. https://doi.org/10.1093/nsr/nwaa041
Zhu J, Kim J, Xiao X, Wang Y, Luo D, Chen R, Xu L, Zhang H, Xiao G, Schoggins J (2020) The immune vulnerability landscape of the 2019 Novel Coronavirus, SARS-CoV-2. bioRxiv. https://doi.org/10.1101/2020.02.08.939553

Steve Kornguth earned his Bachelor of Arts (BA) at Columbia College in New York City and his PhD in biochemistry at the University of Wisconsin-Madison. He was a professor of neurology and biomolecular chemistry at the University of Wisconsin between 1962 and 1998. His research during this period related to neural development, autoimmune diseases, and the development of binding agents for sensors and magnetic resonance image contrast materials. In 1998, Prof. Kornguth relocated to The University of Texas, Austin, and led federally funded programs in biological defense and also on the effects of sleep deprivation on decision-making in soldiers and athletes. His latest research involved head impacts during competitive athletic events on neural structures and protein biomarkers. He served as director of neuroscience programs at the National Science Foundation for 2 years and the Office of Naval Research. Prof. Kornguth also worked as a consultant for Perspecta Engineering in Chantilly, VA, on biological security. Prof. Kornguth passed on September 21, 2020, following a battle with pancreatic cancer.

Clinical and Laboratory Predictors of Severity, Criticality, and Mortality in COVID-19: A Multisystem Disease

22

Bahareh Gholami, Samira Gholami, Amir Hossein Loghman, Behzad Khodaei, Simin Seyedpour, Nasrin Seyedpour, Amene Saghazadeh, and Nima Rezaei

Abstract

Coronavirus disease 2019 (COVID-19) pandemic continues devastating effects on healthcare systems. Such a crisis calls for an urgent need to develop a risk stratification tool. The present chapter aimed to identify laboratory and clinical correlates of adverse outcomes in patients with COVID-19. To this end, we conducted a systematic evaluation of studies that investigated laboratory abnormalities in patients with COVID-19 and compared i. patients with a severe form of disease and patients with a non-severe form of the disease, ii. patients who were in critical condition and patients who were not in critical condition, and iii. patients who survived and patients who died. We included 54 studies in the data synthesis. Compared to patients with a non-severe form of COVID-19, patients

Bahareh Gholami and Samira Gholami contributed equally with all other contributors.

B. Gholami
Systematic Review and Meta-analysis Expert Group (SRMEG), Universal Scientific Education and Research Network (USERN), Tehran, Iran

School of Medicine, Tehran University of Medical sciences, Tehran, Iran

S. Gholami
Systematic Review and Meta-analysis Expert Group (SRMEG), Universal Scientific Education and Research Network (USERN), Tehran, Iran

School of Medicine, Shahid Beheshti University of Medical Sciences, Tehran, Iran

A. H. Loghman
Systematic Review and Meta-analysis Expert Group (SRMEG), Universal Scientific Education and Research Network (USERN), Tehran, Iran

School of Medicine, Kashan University of Medical Sciences, Kashan, Iran

B. Khodaei · S. Seyedpour
Systematic Review and Meta-analysis Expert Group (SRMEG), Universal Scientific Education and Research Network (USERN), Tehran, Iran

School of Medicine, Tehran University of Medical sciences, Tehran, Iran

Nanomedicine Research Association (NRA), Universal Scientific Education and Research Network (USERN), Tehran, Iran

N. Seyedpour
Systematic Review and Meta-analysis Expert Group (SRMEG), Universal Scientific Education and Research Network (USERN), Tehran, Iran

Nanomedicine Research Association (NRA), Universal Scientific Education and Research Network (USERN), Tehran, Iran

Department of Medical Physics and Biomedical Engineering, Tehran University of Medical Sciences, Tehran, Iran

N. Rezaei (ed.), *Coronavirus Disease - COVID-19*, Advances in Experimental Medicine and Biology 1318, https://doi.org/10.1007/978-3-030-63761-3_22

who had a severe form of disease revealed higher values for white blood cells (WBC), polymorphonuclear leukocytes (PMN), total bilirubin, alanine aminotransferase (ALT), creatinine, troponin, procalcitonin, lactate dehydrogenase (LDH), and D-dimer. By contrast, platelet count, lymphocyte count, and albumin levels were decreased in patients with a severe form of COVID-19. Also, patients with a severe phenotype of disease were more likely to have diabetes, chronic heart disease, chronic obstructive pulmonary disease (COPD), cerebrovascular disease, hypertension, chronic kidney disease (CKD), and malignancy. Compared to patients who survived, patients who died had higher WBC, PMN, total bilirubin, ALT, procalcitonin, IL-6, creatinine, PT, lymphocyte count, platelet count, and albumin. Also, non-survivors revealed a higher prevalence of diabetes, chronic heart disease, COPD, cerebrovascular disease, and CKD. Meta-analyses identified several laboratory parameters that might help the prediction of severe, critical, and lethal phenotypes of COVID-19. These parameters correlate with the immune system function, inflammation, coagulation, and liver and kidney function.

A. Saghazadeh
Systematic Review and Meta-analysis Expert Group (SRMEG), Universal Scientific Education and Research Network (USERN), Tehran, Iran

Research Center for Immunodeficiencies, Children's Medical Center, Tehran University of Medical Sciences (TUMS), Tehran, Iran

N. Rezaei (✉)
Research Center for Immunodeficiencies, Children's Medical Center, Tehran University of Medical Sciences (TUMS), Tehran, Iran

Department of Immunology, School of Medicine, Tehran University of Medical Sciences, Tehran, Iran
Network of Immunity in Infection, Malignancy and Autoimmunity (NIIMA), Universal Scientific Education and Research Network (USERN), Tehran, Iran
e-mail: rezaei_nima@tums.ac.ir

Keywords

Biomarker · Clinical · COVID-19 · D-dimer · Diagnostic · Immune system · Inflammation · Interleukin 6 · Laboratory · Mortality · Prognostic · Severity · Systematic review · Meta-analysis

22.1 Introduction

In December 2019, pneumonia cases with unknown etiology emerged in Wuhan, China (Lu et al.). Lower respiratory tract sampling and sequencing process could reveal a novel coronavirus from the family of betacoronavirus, which was subsequently named 2019 novel coronavirus (2019-nCoV) (Zhu et al. 2020). This family also included severe acute respiratory syndrome (SARS) and the Middle East respiratory syndrome (MERS), which have caused outbreaks over the last two decades (Ksiazek et al. 2003; de Wit et al. 2016; Jabbari et al. 2020). Coronavirus disease 2019 (COVID-19) has rapidly spread throughout the world, posing a global emergency (Hanaei and Rezaei 2020; Rezaei 2020a).

The Chinese Centers for Disease Control and Prevention (CDC) has suggested classifying COVID-19 patients according to their clinical state into three categories. Mild disease manifestations include no pneumonia or mild pneumonia and occur in 81% of cases. Severe disease characterized by dyspnea, respiratory frequency ≥ 30/min, blood oxygen saturation (SpO2) $\leq 93\%$, PaO2/FiO2 ratio < 300, and/or lung infiltrates $>50\%$ within 24–48 h is present in 14% of cases. Critical conditions accompanied by respiratory failure, septic shock, and/or multiple organ dysfunction or failure occur in 5% of cases (The Novel Coronavirus Pneumonia Emergency Response Epidemiology 2020). While most patients experience mild respiratory illness, some cases, especially the old population and patients with comorbid conditions, develop a severe form of disease requiring intensive management (Guan et al. 2020b; Lotfi and Rezaei 2020). Chinese CDC has reported an overall case fatality rate of

2.3%, with the majority of deceased patients being 60 years old and above and having preexisting comorbid conditions such as hypertension, cardiovascular disease, and diabetes (The Novel Coronavirus Pneumonia Emergency Response Epidemiology 2020; Shamshirian and Rezaei 2020). Accordingly, the resistance and susceptibility to disease are governed by genetic factors and nongenetic factors such as biological age, pregnancy, and chronic conditions, which affect the expression of the receptors which mediate virus entry, i.e., angiotensin-converting enzyme 2 (ACE2), and the function of the immune system (Yousefzadegan and Rezaei 2020; Ahmadi et al. 2020; Ahanchian et al. 2020; Darbeheshti and Rezaei 2020; Babaha and Rezaei 2020; Mirbeyk and Rezaei 2020).

With the increasing number of people infected with the virus, the burden of COVID-19 on healthcare systems has been overwhelming. Thus, early and efficient detection of high-risk patients is a priority for maximizing the utilization of limited resources, while current diagnostics are not as rapid and as reliable as the current pandemic condition needs (Basiri et al. 2020a). Furthermore, no clinically proven specific antiviral agent has been confirmed for COVID-19. However, approaches targeting the immune system and virus entry, computational methods of drug discovery, regenerative medicine, medical biotechnology, and picotechnology offer hope for the future amid the pandemic (Mohamed et al. 2020b; Rezaei 2020b; Rabiee et al. 2020; Basiri et al. 2020b; Sharifkashani et al. 2020; Lotfi et al. 2020; Mansourabadi et al. 2020; Pashaei and Rezaei 2020; Fathi and Rezaei 2020; Jahanshahlu and Rezaei 2020b). Thus supportive management, including oxygen therapy and conservative fluid management, along with anti-inflammatory and antiviral agents, remains to be the most important management strategy (Cascella et al. 2020; Saghazadeh and Rezaei 2020b). Dynamic changes of laboratory parameters, as well as clinical and imaging findings of patients infected with COVID-19, have been depicted in the literature (Zhou et al. 2020a; Liu et al. 2020a). Results are, however, not consistent. The present chapter aims to identify laboratory and clinical correlates of adverse outcomes in patients with COVID-19. For this purpose, a systematic evaluation of literature was conducted according to the Preferred Reporting Items for Systematic Reviews and Meta-Analyses (PRISMA).

22.2 Methods

22.2.1 Search Strategy

The published literature was searched from December 2019 to April 20, 2020. Databases including EMBASE, MEDLINE via PubMed, and Scopus were searched for observational studies of COVID-19. The keywords for search strategy included "Wuhan seafood market pneumonia virus," "COVID19," "2019-nCoV," "novel coronavirus," "coronavirus disease-19," "severe acute respiratory syndrome coronavirus 2," "SARS-CoV-2," "characteristic," "finding," "lab," "imaging," "comorbidit*," "sever*," "critical," "hospital*," "ICU," "non-ICU," "death," "mortality," and "surviv*." Detailed search strategies are presented in Table 22.1.

22.2.2 Study Selection

The studies identified were imported into an EndNote library. After duplicate removal, the two groups (three reviewers in each group) performed the screening titles and abstracts for inclusion and then reviewing the full text of potentially relevant articles. Any questions or conflicts were resolved through discussion with the members of the other group. Concerning studies with overlapping samples, the study that had the largest sample size was included in the final analysis. Studies that i. included subjects with confirmed COVID-19 based on laboratory testing; ii. evaluated individual laboratory characteristics, symptoms, and comorbidities in predicting severe COVID-19 infection or mortality caused by COVID-19 infection; and iii. were written in English were eligible to be included in the systematic review. If the criteria for disease severity were not determined, then the study was excluded

Table 22.1 Search strategy

Database	Search date	Search strategy	Number of search results
Embase	March 20, 2020	(characterist* OR finding$ OR feature$ OR lab OR labrat* OR imaging? OR radiolog* OR "ct scan" OR "computed tomograph*" OR clinical OR comorbidit* OR symptom* OR discript*) AND (sars2 OR "covid-19" OR "2019-ncov" OR "coronavirus disease 2019" OR "novel coronavirus" OR "coronavirus disease-19" OR "severe acute respiratory syndrome coronavirus 2" OR "sars-cov-2" OR "Wuhan coronavirus" OR "ncov-2019" OR "Wuhan seafood market pneumonia virus" OR "covid19 virus") AND ((sever* OR critical OR hospital* OR icu OR intensive) AND care OR death OR mortality OR surviv* OR fatal* OR "disease progression" OR "invasive management" OR outcome OR predic* OR prognos* OR "risk factor" OR (risk NEAR/2 factor NEAR/2 (mortality OR sever*)) OR correlat* OR (risk NEAR/2 death) OR "non-icu" OR "mechanical ventilation" OR "respiratory failure") AND [english]/lim AND [2019–2020]/py	162
Scopus	March 20, 2020	(characterist* OR finding OR feature OR lab OR labrat* OR imaging OR radiolog* OR ct_scan OR computed_ tomograph* OR clinical OR comorbidit* OR symptom* OR discript*) AND (sars2 OR covid_19 OR 2019_ncov OR coronavirus_disease_2019 OR novel_coronavirus OR coronavirus_disease-19' OR severe_acute_respiratory_syndrome_ coronavirus_2 OR sars_cov_2 OR wuhan_coronavirus OR ncov_2019 OR wuhan_seafood_market_pneumonia_virus OR covid19_virus) AND ((sever* OR critical OR hospital* OR icu OR intensive) AND care OR death OR mortality OR surviv* OR fatal* OR 'disease AND progression' OR invasive_management OR outcome OR predic* OR prognos* OR risk_factor OR (risk W/2 factor W/2 (mortality OR sever*)) OR correlat* OR (risk W/2 death) OR non_icu OR mechanical_ventilation OR respiratory_failure) AND (LIMIT-TO (PUBYEAR, 2020) OR LIMIT-TO (PUBYEAR, 2019)) AND (LIMIT-TO (LANGUAGE, "English")) AND (LIMIT-TO (SUBJAREA, "MEDI"))	195
PubMed	March 20, 2020	((((characteristic) OR (finding) OR (feature) OR (lab) OR (laboratory) OR (imaging) OR (radiolog*) OR (CT scan) OR (computed tomography) OR (clinical) OR (comorbidit*) OR (symptom))) AND ((SARS2) OR (Wuhan seafood market pneumonia virus) OR (COVID19 virus) OR(COVID-19) OR (2019-nCoV) OR (coronavirus disease 2019) OR (novel coronavirus) OR (coronavirus disease-19) OR (severe acute respiratory syndrome coronavirus 2) OR (SARS-CoV-2) OR (Wuhan coronavirus) OR (nCoV-2019))) AND ((sever*) OR (critical) OR (hospital*) OR (ICU) OR (non-ICU) OR (intensive care) OR (death) OR (mortality) OR (surviv*) OR (fatal*) OR (disease progression) OR (invasive management) (predic*) OR (prognosis)OR (risk factor)OR (risk factor n2 mortality) OR (risk factor n2 sever*) OR (correlat*) OR (influencing factors) OR (risk n2 death) OR (outcome) OR (PaO2/FiO) OR (mechanical ventilation) OR (respiratory failure))	104

Table 22.2 The definition of outcome and population

	Definition
Laboratory-confirmed case	The laboratory-confirmed case was defined as the presence of SARS-CoV-2 in respiratory specimens (including nasal and pharyngeal swabs) detected by the reverse-transcriptase polymerase chain reaction (RT-PCR), according to the protocol established by the World Health Organization
Severe disease	The following criteria defined severe disease: 1) respiratory distress (respiratory rate ≥ 30 breaths/minute); 2) an oxygen saturation (SpO2) ≤ 93% in the resting state; 3) hypoxemia defined as an arterial partial pressure of oxygen divided by the fraction of inspired oxygen (PaO2/FiO2 ratio) ≤ 300 mmHg; 4) the presence of radiographic progression, defined as ≥50% increase of target lesion within 24–48 hours. The criterion was based on the Chinese COVID-19 prevention and control program (seventh edition, http://en.nhc.gov.cn/2020-03/29/c_78469.htm) and American Thoracic Society guideline (Mandell et al. 2007) and World Health Organization interim guidance

(Table 22.2). Also, types of articles other than original observational studies, e.g., opinion, editorial, case report, book chapter, review, and meta-analysis, studies confined to the pediatric population or pregnant women, and studies rated as low-quality evidence were not included.

22.2.3 Data Extraction and Quality Assessment

We used the Newcastle-Ottawa scale (NOS) and its modified version for the assessment of the quality of retrospective cohort studies and cross-sectional studies. This scale gives the maximum score of 9 points for the least risk of bias in four domains: the selection of study groups (4 points), comparability of groups (2 points), and ascertainment of exposure and outcomes (3 points). Age and sex were defined as the most relevant covariate for comparability. For the cross-sectional studies, we used the modified NOS, which gives a maximum of 10 scores for the least risk of bias in three domains: the selection of study groups (5 points), comparability of groups (2 points), and ascertainment of exposure and outcomes (3 points).

Three authors independently extracted the data by a predesigned spreadsheet in Excel, including the author's name, publication year, setting (country/hospital name), study design, sample size, clinical and laboratory and comorbidities, and severity criteria.

22.2.4 Data Synthesis

All between-group meta-analyses were performed using Comprehensive Meta-Analysis. For the continuous type of outcome, we entered the number of participants in two groups and the mean and standard deviation (SD) of laboratory data. We used methods described in Luo et al. (2018) and Wan et al. (2014) to estimate mean and SD based on median (IQR). A meta-analysis was conducted for data on laboratory and clinical characteristics, comorbidities, and vital status when there were at least three between-group comparisons for each variable. Because studies used different measurement scales, we employed the standardized mean difference (SMD) as effect size (ES) measure. For categorical data, we used the dichotomous method to calculate the odds ratio (OR) with its 95% confidence interval as ES measure. As explained by Higgins and Green et al., the ES of 0.2, 0.5, and 0.8 represent small, moderate, and large effect estimates, respectively (Higgins and Wells 2011).

We investigated heterogeneity across studies by Cochran's Q test, which is calculated by the weighted sum of squared differences between individual study ES estimates. A P-value of 0.10 or less indicates the presence of heterogeneity. The I2 index was used to get a more precise estimate of heterogeneity. If the I2 is less than 40%, the heterogeneity is considered to be not significant. The fixed-effects model is the preferred method of meta-analysis. However, if I2 estimates exceeded 40%, the random-effects approach was applied as the meta-analysis model. We based our analysis on the different outcome stratification including critical/ICU vs noncritical/non-ICU, severe/critical vs non-severe, and survivors vs non-survivors. Subgroup analysis was performed to explore the potential sources for heterogeneity. We also applied univariate meta-regression analyses to investigate the effect of potential moderators, e.g., sample size, the difference in the mean age of two groups (years), and percentage of female patients on the effect size and heterogeneity. Finally, to identify studies exerting excessive influence on the results or causing high heterogeneity among studies, we conducted a sensitivity analysis by systematically excluding each study at a time and then rerunning the analysis to assess the change in ES.

We assessed publication bias visually by the degree of funnel plot asymmetry and used Begg-Mazumdar Kendall's tau (Begg and Mazumdar 1994) and Egger bias test to quantify it (Duval and Tweedie 2000). A P-value of 0.10 or less suggests the presence of publication bias across studies. The trim and fill method was used to adjust effect sizes in cases that there was evidence of publication bias (Duval and Tweedie 2000).

22.3 Results

22.3.1 Characteristics of Selected Studies

The database search resulted in 461 records. After the removal of duplicates (n = 60) and retrieval of additional 14 studies through hand search of additional references, the title/abstract of 415 search results was screened for potential eligibility. Three hundred forty papers were excluded through screening. Seventy-five full-text articles were reviewed in detail, and 32 met the inclusion criteria. Moreover, an updated search was conducted and resulted in the inclusion of 22 studies. In total, 54 studies were included in the meta-analysis. Figure 22.1 provides an overview of study selection for systematic review and meta-analysis according to PRISMA guidelines.

22.3.2 Study and Patient Characteristics

As summarized in Table 22.3, all included studies were from China, except two studies, which were from Singapore (Young et al. 2020; Fan et al. 2020). All studies were designed as retrospective cohort studies except two with a prospective cohort design (Liu et al. 2020b; Momtazmanesh et al. 2020) and one which was a cross-sectional study (Zhang et al. 2020b). Tables 22.4 and 22.5 show the results of the quality assessment for studies included in the data synthesis.

There were 16 studies that compared critical or severe vs non-severe patients (Zhao et al. 2020; Zhang et al. 2020a; Guan et al. 2020b; Li et al. 2020a; Zhou et al. 2020b; Cai et al.; Tian et al. 2020; Liu et al. 2020b, d; Wu et al. 2020; Feng et al. 2020c; Wan et al. , 2020; Xu et al. 2020; ; Han et al. 2020b; Wang et al. 2020d); 3 studies compared 3 groups of patients, namely, critical, severe, and non-severe (Feng et al. 2020b; Liu et al. 2020c; Han et al. 2020a); 20 studies compared severe vs non-severe patients (Qu et al. 2020; Zhang et al. 2020b, c; Wang et al. 2020e; Cao 2020; Gao et al. 2020; Chen et al. 2020a, e; Liu et al. 2020a; lei and Jian-ya 2020; Lu et al. 2020; Chu et al. 2020; Ji et al. 2020; Xie et al. 2020; Zheng et al. 2020a; c; Li et al. 2020c; Gong et al. 2020; Young et al. 2020; Qin et al. 2020; Nie et al. 2020); 6 studies compared critical/ICU vs noncritical/non-ICU patients (Wang et al. 2020a; Fan et al. 2020; Du et al. 2020; Yan et al. 2020; Huang et al. 2020; Chen et al. 2020b); and 9 studies compared survivor vs non-survivor patients (Tang et al. 2020; Yang et al. 2020; Zhou et al. 2020a; Wu et al. 2020; Momtazmanesh et al. 2020; Chen et al. 2020c, d; Li et al. 2020c; Wang et al. 2020c; Cao et al. 2020). Two studies were not entered into analysis because of the shared samples (Chen et al. 2020c; Huang et al. 2020). A total of 8603 patients were included in the meta-analysis. Tables 22.6, 22.7, 22.8, 22.9, 22.10, 22.11, and 22.12 provide a summary of the

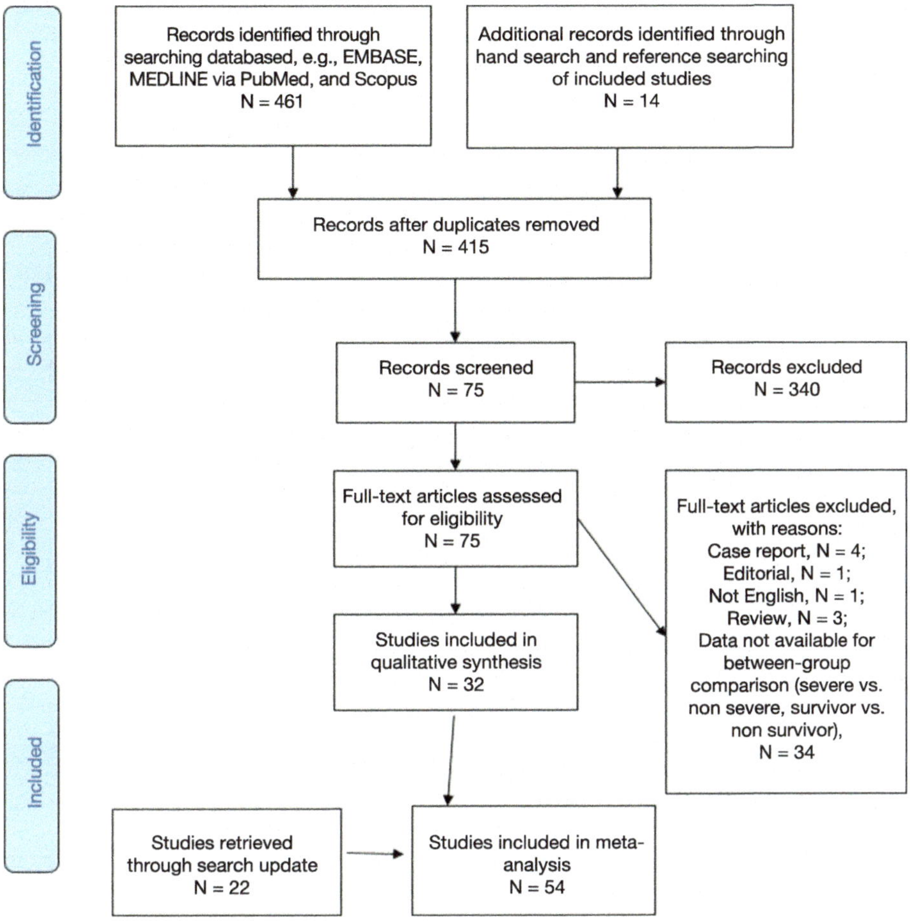

Fig. 22.1 PRISMA flow diagram of study selection

results of meta-analyses, subgroup analyses, and meta-regressions.

22.3.3 Laboratory Findings

22.3.3.1 Severity Analysis

Twenty-seven studies provided data on white blood cell (WBC) values for patients with severe (n = 1131) and non-severe (n = 3293) COVID-19. Meta-analysis showed that patients with non-severe outcome had lower WBC compared to patients with severe COVID-19 (SMD −0.40; 95% CI, −0.58, −0.21; P < 0.0001). There was no evidence of publication bias (Begg's P = 0.77; Egger's P = 0.10). Significant heterogeneity existed across studies (I2 = 84%), and therefore, the random-effects model was applied. Meta-regression analyses demonstrated no significant effect of percentages of female patients, the difference in the mean age, and sample size on the heterogeneity. Neither subgroup meta-analyses nor sensitivity analysis showed a significant effect of specific outcome or study on the effect size and related heterogeneity. Meta-analysis of data on four studies showed that critical COVID-19 patients (n = 131) had higher WBC than noncritical COVID-19 patients (n = 348) (SMD −0.58; 95% CI, −0.79, −0.37; P < 0.0001). There was no significant heterogeneity and evidence of publication bias as well.

Pooled data from 22 studies, including 2215 non-severe and 1109 severe COVID-19 patients, demonstrated that polymorphonuclear (PMN)

Table 22.3 Characteristics of studies included in the systematic review

Study	Setting			Sample size	Outcomes	Adverse outcome			Non-adverse outcome		
	Country	City	Hospital			Patients (%)	Age	Female %	Patients (%)	Age	Female %
Zhao et al.	China	Four institutions in Hunan province		101	Non-severe vs severe	14 (13.86%)	53.71 (12.6)	44.83%	87 (86.14%)	42.94 (11.7)	42.86%
Zhang et al.	China	Wuhan	Zhongnan Hospital of Wuhan University	221	Non-severe vs severe	55 (24.89%)	62.0 (52.0–74.0)	36.36%	166 (75.11%)	51 (36.0–64.3)	56.02%
Guan et al.	China	552 hospitals in 30 provinces		1099	Non-severe vs severe	173 (15.74%)	52.0 (40.0–65.0)	42.20%	926 (84.26%)	45.0 (34.0–57.0)	41.82%
Li et al.	China	Multicenter		83	Non-severe vs severe	25 (30.12%)	53.7 (12.3)	40.00%	58 (69.88%)	41.9 (10.6)	50.00%
Zhou et al.	China	Wuhan	CentralHospital of Wuhan	377	Non-severe vs severe	117 (31.03%)	65.63 (14.03)	41.88%	260 (68.97%)	48.35 (16.17)	60.77%
Cai et al.	China	Shenzhen	Third People's Hospital of Shenzhen	298	Non-severe vs severe	58 (19.46%)	62.5 (56–66)	43.10%	240 (80.54%)	41 (31–56)	53.75%
Tian et al.	China	57 hospitals across Beijing		262	Non-severe vs severe	46 (17.56%)	61.4 (1–94)	43.48%	216 (82.44%)	44.5 (1–93)	53.24%
Liu et al.	China	Wuhan	Central Hospital of Wuhan	109	Non-severe vs severe	53 (48.62%)	61 (52–70)	47.17%	56 (51.38%)	49 (37–59)	44.64%
Liu et al.	China	Beijing	Beijing Ditan Hospital	61	Non-severe vs severe	17 (27.87%)	56 (34–73)	41.18%	44 (72.13%)	41 (1–76)	52.27%
Wu et al.	China	Wuhan	Jinyintan Hospital	201	Non-severe vs severe	84 (41.79%)	58.5 (50.0–69.0)	28.57%	117 (58.21%)	48.0 (40.0–54.0)	41.88%
				84	Survivor vs non-survivor	44 (52.38%)	68.5 (59.3–75.0)	34.09%	40 (47.62%)	50.0 (40.3–56.8)	22.50%
Feng et al.	China	Changsha	Third Xiangya Hospital, Changsha Public Health Treatment Center, Second People's Hospital of Hunan	141	Non-severe vs severe	15 (10.64%)	58 (44–66)	53.33%	126 (89.36%)	41 (33–52)	48.41%

Wan et al.	China	Chongqing	Three Gorges Central Hospital	123	Non-severe vs severe	21 (15.22%)	61.29 ± 15.55	47.62%	102 (73.91%)	43.05 ± 13.12	46.08%
Xu et al.	China	Single center		50	Non-severe vs severe	13 (26%)	60.2 (34–85 range)	46.15%	37 (74%)	36.75 (3–62 range)	40.54%
Wan et al.	China	Chongqing	Three Gorges Central Hospital	135	Non-severe vs severe	40 (29.63%)	56 (52–73)	47.50%	95 (70.37%)	44 (33–49)	45.26%
Han et al.	China	Wuhan	Renmin Hospital of Wuhan University	47	Non-severe vs severe	24 (51.06%)	65.08 (31–87)	29.17%	23 (48.94%)	64.74 (41–81)	60.87%
Qu et al.	China	Huizhou	Huizhou Municipal Central Hospital	30	Non-severe vs severe	3 (10%)	60 ± 5.29	NA	27 (9%)	49.44 ± 14.86	NA
Zhang et al.	China	Wuhan	No. 7 Hospital of Wuhan	140	Non-severe vs severe	58 (41.43%)	64 (25–87)	43.10%	82 (58.57%)	51.5 (26–78)	53.66%
Wang et al.	China	Wuhan	Union Hospital	69	Non-severe vs severe	14 (20.29%)	70.5 (62.0–77.0)	50.00%	55 (79.71%)	37.0 (32.0–51.0)	54.55%
Cao et al.	China	Xiangyang	Xiangyang No 1 Hospital	128	Non-severe vs severe	21 (16.41%)	NA	42.86%	107 (83.59%)	NA	55.14%
Gao et al.	China	Fuyang	Fuyang Second People's Hospital	43	Non-severe vs severe	15 (34.88%)	45.20 ± 7.68	40.00%	28 (65.12%)	42.96 ± 14.00	39.29%
Chen et al.	China	Wuhan	Tongji Hospital	21	Non-severe vs severe	11 (52.38%)	61.0 (56.5–66.0)	9.09%	10 (47.62%)	52.0 (42.8–56.0)	30.00%
Liu et al.	China	Wuhan	Union Hospital	40	Non-severe vs severe	13 (32.5%)	59.7 ± 10.1	46.15%	27 (67.5%)	43.2 ± 12.3	70.37%
Liu et al.	China	Chongqing	Three Gorges Hospital	51	Non-severe vs severe	7 (13.73%)	52 (44–60)	42.86%	44 (86.27%)	44 (33–49)	36.36%
Zhang et al.	China	Chongqing	Chongqing Public Health Medical Center	43	Non-severe vs severe	14 (32.56%)	61.7 ± 9.22	64.29%	29 (67.44%)	44.34 ± 15.84	41.38%
Lu et al.	China	Wuhan	Wuhan Hankou Hospital	577	Non-severe vs severe	100 (17.33%)	61 (49–71)	50.00%	338 (58.58%)	57 (43–65)	54.44%
Chu et al.	China	Wuhan	Tongji Hospital	54	Non-severe vs severe	43 (79.63%)	38 (26–66)	30.23%	11 (20.37%)	47 (36–73)	45.45%

(continued)

Table 22.3 (continued)

Study	Setting			Sample size	Outcomes	Adverse outcome			Non-adverse outcome		
	Country	City	Hospital			Patients (%)	Age	Female %	Patients (%)	Age	Female %
Ji et al.	China	Fuyang and Beijing	Fuyang Second People's Hospital, Fifth Medical Center of Chinese PLA General Hospital	208	Non-severe vs severe	40 (19.23%)	57.7 ± 15.9	30.00%	168 (80.77%)	40.7 ± 14.7	47.02%
Xie et al.	China	Wuhan	Jinyintan Hospital	79	Non-severe vs severe	28 (35.44%)	62.5 (50.5–67.8)	NA	51 (64.56%)	59.0 (46.0–66.0)	NA
Zheng et al.	China	Changsha	North Hospital of Changsha First Hospital (Changsha Public Health Treatment Center)	161	Non-severe vs severe	30 (18.63%)	57 (46.5–66)	50.00%	131 (81.37%)	40 (31–51)	41.22%
Zheng et al.	China	Chengdu	Chengdu Public Health Clinical Medical Center	99	Non-severe vs severe	32 (32.32%)	63.81 ± 16.51	40.63%	67 (67.68%)	42.51 ± 15.11	52.24%
Li et al.	China	Wuhan	Department of Thoracic Surgery of Tongji Hospital	25	Non-severe vs severe	9 (36%)	NA	33.33%	16 (64%)	NA	62.5%
				25	Survivor vs non-survivor	5 (20%)	NA	20%	20 (80%)	NA	60%
Gong et al.	China	Guangzhou and Wuhan	Guangzhou Eighth People's Hospital, Zhongnan Hospital of Wuhan University, and the Third Affiliated Hospital of Sun Yat-sen University, China	372	Non-severe vs severe	28 (7.53%)	63.5 (54.5–72.0)	42.86%	161 (43.28%)	45.0 (33.0–62.0)	55.28%

Young et al.	Singapore	4 hospitals in Singapore		18	Non-severe vs severe	6 (33.33%)	56 (47–73)	66.67%	12 (66.67%)	37 (31–56)	41.67%
Feng et al.	China	Wuhan, Shanghai and Anhui Province	Jinyintan Hospital, Shanghai Public Health Clinical Center and Tongling People's Hospital	476	Non-severe vs severe	124 (26.05%)	58.54 ± 14.38	34.68%	352 (73.94%)	50.29 ± 19.35	46.02%
Liu et al.	China	6 hospitals in Anhui province		67	Non-severe vs severe	24 (32.81%)	46.55 ± 14.30	58.33%	43 (67.19%)	33.4 ± 12.2	34.88%
Han et al.	China	Wuhan	Renmin Hospital of Wuhan University	273	Non-severe vs severe	75 (27.47%)	58.68 ± 14.80	65.33%	198 (72.52%)	58.95 ± 10.8	64.14%
Nie et al.	China	Wuhan	Renmin Hospital	67	Non-severe vs severe	25 (25.77%)	58 (47–67)	48.00%	72 (74.23%)	37 (29–55)	70.83%
Qin et al.	China	Wuhan	Tongji Hospital	452	Non-severe vs severe	286 (63.27%)	61 (51–69)	45.80%	166 (36.73%)	53 (41.25–62)	51.81%
Wang et al.	China	Suzhou	Suzhou Fifth People's Hospital	69	Non-severe vs severe	18 (26.09%)	46.0 (37.3–60.3)	27.78%	51 (73.91%)	41.0 (35.0–57.0)	43.14%
Wang et al.	China	Wuhan	Zhongnan Hospital	138	ICU vs non-ICU	36 (26.09%)	66 (57–78)	38.89%	102 (73.91)	51 (37–62)	48.04%
Fan et al.	Singapore	NA	National Centre for Infectious Diseases (NCID)	67	ICU vs non-ICU	9 (13.43%)	54 (47–62)	33.33%	58 (86.57%)	41 (32–53)	46.55%
Du et al.	China	Wuhan	Wuhan Pulmonary Hospital, Tianyou Hospital, Central Hospital of Wuhan	109	ICU vs non-ICU	51 (46.79%)	68.4 ± 9.7	29.41%	58 (53.21%)	72.7 ± 11.6	34.48%
Yan et al.	China	Hainan province	Hospital of Hainan Medical University	168	Critical vs noncritical	36 (21.43%)	61 (50.3–68)	41.67%	132 (78.57%)	49 (34–60)	54.55%

(continued)

Table 22.3 (continued)

Study	Setting			Sample size	Outcomes	Adverse outcome			Non-adverse outcome		
	Country	City	Hospital			Patients (%)	Age	Female %	Patients (%)	Age	Female %
Tang et al.	China	Wuhan	Tongji Hospital	183	Survivor vs non-survivor	21 (11.48%)	64.0 ± 20.7	23.81%	162 (88.52%)	52.4 ± 15.6	50.62%
Yang et al.	China	Wuhan	Jinyintan Hospital	52	Survivor vs non-survivor	32 (61.54%)	64·6 ± 11·2	34.38%	20 (38.46%)	51·9 ± 12·9	30.00%
Zhou et al.	China	Wuhan	Jinyintan Hospital and Wuhan Pulmonary Hospital	191	Survivor vs non-survivor	54 (28.27%)	69.0 (63.0–76.0)	29.63%	137 (71.23%)	52.0 (45.0–58.0)	40.88%
Du et al.	China	Wuhan	Wuhan Pulmonary Hospital	179	Survivor vs non-survivor	21 (11.73%)	70.2 ± 7.7	52.38%	158 (88.27%)	56.0 ± 13.5	44.94%
Chen et al.	China	Wuhan	Tongji Hospital	274	Survivor vs non-survivor	113 (41.24%)	68.0 (62.0–77.0)	26.55%	161 (58.76%)	51.0 (37.0–66.0)	45.34%
Wang et al.	China	Wuhan	Tongji Hospital	344	Survivor vs non-survivor	133 (38.66%)	70 (62–77)	44.36%	211 (61.34%)	57 (47–69)	50.24%
Cao et al.	China	Wuhan	Wuhan University Zhongnan Hospital	102	Survivor vs non-survivor	17 (16.67%)	72 (63–81)	23.53%	85 (83.33%)	53 (47–66)	52.94%
Huang et al.	China	Wuhan	Jinyintan Hospital	41	ICU vs non-ICU	13 (31.71%)	49·0 (41·0–61·0)	15.38%	28 (68.29%)	49·0 (41·0–57·5)	32.14%
Chen et al.	China	Shanghai	Shanghai Public Health Clinical Center (SPHCC)	249	ICU vs non-ICU	22 (8.84%)	NA	NA	227 (91.16%)	NA	NA
Chen et al.	China	Wuhan	Zhongnan Hospital of Wuhan University	55	Survivor vs non-survivor	19 (34.55%)	77 (median)	15.79%	36 (65.45%)	72 (median)	50.00%
Chen et al.	China	Guangzhou	Guangzhou Eighth People's Hospital	57	Mild vs severe	18 (31.58%)	NA	NA	39 (68.42%)	NA	NA

Table 22.4 The results of quality assessment for cohort studies included in the systematic review

Study	Selection	Comparability	Outcome
Wang et al. (2020a)	***	*	**
Wang et al. (2020d)	***	*	**
Cao (2020)	***	*	*
Huang et al. (2020)	***	**	***
Chen et al. (2020b)	***	**	***
Gao et al. (2020)	****	**	*
Guan et al. (2020a, b)	***	--	**
Yang et al. (2020)	***	--	***
Li et al. (2020a)	***	*	**
Xu et al. (2020)	***	*	---
Fan et al. (2020)	***	*	**
Zhou et al. (2020a, b)	****	**	***
Cai et al.	***	**	**
Liu et al. (2020d)	***	*	***
Tian et al. (2020)	***	*	**
Liu et al. (2020b)	***	**	***
Wu et al. (2020)	***	**	***
Liu et al. (2020a)	***	*	**
Zhang et al. (2020a)	***	--	**
Zhang et al. (2020c)	***	--	**
Qu et al. (2020)	***	*	**
Tang et al. (2020)	****	--	**
Zhao et al. (2020)	***	--	**
Feng et al. (2020b)	****	*	***
Zhang et al. 2020b)	***	**	**
lei and Jian-ya (2020)	***	*	***
Wan et al. (2020)	***	--	**
Momtazmanesh et al. (2020)	****	**	***
Chen et al. (2020e)	***	--	**
Lu et al. (2020)	***	**	***
Chu et al. (2020)	**	*	**
Ji et al. (2020)	****	**	***
Li et al. (2020b)	***	*	**
Liu et al. (2020c)	***	--	**
Wang et al. (2020b)	****	*	***
Xie et al. (2020)	***	**	**
Zheng et al. (2020a)	***	*	**
Chen et al. (2020c)	***	*	**
Chen et al. (2020d)	****	--	*
Han et al. (2020b)	***	*	***
Feng et al. (2020a)	***	**	***
Wan et al.	***	*	**
Du et al. (2020)	***	*	**
Han et al. (2020a)	***	--	**
Zheng et al. (2020b)	***	--	**
Yan et al. (2020)	***	*	**
Cao et al. (2020)	****	*	***
Young et al. (2020)	***	--	***
Gong et al. (2020)	***	*	***

(continued)

Table 22.4 (continued)

Study	Selection	Comparability	Outcome
Nie et al. (2020)	***	*	***
Qin et al. (2020)	***	*	**
Wang et al. (2020c)	***	**	**
Chen et al. (2020a)	***	*	**

Table 22.5 The results of quality assessment for a cross-sectional study included in the systematic review

Author	Selection	Comparability	Outcome
Zhou et al. (2020a, b)	***	*	**

cell counts were increased in severe compared to non-severe patients (SMD −0.72; 95% CI, −0.88, −0.57; P < 0.0001). There was significant heterogeneity among studies (I2 = 64%). No evidence of publication bias was observed (Begg's P = 0.498; Egger's P = 0.496). Meta-analysis for data derived from four studies comparing critical and noncritical COVID-19 patients showed a similar result with the SMD of −0.74 (95% CI, −1.10 to −0.39; P < 0.0001). Critical patients had higher PMN compared to the noncritical group.

Patients with severe outcomes (n = 1890) had lower lymphocyte count compared to non-severe patients (n = 4035) (SMD 0.68; 95% CI, 0.53, 0.84; P < 0.0001). Heterogeneity was high (I2 = 80%). Also, critical patients had lower lymphocyte counts than noncritical patients.

Patients with severe disease (n = 2288) had significantly lower platelets than non-severe cases (n = 627) with an SMD of 0.21 (95% CI, 0.06, 0.35; P = 0.004). Moderate heterogeneity was present (I2 = 50.5%). No evidence of publication bias was observed (Begg's P = 0.707; Egger's P = 0.241).

Eight studies documented CD4 counts for 506 non-severe and 587 severe patients. Because of evidence of publication bias (Begg's P = 0.009), the ES was adjusted using trim and fill method, resulting in an ES of 0.80 (95% CI, −0.06, 1.62), which showed no significant difference between severe and non-severe patients in regard to CD4 counts. Heterogeneity was high (I2 = 96%).

Total bilirubin (TB) was investigated in 15 studies, with a total of 1528 non-severe and 650 severe individuals. Severe patients had higher TB compared to the non-severe group (SMD −0.33; 95% CI, −0.46, −0.21, P < 0.0001), and no substantial heterogeneity was seen (I2 = 31%).

Alanine transaminase (ALT) was documented in 23 studies accounting for a total of 1875 non-severe and 775 severe patients. ALT levels were higher in severe patients than non-severe patients (SMD −0.52; 95% CI, −0.77, −0.27; P < 0.0001). There was evidence of publication bias (Begg's P = 0.009). The ES was adjusted using the trim and fill method and estimated to be about −0.56 with 95% CI, −0.79, −0.32. There was high heterogeneity across studies (I2 = 85.5%). A sensitivity analysis showed that I2 dropped to 49% with the associated ES of −0.36 (95% CI, −0.50, −0.22) when the study (Zheng et al. 2020a) was excluded. Similarly, AST levels were higher in severe patients compared to non-severe patients (SMD −0.79; 95% CI, −1.00, −0.58; P < 0.0001). Heterogeneity was high (I2 = 74.5%). Critical patients had higher AST and ALT levels compared to noncritical patients (SMD −0.78; 95% CI, −1.27, −0.29; and SMD −0.45; 95% CI, −0.68, −0.22), respectively.

Creatinine was increased in severe patients (n = 751) compared to non-severe patients (n = 2320) (SMD −0.21; 95% CI, −0.34, −0.08; P = 0.001). No evidence of publication bias was found (Begg's P = 0.785; Egger's P = 0.282). Moderate heterogeneity was seen (I2 = 52%). Furthermore, BUN was higher in severe patients than non-severe patients (SMD −0.55; 95% CI, −0.93, −0.17; P < 0.0001).

Severe patients showed no difference in potassium levels compared with non-severe patients (SMD = 0.18, p = 0.269). However, the exclusion of a study (Feng et al. 2020b) could resolve heterogeneity (I2 = 0.0), and the ES became significant (SMD 0.25; 95% CI, 0.12, 0.39; P < 0.0001). Sodium levels were not different between severe and non-severe patients, and sensitivity analysis

Table 22.6 Summary of the results of meta-analyses of clinical and laboratory characteristics of patients with COVID-19: patients who survived versus patients who died

Factor	No. of pairwise	No. of subjects		Meta-analysis				Heterogeneity			Publication bias	
		Survived	Died	SMD /OR	95% CI		P-value	I2%	Chi2	P-value	P-value of Begg's test	P-value of Egger's test
WBC	5	707	365	−1.01	−1.33	−0.696	<0.0001	78.57	18.67	0.001	0.327	0.356
PMN	4	570	311	−1.05	−1.53	−0.584	<0.0001	87.76	24.52	<0.0001	0.734	0.191
Lymphocyte	6	727	397	0.97	0.52	1.43	<0.0001	89.79	48.97	<0.0001	0.024	0.0437
Platelet	5	569	376	0.45	0.19	0.70	0.001	67.56	12.33	0.015	0.0864	0.0717
Hb	3	318	199	0.08	0.10	1.26	0.383	0.00	0.59	<0.0001	1	0.713
Ferritin	3	259	223	−1.03	−1.23	−0.83	<0.0001	0.00	22.64	<0.0001	0.296	0.422
TnI	3	415	277	−3.08	−1.041	−0.722	<0.0001	0.00	1.038	<0.0001	0.296	0.149
CRP	4	554	296	−4.09	−1.49	−0.4	0.001	90.47	31.49	<0.0001	0.734	0.171
PCT	4	622	301	−1.05	−1.27	−0.83	<0.0001	48.37	5.81	0.121	0.734	0.320
CK	3	504	298	−0.78	−0.93	−0.63	<0.0001	0.00	1.74	0.418	1.00	0.636
AST	4	570	311	−0.52	−1.066	0.027	0.062	91.46	35.14	<0.0001	0.734	0.115
ALT	5	705	365	−0.36	−0.54	−0.17	<0.0001	44.65	7.22	0.124	0.462	0.236
TB	5	590	343	−0.79	−1.01	−0.56	<0.0001	49.29	7.88	0.096	0.308	0.097
Alb	5	496	232	0.88	0.46	1.30	<0.0001	87.95	33.21	<0.0001	0.086	0.091
PT	7	880	418	−0.78	−1.89	−0.37	<0.0001	88.54	52.36	<0.0001	1.00	0.345
PTT	4	521	199	0.26	−0.62	0.41	0.695	86.19	21.72	<0.0001	0.308	0.903
Cr	5	590	343	−0.34	−0.77	0.09	0.123	87.55	32.13	<0.0001	1.00	0.898
IL-6	4	498	284	−1.22	−1.38	−1.07	<0.0001	0.00	2.80	0.422	0.734	0.541
D-dimer	5	677	349	−0.66	−1.10	−0.22	0.003	89.02	36.44	<0.0001	0.22	0.32
Fever[a]	7	601	551	1.077	0.70	1.65	0.735	0.00	2.71	0.844	0.763	0.437
Cough	7	489	414	0.82	0.26	2.56	0.739	92.58	80.92	<0.0001	1.00	0.464
Fatigue	5	752	338	0.75	0.57	0.90	0.042	17.64	4.85	0.302	0.806	0.549
Dyspnea	5	590	343	0.34	0.25	0.46	<0.0001	33.68	6.03	0.197	1.00	0.732
Diarrhea	4	503	236	1.33	0.85	2.09	0.206	0.00	2.21	<0.0001	0.734	0.369
Nausea/vomiting	4	318	199	1.14	0.54	2.39	0.71	0.00	1.39	<0.0001	1.00	0.79
Myalgia/arthralgia	5	561	237	0.95	0.63	1.42	0.811	32.99	5.97	0.201	0.806	0.552
Headache	3	339	166	0.66	0.20	2.17	0.500	60.59	5.07	0.079	1.00	0.720
Sputum production	4	610	311	0.84	0.50	1.43	0.535	61.83	7.86	0.049	0.730	0.630
DM	7	812	414	0.49	0.36	0.67	<0.0001	14.60	7.02	0.318	0.071	0.098

(continued)

Table 22.6 (continued)

Factor	No. of pairwise	No. of subjects		Meta-analysis				Heterogeneity			Publication bias	
		Survived	Died	SMD /OR	95% CI		P-value	I2%	Chi2	P-value	P-value of Begg's test	P-value of Egger's test
CHD	6	654	393	0.31	0.13	0.72	0.006	61.11	12.85	0.025	1.00	0.651
COPD	3	368	192	0.13	0.052	0.34	<0.0001	0.00	0.38	0.824	1.00	0.669
Cerebrovascular disease	3	266	162	0.125	0.034	0.465	0.002	0.002	0.32	0.852	0.29	0.047
CKD	3	383	184	0.095	0.023	0.391	0.001	0.00	0.49	0.782	1.00	0.747
HTN	6	792	382	0.48	0.22	1.03	0.061	85.87	35.38	<0.0001	0.45	0.63
Malignancy	4	561	237	0.44	0.15	1.3	0.149	0.00	1.34	<0.0001	0.089	0.061
Gender (female)	8	974	435	1.45	0.97	2.16	0.066	51.46	14.42	0.044	0.901	0.825
Age	8	974	435	−1.07	−1.35	−0.78	<0.0001	77.49	31.04	<0.0001	0.710	0.648
PR	4	550	299	−0.037	−0.822	0.066	0.095	86.03	21.48	<0.0001	1.00	0.820

WBC white blood cells, *PMN* polymorphonuclear leukocytes, *Hb* hemoglobin, *TnI* troponin I, *CRP* C-reactive protein, *PCT* procalcitonin, *CK* creatine kinase, *AST* aspartate aminotransferase, *ALT* alanine transaminase, *TB* total bilirubin, *Alb* albumin, *PT* prothrombin time, *PTT* partial thromboplastin time, *Cr* creatinine, *IL-6* interleukin 6, *DM* diabetes mellitus, *CHD* coronary heart disease, *COPD* chronic obstructive pulmonary disease, *CKD* chronic kidney disease, *HTN* hypertension, *PR* pulse rate, *No.* number, *SMD* standardized mean difference, *OR* odds ratio, *CI* confidence interval

[a]Reported as a categorical variable

Table 22.7 Summary of the results of meta-analyses of laboratory measures in patients with COVID-19: patients with a severe form of COVID-19 versus patients with a non-severe form of COVID-19

Factor	No. of pairwise	No. of subjects		Meta-analysis				Heterogeneity			Publication bias	
		Non-severe	Severe	SMD /OR	95%CI		P-value	I2%	Chi2	P-value	P-value of Begg's test	P-value of Egger's test
WBC	27	3293	1131	−0.40	−0.58	−0.21	<0.0001	83.86	161.12	<0.0001	0.77	0.010
PMN	22	2215	1109	−0.72	−0.88	−0.57	<0.0001	63.72	67.07	<0.0001	0.498	0.496
Lymphocyte	31	4035	1890	0.68	0.53	0.84	<0.0001	79.92	149.93	<0.0001	0.586	0.298
Monocytes	11	872	575	0.17	0.57	0.428	0.003	30.70	14.43	0.154	0.640	0.073
Eosinophil	3	164	86	0.23	−0.03	0.50	0.086	33.50	3.00	0.2222	0.296	0.064
CD4 cells	8	506	587	1.80	1.17	2.57	<0.0001	96.17	181.82	<0.0001	0.009	0.062
CD8 cells	8	862	593	1.46	0.81	2.10	<0.0001	95.86	168.91	<0.0001	0.035	0.084
CD3 cells	4	521	176	1.18	0.67	1.69	<0.0001	84.77	19.70	<0.0001	0.308	0.250
Platelet	18	2288	627	0.21	0.06	0.35	0.004	50.47	34.32	0.008	0.707	0.241
Hb	12	1827	472	0.24	0.14	0.34	<0.0001	33.61	16.57	0.121	0.731	0.677
Ferritin	4	470	489	−0.88	−1.26	−0.51	<0.0001	69.82	9.94	0.019	1.00	0.226
D-dimer	17	1954	727	−0.93	−1.12	−0.74	<0.0001	72.77	58.77	<0.0001	0.901	0.619
TnI	3	585	199	−0.98	−1.18	−0.77	<0.0001	0.00	1.31	0.519	1.00	0.370
CRP	23	1677	906	−1.23	−1.48	−0.98	<0.0001	87.20	172.00	<0.0001	0.072	0.004
ESR	7	357	252	−0.71	−1.13	−0.29	0.001	91.74	72.85	<0.0001	0.071	0.054
PCT	13	1170	707	−0.81	−1.07	−0.55	<0.0001	81.08	63.43	<0.0001	0.200	0.202
CK	13	1222	448	−0.50	−0.75	−0.26	<0.0001	74.94	47.88	<0.0001	1.00	0.782
AST	20	1569	630	−0.79	−1.00	−0.58	<0.0001	74.49	74.48	<0.0001	0.922	0.640
ALT	23	1875	775	−0.52	−0.77	−0.27	<0.0001	85.45	151.28	<0.0001	0.009	0.206
TB	15	1528	650	−0.33	−0.46	−0.21	<0.0001	31.00	20.29	0.121	0.766	0.972
Alb	13	1719	868	1.02	0.81	1.23	<0.0001	76.46	50.97	<0.0001	0.760	0.954
Pt	11	796	415	−0.30	−0.64	0.03	0.074	84.18	63.22	<0.0001	0.755	0.689
PTT	9	644	273	0.05	−0.29	0.39	0.772	79.03	38.15	<0.0001	0.251	0.330
LDH	17	1590	587	−1.33	−1.54	−1.17	<0.0001	68.41	50.65	<0.0001	0.433	0.086
Fibrinogen	5	533	193	−0.95	−1.59	−0.31	0.003	87.01	30.79	<0.0001	0.220	0.160
Cr	21	2320	751	−0.21	−0.34	−0.08	0.001	52.31	41.93	0.003	0.785	0.282
BUN	9	821	294	−0.55	−0.93	−0.17	<0.0001	84.97	53.23	<0.0001	0.251	0.143
Sodium	6	1470	381	0.33	−0.05	0.72	0.096	87.33	39.48	<0.0001	1.000	0.413

(continued)

Table 22.7 (continued)

Factor	No. of pairwise	No. of subjects		Meta-analysis				Heterogeneity			Publication bias	
		Non-severe	Severe	SMD /OR	95%CI		P-value	I2%	Chi2	P-value	P-value of Begg's test	P-value of Egger's test
Potassium	6	1473	375	0.18	−0.13	0.49	0.269	79.74	24.68	<0.0001	1.00	0.516
Glucose	5	451	322	−1.02	−1.30	−0.74	<0.0001	44.07	7.15	0.128	0.462	0.608
CK-MB	9	1156	487	−0.44	−0.71	−0.18	0.001	80.40	40.83	<0.0001	0.602	0.590
C3	4	288	384	−0.17	−0.49	0.14	0.291	53.17	6.4	0.093	1.00	0.515
C4	4	288	384	−0.20	−0.79	0.37	0.484	85.58	20.81	<0.0001	0.734	0.709
IgM	5	502	374	0.18	0.02	0.33	0.020	0.00	0.85	0.932	0.220	0.085
IgG	5	502	374	−0.001	−0.15	0.15	0.994	6.66	4.28	0.369	0.462	0.318
IgA	5	502	374	−0.47	−1.04	0.08	0.094	88.30	34.19	<0.0001	0.027	0.282
TNFα	3	310	201	−0.47	−1.00	0.05	0.077	59.53	4.94	0.084	0.296	0.185
IL-6	10	863	532	−1.67	−2.39	−0.95	<0.0001	96.15	234.19	<0.0001	0.371	0.051
IL-10	5	376	355	−3.49	−5.46	−1.47	<0.0001	98.22	224.95	<0.0001	0.220	0.105
IL-4	3	203	60	−0.28	−0.55	0.007	0.056	0.00	0.27	0.871	1.000	0.860

WBC white blood cells, *PMN* polymorphonuclear leukocytes, *Hb* hemoglobin, *TnI* troponin I, *CRP* C-reactive protein, *ESR* erythrocyte sedimentation rate, *PCT* procalcitonin, *CK* creatine kinase, *AST* aspartate aminotransferase, *ALT* alanine transaminase, *TB* total bilirubin, *Alb* albumin, *PT* prothrombin time, *PTT* partial thromboplastin time, *Cr* creatinine, *LDH* lactate dehydrogenase, *BUN* blood urea nitrogen, *CK-MB* creatine kinase-MB, *C3* complement component C3, *C4* complement component C4, *IgM* immunoglobulin M, *IgG* immunoglobulin G, *IgA* immunoglobulin A, *TNFα* tumor necrosis factor-alpha, *IL-6* interleukin 6, *IL-6* interleukin 10, *IL-6* interleukin 4, *No.* number, *SMD* standardized mean difference, *OR* odds ratio, *CI* confidence interval

Table 22.8 Summary of the results of meta-analyses of clinical characteristics and comorbidities in patients with COVID-19: patients with a severe form of COVID-19 versus patients with a non-severe form of COVID-19

Factor	No. of pairwise	No. of subjects		Meta-analysis				Heterogeneity			Publication bias	
		Non-severe	Severe	SMD /OR	95% CI		P-value	I2%	Chi2	P-value	P value of Begg's test	P-value of Egger'stest
Age	28	4006	1493	−0.89	−1.05	−0.72	<0.0001	76.61	115.43	<0.0001	0.797	0.051
Gender (female)	35	4635	1707	1.32	1.17	1.49	<0.0001	18.19	41.56	0.175	0.690	0.799
History of smoking	9	1800	428	0.53	0.38	0.73	<0.0001	0.00	7.045	0.532	0.602	0.795
HTN	25	1787	1322	0.40	0.30	0.52	<0.0001	45.33	43.90	0.008	1.000	0.964
DM	25	3197	1322	0.36	0.26	0.52	<0.0001	47.44	45.66	0.005	0.005	0.038
CHD	20	3036	1259	0.34	0.25	0.45	<0.0001	1.35	19.26	0.440	0.381	0.611
CKD	8	1871	792	0.35	0.18	0.68	0.002	14.09	8.14	0.320	0.901	0.383
Cerebrovascular disease	10	2298	886	0.30	0.18	0.50	<0.0001	0.00	7.41	0.594	0.474	0.892
COPD	15	2607	742	0.20	0.12	0.33	<0.0001	0.00	8.00	0.889	0.022	0.087
Chronic liver disease	12	2126	788	0.71	0.37	1.36	0.309	18.67	13.52	0.260	0.631	0.762
Malignancy	9	1910	748	0.42	0.24	0.72	0.002	0.00	5.58	0.693	0.117	0.480
Immunodeficiency	3	1444	282	0.76	0.15	3.80	0.740	0.00	0.03	0.984	1.000	0.323
Fever[a]	27	3484	1339	0.70	0.41	1.18	0.185	80.74	135.02	<0.0001	0.559	0.097
Fever[b]	8	1579	441	−0.38	−0.67	−0.09	0.010	81.12	37.08	<0.0001	0.386	0.690
RR	6	632	181	−0.78	−1.26	−0.30	0.001	83.29	29.92	<0.0001	1.000	0.825
PR	4	256	107	−0.33	−0.56	−0.09	0.007	0.00	2.26	0.519	0.089	0.025
Chill	5	1037	260	0.62	0.42	0.93	0.023	0.00	2.65	0.616	1.000	0.617
Cough	28	3367	1327	0.78	0.60	1.01	0.064	53.45	58.01	<0.0001	0.441	0.019
Sputum production	18	2247	947	0.66	0.55	0.79	<0.0001	34.46	25.94	0.076	0.939	0.313
Chest pain	10	625	302	0.58	0.28	1.20	0.146	39.37	14.84	0.095	1.000	0.935
Myalgia/arthralgia	15	2248	864	0.79	0.62	1.009	0.060	35.53	21.71	0.085	0.766	0.693
Nausea/vomiting	12	1885	765	0.75	0.51	1.10	0.151	39.55	18.19	0.077	0.837	0.966
Diarrhea	22	2630	1029	0.90	0.57	1.40	0.644	55.48	47.17	0.001	0.283	0.491
Sore throat	16	2422	930	1.01	0.76	1.35	0.903	0.00	11.39	0.724	0.162	0.229
Dyspnea	25	3016	1334	0.22	0.14	0.33	<0.0001	73.50	90.56	<0.0001	0.075	0.083
Headache	17	2500	871	0.83	0.64	1.08	0.178	0.00	11.02	0.808	0.174	0.168
Hemoptysis	4	1519	602	0.25	0.11	0.55	0.001	0.00	0.52	0.914	0.308	0.169
Fatigue	20	2519	981	0.71	0.52	0.97	0.032	54.64	41.89	0.002	0.581	0.273

(continued)

Table 22.8 (continued)

Factor	No. of pairwise	No. of subjects		Meta-analysis				Heterogeneity			Publication bias	
		Non-severe	Severe	SMD /OR	95% CI		P-value	I2%	Chi2	P-value	P value of Begg's test	P-value of Egger'stest
Abdominal pain	6	538	460	0.49	0.24	0.98	0.045	0.00	4.44	0.488	0.259	0.101

DM diabetes mellitus, *CHD* coronary heart disease, *COPD* chronic obstructive pulmonary disease, *CKD* chronic kidney disease, *HTN* hypertension, *RR* respiratory rate, *PR* pulse rate, *No.* number, *SMD* standardized mean difference, *OR* odds ratio, *CI* confidence interval

[a]Reported as a categorical variable

[b]Reported as a continuous variable

Table 22.9 Summary of the results of meta-analyses of clinical and laboratory characteristics of patients with COVID-19: patients who either were in critical condition or required ICU admission versus patients who neither were in critical condition nor required ICU admission

Factor	No. of pairwise	No. of subjects		Meta-analysis				Heterogeneity			Publication bias	
		Noncritical	Critical	SMD /OR	95% CI		P-value	I2%	Chi2	P-value	P-value of Begg's test	P-value of Egger'stest
WBC	4	348	131	−0.58	−0.79	−0.37	<0.0001	15.29	3.54	0.315	1.00	0.806
PMN	4	347	131	−0.74	−1.10	−0.39	<0.0001	60.88	7.66	0.053	1.00	0.854
Lymphocyte	4	346	131	0.53	0.09	0.98	0.018	75.56	12.27	0.006	0.308	0.382
Monocyte	3	346	131	0.16	−0.37	0.69	0.556	74.68	7.9	0.019	0.296	0.160
Platelet	4	350	131	0.15	−0.15	0.46	0.946	50.68	6.08	0.108	0.734	0.555
D-dimer	3	191	102	−0.74	−1.14	−0.34	<0.0001	57.15	4.66	0.097	1.00	0.85
LDH	3	260	76	−1.64	−2.38	−0.89	<0.0001	80.52	10.27	0.006	0.296	0.330
AST	3	235	116	−0.78	−1.27	−0.29	0.002	76.70	8.58	0.014	1.000	0.567
ALT	3	241	118	−0.45	−0.68	−0.22	<0.0001	0.00	0.99	0.608	1.000	0.838
PTT	3	229	114	0.11	−0.12	0.34	0.347	10.73	2.24	0.326	1.000	0.794
PT	3	229	113	−0.27	−0.50	−0.46	0.019	0.00	1.17	0.556	1.000	0.928
Cr	3	248	117	−0.03	−0.75	0.68	0.926	89.91	19.82	<0.0001	1.000	0.879
BUN	3	253	116	−0.42	−1.28	0.42	0.325	92.64	27.19	<0.0001	1.000	0.539
Age	3	292	123	−0.17	−0.85	−0.50	0.614	91.90	24.69	<0.0001	1.000	0.784
Gender (female)	3	292	123	1.46	0.93	2.29	0.094	0.00	0.25	0.879	0.296	0.046
RR	3	292	123	−0.45	−0.99	0.08	0.100	83.58	12.18	0.002	1.000	0.359
PR	3	292	123	−0.55	−0.97	−0.12	0.012	73.51	7.55	0.023	1.000	0.399
Fever[a]	3	292	123	0.41	0.19	0.86	0.019	0.00	0.46	0.793	1.000	0.817
Cough	3	292	123	0.94	0.60	1.48	0.818	0.00	0.25	0.878	1.000	0.534
Fatigue	3	292	123	0.80	0.29	2.21	0.676	78.04	9.11	0.011	1.000	0.600
Dyspnea	3	292	123	0.22	0.06	0.71	0.012	73.46	7.53	0.023	1.000	0.693
Diarrhea	3	292	123	0.59	0.32	1.08	0.088	0.00	0.74	0.700	1.000	0.332
Myalgia/arthralgia	3	292	123	0.81	0.45	1.43	0.475	0.00	1.93	0.380	1.000	0.834
Headache	3	292	123	0.61	0.28	1.32	0.21	0.00	0.51	0.775	1.000	0.208
Sputum production	3	292	123	1.19	0.74	1.91	0.454	0.00	0.15	0.924	1.000	0.341
DM	3	292	123	0.31	0.12	0.82	0.018	59.21	4.90	0.086	0.296	0.005
Chronic liver disease	3	292	123	1.10	0.14	8.34	0.921	49.44	3.95	0.138	0.296	0.012
CKD	3	292	123	0.70	0.06	8.26	0.77	68.62	6.37	0.041	1.000	0.676

(continued)

Table 22.9 (continued)

Factor	No. of pairwise	No. of subjects		Meta-analysis				Heterogeneity			Publication bias	
		Noncritical	Critical	SMD /OR	95% CI		P-value	I2%	Chi2	P-value	P-value of Begg's test	P-value of Egger'stest
HTN	3	292	123	0.398	0.12	1.30	0.128	83.73	12.29	0.002	1.000	0.525
Malignancy	3	292	123	1.02	0.38	2.72	0.961	23.69	2.62	0.270	1.000	0.728

WBC white blood cells, *PMN* polymorphonuclear leukocytes, *LDH* lactate dehydrogenase, *AST* aspartate aminotransferase, *ALT* alanine transaminase, *PT* prothrombin time, *PTT* partial thromboplastin time, *Cr* creatinine, *BUN* blood urea nitrogen, *DM* diabetes mellitus, *CKD* Chronic kidney disease, *HTN* hypertension, *RR* respiratory rate, *PR* pulse rate, *No.* number, *SMD* standardized mean difference, *OR* odds ratio, *CI* confidence interval

[a]Reported as a categorical variable

Table 22.10 Summary of the results of meta-analyses of subgroups of patients with a severe and non-severe form of COVID-19

Factor	Between-group comparison	No. of pairwise	Meta-analysis				Heterogeneity		
			SMD /OR	95% CI		P-value	I2%	Chi2	P-value
WBC	A[a]	12	−0.39	−0.67	−0.12	0.004	88.29	93/98	<0.0001
	B[b]	15	−0.40	−0.66	−0.14	0.001	78.88	66.29	<0.0001
PMN	A	10	−0.74	−0.89	−0.59	<0.0001	45.35	16.17	0.063
	B	12	−0.73	−1.01	−0.45	<0.0001	75.93	45.70	<0.0001
Lymphocyte	A	14	0.81	0.60	1.03	<0.0001	85.09	87.22	<0.0001
	B	17	0.55	0.32	0.79	<0.0001	72.64	58.48	<0.0001
CD4 cells	A	4	2.89	1.35	4.43	<0.0001	98.29	176.20	<0.0001
	B	4	0.79	061	0.96	<0.0001	0.00	2.73	0.434
CD8 cells	A	4	2.24	0.86	3.62	0.001	98.07	155.52	<0.0001
	B	4	0.60	0.31	0.89	<0.0001	35.40	4.64	0.200
Platelet	A	6	0.29	0.04	0.55	<0.0001	72.07	17.90	0.003
	B	12	0.12	−0.002	0.26	0.054	0.00	8.18	0.697
Hb	A	3	0.31	0.17	0.45	<0.0001	0.00	1.13	0.566
	B	8	0.22	0.04	0.41	0.016	36.89	11.09	0.135
CRP	A	7	−1.20	−1.49	−0.91	<0.0001	72.32	21.67	0.001
	B	16	−1.26	−1.61	−0.91	<0.0001	89.02	136.62	<0.0001
PCT	A	4	−1.27	−1.63	−0.62	<0.0001	86.83	22.78	<0.0001
	B	9	−0.61	−0.88	−0.34	<0.0001	61.69	21.04	0.007
Ferritin	B	3	−0.90	−1.49	−0.31	0.003	65.83	5.85	0.054
	A	5	−0.43	−0.75	−0.16	0.008	77.29	17.61	0.001
CK	B	8	−0.56	−0.96	−0.16	0.006	75.37	28.43	<0.0001
	A	7	−0.89	−1.13	−0.66	<0.0001	65.06	17.17	0.009
AST	B	13	−0.70	−1.04	−0.36	<0.0001	78.07	54.72	<0.0001
	A	8	−0.35	−0.61	−0.10	<0.0001	73.13	26.05	<0.0001
ALT	B	15	−0.62	−1.02	−0.22	0.002	88.32	119.86	<0.0001
TB	A	5	−0.36	−0.52	−0.21	<0.0001	0.00	1.29	0.863
	B	8	−0.25	−0.50	−0.01	0.035	48.91	13.07	0.057
Alb	A	6	0.92	0.67	1.17	<0.0001	70.38	16.88	0.005
	B	7	1.14	0.78	1.49	<0.0001	79.04	28.62	<0.0001
PT	A	5	−0.65	−0.91	−0.38	<0.0001	53.90	8.67	0.070
	B	6	−0.001	−0.445	0.442	0.95	76.17	20.98	0.001
PTT	A	4	−0.30	−0.71	0.11	0.151	80.60	15.46	0.001
	B	5	0.45	−0.10	1.02	0.114	75.01	16.01	0.003
LDH	A	7	−1.23	−1.57	−0.90	<0.0001	82.13	33.58	<0.0001
	B	10	−1.41	−1.59	−1.22	<0.0001	20.92	11.38	0.250
Cr	A	9	−0.31	−0.48	−0.22	<0.0001	66.85	24.13	0.002
	B	12	−0.09	−0.23	0.03	0.139	0.00	10.25	0.50
BUN	A	3	−1.04	−1.21	−0.87	<0.0001	0.00	0.50	0.778
	B	6	−0.25	−0.71	0.20	0.270	73.67	18.99	0.002
D-dimer	A	8	−1.05	−1.29	−0.81	<0.0001	75.82	28.95	<0.0001
	B	9	−0.79	−1.10	−0.49	<0.0001	67.66	24.73	0.002
Sodium	A	4	0.46	−0.01	0.95	0.058	91.99	37.48	<0.0001
Potassium	A	4	0.06	−0.28	0.41	0.731	84.66	19.56	<0.0001

(continued)

Table 22.10 (continued)

Factor	Between-group comparison	No. of pairwise	Meta-analysis SMD /OR	95% CI		P-value	Heterogeneity I2%	Chi2	P-value
IL-6	A	3	−3.07	−5.20	0.94	0.005	98.77	162.79	<0.0001
	B	7	−1.11	−1.69	−0.54	<0.0001	86.82	45.54	<0.0001
CK-MB	A	6	−0.58	−0.84	−0.32	<0.0001	75.94	20.78	0.001
	B	3	−0.11	−0.78	0.59	0.746	82.49	11.42	0.003

WBC white blood cells, *PMN* polymorphonuclear leukocytes, *Hb* hemoglobin, *CRP* C-reactive protein, *PCT* procalcitonin, *CK* creatine kinase, *AST* aspartate aminotransferase, *ALT* alanine transaminase, *TB* total bilirubin, *Alb* albumin, *PT* prothrombin time, *PTT* partial thromboplastin time, *LDH* lactate dehydrogenase, *Cr* creatinine, *BUN* blood urea nitrogen, *IL-6* interleukin 6, *CK-MB* creatine kinase-MB, *SMD* standardized mean difference, *OR* odds ratio, *CI* confidence interval

[a]Stduies comparing patients with a severe form of disease or patients who were in a critical condition and patients with a non-severe form of disease or patients who were not in a critical condition

[b]Stduies comparing patients with a severe form of disease and patients with a non-severe form of the disease

Table 22.11 Adjusted ES using trim and fill methods

Factor	Comparison group	ES	95% CI
Lymphocyte	Survived vs died	1.20	1.06 to 1.34
Alb	Survived vs died	1.23	0.79 to 1.69
Cerebrovascular disease	Survived vs died	0.17	0.05 to 0.51
Malignancy	Survived vs died	0.30	0.11 to 0.78
Monocytes	Severe vs non-severe	0.08	−0.020 to 0.185
Eosinophil	Severe vs non-severe	0.05	−0.170 to 0.285
CD4 cells	Severe vs non-severe	0.80	−0.006 to 1.62
CD8 cells	Severe vs non-severe	0.59	−0.15 to 1.33
CRP	Severe vs non-severe	−0.64	−1.02 to -0.25
ESR	Severe vs non-severe	−0.71	−1.13 to −0.29
ALT	Severe vs non-severe	−0.56	−0.79 to −0.32
LDH	Severe vs non-severe	−1.39	−1.62 to −1.16
IgM	Severe vs non-severe	0.20	0.06 to 0.34
IgA	Severe vs non-severe	−0.47	−1.03 to 0.08
IL-6	Severe vs non-severe	−0.53	−1.29 to 0.22
COPD	Severe vs non-severe	0.42	0.27 to 0.64
Fever	Severe vs non-severe	1.36	0.83 to 2.23
Dyspnea	Severe vs non-severe	0.31	0.20 to 0.49
Cough	Severe vs non-severe	1.10	0.84 to 1.43
PR	Severe vs non-severe	−0.39	−0.61 to −0.16
Age	Severe vs non-severe	−0.59	−0.76 to −0.41
Chronic liver disease	Critical vs no-critical	0.25	0.03 to 2.10
DM	Survived vs died	0.59	0.45 to 0.79
	Critical vs noncritical	0.69	0.24 to 2.01
	Severe vs non-severe	0.48	0.32 to 0.71

Alb albumin, *CRP* C-reactive protein, *ESR* erythrocyte sedimentation rate, *ALT* alanine transaminase, *LDH* lactate dehydrogenase, *IgM* immunoglobulin M, *IgA* immunoglobulin A, *IL-6* interleukin 6, *COPD* chronic obstructive pulmonary disease, *PR* pulse rate, *DM* diabetes mellitus, *ES* effect size, *CI* confidence interval

Table 22.12 Summary of the results of meta-regressions

Variable	Moderator	No of pairwise	Meta-regression Coefficient	SE	95% CI		z	P-value	The proportion of total between-studyvariance explained R2 analog
WBC	Sample size	20	0.0006	0.0004	−0.0001	0.0014	1.6	0.1085	0.27
	Age difference	18	0.018	0.0236	−0.0284	0.0643	0.76	0.4467	0
	Gender	20	0.0219	0.0142	−0.0059	0.0497	1.54	0.1226	0
D-dimer	Sample size	12	0.0005	0.0007	−0.0008	0.0018	0.79	0.4324	0
	Age difference	12	−0.0234	0.0179	−0.0585	0.0117	−1.31	0.1908	0.46
	Gender	12	−0.0097	0.0116	−0.0323	0.013	−0.83	0.4039	0.15

No, number, *SE* standard error, *CI* confidence interval

showed no significant effect of any individual study.

Troponin I (TnI) was higher in severe patients (n = 199) than non-severe patients (n = 585) (SMD −0.98; 95% CI, −1.18, −0.77). Thirteen studies reported data on albumin levels, including 868 severe patients and 1719 non-severe patients. Severe patients had lower albumin levels (SMD 1.02; 95% CI, 0.81, 1.23; P < 0.0001). Also, severe (n = 1170) patients had higher procalcitonin levels compared to non-severe patients (n = 707) (SMD −0.81; 95% CI, −1.07, −0.55; P < 0.0001).

A meta-analysis of 11 studies providing data on prothrombin time (PT) showed that patients with severe disease were not different from non-severe patients (SMD −0.30; 95% CI, −0.64, 0.03; P = 0.074). Also, PTT showed no significant difference between the two groups (SMD = 0.05; 95% CI, −0.29 to 0.39; P = 0.772). Critical patients had higher PT compared to non-critical patients (SMD −0.27; 95% CI, −0.50, −0.46, P = 0.019), though PTT was not different between two critical and noncritical groups (SMD 0.11; 95% CI, −0.12, 0.34; P = 0.347).

Seventeen studies reported data on LDH levels, and the meta-analysis revealed that severe patients had higher LDH levels than non-severe patients (SMD −1.33; 95% CI, −1.54, −1.17; P < 0.0001). There was moderate heterogeneity (I2 = 68.41). Also, severe patients had significantly higher D-dimer levels compared to non-severe patients (SMD −0.93; 95% CI, −1.12, −0.74; P < 0.0001). Heterogeneity among studies was relatively high (I2 = 72.77%), but meta-regression analysis could not find any significant potential moderators.

22.3.3.2 Survival Analysis

A total of five studies reported the white blood cell count concerning survival outcomes. Non-survivor patients (n = 707) had significantly higher WBC than survivors (n = 365) with an ES of −1.01 (95% CI, −1.33, −0.69). No evidence of publication bias was observed (Begg's P = 0.327; Egger's P = 0.356). Heterogeneity was high (I2 = 79%). Sensitivity analysis was performed, leaving out a study that only included patients with ARDS (acute respiratory syndrome) (Wu et al. 2020), and as a result, heterogeneity was abolished, and ES became −1.18 (95% CI, −1.33, −1.03).

Analysis of four studies with a total number of 570 survivors and 311 non-survivors revealed significantly higher PMN counts in non-survivor patients compared to survivors (SMD −1.05; 95% CI, −1.53, −0.58; P < 0.0001). Heterogeneity was high (I2 = 87.76%). When removing the abovementioned study (Wu et al. 2020), heterogeneity dropped to 55%, and the ES remained significant −1.32 (SMD −1.32; 95% CI, −1.59, −1.06).

Patients who survived (n = 727) had significantly higher lymphocyte than non-survivors (n = 397) (SMD 0.97; 95% CI, 0.52, 1.42). Significant publication bias was observed (Begg's P = 0.024 Egger's P = 0.043). The trim and fill method was used to adjust the ES (SMD 1.20; 95% CI, 1.06, 1.34). No individual study showed a significant influence on the ES.

Compared to non-survivors (n = 376), survivors had significantly higher platelet counts (SMD 0.45; 95% CI, 0.19, 0.70; P = 0.001). Heterogeneity was relatively high (I2 = 67.56%). When the study (Yang et al. 2020) was removed, there was no heterogeneity across studies further (I2 = 0%), while the pooled effect remained significant.

Non-survivor patients showed higher ALT (SMD −0.36, P < 0.0001) and total bilirubin (SMD = −0.79, p < 0.0001) than survivors. Albumin levels were measured in five studies. Survivor patients showed higher albumin levels compared to non-survivors (SMD = 0.88, p < 0.0001). However, heterogeneity was high (I2 = 87.95). Sensitivity analysis showed a robust ES. Additionally, with the exclusion of two studies (Wu et al. 2020; Momtazmanesh et al. 2020), I2 reduced to 58.41% (Wu et al. 2020; Momtazmanesh et al. 2020).

IL-6 (SMD −1.22; P < 0.0001) and procalcitonin (SMD −1.05; P < 0.0001) levels were significantly higher in non-survivors than survivors. For both analyses, there was no evidence of publication bias and significant heterogeneity.

A meta-analysis of data from five studies showed no difference between non-survivors (n = 343) and survivors (n = 590) regarding creatinine levels (SMD −0.34; 95% CI, −0.77, 0.09; P = 0.123). However, there was high heterogeneity (I2 = 87.55). Sensitivity analysis excluding two studies reduced heterogeneity to I2 = 0%, though the ES remained not significant (SMD 0.016; 95% CI, −0.18, 0.21; P = 0.876).

PT was significantly higher in non-survivors than survivors (SMD −0.78; 95% CI, −1.89, −0.37; P < 0.0001). However, there was no difference regarding PTT between survivors and non-survivors (SMD 0.26; 95% CI, −0.62, 0.41; P = 0.695).

22.3.4 Symptoms and Vital Status

22.3.4.1 Severity Analysis

Compared to severe patients, non-severe patients were less likely to have chills (OR 0.62; 95% CI, 0.42, 0.93), sputum production (OR 0.66; 95% CI, 0.54 to 0.79), dyspnea (OR 0.22; 95% CI, 0.14, 0.33), hemoptysis (OR 0.25; 95% CI, 0.11, 0.55), fatigue (OR 0.71; 95% CI, 0.52, 0.97), and abdominal pain (OR 0.49; 95% CI, 0.24, 0.98). Patients with severe outcomes had significantly higher fever than non-severe patients (SMD −0.38; 95% CI, −0.67, −0.09; P = 0.01).

22.3.4.2 Survival Analysis

The only symptoms that showed a difference between survivors and non-survivors were dyspnea and fatigue, which were less likely to occur in survivors (OR 0.34; 95% CI 0.25, 0.46; and OR 0.75; 95% CI 0.57, 0.90).

22.3.5 Comorbidities and Demographic Characteristics

22.3.5.1 Severity Analysis

Severe patients (n = 4006) were significantly older than non-severe patients (n = 1493) (SMD −0.59; 95% CI, −0.76, −0.41). A meta-analysis of data derived from nine studies showed that non-severe patients were less likely to be smokers than severe patients (OR 0.53; 95% CI, 0.38, 0.73). Male patients had a higher risk of developing the severe disease compared to females (OR 1.32; 95% CI, 1.17, 1.49). Non-severe patients were less likely to have diabetes (OR 0.36; 95% CI, 0.26 to 0.52), chronic heart disease (OR 0.34; 95% CI, 0.25, 0.45), COPD (OR 0.20; 95% CI, 0.12, 0.33), cerebrovascular disease (OR 0.30; 95% CI, 0.18, 0.50), hypertension (OR 0.40; 95% CI, 0.30, 0.52), CKD (OR 0.35; 95% CI, 0.18, 0.68), and malignancy (OR 0.42; 95% CI, 0.24, 0.72) than severe patients. For all analyses, there was low to moderate heterogeneity, with I2 ranging from 0 to 47%.

22.3.5.2 Survival Analysis

A meta-analysis of data from eight studies demonstrated that non-survivors (n = 435) were older than survivors (n = 974) with the SMD of −1.07 (95% CI, −1.35, −0.78; P < 0.0001). Male patients were not different from females concerning survival (p = 0.066; I2 = 51.46).

Patients who survived were less likely to have diabetes (OR 0.49, 0.36 to 0.67), chronic heart disease (OR 0.31; 95% CI, 0.13–0.72), COPD (OR 0.13; 95% CI, 0.052–0.34), cerebrovascular disease (OR 0.125; 95% CI, 0.034–0.46), and CKD (OR 0.095; 95% CI, 0.023–0.39). For all analyses, there was low to moderate heterogeneity, with I2 ranging from 0 to 61%.

22.4 Discussion

Hundreds of thousands of deaths due to COVID-19 during only a few months have placed a heavy burden on the shoulder of each medical researcher by explicitly proving that our knowledge, at the present condition of the pandemic, can seldom offer a useful product to humanity unless all data are collected from other disciplines and then it would be entirely fair to arrive at conclusions from the accumulation of this data (Momtazmanesh et al. 2020; Mohamed et al. 2020a; Rzymski et al. 2020; Moradian et al. 2020; Kafieh et al. 2020). For example, medical doctors initially described COVID-19 as a disease of the respiratory system. It did not take long when a wave of studies reporting anosmia in patients with COVID-19 emerged, putting forward the neurotropism of COVID-19 (Yazdanpanah et al. 2020b; Jahanshahlu and Rezaei 2020a; Saleki et al. 2020). The next wave was actually of two sources of studies in parallel. One was studies that reported stroke events in young patients with COVID-19, and the other was studies that provided evidence of a high incidence of thromboembolic events in patients with COVID-19. Therefore, the second wave strongly confirmed that the involvement of the central nervous system might be why patients with COVID-19 die from acute respiratory failure in a few days as well as suggested coagulopathies as another significant contributor to death from COVID-19. In the meantime, many disciplinary perspectives on the immunology of COVID-19 come into the competition, and the perspective which is already dominant deals inflammatory cells and molecules widely distributed throughout the critical organs to mediate COVID-19 pathogenesis (Pourahmad et al. 2020; Bahrami et al. 2020; Yazdanpanah et al. 2020a; Mansourabadi et al. 2020; Saghazadeh and Rezaei 2020a; Pashaei and Rezaei 2020; Nasab et al. 2020; Saghazadeh and Rezaei 2020b; Rokni et al. 2020). As a result, taking a look at the current evidence links us to a variety of biomarkers and clinical features that might predict the prognosis of COVID-19. The present chapter aimed to offer the most reliable predictors, and for this, we performed a systematic synthesis of the data.

We have stratified patients according to their diverse outcomes: survivor vs non-survivor, critical vs noncritical, and severe vs non-severe. The analysis by severity outcome demonstrated that higher levels of WBC with PMN dominancy might be related to disease severity and subsequent mortality. In the analysis of the WBC in severe patients, a relatively high amount of heterogeneity was observed. Meta-regression of the sample size could only explain 28% of that heterogeneity, and the two other moderators, including the difference in age between two groups and percentages of female patients, could not explain heterogeneity significantly.

We could not find a significant difference regarding CD4+ and CD8+ cell counts between severe and non-severe patients. However, previous investigations have shown that the depletion of T cells occurs during the COVID-19 course (Diao et al. 2020; Fathi and Rezaei 2020). The study (Zheng et al. 2020b) has shown extensive lymphocyte subset reduction in the blood of patients infected with COVID-19 compared to patients with other types of pneumonia. CD45+ lymphocytes, CD3+ lymphocytes, CD4+ T cells, CD8+ T cells, and CD19+ B cells were significantly lower in patients with COVID-19 than those in non-COVID-19-infected pneumonia patients with the same radiological stage.

In line with previous studies, thrombocytopenia was found to be associated with the severity, mortality, and critical outcomes of COVID-19

(Lippi et al. 2020). However, the most remarkable effect size was observed for comparison between survivors and non-survivor patients, which showed moderate ES. Previous literature recognized thrombocytopenia as an indicator that occurs in critically ill patients and is associated with mortality (Williamson et al. 2013). Therefore, COVID-19 can be more challenging in patients whom themselves suffer from immune thrombocytopenia (Sahu et al. 2020).

Acute-phase reactants, including ESR, CRP, ferritin, and procalcitonin, were significantly increased in patients with severe disease compared to patients with the non-severe disease and in patients who died compared to patients who survived. In contrast, negative phase reactants such as albumin, on the other hand, were decreased in patients with severe disease and in patients who died.

Liver enzymes were increased in severe, critical, and deceased patients compared with their counterparts without adverse outcomes. Given the relatively low prevalence of chronic liver disease before infection with COVID-19, studies suggest that increased liver enzymes, as seen in severe patients, might result from a dysregulated immune response against the virus rather than underlying liver comorbidity (Mantovani et al. 2020). Studies postulating the same hypothesis emphasize an inflammatory storm having played a role in immune dysregulation. Other studies believe that respiratory distress syndrome-induced hypoxia and subsequent hepatic ischemia and hypoxia-reperfusion dysfunction might contribute to liver dysfunction in COVID-19 (Feng et al. 2020a).

Our findings pointed to a higher level of TnI in both severe patients and non-survivors. A cohort of 416 hospitalized patients with COVID-19 established that 19.7% of these patients had a cardiac injury and showed an independent correlation between cardiac injury and a higher risk of mortality in patients with COVID-19 (Shi et al. 2020). Several mechanisms could be responsible for cardiac injury, including increased myocardial demand relative to the blood supply and hypoxia, systemic inflammation derived from cytokines, and direct cardiac damage due to COVID-19 (Guo et al. 2020).

On the other hand, there was not a significant difference regarding creatinine neither between severe and non-severe patients nor between survivors and non-survivors. BUN, on the other hand, was significantly increased in severe patients compared to the non-severe group. The authors in Wang et al. (2020b) evaluated the kidney function of patients with COVID-19 over 4 weeks and found that patients with COVID-19 experienced no acute renal injury and only showed a mildly elevated BUN and creatinine. However, there is controversy around acute kidney injury in severe patients.

Electrolytes, including sodium and potassium, showed no significant difference between severe and non-severe patients. However, there was high heterogeneity in studies reporting potassium, and during sensitivity analysis, a significant ES was observed indicating hypokalemia in severe cases. Studies report hypokalemia to be present in a majority of critically ill patients. It can be attributed to increased urine potassium excretion due to the renin-angiotensin system (RAS) disturbances and enhanced delivery of sodium and water to distal tubules (Li et al. 2020b).

Also, PT and PTT in severe patients were not significantly different from non-severe patients. However, PT was higher in critical patients and non-survivors compared to noncritical and survivors, respectively. It might be due to the measurement time of coagulation parameters, which mostly was at the time of admission, while it would undergo significant change over the disease course. D-dimer, on the other hand, was significantly increased in patients with poor outcome.

Patients with poor outcomes, e.g., severe disease and mortality, were more likely to have COPD, CHD, DM, HTN, CKD, cerebrovascular disease, and underlying malignancy compared to their counterparts. A nationwide analysis of 1590 patients in China reported that after adjusting for age and smoking status, COPD, diabetes, hypertension, and malignancy predicted increased admission to ICU, need for invasive ventilation, and death rate (Guan et al. 2020a).

Finally, patients with dyspnea at baseline were about four times more likely to progress to severe or critical condition. Also, hemoptysis increased the likelihood of developing a severe disease by about four times.

22.5 Conclusion

The current pandemic of COVID-19 has affected the healthcare system profoundly (Moazzami et al. 2020), and this effect can take a long time to compensate for it, and as the risk of reinfection is possible (Jabbari and Rezaei 2020), we must work at practice management. Early and efficient detection of high-risk patients is a priority for maximizing the utilization of limited resources. The present chapter aimed to offer the most reliable predictors, and for this, we performed a systematic synthesis of the data. Meta-analyses identified several laboratory parameters (WBC, PMN, IL-6, total bilirubin, ALT, creatinine, troponin, procalcitonin, LDH, and D-dimer), comorbidities (diabetes, chronic heart disease, COPD, cerebrovascular disease, hypertension, CKD, and malignancy), and symptoms (dyspnea and hemoptysis) that might help the prediction of severe, critical, and lethal phenotypes of COVID-19. These parameters correlate with the immune system function, inflammation, coagulation, and liver and kidney function, supporting the view that COVID-19 is a multisystem disorder.

References

Ahanchian H, Moazzen N, Faroughi MSD, Khalighi N, Khoshkhui M, Aelami MH, Haghi NSM, Rezaei N (2020) COVID-19 in a child with primary specific antibody deficiency

Ahmadi M, Saffarzadeh N, Habibi MA, Hajiesmaeili F, Rezaei N (2020) Colon cancer and SARS-CoV-2: impact of ACE2 expression in susceptibility to COVID-19. bioRxiv

Babaha F, Rezaei N (2020) Primary immunodeficiency diseases in COVID-19 pandemic: a predisposing or protective factor? Am J Med Sci. https://doi.org/10.1016/j.amjms.2020.07.027

Bahrami A, Vafapour M, Moazzami B, Rezaei N (2020) Hyperinflammatory shock related to COVID-19 in a patient presenting with multisystem inflammatory syndrome in children: first case from Iran. J Paediatr Child Health. https://doi.org/10.1111/jpc.15048

Basiri A, Heidari A, Nadi MF, Fallahy MTP, Nezamabadi SS, Sedighi M, Saghazadeh A, Rezaei N (2020a) Microfluidic devices for detection of RNA viruses. Rev Med Virol n/a (n/a):e2154. https://doi.org/10.1002/rmv.2154

Basiri A, Pazhouhnia Z, Beheshtizadeh N, Hoseinpour M, Saghazadeh A, Rezaei N (2020b) Regenerative medicine in COVID-19 treatment: real opportunities and range of promises. Stem Cell Rev Rep:1–13. https://doi.org/10.1007/s12015-020-09994-5

Begg CB, Mazumdar M (1994) Operating characteristics of a rank correlation test for publication Bias. Biometrics 50(4):1088–1101. https://doi.org/10.2307/2533446

Cai Q, Huang D, Ou P, Yu H, Zhu Z, Xia Z, Su Y, Ma Z, Zhang Y, Li Z, He Q, Liu L, Fu Y, Chen J COVID-19 in a designated infectious diseases hospital outside Hubei Province, China. Allergy n/a (n/a) https://doi.org/10.1111/all.14309

Cao W (2020) Clinical features and laboratory inspection of novel coronavirus pneumonia (COVID-19) in Xiangyang, Hubei. med Rxiv:2020.2002.2023.20026963. https://doi.org/10.1101/2020.02.23.20026963

Cao J, Tu W-J, Cheng W, Yu L, Liu Y-K, Hu X, Liu Q (2020) Clinical features and short-term outcomes of 102 patients with Corona virus disease 2019 in Wuhan, China. Clin Infect Dis. https://doi.org/10.1093/cid/ciaa243

Cascella M, Rajnik M, Cuomo A, Dulebohn SC, Di Napoli R (2020) Features, evaluation and treatment coronavirus (COVID-19). In: Statpearls [internet]. StatPearls Publishing

Chen G, Wu D, Guo W, Cao Y, Huang D, Wang H, Wang T, Zhang X, Chen H, Yu H, Zhang X, Zhang M, Wu S, Song J, Chen T, Han M, Li S, Luo X, Zhao J, Ning Q (2020a) Clinical and immunologic features in severe and moderate forms of Coronavirus disease 2019. medRxiv:2020.2002.2016.20023903. https://doi.org/10.1101/2020.02.16.20023903

Chen G, Wu D, Guo W, Cao Y, Huang D, Wang H, Wang T, Zhang X, Chen H, Yu H (2020b) Clinical and immunological features of severe and moderate coronavirus disease 2019. J Clin Invest 130(5)

Chen J, Qi T, Liu L, Ling Y, Qian Z, Li T, Li F, Xu Q, Zhang Y, Xu S, Song Z, Zeng Y, Shen Y, Shi Y, Zhu T, Lu H (2020c) Clinical progression of patients with COVID-19 in Shanghai, China. J Infect. https://doi.org/10.1016/j.jinf.2020.03.004

Chen T, Dai Z, Mo P, Li X, Ma Z, Song S, Chen X, Luo M, Liang K, Gao S, Zhang Y, Deng L, Xiong Y (2020d) Clinical characteristics and outcomes of older patients with coronavirus disease 2019 (COVID-19) in Wuhan, China (2019): a single-centered, retrospective study. J Gerontol Ser A. https://doi.org/10.1093/gerona/glaa089

Chen T, Wu D, Chen H, Yan W, Yang D, Chen G, Ma K, Xu D, Yu H, Wang H (2020e) Clinical characteristics of 113 deceased patients with coronavirus disease 2019: retrospective study. BMJ:368

Chu J, Yang N, Wei Y, Yue H, Zhang F, Zhao J, He L, Sheng G, Chen P, Li G, Wu S, Zhang B, Zhang S, Wang C, Miao X, Li J, Liu W, Zhang H (2020) Clinical characteristics of 54 medical staff with COVID-19: a retrospective study in a single center in Wuhan, China. J Med Virol. https://doi.org/10.1002/jmv.25793

Darbeheshti F, Rezaei N (2020) Genetic predisposition models to COVID-19 infection. Med Hypotheses 142:109818. https://doi.org/10.1016/j.mehy.2020.109818

de Wit E, van Doremalen N, Falzarano D, Munster VJ (2016) SARS and MERS: recent insights into emerging coronaviruses. Nat Rev Microbiol 14(8):523

Diao B, Wang C, Tan Y, Chen X, Liu Y, Ning L, Chen L, Li M, Liu Y, Wang G, Yuan Z, Feng Z, Wu Y, Chen Y (2020) Reduction and functional exhaustion of T Cells in patients with Coronavirus disease 2019 (COVID-19). medRxiv:2020.2002.2018.20024364. https://doi.org/10.1101/2020.02.18.20024364

Du R-H, Liu L-M, Yin W, Wang W, Guan L-L, Yuan M-L, Li Y-L, Hu Y, Li X-Y, Sun B (2020) Hospitalization and critical care of 109 decedents with COVID-19 pneumonia in Wuhan, China. Ann Am Thorac Soc

Duval S, Tweedie R (2000) Trim and fill: a simple funnel-plot–based method of testing and adjusting for publication Bias in meta-analysis. Biometrics 56(2):455–463. https://doi.org/10.1111/j.0006-341X.2000.00455.x

Fan BE, Chong VCL, Chan SSW, Lim GH, Lim KGE, Tan GB, Mucheli SS, Kupcran P, Ong KH (2020) Hematologic parameters in patients with COVID-19 infection. Am J Hematol n/a (n/a) https://doi.org/10.1002/ajh.25774

Fathi N, Rezaei N (2020) Lymphopenia in COVID-19: therapeutic opportunities. Cell Biol Int 44(9):1792–1797. https://doi.org/10.1002/cbin.11403

Feng G, Zheng KI, Yan Q-Q, Rios RS, Targher G, Byrne CD, Van Poucke S, Liu W-Y, Zheng M-H (2020a) COVID-19 and liver dysfunction: current insights and emergent therapeutic strategies. J Clin Transl Hepatol 8(1):18

Feng Y, Ling Y, Bai T, Xie Y, Huang J, Li J, Xiong W, Yang D, Chen R, Lu F (2020b) COVID-19 with different severity: a multi-center study of clinical features. Am J Respir Crit Care Med

Feng Z, Yu Q, Yao S, Luo L, Duan J, Yan Z, Yang M, Tan H, Ma M, Li T, Yi D, Mi Z, Zhao H, Jiang Y, He Z, Li H, Nie W, Liu Y, Zhao J, Wang W (2020c) Early prediction of disease progression in 2019 Novel Coronavirus pneumonia patients outside Wuhan with CT and clinical characteristics. https://doi.org/10.1101/2020.02.19.20025296

Gao Y, Li T, Han M, Li X, Wu D, Xu Y, Zhu Y, Liu Y, Wang X, Wang L (2020) Diagnostic utility of clinical laboratory data determinations for patients with the severe COVID-19. J Med Virol. https://doi.org/10.1002/jmv.25770

Gong J, Ou J, Qiu X, Jie Y, Chen Y, Yuan L, Cao J, Tan M, Xu W, Zheng F (2020) Multicenter development and validation of a novel risk nomogram for early prediction of severe 2019-novel coronavirus pneumonia. Available at SSRN 3551365

Guan W-j, Liang W-h, Zhao Y, Liang H-r, Chen Z-s, Li Y-m, Liu X-q, Chen R-c, Tang C-l, Wang T, Ou C-q, Li L, Chen P-y, Sang L, Wang W, Li J-f, Li C-c, Ou L-m, Cheng B, Xiong S, Ni Z-y, Xiang J, Hu Y, Liu L, Shan H, Lei C-l, Peng Y-x, Wei L, Liu Y, Hu Y-h, Peng P, Wang J-m, Liu J-y, Chen Z, Li G, Zheng Z-j, Qiu S-q, Luo J, Ye C-j, Zhu S-y, Cheng L-l, Ye F, Li S-y, Zheng J-p, Zhang N-f, Zhong N-s, He J-x (2020a) Comorbidity and its impact on 1590 patients with Covid-19 in China: a nationwide analysis. Eur Respir J:2000547. https://doi.org/10.1183/13993003.00547-2020

Guan W-J, Ni Z-Y, Hu Y, Liang W-H, Ou C-Q, He J-X, Liu L, Shan H, Lei C-L, Hui DSC, Du B, Li L-J, Zeng G, Yuen K-Y, Chen R-C, Tang C-L, Wang T, Chen P-Y, Xiang J, Li S-Y, Wang J-L, Liang Z-J, Peng Y-X, Wei L, Liu Y, Hu Y-H, Peng P, Wang J-M, Liu J-Y, Chen Z, Li G, Zheng Z-J, Qiu S-Q, Luo J, Ye C-J, Zhu S-Y, Zhong N-S (2020b) Clinical characteristics of coronavirus disease 2019 in China. N Engl J Med. https://doi.org/10.1056/NEJMoa2002032

Guo T, Fan Y, Chen M, Wu X, Zhang L, He T, Wang H, Wan J, Wang X, Lu Z (2020) Cardiovascular implications of fatal outcomes of patients with coronavirus disease 2019 (COVID-19). JAMA Cardiol

Han H, Xie L, Liu R, Yang J, Liu F, Wu K, Chen L, Hou W, Feng Y, Zhu C (2020a) Analysis of heart injury laboratory parameters in 273 COVID-19 patients in one hospital in Wuhan, China. J Med Virol

Han Y, Zhang H, Mu S, Wei W, Jin C, Xue Y, Tong C, Zha Y, Song Z, Gu G (2020b) Lactate dehydrogenase, a risk factor of severe COVID-19 patients. medRxiv

Hanaei S, Rezaei N (2020) COVID-19: developing from an outbreak to a pandemic. Arch Med Res. https://doi.org/10.1016/j.arcmed.2020.04.021

Higgins J, Wells G (2011) Cochrane handbook for systematic reviews of interventions

Huang C, Wang Y, Li X, Ren L, Zhao J, Hu Y, Zhang L, Fan G, Xu J, Gu X, Cheng Z, Yu T, Xia J, Wei Y, Wu W, Xie X, Yin W, Li H, Liu M, Xiao Y, Gao H, Guo L, Xie J, Wang G, Jiang R, Gao Z, Jin Q, Wang J, Cao B (2020) Clinical features of patients infected with 2019 novel coronavirus in Wuhan, China. Lancet 395(10223):497–506. https://doi.org/10.1016/S0140-6736(20)30183-5

Jabbari P, Rezaei N (2020) With risk of reinfection, is COVID-19 here to stay? Disaster Med Public Health Prep:1. https://doi.org/10.1017/dmp.2020.274

Jabbari P, Jabbari F, Ebrahimi S, Rezaei N (2020) COVID-19: a chimera of two pandemics. Disaster Med Public Health Prep:1–2. https://doi.org/10.1017/dmp.2020.223

Jahanshahlu L, Rezaei N (2020a) Central nervous system involvement in COVID-19. Arch Med Res. https://doi.org/10.1016/j.arcmed.2020.05.016

Jahanshahlu L, Rezaei N (2020b) Monoclonal antibody as a potential anti-COVID-19. Biomed Pharmacother Biomed Pharmacotherapie 129:110337. https://doi.org/10.1016/j.biopha.2020.110337

Ji D, Zhang D, Xu J, Chen Z, Yang T, Zhao P, Chen G, Cheng G, Wang Y, Bi J, Tan L, Lau G, Qin E (2020) Prediction for progression risk in patients with COVID-19 pneumonia: the CALL score. Clin Infect Dis. https://doi.org/10.1093/cid/ciaa414

Kafieh R, Arian R, Saeedizadeh N, Minaee S, Yadav SK, Vaezi A, Rezaei N, Javanmard SH (2020) COVID-19 in Iran: a deeper look into the future. medRxiv

Ksiazek TG, Erdman D, Goldsmith CS, Zaki SR, Peret T, Emery S, Tong S, Urbani C, Comer JA, Lim W (2003) A novel coronavirus associated with severe acute respiratory syndrome. N Engl J Med 348(20):1953–1966

Lei I, Jian-Ya G (2020) Clinical characteristics of 51 patients discharged from hospital with COVID-19 in Chongqing, China. med Rxiv:2020.2002.2020.20025536. https://doi.org/10.1101/2020.02.20.20025536

Li K, Wu J, Wu F, Guo D, Chen L, Fang Z, Li C (2020a) The clinical and chest CT features associated with severe and critical COVID-19 pneumonia. Investig Radiol

Li X, Hu C, Su F, Dai J (2020b) Hypokalemia and clinical implications in patients with coronavirus disease 2019 (COVID-19). MedRxiv

Li YK, Peng S, Li LQ, Wang Q, Ping W, Zhang N, Fu XN (2020c) Clinical and transmission characteristics of Covid-19 - a retrospective study of 25 cases from a single thoracic surgery department. Curr Med Sci. https://doi.org/10.1007/s11596-020-2176-2

Lippi G, Plebani M, Henry BM (2020) Thrombocytopenia is associated with severe coronavirus disease 2019 (COVID-19) infections: a meta-analysis. Clin Chim Acta

Liu J, Li S, Liu J, Liang B, Wang X, Wang H, Li W, Tong Q, Yi J, Zhao L, Xiong L, Guo C, Tian J, Luo J, Yao J, Pang R, Shen H, Peng C, Liu T, Zhang Q, Wu J, Xu L, Lu S, Wang B, Weng Z, Han C, Zhu H, Zhou R, Zhou H, Chen X, Ye P, Zhu B, He S, He Y, Jie S, Wei P, Zhang J, Lu Y, Wang W, Zhang L, Li L, Zhou F, Wang J, Dittmer U, Lu M, Hu Y, Yang D, Zheng X (2020a) Longitudinal characteristics of lymphocyte responses and cytokine profiles in the peripheral blood of SARS-CoV-2 infected patients. medRxiv:2020.2002.2016.20023671. https://doi.org/10.1101/2020.02.16.20023671

Liu J, Liu Y, Xiang P, Pu L, Xiong H, Li C, Zhang M, Tan J, Xu Y, Song R, Song M, Wang L, Zhang W, Han B, Yang L, Wang X, Zhou G, Zhang T, Li B, Wang Y, Chen Z, Wang X (2020b) Neutrophil-to-lymphocyte ratio predicts severe illness patients with 2019 novel coronavirus in the early stage. medRxiv:2020.2002.2010.20021584. https://doi.org/10.1101/2020.02.10.20021584

Liu KC, Xu P, Lv WF, Qiu XH, Yao JL, Gu JF, Wei W (2020c) CT manifestations of coronavirus disease-2019: a retrospective analysis of 73 cases by disease severity. Eur J Radiol 126. https://doi.org/10.1016/j.ejrad.2020.108941

Liu Y, Sun W, Li J, Chen L, Wang Y, Zhang L, Yu L (2020d) Clinical features and progression of acute respiratory distress syndrome in coronavirus disease 2019. medRxiv

Lotfi M, Rezaei N (2020) SARS-CoV-2: a comprehensive review from pathogenicity of the virus to clinical consequences. J Med Virol. https://doi.org/10.1002/jmv.26123

Lotfi M, Hamblin MR, Rezaei N (2020) COVID-19: transmission, prevention, and potential therapeutic opportunities. Clin Chim Acta 508:254–266. https://doi.org/10.1016/j.cca.2020.05.044

Lu J, Hu S, Fan R, Liu Z, Yin X, Wang Q, Lv Q, Cai Z, Li H, Hu Y, Han Y, Hu H, Gao W, Feng S, Liu Q, Li H, Sun J, Peng J, Yi X, Zhou Z, Guo Y, Hou J (2020) ACP risk grade: a simple mortality index for patients with confirmed or suspected severe acute respiratory syndrome coronavirus 2 disease (COVID-19) during the early stage of outbreak in Wuhan, China. medRxiv:2020.2002.2020.20025510. https://doi.org/10.1101/2020.02.20.20025510

Lu H, Stratton CW, Tang YW. Outbreak of pneumonia of unknown etiology in Wuhan China: the mystery and the miracle. J Med Virol

Luo D, Wan X, Liu J, Tong T (2018) Optimally estimating the sample mean from the sample size, median, mid-range, and/or mid-quartile range. Stat Methods Med Res 27(6):1785–1805

Mansourabadi AH, Sadeghalvad M, Mohammadi-Motlagh HR, Rezaei N (2020) The immune system as a target for therapy of SARS-CoV-2: a systematic review of the current immunotherapies for COVID-19. Life Sci 258:118185. https://doi.org/10.1016/j.lfs.2020.118185

Mantovani A, Beatrice G, Dalbeni A (2020) Coronavirus disease 2019 (COVID-19) and prevalence of chronic liver disease: a meta-analysis. Liver Int

Mirbeyk M, Rezaei N (2020) The impact of COVID-19 on pregnancy and neonatal health: a systematic review

Moazzami B, Razavi-Khorasani N, Dooghaie Moghadam A, Farokhi E, Rezaei N (2020) COVID-19 and telemedicine: immediate action required for maintaining healthcare providers well-being. J Clin Virol 126:104345. https://doi.org/10.1016/j.jcv.2020.104345

Mohamed K, Rodríguez-Román E, Rahmani F, Zhang H, Ivanovska M, Makka SA, Joya M, Makuku R, Islam MS, Radwan N, Rahmah L, Goda R, Abarikwu SO, Shaw M, Zoghi S, Irtsyan S, Ling I, Cseprekal O, Faten AB, Hazar Sayar E, Soloukey C, Grancini G, Rezaei N (2020a) Borderless collaboration is needed for COVID-19-A disease that knows no borders. Infect

Control Hosp Epidemiol:1–2. https://doi.org/10.1017/ice.2020.162
Mohamed K, Yazdanpanah N, Saghazadeh A, Rezaei N (2020b) Computational drug discovery and repurposing for the treatment of Covid-19: a systematic review. Available at SSRN 3583748
Momtazmanesh S, Ochs HD, Uddin LQ, Perc M, Routes JM, Vieira DN, Al-Herz W, Baris S, Prando C, Rosivall L, Abdul Latiff AH, Ulrichs T, Roudenok V, Aldave Becerra JC, Salunke DB, Goudouris E, Condino-Neto A, Stashchak A, Kryvenko O, Stashchak M, Bondarenko A, Rezaei N (2020) All together to fight COVID-19. Am J Trop Med Hyg 102(6):1181–1183. https://doi.org/10.4269/ajtmh.20-0281
Moradian N, Ochs HD, Sedikies C, Hamblin MR, Camargo CA Jr, Martinez JA, Biamonte JD, Abdollahi M, Torres PJ, Nieto JJ, Ogino S, Seymour JF, Abraham A, Cauda V, Gupta S, Ramakrishna S, Sellke FW, Sorooshian A, Wallace Hayes A, Martinez-Urbistondo M, Gupta M, Azadbakht L, Esmaillzadeh A, Kelishadi R, Esteghamati A, Emam-Djomeh Z, Majdzadeh R, Palit P, Badali H, Rao I, Saboury AA, Jagan Mohan Rao L, Ahmadieh H, Montazeri A, Fadini GP, Pauly D, Thomas S, Moosavi-Movahed AA, Aghamohammadi A, Behmanesh M, Rahimi-Movaghar V, Ghavami S, Mehran R, Uddin LQ, Von Herrath M, Mobasher B, Rezaei N (2020) The urgent need for integrated science to fight COVID-19 pandemic and beyond. J Transl Med 18(1):205. https://doi.org/10.1186/s12967-020-02364-2
Nasab MG, Saghazadeh A, Rezaei N (2020) SARS-CoV-2-a tough opponent for the immune system. Arch Med Res. https://doi.org/10.1016/j.arcmed.2020.05.020
Nie S, Zhao X, Zhao K, Zhang Z, Zhang Z, Zhang Z (2020) Metabolic disturbances and inflammatory dysfunction predict severity of coronavirus disease 2019 (COVID-19): a retrospective study. medRxiv
Pashaei M, Rezaei N (2020) Immunotherapy for SARS-CoV-2: potential opportunities. Expert Opin Biol Ther:1–5. https://doi.org/10.1080/14712598.2020.1807933
Pourahmad R, Moazzami B, Rezaei N (2020) Efficacy of Plasmapheresis and immunoglobulin replacement therapy (IVIG) on patients with COVID-19. SN Compr Clin Med:1–5
Qin C, Zhou L, Hu Z, Zhang S, Yang S, Tao Y, Xie C, Ma K, Shang K, Wang W (2020) Dysregulation of immune response in patients with COVID-19 in Wuhan, China. Clin Infect Dis
Qu R, Ling Y, Zhang Y, Ly W, Chen X, Li X, Xy L, Hm L, Guo Z, Ren H (2020) Platelet-to-lymphocyte ratio is associated with prognosis in patients with Corona virus Disease-19. J Med Virol
Rabiee N, Rabiee M, Bagherzadeh M, Rezaei N (2020) COVID-19 and picotechnology: potential opportunities. Med Hypotheses 144:109917. https://doi.org/10.1016/j.mehy.2020.109917
Rezaei N (2020a) COVID-19 affects healthy pediatricians more than pediatric patients. Infect Control Hosp Epidemiol 1. https://doi.org/10.1017/ice.2020.139
Rezaei N (2020b) COVID-19 and medical biotechnology. Avicenna J Med biotechnol 12(3):139
Rokni M, Hamblin MR, Rezaei N (2020) Cytokines and COVID-19: friends or foes? Hum Vaccin Immunother:1–3
Rzymski P, Nowicki M, Mullin GE, Abraham A, Rodríguez-Román E, Petzold MB, Bendau A, Sahu KK, Ather A, Naviaux A-F (2020) Quantity does not equal quality: scientific principles cannot be sacrificed. Int Immunopharmacol 86:106711
Saghazadeh A, Rezaei N (2020a) Immune-epidemiological parameters of the novel coronavirus - a perspective. Expert Rev Clin Immunol 16(5):465–470. https://doi.org/10.1080/1744666x.2020.1750954
Saghazadeh A, Rezaei N (2020b) Towards treatment planning of COVID-19: rationale and hypothesis for the use of multiple immunosuppressive agents: anti-antibodies, immunoglobulins, and corticosteroids. Int Immunopharmacol 84:106560. https://doi.org/10.1016/j.intimp.2020.106560
Sahu KK, Siddiqui AD, Rezaei N, Cerny J (2020) Challenges for management of immune thrombocytopenia during COVID-19 pandemic. J Med Virol. https://doi.org/10.1002/jmv.26251
Saleki K, Banazadeh M, Saghazadeh A, Rezaei N (2020) The involvement of the central nervous system in patients with COVID-19. Rev Neurosci 31(4):453–456. https://doi.org/10.1515/revneuro-2020-0026
Shamshirian D, Rezaei N (2020) Cardiovascular diseases burden in COVID-19: Systematic review and meta-analysis
Sharifkashani S, Bafrani MA, Khaboushan AS, Pirzadeh M, Kheirandish A, Yavarpour Bali H, Hessami A, Saghazadeh A, Rezaei N (2020) Angiotensin-converting enzyme 2 (ACE2) receptor and SARS-CoV-2: potential therapeutic targeting. Eur J Pharmacol 884:173455. https://doi.org/10.1016/j.ejphar.2020.173455
Shi S, Qin M, Shen B, Cai Y, Liu T, Yang F, Gong W, Liu X, Liang J, Zhao Q (2020) Association of cardiac injury with mortality in hospitalized patients with COVID-19 in Wuhan, China. JAMA Cardiol
Tang N, Li D, Wang X, Sun Z (2020) Abnormal coagulation parameters are associated with poor prognosis in patients with novel coronavirus pneumonia. J Thromb Haemost
The Novel Coronavirus Pneumonia Emergency Response Epidemiology T (2020) The epidemiological characteristics of an outbreak of 2019 novel coronavirus diseases (COVID-19) – China, 2020. China

CDC Wkly 2(8):113–122. https://doi.org/10.46234/ccdcw2020.032

Tian S, Hu N, Lou J, Chen K, Kang X, Xiang Z, Chen H, Wang D, Liu N, Liu D (2020) Characteristics of COVID-19 infection in Beijing. J Infect

Wan X, Wang W, Liu J, Tong T (2014) Estimating the sample mean and standard deviation from the sample size, median, range and/or interquartile range. BMC Med Res Methodol 14(1):135. https://doi.org/10.1186/1471-2288-14-135

Wan S, Yi Q, Fan S, Lv J, Zhang X, Guo L, Lang C, Xiao Q, Xiao K, Yi Z, Qiang M, Xiang J, Zhang B, Chen Y (2020) Characteristics of lymphocyte subsets and cytokines in peripheral blood of 123 hospitalized patients with 2019 novel coronavirus pneumonia (NCP). https://doi.org/10.1101/2020.02.10.20021832

Wan S, Xiang Y, Fang W, Zheng Y, Li B, Hu Y, Lang C, Huang D, Sun Q, Xiong Y, Huang X, Lv J, Luo Y, Shen L, Yang H, Huang G, Yang R Clinical features and treatment of COVID-19 patients in Northeast Chongqing. J Med Virol n/a (n/a) https://doi.org/10.1002/jmv.25783

Wang D, Hu B, Hu C, Zhu F, Liu X, Zhang J, Wang B, Xiang H, Cheng Z, Xiong Y (2020a) Clinical characteristics of 138 hospitalized patients with 2019 novel coronavirus–infected pneumonia in Wuhan, China. JAMA

Wang L, Li X, Chen H, Yan S, Li Y, Li D, Gong Z (2020b) SARS-CoV-2 infection does not significantly cause acute renal injury: an analysis of 116 hospitalized patients with COVID-19 in a single hospital, Wuhan, China medRxiv:20202002.2019.20025288. https://doi.org/10.1101/2020.02.19.20025288

Wang Y, Lu X, Chen H, Chen T, Su N, Huang F, Zhou J, Zhang B, Li Y, Yan F, Wang J (2020c) Clinical course and outcomes of 344 intensive care patients with COVID-19. Am J Respir Crit Care Med. https://doi.org/10.1164/rccm.202003-0736LE

Wang Y, Yao L, Zhang J-P, Tang P-J, Ye Z-J, Shen X-H, Xu J-C, Wu M-Y, Yu X (2020d) Clinical Characteristics and Laboratory Indicator Analysis of 69 COVID-19 Pneumonia Patients in Suzhou, China. China (3/3/2020)

Wang Z, Yang B, Li Q, Wen L, Zhang R (2020e) Clinical features of 69 cases with coronavirus disease 2019 in Wuhan, China. Clin Infect Dis

Williamson DR, Lesur O, Tétrault J-P, Nault V, Pilon D (2013) Thrombocytopenia in the critically ill: prevalence, incidence, risk factors, and clinical outcomes. Can J Anesth 60(7):641–651. https://doi.org/10.1007/s12630-013-9933-7

Wu C, Chen X, Cai Y, Zhou X, Xu S, Huang H, Zhang L, Zhou X, Du C, Zhang Y (2020) Risk factors associated with acute respiratory distress syndrome and death in patients with coronavirus disease 2019 pneumonia in Wuhan, China. JAMA Intern Med

Xie H, Zhao J, Lian N, Lin S, Xie Q, Zhuo H (2020) Clinical characteristics of non-ICU hospitalized patients with coronavirus disease 2019 and liver injury: a retrospective study. Liver Int. https://doi.org/10.1111/liv.14449

Xu YH, Dong JH, An WM, Lv XY, Yin XP, Zhang JZ, Dong L, Ma X, Zhang HJ, Gao BL (2020) Clinical and computed tomographic imaging features of novel coronavirus pneumonia caused by SARS-CoV-2. J Infect. https://doi.org/10.1016/j.jinf.2020.02.017

Yan S, Song X, Lin F, Zhu H, Wang X, Li M, Ruan J, Lin C, Liu X, Wu Q, Luo Z, Fu W, Chen S, Yuan Y, Liu S, Yao J, Lv C (2020) Clinical characteristics of coronavirus disease 2019 in Hainan, China. med Rxiv:2020.2003.2019.20038539. doi:https://doi.org/10.1101/2020.03.19.20038539

Yang X, Yu Y, Xu J, Shu H, Liu H, Wu Y, Zhang L, Yu Z, Fang M, Yu T (2020) Clinical course and outcomes of critically ill patients with SARS-CoV-2 pneumonia in Wuhan, China: a single-centered, retrospective, observational study. Lancet Respir Med

Yazdanpanah F, Hamblin MR, Rezaei N (2020a) The immune system and COVID-19: friend or foe? Life Sci 256:117900. https://doi.org/10.1016/j.lfs.2020.117900

Yazdanpanah N, Saghazadeh A, Rezaei N (2020b) Anosmia: a missing link in the neuroimmunology of coronavirus disease 2019 (COVID-19). Rev Neurosci. https://doi.org/10.1515/revneuro-2020-0039

Young BE, Ong SWX, Kalimuddin S, Low JG, Tan SY, Loh J, Ng O-T, Marimuthu K, Ang LW, Mak TM, Lau SK, Anderson DE, Chan KS, Tan TY, Ng TY, Cui L, Said Z, Kurupatham L, Chen MI-C, Chan M, Vasoo S, Wang L-F, Tan BH, Lin RTP, Lee VJM, Leo Y-S, Lye DC, Team ftSNCOR (2020) Epidemiologic features and clinical course of patients infected with SARS-CoV-2 in Singapore. JAMA. https://doi.org/10.1001/jama.2020.3204

Yousefzadegan S, Rezaei N (2020) Case report: death due to COVID-19 in three brothers. Am J Trop Med Hyg 102(6):1203–1204. https://doi.org/10.4269/ajtmh.20-0240

Zhang G, Hu C, Luo L, Fang F, Chen Y, Li J, Peng Z, Pan H (2020a) Clinical features and outcomes of 221 patients with COVID-19 in Wuhan, China. medRxiv

Zhang H, Wang X, Fu Z, Luo M, Zhang Z, Zhang K, He Y, Wan D, Zhang L, Wang J (2020b) Potential factors for prediction of disease severity of COVID-19 patients

Zhang JJ, Dong X, Cao YY, Yuan YD, Yang YB, Yan YQ, Akdis CA, Gao YD (2020c) Clinical characteristics of 140 patients infected with SARS-CoV-2 in Wuhan, China. Allergy. https://doi.org/10.1111/all.14238

Zhao W, Zhong Z, Xie X, Yu Q, Liu J (2020) Relation between chest CT findings and clinical conditions of coronavirus disease (COVID-19) pneumonia: a multicenter study. Am J Roentgenol:1–6

Zheng F, Tang W, Li H, Huang YX, Xie YL, Zhou ZG (2020a) Clinical characteristics of 161 cases of corona virus disease 2019 (COVID-19) in Changsha. Eur Rev Med Pharmacol Sci 24(6):3404–3410. https://doi.org/10.26355/eurrev_202003_20711

Zheng Y, Huang Z, Ying G, Zhang X, Ye W, Hu Z, Hu C, Wei H, Zeng Y, Chi Y (2020b) Comparative study of the lymphocyte change between COVID-19 and non-COVID-19 pneumonia cases suggesting uncontrolled inflammation might not be the main reason of tissue injury. medRxiv

Zheng Y, Xu H, Yang M, Zeng Y, Chen H, Liu R, Li Q, Zhang N, Wang D (2020c) Epidemiological characteristics and clinical features of 32 critical and 67 non-critical cases of COVID-19 in Chengdu. J Clin Virol 104366

Zhou F, Yu T, Du R, Fan G, Liu Y, Liu Z, Xiang J, Wang Y, Song B, Gu X (2020a) Clinical course and risk factors for mortality of adult inpatients with COVID-19 in Wuhan, China: a retrospective cohort study. Lancet

Zhou Y, Yang Z, Guo Y, Geng S, Gao S, Ye S, Hu Y, Wang Y (2020b) A new predictor of disease severity in patients with COVID-19 in Wuhan, China. medRxiv

Zhu N, Zhang D, Wang W, Li X, Yang B, Song J, Zhao X, Huang B, Shi W, Lu R (2020) A novel coronavirus from patients with pneumonia in China, 2019. N Engl J Med

23 Diagnostic Tests for COVID-19

Tung Phan and Kristin Nagaro

Abstract

The diagnosis of severe acute respiratory syndrome coronavirus 2 (SARS-CoV-2) has ramifications on both an individual level and a public health level. The use of appropriate testing mechanisms is paramount to preventing transmission, along with offering treatment to those who are infected and show appropriate symptomatology. The choice of employing a specific test often relies on laboratory capabilities, including the abilities of the medical technologists, the cost of testing platforms, and the individual quirks of each test. This chapter intends to discuss the relevant issues relating to diagnostic testing for SARS-CoV-2, including specimen types and collection methods, viral detection methods, and serological testing.

Keywords

COVID-19 · Diagnostic testing · Molecular detection · SARS-CoV-2 · Serodiagnosis · Specimen collection · Specimen types · Viral detection

T. Phan
University of Pittsburgh and University of Pittsburgh Medical Center, Pittsburgh, PA, USA

K. Nagaro (✉)
Indiana University School of Medicine, Indianapolis, IN, USA

23.1 Introduction

Identification of severe acute respiratory syndrome coronavirus 2 (SARS-CoV-2) is one of the most critical aspects of care, impacting both treatment of individual patients and preventative recommendations on the scale of public and global health. However, given the rapid and transmittable nature of SARS-CoV-2, along with other complicated factors beyond the scope of medicine, the ability to implement testing protocols has been made more difficult than would be preferred. Also, many companies have created SARS-CoV-2 testing kits, and the choice to use one test over another is dependent on a multitude of factors, including specimen type, costs of labor and kits, and more. In order to organize these issues, this chapter will focus first on the specimens, following the testing mechanisms. It is expected, however, that as research and testing continue to shine a light on this subject, much of this chapter will likely require regular updates in order to keep abreast of recent changes.

23.2 Specimen Collection

As with any infection, the choice of the specimen depends on the most common anatomical sites affected by the virus. Active viral replication has been noted in the upper respiratory tract, discovered through reverse transcription-polymerase

N. Rezaei (ed.), *Coronavirus Disease - COVID-19*, Advances in Experimental Medicine and Biology 1318, https://doi.org/10.1007/978-3-030-63761-3_23

chain reaction (RT-PCR) on subgenomic messenger RNAs (sgRNA), within the first 5 days of symptom presentation (Wolfel et al. 2020). As such, the most reliable recommendation states that a nasopharyngeal (NP) swab performed by a healthcare practitioner should be used as the primary specimen for testing purposes. It has also been accepted that in the instance that an NP swab cannot be collected, nasal swabs, nasal turbinate swabs, or oropharyngeal (OP) swabs may be collected and used for testing. If an OP swab is collected along with an NP swab, it should be placed in the same container (CDC 2020). There is conflicting data on the requirement for an OP swab to be collected concurrently but is generally not suggested that clinicians rely only on an OP swab over an NP given the lower sensitivity of OP swabs (Wang et al. 2020a, b).

Lower respiratory tract specimens are also acceptable, preferably from patients who have already been intubated. The highest rates of positive viral RNA tests are reported from bronchoalveolar lavage (BAL) specimens due to having higher viral loads in comparison to upper respiratory tract specimens (Wang et al. 2020a; Yu et al. 2020). However, there is a danger for droplet dispersion of viral particles causing iatrogenic infections; it is advised that bronchoscopy should be limited to specific therapeutic indications and patients who have already been intubated. Induced sputum is also not recommended for the same reason, but expectorated sputum can be collected from patients with a productive cough (CDC 2020). These recommendations have been expounded upon by the Infectious Diseases Society of America, who has provided a template to help guide healthcare professionals on the choice of testing methodologies (IDSA 2020). Of note, there are no recommendations for serological testing under these guidelines.

There have been multiple issues surrounding the pre-analytical components of laboratory testing. Although NP swabs are the preferred method, there was a global shortage of swabs requiring laboratories to come up with unique ways to address this crisis. One idea was to use additive manufacturing (also known as 3D printing) to produce more swabs. Cox and colleagues used a filament-based printer that allowed the swabs to remain flexible, while releasing the sample from the swab into a proper medium (Cox and Koepsell 2020). Callahan et al. performed a multistep preclinical evaluation on 160 swab designs and 48 materials from 24 companies, laboratories, and individuals. However, none of the prototypes tested were flocked, though their performance was statistically indistinguishable from the flocked control swab (Callahan et al. 2020). Overall, there is a dearth of publications comparing manufactured swabs against 3D-printed ones, leaving questions about the equivocality of the different types of swabs.

Another issue has been the limited quantities of viral transport media or universal transport media. Swabs are placed in the media in order to preserve the cells and potential viruses at room temperature, inhibit bacterial and fungal growth, and allow for long-term preservation when frozen (Diagnostics 2020). The Food and Drug Administration (FDA) has offered a list of acceptable swab types (NP, OP, mid-turbinate, and anterior nares) along with transport media products, including VTM/UTM and phosphate-buffered saline (FDA 2020).

Many additional specimen types have also been evaluated, in part to determine the duration of patient isolation and decision-making regarding the safety of hospital discharge. SARS-CoV-2 RNA has been found in both fecal specimens (Young et al. 2020; Zhang et al. 2020b) and respiratory specimens (Paules et al. 2020) late in the clinical course. One study from Beijing identified 133 SARS-CoV-2-positive patients via quantitative reverse transcription PCR (RT-qPCR) and found 22 patients who had positive sputum or fecal specimens even after pharyngeal swabs tested negative (Chen et al. 2020a). SARS-CoV-2 was detected in a total of 28 (66.7%) fecal specimens originating from 48 Chinese COVID-19 patients (Chen et al. 2020c). A fecal specimen collected from a COVID-19 patient in the United States was tested positive on day 7 of her illness (Holshue et al. 2020). Peng et al. 2020 revealed the presence of SARS-CoV-2's RNA in urine, blood, and anal swabs in some COVID-19 cases using RT-qPCR targeting the N gene (Peng et al. 2020). The cerebrospinal fluid (CSF) from a case of meningitis/encephalitis in Japan tested positive for SARS-CoV-2 (Moriguchi et al. 2020). However, there are currently no clinical testing platforms for non-respiratory specimens, leaving this area within the purview of research use only.

Proper transport of clinical specimens has also been paramount in making sure that test results are as accurate as possible. For bronchoalveolar lavage (BAL) or tracheal aspirates, 2–3 mL of fluid placed in a sterile, leak-proof, screw-cap container is the preferred collection method. This container prevents leakage and contamination of the original specimen or any other specimens that may have been transported concurrently with SARS-CoV-2-positive specimens (CDC 2020). Also, the use of proper technique helps improve the sensitivity of a test, particularly for NP swabs. NP swabs should be inserted into the nostril parallel to the palate, and the swab should come into contact with the pharyngeal mucosa for several seconds before removal (Baden et al. 2009; Miller et al. 2018). CDC also advises that specimens may be stored at 2–8 °C and should be tested within 72 hours of collection.

Lastly, the FDA has issued an Emergency Use Authorization (EUA) to the first stand-alone at-home sample collection kit, the EverlyWell COVID-19 test home collection kit, which can be used by multiple COVID-19 diagnostic tests. They had previously issued a different EUA for the Pixel, a home collection kit by LabCorp for COVID-19, but could only be used with LabCorp's COVID-19 RT-PCR test. Both collection kits contain a nasal swab and a saline-filled container; the patient self-swabs to collect a nasal sample, which is then placed in the container and returned to the specified CLIA-certified laboratories. In the case of the LabCorp test, it is returned to LabCorp. For the EverlyWell test, it may be shipped to either Fulgent Therapeutics or Assurance Scientific Laboratories (FDA 2020).

23.3 Viral Detection

SARS-CoV-2 RNA was first detected in human BAL specimens collected from hospitalized patients with unexplained pneumonia by a real-time RT-PCR assay using pan-betacoronavirus degenerate primers targeting a highly conserved RNA-dependent RNA polymerase (RdRp) region, also named nonstructural protein 12 (NSP12) (Zhu et al. 2020). The complete genome of SARS-CoV-2 was amplified and sequenced using an unbiased viral metagenomic approach (Zhu et al. 2020). An attempt to isolate SARS-CoV-2 was performed in different cell lines, including human airway epithelial cells, Vero E6 cells (African green monkey cells), and Huh-7 cells (Human hepatocarcinomatous cells). It was found that cytopathic effects (CPEs) were seen in human airway epithelial cells 96 hours after inoculation on surface layers of human airway epithelial cells. No CPEs were observed in the Vero E6 and Huh-7 cell lines until 6 days after inoculation (Zhu et al. 2020). Cell culture supernatant from human airway epithelial cell cultures was collected, and viral particles were subsequently observed under transmission electron microscopy (Zhu et al. 2020).

For safety, the CDC recommends that virus isolation in cell culture from clinical specimens collected from COVID-19 patients under investigation should not be attempted due to the possible propagation of SARS-CoV-2. Although cell culture has been considered as the gold standard method for the laboratory diagnosis of viruses for many years, this approach often requires intensive labor and technical expertise (Hsiung 1984; Storch 2000). Also, the lack of a BSL3 biocontainment facility in many clinical laboratories combined with the unavailability of commercial antisera against SARS-CoV-2 for culture confirmation creates significant obstacles. Thus, the rapid and accurate diagnostic tools are essential in early detection and monitoring of diseases, which help to improve overall cost-effectiveness and to reduce mortality (Niesters 2004; Scagnolari et al. 2017).

Genomic analysis revealed that the complete genome of SARS-CoV-2 shared only less than 50% identities at the nucleotide level to those of four common human coronaviruses (hCoV) known as HCoV-229E, -NL63, -OC43, and -HKU1 (Phan 2020). Given the high genomic divergence, commercial multiplex molecular assays for the detection of a panel of respiratory microorganisms, including HCoV-229E, -NL63, -OC43, and -HKU1, showed no evidence of cross-reactivity with SARS-CoV-2 (Phan 2020). In order to screen for SARS-CoV-2 in COVID-19 patients under investigation, two major protocols for laboratory-developed tests (LDTs) were quickly published. The CDC assay known as CDC 2019-nCoV real-time RT-PCR diagnostic panel was designed for specific detection of different regions of the nucleocapsid (N) gene of SARS-CoV-2 (https://www.cdc.gov/coronavirus/2019-ncov/lab/rt-pcr-panel-primer-probes.html). At first, the CDC assay included three separate RT-PCR reactions aimed at universal detection of SARS-like betaCoVs and specific detection of SARS-CoV-2. Later, the CDC assay decided to remove the primer and probe set specific for SARS-like betaCoVs but kept the two primer and probe sets for SARS-CoV-2. The second protocol was also an RT-PCR assay, developed by Corman, and targeted three different regions of the RdRp, N, and envelope (E) genes of SARS-CoV-2 (Corman et al. 2020).

Since SARS-CoV-2 continues to spread to countries and territories around the world, causing a rapid increase in the number of COVID-19 cases, faster and more-accessible testing is a crucial component in managing the pandemic. As of May 2020, there were 311 different molecular assays worldwide in the SARS-CoV-2 pipeline (FIND 2020). However, in the United States, there are only 51 molecular tests that have been authorized by the Emergency Use Authorization (EUA) protocol (https://www.fda.gov/medical-devices/emergency-situations-medical-devices/emergency-use-authorizations#covid19ivd). These tests can qualitatively measure SARS-CoV-2 RNA in clinical specimens collected in the upper and lower respiratory tracts from patients being tested for COVID-19. Commercial tests vary in their sensitivity, as defined with parameters such as detection limit and the amount of time required for testing. Also, these commercial tests vary in targeting different regions of the SARS-CoV-2 genome. While Cepheid Xpert Xpress SARS-CoV-2 amplifies the E and N genes (Loeffelholz et al. 2020), only results from the RdRp and surface (S) genes are used to generate test results in the Simplexa COVID-19 Direct assay (Zhen et al. 2020). The Panther Fusion SARS-CoV-2 assay amplifies and detects two conserved regions of the RdRp gene (Zhen et al. 2020). Lieberman et al. 2020 reported that the CDC-based LDT and Cepheid Xpert Xpress SARS-CoV-2 were the most sensitive assays for SARS-CoV-2. The Panther Fusion SARS-CoV-2, Simplexa COVID-19 Direct, and Cobas SARS-CoV-2 were 100% specific but failed to detect positive samples near the limit of detection of the CDC-based LDT (Lieberman et al. 2020).

One of the emerging molecular detection methods for viral infections is the use of clustered regularly interspaced short palindromic repeats (CRISPR) and CRISPR-associated Cas proteins (Burbelo et al. 2019). Broughton et al. 2020 developed a rapid (under 40 minutes), easy-to-implement, and accurate CRISPR-Cas12-based lateral flow assay to detect SARS-CoV-2. The positive predictive agreement and negative predictive agreement of this assay relative to the CDC-based LDT were 95% and 100%, respectively (Broughton et al. 2020). Another research group also developed a CRISPR/Cas12a-based assay with a naked eye readout called CRISPR/Cas12a-NER. According to this research group, CRISPR/Cas12a-NER can detect as few as ten copies of the virus gene in 45 minutes without a special instrument (Wang et al. 2020c). Lastly, the Sherlock CRISPR SARS-CoV-2 kit is the first CRISPR-based diagnostic assay receiving an EUA by the FDA.

Point-of-care testing (POCT) is essential for the rapid detection of SARS-CoV-2 in the outpatient setting. It allows physicians and medical professionals to accurately receive real-time, lab-quality diagnostic results within minutes rather than hours. The use of a faster COVID-19 diagnosis leads to improved monitoring and management by enabling the early administration of treatment (if necessary), which in turn leads to improved health outcomes for patients. Finally, rapid diagnostics would help to prevent disease spread between close contacts at home and in the community (Goyder et al. 2020; Kankaanpaa et al. 2018; Kankaanpaa et al. 2016). The FDA has approved the emergency use of a portable, rapid swab test for SARS-CoV-2, known as

Abbott ID NOW COVID-19, which is a molecular point-of-care test system for the detection of SARS-CoV-2. The test is an automated assay that takes only 5 minutes to show positive results and only 13 minutes to show negative results, using the ID NOW molecular platform, but accuracy issues are surrounding false-negative results. Zhen et al. 2020 performed a clinical evaluation of three sample-to-answer platforms for the detection of SARS-CoV-2. It was found that the Cepheid Xpert Xpress SARS-CoV-2 had the lowest limit of detection (100 copies/mL), followed by the ePlex SARS-CoV-2 (1000 copies/mL) and the ID NOW COVID-19 (20,000 copies/mL). According to this research group, the ID NOW COVID-19 produced the most rapid time to result per specimen (~17 minutes) as compared to the Cepheid Xpert Xpress SARS-CoV-2 (~46 minutes) and the ePlex SARS-CoV-2 (~1.5 hours) (Zhen et al. 2020). The decision to choose any of these platforms must weigh the benefits of rapid results versus concerns over false negatives, as a positive result is dependent on the patient's viral load or quality of the sample provided for each platform.

One final complicating matter is that SARS-CoV-2 had become pandemic during the winter months when influenza A, influenza B, and a multitude of other upper respiratory viruses are seasonally endemic. From an initial study from Wuhan, China, in January 2020, they performed respiratory pathogen testing on 99 patients who were SARS-CoV-2 positive (amidst other clinical characteristics). No other viral pathogens were detected, while one polymicrobial infection and four cases of fungal infections were noted (Chen et al. 2020b). However, a more recent study looking at 1217 patients in California suggested a higher rate of coinfection, closer to 7.5%, with both SARS-CoV-2 and a second pathogen (Kim et al. 2020). As such, patients tested positive for a non-SARS-CoV-2 pathogen are not necessarily excluded from the possibility of being infected with SARS-CoV-2 superimposed upon initial infection.

Finally, it should be noted that, at this time, there are no at-home PCR testing kits for SARS-CoV-2. It is required that all SARS-CoV-2 testing be performed by laboratories that are Clinical Laboratory Improvement Amendments of 1988 (CLIA'88) certified to perform high complexity tests.

23.4 Serological Tests

Another diagnostic approach is to devise blood tests looking for IgM and IgG antibodies against SARS-CoV-2. Many companies around the world have raced to develop antibody tests, though there are many broader questions beyond the straightforward positive vs. negative implications. As of April 2020, the FDA has not validated any antibody tests for the diagnosis of SARS-CoV-2 infection. However, this does not imply that they have no use; the FDA has stated that they are to be used as confirmatory tests, determining whether the patient is in acute infection (via IgM) or either is recovering or has recovered (via IgG). It may offer insight into the active epidemiology of COVID-19, along with assisting in retrospective analyses of the outbreak.

The biology of human coronavirus (hCoV) has determined the sites for antibody testing. Spike (S) proteins mediate receptor binding and fusion functions, while N proteins form the capsid surrounding hCoV RNA. Cleavage of the S protein creates S1 and S2 subunits; S1 is primarily involved with viral receptors that bind to human ACE2 as the entry point into cells, while S2 is involved with viral fusion (Du et al. 2009). Previous studies showed that antibody-based tests against the N proteins show false positivity when looking at specific hCoVs due to sequence conservation within N proteins among hCoV (Chan et al. 2009). Meyer et al. evaluated the percent of amino acid pairwise-identity between the hCoV to SARS-CoV in both the S and N proteins. It, too, showed an overall lower degree of conservation among S proteins in comparison to the N proteins. Western blots and ELISA techniques that were based on the N proteins were often very sensitive. However, they had more reduced specificity, implying that a combination of both N protein and S protein antibodies is necessary for improving detection without compromising specificity (Meyer et al. 2014). Indeed, previous SARS-CoV infections have provided historical evidence that there is difficulty teasing apart the distinction between antibody evidence indicating SARS-CoV infection and infection with other hCoVs. Previous tests have shown that hCoV OC43 and hCoV 229E have cross-reacted with SARS-CoV in ELISA testing, leading to false-positive results (Woo et al. 2004). Multiple types of serological assays were used during the SARS-CoV outbreak, including IFA, ELISA, and Western blot analysis; IFA and ELISA showed high sensitivity but low specificity, leading to a false-positive rate between 0 and 30% (Infantino et al. 2020).

Another complicating factor is that the timeline from a negative response to IgM or IgG conversion is also not fully defined. One patient from Finland became positive with undetectable IgM and IgG levels on day 4 of symptom presentation and fully detectable for both on Day 9 (Haveri et al. 2020). Zhao J et al. tested SARS-CoV-2 PCR-positive patients and performed ELISA testing for IgM, IgG, and total antibody levels at multiple time points during the patient's hospitalization. This study showed that total antibody against SARS CoV-2 was 100% by day 15–39 post-symptom onset, but the individual sensitivity and specificity of either IgM or IgG were as low as 19% and as high as 99% (Zhao et al. 2020a). An additional study focused on the early humoral response with IgM, IgA, and IgG used as a diagnostic tool. All three could appear as early as day 1 post-symptom onset, though the median time of IgM and IgA detection was around day 5, while IgG detection was more often on day 14 (Guo et al. 2020a). Zhang G et al. followed 112 PCR-positive patients for 50 days and notably showed a statistically significant difference in the IgG titers the longer they were from the onset of infection. However, the time of persistence in the body past this point is still uncertain (Zhang et al. 2020a).

Our understanding of the persistence of humoral immune response from SARS-CoV infection may serve to enhance and support our current knowledge of SARS-CoV-2. Chen et al. found that, in a study of 36 patients with probable SARS-CoV infection and 48 with suspected SARS-CoV, IgG antibody persisted at least for 60 days post-fever onset, while IgM persisted for 28.5 ± 8.7 days (Chen et al. 2004). Further research by this team showed that IgG antibody production seemed to be associated with T-cell immunity but did not progress further than 60-day testing, but this research is complicated by the fact that only nine patients in the probable infection group had been SARS-CoV PCR positive. A long-term prospective cohort followed 34 SARS-CoV-infected healthcare workers from 2002 to 2003 for 13 years with annual serum samples. IgG was detected via ELISA, and indirect ELISA was able to quantify antibodies against the whole virus and N199 antigen (used for detecting a specific antibody). It was found that although IgG levels decreased from 2003 to 2015, they remained higher than levels in non-SARS-CoV-infected healthcare workers (Guo et al. 2020b). It is hoped that the presence of IgG antibodies could provide partial to complete immunity, but as there has been no recurrence of SARS-CoV, we are unable to state one way or another definitively.

As of May 2020, 12 antibody tests have received EUA approval in the United States (https://www.fda.gov/medical-devices/emergency-situations-medical-devices/emergency-use-authorizations#covid19ivd). The qSARS-CoV-2 IgG/IgM rapid test (Cellex Inc.) was the first antibody test to receive EUA approval, intended to qualitatively detect IgG and IgM antibodies in human serum, plasma, and whole blood. Ortho Clinical Diagnostics Inc. has two different tests: the VITROS Anti-SARS-CoV-2 total, which can detect either IgM or IgG but cannot distinguish between them, and Anti-SARS-CoV-2 IgG test to be used on serum or EDTA. The DDP COVID-19 rapid test (Chembio Diagnostic Systems, Inc.) is a 15-minute test IgM/IgG performed from a fingerstick; it is likely a lateral flow assay since its Zika IgM/IgG is patented under the same technology, but it does not formally state otherwise. The Mount Sinai Laboratory has an enzyme-linked immunosorbent assay (ELISA) test that can offer IgG antibody titers in serum and plasma.

Abbott Laboratories has a chemiluminescent microparticle immunoassay for IgG, and DiaSorin Inc. has an S1/S2 IgG antibody test. Autobio Diagnostics detects both IgM/IgG. It should be noted that all of them vary in their format (lateral flow immunoassays (LFAs), ELISAs, and chemiluminescent immunoassays), the antibody classes they detect, the antigens used, and the specimen type. It is paramount that laboratories understand the differences between testing platforms in order to understand the advantages and disadvantages of each (Theel et al. 2020). Once again, all these tests are authorized for use by a CLIA-certified laboratory that can perform moderate- to high-complexity tests. Thus, they are not approved for at-home usage.

LFAs have also found a niche with POC testing platforms, which have been employed in laboratories and clinics worldwide. Many tests (influenza types A and B, adenovirus, respiratory syncytial virus, and others) have received CLIA waivers that allow for POC use (Kozel and Burnham-Marusich 2017). A lateral flow immunoassay requires two major analytes: microbial antigens and antibodies specific for microbial antigens (Qin et al. 2019). The Sofia 2 SARS Antigen FIA is the first COVID-19 antigen test to be granted the EUA by the FDA. It is an LFA sandwich assay that is used with the Sofia 2 instrument intended for the qualitative detection of the nucleocapsid protein antigen from SARS-CoV-2 in NP and nasal (NS) swabs. Again, there are no EUA tests available for non-respiratory specimens.

One study from China looked at 397 SARS-CoV-2 PCR-positive patients in comparison to 128 negative patients through a lab-developed IgM/IgG antibody test, which showed a sensitivity of 88.7% and specificity of 90.6%. The most notable limitation was a lack of data on that time from post-symptom presentation to blood draw (Li et al. 2020). Nine specimens taken from SARS-CoV-2 PCR-positive patients in France were compared with patients who had PCR-positive hCoV/SARS/MERS infections. Although it was a small cohort, ELISA testing revealed that the S1 subunit was more specific for SARS-CoV-2 than S2 or N protein, and there was still noted cross-reactivity with different hCoV strains (Okba et al. 2020). Global studies are looking at the applicability of serology testing as the diagnostic test of choice in order to find patients who may be asymptomatically infected. One study from China compared 412 negative samples against 69 positive samples (per PCR), using an ELISA kit that used the S1 protein as the capturing antigen, and found a sensitivity of 97.1% and specificity of 97.3%. When this was used on 276 asymptomatic healthcare workers, 28 were found to be positive though there was no discussion on whether the SARS-CoV-2 antibody-positive patients were also positive via molecular techniques (Zhao et al. 2020b).

Given the limited time that has progressed since the beginning of the pandemic – December 2019 to May 2020 – systemic studies on diagnostic performance and meaningful predictive value of SARS-CoV-2 serological tests have been lacking. It is in part due to the reality that 6 months does not give a robust view of the long-term human immunological response to SARS-CoV-2 infection nor cement the implication that the presence of IgG is evidence of protection against recurrent infection, as we have only been through one wave as of May 2020. One final area of relevance to the realm of serological diagnostics is the concern that infection with SARS-CoV-2, and the subsequent development of antibodies, could lead to immune-enhanced disease; this is the possibility that repeated exposure to SARS-CoV-2 could cause a more severe response the second time around. It is a topic of particular alarm for vaccine development, whose whole existence is to create antibodies against a specific pathogen. More concerning, infection-induced antibody response would almost certainly be higher than one caused by a vaccine (de Alwis et al. 2020). Although no publications have made a note of any recurrent infections and one study from China showed that reinfection with SARS-CoV-2 did not occur among rhesus macaques that were infected and then rechallenged with the virus (Bao et al. 2020), the possibility of hyperimmune reactions has not been completely ruled out. Until we have additional data, the ability to affect clinical decision-making should be undertaken with important caveats.

23.5 Conclusion

As we continue to progress toward a deeper, more thorough understanding of SARS-CoV-2, the number of diagnostic tests will continue to increase. By that same token, the urge to increase our testing options and capabilities should not undercut the importance of having rigorous studies to prove the efficacy and validation of each test. Longitudinal studies will be necessary to have a better appreciation for the clinical utility of serology tests, and additional epidemiological evaluations via molecular techniques will provide a grander overview of the effect of COVID-19 throughout the world. Through persistent research, we will continue to gain sharper insight into the meaning of these tests and better able to bring this information to both clinicians and the patients.

References

Baden LR, Drazen JM, Kritek PA, Curfman GD, Morrissey S, Campion EW (2009) H1N1 influenza a disease–information for health professionals. N Engl J Med 360:2666–2667

Bao L, Deng Q, Gao H et al (2020) Reinfection could not occur in SARS-CoV-2 infected rhesus macaques. bioRxIV

Broughton JP, Deng X, Yu G, Fasching CL, Servellita V, Singh J, Miao X, Streithorst JA, Granados A, Sotomayor-Gonzalez A, Zorn K, Gopez A, Hsu E, Gu W, Miller S, Pan CY, Guevara H, Wadford DA, Chen JS, Chiu CY (2020) CRISPR-Cas12-based detection of SARS-CoV-2. Nat Biotechnol 38:870

Burbelo PD, Iadarola MJ, Chaturvedi A (2019) Emerging technologies for the detection of viral infections. Future Virol 14:39–49

Callahan CJ, Lee R, Zulauf KE, Tamburello L, Smith KP, Previtera J, Cheng A, Green A, Azim AA, Yano A, Doraiswami N, Kirby JE, Arnaout RA (2020) Open Development and Clinical Validation Of Multiple 3D-Printed Nasopharyngeal Collection Swabs: Rapid Resolution of a Critical COVID-19 Testing Bottleneck. J Clin Microbiol 58(8):e00876-20

CDC, Centers for Disease Control and Prevention (2020) Interim guidelines for collecting, handling, and testing clinical specimens from persons for Coronavirus disease 2019 (COVID-19). Accessed May 24. https://www.cdc.gov/coronavirus/2019-ncov/lab/guidelines-clinical-specimens.html

Chan CM, Tse H, Wong SS, Woo PC, Lau SK, Chen L, Zheng BJ, Huang JD, Yuen KY (2009) Examination of seroprevalence of coronavirus HKU1 infection with S protein-based ELISA and neutralization assay against viral spike pseudotyped virus. J Clin Virol 45:54–60

Chen X, Zhou B, Li M, Liang X, Wang H, Yang G, Wang H, Le X (2004) Serology of severe acute respiratory syndrome: implications for surveillance and outcome. J Infect Dis 189:1158–1163

Chen C, Gao G, Xu Y, Pu L, Wang Q, Wang L, Wang W, Song Y, Chen M, Wang L, Yu F, Yang S, Tang Y, Zhao L, Wang H, Wang Y, Zeng H, Zhang F (2020a) SARS-CoV-2-positive sputum and Feces after conversion of pharyngeal samples in patients with COVID-19. Ann Intern Med 172:832–834

Chen N, Zhou M, Dong X, Qu J, Gong F, Han Y, Qiu Y, Wang J, Liu Y, Wei Y, Xia J, Yu T, Zhang X, Zhang L (2020b) Epidemiological and clinical characteristics of 99 cases of 2019 novel coronavirus pneumonia in Wuhan, China: a descriptive study. Lancet 395:507–513

Chen Y, Chen L, Deng Q, Zhang G, Wu K, Ni L, Yang Y, Liu B, Wang W, Wei C, Yang J, Ye G, Cheng Z (2020c) The presence of SARS-CoV-2 RNA in the feces of COVID-19 patients. J Med Virol 92:833–840

Corman VM, Landt O, Kaiser M, Molenkamp R, Meijer A, Chu DK, Bleicker T, Brunink S, Schneider J, Schmidt ML, Mulders DG, Haagmans BL, van der Veer B, van den Brink S, Wijsman L, Goderski G, Romette JL, Ellis J, Zambon M, Peiris M, Goossens H, Reusken C, Koopmans MP, Drosten C (2020) Detection of 2019 novel coronavirus (2019-nCoV) by real-time RT-PCR. Euro Surveill 25:2000045

Cox JL, Koepsell SA (2020) 3D-printing to address COVID-19 testing supply shortages. Lab Med 51:e45

de Alwis R, Chen S, Gan ES, Ooi EE (2020) Impact of immune enhancement on Covid-19 polyclonal hyperimmune globulin therapy and vaccine development. EBioMedicine 55:102768

Diagnostics, Copan (2020) UTM: viral transport, Accessed May 13. https://www.copanusa.com/wp-content/uploads/2019/08/UTM-Brochure.pdf

Du L, He Y, Zhou Y, Liu S, Zheng BJ, Jiang S (2009) The spike protein of SARS-CoV–a target for vaccine and therapeutic development. Nat Rev Microbiol 7:226–236

FDA, U.S. Food & Drug Administration US (2020) FAQs on testing for SARS-CoV-2, Accessed May 13. https://www.fda.gov/medical-devices/emergency-situations-medical-devices/faqs-testing-SARS-CoV-2

FIND, Foundation for Innovative New Diagnostics (2020) SARS-CoV-2 Diagnostic Pipeline, Accessed May 7. https://www.finddx.org/covid-19/pipeline/

Goyder C, Tan PS, Verbakel J, Ananthakumar T, Lee JJ, Hayward G, Turner PJ, Van Den Bruel A (2020) Impact of point-of-care panel tests in ambulatory care:

a systematic review and meta-analysis. BMJ Open 10:e032132

Guo L, Ren L, Yang S, Xiao M, Chang FY, Dela Cruz CS, Wang Y, Wu C, Xiao Y, Zhang L, Han L, Dang S, Xu Y, Yang Q, Xu S, Zhu H, Xu Y, Jin Q, Sharma L, Wang L, Wang J (2020a) Profiling early humoral response to diagnose novel Coronavirus disease (COVID-19). Clin Infect Dis 71:778

Guo, X, Guo Z, Duan C, Chen Z, Wang G, Lu Y, Li M, Lu J (2020b) Long-term persistence of IgG antibodies in SARS-CoV infected healthcare workers. medRxiv 2020.02.12.20021386. https://doi.org/10.1101/2020.02.12.20021386

Haveri A, Smura T, Kuivanen S, Osterlund P, Hepojoki J, Ikonen N, Pitkapaasi M, Blomqvist S, Ronkko E, Kantele A, Strandin T, Kallio-Kokko H, Mannonen L, Lappalainen M, Broas M, Jiang M, Siira L, Salminen M, Puumalainen T, Sane J, Melin M, Vapalahti O, Savolainen-Kopra C (2020, Mar) Serological and molecular findings during SARS-CoV-2 infection: the first case study in Finland, January to February 2020. Euro Surveill 25(11):2000266. https://doi.org/10.2807/1560-7917.ES.2020.25.11.2000266

Holshue ML, DeBolt C, Lindquist S, Lofy KH, Wiesman J, Bruce H, Spitters C, Ericson K, Wilkerson S, Tural A, Diaz G, Cohn A, Fox L, Patel A, Gerber SI, Kim L, Tong S, Lu X, Lindstrom S, Pallansch MA, Weldon WC, Biggs HM, Uyeki TM, Pillai SK, Case Investigation Team Washington State -nCo V (2020) First Case of 2019 novel Coronavirus in the United States. N Engl J Med 382:929–936

Hsiung GD (1984) Diagnostic virology: from animals to automation. Yale J Biol Med 57:727–733

IDSA, Infectious DIseases Society of America (2020) Infectious Diseases Society of America guidelines on the diagnosis of COVID-19.

Infantino M, Damiani A, Gobbi FL, Grossi V, Lari B, Macchia D, Casprini P, Veneziani F, Villalta D, Bizzaro N, Cappelletti P, Fabris M, Quartuccio L, Benucci M, Manfredi M (2020) Serological assays for SARS-CoV-2 infectious disease: benefits, limitations and perspectives. Isr Med Assoc J 22:203–210

Kankaanpaa M, Raitakari M, Muukkonen L, Gustafsson S, Heitto M, Palomaki A, Suojanen K, Harjola VP (2016) Use of point-of-care testing and early assessment model reduces length of stay for ambulatory patients in an emergency department. Scand J Trauma Resusc Emerg Med 24:1–7

Kankaanpaa M, Holma-Eriksson M, Kapanen S, Heitto M, Bergstrom S, Muukkonen L, Harjola VP (2018) Comparison of the use of comprehensive point-of-care test panel to conventional laboratory process in emergency department. BMC Emerg Med 18:43

Kim D, Quinn J, Pinsky B, Shah NH, Brown I (2020) Rates of co-infection between SARS-CoV-2 and other respiratory pathogens. JAMA 323:2085–2086

Kozel TR, Burnham-Marusich AR (2017) Point-of-care testing for infectious diseases: past, present, and future. J Clin Microbiol 55:2313–2320

Li Z, Yi Y, Luo X, Xiong N, Liu Y, Li S, Sun R, Wang Y, Hu B, Chen W, Zhang Y, Wang J, Huang B, Lin Y, Yang J, Cai W, Wang X, Cheng J, Chen Z, Sun K, Pan W, Zhan Z, Chen L, Ye F (2020) Development and clinical application of a rapid IgM-IgG combined antibody test for SARS-CoV-2 infection diagnosis. J Med Virol 92:1518–1524

Lieberman JA, Pepper G, Naccache SN, Huang ML, Jerome KR, Greninger AL (2020, Jul 23) Comparison of Commercially Available and Laboratory Developed Assays for in vitro Detection of SARS-CoV-2 in Clinical Laboratories. J Clin Microbiol 58(8):821–20. https://doi.org/10.1128/JCM.00821-20

Loeffelholz MJ, Alland D, Butler-Wu SM, Pandey U, Perno CF, Nava A, Carroll KC, Mostafa H, Davies E, McEwan A, Rakeman JL, Fowler RC, Pawlotsky JM, Fourati S, Banik S, Banada PP, Swaminathan S, Chakravorty S, Kwiatkowski RW, Chu VC, Kop J, Gaur R, Sin MLY, Nguyen D, Singh S, Zhang N, Persing DH (2020) Multicenter evaluation of the Cepheid Xpert Xpress SARS-CoV-2 test. J Clin Microbiol 58:e00926-20

Meyer B, Drosten C, Muller MA (2014) Serological assays for emerging coronaviruses: challenges and pitfalls. Virus Res 194:175–183

Miller JM, Binnicker MJ, Campbell S, Carroll KC, Chapin KC, Gilligan PH, Gonzalez MD, Jerris RC, Kehl SC, Patel R, Pritt BS, Richter SS, Robinson-Dunn B, Schwartzman JD, Snyder JW, Telford S 3rd, Theel ES, Thomson RB Jr, Weinstein MP, Yao JD (2018) A guide to utilization of the microbiology laboratory for diagnosis of infectious diseases: 2018 update by the infectious diseases society of America and the American society for microbiology. Clin Infect Dis 67:e1–e94

Moriguchi T, Harii N, Goto J, Harada D, Sugawara H, Takamino J, Ueno M, Sakata H, Kondo K, Myose N, Nakao A, Takeda M, Haro H, Inoue O, Suzuki-Inoue K, Kubokawa K, Ogihara S, Sasaki T, Kinouchi H, Kojin H, Ito M, Onishi H, Shimizu T, Sasaki Y, Enomoto N, Ishihara H, Furuya S, Yamamoto T, Shimada S (2020) A first case of meningitis/encephalitis associated with SARS-Coronavirus-2. Int J Infect Dis 94:55–58

Niesters HG (2004) Molecular and diagnostic clinical virology in real time. Clin Microbiol Infect 10:5–11

Okba NMA, Muller MA, Li W, Wang C, GeurtsvanKessel CH, Corman VM, Lamers MM, Sikkema RS, de Bruin E, Chandler FD, Yazdanpanah Y, Le Hingrat Q, Descamps D, Houhou-Fidouh N, Reusken C, Bosch BJ, Drosten C, Koopmans MPG, Haagmans BL (2020) Severe acute respiratory syndrome Coronavirus 2-specific antibody responses in Coronavirus disease patients. Emerg Infect Dis 26:1478–1488

Paules CI, Marston HD, Fauci AS (2020) Coronavirus infections-more than just the common cold. JAMA 323:707

Peng L, Liu J, Xu W, Luo Q, Chen D, Lei Z, Huang Z, Li X, Deng K, Lin B, Gao Z (2020) SARS-CoV-2 can be detected in urine, blood, anal swabs, and oropharyngeal swabs specimens. J Med Virol 92:1676

Phan T (2020) Novel coronavirus: From discovery to clinical diagnostics. Infect Genet Evol 79:104211

Qin Q, Wang K, Yang J, Xu H, Cao B, Wo Y, Jin Q, Cui D (2019) Algorithms for immunochromatographic assay: review and impact on future application. Analyst 144:5659–5676

Scagnolari C, Turriziani O, Monteleone K, Pierangeli A, Antonelli G (2017) Consolidation of molecular testing in clinical virology. Expert Rev Anti-Infect Ther 15:387–400

Storch GA (2000) Diagnostic virology. Clin Infect Dis 31:739–751

Theel ES, Slev P, Wheeler S, Couturier MR, Wong SJ, Kadkhoda K (2020) The role of antibody testing for SARS-CoV-2: is there one? J Clin Microbiol 58:e00797-20

Wang W, Xu Y, Gao R, Lu R, Han K, Wu G, Tan W (2020a) Detection of SARS-CoV-2 in different types of clinical specimens. JAMA

Wang X, Tan L, Wang X, Liu W, Lu Y, Cheng L, Sun Z (2020b) Comparison of nasopharyngeal and oropharyngeal swabs for SARS-CoV-2 detection in 353 patients received tests with both specimens simultaneously. Int J Infect Dis 94:107–109

Wang X, Zhong M, Liu Y, Ma P, Dang L, Meng Q, Wan W, Ma X, Liu J, Yang G, Yang Z, Huang X, Liu M (2020c) Rapid and sensitive detection of COVID-19 using CRISPR/Cas12a-based detection with naked eye readout, CRISPR/Cas12a-NER. Sci Bull (Beijing) 65:1436–1439

Wolfel R, Corman VM, Guggemos W, Seilmaier M, Zange S, Muller MA, Niemeyer D, Jones TC, Vollmar P, Rothe C, Hoelscher M, Bleicker T, Brunink S, Schneider J, Ehmann R, Zwirglmaier K, Drosten C, Wendtner C (2020) Virological assessment of hospitalized patients with COVID-2019. Nature 581:465–469

Woo PC, Lau SK, Wong BH, Chan KH, Hui WT, Kwan GS, Peiris JS, Couch RB, Yuen KY (2004) False-positive results in a recombinant severe acute respiratory syndrome-associated coronavirus (SARS-CoV) nucleocapsid enzyme-linked immunosorbent assay due to HCoV-OC43 and HCoV-229E rectified by Western blotting with recombinant SARS-CoV spike polypeptide. J Clin Microbiol 42:5885–5888

Young BE, Ong SWX, Kalimuddin S, Low JG, Tan SY, Loh J, Ng OT, Marimuthu K, Ang LW, Mak TM, Lau SK, Anderson DE, Chan KS, Tan TY, Ng TY, Cui L, Said Z, Kurupatham L, Chen MI, Chan M, Vasoo S, Wang LF, Tan BH, Lin RTP, Lee VJM, Leo YS, Lye DC, Research TSNCO (2020) Epidemiologic features and clinical course of patients infected with SARS-CoV-2 in Singapore. JAMA 323:1488

Yu F, Yan L, Wang N, Yang S, Wang L, Tang Y, Gao G, Wang S, Ma C, Xie R, Wang F, Tan C, Zhu L, Guo Y, Zhang F (2020) Quantitative detection and viral load analysis of SARS-CoV-2 in infected patients. Clin Infect Dis 71:793

Zhang G, Nie S, Zhang Z, Zhang Z (2020a) Longitudinal change of SARS-Cov2 antibodies in patients with COVID-19′. In: J infect dis

Zhang W, Du RH, Li B, Zheng XS, Yang XL, Hu B, Wang YY, Xiao GF, Yan B, Shi ZL, Zhou P (2020b) Molecular and serological investigation of 2019-nCoV infected patients: implication of multiple shedding routes. Emerg Microbes Infect 9:386–389

Zhao J, Yuan Q, Wang H, Liu W, Liao X, Su Y, Wang X, Yuan J, Li T, Li J, Qian S, Hong C, Wang F, Liu Y, Wang Z, He Q, Li Z, He B, Zhang T, Fu Y, Ge S, Liu L, Zhang J, Xia N, Zhang Z (2020a) Antibody responses to SARS-CoV-2 in patients of novel coronavirus disease 2019. Clin Infect Dis 71:2027–2034

Zhao R, Li M, Song H, Chen J, Ren W, Feng Y, Gao GF, Song J, Peng Y, Su B, Guo X, Wang Y, Chen J, Li J, Sun H, Bai Z, Cao W, Zhu J, Zhang Q, Sun Y, Sun S, Mao X, Su J, Chen X, He A, Gao W, Jin R, Jiang Y, Sun L (2020b) Early detection of SARS-CoV-2 antibodies in COVID-19 patients as a serologic marker of infection. Clin Infect Dis 71:2066–2072

Zhen W, Smith E, Manji R, Schron D, Berry GJ (2020) Clinical evaluation of three sample-to-answer platforms for the detection of SARS-CoV-2. *J Clin Microbiol* 58:e00783-20

Zhu N, Zhang D, Wang W, Li X, Yang B, Song J, Zhao X, Huang B, Shi W, Lu R, Niu P, Zhan F, Ma X, Wang D, Xu W, Wu G, Gao GF, Tan W, Coronavirus ICN, Team Research (2020) A novel Coronavirus from patients with pneumonia in China, 2019. N Engl J Med 382:727–733

The Role of Medical Imaging in COVID-19

24

Houman Sotoudeh and Masoumeh Gity

Abstract

The coronavirus disease 2019 (COVID-19) pandemic launched in the third decade of the twenty-first century and continued to present time to cause the worst challenges the modern medicine has ever encountered. Medical imaging is an essential part of the universal fight against this pandemic. In the absence of documented treatment and vaccination, early accurate diagnosis of infected patients is the backbone of this pandemic management. This chapter reviews different aspects of medical imaging in the context of COVID-19.

H. Sotoudeh (✉)
Department of Radiology, University of Alabama at Birmingham (UAB),
Birmingham, AL, USA

Systematic Review and Meta-analysis Expert Group (SRMEG), Universal Scientific Education and Research Network (USERN), Birmingham, AL, USA
e-mail: hsotoudeh@uabmc.edu

M. Gity
Advanced Diagnostic and Interventional Radiology Research Center (ADIR), Tehran University of Medical Sciences, Tehran, Iran

Department of Radiology, Medical Imaging Center, Tehran University of Medical Sciences, Tehran, Iran

Keywords

Chest X-ray · COVID-19 · CT · Medical imaging · MRI · Pandemic · SARS-CoV-2 · Ultrasound

24.1 Introduction

Medical imaging crucially helps the diagnosis and management of coronavirus disease 2019 (COVID-19). It is known that the sensitivity of imaging is higher than the reverse transcription-polymerase chain reaction (PCR) test, especially in the early stage of the infection. Chest imaging is the primary diagnostic test in this infection; however, extrapulmonary complications are also diagnosed by imaging. Because of availability and high anatomic resolution, a computed tomography (CT) scan is the most commonly studied medical imaging modality in this infection. CT can diagnose the infection in its earliest stages, estimate the severity of the disease, and diagnose the possible superimposed complications. This chapter reviews the potentials of medical imaging in COVID-19. Chest X-ray (CXR) and thoracic CT scans constitute the main parts of this chapter. Other modalities, such as ultrasound and extrapulmonary imaging, are discussed briefly.

N. Rezaei (ed.), *Coronavirus Disease - COVID-19*, Advances in Experimental Medicine and Biology 1318, https://doi.org/10.1007/978-3-030-63761-3_24

24.2 Pulmonary Imaging

24.2.1 CXR

CXR remains the first Imaging modality for patients with COVID-19 because of low radiation risk and low cost. However, its application is limited because of low sensitivity and specificity (Jajodia et al. 2020; Nair et al. 2020; Revel et al. 2020; Rubin et al. 2020). The sensitivity of the CXR to detect PCR-confirmed COVID-19 cases has been reported as low as 69% (Jajodia et al. 2020). However, CXR is critical for patient follow-up and detection of the superimposed complications during hospitalization in patients admitted to the intensive care unit (ICU) (Rubin et al. 2020). Routine follow-up CXR for stable intubated patients in ICU is not recommended (Rubin et al. 2020). The role of CXR is even more critical in patients with COVID-19, given the fact that CT scanners must be disinfected after scanning each patient, compromising workflow for the next patients (American College of 2020).

The classical presentation of COVID-19 on CXR includes multiple foci of ground-glass opacities (GGOs) (68% of patients) and consolidations, which are located mainly in the peripheral and lower zones of the lungs (Figs. 24.1, 24.2, and 24.3). The pulmonary findings can be bilateral or unilateral (Yoon et al. 2020). The most recent meta-analysis estimated the prevalence of bilateral and unilateral lesions to be about 73% and 25%, respectively (Rodriguez-Morales et al. 2020). The diagnosis of GGO is challenging on CXR and can be better evaluated on a CT scan (Jacobi et al. 2020). Distribution of lesions on CXR can be helpful to differentiate COVID-19 pneumonia from bacterial pneumonia (which is often unilateral and lobar). However, the CXR presentation of COVID-19 has a significant overlap with other viral pneumonia. Pleural effusion is not a common finding on CXR and has been reported in about 3% of patients with COVID-19. Pulmonary cavitation, pneumothorax, and parenchymal calcifications are rarely seen on CXR of COVID-19 patients and require clinical evaluation to exclude other causes (Salehi et al. 2020). Pneumothorax can be seen as a complication of intubation and mechanical ventilation (Wong et al. 2019). In severe form of COVID-19 pneumonia, the entire lungs are involved, and the CXR findings would be similar to acute respiratory distress syndrome (ARDS) secondary to other diseases (Figs. 24.4 and 24.5). These patients are usually critically ill and need intubation and ventilation in the ICU. Typically, the CXR findings peak at 10–12 days after the initial symptoms. About 1–3 weeks after the initial symptoms, the lesions on CXR will evolve to diffuse coalescent or consolidative patterns (Jacobi et al. 2020). Given the

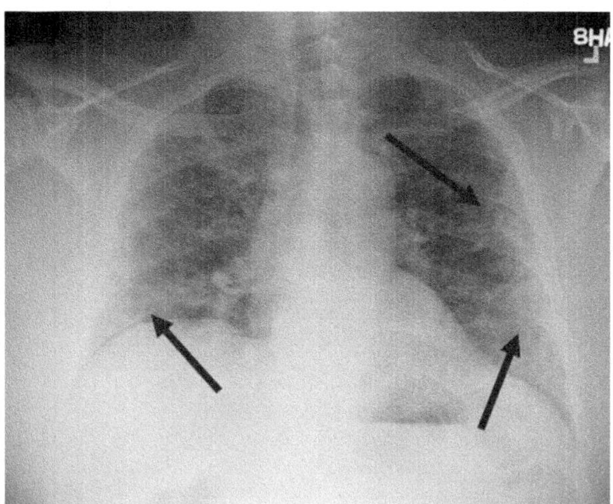

Fig. 24.1 Chest X-ray of a COVID-19 patient. There are peripheral ground-glass opacities in both lungs (arrows). Subsequent chest CT confirmed the peripheral ground-glass opacities (not shown here)

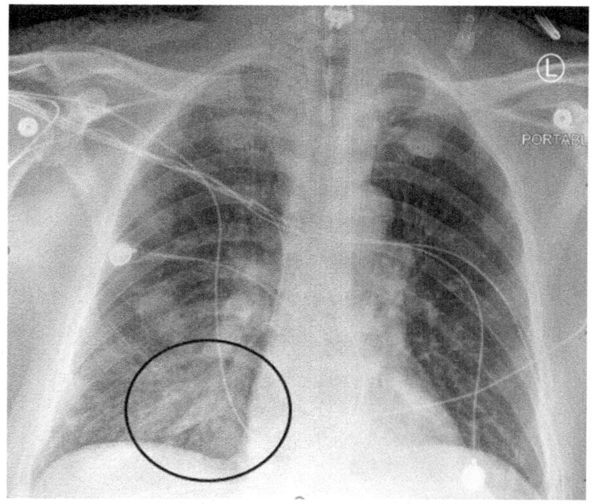

Fig. 24.2 Chest X-ray in COVID-19 patients. There is an area of consolidation in the right lung base (circle)

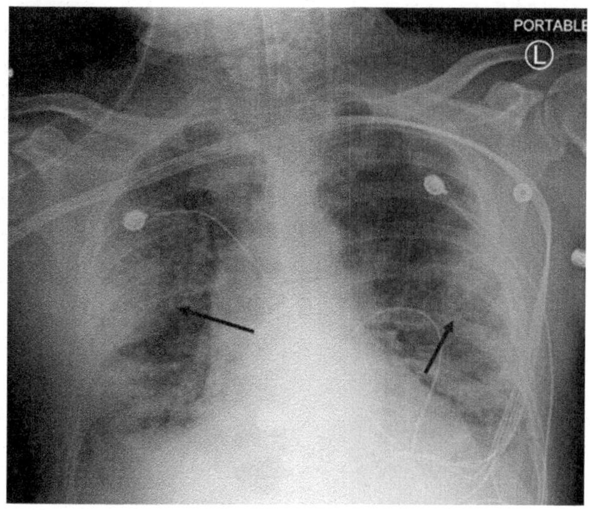

Fig. 24.3 Chest X-ray from COVID-19 pneumonia. There are several areas of peripheral consolidations in both lungs (arrows)

low radiation dose of CXR, this modality has been recommended for serial imaging in hospitalized patients and to exclude superimposed complications (Borghesi and Maroldi 2020). One CXR scoring system has been suggested for COVID-19 pneumonia. In this system, each lung is divided into three zones (upper, middle, and lower) with a total of six zones bilaterally. Then score 0 to 3 is assigned to each zone: score 0, no visible lesion; score 1, interstitial infiltrations; score 2, interstitial more than alveolar infiltrations; and score 3, alveolar more than interstitial infiltrations. In this manner, the CXR SCORE is defined as the sum of all zone score falls between 0 and 18, with the higher the number, the worse the pulmonary involvement is (Borghesi and Maroldi 2020).

24.2.2 CT Scan

Given the higher anatomic resolution and very high sensitivity (about 97%) of CT scan to detect patients with COVID-19, CT scan is the backbone of the imaging in these patients. Because of the radiation risk and the need for disinfecting the CT scanner after scanning the COVID-19 patients, CT scan is not recommended for PCR-positive COVID-19 patients who are asymptomatic or have mild clinical

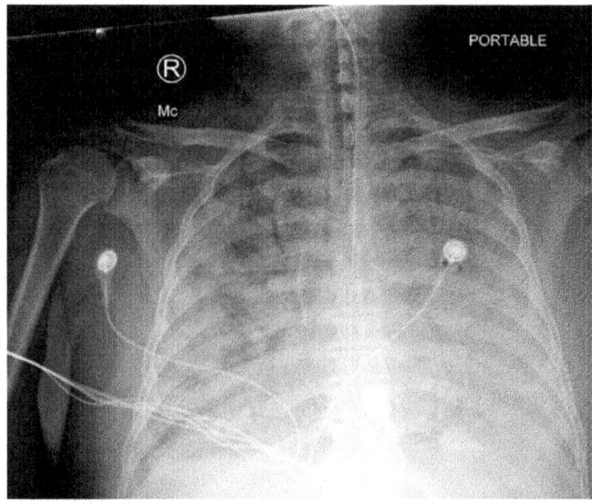

Fig. 24.4 Chest X-ray in a COVID-19 patient with respiratory distress. There are diffuse and extensive pulmonary consolidations with air bronchograms

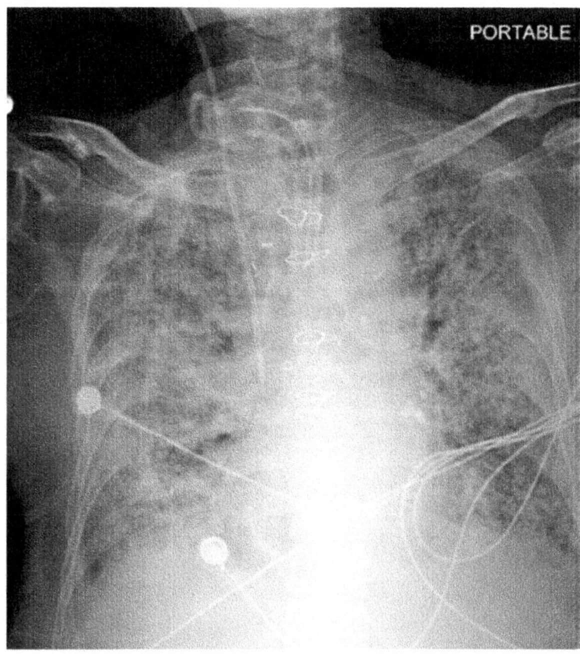

Fig. 24.5 Chest X-ray in a COVID-19 patient with respiratory distress. There are extensive diffuse air space and interstitial opacities in both lungs

symptoms (Nair et al. 2020; Rubin et al. 2020). However, a CT scan is recommended for patients with moderate to severe clinical symptoms as the baseline imaging. Also, it has been recommended for evaluation of possible complications during hospitalization (superimposed bacterial infection, pulmonary embolism, and pleural effusion) (Rubin et al. 2020). CT scan is considered the most effective diagnostic technique in COVID-19 management, especially in early stages (Borghesi and Maroldi 2020).

Despite the abovementioned high sensitivity of the CT scan in COVID-19, it must be noted that a negative CT scan cannot exclude this infection. A negative CT scan has been reported in patients with COVID-19, mainly if it is performed in the first few days after initial symptoms start (Bernheim et al. 2020; Pan et al. 2020a; Xiong et al. 2020). CT findings in COVID-19 are nonspecific but can be helpful when considered in the context of patient exposure and clinical findings. The most common image presentations of COVID-19 on chest CT are listed below.

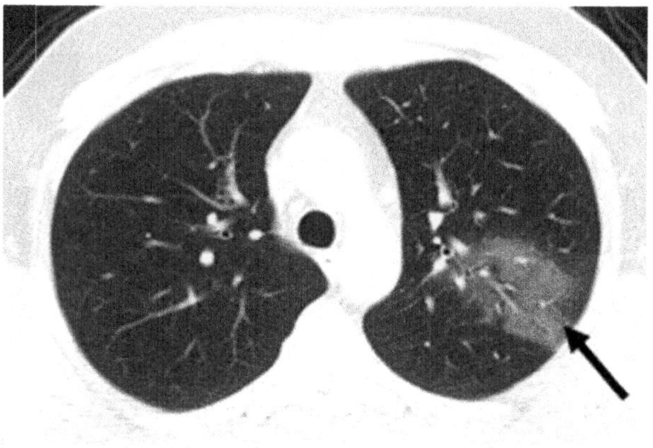

Fig. 24.6 Axial chest CT with lung window. A single patchy area of ground-glass opacity (arrow) in the left lung in a COVID-19-positive patient

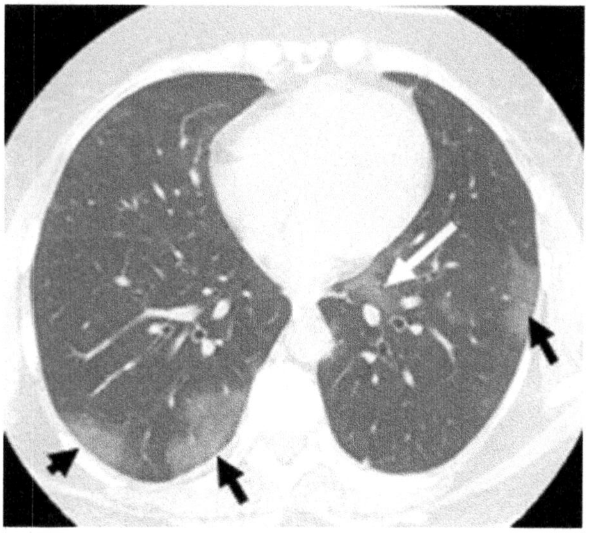

Fig. 24.7 Axial chest CT with lung window. Multiple ground-glass pulmonary opacities in a COVID-19 patient. Lesions are mostly subpleural (black arrows) but with a single central lesion (white arrow)

24.2.2.1 GGO

GGO is consistent with parenchymal lesions causing increased density, while the vascular structures are visible through them (Figs. 24.6 and 24.7). Different mechanisms have been described as the causes of the GGO, including partial filling of alveoli, interstitial lung diseases, and partial alveolar collapse (Hansell et al. 2008; Ufuk and Savas 2020). The most common presentation COVID-19 on CT is multifocal patchy GGO involving the peripheral zone of the lungs, especially in the posterior lung bases (Figs. 24.6 and 24.7). GGOs have been observed in 88% of patients and are among the first radiologic presentations (Li et al. 2020; Guan et al. 2020; Ufuk and Savas 2020; Chung et al. 2020; Pan et al. 2020a; Bernheim et al. 2020; Wu et al. 2020a; Song et al. 2020; Han et al. 2020; Xu et al. 2020a, b; Zhao et al. 2020; Zhou et al. 2020; Li and Xia 2020; Yang et al. 2020; Ai et al. 2020; Xiong et al. 2020; Bai et al. 2020a, b; Cheng et al. 2020; Wang et al. 2020a, b; Colombi et al. 2020; Zhang et al. 2020; Li 2020; Liu et al. 2020b; Grillet et al. 2020; Oudkerk et al. 2020). It is believed that GGO is secondary to the early stages of alveolar damage and pulmonary edema (Jajodia et al. 2020; Xu et al. 2020b). A serial CT scan has also been suggested for the evaluation of GGO progression. The time interval between these follow-up CTs is usually 3–7 days (Borghesi and Maroldi 2020; Bernheim et al. 2020; Pan et al. 2020a).

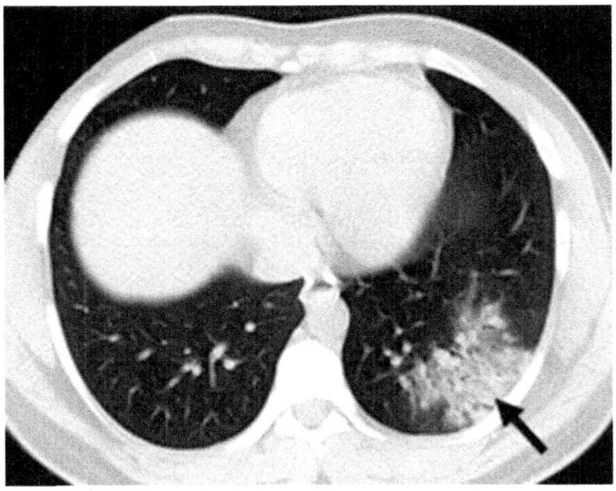

Fig. 24.8 Axial chest CT. A single patchy subpleural consolidation in the left lung obscuring the vascular structures

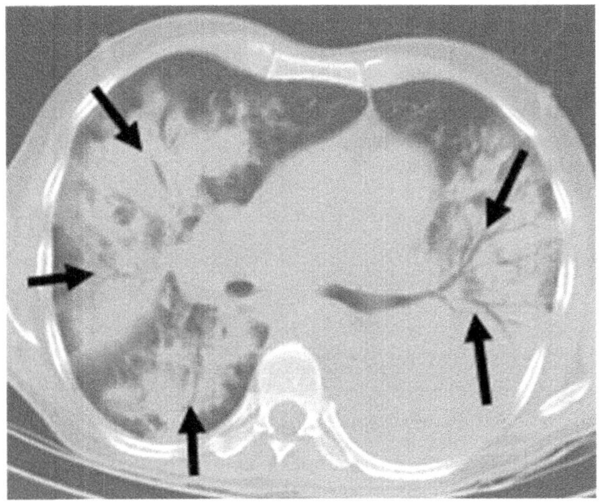

Fig. 24.9 Axial chest CT. Extensive diffuse pulmonary consolidations involving the bilateral lungs. Prominent bilateral air bronchograms (black arrows)

24.2.2.2 Consolidations

Pulmonary consolidations are the parenchymal lesions with increased density, while the vascular structures are not visible through them (Figs. 24.8, 24.9, and 24.10). The mechanisms underlying consolidations are complete filling of air spaces by fluid, blood, neoplastic cells, and pus (Hansell et al. 2008).

Pulmonary consolidations are commonly segmental or subsegmental and located in peripheral and lower zones of the lungs, and they are usually patchy and round in morphology (Figs. 24.8, 24.9, and 24.10). Consolidations are more common in the later stage of COVID-19, in severe cases, and patients older than 50 years old. Linear shape consolidations also have been reported and are likely secondary to evolving organizing pneumonia (Jajodia et al. 2020; Hani et al. 2020; Ufuk and Savas 2020; Chung et al. 2020; Wu et al. 2020a; Huang et al. 2020; Pan et al. 2020a, b; Xiong et al. 2020; Zhou et al. 2020). Consolidations are often associated with GGO (Chung et al. 2020).

24.2.2.3 Crazy-Paving Pattern

Like GGO, a crazy-paving pattern is consistent with increased parenchymal density, while the vascular structures are visible through them. However, in the crazy-paving pattern, there is interlobular septal thickening and enlarged intralobular lines (Fig. 24.11). The likely underlying cause for this pattern is interstitial inflammation

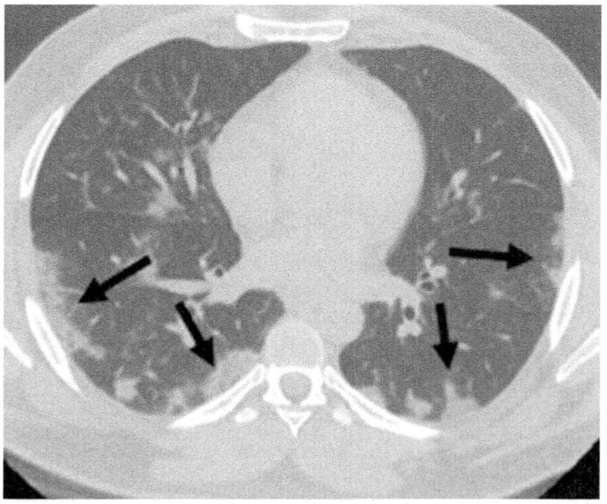

Fig. 24.10 Axial chest CT. Multiple subpleural consolidations in both lungs in a patient with COVID-19 pneumonia

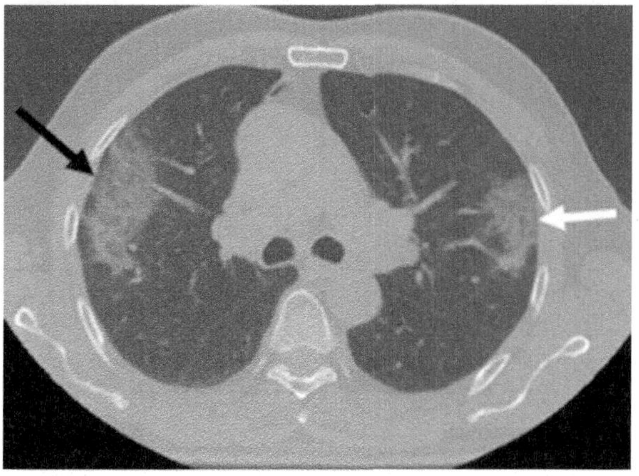

Fig. 24.11 Axial chest CT. A peripheral crazy-paving opacity in the right lung (black arrow) and a subpleural crazy-paving opacity in the left lung (white arrow). In crazy-paving opacities, the density of lesions is increased; however, there is associated interlobular septal thickening and enlarged intralobular lines

in association with alveolar damage (Wu et al. 2020a; Hansell et al. 2008). The crazy-paving pattern is relatively common in COVID-19 and has been reported in up to 90% of patients (Jajodia et al. 2020; Xie et al. 2020; Fang et al. 2020; Bernheim et al. 2020; Wu et al. 2020a; Song et al. 2020; Pan et al. 2020b; Han et al. 2020; Xu et al. 2020a, b; Zhao et al. 2020; Zhou et al. 2020; Li and Xia 2020; Yang et al. 2020; Ai et al. 2020; Xiong et al. 2020; Bai et al. 2020a; Wang et al. 2020a, b ; Colombi et al. 2020; Zhang et al. 2020; Li 2020; Guan et al. 2020; Liu et al. 2020b).

24.2.2.4 Pulmonary Vascular Enlargement (PVE)

Engorged dilated vessel within the pulmonary opacities has been described as a useful diagnostic finding in COVID-19 pneumonia (Lomoro et al. 2020; Ye et al. 2020) (Fig. 24.12). By definition, the PVE is subsegmental pulmonary arteries and veins within the pulmonary lesion with a size of greater than 3 mm (Zhao et al. 2020; Zhou et al. 2020; Li and Xia 2020; Bai et al. 2020a; Wang et al. 2020a). Compared to pneumonia related to other causes, PVE is more common in pneumonia due to COVID-19 (22% versus 59%) and is likely secondary to vascular inflammation.

24.2.2.5 Air Bronchograms

The air bronchogram sign is one of the main presentations of filled air spaces and is consistent with persistent air within bronchioles while there is an increased density of adjacent parenchyma (Fig. 24.13). Air bronchograms occur in up to

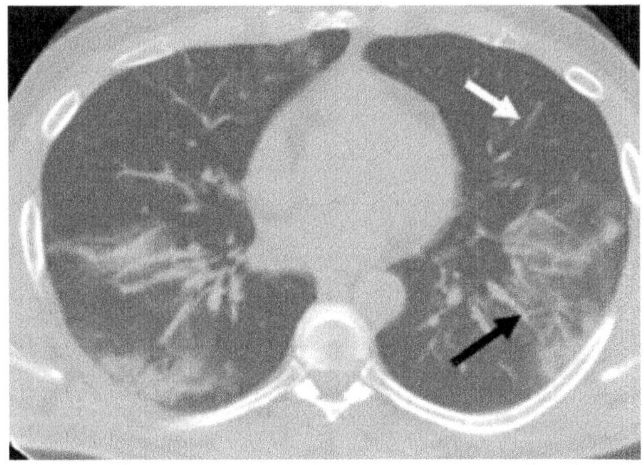

Fig. 24.12 Axial chest CT. Patchy consolidations in both lungs. The vessels within the left-sided opacity (black arrow) are larger than the vessels of the uninvolved areas (white arrow)

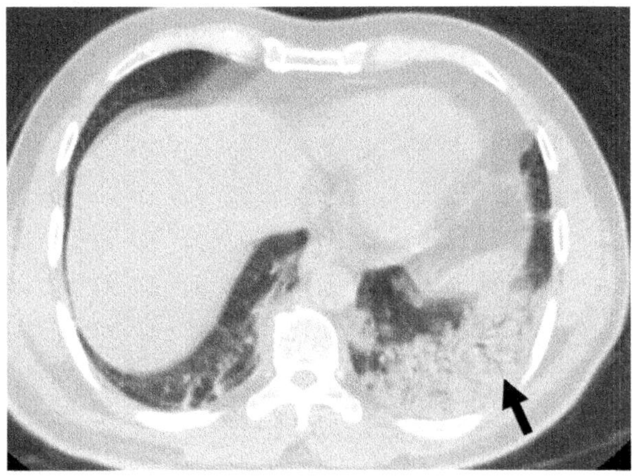

Fig. 24.13 Axial chest CT. A large area of consolidation in the left lung base in a patient with COVID-19 pneumonia. Classical air bronchograms within the lesion (black arrow)

80% of patients with COVID-19 (Jajodia et al. 2020; Song et al. 2020).

24.2.2.6 Reverse Halo Sign (RHS) or the Atoll Sign

The RHS or atoll sign is consistent with a GGO, which has been encased by a rim of pulmonary consolidation (Fig. 24.14). RHS is classically associated with cryptogenic organized pneumonia (COP); however, it has been reported in up to 15% of patients with COVID-19 (Hansell et al. 2008; Bernheim et al. 2020; Wang et al. 2020a; Ufuk and Savas 2020).

24.2.2.7 Thickening of Interlobular Septa

This CT finding is one of the delayed presentations of COVID-19 and usually happens after GGO (Fig. 24.15). It has been reported in up to 80% of patients (Ufuk and Savas 2020; Bernheim et al. 2020; Wu et al. 2020a; Song et al. 2020; Pan et al. 2020b).

24.2.2.8 Diffuse Pulmonary Involvement/White Lungs

In the worst form of pulmonary involvement, COVID-19 causes the involvement of all pulmonary lobes and segments. This pattern occurs in critically ill patients and patients admitted to ICU (Fig. 24.16).

24.2.2.9 Subpleural Curvilinear Lines

The subpleural curvilinear lines are among the late radiologic finings and are because of developing permanent pulmonary fibrosis and have been reported in about 20% of patients (Ufuk and Savas 2020; Wu et al. 2020a; Pan et al. 2020b; Xiong et al. 2020) (Fig. 24.17).

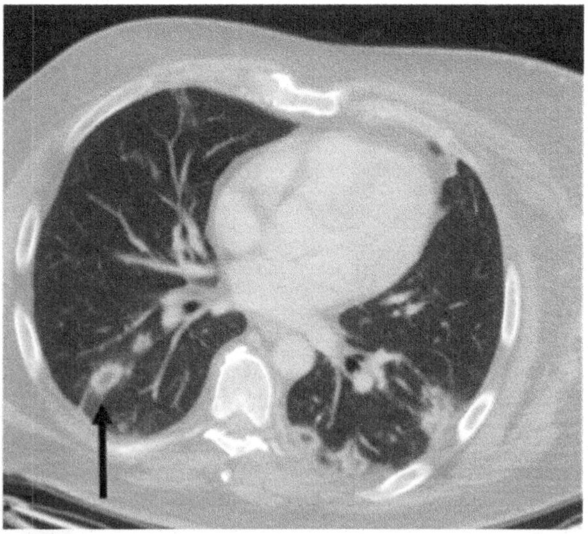

Fig. 24.14 Axial chest CT from a patient suspected of COVID-19 infection. There is an area of ground-glass opacity in the right lung, which is surrounded by a rim of consolidation compatible with the reverse halo sign (RHS) or atoll sign

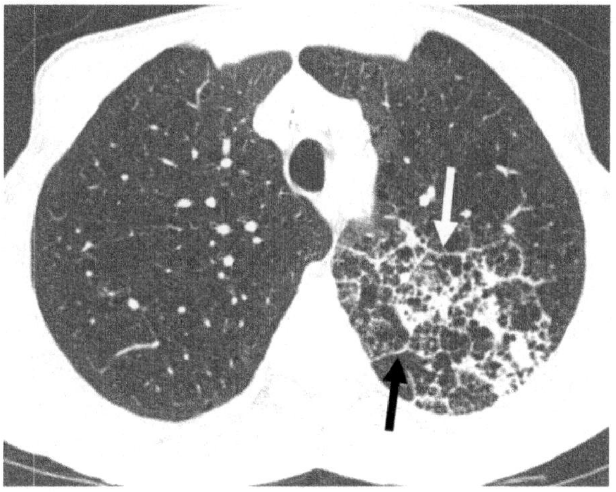

Fig. 24.15 An axial chest CT in a patient with emphysema and superimposed COVID-19 pneumonia. Significant "thickening of interlobular septa" in the left lung (arrows)

24.2.2.10 Pleural Effusion and Thickening

Effusion and thickening of the pleura are not among the classical presentation of COVID-19. They have been reported in a few cases (less than 20%) and late stages of infection (Fig. 24.18). However, the presence of pleural effusions is associated with poor prognosis (Ufuk and Savas 2020; Chung et al. 2020; Bernheim et al. 2020; Wu et al. 2020a; Song et al. 2020; Pan et al. 2020a,b; Han et al. 2020; Xu et al. 2020a; Zhao et al. 2020; Zhou et al. 2020; Li and Xia 2020; Ai et al. 2020; Bai et al. 2020a; Cheng et al. 2020).

24.2.2.11 Lymphadenopathy

Mediastinal lymphadenopathy is not common in COVID-19. Lymphadenopathy, especially in the presence of pleural effusion, micronodules, and tree-in-bud pattern, necessitates evaluation for bacterial infection/superinfection (Fig. 24.19).

24.2.2.12 Lung Cavitary Lesions

The cavitary pulmonary lesions are not a typical presentation of COVID-19 (Fig. 24.20). They have been reported in a few case reports, so the presence of cavitary lesions necessitates the exclusion of other pathologies (Xu et al. 2020c).

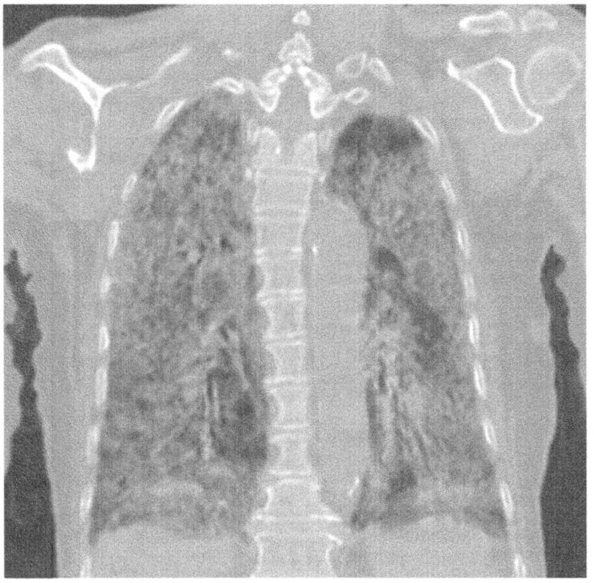

Fig. 24.16 Chest CT with coronal plane reconstruction. There are diffuse extensive pulmonary opacities involving all lobes and segments

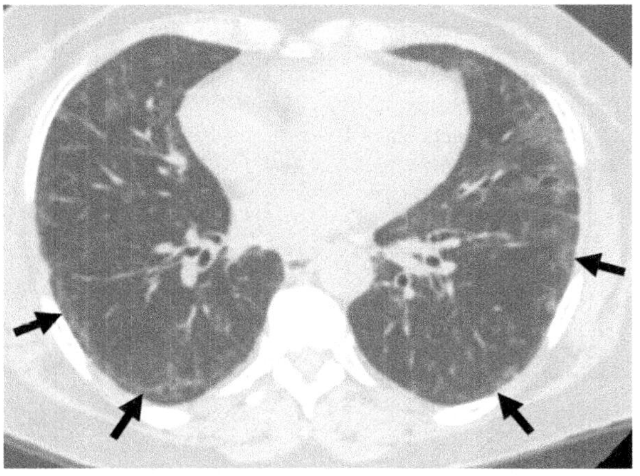

Fig. 24.17 Axial chest CT from a patient with COVID-19 pneumonia. There are subpleural curvilinear lines in both lungs (arrows)

24.2.2.13 Parenchymal Calcification

Pulmonary calcifications are not the classical presentation of COVID-19 pneumonia (Shi et al. 2020), but they are commonly seen in other chronic pulmonary lesions (Fig. 24.21). The presence of parenchymal calcification in confirmed COVID-19 patients can be coincidental.

24.2.2.14 Centrilobular Nodules and Tree-in-Bud Nodularity

By definition, pulmonary nodules are the round pulmonary opacities smaller than 30 mm. The pulmonary nodularity is uncommon in COVID-19 and occurs in about 6–7% of patients (Fig. 24.22). In the case of pulmonary nodularity in a patient suspected to have COVID-19, other etiologies, including bacterial superinfection and aspiration, must be excluded (Ufuk and Savas 2020).

24.2.2.15 Air-Bubble Sign

The air-bubble sign is consistent with the foci of air densities within the pulmonary lesions (Fig. 24.23). They are likely secondary to enlarged air-filled air spaces or bronchioles within the lesions. These lesions have been reported in up to 54% of patients with COVID-19 (Ufuk and Savas 2020; Zhou et al. 2020; Yang et al. 2020; Cheng et al. 2020).

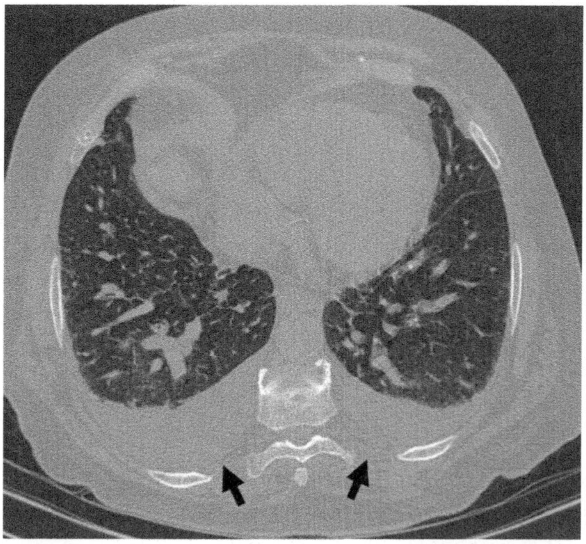

Fig. 24.18 Axial chest CT from a patient with COVID-19 pneumonia shows moderate bilateral pleural effusions. Pleural effusion is not common in COVID-19

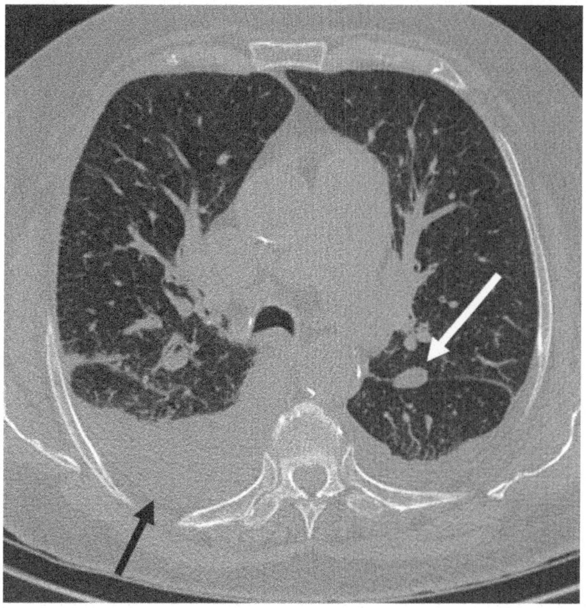

Fig. 24.19 Axial chest CT from a COVID-19 patient. There is a right-sided moderate pleural effusion (black arrow) and left-sided intra-fissure lymphadenopathy (white arrow). Pleural effusion and lymphadenopathy are not common in COVID-19 infection

24.2.2.16 Lesion Locations on Chest CT

Single lung lesion is not typical for COVID-19 pneumonia. More than 70% of lesions are multifocal. The lesions are mainly peripherally located and in lower lung lobes. Lesion distribution along the bronchovascular bundles is not a typical presentation and has been reported only in 12% of patients (Jajodia et al. 2020; Xiong et al. 2020; Pan et al. 2020a; Qin et al. 2020; Wang et al. 2020b; Yoon et al. 2020; Peng et al. 2020; Salehi et al. 2020; Xu et al. 2020d; Hani et al. 2020; Lomoro et al. 2020; Ye et al. 2020; Caruso et al. 2020; Chen et al. 2020b; Wu et al. 2020a; Zhao et al. 2020).

24.2.2.17 CT Scoring

Many researchers suggested and implemented scoring systems based on the volume of lung parenchymal involvement on CT scan. Usually, CT scoring consists of visual evaluation of five different lung lobes. Involvement of each lobe is then visually estimated as follows: score 0, if there is no visible lesion (0% involvement); score

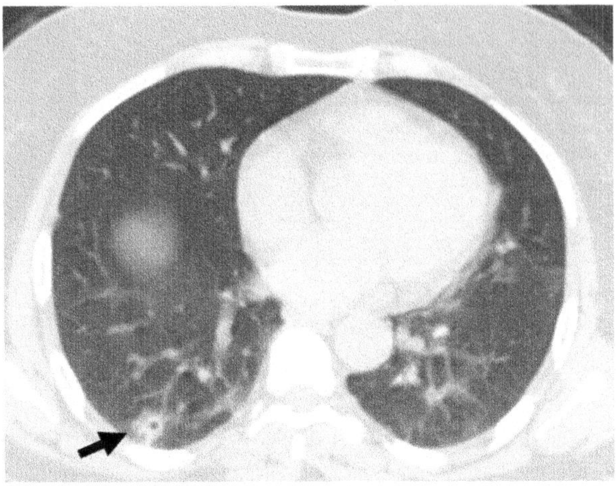

Fig. 24.20 Axial chest CT in COVID-19 pneumonia showing an evolving cavitary lesion in the right lung (arrow)

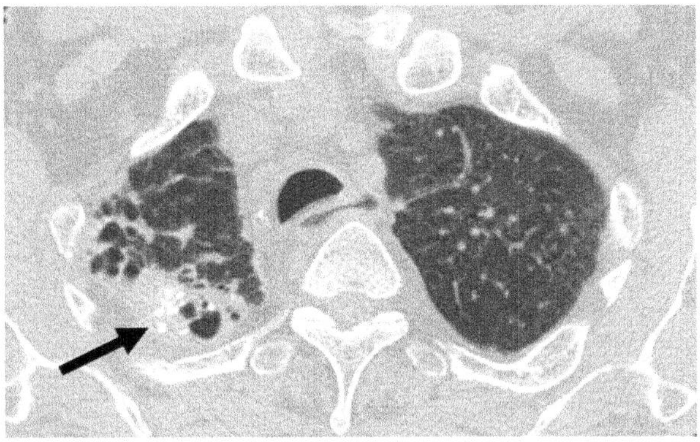

Fig. 24.21 Axial chest CT in a confirmed COVID-19 patient. There are cavitary lesions and foci of parenchymal calcifications in the right lung apex. These lesions are not common in COVID-19 and are likely secondary to underlying pulmonary lesions

1, for minimal lobar involvement (1–25%, of lobe volume); score 2, for mild lobar involvement (26–50% of lobe volume); score 3, for moderate involvement (51–75% of lobe volume); and score 4, for severe involvement (76–100% lobe volume). The overall lung involvement is the sum of five lobar scores. It is a score between 0 and 20, and the higher the score, the worse the pulmonary involvement is (Fu et al. 2020; Chung et al. 2020).

24.2.2.18 The Role of CT Scan in PCR-Negative Patients

PCR tests can be negative even in moderately or severely ill patients. The sensitivity of the nucleic acid test by PCR is about 30–50%. There are many reasons for negative PCR, including low viral load, nonstandard sampling technique, and diagnostic kit failure (Chen et al. 2020a). If the clinical presentation is concerning for COVID-19, CT scan can be performed, and CT scan findings can be used for patient management despite the negative PCR results. A serial PCR test has been proposed for the evaluation of these PCR-negative but CT-positive patients (Rubin et al. 2020).

24.2.2.19 Changes of Imaging Findings Over Time

As mentioned above, the sensitivity of thoracic CT is higher than that of swab and PCR tests for COVID-19 diagnosis. So, it is not uncommon to observe PCR-negative patients who have positive chest CT, especially in the very early stage of the disease. According to the disease stage, CT findings in COVID-19 are clustered into five:

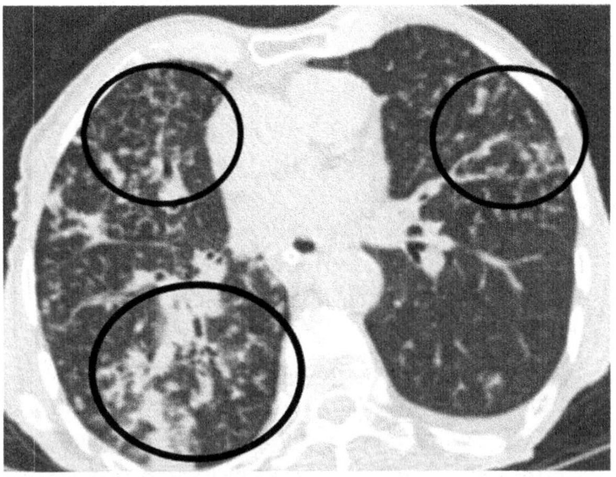

Fig. 24.22 Axial chest CT in COVID-19 patient. Multiple areas of tree-in-bud nodularities in the lung (circles)

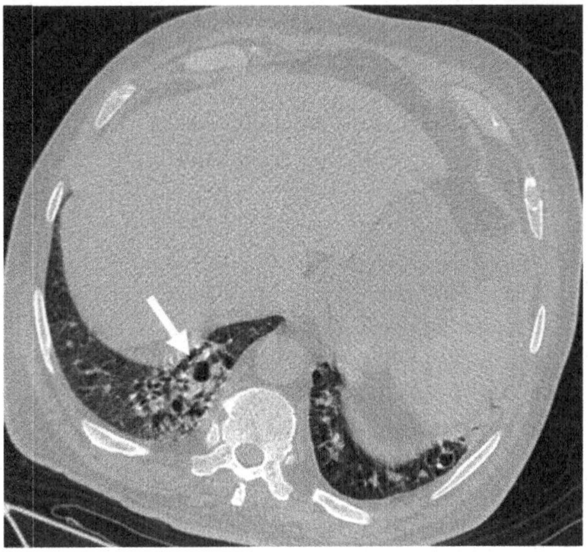

Fig. 24.23 Axial chest CT in COVID-19 patient. There is a right lung base opacity in association with foci of small air densities within the lesion compatible with the "air-bubble sign"

- Ultra-early stage (before symptoms), scattered foci of bilateral subpleural GGOs
- Early stage (1–3 days after initial symptoms), multiple bilateral GGOs and irregular interlobular septa
- Rapid progression stage (3–7 days after initial symptoms), subpleural consolidations with air bronchograms
- Consolidation stage (the second week after initial symptoms), an increase in the size of the consolidations
- Dissipation stage (2–3 after initial symptoms), patchy consolidations with reticular opacities with bronchial and interlobular septal thickening (Fatima et al. 2020; Jin et al. 2020)

The peak of CT findings occurs between 9 and 13 days after initial symptoms (Kanne et al. 2020).

24.2.2.20 Structured Reporting

The structured reporting is a way of standardization of the radiologic reports among different radiologists and medical centers. The structured reporting helps the radiologists to reduce variabilities and uncertainty in reports and improves communication between the radiologists and clinicians. In this context, the Radiology Society of North America (RSNA) has proposed a structured reporting system for COVID-19 patients. In this system, the pulmonary lesions are divided into different classes:

- *Typical appearance of COVID-19*: peripheral bilateral GGO with(out) consolidations or crazy-paving pattern
- *Indeterminate appearance*: the absence of typical findings and presence of multifocal, perihilar, diffuse, or unilateral GGO
- *Atypical appearance*: the absence of typical and indeterminate findings and presence of lobar or segmental consolidation without GGO, pulmonary nodule, cavities, pleural effusions, and smooth interlobular septal thickening
- *Negative appearance*: no evidence of pneumonia (Simpson et al. 2020)

24.2.2.21 Diagnostic Challenges in Patients with COVID-19

As mentioned above, CT findings of COVID-19 are very sensitive but not specific. Viral pneumonia of other causes is the most challenging and common differential diagnosis of COVID-19, characterized by GGO, consolidations, and thickening of interlobular septa. Accurate diagnosis by solo evaluation of the chest imaging is challenging; however, classical chest CT findings during the COVID-19 pandemic in patients with prior exposure to the virus are very suggestive for COVID-19. Several noninfectious diseases can also mimic the radiologic presentation of COVID-19 on CXR and CT. Table 24.1 summarizes the most common types of pneumonia and their similarities and differences in medical imaging.

24.2.3 Ultrasound

During the COVID-19 pandemic, in many centers, ultrasound (US) was used as the first imaging modality to triage, treatment monitoring, complication detection, and evaluation of pleural effusion (Soldati et al. 2019, 2020; Kim et al. 2020). However, given the fact that the ultrasound exam imposes an additional risk of infection to the radiologists and sonographers, it should be limited to the critically ill patients and when the ultrasound result can change the patient management. A fair indication for lung ultrasound is pregnant patients, considering the hazard of CXR and CT scan in this group. It has been reported that the ultrasound can detect the peripheral pulmonary lesion with higher accuracy than the chest X-ray (Soldati et al. 2020).

US findings of the healthy lung include subcutaneous layers, horizontal pleural line with sliding movements during respiratory activity, and horizontal reverberation lines with regular interval lines called A-lines. In contrast, peripheral and subpleural lesions will generate vertical B lines, which are different in thickness and brightness according to the pathology. In the early stages of COVID-19, like other viral pneumonia, the pleural line is usually irregular, thickened, and blurred (Soldati et al. 2019).

24.2.4 Neuroimaging

CNS involvement has been reported in 44% of ICU patients with neurologic symptoms (Kandemirli et al. 2020). The prevalence of neuroimaging abnormalities is higher in patients with more severe respiratory diseases. Common indications for brain imaging in patients with COVID-19 include altered mental status, syncope/fall, and focal neurologic deficit (Radmanesh et al. 2020b). Cortical and lacunar infarctions are among the known CNS complications (Fig. 24.24). Ischemic stroke in young patients without a history of prior cardiovascular disease has been attributed to COVID-19 (Radmanesh et al. 2020b). These observations can be secondary to endothelial damage or coagulopathy associated with this infection. Extracorporeal membrane oxygenation (ECMO) is commonly used in ICU patients with respiratory distress. Anticoagulation medications are used in ECMO patients to avoid clot formation in the connecting tube, and intracranial hemorrhage is a well-known complication of ECMO (Fig. 24.25). However, patients with COVID-19 show an increased risk of intracranial hemorrhages, even without ECMO (Fig. 24.26) (Radmanesh et al.

Table 24.1 Similarities and differences between COVID-19 and other pneumonia on chest imaging

Other pneumonia	Similarity with COVID-19	Differences from COVID-19
Bacterial pneumonia	Consolidation can happen	Bacterial consolidation is usually single and with the lobar distribution. Clinical and laboratory presentation is different than COVID-19
Influenza	Unilateral or bilateral patchy ground-glass opacities+/– consolidations	In addition to the peripherally located lesions, Influenza lesions can be along with the bronchovascular bundles. Also, pulmonary vascular enlargement in COVID-19 may differentiate these two viral pneumonias
Severe acute respiratory syndrome coronavirus (SARS-2003) and Middle East respiratory syndrome coronavirus (MERS-2012)	Peripheral, bilateral, focal,, or multifocal ground-glass opacities and consolidations	Epidemiology is important. SARS and MERS are not active pathogens at the time of writing this manuscript. However, on imaging, there is a significant overlap with COVID-19
Mycoplasma pneumonia	GGO and consolidations can happen	Mycoplasma is common in children and may have a lobular distribution of consolidation. Ill-defined centrilobular nodules, thickening of the bronchovascular bundles, and lymphadenopathy are common in mycoplasma but uncommon in COVID-19
Adenovirus	Multifocal consolidation and ground-glass opacities	Adenovirus pneumonia is more common in children. Atelectasis and right upper lobe collapse are common in adenovirus
Respiratory syncytial virus	GGO and consolidations can happen	Centrilobular nodules and bronchial wall thickening are common in RSV but uncommon in COVID-19. RSV is more common in infants and in patients with immunosuppression and chronic lung disease
Human parainfluenza virus	GGO and consolidations can happen	Lesions of human parainfluenza are mainly centrally located, while COVID-19 lesions are mostly in subpleural areas
Cryptogenic organizing pneumonia	Multiple patchy alveolar opacities. Lesions are often bilateral and subpleural	In cryptogenic organizing pneumonia, there is no predominant craniocaudal distribution, while in COVID-19, lesions are more common in lung bases

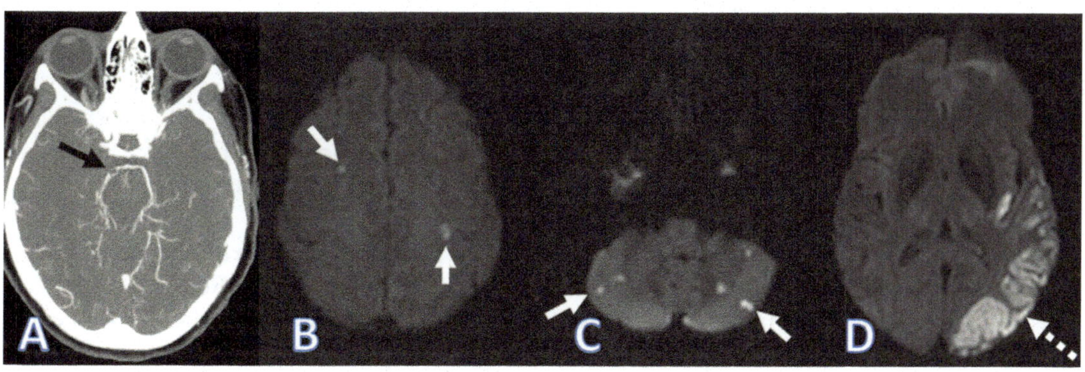

Fig. 24.24 A COVID-19 patient with altered mental status and without a history of cardiovascular disease. Maximum intensity projection (MIP) of CT angiogram shows a cutoff of the right posterior cerebral artery (black arrow **a**). Diffusion-weighted imaging (DWI) shows multiple foci of diffusion restriction and infarctions in bilateral centrum semiovale and cerebellum (white arrows **b** and **c**) and cortical infarction involving the left occipito-temporal lobes and left insula (dash white arrow **d**)

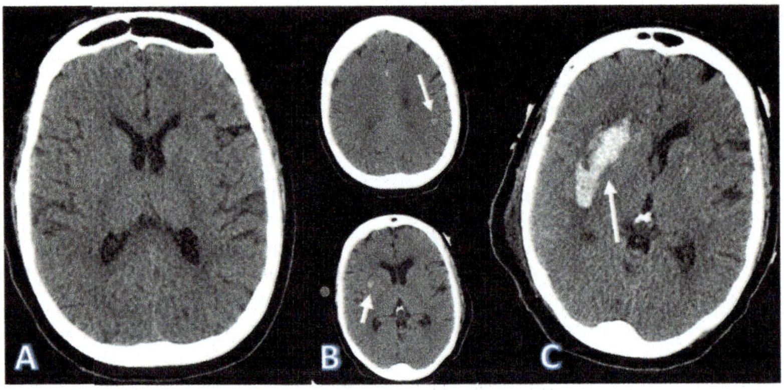

Fig. 24.25 A COVID-19 patient with severe respiratory distress on ECMO. Axial baseline brain CT shows mild microvascular disease; otherwise, no acute intracranial lesion (**a**). Follow-up CT 2 days later shows tiny new foci of new intraparenchymal hemorrhages (arrows, **b**). Follow-up CT 2 days later shows interval enlargement of the right basal ganglia hemorrhage now with mass effect (arrow, **c**)

2020b). Intraventricular hemorrhage (Fig. 24.27) and cortical-based abnormal signal intensity are the other COVID-19 manifestations in neuroimaging (Fig. 24.28). Acute necrotizing encephalopathy (ANE) is the potential life-threatening but rare complication of COVID-19 (Poyiadji et al. 2020). ANE presents on brain CT by bilateral thalami and basal ganglia hypodensities and swelling. The same structures show T2 and FLAIR hyper signal intensities with petechial microhemorrhages on blood-sensitive sequences (GRE and SWI) (Poyiadji et al. 2020). Diffuse leukoencephalopathy and brain parenchymal microhemorrhages are among other CNS complications of COVID-19 (Fig. 24.29) (Radmanesh et al. 2020a).

24.2.5 Imaging Findings in Pediatrics

The prevalence and severity of COVID-19 in children are less than that in adults. Also, the rate of negative chest CT is higher in children than adults, and a negative CT rate of up to 77% has been reported in pediatrics. In positive CT cases, peripheral GGO, crazy-paving pattern, and the halo and reverse halo signs have been reported as dominant lesions. Lesions are more common in the lung bases, and CT lesions are more severe in older children (Steinberger et al. 2020; Bai et al. 2020b).

24.2.6 Imaging of COVID-19 in Pregnancy

Diagnosis of COVID-19 in pregnancy is challenging because clinical and laboratory findings are different from nonpregnant patients. It has been reported that the body temperature, white blood cell, and neutrophil count are higher in pregnant patients. The main CT findings in these patients are the same as nonpregnant cases, and dominant lesions are GGO (81%) and consolidations (17%) (Wu et al. 2020b; Liu et al. 2020a; Elshafeey et al. 2020). On CT scans of pregnant patients with COVID-19, pulmonary lesions are bilateral in 79% and unilateral in 17%. CT scan is negative in 3% of pregnant patients with COVID-19 (Elshafeey et al. 2020).

24.2.7 Imaging of COVID-19 Complications

24.2.7.1 Acute Respiratory Distress Syndrome (ARDS)

ARDS is the common endpoint of many etiologies, causing fluid accumulation in the alveoli and subsequent impaired blood oxygenation. ARDS is common in COVID-19 and on imaging presents as diffuse and bilateral GGO and consolidations (Tang et al. 2020).

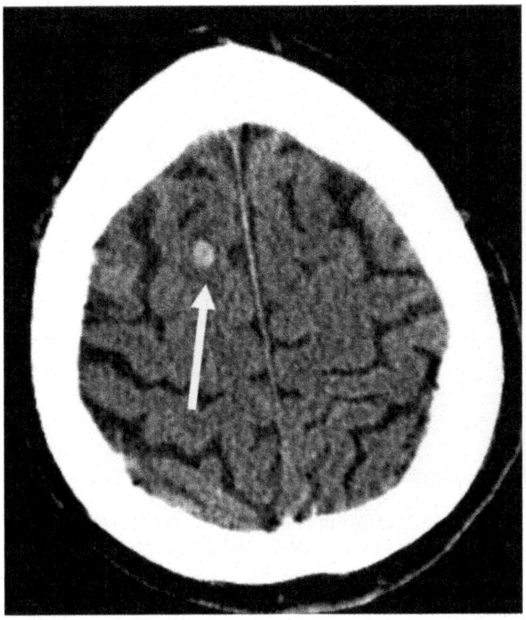

Fig. 24.26 Axial brain CT from a COVID-19 patient shows a small parenchymal hemorrhage in the right frontal lobe. The patient was not on ECMO

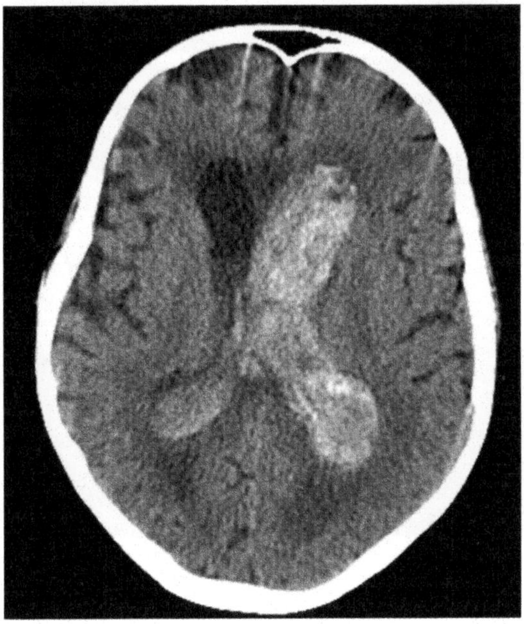

Fig. 24.27 Axial brain CT in a COVID-19 patient with acute altered mental status shows extensive intraventricular hemorrhage and acute hydrocephalus. Periventricular white matter hypodensity is compatible with subependymal interstitial edema secondary to acute hydrocephalus. The patient was not on ECMO

24.2.7.2 Bacterial Superinfection

Pleural effusions, lymphadenopathies, tree-in-bud opacities, and pulmonary cavities are not among the classical presentation of COVID-19. The presence of these conditions in a PCR-positive patient necessitates clinical evaluation to exclude superimposed bacterial infection (Revel et al. 2020).

24.2.7.3 Pulmonary Thromboembolism (PTE)

PTE is very common in critically ill COVID-19 patients and has been reported to happen in 30% of them. Imaging findings are the same as other PTE. They include cardiomegaly, pleural effusions, atelectasis, Hampton hump, and Westermark sign on the CXR, filling defects within the pulmonary arteries on pulmonary CT angiogram (CTA), and V-Q mismatch on ventilation-perfusion nuclear medicine scan (Poyiadji et al. 2020). The diagnosis of PTE usually necessitates CTA after contrast injection. Evaluation of PTE is one of the indications of postcontrast imaging in these patients.

24.2.7.4 Radiomics and Artificial Intelligence (AI)

Radiomics is an evolving research topic in radiology. The radiomics analysis includes feature extraction by computer from the medical images which are not apparent to the human eyes and then using these features for disease classification. The radiomics techniques are often associated with AI models. The most common applications of radiomics are about neoplasms, and very few data are available about infections and pneumonia.

AI now has a significant footprint on COVID-19 imaging. There are several reports regarding the performance of AI in the diagnosis of COVID-19 using CXR and chest CT scans. The most common AI models in COVID-19 imaging are the convolutional neural network (CNN) and related models. Sensitivities of 67–100% and specificities of 81–100% for prediction of COVID-19 pneumonia have been reported in different studies (Ito et al. 2020).

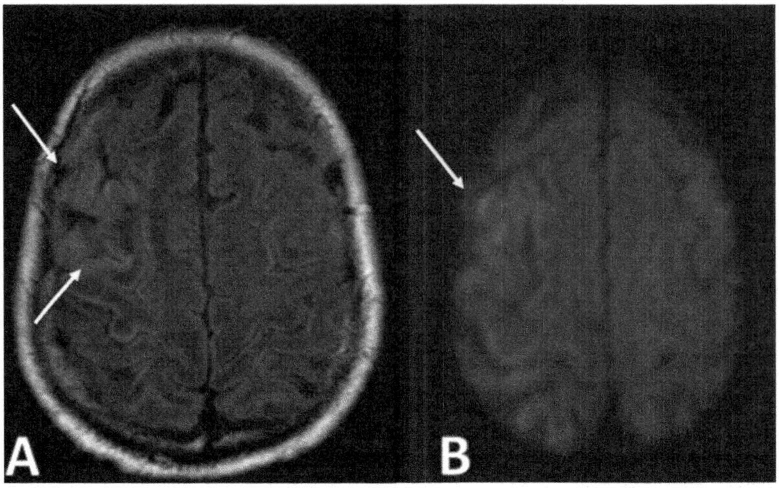

Fig. 24.28 A COVID-19 patient with acute altered mental status. Axial FLAIR MRI shows mild cortical swelling and hyper signal intensity in the right frontal lobe (arrows, **a**). Axial diffusion-weighted imaging (DWI) shows minimal cortical diffusion restriction in this region (arrow, **b**). Follow-up MRI did not demonstrate a change in the signal alteration or evolving encephalomalacia

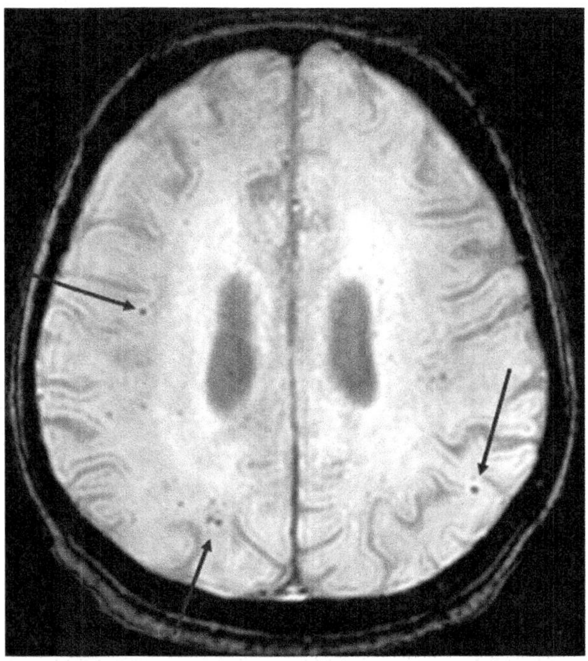

Fig. 24.29 Axial susceptibility-weighted imaging (SWI) in a COVID-19 patient shows multiple tiny foci of parenchymal microhemorrhages (arrows)

24.3 Conclusion

Medical imaging is essential in COVID-19 management. A chest CT scan is more sensitive than the PCR test and can be decisive in the early stage of infection. The main CT presentations are multiple bilateral GGO, consolidations, and crazy-paving patterns in peripheral and subpleural parts of the lungs and sometimes interlobular septal thickening. Pulmonary nodularity, pleural effusions, lymphadenopathy, parenchymal calcifications, and cavitary lesions are not the typical presentations of the COVID-19. Despite the high sensitivity, CT findings are not specific. Similar CT findings can be seen in other types of pneumonia, especially viral infections. CXR is critical for the follow-up of the critically ill and ICU-admitted patients and for the detection of superimposed complications. US can be used as a diagnostic test in COVID-19 patients; however, its application is limited to the peripheral pulmonary lesions and pleural spaces. Imaging also helps the evaluation of the extrapulmonary manifestation of this infection. Imaging scoring and reporting systems have been proposed to improve the communication between the radiologists and clinical experts. Every healthcare worker involved in the COVID-19 pandemic must be familiar with applications, challenges, and pitfalls of the related imaging modalities.

References

Ai T, Yang Z, Hou H, Zhan C, Chen C, Lv W, Tao Q, Sun Z, Xia L (2020) Correlation of chest CT and RT-PCR testing in coronavirus disease 2019 (COVID-19) in China: a report of 1014 cases. Radiology:200642. https://doi.org/10.1148/radiol.2020200642

American College of R (2020) ACR recommendations for the use of chest radiography and computed tomography (CT) for suspected COVID-19 infection. ACR website Advocacy-and Economics/ACR-Position-Statements/Recommendations-for-Chest-Radiography-and-CTfor-Suspected-COVID19-Infection Updated March 22

Bai HX, Hsieh B, Xiong Z, Halsey K, Choi JW, Tran TML, Pan I, Shi LB, Wang DC, Mei J, Jiang XL, Zeng QH, Egglin TK, Hu PF, Agarwal S, Xie F, Li S, Healey T, Atalay MK, Liao WH (2020a) Performance of radiologists in differentiating COVID-19 from viral pneumonia on chest CT. Radiology:200823. https://doi.org/10.1148/radiol.2020200823

Bai K, Liu W, Liu C, Fu Y, Hu J, Qin Y, Zhang Q, Chen H, Xu F, Li C (2020b) Clinical analysis of 25 novel coronavirus infections in children. Pediatr Infect Dis J. https://doi.org/10.1097/INF.0000000000002740

Bernheim A, Mei X, Huang M, Yang Y, Fayad ZA, Zhang N, Diao K, Lin B, Zhu X, Li K, Li S, Shan H, Jacobi A, Chung M (2020) Chest CT findings in coronavirus disease-19 (COVID-19): relationship to duration of infection. Radiology 295(3):200463. https://doi.org/10.1148/radiol.2020200463

Borghesi A, Maroldi R (2020) COVID-19 outbreak in Italy: experimental chest X-ray scoring system for quantifying and monitoring disease progression. Radiol Med 125(5):509–513. https://doi.org/10.1007/s11547-020-01200-3

Caruso D, Zerunian M, Polici M, Pucciarelli F, Polidori T, Rucci C, Guido G, Bracci B, de Dominicis C, Laghi A (2020) Chest CT Features of COVID-19 in Rome, Italy. Radiology:201237. https://doi.org/10.1148/radiol.2020201237

Chen H, Ai L, Lu H, Li H (2020a) Clinical and imaging features of COVID-19. Radiol Infect Dis 7:43. https://doi.org/10.1016/j.jrid.2020.04.003

Chen R, Chen J, Meng QT (2020b) Chest computed tomography images of early coronavirus disease (COVID-19). Can J Anaesth 67(6):754–755. https://doi.org/10.1007/s12630-020-01625-4

Cheng Z, Lu Y, Cao Q, Qin L, Pan Z, Yan F, Yang W (2020) Clinical features and chest CT manifestations of coronavirus disease 2019 (COVID-19) in a Single-Center Study in Shanghai, China. Am J Roentgenol 215:1–6. https://doi.org/10.2214/AJR.20.22959

Chung M, Bernheim A, Mei X, Zhang N, Huang M, Zeng X, Cui J, Xu W, Yang Y, Fayad ZA, Jacobi A, Li K, Li S, Shan H (2020) CT imaging features of 2019 novel coronavirus (2019-nCoV). Radiology 295(1):202–207. https://doi.org/10.1148/radiol.2020200230

Colombi D, Bodini FC, Petrini M, Maffi G, Morelli N, Milanese G, Silva M, Sverzellati N, Michieletti E (2020) Well-aerated lung on admitting chest CT to predict adverse outcome in COVID-19 pneumonia. Radiology:201433. https://doi.org/10.1148/radiol.2020201433

Elshafeey F, Magdi R, Hindi N, Elshebiny M, Farrag N, Mahdy S, Sabbour M, Gebril S, Nasser M, Kamel M, Amir A, Maher Emara M, Nabhan A (2020) A systematic scoping review of COVID-19 during pregnancy and childbirth. Int J Gynaecol Obstet 150:47. https://doi.org/10.1002/ijgo.13182

Fang Y, Zhang H, Xie J, Lin M, Ying L, Pang P, Ji W (2020) Sensitivity of chest CT for COVID-19: comparison to RT-PCR. Radiology:200432. https://doi.org/10.1148/radiol.2020200432

Fatima S, Ratnani I, Husain M, Surani S (2020) Radiological findings in patients with COVID-19. Cureus 12(4):e7651. https://doi.org/10.7759/cureus.7651

Fu F, Lou J, Xi D, Bai Y, Ma G, Zhao B, Liu D, Bao G, Lei Z, Wang M (2020) Chest computed tomography findings of coronavirus disease 2019 (COVID-19) pneumonia. Eur Radiol 30:5489. https://doi.org/10.1007/s00330-020-06920-8

Grillet F, Behr J, Calame P, Aubry S, Delabrousse E (2020) Acute pulmonary embolism associated with COVID-19 pneumonia detected by pulmonary CT angiography. Radiology:201544. https://doi.org/10.1148/radiol.2020201544

Guan CS, Lv ZB, Yan S, Du YN, Chen H, Wei LG, Xie RM, Chen BD (2020) Imaging features of coronavirus disease 2019 (COVID-19): evaluation on thin-section CT. Acad Radiol 27(5):609–613. https://doi.org/10.1016/j.acra.2020.03.002

Han R, Huang L, Jiang H, Dong J, Peng H, Zhang D (2020) Early clinical and CT manifestations of coronavirus disease 2019 (COVID-19) pneumonia. AJR Am J Roentgenol 215:1–6. https://doi.org/10.2214/AJR.20.22961

Hani C, Trieu NH, Saab I, Dangeard S, Bennani S, Chassagnon G, Revel MP (2020) COVID-19 pneumonia: a review of typical CT findings and differential diagnosis. Diagn Interv Imaging 101(5):263–268. https://doi.org/10.1016/j.diii.2020.03.014

Hansell DM, Bankier AA, MacMahon H, McLoud TC, Muller NL, Remy J (2008) Fleischner society: glossary of terms for thoracic imaging. Radiology 246(3):697–722. https://doi.org/10.1148/radiol.2462070712

Huang P, Liu T, Huang L, Liu H, Lei M, Xu W, Hu X, Chen J, Liu B (2020) Use of chest CT in combination with negative RT-PCR assay for the 2019 novel coronavirus but high clinical suspicion. Radiology 295(1):22–23. https://doi.org/10.1148/radiol.2020200330

Ito R, Iwano S, Naganawa S (2020) A review on the use of artificial intelligence for medical imaging of the lungs of patients with coronavirus disease 2019. Diagn Interv Radiol 26:443. https://doi.org/10.5152/dir.2019.20294

Jacobi A, Chung M, Bernheim A, Eber C (2020) Portable chest X-ray in coronavirus disease-19 (COVID-19): a pictorial review. Clin Imaging 64:35–42. https://doi.org/10.1016/j.clinimag.2020.04.001

Jajodia A, Ebner L, Heidinger B, CA K, Prosch H (2020) Imaging in corona virus disease 2019 (COVID-19)-a scoping review. Eur J Radiol Open:100237. https://doi.org/10.1016/j.ejro.2020.100237

Jin YH, Cai L, Cheng ZS, Cheng H, Deng T, Fan YP, Fang C, Huang D, Huang LQ, Huang Q, Han Y, Hu B, Hu F, Li BH, Li YR, Liang K, Lin LK, Luo LS, Ma J, Ma LL, Peng ZY, Pan YB, Pan ZY, Ren XQ, Sun HM, Wang Y, Wang YY, Weng H, Wei CJ, Wu DF, Xia J, Xiong Y, Xu HB, Yao XM, Yuan YF, Ye TS, Zhang XC, Zhang YW, Zhang YG, Zhang HM, Zhao Y, Zhao MJ, Zi H, Zeng XT, Wang YY, Wang XH, Management ftZHoWUNC, Research Team E-BMCoCIE, Promotive Association for M, Health C (2020) A rapid advice guideline for the diagnosis and treatment of 2019 novel coronavirus (2019-nCoV) infected pneumonia (standard version). Mil Med Res 7(1):4. https://doi.org/10.1186/s40779-020-0233-6

Kandemirli SG, Dogan L, Sarikaya ZT, Kara S, Akinci C, Kaya D, Kaya Y, Yildirim D, Tuzuner F, Yildirim MS, Ozluk E, Gucyetmez B, Karaarslan E, Koyluoglu I, Demirel Kaya HS, Mammadov O, Kisa Ozdemir I, Afsar N, Citci Yalcinkaya B, Rasimoglu S, Guduk DE, Kedir Jima A, Ilksoz A, Ersoz V, Yonca Eren M, Celtik N, Arslan S, Korkmazer B, Dincer SS, Gulek E, Dikmen I, Yazici M, Unsal S, Ljama T, Demirel I, Ayyildiz A, Kesimci I, Bolsoy Deveci S, Tutuncu M, Kizilkilic O, Telci L, Zengin R, Dincer A, Akinci IO, Kocer N (2020) Brain MRI findings in patients in the intensive care unit with COVID-19 infection. Radiology:201697. https://doi.org/10.1148/radiol.2020201697

Kanne JP, Little BP, Chung JH, Elicker BM, Ketai LH (2020) Essentials for radiologists on COVID-19: an update-radiology scientific expert panel. Radiology:200527. https://doi.org/10.1148/radiol.2020200527

Kim DJ, Jelic T, Woo MY, Heslop C, Olszynski P (2020) Just the facts: recommendations on point-of-care ultrasound use and machine infection control during the coronavirus disease 2019 pandemic. CJEM 22:1–5. https://doi.org/10.1017/cem.2020.364

Li M (2020) Chest CT features and their role in COVID-19. Radiol Infect Dis 7:51. https://doi.org/10.1016/j.jrid.2020.04.001

Li Y, Xia L (2020) Coronavirus disease 2019 (COVID-19): role of chest CT in diagnosis and management. AJR Am J Roentgenol 214(6):1280–1286. https://doi.org/10.2214/AJR.20.22954

Li K, Fang Y, Li W, Pan C, Qin P, Zhong Y, Liu X, Huang M, Liao Y, Li S (2020) CT image visual quantitative evaluation and clinical classification of coronavirus disease (COVID-19). Eur Radiol 30:4407. https://doi.org/10.1007/s00330-020-06817-6

Liu H, Liu F, Li J, Zhang T, Wang D, Lan W (2020a) Clinical and CT imaging features of the COVID-19 pneumonia: focus on pregnant women and children. J Infect 80(5):e7–e13. https://doi.org/10.1016/j.jinf.2020.03.007

Liu KC, Xu P, Lv WF, Qiu XH, Yao JL, Gu JF, Wei W (2020b) CT manifestations of coronavirus disease-2019: a retrospective analysis of 73 cases by disease severity. Eur J Radiol 126:108941. https://doi.org/10.1016/j.ejrad.2020.108941

Lomoro P, Verde F, Zerboni F, Simonetti I, Borghi C, Fachinetti C, Natalizi A, Martegani A (2020) COVID-19 pneumonia manifestations at the admission on chest ultrasound, radiographs, and CT: single-center study and comprehensive radiologic literature review. Eur J Radiol Open 7:100231. https://doi.org/10.1016/j.ejro.2020.100231

Nair A, Rodrigues JCL, Hare S, Edey A, Devaraj A, Jacob J, Johnstone A, McStay R, Denton E, Robinson G (2020) A British Society of Thoracic Imaging statement: considerations in designing local imaging diagnostic algorithms for the COVID-19 pandemic. Clin Radiol 75(5):329–334. https://doi.org/10.1016/j.crad.2020.03.008

Oudkerk M, Buller HR, Kuijpers D, van Es N, Oudkerk SF, McLoud TC, Gommers D, van Dissel J, Ten Cate H, van Beek EJ (2020) Diagnosis, prevention, and treatment of thromboembolic complications in COVID-19: report of the National Institute for Public Health of the Netherlands. Radiology:201629. https://doi.org/10.1148/radiol.2020201629

Pan F, Ye T, Sun P, Gui S, Liang B, Li L, Zheng D, Wang J, Hesketh RL, Yang L, Zheng C (2020a) Time course of lung changes at chest CT during recovery from coronavirus disease 2019 (COVID-19). Radiology 295(3):715–721. https://doi.org/10.1148/radiol.2020200370

Pan Y, Guan H, Zhou S, Wang Y, Li Q, Zhu T, Hu Q, Xia L (2020b) Initial CT findings and temporal changes in patients with the novel coronavirus pneumonia (2019-nCoV): a study of 63 patients in Wuhan, China. Eur Radiol 30(6):3306–3309. https://doi.org/10.1007/s00330-020-06731-x

Peng QY, Wang XT, Zhang LN, Chinese Critical Care Ultrasound Study G (2020) Findings of lung ultrasonography of novel corona virus pneumonia during the 2019-2020 epidemic. Intensive Care Med 46(5):849–850. https://doi.org/10.1007/s00134-020-05996-6

Poyiadji N, Shahin G, Noujaim D, Stone M, Patel S, Griffith B (2020) COVID-19-associated acute hemorrhagic necrotizing encephalopathy: CT and MRI features. Radiology:201187. https://doi.org/10.1148/radiol.2020201187

Qin C, Liu F, Yen TC, Lan X (2020) (18)F-FDG PET/CT findings of COVID-19: a series of four highly suspected cases. Eur J Nucl Med Mol Imaging 47(5):1281–1286. https://doi.org/10.1007/s00259-020-04734-w

Radmanesh A, Derman A, Lui YW, Raz E, Loh JP, Hagiwara M, Borja MJ, Zan E, Fatterpekar GM (2020a) COVID-19 – associated diffuse leukoencephalopathy and microhemorrhages. Radiology:202040. https://doi.org/10.1148/radiol.2020202040

Radmanesh A, Raz E, Zan E, Derman A, Kaminetzky M (2020b) Brain imaging use and findings in COVID-19: a single academic center experience in the epicenter of disease in the United States. AJNR Am J Neuroradiol 41:1179. https://doi.org/10.3174/ajnr.A6610

Revel MP, Parkar AP, Prosch H, Silva M, Sverzellati N, Gleeson F, Brady A, European Society of R, the European Society of Thoracic I (2020) COVID-19 patients and the radiology department – advice from the European Society of Radiology (ESR) and the European Society of Thoracic Imaging (ESTI). Eur Radiol. https://doi.org/10.1007/s00330-020-06865-y

Rodriguez-Morales AJ, Cardona-Ospina JA, Gutierrez-Ocampo E, Villamizar-Pena R, Holguin-Rivera Y, Escalera-Antezana JP, Alvarado-Arnez LE, Bonilla-Aldana DK, Franco-Paredes C, Henao-Martinez AF, Paniz-Mondolfi A, Lagos-Grisales GJ, Ramirez-Vallejo E, Suarez JA, Zambrano LI, Villamil-Gomez WE, Balbin-Ramon GJ, Rabaan AA, Harapan H, Dhama K, Nishiura H, Kataoka H, Ahmad T, Sah R, Latin American Network of Coronavirus Disease C-REahwlo (2020) Clinical, laboratory and imaging features of COVID-19: a systematic review and meta-analysis. Travel Med Infect Dis 34:101623. https://doi.org/10.1016/j.tmaid.2020.101623

Rubin GD, Ryerson CJ, Haramati LB, Sverzellati N, Kanne JP, Raoof S, Schluger NW, Volpi A, Yim JJ, Martin IBK, Anderson DJ, Kong C, Altes T, Bush A, Desai SR, Goldin J, Goo JM, Humbert M, Inoue Y, Kauczor HU, Luo F, Mazzone PJ, Prokop M, Remy-Jardin M, Richeldi L, Schaefer-Prokop CM, Tomiyama N, Wells AU, Leung AN (2020) The role of chest imaging in patient management during the COVID-19 pandemic: a multinational consensus statement from the Fleischner society. Radiology:201365. https://doi.org/10.1148/radiol.2020201365

Salehi S, Abedi A, Balakrishnan S, Gholamrezanezhad A (2020) Coronavirus disease 2019 (COVID-19): a systematic review of imaging findings in 919 patients. AJR Am J Roentgenol 215:1–7. https://doi.org/10.2214/AJR.20.23034

Shi H, Han X, Jiang N, Cao Y, Alwalid O, Gu J, Fan Y, Zheng C (2020) Radiological findings from 81 patients with COVID-19 pneumonia in Wuhan, China: a descriptive study. Lancet Infect Dis 20(4):425–434. https://doi.org/10.1016/S1473-3099(20)30086-4

Simpson S, Kay FU, Abbara S, Bhalla S, Chung JH, Chung M, Henry TS, Kanne JP, Kligerman S, Ko JP, Litt H (2020) Radiological Society of North America Expert Consensus Statement on Reporting Chest CT Findings Related to COVID-19. Endorsed by the Society of Thoracic Radiology, the American College of Radiology, and RSNA. J Thorac Imaging. https://doi.org/10.1097/RTI.0000000000000524

Soldati G, Demi M, Smargiassi A, Inchingolo R, Demi L (2019) The role of ultrasound lung artifacts in the diagnosis of respiratory diseases. Expert Rev Respir Med 13(2):163–172. https://doi.org/10.1080/17476348.2019.1565997

Soldati G, Smargiassi A, Inchingolo R, Buonsenso D, Perrone T, Briganti DF, Perlini S, Torri E, Mariani A, Mossolani EE, Tursi F, Mento F, Demi L (2020) Is there a role for lung ultrasound during the COVID-19 pandemic? J Ultrasound Med 39:1459. https://doi.org/10.1002/jum.15284

Song F, Shi N, Shan F, Zhang Z, Shen J, Lu H, Ling Y, Jiang Y, Shi Y (2020) Emerging 2019 novel coronavirus (2019-nCoV) pneumonia. Radiology 295(1):210–217. https://doi.org/10.1148/radiol.2020200274

Steinberger S, Lin B, Bernheim A, Chung M, Gao Y, Xie Z, Zhao T, Xia J, Mei X, Little BP (2020) CT features of coronavirus disease (COVID-19) in 30 pediatric patients. AJR Am J Roentgenol:1–9. https://doi.org/10.2214/AJR.20.23145

Tang X, Du RH, Wang R, Cao TZ, Guan LL, Yang CQ, Zhu Q, Hu M, Li XY, Li Y, Liang LR, Tong ZH, Sun B, Peng P, Shi HZ (2020) Comparison of hospitalized patients with ARDS caused by COVID-19 and H1N1. Chest. https://doi.org/10.1016/j.chest.2020.03.032

Ufuk F, Savas R (2020) Chest CT features of the novel coronavirus disease (COVID-19). Turk J Med Sci 50:664. https://doi.org/10.3906/sag-2004-331

Wang J, Xu Z, Wang J, Feng R, An Y, Ao W, Gao Y, Wang X, Xie Z (2020a) CT characteristics of patients infected with 2019 novel coronavirus: association with clinical type. Clin Radiol 75(6):408–414. https://doi.org/10.1016/j.crad.2020.04.001

Wang K, Kang S, Tian R, Zhang X, Zhang X, Wang Y (2020b) Imaging manifestations and diagnostic value of chest CT of coronavirus disease 2019 (COVID-19) in the Xiaogan area. Clin Radiol 75(5):341–347. https://doi.org/10.1016/j.crad.2020.03.004

Wong HYF, Lam HYS, Fong AH, Leung ST, Chin TW, Lo CSY, Lui MM, Lee JCY, Chiu KW, Chung T,

Lee EYP, Wan EYF, Hung FNI, Lam TPW, Kuo M, Ng MY (2019) Frequency and distribution of chest radiographic findings in COVID-19 positive patients. Radiology:201160. https://doi.org/10.1148/radiol.2020201160

Wu J, Wu X, Zeng W, Guo D, Fang Z, Chen L, Huang H, Li C (2020a) Chest CT findings in patients with coronavirus disease 2019 and its relationship with clinical features. Investig Radiol 55(5):257–261. https://doi.org/10.1097/RLI.0000000000000670

Wu X, Sun R, Chen J, Xie Y, Zhang S, Wang X (2020b) Radiological findings and clinical characteristics of pregnant women with COVID-19 pneumonia. Int J Gynaecol Obstet 150:58. https://doi.org/10.1002/ijgo.13165

Xie X, Zhong Z, Zhao W, Zheng C, Wang F, Liu J (2020) Chest CT for typical 2019-nCoV pneumonia: relationship to negative RT-PCR testing. Radiology:200343. https://doi.org/10.1148/radiol.2020200343

Xiong Y, Sun D, Liu Y, Fan Y, Zhao L, Li X, Zhu W (2020) Clinical and high-resolution CT features of the COVID-19 infection: comparison of the initial and follow-up changes. Investig Radiol 55(6):332–339. https://doi.org/10.1097/RLI.0000000000000674

Xu X, Yu C, Qu J, Zhang L, Jiang S, Huang D, Chen B, Zhang Z, Guan W, Ling Z, Jiang R, Hu T, Ding Y, Lin L, Gan Q, Luo L, Tang X, Liu J (2020a) Imaging and clinical features of patients with 2019 novel coronavirus SARS-CoV-2. Eur J Nucl Med Mol Imaging 47(5):1275–1280. https://doi.org/10.1007/s00259-020-04735-9

Xu YH, Dong JH, An WM, Lv XY, Yin XP, Zhang JZ, Dong L, Ma X, Zhang HJ, Gao BL (2020b) Clinical and computed tomographic imaging features of novel coronavirus pneumonia caused by SARS-CoV-2. J Infect 80(4):394–400. https://doi.org/10.1016/j.jinf.2020.02.017

Xu Z, Pan A, Zhou H (2020c) Rare CT feature in a COVID-19 patient: cavitation. Diagn Interv Radiol 26:380. https://doi.org/10.5152/dir.2020.20181

Xu Z, Shi L, Wang Y, Zhang J, Huang L, Zhang C, Liu S, Zhao P, Liu H, Zhu L, Tai Y, Bai C, Gao T, Song J, Xia P, Dong J, Zhao J, Wang FS (2020d) Pathological findings of COVID-19 associated with acute respiratory distress syndrome. Lancet Respir Med 8(4):420–422. https://doi.org/10.1016/S2213-2600(20)30076-X

Yang W, Cao Q, Qin L, Wang X, Cheng Z, Pan A, Dai J, Sun Q, Zhao F, Qu J, Yan F (2020) Clinical characteristics and imaging manifestations of the 2019 novel coronavirus disease (COVID-19):a multi-center study in Wenzhou city, Zhejiang, China. J Infect 80(4):388–393. https://doi.org/10.1016/j.jinf.2020.02.016

Ye Z, Zhang Y, Wang Y, Huang Z, Song B (2020) Chest CT manifestations of new coronavirus disease 2019 (COVID-19): a pictorial review. Eur Radiol 30:4381. https://doi.org/10.1007/s00330-020-06801-0

Yoon SH, Lee KH, Kim JY, Lee YK, Ko H, Kim KH, Park CM, Kim YH (2020) Chest radiographic and CT findings of the 2019 novel coronavirus disease (COVID-19): analysis of nine patients treated in Korea. Korean J Radiol 21(4):494–500. https://doi.org/10.3348/kjr.2020.0132

Zhang R, Ouyang H, Fu L, Wang S, Han J, Huang K, Jia M, Song Q, Fu Z (2020) CT features of SARS-CoV-2 pneumonia according to clinical presentation: a retrospective analysis of 120 consecutive patients from Wuhan city. Eur Radiol 30:4417. https://doi.org/10.1007/s00330-020-06854-1

Zhao W, Zhong Z, Xie X, Yu Q, Liu J (2020) Relation between chest CT findings and clinical conditions of coronavirus disease (COVID-19) pneumonia: a Multicenter study. AJR Am J Roentgenol 214(5):1072–1077. https://doi.org/10.2214/AJR.20.22976

Zhou S, Wang Y, Zhu T, Xia L (2020) CT features of coronavirus disease 2019 (COVID-19) pneumonia in 62 patients in Wuhan, China. AJR Am J Roentgenol 214(6):1287–1294. https://doi.org/10.2214/AJR.20.22975

25 Therapeutic Development in COVID-19

Chan Yang, Yuan Huang, and Shuwen Liu

Abstract

Since the outbreak of coronavirus disease 2019 (COVID-19) caused by the SARS-CoV-2, the disease has spread rapidly worldwide and developed into a global pandemic, causing a significant impact on the global health system and economic development. Scientists have been racing to find effective drugs and vaccines for the treatment and prevention of COVID-19. However, due to the diversity of clinical manifestations caused by COVID-19, no standard antiviral regimen beyond supportive therapy has been established. Ongoing clinical trials are underway to evaluate the efficacy of drugs that primarily act on the viral replication cycle or enhanced immunity of patients. This chapter will summarize the currently used antiviral and adjuvant therapies in clinical practice and provide a theoretical basis for the future treatment of COVID-19.

Keywords

Antiviral · COVID-19 · Drugs · SARS-CoV-2 · Therapy

25.1 Introduction

Since the beginning of the twenty-first century, coronaviruses (CoVs) have caused massive outbreaks of lethal human pneumonia. Severe acute respiratory syndrome coronavirus (SARS-CoV), a virus spread rapidly through the air and readily transmissible between humans, broke out in 2003 and ultimately infected 8096 people with 774 deaths in 26 countries on 5 continents (Stadler and Rappuoli 2005; Peiris et al. 2004). The Middle East respiratory syndrome coronavirus (MERS-CoV) appeared in the Arabian Peninsula in 2012, with a mortality rate of 35.7% (Hui et al. 2018; Chafekar and Fielding 2018). Severe acute respiratory syndrome coronavirus 2 (SARS-CoV-2) is the seventh known human coronavirus identified as the cause of coronavirus disease 2019 (COVID-19). The disease appeared first in

C. Yang · Y. Huang
Guangdong Provincial Key Laboratory of New Drug Screening, School of Pharmaceutical Sciences, Southern Medical University, Guangzhou, China

S. Liu (✉)
Guangdong Provincial Key Laboratory of New Drug Screening, School of Pharmaceutical Sciences, Southern Medical University, Guangzhou, China

State Key Laboratory of Organ Failure Research, Guangdong Provincial Institute of Nephrology, Southern Medical University, Guangzhou, China

Network of Immunity in Infection, Malignancy and Autoimmunity (NIIMA), Universal Scientific Education and Research Network (USERN), Guangzhou, China
e-mail: liusw@smu.edu.cn

N. Rezaei (ed.), *Coronavirus Disease - COVID-19*, Advances in Experimental Medicine and Biology 1318, https://doi.org/10.1007/978-3-030-63761-3_25

Wuhan, China. As of the end of May 2020, a total of 5.56 million cases of SARS-CoV-2 infection (with 351,826 deaths) are recorded worldwide (WHO Web). A high degree of clinical awareness of the possibility of SARS-CoV-2 infection is essential.

To date, the approach to COVID-19 management focuses on symptomatic and supportive caring (Rabaan et al. 2020). Meanwhile, several effective antiviral drugs and drug candidates have been evaluated in vitro, and a few of them are under urgent investigation, including broad-spectrum antiviral and antimalarial drugs that interfere with viral replication (Huang et al. 2020b; Lythgoe and Middleton 2020). Future outbreaks like the ongoing COVID-19 pandemic are likely to continue. Therefore, in addition to containing this outbreak, efforts should be made to develop comprehensive measures to prevent future outbreaks.

25.2 Antiviral Treatments for COVID-19

25.2.1 Virus-Based Drugs

This class of drugs acts directly on the virus. The main antiviral drugs clinically tested so far are 3C-like protease (3CLpro) inhibitors and RNA polymerase (RdRp) inhibitors (Fig. 25.1). The ongoing research and clinical data suggest that there is a potential list of reuse drugs with appropriate pharmacological effects and therapeutic effects in the treatment of COVID-19 patients. However, since most clinical studies are single-arm studies or studies with small sample sizes, further clinical studies are still needed to evaluate the efficacy of antiviral drugs.

25.2.1.1 3CLpro Inhibitors

3CLpro plays an essential role in the autocleavage process of CoVs. Previous studies on other coronaviruses found that the SARS-CoV and MERS-CoV genomes contain two open reading frames (ORF), ORF1a and ORF1b, which are translated via the host ribosomes into two respective viral polyproteins, pp1a and pp1ab. Subsequently, 3CLpro and papain-like protease (PLpro) are encoded by ORF1a. PLpro cleaves the first three cleavage sites of its polyproteins, while 3CLpro is responsible for cleavage of the remaining 11 sites, resulting in the release of a total of 16 nonstructural proteins (nsp) (Anand et al. 2003). Since the autocleavage process is essential for viral transmission, 3CLpro is a crucial drug target against coronavirus infection.

Lopinavir/Ritonavir (LPV/r)

Lopinavir is a novel protease inhibitor (PI) developed from ritonavir, which is mainly used to suppress plasma viral load and enhance immunity in HIV-1-infected patients (Trinh et al. 2012). Co-formulated LPV/r is easy to administer and can be used in combination with other antiretroviral drugs (Yang et al. 2018). It was found that LPV/r is a 3CLpro inhibitor against SARS-CoV by affecting the formation of the viral RNA replicase precursor multi-proteins pp1a and pp1ab and blocking the release of viral polymerase and helical enzymes (Anand et al. 2003). Considering the similarity between SARS-CoV-2 and SARS-CoV, LPV/r was recommended for anti-SARS-CoV-2 treatment at the beginning of the COVID-19 outbreak (Kannan et al. 2020; Martinez 2020). A randomized trial of 127 COVID-19 patients from February 10 to March 20, 2020 showed that triple-antiviral therapy with LPV/r, ribavirin, and interferon beta-1b (IFN-β) could suppress viral load, shorten hospital stay, and reduce mortality in COVID-19 patients compared with LPV/r alone (Hung et al. 2020). However, another randomized trial of LPV/r (400/100 mg, twice daily for 14 days) in 199 severe COVID-19 hospitalized patients showed that LPV/r did not reduce the time of clinical symptom relief (Cao et al. 2020). Based on available data, it is, thus, difficult to assess the therapeutic effect of LPV/r monotherapy or combination therapy on COVID-19 (Ye et al. 2020). Importantly, LPV/r is not likely to develop resistance, but the adverse reactions it causes, e.g., gastrointestinal and metabolic disorders, should not be ignored (Aspiroz et al. 2013). Therefore, it is necessary to implement LPV/r

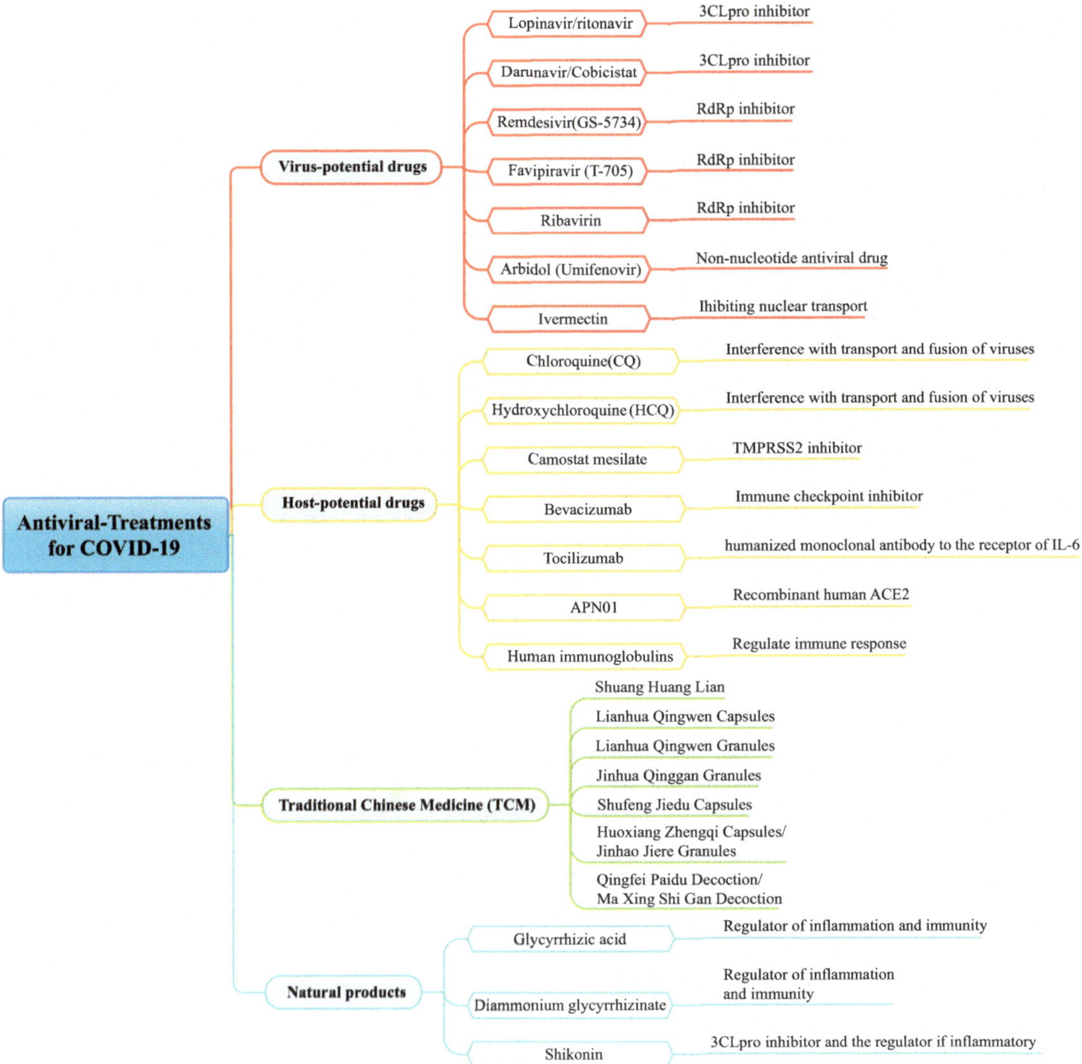

Fig. 25.1 Antiviral treatments for COVID-19. The figure summarizes potential antiviral drugs and traditional Chinese medicine (TCM) for COVID-19 treatment. Targeting the virus (viral-based potential drugs, red border) and host-related factors (host-based potential drugs, yellow border) are two main strategies for the development of antiviral drugs. TCMs (green border) and natural products (blue border) are, in particular, useful in China for the treatment of COVID-19

dosing and monitoring in patients with COVID-19.

Darunavir/Cobicistat

Similar to LPV/r, darunavir/cobicistat has also been used to treat HIV, where cobicistat increases the plasma concentration of darunavir by inhibiting CYP3A-mediated metabolism (Deeks 2018). Recent studies have documented the inhibitory effect of darunavir on 3CLpro (Sk et al. 2020) and identified it as a SARS-CoV-2 3CLpro inhibitor by structure-based virtual screening (Costanzo et al. 2020). However, their therapeutic efficacy requires further study and validation. Clinical trials of darunavir/cobicistat are currently registered and underway.

25.2.1.2 RdRp Inhibitors

RdRp is the central catalytic subunit of all positive-stranded RNA viruses that mediate RNA synthesis. RdRp synthesizes complementary RNA chains using viral mRNA as a template and nucleoside analogs as substrates, but RdRp cannot distinguish between nucleosides and nucleoside analogs. Competitive binding of RdRp inhibitors to RdRp results in blocked steric binding of nucleosides, thus affecting viral RNA synthesis (Wang et al. 2020c). Since the catalytic domain of RdRp is conserved in various RNA viruses (Boonrod et al. 2004), the RdRp of SARS-CoV-2 was modeled, validated, and then targeted using other antiviral drugs currently approved on the market.

Remdesivir (GS-5734)

The first reported COVID-19 patient in the United States was treated with remdesivir (Holshue et al. 2020). Sixty-eight percent of them experienced symptomatic relief after remdesivir treatment, with a morbidity rate of about 18% (Grein et al. 2020). No accelerated recovery of COVID-19 hospitalized patients and no reduction in disease mortality were observed with remdesivir (NCT04257656). However, another study in COVID-19 hospitalized patients (n = 1063) showed that remdesivir was superior to placebo in reducing the recovery time (NCT04280705) (Beigel et al. 2020). Future trials in severe COVID-19 patients may help confirm the efficacy of remdesivir (Mahase 2020).

Favipiravir (T-705)

Favipiravir (6-fluoro-3-hydroxy-2-pyrazinecarboxamide) is an effective selective and potent RdRp inhibitor with a mechanism of action that enhances its broader antiviral activity (Shiraki and Daikoku 2020). Favipiravir shows inhibitory effects on influenza viruses, West Nile virus, yellow fever virus, as well as against flaviviruses, alphaviruses, arenaviruses, bunyaviruses, and SARS-CoV-2 (Fang et al. 2020; Janowski et al. 2020).

In February, favipiravir was declared effective in China for the treatment of COVID-19. An open-label controlled study of 80 patients with COVID-19 in which 35 patients received oral favipiravir plus interferon IFN-α by aerosol inhalation and 45 patients received lopinavir/ritonavir (LPV/r) plus IFN-α by aerosol inhalation showed an advantage of favipiravir in preventing COVID-19 progression and viral clearance. Also, favipiravir treatment led to significant improvement in the chest radiography and was associated with fewer adverse reactions compared to the control group (Cai et al. 2020). On March 8, 2020, another multicenter, randomized, and controlled clinical trial study has recruited 150 patients to investigate the safety of favipiravir in COVID-19 (Pilkington et al. 2020). Currently, several clinical trials are underway to evaluate the safety and efficacy of favipiravir in COVID-19 treatment.

Ribavirin

Ribavirin was first approved for the treatment of respiratory syncytial virus infection and later for the treatment of hepatitis C virus (HCV). As a broad-spectrum antiviral drug, ribavirin has inhibitory effects on a variety of DNA and RNA viruses.

In most cases, ribavirin is combined with IFN for COVID-19 treatment (Khalili et al. 2020). However, during treatment with ribavirin, some patients with SARS-CoV developed hemolytic hemorrhage and decreased hemoglobin levels (Tsang and Zhong 2003). The safety of ribavirin in large doses requires further attention.

25.2.1.3 Other Antiviral Drugs

Arbidol (Umifenovir)

Arbidol is a non-nucleotide antiviral drug developed by the Soviet Research Center for Medicinal Chemistry and is an officially designated anti-influenza drug (Brooks et al. 2012). The use of Arbidol as an anti-COVID-19 drug was first proposed in the New Coronavirus Pneumonia Diagnosis and Treatment Program (China, Trial Sixth Edition) and was subsequently widely used in the clinical treatment of adult COVID-19 patients (200 mg, three times a day, less than 10 days). A case analysis of four

COVID-19 patients showed that Arbidol, in combination with LPV/r and Shufeng Jiedu capsules, significantly relieved the clinical symptoms of pneumonia (fever, cough, and dyspnea) (Deng et al. 2020). Another clinical trial showed that darunavir, in combination with Arbidol, is a viable strategy to combat the SARS-CoV-2 epidemic in China (Costanzo et al. 2020). Also, Arbidol monotherapy is superior to LPV/r in the treatment of COVID-19 (Zhu et al. 2020), with no significant side effects (Wen et al. 2020). However, an observation in non-ICU patients showed that the use of Arbidol might not improve prognosis or accelerate SARS-CoV-2 clearance (Lian et al. 2020). Two clinical trials are currently underway in China to assess the efficacy and safety of Arbidol in the treatment of COVID-19 pneumonia (NCT04252274, NCT04252885).

Ivermectin

Targeting the nuclear transport process is a potential therapeutic approach against RNA viruses. Ivermectin is a Food and Drug Administration (FDA)-approved antiparasitic drug that has also shown antiviral activity against HIV and dengue virus (Grossi et al. 2018; Wagstaff et al. 2012). Recently, a study demonstrated that ivermectin reduced viral RNA by 5000-fold within 48 h after SARS-CoV-2 infection in vivo (Bray et al. 2020; Patri and Fabbrocini 2020). With its established safety, the next step is to prove the efficacy of ivermectin in the treatment of COVID-19 and conduct trials to find the appropriate dosage (NCT04390022).

25.2.2 Host-Based Drugs

Clinical investigations found high concentrations of cytokines in the plasma of patients with severe COVID-19 and suggested that the cytokine storm is associated with disease severity. Below are therapeutic drugs that have shown antiviral effects by inhibiting viral entry and cell-viral membrane fusion during virus invasion and also can act on the host immune and inflammatory responses.

25.2.2.1 Chloroquine (CQ) and Hydroxychloroquine (HCQ)

Chloroquine (CQ) was first recommended as an anti-coronavirus trial drug in the Diagnostic and Treatment Protocol for COVID-19 (Trial Version 6) in China (Wang et al. 2020d). As a classic drug for the treatment of rheumatism and malaria, chloroquine has been used clinically for more than 70 years and has also been effective in the treatment of various viral infections (Rajic et al. 2018; Aguiar et al. 2018). Hydroxychloroquine (HCQ) is an antimalarial drug developed in 1946 based on the CQ structure. In addition to antimalarial properties, the pharmacological effects of HCQ include immunomodulation and antiviral and antibacterial effects (Plantone and Koudriavtseva 2018; Keshavarzi 2016).

Angiotensin-converting enzyme 2 (ACE2) is the critical receptor for SARS-CoV-2 (Ziegler et al. 2020). CQ and HCQ can bind to terminal glycosylation of ACE2 with high affinity and block viral entry into cells (Singh et al. 2020b). On the other hand, CQ and HCQ have inhibitory effects on specific cellular populations, such as neutrophils, and thereby can reduce the production of pro-inflammatory cytokines such as TNF-α, IL-1, and interleukin 6 (IL-6) (Mo and Chen 2019). Therefore, CQ and HCQ act as anti-inflammatory agents to alleviate severe clinical symptoms and may effectively suppress SARS-CoV-2 infection (Klimke et al. 2020; Fantini et al. 2020). Chloroquine phosphate is the phosphate salt of chloroquine. On February 4, 2020, the Institute of Virology, Wuhan, China, jointly published the results of effective in vitro inhibition of SARS-CoV-2 by chloroquine phosphate (Wang et al. 2020b).

In the first clinical trial, CQ effectively reduced viral load, inhibited the progression of pneumonia, improved lung imaging performance, and shortened disease progression. Subsequently, a treatment efficacy study in 80 patients with COVID-19 provided evidence of the effectiveness of coadministration of HCQ and azithromycin and their potential role in the early reduction of infectivity (Gautret et al. 2020). However, the efficacy of HCQ positively correlated with its plasma concentration. Low concentration of HCQ achieved no satisfactory therapeutic effects, while using high concentrations of HCQ caused toxic side effects (Gao et al. 2020). The widespread use of this combination, i.e., HCQ and azithromycin, by clinicians worldwide has raised concerns as both drugs have been independently shown to increase the risk of prolonged QT intervals and sudden pharmacological death (Singh et al. 2020a). HCQ, in the form of a low-dose nonsystemic aerosol, is currently used in high-risk, critically ill, and elderly patients with COVID-19 (Klimke et al. 2020). Ongoing trials are actively recruiting with the hope of validating the potential of HCQ in the treatment and prevention of COVID-19.

25.2.2.2 Camostat Mesilate

Transmembrane protease serine 2 (TMPRSS2) is a serine protease that primes the spike protein of highly pathogenic human coronaviruses. The proteolysis of TMPRSS2 plays an essential role in priming SARS-CoV-2 spike protein for entry (Hoffmann et al. 2020a). Camostat mesilate, an inhibitor of TMPRSS2, can block the entry of SARS-CoV-2 into the lung cells and has been effective in protecting the mice against SARS-CoV infection, with a survival rate of 60% (Zhou et al. 2015).

25.2.2.3 Bevacizumab

Bevacizumab is a recombinant humanized monoclonal antibody that inhibits angiogenesis through specific binding to elevated vascular endothelial growth factor (VEGF) (Ruan et al. 2018). An ongoing trial is now evaluating the efficacy of bevacizumab as a unique approach in the treatment of SARS-CoV-2 infection (NCT04275414).

25.2.2.4 Recombinant Human ACE2 (APN01)

Human recombinant soluble ACE2 significantly blocks the early stages of SARS-CoV-2 infections (Monteil et al. 2020). APN01 was successfully developed in 2010 and can mechanistically prevent viral infections and prevent acute lung injury. A phase II study of acute respiratory distress syndrome has demonstrated a high safety profile of APN01 in reducing angiotensin II (Ang II) and IL-6 levels (Hoffmann et al. 2020b).

25.2.2.5 Tocilizumab

Tocilizumab is a humanized monoclonal antibody (mAb) directed against the IL-6 receptor. It was first approved by the FDA in 2010 for the treatment of rheumatoid arthritis. According to the Diagnosis and Treatment of Pneumonia Caused by COVID-19 (7th edition) of China, tocilizumab can be used to treat severe lung injury in patients with COVID-19 (Colaneri et al. 2020). Tocilizumab has been approved in China for the treatment of COVID-19 pneumonia and IL-6 elevation (ChiCTR2000029765) (Mazzitelli et al. 2020; Antinori et al. 2020). Currently, seven clinical trials of intravenous tocilizumab in COVID-19 patients are underway (NCT04315480, NCT04306705, NCT04332094, NCT04332913, NCT04317092, NCT04335071, NCT04310228).

25.2.2.6 Human Immunoglobulin

Human immunoglobulins are the cornerstone of treatment for patients with primary immunodeficiency diseases affecting the humoral immune system. Immunoglobulins contain IgG antibodies that are broadly resistant to viruses, bacteria, or other pathogens, with dual therapeutic effects of immune replacement and immune regulation (Perez et al. 2017). In patients with severe COVID-19, intravenous immunoglobulin can be given in the early stages of infection, depending on the condition (Moeinzadeh et al. 2020).

25.2.3 Neutralizing Antibodies (nAbs)

nAbs provide specific immune defense against viral infection in patients (Table 25.1). The spike (S) protein of coronaviruses is the primary inducer of nAbs. The S protein of SARS-CoV-2 and SARS-CoV consists of the S1 subunit and the S2 subunit. The virus recognizes the ACE2 receptor through the receptor-binding domain (RBD) located on the S1 subunit, and the S2 subunit mediates viral-cell fusion (Zhou and Zhao 2020).

Table 25.1 Representative SARS-CoV-2 targeting nAbs

nAbs	Source	Target region	Protective efficacy	References
80R scFv	Human	SARS-CoV S1 aa 261–672	Blocking RBD-ACE2 binding and inhibiting syncytium formation	Tian et al. (2020)
47D11	Human	SARS-CoV S1 aa 525–533; SARS-CoV-2 S1 aa 338–506	NA	Ho (2020)
CR3022 CR3014	Human	SARS-CoV RBD aa 318–510; CR3022 binds SARS-CoV-2 RBD	Protect ferrets against SARS-CoV and SARS-CoV-2 infection	Tian et al. (2020)
33G4 35B5 30F9	Mouse	SARS-CoV RBD, blocking RBD-ACE2 receptor binding	NA	Du et al. (2017)
MERS-27 m336 MERS-GD27 MCA1	Human	MERS-CoV RBD, blocking RBD-DPP4 receptor binding	Preventing and treating MERS-CoV (strain EMC2012) challenge in hDPP4-Tg mice, rabbits, or common marmosets	Du et al. (2017) and Zhou et al. (2019)
HCAb-83	Dromedary camel	MERS-CoV RBD aa 539	Prophylactically prevents MERS-CoV (strain EMC2012) challenge in hDPP4-Tg mice	Zhou et al. (2019)
311mab-31B5/32D4	Human	Cloned from single memory B cells separation from recovered COVID-19 patients	Binding to the SARS-CoV-2 S to effectively neutralize the SARS-CoV-2 infection	Wang et al. (2020a) and Chen et al. (2020b)
P2C-1F11/ P2B-2F6	Human	SARS-CoV-2 RBD, derived from single B cells of SARS-CoV-2 infected individuals	Anti-SARS-CoV-2 neutralization activity	Ju et al. (2020)
B38	Human	SARS-CoV-2 RBD 470-loop supported by Q474	Blocking binding between RBD of SARS- CoV-2 and ACE2	Wu et al. (2020)

nAbs neutralizing antibody, *S* spike, *aa* amino acid, *RBD* receptor-binding domain, *ACE2* angiotensin-converting enzyme 2, *NA* not applicable, *hDPP4-Tg mice* human DPP4-transgenic mice

25.2.4 Traditional Chinese Medicine (TCM) and Natural Products

SARS-CoV-2 and SARS-CoV share similarities in epidemiology, genomics, and pathogenesis. Due to these similarities, studies investigating the therapeutic efficacy of TCM in patients with SARS will be summarized below.

25.2.4.1 Chinese Medicine Preparations and Prescriptions

Shuang Huang Lian, a TCM herbal product made from Lonicerae japonicae Flos, Scutellariae radix, and Fructus Forsythiae, purportedly has anti-SARS-CoV-2 activity (Ni et al. 2020). It has been shown that Lianhua Qingwen capsules/granules could inhibit SARS-CoV-2 propagation in vitro and reduce virus-induced gene expression of IL-6, IL-8, TNFα, IP-10, and MCP-1. Jinhua Qinggan granules, Lianhua Qingwen capsules, and Shufeng Jiedu capsules can be used for the prevention and treatment of patients under medical observation (Hu et al. 2020; Runfeng et al. 2020). Huoxiang Zhengqi capsules are recommended for COVID-19 patients with gastrointestinal manifestations. Moreover, the combined use of Huoxiang Zhengqi oral liquid and Jinhao Jiere granules showed efficacy in patients with COVID-19 (Ho et al. 2020). Other Chinese medicines, Qingfei Paidu decoction and Ma Xing Shi Gan decoction, have also been used clinically for the treatment of COVID-19 patients in China (Yang et al. 2020).

25.2.4.2 Natural Products

Glycyrrhizic acid is a triterpene isolated from licorice root that significantly inhibits the entry and replication of SARS-CoV, providing a promising anti-coronavirus compound (Zhao et al. 2020). Clinical trials of diammonium glycyrrhizinate combined with vitamin C for the treatment of COVID-19 are currently underway. Another natural compound, Shikonin, has been identified as a primer inhibitor targeting the 3CLpro of SARS-CoV-2 and has potential for optimization as an antiviral primer (Jin et al. 2020). However, its specific anti-SARS-CoV-2 effect still requires further experimental and clinical verification.

25.3 Supportive Treatment for COVID-19

Similar to SARS-CoV and MERS-CoV, common clinical symptoms of COVID-19 include fever, cough, sore throat, headache, myalgia, and shortness of breath (Huang et al. 2020a). Meanwhile, a small number of patients present with symptoms of gastrointestinal infection and a diagnosis of conjunctivitis (Khavandi et al. 2020; Kim et al. 2020). However, the clinical course of COVID-19 pneumonia shows variable disease severity and rapid progression (Huang et al. 2020c). Mostly, elderly patients and patients with underlying diseases develop ARDS, cytokine storm, acute cardiac injury, arrhythmia, and secondary infections (Liang et al. 2020; Jiang et al. 2020).

When the patient has difficulty breathing, adequate oxygen should be given immediately. Almost all patients received oxygen therapy, and the World Health Organization (WHO) recommends extracorporeal membrane oxygenation (ECMO) in patients with refractory hypoxemia (Thibodeaux et al. 2020; Firstenberg et al. 2020). In patients with hypoxemia, nasal prongs, masks, a high-flow nasal catheter (HFNC), and noninvasive ventilation are recommended. However, mechanical ventilation or even additional porous membrane oxygenation support may be required (Thibodeaux et al. 2020). If there is no sign of inadequate tissue perfusion, fluid resuscitation should be done relatively conservatively. Otherwise, it may cause pulmonary edema and aggravate the hypoxic state. Considering that systemic corticosteroids may delay viral clearance, the use of corticosteroids is also not recommended (Johnson et al. 2020). Nevertheless, corticosteroids can be applied if other reasons justify their use (Zha et al. 2020). Plasma therapy and intravenous immunoglobulin therapy depend on the circumstances (Chen et al. 2020a). Notably, heparin and nafamostat combination therapy has shown potential for the treatment of severe COVID-19 (Alijotas-Reig et al. 2020; Asakura and Ogawa 2020).

25.4 Precautions of COVID-19

Standard precautions, including respiratory and eye protection, are recommended for all healthcare providers caring for patients who are known or suspected to have COVID-19 pneumonia (Gong et al. 2020). Even after two sets of negative test results, patients are still likely to become carriers of the virus (Liu et al. 2020). Therefore, it is recommended to remove precautions based not only on laboratory and radiological findings but also on the clinical evaluation of patients.

25.5 Conclusion

This chapter outlined possible therapeutic options currently under investigation for COVID-19. We summarized clinical trials as of the end of May 2020, which were initiated rapidly after the pandemic emergency. Although trials of specific drugs and therapeutic antibodies based on SARS-CoV-2 are underway, the solution will take longer due to the need for testing of safety. On the other hand, repurposing existing therapeutic agents previously designed for other viruses and diseases is an effective way to respond quickly to pandemics, as most of these agents have already been tested for safety. As the biological properties of SARS-CoV-2 continue to be understood, more potential mechanisms will be revealed and confirmed, and more potential targets and drugs will be developed. However, the development of related vaccines and new drugs and the addition of new indications for "old drugs" require experimental verification and further clinical studies, and drug development often lags far behind the development of the epidemic. The increasing number of recovered patients and the recently recorded higher recovery/mortality ratio seem to favor experimental therapy. Therefore, the most effective response to the ongoing public health emergency caused by SARS-CoV-2 worldwide remains active and strict quarantine measures to avoid the further spread of the virus. Screening and evaluation of old drugs known to be safe are of other effective strategies for timeliness.

References

Agostini ML, Andres EL, Sims AC, Graham RL, Sheahan TP, Lu X, Smith EC, Case JB, Feng JY, Jordan R, Ray AS, Cihlar T, Siegel D, Mackman RL, Clarke MO, Baric RS, Denison MR (2018) Coronavirus susceptibility to the antiviral remdesivir (GS-5734) is mediated by the viral polymerase and the proofreading exoribonuclease. MBio 9(2). https://doi.org/10.1128/mBio.00221-18

Aguiar ACC, Murce E, Cortopassi WA, Pimentel AS, Almeida M, Barros DCS, Guedes JS, Meneghetti MR, Krettli AU (2018) Chloroquine analogs as antimalarial candidates with potent in vitro and in vivo activity. Int J Parasitol Drugs Drug Resist 8(3):459–464. https://doi.org/10.1016/j.ijpddr.2018.10.002

Alijotas-Reig J, Esteve-Valverde E, Belizna C, Selva-O'Callaghan A, Pardos-Gea J, Quintana A, Mekinian A, Anunciacion-Lunell A, Miro-Mur F (2020) Immunomodulatory therapy for the management of severe COVID-19. Beyond the antiviral therapy: a comprehensive review. Autoimmun Rev:102569. https://doi.org/10.1016/j.autrev.2020.102569

Anand K, Ziebuhr J, Wadhwani P, Mesters JR, Hilgenfeld R (2003) Coronavirus main proteinase (3CLpro) structure: basis for design of anti-SARS drugs. Science 300(5626):1763–1767. https://doi.org/10.1126/science.1085658

Antinori S, Bonazzetti C, Gubertini G, Capetti A, Pagani C, Morena V, Rimoldi S, Galimberti L, Sarzi-Puttini P, Ridolfo AL (2020) Tocilizumab for cytokine storm syndrome in COVID-19 pneumonia: an increased risk for candidemia? Autoimmun Rev:102564. https://doi.org/10.1016/j.autrev.2020.102564

Asakura H, Ogawa H (2020) Potential of heparin and nafamostat combination therapy for COVID-19. J Thromb Haemost. https://doi.org/10.1111/jth.14858

Aspiroz EL, Figueroa SEC, Hurle ADG, Sanchez MJG (2013) New strategies in the optimisation of lopinavir/ritonavir doses in human immunodeficiency virus-infected patients. Enferm Infec Micr Cl 31(1):36–43. https://doi.org/10.1016/j.eimc.2011.11.018

Beigel JH, Tomashek KM, Dodd LE, Mehta AK, Zingman BS, Kalil AC, Hohmann E, Chu HY, Luetkemeyer A, Kline S, Lopez de Castilla D, Finberg RW, Dierberg K, Tapson V, Hsieh L, Patterson TF, Paredes R, Sweeney DA, Short WR, Touloumi G, Lye DC, Ohmagari N, Oh MD, Ruiz-Palacios GM, Benfield T, Fatkenheuer G, Kortepeter MG, Atmar RL, Creech CB, Lundgren J, Babiker AG, Pett S, Neaton JD, Burgess TH, Bonnett T, Green M, Makowski M, Osinusi A, Nayak S, Lane HC, Members A-SG (2020) Remdesivir for the treatment of Covid-19 – preliminary report. N Engl J Med. https://doi.org/10.1056/NEJMoa2007764

Boonrod K, Galetzka D, Nagy PD, Conrad U, Krczal G (2004) Single-chain antibodies against a plant viral RNA-dependent RNA polymerase confer virus resistance. Nat Biotechnol 22(7):856–862. https://doi.org/10.1038/nbt983

Bray M, Rayner C, Noel F, Jans D, Wagstaff K (2020) Ivermectin and COVID-19: a report in antiviral research, widespread interest, an FDA warning, two letters to the editor and the authors' responses. Antiviral Res:104805. https://doi.org/10.1016/j.antiviral.2020.104805

Brooks MJ, Burtseva EI, Ellery PJ, Marsh GA, Lew AM, Slepushkin AN, Crowe SM, Tannock GA (2012) Antiviral activity of arbidol, a broad-spectrum drug for use against respiratory viruses, varies according to test conditions. J Med Virol 84(1):170–181. https://doi.org/10.1002/jmv.22234

Brown AJ, Won JJ, Graham RL, Dinnon KH 3rd, Sims AC, Feng JY, Cihlar T, Denison MR, Baric RS, Sheahan TP (2019) Broad spectrum antiviral remdesivir inhibits human endemic and zoonotic deltacoronaviruses with a highly divergent RNA dependent RNA polymerase. Antivir Res 169:104541. https://doi.org/10.1016/j.antiviral.2019.104541

Cai Q, Yang M, Liu D, Chen J, Shu D, Xia J, Liao X, Gu Y, Cai Q, Yang Y, Shen C, Li X, Peng L, Huang D, Zhang J, Zhang S, Wang F, Liu J, Chen L, Chen S, Wang Z, Zhang Z, Cao R, Zhong W, Liu Y, Liu L (2020) Experimental treatment with Favipiravir for COVID-19: an open-label control study. Engineering (Beijing). https://doi.org/10.1016/j.eng.2020.03.007

Cao B, Wang Y, Wen D, Liu W, Wang J, Fan G, Ruan L, Song B, Cai Y, Wei M, Li X, Xia J, Chen N, Xiang J, Yu T, Bai T, Xie X, Zhang L, Li C, Yuan Y, Chen H, Li H, Huang H, Tu S, Gong F, Liu Y, Wei Y, Dong C, Zhou F, Gu X, Xu J, Liu Z, Zhang Y, Li H, Shang L, Wang K, Li K, Zhou X, Dong X, Qu Z, Lu S, Hu X, Ruan S, Luo S, Wu J, Peng L, Cheng F, Pan L, Zou J, Jia C, Wang J, Liu X, Wang S, Wu X, Ge Q, He J, Zhan H, Qiu F, Guo L, Huang C, Jaki T, Hayden FG, Horby PW, Zhang D, Wang C (2020) A trial of lopinavir-ritonavir in adults hospitalized with severe Covid-19. N Engl J Med 382(19):1787–1799. https://doi.org/10.1056/NEJMoa2001282

Chafekar A, Fielding BC (2018) MERS-CoV: understanding the latest human coronavirus threat. Viruses 10(2). https://doi.org/10.3390/v10020093

Chen L, Xiong J, Bao L, Shi Y (2020a) Convalescent plasma as a potential therapy for COVID-19. Lancet Infect Dis 20(4):398–400. https://doi.org/10.1016/S1473-3099(20)30141-9

Chen X, Li R, Pan Z, Qian C, Yang Y, You R, Zhao J, Liu P, Gao L, Li Z, Huang Q, Xu L, Tang J, Tian Q, Yao W, Hu L, Yan X, Zhou X, Wu Y, Deng K, Zhang Z, Qian Z, Chen Y, Ye L (2020b) Human monoclonal antibodies block the binding of SARS-CoV-2 spike protein to angiotensin converting enzyme 2 receptor. Cell Mol Immunol 17(6):647–649. https://doi.org/10.1038/s41423-020-0426-7

Colaneri M, Bogliolo L, Valsecchi P, Sacchi P, Zuccaro V, Brandolino F, Montecucco C, Mojoli F, Giusti EM, Bruno R, The Covid Irccs San Matteo Pavia Task F (2020) Tocilizumab for treatment of severe COVID-19 patients: preliminary results from SMAtteo Covid19 REgistry (SMACORE). Microorganisms 8(5). https://doi.org/10.3390/microorganisms8050695

Costanzo M, De Giglio MAR, Roviello GN (2020) SARS-CoV-2: recent reports on antiviral therapies based on lopinavir/ritonavir, darunavir/umifenovir, hydroxychloroquine, remdesivir, favipiravir and other drugs for the treatment of the new coronavirus. Curr Med Chem. https://doi.org/10.2174/0929867327666200416131117

de Wit E, Feldmann F, Cronin J, Jordan R, Okumura A, Thomas T, Scott D, Cihlar T, Feldmann H (2020) Prophylactic and therapeutic remdesivir (GS-5734) treatment in the rhesus macaque model of MERS-CoV infection. Proc Natl Acad Sci U S A 117(12):6771–6776. https://doi.org/10.1073/pnas.1922083117

Deeks ED (2018) Darunavir/cobicistat/emtricitabine/tenofovir alafenamide: a review in HIV-1 infection. Drugs 78(10):1013–1024. https://doi.org/10.1007/s40265-018-0934-2

Deng L, Li C, Zeng Q, Liu X, Li X, Zhang H, Hong Z, Xia J (2020) Arbidol combined with LPV/r versus LPV/r alone against corona virus disease 2019: a retrospective cohort study. J Inf Secur. https://doi.org/10.1016/j.jinf.2020.03.002

Du L, Yang Y, Zhou Y, Lu L, Li F, Jiang S (2017) MERS-CoV spike protein: a key target for antivirals. Expert Opin Ther Targets 21(2):131–143. https://doi.org/10.1080/14728222.2017.1271415

Fang QQ, Huang WJ, Li XY, Cheng YH, Tan MJ, Liu J, Wei HJ, Meng Y, Wang DY (2020) Effectiveness of favipiravir (T-705) against wild-type and oseltamivir-resistant influenza B virus in mice. Virology 545:1–9. https://doi.org/10.1016/j.virol.2020.02.005

Fantini J, Di Scala C, Chahinian H, Yahi N (2020) Structural and molecular modelling studies reveal a new mechanism of action of chloroquine and hydroxychloroquine against SARS-CoV-2 infection. Int J Antimicrob Agents:105960. https://doi.org/10.1016/j.ijantimicag.2020.105960

Firstenberg MS, Stahel PF, Hanna J, Kotaru C, Crossno J Jr, Forrester J (2020) Successful COVID-19 rescue therapy by extra-corporeal membrane oxygenation (ECMO) for respiratory failure: a case report. Patient Saf Surg 14:20. https://doi.org/10.1186/s13037-020-00245-7

Gao JJ, Tian ZX, Yang X (2020) Breakthrough: chloroquine phosphate has shown apparent efficacy in treatment of COVID-19 associated pneumonia in clinical studies. Biosci Trends 14(1):72–73. https://doi.org/10.5582/bst.2020.01047

Gautret P, Lagier JC, Parola P, Hoang VT, Meddeb L, Sevestre J, Mailhe M, Doudier B, Aubry C, Amrane S, Seng P, Hocquart M, Eldin C, Finance J, Vieira VE, Tissot-Dupont HT, Honore S, Stein A, Million M, Colson P, La Scola B, Veit V, Jacquier A, Deharo JC, Drancourt M, Fournier PE, Rolain JM, Brouqui P, Raoult D (2020) Clinical and microbiological effect of a combination of hydroxychloroquine and azithromycin in 80 COVID-19 patients with at least a six-day follow up: a pilot observational study. Travel Med Infect Dis 34:101663. https://doi.org/10.1016/j.tmaid.2020.101663

Gong K, Xu Z, Cai Z, Chen Y, Wang Z (2020) Internet hospitals help prevent and control the epidemic of COVID-19 in China: multicenter user profiling study. J Med Internet Res 22(4):e18908. https://doi.org/10.2196/18908

Grein J, Ohmagari N, Shin D, Diaz G, Asperges E, Castagna A, Feldt T, Green G, Green ML, Lescure FX, Nicastri E, Oda R, Yo K, Quiros-Roldan E, Studemeister A, Redinski J, Ahmed S, Bernett J, Chelliah D, Chen D, Chihara S, Cohen SH, Cunningham J, D'Arminio Monforte A, Ismail S, Kato H, Lapadula G, L'Her E, Maeno T, Majumder S, Massari M, Mora-Rillo M, Mutoh Y, Nguyen D, Verweij E, Zoufaly A, Osinusi AO, DeZure A, Zhao Y, Zhong L, Chokkalingam A, Elboudwarej E, Telep L, Timbs L, Henne I, Sellers S, Cao H, Tan SK, Winterbourne L, Desai P, Mera R, Gaggar A, Myers RP, Brainard DM, Childs R, Flanigan T (2020) Compassionate use of remdesivir for patients with severe Covid-19. N Engl J Med. https://doi.org/10.1056/NEJMoa2007016

Grossi PA, Lombardi D, Petrolo A, Rovelli C, Di Rosa Z, Perriccioli G, Rossi A, Minoja G, Scaglione F, Dalla Gasperina D (2018) Strongyloides stercoralis hyperinfection in an HIV-infected patient successfully treated with subcutaneous ivermectin. Trop Med Infect Dis 3(2). https://doi.org/10.3390/tropicalmed3020046

Ho M (2020) Perspectives on the development of neutralizing antibodies against SARS-CoV-2. Antib Ther 3(2):109–114. https://doi.org/10.1093/abt/tbaa009

Ho LTF, Chan KKH, Chung VCH, Leung TH (2020) Highlights of traditional Chinese medicine frontline expert advice in the China national guideline for COVID-19. Eur J Integr Med:101116. https://doi.org/10.1016/j.eujim.2020.101116

Hoffmann M, Kleine-Weber H, Schroeder S, Kruger N, Herrler T, Erichsen S, Schiergens TS, Herrler G, Wu

NH, Nitsche A, Muller MA, Drosten C, Pohlmann S (2020a) SARS-CoV-2 cell entry depends on ACE2 and TMPRSS2 and is blocked by a clinically proven protease inhibitor. Cell 181(2):271. https://doi.org/10.1016/j.cell.2020.02.052

Hoffmann M, Kleine-Weber H, Schroeder S, Kruger N, Herrler T, Erichsen S, Schiergens TS, Herrler G, Wu NH, Nitsche A, Muller MA, Drosten C, Pohlmann S (2020b) SARS-CoV-2 cell entry depends on ACE2 and TMPRSS2 and is blocked by a clinically proven protease inhibitor. Cell 181(2):271–280 e278. https://doi.org/10.1016/j.cell.2020.02.052

Holshue ML, DeBolt C, Lindquist S, Lofy KH, Wiesman J, Bruce H, Spitters C, Ericson K, Wilkerson S, Tural A, Diaz G, Cohn A, Fox L, Patel A, Gerber SI, Kim L, Tong S, Lu X, Lindstrom S, Pallansch MA, Weldon WC, Biggs HM, Uyeki TM, Pillai SK, Washington State -nCo VCIT (2020) First case of 2019 novel coronavirus in the United States. N Engl J Med 382(10):929–936. https://doi.org/10.1056/NEJMoa2001191

Hu K, Guan WJ, Bi Y, Zhang W, Li L, Zhang B, Liu Q, Song Y, Li X, Duan Z, Zheng Q, Yang Z, Liang J, Han M, Ruan L, Wu C, Zhang Y, Jia ZH, Zhong NS (2020) Efficacy and safety of Lianhuaqingwen capsules, a repurposed Chinese Herb, in patients with coronavirus disease 2019: a multicenter, prospective, randomized controlled trial. Phytomedicine:153242. https://doi.org/10.1016/j.phymed.2020.153242

Huang C, Wang Y, Li X, Ren L, Zhao J, Hu Y, Zhang L, Fan G, Xu J, Gu X, Cheng Z, Yu T, Xia J, Wei Y, Wu W, Xie X, Yin W, Li H, Liu M, Xiao Y, Gao H, Guo L, Xie J, Wang G, Jiang R, Gao Z, Jin Q, Wang J, Cao B (2020a) Clinical features of patients infected with 2019 novel coronavirus in Wuhan, China. Lancet 395(10223):497–506. https://doi.org/10.1016/S0140-6736(20)30183-5

Huang J, Song W, Huang H, Sun Q (2020b) Pharmacological therapeutics targeting RNA-dependent RNA polymerase, proteinase and spike protein: from mechanistic studies to clinical trials for COVID-19. J Clin Med 9(4). https://doi.org/10.3390/jcm9041131

Huang X, Wei F, Hu L, Wen L, Chen K (2020c) Epidemiology and clinical characteristics of COVID-19. Arch Iran Med 23(4):268–271. https://doi.org/10.34172/aim.2020.09

Hui DS, Azhar EI, Kim YJ, Memish ZA, Oh MD, Zumla A (2018) Middle East respiratory syndrome coronavirus: risk factors and determinants of primary, household, and nosocomial transmission. Lancet Infect Dis 18(8):e217–e227. https://doi.org/10.1016/S1473-3099(18)30127-0

Hung IF, Lung KC, Tso EY, Liu R, Chung TW, Chu MY, Ng YY, Lo J, Chan J, Tam AR, Shum HP, Chan V, Wu AK, Sin KM, Leung WS, Law WL, Lung DC, Sin S, Yeung P, Yip CC, Zhang RR, Fung AY, Yan EY, Leung KH, Ip JD, Chu AW, Chan WM, Ng AC, Lee R, Fung K, Yeung A, Wu TC, Chan JW, Yan WW, Chan WM, Chan JF, Lie AK, Tsang OT, Cheng VC, Que TL, Lau CS, Chan KH, To KK, Yuen KY (2020) Triple combination of interferon beta-1b, lopinavir-ritonavir, and ribavirin in the treatment of patients admitted to hospital with COVID-19: an open-label, randomised, phase 2 trial. Lancet. https://doi.org/10.1016/S0140-6736(20)31042-4

Janowski AB, Dudley H, Wang D (2020) Antiviral activity of ribavirin and favipiravir against human astroviruses. J Clin Virol 123:104247. https://doi.org/10.1016/j.jcv.2019.104247

Jiang F, Deng L, Zhang L, Cai Y, Cheung CW, Xia Z (2020) Review of the clinical characteristics of coronavirus disease 2019 (COVID-19). J Gen Intern Med 35(5):1545–1549. https://doi.org/10.1007/s11606-020-05762-w

Jin Z, Du X, Xu Y, Deng Y, Liu M, Zhao Y, Zhang B, Li X, Zhang L, Peng C, Duan Y, Yu J, Wang L, Yang K, Liu F, Jiang R, Yang X, You T, Liu X, Yang X, Bai F, Liu H, Liu X, Guddat LW, Xu W, Xiao G, Qin C, Shi Z, Jiang H, Rao Z, Yang H (2020) Structure of M(pro) from SARS-CoV-2 and discovery of its inhibitors. Nature. https://doi.org/10.1038/s41586-020-2223-y

Johnson KM, Belfer JJ, Peterson GR, Boelkins MR, Dumkow LE (2020) Managing COVID-19 in renal transplant recipients: a review of recent literature and case supporting corticosteroid-sparing immunosuppression. Pharmacotherapy. https://doi.org/10.1002/phar.2410

Ju B, Zhang Q, Ge J, Wang R, Sun J, Ge X, Yu J, Shan S, Zhou B, Song S, Tang X, Yu J, Lan J, Yuan J, Wang H, Zhao J, Zhang S, Wang Y, Shi X, Liu L, Zhao J, Wang X, Zhang Z, Zhang L (2020) Human neutralizing antibodies elicited by SARS-CoV-2 infection. Nature. https://doi.org/10.1038/s41586-020-2380-z

Kannan S, Ali PSS, Sheeza A, Hemalatha K (2020) COVID-19 (novel coronavirus 2019) – recent trends. Eur Rev Med Pharmacol 24(4):2006–2011

Keshavarzi F (2016) Fungistatic effect of hydroxychloroquine, lessons from a case. Med Mycol Case Rep 13:17–18. https://doi.org/10.1016/j.mmcr.2016.09.003

Khalili JS, Zhu H, Mak NSA, Yan Y, Zhu Y (2020) Novel coronavirus treatment with ribavirin: groundwork for an evaluation concerning COVID-19. J Med Virol. https://doi.org/10.1002/jmv.25798

Khavandi S, Tabibzadeh E, Naderan M, Shoar S (2020) Corona virus disease-19 (COVID-19) presenting as conjunctivitis: atypically high-risk during a pandemic. Cont Lens Anterior Eye. https://doi.org/10.1016/j.clae.2020.04.010

Kim GU, Kim MJ, Ra SH, Lee J, Bae S, Jung J, Kim SH (2020) Clinical characteristics of asymptomatic and symptomatic patients with mild COVID-19. Clin Microbiol Infect. https://doi.org/10.1016/j.cmi.2020.04.040

Klimke A, Hefner G, Will B, Voss U (2020) Hydroxychloroquine as an aerosol might markedly reduce and even prevent severe clinical symptoms after SARS-CoV-2 infection. Med Hypotheses 142:109783. https://doi.org/10.1016/j.mehy.2020.109783

Lian N, Xie H, Lin S, Huang J, Zhao J, Lin Q (2020) Umifenovir treatment is not associated with improved outcomes in patients with coronavirus disease 2019: a retrospective study. Clin Microbiol Infect. https://doi.org/10.1016/j.cmi.2020.04.026

Liang WH, Guan WJ, Li CC, Li YM, Liang HR, Zhao Y, Liu XQ, Sang L, Chen RC, Tang CL, Wang T, Wang W, He QH, Chen ZS, Wong SS, Zanin M, Liu J, Xu X, Huang J, Li JF, Ou LM, Cheng B, Xiong S, Xie ZH, Ni ZY, Hu Y, Liu L, Shan H, Lei CL, Peng YX, Wei L, Liu Y, Hu YH, Peng P, Wang JM, Liu JY, Chen Z, Li G, Zheng ZJ, Qiu SQ, Luo J, Ye CJ, Zhu SY, Cheng LL, Ye F, Li SY, Zheng JP, Zhang NF, Zhong NS, He JX (2020) Clinical characteristics and outcomes of hospitalised patients with COVID-19 treated in Hubei (epicenter) and outside Hubei (non-epicenter): a nationwide analysis of China. Eur Respir J. https://doi.org/10.1183/13993003.00562-2020

Liu X, Na RS, Bi ZQ (2020) Challenges to prevent and control the outbreak of novel coronavirus pneumonia (COVID-19). Zhonghua Liu Xing Bing Xue Za Zhi 41:E029. https://doi.org/10.3760/cma.j.cn112338-20200216-00108

Lythgoe MP, Middleton P (2020) Ongoing clinical trials for the management of the COVID-19 pandemic. Trends Pharmacol Sci 41(6):363–382. https://doi.org/10.1016/j.tips.2020.03.006

Mahase E (2020) Covid-19: Remdesivir is helpful but not a wonder drug, say researchers. Bmj-Brit Med J:369. https://doi.org/10.1136/bmj.m1798

Martinez MA (2020) Compounds with therapeutic potential against novel respiratory 2019 coronavirus. Antimicrob Agents Ch 64 (5):ARTN e00399-20. https://doi.org/10.1128/AAC.00399-20

Mazzitelli M, Arrighi E, Serapide F, Pelle MC, Tassone B, Lionello R, Marrazzo G, Lagana D, Costanzo FS, Matera G, Trecarichi EM, Torti C, Group IUC (2020) Use of subcutaneous tocilizumab in patients with COVID-19 pneumonia. J Med Virol. https://doi.org/10.1002/jmv.26016

Mo S, Chen D (2019) Chloroquine modulates antitumor immune response by resetting tumor-associated macrophages toward M1 phenotype. Eur J Immunol 49:564–564

Moeinzadeh F, Dezfouli M, Naimi A, Shahidi S, Moradi H (2020) Newly diagnosed glomerulonephritis during COVID-19 infection undergoing immunosuppression therapy, a case report. Iran J Kidney Dis 14(3):239–242

Monteil V, Kwon H, Prado P, Hagelkruys A, Wimmer RA, Stahl M, Leopoldi A, Garreta E, Hurtado Del Pozo C, Prosper F, Romero JP, Wirnsberger G, Zhang H, Slutsky AS, Conder R, Montserrat N, Mirazimi A, Penninger JM (2020) Inhibition of SARS-CoV-2 infections in engineered human tissues using clinical-grade soluble human ACE2. Cell 181(4):905–913 e907. https://doi.org/10.1016/j.cell.2020.04.004

Ni L, Zhou L, Zhou M, Zhao J, Wang DW (2020) Combination of western medicine and Chinese traditional patent medicine in treating a family case of COVID-19. Front Med 14(2):210–214. https://doi.org/10.1007/s11684-020-0757-x

Patri A, Fabbrocini G (2020) Hydroxychloroquine and ivermectin: a synergistic combination for COVID-19 chemoprophylaxis and treatment? J Am Acad Dermatol 82(6):e221. https://doi.org/10.1016/j.jaad.2020.04.017

Peiris JS, Guan Y, Yuen KY (2004) Severe acute respiratory syndrome. Nat Med 10(12 Suppl):S88–S97. https://doi.org/10.1038/nm1143

Perez EE, Orange JS, Bonilla F, Chinen J, Chinn IK, Dorsey M, El-Gamal Y, Harville TO, Hossny E, Mazer B, Nelson R, Secord E, Jordan SC, Stiehm R, Vo AA, Ballow M (2017) Update on the use of immunoglobulin in human disease: a review of evidence. J Allergy Clin Immunol 139(3):S1–S46. https://doi.org/10.1016/j.jaci.2016.09.023

Pilkington V, Pepperrell T, Hill A (2020) A review of the safety of favipiravir – a potential treatment in the COVID-19 pandemic? J Virus Erad 6(2):45–51

Plantone D, Koudriavtseva T (2018) Current and future use of chloroquine and hydroxychloroquine in infectious, immune, neoplastic, and neurological diseases: a mini-review. Clin Drug Invest 38(8):653–671. https://doi.org/10.1007/s40261-018-0656-y

Rabaan AA, Al-Ahmed SH, Haque S, Sah R, Tiwari R, Malik YS, Dhama K, Yatoo MI, Bonilla-Aldana DK, Rodriguez-Morales AJ (2020) SARS-CoV-2, SARS-CoV, and MERS-COV: a comparative overview. Infez Med 28(2):174–184

Rajic KPZ, Mlinaric Z, Uzelac L, Kralj M, Zorc B (2018) Chloroquine urea derivatives: synthesis and antitumor activity in vitro. Acta Pharma 68(4):471–483. https://doi.org/10.2478/acph-2018-0039

Ruan G, Ye L, Liu G, An J, Sehouli J, Sun P (2018) The role of bevacizumab in targeted vascular endothelial growth factor therapy for epithelial ovarian cancer: an updated systematic review and meta-analysis. Onco Targets Ther 11:521–528. https://doi.org/10.2147/OTT.S155581

Runfeng L, Yunlong H, Jicheng H, Weiqi P, Qinhai M, Yongxia S, Chufang L, Jin Z, Zhenhua J, Haiming J, Kui Z, Shuxiang H, Jun D, Xiaobo L, Xiaotao H, Lin W, Nanshan Z, Zifeng Y (2020) Lianhuaqingwen exerts antiviral and anti-inflammatory activity against novel coronavirus (SARS-CoV-2). Pharmacol Res 156:104761. https://doi.org/10.1016/j.phrs.2020.104761

Shiraki K, Daikoku T (2020) Favipiravir, an anti-influenza drug against life-threatening RNA virus infections. Pharmacol Ther 209:107512. https://doi.org/10.1016/j.pharmthera.2020.107512

Singh AK, Singh A, Shaikh A, Singh R, Misra A (2020a) Chloroquine and hydroxychloroquine in the treatment of COVID-19 with or without diabetes: a systematic search and a narrative review with a special reference to India and other developing countries. Diabetes Metab Syndr 14(3):241–246. https://doi.org/10.1016/j.dsx.2020.03.011

Singh AK, Singh A, Singh R, Misra A (2020b) Hydroxychloroquine in patients with COVID-19: a systematic review and meta-analysis. Diabetes Metab Syndr 14(4):589–596. https://doi.org/10.1016/j.dsx.2020.05.017

Sk MF, Roy R, Jonniya NA, Poddar S, Kar P (2020) Elucidating biophysical basis of binding of inhibitors to SARS-CoV-2 main protease by using molecular dynamics simulations and free energy calculations. J Biomol Struct Dyn:1–21. https://doi.org/10.1080/07391102.2020.1768149

Stadler K, Rappuoli R (2005) SARS: understanding the virus and development of rational therapy. Curr Mol Med 5(7):677–697. https://doi.org/10.2174/156652405774641124

Thibodeaux K, Speyrer M, Raza A, Yaakov R, Serena TE (2020) Hyperbaric oxygen therapy in preventing mechanical ventilation in COVID-19 patients: a retrospective case series. J Wound Care 29 (Sup5a):S4–S8. https://doi.org/10.12968/jowc.2020.29.Sup5a.S4

Tian X, Li C, Huang A, Xia S, Lu S, Shi Z, Lu L, Jiang S, Yang Z, Wu Y, Ying T (2020) Potent binding of 2019 novel coronavirus spike protein by a SARS coronavirus-specific human monoclonal antibody. Emerg Microbes Infect 9(1):382–385. https://doi.org/10.1080/22221751.2020.1729069

Trinh R, Gathe J, Reynes J, Pulido F, Tian M, Fredrick L, Podsadecki T, Norton M, Nilius A (2012) Lopinavir/ritonavir (LPV/r) combined with raltegravir (RAL) or tenofovir/emtricitabine (TDF/FTC) in antiretroviral-naive subjects: 96-week safety and efficacy results of the PROGRESS study. HIV Med 13:69–69

Tsang K, Zhong NS (2003) SARS: pharmacotherapy. Respirology 8(Suppl):S25–S30. https://doi.org/10.1046/j.1440-1843.2003.00525.x

Wagstaff KM, Sivakumaran H, Heaton SM, Harrich D, Jans DA (2012) Ivermectin is a specific inhibitor of importin alpha/beta-mediated nuclear import able to inhibit replication of HIV-1 and dengue virus. Biochem J 443(3):851–856. https://doi.org/10.1042/BJ20120150

Wang C, Li W, Drabek D, Okba NMA, van Haperen R, Osterhaus A, van Kuppeveld FJM, Haagmans BL, Grosveld F, Bosch BJ (2020a) A human monoclonal antibody blocking SARS-CoV-2 infection. Nat Commun 11(1):2251. https://doi.org/10.1038/s41467-020-16256-y

Wang M, Cao R, Zhang L, Yang X, Liu J, Xu M, Shi Z, Hu Z, Zhong W, Xiao G (2020b) Remdesivir and chloroquine effectively inhibit the recently emerged novel coronavirus (2019-nCoV) in vitro. Cell Res 30(3):269–271. https://doi.org/10.1038/s41422-020-0282-0

Wang M, Li R, Shu B, Jing X, Ye HQ, Gong P (2020c) Stringent control of the RNA-dependent RNA polymerase translocation revealed by multiple intermediate structures. Nat Commun 11(1):2605. https://doi.org/10.1038/s41467-020-16234-4

Wang ML, Cao RY, Zhang LK, Yang XL, Liu J, Xu MY, Shi ZL, Hu ZH, Zhong W, Xiao GF (2020d) Remdesivir and chloroquine effectively inhibit the recently emerged novel coronavirus (2019-nCoV) in vitro. Cell Res 30(3):269–271. https://doi.org/10.1038/s41422-020-0282-0

Wen CY, Xie ZW, Li YP, Deng XL, Chen XT, Cao Y, Ou X, Lin WY, Li F, Cai WP, Li LH (2020) Real-world efficacy and safety of lopinavir/ritonavir and arbidol in treating with COVID-19: an observational cohort study. Zhonghua Nei Ke Za Zhi 59(0):E012. https://doi.org/10.3760/cma.j.cn112138-20200227-00147

Wu Y, Wang F, Shen C, Peng W, Li D, Zhao C, Li Z, Li S, Bi Y, Yang Y, Gong Y, Xiao H, Fan Z, Tan S, Wu G, Tan W, Lu X, Fan C, Wang Q, Liu Y, Zhang C, Qi J, Gao GF, Gao F, Liu L (2020) A noncompeting pair of human neutralizing antibodies block COVID-19 virus binding to its receptor ACE2. Science 368(6496):1274–1278. https://doi.org/10.1126/science.abc2241

Yang JC, Nikanjam M, Best BM, Pinto J, Chadwick EG, Daar ES, Havens PL, Rakhmanina N, Capparelli EV (2018) Population pharmacokinetics of lopinavir/ritonavir: changes across formulations and human development from infancy through adulthood. J Clin Pharmacol 58(12):1604–1617. https://doi.org/10.1002/jcph.1293

Yang R, Liu H, Bai C, Wang Y, Zhang X, Guo R, Wu S, Wang J, Leung E, Chang H, Li P, Liu T, Wang Y (2020) Chemical composition and pharmacological mechanism of Qingfei Paidu decoction and Ma Xing Shi Gan decoction against coronavirus disease 2019 (COVID-19): in silico and experimental study. Pharmacol Res 157:104820. https://doi.org/10.1016/j.phrs.2020.104820

Ye XT, Luo YL, Xia SC, Sun QF, Ding JG, Zhou Y, Chen W, Wang XF, Zhang WW, Du WJ, Ruan ZW, Hong L (2020) Clinical efficacy of lopinavir/ritonavir in the treatment of coronavirus disease 2019. Eur Rev Med Pharmacol Sci 24(6):3390–3396. https://doi.org/10.26355/eurrev_202003_20706

Zha L, Li S, Pan L, Tefsen B, Li Y, French N, Chen L, Yang G, Villanueva EV (2020) Corticosteroid treatment of patients with coronavirus disease 2019 (COVID-19). Med J Aust 212(9):416–420. https://doi.org/10.5694/mja2.50577

Zhao X, Jiang Y, Zhao Y, Xi H, Liu C, Qu F, Feng X (2020) Analysis of the susceptibility to COVID-19 in pregnancy and recommendations on potential drug screening. Eur J Clin Microbiol Infect Dis. https://doi.org/10.1007/s10096-020-03897-6

Zhou GY, Zhao Q (2020) Perspectives on therapeutic neutralizing antibodies against the novel coronavirus SARS-CoV-2. Int J Biol Sci 16(10):1718–1723. https://doi.org/10.7150/ijbs.45123

Zhou Y, Vedantham P, Lu K, Agudelo J, Carrion R, Nunneley JW, Barnard D, Pohlmann S, McKerrow JH, Renslo AR, Simmons G (2015) Protease inhibitors targeting coronavirus and filovirus entry. Antivir Res 116:76–84. https://doi.org/10.1016/j.antiviral.2015.01.011

Zhou Y, Yang Y, Huang J, Jiang S, Du L (2019) Advances in MERS-CoV vaccines and therapeutics based on the receptor-binding domain. Viruses 11(1). https://doi.org/10.3390/v11010060

Zhu Z, Lu Z, Xu T, Chen C, Yang G, Zha T, Lu J, Xue Y (2020) Arbidol monotherapy is superior to lopinavir/ritonavir in treating COVID-19. J Inf Secur. https://doi.org/10.1016/j.jinf.2020.03.060

Ziegler CGK, Allon SJ, Nyquist SK, Mbano IM, Miao VN, Tzouanas CN, Cao Y, Yousif AS, Bals J, Hauser BM, Feldman J, Muus C, Wadsworth MH, 2nd, Kazer SW, Hughes TK, Doran B, Gatter GJ, Vukovic M, Taliaferro F, Mead BE, Guo Z, Wang JP, Gras D, Plaisant M, Ansari M, Angelidis I, Adler H, Sucre JMS, Taylor CJ, Lin B, Waghray A, Mitsialis V, Dwyer DF, Buchheit KM, Boyce JA, Barrett NA, Laidlaw TM, Carroll SL, Colonna L, Tkachev V, Peterson CW, Yu A, Zheng HB, Gideon HP, Winchell CG, Lin PL, Bingle CD, Snapper SB, Kropski JA, Theis FJ, Schiller HB, Zaragosi LE, Barbry P, Leslie A, Kiem HP, Flynn JL, Fortune SM, Berger B, Finberg RW, Kean LS, Garber M, Schmidt AG, Lingwood D, Shalek AK, Ordovas-Montanes J, lung-network@humancellatlas.org HCALBNEa, Network HCALB (2020) SARS-CoV-2 receptor ACE2 is an interferon-stimulated gene in human airway epithelial cells and is detected in specific cell subsets across tissues. Cell. https://doi.org/10.1016/j.cell.2020.04.035

Immune-Based Therapy for COVID-19

26

Abdolreza Esmaeilzadeh, Davood Jafari, Safa Tahmasebi, Reza Elahi, and Elnaz Khosh

Abstract

Severe acute respiratory syndrome coronavirus 2 (SARS-CoV-2) is a novel zoonotic virus identified as the cause of coronavirus disease 2019 (COVID-19) that has crossed species and infected humans. In order to develop new insights on the immune-based treatments against this disease, it is vital to understand the immunopathology of the COVID-19, implications of the immune response to SARS-CoV-2, and immune dysfunction in response to SARS-CoV-2. There is no approved drug for the treatment of COVID-19. It is, thus, promising to design immune-based treatments that inhibit the infectious mechanism of the virus, improve the inadequate immune response, or regulate the hyperactivated immune response in severely ill patients. According to the antiviral immune response against the virus, antibody-based immunotherapies of COVID-19 include injection of convalescent plasma from recovered patients, high-dose intravenous immunoglobulins (IVIG), monoclonal antibodies, and polyclonal antibodies. Also, cell-based treatment, vaccine-based approaches, cytokine-based immunotherapy, immune checkpoint inhibitors, JAK inhibitors, decoy receptors, and immunosuppressive drugs are discussed in this chapter.

Keywords

Acute respiratory distress syndrome · Coronavirus · COVID-19 · Immune system · Immunotherapy · SARS-CoV-2

26.1 Introduction

Severe acute respiratory syndrome coronavirus 2 (SARS-CoV-2) is a novel zoonotic virus that has crossed species and infected humans, causing coronavirus disease 2019 (COVID-19). First cases of COVID-19 were reported in *Wuhan, China*; however, it has now spread worldwide and, thus, has been announced as a global pandemic (Tahmasebi et al. 2020). In order to develop new insights on the immune-based treatments against this disease, it is vital to understand

A. Esmaeilzadeh (✉) · D. Jafari
Department of Immunology, School of Medicine, Zanjan University of Medical Sciences, Zanjan, Iran

Immunotherapy Research and Technology Group, Zanjan University of Medical Sciences, Zanjan, Iran
e-mail: a46reza@zums.ac.ir

S. Tahmasebi
Department of Immunology, School of Public Health, Tehran University of Medical Sciences, Tehran, Iran

R. Elahi · E. Khosh
School of Medicine, Zanjan University of Medical Sciences, Zanjan, Iran

N. Rezaei (ed.), *Coronavirus Disease - COVID-19*, Advances in Experimental Medicine and Biology 1318, https://doi.org/10.1007/978-3-030-63761-3_26

the immunopathology of the novel coronavirus, implications of the immune response to SARS-CoV-2, and immune dysfunction in response to SARS-CoV-2.

Immune-based treatments against COVID-19 are designed to inhibit the infectious mechanism of the virus, improve the inadequate immune response, or regulate the hyperactivated immune response in severely ill patients (Khoury et al. 2020). According to the antiviral immune response against the virus, immune-based therapies of COVID-19 have been described in this chapter. One of the immune-based therapies for COVID-19 is antibody-based treatments, including injection of convalescent plasma from recovered patients, high-dose intravenous immunoglobulins (IVIG), monoclonal antibodies, and polyclonal antibodies (Kumar et al. 2020). Also, since the inflammatory pathways and immune checkpoints have major roles in the uncontrolled immune response against the virus, JAK inhibitors, immune checkpoint inhibitors, and decoy receptors are the other immune-based therapeutic approaches (Tay et al. 2020).

Since the hyper-inflammation caused by the cytokine storm is the major cause of the respiratory and multi-organ failure in severely ill patients, different strategies have shown clinical benefits for the management of the cytokine storm in COVID-19 patients (Wang et al. 2020d). One of these approaches is the suppression of the severe inflammatory response using immunosuppressive drugs, such as cytotoxic chemotherapy and low-dose corticosteroids and nonsteroidal anti-inflammatory drugs (NSAIDs). Given the significant roles of the immune cells in the immune responses against SARS-CoV-2, cell-based immunotherapies have been recently introduced as new treatment options for SARS-CoV-2. The cell-based immunotherapies include the injection of the mesenchymal stem cells (MSCs), natural killer (NK) cells, dendritic cells (DCs), and intrapulmonary macrophages and inhibition of the T-cell exhaustion (Khoury et al. 2020). Also, this chapter includes an overview of the immune-based clinical trials to provide a current picture of the immune-based therapies for COVID-19.

26.2 Immunopathology and Its Implications for Acute Respiratory Distress Syndrome (ARDS) Therapy

Coronaviruses have unsegmented, enveloped, and positive-sense single-stranded RNA genome. Notably, the four crucial proteins, including nucleocapsid (N), envelope (E), membrane (M), and spike (S), are encoded by the coronavirus genome. The abovementioned proteins are necessary for virus replication and pathogenesis. Spike protein is the substantial particle for the entrance of the virus through the cellular receptors on the host cell surface. The releasing and replication processes of the RNA are initiated in the cytoplasm. Eventually, the vesicles composed of the newly formed coronaviruses fuse with the cell membrane and release the viral particles out of the cells (Wang et al. 2020c; Kui et al. 2020; Mousavizadeh and Ghasemi 2020; de Haan et al. 1998; Woo et al. 2010). The most common receptor for the entry of the SARS-CoV-2 to the cells is angiotensin-converting enzyme 2 (ACE2).

Additionally, the SARS-CoV-2 administrates the transmembrane serine protease2 TMPRSS2 and endosomal cysteine proteases cathepsin B and L (CatB/L), while entering the cells through ACE2. The diffuse expression of the ACE2 recommends that multiple organs can be infected by the novel coronavirus (Lukassen et al. 2020; Wambier and Goren 2020; Hoffmann et al. 2020). In addition to ACE2, the CD147 receptor may be involved in the invasion of the novel coronavirus. Based on the expression of CD147 by the inflamed tissues, tumor cells, and infected tissues, the cross-reaction with the healthy cells would be low (Guo et al. 2020; Kosugi et al. 2015).

Of note, the most common clinical manifestations of this highly pathogenic virus are fever, cough, fatigue, and shortness of breath. The high proportion of infected patients experience mild to moderate symptoms of the disease. Besides, they would have a good prognosis. Approximately 10% of the patients may have a severe infection of the novel coronavirus. Additionally, about 5% of the COVID-19 cases would be in the critical condition of the disease. The clinical complications of the disease, such as acute respiratory disease syndrome (ARDS), kidney injury, shock, arrhythmia, and death, may occur in patients with critical condition. It is worth noting that older patients, especially with concomitant medical illnesses such as cardiovascular diseases, cerebrovascular diseases, hypertension, and diabetes, are highly likely to progress to severe and critical conditions (Zhou et al. 2020a; Abdulamir and Hafidh 2020; Prompetchara et al. 2020).

SARS-CoV-2 penetrates to the lungs by passing through the respiratory tract. Then, the virus can appear in the blood and cause the viremic condition. After that, COVID-19 can be in each organ that expresses ACE2 receptors. In addition to the pathogenic features of the virus, the inflammatory response critically contributes to the progression of COVID-19. The immune system alterations during the pathogenesis of the coronavirus would be as follows: decreasing the number of lymphocytes ($CD8^+$, $CD4^+$ T cells, and B cells) and NK cells, increasing the levels of immunosuppressive regulatory T (T reg) cells, and upregulating the proinflammatory cytokines. Innate and adaptive immune responses will destroy the pulmonary tissue and lead toward ARDS in some infected patients. ARDS is a vital clinical feature of the novel coronavirus. High levels of proinflammatory cytokines and chemokines would cause a critical phenomenon, called the cytokine storm. The cytokine storm is a suggested mechanism underlying ARDS conditions and multiple organ failures (Wang et al. 2020c; Lin et al. 2020; Shi et al. 2020). Overall, scientists should become acquainted with the innate and adaptive immune responses of the body against SARS-CoV-2 to know better the COVID-19 immunopathogenesis.

26.3 Immunopathogenesis of COVID-19

The transmission of the SARS-CoV-2 occurs mainly through the droplets (Rabi et al. 2020). COVID-19 causes respiratory symptoms that vary from mild respiratory infection to severe pneumonia, which can lead to respiratory failure. Respiratory failure is the leading cause of death in COVID-19 patients. The underlying cause of respiratory failure is the ARDS, which is, in turn, mediated by the severe inflammatory response or the cytokine storm (Mason 2020). After SARS-CoV-2 infects the lung tissue, the cytotoxic effects of the virus on epithelial cells induce innate and further adaptive immune responses. The interaction between the major pathogenic microbes and subsequent immune response induces inflammation in the respiratory system that contains multiple cytokines and molecules as its core. The cytokine storm is the result of hyperinflammation, caused by a severe uncontrolled immune response to the virus. The histopathological findings of the SARS-CoV-2 include bilateral consolidation, ground-glass opacity, hyaline membrane formation, desquamation of the pneumocytes, and fibromyxoid exudates in the lungs. These findings are consistent with a bilateral diffuse alveolar damage (DAD) of the lungs, which is exerted by severe inflammation and cytokine storm (Tian et al. 2020a). It is, therefore, necessary to understand the anti-SARS-CoV-2 immune mechanisms for controlling the disease and developing effective treatments (Khoury et al. 2020).

The immune system mainly determines the severity of the disease. The highest mortality rate of the COVID-19 has been reported in elderly patients with comorbidities that compromise the antiviral function of the immune system. Since this compromised immune response cannot inhibit the replication of the virus, lower respiratory airway and alveoli are highly susceptible to infection with the SARS-CoV-2. The most critical injury to the lung is caused by the severe inflammation and ARDS, rather than direct cytopathologic effects of the SARS-CoV-2 (Xiao et al. 2020). Both innate and adaptive immunity are involved in the immunopathogenesis of the novel coronavirus (Fig. 26.1).

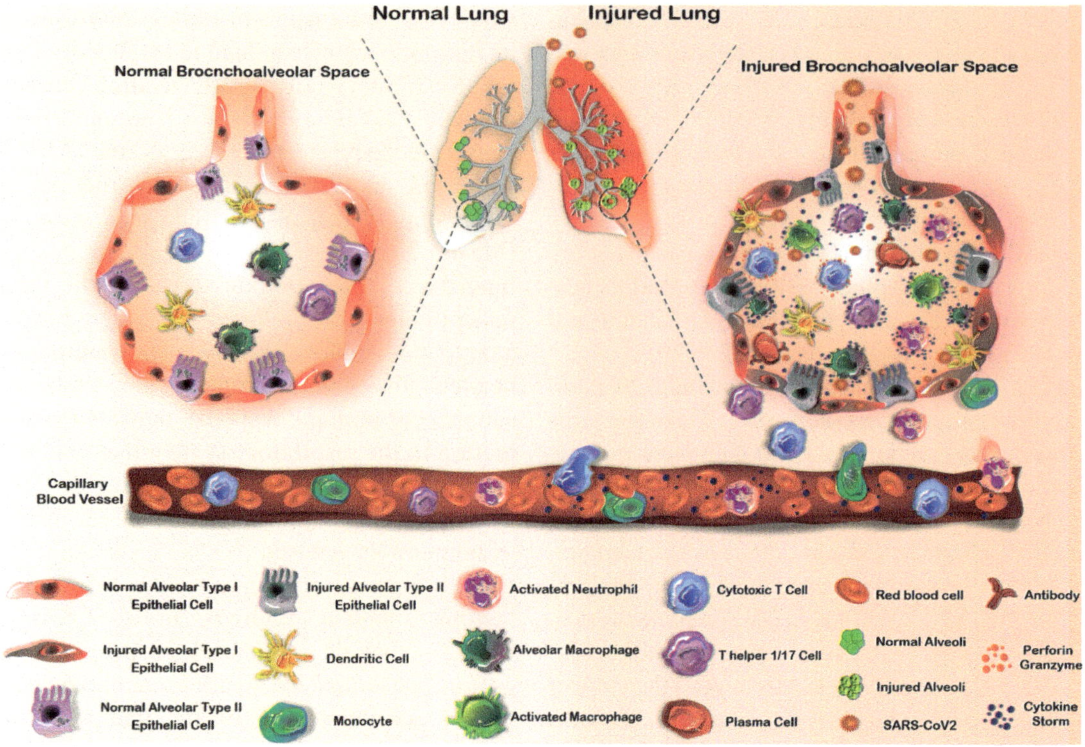

Fig. 26.1 The immunopathogenesis of SARS-CoV-2. Virus antigens are presented to TH1/17 and CTL cells by antigen-presenting cells (alveolar epithelial I/II cells, macrophages, and dendritic cells). On the one hand, TH cells, as well as DCs, activate the CTLs, which attack viruses by releasing perforin and granzymes. On the other hand, TH cells stimulate the activation of neutrophils and macrophages by the cell to cell connection and producing the inflammatory cytokines. Also, TH cells convert the B lymphocytes to plasma cells and lead to the production of antiviral antibodies. Activated neutrophils, monocytes, and lymphocytes also are recruited from the peripheral blood to the lung alveoli by chemotactic factors produced by macrophages and damaged epithelial cells. As a result, the influx of large numbers of immune cells into the bronchoalveolar space and the overproduction of inflammatory cytokines by TH cells activated neutrophils, and macrophages cause overstated immune responses, which elicit the cytokine storms, alveolar epithelial cell damage, ARDS, and other organs

26.3.1 Innate Immune Response to SARS-COV-2

Macrophages and lymphocytes are immune compartments of the alveoli, which initiate the immune response against the virus. Airway epithelial cells can act as immune effector cells, as well. The damage to the pneumocytes induces the presentation of adhesion molecules and the production of the immune mediators that recruit further macrophages and lymphocytes to the alveoli. Macrophages recognize the viral antigens through innate immune receptors, such as toll-like receptors (TLR). The identification of viral antigens would activate the intracellular inflammatory cascade of the macrophages and other inflammatory cells, which leads to the production of the interleukin 1 beta (IL)-1β, IL-6, and tumor necrosis factor alpha (TNF-α). The release of the inflammatory mediators increases vascular permeability and the migration of the immune cells such as neutrophils, lymphocytes, and monocytes to the lungs (Liu et al. 2020b). NK cells are among the immune cells that contribute to the immune response against the virus, as well (Prompetchara et al. 2020).

COVID-19 patients have increased levels of serum acute proinflammatory chemokines and cytokines, such as IL-1β, interferon gamma (IFN-γ), interferon-γ-inducible protein 10 (IP-

10), and monocyte chemoattractant protein 1 (MCP-1). Besides, severe COVID-19 patients who required intensive care unit (ICU) admission had higher levels of TNF-α, granulocyte colony-stimulating factor (G-CSF), MCP-1, macrophage inflammatory protein 1 alpha (MIP-1A), IL-2, 7, 10, and IL-10, indicating severe inflammation in critically ill patients (Huang et al. 2020). Thus, elevated proinflammatory cytokines and chemokines that lead to hyper-inflammation are the underlying cause of the lung tissue damage in COVID-19. Management of the inflammatory response has been shown to increase the outcome of critically ill COVID-19 patients in the clinical setting (Jose and Manuel 2020).

The laboratory findings also show an increased number of neutrophils in COVID-19 patients, which demonstrate the activation of innate immunity. The neutrophil-to-lymphocyte ratio (NLR) is increased in severe cases and has been proposed as a prognostic factor in COVID-19 (Li et al. 2020).

26.3.2 Adaptive Immune Response to SARS-COV-2

B and T lymphocytes have a significant role in the adaptive immune response against SARS-CoV-2. The binding of the viral peptides to the major histocompatibility complex (MHC)-I and II on antigen-presenting cells (APCs) stimulates the $CD8^+$ cytotoxic T cells (CTLs) and $CD4^+$ helper T cells (Th) to exert an adaptive immune response against the novel coronavirus. Helper T cells present the viral antigens to the B lymphocytes. B lymphocytes subsequently neutralize the virus by producing neutralizing antiviral antibodies. These neutralizing antibodies can bind to the S protein of the free extracellular viruses and, thus, can inhibit the infection of the healthy cells. Also, CTLs can directly kill the virus-infected cells that express viral antigens on their cell surface by inducing apoptosis through granzymes and perforin (Tay et al. 2020).

Laboratory results of the COVID-19 patients show a decreased lymphocyte count. The immunological assays have proved that lymphopenia is a common laboratory finding in COVID-19 patients, which mainly contains a reduced number of both CD4+ and CD8+ T cells (Tay et al. 2020). Several reasons might lie in the reduced number of T cells in COVID-19, including apoptosis/pyroptosis of T cells due to the infection by SARS-CoV-2, pulmonary infiltration of T cells, suppression of bone marrow due to severe diffuse inflammation, and margination of T cells into the vessels. SARS-CoV-2 can infect all lymphocyte subsets, including B cells, T cells, and NK cells; however, T lymphocytes are the most common group. Lymphopenia can subsequently increase the risk of secondary bacterial infections and disease severity. The severity of the lymphopenia is associated with poor prognosis in critically ill patients (Liu et al. 2020d).

The adaptive immune system also contributes to the hyper-inflammation and cytokine storm by producing proinflammatory cytokines. The IL-12 produced by immune cells induces the Th1 cells to activate the CTLs. Also, Th1 cells, in combination with monocytes, contribute to produce pro-inflammatory cytokines and growth factors such as IL-6 and GM-CSF, which have critical roles in the progression of mild pulmonary inflammation to severe ARDS (Fu et al. 2020). Type 2 helper (Th2) cells have been reported to be reduced in COVID-19. The effects of the macrophage-associated IL-4 on Th2 cells lead to the release of the inflammatory cytokines by Th2 cells, which are subsequently involved in the severe inflammatory response and cytokine storm (Prompetchara et al. 2020).

26.3.3 Challenges of the Immune Response Against SARS-COV-2

26.3.3.1 Immune-Compromised Patients

The immune system is the most prominent defender of the body against SARS-CoV-2. Immune-competent COVID-19 patients have shown less morbidity and mortality in comparison to immune-compromised patients. Most of the mortalities during the COVID-19 pandemic have been reported in immune-compromised and comorbid patients, such as cancer patients. The insufficient immune response against SARS-CoV-2 leads to increased replication of the virus and the progression of the disease (Wang et al. 2020a). Thus, the prophylactic protection and stimulation of the immune response in immune-compromised patients seem to be necessary.

26.3.3.2 Cytokine Storm

The cytokines and chemokines produced by the epithelial cells and the alveolar macrophages activate the innate and adaptive immune components. The recruitment of these cells to the lung tissue induces the immune responses against SARS-CoV-2. In some severe cases, the immune response can be very severe that can cause cytokine storm and ARDS in the lung. GM-CSF and IL-6 produced by the monocytes and Th cells are the major factors that lead to the inflammation (Wang et al. 2020d). The increased serum level of IL-6 is associated with a more severe form of the disease. Hyper-inflammation causes the alveolus filled with the inflammatory exudates and fibrous tissue, which can finally lead to respiratory failure (Jose and Manuel 2020). Management of this hyperinflammatory response using the immunomodulatory approaches can reduce lung tissue damage and fibrosis. These approaches are the cytokine-targeting therapies using the interventions that mediate the adaptive immune response toward the Th2-mediated pathway and include injection of MSCs as immunomodulatory cells and using the immunomodulatory cytokines and agents (Khoury et al. 2020).

26.3.3.3 Exhaustion of T Cells

Clinical studies on the immune cells of COVID-19 patients have shown that the T cells of these patients have an increased expression of T-cell exhaustion markers, such as programmed death-1 (PD-1) and T-cell immunoglobulin mucin-3 (TIM-3). Exhausted T cells cannot exert appropriate immune responses against the virus (Diao et al. 2020). Inhibiting the exhaustion of T cells could be considered as an immune-enhancing approach in the treatment of SARS-CoV-2 individuals.

26.3.3.4 Dysregulation of Type I IFN

The pathogenesis of SARS-CoV-2 is similar to that of SARS-CoV. The SARS-CoV uses multiple immune-evasion mechanisms. One of the most critical immune pathways that inhibit viral replication is the type I interferon (IFN-I) pathway. SARS-CoV has shown to interfere with host IFN-I production by suppressing the STAT-1, the upstream signaling of the IFN-I. IFN-I depletion dampens the optimum antiviral immune responses, thereby paving the way for the pathogenesis of SARS-CoV (Hu et al. 2017). Accordingly, IFN-based treatments have been proposed for the treatment of COVID-19. IFN-$\alpha 1\beta$, IFN-β, recombinant human IFN-α, IFN-$\beta 1$A, and IFN-$\beta 1$B are such IFN-based immunotherapies currently under investigation by multiple clinical trials. Early administration of IFN-based therapeutics can induce immune responses in the first stages of the disease and prevent the cytokine storm. In contrast, treatment with IFNs at the severe stages might cause an immune response that, in turn, further stimulates cytokine storm and hyper-inflammation in the lungs (Mosaddeghi et al. 2020).

26.3.4 Immunotherapy of COVID-19

While there are no vaccines or specific medicines approved for the prevention and treatment of COVID-19, immunologists offer insights for immunotherapy of COVID-19. Immunotherapy is effective against diseases caused by similar viruses such as SARS-CoV and MERS-CoV. Fig. 26.2 illustrates a graphical overview of the key design characteristics of the different types of immunotherapy-based strategies.

26.3.4.1 Antibody-Based Treatments for COVID-19

Convalescent-Plasma Therapy

Convalescent plasma (CP) or hyper-immune immunoglobulin therapy is a passive-adaptive-immunotherapy used to protect the individuals from clinical infections or to prevent the disease progression (Casadevall et al. 2004; Luke et al.

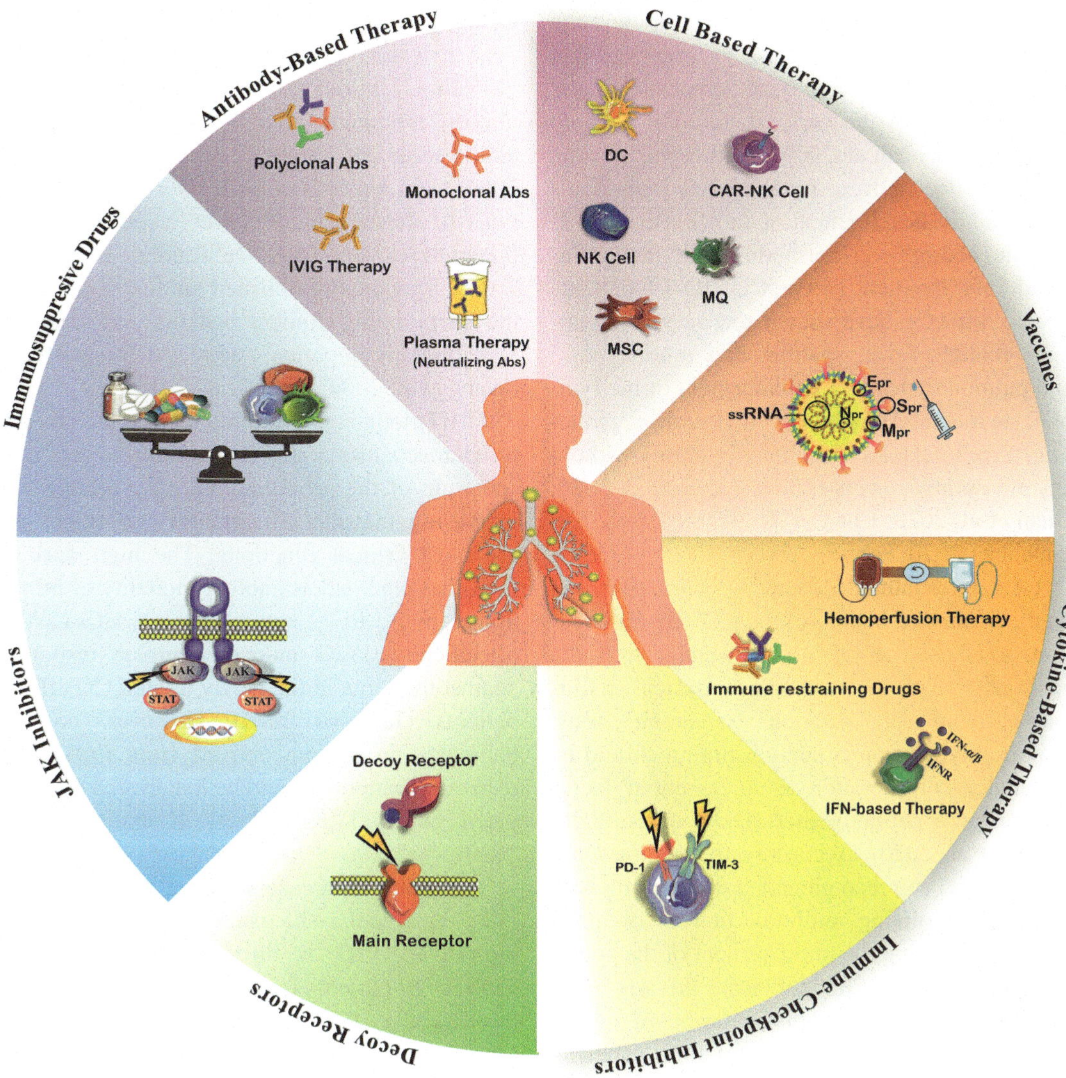

Fig. 26.2 Immune-based therapies at a glance

2006). Plasma therapy has shown promise as postexposure prophylaxis in polio, measles, hepatitis, rabies, and mumps. It is also known as a therapeutic approach with acceptable efficacy and safety for Ebola, influenza, Argentine hemorrhagic fever, SARS, MERS, and COVID-19 (Casadevall and Pirofski 2020; Sahr et al. 2017; Hung et al. 2011). Plasma therapy is particularly useful during outbreaks when time and resources are limited for the immunoglobulin preparations.

In exposed individuals, plasma contains the anti-SARS-CoV-2-specific neutralizing and non-neutralizing antibodies. Antibodies contribute to the elimination of infection by binding to pathogens and neutralizing them, directly or by activating the indirect cytotoxic routes, including antibody-dependent cellular cytotoxicity (ADCC), complement-dependent cytotoxicity (CDC), and phagocytosis (Van Erp et al. 2019; Gunn et al. 2018). As an encouraging report, CP has been approved by the Food and Drug Administration Agency (FDA) for the treatment of COVID-19 (Administration 2020, May 1). CP helps to improve the immunity of SARS-CoV-2-

infected patients, suppress the viremia, clear the infected cells, prevent disease progression, and augment the success rate of supportive care. When applied in the early stages, CP is more effective (Cheng et al. 2005). To this end, plasma can be collected from rehabilitated individuals following the development of anti-SARS-CoV-2-specific neutralizing antibodies and then reinfused into the new cases. COVID-19 patients treated with CP have shown recovery of symptoms (cough, fever, sputum, and muscle pain) and pulmonary lesions, enhancement of the lymphocyte ratio and blood oxygen saturation, reduction of the viral load, and clinical stability without any measurable adverse effects (Shen et al. 2020; Duan et al. 2020; Chen et al. 2020b; Law 2020; xinhuanet 2020, February 28).

Of note, qualified donors are selected based on the confirmed history of the COVID-19 infection, a negative follow-up molecular test, the absence of symptoms, standard physical examination, and pre-donation screening tests. Importantly, plasma collection timing should be put under consideration as an essential factor affecting the antibody titer (Okba et al. 2020; Bloch et al. 2020). According to the investigations, it has been recommended that plasma with a high neutralizing antibody titer is obtained 14 days following the improvement of the symptoms (Bloch et al. 2020). Plasma therapy has a variable dosage of administration based on the prevention or the treatment aims; thus, a specific dose has not been clarified. Nevertheless, some planned clinical trials have proposed the use of one unit of plasma (200 mL) for postexposure prophylaxis and one/two units for therapeutic purposes (29–31). The dosage of CP in COVID-19 has been determined according to the previous experiences of utilizing CP in SARS, which was 5 mL/kg of plasma with ≥1:160 titers (Cheng et al. 2005). In summary, plasma therapy can provide great hope to improve the treatment of COVID-19 (Law 2020).

IVIG Administration

High-dose intravenous immunoglobulin (IVIg) is an effective and safe immune modulator collected from a pool of more than 1000 healthy donors, which contains the polyclonal immunoglobulin G (IgG). IVIG is a potent immunomodulatory treatment with a broad spectrum of antibacterial and antiviral hallmarks (Bayry et al. 2004). As a protective effect, IVIg prevents infections of the upper and lower respiratory tracts (Galeotti et al. 2017; De Ranieri and Fenny 2017). Previously published studies pointed out the encouraging clinical benefits and safety of IVIG therapy in patients with SARS and MERS (Wang et al. 2004; Arabi et al. 2014).

IVIG modulates the immune responses by different mechanisms, of which inhibiting the broad spectrum of the proinflammatory cytokines and increasing the anti-inflammatory cytokines are the most critical functions. The high dose of IVIG has emerged as another modality of immunotherapy to neutralize SARS-CoV-2. Specifically, IVIG induces immune responses against the virus in the newly treated COVID-19 patients (Jawhara 2020). However, possible pathogens in plasma derived from COVID-19 recovered patients should be inactivated or eliminated before IVIG therapy (Caballero et al. 2010).

According to a case series of the COVID-19 IVIG therapy, 0.3–0.5 g per kg weight per day of high-dose IVIG is allowed to prescribe to COVID-19 patients for 5 days (Cao et al. 2020). Successfully, findings have revealed that the administration of high-dose IVIg at the proper point could inhibit the disease progression to severe conditions and increase the patients' recovery in early stages. Patients demonstrated no fever and breathing recuperation in 3–5 days following the IVIG therapy. Also, no adverse effects were observed in all treated patients (Cao et al. 2020; Xie et al. 2020). On account of the ability of IVIG therapy in augmenting the passive immunity and modulating the inflammatory condition with satisfactory efficacy and safety, it could be considered as an alternative treatment for SARS-CoV-2 cases. Moreover, IVIG, in combination with other supportive care, would be a hopeful treatment for the novel coronavirus.

Monoclonal Antibodies (mAbs)

Antibody-based therapy is another practical approach that can be applied to COVID-19. Antibodies can be obtained from infected cases or produced by laboratory manufacturing. mAbs are characterized by high specificity and purity, high safety, and low rate of blood-borne infections that can overcome the restrictions related to IVIG and CP preparations. mAb therapy, as effective passive antibody therapy, has appeared to prevent or to treat the various diseases (Sui et al. 2005; Bayry et al. 2007). In the recent decade, several mAbs have been developed or are under development against different viral infections (Group and Team 2016; Gupta et al. 2016; Prabakaran et al. 2009). Mainly, mAbs can be produced to target the specific epitopes of SARS-CoV-2 surface proteins and inhibit the viral entry to the host cells, replication, and progression.

Promising results have been observed with specific neutralizing mAb, such as m396 and CR3022 against SARS-CoV (Prabakaran et al. 2009). Accordingly, the receptor-binding domain (RBD) of SRAS-CoV-2, which is similar to SARS-CoV, can be targeted by CR3014, a SARS-CoV-specific mAb. Targeting the SARS-CoV-2 RBD by CR3022 resulted in the notable suppression of SARS-CoV-2 infectivity. Thereby, CR3022, as a monotherapy or in combination with other drugs, might be helpful in the treatment of COVID-19-infected patients. In contrast, m396 and CR3014, the other SARS-CoV-specific neutralizing mAbs, were not significantly effective against SARS-CoV-2 due to the presence of no connection between the mentioned antibodies and SARS-CoV-2 spike protein. However, they can be applied as an alternative treatment for SARS-CoV-2 (Tian et al. 2020b). 47D11 is the first mAb introduced to neutralize SRAS-CoV-2. Interestingly, 47D11 cross-neutralizes both SARS-CoV and SARS-CoV-2 by binding to the conserved core domain of the spike RBD (S1B) with identical affinities. 47D11 could powerfully suppress the infection of SARS-CoV and SARS-CoV-2 through the unknown mechanism (Wang et al. 2020b). Also, three human mAbs, including 311mab-31B5, 311mab-32D4, and 311mab-31B9, specifically and strongly bind to SARS-CoV-2 RBD protein. Among these, 311mab-31B5 and 311mab-32D4 could potentially suppress the interaction between the SARS-CoV-2 RBD and hACE2 proteins, leading to the virus neutralization. Encouragingly, two described mAbs could be considered as potent prophylactic and therapeutic agents against COVID-19 (Chen et al. 2020d). Up to now, many clinical trial studies have been initiated to investigate the efficacy of different mAbs against SARS-CoV-2, which may be helpful to disease management.

Polyclonal Antibodies

In addition to mAbs, polyclonal antibodies also can be developed as prophylactic and therapeutic agents against viral infections like the coronaviruses family, mostly MERS-CoV (Sheahan et al. 2017). As an example, SAB-301, a human polyclonal antibody (50 mg/kg), has revealed the potent and well-tolerated antiviral function against MERS-CoV in healthy participants (Jean et al. 2020). Moreover, monoclonal and polyclonal antibodies have also shown powerful prophylactic and therapeutic effects against influenza by inhibiting virus replication (Beigel et al. 2019). Thus far, various polyclonal antibodies have obtained a license to target some viral infections, including varicella-zoster virus (VZV), cytomegalovirus (CMV), hepatitis B virus (HBV), and influenza. These data suggest that polyclonal antibodies as effective therapeutic agents may be useful in COVID-19 as well.

26.3.4.2 Hemoperfusion Therapy

Extracorporeal organ support therapies such as hemoperfusion and hemoadsorption therapies would be beneficial for the critical conditions of COVID-19 (Ronco et al. 2020). Blood purification techniques would assist in eliminating the molecular toxins and extra-produced cytokines from the body. They also play important roles in stabilizing body temperature, volume balance, and hemodynamics. The plasma/whole blood adsorption technologies can be administrated to scavenge the excessive inflammatory cytokines in the course of the novel coronavirus infection (Maoujoud et al. 2020). Since the proinflammatory cytokines are prominent in the early stages of the disease, it is recommended to use plasma/blood adsorption therapies in the initial stages of the infection. These methods can have a therapeutic effect on the pathogenesis of COVID-19. The application of blood purification techniques in COVID-19 should be considered in situations such as ARDS, volume overload, resistance to the diuretics, and continuous inflammatory fever, which cannot be subsided by glucocorticoids (Rangaswami et al. 2019).

Notably, these methods are better to perform every 12 h in the early stages of the disease. After that, the application of adsorption technologies can be reduced to one time in 24 h. The reduction in IL-6/IL-10 ratio represents that anti-inflammatory cytokines have replaced proinflammatory cytokines. Then, the body temperature would return to normal. At this time, the adsorption therapies should be ceased. Extracorporeal membrane oxygenation (ECMO) is one of the respiratory supports in SARS-CoV-2-infected patients. However, it should be considered that ECMO could deteriorate the release of cytokine phenomenon (Al-Fares et al. 2019). So the blood purification techniques would be helpful for patients who underwent ECMO.

When using blood/plasma adsorption techniques, it is necessary to adjust the dosage of the drugs, in particular, albumin (Association 2020).

26.3.4.3 Immune Checkpoint Inhibitor-Based Therapies

$CD4^+$ T cells, $CD8^+$ T cells, and NK cells would become exhausted in the course of the COVID-19. The increased expression of CD94/NK group 2 member A (NKG2A) by NK cells and CTLs plays a role in this functional exhaustion. NKG2A upregulation would reduce the capability of CTLs and NK cells in producing granzyme B, IFN-γ, CD107a, IL-2, and TNF-α. Thus, decreasing the $NKG2A^+$ NK and CTLs may help the management of SARS-CoV-2 infection. Additionally, COVID-19-infected patients have higher serum levels of IL-10, which is an inhibitory and anti-inflammatory cytokine and might affect T-cell exhaustion. Also, PD-1 and TIM-3 are exhaustion markers that have shown to increase in COVID-19. A clinical trial study is ongoing to investigate the effect of PD-1 inhibition on the healing process of COVID-19 patients (NCT04268537). Therefore, inhibition of PD-1, TIM-3, NKG2A, and other involved immune checkpoints may be an efficacious therapy to prevent T-cell exhaustion and to reinforce the immune system against SARS-CoV-2 infection (Zheng et al. 2020a, b; Diao et al. 2020).

26.3.4.4 JAK Inhibitors

The collaboration between the Janus kinase (JAKs) family and signal transducer and activator of transcription proteins (STATs) forms JAK-STAT signaling pathways. This cellular pathway is essential for the immune reactions of the cells to the exogenous signals. The inhibition of JAKs family containing TYK3, JAK1, JAK2, and JAK3 will hinder the release of cytokines. The inhibition of the JAK1 signaling pathway would decrease the expression of Th2 and related cytokines. Additionally, the disruption of STAT3 phosphorylation followed by JAK1 inhibition would reduce the level of IL-6. IL-6 is one of the central destructive cytokines in the cytokine storm of the COVID-19; therefore, JAK inhibitors could be a candidate to control the cytokine storm (Russell et al. 2020a; Napolitano et al. 2020; Peterson et al. 2020). From recent literature, baricitinib (Olumiant), which is a JAK inhibitor drug, has been introduced to diminish the entrance of the virus to the host cells and also inflammation in the COVID-19-infected patients. Baricitinib performs its function through inhibiting the adaptor-associated protein kinase 1 (AAK1) receptor that involves in the virus endocytosis (Richardson et al. 2020).

26.3.4.5 Development of Decoy Receptors

The viral spike (S) protein mediates virus entry into cells through its attachment to an enzyme attached to the surface of cells, called ACE2, and is a primary determinant of cell tropism and pathogenesis (Walls et al. 2020; Belouzard et al. 2012). Immunologists are currently studying the potential therapeutic roles of the Fc variants against the COVID-19 crisis. Thus, it would be desirable to convert the soluble form of ACE2 receptor into an immunoadhesin composition fused to an Fc region of antibody molecules (ACE2-Fc), thereby lengthening the half-life of the circulating fusion protein while also triggering an immune response against the virus (Kruse 2020). This approach has the potential to prevent or to treat not just infection by SARS-CoV-2 but also the other coronaviruses in humans and animals that use ACE2 as the critical receptor to invade the cells. While not tested in an animal model, an in vitro study revealed that a human IgG1 Fc domain fused to the extracellular domain of ACE2 enzyme (known as an ACE2-NN-Ig) effectively neutralized SARS coronavirus, with a 50% inhibitory concentration of approximately 2 nM (Moore et al. 2004). The results of the previous study show promising effects of these agents in the clinical setting and will be ready for testing within months.

26.3.4.6 Management of the Cytokine Storm

Immune-Restraining Approaches

According to the SARS-CoV and MERS-CoV infections, uncontrolled production of the proinflammatory cytokines elicits the cytokine release syndrome (CRS) (Channappanavar and Perlman 2017; Cameron et al. 2008; Min et al. 2016). Based on the investigations in COVID-19 patients, elevated levels of TNF-α, G-CSF, MIP-1A, MCP-1, IP-10, IL-10, and IL-7 in severe conditions have been highlighted the importance of CRS in disease inflammation and progression (Chen et al. 2020c). IL-6, IFN-γ, and GM-CSF have shown strong potential in generating the cytokine storm (Liu et al. 2020c; Zhou et al. 2020b; He et al. 2020).

Plasma therapy, vaccines, MSCs therapy, interferon-based therapy, and inhibitory agents have been developed to block the inflammatory cytokines and manage the cytokine storm (Barrett et al. 2014). Tocilizumab, a humanized mAb, is used to block the IL-6 receptor in SARS-CoV-2 cases, which suppresses the iatrogenic cytokine storm and prevents the respiratory system destruction (Cascella et al. 2020).

Corticosteroids, in particular dexamethasone, are the other drugs used to manage CRS. Of note, the injection of corticosteroids for COVID-19-associated cytokine storm and pneumonia has not been approved due to the increased risk of vascular necrosis and diabetes (Maschalidi et al. 2016).

Hyaluronan (HA) can be considered as another potential cause of ARSD and fatalities in SARS-CoV-2 patients (Xu et al. 2020; Bell et al. 2019). Moreover, studies show that the increased levels of inflammatory cytokines, such as IL-6, IL-1β, and TNF-α in the lungs of COVID-19 cases lead to HA generation by inducing HA synthase-2 (HAS2) (Bell et al. 2019). Therefore, it has been suggested that the prescription of the hyaluronidase and 4-methylumbelliferone (4-MU) (HAS2 inhibitor) may help the rehabilitation of breathing in COVID-19 individuals (Collum et al. 2017). Moreover, targeting inflammatory cytokines can help inhibition of HAS2 and HA production.

Interferon-Based Immunotherapy

The first line of the defense against viruses is the innate immune responses, which mainly involves IFNs (Schneider et al. 2014). IFN-α2b, a member of the type I IFN, is quickly produced in viral infections and can suppress the replication of coronaviruses, including human coronaviruses (Turner et al. 1986). IFN-β, another type I IFN, is efficient in suppressing the SARS-CoV replication (Morgenstern et al. 2005). Additionally, in MERS-CoV infection, IFN-β-1b has led to clinical improvement by inhibiting virus replication (Chan et al. 2015). Based on these results, IFN-based therapies can be suggested for treating COVID-19. In COVID-19 patients, the production of IFN-1 is inhibited by the virus through the suppression of the STAT-1 intracellular signaling pathway (Mosaddeghi et al. 2020).

Interestingly, the low secretion threshold of IFN in children and immediate production of IFN after infection may justify the low infection and mortality rate of the SARS-CoV-2 in children. Contrarily, increased secretion threshold of IFN in elderly patients can be considered as a reason for the higher mortality rate in these cases. It has shown promise that applying the IFN-based immunotherapy would be useful in treating the COVID-19 patients, particularly in early stages (Mosaddeghi et al. 2020). In this context, various clinical trials have been designed to investigate the efficacy of IFN-β, IFN-α2β, and recombinant human IFN-α against SARS-CoV-2. We might expect that IFNs-based therapy can heavily trigger the innate immune responses in addition to the direct suppression of the SARS-CoV-2. However, precautions should be considered in using the high dose of IFNs due to the possibility of cytokine storm occurrence.

26.3.4.7 SARS-CoV-2 Vaccination

The development of a safe and effective vaccine for COVID-19 that can be used globally is a priority for ending the pandemic and preventing its future recurrence (Amanat and Krammer 2020; Yamey et al. 2020). Since the SARS-CoV-2 virus exhibits a high degree of genetic homology with SARS-CoV and MERS-CoV, analyzing the vaccine patents relevant to the SARS-CoV and MERS-CoV viruses may potentially facilitate the design of anti-SARS-CoV-2 vaccines (Ramaiah and Arumugaswami 2020; Lurie et al. 2020). The strategies covered in these patents, which may ultimately offer promise as preventive vaccines against COVID-19, include recombinant spike protein-based vaccines, attenuated and whole inactivated vaccines, virus-like particle (VLP) vaccines, viral vectors, and DNA-based, as well as mRNA-based, vaccines (Liu et al. 2020a; Dhama et al. 2020).

The comparative evaluation performed on full-length S protein sequences identified that the most variable residues among MERS-CoV, SARS-CoV, and SARS-CoV-2 reside within the S1 subunit of the S protein that exposes RBD (Yu et al. 2020). It was reported that viral S protein subunit vaccines induced higher neutralizing antibody titers, T-cell responses, and more induction of protective immunity than other S protein-based strategies, including full-length S protein, live-attenuated SARS-CoV, and DNA-based S protein vaccines (Liu et al. 2020a).

Previous research analyzed the similarity in T-cell epitopes of SARS- and MERS-CoV, revealed the potential cross-reactivity of the coronaviruses, and assessed the possibility of developing universal CoV vaccines (Dhama et al. 2020). Immunoinformatics tools can also be used for the identification of significant epitopes in COVID-19 vaccine candidates. Recently, immunoinformatics was used to predict potential CTL and B-cell epitopes in SARS-CoV-2 spike glycoprotein. Molecular dynamic simulation analysis of interactions between these epitopes and their corresponding MHC class I molecules showed that the CTL epitopes bind with MHC class I peptide-binding grooves via multiple contacts, thus suggesting their potential for eliciting immune responses (Baruah and Bose 2020).

The Coalition for Epidemic Preparedness Innovations (CEPI) has collaborated with the University of Queensland to develop the DNA medicine vaccine platform against COVID-19. The vaccine platform utilizes for delivering synthetic genes into cells for translation into viral antigenic proteins, which generate T-cell and antibody responses. CEPI has also announced funding to Moderna for developing a vaccine candidate expressing SARS-CoV-2 S protein in the mRNA vaccine platform technology. This vaccine is expected to undergo clinical testing in the coming months (Dhama et al. 2020).

Bacillus Calmette-Guérin (BCG) vaccination has been shown to offer broad protection to viral respiratory infections (Moorlag et al. 2019). Recent studies show that national differences in the COVID-19 outbreak may be partly explicable by the different national policies on BCG vaccination in childhood, indicating its long-lasting protection against the current strain of coronavirus (Miller et al. 2020). BCG vaccination significantly elevates the production of proinflammatory cytokines, specifically IFN-γ and IL-1B, which has been demonstrated to play a major role in antiviral defense (Kleinnijenhuis et al. 2014). The phenomenon known as "trained immunity" is proposed to be caused by epigenetic changes leading to the reactivation of the genes encoding for proinflammatory mediators (Netea et al. 2016). Therefore, a combination of reduced morbidity and mortality could make the BCG vaccination a potential new tool in the fight against the COVID-19 pandemic.

26.3.4.8 Cell-Based Immunotherapy

Mesenchymal Stem Cells (MSCs)

In the last decade, immune cell-based therapy, as a promising therapeutic strategy, has demonstrated a significant breakthrough in treating different kinds of diseases. Among the stem cell-based therapies, MSC therapy has revealed the prominent curative potentiality in the most incurable diseases (Golchin and Farahany 2019; Golchin et al. 2019).

From this standpoint, MSC therapy probably reduces the cytokine storm in the SARS-CoV-2 infection by inhibiting the production of TNF-α, IFN-γ, IL-1α, IL-6, and IL-12 cytokines (Uccelli and de Rosbo 2015). Additionally, MSCs may improve the ARDS by secreting IL-10, VEGF, keratinocyte growth factor (KGF), and hepatocyte growth factor (HGF) in COVID-19 cases (Uccelli and de Rosbo 2015). MSCs entrap in the lungs following the intravenous administration into COVID-19 patients. Then, MSCs restore the pulmonary microenvironment, inhibit the alveolar epithelial cells destruction, arrest the pulmonary fibrosis, and recover the dysfunctionality and pneumonia of the lung (Leng et al. 2020).

Based on the above-described details and the experiences of MSC therapy in H5N1 viral infections with similar functions in the lung, the use of this treatment has been suggested in COVID-19 patients (Chen et al. 2020a). The summarized results of human MSC therapy in COVID-19 patients point out that most of the patients can remove from the ventilator and can walk on a few days after treatment. MSC therapy can lead to improvement of symptoms, pulmonary function, and T-cell frequency, induction of T cells toward regulatory phenotypes, inhibition of inflammatory cytokines, and increased secretion of anti-inflammatory cytokines (Liang et al. 2020; Leng et al. 2020).

To increase the effectiveness of anti-inflammatory responses of MSCs, they can be pretreated by IFN-γ, which is absent in severe cases of COVID-19 (Wang et al. 2014). Clinical trials are currently investigating the safety and efficacy of MSC therapy in COVID-19 pneumonia.

Natural Killer (NK) Cells

NK cells are a subgroup of lymphocytes that have an essential role in the initial immune responses against the tumor cells and virus-infected cells. IFNs and macrophage-secreted cytokines can activate the NK cells. Since NK cells are members of the innate immune system, they initially trigger the antiviral immune responses independent of MHC. NK cells have cytoplasmic granules that contain perforin and proteases. Through these granules, NK cells exert their cytotoxic effects on the target cells. The NK-guided immune responses against the virus are independent of the MHC and, thus, are a rapid immune response before the activation of the adaptive immune cells (Tay et al. 2020).

Considering the important role of the NK cells against viral respiratory infections, multiple clinical trials have started to assess the therapeutic efficacy of NK-cell therapy in COVID-19 (NCT04365101 and NCT04280224). Chimeric antigen receptor (CAR)-NK cells have also been among the novel therapeutics for the treatment of the COVID-19. An example is a phase I/II study that is designed for the treatment of SARS-CoV-2 patients with universal off-the-shelf NKG2D-ACE2 CAR-NK-cell therapy (NCT04324996).

T Cells and Chimeric Antigen Receptor (CAR)-T-Cell Therapy

CTLs are the most important effector cells of the adaptive immunity against SARS-CoV-2. As mentioned earlier, studies have exhibited a higher expression of the T-cell exhaustion markers such as PD-1 and Tim-3 (Diao et al. 2020). Thus, inhibiting the exhaustion of the T cells could be considered as a potential therapeutic approach in COVID-19.

CAR-T cells are CTLs engineered to express an antigen-specific receptor. By the expression of the CAR, CAR T cells can recognize and lyse the specific target cells that express the antigens (Elahi et al. 2018; Tahmasebi et al. 2019). CAR-T cells have shown promising results in the treatment of multiple cancers, allergic disorders, and viral infections, such as HIV and hepatitis (Qi et al. 2020; Tahmasebi et al. 2019; Esmaeilzadeh et al. 2020). CAR-T-cell therapy could also be considered as a potential therapeutic approach in the treatment of COVID-19 patients.

Dendritic Cells (DCs) and Engineered DCs

DCs participate in the antiviral immune responses against SARS-CoV-2 by presenting viral antigens to immune cells and inducing the production of proinflammatory cytokines, such as IL-6. DCs are made up of two subsets, including DC-1 and DC-2. DC-1 cells stimulate the T cells to produce antiviral immune responses in COVID-19. Thus, DC-1 cells critically contribute to the progression of the disease to severe respiratory failure (Rajaei and Dabbagh 2020). Management of the inflammatory functions of DCs may be an efficient approach to suppress severe inflammatory responses. Also, engineered DCs can be another potential therapeutic approach in severe SARS-CoV-2.

Modulation of the Intrapulmonary Macrophage

Intrapulmonary macrophages are among the first immune cells that start the inflammatory responses in the alveoli by producing proinflammatory cytokines, such as IL-6 and IL-1β (Tay et al. 2020). Macrophages are divided into two subtypes: M1 macrophages with proinflammatory properties and M2 macrophages with anti-inflammatory properties. In COVID-19, M1 macrophages induce the inflammation by releasing the proinflammatory cytokines (Prompetchara et al. 2020). Thus, it would be beneficial to modulate the intrapulmonary M1 macrophages to reduce their inflammatory cytokine production. Since M2 macrophages have considerable roles in tissue repair and remodeling, modulation of M2 macrophages could be considered as a potential treatment approach to decrease the lung tissue damage and fibrosis.

26.3.4.9 Immunosuppressive Drugs

Cytotoxic Chemotherapy

Cytotoxic chemotherapy drugs are utilized to inhibit pathologic cell divisions in multiple cancers. Also, these types of drugs may have therapeutic effects against the novel coronavirus. An in vitro analysis has shown that 6-thioguanine (6TG) and 6-mercaptopurine (6MP) have the inhibitory impacts on SARS-CoV. Indeed, tacrolimus is one of the immunosuppressor drugs that are involved in the inhibition of calcineurin. Calcineurin has a role in the development process of T cells. Studies have demonstrated that tacrolimus may have antiviral functions in the treatment of human coronaviruses (Carbajo-Lozoya et al. 2012; Carbajo-Lozoya et al. 2014). Moreover, thalidomide would have an impact on the disruption of the messenger RNA (mRNA) in the blood cells and the reduction of the TNF-α. Clinical trials are investigating the therapeutic features of thalidomide in the treatment of SARS-CoV-2 (Medicine 2020 Mar 12; ClinicalTrials.gov 2020 Mar 12). Notably, the administration of mycophenolate mofetil has not been recommended in the treatment of the coronaviruses (Russell et al. 2020a). However, more in vivo and in vitro studies are necessary to define the therapeutic impacts of cytotoxic chemotherapy drugs on COVID-19.

Low-Dose Steroids and NSAIDs

Low-dose steroids and nonsteroidal anti-inflammatory drugs (NSAIDs) might help to control the inflammatory reactions in the course of COVID-19. Indeed, corticosteroids can decrease immunopathological devastations of COVID-19. NSAIDs inhibit cyclooxygenase (COX) enzymes (COX-1 and COX-2), which, in turn, can subside fever and inflammation in COVID-19. Among NSAIDs, naproxen and indomethacin can also disrupt viral replication and therefore have antiviral features. However, more researches are required to determine the definite effects of steroids and NSAIDs on the pathogenesis of COVID-19 (Russell et al. 2020a,b).

26.4 COVID-19 Immune-Based Clinical Trials at a Glance

Given the urgency to find new life-saving therapies, many centers are attempting to identify the potentials of existing immune-based therapies and repurposing them for COVID-19 treatment. Progress has been extraordinary, and hundreds of

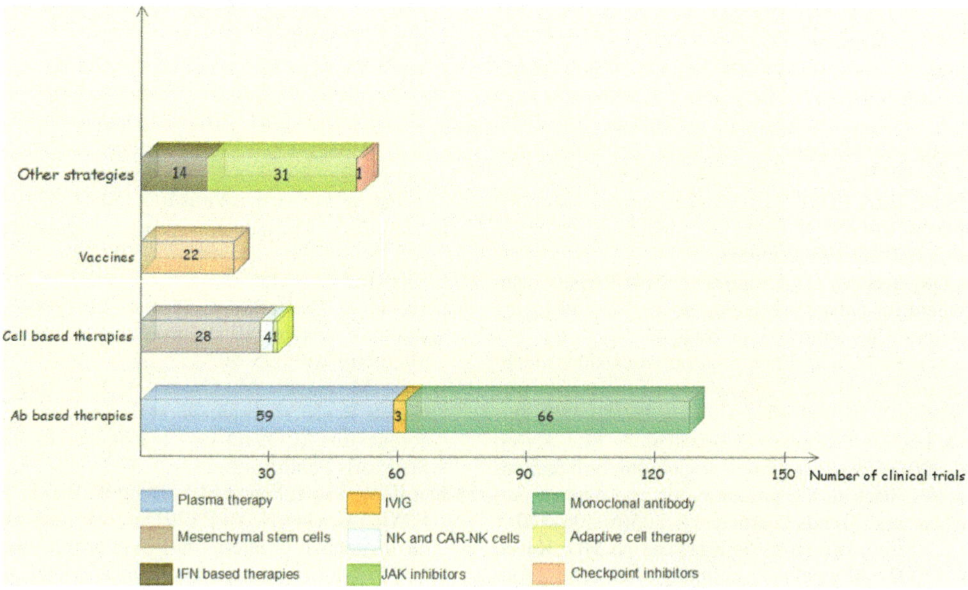

Fig. 26.3 Clinical trials underway on the manipulation of the immune response toward immunotherapy of COVID-19

clinical trials have been initiated. At the time of writing, a search of ClinicalTrials.gov revealed almost 230 results for COVID-19 immune-based clinical trials, with some of them summarized in Fig. 26.3.

26.5 Conclusion

The immunopathology of SARS-CoV-2 critically contributes to the progression of COVID-19, respiratory system injury, and other organ failures. In this context, the most immunopathological challenges include viral evasion from the immune system, hyper-inflammation derived from a cytokine storm along with inflammatory myeloid cells, lymphopenia mostly involving T cells, as well as dysfunction of T and NK cells. Various immune-based therapeutic regimens have been proposed for better disease management. At present, antibody-based therapies, including convalescent plasma therapy (neutralizing Abs), high-dose IVIG, and monoclonal and polyclonal Abs, are of the most commonly used immunotherapy strategies. Further research is necessary to investigate the potential of other immunotherapeutics, e.g., immune checkpoint inhibitors, JAK inhibitors, decoy receptors, cytokine-based therapies (hemoperfusion and interferon-based therapies), vaccines, cell-based treatments (stem cells, NK cells, CAR NK/T cells, DCs, engineered DCs, and modulatory macrophages), and immunosuppressive drugs (cytotoxic chemotherapy, low-dose steroids, and NSAIDs) in COVID-19 treatment and prevention. Meanwhile, the need is still felt to continue the understating of SARS-CoV-2 immunopathogenesis to recommend novel therapeutic approaches.

References

Abdulamir AS, Hafidh RR (2020) The possible immunological pathways for the variable immunopathogenesis of COVID--19 infections among healthy adults, elderly and children. Electron J Gen Med 17(4)

Administration FaD (2020) Recommendations for investigational COVID-19 convalescent plasma. https://www.fda.gov/vaccines-blood-biologics/investigational-new-drug-ind-or-device-exemption-ide-process-cber/recommendations-investigational-covid-19-convalescent-plasma

Al-Fares A, Pettenuzzo T, Del Sorbo L (2019) Extracorporeal life support and systemic inflammation. Intensive Care Med Exp 7(1):46

Amanat F, Krammer F (2020) SARS-CoV-2 vaccines: status report. Immunity

Arabi YM, Arifi AA, Balkhy HH, Najm H, Aldawood AS, Ghabashi A, Hawa H, Alothman A, Khaldi A, Al Raiy B (2014) Clinical course and outcomes of critically ill patients with Middle East respiratory syndrome coronavirus infection. Ann Intern Med 160(6):389–397

Association ERA-EDaT (2020) Expert consensus on the application of special blood purification technology in severe COVID-19 pneumonia

Barrett DM, Teachey DT, Grupp SA (2014) Toxicity management for patients receiving novel T-cell engaging therapies. Curr Opin Pediatr 26(1):43

Baruah V, Bose S (2020) Immunoinformatics-aided identification of T cell and B cell epitopes in the surface glycoprotein of 2019-nCoV. J Med Virol 92(5):495–500

Bayry J, Lacroix-Desmazes S, Kazatchkine MD, Kaveri SV (2004) Intravenous immunoglobulin for infectious diseases: back to the pre-antibiotic and passive prophylaxis era? Trends Pharmacol Sci 25(6):306–310

Bayry J, Lacroix-Desmazes S, Kazatchkine MD, Kaveri SV (2007) Monoclonal antibody and intravenous immunoglobulin therapy for rheumatic diseases: rationale and mechanisms of action. Nat Clin Pract Rheumatol 3(5):262–272

Beigel JH, Nam HH, Adams PL, Krafft A, Ince WL, El-Kamary SS, Sims AC (2019) Advances in respiratory virus therapeutics–a meeting report from the 6th isirv antiviral group conference. Antivir Res

Bell TJ, Brand OJ, Morgan DJ, Salek-Ardakani S, Jagger C, Fujimori T, Cholewa L, Tilakaratna V, Östling J, Thomas M (2019) Defective lung function following influenza virus is due to prolonged, reversible hyaluronan synthesis. Matrix Biol 80:14–28

Belouzard S, Millet JK, Licitra BN, Whittaker GR (2012) Mechanisms of coronavirus cell entry mediated by the viral spike protein. Viruses 4(6):1011–1033

Bloch EM, Shoham S, Casadevall A, Sachais BS, Shaz B, Winters JL, van Buskirk C, Grossman BJ, Joyner M, Henderson JP (2020) Deployment of convalescent plasma for the prevention and treatment of COVID-19. J Clin Invest

Caballero S, Nieto S, Gajardo R, Jorquera JI (2010) Viral safety characteristics of Flebogamma® DIF, a new pasteurized, solvent-detergent treated and Planova 20 nm nanofiltered intravenous immunoglobulin. Biologicals 38(4):486–493

Cameron MJ, Bermejo-Martin JF, Danesh A, Muller MP, Kelvin DJ (2008) Human immunopathogenesis of severe acute respiratory syndrome (SARS). Virus Res 133(1):13–19

Cao W, Liu X, Bai T, Fan H, Hong K, Song H, Han Y, Lin L, Ruan L, Li T (2020) High-dose intravenous immunoglobulin as a therapeutic option for deteriorating patients with coronavirus disease 2019. In: Open forum infectious diseases. vol 3. Oxford University Press US, p ofaa102

Carbajo-Lozoya J, Müller MA, Kallies S, Thiel V, Drosten C, von Brunn A (2012) Replication of human coronaviruses SARS-CoV, HCoV-NL63 and HCoV-229E is inhibited by the drug FK506. Virus Res 165(1):112–117

Carbajo-Lozoya J, Ma-Lauer Y, Malešević M, Theuerkorn M, Kahlert V, Prell E, von Brunn B, Muth D, Baumert TF, Drosten C (2014) Human coronavirus NL63 replication is cyclophilin A-dependent and inhibited by non-immunosuppressive cyclosporine A-derivatives including Alisporivir. Virus Res 184:44–53

Casadevall A, Pirofski L-a (2020) The convalescent sera option for containing COVID-19. J Clin Invest 130(4):1545–1548

Casadevall A, Dadachova E, Pirofski L-a (2004) Passive antibody therapy for infectious diseases. Nat Rev Microbiol 2(9):695–703

Cascella M, Rajnik M, Cuomo A, Dulebohn SC, Di Napoli R (2020) Features, evaluation and treatment coronavirus (COVID-19). In: Statpearls [internet]. StatPearls Publishing

Chan JF-W, Yao Y, Yeung M-L, Deng W, Bao L, Jia L, Li F, Xiao C, Gao H, Yu P (2015) Treatment with lopinavir/ritonavir or interferon-β1b improves outcome of MERS-CoV infection in a nonhuman primate model of common marmoset. J Infect Dis 212(12):1904–1913

Channappanavar R, Perlman S (2017) Pathogenic human coronavirus infections: causes and consequences of cytokine storm and immunopathology. In: Seminars in immunopathology, vol 5. Springer, pp 529–539

Chen J, Hu C, Chen L, Tang L, Zhu Y, Xu X, Chen L, Gao H, Lu X, Yu L (2020a) Clinical study of mesenchymal stem cell treating acute respiratory distress syndrome induced by epidemic Influenza A (H7N9) infection, a hint for COVID-19 treatment. Engineering

Chen L, Xiong J, Bao L, Shi Y (2020b) Convalescent plasma as a potential therapy for COVID-19. Lancet Infect Dis 20(4):398–400

Chen N, Zhou M, Dong X, Qu J, Gong F, Han Y, Qiu Y, Wang J, Liu Y, Wei Y (2020c) Epidemiological and clinical characteristics of 99 cases of 2019 novel coronavirus pneumonia in Wuhan, China: a descriptive study. Lancet 395(10223):507–513

Chen X, Li R, Pan Z, Qian C, Yang Y, You R, Zhao J, Liu P, Gao L, Li Z (2020d) Human monoclonal antibodies block the binding of SARS-CoV-2 spike protein to angiotensin converting enzyme 2 receptor. Cell Mol Immunol:1–3

Cheng Y, Wong R, Soo Y, Wong W, Lee C, Ng M, Chan P, Wong K, Leung C, Cheng G (2005) Use of convalescent plasma therapy in SARS patients in Hong Kong. Eur J Clin Microbiol Infect Dis 24(1):44–46

ClinicalTrials.gov (2020) The efficacy and safety of thalidomide in the adjuvant treatment of moderate new coronavirus (Covid-19) pneumonia. https://clinicaltrials.gov/ct2/show/NCT04273529?term=NCT04273529&draw=2&rank=1

Collum SD, Chen NY, Hernandez AM, Hanmandlu A, Sweeney H, Mertens TC, Weng T, Luo F, Molina JG, Davies J (2017) Inhibition of hyaluronan synthesis attenuates pulmonary hypertension associated with lung fibrosis. Br J Pharmacol 174(19):3284–3301 de Haan CA, Kuo L, Masters PS, Vennema H, Rottier PJ (1998) Coronavirus particle assembly: primary structure requirements of the membrane protein. J Virol 72(8):6838–6850

De Ranieri D, Fenny NS (2017) Intravenous immunoglobulin in the treatment of primary immunodeficiency diseases. Pediatr Ann 46(1):e8–e12

Dhama K, Sharun K, Tiwari R, Dadar M, Malik YS, Singh KP, Chaicumpa W (2020) COVID-19, an emerging coronavirus infection: advances and prospects in designing and developing vaccines, immunotherapeutics, and therapeutics. Hum Vaccin Immunother:1–7

Diao B, Wang C, Tan Y, Chen X, Liu Y, Ning L, Chen L, Li M, Liu Y, Wang G (2020) Reduction and functional exhaustion of T cells in patients with coronavirus disease 2019 (COVID-19). Front Immunol 11:827

Duan K, Liu B, Li C, Zhang H, Yu T, Qu J, Zhou M, Chen L, Meng S, Hu Y (2020) The feasibility of convalescent plasma therapy in severe COVID-19 patients: a pilot study. medRxiv

Elahi R, Khosh E, Tahmasebi S, Esmaeilzadeh A (2018) Immune cell hacking: challenges and clinical approaches to create smarter generations of chimeric antigen receptor T cells. Front Immunol 9:1717

Esmaeilzadeh A, Tahmasebi S, Athari SS (2020) Chimeric antigen receptor -T cell therapy: applications and challenges in treatment of allergy and asthma. Biomed Pharmacother 123:109685. https://doi.org/10.1016/j.biopha.2019.109685

Fu B, Xu X, Wei H (2020) Why tocilizumab could be an effective treatment for severe COVID-19? J Transl Med 18(1):1–5

Galeotti C, Kaveri SV, Bayry J (2017) IVIG-mediated effector functions in autoimmune and inflammatory diseases. Int Immunol 29(11):491–498

Golchin A, Farahany TZ (2019) Biological products: cellular therapy and FDA approved products. Stem Cell Rev Rep 15(2):166–175

Golchin A, Farahany TZ, Khojasteh A, Soleimanifar F, Ardeshirylajimi A (2019) The clinical trials of mesenchymal stem cell therapy in skin diseases: an update and concise review. Curr Stem Cell Res Ther 14(1):22–33

Group TPIW, Team M-NPIS (2016) A randomized, controlled trial of ZMapp for Ebola virus infection. N Engl J Med 375(15):1448

Gunn BM, Yu W-H, Karim MM, Brannan JM, Herbert AS, Wec AZ, Halfmann PJ, Fusco ML, Schendel SL, Gangavarapu K (2018) A role for Fc function in therapeutic monoclonal antibody-mediated protection against Ebola virus. Cell Host Microbe 24(2):221–233. e225

Guo X, Guo Z, Duan C, Wang G, Lu Y, Li M, Lu J (2020) Long-term persistence of IgG antibodies in SARS-CoV infected healthcare workers. medRxiv

Gupta P, Kamath AV, Park S, Chiu H, Lutman J, Maia M, Tan M-W, Xu M, Swem L, Deng R (2016) Preclinical pharmacokinetics of MHAA4549A, a human monoclonal antibody to influenza A virus, and the prediction of its efficacious clinical dose for the treatment of patients hospitalized with influenza A. In: MAbs, vol 5. Taylor & Francis, pp 991–997

He J, Tao H, Yan Y, Huang S-Y, Xiao Y (2020) Molecular mechanism of evolution and human infection with the novel coronavirus (2019-nCoV). bioRxiv

Hoffmann M, Kleine-Weber H, Schroeder S, Krüger N, Herrler T, Erichsen S, Schiergens TS, Herrler G, Wu N-H, Nitsche A (2020) SARS-CoV-2 cell entry depends on ACE2 and TMPRSS2 and is blocked by a clinically proven protease inhibitor. Cell

Hu Y, Li W, Gao T, Cui Y, Jin Y, Li P, Ma Q, Liu X, Cao C (2017) SARS coronavirus nucleocapsid inhibits type I interferon production by interfering with TRIM25-mediated RIG-I ubiquitination. J Virol:02143–02116

Huang C, Wang Y, Li X, Ren L, Zhao J, Hu Y, Zhang L, Fan G, Xu J, Gu X (2020) Clinical features of patients infected with 2019 novel coronavirus in Wuhan, China. Lancet 395(10223):497–506

Hung IF, To KK, Lee C-K, Lee K-L, Chan K, Yan W-W, Liu R, Watt C-L, Chan W-M, Lai K-Y (2011) Convalescent plasma treatment reduced mortality in patients with severe pandemic influenza A (H1N1) 2009 virus infection. Clin Infect Dis 52(4):447–456

Jawhara S (2020) Could intravenous immunoglobulin collected from recovered coronavirus patients protect against COVID-19 and strengthen the immune system of new patients? Int J Mol Sci 21(7):2272

Jean S-S, Lee P-I, Hsueh P-R (2020) Treatment options for COVID-19: the reality and challenges. J Microbiol Immunol Infect

Jose RJ, Manuel A (2020) COVID-19 cytokine storm: the interplay between inflammation and coagulation. Lancet Respir Med

Khoury M, Cuenca J, Cruz FF, Figueroa FE, Rocco PR, Weiss DJ (2020) Current status of cell-based therapies for respiratory virus infections: applicability to COVID-19. Eur Respir J

Kleinnijenhuis J, Quintin J, Preijers F, Benn CS, Joosten LA, Jacobs C, Van Loenhout J, Xavier RJ, Aaby P, Van Der Meer JW (2014) Long-lasting effects of BCG vaccination on both heterologous Th1/Th17 responses and innate trained immunity. J Innate Immun 6(2):152–158

Kosugi T, Maeda K, Sato W, Maruyama S, Kadomatsu K (2015) CD147 (EMMPRIN/Basigin) in kidney diseases: from an inflammation and immune system viewpoint. Nephrol Dial Transplant 30(7):1097–1103

Kruse RL (2020) Therapeutic strategies in an outbreak scenario to treat the novel coronavirus originating in Wuhan, China. F1000Research 9

Kui L, Fang Y-Y, Deng Y, Liu W, Wang M-F, Ma J-P, Xiao W, Wang Y-N, Zhong M-H, Li C-H (2020) Clinical characteristics of novel coronavirus cases in tertiary hospitals in Hubei Province. Chin Med J

Kumar GV, Jeyanthi V, Ramakrishnan S (2020) A short review on antibody therapy for COVID-19. New Microbes New Infect:100682

Law PK (2020) Emergent serum therapy and antibody medicine to counteract sudden attacks of COVID-19 and other pathogenic epidemics. Scientific Research Publishing

Leng Z, Zhu R, Hou W, Feng Y, Yang Y, Han Q, Shan G, Meng F, Du D, Wang S (2020) Transplantation of ACE2-mesenchymal stem cells improves the outcome of patients with COVID-19 pneumonia. Aging Dis 11(2):216–228

Li Z, Yi Y, Luo X, Xiong N, Liu Y, Li S, Sun R, Wang Y, Hu B, Chen W, Zhang Y, Wang J, Huang B, Lin Y, Yang J, Cai W, Wang X (2020) Development and clinical application of a rapid IgM-IgG combined antibody test for SARS-CoV-2 infection diagnosis. https://doi.org/10.1002/jmv.25727

Liang B, Chen J, Li T, Wu H, Yang W, Li Y, Li J, Yu C, Nie F, Ma Z (2020) Clinical remission of a critically ill COVID-19 patient treated by human umbilical cord mesenchymal stem cells. ChinaXiv 2:v1

Lin L, Lu L, Cao W, Li T (2020) Hypothesis for potential pathogenesis of SARS-CoV-2 infection–a review of immune changes in patients with viral pneumonia. Emerg Microbes Infect 9(1):727–732

Liu C, Zhou Q, Li Y, Garner LV, Watkins SP, Carter LJ, Smoot J, Gregg AC, Daniels AD, Jervey S (2020a) Research and development on therapeutic agents and vaccines for COVID-19 and related human coronavirus diseases. ACS Publications

Liu L, Liu W, Wang S, Zheng S (2020b) A preliminary study on serological assay for severe acute respiratory syndrome coronavirus 2 (SARS-CoV-2) in 238 admitted hospital patients. medRxiv:2020.2003.2006.20031856. doi:https://doi.org/10.1101/2020.03.06.20031856

Liu T, Zhang J, Yang Y, Zhang L, Ma H, Li Z, Zhang J, Cheng J, Zhang X, Wu G (2020c) The potential role of IL-6 in monitoring coronavirus disease 2019. Available at SSRN 3548761

Liu Y, Du X, Chen J, Jin Y, Peng L, Wang HH, Luo M, Chen L, Zhao Y (2020d) Neutrophil-to-lymphocyte ratio as an independent risk factor for mortality in hospitalized patients with COVID-19. J Inf Secur

Lukassen S, Chua RL, Trefzer T, Kahn NC, Schneider MA, Muley T, Winter H, Meister M, Veith C, Boots AW (2020) SARS-CoV-2 receptor ACE2 and TMPRSS2 are primarily expressed in bronchial transient secretory cells. EMBO J

Luke TC, Kilbane EM, Jackson JL, Hoffman SL (2006) Meta-analysis: convalescent blood products for Spanish influenza pneumonia: a future H5N1 treatment? Ann Intern Med 145(8):599–609

Lurie N, Saville M, Hatchett R, Halton J (2020) Developing Covid-19 vaccines at pandemic speed. N Engl J Med

Maoujoud O, Asserraji M, Belarbi M, Zemraoui N (2020) Coronavirus disease 19 pandemic and extracorporeal therapies in intensive care:"Bis Peccare in Bello Non Licet". Blood Purif:1–2

Maschalidi S, Sepulveda FE, Garrigue A, Fischer A, de Saint BG (2016) Therapeutic effect of JAK1/2 blockade on the manifestations of hemophagocytic lymphohistiocytosis in mice. Blood J Am Soc Hematol 128(1):60–71

Mason RJ (2020) Pathogenesis of COVID-19 from a cell biology perspective. Eur Respiratory Soc

Medicine NLo (2020) The efficacy and safety of thalidomide combined with low-dose hormones in the treatment of severe Covid-19. https://clinicaltrials.gov/ct2/show/NCT04273581?term=NCT04273581&draw=2&ran50

Miller A, Reandelar MJ, Fasciglione K, Roumenova V, Li Y, Otazu GH (2020) Correlation between universal BCG vaccination policy and reduced morbidity and mortality for COVID-19: an epidemiological study. MedRxiv

Min C-K, Cheon S, Ha N-Y, Sohn KM, Kim Y, Aigerim A, Shin HM, Choi J-Y, Inn K-S, Kim J-H (2016) Comparative and kinetic analysis of viral shedding and immunological responses in MERS patients representing a broad spectrum of disease severity. Sci Rep 6(1):1–12

Moore MJ, Dorfman T, Li W, Wong SK, Li Y, Kuhn JH, Coderre J, Vasilieva N, Han Z, Greenough TC (2004) Retroviruses pseudotyped with the severe acute respiratory syndrome coronavirus spike protein efficiently infect cells expressing angiotensin-converting enzyme 2. J Virol 78(19):10628–10635

Moorlag S, Arts RJ, van Crevel R, Netea MG (2019) Non-specific effects of BCG vaccine on viral infections. Clin Microbiol Infect 25(12):1473–1478

Morgenstern B, Michaelis M, Baer PC, Doerr HW, Cinatl J Jr (2005) Ribavirin and interferon-β synergistically inhibit SARS-associated coronavirus replication in animal and human cell lines. Biochem Biophys Res Commun 326(4):905–908

Mosaddeghi P, Negahdaripour M, Dehghani Z, Farahmandnejad M, Moghadami M, Nezafat N, Masoompour SM (2020) Therapeutic approaches for COVID-19 based on the dynamics of interferon-mediated immune responses

Mousavizadeh L, Ghasemi S (2020) Genotype and phenotype of COVID-19: their roles in pathogenesis. J Microbiol Immunol Infect

Napolitano M, Fabbrocini G, Patruno C (2020) Potential role of Janus kinase inhibitors in COVID-19. J Am Acad Dermatol

Netea MG, Joosten LA, Latz E, Mills KH, Natoli G, Stunnenberg HG, O'Neill LA, Xavier RJ (2016) Trained immunity: a program of innate immune memory in health and disease. Science 352 (6284):aaf1098

Okba NM, Muller MA, Li W, Wang C, GeurtsvanKessel CH, Corman VM, Lamers MM, Sikkema RS, de Bruin E, Chandler FD (2020) SARS-CoV-2 specific antibody responses in COVID-19 patients. medRxiv

Peterson D, Damsky W, King B (2020) The use of Janus kinase inhibitors in the time of severe acute respiratory syndrome coronavirus 2 (SARS-CoV-2). J Am Acad Dermatol 82(6):e223–e226

Prabakaran P, Zhu Z, Xiao X, Biragyn A, Dimitrov AS, Broder CC, Dimitrov DS (2009) Potent human monoclonal antibodies against SARS CoV, Nipah and Hendra viruses. Expert Opin Biol Ther 9(3):355–368

Prompetchara E, Ketloy C, Palaga T (2020) Immune responses in COVID-19 and potential vaccines: lessons learned from SARS and MERS epidemic. Asian Pac J Allergy Immunol 38(1):1–9

Qi J, Ding C, Jiang X, Gao Y (2020) Advances in developing CAR T-cell therapy for HIV cure. Front Immunol 11:361

Rabi FA, Al Zoubi MS, Kasasbeh GA, Salameh DM, Al-Nasser AD (2020) SARS-CoV-2 and coronavirus disease 2019: what we know so far. Pathogens 9(3):231

Rajaei S, Dabbagh A (2020) The immunologic basis of COVID-19: a clinical approach. J Cell Molecul Anestha 5(1):37–42

Ramaiah A, Arumugaswami V (2020) Insights into cross-species evolution of novel human coronavirus 2019-nCoV and defining immune determinants for vaccine development. bioRxiv

Rangaswami J, Bhalla V, Blair JE, Chang TI, Costa S, Lentine KL, Lerma EV, Mezue K, Molitch M, Mullens W (2019) Cardiorenal syndrome: classification, pathophysiology, diagnosis, and treatment strategies: a scientific statement from the American Heart Association. Circulation 139(16):e840–e878

Richardson P, Griffin I, Tucker C, Smith D, Oechsle O, Phelan A, Stebbing J (2020) Baricitinib as potential treatment for 2019-nCoV acute respiratory disease. Lancet (London, England) 395 (10223):e30

Ronco C, Navalesi P, Vincent JL (2020) Coronavirus epidemic: preparing for extracorporeal organ support in intensive care. Lancet Respir Med 8(3):240–241

Russell B, Moss C, George G, Santaolalla A, Cope A, Papa S, Van Hemelrijck M (2020a) Associations between immune-suppressive and stimulating drugs and novel COVID-19 – a systematic review of current evidence. Ecancermedicalscience 14

Russell B, Moss C, Rigg A, Van Hemelrijck M (2020b) COVID-19 and treatment with NSAIDs and corticosteroids: should we be limiting their use in the clinical setting? Ecancermedicalscience 14

Sahr F, Ansumana R, Massaquoi T, Idriss B, Sesay F, Lamin J, Baker S, Nicol S, Conton B, Johnson W (2017) Evaluation of convalescent whole blood for treating Ebola Virus Disease in Freetown, Sierra Leone. J Inf Secur 74(3):302–309

Schneider WM, Chevillotte MD, Rice CM (2014) Interferon-stimulated genes: a complex web of host defenses. Annu Rev Immunol 32:513–545

Sheahan TP, Sims AC, Graham RL, Menachery VD, Gralinski LE, Case JB, Leist SR, Pyrc K, Feng JY, Trantcheva I (2017) Broad-spectrum antiviral GS-5734 inhibits both epidemic and zoonotic coronaviruses. Sci Transl Med 9(396)

Shen C, Wang Z, Zhao F, Yang Y, Li J, Yuan J, Wang F, Li D, Yang M, Xing L (2020) Treatment of 5 critically ill patients with COVID-19 with convalescent plasma. JAMA 323(16):1582–1589

Shi Y, Tan M, Chen X, Liu Y, Huang J, Ou J, Deng X (2020) Immunopathological characteristics of coronavirus disease 2019 cases in Guangzhou, China. medRxiv

Sui J, Li W, Roberts A, Matthews LJ, Murakami A, Vogel L, Wong SK, Subbarao K, Farzan M, Marasco WA (2005) Evaluation of human monoclonal antibody 80R for immunoprophylaxis of severe acute respiratory syndrome by an animal study, epitope mapping, and analysis of spike variants. J Virol 79(10):5900–5906

Tahmasebi S, Elahi R, Esmaeilzadeh A (2019) Solid tumors challenges and new insights of car t cell engineering. Stem Cell Rev Rep 15(5):619–636

Tahmasebi S, Khosh E, Esmaeilzadeh A (2020) The outlook for diagnostic purposes of the 2019-novel coronavirus disease. J Cell Physiol 235:9211–9229. https://doi.org/10.1002/jcp.29804

Tay MZ, Poh CM, Rénia L, MacAry PA, Ng LF (2020) The trinity of COVID-19: immunity, inflammation and intervention. Nat Rev Immunol:1–12

Tian S, Hu W, Niu L, Liu H, Xu H, Xiao S-Y (2020a) Pulmonary pathology of early phase 2019 novel coronavirus (COVID-19) pneumonia in two patients with lung cancer. J Thorac Oncol

Tian X, Li C, Huang A, Xia S, Lu S, Shi Z, Lu L, Jiang S, Yang Z, Wu Y (2020b) Potent binding of 2019 novel coronavirus spike protein by a SARS coronavirus-specific human monoclonal antibody. Emerg Microbes Infect 9(1):382–385

Turner RB, Felton A, Kosak K, Kelsey DK, Meschievitz CK (1986) Prevention of experimental coronavirus colds with intranasal α-2b interferon. J Infect Dis 154(3):443–447

Uccelli A, de Rosbo NK (2015) The immunomodulatory function of mesenchymal stem cells: mode of action and pathways. Ann N Y Acad Sci 1351(1):114–126

Van Erp EA, Luytjes W, Ferwerda G, Van Kasteren PB (2019) Fc-mediated antibody effector functions during respiratory syncytial virus infection and disease. Front Immunol 10

Walls AC, Park Y-J, Tortorici MA, Wall A, McGuire AT, Veesler D (2020) Structure, function, and antigenicity of the SARS-CoV-2 spike glycoprotein. Cell

Wambier CG, Goren A (2020) SARS-COV-2 infection is likely to be androgen mediated. J Am Acad Dermatol

Wang J-T, Sheng W-H, Fang C-T, Chen Y-C, Wang J-L, Yu C-J, Chang S-C, Yang P-C (2004) Clinical manifestations, laboratory findings, and treatment outcomes of SARS patients. Emerg Infect Dis 10(5):818

Wang Y, Chen X, Cao W, Shi Y (2014) Plasticity of mesenchymal stem cells in immunomodulation: pathological and therapeutic implications. Nat Immunol 15(11):1009

Wang B, Li R, Lu Z, Huang Y (2020a) Does comorbidity increase the risk of patients with COVID-19: evidence from meta-analysis. Aging (Albany NY) 12(7):6049

Wang C, Li W, Drabek D, Okba NM, van Haperen R, Osterhaus AD, van Kuppeveld FJ, Haagmans BL, Grosveld F, Bosch B-J (2020b) A human monoclonal antibody blocking SARS-CoV-2 infection. Nat Commun 11(1):1–6

Wang D, Hu B, Hu C, Zhu F, Liu X, Zhang J, Wang B, Xiang H, Cheng Z, Xiong Y (2020c) Clinical characteristics of 138 hospitalized patients with 2019 novel coronavirus–infected pneumonia in Wuhan, China. Jama

Wang W, He J, Wu S (2020d) The definition and risks of cytokine release syndrome-like in 11 COVID-19-infected pneumonia critically ill patients: disease characteristics and retrospective analysis. Medrxiv

Woo PC, Huang Y, Lau SK, Yuen K-Y (2010) Coronavirus genomics and bioinformatics analysis. Viruses 2 (8):1804–1820

Xiao S-Y, Wu Y, Liu H (2020) Evolving status of the 2019 novel coronavirus infection: proposal of conventional serologic assays for disease diagnosis and infection monitoring. J Med Virol 92(5):464–467. https://doi.org/10.1002/jmv.25702

Xie Y, Cao S, Li Q, Chen E, Dong H, Zhang W, Yang L, Fu S, Wang R (2020) Effect of regular intravenous immunoglobulin therapy on prognosis of severe pneumonia in patients with COVID-19. J Inf Secur

xinhuanet (2020) China puts 245 COVID-19 patients on convalescent plasma therapy. http://www.xinhuanet.com/english/2020-02/28/c_138828177.htm

Xu Z, Shi L, Wang Y, Zhang J, Huang L, Zhang C, Liu S, Zhao P, Liu H, Zhu L (2020) Pathological findings of COVID-19 associated with acute respiratory distress syndrome. Lancet Respir Med 8(4):420–422

Yamey G, Schäferhoff M, Hatchett R, Pate M, Zhao F, McDade KK (2020) Ensuring global access to COVID-19 vaccines. Lancet 395(10234):1405–1406

Yu F, Du L, Ojcius DM, Pan C, Jiang S (2020) Measures for diagnosing and treating infections by a novel coronavirus responsible for a pneumonia outbreak originating in Wuhan, China. Microbes Infect

Zheng H-Y, Zhang M, Yang C-X, Zhang N, Wang X-C, Yang X-P, Dong X-Q, Zheng Y-T (2020a) Elevated exhaustion levels and reduced functional diversity of T cells in peripheral blood may predict severe progression in COVID-19 patients. Cell Mol Immunol:1–3

Zheng M, Gao Y, Wang G, Song G, Liu S, Sun D, Xu Y, Tian Z (2020b) Functional exhaustion of antiviral lymphocytes in COVID-19 patients. Cell Mol Immunol:1–3

Zhou P, Yang X-L, Wang X-G, Hu B, Zhang L, Zhang W, Si H-R, Zhu Y, Li B, Huang C-L (2020a) A pneumonia outbreak associated with a new coronavirus of probable bat origin. Nature 579 (7798):270–273

Zhou Y, Fu B, Zheng X, Wang D, Zhao C, Qi Y, Sun R, Tian Z, Xu X, Wei H (2020b) Aberrant pathogenic GM-CSF+ T cells and inflammatory CD14+ CD16+ monocytes in severe pulmonary syndrome patients of a new coronavirus. BioRxiv

Ventilatory Support in Patients with COVID-19

27

Paolo Maria Leone, Matteo Siciliano, Jacopo Simonetti, Angelena Lopez, Tanzira Zaman, Francesco Varone, and Luca Richeldi

Abstract

Severe acute respiratory syndrome coronavirus 2 (SARS-CoV-2) caused the novel coronavirus disease 2019 (COVID-19) pandemic, which spread throughout the world. Acute hypoxemic respiratory failure is the most dangerous complication of COVID-19 pneumonia. To date, no specific therapeutic drugs or vaccines have been proven efficacious. Ventilatory support is still a significant challenge for physicians facing COVID-19. The mechanisms underlying hypoxemia in those patients are not fully understood, but a new physiopathology model has been proposed. Oxygen therapy should be delivered to patients with mild to moderate hypoxemia. More severe patients could benefit from other treatments (high-flow nasal cannula, noninvasive ventilation or intubation, and invasive ventilation). Given the rapid evolution of COVID-19, there has been a paucity of the high-quality data that typically inform clinical practice guidelines from professional societies, and a worldwide consensus is still lacking. This chapter aims to illustrate the potentials of ventilatory support as therapeutic options for adult and pediatric patients affected by COVID-19 pneumonia.

Keywords

Acute respiratory failure · ARDS · COVID-19 · Guidelines · HFNC · Mechanical ventilation · Oxygen therapy · Pediatric

27.1 Introduction

Since the end of 2019, a novel coronavirus disease (COVID-19) has spread from China all over the world (Zhu et al. 2020). Currently, about ten million people have been affected by this disease, of which more than 470,000 dead worldwide. COVID-19 mainly involves the lungs and can cause a severe acute respiratory infection, so the virus causing the disease is referred to as severe acute respiratory syndrome coronavirus 2 (SARS-CoV-2) (Sanders et al. 2020). Here we review the potentials of ventilatory support as therapeutic options for adult and pediatric patients affected by COVID-19 pneumonia.

P. M. Leone (✉) · M. Siciliano · J. Simonetti · F. Varone · L. Richeldi
Fondazione Policlinico A. Gemelli IRCCS, Rome, Italy

Università Cattolica del Sacro Cuore, Rome, Italy
e-mail: paolomaria.leone@policlinicogemelli.it

A. Lopez · T. Zaman
Department of Medicine, Cedars-Sinai Medical Center, Los Angeles, CA, USA

N. Rezaei (ed.), *Coronavirus Disease - COVID-19*, Advances in Experimental Medicine and Biology 1318, https://doi.org/10.1007/978-3-030-63761-3_27

27.2 Respiratory Failure and Oxygen Therapy

Acute hypoxemic respiratory failure is the most dangerous complication of SARS-CoV-2 pneumonia. To date, the true prevalence of hypoxemic respiratory failure in COVID-19 patients is not evident yet. One of the first papers illustrating the clinical characteristics of 72,314 Cases from China reports a prevalence of 14% (Wu and McGoogan 2020). Another paper reports that 26.1% of patients required intensive care unit (ICU) admission, of whom 61.1% related to acute respiratory distress syndrome (ARDS) (Wang et al. 2020a). A more recent report from China shows that 41.3% of hospitalized patients required oxygen therapy, and 6.1% required mechanical ventilation (Guan et al. 2020).

The mechanisms underlying hypoxemia in those patients are not fully understood. Pulmonary edema, loss of hypoxemic vasoconstriction reflex, microvascular and macrovascular thrombosis, alveolar filling, hemoglobinopathies, and a mismatch between ventilation and perfusion are all suspected (Gattinoni et al. 2020b; Tian et al. 2020; Gattinoni et al. 2020a; Alhazzani et al. 2020). A recent guidance paper suggests starting supplemental oxygen if the peripheral oxygen saturation (SpO_2) is <92% and recommends starting supplemental oxygen if SpO2 is <90% (Alhazzani et al. 2020). Since there are no trials on the use of oxygen therapy in COVID-19 patients, indirect evidence should come from a dataset of acutely ill patients. A recent analysis of large ICU datasets had shown the lowest mortality when SpO_2 was maintained between 94 and 98% (van den Boom et al. 2020). A recent open-label randomized trial investigating liberal versus conservative oxygen therapy in ARDS patients (LOCO2 trial) was stopped for the potential increased risk of serious adverse events and futility (Barrot et al. 2020). However, the final analysis showed a significant difference between groups, with a lower probability of survival in the conservative oxygen group.

Oxygen therapy should be delivered through nasal cannulas or Venturi masks for patients with mild to moderate hypoxemia (McDonald et al. 2016). Unstable patients may be managed through other procedures (high-flow nasal cannula and noninvasive ventilation or intubation). The protection of healthcare providers to the patients is a fundamental issue. Exhaled air dispersion during oxygen delivery via nasal cannulas ranges from 66 cm when oxygen flow is 1 L/min to 1 m when the flow is 3–5 L/min. The maximum exhaled air dispersion distance is 40 cm when using a Venturi mask, compared to less than 10 cm when using a non-rebreathing mask (Ferioli et al. 2020).

27.3 High-Flow Nasal Cannula Oxygen

High-flow nasal cannula (HFNC) oxygen therapy is a new oxygen delivery system widely adopted in patients suffering from respiratory failure, including chronic obstructive pulmonary disease (COPD), pulmonary edema, and acute respiratory failure (ARF). It consists of a warmed and humidified (relative humidity: 100%) gas flow delivered via nasal cannula characterized by a flow rate up to 70 L/min and a fraction of inspired oxygen (FiO_2) up to 100%.

HFNC oxygen therapy has several potential advantages: accurate FiO_2 delivery control, washout of upper airways dead space, provision of a small degree of positive airway pressure, improvement of mucociliary motion and sputum clearance, matching patient's inspiratory flow rate with consequently better comfort and compliance, and reducing respiratory work and dyspnea's severity. All these characteristics improve gas exchange via an increment of the partial arterial pressure of oxygen PaO_2/FiO_2 ratio (Nishimura 2016).

Recent studies have dealt with evidence of aerosolization and the risk of infection with HFNC oxygen therapy. Loh and coworkers experimented and simulated the cough of a patient while using HFNC in order to assess the dispersion of droplets in healthy volunteers by assuming a mean distance of 2.48 (SD 1.03) m at baseline and 2.91 (SD 1.09) m with HFNC. The cough of patients using HFNC might reach a maximum distance of 4.50 m (Loh et al. 2020). Hui and coworkers demonstrated a linear incremental in droplet spread with increasing flow rates using a human patient simulator (Hui et al. 2019). Leung and coworkers showed no differences in surface contamination between the HFNC and facemask oxygen groups in patients with bacterial pneumonia (Leung et al. 2019).

Nowadays, there is still no firm evidence about the risk of using HFNC oxygen therapy in COVID-19 or yeast aerosolization of viruses. Nevertheless, its use seems to be feasible.

In March 2020, the World Health Organization (WHO) released an official document on the management of severe acute respiratory infection (SARI) when a SARS-CoV-2 infection is suspected. Regarding the use of HFNC, the WHO guidelines suggest that HFNC oxygen should only be used in selected patients with hypoxemic respiratory failure with airborne precautions and recommend the use of a negative pressure room whenever possible. Assuming that HFNC oxygen therapy involves aerosolization of the particles, it is necessary to understand whether or not this mechanism entails an increased risk of infection (WHO 2020a).

Various causes of lung injury can result in acute respiratory failure (ARF). Hypoxemic ARF is characterized by severe acute hypoxemia (PaO_2/FiO_2 ratio < 300) and causes a high respiratory drive reflected by clinical signs of respiratory distress. In patients with ARF of various origins, HFNC oxygen has been shown to result in better comfort and oxygenation than standard oxygen therapy. It can be considered as an alternative in hypoxemic ARF patients for whom noninvasive ventilation (NIV) indications remain controversial (Frat et al. 2017).

A randomized controlled trial was performed by Frat and coworkers to compare three strategies of oxygenation in 310 subjects: standard oxygen, HFNC, and NIV in patients with nonhypercapnic ARF. The primary outcome, the intubation rate within 28 days after randomization among the three groups, was not met; on the other hand, 90-day mortality was lower in patients treated by HFNC, probably due to a lower intubation rate in this group of patients than in the two other treatment groups (Frat et al. 2015).

Zhao and coworkers conducted a systematic review and meta-analysis to investigate the effect of HFNC oxygen on the rates of intubation, mechanical ventilation, the escalation of respiratory support, and mortality versus conventional oxygen therapy (COT) or NIV in adult patients with respiratory failure. Compared to COT, HFNC was associated with a significant reduction in the intubation rate, mechanical ventilation, and the escalation of respiratory support, but there was no difference in mortality. HFNC and NIV therapy did not differ for any outcomes of interest (Zhao et al. 2017).

Regarding COVID-19 pneumonia, to our knowledge, the experience of HFNC oxygen in this population is poor. Wang and coworkers performed a retrospective observational study; among 27 patients suffering from severe ARF, HFCN oxygen was used as first-line therapy in 17 patients with a success rate of 59% (10 out of 17). Of the 17 patients undergoing HFNC oxygen, 7 (41%) experienced failure. The HFNC oxygen failure rate was 0% (0/6) in patients with PaO2/ FiO2 > 200 mmHg vs 63% (7/11) in those with $PaO_2/FiO_2 \leq 200$ mmHg ($p = 0.04$). The study originally reported that HFNC was the most commonly used ventilation strategy for COVID-19 pneumonia. Patients with lower PaO_2/FiO_2 (<200 mmHg) were more likely to experience HFNC oxygen failure (Wang et al. 2020b). Studies of small series of patients showed an improvement in respiratory and overall clinical conditions after undergoing HFNC oxygen therapy, supporting this therapy as a good option for COVID-19 (Alhazzani et al. 2020; Cook et al. 2020; Geng et al. 2020).

Recently the self-pronation was proposed for patients undergoing treatment with HFNC oxygen, and it was beneficial and led to an improvement of oxygenation and clinical condition (The Critical Care Western Research et al. 2020). It also seemed to be a viable treatment for pregnant women affected by COVID-19 pneumonia (Vibert et al. 2020). Prone positioning (PP) avoided resorting to mechanical ventilation and intubation (The Critical Care Western Research et al. 2020; Vibert et al. 2020; Berlin et al. 2020). More studies in this field are necessary to assess the role of PP while HFNC oxygen therapy is provided.

Multiple guidelines have developed different recommendations on the use of HFNC oxygen in the respiratory management of patients with COVID-19. However, HFNC oxygen therapy seems to be a useful method associated with a significant reduction in the intubation rate, mechanical ventilation, and escalation of respiratory support in COVID-19 (Lyons and Callaghan 2020). HFNC oxygen benefits are well recognized, while evidence is lacking for incremental risk of infection due to aerosolization (Remy et al. 2020; Li et al. 2020). However, the use of appropriate personal protective equipment (PPE) by healthcare workers remains the cornerstone of aerosolization-related infection risk control. For patients who are deteriorating despite standard nasal oxygen or facemask oxygen, HFNC oxygen remains a valid option and should be considered in the active management of patients with COVID-19 pneumonia.

27.4 Noninvasive Ventilation

In the last months, it has been widely suggested that patients with respiratory failure who are either a known or suspected case of COVID-19 should be intubated and ventilated as early as possible in the disease course, giving no much room to less invasive treatments, including continuous positive airway pressure (CPAP) and NIV. However, adopting an early intubation strategy in subjects affected by respiratory distress with COVID-19 pneumonia or highly suspected for this viral condition may lead to a couple of tricky scenarios: subjects with negative swabs for SARS-CoV-2 may undergo unnecessary intubation and also subjects who would have improved on CPAP or NIV would directly undergo invasive mechanical ventilation (IMV). Also, unnecessary intubation and ventilation of one patient might deny what could be a lifesaving treatment for another patient in resource-limited settings. According to the WHO guidelines, healthcare providers are encouraged to consider CPAP and NIV for the management of respiratory failure in COVID-19, though conditioned to wearing appropriate personal protective equipment (PPE) (WHO 2020a).

The LUNG SAFE study, a large multicenter study, demonstrated there is no significant difference in the need for ICU admission and hospital mortality rates of patients with ARDS receiving NIV or IMV, when ARDS severity, demographic characteristics, and associated comorbidities of both treatment groups were matched. This study showed the rate of mortality and NIV failure in the NIV group correlated with the severity of the patient's respiratory failure. An earlier multicenter study by Antonelli and coworkers showed an early 1 h NIV trial in ARDS patients upon ICU admission could be helpful to clinically stratify them and avoid unnecessary intubation in more than half of the population (Antonelli et al. 2007).

The arterial oxygenation improvement experienced by subjects suffering hypoxemic ARF due to atelectasis, pneumonia, or pulmonary edema who undergo CPAP therapy leads to a reduction in both the risk of airway collapse and the work of breathing. The pathophysiologic explanations for this are the improvement of the functional residual capacity and the shift of the tidal volume to a portion of the pressure-volume curve characterized by higher compliance. Also, the application of positive end-expiratory pressure (PEEP) helps to recruit non-aerated alveoli, stabilizes the airways, and reduces the differences of lung volume distribution in the various segments (Navalesi and Maggiore 2013). Subjects with hypoxemic ARF and ARDS (from mild to moderate) seem to benefit from an initial approach with CPAP. NIV with a facial mask might fail because of the lacking of patient compliance and technical problems tied to the interface system (Chiumello et al. 2017).

The helmet had been proposed 20 years ago as a different option to conventional interfaces. The helmet gear may vary among manufacturers for some details, but its core consists of a hood, which is transparent and latex-free, a soft collar, which has different sizes, and a plastic or metal ring, which acts as a connector between the hood and the collar. Two straps, usually applied underarms, are attached to the ring, which keeps it from flying upward when the gas flow inflates it (Chiumello et al. 2003). The average volume of the hood, when containing the patient's head, is about 15 l. On one of the hood's sides, there is the entrance for high flow gas, which flows out on the opposite side of the hood through an expiratory port with an integrated manometer and an adjustable or fixed PEEP valve.

NIV could be a viable treatment for hypoxemic ARF patients, especially when PP is used. Intubated patients with severe ARDS could benefit from early and prolonged (at least 12 h per day) PP that might improve oxygenation and decrease mortality (Munshi et al. 2017). Since ICUs may be overloaded with COVID-19 patients, awake PP may be useful to improve oxygenation and prevent ICU transfers. A recent study (Elharrar et al. 2020) about COVID-19 patients with hypoxemic ARF managed outside the ICU showed that 63% of them were able to tolerate PP for more than 3 h. However, oxygenation increased during PP in only 25% and was not sustained in half of those after resupination. Previous studies about PP in spontaneously breathing non-intubated subjects showed similar findings (Scaravilli et al. 2015). A trial of PP may still be a viable mechanism to select patients who will benefit the most with this kind of ventilation (Berlin et al. 2020), since no sedation is used, and this could cause extreme discomfort to them.

The European Respiratory Society/American Thoracic Society (ERS/ATS) clinical practice guidelines recommend (although not firmly) NIV as a preventive strategy for avoiding intubation in hypoxemic ARF (Rochwerg et al. 2017). However, it has to be performed by experienced teams in highly selected cooperative patients with early-stage ARDS without any associated major organ dysfunction. In patients with de novo ARF undergoing NIV in assisted pressure control mode, large expiratory tidal volumes (VTE) may be generated, and then reliable monitoring of VTE and unintentional leaks is strongly recommended.

One of the most relevant and unfortunately neglected (at least at the beginning) problems is the protection of workers involved in high-risk interventions such as nebulizing therapy, HFNC oxygen therapy, COT, NIV, patient pronation, and chest physiotherapy. According to the available reports, 3.8% of Chinese healthcare workers were infected; 63% of cases occurred in Wuhan (Wu and McGoogan 2020). In Italy, the rate is even worse (14%) (Niederman et al. 2020). The fundamental defense to date is wearing adequate PPE, such as N95 masks, gowns, hair covers, gloves, and eye and face shields (Ferioli et al. 2020). The use of more efficient respirators such as powered air-purifying respirators for high-risk aerosol-generating procedures (like respiratory therapies) has been proposed but rarely used. Bench studies showed that the dispersion of exhaled air is different depending on the respiratory therapies and interfaces involved. NIV is considered to be an aerosol-generating procedure that may increase the risk of disease transmission (Alhazzani et al. 2020). If available, it should be used in a negative pressure room, and all professionals entering the room should utilize airborne precautions with appropriate PPE, as said before. In order to reduce COVID-19 transmission risk, both assuring an appropriate seal on the patient's mask and using a viral filter, before other devices, on the NIV mask play a pivotal role (Whittle et al. 2020). Cohorting patients with COVID-19 who require NIV is also helpful in reducing transmission to providers and other patients. Nevertheless, the risk of COVID-19 transmission could be reduced to a negligible extent through a helmet interface. Large size range droplets are produced when patients undergo NIV with a vented mask compared with the baseline droplet counts. This increase in large droplet count did not occur when using the NIV circuit modification with a non-vented mask and exhalation filter. On the other hand, oxygen therapy seems not to modify droplets' count and mass (Simonds et al. 2010).

In conclusion, neither outcomes nor the decision to treat COVID-19 patients with NIV is clear-cut. Data from each country look heterogenous, resulting in a lack of confidence when it comes to the time to choose the timing, the interface, and the modality of ventilation in these patients. Additional factors such as resource-limited settings and the risk of infection for professionals decrease the possibility of a successful treatment for these patients. Because of this, to recognize factors affecting outcomes is fundamental in order to provide the best ventilatory support in routine care to COVID-19 patients.

27.5 Invasive Ventilation

Severe COVID-19 patients require intensive care surveillance, and this is mostly due to the development of ARDS. ARDS represents the most common and most severe complication of this viral infection, often leading to death. Subjects admitted to the ICU receive ventilatory support, frequently, in the form of IMV. According to different patients' series, the rate of patients invasively ventilated in the ICUs ranges from 29% to 89.9% (Wang et al. 2020c; Richardson et al. 2020; Grasselli et al. 2020), and in the American series, the mortality rate of mechanically ventilated patients reached 88% (Richardson et al. 2020).

In the United Kingdom, two-thirds of COVID-19 patients admitted to ICUs required mechanical ventilation within 24 h (Mahase 2020). Therefore, it is pivotal not to delay the tracheal intubation and mechanical ventilation. Often intubation is seen as a salvage maneuver, but it should be seen as a proactive procedure, particularly in this condition (Wunsch 2020; Phua et al. 2020). Early treatment of these patients may result in a better outcome.

A more aggressive behavior, with early intubation, is concordant with the COVID-19 pneumonia model proposed by Gattinoni and coworkers in order to avoid self-inflicted lung injuries (Gattinoni et al. 2020a; Marini and Gattinoni 2020). Also, IMV is characterized by lower aerosolization and therefore is a relatively safer procedure for both patients and healthcare professionals (Sorbello et al. 2020). However, on the other hand, the staff will be subjected to an increased risk of being infected with the virus while performing the intubation because of the exposure to secretions with high viral load (Tran et al. 2012), and emergency intubations may be at even higher risk (WHO 2020b).

With little knowledge about the physiopathology of COVID-19 ARDS (CARDS), we initially used the well-accepted lung-protective ventilation approach for ARDS characterized by lower tidal volumes (6 mL/kg ideal body weight) and plateau airway pressure of less than 30 cmH_2O using PP (Matthay et al. 2020; Ziehr et al. 2020; Fan et al. 2017). PP seems to be even more beneficial in the case of CARDS than in the usual ARDS (Archer et al. 2020; Roesthuis et al. 2020). It should be maintained for at least 12 h a day, as suggested by international guidelines for ARDS ventilatory treatment (Fan et al. 2017). By redistributing pulmonary perfusion, PP leads to an improvement of the ventilation-perfusion mismatch (Möhlenkamp and Thiele 2020).

Endothelial damage sustained by micro- and macro-thromboses seems to play a pivotal role in COVID-19-related lung injury, leading to alterations of pulmonary vasoregulation and large intrapulmonary shunt. The primary cause of hypoxia in COVID-19 pneumonia seems to be represented by the ventilation-perfusion mismatch (Marini and Gattinoni 2020). This assumption allowed Gattinoni and coworkers to propose a model for COVID-19 pneumonia dividing patients in "type L" and "type H" (Gattinoni et al. 2020a). These two types of patients seem to be susceptible to different ventilatory treatments. The lung of "type L" patients is characterized by low elastance (high compliance), low weight, low PEEP response, and larger tidal volumes (7–8 mL/kg ideal body weight) that occur mainly in the early phases of the disease. The transition to "type H" leads to a lung whose features are more consistent with severe ARDS: high elastance (low compliance), higher weight, high PEEP response, and acceptance of lower tidal volumes. Therefore, "type H" patients will benefit from the usual ARDS approach. However, "type L" and "type H" are just the extremes of this model, and in between, there are several shadows of gray. When lung edema increases in a "type L" lung, its compliance lowers, and the "baby lung" begins to take shape. Many factors could influence this transition, including disease severity, host response, and suboptimal management. Particularly the aggravation of pulmonary edema will shrink the lung and lead to the "baby lung," which is peculiar to the ARDS of other causes.

Liu and coworkers also showed that hypercapnia has a higher prevalence in CARDS patients ventilated with low tidal volume. In their series, an intermediate tidal volume (7–8 mL/kg ideal body weight) successfully led to the resolution of hypercapnia, and the authors suggested this approach for less severe cases with higher compliance (Liu et al. 2020).

The data presented by Beloncle and coworkers are consistent with Gattinoni's phenotype (Beloncle et al. 2020). In CARDS patients PEEP should be titrated after a systematic recruitment-to-inflation (R/I) ratio evaluation. They identified poorly and highly recruitable patients who may benefit from different PEEP levels and ventilatory strategies in order to prevent ventilator-induced lung injury (VILI). They also observed the dynamism of the phenotypes, with patients moving from one to another over time.

On the other hand, Ziehr and coworkers endorsed the usual ARDS ventilation strategy. In their cohort of 66 patients in Boston, they reported physiopathology conditions consistent with ARDS of other causes, and their patients had a positive response to low tidal volume ventilation, conservative fluid therapy, and PP (Ziehr et al. 2020).

Tracheostomy should be considered as an option in those patients who receive IMV for more than 14 days and for those who have difficult weaning. However, there is no worldwide consensus about its timing in this context (Turri-Zanoni et al. 2020; Takhar et al. 2020).

Tailored lung-protective ventilation is mandatory in order to improve COVID-19 outcomes. The challenge of IMV in COVID-19 patients is mostly due to the heterogenous physiopathology of CARDS.

A big challenge in a time of pandemic is the high occupancy rate of the ICUs (Fagiuoli et al. 2020) and the shortage of equipment (i.e., mechanical ventilator). In order to minimize this issue in the areas most affected by COVID-19, the use of shared ventilators has been proposed, but this practice would not be exempt from any risk (Herrmann et al. 2020).

27.6 Actual Guidelines and Recommendation from Scientific Societies

Given the rapid evolution of COVID-19, there has been a paucity of the high-quality data that typically inform clinical practice guidelines from professional societies. To date, only the Society of Critical Care Medicine (SCCM) has published clinical practice guidelines on mechanical ventilation in patients with COVID-19 within the framework "the Surviving Sepsis Campaign Guidelines on the Management of Critically Ill Adults with Coronavirus Disease 2019" (Alhazzani et al. 2020). The guidelines were produced from a panel of 36 experts from 12 countries. The National Institutes of Health (NIH) also published Coronavirus Disease 2019 Treatment Guidelines (COVID-19 Treatment Guidelines Panel 2020). The NIH panel comprised 52 representatives from a total of 14 professional societies and federal agencies. In both guidelines, the respective panels reviewed the literature for direct and indirect evidence and provided recommendations graded by the strength of recommendation and the level of evidence (Table 27.1).

Both guidelines prefer low tidal volume ventilation (4–6 ml/kg ideal body weight) over high tidal volume ventilation (> 8 ml/kg ideal body weight). While there are no controlled studies of differential ventilator management strategies in patients with CARDS, these recommendations are extrapolated from clinical trials that evaluated mechanical ventilation strategy in patients with "usual" ARDS (Walkey et al. 2017; The Acute Respiratory Distress Syndrome 2000).

The SCCM guidelines recommend the use of a high PEEP over a low PEEP strategy based on a meta-analysis that demonstrated a trend toward improved survival in patients with ARDS with $PaO_2/FiO_2 < 200$ (Briel et al. 2010). Importantly, there is neither an agreement upon the definition

Table 27.1 Summary of clinical practice guidelines for mechanical ventilation in COVID-19

Recommendation	SCCM Surviving Sepsis Campaign	SCCM recommendation strength Rated quality of evidence[a]	NIH	NIH recommendation strength Rated quality of evidence[b]
Tidal volume	Low tidal volume (Vt) ventilation (Vt 4–8 mL/kg of ideal body weight)	Strong recommendation Moderate-quality evidence	Low tidal volume (Vt) ventilation (Vt 4–8 mL/kg of ideal body weight)	Strong recommendation Level I evidence
Plateau pressure target	Target plateau pressure of <30 cm H2O	Strong recommendation Moderate-quality evidence	Target plateau pressure of <30 cm H2O	Strong recommendation Level II evidence
PEEP	Higher PEEP strategy over lower PEEP strategy	Weak recommendation Low-quality evidence	High PEEP strategy over low PEEP strategy	Moderate recommendation Level II evidence
Recruitment maneuver	Suggest use	Weak recommendation Low-quality evidence	Suggest use	Moderate recommendation Level II evidence
Prone ventilation	Suggest prone ventilation for 12–16 h for moderate to severe ARDS	Weak recommendation Low-quality evidence	Suggest prone ventilation for 12–16 h for moderate to severe ARDS	Moderate recommendation Level II evidence
Neuromuscular blockade	Suggest intermittent boluses over continuous infusions. If needed, suggest using continuous infusion up to 48 h.	Weak recommendation Low-quality evidence	Intermittent boluses or continuous infusion to facilitate lung-protective ventilation up to 48 h	Moderate recommendation Level III evidence
Inhaled pulmonary vasodilator	Recommend against routine use of inhaled nitric oxide	Strong recommendation Low-quality evidence	Suggest as rescue therapy for hypoxemia despite optimized ventilation	Optional recommendation, Level III evidence
ECMO	Suggest venovenous ECMO for refractory hypoxemia	Weak recommendation Low-quality evidence	Insufficient data for/against routine use	Moderate recommendation Level III evidence

SCCM Society of Critical Care Medicine, *NIH* National Institutes of Health; *Vt* tidal volume, *PEEP* positive end-expiratory pressure, *ECMO* extracorporeal membrane oxygenation
[a]According to the GRADE approach
[b]Level of evidence: I = One or more randomized trials with clinical outcomes and/or validated laboratory endpoints; II = one or more well-designed, nonrandomized trials or observational cohort studies; III = expert opinion

of "high PEEP" nor clinical trials to suggest what PEEP level in ARDS patients may be optimal. Instead, a somewhat arbitrary cutoff of PEEP >10 cmH_2O has been suggested. In clinical practice, optimal PEEP is likely dependent on several factors such as lung compliance and the extent of disease and may change over time due to disease progression and patient positioning.

Recruitment maneuvers, in which high PEEP is held for variable periods on the order of 20–40 s to maximize alveolar recruitment, are often used in ARDS. Both guidelines suggest the use of recruitment maneuvers in patients with persistent hypoxemia despite optimal ventilator settings. An important caveat, however, is a recommendation against the use of staircase recruitment maneuvers, for the evidence of harm due to barotrauma and cardiovascular instability from prior clinical trials in ARDS (Writing Group for the Alveolar Recruitment for Acute Respiratory

Distress Syndrome Trial et al. 2017; Hodgson et al. 2019).

Galiatsou and coworkers reported that PP facilitates recruitment and reduces overinflated lung areas, thereby decreasing VILI by reducing stress and strain throughout the lung (Galiatsou et al. 2006). Both guidelines make the use of prone ventilation as a moderate-weak recommendation.

The use of neuromuscular blockade agents may help facilitate mechanical ventilation in patients when deep sedation alone if significant patient-ventilator asynchrony exists despite deep sedation. Additionally, the use of these agents may decrease overall oxygen consumption (Murray et al. 2016). Both guidelines prefer intermittent boluses over continuous infusions to maintain lung-protective ventilation and recommend the use of continuous paralytics for no longer than 48 h.

Both guidelines recommend against the routine use of nitric oxide; however, both do suggest a trial of inhaled pulmonary vasodilator as a rescue therapy with taper off if a rapid improvement in oxygenation does not occur.

The two clinical practice guidelines diverge slightly concerning recommendations regarding the use of extracorporeal membrane oxygenation (ECMO). While the SCCM recommends utilization of venovenous ECMO in cases of refractory hypoxemia with maximal mechanical ventilatory support, the NIH guidelines defer recommendations citing insufficient data.

In summary, the clinical guidelines from societies on COVID-19 extrapolate existing data of clinical trials in ARDS. Despite controversy regarding whether CARDS represents a spectrum of disease distinct from that of "usual" ARDS, currently-published guidelines for COVID-19 do not deviate from society recommendations for ARDS appreciably.

27.7 Pediatric Patients: A Challenge

Pediatric patients affected by COVID-19 are concerning because of the poor understanding of the epidemiological data and clinical features of the disease in this population (Tezer and Bedir Demirdağ 2020). To our data, pediatric patients might appear rarely diagnosed due to their asymptomatic clinical condition. An active innate immune response, healthier respiratory tracts, and fewer underlying disorders seem to be reasonable causes for their less susceptible condition (Lee et al. 2020). When compared to adults, children with COVID-19 have a relatively mild clinical course with a good prognosis, even though severe or critical cases occurred, particularly in small children, with subsequent admission to pediatric ICUs (less than 10% among infected children who had coexisting conditions) (Jeng 2020). In a report of 2143 pediatric cases, reported by Dong et al., the median age of confirmed cases was 10 years, 56.6% were male, and the interval from symptom onset to diagnosis was 3 days (Dong et al. 2020). The incubation period in children is usually about 2 days, with a range of 2–10 days (Ludvigsson 2020).

Fever and dry cough are the most frequent symptoms; other gastrointestinal symptoms, including nausea; vomiting; abdominal pain; diarrhea; musculoskeletal symptoms, such as fatigue; and nervous system symptoms, as loss of smell and loss of taste, are also reported. A proportion of SARS-CoV-2-infected children, however, has an asymptomatic condition, but they remain potential carriers of the disease (Jeng 2020).

Dyspnea and hypoxia occur when children become severely ill, thus requiring respiratory support. In their report, Sun and coworkers included eight severe or critically ill children with COVID-19 who were treated at the Wuhan Children's Hospital's ICU; among them, six received HFNC oxygen therapy, and two critically ill patients received mechanical ventilation (Sun et al. 2020).

Ground-glass opacities are the most common radiological characteristic at chest tomography, and lung consolidations may occur in severe cases bilaterally. In the series of 171 children diagnosed with COVID-19, 15.8% of patients have no radiological signs of pneumonia (Lu et al. 2020).

Regarding respiratory management, when hypoxia occurs, children must receive standard oxygen therapy delivered by nasal cannula, as it may be better tolerated (WHO 2020a). Despite using COT, in the absence of a resumption of hypoxemia, HFNC oxygen therapy can be useful. The immediate recognition of signs of ARDS is fundamental since children might benefit from NIV.

The European Society of Paediatric and Neonatal Intensive Care suggests to start with a lung-protective ventilation according to the today's recommendation: VTE 5–7 mL/kg ideal body weight, initial PEEP around 10 cmH_2O and driving pressure ≤ 15 cmH_2O with a FiO_2 adopted to maintain SpO_2 in a range going from 92 to 96% (European Society of Paediatric and Neonatal Intensive 2020).

If clinical conditions do not improve after 2 h from starting the NIV or this cannot be tolerated, with increased work of breathing or increased airway secretions, the onset of severe cough, or signs of hemodynamic instability, patients should promptly undergo IMV. If necessary, ECMO may be considered for pediatric patients with severe ARDS (China National Clinical Research Center for Respiratory et al. 2020).

27.8 Conclusion

Ventilatory support of COVID-19 patients is a challenge, and it becomes even more hardened in pediatric patients. The lack of worldwide consensus makes it tricky figuring out how to manage hypoxemic ARF and CARDS properly. During the COVID-19 pandemic, we adopted the actual standard of care for the "usual" hypoxemic ARF and ARDS for COVID-19 patients. However, a new insight into the pathophysiology of COVID-19 pneumonia and CARDS may cast doubt this is the right path to walk. The new model proposed by Gattinoni and coworkers (Gattinoni et al. 2020a) is exciting, and, nowadays, it might be considered the cornerstone for a careful and safe approach to ventilatory support for patients with COVID-19, mostly when mechanical ventilation is necessary. The risk of iatrogenic injuries, such as VILI, is increased in these patients mostly in the early phases of the disease when their pulmonary compliance is higher, and minimizing this kind of injury would lead to a better prognosis. Physicians have several arrows in the quiver to provide ventilatory support for patients with COVID-19 patients (COT, HFNC oxygen, NIV, IMV, and ECMO), and all of them could find some room in the treatment of hypoxemic ARF and CARDS. Several local scientific societies provided their own guidelines and suggestions based on both the availability of the resources and the local situation. However, a worldwide consensus would be desirable in order to standardize the approach to this tricky and potentially life-threatening condition. Although some kind of intervention seemed to be beneficial in these patients (HFNC oxygen therapy, NIV, and the use of PP in patients undergoing HFNC oxygen and NIV), more and bigger studies are necessary to support these assumptions. The role of ECMO in CARDS is far from being defined, but it should be considered for patients with severe CARDS.

Acknowledgments The authors would like to sincerely thank the entire Columbus COVID Team (Fabiana Baldi, Andrea Bellieni, Vincenzo Brandi, Angelo Calabrese, Maria Rosaria Calvello, Arturo Cicculo, Antonella Cingolani, Camilla Cipriani, Giuseppe Maria Corbo, Cinzia Falsiroli, Massimo Fantoni, Enrica Intini, Giovanni Landi, Paolo Maria Leone, Rosa Liperoti, Francesco Macagno, Ilaria Martis, Giuliano Montemurro, Davide Moschese, Rita Murri, Cristina Pais, Giuliana Pasciuto, Annalisa Potenza, Sara Salini, Matteo Siciliano, Jacopo Simonetti, Eleonora Taddei, Matteo Tosato, Anita Tummolo, Francesco Varone, Giulio Ventura) for facing daily COVID-19 side by side on a multidisciplinary basis, and a special thanks goes to the Alpha Team.

References

Alhazzani W, Møller MH, Arabi YM, Loeb M, Gong MN, Fan E, Oczkowski S, Levy MM, Derde L, Dzierba A, Du B, Aboodi M, Wunsch H, Cecconi M, Koh Y, Chertow DS, Maitland K, Alshamsi F, Belley-Cote E, Greco M, Laundy M, Morgan JS, Kesecioglu J, McGeer A, Mermel L, Mammen MJ, Alexander PE, Arrington A, Centofanti JE, Citerio G, Baw B, Memish ZA, Hammond N, Hayden FG, Evans L, Rhodes A (2020) Surviving sepsis campaign: guidelines on the management of critically ill adults with coronavirus disease 2019 (COVID-19). Intensive Care Med 46(5):854–887. https://doi.org/10.1007/s00134-020-06022-5

Antonelli M, Conti G, Esquinas A, Montini L, Maggiore SM, Bello G, Rocco M, Maviglia R, Pennisi MA, Gonzalez-Diaz G, Meduri GU (2007) A multiple-center survey on the use in clinical practice of non-invasive ventilation as a first-line intervention for acute respiratory distress syndrome. Crit Care Med 35(1):18–25. https://doi.org/10.1097/01.CCM.0000251821.44259.F3

Archer SL, Sharp WW, Weir EK (2020) Differentiating COVID-19 pneumonia from acute respiratory distress syndrome (ARDS) and high altitude pulmonary edema (HAPE): therapeutic implications. Circulation 142:101. CIRCULATIONAHA.120.047915. https://doi.org/10.1161/CIRCULATIONAHA.120.047915

Barrot L, Asfar P, Mauny F, Winiszewski H, Montini F, Badie J, Quenot J-P, Pili-Floury S, Bouhemad B, Louis G, Souweine B, Collange O, Pottecher J, Levy B, Puyraveau M, Vettoretti L, Constantin J-M, Capellier G (2020) Liberal or conservative oxygen therapy for acute respiratory distress syndrome. N Engl J Med 382(11):999–1008. https://doi.org/10.1056/NEJMoa1916431

Beloncle FM, Pavlovsky B, Desprez C, Fage N, Olivier P-Y, Asfar P, Richard J-C, Mercat A (2020) Recruitability and effect of PEEP in SARS-Cov-2-associated acute respiratory distress syndrome. Ann Intensive Care 10(1):55. https://doi.org/10.1186/s13613-020-00675-7

Berlin DA, Gulick RM, Martinez FJ (2020) Severe Covid-19. N Engl J Med:NEJMcp2009575. https://doi.org/10.1056/NEJMcp2009575

Briel M, Meade M, Mercat A, Brower RG, Talmor D, Walter SD, Slutsky AS, Pullenayegum E, Zhou Q, Cook D, Brochard L, Richard J-CM, Lamontagne F, Bhatnagar N, Stewart TE, Guyatt G (2010) Higher vs lower positive end-expiratory pressure in patients with acute lung injury and acute respiratory distress syndrome: systematic review and meta-analysis. JAMA 303(9):865. https://doi.org/10.1001/jama.2010.218

China National Clinical Research Center for Respiratory D, National Center for Children's Health BC, Group of Respirology CPSCMA, Chinese Medical Doctor Association Committee on Respirology P, China Medicine Education Association Committee on P, Chinese Research Hospital Association Committee on P, Chinese Non-government Medical Institutions Association Committee on P, China Association of Traditional Chinese Medicine CoCsHaMR, China News of Drug Information Association CoCsSM, Global Pediatric Pulmonology A, Shen K, Yang Y, Wang T, Zhao D, Jiang Y, Jin R, Zheng Y, Xu B, Xie Z, Lin L, Shang Y, Lu X, Shu S, Bai Y, Deng J, Lu M, Ye L, Wang X, Wang Y, Gao L (2020) Diagnosis, treatment, and prevention of 2019 novel coronavirus infection in children: experts' consensus statement. World J Pediatr 16:223. https://doi.org/10.1007/s12519-020-00343-7

Chiumello D, Pelosi P, Carlesso E, Severgnini P, Aspesi M, Gamberoni C, Antonelli M, Conti G, Chiaranda M, Gattinoni L (2003) Noninvasive positive pressure ventilation delivered by helmet vs. standard face mask. Intensive Care Med 29(10):1671–1679. https://doi.org/10.1007/s00134-003-1825-9

Chiumello D, Brochard L, Marini JJ, Slutsky AS, Mancebo J, Ranieri VM, Thompson BT, Papazian L, Schultz MJ, Amato M, Gattinoni L, Mercat A, Pesenti A, Talmor D, Vincent J-L (2017) Respiratory support in patients with acute respiratory distress syndrome: an expert opinion. Crit Care (Lond) 21(1):240. https://doi.org/10.1186/s13054-017-1820-0

Cook TM, El-Boghdadly K, McGuire B, McNarry AF, Patel A, Higgs A (2020) Consensus guidelines for managing the airway in patients with COVID-19: guidelines from the Difficult Airway Society, the Association of Anaesthetists the Intensive Care Society, the Faculty of Intensive Care Medicine and the Royal College of Anaesthetists. Anaesthesia 75(6):785–799. https://doi.org/10.1111/anae.15054

COVID-19 Treatment Guidelines Panel (2020) Coronavirus Disease 2019 (COVID-19) Treatment Guidelines. National Institutes of Health. Available at https://www.covid19treatmentguidelines.nih.gov/

Dong Y, Mo X, Hu Y, Qi X, Jiang F, Jiang Z, Tong S (2020) Epidemiology of COVID-19 among children in China. Pediatrics 145:e20200702. https://doi.org/10.1542/peds.2020-0702

Elharrar X, Trigui Y, Dols A-M, Touchon F, Martinez S, Prud'homme E, Papazian L (2020) Use of prone positioning in nonintubated patients with COVID-19 and hypoxemic acute respiratory failure. JAMA 323:2336. https://doi.org/10.1001/jama.2020.8255

Eurn Society of Paediatric and Neonatal Intensive C (2020) Practice recommendations for managing children with proven or suspected COVID-19

Fagiuoli S, Lorini FL, Remuzzi G (2020) Adaptations and lessons in the Province of Bergamo. N Engl J Med 382:e71.:NEJMc2011599. https://doi.org/10.1056/NEJMc2011599

Fan E, Del Sorbo L, Goligher EC, Hodgson CL, Munshi L, Walkey AJ, Adhikari NKJ, Amato MBP, Branson R, Brower RG, Ferguson ND, Gajic O, Gattinoni L, Hess D, Mancebo J, Meade MO, McAuley DF, Pesenti A, Ranieri VM, Rubenfeld GD, Rubin E, Seckel M, Slutsky AS, Talmor D, Thompson BT, Wunsch H,

Uleryk E, Brozek J, Brochard LJ, American Thoracic Society ESoICMaSoCCM (2017) An official American Thoracic Society/European Society of Intensive Care Medicine/Society of Critical Care Medicine clinical practice guideline: mechanical ventilation in adult patients with acute respiratory distress syndrome. Am J Respir Crit Care Med 195(9):1253–1263. https://doi.org/10.1164/rccm.201703-0548ST

Ferioli M, Cisternino C, Leo V, Pisani L, Palange P, Nava S (2020) Protecting healthcare workers from SARS-CoV-2 infection: practical indications. Eur Respir Rev 29(155):200068. https://doi.org/10.1183/16000617.0068-2020

Frat J-P, Thille AW, Mercat A, Girault C, Ragot S, Perbet S, Prat G, Boulain T, Morawiec E, Cottereau A, Devaquet J, Nseir S, Razazi K, Mira J-P, Argaud L, Chakarian J-C, Ricard J-D, Wittebole X, Chevalier S, Herbland A, Fartoukh M, Constantin J-M, Tonnelier J-M, Pierrot M, Mathonnet A, Béduneau G, Delétage-Métreau C, Richard J-CM, Brochard L, Robert R (2015) High-flow oxygen through nasal cannula in acute hypoxemic respiratory failure. N Engl J Med 372(23):2185–2196. https://doi.org/10.1056/NEJMoa1503326

Frat J-P, Coudroy R, Marjanovic N, Thille AW (2017) High-flow nasal oxygen therapy and non-invasive ventilation in the management of acute hypoxemic respiratory failure. Ann Transl Med 5(14):297–297. https://doi.org/10.21037/atm.2017.06.52

Galiatsou E, Kostanti E, Svarna E, Kitsakos A, Koulouras V, Efremidis SC, Nakos G (2006) Prone position augments recruitment and prevents alveolar Overinflation in acute lung injury. Am J Respir Crit Care Med 174(2):187–197. https://doi.org/10.1164/rccm.200506-899OC

Gattinoni L, Chiumello D, Caironi P, Busana M, Romitti F, Brazzi L, Camporota L (2020a) COVID-19 pneumonia: different respiratory treatments for different phenotypes? Intensive Care Med 46:1099. https://doi.org/10.1007/s00134-020-06033-2

Gattinoni L, Coppola S, Cressoni M, Busana M, Rossi S, Chiumello D (2020b) COVID-19 does not lead to a "typical" acute respiratory distress syndrome. Am J Respir Crit Care Med 201(10):1299–1300. https://doi.org/10.1164/rccm.202003-0817LE

Geng S, Mei Q, Zhu C, Yang T, Yang Y, Fang X, Pan A (2020) High flow nasal cannula is a good treatment option for COVID-19. Heart Lung 49:444.:S0147956320301138. https://doi.org/10.1016/j.hrtlng.2020.03.018

Grasselli G, Pesenti A, Cecconi M (2020) Critical care utilization for the COVID-19 outbreak in Lombardy, Italy: early experience and forecast during an emergency response. JAMA 323:1545. https://doi.org/10.1001/jama.2020.4031

Guan W-j, Ni Z-y, Hu Y, Liang W-h, Ou C-q, He J-x, Liu L, Shan H, Lei C-l, Hui DSC, Du B, Li L-j, Zeng G, Yuen K-Y, Chen R-c, Tang C-l, Wang T, Chen P-y, Xiang J, Li S-y, Wang J-l, Liang Z-j, Peng Y-x, Wei L, Liu Y, Hu Y-h, Peng P, Wang J-m, Liu J-y, Chen Z, Li G, Zheng Z-j, Qiu S-q, Luo J, Ye C-j, Zhu S-y, Zhong N-s (2020) Clinical characteristics of coronavirus disease 2019 in China. N Engl J Med 382(18):1708–1720. https://doi.org/10.1056/NEJMoa2002032

Herrmann J, Fonseca da Cruz A, Hawley ML, Branson RD, Kaczka DW (2020) Shared ventilation in the era of COVID-19: a theoretical consideration of the dangers and potential solutions. Respir Care 65:932. respcare.07919. https://doi.org/10.4187/respcare.07919

Hodgson CL, Cooper DJ, Arabi Y, King V, Bersten A, Bihari S, Brickell K, Davies A, Fahey C, Fraser J, McGuinness S, Murray L, Parke R, Paul E, Tuxen D, Vallance S, Young M, Nichol A (2019) Maximal recruitment open lung ventilation in acute respiratory distress syndrome (PHARLAP). A phase II, multicenter randomized controlled clinical trial. Am J Respir Crit Care Med 200(11):1363–1372. https://doi.org/10.1164/rccm.201901-0109OC

Hui DS, Chow BK, Lo T, Tsang OTY, Ko FW, Ng SS, Gin T, Chan MTV (2019) Exhaled air dispersion during high-flow nasal cannula therapy <i>versus</i> CPAP <i>via</i> different masks. Eur Respir J 53(4):1802339. https://doi.org/10.1183/13993003.02339-2018

Jeng M-J (2020) COVID-19 in children: current status. J Chin Med Assoc:1. https://doi.org/10.1097/JCMA.0000000000000323

Lee P-I, Hu Y-L, Chen P-Y, Huang Y-C, Hsueh P-R (2020) Are children less susceptible to COVID-19? J Microbiol Immunol Infect 53:371.:S1684118220300396. https://doi.org/10.1016/j.jmii.2020.02.011

Leung CCH, Joynt GM, Gomersall CD, Wong WT, Lee A, Ling L, Chan PKS, Lui PCW, Tsoi PCY, Ling CM, Hui M (2019) Comparison of high-flow nasal cannula versus oxygen face mask for environmental bacterial contamination in critically ill pneumonia patients: a randomized controlled crossover trial. J Hosp Infect 101(1):84–87. https://doi.org/10.1016/j.jhin.2018.10.007

Li J, Fink JB, Ehrmann S (2020) High-flow nasal cannula for COVID-19 patients: low risk of bio-aerosol dispersion. Eur Respir J 55(5):2000892. https://doi.org/10.1183/13993003.00892-2020

Liu X, Liu X, Xu Y, Xu Z, Huang Y, Chen S, Li S, Liu D, Lin Z, Li Y (2020) Ventilatory ratio in hypercapnic mechanically ventilated patients with COVID-19-associated acute respiratory distress syndrome. Am J Respir Crit Care Med 201(10):1297–1299. https://doi.org/10.1164/rccm.202002-0373LE

Loh N-HW, Tan Y, Taculod J, Gorospe B, Teope AS, Somani J, Tan AYH (2020) The impact of high-flow nasal cannula (HFNC) on coughing distance: implications on its use during the novel coronavirus disease outbreak. Can J Anesth/J Can Anesth 67:893. https://doi.org/10.1007/s12630-020-01634-3

Lu X, Zhang L, Du H, Zhang J, Li YY, Qu J, Zhang W, Wang Y, Bao S, Li Y, Wu C, Liu H, Liu D, Shao J, Peng X, Yang Y, Liu Z, Xiang Y, Zhang F, Silva RM, Pinkerton KE, Shen K, Xiao H, Xu S, Wong GWK (2020) SARS-CoV-2 infection in children. N Engl J

Med 382(17):1663–1665. https://doi.org/10.1056/NEJMc2005073

Ludvigsson JF (2020) Systematic review of COVID-19 in children shows milder cases and a better prognosis than adults. Acta Paediatr 109(6):1088–1095. https://doi.org/10.1111/apa.15270

Lyons C, Callaghan M (2020) The use of high-flow nasal oxygen in COVID-19. Anaesthesia 75:843.:anae.15073. https://doi.org/10.1111/anae.15073

Mahase E (2020) Covid-19: most patients require mechanical ventilation in first 24 hours of critical care. BMJ 368:m1201. https://doi.org/10.1136/bmj.m1201

Marini JJ, Gattinoni L (2020) Management of COVID-19 respiratory distress. JAMA 323:2329. https://doi.org/10.1001/jama.2020.6825

Matthay MA, Aldrich JM, Gotts JE (2020) Treatment for severe acute respiratory distress syndrome from COVID-19. Lancet Respir Med 8(5):433–434. https://doi.org/10.1016/S2213-2600(20)30127-2

McDonald V, Cousins J, Wark P (2016) Acute oxygen therapy: a review of prescribing and delivery practices. Int J Chron Obstruct Pulmon Dis 11:1067. https://doi.org/10.2147/COPD.S103607

Möhlenkamp S, Thiele H (2020) Ventilation of COVID-19 patients in intensive care units. Herz 45:329. https://doi.org/10.1007/s00059-020-04923-1

Munshi L, Del Sorbo L, Adhikari NKJ, Hodgson CL, Wunsch H, Meade MO, Uleryk E, Mancebo J, Pesenti A, Ranieri VM, Fan E (2017) Prone position for acute respiratory distress syndrome. A systematic review and meta-analysis. Ann Am Thorac Soc 14(Supplement_4):S280–S288. https://doi.org/10.1513/AnnalsATS.201704-343OT

Murray MJ, DeBlock H, Erstad B, Gray A, Jacobi J, Jordan C, McGee W, McManus C, Meade M, Nix S, Patterson A, Sands MK, Pino R, Tescher A, Arbour R, Rochwerg B, Murray CF, Mehta S (2016) Clinical practice guidelines for sustained neuromuscular blockade in the adult critically ill patient. Crit Care Med 44(11):2079–2103. https://doi.org/10.1097/CCM.0000000000002027

Navalesi P, Maggiore SM (2013) Chapter 10. Positive end-expiratory pressure. In: Tobin MJ (ed) Principles and practice of mechanical ventilation, 3e. The McGraw-Hill Companies, New York

Niederman MS, Richeldi L, Chotirmall SH, Bai C (2020) Rising to the challenge of COVID-19: advice for pulmonary and critical care and an agenda for research. Am J Respir Crit Care Med 201(9):1019–1022. https://doi.org/10.1164/rccm.202003-0741ED

Nishimura M (2016) High-flow nasal cannula oxygen therapy in adults: physiological benefits, indication, clinical benefits, and adverse effects. Respir Care 61(4):529–541. https://doi.org/10.4187/respcare.04577

Phua J, Weng L, Ling L, Egi M, Lim C-M, Divatia JV, Shrestha BR, Arabi YM, Ng J, Gomersall CD, Nishimura M, Koh Y, Du B (2020) Intensive care management of coronavirus disease 2019 (COVID-19): challenges and recommendations. Lancet Respir Med 8(5):506–517. https://doi.org/10.1016/S2213-2600(20)30161-2

Remy KE, Lin JC, Verhoef PA (2020) High-flow nasal cannula may be no safer than non-invasive positive pressure ventilation for COVID-19 patients. Crit Care 24(1):169. https://doi.org/10.1186/s13054-020-02892-9

Richardson S, Hirsch JS, Narasimhan M, Crawford JM, McGinn T, Davidson KW, the Northwell C-RC, Barnaby DP, Becker LB, Chelico JD, Cohen SL, Cookingham J, Coppa K, Diefenbach MA, Dominello AJ, Duer-Hefele J, Falzon L, Gitlin J, Hajizadeh N, Harvin TG, Hirschwerk DA, Kim EJ, Kozel ZM, Marrast LM, Mogavero JN, Osorio GA, Qiu M, Zanos TP (2020) Presenting characteristics, comorbidities, and outcomes among 5700 patients hospitalized with COVID-19 in the New York City Area. JAMA 323:2052. https://doi.org/10.1001/jama.2020.6775

Rochwerg B, Brochard L, Elliott MW, Hess D, Hill NS, Nava S, Navalesi P, Antonelli M, Brozek J, Conti G, Ferrer M, Guntupalli K, Jaber S, Keenan S, Mancebo J, Mehta S, Raoof S (2017) Official ERS/ATS clinical practice guidelines: non-invasive ventilation for acute respiratory failure. Eur Respir J 50(2):1602426. https://doi.org/10.1183/13993003.02426-2016

Roesthuis L, van den Berg M, van der Hoeven H (2020) Advanced respiratory monitoring in COVID-19 patients: use less PEEP! Crit Care 24(1):230. https://doi.org/10.1186/s13054-020-02953-z

Sanders JM, Monogue ML, Jodlowski TZ, Cutrell JB (2020) Pharmacologic treatments for coronavirus disease 2019 (COVID-19): a review. JAMA 323:1824. https://doi.org/10.1001/jama.2020.6019

Scaravilli V, Grasselli G, Castagna L, Zanella A, Isgrò S, Lucchini A, Patroniti N, Bellani G, Pesenti A (2015) Prone positioning improves oxygenation in spontaneously breathing nonintubated patients with hypoxemic acute respiratory failure: a retrospective study. J Crit Care 30(6):1390–1394. https://doi.org/10.1016/j.jcrc.2015.07.008

Simonds A, Hanak A, Chatwin M, Morrell M, Hall A, Parker K, Siggers J, Dickinson R (2010) Evaluation of droplet dispersion during non-invasive ventilation, oxygen therapy, nebuliser treatment and chest physiotherapy in clinical practice: implications for management of pandemic influenza and other airborne infections. Health Technol Assess 14(46):131. https://doi.org/10.3310/hta14460-02

Sorbello M, El-Boghdadly K, Di Giacinto I, Cataldo R, Esposito C, Falcetta S, Merli G, Cortese G, Corso RM, Bressan F, Pintaudi S, Greif R, Donati A, Petrini F, Società Italiana di Anestesia Analgesia Rianimazione e Terapia Intensiva Airway Research Group aTEAMS (2020) The Italian coronavirus disease 2019 outbreak: recommendations from clinical practice. Anaesthesia 75(6):724–732. https://doi.org/10.1111/anae.15049

Sun D, Li H, Lu X-X, Xiao H, Ren J, Zhang F-R, Liu Z-S (2020) Clinical features of severe pediatric patients with coronavirus disease 2019 in Wuhan: a single center's observational study. World J Pediatr 16:251. https://doi.org/10.1007/s12519-020-00354-4

Takhar A, Walker A, Tricklebank S, Wyncoll D, Hart N, Jacob T, Arora A, Skilbeck C, Simo R, Surda P (2020) Recommendation of a practical guideline for safe tracheostomy during the COVID-19 pandemic. Eur Arch Otorhinolaryngol 277:2173. https://doi.org/10.1007/s00405-020-05993-x

Tezer H, Bedir Demirdağ T (2020) Novel coronavirus disease (COVID-19) in children. Turk J Med Sci 50(SI-1):592–603. https://doi.org/10.3906/sag-2004-174

The Acute Respiratory Distress Syndrome N (2000) Ventilation with lower tidal volumes as compared with traditional tidal volumes for acute lung injury and the acute respiratory distress syndrome. N Engl J Med 342(18):1301–1308. https://doi.org/10.1056/NEJM200005043421801

The Critical Care Western Research G, Slessarev M, Cheng J, Ondrejicka M, Arntfield R (2020) Patient self-proning with high-flow nasal cannula improves oxygenation in COVID-19 pneumonia. Can J Anesth/J Can Anesth 67:1288. https://doi.org/10.1007/s12630-020-01661-0

Tian S, Hu W, Niu L, Liu H, Xu H, Xiao S-Y (2020) Pulmonary pathology of early-phase 2019 novel coronavirus (COVID-19) pneumonia in two patients with lung cancer. J Thorac Oncol 15(5):700–704. https://doi.org/10.1016/j.jtho.2020.02.010

Tran K, Cimon K, Severn M, Pessoa-Silva CL, Conly J (2012) Aerosol generating procedures and risk of transmission of acute respiratory infections to healthcare workers: a systematic review. PLoS One 7(4):e35797. https://doi.org/10.1371/journal.pone.0035797

Turri-Zanoni M, Battaglia P, Czaczkes C, Pelosi P, Castelnuovo P, Cabrini L (2020) Elective tracheostomy during mechanical ventilation in patients affected by COVID-19: preliminary case series from Lombardy, Italy. Otolaryngol Head Neck Surg 163:135.:019459982092896. https://doi.org/10.1177/0194599820928963

van den Boom W, Hoy M, Sankaran J, Liu M, Chahed H, Feng M, See KC (2020) The search for optimal oxygen saturation targets in critically ill patients. Chest 157(3):566–573. https://doi.org/10.1016/j.chest.2019.09.015

Vibert F, Kretz M, Thuet V, Barthel F, De Marcillac F, Deruelle P, Lecointre L (2020) Prone positioning and high-flow oxygen improved respiratory function in a 25-week pregnant woman with COVID-19. Eur J Obstet Gynecol Reprod Biol 250:257.:S0301211520302736. https://doi.org/10.1016/j.ejogrb.2020.05.022

Walkey AJ, Goligher EC, Del Sorbo L, Hodgson CL, Adhikari NKJ, Wunsch H, Meade MO, Uleryk E, Hess D, Talmor DS, Thompson BT, Brower RG, Fan E (2017) Low tidal volume versus non–volume-limited strategies for patients with acute respiratory distress syndrome. A systematic review and meta-analysis. Ann Am Thorac Soc 14(Supplement_4):S271–S279. https://doi.org/10.1513/AnnalsATS.201704-337OT

Wang D, Hu B, Hu C, Zhu F, Liu X, Zhang J, Wang B, Xiang H, Cheng Z, Xiong Y, Zhao Y, Li Y, Wang X, Peng Z (2020a) Clinical characteristics of 138 hospitalized patients with 2019 novel coronavirus–infected pneumonia in Wuhan, China. JAMA 323(11):1061. https://doi.org/10.1001/jama.2020.1585

Wang K, Zhao W, Li J, Shu W, Duan J (2020b) The experience of high-flow nasal cannula in hospitalized patients with 2019 novel coronavirus-infected pneumonia in two hospitals of Chongqing, China. Ann Intensive Care 10(1):37. https://doi.org/10.1186/s13613-020-00653-z

Wang Y, Lu X, Chen H, Chen T, Su N, Huang F, Zhou J, Zhang B, Li Y, Yan F, Wang J (2020c) Clinical course and outcomes of 344 intensive care patients with COVID-19. Am J Respir Crit Care Med 201:1430. https://doi.org/10.1164/rccm.202003-0736LE

Whittle JS, Pavlov I, Sacchetti AD, Atwood C, Rosenberg MS (2020) Respiratory support for adult patients with COVID-19. J Am Coll Emerg Physicians Open 1(2):95–101. https://doi.org/10.1002/emp2.12071

WHO (2020a) Clinical management of severe acute respiratory infection (SARI) when COVID-19 disease is suspected: interim guidance. https://www.who.int/publications-detail/clinical-management-of-severe-acute-respiratory-infection-when-novel-coronavirus-(ncov)-infection-is-suspected

WHO (2020b) Infection prevention and control during health care when novel coronavirus (nCoV) infection is suspected: interim guidance. https://apps.who.int/iris/rest/bitstreams/1266296/retrieve

Writing Group for the Alveolar Recruitment for Acute Respiratory Distress Syndrome Trial I, Cavalcanti AB, Suzumura ÉA, Laranjeira LN, Paisani DM, Damiani LP, Guimarães HP, Romano ER, Regenga MM, Taniguchi LNT, Teixeira C, Pinheiro de Oliveira R, Machado FR, Diaz-Quijano FA, Filho MSA, Maia IS, Caser EB, Filho WO, Borges MC, Martins PA, Matsui M, Ospina-Tascón GA, Giancursi TS, Giraldo-Ramirez ND, Vieira SRR, Assef MGPL, Hasan MS, Szczeklik W, Rios F, Amato MBP, Berwanger O, Ribeiro de Carvalho CR (2017) Effect of lung recruitment and titrated positive end-expiratory pressure (PEEP) vs low PEEP on mortality in patients with acute respiratory distress syndrome: a randomized clinical trial. JAMA 318(14):1335. https://doi.org/10.1001/jama.2017.14171

Wu Z, McGoogan JM (2020) Characteristics of and important lessons from the coronavirus disease 2019 (COVID-19) outbreak in China: summary of a report of 72 314 cases from the Chinese Center for Disease Control and Prevention. JAMA 323(13):1239. https://doi.org/10.1001/jama.2020.2648

Wunsch H (2020) Mechanical ventilation in COVID-19: interpreting the current epidemiology. Am J Respir Crit Care Med 202:1. https://doi.org/10.1164/rccm.202004-1385ED

Zhao H, Wang H, Sun F, Lyu S, An Y (2017) High-flow nasal cannula oxygen therapy is superior to conventional oxygen therapy but not to non-invasive mechanical ventilation on intubation rate: a systematic review and meta-analysis. Crit Care 21(1):184. https://doi.org/10.1186/s13054-017-1760-8

Zhu N, Zhang D, Wang W, Li X, Yang B, Song J, Zhao X, Huang B, Shi W, Lu R, Niu P, Zhan F, Ma X, Wang D, Xu W, Wu G, Gao GF, Tan W (2020) A novel coronavirus from patients with pneumonia in China, 2019. N Engl J Med 382(8):727–733. https://doi.org/10.1056/NEJMoa2001017

Ziehr DR, Alladina J, Petri CR, Maley JH, Moskowitz A, Medoff BD, Hibbert KA, Thompson BT, Hardin CC (2020) Respiratory pathophysiology of mechanically ventilated patients with COVID-19: a cohort study. Am J Respir Crit Care Med 201:1560. https://doi.org/10.1164/rccm.202004-1163LE

28 Nutrition and Immunity in COVID-19

Marjan Moallemian Isfahani, Zahra Emam-Djomeh, Idupulapati M. Rao, and Nima Rezaei

Abstract

Nutrition can strongly influence infection trajectories by either boosting or suppressing the immune system. During the recently emerged pandemic of coronavirus disease 2019 (COVID-19), individuals who possess diets high in fat, refined carbohydrates, and sugars have shown to be highly prone to the disease and associated adverse outcomes. Both micronutrients and macronutrients provide benefits at different stages of the infection. Thus, using appropriate nutritional recommendations and interventions is necessary to combat the infection in patients with COVID-19 in both outpatient and inpatient settings.

Keywords

COVID-19 · Immunity · Micronutrient · Nutrition · Selenium · Vitamin

28.1 Introduction

In December 2019, a novel coronavirus called severe acute respiratory syndrome coronavirus 2 (SARS-CoV-2) begun rapid transmission among people. As of writing this, the number of confirmed cases is approaching 10 million, with 500,000 deaths.

COVID-19 affects not only the respiratory tract, but it can involve other organs associated with increased morbidity (Yazdanpanah et al. 2020b; Shamshirian and Rezaei 2020; Jahanshahlu

M. Moallemian Isfahani
Dietetics and Nutrition Experts Team (DiNET), Universal Scientific Education and Research Network (USERN), Tehran, Iran

National Nutrition and Food Technology Research Institute, Shahid Beheshti University of Medical Sciences, Tehran, Iran

Z. Emam-Djomeh
Dietetics and Nutrition Experts Team (DiNET), Universal Scientific Education and Research Network (USERN), Tehran, Iran

Department of Food Science and Engineering, University College of Agriculture and Natural Resources, University of Tehran, Karaj, Iran

I. M. Rao
Centro Internacional de Agricultura Tropical, Santiago de Cali, Colombia

N. Rezaei (✉)
Department of Immunology, School of Medicine, Tehran University of Medical Sciences, Tehran, Iran

Research Center for Immunodeficiencies, Pediatrics Center of Excellence, Children's Medical Center, Tehran University of Medical Sciences, Tehran, Iran

Network of Immunity in Infection, Malignancy and Autoimmunity (NIIMA), Universal Scientific Education and Research Network (USERN), Tehran, Iran
e-mail: rezaei_nima@tums.ac.ir

N. Rezaei (ed.), *Coronavirus Disease - COVID-19*, Advances in Experimental Medicine and Biology 1318, https://doi.org/10.1007/978-3-030-63761-3_28

and Rezaei 2020a; Saleki et al. 2020; Lotfi and Rezaei 2020). Older adults, patients with preexisting conditions, in particular cardiovascular diseases, hypertension, diabetes, and cancer, and individuals harboring genes that affect immunity have shown the worst outcomes and highest mortality (Shamshirian and Rezaei 2020; Ahmadi et al. 2020; Yousefzadegan and Rezaei 2020; Darbeheshti and Rezaei 2020). After 6-month efforts to crack the spread of the COVID-19 pandemic, there are still no vaccines or effective medications to protect the body against or to treat the infection (Saghazadeh and Rezaei 2020b).

There is evidence of hyper inflammation and immune system dysregulation in people with COVID-19, especially in cases with severe and critical conditions (Rokni et al. 2020; Bahrami et al. 2020; Yazdanpanah et al. 2020a; Saghazadeh and Rezaei 2020a; Fathi and Rezaei 2020; Nasab et al. 2020). As a result, approaches helping the immune hemostasis come to attention, along with other ones that offer to target the virus-cell interaction (Sharifkashani et al. 2020; Mansourabadi et al. 2020; Pashaei and Rezaei 2020; Fathi and Rezaei 2020; Jahanshahlu and Rezaei 2020b; Basiri et al. 2020b; Saghazadeh and Rezaei 2020b; Mohamed et al. 2020b; Rezaei 2020b; Rabiee et al. 2020; Pourahmad et al. 2020; Lotfi et al. 2020).

Appropriate nutrient intake is necessary for the immune system to function correctly, and nutritional therapy plays a crucial role in the prevention of the disease as well as the recovery process by regulating the immunity system (Gombart et al. 2020; High 2001). The few available studies, which have investigated supportive treatment in COVID-19, suggest that nutritional assessment is necessary for all patients with COVID-19, and, accordingly, nutritional support should be provided to patients as soon as possible. It is of fundamental importance to supply nutritional requirements such as energy, protein, and micronutrients that can enhance the patient immune system (Maggini et al. 2018). In particular, micronutrients are essential to maintain the immune system healthy or improve the immune competence in viral infections such as COVID-19. They include vitamins such as A, B, D, E, and C and trace elements, including selenium, zinc, iron, and magnesium.

28.2 Nutritional Assessment

Nutritional Diagnosis and Intervention describes the critical thinking process from assessment to selection of proper, timely, and measurable diagnosis of nutritional deficiencies. Measurement of the indicators of inflammation is also a critical component of nutritional assessment.

Nutritional assessment is the first step in the nutrition care process. Such an assessment must include critical elements of the dietary patterns, food cravings and eating habits, clinical or medical history, and changes in food intake from the time of infection. Alterations in appetite, food variety, balance, and order of food meals as well as anthropometric measurements when possible and biochemical and laboratory values should be considered (Volkert et al. 2019). Also, food allergies, possible adverse effects of pharmaceutical treatments on nutritional status, and food intake condition, as well as information on the medication and herbal supplement use for potential food-drug interactions, must be considered by interviewing the patient or their live-in companion. Of significance are also the fecal extraction rate and the amount of fluid intake. Moreover, the Nutrition Risk in Critically ILL Score (NUTRIC Score) tool is recommended as a part of nutritional assessment in patients admitted to the intensive care unit (ICU) (Kondrup et al. 2003; Ballmer 2020).

28.2.1 Nutritional Care in Outpatients with COVID-19

With the immune system fighting off coronavirus, energy requirements are increased to meet physiological needs. In this regard, a liquid diet, including higher volumes of water, soups, fruit juices, milk, and tea, which needs less energy to digestion, can be considered. Adequate intake of basic food groups with a diversity of the content would help promote immunity in people with

COVID-19 to combat infection. Increased intake of whole fruits and vegetables, which are either excellent resources of vitamins such as Vit A and Vit C or contain ample amounts of pigments and some precursors of vitamin A and Vit C, play a vital role in improving nutritional status and immunity, and thus well-being.

28.2.2 Hospitalized Patients with COVID-19

Clinical nutritional societies have endorsed the Global Leadership Project on Malnutrition (GLIM) standards for the diagnosis of malnutrition. The GLIM recommends a two-step approach in which an initial screening is performed to determine the at-risk status of the patient by utilizing validated screening methods such as MUST or NRS-2002, followed by a secondary diagnostic evaluation to rate the severity of the malnutrition. According to the GLIM, malnutrition diagnosis requires at least one phenotypic criterion and one etiologic criterion (Cederholm et al. 2019; Stratton et al. 2004). The mentioned criteria can be successfully applied to patients at risk of severe COVID-19 or hospitalized due to COVID-19. Poor COVID-19 outcomes are more likely to occur in patients with a higher risk of malnourishment, such as older adults and individuals with comorbidities (Barazzoni et al. 2020).

When the nutritional status of the patient is suboptimal, the dietitian must decide on the most suitable artificial nutrition (AN) solution, e.g., enteral nutrition (EN) and parenteral nutrition (PN). In this regard, oral nutritional supplementation (ONS), tube feeding, supplementary parenteral nutrition (SPN), and total parenteral nutrition (TPN) are available and can help the recovery of patients. ONS is, however, the first choice of care. When oral intake is insufficient, the dietitian should make recommendations on the use of AN combined with either EN or PN. EN is the first tier of AN. If it were insufficient to provide requirements, it would be substituted by PN.

Effective management of a patient with COVID-19 calls for an effective and timely plan of nutritional therapy adjusted considering the clinical characteristics of the individual. The transition from PN to EN is considered when EN is predicted to meet 50% of requirements. Also, when ONS can meet 50% of requirements, the EN can be progressively reduced and finally terminated. In contrast, if ONS is unable to fulfill 50% of requirements, the EN should be continued. Elderly or malnourished patients with COVID-19 are more likely to require intensive care unit (ICU) admission and prolonged hospitalization. When hospital stay lasts longer than 48 h, nutritional therapy should be started as early as possible and no later than 72 h for patients who are unable to eat or admitted to the ICU (Singer et al. 2019; Ballmer 2020).

28.2.2.1 Nutritional Strategies for Non-ICU Patients Hospitalized for COVID-19

Energy Requirements

Infection with COVID-19 raises the demand for energy. It is, therefore, of importance to provide energy supply to patients with COVID-19 to avoid undesirable loss of weight and muscle mass, which may lead to poor functional results. Reliable estimation of patient's caloric needs is necessary to prevent overfeeding or underfeeding.

The gold standard for estimation of caloric needs is the time-consuming method of indirect calorimetry. The following predictive equations and weight-based formulae are helpful to estimate caloric needs simply (Ballmer 2020; Gomes et al. 2018):

- Total energy expenditure (TEE) for patients with comorbidities aged above 65 years: 27 kcal/kg body weight/day
- TEE for severely underweight patients with comorbidities: 30 kcal/kg body weight/day
- TEE for overweight and obese patients: 20 kcal/kg body weight/day

It is noteworthy to mention that chronically underweight patients should be fed carefully and gradually to arrive at the goal of 30 kcal/kg body weight/day because they are at high risk of refeeding syndrome.

Protein Requirements

Based on growing evidence from experimental and epidemiological research, age at hospitalization, muscle mass, gender, and other essential factors should be considered when selecting an appropriate protein intake for patients with COVID-19. In this regard, intake of at least 1 gram protein/kg body weight/day should be attempted for older people (Ballmer 2020; Volkert et al. 2019), particularly for those who are frail, suffer from comorbidities, or are at risk of malnutrition. In general, recent guidelines recommend maintaining the intake of protein within the range of 1.2–2 grams per kilogram of body weight per day with at least 50% of this amount should be of high biological value (HBV) proteins. The amount of protein that patients with COVID-19 and renal failure need is determined between 1 and 1.2 grams per kg body weight per day. It is decreased to 0.8–1 gram protein per kg body weight and day in patients with chronic kidney disease (CKD).

Fat and Carbohydrate Requirements

A 30:70 fat and carbohydrate ratio in patients with no respiratory impairment and a 50:50 percent for ventilated patients may be considered (9).

28.2.2.2 Nutritional Strategies for ICU Patients with COVID-19

Prolonged ICU stays might initiate or aggravate malnutrition accompanied by severe muscle mass wasting and atrophy, which can, in turn, lead to profound disability and morbidity. The situation gets worse when a controlled diet is not present, and a lack of adequate calorie and protein results in exacerbation of nutritional status. Therefore, the prevention of malnutrition is a critical part of COVID-19 management.

Recommendations for nutritional assessment to derive appropriate nutritional strategies are relevant to all critically ill patients, including patients with COVID-19. For non-intubated ICU patients with COVID-19 not achieving the defined energy target with an oral diet, ONS should be considered following EN. Notably, even for patients who can eat independently and unaided, it might be impossible to meet the increased demand for energy and nutrient requirements associated with metabolic stress and recovery. Moreover, critically ill patients dependent on endotracheal intubation and mechanical ventilation are unable to consume food orally. Impairment of chewing and swallowing and anorexia induced by pain-relieving medications are other factors contributing to the inability of oral feeding. The dietician might consider adding PPN in the case patients exhibit limitations with the enteral route or do not reach the defined energy target by oral or enteral nutrition (Gostynska et al. 2019).

Intubated and Ventilated Patients

Enteral Nutrition

ICU patients with COVID-19 who are intubated and ventilated, EN can be administered via a nasogastric tube (NGT). The prone position in itself does not pose a contraindication for EN. However, in the case that the patient shows gastric intolerance or is at high risk for aspiration, post-pyloric feeding must be considered after prokinetic treatment.

Energy Requirements

Indirect calorimetry, when available, is advised as the first-line method of evaluation of energy expenditure (EE). If not available, the second-line method is to assess EE from respiratory gases, e.g., VO2 (oxygen consumption) and VCO2 (carbon dioxide production), which are recorded by the ventilator (Elamin et al. 2012). Predictive equations are in the last line because they can lead to under or overevaluation of needs (Barazzoni et al. 2020).

Protein Requirements

Intake of 1.3 gram protein/kg body weight/day might help to improve the survival rate in frail patients. Depending on the patient's condition, there are, however, a few considerations to bear in mind. For example, preserving skeletal muscle mass is of particular importance for the patient's recovery. It necessitates additional strategies to enforce skeletal muscle anabolism with regard to the highly catabolic state of patients with COVID-19 during ICU admission (McClave et al. 2016).

Lipid Requirements

When administering intravenous lipids, it is important to adhere to the recommendation of 1 g/kg body weight/day with a tolerance of up to 1.5 g/kg/day. Administration of excess lipids may result in waste, storage, or even toxicity (Santacruz et al. 2015).

Carbohydrate Requirements

A low carbohydrate formula, at the maximum rate of 5 mg/kg weight/min, is prescribed to patients for the reduction of CO_2 production and so an improvement of the respiratory function (Taylor et al. 2016; Singer et al. 2019).

Parenteral Nutrition

PN is indicated when the patient is unable to intake sufficient nutrients orally and enterally. It is recommended to be prescribed by day 3–7 after EN intolerance. For patients with infections, PN should meet the following requirements:

Protein Requirements

1.3-grams protein/kg of ideal body weight per day is given (Romano et al. 2020; Heidegger et al. 2013).

Carbohydrate Requirements

Carbohydrates are the primary source of calories in almost all PN formulations. Glucose is the major fuel for the human body. The brain, peripheral nerves, renal medulla, leukocytes, erythrocytes, and bone marrows use glucose as the primary source of oxidative energy. The minimum daily amount of glucose required is estimated to be 100–120 g to meet the needs of the brain. If this requirement is not satisfied by exogenous nutrition, it will be satisfied by gluconeogenesis, which uses amino acid precursors provided by the skeletal muscle proteolysis. The protein-sparing effect of the parenteral provision of glucose is seen in starvation, ensuring the muscle tissue is conserved. It has remained unclear whether such sparing effectively occurs in the critically ill condition.

However, carbohydrate administration should be limited in critically ill patients with COVID-19 and respiratory failure. The carbohydrate requirement is 2 g/kg/day and must not exceed 150 g per day. The oxidation of a mole of carbohydrates leads to the production of equal amounts of CO_2. In patients with respiratory failure, CO_2 production must be avoided to decrease respiratory quotients (Singer et al. 2009, 2019; Gostynska et al. 2019).

Fat Requirements

In long-term ICU patients, lipid intake analysis is an integral part of the energy maintenance program. Due to changes in fat absorption and fat metabolism in critically ill patients undergoing PN, lipid overload and toxicity can occur with excessive intravenous injection of fat, resulting in hypertriglyceridemia and abnormal liver enzymes.

Studies show the association between glycerol concentrations and COVID-19 outcomes. The recommended daily venous fat is 1 gram/kg body weight, which should not exceed the maximum of 1.5 gram/kg body weight (Singer et al. 2009). However, the dose needs to be adjusted according to individual tolerance. Also, for newly critically ill patients with parenteral nutrition, the use of medium and long-chain fatty acids is preferred. Furthermore, the use of omega-3 fatty acids in critically ill patients has been shown to result in a lower risk of infection and death, as well as a decreased length of hospital stay.

Medium- and long-chain fatty acids and an increase in the proportion of ω-3 and ω-9 fatty acids are considered to be the most important factor. Fatty acids significantly affect the immune system. They take part in constructing cell membranes and cytokines, which are factors secreted during immune cell signaling (Ren et al. 2020; Rutting et al. 2019). As an example, ω-3 fatty acids are well-acknowledged to prevent a hyper-inflammatory state by decreasing the production of eicosanoid and specific cytokines. In particular, the eicosapentaenoic acid (EPA; 20:5 ω-3) and docosahexaenoic acid (DHA; 22:6 ω-3) reduce the production of lipid mediators involved in the induction of the pro-inflammatory state. On the contrary, a ω-6 fatty acid, arachidonic acid, has been one of the most crucial eicosanoid precursors in the composition of prostaglandins and leukotrienes, which critically contribute to inflammation.

Fish oil (FO) is a well-established source for EPA and DHA (Kristine Koekkoek et al. 2019). Thus, the administration of FO might be beneficial in improving the prognosis of the critical patient, for its anti-inflammatory and immunomodulatory characteristics (Lu et al. 2017). Enteral FO feeds are, in certain conditions, considered in patients with ARDS (Singer et al. 2006; Pontes-Arruda et al. 2008). Accordingly, an intake of 0.1–0.2 gram EPA + DHA per kg body weight per day is recommended for patients with COVID-19 (Das 2020).

Post-extubation Period and Dysphagia

In ICU patients with dysphagia, if swallowing is proven unsafe, texture-modified food can be considered after extubation, and EN should be administered (Barazzoni et al. 2020). During swallowing training, once the nasoenteral feeding tube is removed, post-pyloric EN can be performed, mainly where high aspiration risk exists. If post-pyloric EN is not possible, temporary PN can be attempted (Neelemaat et al. 2011; Gomes et al. 2018). A key consideration for COVID-19 patients, particularly in the elderly after prolonged extubation, is that the post-extubation swallowing disorder can result in a prolonged recovery process, lasting up to 21 days in many cases (Barazzoni et al. 2020).

The general principles are followed for fluid therapy of patients with COVID-19, to stabilize patients at 30–40 mL/kg per day. In stable ICU patients with COVID-19, it is recommended to maintain 30 mL/kg/day of fluid for adults and 28 mL/kg/day for the elderly, as well as those with large areas of pulmonary consolidation. For every 1 degree C rise in temperature, there is a need for supplemental fluid therapy calculated at 3–5 mL/kg (4 mL/kg) (Yu and Shi 2020; Singer et al. 2019).

28.3 Role of Micronutrients

Table 28.1 gives an overview of the effects of micronutrient deficiency and supplementation on the immune response. Low levels of certain micronutrients such as vitamins A, D, E, C, and B and minerals such as zinc, selenium, and iron can cause a worsening of viral infections (Gombart et al. 2020; Kakodkar et al. 2020). Patients with COVID-19 are at risk for micronutrient deficiency due to reduced intake and increased demands. Although the prevention and treatment of micronutrient deficiencies are significant, there is no evidence that supraphysiological or supra-therapeutic levels of micronutrients will improve COVID 19 outcomes (Barazzoni et al. 2020). Overall, there should be regular allowances to modify the recommendations over the intake of vitamins and trace elements, considering the patient history and clinical characteristics.

28.3.1 Vitamin A

Vitamin A plays a role against infections. Additionally, vitamin A supplementation has been shown to be effective in reducing morbidity and mortality in infectious diseases such as malaria, measles, lung infections, and measles-related pneumonia (Kantoch et al. 2002). Vitamin A, therefore, seems promising for COVID-19 treatment (Zhang and Liu 2020).

28.3.2 B Vitamins

Vitamin B6 is required in protein metabolism. It participates in over 100 other regulatory reactions in various tissues and also plays an important role in the immune system.

Table 28.1 The impact of micronutrient deficiencies and supplementations on the immune response

Micronutrient	Effects of deficiency	Effects of supplementation
Vitamin A	Increased susceptibility to COVID-19 infection	Enhances innate immune response to viral replication
B Vitamins	Depressed host immune responses	Enhances immune response Vitamin B3 provides robust anti-inflammatory response to respiratory tract infection Recommended daily amount for total parenteral nutrition (TPN): 40 mg of vitamin B3 and 6 mg of vitamin B6
Vitamin C	Increased oxidative damage Increased severity of infection	Antioxidant properties prevent the susceptibility to lower respiratory tract infection of COVID-19 Recommended supplementation dose: 1–2 g/day for patients
Vitamin D	Increased susceptibility to acute respiratory tract infections and the severity of infection Increased morbidity and mortality	Reduced acute respiratory tract infection Recommended supplementation dose: 500,000 IU for all ICU patients in the first week
Selenium	Increased oxidative stress Increased viral virulence Induces impairment of host immune function	Improves immune response to COVID-19 in deficient individuals Suppresses the release of free radicals Reduces oxidative damage Recommended supplementation dose: 100 μg/day
Zinc	Depressed immune responses Increased susceptibility to COVID-19 infectious	Recommended supplementation dose: 20–40 mg/day for adults
Iron	Reduced capacity for adequate immune response Increased recurrent acute respiratory tract infection	Its impact can be a double-edged sword: improves immune response to infectious diseases, while irrational abuse of supplementing may increase the availability of iron for pathogen and enhances viral replication

Hyaluronic acid (HA), an extracellular matrix component, is distributed in the lung parenchyma of humans. Based on several studies, increased levels of HA occur in lung tissue of humans with acute respiratory distress syndrome (ARDS). HA can increase the capacity of absorption of water within the lung up to 1000 times its molecular weight. Patients with COVID-19 show dysregulation in the production and regulation of HA (Esposito et al. 2017). Also, they have shown high levels of inflammatory cytokines, which are potent inducers of HA synthase 2 (HAS2) in the alveolar epithelial cells and fibroblasts, and this might explain difficult breathing in COVID-19 patients. Consequently, reducing or inhibiting the production of HA might be promising to ease breathing in COVID-19 patients (Shi et al. 2020).

Vitamin B3 can inhibit HA synthase and diminish the effects of HA. Y. Vitamin B3 supplementation has been shown to significantly reduce neutrophil infiltration into the lungs and also produce a strong anti-inflammatory effect in ventilator-induced lung injury (Shi et al. 2020).

28.3.3 Vitamin C

It is known as an important antioxidant and enzymatic cofactor for physiological reactions such as hormone production, collagen synthesis, and immune potency. Human beings are unable to synthesize ascorbic acid (vitamin C); as a result, it must be obtained from dietary sources. It has been suggested that supplementation with vitamin C may hinder the susceptibility to respiratory tract infections (Hemilä and Louhiala 2013; Kim et al. 2018; Padhani et al. 2020). Vitamin C (1–2 gram per day) is, therefore, suggested for COVID-19 treatment (Chen et al. 2020; Ran et al. 2018).

Also, studies have shown that high-dose intravenous injection of vitamin C (3–10 gram per day) can substantially improve disease outcomes in critically ill patients. So, in addition to the oral intake of vitamin C, some authors suggest the use of high-dose intravenous vitamin C in critically ill patients (Wang et al. 2019; Boretti and Banik 2020). ICU patients who are in the early stages of COVID-19 are candidates for a brief duration of high-dose intravenous vitamin C therapy (Cheng 2020). High-dose vitamin C therapy can induce an inflammatory response. Adding intravenous (IV) glucocorticoid to high-dose vitamin C can help to diminish the inflammatory effect of high-dose vitamin C therapy (Carr 2020) while keeping the other effects of high-dose vitamin C in places, such as improvement of alveolar fluid clearance and epithelial cell functions (Boretti and Banik 2020).

28.3.4 Vitamin D

Vitamin D receptors exist in a variety of immune cells. As a result, vitamin D can affect immunity responses (Zdrenghea et al. 2017). It has been shown to promote the production of monocyte-derived macrophages and modulate the production of inflammatory cytokines. Also, metabolites of vitamin D tend to control the development of different antimicrobial proteins that specifically kill pathogens and are thus likely to help minimize infection, including those occurring in the lungs (Anderson et al. 2020). Deficiency in vitamin D increases the risk of developing respiratory infections (Martineau et al. 2017; Jolliffe et al. 2013; Pham et al. 2019). Observational studies indicate an association between low levels of 25-hydroxyvitamin D (the primary metabolite of vitamin D) in the blood and vulnerability to acute respiratory tract infections. In agreement with these findings, meta-analyses have concluded that supplementation with vitamin D may reduce the risk of respiratory tract infection in children and adults (Gruber-Bzura 2018). While a substantial number of healthy adults, often at the end of the winter season, present with low vitamin D levels.

COVID-19 was first identified in winter 2019 and has mostly affected elderly and middle-age people. Hence, it may be supposed that the virus has infected people who may already have had vitamin D insufficiency (Grant et al. 2020; Jakovac 2020).

Patients with COVID-19 undergoing EN and PN should be tested for vitamin D deficiency. If the amount of vitamin D is less than 12.5 ng/mL (insufficiency), cholecalciferol is given by intramuscular injection or via the route of administration of EN with a maximum single dose of 100,000 IU solution and 500,000 IU per week.

28.3.5 Selenium

Selenium deficiency may induce oxidative stress, which can cause a mild or moderate pathogenic virus to become highly virulent (Rayman 2012). Selenium deficiency induces not only impairment of the host immune system but also can evoke mutations in the RNA viruses (Singer et al. 2019).

Selenium assists a group of enzymes working in conjunction with vitamin E to prevent the formation of free radicals and oxidative damage to cells and tissues (Calder et al. 2020). Synergistic effects of selenium with some herbal products might help to improve immune response to coronavirus infections. Hence, 100 micrograms of selenium supplementation per day may be an important option for the supplemental treatment of the novel coronavirus infection. These, however, should be recommended based on the Dietary Reference Intakes (DRI) for infected patients with renal failure (Zhang and Liu 2020; Jayawardena et al. 2020).

28.3.6 Zinc

Zinc deficiency induces both humoral and cell-mediated immunity dysfunction and increases the vulnerability to infectious diseases. Intracellular zinc can effectively inhibit the replication of a variety of RNA viruses. Furthermore, the low concentration of zinc and pyrithione prevents the replication of SARS coronavirus (SARS-CoV) (te Velthuis et al. 2010). For adults, a recommended dosage is 20–40 mg/day (Gombart et al. 2020). Zinc supplement can, therefore, be an option for the treatment of COVID-19-related symptoms (Zhang and Liu 2020).

28.3.7 Iron

Iron deficiency was confirmed to be a risk factor for persistent acute respiratory tract infections. Furthermore, iron is essential for viral replication (Wessling-Resnick 2018). In other words, iron is necessary for both the host and pathogen, and iron deficiency can inhibit host immunity, whereas iron excess can allow oxidative stress to spread dangerous viral mutations (Liu et al. 2020).

28.4 Conclusion

Infectious disease outbreaks have ever been challenging for humanity in terms of diagnosis and treatment (Basiri et al. 2020a). It is the COVID-19 case as well, though the origin of COVID-19 comes back to coronaviruses, which previously have caused epidemics (Jabbari et al. 2020; Hanaei and Rezaei 2020). Indeed, COVID-19 caused a more widespread pandemic involving both the society and healthcare system (Rezaei 2020a; Moazzami et al. 2020). It has raised many concerns about the risk of infection and reinfection in frail older people who are malnourished; people in special physiological conditions, for example, pregnant women, neonates, and children; people with medical conditions who suffer from an immune deficiency or are immunocompromised; and patients from lower socioeconomic groups (Ahanchian et al. 2020; Babaha and Rezaei 2020; Mirbeyk and Rezaei 2020; Sahu et al. 2020; Jabbari and Rezaei 2020). During the pandemic, prompt and appropriate nutritional therapy play a critical role in the management of COVID-19 in all phases of care. Nutritional therapy can provide highly effective results, ensuring that all patients have a fair chance to fight COVID-19. The current status is, however, different; research has focused on pharmaceutical treatment of COVID-19, while the importance of nutrition is being overlooked with grave consequences. It calls a need for global recommendations on the role of nutrition in the ongoing and future pandemics (Momtazmanesh et al. 2020; Mohamed et al. 2020a; Rzymski et al. 2020; Moradian et al. 2020; Kafieh et al. 2020).

References

Ahanchian H, Moazzen N, Faroughi MSD, Khalighi N, Khoshkhui M, Aelami MH, Haghi NSM, Rezaei N (2020) COVID-19 in a child with primary specific antibody deficiency

Ahmadi M, Saffarzadeh N, Habibi MA, Hajiesmaeili F, Rezaei N (2020) Colon cancer and SARS-CoV-2: impact of ACE2 expression in susceptibility to COVID-19. bioRxiv

Anderson J, Do LAH, Toh ZQ, Hoe E, Reitsma A, Mulholland K, Licciardi PV (2020) Vitamin D induces differential effects on inflammatory responses during bacterial and/or viral stimulation of human peripheral blood mononuclear cells. Front Immunol 11:602. https://doi.org/10.3389/fimmu.2020.00602

Babaha F, Rezaei N (2020) Primary immunodeficiency diseases in COVID-19 pandemic: a predisposing or protective factor? Am J Med Sci. https://doi.org/10.1016/j.amjms.2020.07.027

Bahrami A, Vafapour M, Moazzami B, Rezaei N (2020) hyper-inflammatory shock related to COVID-19 in a patient presenting with multisystem inflammatory syndrome in children: first case from Iran. J Paediatr Child Health. https://doi.org/10.1111/jpc.15048

Ballmer PE (2020) Individualised nutritional support in medical inpatients – a practical guideline. Swiss Med Wkly 150:w20245. https://doi.org/10.4414/smw.2020.20245

Barazzoni R, Bischoff SC, Breda J, Wickramasinghe K, Krznaric Z, Nitzan D, Pirlich M, Singer P (2020) ESPEN expert statements and practical guidance for nutritional management of individuals with SARS-CoV-2 infection. Clin Nutr 39:1631. https://doi.org/10.1016/j.clnu.2020.03.022

Basiri A, Heidari A, Nadi MF, Fallahy MTP, Nezamabadi SS, Sedighi M, Saghazadeh A, Rezaei N (2020a) Microfluidic devices for detection of RNA viruses. Rev Med Virol n/a (n/a):e2154. https://doi.org/10.1002/rmv.2154

Basiri A, Pazhouhnia Z, Beheshtizadeh N, Hoseinpour M, Saghazadeh A, Rezaei N (2020b) Regenerative medicine in COVID-19 treatment: real opportunities and range of promises. Stem Cell Rev Rep:1–13. https://doi.org/10.1007/s12015-020-09994-5

Boretti A, Banik BK (2020) Intravenous vitamin C for reduction of cytokines storm in acute respiratory distress syndrome. PharmaNutrition 12:100190. https://doi.org/10.1016/j.phanu.2020.100190

Calder PC, Carr AC, Gombart AF, Eggersdorfer M (2020) Optimal nutritional status for a well-functioning immune system is an important factor to protect against viral infections. Nutrients 12(4). https://doi.org/10.3390/nu12041181

Carr AC (2020) A new clinical trial to test high-dose vitamin C in patients with COVID-19. Crit Care 24(1):133. https://doi.org/10.1186/s13054-020-02851-4

Cederholm T, Jensen GL, Correia M, Gonzalez MC, Fukushima R, Higashiguchi T, Baptista G, Barazzoni R, Blaauw R, Coats A, Crivelli A, Evans DC, Gramlich L, Fuchs-Tarlovsky V, Keller H, Llido L, Malone A, Mogensen KM, Morley JE, Muscaritoli M, Nyulasi I, Pirlich M, Pisprasert V, de van der Schueren MAE, Siltharm S, Singer P, Tappenden K, Velasco N, Waitzberg D, Yamwong P, Yu J, Van Gossum A, Compher C (2019) GLIM criteria for the diagnosis of malnutrition – a consensus report from the global clinical nutrition community. Clin Nutr 38(1):1–9. https://doi.org/10.1016/j.clnu.2018.08.002

Chen L, Hu C, Hood M, Zhang X, Zhang L, Kan J, Du J (2020) A novel combination of vitamin C, curcumin and glycyrrhizic acid potentially regulates immune and inflammatory response associated with coronavirus infections: a perspective from system biology analysis. Nutrients 12(4). https://doi.org/10.3390/nu12041193

Cheng RZ (2020) Can early and high intravenous dose of vitamin C prevent and treat coronavirus disease 2019 (COVID-19)? Med Drug Discov 5:100028. https://doi.org/10.1016/j.medidd.2020.100028

Darbeheshti F, Rezaei N (2020) Genetic predisposition models to COVID-19 infection. Med Hypotheses 142:109818. https://doi.org/10.1016/j.mehy.2020.109818

Das UN (2020) Can bioactive lipids inactivate coronavirus (COVID-19)? Arch Med Res 51:282. https://doi.org/10.1016/j.arcmed.2020.03.004

Elamin EM, Miller AC, Ziad S (2012) Immune enteral nutrition can improve outcomes in medical-surgical patients with ARDS: a prospective randomized controlled trial. J Nutr Disord Ther 2:109. https://doi.org/10.4172/2161-0509.1000109

Esposito AJ, Bhatraju PK, Stapleton RD, Wurfel MM, Mikacenic C (2017) Hyaluronic acid is associated with organ dysfunction in acute respiratory distress syndrome. Crit Care 21(1):304. https://doi.org/10.1186/s13054-017-1895-7

Fathi N, Rezaei N (2020) Lymphopenia in COVID-19: therapeutic opportunities. Cell Biol Int 44(9):1792–1797. https://doi.org/10.1002/cbin.11403

Gombart AF, Pierre A, Maggini S (2020) A review of micronutrients and the immune system-working in harmony to reduce the risk of infection. Nutrients 12(1). https://doi.org/10.3390/nu12010236

Gomes F, Schuetz P, Bounoure L, Austin P, Ballesteros-Pomar M, Cederholm T, Fletcher J, Laviano A, Norman K, Poulia KA, Ravasco P, Schneider SM, Stanga Z, Weekes CE, Bischoff SC (2018) ESPEN guidelines on nutritional support for polymorbid internal medicine patients. Clin Nutr 37(1):336–353. https://doi.org/10.1016/j.clnu.2017.06.025

Gostynska A, Stawny M, Dettlaff K, Jelinska A (2019) Clinical nutrition of critically ill patients in the context of the latest ESPEN guidelines. Medicina (Kaunas) 55(12). https://doi.org/10.3390/medicina55120770

Grant WB, Lahore H, McDonnell SL, Baggerly CA, French CB, Aliano JL, Bhattoa HP (2020) Evidence that vitamin D supplementation could reduce risk of influenza and COVID-19 infections and deaths. Nutrients 12(4). https://doi.org/10.3390/nu12040988

Gruber-Bzura BM (2018) Vitamin D and influenza-prevention or therapy? Int J Mol Sci 19(8). https://doi.org/10.3390/ijms19082419

Hanaei S, Rezaei N (2020) COVID-19: developing from an outbreak to a pandemic. Arch Med Res 51:582. https://doi.org/10.1016/j.arcmed.2020.04.021

Heidegger CP, Berger MM, Graf S, Zingg W, Darmon P, Costanza MC, Thibault R, Pichard C (2013) Optimisation of energy provision with supplemental parenteral nutrition in critically ill patients: a randomised controlled clinical trial. Lancet 381(9864):385–393. https://doi.org/10.1016/s0140-6736(12)61351-8

Hemilä H, Louhiala P (2013) Vitamin C for preventing and treating pneumonia. Cochrane Database Syst Rev (8):Cd005532. https://doi.org/10.1002/14651858.CD005532.pub3

High KP (2001) Nutritional strategies to boost immunity and prevent infection in elderly individuals. Clin Infect Dis 33(11):1892–1900. https://doi.org/10.1086/324509

Jabbari P, Rezaei N (2020) With risk of reinfection, is COVID-19 here to stay? Disaster Med Public Health Prep:1. https://doi.org/10.1017/dmp.2020.274

Jabbari P, Jabbari F, Ebrahimi S, Rezaei N (2020) COVID-19: a chimera of two pandemics. Disaster Med Public Health Prep 14:1–2. https://doi.org/10.1017/dmp.2020.223

Jahanshahlu L, Rezaei N (2020a) Central Nervous System Involvement in COVID-19. Arch Med Res 51:721. https://doi.org/10.1016/j.arcmed.2020.05.016

Jahanshahlu L, Rezaei N (2020b) Monoclonal antibody as a potential anti-COVID-19. Biomed Pharmacother 129:110337. https://doi.org/10.1016/j.biopha.2020.110337

Jakovac H (2020) COVID-19 and vitamin D-is there a link and an opportunity for intervention? Am J Physiol Endocrinol Metab 318(5):E589. https://doi.org/10.1152/ajpendo.00138.2020

Jayawardena R, Sooriyaarachchi P, Chourdakis M, Jeewandara C, Ranasinghe P (2020) Enhancing immunity in viral infections, with special emphasis on COVID-19: a review. Diabetes Metab Syndr 14(4):367–382. https://doi.org/10.1016/j.dsx.2020.04.015

Jolliffe DA, Griffiths CJ, Martineau AR (2013) Vitamin D in the prevention of acute respiratory infection: systematic review of clinical studies. J Steroid Biochem Mol Biol 136:321–329. https://doi.org/10.1016/j.jsbmb.2012.11.017

Kafieh R, Arian R, Saeedizadeh N, Minaee S, Yadav SK, Vaezi A, Rezaei N, Javanmard SH (2020) COVID-19 in Iran: a deeper look into the future. medRxiv

Kakodkar P, Kaka N, Baig MN (2020) A comprehensive literature review on the clinical presentation, and management of the pandemic coronavirus disease 2019 (COVID-19). Cureus 12(4):e7560. https://doi.org/10.7759/cureus.7560

Kantoch M, Litwinska B, Szkoda M, Siennicka J (2002) Importance of vitamin A deficiency in pathology and immunology of viral infections. Rocz Panstw Zakl Hig 53(4):385–392

Kim WY, Jo EJ, Eom JS, Mok J, Kim MH, Kim KU, Park HK, Lee MK, Lee K (2018) Combined vitamin C, hydrocortisone, and thiamine therapy for patients with severe pneumonia who were admitted to the intensive care unit: propensity score-based analysis of a before-after cohort study. J Crit Care 47:211–218. https://doi.org/10.1016/j.jcrc.2018.07.004

Kondrup J, Allison SP, Elia M, Vellas B, Plauth M (2003) ESPEN guidelines for nutrition screening 2002. Clin Nutr 22(4):415–421. https://doi.org/10.1016/s0261-5614(03)00098-0

Kristine Koekkoek W, Panteleon V, van Zanten AR (2019) Current evidence on omega-3 fatty acids in enteral nutrition in the critically ill: a systematic review and meta-analysis. Nutrition 59:56–68. https://doi.org/10.1016/j.nut.2018.07.013

Liu W, Zhang S, Nekhai S, Liu S (2020) Depriving iron supply to the virus represents a promising adjuvant therapeutic against viral survival. Curr Clin Microbiol Rep 7:1–7. https://doi.org/10.1007/s40588-020-00140-w

Lotfi M, Rezaei N (2020) SARS-CoV-2: a comprehensive review from pathogenicity of the virus to clinical consequences. J Med Virol 92:1864. https://doi.org/10.1002/jmv.26123

Lotfi M, Hamblin MR, Rezaei N (2020) COVID-19: transmission, prevention, and potential therapeutic opportunities Clinica chimica acta. Int J Clin Chem 508:254–266. https://doi.org/10.1016/j.cca.2020.05.044

Lu C, Sharma S, McIntyre L, Rhodes A, Evans L, Almenawer S, Leduc L, Angus DC, Alhazzani W (2017) Omega-3 supplementation in patients with sepsis: a systematic review and meta-analysis of randomized trials. Ann Intensive Care 7(1):58. https://doi.org/10.1186/s13613-017-0282-5

Maggini S, Pierre A, Calder PC (2018) Immune function and micronutrient requirements change over the life course. Nutrients 10(10). https://doi.org/10.3390/nu10101531

Mansourabadi AH, Sadeghalvad M, Mohammadi-Motlagh HR, Rezaei N (2020) The immune system as a target for therapy of SARS-CoV-2: a systematic review of the current immunotherapies for COVID-19. Life Sci 258:118185. https://doi.org/10.1016/j.lfs.2020.118185

Martineau AR, Jolliffe DA, Hooper RL, Greenberg L, Aloia JF, Bergman P, Dubnov-Raz G, Esposito S, Ganmaa D, Ginde AA, Goodall EC, Grant CC, Griffiths CJ, Janssens W, Laaksi I, Manaseki-Holland S, Mauger D, Murdoch DR, Neale R, Rees JR, Simpson S Jr, Stelmach I, Kumar GT, Urashima M, Camargo CA Jr (2017) Vitamin D supplementation to prevent acute respiratory tract infections: systematic review and meta-analysis of individual participant data. BMJ 356:i6583. https://doi.org/10.1136/bmj.i6583

McClave SA, Taylor BE, Martindale RG, Warren MM, Johnson DR, Braunschweig C, McCarthy MS, Davanos E, Rice TW, Cresci GA, Gervasio JM, Sacks GS, Roberts PR, Compher C (2016) Guidelines for the provision and assessment of nutrition support therapy in the adult critically ill patient: Society of Critical Care Medicine (SCCM) and American Society for Parenteral and Enteral Nutrition (A.S.P.E.N.). JPEN J Parenter Enteral Nutr 40(2):159–211. https://doi.org/10.1177/0148607115621863

Mirbeyk M, Rezaei N (2020) The impact of COVID-19 on pregnancy and neonatal health: a systematic review

Moazzami B, Razavi-Khorasani N, Dooghaie Moghadam A, Farokhi E, Rezaei N (2020) COVID-19 and telemedicine: immediate action required for maintaining healthcare providers well-being. J Clin Virol 126:104345. https://doi.org/10.1016/j.jcv.2020.104345

Mohamed K, Rodríguez-Román E, Rahmani F, Zhang H, Ivanovska M, Makka SA, Joya M, Makuku R, Islam MS, Radwan N, Rahmah L, Goda R, Abarikwu SO, Shaw M, Zoghi S, Irtsyan S, Ling I, Cseprekal O, Faten AB, Hazar Sayar E, Soloukey C, Grancini G, Rezaei N (2020a) Borderless collaboration is needed for COVID-19-a disease that knows no borders. Infect Control Hosp Epidemiol 41:1–2. https://doi.org/10.1017/ice.2020.162

Mohamed K, Yazdanpanah N, Saghazadeh A, Rezaei N (2020b) Computational drug discovery and repurposing for the treatment of Covid-19: a systematic review. Available at SSRN 3583748

Momtazmanesh S, Ochs HD, Uddin LQ, Perc M, Routes JM, Vieira DN, Al-Herz W, Baris S, Prando C, Rosivall L, Abdul Latiff AH, Ulrichs T, Roudenok V, Aldave Becerra JC, Salunke DB, Goudouris E, Condino-Neto A, Stashchak A, Kryvenko O, Stashchak M, Bondarenko A, Rezaei N (2020) All together to fight COVID-19. Am J Trop Med Hyg 102(6):1181–1183. https://doi.org/10.4269/ajtmh.20-0281

Moradian N, Ochs HD, Sedikies C, Hamblin MR, Camargo CA Jr, Martinez JA, Biamonte JD, Abdollahi M, Torres PJ, Nieto JJ, Ogino S, Seymour JF, Abraham A, Cauda V, Gupta S, Ramakrishna S, Sellke FW, Sorooshian A, Wallace Hayes A, Martinez-Urbistondo M, Gupta M, Azadbakht L, Esmaillzadeh A, Kelishadi R, Esteghamati A, Emam-Djomeh Z, Majdzadeh R, Palit P, Badali H, Rao I, Saboury AA, Jagan Mohan Rao L, Ahmadieh H, Montazeri A, Fadini GP, Pauly D, Thomas S, Moosavi-Movahed AA, Aghamohammadi A, Behmanesh M, Rahimi-Movaghar V, Ghavami S, Mehran R, Uddin LQ, Von Herrath M, Mobasher B, Rezaei N (2020) The urgent need for integrated science to fight COVID-19 pandemic and beyond. J Transl Med 18(1):205. https://doi.org/10.1186/s12967-020-02364-2

Nasab MG, Saghazadeh A, Rezaei N (2020) SARS-CoV-2-a tough opponent for the immune system. Arch Med Res 51:589. https://doi.org/10.1016/j.arcmed.2020.05.020

Neelemaat F, Bosmans JE, Thijs A, Seidell JC, van Bokhorst-de van der Schueren MA (2011) Post-discharge nutritional support in malnourished elderly individuals improves functional limitations. J Am Med Dir Assoc 12(4):295–301. https://doi.org/10.1016/j.jamda.2010.12.005

Padhani ZA, Moazzam Z, Ashraf A, Bilal H, Salam RA, Das JK, Bhutta ZA (2020) Vitamin C supplementation for prevention and treatment of pneumonia. Cochrane Database Syst Rev 4(4):Cd013134. https://doi.org/10.1002/14651858.CD013134.pub2

Pashaei M, Rezaei N (2020) Immunotherapy for SARS-CoV-2: potential opportunities. Expert Opin Biol Ther 20:1–5. https://doi.org/10.1080/14712598.2020.1807933

Pham H, Rahman A, Majidi A, Waterhouse M, Neale RE (2019) Acute respiratory tract infection and 25-hydroxyvitamin D concentration: a systematic review and meta-analysis. Int J Environ Res Public Health 16(17). https://doi.org/10.3390/ijerph16173020

Pontes-Arruda A, Demichele S, Seth A, Singer P (2008) The use of an inflammation-modulating diet in patients with acute lung injury or acute respiratory distress syndrome: a meta-analysis of outcome data. JPEN J Parenter Enteral Nutr 32(6):596–605. https://doi.org/10.1177/0148607108324203

Pourahmad R, Moazzami B, Rezaei N (2020) Efficacy of plasmapheresis and immunoglobulin replacement therapy (IVIG) on patients with COVID-19. SN Compr Clin Med:1–5

Rabiee N, Rabiee M, Bagherzadeh M, Rezaei N (2020) COVID-19 and picotechnology: potential opportunities. Med Hypotheses 144:109917. https://doi.org/10.1016/j.mehy.2020.109917

Ran L, Zhao W, Wang J, Wang H, Zhao Y, Tseng Y, Bu H (2018) Extra dose of vitamin C based on a daily supplementation shortens the common cold: a meta-analysis of 9 randomized controlled trials. Biomed Res Int 2018:1837634. https://doi.org/10.1155/2018/1837634

Rayman MP (2012) Selenium and human health. Lancet 379(9822):1256–1268. https://doi.org/10.1016/s0140-6736(11)61452-9

Ren Y, Yao MC, Huo XQ, Gu Y, Zhu WX, Qiao YJ, Zhang YL (2020) Study on treatment of "cytokine storm" by anti-2019-nCoV prescriptions based on arachidonic acid metabolic pathway. Zhongguo Zhong Yao Za Zhi 45(6):1225–1231. https://doi.org/10.19540/j.cnki.cjcmm.20200224.405

Rezaei N (2020a) COVID-19 affects healthy pediatricians more than pediatric patients. Infect Control Hosp Epidemiol 1:1106. https://doi.org/10.1017/ice.2020.139

Rezaei N (2020b) COVID-19 and medical biotechnology. Avicenna J Med Biotechnol 12(3):139

Rokni M, Hamblin MR, Rezaei N (2020) Cytokines and COVID-19: friends or foes? Hum Vaccin Immunother:1–3

Romano L, Bilotta F, Dauri M, Macheda S, Pujia A, De Santis GL, Tarsitano MG, Merra G, Di Renzo L, Esposito E, De Lorenzo A (2020) Short report – medical nutrition therapy for critically ill patients with COVID-19. Eur Rev Med Pharmacol Sci 24(7):4035–4039. https://doi.org/10.26355/eurrev_202004_20874

Rutting S, Zakarya R, Bozier J, Xenaki D, Horvat JC, Wood LG, Hansbro PM, Oliver BG (2019) Dietary fatty acids amplify inflammatory responses to infection through p38 MAPK signaling. Am J Respir Cell Mol Biol 60(5):554–568. https://doi.org/10.1165/rcmb.2018-0215OC

Rzymski P, Nowicki M, Mullin GE, Abraham A, Rodríguez-Román E, Petzold MB, Bendau A, Sahu KK, Ather A, Naviaux A-F (2020) Quantity does not equal quality: scientific principles cannot be sacrificed. Int Immunopharmacol 86:106711

Saghazadeh A, Rezaei N (2020a) Immune-epidemiological parameters of the novel coronavirus – a perspective. Expert Rev Clin Immunol 16(5):465–470. https://doi.org/10.1080/1744666x.2020.1750954

Saghazadeh A, Rezaei N (2020b) Towards treatment planning of COVID-19: rationale and hypothesis for the use of multiple immunosuppressive agents: anti-antibodies, immunoglobulins, and corticosteroids. Int Immunopharmacol 84:106560. https://doi.org/10.1016/j.intimp.2020.106560

Sahu KK, Siddiqui AD, Rezaei N, Cerny J (2020) Challenges for management of immune thrombocytopenia during COVID-19 pandemic. J Med Virol 92:2277. https://doi.org/10.1002/jmv.26251

Saleki K, Banazadeh M, Saghazadeh A, Rezaei N (2020) The involvement of the central nervous system in patients with COVID-19. Rev Neurosci 31(4):453–456. https://doi.org/10.1515/revneuro-2020-0026

Santacruz CA, Orbegozo D, Vincent JL, Preiser JC (2015) Modulation of dietary lipid composition during acute respiratory distress syndrome: systematic review and meta-analysis. JPEN J Parenter Enteral Nutr 39(7):837–846. https://doi.org/10.1177/0148607114562913

Shamshirian D, Rezaei N (2020) Cardiovascular diseases burden in COVID-19: systematic review and meta-analysis

Sharifkashani S, Bafrani MA, Khaboushan AS, Pirzadeh M, Kheirandish A, Yavarpour Bali H, Hessami A, Saghazadeh A, Rezaei N (2020) Angiotensin-converting enzyme 2 (ACE2) receptor and SARS-CoV-2: potential therapeutic targeting. Eur J Pharmacol 884:173455. https://doi.org/10.1016/j.ejphar.2020.173455

Shi Y, Wang Y, Shao C, Huang J, Gan J, Huang X, Bucci E, Piacentini M, Ippolito G, Melino G (2020) COVID-19 infection: the perspectives on immune responses. Cell Death Differ 27(5):1451–1454. https://doi.org/10.1038/s41418-020-0530-3

Singer P, Theilla M, Fisher H, Gibstein L, Grozovski E, Cohen J (2006) Benefit of an enteral diet enriched with eicosapentaenoic acid and gamma-linolenic acid in ventilated patients with acute lung injury. Crit Care Med 34(4):1033–1038. https://doi.org/10.1097/01.Ccm.0000206111.23629.0a

Singer P, Berger MM, Van den Berghe G, Biolo G, Calder P, Forbes A, Griffiths R, Kreyman G, Leverve X, Pichard C, Espen (2009) ESPEN guidelines on parenteral nutrition: intensive care. Clin Nutr 28(4):387–400. https://doi.org/10.1016/j.clnu.2009.04.024

Singer P, Blaser AR, Berger MM, Alhazzani W, Calder PC, Casaer MP, Hiesmayr M, Mayer K, Montejo JC, Pichard C, Preiser JC, van Zanten ARH, Oczkowski S, Szczeklik W, Bischoff SC (2019) ESPEN guideline on clinical nutrition in the intensive care unit. Clin Nutr 38(1):48–79. https://doi.org/10.1016/j.clnu.2018.08.037

Stratton RJ, Hackston A, Longmore D, Dixon R, Price S, Stroud M, King C, Elia M (2004) Malnutrition in hospital outpatients and inpatients: prevalence, concurrent validity and ease of use of the 'malnutrition universal screening tool' ('MUST') for adults. Br J Nutr 92(5):799–808. https://doi.org/10.1079/bjn20041258

Taylor BE, McClave SA, Martindale RG, Warren MM, Johnson DR, Braunschweig C, McCarthy MS, Davanos E, Rice TW, Cresci GA, Gervasio JM, Sacks GS, Roberts PR, Compher C (2016) Guidelines for the provision and assessment of nutrition support therapy in the adult critically ill patient: Society of Critical Care Medicine (SCCM) and American Society for Parenteral and Enteral Nutrition (A.S.P.E.N.). Crit Care Med 44(2):390–438. https://doi.org/10.1097/ccm.0000000000001525

te Velthuis AJ, van den Worm SH, Sims AC, Baric RS, Snijder EJ, van Hemert MJ (2010) Zn(2+) inhibits coronavirus and arterivirus RNA polymerase activity in vitro and zinc ionophores block the replication of these viruses in cell culture. PLoS Pathog 6(11):e1001176. https://doi.org/10.1371/journal.ppat.1001176

Volkert D, Beck AM, Cederholm T, Cruz-Jentoft A, Goisser S, Hooper L, Kiesswetter E, Maggio M, Raynaud-Simon A, Sieber CC, Sobotka L, van Asselt D, Wirth R, Bischoff SC (2019) ESPEN guideline on clinical nutrition and hydration in geriatrics. Clin Nutr 38(1):10–47. https://doi.org/10.1016/j.clnu.2018.05.024

Wang Y, Lin H, Lin BW, Lin JD (2019) Effects of different ascorbic acid doses on the mortality of critically ill patients: a meta-analysis. Ann Intensive Care 9(1):58. https://doi.org/10.1186/s13613-019-0532-9

Wessling-Resnick M (2018) Crossing the iron gate: why and how transferrin receptors mediate viral entry. Annu Rev Nutr 38:431–458. https://doi.org/10.1146/annurev-nutr-082117-051749

Yazdanpanah F, Hamblin MR, Rezaei N (2020a) The immune system and COVID-19: friend or foe? Life Sci 256:117900. https://doi.org/10.1016/j.lfs.2020.117900

Yazdanpanah N, Saghazadeh A, Rezaei N (2020b) Anosmia: a missing link in the neuroimmunology of coronavirus disease 2019 (COVID-19). Rev Neurosci 31:691. https://doi.org/10.1515/revneuro-2020-0039

Yousefzadegan S, Rezaei N (2020) Case report: death due to COVID-19 in three brothers. Am J Trop Med Hyg 102(6):1203–1204. https://doi.org/10.4269/ajtmh.20-0240

Yu KY, Shi HP (2020) Explanation of expert recommendations on medical nutrition for patients with novel coronavirus pneumonia. Zhonghua Yi Xue Za Zhi 100(10):724–728. https://doi.org/10.3760/cma.j.cn112137-20200205-00196

Zdrenghea MT, Makrinioti H, Bagacean C, Bush A, Johnston SL, Stanciu LA (2017) Vitamin D modulation of innate immune responses to respiratory viral infections. Rev Med Virol 27(1). https://doi.org/10.1002/rmv.1909

Zhang L, Liu Y (2020) Potential interventions for novel coronavirus in China: a systematic review. J Med Virol 92(5):479–490. https://doi.org/10.1002/jmv.25707

Dietary Supplements for COVID-19

29

Gerard E. Mullin, Berkeley Limektkai, Lin Wang,
Patrick Hanaway, Loren Marks,
and Edward Giovannucci

Abstract

Coronavirus disease 2019 (COVID-19) is a highly contagious infectious disease that can rapidly escalate to respiratory failure and death. It has infected millions of people worldwide. The trajectory of this disease continues to progress in some areas of the United States and worldwide. The Institute for Health Metrics now predicts a resurgence of infections in the fall of 2020. The pathogenesis of COVID-19 includes an inflammatory phase with either resolution or the potential to accelerate to a cytokine storm, characterized by high interleukin (IL)-6 and other inflammatory markers. COVID-19 is a condition without a gold-standard treatment. The US Federal Drug Administration (FDA) issued an emergency use authorization for remdesivir in severe cases of COVID-19, which shortened the recovery time in hospitalized patients with lower respiratory tract infection in one study. Although several vaccine trials are underway, no vaccines are available for primary prevention of COVID-19 at this time. Dietary supplement sales have dramatically risen during the COVID-19 pandemic despite depressed economic conditions. Commonly used immune-modulating dietary supplements, including vitamin D, ascorbic acid, zinc, and melatonin, are reviewed in this manuscript highlighting biological plausibility for salutary benefit against COVID-19. Ongoing clinical trials recruiting subjects at the time of this writing are provided for each dietary supplement.

Keywords

COVID-19 · Cytokine storm · Inflammasome · Melatonin · SARS-CoV-2 · Vitamin D · Vitamin C · Zinc

G. E. Mullin (✉)
The Johns Hopkins School of Medicine,
Baltimore, MD, USA

Department of Medicine, Johns Hopkins Hospital,
Baltimore, MD, USA
e-mail: gmullin1@jhmi.edu

B. Limektkai
The University of California Los Angeles School of Medicine, Los Angeles, CA, USA

L. Wang
The Johns Hopkins Bloomberg School of Public Health, Baltimore, MD, USA

P. Hanaway
Cleveland Clinic Center for Functional Medicine,
Cleveland, OH, USA

L. Marks
Integrative Nutrition, New York, NY, USA

E. Giovannucci
The Harvard TH Chan School of Public Health,
Boston, MA, USA

N. Rezaei (ed.), *Coronavirus Disease - COVID-19*, Advances in Experimental Medicine and Biology 1318, https://doi.org/10.1007/978-3-030-63761-3_29

29.1 Introduction

29.1.1 COVID-19: Epidemiology and Pathogenesis

Coronavirus disease 2019 (COVID-19) is responsible for the third major coronavirus (CoV) pandemic in recent history (Mahase 2020). There is no definitive treatment for COVID-19, which remains a highly contagious infectious disease that can escalate rapidly to respiratory failure and death (Mahase 2020). The pandemic of COVID-19 has infected over nine million people worldwide, including over two million people in the United States ((JHU); Jin et al. 2015; Urwyler et al. 2019). The trajectory of disease continues to progress in some areas of the United States and worldwide, and the University of Washington Institute for Health Metrics and Evaluation (IHME) now predicts a resurgence of infections in the fall of 2020 ((IHME) 2020). Overall, 81% of adults infected by COVID-19 self-resolve while 19% progress, and many require hospitalization ((JHU)). Older people aged 65 or over and those with comorbidities (immunocompromised, cardiopulmonary disease, cancer, diabetes, obesity, and kidney failure) are most at risk to experience rapid disease progression, respiratory deterioration requiring intensive care unit (ICU) admission, and mortality (Park et al. 2020; Zhou et al. 2020). COVID-19 is classified according to four levels of severity based on symptoms: mild, moderate, severe, and critical (Siddiqi and Mehra 2020). Critical disease has a 49% case fatality rate related to acute respiratory distress syndrome (ARDS), systemic inflammatory response syndrome (SIRS), septic shock, and multiorgan failure) (Wang et al. 2020c).

The pathogenesis of COVID-19 illness proceeds variably according to distinct phases of the disease:

- Incubation with asymptomatic viral replication.
- Symptomatic with constitutional, respiratory, and other systemic symptoms (i.e., gastrointestinal) of variable duration and severity.
- An inflammatory phase with either resolution or acceleration to a cytokine storm, whereby severe illness is associated with increased plasma concentrations of pro-inflammatory cytokines, including interleukin (IL)-6, IL-10, granulocyte colony-stimulating factor (G-CSF), monocyte chemoattractant protein 1 (MCP1), macrophage inflammatory protein (MIP)-1α, and tumor necrosis factor (TNF)-α (Huang et al. 2020; Alunno et al. 2020). Most severe conditions, such as multiorgan failure, occur in those with high IL-6 levels (Han et al. 2020).

The coronavirus responsible for COVID-19, severe acute respiratory syndrome coronavirus 2 (SARS-CoV-2), also attaches to toll-like receptors (TLRs) in pulmonary macrophages, causing IL-1-beta production and inflammasome activation. Once inflammasomes are activated, and injurious pro-inflammatory cytokines are produced, an extensive injury with loss of lung tissue and subsequent fibrosis with permanent respiratory dysfunction may result (Conti et al. 2020). Respiratory decompensation is frequent and can include the development of ARDS requiring mechanical ventilation. Angiotensin-converting enzyme 2 (ACE2) converts angiotensin I to angiotensin (1–7), which regulates blood pressure, systemic vascular resistance, and fluid-electrolyte balance. COVID-19 infects human cells as spike (S) protein enters cells via ACE2 receptors. Once inside the cell, furin facilitates the cleavage of COVID-19 spike protein by transmembrane protease serine 2. The glycoprotein cleavage byproduct of S protein facilitates the entry of viral genetic material into the cell. This step appears to be required for infection of lung tissues and occurs in other aggressive viral infections such as dengue, avian influenza, and anthrax (Shang et al. 2020; Walls et al. 2020). However, furin protease activation has not been seen with prior coronavirus infections or its ancestor viruses (i.e., SARS-CoV-1). Conditions associated with elevated furin, including diabetes, obesity, and hypertension, overlap significantly with vulnerability to the severe form of COVID-19. ACE2 is constitutively expressed in respiratory and oral epithelium, lung parenchyma, and other tissues, including gut epithelium. Increased ACE2 activity under physiologic conditions appears to protect against COVID-19 (Brojakowska et al. 2020). However, higher expression of ACE2 on the surface of cells in COVID-19 is linked to heightened disease severity (Brojakowska et al. 2020). Furthermore, it seems that furin is implicated in the pathogenesis of SARS-CoV-2 infection and potentially in the increased rates of human-to-human transmission. ACE2 expression appears to be associated with COVID-19 severity.

NOD-, LRR-, and pyrin domain-containing protein 3 (*NLRP3*) is an intracellular sensor that detects a broad range of microbial motifs. Inflammasomes are formed by different substances, including lipopolysaccharide (LPS) from bacterial cell walls, pathogen-associated molecular patterns (PAMPs) from viruses, bacteria, and fungi. Inflammasomes can be activated by damage-associated molecular patterns (DAMPs) and pro-inflammatory cytokines (IL-1β, TNFα) (Korakas et al. 2020). Activation of the NLRP3 inflammasome appears to be another critical event in the acceleration of the inflammatory phase of the disease to cytokine storm (Fig. 29.1). Inflammasomes activate caspase-1 leading to pyroptosis, a pathway of cell death is uniquely dependent on caspase-1 (Fink and Cookson 2005), and stimulate maturation and secretion of the pro-inflammatory cytokines, interleukin-1beta (IL-1β) (Parisi and Leosco 2020) and interleukin-18 (IL-18) through nuclear factor kappa-B (NF-κB) signaling. A cytokine storm may occur in COVID-19 patients characterized by a failure of the immune system to counter regulate NLRP3 inflammasome activity

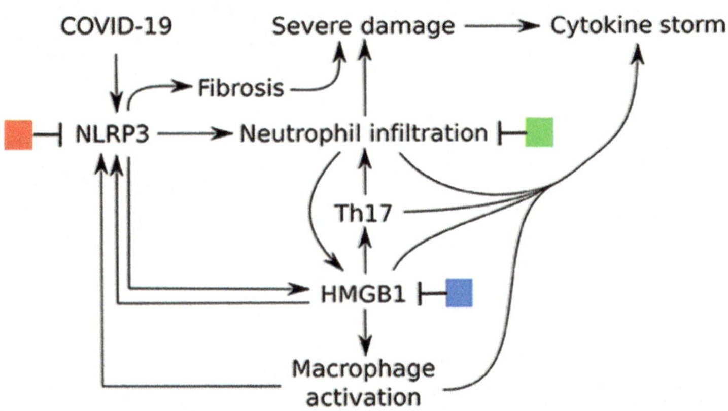

Fig. 29.1 Central role of NLRP3 inflammasome activation in the severe symptomatic phase of COVID-19 and potential options for treatment. The DAMPs released after NLRP3 inflammasome activation have a dual function. In a normal immune reaction, they induce the necessary co-stimulatory activation of the APC, but they also play a role in resolution and tissue regeneration. Only in case of a hyperactivation of the NLRP3 inflammasome DAMPs are released in high concentrations and result in pyroptosis, high-mobility group box 1 (HMGB1) release, activation of macrophages, neutrophil infiltration and reduced apoptosis, excessive cytokine production (IL-1β, IL-2, IL-6, IL-17, TNF-α, GM-CSF, IFN-γ, CXCL10, CCL2, and CCL3, cytokine storm), and fibrosis *DAMPs* damage-associated molecular patterns, *NLRP3* NOD-, LRR- and pyrin domain-containing protein 3, *PAMPs* pathogen-associated molecular patterns, *IL-1β* interleukin-1 beta, *IL-2* interleukin-2, *IL-6* interleukin-6, *IL-7* interleukin-7, *TNF-a* tumor necrosis factor-alpha, *CXCL10* interferon gamma-induced protein 10 or chemokine 10, *GM CSF* granulocyte colony-stimulating factor, *CCL2* C-C motif chemokine ligand 2, *CCL3* C-C motif chemokine ligand 3, *NF-κB* nuclear factor kappa-B, *Th17* T-helper 17 cells, *HMGB1* high-mobility group box 1 Adapted with permission (van den Berg and te Velde 2020)

(Paramo 2020). Modulation of ACE2 expression and furin proteases and prevention of the induction NLRP3 are targets of therapy in COVID-19.

29.1.2 COVID-19: Current Treatment Paradigm

Investigational therapies against COVID-19 that are being tested include antiviral agents, immune modulators, cellular therapies, vaccines, convalescent plasma, traditional Chinese medicines, combination agents, or other medications (Bhagavathula et al. 2020; Guo et al. 2020). Hospitalization for COVID-19-related complications is primarily for supportive care, as therapies to prevent the progression of the respiratory disease remain investigational.

The treatment with the most testing thus far has been the antiviral remdesivir. A double-blind, randomized placebo-controlled trial (RCT) of 1063 patients with COVID-19 and lower respiratory tract involvement compared to remdesivir (200 mg loading on day one, 100 mg daily days 2–10) to placebo and showed benefit in shortening recovery time from 15 to 11 days (rate ratio for recovery, 1.32; 95% CI, 1.12–1.55; $P < 0.001$) (Beigel et al. 2020). The benefit was seen mainly in those with a severity score of 5 (required oxygen), which represents more severe disease. Remdesivir may facilitate quicker recovery for patients who are hospitalized with COVID-19 and require supplemental oxygen therapy.

Wang et al. reported a double-blind RCT of 237 hospitalized patients with COVID-19 lower respiratory tract infections and severe disease, comparing intravenous remdesivir (200 mg on day 1 followed by 100 mg on days 2–10 in single daily infusions) to placebo infusions for 10 days (Wang et al. 2020d). The primary endpoint was clinical improvement up to day 28, defined as the time from randomization to the point of a decline of two levels on a six-point ordinal scale of clinical status (from 1 = discharged to 6 = death) or discharged alive from the hospital, whichever came first. The study was terminated before reaching the anticipated sample size as stringent public health measures in Wuhan created difficulties with enrollment. The intention to treat populations did not show a difference in time to clinical improvement. However, those who had symptom onset in less than 10 days before treatment appeared to trend toward benefit with remdesivir over placebo.

Goldman et al. conducted a randomized non-placebo-controlled clinical trial of remdesivir in 397 patients who underwent 1:1 randomization with intravenous remdesivir for either 5 days or 10 days. A total of 200 subjects were treated for 5 days, and 197 were treated for 10 days. Remdesivir was administered intravenously to subjects at 200 mg on day 1 and then 100 mg once daily after that. The primary endpoint was the clinical status of subjects on day 14, assessed on a 7-point ordinal scale. There was no difference between a 5-day course and a 10-day course of remdesivir in non-ventilated subjects with severe COVID-19. Both groups showed an improvement by reduction of 2 more points in the 7-point ordinal scale [64% (5 days) vs. 54% (10 days), $p = 0.14$] (Goldman et al. 2020). However, a placebo was not utilized, limiting the conclusions drawn from the study.

Treatment to prevent the progression of COVID-19 to severe disease is lacking, and the public is using a variety of nutraceutical agents as a preventative measure to prevent or mitigate the progression of COVID-19 (Table 29.1) (Hemila and Chalker 2013). The nutraceutical agents with biological plausibility (Iddir et al. 2020) with the potential to address in part the pathophysiology of COVID-19 are being explored in clinical trials (clinicaltrials.gov, WHO database) and are reviewed below.

29.2 Dietary Supplements Being Studied for COVID-19

29.2.1 Ascorbic Acid

Ascorbic acid contributes to immune defense by supporting various cellular functions of both the innate and adaptive immune systems (Holmannova et al. 2012). Ascorbic acid accu-

Table 29.1 Summary of dietary supplements being studied in clinical trials

Dietary supplements used by the public and in clinical trials	Dose Oral = PO Intravenous = IV	Mechanism(s) of action Innate and/or adaptive immunity	Antiviral	NLRP3	Cytokines	Adverse events
Ascorbic acid	1–3 g PO daily Intravenous doses (12–24 g daily) used in the critical care setting	xx	xx			Risk of loose stools at high doses >5 ms daily. History of oxalate kidney stones a contraindication
Curcumin	500–1000 mg PO twice daily	xx	xx	xx	xx	GI intolerance in formulations that use black pepper to enhance absorption
Green tea (epigallocatechin gallate; EGCG).	4 cups of tea or 225 mg of EGCG PO daily.	xx	xx	xx	xx	Palpitations from excessive caffeine. Extracts can cause liver injury
Melatonin	5–20 mg PO daily	xx	xx	xx	xxx	Drowsiness.
Quercetin	1000 mg PO twice daily	xx	xx	xx	xx	Generally regarded as safe (GRAS)
Resveratrol	100–150 mg PO daily	xx	xx	xx	xx	NA
Vitamin D	5000–10,000 IU D3 formulation PO daily, emulsified formulation	xx	xx		xx	Risk of hypercalcemia with excessive serum levels [(25(OH) D>100 ng/ml]
Zinc	30–60 mg PO daily, in divided dose	xx	xx			Depletes serum copper. Consider copper supplement

Adapted with permission (Evans et al. 2020)

mulates in phagocytic cells, such as neutrophils, and can enhance chemotaxis, phagocytosis, generation of reactive oxygen species, and ultimately microbial killing (Carr and Maggini 2017). Ascorbic acid has a long history of use for viral respiratory infections, and there is in vitro evidence of its activity against coronavirus in chick embryo ciliated tracheal organ cultures (Atherton et al. 1978). Intravenous ascorbic acid has been shown to reduce the length of stay and ventilator requirements in the critical care unit setting. *Vitamin C* has been shown to shorten the duration of mechanical *ventilation* by about 20% in patients who required mechanical *ventilation* for over a 24 h period (95% CI 7.7% to 27%; p = 0.001) (Hemila and Chalker 2019). In combination with thiamine (Amrein et al. 2011), ascorbic acid and hydrocortisone may improve outcomes among patients with a critical illness such as sepsis and ARDS (Hager et al. 2019, 2020; Fowler 3rd et al. 2019, 2020; Kim et al. 2018, 2019; Hwang et al. 2019; Shin et al. 2019). The putative mechanisms include improvement of pulmonary capillary integrity, correction of sepsis-induced coagulopathy, attenuation of oxidative endothelial injury, and lowering of serum TNF-alpha (Chen et al. 2014).

Vitamin C supplementation could help the prevention and treatment of respiratory and systemic infections (Turski et al. 2020; Carr and Maggini 2017). High doses of ascorbic acid reduce the severity and duration of common cold symptoms caused by rhinovirus, a coronavirus with a typical mild self-limited course if untreated (Hemila and Chalker 2013). The treatment of critically ill patients with ascorbic acid has shown mixed results on mortality, length of intensive care unit stay, and duration of mechanical ventilation (Carr 2019a, b; Hager et al. 2020). There are some clinical trials involving ascorbic acid for COVID-19 underway (Carr 2020). A clinical trial to investigate vitamin C infusion for the treatment of severe 2019-nCoV-infected pneumonia will treat 140 patients with intravenous vitamin C at a dose of 24 g/day versus matching placebo for 7 days. Study endpoints include the need for mechanical ventilation, vasopressor drug requirements, sequential organ failure assessment scores, intensive care unit length of stay, and 28-day mortality (Identifier: NCT04264533na). Other clinical trials involving ascorbic acid for COVID-19 are registered in clinical trials.gov, and the World Health Organization Trial Registry Network (WHO ICTRP database).

29.2.2 Phytochemicals

Phytonutrients are plant-derived biochemicals, some of which impart disease-fighting health-benefits when ingested. Some phytonutrients are powerful anti-inflammatory agents modulating NLRP3 inflammasome activation, thereby mitigating degenerative diseases (Table 29.2 and Fig. 29.2) (Jahan et al. 2017b). The following

Table 29.2 Phytochemicals shown to inhibit NLRP3 inflammasome activation in disease model systems being studies in COVID-19

Phytochemical	Model system	Inflammasome activator	Signaling pathway	Human disease
Catechins (Green tea)	AGS cells, BALB/c mice	*Helicobacter pylori*	Inhibited NLRP3 and IL-1β activation and caspase-1 signaling	*Helicobacter pylori* infection
Curcumin (*Curcuma longa*)	C57BL/6 mice, murine macrophage cells line J774A.1	LPS, cells infected with Salmonella	Modulated TLR4-NF-κB pathway, inhibits NLRP3 including IL-1β- HMGB-1- ROS	Septic shock
Epigallocatehin-3 gallate (*Camellia sinensis*)	Human metastatic melanoma cell lines; 1205Lu & HS294T, female athymic nu/nu mice, New Zealand lupus mice	Melanoma cells, lupus prone mice	Inhibited NLRP3-ROS-NF-κB activation, impairs caspase-1 and IL-1β expression, enhances autophagy, Nrf2 antioxidant signaling pathway and increases Treg cell activity systemically	Melanoma (skin cancer), systemic lupus erythematous, lupus nephritis
Quercetin (*Quercus tinctoria*)	STZ-induced and fructose fed rat models of diabetic complications in rats	STZ, fructose	Inhibited NLRP3 activation, caspase-1, ASC, IL-1β, -6 and TNF-α, mediates JAK2-STAT3- PPAR-γ & IR-IRS1-Akt-ERK1/2 signaling pathways	Diabetic nephropathy, hyperuricemia dyslipidemia
Sulforaphane (broccoli and other cruciferous vegetables	Murine hepatic cells	High-fat diet	Inhibited NLRP3 expression, ASC and caspase-1, activates AMPK and reduced ROS and mitochondrial dysfunction	Non-alcoholic fatty liver disease

STZ streptozocin, *ROS* reactive oxygen species, *TNF* tumor necrosis factor, *PPARs peroxisome proliferator-activated receptors*, *JAK-2* Janus kinase-2, *ASC* apoptosis-associated speck-like protein containing a CARD, *AMPK* AMP-activated protein kinase, *STAT-3* signal transducer and activator of transcription 3, *IL-1β* interleukin-1beta, *IL-6* interleukin-6, *ERK1/2* extracellular signal-regulated protein kinase 1/2, *IRS1* insulin receptor substrate 1, *Treg* T-regulatory cells, *Nrf2* nuclear factor erythroid 2-related factor 2, *TLR4* toll-like receptor 4, *NF-kβ* nuclear factor kappa B, *HMGB1* high-mobility group box 1

Adapted with permission (Jahan et al. 2017a)

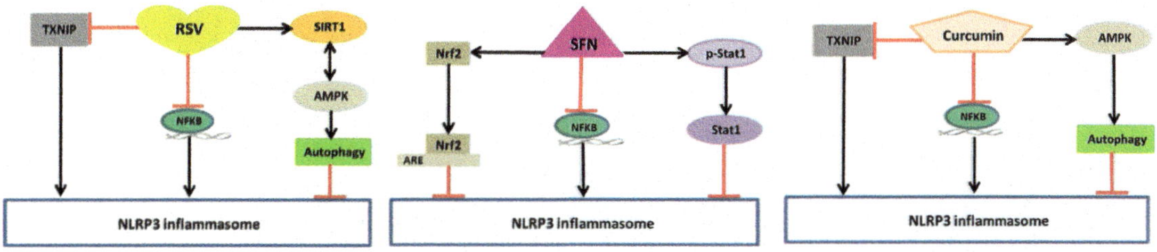

Fig. 29.2 Schematic model illustrating the underlying anti-inflammatory mechanisms of curcumin through regulation of NLRP3 inflammasome activity. Curcumin could suppress the activity of NLRP3 inflammasome through different pathways, including deterrence of K+ efflux, inhibition of ER stress and decreased levels of ROS and TXINP through AMPK activation, prevention of NLRP3 components assembly via blocking the binding of ASC to NLRP3, and suppression of NF-κB signaling pathway, which leads to the prevention of NLRP3 and pro-IL-1β expression
PAMPS pathogen-associated molecular patterns, *DAMPS* damage-associated molecular patterns, *TLR* toll-like receptor, *TNFR* tumor necrosis factor receptor, *NF-κB* nuclear factor kappa-B, *NLRP3* NOD-like receptor pyrin domain-containing 3, *IL-1β* interleukin-1β, *IL-18* interleukin 18, *AMPK* 5′ adenosine monophosphate-activated protein kinase, *ER* endoplasmic reticulum, *ROS* reactive oxygen species, *TXNIP* thioredoxin-interacting protein, *ATP* adenosine triphosphate, *P2 × 7R* purinergic 2 × 7 receptor, *MSU* monosodium urate crystal, *K+* potassium, *Ca+* calcium, *ASC* apoptosis-associated speck-like protein containing a caspase recruitment domain
Adapted with permission (Hassanzadeh et al. 2020)

phytonutrients described here are essential to consider for COVID-19, due to their actions on NLRP3 activation, viral replication, and immunity. Curcumin can exert its anti-inflammatory role mainly by preventing the activation of NLRP3 inflammasomes (Hasanzadeh et al. 2020; Olcum et al. 2020). Curcumin downregulates NF-kappa B (NF-κB) signaling, interrupts IL-1β maturation, and reduces the secretion and release of interleukins. Collectively, these actions are the most prominent mechanisms of curcumin in modulating inflammasomes (Figs. 29.3 and 29.4) (Hasanzadeh et al. 2020). A recent review of phytochemicals and their influence on intracellular signaling mechanisms of action on NLRP3 inflammasome activation focused on sulforaphane (SFN), curcumin, and resveratrol (RSV) (Olcum et al. 2020). SFN, which is present in cruciferous vegetables, was identified as a potent inhibitor of NLRP3 inflammasome activation. SFN is also known as a strong promoter of the Nrf2 transcription factor, which is the primary regulator of numerous cytoprotective, antioxidant, and anti-inflammatory genes in various tissues and cell types, and it has a role in maintaining cellular redox balance (Fig. 29.4). RSV can suppress NLRP3 inflammasome by different mechanisms and pathways, including NAD-dependent deacetylase sirtuin-1 (SIRT1) and autophagy activation. RSV appears to be a potent SIRT1 activator, which alters and inhibits the acetylation of inflammatory proteins (Sui et al. 2016). In most of the studies, RSV inhibits NLRP3 inflammasome via enhancing autophagy by activating SIRT1 (Qi et al. 2019). RSV may also upregulate ACE2 expression and may benefit the host against COVID-19 infection (Horne and Vohl 2020).

Other NLRP3 activation inhibitors include green tea phytochemicals (Zhang et al. 2019; Wang et al. 2020a). Phytochemicals can also prevent and or mitigate COVID-19 by serving as virus main protease (Mpro) inhibitors to replication. A study using molecular docking technology revealed that *Phaseolus vulgaris* phytochemicals had maximum binding with Mpro and ACE2, while quercetin 3-glucuronide-7-glucoside and quercetin 3-vicianoside gave even

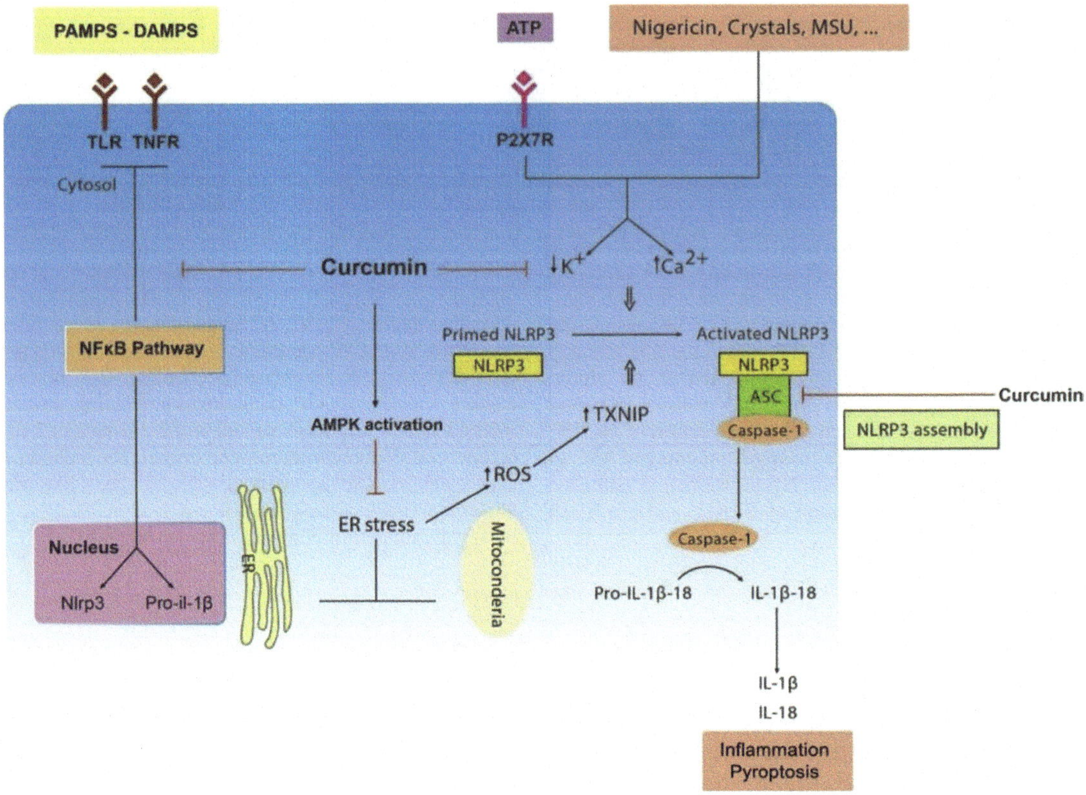

Fig. 29.3 NLRP3 inflammasome suppression mechanisms of sulforaphane (SFN), curcumin, and resveratrol (RSV). All three phytochemicals use NF-kB inhibition to suppress inflammasome activation. Other than this mechanism, SFN leads to inflammasome suppression via Nrf2 activation, STAT-1 activation. RSV and curcumin lead to inflammasome suppression via TXNIP inhibition or leading to AMPK-induced autophagy
Adapted with permission (Olcum et al. 2020)

better binding energy with both the targets (Joshi et al. 2020). Quercetin has been shown to inhibit hepatitis C viral replication (Khan et al. 2013) and modulate NLRP3 inflammasome activation, is antifibrotic by stabilizing mast cell function, and promotes resolvins, which stabilize collateral damage to host tissues. A subsequent study also using molecular docking technology showed that robust one exhibited excellent binding affinity properties against Mpro of SARS-CoV-2 (Rasool et al. 2020). This phytonutrient antioxidant flavonoid compound has also been shown to be an inhibitor against the protease of the dengue virus (Mishra et al. 2016). RSV has also been shown to have in vitro activity against coronaviruses such as MERS-CoV and is effective clinically against rhinovirus (Baldassarre et al. 2020; Lin et al. 2017).

29.2.3 Melatonin

Melatonin is receiving increasing attention in the press as a natural product with a multitude of biological effects that may be relevant to COVID-19 (Reiter et al. 2020). Melatonin attenuates several actions that protect the host against COVID-19, including pro-inflammatory cytokine production, inducible nitric oxide synthases, neuronal nitric oxide synthase, cyclooxygenase-2, high-mobility group box 1 signaling, TLR4 activation, inflammasome NLRP3 activation, and NF-κB activation (El-Missiry et al. 2020). Melatonin induces anti-inflammatory cytokines while having a high antioxidant capacity, which buffers the injurious inflammatory injury during the resolution phase of COVID-19 (El-Missiry et al. 2020). In COVID-19, melatonin may increase resistance to

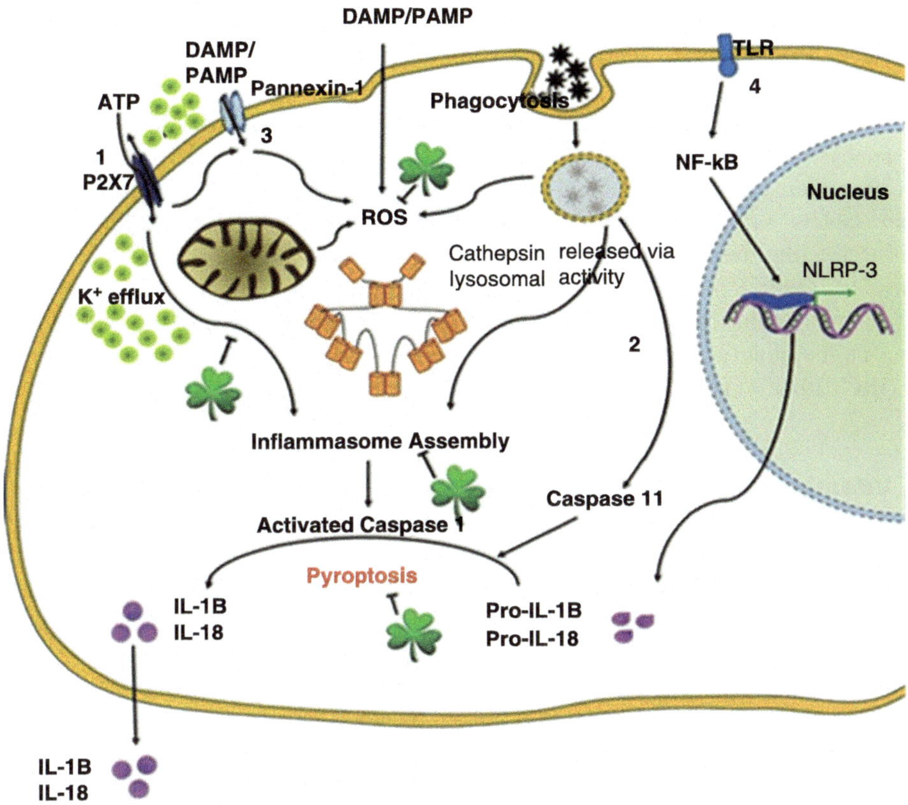

Fig. 29.4 Inflammasome activation pathways concerning disease development and phytonutrients: i, some inflammasome agonists as ATP, triggers P2X7-dependent pore formation by the pannexin-1 hemichannel, allowing extracellular agonists to enter the cytosol and directly trigger inflammasome assembly; ii, crystalline or particulate inflammasome agonists that are engulfed by the cells have characteristic physical properties which lead to lysosomal rupture. The inflammasome senses lysosomal content released in the cytoplasm, for example, via cathepsin B-dependent processing of a direct NLRP3 ligand; iii, all danger-associated molecular patterns (DAMPs) and pathogen-associated molecular patterns (PAMPs), stressed mitochondria including ATP and particulate/crystalline activators, cause the generation of reactive oxygen species (ROS). A ROS-mediated pathway triggers inflammasome complex formation; and iv, toll-like receptor (TLR) senses lipopolysaccharides and prime the NF-κB, a transcription factor which triggers pro-IL-1 and pro-IL-18 expressions which sequentially get converted into IL-1beta and IL-18. Caspase-1 clustering induces autoactivation and caspase-1-dependent maturation and secretion of pro-inflammatory cytokines, such as interleukin-1beta (IL-1b) and IL-18. Against all possible pathways, phytochemicals (shown by green leaf) are used for the therapeutic activity to inhibit inflammasome induced diseases
Adapted with permission (Jahan et al. 2017a)

infection by upregulating ACE2 expression while inhibiting NLRP3 inflammasome activation (Zhang et al. 2019). Multiple actions of melatonin as an anti-inflammatory, antioxidant, and antiviral (against other viruses) make it a reasonable choice for use (Reiter et al. 2020). Mitochondrial intracellular heme oxygenase (HO-1) is low in the elderly, hypertensives, and diabetics, which may relate to COVID-19 susceptibility, and melatonin raises HO-1 (Hooper 2020). Two active COVID-19 clinical trials include melatonin. One clinical trial (NCT04409522) is evaluating the therapeutic effects of high-dose melatonin (9 mg) vs. usual care for 7–10 days in 55 patients with COVID-19 while measuring inflammatory cytokines and pathways to include NLRP3 inflammasome activation. The other study (NCT04353128) involves

450 Spanish healthcare workers who will be randomized to either melatonin 2 mg or placebo before bed for 12 weeks as primary prevention of COVID-19. The World Health Organization (WHO) database lists two melatonin intervention COVID-19 studies: evaluation of the efficacy of melatonin tablets as auxiliary medication in accelerating the improvement of the COVID-19 symptoms and clinical findings (IRCT20200408046988N1) and the effect of melatonin on the quality of sleep in COVID-19 patients (IRCT20200411047030N1).

29.2.4 Vitamin D

Vitamin D, once converted to the active 1,25hydroxyvitamin D (1,25VitD) form, has endocrine and paracrine properties. The paracrine action of 1,25VitD may be most important for immunity. Activated vitamin D augments innate cellular immunity against microbes, protects against bacterial and viral acute respiratory tract infections (including influenza), and regulates adaptive immune responses such as those seen in the COVID-19-associated cytokine storm (Zdrenghea et al. 2017; Martineau et al. 2017). Vitamin D influences T-helper cell (Th) cell differentiation by its effect on antigen-presenting cells (APCs). Vitamin D is involved in APC activity that modulates T-cell differentiation into an effector cell with pro-inflammatory (Th1) or anti-inflammatory (Th2) properties; thus, modulation of APCs is crucial in initiating and maintaining adaptive immune response and self-tolerance. Vitamin D modulates adaptive immunity by suppressing Th1 and Th17 responses that are overactivated in a COVID-19 cytokine storm (Wu and Yang 2020; Miraglia Del Giudice et al. 2018). Vitamin D induces Th2 cytokines such as IL-10, which counterbalance Th1 pro-inflammatory cytokines while increasing T-regulatory cells (Tregs) and their activity (Fawaz et al. 2016). By inducing Tregs and increasing their numbers, vitamin D harmonizes inflammatory responses. In COVID-19, there are reports that Th1/Th17 cytokines are high, Th2 cytokines such as IL-10 are low, and Tregs are low in number and dysfunctional in the setting of a cytokine storm (Wang et al. 2020b; Chen et al. 2020).

Vitamin D enhances innate cellular immunity to prevent and mitigate viral respiratory infections. There are two main mechanisms by which vitamin D has been shown to prevent viral respiratory tract infections. One is the promotion of respiratory epithelial and alveolar barrier function junctions to prevent the infiltration of immune cells in the lungs and other respiratory tissues. The other is an increased immune-enhanced viral killing while avoiding injurious inflammatory response (Grant et al. 2020). Vitamin D enhances the production of β-defensin-2 and LL-37 cathelicidin. Pulmonary epithelial cells have a high expression of 1α-hydroxylase, which produces calcitriol, the active form of vitamin D. Calcitriol inhibits bronchial smooth muscle cell proliferation and elaboration of pro-inflammatory cytokines, chemokines, and matrix metalloproteinases, preventing lung injury (Sandhu and Casale 2010). Vitamin D upregulates cAMP, not only by monocytes and macrophages but also in cells participating in the innate immune system, including the respiratory tract, by increasing their antimicrobial activity and epithelial barrier function (Dhawan et al. 2015).

Vitamin D is well-known to participate in the defense against some respiratory pathogens, including intracellular pathogens and bacterial and viral pathogens (Anderson et al. 2020). Lower-serum vitamin D levels are associated with adult new-onset severe sepsis, including septic shock, and high doses of enteral vitamin D3 (400,000 IU) have been shown to increase circulating cAMP and reduce inflammatory cytokines IL-6 and IL-1 when compared to placebo (Quraishi et al. 2015). In viral respiratory infections, vitamin D metabolites modulate several chemokines and pro-inflammatory cytokines (CXCL8, CXCL10, TNF-α, and IL-6) (Greiller and Martineau 2015). Observational and interventional studies demonstrate that Vitamin D can impart primary and secondary protection against viral respiratory tract illness (Teymoori-

Rad et al. 2019; Beard et al. 2011). A meta-analysis of 25 eligible randomized controlled trials (total 11, 321 participants) concluded that Vitamin D supplementation was safe and protected against acute respiratory tract infection (adjusted odds ratio 0.88, 95% confidence interval 0.81–0.96) (Martineau et al. 2017). Patients who were very vitamin D-deficient as defined by baseline 25-hydroxyvitamin D levels <25 nmol/L (adjusted odds ratio 0.30, 0.17–0.53) derived more protection against respiratory tract infection compared to patients who were not so identified with baseline 25-hydroxyvitamin D levels ≥25 nmol/L (adjusted odds ratio 0.75, 0.60–0.95). Those not receiving bolus doses experienced the most benefit (adjusted odds ratio 0.81, 0.72–0.91).

There is growing speculation that vitamin D deficiency may render hosts vulnerable to COVID-19 infection and that vitamin D may serve a primary and secondary preventative role (Weir et al. 2020; Grant et al. 2020; Wei and Christakos 2015; Dhawan et al. 2015). Countries related to high mortality rates early in the COVID-19 pandemic (Italy, Spain, and the United Kingdom) are more likely to suffer from lower vitamin D levels than countries that were not as severely affected (Grant et al. 2020). An analysis of a world database by investigators at Northwestern University examined severe SARS-CoV-2 illness and vitamin D deficiency prevalence; data revealed that the risk of severe SARS-CoV-2 cases among patients with severe vitamin D deficiency was 17.3%, compared with a risk of 14.6% for patients with normal vitamin D levels (a relative reduction of 15.6%) (Daneshkhah et al. 2020). The authors suggested that the correction of vitamin D deficiency may reduce SARS-CoV-2 severity by suppressing the cytokine storm. The WHO database lists one study that is actively recruiting; IRCT20200401469O9N1 9 (https://en.irct.ir/trial/46875), an RCT to evaluate the efficacy of 1000 IU of vitamin D3 or placebo daily for 8 weeks. The duration of COVID-19 infection is the primary endpoint and the WHO severity scale as the secondary endpoint.

29.2.5 Zinc

Zinc is a micronutrient with an established role in robust and effective immune responses, including antiviral immunity and adaptive immune responses, including antibody formation (Gammoh and Rink 2017). Older adults (65 years and older) are at increased risk of zinc insufficiency or deficiency. Further, because of the high incidence of diarrhea with SARS-CoV2 infection (seen in approximately 20% of patients), it was possible to assume these patients were zinc insufficient (Lee et al. 2020). Zinc is particularly attractive to consider in SARS-CoV2 infection. In vitro studies show zinc to inhibit coronavirus RNA replication (Fig. 29.5) (te Velthuis et al. 2010; Read et al. 2019). Zinc lozenges at symptom onset reduce the duration of symptoms from illness attributed to more innocuous coronavirus infections (i.e., the common cold) (Mossad et al. 1996; Hemila 2017; Hemila et al. 2020). Also, hydroxychloroquine (HCQ) mobilizes zinc into lysosomes suggesting there may be synergy between HCQ and zinc to amplify efficacy (in vitro data support this synergy in cell culture assays of HCQ-induced cytotoxicity and HCQ-induced inhibition of autophagic flux) (Xue et al. 2014). The WHO database lists one study using zinc sulfate (IRCT20180425039414N2; https://en.irct.ir/trial/47516). It investigates the effect of 220 mg of zinc sulfate on the clinical course of 80 inpatients with COVID-19. Forty patients will receive a combination of zinc sulfate 220 mg with hydroxychloroquine 200 mg twice daily and then once for 5 days, and the other 40 patients will receive only hydroxychloroquine in the same manner in a parallel clinical trial at the Esfahan University of Medical Sciences in Iran (https://en.irct.ir/trial/47516).

29.3 Conclusion

There is a dearth of surveys to precisely indicate the usage of dietary supplements by the public for COVID-19. However, the dietary supplement industry has reported a global boost in sales during the COVID-19 pandemic as people sought

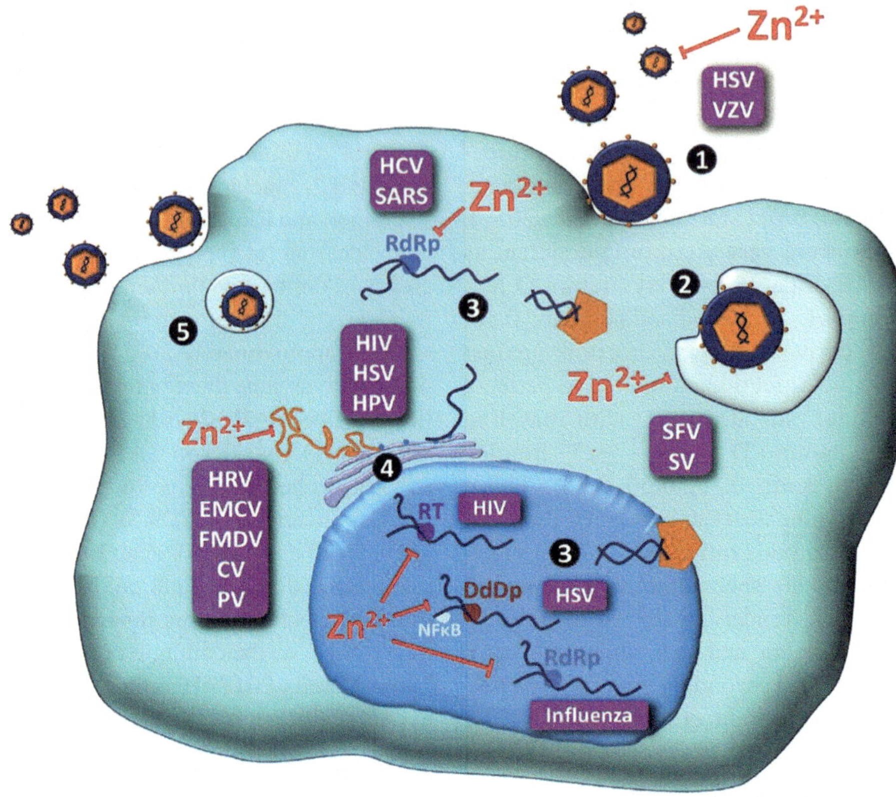

Fig. 29.5 The diverse stages of viral replication cycles that are inhibited by zinc. In vitro studies have demonstrated some mechanisms by which zinc interferes with the viral replication cycle. These include free virus inactivation: i, inhibition of viral uncoating; ii, viral genome transcription; iii and iv, viral protein translation and polyprotein processing. No studies to date, however, have demonstrated zinc-mediated inhibition of virus assembly and/or particle release
CV coronavirus, *DdDp* DNA-dependent DNA polymerase, *EMCV* encephalomyocarditis virus, *FMDV* foot and mouth disease virus, *HCV* hepatitis C virus, *HIV* human immunodeficiency virus, *HPV* human papilloma virus, *HRV* human rhinovirus, *HSV* herpes simplex virus, *PV* polio virus, *RdRp* RNA-dependent RNA polymerase, *RT* reverse transcriptase, *SARS* severe acute respiratory syndrome coronavirus, *SFV* Semliki Forest virus, *SV* sindbis virus, *VZV* varicella-zoster virus, *Zn* zinc
Adapted with permission (Read et al. 2019)

the aid of natural medicines in an attempt to prevent or mitigate COVID-19. There is some evidence to indicate that immune-modulating dietary supplements may play a role in benefiting the public from COVID-19. However, dietary supplements are not without cost or potential harm, albeit low risk in the majority of cases. Any benefit of dietary supplements against COVID-19 depends on biological plausibility, the peer-reviewed literature, not direct studies in humans, preferably RCT. The approach was heuristic and served its place at a time when morbidity and mortality spread across the world in an uncontrolled manner. Ultimately, the many clinical trials underway now that will soon illuminate whether the millions of dollars spent by the public and their actions during the COVID-19 pandemic had any merit. Only time will tell.

Acknowledgments The authors would like to thank the members of the Institute for Functional Medicine Task Force (Drs. Joel Evans, Dan Lukaczer, Robert Luby, and Patrick Hanaway) and Dr. Keith Berkowitz for inspiring this manuscript.

References

Alunno A, Carubbi F, Rodriguez-Carrio J (2020) Storm, typhoon, cyclone or hurricane in patients with COVID-19? Beware of the same storm that has a different origin. RMD Open 6(1):e001295. https://doi.org/10.1136/rmdopen-2020-001295

Amrein K, Ribitsch W, Otto R, Worm HC, Stauber RE (2011) Severe lactic acidosis reversed by thiamine within 24 hours. Crit Care 15(6):457. https://doi.org/10.1186/cc10495

Anderson J, Do LAH, Toh ZQ, Hoe E, Reitsma A, Mulholland K, Licciardi PV (2020) Vitamin D induces differential effects on inflammatory responses during bacterial and/or viral stimulation of human peripheral blood mononuclear cells. Front Immunol 11:602. https://doi.org/10.3389/fimmu.2020.00602

Atherton JG, Kratzing CC, Fisher A (1978) The effect of ascorbic acid on infection chick-embryo ciliated tracheal organ cultures by coronavirus. Arch Virol 56(3):195–199. https://doi.org/10.1007/bf01317848

Baldassarre ME, Di Mauro A, Labellarte G, Pignatelli M, Fanelli M, Schiavi E, Mastromarino P, Capozza M, Panza R, Laforgia N (2020) Resveratrol plus carboxymethyl-beta-glucan in infants with common cold: a randomized double-blind trial. Heliyon 6:e03814. https://doi.org/10.1016/j.heliyon.2020.e03814

Beard JA, Bearden A, Striker R (2011) Vitamin D and the antiviral state. J Clin Virol 50(3):194–200. https://doi.org/10.1016/j.jcv.2010.12.006

Beigel JH, Tomashek KM, Dodd LE, Mehta AK, Zingman BS, Kalil AC, Hohmann E, Chu HY, Luetkemeyer A, Kline S, Lopez de Castilla D, Finberg RW, Dierberg K, Tapson V, Hsieh L, Patterson TF, Paredes R, Sweeney DA, Short WR, Touloumi G, Lye DC, Ohmagari N, Oh MD, Ruiz-Palacios GM, Benfield T, Fatkenheuer G, Kortepeter MG, Atmar RL, Creech CB, Lundgren J, Babiker AG, Pett S, Neaton JD, Burgess TH, Bonnett T, Green M, Makowski M, Osinusi A, Nayak S, Lane HC, Members A-SG (2020) Remdesivir for the treatment of Covid-19 – preliminary report. N Engl J Med 383:993. https://doi.org/10.1056/NEJMoa2007764

Bhagavathula AS, Aldhaleei WA, Rovetta A, Rahmani J (2020) Vaccines and drug therapeutics to lock down novel coronavirus disease 2019 (COVID-19): a systematic review of clinical trials. Cureus 12(5):e8342. https://doi.org/10.7759/cureus.8342

Brojakowska A, Narula J, Shimony R, Bander J (2020) Clinical implications of SARS-CoV-2 interaction with renin angiotensin system: JACC review topic of the week. J Am Coll Cardiol 75(24):3085–3095. https://doi.org/10.1016/j.jacc.2020.04.028

Carr AC (2019a) Duration of intravenous vitamin C therapy is a critical consideration. Crit Care Resusc 21(3):220–221

Carr AC (2019b) Vitamin C administration in the critically ill: a summary of recent meta-analyses. Crit Care 23(1):265. https://doi.org/10.1186/s13054-019-2538-y

Carr AC (2020) A new clinical trial to test high-dose vitamin C in patients with COVID-19. Crit Care 24(1):133. https://doi.org/10.1186/s13054-020-02851-4

Carr AC, Maggini S (2017) Vitamin C and immune function. Nutrients 9(11):1211. https://doi.org/10.3390/nu9111211

Chen Y, Luo G, Yuan J, Wang Y, Yang X, Wang X, Li G, Liu Z, Zhong N (2014) Vitamin C mitigates oxidative stress and tumor necrosis factor-alpha in severe community-acquired pneumonia and LPS-induced macrophages. Mediat Inflamm 2014:426740. https://doi.org/10.1155/2014/426740

Chen G, Wu D, Guo W, Cao Y, Huang D, Wang H, Wang T, Zhang X, Chen H, Yu H, Zhang X, Zhang M, Wu S, Song J, Chen T, Han M, Li S, Luo X, Zhao J, Ning Q (2020) Clinical and immunological features of severe and moderate coronavirus disease 2019. J Clin Invest 130(5):2620–2629. https://doi.org/10.1172/JCI137244

Conti P, Ronconi G, Caraffa A, Gallenga CE, Ross R, Frydas I, Kritas SK (2020) Induction of pro-inflammatory cytokines (IL-1 and IL-6) and lung inflammation by Coronavirus-19 (COVI-19 or SARS-CoV-2): anti-inflammatory strategies. J Biol Regul Homeost Agents 34(2):327. https://doi.org/10.23812/CONTI-E

Daneshkhah A, Agrawal V, Eshein A, Subramanian H, Roy HK, Backman V (2020) The possible role of vitamin D in suppressing cytokine storm and associated mortality in COVID-19 patients. BMJ Yale. https://doi.org/10.1101/2020.04.08.20058578

Dhawan P, Wei R, Sun C, Gombart AF, Koeffler HP, Diamond G, Christakos S (2015) C/EBPalpha and the vitamin D receptor cooperate in the regulation of cathelicidin in lung epithelial cells. J Cell Physiol 230(2):464–472. https://doi.org/10.1002/jcp.24729

El-Missiry MA, El-Missiry ZM, Othman AI (2020) Melatonin is a potential adjuvant to improve clinical outcomes in individuals with obesity and diabetes with coexistence of Covid-19. Eur J Pharmacol 882:173329. https://doi.org/10.1016/j.ejphar.2020.173329

Evans JM, Luby R, Lukaczer D, Rountree R, Stone PM, Guilliams TG, Yanuck S, Messier H, Ramsdell K, Hanaway PJ (2020) The functional medicine approach to COVID-19: virus-specific nutraceutical and botanical agents. Integr Med: Clinician's J 19:34

Fawaz L, Mrad MF, Kazan JM, Sayegh S, Akika R, Khoury SJ (2016) Comparative effect of 25(OH)D3 and 1,25(OH)2D3 on Th17 cell differentiation. Clin Immunol 166–167:59–71. https://doi.org/10.1016/j.clim.2016.02.011

Fink SL, Cookson BT (2005) Apoptosis, pyroptosis, and necrosis: mechanistic description of dead and dying eukaryotic cells. Infect Immun 73(4):1907–1916. https://doi.org/10.1128/IAI.73.4.1907-1916.2005

Fowler AA 3rd, Truwit JD, Hite RD, Morris PE, DeWilde C, Priday A, Fisher B, Thacker LR 2nd, Natarajan R, Brophy DF, Sculthorpe R, Nanchal R, Syed A, Sturgill

J, Martin GS, Sevransky J, Kashiouris M, Hamman S, Egan KF, Hastings A, Spencer W, Tench S, Mehkri O, Bindas J, Duggal A, Graf J, Zellner S, Yanny L, McPolin C, Hollrith T, Kramer D, Ojielo C, Damm T, Cassity E, Wieliczko A, Halquist M (2019) Effect of vitamin C infusion on organ failure and biomarkers of inflammation and vascular injury in patients with sepsis and severe acute respiratory failure: the CITRIS-ALI randomized clinical trial. JAMA 322(13):1261–1270. https://doi.org/10.1001/jama.2019.11825

Fowler AA 3rd, Fisher BJ, Kashiouris MG (2020) Vitamin C for sepsis and acute respiratory failure-reply. JAMA 323(8):792–793. https://doi.org/10.1001/jama.2019.21987

Gammoh NZ, Rink L (2017) Zinc in infection and inflammation. Nutrients 9(6). https://doi.org/10.3390/nu9060624

Goldman JD, Lye DCB, Hui DS, Marks KM, Bruno R, Montejano R, Spinner CD, Galli M, Ahn MY, Nahass RG, Chen YS, SenGupta D, Hyland RH, Osinusi AO, Cao H, Blair C, Wei X, Gaggar A, Brainard DM, Towner WJ, Munoz J, Mullane KM, Marty FM, Tashima KT, Diaz G, Subramanian A, Investigators G-U-540-5773 (2020) Remdesivir for 5 or 10 days in patients with severe Covid-19. N Engl J Med 383:1827. https://doi.org/10.1056/NEJMoa2015301

Grant WB, Lahore H, McDonnell SL, Baggerly CA, French CB, Aliano JL, Bhattoa HP (2020) Evidence that vitamin D supplementation could reduce risk of influenza and COVID-19 infections and deaths. Nutrients 12(4). https://doi.org/10.3390/nu12040988

Greiller CL, Martineau AR (2015) Modulation of the immune response to respiratory viruses by vitamin D. Nutrients 7(6):4240–4270. https://doi.org/10.3390/nu7064240

Guo YR, Cao QD, Hong ZS, Tan YY, Chen SD, Jin HJ, Tan KS, Wang DY, Yan Y (2020) The origin, transmission and clinical therapies on coronavirus disease 2019 (COVID-19) outbreak – an update on the status. Mil Med Res 7(1):1. ARTN 11. https://doi.org/10.1186/s40779-020-00240-0

Hager DN, Hooper MH, Bernard GR, Busse LW, Ely EW, Fowler AA, Gaieski DF, Hall A, Hinson JS, Jackson JC, Kelen GD, Levine M, Lindsell CJ, Malone RE, McGlothlin A, Rothman RE, Viele K, Wright DW, Sevransky JE, Martin GS (2019) The vitamin C, thiamine and steroids in sepsis (VICTAS) protocol: a prospective, multi-center, double-blind, adaptive sample size, randomized, placebo-controlled, clinical trial. Trials 20(1):197. https://doi.org/10.1186/s13063-019-3254-2

Hager DN, Hinson JS, Rothman RE (2020) Vitamin C for sepsis and acute respiratory failure. JAMA 323(8):791–792. https://doi.org/10.1001/jama.2019.21984

Han H, Ma Q, Li C, Liu R, Zhao L, Wang W, Zhang P, Liu X, Gao G, Liu F, Jiang Y, Cheng X, Zhu C, Xia Y (2020) Profiling serum cytokines in COVID-19 patients reveals IL-6 and IL-10 are disease severity predictors. Emerg Microbes Infect 9(1):1123–1130. https://doi.org/10.1080/22221751.2020.1770129

Hasanzadeh S, Read MI, Bland AR, Majeed M, Jamialahmadi T, Sahebkar A (2020) Curcumin: an inflammasome silencer. Pharmacol Res 159:104921. https://doi.org/10.1016/j.phrs.2020.104921

Hassanzadeh S, Read MI, Bland AR, Majeed M, Jamialahmadi T, Sahebkar A (2020) Curcumin: an inflammasome silencer. Pharmacol Res 159:104921

Hemila H (2017) Zinc lozenges and the common cold: a meta-analysis comparing zinc acetate and zinc gluconate, and the role of zinc dosage. JRSM Open 8(5):2054270417694291. https://doi.org/10.1177/2054270417694291

Hemila H, Chalker E (2013) Vitamin C for preventing and treating the common cold. Cochrane Database Syst Rev (1):CD000980. https://doi.org/10.1002/14651858.CD000980.pub4

Hemila H, Chalker E (2019) Vitamin C can shorten the length of stay in the ICU: a meta-analysis. Nutrients 11(4). https://doi.org/10.3390/nu11040708

Hemila H, Haukka J, Alho M, Vahtera J, Kivimaki M (2020) Zinc acetate lozenges for the treatment of the common cold: a randomised controlled trial. BMJ Open 10(1):e031662. https://doi.org/10.1136/bmjopen-2019-031662

Holmannova D, Kolackova M, Krejsek J (2012) Vitamin C and its physiological role with respect to the components of the immune system. Vnitr Lek 58(10):743–749

Hooper PL (2020) COVID-19 and heme oxygenase: novel insight into the disease and potential therapies. Cell Stress Chaperones 25:707. https://doi.org/10.1007/s12192-020-01126-9

Horne JR, Vohl MC (2020) Biological plausibility for interactions between dietary fat, resveratrol, ACE2, and SARS-CoV illness severity. Am J Physiol Endocrinol Metab 318(5):E830–E833. https://doi.org/10.1152/ajpendo.00150.2020

Huang C, Wang Y, Li X, Ren L, Zhao J, Hu Y, Zhang L, Fan G, Xu J, Gu X, Cheng Z, Yu T, Xia J, Wei Y, Wu W, Xie X, Yin W, Li H, Liu M, Xiao Y, Gao H, Guo L, Xie J, Wang G, Jiang R, Gao Z, Jin Q, Wang J, Cao B (2020) Clinical features of patients infected with 2019 novel coronavirus in Wuhan, China. Lancet 395(10223):497–506. https://doi.org/10.1016/S0140-6736(20)30183-5

Hwang SY, Park JE, Jo IJ, Kim S, Chung SP, Kong T, Shin J, Lee HJ, You KM, Jo YH, Kim D, Suh GJ, Kim T, Kim WY, Kim YJ, Ryoo SM, Choi SH, Shin TG, Korean Shock Society I (2019) Combination therapy of vitamin C and thiamine for septic shock in a multicentre, double-blind, randomized, controlled study (ATESS): study protocol for a randomized controlled trial. Trials 20(1):420. https://doi.org/10.1186/s13063-019-3542-x

Iddir M, Brito A, Dingeo G, Fernandez Del Campo SS, Samouda H, La Frano MR, Bohn T (2020) Strengthening the immune system and reducing inflammation and oxidative stress through diet and nutrition: considerations during the COVID-19 crisis. Nutrients 12(6):1562. https://doi.org/10.3390/nu12061562

Jahan S, Kumar D, Chaturvedi S, Rashid M, Wahajuddin M, A Khan Y, N Goyal S, R Patil C, Mohanraj R, Subramanya S (2017a) Therapeutic targeting of NLRP3 inflammasomes by natural products and pharmaceuticals: a novel mechanistic approach for inflammatory diseases. Curr Med Chem 24(16):1645–1670

Jahan S, Kumar D, Chaturvedi S, Rashid M, Wahajuddin M, Khan YA, Goyal SN, Patil CR, Mohanraj R, Subramanya S, Ojha S (2017b) Therapeutic targeting of NLRP3 inflammasomes by natural products and pharmaceuticals: a novel mechanistic approach for inflammatory diseases. Curr Med Chem 24(16):1645–1670. https://doi.org/10.2174/0929867324666170227121619

Jin D, Wu S, Zhang YG, Lu R, Xia Y, Dong H, Sun J (2015) Lack of vitamin D receptor causes dysbiosis and changes the functions of the murine intestinal microbiome. Clin Ther 37(5):996–1009 e1007. https://doi.org/10.1016/j.clinthera.2015.04.004

Joshi T, Joshi T, Sharma P, Mathpal S, Pundir H, Bhatt V, Chandra S (2020) In silico screening of natural compounds against COVID-19 by targeting Mpro and ACE2 using molecular docking. Eur Rev Med Pharmacol Sci 24(8):4529–4536. https://doi.org/10.26355/eurrev_202004_21036

Khan M, Masoud MS, Qasim M, Khan MA, Zubair M, Idrees S, Ashraf A, Ashfaq UA (2013) Molecular screening of phytochemicals from Amelanchier Alnifolia against HCV NS3 protease/helicase using computational docking techniques. Bioinformation 9(19):978–982. https://doi.org/10.6026/97320630009978

Kim WY, Jo EJ, Eom JS, Mok J, Kim MH, Kim KU, Park HK, Lee MK, Lee K (2018) Combined vitamin C, hydrocortisone, and thiamine therapy for patients with severe pneumonia who were admitted to the intensive care unit: propensity score-based analysis of a before-after cohort study. J Crit Care 47:211–218. https://doi.org/10.1016/j.jcrc.2018.07.004

Kim WY, Jung JW, Choi JC, Shin JW, Kim JY (2019) Subphenotypes in patients with septic shock receiving vitamin C, hydrocortisone, and thiamine: a retrospective cohort analysis. Nutrients 11(12):2976. https://doi.org/10.3390/nu11122976

Korakas E, Ikonomidis I, Kousathana F, Balampanis K, Kountouri A, Raptis A, Palaiodimou L, Kokkinos A, Lambadiari V (2020) Obesity and COVID-19: immune and metabolic derangement as a possible link to adverse clinical outcomes. Am J Physiol Endocrinol Metab 319(1):E105–E109. https://doi.org/10.1152/ajpendo.00198.2020

Lee IC, Huo TI, Huang YH (2020) Gastrointestinal and liver manifestations in patients with COVID-19. J Chin Med Assoc 83:521. https://doi.org/10.1097/JCMA.0000000000000319

Lin SC, Ho CT, Chuo WH, Li S, Wang TT, Lin CC (2017) Effective inhibition of MERS-CoV infection by resveratrol. BMC Infect Dis 17(1):144. https://doi.org/10.1186/s12879-017-2253-8

Mahase E (2020) Covid-19: WHO declares pandemic because of "alarming levels" of spread, severity, and inaction. BMJ-Brit Med J 368:m1036. ARTN m1036. https://doi.org/10.1136/bmj.m1036

Martineau AR, Jolliffe DA, Hooper RL, Greenberg L, Aloia JF, Bergman P, Dubnov-Raz G, Esposito S, Ganmaa D, Ginde AA, Goodall EC, Grant CC, Griffiths CJ, Janssens W, Laaksi I, Manaseki-Holland S, Mauger D, Murdoch DR, Neale R, Rees JR, Simpson S Jr, Stelmach I, Kumar GT, Urashima M, Camargo CA Jr (2017) Vitamin D supplementation to prevent acute respiratory tract infections: systematic review and meta-analysis of individual participant data. BMJ 356:i6583. https://doi.org/10.1136/bmj.i6583

Miraglia Del Giudice M, Indolfi C, Strisciuglio C (2018) Vitamin D: immunomodulatory aspects. J Clin Gastroenterol 52 Suppl 1, Proceedings from the 9th Probiotics, Prebiotics and New Foods, Nutraceuticals and Botanicals for Nutrition & Human and Microbiota Health Meeting, held in Rome, Italy from September 10–12, 2017:S86–S88. https://doi.org/10.1097/MCG.0000000000001112

Mishra AK, Upadhyay R, Chaurasia JK, Tiwari KN (2016) Comparative antioxidant study in different flower extracts of Nyctanthes arbor-tristis (L.) (Oleaceae): an important medicinal plant. Braz J Bot 39(3):813–820. https://doi.org/10.1007/s40415-016-0283-x

Mossad SB, Macknin ML, Medendorp SV, Mason P (1996) Zinc gluconate lozenges for treating the common cold. A randomized, double-blind, placebo-controlled study. Ann Intern Med 125(2):81–88. https://doi.org/10.7326/0003-4819-125-2-199607150-00001

Olcum M, Tastan B, Ercan I, Eltutan IB, Genc S (2020) Inhibitory effects of phytochemicals on NLRP3 inflammasome activation: a review. Phytomedicine 75:153238. https://doi.org/10.1016/j.phymed.2020.153238

Paramo JA (2020) Inflammatory response in relation to COVID-19 and other prothrombotic phenotypes. Reumatol Clin. https://doi.org/10.1016/j.reuma.2020.06.004

Parisi V, Leosco D (2020) Precision medicine in COVID-19: IL-1beta a potential target. JACC Basic Transl Sci 5:543. https://doi.org/10.1016/j.jacbts.2020.04.006

Park M, Cook AR, Lim JT, Sun Y, Dickens BL (2020) A systematic review of COVID-19 epidemiology based on current evidence. J Clin Med 9(4):967. https://doi.org/10.3390/jcm9040967

Qi Y, Shang L, Liao Z, Su H, Jing H, Wu B, Bi K, Jia Y (2019) Intracerebroventricular injection of resveratrol ameliorated Abeta-induced learning and cognitive decline in mice. Metab Brain Dis 34(1):257–266. https://doi.org/10.1007/s11011-018-0348-6

Quraishi SA, De Pascale G, Needleman JS, Nakazawa H, Kaneki M, Bajwa EK, Camargo CA Jr, Bhan I (2015) Effect of cholecalciferol supplementation on vitamin D status and cathelicidin levels in sepsis: a randomized, placebo-controlled trial. Crit Care

Med 43(9):1928–1937. https://doi.org/10.1097/CCM.0000000000001148
Rasool N, Akhtar A, Hussain W (2020) Insights into the inhibitory potential of selective phytochemicals against Mpro of 2019-nCoV: a computer-aided study. Struct Chem:1–7. https://doi.org/10.1007/s11224-020-01536-6
Read SA, Obeid S, Ahlenstiel C, Ahlenstiel G (2019) The role of zinc in antiviral immunity. Adv Nutr 10(4):696–710. https://doi.org/10.1093/advances/nmz013
Reiter RJ, Abreu-Gonzalez P, Marik PE, Dominguez-Rodriguez A (2020) Therapeutic algorithm for use of melatonin in patients with COVID-19. Front Med (Lausanne) 7:226. https://doi.org/10.3389/fmed.2020.00226
Sandhu MS, Casale TB (2010) The role of vitamin D in asthma. Ann Allergy Asthma Immunol 105(3):191–199.; quiz 200-192, 217. https://doi.org/10.1016/j.anai.2010.01.013
Shang J, Wan Y, Luo C, Ye G, Geng Q, Auerbach A, Li F (2020) Cell entry mechanisms of SARS-CoV-2. Proc Natl Acad Sci U S A 117(21):11727–11734. https://doi.org/10.1073/pnas.2003138117
Shin TG, Kim YJ, Ryoo SM, Hwang SY, Jo IJ, Chung SP, Choi SH, Suh GJ, Kim WY (2019) Early vitamin C and thiamine administration to patients with septic shock in emergency departments: propensity score-based analysis of a before-and-after cohort study. J Clin Med 8(1):102. https://doi.org/10.3390/jcm8010102
Siddiqi HK, Mehra M (2020) COVID-19 illness in native and immunosuppressed states: a clinical-therapeutic staging proposal. J Heart Lung Transplant 39:405. https://doi.org/10.1016/j.healun.2020.03.012
Sui DM, Xie Q, Yi WJ, Gupta S, Yu XY, Li JB, Wang J, Wang JF, Deng XM (2016) Resveratrol protects against sepsis-associated encephalopathy and inhibits the NLRP3/IL-1beta axis in microglia. Mediat Inflamm 2016:1045657. https://doi.org/10.1155/2016/1045657
te Velthuis AJ, van den Worm SH, Sims AC, Baric RS, Snijder EJ, van Hemert MJ (2010) Zn(2+) inhibits coronavirus and arterivirus RNA polymerase activity in vitro and zinc ionophores block the replication of these viruses in cell culture. PLoS Pathog 6(11):e1001176. https://doi.org/10.1371/journal.ppat.1001176
Teymoori-Rad M, Shokri F, Salimi V, Marashi SM (2019) The interplay between vitamin D and viral infections. Rev Med Virol 29(2):e2032. https://doi.org/10.1002/rmv.2032
Turski WA, Wnorowski A, Turski GN, Turski CA, Turski L (2020) AhR and IDO1 in pathogenesis of Covid-19 and the "systemic AhR activation syndrome" translational review and therapeutic perspectives. Restor Neurol Neurosci 38:343. https://doi.org/10.3233/RNN-201042
Urwyler P, Boesing M, Abig K, Cattaneo M, Dieterle T, Zeller A, Bachler H, Markun S, Senn O, Merlo C, Essig S, Ullmer E, Rutishauser J, Schuurmans MM, Leuppi JD (2019) Reduction of corticosteroid use in outpatient treatment of exacerbated COPD – study protocol for a randomized, double-blind, non-inferiority study, (the RECUT-trial). Trials 20(1):727. https://doi.org/10.1186/s13063-019-3856-8
van den Berg DF, te Velde AA (2020) Severe COVID-19: NLRP3 inflammasome dysregulated. Front Immunol 11:1580
Walls AC, Park YJ, Tortorici MA, Wall A, McGuire AT, Veesler D (2020) Structure, function, and antigenicity of the SARS-CoV-2 spike glycoprotein. Cell 181(2):281–292.e286. https://doi.org/10.1016/j.cell.2020.02.058
Wang D, Gao Q, Wang T, Kan Z, Li X, Hu L, Peng CY, Qian F, Wang Y, Granato D (2020a) Green tea polyphenols and epigallocatechin-3-gallate protect against perfluorodecanoic acid induced liver damage and inflammation in mice by inhibiting NLRP3 inflammasome activation. Food Res Int 127:108628. https://doi.org/10.1016/j.foodres.2019.108628
Wang F, Hou H, Luo Y, Tang G, Wu S, Huang M, Liu W, Zhu Y, Lin Q, Mao L, Fang M, Zhang H, Sun Z (2020b) The laboratory tests and host immunity of COVID-19 patients with different severity of illness. JCI Insight 5(10):e137799. https://doi.org/10.1172/jci.insight.137799
Wang Y, Wang Y, Chen Y, Qin Q (2020c) Unique epidemiological and clinical features of the emerging 2019 novel coronavirus pneumonia (COVID-19) implicate special control measures. J Med Virol 92:568. https://doi.org/10.1002/jmv.25748
Wang Y, Zhang D, Du G, Du R, Zhao J, Jin Y, Fu S, Gao L, Cheng Z, Lu Q, Hu Y, Luo G, Wang K, Lu Y, Li H, Wang S, Ruan S, Yang C, Mei C, Wang Y, Ding D, Wu F, Tang X, Ye X, Ye Y, Liu B, Yang J, Yin W, Wang A, Fan G, Zhou F, Liu Z, Gu X, Xu J, Shang L, Zhang Y, Cao L, Guo T, Wan Y, Qin H, Jiang Y, Jaki T, Hayden FG, Horby PW, Cao B, Wang C (2020d) Remdesivir in adults with severe COVID-19: a randomised, double-blind, placebo-controlled, multicentre trial. Lancet 395(10236):1569–1578. https://doi.org/10.1016/S0140-6736(20)31022-9
Wei R, Christakos S (2015) Mechanisms underlying the regulation of innate and adaptive immunity by vitamin D. Nutrients 7(10):8251–8260. https://doi.org/10.3390/nu7105392
Weir EK, Thenappan T, Bhargava M, Chen Y (2020) Does vitamin D deficiency increase the severity of COVID-19? Clin Med (Lond) 20:E107. https://doi.org/10.7861/clinmed.2020-0301
Wu D, Yang XO (2020) TH17 responses in cytokine storm of COVID-19: an emerging target of JAK2 inhibitor Fedratinib. J Microbiol Immunol Infect 53(3):368–370. https://doi.org/10.1016/j.jmii.2020.03.005
Xue J, Moyer A, Peng B, Wu J, Hannafon BN, Ding WQ (2014) Chloroquine is a zinc ionophore. PLoS One 9(10):e109180. https://doi.org/10.1371/journal.pone.0109180
Zdrenghea MT, Makrinioti H, Bagacean C, Bush A, Johnston SL, Stanciu LA (2017) Vitamin D modulation of innate immune responses to respiratory

viral infections. Rev Med Virol 27(1):e1909. https://doi.org/10.1002/rmv.1909

Zhang J, Lu X, Liu M, Fan H, Zheng H, Zhang S, Rahman N, Wolczynski S, Kretowski A, Li X (2019) Melatonin inhibits inflammasome-associated activation of endothelium and macrophages attenuating pulmonary arterial hypertension. Cardiovasc Res 116:2156. https://doi.org/10.1093/cvr/cvz312

Zhou F, Yu T, Du R, Fan G, Liu Y, Liu Z, Xiang J, Wang Y, Song B, Gu X, Guan L, Wei Y, Li H, Wu X, Xu J, Tu S, Zhang Y, Chen H, Cao B (2020) Clinical course and risk factors for mortality of adult inpatients with COVID-19 in Wuhan, China: a retrospective cohort study. Lancet 395(10229):1054–1062. https://doi.org/10.1016/S0140-6736(20)30566-3

Photobiomodulation and Antiviral Photodynamic Therapy in COVID-19 Management

30

Reza Fekrazad, Sohrab Asefi,
Maryam Pourhajibagher, Farshid Vahdatinia,
Sepehr Fekrazad, Abbas Bahador,
Heidi Abrahamse, and Michael R. Hamblin

Abstract

Coronavirus disease 2019 (COVID-19) has shocked the world by its spread and contagiousness. There is no approved vaccine and no proven treatment for this infection. Some potential treatments that have already been associated with antiviral and anti-inflammatory effects are under investigation. Photobiomodulation therapy (PBMT) is a photon-based therapy that uses light to mediate a variety of metabolic, analgesic, anti-inflammatory, and immunomodulatory effects. Antiviral photodynamic therapy (aPDT) is a branch of photodynamic therapy based on the reaction between a photosensitizing agent and a light source in the presence of oxygen, which can produce oxidative and free radical agents to damage the viral structures such as proteins and nucleic acids. This chapter aims to discuss

R. Fekrazad (✉)
Radiation Sciences Research Center, Laser Research Center in Medical Sciences, AJA University of Medical Sciences, Tehran, Iran

International Network for Photo Medicine and Photo Dynamic Therapy (INPMPDT), Universal Scientific Education and Research Network (USERN), Tehran, Iran

S. Asefi
Department of Orthodontics, School of Dentistry, Tehran University of Medical Sciences, Tehran, Iran

M. Pourhajibagher
Dental Research Center, Dentistry Research Institute, Tehran University of Medical Sciences, Tehran, Iran

F. Vahdatinia
Dental Implants Research Center, School of Dentistry, Hamadan University of Medical Sciences, Hamadan, Iran

S. Fekrazad
International Network for Photo Medicine and Photo Dynamic Therapy (INPMPDT), Universal Scientific Education and Research Network (USERN), Tehran, Iran

School of Medicine, Tehran University of Medical Sciences, Tehran, Iran

A. Bahador
Oral Microbiology Laboratory, Department of Medical Microbiology, School of Medicine, Tehran University of Medical Sciences, Tehran, Iran

H. Abrahamse
Laser Research Centre, Faculty of Health Sciences, University of Johannesburg, Johannesburg, South Africa

M. R. Hamblin
Laser Research Centre, Faculty of Health Sciences, University of Johannesburg, Johannesburg, South Africa

Wellman Center for Photomedicine, Massachusetts General Hospital, Boston, MA, USA

Department of Dermatology, Harvard Medical School, Boston, MA, USA

N. Rezaei (ed.), *Coronavirus Disease - COVID-19*, Advances in Experimental Medicine and Biology 1318, https://doi.org/10.1007/978-3-030-63761-3_30

the potential therapeutic benefit of PBMT and aPDT in the context of the novel coronavirus. Studies indicate that PBMT and aPDT could be useful in many viral and bacterial pulmonary complications like influenza, SARS-CoV, and MERS, but we found no direct study on SARS-CoV-2. With a combination of PBMT and aPDT, we may be able to combat COVID-19 with minimal interference with pharmaceutical agents. It might improve the efficacy of PBMT and aPDT by using monoclonal antibodies and preparing new photosensitizers at the nanoscale that target the lung tissue specifically. More animal and human studies would need to take place to reach an effective protocol. This chapter would encourage other scientists to work on this new platform.

Keywords

Antiviral photodynamic therapy · COVID-19 · Low-level laser therapy · Photobiomodulation · Photodynamic therapy

30.1 Introduction

In late 2019, several patients presenting with symptoms of pneumonia without any identified cause were reported in Wuhan, Hubei Province, China (Li et al. 2020). After analysis of the sequence of development and possible exposure, the new disease was determined to be caused by a new coronavirus (CoV) named 2019-nCoV (Zhu et al. 2020). Subsequently, the World Health Organization (WHO) announced the nomenclature of Coronavirus Disease-2019 (COVID-19), for this novel coronavirus pneumonia-related disease on February 11, 2020 (WHO 2020). At the same time, the International Committee on Taxonomy of Viruses (ICTV) introduced the new coronavirus as SARS-CoV-2 (Gorbalenya et al. 2020).

30.2 Virology

30.2.1 Origin, Classification, and Genetical Features

Clinicians diagnosed the disease as a pneumonia-related infection caused by a virus based on clinical manifestations, chest radiographs, and blood tests. First epidemiological studies concluded that most of the suspected cases had been exposed to a local Huanan Seafood Market. Scientists first suggested that the origin of the outbreak was the market because SARS-CoV-2 had been isolated in environmental samples. However, this conclusion was disputed because the earliest/first case had no reported link to the mentioned market (Huang et al. 2020). Furthermore, at least two different strains of SARS-CoV-2 had been identified a few months earlier before COVID-19 was officially reported (Xiong et al. 2020). At present, the precise origin of SARS-CoV-2 remains unknown.

SARS-CoV-2 was first isolated from the bronchoalveolar lavage fluid (BALF) of three COVID-19 patients in Wuhan Jinyintan Hospital on December 30, 2019 (Zhu et al. 2020). Sequencing and evolutionary tree analysis showed that SARS-CoV-2 is a member of the CoV family (Zhu et al. 2020; Zhou et al. 2020). Coronaviruses are enveloped positive-sense single-stranded RNA viruses. These viruses are associated with respiratory, enteric, hepatic, and neurologic diseases (Weiss and Leibowitz 2011; De Wilde et al. 2018), and both SARS coronavirus (SARS-CoV) and MERS coronavirus (MERS-CoV) are members of CoVs (Weiss and Leibowitz 2011). Genetic studies indicate that SARS-CoV-2 shares 79.5% and 50% sequence identity to SARS-CoV and MERS-CoV, respectively, and belongs to the B (*Sarbecovirus*) of CoVs (Zhu et al. 2020; Zhou et al. 2020; Lu et al. 2020; Wu et al. 2020). Like other CoVs, the SARS-CoV-2 virion has a nucleocapsid composed of genomic RNA and phosphorylated nucleocapsid (N) protein. It has two types of spike proteins: the spike glycoprotein trimer (S)

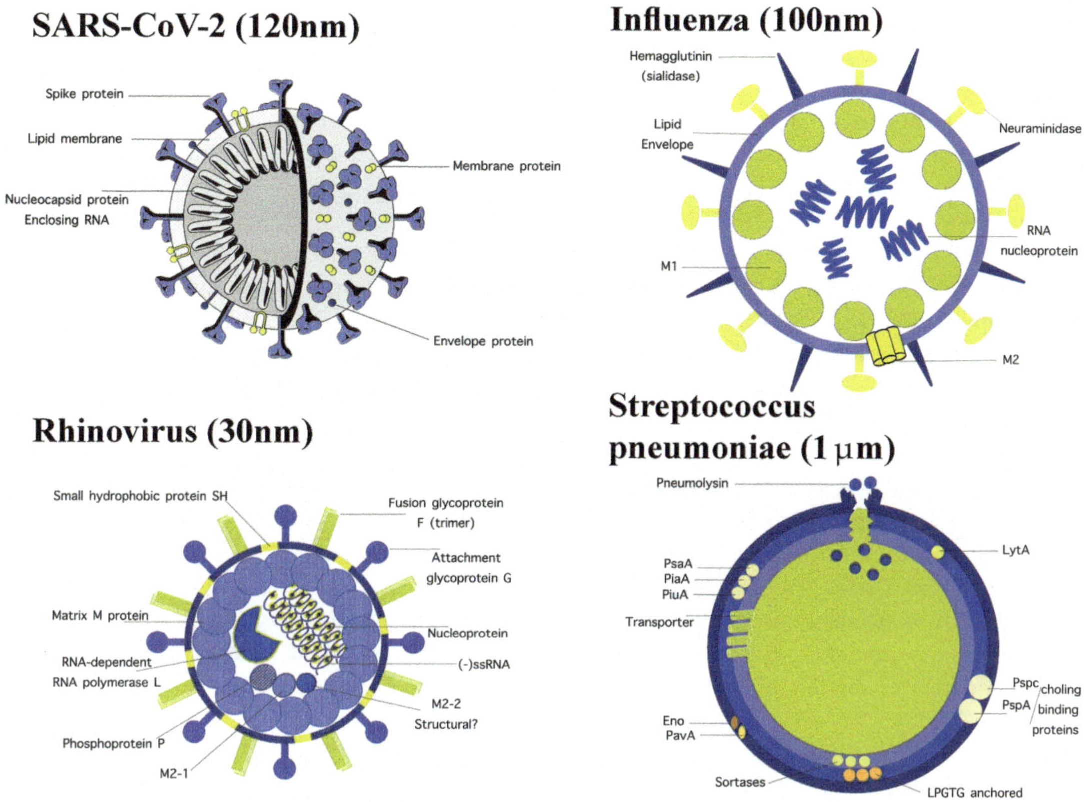

Fig. 30.1 Structural comparison between SARS-Cov-2, influenza, rhinovirus, and *Streptococcus pneumoniae*

and the hemagglutinin-esterase (HE). The former is common to all CoVs, while the latter is shared among certain CoVs. The envelope (E) protein and the membrane (M) protein are among the S proteins in the viral envelope (Wu et al. 2020). Figure 30.1 shows SARs-CoV-2 and some similar viruses and bacteria.

30.2.2 Receptor Interactions and Cell Entry

Human angiotensin-converting enzyme 2 (ACE2) is the receptor that the SARS-CoV-2 uses to enter cells just like SARS-CoV (Zhou et al. 2020; Li et al. 2003). This receptor is a membrane receptor found in the lung, heart, kidney, and intestine and has been associated with cardiovascular diseases. ACE2 possesses a direct binding site for the S proteins of CoVs (Donoghue et al. 2000). The S protein undergoes a structural rearrangement to fuse the viral membrane and the host cell (De Wilde et al. 2018). The S1 subunit triggers this process, which leads to the transition of the S2 subunit to a highly stable post-fusion conformation (De Wilde et al. 2018).

30.3 Epidemiology

30.3.1 Source of Infection

Patients are the primary source of infection, especially those presenting with severe infection and symptoms. Other potential sources of infection are asymptomatic persons or patients who are still within the incubation period and have no signs or symptoms of respiratory infection. These patients help virus shedding (Hoehl et al. 2020). Moreover, it is the first time in the history of human infectious diseases that samples from recovered COVID-19 patients continue to present a positive RT-PCR test (Lan et al. 2020).

30.3.2 Routes of Transmission

Respiratory droplets and contact transition are the main transition routes. Recent reports implied a fecal-oral risk of transition as well (Commission 2020). Virus-contaminated foods are yet to be recognized as a source of infection and transition. Besides, the possibility of mother to baby transmission of SARS-CoV-2 during pregnancy or childbirth is unknown.

30.3.3 High-Risk Population

Age, obesity, smoking as well as underlying disorders such as asthma, diabetes, cardiovascular diseases, and cancer increase the risk and susceptibility to SARS-CoV-2. Additionally, people in close contact with infectious sources, including healthcare workers and family members of infected patients, are the high-risk population (Jin et al. 2020b).

30.3.4 Spectrum of Infection

COVID-19 is considered to be a self-limiting infectious disease, and patients mostly suffer from mild symptoms and recover in 1–2 weeks. SARS-CoV-2 outcomes range from asymptomatic infection (1.2%) and mild to moderate infection (80.9%) to severe infection (13.8%), critical infection (4.7%), and death (2.3%) (Jin et al. 2020b). Figure 30.2 shows the confirmed cases and total deaths related to COVID-19.

30.4 Diagnosis

30.4.1 Nucleic Acid Test

Shortly after the initial outbreak, viral sequence detection diagnostic tests using reverse transcription-polymerase chain reaction (RT-PCR) or next-generation sequencing platforms became available. False negatives remain a problem when using nucleic acid tests. As a solution to the low detection efficiency, improved types of rapid viral nucleic acid diagnostic tests have been invented, especially a nucleic acid paper test, which has made naked-eye detection of SARS-CoV-2 possible within 3 min (Jin et al. 2020b).

30.4.2 Serologic Diagnosis

Studies show that SARS-CoV-2 patients have acute serological responses (Zhou 2020). Immunochromatography, colloidal gold, and other relevant detection methods for serum antibodies became available in a short time (Jin et al. 2020b).

30.4.3 Chest Radiography or Computed Tomography (CT) Scan

It is an important tool for COVID-19 diagnosis in clinical practice. Most COVID-19 patients display similar features on CT Scan images, such as signs of bilateral distribution of patchy shadows and ground-glass opacity (Kanne 2020).

30.5 Pathogenesis

30.5.1 Virus Entry and Spread

SARS-CoV-2 is transmitted via respiratory droplets, contact, and possibly through fecal-oral transmission (Jin et al. 2020b). The mucosal epithelium of the upper respiratory tract (nasal cavity and pharynx) is the place in which the virus is presumed first to enter and replicated. On the other hand, further virus replication takes place in the lower respiratory tract and gastrointestinal mucosa (Xiao et al. 2020), causing a mild viremia. Asymptomatic cases are controlled in this stage. Non-respiratory symptoms were also seen in some patients, such as the acute liver and heart injury, kidney failure, and diarrhea (Wang et al. 2020; Jin et al. 2020b), suggesting multiple organ

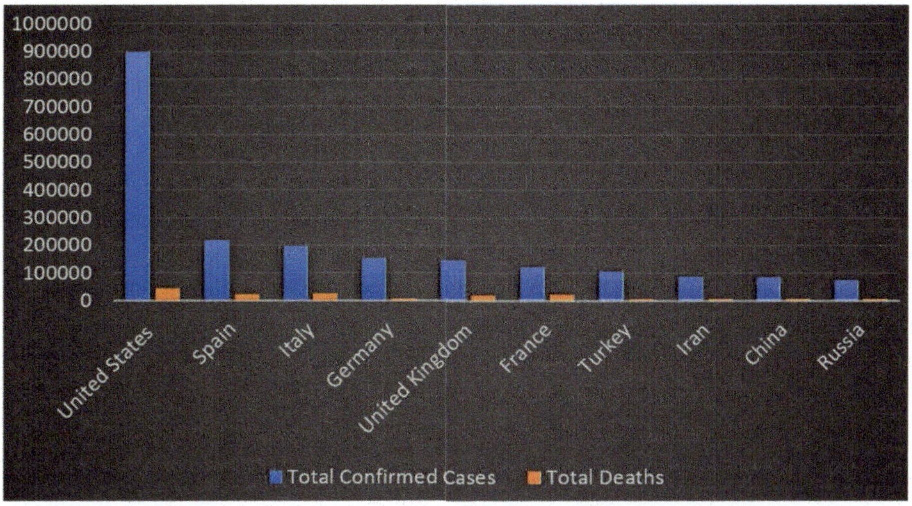

Fig. 30.2 Top ten countries with reported laboratory-confirmed COVID-19 cases and death as of 25 April 2020 (WHO 2020)

involvement. ACE2 is expressed in the nasal mucosa, bronchus, lung, heart, esophagus, kidney, stomach, bladder, and ileum, so all these organs are vulnerable to SARS-CoV-2 infection (Zou et al. 2020). In recent studies, scientists have suggested the potential pathogenicity of the SARS-CoV-2 in testicular tissues raising the concern of fertility effects in younger patients (Fan et al. 2020).

30.5.2 Pathological Findings

The first report (Xu et al. 2020) of pathological findings in a severe COVID-19 patient showed bilateral diffuse alveolar damage (DAD) with cellular fibromyxoid exudates. Tissue from the left lung displayed pulmonary edema with hyaline membrane formation suggesting an early phase of acute respiratory distress syndrome (ARDS). The right lung, on the other hand, demonstrated the desquamation of pneumocytes and hyaline membrane formation, indicating acute respiratory distress syndrome. Both lungs showed interstitial mononuclear inflammatory infiltrates, dominated by lymphocytes. Viral cytopathic-like changes also occur in intra-alveolar spaces. All these pulmonary pathological findings are similar to those of SARS (Ding et al. 2003) and MERS (De Wilde et al. 2018). On the other hand, massive mucus secretion was found in the lungs of patients who died from COVID-19, which is different from SARS and MERS (Jin et al. 2020b).

30.6 Clinical Features

In the first 41 patients (Huang et al. 2020), the most common symptoms were fever (98%), cough (76%), and myalgia or fatigue (44%). Additional symptoms were sputum production (28%), headache (8%), hemoptysis (5%), and diarrhea (3%). Moreover, more than 50% of patients experienced dyspnea. Blood tests showed average or reduced (25%) white blood cell count and lymphopenia (65%) (Huang et al. 2020). Bilateral involvement of the lungs was seen in 98% of patients, as demonstrated in thoracic CT. ICU patients mostly display bilateral multiple lobular involvements and subsegmental areas of consolidation, while non-ICU patients show bilateral ground-glass opacity and subsegmental areas of consolidation (Zhu et al. 2020; Huang et al. 2020).

30.6.1 Acute Respiratory Distress Syndrome (ARDS)

ARDS is a fatal lung condition in which sufficient oxygen is prevented from reaching the lungs and hence the circulation. It is responsible for most deaths in respiratory disorders and acute lung injury (Thompson et al. 2017). Clinically, fatal cases of human SARS-CoV, MERS-CoV, and SARS-CoV-2 infections exhibit severe respiratory distress requiring mechanical ventilation, while the histopathology findings also support ARDS (Xu et al. 2020; Ding et al. 2003; Ng et al. 2016). Based on previous studies, the occurrence of ARDS is closely connected to genetic features and inflammatory findings. There are more than 40 candidate genes related to the development of ARDS and its consequences (Meyer et al. 2013). These genes mainly involve molecular pathways of ACE2 and cytokines (e.g., IL-10, interleukin 10; TNF, tumor necrosis factor (TNF); and VEGF, vascular endothelial growth factor). Different outcomes of ARDS are considered to be associated with increased levels of plasma IL-6 and IL-8 (Thompson et al. 2017). These biomarkers suggest both an inflammatory explanation and possible treatment for ARDS following SARS-CoV-2 infection.

30.6.2 Cytokine Storm Phenomenon

The leading cause of fatality is likely to be uncontrolled pulmonary inflammation, which has been shown to be caused by extreme inflammatory responses during SARS-CoV-2 infection. The result of a recent study concluded that rapid viral replication and cellular damage, virus-induced ACE2 downregulation and shedding, and antibody-dependent enhancement (ADE) are the causes of aggressive inflammation caused by SARS-CoV-2 (Fu et al. 2020). SARS-CoV-2 and SARS-CoV use the same receptor (ACE2) for cell entry, suggesting the same cells are targeted and infected (Gu et al. 2005). Excessive production of pro-inflammatory cytokines and chemokines is triggered by massive epithelial and endothelial cell death and vascular leakage caused by the initial onset of rapid viral replication (Jin et al. 2020b). Pulmonary loss of ACE2 function is considered to be related to acute lung injury (Imai et al. 2008) because it leads to dysfunction of the renin-angiotensin system (RAS), which worsens the inflammation and causes vascular permeability. ADE is a well-known phenomenon in virology, and it has been confirmed in multiple viral infections (Takada 2003). ADE can induce viral cellular uptake of the virus-antibody complexes resulting in enhanced infection of target cells (Takada and Kawaoka 2003).

30.6.3 Cellular Immune Dysfunction

In patients with severe COVID disease, peripheral CD4+ and CD8+ T cells are reduced in counts, but their activity is increased, as indicated by an abundance of cytotoxic granules, including CD8+ T cells and CD4+ T cells (Xu et al. 2020).

30.7 Potential Therapeutics

There is no specific treatment or vaccine for COVID-19, and, still, systematic treatment strategies are recommended for clinical practice (Jin et al. 2020a). Potential therapeutic strategies currently available for the treatment of SARS-CoV-2 are listed below (Mahase 2020).

30.7.1 Chloroquine

It is an approved treatment for malaria and rheumatoid arthritis. It has been tested against SARS-CoV-2 and has been suggested to be highly effective, but the evidence is limited and contradictory. Chloroquine can interfere with viral cell entry by changing the acidity of the endocytosis compartment and inhibition of initial viral replication. Chloroquine may also change the ability of the virus to bind to the outside of the host cell. Finally, this drug has various effects on immune cells, which has made it a possible treatment for autoimmune conditions like lupus and rheumatoid arthritis. However, the side effects of chloroquine, when used in conjunction with additional drugs, do not render it a safe treatment compound.

30.7.2 Kaletra

Kaletra (lopinavir/ritonavir) is usually used to treat HIV. The combination of Kaletra and other antiviral drugs has been tested against SARS. Lopinavir is an inhibitor of HIV protease, which prevents cleavage of the Gag-Pol polyprotein, resulting in the production of immature, noninfectious viral particles. As co-formulated in Kaletra, ritonavir can inhibit the CYP3A-mediated metabolism of lopinavir, thereby providing increased plasma levels of lopinavir.

30.7.3 Interferon

SNG001 is an inhaled form of interferon β, which is a part of the lung's defense system against viruses. It is believed that SNG001 can increase the production of INF-β, which is considered to be suppressed by coronaviruses. This phenomenon could prevent or decrease symptoms of severe respiratory illness, such as pneumonia.

30.7.4 Remdesivir

It is going through clinical trials for potential COVID-19 treatment as a virus replication inhibitor. Wang et al. found that this drug alone cannot provide significant clinical or antiviral effects in severely ill patients with COVID-19. Further studies with larger sample sizes are necessary for a better understanding of this drug's effects. Furthermore, different strategies to improve the antiviral potency of this drug, such as higher dose regimens and combination with antivirals and immunosuppressants targeting pro-inflammatory cytokines, such as IL-6, IL-1, or TNF-α, need to be done (Wang et al. 2020).

30.7.5 Tocilizumab (Actemra)

It is a monoclonal antibody used for rheumatoid arthritis treatment. It blocks the IL-6 signaling pathway.

30.7.6 hrsACE2 (Human Recombinant Soluble Angiotensin-Converting Enzyme 2)

It is a genetically modified type of ACE2 which could significantly prevent SARS-CoV-2 entry (Monteil et al. 2020).

30.7.7 Immunoglobulin Therapy

Convalescent plasma or hyperimmune immunoglobulins may help with both the removal of free virus and infected cell immune clearance (Sanders et al. 2020). All these treatments can be seen in Fig. 30.3. In conclusion, most current treatment regimens are focused on antiviral and anti-inflammatory effects, stifling the cytokine storm and increasing tissue oxygenation.

30.8 Photobiomodulation Therapy

Photobiomodulation therapy refers to the interaction between visible light (usually red) or near-infrared light with low energy density and biological tissue that does not induce thermal effects (tissue temperature does not rise above 98 F). Light sources in PBM can be laser or light-emitting diode (LED) systems that can emit wavelengths between 600 nm–1200 nm with energy densities of 1–20 J/cm^2. Due to their low energy density, this type of irradiation can treat various injuries or pathologies without causing thermal damage in the target tissue (Migliorati et al. 2013; Theocharidou et al. 2017; Huang et al. 2011).

Biomodulation effects of PBMT occur with the interaction of light with photoacceptors,

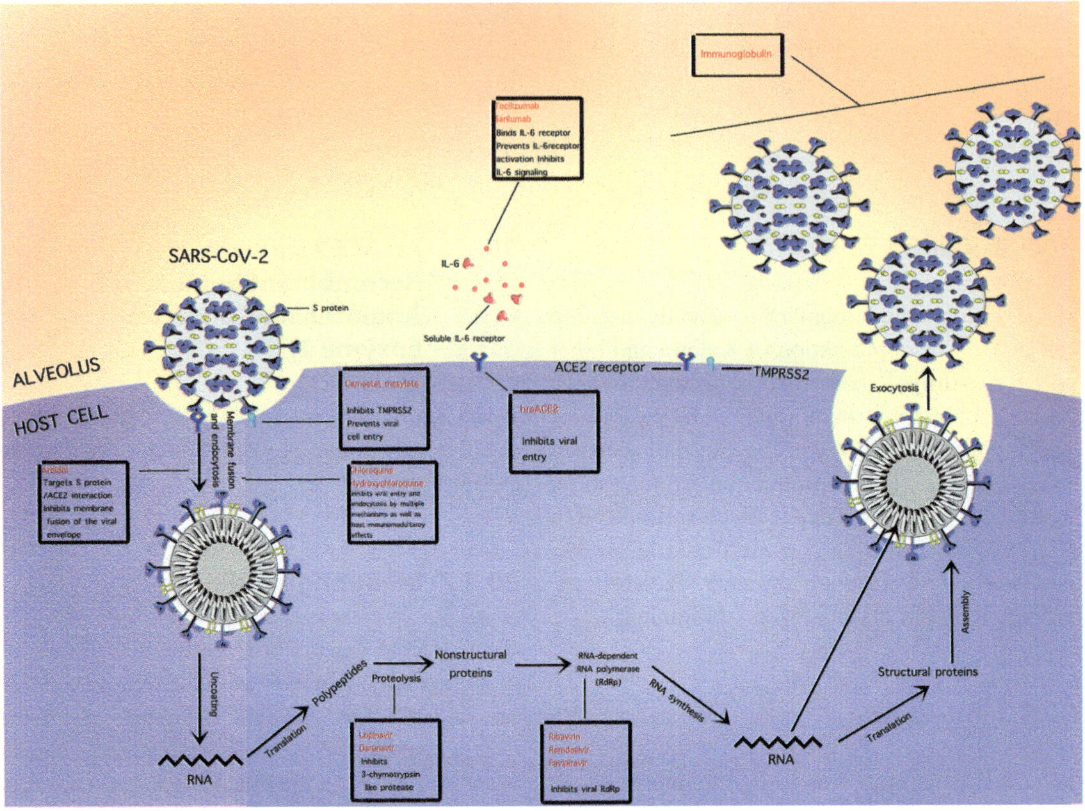

Fig. 30.3 Current SARS-CoV-2 treatment modalities

resulting in mitochondrial signaling events that mediate photophysical, photochemical, and photobiological effects in tissue (Migliorati et al. 2013; Theocharidou et al. 2017). Photochemical reactions can be caused by the effect of visible light or NIR (near-infrared) on mitochondrial photoacceptors and the process of ATP synthesis. The effects of light radiation on Ca ++ channels in cell membranes is another basis for photophysical reactions (Huang et al. 2011; Karu 1989b; Smith 1991; Lopes et al. 2009, 2010; Heidari et al. 2017).

Given the importance of mitochondria in energy production and cellular metabolism, recent studies have paid particular attention to their role in the mechanism of action of PBM (Fig. 30.4). The first PBMT mechanism at the cellular level occurs by the absorption of monochromatic visible and NIR by cellular respiratory chain components. Absorption of R-NIR radiations by mitochondrial photoacceptors activates the respiratory chain, NADH dehydrogenase, cytochrome C reductase, and oxidase, as well as ATP synthase (Karu 2008; Karu and Kolyakov 2005). Accordingly, cytochrome c oxidase (Cco) can be considered the primary photoacceptor that absorbs R-NIR radiations in mammalian cells (Karu 1989a; Huang et al. 2011; Karu and Kolyakov 2005).

The absorption of light by the components of the respiratory chain causes the short-term activation of the respiratory chain and the oxidation of the NADH pool. This stimulation of oxidative phosphorylation leads to changes in mitochondrial redox status as well as cell cytoplasm. The electron transfer chain can provide increased levels of energy to the cell by increasing ATP production, as well as increasing the electrical potential of the mitochondrial membrane, alkalizing the cytoplasm, and activating the synthesis

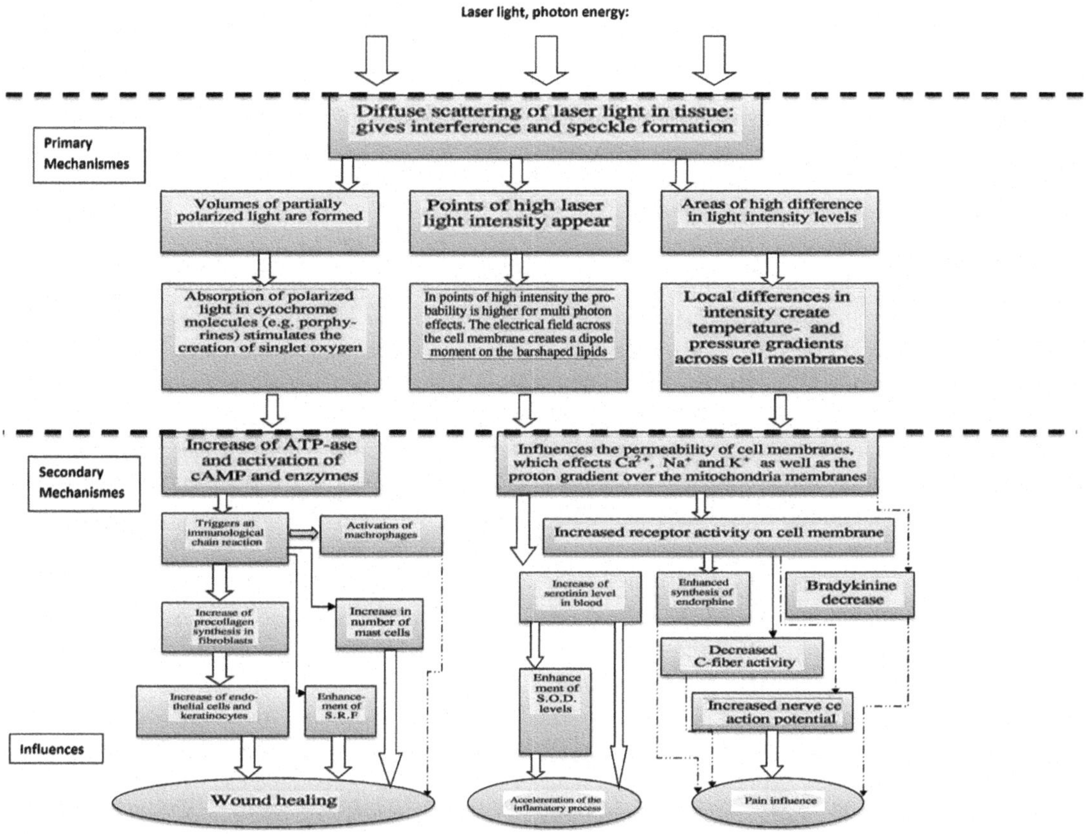

Fig. 30.4 Mechanisms of photobiomodulation

of nucleic acids. Also, infrared radiation transmits biological messages by activating ion channels and affecting the permeability of cell membranes as well as the concentration of Na + -K + flux and Ca + 2 ions (Smith 1991). Because ATP is a common form of energy for the cell, PBMT has the potential to stimulate normal cell functions. By increasing cellular respiratory metabolism, PBMT can also affect the cellular electrical-physiological characteristics (Moslemi et al. 2014; Hamblin and Demidova 2006). Further stimulation of cell proliferation and inhibition of cell death, as well as the effect on the production of extracellular matrix proteins, are among the most critical effects of PBMT on the expression of genes (Theocharidou et al. 2017).

In general, PBMT can be used in both direct and indirect ways. For example, in the transthoracic method, the lung tissue is exposed to direct infrared lasers with wavelengths of 810 nm or 940–970 nm. Intravenous radiation is also a type of direct radiation that can play an important role in the systemic effects of PBMT. The indirect method can be used to irradiate blood cells through the skin adjacent to the blood vessels in the wrist, nasal mucosa, and the back area of the knee, or even sublingual tissues, which is called the transcutaneous technique (Fekrazad 2020).

Among the most important clinical applications of PBMT are accelerating the healing process of the skin or mucosal wounds; reducing acute and chronic pain; regulating the function of the immune system, especially in people with immunodeficiency, anti-inflammatory function by regulating levels of pro-inflammatory cytokines, as well as oxygenation of damaged tissue; and helping tissue regeneration by stimulating stem cell proliferation and differentiation (Fig. 30.4) (Desmet et al. 2006; Chiari 2016; Fujita et al. 2008; Fekrazad 2020).

30.8.1 PBM and Pro-inflammatory Cytokine

Cytokines are hormone-like soluble proteins or glycoproteins that allow communication between cells and the extracellular environment. When an organ or limb is damaged or infected, an immune response is triggered to suppress the infection, in which the release of pro-inflammatory cytokines indicates the body's attempt to respond to the infection. Several cytokines play a key role in the development of acute inflammatory responses, including IL-1, IL-6, IL-8, IL-11, TNF-α, G-CSF (granulocyte colony-stimulating factor), and GM-CSF (granulocyte-macrophage colony-stimulating factor). Among these, IL-1 and TNF (α and β) have the highest power to induce inflammation (Mahnam and Payab 2016).

Studies have shown that one of the important functions of PBM is to reduce all the markers of immune activation such as MHC class II, CD86, and CD11c in inflammatory cells. Also, the IL-12 secretion has been another effect of PBM. Hwang et al. stated the decreased expression of IL-8 and IL-6 (two important pro-inflammatory cytokines in macrophages) with PBM. Chen et al. confirmed the vital role of PBM in reducing inflammatory markers (cyclooxygenase-2 (COX2), prostaglandin E2 (PGE2), granulocyte colony-stimulating factor (GCSF), regulated on activated normal T-cell expressed and secreted (RANTES), and CXCL11) (Chen et al. 2011; Hwang et al. 2015). Yamaura et al. also used PBM (810 nm, 5 or 25 J/cm2) to confirm its role in reducing levels of TNF-α and interleukins (IL)-1beta, and IL-8 (Yamaura et al. 2009).

30.8.2 PBM on Macrophage Phenotype

Another important effect of PBM at the cellular level is its role in altering the phenotype of monocyte or macrophage cells. Monocytes and macrophages critically contribute to innate immunity. These cells detect dangerous and invasive pathogens in the blood and tissues and exert a strong inflammatory immune response that eventually leads to the removal of the pathogens and the establishment of homeostasis in the body. Activated macrophages are differentiated into two different phenotypes, classical macrophages (M1) and alternative macrophages (M2). These phenotypes differ in terms of cytokines, chemokines, ligand receptors, and function. M1 macrophages produce large amounts of lymphokines and are associated with the expression of genes such as IFNγ, TNF-β, and TNF-α. The M2 phenotype shows high production of TGFβ, and IL-10, and exerts anti-inflammatory effects that induce tolerance, immune regulation, and tissue repair. Also, both types of macrophage (M1 and M2 phenotypes) are present in different organs, such as microglia in the brain, alveolar macrophages in the lung, and Kupffer cells in the liver, and are specific to that organ (Hamblin 2017; Jalili et al. 2015).

Fernandes et al. were among the first researchers who found the effect of PBM on reducing TNF-α, COX-2, and iNOS expression in M1 cells. Also, in another study, PBM had a significant effect on reducing CCL3 gene expression in M1 cells. Increased cAMP and decreased TNF-α mRNA are other identified effects following the effect of PBM on alveolar macrophages (Hamblin 2017; Fernandes et al. 2015; Sousa et al. 2017) (Fig. 30.4).

30.8.3 Role of Photobiomodulation in COVID-19 Management

There is not yet any definitive treatment for COVID-19 disease. The main recommended approaches for COVID-19 management are increasing tissue oxygenation, facilitating damaged tissue repair, eliminating the virus, and reducing inflammation. Every treatment, which can promote regeneration of respiratory or other impaired tissues, can be helpful in controlling COVID-19 progression. These may include directly targeted organ or indirect methods as influencing blood flow or oxygenation to damaged tissues. So clinicians usually alleviate symptoms, especially by decreasing inflammation.

One of the innovative approaches is photobiomodulation therapy. Light or low-level lasers irradiate the tissues, and internal or external photoacceptors can absorb the emitted energy. In this noninvasive method, free radicals are produced after light irradiation, which leads to biostimulatory cascades in cellular processes to promote cell proliferation, differentiation, or decrease inflammation (Fekrazad 2020; Tuner and Hode 2004; Khorsandi et al. 2020).

When photoacceptors absorb light or laser, free radicals are produced. Reactive oxygen species (ROS) are of common radicals in this regard. They include singlet oxygen, nitric oxide (NO), or hydrogen peroxide (H_2O_2) (Derr and Fine 1965; Lubart et al. 1990; Mrowiec et al. 1997). They can play the role of secondary messengers at low concentrations, whereas high concentrations can be destructive for cell processes.

Photoacceptors can be divided into internal or external. Internal photoacceptors contain chromophores (like porphyrins) and respiratory chain components in the cell mitochondria, such as NADH oxidase and cytochrome a/a3 (Tuner and Hode 2004; Karu 1996). Photosensitizer drugs are examples of external photoacceptors.

Free radicals can influence cell processes through many mechanisms. Singlet oxygen can promote ATP formation, and ATPs can be used in energy-consuming processes like proliferation or differentiation (Passarella et al. 1984; Hu et al. 2007). Also, low doses of laser irradiation can increase intracellular calcium ions by antiporter process on the cell membrane (Nasu et al. 1989; AlGhamdi et al. 2012) or by increasing the proton-motive force (pmf) in mitochondria (Eduardo et al. 2008) which then affect DNA/RNA synthesis via increasing intracellular pH (alkaline) leading to mitosis and cell proliferation. Also, mitochondrial activation stimulates Na,K-ATPase activity in order to provide more ATP that is necessary for cell division. If the cellular function is impaired (especially by hypoxia), there is more benefit of photobiomodulation compared to situations where the cell functions are intact (Tezel et al. 2009).

Previous studies have shown the benefits of lasers in medical care, including wound healing acceleration, anti-inflammatory effects, and disinfection (Tuner and Hode 2004). One of the advantages of photobiomodulation application is its local effect on the target tissues. General effects of photobiomodulation include metabolic, analgesic, anti-inflammatory, and immunomodulatory functions. The regenerative power of photobiomodulation can increase cellular ability to overcome various problems, such as metabolic and neurological diseases and cardiac or physical rehabilitation. Photobiomodulation is a noninvasive intervention that is local to the target organ and has the least side effects. There have been no adverse systemic effects reported. Based on the reasons mentioned above and the lack of currently available treatment for COVID-19, it is likely that photobiomodulation can be helpful in controlling COVID-19, as an alternative or adjunctive treatment, especially in severe cases with ARDS.

30.8.4 Photobiomodulation of the Lungs

Oliveira et al. designed an in vivo study to evaluate the anti-inflammatory effect of low-level laser therapy on pulmonary and extrapulmonary lipopolysaccharide (LPS)-induced ARDS. They experimentally induced ARDS by intratracheal or intraperitoneal administration of LPS in mice. After 1 h, mice were irradiated by an infrared laser (830 nm wavelength, energy density 3 J/cm^2 per application, power 35mW, continuous mode) in direct contact with the chest skin. They used three points, first at the end part of the trachea, second on the right lung, and third on the left lung. Each irradiation lasted 80 s. 9 J/cm^2 energy was delivered in each application to mice for 240 s. This procedure was repeated three times with a 1 h interval between each application (Oliveira Jr et al. 2014). They found that photobiomodulation could reduce pulmonary inflammation. It was confirmed by a significantly reduced number of total cells and neutrophils in bronchoalveolar lavage (BAL) and neutrophils in the lung parenchyma.

Neutrophil migration to the lungs and secretion of several mediators play important roles in ARDS. These mediators include free radicals, proteases, cytokines, and chemokines (Matute-Bello et al. 2008). The severity and mortality of ARDS have a direct relationship with neutrophil infiltration in the lungs (Fialkow et al. 2006).

Oliveira reported that photobiomodulation could inhibit neutrophil migration to the lungs. They also showed that IL-1 beta, IL-6, IL-8, KC, and TNF-α levels were reduced significantly in BAL fluid and serum after laser irradiation (Oliveira Jr et al. 2014).

IL-1 beta contributes to the initiation of the inflammatory process, and neutrophils are the primary source of IL-1 beta production. It increases the survival rate of neutrophils and is related to the development of ARDS and its poor prognosis. IL-6 is also related to the poor prognosis of ARDS (Meduri et al. 1995; Cho et al. 2012; Rojas et al. 2013; Sharifov et al. 2013) and the prolongation of inflammation (Meduri et al. 1995; Fu et al. 2012). IL-8 stimulates neutrophil chemotaxis and also can increase neutrophil survival in combination with CXCL1/KC. These two functional homologs are related to ARDS severity and morbidity (Meduri et al. 1995; Cho et al. 2012; McGettrick et al. 2006). Therefore, reducing CXCL1/KC by photobiomodulation could be critical in severe ARDS cases. TNF-α affects neutrophil adhesion and activation, and it increases IL-6 generation. TNF-α also induces intravascular coagulation and edema, especially in acute inflammation (Souza et al. 2002; Aimbire et al. 2006).

In one experimental study, Aimbire et al. evaluated Ga-Al-As laser irradiation to reduce airway and lung inflammation induced by gram-negative bacterial LPS in rats (Aimbire et al. 2006). They used a 685 nm wavelength diode laser with 1, 2.5, and 5 J/cm^2 energy densities in continuous mode. They found that 1 and 5 J/cm^2 energy densities were not effective in decreasing rat tracheal hypersensitivity, BAL cellular content, and myeloperoxidase (MPO) activity. Therefore, 2.5 J/cm^2 was selected, the power was 12 mW, the spot size was 0.08 cm^2, the exposure time was 1 min and 20 s, and the irradiation interval was in 90 min, 6, 24, 48 h. Photobiomodulation reduced inflammation in BAL and neutrophil migration to the lungs. PBM also reduced PGE_2 and TXB_2 levels in BAL. PGE_2 has an important role in releasing mediators from mast cells, surveillance and chemotaxis of neutrophils, and mitogenesis (Johnson et al. 1995). TXB_2 mediates inflammation and can stimulate bronchial constriction (Pang and Knox 1997).

De Lima (de Lima et al. 2011a) confirmed the beneficial effects of InGaAlP laser on ARDS induced by LPS in an in vivo study in mice. Laser parameters were 685 nm wavelength, 4.5 J/cm^2 energy density, 17.85 W/cm^2 power density, 35 mW power, 252 s time of irradiation, and 8.82 J energy dose. Irradiation was carried out 15 min after ARDS induction. PBM decreased neutrophil recruitment and TNF production in BAL fluid and also increased cAMP indirectly.

Another study (de Lima et al. 2011b) evaluated the effect of PBM on the acute lung inflammation induced by intestinal ischemia and reperfusion. Diode laser (660 nm, 30 mW, 0.08 cm2 spot size, 5.4 J) was used directly on the rat skin over the right bronchus. Rats were irradiated 1 h after mesenteric artery occlusion. Photobiomodulation reduced inflammation, neutrophil migration, MPO activity, and TNF production. One side effect of laser irradiation was an increase in IL-10. As mentioned before, TNF mediates neutrophil accumulation. IL-10 is associated with a reduction in inflammatory cytokines and the healing of tissue injury. They suggested that acute lung inflammation due to intestinal ischemia may occur with an imbalance between TNF and IL-10 production (de Perrot et al. 2003). So it is promising to promote IL-10 generation and decrease TNF production in order to reduce lung inflammation and subsequent injury.

In addition to TNF-α and IL-1β, another study considered some specific neutrophil chemoattractants (a family of cytokines and neutrophil chemoattractants, CINCs) and macrophage inflammatory protein 2 (MIP-2). Also, CD18 integrin, on polymorphonuclear leukocytes (PMNs), and intracellular adhesion molecules-1 (ICAM-1), on endothelial cells, have important roles in PMN migration into an inflammatory site (de Lima et al. 2010).

In an in vivo study on rats with acute pulmonary inflammation, which was induced by *Escherichia coli* LPS, photobiomodulation reduced TNF-α and IL-1β in BAL fluid. It was suggested that inhibition of TNF-α and IL-1β production might be due to ICAM-1 upregulation. There was no influence on CINC-1, MIP-2, or the expression of IL-10 after laser irradiation. GaAsAl diode laser was used in this regard. Laser parameters were as follows: 650 nm wavelength, power 2.5 mW, power density 31.2 mW/cm2, energy density 1.3 J/cm2, laser spot size 0.08 cm2, and irradiation time 42 s (de Lima et al. 2010).

da Silva achieved similar results in the reduction of lung inflammation by a diode laser (continuous mode, 30 mW, 660 nm, 60 s/point, a spot size of 0.14 cm2, a power density of 210 mW cm2, and energy density of 12.86 J/cm2). Pulmonary inflammation was induced by formaldehyde (%1). The lower number of neutrophils, degranulation of mast cells, and MPO activity were observed along with reduced levels of IL-6 and TNF-α and increased level of IL-10 in the lungs (Miranda da Silva et al. 2015).

Some studies have concluded that light could be useful for COVID-19 symptoms and also suggested that blue and ultraviolet wavelengths light could be beneficial in virus eradication. Further research is necessary to test the effectiveness of light in COVID-19 disease (Enwemeka et al. 2020).

30.8.5 Indirect Photobiomodulation

Lymphopenia is one of the main laboratory findings in COVID-19. These reduced numbers of lymphocytes may represent the lower ability of the immune system to eliminate the virus. (Lu et al. 2019)

Al Musawi et al. (2016) designed an in vitro study and found that photobiomodulation could significantly increase the lymphocyte count in whole blood samples. They used 405, 589, and 780 nm wavelength lasers with 36, 54, 72, and 90 J/cm2 energy densities for all groups. It was found that 598 nm wavelength at 72 J/cm2 fluence had the best effects. There was a significant increase in CD45 lymphocytes and natural killer (CD16 and CD56) cells. However, other cells, like CD3 T lymphocytes, T-suppressor (CD3, CD8) cells, T-helper (CD3, CD4) cells, and CD19 B lymphocytes, did not show any significant changes. As explained before, T-helper cells can influence the inflammatory cascade by releasing mediators. Therefore, photobiomodulation may be able to increase lymphocyte or natural killer cells without any exacerbation in inflammation.

Neutrophils have a critical role in ARDS pathogenesis (Oliveira Jr et al. 2014). Neutrophils undergo an influx into pulmonary tissues, and they release high quantities of cytokines, ROS, and inflammatory mediators. ROS are correlated with the severity of inflammation (de Lima et al. 2011b). Fujimaki et al. (2003) reported that photobiomodulation could attenuate ROS production in neutrophils without influencing cell viability. This effect was seen in both nonsmoker and smoker patients, but it was more pronounced in smoker patients. It was explained that MPO produces the ROS in neutrophils, and laser irradiation can reduce MPO activity. Infrared diode laser (GaAlAs) was used with parameters of 830 nm wavelength, continuous wave mode, power density 150 mW/cm2, power 1000 mW, spot size 6.6 cm^2.

30.9 Photodynamic Therapy (PDT)

The history of photodynamic therapy dates back thousands of years to the ancient civilizations of Egypt, India, and China. The ancient people discovered that administering some plants to patients and exposing them to sunlight could help the treatment of different diseases such as psoriasis, rickets, vitiligo, and skin cancer. However, about 100 years ago, Hermann von Tappeiner and Jesionek were the first to describe the photosensitization reaction and active oxygen formation, based on the accidental findings of his student

Oscar Raab, who was studying the reaction between the fluorescent dyes and microorganisms. They named the technique "photodynamic action" (Dolmans et al. 2003).

The PDT technique was successfully tested in the early twentieth century for the treatment of malignant tumors, especially skin cancer. However, due to the lack of reliable evidence, it was not welcomed by the scientific community until the 1970s. Subsequently, following Dougherty's initial studies using porphyrin derivatives, commercial photosensitizer products, and appropriate light sources were introduced, leading to successful clinical studies (Dolmans et al. 2003; Allison and Moghissi 2013).

PDT works based on the interaction between light, a photosensitizer (PS), and oxygen. ROS production due to the interaction between light radiation and nontoxic PSs can occur either inside the cell or in the immediate environment. Cellular apoptosis or cell necrosis occurs with minimal damage to adjacent tissues. Thus, an advantage of PDT is its dual selectivity, meaning that both PS and light with a specific wavelength are selectively confined to the target cells or tissue (Bargrizan et al. 2019; Saffarpour et al. 2018).

30.9.1 Mechanism of Photodynamic Therapy

Apoptosis or cellular necrosis (cellular PDT), as well as the destruction of blood vessels (vascular-PDT), results from the generation of ROS, which oxidizes biomolecules. It is the basis for the use of PDT in the treatment of various diseases, including cancers and bacterial infections.

In general, photodynamic reactions occur in two ways, one is mediated by ROS and occurs where photosensitizers are exposed to light, and the second group is related to reactions that are not caused by oxygen called photochemotherapy (Dolmans et al. 2003). After absorbing the light energy, the PS is excited to a long-lived excited triplet state. In this case, it either returns to the ground state by losing energy, or it can undergo two types of photochemical reactions called type I and type II. In the type I reaction, the transfer of energy from the PS causes free radicals in adjacent molecules, and during their reaction with oxygen, ROS (superoxide and hydroxyl radicals) are formed. In type II reactions, the PS-excited energy is transferred directly to the ground-state triplet $^3O_2$ oxygen molecule, forming singlet oxygen $^1O_2$, another type of ROS. Photodithazine and phthalocyanine are two types of PSs with the type II mechanism of action. Malachite green from the triarylmethane family is another dominant type of PS that appears to undergo the type I mechanism of action (Reis et al. 2019), as seen in Fig. 30.5.

30.9.2 PDT for Infectious Disease

An important challenge in the treatment of infectious diseases is the emergence of bacteria that are resistant to the effects of antibiotics. Therefore, many efforts have been made in recent decades to find effective treatments for resistant infections. In the case of viral infections, there are several limitations, such as the lack of effective vaccinations for viruses, different immune responses of people to vaccination, limited range of antiviral drugs, and antiviral drug resistance (Shahbaz et al. 2016). Due to the desire for targeted therapy and the failure of antibiotics, the PDT method has been proposed as an effective method in the treatment of microbial infections (Shahbaz et al. 2016; Afrasiabi et al. 2019; Ahrari et al. 2018).

30.9.3 Antimicrobial Photodynamic Therapy

Antimicrobial photodynamic therapy (aPDT), also known as photoactivated disinfection, is a new type of treatment that can inhibit bacterial, viral, fungal, and parasitic infections. aPDT is also known as photodynamic inactivation (PDI), lethal photosensitization, photoactivated disinfection (PAD), or photodynamic antimicrobial chemotherapy (PACT), and, as mentioned above, the most important application of aPDT is the

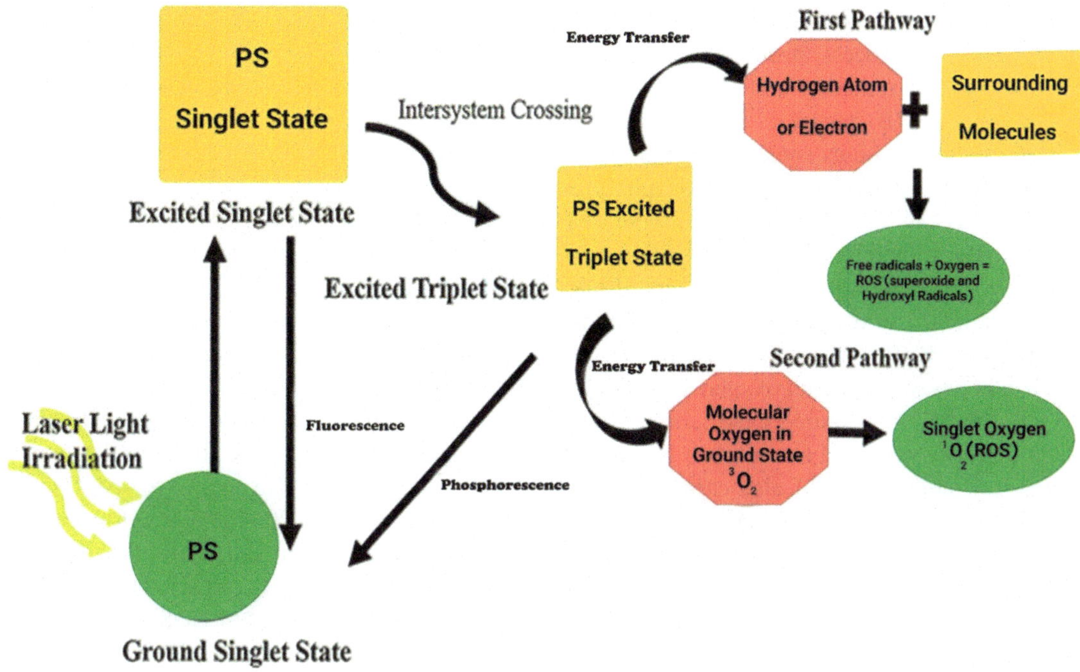

Fig. 30.5 Mechanism of photodynamic therapy

treatment of multidrug-resistant infections. The importance of aPDT treatments is such that no bacterial resistance to this method has been reported so far (Kharkwal et al. 2011).

The most important advantages of using aPDI technique include (Reis et al. 2019) lack of toxicity and gene mutations in the long run; ability to destroy microbes in a short time (a few seconds to a few minutes); no damage to adjacent tissues; the ability to access and affect areas of complex anatomy; decreased risk of bacteremia, in particular in patients with weakened immune systems; and high repeatability without antimicrobial resistance.

30.9.4 Photoinactivation of Viruses

As mentioned, aPDI can also have an antiviral function. Enveloped viruses are the most important group of viruses that are destroyed by aPDI. The presence of lipid and protein structures in the envelope plays a vital role in the possibility of PS binding to the virus envelope, and the most important mechanism for virus inactivation is protein damage (Kashef et al. 2017; Käsermann and Kempf 1998). The PDT virus inactivation mechanism is thought to be due to one or more of the following three possibilities: (i) damage to the cell membrane or virus envelope, (ii) inactivation of proteins or essential enzymes, and (iii) damage to the structure of DNA.

Cellular damage due to the effect of aPDI falls into two groups: morphological and functional. The most important functional damage is the loss of enzymatic activity, oxidation, and denaturation of proteins, as well as inhibition of metabolic processes such as DNA synthesis or glucose transport. Changes in the mesosome structure, as well as direct damage to the cell wall, are the most critical morphological damage. Damage to the cell wall causes the intracellular contents to leak and disrupt the membrane transport system (Kashef et al. 2017; Hamblin 2017).

Methylene blue (MB), or 3,7-bis(dimethylamino)phenothiazine-5-ium chloride, is an important PS that, along with light, can inactivate RNA viruses such as HIV-1 in blood or blood products. The mechanism of action of MB in the inactivation of viruses refers to the chemical process through the stimulation of MB by light radiation and change from the ground state (MB) to an excited singlet state (^{1}MB). ^{1}MB can lose its energy and return to the ground state or become an excited triplet state (^{3}MB). ^{3}MB can also slowly lose its energy and return to the ground state or act according to type I or type II reactions. As mentioned, type II reactions take place in the presence of oxygen, and ^{3}MB is converted to the ground state by converting triplet oxygen to singlet oxygen. Singlet-state oxygen has 23 kcal more energy than ground-state triplet oxygen and has a much higher oxidizing potential. ^{3}MB can also carry out type I reactions in O2-saturated organic solvent or air-saturated aqueous solutions through H atom or electron transfer. MB can make strong bonds to DNA, especially in G-C-rich regions, and break nucleic acid strands in the presence of light (Cadet et al. 1986; Floyd et al. 2004).

30.10 Antiviral Photodynamic Therapy in COVID-19

The initial clinical applications of aPDT as a potentially effective and safe treatment modality for infectious diseases were directed against superficial viral diseases, including herpes genitalis (Chang et al. 1975; Roome et al. 1975). Besides in vitro work on different viruses, in recent years several clinical trials of systemic and topical aPDT as a treatment approach and adjuvant therapy with surgery, cryotherapy, or chemotherapy have been conducted against viral infections (Abramson et al. 1992; Bujia et al. 1993; Shikowitz et al. 1998; Zhou et al. 2014; Wiehe et al. 2019). Currently, there are two main clinical and medical aspects for aPDT of viruses: one is the treatment of local and superficial viral infections, including herpes simplex virus (HSV) and human papillomavirus (HPV) infections, and the other is the area of blood product decontamination. An inhibitory effect of aPDT was observed on human immunodeficiency virus (HIV) through the decrease of viral integrase and protease activities. Antiviral potentials of Buckminsterfullerene and its derivatives have been tested against some viruses such as HIV by several researchers. These compounds can effectively degrade HIV protease under photo-irradiation and inhibit HIV replication in living cells (Friedman et al. 1993). Many reports have also emerged over the years of the use of aPDT as a practical methodology for sterilization of blood or blood products (Lulic et al. 2009; Azarpazhooh et al. 2010). These studies generally involved long known human viruses but have also been carried out on recently emerged viruses, e.g., coronaviruses.

Several studies have revealed that lipid-enveloped viruses are more susceptible to inactivation using aPDT than non-enveloped viruses (Hamblin and Hasan 2004). Although there is no evidence explicitly investigating the effect of aPDT on viral lipids and proteins, there are investigations about the effect of ROS on viral lipids (Girotti 2001; Costa et al. 2012; Baptista et al. 2017). Since there are viral lipids in the envelope of the COVID-19, these should be sensitive to the effects of aPDT. aPDT damage to the structure of the viral envelope can inhibit the binding of the virus to the cells because the envelope plays a vital role in the attachment of viruses to the host cell surface. The effect of aPDT as an adjunct therapy for a viral infection such as MERS and influenza caused by MERS-CoV and orthomyxoviruses, respectively, can be generalized to the COVID-19 due to the similarity of viral pathogenesis to cause primarily mild to severe respiratory infections. In humans, these viruses can be detected with higher viral load and longer duration in the respiratory tract and have also been detected in feces, serum, and urine and blood samples (Ho et al. 2013; Hirose et al. 2016; Al-Abdely et al. 2019). They are highly contagious, causing millions of infections each year.

A recent study by Jin et al. (2020b) conducted aPDT with methylene blue as a photosensitizer (Table 30.1) could be effective against SARS-CoV-2 in plasma without any side effects. Previously, it has proven that methylene blue photochemical technology not only inactivates lipid-enveloped viruses in vitro but also can be applied in clinical treatment without damage to other components of the plasma (Eickmann et al.

Table 30.1 Summary of studies that used either direct or indirect methods for photobiomodulation

No	Study	Year	Induction of lung inflammation	Target tissue/ organ	Light source	Light parameters	Results
1	Aimbire et al. (2005)	2005	Airway and lung inflammation induced by gram-negative bacterial lipopolysaccharide (LPS) intravenous injection	Rat lung	Ga-Al-As diode laser	12 mW, irradiation time = 1 min 20 s, spot size = 0.08 cm2, Continuous wave	Photobiomodulation reduced RTHR, BAL and lung neutrophils influx which led to the anti-inflammatory effect. It is associated with inhibiting of COX-2-derived metabolites
2	Aimbire et al. (2006)	2006	The complex immune reaction by Instillation of Ovalbumin intrabronchial followed by IV injection	Rat lung	Ga-AsI-Al laser	650 nm, 42 s at 5 min after induction, continuous mode 0.04, 0.11, and 0.22 Joules Spot size = 0.08 cm2 Irradiated through the skin over upper bronchi	LLLT significantly reduced TNF-α expression in a dose-dependent manner. 0.11 J had the best efficacy
3	Aimbire et al. (2008)	2008	Acute lung injury induced by IV injection of lipopolysaccharide (LPS) (5 mg/kg)	Rat lung	diode laser	660 nm, 30 mW spot size= 0.785 cm2; skin over the upper bronchus	LLLT decreased lung permeability, neutrophils influx, MPO activity, IL-1β expression, and its mRNA Laser anti-inflammatory effect at 4, 12, and 24 h after LPS exposure
4	Mafra de Lima et al. (2009)	2009	TNF-a-induced acute lung inflammation	Rat-dissected bronchi with or without TNF-α	Ga-As-Al laser	650 nm, 2.5 mW, spot size = 0.08 cm2, 0.44 J, irradiation time = 42 s	Photobiomodulation reduced BSM hyperreactivity or relaxation after acetylcholine and isoproterenol application, respectively Laser irradiation decreased TNF-α mRNA expression
5	de Lima et al. (2010b)	2010	Acute pulmonary inflammation induced by aerosol of lipopolysaccharide from *Escherichia coli* (0.3 mg/ mL)	Rat lung	GaAsAl diode laser	650 nm, 2.5 mW, spot size = 0.08 cm2, irradiation time = 42 s The skin over the upper bronchus	Photobiomodulation reduced pulmonary edema, the neutrophil influx, endothelial cytoskeleton damage, TNF-α, IL-1β, and ICAM-1 expression Levels of CINC-1, MIP-2, and IL-10 did not affect by laser irradiation
6	de Lima et al. (2010b)	2010	4 h incubation with LPS or H2O2 for inducing acute lung inflammation	Rat AM cell line AMJ2-C11	diode laser	660 nm, 30 mW, spot size = 0.785 cm2, irradiation time = 252 s	LLLT decreased MIP-2 mRNA expression and intracellular ROS generation

(continued)

Table 30.1 (continued)

No	Study	Year	Induction of lung inflammation	Target tissue/organ	Light source	Light parameters	Results
7	de Lima et al. (2011b)	2011	Acute lung inflammation induced by intestinal ischemia and reperfusion	Rat lung	diode laser	660 nm, 30 mW, 5.4 J, spot size= 0.08 cm2 Irradiation time= 3 min skin over the upper bronchus	Lung edema, neutrophils influx, MPO activity and TNF expression and production were reduced by LLL irradiation, in contrast, IL-10 production was increased
8	de Lima et al. (2011a)	2011	Acute lung inflammation induced by LPS inhalation or TNF intranasal instillation	Rat-dissected bronchi	InGaAlP laser	660 (685) nm, 8.82 J, irradiation time = 252 s On the skin over the right upper bronchus	Photobiomodulation increased The cAMP indirectly in Alveolar Macrophages by a TNF-dependent mechanism
9	de Lima et al. (2013a)	2013	Acute lung injury induced by gut ischemia and reperfusion	Rat lung	diode laser	660 nm, 30 mW, 5.4 J, continuous mode, irradiation= 180 s, spot size= 0.08 cm2 The skin over the bronchus in the direction of the trachea distal Two-series laser irradiations: (a) 5 min after initial or 5 min before the end of the intestinal reperfusion (b) 30 min after the beginning of the reperfusion	Lung edema, neutrophils influx, MPO activity, and ICAM-1 mRNA expression, ROS formation were reduced by LLL irradiation GSH concentration in the lung, HSP70, and PPARγ expression were increased after LLLT Photobiomodulation reduce acute lung injury induced by gut ischemia and reperfusion
10	de Lima et al. (2013b)	2013	ARDS induced by intestinal ischemia and reperfusion	Rat lung	diode laser	660 nm, 30 mW, 5.4 J, continuous mode, irradiation= 3 min Spot size= 0.08 cm2 Skin over the upper bronchus	Photobiomodulation airway reactivity dysfunction by decreasing lung edema, MPO activity, and TNF-α, iNOS or ICAM-1 expressions LLLT increased the production of IL-10
11	de Lima et al. (2014)	2014	ARDS induced by intestinal ischemia and reperfusion	Rat lung	diode laser	660 nm, 30 mW, 5.4 J, continuous mode, irradiation = 5 min Spot size = 0.08 cm2 skin over the upper bronchus	IL-6 and TNF mRNA expression and proteins were reduced significantly the following photobiomodulation However, IL-1β and MPO activity was reduced by all doses except 1 J/cm^2 Photobiomodulation increased IL-10 protein in 1 J/cm^2
12	Oliveira Jr et al. (2014)	2014	Acute lung inflammation and ARDS induced by LPS	Rat's lung	Infra-red laser	830 nm, 35 mW, 80 s per point, 3 points per application (total 240 s), continuous mode Direct contact to skin Point 1 was in the end part of trachea, point 2 and 3 were in the right and left lungs, respectively	Photobiomodulation significantly inflammation in LPS-induced ARDS and reduced number of total cells, neutrophils in BAL or lung parenchyma and levels of IL-1beta, IL-6, KC and TNF-alpha in BAL fluid and in serum

13	Miranda da Silva et al. (2015)	2015	Lung inflammation induced by formaldehyde (1%) or vehicle inhalation (distilled water)	Rat lung	diode laser	660 nm, 30 mW, spot size = 0.14 cm^2, 60 s/point, 1.8 J for 1 min	Photobiomodulation reduced neutrophil influx (MPO activity), leukocyte number, degranulation of mast cells, and lung microvascular permeability IL-6 and TNF-α production were reduced, but IL-10 generation was increased after laser irradiation
14	de Oliveira Junior et al. (2015)	2015	ARDS induced by LPS intratracheal or intraperitoneal	Rat lung	Infra-red laser	830 nm, 35 mW, 80 s per point 3 points per application) direct contact with the skin	Photobiomodulation significantly reduced ARDS in both inducing ways Photobiomodulation decreased total cell and neutrophil count in the BAL and neutrophil count in lung parenchyma and also reduced levels of IL-1beta, IL-6, KC, and TNF-alpha in BAL and in serum IL-10 level didn't increase by photobiomodulation
15	da Silva Sergio et al. (2018)	2018	Acute lung injury induced by intraperitoneal *Escherichia coli* LPS injection	Alveolar epithelial cells of rat lung	AsGaAl diode laser	808 nm, 100 mW, spot size= 0.028 cm2, 2 and 5 J energy per point, four points of irradiation, time= 2 and 5 s per point Skin over the lung	Photobiomodulation increased Bcl-2 mRNA levels but reduced caspase-3 mRNA levels in acute lung injury Also, photobiomodulation reduced DNA fragmentation in alveolar cells but increased in polymorphonuclear cells which inducing apoptosis in these inflammatory cells
Indirect application of photobiomodulation							
16	Fujimaki et al. (2003)	2003	Isolated neutrophils from peripheral blood samples	Human neutrophils	GaAlAs laser	830 nm continuous wave, 1000 mW, spot size 6.6 cm2, irradiation time = 30 or 60 s	Photobiomodulation attenuated ROS production in neutrophils and can reduce oxidative tissue injury
17	Musawi et al. (2016)	2016	Whole blood sample		Diode laser	wavelengths of 405, 589, and 780 nm, 10 mW, spot size = 0.332 cm2, irradiation time = 20, 30, 40, and 50 min	Photobiomodulatory effects are related to laser parameter. wavelength of 589 nm and fluence of 72 J/cm2 had the best results Photobiomodulation significantly increased CD45 lymphocytes and natural killer (NK) (CD16, CD56) cells, but there was no significant change for CD3 T lymphocytes, T-suppressor (CD3, CD8) cells, T-helper (CD3, CD4) cells, and CD19 B lymphocytes

2018, 2020; Xu et al. 2005). According to the Jin et al. results, 1, 2, and 4 μM of methylene blue photochemical therapy using 630 nm wavelength light for 2 min could completely inactivate the virus, and the viral titer of SARS-CoV-2 decreased to 4.5 $\log_{10}$ median Tissue Culture Infectious Dose (TCID50)/mL.

As commented on by Eickmann et al. (2018, 2020), methylene blue plus visible light using light doses as low as 30 J/cm^2, or 25% of the standard full light dose of 120 J/cm^2, can reduce SARS-CoV and MERS-CoV by more than 3.1 and 3.3 $\log_{10}$ $TCID_{50}$/mL in plasma, respectively.

The inactivation of the MERS-CoV using a riboflavin-based and ultraviolet light-based photochemical treatment in plasma products was evaluated by (Keil et al. 2016). In this study, the treatments were performed using pooled and individual plasma units. The mean reductions in the log titer of MERS-CoV for the pooled and individual donor plasma were ≥ 4.07 and ≥ 4.42, respectively. Also, the results of this study suggested that riboflavin and UVA light might be able to reduce the risk of MERS-CoV transfusion transmission in both platelet and plasma products.

A lipophilic thiazolidine derivative named LJ001, and an arylmethylidene rhodanine derivative were described in 2010 as a new broad-spectrum antiviral compound (Wolf et al. 2010) against more than 15 different enveloped viruses. LJ001 was then introduced by (Vigant et al. 2013) as a new class of antiviral photosensitizer with increased potency, good $^1O_2$ quantum yields, and red-shifted absorption spectra. On the other hand, they determined the effects of treated influenza virus A (A/PR/8/34 H1N1) with five mM of LJ001, exposed to visible light for one h on the phospholipid composition of viral membranes. As previously reported (Lorizate and Kräusslich 2011), the lipid composition can affect the biophysical properties of viral membranes that impact the efficiency of virus-cell fusion. The results of Vigant et al. (2013) showed that LJ001-treated viruses had an up to a 300-fold increase in oxidized forms of unsaturated phospholipids. It was suggested that aPDT by LJ001 targets the viral lipid membrane and can inhibit virus-cell fusion.

In another study by Lenard et al. (1993), hypericin (4,5,7,4′,5′,7′-hexahydroxy-2,2′-dimethyl-naphthodianthrone, an anthraquinone derivative and one of the main active compounds in St. John's wort (*Hypericum perforatum* L.) (Lyles et al. 2017; Klemow et al. 2018)) and rose bengal were used as the photosensitizers during aPDT to inhibit viral fusion of the influenza virus. The finding showed that aPDT using 80 nM hypericin and a standard fluorescent lamp against the influenza virus led to an extensive cross-linking of G and M proteins and may disrupt the capacity of the viruses to attach and penetrate the host cells. Also, aPDT of the influenza virus using rose bengal at a concentration of 50 nM plus a fluorescent lamp modified the HA fusion protein and led to protein cross-links.

Chen et al. investigated the potential antiviral activity of curcumin against influenza virus propagation. The results indicated that curcumin inhibits hemagglutinin as a viral protein in influenza a virus in cell culture. Chen et al. (2010) and Sobotta et al. (2016) evaluated the photodynamic activity of copper (Cu^{2+})- and metal-free phthalocyanine, possessing 1,4,7-trioxanonyl substituents against several viruses such as parainfluenza virus-3, influenza A virus (H1N1, H3N2), and influenza B virus. In their experiments, 100 μM of this compound was exposed to the virus strains, irradiated for 30 min at 735 nm with a light intensity of 4.5 mW/cm^2 and a light dose of 8.1 J/cm^2. The average percentages of virus infectivity versus control were 4.6, <1.6, <3.3, and 30 for parainfluenza virus-3, influenza A virus (H1N1, H3N2), and influenza B virus, respectively. The findings of Sobotta et al. study indicated that photoactivation of Cu^{2+} phthalocyanine could generate various reactive oxygen species (ROS), including singlet oxygen, which leads to decrease infectivity of a wide variety of enveloped viruses. Also, one of the desired properties of phthalocyanine is that it can be activated with 735 nm wavelength in the far-red light spectrum that is not only much less damaging for nearby tissues but also in combination with phthalocyanine affords the local generation of ROS.

Phthalocyanines were used by Ke et al. (2014) due to their absorption in the tissue-penetrating red visible region, as well as production of the highly efficient singlet oxygen. In their study, the efficacy of a series of zinc (II) phthalocyanines conjugated with an oligolysine chain in photodynamic inactivation of influenza A virus (H1N1) was examined. The results showed that influenza A virus (H1N1) was susceptible to photodynamic inactivation with a half-maximal inhibitory concentration (IC_{50}) values as low as1 pM upon irradiation at $\lambda > 610$ nm under a light source of a 300 W halogen lamp with a water tank for cooling at a fluence rate of 40 mW/cm^2 for 20 min. Eventually, it was determined that phthalocyanines are promising photosensitizers for the photoinactivation of influenza A enveloped virus.

Using porphyrin-conjugated multiwalled carbon nanotubes (NT-P), the study (Banerjee et al. 2012) demonstrated the effects of visible light on inactivation of influenza A virus (H3N2). In the first series of experiments, the dependence of viral inactivation on the concentration of NT-P under 90 min light irradiation was evaluated. The virus infectivity percentage was reduced from 86% to 2% as the concentration of NT-P was increased from 0 to 1500 μg/mL. In the second series of experiments, the virus inactivation was examined using 1000 μg/mL of NT-P with different times of light irradiation from 0 to 90 min. The data showed that the percentage of virus infectivity was sharply decreased from 78% to 3% in a dose-dependent manner after increasing time of light irradiation (0–30 min). The results showed that 1000 μg/mL of NT-P and 30 min of irradiation with visible light from a compact fluorescence lamp (350 W) as the optimal dose of aPDT could cause more than a 250-fold reduction (with only 1% of virus infectivity) as the effective infectious viral dose. This inactivation was due to the production of ROS by the protoporphyrin IX (PPIX) part of NT-P in the presence of light. There was no significant reduction in the percentage of infected cells between viruses treated with NT-P in dark conditions without any NT-P under light, so this finding establishes the synergistic effect of light and NT-P. Overall, these results suggest that NT-P with light may be used effectively against influenza viruses without any emergence of resistance to treatment.

In another study, the photosensitizer and light dose dependence of aPDT for virus inactivation was also confirmed by a study from (Belousova et al. 2014). They evaluated the photodynamic inactivation of influenza virus A/Puerto Rico/8/34 (H1N1) by a solid-phase fullerene-based photosensitizer (SPFPS). In this study, an ultrabright diode at a maximum wavelength of 460 nm and a mean power density of 180 mW/cm^2 was used as the source of irradiation. Results demonstrated that the inactivation of the influenza virus depended on the concentration of SPFPS and the dose of irradiation. To confirm the obtained data, two studies were designed: first, a model containing 0.1 g/mL of SPFPS and different time periods of irradiation (10–60 min; dose108-648 J/cm^2) was used. The results showed the infectious titer of the virus was reduced with increasing irradiation dose, but it did not reach the zero level. Interestingly, some infectious virus was still detectable at the highest dose, i.e., 648 J/cm^2 (60 min) irradiation. In the next model, the highest concentration of SPFPS with the lowest dose of radiation was used. The findings showed that the virus was wholly inactivated with 2 mg/mL SPFPS at an irradiation dose of 108 J/cm^2 (30 min) irradiation. According to the data presented by Belousova et al., optimization of the aPDT conditions is important to attain complete inactivation of enveloped viruses in a relatively short time.

On the other hand, the photodynamic inactivation using hematoporphyrin of influenza A/WSN was reported by Perlin et al. (1987). It was shown that hematoporphyrin, as a pigmented, iron-free natural breakdown product of hemoglobin, has good photoactivated antiviral efficacy, which results from ROS generated by energy transfer from the light-excited pigment to oxygen. The influenza virus was inactivated entirely within 15 min of incubation with 2.5 μg/mL hematoporphyrin in the presence of a fluorescent lamp as a visible light source. In addition, it was found that guanosine monophosphate was the only nucleotide to be decomposed in the presence of hematoporphyrin and fluorescent lamp, which could inhibit the influenza virus replication. It was revealed that no cytotoxicity could be detected with five μg/mL of hematoporphyrin against primary chick embryo fibroblasts (CEF). Inhibition of 50% of CEF cells was determined with 25 μg/mL after overnight incubation. No CEF cell destruction was seen at 50 μg/mL within 24 h. Therefore, a selective index of 100 for the antiviral effect of hematoporphyrin was calculated by the ratio of the cytotoxic concentration (50 μg/mL) to the inhibitory concentration of the virus (0.5 μg/ml).

In addition, aPDT has been tested for the treatment of recurrent respiratory papillomatosis. Goon et al. (2008), Venkatesan et al. (2012), and Hu et al. (2018, 2019), in their evaluation of the literature involving the treatment of recurrent respiratory papillomatosis, concluded that aPDT could be applied as an adjuvant treatment for viral infections.

In a randomized prospective trial, Shikowitz et al. (1998) evaluated the immediate and long-term results of aPDT using one of two doses of 3.26 mg/kg or 4.25 mg/kg dihematoporphyrin ether (DHE) for respiratory papillomatosis. The light was administered using an Aurora/M tunable Argon pumped dye laser emitting 630 nm red light through a 400 μm diameter flexible quartz fiber for 100–200 s. According to the results, there was a notable improvement and a more significant decrease in papilloma growth rate in patients receiving 4.25 mg/kg DHE. The photosensitizer DHE is highly lipophilic and binds to the cytoplasmic membrane and slowly diffuses into cytoplasm around the perinuclear area. Also, the 3-year follow-up of patients confirmed that the improvement was maintained.

Shikowitz et al. (2005), in another clinical trial study, investigated the efficacy of aPDT with 0.15 mg/kg meso-tetra(hydroxyphenyl) chlorin (m-THPC, a second-generation photosensitizer molecule based on a chlorin) for respiratory papillomatosis. Diode laser at a 652 nm wavelength as an activating light was irradiated with 80–100 J for adults and 60–80 J for children over an activation time from 200 to 330 s. The results showed there was a significant reduction in lesion severity 6–9 months after aPDT, with 45% of patients in the early treatment group free of laryngeal disease. Successful treatment of recurrent respiratory papillomatosis with aPDT has been reported, probably through an improved immune response.

A notable review by Lieder et al. (2014) provided an overview of aPDT as a relatively powerful method for local destruction of papilloma in the mucosal lining of the upper airway in recurrent respiratory papillomatosis. Although more high-quality randomized controlled trials are required to determine whether aPDT alters the course of the recurrent respiratory papillomatosis, aPDT currently continues to be used in some centers in the USA and Europe to treat recurrent respiratory papillomatosis.

aPDT, as antiviral therapy, was also investigated in a study by Kimberlin (2004), showing promise in controlled trials for juvenile-onset recurrent respiratory papillomatosis. Venkatesan et al. (2012) showed that aPDT might provide benefit by altering the immune response in the case of recurrent respiratory papillomatosis, triggering an immune response to even low concentrations of viral proteins. Moreover, other studies have shown decreased papilloma growth and potential long-term benefits (Abramson et al. 1992, 1994; Borkowski et al. 1999).

In one in vivo study, a murine model of recurrent respiratory papillomatosis was studied by Lee et al. (2010)). It was shown that papilloma completely regressed under 1.0 mg/kg phthalocyanine photosensitizer and 675 nm photoactivated light at 150 J/cm^2. As well, papilloma did not regrow within the observation period of up to 79 days.

Recently, successful treatment of adult-onset recurrent respiratory papillomatosis with CO_2 laser and aPDT was reported by Lu et al. (2019). After the patient was diagnosed with recurrent respiratory papillomatosis, CO_2 laser therapy was applied, followed by aPDT. The neoplasm was removed by CO_2 laser, and aPDT was then performed at the surgical site. During aPDT, 20% aminolaevulinic acid (ALA) as a photosensitizer was excited using irradiation of 635 nm wavelength and 120 J/cm^2 power for 30 min. Since there was no recurrence found during follow-up 15 months after treatment, it was suggested aPDT might be an effective approach in preventing the recurrence of recurrent respiratory papillomatosis.

Altogether, recent experimental and clinical studies have revealed that aPDT is an effective approach for treating pulmonary infections. Geralde et al. (2017) evaluated the efficacy of aPDT with ICG (10 μmol/L) as a photosensitizer and extracorporeal illumination using an infrared light source with a 780 nm laser device for the treatment of pneumonia in an experimental mouse model. The findings confirmed the potential of aPDT to eliminate *Streptococcus pneumoniae* and treat lung infections by administration of the ICG and extracorporeal illumination with infrared light. Geralde et al. (2017) also assessed the effect of aPDT on alveolar macrophages as critical lung phagocytic cells responsible for innate immunity. Notably, the viability of alveolar macrophage was more than 90% following aPDT with light at 850 nm and ICG suggesting that aPDT did not harm the host immune system and could be a safe treatment modality (Geralde et al. 2017).

The penetration depth of the light is the main limiting factor in the widespread clinical acceptability of aPDT in the case of deeper organs and thicker lesions. Among the photosensitizers, the absorption peak of ICG is in near-infrared (810 nm) where penetration into tissues is increased due to low absorption by water, hemoglobin, and melanin, which makes ICG an ideal photosensitizer in extracorporeal aPDT applications for deep-seated lesions, which need light penetration through the lung parenchyma (Urbanska et al. 2002; Crescenzi et al. 2004).

A critical issue in the clinical use of aPDT is the efficient delivery of light. A study by Geralde et al. (2014)) revealed an insignificant reduction in the intensity of light transmittance from an 810 nm laser through the mouse chest in a postmortem model. Based on these results, infrared light with 810 nm may be used effectively for extracorporeal application in aPDT.

Therefore, for the best use of aPDT in clinical application, it is necessary to pay attention to many factors. How the photosensitizers for aPDT are delivered into the lung environment could be via intratracheal or intravenous pathways. It has been demonstrated that these delivery routes of photosensitizers can homogeneously expose all affected areas in the lung without complications. However, the poor penetration of the activating light through to the target tissue is also a limitation in the clinical usage of aPDT (Zhang et al. 2014). According to one recent clinical trial, aPDT can be used at the oropharynx in subclinical infections in the upper respiratory tract. This method would help to reduce the microbial load and infection complications (Fekrazad et al. 2017). Some of the various photosensitizers are used in viral disease are summarized in Table 30.2.

In this chapter, it was our objective to provide an overview of aPDT against viruses, including influenza viruses, MERS-CoV, and orthomyxoviruses over the last decade, to identify current applications and developments that may be useful against COVID-19. The most commonly used clinical application of the aPDT against these viruses is currently the field of viral decontamination of blood products. Looking at this field of aPDT application as an antiviral phototherapy, it becomes evident that there is no "single" specific photosensitizer, which is best suited for aPDT against viruses, but the choice of PS depends on the virus target and specific application. Given the rise of viral resistance against antiviral agents, aPDT, as a minimally invasive treatment modality, could be an alternative treatment to intractable viral diseases. It will also be important to study possible synergistic effects between aPDT and classical antiviral drugs. Although, more studies to confirm the effect of aPDT are still required, especially in photodynamic approaches to the inactivation of enveloped viruses, given this literature review of photoactivated-based treatment in the modern era, and its established safety and efficacy against the viruses that typically infected the respiratory tract, it could be

Table 30.2 Properties of some of the most widely used photosensitizers against viruses

Photosensitizers	Molecular structure	Wavelength (nm)	Target viruses
Curcumins	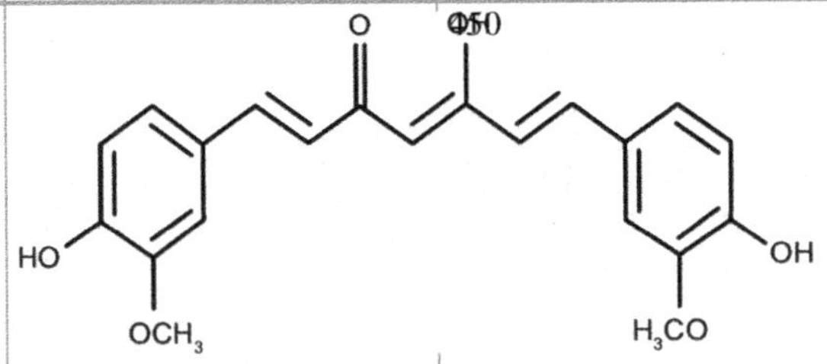	450	VSV FCV FHV HPV IAV HIV-1 HSV-2 HBV HCV
Hypericin	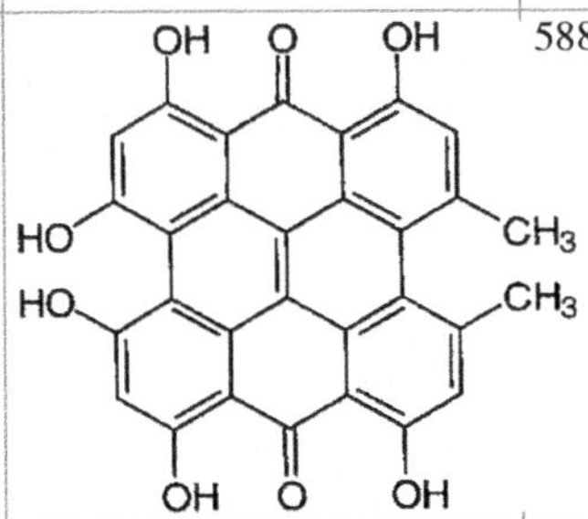	588	HSV HIV IAV VSV HCV
Methylene blue	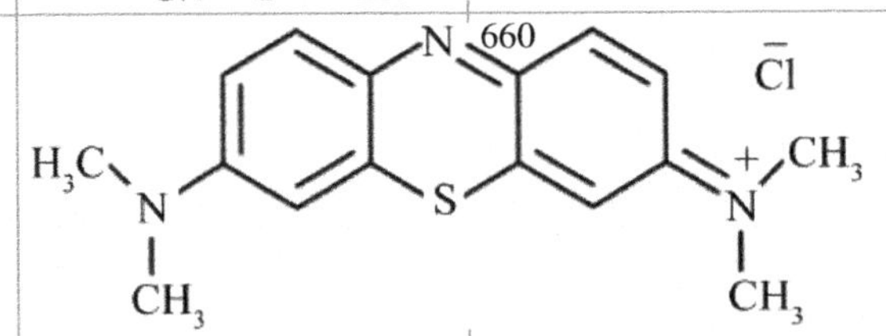	660	HSV HIV IAV VSV HCV SV DENV ZV SARS-CoV-2
Porphyrins	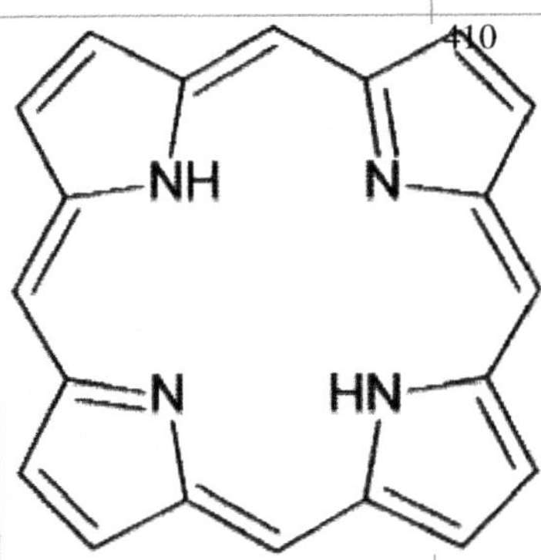	410	HSV-1 HIV-1 BoHV VSV HPV

(continued)

Table 30.2 (continued)

Photosensitizers	Molecular structure	Wavelength (nm)	Target viruses
Phytochlorin	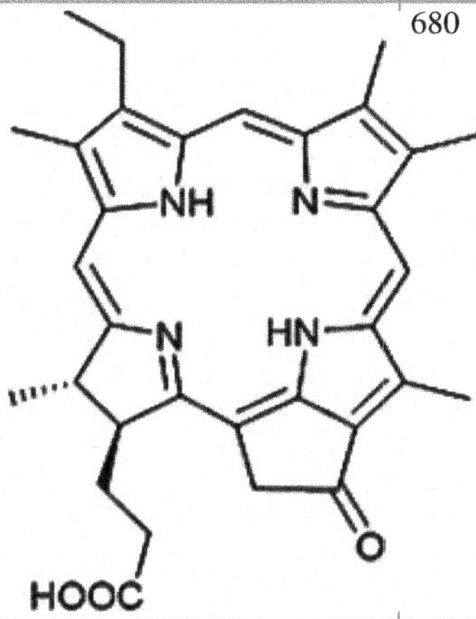	680	HPV BVDV EBV ZV HSV-1
Phthalocyanines		300–400 and 600–700	VSV HSV HIV RV IAV
Riboflavin		365–445	HIV WNV VSV IAV PPV HAV EMV SV MERS-CoV

BoHV Bovine herpesvirus, *BVDV* bovine viral diarrhea virus, *DENV* dengue virus, *EBV* Epstein-Barr virus, *EMV* encephalomyocarditis virus, *FCV Feline coronavirus*, *FHV* feline herpes viruses, *HAV* hepatitis A virus, *HBV* hepatitis B virus, *HCV* hepatitis C virus, *HSV* herpes simplex virus, *IVA* influenza A virus, *MERS-CoV* Middle East respiratory syndrome coronavirus, *HPV* human papillomavirus, *NV Norovirus*, *PPV* porcine parvovirus, *RV* rhinovirus, *SARS-COV-2* severe acute respiratory syndrome coronavirus 2, *SV* sindbis virus, *VSV* vesicular stomatitis virus, *WNV* West Nile virus, *ZV* Zika virus

suggested that aPDT may be effective against COVID-19.

Many experimental and clinical studies have reported that aPDT can be applied in combination with other therapeutic modalities, including supportive therapies, chemotherapy, and surgery. aPDT could be a new adjuvant therapy to the management of viral lung diseases and reduce the proximal extent of the infected tissue. Therefore, the main indications for the use of aPDT in the clinical management of viral lung disease are the early stage of the disease, superficial, local, peripherally and centrally located affected endobronchial tissue, and pleural disease.

30.11 Conclusion

We have discussed the possibility of using PBM and photodynamic therapy for COVID-19. The best approach may be a combination of both methods, as mentioned earlier. PDT would be useful for virus destruction, while PBM would be better for improving tissue oxygenation and reduction or inhibition of the cytokine storm that occurs in severe inflammation. When PDT and PBM combined, we can reach these goals with minimal interference with medications and battle this disease with biophysical methods. The use of aPDT could be improved by using monoclonal antibodies to target lung tissue specifically. PDT can be improved by using nanotechnology to prepare more effective photosensitizers in the nanoscale and more improved targeting of lung tissue to obtain better results. Further animal and human studies are required before we can reach an optimal protocol. This chapter could encourage other scientists to work on this new pandemic problem.

References

Abramson AL, Shikowitz MJ, Mullooly VM, Steinberg BM, Amella CA, Rothstein HR (1992) Clinical effects of photodynamic therapy on recurrent laryngeal papillomas. Arch Otolaryngol Head Neck Surg 118(1):25–29

Abramson AL, Shikowitz MJ, Mullooly VM, Steinberg BM, Hyman RB (1994) Variable light-dose effect on photodynamic therapy for laryngeal papillomas. Arch Otolaryngol Head Neck Surg 120(8):852–855

Afrasiabi S, Pourhajibagher M, Bahador A (2019) The photomodulation activity of metformin against oral microbiome. J Lasers Med Sci 10(3):241

Ahrari F, Shahabi M, Fekrazad R, Eslami N, Mazhari F, Ghazvini K, Emrani N (2018) Antimicrobial photodynamic therapy of Lactobacillus acidophilus by indocyanine green and 810-nm diode laser. Photodiagn Photodyn Ther 24:145–149

Aimbire F, Albertini R, Pacheco M, Castro-Faria-Neto H, Leonardo P, Iversen V, Lopes-Martins R, Bjordal J (2006) Low-level laser therapy induces dose-dependent reduction of TNFα levels in acute inflammation. Photomed Laser Surg 24(1):33–37

Al Musawi MS, Jafar MS, Al-Gailani BT, Ahmed NM, Suhaimi FM, Suardi N (2016) In vitro mean red blood cell volume change induced by diode pump solid state low-level laser of 405 nm. Photomed Laser Surg 34(5):211–214

Al-Abdely HM, Midgley CM, Alkhamis AM, Abedi GR, Lu X, Binder AM, Alanazi KH, Tamin A, Banjar WM, Lester S (2019) Middle East respiratory syndrome coronavirus infection dynamics and antibody responses among clinically diverse patients, Saudi Arabia. Emerg Infect Dis 25(4):753

AlGhamdi KM, Kumar A, Moussa NA (2012) Low-level laser therapy: a useful technique for enhancing the proliferation of various cultured cells. Lasers Med Sci 27(1):237–249

Allison RR, Moghissi K (2013) Photodynamic therapy (PDT): PDT mechanisms. Clin Endosc 46(1):24

Azarpazhooh A, Shah PS, Tenenbaum HC, Goldberg MB (2010) The effect of photodynamic therapy for periodontitis: a systematic review and meta-analysis. J Periodontol 81(1):4–14

Banerjee I, Douaisi MP, Mondal D, Kane RS (2012) Light-activated nanotube–porphyrin conjugates as effective antiviral agents. Nanotechnology 23(10):105101

Baptista MS, Cadet J, Di Mascio P, Ghogare AA, Greer A, Hamblin MR, Lorente C, Nunez SC, Ribeiro MS, Thomas AH (2017) Type I and type II photosensitized oxidation reactions: guidelines and mechanistic pathways. Photochem Photobiol 93(4):912–919

Bargrizan M, Fekrazad R, Goudarzi N, Goudarzi N (2019) Effects of antibacterial photodynamic therapy on salivary mutans streptococci in 5-to 6-year-olds with severe early childhood caries. Lasers Med Sci 34(3):433–440

Belousova I, Kislyakov I, Muraviova T, Starodubtsev A, Kris'ko T, Selivanov E, Sivakova N, Golovanova I, Volkova S, Shtro A (2014) Photodynamic inactivation of enveloped virus in protein plasma preparations by solid-phase fullerene-based photosensitizer. Photodiagn Photodyn Ther 11(2):165–170

Borkowski G, Sommer P, Stark T, Sudhoff H, Luckhaupt H (1999) Recurrent respiratory papillomatosis associated with gastroesophageal reflux disease in children. Eur Arch Otorhinolaryngol 256(7):370–372

Bujia J, Feyh J, Kastenbauer E (1993) Photodynamic therapy with derivatives from hemotoporphyrines for recurrent laryngeal papillomatosis of the children. Early results. In: Anales otorrinolaringologicos ibero-americanos, vol 3. p 251

Cadet J, Berger M, Decarroz C, Wagner J, Van Lier J, Ginot Y, Vigny P (1986) Photosensitized reactions of nucleic acids. Biochimie 68(6):813–834

Chang TW, Fiumara N, Weinstein L (1975) Genital herpes: treatment with methylene blue and light exposure. Int J Dermatol 14(1):69–71

Chen D-Y, Shien J-H, Tiley L, Chiou S-S, Wang S-Y, Chang T-J, Lee Y-J, Chan K-W, Hsu W-L (2010) Curcumin inhibits influenza virus infection and haemagglutination activity. Food Chem 119(4):1346–1351

Chen AC-H, Huang Y-Y, Sharma SK, Hamblin MR (2011) Effects of 810-nm laser on murine bone-marrow-derived dendritic cells. Photomed Laser Surg 29(6):383–389

Chiari S (2016) Photobiomodulation and lasers. In: Tooth movement, vol 18. Karger Publishers, Basel, pp 118–123

Cho JS, Guo Y, Ramos RI, Hebroni F, Plaisier SB, Xuan C, Granick JL, Matsushima H, Takashima A, Iwakura Y (2012) Neutrophil-derived IL-1β is sufficient for abscess formation in immunity against Staphylococcus aureus in mice. PLoS Pathog 8(11):e1003047

Commission GOoNH (2020) General Office of National Administration of Traditional Chinese Medicine. Diagnostic and treatment protocol for Novel Coronavirus Pneumonia

Costa L, Faustino MAF, Neves MGP, Cunha Â, Almeida A (2012) Photodynamic inactivation of mammalian viruses and bacteriophages. Viruses 4(7):1034–1074

Crescenzi E, Varriale L, Iovino M, Chiaviello A, Veneziani BM, Palumbo G (2004) Photodynamic therapy with indocyanine green complements and enhances low-dose cisplatin cytotoxicity in MCF-7 breast cancer cells. Mol Cancer Ther 3(5):537–544

de Lima FM, Villaverde A, Salgado M, Castro-Faria-Neto H, Munin E, Albertini R, Aimbire F (2010) Low intensity laser therapy (LILT) in vivo acts on the neutrophils recruitment and chemokines/cytokines levels in a model of acute pulmonary inflammation induced by aerosol of lipopolysaccharide from Escherichia coli in rat. J Photochem Photobiol B Biol 101(3):271–278

de Lima FM, Moreira LM, Villaverde A, Albertini R, Castro-Faria-Neto HC, Aimbire F (2011a) Low-level laser therapy (LLLT) acts as cAMP-elevating agent in acute respiratory distress syndrome. Lasers Med Sci 26(3):389–400

de Lima FM, Villaverde A, Albertini R, Corrêa J, Carvalho R, Munin E, Araújo T, Silva J, Aimbire F (2011b) Dual Effect of low-level laser therapy (LLLT) on the acute lung inflammation induced by intestinal ischemia and reperfusion: action on anti-and pro-inflammatory cytokines. Lasers Surg Med 43(5):410–420

de Perrot M, Fischer S, Liu M, Imai Y, Martins S, Sakiyama S, Tabata T, Bai X-H, Waddell TK, Davidson BL (2003) Impact of human interleukin-10 on vector-induced inflammation and early graft function in rat lung transplantation. Am J Respir Cell Mol Biol 28(5):616–625

De Wilde AH, Zevenhoven-Dobbe JC, Beugeling C, Chatterji U, De Jong D, Gallay P, Szuhai K, Posthuma CC, Snijder EJ (2018) Coronaviruses and arteriviruses display striking differences in their cyclophilin A-dependence during replication in cell culture. Virology 517:148–156

Derr V, Fine S (1965) Free radical occurrence in some laser-irradiated biologic materials. Fed Proc 24(SUPPL 14):99+

Desmet KD, Paz DA, Corry JJ, Eells JT, Wong-Riley MT, Henry MM, Buchmann EV, Connelly MP, Dovi JV, Liang HL (2006) Clinical and experimental applications of NIR-LED photobiomodulation. Photomed Laser Ther 24(2):121–128

Ding Y, Wang H, Shen H, Li Z, Geng J, Han H, Cai J, Li X, Kang W, Weng D (2003) The clinical pathology of severe acute respiratory syndrome (SARS): a report from China. J Pathol: J Pathol Soc G B Irel 200(3):282–289

Dolmans DE, Fukumura D, Jain RK (2003) Photodynamic therapy for cancer. Nat Rev Cancer 3(5):380–387

Donoghue M, Hsieh F, Baronas E, Godbout K, Gosselin M, Stagliano N, Donovan M, Woolf B, Robison K, Jeyaseelan R (2000) A novel angiotensin-converting enzyme–related carboxypeptidase (ACE2) converts angiotensin I to angiotensin 1–9. Circ Res 87(5):e1–e9

Eduardo FP, Bueno DF, de Freitas PM, Marques MM, Passos-Bueno MR, Eduardo CP, Zatz M (2008) Stem cell proliferation under low intensity laser irradiation: a preliminary study. Lasers Surg Med: Off J Am Soc Laser Med Surg 40(6):433–438

Eickmann M, Gravemann U, Handke W, Tolksdorf F, Reichenberg S, Müller TH, Seltsam A (2018) Inactivation of Ebola virus and middle east respiratory syndrome coronavirus in platelet concentrates and plasma by ultraviolet C light and methylene blue plus visible light, respectively. Transfusion 58(9):2202–2207

Eickmann M, Gravemann U, Handke W, Tolksdorf F, Reichenberg S, Müller TH, Seltsam A (2020) Inactivation of three emerging viruses–severe acute respiratory syndrome coronavirus, Crimean–Congo haemorrhagic fever virus and Nipah virus–in platelet concentrates by ultraviolet C light and in plasma by methylene blue plus visible light. Vox Sang 115(3):146–151

Enwemeka CS, Bumah VV, Masson-Meyers DS (2020) Light as a potential treatment for pandemic coronavirus infections: a perspective. J Photochem Photobiol B Biol 207:111891

Fan C, Li K, Ding Y, Lu WL, Wang J (2020) ACE2 expression in kidney and testis may cause kidney and testis damage after 2019-nCoV infection. MedRxiv. https://doi.org/10.1101/2020.02.12.20022418

Fekrazad R (2020) Photobiomodulation and antiviral photodynamic therapy as a possible novel approach in COVID-19 management. Mary Ann Liebert, Inc., publishers 140 Huguenot Street, 3rd Floor New …

Fekrazad R, Seraj B, Chiniforush N, Rokouei M, Mousavi N, Ghadimi S (2017) Effect of antimicrobial photodynamic therapy on the counts of salivary Streptococcus mutans in children with severe early childhood caries. Photodiagn Photodyn Ther 18:319–322

Fernandes KPS, Souza NHC, Mesquita-Ferrari RA, da Silva DFT, Rocha LA, Alves AN, de Brito Sousa K, Bussadori SK, Hamblin MR, Nunes FD (2015) Photobiomodulation with 660-nm and 780-nm laser on activated J774 macrophage-like cells: effect on M1 inflammatory markers. J Photochem Photobiol B Biol 153:344–351

Fialkow L, Fochesatto Filho L, Bozzetti MC, Milani AR, Rodrigues Filho EM, Ladniuk RM, Pierozan P, de Moura RM, Prolla JC, Vachon E (2006) Neutrophil apoptosis: a marker of disease severity in sepsis and sepsis-induced acute respiratory distress syndrome. Crit Care 10 (6):R155

Floyd RA, Schneider JE Jr, Dittmer DP (2004) Methylene blue photoinactivation of RNA viruses. Antivir Res 61(3):141–151

Friedman SH, DeCamp DL, Sijbesma RP, Srdanov G, Wudl F, Kenyon GL (1993) Inhibition of the HIV-1 protease by fullerene derivatives: model building studies and experimental verification. J Am Chem Soc 115(15):6506–6509

Fu P-K, Wu C-L, Tsai T-H, Hsieh C-L (2012) Anti-inflammatory and anticoagulative effects of paeonol on LPS-induced acute lung injury in rats. Evid Based Complement Alternat Med 2012:1–12

Fu Y, Cheng Y, Wu Y (2020) Understanding SARS-CoV-2-mediated inflammatory responses: from mechanisms to potential therapeutic tools. Virol Sin 35:1–6

Fujimaki Y, Shimoyama T, Liu Q, Umeda T, Nakaji S, Sugawara K (2003) Low-level laser irradiation attenuates production of reactive oxygen species by human neutrophils. J Clin Laser Med Surg 21(3):165–170

Fujita S, Yamaguchi M, Utsunomiya T, Yamamoto H, Kasai K (2008) Low-energy laser stimulates tooth movement velocity via expression of RANK and RANKL. Orthod Craniofacial Res 11(3):143–155

Geralde MC, Leite IS, Inada NM, Grecco C, Medeiros AI, Kurachi C, Bagnato VS (2014) Pulmonary decontamination for photodynamic inactivation with extracorporeal illumination. SPIE BiOS. International Society for Optics and Photonics 8927:89271B

Geralde MC, Leite IS, Inada NM, Salina ACG, Medeiros AI, Kuebler WM, Kurachi C, Bagnato VS (2017) Pneumonia treatment by photodynamic therapy with extracorporeal illumination-an experimental model. Phys Rep 5(5):e13190

Girotti AW (2001) Photosensitized oxidation of membrane lipids: reaction pathways, cytotoxic effects, and cytoprotective mechanisms. J Photochem Photobiol B Biol 63(1–3):103–113

Goon P, Sonnex C, Jani P, Stanley M, Sudhoff H (2008) Recurrent respiratory papillomatosis: an overview of current thinking and treatment. Eur Arch Otorhinolaryngol 265(2):147–151

Gorbalenya AE, Baker SC, Baric R, Groot RJd, Drosten C, Gulyaeva AA, Haagmans BL, Lauber C, Leontovich AM, Neuman BW (2020) Severe acute respiratory syndrome-related coronavirus: the species and its viruses–a statement of the Coronavirus Study Group. BioRxiv. https://doi.org/10.1101/2020.02.07.937862

Gu J, Gong E, Zhang B, Zheng J, Gao Z, Zhong Y, Zou W, Zhan J, Wang S, Xie Z (2005) Multiple organ infection and the pathogenesis of SARS. J Exp Med 202(3):415–424

Hamblin MR (2017) Mechanisms and applications of the anti-inflammatory effects of photobiomodulation. AIMS Biophys 4(3):337

Hamblin MR, Demidova TN (2006) Mechanisms of low level light therapy. Proc SPIE 6140:1–12

Hamblin MR, Hasan T (2004) Photodynamic therapy: a new antimicrobial approach to infectious disease? Photochem Photobiol Sci 3(5):436–450

Heidari M, Paknejad M, Jamali R, Nokhbatolfoghahaei H, Fekrazad R, Moslemi N (2017) Effect of laser photobiomodulation on wound healing and postoperative pain following free gingival graft: a split-mouth triple-blind randomized controlled clinical trial. J Photochem Photobiol B Biol 172:109–114

Hirose R, Daidoji T, Naito Y, Watanabe Y, Arai Y, Oda T, Konishi H, Yamawaki M, Itoh Y, Nakaya T (2016) Long-term detection of seasonal influenza RNA in faeces and intestine. Clin Microbiol Infect 22(9):813. e811–813. e817

Ho Y-L, Yoshino A, Tonacio AC, Zahredine A, Latif A, Caiaffa Filho HH, dos Santos SDS (2013) Detection of pandemic influenza-A (H1N1)-2009 virus in urine. Intensive Care Med 39(6):1168

Hoehl S, Rabenau H, Berger A, Kortenbusch M, Cinatl J, Bojkova D, Behrens P, Böddinghaus B, Götsch U, Naujoks F (2020) Evidence of SARS-CoV-2 infection in returning travelers from Wuhan, China. N Engl J Med 382(13):1278–1280

Hu W-P, Wang J-J, Yu C-L, Lan C-CE, Chen G-S, Yu H-S (2007) Helium–neon laser irradiation stimulates cell proliferation through photostimulatory effects in mitochondria. J Investig Dermatol 127(8):2048–2057

Hu Z, Liu L, Zhang W, Liu H, Li J, Jiang L, Zeng K (2018) Dynamics of HPV viral loads reflect the treatment effect of photodynamic therapy in genital warts. Photodiagn Photodyn Ther 21:86–90

Hu S, Yang Y, Jiang B, Su D, Zhang L, Huang Z, Zhang F (2019) Treatment of condyloma acuminatum using the combination of laser ablation and ALA-PDT. Photodiagn Photodyn Ther 25:193–196

Huang Y-Y, Sharma SK, Carroll J, Hamblin MR (2011) Biphasic dose response in low level light therapy–an update. Dose-Response 9(4):602. dose-response. 11-009. Hamblin

Huang C, Wang Y, Li X, Ren L, Zhao J, Hu Y, Zhang L, Fan G, Xu J, Gu X (2020) Clinical features of patients infected with 2019 novel coronavirus in Wuhan, China. Lancet 395(10223):497–506

Hwang MH, Shin JH, Kim KS, Yoo CM, Jo GE, Kim JH, Choi H (2015) Low level light therapy modulates inflammatory mediators secreted by human annulus fibrosus cells during intervertebral disc degeneration in vitro. Photochem Photobiol 91(2):403–410

Imai Y, Kuba K, Penninger JM (2008) The discovery of angiotensin-converting enzyme 2 and its role in acute lung injury in mice. Exp Physiol 93(5):543–548

Jalili A, Moslemi E, Izadi A, Mosaffa N (2015) Assessing the gene expression of IL4, TNFα, TGFβ and IFNγ in studying M1 and M2 macrophages derived from human monocyte. Res Med 39(1):9–13

Jin C, Yu B, Zhang J, Wu H, Zhou X, Yao H, Liu F, Lu X, Cheng L, Jiang M (2020a) Methylene blue photochemical treatment as a reliable SARS-CoV-2 plasma virus inactivation method for blood safety and convalescent plasma therapy for the COVID-19 outbreak. Res Sq. https://doi.org/10.21203/rs.3.rs-17718/v1

Jin Y, Yang H, Ji W, Wu W, Chen S, Zhang W, Duan G (2020b) Virology, epidemiology, pathogenesis, and control of COVID-19. Viruses 12(4):372

Johnson P, Armour CL, Carey D, Black JL (1995) Heparin and PGE2 inhibit DNA synthesis in human airway smooth muscle cells in culture. Am J Phys Lung Cell Mol Phys 269(4):L514–L519

Kanne JP (2020) Chest CT findings in 2019 novel coronavirus (2019-nCoV) infections from Wuhan, China: key points for the radiologist. Radiology 295:16. Radiological Society of North America

Karu T (1989a) Laser biostimulation: a photobiological phenomenon. J Photochem Photobiol B Biol 3(4):638

Karu T (1989b) Photobiology of low-power laser effects. Health Phys 56(5):691–704

Karu TI (1996) Mechanisms of interaction of monochromatic visible light with cells. Proc Soc Photogr Instrum Eng 2630:2–9

Karu TI (2008) Mitochondrial signaling in mammalian cells activated by red and near-IR radiation. Photochem Photobiol 84(5):1091–1099

Karu TI, Kolyakov S (2005) Exact action spectra for cellular responses relevant to phototherapy. Photomed Laser Ther 23(4):355–361

Käsermann F, Kempf C (1998) Buckminsterfullerene and photodynamic inactivation of viruses. Rev Med Virol 8(3):143–151

Kashef N, Huang Y-Y, Hamblin MR (2017) Advances in antimicrobial photodynamic inactivation at the nanoscale. Nanophotonics 6(5):853–879

Ke MR, Eastel JM, Ngai KL, Cheung YY, Chan PK, Hui M, Ng DK, Lo PC (2014) Oligolysine-conjugated Zinc (II) phthalocyanines as efficient photosensitizers for antimicrobial photodynamic therapy. Chem Asian J 9(7):1868–1875

Keil SD, Bowen R, Marschner S (2016) Inactivation of M iddle E ast respiratory syndrome coronavirus (MERS-C o V) in plasma products using a riboflavin-based and ultraviolet light-based photochemical treatment. Transfusion 56(12):2948–2952

Kharkwal GB, Sharma SK, Huang YY, Dai T, Hamblin MR (2011) Photodynamic therapy for infections: clinical applications. Lasers Surg Med 43(7):755–767

Khorsandi K, Hosseinzadeh R, Abrahamse H, Fekrazad R (2020) Biological responses of stem cells to photobiomodulation therapy. Curr Stem Cell Res Ther 15:400

Kimberlin DW (2004) Current status of antiviral therapy for juvenile-onset recurrent respiratory papillomatosis. Antivir Res 63(3):141–151

Klemow KM, Bartlow A, Crawford J, Kocher N, Shah J, Ritsick M (2018) Herbal medicine: biomolecular and clinical aspects, vol 2(11). CRC Press, Boca Raton, pp 211–228

Lan L, Xu D, Ye G, Xia C, Wang S, Li Y, Xu H (2020) Positive RT-PCR test results in patients recovered from COVID-19. JAMA 323(15):1502–1503

Lee RG, Vecchiotti MA, Heaphy J, Panneerselvam A, Schluchter MD, Oleinick NL, Lavertu P, Alagramam KN, Arnold JE, Sprecher RC (2010) Photodynamic therapy of cottontail rabbit papillomavirus-induced papillomas in a severe combined immunodeficient mouse xenograft system. Laryngoscope 120(3):618–624

Lenard J, Rabson A, Vanderoef R (1993) Photodynamic inactivation of infectivity of human immunodeficiency virus and other enveloped viruses using hypericin and rose bengal: inhibition of fusion and syncytia formation. Proc Natl Acad Sci 90(1):158–162

Li W, Moore MJ, Vasilieva N, Sui J, Wong SK, Berne MA, Somasundaran M, Sullivan JL, Luzuriaga K, Greenough TC (2003) Angiotensin-converting enzyme 2 is a functional receptor for the SARS coronavirus. Nature 426(6965):450–454

Li Q, Guan X, Wu P, Wang X, Zhou L, Tong Y, Ren R, Leung KS, Lau EH, Wong JY (2020) Early transmission dynamics in Wuhan, China, of novel coronavirus–infected pneumonia. N Engl J Med 382:1199

Lieder A, Khan MK, Lippert BM (2014) Photodynamic therapy for recurrent respiratory papillomatosis. Cochrane Database Syst Rev (6):CD009810. https://doi.org/10.1002/14651858

Lopes NNF, Plapler H, Chavantes MC, Lalla RV, Yoshimura EM, Alves MTS (2009) Cyclooxygenase-2 and vascular endothelial growth factor expression in 5-fluorouracil-induced oral mucositis in hamsters: evaluation of two low-intensity laser protocols. Support Care Cancer 17(11):1409–1415

Lopes NNF, Plapler H, Lalla RV, Chavantes MC, Yoshimura EM, da Silva MAB, Alves MTS (2010) Effects of low-level laser therapy on collagen expression and neutrophil infiltrate in 5-fluorouracil-induced oral mucositis in hamsters. Lasers Surg Med 42(6):546–552

Lorizate M, Kräusslich H-G (2011) Role of lipids in virus replication. Cold Spring Harb Perspect Biol 3(10):a004820

Lu S, Liu Y, Shi R, Zhou P (2019) Successful treatment of adult-onset recurrent respiratory papillomatosis with CO2 laser and photodynamic therapy. Case Rep Otolaryngol 2019:7394879

Lu R, Zhao X, Li J, Niu P, Yang B, Wu H, Wang W, Song H, Huang B, Zhu N (2020) Genomic characterisation and epidemiology of 2019 novel coronavirus: implications for virus origins and receptor binding. Lancet 395(10224):565–574

Lubart R, Malik Z, Rochkind S, Fisher T (1990) A possible mechanism of low level laser-living cell interaction. Laser Ther 2(2):65–68

Lulic M, Leiggener Görög I, Salvi GE, Ramseier CA, Mattheos N, Lang NP (2009) One-year outcomes of repeated adjunctive photodynamic therapy during periodontal maintenance: a proof-of-principle randomized-controlled clinical trial. J Clin Periodontol 36(8):661–666

Lyles JT, Kim A, Nelson K, Bullard-Roberts AL, Hajdari A, Mustafa B, Quave CL (2017) The chemical and antibacterial evaluation of St. John's Wort oil macerates used in Kosovar traditional medicine. Front Microbiol 8:1639

Mahase E (2020) Covid-19: what treatments are being investigated? Br Med J Publ Group 368:m1252

Mahnam K, Payab N (2016) Proactive cytokines and their containment methods. J Biosafety 9(2):74–87

Matute-Bello G, Frevert CW, Martin TR (2008) Animal models of acute lung injury. Am J Phys Lung Cell Mol Phys 295(3):L379–L399

McGettrick HM, Lord JM, Wang KQ, Rainger GE, Buckley CD, Nash GB (2006) Chemokine-and adhesion-dependent survival of neutrophils after transmigration through cytokine-stimulated endothelium. J Leukoc Biol 79(4):779–788

Meduri GU, Headley S, Kohler G, Stentz F, Tolley E, Umberger R, Leeper K (1995) Persistent elevation of inflammatory cytokines predicts a poor outcome in ARDS: plasma IL-1β and IL-6 levels are consistent and efficient predictors of outcome over time. Chest 107(4):1062–1073

Meyer NJ, Feng R, Li M, Zhao Y, Sheu C-C, Tejera P, Gallop R, Bellamy S, Rushefski M, Lanken PN (2013) IL1RN coding variant is associated with lower risk of acute respiratory distress syndrome and increased plasma IL-1 receptor antagonist. Am J Respir Crit Care Med 187(9):950–959

Migliorati C, Hewson I, Lalla RV, Antunes HS, Estilo CL, Hodgson B, Lopes NNF, Schubert MM, Bowen J, Elad S (2013) Systematic review of laser and other light therapy for the management of oral mucositis in cancer patients. Support Care Cancer 21(1):333–341

Miranda da Silva C, Peres Leal M, Brochetti RA, Braga T, Vitoretti LB, Saraiva Camara NO, Damazo AS, Ligeiro-de-Oliveira AP, Chavantes MC, Lino-dos-Santos-Franco A (2015) Low level laser therapy reduces the development of lung inflammation induced by formaldehyde exposure. PLoS One 10(11):e0142816

Monteil V, Kwon H, Prado P, Hagelkrüys A, Wimmer RA, Stahl M, Leopoldi A, Garreta E, Del Pozo CH, Prosper F (2020) Inhibition of SARS-CoV-2 infections in engineered human tissues using clinical-grade soluble human ACE2. Cell 181:905

Moslemi N, Heidari M, Fekrazad R, Nokhbatolfoghahaie H, Yaghobee S, Shamshiri A, Paknejad M (2014) Evaluation of the effect of 660nm low power laser on pain and healing in palatal donor site: a randomized controlled clinical trial. J Dent Med 27(1):71

Mrowiec J, Sieron A, Plech A, Cieslar G, Biniszkiewicz T, Brus R (1997) Analgesic effect of low-power infrared laser radiation in rats. Proc SPIE 3198:83–89

Nasu F, Tomiyasu K, Inomata K, Calderhead R (1989) Cytochemical effects of GaAlAs diode laser radiation on rat saphenous artery calcium ion dependent adenosine triphosphatase activity. Laser Ther 1(2):89

Ng DL, Al Hosani F, Keating MK, Gerber SI, Jones TL, Metcalfe MG, Tong S, Tao Y, Alami NN, Haynes LM (2016) Clinicopathologic, immunohistochemical, and ultrastructural findings of a fatal case of Middle East respiratory syndrome coronavirus infection in the United Arab Emirates, April 2014. Am J Pathol 186(3):652–658

Oliveira MC Jr, Greiffo FR, Rigonato-Oliveira NC, Custódio RWA, Silva VR, Damaceno-Rodrigues NR, Almeida FM, Albertini R, Lopes-Martins RÁB, de Oliveira LVF (2014) Low level laser therapy reduces acute lung inflammation in a model of pulmonary and extrapulmonary LPS-induced ARDS. J Photochem Photobiol B Biol 134:57–63

Pang L, Knox AJ (1997) Effect of interleukin-1β, tumour necrosis factor-α and interferon-γ on the induction of cyclo-oxygenase-2 in cultured human airway smooth muscle cells. Br J Pharmacol 121(3):579–587

Passarella S, Casamassima E, Molinari S, Pastore D, Quagliariello E, Catalano I, Cingolani A (1984) Increase of proton electrochemical potential and ATP synthesis in rat liver mitochondria irradiated in vitro by helium-neon laser. FEBS Lett 175(1):95–99

Perlin M, Mao J, Otis E, Shipkowitz N, Duff R (1987) Photodynamic inactivation of influenza and herpes viruses by hematoporphyrin. Antivir Res 7(1):43–51

Reis ACM, Regis WFM, Rodrigues LKA (2019) Scientific evidence in antimicrobial photodynamic therapy: an alternative approach for reducing cariogenic bacteria. Photodiagn Photodyn Ther 26:179–189

Rojas M, Parker RE, Thorn N, Corredor C, Iyer SS, Bueno M, Mroz L, Cardenes N, Mora AL, Stecenko AA (2013) Infusion of freshly isolated autologous bone marrow derived mononuclear cells prevents endotoxin-induced lung injury in an ex-vivo perfused swine model. Stem Cell Res Ther 4(2):1–12

Roome A, Tinkler A, Hilton A, Montefiore D, Waller D (1975) Neutral red with photoinactivation in the treatment of herpes genitalis. Sex Transm Infect 51(2):130–133

Saffarpour A, Nozari A, Fekrazad R, Saffarpour A, Heibati MN, Iranparvar K (2018) Microstructural evaluation of contaminated implant surface treated by laser, photodynamic therapy, and chlorhexidine 2%. Int J Oral Maxillofac Implants 33(5):1019

Sanders JM, Monogue ML, Jodlowski TZ, Cutrell JB (2020) Pharmacologic treatments for coronavirus disease 2019 (COVID-19): a review. JAMA 323(18):1824–1836

Shahbaz B, Norouzi M, Tabatabai H (2016) Mechanism of action and application of virocids in health care-associated viral infections. Tehran Univ Med J TUMS Publ 73(12):837–855

Sharifov OF, Xu X, Gaggar A, Grizzle WE, Mishra VK, Honavar J, Litovsky SH, Palgunachari MN, White CR, Anantharamaiah G (2013) Anti-inflammatory mechanisms of apolipoprotein AI mimetic peptide in acute respiratory distress syndrome secondary to sepsis. PLoS One 8(5):e64486

Shikowitz MJ, Abramson AL, Freeman K, Steinberg BM, Nouri M (1998) Efficacy of DHE photodynamic therapy for respiratory papillomatosis: immediate and long-term results. Laryngoscope 108(7):962–967

Shikowitz MJ, Abramson AL, Steinberg BM, DeVoti J, Bonagura VR, Mullooly V, Nouri M, Ronn AM, Inglis A, McClay J (2005) Clinical trial of photodynamic therapy with meso-tetra (hydroxyphenyl) chlorin for respiratory papillomatosis. Arch Otolaryngol Head Neck Surg 131(2):99–105

Smith KC (1991) The photobiological basis of low level laser radiation therapy. Laser Ther 3(1):19–24

Sobotta L, Wierzchowski M, Mierzwicki M, Gdaniec Z, Mielcarek J, Persoons L, Goslinski T, Balzarini J (2016) Photochemical studies and nanomolar photodynamic activities of phthalocyanines functionalized with 1, 4, 7-trioxanonyl moieties at their non-peripheral positions. J Inorg Biochem 155:76–81

Sousa KB, Araujo LdS, Pedroso NM, Santos DdS, Rodrigues MF, Ferrari RM, Bussadori S, Fernandes K (2017) Photobiomodulation effects on gene and protein expression of proinflammatory chemokines and cytokines by J774 macrophages polarized to M1 phenotype. In: Lasers in surgery and medicine. Wiley 111 River St, Hoboken 07030-5774, NJ USA, pp 36–37

Souza DG, Soares AC, Pinho V, Torloni H, Reis LF, Martins MT, Dias AA (2002) Increased mortality and inflammation in tumor necrosis factor-stimulated gene-14 transgenic mice after ischemia and reperfusion injury. Am J Pathol 160(5):1755–1765

Takada A, Kawaoka Y (2003) Antibody-dependent enhancement of viral infection: molecular mechanisms and in vivo implications. Rev Med Virol 13(6):387–398

Tezel A, Kara C, Balkaya V, Orbak R (2009) An evaluation of different treatments for recurrent aphthous stomatitis and patient perceptions: Nd: YAG laser versus medication. Photomed Laser Surg 27(1):101–106

Theocharidou A, Bakopoulou A, Kontonasaki E, Papachristou E, Hadjichristou C, Bousnaki M, Theodorou G, Papadopoulou L, Kantiranis N, Paraskevopoulos K (2017) Odontogenic differentiation and biomineralization potential of dental pulp stem cells inside Mg-based bioceramic scaffolds under low-level laser treatment. Lasers Med Sci 32(1):201–210

Thompson BT, Chambers RC, Liu KD (2017) Acute respiratory distress syndrome. N Engl J Med 377(6):562–572

Tuner J, Hode L (2004) The laser therapy handbook. Prima Books AB, Tallinn

Urbanska K, Romanowska-Dixon B, Matuszak Z, Oszajca J, Nowak-Sliwinska P, Stochel G (2002) Indocyanine green as a prospective sensitizer for photodynamic therapy of melanomas. Acta Biochim Pol 49(2):387–391

Venkatesan NN, Pine HS, MP U (2012) Recurrent respiratory papillomatosis: review and treatment update. Otolaryngol Clin N Am 45:671–694

Vigant F, Lee J, Hollmann A, Tanner LB, Ataman ZA, Yun T, Shui G, Aguilar HC, Zhang D, Meriwether D (2013) A mechanistic paradigm for broad-spectrum antivirals that target virus-cell fusion. PLoS Pathog 9(4):e1003297

Wang D, Hu B, Hu C, Zhu F, Liu X, Zhang J, Wang B, Xiang H, Cheng Z, Xiong Y (2020) Clinical characteristics of 138 hospitalized patients with 2019 novel coronavirus–infected pneumonia in Wuhan, China. JAMA 323(11):1061–1069

Weiss SR, Leibowitz JL (2011) Coronavirus pathogenesis. In: Advances in virus research, vol 81. Elsevier, London, pp 85–164

World Health Organization (WHO) (2020) World Health Organization Press Conference. The World Health Organization (WHO) Has Officially Named the Disease Caused by the Novel Coronavirus as COVID-19. Available online: https://www.who.int/emergencies/diseases/novel-coronavirus-2019. Accessed on 11 Feb 2020

Wiehe A, O'Brien JM, Senge MO (2019) Trends and targets in antiviral phototherapy. Photochem Photobiol Sci 18(11):2565–2612

Wolf MC, Freiberg AN, Zhang T, Akyol-Ataman Z, Grock A, Hong PW, Li J, Watson NF, Fang AQ, Aguilar HC (2010) A broad-spectrum antiviral targeting entry of enveloped viruses. Proc Natl Acad Sci 107(7):3157–3162

Wu F, Zhao S, Yu B, Chen Y-M, Wang W, Hu Y, Song Z-G, Tao Z-W, Tian J-H, Pei Y-Y (2020) Complete genome characterisation of a novel coronavirus associated with severe human respiratory disease in Wuhan, China. BioRxiv. https://doi.org/10.1101/2020.01.24.919183

Xiao F, Tang M, Zheng X, Liu Y, Li X, Shan H (2020) Evidence for gastrointestinal infection of SARS-CoV-2. Gastroenterology 158(6):1831-1833. e1833

Xiong C, Jiang L, Chen Y, Jiang Q (2020) Evolution and variation of 2019-novel coronavirus. Biorxiv. https://doi.org/10.1101/2020.01.30.926477

Xu J, Duan S, Zhou X, Ma P, Zhao X, Jiang S, Huang J, Zhang Y, Lu L (2005) Inactivation of SARS coronavirus in human plasma by methylene blue/light method. Bull Acad Mil Med Sci 29(2):142–144

Xu Z, Shi L, Wang Y, Zhang J, Huang L, Zhang C, Liu S, Zhao P, Liu H, Zhu L (2020) Pathological findings of COVID-19 associated with acute respiratory distress syndrome. Lancet Respir Med 8(4):420–422

Yamaura M, Yao M, Yaroslavsky I, Cohen R, Smotrich M, Kochevar IE (2009) Low level light effects on inflammatory cytokine production by rheumatoid arthritis synoviocytes. Lasers Surg Med: Off J Am Soc Laser Med Surg 41(4):282–290

Zhang X, Liu T, Li Z, Zhang X (2014) Progress of photodynamic therapy applications in the treatment of musculoskeletal sarcoma. Oncol Lett 8(4):1403–1408

Zhou C, Sun B, Wang F, Dai Z, Han Z, Han J, Chen M, Shen Y (2014) Coblation plus photodynamic therapy (PDT) for the treatment of juvenile onset laryngeal papillomatosis. World J Surg Oncol 12(1):1–6

Zhou P, Yang X-L, Wang X-G, Hu B, Zhang L, Zhang W, Si H-R, Zhu Y, Li B, Huang C-L (2020) A pneumonia outbreak associated with a new coronavirus of probable bat origin. Nature 579(7798):270–273

Zhu N, Zhang D, Wang W, Li X, Yang B, Song J, Zhao X, Huang B, Shi W, Lu R (2020) A novel coronavirus from patients with pneumonia in China, 2019. N Engl J Med 382:727

Zou X, Chen K, Zou J, Han P, Hao J, Han Z (2020) Single-cell RNA-seq data analysis on the receptor ACE2 expression reveals the potential risk of different human organs vulnerable to 2019-nCoV infection. Front Med 14:1–8

The COVID-19 Vaccine Landscape

31

Till Koch, Anahita Fathi, and Marylyn M. Addo

Abstract

The history of vaccine development spans centuries. At first, whole pathogens were used as vaccine agents, either inactivated or attenuated, to reduce virulence in humans. Safety and tolerability were increased by including only specific proteins as antigens and using cell culture methods, while novel vaccine strategies, like nucleic acid- or vector-based vaccines, hold high promise for the future. Vaccines have generally not been employed as the primary tools in outbreak response, but this might change since advances in medical technology in the last decades have made the concept of developing vaccines against novel pathogens a realistic strategy. Wandering the uncharted territory of a novel pathogen, severe acute respiratory syndrome coronavirus 2 (SARS-CoV-2), we can learn from other human *Betacoronaviridae* that emerged in the last decades, SARS-CoV-1 and MERS-CoV. We can identify the most likely target structures of immunity, establish animal models that emulate human disease and immunity as closely as possible, and learn about complex mechanisms of immune interaction such as cross-reactivity or antibody-dependent enhancement (ADE). However, significant knowledge gaps remain. What are the correlates of protection? How do we best induce immunity in vulnerable populations like the elderly? Will the immunity induced by vaccination (or by natural infection) wane over time? To date, at least 149 vaccine candidates against SARS-CoV-2 are under development. At the time of writing, at least 17 candidates have already progressed past preclinical studies (in vitro models and in vivo animal experiments) into clinical development. This chapter will provide an overview of this rapidly developing field.

Keywords

Antibody-dependent enhancement · COVID-19 · Immunity · Safety · SARS-CoV-2 · Vaccine

Till Koch and Anahita Fathi are contributed equally.

T. Koch (✉) · A. Fathi (✉) · M. M. Addo
First Department of Medicine, University Medical Center Hamburg-Eppendorf, Hamburg, Germany

Department for Clinical Immunology of Infectious Diseases, Bernhard Nocht Institute for Tropical Medicine, Hamburg, Germany

German Center for Infection Research, Hamburg-Lubeck-Borstel-Riems, Hamburg, Germany
e-mail: t.koch@uke.de

N. Rezaei (ed.), *Coronavirus Disease - COVID-19*, Advances in Experimental Medicine and Biology 1318, https://doi.org/10.1007/978-3-030-63761-3_31

31.1 Introduction

31.1.1 Genesis of Immunization

Vaccines are one of the most significant public health interventions of all time and have saved countless lives. Despite a lack of knowledge about the underlying mechanisms, protection from infectious diseases by inoculation is a practice that is likely centuries old. One of the deadliest scourges of humanity in centuries past was smallpox, a disease with a 30% mortality rate that often left survivors scarred for life. The fact that survivors of smallpox became immune to the disease was well appreciated and led to the discovery of variolation, namely, the inoculation of a patient with dried and pulverized material from a smallpox lesion (Leung 2011). It conveyed robust protection from deadly smallpox, albeit at the cost of severe side effects such as an about 3% mortality rate and the possibility of smallpox outbreaks following the procedure. Variolation is thought to have originated in the sixteenth century in China, subsequently spreading to India, where different routes of application (nasal, oral, subcutaneous) were tried. In the early eighteenth century, variolation became popular in parts of Europe, especially in the British Empire (Barquet 1997), where King George I had decreed that all citizens could receive variolation for free. Several decades later, at the end of the eighteenth century, the concept of vaccination was introduced by Edward Jenner. He discovered that inoculation with material from a cowpox lesion, a distinct but related disease, inhibited the effects of variolation and subsequently protected the recipient from smallpox. This discovery was especially remarkable as the concepts of human disease were still more or less the same as in the 2000 years before, namely, that disease was caused by imbalances in the four humors, or "bad air" (miasmatic theory). Germ theory and pathophysiology would not be described for decades to come, but the medical community at the time quickly realized the potential of this discovery. Discussion on the worldwide eradication of smallpox arose, and in the first global eradication campaign from 1803 to 1806, called the "Royal Philanthropic Vaccine Expedition," Dr. Francisco Javier de Balmis and his team believed to have vaccinated over 100,000 participants worldwide (Soto-Perez-de-Celis 2008).

31.1.2 Primal Vaccine Designs

It took almost another century for the next vaccine, a live-attenuated cholera vaccine, to be introduced (Lopez et al. 2014). By that time, the entire concept of health and disease had been revolutionized and now resembled theoretical constructs still recognizable today. With the knowledge of microorganisms as the cause of infectious diseases, vaccine development began to proliferate. For decades, vaccine design followed two principal methods, inactivation and attenuation. For inactivation, whole viruses or toxins were treated with chemical agents such as formaldehyde (diphtheria, influenza whole virus vaccine, inactivated polio vaccine) (Glenny and Hopkins 1923), heat, and oxygen (anthrax, rabies) (Pasteur 1885). Another strategy was the attenuation of live pathogens. It was first achieved by serial propagation in animals or growth medium, the latter leading to the Bacillus Calmette-Guerin (BCG) vaccine, which is still used as a tuberculosis vaccine in parts of the world. However, it required the novel technology of cell culture in the 1950s (Taylor 2014) that enabled the propagation of live viruses until they developed key mutations that reduced virulence, to allow for the development of attenuated vaccines against several pathogens. Vaccines against mumps, measles, and rubella (Hilleman et al. 1967; Katz et al. 1960; Plotkin et al. 1969) were created as well as the oral polio vaccine. The latter was described by Sabin in 1961 (Blume 2000) and subsequently almost succeeded in the goal of eliminating a second human disease.

31.1.3 Vaccine Strategies Beyond Whole Pathogens

Since it is not the whole pathogen that induces immunity, the idea quickly arose only to use part of the pathogen in order to increase vaccine safety. In the case of viruses, vaccine strategies mainly focused on the amino acid sequence of capsid proteins which contains specific areas that seem to be most immunogenic. For more complex organisms like bacteria, the outermost membrane-bound polysaccharides were utilized to develop vaccines against meningococcus (Gotschlich et al. 1969), pneumococcus (Heidelberger et al. 1948), and typhoid bacteria (Landy et al. 1954). However, these first polysaccharide vaccines had limited efficacy in children and the elderly, where polysaccharides alone are not sufficient to mount a potent B-cell response. This was overcome by coupling the polysaccharides to more immunogenic proteins (Schneerson et al. 1980), thereby activating T-helper cells to aid the production of antibodies.

The next step was to develop split vaccines, for which viruses like influenza are broken apart by detergents. To avoid complications arising from impurities, only the antigen that infers immunity was purified, creating acellular subunit (Cate et al. 1977) or protein vaccines (Edwards and Karzon 1990). However, the purification process was still relatively cumbersome, and some potentially allergenic substances could not be removed at all, e.g., egg proteins from viruses grown in chicken embryoblasts. The next revolution was, therefore, the production of the antigenic proteins by genetically altering cells in vitro to produce the protein of choice, first achieved in 1982 by expressing the surface antigen of the hepatitis B virus in yeast cells (Valenzuela et al. 1982). It allowed for the rapid generation of large amounts of protein and, therefore, vaccine (Hilleman et al. 1983). Some protein vaccines display unique properties, such as the human papillomavirus vaccine. It is based on the major capsid protein L1, which aggregates and self-assembles to form virus-like particles (VLP), which are highly immunogenic compared to the soluble protein (Kirnbauer et al. 1992).

Peptide vaccines aim to even further reduce the chance of unwanted effects such as allergenicity or reactogenicity, by including only those amino acid sequences that confer immunity. Polypeptides, usually 20–30 amino acids of length, can easily be produced at large scale and offer the advantage of only activating immune responses against the desired epitope of the antigen. In a process called reverse vaccinology, only those peptide sequences are included for which genomic analysis predicts strong immunity (Rappuoli 2000). However, due to their small size, peptides are usually not very immunogenic and require additional molecules as carrier substances and adjuvants (Aguilar and Rodríguez 2007) to increase immunogenicity (Purcell et al. 2007). No peptide vaccines have been licensed to date.

31.1.4 Naked Vaccines

In the early 1990s, an entirely novel approach was discovered. Injection of nucleic acids, i.e., plasmid deoxyribonucleic acid (DNA), elicited an immune response (Tang et al. 1992; Ulmer et al. 1993), a finding later repeated for ribonucleic acid (RNA) (Fleeton et al. 2001). Once the desired sequence of bases is known, these "naked" nucleic acid (DNA/RNA) vaccines are easy and quick to generate in large quantities, and, since a cell-free enzymatic transcription reaction is utilized, quality and safety are improved (Hekele et al. 2013). After cell entry, using the host's intracellular machinery, the desired protein is translated, either directly from RNA or after transcription of messenger RNA (mRNA) from the injected DNA. When antigen-presenting cells like dendritic cells present these peptides via major histocompatibility complex (MHC) molecules, a potent response of both humoral and cellular immunity can be elicited (Condon et al. 1996). Besides, foreign nucleic acids are also recognized by parts of the innate immune system, such as toll-like receptors (TLR) that may also trigger cellular immune responses. Despite these theoretical concepts, DNA-based vaccines have so far proven to be not particularly immunogenic, often requiring multiple booster vaccinations to counter a rapid waning of immunity. Recently, novel delivery technologies like electroporation and activation of innate immunity have increased the effectiveness of DNA vaccines (Grunwald and Ulbert 2015). Another potential drawback is that, at least theoretically, DNA-based vaccines could integrate into the host's genome, causing mutations or new diseases (Nichols et al. 1995).

RNA-based vaccines would not be prone to the latter concern since human cells lack the ability of reverse transcription, and over time, RNA is simply degraded within cells. However, due to ubiquitous RNases, RNA molecules are also rapidly degraded in extracellular surroundings. It necessitates the formulation of target RNA sequences in lipid nanoparticles, which also facilitate cell entry (Geall et al. 2012). However, as with DNA, the immunogenicity induced is relatively low, so large amounts of RNA are needed to establish an adequate immune response. An improvement was proposed by the use of self-amplifying mRNA vaccines, containing both viral RNA and enzymes for its replication (RNA-dependent RNA polymerases). Since these vaccines are lacking the genes encoding for the structural proteins that compose elements like the capsid, no new virions are formed. The accumulation of vast amounts of replicons (strands of target RNA originating from a single founder sequence) in one cell mimics a viral infection and leads to a robust immune response, e.g., via the recognition of RNA by TLR. The amplification in vivo leads to lower dosages, usually in the microgram range. Self-amplifying mRNAs have shown to be safe and immunogenic in various animal models, including mice (Petsch et al. 2012) and nonhuman primates (NHP) (Bogers et al. 2015). However, no nucleic acid-based vaccine has yet been licensed.

31.1.5 Construction Kits for Vaccines

Vector vaccines have solved the problem of delivering the target sequence to a cell since they are making use of the natural ability of viruses to infect cells. First viral vectors were created based on simian virus 40 in the 1970s (Jackson et al. 1972) and vaccinia virus in the 1980s (Mackett et al. 1982), and several types of viral vectors have been developed since then (Ura et al. 2014). After the introduction into the cell, the cell's tools are used to replicate viral DNA/RNA and to translate RNA into structural proteins. Since this not only mimics but is, in fact, a viral infection, immunogenic antigens are synthesized intracellularly and at high levels. It elicits strong humoral and cellular immune responses without an adjuvant, including activation of cytotoxic T cells, which eliminate infected cells. Commonly used viral vectors include, among others, adenovirus (Ad), Modified Vaccinia virus Ankara (MVA), yellow fever virus (YF), and vesicular stomatitis virus (VSV). Depending on the virus it is based on, each viral vector has unique properties and faces unique challenges.

Human Ad vectors are usually replication-defective after deletion of E1 and E3 genes, increasing their safety profile. Ad5, which has been studied extensively, confers strong immunogenicity after the first dose of vaccination but may be hampered by preexisting immunity against the vector (Sekaly 2008). The use of chimpanzee Ad (ChAd), which is nonpathogenic in humans and confers little preexisting immunity, could circumvent this caveat (Morris et al. 2016).

After losing about 15% of the Vaccinia virus genes, MVA is not replication-deficient per se but so highly attenuated that it lost its ability to replicate in mammalian cells. An additional adjuvant effect is conferred by the stimulation of the innate immune system, e.g., via TLR (Zhu et al. 2007a) because the virus has lost genes hampering the innate immune response. MVA, therefore, confers high levels of immune responses (Sutter and Staib 2003). However, in most studies, more than one injection is needed to confer protective immunity. MVA has been safely administered to over 100,000 recipients in the eradication of smallpox (Mayr et al. 1978) and has been manufactured at a large scale for stockpiling in the United States (Frey et al. 2013).

As an example of replication-competent viral vectors, VSV usually causes relevant disease only in animals and is not or only mildly pathogenic in humans (Geisbert and Feldmann 2011). Live vaccines offer the highest degree of immunogenicity, both humoral and cellular, due to the potentiation of immunogenic material through replication. The distribution of the viral vector throughout the host can confer systemic and mucosal immunity. Concerns associated with replication-competent vectors are those of genetic shifts, i.e., the recombination of genetic material from different viruses or viral strains that could allow for an altered phenotype and increased virulence. Since the VSV genome is encoded in a single RNA strand, it lacks the potential for reassortment in vivo. As with adenoviruses, preexisting immunity against the vector could hamper the induced immune response, but VSV seroprevalence in humans is low (Johnson et al. 1966). The second viral vector vaccine to obtain licensure, rVSV-ZEBOV (marketed by MSD as Ervebo®), which was licensed by the US Food and Drug Administration (FDA) in 2019, has shown to be nearly 100% protective against Ebola virus disease (EVD) after a single injection (Henao-Restrepo et al. 2017).

The YF-based, tetravalent dengue vaccine CYD-TDV (marketed by Sanofi Pasteur as Dengvaxia®) was the first viral vector vaccine to obtain licensure (Mexico, 2015). It confers strong immunity (Guy et al. 2011) and is currently licensed in over 20 countries. Different outcomes depending on previous monotypic or multitypic infections with dengue virus (DENV) indicate complex mechanisms of immune interaction and currently limit the use of CYD-TDV to patients with prior dengue infection (Sridhar et al. 2018).

In summary, vector vaccine platforms promise to confer the strongest, longest-lasting immunity among the novel vaccine strategies and have successfully progressed to licensure in two cases but also face challenges in safety and production capacity that need to be addressed for each viral vector.

31.1.6 Vaccines Against Emerging Pathogens

Vaccine development gained substantial significance in the face of the West African Ebola virus epidemic of 2014–2016, when, with no therapeutic or preventive measures at hand, more than 28,000 people were infected and over 11,000 people died (Fathi et al. 2019).

Traditionally, vaccine design had not been an established method of outbreak containment instead of relying solely on non-pharmaceutical interventions (NPI). Even in the case of influenza, where the vaccine is renewed annually, and production pipelines are well-established, the vaccine against the H1N1 strain came too late to have an impact on the so-called swine flu pandemic in 2009 (Girard et al. 2010). However, with novel vaccine development technologies and the threat of a high-consequence pathogen spreading globally, clinical testing of already existing vaccine candidates against EVD was implemented quickly after the beginning of the outbreak. Alas, it also came too late to contain the outbreak. This tragedy underlined the need for swift vaccine development against emerging pathogens. It led the World Health Organization (WHO) to convene a group of experts and establish a list of known pathogens that were likely to cause public health emergencies in the future and for which research and development (R&D) efforts to develop vaccine candidates and causal therapies should, therefore, be prioritized. The regularly updated list, called WHO R&D Blueprint (WHO 2019), contained nine priority pathogens before the COVID-19 pandemic, among them the highly pathogenic coronaviruses (CoV) SARS-CoV-1 and MERS-CoV, as well as a Disease X, a yet unknown pathogen that would need urgent R&D efforts once the outbreak would hit. This pathogen in the current pandemic has turned out to be the newly discovered SARS-CoV-2.

The COVID-19 pandemic caused by SARS-CoV-2 has asserted the need to have vaccines against emerging pathogens at hand as quickly as possible. In this regard, the R&D strategy aims to promote vaccine technologies that can then be adapted to the specific pathogen and manufactured at a large scale. Once a priority pathogen has been identified, the next objective is to develop pathogen-specific vaccines. In the case an outbreak can be contained by NPI, as has been the case for the SARS-CoV-1 outbreak in 2002/2003 and large MERS-CoV outbreaks in 2014 (Saudi-Arabia) and 2015 (South Korea), the goal is to advance the development of multiple vaccine candidates nevertheless. To accomplish this goal, preclinical and early clinical stages are designed to demonstrate the safety and immunogenicity of candidate vaccines, usually involving healthy volunteers (phases 1–2). Ultimately, it is aspired to stockpile thousands of dosages and have these at hand in order to be able to start advanced phase clinical trials as soon as an outbreak of the pathogen arises.

While history has now taught us that this endeavor may prove itself useful for emergency preparedness, there is a little economic incentive for pharmaceutical companies to develop vaccines that may never be used, let alone be licensed. In consequence, powerful funding mechanisms had to be established. One such mechanism is the Coalition for Epidemic Preparedness Innovations (CEPI) (Burki 2017), a public-private partnership supported by multiple governments as well as philanthropic organizations such as the Bill and Melinda Gates Foundation, among others. CEPI plays a significant role in funding current efforts in SARS-CoV-2 vaccine development concerning candidates that can be manufactured swiftly and at scale and will be available globally (Thanh Le et al. 2020). It is currently supporting the development of eight candidates, of which three have already advanced into clinical trials (CEPI 2020).

31.2 SARS-CoV-2 Vaccine Design Challenges

31.2.1 Vaccines Against Respiratory Viruses

Viruses that cause respiratory disease in humans presumably possess the highest potential for pandemic spread. Unfortunately, to date, the development of vaccines against these pathogens has proven immensely difficult. The only respiratory virus that has approved vaccines available is influenza, and the flu vaccine efficacy, changing annually, is comparably low, ranging from 40% to 80% (Hannoun 2013). Other respiratory viruses include parainfluenza, metapneumovirus, adenovirus, rhinovirus, and respiratory syncytial virus (RSV). The latter is especially lethal in vulnerable populations like children and the elderly, accounting for annually over 110,000 deaths globally in children under the age of 5 (Nair et al. 2010). It has been a focus of vaccine research for decades, but research experienced devastating setbacks early on, when a vaccine trial in the 1960s, examining an inactivated whole virus vaccine, led to worse outcomes in vaccinated children compared to placebo (Fulginiti et al. 1969). This observation was likely due to unfavorable immune interactions (see below) such as the induction of non-neutralizing antibodies (Murphy et al. 1986) or a T_H2-dominant cellular response (Openshaw et al. 2001). To date, at least 30 RSV vaccine candidates have entered clinical stages of development; however, none have been licensed. Only one has progressed to phase 3 trials, and the efficacy of candidate vaccines seems to be relatively low (Rossey and Saelens 2019). In light of the, sometimes tragic, history of vaccine development against respiratory viruses, the development of a safe and efficacious vaccine against SARS-CoV-2 can, therefore, be expected to pose a significant challenge. This history also highlights that despite the urgent need for accelerated development of a SARS-CoV-2 vaccine, safety evaluation efforts must not be compromised.

31.2.2 The *Coronaviridae* Family

The RNA sequence of SARS-CoV-2 was released in mid-January 2020. Since then, the discovery of a vaccine has been made a priority by the scientific community worldwide. There are now more than 100 vaccine candidates in the preclinical or clinical stages (as of June 2020), and the landscape is continuously growing (Thanh Le et al. 2020). Keeping in mind that vaccine development usually takes decades rather than years, this evolution certainly is unprecedented and extremely impressive. However, it requires to address many open questions along with the advancement of vaccine candidates. In the quest to find answers to these questions, research on other coronaviruses can guide our way. Including the novel SARS-CoV-2 (Gorbalenya et al. 2020), the family of *Coronaviridae* encompasses 46 species (ICTV 2020), infecting a wide variety of mammals. Most are limited in their host range, and up to date, seven *Coronaviridae* are known to infect humans. Four of these, namely, 229E, NL63, OC43, and HKU1, are considered common cold coronaviruses and cause self-limiting infections of the upper respiratory tract in humans. The genera of *Betacoronaviridae* encompass two of the latter, OC43 and HKU1 (lineage A), as well as the three highly pathogenic coronaviruses SARS-CoV-1, SARS-CoV-2 (both lineage B) and MERS-CoV (lineage C).

31.2.3 Animal Models

One challenge is to establish an appropriate animal model of SARS-CoV-2 infection in order to investigate vaccine immunogenicity and efficacy, which is further complicated by the fact that infection experiments have to be performed under enhanced security measures, i.e., in Biosafety Level 3 laboratories. No animal models have yet been defined that would adequately emulate the natural course disease in humans. Different animal models have been evaluated for SARS-CoV-1 and MERS-CoV, but the animals mostly show a reduced clinical severity of the disease, limiting the evaluation of vaccine efficacy (Sutton and Subbarao 2015).

For MERS-CoV, which uses dipeptidyl peptidase 4 (DPP4) as a receptor, common marmosets, mice (Gretebeck and Subbarao 2015), and dromedary camels have been studied. The latter, which serve as intermediate hosts, can be infected and transmit the virus to humans, allowing studies for immunogenicity (Haagmans et al. 2016), but due to the distribution pattern of DPP4, the disease is limited to the upper respiratory tract. Transgenic mice, which express the human entry receptor (angiotensin-converting enzyme 2 (ACE2) or DPP4 for SARS-CoV-1 and MERS-CoV, respectively), have been established as models (Netland et al. 2010; Munster et al. 2017). They exhibit enhanced infection sensitivity and a more severe course of the disease (high levels of viremia, lower respiratory tract infections, and death), allowing for better evaluation of vaccine efficacy in preclinical studies.

For SARS-CoV-1, rhesus macaques, inbred BALB/c mice, hamsters, and ferrets have been evaluated (Gretebeck and Subbarao 2015). Since SARS-CoV-2 also uses ACE2 as an entry receptor, one might expect a similar host range, and, indeed, SARS-CoV-2 has been shown to replicate in cats and ferrets, but not in dogs, pigs, and chickens (Shi et al. 2020). In line with observations made in SARS-CoV-1 vaccine research, SARS-CoV-2 infection could successfully be established in the NHP model of rhesus macaques when challenged with high amounts of the infectious virus via the intranasal or intratracheal route and the animals furthermore show clinical signs of illness, such as pneumonitis, when challenged (Chandrashekar et al. 2020). The model also allowed to assess the course of infection and established that the macaques were protected from symptomatic disease upon rechallenge.

While NHP often turn out to become the gold standard for infection and vaccine studies, these animal models also harbor quite a few challenges, both for ethical as well as practical reasons, such as availability. The establishment of an appropriate small animal model for toxicology, preclinical immunogenicity, and, ideally, efficacy studies would, therefore, be advantageous.

The ferret model has played a significant role in vaccine research for respiratory infections in general and SARS-CoV-1 research in particular, including the assessment of antibody-dependent enhancement (ADE) of disease (see below). Ferrets can be infected with SARS-CoV-2 and may also transmit the infection to their kin (Kim et al. 2020). While viral antigen can be isolated from tracheal secretions as well as urine and stool, the disease course seems to be relatively mild. Hamsters have previously been established as animal models in SARS-CoV-1 infection (Subbarao and Roberts 2006) and also seem to be an adequate small animal model permissible for SARS-CoV-2 infection (Sia et al. 2020).

Based on experiences made from studies involving SARS-CoV-1 (Song et al. 2019), mice seem to be an adequate model to assess immunogenicity after infection and vaccination, while viral replication seems to be transient and not to induce noticeable disease. In SARS-CoV-2 vaccine research, a BALB/c mouse model and a model using Wistar rats have successfully been established for the assessment of vaccine immunogenicity with comparable binding antibody titers (Gao et al. 2020). Furthermore, the generation of a mouse-adapted SARS-CoV-2 strain seems to be achievable through serial virus passage in aged BALB/c mice. Some analyses, albeit not yet peer-reviewed (Gu et al. 2020), indicate that this strain could be more pathogenic in mice, as it may lead to disease (i.e., pneumonia) and may, therefore, be used as a challenge model in vaccine studies. Taken together, promising small animal models have been described for SARS-CoV-2 alongside the NHP model.

31.2.4 Vaccine Targets

Another question that is needed to be addressed is the definition of an optimal target antigen for vaccine design. The spike (S) glycoprotein is a surface antigen common to all coronaviruses. It consists of two subunits: S1 with a receptor-binding domain (RBD) and S2. These subunits together form one monomer; three of these monomers then make up the S trimer (Lan et al. 2020). It is the S protein that binds to the ACE2 receptor and mediates viral cell entry. S protein is a highly immunogenic antigen in both MERS-CoV and SARS-CoV-1 vaccine studies (Koch et al. 2020; Folegatti et al. 2020; Gao et al. 2003), and neutralizing antibodies against S glycoprotein have been shown to correlate with protection against SARS (Jiang et al. 2005) and MERS (Du et al. 2016).

Potent neutralizing antibodies against S protein, notably the RBD, have also been described in SARS-CoV-2. Unsurprisingly, in some form or other, S protein is incorporated in the design of most candidate vaccines against SARS-CoV-2 (Thanh Le et al. 2020). S protein, however, is a metastable protein, i.e., it undergoes a conformational change upon binding to the cellular ACE receptor and fusion of the virus with the cell membrane. The delivery of a stable prefusion spike protein as a vaccine target will be necessary to induce effective antibody responses while avoiding the generation of ADE (Graham 2020) (see below). Although SARS-CoV-2 diversity is low, the possibility of antigenic drift of the surface S protein under increasing pressure of natural selection has become a growing concern. Currently, the development of most vaccine candidates is based on the sequence of the original Wuhan isolate. Different isolates have been described since, with most being single-nucleotide polymorphisms (SNP). One isolate, featuring a G-to-A base change in the spike antigen (position 23,403 in the Wuhan reference isolate), named D614G, was described as more transmissible in a preprint manuscript (Korber et al. 2020). If this observational data were validated, it would indicate the emergence of a novel SARS-CoV-2 strain, i.e., a virus with phenotypic changes, rather than just a different isotype. However, to date, no genotypic changes leading to different properties in virulence or transmissibility have been demonstrated for other members of the *Coronaviridae* family.

31.2.5 Vulnerable Populations

Another challenge will be the assessment of vaccine candidates in vulnerable populations. From what is known to date, pregnant women and children are not disproportionately affected by COVID-19 (Qiu et al. 2020; Castagnoli et al. 2020; WHO 2020b), and these populations will likely be excluded from early phase vaccine trials, as is common practice. On the contrary, the elderly are at an exceptionally high risk of developing a severe disease course, and the age of ≥ 80 years has been associated with a case fatality rate of up to 15% (Wu and McGoogan 2020). Vaccines are likewise not commonly tested in elderly individuals until late in the development process, as immune senescence challenges the assessment of vaccine immunogenicity. In the case of COVID-19, the protection of older patient populations, however, is a priority, and trials will need to investigate the safety, immunogenicity, and efficacy of vaccine candidates in this group early in the development process. Immunocompromised individuals comprise another vulnerable group, and specific vaccine platforms, i.e., nonreplicating vaccines, may be more suitable for them. Even if an otherwise effective vaccine turns out not to be appropriate for vulnerable populations, however, their indirect protection can, of course, be achieved through widespread vaccination of less affected parts of the society.

31.3 Immunity in Covid-19 Vaccine Design

31.3.1 What Does Immune Protection Consist of: An Open Question

As outlined above, the first vaccines were developed without a clear understanding of immunology or even major concepts of modern medicine like germ theory. However, as our understanding of the body's immune response grew, the question arose as to which measurements signal immunity against a pathogen or protection from a disease, that is, what the correlates of protection (CoP) are. This question proved to be undoubtedly complex (Plotkin 2013). Concepts such as immunity or protection are not binary ("protected" or "not protected"); there are rather many different shades in between. Immune responses after vaccination can lead to less severe disease without preventing infection, can protect from disease while still allowing infection and transmission to others, or can confer sterilizing immunity. In rare cases of failed vaccine development, vaccination has even led to more severe disease (Vennema et al. 1990; Buchbinder et al. 2008; Steinbrook 2007). As two prominent examples of completely different protection conferred, whole virus-inactivated polio vaccine (IPV) protects the host from the disease. Still, it allows infection with the virus, its replication in the human gut, and fecal-oral transmission to others, while live-attenuated oral polio vaccine (OPV) confers sterilizing mucosal immunity that prevents transmission (WHO 2016).

Apart from pathogen factors such as higher or lower infective dose, host factors also play a role in determining an immune response and, thus, measurements of CoP. Host factors include immune status (congenital or acquired immunodeficiencies), coexisting infections or other diseases, age, nutrition, and genetic differences, with the latter primarily occuring in MHC molecules (Plotkin 2013). Likewise, the timing of measurements matters for the CoP at hand. All measurable immune responses wane over time, as active antibody-producing plasmablasts are replaced by memory B cells (Kurosaki et al. 2015). Thus, whether a measurement is performed at the peak of immune responses a few weeks after vaccination/disease or at the beginning of an infection, possibly years after the last vaccination, can change the resulting measurements. It has been proposed to further differentiate CoP into two mutually exclusive categories. Mechanistic correlates of protection (mCoP) are the causal agents of protection, whereas nonmechanistic correlates of protection (nCoP) are merely a predictor of protection without being its causal agent (Plotkin and Gilbert 2012).

For many pathogens, antibodies are dominant humoral CoP. These include binding antibodies, usually measured by enzyme-linked immunosorbent assay (ELISA), as well as neutralizing antibodies, usually measured by plaque or neutralization assays. Antibody assays generally measure immunoglobulin G (IgG) without further subdivision into the four subclasses IgG1–4 (Vidarsson et al. 2014) and generally measure neither serum IgM nor mucosal IgA, with the latter likely being important for protection against noninvasive infections like influenza (Belshe et al. 2000). However, the mere quantity of antibodies may not be enough to describe their effectiveness against a pathogen fully; the quality of the antibody response may be as important. Direct, Fab-mediated antibody functions, like binding or neutralization, are well-established, but indirect, Fc-mediated functions, such as antibody-dependent cellular cytotoxicity (ADCC) or complement-dependent cytotoxicity (CDC), may also have protective effects against different pathogens, particularly HIV (Vargas-Inchaustegui and Robert-Guroff 2013; Smith et al. 2014). After vaccination against *Neisseria meningitidis*, bactericidal antibodies that use Fc-dependent CDC to lyse cells can be considered mCoP, while binding antibodies measured by ELISA are nCoP (Borrow et al. 2005). Currently, a big challenge is the identification and implementation of assays that allow for reliable measurements with reproducible results that are comparable between laboratories (Wines et al. 2017).

Cellular immune responses also play a significant role in protection from pathogens, and their activity is intertwined with humoral immunity. Traditionally, T cells have been divided by the cluster of differentiation (CD) molecules 4 and 8 into CD4+ T-helper cells (T_H) and CD8+ cytotoxic T cells (T_C), a method still used today to describe cellular CoP, e.g., in influenza (Wilkinson et al. 2012). CD4+ T-helper cells can be further subdivided, according to CD molecules and functions, into T_H1, T_H2, T_H17, and Tregs (Zhu and Paul 2008). T_H1 (type 1 helper cells) use IFN-γ and IL-2 and trigger a cell-mediated response by macrophages and CD8 T cells, mostly against intracellular pathogens like bacteria and protozoa. T_H2 (Type 2 helper cells) use cytokines like IL-4 and IL-10 and trigger a humoral response by activating B cells, eosinophils, and basophils. T_H17 (T-helper 17 cells) secrete IL-17 and shape the adaptive immune response, protecting against pathogens like fungi. Regulatory T cells (Tregs) use cytokines like TGF-β (Kretschmer et al. 2005) to modulate the immune system and dampen immune responses where needed, in order to prevent a reaction against self-antigens that could lead to autoimmune disease.

Of paramount interest for vaccine development is the T-cell function of retaining memory, i.e. the ability to elicit a more robust and more rapid immune response when encountering an antigen for the second time after being "primed" through a previous encounter. Flow cytometry based on antibodies with fluorescent dyes is traditionally used to define various subgroups of T cells according to surface molecules. These subsets are transverse with the CD4/CD8 differentiation and include subsets such as effector memory (T_{EM}), central memory (T_{CM}), or tissue-resident memory T cells (Trm). Some of these, like CD8+ T_{CM}, have been proposed as ideal populations for protection because they remain in circulation through lymphoid tissue and have an especially long half-life (Gerlach et al. 2016). However, novel technologies like mass cytometry have led to an even greater plethora of T-cell memory subsets (Newell and Cheng 2016). Some authors, therefore, believe that such minuscule distinctions between subsets according to surface molecules alone are futile and that memory T-cell populations should be viewed more as a continuum (Jameson and Masopust 2018) to describe key characteristics such as localization in different tissues and trafficking between sites.

While antibodies play a role in protection after virtually all vaccines, cellular CoP is more diverse between different pathogens. For some pathogens, exact cutoff values for CoP can be defined. Neutralizing antibodies, the mechanistic CoP against toxin-producing bacteria causing tetanus (Edsall 1959) or diphtheria (Björkholm et al. 1986), are protective above 0.1 IU/ml; for rabies, the value is 0.5 IU/ml (WHO 2007). For herpes zoster, levels of CD4+ T-cell responses correlate with protection, although an exact cut-off has not yet been defined (Weinberg et al. 2009; Gilbert and Luedtke 2018). In the case of EVD, determining the protective immune response is seemingly even more complex, with innate immunity, antibodies (Medaglini et al. 2018) and T cells (Meyer et al. 2019) contributing to immunity.

In animal models, T-cell responses were critical for clearance of both SARS-CoV-1 and MERS-CoV (Zhao et al. 2010; Liu et al. 2017), and in human survivors of SARS or MERS, both humoral and cellular immune responses could be detected (Li et al. 2006; Zhao et al. 2017).

31.3.2 Duration of Immunity

The duration of immune responses is of particular concern with coronaviruses. It stems from the fact that humans can get reinfected with the common cold CoV multiple times. In decades past, challenge models with human volunteers have demonstrated waning protection against hCoV-229E and hCoV-OC43 (the only two human CoV known at that time) despite measurable antibody titers (Bradburne and Somerset 1972; Callow et al. 1990). Serological studies in survivors have demonstrated antibody responses 24 and 36 months after infection with SARS-CoV-1 and MERS-CoV, respectively (Liu et al. 2006; Payne et al. 2016), while memory T-cell responses could be detected up to 6 years after infection in case of SARS-CoV-1 (Tang et al. 2011). How this translates to long-term protection is unknown, especially since memory B cells could not be demonstrated after 6 years (Tang et al. 2011).

31.3.3 Immune Interaction: Cross-Reactivity, ADE, and VAERD

Another concern might be immune interactions within the family of *Coronaviridae*. Specifically, two possibly opposing phenomena come to mind.

On the one hand, cross-reactivity of antibodies might offer a certain degree of protection, particularly antibodies directed against other *Betacoronaviridae*, either lineage A common cold CoV (OC43 and HKU1) or lineage C (MERS-CoV). Cross-reactivity was observed between different isolates of SARS-CoV-1 (Zhu et al. 2007b) and between SARS-CoV-1 and *Betacoronavirus* OC43 as well as, to a lesser degree, *Alphacoronavirus* 229E (Che et al. 2005). CR3022, a human monoclonal antibody directed against the S protein of SARS-CoV-1, showed potent binding of the SARS-CoV-2 spike RBD (dissociation constant (K_d) of 6.3 nM) (Tian et al. 2020). Little is known about the cross-reactivity of MERS-CoV and other coronaviruses. In one study from 2016, only minimal cross-reactivity was observed between a dromedary CoV (DcCoV UAE-HKU23) and MERS-CoV (Woo et al. 2016).

On the other hand, ADE might induce more severe disease. The phenomenon is still incompletely understood, but it is believed that in ADE, antibodies bind non-neutralizing parts of viral surface proteins; in the case of coronaviruses, these are most likely regions of the S protein other than the RBD. Viral entry is then most likely mediated by a different pathway than interaction with the normal host receptor instead relying on viral entry via Fc receptors (FcR) (Jaume et al. 2011). Whether or not ADE occurs depends on several factors such as the antigen targeted by the antibody, the strength of the antigen-antibody interaction, and the antibody concentration (Pierson et al. 2008). A similar phenomenon, called vaccine-associated enhanced respiratory disease (VAERD), has been observed in swine that were vaccinated against influenza A using an inactivated whole virus vaccine (Gauger et al. 2011).

ADE has been described in some viruses, including an *Alphacoronavirus* which causes a fatal disease in cats, called feline infectious peritonitis virus (FIPV) (Hohdatsu et al. 1998). Importantly, a vaccinia virus-based vector vaccine encoding the S protein of FIPV increased mortality in cats (Vennema et al. 1990). In humans, ADE has been well documented for arboviruses of the *Flaviviridae* family, such as the Zika virus (ZIKV) (Khandia et al. 2018) or dengue virus (DENV) (Halstead 2003; Dejnirattisai et al. 2010). It is well documented that DENV infection with one of the four serotypes can increase the severity of disease in subsequent infections by another serotype (Guzman and Vazquez 2010). In human coronaviruses, ADE was observed for SARS-CoV-1 in vitro (Jaume et al. 2011) and in vivo in animal studies (Yasui et al. 2008). In studies on MERS-CoV, an increase in lung pathology has been observed in mice vaccinated with an inactivated MERS-CoV vaccine (Agrawal et al. 2016). In rabbits, animals who developed non-neutralizing antibodies after a first infection with MERS-CoV also showed increased pulmonary inflammation upon reinfection, compared to those that developed neutralizing antibodies (Houser et al. 2017). Whether these findings translate into more severe disease in coronaviruses and, importantly, whether or not ADE can be expected in COVID-19 in humans remain unclear. ADE has been hypothesized to be a possible reason for different clinical outcomes in SARS and COVID-19 (Tetro 2020), but to this point, this remains speculative.

A possible solution to ADE would be to include only those parts of the S protein that elicit neutralizing antibodies, such as the RBD, while excluding other parts that might elicit non-neutralizing immune responses. When such vaccines against CoV were tested in animal models, no signs of ADE were observed (He et al. 2004; Wang et al. 2015). Another strategy to avoid ADE might be the use of adjuvants. In mice vaccinated against SARS-CoV-1, the addition of a polysaccharide-based adjuvant increased the levels of neutralizing antibodies and reduced lung pathology post-challenge (Honda-Okubo et al. 2015). In summary, while ADE is a known phenomenon in humans for some *Flaviviridae* such as DENV, it has only been demonstrated in animal studies for coronaviruses. However, vaccine candidates against SARS-CoV-2 will need to be carefully evaluated for the potential to induce ADE in humans.

31.3.4 Differences in Vaccine-Induced Immunity

The levels of immunity conferred by the different vaccine platforms that are under investigation for a SARS-CoV-2 vaccine vary tremendously (Table 31.1). Increased immunity sometimes comes at the price of side effects, while safer vaccines may be less immunogenic, necessitating adjuvants or repeated booster immunizations. On the one end of the spectrum, live, attenuated yellow fever (YF) 17D vaccine, developed in the 1930s (Theiler and Smith 1937), confers a nearly 100% protective immunity, lasting many years and often even a lifetime after a single injection (Kareko et al. 2019). However, in rare cases, severe adverse events, such as vaccine-associated neurotropic (YEL-AND) and viscerotropic disease (YEL-AVD), can be associated with the vaccine (Silva et al. 2010). On the other end of the spectrum, subunit influenza vaccines confer varying immunity and between 40% and 80% protection, have to be renewed yearly, and have little more side effects than placebo injections (Hannoun 2013). Their low and varying degree of protection is due to the antigenic shifts and drifts that occur, in which reassortment of genetic material between different viruses (antigenic shift) or through accumulated mutations within one virus (antigenic drift) results in a new strain, i.e., a new phenotype of the virus. The immune history of previous exposures to influenza virus strains likely also contributes to variability in protection (Lewnard and Cobey 2018). Inactivated whole virus rabies vaccines elicit high levels of protection but have to be boostered several times for lasting protection and up to five times when used as postexposure prophylaxis (Ertl 2009). It shows that different pathogens require distinct approaches in vaccine development.

31.3.5 Vaccine Development Against SARS-CoV-1 and MERS-CoV

The original vaccine development pipeline against SARS-CoV-1 encompassed different

Table 31.1 Platforms under evaluation for SARS-CoV-2 vaccine development

Vaccine platform	Examples	Previous licensure of the technology?	Advantages	Disadvantages	Example for SARS-CoV-2 candidates[a]	SARS-CoV-2 vaccine data from animal models available?	In clinical testing? (ClinicalTrials.gov identifier)
DNA	DNA plasmids with/without electroporation	No	Easy to produce, thermostable, can be provided at large scale	May lack immunogenicity, needs a special device for injection, boost necessary	INO-4800, multiple variants (full-length spike, RBD, etc.) in preclinical testing	Yes (Yu et al. 2020)	Yes (NCT04336410)
RNA	Lipid-nanoparticle-encapsulated mRNA (multiple targets)	No	Easy to produce, can be provided at large scale	Reactogenic, boost likely necessary	Moderna's mRNA-1273, Pfizer's BNT162b2, multiple others in preclinical testing	Yes (Erasmus et al. 2020)	Yes (NCT04283461)
Replication-deficient viral vectors	Adenovirus (human (type 5, type 26), chimpanzee, simian), MVA	No (MVA has however been licensed as a poxvirus vaccine, and the Ad26.ZEBOV/MVA-BN-Filo vaccine is imminently expected to receive licensure)	Immunogenic (including cellular immunity), can be produced relatively quickly	Preexisting immunity to the vector may be a problem, often has to be stored frozen	ChadOx1 nCoV-19, multiple others in preclinical testing	Yes (van Doremalen et al. 2020)	Yes, results available (Zhu et al. 2020)
Replication-competent viral vectors	Measles, yellow fever (YF17D), vesicular stomatitis virus (VSV), Newcastle disease virus (NDV)	Yes: YF-based, tetravalent dengue vaccine CYD-TDV; VSV-based Ebola vaccine rVSVΔG-ZEBOV-GP	Immunogenic (including cellular immunity), single-shot regimen possible, can be produced relatively quickly	Reactogenic, not suitable for pregnant or immunocompromised individuals, preexisting immunity to the vector may be a problem in some cases, has to be stored frozen	Measles vector expressing N/S antigen (i.e., DZIF, Themis) among others	No	No

(continued)

Table 31.1 (continued)

Vaccine platform	Examples	Previous licensure of the technology?	Advantages	Disadvantages	Example for SARS-CoV-2 candidates[a]	SARS-CoV-2 vaccine data from animal models available?	In clinical testing? (ClinicalTrials.gov identifier)
Subunit vaccines	Protein subunit, conjugate, and polysaccharide vaccines	Yes (i.e., HBV vaccine, pneumococcus vaccines)	Safe and well-tolerated, well-established approach	Booster vaccinations necessary, adjuvants often necessary	Full-length SARS-CoV-2 S glycoprotein, adjuvanted (Novavax), among many others	No	Yes (NCT04368988)
Whole inactivated virus	Chemically or physically inactivated virus, with or without adjuvant	Yes (i.e., IPV)	Usually safe and well-tolerated, well-established approach	Booster vaccinations necessary, adjuvants often necessary	PiCoVacc (Sinovac), inactivated using ß-propriolactone, and many others	Yes (Gao et al. 2020)	Yes, multiple (i.e., NCT04352608)
Virus-like particle (VLP)	Enveloped or naked VLP	Yes, i.e., human papillomavirus vaccine (Gardasil™)	Safe and well-tolerated	May need a booster vaccination, and manufacturing may be challenging and costly	VLP based on RBD	No	No
Live-attenuated virus	Viruses attenuated through multiple passages in nonhuman hosts/cell culture	Yes (i.e., varicella, measles-mumps-rubella vaccination)	Immunogenic (including cellular immunity), well-established approach	Lengthy development process, extensive safety assessment necessary	Serum Institute of India codon-deoptimized live-attenuated vaccine	No	No

DZIF German Center for Infection Research, *HBV* hepatitis B virus, *IPV* inactivated polio vaccine, *MVA* Modified Vaccinia virus Ankara, *RBD* receptor-binding domain

[a]As in DRAFT landscape of COVID-19 candidate vaccines – May 22, 2020

platforms, including inactivated and attenuated whole virus vaccines, vector-based vaccine platforms (parainfluenza virus, VSV, Ad, MVA, and others), recombinant protein, VLP, as well as DNA-based vaccines (Padron-Regalado 2020). As elaborated above, mostly the S glycoprotein was used as the immunogenic target of choice. Either the complete S protein or fragments (S1 subdomain or RBD) were used; only a few vaccines also included the nucleocapsid (N) protein. Two vaccines made it to clinical development stages (phase 1 studies), namely, an inactivated whole virus vaccine (Lin et al. 2007) and a DNA-based vaccine (Martin et al. 2008). However, once non-pharmaceutical interventions stopped the outbreak of SARS in 2003, funding dried up, and the vaccines were not developed further, exemplified by another planned but withdrawn phase 1 trial of a recombinant S protein-based vaccine (clinicaltrials.gov 2011). After more significant outbreaks of MERS-CoV in Saudi-Arabia in 2014 and South Korea in 2015, small-scale "smouldering" transmission is still continuously occurring on the Arabian Peninsula. As with SARS, the preclinical vaccine development landscape spans a wide range of technologies, including whole virus-, vector-, protein-, DNA-, or nanoparticle-based vaccines (Padron-Regalado 2020). In contrast to SARS, however, efforts in vaccine development have not been halted entirely after the initial significant outbreaks were contained. This was due to a shift in policies, exemplified by the WHO's plan for R&D against emerging infections. MERS was placed on the list of priority diseases (WHO 2019), and to date, three candidate vaccines have completed phase 1 clinical trials, namely, one DNA (Modjarrad et al. 2019) and two vector-based vaccines, one ChAd (Folegatti et al. 2019), and one MVA-based (Koch et al. 2020), with preparations for phase 2 studies underway.

31.4 Current Status of SARS-CoV-2 Vaccine Development

An ideal vaccine against SARS-CoV-2 needs to meet quite a few requirements. Some of these apply to vaccines designed against priority pathogens in general, while others are specific to SARS-CoV-2. In a best-case scenario, a vaccine should confer immunity quickly after vaccination and demonstrate a favorable safety profile. Specifically, immune interactions such as ADE and VAERD, as observed in other respiratory virus vaccine trials (Graham 2020), need to be ruled out. It should further be possible to manufacture the vaccine quickly and on a large scale, and it should be storable under simple conditions.

The distinct vaccine platforms are known today to satisfy many of these requirements, although no one platform fulfills them all. It is, therefore, reasonable to develop multiple SARS-CoV-2 vaccines in parallel. To date, more than 100 vaccine candidates are under investigation, and new candidates join the list almost daily. For the most up-to-date information on the current state of vaccine development, an extensive list of vaccine candidates in preclinical and clinical evaluation that is updated regularly has been established by WHO, which is accessible at https://www.who.int/publications/m/item/draft-landscape-of-covid-19-candidate-vaccines. Of the candidates in development as of June 29, 2020, 132 are in the preclinical stage, while 17 vaccines are under clinical evaluation (WHO 2020a).

A human adenovirus type 5 (Ad5) vectored vaccine candidate became the first about which clinical phase 1 test results were reported (Zhu et al. 2020). This clinical trial tested three single doses of recombinant Ad5 expressing full-length SARS-CoV-2 spike glycoprotein (5×10^{10}, 1×10^{11}, and 1.5×10^{11} viral particles) in healthy volunteers previously not exposed to SARS-CoV-2 and assessed safety as well as humoral and cellular immune responses. The vaccine seemed to be well-tolerated, causing mostly mild or moderate adverse events of transient nature. Also, all dose groups developed binding antibodies against the RBD of S glycoprotein, while the level of neutralizing antibodies and interferon-gamma T-cell responses was highest in the high dose cohort. Preexisting immunity to the vector, however, seemed to interfere with the generation of immune responses, a phenomenon that has been a topic of debate in previous Ad5 vectored vaccine candidates and will undoubtedly need to be a focus of further investigation in the currently ongoing phase 2 trial. Another prominent Ad-vectored candidate is the rAd26 SARS-CoV-2 vaccine pursued by Janssen and Janssen, which is currently in the preclinical phase of development. The University of Oxford's Jenner Institute is meanwhile pursuing a nonhuman adenovirus vector for SARS-CoV-2 vaccine development, namely, ChAdOx1, a chimpanzee adenovirus, which is currently in the phase 2/3 clinical stage (WHO 2020a). In mice and NHP, this vaccine was able to elicit humoral and cellular responses against the SARS-CoV-2 spike encoded in the vector. However, in the NHP model that was used to assess efficacy, immunity was not sterile, and vaccinated animals did show symptoms after challenge, albeit those were milder than in mock-vaccinated animals (currently published as a preprint (van Doremalen et al. 2020)).

mRNA vaccines have likewise proceeded to clinical trials. Moderna's mRNA-1273, the first vaccine candidate to have entered clinical trials, is a lipid nanoparticle-encapsulated mRNA encoding the full-length stabilized prefusion spike, which is applied in a prime-boost manner and is currently in phase 2 development (WHO 2020a). Other anti-SARS-CoV-2 mRNA vaccines include BioNTech's/Pfizer's candidate BNT162b2, which is being tested in a phase 1/2 study that is, notably, randomized and placebo-controlled, and compares a prime-only regimen with the more common prime-boost approach. For both companies' mRNA-based vaccine platforms, this is the first time of progressing to clinical stages of vaccine development. Many other mRNA vaccines are in the preclinical stages of vaccine development, among them CureVac's mRNA vaccine candidate, expected to proceed to clinical trials within months. The latter is the only mRNA vaccine platform that has, albeit against another pathogen, completed a phase 1 study and whose data was published (Alberer et al. 2017). As a rabies vaccine, this platform was tested in multiple vaccination regimens and forms of application. It was reported to be generally safe. However, reactogenicity was high, and severe adverse events such as high fever, myalgia, and fatigue did occur in individual cases across dose cohorts. The vaccination appeared to be more immunogenic when given intradermally or with an intramuscular needle-free device, and less mRNA was needed to induce neutralizing antibody titers that are commonly regarded as protective against rabies.

DNA vaccines complement their mRNA counterparts in the field of nucleic acid vectors. To date, Inovio's INO-4800, which is administered via intradermal injection and electroporation, has entered clinical trials. A preclinical study of this candidate in mice and guinea pigs shows promising results with regard to immunogenicity (Smith et al. 2020). Inovio's platform has so far been assessed in a phase 1 clinical trial involving an anti-MERS-CoV vaccine (Modjarrad et al. 2019). This dose escalation study was performed as a prime-boost-boost regimen, and the vaccine demonstrated a favorable safety profile. However, not all of the participants seroconverted, irrespective of dosage. The limited immunogenicity of DNA vaccines remains a concern. Encouragingly, SARS-CoV-2 DNA vaccine candidates studied in rhesus macaques showed promising results. Here, a DNA vaccine expressing the RBD was able to induce antibody and T-cell responses that reduced viral load in the respiratory tract and ameliorated disease (Yu et al. 2020). While sterilizing immunity may not have been achieved in the NHP, the neutralizing antibodies that were induced after vaccination were of comparable magnitude to those observed in convalescent macaques after SARS-CoV-2 infection (Chandrashekar et al. 2020).

In general, both mRNA as well as DNA vaccine candidates have the advantage that they may be produced in large quantities in a short manner of time.

As the third major pole in vaccine platforms, inactivated candidates play a significant role, as they are relatively easy to manufacture, and production capacities for inactivated virus vaccines are distributed worldwide. Currently, multiple inactivated SARS-CoV-2 vaccines, with or without the addition of adjuvants, are under clinical evaluation in phase 1 and 2 trials (WHO 2020a). So far, Sinovac has reported the first preclinical results of its candidate (Gao et al. 2020). The company obtained inactivated SARS-CoV-2 by passaging the virus multiple times, propagating it in cell culture, and inactivating it with propriolactone. Efficacy of the vaccine was assessed using different doses and schedules, with and without adjuvant; it was immunogenic in mice, rats, and NHP. Specifically, vaccination with the highest dose investigated in this study prevented viral replication in NHP, and, while the number of macaques tested in this study was limited, no ADE ($n = 4$), obvious clinical symptoms, or notable histopathological findings ($n = 20$) were noted after vaccination.

31.5 Conclusion

Vaccines are undoubtedly one of the enormous contributions to human health. The history of vaccination spans over hundreds of years. Even though in the eighteenth century, the core foundation of medical understanding differed from today, the observation that inoculation of animals and humans could protect them from smallpox was so striking that it was widely used, even undertaking the challenging endeavor of the elimination of the disease in the early nineteenth century. The first vaccines were based on whole pathogens, but as knowledge and medical understanding grew, more sophisticated methods were developed to create vaccines that offered advantages such as better safety profiles, more accessible means of production, or better immune responses.

Vaccines traditionally took years, if not decades, to develop, so they were naturally unqualified as a tool for outbreak response, i.e., the acute fight against novel or emerging pathogens. In the last decades, global travel and a growing world population have enabled pathogens to cause outbreaks and pandemics at an unprecedented magnitude, and the world vowed to learn from events such as the West African Ebola epidemic in 2014–2016. One lesson learned was that rapidly producible vaccines as tools in an outbreak scenario are urgently needed. Novel technologies such as RNA-based or vector-based vaccines hold promise to make vaccine platforms available that could rapidly be adapted to new pathogens.

The question arises, which pathogen is going to cause the next pandemic. On the one hand, certain known pathogens that have the potential for global spread can and have been identified, and efforts to develop diagnostics, therapeutics, and vaccines against these are ongoing. On the other hand, it is vitally important also to develop strategies that will enable us to combat yet unknown emerging pathogens that might cause what WHO has termed "Disease X". The first Disease X of this millennium has turned out to be COVID-19, even though it technically belongs to a class of already known viruses.

To develop a vaccine to protect against SARS-CoV-2, the causative agent of COVID-19, we can learn from research on its closely related relatives, SARS-CoV-1 and MERS-CoV, which have allowed the identification of target structures and development of animal models that recapitulate human disease. However, significant challenges remain. We need to understand better what constitutes immunity against the virus, how this immunity changes over time, and whether we have to expect phenomena of immune interaction.

In the COVID-19 pandemic, an unprecedented collective effort has been mobilized by the scientific community as well as stakeholders in pharmaceutical, philanthropic, and governmental organizations.

The number of vaccine candidates that have entered clinical trials as of the writing of this chapter will undoubtedly already be outdated once it is published, but some of these promising early candidates will likely progress to later stages of development and might even achieve licensure. While inactivated whole virus vaccines would offer a simple production procedure that could make use of existing facilities worldwide, novel vaccine strategies offer other advantages. mRNA-based and DNA-based vaccine candidates promise rapid production, and viral vector vaccines promise to induce a strong and hopefully lasting immunity. Despite promising progress in SARS-CoV-2 vaccine development, it must be acknowledged that traditionally, the development of vaccines against respiratory viruses has been challenging. Finally, there are no guarantees that the development and licensure of a safe and effective vaccine will be achieved swiftly or – for that matter – at all.

There are still many unknowns in the field of SARS-CoV-2 vaccine research, while, at the same time, the scientific community is generating new information at unprecedented speed. This rapidly generated knowledge will need to be incorporated into ongoing strategic and rational approaches to vaccine design and distribution. To achieve the goal of creating a SARS-CoV-2 vaccine, the distinct aspects of the SARS-CoV-2 vaccine design discussed in this chapter need to be taken into account.

References

Agrawal AS, Tao X, Algaissi A, Garron T, Narayanan K, Peng B-H, Couch RB, Tseng C-TK (2016) Immunization with inactivated Middle East Respiratory Syndrome coronavirus vaccine leads to lung immunopathology on challenge with live virus. Hum Vaccin Immunother 12(9):2351–2356. https://doi.org/10.1080/21645515.2016.1177688

Aguilar JC, Rodríguez EG (2007) Vaccine adjuvants revisited. Vaccine 25(19):3752–3762. https://doi.org/10.1016/j.vaccine.2007.01.111

Alberer M, Gnad-Vogt U, Hong HS, Mehr KT, Backert L, Finak G, Gottardo R, Bica MA, Garofano A, Koch SD (2017) Safety and immunogenicity of a mRNA rabies vaccine in healthy adults: an open-label, non-randomised, prospective, first-in-human phase 1 clinical trial. Lancet 390(10101):1511–1520

Barquet N (1997) Smallpox: the triumph over the most terrible of the ministers of death. Ann Intern Med 127(8_Part_1):635. https://doi.org/10.7326/0003--4819-127-8_Part_1-199710150-00010

Belshe Robert B, Gruber William C, Mendelman Paul M, Mehta Harshvardhan B, Mahmood K, Reisinger K, Treanor J, Zangwill K, Hayden Frederick G, Bernstein David I, Kotloff K, King J, Piedra Pedro A, Block Stan L, Yan L, Wolff M (2000) Correlates of immune protection induced by live, attenuated, cold-adapted, trivalent, intranasal influenza virus vaccine. J Infect Dis 181(3):1133–1137. https://doi.org/10.1086/315323

Björkholm B, Böttiger M, Christenson B, Hagberg L (1986) Antitoxin antibody levels and the outcome of illness during an outbreak of diphtheria among alcoholics. Scand J Infect Dis 18(3):235–239

Blume S (2000) Essay on science and society:a brief history of polio vaccines. Science 288(5471):1593–1594. https://doi.org/10.1126/science.288.5471.1593

Bogers WM, Oostermeijer H, Mooij P, Koopman G, Verschoor EJ, Davis D, Ulmer JB, Brito LA, Cu Y, Banerjee K (2015) Potent immune responses in rhesus macaques induced by nonviral delivery of a self-amplifying RNA vaccine expressing HIV type 1 envelope with a cationic nanoemulsion. J Infect Dis 211(6):947–955

Borrow R, Balmer P, Miller E (2005) Meningococcal surrogates of protection?Serum bactericidal antibody activity. Vaccine 23(17–18):2222–2227. https://doi.org/10.1016/j.vaccine.2005.01.051

Bradburne AF, Somerset BA (1972) Coronavirus antibody titres in sera of healthy adults and experimentally infected volunteers. Epidemiol Infect 70(2):235–244

Buchbinder SP, Mehrotra DV, Duerr A, Fitzgerald DW, Mogg R, Li D, Gilbert PB, Lama JR, Marmor M, del Rio C (2008) Efficacy assessment of a cell-mediated immunity HIV-1 vaccine (the Step Study): a double-blind, randomised, placebo-controlled, test-of-concept trial. Lancet 372(9653):1881–1893

Burki T (2017) CEPI: preparing for the worst. Lancet Infect Dis 17(3):265–266. https://doi.org/10.1016/S1473-3099(17)30062-2

Callow KA, Parry HF, Sergeant M, Tyrrell DAJ (1990) The time course of the immune response to experimental coronavirus infection of man. Epidemiol Infect 105(2):435–446

Castagnoli R, Votto M, Licari A, Brambilla I, Bruno R, Perlini S, Rovida F, Baldanti F, Marseglia GL (2020) Severe acute respiratory syndrome coronavirus 2 (SARS-CoV-2) infection in children and adolescents: a systematic review. JAMA Pediatr

Cate TR, Couch RB, Kasel JA, Six HR (1977) Clinical trials of monovalent influenza A/New Jersey/76 virus vaccines in adults: reactogenicity, antibody response, and antibody persistence. J Infect Dis 136(Suppl_3):S450–S455

CEPI (2020) research potfolio. https://cepi.net/research_dev/our-portfolio/. Accessed 22 May 2020

Chandrashekar A, Liu J, Martinot AJ, McMahan K, Mercado NB, Peter L, Tostanoski LH, Yu J, Maliga Z, Nekorchuk M (2020) SARS-CoV-2 infection protects against rechallenge in rhesus macaques. Science

Che X, Qiu L, Liao Z, Wang Y, Wen K, Pan Y, Hao W, Mei Y, Cheng Vincent CC, Yuen K (2005) Antigenic cross-reactivity between severe acute respiratory syndrome–associated coronavirus and human coronaviruses

229E and OC43. J Infect Dis 191(12):2033–2037. https://doi.org/10.1086/430355

clinicaltrials.gov (2011) Phase I dose escalation SARS-CoV recombinant S protein, with and without adjuvant, vaccine study. https://clinicaltrials.gov/ct2/show/NCT01376765. Accessed 09 June 2020

Condon C, Watkins SC, Celluzzi CM, Thompson K, Falo LD (1996) DNA–based immunization by in vivo transfection of dendritic cells. Nat Med 2(10):1122–1128. https://doi.org/10.1038/nm1096-1122

Dejnirattisai W, Jumnainsong A, Onsirisakul N, Fitton P, Vasanawathana S, Limpitikul W, Puttikhunt C, Edwards C, Duangchinda T, Supasa S, Chawansuntati K, Malasit P, Mongkolsapaya J, Screaton G (2010) Cross-reacting antibodies enhance dengue virus infection in humans. Science 328(5979):745–748. https://doi.org/10.1126/science.1185181

Du L, Tai W, Zhou Y, Jiang S (2016) Vaccines for the prevention against the threat of MERS-CoV. Expert Rev Vaccines 15(9):1123–1134. https://doi.org/10.1586/14760584.2016.1167603

Edsall G (1959) Specific prophylaxis of tetanus. J Am Med Assoc 171(4):417–427

Edwards KM, Karzon DT (1990) Pertussis vaccines. Pediatr Clin N Am 37(3):549–566

Erasmus JH, Khandhar AP, Walls AC, Hemann EA, O'Connor MA, Murapa P, Archer J, Leventhal S, Fuller J, Lewis T, Draves KE, Randall S, Guerriero KA, Duthie MS, Carter D, Reed SG, Hawman DW, Feldmann H, Gale M Jr, Veesler D, Berglund P, Fuller DH (2020) Single-dose replicating RNA vaccine induces neutralizing antibodies against SARS-CoV-2 in nonhuman primates. BioRxiv (preprint).https://doi.org/10.1101/2020.05.28.121640

Ertl HCJ (2009) Novel vaccines to human rabies. PLoS Negl Trop Dis 3(9):e515. https://doi.org/10.1371/journal.pntd.0000515

Fathi A, Dahlke C, Addo MM (2019) Recombinant vesicular stomatitis virus vector vaccines for WHO blueprint priority pathogens. Hum Vaccin Immunother 15(10):2269–2285

Fleeton MN, Chen M, Berglund P, Rhodes G, Parker SE, Murphy M, Atkins GJ, Liljeström P (2001) Self-replicative RNA vaccines elicit protection against influenza A virus, respiratory syncytial virus, and a tickborne encephalitis virus. J Infect Dis 183(9):1395–1398

Folegatti PM, Bellamy D, Flaxman A, Mair C, Ellis C, Ramon RL, Ramos Lopez F, Mitton C, Baker M, Poulton I, Lawrie A, Roberts R, Minassian A, Ewer KJ, Evans TG, Hill AVS, Gilbert SC (2019) Safety and immunogenicity of the heterosubtypic Influenza A vaccine MVA-NP+M1 manufactured on the AGE1.CR.pIX avian cell line. Vaccine 7(1):33. https://doi.org/10.3390/vaccines7010033

Folegatti PM, Bittaye M, Flaxman A, Lopez FR, Bellamy D, Kupke A, Mair C, Makinson R, Sheridan J, Rohde C, Halwe S, Jeong Y, Park Y-S, Kim J-O, Song M, Boyd A, Tran N, Silman D, Poulton I, Datoo M, Marshal J, Themistocleous Y, Lawrie A, Roberts R, Berrie E, Becker S, Lambe T, Hill A, Ewer K, Gilbert S (2020) Safety and immunogenicity of a candidate Middle East respiratory syndrome coronavirus viral-vectored vaccine: a dose-escalation, open-label, non-randomised, uncontrolled, phase 1 trial. Lancet Infect Dis:S1473309920301602. https://doi.org/10.1016/S1473-3099(20)30160-2

Frey SE, Winokur PL, Salata RA, El-Kamary SS, Turley CB, Walter EB, Hay CM, Newman FK, Hill HR, Zhang Y, Chaplin P, Tary-Lehmann M, Belshe RB (2013) Safety and immunogenicity of IMVAMUNE® smallpox vaccine using different strategies for a post event scenario. Vaccine 31(29):3025–3033. https://doi.org/10.1016/j.vaccine.2013.04.050

Fulginiti V, Eller J, Sieber O, Joyner J, Minamitani M, Meiklejohn G (1969) Respiratory virus immunization: a field trial of two inactivated respiratory virus vaccines; an aqueous trivalent parainfluenza virus vaccine and an alum-precipitated respiratory syncytial virus vaccine. Am J Epidemiol 89(4):435–448

Gao W, Tamin A, Soloff A, D'Aiuto L, Nwanegbo E, Robbins PD, Bellini WJ, Barratt-Boyes S, Gambotto A (2003) Effects of a SARS-associated coronavirus vaccine in monkeys. Lancet 362(9399):1895–1896

Gao Q, Bao L, Mao H, Wang L, Xu K, Yang M, Li Y, Zhu L, Wang N, Lv Z (2020) Development of an inactivated vaccine candidate for SARS-CoV-2. Science

Gauger PC, Vincent AL, Loving CL, Lager KM, Janke BH, Kehrli ME Jr, Roth JA (2011) Enhanced pneumonia and disease in pigs vaccinated with an inactivated human-like (δ-cluster) H1N2 vaccine and challenged with pandemic 2009 H1N1 influenza virus. Vaccine 29(15):2712–2719

Geall AJ, Verma A, Otten GR, Shaw CA, Hekele A, Banerjee K, Cu Y, Beard CW, Brito LA, Krucker T (2012) Nonviral delivery of self-amplifying RNA vaccines. Proc Natl Acad Sci 109(36):14604–14609

Geisbert TW, Feldmann H (2011) Recombinant vesicular stomatitis virus–based vaccines against ebola and marburg virus infections. J Infect Dis 204(Suppl_3):S1075–S1081. https://doi.org/10.1093/infdis/jir349

Gerlach C, Moseman EA, Loughhead SM, Alvarez D, Zwijnenburg AJ, Waanders L, Garg R, Juan C, von Andrian UH (2016) The chemokine receptor CX3CR1 defines three antigen-experienced CD8 T cell subsets with distinct roles in immune surveillance and homeostasis. Immunity 45(6):1270–1284

Gilbert PB, Luedtke AR (2018) Statistical learning methods to determine immune correlates of herpes zoster in vaccine efficacy trials. J Infect Dis 218(Suppl_2):S99–S101. https://doi.org/10.1093/infdis/jiy421

Girard MP, Tam JS, Assossou OM, Kieny MP (2010) The 2009 A (H1N1) influenza virus pandemic: a review. Vaccine 28(31):4895–4902

Glenny A, Hopkins BE (1923) Diphtheria toxoid as an immunising agent. Br J Exp Pathol 4(5):283

Gorbalenya AE, Baker SC, Baric RS, de Groot RJ, Drosten C, Gulyaeva AA, Haagmans BL, Lauber C, Leontovich AM, Neuman BW, Penzar D, Perlman S,

LLM P, Samborskiy DV, Sidorov IA, Sola I, Ziebuhr J, Coronaviridae Study Group of the International Committee on Taxonomy of V (2020) The species severe acute respiratory syndrome-related coronavirus: classifying 2019-nCoV and naming it SARS-CoV-2. Nat Microbiol 5(4):536–544. https://doi.org/10.1038/s41564-020-0695-z

Gotschlich EC, Liu TY, Artenstein MS (1969) Human immunity to the meningococcus: III. Preparation and immunochemical properties of the group A, group B, and group C meningococcal polysaccharides. J Exp Med 129(6):1349–1365

Graham BS (2020) Rapid COVID-19 vaccine development. Science 368(6494):945–946

Gretebeck LM, Subbarao K (2015) Animal models for SARS and MERS coronaviruses. Curr Opin Virol 13:123–129. https://doi.org/10.1016/j.coviro.2015.06.009

Grunwald T, Ulbert S (2015) Improvement of DNA vaccination by adjuvants and sophisticated delivery devices: vaccine-platforms for the battle against infectious diseases. Clin Exp Vaccin Res 4(1):1. https://doi.org/10.7774/cevr.2015.4.1.1

Gu H, Chen Q, Yang G, He L, Fan H, Deng Y-Q, Wang Y, Teng Y, Zhao Z, Cui Y (2020) Rapid adaptation of SARS-CoV-2 in BALB/c mice: Novel mouse model for vaccine efficacy. bioRxiv

Guy B, Barrere B, Malinowski C, Saville M, Teyssou R, Lang J (2011) From research to phase III: preclinical, industrial and clinical development of the Sanofi Pasteur tetravalent dengue vaccine. Vaccine 29(42):7229–7241. https://doi.org/10.1016/j.vaccine.2011.06.094

Guzman MG, Vazquez S (2010) The complexity of antibody-dependent enhancement of dengue virus infection. Viruses 2(12):2649–2662. https://doi.org/10.3390/v2122649

Haagmans BL, van den Brand JMA, Raj VS, Volz A, Wohlsein P, Smits SL, Schipper D, Bestebroer TM, Okba N, Fux R, Bensaid A, Solanes Foz D, Kuiken T, Baumgartner W, Segales J, Sutter G, Osterhaus ADME (2016) An orthopoxvirus-based vaccine reduces virus excretion after MERS-CoV infection in dromedary camels. Science 351(6268):77–81. https://doi.org/10.1126/science.aad1283

Halstead SB (2003) Neutralization and antibody-dependent enhancement of dengue viruses. Adv Virus Res 60:421–467. https://doi.org/10.1016/s0065-3527(03)60011-4

Hannoun C (2013) The evolving history of influenza viruses and influenza vaccines. Expert Rev Vaccines 12(9):1085–1094. https://doi.org/10.1586/14760584.2013.824709

He Y, Zhou Y, Liu S, Kou Z, Li W, Farzan M, Jiang S (2004) Receptor-binding domain of SARS-CoV spike protein induces highly potent neutralizing antibodies: implication for developing subunit vaccine. Biochem Biophys Res Commun 324(2):773–781. https://doi.org/10.1016/j.bbrc.2004.09.106

Heidelberger M, MacLeod CM, Di Lapi MM (1948) The human antibody response to simultaneous injection of six specific polysaccharides of pneumococcus. J Exp Med 88(3):369–372

Hekele A, Bertholet S, Archer J, Gibson DG, Palladino G, Brito LA, Otten GR, Brazzoli M, Buccato S, Bonci A, Casini D, Maione D, Qi Z-Q, Gill JE, Caiazza NC, Urano J, Hubby B, Gao GF, Shu Y, De Gregorio E, Mandl CW, Mason PW, Settembre EC, Ulmer JB, Craig Venter J, Dormitzer PR, Rappuoli R, Geall AJ (2013) Rapidly produced SAM® vaccine against H7N9 influenza is immunogenic in mice. Emerg Microbes Infect 2(1):1–7. https://doi.org/10.1038/emi.2013.54

Henao-Restrepo AM, Camacho A, Longini IM, Watson CH, Edmunds WJ, Egger M, Carroll MW, Dean NE, Diatta I, Doumbia M, Draguez B, Duraffour S, Enwere G, Grais R, Gunther S, Gsell P-S, Hossmann S, Watle SV, Kondé MK, Kéïta S, Kone S, Kuisma E, Levine MM, Mandal S, Mauget T, Norheim G, Riveros X, Soumah A, Trelle S, Vicari AS, Røttingen J-A, Kieny M-P (2017) Efficacy and effectiveness of an rVSV-vectored vaccine in preventing Ebola virus disease: final results from the Guinea ring vaccination, open-label, cluster-randomised trial (Ebola Ça Suffit!). Lancet 389(10068):505–518. https://doi.org/10.1016/S0140-6736(16)32621-6

Hilleman MR, Weibel RE, Buynak EB, Stokes J Jr, Whitman JE Jr (1967) Live, attenuated mumps-virus vaccine: protective efficacy as measured in a field evaluation. N Engl J Med 276(5):252–258

Hilleman MR, McAleer WJ, Buynak EB, McLean AA (1983) The preparation and safety of hepatitis B vaccine. J Infect 7:3–8. https://doi.org/10.1016/S0163-4453(83)96465-4

Hohdatsu T, Yamada M, Tominaga R, Makino K, Kida K, Koyama H (1998) Antibody-dependent enhancement of feline infectious peritonitis virus infection in feline alveolar macrophages and human monocyte cell line U937 by serum of cats experimentally or naturally infected with feline coronavirus. J Vet Med Sci 60(1):49–55. https://doi.org/10.1292/jvms.60.49

Honda-Okubo Y, Barnard D, Ong CH, Peng B-H, Tseng C-TK, Petrovsky N (2015) Severe acute respiratory syndrome-associated coronavirus vaccines formulated with delta inulin adjuvants provide enhanced protection while ameliorating lung eosinophilic immunopathology. J Virol 89(6):2995–3007. https://doi.org/10.1128/JVI.02980-14

Houser KV, Broadbent AJ, Gretebeck L, Vogel L, Lamirande EW, Sutton T, Bock KW, Minai M, Orandle M, Moore IN, Subbarao K (2017) Enhanced inflammation in New Zealand white rabbits when MERS-CoV reinfection occurs in the absence of neutralizing antibody. PLoS Pathog 13(8):e1006565. https://doi.org/10.1371/journal.ppat.1006565

ICTV (2020) International Committee on Taxonomy of Viruses – Virus Taxonomy: 2019 Release. EC 51, Berlin, Germany, July 2019, Email ratification March

2020 (MSL #35). https://talk.ictvonline.org/taxonomy/. Accessed 10 June 2020

Jackson DA, Symons RH, Berg P (1972) Biochemical method for inserting new genetic information into DNA of simian virus 40: circular SV40 DNA molecules containing lambda phage genes and the galactose operon of Escherichia coli. Proc Natl Acad Sci 69(10):2904–2909. https://doi.org/10.1073/pnas.69.10.2904

Jameson SC, Masopust D (2018) Understanding subset diversity in T cell memory. Immunity 48(2):214–226. https://doi.org/10.1016/j.immuni.2018.02.010

Jaume M, Yip MS, Cheung CY, Leung HL, Li PH, Kien F, Dutry I, Callendret B, Escriou N, Altmeyer R, Nal B, Daeron M, Bruzzone R, Peiris JSM (2011) Anti-severe acute respiratory syndrome coronavirus spike antibodies trigger infection of human immune cells via a pH- and cysteine protease-independent Fc R pathway. J Virol 85(20):10582–10597. https://doi.org/10.1128/JVI.00671-11

Jiang S, He Y, Liu S (2005) SARS vaccine development. Emerg Infect Dis 11(7):1016

Johnson KM, Peralta PH, Vogel JE (1966) Clinical and serological response to laboratory-acquired human infection by Indiana type vesicular stomatitis virus (VSV). Am J Trop Med Hyg 15(2):244–246. https://doi.org/10.4269/ajtmh.1966.15.244

Kareko BW, Booty BL, Nix CD, Lyski ZL, Slifka MK, Amanna IJ, Messer WB (2019) Persistence of neutralizing antibody responses among yellow fever virus 17D vaccinees living in a nonendemic setting. J Infect Dis. https://doi.org/10.1093/infdis/jiz374

Katz SL, Kempe CH, Black FL, Lepow ML, Krugman S, Haggerty RJ, Enders JF (1960) Studies on an attenuated measles-virus vaccine: VIII. General summary and evaluation of the results of vaccination. Am J Dis Child 100(6):942–946

Khandia R, Munjal A, Dhama K, Karthik K, Tiwari R, Malik YS, Singh RK, Chaicumpa W (2018) Modulation of dengue/zika virus pathogenicity by antibody-dependent enhancement and strategies to protect against enhancement in zika virus infection. Front Immunol 9:597. https://doi.org/10.3389/fimmu.2018.00597

Kim Y-I, Kim S-G, Kim S-M, Kim E-H, Park S-J, Yu K-M, Chang J-H, Kim EJ, Lee S, Casel MAB, Um J, Song M-S, Jeong HW, Lai VD, Kim Y, Chin BS, Park J-S, Chung K-H, Foo S-S, Poo H, Mo I-P, Lee O-J, Webby RJ, Jung JU, Choi YK (2020) Infection and rapid transmission of SARS-CoV-2 in ferrets. Cell Host Microbe 27(5):704–709.e702. https://doi.org/10.1016/j.chom.2020.03.023

Kirnbauer R, Booy F, Cheng N, Lowy DR, Schiller JT (1992) Papillomavirus L1 major capsid protein self-assembles into virus-like particles that are highly immunogenic. Proc Natl Acad Sci 89(24):12180–12184

Koch T, Dahlke C, Fathi A, Kupke A, Krähling V, Okba NMA, Halwe S, Rohde C, Eickmann M, Volz A, Hesterkamp T, Jambrecina A, Borregaard S, Ly ML, Zinser ME, Bartels E, Poetsch JSH, Neumann R, Fux R, Schmiedel S, Lohse AW, Haagmans BL, Sutter G, Becker S, Addo MM (2020) Safety and immunogenicity of a modified vaccinia virus Ankara vector vaccine candidate for Middle East respiratory syndrome: an open-label, phase 1 trial. Lancet Infect Dis:S1473309920302486. https://doi.org/10.1016/S1473-3099(20)30248-6

Korber B, Fischer W, Gnanakaran SG, Yoon H, Theiler J, Abfalterer W, Foley B, Giorgi EE, Bhattacharya T, Parker MD (2020) Spike mutation pipeline reveals the emergence of a more transmissible form of SARS-CoV-2. bioRxiv

Kretschmer K, Apostolou I, Hawiger D, Khazaie K, Nussenzweig MC, von Boehmer H (2005) Inducing and expanding regulatory T cell populations by foreign antigen. Nat Immunol 6(12):1219–1227

Kurosaki T, Kometani K, Ise W (2015) Memory B cells. Nat Rev Immunol 15(3):149–159

Lan J, Ge J, Yu J, Shan S, Zhou H, Fan S, Zhang Q, Shi X, Wang Q, Zhang L (2020) Structure of the SARS-CoV-2 spike receptor-binding domain bound to the ACE2 receptor. Nature:1–6

Landy M, Gaines S, Seal JR, Whiteside JE (1954) Antibody responses of man to three types of antityphoid immunizing agents: heat-phenol fluid vaccine, acetone-dehydrated vaccine, and isolated Vi and 0 antigens. Am J Public Health Nations Health 44(12):1572–1579

Leung AKC (2011) "Variolation" and vaccination in late imperial China, Ca 1570–1911. In: Plotkin SA (ed) History of vaccine development. Springer New York, New York, pp 5–12

Lewnard J, Cobey S (2018) Immune history and influenza vaccine effectiveness. Vaccine 6(2):28. https://doi.org/10.3390/vaccines6020028

Li T, Xie J, He Y, Fan H, Baril L, Qiu Z, Han Y, Xu W, Zhang W, You H, Zuo Y, Fang Q, Yu J, Chen Z, Zhang L (2006) Long-term persistence of robust antibody and cytotoxic T cell responses in recovered patients infected with SARS coronavirus. PLoS One 1(1):e24. https://doi.org/10.1371/journal.pone.0000024

Lin J-T, Zhang J-S, Su N, Xu J-G, Wang N, Chen J-T, Chen X, Liu Y-X, Gao H, Jia Y-P, Liu Y, Sun R-H, Wang X, Yu D-Z, Hai R, Gao Q, Ning Y, Wang H-X, Li M-C, Kan B, Dong G-M, An Q, Wang Y-Q, Han J, Qin C, Yin W-D, Dongs X-P (2007) Safety and immunogenicity from a phase I trial of inactivated severe acute respiratory syndrome coronavirus vaccine. Antivir Ther (Lond) 12(7):1107–1113

Liu W, Fontanet A, Zhang PH, Zhan L, Xin ZT, Baril L, Tang F, Lv H, Cao WC (2006) Two-year prospective study of the humoral immune response of patients with severe acute respiratory syndrome. J Infect Dis 193(6):792–795. https://doi.org/10.1086/500469

Liu WJ, Lan J, Liu K, Deng Y, Yao Y, Wu S, Chen H, Bao L, Zhang H, Zhao M, Wang Q, Han L, Chai Y, Qi J, Zhao J, Meng S, Qin C, Gao GF, Tan W (2017) Protective T cell responses featured by concordant recognition of middle east respiratory syndrome coronavirus–derived CD8 ⁺ T cell epitopes and host MHC. J Immunol 198(2):873–882. https://doi.org/10.4049/jimmunol.1601542

Lopez AL, Gonzales MLA, Aldaba JG, Nair GB (2014) Killed oral cholera vaccines: history, development and implementation challenges. Ther Adv Vaccines 2(5):123–136. https://doi.org/10.1177/2051013614537819

Mackett M, Smith GL, Moss B (1982) Vaccinia virus: a selectable eukaryotic cloning and expression vector. Proc Natl Acad Sci 79(23):7415–7419. https://doi.org/10.1073/pnas.79.23.7415

Martin JE, Louder MK, Holman LA, Gordon IJ, Enama ME, Larkin BD, Andrews CA, Vogel L, Koup RA, Roederer M, Bailer RT, Gomez PL, Nason M, Mascola JR, Nabel GJ, Graham BS (2008) A SARS DNA vaccine induces neutralizing antibody and cellular immune responses in healthy adults in a phase I clinical trial. Vaccine 26(50):6338–6343. https://doi.org/10.1016/j.vaccine.2008.09.026

Mayr A, Stickl H, Müller H, Danner K, Singer H (1978) The smallpox vaccination strain MVA: marker, genetic structure, experience gained with the parenteral vaccination and behaviour in organisms with a debilitated defence mechanism. Zentralblatt fur Bacteriologie, Parasitenkunde, Infektionskrankheiten und Hygiene 1 (5/6):375-390. PMID: 219640

Medaglini D, Santoro F, Siegrist C-A (2018) Correlates of vaccine-induced protective immunity against Ebola virus disease. Semin Immunol 39:65–72. https://doi.org/10.1016/j.smim.2018.07.003

Meyer M, Malherbe DC, Bukreyev A (2019) Can ebola virus vaccines have universal immune correlates of protection? Trends Microbiol 27(1):8–16. https://doi.org/10.1016/j.tim.2018.08.008

Modjarrad K, Roberts CC, Mills KT, Castellano AR, Paolino K, Muthumani K, Reuschel EL, Robb ML, Racine T, Oh M-d, Lamarre C, Zaidi FI, Boyer J, Kudchodkar SB, Jeong M, Darden JM, Park YK, Scott PT, Remigio C, Parikh AP, Wise MC, Patel A, Duperret EK, Kim KY, Choi H, White S, Bagarazzi M, May JM, Kane D, Lee H, Kobinger G, Michael NL, Weiner DB, Thomas SJ, Maslow JN (2019) Safety and immunogenicity of an anti-Middle East respiratory syndrome coronavirus DNA vaccine: a phase 1, open-label, single-arm, dose-escalation trial. Lancet Infect Dis. https://doi.org/10.1016/S1473-3099(19)30266-X

Morris SJ, Sebastian S, Spencer AJ, Gilbert SC (2016) Simian adenoviruses as vaccine vectors. Futur Virol 11(9):649–659. https://doi.org/10.2217/fvl-2016-0070

Munster VJ, Wells D, Lambe T, Wright D, Fischer RJ, Bushmaker T, Saturday G, van Doremalen N, Gilbert SC, de Wit E, Warimwe GM (2017) Protective efficacy of a novel simian adenovirus vaccine against lethal MERS-CoV challenge in a transgenic human DPP4 mouse model. npj Vaccines 2(1):28. https://doi.org/10.1038/s41541-017-0029-1

Murphy B, Prince G, Walsh E, Kim H, Parrott R, Hemming V, Rodriguez W, Chanock R (1986) Dissociation between serum neutralizing and glycoprotein antibody responses of infants and children who received inactivated respiratory syncytial virus vaccine. J Clin Microbiol 24(2):197–202

Nair H, Nokes DJ, Gessner BD, Dherani M, Madhi SA, Singleton RJ, O'Brien KL, Roca A, Wright PF, Bruce N (2010) Global burden of acute lower respiratory infections due to respiratory syncytial virus in young children: a systematic review and meta-analysis. Lancet 375(9725):1545–1555

Netland J, DeDiego ML, Zhao J, Fett C, Álvarez E, Nieto-Torres JL, Enjuanes L, Perlman S (2010) Immunization with an attenuated severe acute respiratory syndrome coronavirus deleted in E protein protects against lethal respiratory disease. Virology 399(1):120–128. https://doi.org/10.1016/j.virol.2010.01.004

Newell EW, Cheng Y (2016) Mass cytometry: blessed with the curse of dimensionality. Nat Immunol 17(8):890

Nichols WW, Ledwith BJ, Manam SV, Troilo PJ (1995) Potential DNA vaccine integration into host cell genome. Ann N Y Acad Sci 772(1 DNA Vaccines):30–39. https://doi.org/10.1111/j.1749-6632.1995.tb44729.x

Openshaw PJ, Culley FJ, Olszewska W (2001) Immunopathogenesis of vaccine-enhanced RSV disease. Vaccine 20:S27–S31

Padron-Regalado E (2020) Vaccines for SARS-CoV-2: lessons from other coronavirus strains

Pasteur L (1885) Méthode pour prévenir la rage après morsure

Payne DC, Iblan I, Rha B, Alqasrawi S, Haddadin A, Al Nsour M, Alsanouri T, Ali SS, Harcourt J, Miao C, Tamin A, Gerber SI, Haynes LM, Al Abdallat MM (2016) Persistence of antibodies against middle east respiratory syndrome coronavirus. Emerg Infect Dis 22(10):1824–1826. https://doi.org/10.3201/eid2210.160706

Petsch B, Schnee M, Vogel AB, Lange E, Hoffmann B, Voss D, Schlake T, Thess A, Kallen K-J, Stitz L, Kramps T (2012) Protective efficacy of in vitro synthesized, specific mRNA vaccines against influenza A virus infection. Nat Biotechnol 30(12):1210–1216. https://doi.org/10.1038/nbt.2436

Pierson TC, Fremont DH, Kuhn RJ, Diamond MS (2008) Structural insights into the mechanisms of antibody-mediated neutralization of flavivirus infection: implications for vaccine development. Cell Host Microbe 4(3):229–238

Plotkin SA (2013) Complex correlates of protection after vaccination. Clin Infect Dis 56(10):1458–1465. https://doi.org/10.1093/cid/cit048

Plotkin SA, Gilbert PB (2012) Nomenclature for immune correlates of protection after vaccination. Clin Infect Dis 54(11):1615–1617. https://doi.org/10.1093/cid/cis238

Plotkin SA, Farquhar JD, Katz M, Buser F (1969) Attenuation of RA 27/3 rubella virus in WI-38 human diploid cells. Am J Dis Child 118(2):178–185

Purcell AW, McCluskey J, Rossjohn J (2007) More than one reason to rethink the use of peptides in vaccine design. Nat Rev Drug Discov 6(5):404–414. https://doi.org/10.1038/nrd2224

Qiu H, Wu J, Hong L, Luo Y, Song Q, Chen D (2020) Clinical and epidemiological features of 36 children with coronavirus disease 2019 (COVID-19) in Zhejiang, China: an observational cohort study. Lancet Infect Dis

Rappuoli R (2000) Reverse vaccinology. Curr Opin Microbiol 3(5):445–450. https://doi.org/10.1016/S1369-5274(00)00119-3

Rossey I, Saelens X (2019) Vaccines against human respiratory syncytial virus in clinical trials, where are we now? Expert Rev Vaccines 18(10):1053–1067

Schneerson R, Barrera O, Sutton A, Robbins JB (1980) Preparation, characterization, and immunogenicity of Haemophilus influenzae type b polysaccharide-protein conjugates. J Exp Med 152(2):361–376

Sekaly R-P (2008) The failed HIV Merck vaccine study: a step back or a launching point for future vaccine development? J Exp Med 205(1):7–12. https://doi.org/10.1084/jem.20072681

Shi J, Wen Z, Zhong G, Yang H, Wang C, Huang B, Liu R, He X, Shuai L, Sun Z, Zhao Y, Liu P, Liang L, Cui P, Wang J, Zhang X, Guan Y, Tan W, Wu G, Chen H, Bu Z (2020) Susceptibility of ferrets, cats, dogs, and other domesticated animals to SARS–coronavirus 2. Science:eabb7015. https://doi.org/10.1126/science.abb7015

Sia SF, Yan L-M, Chin AW, Fung K, Choy K-T, Wong AY, Kaewpreedee P, Perera RA, Poon LL, Nicholls JM (2020) Pathogenesis and transmission of SARS-CoV-2 in golden hamsters. Nature:1–7

Silva ML, Espírito-Santo LR, Martins MA, Silveira-Lemos D, Peruhype-Magalhães V, Caminha RC, de Andrade Maranhão-Filho P, Auxiliadora-Martins M, de Menezes Martins R, Galler R, da Silva Freire M, Marcovistz R, Homma A, Teuwen DE, Elói-Santos SM, Andrade MC, Teixeira-Carvalho A, Martins-Filho OA (2010) Clinical and immunological insights on severe, adverse neurotropic and viscerotropic disease following 17D yellow fever vaccination. Clin Vaccine Immunol 17(1):118–126. https://doi.org/10.1128/CVI.00369-09

Smith AJ, Wietgrefe SW, Shang L, Reilly CS, Southern PJ, Perkey KE, Duan L, Kohler H, Müller S, Robinson J, Carlis JV, Li Q, Johnson RP, Haase AT (2014) Live simian immunodeficiency virus vaccine correlate of protection: immune complex–inhibitory fc receptor interactions that reduce target cell availability. J Immunol 193(6):3126–3133. https://doi.org/10.4049/jimmunol.1400822

Smith TR, Patel A, Ramos S, Elwood D, Zhu X, Yan J, Gary EN, Walker SN, Schultheis K, Purwar M (2020) Immunogenicity of a DNA vaccine candidate for COVID-19. Nat Commun 11(1):1–13

Song Z, Xu Y, Bao L, Zhang L, Yu P, Qu Y, Zhu H, Zhao W, Han Y, Qin C (2019) From SARS to MERS, thrusting coronaviruses into the spotlight. Viruses 11(1):59

Soto-Perez-de-Celis E (2008) The royal philanthropic expedition of the vaccine: a landmark in the history of public health. Postgrad Med J 84(997):599–602. https://doi.org/10.1136/pgmj.2008.069450

Sridhar S, Luedtke A, Langevin E, Zhu M, Bonaparte M, Machabert T, Savarino S, Zambrano B, Moureau A, Khromava A, Moodie Z, Westling T, Mascareñas C, Frago C, Cortés M, Chansinghakul D, Noriega F, Bouckenooghe A, Chen J, Ng S-P, Gilbert PB, Gurunathan S, DiazGranados CA (2018) Effect of dengue serostatus on dengue vaccine safety and efficacy. N Engl J Med 379(4):327–340. https://doi.org/10.1056/NEJMoa1800820

Steinbrook R (2007) One step forward, two steps back – will there ever be an AIDS vaccine? N Engl J Med 357(26):2653–2655

Subbarao K, Roberts A (2006) Is there an ideal animal model for SARS? Trends Microbiol 14(7):299–303

Sutter G, Staib C (2003) Vaccinia vectors as candidate vaccines: the development of modified vaccinia virus ankara for antigen delivery. CDTID 3(3):263–271. https://doi.org/10.2174/1568005033481123

Sutton TC, Subbarao K (2015) Development of animal models against emerging coronaviruses: from SARS to MERS coronavirus. Virology 479–480:247–258. https://doi.org/10.1016/j.virol.2015.02.030

Tang D-C, DeVit M, Johnston SA (1992) Genetic immunization is a simple method for eliciting an immune response. Nature 356(6365):152–154

Tang F, Quan Y, Xin Z-T, Wrammert J, Ma M-J, Lv H, Wang T-B, Yang H, Richardus JH, Liu W, Cao W-C (2011) Lack of peripheral memory B cell responses in recovered patients with severe acute respiratory syndrome: a six-year follow-up study. J Immunol 186(12):7264–7268. https://doi.org/10.4049/jimmunol.0903490

Taylor MW (2014) A history of cell culture. In: Viruses and man: a history of interactions. Springer, Cham, pp 41–52

Tetro JA (2020) Is COVID-19 receiving ADE from other coronaviruses? Microbes Infect 22(2):72–73. https://doi.org/10.1016/j.micinf.2020.02.006

Thanh Le T, Andreadakis Z, Kumar A, Gomez Roman R, Tollefsen S, Saville M, Mayhew S (2020) The COVID-19 vaccine development landscape. Nat Rev Drug Discov 19(5):305–306. https://doi.org/10.1038/d41573-020-00073-5

Theiler M, Smith HH (1937) The use of yellow fever virus modified by in vitro cultivation for human immunization. J Exp Med 65(6):787–800. https://doi.org/10.1084/jem.65.6.787

Tian X, Li C, Huang A, Xia S, Lu S, Shi Z, Lu L, Jiang S, Yang Z, Wu Y, Ying T (2020) Potent binding of 2019 novel coronavirus spike protein by a SARS coronavirus-specific human monoclonal antibody. Emerg Microbes Infect 9(1):382–385. https://doi.org/10.1080/22221751.2020.1729069

Ulmer JB, Donnelly JJ, Parker SE, Rhodes GH, Felgner PL, Dwarki VJ, Gromkowski SH, Deck RR, DeWitt CM, Friedman A (1993) Heterologous protection against influenza by injection of DNA encoding a viral protein. Science 259(5102):1745–1749

Ura T, Okuda K, Shimada M (2014) Developments in viral vector-based vaccines. Vaccine 2(3):624–641. https://doi.org/10.3390/vaccines2030624

Valenzuela P, Medina A, Rutter WJ, Ammerer G, Hall BD (1982) Synthesis and assembly of hepatitis B virus surface antigen particles in yeast. Nature 298(5872):347–350. https://doi.org/10.1038/298347a0

van Doremalen N, Lambe T, Spencer A, Belij-Rammerstorfer S, Purushotham J, Port J, Avanzato V, Bushmaker T, Flaxman A, Ulaszewska M (2020) ChAdOx1 nCoV-19 vaccination prevents SARS-CoV-2 pneumonia in rhesus macaques. bioRxiv

Vargas-Inchaustegui D, Robert-Guroff M (2013) Fc receptor-mediated immune responses: new tools but increased complexity in HIV prevention. CHR 11(5):407–420. https://doi.org/10.2174/1570162X113116660063

Vennema H, de Groot RJ, Harbour DA, Dalderup M, Gruffydd-Jones T, Horzinek MC, Spaan WJ (1990) Early death after feline infectious peritonitis virus challenge due to recombinant vaccinia virus immunization. J Virol 64(3):1407–1409

Vidarsson G, Dekkers G, Rispens T (2014) IgG subclasses and allotypes: from structure to effector functions. Front Immunol 5:520. https://doi.org/10.3389/fimmu.2014.00520

Wang L, Shi W, Joyce MG, Modjarrad K, Zhang Y, Leung K, Lees CR, Zhou T, Yassine HM, Kanekiyo M, Yang Z-y, Chen X, Becker MM, Freeman M, Vogel L, Johnson JC, Olinger G, Todd JP, Bagci U, Solomon J, Mollura DJ, Hensley L, Jahrling P, Denison MR, Rao SS, Subbarao K, Kwong PD, Mascola JR, Kong W-P, Graham BS (2015) Evaluation of candidate vaccine approaches for MERS-CoV. Nat Commun 6:7712. https://doi.org/10.1038/ncomms8712

Weinberg A, Zhang JH, Oxman MN, Johnson GR, Hayward AR, Caulfield MJ, Irwin MR, Clair J, Smith JG, Stanley H (2009) Varicella-zoster virus–specific immune responses to herpes zoster in elderly participants in a trial of a clinically effective zoster vaccine. J Infect Dis 200(7):1068–1077

WHO (2007) Rabies vaccines WHO position paper. https://www.who.int/wer/2010/wer8532.pdf?ua=1. Accessed 09 June 2020

WHO (2016) Polio vaccines: WHO position paper – March, 2016. https://www.who.int/wer/2016/wer9112.pdf. Accessed 09 June 2020

WHO (2019) List of Blueprint priority diseases. http://www.who.int/blueprint/priority-diseases/en/. Accessed 09 June 2020

WHO (2020a) DRAFT landscape of COVID-19 candidate vaccines. https://www.who.int/publications/m/item/draft-landscape-of-covid-19-candidate-vaccines. Accessed 29 June 2020

WHO (2020b) Report of the WHO-China joint mission on coronavirus disease 2019 (COVID-19). Geneva. https://www.who.int/docs/default-source/coronaviruse/who-china-joint-mission-on-covid-19-final-report.pdf. Accessed 09 June 2020

Wilkinson TM, Li CKF, Chui CSC, Huang AKY, Perkins M, Liebner JC, Lambkin-Williams R, Gilbert A, Oxford J, Nicholas B, Staples KJ, Dong T, Douek DC, McMichael AJ, Xu X-N (2012) Preexisting influenza-specific CD4+ T cells correlate with disease protection against influenza challenge in humans. Nat Med 18(2):274–280. https://doi.org/10.1038/nm.2612

Wines BD, Billings H, McLean MR, Kent SJ, Hogarth PM (2017) Antibody functional assays as measures of fc receptor-mediated immunity to HIV – new technologies and their impact on the HIV vaccine field. CHR 15(3). https://doi.org/10.2174/1570162X15666170320112247

Woo P, Lau S, Fan R, Lau C, Wong E, Joseph S, Tsang A, Wernery R, Yip C, Tsang C-C, Wernery U, Yuen K-Y (2016) Isolation and characterization of dromedary camel coronavirus UAE-HKU23 from dromedaries of the middle east: minimal serological cross-reactivity between MERS coronavirus and dromedary camel coronavirus UAE-HKU23. IJMS 17(5):691. https://doi.org/10.3390/ijms17050691

Wu Z, McGoogan JM (2020) Characteristics of and important lessons from the coronavirus disease 2019 (COVID-19) outbreak in China: summary of a report of 72 314 cases from the Chinese Center for Disease Control and Prevention. JAMA 323(13):1239–1242

Yasui F, Kai C, Kitabatake M, Inoue S, Yoneda M, Yokochi S, Kase R, Sekiguchi S, Morita K, Hishima T, Suzuki H, Karamatsu K, Yasutomi Y, Shida H, Kidokoro M, Mizuno K, Matsushima K, Kohara M (2008) Prior immunization with severe acute respiratory syndrome (SARS)-associated coronavirus (SARS-CoV) nucleocapsid protein causes severe pneumonia in mice infected with SARS-CoV. J Immunol 181(9):6337–6348. https://doi.org/10.4049/jimmunol.181.9.6337

Yu J, Tostanoski LH, Peter L, Mercado NB, McMahan K, Mahrokhian SH, Nkolola JP, Liu J, Li Z, Chandrashekar A (2020) DNA vaccine protection against SARS-CoV-2 in rhesus macaques. Science

Zhao J, Zhao J, Perlman S (2010) T cell responses are required for protection from clinical disease and for virus clearance in severe acute respiratory syndrome coronavirus-infected mice. J Virol 84(18):9318–9325. https://doi.org/10.1128/JVI.01049-10

Zhao J, Alshukairi AN, Baharoon SA, Ahmed WA, Bokhari AA, Nehdi AM, Layqah LA, Alghamdi MG, Al Gethamy MM, Dada AM, Khalid I, Boujelal M, Al Johani SM, Vogel L, Subbarao K, Mangalam A, Wu C, Ten Eyck P, Perlman S, Zhao J (2017) Recovery from the Middle East respiratory syndrome is associated with antibody and T-cell responses. Sci Immunol 2(14). https://doi.org/10.1126/sciimmunol.aan5393

Zhu J, Paul WE (2008) CD4 T cells: fates, functions, and faults. Blood 112(5):1557–1569. https://doi.org/10.1182/blood-2008-05-078154

Zhu J, Martinez J, Huang X, Yang Y (2007a) Innate immunity against vaccinia virus is mediated by TLR2 and requires TLR-independent production of IFN-beta. Blood 109(2):619–625. https://doi.org/10.1182/blood-2006-06-027136

Zhu Z, Chakraborti S, He Y, Roberts A, Sheahan T, Xiao X, Hensley LE, Prabakaran P, Rockx B, Sidorov IA, Corti D, Vogel L, Feng Y, Kim JO, Wang LF, Baric R, Lanzavecchia A, Curtis KM, Nabel GJ, Subbarao K, Jiang S, Dimitrov DS (2007b) Potent cross-reactive neutralization of SARS coronavirus isolates by human monoclonal antibodies. Proc Natl Acad Sci 104(29):12123–12128. https://doi.org/10.1073/pnas.0701000104

Zhu F-C, Li Y-H, Guan X-H, Hou L-H, Wang W-J, Li J-X, Wu S-P, Wang B-S, Wang Z, Wang L (2020) Safety, tolerability, and immunogenicity of a recombinant adenovirus type-5 vectored COVID-19 vaccine: a dose-escalation, open-label, non-randomised, first-in-human trial. Lancet

Prevention of COVID-19: Preventive Strategies for General Population, Healthcare Setting, and Various Professions

32

Shirin Moossavi, Kelsey Fehr, Hassan Maleki, Simin Seyedpour, Mahdis Keshavarz-Fathi, Farhad Tabasi, Mehrdad Heravi, Rayka Sharifian, Golnaz Shafiei, Negin Badihian, Roya Kelishadi, Shahrzad Nematollahi, Majid Almasi, Saskia Popescu, Mahsa Keshavarz-Fathi, and Nima Rezaei

Abstract

The disease 2019 (COVID-19) made a public health emergency in early 2020. Despite attempts for the development of therapeutic modalities, there is no effective treatment yet. Therefore, preventive measures in various settings could help reduce the burden of disease. In this chapter, the transmission of severe acute respiratory syndrome coronavirus 2 (SARS-CoV-2) causing COVID-19, non-pharmaceutical approaches at individual and population level, chemoprevention, immunoprevention, preventive measures in different healthcare settings and other professions, special considerations in high-risk groups, and

S. Moossavi
Department of Physiology and Pharmacology, Cumming School of Medicine, University of Calgary, Calgary, Canada

Digestive Oncology Research Center, Digestive Disease Research Institute, Tehran University of Medical Sciences, Tehran, Iran

Microbiome and Microbial Ecology Interest Group (MMEIG), Universal Scientific Education and Research Network (USERN), Calgary, Canada

K. Fehr
Children's Hospital Research Institute of Manitoba, Winnipeg, MB, Canada

Department of Pediatrics and Child Health, University of Manitoba, Winnipeg, MB, Canada

Microbiome and Microbial Ecology Interest Group (MMEIG), Universal Scientific Education and Research Network (USERN), Winnipeg, MB, Canada

H. Maleki
Department of Medical Nanotechnology and Nano Drug Delivery Research Center, Kermanshah University of Medical Sciences, Kermanshah, Iran

Nanomedicine Research Association (NRA), Universal Scientific Education and Research Network (USERN), Kermanshah, Iran

S. Seyedpour
School of Medicine, Tehran University of Medical Sciences, Tehran, Iran

Nanomedicine Research Association (NRA), Universal Scientific Education and Research Network (USERN), Tehran, Iran

M. Keshavarz-Fathi
School of Pharmacy, Shahid Beheshti University of Medical Sciences, Tehran, Iran

Cancer Immunology Project (CIP), Universal Scientific Education and Research Network (USERN), Tehran, Iran

N. Rezaei (ed.), *Coronavirus Disease - COVID-19*, Advances in Experimental Medicine and Biology 1318, https://doi.org/10.1007/978-3-030-63761-3_32

the role of organizations to hamper the psychosocial effects will be discussed.

Keywords

COVID-19 · Healthcare · Nosocomial infection · Personal protective equipment · Preventive measures · Psychological effects · Vaccine

32.1 Introduction

In December 2019, a new outbreak of an unknown acute respiratory tract infection and pneumonia caused by a novel coronavirus was reported in Wuhan, China (Huang et al. 2020). The disease spread quickly among other world regions, and by late January 2020, the World Health Organization (WHO) officially introduced the severe acute respiratory syndrome coronavirus 2 (SARS-CoV2), also known as COVID-19, as an international public health emergency (Mallineni

F. Tabasi
School of Medicine, Zahedan University of Medical Sciences, Zahedan, Iran

Department of Physiology, Faculty of Medical Sciences, Tarbiat Modares University, Tehran, Iran

Medical Education Development Network (MED_NET), Universal Scientific Education and Research Network (USERN), Zahedan, Iran

M. Heravi
School of Medicine, Zahedan University of Medical Sciences, Zahedan, Iran

Medical Education Development Network (MED_NET), Universal Scientific Education and Research Network (USERN), Zahedan, Iran

R. Sharifian
Men's Health and Reproductive Health Research Center, Shahid Beheshti University of Medical Sciences, Tehran, Iran

Iranian Association of Urology, Research Committee (IUA-RC), Tehran, Iran

Urology Research Taskforce (URT), Universal Scientific Education and Research Network (USERN), Tehran, Iran

G. Shafiei
Gametogenesis Research Center, Kashan University of Medical Sciences, Kashan, Iran

Department of Anatomy, Afzalipour Faculty of Medicine, Kerman University of Medical Sciences, Kerman, Iran

Brain Cancer Research Core (BCRC), Universal Scientific Education and Research Network (USERN), Kashan, Iran

N. Badihian · R. Kelishadi
Child Growth and Development Research Center, Research Institute for Primordial Prevention of Non-communicable Disease, Isfahan University of Medical Sciences, Isfahan, Iran

Primordial Prevention of Non Communicable Disease Group (PPNCDG), Universal Scientific Education and Research Network (USERN), Isfahan, Iran

S. Nematollahi
Urology Research Taskforce (URT), Universal Scientific Education and Research Network (USERN), Tehran, Iran

Men's Health and Reproductive Health Research Center, Shahid Beheshti University of Medical Sciences, Tehran, Iran

M. Almasi
Gametogenesis Research Center, Kashan University of Medical Sciences, Kashan, Iran

Brain Cancer Research Core (BCRC), Universal Scientific Education and Research Network (USERN), Kashan, Iran

S. Popescu
HonorHealth, Phoenix, AZ, USA

Mel and Enid Zuckerman College of Public Health, University of Arizona, Tucson, AZ, USA

M. Keshavarz-Fathi (✉)
School of Medicine, Tehran University of Medical Sciences, Tehran, Iran

Cancer Immunology Project (CIP), Universal Scientific Education and Research Network (USERN), Tehran, Iran

N. Rezaei
Research Center for Immunodeficiencies, Children's Medical Center, Tehran University of Medical Sciences, Tehran, Iran

Department of Immunology, School of Medicine, Tehran University of Medical Sciences, Tehran, Iran

Network of Immunity in Infection, Malignancy and Autoimmunity (NIIMA), Universal Scientific Education and Research Network (USERN), Stockholm, Sweden
e-mail: rezaei_nima@tums.ac.ir; m-keshavarz@student.tums.ac.ir

et al. 2020; He et al. 2020). The situation got worse rapidly, and on March 11, 2020, COVID-19 was declared as a pandemic affecting almost all countries of the globe (Rodriguez-Morales et al. 2020; Hanaei and Rezaei 2020).

SARS-CoV2 is an enveloped betacoronavirus with a positive-sense single-stranded RNA genome (30 kb) encoding 27 proteins, including the spike protein and an RNA-dependent RNA polymerase. The spike protein mediates viral attachment to angiotensin-converting enzyme 2 (ACE2) receptor on epithelial cells leading to virus internalization during infection (Zhou et al. 2020b; Zhang and Liu 2020; Rezaei 2020b; Saghazadeh and Rezaei 2020b). ACE2 is highly expressed on lung alveolar epithelial cells and other tissues, including the gastrointestinal tract and liver (Muus et al. 2020; Sharifkashani et al. 2020). Following the entry, the virus life cycle is completed within the host cells, and the virions are assembled via the interaction of viral RNA and protein at the endoplasmic reticulum and Golgi complex (Hoffmann et al. 2020) (Fig. 32.1).

The SARS-CoV-2 infection leads to asymptomatic or mild disease in the majority of people (Weiss and Murdoch 2020; Lotfi and Rezaei 2020). In symptomatic cases, the most common clinical symptoms of COVID-19 include fever (87.9%), cough (67.7%), and fatigue (38.1%). However, the disease can involve other systems and cause gastrointestinal symptoms and loss of sense of smell (Yazdanpanah et al. 2020b;

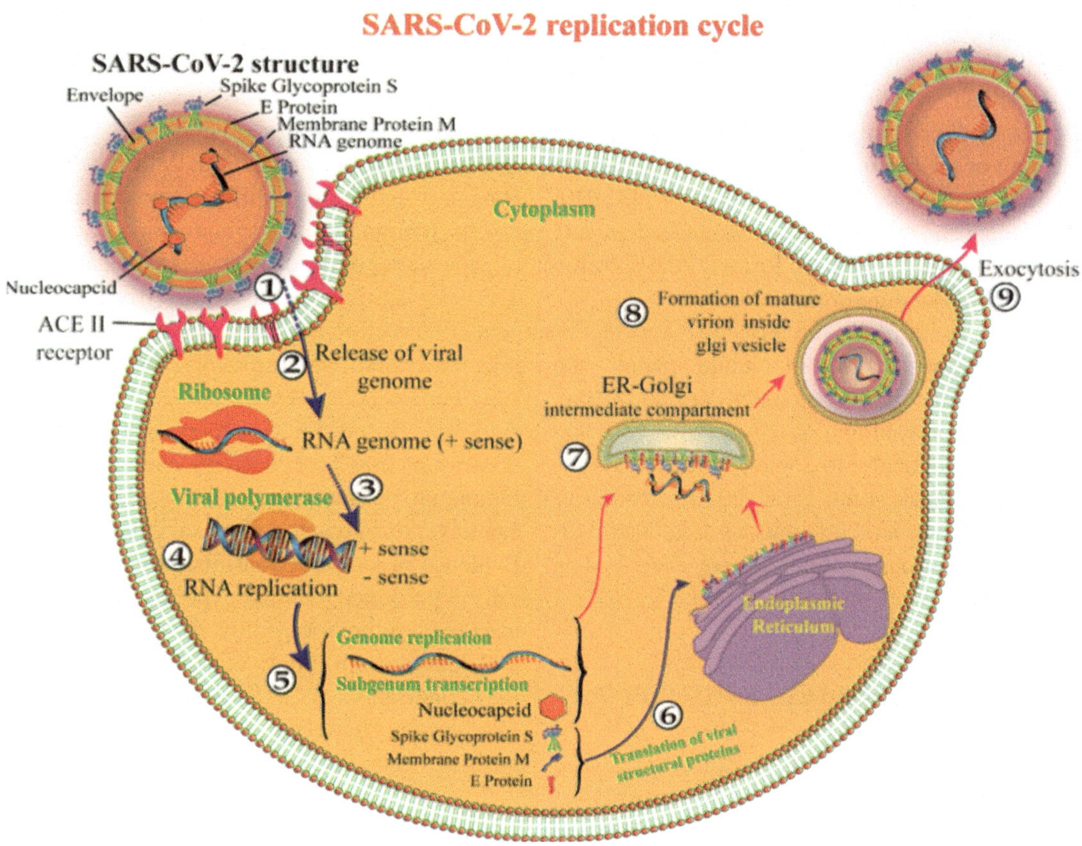

Fig. 32.1 The replication cycle of SARS-CoV2 in host cells: 1, binding and viral entry via membrane fusion or endocytosis; 2, release of viral genome; 3, translation of viral polymerase protein; 4, RNA replication; 5, genome replication and subgenomic transcription; 6, translation of viral structural protein; 7, S, E, M proteins combine with nucleocapsid; 8, formation of mature virion; and 9, exocytosis

Jahanshahlu and Rezaei 2020a; Saleki et al. 2020). Acute respiratory distress syndrome (ARDS) and death occur in 3.4% and 1.4% of diagnosed patients, respectively (Guan et al. 2020). Atypical manifestation can be observed in elderly or immunocompromised patients (Godaert et al. 2020). The average incubation period for COVID-19 is estimated to be 5 days, and an initial estimate of the basic reproduction number (R_0, the average number of cases directly infected by one individual) was 2·2 (95% CI 1·4–3·9) (Wilder-Smith et al. 2020). In addition to genetic factors, older age, cardiovascular disease, hypertension, diabetes, chronic respiratory disease, cancer, smoking, and obesity are predisposing factors to severe/fatal disease (Yang et al. 2020a; Shamshirian and Rezaei 2020; Ahmadi et al. 2020; Yousefzadegan and Rezaei 2020; Darbeheshti and Rezaei 2020). If or not immune deficiency disorders make individuals more susceptible to COVID-19 remained a mystery (Ahanchian et al. 2020; Babaha and Rezaei 2020), as the role of the immune system in conducting or misconducting in the fight against the disease is open to question (Rokni et al. 2020; Yazdanpanah et al. 2020a; Saghazadeh and Rezaei 2020a; Lotfi and Rezaei 2020; Nasab et al. 2020; Bahrami et al. 2020).

There is currently no effective vaccine or treatment for COVID-19, while people are affected by the disease globally (Rezaei 2020a). Therefore, prevention strategy is the best and most effective action to contain and mitigate the disease outbreak (Lotfi et al. 2020). Preventive strategies for general populations, healthcare settings, and various professions are described in this chapter.

32.2 Transmission of SARS-CoV-2

Identification of transmission routes of SARS-CoV-2 is imperative for the development of effective preventive strategies at the individual and population levels. Current evidence suggests the virus could be transmitted from wild animals to humans, from humans to humans, and potentially from humans to pet animals.

32.2.1 Animal-to-Human Transmission

The first COVID-19 cases were identified among workers at the South China Seafood Wholesale Market in Wuhan, China. Phylogenetic analyses indicated that SARS-CoV-2 has the highest genome sequence similarity to a bat coronavirus Bat-CoV-RaTG13 from *Rhinolophus affinis* (Hu et al. 2018), suggesting SARS-CoV2 may have crossed species from bats to wild mammals traded at the Wuhan seafood market. Additionally, SARS-CoV-2 was closely related to Pangolin-CoV isolated from Malayan pangolin (*Manis javanica*) (Zhang et al. 2020).

32.2.2 Human-to-Human Transmission

Human-to-human transmission immediately followed the initial animal-to-human transmission resulting in the rapid spread of the virus in China and beyond (Lu et al. 2020a; McBryde 2020). Human-to-human transmission could occur during asymptomatic, pre-symptomatic, or symptomatic stages of the disease (Ferretti et al. 2020). The virus is primarily transmitted from person to person directly through close contact with infected people via respiratory droplets from a cough or sneezing (Berhe et al. 2020). The virus could also be transmitted via aerosols during procedures such as bronchoscopy and endotracheal intubation (Organization 2020d). The aerosols can travel to a greater distance and persist longer. Evidence of airborne transmission must be carefully interpreted as the experimental systems used (e.g., a three-jet Collison nebulizer fed into a Goldberg drum) might not accurately simulate the conditions of a normal human cough. Furthermore, the clinical validity of the experiment is yet undetermined (van Doremalen et al. 2020).

While the virus is primarily detected in the upper respiratory tract, the virus has also been found in fecal samples from patients. Evidence is accumulating on the presence of the virus in sewage, wastewater, and river (Bar-Or et al. 2020), indicating a potential fecal-oral route of transmission. Also, the virus has been found in the breast milk (Groß et al. 2020), yet the vertical transmission of the virus is not yet established. So far, no evidence of congenital transmission has been found (Chen et al. 2020b).

32.2.3 Environment-to-Human Transmission

Contact with infected inanimate surfaces and human samples could also lead to a viral transmission. Inanimate surfaces could include healthcare equipment such as thermometers and stethoscopes or general household and public surfaces such as doorknobs and supermarket baskets (Liu et al. 2020; Chan et al. 2020; Huang et al. 2020; Ong et al. 2020; Ferretti et al. 2020).

32.2.4 Human-to-Animal Transmission

Despite the limited evidence on human-to-animal transmission of SARS-CoV-2, case reports have identified several pet and zoo animals infected with the virus (Belgium 2020). ACE2 orthologs are present in different wild and domesticated animal species, including cats and dogs, and thus provide a mechanism for these animals to become infected (Li et al. 2020; Shi et al. 2020a).

32.3 Non-pharmaceutical Preventive Interventions

Strategies to tackle an infectious disease outbreak are broadly categorized as non-pharmaceutical and pharmaceutical approaches. Non-pharmaceutical approaches are implemented on the community/country level and involve actions within the responsibility of individuals and organizations. In the absence of effective pharmaceutical prevention or treatment for the SARS-CoV-2, the most effective approach to containing the disease and mitigating the health impacts is non-pharmaceutical interventions.

32.3.1 Preventive Measures at the Individual Level

During the SARS-CoV2 pandemic, the main focus of the public health authorities was to minimize the risk of virus transmission and thus to minimize the burden on the healthcare system, in other words, to flatten the epidemiologic curve of the number of infected/hospitalized patients at any given time. Regardless of the success of the countries and jurisdictions in successfully containing the spreading virus, several interventions directed at the individual level highlight the role that every single one had to play during the crisis. Individual-level interventions mainly aimed to minimize the spread of the virus within the community and reduce the likelihood of the colonization of the virus on the mucus membrane.

The rapidly evolving knowledge about the potential routes of the virus transmission informs the preventive measures at the individual level. The virus is thought to have zoonotic origins, with the first cases linked to a live animal market in Wuhan, China. As a result, individuals were advised to abstain from trading and consuming live/wild animals. Additionally, individuals were advised to minimize contact with wild animals co-habiting residential areas (Hou et al. 2020). There are a few reported cases of confirmed COVID-19 in pet cats and dogs (Belgium 2020). The extent to which pet animals are susceptible to SARS-CoV-2 infection and transmission is currently unclear. However, pet owners were advised to minimize their pet contact with other humans, pets, or wild animals.

While the initial transmission of SARS-CoV-2 was zoonotic, the large-scale community transfer occurred through human-to-human transmission. In accordance, public health recommendations were focused on reducing the probability of spreading or acquiring the virus through respiratory droplets. These measures include (i) proper hand hygiene including frequent handwashing for at least 20 s with soap and water or alcohol-based hand sanitizers; (ii) proper sneezing/coughing hygiene by sneezing or coughing into the elbow; and (iii) physical distancing by avoiding non-essential activities, maintaining at least

2 m distance, and avoiding large gatherings including family gatherings where proper physical distancing is not feasible. The virus can survive on surfaces, and thus frequently cleaning the household surfaces is recommended. Further, into the pandemic, it was demonstrated that viruses exist in aerosols and thus could travel longer than initially projected (van Doremalen et al. 2020). Consequently, wearing a face mask (including homemade face masks) by individuals in situations where physical distancing is not possible was recommended.

Contact tracing is a cornerstone of infectious outbreak containment (Hellewell et al. 2020). Comprehensive and stringent contact tracing for patients could be done through the use of spatial surveillance technology and mobile phone applications (Ferretti et al. 2020). Countries such as Taiwan, Singapore, and Iceland that implemented stringent and innovative contact tracing very early on during the pandemic were among the most successful in containing the virus (Wang et al. 2020a; Lee et al. 2020; Gudbjartsson et al. 2020).

32.3.2 Surveillance, Patient Identification, and Contact Tracing

While public health recommendations targeting uninfected/asymptomatic individuals aim to disrupt the widespread community transmission of the virus, an infected individual must be promptly diagnosed and isolated. Therefore, diagnostic criteria for the definite and suspected case should be developed based on the common presentation of the disease while modifying the case definition by active monitoring of the patients, including those with unusual presentations in the elderly and immunocompromised people (2010).

Active surveillance requires systematic screening of the population for patient identification (2010). Many countries adopted the passive surveillance strategy that identifies patients through the established healthcare system. This strategy results in the classical iceberg phenomenon whereby the majority of infected individuals not requiring (or not seeking) medical attention will not be identified (2010). Thus, active surveillance via mass testing is the preferred strategy to mitigate the health and economic consequences of COVID-19 pandemics. For an outbreak as widespread as the SARS-CoV-2 pandemics, spatial surveillance using novel technologies including GPS tracking, phone records, and travel history has proven useful in some countries (2010; Wang et al. 2020a; Lee et al. 2020); however, social acceptability of these measures might vary in different countries.

32.3.3 Population-Level Strategies

While the majority of countries are dealing with widespread SARS-CoV-2 outbreak, two broad population-level strategies have been undertaken for containment of the virus. The first strategy exemplified by Iceland aimed to contain the disease and was based on active patient identification by mass testing and stringent contact tracing (Gudbjartsson et al. 2020). While this strategy has proven effective, it is contingent on active surveillance and increasing testing capacity long before the community spread occurs in a nation. When there is evidence of widespread community spread resulting in patients without a visible link to travel history to the outbreak epicenters, the main focus would be shifted toward mitigation of the situation by disrupting the community transmission chain. In this approach, large-scale physical distancing and stay and shelter measures are implemented. China, many of the European countries, Canada, and parts of the United States have adopted the latter strategy (Adhikari et al. 2020). These approaches mainly aiming to disrupt the community spread of the virus include the prohibition of gathering (including funeral services), implementing physical distancing (e.g., in the supermarkets), and changes to the public transport maintenance and operations (Ebrahim et al. 2020).

High-level strategies and guidelines to contain or mitigate the spread of the virus are developed and disseminated by the national or local jurisdictional authorities. These measures include limiting international travel and point-of-entry surveillance to prevent cross-regional transmission of the virus (Lin et al. 2020); surveillance, identification, and contact tracing; and procurement and manufacturing of personal protective equipment through the repurposing of existing local businesses. The authorities are also instrumental in exhibiting leadership and promoting community engagement and unity in the face of an unprecedented challenge. Several additional initiatives and emergency measures to address the financial and physiological needs of the population are also achieved through high-level leadership (Anderson et al. 2020).

32.3.4 The Rapid Development of Diagnostic Capabilities

Developing a reliable diagnostic test is a priority in response to the outbreak of an emergent pathogen. In order to acquire accurate population-level information regarding the epidemiology of the COVID-19, several questions need to be answered: the transmission rate, infectivity, the percentage of infected, the percentage of asymptomatic carriers, and the percent requiring hospitalization and intensive care. Mass testing is, therefore, required, and for this, there is an acute need for developing and expanding testing capabilities with rapid turnaround, high sensitivity, and specificity.

Currently, the gold standard for patient identification is reverse transcription polymerase chain reaction (RT-PCR) of the viral RNA in the upper respiratory samples (Corman et al. 2020). RT-PCR is a simple molecular method routinely done in clinical and research laboratories. Specifically, RT-PCR is a common method of identification of viral pathogens in upper respiratory infections. Initial diagnostic assays were developed based on the publicly available genome sequences of naturally occurring SARS-CoV (Corman et al. 2020). Following the successful isolation of the virus, tests were developed to target the envelope protein gene followed by confirmation by RNA-dependent RNA polymerase gene assay (Corman et al. 2020). Drive-through screening centers are advantageous to maximize the accessibility of the test for mass testing (Kwon et al. 2020).

Additional available laboratory diagnostic techniques include serological testing of antibodies against the virus indicative of the previous infection (Jabbari and Rezaei 2020). Understanding the immunity landscape of the population is crucial for accurate epidemiological modeling, public health monitoring, and devising strategies to ease population-level physical distancing measures (2010). Virus whole-genome sequencing is required to map the evolution pattern of the virus, while isolation of the virus in culture is required for research (Berhe et al. 2020; Moore et al. 2020).

Unfortunately, standardized SARS-CoV-2 test kits are in short supply for many countries, limiting the ability to estimate prevalence rates accurately and use large-scale testing methods (Eberhardt et al. 2020). A few protocols and options to accelerate testing and make it more affordable have been proposed and include repurposing laboratory equipment, repurposing government and private laboratories, automated RNA extraction equipment, pooling samples such that one reaction includes multiple individuals, and optimizing the testing scheme based on the prevalence of COVID-19 in the area (Eberhardt et al. 2020; Hogan et al. 2020). While centralized laboratories can oversee the quality assessments, additional testing sites, especially in remote areas, can significantly enhance the testing capabilities (Hockemeyer et al. 2020).

32.4 Pharmaceutical Prevention

32.4.1 Chemoprevention

Despite the lack of an active pharmaceutical preventive measure, several promising candidates, including chloroquine/hydroxychloroquine and remdesivir, are being investigated in international clinical trials.

32.4.1.1 Chloroquine and Derivatives

Chloroquine and its derivative, hydroxychloroquine, are a group of antimalarial drugs, which are also commonly used in autoimmune diseases such as systemic lupus erythematosus and rheumatoid arthritis. They interfere with the glycosylation of cellular receptors and increase endosomal pH, thus disrupting the virus life cycle (Al-Bari 2017). Chloroquine has exhibited inhibitory effects against SARS-CoV-2 in vitro (Vincent et al. 2005; Wang et al. 2020d), while both chloroquine and hydroxychloroquine have shown some preventive efficacy against SARS-CoV-2 when administered before or after the infection in patients (Vincent et al. 2005; Gautret et al. 2020). Although chloroquine and hydroxychloroquine are inexpensive and have a relatively good safety profile, it is noteworthy to highlight the potential fatal cardiac side effects, including ventricular arrhythmias, QT prolongation, and cardiac toxicity (Shah et al. 2020; https://clinicaltrials.gov/ct2/show/NCT04308668). Therefore, self-medication is strongly prohibited, and administration is recommended in the context of a clinical trial.

32.4.1.2 Remdesivir

Remdesivir, produced initially to treat the Ebola virus infection, is a nucleotide analog pro-drug with a broad-spectrum antiviral effect against RNA viruses through inhibition of viral RNA-dependent RNA polymerase (Gordon et al. 2020; Tchesnokov et al. 2019; Mohamed et al. 2020b). Remdesivir has previously shown some efficacy against the other major coronaviruses, e.g., MERS-CoV and SARS-COV-1 (Agostini et al. 2018). Given the high degree of sequence similarity between the RNA-dependent RNA polymerases of SARS-CoV-1 and SARS-CoV-2, remdesivir became a promising candidate for further investigation (Hillaker et al. 2020). Remdesivir has demonstrated efficacy against SARS-CoV-2 infection in vitro and in vivo (Wang et al. 2020d). Currently, remdesivir is recommended for severe COVID-19 requiring invasive mechanical ventilation. However, multiple clinical trials are currently ongoing to confirm its prophylactic effectiveness.

32.4.2 Immunoprevention

Active immune-based interventions are one of the main approaches for the prevention of infectious diseases exemplified by vaccine as an effective preventive strategy for existing and emerging viral infections (Gross et al. 2018; Rauch et al. 2018; Ahmed et al. 2020). Alternatively, passive immunity approaches using adoptive cell transfer and antibodies against the virus are useful before the availability of the vaccine (Luke et al. 2010; Yan et al. 2020; Mansourabadi et al. 2020; Pashaei and Rezaei 2020; Fathi and Rezaei 2020; Jahanshahlu and Rezaei 2020b; Basiri et al. 2020b).

32.4.2.1 Vaccine

Currently, many investigators are working on developing and testing vaccines against COVID-19. Different approaches are being undertaken, including attenuated live virus, inactivated virus, recombinant protein, peptide-based subunit vaccine, DNA vaccine, mRNA vaccine, and adenovirus vaccine (Gao et al. 2020; Chen et al. 2020d; Kalita et al. 2020a; Dilucca et al. 2020; Wang et al. 2020b; Thanh Le et al. 2020) (Table 32.1). Many of the investigational vaccines target spike protein, which is essential for the pathogenicity of the virus (Ahmed et al. 2020). S protein includes S1 and S2 subunits. S1 subunit has a receptor-binding domain (RBD) that binds to angiotensin-converting enzyme 2 (ACE2) receptor, with very high affinity, to enter the host cell. S2 plays a role in the fusion of the virus and the host membrane (Tai et al. 2020). Other components of SARS-CoV-2, including membrane (M) protein and nucleocapsid (N) protein, have the potential to be targeted by vaccines as well (Kalita et al. 2020a).

Subunit vaccines target several immunogenic epitopes by using recombinant protein (Kalita et al. 2020b). Antigen-presenting cells (APC) such as dendritic cells (DC) or artificial APCs (aAPC) loaded with selected peptides are the next approach for immunization, which could activate a robust T-cell response (Ji et al. 2020; Neal et al. 2017). Viral or bacterial vectors encoding S protein, at the high expression level, have also been used for stimulation of an immune response against COVID-19 (Thanh Le et al. 2020). Adenovirus as a vector induces long-term and robust immunity and has more potential for stimulation of $CD4^+$ and $CD8^+$ immune responses. This vector has been used for vaccination against other respiratory viruses, including influenza (Moraes et al. 2011). *Bifidobacterium longum*, a commensal gut bacterium, is also used as a vector. This novel vaccine, which is orally administered and leads to the uptake of the antigen by the gut immune system, has been previously tested for cancer vaccine and showed safety and high immunogenicity (Shirakawa and Kitagawa 2018). Genetic vaccines, including DNA- and mRNA-based vaccines delivered through various platforms such as nanoparticle-based strategies or direct delivery, are also being developed for COVID-19 (Wang et al. 2020b; Thanh Le et al. 2020).

32.4.2.2 Convalescent Plasma

Convalescent plasma has been previously used to prevent or treat viral infections such as Ebola, influenza, Middle East respiratory syndrome (MERS), and severe acute respiratory syndrome (SARS) (Luke et al. 2010; Yan et al. 2020; Stokes et al. 1935; Sahr et al. 2017; Zhou et al. 2007; Ko et al. 2018; Cheng et al. 2005). The convalescent plasma of patients recovered from COVID-19 was found to be useful for the treatment and prevention of COVID-19 (Chen et al. 2020c; Bloch et al. 2020) (Shen et al. 2020). The donors are selected if the following criteria are met: (i) history of a confirmed diagnosis of COVID-19; (ii) a negative result for SARS-CoV-2 molecular tests on the nasopharyngeal or blood sample after 14 days of recovery; (iii) no current clinical symptoms; and (iv) anti-SARS-CoV-2 antibodies at the optimal titer (>1:320) (Bloch et al. 2020; Langhi et al. 2020). The convalescent plasma obtained from these patients contains protective antibodies against the virus. In the recipient, the antibodies bind to and neutralize the virus and thus prevent its pathogenicity (Chen et al. 2020c; Pourahmad et al. 2020). Also, some other pathways of the immune system, including complement activation, phagocytosis, and antibody-dependent cellular cytotoxicity (ADCC), can be activated through this method (van Erp et al. 2019). Clinical trials are ongoing to assess the effectiveness of convalescent plasma in high-risk exposure groups (NCT04323800, NCT04390503) (Table 32.1).

32.4.2.3 Non-specific Immunoprevention

While the research community is racing to develop a safe and effective vaccine, inducing non-specific immunity is suggested to be an attractive alternative (Miyasaka 2020; Franklin et al. 2020). There is some evidence that a history of BCG vaccination might confer non-specific protection against COVID-19 (Miyasaka 2020; Miller et al. 2020). However, other population characteristics of the countries with routine BCG vaccination, such as age, could compound the observation (Hensel et al. 2020). Measles, mumps, and rubella share protein homology with SARS-CoV-2 (Franklin et al. 2020), and thus measles-mumps-rubella (MMR) vaccine is the other non-specific approach currently under investigation in clinical trials to prevent COVID-19 (Table 32.1).

Type I interferons (IFNs), including IFN-α and IFN-β, are used to enhance immunity against viral infections (Teijaro 2016; Shi et al. 2020b). IFN-α was shown to be effective against SARS-CoV in vitro (Ströher et al. 2004). A clinical trial is ongoing to assess the role of recombinant human IFNα-1b nasal drop for the prevention of COVID-19 in healthcare workers (NCT04320238) (Table 32.1).

32.4.3 Traditional Medicine

Traditional medicine is a broad term used to refer to treatments, including traditional Chinese medicine, Indian Ayurveda, Arabic Unani medicine, and various forms of local medicine (Alves and Rosa 2007). Traditional medicine approaches are also being studied for COVID-19 in China (Chen and Du 2020; Yang et al. 2020c).

Table 32.1 Immune-based modalities for prevention of COVID-19 under investigation in ongoing clinical trials

Modality	Type of modality	Constituent	Trial's phase	Participants	Reference
Ad5-nCoV vaccine	Active specific	Adenovirus type 5 vector encoding spike protein	II	500	NCT04341389
Ad5-nCoV vaccine	Active specific	Adenovirus type 5 vector encoding spike protein	I	108	NCT04313127
Inactivated SARS-CoV-2 vaccine	Active specific	Inactivated SARS-CoV-2	I-II	422	NCT04383574
Inactivated SARS-CoV-2 vaccine	Active specific	Inactivated SARS-CoV-2	I-II	744	NCT04352608
LV-SMENP-DC vaccine and antigen-specific CTLs	Active specific	Dendritic cells modified by lentiviral vector encoding COVID-19 minigene SMENP	I-II	100	NCT04276896
aAPC vaccine	Active specific	Artificial antigen-presenting cells modified by a lentiviral vector encoding COVID-19 minigene	I	100	NCT04299724
BNT162 (a1, b1, b2, c2) vaccine	Active specific	SARS-CoV-2 RNA	I-II	7600	NCT04368728
BNT162 (a1, b1, b2, c2) vaccine	Active specific	SARS-CoV-2 RNA	I-II	200	NCT04380701
NVX-CoV2373 vaccine	Active specific	SARS-CoV-2 recombinant spike protein nanoparticle vaccine ±MATRIX-M™ as adjuvant	I	131	NCT04368988
ChAdOx1 nCoV-19 vaccine	Active specific	Adenovirus vector encoding spike protein	I-II	1090	NCT04324606
bacTRL-spike vaccine	Active specific	Bacterial vaccine delivering spike protein plasmid (oral)	I	84	NCT04334980
AV-COVID-19 vaccine	Active specific	Autologous DC loaded with SARS-CoV-2 antigens, ± GM-CSF	I-II	180 HCW	NCT04386252
INO-4800 vaccine	Active specific	DNA vaccine injected by electroporation	I	40	NCT04336410
mRNA-1273 vaccine	Active specific	LNP-encapsulated mRNA-based vaccine encoding spike protein	I	105	NCT04283461
BCG vaccine	Active non-specific	Attenuated *Mycobacterium bovis*	III	1500 HCW	NCT04328441
BCG vaccine	Active non-specific	Attenuated *Mycobacterium bovis*	III	900 HCW	NCT04350931
BCG vaccine	Active non-specific	Attenuated *Mycobacterium bovis*	III	1000 HCW	NCT04362124
BCG vaccine	Active non-specific	Attenuated *Mycobacterium bovis*	III	4170 HCW	NCT04327206
BCG vaccine	Active non-specific	Attenuated *Mycobacterium bovis*	III	500 HCW	NCT04379336
BCG vaccine	Active non-specific	Attenuated *Mycobacterium bovis*	III	1500 HCW	NCT04373291
BCG vaccine	Active non-specific	Attenuated *Mycobacterium bovis*	IV	1800 HCW	NCT04348370
BCG vaccine	Active non-specific	Attenuated *Mycobacterium bovis*	III	1120 HCW	NCT04384549
VPM1002 vaccine	Active non-specific	Recombinant BCG vaccine	III	1200 HCW	NCT04387409
MMR vaccine	Active non-specific	Attenuated MMR viruses	III	200 HCW	NCT04357028
rhIFNα nasal drop	Passive non-specific	Recombinant human IFNα-1b	III	2944 HCW	NCT04320238
HB-adMSCs	Active specific	Stem cell-based intervention	II	100	NCT04348435

Anti-SARS-CoV-2 convalescent plasma	Passive specific	Convalescent plasma obtained from patients, who recovered from COVID-19	II	200 individuals with close contacts	NCT04390503
Anti-SARS-CoV-2 convalescent plasma	Passive specific	Convalescent plasma obtained from patients, who recovered from COVID-19	II	150 individuals with close contact or high risk of severe disease	NCT04323800
Anti-Corona VS2 immunoglobulins	Passive specific	Purified hyper immunoglobulins containing anti-Corona VS2 immunoglobulins from convalescent plasma obtained from COVID-19 patients	–	100	NCT04383548

HB-adMSCs Hope Biosciences adipose-derived mesenchymal stem cells, *MMR* measles-mumps-rubella, *BCG* Bacillus Calmette–Guérin, *aAPC* artificial antigen-presenting cells, *HCW* healthcare worker, *LNP* lipid nanoparticle, *IFNα* interferon-alpha

32.5 Prevention in Healthcare Settings

Hospital-associated transmission has been reported as one of the important routes for COVID-19 transmission (Zhou et al. 2020a). Specifically, healthcare workers are at increased risk of contracting the disease (Wang et al. 2020g). Based on a report by the Centers for Disease Control and Prevention (CDC), in a sample of 1432 US healthcare personnel diagnosed with COVID-19, 55% reported contact with a laboratory-confirmed COVID-19 case only in a healthcare setting (Basiri et al. 2020a). Moreover, based on the statistics from those states of the United States with a complete report on the occupational status of the recognized patients, almost 11% of the reported COVID-19 cases were healthcare personnel (Basiri et al. 2020a). Healthcare providers (HCP) should be monitored routinely for symptoms suggestive of COVID-19 and stay at home if they are ill. It is essential to exclude HCP at a higher risk of the severe form of COVID-19 (e.g., with underlying disease, older age, pregnancy) during the crisis and severe resource limitations, from the care of or contact with patients suspected or confirmed COVID-19 (Xu et al. 2020).

The healthcare environment poses unique challenges for personal protection as physical distancing is not feasible, and many life-saving procedures can enhance the dissemination of the virus. An important aspect in healthcare settings is the risk of COVID-19 transmission from patients undergoing oxygen delivery through different methods. The exhaled air during oxygen delivery through a nasal cannula spreads based on the amount of oxygen flow. Air dispersion in the direction of the end of the patient's bed is almost 66 cm, 70 cm, and 1 m when the oxygen flow setting is 1 L/min, 3 L/min, and 3–5 L/min, respectively (Hui et al. 2011). Using an oronasal mask with an oxygen flow of 4 L/min would decrease the air dispersion to 40 cm (Hui et al. 2006). The amount of air leakage in noninvasive ventilation via a full face mask can also be considered if the inspiratory positive airway pressure (IPAP) increases to 18 cm H2O (Hui et al. 2015). The least air dispersion is when the oxygen is delivered using a CPAP via an oronasal mask (20 cmH2O) or during noninvasive ventilation via a helmet with tight air cushion (IPAP 20 cmH2O, EPAP 10 cmH2O) (Hui et al. 2015). In these two methods, the exhaled air dispersion is considered negligible. Other oxygen delivery methods, including venturi mask, nonrebreathing mask, and CPAP via nasal pillows, have an air dispersion distance of less than 35 cm if used correctly (Hui et al. 2011; Hui et al. 2014). The nebulization of drugs in a COVID-19 patient can also cause the dissemination of the virus (Hui et al. 2014). Using a jet nebulizer would cause some sideway air leakages. Dispersion distance would increase with the level of lung injury and reaches almost 80 cm in a patient with severe lung injury (oxygen consumption of 500 ml/min, lung compliance of 10 ml/cm H2O) (Hui et al. 2014; Ferioli et al. 2020).

32.5.1 Patient Assessment and Transport

It is essential to screen and triage all patients for clinical symptoms of COVID-19 before healthcare visits to reduce unnecessary exposures. If possible, entry into the healthcare facility should be avoided, and telemedicine should be used for initial screening to determine whether a patient may have symptoms consistent with COVID-19 and whether a healthcare visit is needed (Zhang et al. 2020; Moazzami et al. 2020). A patient suspected of having COVID-19 should follow the instructions of the local public health officials.

All HCP responsible for transportation of patients with suspected COVID-19 should be trained on infection control and prevention strategies and use proper PPE (Outpatient and Ambulatory Care Settings: Responding to Community Transmission of COVID-19 in the United States 2020 (Reviewed April 7, 2020)). All patients should be assessed for respiratory symptoms upon entry into healthcare facilities, and patients in waiting areas should be spaced at least 2 m apart. Also, contact should be limited to only what is necessary for proper care (Zhang et al. 2020).

32.5.2 Personnel Protective Equipment

The use and disposal of personal protective equipment (PPE) are the cornerstone of protection against SARS-CoV-2 (Xu et al. 2020). PPE includes eye/face protection, face mask, N95 respirator, and gowns.

32.5.2.1 Eye/Face Protection

Eye protection should be prioritized for prolonged close contact of probably infected individuals and aerosol-generating procedures (Xu et al. 2020). Reusable equipment such as face shields and goggles are preferred. Using powered air-purifying respirators (PAPRs) or full face respirator (with built-in eye protection) could be considered. If multiple users use them, proper disinfection should be considered (Xu et al. 2020).

32.5.2.2 Face Masks

Surgical masks could protect against respiratory viruses. HCP must avoid touching face masks, and the facemask should be removed once soiled or damaged. Extended use of face masks, i.e., wearing the same mask for repeated contacts or with different patients without removing between visits, reduces their effectiveness for healthcare workers and increases the risk of transmission to patients (Wan et al. 2020). Face shield and homemade masks could also be used if resources are severely limited (Wan et al. 2020).

32.5.2.3 N95 Respirators

N95 respirators are the most effective PPE against airborne transmissible pathogens and are recommended for aerosol-generating procedures and fluid hazards. The optimum effectiveness of N95 respirators depends on the fit and proper use. Extended use of these respirators for 8–12 h is acceptable and does not put the patients and personnel at a higher risk of exposure (Zhang et al. 2020). N95 respirators should not be shared among different HCPs. If resources are extremely limited during the COVID-19 pandemic, N95 respirator reuse is allowed to no more than five times (Hick et al. 2009; recommended guidance for extended use and limited reuse of N95 filtering facepiece respirators in healthcare settings 2014 (Reviewed March 27, 2020)).

32.5.2.4 Gowns

A gown protects HCPs by providing a physical barrier against microorganisms, body fluids, and particulate material (Kilinc 2015). Gowns should be prioritized for care activities where splashes or aerosol-generating procedures are anticipated (Zu et al. 2020). Isolation gowns should be reserved for critical situations, and alternation with at least equivalent protection (like surgical gowns or other impermeable protective clothing) should be prioritized (Zu et al. 2020). In a crisis, gown use should shift toward isolation gown and extended use of isolation gown. Extended use must be considered when the same gown is worn by the same HCP during interaction with patients with the same infectious disease in the same location, without any additional co-infection, which is transmissible by contact. Gowns made of polyester or polyester-cotton are considered washable and can be safely laundered and reused (Zu et al. 2020). In extreme limitation, other clothing (e.g., disposable/reusable laboratory coats, aprons, or patient gowns) that have not been evaluated for or have unknown effectiveness can be used, but not considered as PPE (Zu et al. 2020).

32.5.3 Prevention of Nosocomial Infection

Several measures are recommended to minimize virus transmission within the healthcare setting, including following preventive protocols before patient arrival and during visit and assessment, thorough disinfection of patient areas following each visit of suspected or confirmed cases, separation of confirmed COVID-19 patients from uninfected patients as much as possible, and prioritizing airborne isolation rooms for patients undergoing aerosol-generating procedures

(Shereen et al. 2020). Pre-symptomatic or asymptomatic carriers are an important source of community transmission. It is especially important in healthcare units caring for non-COVID-19 patients (Ong et al. 2020). While general recommendations are provided for healthcare workers, specialized units require additional context-specific precautions.

Within the healthcare setting, education and training of healthcare workers and ancillary staff are critical. As the PPE and processes required for COVID-19 are unique, it is important to ensure all staff coming into contact with patients and their environments have adequate training to ensure safety. It is recommended that efforts focus on PPE donning, doffing, and disinfection practices but also encompass those ancillary departments such as environmental services and laboratory workers (Popescu 2020).

Prevention of nosocomial infections for patients and staff alike should include the utilization of proper disinfection products. It is recommended that those disinfectants meet national or World Health Organization standards for efficacy against human coronaviruses (WHO. Cleaning and disinfection of environmental surfaces in the context of COVID-19. Geneva: World Health Organization, May 16, 2020. https://www.who.int/publications-detail/cleaning-and-disinfection-of-environmental-surfaces-inthe-context-of-covid-19 (Accessed May 29, 2020)). Care should be taken to ensure manufacturer recommendations, including wet/kill times, are followed. Staff should also be trained on the importance of using environmental disinfectants when caring for COVID-19 patients.

Rapid identification and isolation of patients under investigation for COVID-19 are necessary. These efforts should include screening processes early in the patient's registration to a healthcare facility. Asking patients about their symptoms and potential exposures should be considered for the triage and registration process (CDC. Screening Dialysis Patients for COVID-19. Centers for Disease Control and Prevention 2020. https://www.cdc.gov/coronavirus/2019-ncov/hcp/dialysis/screening.html (Accessed May 29, 2020)). Signage at entrances is also recommended to encourage patients to self-identify and ask for a mask if they are experiencing symptoms. All patients with COVID-19-like symptoms should be given a mask and appropriately isolated until they can be evaluated. If possible, the utilization of the electronic medical record is recommended to ensure rapid alerts and patient tracking. Proper isolation also includes isolation precaution signage outside the patient's room, but also efforts to cohort confirmed cases in units that have appropriate signage to all entering the space. The rapid identification and isolation of potential COVID-19 patients are critical in reducing secondary cases (Heinzerling et al. 2020).

Lastly, hospital-wide efforts can be utilized to prevent the spread of COVID-19. These include universal masking for all staff during their shifts, closure to patient visitors, and screening of employees upon entrance to the hospital. Cancellation of elective surgeries has also been suggested to help reduce healthcare burden during periods of high COVID-19 community transmission. Established processes for contact tracing when patients or staff are identified as having COVID-19 are recommended to help reduce secondary cases. These additional COVID-19 efforts are recommended during periods of sustained community transmission and can be de-escalated as cases decline (Popescu 2020).

32.5.3.1 Pediatric Healthcare Settings

Healthcare providers are encouraged to maintain immunization during the pandemic, if possible (Guidance on Providing Pediatric Ambulatory Services via Telehealth During COVID-19 2020 (Updated April 15, 2020)). As children are less commonly presenting with COVID-19 symptoms, efforts should be made to minimize the risk of virus transmission from a child to others. Outpatient or inpatient management should be decided on a case-by-case basis and based on the severity of the illness and medical condition, presentation, and care ability of parents or guardians (Information for Pediatric Healthcare Providers 2020 (Updated April 17, 2020)).

32.5.3.2 Obstetric Healthcare Settings

Pregnant women should be assessed for COVID-19 symptoms before arrival, while pregnant women with suspected or confirmed COVID-19 should notify obstetric facilities as soon as possible. Pregnant women suspected of COVID-19 should be prioritized for testing upon admission (Chen and Du 2020). If the mother is a confirmed case of COVID-19, decisions regarding mother/infant contact should be made on a case-by-case basis and involve the parents in the decision-making process. The presence of the partner during the delivery should be discussed with the pregnant woman and should be allowed following precautions (Chen and Du 2020).

32.5.3.3 Hemodialysis Settings

Patients with symptoms presumptive of COVID-19 should be identified before entering the treatment area. It is recommended that all patients be provided with face masks at entry, regardless of symptoms. Instructions about source control, distancing, and respiratory hygiene should be provided to patients and HCP by facilities. During dialysis, patients should be separated at least 2 m from other patients (Ikizler and Kliger 2020). As in other healthcare settings, suspected or confirmed COVID-19 patients should be treated in a separate, isolated room. If this is not possible, the patient should be placed at a corner (away from traffic) at least 2 m from the closest patient. Routine protocols should be used for the disinfection of dialysis stations using products effective against SARS-CoV-2 (control and prevention).

32.5.3.4 Laboratory Settings

All precautions should be undertaken when working with and handling specimens suspected or confirmed for SARS-CoV-2. Risk assessments should be performed based on the procedures, hazards of processes, equipment, and available resources. Standard precautions should be considered, including PPE, hand hygiene, waste management, and routine decontamination interventions with disinfectants effective against SARS-CoV-2. Class II biological safety cabinets (BSC) should be used for producers with high potential of aerosols or droplet generation. Any laboratory activities involving SARS-CoV-2 should be conducted in biosafety level 3 (BSL-3) laboratories and using BSL-3 practices (Interim Laboratory Biosafety Guidelines for Handling and Processing Specimens Associated with Coronavirus Disease 2019 (COVID-19) 2020 (Updated March 31, 2020)).

32.6 Specific Guidelines for Other Professions

In pandemic situations, all occupational fields are encouraged to use the recommended preventative measures such as the use of masks and gloves, limiting the number of on-site staff, and working from home whenever possible. However, careers come with different levels of occupational exposure when it comes to COVID-19.

32.6.1 Forensic Medicine

It is recommended that autopsies on suspected or confirmed cases of COVID-19 be performed in airborne infection isolation rooms. These rooms must have three components: (i) negative pressure; (ii) airflow rate of 6–12 ACH (air changes per hour); and (iii) air exhaustion to either unoccupied places or the use of a high-efficacy particulate air (HEPA) filter to remove contaminants from the exhaust air (Centers for Disease Control and Prevention 2015, November 5). Vacuum tools and the use of biosafety cabinets are also recommended, along with minimizing the number of personnel present at a time (Occupational Safety and Health Administration).

32.6.2 Funeral Homes

Mortuary staff must take precautions to prevent injuries when using sharps and dispose of them properly. In addition, face touching should be avoided during tasks, and proper handwashing and PPE disposal procedures followed. Besides general autopsy PPE and special considerations like using NIOSH-certified disposable N95 or powered, air-purifying respirators (PARPs) with HEPA filters must be considered due to the high chance of aerosol production. Fast disposal of corpses is also essential to prevent exposure. However, whether a specific manner of disposal prevents exposure more than others is unknown (Occupational Safety and Health Administration).

32.6.3 Airline Industry

Alongside the other industries, airlines have been greatly affected during the current pandemic, and they need to apply preventive strategies for vital flights necessary for COVID-19 support (Administration 2020). If travelers are suspected of having COVID-19 during flights, it is recommended that the situation be reported before landing (Prevention 2020). For crew members who anticipate potential contact with symptomatic patients or their body fluids, it is necessary to wear disposable gloves, and a face mask is necessary (Association 2017). After the landing, both the hard and soft portions of the symptomatic travelers' seat must be cleaned and disinfected within 2 m (Prevention 2020i).

32.6.4 Border Control

For border security officers, frequent contact with travelers increases the risk of contracting the disease. Prevention strategies begin with applying physical shields between officers and passengers, isolating passengers suspected of having the disease, and providing waiting spots a distance of 2 m from stations (Occupational Safety and Health Administration 2020).

32.7 Considerations in Special Groups

32.7.1 High-Risk Groups and Elderly

Some groups of individuals are more susceptible to developing severe COVID-19, warranting specific attention to the high-risk group. The high-risk patients include the elderly (age $\geq$ 65 years) and those with one or more comorbidities including chronic pulmonary conditions (including moderate to severe asthma, chronic obstructive pulmonary disease), cardiovascular diseases (including heart failure, coronary artery disease, congenital heart disease, cardiomyopathies, and pulmonary hypertension), immunocompromised states (including immunodeficiency, using immunosuppressive medications, organ transplantation, chemotherapy, and radiotherapy), obesity, diabetes, liver, and kidney failure (Prevention 2020a).

Specific preventive measures are recommended for each group. They should rigorously implement hand hygiene, social distancing, and other general measures, provide at least a 2-week supply of drugs and other utilities to minimize home leaving, reschedule medical appointments with their caregiver, and continue their current medications according to their physician's advice (Prevention 2020g). CDC also recommends people over 65 years, immunocompromised, and patients with liver disease should be vaccinated against influenza and pneumococcal disease (Prevention 2020a, g). Recently, CDC has recommended high-risk and elderly populations to develop care plans, which are a summary of each person's previous health condition, medications, and treatment plans. Care plans reduce hospitalization and provide valuable data for better management of patients during emergencies (Prevention 2020b). Nursing home and long-term care facilities also face a high risk of transmission. CDC provides instructions for these centers, such as limiting visitors and regular surveillance of staff and residents to keep them safe (Prevention 2020h).

32.7.2 Pregnancy and Breastfeeding

Pregnant women might be more susceptible to COVID-19 as a result of the physiologic alteration during pregnancy. However, no sufficient evidence is available whether pregnancy increases the risk of severe COVID-19 (Kourtis et al. 2014; Nelson-Piercy 2015). Pregnant women infected with COVID-19 demonstrated the same symptoms as others. Fever and respiratory symptoms (cough, dyspnea) were the most frequent ones (Zaigham and Andersson 2020; Di Mascio et al. 2020; Della Gatta et al. 2020; Monteleone et al. 2020). Also, regarding reports of fetal growth restriction in pregnant women with SARS infection, follow-up ultrasonography is recommended for patients who recovered from COVID-19 to monitor the probable effects of infection on fetal growth (Di Mascio et al. 2020).

According to a meta-analysis, spontaneous abortion, preeclampsia, perinatal death, and cesarean were more prevalent in pregnant women infected with SARS, MERS, or COVID-19 compared to healthy women (Di Mascio et al. 2020). However, most pregnant patients developed mild to moderate COVID-19 with no increase in pregnancy complications compared to the healthy population (Della Gatta et al. 2020; Zaigham and Andersson 2020; Wang et al. 2020f).

Preterm labor was the most prevalent complication (41–42%) of pregnancies with COVID-19 (Di Mascio et al. 2020; Zaigham and Andersson 2020). Fetal death was estimated to be about 7.0% in a study on 41 cases (Di Mascio et al. 2020). Respiratory distress, fever, tachycardia, vomiting, and pneumothorax have been reported among complicated neonates of mothers with COVID-19 (Zaigham and Andersson 2020). Early data failed to demonstrate an increased risk of developing severe disease in mothers and their fetuses with COVID-19; however, regarding the ongoing outbreak and limited studies on pregnant populations, more high qualitative prospective studies are needed to be conducted (Monteleone et al. 2020). Up to now, no evidence supported the vertical transmission of COVID-19 (Di Mascio et al. 2020; Monteleone et al. 2020), and SARS-COV-2 has not been isolated from the blood, placenta, umbilical cord blood, amniotic fluid, pharyngeal swab, and breast milk samples (Wang et al. 2020c; Chen et al. 2020b).

Some modification in prenatal care during the outbreak has been suggested including restricting the number and duration of in-person visits, scheduling laboratory, non-stress test, and ultrasonography to minimize patient referral, and where possible taking advantage of telemedicine (Chen et al. 2020a). Similar to other COVID-19 cases, pregnant women with confirmed or suspected disease should be isolated in a specific room for delivery (Chen et al. 2020a). However, the use of a surgical mask might interfere with mothers' forceful ventilation during the active phase of delivery, which can be prevented by appropriate use and proper choice of analgesia. COVID-19 is not considered an indication for cesarean delivery (Organization 2020a). However, if possible, rescheduling delivery after recovery can prevent virus transmission to the neonate (Chen et al. 2020a). Assisted reproductive treatment (ART) is recommended to be continued after adequate evidence provided regarding the effect of COVID-19 on both mother and fetus (Monteleone et al. 2020).

The virus can readily transfer through droplets; therefore, neonates are recommended to be separated from infected mothers. CDC recommends the following criteria for the duration of isolation:

- If she has confirmed COVID-19, she can discontinue isolation at least 7 days after initiation of symptoms and should remain afebrile for 72 h without consumption of antipyretics.
- If her infant needs supportive care in the hospital, she can discontinue isolation if she remains afebrile for 72 h without consumption of antipyretics with improvement in her respiratory symptoms with two continuous negative nasopharyngeal swab results.
- Asymptomatic mothers with a positive test for COVID-19 can discontinue isolation 7 days after their first positive result.

It is recommended that mothers wear face masks and keep social distancing 3 days after isolation (Prevention 2020c). The suspected infants should be isolated from other neonates. Those who take care of the neonate should use proper PPE. If separation was not feasible, other strategies should be implemented to reduce the risk of transmission, such as wearing masks while breastfeeding, keeping infants in at least 6 ft distance from mother, or using donated breast milk from other healthy mothers (Chen et al. 2020a; Stuebe 2020; Organization 2020a; Davanzo 2020).

The SARS-COV-2 has not been isolated from breast milk (Chen et al. 2020b). WHO recommends cautious breastfeeding due to its beneficial effect on infant's health. However, the effect of breastfeeding during antiviral treatment is not apparent. Breast pumping should be safely conducted by cleaning hands, washing breast and pump with soap and water, and utilizing respiratory hygiene (Berveiller et al. 2020; WHO 2020).

32.7.3 Children

One to five percent of COVID-19 patients are children with a mean age of 7 years old (Dong et al. 2020; Lu et al. 2020b; Wei et al. 2020; Wu and McGoogan 2020). Boys seem to be more affected than girls (COVID et al. 2020; Dong et al. 2020). However, the true prevalence of COVID-19 in children is unclear. Reports from SARS and MERS infection indicated that most children might remain asymptomatic (Hon et al. 2003; Bitnun et al. 2003; Al-Tawfiq et al. 2016) while still might be able to transfer the infection to others, including high-risk populations and the elderly (Zhou et al. 2020b; Ludvigsson 2020; Chan et al. 2020). Despite differences in the immune system of children and adults, the symptoms are similar (Ludvigsson 2020).

The outbreak can affect children with other medical conditions. They might be at higher risk of developing severe infections (Hagmann 2020). On the other hand, it might restrict their access to the healthcare system and supporting facilities. No specific guideline has been published to address these needs yet (Dayal). In a study in China on 171 pediatric patients, all three children who developed severe COVID-19 had other comorbidities, including leukemia, intussusception, and bilateral hydronephrosis (Yang et al. 2020b). Children with pulmonary dysfunction; immunocompromised state; renal, hematologic, and liver disease; obesity; and endocrine disorders are also more vulnerable to severe infection (Hagmann 2020; Wang et al. 2020h). High-risk children should also be trained to use general preventive measures such as hand hygiene and social distancing. No specific treatment has been recommended for infected children (Control and Prevention 2020). Parents should contact children's physician, planning for future appointments as well as providing a sufficient supply of drugs and equipment (Organization 2020e).

CDC recommends several measures to prevent children from COVID-19. Different guidelines are provided regarding the degree of transmission in an area (low, moderate, high), which is determined by local health services. Schools have a fundamental role in providing a healthy educational environment for both staff and children and collaborating with healthcare offices to develop a comprehensive preventive strategy based on the condition of COVID-19 in their region. They should provide adequate amounts of sanitary supplies, monitor the school absence, and implement social distancing when needed. Schools should also prepare an electronic-aid curriculum beforehand to avoid interruption in educational programs (Prevention 2020e).

Parents should regularly clean and disinfect most touched areas of the home and should avoid their children from in-person contact with other children and high-risk adults. Recently, CDC has recommended a cloth face mask as a new measure for individuals above 2 years old. The cloth mask might be useful in the prevention of the spread of disease to others (Prevention 2020i). However, people who are in close contact with patients still should use N95 or surgical mask (Prevention 2020e).

Children with mild symptoms can be managed at home (Zheng et al. 2020). Sick children should be kept in an isolated room with separate bath and equipment and monitored for any sign of deterioration. In case of dyspnea, chest discomfort, cyanosis, shock, poor feeding, or any deterioration in a child's health condition, they should be visited by a doctor (Greenhalgh et al. 2020; Organization 2020a). Other family members should keep at least 6 ft distance, wear a mask while contacting the child, and adequately disinfect households. It is also recommended for the ill child caregivers to quarantine themselves (Gandhi et al. 2020; Organization 2020b, f; Prevention 2020d).

Healthcare providers should properly use PPE, prioritize routine child screening visits, and prepare separate room or time to visit sick and asymptomatic children. Where possible, the American Psychological Association (APA) recommends taking advantage of telemedicine (Pediatrics 2020). Screening and vaccination of infants and children under 2 years old should be continued as routine (Prevention 2020f).

32.8 Prevention of Social Crisis

32.8.1 Education, Knowledge, and Awareness

Lockdowns including school closure can influence the education of 80% of children all around the world. Approximately 140 countries have had nationwide school closure, and some others have had local closures. These closures can ultimately lead to adverse effects on educational, social, physical, mental, and food accessibility, especially in low socioeconomic regions (Bitler MP, Seifoddini A. Health impacts of food assistance: evidence from the United States. Annu Rev. Resour Economics 2019; 11: 261–87). Inequalities in educational outcomes following school disclosure are also attributed to non-school factors, which has a tremendous burden on low socioeconomic households (Wim-Van and Parolin 2020). Specific attentions to low socioeconomic regions are required to reduce the global challeges of the pandemic.

In order to increase the awareness to COVID-19, UNICEF has launched a Knowledge, Attitude, and Practice (KAP) survey on COVID-19 on May 01, 2020, to improve efforts addressing the COVID-19 pandemic. The survey assesses the comprehension, knowledge, and behavior of people in response to COVID-19 (World Health Organization, UNICEF: Knowledge, Attitude and Practice Survey on COVID-19 response. Available at https://www.unicef.org/guyanasuriname/press-releases/knowledge-attitude-and-practice-survey-covid-19-response. Accessed on 12 May 2020). Increased level of knowledge and awareness can lead to prevention of the disease.

32.8.2 Psychological Aspects

COVID-19 pandemic influences many psychological aspects and induces stressful situation and panic. WHO has released practical instructions for the general population and governments to improve mental health status during the COVID-19 outbreak. Some special groups, including children, elderlies, and people with a history of mental or physical disease are at higher risk of being affected by adverse effects of COVID-19 on mental health. Anxiety, stress, fear, and interpersonal violence are some examples of these adverse effects (World Health Organization, Regional Office for Europe. Mental health and psychological resilience during the COVID-19 pandemic. Available at http://www.euro.who.int/en/health-topics/health-emergencies/coronavirus-covid-19/news/news/2020/3/mental-health-and-psychological-resilience-during-the-covid-19-pandemic, accessed on 12 May 2020).

32.8.3 Economic Aspects

The COVID-19 outbreak has disrupted the worldwide economy. The economic impact of COVID-19 is too complex and uncertain to be responded by an appropriate policy (McKibbin, Warwick J. and Fernando, Roshen. The Global Macroeconomic Impacts of COVID-19: Seven Scenarios (March 2, 2020). CAMA Working Paper No. 19/2020. Available at SSRN: https://

ssrn.com/abstract=3547729 or https://doi.org/10.2139/ssrn.3547729). The economic effects of the COVID-19 outbreak might be underestimated if the historical effects of the SARS crisis are considered (Fernandes, Nuno, Economic Effects of Coronavirus Outbreak (COVID-19) on the World Economy (March 22, 2020). Available at SSRN: https://ssrn.com/abstract=3557504 or https://doi.org/10.2139/ssrn.3557504). It has been estimated that even a contained outbreak could lead to a significant influence on the world economy (McKibbin, Warwick J. and Fernando, Roshen. The Global Macroeconomic Impacts of COVID-19: Seven Scenarios (March 2, 2020). CAMA Working Paper No. 19/2020. Available at SSRN: https://ssrn.com/abstract=3547729 or https://doi.org/10.2139/ssrn.3547729). During the COVID-19 outbreak, a 3–15% decline in the gross domestic product (GDP) could be anticipated, and countries that are more dependent on tourism or foreign trade will be more affected by the pandemic (Fernandes, Nuno, Economic Effects of Coronavirus Outbreak (COVID-19) on the World Economy (March 22, 2020). Available at SSRN: https://ssrn.com/abstract=3557504 or https://doi.org/10.2139/ssrn.3557504). As a solution, investments in public health could prevent disastrous expenses during similar outbreaks especially in developing countries (McKibbin, Warwick J. and Fernando, Roshen. The Global Macroeconomic Impacts of COVID-19: Seven Scenarios (March 2, 2020). CAMA Working Paper No. 19/2020. Available at SSRN: https://ssrn.com/abstract=3547729 or https://doi.org/10.2139/ssrn.3547729).

32.8.4 Role of International and Non-governmental Organizations

International organizations should respond to the COVID-19 pandemic through two main roles: solving the current problem and preventing the next wave. They should accelerate both treatment development and prevention of disease transmission (Gates 2018). Establishment of databases and setting rules for providing instant scientific information are required. International organizations such as WHO and the Global Research Collaboration for Infectious Disease Preparedness can facilitate international collaborations for sharing data, prioritizing research scope, running clinical trials, and getting regulatory approval (Gates 2020).

Non-profit and non-governmental organizations can also play an essential role in expanding budgets and funds to develop appropriate vaccine and treatment (Thanh Le et al. 2020; Mukherjee 2020). However, governmental funding can minimize the risk of conflicts for pharmaceutical companies. Finally, governmental and non-governmental organizations should support the preparation and distribution of therapeutics and vaccines and provide an accessible and equitable way to reach vaccines to the target population (Bollyky et al. 2020).

32.8.5 Regulation

On January 30, 2020, WHO announced the outbreak of COVID-19 as a public health emergency of international concern (PHEIC) (WHO. WHO Director-General's statement on IHR Emergency Committee on Novel Coronavirus (2019-nCoV). Geneva: World Health Organization, Jan 30, 2020. https://www.who.int/dg/speeches/detail/who-director-general-s-statement-on-ihr-emergencycommittee-on-novel-coronavirus-(2019-ncov) (accessed Jan 30, 2020)). International Health Regulations (IHR 2005), which were released by WHO in 2007, aimed at the empowerment and preparedness of countries in response to public health emergencies and epidemics. Implementation and regular assessment of IHR in countries lead to responding to epidemics and pandemics by prevention, surveillance, detection, management, and risk assessment (Global Preparedness Monitoring Board. Annual report on global preparedness for health emergencies. Geneva: World Health Organization, 2019. https://apps.who.int/gpmb/assets/annual_report/GPMB_annualreport_2019.pdf (accessed Jan 25, 2020)). The criteria of WHO for IHR capacities consist of actions that improve the pre-

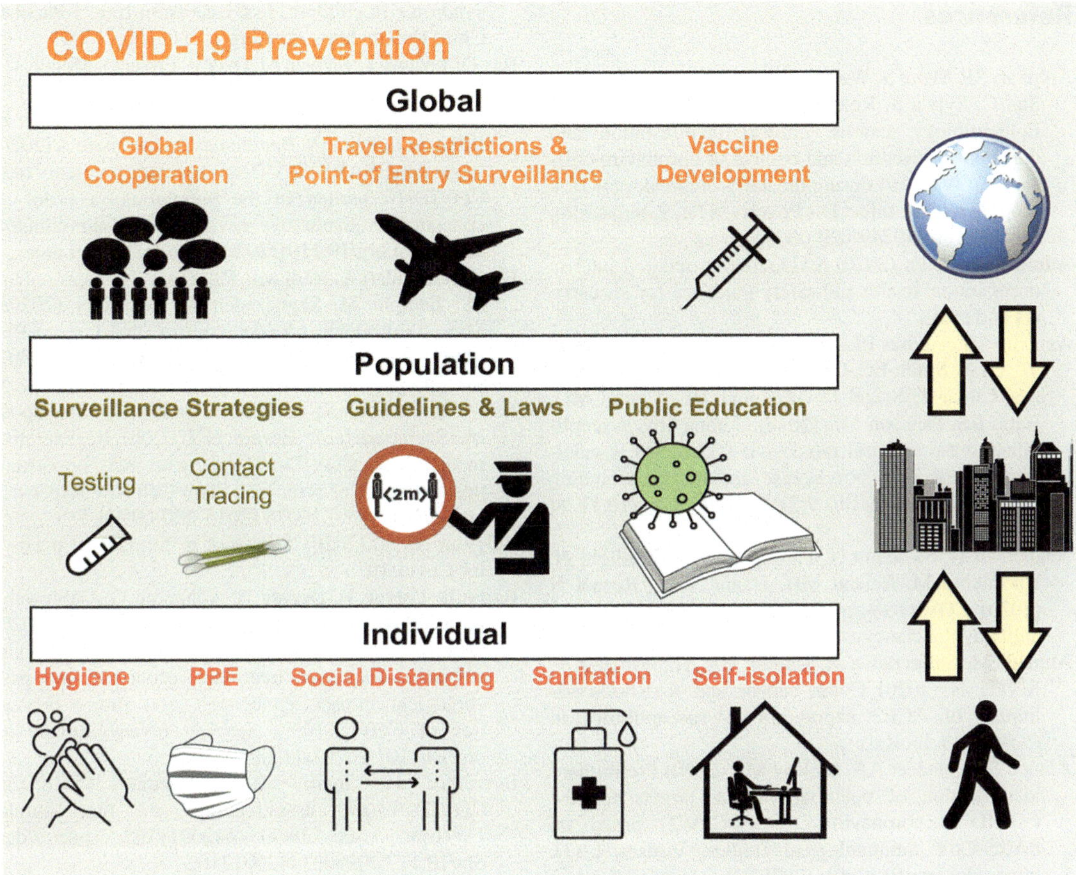

Fig. 32.2 COVID-19 prevention at the global, population, and individual level

paredness of countries in response to health emergencies and empower international health security (WHO. WHO benchmarks for international health regulations (IHR) capacities. Geneva: World Health Organization, 2019. www.who.int/ihr/publications/9789241515429/en/ (accessed Jan 26, 2020); Kandel et al. 2020). Applying these regulations could help to mitigate the pandemic's damages.

32.9 Conclusion

During the last two decades, coronaviruses have continued to rise epidemics and thereby examining the lessons we have learned from the previous epidemics (Jabbari et al. 2020). Nevertheless, the ongoing COVID-19 pandemic took the global community by surprise and resulted in unprecedented preventive measures, including nationwide lockdowns, as summarized in Fig. 32.2. It is promising that public health officials, biomedical researchers, the research and development sector, and nongovernmental organizations made a unified response (Momtazmanesh et al. 2020; Mohamed et al. 2020a; Rzymski et al. 2020; Moradian et al. 2020), resulting in the proposals for early diagnosis of disease (Rabiee et al. 2020; Basiri et al. 2020a) and development of more than 100 experimental vaccines and international clinical trials.

References

Adhikari SP, Meng S, Wu YJ, Mao YP, Ye RX, Wang QZ, Sun C, Sylvia S, Rozelle S, Raat H, Zhou H (2020) Epidemiology, causes, clinical manifestation and diagnosis, prevention and control of coronavirus disease (COVID-19) during the early outbreak period: a scoping review. Infect Dis Poverty 9(1):29. https://doi.org/10.1186/s40249-020-00646-x

Administration Fa (2020) SAFO20009: updated interim occupational health and safety guidance for air carriers and crews

Agostini ML, Andres EL, Sims AC, Graham RL, Sheahan TP, Lu X, Smith EC, Case JB, Feng JY, Jordan R, Ray AS, Cihlar T, Siegel D, Mackman RL, Clarke MO, Baric RS, Denison MR (2018) Coronavirus susceptibility to the antiviral remdesivir (GS-5734) is mediated by the viral polymerase and the proofreading exoribonuclease. MBio 9(2). https://doi.org/10.1128/mBio.00221-18

Ahanchian H, Moazzen N, Faroughi MSD, Khalighi N, Khoshkhui M, Aelami MH, Haghi NSM, Rezaei N (2020) COVID-19 in a child with primary specific antibody deficiency

Ahmadi M, Saffarzadeh N, Habibi MA, Hajiesmaeili F, Rezaei N (2020) Colon cancer and SARS-CoV-2: impact of ACE2 expression in susceptibility to COVID-19. bioRxiv

Ahmed SF, Quadeer AA, McKay MR (2020) Preliminary identification of potential vaccine targets for the COVID-19 coronavirus (SARS-CoV-2) based on SARS-CoV immunological studies. Viruses 12(3). https://doi.org/10.3390/v12030254

Al-Bari MAA (2017) Targeting endosomal acidification by chloroquine analogs as a promising strategy for the treatment of emerging viral diseases. Pharmacol Res Perspect 5(1):e00293–e00293. https://doi.org/10.1002/prp2.293

Al-Tawfiq JA, Kattan RF, Memish ZA (2016) Middle East respiratory syndrome coronavirus disease is rare in children: an update from Saudi Arabia. World J Clin Pediatr 5(4):391

Alves RR, Rosa IM (2007) Biodiversity, traditional medicine and public health: where do they meet? J Ethnobiol Ethnomed 3(1):14

Anderson RM, Heesterbeek H, Klinkenberg D, Hollingsworth TD (2020) How will country-based mitigation measures influence the course of the COVID-19 epidemic? Lancet 395(10228):931–934. https://doi.org/10.1016/s0140-6736(20)30567-5

Association Iat (2017) Universal precaution kit

Babaha F, Rezaei N (2020) Primary immunodeficiency diseases in COVID-19 pandemic: a predisposing or protective factor? Am J Med Sci. https://doi.org/10.1016/j.amjms.2020.07.027

Bahrami A, Vafapour M, Moazzami B, Rezaei N (2020) Hyperinflammatory shock related to COVID-19 in a patient presenting with multisystem inflammatory syndrome in children: first case from Iran. J Paediatr Child Health. https://doi.org/10.1111/jpc.15048

Bar-Or I, Yaniv K, Shagan M, Ozer E, Erster O, Mendelson E, Mannasse B, Shirazi R, Kramarsky-Winter E, Nir O, Abu-Ali H, Ronen Z, Rinott E, Lewis YE, Friedler E, Bitkover E, Paitan Y, Berchenko Y, Kushmaro A (2020) Regressing SARS-CoV-2 sewage measurements onto COVID-19 burden in the population: a proof-of-concept for quantitative environmental surveillance. https://doi.org/10.1101/2020.04.26.20073569.this

Basiri A, Heidari A, Nadi MF, Fallahy MTP, Nezamabadi SS, Sedighi M, Saghazadeh A, Rezaei N (2020a) Microfluidic devices for detection of RNA viruses. Rev Med Virol n/a (n/a):e2154. https://doi.org/10.1002/rmv.2154

Basiri A, Pazhouhnia Z, Beheshtizadeh N, Hoseinpour M, Saghazadeh A, Rezaei N (2020b) Regenerative medicine in COVID-19 treatment: real opportunities and range of promises. Stem Cell Rev Rep:1–13. https://doi.org/10.1007/s12015-020-09994-5

Belgium NVSo (2020) A case of a Belgian cat positive for Covid-19

Berhe B, Legese H, Degefa H, Adhanom G, Gebrewahd A, Mardu F, Tesfay K, Welay M, Negash H (2020) Global epidemiology, pathogenesis, immune response, diagnosis, treatment, economic and psychological impact, challenges, and future prevention of COVID-19: a Scoping review. https://doi.org/10.1101/2020.04.02.20051052

Berveiller P, Guerby P, Garabedian C (2020) COVID19 and breastfeeding: not that simple. J Hum Lact: 0890334420917102. https://doi.org/10.1177/0890334420917102

Bitler MP, Seifoddini A (2019) Health impacts of food assistance: evidence from the United States. Annu Rev Resour Econo 11:261–287

Bitnun A, Allen U, Heurter H, King SM, Opavsky MA, Ford-Jones EL, Matlow A, Kitai I, Tellier R, Richardson S (2003) Children hospitalized with severe acute respiratory syndrome-related illness in Toronto. Pediatrics 112(4):e261–e261

Bloch EM, Shoham S, Casadevall A, Sachais BS, Shaz B, Winters JL, van Buskirk C, Grossman BJ, Joyner M, Henderson JP, Pekosz A, Lau B, Wesolowski A, Katz L, Shan H, Auwaerter PG, Thomas D, Sullivan DJ, Paneth N, Gehrie E, Spitalnik S, Hod E, Pollack L, Nicholson WT, Pirofski L-A, Bailey JA, Tobian AAR (2020) Deployment of convalescent plasma for the prevention and treatment of COVID-19. J Clin Invest doi:https://doi.org/10.1172/JCI138745

Bollyky TJ, Gostin LO, Hamburg MA (2020) The equitable distribution of COVID-19 therapeutics and vaccines. JAMA

CDC (2020) Screening dialysis patients for COVID-19. Centers for Disease Control and Prevention. https://www.cdc.gov/coronavirus/2019-ncov/hcp/dialysis/screening.html. Accessed 21 May 2020

Centers for disease control and prevention (2015, November 5) Airborne infection isolation room

(AIIR). https://www.cdc.gov/infectioncontrol/guidelines/isolation/glossary.html

Chan JF-W, Yuan S, Kok K-H, To KK-W, Chu H, Yang J, Xing F, Liu J, Yip CC-Y, Poon RW-S (2020) A familial cluster of pneumonia associated with the 2019 novel coronavirus indicating person-to-person transmission: a study of a family cluster. Lancet 395(10223):514–523

Chen H, Du Q (2020) Potential natural compounds for preventing 2019-nCoV infection. Preprints:2020010358 (v2020010353)

Chen D, Yang H, Cao Y, Cheng W, Duan T, Fan C, Fan S, Feng L, Gao Y, He F (2020a) Expert consensus for managing pregnant women and neonates born to mothers with suspected or confirmed novel coronavirus (COVID-19) infection. Int J Gynecol Obstet

Chen H, Guo J, Wang C, Luo F, Yu X, Zhang W, Li J, Zhao D, Xu D, Gong Q, Liao J, Yang H, Hou W, Zhang Y (2020b) Clinical characteristics and intrauterine vertical transmission potential of COVID-19 infection in nine pregnant women: a retrospective review of medical records. Lancet 395(10226):809–815. https://doi.org/10.1016/s0140-6736(20)30360-3

Chen L, Xiong J, Bao L, Shi Y (2020c) Convalescent plasma as a potential therapy for COVID-19. Lancet Infect Dis 20(4):398–400. https://doi.org/10.1016/S1473-3099(20)30141-9

Chen WH, Hotez PJ, Bottazzi ME (2020d) Potential for developing a SARS-CoV receptor-binding domain (RBD) recombinant protein as a heterologous human vaccine against coronavirus infectious disease (COVID)-19. Hum Vaccin Immunother:1–4. https://doi.org/10.1080/21645515.2020.1740560

Cheng Y, Wong R, Soo YO, Wong WS, Lee CK, Ng MH, Chan P, Wong KC, Leung CB, Cheng G (2005) Use of convalescent plasma therapy in SARS patients in Hong Kong. Eur J Clin Microbiol Infect Dis 24(1):44–46. https://doi.org/10.1007/s10096-004-1271-9

Control CfD, Prevention Interim Additional Guidance for Infection Prevention and Control Recommendations for Patients with Suspected or Confirmed COVID-19 in Outpatient Hemodialysis Facilities, 2020. Full text

Control CfD, Prevention (2020) Interim guidelines for collecting, handling, and testing clinical specimens from Persons Under Investigation (PUIs) for coronavirus disease 2019 (COVID-19). February 14, 2020

Corman VM, Landt O, Kaiser M, Molenkamp R, Meijer A, Chu DK, Bleicker T, Brunink S, Schneider J, Schmidt ML, Mulders DG, Haagmans BL, van der Veer B, van den Brink S, Wijsman L, Goderski G, Romette JL, Ellis J, Zambon M, Peiris M, Goossens H, Reusken C, Koopmans MP, Drosten C (2020) Detection of 2019 novel coronavirus (2019-nCoV) by real-time RT-PCR. Euro Surveill 25(3). https://doi.org/10.2807/1560-7917.ES.2020.25.3.2000045

COVID C, COVID C, COVID C, Bialek S, Gierke R, Hughes M, McNamara LA, Pilishvili T, Skoff T (2020) Coronavirus disease 2019 in children—United States, February 12–April 2, 2020. Morb Mortal Wkly Rep 69(14):422

Dalton L, Rapa E, Stein A (2020) Protecting the psychological health of children through effective communication about COVID-19. Lancet Child Adolesc Health 4(5):346–347. https://doi.org/10.1016/S2352-4642(20)30097-3

Darbeheshti F, Rezaei N (2020) Genetic predisposition models to COVID-19 infection. Med Hypotheses 142:109818. https://doi.org/10.1016/j.mehy.2020.109818

Davanzo R (2020) Breast feeding at the time of COVID-19: do not forget expressed mother's milk, please. Arch Dis Child Fetal Neonatal Ed. https://doi.org/10.1136/archdischild-2020-319149

Dayal D (2020) We urgently need guidelines for managing COVID-19 in children with comorbidities. Acta Paediatr 109(7):1497–1498. https://doi.org/10.1111/apa.15304

Della Gatta AN, Rizzo R, Pilu G, Simonazzi G (2020) COVID19 during pregnancy: a systematic review of reported cases. Am J Obstet Gynecol

Di Mascio D, Khalil A, Saccone G, Rizzo G, Buca D, Liberati M, Vecchiet J, Nappi L, Scambia G, Berghella V (2020) Outcome of Coronavirus spectrum infections (SARS, MERS, COVID 1-19) during pregnancy: a systematic review and meta-analysis. Am J Obstet Gynecol MFM:100107

Dilucca M, Forcelloni S, Georgakilas AG, Giansanti A, Pavlopoulou A (2020) Codon usage and phenotypic divergences of SARS-CoV-2 genes. Viruses 12(5). https://doi.org/10.3390/v12050498

Dong Y, Mo X, Hu Y, Qi X, Jiang F, Jiang Z, Tong S (2020) Epidemiological characteristics of 2143 pediatric patients with 2019 coronavirus disease in China. Pediatrics

Eberhardt JN, Breuckmann NP, Eberhardt CS (2020) Multi-stage group testing optimizes COVID-19 mass population testing. medRxiv. https://doi.org/10.1101/2020.04.10.20061176

Ebrahim SH, Ahmed QA, Gozzer E, Schlagenhauf P, Memish ZA (2020) Covid-19 and community mitigation strategies in a pandemic. BMJ:m1066. https://doi.org/10.1136/bmj.m1066

Fathi N, Rezaei N (2020) Lymphopenia in COVID-19: therapeutic opportunities. Cell Biol Int 44(9):1792–1797. https://doi.org/10.1002/cbin.11403

Ferioli M, Cisternino C, Leo V, Pisani L, Palange P, Nava S (2020) Protecting healthcare workers from SARS-CoV-2 infection: practical indications. European respiratory review: an official journal of the European respiratory. Society 29(155). https://doi.org/10.1183/16000617.0068-2020

Fernandes, Nuno, Economic Effects of Coronavirus Outbreak (COVID-19) on the World Economy (March 22, 2020). Available at SSRN: https://ssrn.com/abstract=3557504 or https://doi.org/10.2139/ssrn.3557504

Ferretti L, Wymant C, Kendall M, Zhao L, Nurtay A, Abeler-Dorner L, Parker M, Bonsall DG, Fraser C (2020) Quantifying dynamics of SARS-CoV-2 transmission suggests that epidemic control and avoidance is feasible through instantaneous digital contact tracing. https://doi.org/10.1101/2020.03.08.20032946

Franklin R, Young A, Neumann B, Fernandez R, Joannides A, Reyahi A, Modis Y (2020) Homologous protein domains in SARS-CoV-2 and measles, mumps and rubella viruses: preliminary evidence that MMR vaccine might provide protection against COVID-19. medRxiv:2020.2004.2010.20053207. https://doi.org/10.1101/2020.04.10.20053207

Gail M, Krickeberg K, Samet JM, Tsiatis A, Wong W (2010) Modern infectious disease epidemiology. Springer, Statistics for Biology and Health

Gandhi RT, Lynch JB, del Rio C (2020) Mild or moderate Covid-19. N Engl J Med

Gao Q, Bao L, Mao H, Wang L, Xu K, Yang M, Li Y, Zhu L, Wang N, Lv Z, Gao H, Ge X, Kan B, Hu Y, Liu J, Cai F, Jiang D, Yin Y, Qin C, Li J, Gong X, Lou X, Shi W, Wu D, Zhang H, Zhu L, Deng W, Li Y, Lu J, Li C, Wang X, Yin W, Zhang Y, Qin C (2020) Rapid development of an inactivated vaccine candidate for SARS-CoV-2. Science. https://doi.org/10.1126/science.abc1932

Gates B (2018) Innovation for pandemics. N Engl J Med 378:2057–2060

Gates B (2020) Responding to Covid-19 — a once-in-a-century pandemic? N Engl J Med 382(18):1677–1679. https://doi.org/10.1056/NEJMp2003762

Gautret P, Lagier J-C, Parola P, Meddeb L, Mailhe M, Doudier B, Courjon J, Giordanengo V, Vieira VE, Dupont HT (2020) Hydroxychloroquine and azithromycin as a treatment of COVID-19: results of an open-label non-randomized clinical trial. Int J Antimicrob Agents:105949

Global Preparedness Monitoring Board (2019) Annual report on global preparedness for health emergencies. World Health Organization, Geneva. https://apps.who.int/gpmb/assets/annual_report/GPMB_annualreport_2019.pdf. Accessed 25 Jan 2020

Godaert L, Proye E, Demoustier-Tampere D, Coulibaly PS, Hequet F, Drame M (2020) Clinical characteristics of older patients: The experience of a geriatric short-stay unit dedicated to patients with COVID-19 in France. J Infect. https://doi.org/10.1016/j.jinf.2020.04.009

Gordon CJ, Tchesnokov EP, Feng JY, Porter DP, Götte M (2020) The antiviral compound remdesivir potently inhibits RNA-dependent RNA polymerase from Middle East respiratory syndrome coronavirus. J Biol Chem 295(15):4773–4779

Greenhalgh T, Koh GCH, Car J (2020) Covid-19: a remote assessment in primary care. BMJ:368

Gross L, Lhomme E, Pasin C, Richert L, Thiebaut R (2018) Ebola vaccine development: systematic review of pre-clinical and clinical studies, and meta-analysis of determinants of antibody response variability after vaccination. Int J Infect Dis 74:83–96. https://doi.org/10.1016/j.ijid.2018.06.022

Groß R, Conzelmann C, Müller JA, Stenger S, Steinhart K, Kirchhoff F, Münch J (2020) Detection of SARS-CoV-2in human breastmilk. Lancet (London, England): https://doi.org/10.1016/S0140-6736(1020)31181-31188

Guan W-J, Ni Z-y, Hu Y, Liang W-h, Ou C-q, He J-x, Liu L, Shan H, Lei C-l, Hui DS (2020) Clinical characteristics of coronavirus disease 2019 in China. N Engl J Med 382:1708–1720

Gudbjartsson DF, Helgason A, Jonsson H, Magnusson OT, Melsted P, Norddahl GL, Saemundsdottir J, Sigurdsson A, Sulem P, Agustsdottir AB, Eiriksdottir B, Fridriksdottir R, Gardarsdottir EE, Georgsson G, Gretarsdottir OS, Gudmundsson KR, Gunnarsdottir TR, Gylfason A, Holm H, Jensson BO, Jonasdottir A, Jonsson F, Josefsdottir KS, Kristjansson T, Magnusdottir DN, le Roux L, Sigmundsdottir G, Sveinbjornsson G, Sveinsdottir KE, Sveinsdottir M, Thorarensen EA, Thorbjornsson B, Löve A, Masson G, Jonsdottir I, Möller AD, Gudnason T, Kristinsson KG, Thorsteinsdottir U, Stefansson K (2020) Spread of SARS-CoV-2 in the Icelandic population. N Engl J Med. https://doi.org/10.1056/NEJMoa2006100

Guidance on Providing Pediatric Ambulatory Services via Telehealth During COVID-19 (2020 (Updated April 15, 2020))

Hagmann SH (2020) COVID-19 in children: more than meets the eye. Elsevier

Hanaei S, Rezaei N (2020) COVID-19: developing from an outbreak to a pandemic. Arch Med Res. https://doi.org/10.1016/j.arcmed.2020.04.021

He F, Deng Y, Li W (2020) Coronavirus disease 2019: what we know? J Med Virol. https://doi.org/10.1002/jmv.25766

Heinzerling A, Stuckey MJ, Scheuer T, Xu K, Perkins KM, Resseger H, Magill S, Verani JR, Jain S, Acosta M, Epson E (2020) Transmission of COVID-19 to health care personnel during exposures to a hospitalized patient – Solano County, California, February 2020. MMWR Morb Mortal Wkly Rep 69(15):472–476. https://doi.org/10.15585/mmwr.mm6915e5

Hellewell J, Abbott S, Gimma A, Bosse NI, Jarvis CI, Russell TW, Munday JD, Kucharski AJ, Edmunds WJ, Funk S, Eggo RM, Sun F, Flasche S, Quilty BJ, Davies N, Liu Y, Clifford S, Klepac P, Jit M, Diamond C, Gibbs H, van Zandvoort K (2020) Feasibility of controlling COVID-19 outbreaks by isolation of cases and contacts. Lancet Glob Health 8(4):e488–e496. https://doi.org/10.1016/s2214-109x(20)30074-7

Hensel J, McGrail DJ, McAndrews KM, Dowlatshahi D, LeBleu VS, Kalluri R (2020) Exercising caution in correlating COVID-19 incidence and mortality rates with BCG vaccination policies due to variable rates of SARS CoV-2 testing. medRxiv:2020.2004.2008.20056051. https://doi.org/10.1101/2020.04.08.20056051

Hick JL, Barbera JA, Kelen GD (2009) Refining surge capacity: conventional, contingency, and crisis capacity. Disaster Med Public Health Prep 3(S1):S59–S67

Hillaker E, Belfer JJ, Bondici A, Murad H, Dumkow LE (2020) Delayed initiation of remdesivir in a COVID-19 positive patient. Pharmacotherapy: The Journal of Human Pharmacology and Drug Therapy

Hockemeyer D, Urnov F, Doudna JA (2020) Blueprint for a pop-up SARS-CoV-2 testing lab. medRxiv. https://doi.org/10.1101/2020.04.11.20061424

Hoffmann M, Kleine-Weber H, Schroeder S, Krüger N, Herrler T, Erichsen S, Schiergens TS, Herrler G, Wu N-H, Nitsche A, Müller MA, Drosten C, Pöhlmann S (2020) SARS-CoV-2 cell entry depends on ACE2 and TMPRSS2 and is blocked by a clinically proven protease inhibitor. Cell 181(2):271–280.e278. https://doi.org/10.1016/j.cell.2020.02.052

Hogan CA, Sahoo MK, Pinsky BA (2020) Sample pooling as a strategy to detect community transmission of SARS-CoV-2. JAMA. https://doi.org/10.1001/jama.2020.5445

Hon K, Leung C, Cheng W, Chan P, Chu W, Kwan Y, Li A, Fong N, Ng P, Chiu M (2003) Clinical presentations and outcome of severe acute respiratory syndrome in children. Lancet 361(9370):1701–1703

Hou Z, Lin L, Lu L, Du F, Qian M, Liang Y, Zhang J, Yu H (2020) Public exposure to live animals, behavioural change, and support in containment measures in response to COVID-19 Outbreak: a population-based cross sectional survey in China. https://doi.org/10.1101/2020.02.21.20026146

https://clinicaltrials.gov/ct2/show/NCT04308668

Hu D, Zhu C, Ai L, He T, Wang Y, Ye F, Yang L, Ding C, Zhu X, Lv R (2018) Genomic characterization and infectivity of a novel SARS-like coronavirus in Chinese bats. Emerg Microbes Infect 7(1):1–10

Huang C, Wang Y, Li X, Ren L, Zhao J, Hu Y, Zhang L, Fan G, Xu J, Gu X (2020) Clinical features of patients infected with 2019 novel coronavirus in Wuhan, China. Lancet 395(10223):497–506

Hui DS, Ip M, Tang JW, Wong AL, Chan MT, Hall SD, Chan PK, Sung JJ (2006) Airflows around oxygen masks: a potential source of infection? Chest 130(3):822–826. https://doi.org/10.1378/chest.130.3.822

Hui DS, Chow BK, Chu L, Ng SS, Lai ST, Gin T, Chan MT (2011) Exhaled air dispersion and removal is influenced by isolation room size and ventilation settings during oxygen delivery via nasal cannula. Respirology (Carlton, Vic) 16(6):1005–1013. https://doi.org/10.1111/j.1440-1843.2011.01995.x

Hui DS, Chan MT, Chow B (2014) Aerosol dispersion during various respiratory therapies: a risk assessment model of nosocomial infection to health care workers. Hong Kong Med J = Xianggang yi xue za zhi 20(Suppl 4):9–13

Hui DS, Chow BK, Lo T, Ng SS, Ko FW, Gin T, Chan MTV (2015) Exhaled air dispersion during noninvasive ventilation via helmets and a total facemask. Chest 147(5):1336–1343. https://doi.org/10.1378/chest.14-1934

Ikizler TA, Kliger AS (2020) Minimizing the risk of COVID-19 among patients on dialysis. Nat Rev Nephrol 16(6):311–313. https://doi.org/10.1038/s41581-020-0280-y

Information for Pediatric Healthcare Providers (2020 (Updated April 17, 2020))

Interim Laboratory Biosafety Guidelines for Handling and Processing Specimens Associated with Coronavirus Disease 2019 (COVID-19) (2020 (Updated March 31, 2020))

Jabbari P, Rezaei N (2020) With risk of reinfection, is COVID-19 here to stay? Disaster Med Public Health Prep:1. https://doi.org/10.1017/dmp.2020.274

Jabbari P, Jabbari F, Ebrahimi S, Rezaei N (2020) COVID-19: a chimera of two pandemics. Disaster Med Public Health Prep:1–2. https://doi.org/10.1017/dmp.2020.223

Jahanshahlu L, Rezaei N (2020a) Central Nervous System Involvement in COVID-19. Arch Med Res. https://doi.org/10.1016/j.arcmed.2020.05.016

Jahanshahlu L, Rezaei N (2020b) Monoclonal antibody as a potential anti-COVID-19. Biomed Pharmacother 129:110337. https://doi.org/10.1016/j.biopha.2020.110337

Ji H, Yan Y, Ding B, Guo W, Brunswick M, Niethammer A, SooHoo W, Smith R, Nahama A, Zhang Y (2020) Novel decoy cellular vaccine strategy utilizing transgenic antigen-expressing cells as immune presenter and adjuvant in vaccine prototype against SARS-CoV-2 virus. Med Drug Discov 5:100026. https://doi.org/10.1016/j.medidd.2020.100026

Kalita P, Padhi AK, Zhang KYJ, Tripathi T (2020a) Design of a peptide-based subunit vaccine against novel coronavirus SARS-CoV-2. Microb Pathog 145:104236. https://doi.org/10.1016/j.micpath.2020.104236

Kalita P, Padhi AK, Zhang KYJ, Tripathi T (2020b) Design of a peptide-based subunit vaccine against novel coronavirus SARS-CoV-2. Microb Pathog 145:104236–104236. https://doi.org/10.1016/j.micpath.2020.104236

Kandel N, Chungong S, Omaar A, Xing J (2020) Health security capacities in the context of COVID-19 outbreak: an analysis of international health regulations annual report data from 182 countries. Lancet (London, England) 395:1047–1053

Kilinc FS (2015) A review of isolation gowns in healthcare: fabric and gown properties. J Eng Fibers fabr 10(3):155892501501000313

Ko JH, Seok H, Cho SY, Ha YE, Baek JY, Kim SH, Kim YJ, Park JK, Chung CR, Kang ES, Cho D, Muller MA, Drosten C, Kang CI, Chung DR, Song JH, Peck KR (2018) Challenges of convalescent plasma infusion therapy in Middle East respiratory coronavirus infection: a single centre experience. Antivir Ther 23(7):617–622. https://doi.org/10.3851/imp3243

Kourtis AP, Read JS, Jamieson DJ (2014) Pregnancy and infection. N Engl J Med 370(23):2211–2218. https://doi.org/10.1056/NEJMra1213566

Kwon KT, Ko J-H, Shin H, Sung M, Kim JY (2020) Drive-through screening center for COVID-19: a safe and efficient screening system against massive community outbreak. J Korean Med Sci 35(11). https://doi.org/10.3346/jkms.2020.35.e123

Langhi DM, Santis GCD, Bordin JO (2020) COVID-19 convalescent plasma transfusion. Hematology, transfusion and cell therapy. https://doi.org/10.1016/j.htct.2020.04.003

Lee VJ, Chiew CJ, Khong WX (2020) Interrupting transmission of COVID-19: lessons from containment efforts in Singapore. J Travel Med:taaa039. https://doi.org/10.1093/jtm/taaa039/5804843

Li Y, Wang H, Tang X, Ma D, Du C, Wang Y, Pan H, Zou Q, Zheng J, Xu L, Farzan M, Zhong G (2020) Potential host range of multiple SARS-like coronaviruses and an improved ACE2-Fc variant that is potent against both SARS-CoV-2 and SARS-CoV-1. https://doi.org/10.1101/2020.04.10.032342

Lin S, Qiao Y, Huang J, Yan N (2020) Research on the Influence of Effective Distance Between Cities on the Cross-regional Transmission of COVID-19. https://doi.org/10.1101/2020.03.27.20044958

Liu J, Liao X, Qian S, Yuan J, Wang F, Liu Y, Wang Z, Wang F-S, Liu L, Zhang Z (2020) Community transmission of severe acute respiratory syndrome coronavirus 2, Shenzhen, China, 2020. Emerg Infect Dis 26(6)

Lotfi M, Rezaei N (2020) SARS-CoV-2: a comprehensive review from pathogenicity of the virus to clinical consequences. J Med Virol. https://doi.org/10.1002/jmv.26123

Lotfi M, Hamblin MR, Rezaei N (2020) COVID-19: transmission, prevention, and potential therapeutic opportunities. Clin Chim Acta; Int J Clin Chem 508:254–266. https://doi.org/10.1016/j.cca.2020.05.044

Lu R, Zhao X, Li J, Niu P, Yang B, Wu H, Wang W, Song H, Huang B, Zhu N (2020a) Genomic characterisation and epidemiology of 2019 novel coronavirus: implications for virus origins and receptor binding. Lancet 395(10224):565–574

Lu X, Zhang L, Du H, Zhang J, Li YY, Qu J, Zhang W, Wang Y, Bao S, Li Y (2020b) SARS-CoV-2 infection in children. N Engl J Med

Ludvigsson JF (2020) Systematic review of COVID-19 in children show milder cases and a better prognosis than adults. Acta Paediatrica

Luke TC, Casadevall A, Watowich SJ, Hoffman SL, Beigel JH, Burgess TH (2010) Hark back: passive immunotherapy for influenza and other serious infections. Crit Care Med 38(4 Suppl):e66–e73. https://doi.org/10.1097/CCM.0b013e3181d44c1e

Mallineni SK, Innes NP, Raggio DP, Araujo MP, Robertson MD, Jayaraman J (2020) Coronavirus disease (COVID-19): characteristics in children and considerations for dentists providing their care. Int J Paediatr Dent. https://doi.org/10.1111/ipd.12653

Mansourabadi AH, Sadeghalvad M, Mohammadi-Motlagh HR, Rezaei N (2020) The immune system as a target for therapy of SARS-CoV-2: a systematic review of the current immunotherapies for COVID-19. Life Sci 258:118185. https://doi.org/10.1016/j.lfs.2020.118185

McBryde E (2020) The value of early transmission dynamic studies in emerging infectious diseases. Lancet Infect Dis

McKibbin WJ, Fernando R. The global macroeconomic impacts of COVID-19: seven scenarios (March 2, 2020). CAMA working paper no. 19/2020. Available at SSRN: https://ssrn.com/abstract=3547729 or https://doi.org/10.2139/ssrn.3547729

Miller A, Reandelar MJ, Fasciglione K, Roumenova V, Li Y, Otazu GH (2020) Correlation between universal BCG vaccination policy and reduced morbidity and mortality for COVID-19: an epidemiological study. medRxiv:2020.2003.2024.20042937. https://doi.org/10.1101/2020.03.24.20042937

Miyasaka M (2020) Is BCG vaccination causally related to reduced COVID-19 mortality? EMBO Mol Med. https://doi.org/10.15252/emmm.202012661

Moazzami B, Razavi-Khorasani N, Dooghaie Moghadam A, Farokhi E, Rezaei N (2020) COVID-19 and telemedicine: immediate action required for maintaining healthcare providers well-being. J Clin Virol 126:104345. https://doi.org/10.1016/j.jcv.2020.104345

Mohamed K, Rodríguez-Román E, Rahmani F, Zhang H, Ivanovska M, Makka SA, Joya M, Makuku R, Islam MS, Radwan N, Rahmah L, Goda R, Abarikwu SO, Shaw M, Zoghi S, Irtsyan S, Ling I, Cseprekal O, Faten AB, Hazar Sayar E, Soloukey C, Grancini G, Rezaei N (2020a) Borderless collaboration is needed for COVID-19-A disease that knows no borders. Infect Control Hosp Epidemiol:1–2. https://doi.org/10.1017/ice.2020.162

Mohamed K, Yazdanpanah N, Saghazadeh A, Rezaei N (2020b) Computational drug discovery and repurposing for the treatment of Covid-19: a systematic review. Available at SSRN 3583748

Momtazmanesh S, Ochs HD, Uddin LQ, Perc M, Routes JM, Vieira DN, Al-Herz W, Baris S, Prando C, Rosivall L, Abdul Latiff AH, Ulrichs T, Roudenok V, Aldave Becerra JC, Salunke DB, Goudouris E, Condino-Neto A, Stashchak A, Kryvenko O, Stashchak M, Bondarenko A, Rezaei N (2020) All together to fight COVID-19. Am J Trop Med Hyg 102(6):1181–1183. https://doi.org/10.4269/ajtmh.20-0281

Monteleone PA, Nakano M, Lazar V, Gomes AP, de Martin H, Bonetti TC (2020) A review of initial data on pregnancy during the COVID-19 outbreak: implications for assisted reproductive treatments. JBRA Assist Reprod 24 (2):219

Moore SC, Penrice-Randal R, Alruwaili M, Dong X, Pullan ST, Carter D, Bewley K, Zhao Q, Sun Y, Hartley C, Zhou E-m, Solomon T, Beadsworth MBJ, Cruise J, Bogaert D, Crook DWT, Matthews DA, Davidson AD, Mahmood Z, Aljabr W, Druce J, Vipond RT, Ng L, Renia L, Openshaw P, Baillie JK, Carroll MW, Semple C, Turtle L, Hiscox JA (2020) Amplicon based MinION sequencing of SARS-CoV-2 and metagenomic characterisation of nasopharyngeal swabs from patients with COVID-19. https://doi.org/10.1101/2020.03.05.20032011

Moradian N, Ochs HD, Sedikies C, Hamblin MR, Camargo CA Jr, Martinez JA, Biamonte JD, Abdollahi M, Torres PJ, Nieto JJ, Ogino S, Seymour JF, Abraham

A, Cauda V, Gupta S, Ramakrishna S, Sellke FW, Sorooshian A, Wallace Hayes A, Martinez-Urbistondo M, Gupta M, Azadbakht L, Esmaillzadeh A, Kelishadi R, Esteghamati A, Emam-Djomeh Z, Majdzadeh R, Palit P, Badali H, Rao I, Saboury AA, Jagan Mohan Rao L, Ahmadieh H, Montazeri A, Fadini GP, Pauly D, Thomas S, Moosavi-Movahed AA, Aghamohammadi A, Behmanesh M, Rahimi-Movaghar V, Ghavami S, Mehran R, Uddin LQ, Von Herrath M, Mobasher B, Rezaei N (2020) The urgent need for integrated science to fight COVID-19 pandemic and beyond. J Transl Med 18(1):205. https://doi.org/10.1186/s12967-020-02364-2

Moraes TJ, Lin GH, Wen T, Watts TH (2011) Incorporation of 4-1BB ligand into an adenovirus vaccine vector increases the number of functional antigen-specific CD8 T cells and enhances the duration of protection against influenza-induced respiratory disease. Vaccine 29(37):6301–6312. https://doi.org/10.1016/j.vaccine.2011.06.022

Mukherjee R (2020) Global efforts on vaccines for COVID-19: since, sooner or later, we all will catch the coronavirus. J Biosci 45(1):68. https://doi.org/10.1007/s12038-020-00040-7

Muus C, Luecken MD, Eraslan G, Waghray A, Heimberg G, Sikkema L, Kobayashi Y, Vaishnav ED, Subramanian A, Smilie C, Jagadeesh K, Duong ET, Fiskin E, Triglia ET, Ansari M, Cai P, Lin B, Buchanan J, Chen S, Shu J, Haber AL, Chung H, Montoro DT, Adams T, Aliee H, Samuel J, Andrusivova AZ, Angelidis I, Ashenberg O, Bassler K, Bécavin C, Benhar I, Bergenstråhle J, Bergenstråhle L, Bolt L, Braun E, Bui LT, Chaffin M, Chichelnitskiy E, Chiou J, Conlon TM, Cuoco MS, Deprez M, Fischer DS, Gillich A, Gould J, Guo M, Gutierrez AJ, Habermann AC, Harvey T, He P, Hou X, Hu L, Jaiswal A, Jiang P, Kapellos T, Kuo CS, Larsson L, Leney-Greene MA, Lim K, Litviňuková M, Lu J, Ludwig LS, Luo W, Maatz H, Madissoon E, Mamanova L, Manakongtreecheep K, Marquette C-H, Mbano I, McAdams AM, Metzger RJ, Nabhan AN, Nyquist SK, Penland L, Poirion OB, Poli S, Qi C, Queen R, Reichart D, Rosas I, Schupp J, Sinha R, Sit RV, Slowikowski K, Slyper M, Smith N, Sountoulidis A, Strunz M, Sun D, Talavera-López C, Tan P, Tantivit J, Travaglini KJ, Tucker NR, Vernon K, Wadsworth MH, Waldman J, Wang X, Yan W, Zhao W, Ziegler CGK (2020) Integrated analyses of single-cell atlases reveal age, gender, and smoking status associations with cell type-specific expression of mediators of SARS-CoV-2 viral entry and highlights inflammatory programs in putative target cells. bioRxiv (2020.04.19.049254). https://doi.org/10.1101/2020.04.19.049254

Nasab MG, Saghazadeh A, Rezaei N (2020) SARS-CoV-2-a tough opponent for the immune system. Arch Med Res. https://doi.org/10.1016/j.arcmed.2020.05.020

Neal LR, Bailey SR, Wyatt MM, Bowers JS, Majchrzak K, Nelson MH, Haupt C, Paulos CM, Varela JC (2017) The basics of artificial antigen presenting cells in T cell-based Cancer immunotherapies. J Immunol Res Ther 2(1):68–79

Nelson-Piercy C (2015) Handbook of obstetric medicine. CRC Press, Boca Raton

Occupational Safety and Health Administration (2020) Border protection and transportation security workers and employers

Occupational safety and health administration Postmortem care workers and employers. https://www.osha.gov/SLTC/covid-19/controlprevention.html#interim_increased_risk

Ong SWX, Tan YK, Chia PY, Lee TH, Ng OT, Wong MSY, Marimuthu K (2020) Air, surface environmental, and personal protective equipment contamination by severe acute respiratory syndrome coronavirus 2 (SARS-CoV-2) from a symptomatic patient. Jama

Organization WH (2020a) Clinical management of severe acute respiratory infection (SARI) when COVID-19 disease is suspected: interim guidance, 13 March 2020. World Health Organization

Organization WH (2020b) Home care for patients with COVID-19 presenting with mild symptoms and management of their, Interim guidance, 17 February 2020

Organization WH (2020c) Mental health and psychosocial considerations during the COVID-19 outbreak, 18 March 2020. World Health Organization

Organization WH (2020d) Modes of transmission of virus causing COVID-19: implications for IPC precaution recommendations: scientific brief, 27 March 2020. World Health Organization

Organization WH (2020e) WHO interim guidance on clinical management of severe acute respiratory infection when novel coronavirus (nCoV) infection is suspected

Organization WH (2020f) WHO interim guidance on clinical management of severe acute respiratory infection when novel coronavirus (nCoV) infection is suspected

Outpatient and Ambulatory Care Settings: Responding to Community Transmission of COVID-19 in the United States (2020 (Reviewed April 7, 2020))

Pashaei M, Rezaei N (2020) Immunotherapy for SARS-CoV-2: potential opportunities. Expert Opin Biol Ther:1–5. https://doi.org/10.1080/14712598.2020.1807933

Pediatrics AAo (2020) Guidance on providing pediatric ambulatory services via telehealth during COVID-19

Popescu S (2020) Roadblocks to infection Prevention efforts in healthcare SARS-CoV-2/COVID-19 response. Disaster Med Public Health Prep:1–7. https://doi.org/10.1017/dmp.2020.55

Pourahmad R, Moazzami B, Rezaei N (2020) Efficacy of Plasmapheresis and immunoglobulin replacement therapy (IVIG) on patients with COVID-19. SN Compre Clin Med:1–5

Prevention CfDCa (2020a) COVID-19 Guidance for Older Adults

Prevention CfDCa (2020b) Develop a Care Plan

Prevention CfDCa (2020c) Discontinuation of Isolation for Persons with COVID-19 Not in Healthcare Settings (Interim Guidance)

Prevention CfDCa (2020d) Interim clinical guidance for management of patients with confirmed 2019 novel coronavirus (2019-nCoV) infection

Prevention CfDCa (2020e) K-12 Schools and Child Care Programs

Prevention CfDCa (2020f) Information for pediatric healthcare providers

Prevention CfDCa (2020g) People Who Are at Higher Risk for Severe Illness

Prevention CfDCa (2020h) Preparing for COVID-19: long-term care facilities. Homes, Nursing

Prevention CfDCa (2020i) Updated Interim Guidance for Airlines and Airline Crew: Coronavirus Disease 2019 (COVID-19)

Prevention II (2020) Control recommendations for patients with suspected or confirmed coronavirus disease 2019 (COVID-19) in healthcare settings. Centers for Disease Control and Prevention

Rabiee N, Rabiee M, Bagherzadeh M, Rezaei N (2020) COVID-19 and picotechnology: potential opportunities. Med Hypotheses 144:109917. https://doi.org/10.1016/j.mehy.2020.109917

Rauch S, Jasny E, Schmidt KE, Petsch B (2018) New vaccine technologies to combat outbreak situations. Front Immunol 9:1963. https://doi.org/10.3389/fimmu.2018.01963

Recommended guidance for extended use and limited reuse of N95 filtering facepiece respirators in healthcare settings (2014 (Reviewed March 27, 2020)). National Institute for Occupational Safety and Health (NIOSH)

Rezaei N (2020a) COVID-19 affects healthy pediatricians more than pediatric patients. Infect Control Hosp Epidemiol 1. https://doi.org/10.1017/ice.2020.139

Rezaei N (2020b) COVID-19 and medical biotechnology. Avicenna J Med Biotechnol 12(3):139

Rodriguez-Morales AJ, Cardona-Ospina JA, Gutierrez-Ocampo E, Villamizar-Pena R, Holguin-Rivera Y, Escalera-Antezana JP, Alvarado-Arnez LE, Bonilla-Aldana DK, Franco-Paredes C, Henao-Martinez AF, Paniz-Mondolfi A, Lagos-Grisales GJ, Ramirez-Vallejo E, Suarez JA, Zambrano LI, Villamil-Gomez WE, Balbin-Ramon GJ, Rabaan AA, Harapan H, Dhama K, Nishiura H, Kataoka H, Ahmad T, Sah R (2020) Clinical, laboratory and imaging features of COVID-19: A systematic review and meta-analysis. Travel Med Infect Dis:101623. https://doi.org/10.1016/j.tmaid.2020.101623

Rokni M, Hamblin MR, Rezaei N (2020) Cytokines and COVID-19: friends or foes? Hum Vaccin Immunother:1–3

Rzymski P, Nowicki M, Mullin GE, Abraham A, Rodríguez-Román E, Petzold MB, Bendau A, Sahu KK, Ather A, Naviaux A-F (2020) Quantity does not equal quality: scientific principles cannot be sacrificed. Int Immunopharmacol 86:106711

Saghazadeh A, Rezaei N (2020a) Immune-epidemiological parameters of the novel coronavirus - a perspective. Expert Rev Clin Immunol 16(5):465–470. https://doi.org/10.1080/1744666x.2020.1750954

Saghazadeh A, Rezaei N (2020b) Towards treatment planning of COVID-19: rationale and hypothesis for the use of multiple immunosuppressive agents: anti-antibodies, immunoglobulins, and corticosteroids. Int Immunopharmacol 84:106560. https://doi.org/10.1016/j.intimp.2020.106560

Sahr F, Ansumana R, Massaquoi TA, Idriss BR, Sesay FR, Lamin JM, Baker S, Nicol S, Conton B, Johnson W, Abiri OT, Kargbo O, Kamara P, Goba A, Russell JB, Gevao SM (2017) Evaluation of convalescent whole blood for treating Ebola Virus Disease in Freetown, Sierra Leone. J Infect 74(3):302–309. https://doi.org/10.1016/j.jinf.2016.11.009

Saleki K, Banazadeh M, Saghazadeh A, Rezaei N (2020) The involvement of the central nervous system in patients with COVID-19. Rev Neurosci 31(4):453–456. https://doi.org/10.1515/revneuro-2020-0026

Shah S, Das S, Jain A, Misra DP, Negi VS (2020) A systematic review of the prophylactic role of chloroquine and hydroxychloroquine in coronavirus Disease-19 (COVID-19). Int J Rheum Dis

Shamshirian D, Rezaei N (2020) Cardiovascular diseases burden in COVID-19: Systematic review and meta-analysis

Sharifkashani S, Bafrani MA, Khaboushan AS, Pirzadeh M, Kheirandish A, Yavarpour Bali H, Hessami A, Saghazadeh A, Rezaei N (2020) Angiotensin-converting enzyme 2 (ACE2) receptor and SARS-CoV-2: potential therapeutic targeting. Eur J Pharmacol 884:173455. https://doi.org/10.1016/j.ejphar.2020.173455

Shen C, Wang Z, Zhao F, Yang Y, Li J, Yuan J, Wang F, Li D, Yang M, Xing L, Wei J, Xiao H, Yang Y, Qu J, Qing L, Chen L, Xu Z, Peng L, Li Y, Zheng H, Chen F, Huang K, Jiang Y, Liu D, Zhang Z, Liu Y, Liu L (2020) Treatment of 5 critically ill patients with COVID-19 with convalescent plasma. JAMA 323(16):1582–1589. https://doi.org/10.1001/jama.2020.4783

Shereen MA, Khan S, Kazmi A, Bashir N, Siddique R (2020) COVID-19 infection: origin, transmission, and characteristics of human coronaviruses. J Adv Res

Shi J, Wen Z, Zhong G, Yang H, Wang C, Liu R, He X, Shuai L, Sun Z, Zhao Y, Liang L, Cui P, Wang J, Zhang X, Guan Y, Chen H, Bu Z (2020a) Susceptibility of ferrets, cats, dogs, and different domestic animals to SARS-coronavirus-2. https://doi.org/10.1101/2020.03.30.015347

Shi Y, Wang Y, Shao C, Huang J, Gan J, Huang X, Bucci E, Piacentini M, Ippolito G, Melino G (2020b) COVID-19 infection: the perspectives on immune responses. Cell Death Differ 27(5):1451–1454. https://doi.org/10.1038/s41418-020-0530-3

Shirakawa T, Kitagawa K (2018) Antitumor effect of oral cancer vaccine with Bifidobacterium delivering WT1 protein to gut immune system is superior to WT1 peptide vaccine. Hum Vaccin Immunother 14(1):159–162. https://doi.org/10.1080/21645515.2017.1382787

Stokes JJ, Wolman IJ, Carpenter HC, Margolis J (1935) Prophylactic use of parents' whole blood in anterior poliomyelitis: philadelphia epidemic of 1932. Am J Dis Child 50(3):581–595. https://doi.org/10.1001/archpedi.1935.01970090011002

Ströher U, DiCaro A, Li Y, Strong JE, Aoki F, Plummer F, Jones SM, Feldmann H (2004) Severe acute respiratory syndrome-related coronavirus is inhibited by interferon-α. J Infect Dis 189 (7):1164–1167. doi:https://doi.org/10.1086/382597

Stuebe A (2020) Should infants be separated from mothers with COVID-19? first, do no harm. Breastfeed Med

Tai W, He L, Zhang X, Pu J, Voronin D, Jiang S, Zhou Y, Du L (2020) Characterization of the receptor-binding domain (RBD) of 2019 novel coronavirus: implication for development of RBD protein as a viral attachment inhibitor and vaccine. Cell Mol Immunol:1–8

Tchesnokov EP, Feng JY, Porter DP, Götte M (2019) Mechanism of inhibition of Ebola virus RNA-dependent RNA polymerase by remdesivir. Viruses 11(4):326

Teijaro JR (2016) Type I interferons in viral control and immune regulation. Curr Opin Virol 16:31–40. https://doi.org/10.1016/j.coviro.2016.01.001

Thanh Le T, Andreadakis Z, Kumar A, Gomez Roman R, Tollefsen S, Saville M, Mayhew S (2020) The COVID-19 vaccine development landscape. Nat Rev Drug Discov 19(5):305–306. https://doi.org/10.1038/d41573-020-00073-5

van Doremalen N, Bushmaker T, Morris DH, Holbrook MG, Gamble A, Williamson BN, Tamin A, Harcourt JL, Thornburg NJ, Gerber SI (2020) Aerosol and surface stability of SARS-CoV-2 as compared with SARS-CoV-1. N Engl J Med

van Erp EA, Luytjes W, Ferwerda G, van Kasteren PB (2019) Fc-mediated antibody effector functions during respiratory syncytial virus infection and disease. Front Immunol 10:548. https://doi.org/10.3389/fimmu.2019.00548

Vincent MJ, Bergeron E, Benjannet S, Erickson BR, Rollin PE, Ksiazek TG, Seidah NG, Nichol ST (2005) Chloroquine is a potent inhibitor of SARS coronavirus infection and spread. Virol J 2(1):69

Wan Y, Shang J, Graham R, Baric RS, Li F (2020) Receptor recognition by the novel coronavirus from Wuhan: an analysis based on decade-long structural studies of SARS coronavirus. J Virol 94(7)

Wang CJ, Ng CY, Brook RH (2020a) Response to COVID-19 in Taiwan: big data analytics, new technology, and proactive testing. JAMA. https://doi.org/10.1001/jama.2020.3151

Wang F, Kream RM, Stefano GB (2020b) An evidence based perspective on mRNA-SARS-CoV-2 vaccine development. Med Sci Monit 26:e924700. https://doi.org/10.12659/msm.924700

Wang J, Zhou M, Liu F (2020c) Reasons for healthcare workers becoming infected with novel coronavirus disease 2019 (COVID-19) in China. J Hosp Infect. https://doi.org/10.1016/j.jhin.2020.03.002

Wang M, Cao R, Zhang L, Yang X, Liu J, Xu M, Shi Z, Hu Z, Zhong W, Xiao G (2020d) Remdesivir and chloroquine effectively inhibit the recently emerged novel coronavirus (2019-nCoV) in vitro. Cell Res 30(3):269–271

Wang S, Tu J, Sheng Y (2020e) Clinical characteristics and fecal–oral transmission potential of patients with COVID-19. medRxiv. https://doi.org/10.1101/2020.05.02.20089094.this

Wang S, Wen X, Liu B, Dong Y, Cui M (2020f) Psychological influence of Coronovirus disease 2019 (COVID-19) pandemic on the general public, medical workers and patients with mental disorders and its countermeasures. Psychosomatics. https://doi.org/10.1016/j.psym.2020.05.005

Wang W, Tang J, Wei F (2020g) Updated understanding of the outbreak of 2019 novel coronavirus (2019-nCoV) in Wuhan. China 92(4):441–447. https://doi.org/10.1002/jmv.25689

Wang Y, Liu Y, Liu L, Wang X, Luo N, Ling L (2020h) Clinical outcome of 55 asymptomatic cases at the time of hospital admission infected with SARS-Coronavirus-2 in Shenzhen, China. J Infect Dis

Wei M, Yuan J, Liu Y, Fu T, Yu X, Zhang Z (2020) Novel coronavirus infection in hospitalized infants under 1 year of age in China. JAMA. Published online February 14, 2020: E1-E2

Weiss P, Murdoch DR (2020) Clinical course and mortality risk of severe COVID-19. Lancet 395(10229):1014–1015. https://doi.org/10.1016/S0140-6736(20)30633-4

WHO. WHO benchmarks for international health regulations (IHR) capacities. Geneva: World Health Organization, 2019. www.who.int/ihr/publications/9789241515429/en/. Accessed 26 Jan 26 2020

WHO (2020) Breastfeeding advice during the COVID-19 outbreak

WHO. Cleaning and disinfection of environmental surfaces in the context of COVID-19. Geneva: World Health Organization, May 16, 2020. https://www.who.int/publications-detail/cleaning-and-disinfection--of-environmental-surfaces-inthe-context-of-covid-19. Accessed 29 May 2020

WHO. WHO Director-General's statement on IHR Emergency Committee on Novel Coronavirus (2019-nCoV). Geneva: World Health Organization, Jan 30, 2020. https://www.who.int/dg/speeches/detail/who-director-general-s-statement-on-ihr-emergencycommittee-on-novel-coronavirus-(2019-ncov). Accessed 30 Jan 2020

Wilder-Smith A, Chiew CJ, Lee VJ (2020) Can we contain the COVID-19 outbreak with the same measures as for SARS? Lancet Infect Dis 20(5):e102–e107. https://doi.org/10.1016/s1473-3099(20)30129-8

Wim-Van L, Parolin Z (2020) COVID-19, school closures, and child poverty: a social crisis in the making. Lancet Public Health

World Health Organization, Regional Office for Europe. Mental health and psychological resilience during the COVID-19 pandemic. Available at:http://www.euro.who.int/en/health-topics/health-emergencies/coronavirus-covid-19/news/news/2020/3/mental-health-and-psychological-resilience-during-the-covid-19-pandemic. Accessed 12 May 2020

World Health Organization, UNICEF: Knowledge, attitude and practice survey on COVID-19 response. Available at:https://www.unicef.org/guyanasuriname/press-releases/knowledge-attitude-and-practice-survey-covid-19-response. Accessed 12 May 2020

Wu Z, McGoogan JM (2020) Characteristics of and important lessons from the coronavirus disease 2019 (COVID-19) outbreak in China: summary of a report of 72 314 cases from the Chinese Center for Disease Control and Prevention. JAMA 323(13):1239–1242

Xu J, Zhao S, Teng T, Abdalla AE, Zhu W, Xie L, Wang Y, Guo X (2020) Systematic comparison of two animal-to-human transmitted human coronaviruses: SARS-CoV-2 and SARS-CoV. Viruses 12(2):244

Yan CX, Li J, Shen X, Luo L, Li Y, Li MY (2020) [Biological Product Development Strategies for Prevention and Treatment of Coronavirus Disease 2019]. Sichuan Da Xue Xue Bao Yi Xue Ban 51 (2):139–145. https://doi.org/10.12182/20200360506

Yang J, Zheng Y, Gou X, Pu K, Chen Z, Guo Q, Ji R, Wang H, Wang Y, Zhou Y (2020a) Prevalence of comorbidities and its effects in patients infected with SARS-CoV-2: a systematic review and meta-analysis. Int J Infect Dis 94:91–95

Yang P, Liu P, Li D, Zhao D (2020b) Corona Virus Disease 2019, a growing threat to children? J Infect

Yang Y, Islam MS, Wang J, Li Y, Chen X (2020c) Traditional Chinese medicine in the treatment of patients infected with 2019-new coronavirus (SARS-CoV-2): a review and perspective. Int J Biol Sci 16(10):1708

Yazdanpanah F, Hamblin MR, Rezaei N (2020a) The immune system and COVID-19: friend or foe? Life Sci 256:117900. https://doi.org/10.1016/j.lfs.2020.117900

Yazdanpanah N, Saghazadeh A, Rezaei N (2020b) Anosmia: a missing link in the neuroimmunology of coronavirus disease 2019 (COVID-19). Rev Neurosci. https://doi.org/10.1515/revneuro-2020-0039

Yousefzadegan S, Rezaei N (2020) Case report: death due to COVID-19 in three brothers. Am J Trop Med Hyg 102(6):1203–1204. https://doi.org/10.4269/ajtmh.20-0240

Zaigham M, Andersson O (2020) Maternal and perinatal outcomes with COVID-19: a systematic review of 108 pregnancies. Acta Obstetricia et Gynecologica Scandinavica 99(7). https://doi.org/10.1111/aogs.13867

Zhang L, Liu Y (2020) Potential interventions for novel coronavirus in China: a systematic review. https://doi.org/10.1002/jmv.25707

Zhang Z, Wu Q, Zhang T (2020) Pangolin homology associated with 2019-nCoV. bioRxiv

Zheng F, Liao C, Fan Q-h, Chen H-b, Zhao X-g, Xie Z-g, Li X-l, Chen C-x, Lu X-x, Liu Z-s (2020) Clinical characteristics of children with coronavirus disease 2019 in Hubei, China. Curr Med Sci:1–6

Zhou B, Zhong N, Guan Y (2007) Treatment with convalescent plasma for influenza a (H5N1) infection. N Engl J Med 357(14):1450–1451. https://doi.org/10.1056/NEJMc070359

Zhou P, Huang Z, Xiao Y, Huang X, Fan XG (2020a) Protecting Chinese healthcare workers while combating the 2019 novel coronavirus. Infect Control Hosp Epidemiol:1–4. https://doi.org/10.1017/ice.2020.60

Zhou P, Yang X-L, Wang X-G, Hu B, Zhang L, Zhang W, Si H-R, Zhu Y, Li B, Huang C-L (2020b) A pneumonia outbreak associated with a new coronavirus of probable bat origin. Nature 579(7798):270–273

Zu ZY, Jiang MD, Xu PP, Chen W, Ni QQ, Lu GM, Zhang LJ (2020) Coronavirus disease 2019 (COVID-19): a perspective from China. Radiology:200490

Pharmacist's Role and Pharmaceutical Care During the COVID-19 Pandemic

33

Pedro Amariles, Mónica Ledezma-Morales, Andrea Salazar-Ospina, and Jaime Alejandro Hincapié-García

Abstract

Pharmacists play a vital role in public health during the COVID-19 pandemic. In this chapter, we present the most significant contributions that pharmacists could make to communitarian and hospital settings. It includes the provision of essential services, such as continuous access to medication and supplies, support to medical services, education and information to the population and the health team to reduce the misuse of medications, patient monitoring and follow-up, and the detection and referral of suspected cases of COVID-19. The chapter ends with a discussion over certain elements related to innovation needs, such as telepharmacy services.

Keywords

Coronavirus infections · COVID-19 · Health services · Pharmaceutical services · Pharmacists

33.1 Introduction

SARS-CoV-2 causes coronavirus disease (COVID-19). Due to its high contagious and relatively high mortality, the World Health Organization (WHO) declared it a global health emergency (Rothan and Byrareddy 2020). The lack of similar crisis references in the recent past limits the assessment of what might happen in the immediate future. As a consequence, to control the outbreak advance of the COVID-19 pandemic, the situation demands that information be balanced, coherent, in harmony to public health communication, and based on either science (The Lancet Respiratory Medicine 2020) or the best practice. Healthcare professionals across all settings are now caring for and communicating with patients in a context of high uncertainty.

Pharmacists, as healthcare practitioners, are essential in preventing the spread of COVID-19 and can be active participants in national efforts to fight and contain this outbreak (Al-Quteimat and Amer 2020). In the community settings, pharmacists are likely to be the first option for many patients, refills of medicines, and supply of

P. Amariles (✉) · M. Ledezma-Morales · A. Salazar-Ospina · J. A. Hincapié-García
Research Group on Pharmaceutical Promotion and Prevention, University of Antioquia, Medellín, Colombia

Network of Immunity in Infection, Malignancy and Autoimmunity (NIIMA), Universal Scientific Education and Research Network (USERN), Medellín, Colombia
e-mail: pedro.amariles@udea.edu.co

N. Rezaei (ed.), *Coronavirus Disease - COVID-19*, Advances in Experimental Medicine and Biology 1318, https://doi.org/10.1007/978-3-030-63761-3_33

medical devices. Also, they have a great responsibility by informing the public about preventative measures and advising about behavioral precautions and in the risk assessment, early identification, and referral of individuals suspected to be at a higher risk of being infected, also providing detailed information on pharmacotherapy for COVID-19 so that physicians could better master the characteristics and instructions for the medication usage (Aruru et al. 2020). Pharmacists must be equipped with knowledge and tools to communicate clearly and effectively with patients and healthcare teams on how to limit the impact of COVID-19 (Carico et al. 2020; Li et al. 2020) and recognize all the services that can support during the pandemic. Therefore, this chapter aims to describe the pharmacist's role across communitarian and hospital settings and presents essential information to provide pharmacy services, including pharmaceutical care (PC), in the context of the COVID-19 pandemic, and also to comment on some pharmaceutical practice innovations to provide PC services during the COVID-19 pandemic.

33.2 The Role of a Pharmacist During the COVID-19 Pandemic

Pharmacists, according to specialty and scope of practice, could make substantial contributions in the control, prevention, and treatment of COVID-19, providing essential services that include (Al-Quteimat and Amer 2020; Amariles et al. 2020b; Aruru et al. 2020; Bukhari et al. 2020; Cadogan and Hughes 2020; Carico et al. 2020; Tan et al. 2020):

- Providing continued access to medications and supplies
- Managing minor ailments
- Supporting healthcare needs and services such as triage services
- Educating the public to reduce medication misinformation
- Monitoring and following up patients in coordination with their medical practitioner, focusing on identifying, preventing, and resolving drug-related problems (DRPs)
- Detecting possible cases of the COVID-19 and referring to medical institutions
- Developing workplace protocols
- And adjusting to critical personal protection equipment (PPE) shortages and serving as trustworthy drug information sources.

Pharmacists across the world have a central role in putting each of these measures into action to ensure that their countries emerge safely from the current state of lockdown and should be supported to do so (Dawoud 2020).

33.2.1 Community Pharmacist Support in the COVID-19 Pandemic

In some countries, as the United States, England, or Canada, community pharmacies include both retail and outpatient pharmacies as part of the primary care system. In other countries, retail pharmacies can provide some prescription medications, over-the-counter medications, supplements, and health-related products and devices for sale; however, they are not considered part of the healthcare system (Zheng et al. 2020).

Pharmacists in primary care are the most accessible healthcare professionals. Usually, they are the initial connection with the healthcare system for patients with health-related issues or drug information/healthcare advice requirements (Hedima et al. 2020; International Pharmaceutical Federation 2020; Zheng et al. 2020).

The current COVID-19 pandemic has shown the necessity to link community pharmacy's action to the healthcare process (Amariles et al. 2020b), to expand the capacity to provide healthcare services (Kretchy et al. 2020) and reduce patients' burden on health facilities, such as hospitals and general practices (Bukhari et al. 2020;

Hedima et al. 2020), and, moreover, to accurately detect possible cases of COVID-19 and refer them to healthcare (Amariles et al. 2020b). Community pharmacists play a crucial role in patient education, increasing community awareness, advising on self-care measures to prevent virus transmission, and offering help to individuals with suspected COVID-19 (Al-Quteimat and Amer 2020; Amariles et al. 2020b). Also, community pharmacists should advise people not suspected of having COVID-19 to practice social distancing and avoid enclosed, crowded spaces and inform them to maintain a protective distance of at least 2 meters from any individuals suspected of having COVID-19 (Al-Quteimat and Amer 2020).

Community pharmacists could recommend symptom management for mild conditions, ensure medication refill on schedule, and counsel patients about using over-the-counter medications in case of minor symptoms. Those measures can reduce unnecessary hospital visits, where individuals might be exposed to COVID-19 (Al-Quteimat and Amer 2020; Dawoud 2020). Also, they should be prepared to continue with the provision of skilled PC services, to ensure medication safety through the identification, resolution, and prevention of DRP (Al-Quteimat and Amer 2020; International Pharmaceutical Federation 2020; Zheng et al. 2020).

As the International Pharmaceutical Federation (FIP) stated, community pharmacists have a responsibility of ensuring appropriate storage and supply of pharmaceutical products and medical devices, such as medicines and masks; informing, educating, and counseling the public; referring of suspected COVID-19 cases; and promoting disease prevention and infection control (International Pharmaceutical Federation 2020).

33.2.1.1 Community Pharmacist in Rural and Underserved Areas

The public health contributions of community pharmacists are even more significant in those rural and underserved communities that are experiencing primary care provider shortages because pharmacists become the only accessible healthcare provider in the area (Adunlin et al. 2020; Hedima et al. 2020). When patients experience mild symptoms related to a common cold or seasonal allergies, they often seek advice in their community pharmacies. Both the use of OTC medications and non-pharmacology treatments could be suggested. Pharmacists can refer patients to medical institutions for testing, encourage them to seek treatment more rapidly, and provide information to limit the disease spread (Adunlin et al. 2020).

33.2.1.2 Community Pharmacist to Support Industrial Workers

Industrial workers commonly have underlying health diseases such as respiratory system problems and can suffer occupation-related injuries. They usually self-medicate and visit commune health centers, resulting in a higher risk of getting COVID-19 and spreading it in the communities. Their economic vulnerabilities may cause a reluctance to take time off of work to attend the hospital/clinic. Therefore, it is necessary to involve local pharmacies, community health centers, and health providers or village health collaborators to act as the first point for detection and report of suspected COVID-19 cases, as well as a channel for accurate information about COVID-19 and safety measures (Tran et al. 2020).

33.2.2 Clinical Pharmacist's Role in Hospital Settings During the COVID-19 Pandemic

Clinical pharmacists can select, prepare, or deliver drugs to the patient's bedside (Gross and MacDougall 2020) and optimize COVID-19 patient care through the identification, prevention, and resolution of DRP (Al-Quteimat and Amer 2020). For critically ill patients with COVID-19, who often have comorbidities as renal and hepatic dysfunction or bacterial infections, pharmacists can actively participate in multidisciplinary teams and contribute to drug therapy plans, together with physicians. They also can carefully monitor adverse drug reactions (ADR) and drug-drug interactions and evaluate drug treatments for COVID-19 (Aruru et al. 2020).

Clinical pharmacists should be joined to the multidisciplinary team to improve COVID-19 patients' outcomes, reduce mortality, and facilitate the pandemic control (Song et al. 2020). This work can be done through participation in making evidence-based decisions for medication, monitoring, and evaluating medication safety and efficacy and providing scientific information about COVID-19 vaccines. The involvement of pharmacists in information technology, general and critical care, medicine distribution, and management can play a meaningful role in speeding the time to therapy for patients (Gross and MacDougall 2020).

Clinical pharmacists' participation in the hospital's infectious disease team can help to coordinate actions for COVID-19 patients as "monitor compliance with institutional guidelines, prepare and implement local treatment protocols, monitor and manage drug shortages and help in investigations of new drug applications and uses" (Stevens et al. 2020). It is important to note that using drugs with unproven benefits may represent experimental use and that institutional review boards should approve intentional research using such agents with the application of informed consent. Clinical pharmacists can play a significant role in getting individuals enrolled in such ongoing studies (Al-Quteimat and Amer 2020).

There are controversial opinions on the efficacy and safety of COVID-19 pharmacotherapy; hence, clinical pharmacists must be well-prepared to provide the best available recommendations to healthcare practitioners in order to establish standardized treatments. Pharmacists need to be updated with complete information on drugs used as COVID-19 treatment (dosing and dose adjustment, drug-drug interactions, drug-food interactions, and adverse effects) (Al-Quteimat and Amer 2020).

33.3 Pharmacy Services During COVID-19 Pandemic

Pharmacy services are an essential pillar in public health to prevent and contain the COVID-19 pandemic (Liu et al. 2020). The timely supply of preventive and treatment medications, the emergency plans for the pharmacy workforce, and the provision of pharmacy services are critical to support public health emergency operations (Li et al. 2020).

The pharmacy services provided in community and hospital pharmacies to promote COVID-19 control, and ensure the effective and safe medication use, include supply of medications and products, home care guidance, psychological support, and PC (drug dispensing, pharmacotherapy follow-up, patient education, and counseling) (Zheng et al. 2020). In community pharmacy, other services could be implemented as drug indication (Al-Quteimat and Amer 2020; Cadogan and Hughes 2020; Faus Dáder et al. 2018), testing and immunization (Dawoud 2020; Hedima et al. 2020), consulting and referrals (Amariles et al. 2020b; Zheng et al. 2020), and promotion of safe use of infusions (Zheng et al. 2020). In hospital settings, pharmacy services can include evidence-based drug evaluation, guideline and the establishment of drug registration research, and management of off-label (Li et al. 2020; Liu et al. 2020; Song et al. 2020) (Fig. 33.1).

33.3.1 Supply of Medication and Products for COVID-19 Prevention

Medicine supplies are essential elements of all functioning health systems (World Health Organization 2018). During the pandemic, supply and dispensing of medicines and medical devices should be prioritized over non-essential products (International Pharmaceutical Federation 2020) to guarantee an adequate supply to the public of their daily medications and the products to prevent COVID-19 (masks and alcohol-based hand rubs) (Aruru et al. 2020;

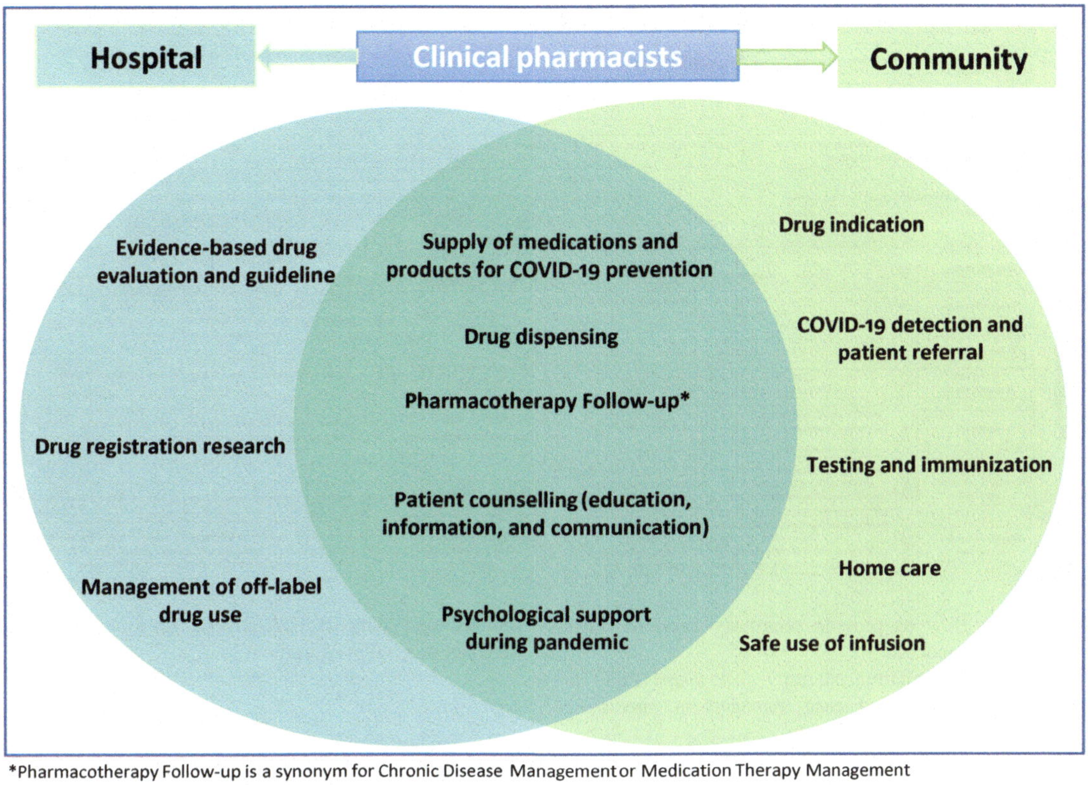

Fig. 33.1 Pharmacy services during the COVID-19 pandemic
*Pharmacotherapy follow-up is a synonym for chronic disease management or medication therapy management

Zheng et al. 2020). Therefore, focused actions such as establishing remote pharmacy services should be considered, and national emergency drug formularies and COVID-19 therapeutic drug list should be created so pharmacists can monitor and resolve potential drug shortages associated with the pandemic (Liu et al. 2020; Ying et al. 2020). It is also essential to be prepared to manage any potential pharmacy consumable shortages (such as PPEs, intravenous bags, and syringes) (Al-Quteimat and Amer 2020).

33.3.2 Pharmaceutical Care

PC refers to the responsibility that pharmacists have to contribute to get the best health outcomes in patients receiving pharmacologic and non-pharmacologic treatments, focusing on solving patient medication needs (Faus Dáder et al. 2018; Foro de Atención Farmacéutica-Farmacia 2019).

In the COVID-19 pandemic, PC is an essential complement for clinical management. It advances the level of pharmacotherapy, improves patient outcomes and medication compliance, and reduces the drug-induced risk through medication review and medication guidance (Hua et al. 2020; International Pharmaceutical Federation 2020). PC must be done according to the patient's needs, providing updated treatment plans, monitoring potential drug interactions, and focusing on particular population medication. The implementation of remote pharmaceutical service can be required to provide PC services during the pandemic (Ying et al. 2020).

Some PC services can be implemented or strengthened during the COVID-19 pandemic in community and hospital settings (Fig. 33.2).

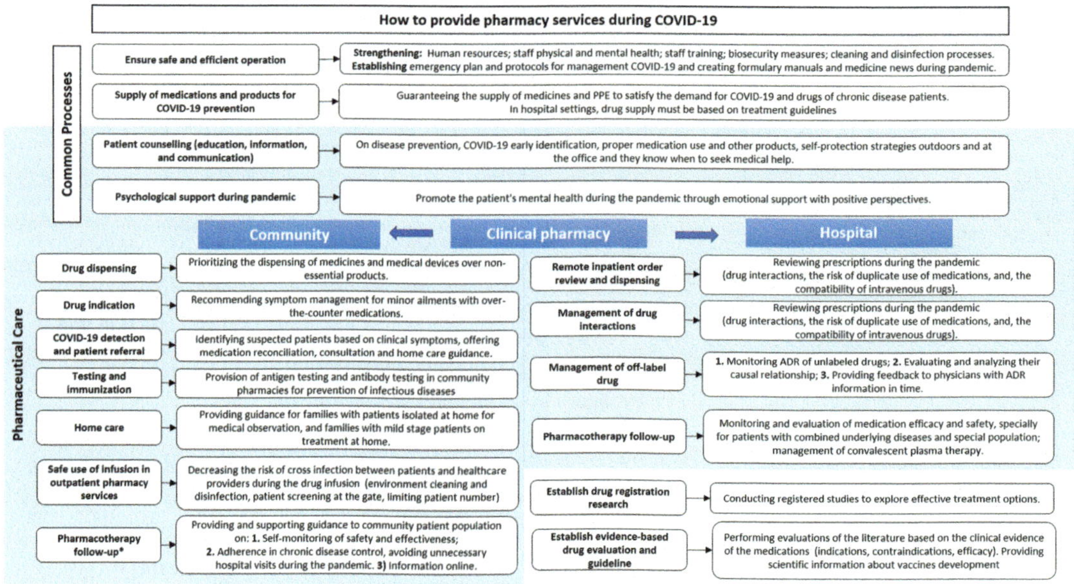

Fig. 33.2 How to provide pharmacy services during COVID-19 pandemic
*Synonyms of pharmacotherapy follow-up (PFT) are assumed as chronic disease management, medication therapy management. *ADR* adverse drug reaction, *PPE* personal protective equipment

33.3.2.1 Drug Dispensing

Dispensing is a central service of the pharmacist's professional exercise, which guarantees access to medicines for patients, including the necessary information, so patients know the correct use of medicines, as well as avoid and solve some DRPs. Also, dispensing contributes to the strength of the patient-pharmacist relationship (Faus Dáder et al. 2018; Foro de Atención Farmacéutica-Farmacia 2019; Hua et al. 2020). When dispensing new medications to patients, the pharmacist must assess the current medication list to identify duplicate therapy and provide medication education to ensure the safe use of dispensed medications (Zheng et al. 2020).

Community pharmacies are likely to become significant outlets where medical masks have to be available for the public. Thus, pharmacists will need to not only ensure its supply but also provide appropriate instructions on how to use these masks (Dawoud 2020) correctly.

Mail order or home delivery services should be implemented to promote door-step prescription refills and other non-prescription supplies (Kretchy et al. 2020) to guarantee access to medicines. Besides, the provision of online or phone counseling by the pharmacist should be considered, whenever feasible, in order to promote social distancing and self-isolation and to help patients who have limited access to healthcare, especially high-risk individuals (Al-Quteimat and Amer 2020; Kretchy et al. 2020). These services can be offered by working with social workers, volunteers, care coordinators, or drug companies (Zheng et al. 2020) and can be supported using information network technology. In this way, it is possible to reduce infectious risk caused by cross-contact in the dispensing process (Ying et al. 2020).

In some cases, the autonomy of pharmacists must be encouraged. It can help pharmacists to support clinicians and to ensure timely access to medicines during the COVID-19 pandemic (Iacobucci 2020). In this sense, countries as France and Portugal have removed some administrative barriers and have authorized pharmacist to repeat dispensing of prescribed medicines for patients with long-term conditions, to reduce the need for medical appointments and release resources. It is fundamental to remove/waive any administrative barriers to improve access to pharmaceutical products and assure the quality, safety, and efficacy of both medicines and medical devices (International Pharmaceutical Federation 2020).

33.3.2.2 Medication Reconciliation Service

It allows identifying, classifying, evaluating, and resolving possible discrepancies in the pharmacological treatment generated in transit between the care levels of the health system. Through this service, pharmacists contribute to patient safety, avoiding possible damage that involves the use of medicines during patient care transitions and hospital discharge (Faus Dáder et al. 2018; Foro de Atención Farmacéutica-Farmacia 2019).

33.3.2.3 Pharmacotherapy Follow-Up (PTF)

PTF, also known as medical review management or medication therapy management, is the service aimed at ensuring the appropriate use of medicines and medical devices according to each patient's clinical needs (Faus Dáder et al. 2018; Foro de Atención Farmacéutica-Farmacia 2019). During PTF, the pharmacist can determine and recommend individualized drug regimens and improve pharmacotherapy effectiveness and safety according to patient clinical conditions (Li et al. 2020).

PTF involves detection, prevention, and resolution of DRPs or negative outcome related to medicine (NOM). It comprises continuous attention from the pharmacist and collaboration with patients and other healthcare professionals to improve outcomes and patient quality of life (Faus Dáder et al. 2018; Foro de Atención Farmacéutica-Farmacia 2019).

There are multiple methods to provide PTF, such as the Dader Method, that can help pharmacists to achieve the most effective and safe pharmacotherapy as possible for patients. According to this method, a pharmacist conducts a pharmaceutical interview, where drug information and patient health problems are recorded to create an assessment form. After the pharmaceutical interview, the pharmacist makes an assessment and identifies any DRPs and NOMs resulting from the patient's pharmacotherapy. The pharmacist then suggests any necessary interventions, verifying and reassessing the outcomes of the interventions (Pharmaceutical Care Research Group 2006). These interventions can address drug quantity (for instance, modifying dosage or frequency), pharmacological strategy (for instance, adding or changing any drug), and patient education (Faus Dáder et al. 2007). During the COVID-19 pandemic, pharmacists must educate patients to perform self-monitoring on disease control and how to evaluate the appearance of ADR (Zheng et al. 2020) in order to avoid unnecessary hospital visits (Cadogan and Hughes 2020).

33.3.2.4 Patient Counseling (Education, Information, and Communication About COVID-19)

Patient education and counseling are necessary to prepare and motivate patients to follow their therapeutic regimens and monitoring plans. Pharmacists and other healthcare professionals' coordinated efforts will enhance patients' adherence to pharmacotherapy regimens to achieve chronic disease control and the adoption of hygienic-dietary and lifestyle change recommendation (Faus Dáder et al. 2018).

The WHO informs that the "COVID-19 outbreak has been accompanied by a massive 'infodemic' – an over-abundance of information" (World Health Organization 2020), and it represents a severe challenge for health professionals. The current rush to search an effective drug or vaccine, even without adequate safety guarantees, has already become a source of medication-related misinformation, causing public confusion and panic.

Pharmacists have a reliable drug information resource for patients and healthcare providers (Erku et al. 2020). They have an emerging role during the pandemic preventing misinformation surrounding COVID-19 and providing patient counseling and training in the rational use of medicines (Kretchy et al. 2020).

Some topics can be included in patient's education to reduce misinformation around COVID-19, the promotion of rational use of medicines, and self-care measures to prevent COVID-19 transmission (Adams et al. 2020; Al-Quteimat and Amer 2020; Amariles et al. 2020b; Erku et al. 2020; Kretchy et al. 2020; Zheng et al. 2020):

- Mode of transmission and how to prevent it
- Proper hand hygiene
- Selection and proper use of masks
- Evaluation of COVID-19 symptoms and personal contact histories
- Importance of prevention of self-medication and promotion of rational use of medicines, especially of medications used as COVID-19 treatments
- Monitoring the appearance of ADR of prescribed COVID-19 treatments
- Contraindications or drug-disease state interactions with COVID-19 medicines
- Therapy and medications for the treatment of minor symptoms associated with COVID-19
- Risk of procuring drugs from unregistered Internet retail shops, in particular with the risks of falsified medicines
- How to reduce the risk of transmission of emerging pathogens from animals to humans in live animal markets or animal product markets, including consumption of raw meat, milk, or animal organs
- Selection and safe use of disinfection products, proper preparation of sanitizers or antiseptic hand-rub products from locally available chemicals, and promote their appropriate application and usage
- And self-protection strategies outdoors and at work

Pharmacists must collaborate with other health professionals, health professional societies, and associations to facilitate educational and behavioral interventions that will assist the community in complying with standards/procedures aimed at combating the spread of COVID-19 and misleading information. They also have to (Cadogan and Hughes 2020; Erku et al. 2020; Kretchy et al. 2020):

- Provide updated and evidence-based scientific information on vaccines or treatments under investigation.
- Participate actively in COVID-19 hygiene and infection control initiatives or strategies.
- Teach how to examine products for suspicious appearance and continuously monitor medicine product alerts.
- And promote medication adherence and lifestyle habit changes, supported by the application of technology to remind patients of their medications and lifestyle regimen to enhance adherence. Pharmacists can frequently interact with patients using phone calls, SMS text messaging, and social media to educate, especially within this pandemic.

Pharmacists and pharmaceutical associations can prepare materials for communicating information with the community or inside hospitals such as posters, leaflets, websites, text messages, and app alerts, to clarify government guidelines and any other information regarding COVID-19 and the rational use of medicines.

33.3.2.5 Psychological Support During the Pandemic

During drug dispensing, patient interaction, and communication, pharmacy staff must pay attention to the patient's emotional situation to identify patients with excessive anxiety, concern, fear, or blind optimism. In such cases, psychological or emotional support needs to be provided (Zheng et al. 2020).

In hospital settings, if the patient is considered having a psychological problem that needs assessment or treatment, the clinical pharmacist can refer the patient to a psychiatrist. In the community settings, it is crucial to establish a collaborative relationship with its surrounding clinics/hospitals or other medical institutions for mutual patient information sharing and proper referral as needed (Zheng et al. 2020).

All pharmacy staff should encourage the public to combine regular work with adequate rest and exercise to enhance the immune system and relieve negative emotions at the same time (Zheng et al. 2020).

33.3.3 PC Services in Community Pharmacies During the COVID-19 Pandemic

Patient-centered PC should be provided by community pharmacists to promote COVID-19 prevention and control and to ensure a safe and effective treatment (Zheng et al. 2020). It becomes even more relevant in patients with chronic diseases because they may have difficulties accessing health services due to the increased probability of infection in hospitals (Kretchy et al. 2020). The contribution of community pharmacists will facilitate the timely response to acute diseases, as well as the effective control of chronic health problems by ensuring that these drugs are available and promoting adherence to pharmacotherapy during the pandemic (Cadogan and Hughes 2020; Dawoud 2020). Additionally, it will help to alleviate the burden on already strained health systems (Cadogan and Hughes 2020). In this sense, PC services are in line with the WHO initiative to maintain essential services during the pandemic to prevent non-COVID disease burden on healthcare systems, particularly in low- and middle-income countries (Kretchy et al. 2020).

33.3.3.1 Drug Indication

It is a service usually requested in the community pharmacies, in which the pharmacist develops a central role to improve/remove self-limited diseases by indicating non-pharmacology interventions or a drug that does not require a medical prescription. This service allows pharmacists to contribute to medication optimization (Faus Dáder et al. 2018; Foro de Atención Farmacéutica-Farmacia 2019).

Community pharmacists can participate in recommending symptom management for minor ailments such as allergies and skin rashes, coughs and colds, and gastrointestinal complaints, among others (Al-Quteimat and Amer 2020; Cadogan and Hughes 2020). In the United Kingdom and Canada, some schemes enable community pharmacists to assess individuals who present with particular minor ailments and offer them appropriate self-care advice or treatment options with over-the-counter medications or certain prescription medications from agreed formularies (Cadogan and Hughes 2020). In the context of the COVID-19 pandemic, this can help to reduce unnecessary hospital visits, where individuals might be exposed to the virus transmission (Al-Quteimat and Amer 2020) and can be even more critical in those cases when physicians cancel non-urgent appointments. Community pharmacist becomes the single point of in-person contact with the healthcare system (Carico et al. 2020).

33.3.3.2 COVID-19 Detection and Patient Referral

Given the proximity and availability of the community pharmacist and the relationships established with their patients, they can assist in identifying suspected cases of COVID-19 by following established protocols and act as a point of referral to the healthcare system at the community level (Amariles et al. 2020b; Kretchy et al. 2020; Zheng et al. 2020). These actions can help to prevent the spread of the pandemic within communities (Kretchy et al. 2020). Also, it requires some policies to authorize the community pharmacies as a point-of-care testing and provide access to COVID-19 rapid test (Pharmacist as front-line responders for COVID-19 patient care 2020). For instance, in March 2020, in the United States, licensed pharmacists were authorized to order and administer COVID-19 tests, including serology tests (US Department of Health and Human Services 2020).

Amariles et al. proposed a route to detect and counsel possible patients with suspicious COVID-19 from the community pharmacies to link them to the health system. This route has three possible entrances depending on the user's needs: anti-flu drugs, symptoms related to COVID-19 (such as fever, cough, and fatigue), or the request for hygiene products as 70% alcohol or face masks. After a quick interview, the pharmacist evaluates symptoms related to COVID-19 and the history of contact with people with suspected or confirmed COVID-19. According to the answers, the pharmacist should provide self-care education or manage the case as a suspected COVID-19 case (Amariles et al. 2020b). They would recommend immediate isolation in a single room and refer to the designated institutions (Zheng et al. 2020).

Those strategies can expand access to care in underserved areas, reduce unnecessary burdens on emergency departments, and reduce community exposure to the hospital's environment (Hedima et al. 2020).

33.3.3.3 Testing and Immunization

Pharmacists can help expand access to COVID-19 testing through antigen testing and antibody testing in community pharmacies. Antibody testing carries fewer risks, as it should be conducted following recovery from the infection (Dawoud 2020; Hedima et al. 2020). Also, community pharmacies can act as a point of immunization to administer vaccines to prevent infectious diseases (Adunlin et al. 2020). Converting pharmacies as a point of immunization can increase the number of vaccinated patients in the community (Hedima et al. 2020). For instance, in the United States, pharmacies have become typical locations for the flu vaccination, given its convenience and low costs (Adunlin et al. 2020).

33.3.3.4 Home Care

Pharmacists can refer to home care those suspected cases of COVID-19 with the guidance on proper isolation at home and the requirement of medical observation. The home environment must be well-prepared, adequately cleaned, and disinfected, including the suspected patients' tableware and articles for daily use (Zheng et al. 2020).

33.3.3.5 Safe Use of Infusions in Outpatient Pharmacy Services

Additional infusions should be avoided during the COVID-19 pandemic, considering the increased risk of cross-infection among patients and healthcare providers. Therefore, some strategies should be implemented as environment cleaning and disinfection, patient screening at the front door, limiting patient numbers, and patient's separation with a safe distance when lining up or when infusing drugs. It could prevent cross-infection in the healthcare institution and during the drug infusion (Zheng et al. 2020).

33.3.4 PC Services in COVID-19 Hospitalized Patients

According to current information, older COVID-19 patients with underlying diseases (diabetes, cardiovascular disease, chronic obstructive pulmonary disease, malignancy, bacterial or fungal infections) are more likely to develop severe illness. Moreover, the use of multiple medications can lead to potential drug interactions and increase the risk of appearance of ADR. For instance, chloroquine phosphate can cause cardiac arrest as a severe ADR, and it is contraindicated in combination with drugs that cause Q-T prolongation (Sun et al. 2020).

Hospital pharmacists should strengthen PTF for those COVID-19 patients with combined underlying diseases or special populations, such as pregnant or lactating women, children, adolescents, and elderly patients. Therefore, individualized PTF must be done in order to provide close monitoring and evaluation of medication safety and efficacy, avoid the risk of duplicate use of medications, adjust therapy dosages according to their liver and kidney function, manage drug interactions, and monitor convalescent plasma therapy (CPT) (Song et al. 2020; Ying et al. 2020).

33.3.4.1 Remote Inpatient Order Review and Dispensing

In order to minimize any potential exposure, direct contact with COVID-19 patients must be restricted to the caregivers that provide essential health services in a hospital context, including nurses, respiratory therapists, and physicians. In this sense, it is recommended to use innovative methods to dispense drugs to inpatients to reduce the risk of infection. Those methods can include automated dispensers and intelligent drug storage cabinets (Li et al. 2020). Intravenous drugs can be centrally deployed in the Pharmacy Intravenous Admixture Services (Li et al. 2020). Other hospital pharmacy services as patient counseling, medication reconciliation, and order review should be performed remotely with the use of an individual's room phone or via the primary nurse taking care of such individuals (Al-Quteimat and Amer 2020).

33.3.4.2 Management of Drug Interactions

The potential medications to treat patients with COVID-19 can be related to drug interactions. Also, considering that COVID-19 patients could have underlying diseases and, therefore, multiple medicines, it is essential to perform drug interaction evaluations to avoid symptom aggravation. Pharmacists should evaluate potential drug-drug, drug-food, or drug-disease interactions and strengthen the management of patients' medication and diet. Pharmacists can provide the medical team with a list of common risk warnings of potential drug interactions and reactions according to COVID-19 diagnosis and treatment plans according to the literature (Ying et al. 2020). Also, for better assessment, the drug interaction checker of the University of Liverpool can be used: https://www.covid19-druginteractions.org/ (Song et al. 2020).

33.3.4.3 Monitoring and Management of Convalescent Plasma Therapy

CPT can be applied in COVID-19 patients with rapid disease progression, severe or critical condition. This therapy consists of the plasma infusion with a specific titer of virus-specific antibodies, which patients receive to obtain passive immunity and remove pathogens in circulation. CPT therapy has been successfully applied to the treatment of SARS and H1N1 influenza. Before administering CPT, pharmacists should assist clinicians in assessing patients' indications of CPT, combining with allergy history, contraindications of plasma transfusion, and individual clinical conditions. Pharmacists must closely monitor and record adverse effects of plasma transfusion before, during, and after the administration and cooperate with clinicians to manage adverse effects. Common adverse transfusion reactions include transfusion-related circulation overload; acute lung injury; dyspnea; allergy; hypotension; non-hemolytic febrile, acute hemolytic transfusion reactions; delayed hemolytic transfusion reaction; infectious transfusion reaction; etc. (Song et al. 2020).

33.3.4.4 Management of Off-Label Drug Use

Currently, antiviral drugs for COVID-19 have not been approved for marketing, and the safety and efficacy of the medications that have been used for treatment remain unclear to date. This situation has increased the off-label drug use with the following increment of potential drug use risks. Some drugs may cause severe ADR, including hemolytic anemia, cardiotoxicity, infection, and teratogenic effects. Therefore, pharmacists should cooperate with clinicians to formulate medication regimens to avoid inappropriate high-dose medications or drug combinations and to monitor the appearance of ADR. Through their pharmacovigilance systems, they can assess the safety of medicines used to treat COVID-19, evaluate the causal relationship of ADR and present feedback of the results to the healthcare team (Song et al. 2020; Ying et al. 2020), and report the ADR accurately in strict accordance with the regulatory agencies (Sun et al. 2020).

33.3.5 Other Hospital Pharmacy Services During the COVID-19 Pandemic

33.3.5.1 Establish and Participate in Evidence-Based Drug Evaluation and Guidelines

Pharmacists should actively participate in evidence-based evaluations of the safety and effectiveness of COVID-19 drug therapy to assist clinicians in formulating and adjusting drug regimens (Li et al. 2020; Song et al. 2020). It implies a critical appraisal of the literature and decision-making process to select the proper treatments that could be part of guidelines and what has to be available in hospitals. For example, while hydroxychloroquine has been considered to have great promise, a recent trial was unable to show a clinical benefit (Geleris et al. 2020). In the same way, lopinavir/ritonavir was very promising, but the clinical trials could not demonstrate its benefits (Cao et al. 2020).

Clinical pharmacists in all medical institutions should collaborate to establish and optimize prescription review rules and knowledge databases, using or developing a clinical decision support system to review prescriptions in order to reduce the irrational prescribing and medication errors and to ensure medication safety (Li et al. 2020).

To assist clinicians in better understanding and prescribing of COVID-19 treatments, clinical pharmacists can create a rational drug use manual that includes usage and dosage, solvents, precautions, ADR, and dose adjustment for special populations (Li et al. 2020).

33.3.5.2 Establishment of Drug Registration Research

Pharmacists can participate in the proposal and development of registered clinical studies to evaluate the effectiveness and safety of medication regimens in hospitalized patients with COVID-19 (Liu et al. 2020; Song et al. 2020). Besides, pharmacists can cooperate with research sponsors in the management of clinical trials, including the proper supply, use, storage, and disposal of experimental drugs in compliance with relevant clinical trial regulations (Liu et al. 2020).

In the case of vaccine-related clinical trials, hospital pharmacists should assist clinicians and immunologists in closely monitoring patients' clinical manifestations and ADR after vaccination. Pharmacists should also collect data about the effectiveness and safety of the vaccines under evaluation in clinical trials (Song et al. 2020). Short-term efforts to quickly develop lifesaving vaccines and therapeutics are of the highest importance (Amariles et al. 2020a).

33.4 Ensuring a Safe and Efficient Operation for Providing Pharmacy Services

It is necessary to train all pharmacy staff and to establish some protocols to minimize the risk of virus transmission, regulate operating procedures, and ensure an appropriate environment control, staff protection, and emergency plan establishment to decrease occupational exposure risk (Zheng et al. 2020).

33.4.1 Pharmacy Staff Training

Under the COVID-19 epidemic, the staff of community and retail pharmacies are at the front line, providing drugs and protective/hygiene supplies and offering information and education to consumers (Chinese Pharmaceutical Association 2020). Therefore, pharmacists should stay up to date and be familiar with the knowledge latest regarding the COVID-19 (Ung 2020). Besides, in contrast with high-income countries, community pharmacy staff of low- and middle-income countries may not be sufficiently trained and prepared to afront this situation. Hence, it is essential to provide comprehensive and prompt training about virus transmission prevention and control, regulate operating procedures, detect and refer suspected cases, properly handle medical waste, and decrease occupational exposure risk (Chinese Pharmaceutical Association 2020).

The FIP (International Pharmaceutical Federation 2020) and pharmacist researchers (Amariles et al. 2020b; Ung 2020; Zheng et al. 2020) suggest training the entire pharmacy staff in:

- Technical and scientific information on COVID-19 (symptoms, incubation period, and modes of transmission of the virus)
- Epidemiological information on affected areas
- Information and counseling on preventive actions to reduce virus transmission, for instance, the use of mask and other PPE, personal hygiene, social distancing, pharmacy

cleaning, and use of disinfectants, among others

- Pharmacy environment control
- Pharmacology and non-pharmacology treatments of COVID-19
- Criteria for identification and referral of suspected cases of COVID-19, including strategies that each community pharmacy should implement
- Psychological support
- And materials to support the intervention (information brochures, intervention flow charts, and accurate information websites)

In the context of the pandemic, training programs can be developed by universities or pharmacist organizations, and they can be offered to pharmacists in virtual modality. For instance, pharmacists from the Research Group on Pharmaceutical Promotion and Prevention from the University of Antioquia in Colombia developed a free online course, available for all pharmacy staff, especially the community pharmacists. The course aims to teach the different terms, concepts, and practices to provide care and identify suspected cases of COVID-19 from retail pharmacies (Universidad de Antioquia 2020).

Community pharmacy protocols should include some of these recommendations (Consejo General de Colegios Farmacéuticos 2020; International Pharmaceutical Federation 2020; Center for Disease Control and Prevention 2020):

- Contact with consumers/patients should be minimized by dispensing medications through a small window on the door. A plastic protector can be placed on the distribution area, and marks can be placed on the ground to indicate the distance of approximately 2 meters between customers and staff.
- Consumers should be asked to wait their turn outside the pharmacy keeping 2 meters between them.
- Medicine supply to pharmacies should be done without any patient/customer entering the pharmacy (or at least the non-public areas of the pharmacy).
- All the medicines received from wholesale distributors should be cleaned and disinfected before they are taken inside the pharmacy.
- Delimit with tape or some sign those areas of consumer's attention.
- There should be, if possible, isolated and reserved space in the pharmacy where a consumer with suspected COVID-19 can be located. If the pharmacy does not have a suitable isolation room, an area in which individuals can stay at least 2 meters away from the staff and customers is recommended (Al-Quteimat and Amer 2020).
- And frequently clean and sanitize the care areas, mainly when serving a consumer with suspected COVID-19.

33.4.2 Biosecurity Measures

Pharmacy staff can be at medium exposure risk for COVID-19 (International Pharmaceutical Federation 2020) because they are likely to encounter symptomatic or asymptomatic carriers of SARS-CoV-2, and then, the risk of getting infected increases. Since infected individuals may be asymptomatic at least during the incubation period, there is an increasing probability of further transmission of SARS-CoV2 to the general public visiting pharmacies (Hasan et al. 2020). Therefore, all pharmacy staff shall be provided with adequate and sufficient PPEs (Zheng et al. 2020) to ensure their safety and to increase the public's trust in the safety of visiting the pharmacy and using its services (Dawoud 2020).

The WHO and FIP recommend the use of long-sleeved gowns, appropriate eye protection (face shield or goggles), and surgical masks. Those are measures to protect pharmacists from infection and to avoid further dissemination – in case the pharmacy personnel becomes infected themselves (Al-Quteimat and Amer 2020; International Pharmaceutical Federation 2020). Also, it is recommended to wear gloves when entering a room containing suspected or confirmed COVID-19-infected individuals (Al-Quteimat and Amer 2020).

Besides the use of PPEs, some practices must be included during consumer and patient service in community pharmacies (Center for Disease Control and Prevention 2020; Consejo General de Colegios Farmacéuticos 2020; International Pharmaceutical Federation 2020):

- Perform hand hygiene frequently (patient by patient), using an alcohol-based hand sanitizer or soap and water.
- Clean and disinfect objects and surfaces which are touched by people frequently, e.g., telephones, pens, workstations, keyboards, handrails, and doorknobs.
- When it is possible, avoid using phones, desks, offices, or other tools and work equipment of other employees. Always perform cleaning and disinfection before and after using them.
- Respect the safety distance of 1–2 meters.
- In case of respiratory symptoms, notify the boss or supervisor, use the mask, contact the reporting lines, initiate voluntary isolation at home, and avoid close contact with other people. Employees should not go back to work until the criteria for interrupting insulation at home are met.
- And pharmaceutical staff who do not have symptoms but who have a relative at home with COVID-19 should notify their supervisor and follow all precautions to avoid transmission.

33.4.3 Cleaning and Disinfection

COVID-19 transmission occurs mainly through droplets and contact; therefore, it is crucial to promote healthcare professionals and patients' safety during the distribution and supply chain to ensure that medications or pharmacy areas are not contaminated (Aruru et al. 2020). Therefore, it is necessary to carry out proper cleaning and disinfection, especially if a customer or patient suspected of having COVID-19 spent time in any public areas of the pharmacy, such as a waiting area or bathroom, because those areas may have been contaminated with the SARS-CoV-2 (American Pharmacists Association 2020; International Pharmaceutical Federation 2020).

The Chinese Pharmaceutical Association recommends the disinfection of the working environment and associated articles and equipment (Chinese Pharmaceutical Association 2020):

- Items in the public areas as phones, computers, stationery, cash register, balance, table and chair, and door handle: wipe the surface for sterilization with 75% alcohol or 5000 mg/L peracetic acid every 2 h.
- Extensive facilities as shelves, counters, refrigerators, lockers, and air conditioners: wipe for sterilization with 5000 mg/L peracetic acid or 500 mg/L chlorine-containing disinfectant once a day.
- Work clothes: sterilize twice a week (replace when contaminated) by hot washing method. Sterilize at 75 °C for more than 30 min or at 80 °C for more than 10 min (the sterilization time can be extended according to the degree of dirt); or with clothing disinfectant (250–500 mg/L chlorine-containing disinfectant can be used for white clothes) for 30 min, rinse with water repeatedly.
- And cleaning utensils as mops and cleaning rags: should be dedicated to the particular area, rinse with water after each use, soak and sterilize with 500 mg/L chlorine-containing disinfectant for 30 min, rinse again with water, and then dry.

33.4.4 Waste Management

Pharmacy staff should avoid secondary contamination. It is recommended to have a dedicated container/bag to throw discarded masks/gloves. Before disposal, the bag must be sealed and sterilized with 500 mg/L chlorine-containing disinfectant. The sealed bags should be marked with "discarded masks/gloves" (Chinese Pharmaceutical Association 2020).

33.5 Pharmaceutical Practice Innovations in COVID-19 Pandemic

When facing public health emergencies, pharmacists can contribute to analyzing the current situation, working in collaboration, formulating telehealth strategies swiftly, providing information to medical staff, and developing innovative pharmacy services to ensure rational use of medicine (Al-Quteimat and Amer 2020; Li et al. 2020). Thus, it is necessary to establish remote pharmacy services to reduce human-to-human infections. Some remote pharmacy services such as online drug prescribing, drug consultation, and drug delivery services can be implemented to provide out-of-hospital PC access and to help to reduce the risk of cross-infection during unnecessary hospital visits (Liu et al. 2020).

33.5.1 Use of Technology to Provide Information About Medication Availability

Community pharmacist can use apps, web pages, or social networks to provide information of drug availabilities and its store locations to guide the public and patients when they need to buy medications (Zheng et al. 2020).

33.5.2 Telepharmacy, Telepharmaceutical Care, and Innovative Telehealth-Based Practices

In the United States, rural communities have been providing pharmacy services using technology – commonly known as telepharmacy – expanding access to healthcare. This modality has demonstrated value for patient counseling, monitoring of pharmacotherapy effectiveness and safety, medication selection, and the provision of clinical services (Li et al. 2020). With the increased number of COVID-19 cases, telepharmacy can be implemented for providing those services, to remote order entry and medication approval, thereby decreasing the time to get needed medications to ill patients (Adunlin et al. 2020) and to reduce unnecessary visits to the pharmacy (Zheng et al. 2020).

Given the importance of the pharmaceutical interventions discussed before in this chapter, it is necessary to design and develop a remote or an online pharmacotherapy follow-up service, using equipment and software according to the context of each country and institution. This service can be provided using just a telephone or more complex systems such as video chats or mobile apps in places where the Internet is available. The service model should contribute to reducing hospital-acquired infections, to increase the efficiency of pharmacy services, and to perform effective professional services for patients included in the PC process.

Examples of telepharmacy practices are the services offered during the pandemic by the Chinese pharmacists in "Fangcang shelter hospitals" – a novel public health concept for quarantining and treating infected patients with mild symptoms. They have developed novel approaches to introducing drug information, answering patients' questions, and relieving patients' anxiety through broadcast and telephone services and mobile phone apps. Besides, pharmacists have also conducted online clinics or used Internet tools to support these patients (Tan et al. 2020).

33.5.3 Telehealth Counseling and Patient Education

Pharmacists can educate the public, focusing on infection prevention and disease management (Liu et al. 2020). Considering that patients in the hospital and at home need pharmacotherapy counseling, pharmacists should launch a variety of innovative remote pharmaceutical services (Erku et al. 2020; Gamble et al. 2020; Li et al. 2020; Tan et al. 2020):

- Creation or use of radio stations to develop educational programs regarding COVID-19, including prevention and control and medication knowledge, and solve concerns about the disease. Radio shows can include lectures related to patients' therapeutic drugs, COVID-19 reasonable nutrition, dietary advice, self-protection, and medication instructions after discharge, targeting patients' medication needs at admission, during hospitalization, and upon discharge.

- Use of TV programs, the Internet, and social media (Facebook, Instagram, YouTube, or TikTok) to promote medical knowledge through videos or live broadcast. It could be useful to spread accurate information about the epidemic and related pharmacotherapy and to encourage the public to seek advice or clarifications on COVID-19 preventive and therapeutic measures. It is particularly worth mentioning that through relevant science popularization, the public has learned to wash hands and wear masks properly.
- Publication of articles about personal and environmental hygiene and the importance of wearing a surgical mask: these activities can help alleviate public panic and prevent the spread of rumors.
- And the creation of multi-media and other materials with information related to chronic home management, including text, pictures, animated cartoons, and videos, can be used with patients with COVID-19 or assist healthcare at-risk populations, for instance, patients with diabetes.

33.6 Conclusion

In public health emergencies, pharmacists can play an essential role in community and hospital pharmacy. Their pharmacological knowledge, training in pharmaceutical care, and active involvement in educational activities could be valuable in providing medical advice to staff and the public and in ensuring the supply and rational use of medicines. We found significant roles for pharmacists that the epidemic has driven, with particular emphasis on team-based activities, pharmacy management, and pharmaceutical care. We also find innovative services, such as telepharmacy and various advanced practices. These reveal the need to continually update practice and push the boundaries of pharmaceutical services, especially during the outbreak.

References

Adams KK, Baker WL, Sobieraj DM (2020) Myth busters: dietary supplements and COVID-19. Ann Pharmacother:106002802092805. https://doi.org/10.1177/1060028020928052

Adunlin G, Murphy PZ, Manis M (2020) COVID-19: how can rural community pharmacies respond to the outbreak? J Rural Health. https://doi.org/10.1111/jrh.12439

Al-Quteimat OM, Amer AM (2020) SARS-CoV-2 outbreak: how can pharmacists help? Res Soc Adm Pharm:S1551741120302382. https://doi.org/10.1016/j.sapharm.2020.03.018

Amariles P, Hincapie-García J, Julio Montoya C (2020a) Pharmacotherapy for hospitalized patients with COVID-19: Waiting or doing? Res Soc Adm Pharm:S1551741120305301. https://doi.org/10.1016/j.sapharm.2020.05.010

Amariles P, Ledezma-Morales M, Salazar-Ospina A, Hincapié-García JA (2020b) How to link patients with suspicious COVID-19 to health system from the community pharmacies? A route proposal. Res Soc Adm Pharm:S1551741120302485. https://doi.org/10.1016/j.sapharm.2020.03.007

American Pharmacists Association (2020) Coronavirus: resources for pharmacists and an interview with a virus expert

Aruru M, Truong H-A, Clark S (2020) Pharmacy Emergency Preparedness and Response (PEPR): a proposed framework for expanding pharmacy professionals' roles and contributions to emergency preparedness and response during the COVID-19 pandemic and beyond. Res Soc Adm Pharm: S1551741120303235. https://doi.org/10.1016/j.sapharm.2020.04.002

Bukhari N, Rasheed H, Nayyer B, Babar Z-U-D (2020) Pharmacists at the frontline beating the COVID-19 pandemic. J Pharm Policy and Pract 13(1):8,-s40545-40020-00210-w. https://doi.org/10.1186/s40545-020-00210-w

Cadogan CA, Hughes CM (2020) On the frontline against COVID-19: community pharmacists' contribution during a public health crisis. Res Soc Adm Pharm:S1551741120302928. https://doi.org/10.1016/j.sapharm.2020.03.015

Cao B, Wang Y, Wen D, Liu W, Wang J, Fan G, Ruan L, Song B, Cai Y, Wei M, Li X, Xia J, Chen N, Xiang J, Yu T, Bai T, Xie X, Zhang L, Li C, Yuan Y, Chen H, Li H, Huang H, Tu S, Gong F, Liu Y, Wei Y, Dong C, Zhou F, Gu X, Xu J, Liu Z, Zhang Y, Li H, Shang L, Wang K, Li K, Zhou X, Dong X, Qu Z, Lu S, Hu X, Ruan S, Luo S, Wu J, Peng L, Cheng F, Pan L, Zou

J, Jia C, Wang J, Liu X, Wang S, Wu X, Ge Q, He J, Zhan H, Qiu F, Guo L, Huang C, Jaki T, Hayden FG, Horby PW, Zhang D, Wang C (2020) A trial of lopinavir–ritonavir in adults hospitalized with severe Covid-19. N Engl J Med 382(19):1787–1799. https://doi.org/10.1056/NEJMoa2001282

Carico RR, Sheppard J, Thomas CB (2020) Community pharmacists and communication in the time of COVID-19: applying the health belief model. Res Soc Adm Pharm:S155174112030293X. https://doi.org/10.1016/j.sapharm.2020.03.017

Center for Disease Control and Prevention (2020) Coronavirus disease 2019 (COVID-19) – interim guidance for businesses and employers

Chinese Pharmaceutical Association (2020) SARS-CoV-2 infection: expert consensus on guidance and prevention and control strategies for retail pharmacy workforce (1st edition)

Consejo General de Colegios Farmacéuticos (2020) Procedimiento de actuación en la farmacia comunitaria en casos de infección por COVID-19

Dawoud D (2020) Emerging from the other end: key measures for a successful COVID-19 lockdown exit strategy and the potential contribution of pharmacists. Res Soc Adm Pharm:S1551741120305313. https://doi.org/10.1016/j.sapharm.2020.05.011

Erku DA, Belachew SA, Abrha S, Sinnollareddy M, Thomas J, Steadman KJ, Tesfaye WH (2020) When fear and misinformation go viral: Pharmacists' role in deterring medication misinformation during the 'infodemic' surrounding COVID-19. Res Soc Adm Pharm:S1551741120304551. https://doi.org/10.1016/j.sapharm.2020.04.032

Faus Dáder MJ, Sabater-Hernández D, Amariles P (2007) Types of pharmacist interventions intended to prevent and solve negative outcomes associated with medication. Pharmacotherapy 7:e51–e52

Faus Dáder MJ, Amariles P, Martinez-Martinez F (2018) Atención farmacéutica servicios farmacéuticos orientados al paciente. Avicam, Armilla

Foro de Atención Farmacéutica-Farmacia C (2019) Guía práctica para los Servicios Profesionales Farmacéuticos Asistenciales en la Farmacia Comunitaria. Consejo General de Colegios Oficiales de Farmacéuticos, Madrid, España

Gamble A, Pham Q, Goyal S, Cafazzo JA (2020) The challenges of COVID-19 for people living with diabetes: considerations for digital health. JMIR Diabetes 5(2):e19581. https://doi.org/10.2196/19581

Geleris J, Sun Y, Platt J, Zucker J, Baldwin M, Hripcsak G, Labella A, Manson DK, Kubin C, Barr RG, Sobieszczyk ME, Schluger NW (2020) Observational Study of Hydroxychloroquine in Hospitalized Patients with Covid-19. New England Journal of Medicine 0 (0): null. https://doi.org/10.1056/NEJMoa2012410

Gross AE, MacDougall C (2020) Roles of the clinical pharmacist during the COVID-19 pandemic. JACCP: JOURNAL OF THE AMERICAN COLLEGE OF. Clin Pharm 3(3):564–566. https://doi.org/10.1002/jac5.1231

Hasan SS, Kow CS, Zaidi STR (2020) Social distancing and the use of PPE by community pharmacy personnel: does evidence support these measures? Res Soc Adm Pharm:S1551741120304903. https://doi.org/10.1016/j.sapharm.2020.04.033

Hedima EW, Adeyemi MS, Ikunaiye NY (2020) Community pharmacists: on the frontline of health service against COVID-19 in LMICs. Res Soc Adm Pharm:S1551741120303818. https://doi.org/10.1016/j.sapharm.2020.04.017

Hua X, Gu M, Zeng F, Hu H, Zhou T, Zhang Y, Shi C (2020) Pharmacy administration and pharmaceutical care practice in a module hospital during the COVID-19 epidemic. J Am Pharm Assoc:S1544319120301539. https://doi.org/10.1016/j.japh.2020.04.006

Iacobucci G (2020) Covid-19: allow pharmacists to dispense controlled drugs without prescription, urge specialists. BMJ:m1868. https://doi.org/10.1136/bmj.m1868

International Pharmaceutical Federation (2020) COVID-19: Guidelines for pharmacists and the pharmacy workforce

Kretchy IA, Asiedu-Danso M, Kretchy J-P (2020) Medication management and adherence during the COVID-19 pandemic: perspectives and experiences from low-and middle-income countries. Res Soc Adm Pharm:S1551741120303326. https://doi.org/10.1016/j.sapharm.2020.04.007

Li H, Zheng S, Liu F, Liu W, Zhao R (2020) Fighting against COVID-19: innovative strategies for clinical pharmacists. Res Soc Adm Pharm:S1551741120303284. https://doi.org/10.1016/j.sapharm.2020.04.003

Liu S, Luo P, Tang M, Hu Q, Polidoro JP, Sun S, Gong Z (2020) Providing pharmacy services during the coronavirus pandemic. Int J Clin Pharm 42(2):299–304. https://doi.org/10.1007/s11096-020-01017-0

Pharmaceutical Care Research Group UoG (2006) Pharmacotherapy follow-up: the Dader method (3rd revision: 2005). Pharm Pract 4(1):44–53

Pharmacist as front-line responders for COVID-19 patient care (2020)

Rothan HA, Byrareddy SN (2020) The epidemiology and pathogenesis of coronavirus disease (COVID-19) outbreak. J Autoimmun 109:102433. https://doi.org/10.1016/j.jaut.2020.102433

Song Z, Hu Y, Zheng S, Yang L, Zhao R (2020) Hospital pharmacists' pharmaceutical care for hospitalized patients with COVID-19: Recommendations and guidance from clinical experience. Research in Social and Administrative Pharmacy: S1551741120303144. https://doi.org/10.1016/j.sapharm.2020.03.027

Stevens MP, Patel PK, Nori P (2020) Involving antimicrobial stewardship programs in COVID-19 response efforts: all hands on deck. Infect Control Hosp Epidemiol 41(6):744–745. https://doi.org/10.1017/ice.2020.69

Sun X, Li S, Li K, Hu X (2020) Pharmaceutical care of chloroquine phosphate in elderly patients with coronavirus pneumonia (COVID-19). Aging Med 3:98–101. https://doi.org/10.1002/agm2.12104

Tan S-L, Zhang B-K, Xu P (2020) Chinese pharmacists' rapid response to the COVID-19 outbreak. American Journal of Health-System Pharmacy:zxaa120. https://doi.org/10.1093/ajhp/zxaa120

The Lancet Respiratory Medicine (2020) COVID-19: delay, mitigate, and communicate. Lancet Respir Med 8(4):321. https://doi.org/10.1016/S2213-2600(20)30128-4

Tran BX, Vu GT, Latkin CA, Pham HQ, Phan HT, Le HT, Ho RCM (2020) Characterize health and economic vulnerabilities of workers to control the emergence of COVID-19 in an industrial zone in Vietnam. Saf Sci 129:104811. https://doi.org/10.1016/j.ssci.2020.104811

U.S. Department of Health and Human Services (2020) Guidance for licensed pharmacists, COVID-19 testing, and immunity under the PREP Act.

Ung COL (2020) Community pharmacist in public health emergencies: quick to action against the coronavirus 2019-nCoV outbreak. Res Soc Adm Pharm 16(4):583–586. https://doi.org/10.1016/j.sapharm.2020.02.003

Universidad de Antioquia (2020) Cómo actuamos frente al COVID-19 desde las droguerías

World Health Organization (2018) WHO medicines, vaccines and pharmaceutical. WHO

World Health Organization (2020) Novel Coronavirus (2019-nCoV). Situation Report – 13

Ying W, Qian Y, Kun Z (2020) Drugs supply and pharmaceutical care management practices at a designated hospital during the COVID-19 epidemic. Research in Social and Administrative Pharmacy: S1551741120303259. https://doi.org/10.1016/j.sapharm.2020.04.001

Zheng S-Q, Yang L, Zhou P-X, Li H-b, Liu F, Zhao R-S (2020) Recommendations and guidance for providing pharmaceutical care services during COVID-19 pandemic: a China perspective. Res Soc Adm Pharm. https://doi.org/10.1016/j.sapharm.2020.03.012

Impact of COVID-19 on Dentistry

34

Arghavan Tonkaboni,
Mohammad Hosein Amirzade-Iranaq,
Heliya Ziaei, and Amber Ather

Abstract

The coronavirus disease 2019 (COVID-19) pandemic has affected the health-care system drastically, including dental care practice. COVID-19, caused by severe acute respiratory syndrome coronavirus 2 (SARS-CoV-2), is thought to spread via close contact through respiratory droplets and aerosols. Owing to specific characteristics of dental care such as aerosol generation as well as close proximity to patients, dentistry is thought to be associated with the nosocomial spread of infection. The risk of bidirectional spread of infection between patient and dental care providers makes it critical to take additional precautionary measures to mitigate the spread of COVID-19. It is essential to understand that the guidelines for providing dental treatment during the COVID-19 pandemic will vary across the globe, and dental practices should be in compliance with their regional guidelines. This chapter aims to present an overview of the dynamics of COVID-19 transmission and its impact on dentistry and discuss measures to provide dental care during the time of the COVID-19 outbreak effectively.

Keywords

Aerosols · COVID-19 · Dental · Dentistry · Oral · Pandemic · Saliva · SARS-CoV-2

A. Tonkaboni (✉)
Oral and Maxillofacial Medicine Department, School of Dentistry, Tehran University of Medical Sciences, Tehran, Iran

Grupo de investigación en patología oral médico quirúrgica, University of Santiago de Compostela, Santiago, Spain
e-mail: a-tonkaboni@sina.tums.ac.ir

M. H. Amirzade-Iranaq
Universal Network of Interdisciplinary Research in Oral and Maxillofacial Surgery (UNIROMS), Universal Scientific Education and Research Network (USERN), Tehran, Iran

International Otorhinolaryngology Research Association (IORA), Universal Scientific Education and Research Network (USERN), Tehran, Iran

H. Ziaei
International Otorhinolaryngology Research Association (IORA), Universal Scientific Education and Research Network (USERN), Tehran, Iran

School of Dentistry, Tehran University of Medical Sciences, Tehran, Iran

A. Ather (✉)
Department of Endodontics, University of Texas Health Science Center at San Antonio, San Antonio, TX, USA
e-mail: ather@livemail.uthscsa.edu

N. Rezaei (ed.), *Coronavirus Disease - COVID-19*, Advances in Experimental Medicine and Biology 1318, https://doi.org/10.1007/978-3-030-63761-3_34

34.1 Introduction

The outbreak of coronavirus disease 2019 (COVID-19) in Wuhan, China, spread to the rest of the world in an exponential manner (Chen and Yu 2020). COVID-19, a febrile respiratory illness, is caused by a virus bearing close genomic resemblance to severe acute respiratory syndrome coronavirus (SARS-CoV) (Chen and Yu 2020). As a result, this novel coronavirus was named severe acute respiratory syndrome coronavirus 2 (SARS-CoV-2) (of the International CSG 2020). SARS-CoV-2 enters the host cells by efficient binding of the spike (S) viral protein to the angiotensin-converting enzyme 2 (ACE-2) receptors present in several tissues such as the heart, kidney, lung, and brain (Walls et al. 2020; Zhang et al. 2020).

Patients with SARS-CoV-2 infection typically present with fever, myalgia, and difficulty in breathing (Fu et al. 2020). Individuals with comorbidities such as diabetes mellitus, preexisting respiratory disease, or immunocompromised state often present with a severe form of this infection, notably with pneumonia or acute respiratory distress syndrome (Fu et al. 2020). However, recent evidence has also shown the presence of some atypical symptoms in COVID-19 patients such as gastrointestinal upset, anosmia, dysgeusia, pink eye, and even inflammatory symptoms in young children (Fu et al. 2020; Giacomelli et al. 2020; Wu et al. 2020; Verdoni et al. 2020). Interestingly, a subset of COVID-19-affected patients are asymptomatic; these patients often go undiagnosed and can lead to untraceable transmission to multiple individuals (Gandhi et al. 2020).

34.2 Routes of Transmission

The high spread of this infection can also be partly explained by the basic reproduction number (R0) of SARS-CoV-2. R0, defined as the average number of people one infected person will pass the virus to, is relatively high, with an estimate of 1.4–5.5 (Chen 2020).

The mode of transmission of SARS-CoV-2 has been a matter of debate owing to limited scientific evidence. The most commonly proposed routes for SARS-CoV-2 transmission are through inhalation of respiratory droplets, close contact, or fomites (Ong et al. 2020). One of the first studies demonstrating the person-to-person transmission of SARS-CoV-2 was done by Chan et al., who studied five patients in a family cluster presented with unexplained pneumonia (Chan et al. 2020). Also, there have been reports of indirect transmission of COVID-19. Cai et al. investigated a cluster of COVID-19 cases in a shopping mall in China and reported indirect transmission of SARS-CoV-2 via fomites or aerosolization (Cai et al. 2020). The possibility of indirect transmission has been further corroborated by results of a study by Van Doremalen et al., wherein they showed that SARS-CoV-2 could remain viable on surfaces such as stainless steel, glass, or plastic up to 9 days (Van Doremalen et al. 2020).

Another important consideration is the transmission from asymptomatic carriers or patients in their incubation period. The incubation period is the time between infection and onset of symptoms. It ranges from 0 to 24 days (Guan et al. 2020), with a mean value of 5.1 days (Lauer et al. 2020). It has critical implications as disease transmission can occur before any symptoms are apparent (Rothe et al. 2020). To better understand this, Ferretti et al. have outlined four different modes of transmission which are as follows: i, symptomatic transmission, direct transmission from a confirmed COVID-19 individual; ii, presymptomatic transmission, direct transmission from a person before the development of any apparent signs and symptoms; iii, asymptomatic transmission, direct transmission from individuals who never experience noticeable symptoms; and iv, environmental transmission, indirect transmission via contamination, which is difficult to trace back to index patient (Ferretti et al. 2020).

Transmission of COVID-19 to health-care workers in hospitals, even in low contagion areas, has already been reported, and dental care providers are no exception (Sikkema et al. 2020). Dental practitioners and their patients might be at high risk of contracting infections because of several reasons (Ather et al. 2020). First, provision of dental care demands to work in close vicinity to the patient's oropharyngeal region, which has been shown to contain high viral loads of SARS-CoV-2 (To et al. 2020). Second, dental procedures usually involve the handling of sharps, needles, and instruments contaminated with blood and/or saliva. Third the use of a high-speed handpiece, ultrasonic instrument, and air-water syringe generates high levels of aerosol or splatter (Zemouri et al. 2017). Inevitably, the Occupational Safety and Health Administration (OSHA) has categorized the dental profession as a "very high-risk job," especially if it involves performing aerosol-generating procedures (Dentistry Workers and Employers 2020).

34.3 Importance of Oral Health and Impact of COVID-19 on Dental Practice

Oral health critically contributes to the overall well-being of an individual. In a systematic analysis for the Global Burden of Disease Study 2017, it was shown that oral diseases affect 3.5 billion people worldwide (Dye 2017). Oral diseases pose a significant health burden and can be a cause for pain, discomfort, disfigurement, and even death (Oral health 2020). One of the most prevalent oral diseases worldwide is untreated dental caries in permanent teeth. Periodontal (gum) disease, maxillofacial trauma, oral cancer, and dental infections, including abscess, are some other examples of oral diseases that can have a significant impact on the quality of life. Access to oral health-care services is still a significant problem across many countries with challenges such as the unequal distribution of oral health professionals and a lack of appropriate health facilities.

COVID-19 and its rapid spread have drastically affected the dental community worldwide. Various dental organizations responded by specifying guidelines for the provision of dental care during the pandemic. Sets of recommendations emerged. Accordingly, some regions implemented a complete lockdown of dental practices compared to specific areas where dentists continued to provide care for emergency patients (Mallineni et al. 2020). In some regions, dental care professionals were at the forefront, providing care to COVID-19-affected individuals (Cagetti et al. 2020). Despite the disparity in recommendations across different regions in the world, the main goal has been to treat dental patients in need to reduce the burden on the emergency room.

A report published by the American Dental Association (ADA) on April 10, 2020, showed that four of five dentists had closed their practices and were only providing emergency care. It is interesting that dentists and their team joined public health settings, worked with other departments, and contributed to screening and testing patients.

The pandemic also had financial implications. Even though dentists in the public health settings were still being paid during the lockdown, their counterparts employed in private care settings were not being paid, especially if they were non-owners of practices.

34.4 Utilization of Dental Services During COVID-19 Pandemic

The pattern of utilization of dental services is greatly affected by the COVID-19 pandemic. A report published by Guo et al. revealed that the proportion of patients with dental infection increased by 20% during the pandemic, whereas dental trauma cases decreased from 14.2% to 10.5% (Guo et al. 2020a). There was also a significant decrease in patients presenting with non-urgent conditions (Guo et al. 2020a). In another study, collected data from Brazil showed that cases of facial trauma had declined as a result of social distancing practices and lockdown (Maffia et al. 2020).

Meng et al. treated more than 700 patients and provided online consults to over 1600 patients. Although they did not report on the characteristics of all the emergency patients, the recommendation was made to treat a COVID-19 patient at the end of the day in an isolated or negative pressure room (Meng et al. 2020).

Yu et al. reported on the characteristics of dental emergency patients visiting the School and Hospital of Stomatology, Wuhan University (WHUSS), during the COVID-19 outbreak and found that the proportion of endodontic emergencies was 51% (Yu et al. 2020). These patients are usually presented with symptomatic irreversible pulpitis, symptomatic apical periodontitis, and acute apical abscess. The authors also proposed the use of vital pulp therapy procedures such as pulpotomy to relieve pain and, at the same time, reduce treatment duration (Yu et al. 2020).

34.5 Tele-dentistry and Triaging

An exponential increase in the number of COVID-19 cases and its community spread pattern led to the introduction of the "social distancing" concept to flatten the curve (Maharaj and Kleczkowski 2012). During the pandemic, several state and national associations recommended restrictions on dental practices to curb the spread of COVID-19. A sudden increase in the use of teledentistry occurred in order to continue providing care to the dental patient population (Giudice et al. 2020). Teledentistry, similar to telemedicine, makes use of telecommunication modalities to enable screening and triaging of patients and can serve as an alternative to face-to-face visits to evaluate several dental and mucosal conditions (Jampani et al. 2011). It has been reported that in China, 69% of the tertiary dental hospitals offered free dental consultations (Yang et al. 2020). Patients with urgent and emergent needs can be identified and either managed pharmacologically or scheduled to have a face-to-face visit. It is critical to prioritize the follow-up of these patients and schedule them as soon as routine dental treatment resumes. However, teledentistry is an evolving field that may not be routinely practiced worldwide, and there are several inherent limitations which still need to be addressed (Maret et al. 2020). One of the foremost challenges is to maintain the confidentiality of patient data. Therefore, the use of secure and encrypted platforms for communication is highly encouraged. It is also a limitation to perform a physical and detailed visual examination, which might be critical to diagnosing several dental conditions, as a result of which several states or countries have prohibited the routine use of teledentistry (Maret et al. 2020). However, in particular circumstances such as during a public health emergency, guidelines on telemedicine and teledentistry should be revised to facilitate the provision of care for affected or susceptible individuals. For example, in response to the COVID-19 pandemic, the United States Department of Health and Human Services has eased the Health Insurance Portability and Accountability Act (HIPAA) regulations in order to enable free telehealth services to patients (Notification of Enforcement Discretion for Telehealth 2020). In a survey conducted by ADA, it was shown that 58% of dentists in public health settings and 25% of dentists in private practice were utilizing virtual technology/telecommunications to conduct remote problem-focused evaluations (COVID-19 Impact on Dental Practices 2020).

34.6 Pharmacologic Management

During the state of lockdown and restrictions imposed on face-to-face visits, pharmacological management can be a useful alternative to manage several dental complaints. These might include the use of analgesics, antibiotics, and steroids to provide palliative care for pain, swelling, and infection (Ather et al. 2020). It is also a useful strategy to provide symptomatic relief for patients with suspected or confirmed COVID-19 infection. This approach will also provide the dentist with some time to identify centers with adequate resources where dental care can be provided safely to COVID-19-positive patients (Ather et al. 2020).

Ibuprofen, a commonly used non-steroidal anti-inflammatory drug (NSAID) in dental practice, alone or in combination with acetaminophen, may be prescribed for pain relief (Fu et al. 2020). There were some initial concerns about the use of ibuprofen in COVID-19 patients. It might prolong the illness and worsen respiratory problems during the infection (Day 2020). However, the World Health Organization (WHO) in a commentary refuted these findings (The use of non-steroidal anti-inflammatory drugs (NSAIDs) in patients with COVID-19 2020). It stated, "there is no evidence of severe adverse events, acute health care utilization, long-term survival, or quality of life in patients with COVID-19, as a result of the use of NSAIDs."

Corticosteroids may be prescribed for managing dental inflammatory conditions such as post-operative swelling or endodontic flare-up (Ather et al. 2020). There have been conflicting reports on the use of steroids in COVID-19-positive patients. Initial reports raised concerns about prescribing steroids in COVID-19 patients (Li et al. 2020). However recent data has shown that low-dose dexamethasone at 6 mg for up to 10 days can lead to a reduced rate of mortality. Therefore, the prescription of drugs should be assessed on a case-by-case basis and after consultation with the patient's primary care physician.

34.7 Screening of Patients

Before scheduling any patient for an in-office visit, the medical and dental history can be obtained via a telephone call (Ather et al. 2020). Patients should be screened to determine the nature of the chief complaint, as it is crucial to differentiate between routine and urgent dental conditions. Table 34.1 provides a list of commonly encountered urgent and emergent dental conditions. Also, a COVID-19 screening questionnaire should be utilized as a part of a dual-phase triaging (telephonic consultation followed by in-person at the clinic) to evaluate the patient's health status better (Fig. 34.1). Regardless of this primary tele-screening, the patient's body temperature and other symptoms must be evaluated on the day of the appointment (Ather et al. 2020). If they had a fever (over 37 degrees) or considered as a suspected case of COVID-19, deferring the appointment to another date (at least 2 weeks later) is suggested (Giacomelli et al. 2020).

Table 34.1 List of commonly encountered urgent and emergent dental conditions

Dental emergencies	Uncontrolled post-extraction hemorrhage
	Facial trauma with potential to compromise the airway
	Acute pericoronitis/cellulitis/ spreading infection with the potential to compromise the airway and/or cause severe trismus
	Inhalation or ingestion of fractured/ displaced orthodontic appliances or pieces of restoration or tooth
	Oral ulceration with severe dehydration
Urgent dental care	Dental trauma: tooth fracture/ displacement/avulsion with or without soft tissue injury
	Severe dental pain from pulpal inflammation
	Dry socket
	Pericoronitis
	Biopsy
	Oral ulcers of >3 weeks of history
	Abscess or localized bacterial infection
	Dental treatment required before critical medical procedures
	Loss of temporary restoration: the need for a final crown and/or fixed dental prosthesis (FDP)
Other urgent dental care	Extensive dental caries or defective restorations causing pain
	Suture removal
	Denture adjustment on radiation/ oncology patients or when the function is impeded
	Lost temporary restoration on endodontic access openings in patients experiencing pain
	Orthodontic wire/appliance adjustment

Adapted and modified from the National Health Service (NHS), Scottish Dental Clinical Effectiveness Programme, and American Dental Association guidelines

34.8 Considerations for In-Office Dental Visit

Because of the evolving understanding of the disease and its transmission dynamics, provid-

- Any sign of fever, cough, unexplained fatigue or respiratory problem during the last 2 weeks?
- Any contact with suspected or confirmed COVID-19 patients during the last 2 weeks?
- Any travel history to COVID-19 hot spot region during the last 2 weeks?
- Participation in any crowded meeting during the last 2 weeks?

Fig. 34.1 COVID-19 screening questionnaire

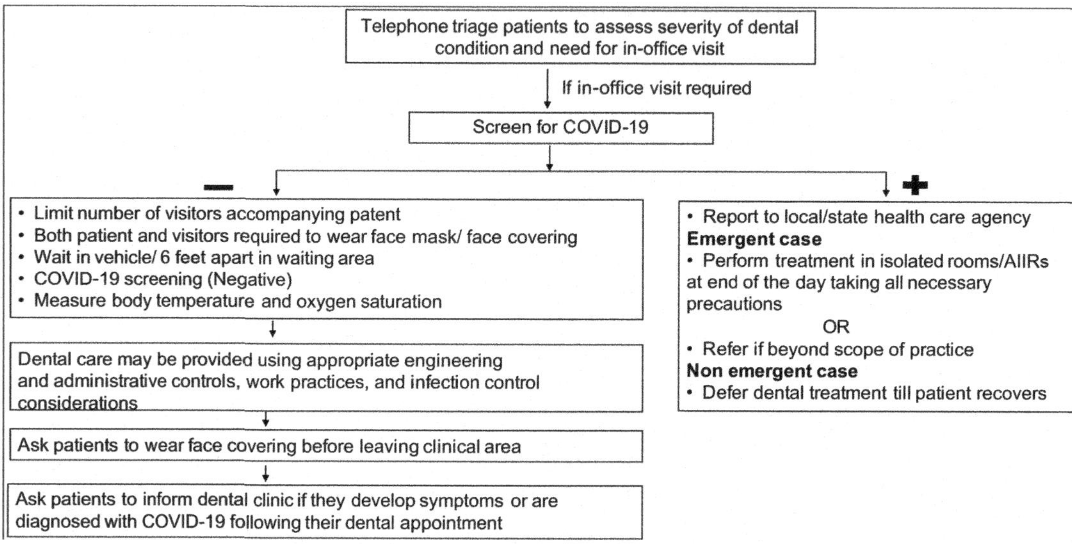

Fig. 34.2 Flowchart for patient triaging and dental management

ing definitive universal recommendations about scheduling an in-office visit is beyond the scope of this text. An in-office dental visit should be scheduled at the discretion of the dental health-care provider and in compliance with local/state/national guidelines. Before scheduling, it is crucial to use professional judgment to carefully assess the risk to benefit ratio, as a significant subset of COVID-19-affected patients might be asymptomatic and can serve as carriers of infection (Rothe et al. 2020). Amid the pandemic, all the patients should be treated as if they are COVID-19 positive. Therefore standard, contact, and air-borne precautions should be in place (Ather et al. 2020). Figure 34.2 is a brief overview of the screening protocol and dental management.

34.8.1 Waiting Area

If possible, patients should be instructed to come alone to minimize the number of people in the waiting area. As an additional measure to avoid crowding, patients can be asked to wait in the vehicle or a designated common area and then be notified via telephone or RFID to come in for the appointment. In order to avoid cross-contamination, a minimum of 6 feet or 1 meter distance between patients should be maintained in the waiting area, and infographics regarding appropriate hand hygiene and respiratory etiquettes such as for coughing and sneezing should be put up. Alcohol-based hand rubs (60–95%) should also be provided in the waiting area and front desk. Patients need to wear a face mask at all times. As the waiting area usually has a lot of flat surfaces, and the fact that coronavirus can persist on inanimate surfaces for several days, frequent disinfection with a neutral pH detergent should be performed (Jamal et al. 2020). Emphasis should be put to reduce the number of flat surfaces as much as possible and unnecessary items from the waiting area, such as magazines and toys (Jamal et al. 2020).

Upon patient arrival, the COVID-19 screening questionnaire should be filled out again to determine any changes in the patient's COVID-19 status (Ather et al. 2020). Dental staff should be trained to measure the patient's temperature using a non-touch thermometer or infrared thermal scanner and oxygen saturation with a pulse oximeter (Ather et al. 2020). These vitals should be recorded for every patient, as silent hypoxemia or low blood oxygen saturation level has been reported as a presenting sign in some asymptomatic COVID-19 patients (Wilkerson et al. 2020). Presence of fever (>100.4 °F = 38 °C), low oxygen saturation (<94%), and symptoms related to COVID-19 should be recorded, and elective care should be deferred for at least 2 weeks. These patients should be asked to self-quarantine themselves, and the findings should be reported to local health authorities and to patient's primary care physician.

34.8.2 Hand Hygiene

It is a simple and effective method to prevent the spread of infectious diseases. Unfortunately, hand hygiene compliance rates have been reported to be as low as 51% (Pettinger and Nettleman 1991). It has also been reported that people touch their face on an average of 23 times per hour, with 44% of these occurrences involving the mucous membranes of the mouth and nose (Kwok et al. 2015). Therefore, a two-before and three-after hand hygiene protocol should be followed in order to reinforce the compliance of handwashing (Safety and World Health 2009). Specifically, the dental care providers should wash their hands before examining a patient, before performing any dental procedures, after touching the patient, after touching the surroundings and equipment without disinfection, and after touching the oral mucosa, blood, damaged skin, or wound (Peng et al. 2020). CDC recommends using alcohol-based hand rubs (ABHRs) with 60% ethanol or 70% isopropyl alcohol as an alternative to washing hands with soap and water for 20 s as the use of ABHRs has been shown to have better compliance among health-care workers (Kratzel et al. 2020). However, if hands are visibly soiled, handwashing with soap and water for at least 20 s is recommended (Kratzel et al. 2020).

34.8.3 Personal Protective Equipment (PPE)

PPE should be worn in order to have an effective barrier against the cross-transmission of infectious agents. The judicious use of PPE is encouraged owing to the overwhelming spread of COVID-19 infection and to prevent the shortage of PPE supply (Using Personal Protective Equipment (PPE) 2020). Therefore, PPE should be chosen depending on the planned procedure and the infection status of the patient. PPE includes the use of gowns, gloves, protective eyewear, surgical facemask with a shield, surgical head cap, and impervious shoe cover (Using Personal Protective Equipment (PPE) 2020; Jamal et al. 2020). Depending on the logistics, infection status of patient, and procedure, more stringent practices such as the use of hazmat suits, National Institute for Occupational Safety and Health (NIOSH)-certified N95, European Standard Filtering Face Piece 2 (EU FFP2), or EU FFP3 respirators conforming to European Standard 149 (EN149) might be needed (Jamal et al. 2020). OSHA recommends using powered air-purifying respirators or supplied-air respirators for aerosol-generating procedures. Furthermore, it is critical to ensure the fit of the masks and respirators. Therefore, masks and respirators should be fit checked and always discarded if soiled (Using Personal Protective Equipment (PPE) 2020).

Guidelines for donning and doffing of PPE must be followed (Using Personal Protective

Equipment (PPE) 2020), as it has been shown that incorrect doffing is a frequent occurrence and can lead to contamination of health-care workers after the patient encounter (Okamoto et al. 2016).

34.8.4 Radiographs

Dental radiography is an integral component of establishing the diagnosis in several dental conditions. The use of intra-oral radiographic films or sensors carries the potential to induce cough or gag reflex (Ather et al. 2020). If needed, films or sensors should be a double barrier to reduce the chances of contamination with the patient's saliva (Hokett et al. 2000). As an alternative, extraoral radiographic techniques such as panoramic radiograph or limited field of view CBCT can be used to obtain diagnostic information (Ather et al. 2020).

34.8.5 Pre-procedural Mouth Rinse

Oral rinses targeting the viral envelope represent a potential method to reduce or inactivate infective viral particles (O'Donnell et al. 2020). Povidone-iodine (PVP-I) mouthwash in a concentration of 0.23% has been shown to inactivate both SARS-CoV and MERS-CoV following a 15-second exposure (Eggers et al. 2018). An in vitro study published in June 2020 confirmed the rapid SARS-CoV-2 viricidal activity of PVP-I [0.50–1.5%%] (Bidra et al. 2020). However, caution should be exercised while using high concentrations, as they have been shown to possess ciliotoxicity on human respiratory tissues (O'Donnell et al. 2020). Rare reports of an allergic reaction to PVP-I have also been reported in the literature (O'Donnell et al. 2020). Another alternative is an oxidizing agent, hydrogen peroxide, which should be diluted to 1–1.5% for oral use, and the patient should gargle with the solution for 30–60 s (Peng et al. 2020). Data for hydrogen peroxide comes from an in vitro study confirming the efficacy of 0.5% hydrogen peroxide against SARS-CoV-2 on inanimate surfaces (Kampf et al. 2020). Conflicting results are available in the literature on the efficacy of chlorhexidine mouth rinse against SARS-CoV-2. Although Meng et al. reported against the use of chlorhexidine rinse for SARS-CoV-2 (Meng et al. 2020), another study showed the efficacy of chlorhexidine mouth rinse in reducing SARS-CoV-2 load in saliva (Yoon et al. 2020).

34.8.6 Aerosol-Generating Procedures (AGP)

Aerosol-generating procedures (AGPs) are defined as any medical and patient care procedure that results in the production of airborne particles (aerosols) (Bolton et al. 2020). Aerosols are defined as particles less than 50 micrometers in diameter compared to splatter characterized by having a diameter of greater than 50 micrometers (Harrel and Molinari 2004). Aerosols have been suggested as the greatest airborne infection threat in dentistry due to their ability to stay airborne and the potential to enter respiratory passages (Harrel and Molinari 2004). The potential of aerosols to contain saliva, nasopharyngeal secretions, blood, and microorganisms can pose a risk for nosocomial spread of infection (Zemouri et al. 2017). It is noteworthy that aerosols from severe acute respiratory syndrome coronavirus (SARS-CoV) can travel more than 6 feet (Kutter et al. 2018). Studies have demonstrated high viral load in nasopharyngeal secretions and saliva of COVID-19-affected individuals (Xu et al. 2020; To et al. 2020). In addition to lungs, salivary glands have also been shown to possess human angiotensin-converting enzyme 2 receptors (Liu et al. 2011), which can act as the binding site for SARS-CoV-2 (Using Personal Protective Equipment (PPE) 2020).

Interestingly, high concentrations of virions have also been reported on oral mucosa and tongue of affected patients (Xu et al. 2020). However, the transmission of SARS-CoV-2 via aerosols is still to be proven. There have been reports of the ability of SARS-CoV-2 to be aerosolized and transmitted to other individuals. A recent study by Guo et al. showed the potential for SARS-CoV-2 to be aerosolized and reported on-air and surface contamination in COVID-19 hospital wards (Guo et al. 2020b). This study also highlighted the potential for this infection to spread to health-care professionals. Even after completing treatment, aerosols are suspended in the air within the treatment room, with more substantial and larger particles settling faster (Bennett et al. 2000; Nikitin et al. 2014). Microbial contamination can be detected as long as 30 min after an aerosol-generating procedure (Nikitin et al. 2014; Bennett et al. 2000). Therefore, to avoid cross-transmission, a waiting time of at least 30 min between patients is recommended along with adequate ventilation of the operatory (coronavirus Sars-CoV 2020). Settling occurs on all horizontal surfaces, after which these can act as a vehicle for transmission of the SARS-CoV-2 virus via indirect contact. Also, the viability of SARS-CoV-2 in aerosols can be up to 3 h, while it can maintain up to 3 days on surfaces such as stainless steel and glass (Van Doremalen et al. 2020).

In dentistry, the most common sources for aerosol generation are high-speed handpieces, ultrasonic units, air abrasion devices, and three-way air-water syringes (Harrel and Molinari 2004). Depending on the status of the COVID-19 pandemic in a specific region, recommendations are put forward to restrict the use of aerosol-generating devices. Alternative measures to AGP can be undertaken to achieve the same treatment goal, e.g., scaling and root planing can be performed manually instead of using ultrasonic units. Chemo-mechanical caries removal (atraumatic restorative technique) can be used to avoid aerosol generation from high-speed rotary handpieces. However, in situations where AGP needs to be undertaken, a four-handed dentistry technique with rubber dam application and high-volume evacuation should be utilized (Ather et al. 2020). High-speed handpieces should also have anti-retraction valves to prevent the aspiration of contaminants. It is also advisable to perform AGP in isolated closed rooms during the pandemic (Peng et al. 2020).

Also, the use of lasers has been implicated in the spread of viral particles via plumes. Ziegler et al. investigated that application of Er:YAG laser can enhance the virus spread through longer distances and promote its transmission significantly in the dental setting (Ziegler et al. 1998).

34.8.7 Use of Rubber Dam and High-Volume Evacuation

Rubber dam isolation involves the application of a latex/non-latex sheet in a frame over the tooth/teeth and isolates the site from the rest of the oral cavity (Ahmed et al. 2014). One of the purposes of a rubber dam is to prevent the seepage of saliva into the field of operation (Ahmed et al. 2014). It has been reported that if applied appropriately and further sealed with a barrier such as OpalDam/OraSeal (Ultradent Products Inc., South Jordan UT, USA), the only source of contamination would be the tooth that is being treated (Diegritz et al. 2020; Li et al. 2004). After the rubber dam application, it is advisable to disinfect the dam and tooth with 0.5–5.25% sodium hypochlorite solution, which has shown to possess virucidal activity against SARS-CoV-2 (Kampf et al. 2020). Cochran et al. reported a significant reduction in atmospheric microbial contamination when rubber dams were used (Cochran et al. 1989). A further step to reduce microbial contamination and aerosols is to routinely use high-volume evacuation during aerosol-generating procedures (Jamal et al. 2020).

34.8.8 Negative-Pressure Treatment Room/Airborne Infection Isolation Rooms (AIIRs)

For suspected or confirmed patients with COVID-19 or individuals meeting any criteria outlined in Fig. 34.2, the dental procedure should be performed at a center with a provision for a negative-pressure room (Ather et al. 2020; Jamal et al. 2020). AIIRs are rooms at negative pressure relative to the surrounding areas and with a minimum of six air changes per hour (Ather et al. 2020). Anticipatory knowledge of centers with provision for AIIRs will help dentists to provide emergent dental care if the need arises (Ather et al. 2020). By considering the shortage of negative-pressure rooms and the overwhelmed hospitals, there should be another alternative to provide care in an isolated room. Also, the appointment should be scheduled as the last patient of the day (Meng et al. 2020), and if possible, it is advisable to defer the treatment for at least 2 weeks (Ather et al. 2020). However, if these patients have an urgent dental problem, palliative care such as prescribing antibiotics and analgesics should be considered.

34.8.9 Disinfection of Instruments, Surfaces, and Operatory

The use of single-use disposable instruments and devices such as mouth mirrors and syringes is recommended (Peng et al. 2020). For reusable patient care items, procedures for cleaning, disinfection, and sterilization can be performed according to the CDC guidelines.

The workspace environment can often get contaminated due to the microbial-laden aerosols generated during dental procedures. These workspace surfaces can be separated into housekeeping surfaces (such as floors and walls) and clinical contact surfaces (Schneiderman and Cartee 2020). Clinical contact surfaces include the dental radiographic equipment, knobs, switches, light handles, drawer handles, chairside computers, countertops, nitrous oxide equipment, and digital impression devices (Schneiderman and Cartee 2020). Impervious barriers should be placed on clinical contact surfaces and should be replaced between patients (Schneiderman and Cartee 2020). Clinical contact surfaces require disinfection between patients. The United States Environmental Protection Agency (EPA) provides a list of approved disinfectants for SARS-CoV-2 disinfection and specifies the contact time required for adequate disinfection.

Regarding disinfectants, the WHO has stated, "The recommendation of 0.1% (1000 ppm) in the context of COVID-19 is a conservative concentration that will inactivate the vast majority of other pathogens that may be present in the health care setting. However, for blood and body fluids, large spills (i.e., more than about 10 mL) a concentration of 0.5% (5000 ppm) is recommended." A review of literature done by Kampf et al. showed that coronaviruses could persist on inanimate surfaces for up to 9 days and surface disinfection with 0.1% sodium hypochlorite or 62–71% ethanol significantly reduces coronavirus infectivity on surfaces within 1 min exposure time (Kampf et al. 2020). Specific research on SARS-CoV-2 done by Van Doremalen et al. revealed that it could persist on inanimate surfaces for up to 3 days with more stability on surfaces such as plastic and stainless steel (Van Doremalen et al. 2020). Therefore, adequate disinfection of surfaces should be carried out to mitigate the risk of indirect transmission of SARS-CoV-2.

In addition to chemical disinfectants, UV-C might have a potential role in the disinfection of operatories and workspace clinical surfaces (Casini et al. 2019). However, the data on UV-C light disinfection is mostly against SARS-CoV-1 and has conflicting results (Kariwa et al. 2006; Darnell et al. 2004). No conclusive data on the efficacy of UV disinfection against SARS-CoV-2 has been published to date in the scientific literature.

Ozone gas is also a potential agent that has been tested for eliminating various types of viruses (Roy et al. 1981). It has been reported that SARS-CoV is also sensitive to ozone gas (Schentag et al. 2004). Enveloped viruses are vulnerable to inactivation in the presence of ozone gas, which has an oxidizing effect; reactive oxygen species can have negative effects on the integrity of the envelope because of their oxidation (Kekez and Sattar 1997). This inactivation may work on the novel coronavirus, which possesses a lipid envelope, like the SARS-CoV. However, the concentration of more than 1 ppm is not recommended in an occupied room because of its toxicity (Paffett et al. 2015). Another study investigated that virus in the aerosols was inactivated in a few seconds after exposure to ozone gas (Kekez and Sattar 1997).

34.8.10 Workspace Considerations

As part of the COVID-19 outbreak preparedness, clinics were shut down, and elective procedures were withheld. As a result, patient volume across most health systems declined, helping reduce the nosocomial spread of the virus. Moving forward, several regions across the world have moved toward a phased opening of the economy and reopening of dental practices for elective care. However, a shift from emergent care to elective care will increase patient flow and subsequently challenge the feasibility of "social distancing" practices, which is the bedrock for continuing to open the country.

Unfortunately, several modern health-care facilities have open floor plan practices and "densified" waiting rooms, especially for outpatient care. Such proximity of patients, family members, and health-care workers, combined with the high infectivity of SARS CoV-2, poses a high risk for cross-contamination and limits the ability of facilities to prevent COVID-19 spread.

As we move toward possibly providing elective care, changes in the architectural design and layout of the work environment should be considered. Design interventions such as placing glass barriers at the front desk, automated doors, and minimizing flat surfaces as much as possible to reduce fomite transmission of microbes are all easy fixes that can be incorporated into any work environment. Also, optimizing the design of operatories in terms of creating separate rooms or cubicles with automated doors would also be a critical step. Even though high-efficiency particulate air (HEPA) filters can remove particles as small as 0.3 microns, it is still ineffective for coronaviruses (Bauchner et al. 2020). Therefore, increasing outside air fractions in the ventilation systems or having a window is a simple step to dilute the contaminants. Materials such as copper may find increased use as it has shown to be better than steel and plastic in terms of resisting SARS-CoV-2 survival (Van Doremalen et al. 2020).

34.9 Conclusion

Infectious disease outbreaks have been on the rise for the past few decades. The COVID-19 pandemic has been one of the most humbling experiences for the health-care system. Although the topic of COVID-19 transmission via aerosols in dental practices has been controversial, the possibility still exists and calls for an immediate action by the dental community to play a dual role by not only providing adequate care to the patients but also making sure that appropriate measures are in place to contain the spread of this infection. The practice of dentistry might experience specific changes such as careful screening of patients, strict adherence to safety protocols, and modification of workspaces. It is essential to understand that the guidelines and recommendations provided in this chapter might change over a while as the understanding of COVID-19 matures. Therefore, dentists need to keep themselves abreast of the dynamic trends in the understanding of COVID-19 and should follow the local or regional dental guidelines at all times.

References

Ahmed HMA, Cohen S, Lévy G, Steier L, Bukiet F (2014) Rubber dam application in endodontic practice: an update on critical educational and ethical dilemmas. Aust Dent J 59(4):457–463

Ather A, Patel B, Ruparel NB, Diogenes A, Hargreaves KM (2020) Coronavirus disease 19 (COVID-19): implications for clinical dental care. J Endod 46:584–595

Bauchner H, Fontanarosa PB, Livingston EH (2020) Conserving supply of personal protective equipment—a call for ideas. JAMA 323(19):1911–1911

Bennett AM, Fulford MR, Walker JT, Bradshaw DJ, Martin MV, Marsh PD (2000) Microbial aerosols in general dental practice. Br Dent J 189(12):664–667

Bidra AS, Pelletier JS, Westover JB, Frank S, Brown SM, Tessema B (2020) Rapid in-vitro inactivation of severe acute respiratory syndrome coronavirus 2 (SARS-CoV-2) using povidone-iodine oral antiseptic rinse. J Prosthodont. https://doi.org/10.1111/jopr.13209

Bolton L, Mills C, Wallace S, Brady MC, of Speech RC, Covid LTR, Group A (2020) Aerosol generating procedures, dysphagia assessment and COVID-19: a rapid review. Int J Lang Commun Disord 55(4):629–636

Cagetti MG, Cairoli JL, Senna A, Campus G (2020) COVID-19 outbreak in North Italy: an overview on dentistry. A questionnaire survey. Int J Environ Res Public Health 17(11):3835

Cai J, Sun W, Huang J, Gamber M, Wu J, He G (2020) Indirect virus transmission in cluster of COVID-19 cases, Wenzhou, China. Emerg Infect Dis. https://doi.org/10.3201/eid2606.200412

Casini B, Tuvo B, Cristina ML, Spagnolo AM, Totaro M, Baggiani A, Privitera GP (2019) Evaluation of an ultraviolet C (UVC) light-emitting device for disinfection of high touch surfaces in hospital critical areas. Int J Environ Res Public Health 16(19):3572

Chan JF-W, Yuan S, Kok K-H, To KK-W, Chu H, Yang J, Xing F, Liu J, Yip CC-Y, Poon RW-S (2020) A familial cluster of pneumonia associated with the 2019 novel coronavirus indicating person-to-person transmission: a study of a family cluster. Lancet 395(10223):514–523

Chen J (2020) Pathogenicity and transmissibility of 2019-nCoV—a quick overview and comparison with other emerging viruses. Microbes and infection 22(2):69–71

Chen X, Yu B (2020) First two months of the 2019 Coronavirus Disease (COVID-19) epidemic in China: real-time surveillance and evaluation with a second derivative model. Glob Health Res Policy 5(1):1–9

Cochran MA, Miller CH, Sheldrake MA (1989) The efficacy of the rubber dam as a barrier to the spread of microorganisms during dental treatment. J Am Dent Assoc 119(1):141–144

coronavirus Sars-CoV (2020) H Infectiepreventie in mondzorgpraktijken

COVID-19 Impact on Dental Practices (2020) ADA Health Policy Institute. https://www.ada.org/en/science-research/health-policy-institute/covid-19-dentistseconomic-impact. Accessed 11 July 2020

Darnell MER, Subbarao K, Feinstone SM, Taylor DR (2004) Inactivation of the coronavirus that induces severe acute respiratory syndrome, SARS-CoV. J Virol Methods 121(1):85–91

Day M (2020) Covid-19: ibuprofen should not be used for managing symptoms, say doctors and scientists. Br Med J Publ Group 368:m1086

Dentistry Workers and Employers (2020) Occupational Safety and Health Administration. https://www.osha.gov/SLTC/covid-19/dentistry.html

Diegritz C, Manhart J, Bücher K, Grabein B, Schuierer G, Kühnisch J, Kunzelmann K-H, Hickel R, Fotiadou C (2020) A detailed report on the measures taken in the Department of Conservative Dentistry and Periodontology in Munich at the beginning of the COVID-19 outbreak. Clin Oral Investig:1–11

Dye BA (2017) The global burden of oral disease: research and public health significance. J Dent Res 96(4):361–363

Eggers M, Koburger-Janssen T, Eickmann M, Zorn J (2018) In vitro bactericidal and virucidal efficacy of povidone-iodine gargle/mouthwash against respiratory and oral tract pathogens. Infect Dis Ther 7(2):249–259

Ferretti L, Wymant C, Kendall M, Zhao L, Nurtay A, Abeler-Dörner L, Parker M, Bonsall D, Fraser C (2020) Quantifying SARS-CoV-2 transmission suggests epidemic control with digital contact tracing. Science 368(6491):eabb6936

Fu L, Wang B, Yuan T, Chen X, Ao Y, Fitzpatrick T, Li P, Zhou Y, Lin Y, Duan Q (2020) Clinical characteristics of coronavirus disease 2019 (COVID-19) in China: a systematic review and meta-analysis. J Infect 80(6):656–665

Gandhi M, Yokoe DS, Havlir DV (2020) Asymptomatic transmission, the Achilles' heel of current strategies to control COVID-19. Mass Medical Soc 382(22):2158–2160

Giacomelli A, Pezzati L, Conti F, Bernacchia D, Siano M, Oreni L, Rusconi S, Gervasoni C, Ridolfo AL, Rizzardini G (2020) Self-reported olfactory and taste disorders in patients with severe acute respiratory coronavirus 2 infection: a cross-sectional study. Clin Infect Dis 71(15):889–890

Giudice A, Barone S, Muraca D, Averta F, Diodati F, Antonelli A, Fortunato L (2020) Can teledentistry improve the monitoring of patients during the covid-19 dissemination? a descriptive pilot study. Int J Environ Res Public Health 17(10):3399

Guan W-j, Ni Z-y, Hu Y, Liang W-h, Ou C-q, He J-x, Liu L, Shan H, Lei C-l, Hui DSC (2020) Clinical characteristics of coronavirus disease 2019 in China. N Engl J Med 382(18):1708–1720

Guo H, Zhou Y, Liu X, Tan J (2020a) The impact of the COVID-19 epidemic on the utilization of emergency dental services. J Dental Sci. https://doi.org/10.1016/j.jds.2020.02.002

Guo Z-D, Wang Z-Y, Zhang S-F, Li X, Li L, Li C, Cui Y, Fu R-B, Dong Y-Z, Chi X-Y (2020b) Aerosol and surface distribution of severe acute respiratory syndrome coronavirus 2 in hospital wards, Wuhan, China, 2020. Emerg Infect Dis 26(7):10–3201

Harrel SK, Molinari J (2004) Aerosols and splatter in dentistry: a brief review of the literature and infection control implications. J Am Dent Assoc 135(4):429–437

Hokett SD, Honey JR, Ruiz F, Baisden MK, Hoen MM (2000) Assessing the effectiveness of direct digital: radiography barrier sheaths and finger cots. J Am Dent Assoc 131(4):463–467

Jamal M, Shah M, Almarzooqi SH, Aber H, Khawaja S, El Abed R, Alkhatib Z, Samaranayake LP (2020) Overview of transnational recommendations for COVID-19 transmission control in dental care settings. Oral Dis. https://doi.org/10.1111/odi.13431

Jampani ND, Nutalapati R, Dontula BSK, Boyapati R (2011) Applications of teledentistry: a literature review and update. J Int Soc Prev Commun Dent 1(2):37

Kampf G, Todt D, Pfaender S, Steinmann E (2020) Persistence of coronaviruses on inanimate surfaces and their inactivation with biocidal agents. J Hosp Infect 104(3):246–251

Kariwa H, Fujii N, Takashima I (2006) Inactivation of SARS coronavirus by means of povidone-iodine, physical conditions and chemical reagents. Dermatology 212(Suppl. 1):119–123

Kekez MM, Sattar SA (1997) A new ozone-based method for virus inactivation: preliminary study. Phys Med Biol 42(11):2027

Kratzel A, Todt D, V'kovski P, Steiner S, Gultom M, Thao TTN, Ebert N, Holwerda M, Steinmann J, Niemeyer D (2020) Inactivation of severe acute respiratory syndrome coronavirus 2 by WHO-recommended hand rub formulations and alcohols. Emerg Infect Dis 26(7):1592

Kutter JS, Spronken MI, Fraaij PL, Fouchier RAM, Herfst S (2018) Transmission routes of respiratory viruses among humans. Curr Opin Virol 28:142–151

Kwok YLA, Gralton J, McLaws M-L (2015) Face touching: a frequent habit that has implications for hand hygiene. Am J Infect Control 43(2):112–114

Lauer SA, Grantz KH, Bi Q, Jones FK, Zheng Q, Meredith HR, Azman AS, Reich NG, Lessler J (2020) The incubation period of coronavirus disease 2019 (COVID-19) from publicly reported confirmed cases: estimation and application. Ann Intern Med 172(9):577–582

Li RWK, Leung KWC, Sun FCS, Samaranayake LP (2004) Severe acute respiratory syndrome (SARS) and the GDP. Part II: implications for GDPs. Br Dent J 197(3):130–134

Li H, Chen C, Hu F, Wang J, Zhao Q, Gale RP, Liang Y (2020) Impact of corticosteroid therapy on outcomes of persons with SARS-CoV-2, SARS-CoV, or MERS-CoV infection: a systematic review and meta-analysis. Leukemia 34:1–9

Liu L, Wei Q, Alvarez X, Wang H, Du Y, Zhu H, Jiang H, Zhou J, Lam P, Zhang L (2011) Epithelial cells lining salivary gland ducts are early target cells of severe acute respiratory syndrome coronavirus infection in the upper respiratory tracts of rhesus macaques. J Virol 85(8):4025–4030

Maffia F, Fontanari M, Vellone V, Cascone P, Mercuri LG (2020) Impact of COVID-19 on maxillofacial surgery practice: a worldwide survey. Int J Oral Maxillofac Surg 49(6):827–835

Maharaj S, Kleczkowski A (2012) Controlling epidemic spread by social distancing: do it well or not at all. BMC Public Health 12(1):679

Mallineni SK, Innes NP, Raggio DP, Araujo MP, Robertson MD, Jayaraman J (2020) Coronavirus disease (COVID-19): characteristics in children and considerations for dentists providing their care. Int J Paediatr Dent 30(3):245–250

Maret D, Peters OA, Vaysse F, Vigarios E (2020) Integration of telemedicine into the public health response to COVID-19 must include dentists. Int Endod J 53(6):880–881

Meng L, Hua F, Bian Z (2020) Coronavirus disease 2019 (COVID-19): emerging and future challenges for dental and oral medicine. J Dent Res 99(5):481–487

Nikitin N, Petrova E, Trifonova E, Karpova O (2014) Influenza virus aerosols in the air and their infectiousness. Adv Virol 2014:859090

Notification of Enforcement Discretion for Telehealth (2020) HHS.gov. https://www.hhs.gov/hipaa/for--professionals/special-topics/emergencypreparedness/notification-enforcement-discretion-telehealth/index.html. Accessed 11 July 2020

O'Donnell VB, Thomas D, Stanton R, Maillard J-Y, Murphy RC, Jones SA, Humphreys I, Wakelam MJO, Fegan C, Wise MP (2020) Potential role of oral rinses targeting the viral lipid envelope in SARS-CoV-2 infection. Function 1(1):zqaa002

Of the International CSG (2020) The species severe acute respiratory syndrome-related coronavirus: classifying 2019-nCoV and naming it SARS-CoV-2. Nat Microbiol 5(4):536

Okamoto K, Rhee Y, Schoeny M, Lolans K, Cheng J, Reddy S, Salazar E, Warpinski G, Carr M, McIlroy G (2016) Impact of doffing errors on healthcare worker self-contamination when caring for patients on contact precautions. Oxford University Press, Oxford, p 1388

Ong SWX, Tan YK, Chia PY, Lee TH, Ng OT, Wong MSY, Marimuthu K (2020) Air, surface environmental, and personal protective equipment contamination by severe acute respiratory syndrome coronavirus 2 (SARS-CoV-2) from a symptomatic patient. JAMA 323(16):1610–1612

Oral health (2020) World Health Organization. https://www.who.int/news-room/fact-sheets/detail/oral-health

Paffett ML, Zychowski KE, Sheppard L, Robertson S, Weaver JM, Lucas SN, Campen MJ (2015) Ozone inhalation impairs coronary artery dilation via intracellular oxidative stress: evidence for serum-borne factors as drivers of systemic toxicity. Toxicol Sci 146(2):244–253

Peng X, Xu X, Li Y, Cheng L, Zhou X, Ren B (2020) Transmission routes of 2019-nCoV and controls in dental practice. Int J Oral Sci 12(1):1–6

Pettinger A, Nettleman MD (1991) Epidemiology of isolation precautions. Inf Control Hospital Epidemiol 12(5):303–307

Rothe C, Schunk M, Sothmann P, Bretzel G, Froeschl G, Wallrauch C, Zimmer T, Thiel V, Janke C, Guggemos W (2020) Transmission of 2019-nCoV infection from an asymptomatic contact in Germany. N Engl J Med 382(10):970–971

Roy D, Wong PK, Engelbrecht RS, Chian ES (1981) Mechanism of enteroviral inactivation by ozone. Appl Environ Microbiol 41(3):718–723

Safety WHOP, World Health O (2009) A guide to the implementation of the WHO multimodal hand hygiene improvement strategy. World Health Organization, Geneva

Schentag JJ, Akers C, Campagna P, Chirayath P (2004) SARS: clearing the air. National Academies Press, New York

Schneiderman MT, Cartee DL (2020) Surface disinfection. In: Infection control in the dental office. Springer, Cham, pp 169–191

Sikkema RS, Pas SD, Nieuwenhuijse DF, O'Toole Á, Verweij J, van der Linden A, Chestakova I,

Schapendonk C, Pronk M, Lexmond P (2020) COVID-19 in health-care workers in three hospitals in the south of the Netherlands: a cross-sectional study. Lancet Infect Dis 20(11):1273–1280

The use of non-steroidal anti-inflammatory drugs (NSAIDs) in patients with COVID-19 (2020). https://www.who.int/news-room/commentaries/detail/the-use-of-nonsteroidal-anti-inflammatory-drugs-(nsaids)-in-patients-with-covid-19. Accessed 11 July 2020

To KK-W, Tsang OT-Y, Leung W-S, Tam AR, Wu T-C, Lung DC, Yip CC-Y, Cai J-P, Chan JM-C, Chik TS-H (2020) Temporal profiles of viral load in posterior oropharyngeal saliva samples and serum antibody responses during infection by SARS-CoV-2: an observational cohort study. Lancet Infect Dis 20(5):565

Using Personal Protective Equipment (PPE) (2020) Centers for Disease Control and Prevention. https://www.cdc.gov/coronavirus/2019-ncov/hcp/using-ppe.html. Accessed 11 July 2020

Van Doremalen N, Bushmaker T, Morris DH, Holbrook MG, Gamble A, Williamson BN, Tamin A, Harcourt JL, Thornburg NJ, Gerber SI (2020) Aerosol and surface stability of SARS-CoV-2 as compared with SARS-CoV-1. N Engl J Med 382(16):1564–1567

Verdoni L, Mazza A, Gervasoni A, Martelli L, Ruggeri M, Ciuffreda M, Bonanomi E, D'Antiga L (2020) An outbreak of severe Kawasaki-like disease at the Italian epicentre of the SARS-CoV-2 epidemic: an observational cohort study. Lancet 395:1771

Walls AC, Park Y-J, Tortorici MA, Wall A, McGuire AT, Veesler D (2020) Structure, function, and antigenicity of the SARS-CoV-2 spike glycoprotein. Cell 181(2):281–292

Wilkerson RG, Adler JD, Shah NG, Brown R (2020) Silent hypoxia: a harbinger of clinical deterioration in patients with COVID-19. Am J Emerg Med. https://doi.org/10.1016/j.ajem.2020.05.044

Wu P, Duan F, Luo C, Liu Q, Qu X, Liang L, Wu K (2020) Characteristics of ocular findings of patients with coronavirus disease 2019 (COVID-19) in Hubei Province, China. JAMA ophthalmology 138(5):575–578

Xu J, Li Y, Gan F, Du Y, Yao Y (2020) Salivary glands: potential reservoirs for COVID-19 asymptomatic infection. J Dental Res. https://doi.org/10.1177/0022034520918518

Yang Y, Zhou Y, Liu X, Tan J (2020) Health services provision of 48 public tertiary dental hospitals during the COVID-19 epidemic in China. Clin Oral Investig:1

Yoon JG, Yoon J, Song JY, Yoon S-Y, Lim CS, Seong H, Noh JY, Cheong HJ, Kim WJ (2020) Clinical significance of a high SARS-CoV-2 viral load in the saliva. J Korean Med Sci 35(20):e195

Yu J, Zhang T, Zhao D, Haapasalo M, Shen Y (2020) Characteristics of endodontic emergencies during COVID-19 outbreak in Wuhan. J Endod 46(6):730

Zemouri C, de Soet H, Crielaard W, Laheij A (2017) A scoping review on bio-aerosols in healthcare and the dental environment. PLoS One 12(5):e0178007

Zhang H, Penninger JM, Li Y, Zhong N, Slutsky AS (2020) Angiotensin-converting enzyme 2 (ACE2) as a SARS-CoV-2 receptor: molecular mechanisms and potential therapeutic target. Intensive Care Med 46(4):586–590

Ziegler BL, Thomas CA, Meier T, Müller R, Fliedner TM, Weber L (1998) Generation of infectious retrovirus aerosol through medical laser irradiation. Lasers Surg Med Off J Am Soc Laser Med Surg 22(1):37–41

The Implications of COVID-19 to Ophthalmology

35

Tracy H. T. Lai, Emily W. H. Tang, and Kenneth K. W. Li

Abstract

The coronavirus disease 2019 (COVID-19) pandemic has brought unprecedented challenges to ophthalmology. At least 16 ophthalmologists worldwide have succumbed to COVID-19. It reflects the susceptibility of ophthalmologists to COVID-19 infection as they are in close proximity to patients. This chapter provides an overview of the ocular manifestations of COVID-19, risks of COVID-19 to ophthalmologists and patients, clinical service adjustments due to COVID-19, and infection control measures to minimize the transmission of COVID-19 in ophthalmic practice.

Keywords

Conjunctivitis · Coronavirus · COVID-19 · Infection control · Ophthalmology · SARS-CoV-2

T. H. T. Lai · E. W. H. Tang · K. K. W. Li (✉)
Department of Ophthalmology, United Christian Hospital, Kowloon, Hong Kong

Department of Ophthalmology, Tseung Kwan O Hospital, Tseung Kwan O, Hong Kong
e-mail: lkw856@ha.org.hk

35.1 Introduction

The coronavirus disease 2019 (COVID-19) is a highly infectious disease caused by severe acute respiratory syndrome coronavirus 2 (SARS-CoV-2). It quickly evolved into a global threat since the first reported case in December 2019. The World Health Organization (WHO) officially declared COVID-19 a global pandemic on March 11, 2020 (WHO Director-General's opening remarks at the media briefing on COVID-19 2020). Since then, there has been continuing exponential growths in both the newly confirmed COVID-19 cases and the death toll. At the time of writing, more than 71 million people worldwide are confirmed to be infected with COVID-19, and the number of deaths has already surpassed 16 million (COVID-19 Coronavirus Pandemic 2020). Border control, quarantine policies, lockdown, and social distancing measures have been implemented in various parts of the world to contain the virus and mitigate the spread of infection. Some localities, such as Hong Kong, employed a containment approach, which involves the use of extensive testing of COVID-19, tracing of all suspected cases, quarantining of all close contacts, and admission of all COVID-19 patients. Some countries, such as the United Kingdom and the United States, adopted a mitigation approach, in which only severe COVID-19 cases would be admitted and mild COVID-19 cases would be sent home to self-isolate.

N. Rezaei (ed.), *Coronavirus Disease - COVID-19*, Advances in Experimental Medicine and Biology 1318, https://doi.org/10.1007/978-3-030-63761-3_35

35.1.1 Hospital-Wide Contingency Measures

During the COVID-19 pandemic, hospitals worldwide have implemented contingency measures (Adlhoch et al. 2020). Triage systems were set up at hospital entrances to identify suspected COVID-19 cases who are then diverted to the accident and emergency department or tents outside the hospital for medical care and testing. Due to the overwhelming number of patients, operating theatres were converted into intensive care units (ICU) for COVID-19 patients in numerous countries. Regular inpatient wards have been equipped with air filtration systems to convert them into isolation wards. When the capacity of the hospital was overwhelmed, prefabricated buildings and mobile military hospitals were set up. To reduce the risk of cross-infection and conserve resources, such as human resources, inpatient beds, and personal protective equipment (PPE), many nonurgent outpatient clinics, elective investigations, and operations were suspended. Guidelines were established to conserve and optimize the use of PPE, and visitors and accompanying persons were kept to a minimum.

The policy on the use of surgical masks by health-care workers and patients within the hospital varies among different countries (Tang et al. 2020). Public Health England (PHE) does not recommend all health-care workers to wear surgical masks in hospitals (COVID-19 personal protective equipment (PPE) 2020). Contrarily, in Hong Kong, under the emergency response level, everyone that enters the hospital premises (health-care worker, patient, or visitor) is required to wear mask (Recommended Personal Protective Equipment (PPE) in hospitals/clinics under Serious/Emergency Response Level. Coronavirus disease (COVID-19) 2020).

35.1.2 Deployment of Ophthalmologists to Emergency Rooms and Medical Departments

With the increase in clinical workload due to COVID-19, numerous ophthalmologists worldwide have been deployed to medicine departments, ICUs, and accident and emergency departments. For example, ophthalmologists from New York (Matossian 2020) and Singapore (Li et al. 2020b) were deployed to emergency rooms and screening centers to conduct examination and screening on suspected COVID-19 patients. In the United Kingdom, ophthalmologists were deployed to acute wards and ICUs (Attzs and Lakhani 2020; Yang et al. 2020). In Hong Kong, ophthalmologists were deployed to non-COVID medical wards and medical outpatient clinics to relieve the workload of medical colleagues who had to take care of COVID-19 patients. To prepare medical staff for deployment, it is of paramount importance to equip them with the latest knowledge in both clinical medicine and infection control.

35.2 Ocular Manifestations and COVID-19

35.2.1 Conjunctivitis and COVID-19

35.2.1.1 Prevalence of Conjunctivitis and SARS-CoV-2 in Conjunctival Swabs

The prevalence of ocular involvement in COVID-19 patients varies widely among studies. In a study of 1099 COVID-19 cases, only nine (0.8%) had conjunctival congestion (Guan et al. 2020). In another case series of 535 COVID-19 patients in Wuhan, 27 (5.0%) had conjunctival congestion, of which 4 (0.75%) had conjunctival congestion as the initial symptom (Chen et al. 2020a). Hong et al. administered the Ocular Surface Disease Index (OSDI) and Salisbury Eye Evaluation Questionnaire (SEEQ) to 56 COVID-19 cases. A total of 15 patients (26.8%) reported aggravation of ocular symptoms, 6 (10.7%) had prodromal ocular symptoms before the disease onset, and only 2 (3.6%) had conjunctivitis (Hong et al. 2020).

SARS-CoV-2 transmission through conjunctival secretions and the tear fluid remained controversial. Zhang et al. from Wuhan, which was the epicenter of COVID-19 in China, reported conjunctivitis in only 2 out of 72 patients (2.8%)

(Zhang et al. 2020). Only one patient (1.4%) had a positive conjunctival swab. They found that out of 121 COVID-19 patients, 8 (6.6%) had ocular symptoms, with only 3 (2.5%) who had a positive conjunctival swab reverse transcription polymerase chain reaction (RT-PCR) for SARS-CoV-2. 2 of these 3 patients were in critical condition (Zhou et al. 2020). In a Singaporean study of 64 tear samples, which were collected from 17 COVID-19 patients using Schirmer's tests, all samples tested negative for SARS-CoV-2 ribonucleic acid (RNA) (Seah et al. 2020). Out of the 17 patients, only 1 (5.9%) had conjunctival injection and chemosis. Another study conducted by Xia et al. included 30 COVID-19 patients (Xia et al. 2020). Some of these patients overlapped with patients who were included in the study by Hong et al., which was previously mentioned. Only 1 patient (3.3%) had conjunctivitis whose conjunctival swab tested positive for SARS-CoV-2. All the other patients had negative conjunctival swabs. Karimi et al. from Iran collected tear samples from 43 COVID-19 patients, of which only 1 (2.3%) had conjunctivitis; 3 (7%) samples tested positive for SARS-CoV-2 (Karimi et al. 2020). Most studies did not report on how conjunctival swabs were taken, but Karimi et al. mentioned that they obtained all eye swabs on the first day of admission and placed them in the lower fornix for a longer duration (at least 10 seconds). Such a method of sample collection may explain the slightly higher positive rate of tear RT-PCR for SARS-CoV-2 in their study compared with that in other studies. A study by Wu et al. included 38 COVID-19 patients (Wu et al. 2020). Among them, 12 (31.6%) had ocular manifestations compatible with conjunctivitis (conjunctival hyperemia, chemosis, epiphora, or increased secretions), and 8 were in critical condition. Only 2 (5.3%) had positive conjunctival swabs. The percentage of patients with ocular manifestations in this study was much higher than those in other studies. However, we need to bear in mind that such ocular manifestations are common in ventilated patients and may not necessarily imply COVID-19-related conjunctivitis (Stevens et al. 2020). In addition to the conventional treatment of conjunctivitis, some patients were treated with ganciclovir eye ointment, azithromycin, or ribavirin eye drops. The efficacy of these drugs for COVID-19-associated conjunctivitis will be the focus of future research.

In summary, out of 1981 patients reported, 75 (3.8%) had ocular manifestations and 10 (0.5%) had positive conjunctival swabs. The WHO-China Joint Mission on COVID-19 estimated the incidence of conjunctival congestion to be about 0.8%, based on a study on 55,924 laboratory-confirmed cases (Report of the WHO-China Joint Mission on Coronavirus Disease 2019 (COVID-19) 2020). The rate of SARS-CoV-2 RNA detection in tears and conjunctival swabs may be influenced by the method of sampling and the definition (clinical vs. laboratory-confirmed diagnosis) of COVID-19 and severity classes. With regard to conjunctival swab, the rate of SARS-CoV-2 RNA detection was affected by the number of samples taken from each patient, the amount of ocular discharge collected, and the time from the date of illness onset to the date of conjunctival sampling. A single sample, an inadequate sampling, and a sample taken after the disappearance of ocular symptoms may have led to a false-negative result.

35.2.1.2 The Conjunctiva as a Portal for SARS-CoV-2 Infection and the Importance of Eye Protection

The presence of SARS-CoV-2 in the conjunctiva can be caused by direct inoculation, migration of the virus from the respiratory tract through the nasolacrimal duct, or infection of the lacrimal gland and consequent shedding of the virus (Seah et al. 2020). Whether the conjunctiva is a possible site of entry for SARS-CoV-2 remains controversial (Napoli et al. 2020). Ma et al. suggested that the conjunctiva would be less likely to be infected by SARS-CoV-2 due to the presence of angiotensin-converting enzyme 2 (ACE2) and the absence of type II transmembrane serine protease (TMPRSS2), which are both required for the occurrence of SARS-CoV-2 infection (Ma et al. 2020). However, anecdotal reports suggested that conjunctivitis may be the only initial symptom of COVID-19 before the appearance of systemic symptoms, and there have been reports of ocular transmission of COVID-19 to health-care workers due to the lack of eye protection.

These are pieces of evidence proving that the conjunctiva could potentially be a portal for SARS-CoV-2 infection, either by droplets or by direct contact with secretions from infected cases, followed by subsequent inoculation into the mucous membranes. Dr. Guangfa Wang, a member of the Chinese national expert panel on pneumonia, experienced conjunctivitis in the left eye 3 hours before the onset of cough (Lu et al. 2020). He suspected that the SARS-CoV-2 virus entered his body through his eyes as he wore an N95 respirator but did not wear any eye protection when he visited COVID-19 patients in Wuhan. A nurse working in the emergency department developed conjunctivitis and fever without any other systemic symptoms of COVID-19 (Zhang et al. 2020). She reported that her protective goggles occasionally dislocated and touched her eyelids. She later tested positive for COVID-19. Hui et al. from the University of Hong Kong studied the infectivity of SARS-CoV-2 and SARS-CoV using ex vivo human conjunctival tissues (Hui et al. 2020). Conjunctival cultures were more extensively infected by SARS-CoV-2 than by SARS-CoV, as indicated by immunohistochemical staining, and had higher infectious viral titers. The demonstration of infection and replication of SARS-CoV-2 in the conjunctiva suggested that the conjunctiva could be a possible portal for SARS-CoV-2 infection and supported the important practice of wearing eye protection by health-care workers (Lai et al. 2020b).

35.2.1.3 Conjunctivitis May Present as an Initial Symptom or May Occur in the Later Stages of COVID-19

While conjunctivitis could present as an initial symptom of COVID-19, it may also occur during the later stages of COVID-19. A 30-year-old man developed acute bilateral conjunctivitis 13 days after the disease onset, and his conjunctival swab tested positive for SARS-CoV-2 RNA (Chen et al. 2020b). He suffered from eye redness, tearing, and foreign body sensation. Through the physical examination, bilateral conjunctival injection, follicles, and preauricular lymph nodes were observed. Another 63-year-old male developed hemorrhagic pseudomembranous conjunctivitis on day 17 after the disease onset (Navel et al. 2020). His conjunctival swab and scraping tested negative for any bacteria or virus. It suggests that conjunctivitis may reflect SARS-CoV-2 infection or be secondary to other infections.

35.2.1.4 Severe Keratoconjunctivitis in COVID-19

The severity of conjunctivitis in patients with COVID-19 varies from mild to severe. While some only experienced eye redness and tearing, there have been reports of cases of severe conjunctivitis with corneal involvement causing visual impairment. A 29-year-old woman in Canada developed keratoconjunctivitis with a swollen eyelid, photophobia, and eye discharge (Cheema et al. 2020). The slit-lamp examination revealed conjunctival injection, follicles, and corneal involvement with pseudodendrites and subepithelial infiltrates with overlying corneal defects with a drop in vision from 20/20 to 20/40. A tender preauricular lymph node was palpable. The conjunctival swab of the patient tested positive for SARS-CoV-2 RNA.

35.2.2 Neuro-ophthalmological manifestations of COVID-19

Miller Fisher syndrome, a variant of Guillain–Barre Syndrome, was reported in one case of COVID-19 (Gutierrez-Ortiz et al. 2020). This syndrome is characterized by ophthalmoplegia, ataxia, and areflexia. A 50-year-old man presented with vertical diplopia 5 days after the onset of cough, fever, anosmia, and ageusia. The physical examination revealed right hypertropia, limited adduction of the right eye, and abduction nystagmus of the left eye, compatible with right internuclear ophthalmoplegia and oculomotor nerve palsy. An ataxic gait was detected, whereas deep tendon reflexes in the upper and lower limbs were absent. He tested positive for SARS-CoV-2 RNA in the nasopharyngeal swab and antibodies to the ganglioside GD1b in the blood. Following treatment with intravenous immunoglobulin, his neurological symptoms resolved over 2 weeks.

Bilateral abducens nerve palsy was also reported in a COVID-19 patient (Gutierrez-Ortiz et al. 2020). The patient presented with acute-onset diplopia. Three days before, he suffered from fever and diarrhea without any respiratory symptoms. The physical examination revealed bilateral abduction deficit and esotropia, which was more marked at a distance than near. All deep tendon reflexes were absent, but there was no ataxia. He tested positive for SARS-CoV-2 RNA in the oropharyngeal swab. The anti-ganglioside antibody test was not performed for this case. He completely recovered after 2 weeks.

The pathogenesis of Miller Fisher syndrome and cranial nerve palsy in COVID-19 patients may be due to either an aberrant immune response to SARS-CoV-2 or direct pathogenic effects of SARS-CoV-2. The former seems to be more likely as there have been reports of Miller Fisher syndrome after different bacterial and viral infections, suggesting a para- or post-viral immune response. Moreover, SARS-CoV-2 was not detected in the cerebrospinal fluid of both patients, making the latter less likely to be true.

35.2.3 Retinal Changes in COVID-19 Patients

Asymptomatic retinal changes were detected in the retinas of COVID-19 patients. A study from Brazil reported optical coherence tomography findings of 12 COVID-19 infected adults 11–33 days after the onset of symptoms (Marinho et al. 2020). All patients showed hyper-reflective lesions in the ganglion cells and inner plexiform layers, especially at the papillomacular bundle. Four patients had cotton wool spots and retinal hemorrhages along the retinal arcades. Optical coherence tomography angiograms were normal, and the vision was not affected. However, it was not mentioned whether the patients had any pre-existing ocular pathologies, and no follow-up scans were performed. These retinal changes may be the result of microvascular occlusion secondary to COVID-19.

35.2.4 Hydroxychloroquine and ocular toxicity

Both chloroquine and hydroxychloroquine bind with melanin in the retinal pigment epithelium, which could lead to damage in the photoreceptors and outer nuclear layers, thus resulting in a classical "bull's eye" maculopathy after many years of drug use (Ruamviboonsuk et al. 2020; Yusuf et al. 2017). Chloroquine and hydroxychloroquine were postulated to reduce the viral replication of coronavirus, though their effectiveness against COVID-19 has not been proven yet. At the time of writing, more than 80 clinical trials are currently underway (Ferner and Aronson 2020; Gautret et al. 2020; Gao et al. 2020). However, the WHO has halted its clinical trials involving the use of hydroxychloroquine as it has no proven effect in the reduction of mortality ("Solidarity" clinical trial for COVID-19 treatments 2020). The standard dosage prescribed in clinical trials was 800 mg per day for the first day, followed by 400 mg per day for 5 days. Although this is higher than the dosage of <5mg/kg recommended by the American Academy of Ophthalmology (Marmor et al. 2016), retinopathy is rarely observed before 10 or more years of hydroxychloroquine usage. The risk associated with high-dose hydroxychloroquine over a concise period is very low, and an eye exam is not likely to be necessary (Marmor 2020).

35.3 The Risks of COVID-19 to Ophthalmologists and Patients

According to the WHO, during previous coronavirus outbreaks, such as SARS, human-to-human transmission occurred through droplets, contacts, and fomites. Evidence shows similar modes of transmission for COVID-19 (Coronavirus disease 2019 Situation Report – 25 2020). More than 15,000 health-care workers in Italy were infected with COVID-19, and the latest estimates from Lombardy suggested that 20% of the health-care professionals were positive for SARS-CoV-2 RNA (Veritti et al. 2020). The Centers for Disease Control and Prevention (CDC) reported that at least 9,282 health-care workers in the United States had been infected with COVID-19 (Characteristics of Health Care Personnel with COVID-19 – United States, February 12-April 9, 2020). Moreover, more than 1,800 health-care workers worldwide have died from COVID-19 (In Memoriam: Healthcare Workers Who Have Died of COVID-19 2020). According to a survey conducted on 2,306 residents in New York City, anesthesiology, emergency medicine, and ophthalmology clustered as high-risk specialties, as indicated by the proportion of residents infected with COVID-19 (Breazzano et al. 2020). In Wuhan, otolaryngologists and ophthalmologists had higher rates of infection compared with their colleagues in other specialties (Patel et al. 2020). Dr. Li Wenliang, an ophthalmologist at Wuhan Central Hospital, contracted COVID-19 when he visited an initially asymptomatic glaucoma patient. He succumbed to the disease 1 month later (Parrish et al. 2020). His colleagues, Drs Mei Zhongming and Zhu Heping, also contracted the virus at the same hospital and died in March 2020 (Ting et al. 2020). At the time of writing, 16 ophthalmologists have already died due to COVID-19 (In Memoriam: Ophthalmologist Deaths from COVID-19 2020). A survey conducted across hospitals in Wuhan reported that 14 ophthalmologists, 12 ophthalmic nurses, and 2 ophthalmic technicians were infected with COVID-19 (Qiao et al. 2020). Compared with the control group of eye professionals who did not have COVID-19, the eye professionals in the case group were older ($p = 0.01$), had been in practice longer ($p = 0.001$), had higher rates of contact with confirmed or suspected cases ($p = 0.002$), and reported higher rates of lack of sleep ($p = 0.008$) and lack of PPE ($p = 0.02$). Also, the control group participants more frequently avoided direct skin contact with patients by using gloves or cotton tips ($p = 0.03$). Ophthalmologists are particularly at risk for COVID-19 due to the proximity to the patient's upper respiratory tract during the slit-lamp examination and direct ophthalmoscopy, contact with ocular secretions, and seeing a large number of patients (Breazzano et al. 2020). Anecdotal reports suggest that patients with subclinical infection can transmit infection (Chang et al. 2020) and that when no eye protection is worn, the transmission of the virus can occur via aerosol contact with the conjunctiva (Lu et al. 2020).

Ophthalmologists may also be caught off guard because conjunctivitis, though uncommon, could be the first presenting symptom of COVID-19 before the appearance of common ones, such as cough and fever (Lu et al. 2020). The American Academy of Ophthalmology has issued an alert advising ophthalmologists to wear masks and eye protection when seeing patients with conjunctivitis, respiratory symptoms, and a recent history of international travel (Alert: Important coronavirus context for ophthalmologists 2020).

The world's population is aging. Given the direct association between aging and cataract, most ophthalmic clinics are extremely busy and crowded. Elderly patients also appear to be at increased risk of severe COVID-19 infection and mortality (Wang et al. 2020; Zhang 2020). Furthermore, since ophthalmic consultations often involve multiple investigations, including visual acuity, intraocular pressure measurement, and pupillary dilatation, it is not uncommon that patients have prolonged stay in the clinic to complete the whole examination (What Happens at an Eye Exam? 2019). All these factors potentially increase the risk of cross-infection, between patients and between health-care workers and patients, in ophthalmology outpatient clinics.

35.4 Strategies to Prevent COVID-19 Transmission in Ophthalmology Outpatient Clinics

We previously described the use of a three-level hierarchy approach as a framework to minimize the COVID-19 transmission in ophthalmology: (i) administrative control, (ii) environmental control, and (iii) the use of PPE (Lai et al. 2020a).

35.4.1 Administrative Control

It is the first and most important level of the hierarchy, which involves the most significant number of people. It consists of the following measures aimed at reducing the risk of exposure of uninfected people to people infected with COVID-19.

35.4.1.1 Lowering Patient Attendance

With many countries facing an aging population, most ophthalmic clinics are crowded with elderly patients. It is important to reduce the number of outpatient attendance to prevent infection in this vulnerable group of patients. In countries under lockdown due to severe COVID-19 outbreak, elective clinic visits are suspended, and only cases with emergency ophthalmic conditions would be seen. On March 18, 2020, the American Academy of Ophthalmology recommended all ophthalmologists in the United States to immediately cease the provision of any treatment other than urgent or emergent care (Recommendations for urgent and nonurgent patient care 2020). The Royal College of Ophthalmologists in the United Kingdom made similar recommendations on March 30, 2020. They suggested that all face-to-face outpatient activities should be postponed unless patients are at a high risk of rapid, significant harm if their appointments are delayed (Protecting Patients, Protecting Staff 2020). Such conditions included acute glaucoma, wet active age-related macular degeneration, sight-threatening proliferative diabetic retinopathy, ischemic central retinal vein occlusion, acute retinal detachment, severe active uveitis (Hung and Li 2020), active ocular and adnexal cancers, retinopathy of prematurity, endophthalmitis, sight-threatening trauma or orbital disease, and giant cell arteritis affecting the vision (Protecting Patients, Protecting Staff 2020). International vitreoretinal experts have also published risk stratification guidelines for patients receiving anti-vascular endothelial growth factor (anti-VEGF) injections (Korobelnik et al. 2020). Patients with a recent onset of significant visual loss, neovascular age-related macular degeneration, neovascular glaucoma, new central retinal occlusion, and only eye would be prioritized over patients with diabetic macular edema and branch retinal vein occlusion. In Northern Carolina and Maryland in the United States, drive-through clinics were set up such that the patient would sit in his/her car and an ophthalmologist wearing a face shield and surgical mask would examine the patient using portable equipment. It reduces the risk of cross-infection among patients in the waiting hall (Linnehan 2020; Robinson 2020). Any talking between the ophthalmologist and the patient was discouraged, and any communication would be made via telemedicine after the consultation. To maintain social distancing, some clinics advised patients to wait in their motor vehicles, and they would be called upon to enter the clinic when it was their turn.

Telemedicine has been widely implemented during the COVID-19 pandemic, especially in countries with a more severe outbreak. Virtual consultations are conducted using video-conferencing applications. For example, at the New York University Langone Health Medical Center, over 5,000 video visits were conducted within 1 month of the pandemic onset (Grossman et al. 2020). Oculoplastic surgery is a subspecialty particularly suitable for teleconsultation. Websites and apps for checking visual acuity are widely available, and external appearance and extraocular movements can be checked during the video consultation. Patients with ptosis, dermatochalasis, ectropion, entropion, eyelid retraction, epiphora, and other conditions that rarely threaten vision can be readily identified using telemedicine (Langer and Bernardini 2020). By using telemedicine for pre-screening, both ophthalmologists and patients would feel more comfortable to postpone appointments safely and reduce the number of face-to-face consultations. Follow-up visits for preseptal cellulitis and stable thyroid eye disease could also be conducted. With the use of absorbable sutures in blepharoplasty, entropion/ectropion correction, and lid mass excisional biopsy, postoperative visits can also be conducted using telemedicine. Our hospital has also successfully conducted pilot teleconsultations for stable oculoplastic patients. Nevertheless, virtual consultations would be more difficult for subspecialties requiring intraocular examination or more sophisticated investigations, such as visual field and optical coherence tomography. Developing countries may not have sufficient information technology support and resources to conduct telemedicine. On the other hand, developed countries also face issues of legislation, coding, and billing that need to be addressed.

In some parts of the world where the COVID-19 outbreak is relatively less severe, such as Hong Kong, ophthalmic clinics continued their activities while giving the patients an option to postpone their appointments. Patients with fever, flu symptoms, or recent travel were advised to avoid ophthalmic clinic attendance. The number of accompanying persons was limited to one for each patient. Despite active solicitation of patients at the entrance of the clinic to reschedule their appointments, it would be more desirable if patients are informed of this option well before their scheduled appointments. Conveying this message to a large number of outpatients is often challenging. Our hospital took advantage of short message service (SMS) to send information to patients through their electronic mobile devices, at least 5 days in advance before their scheduled appointments. The content of the SMS included (i) the option of rebooking and drug refill via telephone hotline and (ii) the suggestion for patients with fever, flu symptoms, or recent travel history to seek medical advice and avoid ophthalmic clinic attendance. An inquiry hotline was set up, and an ophthalmic specialist nurse answered all calls. The medical records of patients who requested postponement of their appointments were screened by ophthalmologists, who would decide whether to accept the request for postponement, the duration of postponement, and the need for drug-refill based on the clinical conditions at the most recent follow-up, as documented in the electronic medical record. The rescheduling was based on the following agreed protocol with one overarching principle that only patients with stable clinical conditions would be permitted to postpone their appointments. Patients with the following clinical conditions would need to keep their original appointment: (i) cases on systemic or topical steroids, (ii) cases within 2 weeks postoperatively, (iii) cases with uncontrolled intraocular pressure in the previous visits, and (iv) cases with conditions requiring frequent follow-ups. In our hospitals, a total of 17,028 SMS were sent from February to April 2020. The overall response rate was 23.6% (4,011/17,028 patients). A total of 14.3% (2,439/17,028) of patients postponed their appointments, which led to an overall reduction of 13.9% in clinic attendance. Two hundred patients were invited to answer a questionnaire regarding satisfaction level and reason(s) for appointment rescheduling. The overall satisfaction was high (96%). The main reason for postponing appointments was worries about infection risk (93.1%) (Lai et al. 2020d).

The elderly with comorbidities are particularly vulnerable to COVID-19 (Li et al. 2020c). Outbreaks in elderly homes or long-term facilities for dependent elderly can be disastrous (McMichael et al. 2020; Etard et al. 2020). To reduce the risk of infection in the elderly, we proactively called nursing homes in our catchment area to encourage their residents to utilize our drug refill service and consider postponing their appointments. The nursing home staff could collect drugs, on the patients' behalf, at either our hospital pharmacy or nearby community pharmacies.

35.4.1.2 Suspension of Elective Surgeries and Clinical Services

In numerous countries, nonurgent operations were suspended to reduce the risk of transmission due to cross-infection and to conserve resources, such as inpatient beds, human resources, and PPE used to combat the COVID-19 outbreak (Recommendations for urgent and nonurgent patient care 2020; Protecting Patients, Protecting Staff 2020). Emergency surgeries, such as ruptured globe repair, lid laceration repair, retinal detachment repair, and vitrectomy for endophthalmitis, were continued (Tang et al. 2020a). Some countries have suspended solicitation of eye tissue donation during the period of the outbreak. Before harvesting corneal grafts (Busin et al. 2020) and amniotic membranes (Yeung et al. 2020), all donors were required to undergo PCR testing for SARS-CoV-2. Other clinical services, such as electrodiagnostic studies, botulinum toxin injections, and contact lens clinics, were suspended.

35.4.1.3 Patient Triage

Up to 89% of COVID-19-infected cases suffer from fever (Guan et al. 2020). Therefore, it is necessary to ensure that patients with fever are identified before they enter the clinical area. Many ophthalmology clinics worldwide have set up triage stations to screen for high-risk patients suspected of having COVID-19. At our hospital, thermal infrared cameras were set up at the hospital entrance. All patients and their accompanying persons were screened. Anyone not wearing masks would not be allowed to enter the hospital premises. Those with fever were prohibited from entering the eye clinic and were diverted to the Accident and Emergency Department by healthcare workers wearing appropriate PPE. Those with urgent eye conditions and fever were seen by the on-call ophthalmologist either at the Accident and Emergency Department, inpatient isolation rooms, or inpatient wards after admission.

For afebrile patients, questionnaires to screen for epidemiological history (**T**ravel to affected areas during the incubation period, **O**ccupation, **C**ontact of a suspected or confirmed case, and **C**luster of cases (TOCC)) were administered by triage nurses. Besides, patients fulfilling the following criteria were also identified, and their clinic appointments were postponed for at least 14 days, which was the incubation period of COVID-19 (Symptoms of Coronavirus Disease 2019 2020): (i) patients (or patients who have family members or accompanying persons) who travelled to outbreak areas within 14 days; (ii) patients with symptoms of upper respiratory tract infection; and (iii) patients with acute conjunctivitis. Any patient fulfilling any of the above criteria but requiring urgent ophthalmic attention would be diverted to a separate waiting area and seen by a designated ophthalmologist in a particular room. Designated equipment and instruments were used and not shared with other patients. Figure 35.1 presents the patient triage workflow.

Inpatient consultations from other specialties were seen in their respective parent wards rather than in the outpatient clinic, thereby mitigating the risk of nosocomial transmission of COVID-19.

35.4.1.4 Reduction of Aerosol and Droplet Generation in Ophthalmology

COVID-19 transmission can occur via aerosols (Zhou 2020). Therefore, routine aerosol-generating procedures (AGP) in ophthalmic practice should be suspended.

Non-contact tonometry (NCT) is a potential source for micro aerosol (Britt et al. 1991). Using a camera and flash electrically coupled to an NCT machine, Britt et al. studied the disruption of tear film when a pulse of pressurized air was blown towards the eyes. They reported tear film dehiscence and micro aerosol formation. Therefore, it is prudent to suspend the use of NCT in outbreak areas. Other ways of intraocular pressure measurement, such as i-Care tonometry or Goldmann applanation tonometry, can be employed instead (Ng et al. 2020). With the use of disposable tips in tonometry, the risk of cross-infection is minimized.

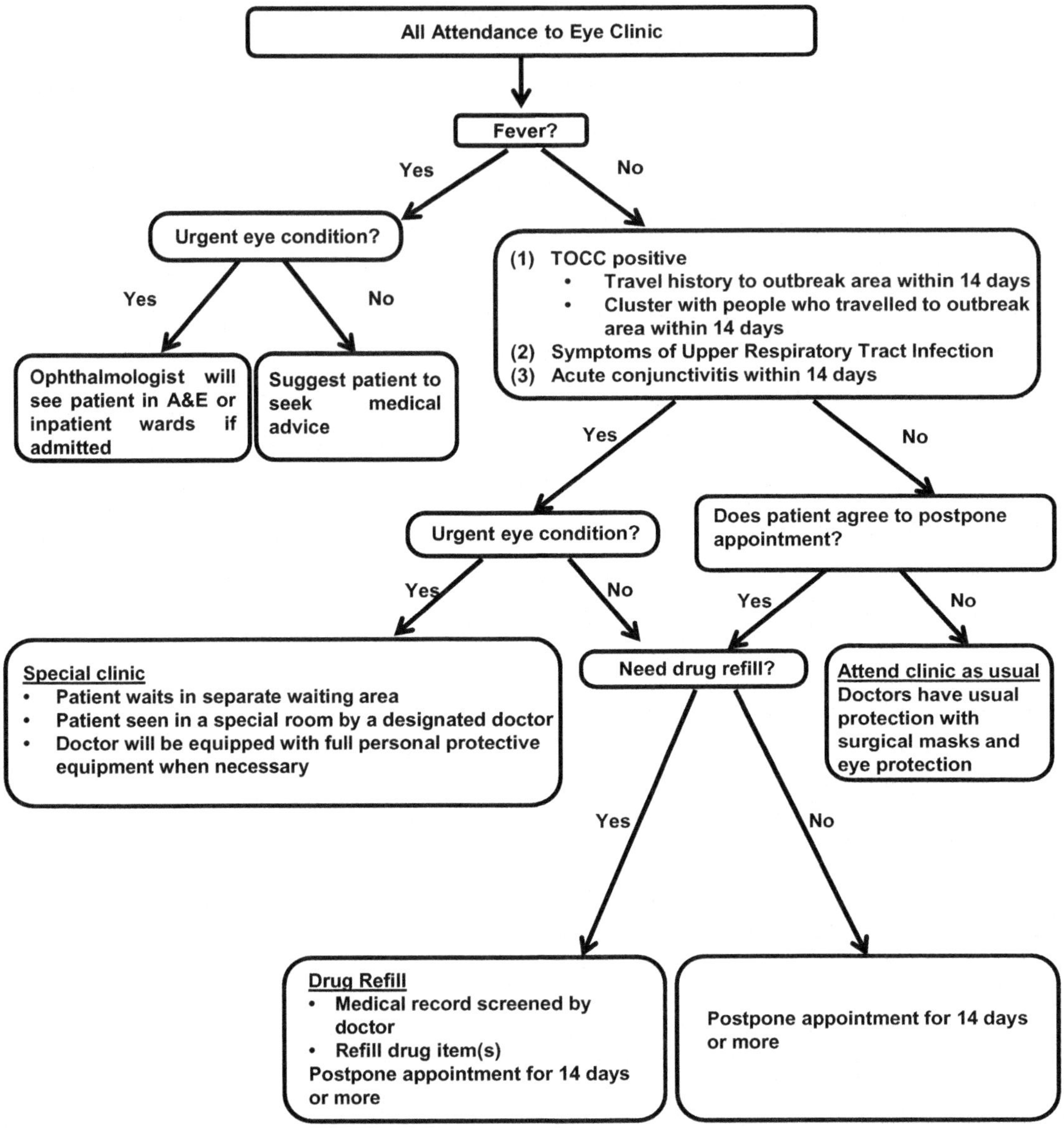

Fig. 35.1 Patient triage in the ophthalmology outpatient clinic. Abbreviations: *A&E* Accident and Emergency Department, *TOCC* Travel, Occupation, Contact, and Clustering. (Adapted with permission (Lai THT et al. Stepping up infection control measures in ophthalmology during the novel coronavirus outbreak: an experience from Hong Kong. Graefe's Archive for Clinical and Experimental Ophthalmology 258 (5):1049–1055. https://doi.org/10.1007/s00417-020-04641-8))

Oculoplastic surgeons often utilize nasal endoscopy in the outpatient clinic and during endonasal surgery. Workman et al. from Harvard Medical School simulated sneezing during endoscopy using an intranasal atomizer placed in a cadaver head and performed suction and high-speed (75,000 rpm) drilling of a cadaver head (Workman et al. 2020). If the cadaver was unmasked, a simulated sneeze event generated aerosol distribution of up to 66 cm from the nares. Suction did not produce aerosols, but high-speed drills caused significant aerosol contamination up to 30 cm away from the nares. A recent study demonstrated that aerosolized particles of SARS-CoV-2 of <5 μm remain viable in air for at least 3 hours (van Doremalen et al. 2020). As nasal endoscopy may irritate the nasal mucosa and provoke sneezing and dacryocystorhinostomy and medial wall decompression often involve the use of high-speed drills, they pose risk of infection to the operator. Thus, they should be avoided as much as possible (Mak and Yuen 2020). If endoscopy is unavoidable, appropriate PPE, including N95/FFP3 respirators, should be worn (Tran et al. 2012).

As vitrectomy operates at a speed of 10,000 CPM, it is also potentially generates aerosol. Both the British and Eire Association of Vitreoretinal Surgeons and the Royal College of Ophthalmologists in the United Kingdom have recently recommended the use of FFP3 respirators and eye protection for vitrectomy surgeries owing to their potential to cause aerosol generation (Vitreoretinal surgery during the COVID-19 pandemic 2020). Phacoemulsification has an even higher frequency at 50,000 Hz, but whether it generates aerosol remains to be investigated. Even if both procedures are, in fact, AGPs and unless the viral carriage of intraocular fluid is exceptionally high, the amount of released aerosol should be minimal (Li et al. 2020a).

If ophthalmic surgery needs to be performed during this pandemic, it should preferably be performed under local anesthesia and not general anesthesia, as endotracheal intubation generates aerosol (Tran et al. 2012). If emergency operation under general anesthesia is inevitable (e.g., repair of the ruptured globe or macula-on retinal detachment), ophthalmologists should work closely with anesthetists and internists to ensure COVID-19 rapid test is conducted and identify TOCC positive cases before undergoing general anesthesia (CDC Tests for COVID-19 2020). For patients who tested positive for COVID-19, for patients who tested negative but considered at a high risk, or in ultra-urgent situations in which it was not feasible to wait for the test results, operations should be performed in negative-pressure isolation operating theatres, and all staff should wear isolation gowns, N95/FFP3 respirators, and protective eyewear. Our department, in collaboration with anesthetists and other surgical departments, published a risk stratification protocol for emergency operations under general anesthesia (Fig. 35.2) (Wong et al. 2020b). The protocol was established to protect health-care workers from potential COVID-19 patients by providing a clear guideline during emergency surgical situations while simultaneously conserving PPE.

35.4.1.5 Infection Control Training and Staff Monitoring

All health-care workers should undergo infection control training to update them with the knowledge of COVID-19. Such training is also provided to educate them on the proper technique of hand hygiene and donning and doffing of PPE. Staff are instructed to be extra careful when applying multidose eye drops and to avoid contact between the patients' eyelashes and the bottle tip. All staff are required to measure and report their body temperatures before work and report any symptoms, such as fever, chills, myalgia, sore throat, runny nose, cough, vomiting, diarrhea, or pneumonia. Staff are discouraged from non-essential travels and are required to report their travel histories after returning from overseas. Certain countries have imposed compulsory quarantine measures after international travel, which should be strictly followed.

35.4.1.6 Team Splitting to Conserve PPE and Prevent Cross-infection

Due to the cutdown of elective clinical services, the number of ophthalmologists required to provide clinical service was reduced. Some ophthalmology departments divided their workforce into two separate teams. One team provided clinical services in the hospital, and the other team worked from home and/or conducted teleconsultations for nonurgent cases. Their roles would be interchanged every 1 to 2 weeks. The rationale of this arrangement was twofold: first, to only keep clinically essential staff in the hospital so that the other team of staff could work from home, reducing the risk of getting infected and lowering the consumption of PPE, and, second, to prevent

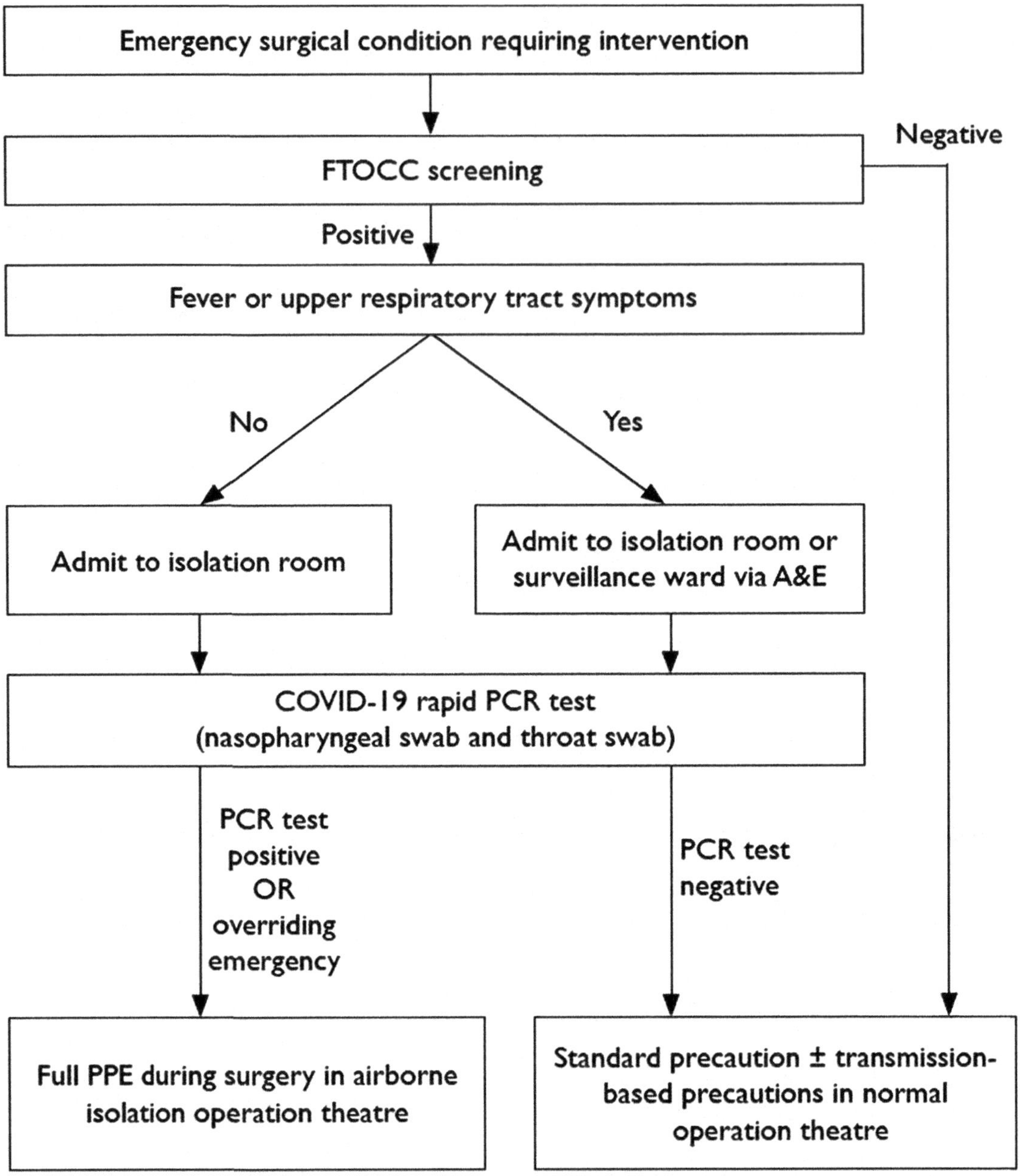

FIG. Risk stratification protocol for arrangement of emergency operation
Abbreviations: A&E = accident and emergency; COVID-19 = coronavirus disease 2019; FTOCC = fever, travel history to areas with active community transmission of COVID-19 within the past 14 days, occupation, contact, and clustering; Full PPE = full personal protective equipment, including splash resistant gown, disposable gloves, N95 respirator, and goggles or face shield; PCR = polymerase chain reaction

Fig. 35.2 Risk stratification protocol for the arrangement of emergency operation. A&E, Accident and Emergency Department; COVID-19, coronavirus disease 2019; FTOCC, Fever, Travel, Occupation, Contact, and Clustering; Full PPE, full personal protective equipment, including splash-resistant gown, disposable gloves, N95 respirator, and goggles or face shield; PCR, polymerase chain reaction. (Adapted with permission (Wong DHT et al. Risk stratification protocol to reduce consumption of personal protective equipment for emergency surgeries during COVID-19 pandemic. Hong Kong Med J 2020 26 (3):252–254. https://doi.org/10.12809/hkmj208533))

cross-infection between the members of the two teams and to ensure that essential clinical services are maintained. In the worst-case scenario that a member of staff tested positive for COVID-19, and the whole team needed to be quarantined, the other team could fill in to provide clinical service.

35.4.2 Environmental Control

Environmental control aims to prevent the spread of SARS-CoV-2 and reduce the concentration of infectious droplets in ambient air. Air ventilation in the waiting areas should be enhanced by opening the fresh air dampers in the air handling equipment to achieve a higher fresh air rate with improved air dilution. Mobile high-efficiency particulate air (HEPA) units could be added to augment the total air change rates in waiting areas where necessary. Patients were advised to practice social distancing from other patients wherever possible and were required to sit in every other seat in the waiting areas. Signage and broadcast advising respiratory hygiene and cough etiquette should be set up. Hand sanitizer dispensers should be installed in waiting areas, and patients are required to apply disinfection hand rub before entering consultation rooms.

The proximity between ophthalmologists and patients during slit-lamp examination puts ophthalmologists at risk of infection, as droplets from a cough or sneeze can travel up to 6 feet (How Flu Spreads 2020). Protective plastic shields should be installed on slit lamps to lower the risk of transmission via droplets (Wong et al. 2020a) (Fig. 35.3). These shields act as barriers to droplets (Liu et al. 2020; Poostchi et al. 2020), and they should be cleaned and disinfected after every clinic session and when the shields are visibly soiled or contaminated. The consultation time should be kept as short as possible, and patients are advised to refrain from talking while having examination at the slit lamp. Equipment such as slit lamps, indirect binocular ophthalmoscopes, visual field analyzers, and environmental surfaces that are frequently touched by health-

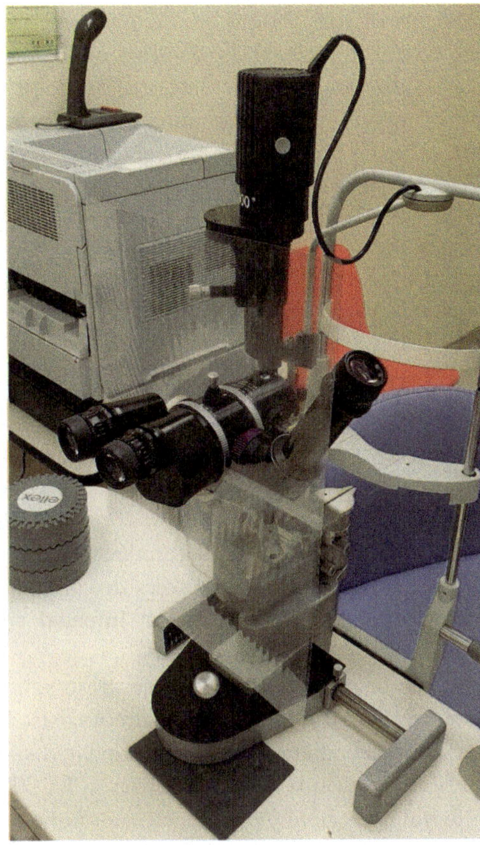

Fig. 35.3 Protective shield installed on a slit lamp. (Adapted with permission (Lai THT et al. Stepping up infection control measures in ophthalmology during the novel coronavirus outbreak: an experience from Hong Kong. Graefe's Archive for Clinical and Experimental Ophthalmology 258 (5):1049–1055. https://doi.org/10.1007/s00417-020-04641-8))

care workers and patients (such as doorknobs, phones, and computer keyboards) should be disinfected according to the manufacturer and local disinfection guidelines. Ultraviolet disinfection machines could be set up for staff to disinfect their mobile phones, staff cards, and other small items.

The use of video conferences through personal mobile devices is helpful to minimize the gathering of staff for situation updates or academic meetings. Virtual symposiums and webinars are excellent platforms for knowledge exchange, allowing ophthalmologists to globally attend and acquire the most updated knowledge. To minimize droplet transmission, the breaktime of staff should be staggered such that staff would not gather in pantries at the same time. Plastic barriers are installed in pantries, and signage is put up to remind staff to refrain from chatting while having meals. Appropriate distancing between diners in canteens should be adopted. For instance, the staff is recommended to sit in a one-way direction behind plastic barriers.

35.4.3 Use of PPE

The first two levels of the hierarchy approach reduce exposure to SARS-CoV-2 but do not eliminate the risk in clinics and wards. The use of PPE in these situations could further reduce the risk of exposure of health-care workers to infectious droplets expelled from a patient infected with COVID-19 (Chan and Yuen 2020).

Some countries initially had reservations about the use of PPE by ophthalmologists, but their recommendations have been updated since more evidence came to light. On March 30, 2020, the Royal College of Ophthalmologists published that "clinicians may wish to wear standard surgical masks while recognizing that they are of uncertain benefit. Gowns and gloves are not recommended when seeing patients without respiratory symptoms" (Protecting Patients, Protecting Staff 2020). Due to the spread of COVID-19 in the United Kingdom, the Royal College of Ophthalmologists updated its PPE guidelines in April 2020 (PPE and Staff Protection Requirements for Ophthalmology 2020; PPE requirements for ophthalmology 2020). Accordingly, all ophthalmologists need to wear surgical masks and gloves. Also, the American Academy of Ophthalmology suggested that when seeing low-risk patients, the use of surgical masks and eye protection for clinicians as well as surgical masks for the patients may reduce asymptomatic and presymptomatic transmission (Alert: Important coronavirus context for ophthalmologists 2020).

Apart from practicing universal masking (Naveed et al. 2020; Tang et al. 2020b), eye protection with either visors or protective eyewear should be provided to all ophthalmologists (Lai et al. 2020b). Instead of directly touching patients, ophthalmologists could wear gloves or use cotton tips to lift the patients' eyelids during the examination. Ophthalmologists attending higher-risk patients at designated areas should take extra precautions and wear full PPE, including isolation gown, gloves, cap, eye protection, and a surgical mask (or N95/FFP3 respirator when necessary). Hand hygiene is particularly important, and ophthalmologists should practice hand hygiene using the WHO formula alcohol hand rub or hand washing before and after every patient encounter (Lai et al. 2020c). If gloves are worn, they should be removed, followed by hand hygiene between patients.

35.5 Conclusion

At the time of writing, COVID-19 has taken the lives of more than 16 million people worldwide, including more than 1,800 health-care workers and 16 ophthalmologists (COVID-19 Coronavirus Pandemic 2020; In Memoriam: Healthcare Workers Who Have Died of COVID-19 2020; In Memoriam: Ophthalmologist Deaths From COVID-19 2020). Ocular manifestations of COVID-19 include conjunctivitis (Hong et al. 2020; Zhang et al. 2020; Seah et al. 2020; Xia et al. 2020; Karimi et al. 2020), retinal hemorrhages, cotton wool spots (Marinho et al. 2020), internuclear ophthalmoplegia, and abducens nerve palsy (Gutierrez-Ortiz et al. 2020). Conjunctivitis may occur as the initial presentation before the appearance of other symptoms. Ophthalmologists are at risk of infection due to the proximity between ophthalmologists and patients during examination (Lu et al. 2020). Ophthalmologists need to stay vigilant, use appropriate PPE, and practice frequent hand hygiene during their clinical work. Moreover, the use of a three-level hierarchy approach, including administrative control, environmental control, and use of PPE, provides an excellent framework to minimize the transmission of COVID-19 in ophthalmology.

References

"Solidarity" clinical trial for COVID-19 treatments (2020) World Health Organization. https://www.who.int/emergencies/diseases/novel-coronavirus-2019/global-research-on-novel-coronavirus-2019-ncov/solidarity-clinical-trial-for-covid-19-treatments. Accessed 29 May 2020

Adlhoch C, Baka A, Cenciarelli O, Ciotti M, Riley P, Spiteri G, van der Werf M, Wiltshire E (2020) Guidance for health system contingency planning during widespread transmission of SARS-CoV-2 with high impact on healthcare services. European Centre for Disease Prevention and Control. https://www.ecdc.europa.eu/sites/default/files/documents/COVID-19-guidance-health-systems-contingency-planning.pdf. Accessed 10 May 2020

Alert: Important coronavirus context for ophthalmologists (2020) American academy of ophthalmology. https://www.aao.org/headline/alert-important-coronavirus-context. Accessed 10 May 2020

Attzs MS, Lakhani BK (2020) Redeployment: a new paradigm for ophthalmology professionals. Eye (Lond) 34(7):1239–1240. https://doi.org/10.1038/s41433-020-0915-z

Breazzano MP, Shen J, Abdelhakim AH, Dagi Glass L, Horowitz J, Xie SX, De Moraes CG, Chen-Plotkin A, Chen RW (2020) New York City COVID-19 resident physician exposure during exponential phase of pandemic. J Clin Invest 30(9):4726–473. https://doi.org/10.1172/jci139587

Britt JM, Clifton BC, Barnebey HS, Mills RP (1991) Microaerosol formation in non-contact 'air-puff' tonometry. Arch Ophthalmol (Chicago, Ill : 1960) 109(2):225–228. https://doi.org/10.1001/archopht.1991.01080020071046

Busin M, Yu AC, Ponzin D (2020) Coping with COVID-19: An Italian perspective on corneal surgery and eye banking in the time of a pandemic and beyond. Ophthalmology 127(9):e68–e69. https://doi.org/10.1016/j.ophtha.2020.04.031

CDC Tests for COVID-19 (2020) Centers for disease control and prevention. https://www.cdc.gov/coronavirus/2019-ncov/about/testing.html. Accessed 16 Feb 2020

Chan KH, Yuen KY (2020) COVID-19 epidemic: disentangling the re-emerging controversy about medical facemasks from an epidemiological perspective. Int J Epidemiol 49(4):1063–1066. https://doi.org/10.1093/ije/dyaa044

Chang D, Xu H, Rebaza A, Sharma L, Dela Cruz CS (2020) Protecting health-care workers from subclinical coronavirus infection. Lancet Respir Med 8(3):e13. https://doi.org/10.1016/s2213-2600(20)30066-7

Characteristics of Health Care Personnel with COVID-19 – United States, February 12-April 9, 2020 (2020) MMWR Morbidity and mortality weekly report 69(15):477–481. https://doi.org/10.15585/mmwr.mm6915e6

Cheema M, Aghazadeh H, Nazarali S, Ting A, Hodges J, McFarlane A, Kanji JN, Zelyas N, Damji KF, Solarte C (2020) Keratoconjunctivitis as the initial medical presentation of the novel coronavirus disease 2019 (COVID-19). Can J Ophthalmol Journal Canadien d'ophtalmologie 55(4):e125–e129. https://doi.org/10.1016/j.jcjo.2020.03.003

Chen L, Deng C, Chen X, Zhang X, Chen B, Yu H, Qin Y, Xiao K, Zhang H, Sun X (2020a) Ocular manifestations and clinical characteristics of 535 cases of COVID-19 in Wuhan, China: a cross-sectional study. Acta Ophthalmol. https://doi.org/10.1111/aos.14472

Chen L, Liu M, Zhang Z, Qiao K, Huang T, Chen M, Xin N, Huang Z, Liu L, Zhang G, Wang J (2020b) Ocular manifestations of a hospitalised patient with confirmed 2019 novel coronavirus disease. Br J Ophthalmol 104(6):748–751. https://doi.org/10.1136/bjophthalmol-2020-316304

Coronavirus disease 2019 Situation Report – 25. (2020) World Health Organization. https://www.who.int/docs/default-source/coronaviruse/situation-reports/20200214-sitrep-25-covid-19.pdf?sfvrsn=61dda7d_2. Accessed 15 Feb 2020

COVID-19 Coronavirus Pandemic (2020) https://www.worldometers.info/coronavirus/. Accessed 12 Dec 2020

COVID-19 personal protective equipment (PPE) (2020) Public health England. https://www.gov.uk/government/publications/wuhan-novel-coronavirus-infection-prevention-and-control/covid-19-personal-protective-equipment-ppe. Accessed 10 May 2020

Etard JF, Vanhems P, Atlani-Duault L, Ecochard R (2020) Potential lethal outbreak of coronavirus disease (COVID-19) among the elderly in retirement homes and long-term facilities, France, March 2020. Euro Surveillance: bulletin Europeen sur les maladies transmissibles = European communicable disease bulletin 25 (15). doi:https://doi.org/10.2807/1560-7917.es.2020.25.15.2000448

Ferner RE, Aronson JK (2020) Chloroquine and hydroxychloroquine in covid-19. BMJ (Clinical research ed) 369:m1432. https://doi.org/10.1136/bmj.m1432

Gao J, Tian Z, Yang X (2020) Breakthrough: Chloroquine phosphate has shown apparent efficacy in treatment of COVID-19 associated pneumonia in clinical studies. Biosci Trends 14(1):72–73. https://doi.org/10.5582/bst.2020.01047

Gautret P, Lagier JC, Parola P, Hoang VT, Meddeb L, Mailhe M, Doudier B, Courjon J, Giordanengo V, Vieira VE, Dupont HT, Honoré S, Colson P, Chabrière E, La Scola B, Rolain JM, Brouqui P, Raoult D (2020) Hydroxychloroquine and azithromycin as a treatment of COVID-19: results of an open-label non-randomized clinical trial. Int J Antimicrob Agents:105949. https://doi.org/10.1016/j.ijantimicag.2020.105949

Grossman SN, Calix R, Tow S, Odel JG, Sun LD, Balcer LJ, Galetta SL, Rucker JC (2020) Neuro-ophthalmology in the Era of COVID-19: Future implications of a public health crisis. Ophthalmology. https://doi.org/10.1016/j.ophtha.2020.05.004

Guan WJ, Ni ZY, Hu Y, Liang WH, Ou CQ, He JX, Liu L, Shan H, Lei CL, Hui DSC, Du B, Li LJ, Zeng G, Yuen KY, Chen RC, Tang CL, Wang T, Chen PY, Xiang J, Li SY, Wang JL, Liang ZJ, Peng YX, Wei L, Liu Y, Hu YH, Peng P, Wang JM, Liu JY, Chen Z, Li G, Zheng ZJ, Qiu SQ, Luo J, Ye CJ, Zhu SY, Zhong NS (2020) Clinical characteristics of coronavirus disease 2019 in China. N Engl J Med 382(18):1708–1720. https://doi.org/10.1056/NEJMoa2002032

Gutierrez-Ortiz C, Mendez A, Rodrigo-Rey S, San Pedro-Murillo E, Bermejo-Guerrero L, Gordo-Manas R, de Aragon-Gomez F, Benito-Leon J (2020) Miller Fisher Syndrome and polyneuritis cranialis in COVID-19. Neurology doi:https://doi.org/10.1212/wnl.0000000000009619

Hong N, Yu W, Xia J, Shen Y, Yap M, Han W (2020) Evaluation of ocular symptoms and tropism of SARS-CoV-2 in patients confirmed with COVID-19. Acta Ophthalmol. https://doi.org/10.1111/aos.14445

How Flu Spreads (2020) Centers for disease control and prevention. https://www.cdc.gov/flu/about/disease/spread.htm. Accessed 15 Feb 2020

Hui KPY, Cheung MC, Perera R, Ng KC, Bui CHT, Ho JCW, Ng MMT, Kuok DIT, Shih KC, Tsao SW, Poon LLM, Peiris M, Nicholls JM, Chan MCW (2020) Tropism, replication competence, and innate immune responses of the coronavirus SARS-CoV-2 in human respiratory tract and conjunctiva: an analysis in ex-vivo and in-vitro cultures. Lancet Respir Med 8(7):687–695. https://doi.org/10.1016/s2213-2600(20)30193-4

Hung JCH, Li KKW (2020) Implications of COVID-19 for uveitis patients: perspectives from Hong Kong. Eye (Lond) 34(7):1163–1164. https://doi.org/10.1038/s41433-020-0905-1

In Memoriam: Healthcare Workers Who Have Died of COVID-19 (2020) https://www.medscape.com/viewarticle/927976. Accessed 12 Dec 2020

In Memoriam: Ophthalmologist Deaths From COVID-19 (2020) American academy of ophthalmology. https://www.aao.org/coronavirus/deaths. Accessed 12 Dec 2020

Karimi S, Arabi A, Shahraki T, Safi S (2020) Detection of severe acute respiratory syndrome Coronavirus-2 in the tears of patients with Coronavirus disease 2019. Eye (Lond) 34(7):1220–1223. https://doi.org/10.1038/s41433-020-0965-2

Korobelnik JF, Loewenstein A, Eldem B, Joussen AM, Koh A, Lambrou GN, Lanzetta P, Li X, Lövestam-Adrian M, Navarro R, Okada AA, Pearce I, Rodríguez FJ, Wong DT, Wu L (2020) Guidance for anti-VEGF intravitreal injections during the COVID-19 pandemic. Graefe's Arch Clin Exp Ophthalmol = Albrecht von Graefes Archiv fur klinische und experimentelle Ophthalmologie 258(6):1149–1156. https://doi.org/10.1007/s00417-020-04703-x

Lai THT, Tang EWH, Chau SKY, Fung KSC, Li KKW (2020a) Stepping up infection control measures in ophthalmology during the novel coronavirus outbreak: an experience from Hong Kong. Graefes Arch Clin Exp Ophthalmol 258(5):1049–1055. https://doi.org/10.1007/s00417-020-04641-8

Lai THT, Tang EWH, Chau SKY, Li KKW (2020b) Reply to ocular manifestation, eye protection, and COVID-19. Graefe's Arch Clin Exp Ophthalmol = Albrecht von Graefes Archiv fur klinische und experimentelle Ophthalmologie 258(6):1341. https://doi.org/10.1007/s00417-020-04663-2

Lai THT, Tang EWH, Fung KSC, Li KKW (2020c) Reply to "Does hand hygiene reduce SARS-CoV-2 transmission?". Graefe's Arch Clin Exp Ophthalmol = Albrecht von Graefes Archiv fur klinische und experimentelle Ophthalmologie 258(5):1135. https://doi.org/10.1007/s00417-020-04653-4

Lai THT, Lee M, Au AKH, Tang HHY, Tang EWH, Li KKW (2020d) The use of short message service (SMS) to reduce outpatient attendance in ophthalmic clinics during the coronavirus pandemic. International ophthalmology:1–8. https://doi.org/10.1007/s10792-020-01616-w

Langer PD, Bernardini FP (2020) Oculofacial plastic surgery and the COVID-19 pandemic: current reactions and implications for the future. Ophthalmology. https://doi.org/10.1016/j.ophtha.2020.04.035

Li KKW, Joussen AM, Kwan JKC, Steel DHW (2020a) FFP3, FFP2, N95, surgical masks and respirators: what should we be wearing for ophthalmic surgery in the COVID-19 pandemic? Graefe's Arch Clin Exp Ophthalmol = Albrecht von Graefes Archiv fur klinische und experimentelle Ophthalmologie 258(8):1587–1589. https://doi.org/10.1007/s00417-020-04751-3

Li KZ, Yong VKY, Lee LKM, Chin CF, Yip LWL (2020b) When ophthalmologists step up to the COVID-19 frontlines. Eye (Lond) 34(7):1237–1238. https://doi.org/10.1038/s41433-020-0918-9

Li Q, Guan X, Wu P, Wang X, Zhou L, Tong Y, Ren R, Leung KSM, Lau EHY, Wong JY, Xing X, Xiang N, Wu Y, Li C, Chen Q, Li D, Liu T, Zhao J, Liu M, Tu W, Chen C, Jin L, Yang R, Wang Q, Zhou S, Wang R, Liu H, Luo Y, Liu Y, Shao G, Li H, Tao Z, Yang Y, Deng Z, Liu B, Ma Z, Zhang Y, Shi G, Lam TTY, Wu JT, Gao GF, Cowling BJ, Yang B, Leung GM, Feng Z (2020c) Early transmission dynamics in Wuhan, China, of novel Coronavirus-infected pneumonia. N Engl J Med 382(13):1199–1207. https://doi.org/10.1056/NEJMoa2001316

Linnehan R (2020) Innovative drive-through clinic offers flexibility for patients with glaucoma. https://www.healio.com/ophthalmology/practice-management/news/online/%7Bce6723c1-f462-4ee6-ab72-770b72cf7d2b%7D/innovative-drive-through-clinic-offers-flexibility-for-patients-with-glaucoma. Accessed 9 May 2020

Liu J, Wang AY, Ing EB (2020) Efficacy of slit lamp breath shields. Am J Ophthalmol. https://doi.org/10.1016/j.ajo.2020.05.005

Lu CW, Liu XF, Jia ZF (2020) 2019-nCoV transmission through the ocular surface must not be ignored. Lancet (London, England) 395(10224):e39. https://doi.org/10.1016/s0140-6736(20)30313-5

Ma D, Chen CB, Jhanji V, Xu C, Yuan XL, Liang JJ, Huang Y, Cen LP, Ng TK (2020) Expression of SARS-CoV-2 receptor ACE2 and TMPRSS2 in human primary conjunctival and pterygium cell lines and in mouse cornea. Eye (Lond) 34(7):1212–1219. https://doi.org/10.1038/s41433-020-0939-4

Mak ST, Yuen HK (2020) Oculoplastic surgery practice during the COVID-19 novel coronavirus pandemic: experience sharing from Hong Kong. Orbit:1–3. https://doi.org/10.1080/01676830.2020.1754435

Marinho PM, Marcos AAA, Romano AC, Nascimento H, Belfort R Jr (2020) Retinal findings in patients with COVID-19. Lancet (London, England) 395(10237):1610. https://doi.org/10.1016/s0140-6736(20)31014-x

Marmor MF (2020) COVID-19 and Chloroquine/Hydroxychloroquine: is there ophthalmological concern? Am J Ophthalmol 213:A3–a4. https://doi.org/10.1016/j.ajo.2020.03.028

Marmor MF, Kellner U, Lai TY, Melles RB, Mieler WF (2016) Recommendations on screening for Chloroquine and Hydroxychloroquine Retinopathy (2016 revision). Ophthalmology 123(6):1386–1394. https://doi.org/10.1016/j.ophtha.2016.01.058

Matossian CA (2020) VIDEO: Ophthalmologist shares 'deployment' to ER in fight against coronavirus. https://www.healio.com/ophthalmology/practice-management/news/online/%7B40e41819-067c-4c76-8277-6ac7f674719f%7D/video-ophthalmologist-shares-deployment-to-er-in-fight-against-coronavirus. Accessed 10 May 2020

McMichael TM, Clark S, Pogosjans S, Kay M, Lewis J, Baer A, Kawakami V, Lukoff MD, Ferro J, Brostrom-Smith C, Riedo FX, Russell D, Hiatt B, Montgomery P, Rao AK, Currie DW, Chow EJ, Tobolowsky F, Bardossy AC, Oakley LP, Jacobs JR, Schwartz NG, Stone N, Reddy SC, Jernigan JA, Honein MA, Clark TA, Duchin JS (2020) COVID-19 in a long-term care facility – King County, Washington, February 27-March 9, 2020. MMWR Morb Mortal Wkly Rep 69(12):339–342. https://doi.org/10.15585/mmwr.mm6912e1

Napoli PE, Nioi M, d'Aloja E, Fossarello M (2020) The Ocular Surface and the Coronavirus disease 2019: does a dual 'Ocular Route' exist? J Clin Med 9(5). https://doi.org/10.3390/jcm9051269

Naveed H, Scantling-Birch Y, Lee H, Nanavaty MA (2020) Controversies regarding mask usage in ophthalmic units in the United Kingdom during the COVID-19 pandemic. Eye (Lond) 34(7):1172–1174. https://doi.org/10.1038/s41433-020-0892-2

Navel V, Chiambaretta F, Dutheil F (2020) Haemorrhagic conjunctivitis with pseudomembranous related to SARS-CoV-2. Am J Ophthalmol Case Rep 19:100735. https://doi.org/10.1016/j.ajoc.2020.100735

Ng VWK, Tang GCH, Lai THT, Tang EWH, Li KKW (2020) Intraocular pressure measurement during COVID pandemic. Indian J Ophthalmol 68(5):950–951. https://doi.org/10.4103/ijo.IJO_723_20

Parrish RK 2nd, Stewart MW, Duncan Powers SL (2020) Ophthalmologists Are More Than Eye doctors-in memoriam Li Wenliang. Am J Ophthalmol 213:A1–a2. https://doi.org/10.1016/j.ajo.2020.02.014

Patel ZM, Fernandez-Miranda J, Hwang PH, Nayak JV, Dodd R, Sajjadi H, Jackler RK (2020) Letter: precautions for endoscopic transnasal skull base surgery during the COVID-19 pandemic. Neurosurgery 87(1):E66–e67. https://doi.org/10.1093/neuros/nyaa125

Poostchi A, Kuet ML, Pegg K, Wilde C, Richardson PS, Patel MK (2020) Efficacy of slit lamp breath shields. Eye (Lond) 34(7):1185–1186. https://doi.org/10.1038/s41433-020-0940-y

PPE and Staff Protection Requirements for Ophthalmology (2020). https://www.rcophth.ac.uk/wp-content/uploads/2020/04/UPDATED-RCOphth-PPE-for-ophthalmology-090420.pdf. Accessed 10 May 2020

PPE requirements for ophthalmology (2020) The royal college of ophthalmologists. https://www.rcophth.ac.uk/wp-content/uploads/2020/04/UPDATED-RCOphth-PPE-Principles-for-ophthalmic-staff-protection-090420.pdf. Accessed 10 May 2020

Protecting Patients, Protecting Staff (2020) The royal college of ophthalmologists. https://www.rcophth.ac.uk/wp-content/uploads/2020/03/Protecting-Patients-Protecting-Staff-UPDATED-300320.pdf. Accessed 9 May 2020

Qiao C, Zhang H, He M, Ying G, Chen C, Song Y, Oatts J, Liu Z, Xing Y, Xiao Z, Ke M, Ye Y, Chen X, Luo J, Han Y, Wang N (2020) Symptomatic COVID-19 in eye professionals in Wuhan, China. Ophthalmology 127(9):1268–1270. https://doi.org/10.1016/j.ophtha.2020.04.026

Recommendations for urgent and non-urgent patient care (2020) American academy of ophthalmology. https://www.aao.org/headline/new-recommendations-urgent-nonurgent-patient-care. Accessed 9 May 2020

Recommended Personal Protective Equipment (PPE) in hospitals/clinics under Serious/Emergency Response

Level. Coronavirus disease (COVID-19) (2020) Centre for health protection, Hong Kong. https://www.chp.gov.hk/files/pdf/recommended_ppe_for_nid_eng.pdf. Accessed 24 May 2020

Report of the WHO-China Joint Mission on Coronavirus Disease 2019 (COVID-19) (2020) World Health Organization. https://www.who.int/docs/default--source/coronaviruse/who-china-joint-mission-on-covid-19-final-report.pdf. Accessed 15 Mar 2020

Robinson L (2020) Patients receive eye service via drive-thru clinic to avoid spread of coronavirus. https://www.wbaltv.com/article/patients-receive-eye-service-drive-thru-clinic-to-avoid-spread-of-coronavirus/32407510. Accessed 9 May 2020

Ruamviboonsuk P, Lai TYY, Chang A, Lai CC, Mieler WF, Lam DSC (2020) Chloroquine and Hydroxychloroquine retinal toxicity consideration in the treatment of COVID-19. Asia-Pac J Ophthalmol (Philadelphia, Pa) 9(2):85–87. https://doi.org/10.1097/apo.0000000000000289

Seah IYJ, Anderson DE, Kang AEZ, Wang L, Rao P, Young BE, Lye DC, Agrawal R (2020) Assessing viral shedding and infectivity of tears in Coronavirus disease 2019 (COVID-19) patients. Ophthalmology 127(7):977–979. https://doi.org/10.1016/j.ophtha.2020.03.026

Stevens DV, Tran AQ, Kim E (2020) Complications of orbital emphysema in a COVID-19 patient. Ophthalmology 127(7):990. https://doi.org/10.1016/j.ophtha.2020.05.011

Symptoms of Coronavirus Disease 2019 (2020) Centers for disease control and prevention. https://www.cdc.gov/coronavirus/2019-ncov/about/symptoms.html. Accessed 16 Feb 2020

Tang EWH, Wong DHT, Chan YYY, Li KKW (2020a) Emergency ophthalmic surgeries during COVID-19-a Hong Kong perspective. Graefe's Archive for Clinical and Experimental Ophthalmology = Albrecht von Graefes Archiv fur klinische und experimentelle Ophthalmologie 258(12):2867–2868. https://doi.org/10.1007/s00417-020-04768-8

Tang GCH, Steel DHW, Li KKW (2020b) Why ophthalmologists should mask: a perspective from Hong Kong. Eye (Lond) 34(7):1168–1169. https://doi.org/10.1038/s41433-020-0916-y

Tang SW, Romano MR, Wong DH, Montericio A, Yip NK, Moltalbano C, Raimondi R, Li KK (2020c) The use of personal protective equipment in clinical ophthalmology during corona virus disease-2019: a review of international guidelines and literature. Curr Opin Ophthalmol 31 (5):435–446. https://doi.org/10.1097/icu.0000000000000691

Ting D, Yip MYT, Li JPO, Shantha J, Yeh S, Chodosh J, Fung AT, Lin H, Mehta J (2020) The complexities of COVID-19 in ophthalmology. https://theophthalmologist.com/subspecialties/the-complexities-of-covid-19-in-ophthalmology. Accessed 8 May 2020

Tran K, Cimon K, Severn M, Pessoa-Silva CL, Conly J (2012) Aerosol generating procedures and risk of transmission of acute respiratory infections to healthcare workers: a systematic review. PLoS One 7(4):e35797. https://doi.org/10.1371/journal.pone.0035797

van Doremalen N, Bushmaker T, Morris DH, Holbrook MG, Gamble A, Williamson BN, Tamin A, Harcourt JL, Thornburg NJ, Gerber SI, Lloyd-Smith JO, de Wit E, Munster VJ (2020) Aerosol and surface stability of SARS-CoV-2 as compared with SARS-CoV-1. N Engl J Med 382(16):1564–1567. https://doi.org/10.1056/NEJMc2004973

Veritti D, Sarao V, Bandello F, Lanzetta P (2020) Infection control measures in ophthalmology during the COVID-19 outbreak: a narrative review from an early experience in Italy. Eur J Ophthalmol 30(4):621–628. https://doi.org/10.1177/1120672120927865

Vitreoretinal surgery during the COVID-19 pandemic (2020) The royal college of ophthalmologists. https://rcophth.wpengine.com/wp-content/uploads/2020/04/Vitreoretinal-surgery-management-guidance-070420-1.pdf. Accessed 24 May 2020

Wang D, Hu B, Hu C, Zhu F, Liu X, Zhang J, Wang B, Xiang H, Cheng Z, Xiong Y, Zhao Y, Li Y, Wang X, Peng Z (2020) Clinical characteristics of 138 hospitalized patients with 2019 novel Coronavirus-infected pneumonia in Wuhan, China. JAMA 323(11):1061–1069. https://doi.org/10.1001/jama.2020.1585

What Happens at an Eye Exam? (2019) WebMD. https://www.webmd.com/eye-health/what-to-expect--checkup-eye-exam-adults#2. Accessed 16 Feb 2020

WHO Director-General's opening remarks at the media briefing on COVID-19 (2020) World Health Organization. https://www.who.int/dg/speeches/detail/who-director-general-s-opening-remarks-at-the-media-briefing-on-covid-19%2D%2D-11-march-2020. Accessed 23 Apr 2020

Wong DHT, Mak ST, Yip NKF, Li KKW (2020a) Protective shields for ophthalmic equipment to minimise droplet transmission of COVID-19. Graefe's Arch Clin Exp Ophthalmol = Albrecht von Graefes Archiv fur klinische und experimentelle Ophthalmologie 258(7):1571–1573. https://doi.org/10.1007/s00417-020-04683-y

Wong DHT, Tang EWH, Njo A, Chu CKL, Chau SKY, Chow TL, Lim HS, Fung KSC, Li KKW (2020b) Risk stratification protocol to reduce consumption of personal protective equipment for emergency surgeries during COVID-19 pandemic. Hong Kong Med J = Xianggang yi xue za zhi 26(3):252–254. https://doi.org/10.12809/hkmj208533

Workman AD, Welling DB, Carter BS, Curry WT, Holbrook EH, Gray ST, Scangas GA, Bleier BS (2020) Endonasal instrumentation and aerosolization risk in the era of COVID-19: simulation, literature review, and proposed mitigation strategies. Int Forum Allergy Rhinol 10(7):798–805. https://doi.org/10.1002/alr.22577

Wu P, Duan F, Luo C, Liu Q, Qu X, Liang L, Wu K (2020) Characteristics of ocular findings of patients with Coronavirus disease 2019 (COVID-19) in Hubei

Province, China. JAMA Ophthalmol 138(5):575–578. https://doi.org/10.1001/jamaophthalmol.2020.1291

Xia J, Tong J, Liu M, Shen Y, Guo D (2020) Evaluation of coronavirus in tears and conjunctival secretions of patients with SARS-CoV-2 infection. J Med Virol 92(6):589–594. https://doi.org/10.1002/jmv.25725

Yang E, Sathianathan V, Bourkiza R, Mensah E (2020) Sorry ICU, I'm an Ophthalmologist! Eye (Lond) 34(7):1170–1171. https://doi.org/10.1038/s41433-020-0914-0

Yeung TLM, Liu S, Li BCY, Mok KM, Li KKW (2020) Amniotic membrane harvesting during COVID-19 pandemic. Eye (Lond) 1. https://doi.org/10.1038/s41433-020-0947-4

Yusuf IH, Sharma S, Luqmani R, Downes SM (2017) Hydroxychloroquine retinopathy. Eye 31(6):828–845. https://doi.org/10.1038/eye.2016.298

Zhang Y (2020) The epidemiological characteristics of an outbreak of 2019 novel Coronavirus diseases (COVID-19). China CDC Wkly 2020

Zhang X, Chen X, Chen L, Deng C, Zou X, Liu W, Yu H, Chen B, Sun X (2020) The evidence of SARS-CoV-2 infection on ocular surface. Ocul Surf 18(3):360–362. https://doi.org/10.1016/j.jtos.2020.03.010

Zhou W (2020) Shanghai officials reveal novel coronavirus transmission modes. https://www.chinadaily.com.cn/a/202002/08/WS5e3e7d97a310128217275fc3.html. Accessed 15 Feb 2020

Zhou Y, Duan C, Zeng Y, Tong Y, Nie Y, Yang Y, Chen Z, Chen C (2020a) Ocular findings and proportion with conjunctival SARS-COV-2 in COVID-19 patients. Ophthalmology 127(7):982–983. https://doi.org/10.1016/j.ophtha.2020.04.028

Challenges of Cellular Therapy During the COVID-19 Pandemic

36

Kamal Kant Sahu, Sikander Ailawadhi, Natalie Malvik, and Jan Cerny

Abstract

Currently, coronavirus disease 2019 (COVID-19) has spread worldwide and continues to rise. There remains a significant unmet need for patients with hematological malignancies requiring specialized procedures and treatments, like cellular therapy to treat or cure their disease. For instance, chimeric antigen receptor T (CAR-T) cell therapy is approved for relapsed/refractory (after two or more lines of therapy) diffuse large B cell lymphoma and B cell acute lymphoblastic leukemia that is refractory or in the second relapse in patients younger than 25 years of age. Similarly, hematopoietic stem cell transplantation (HSCT) can be a lifesaving procedure for many patients, such as those with acute myeloid leukemia with high-risk cytogenetics. Unfortunately, the COVID-19 pandemic has thrust upon the hematologists and transplant specialists' unique challenges with the implementation and management of cellular therapy. One of the significant concerns regarding this immunocompromised patient population is the significant risk of acquiring SARS-CoV-2 infection due to its highly contagious nature. Experts have recommended that if medically indicated, especially in high-risk disease (where chemotherapy is unlikely to work), these lifesaving procedures should not be delayed even during the COVID-19 pandemic. However, proceeding with CAR-T cell therapy and HSCT during the pandemic is a considerable task and requires dedication from the transplant team and buy-in from the patients and their family or support system. Open conversations should be held with the patients about the risks involved in undergoing cellular therapies during current times and the associated future uncertainties.

Keywords

Cellular therapy · Chimeric antigen receptor · Coronavirus · COVID-19 · Pandemic · Stem cells · T cells

K. K. Sahu (✉)
Department of Medicine, Saint Vincent Hospital, Worcester, MA, USA

Systematic Review and Meta-analysis Expert Group (SRMEG), Universal Scientific Education and Research Network (USERN), Worcester, MA, USA

S. Ailawadhi
Division of Hematology-Oncology, Department of Medicine, Mayo Clinic, Jacksonville, FL, USA

N. Malvik
Division of Transfusion Medicine, University of Massachusetts Medical Center, Worcester, MA, USA

J. Cerny
Division of Hematology/Oncology, Department of Medicine, University of Massachusetts Medical Center, Worcester, MA, USA

N. Rezaei (ed.), *Coronavirus Disease - COVID-19*, Advances in Experimental Medicine and Biology 1318, https://doi.org/10.1007/978-3-030-63761-3_36

36.1 Introduction

Currently, the world is facing an unseen health crisis in the form of coronavirus disease 2019 (COVID-19) pandemic (Sahu et al. 2020i, j, l). With 9,170,913 confirmed cases and 473,277 deaths due to COVID-19 as of June 22, 2020, this pandemic has affected the physical, mental, and economic well-being of humanity (Mishra et al. 2020a). The ongoing pandemic presents a unique challenge to deliver cellular therapy to patients with benign and malignant hematological disorders (Block 2020; Sahu and Cerny 2020; Jindal et al. 2020; Sahu et al. 2020a, s, r). It has extended beyond the routine patient care and involves other aspects like logistics, transportation, infusion, administration, distribution, and resource allocation (Broxmeyer and Parker 2020; Dholaria and Savani 2020). Hematopoietic stem cell transplantation (HSCT) and chimeric antigen receptor T (CAR-T) cell therapy are the two major forms of cellular therapy used for the patients suffering from various hematological malignancies (Ardura et al. 2020; Weinkove et al. 2020; Hunter et al. 2019). These treatments and their associated procedures are exceedingly complex. They require a great deal of coordination, communication, and a well-structured network of teams to ensure their successful implementation, and this process is TYPICALLY initiated weeks or months before the actual cell infusion itself. Unfortunately, like many other healthcare sectors, the management of cellular therapies has its practical challenges during the COVID-19 pandemic that need immediate attention and intervention to ensure delivery of the best and safest care possible (Hunter et al. 2019; Mishra et al. 2020c). We now review the numerous facets of cellular therapy, including patient and donor selection, travel and logistic hurdles, implementation, and delivery, scarcity, and prioritization of available resources during the ongoing COVID-19 pandemic (Sahu et al. 2020q).

36.1.1 Current Cellular Therapy Types

For the treatment of various malignancies and degenerative disorders, different sources of stem cells have been used. The main types of cell therapy are as follows: hematopoietic stem cell transplantation (bone marrow, mobilized peripheral blood, or umbilical cord blood); genetically engineered cellular therapies, chimeric antigen receptor T cell therapies, or pluripotent stem cell-based therapies; and others, like mesenchymal stem/stromal cells.

The current COVID-19 pandemic has significantly disrupted the regular functioning of most of the cellular therapy programs. It has caused a delay in the treatment of various disorders and, as such, put potentially patients with advanced malignancies at the risk of disease progression. On the other hand, proceeding with cellular therapy poses its specific challenges, such as placing the recipients at high risk of infection, potentially for prolonged periods, due to immunocompromised states. COVID-19 infection in the post-cellular therapy setting could be particularly challenging due to the paucity of experience of the transplant specialists and infectious disease physicians with COVID-19 management. Additionally, unlike what is typically expected and available, prompt intensive care services might not be fully assessable to post-cellular therapy patients due to the potential saturation of healthcare services. Therefore, it is necessary to discuss the significant challenges (and potential solutions) being faced by cellular therapists, transplant physicians, and various stem cell programs during this current situation.

36.1.2 General Rules and Policies to Prevent Infection with SAR-CoV-2 Causing COVID-19

As the incidence of COVID-19 continues to rise, almost all hospitals across the globe have adopted standard guidelines for good hygiene practices based on the recommendations suggested by the World Health Organization (WHO), Centers for Disease Control and Prevention (CDC), and other government and local public health agencies (Sahu and Kumar 2020a). Also, there are institutional guidelines, which are tailored as per the local COVID-19 disease burden state (county, city) laws and regulations (Samaha and Kattan

2020). It is because the COVID situation is hugely variable between different countries, and hence the guidelines and recommendations suggested by national and international authorities need to be adjusted according to the local needs (Sahu et al. 2020d; Samaha and Kattan 2020). In particular, pregnant women, patients with HIV/AIDS, cancer patients, elderly individuals with multiple comorbidities, and other immunocompromised individuals are at the highest risk of a dismal outcome if infected with the SARS-CoV-2 (Sahu et al. 2020c, k, m, r; Liang et al. 2020).

Among the cancer patients, those awaiting HSCT or CAR-T cell therapy require special attention and planning due to the unique complexities of cellular therapy. These procedures are complex to execute due to the various factors like cell acquisition, processing, procedural guidelines, transportation (for HSCT, donor center to transplant center; for CAR-T cells, collection unit/center to the manufacturing facility and then back to the center that infuses cells), genetic engineering, international travel restrictions, staffing issues (need for quarantine, lack of child care, COVID illness transmission), availability of cryopreservation, and donor/recipient eligibility criteria (Neelapu et al. 2017) (Table 36.1).

Table 36.1 Areas of potential disruptions in cellular therapy during the COVID-19 pandemic

Apheresis procedure and availability of efficient staff
Stem cell or cell processing lab
Manufacturing site vicinity
Drug shortage: concern over nonavailability of tocilizumab for CAR-T cell recipients due to its current use in critically ill COVID-19 patients. Also, a shortage of other vital drugs like azithromycin, IVIG, and albumin
Shipping and transportation due to international border closures and air travel restrictions: based on the local, regional rules and regulations both at the place of collection center and the infusion center
Referring bias (either patients are hesitating to go to hospital or primary referring oncologists are not referring the patients)
Hospital and ICU capacity (lack of availability due to surge in COVID-19 patients' related occupancy)
Blood bank (blood products and platelet shortages due to lack of donation drives)
Laboratory testing and result interpretation (staff and reagent shortages)
Radiology (staff shortages, nonavailability, need for additional visits)
Pathology (staff shortages, lack of antigens and sparse supply of essential dyes, specimen processing)
Transfusion medicine (lack of adequate and expert staff)
Absence of caregivers (caregivers may not be able to travel or are unavailable, restrictive hospital visitor policy)
Overall staffing and nursing issues (due to the reallocation of the CAR-T specialized nurses to non-CAR-T floors/ICUs)
Lodging and housing services (local housing closures): Hope Lodge has been temporarily suspended

36.2 Chimeric Antigen Receptor T (CAR-T) Cell Therapy

36.2.1 Currently Approved CAR-T Cell Therapeutics

Chimeric antigen receptor (CAR) T cell therapy has led to a paradigm shift in the management of patients suffering from recurrent B cell acute lymphoblastic leukemia (ALL) up to 25 years of age and adult recurrent B cell non-Hodgkin's lymphomas (NHL) (Schuster et al. 2019; Maude et al. 2018). Currently, there are two anti-CD19 CAR-T cell products approved by the United States Food and Drug Administration (FDA): axicabtagene ciloleucel (Yescarta®; Kite/Gilead) for the treatment of adult relapsed/refractory (R/R) diffuse large B cell lymphoma (DLBCL) and tisagenlecleucel (Kymriah®; Novartis), both for adult R/R DLBCL and R/R B cell ALL.

In the pre-COVID-19 era, outpatient CAR-T cell therapy had been gaining popularity, as most centers found it easy and more cost-effective to arrange for post-infusion lodging than inpatient stays, and follow-up plans were generally well-executed and followed. The exception to this is axicabtagene ciloleucel, which requires post-infusion therapy hospitalization due to higher rates of toxicities, especially the cytokine release syndrome (CRS) and neurotoxicity. These side effects may add to the ongoing disease process in a patient with COVID-19 (Mishra et al. 2020b, d).

Now, most centers are hospitalizing all CAR-T cell therapy recipients, irrespective of the type of CAR-T cell therapy received. It is presumably to ensure a safe environment for the patient during and immediately after the CAR-T cell infusion. However, this stance is debatable, as some cellular therapists may argue that patients would experience far fewer potential SARS-CoV-2 exposures at home than as an inpatient in the hospital. We believe that the risk of individual exposure is extremely variable and depends on the cellular therapy center/hospital measures that are in place to prevent SARS-CoV-2 transmission; adherence of patients (and their families) to these measures, so they protect themselves from disease; and last but not least the level of community transmission of the virus itself.

36.2.2 Patient Selection for the Receipt of CAR-T Cell Therapy During the COVID-19 Pandemic

In general, experts are usually in favor of proceeding with treatment whenever a patient needs CAR-T cell therapy. This suggestion is based on the results of multiple studies showing positive outcomes of CAR-T cell therapy in refractory/relapsed cases, which in the pre-CAR-T cell era had otherwise very dismal outcomes (Romero 2018; Bachanova et al. 2020).

Treatment during the COVID-19 pandemic has raised nearly innumerable challenges to the successful delivery of potentially lifesaving CAR-T cell therapy. Recently, experts from eight US academic institutions created a "CAR T-cell Consortium" to assimilate the data and resources to aid hematologists and transplant specialists in the optimization of cellular therapy care.

CAR-T cell therapy is a form of individualized therapy that involves series of complicated steps requiring professional expertise in the cell manufacturing process and handling (e.g., T cell collection/selection, transduction, gene modification, and washing). Every cellular therapy program should have its schema or protocol that takes into consideration the various aspects of logistics, resource scarcities, local COVID-19 burden, and urgency of the CAR-T cell therapy/HSCT (Fig. 36.1) (Dave et al. 2019). Bachanova et al. recommended establishing a triage algorithm for assessing and prioritizing patients for urgent versus potentially delaying the procedure (Bachanova et al. 2020; Paul et al. 2020).

Choosing a suitable candidate for CAR-T cell therapy or HSCT is based on the significant factors, such as pretreatment disease bulk, performance status as defined by the Eastern Cooperative Oncology Group (ECOG), preexisting comorbidities and organ function status, and serum biomarkers such as serum lactate dehydrogenase (LDH) (Brudno and Kochenderfer 2019).

In the case of adult relapsed/refractory (R/R) diffuse large B cell lymphoma (DLBCL), multivariate analyses have shown ECOG performance status >2 and elevated LDH before lymphodepleting chemotherapy as the two significant predictors of poor outcomes in CAR-T cell therapy. Also, a more considerable tumor bulk has been found to be associated with increased acute toxicity. It may be prudent to defer CAR-T cell therapy for these patients due to their poor prognostic factors and considering the constrained resources and unpredictable environment during the current COVID-19 crisis (Table 36.2).

In the case of cellular therapy in R/R ALL, as mentioned above, tisagenlecleucel has been FDA approved. For R/R ALL, experts have suggested that rates of CRS, availability of intensive care unit (ICU) beds, and access to tocilizumab could be major deciding factors (Ahmad et al. 2020).

In a recent webinar, Dr. Schuster from the University of Pennsylvania suggested evaluating patients with relapsed/refractory (R/R) B-NHL for CAR-T cell therapy based on the availability of a bridging therapy (lenalidomide, monoclonal antibodies), which could be used as an alternative treatment during the COVID-19 crisis.

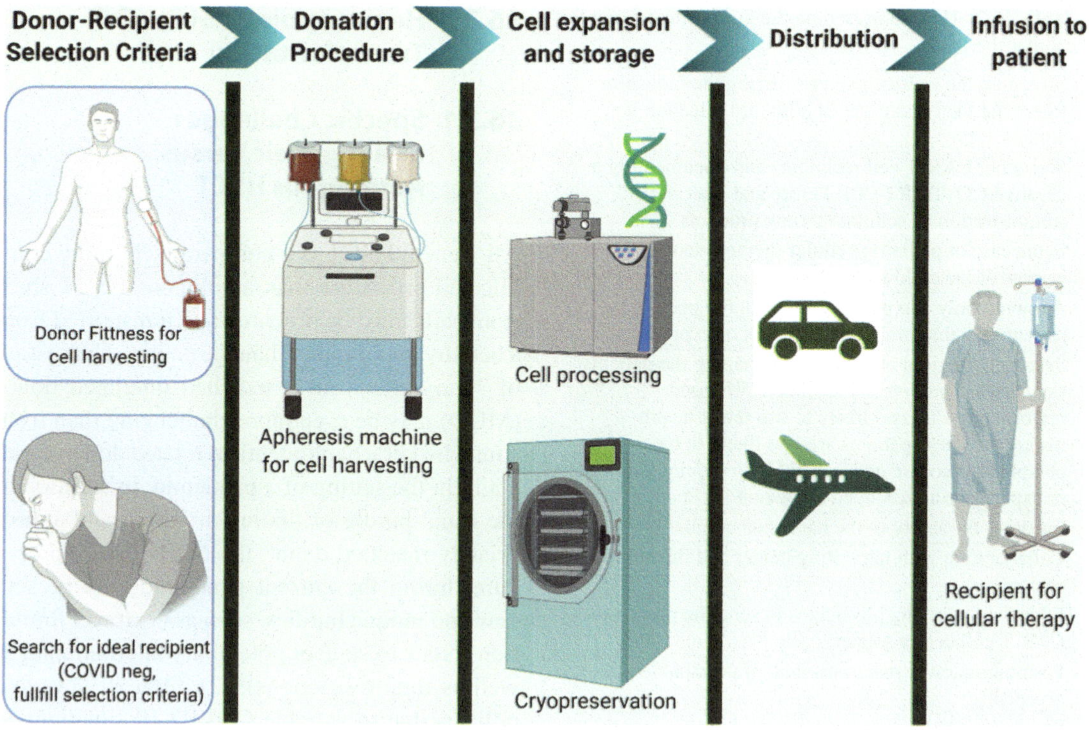

Fig. 36.1 Various steps and procedures from the collection center to the infusion center, which can be hampered during the COVID-19 pandemic

36.2.3 Managing Toxicities of CAR-T Cell Therapy During the COVID-19 Pandemic

CAR-T cell therapy requires the infusion of genetically engineered cellular product, which can lead to an immediate hypersensitivity reaction. Other common side effects include CRS, neurotoxicity, hypotension, hypogammaglobulinemia, and infections in the short and long term (Ahmad et al. 2020; Boyiadzis et al. 2018). CRS rates occur more in patients with ALL compared to patients with DLBCL.

Tocilizumab, a recombinant monoclonal antibody to the IL-6 receptor, is currently being explored for its clinical efficacy in critically ill patients with COVID-19 (Sahu and Kumar 2020b). Early reports from China showed tocilizumab had a positive effect with regard to reduced disease severity and the ability to wean patients with cytokine storm from ventilatory support. Tocilizumab has been the mainstay of the management of recipients of CAR-T cell therapy who experience advanced (grade > 2) CRS (Hassoun et al. 2020). Since its approval for use in CRS, there has been a gradual trend toward an earlier intervention with tocilizumab based on the encouraging data showing a decrease in the frequency, severity, and duration of CRS. Preemptive use of tocilizumab in patients with a high disease burden has also been studied recently, which showed a one-third reduction in grade 4 CRS, thereby resulting in lower ICU admission rates and critical care bed utilization (Gardner et al. 2019). This option may also be explored for CAR-T cell recipients during the current scenario so that critical care beds could be saved for patients with severe COVID-19. From a practical standpoint, many centers have already modified their

Table 36.2 How to overcome the hurdles to accomplish cellular therapy?

Screening the waitlist and prioritizing the indication where the likely cure rate is high, or indication is urgent)
Screening CAR-T cell recipients and documenting a negative COVID RT-PCR before and after the administration of cellular therapy products
In the case of pediatric cellular therapy, screening the parents of the child for COVID
Allowing only one parent to be with the pediatric patient, thereby minimizing the risk of exposure
Selecting the right candidate and triaging the patients (low disease burden and normal LDH, good performance disease) likely to use fewer health resources. These patients are less likely to have a complicated course nad less likely to require ICU beds or experience a prolonged cytopenia
Ensuring the safety of the harvested cellular products
Collaboration with harvesting center and infusion center
Educating referring physicians to send the patient to CAR-T cell center sooner
Postponing lower risk transplants if possible/acceptable
Stringent social isolation policy creating a separate cancer clinic space dedicated to CAR-T cell patients with the provision of separate elevators, pathways, and areas for waiting
Assessing the local COVID-19 disease burden within the patient's community
Patients with high tumor load and high LDH with a high likelihood of going to CRS are deferred from CAR-T cell therapy and providing them alternative treatment with readdressing the possibility of CAR-T cell later
Frank and timely discussion of goals of care with patients and caregivers during the early phase of treatment to avoid last-minute discussion
Early use of tocilizumab and steroids in CRS
The transition from outpatient to inpatient in case of high community transmission of COVID-19
House availability
Virtual or telehealth visits for the patients to minimize hospital visits
Lab draws at home to minimize hospital visits and associated exospores

cellular therapy treatment strategies and have lowered the threshold for initiating steroids and tocilizumab in patients with early CRS, in an attempt to free up the critical care beds for potential COVID-19 patients (Hassoun et al. 2020).

36.3 Hematopoietic Stem Cell Transplantation

36.3.1 Specific Challenges of Allogeneic Versus Autologous HSCT

Allogeneic HSCT is a comparatively more complicated procedure than autologous HSCT since stem cells must be procured and transported from a healthy donor rather than the patient. Harvesting of stem cells from a matched unrelated donor (MUD) may be even more challenging than from a matched or a haploidentical related donor, especially in the setting of a pandemic. In addition to the usual hurdle of finding a willing and appropriately matched donor, the MUD HSCT procedure during the current pandemic may present new and unique hurdles, such as (1) travel limitations (state to state or potentially international, as well as time restrictions), (2) different laws and policies due to variable COVID-19 situations in different regions or countries (donor center impact vs. transplant center impact), (3) willingness of donor to go to the donor center (hospital) during this infectious disease pandemic and health crisis, and (4) the risk of being an asymptomatic SARS-CoV-2 carrier or becoming infected at the collection facility (applicable to both the MUD donor and recipient).

36.3.2 Triaging the Indication for HSCT

It is becoming increasingly apparent that no one-size-fits-all set of guidelines for managing cellular therapy amid the current COVID-19 pandemic. It is crucial to consider all the variables when deciding whether to proceed with a transplant. The availability of adequately trained transplant staff (in some hospitals, the team members may be required to COVID-19 specific units), number of open critical care beds and ventilators, access to the stem cell products, location of the cryopreservation and processing facility, local and institutional guidelines, travel restrictions, patient disease status, the urgency of transplant, availability of alternative bridging therapies, unrelated versus related versus autologous transplantation, patient preference, risk of contracting COVID-19 infection, availability of reliable COVID-19 screening tests for both donor and recipient, and numerous other factors need to be weighed. According to the European Society for Blood and Marrow Transplantation (EBMT), it is impossible to provide universal guidelines that are clear and definitive, regarding which patients should move forward with transplants and which can be safely delayed (Ljungman et al. 2020). In general, a multidisciplinary team approach should be employed whenever possible, and every patient should be assessed by an institutional multidisciplinary tumor board to discuss the indications for urgent versus delayed transplant. Additionally, delay of allogeneic HSCT may be possible in certain situations, especially for nonmalignant disorders.

36.4 Recommendations from Transplant Societies

Continuing to provide HSCT services and CAR-T cell therapy during the COVID-19 pandemic would prove challenging for any cellular therapy and transplant programs. The current COVID-19 situation is fluid and rapidly evolving, with variable disease burden in different countries. Most patients slated to receive CAR-T cell therapy cannot interrupt or delay treatment, as they have no further treatment options. The same is also true for many HSCT patients. It makes for a unique group of patients with a compelling reason to push forward with treatment despite pandemic constraints. Continuing with these treatments should be considered if centers can accommodate the patients while considering national/state/institutional regulations as well.

In March 2020, the European Society for Blood and Marrow Transplantation (EBMT) led by Dr. Ljungman and colleagues published interim guidelines and recommendations for transplant centers, recipients, and donors (Ljungman et al. 2020). Since then, the recommendations have undergone a series of modifications, the latest of which is version 8, dated May 18, 2020 (Mahmoudjafari et al. 2020; Ljungman et al. 2020). Similarly, the American Society for Transplantation and Cellular Therapy (ASTCT) has also issued interim guidelines, the majority of which were adopted from EBMT (Mahmoudjafari et al. 2020).

In the setting of this COVID-19 crisis, it is imperative to balance the need to limit the spread of the virus in the community while protecting the patients from infection, with ensuring their access to this last line of potentially lifesaving treatment. The guidelines provided by these transplant societies will be most helpful to the clinicians tasked with determining and discussing with the patients whether they were more at risk of dying from their malignancy or of a potential SARS-CoV-2 infection (Fig. 36.2).

36.5 More on Selecting the Ideal Donor and Recipient

The pandemic may affect donors and the standard procedure of allogeneic stem cell transplantation. Under the current condition, it is even more critical to assess the health condition of both the donor (related or unrelated) and the recipient before proceeding with harvesting donor cells or infusing the product to the recipient.

In this situation, donors might be concerned about leaving their house and being infected with COVID-19 at the harvesting/donor center. It is a practical and genuine concern from the donor's perspective, especially when it is regularly advised by the world authorities to stay at home, and the act of donating cells is not imperative to improving their own health. Similarly, every donor requires a health and eligibility check before being accepted as a donor (Table 36.3). However, with many outpatient clinics closed or are open for only limited hours, donors may find it challenging to complete their pre-donation workup and screening tests. From the clinician's perspective, other concerns include the donor participating in "super spreader events" or having recently traveled to or from a geographically

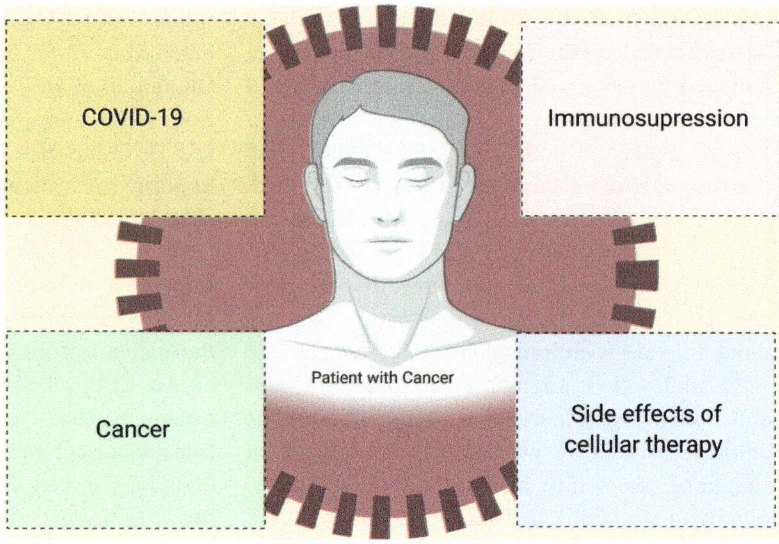

Fig. 36.2 Patient concerns while considering cellular therapy (CAR-T cell and HSCT) during the COVID-19 pandemic

Table 36.3 Recommendations for donor

Scenario	Recommendations
If a donor gets a COVID-19 disease	To be excluded from stem cell donation
	At least 3 months deferral to be considered unless the need for donation is urgent when individual consideration should be made
If a donor gets a close contact with a COVID-19 confirmed case	**In the case of nonurgent transplant**: Stem cell harvest shall be deferred for at least 28 days, and donor should be watched closely for any symptoms
	In the case of urgent need of transplant: If the donor is entirely well, and a COVID-19 test is negative, and there are no suitable alternative donors, an earlier collection may be considered subject to careful risk assessment if local quarantine requirements permit
If a donor has traveled to high-risk COVID-19 areas or came to close contact with an individual traveling from such areas	The donor should be excluded from donation for at least 28 days

Prepared with data from EBMT, version 4.3, dated March 23, 2020

high-risk area, experiencing active symptoms, or having been recently diagnosed with COVID-19.

Similarly, while considering recipients for cellular therapy, the candidate is recommended to isolate him-/herself for at least 14 days, which they may have been doing already due to their immunosuppressive therapies. Also, all recipients should have a negative SARS-CoV-2 RT-PCR test before starting the conditioning regimen. If the recipient tested positive for SARS-CoV-2, then delaying the HSCT depends mainly on the risk of relapsed or progressive cancer (Table 36.4).

36.6 COVID-19 and Stem Cell Graft Cryopreservation

Numerous countries have implemented various travel restrictions to contain the spread of COVID-19. Because of these restrictions, the timely delivery of donor stem cells from the site of harvesting to the site of transfusion may be jeopardized. As discussed above, this, in particular, occurs in the case of MUD HSCT. To ensure that harvested donor cells are available for infusion on the scheduled date, most international registries are strongly favoring cryopreservation

Table 36.4 Recommendations for transplant candidate patients (HSCT and cellular therapy recipients)

Scenario	Recommendations
Before planned admission for HSCT/cellular therapy	Home isolation for 14 days or more before starting the transplant conditioning
	To avoid the non-necessary clinic visits to reduce the risk to a minimum and use telehealth where able
Before starting conditioning regimen	All recipients should have COVID-19 RT-PCR tested negative before the starting conditioning regimen irrespective of respiratory symptoms
In case, a potential recipient gets a close contact with COVID-19 confirmed case	Transplantation should not be performed within at least 14 days and preferably 21 days from the last date of contact, and the recipient should be closely monitored for COVID-19 symptoms and COVID-19 RT-PCR negativity to be confirmed before undertaking transplant procedure again
If a recipient gets a COVID-19 disease	**High-risk cancer disease:** Deferral until the recipient COVID-19 patient is asymptomatic and has two repeated virus PCR negativity at least 1 week apart (deferral of 14 days minimum)
	Low-risk cancer disease: Deferral for up to 3 months

Prepared with data from EBMT, version 4.3, dated March 23, 2020

of the donated grafts before starting the transplant conditioning (Hamadani et al. 2020; Kao et al. 2011). Cryopreserving the graft guarantees the availability of stem cell products on the intended day of transplantation. Since March 30, 2020, the National Marrow Donor Program (NMDP) has implemented a mandatory step for all transplant programs to cryopreserve harvests before transfusion. Previously, cryopreservation posed a potential (ethical) dilemma as the harvested cells put the healthy volunteer donor at some (THOUGH typically minor) risk, and the cryopreserved cells may not have been ultimately used due to recipient's condition, i.e., disease progression, uncontrolled infection, or organ failure, leading to disqualify a transplant candidate.

Similarly, centers that have traditionally used only fresh cells were concerned about any changes to the usual function of the graft, especially graft viability and its impact on engraftment issues (graft failure or delayed engraftment). However, these concerns have been eliminated and addressed by a recently published CIBMTR study (Antonenas et al. 2006; Yu et al. 2018; Guttridge et al. 2006). This strategy attempts to ensure patient safety so that no patient receives a conditioning regimen in the potential absence of unrelated donor cell infusion on the transplant day.

Ideally, infusion of marrow and peripheral blood stem cells (PBSC) should occur within 48 h of collection. However, this is often not possible during the current COVID-19 situation, and fear of unintended delays during the transit from the collection center to the infusion center has become real. As mentioned above, NMDP recently addressed this by issuing a statement that harvested products have no expiration date, and hence infusion and cryopreservation at greater than 72 h would not be considered as a deviation from NMDP standards or against FDA regulations. The mean CD34+ cell viability for PBSC and bone marrow products, when stored over 72 h at 4 °C, was found to be higher than 75% (Fournier et al. 2020; Kao et al. 2011). Hence, this optimal cell viability and recovery allow for flexibility in the time duration from the point of harvest to the infusion. Hamadani et al. recently compared fresh versus cryopreserved grafts and found no difference in the time to hematopoietic recovery, risk of acute graft-versus-host disease (GVHD), non-relapse mortality (NRM), or overall survival (OS) (Hamadani et al. 2020; Sahu and Siddiqui 2020; Eapen et al. 2020). These studies provide us the evidence that during the current COVID-19 situation, having the option of stem cell graft cryopreservation can provide extra assurance and confidence for completing MUD HSCTs.

36.7 A Possible Cargo Solution as an Alternative to Cryopreservation

Cryopreservation is an effective way to preserve stem cell products, but it utilizes a good deal of resources. Recently, a multi-agency pilot project in which NMDP is collaborating with the German National Bone Marrow Donor Registry (ZKRD, Zentrales Knochenmarkspender-Register Deutschland), the German Bone Marrow Donor File (Deutsche KnochenMarkSpenderdatei, DKMS), and the World Marrow Donor Association (WMDA) was created to test the transportation of fresh stem cell products using cargo services. The successful shipment of two test blood samples via this method from Frankfurt, Germany, to the United States was recently completed. With these encouraging results, staring April 6, 2020, both NMDP and DKMS have agreed to use cargo rather than passenger planes for all stem cell products. While this seems to be a practically possible and logical solution, it requires additional exploration before being incorporated into guidelines for widespread use until air travels return the normal situations.

36.8 Searching for the Best Stem Cell Source

For HSCT, only three graft sources exist: bone marrow, mobilized peripheral blood cells, and cord blood. PBSC has become the preferred source for its easily accessible site, typically high yield, and higher graft versus leukemia effect. Conversely, cord blood has been less often preferred due to the main disadvantages of insufficient cord blood stem cell dose for adult transplants, slower engraftment, and immune recovery. Recently, there has been an upturn in the search trend for the grafts. Dr. Bronwen Shaw, the president-elect of the WMDA, shared data in a webinar and showed that the number of donor searches has gone own down by 20% and, instead, that of cord blood rose by 58% during March 2020. It may indicate that transplant specialists are trying to overcome the logistic hurdles associated with unrelated related grafts by exploring the possibility of cord blood as the alternative source for HSCT. One expects that a similar trend exists for haploidentical donors and grafts.

36.9 Supportive Care and Services for Cellular Therapy During the COVID-19 Pandemic

CAR-T cell therapy with tisa-cel in many centers is handled as an outpatient procedure with possible admission for complications, while axi-cel remains full hospitalization. Healthcare infrastructures worldwide are overwhelmed (or nearly so) by the gigantic load of an exponential increase in the number of patients with COVID-19. Critically ill patients are the most challenging to treat and require multidisciplinary teams. The Hospital Incident Command Systems (HICS) are currently busy dealing with the clinical and logistical complexities of staff shortages; limited resources; manufacturing issues; scarcity of personal protective equipment (PPEs), hospital beds, ICU beds, NIV/BiPAP machines, and mechanical ventilators; medication shortages; and blood product shortages (Eapen et al. 2020; Sahu et al. 2020h; Gehrie et al. 2020).

A unique deficiency noted is the scarcity of cryopreservation bags during the COVID-19 era, most probably due to the increased practice of cryopreservation and delayed allogeneic HSCTs (Tanhehco and Schwartz 2020). To overcome this challenge, it is recommended to communicate with respective primary vendors and order for an estimated supply for approximately the next 6 months, be sure to have an alternate backup vendor, and network with other cell therapy labs to share these suddenly scarce resources.

Even the American Cancer Society has made concessions. Its "Hope lodge" initiative, which offers free lodging for cancer patients and their caregivers during treatment, has temporarily suspended its operations.

Cellular therapy is usually an expensive procedure, and much of the financial burden falls on the recipients. International registries surely recognize that during this COVID-19 crisis, patients may be left financially vulnerable. NMDP has decided to assist by reducing the "Related Donor Workup and Collection" fee to $15,000 for all fully matched and haploidentical related donors in the United States and Mexico. Also, NMDP has agreed to provide in-home swab collections for patients and related donor HLA typing along with a full waiver of the fee.

Under normal circumstances, transplant and cellular therapy programs use a significant number of PPEs (depending on institutional protocols and guidelines). During COVID-19 times, limited availability of PPE restricts the liberal use of PPE while caring for oncology patients and those undergoing cellular therapy, as well as when processing cellular products in the laboratory. It could potentially place the patients at increased risk of acquiring iatrogenic or nosocomial infections. These resource limitations pose an unprecedented challenge to cellular therapy programs to provide standard care.

To avoid a potential delay in the shipment of cellular therapy products, it requires transparent cross talk and a coordinated effort between the cellular therapy program team and the hospital's emergency response team. There should be an open discussion regarding the patients who have an urgent need for HSCT or CAR-T cell therapy so that these patients and their corresponding teams can be provided with the necessary resources for the successful implementation of cellular therapy.

Cellular therapists have expressed their concerns over the reduction in the number of referrals for transplants and CAR-T cell therapy since the appearance of COVID-19. Possible explanations for this include patient reluctance to go to office/hospital visits for CAR-T cell therapy or HSCT suitability evaluations referrals, as well as the reluctance of primary care providers (or even oncologists) to send patients to larger centers during the COVID-19 outbreak. The delay and reduction in referrals could have two significant impacts. One is that delays in initiating CAR-T cell therapy may lead to progression of the disease and, therefore, the loss of precious time, and the other is that in the absence of appropriate cellular therapy, malignancies and tumors may progress, leading to a decline in performance status. Ultimately, these could either make cancer patients no longer suitable for the treatment or place them at higher risk for complications if CAR-T cell therapy is pursued at a later date. Additionally, there may be a post-COVID-19 referral surge which exceeds the capacity of cellular programs. The cellular therapy program is a new addition to the treatment facility, and hence these challenges were either not faced or remained unattended during the previous outbreaks.

36.10 Challenges During Post-transplant Setting in COVID-19 Era

During the post-transplant period, recipients are always faced with a hefty set of challenges. These include, but not limited to, relapse of disease, superadded infections, drug toxicities, and graft-vs.-host disease (GVHD) (Sahu et al. 2016, 2020f, g). Immunosuppression and delayed immune reconstitution play a significant role in predisposing transplant recipients to infection. In the case of allogeneic HSCT, the risk further increases owning to the immunosuppressant drugs, which improve graft survival but, in turn, make the patient susceptible to various infectious complications (Sahu et al. 2016; Sheshadri et al. 2019; Duncan and Wilkes 2005; Heeger and Dinavahi 2012).

36.11 COVID-19 in Post-transplant/CAR-T Cell Therapy Setting

There are guidelines for reporting infections in the post-transplant setting. The international stem cell registries like the Center for International Blood and Marrow Transplant Research (CIBMTR) keep a close track of the incidence of opportunistic infections via effective data collection systems such as the Transplant Essential Data (TED) track that collects minimal data and a Comprehensive Report Form (CRF) track that requires more comprehensive data gathering (D'Souza et al. 2017).

Due to the novelty of COVID-19, data is scarce on the potential effects of COVID-19 on the post-transplant setting. It is especially the case during the first few months of the pandemic when much of the relevant data could not be captured due to the lack of a set format, absence of universal diagnostic tools, deficiencies in data gathering, and unstructured networking. CIBMTR has acted promptly to address these issues and has modified the post-HSCT follow-up data form (2100) and cellular therapy necessary data follow-up data form (4100) in order to incorporate the new pathogen of SARS-CoV-2. These changes were implemented on March 27, 2020, and with this now effective data gathering and networking, a useful data pool is being generated that could prove to be a cornerstone for establishing future recommendations. As of May 29, 2020, 122 COVID-19 infections in the post cellular therapy stage have been reported. Most of the reporting centers are in the United States. It again suggests the importance of increasing the awareness of reporting among the non-US-based transplant programs (48 out of 56 reporting centers). The majority of the patients received allogeneic HSCT (74/122), followed by autologous HSCT (40/122) and other cellular therapies (6/122). Most patients belonged to the age group 60 and above (50/122). An interesting fact to note is that most of the patients confirmed to have COVID-19 were within the first year (29 patients) or second year (38 patients) post-infusion. The main problem with the data is the lack of testing across the board as we do not know the denominator for the COVID positivity. As more data will come into the databases (CIBMTR and EBMT), we hope we would have a better understanding of the risk predisposition, susceptible age, and outcome of these patients. Reporting from developing countries, where the resources are already sparse for management of both COVID-19 and cancer patients, should also be encouraged (Sahu et al. 2020n; The 2020).

Evaluating a patient for respiratory symptoms during the post-HSCT setting is challenging. SARS-CoV-2 is just an addition to the already identified differentials of opportunistic infections, thromboembolic events, fungal infections, GvHD, and idiopathic pneumonia syndrome (Sahu et al. 2020o, p). Appropriate imaging and laboratory investigations would help in evaluating such patients (Sahu et al. 2020e; Lal et al. 2020). If confirmed, COVID-19 positivity in a post-HSCT setting would need aggressive management and follow-up. The use of convalescent plasma therapy, remdesivir, tocilizumab, and other investigational drugs should be assessed with assistance from the infectious disease colleagues (Sahu et al. 2020b). In addition to providing appropriate medical therapy, it is also essential to ensure that adequate nutrition is maintained in cancer patients, especially in post-transplant settings (Gibbs et al. 2020).

36.12 Impact of the Clinical Trials Involving CAR-T Cell Therapies and Its Implications

Randomized clinical trials (RCT) focusing on the various aspects of cellular therapy have been profoundly affected and have been put on hold. Newer requirements, especially for the high-risk protocols, are on hold, especially in places where the health resources are scarce. Some centers (without severe resource constrains) continue to enroll on to trials which offer survival benefit to patients who would not have otherwise good alternatives. So, some continue with the most impactful RCT and phase II trials that could reveal lifesaving outcomes. Similarly, clinical trial monitoring is an essential aspect of clinical trial conduct to ensure all required guidelines are adequately followed. This aspect is difficult to fulfill due to current regulations that debar all the non-hospital personnel like clinical trial monitors or auditors from visiting hospitals.

36.13 Future Directions

The current pandemic has brought challenges but also opened new gateways for opportunities at the same time. Due to lack of data, the cellular therapists and transplant physicians so far have innovated their own ways to deal with the COVID-19-related challenges (from streamlining of protocols and communication to establishing new standards for cryopreservation of unrelated allogeneic stem cells). We believe that the use of telehealth visits, excellent communication with the referring providers, the meticulous selection of right recipient candidates, and timely goals of care discussion would hold critical importance for cellular therapy during the COVID-19 pandemic. We also emphasize a well-structured contingency plan should be provided in place to ensure the preparedness for post-pandemic challenges like the referral surge and the possibility of the second wave of COVID-19 (Al-Shamsi et al. 2020). Also, it is crucial to understand that there is much information available in the literature since the outbreak started. However, clinicians and specialists need to study them diligently before coming to any conclusion (Rzymski et al. 2020).

36.14 Conclusion

The COVID-19 pandemic is an ongoing and evolving healthcare crisis. Because the situation is fluid and expected to change further, we welcome that hematological societies have been forming universal portals that are easily accessible for the specialists for guidance, opinion, and data sharing. According to the current status of COVID-19, some predict that this pandemic may last until the end of 2020. Many patients are struggling with relapsed/refractory hematological diseases, and we ought to strive to enhance their survival and quality of life by providing advanced cellular therapeutics to the best of our ability while maximizing available resources and preventing further morbidity and mortality caused by the spread of COVID-19.

Acknowledgments The authors acknowledge image construction provided by Biorender.com.

References

Ahmad A, Uddin S, Steinhoff M (2020) CAR-T cell therapies: an overview of clinical studies supporting their approved use against acute lymphoblastic leukemia and large B-cell lymphomas. Int J Mol Sci 21(11). https://doi.org/10.3390/ijms21113906

Al-Shamsi HO, Alhazzani W, Alhuraiji A, Coomes EA, Chemaly RF, Almuhanna M, Wolff RA, Ibrahim NK, Chua MLK, Hotte SJ, Meyers BM, Elfiki T, Curigliano G, Eng C, Grothey A, Xie C (2020) A practical approach to the management of cancer patients during the novel coronavirus disease 2019 (COVID-19) pandemic: an International Collaborative Group. Oncologist 25(6):e936–e945. https://doi.org/10.1634/theoncologist.2020-0213

Antonenas V, Garvin F, Webb M, Sartor M, Bradstock KF, Gottlieb D (2006) Fresh PBSC harvests, but not BM, show temperature-related loss of CD34 viability during storage and transport. Cytotherapy 8(2):158–165. https://doi.org/10.1080/14653240600620994

Ardura M, Hartley D, Dandoy C, Lehmann L, Jaglowski S, Auletta JJ (2020) Addressing the impact of the coronavirus disease 2019 (COVID-19) pandemic on hematopoietic cell transplantation: learning networks as a means for sharing best practices. Biol Blood Marrow Transplant 26(7):e147–e160. https://doi.org/10.1016/j.bbmt.2020.04.018

Bachanova V, Bishop MR, Dahi P, Dholaria B, Grupp SA, Hayes-Lattin B, Janakiram M, Maziarz RT, McGuirk JP, Nastoupil LJ, Oluwole OO, Perales MA, Porter DL, Riedell PA (2020) Chimeric antigen receptor T cell therapy during the COVID-19 pandemic. Biol Blood Marrow Transplant 26(7):1239–1246. https://doi.org/10.1016/j.bbmt.2020.04.008

Block KI (2020) Integrative cancer therapies: learning from COVID-19. Integr Cancer Ther 19:1534735420932652. https://doi.org/10.1177/1534735420932652

Boyiadzis MM, Dhodapkar MV, Brentjens RJ, Kochenderfer JN, Neelapu SS, Maus MV, Porter DL, Maloney DG, Grupp SA, Mackall CL, June CH, Bishop MR (2018) Chimeric antigen receptor (CAR) T therapies for the treatment of hematologic malignancies: clinical perspective and significance. J Immunother Cancer 6(1):137. https://doi.org/10.1186/s40425-018-0460-5

Broxmeyer HE, Parker GC (2020) Impact of COVID-19 and future emerging viruses on hematopoietic cell transplantation and other cellular therapies. Stem Cells Dev 29(10):625–626. https://doi.org/10.1089/scd.2020.0064

Brudno JN, Kochenderfer JN (2019) Recent advances in CAR T-cell toxicity: mechanisms, manifestations and management. Blood Rev 34:45–55. https://doi.org/10.1016/j.blre.2018.11.002

Dave H, Jerkins L, Hanley PJ, Bollard CM, Jacobsohn D (2019) Driving the CAR to the bone marrow transplant program. Curr Hematol Malig Rep 14(6):561–569. https://doi.org/10.1007/s11899-019-00544-6

Dholaria B, Savani BN (2020) How do we plan hematopoietic cell transplant and cellular therapy with the looming COVID-19 threat? Br J Haematol 189(2):239–240. https://doi.org/10.1111/bjh.16597

D'Souza A, Lee S, Zhu X, Pasquini M (2017) Current use and trends in hematopoietic cell transplantation in the United States. Biol Blood Marrow Transplant 23(9):1417–1421. https://doi.org/10.1016/j.bbmt.2017.05.035

Duncan MD, Wilkes DS (2005) Transplant-related immunosuppression: a review of immunosuppression and pulmonary infections. Proc Am Thorac Soc 2(5):449–455. https://doi.org/10.1513/pats.200507-073JS

Eapen M, Zhang MJ, Tang XY, Lee SJ, Fei MW, Wang HL, Hebert KM, Arora M, Chhabra S, Devine SM, Hamadani M, D'Souza A, Pasquini MC, Phelan R, Rizzo JD, Saber W, Shaw BE, Weisdorf DJ, Horowitz MM (2020) Hematopoietic cell transplantation with cryopreserved grafts for severe aplastic anemia. Biol Blood Marrow Transplant 26(7):e161–e166. https://doi.org/10.1016/j.bbmt.2020.04.027

Fournier D, Lewin A, Simard C, Trépanier P, Néron S, Ballerini L, Codinach M, Elmoazzen H, Halpenny M, Kogler G, Liedtke S, Louis I, Molluna CA, Pineault N, Prasath A, Querol S, Saccardi R, Sutherland DR, Thérien C, Urbani S (2020) Multi-laboratory assay for harmonization of enumeration of viable CD34+ and CD45+ cells in frozen cord blood units. Cytotherapy 22(1):44–51. https://doi.org/10.1016/j.jcyt.2019.10.009

Gardner RA, Ceppi F, Rivers J, Annesley C, Summers C, Taraseviciute A, Gust J, Leger KJ, Tarlock K, Cooper TM, Finney OC, Brakke H, Li DH, Park JR, Jensen MC (2019) Preemptive mitigation of CD19 CAR T-cell cytokine release syndrome without attenuation of antileukemic efficacy. Blood 134(24):2149–2158. https://doi.org/10.1182/blood.2019001463

Gehrie E, Tormey CA, Sanford KW (2020) Transfusion service response to the COVID-19 pandemic. Am J Clin Pathol. https://doi.org/10.1093/ajcp/aqaa111

Gibbs L, Minnelli N, Larrivee J, Sahu KK, Siddiqui AD (2020) Oncology dietitians sound alarm in key nutrition needs of cancer patients during COVID-19 pandemic. JCO Oncol Pract 2020:Op2000349. https://doi.org/10.1200/op.20.00349

Guttridge MG, Sidders C, Booth-Davey E, Pamphilon D, Watt SM (2006) Factors affecting volume reduction and red blood cell depletion of bone marrow on the COBE spectra cell separator before haematopoietic stem cell transplantation. Bone Marrow Transplant 38(3):175–181. https://doi.org/10.1038/sj.bmt.1705420

Hamadani M, Zhang MJ, Tang XY, Fei M, Brunstein C, Chhabra S, D'Souza A, Milano F, Phelan R, Saber W, Shaw BE, Weisdorf D, Devine SM, Horowitz MM (2020) Graft cryopreservation does not impact overall survival after allogeneic hematopoietic cell transplantation using post-transplantation cyclophosphamide for graft-versus-host disease prophylaxis. Biol Blood Marrow Transplant 26(7):1312–1317. https://doi.org/10.1016/j.bbmt.2020.04.001

Hassoun A, Thottacherry ED, Muklewicz J, Aziz QU, Edwards J (2020) Utilizing tocilizumab for the treatment of cytokine release syndrome in COVID-19. J Clin Virol 128:104443. https://doi.org/10.1016/j.jcv.2020.104443

Heeger PS, Dinavahi R (2012) Transplant immunology for non-immunologist. Mt Sinai J Med 79(3):376–387. https://doi.org/10.1002/msj.21314

Hunter BD, Rogalski M, Jacobson CA (2019) Chimeric antigen receptor T-cell therapy for the treatment of aggressive B-cell non-Hodgkin lymphomas: efficacy, toxicity, and comparative chimeric antigen receptor products. Expert Opin Biol Ther 19(11):1157–1164. https://doi.org/10.1080/14712598.2019.1644316

Jindal V, Sahu KK, Gaikazian S, Siddiqui AD, Jaiyesimi I (2020) Cancer treatment during COVID-19 pandemic. Med Oncol 37(7):58. https://doi.org/10.1007/s12032-020-01382-w

Kao GS, Kim HT, Daley H, Ritz J, Burger SR, Kelley L, Vierra-Green C, Flesch S, Spellman S, Miller J, Confer D (2011) Validation of short-term handling and storage conditions for marrow and peripheral blood stem cell products. Transfusion 51(1):137–147. https://doi.org/10.1111/j.1537-2995.2010.02758.x

Lal A, Mishra AK, Sahu KK (2020) CT chest findings in coronavirus disease-19 (COVID-19). J Formos Med Assoc 119(5):1000–1001. https://doi.org/10.1016/j.jfma.2020.03.010

Liang W, Guan W, Chen R, Wang W, Li J, Xu K, Li C, Ai Q, Lu W, Liang H, Li S, He J (2020) Cancer patients in SARS-CoV-2 infection: a nationwide analysis in China. Lancet Oncol 21(3):335–337. https://doi.org/10.1016/s1470-2045(20)30096-6

Ljungman P, Mikulska M, de la Camara R, Basak GW, Chabannon C, Corbacioglu S, Duarte R, Dolstra H, Lankester AC, Mohty M, Montoto S, Murray J, Peffault de Latour R, Snowden JA, Yakoub-Agha I, Verhoeven B, Kröger N, Styczynski J (2020) The challenge of COVID-19 and hematopoietic cell transplantation; EBMT recommendations for management of hematopoietic cell transplant recipients, their donors, and patients undergoing CAR T-cell therapy. Bone Marrow Transplant:1–6. https://doi.org/10.1038/s41409-020-0919-0

Mahmoudjafari Z, Alexander M, Roddy J, Shaw R, Shigle TL, Timlin C, Culos K (2020) American Society for Transplantation and Cellular Therapy Pharmacy Special Interest Group Position Statement on Pharmacy Practice Management and Clinical Management for COVID-19 in Hematopoietic Cell Transplantation and Cellular Therapy Patients in the United States. Biol Blood Marrow Transplant 26(6):1043–1049. https://doi.org/10.1016/j.bbmt.2020.04.005

Maude SL, Laetsch TW, Buechner J, Rives S, Boyer M, Bittencourt H, Bader P, Verneris MR, Stefanski HE,

Myers GD, Qayed M, De Moerloose B, Hiramatsu H, Schlis K, Davis KL, Martin PL, Nemecek ER, Yanik GA, Peters C, Baruchel A, Boissel N, Mechinaud F, Balduzzi A, Krueger J, June CH, Levine BL, Wood P, Taran T, Leung M, Mueller KT, Zhang Y, Sen K, Lebwohl D, Pulsipher MA, Grupp SA (2018) Tisagenlecleucel in children and young adults with B-cell lymphoblastic leukemia. N Engl J Med 378(5):439–448. https://doi.org/10.1056/NEJMoa1709866

Mishra AK, Sahu KK, George AA, Lal A (2020a) A review of cardiac manifestations and predictors of outcome in patients with COVID-19. Heart Lung. https://doi.org/10.1016/j.hrtlng.2020.04.019

Mishra AK, Sahu KK, George AA, Sargent J, Lal A (2020b) Cerebrovascular events in COVID-19 patients. Monaldi Arch Chest Dis 90(2). https://doi.org/10.4081/monaldi.2020.1341

Mishra AK, Sahu KK, Lal A, Sargent J (2020c) Mechanisms of stroke and the role of anticoagulants in COVID-19. J Formos Med Assoc. https://doi.org/10.1016/j.jfma.2020.06.026

Mishra AK, Sahu KK, Lal A, Sargent J (2020d) Patterns of heart injury in COVID-19 and relation to outcome. J Med Virol. https://doi.org/10.1002/jmv.25847

Neelapu SS, Locke FL, Bartlett NL, Lekakis LJ, Miklos DB, Jacobson CA, Braunschweig I, Oluwole OO, Siddiqi T, Lin Y, Timmerman JM, Stiff PJ, Friedberg JW, Flinn IW, Goy A, Hill BT, Smith MR, Deol A, Farooq U, McSweeney P, Munoz J, Avivi I, Castro JE, Westin JR, Chavez JC, Ghobadi A, Komanduri KV, Levy R, Jacobsen ED, Witzig TE, Reagan P, Bot A, Rossi J, Navale L, Jiang Y, Aycock J, Elias M, Chang D, Wiezorek J, Go WY (2017) Axicabtagene ciloleucel CAR T-cell therapy in refractory large B-cell lymphoma. N Engl J Med 377(26):2531–2544. https://doi.org/10.1056/NEJMoa1707447

Paul S, Rausch CR, Jain N, Kadia T, Ravandi F, DiNardo CD, Welch MA, Dabaja BS, Daver N, Garcia-Manero G, Wierda W, Pemmaraju N, Montalban Bravo G, Thompson P, Verstovsek S, Konopleva M, Kantarjian H, Jabbour E (2020) Treating leukemia in the time of COVID-19. Acta Haematol:1–13. https://doi.org/10.1159/000508199

Romero D (2018) Haematological cancer: Favourable outcomes with CAR T cells. Nat Rev Clin Oncol 15(2):65. https://doi.org/10.1038/nrclinonc.2017.208

Rzymski P, Nowicki M, Mullin GE, Abraham A, Rodríguez-Román E, Petzold MB, Bendau A, Sahu KK, Ather A, Naviaux AF, Janne P, Gourdin M, Delanghe JR, Ochs HD, Talmadge JE, Garg M, Hamblin MR, Rezaei N (2020) Quantity does not equal quality: scientific principles cannot be sacrificed. Int Immunopharmacol 86:106711. https://doi.org/10.1016/j.intimp.2020.106711

Sahu KK, Cerny J (2020) Managing patients with hematological malignancies during COVID-19 pandemic. Expert Rev Hematol. https://doi.org/10.1080/17474086.2020.1787147

Sahu K, Kumar R (2020a) Current perspective on pandemic of COVID-19 in the United States. J Family Med Primary Care 9(4):1784–1791. https://doi.org/10.4103/jfmpc.jfmpc_424_20

Sahu K, Kumar R (2020b) Preventive and treatment strategies of COVID-19: from community to clinical trials. J Family Med Primary Care 9(5):2149–2157. https://doi.org/10.4103/jfmpc.jfmpc_728_20

Sahu KK, Siddiqui AD (2020) From Hematologist's desk: the effect of COVID-19 on the blood system. Am J Hematol. https://doi.org/10.1002/ajh.25849

Sahu KK, Prakash G, Khadwal A, Varma SC, Malhotra P (2016) A rare case of hemorrhagic cystitis in allogeneic hematopoietic stem cell transplant patient. Indian J Hematol Blood Transfus 32(Suppl 1):196–200. https://doi.org/10.1007/s12288-015-0530-5

Sahu KK, Jindal V, Siddiqui AD (2020a) Managing COVID-19 in patients with Cancer: a double blow for oncologists. JCO Oncol Pract 16(5):223–225. https://doi.org/10.1200/op.20.00167

Sahu KK, Jindal V, Siddiqui AD, Cerny J, Gerber JM (2020b) Convalescent plasma therapy: a passive therapy for an aggressive COVID-19. J Med Virol. https://doi.org/10.1002/jmv.26047

Sahu KK, Lal A, Mishra AK (2020c) COVID-2019 and pregnancy: a plea for transparent reporting of all cases. Acta Obstet Gynecol Scand 99(7):951. https://doi.org/10.1111/aogs.13850

Sahu KK, Lal A, Mishra AK (2020d) Latest updates on COVID-2019: a changing paradigm shift. J Med Virol 92(6):533–535. https://doi.org/10.1002/jmv.25760

Sahu KK, Lal A, Mishra AK (2020e) An update on CT chest findings in coronavirus disease-19 (COVID-19). Heart Lung. https://doi.org/10.1016/j.hrtlng.2020.03.007

Sahu KK, Lal A, Mishra AK, George SV (2020f) Methemoglobinemia in transplant recipient patients: future of dapsone as pneumocystis carinii pneumonia prophylaxis. Prog Transplant 30(1):71–72. https://doi.org/10.1177/1526924819893316

Sahu KK, Mahagaokar K, Patel B, Winokur D, Suzuki S, Daly JS, Ramanathan M, Cerny J (2020g) Strongyloides stercoralis hyperinfection syndrome in mantle cell lymphoma in post-transplant setting. Ann Hematol. https://doi.org/10.1007/s00277-020-04049-8

Sahu KK, Mishra AK, Lal A (2020h) COIVD 19 disease: tackling a pandemic in 21st century. QJM. https://doi.org/10.1093/qjmed/hcaa117

Sahu KK, Mishra AK, Lal A (2020i) Comprehensive update on current outbreak of novel coronavirus infection (2019-nCoV). Ann Transl Med 8(6):393. https://doi.org/10.21037/atm.2020.02.92

Sahu KK, Mishra AK, Lal A (2020j) COVID-2019: update on epidemiology, disease spread and management. Monaldi Arch Chest Dis 90(1). https://doi.org/10.4081/monaldi.2020.1292

Sahu KK, Mishra AK, Lal A (2020k) Re: from the frontlines of COVID-19-how prepared are we as obstetricians? a commentary. Bjog 127(8):1041. https://doi.org/10.1111/1471-0528.16257

Sahu KK, Mishra AK, Lal A (2020l) Trajectory of the COVID-19 pandemic: chasing a moving target. Ann Transl Med 8(11):694. https://doi.org/10.21037/atm-20-2793

Sahu KK, Mishra AK, Lal A (2020m) A twin challenge to handle: COVID-19 with pregnancy. J Med Virol. https://doi.org/10.1002/jmv.25784

Sahu KK, Mishra AK, Lal A, Sahu SA (2020n) India fights Back: COVID-19 pandemic. Heart Lung. https://doi.org/10.1016/j.hrtlng.2020.04.014

Sahu KK, Mishra AK, Martin K, Chastain I (2020o) COVID-19 and clinical mimics. Correct diagnosis is the key to appropriate therapy. Monaldi Arch Chest Dis 90(2). https://doi.org/10.4081/monaldi.2020.1327

Sahu KK, Mishra AK, Martin K, Chastain I (2020p) COVID-19 and restrictive lung disease: a deadly combo to trip off the fine balance. Monaldi Arch Chest Dis 90(2). https://doi.org/10.4081/monaldi.2020.1346

Sahu KK, Siddiqui AD, Cerny J (2020q) COVID-19 pandemic and impact on hematopoietic stem cell transplantation. Bone Marrow Transplant:1–3. https://doi.org/10.1038/s41409-020-0913-6

Sahu KK, Siddiqui AD, Cerny J (2020r) Managing sickle cell patients with COVID-19 infection: the need to pool our collective experience. Br J Haematol. https://doi.org/10.1111/bjh.16880

Sahu KK, Siddiqui AD, Rezaei N, Cerny J (2020s) Challenges for management of immune thrombocytopenia during COVID-19 pandemic. J Med Virol. https://doi.org/10.1002/jmv.26251

Samaha R, Kattan J (2020) Hematopoietic stem cell transplantation dilemma during the COVID-19 era. Future Oncol. https://doi.org/10.2217/fon-2020-0414

Schuster SJ, Bishop MR, Tam CS, Waller EK, Borchmann P, McGuirk JP, Jäger U, Jaglowski S, Andreadis C, Westin JR, Fleury I, Bachanova V, Foley SR, Ho PJ, Mielke S, Magenau JM, Holte H, Pantano S, Pacaud LB, Awasthi R, Chu J, Anak Ö, Salles G, Maziarz RT (2019) Tisagenlecleucel in adult relapsed or refractory diffuse large B-cell lymphoma. N Engl J Med 380(1):45–56. https://doi.org/10.1056/NEJMoa1804980

Sheshadri A, Chemaly RF, Alousi AM, Shah PK, Rondon G, Bashoura L, Kmeid J, Azzi J, Blanco DW, Kaous M, Dickey BF, Champlin RE, Shah DP (2019) Pulmonary impairment after respiratory viral infections is associated with high mortality in allogeneic hematopoietic cell transplant recipients. Biol Blood Marrow Transplant 25(4):800–809. https://doi.org/10.1016/j.bbmt.2018.11.022

Tanhehco YC, Schwartz J (2020) How the COVID-19 pandemic changed cellular therapy at Columbia University Irving Medical Center/NewYork-Presbyterian Hospital. Transfusion. https://doi.org/10.1111/trf.15956

The L (2020) India under COVID-19 lockdown. Lancet 395(10233):1315. https://doi.org/10.1016/s0140-6736(20)30938-7

Weinkove R, McQuilten ZK, Adler J, Agar MR, Blyth E, Cheng AC, Conyers R, Haeusler GM, Hardie C, Jackson C, Lane SW, Middlemiss T, Mollee P, Mulligan SP, Ritchie D, Ruka M, Solomon B, Szer J, Thursky KA, Wood EM, Worth LJ, Yong MK, Slavin MA, Teh BW (2020) Managing haematology and oncology patients during the COVID-19 pandemic: interim consensus guidance. Med J Aust 212(10):481–489. https://doi.org/10.5694/mja2.50607

Yu NH, Chun SY, Ha YS, Kim HT, Kim DH, Kim J, Chung JW, Lee JN, Song PH, Yoo ES, Kim BS, Kwon TG (2018) Optimal stem cell transporting conditions to maintain cell viability and characteristics. Tissue Eng Regen Med 15(5):639–647. https://doi.org/10.1007/s13770-018-0133-y

37 COVID-19 Amid Rumors and Conspiracy Theories: The Interplay Between Local and Global Worlds

Inayat Ali

Abstract

Stories and narratives are part of our human sociocultural history, which are always preserved in what I call "societal memory." We construct stories to weave meanings that help us make sense of our lifeworlds. Like stories, rumors and conspiracy theories can offer deep meanings when analyzed in specific contexts. Such narratives become most prominent in times of looming uncertainties, anxieties, and fears. Thus, the challenging coronavirus disease 2019 (COVID-19) pandemic has become surrounded by plentiful rumors and conspiracy theories. These narratives reveal geopolitics when they code the pandemic as "bioengineered." They also demonstrate local concerns, as in Pakistan, people started drinking "miraculous" tea as a form of prevention, shaving their heads, and/or praying to God to undo his "punishment." Some conceptualized the pandemic as an invented "plot." These narratives seem to empower individuals to make sense of this pandemic and to deal with its multidimensional effects: they allow them to feel confident enough to go outside and earn their livelihood. In this chapter, the author builds on his long-term ethnographic fieldwork on infectious diseases, recent telephone interviews, and content analysis of the media to discuss narratives revolving around COVID-19 in Pakistan. The author argues that these rumors and conspiracy theories are social phenomena pregnant with multiple meanings that deserve to be thoroughly explored, especially by anthropologists. A dearth of understanding about COVID-19 and narratives surrounding it would substantially impede the strategies to deal with this ongoing pandemic.

Keywords

Conspiracy theories · Coronavirus · COVID-19 · Narratives · Pakistan · Public anthropology · Rumors, Disparities, Societal memory

A different and shorter version of this paper was first published in *Disaster Medicine and Public Health Preparedness* in September 2020. It is republished here with Creative Common permission: Ali, I. 2020. "Impacts of Rumors and Consipriacy Theories Surrounding COVID-19 on Preparedness Programs." Disaster Medicine and Public Health Preparedness. 1–6.

I. Ali, (✉)
Department of Social and Cultural Anthropology, University of Vienna, Vienna, Austria

N. Rezaei (ed.), *Coronavirus Disease - COVID-19*, Advances in Experimental Medicine and Biology 1318, https://doi.org/10.1007/978-3-030-63761-3_37

37.1 Introduction

Stories and narratives are part of our human sociocultural history. From our early beginnings as foragers, we humans have created, told, and retold myths, legends, folktales, and other types of stories both to help us make sense of our life-world and to encode knowledge and wisdom learned over generations, thereby passing it on. Before the invention of writing, tales were the best sources to preserve and share specific meanings and lessons. Generations to generations, narratives have continued flowing, as we are "story-creating and story-telling animals," who live, remember, and dream by stories: we "domesticate this wild world (…) by narrative" (Reck 1983).

Narratives are social phenomena, and many are created to make sense of and to deal with the unprecedented uncertain and challenging situations. These narratives are always preserved in what I call elsewhere "societal memory" of every society (Ali 2020b), which should best be understood in light of sociocultural, economic, and (geo-)political contexts. The rapidity of their spread reveals the intensity of the challenge faced. Natural disasters, wars, and contagious outbreaks, epidemics, and pandemics, such as the extraordinary challenging 2020 coronavirus disease 2019 (COVID-19), have accumulated multiple narratives. After the rapid escalation of disease transmission (as of 31 May 2020, there were over 6 million cases worldwide and around 370,000 deaths) (Johns Hopkins University 2020), narratives steadfastly spread at local, national, and global levels (Ali 2020c). At local levels, people brainstormed probable causes of COVID-19 and circulated advice about preventions and "cures"—such as "miraculous" tea, shaving their heads, and/or praying to God to undo his "punishment" in Pakistan. At all levels, rumors and conspiracy theories spread about the potential agents behind this virus. These narratives demonstrate geopolitics when some conceptualize and code the pandemic as "bioengineered" or an invented "plot." When these narratives take the form of rumors and conspiracy theories, it is indispensable to document them for a thorough understanding of how people make sense of (what seems) a looming threat and most crucial to question the underlying reasons for these speculations: Why have these narratives started and what do they reveal? I have aimed to come to comprehend how different stakeholders located in diverse places construct, generate meanings, and negotiate outbreaks of disease, notably the 2020 COVID-19 pandemic. In the following section, I begin by exploring the many similarities between viruses and rumors.

37.2 Viral Rumors: Comparing Viruses and Rumors

Although COVID-19 and rumors and conspiracy theories circulating in response to it appear to lie at two different poles, they have significant similarities. As viruses are a reality, so are these competing narratives.

COVID-19 is a viral infection, and viruses always have an origin: they emerge, spread, and sometimes cause a major outbreak. Such is the case with COVID-19, which appeared to have originated from the Huanan Seafood Wholesale Market in Wuhan, China and then was rapidly transmitted from person to person and then countries to continents (Ali 2020c). It has exerted profound effects on people's sociocultural patterns, economic situations, and physical and mental health. It has considerably challenged governments and global stakeholders, such as the World Health Organization (WHO), to deal with it effectively.

For studying the behaviors and patterns of viruses, we have experts such as virologists, microbiologists, and epidemiologists who appropriately investigate them. They explore where exactly a virus broke out, why and how it emerged as well as spread so rapidly, and what are its modes of transmission and patterns of spread. Under what circumstances can it breed further?

Correspondingly, rumors and conspiracy theories also have specificities that need to be critically studied, adequately understood, and amply illustrated. These narratives also emerge in specific situations and diverse places and travel at a pace that is almost as rapid as the virus itself. Like the outbreak of COVID-19, these narratives significantly affect multiple aspects of life and have their own set of researchers like social scientists, especially anthropologists, who can systematically study the patterns, modes, and effects of these narratives (Ali 2020a). Unlike a viral outbreak, it is usually not possible to trace the origins of these narratives; nonetheless, it is perfectly possible to locate in which area they started, under what circumstances they broke out, what meanings they are pregnant with, and what are their short-term and long-term consequences.

Another important commonality shared by viruses and rumors is the near-impossibility of stopping their spread or of completely eradicating them, as indicated in the saying, "The rumor went viral." Various old viruses re-emerge (e.g., measles) from time to time and place to place, and new viruses emerge (e.g., COVID-19). Viruses can be classified into categories due to their similarities, such as the severe acute respiratory syndrome coronavirus (SARS-CoV) family of viruses, of which the COVID-19 is a part. Similarly, rumors can be categorized in terms of scale, e.g., local, national, or global, and the arenas such as health/medical-related. Viruses generate rumors and conspiracy theoriey, and those narratives can affect how people deal with those viruses. During pandemics, rumors and conspiracy theories can affect preparedness programs both adversely and positively.

Two additional commonalities between viruses and these narratives are that both are highly contagious and affect "at-risk" populations. A virus severely affects those who are "weak" in terms of their immune systems, and that "weakness" often stems from their socioeconomic and political situations. That same population is especially vulnerable to the detrimental effects of rumors and conspiracy theories. These factors are critically interdependent and intricately interrelated.

Just as virologists, microbiologists, and epidemiologists can help to understand a virus to contain and deal properly with it, social scientists, especially anthropologists, can help to meticulously understand viral rumors and conspiracy theories, which is necessary to aptly contain, and deal with them. The latter, in particular, involves anthropologists who are highly equipped, well-positioned, and appropriately skilled to study micro-level phenomena (e.g., rumors) and link them to macro-level structures (e.g., economics and politics) (Ali 2020a).

Having described the general interlinkages between viruses and narratives, in the following sections, I briefly mention about methods and then describe and analyze some of the specific narratives surrounding COVID-19 and offer a general and historical overview with an emphasis on health-related narratives.

37.3 Materials and Methods

Concerning cultural understanding of and dealing with health and illnesss, I draw on my previous ethnographic fieldwork that I commenced as early as in 2008. This background data are mainly based on my PhD research that focused on measles and vaccination in Pakistan. Moreover, the data for COVID-19 is based on my project on COVID-19 in Pakistan that is approved by Pakistan's National Bioethics Committee of Pakistan (reference No.4-87/NBC-471-COVID-19-09/20/). Under this project, I did telephone interviews to explore how different stakeholdrs perceive, negotiate, and dealt with COVID-19 at village, provincial, and national levels. The data I analyze herein also come from the content analysis of mainstream social media reports that I began to collect and undertake in January 2020. Moreover, I also have borrowed from the relevant and extensive body of literature, mainly from psychology and anthropology, to provide a historical outline of rumors and conspiracy theories.

37.4 The COVID-19 Pandemic and Infodemic

As the number of COVID-19 cases swiftly escalated, so did the rumors and conspiracy theories (Ali 2020c, d) that led the WHO (2020) to coin a new metaphor "Infodemic" to encapsulate the pandemic of "(mis)information" that rapidly spread via social media and word of mouth. As stated earlier, these narratives give deep meansing if analyzed in a context. Globally, these narratives reveal geopolitics. An Iranian-owned media group, Press TV, publicized a conspiracy theory while disseminating an interview with Dennis Etler, a political analyst and anthropologist. Etler argued that although there is no evidence that the United States has bioengineered the virus, the country has deep interests in affecting China's economy (Ali 2020c). Merely saying that the United States did not bioengineer the virus on media can immediately start a plethora of narratives that it did!

Likewise, China and Russia pointed the finger at the United States (Rogers et al. 2020). For instance, Chinese officials claimed that the United States army had introduced the virus to China (Myers 2020). The Venezuelan President also argued that the United States used the virus as a "bioweapon" for targeting China (Fisher 2020). Turkmenistan and Tajikistan asked their citizens to continue working after calling the virus and its treatment "bogus" (Fisher 2020).

In contrast, Matteo Salvini, the leader of Italy's anti-migrant League Party, contended that the outbreak of the virus is China's doing, insisting that the Chinese have deliberately cultivated a "lung supervirus" from "bats and rats" (Fisher 2020). Analogously, the United States blamed China for the origin and early transmission of the virus: United States Senator Tom Cotton suspected that perhaps China produced the virus in its weapons lab (Fisher 2020)—a geopolitically loaded rumor that went so far that it was being addressed on United States National Public Radio. President Trump started calling the virus a "foreign virus" and "the China Virus" (Rogers et al. 2020), 2020), warned WHO against being so "China-centric," and is cutting funding to the organization, though WHO firmly rejected that criticism (Reuters 2020a, b).

These news reports, social media videos, and stories have accumulated a plethora of comments that discuss, argue, and counter-argue these narratives. For example, the news report on the WHO's rejection of the United States allegations received many online comments (Reuters 2020b). These comments demonstrated how people negotiated the pandemic by finding a "scapegoat." Undoubtedly, people often need somewhere to go with their anger and confusion; blaming and shaming serve that purpose and make the people doing it feel sturdy. The commentators often anonymized their personal threats yet make their points vividly. Needless to say, their accuracy is debatable, which begs the question, who is behind them, and what is their political agenda?

Furthermore, on social media, especially on Facebook and Instagram, people shared speculations that Bill Gates, on behalf of "Big Pharma," is behind the emergence of the 2020 COVID-19 pandemic. People refered to his Technology, Entertainment, and Design (TED) talk of 2015, in which he argued that the next viral outbreak might prove more deadly than war (Zadrozny 2020). Predicting the occurrence of something may lead people to believe that you created that something, perhaps just to prove yourself right.

37.5 Rumors and Conspiracy Theories at Local Levels

In Pakistan, various competing narratives with unknown and untraceable sources emerged, specifically in the villages. These narratives about COVID-19 in that country were not exceptional, given that numerous tales regarding outbreaks and epidemics have long been prevailing there. Various rumors and conspiracy theories prevail about vaccination campaigns—that they are a means of "depopulation" via the sterilization of Muslim women and that hidden stakeholders are behind these campaigns (Ali 2020b, c). About COVID-19, comparable stories surfaced to trace a "hidden" agent: the pandemic was "bioengineered" either by the United States or by "Big Pharma." Speculations also contained home remedies, such as drinking garlic water (which might help) or "blowing hot air from a hairdryer through your nostrils"; the country's health ministry suggests keeping your throat moist (Dawn 2020). Many people followed these (relatively "harmless") suggestions.

In the following section, I describe and analyze six specific narratives that circulated in Pakistan, mainly during the start of the pandemic, from killing the infected person to a talking infant:

- In the early days of the COVID-19 pandemic, in a small village in Sindh Province, out of extreme anxiety and uncertainty was born the rumor that to eliminate the virus, the government is setting infected people ablaze (Ali 2020c).
- In the small towns of Azad Jammu and Kashmir (AJK), a similar rumor surfaced that the government shoots the infected persons.

After hearing these rumors, family members, particularly (older) parents, became extremely concerned about the mobility of the younger generation and sought to ensure that they all stayed at home. In this specific case, we can see that rumors can have positive, instead of solely adverse effects, to deal with the pandemic. Furthermore, as the pandemic grew extraordinary, narratives emerged that linked it to the supernatural scale.

- In a town of Punjab province, the narrative was that the virus was nothing but the punishment of God. This punishment was said to be a result of the opening of cinemas in Saudi Arabia and general disbelief in God in the Global North, where everything is "too open," especially romance. God has shown us how powerful he is in that science is unable to deal with the ongoing pandemic. Because Pakistanis believe in and worship God, their number of cases is far less than in the "Godforsaken" countries of the United States and Europe.
- A vastly different rumor spread mainly in Sindh and Punjab provinces that there is no coronavirus: how would it be possible that one gets infected via shaking hands or standing close to someone? The rumor continued that the government was only imposing lockdowns to receive global attention for potential foreign aid and that there is no danger.
- A widespread rumor broke out in Sindh province, from its district Hyderabad to district Ddaharki, that shaving one's head is a great preventive measures that protects against the virus. As soon as this rumor traveled, many men (the shaving of a woman's head is considered shameful) immediately shaved their heads, considering it a very affordable preventive measure that costs only US5–10 cents. In one village of Sindh, over 50 men shaved their heads.
- An interesting and unconfirmed rumor about an infant emerged in the northern part of Sindh Province. In a matter of hours, it spread in many districts (as I confirmed while talking to my family and friends). The source of the rumor was untraceable, and the forms it took varied by region, although with the same central idea. In district Khairpur, the story was that an infant was born in a nearby village of Gambat to a Syed family (believed to be descendants of the Prophet Muhammad). In Shikarpur district, a boy was born in an unnamed village. In Naushahro Feroze District, a boy was born there. Here is the general content of this rumor:

> After his miraculous birth, the boy started talking. "I will not survive. I am here to tell you something important about the current coronavirus that the disease is deadly. I will die at noon and will bring the coronavirus with me", said the child. It could kill everyone if a recommended measure were not taken. The measure was brewing green tea, and every person should drink five sips. The one who would drink these five sips would survive; the rest would die. "As long as my heart beats, I ask you to please drink tea". After conveying this message, the child died.

After this rumor spread, people started making calls and sending messages via their smart mobile phones to their families and friends. Taking the virus and the cure seriously, people paid in-person visits to their neighbors to convey the message. They woke those who were sleeping and got them to start brewing the tea and drinking five sips.

These six narratives surrounding COVID-19 have actual effects on behavior, both negative and positive. The first two rumors about the government burning or shooting infected people can compel people to refrain from being tested for COVID-19, as well as to take preventive measures to avoid infection.

The third rumor, "an act of God," indexes the religious belief system and gives people hope that the pandemic can be supernaturally dealt with through worshiping God. The fourth rumor that there is no coronavirus reveals the scale of the mistrustful relationship between people and the government. These third and fourth rumors can be seen against the historical background of the country and region: colonization, poverty, and aid dependency.

The fifth rumor, "shaving one's head," is a good example of how a society can come up with an easily accessible "cure" in the absence of an effective healthcare system. It is relatively harmless and seems to index people's desire for disease prevention by any means suggested, as does the sixth rumor "drinking five sips of green tea." This rumor has some basis in fact since studies have shown the positive effects of black or green tea (as well as garlic) on the human immune system, although solely five sips are unlikely to help much.

Moreover, the number "five" has a special meaning within the local belief system, as this digit carries a vital meaning within the local numerology. Many people across the country, specifically those who practice Sufism and Shia-Islam, regard numeral five as lucky and sacred. They link this number to Panjtan Pak (literally, "the Holy Five"), who include the Prophet Muhammad, his cousin and son-in-law Ali, his daughter Fatima, and his grandsons Hassan and Hussain. People believe in their supernatural powers to be advocates to God for seeking help and blessings in normal and especially during extraordinary events. In this way, "drinking five sips of tea" has physical and symbolic powers. Both the properties of tea and the symbolic power of five can exert positive impacts on people's physical and mental bodies.

I argue that the specificity of the instructions can make them seem more important and more valid, perhaps the mark of a society that is pooling its resources in an egalitarian way against a deadly and overwhelming threat via a simple treatment that is available to everyone, including the marginalized and the poor.

37.6 Rumors Revisited

Rumors and conspiracy theories constitute normal social phenomena (Bonhomme 2016). These narratives are universal (Van Prooijen 2018) and part of our sociocultural history and modes of enculturation, especially around threatening situations. Rumors can be interpreted as "brief, oral, non-narrative statement[s] based on hearsay" (Turner 1993) with unknown reliability (VandenBos 2007); as "improvised news" (Shibutani 1966); and as symptoms of psychological diseases such as paranoia, schizophrenia, a phobia, or hysteria. Thus, their study can be based on the principles of social pathology (Reumaux 1994). And, conspiracy theories constitute an extreme version of rumors (Taylor 2019).

Through rumors and conspiracy theories, people link threatening critical events to specific actors who devise "secret plots" for achieving some "vested interests" (Douglas et al. 2017; Van Prooijen 2018). Again, attributing agency gives people someone or something to hold accountable. Although during their continuous transmission processes, the content of these narratives continually changes, the underlying purposes remain to a great extent the same (DiFonzo and Bordia 2007; Allport and Postman 1947). These narratives explain a cause-and-effect relationship: people weigh the situation, explore its leading causes, assess the risks, decide its potential benefits, and make a final decision. Based on their conclusions, people adapt to the required measures.

Rumors and conspirarcy theories play a profound role in dealing with uncertainties and extraordinary events. They can cause further anxieties, fear, suspicions, and social disruption, as well as make more positive effects while helping people to cope with those connotations. In the following section, I present a brief overview of worldwide rumors and conspiracy theories regarding health and illness.

37.6.1 The Outbreak of Contagious Rumors: A Global Overview

As previously noted, rumors and conspiracy theories emerge and rapidly travel during challenging outbreaks, epidemics, and pandemics to help people make sense of them and/or to feel empowered to deal with them.

During the bubonic plague outbreak in 1576, in the Italian city of Padua, a pamphlet was widely spread that said: "Wicked ones (sciagurati) were spreading the disease intentionally with infected clothing, and poisonous ointments (untioni) rubbed on door handles and knockers" (Cohn 2010).

During the Spanish flu pandemic in 1918, several narratives emerged to trace the probable cause. In the United States, a rumor had it that a Bayer pharmaceutical company poisoned people with flu while testing aspirin tablets, resulting in the viral outbreak (Crosby 2003), or a "nurse of German extraction" supposedly distributed "germs in the hospital wards of Camp Meade and was later executed" (Schoch-Spana 2004). To express anti-German sentiments, people regarded the pandemic "as an immoral, murderous 'Hun of a disease' that German spies spread in the United States." In England, the rumor was similar that the Spanish flu was "directly traceable to the German use of poison gas, the after-effects of which have induced the growth of [a] new type of streptococcus" (Johnson 2006).

In 1987, speculations emerged that the AIDS pandemic was a bioengineered weapon of the Central Intelligence Agency (CIA) to eliminate homosexuality or to commit racial genocide by eradicating African-Americans (Turner 1993; Heller 2015). In Indonesia, widespread speculations arose that people in crowded malls and elevators were attacked with a needle and left with a note, "Welcome to the AIDS Club" (Kroeger 2003). In New York City's Chinatown during the 2003 SARS epidemic, speculation circled that the virus had infected Americans, including many in Chinatown (Eichelberger 2007), and in Hong Kong, one news item emerged that the region soon would be declared SARS-infected and shut off from the world (Cheng and Cheung 2005). In Australia, health officials were compelled to circumvent conspiracy theories about the SARS outbreak that it was an act of terrorism or biological warfare executed by some malevolent country or agency (Lee 2014). Similarly, in 2009, regarding the swine flu, a rumor held that terrorists used infected Mexicans to immigrate to the United States as "walking germ warfare weapons" (Smallman 2015). Another rumor held that a biotech company created the outbreak of the Zika virus in 2016 after genetically modifying mosquitoes for combating dengue fever (Jacobs 2016).

Moreover, multiple rumors exist about human bodies across the world: collecting human heads as forms of sacrifice in Eastern Indonesia (Stewart and Strathern 2004); stealing organs after killing people in Brazil (Scheper-Hughes 1996, 2000); foreigners supposedly abducting local children to make them sex slaves or to steal their organs in Guatemala (Dubinsky 2010); "whites" killing volunteers while conducting clinical trials to steal their blood in South Africa (Saethre and Stadler 2013); and "genital thieves" or "sex thieves" who were stealing genitals in the streets while working via witchcraft—once found, they were publicly attacked and lynched (Bonhomme 2016).

Family planning programs have also attracted numerous rumors and dconspiracy theories worldwide. For instance, rumors proliferated across the world that some stakeholders want to eliminate "undesirable" populations (Ali 2020b). The medical/health sector is said to play a pivotal role in that population control scheme via producing new viruses and vaccines to achieve the stated goals (Horowitz 1997). This perception has roots in the notion that economically underprivileged people refuse to engage in family planning and could destroy the environment via their explosive population growth (Hardin 1968). After the support of the United States president, Lyndon Johnson, in 1965, a wide range of family planning programs was initiated, particularly in Asia (Lock and Nguyen 2018), on a massive scale, in a "Cold War" approach which held that the populations of the then-called "Third World" countries might take over the entire world (Lock and Nguyen 2018).

This narrative further paved the way for various narratives about a "New World Order." For instance, during the Zika virus outbreak in 2016, a rumor held that a global elite, members of a mysterious organization called the "New World Order," designed a "plot" to control the population (Jacobs 2016). The SARS outbreak was linked to this "plot" (Lee 2014). People often link vaccination with infertility in women (Feldman-Savelsberg et al. 2000; Andrade and Hussain 2018; Renne 2010), vaccination as a "Western plot" or "conspiracy" to sterilize Muslim women (Hussain et al. 2016; Ali 2020b, c), and vaccine side effects causing deaths of children (Ali 2020b). In the United States, mass sterilizations of Puerto Rican and Native American women have occurred (Lopez 1993). Another rumor that vaccination campaigns were a "Western plot" turned into "reality" when a fake vaccinator was sent to discover the hideout of Osama bin Laden (Ali 2020b, c). Since this was a covert vaccination drive with ulterior motives, people began suspecting vaccination teams as "spies" and started attacking and killing them (Andrade and Hussain 2018; McGirk 2015).

37.6.2 Rumors and Conspiracy Theories: Reasons and Modes of Travel

While devoting considerable attention to document rumors and conspiracy theories, it is crucial to ask a simple question: why do these stories emerge and receive wide and significant attention? The simple answer is because, as previously noted, these narratives provide meanings of an uncertain phenomenon and empower people to feel some sense of control in the face of frightening situations. Besides, these narratives are mysterious and captivating (Van Prooijen 2018) and can bring excitement to people's lives. More complex answers involve specific motives that lie behind people's desire to believe. These encompass epistemic factors that are required to understand one's environment sufficiently; the existential need to feel safe and in control of one's environment; and the social need to maintain a positive image of oneself and one's in-group (Douglas et al. 2017).

These properties lead narratives to travel rapidly via major three modes: through the transmission of information via social media or word of mouth, personal experiences, and observational learning (Goumon and Špinka 2016; Barlow 2004). Prior to the presence of mass media and social media, word of mouth was the primary source of rumor transmission. This historical source of circulation still prevails in those societies where access to media is low (Ali 2020b).

These narratives may make people feel special that they possess some critical and rare information that enhances their self-esteem (Taylor 2019). Some people deliberately craft specific rumors and conspiracy theories, and others produce counter-narratives; hence, in a way, both equally participate in their spread (Douglas et al. 2017); to generate a counter-narrative to a given conspiracy theory is to give attention to and to spread that theory further. Some people are more prone to such speculations, especially those who believe in other similar narratives (Bruder et al. 2013; Douglas et al. 2019; Galliford and Furnham 2017; Lewandowsky et al. 2013). For example, those who believe that Zika was a scheme of Monsanto were likely also to believe that someone from inside was involved in 9/11 and that the National Aeronautics and Space Administration (NASA) moon landing never happened (Taylor 2019). Analogously, in Pakistan, those who believed in rumors and conspiracy theories about family planning and vaccination are likely to also believe in rumors related to COVID-19.

Rumors and conspiracy theories index ambiguities (Butt 2005), convey coalescences of meaning, and reveal interconnections and interplays between local and global contexts. The speculations grow successfully in specific sociocultural, economic, and political contexts (Douglas et al. 2017). Their circulation continues in waves of what I term "the provision of meaning." Rumors and conpiracy theories tend to occur in multiple versions and can constitute what Scheper-Hughes (1996) refers to as "global mass hysteria." They describe the relationships among (geo) political and economic low-power and high-power groups. These narratives can explicate and indicate the politics of inequality (Butt 2005).

Therefore, their meanings must be deciphered within their historical, sociocultural, economic, and (geo)political contexts to be fully understood (Kingori et al. 2010). For example, refusals of anti-tetanus vaccination in Cameroon have roots in colonization and mistrust in governments (Feldman-Savelsberg et al. 2000). The emergence of the "AIDS Club" rumor in Indonesia has roots in the sociocultural and political realities of the 1990s (Kroeger 2003). Suspicions of vaccination in Pakistan are entrenched in sociocultural, economic, historical, and (geo)political contexts (Ali 2020b, c).

Rumors and cospiracy theories can illuminate a lack of power and can allow the less powerful to express their fears. When local people see themselves as powerless against the dominant institutions, e.g., the government or international non-governmental organizations (INGOs), they spin different narratives (Bass 2008) to meet an important objective. Such narratives assist people to "create an alternative public sphere and maneuver to speak in oppositional voices" (Perice 1997).

Conversely, rumors and conspiracy theories can be created and spread by people who enjoy a great deal of power and wish to use that power to promote their agendas. For instance, during this ongoing global coronavirus pandemic, Donald Trump considered the coronavirus as a "foreign virus" that "will not have a chance against us" (Woodward 2020). He criticized Europe for failing to control the outbreak and argued that the United States' ban on China plays a role in containing the virus. Some regarded his views as xenophobic and as excuses to shift the blame. The controversies and criticisms grew over time as cases and deaths caused by the COVID-19 startlingly escalated.

In April 2020, due to his approach and remarks, many criticized Trump for his self-promotion when he claimed: "When somebody is the president of the United States, the authority is total" (Parker 2020). After this media briefing—in which he made claims like "Everything we did was right"—media analysts competed in insisting that Trump's paramount concern is not the virus but himself (Parker 2020). Afterward, Trump played a "propaganda-style" biographic video, in which his supporters were praising him: "The president has been outstanding through all this," stated Republican Florida Governor Ron DeSantis (Parker 2020).

In addition to giving power to the powerless and more power to those already enjoying it, such narratives can help to negotiate a given phenomenon. For example, in South Africa, trial staff, community, and trial participants negotiated clinical trials for assessing a microbicide vaginal gel to prevent HIV in women: white kill the participants and sell their blood; women participate for the financial gains, and volunteers participate for health and aid (Stadler and Saethre 2010). In South Africa, while defying the dominant epistemologies of HIV/AIDS, people articulate it through rumors when elders consider the disease as curable, and the younger generation regards it as "foreign diseases" (Stadler 2003). People contested Ghanaian national identity through the spread of the Sakawa rumor, in which Sakawa boys were said to have manipulated evil occult powers to enter into the Internet and perform fraud (Armstrong 2016). This act of observing rituals to commit fraud was negotiated by the Ghanaians while linking to the national identity. The rumor was:

> Sakawa boys sleep in coffins or do not wash for weeks; some even kill a small girl and eat her like fufu (a national dish)! They do whatever these Sakawa leaders tell them to do, and then they have the ju-ju power to do their evil tricks. Then their spirit can enter the internet, possessing the obruni (White person) to get their money! It is evil, selfish behavior, and they bring shame to Ghana. Kwame, Abiriw, Ghana, 07 August 2009 (Armstrong 2016)

Although rumors and conspiracy theories can be true or false, or half truth and half lie (Taylor 2019), they are always a form of communication, and they can convey hidden and vital meanings, reflecting beliefs and perceptions regarding the workings of the world, and collective explanations and illustrations about complex circumstances (Kroeger 2003).

These narratives often have their roots in social institutions (Heath et al. 2005). They can function to mobilize crowds (Rudé 1981; Thompson 1971) and to unite rioting masses (Guha 1999). Rumors differ from other types of communication because of their enunciative and performative functions (Bhabha 1994). Due to their flexibility, rumors work as a practical mode to reflect upon hard economic and political situations (Butt 2005). In rumors (and conspiracy theories), words transform from a medium of communication to an apparatus of force (Das 1998). An example is of the 900,000 people in Brazil who became a protesting crowd when a rumor spread about the closure of the Bolsa Familia Program (the Brazilian government's social program that provides financial support to economically poor people) (Morton 2014).

Raymond Firth (1955) explored two Tikopian rumors, one attributing extraordinary power to a person after death and another about the destruction of Honiara by a hurricane. These rumors functioned as a prototype of the cargo-cult in Tikopia and succeeded promptly in the case of severe dangers, e.g., famine. Rumors can disappear and reappear during similar circumstances but with prominent modifications, e.g., additions and subtractions, such as in Indonesia (Forth 2009).

For a rumor or a conspiracy theory to have powerful implications, it does not matter whether it is true or false. A rumor containing private information can be false, and so can "official" information (Bonhomme 2016). For their believers, a rumor or a conspiracy theory always counts as real and meaningful. Perhaps their credibility is unrelated to these narratives but rather to the credibility of the tellers. The authentication of anything can be challenged.

37.6.3 (Mis)Trust as Pretext

Rumors and conspiracy theories also serve as analytical windows into institutionalized forms of inequalities or what Farmer (1996) terms "structural violence" and what Kleinman et al. (1997) call "social suffering." In Nigeria as elsewhere, rumors help to understand "the origins of inequality and the structure of power" existing in their society (Smith 2001). Behind these narratives, there is always an invisible interplay between trust and mistrust. This trust deficit may occur between citizens and governments, between people and global stakeholders (e.g., big corporations), and between or among countries. People suspect that the government withholds valuable information to make secret deals or that power strings are in invisible hands and encode such suspicions in rumors (Kroeger 2003).

Hence, the next question is what lies underneath these phenomena of (mis)trust? Following this line of inquiry, we should explore several economic and political factors that have occurred throughout our social and biological history. For example, the suspicions surrounding vaccination programs in Pakistan have roots in regional history, British colonization, the Cold War, the USSR invasion, and contemporary geopolitics (Ali 2020a, b). These "critical events" (in the terminology of Veena Das (1995)) have germinated the seeds of suspicions about the vaccine, the government, and global actors (Ali 2020b, c; McGirk 2015). Sentiments of anti-tetanus vaccination in Cameroon show the links with colonization and mistrust in the government (Feldman-Savelsberg et al. 2000); narratives concerning the "AIDS Club" in Indonesia have roots in the sociocultural and political realities of the 1990s (Kroeger 2003). Likewise, local resentment against polio eradication efforts in Northern Nigeria has a sociocultural, political, and historical context, which shaped the local response to suspect the vaccine's quality, mistrust the government, and regard politicians as corrupt (Renne 2010; Ali 2020c).

Analogously, the current rumors and conspiracies surrounding COVID-19 and spinning at the global level reveal unceasing geopolitics. As we have seen, the big superpowers like China, the United States, and Russia are actively engaged in generating and spreading rumors and conspiracy theories. Other countries, such as Iran, Italy, and some Latin American countries, are also participating, either as supporters or adversaries. This scale of negotiation and contestation embodies the historical processes of the World Wars and the Cold Wars. While China and Russia are blaming the United States, the United States is blaming China for the origin and spread of this virus. This embodiment signifies a "win-or-lose" situation. Countries strategize to win their old battles.

37.7 Summary

Whenever extraordinary uncertainties and threats to life surface, rumors and conspiracy theories quickly spread. If the new phenomena resemble an old one, then the new narratives also resemble the older versions, but with specific modifications. These narratives are always preserved in every society's "societal memory" (Ali 2020b), which should best be seen within sociocultural, economic, and (geo-)political contexts. Each society accumulates information—about every phenomenon, including the potential threats, enemies, counterstrategies, and solutions—and converts it into collective knowledge. As soon as similar circumstances and reasons emerge, society releases its "antibodies," i.e., rumors, to deal with them.

Throughout history, rumors and conspiracy theories have surrounded pandemics and medical interventions. The vaccination programs are good examples to cite, as they have remained the center of multiple rumors and conspiracy theories rooted in historical events such as the "Cold War" and colonization.

COVID-19 is no exception, as a novel coronavirus caused it, and by the writing of this chapter, researchers were still unraveling its characteristics. The escalation worldwide was so rapid and worrying, and the media presented it as "an invisible enemy" that has severely "threatened" the world, which seems at "war." These unusual metaphors demonstrated overwhelming circumstances, which have generated massive uncertainties and anxieties and sparked a plethora of narratives. The narratives either were pregnant with a potential solution or probable "enemy" behind it while causing further panic amid uncertainty.

In Pakistan, narratives surrounding the 2020 COVID-19 pandemic under circulation at local levels might qualify as societal coping mechanisms to avoid the threat of becoming either infected or deceived. Most notably, these rumors and conspiracy theories ran rampant in those places where structural inequalities prevail. Considering shaving heads, drinking five sips of green tea, and worshipping God as preventive measures or potential cures decreases people's fear of infection and made them feel that if they were infected, they could easily be treated. Regarding the virus as a conspiracy and not something real, like interpreting the pandemic as the government's ticket to receiving foreign aid, might have removed the massive burden of fear and decreases people's worries that there was no danger in going out to earn their livelihoods. In sum, almost all of the above narratives give power back to people to do something that can turn away an extraordinary existential and physical threat.

37.8 Conclusion

Stories and narratives are part of our human sociocultural history. It is indispensable to understand that constructing stories, weaving meanings, and deciphering them are the long-standing human social phenomena. We cannot survive without narratives. Functioning as building blocks of society and as means of enculturation, these narratives help us to encode crucial information that is necessary for biological, sociocultural, emotional, psychological, economic, and—the most important—political survival. Such narratives rapidly surround a new and uncertain phenomenon, as it has happened in the case of COVID-19 that has caused exceptional circumstances. Considerable efforts were taken and (are still underway) to understand the virus better. Specialized scientists unceasingly unravelled its characteristics, origin, mechanism of spread, and "high-risk" populations, to counter it via effective preventive measures and develop medicines, including an appropriate vaccine.

Similarly, it was revealed that studying viral rumors and conspiracy theories is as crucial as researching the coronavirus to lucidly comprehend the content, sources, modes of spread, and impacts of these narratives. Focus can be on sociocultural, economic, and (geo-)political factors that lie underneath these stories. One can ask why do they spread, in which circumstances and in which parts of the world, and who are their beneficiaries?

As an inadequate understanding of COVID-19 hinders effectively dealing with it at the medical level, so does lack of understanding of the rumors and conspiracy theories surrounding it impede dealing with this pandemic at societal levels. Just as the scientific inquiries of the virus, these narratives beg the same paramount attention. To study and analyzed their causes and implications, anthropologists are well-positioned who need to be engaged by governments and global stakeholders, such as WHO.

It is hoped that the present chapter of rumors and conspiracy theories surrounding COVID-19 will empower concerned stakeholders to understand these narratives' sources and effects, both negative and positive, and to find more effective means of spreading truths that can also be encoded in narratives. Furthermore, this chapter will generate debates on narratives—our most popular and long-standing form of information dissemination.

Acknowledgments The author would like to thank Robbie Davis-Floyd for her immense help in terms of thorough reading, critical comments, and invaluable suggestions. Although the author received no specific funding for this article, he acknowledges the Higher Education Commission (HEC) of Pakistan's grant (PD/OSS-II/Batch-IV/Austria/2012/9903), which supported Ph.D. work that has significantly informed this article.

References

Ali I (2020a) Anthropology in emergencies: the roles of anthropologists during the COVID-19 pandemic. Pract Anthropol 42(3):16–22

Ali I (2020b) Constructing and negotiating measles: the case of Sindh Province of Pakistan. University of Vienna, Vienna

Ali I (2020c) The COVID-19 pandemic: making sense of rumor and fear. Med Anthropol 39(5):376–379. https://doi.org/10.1080/01459740.2020.1745481

Ali I (2020d) COVID-19: are we ready for the second wave? Disaster Med Public Health Prep:1–3. https://doi.org/10.1017/dmp.2020.149

Allport GW, Postman L (1947) The psychology of rumor. Holt, New York

Andrade GE, Hussain A (2018) Polio in Pakistan: political, sociological, and epidemiological factors. Cureus 10(10):e3502–e3502. https://doi.org/10.7759/cureus.3502

Armstrong A (2016) 'Sakawa' Rumours: occult Internet fraud and Ghanaian identity. In: Littlewood R, Lynch R (eds) Cosmos, Gods and madmen: frameworks in the anthropologies of medicine. Berghahn, New York, pp 67–92

Barlow DH (2004) Anxiety and its disorders: the nature and treatment of anxiety and panic. Guilford Press, New York

Bass D (2008) Paper tigers on the prowl: rumors, violence and agency in the up-country of Sri Lanka. Anthropol Q 81(1):269–295

Bhabha HK (1994) By bread alone: signs of violence in the mid-nineteenth century. In: Bhabha HK (ed) The location of culture. Routledge, London, pp 198–211

Bonhomme J (2016) The sex thieves: the anthropology of a rumor (trans: Horsfall D). Hau Books, Chicago

Bruder M, Haffke P, Neave N, Nouripanah N, Imhoff R (2013) Measuring individual differences in generic beliefs in conspiracy theories across cultures: conspiracy mentality questionnaire. Front Psychol 4:225

Butt L (2005) "Lipstick Girls" and "Fallen Women": AIDS and conspiratorial thinking in Papua, Indonesia. Cult Anthropol 20(3):412–441

Cheng C, Cheung MW (2005) Psychological responses to outbreak of severe acute respiratory syndrome: a prospective, multiple time-point study. J Pers 73(1):261–285

Cohn SK (2010) Cultures of plague: medical thinking at the end of the renaissance. Oxford University Press, Oxford

Crosby AW (2003) America's forgotten pandemic: the influenza of 1918. Cambridge University Press, Cambridge

Das V (1995) Critical events: an anthropological perspective on contemporary India. Oxford University Press, New Delhi

Das V (1998) Specificities: official narratives, rumour, and the social production of hate. Soc Identities 4(1):109–130. https://doi.org/10.1080/13504639851915

Dawn (2020) Covid-19 misinformation Dawn, 31 March, p 06

DiFonzo N, Bordia P (2007) Rumor psychology: social and organizational approaches. American Psychological Association, Washington, DC

Douglas KM, Sutton RM, Cichocka A (2017) The psychology of conspiracy theories. Curr Dir Psychol Sci 26(6):538–542

Douglas KM, Uscinski JE, Sutton RM, Cichocka A, Nefes T, Ang CS, Deravi F (2019) Understanding conspiracy theories. Polit Psychol 40:3–35

Dubinsky K (2010) Babies without borders: adoption and migration across the Americas. University of Toronto Press, Toronto

Eichelberger L (2007) SARS and New York's Chinatown: the politics of risk and blame during an epidemic of fear. Soc Sci Med 65(6):1284–1295

Farmer P (1996) On suffering and structural violence: a view from below. Race Ethn Multidiscipl Glob Contexts 3(1):11–28

Feldman-Savelsberg P, Ndonko FT, Schmidt-Ehry B (2000) Sterilizing vaccines or the politics of the womb: retrospective study of a rumor in Cameroon. Med Anthropol Q 14(2):159–179

Firth R (1955) The theory of 'Cargo' cults: a note on Tikopia. Man 55(9):130–132

Fisher M (2020) Why coronavirus conspiracy theories flourish. And why it matters. New York Times, 08 April

Forth G (2009) Heads under bridges or in mud: reflections on a southeast Asian 'Diving Rumour'. Anthropol Today 25(6):3–6

Galliford N, Furnham A (2017) Individual difference factors and beliefs in medical and political conspiracy theories. Scand J Psychol 58(5):422–428

Goumon S, Špinka M (2016) Emotional contagion of distress in young pigs is potentiated by previous exposure to the same stressor. Anim Cogn 19(3):501–511

Guha R (1999) Elementary aspects of peasant insurgency in colonial India. Duke University Press, Duke

Hardin G (1968) The tragedy of the commons. Science 162(3859):1243–1248

Heath C, Campion-Vincent V, Fine GA (2005) Rumor and legend: seven questions. In: Fine GA, Vincent VC, Heath C (eds) The social impact of rumor and legend. Transaction Press, Rumor Mills, pp 255–265

Heller J (2015) Rumors and realities: making sense of HIV/AIDS conspiracy narratives and contemporary legends. Am J Public Health 105(1):e43–e50

Horowitz LG (1997) Emerging viruses: AIDS and Ebola: nature, accident, or intentional? Healthy World, Dist

Hussain SF, Boyle P, Patel P, Sullivan R (2016) Eradicating polio in Pakistan: an analysis of the challenges and solutions to this security and health issue. Glob Health 12(1):63

Jacobs A (2016) Conspiracy theories about Zika spread through Brazil with the virus. The New York Times, 16 February

Johns Hopkins University (2020) Coronavirus COVID-19 global cases. Johns Hopkins University. https://gisanddata.maps.arcgis.com/apps/opsdashboard/index.html#/bda7594740fd40299423467b48e9ecf6. Accessed 22 Apr 2020

Johnson N (2006) Britain and the 1918–19 influenza pandemic: a dark epilogue. Routledge, New York

Kingori P, Muchimba M, Sikateyo B, Amadi B, Kelly P (2010) 'Rumours' and clinical trials: a retrospective examination of a paediatric malnutrition study in Zambia, southern Africa. BMC Public Health 10(1):556. https://doi.org/10.1186/1471-2458-10-556

Kleinman A, Das V, Lock M (1997) Social suffering. University of California press, Berkeley

Kroeger KA (2003) AIDS rumors, imaginary enemies, and the body politic in Indonesia. Am Ethnol 30(2):243–257

Lee JD (2014) An epidemic of rumors: how stories shape our perception of disease. University Press of Colorado, Boulder

Lewandowsky S, Oberauer K, Gignac GE (2013) NASA faked the moon landing—therefore,(climate) science is a hoax: an anatomy of the motivated rejection of science. Psychol Sci 24(5):622–633

Lock M, Nguyen V-K (2018) An anthropology of biomedicine. John Wiley & Sons, New York

Lopez I (1993) Agency and constraint: sterilization and reproductive freedom among Puerto Rican women in New York City. Urban Anthropol Stud Cult Syst World Econ Dev 22(3/4):299–323

McGirk T (2015) How the bin laden raid put vaccinators under the gun in Pakistan Polio's surprising comeback:1

Morton GD (2014) Protest before the protests: the unheard politics of a welfare panic in Brazil. Anthropol Q 87(3):925–933

Myers SL (2020) China spins tale that the U.S. Army started the Coronavirus Epidemic. New York Times, 13 March

Parker A (2020) 'Everything we did was right': trump accused of using coronavirus briefing to focus on himself. The Independent, 14 April

Perice GA (1997) Rumors and politics in Haiti. Anthropol Q 70(1):1–10. https://doi.org/10.2307/3317797

Reck GG (1983) Narrative anthropology. Anthropol Humanism 8(1):8–12

Renne EP (2010) The politics of polio in Northern Nigeria. Indiana University Press, Bloomington

Reumaux F (1994) Toute la ville en parle: Esquisse d'une théorie des rumeurs. Editions L'Harmattan, Paris

Reuters (2020a) Trump says WHO is China-centric, 'really blew it' on coronavirus; threatens to freeze US funding Dawn, 08 April

Reuters (2020b) WHO denies being 'China-centric' after Trump comment, threat to cut funding Dawn, 08 April

Rogers K, Jakes L, Swanson A (2020) Trump defends using 'Chinese Virus' label, ignoring growing criticism. New York Times, 18 March

Rudé GF (1981) The crowd in history: a study of popular disturbances in France and England, 1730–1848. Lawrence & Wishart Limited, London

Saethre E, Stadler J (2013) Malicious whites, greedy women, and virtuous volunteers: negotiating social relations through clinical trial narratives in South Africa. Med Anthropol Q 27(1):103–120

Scheper-Hughes N (1996) Theft of life: the globalization of organ stealing rumours. Anthropol Today 12(3):3–11. https://doi.org/10.2307/2783143

Scheper-Hughes N (2000) The global traffic in human organs. Curr Anthropol 41(2):191–224. https://doi.org/10.1086/300123

Schoch-Spana M (2004) Lessons from the 1918 pandemic influenza: psychosocial consequences of a catastrophic outbreak of disease. In: Ursano RJ, Norwood AE, Fullerton CS (eds) Bioterrorism: psychological and public health interventions. Cambridge University Press, New York, pp 38–55

Shibutani T (1966) Improvised news: a sociological study of rumor. Ardent Media, London

Smallman S (2015) Whom do you trust? doubt and conspiracy theories in the 2009 influenza pandemic. J Int Glob Stud 6(2):1–24

Smith DJ (2001) Ritual killing, 419, and fast wealth: inequality and the popular imagination in Southeastern Nigeria. Am Ethnol 28(4):803–826

Stadler J (2003) Rumor, gossip and blame: implications for HIV/AIDS prevention in the South African lowveld. AIDS Educ Prev 15 (4: Special issue):357–368

Stadler J, Saethre E (2010) Rumours about blood and reimbursements in a microbicide gel trial. Afr J AIDS Res 9(4):345–353. https://doi.org/10.2989/16085906.2010.545636

Stewart PJ, Strathern A (2004) Witchcraft, sorcery, rumors and gossip. Cambridge University Press, Cambridge

Taylor S (2019) The psychology of pandemics: preparing for the next global outbreak of infectious disease. Cambridge Scholars Publishing, Cambridge

Thompson EP (1971) The moral economy of the English crowd in the eighteenth century. Past Present 50(1):76–136

Turner PA (1993) I heard it through the grapevine: rumor in African-American culture. University of California Press, Berkley

Van Prooijen J-W (2018) The psychology of conspiracy theories. Routledge, Abingdon

VandenBos GR (2007) APA dictionary of psychology. American Psychological Association, Washington, DC

WHO (2020) Coronavirus disease 2019 (COVID-19) Situation Report – 13. Coronavirus disease 2019 (COVID-19). World Health Organisation, Geneva

Woodward A (2020) 'The virus is already here': Trump accused of xenophobia and shifting blame in historic coronavirus address. Independent, 12 March

Zadrozny B (2020) Coronavirus conspiracy video spreads on Instagram among black celebrities. National Broadcasting Company (NBC) News, 14 March

Exploration of the Epidemiological and Emotional Impact of Quarantine and Isolation During the COVID-19 Pandemic

38

Helia Mojtabavi, Nasirudin Javidi, Anne-Frédérique Naviaux, Pascal Janne, Maximilien Gourdin, Mahsa Mohammadpour, Amene Saghazadeh, and Nima Rezaei

Abstract

Starting in December 2019 in Wuhan Municipal Health Commission, the coronavirus disease 2019 (COVID-19) has crossed the borders forming a pandemic in 2020. The absence of pharmacological interventions has pushed governments to apply different sets of old, non-pharmacological interventions, which are, though temporary, helpful to prevent further pandemic propagation. In the context of COVID-19, research confirms that quarantine is useful, mainly if applied early and if combined with other public health measures. However, the efficacy of quarantine and isolation is limited in many ways, ranging from legal issues and suspension of economic activities to mental health considerations. This chapter is an exploration of (i) epidemiological impact of isolation and quarantine; (ii) emotional impact of isolation and quarantine; and (iii) the possible effect of culture on quarantine experience.

H. Mojtabavi
Research Center for Immunodeficiencies, Tehran University of Medical Sciences, Tehran, Iran

MetaCognition Interest Group (MCIG), Universal Scientific Education and Research Network (USERN), Tehran, Iran

N. Javidi
Behavioral Sciences Research Center, Life style institute, Baqiyatallah University of Medical Sciences, Tehran, Iran

Clinical Psychology and Psychotherapy Studies (CPPS), Universal Scientific Education and Research Network (USERN), Tehran, Iran

A.-F. Naviaux
College of Psychiatrists of Ireland and Health Service Executive (HSE), Summerhill Community Mental Health Service, Summerhill, Wexford, Ireland

Faculty of Medicine, Université catholique de Louvain, Woluwé-Saint-Lambert, Belgium

P. Janne
Université catholique de Louvain, CHU UCL Namur, Avenue Dr. G. Thérasse, Yvoir, Belgium

Faculty of Psychology, Université catholique de Louvain, Ottignies-Louvain-la-Neuve, Belgium

M. Gourdin
Faculty of Medicine, Université catholique de Louvain, Woluwé-Saint-Lambert, Belgium

Université catholique de Louvain, CHU UCL Namur, Avenue Dr. G. Thérasse, Yvoir, Belgium

M. Mohammadpour
MetaCognition Interest Group (MCIG), Universal Scientific Education and Research Network (USERN), Tehran, Iran

N. Rezaei (ed.), *Coronavirus Disease - COVID-19*, Advances in Experimental Medicine and Biology 1318, https://doi.org/10.1007/978-3-030-63761-3_38

Keywords

Anxiety · COVID-19 · Emotional regulation · Iran · Quarantine · Social isolation

38.1 Introduction

In late December 2019, the world faced the problem of a novel coronavirus disease (COVID-19); its spread was as rapid as it was declared as a pandemic by the World Health Organization (WHO) on March 11, 2020 (Hanaei and Rezaei 2020). As of writing this chapter, all countries worldwide have experienced the first wave, some are undergoing the second wave, and the others are aware and possibly preparing for the second wave of the COVID-19 pandemic. About 6 months have passed since its first report, and it has affected nine million cases and claimed more than 460,000 lives worldwide. Such a high number of deaths state explicitly how much the SARS-CoV-2 is violent, and multiple routes of transmission might explain its violence (Yazdanpanah et al. 2020b; Jahanshahlu and Rezaei 2020a; Lotfi et al. 2020; Saleki et al. 2020). Genetic factors and medical conditions can intensify infection severity as well (Yousefzadegan and Rezaei 2020; Darbeheshti and Rezaei 2020; Shamshirian and Rezaei 2020; Ahmadi et al. 2020; Sahu et al. 2020). The current literature provides a library of evidence of inflammation in people with severe COVID-19 along with reports of a relatively lower susceptibility to develop COVID-19 in people lacking immunocompetence (Ahanchian et al. 2020; Bahrami et al. 2020; Babaha and Rezaei 2020). Totally, these findings have made doubt on the role the immune system truly plays against COVID-19 (Rokni et al. 2020; Yazdanpanah et al. 2020a).

There is neither vaccine for prevention of disease nor a specific treatment, other than supportive care, and previously identified options for dealing with viral pneumonia that causes immune dysregulation (Rokni et al. 2020; Bahrami et al. 2020; Yazdanpanah et al. 2020a; Saghazadeh and Rezaei 2020a; Fathi and Rezaei 2020; Lotfi and Rezaei 2020; Nasab et al. 2020) progressing to acute respiratory distress syndrome and multiorgan failure, e.g., antivirals (Mohamed et al. 2020b) and immunosuppressants (monoclonal antibodies, corticosteroids, plasma therapy, and intravenous immunoglobulin) (Saghazadeh and Rezaei 2020b; Mohamed et al. 2020b; Lotfi et al. 2020; Pourahmad et al. 2020; Jahanshahlu and Rezaei 2020b). However, a number of promising opportunities appear for specific treatment of COVID-19. They include therapies that can precisely target signaling pathways engaged in hyper-inflammation and viral entry into the cells as well as cell therapies that can help enhance antiviral immunity while attenuating inflammatory responses (Sharifkashani et al. 2020; Mansourabadi et al. 2020; Pashaei and Rezaei 2020; Basiri et al. 2020b; Rezaei 2020b).

The absence of pharmacological interventions has made the health-care system to rely on early diagnosis of infection, which is, in turn, challenging and needs more reliable techniques (Basiri et al. 2020a; Rabiee et al. 2020), to provide an extended window for care delivery. However, COVID-19 has affected the health-care providers as well (Rezaei 2020a; Moazzami et al. 2020), and this has pushed governments to apply different sets of old, non-pharmacological interventions, based on societal measures and behavior

A. Saghazadeh
Research Center for Immunodeficiencies, Tehran University of Medical Sciences, Tehran, Iran

Systematic Review and Meta-analysis Expert Group (SRMEG), Universal Scientific Education and Research Network (USERN), Tehran, Iran

N. Rezaei (✉)
Research Center for Immunodeficiencies, Tehran University of Medical Sciences, Tehran, Iran

Department of Immunology, School of Medicine, Tehran University of Medical Sciences, Tehran, Iran

Network of Immunity in Infection, Malignancy and Autoimmunity (NIIMA), Universal Scientific Education and Research Network (USERN), Tehran, Iran
e-mail: rezaei_nima@tums.ac.ir

adjustment, including cordon sanitaire, traffic restriction, social distancing, home confinement, centralized quarantine, and universal symptom survey, which are, though temporary, helpful to prevent infection and re-infection with the SARS-CoV-2 pandemic (Pan et al. 2020; Jabbari and Rezaei 2020; Jabbari et al. 2020).

Quarantine refers to the restriction of the movement and separation of people who are not ill and may have been exposed to an infectious disease, while isolation means separation of people who are ill and have been infected. Both quarantine and isolation aim at controlling the spread of communicable diseases, such as cholera, diphtheria, tuberculosis, plague, smallpox, yellow fever, and viral hemorrhagic fevers (Parmet and Sinha 2020). In the context of COVID-19, research confirms that quarantine is useful, mainly if applied earlier and combined with other public health measures (Nussbaumer-Streit et al. 2020). Thus, several strategies were implemented, including cocooning, social distancing, movement restricting, and self-isolating. However, the efficacy of quarantine and isolation is limited in numerous ways, ranging from legal issues and suspension of economic activities – which will be discussed in Chaps. 40, 41, 42 – to mental health considerations. This chapter is an exploration of (i) epidemiological impact of isolation and quarantine; (ii) emotional impact of isolation and quarantine; and (iii) the possible effect of culture on quarantine experience.

38.2 Factors Determining the Epidemiological Impact of Quarantine and Isolation During the COVID-19 Outbreak

38.2.1 Combination with Other Testing and Tracing

Using mathematical modeling of data from UK participants ($n = 40{,}162$), the study estimated that each of random mass testing of 5% of the population each week, self-isolation of symptomatic cases within the household, and self-isolation outside the household would reduce mean transmission of disease by about 2%, 29%, and 35%, respectively. Self-isolation effectiveness increased to 37%, 47%, and 57% when combined with household quarantine, manual contact tracing of acquaintances, and app-based contact tracing. Furthermore, it reached a high value of 64% in combination with both household quarantine and manual contact tracing of acquaintances (Kucharski et al. 2020).

As summarized in a systematic review (Lahiri et al. 2020), the most commonly used models for prediction of the behavior of COVID-19 pandemic included susceptible exposed infected recovery (SEIR), susceptible infected recovered (SIR), and neural network models and their modified versions. The review highlighted that control measures, ranging from isolation, social distancing, lockdown, and quarantine to port-of-entry screening, testing, tracing, personal hygiene, and the increase in the number of hospital beds, should be combined to enhance each other's strengths and also to compensate for each other's shortcomings.

In the study (Wilasang et al. 2020), the authors compared ten countries, e.g., Belgium, China, France, Germany, Iran, South Korea, Spain, Thailand, the United States, and the United Kingdom, in terms of reduction of the reproduction number, starting from the date when the reported number of cases exceeded 100 to 3 weeks after control measures were applied. As of April 07, 2020, two countries could achieve and maintain a reproduction number of about one to below one after implementation of control measures for 3–4 weeks: China and South Korea. In China, control measures include a strict lockdown, total standstill, stay-at-home order, contact tracing, and quarantine, reducing the reproduction number from 3.10 to about 1. In South Korea, these measures were a strict lockdown and active case finding, leading to a decrease in reproduction number from about three to below one. The other eight countries applied only one of quarantine (lockdown) and active case finding and failed to control the COVID-19 as of study completion date.

In summary, the findings of the models are consistent with an analysis of real-time data demonstrating the more effectiveness of comprehensive attempts, comprised of early isolation, contact tracing, and quarantine, to bring down the reproduction number of COVID-19 (Wilasang et al. 2020).

38.2.2 Digital Contact Tracing

As above described, self-isolation alone could reduce transmission by 37%, and this reduction increased to 47% when combined with the manual tracing of contacts compared to 57% when combined with the app-based tracing of contacts (Kucharski et al. 2020). Telehealth or telemedicine offers digital technologies that not only accelerate screening and contact tracing but also facilitate treatment and surveillance. Interestingly, the use of telemedicine assists clinicians with holding virtual visits while being quarantined at home (Hollander and Carr 2020). In this manner, quarantine and isolation will work more effectively employing digitalization, a concept referred to as "Digital herd immunity" by Bulchandani et al. (2020) and Caetano et al. (2020).

38.2.3 Combination with Travel Restrictions and School Closures

When combined with travel restrictions, a review of studies suggests that isolation and quarantine could control the outbreak better in terms of reduction in the number of confirmed cases and duration of the outbreak of COVID-19 (Nussbaumer-Streit et al. 2020). Models, however, forecast that travel restrictions, even if strictly sustained, reveal more than a moderate effect on the trajectory of COVID-19, if and only if combined with other control measures that can reduce disease transmission by at least half (Chinazzi et al. 2020).

38.2.4 Degree of Adherence to Quarantine

The study (Sjödin et al. 2020) investigated the correlation between the degree of adherence to quarantine – defined as the degree to which the community correctly follows quarantine laws and limits – and the effectiveness of a 2-week lockdown with community quarantine in Italy. The four different degrees of adherence to community quarantine were complete noncompliance to community quarantine, medium adherence to community quarantine, near-complete community quarantine, and a complete community quarantine, on the average of 10, 5, 1, and 0 h out-of-household activities per day, correspondingly. The more time spent out of the house (the less the community tended to adhere to the quarantine), the more the number of secondary cases increased by the end of quarantine.

38.2.5 Response System

Based on mobility data from mobile devices in the Boston metropolitan area, an agent-based model predicted that it would suffice to keep a potential second wave of disease at the place within the capacity of the health-care system, if a response system using social distancing combined with testing, contact tracing, and household quarantine permits detection of 50% of symptomatic cases, tracing of 40% of their contacts and households, and 9% of individuals quarantined (Aleta et al. 2020). Under such a system, economic activities can be re-opened while preventing the oversaturation of the health-care system.

38.2.6 Early Implementation

The study (Moris and Schizas 2020) using data from the Worldometer compared 22 countries in terms of delta days defined as the gap between the first-time implementation of control measures and the date of the first case per million population (PMP) confirmed and its impact on 30-day mortality and incidence outcomes. Among the countries included in the analysis, only Iceland and Switzerland implemented large-scale sampling, while the remaining 20 countries relied on lockdown as the control measure. Delta days ranged from −1 day for Hong Kong to 29 days for Sweden, and the less the delta days (control measures occur earlier), the lower the 30-day incidence of COVID-19. By observing no significant association between gross domestic product (GDP) health expenditure and disease outcomes, control measures and their timing appeared to be a more critical element of success.

38.2.7 Duration of Quarantine

Studies using different epidemiological models make different recommendations on a preferred duration of lockdown. Other factors that play a role in determining a preferred duration of lockdown include the degree of adherence to quarantine and the population and household size. For India, studies mostly conclude on 3 weeks of lockdown for adequate detection of cases and controlling the disease spread in the country, while other studies call the need to adopt the lockdown for an extended duration of 42–56 days as reviewed in Lahiri et al. (2020). For Italy, modeling data related to a 2-week community quarantine estimated that for medium average household size (three persons), a 1-month duration of quarantine is required for adequate control of outbreak when the degree of adherence to community quarantine is near complete (Sjödin et al. 2020). The duration increased to 54 days under medium adherence to community quarantine.

38.2.8 Individuals Quarantined

Using a SEIRQ epidemic model with high accuracy for COVID-19 predictions, the study (Cui et al. 2020) evaluated the comparative effect of quarantine of different populations, including susceptible individuals, exposed individuals, and infected individuals, on COVID-19 variations. The model predicted that quarantine of susceptible individuals is more important for controlling the spread of disease as indicated by the number of confirmed cases when compared to the quarantine of infected and exposed individuals. No direct evidence exists regarding the effectiveness of quarantine of travelers on controlling the spread of COVID-19 outbreak. There is only a little indirect evidence on the issue regarding the outbreak of SARS, as reviewed in Nussbaumer-Streit et al. (2020).

38.2.9 Household Size

The study (Sjödin et al. 2020) addressed whether household size would influence the effectiveness of community quarantine in Italy concerning the degree of adherence to community quarantine. There were four different average household sizes, larger (six persons), medium (three persons), small (two persons), and single-person, and four degrees of adherence to community quarantine, no, medium, near-complete, and complete community quarantine. Considering a town of 5000 persons, the model provided estimations of the number of secondary cases under the quarantine scenarios mentioned for different household sizes as follows: larger, 43, 29, 19, 16; medium, 20, 12, 8, 7; and small, 11, 7, 4, 3. It shows the interaction between the household size and the degree of adherence to quarantine and its impact on the effectiveness of community quarantine.

38.2.10 Population Size

The study mentioned above also found a direct relationship between the population size and the number of secondary cases. For a town of 50,000 people, the number of secondary cases was estimated to be 110, 70, 40, and 30 for a small household size under no, medium, near-complete, and complete community quarantine scenarios.

38.3 COVID-19: Psychological Challenges and Opportunities Ahead

38.3.1 COVID-19: The Trauma

COVID-19 is a severe threat to global health and is now recognized as one of the significant human concerns internationally. The impact of COVID-19 on mental health cannot be dismissed even though there are so many things to tackle at the same time, like stopping the virus and helping those struggling financially. The cost of psychological support skyrockets when left unaddressed. We are all affected by the pandemic, both emotionally and mentally, but to different degrees.

As such, COVID-19 is traumatic in many ways. From a psychological point of view, a trauma is a response to a deeply distressing or disturbing event that overwhelms an individual's capacity or even ability to cope. It generates feelings of helplessness/hopelessness and reduces the sense of self and the access to the full range of emotions and experiences in an individual. Medically, trauma can also refer to physical damage (which is evident in this pandemic). COVID-19 is to be considered as complex trauma, being exposed to varied and multiple traumatic events of an invasive and interpersonal nature.

In such a context, the most critical issues related to COVID-19 are stress, anxiety, and their consequences, which can also be worsened by necessary public health actions to stop the spread of the disease (such as quarantine and isolation). These can make people feel lonely, powerless, and isolated.

38.3.2 Stress: A Double Impact

Stress is a normal physiological response to an abnormal situation. It is a fundamental component of our lives. It enables us (body and soul) to adapt and adjust to both positive and negative situations through increased cortisol levels. Stress is multifactorial, not constant, and varies from one individual to another. Stress can affect how people feel emotionally, mentally, and physically. It can also affect how people behave.

During the COVID-19 outbreak, stress has caused fear and worry (e.g., about work, health, support, finances); changed eating or sleep patterns; worsened focus, chronic health problems, and mental health conditions; and increased our use of alcohol/tobacco and other substances. From this point of view, stress appeared to make people vulnerable and to possibly aggravate a situation already dramatic by disabling emotional management and regulation. In that sense, it can contribute to intensifying the damages of the coronavirus, causing confusion, discomfort, and chaos. However, stress has also been extremely beneficial during the COVID-19 crisis. Thanks to stress, we have been able to react to the threat of COVID-19 and take action to protect ourselves. Stress is crucial to raise our attention, focus, and alertness in a situation of insecurity. It helps behavioral change or adaptation and enables planning in order to reduce the risks and the dangers of our environment. Stress is essential for an organism to initiate and activate a cohesive defense while exposed to a threat. Stress increases the body's ability and energy (via cortisol secretion) in order to preserve personal and environmental health. During the COVID-19 crisis, stress helped make people aware of the gravity of the situation, to take things seriously, and attempt to limit the spread of coronavirus. Stress will become a problem when affecting how people cope with day-to-day life, and both beneficial and destructive aspects of stress have been observed during the pandemic: a double impact.

38.3.3 Anxiety

Contrary to fear, which is a response to a well-defined and very real and concrete danger, anxiety is a response to a vague or unknown threat. It occurs when we believe that a menacing or unfortunate event could happen and are expecting it. Like stress, anxiety is experienced at various individual degrees and intensity. The extent of the anxiety manifestations greatly depends on how we perceive the anticipated event.

Anxiety appears directly connected to people's uncertainty, cognitive ambiguity, and lack of scientific information about the virus. As per the DSM-5 (2013), anxiety can present in many ways and forms: generalized anxiety disorder (GAD), obsessive-compulsive disorder (OCD), post-traumatic stress disorder (PTSD), and phobia (including social phobia), with or without a panic attack. Levels of anxiety, depression, and PTSD have doubled in society since the onset of COVID-19, with as many as 4/10 people experiencing symptoms (Lucey 2020).

38.3.4 Challenges and Behaviors During the COVID-19 Crisis

During the lockdown, challenges were legion; some people reported suffering from quarantine and isolation measures, as they felt lonely, restricted, and withdrawn from the environment. Self-isolation reduced drastically social contact, and many people felt trapped. Other people felt at risk of losing their livelihood, especially the ones who had a horrific experience (like being on a respirator or extended stay in ICU). The uncertainty status over the disease, coupled with constant exposure to harmful and frightening news, made life extremely difficult. The decimation of normal routine and structure, the loss of freedoms, and the boredom associated with quarantine made the whole process very lengthy for the population. Some individuals were particularly sensitive to the changes in work circumstances, the financial insecurity (including bankruptcy, financial loss), inadequate supplies and infrastructure, and the fact that their life was "on pause." It was very hard to adjust and obey societal changes that were not always well explained nor well understood and which heavily impacted families (as weddings and funerals, e.g., had to follow a different protocol). Prolonged uncertainty and impossibility to plan the future were probably the hardest challenges population faced during the COVID-19 crisis. Of course, while confined, many had to closely face the cracks, the conflicts, and the failures of their couple/family, without being able to escape. Similarly, being separated from loved ones led to worrying for their safety, well-being, access to medical assistance, mental health, and daily services and supports and to wondering about how their loved ones' condition would evolve without those regular supports.

In terms of behaviors, alarm, panic, and terror appeared first, but also anger and confusion/delirium. Maximum individualism was illustrated by food hoarding and compulsive buying. There were thefts in shops, pharmacies, and hospitals to the point that hydro-alcoholic gel distributors had to be removed from the wards. Some aggressive behaviors also emerged, either via assaults of various types (from threats to gun attacks), including against health-care professionals or in a domestic context with a raise in arguments, abuse, and violence within the home (couple and family crises). If insecurity certainly motivated these manifestations, increased consumption of alcohol and drugs associated with boredom and despair is also to be taken into account.

38.3.5 Mental Health Alterations During the COVID-19 Crisis

Millions of people are suffering from psychological repercussions, especially those closest to the pandemic. In patients acutely ill with severe COVID-19, available data suggest (Rogers et al. 2020) that 65% experience delirium, 69% have agitation after the withdrawal of sedation, and 21% have altered consciousness. In the general population under lockdown, not only anxiety increased, but depression also intensified; depression is a mood disorder that causes, among other symptoms, a persistent feeling of sadness and loss of interest. Hypochondria is not to be forgotten in the picture. It is also called health anxiety and consists of an excessive preoccupation with one's health. Besides cyberchondria otherwise known as "compuchondria," COVID-chondria (excessive preoccupation with one's health around the coronavirus), COVID-phobia (extreme or irrational fear of or aversion to COVID-19), and paranoid states (in which persecutory and grandiose ideas and delusions constitute a significant part of the symptoms) were also observed. Altogether, anxiety appears to be the most substantial mental health alteration and most common psychiatric disorder to affect all individuals (in the general population but also health-care professionals) during the COVID-19 crisis. It only highlights the necessity to invest in mental health services, hotlines, counseling, guidance, information, and online help and support groups.

38.3.6 Psychotherapy Approaches to Anxiety

Psychologists use psychotherapeutic approaches such as cognitive behavioral therapy (CBT) approaches and emotionally focused therapy (EFT) approaches to help regulate anxiety levels in order to treat individuals (Najafi et al. 2015a). In the CBT approach, harmful and wrong beliefs are considered as the basis of negative emotions. Healthy beliefs lead to healthy emotions and, ultimately, to healthy functioning. This approach also helps to improve anxiety through the use of psychological training techniques, cognitive reconstruction, coping, and role-playing. The EFT approach is also a short-term approach, which is an empiric humanistic approach to rebuilding the emotional business and uses a structural, systemic approach to rebuilding interactions (Javidi et al. 2013; Soleimani et al. 2015; Najafi et al. 2015b). This approach focuses on emotions and helps people to achieve emotional self-regulation. It uses the components of focusing on positive emotions, emotional reconstruction, and finding new meanings to better communicate with others and better adapt to the environment (Madahi et al. 2013). In this approach, the therapist pays special attention to attachment styles and the communication patterns of individuals (Javidi et al. 2001; Najafi et al. 2015b). On this basis, the therapist regulates the treatment process in such a way as to lead to a secure attachment, so that people can move more confidently toward a positive and constructive relationship in order to be able to express their emotions constructively (Javidi et al. 2012, 2013). The precarious feeling of anxiety is being treated in this approach. It is because the main focus of this approach is to create a safe, emotional, and supportive environment. Both approaches help to control and manage anxiety by managing and regulating emotions.

38.3.7 The Notion of Locus of Control and the Response Ability

The notion of "locus of control" first appeared in 1966, thanks to Julian Rotter. It was formulated after observing recurrent patterns in how people perceive what is happening to them (and to their health). On the one hand, he observed that some individuals believe they are directly responsible for what is happening to them, as directly resulting from their actions or behaviors (internal locus of control). On the other hand, Rotter noticed that some individuals consider that the "reinforcements" they receive do not result from their actions but are depending on external factors (luck, fate, others, external locus of control). More specifically, this theory aims to predict which behavior an individual will adopt, knowing different reactions are possible. Multiple studies have applied the concept to the health sector and have identified three subtypes in terms of beliefs, thinking, and behaviors: (i) internal locus of control; (ii) external "powerful others" locus of control; and (iii) external "chance" locus of control. There are some obvious connections between COVID-19 and the health locus of control (Naviaux et al. 2020).

Indeed, people who have a rather internal locus of control will look after themselves and self-observe (fever, cough, shortness of breath, sore throat, headaches), while individuals with an external locus of control will consider that the course of the coronavirus depends either on luck, destiny, or "the others." The health locus of control will also predict how people will handle prevention measures, quarantine, and isolation; knowing that people with an internal locus of control will better comply (than people with an external locus of control) because they feel responsible ("response-able" or "able to provide a proper response" in order to protect their health). Therefore, opting for an internal locus of control during the COVID-19 and choosing response ability is not only salutary but also sanitary.

38.3.8 Vulnerable Groups

During the pandemic, children and young people, older adults, health-care providers, and people with comorbid conditions appear as the most psychologically vulnerable populations. However, pregnant women are the same even if they are not infected (Mirbeyk and Rezaei 2020).

38.3.8.1 Children and Young People

Schools and colleges have been closed, so children and young people are staying at home during the pandemic. Parenting is the key. It is important to help young ones find a positive way to express their feelings, such as sadness, boredom, or fear. For example, engaging in creative activities such as drawing or clay modeling can prove extremely helpful in that regard, as it allows us to express and communicate feelings and emotions. Isolation can be challenging for little ones who miss their friends and their activities. It is essential to keep children close to their parents and family if safe for the child and to maintain regular contact with parents and caregivers in case of separation. It is crucial to make time for younger people to check in with friends and family they might be worried about, using phone or video calls, so that they can see they are okay. Young people should be informed about what is happening in the world, using words and language they can access, and not overwhelming them with unnecessary information. Adults need to be honest with them and address their concerns in a reassuring way to ease anxiety. Maintaining familiar routines (such as bedtime rituals, hygiene routines, and a healthy diet) increases the feeling of safety. Creating new routines and dedicating some time to daily physical exercise are also essential. Children must be encouraged to pursue learning and age-appropriate activities, to continue to play, to have fun, and to socialize with others if only within the family. They might seek more attachment and be more demanding to parents. Intuitively they will get cues on how to manage difficult times by observing adults' behaviors and emotions. Supervised and limited access to (social) media is given. Just like for any vulnerable group, we need to help, support, and stand by our children.

38.3.8.2 Elderly People

Older adults and those with cognitive decline or dementia can be deeply affected by staying at home as they can experience an increased feeling of isolation and loneliness. The lack of daily tasks and their physical vulnerability can make them feel more anxious, angry, or even agitated. Adjusting to a new situation can take a while, and do so, it is crucial to develop a regular structure to the day as routines make the world safer and more predictable. Staying active, eating and drinking healthily, and receiving practical and emotional support through family, friends, and professionals will help maintain good mental health. Participating in groups or events online and developing new interests will also be beneficial. Information should be provided clearly and simply so that people with cognitive impairment can understand. Information should be repeated if necessary.

38.3.8.3 People with Pre-existing Conditions

People with pre-existing conditions might experience increased stress over the threat of the infection. They should activate their social contacts for assistance and make sure they have access to medication, food, services, and medical assistance.

38.3.8.4 People with Pre-existing Mental Health Issues

There is a risk of exacerbation and deterioration of pre-existing mental health issues. Physical exercise, maintaining daily routines, and a healthy diet, as well as regular activities, are helpful to cope. Having access to appropriate (and not overloading) information is essential, just like it is to keep in touch with family, friends, and professionals. Access to medication, food, services, and medical assistance should be available.

38.3.9 Health-Care Professionals (HCP) During the COVID-19 Crisis

HCP are disproportionally affected by work-related stress because of excessive workloads, working in an emotionally charged environment where demand outweighs capacity and facing a lack of management and government support and unrealistic public expectations and perception of stigma. HCP are at high risk of decompensation regarding both physical and mental health (burn-out in essential workers), and access to appropriate support should be put in place. HCP can also experience cynicism regarding a possible change as they face (unprecedented) high levels of mortality and extremely distressed patients (including treating infected colleagues). No child care arrangements have been made for HCP who have to deal with social isolation and interpersonal distancing, infection control procedures, diminished collegial social interaction, and re-deployment in unfamiliar environments or with unfamiliar colleagues. They often do not dispose of adequate supplies and equipment, rationing personal protective equipment and facing ethical dilemmas of rationing access to ventilators and other essential health supplies. HCP are, of course, facing a higher risk of contamination. Personal worries include infection risk to self and others, and concerns regarding the well-being of family members who are “home-schooled,” quarantined, or infected are real (Gavin et al. 2020).

The only study to date on the psychological impact on COVID-19 frontline HCP reports high rates of depression, distress, anxiety, and insomnia (Lai et al. 2020). Ensuring that HCP is safe and able to carry out their job is crucial. HCP are per definition exposed to the virus in their life-saving activity and should not be stigmatized for this.

Unfortunately, the sad reality is that HCP have been subject to many shocking forms of violence. It includes harassment, stigmatization, and physical violence. In response to this concern, on May 26, 2020, 13 medical and humanitarian global organizations representing more than 30 million HCP have issued a declaration condemning increasing incidents of attacks against health workers and facilities (healthcareindanger.org 2020).

38.3.10 Opportunities

Along with the damage it has caused, the COVID-19 pandemic has also provided opportunities for people to develop their existential and skill-building capabilities. It has led to achievements, including hardware and software upgrades, improved knowledge of specialists and public adherence to the principles of health and hygiene, and an increase in the spirit of sympathy between people and the authorities. It would also help to understand the importance of health-care systems, education, and research and identify strengths and weaknesses in various areas, in particular, in the field of crisis intervention. Its psychological achievements also include improving individual emotional regulation, identifying individual existential capacities in times of crisis, controlling anxious beliefs, improving psychological adjustment, recalling anxiety about death, enhancing creativity and innovation, individual self-assessment in the field of courage/fear and indifference, a better understanding of the value of health, a distinction between opportunism and profit- and fame-seeking and truth-seeking, human attention to family and friends, the empowerment of sympathy, and the creation of home opportunities and home quarantine jobs.

38.4 True Face of Quarantine and Social Isolation in Iran

One of the primary methods in breaking the chain of infection is the separation of people exposed to a contagious disease. Three main methods could achieve this: (i) isolation, separating the affected individuals from the rest of the population; (ii) quarantine, applying distance on any suspected member; and (iii) social distancing (Huremović 2019). Although quarantine is considered a passable action in outbreak management (quarantine and isolation), many governments, e.g., the United States, Singapore, and Iran, doubted about its actual effectiveness in the current COVID-19 outbreak, mainly due to socioeconomic burdens. However, social distancing described as canceling mass gatherings and closing schools has been the preferred measure for the abovementioned governmental health authorities so far.

According to the official reports from Iran in May 19, there are 122,492 confirmed cases; and the number may be underestimated due to a limited number of COVID-19 PCR diagnostic kits, hence the US ban (Retrieved from worldometers on May 19, 2020). Indeed, there might be several cases with milder symptoms (or even asymptomatic) who were not included. On February 20, 2020, the National Headquarters for Fighting Coronavirus in Iran has made a decision for closing schools and universities until the further announcement, asked people to house-quarantine themselves and avoid unnecessary travel, and dedicated specific hospitals merely to manage COVID-19 in isolation. From March 6, 2020, province-wide quarantine was applied to two Northern provinces of Iran (Mazandaran and Gilan), which did not exceed 3 weeks. Although only two mentioned provinces have gone under the official mass quarantine, remarkable portion from many other metropolitan residents, e.g., Tehran, Karaj, Mashhad, Isfahan, and Shiraz, have voluntarily house-quarantined themselves. Moreover, most airlines stopped traveling to and from Iran, which brings a sense of entrapment within their own geographical borders in the population. These measures were applied a week before the pandemic declaration (https://www.who.int/dg/speeches/detail/who-director-general-s-opening-remarks-at-the-media-briefing-on-covid-19%2D%2D-11-march-2020 accessed on April 16th).

38.4.1 Review of the Literature

Outbreaks themselves attenuate community-wide anxiety, starting as the first death being reported through the increasing reports in media and by the growing number of newly infected cases. The mass quarantine can raise this imposed anxiety due to several reasons. It represents the authorities' belief that not only is the situation difficult, but it also is liable to worsen. Also, it brings the feeling of being trapped and having lost control among the population under quarantine. Besides, the absence of clear messages and facts on the ongoing situation escalates to fear, and as regards, people turn to less reliable sources for the information, which makes the possible ground to plant rumor (Rubin and Wessely 2020).

Quarantine forces people to separate from loved ones, brings loss of freedom and uncertainty over disease course, and has the potential to expose individuals to profound boredom. The latter could lead to suicide, as previously been reported. Consequently, successful mandatory mass quarantine application is not reached unless we are aware of the potential psychological costs that we have to pay and do our best to reduce it (Brooks et al. 2020; Barbisch et al. 2015). However, social isolation describing the loss of social integrity with the loved ones has severe health damages as reducing immune responses to viral infections and inducing inflammatory cytokines (Leschak and Eisenberger 2019).

Although we could not find strong evidence to back up our claim, Persians are considered a friendly social community that relies intensely on their social network and prioritizes the family gatherings to any other personal activity. Hence, we presume the Iranians to be much vulnerable to this house confinement. Moreover, this quarantine is concurrent with the Persian celebration of New Year, anciently called "Nowrouz," which directly translates to the new day. Nowrouz marks the beginning of spring in the Northern Hemisphere and marks the first day of the first month (Farvardin) of the Iranian calendars, which usually is in March 21. It is the most important national holiday for the Persian population with the essential tradition of making short visits to the homes of family, friends, and neighbors. This concurrency could also be considered a mental burden for the population.

Reviewing the literature revealed the null experience of previous quarantine in the Iranian population. Thus no evidence is available on the community acceptance, the effectiveness of the modality based on Persian culture, and their mental health within and afterward. It raises the need to learn from different nations' experiences in order to apply better facilities and measures during the quarantine period and to minimize long-term psychosociophysical burdens on this huge quarantined population.

It is crucial to plan mitigation measures along the way since the psychophysical burden could be troublesome in the future. The only proven risk factor is pre-existing poor mental health. Coping mechanisms focus on four main groups, including the role of government, how the individuals could cope, and the impact on health-care workers and among youngsters. Lessons learned from previous severe acute respiratory syndrome coronavirus (SARS-COV) and Ebola outbreak alongside with author's recommendation led to the following coping strategies in brief (Robertson et al. 2004; Brooks et al. 2020).

38.4.2 Recommended Strategies

38.4.2.1 The Government

For the government, clarity is the key. That is what makes people believe them and listen to their instructions and reduces panic among the public. Governments should propose an exact extent of quarantine relevant to the incubation period and stick to it, inform the population about the disease and the reasons of the quarantine, and do it through altruism rather than compulsion. It is also crucial that governments plan to provide the supplies and the resources for the needs of people during the quarantine. One of the essential needs is communication. Lack of proper deeds to communicate through social networks can make people anxious and should be avoided. Also, it is necessary to build particular forms of communication such as support groups and special communicating lines that ease the connection between quarantined people and other affected ones and also health-care providers, such as psychiatrists. Lastly, health authorities should provide a mechanism by which people with the symptoms of the disease can connect to the health-care workers and learn what to do if they develop illness symptoms (Brooks et al. 2020). Currently, several online platforms and hotlines are provided in Iran to reassure people about the symptoms they are experiencing in order to ease the corona phobia.

38.4.2.2 The Individual

The individual responsibility is to trust the government and do as they are asked to, pay attention to the instructions, maintain the quarantine, and avoid stigmatizing the affected people instead of supporting them. It is crucial only to pass valid information and avoid spreading fear and fake news about the disease. People should only buy what they need instead of buying loads of supplies out of panic. They should help each other to overcome the income reduction during quarantine (which is the reason for many low-income families to break the quarantine). Also, they should avoid overloading hospitals out of fear, so that there would be enough time and resources remained for the people in need (Robertson et al. 2004; Brooks et al. 2020).

People experience boredom, loneliness, fear, and anxiety at this time. According to previous studies, the most useful way to cope with these feelings is the stress adaptation model. In this model, you identify, articulate, and normalize the stressors as much as possible. Prolonged home-stays lead to severe inactivity, which contributes to anxiety and depression and could cause a range of chronic health conditions. Maintaining regular physical activity at the house, using a smartphone application or other multimedia, can boost the energy, lower the stress, and enhance the immune system (Chen et al. 2020; Sellami et al. 2018). Examples of home exercises include walking in the house; alternating leg lunges; stair climbing;

stand-to-sit and sit-to-stand, using a chair and from the floor; chair squats; sit-ups and pushups; and yoga (National Center for Complementary and Integrative Health. Yoga: what you need to know. Available at:).

We also recommend the practice of mindfulness meditation to the people under quarantine. Mindfulness meditation mainly represents a mental training framework for cultivating a state of mindful awareness in daily life. Besides the known role of stress management, mindfulness has shown to enhance cell-mediated immune functions by increasing the number of $CD4^+$ T lymphocyte and their activity. It also reduces inflammatory response by reductions in the activity of the cellular transcription factor NF-kB and reductions in circulating levels of CRP (Black and Slavich 2016; Bower and Irwin 2016). In this matter, several online sessions could be found, and the most accessible professional application, which authors recommend, is Headspace (Headspace).

38.4.2.3 Health-Care Providers

HCP need special attention during this time. Not only should they be provided with organizational support, but also the affected staff should receive proper support from their colleagues, so that they will not feel guilty for causing more work for them. It is crucial to forming a leadership group from different specialties, for example, psychiatrists, internists, immunologists, virologists, and infectious disease specialists among epidemiologists and managers. They should keep in mind that it is their job to provide transparent information and supportive atmosphere among the colleagues and that the health-care workers can be more vulnerable because they get isolated from their loved ones, they cannot work when they are most needed, and all of this makes them prone to more prolonged adverse psychological effects. Useful measurement is to link psychiatrists with other colleagues to inform ease the mental support. Senior staff should act as role models, for example, in using the support system and medical care advice (Brooks et al. 2020; Robertson et al. 2004).

38.4.2.4 Children

They are a vulnerable group that could easily be neglected during the home confinement. Stressors such as prolonged duration; fears of being infected; fear of death; inability to perceive real information; frustration and boredom; losing their social network of classmates, friends, and teachers; and familial financial loss can raise problematic psychological issues among children (Wang et al. 2020). Based on a study by Sprang and Silman, children are more prone to develop PTSD, when exposed to quarantine, compared to their parents (Sprang and Silman 2013). An integrated approach is necessary for this matter. Parents can build committees to work together and bridge the needs of students with school requirements. Psychologists and social workers can offer online services for children to learn how to cope with mental health issues caused by domestic conflicts, tension with parents, and anxiety from becoming infected. Lastly, the role of schools is more critical, not only to deliver educational materials to children but also to offer an opportunity for students to interact with teachers and obtain psychological counseling. They actively promote a health-conscious schedule as an obligatory school curriculum, encouraging the children to build personal hygiene habits, encouraging physical activities, and teaching healthy diet contents and sleep hygiene (Wang et al. 2020).

38.5 Conclusion

Currently, the COVID-19 outbreak has put many nations under serious psychological stress due to quarantine. Social relationships are described as what exists between two people when each can influence the other's thoughts, feelings, and/or behavior has lost (Clark et al. 2015). Loneliness and decreased social interactions affect individuals' health drastically. Apart from depression and anxiety, a lack of proper social contact exposes individuals to a higher risk of coronary heart disease and stroke, suppresses antiviral immunity, and increases the inflammatory responses (Valtorta et al. 2016b; Valtorta et al. 2016a; Wang et al. 2018). Due to a lack of a unified framework within studies that score social isolation, the existing data is diverse (Valtorta et al. 2016a). Commonly nations are believed to be different in their social attachments. Ultimately without a comparable social connectivity baseline score plus zero data on the national social connectivity diversity, a scientific comparison is impracticable. It is presumed that prior enriched social connection could be a possible threatening factor for the within and post lockdown psychophysical health. Thus, different measurements should be directed to address the impact of personality traits and cultural differences on the psychological experience of confinement. It is also crucial to address the early and late psychosociophysical impact of quarantine on people's lives and to reduce the costs as much as possible before it becomes too late. This universal house confinement makes the ground for further research on how various nations react to the imposed quarantine, hopefully preparing for international collaboration and interdisciplinary research as a very strong response to this pandemic (Momtazmanesh et al. 2020; Mohamed et al. 2020a; Rzymski et al. 2020; Moradian et al. 2020; Kafieh et al. 2020).

References

Ahanchian H, Moazzen N, Faroughi MSD, Khalighi N, Khoshkhui M, Aelami MH, Haghi NSM, Rezaei N (2020) COVID-19 in a child with primary specific antibody deficiency

Ahmadi M, Saffarzadeh N, Habibi MA, Hajiesmaeili F, Rezaei N (2020) Colon cancer and SARS-CoV-2: impact of ACE2 expression in susceptibility to COVID-19. bioRxiv

Aleta A, Martin-Corral D, Pastore YPA, Ajelli M, Litvinova M, Chinazzi M, Dean NE, Halloran ME, Longini IM, Merler S, Pentland A, Vespignani A, Moro E, Moreno Y (2020) Modeling the impact of social distancing, testing, contact tracing and household quarantine on second-wave scenarios of the COVID-19 epidemic. medRxiv: the preprint server for health sciences. https://doi.org/10.1101/2020.05.06.20092841

Babaha F, Rezaei N (2020) Primary immunodeficiency diseases in COVID-19 pandemic: a predisposing or protective factor? Am J Med Sci. https://doi.org/10.1016/j.amjms.2020.07.027

Bahrami A, Vafapour M, Moazzami B, Rezaei N (2020) Hyperinflammatory shock related to COVID-19 in a patient presenting with multisystem inflammatory syndrome in children: first case from Iran. J Paediatr Child Health. https://doi.org/10.1111/jpc.15048

Barbisch D, Koenig KL, Shih FY (2015) Is there a case for quarantine? Perspectives from SARS to Ebola. Disaster Med Public Health Prep 9(5):547–553. https://doi.org/10.1017/dmp.2015.38

Basiri A, Heidari A, Nadi MF, Fallahy MTP, Nezamabadi SS, Sedighi M, Saghazadeh A, Rezaei N (2020a) Microfluidic devices for detection of RNA viruses. Rev Med Virol n/a (n/a):e2154. https://doi.org/10.1002/rmv.2154

Basiri A, Pazhouhnia Z, Beheshtizadeh N, Hoseinpour M, Saghazadeh A, Rezaei N (2020b) Regenerative medicine in COVID-19 treatment: real opportunities and range of promises. Stem Cell Rev Rep:1–13. https://doi.org/10.1007/s12015-020-09994-5

Black DS, Slavich GM (2016) Mindfulness meditation and the immune system: a systematic review of randomized controlled trials. Ann N Y Acad Sci 1373(1):13–24. https://doi.org/10.1111/nyas.12998

Bower JE, Irwin MR (2016) Mind-body therapies and control of inflammatory biology: a descriptive review. Brain Behav Immun 51:1–11. https://doi.org/10.1016/j.bbi.2015.06.012

Brooks SK, Webster RK, Smith LE, Woodland L, Wessely S, Greenberg N, Rubin GJ (2020) The psychological impact of quarantine and how to reduce it: rapid review of the evidence. Lancet. https://doi.org/10.1016/S0140-6736(20)30460-8

Bulchandani VB, Shivam S, Moudgalya S, Sondhi SL (2020) Digital herd immunity and COVID-19. arXiv preprint arXiv:200407237

Caetano R, Silva AB, Guedes A, Paiva CCN, Ribeiro GDR, Santos DL, Silva RMD (2020) Challenges and opportunities for telehealth during the COVID-19 pandemic: ideas on spaces and initiatives in the Brazilian context. Cad Saude Publica 36(5):e00088920. https://doi.org/10.1590/0102-311x00088920

Chen P, Mao L, Nassis GP, Harmer P, Ainsworth BE, Li F (2020) Wuhan coronavirus (2019-nCoV): the need to maintain regular physical activity while taking precautions. J Sport Health Sci 9(2):103–104. https://doi.org/10.1016/j.jshs.2020.02.001

Chinazzi M, Davis JT, Ajelli M, Gioannini C, Litvinova M, Merler S, Piontti y AP, Mu K, Rossi L, Sun K (2020) The effect of travel restrictions on the spread of the 2019 novel coronavirus (COVID-19) outbreak. Science 368(6489):395–400

Clark M, Aragón O, Hirsch J (2015) Social relationships in adulthood. In. https://doi.org/10.1016/B978-0-08-097086-8.34026-0

Cui Q, Hu Z, Li Y, Han J, Teng Z, Qian J (2020) Dynamic variations of the COVID-19 disease at different quarantine strategies in Wuhan and mainland China. J Infect Public Health 13(6):849–855. https://doi.org/10.1016/j.jiph.2020.05.014

Darbeheshti F, Rezaei N (2020) Genetic predisposition models to COVID-19 infection. Med Hypotheses 142:109818. https://doi.org/10.1016/j.mehy.2020.109818

Fathi N, Rezaei N (2020) Lymphopenia in COVID-19: therapeutic opportunities. Cell Biol Int 44(9):1792–1797. https://doi.org/10.1002/cbin.11403

Gavin B, Hayden J, Adamis D, McNicholas F (2020) Caring for the psychological well-being of healthcare professionals in the Covid-19 pandemic crisis. Irish Medical J 113(4):51.

Hanaei S, Rezaei N (2020) COVID-19: developing from an outbreak to a pandemic. Arch Med Res

Headspace. https://www.headspace.com/headspace-meditation-app

Hollander JE, Carr BG (2020) Virtually perfect? Telemedicine for Covid-19. N Engl J Med 382 (18):1679–1681. https://doi.org/10.1056/NEJMp2003539. https://www.who.int/dg/speeches/detail/who-director-general-s-opening-remarks-at- -media-briefing-on-covid-19%2D%2D-11-march-2020. Accessed on April 16th

Huremović D (2019) Psychiatry of pandemics a mental health response to infection outbreak: a mental health response to infection outbreak. https://doi.org/10.1007/978-3-030-15346-5

Jabbari P, Rezaei N (2020) With risk of reinfection, is COVID-19 here to stay? Disaster Med Public Health Prep:1. https://doi.org/10.1017/dmp.2020.274

Jabbari P, Jabbari F, Ebrahimi S, Rezaei N (2020) COVID-19: a chimera of two pandemics. Disaster Med Public Health Prep:1–2. https://doi.org/10.1017/dmp.2020.223

Jahanshahlu L, Rezaei N (2020a) Central Nervous System Involvement in COVID-19. Arch Med Res. https://doi.org/10.1016/j.arcmed.2020.05.016

Jahanshahlu L, Rezaei N (2020b) Monoclonal antibody as a potential anti-COVID-19. Biomed Pharmacother 129:110337. https://doi.org/10.1016/j.biopha.2020.110337

Javidi N, Soleimani A, Ahmadi K (2001) The effectiveness of emotionally focused couples therapy (EFT) to improve communication patterns in couples

Javidi N, Soleimani A, Ahmadi Kh SM (2012) The effectiveness of training of emotion management strategies according to emotionally focused couple therapy (EFT) to increase sexual satisfaction in couples. J Health Psychol 1(3):5–18

Javidi N, Soleimani A, Ahmadi Kh SM (2013) Efficacy of emotionally-focused couples therapy (EFCT) to improve family functioning. J Behav Sci 7(2):181–188

Kafieh R, Arian R, Saeedizadeh N, Minaee S, Yadav SK, Vaezi A, Rezaei N, Javanmard SH (2020) COVID-19 in Iran: a deeper look into the future. medRxiv

Kucharski AJ, Klepac P, Conlan AJK, Kissler SM, Tang ML, Fry H, Gog JR, Edmunds WJ (2020) Effectiveness of isolation, testing, contact tracing, and physical distancing on reducing transmission of SARS-CoV-2 in different settings: a mathematical modelling study. Lancet Infect Dis. https://doi.org/10.1016/s1473-3099(20)30457-6

Lahiri A, Jha SS, Bhattacharya S, Ray S, Chakraborty A (2020) Effectiveness of preventive measures against COVID-19: a systematic review of in Silico modeling studies in Indian context. Indian J Public Health 64(Supplement):S156–s167. https://doi.org/10.4103/ijph.IJPH_464_20

Lai J, Ma S, Wang Y, Cai Z, Hu J, Wei N, Wu J, Du H, Chen T, Li R, Tan H, Kang L, Yao L, Huang M, Wang H, Wang G, Liu Z, Hu S (2020) Factors associated with mental health outcomes among health care workers exposed to Coronavirus disease 2019. JAMA Network Open 3(3):e203976. https://doi.org/10.1001/jamanetworkopen.2020.3976

Leschak CJ, Eisenberger NI (2019) Two distinct immune pathways linking social relationships with health: inflammatory and antiviral processes. Psychosom Med 81(8):711–719. https://doi.org/10.1097/psy.0000000000000685

Lotfi M, Rezaei N (2020) SARS-CoV-2: a comprehensive review from pathogenicity of the virus to clinical consequences. J Med Virol. https://doi.org/10.1002/jmv.26123

Lotfi M, Hamblin MR, Rezaei N (2020) COVID-19: transmission, prevention, and potential therapeutic opportunities. Clin Chim Acta 508:254–266. https://doi.org/10.1016/j.cca.2020.05.044

Lucey J (2020, 15th May) https://www.rte.ie/news/world/2020/0515/1138301-pandemic-mental-health-crisis/

Madahi ME, Samadzadeh M, Javidi N (2013) The communication patterns & satisfaction in married students. Procedia Soc Behav Sci 84:1190–1193

Mansourabadi AH, Sadeghalvad M, Mohammadi-Motlagh HR, Rezaei N (2020) The immune system as a target for therapy of SARS-CoV-2: a systematic review of the current immunotherapies for COVID-19. Life Sci 258:118185. https://doi.org/10.1016/j.lfs.2020.118185

Mirbeyk M, Rezaei N (2020) The impact of COVID-19 on pregnancy and neonatal health: a systematic review

Moazzami B, Razavi-Khorasani N, Dooghaie Moghadam A, Farokhi E, Rezaei N (2020) COVID-19 and telemedicine: immediate action required for maintaining healthcare providers well-being. J Clin Virol 126:104345. https://doi.org/10.1016/j.jcv.2020.104345

Mohamed K, Rodríguez-Román E, Rahmani F, Zhang H, Ivanovska M, Makka SA, Joya M, Makuku R, Islam MS, Radwan N, Rahmah L, Goda R, Abarikwu SO, Shaw M, Zoghi S, Irtsyan S, Ling I, Cseprekal O, Faten AB, Hazar Sayar E, Soloukey C, Grancini G, Rezaei N (2020a) Borderless collaboration is needed for COVID-19-a disease that knows no borders. Infect Control Hosp Epidemiol:1–2. https://doi.org/10.1017/ice.2020.162

Mohamed K, Yazdanpanah N, Saghazadeh A, Rezaei N (2020b) Computational drug discovery and repurposing for the treatment of Covid-19: a systematic review. Available at SSRN 3583748

Momtazmanesh S, Ochs HD, Uddin LQ, Perc M, Routes JM, Vieira DN, Al-Herz W, Baris S, Prando C, Rosivall L, Abdul Latiff AH, Ulrichs T, Roudenok V, Aldave Becerra JC, Salunke DB, Goudouris E, Condino-Neto A, Stashchak A, Kryvenko O, Stashchak M, Bondarenko A, Rezaei N (2020) All together to fight COVID-19. Am J Trop Med Hyg 102(6):1181–1183. https://doi.org/10.4269/ajtmh.20-0281

Moradian N, Ochs HD, Sedikies C, Hamblin MR, Camargo CA Jr, Martinez JA, Biamonte JD, Abdollahi M, Torres PJ, Nieto JJ, Ogino S, Seymour JF, Abraham A, Cauda V, Gupta S, Ramakrishna S, Sellke FW, Sorooshian A, Wallace Hayes A, Martinez-Urbistondo M, Gupta M, Azadbakht L, Esmaillzadeh A, Kelishadi R, Esteghamati A, Emam-Djomeh Z, Majdzadeh R, Palit P, Badali H, Rao I, Saboury AA, Jagan Mohan Rao L, Ahmadieh H, Montazeri A, Fadini GP, Pauly D, Thomas S, Moosavi-Movahed AA, Aghamohammadi A, Behmanesh M, Rahimi-Movaghar V, Ghavami S, Mehran R, Uddin LQ, Von Herrath M, Mobasher B, Rezaei N (2020) The urgent need for integrated science to fight COVID-19 pandemic and beyond. J Transl Med 18(1):205. https://doi.org/10.1186/s12967-020-02364-2

Moris D, Schizas D (2020) Lockdown during COVID-19: the Greek success. Vivo (Athens, Greece) 34(3 Suppl):1695–1699. https://doi.org/10.21873/invivo.11963

Najafi M, Soleimani A, Ahmadi K, Javidi N, Hoseni E, Pirjavid F (2015a) The study of the effectiveness of couple emotionally focused therapy (EFT) on increasing marital adjustment and improving the physical and psychological health of the infertile couples. Iran J Obstet Gynecol Infertil 17(133):8–21

Najafi M, Soleimani AA, Ahmadi K, Javidi N, Kamkar EH (2015b) The effectiveness of emotionally focused therapy on enhancing marital adjustment and quality of life among infertile couples with marital conflicts. Int J Fertil Steril 9(2):238

Nasab MG, Saghazadeh A, Rezaei N (2020) SARS-CoV-2-a tough opponent for the immune system. Arch Med Res. https://doi.org/10.1016/j.arcmed.2020.05.020

National Center for Complementary and Integrative Health. Yoga: what you need to know. Available at. https://nccih.nih.gov/health/yoga/introduction.htm. Accessed 3 Aug 2020

Naviaux A, Janne P, Gourdin M (2020) Response-ability facing Covid-19: a matter of psyche, behavior and civism. HSOA Journal of Clinical Immunology and Immunotherapy 6:021. https://doi.org/10.24966/CIIT-8844/1000021

Nussbaumer-Streit B, Mayr V, Dobrescu AI, Chapman A, Persad E, Klerings I, Wagner G, Siebert U, Christof C, Zachariah C (2020) Quarantine alone or in combination with other public health measures to control COVID-19: a rapid review. Cochrane Database Syst Rev 4:CD013574

Pan A, Liu L, Wang C, Guo H, Hao X, Wang Q, Huang J, He N, Yu H, Lin X (2020) Association of public health interventions with the epidemiology of the COVID-19 outbreak in Wuhan, China. JAMA 323(19):1915–1923

Parmet WE, Sinha MS (2020) Covid-19—the law and limits of quarantine. N Engl J Med 382(15):e28

Pashaei M, Rezaei N (2020) Immunotherapy for SARS-CoV-2: potential opportunities. Expert Opin Biol Ther:1–5. https://doi.org/10.1080/14712598.2020.1807933

Pourahmad R, Moazzami B, Rezaei N (2020) Efficacy of plasmapheresis and immunoglobulin replacement therapy (IVIG) on patients with COVID-19. SN Compr Clin Med 2:1–5

Quarantine and Isolation. https://www.cdc.gov/quarantine/index.html

Rabiee N, Rabiee M, Bagherzadeh M, Rezaei N (2020) COVID-19 and picotechnology: potential opportunities. Med Hypotheses 144:109917. https://doi.org/10.1016/j.mehy.2020.109917. Retrieved from worldometers on May 19th 2020. https://www.worldometers.info/coronavirus/

Rezaei N (2020a) COVID-19 affects healthy pediatricians more than pediatric patients. Infect Control Hosp Epidemiol:1. https://doi.org/10.1017/ice.2020.139

Rezaei N (2020b) COVID-19 and medical biotechnology. Avicenna J Med Biotechnol 12(3):139

Robertson E, Hershenfield K, Grace SL, Stewart DE (2004) The psychosocial effects of being quarantined following exposure to SARS: a qualitative study of Toronto health care workers. Can J Psychiatr 49(6):403–407. https://doi.org/10.1177/070674370404900612

Rogers JP, Chesney E, Oliver D, Pollak TA, McGuire P, Fusar-Poli P, Zandi MS, Lewis G, David AS (2020) Psychiatric and neuropsychiatric presentations associated with severe coronavirus infections: a systematic review and meta-analysis with comparison to the COVID-19 pandemic. The Lancet Psychiatry 7(7):611–627. https://doi.org/10.1016/S2215-0366(20)30203-0

Rokni M, Hamblin MR, Rezaei N (2020) Cytokines and COVID-19: friends or foes? Hum Vaccin Immunother 16:1–3

Rubin GJ, Wessely S (2020) The psychological effects of quarantining a city. BMJ (Clin Res Ed) 368:m313. https://doi.org/10.1136/bmj.m313

Rzymski P, Nowicki M, Mullin GE, Abraham A, Rodríguez-Román E, Petzold MB, Bendau A, Sahu KK, Ather A, Naviaux A-F (2020) Quantity does not equal quality: scientific principles cannot be sacrificed. Int Immunopharmacol 86:106711

Saghazadeh A, Rezaei N (2020a) Immune-epidemiological parameters of the novel coronavirus – a perspective. Expert Rev Clin Immunol 16(5):465–470. https://doi.org/10.1080/1744666x.2020.1750954

Saghazadeh A, Rezaei N (2020b) Towards treatment planning of COVID-19: Rationale and hypothesis for the use of multiple immunosuppressive agents: anti-antibodies, immunoglobulins, and corticosteroids. Int Immunopharmacol 84:106560

Sahu KK, Siddiqui AD, Rezaei N, Cerny J (2020) Challenges for management of immune thrombocytopenia during COVID-19 pandemic. J Med Virol. https://doi.org/10.1002/jmv.26251

Saleki K, Banazadeh M, Saghazadeh A, Rezaei N (2020) The involvement of the central nervous system in patients with COVID-19. Rev Neurosci 31(4):453–456. https://doi.org/10.1515/revneuro-2020-0026

Sellami M, Gasmi M, Denham J, Hayes LD, Stratton D, Padulo J, Bragazzi N (2018) Effects of acute and chronic exercise on immunological parameters in the elderly aged: can physical activity counteract the effects of aging? Front Immunol 9:2187. https://doi.org/10.3389/fimmu.2018.02187

Shamshirian D, Rezaei N (2020) Cardiovascular diseases burden in COVID-19: systematic review and meta-analysis

Sharifkashani S, Bafrani MA, Khaboushan AS, Pirzadeh M, Kheirandish A, Yavarpour Bali H, Hessami A, Saghazadeh A, Rezaei N (2020) Angiotensin-converting enzyme 2 (ACE2) receptor and SARS-CoV-2: potential therapeutic targeting. Eur J Pharmacol 884:173455. https://doi.org/10.1016/j.ejphar.2020.173455

Sjödin H, Wilder-Smith A, Osman S, Farooq Z, Rocklöv J (2020) Only strict quarantine measures can curb the coronavirus disease (COVID-19) outbreak in Italy, 2020. Eur Secur 25(13):2000280

Soleimani AA, Najafi M, Ahmadi K, Javidi N, Kamkar EH, Mahboubi M (2015) The effectiveness of emotionally focused couples therapy on sexual satisfaction and marital adjustment of infertile couples with marital conflicts. Int J Fertil Steril 9(3):393

Sprang G, Silman M (2013) Posttraumatic stress disorder in parents and youth after health-related disasters. Disaster Med Public Health Prep 7(1):105–110. https://doi.org/10.1017/dmp.2013.22

Valtorta NK, Kanaan M, Gilbody S, Hanratty B (2016a) Loneliness, social isolation and social relationships: what are we measuring? A novel framework for classifying and comparing tools. BMJ Open 6(4):e010799. https://doi.org/10.1136/bmjopen-2015-010799

Valtorta NK, Kanaan M, Gilbody S, Ronzi S, Hanratty B (2016b) Loneliness and social isolation as risk factors for coronary heart disease and stroke: systematic review and meta-analysis of longitudinal observational studies. Heart 102(13):1009–1016. https://doi.org/10.1136/heartjnl-2015-308790

Wang J, Mann F, Lloyd-Evans B, Ma R, Johnson S (2018) Associations between loneliness and perceived social support and outcomes of mental health problems: a systematic review. BMC Psychiatry 18(1):156. https://doi.org/10.1186/s12888-018-1736-5

Wang G, Zhang Y, Zhao J, Zhang J, Jiang F (2020) Mitigate the effects of home confinement on children during the COVID-19 outbreak. Lancet. https://doi.org/10.1016/S0140-6736(20)30547-X

Wilasang C, Sararat C, Jitsuk NC, Yolai N, Thammawijaya P, Auewarakul P, Modchang C (2020) Reduction in effective reproduction number of COVID-19 is higher in countries employing active case detection with prompt isolation. J Travel Med. https://doi.org/10.1093/jtm/taaa095

Yazdanpanah F, Hamblin MR, Rezaei N (2020a) The immune system and COVID-19: friend or foe? Life Sci 256:117900. https://doi.org/10.1016/j.lfs.2020.117900

Yazdanpanah N, Saghazadeh A, Rezaei N (2020b) Anosmia: a missing link in the neuroimmunology of coronavirus disease 2019 (COVID-19). Rev Neurosci. https://doi.org/10.1515/revneuro-2020-0039

Yousefzadegan S, Rezaei N (2020) Case report: death due to COVID-19 in three brothers. Am J Trop Med Hyg 102(6):1203–1204. https://doi.org/10.4269/ajtmh.20-0240

The Main Sources and Potential Effects of COVID-19-Related Discrimination

39

Piotr Rzymski, Hanna Mamzer,
and Michał Nowicki

Abstract

The outbreak of a new coronavirus disease (COVID-19), which appeared in late 2019 and eventually resulted in the announcement of a pandemic by the World Health Organization, led to global fear and panic as well as the spread of false information and fake news from different sources. As a result, a sharp increase in prejudice, discrimination, and xenophobia against different groups of people was observed in different geographical locations. This chapter presents the psychological and social sources of stereotypes and prejudices that take forms in the COVID-19 pandemic. These sources can be located in psychosocial processes, such as (i) socially generated and reinforced fears; (ii) human responses to stress induced by certain types of stimuli; (iii) sense of helplessness based on the lack of control over reality; (iv) psychological responses reinforced by conformism (crowd psychology); and (v) the stigmatization process. The chapter also presents the main groups of increased risk of experiencing prejudice and discrimination during the COVID-19 pandemic (Asians, health-care workers, COVID-19 patients, and their relatives). Moreover, it provides a documented example of such behaviors. The groups at higher risk of more adverse effects of COVID-19 due to pre-pandemic discrimination are also discussed. Finally, initiatives taken to mitigate the discrimination associated with COVID-19 are presented, as well as the recommendations and good practices for preventing these behaviors during future outbreaks and for limiting discrimination against COVID-19 until the disease can be contained.

Keywords

COVID-19 · Discrimination · Fear · Prejudice · SARS-CoV-2 · Stigma · Xenophobia

P. Rzymski (✉)
Department of Environmental Medicine, Poznan University of Medical Sciences, Poznań, Poland

Network of Immunity in Infection, Malignancy and Autoimmunity (NIIMA), Universal Scientific Education and Research Network (USERN), Poznań, Poland
e-mail: rzymskipiotr@ump.edu.pl

H. Mamzer
Faculty of Sociology Adam Mickiewicz University, Szamarzewskiego 89 c, 60-568, Poznań, Poland

M. Nowicki
Chair of Histology and Embryology, Poznan University of Medical Sciences, Poznań, Poland

N. Rezaei (ed.), *Coronavirus Disease - COVID-19*, Advances in Experimental Medicine and Biology 1318, https://doi.org/10.1007/978-3-030-63761-3_39

39.1 Introduction

Although modern human history has already dealt with the epidemic and pandemic of infectious diseases, the pandemic of the new coronavirus disease (COVID-19) associated with the beta coronavirus or the severe acute respiratory syndrome coronavirus 2 (SARS-CoV-2) has met with an unprecedented response on many levels. Firstly, it has received an extraordinary joint reaction from the global scientific community (Deng 2020; Yi et al. 2020). The Chinese Center for Disease Control and Prevention (China CDC) confirmed a cluster of severe acute pneumonia of unknown etiology on 31 December 2019 (Chen et al. 2020a). Complete genome sequences of the causative agent (provisionally named "2019n-CoV" and later officially classified as SARS-CoV-2 by the International Committee on Taxonomy of Viruses (Gorbalenya et al. 2020)) were made publicly available by Chinese researchers 11 days later (Zhang 2020). At the end of January 2020, a team of Australian scientists isolated the virus and maintained its passage in cell culture (Caly et al. 2020). For comparison, SARS-CoV, the causative agent of the severe acute respiratory syndrome (SARS), was identified in late March 2003 with full-genome sequences made available after 5 months following the identification of the first case in November 2002 (Hawkey et al. 2003; Marra et al. 2003; Rota et al. 2003; Zeng et al. 2003). All these rapid achievements in January 2020 allowed not only to develop diagnostic tools of SARS-CoV-2 infection (Corman et al. 2020; Ai et al. 2020) but also to study the viral mechanism of cell entry and propagation (Hoffmann et al. 2020; Ou et al. 2020) and to initiate first experimental and clinical studies on drugs and the development of vaccine candidates (Rismanbaf 2020; Zhang and Wang 2020; Yao et al. 2020). At the end of April 2020, over 90 candidates were tested, with 8 entering a phase I clinical trial (Thanh Le et al. 2020; Callaway 2020). The number of pharmaceuticals has been repurposed to evaluate them as a treatment option of COVID-19 (Rosa and Santos 2020). *An interactive online dashboard updated by Johns Hopkins University was developed to track the daily global COVID-19 situation* (Dong et al. 2020). Within the first 4 months following an outbreak, over 8000 papers – original research, reviews, and commentaries – were searchable via PubMed/Medline database with key terms "SARS-CoV-2" and "COVID-19." These papers covered and discussed all the essential aspects of the pandemic and its etiological factor, encompassing molecular biology and pathogenicity of SARS-CoV-2, diagnostics of infection and its clinical manifestations, potential treatment options, epidemiological data, mathematical modeling of disease transmission and mitigation, and preventative measures (Rothan and Byrareddy 2020; Huang et al. 2020; Cao et al. 2020; Tang et al. 2020; Andersen et al. 2020; Wu et al. 2020a; Ozma et al. 2020). At the same time, authors were increasingly posting online preprints to facilitate the swift distribution of scientific information (Kupferschmidt 2020; Majumder and Mandl 2020). A number of publishers have taken an effort to create collections of articles devoted to COVID-19, ensure timely peer-review process, and provide an online review forum for preprints (Johansson and Saderi 2020; Wellcome Trust 2020; Eisen and Akhmanova 2020; Mamzer 2020). One could, therefore, potentially assume that the size of scientific response seen during the COVID-19 outbreak should effectively prevent a spread of misinformation, counteract fake news, and positively mitigate prejudice, discrimination, and xenophobia related to COVID-19. However, at the same time, an ongoing outbreak has received enormous coverage from worldwide mass media. As shown by LexisNexis, only during January, over 41,000 English language print news articles on coronavirus were published, and over 45% of them were included in the headlines (Ducharme 2020). The print and online articles were heavily accompanied by graphical content, often images of medical staff in special biohazard protective clothing as well as people, in early outbreak stages those of Asian origin, wearing a facemask in public spaces. These masks, mostly surgical ones, have quickly become a symbol of the COVID-19 outbreak. However, contrary to Australia, Africa, Europe, South America, and

North America, masks were already widely used in various Asian countries for various purposes – not only to prevent infections but also for the protection from allergens and pollution, as well as a "social firewall" (Burgess and Horii 2012; Wada et al. 2012). This cultural discrepancy could, therefore, create an unsupported notion that all individuals of Asian origin, located outside the Asian continent, particularly those wearing face masks during autumn to winter season, are maybe a source of COVID-19 infection (Stoye 2020; Rzymski and Nowicki 2020b). Furthermore, a number of studies clearly show that mass media are often dominated by sensationalism (Ransohoff and Ransohoff 2001), negative news are reported more frequently and receive more air time (Johnson 1996; Garz 2014), and "clickbait" techniques (that means content, most often a headline and a related image, is purposely designed to generate the highest revenue) effectively attract attention and encourage forwarding of the material over online social networks at the expenses of accuracy and quality of information (Burger et al. 2019; Bolton and Yaxley 2017). All in all, this has partially promoted misinformation and misconceptions on COVID-19 (Shimizu 2020; Bastani and Bahrami 2020; Kouzy et al. 2020) and could have added to prejudice and xenophobic reactions in different geographical locations. The media were reporting on "deadly virus" (Wong 2020), "killer coronavirus" (Matthews et al. 2020), "alarming spread" (Hindu 2020), and "highly contagious disease" (Kasraoui 2020) and provided daily updates on the number of confirmed cases and death toll. Compared with the previous epidemic outbreaks, COVID-19 has emerged during a time of increasing global use of social media that plays an important role in the dissemination of various information. Social media, such as Facebook or Twitter, were forced to undertake actions to combat fake news related to the COVID-19 outbreak, although their users are still able to share rumors, fake stories, and unsupported advice directly with each other. The modern social networks have also provided a platform for a rapid spread of conspiracy theories, such as that a SARS-CoV-2 was purposely designed as a biological weapon (Calisher et al. 2020; Mian and Khan 2020; Ioannidis 2020). As highlighted, social media played an important role in the spread of fear and panic (Depoux et al. 2020). All of this represented a serious threat that evidence-based information is not reaching the public in a sufficient manner despite efforts undertaken by health authorities and the academia. The World Health Organization (WHO) has called the spread of misinformation and fake news related to COVID-19 and SARS-CoV-2 as "infodemic" (Zarocostas 2020).

COVID-19 outbreak has also surged global political tensions with different leaders who made public accusations for the global health threat. These accusations were most often directed towards China or Asian countries in general and could have also fueled xenophobia. The WHO issued a guideline in 2015 that explicitly states to avoid a previous practice of terming diseases and their etiological agents with references to geographical locations (e.g., Middle East Respiratory Syndrome), people's names (e.g., Creutzfeldt-Jakob disease), species of animal (e.g., swine flu), cultural, population, occupational, and industrial references. It also clearly highlights that terms inciting undue fear should not be used (Fukuda et al. 2015). That, however, has been violated not only by the mass media, e.g., referencing to SARS-CoV-2 as "deadly virus," but also by selected prominent political leaders who made statements with terms, such as "Chinese virus." The latter, massively repeated through social media (Budhwani and Sun 2020), possibly contributed to xenophobia towards the people of Chinese origin, while the former could potentially add to fears and prejudice not only towards infected people and their relatives but also medical staff.

In consequence, the COVID-19 outbreak has increased the magnitude of the backlash of prejudiced and discriminatory reactions towards different groups of people. Such behaviors are not uncommon during outbreaks and have been seen before (Person et al. 2004; White 2020; O'Shea et al. 2020); however, the COVID-19 pandemic has happened during unprecedented times in terms of global information spread and sharing, and the importance of social and worldwide mass media in this process is more than ever. Therefore, some of these reactions may be further magnified as compared with other outbreaks. Moreover, certain specific features of COVID-19 may also add to the prejudice and discrimination. Firstly, the disease and its etiological factors have been frequently referred to as "novel" (Jiang et al. 2020), which could promote the fears of the danger of the unknown. Secondly, evidence of the presymptomatic spread of SARS-CoV-2 had been accumulating with some studies showing that the virus can be infectious already 2–3 days before the onset of symptoms, reaching its peak half a day before their occurrence. Although such data is important to model the disease transmission and incorporate effective mitigation strategies, it can also drive fear and anxiety over contracting the SARS-CoV-2 and increase a level of suspicion that every encountered subject can be a source of the infection. Apart from the hygiene regime, social distancing interventions are effective in lowering the spread of infectious agents, such as SARS-CoV-2 (Ferguson et al. 2009; Matrajt and Leung 2020). However, one should note that they also have various unintended effects. Besides mental health effects, they can reinforce prejudice towards other individuals, which may persist even after lessening the restriction measures and reopening after lockdowns (Venkatesh and Edirappuli 2020). Moreover, at the beginning of the outbreak, the stability of SARS-CoV-2 outside a host was unknown and could only be hypothesized and partially predicted from experiences with other coronaviruses (Kampf et al. 2020). The experimental studies have, however, shown that it can remain viable for a few hours on copper surfaces, tissues, and print paper; 3 days on plastic, stainless steel, and banknotes; and up to a week on a surgical face mask (Chin et al. 2020; van Doremalen et al. 2020). These investigations were conducted under immobilized and stable conditions; therefore, they should be interpreted with caution as regards extrapolation to the realistic scenario under which both temperature and humidity are changed, while objects are subject to movement. Nevertheless, the possibility of SARS-CoV-2 to survive longer on plastic and stainless steel has added to anxiety and fear that a virus can be contracted during different daily activities shared with other people (e.g., shopping).

This chapter reviews the psychological and social sources of stereotypes and prejudices that take strong forms in the situation of the COVID-19 pandemic. As outlined, these sources can lie in the following psychosocial processes: i. socially generated and reinforced fears; ii. human responses to stress induced by certain types of stimuli; iii. sense of helplessness based on the lack of control over reality; iv. psychological responses reinforced by conformism (crowd psychology); and v. the stigmatization process. The chapter also presents selected examples of behaviors, including those documented during the COVID-19 outbreak, to illustrate the abovementioned mechanisms better. The main initiatives to prevent and minimize discrimination during the pandemic are also given. Finally, a list of recommendations on preventative measures to be developed and implemented before future outbreaks is provided.

39.2 Basic Functions of Stereotypes and Prejudices

Stereotypes and prejudices have important pragmatic roles in the functioning of a society. Their most in-depth, basically biological background refers to primary, evolutionary old reactions resulting from physiologically controlled behavior associated with deep brain structures. Their task is to provide behavioral strategies that maximize the chance of survival. Currently, in the era of living in post-modern societies based on developed cultures, the functions of stereotypes and prejudices are limited to the right selection of information and the elimination of unnecessary data. Judgments based on stereotypes and prejudices may be misleading, but they are quick (Aronson et al. 1994). It allows for equally fast reactions. In prehistoric contexts, such rapid reactions, especially those related to fleeing, it repeatedly saved lives and therefore constituted the quintessence of the self-preservation instinct. At present, their functions are less critical in avoiding physical threats, but the selection of information is adequate. The functioning of stereotypes results from the specificity of human brain functioning, categorization of information, and it is impossible to disable this process. As shown in neurobiological studies, the stereotypic attitudes can only be moderated and inhibited, a process that employs specific areas of the frontal lobe, such as the dorsolateral prefrontal cortex that is involved in higher-order processing (Kadota et al. 2010; Knutson et al. 2007).

In the social sciences and humanities, there is a consensus on how to define stereotypes and prejudices (Zimbardo 1994). It is assumed that the stereotype is an exaggerated generalization, ascribing identical features to each person belonging to a given social category, without taking into account the real differences between individuals. In this sense, stereotypes are an element of larger units. They are cognitive elements of attitudes or permanent structures made up of emotional, behavioral, and cognitive elements which express the attitude of its host to an object, a person or a group of people, or a phenomenon. By contrast, prejudice is a much broader concept, defined as an attitude towards a social category based on false or incomplete information.

It is common for both concepts that in today's Western civilization, judging other people through stereotypes and social prejudices is an undesirable and negative behavior (Angermeyer and Matschinger 2005; Baumann 2007; Rössler 2016; Puddifoot 2019).

It might be due to changes in the realities we live in today, compared with the realities our ancestors lived in before. In the absence of direct threats from competition for the primary resources indispensable for survival, Western cultures assume that the use of stereotypes is not justified. Nevertheless, it is still challenging to stay distant from stereotypes, because, as mentioned earlier, their origins remain in atavistic physiological mechanisms (Link et al. 1999; Marcelin et al. 2019; Pascoe and Smart Richman 2009).

39.3 Psychosocial Sources of Discrimination, Stereotypes, and Prejudices

Some specific circumstances increase the probability of triggering stereotypes and prejudices; these are all situations that generate a sense of threat. Appearing fears cause an increase in the level of mistrust and suspicion towards other people (Giddens 1991). Again, from effective behavioral strategies aimed at maximizing the chances of survival, this is a highly efficient approach. In the cultural world of a man, however, moral and ethical judgments that require the elimination of stereotypes are involved in the whole process. Hence, in situations commonly considered by people as threatening, many emotional tensions appear, as well as emotional–rational thinking tensions.

The outbreak of the COVID-19 pandemic is a phenomenon that people have not experienced in the last century (except for the Spanish Flu pandemic). The memory of such situations is therefore no longer appropriately embedded socially; hence it is difficult to refer to the knowledge retained on this subject, which might be based on direct experience. Such knowledge has, of course, been recorded in other forms, in the form of notes, chronicles, and diaries or fiction. Excellent examples of this kind of publication are Albert Camus "Plague" and its prototype – Daniel Defoe's "A Journal of the Plague Year." Reading these books and combining them with the observations of the surrounding reality reveal the existence of universal regularities in human behavior. The COVID-19 pandemic generates many fears leading to the appearance of intensified stigmatizing reactions, which are based on stereotypes. Unfortunately, these reactions are reinforced by messages conveying political propaganda and by the media's search for sensations and the focus on presenting exaggerated reactions. The worldwide COVID-19 pandemic is a very powerful stressor-like stimulus, and as Selye (1976) points out in his biological concept of stress, it is a kind of stimulus that knocks the body out of the homeostatic balance. Moreover, we encounter here an imbalance in both types of homeostasis, psychological and biological (somatic).

Psychological homeostasis is mainly based on the sense of security, and security is embedded in the predictability of events. Once the events are predictable, the situation is perceived as safe and controllable by people. These perceptions might be delusional, as the events may not necessarily be predictable, but if people perceive them as such, they are accompanied by a sense of acceptance, peace, and precisely a sense of situational awareness. Moreover, biological homeostasis results from somatic integrity and lack of health disorders. The loss of health (particularly broadly understood – as proposed by the WHO[1]) is one of the most extreme stressors disrupting homeostasis[2] and activating all possible actions as well as mobilizing resources needed to restore the lost balance. In this sense, stimuli causing somatic disorders are perceived with increased sensitivity and are accompanied by high levels of emotional tension.

There is another factor that has a significant impact on human behavior and well-being; however, this impact is almost completely ignored by the Western model public health system. It is the sense of self-efficacy. Seligman (1975) classic experiments showed that a profound lack of control over reality leads to mental disorders ranging from depressed moods to advanced forms of exogenous depression (Rotter 1966, 1990). It is also confirmed by an animal model research where the generation of a sense of stressor control was associated with reduced glucocorticosteroid levels and pain – despite the absence of corresponding modulations of the intensity of a substantial stressor (Maier et al. 1986; Dess et al. 1983; Maier et al. 1982). The influence of the sense of control over the stressor on psychophysiological reactions was also proven by an experimental research involving humans (Lundberg and Frankenhaeuser 1978; Houston 1972). During a pandemic, where an individual has very little or no influence on the threat, it is particularly essential to generate a sense of self-efficacy. A type of a positive psychological influence on society is the widely used encouragement to maintain basic hygiene principles, such as handwashing and disinfection of objects. Citizens, therefore, have a sense of influence on the situation, as they are personally responsible for these principles. Paradoxically, in a situation of danger and lack of control and influence on the environment, such a mechanism may provide a kind of psychological support – by being able to control their own behavior, people get the impression of controlling the danger, thus reducing, at least partially, the level of stress accompanying the epidemiological situation.

In the nineteenth century, French social psychologist and sociologist Gustav Le Bon (1897) described the mechanisms of social functioning in an exceptionally adequate manner. His classic work "Psychology of Crowds" is a universal description of human behavior in large and anonymous groups. Le Bon points to the irrationality of the "crowd," the vehemence of its behaviors, unidirectionality and exaggeration, the use of emotions, and the impossibility of rational control over it. Such behaviors of the "crowd" (in current post-modern societies, we would say the "mass audience") are very dangerous; they threaten the safety of the people and contribute to causing harm, particularly to those social categories that are weak.

39.4 Discrimination and Prejudice Under a Pandemic Scenario

Discrimination is a process where five consecutive and overlapping stages can be distinguished, passing smoothly from one stage to another. The initial stage is described as a psychological distance between an individual and the others.

[1] WHO's Definition of Health; https://www.who.int/about/who-we-are/frequently-asked-questions (Accessed 2-April 2020)

[2] In 1970 Thomas H. Holmes and his colleagues constructed a list of stressful events. This is a scale that measures stress in the Life Change Units (LCU). According to Holmes, people who scored more than 300 LCU in total were prone to suffer from a serious disease within 2 years (Zimbardo 1994). Loss of one' s health is presented as the sixth most stressful life situation (after death of a spouse, divorce, separation, or death of a close family member) on the list consisting of 43 different situations.

Distancing occurs in the absence or exclusion of empathy. It rapidly turns into devaluation and thus depreciation of the value of an individual. Pejorative terms are introduced to social practice, such as "animals," "subhumans," "others," "barbarians," "they," and even "cockroaches," "rats," or "plague" (Sontag 1979, 1989). People described with these names lose their humanity in the eyes of a person who makes such a description. It activates the possibility of proceeding to the next stage of the process, the delegitimization and thus the sanctioning of the deprivation of rights of a given social category (e.g., the virus carriers). At this stage, the psychological process is transformed into a social process and leads to concrete actions towards other people (e.g., labeling of sick people's homes). The fourth stage is physical segregation (creation of ghettos, places of isolation and closed camps, or, e.g, of "total institutions" as described by (Goffman 1986), such as prisons, hospitals, or infected and contaminated zones, and even "contaminated" houses' labeling.[3]). The most drastic is, however, the fifth stage, which can assume the form of physical extermination and therefore taking the lives of people. The gradual nature of the phenomenon of discrimination and the swift transformation from one stage to another make it difficult to feel and notice how, out of the seemingly innocent phenomenon of emotional detachment, reactions take the form of segregation leading to extermination.

What is very perilous is the fact that discriminatory processes occur at different levels simultaneously and concern different social categories. In a pandemic, people with more resources (material and financial, but above all, the resources in the form of social and cultural capital) have better abilities to cope with the challenge of the disease. People with no resources have significantly less access to these opportunities,[4] which can cause an increase in fear and anxiety. The mechanism of combining discrimination with stereotypes consists in attributing the act of doing (the guilt) to some social category based on beliefs (faith) and not the facts. Most often, as Sontag (1979, 1989) points out, people that are accused of carrying the disease are "strangers," i.e., people who differ from the dominant majority (a distinctive feature may be any possible trait: gender, age, and ethnicity or religion, as well as lifestyle with dietary, sexual, or other preferences). The psychological mechanism of the "scapegoat," which otherwise performs important social functions, works precisely in this way. The consequences of the discriminatory processes are always negative, both socially and psychologically. For an effectively, rationally, and functionally well-managed community, they should be eliminated through systemic, educational, and grassroots solutions.

Activities that may seem rational in the pandemic are potentially burdened with a high risk of negative social consequences – the initiation of stereotyping and stigmatization processes, i.e., psychological and social processes based on emotions. This kind of behavior (as Le Bon pointed out, by the way) is potentially very dangerous and, in reality, very difficult to eradicate and control.

They can occur at the following levels:

- Individual activities – stigmatizing sick people, suspected of infection, and blaming people of foreign origin for bringing the disease to Poland – scapegoat mechanism;
- Individual actions, but grouped in the form of collective actions, such as the crowd behavior described by Le Bon (hate and aversion, rejection and labeling of medical personnel and

[3]https://tvn24.pl/bialystok/wysokie-mazowieckie-starosta-bogdan-zielinski-proponuje-oznaczanie-domow-osob-zakazonych-koronawirusem-4559094 [accessed on April 20.2020].

[4]For example, without savings, they cannot accumulate food stocks. They have to buy products on a daily basis, and this exposes them to the virus and disease. Lack of knowledge, awareness of the mechanisms of a virus's impact and transmission, etc. makes it impossible to protect effectively against the virus due to a lack of understanding of the importance of certain procedures. For example, wealthier people can buy tests to identify the virus in their bodies. Many employers are beginning to verbalize expectations that they will only employ healthy people. Therefore, poorer people are losing equal opportunities on the labor market.

patients and their families as potential COVID-19 vectors);

- The activities of enterprises, which consist of refusing to serve selected groups of people due to fear of transmission of infection;
- Local administration initiatives – here the examples are the proposals to label houses where people are either infected with SARS-CoV-2 or in quarantine;
- And governments' measures – for example, introducing mechanisms to control citizens' behavior.

39.5 The Main Targets of COVID-19-Related Prejudice and Discrimination

Considering that the COVID-19 pandemic is a complicated situation affecting various sectors (e.g., health and economy), the associated prejudice and discrimination may be faced by different, often unrelated, groups of individuals. As opposed to previous outbreaks of discrimination that were solely based on skin color, sexual orientation, or some form of disability (Casey et al. 2019; Krnjacki et al. 2018; Monk 2015), the COVID-19-related discrimination may have sources in geographical origin, health status, and specific occupation. It has been predominantly affecting the following groups of individuals: (i) Chinese citizens; (ii) Asian descents; (iii) patients suffering from COVID-19; (iv) relatives of COVID-19 patients; and (v) health-care workers (Fig. 39.1). The sources, examples, and potential outcomes of COVID-19 discrimination in these groups are provided in the subsequent subsections. Moreover, certain minorities – Native Americans and African Americans – suffered from pre-pandemic effects of discrimination that, at least partially, resulted in an increased rate of infectivity and mortality due to COVID-19. This aspect has been discussed in the subsection "Other groups."

39.5.1 Chinese Citizens

Chinese citizens located in China and the Asian continent were the target of discrimination due to accusations for being responsible for the COVID-19 outbreak and its consequences. It may be further magnified by media reports and political leaders deliberately using references, such as "Chinese virus," effectively demonizing individuals inhabiting China (Budhwani and Sun 2020). Even within China, the stigma related to COVID-19 was recorded; e.g., hotels in Yunnan province declined to fulfill the pre-booked reservations from guests coming from Hubei, regardless of their health conditions (He et al. 2020). Some accommodation places had reportedly

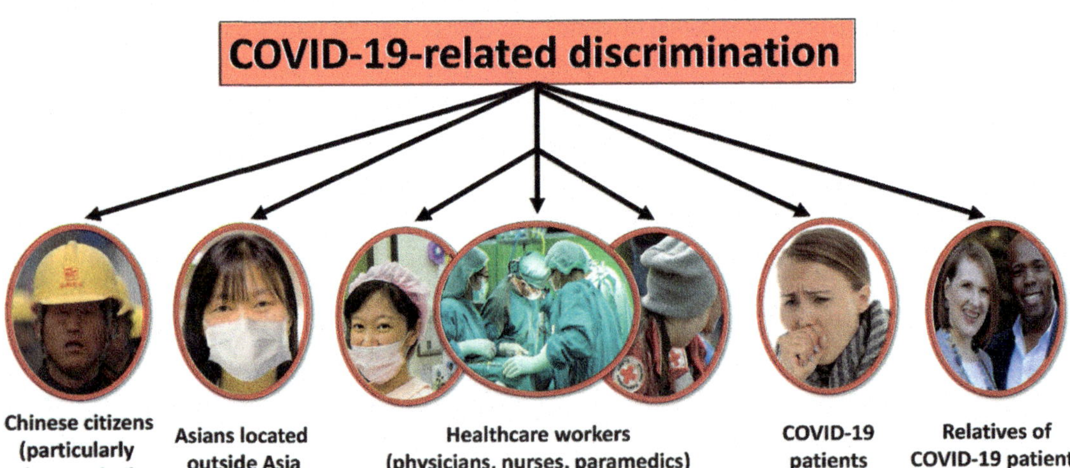

Fig. 39.1 The groups of individuals with increased risk of prejudice and discrimination related to COVID-19

hung signs stating that "people from Wuhan and cars from Hubei are not welcomed here," while travelers from this province had their personal data, including ID number and address, posted online (World Journal 2020). Visitors from Hubei province were also facing blockades set by rural and urban communities that effectively stopped them from further traveling. Importantly, these blockades did not discriminate visitors based on medical checks, which were not performed and not verified. Cars with Hubei's registration numbers have been subject to attacks in various Chinese provinces (He et al. 2020). As shown via a survey performed across 31 provinces in Mainland China, over 15% of respondents declared to expel all individuals from Hubei from their own communities actively and 50% reported that they would avoid people associated with Hubei. In contrast, nearly 90% were committed to reporting to local authorities the appearance of anyone from Hubei in their community (He et al. 2020).

39.5.2 Asians Located Outside Asia

Individuals of Asian origin located outside an Asian continent (students and workers) were subject to discrimination related to COVID-19 due to: (i) the emergence of this disease in an Asian country (China) and (ii) the culture of wearing face masks in public spaces, (a practice not commonly practiced in Western countries) which due to media-coverage has quickly become one of the symbols of the COVID-19 pandemic. All in all, these subjects may be victims of unsupported assumptions that every Asian, especially the one wearing a face mask in public, is a potential source of SARS-CoV-2 infection, as well as victims of accusations to be responsible for an outbreak and its consequences (Rzymski and Nowicki 2020a; He et al. 2020). The COVID-19-related discrimination in this regard may also be associated with the views on Chinese dietary habits, particularly the consumption of wild animals sold at wet markets. It is due to the zoonotic origin of COVID-19, the emergence of SARS-CoV-2 in bats, the potential role of the intermediate host (e.g., pangolins) in transmission to human, and the reported association of initial COVID-19 cases with the Huanan Seafood Wholesale Market in Wuhan (Andersen et al. 2020; Ji et al. 2020; Wan et al. 2020). These practices are often considered unacceptable by members of other cultures, although they are also not revealed by people of Asian descent. To work or study abroad, particularly outside the Asian continent, these individuals must travel far from home, adapt to another culture and traditions, and overcome a language barrier. Therefore, experiencing COVID-19-related prejudice/xenophobia may add to the feeling of isolation, jeopardize the adaptation to a new culture, and affect mental health. Finally, facing discrimination may adversely affect the care-seeking behaviors of individuals of Asian descent, as seen during the SARS outbreak in 2003 (Person et al. 2004; Bruns et al. 2020).

As the epidemic COVID-19 situation evolved, the prejudice and discrimination behaviors towards Asian descents were numerously reported by mass media from different parts of the world. In the USA, the University of California, Berkeley, issued an online infographic, for which it later apologized, listing the normal reactions that one can experience during the COVID-19 pandemic and including the following on the list: "Xenophobia: fears about interacting with those who might be from Asia and guilt about these feelings" (Asmelash 2020). The range of discriminatory acts varied from verbal abuse to violent attacks and occurred in public spaces, such as on the streets, in public transportation, and on campuses as well as in shops (Campbell 2020; Sosa and Brown 2020; Zhao 2020). The research surveying Asian students in Poland found that over 60% had faced some form of prejudice and discrimination related to COVID-19, even though the study took place before the first case was confirmed in this country (Rzymski and Nowicki 2020a). A survey of Chinese residents overseas across 70 countries has found that 25% of surveyed had experienced discrimination related to COVID-19. Its most commonly reported manifestations included being laid-off without proper cause, abuses in the public spaces, and rejection of rental housing. Interestingly, respondents living in countries with a higher number of confirmed cases of COVID-19 were less likely to report cases of discrimination (He et al. 2020). It points out that disease-related discrimination rate may be higher before the arrival of an infectious agent and underlines the need for preventive measures to be implemented as early as possible during future outbreaks. The discriminatory behaviors appeared to affect more likely those individuals of Asian origin who tended to wear face masks in public (Rzymski and Nowicki 2020a). Before implementation of their universal use, wearing face masks implied sickness in various regions outside Eastern Asian countries (in which their use is frequent and linked to a broad range of motivations). These cultural differences contributed to the increased discrimination under the circumstances of novel infectious disease that emerged in China.

39.5.3 Health-Care Workers

Health-care workers, such as physicians, nurses, and paramedics, were at increased risk of prejudice and discrimination during the COVID-19 outbreak. This group constitutes an essential front line in the identification and treatment of patients with COVID-19, and this may result in public fears as well as fears of their relatives that they can represent a source of SARS-CoV-2 infection (Rana et al. 2020). That is due to their direct contact with COVID-19 patients, including severe cases, and increased risk of infection, despite protective equipment and other measures undertaken to prevent transmissions in various regions. Health-care workers revealed increased rates of infection in various world regions (The Lancet 2020; International Council of Nurses 2020; Folgueira et al. 2020), and these statistics were immediately covered by mass media, adding to fears of contracting the disease (Hirji 2020).

Some prejudice and discrimination behaviors related to COVID-19 and aimed at medical personnel were reported, including highly violent acts. For example, in the Philippines, an ambulance driver was shot for parking his vehicle in a residential area, a hospital utility worker was flushed with bleach, a nurse was splattered with chlorine, residents were protesting against a return of some health-care workers to their home (Rubrico 2020). The verbal and physical assaults of health-care workers were also reported in other countries, e.g., the United Kingdom (Coates 2020). The health-care workers were also targeted in Mexico by those who feared the spread of the virus; the reported incidences included physical assaults in the streets, flushing with bleach, and banning from the use of public transportation (Rios 2020). Similarly, in Argentina, incidences of denying public transportation to health-care workers on their way to the hospital were recorded (Universidad 2020). In India, medical doctors wearing personal protective equipment were assaulted by a mob throwing stones (Pandey 2020). Attacks on ambulances were also reported (The Economic Times 2020). In Colombia, a pediatrician and his family first received death threats, and later their apartment in Bogotá was attacked (Cabrera 2020). The refusals to service medical doctors in shops, bakeries, gas stations were also reported from various world regions (D'Angelo 2020; Wroński 2020).Moreover, the stigmatization of health-care workers may also result in prejudice directed towards their families. As reported in Japan, children of medical professionals were shut out from daycare centers or asked to provide proof that they are not infected with SARS-CoV-2. In India, families of physicians have been ostracized by their neighbors, fearing that they may be a source of SARS-CoV-2 (Pandey 2020).

One should note that medical workers working on the "front line" are by itself more prone to burnout feelings, stress, and trauma due to overwhelmed health system, equipment shortages, and often challenging decisions to be undertaken (Wu et al. 2020b; Zhang et al. 2020; Tsamakis et al. 2020; Chen et al. 2020b). Facing prejudice and stigma, also by their families, may further add to the burdens of an already stretched workforce and drive the burnout feeling, decrease the level of job satisfaction, and lead to mental health issues.

39.5.4 COVID-19 Patients and Their Relatives

Patients who suffer from COVID-19 could face discrimination due to fears exhibited by family, relatives, and society. The stigmatization related to infectious diseases (e.g., Ebola, HCV, and HIV) has been numerously observed in the past (Pellecchia et al. 2015; Zacks et al. 2006; Crandall and Coleman 1992). Facing prejudice may add to stress and the feeling of isolation already generated by quarantine/hospitalization and affect the mental health of patients, including convalescents. During the COVID-19 pandemic number of discrimination behaviors towards infected patients was seen. In Colombia, a mob was throwing stones at houses of hospitalized patients (Semana 2020). In Russia, COVID-19 patients reported online bullying, threats, and intimidation, while the deputy health minister of the Bashkortostan publicly admitted that "there is such an aggressive stigma that people cannot even leave the hospital" (Cordell 2020). While in Kenya, a man was beaten to death by a mob after being suspected to be SARS-CoV-2-positive (Dyer 2020). As reported by the Argentinian National Institute Against Discrimination (Instituto Nacional contra la Discriminación), the number of reported cases of discrimination acts towards patients was outnumbering this towards health-care workers and included, but not limited to, harassments from the neighborhood, online publications of pictures of COVID-19 patients, and placing warning signs at their homes (Universidad 2020).

The prejudice and discrimination aimed at COVID-19 patients may be magnified by insufficient measures undertaken to ensure patient confidentiality; this is challenging since authorities need to take steps to identify those who maybe have been exposed to SARS-CoV-2. Not obeying the law of self-quarantine of patients with mild symptoms may also add to discrimination – as reported in Vietnam, a case of two Muslim COVID-19 attending public events have resulted in demands to imprison the entire Muslim population in the country (Vietgiaitri 2020).

The discrimination of COVID-19 patients could, to some extent, play a role in health outcomes for infected individuals, as it could effectively delay reporting the onset of clinical symptoms to emergency and subsequently result in worsening the disease outcomes (Bruns et al. 2020). Such delays were observed, e.g., in the United States (McNamara 2020), also in case of clinically disturbing symptoms, such as COVID-19-associated large-vessel strokes (Oxley et al. 2020).

Moreover, the family of COVID-19 patients was also at higher risk of facing accusations of spreading disease and related prejudice and stigma due to societal pressures and fears. For example, it was reported that one family in Poland experienced denials to shop in stores located in their hometown – a case that forced local authorities to issue a statement condemning such behaviors (Partyła 2020). In Ghana, a family of deceased COVID-19 patient, whose name and the family name was announced on the local radio, had to face stigmatization from the local community (GhanaWeb 2020). The COVID-19-related prejudice and discrimination towards relatives of COVID-19 patients may disturb the social functioning and generate the feeling of exclusion.

39.5.5 Other Groups

Particularly vulnerable to stigmatization in times of increased anxiety caused by a pandemic are the "Others,"that is, those whose behaviors, actions, and appearance are significantly different from the dominant majority. An example is the Navajo Indian community in the states of Utah, Arizona, and New Mexico in the United States. As reported, the infectivity of Navahos on the reservation is 22-fold higher compared with the national average with a death rate of approximately 4% (Soto and Hakim 2020). It can be hypothesized that this group may reveal some specific biological factors essential for SARS-CoV-2 cell entry, especially higher expression of factors, such as angiotensin-converting enzyme 2 (ACE-2). Considering that increased tissue ACE-2 levels have been evidenced for selected populations, i.e., East Asians (Cao et al. 2020), it remains unknown whether this is a case for the Navajo tribe. Besides, it is also unclear whether this could explain increased mortality. In turn, it is highly plausible that pre-pandemic discrimination has contributed to increased infectivity and mortality rate within this group. Native Americans are an example of chronically discriminated and stigmatized communities, a phenomenon that led to lower-income, wealth, poverty, and high rates of unemployment and addictions. It, in turn, would contribute to disparities in health status that result in age-adjusted death rate for adults exceeding by nearly 40% that of the general population and higher rate of deaths to cardiovascular diseases, chronic liver disease, cirrhosis, diabetes, influenza, pneumonia, and tuberculosis (Jones 2006; Whitesell et al. 2012; Sarche and Spicer 2008). Thus, it is not surprising that Native Americans experience more significant effects of the COVID-19 pandemic.

A similar issue of pre-pandemic discrimination resulting in worse health outcomes under the COVID-19 pandemic situation could also be seen in the case of African Americans. As reported in the United States, the rate of infection and death related to COVID-19 in predominantly black counties is over three-fold and six-fold higher compared with predominantly white counties. Moreover, the black population represents as little as approx. 13% in the United States, yet African Americans represent over 30% of COVID-19 hospitalization in this country (Yancy 2020). However, some biological hypotheses could be put forward to, at least partially, explain this phenomenon; one cannot disregard the long-term effect of racism experienced by the black community on its situation during the COVID-19 pandemic. Racism has a profound effect on social and health disparities (Nelson 2016; Brondolo et al. 2009), and these effects are further highlighted and magnified under the pandemic situation. Living in the racialized environment resulted in blacks having lower income, less wealth, worse (if any) health insurance, and poor access to education when compared with whites (Sohn 2017; Donnor 2013; Patten 2016). Under such circumstances, it comes as no surprise that long-standing discrimination of the black population in the United States and associated trauma and stress from poverty (Himmelstein et al. 2015; Britt-Spells et al. 2018) could have contributed to the higher vulnerability to COVID-19. Of note, during the early stages of the COVID-19 outbreak, a false claim that black people are immune to SARS-CoV-2 infection, which may also have a devastating effect on health outcomes for this community. The misinformation has likely arisen from the news that an individual of Cameroon-origin, who represented an initial group of COVID-19 cases in China, has responded well to treatment (Padayachee and du Toit 2020). This misinformation brought memories of the yellow fever epidemic in 1793; during which 400 blacks lost their lives due to false susceptibility claims (Hogarth 2019).

39.5.6 Future Groups

At the moment of preparation of this chapter, a vaccine for COVID-19 is not yet available, the pandemic is not yet contained, and it is expected that the number of infected people will experience seasonal rises. It may, therefore, create a situation under which new groups will face prejudice, discrimination, and stigma. It may particularly affect those traveling from regions temporarily facing worsening of the epidemiological situation. For example, it has already been reported that individuals of African descent are facing discrimination in China due to fears of imported transmissions of SARS-CoV-2 – Nigerians, Ghanaians, and Ugandans in Guangzhou have been rejected by hotels, evicted by landlords, and, in some cases, left homeless (Hood 2020).

39.6 Initiatives Undertaken to Decrease Discrimination Related to COVID-19

The discriminatory behaviors reported during the COVID-19 pandemic were met with some reactions and initiatives from different authorities and individuals. The most notable included:

- On 30 January 2020, WHO advised all countries to act against actions to promote stigma and discrimination during the COVID-19 outbreak (WHO 2020).
- Statements of political leaders aiming to prevent discrimination and show support to those at risk. The most prominent politicians involved in such actions included: Canadian prime minister Justin Trudeau, Italian president Sergio Mattarella, and mayor of Toronto John Tory (Global News 2020; NBC News 2020b; Il Messaggero 2020).
- Statement of the United Nations High Commissioner for Human Rights Michelle Bachelet on 24 February 2020 calling against a disturbing wave of prejudice against people of Chinese and East Asian ethnicity (Reuters 2020).
- Global online initiatives providing (besides other aims) support to health-care workers engaged in combating COVID-19, e.g., Global Citizen's "Together at Home."
- Various local and national initiatives aiming to support health-care workers (Jeffrey 2020; Hess 2020; Deshpande 2020).
- Statements supportive to health-care workers published by various authors in the peer-reviewed journals (Xiang et al. 2020; Adams and Walls 2020; Calisher et al. 2020).
- A website hosted by the Asian Pacific Policy and Planning Council and Chinese for Affirmative Action that allowed reporting the discrimination related to COVID-19 (CAASF 2020).
- A hotline launched for inhabitants of New York that allowed reporting discrimination related to COVID-19 (NBC News 2020a).
- And a call-to-action published in Science that urged the academic world to proactively develop policies and implement measures preventing discrimination against Asian students and staff located outside an Asian continent (Rzymski and Nowicki 2020b).

39.7 Recommendations to Decrease Discrimination During Future Outbreaks

Considering that prejudice and discrimination during outbreaks of infectious diseases are not uncommon and are also present during the COVID-19 pandemic, affecting different groups of individuals, there is a need to develop recommendations aiming at the prevention of this phenomenon during future epidemic. Practical implementations of preventive measures in this regard require commitment from various parties, such as political leaders, health authorities, academics, and non-governmental institutions. Mass and social media must be employed to increase the output of actions undertaken and their promotion (Parke 2007). Likely, future outbreaks will also result in epidemiological situations with global impacts. Therefore, the successful implementation of anti-discriminatory practices requires to follow the advice on intercultural communication that also regards reporting and visualizing public health risks (Grabill and Simmons 1998; Welhausen 2015).

Learning from the COVID-19 pandemic, we suggest 13 recommendations that may be useful in the prevention of future outbreak-related discrimination as well as in mitigating COVID-19 discrimination until the containment of this disease will become possible:

- Development and enforcement of international and national regulations preventing the promotion of pseudoscience and scientifically unsupported claims as both fuel fear and panic, and magnify prejudice and discrimination;
- Development of effective measures to maintain the confidentiality of those who are seek-

ing health care due to outbreak and all individuals who may be part of contact investigation;

- Development of guidelines on how to report and visualize health data to decrease adverse impacts on public perceptions, panic promotion, and fueling fear and stigma;
- Advising political leaders to follow the WHO guidelines when reporting on emerging public health threat and avoid referencing to its etiological factor by country of origin as this promotes unwanted prejudice and discrimination;
- Political leaders and health authorities should be encouraged to speak out against disease-related discrimination proactively and to release statements on "zero-tolerance" to such behavior, both in public as well as online space;
- Engagement of political leaders and national authorities in the proactive reduction of health disparities in minorities as these disparities magnify the adverse outcome of the pandemic;
- Engagement of social media to effectively block a material that contains signs of disease-related discrimination from online publication;
- Engagement of social media in supporting patients and health workers via reassuring posts and infographics;
- The strong commitment of academics in the prevention of the spread of fake news and disease-related discrimination – as seen during COVID-19 pandemic;
- Sharing the accurate information and raising awareness on the disease and its etiological factor in a non-emotional manner with particular attention to avoid increasing tension and fear as well as ensuring that shared information (e.g., in the form of infographics) does not reinforce stereotypes;
- Involvement of non-governmental organizations in monitoring and reporting of disease-related discrimination;
- Statements from international and national authorities to support health workers during outbreaks as they face high levels of stress and trauma;
- Promotion of social actions to thank and support health-care workers during their battle against outbreak;
- And the involvement of individuals associated with the management of international student exchange programs in the prevention of prejudice and discrimination that studies from specific geographical locations may face during the time of the outbreak.

39.8 Conclusion

The present chapter demonstrates that the COVID-19 pandemic has, predictably, resulted in the backlash of fear-driven prejudice and discrimination, which can lie in the standard psychological context. These behaviors affected different groups of individuals with Chinese citizens, people of Asian descent, COVID-19 patients, their families and relatives, and health-care workers being at the highest risk of experiencing them. At the same time, the number of initiatives has been undertaken to reduce the effects of discrimination. Future outbreaks will require the swift implementation of measures preventing prejudice and discrimination and to react a priori rather than a posteriori. It requires not only action from international and national authorities, academics, and non-governmental organizations but also the commitment of mass and social media to report on the epidemiological situation in a responsible manner and to prevent a spread of fear- and panic-fueling misinformation and fake news.

References

Adams JG, Walls RM (2020) Supporting the health care workforce during the COVID-19 global epidemic. JAMA 323(15):1439–1440. https://doi.org/10.1001/jama.2020.3972

Ai T, Yang Z, Hou H, Zhan C, Chen C, Lv W, Tao Q, Sun Z, Xia L (2020) Correlation of Chest CT and RT-PCR Testing in Coronavirus Disease 2019 (COVID-19) in China: A Report of 1014 Cases.

Radiology 0 (0):200642. doi:https://doi.org/10.1148/radiol.2020200642

Andersen KG, Rambaut A, Lipkin WI, Holmes EC, Garry RF (2020) The proximal origin of SARS-CoV-2. Nat Med. https://doi.org/10.1038/s41591-020-0820-9

Angermeyer MC, Matschinger H (2005) Labeling—stereotype—discrimination. Soc Psychiatry Psychiatr Epidemiol 40(5):391–395. https://doi.org/10.1007/s00127-005-0903-4

Aronson E, Wilson TD, Akert RM (1994) Social psychology: the heart and the mind. HarperCollins College Publishers

Asmelash L (2020) UC Berkeley faces backlash after stating 'xenophobia' is 'common' or 'normal' reaction to coronavirus. CNN

Bastani P, Bahrami MA (2020) COVID-19 related misinformation on social media: a qualitative study from Iran. J Med Internet Res. https://doi.org/10.2196/18932

Baumann AE (2007) Stigmatization, social distance and exclusion because of mental illness: the individual with mental illness as a 'stranger'. Int Rev Psychiatry 19(2):131–135. https://doi.org/10.1080/09540260701278739

Bolton DM, Yaxley J (2017) Fake news and clickbait – natural enemies of evidence-based medicine. BJU Int 119(Suppl 5):8–9. https://doi.org/10.1111/bju.13883

Britt-Spells AM, Slebodnik M, Sands LP, Rollock D (2018) Effects of perceived discrimination on depressive symptoms among black men residing in the United States: a meta-analysis. Am J Mens Health 12(1):52–63. https://doi.org/10.1177/1557988315624509

Brondolo E, Gallo LC, Myers HF (2009) Race, racism and health: disparities, mechanisms, and interventions. J Behav Med 32(1):1–8. https://doi.org/10.1007/s10865-008-9190-3

Bruns DP, Kraguljac NV, Bruns TR (2020) COVID-19: facts, cultural considerations, and risk of stigmatization. J Transcult Nurs 1043659620917724. https://doi.org/10.1177/1043659620917724

Budhwani H, Sun R (2020) Referencing the novel coronavirus as the "Chinese virus" or "China virus" on twitter: COVID-19 stigma. J Med Internet Res. https://doi.org/10.2196/19301

Burger P, Kanhai S, Pleijter A, Verberne S (2019) The reach of commercially motivated junk news on Facebook. 14(8):e0220446. https://doi.org/10.1371/journal.pone.0220446

Burgess A, Horii M (2012) Risk, ritual and health responsibilisation: Japan's 'safety blanket' of surgical face mask-wearing. Sociol Health Illn 34(8):1184–1198. https://doi.org/10.1111/j.1467-9566.2012.01466.x

CAASF (2020) Report COVID-19 Related Harassment, Discrimination, and Bigotry. https://caasf.org/2020/03/fight-covid-19-xenophobia-help-stopaapihate/. Accessed 2 May 2020

Cabrera D (2020) Investigan agresión al apartamento del médico pediatra amenazado en Bogotá. https://www.rcnradio.com/bogota/investigan-agresion-al-apartamento-del-medico-pediatra-amenazado-en-bogota. Accessed 2 May 2020

Calisher C, Carroll D, Colwell R, Corley RB, Daszak P, Drosten C, Enjuanes L, Farrar J, Field H, Golding J, Gorbalenya A, Haagmans B, Hughes JM, Karesh WB, Keusch GT, Lam SK, Lubroth J, Mackenzie JS, Madoff L, Mazet J, Palese P, Perlman S, Poon L, Roizman B, Saif L, Subbarao K, Turner M (2020) Statement in support of the scientists, public health professionals, and medical professionals of China combatting COVID-19. Lancet 395(10226):e42–e43. https://doi.org/10.1016/S0140-6736(20)30418-9

Callaway E (2020) Scores of coronavirus vaccines are in competition — how will scientists choose the best? Nature. https://doi.org/10.1038/d41586-020-01247-2

Caly L, Druce J, Roberts J, Bond K, Tran T, Kostecki R, Yoga Y, Naughton W, Taiaroa G, Seemann T, Schultz MB, Howden BP, Korman TM, Lewin SR, Williamson DA (2020) Isolation and rapid sharing of the 2019 novel coronavirus (SARS-CoV-2) from the first patient diagnosed with COVID-19 in Australia. doi:https://doi.org/10.5694/mja2.50569

Campbell L (2020) Chinese in UK report 'shocking' levels of racism after coronavirus outbreak. https://www.theguardian.com/uk-news/2020/feb/09/chinese-in-uk-report-shocking-levels-of-racism-after-coronavirus-outbreak. Accessed 2 May 2020

Cao Y, Li L, Feng Z, Wan S, Huang P, Sun X, Wen F, Huang X, Ning G, Wang W (2020) Comparative genetic analysis of the novel coronavirus (2019-nCoV/SARS-CoV-2) receptor ACE2 in different populations. Cell Discov 6(1):11. https://doi.org/10.1038/s41421-020-0147-1

Casey LS, Reisner SL, Findling MG (2019) Discrimination in the United States: Experiences of lesbian, gay, bisexual, transgender, and queer Americans. 54 Suppl 2:1454–1466. https://doi.org/10.1111/1475-6773.13229

Chen N, Zhou M, Dong X, Qu J, Gong F, Han Y, Qiu Y, Wang J, Liu Y, Wei Y, Xia Ja YT, Zhang X, Zhang L (2020a) Epidemiological and clinical characteristics of 99 cases of 2019 novel coronavirus pneumonia in Wuhan, China: a descriptive study. Lancet 395(10223):507–513. https://doi.org/10.1016/S0140-6736(20)30211-7

Chen Q, Liang M, Li Y, Guo J, Fei D, Wang L, He L, Sheng C, Cai Y, Li X, Wang J, Zhang Z (2020b) Mental health care for medical staff in China during the COVID-19 outbreak. Lancet Psychiatry 7(4):e15–e16. https://doi.org/10.1016/s2215-0366(20)30078-x

Chin AWH, Chu JTS, Perera MRA, Hui KPY, Yen H-L, Chan MCW, Peiris M, Poon LLM (2020) Stability of SARS-CoV-2 in different environmental conditions. Lancet Microbe. https://doi.org/10.1016/S2666-5247(20)30003-3

Coates M (2020) Covid-19 and the rise of racism. BMJ 369:m1384. https://doi.org/10.1136/bmj.m1384

Cordell J (2020) In Russia, Coronavirus Patients Fight Infection, Stigma and Harassment. The Moscow Times. https://www.themoscowtimes.com/2020/04/15/in-russia-coronavirus-patients-fight-infection-stigma-and-harassment-a69993. Accessed 2 May 2020

Corman VM, Landt O, Kaiser M, Molenkamp R, Meijer A, Chu DK, Bleicker T, Brunink S, Schneider J, Schmidt ML, Mulders DG, Haagmans BL, van der Veer B, van den Brink S, Wijsman L, Goderski G, Romette JL, Ellis J, Zambon M, Peiris M, Goossens H, Reusken C, Koopmans MP, Drosten C (2020) Detection of 2019 novel coronavirus (2019-nCoV) by real-time RT-PCR. Euro Surveillance: bulletin Europeen sur les maladies transmissibles = European communicable disease bulletin 25(3). https://doi.org/10.2807/1560-7917.es.2020.25.3.2000045

Crandall CS, Coleman R (1992) Aids-related stigmatization and the disruption of social relationships. J Soc Pers Relat 9(2):163–177. https://doi.org/10.1177/0265407592092001

D'Angelo B (2020) Coronavirus: Indiana gas station owner apologizes after clerk kicks out doctor for being Asian Fox13. https://www.fox13memphis.com/news/trending/coronavirus-indiana-gas-station-owner-apologizes-after-clerk-kicks-out-doctor-being-asian/TNT2MVQEFZGZTI2QZWT6Q5RRZE/. 2 May 2020

Deng CX (2020) The global battle against SARS-CoV-2 and COVID-19. Int J Biol Sci 16(10):1676–1677. https://doi.org/10.7150/ijbs.45587

Depoux A, Martin S, Karafillakis E, Bsd RP, Wilder-Smith A, Larson H (2020) The pandemic of social media panic travels faster than the COVID-19 outbreak. J Travel Med. https://doi.org/10.1093/jtm/taaa031

Deshpande S (2020) How the world is thanking doctors. Conde Nast Traveller. https://www.cntraveller.in/story/photos-world-thanking-doctors-covid-19-pandemic-images-india-italy-london-australia/

Dess NK, Linwick D, Patterson J, Overmier JB, Levine S (1983) Immediate and proactive effects of controllability and predictability on plasma cortisol responses to shocks in dogs. Behav Neurosci 97(6):1005–1016. https://doi.org/10.1037//0735-7044.97.6.1005

Dong E, Du H, Gardner L (2020) An interactive web-based dashboard to track COVID-19 in real time. Lancet Infect Dis. https://doi.org/10.1016/S1473-3099(20)30120-1

Donnor J (2013) Education as the property of whites: African Americans' continued quest for good schools. In: Lynn M, Dixson AD (eds) Handbook of critical race theory in education. Routledge, pp 195–203

Ducharme J (2020) News Coverage of Coronavirus in 2020 is Very Different Than it Was For Ebola in 2018. Time. https://time.com/5779872/coronavirus-ebola-news-coverage/. Accessed 2 May 2020

Dyer C (2020) Kenyan man is beaten to death by a mob of youths after they suspected him of having coronavirus. Daily Mail. https://www.dailymail.co.uk/news/article-8129791/Kenyan-man-beaten-death-mob-youths-suspected-having-coronavirus.html. Accessed 2 May 2020

Eisen MB, Akhmanova A (2020) Publishing in the time of COVID-19. 9. https://doi.org/10.7554/eLife.57162

Ferguson N, Nicoll A, Schwartz B (2009) SIP 5: social distancing during a pandemic. Not sexy, but sometimes effective: social distancing and non-pharmaceutical interventions. Vaccine 27(45):6383–6386. https://doi.org/10.1016/j.vaccine.2009.08.010

Folgueira MD, Munoz-Ruiperez C, Alonso-Lopez MA, Delgado R (2020) SARS-CoV-2 infection in Health Care Workers in a large public hospital in Madrid, Spain, during March 2020. medRxiv:2020.2004.2007.20055723. doi:https://doi.org/10.1101/2020.04.07.20055723

Fukuda K, Wang R, Vallat B (2015) Naming diseases: first do no harm. Science 348(6235):643–643. https://doi.org/10.1126/science.348.6235.643

Garz M (2014) Good news and bad news: evidence of media bias in unemployment reports. Public Choice 161(3):499–515. https://doi.org/10.1007/s11127-014-0182-2

GhanaWeb (2020) Family of deceased Coronavirus patient faces stigma. https://www.ghanaweb.com/GhanaHomePage/NewsArchive/Family-of-deceased-Coronavirus-patient-faces-stigma-923908. Accessed 2 May 2020

Giddens A (1991) Modernity and self-identity: self and Society in the Late Modern age. Stanford University Press, Stanford

Global News (2020) Trudeau condemns racism linked to coronavirus outbreak. https://globalnews.ca/video/6494388/trudeau-condemns-racism-linked-to-coronavirus-outbreak/. Accessed 2 May 2020

Goffman E (1986) Stigma: notes on the Management of Spoiled Identity. Simon & Schuster, New York

Gorbalenya AE, Baker SC, Baric RS, de Groot RJ, Drosten C, Gulyaeva AA, Haagmans BL, Lauber C, Leontovich AM, Neuman BW, Penzar D, Perlman S, LLM P, Samborskiy DV, Sidorov IA, Sola I, Ziebuhr J, Coronaviridae Study Group of the International Committee on Taxonomy of V (2020) The species severe acute respiratory syndrome-related coronavirus: classifying 2019-nCoV and naming it SARS-CoV-2. Nat Microbiol 5(4):536–544. https://doi.org/10.1038/s41564-020-0695-z

Grabill JT, Simmons WM (1998) Toward a critical rhetoric of risk communication: producing citizens and the role of technical communicators. Tech Commun Q 7(4):415–441. https://doi.org/10.1080/10572259809364640

Hawkey PM, Bhagani S, Gillespie SH (2003) Severe acute respiratory syndrome (SARS): breath-taking progress. J Med Microbiol 52(Pt 8):609–613. https://doi.org/10.1099/jmm.0.05321-0

He J, He L, Zhou W, Nie X, He M (2020) Discrimination and social exclusion in the outbreak of COVID-19. Int J Environ Res Public Health 17(8). https://doi.org/10.3390/ijerph17082933

Hess A (2020) In Praise of Quarantine Clapping. New York Times. https://www.nytimes.com/2020/04/09/arts/virus-quarantine-clapping.html. Accessed 2 May 2020

Himmelstein MS, Young DM, Sanchez DT, Jackson JS (2015) Vigilance in the discrimination-stress model for Black Americans. Psychol Health

30(3):253–267. https://doi.org/10.1080/08870446.2014.966104

Hindu (2020) Alarming spread: on novel coronavirus outbreak The Hindu. https://www.thehindu.com/opinion/editorial/alarming-spread/article30677660.ece. Accessed 2 May 2020

Hirji Z (2020) Thousands of US health care workers have been infected by the coronavirus. This Is How Each State Stacks Up, Buzz Feed News. https://www.buzzfeednews.com/article/zahrahirji/us-health-care-workers-coronavirus. Accessed 2 May 2020

Hoffmann M, Kleine-Weber H, Schroeder S, Kruger N, Herrler T, Erichsen S, Schiergens TS, Herrler G, Wu NH, Nitsche A, Muller MA, Drosten C, Pohlmann S (2020) SARS-CoV-2 cell entry depends on ACE2 and TMPRSS2 and is blocked by a clinically proven protease inhibitor. Cell. https://doi.org/10.1016/j.cell.2020.02.052

Hogarth RA (2019) The myth of innate racial differences between White and black People's bodies: lessons from the 1793 yellow fever epidemic in Philadelphia, Pennsylvania. Am J Public Health 109(10):1339–1341. https://doi.org/10.2105/AJPH.2019.305245

Hood L (2020) Mistreatment of Africans in Guangzhou threatens China's coronavirus diplomacy The Conversation. https://theconversation.com/mistreatment-of-africans-in-guangzhou-threatens-chinas-coronavirus-diplomacy-136348. Accessed 2 May 2020

Houston BK (1972) Control over stress, locus of control, and response to stress. J Pers Soc Psychol 21(2):249–255. https://doi.org/10.1037/h0032328

Huang C, Wang Y, Li X, Ren L, Zhao J, Hu Y, Zhang L, Fan G, Xu J, Gu X, Cheng Z, Yu T, Xia J, Wei Y, Wu W, Xie X, Yin W, Li H, Liu M, Xiao Y, Gao H, Guo L, Xie J, Wang G, Jiang R, Gao Z, Jin Q, Wang J, Cao B (2020) Clinical features of patients infected with 2019 novel coronavirus in Wuhan, China. Lancet (London, England) 395(10223):497–506. https://doi.org/10.1016/s0140-6736(20)30183-5

Il Messaggero (2020) Coronavirus, Mattarella visita a sorpresa la scuola multietnica all'Esquilino. La preside: «Ha voluto stringere le mani ai bambini». https://www.ilmessaggero.it/roma/news/coronavirus_mattarella_esquilino_scuola_manin_cinesi_news-5033113.html. Accessed 2 May 2020

International Council of Nurses (2020) High proportion of healthcare workers with COVID-19 in Italy is a stark warning to the world: protecting nurses and their colleagues must be the number one priority. https://www.icn.ch/news/high-proportion-healthcare-workers-covid-19-italy-stark-warning-world-protecting-nurses-and. 2 May 2020

Ioannidis JPA (2020) Coronavirus disease 2019: the harms of exaggerated information and non-evidence-based measures. Eur J Clin Investig 50(4):e13222. https://doi.org/10.1111/eci.13222

Jeffrey A (2020) US cities light up landmarks blue in support of health-care workers. CNBC. https://www.cnbc.com/2020/04/10/in-photos-us-cities-light-up-landmarks-blue-in-support-of-health-care-workers.html. Accessed 2 May 2020

Ji W, Wang W, Zhao X, Zai J, Li X (2020) Cross-species transmission of the newly identified coronavirus 2019-nCoV. J Med Virol 92(4):433–440. https://doi.org/10.1002/jmv.25682

Jiang S, Xia S, Ying T, Lu L (2020) A novel coronavirus (2019-nCoV) causing pneumonia-associated respiratory syndrome. Cell Mol Immunol. https://doi.org/10.1038/s41423-020-0372-4

Johansson MA, Saderi D (2020) Open peer-review platform for COVID-19 preprints. Nature 579(29). https://doi.org/10.1038/d41586-020-00613-4

Johnson RN (1996) Bad News revisited: the portrayal of violence, conflict, and suffering on television News. Peace Conflict 2(3):201–216. https://doi.org/10.1207/s15327949pac0203_2

Jones DS (2006) The persistence of American Indian health disparities. Am J Public Health 96(12):2122–2134. https://doi.org/10.2105/AJPH.2004.054262

Kadota H, Sekiguchi H, Takeuchi S, Miyazaki M, Kohno Y, Nakajima Y (2010) The role of the dorsolateral prefrontal cortex in the inhibition of stereotyped responses. Exp Brain Res 203(3):593–600. https://doi.org/10.1007/s00221-010-2269-4

Kampf G, Todt D, Pfaender S, Steinmann E (2020) Persistence of coronaviruses on inanimate surfaces and their inactivation with biocidal agents. J Hosp Infect 104(3):246–251. https://doi.org/10.1016/j.jhin.2020.01.022

Kasraoui S (2020) Health Officials Deny Rumors of Coronavirus in Fez. Morocco World News. https://www.moroccoworldnews.com/2020/01/292348/health-officials-deny-rumors-of-coronavirus-in-fez/. Accessed 2 May 2020

Knutson KM, Mah L, Manly CF, Grafman J (2007) Neural correlates of automatic beliefs about gender and race. Hum Brain Mapp 28(10):915–930. https://doi.org/10.1002/hbm.20320

Kouzy R, Abi Jaoude J, Kraitem A, El Alam MB, Karam B, Adib E, Zarka J, Traboulsi C, Akl EW, Baddour K (2020) Coronavirus Goes viral: quantifying the COVID-19 misinformation epidemic on twitter. Cureus 12(3):e7255. https://doi.org/10.7759/cureus.7255

Krnjacki L, Priest N, Aitken Z, Emerson E, Llewellyn G, King T, Kavanagh A (2018) Disability-based discrimination and health: findings from an Australian-based population study. Aust N Z J Public Health 42(2):172–174. https://doi.org/10.1111/1753-6405.12735

Kupferschmidt K (2020) Preprints bring 'firehose' of outbreak data. Science 367(6481):963–964. https://doi.org/10.1126/science.367.6481.963

Link BG, Phelan JC, Bresnahan M, Stueve A, Pescosolido BA (1999) Public conceptions of mental illness: labels, causes, dangerousness, and social distance. Am J Public Health 89(9):1328–1333. https://doi.org/10.2105/ajph.89.9.1328

Le Bon G (1897) The crowd: a study of the popular mind. T. Fisher Unwin.

LundbergU,FrankenhaeuserM(1978)Psychophysiological reactions to noise as modified by personal control over noise intensity. Biol Psychol 6(1):51–59. https://doi.org/10.1016/0301-0511(78)90006-6

Maier SF, Drugan RC, Grau JW (1982) Controllability, coping behavior, and stress-induced analgesia in the rat. Pain 12(1):47–56. https://doi.org/10.1016/0304-3959(82)90169-5

Maier SF, Ryan SM, Barksdale CM, Kalin NH (1986) Stressor controllability and the pituitary-adrenal system. Behav Neurosci 100(5):669–674. https://doi.org/10.1037/0735-7044.100.5.669

Majumder MS, Mandl KD (2020) Early in the epidemic: impact of preprints on global discourse about COVID-19 transmissibility. Lancet Glob Health 8(5):e627–e630. https://doi.org/10.1016/s2214-109x(20)30113-3

Mamzer H (2020) Postmodern society and Covid-19 pandemic: old, new and scary. Society Register 4(2):7–18. https://doi.org/10.14746/sr.2020.4.2.01

Marcelin JR, Siraj DS, Victor R, Kotadia S, Maldonado YA (2019) The impact of unconscious Bias in healthcare: how to recognize and mitigate it. J Infect Dis 220(Supplement_2):S62–S73. https://doi.org/10.1093/infdis/jiz214

Marra MA, Jones SJ, Astell CR, Holt RA, Brooks-Wilson A, Butterfield YS, Khattra J, Asano JK, Barber SA, Chan SY, Cloutier A, Coughlin SM, Freeman D, Girn N, Griffith OL, Leach SR, Mayo M, McDonald H, Montgomery SB, Pandoh PK, Petrescu AS, Robertson AG, Schein JE, Siddiqui A, Smailus DE, Stott JM, Yang GS, Plummer F, Andonov A, Artsob H, Bastien N, Bernard K, Booth TF, Bowness D, Czub M, Drebot M, Fernando L, Flick R, Garbutt M, Gray M, Grolla A, Jones S, Feldmann H, Meyers A, Kabani A, Li Y, Normand S, Stroher U, Tipples GA, Tyler S, Vogrig R, Ward D, Watson B, Brunham RC, Krajden M, Petric M, Skowronski DM, Upton C, Roper RL (2003) The genome sequence of the SARS-associated coronavirus. Science 300(5624):1399–1404. https://doi.org/10.1126/science.1085953

Matrajt L, Leung T (2020) Evaluating the effectiveness of social distancing interventions to delay or flatten the epidemic curve of coronavirus disease. Emerg Infect Dis 26(8). https://doi.org/10.3201/eid2608.201093

Matthews S, Blanchard S, Boyd C (2020) Russia closes its eastern border as killer coronavirus spreads to EVERY region of China: Death toll jumps to 170 and almost 8,300 people have caught the deadly SARS-like infection worldwide. DailyMail. https://www.dailymail.co.uk/health/article-7946387/Killer-coronavirus-spreads-region-China-death-toll-jumps-170.html. Accessed 2 May 2020

McNamara D (2020) COVID-19: Are Acute Stroke Patients Avoiding Emergency Care? Medscape. https://www.medscape.com/viewarticle/928337. Accessed 2 May 2020

Mian A, Khan S (2020) Coronavirus: the spread of misinformation. BMC Med 18(1):89. https://doi.org/10.1186/s12916-020-01556-3

Monk EP Jr (2015) The cost of color: skin color, discrimination, and health among African-Americans. AJS 121(2):396–444. https://doi.org/10.1086/682162

NBC News (2020a) NY attorney general launches hotline for people to report discrimination related to coronavirus. https://www.nbcnews.com/news/asian-america/ny-attorney-general-launches-hotline-asian-americans-report-discrimination-n1167766. Accessed 2 May 2020

NBC News (2020b) Toronto mayor denounces xenophobia against Chinese community amid coronavirus fears. https://www.nbcnews.com/news/asian-america/toronto-mayor-denounces-xenophobia-against-chinese-community-amid-coronavirus-fears-n1125991. Accessed 2 May 2020

Nelson S (2016) Race, racism, and health disparities: what can I do about it? Creat Nurs 22(3):161–165. https://doi.org/10.1891/1078-4535.22.3.161

O'Shea BA, Watson DG, Brown GDA, Fincher CL (2020) Infectious disease prevalence, not race exposure, predicts both implicit and explicit racial prejudice across the United States. Soc Psycho Personal Sci 0(0):1948550619862319. https://doi.org/10.1177/1948550619862319

Ou X, Liu Y, Lei X, Li P, Mi D, Ren L, Guo L, Guo R, Chen T, Hu J, Xiang Z, Mu Z, Chen X, Chen J, Hu K, Jin Q, Wang J, Qian Z (2020) Characterization of spike glycoprotein of SARS-CoV-2 on virus entry and its immune cross-reactivity with SARS-CoV. Nat Commun 11(1):1620. https://doi.org/10.1038/s41467-020-15562-9

Oxley TJ, Mocco J, Majidi S, Kellner CP, Shoirah H, Singh IP, De Leacy RA, Shigematsu T, Ladner TR, Yaeger KA, Skliut M, Weinberger J, Dangayach NS, Bederson JB, Tuhrim S, Fifi JT (2020) Large-vessel stroke as a presenting feature of Covid-19 in the Young. N Engl J Med:e60. https://doi.org/10.1056/NEJMc2009787

Ozma MA, Maroufi P, Khodadadi E, Kose S, Esposito I, Ganbarov K, Dao S, Esposito S, Dal T, Zeinalzadeh E, Kafil HS (2020) Clinical manifestation, diagnosis, prevention and control of SARS-CoV-2 (COVID-19) during the outbreak period. Infez Med 28(2):153–165

Padayachee N, du Toit LC (2020) Debunking 9 popular myths doing the rounds in Africa about the coronavirus The Conversation. https://theconversation.com/debunking-9-popular-myths-doing-the-rounds-in-africa-about-the-coronavirus-135580. Accessed 3 May 2020

Pandey V (2020) Coronavirus: India doctors 'spat at and attacked'. BBC. https://www.bbc.com/news/world-asia-india-52151141. Accessed 2 May 2020

Parke J (2007) Getting the message across: using media to reduce racial prejudice and discrimination

Partyła M (2020) Sulmierzyce: Szykanowana rodzina chorego na Covid-19. https://www.rmf24.pl/raporty/raport-koronawirus-z-chin/polska/news-sulmierzyce-szykanowana-rodzina-chorego-na-covid-19,nId,4421560. Accessed 2 May 2020

Pascoe EA, Smart Richman L (2009) Perceived discrimination and health: a meta-analytic review. Psychol Bull 135(4):531–554. https://doi.org/10.1037/a0016059

Patten E (2016) Racial, gender wage gaps persist in U.S. despite some progress. Pew Research. https://www.pewresearch.org/fact-tank/2016/07/01/racial-gender-wage-gaps-persist-in-u-s-despite-some-progress/. Accessed 3 May 2020

Pellecchia U, Crestani R, Decroo T, Van den Bergh R, Al-Kourdi Y (2015) Social consequences of Ebola containment measures in Liberia. PLoS One 10(12):e0143036. https://doi.org/10.1371/journal.pone.0143036

Person B, Sy F, Holton K, Govert B, Liang A, National Center for Inectious Diseases SCOT (2004) Fear and stigma: the epidemic within the SARS outbreak. Emerg Infect Dis 10 (2):358–363. doi:https://doi.org/10.3201/eid1002.030750

Puddifoot K (2019) Stereotyping patients. J Soc Philos 50(1):69–90. https://doi.org/10.1111/josp.12269

Rana W, Mukhtar S, Mukhtar S (2020) Mental health of medical workers in Pakistan during the pandemic COVID-19 outbreak. Asian J Psychiatr 51:102080–102080. https://doi.org/10.1016/j.ajp.2020.102080

Ransohoff DF, Ransohoff RM (2001) Sensationalism in the media: when scientists and journalists may be complicit collaborators. Eff Clin Pract 4

Reuters (2020) U.N. asks world to fight virus-spawned discrimination. https://www.reuters.com/article/us-china-health-rights/u-n-asks-world-to-fight-virus-spawned-discrimination-idUSKCN20L16B. Accessed 2 May 2020

Rios L (2020) Mexican Doctors Attacked With Bleach, Assaulted Over Coronavirus Transmission Fears. Time. https://time.com/5827496/mexico-doctors-coronavirus-attacks/. Accessed 2 May 2020

Rismanbaf A (2020) Potential treatments for COVID-19; a narrative literature review. Arch Acad Emerg Med 8(1):e29

Rosa SGV, Santos WC (2020) Clinical trials on drug repositioning for COVID-19 treatment. Rev Panam Salud Publica 44:e40–e40. https://doi.org/10.26633/RPSP.2020.40

Rössler W (2016) The stigma of mental disorders. EMBO Rep 17(9):1250–1253. https://doi.org/10.15252/embr.201643041

Rota PA, Oberste MS, Monroe SS, Nix WA, Campagnoli R, Icenogle JP, Penaranda S, Bankamp B, Maher K, Chen MH, Tong S, Tamin A, Lowe L, Frace M, DeRisi JL, Chen Q, Wang D, Erdman DD, Peret TC, Burns C, Ksiazek TG, Rollin PE, Sanchez A, Liffick S, Holloway B, Limor J, McCaustland K, Olsen-Rasmussen M, Fouchier R, Gunther S, Osterhaus AD, Drosten C, Pallansch MA, Anderson LJ, Bellini WJ (2003) Characterization of a novel coronavirus associated with severe acute respiratory syndrome. Science 300(5624):1394–1399. https://doi.org/10.1126/science.1085952

Rothan HA, Byrareddy SN (2020) The epidemiology and pathogenesis of coronavirus disease (COVID-19) outbreak. J Autoimmun 109:102433. https://doi.org/10.1016/j.jaut.2020.102433

Rotter JB (1966) Generalized expectancies for internal versus external control of reinforcement. Psychol Monogr Gen Appl 80(1):1–28. https://doi.org/10.1037/h0092976

Rotter JB (1990) Internal versus external control of reinforcement: a case history of a variable. Am Psychol 45(4):489–493. https://doi.org/10.1037/0003-066X.45.4.489

Rubrico JGU (2020) Covid-19 and the discrimination of medical personnel in the Philippines. Italian Institute for International Political Studies

Rzymski P, Nowicki M (2020a) COVID-19-related prejudice towards Asian medical students: a consequence of SARS-CoV-2 fears in Poland. J Infect Public Health. https://doi.org/10.1016/j.jiph.2020.04.013

Rzymski P, Nowicki M (2020b) Preventing COVID-19 prejudice in academia. Science 368(6484):1313. https://doi.org/10.1126/science.abb4870

Sarche M, Spicer P (2008) Poverty and health disparities for American Indian and Alaska native children: current knowledge and future prospects. Ann N Y Acad Sci 1136:126–136. https://doi.org/10.1196/annals.1425.017

Seligman MEP (1975) Helplessness: On depression, development, and death. On depression, development, and death. W H Freeman/Times Books/ Henry Holt & Co, New York, NY, US, Helplessness

Selye H (1976) Stress without distress. In: Serban G (ed) Psychopathology of human adaptation. Springer US, Boston, MA, pp 137–146. https://doi.org/10.1007/978-1-4684-2238-2_9

Semana (2020) Intolerancia en Neiva: atacan a piedra casa de pacientes con coronavirus https://www.semana.com/nacion/articulo/intolerancia-en-neiva-atacan-a-piedra-casa-de-pacientes-con-coronavirus/657040. Accessed 2 May 2020

Shimizu K (2020) 2019-nCoV, fake news, and racism. Lancet 395(10225):685–686. https://doi.org/10.1016/S0140-6736(20)30357-3

Sohn H (2017) Racial and ethnic disparities in health insurance coverage: dynamics of gaining and losing coverage over the life-course. Popul Res Policy Rev 36(2):181–201. https://doi.org/10.1007/s11113-016-9416-y

Sontag S (1979) Illness as metaphor. First Vintage Books Edition

Sontag S (1989) AIDS and its metaphors. Allen Lane

Sosa A, Brown L (2020) Woman wearing face mask attacked in possible coronavirus hate crime New York

Post. https://nypost.com/2020/02/05/woman-wearing-face-mask-attacked-in-possible-coronavirus-hate-crime/. Accessed 2 May 2020

Soto J, Hakim S (2020) Medical basis for increased susceptibility of COVID-19 among the Navajo and other indigenous tribes. doi:https://doi.org/10.20944/preprints202004.0217.v1

Stoye E (2020) 'No one is allowed to go out': your stories from the coronavirus outbreak. Nature. https://doi.org/10.1038/d41586-020-00478-7

Tang B, Wang X, Li Q, Bragazzi NL, Tang S, Xiao Y, Wu J (2020) Estimation of the transmission risk of the 2019-nCoV and its implication for public health interventions. J Clin Med 9(2). https://doi.org/10.3390/jcm9020462

Thanh Le T, Andreadakis Z, Kumar A, Gomez Roman R, Tollefsen S, Saville M, Mayhew S (2020) The COVID-19 vaccine development landscape. Nat Rev Drug Discov. https://doi.org/10.1038/d41573-020-00073-5

The Economic Times (2020) Stones pelted at medical team, police in UP's Moradabad. https://economictimes.indiatimes.com/news/politics-and-nation/stones-pelted-at-medical-team-police-in-ups-moradabad/articleshow/75161729.cms?from=mdr

The Lancet (2020) COVID-19: protecting health-care workers. Lancet (London, England) 395(10228):922–922. https://doi.org/10.1016/S0140-6736(20)30644-9

Tsamakis K, Rizos E, JM A, Chaidou S, Kympouropoulos S, Spartalis E, Spandidos DA, Tsiptsios D, Triantafyllis AS (2020) COVID-19 pandemic and its impact on mental health of healthcare professionals. Exp Ther Med 19(6):3451–3453. https://doi.org/10.3892/etm.2020.8646

Universidad (2020) Coronavirus: crecen las denuncias por discriminación a médicos y pacientes. http://www.unidiversidad.com.ar/coronavirus-crecen-la-denuncias-por-discriminacion-a-medicos-y-pacientes. Accessed 2 May 2020

van Doremalen N, Bushmaker T, Morris DH, Holbrook MG, Gamble A, Williamson BN, Tamin A, Harcourt JL, Thornburg NJ, Gerber SI, Lloyd-Smith JO, de Wit E, Munster VJ (2020) Aerosol and surface stability of SARS-CoV-2 as compared with SARS-CoV-1. N Engl J Med 382(16):1564–1567. https://doi.org/10.1056/NEJMc2004973

Venkatesh A, Edirappuli S (2020) Social distancing in covid-19: what are the mental health implications? BMJ 369:m1379. https://doi.org/10.1136/bmj.m1379

Vietgiaitri (2020) Nguyễn Sin tức giận đòi bỏ tù bệnh nhân thứ 100. https://vietgiaitri.com/nguyen-sin-tuc-gian-doi-bo-tu-benh-nhan-thu-100-20200323i4781922/. Accessed 2 May 2020

Wada K, Oka-Ezoe K, Smith DR (2012) Wearing face masks in public during the influenza season may reflect other positive hygiene practices in Japan. BMC Public Health 12:1065. https://doi.org/10.1186/1471-2458-12-1065

Wan Y, Shang J, Graham R, Baric RS, Li F (2020) Receptor recognition by the novel coronavirus from Wuhan: an analysis based on decade-long structural studies of SARS coronavirus. J Virol 94(7):e00127–e00120. https://doi.org/10.1128/jvi.00127-20

Welhausen C (2015) Visualizing a non-pandemic: considerations for communicating public health risks in intercultural. Contexts 62:244–257

Wellcome Trust (2020) Sharing research data and findings relevant to the novel coronavirus (COVID-19) outbreak. Wellcome Trust. https://wellcome.ac.uk/coronavirus-covid-19/open-data. Accessed 25 Apr 2020

White AIR (2020) Historical linkages: epidemic threat, economic risk, and xenophobia. Lancet. https://doi.org/10.1016/S0140-6736(20)30737-6

Whitesell NR, Beals J, Crow CB, Mitchell CM, Novins DK (2012) Epidemiology and etiology of substance use among American Indians and Alaska natives: risk, protection, and implications for prevention. Am J Drug Alcohol Abuse 38(5):376–382. https://doi.org/10.3109/00952990.2012.694527

WHO (2020) Statement on the second meeting of the International Health Regulations (2005) Emergency Committee regarding the outbreak of novel coronavirus (2019-nCoV). https://www.who.int/news-room/detail/30-01-2020-statement-on-the-second-meeting-of-the-international-health-regulations-(2005)-emergency-committee-regarding-the-outbreak-of-novel-coronavirus-(2019-ncov). Accessed 2 May 2020

Wong SL (2020) China warns that spread of deadly virus will accelerate. Financial Times. https://www.ft.com/content/7bb597c8-3ff7-11ea-a047-eae9bd51ceba. Accessed 2 May 2020

World Journal (2020) 恐慌擴散!機上有武漢人...上海旅客拒登機

Wroński M (2020) Piekarnie nie chcą obsługiwać lekarzy. Interwencja Rzecznika Praw Obywatelskich. Puls HR. https://www.pulshr.pl/rynek-zdrowia/piekarnie-nie-chca-obslugiwac-lekarzy-interwencja-rzecznika-praw-obywatelskich,73305.html. Accessed 2 May 2020

Wu D, Wu T, Liu Q, Yang Z (2020a) The SARS-CoV-2 outbreak: what we know. IJID. https://doi.org/10.1016/j.ijid.2020.03.004

Wu Y, Wang J, Luo C, Hu S, Lin X, Anderson AE, Bruera E, Yang X, Wei S, Qian Y (2020b) A comparison of burnout frequency among oncology physicians and nurses working on the front lines and usual wards during the COVID-19 epidemic in Wuhan. China J Pain Symp Manag. https://doi.org/10.1016/j.jpainsymman.2020.04.008

Xiang YT, Jin Y, Wang Y, Zhang Q, Zhang L, Cheung T (2020) Tribute to health workers in China: a group of respectable population during the outbreak of the COVID-19. Int J Biol Sci 16(10):1739–1740. https://doi.org/10.7150/ijbs.45135

Yancy CW (2020) COVID-19 and African Americans. JAMA. https://doi.org/10.1001/jama.2020.6548

Yao TT, Qian JD, Zhu WY, Wang Y, Wang GQ (2020) A systematic review of lopinavir therapy for SARS coronavirus and MERS coronavirus-A possible reference for coronavirus disease-19 treatment option. https://doi.org/10.1002/jmv.25729

Yi Y, Lagniton PNP, Ye S, Li E, Xu RH (2020) COVID-19: what has been learned and to be learned about the novel coronavirus disease. Int J Biol Sci 16(10):1753–1766. https://doi.org/10.7150/ijbs.45134

Zacks S, Beavers K, Theodore D, Dougherty K, Batey B, Shumaker J, Galanko J, Shrestha R, Fried MW (2006) Social stigmatization and hepatitis C virus infection. J Clin Gastroenterol 40(3)

Zarocostas J (2020) How to fight an infodemic. Lancet (London, England) 395(10225):676. https://doi.org/10.1016/s0140-6736(20)30461-x

Zeng FY, Chan CW, Chan MN, Chen JD, Chow KY, Hon CC, Hui KH, Li J, Li VY, Wang CY, Wang PY, Guan Y, Zheng B, Poon LL, Chan KH, Yuen KY, Peiris JS, Leung FC (2003) The complete genome sequence of severe acute respiratory syndrome coronavirus strain HKU-39849 (HK-39). Exp Biol Med (Maywood) 228(7):866–873. https://doi.org/10.1177/15353702-0322807-13

Zhang Y-Z (2020) Novel 2019 coronavirus genome. Virological. http://virological.org/t/novel-2019-coronavirus-genome/319. Accessed 12 Apr 2002 2020

Zhang Q, Wang Y (2020) Clinical trial analysis of 2019-nCoV therapy registered in China. https://doi.org/10.1002/jmv.25733

Zhang C, Yang L, Liu S, Ma S, Wang Y, Cai Z, Du H, Li R, Kang L, Su M, Zhang J, Liu Z, Zhang B (2020) Survey of insomnia and related social psychological factors among medical staff involved in the 2019 novel coronavirus disease outbreak. Front Psych 11:306. https://doi.org/10.3389/fpsyt.2020.00306

Zhao I (2020) Coronavirus has sparked racist attacks on Asians in Australia — including me. ABC. https://www.abc.net.au/news/2020-02-01/coronavirus-has-sparked-racist-attacks-on-asian-australians/11918962. Accessed 2 May 2020

Zimbardo P (1994) Psychology. Harpercollins College Div

40 Potential Mechanisms of COVID-19-Related Psychological Problems and Mental Disorders

Alfred Shaw

Abstract

As the coronavirus disease 2019 (COVID-19) pandemic has spread, so has the psychological impact of the disease been felt worldwide. Despite this, the mechanisms of COVID-19-related psychological problems and mental disorders remain unclear. As such, effective therapeutic schemes or intervention strategies cannot be developed. It is, therefore, necessary to establish a theoretical basis of psychological problems and mental disorders related to public health emergencies such as COVID-19. Herein, the potential mechanisms of occurrence and development of COVID-19-related psychological problems and mental disorders have been discussed from two angles: the pandemic as a public health emergency itself and the extensive quarantine situation during the pandemic.

A. Shaw (✉)
Network of Immunity in Infection, Malignancy and Autoimmunity (NIIMA), Universal Scientific Education and Research Network (USERN), Beijing, China

Keywords

COVID-19 · Mechanism · Mental disorders · Pandemic · Psychology · Quarantine

40.1 Introduction

Beginning early December 2019, the COVID-19 outbreak spread globally within 3 months to become a full-blown pandemic. It resulted in substantial social impacts and economic losses. COVID-19 is an infectious respiratory disease that seriously threatens patients' health and life. Its potent infectivity, disability, and fatality rates have caused widespread public panic. Many governments have formulated and implemented strict quarantine measures to prevent further spread of the disease. Public health emergencies such as the COVID-19 pandemic have severe negative psychological impacts on individuals both randomly and spatially within the population.

In contrast with typical social emergencies that have more simple stressors, the strict quarantine measures imposed to prevent further COVID-19 spread across communities further increase individual susceptibility to psychological problems and mental disorders. As such, these conditions have been on the rise. Studies focusing on the correlation between public health emergencies and related psychological problems

N. Rezaei (ed.), *Coronavirus Disease - COVID-19*, Advances in Experimental Medicine and Biology 1318, https://doi.org/10.1007/978-3-030-63761-3_40

and mental disorders have confirmed that there is a close relationship between these events. Research has also confirmed that employing psychological crisis intervention strategies during social emergencies is effective. However, the current understanding of the mechanisms of COVID-19-related psychological problems and mental disorders is unclear. Cognizant to this, effective therapeutic schemes or intervention strategies cannot be developed. It is, therefore, necessary to develop a theoretical basis of psychological problems and mental disorders related to public health emergencies such as COVID-19.

40.2 Psychological Stress

Stress is an adaptive and instinctive response by the body when the mind perceives that environmental changes pose a threat or challenge to the body (Tachè 2014; Rom and Reznick 2016; Otto et al. 1990). Hans Selye was the first physiologist to systematically use the concept of stress to describe the regulatory responses which occur when the body is threatened. He introduced the core concepts of stress and stressor. He interpreted stress as any stimulus that seriously threatens the body's homeostasis and stressor as the stimulation which causes stress (Szabo et al. 2012, 2017; Szabó and Gyires 2015). Currently, the meaning of the term stressor has expanded from a single external physical or chemical factor to a whole variety of external stimulus factors, including psychological stimulus owing to the deepening understanding of this concept (Koffer et al. 2016; Leger et al. 2016). Different stress theories have been developed in the last century. Currently, there are three commonly recognized stress theoretical models: the cognitive appraisal model, the system model, and the process model. As a result, the definition of psychological stress has changed considerably.

Proponents of stress theories define various biological, psychological, social, and behavioral changes caused by stressors as psychosomatic reactions of stress. These include physiological and psychological reactions to stress. On the one hand, psychological reactions to stress include emotional, cognitive, and behavioral stress responses. On the other hand, physiological reactions to stress can induce related psychological reactions through pathways such as the mediating mechanism. Based on existing studies, stress can lead to mental and psychological disorders such as acute stress disorder (Bryant 2017, 2018; Santana et al. 2017), post-traumatic stress disorder (PTSD) (Jakovljević et al. 2012; Qi et al. 2016; Bryant 2019), and adjustment disorder (Appart et al. 2017; Strain 2018; Gradus et al. 2014). The COVID-19 pandemic is a typical health emergency that is both a social and environmental event stressor based on the phenomenological classification of life events. As such, it can lead to a series of psychological problems and mental disorders. Given the implementation of strict quarantine measures imposed by various governments, individual support systems are also affected, thus further increasing their susceptibility to psychological problems and mental disorders.

40.3 The Zeitgebers and Zeitstörers Theory

Proponents of the zeitgebers and zeitstörers theory believe that the basis for individual biological rhythms is a timer composed of stable biological and environmental factors (Kupfer and Reynolds 1983; Ehlers et al. 1988). The specific physical, chemical, and psychosocial events that interfere with an individual biological clock are considered to be zeitstörers. They are the critical risk factors for the occurrence and development of psychological problems and mental disorders (Frank 2007). For example, changes in light exposure can affect the establishment of an individual biological clock. This phenomenon is caused by the impact of sudden changes in light exposure, meal times, and sleep time as a result of flying across different time zones. It can cause jet lag responses and mental health symptoms in susceptible populations (Finkelstein 1989; Patterson 1989). Wever suggested that social and natural zeitstörers have the same effects on the biological clock (Wever 1979). These zeitstörers include adverse events such as divorce, bankruptcy, and widowhood. They lead to psychological problems and mental disorders by breaking individual circadian rhythms and sleep-awakening cycles. As the understanding of the zeitgeber and zeitstörers theory deepens, psychologists can further explore the mechanisms of adverse events that lead to psychological problems and mental disorders. Currently, it is believed that adverse events initially cause social rhythm disturbance by breaking social zeitgebers. They then affect individual biological rhythms, consequently leading to psychological problems and mental disorders. The COVID-19 pandemic is a typical public health emergency that can be regarded as a type of zeitstörer. As such, its related adverse socio-psychological effects can be explained by the zeitgeber and zeitstörers theory.

40.4 The Shutdown of Normal Reaction to Fear

People fear when they perceive a threat. It leads to a series of physiological changes in both the endocrine and autonomic nervous systems. On the one hand, these physiological changes can help individuals to deal with the threats better. On the other hand, they may lead to a series of physical and psychological symptoms (Garcia 2017; Seligman 2016). Psychologists postulate that fear stimulates four reactions: fight, flight, freeze, and tend-and-befriend. Tend-and-befriend can be to seek help from others or trying to make the situation less dangerous (Jones and Monfils 2016; Bracha 2004; Taylor et al. 2000). The interference of typical psychological responses to fear may lead to psychological problems and mental disorders (Garcia 2017). During the COVID-19 pandemic, quarantine likely interferes with the psychological responses to fear, possibly leading to the occurrence of psychological problems and mental disorders. Individuals pay attention to the spread of the pandemic every day because of its status as a threat event. Any negative news can be a stimulator that gradually increases public fears. Cognizant to this, the mechanisms of abnormal psychological responses to fear that lead to the occurrence of psychological problems and mental disorders can be said to be the shutdown of normal reaction to fear.

40.5 Potential Quarantine-Related Mechanisms

Although the COVID-19 pandemic has significant psychological impacts on the majority of the population, only a minority of the population is under the standard of the theories mentioned above. As such, it is worth noting that the impact of the stressor of the outbreak itself is based on the psychological stress theory, and the nature of social zeitstörers events based on the zeitgeber and zeitstörers theory should not be underestimated. It is, therefore, necessary to independently analyze the individual social and psychological impacts of quarantine to understand the social-psychological impact of the COVID-19 pandemic comprehensively.

40.5.1 Interpersonal and Social Rhythm Disturbance

The term interpersonal and social rhythm refers to a series of behavioral patterns and regulations such as eating, work, rest times, and social tendencies (Ehlers et al. 1988; Sylvia et al. 2009). Previous studies have confirmed that interpersonal and social rhythm disturbance is closely associated with the occurrence and development of various psychological problems and mental disorders. It is mainly the case in the occurrence and development of the bipolar disorder (Grandin et al. 2006; Lieverse et al. 2013; Meyer and Maier 2006; Margraf et al. 2016; Shen et al. 2008). In recent years, several studies have further confirmed that there is indeed an association between interpersonal and social rhythm disturbance and the onset of mental disorders. Moreover, other studies have shown that interpersonal and social disturbance is associated not only with the early onset of depression, anxiety, mania, and other mental disorders but also with the severity of these disorders (Goldstein et al. 2018; Boyce and Barriball 2010; Coles et al. 2015).

Currently, psychologists are exploring the potential physiological and pathological mechanisms of psychological problems and mental disorders caused by interpersonal and social rhythm disturbance. Most psychologists postulate that interpersonal and social disturbance can lead to biological disturbance (Grandin et al. 2006). Previous studies have shown that the biological circadian rhythm and the sleep-awakening cycle positively correlate with emotion (Kim et al. 2017; Yan et al. 2019). For example, the occurrence and development of sleep disorder, depression, and anxiety are closely associated with the disturbance of a biological rhythm. As such, individuals living with mental disorders can experience aggravated symptoms as a result of biological rhythm disturbance (Lanfumey et al. 2013; Monteleone et al. 2011). Cognizant to this, it is generally believed that interpersonal and social rhythm disturbance can lead to psychological problems and mental disorders by affecting an individual biological rhythm.

Strict quarantine measures have led to the suspension of work, production, and schools, thus directly impacting negatively on the working and living habits of many people. Interruption of typical social activities, staying at home for long periods, and lack of physical exercise may lead to interpersonal and social rhythm disturbance, consequently leading to psychological problems and mental disorders. For instance, a student who would generally maintain a sleep pattern from ten in the evening to seven in the morning may find his sleep affected by staying at home because of the pandemic. If the student is unable to adapt to the rhythm of home quarantine, she/he is susceptible to emotional problems such as apathy and irritability and mental health challenges such as depression and anxiety. In this example, only the sleep rhythm has been considered to show the positive correlation between interpersonal and social rhythm disturbance and the associated psychological problems and mental disorders. However, multiple factors are affected by home quarantine. The synergistic effect of these factors can lead to psychological problems and mental disorders in vulnerable populations. It may, therefore, be inappropriate to include quarantine as a social zeitstörer event in the zeitgeber and zeitstörers theory owing to the multifaceted and complex nature of the impacts of quarantine. Cognizant to this, quarantine is an independent topic whose psychological impacts need to be comprehensively discussed and analyzed.

40.5.2 Transient Unmet Demands for Group Belongings

Group belonging is a primary human behavioral character. Individuals establish stable emotional and positive behavioral interactions among themselves by forming long-term caring relationships (Baumeister and Leary 1995; Carvallo and Gabriel 2006; Lavigne et al. 2011). Group behavior and intimate relationships are formed to meet this behavioral character. Also, self-presentation, correction, interpretation, and group integration are behaviors that help individuals integrate into groups (Baumeister and Roy 1982; Scott and Schlenker 1981). Striving to fit in a group is a basic and natural motivation of human beings. As such, lack of a need to belong to certain groups may lead to psychological challenges. Some psychologists postulate that psychological problems are a result of negative emotional and behavioral responses by individuals when social relationships are threatened. Common emotional problems such as anxiety, depression, grief, loneliness, and relationship problems, among others, can be attributed to the lack of group belongings. Many neurotic, non-adaptive, and destructive behaviors are individual reactions as they try to establish or maintain social relationships or to express frustration and confusion (Baumeister and Leary 1995). During the COVID-19 pandemic, most people, especially those who are isolated outside their hometown, find it hard to meet their demand of belonging to a group. Though most family support systems are functional, social support systems have been affected by mass quarantine. Cognizant to this, the lack of group belongings during the pandemic could become a universal psychological event leading to psychological problems and mental disorders. However, group belonging is only transiently missed during the pandemic period. A majority of the support systems will be restored and improved afterward, thereby preventing long-term impacts.

40.6 The Gut-Brain Axis

Emotional and psychological stress affects the composition of the gut microflora. It not only affects digestive, metabolic, and immune functions but also regulates sleep and mental states through the gut-brain axis (Li et al. 2018). There is an interaction between emotional and psychological stress and gut microbial environment (Eriksson et al. 2015; Gracie et al. 2018). Changes to the proportion or composition of gut microflora influence the pathogenesis of insomnia, circadian rhythm, and affective and metabolic disorders. In the same line, emotional and psychological stress can affect the proportion or composition of gut microbes, thus further affecting individual mental state through the gut-brain axis. During COVID-19 quarantine, individual negative emotions can positively correlate with abnormal changes to the gut microbial environment, thereby potentially leading to an increase in symptoms of psychological problems and mental disorders.

40.7 Discussion

The psychological impact of COVID-19 is extensive. It is similar to that of any public health emergency, especially for special groups such as frontline medical professionals (Shigemura et al. 2020), the elderly (Lima et al. 2020), and children confined to their homes (Wang et al. 2020). Public health emergencies are different from each other. As such, sudden psychological stress events, as well as the general psychological crisis intervention strategies, vary. That makes it challenging to get the best results. From this point of view, it is, therefore, necessary to develop new targeted psychological intervention strategies to help people cope with COVID-19-related psychological problems and mental disorders. It requires in-depth knowledge of the mechanisms of COVID-19-related psychological problems and mental disorders for it to be a success. Herein, the potential mechanisms of occurrence and development of COVID-19-related psychological problems and mental disorders have been discussed from two angles: the pandemic as a public health emergency itself and the extensive quarantine situation during the pandemic (Fig. 40.1). Besides these mechanisms, there are other complex potential mechanisms of psychological problems and mental disorders related to COVID-19. For instance, the psychological problems of people living with chronic diseases can be attributed to the existing burden of physical disease as well as their physiological characteristics. Cognizant to this, potential mechanisms of psychological problems and mental disorders associated with COVID-19 are unique to specific populations.

Several potential mechanisms discussed in this article focus on the specific processes of COVID-19-related psychological problems and mental disorders (Fig. 40.2). The essential characteristics of these specific processes can be summarized in three points: differences in process performance, the process of action varies with people because it is closely related to individual psychological and physiological characteristics; the interaction of several pathways, one symptom may be related to multiple pathways because pathways interact in an interrelated manner, and this can further be divided into single and mixed modes of action depending on the existence of a dominant way of action; and the predictability of the mechanisms involved, the potential mechanism involved can be inferred from the individual psychological and physiological characteristics and the patient symptoms as well. These three characteristics are crucial to understanding the potential mechanisms of COVID-19-related psychological problems and mental disorders. For better and more complete results, a particular focus needs to be put in combining individual psychological and physiological characteristics in clinical practice.

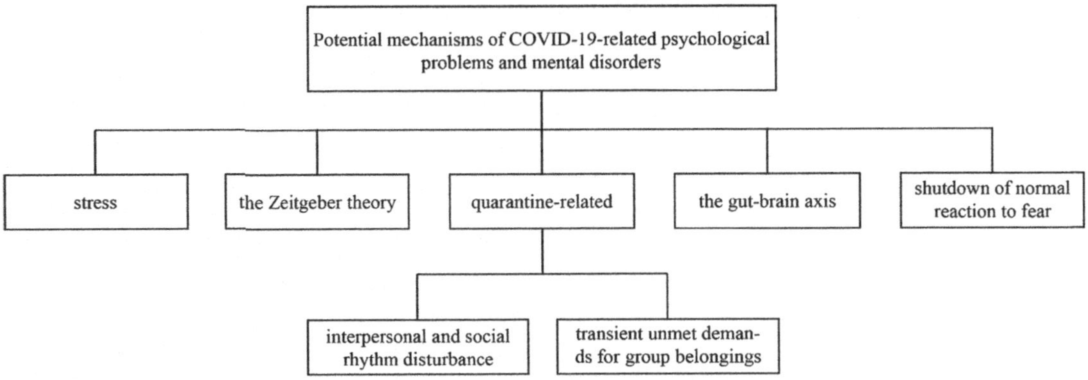

Fig. 40.1 Potential mechanisms of COVID-19-related psychological problems and mental disorders

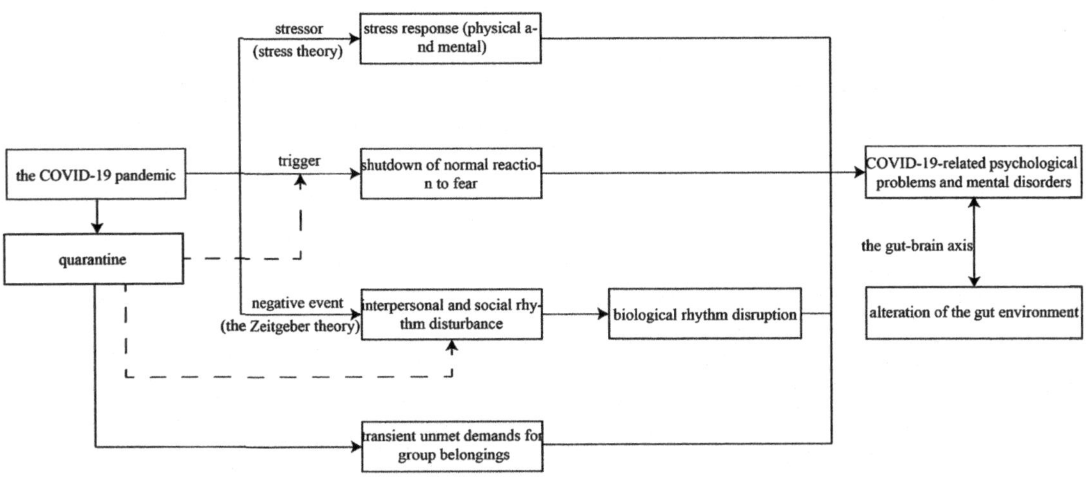

Fig. 40.2 A flowchart of potential mechanisms contributing to the relationship between the COVID-19 pandemic and related psychological problems and mental disorders

Based on the potential mechanisms discussed in this article, some existing intervention strategies may be useful in the management and treatment of COVID-19-related psychological problems and mental disorders. For example, psychological problems and mental disorders caused by interpersonal and social rhythm disturbances can be managed and treated using the interpersonal and social rhythm therapy (IPSRT) (Bottai et al. 2010; Goldstein et al. 2018; Frank et al. 2007). In the same line, patients with intestinal dysfunctions can drink probiotics reasonably to prevent further development of the disease (Bottai et al. 2010; Frank et al. 2005; Dickerson et al. 2018; Butler et al. 2019). Nonetheless, further studies focusing on the mechanisms of COVID-19-related psychological problems and mental disorders are needed to provide a more solid theoretical basis for clinical practice.

40.8 Conclusion

The potential mechanisms of occurrence and development of COVID-19-related psychological problems and mental disorders have been discussed from two angles: the pandemic as a public health emergency itself and the extensive quarantine situation during the pandemic. The basic characteristics of the specific processes of COVID-19-related psychological problems and mental disorders can be summarized in three points: i, differences in process performance; ii, the interaction of several pathways; and iii, the predictability of the mechanisms involved. Further research is necessary to investigate other potential mechanisms of psychological problems and mental disorders related to COVID-19.

References

Appart A, Lange AK, Sievert I, Bihain F, Tordeurs D (2017) Adjustment disorder and DSM-5: a review. Encéphale 43(1):41–46. https://doi.org/10.1016/j.encep.2015.06.007

Baumeister RF (1982) A self-presentational view of social phenomena. Psychol Bull 91(1):3–26

Baumeister RF, Leary MR (1995) The need to belong: desire for interpersonal attachments as a fundamental human motivation. Psychol Bull 117(3):497–529

Bottai T, Biloa-Tang M, Christophe S, Dupuy C, Jacquesy L, Kochman F, Meynard JA, Papeta D, Rahioui H, Adida M, Fakra E, Kaladjian A, Pringuey D, Azorin JM (2010) Interpersonal and social rhythm therapy (IPSRT). Encéphale 36(6):S206–S217. https://doi.org/10.1016/s0013-7006(10)70059-2

Boyce P, Barriball E (2010) Circadian rhythms and depression. Aust Fam Physician 39(5):307–310

Bracha HS (2004) Freeze, flight, fight, fright, faint: adaptationist perspectives on the acute stress response spectrum. CNS Spectr 9(9):679–685. https://doi.org/10.1017/s1092852900001954

Bryant RA (2017) Acute stress disorder. Curr Opin Psychol 14:127–131. https://doi.org/10.1016/j.copsyc.2017.01.005

Bryant RA (2018) The current evidence for acute stress disorder. Curr Psychiatry Rep 20(12):111. https://doi.org/10.1007/s11920-018-0976-x

Bryant RA (2019) Post-traumatic stress disorder: a state-of-the-art review of evidence and challenges. World Psychiatry 18(3):259–269. https://doi.org/10.1002/wps.20656

Butler MI, Mörkl S, Sandhu KV, Cryan JF, Dinan TG (2019) The gut microbiome and mental health: what should we tell our patients?: Le microbiote intestinal et la Santé Mentale : que Devrions-nous dire à nos patients? Can J Psychiatr 64(11):747–760. https://doi.org/10.1177/0706743719874168

Carvallo M, Gabriel S (2006) No man is an island: the need to belong and dismissing avoidant attachment style. Personal Soc Psychol Bull 32(5):697–709. https://doi.org/10.1177/0146167205285451

Coles ME, Schubert JR, Nota JA (2015) Sleep, circadian rhythms, and anxious traits. Curr Psychiatry Rep 17(9):73. https://doi.org/10.1007/s11920-015-0613-x

Dickerson F, Adamos M, Katsafanas E, Khushalani S, Origoni A, Savage C, Schweinfurth L, Stallings C, Sweeney K, Goga J, Yolken RH (2018) Adjunctive probiotic microorganisms to prevent rehospitalization in patients with acute mania: a randomized controlled trial. Bipolar Disord 20(7):614–621. https://doi.org/10.1111/bdi.12652

Ehlers CL, Frank E, Kupfer DJ (1988) Social zeitgebers and biological rhythms. A unified approach to understanding the etiology of depression. Arch Gen Psychiatry 45(10):948–952. https://doi.org/10.1001/archpsyc.1988.01800340076012

Eriksson EM, Andrén KI, Kurlberg GRK, Eriksson HT (2015) Aspects of the non-pharmacological treatment of irritable bowel syndrome. World J Gastroenterol 40:249–259

Finkelstein J (1989) Social zeitgebers and biological rhythms. Arch Gen Psychiatry 46(9):859

Frank E (2007) Interpersonal and social rhythm therapy: a means of improving depression and preventing relapse in bipolar disorder. J Clin Psychol 63(5):463–473. https://doi.org/10.1002/jclp.20371

Frank E, Kupfer DJ, Thase ME, Mallinger AG, Swartz HA, Fagiolini AM, Grochocinski V, Houck P, Scott J, Thompson W, Monk T (2005) Two-year outcomes for interpersonal and social rhythm therapy in individuals with bipolar I disorder. Arch Gen Psychiatry 62(9):996–1004. https://doi.org/10.1001/archpsyc.62.9.996

Frank E, Swartz HA, Boland E (2007) Interpersonal and social rhythm therapy: an intervention addressing rhythm dysregulation in bipolar disorder. Dialogues Clin Neurosci 9(3):325–332

Garcia R (2017) Neurobiology of fear and specific phobias. Learn Mem 24(9):462–471. https://doi.org/10.1101/lm.044115.116

Goldstein TR, Merranko J, Krantz M, Garcia M, Franzen P, Levenson J, Axelson D, Birmaher B, Frank E (2018) Early intervention for adolescents at-risk for bipolar disorder: a pilot randomized trial of interpersonal and social rhythm therapy (IPSRT). J Affect Disord 235:348–356. https://doi.org/10.1016/j.jad.2018.04.049

Gracie DJ, Hamlin P, John F, Alexander C (2018) Longitudinal impact of IBS-type symptoms on disease activity, healthcare utilization, psychological health, and quality of life in inflammatory bowel disease. Am J Gastroenterol 113:702–712

Gradus JL, Bozi I, Antonsen S, Svensson E, Lash TL, Resick PA, Hansen JG (2014) Severe stress and adjustment disorder diagnoses in the population of Denmark. J Trauma Stress 27(3):370–374. https://doi.org/10.1002/jts.21926

Grandin LD, Alloy LB, Abramson LY (2006) The social zeitgeber theory, circadian rhythms, and mood disorders: review and evaluation. Clin Psychol Rev 26(6):679–694. https://doi.org/10.1016/j.cpr.2006.07.001

Jakovljević M, Brajković L, Lončar M, Cima A (2012) Post-traumatic stress disorders (PTSD) between fallacy and facts: what we know and what we don't know? Psychiatr Danub 24(3):241–245

Jones CE, Monfils MH (2016) Fight, flight, or freeze? The answer may depend on your sex. Trends Neurosci 39(2):51–53. https://doi.org/10.1016/j.tins.2015.12.010

Kim J, Jang S, Choe HK, Chung S, Son GH, Kim K (2017) Implications of circadian rhythm in dopamine and mood regulation. Mol Cells 40(7):450–456. https://doi.org/10.14348/molcells.2017.0065

Koffer RE, Ram N, Conroy DE, Pincus AL, Almeida DM (2016) Stressor diversity: introduction and empirical integration into the daily stress model. Psychol Aging 31(4):301–320. https://doi.org/10.1037/pag0000095

Kupfer DJ, Reynolds CF 3rd (1983) A critical review of sleep and its disorders from a developmental perspective. Psychiatr Dev 1(4):367–386

Lanfumey L, Mongeau R, Hamon M (2013) Biological rhythms and melatonin in mood disorders and their treatments. Pharmacol Ther 138(2):176–184. https://doi.org/10.1016/j.pharmthera.2013.01.005

Lavigne GL, Vallerand RJ, Crevier-Braud L (2011) The fundamental need to belong: on the distinction between growth and deficit-reduction orientations. Personal Soc Psychol Bull 37(9):1185–1201. https://doi.org/10.1177/0146167211405995

Leger KA, Charles ST, Turiano NA, Almeida DM (2016) Personality and stressor-related affect. J Pers Soc Psychol 111(6):917–928. https://doi.org/10.1037/pspp0000083

Li Y, Hao Y, Fan F, Zhang B (2018) The role of microbiome in insomnia, circadian disturbance and depression. Front Psych 9:669. https://doi.org/10.3389/fpsyt.2018.00669

Lieverse R, de Vries R, Hoogendoorn AW, Smit JH, Hoogendijk WJ (2013) Social support and social rhythm regularity in elderly patients with major depressive disorder. Am J Geriatr Psychiatry 21(11):1144–1153. https://doi.org/10.1016/j.jagp.2013.01.052

Lima CKT, Carvalho PMM, Lima I, Nunes J, Saraiva JS, de Souza RI, da Silva CGL, Neto MLR (2020) The emotional impact of coronavirus 2019-nCoV (new coronavirus disease). Psychiatry Res 287:112915. https://doi.org/10.1016/j.psychres.2020.112915

Margraf J, Lavallee K, Zhang X, Schneider S (2016) Social rhythm and mental health: a cross-cultural comparison. PLoS One 11(3):e0150312. https://doi.org/10.1371/journal.pone.0150312

Meyer TD, Maier S (2006) Is there evidence for social rhythm instability in people at risk for affective disorders? Psychiatry Res 141(1):103–114. https://doi.org/10.1016/j.psychres.2005.07.023

Monteleone P, Martiadis V, Maj M (2011) Circadian rhythms and treatment implications in depression. Prog Neuro-Psychopharmacol Biol Psychiatry 35(7):1569–1574. https://doi.org/10.1016/j.pnpbp.2010.07.028

Otto E, Ullsperger P, Metz AM (1990) The history of the concept stress. Z Gesamte Hyg 36(7):363–365

Patterson CW (1989) Social zeitgebers, biological rhythms, and transactional analysis. Arch Gen Psychiatry 46(9):859–860. https://doi.org/10.1001/archpsyc.1989.01810090101018

Qi W, Gevonden M, Shalev A (2016) Prevention of post-traumatic stress disorder after trauma: current evidence and future directions. Curr Psychiatry Rep 18(2):20. https://doi.org/10.1007/s11920-015-0655-0

Rom O, Reznick AZ (2016) The stress reaction: a historical perspective. Adv Exp Med Biol 905:1–4. https://doi.org/10.1007/5584_2015_195

Santana MRM, Zatti C, Spader ML, Malgarim BG, Salle E, Piltcher R, Cerésér KMM, Bastos AG, Freitas LH (2017) Acute stress disorder and defense mechanisms: a study of physical trauma patients admitted to an emergency hospital. Trends Psychiatry Psychother 39(4):247–256. https://doi.org/10.1590/2237-6089-2016-0071

Scott M, Schlenker B (1981) Impression management: the self-concept, social identity, and interpersonal relations. Contemp Sociol 10:582. https://doi.org/10.2307/2067758

Seligman ME (2016) Phobias and preparedness – republished article. Behav Ther 47(5):577–584. https://doi.org/10.1016/j.beth.2016.08.006

Shen GH, Alloy LB, Abramson LY, Sylvia LG (2008) Social rhythm regularity and the onset of affective episodes in bipolar spectrum individuals. Bipolar Disord 10(4):520–529. https://doi.org/10.1111/j.1399-5618.2008.00583.x

Shigemura J, Ursano RJ, Morganstein JC, Kurosawa M, Benedek DM (2020) Public responses to the novel 2019 coronavirus (2019-nCoV) in Japan: mental health consequences and target populations. Psychiatry Clin Neurosci 74(4):281–282. https://doi.org/10.1111/pcn.12988

Strain JJ (2018) The psychobiology of stress, depression, adjustment disorders and resilience. World J Biol Psychiatry 19(sup1):S14–s20. https://doi.org/10.1080/15622975.2018.1459049

Sylvia LG, Alloy LB, Hafner JA, Gauger MC, Verdon K, Abramson LY (2009) Life events and social rhythms in bipolar spectrum disorders: a prospective study. Behav Ther 40(2):131–141. https://doi.org/10.1016/j.beth.2008.04.003

Szabó S, Gyires K (2015) Naming and classification of steroids and human stress ulcers. Articles of historic significance published by Hans Selye 70 years ago. Orv Hetil 156(35):1406–1414. https://doi.org/10.1556/650.2015.30251

Szabo S, Tache Y, Somogyi A (2012) The legacy of Hans Selye and the origins of stress research: a retrospective 75 years after his landmark brief "letter" to the editor# of nature. Stress 15(5):472–478. https://doi.org/10.3109/10253890.2012.710919

Szabo S, Yoshida M, Filakovszky J, Juhasz G (2017) "stress" is 80 years old: from Hans Selye original paper in 1936 to recent advances in GI ulceration. Curr Pharm Des 23(27):4029–4041. https://doi.org/10.2174/1381612823666170622110046

Tachè Y (2014) Hans Selye and the stress response: from "the first mediator" to the identification of the hypothalamic corticotropin-releasing factor. Ideggyogy Sz 67(3–4):95–98

Taylor SE, Klein LC, Lewis BP, Gruenewald TL, Gurung RA, Updegraff JA (2000) Biobehavioral responses to stress in females: tend-and-befriend, not fight-or-flight. Psychol Rev 107(3):411–429. https://doi.org/10.1037/0033-295x.107.3.411

Wang G, Zhang Y, Zhao J, Zhang J, Jiang F (2020) Mitigate the effects of home confinement on children during the COVID-19 outbreak. Lancet 395(10228):945–947. https://doi.org/10.1016/s0140-6736(20)30547-x

Wever RA (1979) The circadian system of man: results of experiments under temporal isolation. Springer, New York

Yan L, Lonstein JS, Nunez AA (2019) Light as a modulator of emotion and cognition: lessons learned from studying a diurnal rodent. Horm Behav 111:78–86. https://doi.org/10.1016/j.yhbeh.2018.09.003

41 Mental Health in Health Professionals in the COVID-19 Pandemic

Antonia Bendau, Andreas Ströhle, and Moritz Bruno Petzold

Abstract

The coronavirus disease 2019 (COVID-19) pandemic has posed enormous challenges to the healthcare systems worldwide, which are mainly shouldered by healthcare workers from all professions. This chapter outlines the potential stressors of the COVID-19 pandemic for health professionals and describes possible consequences for their mental health as well as potential interventions and coping strategies. The chapter is based on preliminary research on the psychosocial implications of the COVID-19 pandemic in health professionals and is complemented by findings from previous outbreaks of high-risk infectious diseases. High proportions of healthcare workers report acute symptoms of anxiety, depression, high psychological stress, and insomnia in the context of the COVID-19 pandemic. Coping strategies and self-care on an individual level, interventions on an institutional level such as specific training and institutional support, as well as social and psychological support can help to mitigate psychological strain. Further reliable and prospective studies regarding the mental health of health professionals, as well as further measures to protect their short- and long-term mental health, are required.

A. Bendau · A. Ströhle · M. B. Petzold (✉)
Department of Psychiatry and Psychotherapy, Charité Universitätsmedizin Berlin, corporate member of Freie Universität Berlin, Humboldt Universität zu Berlin and Berlin Institute of Health, Berlin, Germany
e-mail: moritz.petzold@charite.de

Keywords

Coronavirus · COVID-19 · Healthcare · Nurses · Physicians · Psychological distress · Stress

41.1 Introduction

Since the first reports of its occurrence in December of 2019 in Wuhan, China, the severe acute respiratory syndrome coronavirus 2 (SARS-CoV-2), which can cause the disease named COVID-19, has grown rapidly to one of the most severe international health issues in the last decades and posed an immense threat to the healthcare systems worldwide (Zhu et al. 2020a). Communities worldwide suffer directly from high infection rates associated with increased mortality, and additionally, the measures to contain the spreading of the virus are associated with further indirect physical and mental health consequences (Bohlken et al. 2020; Helmy et al. 2020; Wang et al. 2020a; Xiang et al. 2020). Health professionals of all professions play a

N. Rezaei (ed.), *Coronavirus Disease - COVID-19*, Advances in Experimental Medicine and Biology 1318, https://doi.org/10.1007/978-3-030-63761-3_41

crucial role in facing and overcoming the enormous challenges of the COVID-19 pandemic (Xiong and Peng 2020). Dealing with the pandemic comes with multiple stressors on general and profession-specific levels for health professionals (Kang et al. 2020a). Being confronted with those stressors bears the risk of the development of new or the aggravation of pre-existing mental health issues such as symptoms of anxiety, depression, and high psychological stress (Kang et al. 2020a). These mental health strains might not only affect the performance of health professionals in managing the challenges of the COVID-19 pandemic but could also impair their overall well-being and mental health in the long run (Kang et al. 2020a; Mehta et al. 2018). Therefore it should be of top priority to protect the mental health of health professionals as extensively as possible to enable the control of the pandemic as well as to protect the short- and long-term mental health of professionals (Kang et al. 2020a).

In the following, potential common stressors and psychological reactions of health professionals, as well as possible coping and intervention strategies, are described – based on first insights from preliminary research on the psychosocial consequences of the COVID-19 pandemic in health professionals. Evidence from these first studies is complemented with findings from previous outbreaks of high-risk infectious diseases.

41.2 Stress Factors

41.2.1 Stressors Concerning the General Population

Some of the potential adverse and stress-inducing factors of the pandemic are shared by health professionals and the general population (International Federation of Red Cross and Red Crescent Societies 2020; Petzold et al. 2020d). Those factors can relate to both the anticipation of negative impacts and occurring adverse outcomes. Below outlines these potential stressors.

41.2.1.1 Fear of Being Infected with COVID-19

As for the general population, it is likely for health professionals to be afraid of being infected with COVID-19 and to fear the potentially resultant severe illness outcomes for themselves or loved ones (Wu et al. 2020a; IASC 2020). For example, in a study with 1357 healthcare workers from Henan, China, 85% of them reported a pronounced fear of a potential self-infection with the virus (Zhang et al. 2020b). The risk of infecting oneself and others is significantly adverse in a situation where the transmission pathways and consequences of the virus are not yet fully understood (Petzold et al. 2020d; IASC 2020).

41.2.1.2 Misinterpretation of Symptoms of Other Diseases

As COVID-19 is accompanied by symptoms that are present in a wide range of other diseases like cold or influenza (Wang et al. 2020b), which are present in large parts of the populations, at times of the pandemic, the misinterpretation of symptoms of such other diseases as symptoms of COVID-19 may result in pronounced fears of being infected (Petzold et al. 2020d; IASC 2020).

41.2.1.3 Consequences of the Measures to Contain the Spreading of the Virus

Vast parts of the societies worldwide need to cope with the measures that have been installed to reduce the spread of COVID-19 (IASC 2020). Among those are, for example, the caring for family members at home due to the suspension of schools and childcare services; the omission of cultural, sportive, and leisure activities; and the cancellation of vacations or occupational formation (Wu et al. 2020a; IASC 2020). Moreover, feelings of helplessness, boredom, and depressive symptoms due to isolation or quarantine are possible (IASC 2020).

41.2.1.4 Fear of Social Consequences

The fear of social isolation, when associated with the disease and the fear of separation and losing contact with family and friends due to distancing measurements or isolation/quarantine, is also among the common stressors (Wu et al. 2020a; IASC 2020).

41.2.1.5 Fear of Economic Consequences

Also, potential negative economic consequences due to the pandemic often expose people to fears and worrying (IASC 2020).

41.2.1.6 Unpredictability and Feelings of Helplessness

Like the general population, health professionals need to cope with the highly unpredictable and dynamic situation due to the novelty and lack of information about the virus (Sun et al. 2020; IASC 2020). The pandemic is particularly threatening as to present, there is no vaccine, and COVID-19 can be only treated symptomatically (Shuja et al. 2020; Xiao et al. 2020). Furthermore, the circumstances may require individuals to deal with feelings of helplessness and the inability to protect loved ones from COVID-19 (IASC 2020).

41.2.1.7 Worsening of Pre-Existing Strains

Among individuals with pre-existing health issues or risk factors, the pandemic may put particular strains on physical or mental health (Adams and Walls 2020; Petzold et al. 2020d). For example, the pandemic may lead to a worsening of symptoms of psychological distress or pre-existing mental disorders or results in a reactivation of threatening experiences from previous epidemics or health crises (Adams and Walls 2020; Petzold et al. 2020d).

41.2.2 Stressors Concerning Health Professionals in Particular

Additionally, to those common stressors which apply to the general population, health professionals are confronted with profession-specific stress factors (International Federation of Red Cross and Red Crescent Societies 2020). Besides stressors that already existed before the pandemic (e.g., high workload and high responsibilities), the COVID-19 pandemic may come with new burdening factors and aggravate pre-existing ones. The following presents some factors, especially concerning health professionals working in hospitals.

41.2.2.1 High Occupational Stress

Health professionals report a massive workload, such as longer working hours, high pressure, and more number of patients than usual due to the pandemic as a primary source of physical and psychological strain (Adams and Walls 2020; Cao et al. 2020; El-Hage et al. 2020; IASC 2020).

41.2.2.2 High Exposure Risk

Even with strict safety measures, the infection risk of health professionals working with COVID-19 patients is, on average, higher than that of the general population (Heymann 2020; Strametz et al. 2020; Wang et al. 2020a; Wu et al. 2020a; Xu et al. 2020). It can result in stronger worrying about consequences for one's well-being in case of an infection as well as worrying about infecting family and loved ones with COVID-19 (Adams and Walls 2020; El-Hage et al. 2020; Tsamakis et al. 2020; Wu et al. 2020a) and was the most common reason for health workers to report unwillingness to care for patients diagnosed with COVID-19 in a study with medical staff in Chinese psychiatric hospitals (Shi et al. 2020). Furthermore, insufficient information on the consequences of long-term exposure to COVID-19-infected patients adds strain to this factor (IASC 2020).

41.2.2.3 Stigma and Social Exclusion

Health professionals are confronted (e.g., through verbally and physically assaults and attacks) with the stigmatization attached to people working with COVID-19 patients, e.g., because of concerns of others that the health professionals may be infected and therefore constitute a risk (McKay et al. 2020; IASC 2020). In previous outbreaks of high-risk respiratory infectious diseases (SARS, MERS), perceived stigma also was reported to occur frequently and was a predictor of mental health strains in health professionals (Bell and Wade 2020; Koh et al. 2005; Maunder et al. 2003; Maunder et al. 2006).

41.2.2.4 Strict Safety Measures

The wearing of protective clothing, the permanent need for concentration and vigilance, as well as highly regulated procedural instructions limit spontaneity and autonomy and strains the physical and mental well-being (Adams and Walls 2020; IASC 2020; Petzold et al. 2020d; Poon and Liu 2004). The reduction of physical contact hinders the ability to spend consolidation for patients or colleagues, which might reduce social support that represents a critical factor for mental health (Petzold et al. 2020d; IASC 2020).

41.2.2.5 Separation from Loved Ones

Due to the higher exposure risk and the high workload, healthcare workers often have to keep a distance from loved ones (WHO 2020b; Wu et al. 2020a). For example, health professionals in Chinese hospitals often slept over weeks in the hospital and have been quarantined afterward to avoid cross infections of family members and patients (Wu et al. 2020a). It resulted in separation from their families for weeks and was a commonly reported stress factor (Wu et al. 2020a) and the main reason for negative emotions (Cao et al. 2020).

41.2.2.6 Reduced Social Support

Due to long working hours and stigmatization of health professionals when dealing with COVID-19 patients, health professionals may face a lack of social support (Petzold et al. 2020d; IASC 2020). In the work environment, social support by colleagues might be reduced because of a high workload and transfers to different working environments without the familiar colleagues, which could additionally result in feelings of isolation (Petzold et al. 2020d; Wu et al. 2020a).

41.2.2.7 Reduced Self-Care

Due to a lack of time and energy, healthcare workers might not be able to maintain their self-care (IASC 2020; Petzold et al. 2020d). Mainly a lack of sleep and adequate recovery times are a problem that health professionals are facing in the current pandemic (National Center for PTSD 2020).

41.2.2.8 Dealing with the Intense Emotions of the Patients

Health professionals may be confronted with anger and rage against the government or the healthcare system by patients or their relatives (Chen et al. 2020; McKay et al. 2020; Petzold et al. 2020d). Additionally, healthcare workers might have to deal with an increased rate of patients who die or suffer from severe outcomes of COVID-19 and need to handle the fear and grief of patients and their relatives (Chen et al. 2020; Kang et al. 2020b; National Center for PTSD 2020; Xiong and Peng 2020).

41.2.2.9 Inadequate Preparedness

The rapidly spreading pandemic has surprised the health systems of many countries (WHO 2020c). Therefore, a resource-limited setting due to, e.g., lack of staff, medical devices, and protective equipment, was a common source for strain in healthcare workers (Chung and Yeung 2020; International Council of Nurses 2020). In a study with nurses from China, a lack of knowledge about coping with infectious diseases and respiration devices was reported as a stress factor as well as the pressure to obtain this expertise in a short time (Wu et al. 2020a).

41.2.2.10 Conflicts of Roles and Feelings of Responsibility

The measurements to contain the spreading of the virus, e.g., the closure of schools and daycare institutions for children, might have given rise to role conflicts between family and occupational duties for medical staff because they might have to provide care and home tutoring for their children in addition to their anyhow high workload (Wu et al. 2020a).

41.3 Psychological Strain and Risk and Protective Factors

About the multiple and far-reaching adverse impacts of the pandemic, the occurrence of overburdening, distress, and psychological strain has to be seen as a largely normal reaction to an exceptional event for the general population as well as for health professionals (IASC 2020). Nevertheless, it is crucial to obtain an overview of how many health professionals are affected by such symptoms, which are the most frequent and most affecting symptoms, and to which extent they exceed the limits to pathological reactions (Pothiawala 2020). Currently, substantial research on the psychological consequences of the COVID-19 pandemic on health professionals is lacking. In the following, we summarize insight from the first studies on this subject. Most of the studies are cross-sectional, which comes with several methodological limitations. Table 41.1 shows an overview of the currently existing studies on the mental health of health professionals in the COVID-19 pandemic. Most of the recent studies concerning those topics have been conducted in China (Bohlken et al. 2020). It remains unclear to which extent these results can be generalized to other populations, and further research on this topic seems urgently needed.

41.3.1 Frequently Occurring Symptoms of Psychological Distress

41.3.1.1 Acute Symptoms

In a study with 911 hospital workers in Italy, 86% reported moderate to severe psychological distress during the COVID-19 outbreak (Bettinsoli et al. 2020). Several surveys with health professionals in China showed consistently high rates of symptoms of psychological strain in response to the COVID-19 pandemic (Bohlken et al. 2020). In a study with medical staff in a specially equipped COVID-19 emergency clinic in Peking (Cao et al. 2020), 19% of the staff reported moderate or severe depressive symptoms. Similar results were obtained in clinics in Hong Kong, with 20% of the staff fulfilling the questionnaire criteria of mild and 15% the criteria for moderate depression (Chung and Yeung 2020). Even higher rates occurred in hospitals in and around Wuhan, the epicenter of the pandemic (Kang et al. 2020b; Lai et al. 2020). Mild depressive symptoms occurred in 34% to 36%, moderate symptoms in 9% up to 22%, and severe symptoms in 6% of the medical and nursery staff.

Besides depressive symptoms, symptoms of anxiety were the most commonly reported symptoms among hospital staff. In a study with clinical first-line medical staff in China (Huang et al. 2020), 23% showed symptoms of anxiety (16% mild, 2% moderate, 5% severe). In Wuhan, the rates have again been even higher, with 44% reported anxiety (32% mild, 7% moderate, and 5% severe symptoms) (Lai et al. 2020). These findings are in line with previous epidemics of infectious diseases with high rates of anxiety in reaction to the epidemics (Xiao et al. 2020).

On scales targeting symptoms of post-traumatic stress, 27% of the surveyed frontline staff from China exceeded the cut-off criteria for post-traumatic stress disorder (National Center for PTSD) during the COVID-19 pandemic (Huang et al. 2020). In a study in Wuhan, China, 37% of the healthcare workers scored on mild, 25% on moderate, and 17% on severe extents of symptoms of post-traumatic stress (Lai et al. 2020).

Problems of sleep have been another frequently reported issue. In qualitative interviews, 30% of the health professionals from Wuhan reported suffering from sleep problems (Wu et al. 2020a). In quantitative surveys, 34–36% of the staff showed symptoms of insomnia (26% mild, 7% moderate, and 1% severe) (Lai et al. 2020; Zhang et al. 2020a). Complaints about loss of appetite (22%) occurred, too, as well as other somatic strains (Wu et al. 2020a).

To sum up, on average, approximately 2–15% of the health professionals reported suffering from severe and urgent treatment requiring symptoms, and more than one third reported to be affected with mild to moderate acute impairments in mental well-being (Bohlken et al. 2020). Those proportions seem to be comparable with first reports regarding the general populations (Li et al. 2020; Sun et al. 2020; Tan et al. 2020). These results stem from cross-sectional studies, which suffer from a lack of baseline assessment before the current pandemic and, therefore, should be interpreted with great caution.

Table 41.1 Overview of studies on the mental health of health professionals in the COVID-19 pandemic

Study	Country	Population	Recruitment period	Sample size	Design	Outcome measures
Bettinsoli et al. (2020)	Italy	Hospital staff, different regions of Italy	March–April 2020	580	Cross-sectional, online questionnaire	BRCS, CSES, GHQ-12, and self-developed items
Cao et al. (2020)	China	Medical workers in a fever clinic in Beijing	?	37	Cross-sectional, interview and questionnaire	Qualitative interview, PHQ-9, and MBI
Chung and Yeung (2020)	China	Hospital staff in Hong Kong	February 2020	69	Cross-sectional, online questionnaire	PHQ-9 and free text items
Huang et al. (2020)	China	Hospital staff in an infectious disease hospital	February 2020	246	Cross-sectional, online questionnaire	SAS and PTSD-SS
Kang et al. (2020b)	China	Doctors and nurses in Wuhan	January – February 2020	994	Cross-sectional, online questionnaire	PHQ-9, GAD-7, ISI, IES-R, and self-developed items
Lai et al. (2020)	China	Health workers in different regions of China	January – February 2020	1257	Cross-sectional, online questionnaire	PHQ-9, GAD-7, ISI, and IES-R
Li et al. (2020)	China	Nurses + general public	February 2020	526 nurses	Cross-sectional, app-based questionnaire	The Chinese version of the vicarious traumatization Questionnaire
Liang et al. (2020)	China	Health workers of a hospital in Guangdong	February 2020	59	Cross-sectional, self-report questionnaire	SDS and SAS
Lu et al. (2020)	China	Health workers of a hospital in Fujian	February 2020	2299	Cross-sectional, online questionnaire	NRS on fear, HAMA, and HAMD
Mo et al. (2020)	China	Nurses in Guangxi	February 2020	180	Cross-sectional, online questionnaire	SOS and SAS
Sun et al. (2020) Preprint	China	Nurses in a Hospital in Henan	January – February 2020	20	Cross-sectional, qualitative interview	Qualitative interview analyzed by Colaizzi's 7-step method
Tan et al. (2020)	Singapore	Health workers from two tertiary institutions in Singapore	February – March 2020	470	Cross-sectional, self-report questionnaire	DASS-21 and IES-R
Wu et al. (2020b)	China	Physicians and nurses in frontline vs. usual wards in Wuhan	March 2020	190	Cross-sectional	MBI
Xiao et al. (2020)	China	Medical staff that treated patients with COVID-19	January – February 2020	180	Cross-sectional, self-report questionnaire	SAS, GSES, SASR, PSQI, and SSRS

Xu et al. (2020)	China	Surgical medical staff at a hospital in Shanghai	January – March 2020	120	Cross-sectional	Anxiety scale, depression score, dream anxiety score, and SF-36
Zhang et al. (2020a)	China	Medical staff from China	January – February 2020	1563	Cross-sectional, app-based questionnaire	ISI, PHQ-9, GAD-7, and IES-R
Zhu et al. (2020b) Preprint	China	Medical staff at Tongji hospital	February 2020	5062	Cross-sectional, online questionnaire	IES-R, PHQ-9, and GAD-7

BRCS Brief Resilience Coping Scale, *CSES* Coping Self-Efficacy Scale, *GHQ-12* General Health Questionnaire-12, *PHQ-9* Patient Health Questionnaire, *MBI* Maslach Burnout Inventory, *SAS* Self-rating Anxiety Scale, *PTSD-SS* Post-traumatic Stress Disorder Scale, *GAD-7* General Anxiety Disorder Questionnaire, *ISI* Insomnia Severity Index, *IES-R* Impact of Event Scale-Revised, *SDS* Zung's Self-Rating Depression Scale, *SAS* Zung's Self-Rating Anxiety Scale, *NRS on fear* Numeric Rating Scale on fear, *HAMA* Hamilton Anxiety Scale, *HAMD* Hamilton Depression Scale, *SOS* Stress Overload Scale, *DASS-21* Depression, Anxiety, and Stress Scales, *GSES* General Self-Efficacy Scale, *SASR* Stanford Acute Stress Reaction Questionnaire, *PSQI* Pittsburgh Sleep Quality Index, *SSRS* Social Support Rate Scale, *SF-36* Short Form Health Survey

41.3.1.2 Long-Term Consequences

Due to the dynamic situation in the current pandemic, research on long-term consequences for health professionals is not available. However, all those reactions and underlying stressors among the COVID-19 pandemic described above are in line with the results from previous outbreaks of infectious diseases (Bai et al. 2004; Bell and Wade 2020; Chen et al. 2006; Chew et al. 2020; Maunder et al. 2003). In interpreting these findings, it is highly relevant to emphasize that the development of a wide range of psychological reactions regarding the circumstances of the COVID-19 pandemic represents a mostly normal response after experiencing an exceptional and straining event (WHO and International Labour Organization 2018; Petzold et al. 2020d) and does not automatically lead to mental disorders.

Longitudinal observations of the psychological reactions to previous outbreaks of high-risk infectious diseases indicate that symptoms of psychological strain, such as anxiety and distress, tend to peak early during outbreaks and are transient for most responders, which means that for the majority, the symptoms resolve as time passes (Bell and Wade 2020; WHO and International Labour Organization 2018; Wu et al. 2008). Several qualitative and cross-sectional quantitative studies underpinned this pattern of initially high levels of psychological strain and decreased levels after the outbreak (Bell and Wade 2020; Lu et al. 2006; Xu et al. 2020). It seems plausible because, during the outbreak, the workload and multiple other factors are probably more prevalent and adverse than in a non-outbreak moment (Xu et al. 2020). Nevertheless, the burden faced by health professionals while working during the COVID-19 pandemic has the potential to result in long-lasting psychological strain and an increased occurrence of mental illnesses such as anxiety disorders, depression, or PTSD in the future (Bell and Wade 2020; Petzold et al. 2020d; Wu et al. 2008). The evidence of previous SARS outbreaks supports this potential risk. SARS outbreak correlated with an increased prevalence of PTSD among hospital staff in Singapore from before to 2 months after the outbreak (from approximately 2% to 8%) (Chan and Huak 2004). One to two years after the resolution of the outbreak, health professionals who worked in SARS-affected hospitals in Toronto, Canada, in 2003, reported significantly more psychological strain than hospital staff without exposure to SARS (International Federation of Red Cross and Red Crescent Societies 2020; Maunder et al. 2003; Maunder et al. 2006). Health professionals of SARS-affected hospitals had elevated signs of chronic stress (30% vs. 19%), depressive and anxiety symptoms (45% vs. 30%), and problematic behavior such as increased consumption of tobacco and alcohol (21% vs. 13%). Moreover, the study showed long-term influences on their daily work, e.g., shortening of work hours (9% vs. 2%). However, the rates of diagnosed mental diseases such as depression or PTSD were not significantly elevated (Maunder et al. 2008; Maunder et al. 2006). It indicates the conclusion that long-term effects of outbreaks of infectious diseases tend to be common but often stay in the range of subsyndromal stress response syndromes (Lancee et al. 2008; Maunder et al. 2008). Furthermore, common symptoms seem to change over time. A study in Taiwan found in the aftermath of the SARS outbreak symptoms of depression and avoidance predominantly, whereas during the initial phase of the outbreak, somatic and cognitive symptoms of anxiety and feelings of extreme vulnerability and distress dominated (Chong et al. 2004).

41.3.2 Risk and Protective Factors

On average, female health professionals show a significantly higher symptom severity and frequency than males in several studies (Barati et al. 2020; Bettinsoli et al. 2020; Huang et al. 2020; Kang et al. 2020b; Kisely et al. 2020; Lai et al. 2020; Zhu et al. 2020b). A preprint reported, for example, a significant *hazard ratio* (HR) = 1.32 of women compared to men among the COVID-19 pandemic (Zhu et al. 2020b). Due to the vast body of evidence that shows that women tend to be more vulnerable for affective and anxiety disorders in general, those results may be explained by pre-existing differences rather than by differences in pandemic-related reactions (Wang et al. 2020a). Furthermore, women may be stronger affected by the consequences of the public measures to contain the spreading of the virus (such as school closures) because they seem to be more often in charge of caring for their family members at home than men, regardless of their occupational workload (IASC 2020).

According to the results of studies in China and Italy, the younger age of health professionals seems to be a risk factor for psychological strain (Bettinsoli et al. 2020; Kang et al. 2020b). Those findings are in line with the results of previous outbreaks of infectious diseases (Matsuishi et al. 2012; Ricci-Cabello et al. 2020; Sim et al. 2004; Su et al. 2007). Less working experience and a lower sense of preparedness in young health professionals could be a possible partially mediating variable of those results (Matsuishi et al. 2012; Ricci-Cabello et al. 2020).

Nursery staff and medical technicians reported, on average, higher levels of anxiety, depression, and post-traumatic stress than doctors (Huang et al. 2020; Zhu et al. 2020b). These findings are in line with previous studies in hospitals affected with SARS, MERS, or H1N1 influenza which reported elevated levels of psychological strain in nurses (Bell and Wade 2020; Brooks et al. 2018; Matsuishi et al. 2012; Nickell et al. 2004; Phua et al. 2005). These findings may be explained due to nursery staff (and medical technicians) having more and closer contact with patients than doctors, on average, more night-shifts, and maybe other more overburdening working conditions (Bell and Wade 2020; Nickell et al. 2004; Zhang et al. 2020a; Zhu et al. 2020b). Furthermore, doctors expressed, on average, more confidence in their ability to protect themselves and their patients than nurses (Shi et al. 2020), which could be another possible mediator of those findings.

Compared to administration personnel, medical personnel scored on average significantly higher in symptoms of fear, anxiety, and depression in a study in China (Lu et al. 2020). The results of a study in Singapore are standing in contrast to these findings: non-medical personnel in the health system reported more definite symptoms of depression, anxiety, and (post-traumatic) stress than medical staff (Tan et al. 2020). Possible reasons for the latter result may include reduced accessibility of non-medicals to first-hand medical information, adequate training, protective equipment, and formal psychological support (Tan et al. 2020).

Health professionals with more frequent direct exposure to infected patients showed in two studies in Wuhan higher amounts of mental health disturbances than professionals with less exposure (Kang et al. 2020b; Lai et al. 2020; Wu et al. 2009; Zhu et al. 2020b) which is in line with findings of previous epidemic outbreaks (Bell and Wade 2020; Grace et al. 2005; Koh et al. 2005; McAlonan et al. 2007; Ricci-Cabello et al. 2020; Styra et al. 2008; Wu et al. 2009). In contrast, no significant differences in depressive and anxiety symptoms between health professionals working in units for COVID-19 and medical health workers without direct contact with infected patients became apparent in the context of another study conducted in China (Liang et al. 2020). Moreover, in two other studies in the Wuhan area, the frontline workforce showed, on average, fewer symptoms of traumatization, emotional exhaustion, and depersonalization than non-frontline physicians and nursery staff on general wards (Li et al. 2020; Wu et al. 2020b) which is also congruent with some previous findings regarding a SARS outbreak in Hong Kong (Chan et al. 2005). Those latter findings could be due to potentially better support of the special wards, whereas the general wards may somewhat be neglected by the hospital's administration and the government (Wu et al. 2020b). Therefore, it seems important to avoid an underestimation of the mental health burden for health professionals who are not in direct contact with COVID-19 patients and to address them with adequate support, too.

In studies regarding previous outbreaks of highly infectious diseases, the voluntariness of work in high-risk units seems to be relevant. The clinical staff that was conscripted to such units because of human resources showed particularly low levels of mental health (Bell and Wade 2020; Chen et al. 2005).

Concomitant chronic diseases (HR = 1.51) and a history of mental disorders (HR = 3.27) occurred to be substantial risk factors for stress, anxiety, and depression (Wang et al. 2020a; Zhu et al. 2020b), which is congruent with the findings of previous epidemics (Kisely et al. 2020; Styra et al. 2008). Health workers with a confirmed COVID-19 infection reported more severe depressive symptoms than those without infection (Zhu et al. 2020b). Especially being quarantined because of a suspected or confirmed infection seems to be a predictor of psychological distress (Bell and Wade 2020; Marjanovic et al. 2007). Having family members or relatives proved or suspected with COVID-19 was also significantly associated with a substantially higher risk for the psychological strain of health workers (HR = 1.23) (Zhu et al. 2020b) as well as seeing infected colleagues (Bell and Wade 2020).

In studies in previous SARS outbreaks, working experience seems to be a protective factor against psychological strain due to an outbreak (Kisely et al. 2020; Maunder et al. 2008). In contrast, in a study in China, health professionals with more than 10 years of working experience scored higher on stress, anxiety, and depression in the COVID-19 pandemic than those with less experience (HR = 2.02) (Zhu et al. 2020b). This result may be explained by less occupational exhaustion, a higher proportion of vulnerability for severe outcomes of COVID-19, or family status: the majority of healthcare workers with shorter working experience was single and had to face fewer family responsibilities in addition to the high workload (Zhu et al. 2020b). Health workers with children showed higher levels of stress – probably due to more intense family responsibilities, which would be in line with results from previous SARS outbreaks (Kisely et al. 2020; Maunder et al. 2008; Zhu et al. 2020b).

Care provided by the hospital and department administrators was associated with a reduced risk of showing adverse psychological reactions to the pandemic (*odds ratio* (OR) = 0.76) as reported in a study from China (Zhu et al. 2020b). Protective safety measures seem to be an essential protective factor: full coverage of all hospital units with protection equipment and precautionary measures was associated with reduced levels of psychological strain (OR = 0.69) (Zhu et al. 2020b). Both findings are in line with results from previous SARS outbreaks in which support of colleagues and supervisors and specialized training and safety measures were associated with lower psychological distress (Huremović 2019; Ricci-Cabello et al. 2020). Preparedness in terms of safety equipment, knowledge about the virus and safety measures, appropriate training in hospitals, and the confidence to be familiar with the risks and the protection measures were furthermore associated with a higher willingness to care for patients with COVID-19 (Ricci-Cabello et al. 2020; Shi et al. 2020).

The access to supporting psychological material seems to be a protective factor: severe mental health disturbances were associated with less access to online psychological resources (Kang et al. 2020b). Although distressed health professionals' access to mental healthcare services is limited, the majority evaluated such services as important resources to alleviate acute health disturbances (Kang et al. 2020b).

In a study with hospital staff in Wuhan, regular exercise seemed to help to alleviate adverse psychological impact caused by catastrophic events as the exercise was associated with a lower risk for anxiety symptoms (OR = 0.71) (Zhu et al. 2020b). It is congruent with previous research, which emphasized the anxiolytic and stress-buffering effect of physical activity (Petzold et al. 2020b).

Self-efficacy negatively correlated with anxiety and insomnia among healthcare workers in Chinese hospitals (Xiao et al. 2020). Self-efficacy, therefore, seems to be of help to cope with massive workloads with high risks for physical and mental strains (Xiao et al. 2020).

An overview of potential risk and protective factors regarding the psychological consequences of the COVID-19 pandemic on the mental health of health professionals is given in Fig. 41.1.

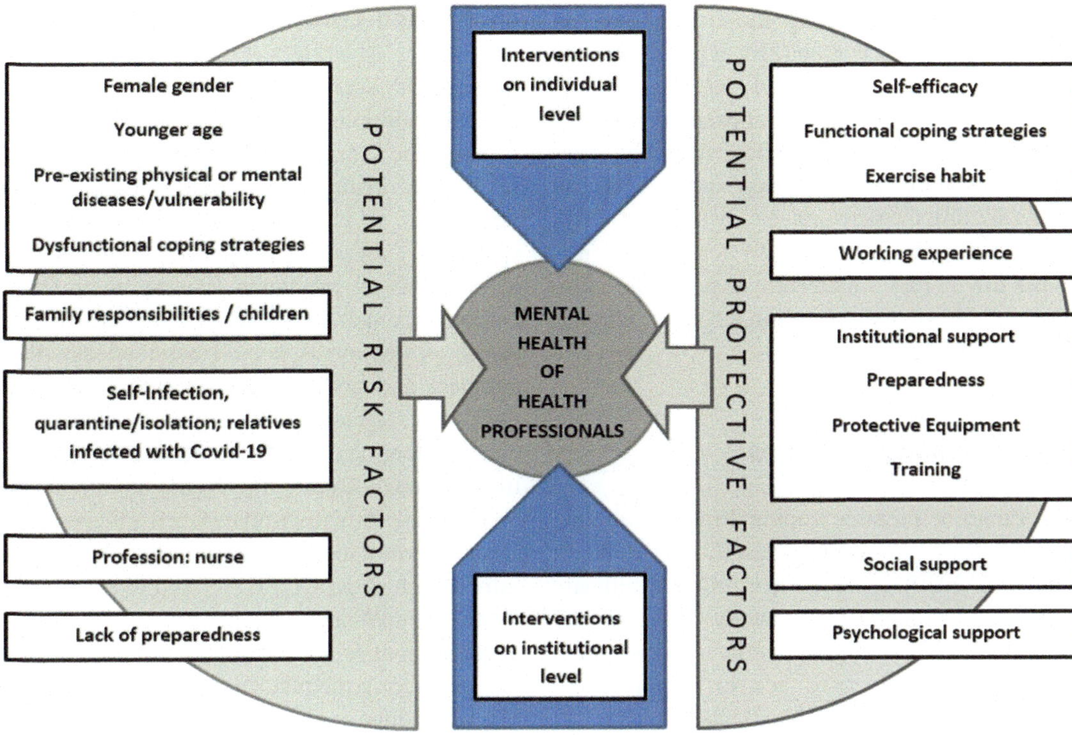

Fig. 41.1 Potential risk and protective factors for the mental health of health professionals

41.4 Coping and Interventions

To maintain essential healthcare services during the COVID-19 pandemic and to protect the short- and long-term mental health of health professionals, adequate coping and intervention strategies are required (Bao et al. 2020; Chung and Yeung 2020; Kang et al. 2020a; Pothiawala 2020). Self-care strategies on an individual level, as well as interventions on a macro-level such as institutional support, should be implemented as early as possible (Maunder et al. 2008; National Center for PTSD 2020; Pothiawala 2020; Xu et al. 2020). In the following, some potentially adaptive strategies are described.

41.4.1 Consideration of Basic Needs

Despite all the demands coming with the COVID-19 pandemic, the basic needs of the health professionals should be considered and addressed by the healthcare workers themselves as well as by their leaders (Adams and Walls 2020; Petzold et al. 2020d). Especially concerning the highly demanding conditions, it is even more critical for health professionals to take rest times sufficiently and regularly, to sleep enough, and to eat sufficiently and healthily (Petzold et al. 2020d). Organizational institutions should ensure that employees have adequate opportunities to take breaks and recover and to carry out necessary self-care measures, even in phases of high workload, and employees should make use of such opportunities (IASC 2020; Petzold et al. 2020c). About the frequent occurrence of insomnia and the importance of sleep quality as a critical indicator of physical and mental health, among other interventions, adaptive shift planning to facilitate regular sleep times should be taken into account (Xiao et al. 2020).

Additionally, physical activity is an important, effective, and frequently underestimated tool to facilitate coping with straining situations (Zhu et al. 2020b). Concerning its stress-alleviating effect in a study in Wuhan (Zhu et al. 2020b) and previous research (Petzold et al. 2020b; Tamminen et al. 2020), regular exercise seems to be an easily affordable and effective coping strategy. In the study, as mentioned earlier from Zhu et al. (2020b), the percentage of healthcare workers who performed regular physical activity was rather low (17%) – therefore, health professionals should be motivated to implement or to maintain physical activity.

41.4.2 Coping Strategies

The range of possible coping strategies among the COVID-19 pandemic is vast (Cao et al. 2020; Sun et al. 2020). It can be helpful for healthcare workers to remember personally adequate and adaptive coping strategies that have helped them to overcome past crises and to reapply them (Petzold et al. 2020d; WHO 2020b).

Dysfunctional and unhealthy coping strategies such as the use of tobacco, alcohol, illicit drugs, or excessive amounts of prescription drugs should be avoided (National Center for PTSD 2020; Petzold et al. 2020d; WHO 2020b). Especially concerning the high proportion of health professionals who tend to use dysfunctional strategies to cope with potential traumatizing events, it becomes even more important to emphasize this aspect and to enhance the usage of adaptive coping strategies (Strametz et al. 2020).

41.4.3 Restriction of Worrying Media Coverage

Rest times should be used for recovering and getting some distance from the omnipresent topic of COVID-19 (WHO 2020b). Keeping oneself informed in regular intervals is important, but, for example, excessive consumption of media and especially the confrontation with misinformation and rumors is counterproductive for mental health (Ullah and Amin 2020; WHO 2020b).

41.4.4 Self-Efficacy

Self-efficacy seems to be an essential tool in coping with multiple stressors (Xiao et al. 2020). To increase the sense of self-efficacy of health professionals, institutions should strengthen the training of psychological skills in general and specifically regarding the particular demands of the COVID-19 pandemic (Huang et al. 2020; National Center for PTSD 2020). Furthermore, health professionals should acknowledge their personal and occupational strengths and avoid negatively assessing and degrading their work contributions and their functioning in personal life (National Center for PTSD 2020). About enhancing occupation-related self-efficacy, it is furthermore important to provide all relevant information about COVID-19 and to ensure adequate occupational training, e.g., the correct application of safety measures or the handling of patients during the pandemic (Barati et al. 2020; Pincha Baduge et al. 2018).

Concerning previous epidemics (H1N1 influenza), a study in Toronto succeeded to enhance the self-efficacy and other factors of resilience and preparedness of healthcare workers with an intervention (from 35% of the staff feeling able to cope with influenza before the training to 76% afterward) (Aiello et al. 2011). A similar program could be envisaged for COVID-19.

41.4.5 Acceptance of Negative Emotions

It appears to be of great importance that health professionals accept their negative emotions like fear, sadness, or anger and attribute them as a normal reaction to an extraordinary situation rather than trying to suppress them (Petzold et al. 2020d; WHO 2020b). Avoidance of negative emotions and the use of coping strategies based on avoidance rather than the problem or emotional-based coping have shown to be associated with adverse long-term consequences in health workers (Edwards et al. 2002; Sim et al. 2004). Furthermore, specific fear regarding infectious diseases might increase when health workers are trying to suppress negative emotions instead of accepting them. A longitudinal study during the Zika outbreak in the USA in 2016 from Dillard et al. (2018) showed in a sample of 561 women in the USA that the suppression of fear might contribute to a future increase in fear, which might constitute a self-reinforcing vicious cycle of fear and suppression (Fig. 41.2).

41.4.6 Self-Concept of Health Professionals in the Pandemic

The COVID-19 pandemic might result in psychological distress and negative emotions like helplessness, sadness, or anger, which in turn might lead to cognitive dissonance in the self-concept of the health professionals. Health professionals might have a strong perception of themselves as a helper, being mentally stable, and having solutions for the problems of other people. In the current circumstances, feelings of helplessness, being overwhelmed, or the wish for being helped might arise. This might lead health professionals to question their self-concept as a health professional and results in psychological distress and reduced self-confidence (Petzold et al. 2020a). In this regard, it seems to be of utmost importance for health professionals to understand that feelings of stress and other negative emotions are not a reflection of a lack of competence concerning the fulfillment of their duties (IASC 2020). In this regard, it is essential for health professionals to engage in self-care and attach great importance to the fulfillment of their own basic needs, especially in times with increased workload, psychological distress, and reduced continuity (WHO 2020b). Self-care reducing thoughts like: "It would be selfish to take time to rest" or "The needs of survivors are more important than the needs of helpers" should be avoided (National Center for PTSD 2020).

41.4.7 Social Support

Social support can occur on different levels, for example, family members, friends, colleagues, supervisors, as well as the society, in general, can deploy social support (Petzold et al. 2020c). Social support through family and friends was found to be of great help to cope with the strains and challenges of the pandemic (Su et al. 2007; Xiao et al. 2020). In a cross-sectional observational study with medical staff treating patients with COVID-19 infection by Xiao et al. (Xiao et al. 2020), the amount of social support was significantly positively associated with the levels of self-efficacy and sleep quality and negatively associated with anxiety and distress. In qualitative interviews, contact with family members via video chat or telephone was named as an efficient coping strategy (Cao et al. 2020). Conversations with colleagues were also reported to be helpful (Cao et al. 2020). These findings are in line with

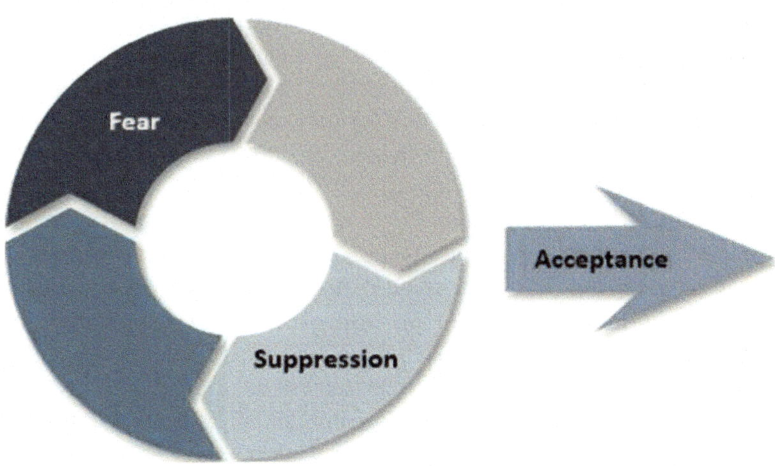

Fig. 41.2 Self-reinforcing vicious cycle of fear and suppression

previous research, which underpins the protective role of social support (Mo et al. 2020).

Thus, it is vital to mobilize and strengthen the social support system of health professionals and to animate them to use it actively (Mo et al. 2020). In terms of work schedules, enough time should be provided for healthcare workers to stay in contact with their families and friends, especially via telephone and web-based communication during isolation (Mo et al. 2020; Petzold et al. 2020c).

Furthermore, the broader society should try to support health professionals. For example, community staff may assist the healthcare workers in terms of childcare services or the completion of everyday tasks (Wu et al. 2020a). Furthermore, society might be able to provide support through showing appreciation for the work of health professionals and through following rules and recommendations to contain the spreading of the virus and to prevent an overburdening of the healthcare system (Kang et al. 2020a; WHO 2020b).

41.4.8 Seeking Professional Support

As health professionals are very likely to experience psychological distress in connection with their work during the COVID-19 pandemic, it seems to be of particular importance for them to seek social and professional support when needed (IASC 2020; Petzold et al. 2020d; WHO 2020b). It might be especially difficult for health professionals as there is still stigmatization toward mental health problems in health professionals (Horsfall et al. 2010; Thornicroft et al. 2007). Since stigmatization toward mental health problems can prevent individuals from seeking support (Clement et al. 2015; Pothiawala 2020), this might hinder health professionals from seeking support from their colleagues, supervisors, or professional treatment, when needed. Therefore, the normalization of the emotional responses in health professionals during the COVID-19 pandemic plays an essential role in this context as well (Petzold et al. 2020d). Fostering self-acceptance and self-forgiving in health professionals in this context might, therefore, be of central importance, as these factors were shown to be associated with reduced stigma and increased mental health, e.g., in firefighters exposed to traumatic events (Carpenter et al. 2020).

Psychological crisis interventions inside organizational institutions (e.g., psychological response teams in the hospital's units) should be provided to health professionals (Chung and Yeung 2020) as well as the access to outside psychological or psychosocial support opportunities (IASC 2020). Furthermore, in this context, anonym and easily accessible psychological support via apps, chat, or telephone hotlines seems to be a promising option (Bao et al. 2020; Cao et al. 2020; Chung and Yeung 2020). For example, only a relatively small amount of healthcare workers in Wuhan participated in counseling or psychotherapy (18%), but half of the staff had accessed online psychological resources (e.g., regarding mental health self-help strategies) (Kang et al. 2020b).

41.4.9 Continuity and Sense of Coherence

Previous research on disaster management resilience and the coping of individuals with psychological distress argues that the concept of continuity plays an important role in maintaining mental health (Omer and Alon 1994). The perception of meaningful continuity includes several aspects in the life of a person which seem stable and predictable for the individual, for example, functional continuity (the conception of being able to maintain routines in one's daily life) or social continuity (stability in social relationships and social roles) (Lahad and Rogel 2004). The concept of continuity is somewhat related to the concept of sense of coherence (Antonovsky 1987). A sense of coherence means the ability of an individual to "make sense of the world" (Antonovsky 1987). The current COVID-19 pandemic has the potential to reduce the sense of coherence in health professionals due to the rapidly changing situation with a lack of scientific evidence regarding the virus, its treatment, and the consequences for the health professionals. It also might disrupt the perception of continuity, for example, through changing teams and roles, reduced social contacts, and changes in the work environment. It might result in psychological distress among health professionals. Therefore, approaches to foster a sense of coherence and perception of continuity among health professionals might play a vital role in strengthening their resilience and coping with the pandemic situation. In this context, fostering information and communication on the current scientific evidence (to enhance a sense of coherence) and focusing on maintaining or re-establishing as much predictability in the everyday work of health professionals as possible (to strengthen the perception of continuity) might be a promising approach to help them cope with the situation.

41.4.10 Preparedness

As a lack of preparedness seems to be a factor of risk and a reason for strain, there should be organizational effort to provide sufficient formal and informal training (Mak et al. 2009; National Center for PTSD 2020; Olesen et al. 2020; Rubin et al. 2016), for example, on specific details about the virus, its transmission, the use of protective equipment, and ethical decision-making, e.g., regarding surge capacity issues (Greenberg et al. 2020; National Center for PTSD 2020).

Moreover, sufficient protective equipment should be provided to reduce the objective infection risk of the healthcare workers as well as the subjective risk perception and the associated psychological strain (Gee and Skovdal 2017; National Center for PTSD 2020; Zhu et al. 2020b). For example, a SARS prevention program in Taiwan succeeded in reducing symptoms of anxiety, depression, and insomnia in nursery staff by providing a higher preparedness (Chen et al. 2006).

Accurate information updates about all relevant aspects and precise and unambiguous communication should be provided to all staff to avoid worries due to uncertainties and to create a sense of control (Maunder et al. 2003; Petzold et al. 2020d; WHO 2020b).

41.4.11 Institutional Support

Besides the already mentioned contributions of institutions and team leaders in health facilities, leadership should continuously keep an eye on the mental distress of health professionals. They should create an atmosphere in which they can communicate about experienced psychological strain (Adams and Walls 2020; WHO 2020b). Enabling collegial support and reciprocal communication opportunities are necessary for this context as well as providing access to mental health and psychosocial support services (Greenberg et al. 2020; Greenberg et al. 2015; Lee et al. 2005; WHO 2020b). It is essential to prevent the stigmatization of psychological support and to emphasize the experience of psychological strain as a mostly normal reaction to the high and adverse demands of the COVID-19 pandemic (International Federation of Red Cross and Red Crescent Societies 2020). Stress and psychological strain should be taken seriously, and protecting health professionals from these strains should be given high priority (Petzold et al. 2020d).

Moreover, leaders should be role-models for self-care and should emphasize the importance of self-care strategies to mitigate stress (Adams and Walls 2020; WHO 2020b). They have much responsibility and are also subject to multiple stressors due to the pandemic and should, therefore, pay attention to their own mental well-being and adequate dealing with stress (Petzold et al. 2020d).

Besides the desirable strive of health professionals to use somewhat functional strategies, the organizational institutions on the macro-level may be able to facilitate this. As enhanced organizational support and sufficient and trustworthy safety equipment predicted lower levels of dysfunctional coping strategies such as avoidance behavior, emotional exhaustion, and excessive anger in a previous SARS outbreak, addressing these issues on the institutional level seems to be of importance (Marjanovic et al. 2007; Phua et al. 2005).

In a study in China, a heavy workload was named as the main reason for psychological strain due to the pandemic, and doctors and nursery staff sought additional staff to reduce their overburdening workload (Cao et al. 2020). Therefore, organizational and governmental institutions should try to identify additional workforce capacity to reduce the workload by distributing it on more individuals (WHO 2020a). Transparent allocation of roles and tasks should be ensured and regularly adapted to changing circumstances (Petzold et al. 2020d). For example, nursery staff in a study in China reported various and unclear duties as a stressor and sought for a more specified task division (Cao et al. 2020).

Team leaders and healthcare institutions, in general, should convey appreciation for the exceptional and important performance of health professionals during the pandemic, as this can play an essential protective role concerning mental health in times of crisis (International Federation of Red Cross and Red Crescent Societies 2020; Khalid et al. 2016; Petzold et al. 2020d; Wong et al. 2012; Wu et al. 2020a).

Especially healthcare workers with an elevated risk for psychological strain should be in the focus of attention for interventions to reduce strain (Adams and Walls 2020; Huang et al. 2020; Lai et al. 2020), but all health professionals should be taken into account (WHO 2020b). Keeping all staff protected from elevated mental health burden and chronic stress to the highest degree possible will maintain and enhance their capacity to fulfill their roles (WHO 2020b). In this context, one should not forget that the COVID-19 pandemic and the related exceptional demands for the healthcare systems worldwide will not disappear overnight, and governments and societies should focus more on long-term solutions for upholding a sufficient occupational capacity than concentrating on minor, short-term responses to the crisis (WHO 2020b).

41.5 Conclusion

To sum up, health professionals have to face multiple stressors on general and profession-specific levels due to the COVID-19 pandemic. High proportions of healthcare workers reported acute symptoms of anxiety, depression, high psychological stress, and insomnia in the context of the pandemic. Studies of the long-term consequences are lacking, but a transient trend can be expected for most healthcare workers. Several coping strategies and self-care on an individual level, interventions on an institutional level such as specific training and institutional support, as well as social support and psychological support seem to be of help to mitigate mental strain.

The generalizability of the results has some limitations. For example, a limitation can be inferred from the fact that the majority of studies regarding the mental health of health professionals rely on observational cross-sectional data, which does not allow any causal conclusions. Especially the frequent lack (or only retrospective collection) of data before the outbreak of COVID-19 and the lack of adequate control groups reduce generalizability. Moreover, not all studies provide sufficient sample sizes for reliable interpretations, and most studies rely on data gained from self-report questionnaires, an approach that bears an additional risk for bias. Because the healthcare systems and the exposure/affectedness with COVID-19 and the governmental measures differ remarkably from country to country, it is essential to keep in mind that the universal generalization of findings of national studies is limited. Moreover, there is only a low number of studies on this topic in general and also a particular lack of studies regarding ambulatory health professionals in the pandemic (Bohlken et al. 2020). Moreover, in general, more generalizable, reliable, and longitudinal studies are required in the future to assess the consequences of the COVID-19 pandemic on the mental health of health professionals more reliably and comprehensively.

References

Adams JG, Walls RM (2020) Supporting the health care workforce during the COVID-19 global epidemic. JAMA. https://doi.org/10.1001/jama.2020.3972

Aiello A, Khayeri MY-E, Raja S, Peladeau N, Romano D, Leszcz M, Maunder RG, Rose M, Adam MA, Pain C, Moore A, Savage D, Schulman RB (2011) Resilience training for hospital workers in anticipation of an influenza pandemic. J Contin Educ Health Prof 31(1):15–20. https://doi.org/10.1002/chp.20096

Antonovsky A (1987) Unraveling the mystery of health: how people manage stress and stay well

Bai Y, Lin C-C, Lin C-Y, Chen J-Y, Chue C-M, Chou P (2004) Survey of stress reactions among health care workers involved with the SARS outbreak. Psychiatric serv (Washington, DC) 55(9):1055–1057. https://doi.org/10.1176/appi.ps.55.9.1055

Bao Y, Sun Y, Meng S, Shi J, Lu L (2020) 2019-nCoV epidemic: address mental health care to empower society. Lancet 395(10224):e37–e38. https://doi.org/10.1016/s0140-6736(20)30309-3

Barati M, Bashirian S, Jenabi E, Khazaei S, Karimi-Shahanjarini A, Zareian S, Rezapur-Shahkolai F, Moeini B (2020) Factors associated with preventive Behaviours of COVID-19 among hospital staff in Iran in 2020: an application of the protection motivation theory. J Hosp Infect. https://doi.org/10.1016/j.jhin.2020.04.035

Bell V, Wade D (2020) Mental health of clinical staff working in high-risk epidemic and pandemic health emergencies: a rapid review of the evidence and meta-analysis. Soc Psychiatry Psychiatr Epidemiol. https://doi.org/10.1101/2020.04.28.20082669

Bettinsoli ML, Di Riso D, Napier J, Moretti L, Bettinsoli P, Delmedico M, Piazzolla A, Moretti B (2020) Psychological impact and contextual factors associated with physical and mental health conditions of Italian healthcare professionals during the Covid-19 disease outbreak. https://doi.org/10.31234/osf.io/w89fz

Bohlken J, Schömig F, Lemke MR, Pumberger M, Riedel-Heller SG (2020) COVID-19-Pandemie: Belastungen des medizinischen personals (COVID-19 pandemic: stress experience of healthcare workers - a short current review). Psychiatr Prax 47(4):190–197. https://doi.org/10.1055/a-1159-5551

Brooks SK, Dunn R, Amlôt R, Rubin GJ, Greenberg N (2018) A systematic, thematic review of social and occupational factors associated with psychological outcomes in healthcare employees during an infectious disease outbreak. J Occup Environ Med 60(3):248–257. https://doi.org/10.1097/jom.0000000000001235

Cao J, Wei J, Zhu H, Duan Y, Geng W, Hong X, Jiang J, Zhao X, Zhu B (2020) A study of basic needs and psychological wellbeing of medical Workers in the Fever Clinic of a tertiary general Hospital in Beijing during the COVID-19 outbreak. Psychother Psychosom:1–3. https://doi.org/10.1159/000507453

Carpenter TP, Pennington ML, Seebeck J, Gomez DR, Denman TC, Kimbrel NA, Cammarata CM, Leto F, Ostiguy WJ, Gulliver SB (2020) Dispositional self-forgiveness in firefighters predicts less help-seeking stigma and fewer mental health challenges. Stigma Health 5(1):29–37. https://doi.org/10.1037/sah0000172

Chan AOM, Huak CY (2004) Psychological impact of the 2003 severe acute respiratory syndrome outbreak on health care workers in a medium size regional general hospital in Singapore. Occup Med (Oxford, England) 54(3):190–196. https://doi.org/10.1093/occmed/kqh027

Chan SSC, Leung GM, Tiwari AFY, Salili F, Leung SSK (2005) The impact of work-related risk on nurses during the SARS outbreak in Hong Kong. Fam Community Health 28(3):274–287

Chen C-S, Wu H-Y, Yang P, Yen C-F (2005) Psychological distress of nurses in Taiwan who worked during the outbreak of SARS. Psychiatric Serv (Washington, DC) 56(1):76–79. https://doi.org/10.1176/appi.ps.56.1.76

Chen R, Chou K-R, Huang Y-J, Wang T-S, Liu S-Y, Ho L-Y (2006) Effects of a SARS prevention programme in Taiwan on nursing staff's anxiety, depression and sleep quality: a longitudinal survey. Int J Nurs Stud 43(2):215–225. https://doi.org/10.1016/j.ijnurstu.2005.03.006

Chen Q, Liang M, Li Y, Guo J, Fei D, Wang L, He L, Sheng C, Cai Y, Li X, Wang J, Zhang Z (2020) Mental health care for medical staff in China during the COVID-19 outbreak. Lancet 7

Chew QH, Wei KC, Vasoo S, Chua HC, Sim K (2020) Narrative synthesis of psychological and coping responses towards emerging infectious disease outbreaks in the general population: practical considerations for the COVID-19 pandemic. Singap Med J. https://doi.org/10.11622/smedj.2020046

Chong M-Y, Wang W-C, Hsieh W-C, Lee C-Y, Chiu N-M, Yeh W-C, Huang O-L, Wen J-K, Chen C-L (2004) Psychological impact of severe acute respiratory syndrome on health workers in a tertiary hospital. British J Psychiatry J Mental Sci 185:127–133. https://doi.org/10.1192/bjp.185.2.127

Chung JPY, Yeung WS (2020) Staff mental health self-assessment during the COVID-19 outbreak. East Asian Archiv Psychiatry Off J Hong Kong College Psychiatrists = Dong Ya jing shen ke xue zhi : Xianggang jing shen ke yi xue yuan qi kan 30(1):34. https://doi.org/10.12809/eaap2014

Clement S, Schauman O, Graham T, Maggioni F, Evans-Lacko S, Bezborodovs N, Morgan C, Rüsch N, Brown JSL, Thornicroft G (2015) What is the impact of mental health-related stigma on help-seeking? A systematic review of quantitative and qualitative studies. Psychol Med 45(1):11–27. https://doi.org/10.1017/s0033291714000129

Dillard JP, Yang C, Li R (2018) Self-regulation of emotional responses to Zika: spiral of fear. PLoS One 13(7):e0199828. https://doi.org/10.1371/journal.pone.0199828

Edwards D, Hannigan B, Fothergill A, Burnard P (2002) Stress management for mental health professionals: a review of effective techniques. Stress Health 18(5):203–215. https://doi.org/10.1002/smi.947

El-Hage W, Hingray C, Lemogne C, Yrondi A, Brunault P, Bienvenu T, Etain B, Paquet C, Gohier B, Bennabi D, Birmes P, Sauvaget A, Fakra E, Prieto N, Bulteau S, Vidailhet P, Camus V, Leboyer M, Krebs M-O, Aouizerate B (2020) C (health professionals facing the coronavirus disease 2019 (COVID-19) pandemic: what are the mental health risks?). L'Encephale. https://doi.org/10.1016/j.encep.2020.04.008

Gee S, Skovdal M (2017) Navigating 'riskscapes': the experiences of international health care workers responding to the Ebola outbreak in West Africa. Health Place 45:173–180. https://doi.org/10.1016/j.healthplace.2017.03.016

Grace SL, Hershenfield K, Robertson E (2005) The occupational and psychosocial impact of SARS on academic physicians in three affected hospitals. Psychosomatics 46(5):385–391

Greenberg N, Wessely S, Wykes T (2015) Potential mental health consequences for workers in the Ebola regions of West Africa--a lesson for all challenging environments. J Mental Health (Abingdon, England) 24(1):1–3. https://doi.org/10.3109/09638237.2014.1000676

Greenberg N, Docherty M, Gnanapragasam S, Wessely S (2020) Managing mental health challenges faced by healthcare workers during covid-19 pandemic. BMJ (Clinical research ed) 368:1–4. https://doi.org/10.1136/bmj.m1211

Helmy YA, Fawzy M, Elaswad A, Sobieh A, Kenney SP, Shehata AA (2020) The COVID-19 pandemic: a comprehensive review of taxonomy, genetics, epidemiology, diagnosis, treatment, and control. J Clin Med 9(4). https://doi.org/10.3390/jcm9041225

Heymann DL (2020) Data sharing and outbreaks: best practice exemplified. Lancet 395(10223):514–523. https://doi.org/10.1016/s0140-6736(20)30154-9

Horsfall J, Cleary M, Hunt GE (2010) Stigma in mental health: clients and professionals. Issues Ment Health Nurs 31(7):450–455. https://doi.org/10.3109/01612840903537167

Huang JZ, Han MF, Luo TD, Ren AK (2020) Mental health survey of medical staff in a tertiary infectious disease hospital for COVID-19

Huremović D (2019) Psychiatry of pandemics: a mental health response to infection outbreak. Springer, Cham. https://doi.org/10.1007/978-3-030-15346-5

International Council of Nurses (2020) ICN Call to Action: Covid-19

International Federation of Red Cross and Red Crescent Societies (2020) Mental Health and Psychosocial Support for Staff, Volunteers and Communities in an Outbreak of Novel Coronavirus

Inter-Agency Standing Committee (IASC) (2020) Briefing note on addressing mental health and psychosocial aspects of COVID-19 Outbreak- Version 1.1

Kang L, Li Y, Hu S, Chen M, Yang C, Yang BX, Wang Y, Hu J, Lai J, Ma X, Chen J, Guan L, Wang G, Ma H, Liu Z (2020a) The mental health of medical workers in Wuhan, China dealing with the 2019 novel coronavirus. Lancet Psychiatry 7(3):e14. https://doi.org/10.1016/s2215-0366(20)30047-x

Kang L, Ma S, Chen M, Yang J, Wang Y, Li R, Yao L, Bai H, Cai Z, Xiang Yang B, Hu S, Zhang K, Wang G, Ma C, Liu Z (2020b) Impact on mental health and perceptions of psychological care among medical and nursing staff in Wuhan during the 2019 novel coronavirus disease outbreak: a cross-sectional study. Brain Behav Immun. https://doi.org/10.1016/j.bbi.2020.03.028

Khalid I, Khalid TJ, Qabajah MR, Barnard AG, Qushmaq IA (2016) Healthcare workers emotions, perceived stressors and coping strategies during a MERS-CoV outbreak. Clin Med Res 14(1):7–14. https://doi.org/10.3121/cmr.2016.1303

Kisely S, Warren N, McMahon L, Dalais C, Henry I, Siskind D (2020) Occurrence, prevention, and management of the psychological effects of emerging virus outbreaks on healthcare workers: rapid review and meta-analysis. BMJ (Clinical Research ed) 369:m1642. https://doi.org/10.1136/bmj.m1642

Koh D, Lim MK, Chia SE, Ko SM, Qian F, Ng V, Tan BH, Wong KS, Chew WM, Tang HK, Ng W, Muttakin Z, Emmanuel S, Fong NP, Koh G, Kwa CT, Tan KB-C, Fones C (2005) Risk perception and impact of severe acute respiratory syndrome (SARS) on work and personal lives of healthcare workers in Singapore: what can we learn? Med Care 43(7):676–682. https://doi.org/10.1097/01.mlr.0000167181.36730.cc

Lahad M, Rogel R (2004) The need for ER protocol in the treatment of public manifesting ASR symptoms following disaster. Aus J Disaster Trauma Stud

Lai J, Ma S, Wang Y, Cai Z, Hu J, Wei N, Wu J, Du H, Chen T, Li R, Tan H, Kang L, Yao L, Huang M, Wang H, Wang G, Liu Z, Hu S (2020) Factors associated with mental health outcomes among health care workers exposed to coronavirus disease 2019. JAMA Netw Open 3(3):e203976. https://doi.org/10.1001/jamanetworkopen.2020.3976

Lancee WJ, Maunder RG, Goldbloom DS (2008) Prevalence of psychiatric disorders among Toronto hospital workers one to two years after the SARS outbreak. Psychiatric serv (Washington, DC) 59(1):91–95. https://doi.org/10.1176/ps.2008.59.1.91

Lee S-H, Juang Y-Y, Su Y-J, Lee H-L, Lin Y-H, Chao C-C (2005) Facing SARS: psychological impacts on SARS team nurses and psychiatric services in a Taiwan general hospital. Gen Hosp Psychiatry 27(5):352–358. https://doi.org/10.1016/j.genhosppsych.2005.04.007

Li Z, Ge J, Yang M, Feng J, Qiao M, Jiang R, Bi J, Zhan G, Xu X, Wang L, Zhou Q, Zhou C, Pan Y, Liu S, Zhang H, Yang J, Zhu B, Hu Y, Hashimoto K, Jia Y, Wang H, Wang R, Liu C, Yang C (2020) Vicarious traumatization in the general public, members, and non-members of medical teams aiding in COVID-19 control. Brain Behav Immun. https://doi.org/10.1016/j.bbi.2020.03.007

Liang Y, Chen M, Zheng X, Liu J (2020) Screening for Chinese medical staff mental health by SDS and SAS during the outbreak of COVID-19. J Psychosom Res 133:110102. https://doi.org/10.1016/j.jpsychores.2020.110102

Lu Y-C, Shu B-C, Chang Y-Y, Lung F-W (2006) The mental health of hospital workers dealing with severe acute respiratory syndrome. Psychother Psychosom 75(6):370–375. https://doi.org/10.1159/000095443

Lu W, Wang H, Lin Y, Li L (2020) Psychological status of medical workforce during the COVID-19 pandemic: a cross-sectional study. Psychiatry Res 288:112936. https://doi.org/10.1016/j.psychres.2020.112936

Mak IWC, Chu CM, Pan PC, Yiu MGC, Chan VL (2009) Long-term psychiatric morbidities among SARS survivors. Gen Hosp Psychiatry 31(4):318–326. https://doi.org/10.1016/j.genhosppsych.2009.03.001

Marjanovic Z, Greenglass ER, Coffey S (2007) The relevance of psychosocial variables and working conditions in predicting nurses' coping strategies during the SARS crisis: an online questionnaire survey. Int J Nurs Stud 44(6):991–998. https://doi.org/10.1016/j.ijnurstu.2006.02.012

Matsuishi K, Kawazoe A, Imai H, Ito A, Mouri K, Kitamura N, Miyake K, Mino K, Isobe M, Takamiya S, Hitokoto H, Mita T (2012) Psychological impact of the pandemic (H1N1) 2009 on general hospital workers in Kobe. Psychiatry Clin Neurosci 66(4):353–360. https://doi.org/10.1111/j.1440-1819.2012.02336.x

Maunder R, Hunter J, Vincent L, Bennett J, Peladeau N, Leszcz M, Sadavoy J, Verhaeghe LM, Steinberg R, Mazzulli T (2003) The immediate psychological and occupational impact of the 2003 SARS outbreak in a teaching hospital. CMAJ : Can Med Assoc J = journal de l'Association medicale canadiene 168(10):1245–1251

Maunder RG, Lancee WJ, Balderson KE, Bennett JP, Borgundvaag B, Evans S, Fernandes CMB, Goldbloom DS, Gupta M, Hunter JJ, McGillis Hall L, Nagle LM, Pain C, Peczeniuk SS, Raymond G, Read N, Rourke SB, Steinberg RJ, Stewart TE, VanDeVelde-Coke S, Veldhorst GG, Wasylenki DA (2006) Long-term psychological and occupational effects of providing hospital healthcare during SARS outbreak. Emerg Infect Dis 12(12):1924–1932. https://doi.org/10.3201/eid1212.060584

Maunder R, Leszcz M, Savage D, Adam MA (2008) Applying the lessons of SARS to pandemic influenza: an evidence-based approach to mitigating the stress experience by healthcare workers. Revue Canadienne de Santé Publique 99(6):487–488

McAlonan GM, Lee AM, Cheung V, Cheung C, Tsang KWT, Sham PC, Chua SE, Wong JG (2007) Immediate and sustained psychological impact of an emerging infectious disease outbreak on health care workers. Can J Psychiatry 52(4):241–247

McKay D, Heisler M, Mishori R, Catton H, Kloiber O (2020) Attacks against health-care personnel must stop, especially as the world fights COVID-19. Lancet. https://doi.org/10.1016/s0140-6736(20)31191-0

Mehta M, Singh MM, Gupta SK, Kushal A (2018) Study of stress among health care professionals: a systemic review. Int J Res Found Hospital Healthcare Admin 6(1):6–11. https://doi.org/10.5005/jp-journals-10035-1084

Mo Y, Deng L, Zhang L, Lang Q, Liao C, Wang N, Qin M, Huang H (2020) Work stress among Chinese nurses to support Wuhan for fighting against the COVID-19 epidemic. J Nurs Manag. https://doi.org/10.1111/jonm.13014

National Center for PTSD (2020) Managing healthcare workers' stress associated with the COVID-19 virus outbreak

Nickell LA, Crighton EJ, Tracy CS, Al-Enazy H, Bolaji Y, Hanjrah S, Hussain A, Makhlouf S, Upshur REG (2004) Psychosocial effects of SARS on hospital staff: survey of a large tertiary care institution. CMAJ Can Med Assoc J = Journal de l'Association medicale canadiene 170(5):793–798. https://doi.org/10.1503/cmaj.1031077

Olesen B, Gyrup HB, Troelstrup MW, Marloth T, Mølmer M (2020) Infection prevention partners up with psychology in a Danish hospital successfully addressing staffs fear during the COVID-19 pandemic. J Hosp Infect. https://doi.org/10.1016/j.jhin.2020.04.033

Omer H, Alon N (1994) The continuity principle: a unified approach to disaster and trauma. Am J Community Psychol 22(02):273–287. https://doi.org/10.1007/bf02506866

Petzold M, Plag J, Ströhle A (2020a) Führen in stressigen Zeiten. Pflege Z 73(6):14–17. https://doi.org/10.1007/s41906-020-0713-z

Petzold MB, Bendau A, Ströhle A (2020b) Körperliche Aktivität in der Prävention und Behandlung von Angsterkrankungen. Psychotherapeut. https://doi.org/10.1007/s00278-020-00414-0

Petzold MB, Plag J, Ströhle A (2020c) Psychische Belastungen können reduziert werden. Deutsches Ärzteblatt Jg 117(13)

Petzold MB, Plag J, Ströhle A (2020d) Umgang mit psychischer Belastung bei Gesundheitsfachkräften im Rahmen der Covid-19-Pandemie (dealing with psychological distress by healthcare professionals during the COVID-19 pandemia). Nervenarzt 91(5):417–421. https://doi.org/10.1007/s00115-020-00905-0

Phua DH, Tang HK, Tham KY (2005) Coping responses of emergency physicians and nurses to the 2003 severe acute respiratory syndrome outbreak. Acad Emerg Med Off J Soc Acad Emerg Med 12(4):322–328. https://doi.org/10.1197/j.aem.2004.11.015

Pincha Baduge MS, Morphet J, Moss C (2018) Emergency nurses' and department preparedness for an ebola outbreak: a (narrative) literature review. Int Emerg Nurs 38:41–49. https://doi.org/10.1016/j.ienj.2017.12.002

Poon E, Liu KS (2004) Impact of severe acute respiratory syndrome on anxiety levels of front-line health care workers. Hong Kong Med 1(10)

Pothiawala S (2020) Psychological impact of the COVID-19 on health care workers in the emergency department. Adv J Emerg Med 4(2s). https://doi.org/10.22114/ajem.v0i0.397

Ricci-Cabello I, Meneses-Echavez JF, Serrano-Ripoll MJ, Fraile-Navarro D, Fiol de Roque MA, Pastor Moreno G, Castro A, Ruiz-Perez I, Zamanillo Campos R, Goncalves-Bradley D (2020) Impact of viral epidemic outbreaks on mental health of healthcare workers: a rapid systematic review. J Affect Disord 277:347–357. https://doi.org/10.1101/2020.04.02.20048892

Rubin GJ, Harper S, Williams PD, Öström S, Bredbere S, Amlôt R, Greenberg N (2016) How to support staff deploying on overseas humanitarian work: a qualitative analysis of responder views about the 2014/15 West African Ebola outbreak. Eur J Psychotraumatol 7:30933. https://doi.org/10.3402/ejpt.v7.30933

Shi Y, Wang J, Yang Y, Wang Z, Wang G, Hashimoto K, Zhang K, Liu H (2020) Knowledge and attitudes of medical staff in Chinese psychiatric hospitals regarding COVID-19. Brain Behav Immun Health 4:100064. https://doi.org/10.1016/j.bbih.2020.100064

Shuja KH, Aqeel M, Jaffar A, Ahmed A (2020) COVID-19 pandemic and impending global mental health implications. Psychiatr Danub 32(1):32–35. https://doi.org/10.24869/psyd.2020.32

Sim K, Chong PN, Chan YH, Soon WSW (2004) Severe acute respiratory syndrome-related psychiatric and posttraumatic morbidities and coping responses in medical staff within a primary health care setting in Singapore. J Clin Psychiatry 65(8):1120–1127. https://doi.org/10.4088/jcp.v65n0815

Strametz R, Raspe M, Ettl B, Huf W, Pitz A (2020) Handlungsempfehlung zu Stärkung der Resilienz von Behandelnden und Umgang mit Second Victims im Rahmen der Covid-19-Pandemie zur Sicherung der Leistungsfähigkeit des Gesundheitswesens. https://doi.org/10.21960/202003

Styra R, Hawryluck L, Robinson S, Kasapinovic S, Fones C, Gold WL (2008) Impact on health care workers employed in high-risk areas during the Toronto SARS outbreak. J Psychosom Res 64(2):177–183. https://doi.org/10.1016/j.jpsychores.2007.07.015

Su T-P, Lien T-C, Yang C-Y, Su YL, Wang J-H, Tsai S-L, Yin J-C (2007) Prevalence of psychiatric morbidity and psychological adaptation of the nurses in a structured SARS caring unit during outbreak: a prospective and periodic assessment study in Taiwan. J Psychiatr Res 41(1–2):119–130. https://doi.org/10.1016/j.jpsychires.2005.12.006

Sun N, Wei L, Shi S, Jiao D, Song R, Ma L, Wang H, Wang C, Wang Z, You Y, Liu S, Wang H (2020) A qualitative study on the psychological experience of caregivers of COVID-19 patients. Am J Infect Control. https://doi.org/10.1016/j.ajic.2020.03.018

Tamminen N, Reinikainen J, Appelqvist-Schmidlechner K, Borodulin K, Mäki-Opas T, Solin P (2020) Associations of physical activity with positive mental health: a population-based study. Ment Health Phys Act 18:100319. https://doi.org/10.1016/j.mhpa.2020.100319

Tan BYQ, Chew NWS, Lee GKH, Jing M, Goh Y, Yeo LLL, Zhang K, Chin H-K, Ahmad A, Khan FA, Shanmugam GN, Chan BPL, Sunny S, Chandra B, Ong JJY, Paliwal PR, Wong LYH, Sagayanathan R, Chen JT, Ying Ng AY, Teoh HL, Ho CS, Ho RC, Sharma VK (2020) Psychological impact of the COVID-19 pandemic on health care workers in Singapore. Ann Intern Med. https://doi.org/10.7326/m20-1083

Thornicroft G, Rose D, Kassam A (2007) Discrimination in health care against people with mental illness. Int Rev Psychiat (Abingdon, England) 19(2):113–122. https://doi.org/10.1080/09540260701278937

Tsamakis K, Rizos E, Manolis A, Chaidou S, Kympouropoulos S, Spartalis E, Spandidos D, Tsiptsios D, Triantafyllis A (2020) COVID-19 pandemic and its impact on mental health of healthcare professionals. Exp Ther Med 19(6):3451–3453. https://doi.org/10.3892/etm.2020.8646

Ullah R, Amin S (2020) The psychological impact of COVID-19 on medical students letter. Psychiatry Res 288:113020. https://doi.org/10.1016/j.psychres.2020.113020

Wang C, Pan R, Wan X, Tan Y, Xu L, Ho CS, Ho RC (2020a) Immediate psychological responses and associated factors during the initial stage of the 2019 coronavirus disease (COVID-19) epidemic among the general population in China. Int J Environ Res Public Health 17(5). https://doi.org/10.3390/ijerph17051729

Wang L, Wang Y, Ye D, Liu Q (2020b) Review of the 2019 novel coronavirus (SARS-CoV-2) based on current evidence. Int J Antimicrob Agents 105948. https://doi.org/10.1016/j.ijantimicag.2020.105948

Wong ELY, Wong SYS, Lee N, Cheung A, Griffiths S (2012) Healthcare workers' duty concerns of working in the isolation ward during the novel H1N1 pandemic. J Clin Nurs 21(9–10):1466–1475. https://doi.org/10.1111/j.1365-2702.2011.03783.x

World Health Organization and International Labour Organization (2018) Occupational safety and health in public health emergencies: A manual for protecting health workers and responders. Geneva

World Health Organization (WHO) (2020a) COVID-19: Operational guidance for maintaining essential health services during an outbreak

World Health Organization (WHO) (2020b) Mental Health Considerations during COVID-19 Outbreak. Geneva

World Health Organization (WHO) (2020c) Situation Report- 119

Wu P, Liu X, Fang Y, Fan B, Fuller CJ, Guan Z, Yao Z, Kong J, Lu J, Litvak IJ (2008) Alcohol abuse/dependence symptoms among hospital employees exposed to a SARS outbreak. Alcohol Alcoholism (Oxford, Oxfordshire) 43(6):706–712. https://doi.org/10.1093/alcalc/agn073

Wu P, Fang Y, Guan Z, Fan B, Kong J, Yao Z, Liu X, Fuller CJ, Susser E, Lu J, Hoven CW (2009) The psychological impact of the SARS epidemic on hospital employees in China: exposure, risk perception, and altruistic acceptance of risk. Can J Psychiatry 54(5):302–311. https://doi.org/10.1177/070674370905400504

Wu D, Jiang C, He C, Li C, Yang L, Yue Y (2020a) Stressors of nurses in psychiatric hospitals during the COVID-19 outbreak. Psychiatry Res 288:112956. https://doi.org/10.1016/j.psychres.2020.112956

Wu Y, Wang J, Luo C, Hu S, Lin X, Anderson AE, Bruera E, Yang X, Wei S, Qian Y (2020b) A comparison of burnout frequency among oncology physicians and nurses working on the frontline and usual wards during the COVID-19 epidemic in Wuhan, China. J Pain Symptom Manag. https://doi.org/10.1016/j.jpainsymman.2020.04.008

Xiang Y-T, Yang Y, Li W, Zhang L, Zhang Q, Cheung T, Ng CH (2020) Timely mental health care for the 2019 novel coronavirus outbreak is urgently needed. Lancet Psychiatry 7(3):228–229. https://doi.org/10.1016/s2215-0366(20)30046-8

Xiao H, Zhang Y, Kong D, Li S, Yang N (2020) The effects of social support on sleep quality of medical staff treating patients with coronavirus disease 2019 (COVID-19) in January and February 2020 in China. Med Sci Monitor Int Med J Exp Clin Res 26:e923549. https://doi.org/10.12659/msm.923549

Xiong Y, Peng L (2020) Focusing on health-care providers' experiences in the COVID-19 crisis. Lancet Glob Health. https://doi.org/10.1016/s2214-109x(20)30214-x

Xu J, Xu Q-H, Wang C-M, Wang J (2020) Psychological status of surgical staff during the COVID-19 outbreak. Psychiatry Res 288:112955. https://doi.org/10.1016/j.psychres.2020.112955

Zhang C, Yang L, Liu S, Ma S, Wang Y, Cai Z, Du H, Li R, Kang L, Su M, Zhang J, Liu Z, Zhang B (2020a) Survey of insomnia and related social psychological factors among medical staff involved in the 2019 novel coronavirus disease outbreak. Front Psych 11:306. https://doi.org/10.3389/fpsyt.2020.00306

Zhang M, Zhou M, Tang F, Wang Y, Nie H, Zhang L, You G, Zhang M (2020b) Knowledge, attitude and practice regarding COVID-19 among health care workers in Henan, China. J Hospital Infect. https://doi.org/10.1016/j.jhin.2020.04.012

Zhu N, Zhang D, Wang W, Li X, Yang B, Song J, Zhao X, Huang B, Shi W, Lu R, Niu P, Zhan F, Ma X, Wang D, Xu W, Wu G, Gao GF, Tan W (2020a) A novel coronavirus from patients with pneumonia in China, 2019. N Engl J Med 382(8):727–733. https://doi.org/10.1056/NEJMoa2001017

Zhu Z, Xu S, Wang H, Liu Z, Wu J, Li G, Miao J, Zhang C, Yang Y, Sun W, Zhu S, Fan Y, Hu J, Liu J, Wang W (2020b) COVID-19 in Wuhan: Immediate Psychological Impact on 5062 Health Workers. https://doi.org/10.1101/2020.02.20.20025338

Treatment of Patients with Mental Illness Amid A Global COVID-19 Pandemic

42

Ankit Jain, Kamal Kant Sahu, and Paroma Mitra

Abstract

A newly discovered coronavirus called severe acute respiratory syndrome coronavirus 2 (SARS-CoV-2) has caused the ongoing pandemic of coronavirus disease 2019 (COVID-19), which is not only physically challenging but also has many subtle and overt mental impacts. The concern of being infected, lack of antiviral agents, preventive strategies of social distancing, and home isolation have created unrest in the society. The way of reacting to emergencies varies from individual to individual, and that this variability lies in our unique personality traits. The COVID-19 pandemic is testing the mental stability of all of us, and hence it is crucial to recognize the vulnerable population and support them to prevent or minimize the catastrophe like post-traumatic stress disorder (PTSD), emotional trauma, and suicides. In this context, the role of psychiatrists, psychotherapists, and other mental healthcare providers is indispensable.

Keywords

Coronavirus · COVID-19 · Mental health · Mental illness · Personality · Psychiatry

42.1 Introduction

Coronavirus disease 2019 (COVID-19) has posed challenges to global citizens and significantly affected their normal functioning (Sahu et al. 2020h). As of May 30, 2020, there are 6,102,522 confirmed COVID-19 cases, with 369,127 deaths (Sahu et al. 2020j). COVID crisis affects various aspects of normal life, including physical, social, economic, and mental wellbeing (Sahu et al. 2020g; Shigemura et al. 2020).

Individual, national, and global efforts are underway to minimize the influence (Sohrabi et al. 2020). Economic packages, volunteer services, tax rebates, and relaxation in formal schooling are only a few examples among the many measures being implemented by various federal governments and national agencies. At an

A. Jain
Department of Psychiatry, Penn State College of Medicine, Milton S. Hershey Medical Center, Hershey, Pennsylvania, USA

K. K. Sahu (✉)
Department of Medicine, Saint Vincent Hospital, Worcester, MA, USA

Systematic Review and Meta-analysis Expert Group (SRMEG), Universal Scientific Education and Research Network (USERN), Worcester, MA, USA

P. Mitra
Department of Psychiatry, Bellevue Hospital Center, New York University School of Medicine, New York, NY, USA

N. Rezaei (ed.), *Coronavirus Disease - COVID-19*, Advances in Experimental Medicine and Biology 1318, https://doi.org/10.1007/978-3-030-63761-3_42

individual level, staying at home, maintaining social distancing, isolation, distance learning, and virtual meetings are components of a new healthy lifestyle.

Almost all nations, international agencies, and non-government organizations are endorsing the lifestyle mentioned above to mitigate community transmission (Sahu et al. 2020g). However, there is an urgent need to study the negative effect of the COVID-19 preventive/isolation practices on our mental health (Dong and Bouey 2020). Isolation, social withdrawal, and lack of physical contact have made people vulnerable to an emotional breakdown, which has been particularly challenging for individuals with known mental health conditions. This chapter provides an overview of changes to the delivery of mental health services with a focus on services in the United States during the COVID-19 pandemic.

42.2 Background

According to the National Alliance on Mental Illness (NAMI), 1 out of 5 adults in the United States is assumed to suffer from some form of mental illness in their lifetime. Moreover, about 1 in 25 adults in the United States experiences serious mental illness such as schizophrenia, bipolar disorder, and substance use disorders (Becker et al. 2019). Mental illness costs the United States about 190 billion in lost earnings every year. About 90% of people who died by suicide have an underlying mental illness. Suicide is considered one of the leading causes of death in the United States. Similar problems do exist in developing countries, but we lack a good insight into the grieve situation due to the lack of a sound reporting system (Slewa-Younan et al. 2020). The ongoing pandemic has different challenges for nationals from different countries (Sahu et al. 2020l).

Studies have shown that approximately one-fourth of homeless adults living have a mental illness, and approximately another one-fourth of incarcerated individuals have a background history of mental illness. About 10.2 million US adults have both substance use and mental health disorders. It leads to a multidimensional health hazard, increased mortality, and economic burden (Sahu et al. 2019; Sahu et al. 2020p).

The current pandemic has impacted pre-existing mental health symptoms in many ways. A more noticeable effect of the COVID-19 pandemic and country-wide lockdown would be the new onset of anxiety and to cause exacerbation of symptoms in diagnosed cases. Any naïve flu-like symptom increases anxiety, and under present circumstances, it is expected that COVID-19 would have a more severe impact. Patients with obsessive-compulsive disorder, especially those who have checking, hoarding, and washing compulsion, are at a higher risk. Advice on improving personal hygiene measures could potentially exaggerate the obsessions and compulsions for frequent washing practices. Another challenge is that as many countries are in various phases of lockdown, many patients might route to panic buying and unwarranted hoarding of essential items. Some patients with special needs might struggle with the specific diet which they are prescribed by the specialists and nutritionists (Gibbs et al. 2020). About the individuals with major depressive disorder (MDD), the practices of social distancing and isolation are a disruption in the daily routine. They thereby can provoke stress, with a vicious worsening of symptoms related to depression, anxiety, and sleep deprivation (Dong and Bouey 2020). Patients during their peak of stress and anxiety can even take extreme steps of attempts to commit suicide (Goyal et al. 2020).

There has been limited data on psychosocial factors that impact mood and anxiety. However, a small study from Iran showed that the inability to join work, dwindling finances, and the long-term impact on the economy are likely to affect new and pre-existing common mental health disorders (Zandifar and Badrfam 2020).

Quarantine can precipitate feelings of fear, anger, anxiety, and panic about the worst possible outcomes of boredom, loneliness, and guilt about not being able to do anything worthwhile for the family and community. Having a history of diagnosed mental illness is a significant risk factor, and all these issues generated out of isolation practices can spark symptoms with renewed severity, and even suicidal thoughts and attempts. In some cases, anxiety can be so overwhelming that it can cause paranoia and nihilistic delusions (Brooks et al. 2020). Patients with mental illness have their own needs. They come importantly with chronic physical conditions and have decreased life expectancy (Walker et al. 2015).

The other important note is that patients with mental illness have higher incidences of cardiac disease, respiratory disease, and obesity – comorbidities, which are known to increase susceptibility to COVID-19 and related adverse outcomes. In particular, persons with schizophrenia have a shortened life span secondary to chronic medical factors. Pregnant women with or without mental disorders are also likely to be affected for fear of possible maternal-fetal transmission of disease, risk of infection to the newborn, and concerns over breastfeeding (Janssen et al. 2015; Sahu et al. 2020d, i; k; Egloff et al. 2020; Jha et al. 2018).

More important characteristics include challenges not only from a physical health perspective but also from a socioeconomic perspective. In the United States, there is an increased incidence of homelessness and limited access to care and resources within this population (Sahu and Kumar 2020a). Similarly, the current pandemic has exposed the health disparities and inequities among the African American and Latino individuals. For instance, in Chicago, rates of COVID-19 cases per 100,000 (as of May 30, 2020) are highest among Latino (2064.1), followed by African American/black (1326.0) as compared to white (568.2) residents. A higher prevalence of certain diseases in African Americans (e.g., sickle cell disease) further increases the COVID-19 severity (Sahu et al. 2020o). These factors increase the risk of contracting COVID-19 and limits seeking care when unwell (Fazel et al. 2008).

Patients with mental illness are also at risk of being non-compliant with their medications, which leads to uncontrolled medical illness and, therefore, more complications (Dolder et al. 2003; Mishra et al. 2020b, d).

Psychiatric care is delivered in many ways in the United States. These include inpatient services, outpatient services, and emergency services. As of May 30, 2020, 105,521 deaths have been recorded across the United States. The severity of the COVID-19 pandemic has an impact on access, delivery, and quality of care. Even legal guidelines and confidentiality rules have been sidestepped to deliver care in a different environment. The delivery of psychiatric care has transformed depending on the setting the psychiatric physician is working. The American Psychiatric Association has issued national COVID-19 practice guidance to help physicians understand the many changes in the delivery of care. In outpatient settings, usually, most patients are seen face to face in the offices weekly, biweekly, monthly, bi-monthly, or once in 3 months. In the United States, there are dedicated psychiatric emergency rooms, separate from medical emergency rooms such as CPEP (Comprehensive Psychiatric Emergency Program) and PES (Psychiatric Emergency Services) units. In these dedicated psychiatric emergency rooms, patients are triaged and assessed, and disposition to self-care, extended observation rooms, or inpatient psychiatry unit is done depending upon the assessment and clinical state of the patient. These services have transitioned to the use of telehealth, which will be described in individual sections below.

During the COVID-19 pandemic, psychiatric physicians in these settings are providing care through a variety of modalities, including using a telephone or video camera, or in-person with personal protective equipment (PPE). At many hospitals, patients considered for inpatient psychiatric admission are now also subjected to rapid testing to determine COVID-19 status. COVID-19-positive and COVID-19-negative inpatient units have since been established. Visitors for patients have been asked to refrain from personal meetings to avoid the risk of spread across most emergency rooms. The role of telemedicine has increasingly been recognized in immunocompromised individuals like patients with cancer and transplant recipients who are advised to avoid unnecessary hospital visits (Sahu et al. 2020a, b, c, n; Jindal et al. 2020).

Psychiatry Consultation Service, which was formerly called Psychosomatic Medicine Services, attends to consultation calls from non-psychiatric departments like medicine, surgery, hematology, radiology, transfusion medicine, obstetrics and gynecology, physical rehabilitation, and medicine, on an inpatient basis (Sahu et al. 2020f; Sahu and Siddiqui 2020; Sahu and Cerny 2020). When a consult is generated by the departments mentioned above, the psychiatry consultation-liaison team evaluates the patient in person and provides recommendations to the primary team. In several hospital systems, especially with large hospital systems, in psychiatric consultation service, a hybrid model has been implemented, utilizing available resources, triaging, and deciding whether a patient is seen in person, via telephone, or video. In many hospitals, consultations are conducted remotely in another location.

Child Psychiatry is a subdivision of the psychiatry department where patients of age less than 18 are provided care in inpatient, outpatient, day programs, partial hospitalization, and substance use programs. These services have also changed in the context of the current pandemic.

Addiction services in the United States encompass both inpatient and outpatient services and services in the community. Inpatient services involve detoxification services from substances such as alcohol, opiates, and benzodiazepines. There are drug rehabilitation services of both short term and long term, which are utilized by patients in maintenance treatment. Other services include community programs for methadone maintenance, buprenorphine clinics, and dual mental health and addiction clinics. Community-based services also include alcoholics anonymous (AA) groups and narcotics addiction (NA) groups.

Across all these settings, substantial concerns have appeared about the availability and use of proper PPE. These worries can evolve into fears and anxieties about falling sick, spreading the virus to loved ones, and dying. On psychiatric units, there can be the fear that PPE "is not patient-friendly," and also, it is a challenge to get psychotic patients to wear masks (Bojdani et al. 2020). Some psychiatric patients have difficulty managing personal hygiene, increasing the risk of contracting the virus. In general, the psychiatric staff is not trained in infectious disease protocols, putting these employees at higher risk than other hospital employees.

The pandemic of COVID-19 has affected the delivery of psychiatric care across the world, including China and Italy, among others. In China, the challenges that the disease presents to the psychiatric services were issued by the government of the People's Republic of China for strengthening the management of treatment of patients with severe mental disorders during the outbreak of COVID-19 (Cui et al. 2020).

Italy is one of the countries badly hit by the pandemic with widespread casualties (Sahu and Kumar 2020b; Day 2020). Lombardy region in Italy, which has one of the most extensive psychiatric services in Lombardy, has described the changes to their system and distancing guidelines while treating patients with mental illness (Sahu et al. 2020e; Percudani et al. 2020).

Given the recent outbreak of the pandemic, there are few studies on the incidence of an ongoing pandemic on persons with mental illness, especially in the closed facility, which is akin to a cluster. A study in Wuhan showed about 40 patients being infected initially, which increased to more than 300 patients in 2 weeks (Li et al. 2020).

The pandemic has disrupted the training of psychiatry residents and medical students. Residents and medical students often perform a vital part in the treatment and care of patients with mental illness. Psychiatric residents and other rotators in medicine have been deployed to medicine services disrupting services in mental health (Bojdani et al. 2020). Major educational conferences have been canceled and postponed (Asmundson and Taylor 2020).

42.3 Mental Health and COVID-19

Databases in Pubmed, OVID, PsycInfo, and APA (American Psychiatric Association) were searched. Specific terms such as "COVID-19," "coronavirus," and "mental health," "psychiatry," and "addiction" were used. The search yielded 494 results, including 29 systematic reviews, and out of these 5 related to non-psychiatric issues. In addition to the information collected from the relevant systematic reviews, the knowledge of clinicians working through the current pandemic is applied to the narrative synthesis below.

42.3.1 Impact of COVID-19 on Emergency Room Services

When one talks about emergency room services in psychiatry, there are several scenarios under which these occur. These situations include when persons are deemed at risk to selves or others or with poor self-care. Sometimes patients walk into services themselves, but often, family members bring patients in asking for services. In states like New York, patients may be brought in for emergent services by the authority if deemed to be unfit to care for themselves (Sullivan and Rivera 2000).

Often patients brought into the emergency room services are persons with severe mental illness, and as discussed in the background above, the current COVID 19 crisis has impacted care more than ever. One of the critical factors is that persons with severe mental illness are often homeless (Folsom et al. 2005; Ayano et al. 2019).

Homelessness within itself leads to several issues, which include being unable to adequately social distance, which is the critical measure to lower the rate of infection. The other issue with persons with mental illness is the ability to absorb and process the necessary information required for maintaining sanitary measures that would prevent COVID-19 infections (Sheffield et al. 2018).

These factors explain that persons with serious mental illnesses who come into the emergency room suffer from different emergent related conditions. Unfortunately, social distancing measures have impacted the delivery of care in the community, and persons with mental illness often have limited resources, including smartphones or the ability to access telehealth given their limitations, as described above, leading to non-compliance with medications. It often leads to exacerbation of an illness and hence patients are brought into the emergency room. An interesting phenomenon is now the formation of delusions around COVID-19 – including grandiose delusions: an example being "the cure for coronavirus" or persecutory delusions such as "the FBI caused coronavirus and are tracking people."

Another phenomenon with persons with depressive and anxiety disorders is social isolation, which leads to further anhedonia and mood dysregulation. Emergency rooms report increased cases of suicidal thoughts among persons with pre-existing mood disorders (Kawohl and Nordt 2020).

COVID-19 pandemic has changed the delivery of care in the emergency rooms, especially with providers using PPE. In the world of psychiatry, it has affected the dynamic of establishing a rapport with a patient tremendously. In behavioral emergencies where a patient is verbally or physically aggressive, the most effective way to assist a patient is the use of verbal de-escalation. It has been relatively challenging with providers, and patients might be unable to connect with them. Another challenge in the emergency room is working with COVID-19-positive patients and the treatment of their urgent psychiatric needs.

42.3.2 Impact of COVID-19 on Inpatient Care

Providing care for patients in the inpatient psychiatric setting has become a challenge in the times when there are protocols in place to maintain social distancing. During the COVID-19 pandemic, inpatient psychiatric units across the country have adapted to the change and faced numerous challenges like patients not adhering to the protocols, inadequate spacing or design of units, and staff not adequately trained to follow protocols. Unsurprisingly, control measures in inpatient units can be a daunting task, mainly when treatment traditionally includes close contact, for example, frequent safety monitoring, group therapy sessions, and eating meals together. Additionally, there are logistical factors to consider, such as shared rooms and bathrooms and tables where food is shared (Li 2020).

Some protocols have been placed: (i) disease precautions (screening, re-screening, PPE, use of designated clothing, cleaning of surfaces, and disinfection), (ii) restriction of visitors and minimization of non-essential contacts (e.g., medical students, physician assistant students, observers, and psychology externs), (iii) physician workforce rescheduling and considerations (creating back-up pools, job reassignment), (iv) operational adjustments (creating isolation rooms in psychiatry inpatient units, tightening criteria for psychiatric admission), and (v) group therapy changes (limiting the number of participants, utilizing physical distancing). It is also fundamentally important to have structured leadership, clear communication, and involvement of all disciplines in decision making (Li 2020).

Inpatient psychiatric units have faced unique challenges in terms of exposure to other patients, staff, and visitors (Benson et al. 2020). In the usual time of functioning, patients share bedrooms with others in the inpatient psychiatry unit, eat in a group in the dining room, and participate in various group activities like play therapy, activity therapy, dance therapy, and group psychotherapy. Some inpatient units have implemented no visitor policies and suspension of group activities. Akin to other countries, as demonstrated by China, some hospitals have enforced the tightening of admission criteria, especially for voluntary admissions for specific substance use disorders (Li 2020). Some hospitals have reduced the capacity of inpatient units to facilitate social distancing norms, while others have converted all shared rooms to single occupancy rooms. The policy now in most inpatient units is to screen patients for COVID-19 and place them in specific units after multiple calls were made for the same by the psychiatric community.

At various hospitals in New York City and most parts of the United States, psychiatric units have opened to provide acute inpatient care for psychiatric patients presently infected with the novel coronavirus, but not medically ill enough to be hospitalized on an infectious disease unit. One challenge in providing this level of care is the difficult-to-predict course of infection. While patients admitted to this unit are discussed with internal medicine colleagues, seen by internists on the psychiatric floor and determined to be medically stable on admission, some patients develop complications and require transfer to a medical unit. It is especially important to be considered in psychiatric facilities where access to emergent interventions, such as intubation, is limited, and patients must be brought by ambulance to the nearest medical facility. Another challenge is disposition planning, particularly for patients experiencing housing instability. Programs, such as the much-needed partial hospitalizations or day hospitals, are mostly not available. There also may be challenges in engaging families to participate in discharge planning for patients infected with the novel coronavirus due to fear.

Some inpatient psychiatry units have reduced their capacities to half to incorporate social distancing measures. Group activities, including various forms of nonpharmacological interventions, including play, artwork, dancing, and group psychotherapy, are currently on hold in most places to ensure that everyone is practicing social distancing measures. Team meetings involving various disciplines to discuss the new admissions, daily progress, and planning disposition of the patients are either conducted in tiny groups or are being done over the phone or video conferencing. Administrative hearings, retention hearings, and medication over objection hearings are being done over video conferencing to continue the treatment for the patients.

The changes made above, though vital for infection control purposes, have formed an inpatient setting that is more detaching for both patients and clinicians. For example, a patient with a major depressive episode who experiences a striking decrease in appetite and difficulty getting out of bed may benefit principally from attending group meals and socializing with others on the unit as forms of behavioral stimulation. COVID-19 pandemic has changed the inpatient environment so that it can prohibit this treatment. Overall, the combination of less face to face interaction, decreased in-person programming, restricted dining to in-room meals instead of shared dining, and the "no visitors policy" is likely contributing to a decrease in the efficacy of a therapeutic milieu on the inpatient psychiatric unit. There are also concerns about a pandemic of loneliness and social disconnection following the resolution of the present pandemic. That may fuel an uptick in psychiatric hospitalizations in the coming months.

42.3.3 Impact of COVID-19 on Outpatient Care

In the outpatient world of psychiatry, care is delivered in many settings, including traditional community clinics, intensive outpatient clinics, day treatment programs, and community teams. In each of the settings, the delivery of service works on personal interaction between providers and patients. In most intensive outpatient programs and day treatment programs, persons with mental illness often stay all day and attend various groups. Day programs often point to the essential backbone of treatment where persons with mental illness remain socially connected to both their peers and treatment. It has been disrupted in the recent pandemic where telehealth has become the norm. Evidence has shown that telehealth is an effective way of treating persons with mental illness (Bashshur et al. 2016).

Again, some of the limitations now include persons with serious mental illness who have difficulty in accessing software technology necessary for telehealth. Another downfall is the loss of social structure for persons with mental illness who often rely on programs for routine and care.

COVID-19 has led to the expansion of telepsychiatry, which had begun to burgeon. During this pandemic, to improve access to care and reduce transmission risk by in-person visits, the US government temporarily waived numerous regulations around telehealth (Torous et al. 2020). Monitoring barriers to telemedicine were also relaxed to allow for various platforms to be used across state lines (Freeman 2020). Remote consultation via telemedicine has been described in other countries, and community hospitals throughout the United States and VA healthcare systems like South Carolina and VA Boston have swiftly incorporated telehealth (Myers et al. 2020).

There are limited studies conducted on the effectiveness of therapy using telehealth services. Cognitive-behavioral therapy (CBT) via telehealth has been studied in older persons and found to be effective (Lichstein et al. 2013).

42.3.4 Impact of COVID-19 on Addiction Services

In the world of addiction, access to substances has decreased significantly due to ongoing social distancing measures and restrictions to pubs, restaurants, and gatherings. It has led to isolation, which exacerbates other symptoms like mood and anxiety. If there is increased use of substances in isolation, this may lead to severe withdrawal symptoms like delirium tremens, which is a life-threatening condition where patients often find themselves in intensive care units if left untreated (Kar et al. 2020). The delivery of services for addiction psychiatry is on a multi-model structure. Addiction services comprise inpatient detoxification and rehabilitation services and outpatient treatment programs. Patients suffering from substance use disorder might struggle during this time period due to the non-availability of substance or medicines like methadone and suboxone (Xiang et al. 2020).

Per the national institute of drug abuse, persons with substance use are at higher risk of developing COVID-19 given that it attacks the respiratory system, and this aspect is of particular importance to people with opioid use and methamphetamine use. Many prominent substance use associations such as the Substance Abuse and Mental Health Services Administration (SAMHSA) and the Drug Enforcement Agency (DEA) have been instrumental in issuing guidelines for continued treatment of those with an addiction disorder.

Inpatient detox programs have remained functional for those in need, though provided with minimal staff and limited patients to manage social distancing.

Most addiction services are served in the community. Addiction services are in many forms, e.g., joint mental health and addiction services, stand-alone methadone programs, and day treatment programs. The backbone of addiction services includes AA and NA. Most have transitioned to online groups and virtual meetings. Patients with addiction often are neglected during disasters and are at higher risk of relapse during this period without adequate monitoring and support.

Methadone programs that rely on testing their clients for illicit substance use and daily dosing have now resorted to giving medications for the month and having virtual meetings. Buprenorphine clinics have transitioned to virtual settings, including the induction phase, which is traditionally done in the clinic to monitor clinical withdrawal.

42.3.5 Impact of COVID-19 on Consultation-Liaison Services

Consultation-liaison (C-L) psychiatrists stand at the junction of psychiatry and the rest of medicine. C-L service, when at its peak, is a crucial entity in the hospital. C-L team provides care for extraordinarily sick patients based on physical examination findings, including a thorough neurological as well as a psychiatric examination, and manages conditions, such as hypoactive or hyperactive delirium, catatonia, and severe alcohol withdrawal, which require frequent visual assessments (De Giorgio et al. 2015).

As of April 9, 2020, over 9000 healthcare workers in the United States have been confirmed to have COVID-19, with at least 27 confirmed deaths (Characteristics of Health Care Personnel with COVID-19 - United States, February 12–April 9, 2020 2020). Because of the risk of sustained transmission of the virus causing COVID-19, like most medical specialties, C-L psychiatry is strained to adjust during the pandemic, possibly bringing permanent changes to this service (Funk et al. 2020).

Owing to nearly universal lack of adequate PPE in the healthcare facilities of the United States, as well as inadequate testing for COVID-19, many C-L psychiatry services have implemented telepsychiatry routes, usually reserved for under-resourced areas. Usually the gold standard of consultation involves in-person evaluation of the patient; however, this crisis calls for the use of technology to decrease viral transmission. Guidelines from the Centers for Disease Control and Prevention recommend the use of telemedicine when possible, the limit on the number of staff providing care to COVID-19 patients, and the use of dedicated healthcare providers who care only for COVID-19 patients during a shift (Sahu et al. 2020a, e; Chang et al. 2020).

Most hospitals in the United States are utilizing the hybrid model of patient care where the consults are usually answered by telepsychiatry-video conferencing interview and assessment of the patient, review of chart, and contacting collateral contacts through telephone unless an in-person patient interview is deemed necessary in a few cases.

In one of the reports from psychiatry consult service in a large general hospital in the United States, there were 19 patients documented with delirium in the setting of COVID-19 as of May 1, 2020. Some patients were admitted with confusion and agitation as their presenting symptom in the absence of respiratory symptoms or other signs of infection (Baller et al. n.d.). In addition to significant agitation and attentional impairment, exams were notable in varying degrees for myoclonus, rigidity, alogia, and abulia (Beach et al. 2020). Some of the treatment recommendations are based on anecdotal observations that some patients with COVID-19 and delirium appear to have increased rates of myoclonus, rigidity, alogia, and abulia, suggesting a dopamine depletion state or catatonia-spectrum condition. When there are no absolute contraindications to any class of medication, it has been recommended to preferentially use alpha-2 agonists and low-potency antipsychotics to manage behavioral disturbance. Universally a thoughtful, individualized, step-based approach to management is vital to the management of delirium associated with COVID-19 while keeping in mind whether there is pre-existing mental health or neurological diagnosis (Rogers et al. 2020). A recent review has described the initial picture of delirium in patients with COVID-19, which is similar to that in patients with SARS (severe acute respiratory syndrome) and MERS (Middle East respiratory syndrome). It might be presented with a variety of symptoms, including agitation, confusion, and to a lesser extent, psychosis and mania like euphoria and psychomotor agitation. Also, long-lasting effects mainly include depression and post-traumatic stress disorder (Rogers et al. 2020).

42.3.6 Impacts of COVID-19 on Healthcare Professionals

Healthcare workers are exposed to mental challenges during the current pandemic. There could be multiple reasons which can cause a healthcare provider prone to an emotional breakdown, including the worsening situation with each passing day, fear of being infected, and subsequently, other family members, prolonged shifts, witnessing critically ill patients in the intensive care unit, and the frustration of not being able to save lives (Wu et al. 2020). The majority of the studies have been reported from China, and similar trends seem to be present in the United States as declared by the healthcare providers. The development of mental issues has also been noted to have a socio-demographic graphic variability with younger aged females (< 30 years) being at the highest risk (Lai et al. 2020; Liang et al. 2020). Cai et al. reported middle-aged medical workers (31–40 years) to be worrying about the family members having the risk of getting cross-infection. The study also showed that health workers >50 years expressed the death of the patients as the primary cause of their stress (Cai et al. 2020).

Healthcare professionals are also subjected to concerns that they may not have adequate PPE, which is instrumental in preventing infection (Rowan and Laffey 2020). There is increasing data that adequate PPE prevents cross-infection between patients. Specifically, for mental health professionals in the inpatient setting, there are ongoing concerns about being part of a "cluster contagion." Another concern has been ethical concerns around psychiatric inpatients who contract COVID-19 and wish to be released into the community without concern for maintaining adequate social distancing measures. Finally, healthcare professionals, like others in the country, are plagued by financial downfall.

42.4 Discussion

The current COVID-19 pandemic has brought about several changes to the delivery of healthcare in the United States (Sahu et al. 2020m). Individuals with a known psychiatric disorder are incredibly prone to disease and related consequences. Many social factors play a role in this context, including homelessness, chronic medical issues, and lack of access to care and also the increased prevalence of diabetes, hypertension, and cigarette smoking, which are of known risk factors for COVID-19 (Dickerson et al. 2013; Mishra et al. 2020a).

On the positive side, however, systems have shown that they can gradually move to telehealth and work with patients with some amount of effectiveness. Also, many hospital systems have worked with providers and patients to provide PPE and come up with innovative methods to work on anxiety management. To cope with COVID-19-related stress, the World Health Organization has issued guidelines and advisories, which apply to outpatient services. Many studies published recently have received criticism worldwide for lacking authenticity. Hence, it is also essential to recognize the importance of educating our patients about the right source of authentic information (Rzymski et al. 2020).

A substantial downside is the increase of mental health symptoms in both people with and without previous mental illness. There has continued to be an increased need for behavioral health services. Symptoms such as anxiety due to loss of jobs, financial insecurity, and fear of death have become prominent in all spheres of life. With each passing day, our understanding of COVID-19 and its effect on the various organs and functioning is increasing (Lal et al. 2020; Mishra et al. 2020d, c). However, at present, it calls for a more dedicated study to understand the impact of COVID-19 on emotional wellbeing and mental health.

42.5 Conclusion

As described above, the current COVID-19 pandemic has affected psychiatric services. People experience bereavement, loss of self, and financial security worldwide, while those who have a history of mental illness are more susceptible to contract the illness and have new challenges to access to care. We believe that all of us have our own narrative of the experience of living during the COVID-19 pandemic; the trauma experienced will be individual. Long-term studies are necessary to understand the full impact of the pandemic on mental health with an emphasis on trauma. Delivery of care is a continually evolving process, and sharing of data and knowledge will lead to more innovative care for our vulnerable and disenfranchised in the face of a changing world.

References

Asmundson GJG, Taylor S (2020) How health anxiety influences responses to viral outbreaks like COVID-19: what all decision-makers, health authorities, and health care professionals need to know. J Anxiety Disord 71:102211. https://doi.org/10.1016/j.janxdis.2020.102211

Ayano G, Tesfaw G, Shumet S (2019) The prevalence of schizophrenia and other psychotic disorders among homeless people: a systematic review and meta-analysis. BMC Psychiatry 19(1):370. https://doi.org/10.1186/s12888-019-2361-7

Baller EB, Hogan CS, Fusunyan MA, Ivkovic A, Luccarelli JW, Madva E, Nisavic M, Praschan N, Quijije NV, Beach SR, Smith FA (n.d.) Neurocovid: pharmacological recommendations for delirium associated with COVID-19. Psychosomatics. https://doi.org/10.1016/j.psym.2020.05.013

Bashshur RL, Shannon GW, Bashshur N, Yellowlees PM (2016) The empirical evidence for telemedicine interventions in mental disorders. Telemed J E Health 22(2):87–113. https://doi.org/10.1089/tmj.2015.0206

Beach SR, Praschan NC, Hogan C, Dotson S, Merideth F, Kontos N, Fricchione GL, Smith FA (2020) Delirium in COVID-19: a case series and exploration of potential mechanisms for central nervous system involvement. Gen Hosp Psychiatry 65:47–53. https://doi.org/10.1016/j.genhosppsych.2020.05.008

Becker M, Cunningham CE, Christensen BK, Furimsky I, Rimas H, Wilson F, Jeffs L, Madsen V, Bieling P, Chen Y, Mielko S, Zipursky RB (2019) Investigating service features to sustain engagement in early intervention mental health services. Early Interv Psychiatry 13(2):241–250. https://doi.org/10.1111/eip.12470

Benson NM, Öngür D, Hsu J (2020) COVID-19 testing and patients in mental health facilities. Lancet Psychiatry 7(6):476–477. https://doi.org/10.1016/S2215-0366(20)30198-X

Bojdani E, Rajagopalan A, Chen A, Gearin P, Olcott W, Shankar V, Cloutier A, Solomon H, Naqvi NZ, Batty N, Festin FED, Tahera D, Chang G, DeLisi LE (2020) COVID-19 pandemic: impact on psychiatric care in the United States. Psychiatry Res 289:113069. https://doi.org/10.1016/j.psychres.2020.113069

Brooks SK, Webster RK, Smith LE, Woodland L, Wessely S, Greenberg N, Rubin GJ (2020) The psychological impact of quarantine and how to reduce it: rapid review of the evidence. Lancet 395(10227):912–920. https://doi.org/10.1016/s0140-6736(20)30460-8

Cai H, Tu B, Ma J, Chen L, Fu L, Jiang Y, Zhuang Q (2020) Psychological impact and coping strategies of frontline medical staff in Hunan between January and march 2020 during the outbreak of coronavirus disease 2019 (COVID-19) in Hubei, China. Med Sci Monit 26:e924171–e924171. https://doi.org/10.12659/MSM.924171

Chang D, Xu H, Rebaza A, Sharma L, Dela Cruz CS (2020) Protecting health-care workers from subclinical coronavirus infection. Lancet Respir Med 8(3):e13. https://doi.org/10.1016/s2213-2600(20)30066-7

Characteristics of Health Care Personnel with COVID-19 - United States, February 12-April 9, 2020 (2020) MMWR Morb Mortal Wkly Rep 69 (15):477–481. https://doi.org/10.15585/mmwr.mm6915e6

Cui LB, Wang XH, Wang HN (2020) Challenges of facing coronavirus disease 2019: psychiatric services for patients with mental disorders. Psychiatry Clin Neurosci 74(6):371–372. https://doi.org/10.1111/pcn.13003

Day M (2020) Covid-19: surge in cases in Italy and South Korea makes pandemic look more likely. BMJ 368:m751. https://doi.org/10.1136/bmj.m751

De Giorgio G, Quartesan R, Sciarma T, Giulietti M, Piazzoli A, Scarponi L, Ferrari S, Ferranti L, Moretti P, Piselli M (2015) Consultation-liaison psychiatry—from theory to clinical practice: an observational study in a general hospital. BMC Res Notes 8(1):475. https://doi.org/10.1186/s13104-015-1375-6

Dickerson F, Stallings CR, Origoni AE, Vaughan C, Khushalani S, Schroeder J, Yolken RH (2013) Cigarette smoking among persons with schizophrenia or bipolar disorder in routine clinical settings, 1999-2011. Psychiatr Serv 64(1):44–50. https://doi.org/10.1176/appi.ps.201200143

Dolder CR, Lacro JP, Jeste DV (2003) Adherence to antipsychotic and nonpsychiatric medications in middle-aged and older patients with psychotic disorders. Psychosom Med 65(1):156–162. https://doi.org/10.1097/01.psy.0000040951.22044.59

Dong L, Bouey J (2020) Public mental health crisis during COVID-19 pandemic, China. Emerg Infect Dis 26(7):1616–1618. https://doi.org/10.3201/eid2607.200407

Egloff C, Vauloup-Fellous C, Picone O, Mandelbrot L, Roques P (2020) Evidence and possible mechanisms of rare maternal-fetal transmission of SARS-CoV-2. J Clin Virol 128:104447. https://doi.org/10.1016/j.jcv.2020.104447

Fazel S, Khosla V, Doll H, Geddes J (2008) The prevalence of mental disorders among the homeless in western countries: systematic review and meta-regression analysis. PLoS Med 5(12):e225. https://doi.org/10.1371/journal.pmed.0050225

Folsom DP, Hawthorne W, Lindamer L, Gilmer T, Bailey A, Golshan S, Garcia P, Unützer J, Hough R, Jeste DV (2005) Prevalence and risk factors for homelessness and utilization of mental health services among 10,340 patients with serious mental illness in a large public mental health system. Am J Psychiatry 162(2):370–376. https://doi.org/10.1176/appi.ajp.162.2.370

Freeman MP (2020) COVID-19 from a psychiatry perspective: meeting the challenges. J Clin Psychiatry 81(2). https://doi.org/10.4088/JCP.20ed13358

Funk MC, Beach SR, Shah SB, Boland R (2020) Consultation-liaison psychiatry in the age of COVID-19: reaffirming ourselves and our worth. Psychosomatics. https://doi.org/10.1016/j.psym.2020.04.013

Gibbs L, Minnelli N, Larrivee J, Sahu KK, Siddiqui AD (2020) Oncology dietitians sound alarm in key nutrition needs of Cancer patients during COVID-19 pandemic. JCO Oncol Pract:Op2000349. https://doi.org/10.1200/op.20.00349

Goyal K, Chauhan P, Chhikara K, Gupta P, Singh MP (2020) Fear of COVID 2019: first suicidal case in India ! Asian J Psychiatr 49:101989. https://doi.org/10.1016/j.ajp.2020.101989

Janssen EM, McGinty EE, Azrin ST, Juliano-Bult D, Daumit GL (2015) Review of the evidence: prevalence of medical conditions in the United States population with serious mental illness. Gen Hosp Psychiatry 37(3):199–222. https://doi.org/10.1016/j.genhosppsych.2015.03.004

Jha S, Salve HR, Goswami K, Sagar R, Kant S (2018) Burden of common mental disorders among pregnant women: a systematic review. Asian J Psychiatr 36:46–53. https://doi.org/10.1016/j.ajp.2018.06.020

Jindal V, Sahu KK, Gaikazian S, Siddiqui AD, Jaiyesimi I (2020) Cancer treatment during COVID-19 pandemic. Med Oncol 37(7):58. https://doi.org/10.1007/s12032-020-01382-w

Kar SK, Arafat SMY, Sharma P, Dixit A, Marthoenis M, Kabir R (2020) COVID-19 pandemic and addiction: current problems and future concerns. Asian J Psychiatr 51:102064–102064. https://doi.org/10.1016/j.ajp.2020.102064

Kawohl W, Nordt C (2020) COVID-19, unemployment, and suicide. Lancet Psychiatry 7(5):389–390. https://doi.org/10.1016/s2215-0366(20)30141-3

Lai J, Ma S, Wang Y, Cai Z, Hu J, Wei N, Wu J, Du H, Chen T, Li R, Tan H, Kang L, Yao L, Huang M, Wang H, Wang G, Liu Z, Hu S (2020) Factors associated with mental health outcomes among health care workers exposed to coronavirus disease 2019. JAMA Netw Open 3(3):e203976. https://doi.org/10.1001/jamanetworkopen.2020.3976

Lal A, Mishra AK, Sahu KK (2020) CT chest findings in coronavirus disease-19 (COVID-19). J Formos Med Assoc 119(5):1000–1001. https://doi.org/10.1016/j.jfma.2020.03.010

Li L (2020) Challenges and priorities in responding to COVID-19 in inpatient psychiatry. Psychiatr Serv 71(6):624–626. https://doi.org/10.1176/appi.ps.202000166

Li W, Yang Y, Liu ZH, Zhao YJ, Zhang Q, Zhang L, Cheung T, Xiang YT (2020) Progression of mental health services during the COVID-19 outbreak in China. Int J Biol Sci 16(10):1732–1738. https://doi.org/10.7150/ijbs.45120

Liang Y, Chen M, Zheng X, Liu J (2020) Screening for Chinese medical staff mental health by SDS and SAS during the outbreak of COVID-19. J Psychosom Res 133:110102. https://doi.org/10.1016/j.jpsychores.2020.110102

Lichstein KL, Scogin F, Thomas SJ, DiNapoli EA, Dillon HR, McFadden A (2013) Telehealth cognitive behavior therapy for co-occurring insomnia and depression symptoms in older adults. J Clin Psychol 69(10):1056–1065. https://doi.org/10.1002/jclp.22030

Mishra AK, Sahu KK, George AA, Lal A (2020a) A review of cardiac manifestations and predictors of outcome in patients with COVID - 19. Heart Lung. https://doi.org/10.1016/j.hrtlng.2020.04.019

Mishra AK, Sahu KK, Lal A (2020b) Reporting of all cardiac medications and their outcome in COVID-19. J Med Virol. https://doi.org/10.1002/jmv.25843

Mishra AK, Sahu KK, Lal A, Sargent J (2020c) Mechanisms of stroke and the role of anticoagulants in COVID-19. J Formos Med Assoc. https://doi.org/10.1016/j.jfma.2020.06.026

Mishra AK, Sahu KK, Lal A, Sargent J (2020d) Patterns of heart injury in COVID-19 and relation to outcome. J Med Virol. https://doi.org/10.1002/jmv.25847

Myers US, Birks A, Grubaugh AL, Axon RN (2020) Flattening the curve by getting ahead of it: how the VA healthcare system is leveraging telehealth to provide continued access to care for Rural Veterans. J Rural Health. https://doi.org/10.1111/jrh.12449

Percudani M, Corradin M, Moreno M, Indelicato A, Vita A (2020) Mental health services in Lombardy during COVID-19 outbreak. Psychiatry Res 288:112980. https://doi.org/10.1016/j.psychres.2020.112980

Rogers JP, Chesney E, Oliver D, Pollak TA, McGuire P, Fusar-Poli P, Zandi MS, Lewis G, David AS (2020) Psychiatric and neuropsychiatric presentations associated with severe coronavirus infections: a systematic review and meta-analysis with comparison to the COVID-19 pandemic. Lancet Psychiatry 7(7):611–627. https://doi.org/10.1016/s2215-0366(20)30203-0

Rowan NJ, Laffey JG (2020) Challenges and solutions for addressing critical shortage of supply chain for personal and protective equipment (PPE) arising from coronavirus disease (COVID19) pandemic - case study from the Republic of Ireland. Sci Total Environ 725:138532. https://doi.org/10.1016/j.scitotenv.2020.138532

Rzymski P, Nowicki M, Mullin GE, Abraham A, Rodríguez-Román E, Petzold MB, Bendau A, Sahu KK, Ather A, Naviaux AF, Janne P, Gourdin M, Delanghe JR, Ochs HD, Talmadge JE, Garg M, Hamblin MR, Rezaei N (2020) Quantity does not equal quality: scientific principles cannot be sacrificed. Int Immunopharmacol 86:106711. https://doi.org/10.1016/j.intimp.2020.106711

Sahu KK, Cerny J (2020) Managing patients with hematological malignancies during COVID-19 pandemic. Expert Rev Hematol. https://doi.org/10.1080/17474086.2020.1787147

Sahu K, Kumar R (2020a) Current perspective on pandemic of COVID-19 in the United States. J Family Med Primary Care 9(4):1784–1791. https://doi.org/10.4103/jfmpc.jfmpc_424_20

Sahu K, Kumar R (2020b) Preventive and treatment strategies of COVID-19: from community to clinical trials. J Family Med Primary Care 9(5):2149–2157. https://doi.org/10.4103/jfmpc.jfmpc_728_20

Sahu KK, Siddiqui AD (2020) From Hematologist's desk: the effect of COVID-19 on the blood system. Am J Hematol. https://doi.org/10.1002/ajh.25849

Sahu KK, Mishra A, Naraghi L (2019) Erythema ab igne as a complication of cannabinoid hyperemesis syndrome. BMJ Case Rep 12(1). https://doi.org/10.1136/bcr-2018-227836

Sahu KK, Jindal V, Siddiqui AD (2020a) Managing COVID-19 in patients with Cancer: a double blow for oncologists. JCO Oncol Pract 16(5):223–225. https://doi.org/10.1200/op.20.00167

Sahu KK, Jindal V, Siddiqui AD, Cerny J (2020b) Facing COVID-19 in the hematopoietic cell transplant setting: a new challenge for transplantation physicians. Blood Cells Mol Dis 83:102439. https://doi.org/10.1016/j.bcmd.2020.102439

Sahu KK, Jindal V, Siddiqui AD, Cerny J, Gerber JM (2020c) Convalescent plasma therapy: a passive therapy for an aggressive COVID-19. J Med Virol. https://doi.org/10.1002/jmv.26047

Sahu KK, Lal A, Mishra AK (2020d) COVID-2019 and pregnancy: a plea for transparent reporting of all cases. Acta Obstet Gynecol Scand 99(7):951. https://doi.org/10.1111/aogs.13850

Sahu KK, Lal A, Mishra AK (2020e) Latest updates on COVID-2019: a changing paradigm shift. J Med Virol 92(6):533–535. https://doi.org/10.1002/jmv.25760

Sahu KK, Lal A, Mishra AK (2020f) An update on CT chest findings in coronavirus disease-19 (COVID-19). Heart Lung. https://doi.org/10.1016/j.hrtlng.2020.03.007

Sahu KK, Mishra AK, Lal A (2020g) Comprehensive update on current outbreak of novel coronavirus infection (2019-nCoV). Ann Transl Med 8(6):393. https://doi.org/10.21037/atm.2020.02.92

Sahu KK, Mishra AK, Lal A (2020h) COVID-2019: update on epidemiology, disease spread and management. Monaldi Arch Chest Dis 90(1). https://doi.org/10.4081/monaldi.2020.1292

Sahu KK, Mishra AK, Lal A (2020i) Re: from the frontlines of COVID-19-how prepared are we as obstetricians? A commentary. BJOG 127(8):1041. https://doi.org/10.1111/1471-0528.16257

Sahu KK, Mishra AK, Lal A (2020j) Trajectory of the COVID-19 pandemic: chasing a moving target. Ann Transl Med 8(11):694. https://doi.org/10.21037/atm-20-2793

Sahu KK, Mishra AK, Lal A (2020k) A twin challenge to handle: COVID-19 with pregnancy. J Med Virol. https://doi.org/10.1002/jmv.25784

Sahu KK, Mishra AK, Lal A, Sahu SA (2020l) India fights Back: COVID-19 pandemic. Heart Lung. https://doi.org/10.1016/j.hrtlng.2020.04.014

Sahu KK, Mishra AK, Martin K, Chastain I (2020m) COVID-19 and clinical mimics. Correct diagnosis is the key to appropriate therapy. Monaldi Archiv Chest Dis 90(2). https://doi.org/10.4081/monaldi.2020.1327

Sahu KK, Siddiqui AD, Cerny J (2020n) COVID-19 pandemic and impact on hematopoietic stem cell transplantation. Bone Marrow Transplant:1–3. https://doi.org/10.1038/s41409-020-0913-6

Sahu KK, Siddiqui AD, Cerny J (2020o) Managing sickle cell patients with COVID-19 infection: the need to pool our collective experience. Br J Haematol. https://doi.org/10.1111/bjh.16880

Sahu KK, Tsitsilianos N, Mishra AK, Suramaethakul N, Abraham G (2020p) Neck abscess secondary to pocket shot intravenous drug abuse. BMJ Case Rep 13(3). https://doi.org/10.1136/bcr-2019-234033

Sheffield JM, Karcher NR, Barch DM (2018) Cognitive deficits in psychotic disorders: a lifespan perspective. Neuropsychol Rev 28(4):509–533. https://doi.org/10.1007/s11065-018-9388-2

Shigemura J, Ursano RJ, Morganstein JC, Kurosawa M, Benedek DM (2020) Public responses to the novel 2019 coronavirus (2019-nCoV) in Japan: mental health consequences and target populations. Psychiatry Clin Neurosci 74(4):281–282. https://doi.org/10.1111/pcn.12988

Slewa-Younan S, Nguyen TP, Al-Yateem N, Rossiter RC, Robb W (2020) Causes and risk factors for common mental illnesses: the beliefs of paediatric hospital staff in the United Arab Emirates. Int J Ment Health Syst 14:35. https://doi.org/10.1186/s13033-020-00367-6

Sohrabi C, Alsafi Z, O'Neill N, Khan M, Kerwan A, Al-Jabir A, Iosifidis C, Agha R (2020) World health organization declares global emergency: a review of the 2019 novel coronavirus (COVID-19). Int J Surg 76:71–76. https://doi.org/10.1016/j.ijsu.2020.02.034

Sullivan AM, Rivera J (2000) Profile of a comprehensive psychiatric emergency program in a new York City municipal hospital. Psychiatry Q 71(2):123–138. https://doi.org/10.1023/a:1004624319072

Torous J, Jän Myrick K, Rauseo-Ricupero N, Firth J (2020) Digital mental health and COVID-19: using technology today to accelerate the curve on access and quality tomorrow. JMIR Ment Health 7(3):e18848. https://doi.org/10.2196/18848

Walker ER, McGee RE, Druss BG (2015) Mortality in mental disorders and global disease burden implications: a systematic review and meta-analysis. JAMA Psychiat 72(4):334–341. https://doi.org/10.1001/jamapsychiatry.2014.2502

Wu PE, Styra R, Gold WL (2020) Mitigating the psychological effects of COVID-19 on health care workers. CMAJ 192(17):E459–e460. https://doi.org/10.1503/cmaj.200519

Xiang Y-T, Yang Y, Li W, Zhang L, Zhang Q, Cheung T, Ng CH (2020) Timely mental health care for the 2019 novel coronavirus outbreak is urgently needed. Lancet Psychiatry 7(3):228–229. https://doi.org/10.1016/S2215-0366(20)30046-8

Zandifar A, Badrfam R (2020) Iranian mental health during the COVID-19 epidemic. Asian J Psychiatr 51:101990. https://doi.org/10.1016/j.ajp.2020.101990

A Shift in Medical Education During the COVID-19 Pandemic

43

Farida Nentin, Nagaraj Gabbur, and Adi Katz

Abstract

The COVID-19 pandemic has abruptly affected every aspect of people's daily lives worldwide. Just like every other area, the medical field has been dramatically impacted by the need to care for a large number of patients while at the same time protecting staff, patients, and their families. Changes in the wake of the pandemic called for the prompt and extensive rechanneling and re-organization of resources. The pandemic has opened challenges and concerns for patient safety, starting with the early recognition that individuals, including medical staff, may spread the virus during the asymptomatic phase. Many healthcare facilities faced resource-limited settings, including challenges in the availability of personal protective equipment for healthcare providers. Additionally, the pandemic has disrupted medical education, both at the undergraduate and at the graduate levels, and according to many predictions, its effects may forever transform the ways medical education is delivered. In this chapter, we are exploring the history of medical education, describe changes in medical education experienced during the COVID-19 pandemic, and predict some of the considerations worth taking into account when envisioning the future of medical education.

Keywords

COVID-19 · Education · Pandemic

43.1 Introduction

The COVID-19 pandemic, caused by the novel SARS-CoV-2 (Arun Krishnan et al. 2020), has abruptly affected virtually every aspect of people's daily lives on a global scale. The medical field is one of the areas that was dramatically impacted by the need to care for a large number of patients while at the same time protecting staff, patients, and their families. The changes in the wake of the COVID-19 pandemic opened the need to re-think the organization and allocation of the available resources, while at the same time

F. Nentin
Department of Obstetrics, Gynecology and Reproductive Science, Icahn School of Medicine at Mount Sinai, New York, NY, USA

N. Gabbur
Department of Obstetrics and Gynecology, Zucker School of Medicine at Hofstra/Northwell, North Shore University Hospital and Long Island Jewish Medical Centers, Manhasset, NY, USA

A. Katz (✉)
Department of Obstetrics and Gynecology, Zucker School of Medicine at Hofstra/Northwell, Lenox Hill Hospital, New York, NY, USA
e-mail: adkatz@northwell.edu

N. Rezaei (ed.), *Coronavirus Disease - COVID-19*, Advances in Experimental Medicine and Biology 1318, https://doi.org/10.1007/978-3-030-63761-3_43

providing practical and professional healthcare and ensuring the wellbeing of patients, healthcare providers, and families. Therefore, the COVID-19 pandemic opened challenges in terms of the delivery of healthcare and patient care and safety. The pandemic also created unprecedented and unpredictable challenges for medical education. Both undergraduate and graduate medical education were profoundly impacted in several medical fields and multiple countries (Chang et al. 2020; Chatziralli et al. 2020; Virarkar et al. 2020; Go and Rajasekaran 2020; Wang et al. 2020).

Considering the impact of the pandemic on medical education, and the relative scarcity of knowledge in terms of the best approaches to reshape and adapt educational resources to an ongoing and unpredictable public health emergency, we want to present the changes in medical education that our institution has adopted during the pandemic. We hope that our experience will complement the teachings that other medical education programs and other institutions gained during these challenging times, and contribute to a framework that will be most certainly needed during future healthcare crises.

43.2 The History of Medical Education: Where Have We Been?

In the nineteenth century, after the Civil War, a marked expansion of knowledge in medical sciences led to a revolution in the practice of medicine (Blaisdell 1988). Before that time, there were no standardized requirements for medical school admission or graduation, nor was there a consistent curriculum. Students during those times were merely observers, passively learning information and techniques. Lectures were given to overcrowded classes, frequently with no clinical experience or hands-on instruction. Students or apprentices went on to practice these skills that they learned without questioning reason (Halperin et al. 2010; Ludmerer 2011).

Accordingly, medical educators recognized a critical need for reform in the medical education system. Between January 1909 and April 1910, Abraham Flexner, a high school principal supported by the Carnegie Foundation, visited all 155 medical colleges that were at the time in operation across the United States and Canada. After visiting, he published in 1910 a report that was primarily addressed to the public, in which he outlined how to reconstruct medical education (Cooke et al. 2006; Mitka 2010). When assessing the medical schools, Flexner evaluated them as an educator, and not as a medical practitioner (Duffy 2011). One of the findings of the report was a high number of substandard medical schools; many of them run for profit by private physicians (Mitka 2010). Based on his report, medical schools were assigned to one of three categories. The first group included medical schools that were compared with Johns Hopkins University; the second group included schools that were viewed as substandard, but possible to be salvaged by providing financial assistance and correcting shortcomings; and the third group included medical schools that were thought to be of such poor quality, that their closure was recommended (Duffy 2011).

One of the significant foundational principles that the Flexner Report was built on is that medical knowledge, as opposed to being static, continually grows and evolves (Ludmerer 2011). The report was particularly valuable for three reasons: its very comprehensive nature, the emphasis on science in the practice of medicine, and its appeal to the general public (Cooke et al. 2006) Additionally, the report also established the biomedical model of medical training as the gold standard (Duffy 2011). Flexner's report was disapproving of the traditional lecture-driven curriculum and, instead, it promoted active learning through various hands-on methods, such as case studies and bedside teaching. Flexner also supported teaching students to become lifelong learners, using research and medical literature to support their decision-making and supplement their medical knowledge and skills. This concept, described in a 100-year-old report, is now one of the Accreditation Council for Graduate Medical Education core competencies by which we measure medical residents' success: practice-based learning and improvement.

The Flexner Report is considered iconic. It accelerated leaders in medical education to pursue a new mission and provide young doctors with the necessary tools to help them mature into problem-solvers and critical thinkers. Less emphasis was placed on the traditional pedagogy of the nineteenth century, and so began an era marked with students as active participants, through clinical clerkships, small-group teaching sessions, and self-directed education, rather than moving through medical school as passive observers (Ludmerer 2011; Halperin et al. 2010).

43.3 Today's Medical Education: How Far Have We Come?

For most of the twentieth century, lectures have provided an efficient way for teachers to transfer knowledge to learners. However, numerous studies have shown that learners do not engage or retain information with the traditional PowerPoint format as compared to active learning educational sessions. Educators are increasingly looking for ways to move away from the podium and incorporate more active learning into their curriculum. The techniques described below allow for a more interactive learning experience. Improving the quality of educational programs in graduate medical education requires the utilization of all instructional methods. Recognizing the benefits of the different educational techniques and utilizing various options throughout an educational conference may better convey educational content and increase retention. With time constraints, didactic lectures are still the optimal way to communicate information to a broad audience. Active learning sessions, such as the flipped classroom, team-based learning, small-group sessions, case-based learning, simulation, and role-playing/standardized patients, are all tools used in both undergraduate and graduate medical education programs to enhance the educational experience.

43.3.1 Flipped Classrooms

As part of the shift from the traditional, lecture-based teaching methodology in medical education to approaches that involve more active participation from students, the flipped classroom (FC) model has received increasing attention (Chen et al. 2017). The first FC experience was reported in late 2012 and subsequently became increasingly popular worldwide (Young et al. 2014; Chen et al. 2017) in both undergraduate and graduate education (Young et al. 2014). The FC model offers educational materials and provides knowledge independently before the classroom meeting (Kraut et al. 2019). While in traditional didactic methods content is delivered synchronously in a classroom, and students may complete assignments at home, in FC the content is delivered at home, before the lecture, asynchronously, and students can later work on assignments in a classroom setting and receive support from an expert in the field (Young et al. 2014). If the asynchronous learning activity that precedes the classroom meeting cannot be accomplished due to time constraints, the facilitator can use 5 to 10 minutes of the session to review the educational materials for the class (Riddell et al. 2017). Class sessions can then focus on simulation, case-based learning, or small-group learning with an instructor who facilitates this process. This format allows learners to apply their acquired knowledge in collaboration with peers while receiving immediate feedback and guidance from teachers (Sait et al. 2017; Wray et al. 2020). The FC approach was reported to significantly enhance student learning in health professions, as compared to the traditional teaching methods (Hew and Lo 2018).

43.3.2 Team-Based Learning

Team-based learning (TBL) was initially introduced for business courses and was later adopted by medical schools, for pre-clinical courses, internships, residency programs, and continuing medical education (Faezi et al. 2018). TBL focuses on applying knowledge in a collaborative environment, improves creative thinking, and was shown to particularly benefit students with a lower academic performance during their final exams (Koles et al. 2005; Inuwa et al. 2012; Faezi et al. 2018). The seven core elements of TBL are team formation, readiness assurance, immediate feedback, sequencing of in-class problem-solving, team application exercises, incentive structure, and peer evaluation (Farland et al. 2013). Building on the flipped classroom model, this approach allows for further advancement in several Accreditation Council for Graduate Medical Education core competencies, such as medical knowledge, professionalism, and interpersonal and communication skills. A TBL session has three segments:

- Review of assigned materials before the knowledge acquired over meeting sessions.
- Followed by working in teams to address the same questions (group readiness assurance test (GRAT)).
- And finally, discussion of answers amongst teams. As there is no single correct answer, mimicking real-life clinical scenarios, this segment encourages peer learning, challenges, and constructive debate.

TBL, similar to flipped classrooms, allows for immediate feedback from instructors as well as peers (Poeppelman et al. 2016). Both flipped classrooms and TBL were shown to increase knowledge. However, they also require increased learner workload and, therefore, they have received mixed reviews over time (Fatmi et al. 2013).

43.3.3 Simulation

Simulation-based medical education refers to any activity that uses simulation aides, usually simplified, to replicate complex clinical scenarios artificially. Simulation represents an alternative to real-life patients (Al-Elq 2010; Datta et al. 2012). The introduction of simulation activities into medical education dates back to the early 1990s (Thomson 1990; Luketich 1990). They subsequently became increasingly used in medical education activities (Zhang et al. 2015). Simulation can be used to teach learners new or challenging procedures. Simulation technologies, including task trainers, computer-based programs, and mannequins, serve as tools for hands-on learning. These activities can be incorporated into education programs as small-group sessions with a single station on a regular basis or as a simulation conference with multiple trainers. Learners can be taught at skills-based stations, while senior faculty or program leadership can assist or observe the simulation, and this allows them to assess the activities in addition to opportunities to provide immediate and meaningful feedback. Interdisciplinary simulation drills have been filmed for immediate debriefing and feedback as an aide for improving communication skills and team-building activities (Haas et al. 2019).

43.3.4 Case-Based and Problem-Based Learning

Problem-based learning (PBL) was initially developed over 50 years ago, to teach basic sciences, and was later adopted as an effective method for delivering postgraduate and continuing medical education (Al-Azri and Ratnapalan 2014). This educational strategy provides opportunities to solve real problems and scenarios (Pilcher 2014). In PBL, learners are given a case as a focal point to direct their learning objectives and to apply their knowledge to solve the case. Learners work in small groups, and the instructor serves as a passive facilitator, guiding student thinking and answering questions as needed. The goal of PBL is to promote self-directed learning through critical thinking and problem-solving skills. Studies have shown that incorporating PBL into learning activities improved these skills, enhanced the ability to analyze complex datasets, and fostered teamwork (Pilcher 2014; Srinivasan et al. 2007).

Case-based learning (CBL), also known as case study teaching or case method learning, has different definitions depending on the discipline

(Thistlethwaite et al. 2012). In the health professions, these CBL activities usually involve patient cases, in which patients may be real, simulated, virtual, or text-based (Thistlethwaite et al. 2012). CBL is a form of PBL, and while the differences between the two are not always obvious, CBL uses a more structured and guided approach than PBL (Thistlethwaite et al. 2012; Suliman et al. 2019). CBL is sometimes called "guided inquiry approach" (Srinivasan et al. 2007). Group discussion is directed toward the intended learning objectives set by an active facilitator. CBL enhances clinical skills and is a more structured learning modality with assigned reading and more guidance from an instructor (Carder et al. 2001; Hansen et al. 2005). CBL was shown to significantly improve knowledge in reproductive endocrinology and infertility, and it was possible to successfully integrate it into residency training (Goldman et al. 2017).

43.4 Disruption Brings Disarray, Turmoil, and Change

The COVID-19 pandemic has challenged the education of medical students, residents, and fellows, by interrupting their education in unprecedented ways, and placed them at risk for gaps in their education (Hoopes et al. 2020). For residents, one of the dangers was the possibility of surgical-skill decay, which is defined as the loss of acquired skills after a period of inactivity (Hoopes et al. 2020).

Since the foundation of ACGME in 1981 (Sofka 2014), and the establishment of organized medical education, including the formation of the American Board of Medical Specialties (ABMS) in 1933 (Nelson 2014), nothing that we can think of has disrupted graduate education so abruptly and to such an extent. Although there is a wide variety of instructional methods available, residency programs typically deliver education primarily through lecture-based strategies (Wray et al. 2020). The pandemic has disrupted medical education and has forced residency programs to re-think the delivery of their educational activities. The impact of these limitations in graduate medical education is predicted to be most strongly felt by medical residents and fellows in their ultimate or penultimate years during their respective training programs (Potts 2020).

Disruption can be inconvenient and cause hardship, but it can also spur innovation and positive change in society. How disruption in the wake of the pandemic caused by COVID-19 has changed the world of GME is a case study example. On January 30, 2020, the WHO declared COVID-19 a global public health emergency (Sohrabi et al. 2020), and on March 4, the governor of California declared a state of emergency (Newsom 2020). On March 6, Stanford was the first major educational institution to go entirely online. On March 13, the President of the United States declared a national emergency, and on March 15, the governor of NY closed schools and all non-essential businesses in the state; subsequently, the ACGME suspended all activities and allowed institutions to declare a Level III Pandemic Emergency (ACGME 2020a; Brooks 2020). This declaration allowed residency programs to suspend all Common Program Requirements (CPR), yet it contained several main tenants (ACGME 2020b):

- Adequate resources and training: all residents and fellows must be trained in and be provided with appropriate infection protection for the clinical setting and the situation.
- Adequate supervision: any resident or fellow who provides care to patients will do so under appropriate supervision for the clinical circumstance and the level of education and their experience.
- Working hour requirements: the ACGME Common Program Requirements addressing work hours remained unchanged.
- Fellows working in a core specialty: fellows in ACGME-accredited programs were allowed to work in their core specialties if several requirements were met. These include being ABMS or AOA board-eligible or board-certified, being appointed to the medical staff of the institution, and maintaining the time they spent on their core specialty at 20% or

less of their annual time for any academic year.

The overall objective of the ACGME was to ensure the safety and supervision of the residents, yet allow residents to care for patients safely. Customarily, ACGME, in general, seeks, through its CPR, to achieve a balance between academic education and clinical service (ACGME). In other words, resident physicians are not to be viewed as cheap labor. In many instances, residents are less expensive to have as employees than House Officers, NPs, and PAs, which command higher salaries and only work about 40 hours a week. Residents can work up to 80 hours per week on average and are usually paid a lower overall salary. Part of having residents as a workforce includes the requirement of protected time each week for didactics. However, with the declaration of a Stage III Pandemic Emergency, ACGME allowed clinical service to be favored over academics in order to attend to the needs of the overwhelming number of patients. As a result, all academic activities, including Grand Rounds, Resident Lectures, Simulation Sessions, Mortality and Morbidity Conferences, and Bedside Teaching Rounds, were canceled overnight.

The Stage III declaration is only allowed to last for 30 days unless requested by the Designated Institutional Official. At the end of this time, programs were required to follow all CPRs. However, during this short time, the dynamics of daily life in the hospital has changed. What educators had taken for granted was no longer possible. The most significant change was gathering in groups. No longer could people gather in groups for a lecture or even Grand Rounds. No matter the educational construct, e.g., TBL, Case Discussions, and Simulations, the main prerequisite was to gather together to participate in the learning activity.

Residency Programs now had a new issue with which to contend. The major question became how to deliver educational activities without meeting in person. Programs turned to technology. Coincidentally, some of the medical schools have been beset with this problem for many years. Some schools have remote learning sites but were responsible for ensuring comparability of education no matter the location of the student (Parisky et al. 2009; Kahn et al. 2014; Winn et al. 2015). For them, it is not practical to bring students back to the main campus for an in-person lecture. Therefore, although the technology has been available for several years, residency programs rarely, if ever, took advantage of this modality. Remote-distance learning technology has mainly consisted of streaming a lecture that allows for live lectures but also allows the learners to ask questions of the lecturer immediately.

The platforms that are used for distance learning provide several advantages. One of the most obvious advantages is that web-based education overcomes physical distances. It provides some convenience for the lecturer, who can be anywhere to deliver the lecture, including home, office, or a different state or country. Learners also have the opportunity to attend, irrespective of their location. There is no need to travel, which sometimes may be a reason why a lecturer might not be able to deliver a lecture. The lecturer does not have to "dress up" for their lecture. There is the added convenience for the lecturer to read from a script. The lecturer may be less nervous when presenting online as compared to when they would present in front of a live audience. Finally, the lecturer may be in almost any setting and be as comfortable as possible (Lahaie 2007; Cook 2007).

Some of the disadvantages include loss of audience contact. Sometimes giving a lecture on a web-based platform may make the lecturer feel as if they are talking to themselves during the video conference. It is very challenging to gauge the audience's nonverbal behavior, receptiveness, mood, and interest. The interaction becomes less individualized. Experienced lecturers can shift gears when sensing how the audience is reacting to the didactic material. They can make the material more interesting if sensing boredom or spend more time on a concept if sensing confusion. The art of teaching, in some ways, is blunted. Online presentations are sometimes plagued by poor internet connection and technical malfunction, or by the speaker's lack of familiarity with the program platform, or features may be problematic. Finally, the audience may feel inhibited to ask questions (Cook 2007; Lahaie 2007).

This crisis also changed resident schedules. Non-urgent services were temporarily suspended, and all clinics and elective surgeries were stopped. As a result, most resident schedules changed to a rotating schedule of day or night shifts and a back-up team in case a resident became sick. Rotations were changed every 3–7 days to decrease the exposure of the residents and allow rest and recovery during these times of increased stress. Residents from other services, such as Dermatology, Orthopedics, Surgery, and Physical Medicine and Rehabilitation, were deployed to make-shift ICU units, newly created Medicine wards, and to the Emergency Room to help with the sudden and ever-increasing influx of patients diagnosed with coronavirus.

With the residents subdivided into groups, all with different schedules, an effort to have protected time became a logistical nightmare, if not an impossibility. Furthermore, increased numbers of faculty were pressed into service as faculty time became stretched as well. Lectures were given sporadically online or not at offered all, and the education mission appeared to have collapsed due to overwhelming clinical demands.

43.5 Disruption Brings Innovation and Adaptation

Web-based delivery of didactics was an adaptation to the pandemic. The next challenge was to determine the best option for the time to deliver these educational activities. As part of that challenge, one of the considerations was how to have faculty participation when faculty time has also become so limited and identifying the individual faculty members that will give the respective presentations. Although there have been several instances of Innovation and Adaptation throughout the country, during these trying times, we want to highlight a case study of one of them, the Education for All (EFA) initiative.

As of 2019, New York had the highest concentration of residency programs in the United States (544) and, as a result, the highest concentration of residents (12%) in the country (AMA). The state with the next largest number of programs, California, has 431 residency programs as of 2019 (AMA). Similarly, there are 35 Ob/Gyn residency programs in New York State. Of these, 31 are located closely together in NYC, Westchester, and Long Island.

Northwell is a hospital system consisting of 23 major hospitals and 4 independent Ob/Gyn residency programs located in Staten Island (1), Manhattan (1), and Long Island (2; 1 in Nassau and 1 in Suffolk County). All four of these programs were struggling with the issues outlined above. The programs decided to share didactics and links between programs. Didactics were given in each program on different days. It alleviated the issue of when to perform those activities. This process started in mid-March 2020. After this initiative had a successful launch, a decision was made to expand the model to the other university programs in NYC and Long Island. An e-mail was subsequently sent out to the other program directors, and their response was immediate; and all of them immediately agreed to participate and to share didactics in their respective hospitals and programs. Subsequently, e-mails were sent out daily with lecture schedules and links to live lectures. A few days later, a decision was made to expand this initiative to all of NY State. An e-mail was sent out to the remainder of the programs, and they all agreed to participate. By word of mouth, information spread, and programs in neighboring states, including Connecticut and Vermont, asked to be included in this initiative. Soon after that, all programs in seven states from the Northeast were apprised of the program.

However, meeting the demand for e-mails and communications that were involved in this process soon became overwhelming for the organizers. Additionally, for residents who were unable to attend a live-streamed lecture, it became impossible to see that lecture at a later date. This is how the Education For All (EFA) initiative was born (All 2020). It involved creating a website to host the lecture schedule with links to live lectures and to also act as a repository for previously recorded lectures. The creation of this platform alleviated the crush of e-mails and constant changes to the lecture schedule that was being sent out. It also allowed the viewing of missed lectures and provided opportunities for residents on night rotations to partake in didactics.

The purpose of EFA was threefold: to support residency didactics across residency programs through sharing of resources, to promote faculty development, and to promote networking among faculty and residents across residency programs.

The Chair of the National Ob/Gyn Education Association, Council on Resident Education in Obstetrics and Gynecology (CREOG), learned about this program and encouraged and promoted national publicity for it. As a result, our EFA initiative became known across the country overnight. Website analytics revealed an increase of >4000% in the visits to the site, with over 6300-page visits by mid-April, when the site was only about 3 weeks old. Analysis of the data also revealed that knowledge about the site had spread internationally and to other continents, including Europe, South America, Africa, and Asia.

EFA accomplished several objectives. It helped alleviate the issue of not having enough faculty to deliver didactics. As a result, live lectures are currently available every morning and afternoon and are also stored on the website for later viewing. It helped promote those "unsung" heroes of every department; those teachers who are not getting NIH grants or are not on the cover of medical journals with the latest research, but who can really teach and teach well; and the teachers that are perennial candidates for the departmental teaching award; every department has one or two. Finally, it helped promote junior faculty by giving them exposure and an opportunity to add activities to their résumés.

Crowdsourcing has been one of the hallmarks of the twenty-first century. YouTube and Wikipedia are prime examples of this trend. These websites utilize the "wisdom" of the crowd. Examples of crowdsourcing can be found in medical education as well, in attempts to solve specific problems (Blackwell et al. 2016; Dai et al. 2017). A prevailing thought in medicine for decades is that the amount of research is proportional to the effectiveness of a teacher. However, actual teaching effectiveness is not measured by the number of publications that a faculty member may have. EFA utilizes the wisdom of all faculty across the United States. It helps smaller programs that may not have enough resources in general (smaller faculty size), and it helps more extensive programs by alleviating the burden of lecture preparation in this stressful time. EFA has helped medical education in the GME space to move from the "sage on the stage" to the democratization of education.

43.6 Medical Education in the Post-COVID-19 Era

Historically, UME and GME have been slow in adopting changes. Traditionally, incorporating new technology into medical education has been a slow and tedious process. For example, despite the availability of resources, few medical schools are using virtual anatomy with three-dimensional imaging, computed tomography scans, and ultrasound. During the COVID-19 pandemic, some medical schools removed students from their clinical rotation, and changes were proposed to the residency interview process. Telemedicine has become established as a viable approach to provide healthcare (Smith et al. 2020). For example, during COVID-19, chatbots provided valuable information through technology (Smith et al. 2020; Priya 2020). In particular, telemedicine reduces exposure to other patients with acute disease, and, as relevantly expressed in a recent article, the only infection that one can catch during a telemedicine consultation is "a computer virus" (Portnoy et al. 2020). In the United States, during the COVID-19 pandemic, telemedicine played a transformative role in healthcare during three phases: to help with stay-at-home patient care, to address the initial COVID-19 hospital surge, and to assist during post-pandemic recovery (Wosik et al. 2020). Virtual patients and simulation replaced the clinical rotation for some students.

The question is whether these UME students will find themselves at a disadvantage and whether their educational experience may be diminished. Residents in a surgical specialty who had significantly less surgical experience secondary to regulations limiting elective surgery and patient may develop a fear of hospitals and medical facilities. Another conundrum is whether virtual reality or augmented reality could replace resident experiences and whether the simulation is sufficient. Historically, simulators have been a costlier option, and they are not available at every teaching institution (Siddiqui and Aslanian 2020). Using simulation before the surgical procedure can improve performance during the case, but it is questionable whether it can also replace the experience with a patient. Would a patient feel comfortable seeing physicians that had the majority of their training performed using virtual situations? Pilots use simulations for many hours before they are allowed to fly a plane, and arguably, it is critical to adopt this to medical education. The criticism for this approach in medicine has been that students will not develop a connection to their patients and will not fully appreciate the variations of the human body. The solution most likely lies somewhere in the middle (Prober and Heath 2012; Moran et al. 2018; Guze 2015).

43.7 Conclusion

The Flexner Report of 1910 prompted a significant reform in medical education. Since it was written, society has seen a boom in medical advancements, the expansion of healthcare systems, the introduction of ACGME Work Hour Requirements, and the advent of the "World Wide Web." Flexner emphasized the importance of active learning in medical training, yet students and trainees often still sit in lecture halls, being talked to while information is projected on PowerPoint slides. Medical training covers a finite period, and it is incumbent on medical educators to maximize this precious time with more productive educational opportunities. In this tech-savvy era of TED Talks, YouTube, podcasts, social media, and innumerous iPhone or Android medical apps, the possibilities for revolutionizing education programs are exciting and endless. The COVID-19 pandemic has impacted humanity in virtually every aspect of livelihood. It has overwhelmed healthcare systems, devastated global economies, and led to surging unemployment and poverty. It is highly likely to create a shadow pandemic of emotional distress and mental health as well. It has impacted education by closing everything from daycare centers to medical schools, leaving parents to homeschool their children, and pushing educators to innovate with the use of available technology to create a new curriculum. The question remains whether we can continue these innovative changes to medical education curricula with the use of technology and increased active learning into the post-COVID world, and the challenge will be to find strategies to do this in an ideal way.

References

ACGME (2020a) ACGME response to the coronavirus (COVID-19). Available at: https://acgme.org/Newsroom/Newsroom-Details/ArticleID/10111/ACGME-Response-to-the-Coronavirus-COVID-19. Last accessed on 7 June 2020

ACGME (2020b) Stage 3: Pandemic emergency status guidance. Available at: https://acgme.org/COVID-19/Three-Stages-of-GME-During-the-COVID-19-Pandemic/Stage-3-Pandemic-Emergency-Status-Guidance. Last accessed on 9 June 2020

ACGME Common Program Requirements (2020). Available at: https://www.acgme.org/What-We-Do/Accreditation/Common-Program-Requirements. Last accessed on 10 June 2020

Al-Azri H, Ratnapalan S (2014) Problem-based learning in continuing medical education: review of randomized controlled trials. Can Fam Physician 60(2):157–165

Al-Elq AH (2010) Simulation-based medical teaching and learning. J Family Community Med 17(1):35–40. https://doi.org/10.4103/1319-1683.68787

All EF (2020) Education for all. Available at: www.efaobgyn.com

AMA Most popular medical residency programs in California (2020). Available at: https://www.ama-assn.org/residents-students/residency/most-popular-medical-residency-programs-california. Last accessed on 10 June 2020

AMA Most popular medical residency programs in New York (2020). Available at: https://www.ama-assn.org/residents-students/residency/most-popular-medical-residency-programs-new-york. Last accessed on 7 June 2020

Arun Krishnan R, Elizabeth Thomas R, Sukumaran A, Paul JK, Vasudevan DM (2020) COVID-19: current trends in Invitro diagnostics. Indian J Clin Biochem 35(3):285–289. https://doi.org/10.1007/s12291-020-00906-5

Blackwell KA, Travis MJ, Arbuckle MR, Ross DA (2016) Crowdsourcing medical education. Med Educ 50(5):576. https://doi.org/10.1111/medu.13010

Blaisdell FW (1988) Medical advances during the civil war. Arch Surg 123(9):1045–1050. https://doi.org/10.1001/archsurg.1988.01400330021001

Brooks OT (2020) What if they threw a pandemic and nobody came? J Natl Med Assoc 112(2):115–116. https://doi.org/10.1016/j.jnma.2020.04.005

Carder L, Willingham P, Bibb D (2001) Case-based, problem-based learning: information literacy for the real world. Res Strategies 18(3):181–190. https://doi.org/10.1016/S0734-3310(02)00087-3

Chang DG, Park JB, Baek GH, Kim HJ, Bosco A, Hey HWD, Lee CK (2020) The impact of COVID-19 pandemic on orthopaedic resident education: a nationwide survey study in South Korea. Int Orthop:1–8. https://doi.org/10.1007/s00264-020-04714-7

Chatziralli I, Ventura CV, Touhami S, Reynolds R, Nassisi M, Weinberg T, Pakzad-Vaezi K, Anaya D, Mustapha M, Plant A, Yuan M, Loewenstein A (2020) Transforming ophthalmic education into virtual learning during COVID-19 pandemic: a global perspective. Eye (Lond):1–8. https://doi.org/10.1038/s41433-020-1080-0

Chen F, Lui AM, Martinelli SM (2017) A systematic review of the effectiveness of flipped classrooms in medical education. Med Educ 51(6):585–597. https://doi.org/10.1111/medu.13272

Cook DA (2007) Web-based learning: pros, cons and controversies. Clin Med (Lond) 7(1):37–42. https://doi.org/10.7861/clinmedicine.7-1-37

Cooke M, Irby DM, Sullivan W, Ludmerer KM (2006) American medical education 100 years after the Flexner report. N Engl J Med 355(13):1339–1344. https://doi.org/10.1056/NEJMra055445

Dai JC, Lendvay TS, Sorensen MD (2017) Crowdsourcing in surgical skills acquisition: a developing Technology in Surgical Education. J Grad Med Educ 9(6):697–705. https://doi.org/10.4300/jgme-d-17-00322.1

Datta R, Upadhyay K, Jaideep C (2012) Simulation and its role in medical education. Med J Armed Forces India 68(2):167–172. https://doi.org/10.1016/s0377-1237(12)60040-9

Duffy TP (2011) The Flexner report--100 years later. Yale J Biol Med 84(3):269–276

Faezi ST, Moradi K, Ghafar Rahimi Amin A, Akhlaghi M, Keshmiri F (2018) The effects of team-based learning on learning outcomes in a course of rheumatology. J Adv Med Educ Prof 6(1):22–30

Farland MZ, Sicat BL, Franks AS, Pater KS, Medina MS, Persky AM (2013) Best practices for implementing team-based learning in pharmacy education. Am J Pharm Educ 77(8):177. https://doi.org/10.5688/ajpe778177

Fatmi M, Hartling L, Hillier T, Campbell S, Oswald AE (2013) The effectiveness of team-based learning on learning outcomes in health professions education: BEME guide no. 30. Med Teach 35(12):e1608–e1624. https://doi.org/10.3109/0142159x.2013.849802

Go B, Rajasekaran K (2020) Effect of COVID-19 in selecting otolaryngology as a specialty. Head Neck 42(7):1409–1410. https://doi.org/10.1002/hed.26251

Goldman KN, Tiegs AW, Uquillas K, Nachtigall M, Fino ME, Winkel AF, Lerner V (2017) Interactive case-based learning improves resident knowledge and confidence in reproductive endocrinology and infertility(). Gynecol Endocrinol 33(6):496–499. https://doi.org/10.1080/09513590.2017.1290075

Guze PA (2015) Using technology to meet the challenges of medical education. Trans Am Clin Climatol Assoc 126:260–270

Haas MRC, Munzer BW, Santen SA, Hopson LR, Haas NL, Overbeek D, Peterson WJ, Cranford JA, Huang RD (2019) #DidacticsRevolution: applying Kotter's 8-step change management model to residency didactics. West J Emerg Med 21(1):65–70. https://doi.org/10.5811/westjem.2019.11.44510

Halperin EC, Perman JA, Wilson EA (2010) Abraham Flexner of Kentucky, his report, medical education in the United States and Canada, and the historical questions raised by the report. Acad Med 85(2):203–210. https://doi.org/10.1097/ACM.0b013e3181c88506

Hansen WF, Ferguson KJ, Sipe CS, Sorosky J (2005) Attitudes of faculty and students toward case-based learning in the third-year obstetrics and gynecology clerkship. Am J Obstet Gynecol 192(2):644–647. https://doi.org/10.1016/j.ajog.2004.10.595

Hew KF, Lo CK (2018) Flipped classroom improves student learning in health professions education: a meta-analysis. BMC Med Educ 18(1):38. https://doi.org/10.1186/s12909-018-1144-z

Hoopes S, Pham T, Lindo FM, Antosh DD (2020) Home surgical skill training resources for obstetrics and gynecology trainees during a pandemic. Obstet Gynecol. https://doi.org/10.1097/aog.0000000000003931

Inuwa IM, Al-Rawahy M, Roychoudhry S, Taranikanti V (2012) Implementing a modified team-based learning strategy in the first phase of an outcome-based curriculum--challenges and prospects. Med Teach 34(7):e492–e499. https://doi.org/10.3109/0142159x.2012.668633

Kahn MJ, Maurer R, Wartman SA, Sachs BP (2014) A case for change: disruption in academic medicine. Acad Med 89(9):1216–1219. https://doi.org/10.1097/acm.0000000000000418

Koles P, Nelson S, Stolfi A, Parmelee D, Destephen D (2005) Active learning in a year 2 pathology curriculum. Med Educ 39(10):1045–1055. https://doi.org/10.1111/j.1365-2929.2005.02248.x

Kraut AS, Omron R, Caretta-Weyer H, Jordan J, Manthey D, Wolf SJ, Yarris LM, Johnson S, Kornegay J (2019) The flipped classroom: a critical appraisal. West J

Emerg Med 20(3):527–536. https://doi.org/10.5811/westjem.2019.2.40979

Lahaie U (2007) Web-based instruction: getting faculty onboard. J Prof Nurs 23(6):335–342. https://doi.org/10.1016/j.profnurs.2007.01.024

Ludmerer KM (2011) Abraham Flexner and medical education. Perspect Biol Med 54:8–16

Luketich GF (1990) A clinical simulations laboratory ten years later. J Biocommun 17(3):26–29

Mitka M (2010) The Flexner report at the century mark: a wake-up call for reforming medical education. JAMA 303(15):1465–1466. https://doi.org/10.1001/jama.303.15.1465

Moran J, Briscoe G, Peglow S (2018) Current Technology in Advancing Medical Education: perspectives for learning and providing care. Acad Psychiatry 42(6):796–799. https://doi.org/10.1007/s40596-018-0946-y

Nelson RN (2014) Demystifying maintenance of certification. Ann Emerg Med 63(4):467–470. https://doi.org/10.1016/j.annemergmed.2013.01.025

Newsom OoGG (2020) Governor newsom declares state of emergency to help state prepare for broader spread of COVID-19. Available at: https://www.gov.ca.gov/2020/03/04/governor-newsom-declares-state-of-emergency-to-help-state-prepare-for-broader-spread-of-covid-19/. Last accessed on 9 June 2020

Parisky A, Ortiz T, McCann K, Hoffmann E, Boulay R (2009) How top US medical schools are using distance learning resources: an exploratory study of four institutions. In: Bastiaens T et al (eds) Proceedings of world conference on E-Learning in corporate, government, healthcare, and higher education; Chesapeake, VA Association for the Advancement of Computing in Education; 2009 pp 2930–2935

Pilcher J (2014) Problem-based learning in the NICU. Neonatal Netw 33(4):221–224. https://doi.org/10.1891/0730-0832.33.4.221

Poeppelman RS, Liebert CA, Vegas DB, Germann CA, Volerman A (2016) A narrative review and novel framework for application of team-based learning in graduate medical education. J Grad Med Educ 8(4):510–517. https://doi.org/10.4300/jgme-d-15-00516.1

Portnoy J, Waller M, Elliott T (2020) Telemedicine in the era of COVID-19. J Allergy Clin Immunol Pract 8(5):1489–1491. https://doi.org/10.1016/j.jaip.2020.03.008

Potts JR 3rd (2020) Residency and fellowship program accreditation: effects of the novel coronavirus (COVID-19) pandemic. J Am Coll Surg 230(6):1094–1097. https://doi.org/10.1016/j.jamcollsurg.2020.03.026

Priya S (2020) Priya S. Singapore government launches COVID-19 chatbot, https://www.opengovasia.com/singapore-government-launches-covid-19-chatbot/ . Last accessed 8 June 2020

Prober CG, Heath C (2012) Lecture halls without lectures--a proposal for medical education. N Engl J Med 366(18):1657–1659. https://doi.org/10.1056/NEJMp1202451

Riddell J, Jhun P, Fung CC, Comes J, Sawtelle S, Tabatabai R, Joseph D, Shoenberger J, Chen E, Fee C, Swadron SP (2017) Does the flipped classroom improve learning in graduate medical education? J Grad Med Educ 9(4):491–496. https://doi.org/10.4300/jgme-d-16-00817.1

Sait MS, Siddiqui Z, Ashraf Y (2017) Advances in medical education and practice: student perceptions of the flipped classroom. Adv Med Educ Pract 8:317–320. https://doi.org/10.2147/amep.S133328

Siddiqui UD, Aslanian HR (2020) The new virtual reality: advanced endoscopy education in the COVID-19 era. Dig Dis Sci:1–4. https://doi.org/10.1007/s10620-020-06386-0

Smith AC, Thomas E, Snoswell CL, Haydon H, Mehrotra A, Clemensen J, Caffery LJ (2020) Telehealth for global emergencies: implications for coronavirus disease 2019 (COVID-19). J Telemed Telecare:1357633x20916567. https://doi.org/10.1177/1357633x20916567

Sofka CM (2014) Developments and innovations in resident and fellowship education: review article. HSS J 10(3):225–229. https://doi.org/10.1007/s11420-014-9396-6

Sohrabi C, Alsafi Z, O'Neill N, Khan M, Kerwan A, Al-Jabir A, Iosifidis C, Agha R (2020) World Health Organization declares global emergency: a review of the 2019 novel coronavirus (COVID-19). Int J Surg 76:71–76. https://doi.org/10.1016/j.ijsu.2020.02.034

Srinivasan M, Wilkes M, Stevenson F, Nguyen T, Slavin S (2007) Comparing problem-based learning with case-based learning: effects of a major curricular shift at two institutions. Acad Med 82(1):74–82. https://doi.org/10.1097/01.ACM.0000249963.93776.aa

Suliman S, Al-Mohammed A, Al Mohanadi D, Karim H, Elbuzidi A, Mubasher M, Rahil A (2019) It is all about patients' stories: case-based learning in residents' education. Qatar Med J 2019(3):17. https://doi.org/10.5339/qmj.2019.17

Thistlethwaite JE, Davies D, Ekeocha S, Kidd JM, MacDougall C, Matthews P, Purkis J, Clay D (2012) The effectiveness of case-based learning in health professional education. A BEME systematic review: BEME guide no. 23. Med Teach 34(6):e421–e444. https://doi.org/10.3109/0142159x.2012.680939

Thomson AN (1990) Case specificity of performance with simulated patients. N Z Med J 103(895):372–374

Virarkar M, Jensen C, Javadi S, Saleh M, Bhosale PR (2020) Radiology education amid COVID-19 pandemic and possible solutions. J Comput Assist Tomogr. https://doi.org/10.1097/rct.0000000000001061

Wang DC, Emmad Q, Jiayi TM, Eric B (2020) COVID-19 pandemic on Canadian radiology residency education-impact and solutions: university of Toronto perspective. Can Assoc Radiol J:846537120940299. https://doi.org/10.1177/0846537120940299

Winn S, McKeown P, Lotfipour S, Maguire GA, Youm JH, Wiechmann W, Fox JC (2015) Remote, synchronous, hands-on ultrasound education. Telemed J E Health 21(7):593–597. https://doi.org/10.1089/tmj.2014.0050

Wosik J, Fudim M, Cameron B, Gellad ZF, Cho A, Phinney D, Curtis S, Roman M, Poon EG, Ferranti J, Katz JN, Tcheng J (2020) Telehealth transformation: COVID-19 and the rise of virtual care. J Am Med Inform Assoc 27(6):957–962. https://doi.org/10.1093/jamia/ocaa067

Wray A, Wolff M, Boysen-Osborn M, Wiechmann W, Paradise S, Runcie E, Sudario G, Toohey S (2020) Not another boring resident didactic conference. AEM Educ Train 4(Suppl 1):S113–s121. https://doi.org/10.1002/aet2.10367

Young TP, Bailey CJ, Guptill M, Thorp AW, Thomas TL (2014) The flipped classroom: a modality for mixed asynchronous and synchronous learning in a residency program. West J Emerg Med 15(7):938–944. https://doi.org/10.5811/westjem.2014.10.23515

Zhang MY, Cheng X, Xu AD, Luo LP, Yang X (2015) Clinical simulation training improves the clinical performance of Chinese medical students. Med Educ Online 20:28796. https://doi.org/10.3402/meo.v20.28796

Reopening Schools After a Novel Coronavirus Surge

44

Dan Li, Elizabeth Z. Lin, Marie A. Brault, Julie Paquette, Sten H. Vermund, and Krystal J. Godri Pollitt

Abstract

The coronavirus disease 2019 (COVID-19) pandemic shook the world in ways not seen since the pandemic influenza of 1918–1919. As of late August 2020, over 25 million persons had been infected, and we will see the global death toll exceed one million by the end of 2020. Both are minimum estimates. All segments of society have been drastically affected. Schools worldwide have been forced to close due to illness and absenteeism, transmission and risk to vulnerable members of the school community, and community concerns. The decision to reopen school during a pandemic will have a tremendous impact on children's safety, growth, and well-being. Not opening invites social isolation and suboptimal educational experiences, especially for youth whose computing assets and online access are limited and those with special needs. The opening has hazards as well, and the mitigation of these risks is the topic of this chapter. Opening schools requires careful considerations of benefits, risks, and precautions. Guiding principles for safety and strategic application of the principles in each educational niche are critical issues to consider during school reopening. The fundamental principles of disease control involve school-directed initiatives (physical distancing and mask use, hand/face and surface cleansing, administrative controls, engineering controls) and individual-level risk reduction approaches to maximize adherence to new guidelines. The school-initiated "top-down" approaches and the individual-level "bottom-up" approaches must be synergized, as no single method will ensure safety. We discuss how to effectively layer strategies in each educational space to increase safety. Since the vulnerability of children has been heightened during this pandemic crisis, we highlight the special considerations for mental health support that should be considered by schools. The safety principles, disease control strategies, and other critical issues discussed here will serve as a starting point for developing a safe, comprehensive, and feasible reopening plan.

Keywords

Coronavirus · COVID-19 · Education · Pandemic · Quarantine · Reopening · School · Disease control strategies · Safety

D. Li · E. Z. Lin · M. A. Brault · S. H. Vermund (✉) · K. J. Godri Pollitt (✉)
Yale School of Public Health, New Haven, CT, USA
e-mail: sten.vermund@yale.edu; krystal.pollitt@yale.edu

J. Paquette
Office of Facilities, Yale University, New Haven, CT, USA

N. Rezaei (ed.), *Coronavirus Disease - COVID-19*, Advances in Experimental Medicine and Biology 1318, https://doi.org/10.1007/978-3-030-63761-3_44

44.1 Introduction

> "Plans are worthless, but planning is everything."
> (*US President Dwight Eisenhower*)

Given the impact of the coronavirus disease 2019 (COVID-19) pandemic on every facet of society, the quote from the late President Eisenhower speaks to the need to think through operations and be adaptable rigorously. Kindergarten through Grade 12 (K-12) schools were closed in early March across the United States, recapitulating the experience of China, South Korea, Iran, Italy, Spain, Britain, and many other nations. Given the high individual and societal costs of these closures, as well as the adverse impact of school closure on children's growth and mental health, nations have placed a high priority on reopening schools. Planning for school reopening as COVID-19 case numbers have remained high in many US states at the start of the 2020–2021 school year presents challenges. The unique nature of each school must be considered and will require a tailored solution for each geographic zone. This chapter aims to present general guidance that can be adapted by K-12 schools as they develop a safe reopening plan that will consider the needs of the children, their families, and staff at the school. Information provided in this chapter is not a comprehensive guide nor are requirements or standards presented.

COVID-19 is an infectious disease caused by a coronavirus identified in late 2019, the so-called severe acute respiratory syndrome coronavirus 2 (SARS-CoV-2). The SARS-CoV-2 virus is the seventh zoonotic human coronavirus to be identified (Banerjee et al. 2019). Four of these coronaviruses cause approximately one-third of common cold cases.

Much concerning COVID-19 transmission, susceptibility, and treatment remain unknown due to the novelty of the disease. COVID-19 may occur without symptoms or cause symptoms that can range from mild to very severe. The most common symptoms are fever, dry cough, and difficulty breathing. Symptoms in severe cases include chest pain or pressure, shortness of breast, and loss of speech or movement, while milder symptoms are muscle aches, headaches, sore throat, diarrhea, conjunctivitis, rash, and loss of taste or smell. The incubation period ranges from 2 to 14 days, with rare later cases (Oran and Topol 2020 #16).

People in prolonged close contact are at the highest risk of COVID-19 infection related to virus-laden droplets and aerosols expelled through coughing, sneezing, talking, singing, or simply breathing (Fennelly 2020 #3). Respiratory aerosols are characterized by a diameter of fewer than 5 micrometers (μm), while droplets are larger (Fennelly 2020 #57). Evidence strongly suggests that the COVID-19 virus can be transmitted through fomites and contact with contaminated surfaces (Santarpia et al. 2020 #22). A growing body of epidemiological studies supports aerosol transmission (Fennelly 2020 #3).

Approximately 15 to 45% of SARS-CoV-2 infections are asymptomatic cases (Nishiura et al. 2020), though one review cites 40–45% to be a better estimate (Oran and Topol 2020). Individuals infected before being symptomatic are termed "pre-symptomatic." They may harbor high viral loads before they feel ill, potentially transmitting viruses efficiently. The role that asymptomatic cases play in the transmission of SARS-CoV-2 is unclear, given their typically lower viral loads (Vermund and Pitzer 2020). One may speculate that lower efficiency of transmission per encounter could still add up to a high transmission burden due to more encounters taking place since an asymptomatic person is less likely to stay-in-residence compared to a person who feels ill.

Currently, there are no effective methods to treat or cure COVID-19, though the antiviral drug remdesivir, the monoclonal antibody tocilizumab, and the steroids methylprednisolone or dexamethasone all have benefits for some patients. Patients with mild or moderate symptoms are asked to self-isolate at home for at least 10 days to prevent community spread. Contacts of cases are asked to self-quarantine for at least 14 days since the time of the encounter with the potentially infectious person. Supportive care is given to individuals with severe symptoms. In severe cases, hospitalization, ventilator treatment, and intensive care unit (Connecticut State Department of Education) support (World Health Organization 2020b) might be required as well.

There is no vaccination or prophylactic pharmacological method currently approved (World Health Organization 2020b). Thus, the risk of subsequent waves of infection remains very high, and disease control in schools is dependent on preventative public health practices on multiple fronts. Effective planning of administrative and engineering controls for risk reduction, communication across the school community, and partnerships with local public health organizations are all key to effective disease prevention.

While children can be infected easily with SARS-CoV-2, the incidence of COVID-19 in children is lower compared to adults; children who contract COVID-19 are more likely to remain without symptoms or have moderate symptoms (Dong et al. 2020). Youth deaths are less frequent than they are represented in the general population (Dong et al. 2020). One hypothesis suggests that milder cases of this disease in children may be attributable to lower levels of angiotensin-converting enzyme 2 (ACE2) gene expression in children compared to adults. Rarely, children may present with the multisystem inflammatory syndrome in children (MIS-C) (Henderson et al. 2020). MIS-C is a condition where different organs become inflamed, including the heart, lungs, kidneys, brain, skin, eyes, or gastrointestinal organs. Fortunately, COVID-19-related MIS-C is extremely rare, with an incidence of 2 per 100,000, and it is treatable with symptomatic and anti-inflammatory measures.

Children tend to have closer interactions with each other compared to older age groups and may be less likely to follow guidance on mask use and proper hand hygiene. Adult supervision and regular positive encouragement of safe personal behaviors are required to promote adherence to recommended COVID-19 public health practices. Children are likely to spread the virus to more vulnerable individuals in their close contacts or households (Somekh et al. 2020). Controlling disease outbreaks in children is likely crucial to protect the well-being of the community as a whole.

In this chapter, we introduce key issues in COVID-19 and school reopening. We will discuss guiding principles and the hierarchy of disease control related to school reopening in a respiratory viral pandemic. The application of these safety principles will also be discussed for specific school spaces. Finally, we will highlight special considerations for vulnerable student populations and school mental health support in a pandemic.

44.2 Fundamental Principles of Infectious Disease Control

44.2.1 R_0 and Its Relevance

The basic reproduction number (R_0) measures the transmission potential of an infectious disease. It is the average number of cases produced by a single case of infection in a given population under theoretical conditions. A variation of the R_0 is the R_t, the effective reproduction number, the $R_{0,}$ as seen in real-world conditions. Both are measures of the transmissibility of an infectious agent – in the current case, SARS-CoV-2. R_0 is influenced by biological, social, behavioral, and environmental factors; control measures can reduce $R_{0,}$ and this is estimated by R_t. Our approach is organized by the basic principles of public health to do what we can to keep R_t below one so that transmission likelihood is low. Our goal is to protect the capacity of a K-12 school to continue to function safely and effectively through the academic year (Delamater et al. 2019; Barratt et al. 2010; Lemecha Obsu and Feyissa Balcha 2020).

44.2.2 Disease Control Principles

Disease control can be categorized as approaches that are directed by the school administration (top-down) and those that are dependent on the personal behaviors of students, teachers, and

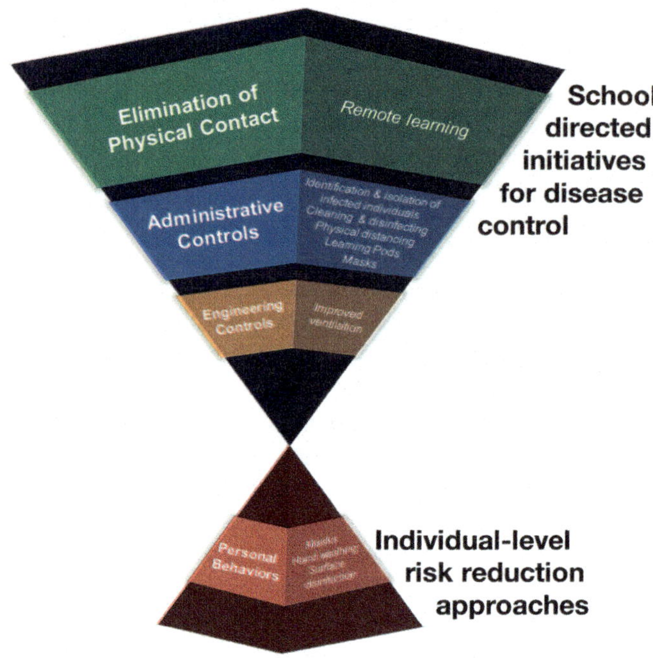

Fig. 44.1 School directed (top-down) and student/teacher (bottom-up) risk reduction strategies to decrease disease transmission

administrative staff in the school (bottom-up) (Fig. 44.1).

Top-down approaches include the elimination of physical contact, administrative controls, and engineering controls, ranked in decreasing order of effectiveness of reducing transmission risks. Bottom-up approaches are focused on individual-level actions such as face masks use and good hand hygiene, cleaning, and disinfection behaviors. No single control measure will eliminate infectious disease risk. Layering multiple strategies from both approaches is critical in creating a safer school environment during the COVID-19 pandemic.

44.3 School-Directed Initiatives for Disease Control

44.3.1 Elimination of Physical Contact

Eliminating physical contact is the most effective method to reduce the risk of disease transmission (National Institute for Occupational Safety and Health 2020a; American Society of Safety Pofessionals 2020). This strategy for disease control shifts learning to a virtual format, which may not be feasible for some schools and presents challenges for successful implementation. Even for schools offering full-time in-person learning, the development of a virtual learning program is essential to accommodate students and teachers who are at high risk for COVID-19 infection due to older age or prior risk profiles, as well as for students or teachers who are in quarantine or isolation. A remote learning platform is also critical in the event of a school closure. In-person courses should be prioritized for academic experiences that cannot be achieved virtually, like hands-on learning experiences. Teachers may benefit from training sessions discussing effective remote learning techniques. Tutorials on their virtual platforms would also likely be useful for most students and their parents. The school should be aware of families where students may be without access to a computer with high-speed internet to participate in a virtual learning environment. A remote option can be a powerful tool that offers students and teachers the safety of learning from home while mitigating the harms of foregoing the in-class academic experience.

44.3.2 Administrative Controls

The risk of infection can be minimized through the implementation of school-initiated administrative controls. These control measures as related to academics, logistics, and school staff/teachers/visitors are summarized in Fig. 44.2.

44.3.2.1 Academics

The academic calendar and class schedule should be adjusted to maximize physical distancing at all times. Staggering arrival and departure times, recess, lunchtimes, and times for locker access for different learning pods is a potential strategy to facilitate physical distancing between students. The time allotted for recess, lunchtime, and locker access can also be extended to give students sufficient time to travel or wash their hands without generating high traffic in the hallways or other spaces. The extra time can also allow staff to provide air ventilation time between different groups of students in a given classroom. Having a hybrid schedule in which students take virtual classes for a portion of the week and in-personal classes for the remainder of time may be useful to reduce occupancy of the spaces. Holidays like fall breaks can be readjusted to avoid travel during the semester.

"Pod" Model

Smaller cohorts or "pods" can be created to limit class sizes (Connecticut State Department of Education 2020; Moroney 2020; North 2020; Cullotta 2020; Moyer 2020). Students within each pod are recommended to eat together, have classes together, and perform school activities together. If a student within a pod becomes sick, this model reduces the risk of a larger outbreak across the wider school community. Pod size can be strategically adjusted based on the size of the school, the number of available teachers and classrooms, and the different social and developmental needs of different age groups. The larger the pod, the easier it is to maintain social cohesion as two students are very unlikely to socialize just by themselves. However, large pods may have a higher risk of inter-pod disease transmission. The number of each pod must balance both adherence to guidelines and also transmission control.

Setting up different entrances and exits for different pods, whenever possible, can help to mitigate risk further. Arrows and signs are helpful to indicate directions of travel and encourage physical distancing. It may be beneficial to allow students from the same pod to use the same classroom throughout the school day and ask teachers to rotate between spaces. This method can minimize item sharing and contact in the

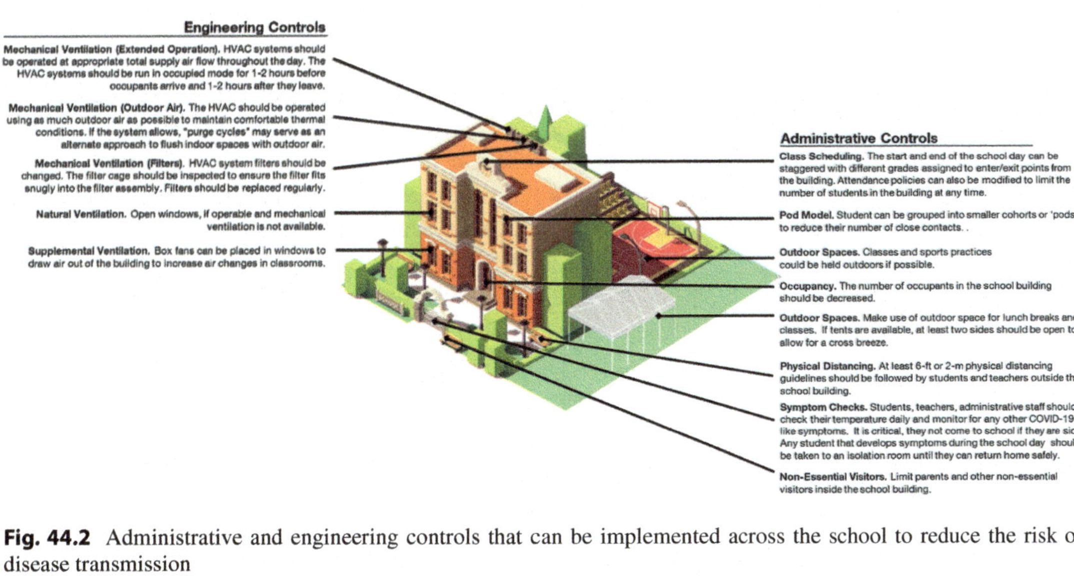

Fig. 44.2 Administrative and engineering controls that can be implemented across the school to reduce the risk of disease transmission

hallways/stairwells. For young children who will find it challenging to wear a mask for a full school day and to remain in the same classroom for the entire day, it has been suggested that students remove their masks while with their pod indoors; this will increase within pod transmission risk, however. A better idea to combat "mask fatigue" is that of "mask breaks" where a student removes a mask for a brief time of 30 seconds or so. Other proximate students should retain their masks if such a break time were contemplated; 2-meter (6-feet) physical distancing should be maintained.

Activities

Before- and after-school programs have traditionally been offered to enrich the learning environment with clubs or sports and provide additional child care support for parents with external commitments. These programs present challenges during the COVID-19 pandemic as students from across the school community have the potential to interact, contravening the core directive of the pod model. These programs are recommended to proceed only if the teachers and students can adhere to public health guiding principles.

Other events also have the potential to be high risk if they include large student gatherings with prolonged close contact in indoor spaces that cannot be well supervised by teachers or parents. Some examples include school assemblies, field trips, arts events (e.g., drama, music, dance), spirit events, sports events, and "homecoming" (common in the United States). These higher-risk gatherings should be held in a virtual online format to maintain the students' social networks while reducing risks to the school community.

44.3.2.2 Logistics

Physical Distancing

Physical distancing requires individuals to maintain a safe distance between themselves and others outside their household to limit disease transmission (Centers for Disease Control and Prevention 2020a). This has also been called "social distancing," but we believe this term to be imprecise, with unfavorable connotations of social isolation. As COVID-19 transmission is currently understood to occur through prolonged close contact with respiratory droplets and aerosols, physical distancing is a critical practice for reducing transmission risk. Approximately 2 meters (6 feet) of distancing has been suggested following the dynamics of droplet deposition (Santarpia et al. 2020; Bi et al. 2020). Airplane, train, and restaurant outbreaks are also supportive of the 2-meter guideline. Accounting for aerosol transmission necessitates increased physical distancing, in particular for higher-risk indoor environments (i.e., music rooms, cafeterias) (Prather et al. 2020). A variety of strategies can be implemented by schools to ensure that students and teachers in indoor spaces can achieve physical distancing guidelines. Teachers can consider removing extra furniture from their classrooms to free floor space and storing less frequently used items that might crowd desks and seats. In cases where physical distancing is not feasible, additional controls, such as plexiglass barriers, can be used. Even low-cost clear shower curtains hung from rods or ceilings can be used as physical barriers.

Cleaning and Disinfecting

Regular cleaning of classrooms should be arranged by the school's administration. Commonly touched surfaces and items in these shared spaces should be periodically and more frequently cleaned *and* disinfected; tabletops, chairs, doorknobs, electronics, and banisters are of particular concern. Surfaces and items that appear to be dirty must be cleaned before being disinfection. Recommended disinfectants for use against SARS-CoV-2 have been compiled by the US Environmental Protection Agency (EPA) (United States Environmental Protection Agency 2020) as well as by other national agencies. No or reduced touch options (i.e., sensor-activated, foot-operated pedals) can also be installed to further reduce the risk of disease transmission through contact of contaminated surfaces.

In the event that a student or teacher is suspected to be COVID-19 positive, it is recommended that access to any spaces recently used by this individual be restricted until after thorough cleaning and disinfection is complete.

To ensure cleaning and disinfection is accomplished thoroughly, schools must have available stock of required supplies and personal protective equipment for custodial staff. It is recommended that custodial staff receive training on the proper use of disinfectants (i.e., sufficient contact time) and personal protective equipment. It may be useful to designate administrative personnel to monitor and maintain the school's inventory of personal protective equipment, hand sanitizers, disinfection wipes, soap, paper towels, and other cleaning and disinfection products.

Noise

The sound level can be used as a proxy for the extent of aerosol release through vocalization; speaking loudly, yelling, or shouting can release more aerosols (Asadi et al. 2020; Asadi et al. 2019). The NIOSH Sound Level Meter is a freely available app that can be used to indirectly inform and remind the students and teachers to keep sound levels low (National Institute for Occupational Safety and Health 2020b). Visual displays of sound levels are also available in the form of a traffic light, which serves as a useful tool for younger students.

Symptom Checks

COVID-19 symptom-check apps can be a useful tool that can be implemented by the school to remind students and teachers to report symptoms and check their daily temperature. The apps will notify the school officials as soon as a COVID-19-like symptom arises. The schools can then take immediate action to encourage students to stay at home, conduct a test, and conduct contact tracing. These apps can also provide additional information regarding student attendance, movement, family member attendance, and parents and teacher communications. These data can allow schools to identify high-risk individuals and families and allow for immediate action. While convenient, the use of these apps requires access to a digital oral thermometer, which present challenges for some low-resource school communities. For schools without apps, conducting non-touch temperature checks before students enter the school building may help mitigate risks. However, a warmly bundled child in a cold winter may run a transient higher fever simply from warming; similarly, a child having walked to school on a hot day may also have a transient low-grade elevated temperature. Sending such healthy students ("false positives") home will be disruptive for both families and schools, so sitting a child in a quiet, temperature-controlled setting for 10 minutes and then rechecking the temperature is advisable.

Testing and Quarantine

Schools are recommended to have the necessary devices and medical supplies to monitor symptoms. If a student gets sick while she or he is in school or shows any COVID-19 symptom upon arrival, the student must be isolated until a parent or guardian comes to guide the student off the school property. Schools are recommended to set up an isolation room where students can stay until they are able to leave the school, except to go to a dedicated restroom or in case of an emergency. Access to personal protective equipment (PPE) is crucial for school staff when interacting with a potentially ill student.

Anyone who might have been exposed must be notified and quarantined immediately. It is debatable whether a class or pod is considered exposed if there has been mask use, physical distancing, hand and surface hygiene, and optimized indoor air management. The US Centers for Disease Control and Prevention considers contact to be the proximity of fewer than 2 meters (6 feet) for at least 15 minutes. If someone in the class qualifies as a contact using this criterion (e.g., friends sharing lunch in close proximity), then they must be quarantined. The rest of the class may not meet this contact definition. If a case is confirmed, the responsible healthcare worker or teacher must file a report to the local health department. Parents or guardians and the student's pediatrician must be notified immediately so that an ill child can return home for 10 days of isolation. Medical attention is needed for children who are more severely ill. Parents or guardians of identified contacts must be advised to pick their children up immediately and keep their children at home for 14 days of quarantine. Furthermore, responsible adults must be educated on prevention strategies to avoid a family cluster outbreak.

For boarding schools and colleges/universities, all students and staff arriving on the school campus are recommended to be "gateway" tested. Ideally, a test within 4 days prior to arrival followed by 14 days of quarantine and repeat testing at day 12–13 or so would help eliminate false-negative cases and those that are incubating. Students and faculties should not be allowed to circulate freely on the school campus until the test results are confirmed as negative. The schools may need to quarantine students in their rooms (ideally without roommates, but at the very least, with roommates who are on the same quarantine schedule). Food must be delivered, and laundry facilitated. Particularly limited use schedules for bathroom and shower use, with students or staff helping clean bathrooms after use, is advised. Schools can also consider "gateway" testing after return home for holidays. Less rigorous approaches may be indicated if the background epidemiologic circumstances suggest very low infection risk and if students are all coming from these low-risk venues.

Any student must be re-tested if presenting with any symptom or exposure (contact definition of fewer than 2 meters [or 6 feet] for at least 15 min) to infected individuals. Boarding schools and colleges/universities can also track epidemiologic trends across the county to judge the needed frequency of routine testing (Centers for Disease Control and Prevention 2020b). With viral background circulation, some modelers advocate frequent routine testing to maintain a virus-free educational environment, though this will be unaffordable for most school districts (Paltiel et al. 2020).

44.3.2.3 Teachers, Staff, and Visitors

Workforce and Communications

School operation is contingent on faculty and staff resuming their teaching and administrative responsibilities. These individuals also play a critical role in mitigating risks through continued communication of disease control strategies to students and the supervision of students to ensure safe personal behavior. Being good role models and communicating with all school personnel daily can help reinforce health messages. Daily briefings help reinforce safety principles, informing adjustments in strategies and community/school health status updates. Regular meetings can be held to evaluate the intervention plans and strategies. It may be helpful for school staff to send regular reminders to students and their parents/guardians of their responsibilities for safe personal behaviors to reduce risks. It can also build consumer confidence in the efforts being made to keep children and staff safe.

Teachers and staff need to stay home if they feel sick. COVID-19 symptoms are so nonspecific that any respiratory symptom can qualify one to stay home; even diarrhea and abdominal pain have been cited as symptoms. Schools should anticipate that the number of days taken for sick leave over the academic year is likely to be higher than in the past. All students and staff in the school should receive influenza vaccines; this will diminish the number of COVID-19-like symptoms and signs that manifest, reducing the year-round (in the tropics) and seasonal (in temperate zones) burden on school nurses, community pediatricians, and families. It may be necessary, when possible, to remove any barriers for taking additional leave from school, such as medical documentation, if local health services are overwhelmed (Edwards et al. 2019). It will be important for schools to build the necessary staffing capacity to fulfill these responsibilities and to fill any staffing gaps that may arise without exacerbating crowding.

Non-essential Visitors

A strict visitor guideline can help schools mitigate risks of disease transmission associated with outside visitors. It is recommended that visitor access be restricted inside the school building. A designated outside pick-up area can be set up for parents/guardians. In case of an emergency or other exceptional circumstance, all visitors need to wear a face mask that covers their nose and mouth. In the case of deliveries or maintenance, physical distancing is prudent, as is excellent hand hygiene and outdoor "hand-off." Establishing a time limit for visitors can also be useful in reducing risk.

44.3.3 Engineering Controls

Establishing well-ventilated school buildings is essential in reducing risks associated with the airborne transmission of SARS-CoV-2. Poorly ventilated indoor spaces can lead to increased exposure of virus-laden aerosols. Engineering controls can be implemented in a school to improve ventilation through (i) increasing the amount of outdoor air introduced into the building; (ii) improving filtration of recirculated air; (iii) increased the volume of air exchange per hour; and (iv) use of supplementation controls in individual classrooms, such as portable air cleaners and window-mounted box fans. When generating a negative pressure in the classroom, fans should be pointed outward, drawing air outdoors. Fans and portable air cleaners can be most useful in older buildings that do not have central mechanical air systems. The optimal configuration of controls will be unique to each school. Figure 44.3 presents a flow diagram to evaluate potential risk reduction controls that may be feasible based on the school's existing infrastructure.

44.3.3.1 Mechanical Ventilation

School buildings that are equipped with heating, ventilation, and air conditioning (HVAC) system are recommended to discuss current operation protocols and system capabilities with facility managers or an HVAC professional. The HVAC system must be commissioned before school reopening to verify the expected performance: commissioning will confirm that (i), filters, dampers, and economizer seals and frames are intact, clean, functional, and responsive to control signals; (ii) temperature and relative sensors are appropriately calibrated and communicating with the building automation system; and (iii) air handling systems are providing sufficient airflow to individual rooms and exhaust fans are functional and venting outdoors.

The operation of ventilation systems with 100% outdoor air, eliminating recirculation of viral material in the building, offers the most significant risk reduction concerning engineering controls. Increased outdoor air use may be feasible for some buildings when ambient temperatures and humidity levels are mild but will present challenges for regions with extreme hot or cold temperatures. Maintaining a building at an optimal indoor temperature during these hot or cold periods will require increased recirculation to avoid damaging the HVAC equipment. While thermal comfort for school occupants is important, adjusting building temperature set points to parallel outdoor conditions will enable effective use of outdoor air and provide an additional option for disease control.

For many buildings, continuous operation with outdoor air will not be feasible. Schools may decide to run a short (10–15 min) 100% outdoor air purge cycle during the day (i.e., after or during the lunch break). Schools running their HVAC system using less than 100% outdoor air will recirculate some amount of air.

In the HVAC's air handling unit, filters can be used as the first line of defense in removing virus-laden aerosol from recirculated air. Filters are rated for their efficiency at airborne contaminant removal. High-efficiency particulate air (HEPA) filters have a single-pass removal efficiency of 99.9% for aerosol sized between 0.3 and 1 μm diameter. Minimum Efficiency Reporting Value (MERV)-rated filters are also effective at removing aerosol and range from MERV1 through to MERV16. MERV13-rated filters have been recommended for use in HVAC systems as a risk reduction strategy for COVID-19 (American Society of Heating Refrigerating and Air-Conditioning Engineers 2020a). In contrast to HEPA filters, the single-pass efficiency of a MERV13-rated filter is 90% for similarly sized aerosols (Azimi and Stephens 2013), a bit lower for the smaller aerosols at 70%. If possible, filters should be upgraded to the highest-rated filter that can be accommodated by the system. While a MERV13 filter is recommended for use, it is

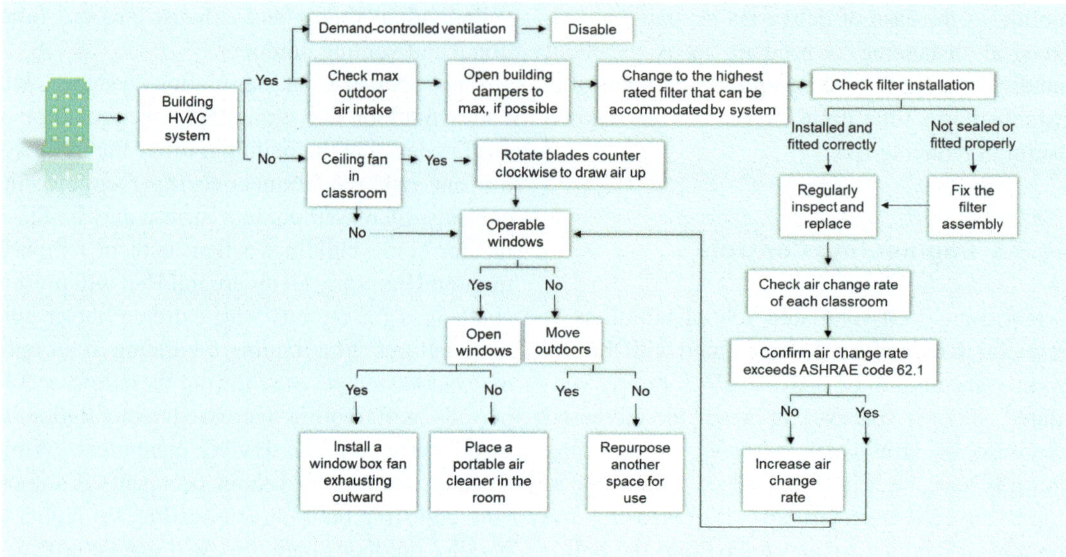

Fig. 44.3 Engineering control flow diagram for optimizing ventilation in schools

important to ensure that a building's HVAC system can accommodate the necessary flow required by a higher-rated filter.

Beyond simply upgrading and periodically changing the system's filter, it is also critical to ensure the filter cage is maintained; any gaps in the housing assembly will impair filter efficiency. When handing these filters, it is important to recognize that they are delicate and require careful handling to preserve performance. Regular inspection of the filter and filter assembly, as well as following manufacturer recommended maintenance, are critical to ensure proper operation. Filter changes are recommended to be carried after a period with no or less building occupancy (i.e., weekend, holiday) following findings from studies evaluating the stability of SARS-CoV-2 on surfaces (van Doremalen et al. 2020). When it is possible, a 10% bleach solution or another appropriate disinfectant can be used for disinfecting filters before removal. Once removed, filters can be placed into a bag for disposal as general waste.

Increasing the number of air changes in a classroom will allow for sequential removal of airborne viral material on each pass through the filter. The number of air changes per hour (ACH) in an indoor space can also be increased to enhance ventilation. ACH should be evaluated for each classroom in relation to the size of the room and occupants' activities. The American Society of Heating, Refrigerating, and Air-Conditioning Engineers (ASHRAE) has established ventilation standards for different types of school spaces (ASHRAE 62.12019 – Ventilation for Acceptable Indoor Air Quality Standards) (American Society of Heating Refrigerating Air-Conditioning Engineers 2019). The amount of airflow into each schoolroom should ideally exceed these minimum standards with a goal of 5–6 ACH.

Enhanced ventilation can additionally be achieved through the extended operation of the HVAC system. It is recommended that the system be turned on in occupied operation at least 2 h before the arrival of staff, teachers, and students and remain running until at least 1 h after all occupants have departed. The system should also be operated at a maximum total airflow to the extent possible, assuming full building occupancy.

HVAC systems in some school buildings may have displacement ventilation capabilities. This system feature introduces conditioned air at a low flow from supply diffusers positioned near the floor and exhaust air near the ceiling to provide enhanced air mixing. The use of displacement ventilation in classrooms and offices is another effective engineering control that provides enhanced air mixing, moving potential contaminant air from students and teachers.

44.3.3.2 Natural Ventilation

For buildings using natural ventilation, operable windows, doors, skylight, or roof ventilators can be opened when outside temperatures and humidity levels permit. In buildings with an HVAC system, the interlock with the windows should be disabled in the building automation system. Portable air filters can be used in lieu of HVAC systems when the outside air is too hot or too cold.

44.3.3.3 Supplemental Controls

Air purifiers, window-mounted box fans, and ultraviolet germicidal irradiation (UV-GI) are supplemental controls that have been used in various indoor environments to reduce virus-laden aerosol. Supplemental controls can be used in combination with other risk reduction strategies, as guided by the flow diagram shown in Fig. 44.3.

Box fans can be placed in the window of a room in the school to enhance the exhaust of indoor air outside. This engineering control offers schools in a naturally ventilated building a low-cost option to increase airflows. Efficiency can be maximized by blocking open-air cavities on the side of the fan. Box fans can also be fitted with filters to create a closed-loop filtration system analogous to filtration via HVAC systems.

Air purifiers with HEPA filters can remove airborne chemicals (e.g., particulate matter) and biological contaminants (e.g., virus, bacteria, fungal spores) (Kettleson et al. 2009; Foarde 2010; Miller-Leiden et al. 1996; Shaughnessy et al. 1994; Kujundzic et al. 2006). Portable air purifiers floor units, as well as ceiling and wall-mounted systems, are available with a range of fan sizes and filter ratings. The optimal size of an air purifier for space can be determined by the size of the classroom and targeted ACH. All air purifiers certified by the Association of Home Appliance Manufacturers (AHAM) have a clean air delivery rate (CADR) that can be used to evaluate performance (Association of Home Appliance Manufacturers 2020). The CADR rating reflects the volume of air the device is capable of filtering per unit time (measured in cubic feet per minute, CFM) and the particle removal efficiency of the filter. For each air purifier, this rating is evaluated for airborne pollen, smoke, and dust; the CADR rating reported for smoke is the most comparable to the size of SARS-CoV-2 aerosol. Airflow pattern (i.e., location of supply and exhaust air, windows) and the configuration of students in a classroom can be used to guide the placement of the device – this is of particular importance for the floor units.

UV-GI systems are capable of inactivating airborne viruses and have been installed in high-risk settings, such as healthcare facilities. These systems use low-pressure, mercury lamps which emit non-ionizing electromagnetic radiation, primarily in the UV-C range (Reed 2010). Doped quartz glass is used in UVGI installations to block transmission of 185 nm wavelength, which is where ozone is generated. Multiple installation formats are available, including upper-air, in-duct, and coil systems. Upper-room air disinfection systems achieve maximal efficiency in spaces with an appropriate fixture and sufficient vertical air exchange in high-ceiling rooms (Xu et al. 2003). In-duct systems can also be a useful option if designed to provide sufficient UV dose, which is determined by exposure time and irradiance. The rapid movement of air through ducts requires in-duct UV lights to operate at a high intensity (Kujundzic et al. 2007). Even if UV dose is optimized, the single-pass airborne inactivation efficiency is low due to the short exposure time (order of seconds) in this UV-GI format. Installation of in-duct systems in spaces with high ACH can improve efficiency (Beggs et al. 2000; Kujundzic et al. 2007).

Coil cleanings systems can be used to decrease the microbial surface loading on cooling coils. This system format has been shown to have maximum effectiveness in climatic regions that experience coil condensation (high outdoor relative humidity) (Luongo and Miller 2016; American Society of Heating Refrigerating and Air-Conditioning Engineers 2019). Installation of upper-air or coil cleaning UV-GI systems may be feasible for some schools as an engineering control strategy. However, installation often requires upgrades to the building's HVAC system, and the high installation and maintenance costs should be considered when evaluating the feasibility of use. In-duct UV-GI systems are unlikely to be effective in a school building where classroom ACHs are low.

44.3.3.4 Unintended Exposures from Outdoor Environments

While maximizing outdoor airflow is recommended in indoor spaces, it is important to be aware of potential contaminants in outdoor air and take precautions that minimize/prevent school occupants from exposure. For buildings with mechanical ventilation, outdoor air inlets should not be positioned near exhaust air outlets to avoid re-entry of potentially contaminated air. In schools using natural ventilation, outdoor airborne contaminants can readily enter the building spaces through open windows, including air pollutants derived from traffic and industrial sources, allergens, pests, and insects. It will be important to consider potential exposure to these contaminants and be selective on which operable windows are used. Considering the potential exposure of students and teachers, exposure to these outdoor factors is also important if scheduling outdoor classes.

44.3.3.5 Sensors to Monitor Building Performance

Achieving the optimal performance of an HVAC system requires regular commissioning. Indirect measures of system performance can also be evaluated using air pollutant sensors. A well-commissioned HVAC system will provide a sufficient number of air changes for space based on the size and number of occupants. Some HVAC systems are designed to adjust air flows in individual classrooms automatically, and offices based on occupancy load monitored using carbon dioxide sensors. Demand-controlled ventilation sequences need to be disabled to maintain HVAC operation at maximum air flows independent of the number of students, teachers, and administrative staff in the school building.

While demand-controlled ventilation is recommended to be disabled, these carbon dioxide sensors can provide useful information about occupancy load and activity levels in relation to the ventilation capacity of the classroom. Airflow may require adjustments if levels of exhaled carbon dioxide steadily increase while occupants are in the space. For classrooms not already equipped with these sensors, low-cost models are commercially available and may provide school facility operators with useful insight on under-ventilated classrooms. Additional sensors are available to evaluate ventilation of air pollutants from other sources such as particulate matter 2.5 μm or less in diameter and volatile organic compound sensors.

Unfortunately, nearly all these strategies increase energy utilization and cost. However, we hope that building air quality can return to normal when COVID-19 can be treated more successfully and safe and effective vaccines are available and deployed universally.

44.4 Student- and Teacher-Directed Risk Reduction Strategies

COVID-19 public health guiding principles include physical distancing, hand/face/surface hygiene, use of face masks covering the nose and mouth, improved air quality and ventilation, and appropriate testing and follow-up (isolation, quarantine, and contract tracing). Adherence to all of these is key to improving the safety of the school community. These principles may become the new normal for daily practice until we have excellent COVID-19 therapeutics and vaccines; school administrative staff and teachers need to communicate and reinforce their importance. This section will discuss how these public health principles may be applied and special considerations for different settings and populations.

44.4.1 Masks

There is strong evidence from epidemiology and economics that supports the use of masks to limit community spread of COVID-19. Face masks provide inward and outward protection. All students, teachers, and administrative staff are recommended to wear face masks throughout the day while on school property (Centers for Disease Control and Prevention 2020a). Education on how to safely choose, wear, care for, clean or discard, and store masks can be useful to encourage adherence. Promoting personalized masks with the school logo or creating masks for students in different classes, houses, clubs, or sports teams may also encourage use. Fun masks for younger children may include favorite cartoon characters or colors. Masks made from cotton fabrics with high thread counts and random fiber orientation seem promising for preventing release and exposure to droplets and aerosols (Dbouk and Drikakis 2020). They are also highly cost-effective (Abaluck et al. 2020).

In contrast, elastic fabric materials are not recommended due to the large pore size, resulting in lower filtration efficiency. Masks should be appropriately sized and provide coverage of both the nose and mouth (Centers for Disease Control and Prevention 2020c). Face masks with an exhalation valve are not recommended as they offer inward protection but no outward protection; thus, they will not prevent the spread of the virus from the mask wearer (Ippolito et al. 2020). Reusable masks can be washed or stored 2–3 days between wear (Hao et al. 2020; Yang 2020; World Health Organization 2020a). Rotating between at least three masks will allow for safer reuse of masks.

In situations where masks cannot be worn (i.e., eating, playing a wind or a brass musical instrument), other precautions can be taken to increase safety. Alternative arrangements should also be made for students who are unable to wear a face mask due to developmental, respiratory, young age, or other physical conditions. Members of the school community must understand that wearing a mask is not a replacement for physical distancing or other guiding public health principles.

Face shields are not a replacement for masks. Masks and face shields apply to different routes of transmission. Face shields can effectively limit the spread of droplets released when coughing or sneezing; less effective protection is offered toward aerosols (Lindsley et al. 2014). Face shields will have limited utility for reducing risk toward airborne transmission (Perencevich et al. 2020). A face shield could be worn in addition to a mask, but face shields should not be worn without a mask unless substantial physical distancing is also possible.

Masks can create challenges in certain pedagogical cases, such as language classes. Foreign language instructors may need to be seen by students to appreciate mouth movements needed to make certain sounds. Video assistance should be considered, or a shield combined with an increase in their physical distance to the students to 12 feet (4 meters) or combined with a plexiglass barrier may be considered so that students can see the instructor mouthing the words to be learned. In these cases where teachers do not wear masks, the duration of the class time without a mask could be minimized, and additional engineering controls for ventilation can be used in the classroom to offer enhanced risk reduction approaches. In situations where lip-reading is required, and physical distancing may make it impossible for the deaf or hard of hearing student read lips, communication may be possible using a mask with a transparent mouth window. School policy should be developed to accommodate members of the school community who cannot wear masks. For instance, certain states are requesting all students that cannot wear masks to develop a particular educational plan with the teachers.

While not ideal, teachers using only face shields can stand near an air purifier or window-mounted box fan. These lessons can be held outdoors if weather permits. Also, lessons can go for 15 minutes with a 10-min non-talking period, limiting the potential for high air aerosol volumes and maximizing the ability of the HVAC or fan or air purifier to cleanse the air.

44.4.2 Hand Hygiene

All members should be consistently reminded to wash their hands as frequently as possible. Hands should be washed with soap and water for at least 20–30 seconds. That is longer than most people realize, the time it takes to sing "Happy Birthday to You" twice. When lacking soap and water, hand sanitizer that contains at least 60% alcohol can be used. For hand sanitizer to be effective, all surfaces of the hands must be covered and then rubbed until dry. It is especially important for students, teachers, and administrative staff to wash their hands after touching common surfaces or their face after they cough or sneeze and after using the restroom (Pradhan et al. 2020). Also, even adults may need a "hand-washing lesson" since, commonly, the hands are not thoroughly washed in everyday practice. Touching one's face should trigger additional hand hygiene, as would touching of potentially contaminated surfaces. Courtesy greetings such as shaking of hands should not be conducted as long as COVID-19 is not well controlled with medications for infected persons and vaccines for the uninfected.

44.5 Case Studies of Specific School Spaces

Approaches to reduce transmission risk through the elimination of physical contact, various administrative and engineering controls, and personal behaviors have been discussed. Achieving maximal disease control requires a layered strategy that includes multiple controls. The optimal combination of risk reduction approaches implemented should be guided by occupancy patterns and activities of students, teachers, staff, and administrators in the space. We will discuss the application of administrative and engineering controls in specific school spaces.

44.5.1 Upper-grade Classrooms

As students spend much of the school day in a classroom, disease control measurements must be implemented in these spaces (Fig. 44.4). Older students are likely to be more adherent to guidelines than lower-grade school students, but teachers and school administrative staff need to regularly reinforce the benefits of adequately wearing a face mask, hand hygiene, and physical distancing.

Schools should be aware of "mask fatigue" with children, as mentioned earlier in the chapter. "Free Mask" breaks can be scheduled during the school day, where students can take off their mask under teacher supervision while physically distanced (>12 ft. or 4 meters apart) outdoors. If this is done indoors, the break should be a few students at a time, maintaining mask use by those students most proximate.

Teachers need to arrange desks such that students are physically distanced at least 2 meters or 6 feet apart. Zig-zag configurations can be used to maximize the number of students that can be

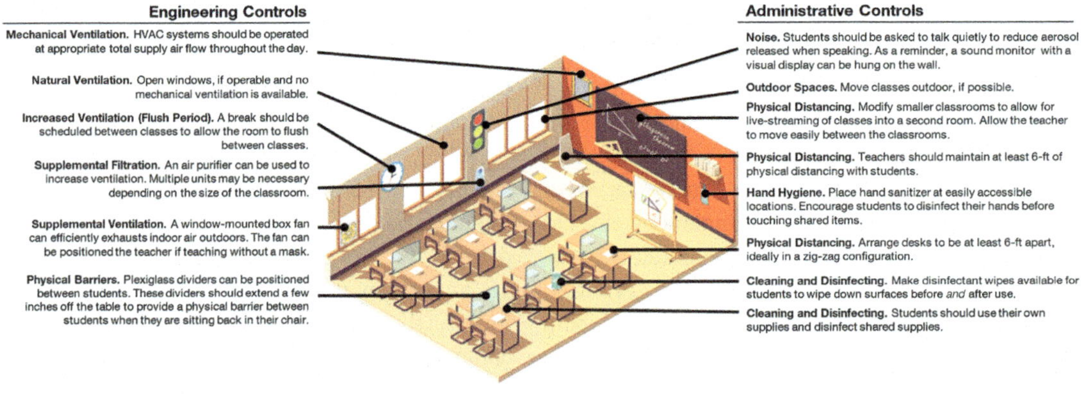

Fig. 44.4 Risk reduction strategies in upper-grade classrooms

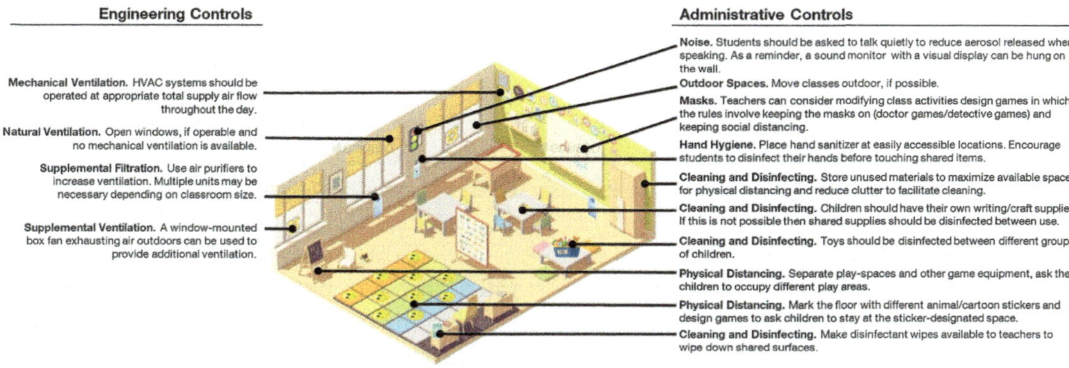

Fig. 44.5 Risk reduction strategies in kindergarten and lower-grade classrooms

accommodated in the classroom. Plexiglass barriers can be used to limit contact between students when 2-meter (6-foot) distancing is not possible. It may also be possible to repurpose larger spaces available in the school, such as gymnasiums, common rooms, and auditorium as additional classrooms.

Also, teachers are recommended to rotate between classrooms rather than students. This strategy can limit occupant movement in the school building and reduce potential contamination of shared surfaces, contact with high-touch items, and close contact in the hallway/stairwells.

Increased ventilation in classrooms is a practical approach to reduce risk to virus-laden aerosols. Schools with mechanical ventilation should operate HVAC at appropriate total supply airflow throughout the day. Naturally ventilated classrooms can make use of operable windows and consider the use of air purifiers or window-mounted box fans for greater airflow.

44.5.2 Kindergarten and Lower grade Classes

Activity patterns of students and teachers in lower-grade classrooms (kindergarten to grade 2) are distinct from the upper-grade classes that were previously discussed – students tend to be more mobile throughout the day in the classroom space, and much of the day is spent near or on the floor. Young students in these classrooms are also likely to have lower adherence to preventive behaviors, including the use of face masks, hand hygiene, and physical distancing. Teachers can consider implementing various creative strategies to overcome age-related challenges, as summarized in Fig. 44.5.

To encourage preventive behaviors, teachers can play health education games with students. These games could cover topics including sneezing into their elbow ("be a vampire"), effective hand hygiene ("wash off the paint or glitter"), and mask use ("decorate your mask").

While physical distancing younger students may prove challenging, classroom furniture can be rearranged with activity stations spaced at least 2 meters or 6 feet apart. Toys, shared supplies, and high-touch surfaces are recommended to be cleaned and disinfected between groups of students. The supply of disinfectant wipes in the classroom will facilitate frequent use by teachers on contaminated surfaces and items. Placing a hand sanitizer bottle in the classroom can encourage use by students, and students can be taught to remind each other and be reminded by the teacher.

Increased ventilation can also be effective in reducing airborne transmission risk. Classes can be held in outdoor spaces if weather permits. Available mechanical and natural ventilation options should also be maximized to increase ACH in the classroom. Supplemental ventilation can be achieved in the classroom through the use of air purifiers and window-mounted box fans. Depending on the size of the classroom, multiple units may be necessary.

44.5.3 Music Rooms

Music rooms are high-risk spaces in schools as singing and wind or brass instruments can generate aerosols. Administrative and engineering controls can be employed to decrease risks, as summarized in Fig. 44.6. Schools should consider holding classes and rehearsals that include singing or wind and brass instruments outside.

To limit exposure, students playing the piano, string instrument, or percussion should wear a face mask over their nose and mouth. Students playing a wind or brass instrument can also be encouraged to wear an "instrument mask," which has an overlapping, layered opening that can be pulled back to access the mouthpiece. While playing, this mask design allows their nose to be covered. While masks offer protection from aerosols and droplets, vulnerable students should be notified of the increased risk of participating in music classes.

Indoor class times should be limited depending on the size of the room and allow for a flush period before the next group of students enters the room. Teachers can control aerosol levels by scheduling 5–10-min breaks during the rehearsal session to permit HVAC or fans to cleanse the room air. During these breaks, all students are recommended to wear masks. These breaks may also serve as an opportune time for teachers to introduce digital platforms in the event at-home learning is necessary.

Ventilation in music rooms should be increased during and after rehearsals. Schools with HVAC systems are recommended to be operated at total supply air flows throughout the day. If continuous use of outdoor air is not possible during the rehearsal time, facilities managers could consider a purge cycle before the next group of students enters the room. Flutes in which forced breaths are passed over the instrument rather than blown into it should be segregated behind plexiglass and/or physically separated by more than 12 feet or 4 meters from other instrumentalists and the conductor.

Each student is recommended to have her or his own music stand. Students who can wear a mask while playing their instrument can be positioned 2 meters or 6 feet apart facing the force. Increased physical distancing is recommended for wind and brass instruments and singers. The occupancy in smaller-sized classrooms should be reduced. A larger ensemble alternatively can be spread across multiple rooms and connected by live stream, if available. Classes may consider meeting in larger, well-ventilated school spaces (i.e., assembly hall, auditorium, gymnasium, etc.) or outdoor locations to achieve the recommended physical distancing. Tents can increase flexibility with outdoor spaces but should be open sided and high pitched.

To further limit exposure of other students to the enhanced aerosols released by these instruments, wind and brass players can be positioned by a ventilation source (i.e., open window, air purifier, window-mounted box fan) surrounded by physical barriers. Students can also use nylon or cloth bell coverings on wind instruments to

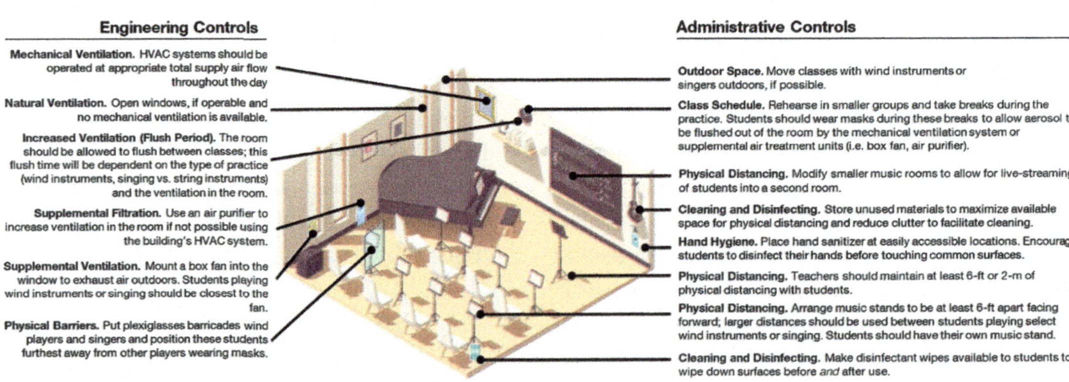

Fig. 44.6 Risk reduction strategies in music rooms

reduce aerosol emissions. Students should not discharge water valves on the floor; students should have their own absorbent pads or container.

If students are obtaining instruments from the school, each student should be assigned to avoid sharing. Similarly, percussionists should be assigned their own stick/mallet.

44.5.4 Art Rooms

Art rooms can be viewed as lower risk as students can wear masks throughout the class session. An overview of approaches to reduce transmission risk reduction in this space is summarized in Fig. 44.7. The risk of COVID-19 transmission by fomites is of relevance when art tools are shared between students. SARS-CoV-2 can remain viable on various surfaces, ranging from 4 h on copper to a maximum of 3 days on plastic and stainless-steel surfaces (van Doremalen et al. 2020). Providing each student with a set of essential art tools can minimize shared objects but may not be feasible in many cases. Any shared supplies and surfaces are recommended to be cleaned and disinfected after use. A block of time can be scheduled near the end of each class allocated to cleaning and disinfecting art supplies. Disinfectant wipes and hand sanitizers are recommended to be accessible at multiple locations around the art room to encourage use during class. Students and teachers should maintain at least 2 meters or 6 feet of physical distance; workstations can be reconfigured to promote adherence to guidelines. When proximity is needed for teaching, this should be very brief. To maximize available space for physical distancing and to facilitate cleaning, unused tools or materials can be organized or stored. Increased ventilation can further decrease risks from the airborne transmission, either via the school's HVAC system or natural ventilation strategies. Supplemental controls (box fans, air purifiers) may also be used if other ventilation options are not available. If the weather permits, teachers can consider holding outdoor classes.

44.5.5 Computer Labs

School computer labs present similar risks as other classroom spaces but also present a risk for transmission of SARS-CoV-2 via the fomites, namely, shared computer peripherals (i.e., mouse, keyboard, screen). Strategies for risk reduction are summarized in Fig. 44.8. Computer peripherals are made from plastics where SARS-CoV-2 can remain active or infectious for approximately 3 days (van Doremalen et al. 2020). Cleaning and disinfection between groups of students are recommended. The availability of hand sanitizer and disinfectant wipes around the computer lab would help to promote good hand hygiene as well as cleaning and disinfection behavior by students. Teachers and students should maintain at

Fig. 44.7 Risk reduction strategies in art classrooms

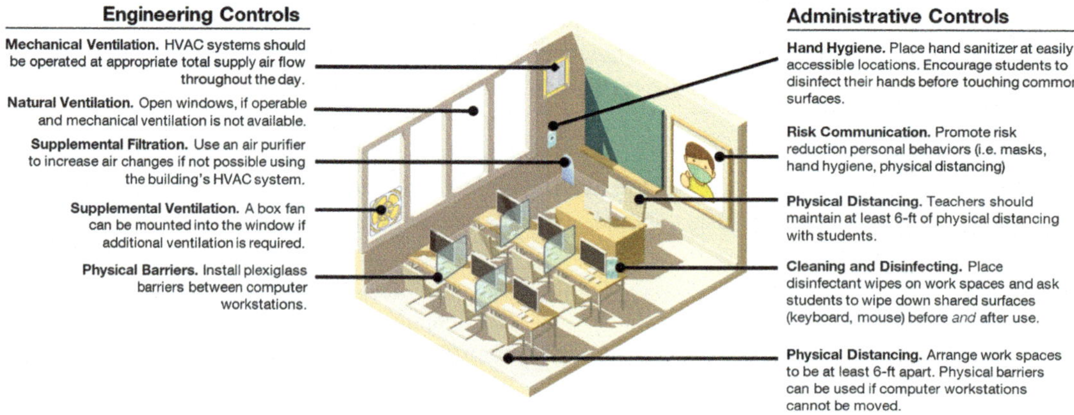

Fig. 44.8 Risk reduction strategies in computer labs

least 2 meters or 6 feet of physical distance while wearing masks, but this may present challenges with computer workstations that cannot be easily moved. Plexiglass barriers can be positioned between workstations to minimize prolonged close contact of students during class times. Ventilation can be increased further in computer classrooms, as with other spaces, to reduce airborne transmission risks through mechanical or natural ventilation options. Supplemental controls like fans and air purifiers can also be used if other ventilation options are not available.

44.5.6 Gymnasiums

While athletics programs are a significant component of school life for many students, COVID-19 presents a high risk of transmission while participating in contact team sports. Sport-related impediments to personal distancing are unavoidable in many teams and contact sports like the US or Australian football, wrestling, football/soccer, field hockey, handball, lacrosse, water polo, and basketball. Contact ball sports such as baseball, softball, or basketball often involve ball sharing and ball transfer, which increases the risk of transmission through fomites. Individual sports are relatively easier to keep physical distancing – swimming with adequately spaced lanes, golf, and tennis. Schools should assess the risks of disease transmission in each sport. Athletes can be physically distanced when playing contact sports. Swimmers in adjacent lanes can start at opposite ends of the pool so that they are not swimming together and instead pass each other. Decisions regarding phased return must be made for each sport. Schools should consider canceling inter-school athletics competitions to limit interaction between students and teachers from different schools. Division get-togethers, team meetings, and team demonstrations are recommended to be held virtually.

Precautions should be taken for sports activities held in indoor gymnasium spaces. An overview of approaches to reduce transmission risk reduction in gymnasiums are summarized in Fig. 44.9. All students on sports teams would likely benefit from education on disease control and discussion of the importance of a "shared responsibility" mentality. The symptoms of students on sports teams should be checked regularly. Non-touch temperature checks before training, practices, and competitions may help mitigate risks. Outdoor practice sessions are recommended to be arranged as much as possible.

Masks are recommended while playing sports, in particular, when indoors. Shared items and surfaces need to be cleaned and disinfected between student groups. Good hand hygiene can be promoted by placing hand sanitizer around the gymnasium. Many gymnasiums have high ceilings and are equipped with dedicated mechanical ventilation systems that have the potential to provide adequate levels of ventilation. To ensure increased ventilation is provided for students and

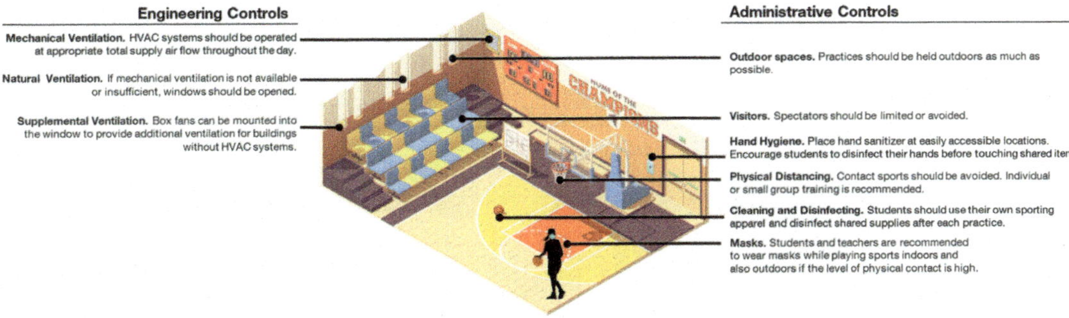

Fig. 44.9 Risk reduction strategies in gyms

teachers in this space for physical education classes or sports practices, available HVAC systems should be operated with demand-controlled ventilation disabled to allow the highest levels of outdoor air to be brought in while the gymnasium occupancy levels are relatively low. Gymnasium spaces may have extensive HVAC capabilities that could be repurposed by the school to hold other classes or events. In gymnasiums that are mechanically ventilated, operable windows and window-mounted box fans may be used to increase ventilation levels further.

44.5.7 Restrooms

Restrooms may impart additional risk for viral transmission. In restrooms, aerosols do not only transmit from human to human or objects but also can be generated and spread by flushing toilets (Knowlton et al. 2018). SARS-CoV-2 has been found in fecal samples of patients with COVID-19 (Knowlton et al. 2018; McDermott et al. 2020). There is no evidence that there is any substantial degree of transmission via the fecal-oral route. Schools still need to be cautious and minimize any risk of fecal bio-aerosol transmission by limiting bathroom crowding and maximizing handwashing and surface cleaning opportunities.

Administrative and engineering controls can be used to decrease the risk of disease transmission, as summarized in Fig. 44.10. Occupancy limits should be set for each restroom. While waiting in a line for the restroom, students must keep at least 2 meters or 6 feet of physical distance. Markers can be used to indicate 2 meters or 6 feet spacing. Frequent cleaning and disinfecting during the school day are recommended to minimize risks of surface contamination. Automated infrastructure in restrooms (automatic doors, touchless faucets, soap dispensers, towel dispensers, and toilet flushers) can be used to reduce touch of high-use surfaces.

Instructions on proper hand-washing techniques can be posted by the sink area together with a clock that students can use to time themselves. If students follow the 20-second guidance for hand-washing, aerosol and droplets sprayed by hand dryers will not likely present disease risk; however, paper towels are recommended given that many younger and even older students may not adhere to recommended procedures. Placement of hand sanitizer by doors may also encourage good hand hygiene.

As most restroom spaces are small, physical barriers such as plexiglass can be positioned between sinks and urinals to lower contact between occupants. Toilet lids can also be installed as another physical barrier to limit exposure to aerosolized fecal material, though this may result in more touching of the lids. Increased ventilation in restrooms will also be critical in limiting airborne transmission; exhaust ventilation should be confirmed to be functioning correctly and should be left on at all times when the school is occupied. Due to the bathroom exhaust, these spaces are negatively pressurized relative to common areas (i.e., airflow is moving into the

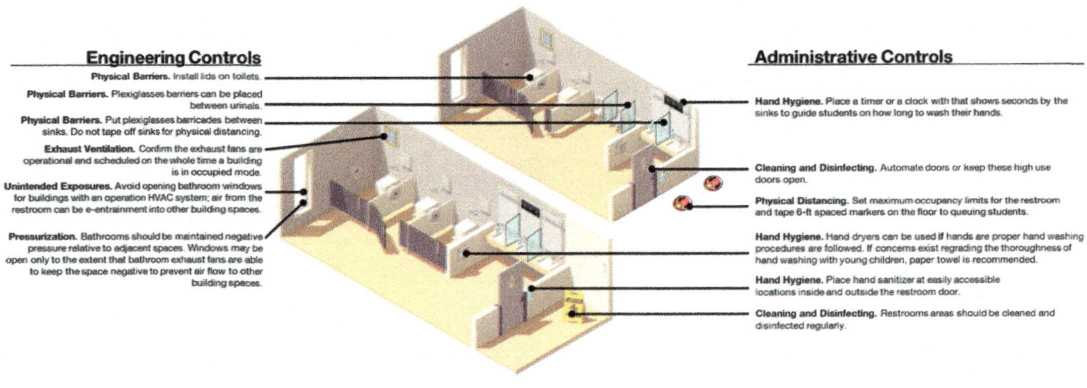

Fig. 44.10 Risk reduction strategies in female (left) and male (right) restrooms

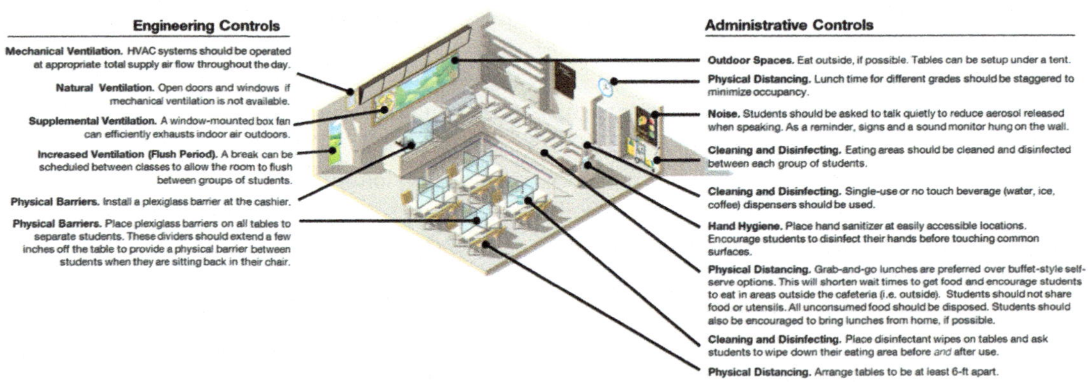

Fig. 44.11 Risk reduction strategies in cafeterias

restroom rather than from the bathroom into adjacent spaces).

44.5.8 Cafeteria

Cafeterias are viewed as high-risk spaces in schools. Wearing face masks is one of the most significant disease controls; however, the mask usage cannot be enforced due to the need to eat. Administrative and engineering controls should be increased to account for the increased release of potentially virus-laden droplets and aerosols (Fig. 44.11).

Lunch periods can be staggered to limit the number of occupants in the cafeteria. Students should be encouraged to eat their lunches with their pod outside (weather permitting) or in their assigned classroom. Students can be encouraged to bring a packed lunch from home if possible. Alternatively, a "grab-and-go" style lunch can be offered by the school to avoid long queues that extend contact times between students and between students and food service staff.

Also, students can be asked to eat lunch at tables that are physically distanced 2 meters or 6 feet apart. Multiple students can be accommodated at a table using plexiglass barriers. Aerosols will be released by students while speaking during lunch. The release will be great with vocalization. Students should be asked to speak quietly, which is recognized to be potentially challenging. A noise monitor with a visual indicator can be placed in the cafeteria to monitor sound levels.

Placement of hand sanitizer bottles at accessible locations in the cafeteria will facilitate good hand hygiene. Access to disinfectant wipes on tables will further encourage students to wipe down surfaces before and after eating.

Evaluating ventilation options will be critical in this space. Mechanical ventilation is recommended to be operated at appropriate total supply airflow throughout the day. Since many building HVAC systems cannot be continuously operated at 100% outdoor air, a short purge cycle can be run between groups of students to bring in more fresh air between use. For naturally ventilated spaces, operable windows should be used. Air can additionally be exhausted outdoors using outward-facing window-mounted box fans. Multiple units can be installed depending on the size of the cafeteria.

44.5.9 Hallways and Stairwells

Although hallways and stairwells are only intermittently occupied continuously during the school day, they are still considered to be higher-risk areas due to high time-limited occupancy, limited space for physical distancing, and potentially lower levels of ventilation. An overview of approaches to reduce transmission risk reduction is summarized in Fig. 44.12. Hallways and stairwells should be marked with tapes to show directions of flow (similar to a road traffic pattern with lanes to traffic in one direction), to avoid congestion, and to remind people of keeping at least 2 meters or 6 feet of physical distance. Also, classes can be staggered to minimize traffic in hallways. The concentrated flow of students and teachers in these spaces will create many high-touch surfaces, such as doorknobs and light switches. Disinfection of these surfaces after class changes can help minimize transmission risks via fomites. Sensor automated, foot-operated, or pop-open doors can also be used to reduce touch of high-use doors. Signs can be placed in these spaces to alert students, teachers, and staff of these non-touch features and encourage opening other doors using elbows, keys, pens, or other tools. To increase ventilation within hallways and stairwells, windows, if operable, can be opened during and after times of high traffic.

44.5.10 Library

School libraries provide important learning resources to students. Beyond access to books, libraries offer a safe and quiet study environment as well as access to computers and the internet. These facilities and amenities are especially important for low-income communities where analogous services may be unavailable at homes. In planning for school reopening, librar-

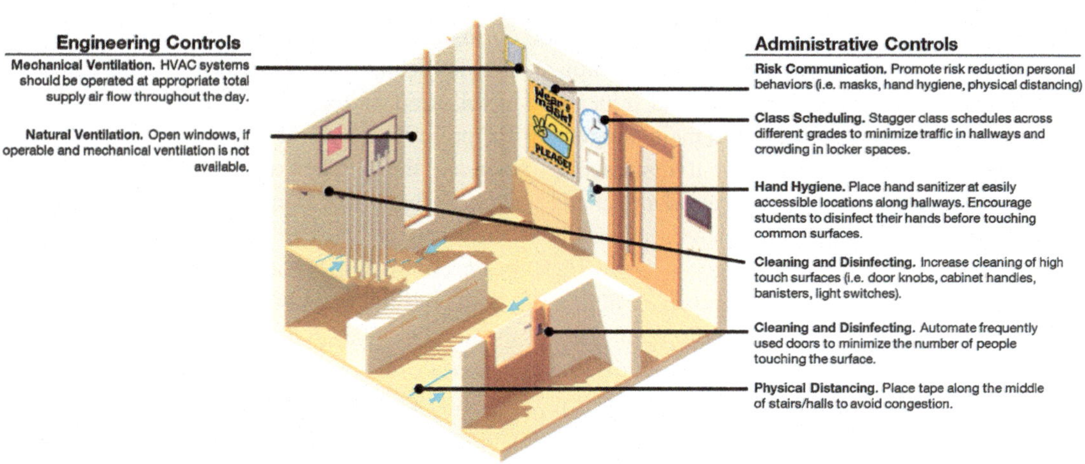

Fig. 44.12 Risk reduction strategies in hallways and stairwells

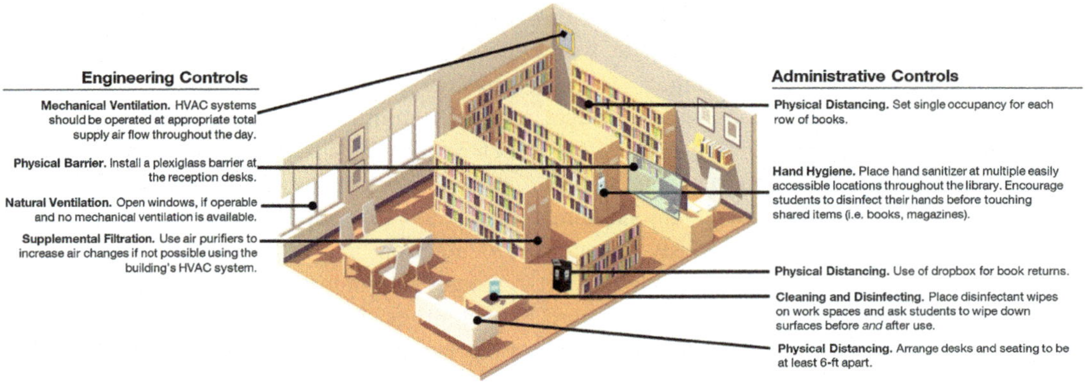

Fig. 44.13 Risk reduction strategies in a library space

ies are of potential concern because of the unstructured time students may spend in this space before, during, and after school. As this is a common space, students will likely interact with others outside of their class pod, which may warrant additional administrative and engineering controls to be implemented. These strategies for risk reduction are summarized in Fig. 44.13.

Building HVAC systems with maximized air exchange or air purifiers should be implemented in the workspace to increase air exchanges. Given the size of most library spaces, schools should consider placing multiple air purifiers with the same CADR rating. Protocols to facilitate physical distancing should be established, such as 2 meters or 6 feet spacing between study/reading spaces and setting single occupancy in bays of books. Placing a plexiglass barrier at the information desk and use of a drop box for book returns can further minimize interactions between librarians and students. Students can also be encouraged to maintain good hand hygiene as well as contribute to disinfecting efforts through the placement of hand sanitizer and disinfectant wipes at multiple locations across the library space.

Issues for libraries may also apply to audio-video studios, radio studios, computer labs, and other specialized areas of a school that may represent closed spaces and multiple persons using the space. Spacing, plexiglass or clear plastic curtain separation, frequent surface hygiene, and maximizing distancing, mask use, and indoor air quality can mitigate risks.

44.5.11 Playgrounds

Schools should plan to maximize students' amount of outdoor recess breaks and time in playgrounds. The risk of transmission in these outdoor spaces can be minimized through various strategies that allow for physical distancing as well as good hand hygiene and cleaning and disinfecting protocols (Fig. 44.14). Recess times can be staggered to reduce the total number in the play area and interaction between students in different pods. Students should wear face masks during recess breaks. Increased teacher supervision is recommended to ensure proper mask use is followed and encourage physical distancing. To promote physical distancing, schools can increase the number of available portable toys, such as balls and sand shovels/buckets. These items can be cleaned and disinfected between groups of students. Play structures are not likely a route of high transmission due to sunlight inactivation. Regularly disinfecting high-touch surfaces (i.e., slides, swings) and good hand hygiene can further lower risks. Hand sanitizer stations can be placed around the play area to encourage

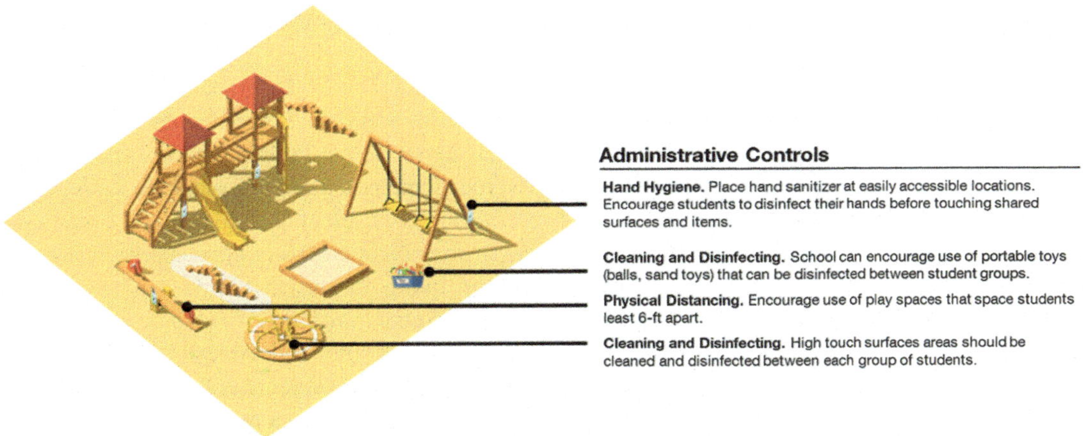

Fig. 44.14 Risk reduction strategies in playgrounds

use, and students can wash their hands before and after recess.

44.5.12 Transportation

Many students rely on school buses for transportation to and from school. Several administrative controls can be implemented to decrease the risks of disease transmission. It is recommended that bus drivers be screened for COVID-19 symptoms and temperature daily before starting their morning route. The maximum occupancy of students on the bus should be decreased to allow for only one child per seat, also seated in a zig-zag pattern to maximize distancing. Cost and time considerations may not make this feasible, but reduced crowding should still be a goal. Hand sanitizer placed at the entrance of the bus can encourage children to disinfect their hands while boarding and exiting the bus. Increased cleaning and disinfection protocols should be followed; seats, seatbelts, and windows should be disinfected between each group of students on the bus. Windows should be kept open as an engineering control whenever possible to promote increased airflow through the bus. Weather will guide the extent to which windows can be opened; however, it is recommended that some windows along the length of the bus be kept open approximately 3 inches irrespective of weather. Similar to the personal behavior recommendations made for all students and teachers on school property, all students and the bus driver are recommended to wear a mask while on the bus or queuing to board the bus. Departure should be from front to back and should involve physical distancing. We appreciate the cost implications.

44.5.13 Isolation Room

Schools should arrange to dedicate rooms to be separate isolation areas. Students that develop COVID-19-like symptoms during the school day can be taken to this space until they can return home safely. Isolation rooms are recommended to be located on an exterior wall to maximize ventilation options and have exterior exits to eliminate infected individuals from mixing with healthy staff and students. It is recommended that this space include a bed if the child needs to lie down and all other furniture be removed to facilitate cleaning and disinfecting protocols. Ideally, a dedicated bathroom will be available for students requiring the use of this space. For contact tracing, anyone who enters the isolation room should be logged. Anyone who enters the designated isolation room must use appropriate PPE. Efficient ventilation systems are critical for isolation rooms. Negative pressure should be maintained in the isolation room to limit the dispersion of air into adjoining spaces. Schools should target to have 6–10 air changes per hour of total air in this space (American Society of Heating Refrigerating and Air-Conditioning Engineers 2020b). For schools with mechanical ventilation, return air from isolation rooms should be exhausted directly outdoors. Schools with natural ventilation can use a window-mounted box fan to exhaust air outdoors. If weather permits, this isolation area can be outdoors in a tent.

44.6 Considerations for Vulnerable Populations

44.6.1 Physically Vulnerable Students

It is especially important to protect students, staff, and faculty at increased risk for severe illnesses from COVID-19. People at increased risk are those with pre-existing conditions, immunocompromised, older age, or those with family members who are sick (Jordan et al. 2020). Students with asthma, respiratory problems, immunocompromised, other underlying neurologic, genetic, metabolic, and developmental conditions are at increased risk for severe illness from COVID-19 than other children (Centers for Disease Control and Prevention 2020c). Students and faculties who live with at-risk individuals must be protected with extra precautions.

All members of the school community should be instructed to discuss their risks and concerns with school nurses/healthcare personnel. Any individual who falls within categories of high risk should be permitted to work/study remotely. Necessary supplies, such as laptops and high-speed internet, should be provided. If needed, the school administration can reassign roles to allow at-risk faculties to work while staying safe. Schools should provide adequate PPE, supervision, and other protections based on each individual's risk. Accommodations can be granted case-by-case upon request. Extra precautions must be taken to protect the privacy of high-risk individuals.

44.6.2 Socially Vulnerable Students

Schools should also be prepared to help socially vulnerable students. Socially vulnerable students include but are not limited to those who are uninsured, homeless, those in foster care and/or group homes, undocumented students (i.e., immigrants who may be in camps or considered illegal), and those living in unstable environments. In the United States, one can add persons with Deferred Action for Childhood Arrival (DACA) status. Protecting these students is both a moral and an ethical obligation. Students with limited resources must be provided with the necessary supplies, technology, and support to ensure a safe and equitable academic environment.

44.6.3 Academically Vulnerable Students

Some students with learning disabilities may encounter difficulties with virtual learning platforms. Extra tutoring and academic resources may be beneficial to these children. Teachers and guardians must work together to provide the best learning environment for these students. If limited face-to-face capacity is available in a re-engineered school environment, the academic vulnerable and special needs students should be prioritized for classroom engagement.

44.6.4 Medically Vulnerable Staff and Teachers

Schools should be safe for teachers and other staff. Persons who are over the age of 60–65 years who have underlying medical conditions like asthma, a chronic obstructive pulmonary disorder, diabetes, uncontrolled hypertension, and others are advised to teach virtually or, at the very least, teach with extra precautions. One of the authors (SHV) of this chapter fits this category, and he will be teaching with universal classroom mask use, maximum physical distancing, abundant hand hygiene, and optimized indoor air quality made possible by his university's investments (Vermund 2020). Not all teachers will be so fortunate, given resource constraints and/or a lack of administrative or political support.

44.7 Mental Health Among Students, Teachers, and Staff

The COVID-19 pandemic has shaken many people's lives. Many students are experiencing worry, sadness, anxiety, and trauma related to the upheaval. Students from minority and marginalized communities already experience higher rates of anxiety and depression, while also having limited access to behavioral healthcare. Parents or guardians may have lost jobs, businesses, or other income. The pandemic will only exacerbate the needs of families.

There is a growing literature on student mental health amid the COVID-19 pandemic. Much research has been conducted with university students and those in countries like China, Italy, and Spain that experienced the early surge of infections. During the pandemic and the resultant school closures, several factors converged to contribute to student mental health and psychological distress (Courtney et al. 2020; Duan et al. 2020; Huckins et al. 2020; Zhou et al. 2020; Cao et al. 2020). The uncertainty and fear associated with the pandemic continue to harm students' well-being. Quarantine and isolation result in disturbing insecurity among students. The media is generating heightened stress. Many students are experiencing anxiety, depression, suicidal ideation, substance abuse, and domestic violence. In China, a study found higher rates of post-COVID anxiety in students than the general population (Phelps and Sperry 2020; Wang and Zhao 2020). Adolescents and young adults may suffer more acutely from loneliness and the loss of support from peers than younger children, contributing to additional distress in this population (Loades et al. 2020). Further, there are early indications that the stressors of the pandemic and lockdown may cause post-traumatic stress among some youth (Loades et al. 2020).

Teachers' and school staff's mental health is also critical in reopening schools and providing support to students. Data for these groups are lacking, though studies thus far indicate that many teachers are struggling to balance their commitments to their students with fears of their personal and family's risks of infection (Odriozola-Gonzalez et al. 2020). The sudden switch to online learning has created new burdens on teachers. Teachers and school staff, similar to their students, also face anxieties related to the uncertainty of future disruptions to education and employment (Araujo et al. 2020). Given the importance of teachers beyond their role as educators, any approaches to incorporate mental health and behavioral interventions into school reopening must also consider the needs of teachers and staff (Zhou 2020).

Despite frequent discussions of the stress that teachers and school staff face with the reopening of schools in the media, the mental health implications of school reopening for students are unclear. One study conducted with US college students found that students were amenable to keeping colleges closed until a COVID vaccine was available, but it is unclear how younger adolescents and children might feel about continued distance learning (Cohen et al. 2020). Further research is needed to explore students' expectations and feelings concerning the reopening of schools amid the pandemic, and this information should be incorporated into reopening plans.

It is crucial to provide mental health services to students, teachers, and school staff during this crisis. Tele-mental health services or virtual well-being sessions are effective ways to make mental health and behavioral interventions accessible and safe. Mental health staff should receive training focusing specifically on telehealth, virtual counseling, and pandemic- and trauma-specific stress counseling. Training is necessary to provide the needed education, confidence, and skills required among mental health providers. Virtual communication is an accessible, valid, and responsive venue for student support. However, in-person arrangements should be made in case of a psychological emergency.

Schools should set up meditation, mindfulness sessions, mental health workshops, and therapeutic group discussion opportunities for students and employees with different needs.

Coordinated care that addresses the needs of and incorporates resources for both students and their supports (teachers, family) will be critical (Zhou 2020). Self-care education can also be integrated into daily teaching and activities. Teachers and schools should promote "physical distancing, NOT social distancing" messaging. Teachers can encourage students to replace high-fives and hugs with "virtual hugs" or "flying hearts." Teachers must be vigilant for signs of depression, anxiety, or burnout and bridge students to necessary support. A high degree of knowledge is not necessarily protective of depression and anxiety, as demonstrated in a study of Chinese medical students who were studying public health during the pandemic shutdown (Chen et al. 2020). The socialization, even limited, provided by the school may help avoid mental health problems for some youth and may also enable parents and guardians the opportunity to make a living. Offering mental health support will be crucial to the well-being of students and teachers as they seek a return to school and a new normal.

44.8 Conclusion

Fundamental principles to keep schools safe are similar to those for any indoor environment during the influenza season, but the needs and proclivities of children must be brought to the fore in this more urgent time or greater risk. Fundamental elements are universal mask use, 2 meters or 6 feet distancing, assertive hygiene for hands, face, and surfaces, optimized air quality, outdoor teaching and activities whenever possible, limiting crowds, and abundant testing. A universal influenza vaccine is highly advisable. Provision for temporary isolation or quarantine, along with contact tracing, is needed. Boarding schools, colleges, and universities that host students from far away may need to invoke their *locum parentis* and provide healthcare, quarantine, isolation, and contact tracing on behalf of distanced parents and guardians. Long-distance travel is not safe for a known or potential SARS-CoV-2 carrier.

The feasibility of maximum safety will vary with local attitudes and resources. If political leaders, parents/guardians, educational administrators, or school staff do not take COVID-19 control seriously, much damage can occur if necessary precautions are not taken appropriately. Improved air quality, masks, hand sanitizers, plexiglass barriers, grab-and-go meals, outdoor tents, and smaller class sizes all take extra resources. Governments must acknowledge this and provide additional resources in this time of COVID-19 before the availability of excellent treatments and vaccines.

The current crisis calls for a shift in culture that recommends the universal, routine mask use. Behavioral adherence must be reinforced and incentivized. Vulnerable groups must be protected, whether students, vulnerable persons living with the students, teachers, or other staff. Mitigation measures hopefully enable us to open up schools in lower-incidence settings around the world. Virtual online learning becomes the preferred method of education in the face of high viral transmission in a school's particular community.

References

Abaluck J, Chevalier J, Christakis NA et al (2020) The case for universal cloth mask adoption and policies to increase supply of medical masks for health workers. Covid Econ (5):147–159. https://cepr.org/sites/default/files/news/CovidEconomics5.pdf

American Society of Heating Refrigerating Air-Conditioning Engineers (2019) Ventilation for acceptable indoor air quality in residential buildings. vol ANSI/ASHRAE Standard 62.1-2019 ANSI/ASHRAE, Atlanta

American Society of Heating Refrigerating and Air-Conditioning Engineers (2019) Ultraviolet air and surface treatment. ASHRAE Handbook, Atlanta

American Society of Heating Refrigerating and Air-Conditioning Engineers (2020a) Reopening of schools and universities. American Society of Heating, Refrigerating and Air-Conditioning Engineers https://www.ashrae.org/technical-resources/reopening-of-schools-and-universities

American Society of Heating Refrigerating and Air-Conditioning Engineers (2020b) Reopening schools and universities guidance. American Society of Heating, Refrigerating and Air-Conditioning Engineers, Atlanta

American Society of Safety Pofessionals (2020) How to apply the hierarchy of controls in a pandemic.

American Society of Safety Pofessionals https://www.assp.org/news-and-articles/2020/03/31/how-to-apply-the-hierarchy-of-controls-in-a-pandemic

Araujo FJO, de Lima LSA, Cidade PIM, Nobre CB, Neto MLR (2020) Impact of Sars-Cov-2 and its reverberation in global higher education and mental health. Psychiatry Res 288:112977. https://doi.org/10.1016/j.psychres.2020.112977

Asadi S, Wexler AS, Cappa CD, Barreda S, Bouvier NM, Ristenpart WD (2019) Aerosol emission and superemission during human speech increase with voice loudness. Sci Rep 9(1):2348. https://doi.org/10.1038/s41598-019-38808-z

Asadi S, Bouvier N, Wexler AS, Ristenpart WD (2020) The coronavirus pandemic and aerosols: does COVID-19 transmit via expiratory particles? Aerosol Sci Technol 0(0):1–4. https://doi.org/10.1080/02786826.2020.1749229

Association of Home Appliance Manufacturers (2020) Air filtration standards association of home appliance manufacturers. https://ahamverifide.org/ahams-air-filtration-standards/

Azimi P, Stephens B (2013) HVAC filtration for controlling infectious airborne disease transmission in indoor environments: predicting risk reductions and operational costs. Build Environ 70:150–160. https://doi.org/10.1016/j.buildenv.2013.08.025

Banerjee A, Kulcsar K, Misra V, Frieman M, Mossman K (2019) Bats and coronaviruses. Viruses 11(1). https://doi.org/10.3390/v11010041

Barratt H, Kirwan M, Shantikumar S (2010) Epidemiology. Public health textbook. Public Health Action Support Team, UK

Beggs CB, Kerr KG, Donnelly JK, Sleigh PA, Mara DD, Cairns G (2000) The resurgence of tuberculosis in the tropics. An engineering approach to the control of mycobacterium tuberculosis and other airborne pathogens: a UK hospital based pilot study. Trans R Soc Trop Med Hyg 94(2):141–146. https://doi.org/10.1016/s0035-9203(00)90250-5

Bi Q, Wu Y, Mei S, Ye C, Zou X, Zhang Z, Liu X, Wei L, Truelove SA, Zhang T, Gao W, Cheng C, Tang X, Wu X, Wu Y, Sun B, Huang S, Sun Y, Zhang J, Ma T, Lessler J, Feng T (2020) Epidemiology and transmission of COVID-19 in 391 cases and 1286 of their close contacts in Shenzhen, China: a retrospective cohort study. Lancet Infect Dis 20(8):911–919. https://doi.org/10.1016/s1473-3099(20)30287-5

Cao W, Fang Z, Hou G, Han M, Xu X, Dong J, Zheng J (2020) The psychological impact of the COVID-19 epidemic on college students in China. Psychiatry Res 287:112934. https://doi.org/10.1016/j.psychres.2020.112934

Centers for Disease Control and Prevention (2020a) How to protect yourself & others. Coronavirus disease 2019 (COVID-19). Centers for disease control and prevention, Atlanta

Centers for Disease Control and Prevention (2020b) Interim considerations for K-12 school administrators for SARS-CoV-2 testing. Coronavirus disease 2019 (COVID-19). Centers for Disease Control and Prevention, Atlanta

Centers for Disease Control and Prevention (2020c) Use of masks to help slow the spread of COVID-19. Coronavirus disease 2019 (COVID-19). Centers for Disease Control and Prevention, Atlanta

Chen S, Li F, Lin C, Han Y, Nie X, Portnoy RN, Qiao Z (2020) Challenges and recommendations for mental health providers during the COVID-19 pandemic: the experience of China's first university-based mental health team. Glob Health 16(1):59. https://doi.org/10.1186/s12992-020-00591-2

Cohen AK, Hoyt LT, Dull B (2020) A descriptive study of coronavirus disease 2019-related experiences and perspectives of a National Sample of college students in spring 2020. J Adolesc Health. https://doi.org/10.1016/j.jadohealth.2020.06.009

Connecticut State Department of Education (2020) Adapt, advance, achieve: Connecticut's plan to learn and grow together. Connecticut State Department of Education, CT, USA

Courtney D, Watson P, Battaglia M, Mulsant BH, Szatmari P (2020) COVID-19 impacts on child and youth anxiety and depression: challenges and opportunities. Can J Psychiatr 706743720935646. https://doi.org/10.1177/0706743720935646

Cullotta KA (2020) 'Pandemic pods' and 'micro-schools': how parents are finding ways to help their kids — and themselves — manage schooling at home. Chicago Tribune, 2020/07/28

Dbouk T, Drikakis D (2020) On respiratory droplets and face masks. Phys Fluids (1994) 32(6):063303. https://doi.org/10.1063/5.0015044

Delamater PL, Street EJ, Leslie TF, Yang YT, Jacobsen KH (2019) Complexity of the basic reproduction number (R0). Emerg Infect Dis 25(1):1–4. https://doi.org/10.3201/eid2501.171901

Dong Y, Mo X, Hu Y, Qi X, Jiang F, Jiang Z, Tong S (2020) Epidemiology of COVID-19 among children in China. Pediatrics 145(6). https://doi.org/10.1542/peds.2020-0702

Duan L, Shao X, Wang Y, Huang Y, Miao J, Yang X, Zhu G (2020) An investigation of mental health status of children and adolescents in China during the outbreak of COVID-19. J Affect Disord 275:112–118. https://doi.org/10.1016/j.jad.2020.06.029

Edwards CH, Tomba GS, Sonbo Kristiansen I, White R, de Blasio BF (2019) Evaluating costs and health consequences of sick leave strategies against pandemic and seasonal influenza in Norway using a dynamic model. BMJ Open 9(4):e027832. https://doi.org/10.1136/bmjopen-2018-027832

Fennelly KP (2020) Particle sizes of infectious aerosols: implications for infection control. Lancet Respir Med 8(9):914–924. https://doi.org/10.1016/S2213-2600(20)30323-4

Foarde KK (2010) Development of a method for measuring single-pass bioaerosol removal efficiencies of a room air cleaner. Aerosol Sci Technol 30(2):223–234. https://doi.org/10.1080/027868299304804

Hao W, Parasch A, Williams S, Li J, Ma H, Burken J, Wang Y (2020) Filtration performances of non-medical materials as candidates for manufacturing facemasks and respirators. Int J Hyg Environ Health 229:113582. https://doi.org/10.1016/j.ijheh.2020.113582

Henderson LA, Canna SW, Friedman KG, Gorelik M, Lapidus SK, Bassiri H, Behrens EM, Ferris A, Kernan KF, Schulert GS, Seo P, MB FS, Tremoulet AH, Yeung RSM, Mudano AS, Turner AS, Karp DR, Mehta JJ (2020) American college of rheumatology clinical guidance for pediatric patients with multisystem inflammatory syndrome in children (MIS-C) associated with SARS-CoV-2 and hyperinflammation in COVID-19. Version 1. Arthritis Rheumatol. https://doi.org/10.1002/art.41454

Huckins JF, daSilva AW, Wang W, Hedlund E, Rogers C, Nepal SK, Wu J, Obuchi M, Murphy EI, Meyer ML, Wagner DD, Holtzheimer PE, Campbell AT (2020) Mental health and behavior of college students during the early phases of the COVID-19 pandemic: longitudinal smartphone and ecological momentary assessment study. J Med Internet Res 22(6):e20185. https://doi.org/10.2196/20185

Ippolito M, Iozzo P, Gregoretti C, Grasselli G, Cortegiani A (2020) Facepiece filtering respirators with exhalation valve should not be used in the community to limit SARS-CoV-2 diffusion. Infect Control Hosp Epidemiol:1–2. https://doi.org/10.1017/ice.2020.244

Jordan RE, Adab P, Cheng KK (2020) Covid-19: risk factors for severe disease and death. BMJ 368:m1198. https://doi.org/10.1136/bmj.m1198

Kettleson EM, Ramaswami B, Hogan CJ Jr, Lee MH, Statyukha GA, Biswas P, Angenent LT (2009) Airborne virus capture and inactivation by an electrostatic particle collector. Environ Sci Technol 43(15):5940–5946. https://doi.org/10.1021/es803289w

Knowlton SD, Boles CL, Perencevich EN, Diekema DJ, Nonnenmann MW (2018) Bioaerosol concentrations generated from toilet flushing in a hospital-based patient care setting. Antimicrob Resist Infect Control 7:16. https://doi.org/10.1186/s13756-018-0301-9

Kujundzic E, Matalkah F, Howard CJ, Hernandez M, Miller SL (2006) UV air cleaners and upper-room air ultraviolet germicidal irradiation for controlling airborne bacteria and fungal spores. J Occup Environ Hyg 3(10):536–546. https://doi.org/10.1080/15459620600909799

Kujundzic E, Hernandez M, Miller SL (2007) Ultraviolet germicidal irradiation inactivation of airborne fungal spores and bacteria in upper-room air and HVAC in-duct configurations. J Environ Eng Sci 6(1):1–9. https://doi.org/10.1139/s06-039

Lemecha Obsu L, Feyissa Balcha S (2020) Optimal control strategies for the transmission risk of COVID-19. J Biol Dyn 14(1):590–607. https://doi.org/10.1080/17513758.2020.1788182

Lindsley WG, Noti JD, Blachere FM, Szalajda JV, Beezhold DH (2014) Efficacy of face shields against cough aerosol droplets from a cough simulator. J Occup Environ Hyg 11(8):509–518. https://doi.org/10.1080/15459624.2013.877591

Loades ME, Chatburn E, Higson-Sweeney N, Reynolds S, Shafran R, Brigden A, Linney C, McManus MN, Borwick C, Crawley E (2020) Rapid systematic review: the impact of social isolation and loneliness on the mental health of children and adolescents in the context of COVID-19. J Am Acad Child Adolesc Psychiatry. https://doi.org/10.1016/j.jaac.2020.05.009

Luongo JC, Miller SL (2016) Ultraviolet germicidal coil cleaning: decreased surface microbial loading and resuspension of cell clusters. Build Environ 105:50–55. https://doi.org/10.1016/j.buildenv.2016.05.024

McDermott CV, Alicic RZ, Harden N, Cox EJ, Scanlan JM (2020) Put a lid on it: are faecal bio-aerosols a route of transmission for SARS-CoV-2? J Hosp Infect. https://doi.org/10.1016/j.jhin.2020.04.024

Miller-Leiden S, Lobascio C, Nazaroff WW, Macher JM (1996) Effectiveness of in-room air filtration and dilution ventilation for tuberculosis infection control. J Air Waste Manag Assoc 46(9):869–882. https://doi.org/10.1080/10473289.1996.10467523

Moroney M (2020) More parents are considering microschools amid COVID-19: here's what they entail. PopSugar Inc. https://www.popsugar.com/family/what-are-coronavirus-school-pods-47639027

Moyer MW (2020) Pods, microschools and tutors: can parents solve the education crisis on their own? The New York Times, 2020/7/22

National Institute for Occupational Safety and Health (2020a) Hierarchy of controls. Workplace safety & health topics. National Institute for Occupational Safety and Health, Atlanta

National Institute for Occupational Safety and Health (2020b) Noise and hearing loss prevention. Workplace safety and health topics. National Institute for Occupational Safety and Health, Atlanta

Nishiura H, Kobayashi T, Miyama T, Suzuki A, Jung SM, Hayashi K, Kinoshita R, Yang Y, Yuan B, Akhmetzhanov AR, Linton NM (2020) Estimation of the asymptomatic ratio of novel coronavirus infections (COVID-19). Int J Infect Dis 94:154–155. https://doi.org/10.1016/j.ijid.2020.03.020

North A (2020) Pandemic learning "pods" don't have to be just for the rich. Vox.com. https://www.vox.com/2020/7/28/21340222/learning-pods-covid-private-pandemic-education-school

Odriozola-Gonzalez P, Planchuelo-Gomez A, Irurtia MJ, de Luis-Garcia R (2020) Psychological effects of the COVID-19 outbreak and lockdown among students and workers of a Spanish university. Psychiatry Res 290:113108. https://doi.org/10.1016/j.psychres.2020.113108

Oran DP, Topol EJ (2020) Prevalence of asymptomatic SARS-CoV-2 infection: a narrative review. Ann Intern Med:M20–M3012. https://doi.org/10.7326/M20-3012. Epub ahead of print

Paltiel AD, Zheng A, Walensky RP (2020) Assessment of SARS-CoV-2 screening strategies to permit the safe reopening of college campuses in the United States. JAMA Netw Open 3(7):e2016818

Perencevich EN, Diekema DJ, Edmond MB (2020) Moving personal protective equipment into the community: face shields and containment of COVID-19. JAMA. https://doi.org/10.1001/jama.2020.7477

Phelps C, Sperry LL (2020) Children and the COVID-19 pandemic. Psychol Trauma 12(S1):S73–S75. https://doi.org/10.1037/tra0000861

Pradhan D, Biswasroy P, Kumar Naik P, Ghosh G, Rath G (2020) A review of current interventions for COVID-19 prevention. Arch Med Res 51(5):363–374. https://doi.org/10.1016/j.arcmed.2020.04.020

Prather KA, Wang CC, Schooley RT (2020) Reducing transmission of SARS-CoV-2. Science 368(6498):1422–1424. https://doi.org/10.1126/science.abc6197

Reed NG (2010) The history of ultraviolet germicidal irradiation for air disinfection. Public Health Rep 125(1):15–27. https://doi.org/10.1177/003335491012500105

Santarpia JL, Rivera DN, Herrera V, Morwitzer MJ, Creager H, Santarpia GW, Crown KK, Brett-Major D, Schnaubelt E, Broadhurst MJ, Lawler JV, Reid SP, Lowe JJ (2020) Aerosol and surface transmission potential of SARS-CoV-2. med Rxiv:2020.2003.2023.20039446. https://doi.org/10.1101/2020.03.23.20039446

Shaughnessy RJ, Levetin E, Blocker J, Sublette KL (1994) Effectiveness of portable indoor air cleaners: sensory testing results. Indoor Air 4(3):179–188. https://doi.org/10.1111/j.1600-0668.1994.t01-1-00006.x

Somekh I, Yakub Hanna H, Heller E, Bibi H, Somekh E (2020) Age-dependent sensory impairment in Covid-19 infection and its correlation with Ace2 expression. Pediatr Infect Dis J. https://doi.org/10.1097/INF.0000000000002817

United States Environmental Protection Agency (2020) List N: disinfectants for use against SARS-CoV-2 (COVID-19). United States Environmental Protection Agency, Washington, DC

van Doremalen N, Bushmaker T, Morris DH, Holbrook MG, Gamble A, Williamson BN, Tamin A, Harcourt JL, Thornburg NJ, Gerber SI, Lloyd-Smith JO, de Wit E, Munster VJ (2020) Aerosol and surface stability of SARS-CoV-2 as compared with SARS-CoV-1. N Engl J Med 382(16):1564–1567. https://doi.org/10.1056/NEJMc2004973

Vermund SH (2020) Is classroom teaching advisable during COVID-19? With precautions, absolutely. The Connecticut Mirror, 2020/7/17, https://ctmirror.org/category/ct-viewpoints/is-classroom-teaching-advisable-during-covid-19-with-precautions-absolutely/

Vermund SH, Pitzer VE (2020) Asymptomatic transmission and the infection fatality risk for COVID-19: implications for school reopening. Clin Infect Dis. https://doi.org/10.1093/cid/ciaa855

Wang C, Zhao H (2020) The impact of COVID-19 on anxiety in Chinese university students. Front Psychol 11:1168. https://doi.org/10.3389/fpsyg.2020.01168

World Health Organization (2020a) Advice on the use of masks in the context of COVID-19. World Health Organization, Geneva

World Health Organization (2020b) Clinical management of COVID-19. COVID-19: clinical care. World Health Organization, Geneva

Xu P, Peccia J, Fabian P, Martyny JW, Fennelly KP, Hernandez M, Miller SL (2003) Efficacy of ultraviolet germicidal irradiation of upper-room air in inactivating airborne bacterial spores and mycobacteria in full-scale studies. Atmos Environ 37(3):405–419. https://doi.org/10.1016/s1352-2310(02)00825-7

Yang W (2020) Some new results on mask testing, thanks to @JTdavidhao It is reassuring the filtration efficiency didn't change much for fabrics after washing/drying cycles. N95 & surgical masks are not for wash. Potential triboelectric charging in dried fabrics. Data: http://tinyurl.com/ve28f3l. Twitter

Zhou X (2020) Managing psychological distress in children and adolescents following the COVID-19 epidemic: a cooperative approach. Psychol Trauma 12(S1):S76–S78. https://doi.org/10.1037/tra0000754

Zhou SJ, Zhang LG, Wang LL, Guo ZC, Wang JQ, Chen JC, Liu M, Chen X, Chen JX (2020) Prevalence and socio-demographic correlates of psychological health problems in Chinese adolescents during the outbreak of COVID-19. Eur Child Adolesc Psychiatry 29(6):749–758. https://doi.org/10.1007/s00787-020-01541-4

45 COVID-19 and Its Impact on Tourism Industry

Dimitrios G. Lagos, Panoraia Poulaki, and Penny Lambrou

Abstract

The current COVID-19 pandemic is leading to significant changes in terms of people's economic behavior, which will inevitably impact the tourism industry and tourism activity both worldwide and in tourism host countries. Immediate control measures, such as necessary restrictions on travel, avoiding physical contact, social distancing, as well as tourists' and patients' changes in priority making, have vanished interest in traveling away from the place of usual residence and seeking to receive tourism services. COVID-19 pandemic has caused immediate impacts across the whole spectrum of economic and social activity. The duration and intensity of the arising malfunction in tourism are not yet known; thus, it is too early to make any assessments of the financial losses that will be recorded on an annual basis. However, an initial approach is necessary in order to assess the range of to date impacts, aiming at a critical appraisal of the current situation. It will mainly help in making the appropriate pandemic management plan in the tourism industry.

Keywords

COVID-19 · Economy · Pandemic · Tourism industry · Tourists

45.1 Introduction

The current COVID-19 pandemic has caused immediate impacts across the whole spectrum of economic and social activity of the global community. The whole world is faced with a health crisis that bears a tremendous impact on our societies and people's lives. The ongoing crisis has affected many sectors, among which travel and tourism have been negatively affected to the greatest extent. Taking into account that millions of jobs are at risk, urgent support must be offered. The COVID-19 outbreak has paralyzed the tourism industry, leaving travelers scrambling to return home and devastating economies that are mostly dependent on tourism (Niestadt 2020).

The duration and intensity of the arising malfunction in tourism are not yet known; thus, it is too early to make any specific predictions on the financial losses that will be eventually recorded.

D. G. Lagos (✉)
Department of Business Administration, University of the Aegean, Chios, Greece
e-mail: d.lagos@aegean.gr

P. Poulaki
Department of Economics and Management of Tourism University of the Aegean, Chios, Greece

P. Lambrou
Hospital Manager – Health Management Consultant, Athens, Greece

N. Rezaei (ed.), *Coronavirus Disease - COVID-19*, Advances in Experimental Medicine and Biology 1318, https://doi.org/10.1007/978-3-030-63761-3_45

Nevertheless, it offers a rare opportunity for reflection and recalibration to grow again and to grow better for the planet and people (Organization 2020). In many popular tourism destinations, hotels have been deserted, and restaurants, bars, tourism attractions, theme parks, and museums are closed.

An initial approach is necessary to assess the range of to date impacts of the crisis due to the COVID-19 pandemic, aiming at a critical appraisal of the situation. It will mainly help in making the appropriate management plan in the global tourism industry economy.

Global international tourism arrivals declined by 20–30% in 2020 as compared to 2019. It can be translated into a loss of US$300–450 billion (€270–407 billion) in spending by international visitors (international tourism receipts) (Organization 2020).

According to the Organisation for Economic Co-operation and Development (OECD), the shock that is implied could bring about a "45–70% decline in the international tourism economy in 2020" (Assessment 2020). The prediction of the World Travel and Tourism Council (WTTC) for the year 2020 is that more than 70 million jobs worldwide and 6 million jobs in the European Union will be lost in the travel and tourism market. The WTTC managing director believes that "once the outbreak is under control, it would take up to 10 months for the tourism sector to return to its normal levels" (WTTC 2020).

In most countries, a considerable number of lives have been lost, with the United States, United Kingdom, France, Italy, Spain, China, Belgium, Brazil, and Germany being at the top of this list (Buhalis 2020). The situation is particularly tricky in several European Union countries that are key tourism destinations, such as Italy, Spain, and France. According to the Italian Tourism Federation (Assoturismo), it is estimated that Italy stands to lose around 60% of its tourists this year (Niestadt 2020).

Urgent measures must be taken to ensure the restarting of the travel and tourism sector, which in Italy represents 10.4% of the total GDP (Valeri 2020). Only the trade associations are trying to create a common technical table to advance proposals aimed at buffering the crisis that is affecting the sector.

According to the World Health Organization (WHO), tourism arrivals in 2019, based on data from tourism destinations worldwide, reached close to 1.5 billion, indicating a 3.8% (54 million arrivals) increase compared to 2018, a steady increase over the last 10 years (Organization 2020). According to WTTC, tourism contributed 8.8 trillion dollars in the global economy in 2018, a figure corresponding to 10.4% of world GDP (WTTC 2020).

Beginning from the end of World War II, tourism has shown impressive growth internationally, and it has been slightly affected by events such as terrorism, wars, and natural disasters. Such factors have been overcome, and the lost ground has been gained in a pretty short time. It constitutes common knowledge deriving from global tourism industry statistics that numbers of international tourism arrivals have been continuously increasing over time since 1950 to the present. In particular, the increase of international tourism arrivals has been observed to grow in numbers in the last 70 years, and only a period of 2 years has seen a decrease in arrivals compared to the previous year. Specifically, the largest decrease was recorded in the period 2008–2009, due to the global economic crisis, which affected international tourism, causing a 4% decrease in 2009 (−37 million arrivals) and a 6% decrease in international tourism revenues. In 2003, the severe acute respiratory syndrome (SARS) epidemic, which also started in a Chinese province, had little effect on world tourism, and international tourism arrivals fell by just 0.4% (− 3 million arrivals) compared to 2002. According to international estimations, it is expected that 290 to 440 million fewer arrivals will occur, due to the COVID-19 pandemic, corresponding to a percentage reduction by 20% to 30%.

According to the OECD, the prospects for global growth remain incredibly uncertain. As a result, annual GDP change is projected to be negative and decline estimated by 2.4% in 2020 as a whole, with growth likely to be negative in the first quarter of 2020. The negative impact on confidence in financial markets in the travel sector and the cessation of supply chains contribute to the downturn in all G20 (Group of 20) economies in 2020, especially those closely linked to Australia, China, such as Japan, and Korea.

The World Tourism Organization (UNWTO) points out that today more than ever, the focus of global interest is on man. In this context, estimates that the sudden and unexpected drop in tourism demand caused by COVID-19 puts in jeopardy millions of jobs and the livelihood of workers, while at the same time jeopardizing the progress made in sustainable development and equality in recent years.

UNWTO analyzes the type of travel restrictions introduced by destinations around the world when the WHO declared COVID-19 an emergency international concern for public health. UNWTO estimates that in the first phase of the pandemic, 100% of destinations worldwide are subject to COVID-19 travel restrictions. In the past, international travel has never been limited so extremely. One hundred eighty destinations, or 83% of all destinations worldwide, now have relevant COVID-19 travel restrictions for more than 4 weeks. Of these, 107 destinations, or 49% of all destinations worldwide, have closed their borders or suspended flights for more than 4 weeks. Of the 217 destinations worldwide, 97 destinations (45%) have fully or partially closed their borders to tourists, 65 destinations (30%) have been entirely or partially suspended internationally, 39 destinations (18%) are closing borders in a more differentiated way prohibiting the entry of passengers from specific countries of origin. In comparison, the remaining 16 destinations (7%) apply different measures, such as quarantine or self-isolation for 14 days or visa measures.

Air travel has been hit particularly hard. Some airlines have ceased operations, while others have temporarily canceled all flights (Niestadt 2020). A significant number of flights have been suspended, and many airports have been closed, preventing the departure of flights. Governments have paid significant amounts of money for citizens' repatriation, testing, and providing hospitality in hotels for quarantined travelers.

Within the hospitality sector, small- and medium-sized enterprises have been particularly severely affected. The drop in demand has led many hotels to the decision to close down. As an indicative example, one could mention the country of France on March 17, 2020, when the hotel occupancy rate was reaching a low percentage of 3.3% (in comparison to 65.3% occupancy rate on February 26). By March 30, restaurants and bars had been closed in almost all member states of the European Union (EU), except for Sweden. In many hotels, restaurants, and bars, thousands of staff layoffs or even redundancies have occurred worldwide.

In the second phase of returning to the new normalcy, the global community is concerned. The nature of the pandemic has divided the international scientific community over the measures to be taken. Thus, each country forms its own strategy and deals with the health crisis according to its own health indicators. However, the global community expects unanimity on the ecosystem of transport, which is an essential factor in the travel and tourism industry.

This chapter reviews major global pandemics over time, to assess the effects of the COVID-19 pandemic on the global tourism industry and to make proposals that can help shape a commonly accepted pandemic management plan.

45.2 Basic Definitions and the History of Major Global Pandemics

Looking at the topic of a pandemic, for a better understanding of its impacts on the global level, it is useful to take into account the etymology of this compound word, with the ancient Greek origin, which consists of the prefix "pan-," which means "all," and the word synthetic "demic," deriving from the word "demos," meaning "people." In other words, it is "an epidemic that spreads through human populations over a vast area, crossing international boundaries and usually affecting a large number of people" (Hoad 1993). According to the WHO, the classical definition of a pandemic is: "the worldwide spread of a new disease" (WHO 2020). At first, a pandemic is classified as an epidemic, which is the rapid spread of an emerging disease across a particular region or more regions or defined as "the occurrence in a community or region of cases of an illness … clearly in excess of normal expectancy" (Porta 2014), and then it is classified as a pandemic when there is a global outbreak of the emerging disease. Therefore, the difference between a pandemic and an epidemic is that while an epidemic may affect just one or a few regions, a pandemic affects the entire world. Pandemics are, therefore, identified by their geographic scale rather than the severity of illness (Porta 2014).

The outbreak of COVID-19 was declared as such a pandemic by the WHO on March 12, 2020. When thinking of a new disease, one bears in mind infectious communicable diseases, such as those that emerged in the past and have either been eradicated or even those that have re-emerged, namely, influenza (flu), plague (bubonic plague, respiratory plague, septic plague), smallpox, tuberculosis, cholera, HIV (AIDS), SARS, and the most recent emerged pandemic of COVID-19. Several categories of factors play a central role in the occurrence of a pandemic. According to the WHO, some environmental factors influence the spreading of infectious diseases. The most important of these are water supply, sanitary facilities, eating habits, and climate conditions. In the context of travel and tourism, several factors affect the occurrence of pandemics, in particular, the mobility of populations, new transport links, and traveling. Another category of factors that we could distinguish is public health policy factors, including the influence of civil liberties on public health policy and pandemic preparedness frameworks on policy levels per country through defined standards and health capacity (Madhav et al. 2017).

As far as the first time of epidemics/pandemics occurrence is concerned, according to archaeological findings and historical surveys, researchers' indication of epidemics and pandemics start from the prehistoric era several thousands of years ago (Jarus 2020). Concerning ancient times, although it does not fall under the definition of the pandemic, as indicated by WHO, it is worth mentioning because it was perhaps the most disastrous and fateful epidemic of recorded ancient history (Langmuir et al. 1985). It is the so-called "Plague of Athens" in ancient Greece, which emerged at 430 B.C. not long after a war between Athens and Sparta had begun. It was an epidemic that ravaged the people of Athens, devastating a significant part of the city population, and lasted for years (430–427 B.C.) (Langmuir et al. 1985). According to several estimations, the death toll was then as high as 100,000 people. According to the Greek historian Thucydides (460–400 B.C.), "*people in good health were all of a sudden attacked by violent heats in the head, and redness and inflammation in the eyes, the inward parts, such as the throat or tongue, becoming bloody and emitting an unnatural and fetid breath*" (Crawley 1914). Initially, it was characterized as the "Thucydides syndrome" by many researchers. However, a later hypothesis showed that the pathogenic cause of the plague of Athens might have been different from the cause of the described syndrome (Langmuir et al. 1985).

Throughout the history of pandemics, the one that has been characterized as the "mother of all pandemics" was the Spanish flu, which was the deadliest influenza pandemic (Taubenberger and Morens 2006). An estimation of 500 million people worldwide has fallen victim to the Spanish flu. The death toll was one-fifth of those, namely, an estimation of 100 million deaths. Some indigenous communities have become extinct because of specific pandemic. During World War I, a combination of significant factors concerning the majority of people led the flu to spread and become even more lethal. Several of those factors include soldiers' limited conditions, lack of sanitation (conditions and facilities), and poor nutrition due to war. Despite the name it was given as "Spanish Flu," the disease was likely not to have started in Spain initially. However, the fact that Spain was a neutral nation during the war led to less strict press censorship, so early numbers of the illness victims could be freely published. So, the first report of the flu case was in Spain. As a result, people probably falsely believed the illness was limited to Spain, and the name Spanish Flu stuck. Since it emerged during the war, poor nutrition of the population played an important role in the emergence of the pandemic as a direct result of weakened immune systems from malnourishment (Jarus 2020).

Table 45.1 provides a synopsis of a subset of major global pandemics, having played a significant role in the history of humanity.

Concerning the influenza viral pathogens, one must note that the threat of an influenza pandemic is ever-present. Because of the constant mutations and the emergence of new influenza strains, we can never be sure about the time and the region where the next probable pandemic will arise. In those terms, another new influenza pandemic is inevitable, and the question that is still pending is when another influenza pandemic will arise in the future. In contrast to annual seasonal influenza epidemics, pandemic influenza is defined as "when a new influenza virus emerges and spreads around the world, and most people do not have immunity" by WHO.

Concerning the most recent and still ongoing crisis due to the COVID-19 emergency, caused by the novel strain of coronavirus of 2019, everything is still under uncertainty, and the situation is new and rapidly evolving. Thus, it is difficult to estimate and predict the real impact of the COVID-19 emergency, as the outbreak is still ongoing to the present day and scientists and researchers are still investigating and learning about this novel strain of the coronavirus. Future research will shed light on the "truth" of this ongoing coronavirus crisis.

Throughout history, pandemics have had a tremendous impact on all facets of human life (health, economy, and society), with disease out-

Table 45.1 Subset of major global pandemics in the history of humanity

Occurrence time	Pandemic disease Etiologic agent/pathogen causing the pandemic	Initiation and spread regions	Death toll estimation (subject to debate upon new evidence)
1346–1353	The black death (bubonic plague) The bacterial **pathogen,** *Yersinia pestis*	Europe Globally	Over 50 million people in Europe
1889–1890	The influenza (flu) Viral **pathogens**, influenza virus	From Russia to Europe and globally	1 million people
1918–1920	The Spanish flu is known as the mother of all pandemics Viral **pathogens**, influenza virus	Probably China Globally Not originally from Spain, despite its name	50 to 100 million people
1957–1958	Asian flu pandemic Viral **pathogens,** influenza virus	Asia Globally	1.1 million people
1981-present	AIDS pandemic and epidemic The viral **pathogen,** HIV	Africa Globally	25–35 million people
2009–2010	H1N1 swine flu pandemic Viral **pathogens**, influenza virus A, (H1N1) 2009 virus	Mexico Globally	Between 151,700 and 575,400 people
2019-present	COVID-19 (the novel coronavirus of December 2019) pandemic, caused by a novel strain of coronavirus, called SARS-CoV-2	China Globally	Not yet finally estimated

Prepared with data from:
(1) https://www.who.int/health-topics/plague#tab=tab_1
(2) https://www.visualcapitalist.com/history-of-pandemics-deadliest/
(3) https://www.visualcapitalist.com/history-of-pandemics-deadliest/; https://www.who.int/news-room/detail/08-04-2020-who-timeline%2D%2D-COVID-19; https://www.who.int/diagnostics_laboratory/EUL/en/
(4) https://www.nih.gov/health-information/coronavirus
Note that many of the death toll numbers listed in table are best estimates based on available research. Some, such as the swine flu, *are subject to debate based on new evidence*

breaks having played a significant role in shaping politics, in crushing revolutions, and in establishing racial and economic discrimination (Snowden 2019). In other words, infectious diseases are as crucial in understanding societal development as economic crises, wars, revolutions, and demographic change. Looking at history, one should note that pandemics have ravaged humanity, changing even the route of history (Jarus 2020). On occasions, this has occurred even more instantaneously than wars or natural disasters. On the global level, one area of significant impact is in the tourism industry, as we are going to indicate and discuss in the following sections since tourism is especially susceptible to measures to counteract pandemics because of restricted mobility and (imposed) social distancing.

45.3 The Effects of the COVID-19 Pandemic on Global Tourism Industry

The pandemic of the COVID-19 has caused significant changes in the economic behavior of people, and this, in turn, will have inevitable negative impacts on global tourism activity, especially in the host countries. Immediate measures are taken, such as the necessary restrictions on travel, fear of physical contact, and changing priorities for patients and travelers, which have almost nullified citizens' interest in moving to receive tourism services away from their place of permanent residence.

As of writing this chapter, one-third of the world's countries are in quarantine. The world community is going through one of the worst periods in its history since World War II. The blow to world tourism is enormous and will last. The WTTC estimates that 75 million jobs are in immediate danger in the travel and tourism sector worldwide because of the pandemic, while the UNWTO estimates that international arrivals and receipts of 2020 will be declining.

The effects on the world economy are estimated to be very damaging both in terms of supply and demand, as tourism accounts for 10.4% of GDP and has a multiplier effect on other activities of the world economy. The vertical reduction in tourism flows has also dragged on other sectors of economic activity, such as air travel, car rental, travel agencies, transportation, tour guides, and food business, especially in regions that demand complementaries.

The United States includes 50 different states; each of them makes its own decisions on the measures of social distancing that are being imposed. The relationship between States and the Federal Government is similar to that of Member States with the European Union.

In the short term and as long as there is an escalation of the crisis, tackling the health crisis in countries in which a health system crisis has occurred must be addressed with targeted measures of support from the states of the world community in the context of a creative intervention policy.

The return to new normalcy should be gradual and in line with the development of pandemic cases. It is the first step that can serve as a psychological means to change the behavior of travelers and their expectation of vacationing in low-risk countries. There is no easy way forward until a vaccine is produced and full medical treatment can be developed (Buhalis 2020).

The forecast for the current tourism season is that there will be a large number of last-minute arrivals; thus, all businesses in the tourism industry must be on standby. The model of mass tourism that has supported the main island destinations for many decades is now becoming insecure, and there is limited scope for optimism. The excellent scenario, especially for the countries of solar tourism, is that the tourism season starts in July and lasts for 3–4 months so that there are employment and tourism and related companies can cover part of their operating expenses. It, however, depends on the condition that an international health protocol is implemented so that travelers can move in safety. In no case, domestic tourism can compensate for the short-term economic losses of the global tourism industry. The so-far significant progress in dealing with the crisis due to the COVID-19 emergency in various countries of the world community is a positive, useful example. It is too early to make serious predictions about what "the next day" will be like, not only for tourism but for the lifestyle that will follow in general. However, at the same time, with the resumption of international travel, it is expected to be much pressure from the major tour operators and airlines for significantly lower prices as part of the standard all-inclusive tour packages offered. To this end, the aviation ecosystem will have to adapt to the new data and take measures to prevent the health crisis. For example, airlines will be forced to reduce the completeness of their aircraft to the logic of increasing the distance that separates passengers from each other. The adoption of travel safety measures will inevitably be accompanied by high operating costs that will lead to forced increases in fares. Respectively, airlines should work with airports and ground service companies to increase sanitation in the reception and waiting areas for passengers. The issuance of health passports similar to those issued by the WHO on yellow fever could be a possible solution for travelers until a safe and effective COVID-19 vaccine is produced. Several of the above measures will not only concern the aviation sector but the entire ecosystem of tourism, which, in any case, will have to adapt to the new condition.

In particular, the United States is intervening financially and subsidizing four of the largest airlines to avoid bankruptcies, acquisitions, or mergers, provided they do not dismiss their employees by September. The EU has abandoned aid initiatives in the Member States and simply expects to consider the legitimacy of each case. In any case, the demand for tourism packages, even with significant discounts, will decrease more than expected.

The above return framework to the new normalcy allows us mainly in the host countries to build a new identity (rebranding), a new tourism image based on a new narrative, and a new approach to tourism that creates a high-value chain. The model of mass tourism, which considers the number of arrivals with a small profit margin and relies on low-cost packages and tickets as an indicator of success, must be abandoned. The one-dimensional tourism product "sun and sea" that many countries of the world have developed has long since completed its life cycle and is now following a slowdown. The new model of tourism development that is now emerging must be based on quality in tourism, where potential visitors will have the opportunity for quality tourism services at a reasonable price (value for money).

There are different levels of challenges across each destination; thus, each destination needs to be treated differently. Regions such as Greece, Bulgaria, Cyprus, and Portugal that have performed well may be considered safer than other countries, and in these countries, more facilities will be able to open up and to be made available to visitors again. Scientists/epidemiologists, in cooperation with local authorities, should make a decision once the parameters for each particular place can be defined (Buhalis 2020). For instance, Italy has implemented a tourism-specific support package. Some countries have re-nationalized coronavirus-hit companies (Niestadt 2020). Travel companies could consider six actions:

- Care for customers and employees to keep them safe and support them through the crisis
- Uncertainty management by setting up a nerve center with a custom, real-time dashboard
- Preserving and optimizing liquidity to ensure access to cash and thereby maintaining critical operations
- Preparedness for recovery by determining when, where, and how to ramp up commercial activity
- Planning to compete in the new world, with changing customer behavior and industry landscape
- Planning to return assets to service and reintegrate employees into the workforce

The recommendations are divided into three key areas, including the management of the crisis and minimization of the impact, offering stimulus and acceleration of recovery, and preparing for the future (Organization 2020). The world is currently faced not only with the dilemma of balancing a trade-off between human health and economic prosperity but also with the question of whether the economy will be relieved directly with liquid funds or whether those funds, even if they are granted on very favorable terms, will have to be repaid in the future. Many of the measures taken by governments of different countries are quite similar (Gallen 2020). The most popular ones include tax/fee deferrals and in some instances also waivers; general loan facilitation, either by government or government-backed or backed by special banking institutions; deferrals of payments, such as for principals, and interest for loans ("credit holiday"); short-time work compensation and wage subsidies; and hardship relief, for different settings, sectors, and geographic perimeters. Tourism contributes both directly and via its multiplier effect indirectly to create jobs globally and economic recovery after crises. Post-crisis periods have shown the capacity of tourism to bounce back strongly and quickly after external shocks (Organization 2020).

There are, of course, several questions that still need to be answered, such as:

I. How long will it take to restore the full operation of the global tourism market?
II. Will the problems of the tourism market solved at the national level, or will there be differences per country?
III. What will the travelers' behavior be, and how is the psychology of the population of affected countries going to be?
IV. What will the structural changes that take place in the global tourism industry look like?

In the first question, we estimate that it will take at least 2 to 4 years to be made feasible for the economy gradual return to the new normalcy and for the economy to restart. The tourism industry needs to regain the trust of travelers/tourists in a relatively short time with healthcare measures that need to be agreed upon internationally.

Concerning the second question, we estimate that each country will implement its own strategic plan concerning health indicators and the magnitude of the pandemic crisis. Reducing tourism activity will act as a "domino effect" and will harm other economic activities that are directly or indirectly linked to the production and distribution of the tourism product.

In regard to the third question, we estimate that the negative conjuncture of international dimensions, which is accompanied by negative psychology that limits the intention of people willing to travel, either turns them into cheaper vacation options or shapes a waiting attitude. The shrinking income and the high rate of insecurity and uncertainty of the inhabitants, for the developed traditional countries of origin of tourism as well as for the emerging markets at the heart of the economic crisis with rising unemployment and shrinking, are all parameters that must be taken into account in planning the tourism crisis.

Finally, given that many countries have significant comparative advantages that can be taken into account (such as cultural elements, natural wealth, enormous historical and architectural heritage) and have the potential for multifaceted tourism development, we estimate the occurrence of modern developments in the tourism industry worldwide. A new perception of the holidays, sustainable tourism development, and changes in tourism standards, forcing a redefinition of tourism policy, need to meet the expectations of the global community.

Each pandemic is unique and part of a specific social, scientific, and historical context. How the global community deals with these issues may well be an important factor in determining the survival of our society and perhaps even our species (Snowden 2019). In particular, the ongoing crisis of COVID-19, in addition to its particular characteristics, differs in powerful countries such as the United States and the EU, which are expected to lead the efforts. However, they have so far shown almost no social solidarity toward the global community.

45.4 The Framework of Policy Proposals to Address the Global Tourism Crisis

Targeted measures are needed in the economies most affected by the coronavirus outbreak. A series of such measures are listed below:

- Establishing standard regulations for the reopening of air, sea, and road transport to ensure the safe and uninterrupted movement of travelers around the world
- Formulation and implementation of a unified strategy for reheating demand and restarting the global tourism industry
- Strengthening businesses and employees affected throughout the crisis due to the COVID-19 emergency by providing substantial liquidity and other interventions
- Creation of a World Tourism Recovery Fund
- Issuance of bonds with a growth clause, where the "coupon" and the debt repayment capital are linked to the economic development of the issuing country
- Strengthening public health infrastructure to address possible cases of COVID-19 among travelers and to grow a sense of security and confidence that activates tourism demand again
- Encouragement to establish a body consisting of specialized executives/experts who will deal with the management of the tourism crisis
- Enhancing fiscal support via more robust public investment

We believe that the governments of the wealthiest countries should be of raised awareness given the COVID-19 pandemic and its impacts and take initiatives toward the support of the entire ecosystem of tourism and aiming at ensuring the smooth running of the competitive tourism market. In this regard, central banks, such as the World Bank and the European Investment Bank, should take immediate action to support the global tourism industry.

It is time to **r**eignite, **r**efocus, **r**edesign, and **r**eengineer the global tourism industry: the new 4Rs of tourism! For immediate recovery, the global tourism industry has several critical conditions that need to apply in order to reignite tourism. Also, *the EU is working on many levels to fight impacts due to the COVID-19 crisis* (Buhalis 2020).

45.5 Conclusion

In addition to being a critical threshold for its development, the model of mass tourism conflicts with two critical issues related to the environment and productivity. One is the function of tourism products based on the exploitation of natural resources, characterized by declining scale yields and consequently reduced average labor productivity. The other is the expansion of tourism demand only for low-income social groups which enter in the tourism market.

The ongoing crisis due to the COVID-19 emergency is exacerbating the problem of declining productivity and is causing concern on the direction of tourism and the role it is going to play in terms of national economies worldwide. This issue is already at the center of discussions not only by the stakeholders in the tourism circuit but also by the bodies pursuing state policy in the various countries.

Various countries, concerning their comparative advantage, have many opportunities to develop unique experiences and products that offer a significant competitive advantage, as long as, gradually, they escape the vicious circle of the masses. Therefore, in the current context of the crisis due to the COVID-19 emergency, the proposed tourism policy framework should focus on the qualitative reconstruction (i.e., differentiation and enrichment) of the tourism product and the effective functioning of the tourism industry in the context of endogenous integrated and sustainable tourism development.

Monoculture, especially in times of crisis, creates multiple problems in tourism destinations where tourism is the main economic activity. We believe that the ongoing crisis allows us to reflect on the current model of tourism development. Any changes should be made gradually throughout the range of activities, as they shake the heliotrope pattern based on the steady increase in arrivals from tourism packages promoted by traditional tour operators, often in the logic of low-quality products with all-inclusive status and with low efficiency.

It is a necessity that governments act quickly and dynamically to overcome the COVID-19 pandemic and the economic impact worldwide. It is necessary to allocate financial resources to cover the deficiencies of public health policies/systems, to prevent infection and reduce transmission of the virus. Well-targeted policies also need to focus on supporting healthcare systems and health professionals as well as employees of the tourism sector and protecting the incomes of vulnerable social groups and businesses during the virus outbreak. It is now clear that we need to trust the welfare state more and invest in public health systems, as well as investing in research and (health) education so that we can better deal with unforeseen future health crises.

The macroeconomic policies adopted can help restore confidence and recover tourism demand, but they cannot offset the immediate disruptions that result from forced travel restrictions. In this context, there must be a significant, credible, and internationally coordinated effort to provide the necessary financial resources for the immediate treatment of public health emergencies, the relief from economic shock, and the development of a path to recovery. In this regard, governments need to work to ensure international cooperation to address the challenge and promote standard policies, through direct funding, to mitigate the effects and speed up recovery.

The COVID-19 crisis has led to economic disruption. It, in turn, is affecting the functioning of financial markets and banks in all countries. It is a severe crisis of the capitalist system calling for international multilateral action to manage this issue effectively. Finally, we believe that for reasons of global solidarity, the establishment of a single global health crisis management plan is more necessary than ever because it will provide guidelines for all states to return to the new normalcy and to restart their economies.

References

Assessment OIE (2020) Coronavirus: The world economy at risk

Buhalis D (2020) Brace the future and time to Reignite, Refocus, Redegine, Reengineer our global tourism industry: the 4Rs of tourism

Crawley R (1914) History of the Peloponnesian war done into English. JM Dent-sons, Limited, London

Gallen S (2020) The response of governments vis-à-vis the economic ramifications of SARS-CoV-2. AIEST, St Gallen

Hoad TF (1993) The concise Oxford dictionary of English etymology. Oxford University Press, Oxford

Jarus O (2020) 20 of the worst epidemics and pandemics in history

Langmuir AD, Worthen TD, Solomon J, Ray CG, Petersen E (1985) The Thucydides syndrome: a new hypothesis for the cause of the plague of Athens. N Engl J Med 313(16):1027–1030

Madhav N, Oppenheim B, Gallivan M, Mulembakani P, Rubin E, Wolfe N (2017) Pandemics: risks, impacts, and mitigation. In

Niestadt M (2020) COVID-19 and the tourism sector

Organization WT (2020) Supporting Jobs and Economies through Travel & Tourism – A Call for Action to Mitigate the Socio-Economic Impact of COVID-19 and Accelerate Recovery, UNWTO, Madrid.40. https://doi.org/10.18111/9789284421633

Porta M (2014) A dictionary of epidemiology. Oxford University Press, Oxford

Snowden FM (2019) Epidemics and society: from the black death to the present. Yale University Press, New Haven

Taubenberger JK, Morens DM (2006) 1918 influenza: the mother of all pandemics. Rev Biomed 17(1):69–79

Valeri (2020) Coronavirus abbatte il turismo

WHO (2020) Coronavirus disease (COVID-19) travel advice

WTTC (2020) Travel & Tourism Global Economic Impacts & Trends 2019

COVID-19 and Its Global Economic Impact

46

Zahra Kolahchi, Manlio De Domenico, Lucina Q. Uddin, Valentina Cauda, Igor Grossmann, Lucas Lacasa, Giulia Grancini, Morteza Mahmoudi, and Nima Rezaei

Abstract

Pandemics are enormous threats to the world that impact all aspects of our lives, especially the global economy. The COVID-19 pandemic has emerged since December 2019 and has affected the global economy in many ways. As the world becomes more interconnected, the economic impacts of the pandemic become more serious. In addition to increased health expenditures and reduced labor force, the pandemic has hit the supply and demand chain massively and caused trouble for manufacturers who have to fire some of their employees or delay their economic activities to prevent more loss. With the closure of manufacturers and companies and reduced travel rates, usage of oil after the beginning of the pandemic has decreased significantly that was unprecedented in the last 30 years. The mining industry is a critical sector in several developing countries, and the COVID-19 pandemic has hit this industry too. Also, world stock markets declined as investors started to

Z. Kolahchi
Universal Scientific Education and Research Network (USERN), Tehran, Iran

M. De Domenico
Max Planck Institute for the Physics of Complex Systems, Dresden, Germany

Universal Scientific Education and Research Network (USERN), Dresden, Germany

L. Q. Uddin
University of Miami, Coral Gables, FL, USA

Universal Scientific Education and Research Network (USERN), Miami, FL, USA

V. Cauda
Department of Applied Science and Technology, Politecnico di Torino, Turin, Italy

Universal Scientific Education and Research Network (USERN), Turin, Italy

I. Grossmann
University of Waterloo, Waterloo, Canada

Universal Scientific Education and Research Network (USERN), Waterloo, Canada

L. Lacasa
Universal Scientific Education and Research Network (USERN), Waterloo, Canada

Queen Mary, University of London, London, UK

G. Grancini
University of Pavia, Pavia, Italy

Universal Scientific Education and Research Network (USERN), Pavia, Italy

M. Mahmoudi
Michigan State University, East Lansing, MI, USA

Universal Scientific Education and Research Network (USERN), East Lansing, MI, USA

N. Rezaei (✉)
Universal Scientific Education and Research Network (USERN), Tehran, Iran

Research Center for Immunodeficiencies, Children's Medical Center, Tehran University of Medical Sciences, Tehran, Iran
e-mail: rezaei_nima@tums.ac.ir

N. Rezaei (ed.), *Coronavirus Disease - COVID-19*, Advances in Experimental Medicine and Biology 1318, https://doi.org/10.1007/978-3-030-63761-3_46

become concerned about the economic impacts of the COVID-19 pandemic. The tourism industry and airlines have also experienced an enormous loss too. The GDP has reduced, and this pandemic will cost the world more than 2 trillion at the end of 2020.

Keywords

COVID-19 · Economic impacts · Global economy · Market · Oil · Pandemic

46.1 Introduction

History reminds us that pandemics go beyond health impact and leaves tracks on the economy of nations and the world in direct and indirect matters; the fourteenth-century pandemic, named the Black Death, changed the direction of the global economy as it brought feudalism to an end and created capitalism (Bobdey and Ray 2020). According to the World Health Organization (WHO), viral infections keep appearing and can emerge as a serious problem for global public health. It is of great importance to learn from previous occasions to not only identify the probable economic impacts of the outbreaks but also to utilize previous experiences in dealing and managing the adverse economic effects effectively, efficiently, and in time (Jabbari et al. 2020). Numerous viral epidemics, including the severe acute respiratory coronavirus syndrome (SARS-CoV) from 2002 to 2003 and H1N1 influenza in 2009, have been reported in the last two decades. Not long ago, Saudi Arabia declared an outbreak of the Middle East Respiratory Coronavirus Syndrome (MERS-CoV) in 2012 (Cascella et al. 2020). In December 2019, an outbreak of unexplained pneumonia in Wuhan attracted massive attention on a global scale (Hanaei and Rezaei 2020). After investigating the etiology, it was revealed that a new coronavirus, severe acute respiratory syndrome coronavirus 2 (SARS-CoV-2), is the causative microorganism (Lotfi and Rezaei 2020) of the so-called coronavirus disease 2019 (COVID-19).

After months, COVID-19 is still a living challenge to medical society (Sahu et al. 2020; Moazzami et al. 2020). It has already transmitted between people globally (Lotfi et al. 2020; Rezaei 2020a), though the behavior of disease is variable depending on numerous factors, including age, sex, medical conditions, immunogenetic background, and the organs affected by the disease (Hessami et al. 2020; Yousefzadegan and Rezaei 2020; Ahmadi et al. 2020; Rokni et al. 2020; Darbeheshti and Rezaei 2020; Bahrami et al. 2020; Yazdanpanah et al. 2020a; Babaha and Rezaei 2020; Lotfi and Rezaei 2020; Nasab et al. 2020; Yazdanpanah et al. 2020b; Jahanshahlu and Rezaei 2020a; Saleki et al. 2020; Ahanchian et al. 2020; Saghazadeh and Rezaei 2020a). There is a widespread effort (Momtazmanesh et al. 2020; Mohamed et al. 2020a; Rzymski et al. 2020; Moradian et al. 2020) to meet the need of developing diagnostic, prophylactic, and therapeutic methods specified for the COVID-19 (Sharifkashani et al. 2020; Mohamed et al. 2020b; Rezaei 2020b; Rabiee et al. 2020; Lotfi et al. 2020; Mansourabadi et al. 2020; Pashaei and Rezaei 2020; Fathi and Rezaei 2020; Basiri et al. 2020a; Basiri et al. 2020b), while old antiviral agents, immunomodulatory drugs, monoclonal antibodies, intravenous immunoglobulin therapy, and plasmapheresis remain as supportive care (Pourahmad et al. 2020; Jahanshahlu and Rezaei 2020b; Saghazadeh and Rezaei 2020b).

The increasingly frequent crises and outbreaks of novel forms of coronaviruses are warning us that diseases like COVID-19 should be expected to be a significant threat to public health. Indeed, novel CoV outbreaks in the upcoming years seem inevitable considering climatological and ecological changes and growing human and animal contacts (Chen et al. 2020). In the first month of the beginning of this outbreak, it was limited to China, and its impact on China's economy caused interruptions and a decrease in productions. Consequently, the worldwide supply chain was interrupted. Activities of thousands of firms and corporations stopped because their necessary equipment and material were provided by China (Fernando 2020). The impact of COVID-19 on the global economy is estimated to be far higher than that of SARS or other outbreaks in the last few decades. In 2002, China had a less critical role in the global economy than nowadays, where it has become an economic superpower (Bobdey and Ray 2020). Despite the restriction of infected cities, just a few months later, the outbreak spread around the world, and, eventually, WHO declared that the situation was a pandemic (Organization). In a situation like this, governments are unable to simultaneously mitigate both the outbreak mortality and the higher-order effects of the viral spread to the highly interdependent socio-economic system. At the beginning of a large-scale epidemic like this pandemic, the short-term goal is controlling the contagion spread and keeping death rates as low as possible. However, avoiding the collapse of national public health systems comes with other costs in terms of both the social and the economic systems, forcing policymakers to take specific measures to mitigate the unavoidable economic downturn in the mid- and long term.

Until now, in the absence of a tailored cure and a vaccine, the most effective strategy to contain the COVID-19 outbreak has relied on non-pharmaceutical interventions, from quarantining the contacts of all people by isolating them in their homes (Burns et al. 2008; Anderson et al. 2020) to physical distancing, changing contact patterns, restricting human mobility, and enhancing individual hygiene (Lai et al. 2020; Chinazzi et al. 2020; Zhang et al. 2020b). Studies estimate that nearly 80% of COVID-19 patients are asymptomatic or have a mild form of the disease. Thus, by symptom-based isolating, containing the outbreak is impossible (Anderson et al. 2020). As long as the quarantine status continues, people cannot go to work, and many firms and companies are unable to operate. Global and domestic transports will continue to be restricted. Inactivity of companies and global transports will reduce oil prices.

Additionally, some consumers and companies panic, show distorted typical patterns of consumption, and generate anomalies in the market. Also, these changes affect global financial markets, and stock market indices around the world drop significantly. Rising healthcare costs put the government under severe financial pressure (Fernando 2020). The more the virus spreads, the more effects it will have on economic activities globally. Government policymakers must provide synchronized and organized responses to mitigate the economic downturn.

From the perspective of complexity science, a pandemic like the COVID-19 is a shock that propagates through our social contacts, favored by the high interconnectivity of our transportation infrastructures. This type of shocks quickly spread through the interdependencies (Gao et al. 2012) of our multilayered socioeconomic system (De Domenico et al. 2013, 2016; Kivelä et al. 2014), and it is known that the robust-yet-fragile feature of real complex systems (Albert et al. 2000; Vespignani 2010; Carlson and Doyle 2002) has the potential to cause an abrupt collapse of a given system (Radicchi and Arenas 2013; Radicchi 2014).

Based on available information about the COVID-19 pandemic and previous literature on viral epidemics, in the present chapter, we outline the impacts of COVID-19 on the global economy.

46.2 Why the Pandemic Affects the Global Economy

Studies strongly link public health to economic well-being and development. Pandemics can impact the global economy in many ways (Fernando 2020; Haacker 2004; Maliszewska 2014; Medicine 2004). A disease usually affects the economy either directly or indirectly. To obtain an estimation of the economic burden of an infectious outbreak, a conventional way to evaluate the reduction of potential revenue from death and disability is using morbidity and mortality rates in addition to healthcare expenditure and related services. There is no doubt that this conventional way underestimates the actual economic burden of those infectious outbreaks, which are incredibly contagious and spreading rapidly around the world. For instance, SARS, HIV/AIDS, and pandemic influenza are previous epidemics that can teach us valuable lessons to estimate and respond to the COVID-19 outbreak appropriately (Fernando 2020).

46.2.1 Behavioral Impact

One of the influential factors of the economic impact of the SARS outbreak was fear. The fear factor is also expressing in the COVID-19 pandemic. Psychological impacts of the fear of an unidentified lethal virus resemble the response to a biological threat or terrorism dangers and lead to anxiety and stress, usually with long-term consequences. At the beginning of a pandemic, a large percentage of the population will feel a great danger, even though their real increased risk of infection is small (Fernando 2020). This fear factor results in three outcomes: first, a significant decline in market demand, especially for transportation services and retail trade. The rapid rate of epidemics makes people refuse social interactions in affected countries. In regions with higher population densities, the unfavorable demand shock is more severe. The psychological shock is raging all across the world, not only with the countries where COVID-19 can spread locally since the world is so intimately connected through international travels (Medicine 2004). Second, operating costs in businesses rise, and based on the region risk, the acceleration of premium costs would occur (Fernando 2020). Third, the labor force is reluctant to get involved in group activities; therefore, a fear factor also reduces the labor force participation rate (Maliszewska 2014). Besides, fear is not the only way how a pandemic can impact behavior. There are numerous theories on the impacts of infectious decease threat, linking it not only to mental health but also to prejudice toward outgroups, more conservative attitudes, and less risk-taking (and less creativity). For instance, behavioral immune system theories suggest that intergroup processes are affected by evolutionary and ontogenetic pressures related to pathogen stress (Faulkner et al. 2004; Fincher et al. 2008; Murray et al. 2011; Schaller and Park 2011; Tybur et al. 2016), suggesting possible effects of pathogen threat to outgroup prejudice, political polarization, cultural values related to traditionalism and individualism, and prosocial and antisocial behavior. Moreover, life history theories in psychology and evolutionary biology suggest that organisms should alter future-oriented behavior and reproductive activities in response to environmental unpredictability (Griskevicius et al. 2011; Varnum and Grossmann 2016, 2017) and suggest possible consequences of the pandemic-induced uncertainty for delay of gratification and birth rates.

By evaluating the last pandemics in 2003 (SARS) and 2009 (H1N1 flu), it is estimated that about 80–90% of the total economic impact of them is because of behavioral impact (Maliszewska 2014).

46.2.2 GDP Loss

To realize the impact of economic growth and GDP development, we start by reviewing the US economy's growth over a long period (Maliszewska 2014). The real per capita of the United States grew from $3340 in 1870 to $33,330 in 2000. The growth rate of per capita GDP was 1.8% per year. Now suppose that the growth rate of the United States decreases by 0.8% per year from 1870 to 2000. In this condition, the real per capita gross domestic product (GDP) in 2000 would be $9450, approximately just 28% of $33,330; a tiny difference in GDP results in significant consequences (Robert J. Barro 2004).

Low- and middle-income countries have considerably higher rates of death during pandemics due to inadequate appropriate medical services, greater dissemination rates of diseases, and higher amounts of malnutrition and comorbidity. Consequently, a more significant economic impact happens in these countries and then affects the whole world (Jamison et al. 2017). Also, open economies seem to be more susceptible to disruptions in pandemic situation, and countries which have large-scale exports are the most affected (Chang et al. 2007). Furthermore, as much as the monetary policy of a country resembles the fixed exchange rate regime, the pandemic costs will be more than countries with floating exchange rate regimes (McKibbin et al. 2006).

Several studies have estimated the possible economic consequences of a pandemic (Chang et al. 2007). Bloom and colleagues assessed the economic impacts of a pandemic by avian flu virus with 0.5% CFR and a 20% attack rate. In this case, through infecting East and South Asia, excluding Japan, a loss of around 6.5% GDP (282.7 US$ billion) will occur (Bloom et al. 2005). Also, global GDP will decline by 0.6 percent (Chang et al. 2007). Burns et al. reported that an avian flu pandemic similar to 1918–1919 Spanish flu can reduce global GDP even by 5% (Burns et al. 2008).

Additionally, McKibbin and Sidorenko simulate an ultra-scenario for a global pandemic, which was similar to 1918–1919 Spanish flu, but with the excess mortality rate for older people. They estimated that in such a situation, the GDP would decrease by 5.5% in the United States and 8% in Europe. Also, 142.2 million people would die, and global GDP would decline by 12.6% (McKibbin et al. 2006). Altogether, evidence shows that the range of economic impact of the pandemic is extensive, although even considering a mild pandemic, the economic consequences may be noticeable.

46.2.3 The Decrease in Tax Revenue and Fiscal Shocks

A decrease in consumption and income would result in the reduction of tax revenue, particularly in low- and middle-income countries with more restrictions on fiscal policy and weak tax systems. It usually coincides with raised health spending and public expenditure, which puts governments under double pressure (Maynard and Bloor 2009; Jamison et al. 2017). When the pandemic is mild to moderate, high-income countries are not affected, and support influenced low- and middle-income countries to mitigate a fiscal crisis. However, in stressful situations, high-income countries are also under pressure and usually reluctant to support others. So, low- and middle-income countries may experience more significant fiscal problems, eventually resulting in reduced appropriate actions to public health and perhaps being forced to change other government expenditures (Jamison et al. 2017).

46.2.4 Reduction of the Labor Force and Forgone Income

Bloom et al. reported that improvement in health capital has a positive correlation with GDP. A 1-year rise in the life expectancy of a community can result in productivity enhancement of about 4%. It is a pretty significant impact, suggesting that higher healthcare spending may be supported solely based on its effects on the output of labor (David et al. 2001).

Forgone income, typically measured by the amount of missed working days due to the disease, is another factor that must be considered while calculating the costs of a pandemic. When somebody dies due to the disease, forgone income is obtained by the determined value of the expected lifetime that has been gone, predicted on estimated income for various age categories and survival estimates. Thus, in the case of deadly outbreaks, the forgone income is significant. For example, over 1 million people die due to malaria every year (Medicine 2004), and approximately 770,000 people died from HIV around the world in 2018 (Organization 2020).

High mortality rates of pandemics also lead to negative population shock, although the impacts on GDP are divergent based on affecting different age categories. An outbreak of an infectious decease, which mainly results in the death of children and the elderly, despite no effect on the economically productive people between 15 and 54 years of age, will enhance the GDP. On the contrary, a decrease in the labor force increases the capital-labor proportion and reduces the return rate of capital, resulting in slower accumulation of capital and, therefore, lower growth rate (Medicine 2004). Noticeably, death is not the only cause of a decline in the labor force; as long as pandemic continues, governments quarantine all people at home and prevent them from going to work. At the same time, schools are closed, so taking care of school-aged children is another reason which gives rise to absenteeism from work. Even people that have the necessary jobs to provide for the basic needs of others sometimes have to stop working because of concerns regarding catching the disease and recovery time (Jamison et al. 2017). It is also evident that in epidemic situations, laborers who need to move between various regions to seek work face many problems. Preventing individuals from working in a workplace with much productivity is another cause of a decline in economic growth (Medicine 2004). If one accounts for the existing social and economic inequality, it is straightforward to realize that the effects of a declining economic growth will not affect all socioeconomic classes uniformly, with the poorest and most vulnerable ones being more exposed to the dramatic effects of financial shocks and recessions.

It has been discussed whether morbidity or mortality factors are the leading cause of the economic costs of a pandemic. Both morbidity and mortality factors are important, but their effects unfold differently over time. In the short term, the morbidity rate through absenteeism, closure of schools, decreasing production, and raised hospital expenditures affects more than the mortality factor. Conversely, the lost labor force due to mortality, and the influence of foregone earnings account for pandemic outcomes in the long term (McKibbin et al. 2006).

An important factor influencing the impact of the reduction of the labor force on the global economy due to absence from work is the duration of the pandemic. It is still unknown whether the COVID-19 pandemic will be mitigated through summer or not. A convenient way to find the answer is to compare COVID-19 to influenza A epidemic. Because of school closure and higher temperature, the transmission of influenza A virus with a basic reproduction number (R_0) of 1.1–1.5 decreases significantly during summer. However, based on mathematical modeling, R_0 of COVID-19 is around 2–3 or even can be up to 6 (Hellewell et al. 2020; Kucharski et al. 2020; Zhang et al. 2020c). Thus its prevalence may not significantly decrease through different seasons as much as influenza A. So, it is probable that COVID-19 pandemic lasts more than one or two seasons, which can increase its impact on the global economy (Anderson et al. 2020; Guo et al. 2020).

Epidemics also can have indirect consequences in the long term. As stated by the child-survivor hypothesis, parents must be sure that a definite number of their children will survive. A life-threatening outbreak, especially if it becomes endemic in some regions, will increase the fertility rate in the future. Therefore, the growth rate of a population is also increased in those regions. According to the quantity–quality trade-off, an increased number of children in a family has adverse effects on their education level. Low levels of education and increased population can significantly reduce GDP and income per capita. For example, the rapid growth of population and an extraordinary fertility rate are correlated with high mortality in some African regions with a high prevalence of malaria (Sachs and Malaney 2002).

46.2.5 Abnormal Consumption Patterns

In addition to the issues mentioned above, fear influences consumption patterns among people, and sometimes economic strategies of firms change also. During the SARS outbreak, it was observed that as community concern becomes challenging, the willingness of people to pay for eliminating the likelihood of infection increases. All of these reasons contribute to generating anomalies in the market (Fernando 2020; Liu et al. 2005).

46.2.6 Increased Healthcare Expenditure

Containment of infection spread in hospitals requires a set of wide-ranging and expensive measures. The staff must use gloves, N95 masks, gowns, and goggles during any contact with patients or procedures. Additionally, hospitals must limit the working staff daily, the number of patients, and the number of visitors as much as possible. Despite a lower number of patients, the cost of supplies will increase significantly (Achonu et al. 2005). Due to reduced patient burden and to prevent more exposures, several employees are sent home during the pandemic situation although they are paid. On the other hand, the working staff is paid more than usual because of the condition. Retail sales of the hospital will close considerably. Thus the revenue will decrease too because part of hospital revenue is obtained from retail sales (Achonu et al. 2005). During the COVID-19 pandemic, while the healthcare systems across the world are subjected to stress, they can even get overwhelmed by high incidence numbers during various epidemic waves, and resources (such as ventilators) and demand (such as ICU patients) could be load-balanced.

46.3 Possible and Definite Impacts of the COVID-19 Pandemic on the Global Economy

As a consequence of the SARS pandemic in 2003 and the COVID-19 pandemic in 2020, the recurrent coronavirus outbreaks which are happening in China may have negative impacts on foreign investors in China or even countries with low public health. Its impact is not limited to these countries; other countries that are also competing for foreign direct investment (FDI) may be influenced (Medicine 2004).

Although the mortality rate of COVID-19 is approximately 4% (Zhang et al. 2020a), studies suggest that there is no strong correlation between the mortality rate of the COVID-19 and its economic impacts. COVID-19 pandemic influences the global economy via other channels (Fernandes 2020).

COVID-19 pandemic started affecting China from December 2019, and its growth prospect decreased to less than 5% in 2020, compared to 6.1% in 2019. It is also estimated that global economic growth will decrease by 2.4%, or even, as a result of the international decline of economic activities, it can decrease by 1.5% in 2020, while it was 2.9% in 2019 (Development 2 March 2020).

As reported by the World Travel and Tourism Council (WTTC), the tourism industry is an enormous economic sector comprising 10.4% of the GDP of the world and 10% of worldwide employment. After the onset of the COVID-19 pandemic, one of the most affected sectors was tourism. Chinese people alone comprise one-third of all tourists around the world; the declined rate of their activity caused a considerable demand drop across the world. Also, when the virus spread in other parts of the world, its effect became larger. It is estimated that a drop rate of tourists around 20%–30% will occur, costing approximately 300 to 450 billion dollars. Furthermore, the International Air Transport Association (IATA) evaluated the airline industry lost nearly $252 billion, with also a 44% fall in income. Besides, WTTC has predicted that about 50 million employees will be at risk of losing their job in the tourism and travel industry as a result of the COVID-19 pandemic (Açikgöz and Günay 2020; Council 2019; Nicola et al. 2020; Chakraborty and Maity 2020).

A British Plastics Federation (BPF) study aims to look at the effect of COVID-19 on UK manufacturing companies. More than 80% of surveyed persons predicted a drop in turnover within the next 6 months, and 98% have the opinion that there are adverse effects of COVID-19 on business activities (Federation 20 March 2020). On the other hand, many countries, including China, have locked their borders to try to contain the infection. The pandemic has hit the supply and demand chain massively and caused trouble for manufacturers who have to fire some of their employees or delay their economic activities to prevent more loss. It is a very terrible and long-lasting consequence of the pandemic because of a dramatic increase in the unemployment rate. Additionally, some of these closed manufacturers may never re-open again. For example, in China, manufacturing output dropped by around 13.5%. Multiple industries practically collapsed entirely. Restaurant sales decreased by 95%, and car sales declined by 92% (McKee and Stuckler 2020). Based on the prediction of Goldman Sachs, the US economy will contract by 24% in the middle of 2020, which is unprecedented in US history (Reinicke March 2020). Its declined business activity leads to the reduction of money circulation and, thus, tax income (McKee and Stuckler 2020). As tax income decrease, less money will be available to support public health, which is vital in containing the infection. Therefore, it generates a vicious cycle that helps to spread the effects of the pandemic.

On the contrary, there is some benefit to adapting to this pandemic. For instance, disrupting the supply chain may force industries to provide their needs from new routes of supply chains and make them find new markets for their products. Furthermore, working from home, online educational sessions, and online shopping could lead to an improvement in digital technology (Açikgöz and Günay 2020).

With the closure of manufacturers and companies and reduced travel rates, usage of oil after the beginning of the pandemic has decreased significantly. It caused a reduced oil price and led to a negotiation between oil producer countries to stabilize the price. However, the argument between two significant oil producers in the world, Russia and Saudi Arabia, prompted Saudi to produce a tremendous amount of oil, which caused more imbalance between the supply and demand of oil. It dramatically decreased the oil price, which was unprecedented in the last 30 years. It is getting back to average price slowly, but it takes a long time to become stable (Nicola et al. 2020). Theoretically, declining oil prices would have a positive effect on growth, as the cheaper fuel becomes, the less cost for business would be. Therefore, the prices of products will be less, too, and the purchasing power of consumers will enhance, although this is only true for importer countries and can hit the economy of exporter countries (Açikgöz and Günay 2020).

The mining industry is a critical sector in several developing countries, and the COVID-19 pandemic has hit this industry too. With the closure of many manufactures and companies worldwide, the demand for materials and metals has decreased considerably. Thus, alongside reduced exportation, the prices dropped dramatically, especially for aluminum and copper. This condition is very similar to the global financial crisis (GFC) of 2008–2009 for the mining industry, but there are some differences too. For example, during the GFC, the gold industry grew conversely. Gold prices increased twice compared to the time before the GFC. It assures investors to take their money and invest it in the gold industry as a safe market. On the contrary, the new pandemic has affected the gold industry too, and some gold companies are experiencing a dramatically negative growth rate. Some of them also had to furlough their employees in order to save the company from more significant loss (Laing 2020).

World stock markets declined as investors started to become concerned about the economic impacts of the COVID-19 pandemic. In order to stabilize markets, governments started to cut interest rates, thus providing cheaper loans (Açikgöz and Günay 2020).

The COVID-19 pandemic will probably cost the world approximately $2 trillion just in 2020, estimated by the United Nations Conference on Trade and Development (UNCTAD) (Development 9 March 2020). The April report of World Economic Prospects predicted that the GDP would decline as much as 7% in the first half of 2020. However, in the second half, some of the business activities may begin again and prevent more loss of GDP. Although if another outbreak of the virus occurs in the second half, the GDP loss will get to 8% (Prospects April 2020).

Physicians are also affected. Recently, a study surveyed 724 physicians and showed that the COVID-19 pandemic negatively impacted 97% of them (Association. 2020). A decline of 70% in surgical interventions and a decline of 33% of in-office visits were mentioned. Research by the HealthLandscape and the American Academy of Family Physicians predicted that approximately 60,000 family practitioners would close or reduced their working considerably. The layoff of nearly 800,000 workers will be inevitable if the condition remains unchanged (Satiani et al. 2020; O'Donnell April 2 2020).

The overcoming of the COVID-19 outbreak can enable new economic and social patterns in the direction of sustainable development in the future, as highlighted in the European Green Deal project (Deal 2019) and also supported by the major universities throughout the world. The health emergency has not canceled the environmental world crisis, but it has, in contrast, worsen it.

To face the health emergency, most of the university institutes have made substantial technological and organizational advancements in few days, delivering lessons virtually, using online platforms that allowed the connection of hundreds of students. Similarly, smart work organizations in most companies and public entities also applied. Such exceptional digital transformation and conversion of the work roles and dues must not be lost after the overcome of the pandemic period but should be valorized to empower workers and companies of flexibility and work delocalization. It will also positively impact the workers' mobility, for example, reducing the need to travel day by day to work and ultimately contributing to the reduction of pollution by own car traveling. It is also a cultural transformation. We all have to occasion to rethink and reshape our lifestyles, methods and processes of education, and production and consumption of goods. Also, we need to redefine environmental policies, promote policies aware to population needs, fight climate changes and biodiversity loss, and sustain the development of highly innovative companies for high-tech innovation and rapid time-to-market delivery of innovative products. It requires obtaining the broad promotion and economic support of fundamental and applied research to solve the real world's problem.

Since the beginning of COVID-19 pandemic, the daily discharge of carbon dioxide (CO2) gas has been reduced drastically by 17% worldwide, as the most significant drop in CO2 emissions since World War II, albeit for unsustainable and unwanted reasons, although the energy consumption will get back to its previous amounts soon, as we see it is returning in China now (Le Quéré et al. 2020; Rugani and Caro 2020; Tollefson 2020).

Carbon emissions usually are reported annually, but this time, after the beginning of the COVID-19 pandemic, the dramatic reduction in the daily discharge of the carbon dioxide was reason enough to attract attention. It was a fantastic result for people how rapid air pollution could disappear by stoppage of driving and using fossil fuels. Such an event has occurred in 2008 and 1970; in those times, it results in paying more attention to the solar and renewable alternative ways to make energy. There is hope now that the COVID-19 pandemic reminds us of the possibility of living in a cleaner world (Le Quéré et al. 2020; Rugani and Caro 2020; Tollefson 2020).

46.4 Conclusion

To summarize, pandemics are enormous threats to the world that impact all aspects of our lives, especially the global economy. Its effects are short term and long term, including reduced human capital, reduced GDP, shocks of stock markets, abnormal patterns of markets, decreased tax revenue, and delayed business activities. For this reason, health and the economy are correlated strictly. The more information we acquire about its mechanisms, the more we can do to handle the situation. Contrary to decades ago, nowadays, the deluge of multidimensional and multilevel data, together with the adequate mathematical and computational methods of analysis, e.g., based on the science of complexity, provides a suitable starting point to map the propagation of shocks through our socioeconomic system. Also, they provide the ability to design a more resilient society. Beyond everything, the most important goal must be preventing another pandemic in the future (Kafieh et al. 2020), and this will be difficult to achieve while the risk of re-infection with COVID-19 is possible (Jabbari and Rezaei 2020).

References

Achonu C, Laporte A, Gardam MA (2005) The financial impact of controlling a respiratory virus outbreak in a teaching hospital: lessons learned from SARS. Canadian J Public Health (Revue canadienne de sante publique) 96(1):52–54. https://doi.org/10.1007/bf03404018

Açikgöz Ö, Günay A (2020) The early impact of the Covid-19 pandemic on the global and Turkish economy. Turk J Med Sci 50(SI-1):520–526. https://doi.org/10.3906/sag-2004-6

Ahanchian H, Moazzen N, Sezavar Dokht Faroughi M, et al (2020) COVID-19 in a child with primary antibody deficiency. Clin Case Rep 00:1–4. https://doi.org/10.1002/ccr3.3643

Ahmadi M, Saffarzadeh N, Habibi MA, Hajiesmaeili F, Rezaei N (2020) Colon cancer and SARS-CoV-2: impact of ACE2 expression in susceptibility to COVID-19. bioRxiv. 2020.06.11.146878

Albert R, Jeong H, Barabási A-L (2000) Error and attack tolerance of complex networks. Nature 406(6794):378–382. https://doi.org/10.1038/35019019

Anderson RM, Heesterbeek H, Klinkenberg D, Hollingsworth TD (2020) How will country-based mitigation measures influence the course of the COVID-19 epidemic? Lancet (London) 395(10228):931–934. https://doi.org/10.1016/s0140-6736(20)30567-5

Association. MGM (2020) Covid-19 financial impact on medical practices. https://www.mgma.com/getattachment/9b8be0c2-0744-41bf-864f-04007d6adbd2/2004-G09621D-COVID-Financial-Impact-One-Pager-8-5x11-MW-2.pdf.aspx?lang¼en-US&ext¼.pdf

Babaha F, Rezaei N (2020) Primary immunodeficiency diseases in COVID-19 pandemic: a predisposing or protective factor? Am J Med Sci. https://doi.org/10.1016/j.amjms.2020.07.027

Bahrami A, Vafapour M, Moazzami B, Rezaei N (2020) Hyperinflammatory shock related to COVID-19 in a patient presenting with multisystem inflammatory syndrome in children: first case from Iran. J Paediatr Child Health. https://doi.org/10.1111/jpc.15048

Barro RJ, Sala-i-Martin X (2004) Economic growth, 2nd edn. MIT Press, Cambridge MA

Basiri A, Heidari A, Nadi MF, Fallahy MTP, Nezamabadi SS, Sedighi M, Saghazadeh A, Rezaei N (2020a) Microfluidic devices for detection of RNA viruses. Reviews in Medical Virology n/a (n/a): e2154. https://doi.org/10.1002/rmv.2154

Basiri A, Pazhouhnia Z, Beheshtizadeh N, Hoseinpour M, Saghazadeh A, Rezaei N (2020b) Regenerative medicine in COVID-19 treatment: real opportunities and range of promises. Stem Cell Rev Rep:1–13. https://doi.org/10.1007/s12015-020-09994-5

Bloom EA, de Witt V, Jose MJC-S (2005) Potential Economic Impact of an Avian Flu Pandemic on Asia. Asian Development Bank. https://www.adb.org/publications/potential-economic-impact-avian-flu-pandemic-asia

Bobdey S, Ray S (2020) Going viral – Covid-19 impact assessment: a perspective beyond clinical practice. J Marine Med Soc 22:9. https://doi.org/10.4103/jmms.jmms_12_20

Burns A, Van der Mensbrugghe D, Timmer H (2008) Evaluating the economic consequences of avian influenza. World Bank, Washington, DC

Carlson JM, Doyle J (2002) Complexity and robustness. Proc Natl Acad Sci 99(suppl 1):2538. https://doi.org/10.1073/pnas.012582499

Cascella M, Rajnik M, Cuomo A, Dulebohn SC, Di Napoli R (2020) Features, evaluation and treatment coronavirus (COVID-19). StatPearls Publishing, Treasure Island

Chakraborty I, Maity P (2020) COVID-19 outbreak: migration, effects on society, global environment and prevention. Sci Total Environ 728:138882. https://doi.org/10.1016/j.scitotenv.2020.138882

Chang C-C, Lee D-H, Lin H-C, Hsu S-H (2007) The potential economic impact of avian flu pandemic on Taiwan

Chen Y, Liu Q, Guo D (2020) Emerging coronaviruses: genome structure, replication, and pathogenesis. J Med Virol 92(4):418–423. https://doi.org/10.1002/jmv.25681

Chinazzi M, Davis JT, Ajelli M, Gioannini C, Litvinova M, Merler S, Pastore Y, Piontti A, Mu K, Rossi L, Sun K, Viboud C, Xiong X, Yu H, Halloran ME, Longini IM, Vespignani A (2020) The effect of travel restrictions on the spread of the 2019 novel coronavirus (COVID-19) outbreak. Science 368(6489):395. https://doi.org/10.1126/science.aba9757

Council WTT (2019) Travel & tourism economic impact 2019 world. London, UK

Darbeheshti F, Rezaei N (2020) Genetic predisposition models to COVID-19 infection. Med Hypotheses 142:109818. https://doi.org/10.1016/j.mehy.2020.109818

David E. Bloom DC, Sevilla J (2001) The effect of health on economic growth: theory and evidence. https://doi.org/10.3386/w8587

De Domenico M, Solé-Ribalta A, Cozzo E, Kivelä M, Moreno Y, Porter MA, Gómez S, Arenas A (2013) Mathematical formulation of multilayer networks. Physic Rev X 3(4):041022. https://doi.org/10.1103/PhysRevX.3.041022

De Domenico M, Granell C, Porter MA, Arenas A (2016) The physics of spreading processes in multilayer networks. Nat Phys 12(10):901–906. https://doi.org/10.1038/nphys3865

Deal TEG (2019) Communication from the commission. https://eur-lex.europa.eu/legal-content/EN/TXT/?qid=1588580774040&uri=CELEX:52019DC0640

Development OfEC-oa (2 March 2020) OECD interim economic assessment coronavirus: the world economy at risk. Paris, France

Development UNCoTa (9 March 2020) The coronavirus shock: a story of another global crisis, foretold and policymaker should be doing about it. Geneva, Switzerland

Fathi N, Rezaei N (2020) Lymphopenia in COVID-19: therapeutic opportunities. Cell Biol Int 44(9):1792–1797. https://doi.org/10.1002/cbin.11403

Faulkner J, Schaller M, Park J, Duncan L (2004) Evolved disease-avoidance mechanisms and contemporary xenophobic attitudes. Group Proc & Amp Intergroup Relation 7:333–353. https://doi.org/10.1177/1368430204046142

Federation BP (20 March 2020) Survey on the effects of coronavirus on the UK plastics industry. London

Fernandes N (2020) Economic effects of coronavirus outbreak (COVID-19) on the world economy SSRN: https://ssrn.com/abstract=3557504 or https://doi.org/10.2139/ssrn.3557504

Fernando WJMaR (2020) The global macroeconomic impacts of COVID-19: Seven scenarios. https://www.brookings.edu/research/the-global-macroeconomic-impacts-of-covid-19-seven-scenarios/

Fincher CL, Thornhill R, Murray DR, Schaller M (2008) Pathogen prevalence predicts human cross-cultural variability in individualism/collectivism. Proc Biol Sci 275(1640):1279–1285. https://doi.org/10.1098/rspb.2008.0094

Gao J, Buldyrev SV, Stanley HE, Havlin S (2012) Networks formed from interdependent networks. Nat Phys 8(1):40–48. https://doi.org/10.1038/nphys2180

Griskevicius V, Delton AW, Robertson TE, Tybur JM (2011) Environmental contingency in life history strategies: the influence of mortality and socioeconomic status on reproductive timing. J Pers Soc Psychol 100(2):241–254. https://doi.org/10.1037/a0021082

Guo Y-R, Cao Q-D, Hong Z-S, Tan Y-Y, Chen S-D, Jin H-J, Tan K-S, Wang D-Y, Yan Y (2020) The origin, transmission and clinical therapies on coronavirus disease 2019 (COVID-19) outbreak – an update on the status. Mil Med Res 7. https://doi.org/10.1186/s40779-020-00240-0

Haacker M (2004, November 19) The macroeconomics of HIV/AIDS: international monetary fund. https://www.imf.org/external/pubs/ft/aids/eng/

Hanaei S, Rezaei N (2020) COVID-19: developing from an outbreak to a pandemic. Arch Med Res. https://doi.org/10.1016/j.arcmed.2020.04.021

Hellewell J, Abbott S, Gimma A, Bosse NI, Jarvis CI, Russell TW, Munday JD, Kucharski AJ, Edmunds WJ, Funk S, Eggo RM (2020) Feasibility of controlling COVID-19 outbreaks by isolation of cases and contacts. Lancet Glob Health 8(4):e488–e496. https://doi.org/10.1016/s2214-109x(20)30074-7

Hessami A, Shamshirian A, Heydari K, Pourali F, Alizadeh-Navaei R, Moosazadeh M, et al (2020) Cardiovascular diseases burden in COVID-19: systematic review and meta-analysis. Am J Emerg Med S0735–6757(20)30908–6

Jabbari P, Rezaei N (2020) With risk of reinfection, is COVID-19 Here to stay? Disaster Med Public Health Prep 1. https://doi.org/10.1017/dmp.2020.274

Jabbari P, Jabbari F, Ebrahimi S, Rezaei N (2020) COVID-19: a chimera of two pandemics. Disaster Med Public Health Prep:1–2. https://doi.org/10.1017/dmp.2020.223

Jahanshahlu L, Rezaei N (2020a) Central nervous system involvement in COVID-19. Arch Med Res. https://doi.org/10.1016/j.arcmed.2020.05.016

Jahanshahlu L, Rezaei N (2020b) Monoclonal antibody as a potential anti-COVID-19. Biomedicine & pharmacotherapy. Biomed Pharmacother 129:110337. https://doi.org/10.1016/j.biopha.2020.110337

Jamison DT, Gelband H, Horton S, Jha P, Mock CN, Nugent R (2017) Disease control priorities, third edition: volume 9. Improving health and reducing poverty, 3rd edn. World Bank Publications, Washington DC

Kafieh R, Arian R, Saeedizadeh N, Minaee S, Yadav SK, Vaezi A, Rezaei N, Javanmard SH (2020) COVID-19 in Iran: a deeper look into the future. medRxiv

Kivelä M, Arenas A, Barthelemy M, Gleeson JP, Moreno Y, Porter MA (2014) Multilayer networks. J Complex Network 2(3):203–271. https://doi.org/10.1093/comnet/cnu016

Kucharski AJ, Russell TW, Diamond C, Liu Y, Edmunds J, Funk S, Eggo RM (2020) Early dynamics of transmission and control of COVID-19: a mathematical modelling study. Lancet Infect Dis 20(5):553–558. https://doi.org/10.1016/s1473-3099(20)30144-4

Lai S, Ruktanonchai NW, Zhou L, Prosper O, Luo W, Floyd JR, Wesolowski A, Santillana M, Zhang C, Du X, Yu H, Tatem AJ (2020) Effect of non-pharmaceutical interventions to contain COVID-19 in China. Nature. https://doi.org/10.1038/s41586-020-2293-x

Laing T (2020) The economic impact of the coronavirus 2019 (Covid-2019): implications for the mining industry. Extractive Indus & Soc. https://doi.org/10.1016/j.exis.2020.04.003

Le Quéré C, Jackson RB, Jones MW, Smith AJP, Abernethy S, Andrew RM, De-Gol AJ, Willis DR, Shan Y, Canadell JG, Friedlingstein P, Creutzig F, Peters GP (2020) Temporary reduction in daily global CO2 emissions during the COVID-19 forced confinement. Nat Clim Chang. https://doi.org/10.1038/s41558-020-0797-x

Liu JT, Hammitt JK, Wang JD, Tsou MW (2005) Valuation of the risk of SARS in Taiwan. Health Econ 14(1):83–91. https://doi.org/10.1002/hec.911

Lotfi M, Rezaei N (2020) SARS-CoV-2: a comprehensive review from pathogenicity of the virus to clinical consequences. J Med Virol. https://doi.org/10.1002/jmv.26123

Lotfi M, Hamblin MR, Rezaei N (2020) COVID-19: transmission, prevention, and potential therapeutic opportunities. Clinica chimica acta. Int J Clinic Chem 508:254–266. https://doi.org/10.1016/j.cca.2020.05.044

Maliszewska M (2014) The economic impact of the 2014 Ebola epidemic: short and medium term estimates for West Africa. World Bank, Washington, DC

Mansourabadi AH, Sadeghalvad M, Mohammadi-Motlagh HR, Rezaei N (2020) The immune system as a target for therapy of SARS-CoV-2: a systematic review of the current immunotherapies for COVID-19. Life Sci 258:118185. https://doi.org/10.1016/j.lfs.2020.118185

Maynard A, Bloor K (2009) The economic impact of pandemic influenza. Br Med J 339(7733). https://doi.org/10.1136/bmj.b4888

McKee M, Stuckler D (2020) If the world fails to protect the economy, COVID-19 will damage health not just now but also in the future. Nat Med 26(5):640–642. https://doi.org/10.1038/s41591-020-0863-y

McKibbin W, Sidorenko A (2006) Global macroeconomic consequences of pandemic influenza. Centre for Applied Macroeconomic Analysis, Crawford School of Public Policy, The Australian National University

Medicine Io (2004) Learning from SARS: preparing for the next disease outbreak: workshop summary. The National Academies Press, Washington, DC. https://doi.org/10.17226/10915

Moazzami B, Razavi-Khorasani N, Dooghaie Moghadam A, Farokhi E, Rezaei N (2020) COVID-19 and telemedicine: immediate action required for maintaining healthcare providers well-being. J Clinic Virol: Official Pub Pan Am Soc Clinic Virol 126:104345. https://doi.org/10.1016/j.jcv.2020.104345

Mohamed K, Rodríguez-Román E, Rahmani F, Zhang H, Ivanovska M, Makka SA, Joya M, Makuku R, Islam MS, Radwan N, Rahmah L, Goda R, Abarikwu SO, Shaw M, Zoghi S, Irtsyan S, Ling I, Cseprekal O, Faten AB, Hazar Sayar E, Soloukey C, Grancini G, Rezaei N (2020a) Borderless collaboration is needed for COVID-19 – a disease that knows no borders. Infection control and hospital epidemiology: 1–2. doi:https://doi.org/10.1017/ice.2020.162

Mohamed K, Yazdanpanah N, Saghazadeh A, Rezaei N (2020b) Computational drug discovery and repurposing for the treatment of Covid-19: a systematic review. Available at SSRN 3583748

Momtazmanesh S, Ochs HD, Uddin LQ, Perc M, Routes JM, Vieira DN, Al-Herz W, Baris S, Prando C, Rosivall L, Abdul Latiff AH, Ulrichs T, Roudenok V, Aldave Becerra JC, Salunke DB, Goudouris E, Condino-Neto A, Stashchak A, Kryvenko O, Stashchak M, Bondarenko A, Rezaei N (2020) All together to fight COVID-19. Am J Trop Med Hyg 102(6):1181–1183. https://doi.org/10.4269/ajtmh.20-0281

Moradian N, Ochs HD, Sedikies C, Hamblin MR, Camargo CA Jr, Martinez JA, Biamonte JD, Abdollahi M, Torres PJ, Nieto JJ, Ogino S, Seymour JF, Abraham A, Cauda V, Gupta S, Ramakrishna S, Sellke FW, Sorooshian A, Wallace Hayes A, Martinez-Urbistondo M, Gupta M, Azadbakht L, Esmaillzadeh A, Kelishadi R, Esteghamati A, Emam-Djomeh Z, Majdzadeh R, Palit P, Badali H, Rao I, Saboury AA, Jagan Mohan Rao L, Ahmadieh H, Montazeri A, Fadini GP, Pauly D, Thomas S, Moosavi-Movahed AA, Aghamohammadi A, Behmanesh M, Rahimi-Movaghar V, Ghavami S, Mehran R, Uddin LQ, Von Herrath M, Mobasher B, Rezaei N (2020) The urgent need for integrated science to fight COVID-19 pandemic and beyond. J Transl Med 18(1):205. https://doi.org/10.1186/s12967-020-02364-2

Murray DR, Trudeau R, Schaller M (2011) On the origins of cultural differences in conformity: four tests of the pathogen prevalence hypothesis. Pers Soc Psychol Bull 37(3):318–329. https://doi.org/10.1177/0146167210394451

Nasab MG, Saghazadeh A, Rezaei N (2020) SARS-CoV-2-a tough opponent for the immune system. Arch Med Res. https://doi.org/10.1016/j.arcmed.2020.05.020

Nicola M, Alsafi Z, Sohrabi C, Kerwan A, Al-Jabir A, Iosifidis C, Agha M, Agha R (2020) The socio-economic implications of the coronavirus and COVID-19 pandemic: a review. Int J Sur g:S1743–9191(1720)30316–30312. https://doi.org/10.1016/j.ijsu.2020.04.018

O'Donnell KNAJ (April 2 2020) Thousands of US medical workers furloughed, laid off as routine patient visits drop during coronavirus pandemic. USA TODAY NETWORK. https://eu.usatoday.com/story/news/health/2020/04/02/coronavirus-pandemic-jobs-us-health-care-workers-furloughed-laid-off/5102320002/. Accessed 20 May 2020

Organization WH (2020) Global Health Observatory (GHO) data. https://www.who.int/gho/hiv/epidemic_status/deaths_text/en/

Organization WH WHO announces COVID-19 outbreak a pandemic (n.d.). http://www.euro.who.int/en/health-topics/health-emergencies/coronavirus-covid-19/news/news/2020/3/who-announces-covid-19-outbreak-a-pandemic

Pashaei M, Rezaei N (2020) Immunotherapy for SARS-CoV-2: potential opportunities. Expert Opin Biol Ther:1–5. https://doi.org/10.1080/14712598.2020.1807933

Pourahmad R, Moazzami B, Rezaei N (2020) Efficacy of Plasmapheresis and immunoglobulin replacement therapy (IVIG) on patients with COVID-19. SN Comprehen Clinic Med:1–5

Prospects WE (April 2020) World GDP to fall 2.8% in 2020, exceeding financial crisis toll. https://resources.oxfordeconomics.com/hubfs/WEP/WEPM%20April%202020%20pdf.pdf

Rabiee N, Rabiee M, Bagherzadeh M, Rezaei N (2020) COVID-19 and picotechnology: potential opportunities. Med Hypotheses 144:109917. https://doi.org/10.1016/j.mehy.2020.109917

Radicchi F (2014) Driving interconnected networks to supercriticality. Physic Rev X 4(2):021014. https://doi.org/10.1103/PhysRevX.4.021014

Radicchi F, Arenas A (2013) Abrupt transition in the structural formation of interconnected networks. Nat Phys 9(11):717–720. https://doi.org/10.1038/nphys2761

Reinicke CMI (March 2020) Goldman Sachs now says US GDP will shrink 24% next quarter amid the coronavirus pandemic – which would be 2.5 times bigger than any decline in history. https://markets.businessinsider.com/news/stocks/us-gdp-drop-record-2q-amid-coronavirus-recession-goldman-sachs-2020-3-1029018308

Rezaei N (2020a) COVID-19 affects healthy pediatricians more than pediatric patients. Infect Control Hosp Epidemiol 1. https://doi.org/10.1017/ice.2020.139

Rezaei N (2020b) COVID-19 and medical biotechnology. Avicenna J Med Biotechnol 12(3):139

Rokni M, Hamblin MR, Rezaei N (2020) Cytokines and COVID-19: friends or foes? Hum Vaccin Immunother 16:2363–2365

Rugani B, Caro D (2020) Impact of COVID-19 outbreak measures of lockdown on the Italian carbon footprint. Sci Total Environ 737:139806. https://doi.org/10.1016/j.scitotenv.2020.139806

Rzymski P, Nowicki M, Mullin GE, Abraham A, Rodríguez-Román E, Petzold MB, Bendau A, Sahu KK, Ather A, Naviaux A-F (2020) Quantity does not equal quality: scientific principles cannot be sacrificed. Int Immunopharmacol 86:106711

Sachs J, Malaney P (2002) The economic and social burden of malaria. Nature 415(6872):680–685. https://doi.org/10.1038/415680a

Saghazadeh A, Rezaei N (2020a) Immune-epidemiological parameters of the novel coronavirus – a perspective. Expert Rev Clin Immunol 16(5):465–470. https://doi.org/10.1080/1744666x.2020.1750954

Saghazadeh A, Rezaei N (2020b) Towards treatment planning of COVID-19: rationale and hypothesis for the use of multiple immunosuppressive agents: anti-antibodies, immunoglobulins, and corticosteroids. Int Immunopharmacol 84:106560. https://doi.org/10.1016/j.intimp.2020.106560

Sahu KK, Siddiqui AD, Rezaei N, Cerny J (2020) Challenges for management of immune thrombocytopenia during COVID-19 pandemic. J Med Virol. https://doi.org/10.1002/jmv.26251

Saleki K, Banazadeh M, Saghazadeh A, Rezaei N (2020) The involvement of the central nervous system in patients with COVID-19. Rev Neurosci 31(4):453–456. https://doi.org/10.1515/revneuro-2020-0026

Satiani B, Zigrang TA, Bailey-Wheaton JL (2020) COVID-19 financial resources for physicians. J Vasc Surg. https://doi.org/10.1016/j.jvs.2020.04.482

Schaller M, Park J (2011) The behavioral immune system (and why it matters). Curr Dir Psychol Sci 20:99–103. https://doi.org/10.1177/0963721411402596

Sharifkashani S, Bafrani MA, Khaboushan AS, Pirzadeh M, Kheirandish A, Yavarpour Bali H, Hessami A, Saghazadeh A, Rezaei N (2020) Angiotensin-converting enzyme 2 (ACE2) receptor and SARS-CoV-2: potential therapeutic targeting. Eur J Pharmacol 884:173455. https://doi.org/10.1016/j.ejphar.2020.173455

Tollefson J (2020) How the coronavirus pandemic slashed carbon emissions – in five graphs. Nature 582(7811):158–159. https://doi.org/10.1038/d41586-020-01497-0

Tybur JM, Inbar Y, Aarøe L, Barclay P, Barlow FK, de Barra M, Becker DV, Borovoi L, Choi I, Choi JA, Consedine NS, Conway A, Conway JR, Conway P, Adoric VC, Demirci DE, Fernández AM, Ferreira DC, Ishii K, Jakšić I, Ji T, van Leeuwen F, Lewis DM, Li NP, McIntyre JC, Mukherjee S, Park JH, Pawlowski B, Petersen MB, Pizarro D, Prodromitis G, Prokop P, Rantala MJ, Reynolds LM, Sandin B, Sevi B, De Smet D, Srinivasan N, Tewari S, Wilson C, Yong JC, Žeželj I (2016) Parasite stress and pathogen avoidance relate to distinct dimensions of political ideology across 30 nations. Proc Natl Acad Sci U S A 113(44):12408–12413. https://doi.org/10.1073/pnas.1607398113

Varnum MEW, Grossmann I (2016) Pathogen prevalence is associated with cultural changes in gender equality. Nat Hum Behav 1(1):0003. https://doi.org/10.1038/s41562-016-0003

Varnum MEW, Grossmann I (2017) Cultural change: the how and the why. Perspect Psychol Sci 12(6):956–972. https://doi.org/10.1177/1745691617699971

Vespignani A (2010) The fragility of interdependency. Nature 464(7291):984–985. https://doi.org/10.1038/464984a

Yazdanpanah F, Hamblin MR, Rezaei N (2020a) The immune system and COVID-19: friend or foe? Life Sci 256:117900. https://doi.org/10.1016/j.lfs.2020.117900

Yazdanpanah N, Saghazadeh A, Rezaei N (2020b) Anosmia: a missing link in the neuroimmunology of coronavirus disease 2019 (COVID-19). Rev Neurosci. https://doi.org/10.1515/revneuro-2020-0039

Yousefzadegan S, Rezaei N (2020) Case report: death due to COVID-19 in three brothers. Am J Trop Med Hyg 102(6):1203–1204. https://doi.org/10.4269/ajtmh.20-0240

Zhang C, Wu Z, Li JW, Zhao H, Wang GQ (2020a) Cytokine release syndrome in severe COVID-19: interleukin-6 receptor antagonist tocilizumab may be the key to reduce mortality. Int J Antimicrob Agents 55(5):105954. https://doi.org/10.1016/j.ijantimicag.2020.105954

Zhang J, Litvinova M, Liang Y, Wang Y, Wang W, Zhao S, Wu Q, Merler S, Viboud C, Vespignani A, Ajelli M, Yu H (2020b) Changes in contact patterns shape the dynamics of the COVID-19 outbreak in China. Science: eabb8001. https://doi.org/10.1126/science.abb8001

Zhang J, Litvinova M, Wang W, Wang Y, Deng X, Chen X, Li M, Zheng W, Yi L, Chen X, Wu Q, Liang Y, Wang X, Yang J, Sun K, Longini IM Jr, Halloran ME, Wu P, Cowling BJ, Merler S, Viboud C, Vespignani A, Ajelli M, Yu H (2020c) Evolving epidemiology and transmission dynamics of coronavirus disease 2019 outside Hubei province, China: a descriptive and modelling study. Lancet Infect Dis. https://doi.org/10.1016/s1473-3099(20)30230-9

47 Retrieval and Investigation of Data on SARS-CoV-2 and COVID-19 Using Bioinformatics Approach

Muhamad Fahmi, Viol Dhea Kharisma, Arif Nur Muhammad Ansori, and Masahiro Ito

Abstract

Sudden emergence and a rapid outbreak of SARS-CoV-2 accompanied by a devastating impact on the economy and public health has driven extensive scientific mobilization to study and elucidate the various associated concerns about SARS-CoV-2. Bioinformatics plays a crucial role in addressing and providing solutions to questions about SARS-CoV-2. It helps shorten the duration for the vaccine development process and the discovery of potential clinical interventions through the simulation and information retrieval, and the development of well-ordered information hubs and resources, which are essential to derive data and meaningful findings from the current massive information about SARS-CoV-2. Advanced algorithms in this field also provide approaches that are essential to elucidate the relationship, origin, and evolutionary process of SARS-CoV-2. Here, we report essential bioinformatics entities, such as database and platform development, molecular evolution and phylogenetic analyses, and vaccine designs, that are useful to solve the SARS-CoV-2 conundrum.

Keywords

COVID-19 · Databases · Information hubs · Molecular evolution · Phylogenetic inference · SARS-CoV-2 · Vaccine design

M. Fahmi · M. Ito (✉)
Advanced Life Sciences Program, Graduate School of Life Sciences, Ritsumeikan University, Kusatsu, Shiga, Japan

Systematic Review and Meta-analysis Expert Group (SRMEG), Universal Scientific Education and Research Network (USERN), Kusatsu, Japan
e-mail: maito@sk.ritsumei.ac.jp

V. D. Kharisma
Master Program in Biology, Department of Biology, Faculty of Mathematic and Natural Sciences, Universitas Brawijaya, Malang, Indonesia

Computational Virology and Complexity Science Research Unit, Division of Molecular Biology and Genetics, Generasi Biologi Indonesia (GENBINESIA) Foundation, Gresik, Indonesia

Systematic Review and Meta-analysis Expert Group (SRMEG), Universal Scientific Education and Research Network (USERN), Malang, Indonesia

A. N. M. Ansori
Doctoral Program in Veterinary Science, Faculty of Veterinary Medicine, Universitas Airlangga, Kampus C Universitas Airlangga, Surabaya, Indonesia

Systematic Review and Meta-analysis Expert Group (SRMEG), Universal Scientific Education and Research Network (USERN), Surabaya, Indonesia

N. Rezaei (ed.), *Coronavirus Disease - COVID-19*, Advances in Experimental Medicine and Biology 1318, https://doi.org/10.1007/978-3-030-63761-3_47

47.1 Introduction

Viruses have been characterized as distinct biological entities for more than 100 years. They were initially discovered in 1892 when Dmitri Ivanovsky observed that the tobacco mosaic disease was caused by tiny rod-shaped infectious particles smaller than any known bacteria; later, these particles were named as "virus" by Martinus Beijerinck in 1898. Since then, they have been isolated from multiple organisms and even the virus itself. In this manner, viruses have been assigned to a unique taxonomic position. The abundance of viruses in the biosphere is estimated to be markedly high and approximately ten times greater than that of bacteria (Breitbart and Rohwer 2005; Suttle 2005). They have been studied extensively owing to their infectivity in humans and other organisms, which causes several diseases, as well as their ability to maintain ecosystem homeostasis and exhibit pathogen control (Flint et al. 2015). Several attributes are considered for the characterization and identification of viruses, including the clinical attributes, pathogenic properties, measurement of the physical structures, and the comparison of genetic material (Flint et al. 2015). Viral genomic sequences were some of the first available genomic information. The development of a more efficient method for virus attribute retrieval requires sophisticated computational tools and techniques for analyzing the extensive data available. Bioinformatics plays a crucial role in contending this challenge, thereby providing substantial support in addressing several common questions pertinent in virology (Chang 2015; Marz et al. 2014).

Severe acute respiratory syndrome coronavirus 2 (SARS-CoV-2) is the seventh coronavirus that has crossed the species barrier to infect the human population. It is a rapidly spreading virus that has posed a significant public threat and proved to have a considerable burden on the global economy and human health. A form of severe respiratory disease caused by this agent, known as the coronavirus disease (COVID-19), was first detected in late December 2019 (Zhu et al. 2020). Later on, it attained a pandemic status, as declared by the World Health Organization (WHO) in early March 2020, and the virus has spread to more than 150 countries and infected over 5.8 million people by the latter part of May 2020 (World Health Organization 2020a, b). The devastating effects exerted by this virus have motivated multiple researchers to study and elucidate the various associated concerns that may help strategize the process of obtaining a solution. Here, we discuss several state-of-the-art computational approaches that have been used to address questions on SARS-CoV-2, including the development of databases and platforms, studies on structural and evolutionary relationships, and designing potential interventions through virtual screening.

47.2 SARS-CoV-2-Related Databases and Platforms

The sudden outbreak and rapid spread of COVID-19 have endangered global health and economy (Li et al. 2020a; Huang et al. 2020). This crisis has called for extensive scientific mobilization for studies on SARS-CoV-2 concerning its characteristics, the mechanism of transmission, clinical aspects of the disease, and management and prevention strategies, with the ultimate aim of countering the devastating outcomes. From January to May 2020, an extensive body of literature and data consisting of several thousand articles has been published. As a result, such vast numbers of publications have overwhelmed researchers in their effort to derive meaningful findings or to track novel information simply. This "explosion" of data calls for novel features that would help establish well-ordered information and advanced computational tools, which will help create a more user-friendly medium for the accession and analysis of "big data" from the literature on SARS-CoV-2. Additionally, several common and virus-specific databases that had been developed before the emergence of SARS-CoV-2 continue to be relevant and useful, especially those that process sequence and structural data. Intelligent use of these media would expedite knowledge discovery, which is essential in coping with the urgency of the crisis (Table 47.1).

47.2.1 Platforms and Tools for SARS-CoV-2 Literature

Scientists are overburdened with the vast array of papers on COVID-19. In the 5 months since the outbreak, more than 23,000 papers on SARS-CoV-2 and COVID-19 have been published in public domains. This repository is proliferating and has doubled in approximately every 20 days within this period (Brainard 2020). Journals and scholarly publishers have provided outstanding support to such efforts for disseminating scientific contributions rapidly by accelerating their editorial assessment, peer review, and publication processes, which has raised concerns about the quality of the resulting publications (Horbach 2020). As a result, the concept of a curated body of literature is integral in dealing with information overload and deriving meaningful patterns from findings across published papers. Currently, several groups have commenced the utilization of state-of-the-art computational approaches, such as artificial intelligence, for curating hubs for easy and effective access or simply filtering papers with specific focus on high-quality publications (Chen et al. 2020; Lorenc et al. 2020; Johns Hopkins Bloomberg School of Public Health 2020; Balakrishnan 2020).

LitCovid is one such prominent hub of curated literature for tracking recent scientific information on COVID-19 (https://www.ncbi.nlm.nih.gov/research/coronavirus/). This open-resource literature hub was developed by researchers under the aegis of the National Center for Biotechnology Information (NCBI). It assists the retrieval of relevant literature using sophisticated search functions and categorized topics, including an overview, disease mechanism, transmission dynamics, treatment, case reports, and epidemic forecasting (Chen et al. 2020). The articles are also categorized based on geographic location with visualization for better access (Chen et al. 2020). Additionally, a platform developed by the Evidence for Policy and Practice Information and Co-ordinating Centre (EPPI-Centre) at the University College London is an alternative resource for accessing the overview and distribution of articles related to COVID-19 (http://eppi.ioe.ac.uk/COVID19_MAP/covid_map_v11.html) (Lorenc et al. 2020). This platform has been built from the Collaborative Approach to Meta-Analysis and Review of Animal Data from Experimental Studies (CAMARADES) infrastructure that had been developed previously, the infrastructure includes systematic review facility, crowdsourcing for screening, annotation and automation techniques, and a web-based portal. This platform provides an updated map of the current evidence on COVID-19 by categorizing publications based on the study type (Lorenc et al. 2020).

The 2019 Novel Coronavirus Research Compendium (NCRC) is an alternative centralized platform available to the public that rapidly curates and reviews emerging scientific evidence on SARS-CoV-2 and COVID-19, with a specific focus on providing accurate and relevant information from original, high-quality, and trending research (https://ncrc.jhsph.edu/) (Johns Hopkins Bloomberg School of Public Health 2020). The demand for reliable and rapidly curated evidence from the public, programs, policies, and researchers amid the fast growth in the body of literature on COVID-19 has driven numerous researchers from Johns Hopkins Schools of Public Health, Johns Hopkins School of Medicine, and other institutions worldwide to develop this platform (Johns Hopkins Bloomberg School of Public Health 2020). This group rapidly evaluates COVID-19-related publications and strictly selects high-quality articles to be categorized under several topics listed in this platform, including diagnostics, modeling, epidemiology, pharmaceutical interventions, non-pharmaceutical interventions, clinical presentation and prognostic risk factors, vaccines, and ecology and spillover (Johns Hopkins Bloomberg School of Public Health 2020).

Table 47.1 Essential databases and platforms for SARS-CoV-2 literature

Platform and tool	Description	Web site	References
LitCovid	A curated literature hub for tracking recent scientific information about SARS-CoV-2	www.ncbi.nlm.nih.gov/research/coronavirus	Chen et al. (2020)
COVID-19: living map of the evidence	Map of the current evidence on COVID-19 by categorizing publications based on the study type	eppi.ioe.ac.uk/COVID19_MAP/covid_map_v11.html	Lorenc T et al. (2020)
Novel Coronavirus Research Compendium (NCRC)	A centralized, publicly available resource that rapidly curates and reviews the emerging scientific evidence about SARS-CoV-2 and COVID-19	ncrc.jhsph.edu/	Johns Hopkins Bloomberg School of Public Health (2020)
COVID-evidence	Continuously updated database of the available worldwide evidence on interventions for COVID-19	covid-evidence.org	COVID-evidence (2020)
Global Coronavirus COVID-19 Clinical Trial Tracker	Map of COVID-19 trials according to geographical, trial, patient, and intervention characteristics	covid-trials.org	Thorlund et al. (2020)
ClinicalTrials.gov	Database of privately and publicly funded clinical studies conducted around the world	clinicaltrials.gov	ClinicalTrials.gov (2000)
International Clinical Trials Registry Platform	Clinical trials registry database by WHO	www.who.int/ictrp/en/	World Health Organization (2005)
COVID-19 Open Research Dataset (CORD-19)	Resource of more than 130,000 scholarly articles about the novel coronavirus for use by the global research community	www.semanticscholar.org/cord19	Wang et al. (2020)
SciSight	A tool for exploring the evolving network of science in the CORD-19	scisight.apps.allenai.org	SciSight (2020)
SPIKE-CORD	A powerful set of tools for effectively searching and interacting with the CORD-19 data	spike.covid-19.apps.allenai.org/search/covid19	SPIKE-COVID (2020)
ViralZone	A knowledge resource to understand virus diversity	www.expasy.org/viralzone/	Hulo et al. (2011)

The WHO initiated an international clinical trial called SOLIDARITY a week after the declaration of the COVID-19 pandemic, and over 100 countries have joined the SOLIDARITY trial and are working simultaneously to solve the issue at hand (Balakrishnan 2020). A clinical trial is a type of research methodology in which tests and treatments for the disease are evaluated by assessing the safety and efficacy of clinical candidate interventions (Friedman et al. 2010). It relies on a foundation of evidence to improve the quality of health care and help stakeholders control the costs involved through careful comparison with alternative interventions (Friedman et al. 2010). Amid the rapid growth in the number of COVID-19 patients globally and the absence of a confirmed drug or treatment method, the urgency to develop or discover an efficient therapeutic strategy for COVID-19 has triggered a spike in clinical trial research and unprecedented growth in the number of findings from studies on a specific disease within a brief period (Thorlund et al. 2020). Hence, an easily accessible platform that summarizes findings from COVID-19 clinical trials is necessary for convenient tracking of relevant information without including irrelevant data in the findings. Several groups of researchers have taken the initiative to develop media that is focused on the mapping and summarization of clinical trial research instead of providing an overview of all articles related to COVID-19. It includes COVID-evidence (covid-evidence.org) and Global Coronavirus COVID-19 Clinical Trial Tracker (covid-trials.org) (Thorlund et al. 2020; COVID-evidence 2020; Ruano et al. 2020). Both platforms use automatic search and expert manual extraction strategies for retrieving data from several sources, including published articles and the International Clinical Trials Registry Platforms, to minimize duplicated entries (Thorlund et al. 2020; COVID-evidence 2020). Additionally, ClinicalTrials.gov (clinicaltrials.gov) and the International Clinical Trials Registry Platform (ICTRP; www.who.int/ictrp/en/) by WHO are relevant resources for tracking clinical trials, although these platforms provide data without recapitulation to avoid unnecessary duplication (Thorlund et al. 2020; World Health Organization 2005; ClinicalTrials.gov 2000).

Additionally, the development of text mining and information retrieval systems is the key to the dynamic discovery and extraction of relevant, nontrivial information from the massive collection of literature. To accommodate such advancements for use in SARS-CoV-2- and COVID-19-related studies, the Allen Institute for AI (AI2) released a COVID-19 Open Research Dataset (CORD-19; www.semanticscholar.org/cord19) in collaboration with governments' several leading institutions (Wang et al. 2020). This resource contains more than 128,000 scholarly articles on COVID-19, SARS-CoV2, and related historical research on coronavirus, which are assembled in a machine-readable format to provide an accessible structure and system for the retrieval process using computational methods such as data mining, machine learning, and natural language processing (Wang et al. 2020). The development and availability of this resource are provided to the global community to apply recent advances of AI techniques and to develop robust and user-friendly computational tools that help researchers find answers to their questions from published studies. The CORD-19 itself consists of several built-in tools such as SciSight and SPIKE-CORD for exploring the evolving network of scientific information and for effectively searching data, respectively (scisight.apps.allenai.org; spike.covid-19.apps.allenai.org/search/covid19) (SciSight 2020; SPIKE-COVID 2020). SciSight helps retrieve a comprehensive network of information using an AI-powered visualization framework. This tool has several features for exploring researcher working networks (the network of science), searching key facets (Faceted Search), and exploring the association between proteins, genes, and cells (proteins/genes/cells), as well as between diseases and chemicals (Diseases/Chemicals) (SciSight 2020). Conversely, SPIKE-CORD offers user-friendly text mining options for effective search and interaction with CORD-19 data. That is an advanced tool and comprises three query modes, including Boolean queries, sequential queries, and structured queries (SPIKE-COVID 2020).

ViralZone (www.expasy.org/viralzone/) is another important resource that provides a clear view of the biological processes of the complete identified virosphere (Hulo et al. 2011). This web resource provides fact sheets with information on sequence information, replication cycles, taxonomy, and epidemiology, as well as graphics describing virion organization, genome transcription, and translation strategies for every identified virus families/genera, thereby creating an accurate and concise information platform to improve the understanding of the complications associated with the massive diversity in viruses. The ViralZone platform has a dedicated resource for SARS-CoV-2 with additional information on antiviral drug development and COVID-19 treatment strategies (Hulo et al. 2011).

To deal with the devastating effects of SARS-CoV-2 infection, the global scientific community has made an effort to contribute to the current information on the disease, which has created a massive collection of readily accessible literature. A challenge involving the tracking of novel findings and the construction of a meaningful network from such a massive data repository awaits clinicians, researchers, and policymakers. Fortunately, several research groups are addressing the challenge and have attempted to create sophisticated platforms and tools that focus on providing user-friendly media. Given these efforts, the utilization of such media aids active tracking of important findings on COVID-19.

47.2.2 Sequence and Structure Database of SARS-CoV-2

The ability to sequence DNA at higher throughput and lower costs leads to the rapid increase in the available sequence data. Sequence analysis plays a vital role in viral surveillance, host reservoir identification, and public health policy debates (Brister et al. 2015). The development of a sequence database is essential in current bioinformatics research and applications for storage and retrieval locations of sequence data and annotation. Generally, three primary public nucleotide sequence databases facilitate the availability of DNA sequence data to the public: NCBI, the European Bioinformatics Institute (EBI), and the DNA Data Bank of Japan (DDBJ). Over the years, the repository of sequences deposited in these databases has grown at an increasing rate. Since these databases form the International Nucleotide Sequence Database Collaboration and exchange updates daily, they follow certain common principles for the arrangement of sequence data and mainly provide the same accession number (Koonin and Galperin 2003).

Additionally, scientific journals generally specify that the sequence data should be cited with the accession number of information from these databases in a paper that describes a nucleotide or protein sequence (Brister et al. 2015). Following the growth in the number of viral genome sequences and the challenges of implementing a purely well-annotated representation of viral genome sequences, NCBI has a unique feature known as the NCBI Viral Genome Resource (www.ncbi.nlm.nih.gov/genome/viruses/) to serve better the needs of the community of virologists. This resource has a central browser for viral and viroid genomes, which lists all viral and viroid species indicated by the corresponding reference sequence and includes links to genome neighbor sequences (Brister et al. 2015).

Specific databases for viral sequences have also been established; this is an alternative of the general public nucleotide sequence databases. The development of a specific database for viruses is essential to avoid the challenges arising from the sharing of viral sequence information. For example, the Global Initiative on Sharing All Influenza Data (GISAID) EpiFlu (https://www.gisaid.org/) was initially developed to address problems experienced while sharing virus sequences, such as the potential intervention by government agencies for the international exchange of information and the hesitation of researchers to share data on sequences of lethal viruses (Elbe and Buckland-Merrett 2017). This database provides protection and assurances to data contributors through a unique data access agreement. Users must register themselves with official affiliation to access the data in this database and comply with several agreements, including the acknowledgment of the contributor while using their data in publications and refraining from data sharing with third parties outside the GISAID community (Elbe and Buckland-Merrett 2017). This mechanism allows researchers and the government to share their sequence data promptly and is important because the immediate availability of viral genomic data helps expedite the processes involved in tracking and recognizing emergent epidemics or pandemics (Elbe and Buckland-Merrett 2017). The GISAID EpiFlu system itself contains above 27,000 partial and complete genomic sequences of SARS-CoV-2 that have been contributed by clinicians and researchers worldwide, as of mid-May 2020.

Given the extent of available sequence information, the number of structurally characterized proteins, nucleic acids, and other biological macromolecules is relatively low. Acquiring three-dimensional molecular structure information helps characterize the molecular function and interaction. Additionally, elucidating the viral protein structure and interaction is fundamental to comprehending the virus receptor recognition mechanism, which is associated with its infectivity, pathogenesis, and host range, and is a requisite in the effort to develop structure-based drugs. The Protein Data Bank (PDB) is a prominent global repository of experimentally determined three-dimensional structures (Sussman et al. 1998). It is one of the first open-access digital resources for biological sciences established in 1971. The Worldwide Protein Data Bank currently manages this database (wwPDB; http://www.wwpdb.org/), which includes the RCSB Protein Data Bank (RCSB PDB; https://www.rcsb.org/), the Protein Data Bank Japan (PDBj; https://pdbj.org/), the Protein Data Bank in Europe (PDBe; https://www.ebi.ac.uk/pdbe/), and the Biological Magnetic Resonance Data Bank (BMRB; http://www.bmrb.wisc.edu/) (Burley et al. 2017). This community manages annotated protein structure information and provides convenient access to experimental data to the community of researchers and students of biological sciences. In response to the COVID-19 pandemic, RCSB PDB has developed COVID-19/SARS-CoV-2 Resources, which provide quick access to all PDB structures of SARS-CoV-2.

47.3 Molecular Evolution and Phylogenetic Analysis of SARS-CoV-2

Recognizing the relationship, origin, and evolutionary process of emerging outbreaks of zoonotic pathogens may help health authorities develop strategies to prevent cross-species transmissions and control outbreaks in the future (Boni et al. 2020; Li et al. 2020b). Evolution is arguably one of the most important aspects to comprehend, especially for rapidly evolving viruses such as RNA viruses. When the genome sequence of the etiological agent of an epidemic or pandemic is available, comparative genomic and phylogenetic analyses are powerful tools to address the abovementioned queries. Next-generation sequencing (NGS) and advanced algorithms in phylogenetics and comparative genomic analyses provide genome-based methods that play a major role in the characterization and identification of viruses, especially in case of rapidly evolving viruses such as RNA viruses (Delwart 2007). In addition to identifying the origins and evolutionary history, molecular phylogenetics can apply to the recognition of transmission routes and source populations of epidemic outbreaks and seasonal diseases (Kühnert et al. 2011) (Fig. 47.1).

47.3.1 Comparative Genomics of SARS-CoV-2

Immediately after the initial report on 27 unusual cases of pneumonia in Wuhan on December 31, 2019, the complete genome sequence of the causative agent, SARS-CoV-2, was made available on January 10, 2020 (Wu et al. 2020; Lu et al. 2020). Measuring the genetic distance is a simple com-

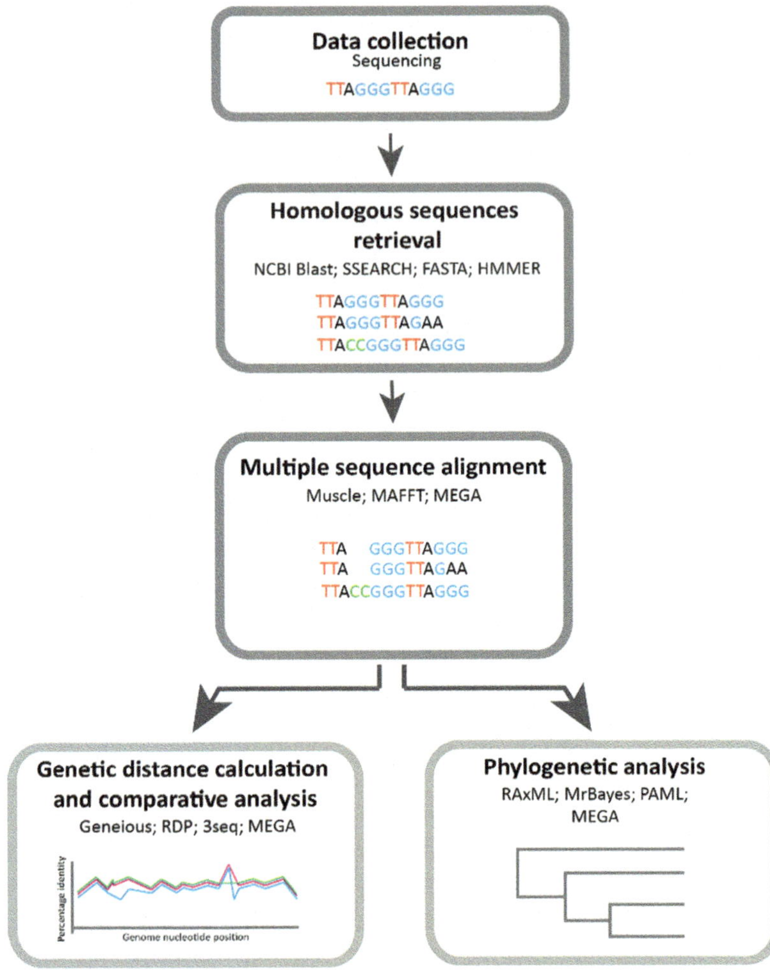

Fig. 47.1 The general workflow of molecular evolution and phylogenetic inferences

putational approach for understanding the origin of biodiversity. For this purpose, the detection of candidate homologous sequences is a fundamental step, which is achieved by searching sequence similarity in sequence databases or GenBank (Pearson 2013). Several available programs statistically estimate sequence similarity against existing sequences in databases, such as Basic Local Alignment Search Tool (BLAST), SSEARCH, FASTA, and HMMER (Altschul et al. 1997; Smith and Waterman 1981; Pearson and Lipman 1988; Johnson et al. 2010). Sequence candidates with significant similarity can be obtained using these programs, as the sequence data in GenBank is becoming increasingly comprehensive, based on the principle that scientific journals that require the data of sequences mentioned in the article should be registered with this platform (Pearson 2013). Once the homologs are detected, building a more accurate alignment model using multiple sequence alignments forms the basis for the calculation of genetic distance and determining genome organization, as well as developing other evolutionary analyses, such as the construction of phylogenetic trees and recombination detection. As computation has become more accessible recently, utilizing a more rigorous multiple sequence alignment method that employs iterative approaches such as MAFFT and MUSCLE is recommended for acquiring high throughput and high accuracy results (Pearson 2013; Katoh et al. 2002; Edgar 2004). Through these steps, the pairwise sequence identities of SARS-CoV-2 initially isolated from several patients were analyzed and were observed to share a nearly identical similarity of above 99.9%, and this indicates a recent host shift into the human population (Lu et al. 2020). Several groups have confirmed that SARS-CoV-2 belongs to the genus *Betacoronavirus*, with the bat SARS-related coronavirus (SARSr-CoV RaTG13) as the most closely related member (96.2% similarity). Human-infecting coronaviruses that are closely related to this virus are SARS-CoV (approximately 79% similarity) and MERS-CoV (approximately 50% similarity) (Lu et al. 2020; Fahmi et al. 2020; Zhou et al. 2020; Paraskevis et al. 2020; Chan et al. 2020).

As SARS-CoV-2 was observed to be related to the genus *Betacoronavirus* and the genome organization of betacoronaviruses has been elucidated previously, the genome organization of SARS-CoV-2 can be determined simply by sequence alignment with the genomes of other members of *Betacoronavirus* (Kim et al. 2020; Marra et al. 2003; Rota et al. 2003). It was performed by alignment to two representative members of *Betacoronavirus*, SARS-CoV Tor2 (GenBank accession number AY274119), and bat SL-CoVZC45 (GenBank accession number MG772933), which are associated with humans and bats, respectively (Wu et al. 2020). The SARS-CoV-2 genome contains a positive-sense single-stranded RNA with a 5′ cap structure and a poly-A 3′ tail, and it has ~29,900 nt encoding ~9860 amino acids (Marra et al. 2003; Rota et al. 2003). The 5′ end of the SARS-CoV-2 genome contains a predicted RNA leader sequence of ~70 nt that is fused with two open reading frames (ORF1a and ORF1b) through short motif transcription regulatory sequences (TRSs) (Wu et al. 2020). The TRSs precede each structural or accessory gene, which aids gene expression (Fehr and Perlman 2015). Usually, the ORF1a and ORF1b overlap in betacoronaviruses, and the region occupies two-thirds of the genome. This region encodes the viral replicase and a translational read-through comprising a − 1 ribosomal frameshift, which allows the translation of this overlapping reading frame into a single polyprotein (Fehr and Perlman 2015; Thiel et al. 2003). Subsequently, upon infection of an appropriate host cell and translation, this large polyprotein commonly undergoes proteolytic processing and is cleaved by virus-encoded proteases into several nonstructural proteins (nsps), including viral papain-like protease (PLpro), main protease (3CLpro, also known as 3-chymotrypsin-like protease), RNA-dependent RNA polymerase (RdRp), and helicase (Hel) (Marra et al. 2003). These proteins mediate the replication process of the viral genome and generate nested transcripts, which are essential for the synthesis of viral proteins (Marra et al. 2003). There are structural and accessory genes downstream of the ORF1a and ORF1b sequences that encode the spike glycoprotein (S), ORF3a, ORF3b, envelope (E), membrane (M), ORF6, ORF7a, ORF7b, ORF8a, ORF8b, and nucleocapsid (N), which are interspersed with the TRS motifs (Fehr and Perlman 2015; Thiel et al. 2003). The structural proteins are responsible for packaging the viral particles and the entry process in host cells, while the accessory proteins are likely to play vital roles in viral pathogenesis (Marra et al. 2003).

A notable feature within the specific region of SARS-CoV-2 can be identified by comparative analysis once the genome organization is known. The S protein is one of the most variable regions within the genomic sequence of SARS-CoV-2. It displays the lowest sequence identity of around 75% to bat-SL-CoVZC45 and bat-SL-CoVZXC21 (Lu et al. 2020). The trimeric S proteins of CoVs mediate binding to the cell receptor ACE2 and membrane fusion in the viral entry process; understanding this protein is compulsory in vaccine design as it elicits an antibody response (Li et al. 2005a; Belouzard et al. 2012; Babcock et al. 2004). The core structure and receptor-binding motif of S protein to the claw-like structure of ACE2 receptors occur in the receptor-binding domain (RBD) (Li 2015, 2008). There are six vital amino acids in the RBD for enhancing the viral binding of SARS-CoV to human ACE2 receptors. They are Y442, L472, N479, D480, T487, and Y491 based on S protein sequence (Wan et al. 2020). With coordinates based on SARS-CoV, those amino acids are L455, F486, Q493, S494, N501, and Y505 in SARS-CoV-2, of which is only one amino acid identical to SARS-CoV (Wan et al. 2020; Andersen et al. 2020). Another notable feature in the S protein of SARS-CoV-2 that has been identified is the presence of polybasic cleavage site (RRAR) at the junction of the two subunits of the S protein: S1 and S2 (Walls et al. 2020). This cleavage site is preceded by inserted proline, which is predicted to result in the o-linked glycosylation of S673, T678, and S686 in the S protein sequence (Andersen et al. 2020). Furin proteases cleave this site during biosynthesis; this process is essential to enable exposure of the fusion sequence for cell membranes, which crucially mediates the cell entry process (Walls et al. 2020). The fact that the cleaving process varies from coronavirus to coronavirus implies that the cleavage site is associated with transmissibility and pathogenesis in the host animal.

CoVs have developed several genetic mechanisms to control replication errors as they have an extraordinarily large RNA genome. Recombination is the capacity to create chimeric molecules from two parental genomes of different origins during coinfection and is one such mechanism that contributes to the genetic stability and diversity of CoVs (Simon-Loriere and Holmes 2011; Lai et al. 1994). Deciphering recombination event that contributes to the virus emergence hence is one way to identify SARS-CoV-2 origins. This analysis is also important to identify virus expansion of viral host range and the evolution of resistance to antivirals (Brown 1997; Gibbs and Weiller 1999; Nora et al. 2007). Numerous tools for examining recombination have been developed within the last two decades, such as SIMPLOT, RDP, TOPALi, and 3seq, all of which are optimized to detect recombination in different ways (Lole et al. 1999; Milne et al. 2008; Martin and Rybicki 2000; Lam et al. 2017). The comprehensive list of tools for analyzing recombination events can be found at http://bioinf.man.ac.uk/robertson/recombination/programs.shtml. Even though CoVs have been reported to be a highly recombinogenic group of viruses, detecting recombination event of the recently emerging CoV from the ancestor that likely develops multiple recombination events, however, is not trivial (Zhang and Holmes 2020; Forni et al. 2017; Hon et al. 2008). It has been reported that SARS-CoV-2 is not a recombinant of any viruses detected to date (Tang et al. 2020). This indicates that the sampled diversity of CoVs, especially within the subgenus sarbecovirus, is massively inefficient.

47.3.2 Molecular Phylogenetics

Elucidating the origin, relationships, and transmission routes of emerging infectious agents is a key to conveniently understanding their biological processes and the potential methods of intervention. These functions are related to molecular evolution, which can be elegantly illustrated by molecular phylogenetic analysis (Kühnert et al. 2011; Lemey et al. 2009). The phylogenetic tree has played a fundamental role in the efforts of inferring the evolutionary history of several infectious agents such as HIV, HCV, and SARS-CoV (Kühnert et al. 2011; Gao et al. 1999; Santiago et al. 2002; Pybus et al. 2009; Markov et al. 2009; Li et al. 2005b). The phylogenetic methods are pertinent for detecting orthology and paralogy, estimating divergence times, reconstructing ancient proteins, finding the important residues to natural selection, identifying recombination points, determining the identity on new pathogens, and identifying mutations likely to be associated with disease (Holder and Lewis 2003). Once the predetermined sequences are aligned, a wide range of methods and software packages are available for reconstructing the phylogenetic tree, which can, at times, make it challenging for researchers to choose the well-suited ones. Each method relies on a different algorithm and has its own advantages and limitations. There are two types of reconstruction methods: distance-based methods and character-based methods (Lemey et al. 2009). Distance-based methods reconstruct trees from the calculated distance matrix of every pair of sequences. These techniques can also be classified into two categories: clustering-based (UPGMA and neighbor-joining) and optimality-based (Fitch-Margoliash and minimum evolution) algorithms (Lemey et al. 2009). These methods generally have a rapid calculation time and a large number of applicable substitution models for calculating the genetic distance scores; however, their applicability in cases with divergent sequences and high variance of large distance estimates is poor (Lemey et al. 2009; Yang and Rannala 2012). Conversely, character-based methods are based on the mutational event scoring of given aligned characters (sites in alignment), which prevents information loss in pairwise distance calculation (Yang and Rannala 2012; Xiong 2006). Character-based methods include maximum likelihood (ML), maximum parsimony (MP), and Bayesian inference (BI). MP follows the "Ockham's razor" principle by selecting the proposed tree with a minimum score of discrete changes to the given alignment data (Lemey et al. 2009). The resulting tree in MP generally has the minimum instances of homoplasy (Lemey et al. 2009). Owing to its simplicity, the computations in MP are less extensive than those in other character-based methods, although these lack explicit assumptions. The ML method estimates trees with the highest likelihood of development using discrete change data of alignment based on multiple proposed trees depicting different evolutionary hypotheses (topologies, branch lengths, and sequence substitution models) of taxa in an alignment. Conversely, BI calculates the posterior probability distribution of the proposed trees based on the model, prior probability distribution, and data. When the data are informative, the majority of the posterior probability is typically concentrated in one tree (or a small subset of trees in an ample tree space) (Lemey et al. 2009; Xiong 2006). At present, ML and BI are considered to represent the best accurate methods since they incorporate the most complex statistical models.

Several authors have constructed the phylogenetic trees of SARS-CoV-2 with that of previous CoVs using the whole genome and specific region sequences to understand its evolutionary history and recombination event (Li et al. 2020b; Wu et al. 2020; Lu et al. 2020; Tang et al. 2020). Based on the currently available genome samples of CoVs, the whole-genome phylogenetic tree indicates that SARS-CoV-2 is closest to SARSr-CoV RaTG13, followed by Pangolin SARSr-CoVs. Two possible scenarios can plausibly explain the origin of SARS-CoV2: natural selection in an animal host prior to zoonotic transfer and natural selection in humans following zoonotic transfer (Andersen et al. 2020). With the first scenario, the genuine progenitor of SARS-CoV2, however, is unrevealed; this uncertainty is attributed to the absence of animal coronaviruses that have the highest similarity on the entire genome location to SARS-CoV-2; this also indicates that subgenus sarbecovirus is likely to have massive hidden diversity (Boni et al. 2020; Andersen et al. 2020). Even though the RaTG13 bat coronavirus has been identified as the virus with the highest similarity to SARS-CoV-2, the RBD in the S protein of SARS-CoV2 is significantly related to that in the Pangolin (*Manis javanica*) coronavirus, which belongs to the sister lineage of the RaTG13 bat coronavirus (Wan et al. 2020; Andersen et al. 2020; Zhang et al. 2020). RBD is a region critical in SARS-CoV-2 transmission in humans because it has a high affinity for human ACE2 (Wan et al. 2020). This binding is likely a result of natural selection of human or human-like ACE2, as a novel binding pattern distinct from those previously predicted was observed (Wan et al. 2020; Andersen et al. 2020). This complication highlights the uncertainty involved in the identification of the direct ancestor of SARS-CoV-2. Conversely, pangolin SARSr-CoVs are likely to be the progenitor of SARS-CoV-2 if the second scenario is applied, following to their significant similarities of the RBD region (Andersen et al. 2020). Since pangolin SARSr-CoVs have RBD that binds with high affinity to humans ACE2, it is likely that this virus jumped to the human population and acquired genomic features that give rise to SARS-CoV-2 (Andersen et al. 2020).

47.4 Computational Approach for One Step of SARS-CoV-2 Vaccine Design

The spike glycoprotein has an important role in binding with host cell receptors; this protein is found on the surface of virions and allows it to be neutralized by antibodies. Vaccines have been proven effective in handling an infectious disease and aims to reduce the deterioration effect caused by that of the etiological agent. Currently, around 90 vaccines have been developed for the clinical trial of SARS-CoV-2 carried out by universities and companies across the world. A variety of vaccines with a distinct approach is tried against SARS-CoV-2, such as virus vaccines (weakened or inactivated virus), viral-vector vaccines (using adenovirus), nucleic-acid vaccines (DNA or RNA vaccines), and protein-based vaccines (protein subunits or virus-like particles) (Sun and Zhang 2014). The development of an effective vaccine against SARS-CoV-2 infection is urgently needed, and it is unlikely to use classical methods in vaccine design under current conditions (Doytchinova and Flower 2007; Lamiable et al. 2016). Currently, there are no inactivated and live-attenuated vaccines efficient enough for providing extensive protection against SARS-CoV-2 infection (Kharisma and Ansori 2020). During the current critical period, the immunoinformatics-based approach could help shorten the duration of the experiment for discovering potential SARS-CoV-2 vaccine candidates. Also, accumulated releases of SARS-CoV-2 genomes in GenBank, NCBI, and GISAID EpiCoV facilitate the immunoinformatics-based approach for the development of virus-based subunit vaccines. The subunit candidate, such as S1 protein or the RBD element of SARS-CoV-2, is one of the precious targets for vaccine development design (Jespersen et al. 2017).

The humoral immune response involves antigen-antibody interactions for pathogen recognition. A specific antibody recognizes antigen through an area called antigenic determinants or B cell epitopes. B cell epitopes can be referred to as surfaces that have amino acid groups accessible for recognizing antibodies or B cell receptors (BCR) and making it possible to trigger a humoral immune response. The introduction of B cells to epitopes is distinct from that of T cells, which have a mechanism through direct binding of antigens and activation through T helper (Th) 2 cells (Sun and Zhang 2014). In addition, on the T-cell surface, there are specific receptors called T-cell receptors, which allow the recognition of antigens displayed by antigen-presenting cells (APCs) through binding to the major histocompatibility complex (MHC) molecules. T-cell epitopes are represented by MHC-1 and MHC-2, both of which are recognized by CD8 and CD4 T cells. CD8 will become cytotoxic T lymphocytes (CTL) following the introduction of the CD8 T epitope, whereas CD4 forms Th or T regulatory cells (Doytchinova and Flower 2007). This is divided into Th1 (plays a role in the cell-mediated immune response), Th2 (forms antibodies through mediated immunity), and Th17 (activates the inflammatory response and protects against extracellular bacteria) (Lamiable et al. 2016; Kharisma and Ansori 2020). Currently, there are numerous immunoinformatics tools for B-cell and T-cell epitope predictions (Table 47.2).

Prediction of B cell epitopes allows us to gain insight into the important position on the surface of antigen that can be recognized by BCR in the adaptive immune response. Immune Epitope Database and Analysis Resource (IEDB) (http://tools.iedb.org/main/) is one such prominent platform for epitope prediction, the approach for epitope prediction can be divided into linear and discontinuous predictions (Sun and Zhang 2014). Linear prediction works by introducing antibodies to the primary structure of the amino acid residue of the antigen. It is distinct to discontinuous prediction, which utilizes the 3D form of an epitope for the introduction. The methods of linear epitope prediction for B cells use the sequence characteristics of antigens through the amino acid scale and hidden Markov models (HMMs) (http://tools.iedb.org/bcell/), which consist of prediction of hydrophilicity, flexibility, accessibility, surface, and antigenic tendencies in polypeptide chains (Sun and Zhang 2014; Doytchinova and Flower 2007). BepiPred is one of methods to predict B cell epitopes in IEDB platform, this method is based on a random forest algorithm trained on epitopes annotated from antibody-antigen protein structures (Sun and Zhang 2014). The other methods include DiscoTope and ElliPro, these methods determine B-cell epitopes in 3D antigen structures based on solvent accessibility and flexibility (Sun and Zhang 2014).

Table 47.2 Immunoinformatics tools of B-cell and T-cell epitope prediction

Prediction	Tool	URL	References
B-cell epitope	BepiPred	http://tools.iedb.org/bcell/	Jespersen et al. (2017)
	DiscoTope	http://tools.iedb.org/discotope/	Kringelum et al. (2012)
	ElliPro	http://tools.iedb.org/ellipro/	Ponomarenko et al. (2008)
	ABCpred	http://crdd.osdd.net/raghava/abcpred/	Saha and Raghava (2006)
	BCPred	http://crdd.osdd.net/raghava/bcepred/	El-Manzalawy et al. (2008)
	COBEpro	http://scratch.proteomics.ics.uci.edu/	Sweredoski and Baldi (2009)
	SVMTriP	http://sysbio.unl.edu/SVMTriP/	Yao et al. (2012)
	LBtope	http://crdd.osdd.net/raghava/lbtope/	Singh et al. (2013)
	EpiPred	http://opig.stats.ox.ac.uk/webapps/newsabdab/sabpred/epipred/	Dunbar et al. (2016)
	Pepitope	http://pepitope.tau.ac.il/	Mayrose et al. (2007)
T-cell epitope	IEDB-MHCI	http://tools.iedb.org/mhci/	Nielsen and Andreatta (2016)
	IEDB-MHCII	http://tools.iedb.org/mhcii/	Nielsen et al. (2007)
	NetMHC	https://services.healthtech.dtu.dk/service.php?NetMHC-4.0	Lundegaard et al. (2008)
	NetMHCII	https://services.healthtech.dtu.dk/service.php?NetMHCII-2.3	Jensen et al. (2018)
	NetMHCIIpan	https://services.healthtech.dtu.dk/service.php?NetMHCIIpan-4.0	Karosiene et al. (2013)
	IL4pred	https://webs.iiitd.edu.in/raghava/il4pred/index.php	Dhanda et al. (2013)
	ProPred	http://crdd.osdd.net/raghava/propred/	Singh and Raghava (2001)
	EpiTOP	http://www.ddg-pharmfac.net/EpiTOP3/	Dimitrov et al. (2010)
	MHCPred	http://www.ddg-pharmfac.net/mhcpred/	Guan et al. (2003)
	PEPVAC	http://imed.med.ucm.es/PEPVAC/	Reche and Reinherz (2005)

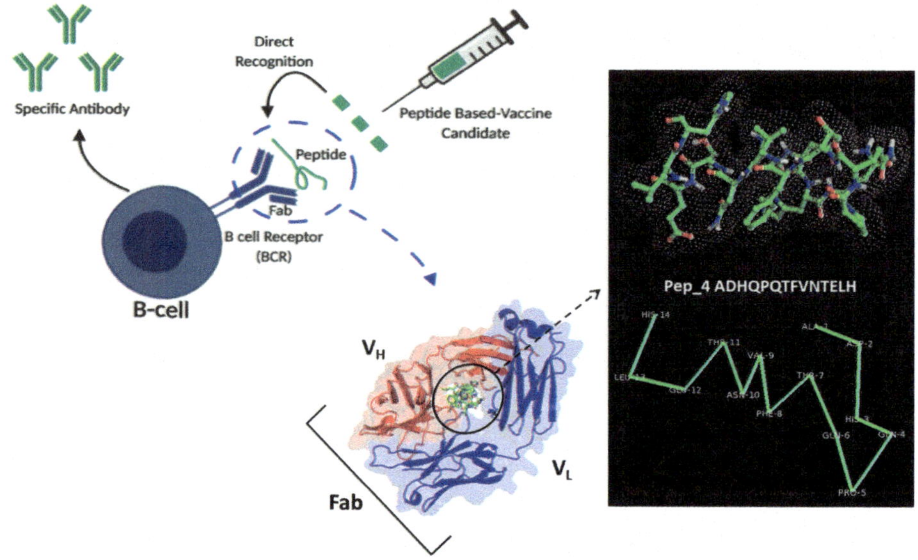

Fig. 47.2 Schematic representation of peptide epitope based-vaccine candidate (green) from spike glycoprotein recognized by B-cell immune response to produce specific antibody against SARS-CoV-2 (Kharisma and Ansori 2020)

Conversely, T-cell epitope prediction aims to identify short peptides in antigens that allow CD4 or CD8 cell stimulation. There are three steps for the prediction of T-cell epitopes, namely, antigen processing, binding of peptides to MHC molecules, and recognition of T-cell receptor.

The predicted results of epitope mapping consist of peptides that are potentially recognized by the immune system. The peptide can be selected based on the antigenicity or the ability of an antigen to trigger the activation of immune cell receptors that recognize it. This prediction is based on linear protein sequence alignment and physicochemical properties of the protein so that it can calculate antigenicity scores on the peptide. VaxiJen v2.0 server http://www.ddg-pharmfac.net/vaxijen/VaxiJen/VaxiJen.html is a platform that can be used to predict the antigenicity of peptides (Sun and Zhang 2014; Doytchinova and Flower 2007). After determination of the epitopes candidates, the next steps include peptide structure prediction and molecular docking with specific receptor to identify the potential of candidates (Fig. 47.2). PEP-FOLD (http://mobyle.rpbs.univ-paris-diderot.fr/cgi-bin/portal.py#forms::PEP-FOLD3) and CLusPro are two examples of tools to predict peptides structures and to perform molecular docking, respectively (Lamiable et al. 2016; Kozakov et al. 2007).

Previously, we characterized the spike glycoprotein of SARS-CoV-2 to obtain epitope-based peptide vaccine against SARS-CoV-2. In that study, SARS-CoV-2 isolates were retrieved from the GISAID EpiCoV and NCBI and then aligned to obtain the conserved region of SARS-CoV-2 spike glycoprotein. We identified Pep_4 ADHQPQTFVNTELH as potential B-cell epitope vaccine candidates to overcome the SARS-CoV-2 outbreak. Pep_4 vaccine candidates are predicted to trigger B-cell immune response through direct binding to the BCR/FAB molecule (Kharisma and Ansori 2020).

47.5 Conclusion

Computational and statistical approaches are essential entities in virology. Here, we reported numerous tools and platforms that have been developed to facilitate global research to study and elucidate the various associated concerns which may help strategize the process of obtaining a solution for SARS-CoV-2 devastating effects. The utilization of such media aids active tracking of important findings and also helps health authorities develop strategies to prevent cross-species transmissions and control outbreaks in the future. Additionally, advanced algorithms and databases facilitate the urgency to find potential interventions such as vaccine design through experimental simulation.

References

Altschul SF, Madden TL, Schäffer AA, Zhang J, Zhang Z, Miller W, Lipman DJ (1997) Gapped BLAST and PSI-BLAST: a new generation of protein database search programs. Nucleic Acids Res. 25(17):3389–3402. https://doi.org/10.1093/nar/25.17.3389

Andersen KG, Rambaut A, Lipkin WI, Holmes EC, Garry RF (2020) The proximal origin of SARS-CoV-2. Nat. Med. 26(4):450–452. https://doi.org/10.1038/s41591-020-0820-9

Babcock GJ, Esshaki DJ, Thomas WD, Ambrosino DM (2004) Amino acids 270 to 510 of the severe acute respiratory syndrome coronavirus spike protein are required for interaction with receptor. J. Virol. 78(9):4552–4560. https://doi.org/10.1128/jvi.78.9.4552-4560.2004

Balakrishnan VS (2020) Increasing accessibility in COVID-19 clinical trials. The Lancet Microbe 1(1):e13

Belouzard S, Millet JK, Licitra BN, Whittaker GR (2012) Mechanisms of coronavirus cell entry mediated by the viral spike protein. Viruses 4(6):1011–1033. https://doi.org/10.3390/v4061011

Boni MF, Lemey P, Jiang X, Lam TT-Y, Perry B, Castoe T, Rambaut A, Robertson DL (2020) Evolutionary origins of the SARS-CoV-2 sarbecovirus lineage responsible for the COVID-19 pandemic. bioRxiv:2020.2003.2030.015008. https://doi.org/10.1101/2020.03.30.015008

Brainard J (2020) Scientists are drowning in COVID-19 papers. Can new tools keep them afloat? https://www.sciencemag.org/news/2020/05/scientists-are-drowning-covid-19-papers-can-new-tools-keep-them-afloat. Accessed 15 May 2020

Breitbart M, Rohwer F (2005) Here a virus, there a virus, everywhere the same virus? Trends Microbiol. 13(6):278–284. https://doi.org/10.1016/j.tim.2005.04.003

Brister JR, Ako-Adjei D, Bao Y, Blinkova O (2015) NCBI viral genomes resource. Nucleic Acids Res. 43(Database issue):D571–D577. https://doi.org/10.1093/nar/gku1207

Brown DWG (1997) Threat to humans from virus infections of non-human primates. Rev. Med. Virol. 7(4):239–246. https://doi.org/10.1002/(sici)1099-1654(199712)7:4<239::Aid-rmv210>3.0.Co;2-q

Burley SK, Berman HM, Kleywegt GJ, Markley JL, Nakamura H, Velankar S (2017) Protein data Bank (PDB): the single global macromolecular structure archive. Methods Mol. Biol. 1607:627–641. https://doi.org/10.1007/978-1-4939-7000-1_26

Chan JF, Kok KH, Zhu Z, Chu H, To KK, Yuan S, Yuen KY (2020) Genomic characterization of the 2019 novel human-pathogenic coronavirus isolated from a patient with atypical pneumonia after visiting Wuhan. Emerg Microbes Infect 9(1):221–236. https://doi.org/10.1080/22221751.2020.1719902

Chang J (2015) Core services: reward bioinformaticians. Nature Nature 520(7546):151–152

Chen Q, Allot A, Lu Z (2020) Keep up with the latest coronavirus research. Nature 579(7798):193. https://doi.org/10.1038/d41586-020-00694-1

ClinicalTrials.gov (2000). ClinicalTrials.gov. https://clinicaltrials.gov/. Accessed 1 April 2020

COVID-evidence (2020) Find evidence on interventions for COVID-19. https://covid-evidence.org/. Accessed 1 May 2020

Delwart EL (2007) Viral metagenomics. Rev. Med. Virol. 17(2):115–131. https://doi.org/10.1002/rmv.532

Dhanda SK, Gupta S, Vir P, Raghava GPS (2013) Prediction of IL4 inducing peptides. Clin Dev Immunol 2013:263952

Dimitrov I, Garnev P, Flower DR, Doytchinova I (2010) EpiTOP—a proteochemometric tool for MHC class II binding prediction. Bioinformatics 26(16):2066–2068

Doytchinova IA, Flower DR (2007) VaxiJen: a server for prediction of protective antigens, tumour antigens and subunit vaccines. BMC Bioinformatics 8(1):4. https://doi.org/10.1186/1471-2105-8-4

Dunbar J, Krawczyk K, Leem J, Marks C, Nowak J, Regep C, Georges G, Kelm S, Popovic B, Deane CM (2016) SAbPred: a structure-based antibody prediction server. Nucleic Acids Res 44(W1):W474–8

Edgar RC (2004) MUSCLE: multiple sequence alignment with high accuracy and high throughput. Nucleic Acids Res. 32(5):1792–1797. https://doi.org/10.1093/nar/gkh340

Elbe S, Buckland-Merrett G (2017) Data, disease and diplomacy: GISAID's innovative contribution to global health. Glob Chall 1(1):33–46. https://doi.org/10.1002/gch2.1018

EL-Manzalawy Y, Dobbs D, Honavar V (2008) Predicting linear B-cell epitopes using string kernels. J Mol Recognit 21(4):243–255

Fahmi M, Kubota Y, Ito M (2020) Nonstructural proteins NS7b and NS8 are likely to be phylogenetically associated with evolution of 2019-nCoV. Infect. Genet. Evol. 81:104272. https://doi.org/10.1016/j.meegid.2020.104272

Fehr AR, Perlman S (2015) Coronaviruses: an overview of their replication and pathogenesis. Methods Mol. Biol. 1282:1–23. https://doi.org/10.1007/978-1-4939-2438-7_1

Flint SJ, Racaniello VR, Rall GF, Skalka AM, Enquist LW (2015) Principles of virology, vol 1, 4th edn. ASM Press, Washington, DC

Forni D, Cagliani R, Clerici M, Sironi M (2017) Molecular evolution of human coronavirus genomes. Trends Microbiol. 25(1):35–48. https://doi.org/10.1016/j.tim.2016.09.001

Friedman LM, Furberg C, DeMets DL, Reboussin DM, Granger CB (2010) Fundamentals of clinical trials, vol 4. Springer International Publishing, Cham. https://doi.org/10.1007/978-3-319-18539-2

Gao F, Bailes E, Robertson DL, Chen Y, Rodenburg CM, Michael SF, Cummins LB, Arthur LO, Peeters M, Shaw GM, Sharp PM, Hahn BH (1999) Origin of HIV-1 in the chimpanzee Pan troglodytes troglodytes. Nature 397(6718):436–441. https://doi.org/10.1038/17130

Gibbs MJ, Weiller GF (1999) Evidence that a plant virus switched hosts to infect a vertebrate and then recombined with a vertebrate-infecting virus. Proc. Natl. Acad. Sci. 96(14):8022–8027. https://doi.org/10.1073/pnas.96.14.8022

Guan P, Doytchinova IA, Zygouri C, Flower DR (2003) MHCPred: a server for quantitative prediction of peptide–MHC binding. Nucleic Acids Res 31:3621–3624

Holder M, Lewis PO (2003) Phylogeny estimation: traditional and Bayesian approaches. Nat. Rev. Genet. 4(4):275–284. https://doi.org/10.1038/nrg1044

Hon C-C, Lam T-Y, Shi Z-L, Drummond AJ, Yip C-W, Zeng F, Lam P-Y, Leung FC-C (2008) Evidence of the recombinant origin of a bat severe acute respiratory syndrome (SARS)-like coronavirus and its implications on the direct ancestor of SARS coronavirus. J. Virol. 82(4):1819–1826. https://doi.org/10.1128/jvi.01926-07

Horbach SPJM (2020) Pandemic publishing: medical journals drastically speed up their publication process for Covid-19. bioRxiv:2020.2004.2018.045963. https://doi.org/10.1101/2020.04.18.045963

Huang C, Wang Y, Li X, Ren L, Zhao J, Hu Y, Zhang L, Fan G, Xu J, Gu X, Cheng Z, Yu T, Xia J, Wei Y, Wu W, Xie X, Yin W, Li H, Liu M, Xiao Y, Gao H, Guo L, Xie J, Wang G, Jiang R, Gao Z, Jin Q, Wang J, Cao B (2020) Clinical features of patients infected with 2019 novel coronavirus in Wuhan, China. Lancet 395(10223):497–506. https://doi.org/10.1016/s0140-6736(20)30183-5

Hulo C, de Castro E, Masson P, Bougueleret L, Bairoch A, Xenarios I, Le Mercier P (2011) ViralZone: a knowledge resource to understand virus diversity. Nucleic Acids Res. 39(Database issue):D576–D582. https://doi.org/10.1093/nar/gkq901

Jensen KK, Andreatta M, Marcatili P, Buus S, Greenbaum JA, Yan Z, Sette A, Peters B, Nielsen M (2018) Improved methods for predicting peptide binding affinity to MHC class II molecules. Immunology 154(3):394–406

Jespersen MC, Peters B, Nielsen M, Marcatili P (2017) BepiPred-2.0: improving sequence-based B-cell epitope prediction using conformational epitopes. Nucleic Acids Res. 45(W1):W24–w29. https://doi.org/10.1093/nar/gkx346

Johns Hopkins Bloomberg School of Public Health (2020) 2019 novel coronavirus research compendium (NCRC). https://ncrc.jhsph.edu Accessed 8 May 2020

Johnson LS, Eddy SR, Portugaly E (2010) Hidden Markov model speed heuristic and iterative HMM search procedure. BMC Bioinformatics 11:431. https://doi.org/10.1186/1471-2105-11-431

Karosiene E, Rasmussen M, Blicher T, Lund O, Buus S, Nielsen M (2013) NetMHCIIpan-3. 0, a common pan-specific MHC class II prediction method including all three human MHC class II isotypes, HLA-DR, HLA-DP and HLA-DQ. Immunogenetics 65(10):711–724

Katoh K, Misawa K, Kuma K, Miyata T (2002) MAFFT: a novel method for rapid multiple sequence alignment based on fast Fourier transform. Nucleic Acids Res. 30(14):3059–3066. https://doi.org/10.1093/nar/gkf436

Kharisma V, Ansori A (2020) Construction of epitope-based peptide vaccine against SARS-CoV-2: Immunoinformatics study. J Pure Appl Microbiol 14(suppl 1):999–1005

Kim D, Lee JY, Yang JS, Kim JW, Kim VN, Chang H (2020) The architecture of SARS-CoV-2 transcriptome. Cell 181(4):914–921.e910. https://doi.org/10.1016/j.cell.2020.04.011

Koonin EV, Galperin MY (2003) Sequence – evolution – function: computational approaches in comparative genomics. Springer US, Boston. https://doi.org/10.1007/978-1-4757-3783-7

Kozakov D, Hall DR, Xia B, Porter KA, Padhorny D, Yueh C, Beglov D, Vajda S (2007) The ClusPro web server for protein-protein docking. Nat Protoc 12(2):255–278. https://doi.org/10.1038/nprot.2016.169

Kringelum JV, Lundegaard C, Lund O, Nielsen M (2012) Reliable B cell epitope predictions: impacts of method development and improved benchmarking. PLoS Comput Biol 8(12):e1002829

Kühnert D, Wu CH, Drummond AJ (2011) Phylogenetic and epidemic modeling of rapidly evolving infectious diseases. Infect. Genet. Evol. 11(8):1825–1841. https://doi.org/10.1016/j.meegid.2011.08.005

Lai MM, Liao CL, Lin YJ, Zhang X (1994) Coronavirus: how a large RNA viral genome is replicated and transcribed. Infect. Agents Dis. 3(2–3):98–105

Lam HM, Ratmann O, Boni MF (2017) Improved algorithmic complexity for the 3SEQ recombination detection algorithm. Mol. Biol. Evol. 35(1):247–251. https://doi.org/10.1093/molbev/msx263

Lamiable A, Thévenet P, Rey J, Vavrusa M, Derreumaux P, Tufféry P (2016) PEP-FOLD3: faster de novo structure prediction for linear peptides in solution and in complex. Nucleic Acids Res. 44(W1):W449–W454. https://doi.org/10.1093/nar/gkw329

Lemey P, Salemi M, Vandamme A-M (2009) The phylogenetic handbook: a practical approach to phylogenetic analysis and hypothesis testing. Cambridge University Press

Li F (2008) Structural analysis of major species barriers between humans and palm civets for severe acute respiratory syndrome coronavirus infections. J. Virol. 82(14):6984–6991. https://doi.org/10.1128/jvi.00442-08

Li F (2015) Receptor recognition mechanisms of coronaviruses: a decade of structural studies. J. Virol. 89(4):1954–1964. https://doi.org/10.1128/jvi.02615-14

Li F, Li W, Farzan M, Harrison SC (2005a) Structure of SARS coronavirus spike receptor-binding domain complexed with receptor. Science 309(5742):1864–1868. https://doi.org/10.1126/science.1116480

Li W, Shi Z, Yu M, Ren W, Smith C, Epstein JH, Wang H, Crameri G, Hu Z, Zhang H, Zhang J, McEachern J, Field H, Daszak P, Eaton BT, Zhang S, Wang LF (2005b) Bats are natural reservoirs of SARS-like coronaviruses. Science 310(5748):676–679. https://doi.org/10.1126/science.1118391

Li Q, Guan X, Wu P, Wang X, Zhou L, Tong Y, Ren R, Leung KSM, Lau EHY, Wong JY, Xing X, Xiang N, Wu Y, Li C, Chen Q, Li D, Liu T, Zhao J, Liu M, Tu W, Chen C, Jin L, Yang R, Wang Q, Zhou S, Wang R, Liu H, Luo Y, Liu Y, Shao G, Li H, Tao Z, Yang Y, Deng Z, Liu B, Ma Z, Zhang Y, Shi G, Lam TTY, Wu JT, Gao GF, Cowling BJ, Yang B, Leung GM, Feng Z (2020a) Early transmission dynamics in Wuhan, China, of novel coronavirus infected pneumonia. N Engl J Med 382(13):1199–1207

Li X, Giorgi EE, Marichann MH, Foley B, Xiao C, Kong X-P, Chen Y, Korber B, Gao F (2020b) Emergence of SARS-CoV-2 through recombination and strong purifying selection. bioRxiv:2020.2003.2020.000885. doi:https://doi.org/10.1101/2020.03.20.000885

Lole KS, Bollinger RC, Paranjape RS, Gadkari D, Kulkarni SS, Novak NG, Ingersoll R, Sheppard HW, Ray SC (1999) Full-length human immunodeficiency virus type 1 genomes from subtype C-infected seroconverters in India, with evidence of intersubtype recombination. J. Virol. 73(1):152–160. https://doi.org/10.1128/jvi.73.1.152-160.1999

Lorenc T, Khouja C, Raine G, Sutcliffe K, Wright K, Sowden A, Thomas J (2020) COVID-19: living map of the evidence. http://eppi.ioe.ac.uk/COVID19_MAP/covid_map_v11.html. Accessed 15 May 2020

Lu R, Zhao X, Li J, Niu P, Yang B, Wu H, Wang W, Song H, Huang B, Zhu N, Bi Y, Ma X, Zhan F, Wang L, Hu T, Zhou H, Hu Z, Zhou W, Zhao L, Chen J, Meng Y, Wang J, Lin Y, Yuan J, Xie Z, Ma J, Liu WJ, Wang D, Xu W, Holmes EC, Gao GF, Wu G, Chen W, Shi W, Tan W (2020) Genomic characterisation and epidemiology of 2019 novel coronavirus: implications for virus origins and receptor binding. Lancet 395(10224):565–574. https://doi.org/10.1016/s0140-6736(20)30251-8

Lundegaard C, Lamberth K, Harndahl M, Buus S, Lund O, Nielsen M (2008) NetMHC-3.0: accurate web accessible predictions of human, mouse and monkey MHC class I affinities for peptides of length 8–11. Nucleic Acids Res 1:W509–12

Markov PV, Pepin J, Frost E, Deslandes S, Labbé AC, Pybus OG (2009) Phylogeography and molecular epidemiology of hepatitis C virus genotype 2 in Africa. J. Gen. Virol. 90(Pt 9):2086–2096. https://doi.org/10.1099/vir.0.011569-0

Marra MA, Jones SJ, Astell CR, Holt RA, Brooks-Wilson A, Butterfield YS, Khattra J, Asano JK, Barber SA, Chan SY, Cloutier A, Coughlin SM, Freeman D, Girn N, Griffith OL, Leach SR, Mayo M, McDonald H, Montgomery SB, Pandoh PK, Petrescu AS, Robertson AG, Schein JE, Siddiqui A, Smailus DE, Stott JM, Yang GS, Plummer F, Andonov A, Artsob H, Bastien N, Bernard K, Booth TF, Bowness D, Czub M, Drebot M, Fernando L, Flick R, Garbutt M, Gray M, Grolla A, Jones S, Feldmann H, Meyers A, Kabani A, Li Y, Normand S, Stroher U, Tipples GA, Tyler S, Vogrig R, Ward D, Watson B, Brunham RC, Krajden M, Petric M, Skowronski DM, Upton C, Roper RL (2003) The genome sequence of the SARS-associated coronavirus. Science 300(5624):1399–1404. https://doi.org/10.1126/science.1085953

Martin D, Rybicki E (2000) RDP: detection of recombination amongst aligned sequences. Bioinformatics 16(6):562–563. https://doi.org/10.1093/bioinformatics/16.6.562

Marz M, Beerenwinkel N, Drosten C, Fricke M, Frishman D, Hofacker IL, Hoffmann D, Middendorf M, Rattei T, Stadler PF, Töpfer A (2014) Challenges in RNA virus bioinformatics. Bioinformatics 30(13):1793–1799. https://doi.org/10.1093/bioinformatics/btu105

Mayrose I, Penn O, Erez E, Rubinstein ND, Shlomi T, Freund NT, Bublil EM, Ruppin E, Sharan R, Gershoni JM, Martz E, Pupko T (2007) Pepitope: epitope mapping from affinity-selected peptides. Bioinformatics 23(23):3244–3246

Milne I, Lindner D, Bayer M, Husmeier D, McGuire G, Marshall DF, Wright F (2008) TOPALi v2: a rich graphical interface for evolutionary analyses of multiple alignments on HPC clusters and multi-core desktops. Bioinformatics 25(1):126–127. https://doi.org/10.1093/bioinformatics/btn575

Nielsen M, Andreatta M (2016) NetMHCpan-3.0; improved prediction of binding to MHC class I molecules integrating information from multiple receptor and peptide length datasets. Genome Med 8(1):1–9

Nielsen M, Lundegaard C, Lund O (2007) Prediction of MHC class II binding affinity using SMM-align, a novel stabilization matrix alignment method. BMC bioinform 8(1):238

Nora T, Charpentier C, Tenaillon O, Hoede C, Clavel F, Hance AJ (2007) Contribution of recombination to the evolution of human immunodeficiency viruses expressing resistance to antiretroviral treatment. J. Virol. 81(14):7620–7628. https://doi.org/10.1128/jvi.00083-07

Paraskevis D, Kostaki EG, Magiorkinis G, Panayiotakopoulos G, Sourvinos G, Tsiodras S (2020) Full-genome evolutionary analysis of the novel corona virus (2019-nCoV) rejects the hypothesis of emergence as a result of a recent recombination event. Infect. Genet. Evol. 79:104212. https://doi.org/10.1016/j.meegid.2020.104212

Pearson WR (2013) An introduction to sequence similarity ("homology") searching. Curr. Protoc. Bioinformatics. Chapter 3:Unit3.1. https://doi.org/10.1002/0471250953.bi0301s42

Pearson WR, Lipman DJ (1988) Improved tools for biological sequence comparison. Proc. Natl. Acad. Sci. U. S. A. 85(8):2444–2448. https://doi.org/10.1073/pnas.85.8.2444

Ponomarenko J, Bui HH, Li W, Fusseder N, Bourne PE, Sette A, Peters B (2008) ElliPro: a new structure-based tool for the prediction of antibody epitopes. BMC Bioinform 9(1):514

Pybus OG, Barnes E, Taggart R, Lemey P, Markov PV, Rasachak B, Syhavong B, Phetsouvanah R, Sheridan I, Humphreys IS, Lu L, Newton PN, Klenerman P (2009) Genetic history of hepatitis C virus in East Asia. J. Virol. 83(2):1071–1082. https://doi.org/10.1128/jvi.01501-08

Reche PA, Reinherz EL (2005) PEPVAC: a web server for multi-epitope vaccine development based on the prediction of supertypic MHC ligands. Nucleic Acids Res 33(1):W138–42

Rota PA, Oberste MS, Monroe SS, Nix WA, Campagnoli R, Icenogle JP, Peñaranda S, Bankamp B, Maher K, Chen MH, Tong S, Tamin A, Lowe L, Frace M, DeRisi JL, Chen Q, Wang D, Erdman DD, Peret TC, Burns C, Ksiazek TG, Rollin PE, Sanchez A, Liffick S, Holloway B, Limor J, McCaustland K, Olsen-Rasmussen M, Fouchier R, Günther S, Osterhaus AD, Drosten C, Pallansch MA, Anderson LJ, Bellini WJ (2003) Characterization of a novel coronavirus associated with severe acute respiratory syndrome. Science 300(5624):1394–1399. https://doi.org/10.1126/science.1085952

Ruano J, Gómez-García F, Pieper D, Puljak L (2020) What evidence-based medicine researchers can do to help clinicians fighting COVID-19? J. Clin. Epidemiol. https://doi.org/10.1016/j.jclinepi.2020.04.015

Saha S, Raghava GPS (2006) Prediction of continuous B-cell epitopes in an antigen using recurrent neural network. Proteins 65(1):40–48

Santiago ML, Rodenburg CM, Kamenya S, Bibollet-Ruche F, Gao F, Bailes E, Meleth S, Soong SJ, Kilby JM, Moldoveanu Z, Fahey B, Muller MN, Ayouba A, Nerrienet E, McClure HM, Heeney JL, Pusey AE, Collins DA, Boesch C, Wrangham RW, Goodall J, Sharp PM, Shaw GM, Hahn BH (2002) SIVcpz in wild chimpanzees. Science 295(5554):465. https://doi.org/10.1126/science.295.5554.465

SciSight (2020) SciSight. https://scisight.apps.allenai.org/. Accessed 1 May 2020

Simon-Loriere E, Holmes EC (2011) Why do RNA viruses recombine? Nat. Rev. Microbiol. 9(8):617–626. https://doi.org/10.1038/nrmicro2614

Singh H, Raghava GPS (2001) ProPred: prediction of HLA-DR binding sites. Bioinformatics 17(12):1236–1237

Singh H, Ansari HR, Raghava GPS (2013) Improved method for linear B-cell epitope prediction using antigen's primary sequence. PLOS ONE 8(5):e62216

Smith TF, Waterman MS (1981) Identification of common molecular subsequences. J. Mol. Biol. 147(1):195–197. https://doi.org/10.1016/0022-2836(81)90087-5

SPIKE-COVID (2020) SPIKE-COVID: Extractive search over CORD 19. https://spike.covid-19.apps.allenai.org/search/covid19. Accessed 2 April 2020

Sun B, Zhang Y (2014) Overview of orchestration of CD4+ T cell subsets in immune responses. Adv. Exp. Med. Biol. 841:1–13. https://doi.org/10.1007/978-94-017-9487-9_1

Sussman JL, Lin D, Jiang J, Manning NO, Prilusky J, Ritter O, Abola EE (1998) Protein Data Bank (PDB): database of three-dimensional structural information of biological macromolecules. Acta Crystallogr. D Biol. Crystallogr. 54(Pt 6 Pt 1):1078–1084. https://doi.org/10.1107/s0907444998009378

Suttle CA (2005) Viruses in the sea. Nature 437(7057):356–361. https://doi.org/10.1038/nature04160

Sweredoski MJ, Baldi P (2009) COBEpro: a novel system for predicting continuous B-cell epitopes. Protein Eng Des Sel 22(3):113–120

Tang X, Wu C, Li X, Song Y, Yao X, Wu X, Duan Y, Zhang H, Wang Y, Qian Z, Cui J, Lu J (2020) On the origin and continuing evolution of SARS-CoV-2. Natl. Sci. Rev. https://doi.org/10.1093/nsr/nwaa036

Thiel V, Ivanov KA, Putics Á, Hertzig T, Schelle B, Bayer S, Weißbrich B, Snijder EJ, Rabenau H, Doerr HW, Gorbalenya AE, Ziebuhr J (2003) Mechanisms and enzymes involved in SARS coronavirus genome expression. J. Gen. Virol. 84(Pt 9):2305–2315. https://doi.org/10.1099/vir.0.19424-0

Thorlund K, Dron L, Park J, Hsu G, Forrest JI, Mills EJ (2020) A real-time dashboard of clinical trials for COVID-19. Lancet Digit Health 2(6):e286–e287. https://doi.org/10.1016/s2589-7500(20)30086-8

Walls AC, Park YJ, Tortorici MA, Wall A, McGuire AT, Veesler D (2020) Structure, function, and antigenicity of the SARS-CoV-2 spike glycoprotein. Cell 181(2):281–292.e286. https://doi.org/10.1016/j.cell.2020.02.058

Wan Y, Shang J, Graham R, Baric RS, Li F (2020) Receptor recognition by the novel coronavirus from Wuhan: an analysis based on decade-long structural studies of SARS coronavirus. J. Virol. 94(7):e00127–e00120. https://doi.org/10.1128/jvi.00127-20

Wang LL, Lo K, Chandrasekhar Y, Reas R, Yang J, Eide D, Funk K, Kinney R, Liu Z, Merrill W (2020) CORD-19: the Covid-19 open research dataset. arXiv preprint arXiv:200410706

World Health Organization (2005) The international clinical trials registry platform – ICTRP. https://www.who.int/ictrp/en/. Accessed 2 April 2020

World Health Organization (2020a) Coronavirus disease 2019 (COVID-19) situation report – 51, 11 March

2020. https://www.who.int/emergencies/diseases/novel-coronavirus-2019/situation-reports. Accessed 15 April 2020

World Health Organization (2020b) WHO coronavirus disease (COVID-19) dashboard. https://covid19.who.int/. Accessed 1 May 2020

Wu F, Zhao S, Yu B, Chen YM, Wang W, Song ZG, Hu Y, Tao ZW, Tian JH, Pei YY, Yuan ML, Zhang YL, Dai FH, Liu Y, Wang QM, Zheng JJ, Xu L, Holmes EC, Zhang YZ (2020) A new coronavirus associated with human respiratory disease in China. Nature 579(7798):265–269. https://doi.org/10.1038/s41586-020-2008-3

Xiong J (2006) Essential bioinformatics. Cambridge University Press, New York

Yang Z, Rannala B (2012) Molecular phylogenetics: principles and practice. Nat. Rev. Genet. 13(5):303–314. https://doi.org/10.1038/nrg3186

Yao B, Zhang L, Liang S, Zhang C (2012) SVMTriP: a method to predict antigenic epitopes using support vector machine to integrate tri-peptide similarity and propensity. PLOS ONE 7(9):e45152

Zhang Y-Z, Holmes EC (2020) A genomic perspective on the origin and emergence of SARS-CoV-2. Cell 181(2):223–227. https://doi.org/10.1016/j.cell.2020.03.035

Zhang T, Wu Q, Zhang Z (2020) Pangolin homology associated with 2019-nCoV. bioRxiv:2020.2002.2019.950253. doi:https://doi.org/10.1101/2020.02.19.950253

Zhou P, Yang XL, Wang XG, Hu B, Zhang L, Zhang W, Si HR, Zhu Y, Li B, Huang CL, Chen HD, Chen J, Luo Y, Guo H, Jiang RD, Liu MQ, Chen Y, Shen XR, Wang X, Zheng XS, Zhao K, Chen QJ, Deng F, Liu LL, Yan B, Zhan FX, Wang YY, Xiao GF, Shi ZL (2020) A pneumonia outbreak associated with a new coronavirus of probable bat origin. Nature 579(7798):270–273. https://doi.org/10.1038/s41586-020-2012-7

Zhu N, Zhang D, Wang W, Li X, Yang B, Song J, Zhao X, Huang B, Shi W, Lu R, Niu P, Zhan F, Ma X, Wang D, Xu W, Wu G, Gao GF, Tan W (2020) A novel coronavirus from patients with pneumonia in China, 2019. N Engl J Med 382(8):727–733

Answering the Challenge of COVID-19 Pandemic Through Innovation and Ingenuity

48

Kathryn Clare Kelley, Jonathan Kamler, Manish Garg, and Stanislaw P. Stawicki

Abstract

The novel coronavirus disease 2019 (COVID-19) pandemic has created a maelstrom of challenges affecting virtually every aspect of global healthcare system. Critical hospital capacity issues, depleted ventilator and personal protective equipment stockpiles, severely strained supply chains, profound economic slowdown, and the tremendous human cost all culminated in what is questionably one of the most profound challenges that humanity faced in decades, if not centuries. Effective global response to the current pandemic will require innovation and ingenuity. This chapter discusses various creative approaches and ideas that arose in response to COVID-19, as well as some of the most impactful future trends that emerged as a result. Among the many topics discussed herein are telemedicine, blockchain technology, artificial intelligence, stereolithography, and distance learning.

Keywords

Artificial intelligence · Blockchain technology · COVID-19 · Healthcare innovation · Pandemic · SARS-CoV-2 · Technology · Telemedicine

48.1 Introduction

During the unprecedented time of the novel coronavirus disease 2019 (COVID-19) pandemic, the medical community, and the world as a whole, has witnessed numerous and often unexpected innovations. The development of these innovations has taken place to provide necessary medical care and education, to create much-needed personal protective equipment (PPE) and other medical devices in light of worldwide shortages, to improve methods of testing, and, finally, to prevent further spread of the highly infectious virus (Chauhan et al. 2020b). This chapter will serve as an overview of the various forms of ingenuity that emerged worldwide to keep vital health and education systems functioning and, of equal importance, to preserve a sense of normalcy in an extraordinary time

K. C. Kelley · S. P. Stawicki (✉)
Department of Surgery, University Campus, St. Luke's University Health Network, Bethlehem, PA, USA

J. Kamler
Departments of Emergency Medicine, NewYork-Presbyterian Health System, New York City, NY, USA

M. Garg
Departments of Emergency Medicine, Weill Cornell Medicine, Columbia University Vagelos College of Physicians and Surgeons, New York City, NY, USA

N. Rezaei (ed.), *Coronavirus Disease - COVID-19*, Advances in Experimental Medicine and Biology 1318, https://doi.org/10.1007/978-3-030-63761-3_48

where quarantines and lockdowns continue to be in effect. Some of the advances and innovations highlighted include the implementations of telemedicine and virtual learning, three-dimensional (3D) printing of medical equipment and devices, artificial intelligence (AI), flexible supply chain adaptations, novel methods of testing for COVID-19, and other 'future trends' of health science and technology.

48.2 Innovation During the COVID-19 Pandemic: Recognizing Limitations and Optimizing Possibilities

Although there is no such thing as the wrong place or wrong time for innovation and ingenuity, there are significant limitations that one must recognize in the context of a truly global event of transformational change. The emergence of the severe acute respiratory syndrome coronavirus 2 (SARS-CoV-2) represents such an event and therefore calls for special measures and considerations when it comes to innovative thinking (World Health Organization 2020a; Bruce 2020). Given the profound and acute economic crisis that is inherent to the life-saving physical distancing measures to mitigate COVID-19, any innovative ideas and their implementations in the face of this great challenge have to occur 'under budget' and with little room or time for experimentation. In this chapter, we will focus on broadly understood technological innovation and, more specifically, on information technology (IT) innovation.

First, it is crucial to recognize that innovation does not necessarily require invention or discovery. Thus, merely re-allocating and re-focusing our energies to existing ways of doing things may bring about incremental benefits well beyond those achieved through discoveries. One example of this general approach is the true embrace of telemedicine as a response to the current pandemic (Chauhan et al. 2020b). Considered by some in its early days to be a tool looking for applications, telemedicine enabled the medical community to address the urgent problem of lack of capacity by allowing patient encounters to be delivered at the point-of-care (Chauhan et al. 2020a; Stawicki et al. 2020b). Related to the area of telemedicine is the much broader domain of telepresence. Here, opportunity abounds in education, including medical school and residency didactics (Chick et al. 2020; Schwartz et al. 2020).

Second, some innovative approaches may appear a bit clumsy or even illogical at first. It is an expected norm in the area of innovation, where significant initial resistance often needs to be overcome before an invention, or new way of doing things is accepted first by early adopters and ultimately by the general public (Kee 2017). Thus, the 3D printing community came together synthesizing face masks, face shields, and ventilator parts initially at a snail's pace only to quickly achieve economies of scale and make a dramatic difference shortly after that for the frontline healthcare workers who were in severe need of high-quality personal protective equipment and hard-to-find equipment replacement (e.g., ventilator) parts (Ishack and Lipner 2020; Kaza et al. 2018; Cavallo et al. 2020; Larrañeta et al. 2020). Other manufacturers, such as those producing sports gear or clothing, have also flexed and adapted their capabilities to help address various acute shortages (Elliott 2020; Newcomb 2020).

Third, extraordinary circumstances lead people to connect the dots in truly unique and innovative ways. For example, when faced with acute shortages of COVID-19 testing kits and supplies, clinical laboratory scientists promptly devised a "sample pooling" testing method, whereby specimens from several patients are processed simultaneously using the same set of reagents. The downside of this approach is that when a positive result is reported within a testing pool, individual specimens from the pooled sample (typically five to ten patients per pool) require dedicated re-testing. Significant efficiencies, both in terms of resources and money saved, and the ability to screen large numbers of patients (especially in areas with lower disease incidence or before community spread) were realized using such implementations (Stawicki et al. 2020b; Hogan et al. 2020).

Finally, nonmedical innovation affects healthcare more profoundly than many appreciate. With global supply chains disrupted and the emergence of regional shortages of critical nonmedical frontline personnel (i.e., truck drivers), novel approaches of addressing logistical challenges arose. Simultaneously, air transportation came to a standstill, further compounding the situation and leaving few viable supply links in our highly interconnected world (Shepard 2020). In one original solution, detached semi-trailers were transported using rail cars, providing much needed and timely supplies to locations where direct truck delivery was either not possible or would have been severely delayed (Stawicki et al. 2020b). In another example, driverless vans were deployed to collect laboratory specimens from patients who required diagnostic testing. Here, a synergy between AI and the need of the moment created a solution that will likely stay with us for a long time (Business Insider 2020).

Below is a case-by-case summary of key developments in each of the above areas of innovation during the COVID-19 pandemic. Also, a section on other potentially novel approaches is provided at the end of the chapter, with a focus squarely on innovative ways and innovative thinking rather than actual developments in the space.

48.2.1 Telemedicine

By the time we completed this manuscript, there have been over 5.9 million confirmed cases of COVID-19 worldwide, with almost 1.8 million cases in the United States alone (World Health Organization 2020c). During this challenging time in history, the emergence of innovative solutions and technological advances has provided humanity with many novel, previously unavailable options (Chauhan 2020). While the concept of telemedicine itself is relatively well established, the applications and widespread utility during the COVID-19 pandemic certainly have been unparalleled. When the movement of individuals is restricted, and quarantines are in place, worldwide populations are at risk for increased daily stressors, various communicable and non-communicable diseases, and undiagnosed mental health disorders (Chauhan 2020). The use of telemedicine service(s) (TMS) has become vital, allowing access to care while decreasing the risk of person-to-person disease transmission (Chauhan 2020).

Telemedicine allows for non-urgent virtual patient assessments, such as focused medical consultations, follow-up encounters, and mental health visits (Chauhan 2020). The service also importantly protects individuals at high risk of contracting COVID-19 (e.g., the elderly, immunosuppressed patients, and those with severe medical comorbidities) (Mehrotra et al. 2020). TMS can serve as a "gatekeeper" to safely identify patients who require escalation of care and who can be safely discharged with remote follow-up, thereby preventing emergency departments and hospitals from becoming overwhelmed with high patient volumes (Chauhan 2020). In one such example, TMS was utilized to care for 12 COVID-19 patients who were received from a quarantined cruise ship (Chauhan 2020). Various modalities for TMS have been employed, allowing for virtual patient assessment and examination. For example, robotic telemedicine carts featuring cameras are devices with screens used for communication between patients and clinicians (known as the "two-tablet approach"), including remote vital signs monitoring enable clinicians to view temperatures, heart rates, blood pressure readings, and pulse oximetry data without any additional human contact (Chauhan 2020). Electronic monitoring in critical care units, which can allow remote surveillance of up to 60–100 ill patients at a time, is being employed by a growing number of institutions (Goh et al. 2020; Ross et al. 2020). Of importance, a significant barrier to maintaining continuity of care via TMS was recently removed by the Centers for Medicare and Medicaid Services (CMS), with the implementation of payment parity between telemedicine and in-person medical visits

(Stawicki et al. 2020b). One area of critical need, especially during the highly stressful COVID-19 pandemic, is mental health and addiction medicine (Ganesh et al. 2020; Zhou et al. 2020; Wright and Caudill 2020; Knopf 2020). The availability of telepresence and "mobile health" is a true game changer for those with mental illness who currently may have only limited access to much-needed care (Torous et al. 2020; Torous and Keshavan 2020). Another area where great strides in innovation have been noted is in education with various modes of virtual learning for all ages.

48.2.2 Virtual Learning

In response to COVID-19, education at all levels has been forced to transition to a largely virtual format. In the United States, most elementary grade students have been effectively moved to either home school or virtual school environments (Butcher 2020). Instead of creating new virtual platforms, some school districts have partnered with existing online learning companies, sharing online video content and webinars to ease the transition to online instruction for traditional classroom teachers (Butcher 2020). The United States' largest state-based virtual school, the Florida Virtual School, has offered customized training programs for its state teachers (Butcher 2020). Other existing online platforms have also proven useful for sharing ideas and generating collective learning (Greenhalgh et al. 2020; Almarzooq et al. 2020; Jayaraman and Jothiswaran 2020; Langford and Damşa 2020).

Concerning medical education, there are certain unique challenges brought on by the COVID-19 pandemic for both medical students and physicians-in-training. Medical schools traditionally focus the first 12–24 months of their basic science curriculum on a lecture-based programmatic format, followed by highly structured clinical medicine clerkships in the later years of training (Rose 2020). The COVID-19 pandemic has particularly affected those students who are now effectively displaced from their clinical clerkships. It forced the development of new methods to attempt to mimic the hospital environment as much as possible. Options currently being employed include high-fidelity immersive simulation featuring virtual clinical cases, making the telemedicine environment available to medical students in a fashion similar to traditional clinics and moving clinical didactic sessions online earlier in the curriculum to prepare for later entry into the clerkship environment (Rose 2020). For physicians-in-training, many in-person academic activities, especially those involving >10 people, have been severely restricted and mainly moved to online settings (Stawicki and Galwankar 2020; Ashokka et al. 2020). This includes various didactic sessions, morbidity and mortality conferences, simulation labs, and other activities that are vital to medical education (Chick 2020). Institutions such as Brooke Army Medical Center have developed innovative solutions to those problems. For General Surgery residents, these solutions include implementation of a flipped virtual classroom model, in which learners are provided with didactic material before the scheduled event, and the online conference session is held with a focus on case-based discussions (Chick 2020). Other methods utilized were online practice questions, teleconferences, telehealth clinics with resident involvement, and facilitated use of surgical videos in order to mitigate the loss of in-person opportunities in the operating room (Chick 2020). Virtual reality immersion, in which multidirectional filming is used to produce 3D video, has successfully been used to support emergency medicine training at the University of Pennsylvania and Cornell University, through the production of realistic walk-throughs of critical airway procedures (University of Pennsylvania 2020). In addition to bringing the bedside to the student, these technologies facilitate telementoring, teleproctoring, coaching, and other less formal but equally vital educational activities (Amparore et al. 2020; Karim et al. 2020).

48.2.3 3D Printing of Personal Protective Equipment (PPE) and Other Essential Medical Supplies

Many hospitals, companies, and researchers have utilized 3D printing as a new manufacturing resource during the COVID-19 pandemic (Chauhan et al. 2020b). COVID-19 quarantine restrictions have instilled a sense of fear and even some degree of hysteria in the general public, leading to panic buying of items such as PPE, in turn leaving those in need of PPE, such as essential healthcare employees, in short supply (COVID-19 Supply Chain Response 2020). An entire movement to supply home-made and improvised PPE developed (Chauhan et al. 2020b). In response, the global 3D printing community has devised numerous reusable PPE devices featuring insertable filters, which are easily made using low-cost desktop filament extrusion printers (Tino et al. 2020). For example, the Food and Drug Administration (FDA), United States Veterans Association, and National Institutes of Health (NIH) 3D Print Exchange are collaborating in developing prototypes for N95 masks, face shields, and ear saver mask holders (Tino et al. 2020; COVID-19 Supply Chain Response 2020).

The COVID-19 pandemic has also led to shortages in essential medical equipment. Such shortages include the lack of masks and hoods required for the proper use of continuous positive airway pressure/positive end-expiratory pressure (CPAP/PEEP) noninvasive ventilatory support (Tino et al. 2020). In Italy, a country significantly impacted by the pandemic, it has been challenging to obtain venturi valves, which are critical components of respiratory support equipment, due to acute supply-demand imbalance (Tino et al. 2020). An Italian startup called Isinnova, with its 3D printing community of physicians and engineers, successfully developed ways of manufacturing Venturi valves in order to boost local supply (Tino et al. 2020). Another novel invention spurred by equipment shortages is the 3D-printed ventilator splitter, a device that allows the use of a single ventilator for multiple patients (Tino et al. 2020). The US FDA has issued a no-objection policy for the creation of such devices, including T-connectors, which meet specifications for placing more than one patient on a single ventilator when the number of patients in need exceeds the available supply of ventilators (Tino et al. 2020).

48.2.4 Innovations in COVID Testing

Essential for controlling the spread of COVID-19 is the ability to accurately test individuals, especially given the number of asymptomatic people who can potentially transmit the virus (Stawicki et al. 2020b). Group testing, or pooled testing, is a quicker, more resource-efficient alternative to individual testing, and its application to the COVID-19 pandemic has proven an ingenious idea (De Wolff 2020). The use of group testing, which originated in the 1940s with syphilis detection in the US military, could potentially save as many as 85–95% of tests that would be otherwise administered (Gollier 2020). The concept in its simplest form is that multiple samples from a group of individuals are pooled together and tested (e.g., five pooled samples of ten individuals) (Bertolotti et al. 2020). If the pooled samples are negative, all of the individuals in the group are deemed negative for the virus (Bertolotti et al. 2020). If a pooled sample is positive, then all individual samples from the group are tested separately (Bertolotti et al. 2020). This approach thus uses much fewer tests than universally applied individual testing. Numerous countries, including the United States, Germany, India, and Israel, have been employing pooled testing and demonstrated its efficacy (Bertolotti et al. 2020; Yelin et al. 2020; Stawicki et al. 2020b).

48.2.5 Nonmedical Innovation

Innovation outside of healthcare has played a significant role in addressing the challenges of COVID-19 pandemic, in part due to the large number of individuals affected directly, as well as the significant impact of the society's collective behavior and attitudes, on the spread of disease. Supply chains have been particularly strained, with elevated demand for PPE, ventilators, test kits, and other essential tools, as well as reduced supply impacted by fragmented transportation networks and worker quarantines, leading to the creative use of available vehicles and broader implementation of driverless vehicles (Stawicki et al. 2020b; Dizikes 2020; Ivanov 2020). The need to mitigate community spread has led to an array of changes to routine practices, known as nonpharmaceutical interventions, including the closure of nonessential businesses and limitations to social gatherings (Stawicki et al. 2020b). Though the effectiveness of these measures in the United States is to be confirmed, evidence from theoretical and historical models, and data from early COVID-19 pandemic experiences in China, suggests they can help control local outbreaks (Hartley and Perencevich 2020). In Japan, lockdowns were avoided largely with the innovative, population-wide application of a simple yet very effective three C's paradigm (Feder 2020). Instead of implementing costly, across-the-board shutdowns, citizens were asked to wear masks, avoid closed spaces, crowded places, and close-contact settings (Feder 2020). Although not perfect, the paradigm appears to be comparably useful to more restrictive measures. Factors that may play a role in the effectiveness of the three C's in Japan may be the existing culture of mask-wearing (Feder 2020).

Local innovation initiatives, primarily initiated at single-institution or hospital network level, include PPE sterilization using hydrogen peroxide vapor and ultraviolet (UV) light application (Nogee and Tomassoni 2020; Kenney et al. 2020; Rubio-Romero et al. 2020). In the setting of acute PPE shortages, the ability to provide frontline healthcare workers with safe and effective means of personal protection is critical to the overall pandemic response (Organization 2020b; Rubio-Romero et al. 2020). Although important in ensuring PPE availability during the acute surge phase of the epidemic, sterilization and reuse of equipment designed specifically for single use should be considered a temporary bridging measure. Also, specific limits and guidelines should be set regarding the number of times PPE is reused.

48.3 Innovative Ways and Innovative Thinking: The Genesis of Future Trends

The numerous scenarios and use cases summarized above illustrate only a small proportion of approaches to innovation in the face of an overwhelming challenge. Optimization of existing technologies, rather than discovery and invention, can be truly transformational and bring about widespread benefit, as demonstrated by telemedicine and other virtual platforms. 3D printing of PPE and various medical equipment parts demonstrates how innovation can leverage creative adaptations toward repurposing existing capabilities. Reevaluation of historical strategies and applying them to new situations is demonstrated with pooled testing, initially developed decades ago. Finally, nonmedical innovation contributes to the overall pandemic response by supporting healthcare providers and patients and minimizing community spread.

The COVID-19 pandemic has shown that a rapidly evolving public health crisis, with multiple emergent needs requiring different solutions, necessitates multi-pronged innovation across many domains beyond healthcare, including technology development, clinical resource allocation, and communal risk mitigation. As the acute phases of the pandemic subside across the various global hotspots, increasing efforts can be directed toward assessing, improving, and broadly deploying the most effective techniques, practices, and approaches in the future. The sections below review these emerging trends.

48.3.1 Technology Development

As of the time of this writing, no therapy has been proven effective in preventing or treating COVID-19 outside of the investigational setting, and novel approaches to technological innovation will continue to provide critical adjuncts to clinical care for the pandemic response (Stawicki et al. 2020b). Technologies using readily available, inexpensive components are likely to have a great impact in the early stages of an outbreak, particularly in resource-limited environments. The increased use of telehealth will help address present needs but also reveal opportunities for broadening the clinical information that can be communicated and acted upon remotely. Telemedicine will also accelerate the transfer of technologies typically found in supervised clinical settings to the home environment, as well as the development of devices capable of automated clinical assessment and intervention (Chauhan et al. 2020a, 2020b; Stawicki et al. 2020b).

Novel technologies derived from inexpensive, readily available materials have been directed toward limiting person-to-person transmission of SARS-CoV-2. These include an intubation tent, made of plastic sheet and tubing, to isolate the heavy aerosol load produced during this high-risk procedure and thus protect nearby healthcare workers (Henneman 2020). A nasal spray delivering an antiviral salt formulation is being developed to potentially reduce the generation and transmission of infectious bioaerosols (Edwards et al. 2004; Pulmatrix, Inc. 2020). If successful, these kinds of innovations can be manufactured easily and distributed widely, helping to more effectively (and less expensively) contain new outbreaks even in resource-constrained settings.

As state-of-the-art technologies gain wider acceptance and use, their potential application beyond current specifications can become more apparent and motivate new research and development. Current telemedicine technologies, for example, communicate real-time audio and visual information but generally do not make use of touch, which is an essential component of a complete in-person physical examination. Haptic interfaces, which enable manual interaction with virtual environments, remote systems, or robots (National Research Council 1994), have found significant medical applications in robotic surgery (e.g., providing tactile feedback to surgeons (Abiri et al. 2019)) and brain-machine interfaces (e.g., enabling sensory feedback from prosthetic limbs (Flesher et al. 2016)). Increased demand for telemedicine patient evaluations is likely to accelerate efforts to add the sense of touch into routine virtual encounters. Similar applications may also be critical in the setting of physical isolation of patients affected by a highly contagious pathogen, and maybe especially applicable in protecting frontline healthcare personnel while enabling the preservation of excellent bedside care (e.g., in intensive care or emergency department setting) (Yang et al. 2020; Javaid et al. 2020).

Telemedicine is also incorporating devices that collect more comprehensive audio and visual data which is traditionally typical of in-person visits, including cardiopulmonary auscultation, otoscopy, and high-resolution visualization of dermatologic lesions (TytoHome 2019). One startup developing tele-auscultation is deploying a device and algorithm to assess lung sounds for fluid buildup, which could be useful in the detection of acute respiratory distress syndrome in COVID-19, as well as chronic conditions such as heart failure (Perry 2020b). Another venture is releasing software that calculates heart rate, oxygen saturation, and respiratory rate using a standard video of the user's cheeks, via the detection of changes in microvascular blood volume known as photoplethysmography (Allen 2007). Examination techniques that can be performed remotely but require some level of clinical expertise or complex tools can be made available more efficiently by being grouped at local clinics teleconnected to the relevant specialists, as is already the case for retinal evaluations in primary care settings or tele-neurology in stroke evaluations.

The increased availability of telemedicine has also accelerated the transition of established diagnostic and therapeutic tools to the home environment. Hospitals attempting to optimize care for COVID-19 patients, while minimizing the risk of nosocomial infection and managing capacity limitations, have developed protocols to discharge lower-risk patients meeting certain criteria with portable pulse oximeters and oxygen concentrators, as well as close telemedicine follow-up plans and strict return precautions (Allen 2020). There has also been growing interest in home electrocardiography devices to monitor COVID-19 patients for cardiac arrhythmias. One such device produces a six-lead electrocardiogram that can be transmitted immediately to a clinician; it includes measurement of the QTc interval, a lengthening of which is one indication of increased arrhythmia risk (American College of Cardiology 2019; Giudicessi et al. 2020). Realizing the full potential of telemedicine may thus enable, in the long-term, improved access to both general and specialty care.

Remote diagnosis and treatment technologies incorporating increasingly sophisticated AI deployments may ultimately reduce the need for synchronous clinician involvement and more advanced point-of-care applications (Vashistha et al. 2018; Attia et al. 2019). This principle is already apparent in automated insulin pumps for diabetics, which measure glucose levels to calculate, and then inject, appropriate doses (FDA 2016). A bracelet-like device, using more complex algorithms, is currently being developed as a potential treatment for trichotillomania and other compulsive behaviors. After being trained to recognize hand movements associated with the pathologic behavior, it alerts the user when these occur and thus allows for conscious redirection (St. Anthony 2019). During the COVID-19 pandemic, its makers suggested the bracelet may be further useful in detecting and minimizing face touching, which may reduce the risk of viral infection (Perry 2020a). Thus, personal devices capable of immediate clinical assessment and intervention will likely expand the reach of medical care and encourage patients to become more intimately aware of their conditions and be more engaged in the management of their own health and well-being. Finally, various technology companies are developing novel ways to keep the public safe and informed through sharing real-time COVID-19 awareness data through innovative Internet- and smartphone-based applications (Abeler et al. 2020; Oliver et al. 2020). Patients and physicians can utilize these apps to monitor health and access additional resources (information, supplies, and personnel). Finally, the use of drones with artificial intelligence can assist in supply delivery and theoretical protective surveillance of large, highly dense populations (Abeler et al. 2020; Oliver et al. 2020; Ichino et al. 2020). Implementations of such systems must be meticulously thought out to avoid any misuse of sensitive information or unwanted manipulation of social discourse (Stawicki et al. 2020a).

48.3.2 Clinical Resource Allocation

A crisis like the COVID-19 pandemic inevitably leads to profound questions regarding the distribution of limited clinical resources, spurring innovative approaches to the allocation of medical devices, physical space, and providers themselves. It is generally understood that the overarching goal is to provide the most benefit to most patients (Christian et al. 2014; Ornelas et al. 2014; Joynt et al. 2019). In some cases, this implies prioritizing care of patients who are less ill and may have a higher chance of survival. Influenza pandemics have prompted attempts to standardize protocols for the allocation of intensive-care resources, but a broad consensus has yet to be reached (Christian et al. 2013). The COVID-19 pandemic, furthermore, has strained human resources in addition to physical facilities and clinical tools. Thus, hospitals, in conjunction with government guidance, have had to devise mechanisms for allocating available capacity as needs have emerged.

For certain clinical interventions, a tradeoff may be required between optimal use and the number of patients eligible to receive such treatments. For example, shortages of a clinical trial drug can prompt efforts to determine whether reduced or otherwise modified treatment courses are also useful (Carlson 2020). The splitting of ventilators, to connect each machine to two or more patients, can present a similar choice (Englisz and Darowski 2000; Neyman and Irvin 2006; Paladino et al. 2008) and has been performed during the COVID-19 pandemic (Allen and Brown 2020). Further evaluation of the risks and benefits of this approach is necessary, however, as it decreases the ability to optimize settings for each patient, and there has been disagreement regarding whether the potential risks are acceptable (Stawicki et al. 2020b).

The increased need for high-level clinical resources during the COVID-19 pandemic has spurred efforts to repurpose adaptable spaces into makeshift intensive care units (ICUs). The broad cancelation of elective surgical procedures, which reduced potential transmission and PPE use, has also enabled the use of underutilized operating rooms for critical COVID-19 patients (Stawicki et al. 2020b). Invasive ventilatory support has been provided with anesthesia machines typically used to sustain patients through general anesthesia during surgery. Though these machines do not provide the same range of settings as standalone ventilators, their use has decreased the need to seek new supplies of ventilators (ASA 2020). Moving generally lower-acuity patients to makeshift spaces, including outdoor tents, convention halls, hotels, and Navy hospital ships, has also provided additional space and off-loaded critical hospital capacity during the surge phase across various COVID-19 hotspots (Stawicki et al. 2020b).

Clinical administrators have shown adaptability in managing provider shortages during times of surging patient volume. With testing capacity for SARS-CoV-2 initially limited and unavailable to even symptomatic healthcare workers, guidelines for quarantine have included nonspecific viral symptoms to maximize sensitivity in identifying potentially infected clinicians and minimize the risk of transmission to colleagues or potentially vulnerable patients. It led to a reduced patient-facing workforce. In parallel, there was the increased availability of clinicians from services whose patient volumes decreased, including anesthesia, surgery, surgical subspecialties, and interventional radiology. Thus ophthalmologists, urologists, and neurosurgeons found themselves working in emergency departments and ICUs under the guidance of attendings in those fields (Miller 2020). Also, in some hospitals, specialized teams could be established to perform specific procedures in these acute settings and reduce the burden on the native staff, including anesthesia teams for intubation and surgery or interventional radiology teams for central line placements.

Finally, novel technologies that are beginning to change how both medicine and public health are practiced extend well beyond telemedicine and 3D printing. For example, implementations of blockchain technology are making it easier and more efficient for patients to receive home medication deliveries (Ting et al. 2020), for public health officials to track cases (Nguyen et al. 2020), and improve supply chain functioning (Pal 2020). Innovative uses of both AI and blockchain technologies, especially when combined with existing data management capacity, enable a truly transformative increase in process efficiency, transparency, and speed while at the same time reducing costs (Nguyen et al. 2020; Stawicki et al. 2020b; Mashamba-Thompson and Crayton 2020).

48.3.3 Communal Risk Mitigation

Mitigating the pandemic has also required innovative strategies among community institutions and government authorities. It is especially true regarding quarantines for suspected or confirmed COVID-19 patients. Isolation guidelines are not always practical, especially among patients who are undomiciled or live in group facilities (Leung et al. 2008; Tsai and Wilson 2020; Culhane et al. 2020). The situation was further complicated by the lack of widespread testing early in the pandemic. Some undomiciled patients with milder symptoms have been considered for hospital admission to reduce potential community transmission (Bond 2020). In New York City, sleeping arrangements at shelters were modified to increase spacing, and additional spaces, including hotel rooms, were also made available (Gilman 2020). Similar arrangements were made in nursing homes and assisted living facilities. Along with increased testing, appropriately structured changes have made it possible to discharge patients to these sites while reducing their risk of transmitting or becoming infected by the virus. On a broader communal scale, testing of sewage to detect SARS-CoV-2 shed in the stool may facilitate early detection of new outbreaks even among asymptomatic carriers, as well as the dynamic measurement of outbreak severity, which can be used to guide the timing and intensity of public health interventions (CBC 2020).

In the United States, the pandemic has required broad public adaptation to guidelines regarding working and studying from home, social distancing, the use of facial coverings, and key changes across other aspects of daily life, which has led to an unprecedented debate on public health and economic risks. Government agencies assess whether to mandate a large-scale safety measure and, for this, compare its cost to a calculated monetary value deemed representative of the impact of the statistically preventable deaths. Before the early 1980s, this value was based primarily on estimates of lifetime earnings; newer models have used the increase in wages for workers choosing high-risk jobs (NPR 2020). Such models, however, have important limitations, including the unclear validity of assumptions about individual risk acceptance, the possible undervaluation of an individual's impact on others, and, of course, the exclusion of potentially overriding ethical considerations regarding preventable harm (NPR 2020). These concerns notwithstanding, the models have been used to calculate a value per statistical life of approximately $7.9 million by the Food and Drug Administration and $9.1 million by the Environmental Protection Agency (Appelbaum 2020). Public policy experts, as of mid-April 2020, thus estimated that avoiding potentially hundreds of thousands or millions of deaths from COVID-19 far outweighed, in statistical terms, the economic activity lost due to societal restrictions (Fink 2020). As these restrictions continue to be re-evaluated, thoughtful and creative approaches to analyzing the underlying issues, and strong patient advocacy, will be necessary to ensure the interests of both individuals and the common good are represented appropriately.

Innovative coordination among manufacturers, governments, and humanitarian organizations will promote the ongoing development of vaccines and treatments for COVID-19 and their fair distribution worldwide. Gavi, an international alliance that has sought to bring vaccines to underserved communities, has created adaptive partnerships with vaccine producers during past outbreaks, including provisions for immediate availability of formulations under development. It enabled pre-license use of the rVSV-ZEBOV vaccine, which was considered essential in containing Ebola outbreaks in the Democratic Republic of the Congo in 2017–2018 (Berkley 2020). Coordination challenges apparent during the Ebola response led, in part, to the establishment in 2016 of the Coalition for Epidemic Preparedness Innovations (CEPI), a global partnership that facilitates cooperation among researchers, manufacturers, and regulators to accelerate the development and deployment of vaccines for emerging threats. Its portfolio includes five vaccine candidates for MERS-CoV; an adaptable platform technology, similar to influenza vaccine platforms, that may permit the rapid development of new vaccines and immunoprophylaxis for multiple diseases; and ongoing support for SARS-CoV-2 vaccine and adjuvant technologies (Coy 2020; CEPI 2020). Novel organizational approaches like these, aligning emerging capabilities with pressing needs across borders, will ensure our collective efforts have the most impact on all patients affected by COVID-19 and future pandemics.

48.4 Conclusion

In summary, the COVID-19 pandemic has presented the global community with formidable challenges, driving innovation and ingenuity in many disciplines. Among the topics discussed in this chapter were unique ways of employing existing resources and technologies to meet the challenge that is upon our planetary community. From stereolithography to new supply chain logistics approaches, the world is uniting in finding innovative strategies against a common enemy, the SARS-CoV-2 pathogen. As the pandemic is still currently ongoing, there will assuredly be more novel ideas and developments, including some directed toward addressing the changes in our daily routines. It may involve highly customized assistive technologies designed to facilitate some of the aspects of physical distancing while overcoming the deleterious aspects of social isolation. Redesign of our current work environments (e.g., decrease in worker density and increase in physical barriers) combined with the evolution of work-at-home paradigms, both beneficial toward reducing viral transmission, will also require ongoing innovation. This chapter serves as a snapshot of the progress made thus far, with possible inspiration for future directions and advancements.

References

Abeler J, Bäcker M, Buermeyer U, Zillessen H (2020) COVID-19 contact tracing and data protection can go together. JMIR Mhealth Uhealth 8(4):e19359

Abiri A, Juo Y-Y, Tao A, Askari SJ, Pensa J, Bisley JW, Dutson EP, Grundfest WS (2019) Artificial palpation in robotic surgery using haptic feedback. Surg Endosc 33(4):1252–1259

Allen J (2007) Photoplethysmography and its application in clinical physiological measurement. Physiol Meas 28(3):R1

Allen J (2020) New York hospital sends some 'borderline' COVID-19 patients home with oxygen monitors. https://www.reuters.com/article/us-health-coronavirus-usa-home/new-york-hospital-sends-some-borderline-covid-19-patients-home-with-oxygen-monitors-idUSKCN21Q3F0. Accessed 30 May 2020

Allen J, Brown N (2020) One ventilator, two patients: New York hospitals shift to crisis mode. https://www.reuters.com/article/us-health-coronavirus-usa-ventilators-idUSKBN21D3M1. Accessed 31 May 2020

Almarzooq Z, Lopes M, Kochar A (2020) Virtual learning during the COVID-19 pandemic: a disruptive technology in graduate medical education. J Am Coll Cardiol 75:2635–2638

American College of Cardiology (2019) FDA approves first six-lead personal ECG device. https://www.acc.org/latest-in-cardiology/articles/2019/05/16/13/02/fda-approves-first-six-lead-personal-ecg-device. Accessed 30 May 2020

Amparore D, Claps F, Cacciamani GE, Esperto F, Fiori C, Liguori G, Serni S, Trombetta C, Carini M, Porpiglia F (2020) Impact of the COVID-19 pandemic on urology residency training in Italy. Minerva Urol Nefrol 72:03868–03860

Appelbaum B (2020) As U.S. Agencies put more value on a life, businesses fret. https://www.nytimes.com/2011/02/17/business/economy/17regulation.html?_r=0&pagewanted=all. Accessed 31 May 2020

ASA (2020) APSF/ASA guidance on purposing anesthesia machines as ICU ventilators. https://www.asahq.org/in-the-spotlight/coronavirus-covid-19-information/purposing-anesthesia-machines-for-ventilators. Accessed 5 Apr 2020

Ashokka B, Ong SY, Tay KH, Loh NHW, Gee CF, Samarasekera DD (2020) Coordinated responses of academic medical centres to pandemics: sustaining medical education during COVID-19. Med Teach 5:1–10

Attia ZI, Noseworthy PA, Lopez-Jimenez F, Asirvatham SJ, Deshmukh AJ, Gersh BJ, Carter RE, Yao X, Rabinstein AA, Erickson BJ (2019) An artificial intelligence-enabled ECG algorithm for the identification of patients with atrial fibrillation during sinus rhythm: a retrospective analysis of outcome prediction. Lancet 394(10201):861–867

Berkley S (2020) We finally have a life-saving vaccine for Ebola. https://www.weforum.org/agenda/2018/05/gavi-life-saving-vaccine-ebola-rVSV-ZEBOV/. Accessed 31 May 2020

Bertolotti P et al (2020) Pooled sample testing for SARS-CoV-2. https://idss.mit.edu/wp-content/uploads/2020/04/04.28.2020pooled_testing.pdf. Accessed 30 Nov 2020

Bond A (2020) As Covid-19 surges among San Francisco's homeless, doctors face difficult choices. https://www.statnews.com/2020/04/11/coronavirus-san-francisco-homeless-doctors-difficult-choices/. Accessed 31 May 2020

Bruce T (2020) Corona Virus Disease (COVID 19): Rapid human ingenuity through innovation and technology as a proactive and admirable response in the storm of a global pandemic. https://www.researchgate.net/profile/Tor_Alexander_Bruce/publication/340253741_COVID_19_RAPID_HUMAN_INGENUITY/links/5e806845458515efa0b53137/COVID-19-RAPID-HUMAN-INGENUITY. Accessed 30 Nov 2020

Business Insider (2020) A Florida lab is using self-driving vans to ferry coronavirus tests. https://www.yahoo.com/news/florida-lab-using-self-driving-091701923.html. Accessed 8 Apr 2020

Butcher J (2020) Public-private virtual-school partnerships and Federal Flexibility for schools during COVID-19. Mercatus Center-George Mason University, Boulder, pp 1–5

Carlson J (2020) Minnesota grapples with promise of COVID-19 drug in limited supply, unknown cost.

https://www.startribune.com/minnesota-grapples-with-promise-of-covid-19-drug-in-limited-supply-unknown-cost/570549072/. Accessed 31 May 2020

Cavallo L, Marcianò A, Cicciù M, Oteri G (2020) 3D printing beyond dentistry during COVID 19 epidemic: a technical note for producing connectors to breathing devices. PRO 2(2):46–52

CBC (2020) Researchers explore whether human poop can help track COVID-19, and curb a second wave. https://www.cbc.ca/radio/thecurrent/the-current-for-may-5-2020-1.5555757/researchers-explore-whether-human-poop-can-help-track-covid-19-and-curb-a-second-wave-1.5556579. Accessed 31 May 2020

CEPI (2020) Coalition for Epidemic Preparedness Innovation: new vaccines for a safer World. https://cepi.net/. Accessed 31 May 2020

Chauhan EA (2020) Novel coronavirus (COVID-19): leveraging telemedicine to optimize care while minimizing exposures and viral transmission. Journal of emergencies. J Emerg Trauma Shock 13(1):20–24

Chauhan V, Galwankar S, Arquilla B, Garg M, Di Somma S, El-Menyar A, Krishnan V, Gerber J, Holland R, Stawicki SP (2020a) Novel coronavirus (COVID-19): leveraging telemedicine to optimize care while minimizing exposures and viral transmission. J Emerg Trauma Shock 13(1):20

Chauhan V, Galwankar SC, Yellapu V, Perez-Figueroa IJ, Stawicki SP (2020b) State of the globe: the trials and tribulations of the COVID-19 pandemic: separated but together, telemedicine revolution, frontline struggle against "silent hypoxia," the relentless search for novel therapeutics and vaccines, and the daunting prospect of "COVIFLU". J Global Infect Dis 12(2):39–43

Chick RC, Clifton GT, Peace KM, Propper BW, Hale DF, Alseidi AA, Vreeland TJ (2020) Using technology to maintain the education of residents during the COVID-19 pandemic. J Surg Educ 77(4): 729–732

Christian MD, Fowler R, Muller MP, Gomersall C, Sprung CL, Hupert N, Fisman D, Tillyard A, Zygun D, Marshal JC (2013) Critical care resource allocation: trying to PREEDICCT outcomes without a crystal ball. Crit Care 17:107

Christian MD, Sprung CL, King MA, Dichter JR, Kissoon N, Devereaux AV, Gomersall CD (2014) Triage: care of the critically ill and injured during pandemics and disasters: CHEST consensus statement. Chest 146(4):e61S–e74S

COVID-19 Supply Chain Response (2020) NIH 3D print exchange. https://3dprint.nih.gov/collections/covid-19-response . Accessed 30 Nov 2020

Coy P (2020) The road to a Coronavirus vaccine runs through Oslo. https://www.bloomberg.com/news/articles/2020-02-13/this-oslo-facility-may-be-the-key-to-the-coronavirus-vaccine. Accessed 31 May 2020

Culhane D, Treglia D, Steif K, Kuhn R, Byrne T (2020) Estimated emergency and observational/quarantine capacity need for the US homeless population related to COVID-19 exposure by county; projected hospitalizations, intensive care units and mortality. https://escholarship.org/content/qt9g0992bm/qt9g0992bm.pdf. Accessed 30 Nov 2020

De Wolff TD, Bittner M-I, Pfluger D, Rehme M, Heuer J, Bittner M-I (2020) Evaluation of pool-based testing approaches to enable population-wide screening for COVID-19. arXiv 2004:11851v2

Dizikes P (2020) Supply chain outlook: the timing of the slowdown. https://news.mit.edu/2020/simchi-levi-supply-chain-covid-19-0325. Accessed 28 May 28 2020

Edwards DA, Man JC, Brand P, Katstra JP, Sommerer K, Stone HA, Nardell E, Scheuch G (2004) Inhaling to mitigate exhaled bioaerosols. Proc Natl Acad Sci 101(50):17383–17388

Elliott H (2020) Column: Bauer's line shift from hockey equipment to medical face shields is inspiring. Los Angeles Times. https://www.latimes.com/sports/hockey/story/2020-05-14/bauer-switch-from-hockey-gear-to-medical-face-shields-inspiring-coronavirus. Accessed 27 May 2020

Englisz M, Darowski M (2000) Artificial ventilation of the lungs for emergencies. Front Med Biol Eng 10(3):177–183

FDA (2016) FDA approves first automated insulin delivery device for type 1 diabetes. https://www.fda.gov/news-events/press-announcements/fda-approves-first-automated-insulin-delivery-device-type-1-diabetes. Accessed 31 May 2020

Feder S (2020) Japan avoided a lockdown by telling everyone to steer clear of the 3 C's. Here's what that means. https://www.businessinsider.com/how-japan-tackled-coronavirus-without-a-lockdown-2020-5. Accessed 30 May 2020

Fink S (2020) Worst-case estimates for U.S. Coronavirus deaths. https://www.nytimes.com/2020/03/13/us/coronavirus-deaths-estimate.html. Accessed 31 May 2020

Flesher SN, Collinger JL, Foldes ST, Weiss JM, Downey JE, Tyler-Kabara EC, Bensmaia SJ, Schwartz AB, Boninger ML, Gaunt RA (2016) Intracortical microstimulation of human somatosensory cortex. Sci Transl Med 8(361):361ra141

Ganesh A, Sahu P, Nair S, Chand P (2020) A smartphone based e-consult in addiction medicine: an initiative in COVID lockdown. Asian J Psychiatr 51:102120

Gilman A (2020) New York City's homeless face additional challenges during COVID-19. https://www.marketwatch.com/story/new-york-citys-homeless-face-additional-challenges-during-covid-19-2020-04-13. Accessed 31 May 2020

Giudicessi JR, Noseworthy PA, Friedman PA, Ackerman MJ (2020) Urgent guidance for navigating and circumventing the QTc-prolonging and torsadogenic potential of possible pharmacotherapies for coronavirus disease 19 (COVID-19). In: Mayo clinic proceedings. Elsevier

Goh KJ, Wong J, Tien J-CC, Ng SY, Duu Wen S, Phua GC, Leong CK-L (2020) Preparing your intensive care unit for the COVID-19 pandemic: practical considerations and strategies. Crit Care 24:1–12

Gollier C, Gossner O (2020) Group testing against COVID-19. Working papers-center for research in economics and statistics. http://gossner.me/wp-content/uploads/2020/03/group-testing20202328.pdf. Accessed 30 Nov 2020

Greenhalgh T, Wherton J, Shaw S, Morrison C (2020) Video consultations for covid-19. BMJ 368:m998. https://doi.org/10.1136/bmj.m998

Hartley DM, Perencevich EN (2020) Public health interventions for COVID-19: emerging evidence and implications for an evolving public health crisis. JAMA 323(19):1908–1909

Henneman J (2020) Build, assembly, use of COVID intubation Tent/Shield. https://vimeo.com/400428231. Accessed 30 May 2020

Hogan CA, Sahoo MK, Pinsky BA (2020) Sample pooling as a strategy to detect community transmission of SARS-CoV-2. JAMA 323(19):1967–1967

Ichino A, Calzolari G, Mattozzi A, Rustichini A, Zanella G, Anelli M (2020) Transition steps to stop COVID-19 without killing the world economy. VoxEU org, 25

Ishack S, Lipner SR (2020) Applications of 3D printing technology to address COVID-19 related supply shortages. Am J Med 133:771–773

Ivanov D (2020) Predicting the impacts of epidemic outbreaks on global supply chains: a simulation-based analysis on the coronavirus outbreak (COVID-19/SARS-CoV-2) case. Transport Res E Log Transport Rev 136:101922

Javaid M, Haleem A, Vaishya R, Bahl S, Suman R, Vaish A (2020) Industry 4.0 technologies and their applications in fighting COVID-19 pandemic. Diabetes Metab Syndr Clin Res Rev 14:419–422

Jayaraman R, Jothiswaran V (2020) Web-based platforms for virtual learning. Biotica Res Today 2(5 Spl):184–186

Joynt G, Gopalan P, Argent A, Chetty S, Wise R, Lai V, Hodgson E, Lee A, Joubert I, Mokgokong S (2019) The critical Care Society of Southern Africa Consensus Guideline on ICU triage and rationing (ConICTri). S Afr J Crit Care (Online) 35(1):53–65

Karim JS, Hachach-Haram N, Dasgupta P (2020) Bolstering the surgical response to COVID-19: how virtual technology will save lives and safeguard surgical practice. BJU Int 125:E18–E19

Kaza A, Rembalsky J, Roma N, Yellapu V, Delong WG, Stawicki SP (2018) Medical applications of stereolithography: an overview. Int J Acad Med 4(3):252

Kee KF (2017) Adoption and diffusion. In: The international encyclopedia of organizational communication. UK Wiley Blackwell, Chichester, pp 1–14

Kenney P, Chan BK, Kortright K, Cintron M, Havill N, Russi M, Epright J, Lee L, Balcezak T, Martinello R (2020) Hydrogen peroxide vapor sterilization of N95 respirators for reuse. medRxiv. https://www.medrxiv.org/content/medrxiv/early/2020/03/27/2020.03.24.20041087.full.pdf. Accessed 30 Nov 2020

Knopf A (2020) Addiction telemedicine comes into its own with COVID-19. Alcohol Drug Abuse Wkly 32(13):5–6

Langford M, Damşa C (2020) Online teaching in the Time of COVID-19. http://iropine.ouhk.edu.hk/file/Online%20Teaching%20in%20the%20Time%20of%20COVID-19.pdf. Accessed 30 Nov 2020

Larrañeta E, Dominguez-Robles J, Lamprou DA (2020) Additive manufacturing can assist in the fight against COVID-19 and other pandemics and impact on the global supply chain. 3D Print Addit Manuf 7:100–103

Leung CS, Ho MM, Kiss A, Gundlapalli AV, Hwang SW (2008) Homelessness and the response to emerging infectious disease outbreaks: lessons from SARS. J Urban Health 85(3):402–410

Mashamba-Thompson TP, Crayton ED (2020) Blockchain and artificial intelligence technology for Novel Coronavirus disease-19 self-testing. Multidisciplinary Digital Publishing Institute. https://www.mdpi.com/2075-4418/10/4/198/pdf. Accessed 30 Nov 2020

Mehrotra A, Ray K, Brockmeyer DM, Barnett ML, Bender JA (2020) Rapidly converting to "virtual practices": outpatient care in the Era of Covid-19. NEJM Catal Innov Care Deliv 1(2):1–5

Miller RW (2020) At NYC hospital treating coronavirus, doctors 'use their MD license like they haven't had to before'. https://www.usatoday.com/story/news/health/2020/04/03/coronavirus-inside-newyork-presbyterians-battle-against-covid-19/5111931002/. Accessed 31 May 2020

National Research Council (1994) Virtual reality: scientific and technological challenges. National Academy Press, Washington, DC

Newcomb A (2020) Gap, Zara start producing masks and other supplies for first responders. https://www.today.com/style/hanes-zara-start-producing-masks-other-supplies-first-responders-t176595. Accessed 27 May 2020

Neyman G, Irvin CB (2006) A single ventilator for multiple simulated patients to meet disaster surge. Acad Emerg Med 13(11):1246–1249

Nguyen D, Ding M, Pathirana PN, Seneviratne A (2020) Blockchain and AI-based solutions to combat coronavirus (COVID-19)-like epidemics: a survey. https://www.techrxiv.org/articles/Blockchain_and_AI-based_Solutions_to_Combat_Coronavirus_COVID-19_-like_Epidemics_A_Survey/12121962/files/22292202.pdf. Accessed 30 Nov 2020

Nogee D, Tomassoni AJ (2020) Covid-19 and the N95 respirator shortage: closing the gap. Infect Control Hosp Epidemiol 41:1

NPR (2020) Planet Money, Episode 991: Lives Vs. The Economy. https://www.npr.org/transcripts/835571843. Accessed 31 May 2020

Oliver N, Lepri B, Sterly H, Lambiotte R, Delataille S, De Nadai M, Letouzé E, Salah AA, Benjamins R, Cattuto C (2020) Mobile phone data for informing public health actions across the COVID-19 pandemic life cycle. American Association for the Advancement of Science

Ornelas J, Dichter JR, Devereaux AV, Kissoon N, Livinski A, Christian MD (2014) Methodology: care of the critically ill and injured during pandemics and disasters: CHEST consensus statement. Chest 146(4):35S–41S

Pal K (2020) Information sharing for manufacturing supply chain management based on Blockchain technology. In: Cross-industry use of Blockchain technology and opportunities for the future. IGI Global, pp 1–17

Paladino L, Silverberg M, Charchaflieh JG, Eason JK, Wright BJ, Palamidessi N, Arquilla B, Sinert R, Manoach S (2008) Increasing ventilator surge capacity in disasters: ventilation of four adult-human-sized sheep on a single ventilator with a modified circuit. Resuscitation 77(1):121–126

Perry TS (2020a) IndieBio to Give 8 Coronavirus-Tech Startups at Least $250,000 Each. https://spectrum.ieee.org/view-from-the-valley/at-work/start-ups/indiebio-eight-coronavirus-tech-startups-funding. Accessed 31 May 2020

Perry TS (2020b) Startups unveil tech to fight coronavirus; the Challenge may be getting enough engineers. https://spectrum.ieee.org/view-from-the-valley/at-work/start-ups/startups-unveil-tech-fight-coronavirus-challenge-engineers. Accessed 30 May 2020

Pulmatrix_Inc. (2020) Pulmatrix and sensory cloud announce partnership to develop and commercialize a potential nasal prophylactic and Anticontagion product for COVID-19. https://www.prnewswire.com/news-releases/pulmatrix-and-sensory-cloud-announce-partnership-to-develop-and-commercialize-a-potential-nasal-prophylactic-and-anticontagion-product-for-covid-19-301041548.html. Accessed 30 May 2020

Rose S (2020) Medical student education in the time of COVID-19. JAMA. https://doi.org/10.1001/jama.2020.5227

Ross SW, Lauer CW, Miles WS, Green JM, Christmas AB, May AK, Matthews BD (2020) Maximizing the calm before the storm: tiered surgical response plan for novel coronavirus (COVID-19). J Am Coll Surg 230:1080–1091.e3

Rubio-Romero JC, del Carmen Pardo-Ferreira M, García JAT, Calero-Castro S (2020) Disposable masks: disinfection and sterilization for reuse, and non-certified manufacturing, in the face of shortages during the COVID-19 pandemic. Saf Sci 129:104830

Schwartz AM, Wilson JM, Boden SD, Moore TJ Jr, Bradbury TL Jr, Fletcher ND (2020) Managing resident workforce and education during the COVID-19 pandemic: evolving strategies and lessons learned. JBJS Open Access 5(2):e0045

Shepard W (2020) China-Europe rail is set to boom as COVID-19 chokes air, Sea and road transport. https://www.forbes.com/sites/wadeshepard/2020/03/31/china-europe-rail-is-set-to-boom-as-covid-19-chokes-air-sea-and-road-transport/#64f9c6f437db. Accessed 10 May 2020

St. Anthony N (2019) HabitAware anti-compulsive behavior firm wins SBIR grant. https://www.startribune.com/habitaware-anti-compulsive-behavior-firm-wins-sbir-grant/566102111/?refresh=true. Accessed 31 May 2020

Stawicki SP, Galwankar SC (2020) Winning together: novel coronavirus (COVID-19) infographic. J Emerg Trauma Shock 13(1):103

Stawicki SP, Firstenberg MS, Papadimos TJ (2020a) The growing role of social Media in International Health Security: the good, the bad, and the ugly. In: Global Health security. Springer, pp 341–357

Stawicki SP, Jeanmonod R, Miller AC, Paladino L, Gaieski DF, Yaffee AQ (2020b) The 2019–2020 novel coronavirus (severe acute respiratory syndrome coronavirus 2) pandemic: a joint American College of Academic International Medicine-World Academic Council of emergency medicine multidisciplinary COVID-19 working group statement. J Global Infect Dis 12(2):47–93

Ting DSW, Carin L, Dzau V, Wong TY (2020) Digital technology and COVID-19. Nat Med 26(4):459–461

Tino R, Moore R, Antoline S, Ravi P, Wake N, Ionita CN, Morris JM, Decker SJ, Sheikh A, Rybicki FJ, Chepelev LL (2020) COVID-19 and the role of 3D printing in medicine. 3D Print Med 6(1):11. https://doi.org/10.1186/s41205-020-00064-7

Torous J, Keshavan M (2020) COVID-19, mobile health and serious mental illness. Schizophr Res 218:36–37

Torous J, Myrick KJ, Rauseo-Ricupero N, Firth J (2020) Digital mental health and COVID-19: using technology today to accelerate the curve on access and quality tomorrow. JMIR Ment Health 7(3):e18848

Tsai J, Wilson M (2020) COVID-19: a potential public health problem for homeless populations. Lancet Public Health 5(4):e186–e187

TytoHome (2019) An easy-access doctor. https://time.com/collection/best-inventions-2019/5733103/tytohome/. Accessed 30 May 2020

University of Pennsylvania (2020) 360° video: resuscitation of a COVID-19 patient with respiratory failure best practices demonstration. https://www.youtube.com/watch?v=nj18ZWnprBs&feature=youtu.be. Accessed 28 May 2020

Vashistha R, Dangi AK, Kumar A, Chhabra D, Shukla P (2018) Futuristic biosensors for cardiac health care: an artificial intelligence approach. 3 Biotech 8(8):358

World Health Organization (2020a) COVID 19 Public Health Emergency of International Concern (PHEIC). Global research and innovation forum: towards a research roadmap

World Health Organization (2020b) Rational use of personal protective equipment for coronavirus disease (COVID-19) and considerations during severe shortages: interim guidance, 6 April 2020. World Health Organization

World Health Organization (2020c) Coronavirus disease (COVID-19) situation report– 104. https://

www.who.int/docs/default-source/coronaviruse/situation-reports/20200503-covid-19-sitrep-104.pdf?sfvrsn=53328f46_4. Accessed 3 May 2020

Wright JH, Caudill R (2020) Remote treatment delivery in response to the COVID-19 pandemic. Psychother Psychosom 89(3):1

Yang G-Z, Nelson BJ, Murphy RR, Choset H, Christensen H, Collins SH, Dario P, Goldberg K, Ikuta K, Jacobstein N (2020) Combating COVID-19—the role of robotics in managing public health and infectious diseases. Sci Robot 5(40): eabb5589

Yelin I, Aharony N, Shaer Tamar E, Argoetti A, Messer E, Berenbaum D, Shafran E, Kuzli A, Gandali N, Shkedi O, Hashimshony T, Mandel-Gutfreund Y, Halberthal M, Geffen Y, Szwarcwort-Cohen M, Kishony R (2020) Evaluation of COVID-19 RT-qPCR test in multi-sample pools. Clin Infect Dis:ciaa531. https://doi.org/10.1093/cid/ciaa531

Zhou X, Snoswell CL, Harding LE, Bambling M, Edirippulige S, Bai X, Smith AC (2020) The role of telehealth in reducing the mental health burden from COVID-19. Telemed e-Health 26(4):377–379

COVID-19 Pandemic: The Influence of Culture and Lessons for Collaborative Activities

49

Linda Simon Paulo, George M. Bwire, Xingchen Pan, Tianyue Gao, Amene Saghazadeh, and Chungen Pan

Abstract

The rapid epidemiological shift from an epidemic/outbreak in Wuhan, China, to a global pandemic of COVID-19 in less than 3 months came with lessons the world's health system should learn to prepare for the future outbreaks. Since February 20, 2020, the total number of confirmed cases of COVID-19 has been increased very slowly in the countries of East Asia, including Japan, South Korea, and China, when compared with those in the Western countries. This chapter begins with an overview of the impact of COVID-19 on healthcare workers and public health facilities, followed by immediate global actions and research in response to the newly emerged pandemic. It includes an evaluation of the potential influence of culture on the implementation of different protective measures to combat the COVID-19 pandemic while at the same time offering suggestions that will make it easier for all populations to adapt protective steps against COVID-19 and other respiratory infectious diseases. Finally, the chapter provides a detailed discussion of lessons we have learned from the pandemic, leading to the conclusion that the transition from individualism to collaborative efforts is the treatment of universal pandemics.

Keywords

Collaboration · Collectivism · COVID-19 · Crisis · Culture · Globalization · Healthcare · Individualism · Transmission

L. S. Paulo
Department of Development Studies, Muhimbili University of Health and Allied Sciences, Dar es Salaam, Tanzania

G. M. Bwire
Department of Pharmaceutical Microbiology, Muhimbili University of Health and Allied Sciences, Dar es Salaam, Tanzania

X. Pan
Department of Human Resources, Shanghai University of Finance and Economics, Shanghai, China

T. Gao
Earl Haig Secondary School, North York, Ontario, Canada

A. Saghazadeh
Research Center for Immunodeficiencies, Children's Medical Center, Tehran University of Medical Sciences, Tehran, Iran

Systematic Review and Meta-analysis Expert Group (SRMEG), Universal Scientific Education and Research Network (USERN), Tehran, Iran

C. Pan (✉)
Laboratory of Molecular Virology & Immunology, Technology Innovation Center, Haid Research Institute, Guangdong Haid Group Co., Ltd., Guangzhou, China

N. Rezaei (ed.), *Coronavirus Disease - COVID-19*, Advances in Experimental Medicine and Biology 1318, https://doi.org/10.1007/978-3-030-63761-3_49

49.1 Introduction

On December 31, 2019, a severe acute respiratory syndrome coronavirus 2 (SARS-CoV-2) first appeared as the cause of Coronavirus Disease 2019 (COVID-19) in Wuhan, Hubei Province, People's Republic of China. On January 13, 2020, the World Health Organization (WHO) reported a case in Thailand, the first outside of China, a woman who had arrived from Wuhan City. In less than a month, since its identification in China, different countries from three continents reported confirmed cases of COVID-19, including Russia, Spain, Sweden, and the United Kingdom. A disease was declared a Public Health Emergency of International Concern by WHO on January 30, 2020. In Africa, Egypt was the first country to confirm the COVID-19 case from a foreigner who presented with no severe symptoms, while in Latin America, Brazil was the first country to confirm the case of COVID-19 in a 61 year old who had visited Italy. In this regard, all six WHO regions reported confirmed COVID-19 cases within 2 months since the first reported case in Wuhan (Fig. 49.1).

Accordingly, in just a few months, COVID-19 has caused over 10 million confirmed cases worldwide and dramatically affected every aspect of people's lives. Every country has to implement multiple precautions to control the spread of COVID-19. For example, as the first profoundly affected country, China has carried out strict surveillance on virus transmission for 5 months since January of 2020 (Pan et al. 2020). Meanwhile, cases are increasing at a fast pace in many other countries. As we have learned more about the virus, we have also learned more about the strategies required to protect against it in the absence of vaccines or therapies.

This chapter would first take a brief look at cultural diversity and how it might complicate the implementation of protective measures against the disease while, at the same time, offering suggestions that will make it easier for all populations to adapt protective steps against COVID-19 and other respiratory infectious diseases. Then, the chapter provides a detailed discussion of lessons we have learned from the pandemic. It leads to the conclusion that the transition from individualism to collaborative efforts is the treatment of universal pandemics.

49.2 Healthcare Workers and Public Health Facilities During the COVID-19 Pandemic

The World Health Assembly designated the 200th birthday of Florence Nightingale in 2020 to be the international year of the nurse and the midwife. The efforts and dedication of this cadre and other healthcare professionals were conspicuously demonstrated during the COVID-19 outbreak. When the rest of the world closed behind, they rushed to help those in need with little or no compensation or personal protective equipment, especially in developing countries. Despite

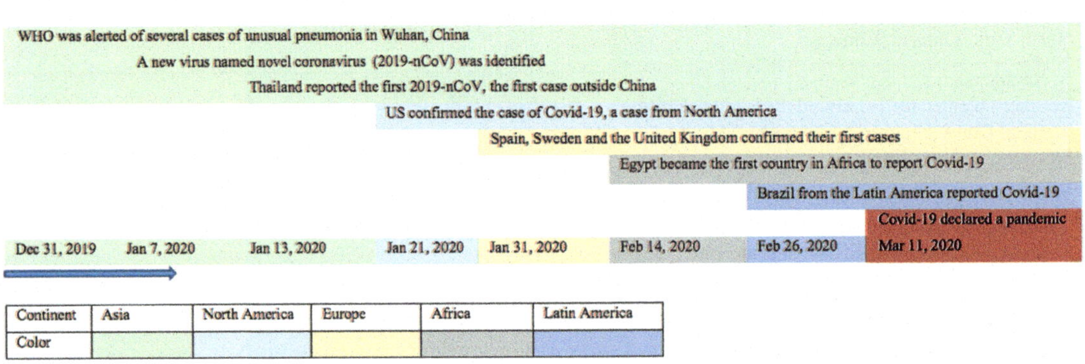

Fig. 49.1 COVID-19 timeline for the selected events. Time series in horizontal against COVID-19 reported cases in vertical across the globe

losses, they diligently kept serving the community. Although they were embraced and celebrated in most countries, some faced harassment, stigma, and violence in several communities (Amon 2020). These injustices transcended to journalists and activists as some governments censored information about COVID-19 to the public. Safeguarding human rights was again highlighted in this pandemic as a critical aspect of the national and international agenda. Authorities should increase efforts to improve the structure of the health system to offer physical and emotional protection to healthcare workers (HCWs). Emotional and psychological support for HCWs is missed in several countries and many fight occupational hazards without receiving proper compensation.

HCWs rely on personal protective equipment (PPE) to protect themselves and their patients from being infected and infecting others. However, there has been some reports of PPE-associated deaths of HCWs across the globe. For example, WHO reported that, since the start of the COVID-19 outbreak, prices have surged, medical masks have seen a sixfold increase, N95 respirators have trebled, and gowns have doubled. In European countries, the region believed to have sound healthcare systems, the health authorities declared to be overwhelmed with the large number of patients requiring intensive care. In these countries with a high number of incidences, hotels are turning into hospitals, and elective and semi-elective operative procedures are canceled to provide enough spaces to save COVID-19 patients. Similar to what developed and developing countries are doing in preparation for the possible wars, the same preparedness and investments should be made in health systems even more. The reason for this is that COVID-19 has proved the world that an infectious disease outbreak is a dangerous invisible enemy both to human health and the world economy.

COVID-19 has reminded the world of the need to invest in basic health systems such as primary healthcare facilities, laboratories, and critical care facilities. Strengthening healthcare systems will both promote health in a broad range and improve the national capacity of dealing with epidemics. Furthermore, improving the health system will play a significant role in life-saving and promote better community health, fundamental to development. Therefore, investing in the health system is necessary as a life-saving investment regardless of the threat of another epidemic. Once it became clear that serious emergency of COVID-19 is underway from China, countries started recruiting and training personnel but not in the speed as the virus did. The world needs trained personnel such as experts in disease surveillance and epidemiology ready to confront and contain the epidemic quickly, but this was not the case not only in developing countries but for developed countries. During the epidemic, it is also crucial to have useful data about what is going on. Unfortunately, this has remained a challenge in most developing countries. An essential part of human resources is model experts for predicting what might happen and what interventions must be in place and prioritized. Additionally, public health centers should have a high-bandwidth internet capacity to help with reporting data and coordinating personnel.

49.3 Immediate Global Action and Research in Response to the COVID-19 Pandemic

During previous outbreaks like Ebola, countries with weak health systems were more vulnerable, while during the COVID-19 pandemic, countries reported having better health systems such as European and North American countries were hardest hit with the virus. The message to the world is that infectious outbreaks have limited containment; what is happening in sub-Saharan countries should keep an alert to other unaffected regions and call for global actions. On the other hand, series of delays were observed in responding to COVID-19 outbreak; for example, on January 23, 2020, WHO reported that "the outbreak was yet to constitute a public emergency of international concern (PHEIC) and there was insufficient evidence of the virus spreading between humans outside of China." Immediate global actions like intensive screening for the returning travelers, travel restrictions to and from the high-risk areas, and a mandatory 14-day quarantine of travelers could have slowed the rapid importations of this virus. Nonetheless, most of the countries closed borders and imposed travel restrictions when it was already late, leading to local transmission originating from the unidentified or late identified imported cases. Several candidate vaccines and drugs for COVID-19 are going through different clinical phases. Customary, vaccine/drug research and development take several years to complete (Gates 2020). For COVID-19, the timeline has been shortened not at the expenses of quality, safety, and effectiveness but under the expedited bureaucracy and globally coordinated manner. It should not end on COVID-19 pandemic but be the direction to take for much other biomedical research. Then there is the question of funding. Budgets for biomedical research need to be expanded over and over. Billions of dollars are usually needed to complete clinical trial phases and secure regulatory approval for vaccines and drugs. It may need the government, public, and pharmaceutical companies funding.

49.4 Health and Disease Surveillance During the COVID-19 Pandemic

COVID-19 is added to the list of zoonotic coronavirus disease after Severe Acute Respiratory Syndrome Coronavirus (SARS-CoV) and the Middle East Respiratory Syndrome Coronavirus (MERS-CoV). Although the information on the sources of SARS-CoV-2 is limited, evidence reveals that SARS-CoV-2 could have been originated from bats (Sun et al. 2020). SARS-CoV-2 outbreak came as of no surprise in the scientific community as researchers previously cautioned about the possible outbreaks from coronavirus after discovering that, still, there is a reservoir of mutating coronavirus in bat (Li et al. 2005). The conventional surveillance system in one health fashion (animal and human health together) is vital in fighting to emerge and re-emerging diseases. One health approach and continued surveillance will keep us alerted on where the epidemic seems most likely to break out. Global investment in better disease surveillance is needed now for both normal situations and epidemics. If done routinely, a well-designed surveillance system can detect early signs of an outbreak.

49.5 Measures to Prevent COVID-19 Transmission

Containing COVID-19 transmission is hampered by the ability of the virus to spread through air droplets during the incubation period when individuals are asymptomatic (He et al. 2020; Stadnytskyi et al. 2020). It complicates the identification and tracking of infectious sources. In contrast, both SARS-CoV and highly pathogenic influenza A virus, H5N1 or H7N9, usually spread after patients show clinical symptoms, making it easier to identify the source of infection and control it at the beginning of an outbreak (C. Jiang et al. 2020).

Up to now, various measures have been adopted against COVID-19, such as immediate and accurate testing of the virus, travel restrictions, social distancing, washing hands frequently, and wearing a mask (Xiao and Torok 2020). Whereas some Eastern Asian countries (China, Japan, and South Korea) have successfully controlled the pandemic, the virus continues to transmit rapidly in Western countries, as shown in Fig. 49.2 (Baidu). Therefore, we asked what would account for the ability of these three East Asian countries to control the spread of the disease, while, at the same time, Western countries could not.

49.6 Collectivism and Individualism Play Roles in the COVID-19 Pandemic

One difference between East and West is the culture of collectivism vs. individualism and the role that each plays in the perception of a public crisis (Hamamura 2012). When a crisis breaks out, it is

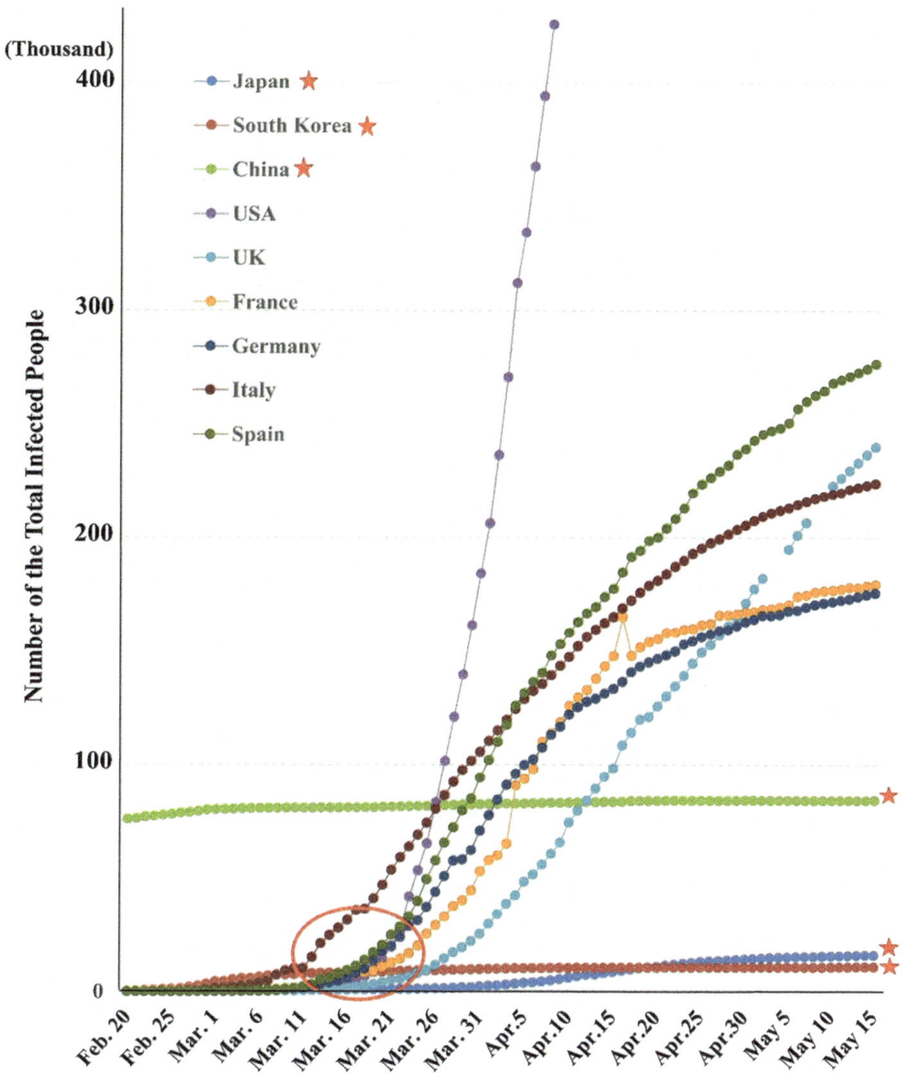

Fig. 49.2 Infection curves of several countries from Eastern Asia, North America, and Europe, respectively, between February 20 and May 15. The United States data were collected from February 20 to April 8. Red stars indicate the three East Asian countries. A red circle indicates the times when most countries carried out travel restrictions or set up temporary hospitals in the hard-hit epidemic areas. The breaks in the United Kingdom curve mean unavailable data

expected that the government will take the lead in coordinating experts and messaging (Pan et al. 2020). However, in Western countries, individualism can make it challenging to achieve unified national action (Rynn 2005). Individualism emphasizes individual responsibility and the pursuit of independence and freedom (Kim et al. 1994). Individualism supposedly inspires initiative, independence, and creativity (Triandis 2001). However, public crises are characterized by a set of strong negative externalities. These are also known as external costs or external diseconomies, referring to the phenomenon that certain behaviors (people or nature) affect other people or enterprises and make them pay extra costs, but the latter cannot obtain corresponding compensation (Rynn 2005). It could account for the relative slowness of (Iii 1995) individualist societies to respond when compared to collectivist societies (Triandis 1990). In the COVID-19 pandemic, however, we have seen that coordination from the public is crucial for stopping the spread of the virus (Kickbusch et al. 2020; Kokudo and Sugiyama 2020; Pan et al. 2020).

For example, we can compare the response to the current COVID-19 to that of the 2008 economic crisis (Shablovsky 2017; Stadler et al. 2003; Taylor 2012). First, both exceeded the intellectual grasp of the population in the breadth and depth of the destruction. While a purely economic crisis goes directly to financial security, the pandemic, as we have seen, affects financial security, food security, and health security. The contagion of COVID-19 is inextricably linked to multinationalism and the ease of travel, while the 2008 recession ushered in a years-long economic contraction in most countries (World Bank). Some have estimated that the effects of COVID-19 will last 1 to 2 years (Michie 2020), implying the strong negative externality of each event. As shown in Table 49.1 and Fig. 49.3, panic in both events caused significant disruption in the US stock market, mostly arising from uncertainty and the Centers for Disease Control and Prevention (CDC) and WHO predictions (Luchtenberg and Vu 2015; Yahoo Finance).

The core problem behind the economic crisis, also known as the subprime mortgage crisis, was the loan rate (Boubakri et al. 2017). In the collectivistic East Asian culture, people are more willing to deposit than to lend. In contrast, the advanced consumption in the Western individualistic culture makes people reluctant to deposit and more willing to lend, as in, for example, the use of credit cards (Boubakri et al. 2017). Banks were simply willing to loan out more than they knew borrowers could payback, but borrowers will be willing to take the "subprime" offers. Some scholars believe that South Korea was more affected by the economic crisis because of the transition from collectivism to individualism

Table 49.1 US stock markets in 2008 and 2020

	2008		2020	
Stock market of	Quote increase (%)	Rank	Quote increase (%)	Rank
Energy	−47	7	−46	1
Finance	−83	1	−35	2
Industry	−65	2	−31	3
Material	−60	3	−28	4
Nonessential	−58	4	−27	5
IT	−53	5	−27	6
Communication devices	−51	6	−25	7
Public utilities	−46	8	−23	8
Necessities	−31	10	−19	9
Medicine	−40	9	−19	10

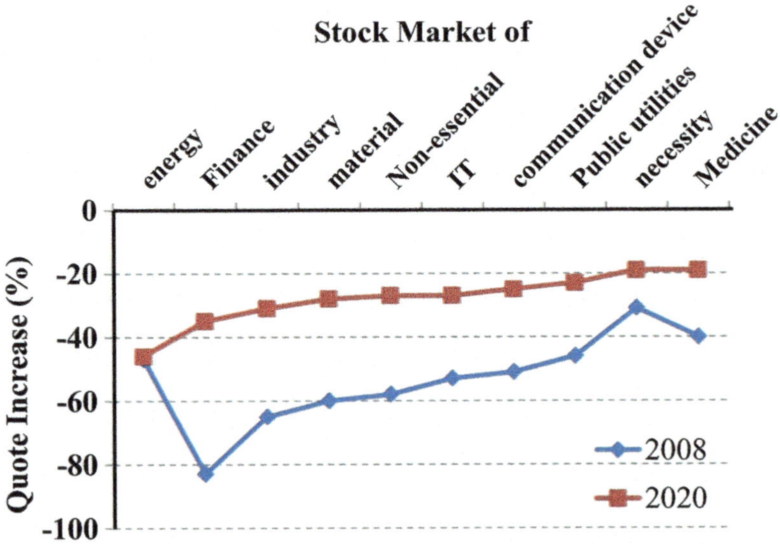

Fig. 49.3 Diagram of the US stock market in 2008 and 2020

(Lee and McNulty 2003). Also, China's economic system is different from that of the West. During the economic crisis, China had more direct government intervention (Lardy 2011). Meanwhile, Japan showed that it had learned the lessons from the 1990s economic crisis (Lardy 2011; Siami Namini 2017), directly introducing a national stimulus program of 75 trillion Japanese yen, leading to a smooth ride through the 2008 economic crisis (Hongzhou 2009). The data in Table 49.2 and Fig. 49.4 show that the GDP (gross domestic product) of Japan and China from 2008 to 2010 increased over that of other countries, except South Korea (World Bank).

Individualism typically focuses on the achievement of personal goals and the direct expression of opinions. During the COVID-19 pandemic, this explains why some in the West spontaneously protested the quarantine policy. In contrast, collectivistic societies respect higher-order decision-making, and people are more likely to accept quarantine orders, stay at home, wear a mask, and practice social distancing (Phuong-Mai et al. 2005). These steps dramatically speeded up the implementation of preventive measures against COVID-19 in East Asian countries.

49.7 Inconsistent Acceptance of Face Covering: An Important Measure for the Prevention of COVID-19 Transmission

Another difference is that the West and the East still have different opinions on the use of face masks, which may play an essential role in controlling virus transmission (S. Jiang and Pan 2020). As shown in Table 49.3, Eastern countries started to wear facial masks when the reported case was only in the hundreds, while it was not until more than hundreds of thousands of cases in the Western countries were found that the strict mask policies there were implemented. These differences are partly due to history and partly due to culture.

The practice of wearing masks against airborne disease began in China during the Manchurian plague, which broke out in the autumn of 1910 in the town of Manzhouli on the Chinese-Russian border (Lynteris 2018). A Chinese scientist named Wu Lien-Teh deduced that it was an airborne disease directly transmissible between humans. Based on this conclusion, he introduced a new type of mask, adapted from existing surgical masks used in Europe, for anti-plague purposes in China. It led to the dissemination of the behavior of mask-wearing, and 60,000 masks were distributed to the general population (Tendfonline).

In Japan, face masks are an especially popular accessory. It started with the Spanish flu pandemic from 1918 to 1920. In 1919, the Japanese government recommended wearing masks along with vaccination and gargling as measures against the pandemic. After the pandemic ended, the ritual of wearing masks went away in other places, but it persisted in Japan (Burgess and Horii 2012).

Wearing masks continued to be shared among East Asian countries up to the twenty-first century. During the SARS epidemic in 2002 to 2003, more than 90% of people in Hong Kong wore face masks (Lau et al. 2004). In a 2011 survey of 120 Tokyo passers-by, 39 identified themselves as regular mask wearers (32.5%). Almost all 120 reported wearing a mask as a general protective device. For the Japanese, wearing a mask had become ritualized, with 62 people in the survey reporting that they had been pressured or told to wear one (Burgess and Horii 2012). Cultural factors also affect mask-wearing in East Asian countries. One of these cultural factors is collectivism. Many people in East Asian countries see wearing masks as a collective responsibility for mitigating viral transmission (SCMP). Wearing masks in these countries fosters a sense of solidarity and mutual obligation and shows that one has civic responsibility (NY times). It

Table 49.2 The GDP of several countries from three different regions of the world from 2007 to 2010

Region	Country	2007–2010 GDP (USD in billions)			
		2007	2008	2009	2010
North America	USA	14452	14713	14449	14992
	Canada	1465	1549	1371	1614
Eastern Asia	Japan	4515	5038	5231	5700
	South Korea	1123	1002	902	1094
	China	3550	4594	5102	6087
Europe	England	3101	2923	2411	2475
	France	2657	2918	2690	2643
	Germany	3421	3730	3398	3396
	Italy	2210	2399	2191	2134

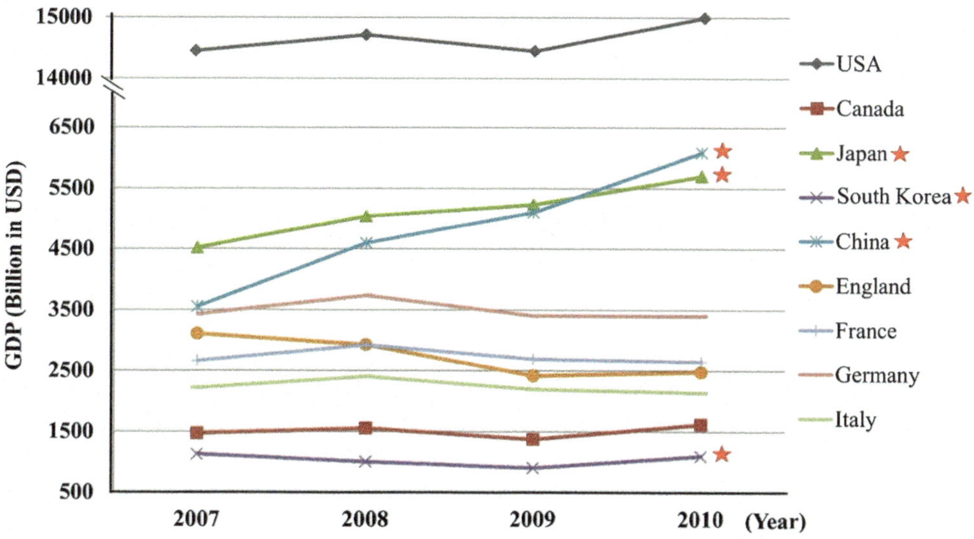

Fig. 49.4 GDP growth curves from 2007 to 2010. Red stars indicate the three Eastern Asia countries

Table 49.3 Mask-wearing by country

Region	Country	Rules	Reported or implemented on	When total cases were[a]
Eastern Asia	Japan	Not mandatory. Two cloth masks delivered to 50 million households	Mar. 4[b]	329
	South Korea	Not mandatory	N/A	
	China	Mandatory in some regions; recommended in others	Jan. 22[c]	571
North America	United States	Recommended for the public, but not mandatory	Apr. 2[d]	236,339
Europe	United Kingdom	Recommended by the government, but no mention of being compulsory	May 12[e]	229,705
	France	Compulsory on public transport and in secondary schools	May 11[f]	177,423
	Germany	Mandatory across all states. Some states impose fines if no mask worn	Apr. 22[g]	150,062
	Italy	Mandatory on public transport and in stores	Apr. 28[h]	201,515
	Spain	Mandatory on public transport	May 2[i]	245,589

Prepared with data from:
[a]https://voice.baidu.com/act/newpneumonia/newpneumonia
[b]https://english.kyodonews.net/news/2020/04/67ad0dfcd954-delivery-of-cloth-masks-from-govt-starts.html;
[c]http://www.wuhan.gov.cn/zwgk/xxgk/zfwj/gfxwj/202003/t20200316_973437.shtml;
[d]https://www.usatoday.com/story/news/politics/2020/04/03/coronavirus-trump-says-cdc-recommending-voluntary-use-face-masks/2938705001/;
[e]https://www.bbc.co.uk/newsround/52200989;
[f]https://www.bbc.com/news/world-europe-52529981;
[g]https://www.cnn.com/2020/04/27/europe/germany-face-mask-mandatory-grm-intl/index.html;
[h]https://www.thelocal.it/20200428/coronavirus-where-should-you-wear-a-face-mask-in-italy; and
[i]https://www.bbc.com/news/world-europe-52513516

also serves to impart a sense of safety for the mask wearer, to reassure the person that infection is unlikely (NY times).

Interestingly, young people in Japan sometimes wear masks to avoid interaction with others, showing disinterest in socializing (JPtoday).

They also wear masks to cover facial deformities and to keep warm in cold weather (SCMP). Some in Japan even see masks as an article of fashion, as indicated by the number of companies offering masks with various colored patterns like polka dots (JPtoday). Masks are also worn in smoggy weather in countries like South Korea and China to keep out harmful air particles (VOAnews).

In the West, meanwhile, wearing masks is stigmatized. Another theory involves the importance of eye contact and facial expressions in Western social interactions, which would preclude the full acceptance of facial masks (TIME). The practice of wearing masks is also negatively associated with East Asian people in Western countries (theAtlantic). It is seen as part of Asian otherness, as indicated by reports of Asian mask wearers in New York receiving odd stares from other commuters (TIME). In East Asian countries, wearing masks in public places is seen as a sign of civic responsibility and conscientiousness, whereas in Western countries, wearing masks is shunned as it gives the impression that one is ill and spreading contagion or that one is spreading unnecessary panic (theAtlantic). All these reasons have prevented the full acceptance of facial masks during the pandemic.

49.8 COVID-19 Pandemic: Lessons for Globalization

Compared to the previous respiratory virus outbreaks such as Spanish Flu and SARS, COVID-19 stroke at the peak of globalization. It is also the time with the highest levels of population and interconnectedness in the world's history. Life expectancy had risen for most countries except in disparate regions with civil unrest and wars. With high interconnectedness, countries relied on others for the importation of manufactured goods and exported manufacturing to other countries. Over the past two decades, manufacturing industries migrated from West to Eastern countries, specifically to China. With the recent formation of the Association of Southeast Asian Nations (ASEAN community), more factories have expanded to Southeast Asian countries, including Vietnam, Cambodia, Indonesia, Malaysia, the Philippines, Myanmar, Laos, Brunei, Singapore, and Thailand. Lenient manufacturing environment that increases profit margins has made these countries suitable for the mass production of goods distributed worldwide. The manufacturing environment in these countries allows compromises with the quality and regulatory standards that are not possible in developed nations, including Europe and the United States.

COVID-19 made the effects of the lenient regulation of manufacturing industries felt throughout the world as countries were flooded with counterfeit medical equipment in the time of dire needs. Spain procured test kits that had an accurate detection rate of 30% significantly low compared to other reviewed tests that had a sensitivity of 71–98% (Jones 2020; Watson et al. 2020). The faulty equipment comprised genetic and antibody tests, face masks, and other protective gears that were reported by several other countries, including the United States, Nepal, Australia, the Philippines and Netherlands (AsiaNews 2020; Chain 2020; Euronews 2020; Loh 2020; Tibetan Review 2020). Thus, the COVID-19 outbreak calls for further efforts to strengthen the regulation of manufacturing in the global supply chain.

49.8.1 Lessons for Leadership and Preparedness Policies

The response to COVID-19 demanded leadership from each sector of any magnitude. It touched people in their capacity as inhabitants of the planet earth, and each one had a role to play to curb the pandemic. During the pandemic, the World Health Organization (WHO) appears as the custodian of the roadmap to a successful exist. Together with coordinating the global efforts to manage the pandemic, the WHO has set out an initiative to ensure equitable access to COVID-19 tools (WHO 2020). However, for some countries, the declaration of COVID-19 pandemic and management guidelines were likely to be delayed and unclear. On May 29, 2020, President Trump announced to terminate

US ties with the WHO, halting a substantial amount of funding to the organization. This controversial decision in the middle of an outbreak threatens to weaken global efforts to prevent and control diseases as well as mitigating other socioeconomic effects on the communities.

The COVID-19 outbreak caused countries to have embraced individual responses, becoming part of the idea of nationalism, which has strongly resurfaced in the past decade. This approach threatened not only global solidarity, but also the regional cooperation showed existing weak links. In Africa, each country chose its way of handling the outbreak with no clear evidence of collaboration or exchange of best practices. The spirit of global solidarity lies in alleviating human suffering. It led to the formation of the United Nations and its branches, including WHO, in 1948. For many decades, the WHO has successfully led global initiatives in eradicating diseases and combating other conditions in collaboration with other entities, including smallpox, polio, women's health, and nutrition. Hence, this is an opportunity to strengthen leadership and preparedness policies at local and global levels for effective management of disease outbreaks and other health conditions.

The gender disparity in the successes of COVID-19 was highlighted by comparing the effective management of the outbreak in female-led countries. However, this success might be attributed to the general scores of equalities within these countries as opposed to individual leaders. It reminds countries on the importance of promoting gender equality and equity in the social, economic, and political development for universal health.

49.8.2 Lessons for the Application of Digital Technologies

The digital solutions have been available for several years; however, their adoption was escalated to a magnitude that was unforeseeable before the outbreak. New technologies call for new policies, leadership, and management strategies. Some occupations and companies may never make their way back after the pandemic requiring to restructure and set new strategies. The use of technology eased social, economic, and public health efforts during the COVID-19 outbreak. Through the use of technology, South Korea tracked cases and managed to isolate suspects without instituting lockdown measures. Although the context may differ between countries, documentation and wide adoption of practical digital measures should be embraced early.

Africa has been intellectually dependent on other countries for more than five decades post-independence. However, in the public health realm, Africa has developed resilience from fighting several infectious diseases and more deadly outbreaks such as Ebola in DRC and West African countries, Lassa fever endemic in Nigeria, and HIV, which continues to ravage many lives. The technologies used in the development of affordable testing kits for Ebola were repurposed to creating affordable test kits for COVID-19 in Senegal and Nigeria. Several African and developing countries incorporated digital and other technologies during COVID-19, which should not cease during the post-COVID-19 outbreak. They should use this opportunity to strengthen innovation capabilities and strengthen their scientific expertise level.

49.8.3 Lessons from Lockdowns

For economies that are yet to graduate from the hand to mouth means of survival, lockdowns proved to be a painful experience for a fraction of the population. The decision to lift lockdowns in these countries was based on economic rather than epidemiological/health reasons. While it is possible to live with the fear of the Coronavirus, this fear is quickly forgotten in the face of the hunger virus. It is not easy to sustain the pains of hunger, and every scientific fact vanishes as food is a basic human need. Based on the economic level classification proposed by Hans Rosling, the author of the book *Factfulness*, each nation has people spread across four economic levels (Hans Rosling 2020). The lowest level one is extremely poor, who have to work each day to eat. They are the most vulnerable in the face of diseases, especially in countries without universal access to healthcare. They are likely to spend up to 40% of their household income on medical expenses in case of significant illnesses.

The second level is a step above level one with a stable income, more belongings, but an easy fall back to level one in case of illnesses or significant catastrophic events such as COVID-19. With hard work and sacrifice level, level three has managed to save enough to absorb shocks from medical expenses with stable access to nutrition, education, and other social services. However, with prolonged lockdowns, for example, in the United States, residents from this level were also challenged into lining up for assistance and filing for unemployment benefits (NY Times 2020).

Level four comprises those in the high-income group on the verge of self-actualization with power, enormous savings, and networks. This classification transcends from the remotest village of a developing country to the middle of a metropolis in the most developed nation. Although there is a significant contextual difference for the four levels in lower- versus middle- and high-income countries as classified by the World Bank, the struggle falls into the four categories at individual levels in each country.

Historically, the lockdown has been employed as the first step to contain and prevent the initial spread of the disease. However, balancing these dynamics to flatten the curve during COVID-19 has spiked debates and paused a moral question without a clear answer. China instituted this measure first, and the rest of the world followed. For developing countries such as those in Africa, it was challenging to follow these measures as those in levels one and two had no means of survival. Those in rural areas who survive from subsistence farming experienced little changes as they continued with regular life routines. The governments of developing countries also faced with choices to procure emergency loans to cover the lockdown expenses, which further intensified their economic struggles. However, it did not sustain them as within 30 days amid increasing cases; the lockdown was lifted to save the economy in several countries, including Ghana, Nigeria, and Kenya. For instance, the lockdown in Ghana lasted for 3 weeks. After which the number of cases doubled from 1200 to more than 2000 (Guardian 2020). South Africa had the most prolonged lockdown with registered higher deaths compared to other countries. After 10 weeks of strict lockdown, it planned to move to level three lockdown on June 1, 2020. The government affirmed into reverting to level four and five lockdowns if cases increase to avoid overloading of hospitals.

49.8.4 Lessons from Mathematical Models

Mathematical models to predict disease behavior were widely used to inform nonpharmaceutical interventions (NPI) during the COVID-19 pandemic, such as the decisions on lockdown and when to lift the lockdown. NPI aims at reducing the chances of disease transmission between individuals by predicting the extreme scenarios with the maximum number of severe cases. Early models were put forward to inform choices such as investing in vaccines or the use of antivirals using data from past influenza pandemics, namely, Spanish flu 1918, Asian flu 1957/1958, Hong Kong flu 1967/1968, and Swine flu 2009. However, the world is devoid of effective vaccines and antivirals against influenza viruses. It calls for more allocation of resources to develop effective vaccines and antivirals for strains that have led to pandemics as well as more research into the unknown potential strains.

The statistical aphorism that all models are wrong, but some are useful, was proven in the COVID-19 outbreak. Some models were thought to be too optimistic about the number of deaths, while others predicted the spread of COVID to have started many days before official announcements. They were questioning if lockdowns were warranted in the first place.

An example of a model that was put forward early during the pandemic predicted 0.3 million deaths in the presence of extreme social distancing and 3.3 million deaths in the absence of any interventions in Africa (UN 2020). Up to

2 months after the first confirmed case in Africa, the total number of deaths from 54 countries was 3926, which is 1% of the total global deaths by the end of May. In the absence of thorough testing, it is not easy to conclude if the number of reported deaths in Africa is accurate. However, anecdotal evidence points toward an exceptional escape of the continent from the disease without apparent factors to support their success.

Except for South Africa, mathematical models for biological processes were not useful in informing decisions to lift lockdowns in sub-Saharan African countries. Most countries lifted lockdowns amid rising cases of COVID-19 due to economic reasons. Models become better with an increasing amount of accurate data. Hence, investing in developing their models by incorporating local factors will help countries make better decisions in the future.

49.9 Conclusion

Frequent crises are usual in human society. Consequently, it is necessary to learn the lessons from the last crisis to handle the next one better. The COVID-19 event is ongoing and likely to be ongoing for an indeterminate time; therefore, it is still too early to draw any conclusions. Here, we have learned that different countries have used similar countermeasures against the virus, but with different results, and we have tried to explain why in the context of cultural differences. However, since the fight is not over, it is time that nations unite with a single purpose, learn from each other, collaborate in finding vaccines and therapies, and adhere to the recommendations of CDC and WHO, as well as local and other medical practitioners and officials.

References

Amon JJ (2020) Human rights protections are needed alongside PPE for health-care workers responding to COVID-19. Lancet Glob Health 8(7):e896. Retrieved from https://doi.org/10.1016/s2214-109x(20)30252-7. https://doi.org/10.1016/s2214-109x(20)30252-7

AsiaNews (2020) COVID-19: Chinese companies selling defective test kits abroad. Retrieved from http://www.asianews.it/news-en/COVID-19:-Chinese-companies-selling-defective-test-kits-abroad-49698.html

Baidu. Retrieved from https://voice.baidu.com/act/newpneumonia/newpneumonia

Boubakri N, Mirzaei A, Samet A (2017) National culture and bank performance: Evidence from the recent financial crisis. J Financ Stab 29:36–56. https://doi.org/10.1016/j.jfs.2017.02.003

Burgess A, Horii M (2012) Risk, ritual and health responsibilisation: Japan's 'safety blanket' of surgical face mask-wearing. Sociol Health Illn 34(aop):1184–1198

Chain B (2020, 2020-05-13) Australia buys 1.5 million faulty coronavirus antibody test kits. Retrieved from https://www.dailymail.co.uk/news/article-8313343/Australia-buys-1-5million-faulty-coronavirus-antibody-test-kits.html

Euronews (2020, 2020-03-29) COVID-19: Netherlands recalls defective Chinese face masks. Retrieved from https://www.euronews.com/2020/03/29/netherlands-recalls-hundreds-of-thousands-of-defective-chinese-face-masks

Gates B (2020) Responding to Covid-19 — A Once-in-a-Century Pandemic? N Engl J Med 382(18):1677–1679. https://www.nejm.org/doi/10.1056/NEJMp2003762

Guardian (2020, 2020-05-03) Easing of lockdown a relief to Ghana's poor – despite fears it is premature. Retrieved from http://www.theguardian.com/global-development/2020/may/03/coronavirus-easing-of-lockdown-a-relief-to-ghanas-poor-despite-fears-it-is-premature

Hamamura T (2012) Are Cultures Becoming Individualistic? A Cross-Temporal Comparison of Individualism-Collectivism in the United States and Japan. Personal Soc Psychol Rev 16(1):3–24

Hans Rosling OR, Anna Rosling Rönnlund. (2020) Factfulness: @goodreads

He D, Zhao S, Lin Q, Zhuang Z, Cao P, Wang MH, Yang L (2020) The relative transmissibility of asymptomatic COVID-19 infections among close contacts. Int J Infect Dis 94:145–147. Retrieved from http://www.sciencedirect.com/science/article/pii/S1201971220302502. https://doi.org/10.1016/j.ijid.2020.04.034

Hongzhou LI (2009) Comments on Japan's policy responses to the financial crisis. Contemporary economy of Japan

Iii JAW (1995) Studies of Individualism-Collectivism: Effects on Cooperation in Groups. Acad Manag J 38(1):152–172

Jiang S, Pan C (2020) Wearing a mask in the era of COVID-19 makes all the difference. Retrieved from https://science.sciencemag.org/content/368/6498/1422/tab-e-letters

Jiang C, Yao X, Zhao Y, Wu J, Huang P, Pan C et al (2020) Comparative review of respiratory diseases caused by coronaviruses and influenza A viruses during epidemic season. Microbes Infect. https://doi.org/10.1016/j.micinf.2020.05.005

Jones S (2020, 2020-03-27) Coronavirus test kits withdrawn in Spain over poor accuracy rate. Retrieved from http://www.theguardian.com/world/2020/mar/27/coronavirus-test-kits-withdrawn-spain-poor-accuracy-rate

JPtoday. Retrieved from https://japantoday.com/category/features/lifestyle/why-do-japanese-people-wear-surgical-masks-its-not-always-for-health-reasons

Kickbusch I, Leung GM, Bhutta ZA, Matsoso MP, Ihekweazu C, Abbasi K (2020) Covid-19: how a virus is turning the world upside down. BMJ 369:m1336

Kim U, Triandis HC, Kâgitcibasi C, Choi S, Yoon G (1994) Individualism and collectivism: theory, method, and applications. Psyccritiques. pp 81–119

Kokudo N, Sugiyama H (2020) Call for international cooperation and collaboration to effectively tackle the COVID-19 pandemic. Glob Health Med. https://doi.org/10.35772/ghm.2020.01019

Lardy N (2011) Sustaining China's economic growth after the global financial crisis. Peterson Institute for International Economics, Washington, DC

Lau J, Tsui HY, Lau MMC, Yang X (2004) SARS transmission, risk factors, and prevention in Hong Kong. Emerg Infect Dis 10(4):587–592

Lee H-C, McNulty M (2003) Korea's Economic crisis and cultural transition toward individualism. ESRI discussion paper series, 71

Li W, Shi Z, Yu M, Ren W, Smith C, Epstein JH, Wang H, Crameri G, Hu Z, Zhang H, Zhang J, McEachern J, Field H, Daszak P, Eaton BT, Zhang S, Wang LF (2005) Bats are natural reservoirs of SARS-like coronaviruses. Science 310(5748):676–9. https://science.sciencemag.org/content/310/5748/676

Loh T (2020, 2020-04-07T09:31:02.505Z) New test hopes dashed as U.K. finds antibody kits don't deliver. Retrieved from https://www.bloomberg.com/news/articles/2020-04-07/new-test-hopes-dashed-as-u-k-finds-antibody-kits-don-t-deliver

Luchtenberg KF, Vu QV (2015) The 2008 financial crisis: Stock market contagion and its determinants. Res Int Bus Financ 33:178–203

Lynteris C (2018) Plague masks: the visual emergence of anti-epidemic personal protection equipment. Med Anthropol 37(6):442–457

Michie J (2020) The covid-19 crisis – and the future of the economy and economics. Int Rev Appl Econ 34(3):301–303. Retrieved from https://doi.org/10.1080/02692171.2020.1756040. https://doi.org/10.1080/02692171.2020.1756040

NY Times (2020, 20200412) It's 'people, people, people' as lines stretch Across America. Retrieved from https://www.nytimes.com/2020/04/12/us/coronavirus-long-lines-america.html

NY times. Retrieved from https://www.nytimes.com/2020/04/09/world/europe/virus-mask-wearing.html

Pan X, Ojcius DM, Gao T, Li Z, Pan C, Pan C (2020) Lessons learned from the 2019-nCoV epidemic on prevention of future infectious diseases. Microbes Infect 22(2):86–91. Retrieved from https://www.ncbi.nlm.nih.gov/pubmed/32088333. https://doi.org/10.1016/j.micinf.2020.02.004

Phuong-Mai N, Terlouw C, Pilot A (2005) Cooperative learning vs Confucian heritage culture's collectivism: confrontation to reveal some cultural conflicts and mismatch. Asia Eur J 3(3):403–419. Retrieved from https://doi.org/10.1007/s10308-005-0008-4. https://doi.org/10.1007/s10308-005-0008-4

Rynn MY (2005) Cultures and organizations: software of the mind. Choice Curr Rev Acad Libr 42(10):1860–1861

SCMP. Retrieved from https://www.scmp.com/news/china/society/article/3075211/face-masks-and-coronavirus-how-culture-affects-your-decision

Shablovsky S (2017) The legacy of the Spanish flu Nearly a century after it killed millions, a journalist reflects on how a pandemic changed the world. Science 357(6357):1245–1245. Retrieved from <Go to ISI>://WOS:000411337700028. https://doi.org/10.1126/science.aao4093

Siami Namini S (2017) China's economy and the global financial crisis. Int J Econ Financ Issues 7:259–265

Stadler K, Masignani V, Eickmann M, Becker S, Abrignani S, Klenk HD, Rappuoli R (2003) SARS - Beginning to understand a new virus. Nat Rev Microbiol 1(3):209–218. Retrieved from <Go to ISI>://WOS:000220431600014

Stadnytskyi V, Bax CE, Bax A, Anfinrud P (2020) The airborne lifetime of small speech droplets and their potential importance in SARS-CoV-2 transmission. Proc Natl Acad Sci:202006874. Retrieved from https://www.pnas.org/content/pnas/early/2020/05/12/2006874117.full.pdf. https://doi.org/10.1073/pnas.2006874117

Sun P, Lu X, Xu C, Sun W, Pan B (2020) Understanding of COVID-19 based on current evidence. J Med Virol 92(6):548–551. https://onlinelibrary.wiley.com/doi/full/10.1002/jmv.25722

Taylor AM (2012) Global financial stability and the lessons of history: a review of Carmen M. Reinhart and Kenneth S. Rogoff's this time is different: eight centuries of financial folly. J Econ Lit 50(4):1092–1105. Retrieved from <Go to ISI>://WOS:000312738600005. https://doi.org/10.1257/jel.50.4.1092

Tendfonline. Retrieved from https://www.tandfonline.com/doi/full/10.1080/01459740.2017.1423072

The Atlantic. Retrieved from https://www.theatlantic.com/politics/archive/2020/04/america-asia-face-mask-coronavirus/609283/

Tibetan Review (2020) Nepal, Czech Republic and Philippines also find Covid-19 test kits from China faulty | Tibetan Review. Retrieved from https://www.tibetanreview.net/nepal-czech-republic-and-philippines-also-find-covid-19-test-kits-from-china-faulty/

TIME. Retrieved from https://time.com/5799964/coronavirus-face-mask-asia-us/

Triandis H (1990) Cross-cultural studies of individualism and collectivism. Neb Symp Motiv 37(11):41–133

Triandis HC (2001) Individualism-Collectivism and Personality. J Pers 69(6):907–924

UN (2020) COVID-19 in Africa: protecting lives and economies | United Nations economic commission for Africa. Retrieved from https://www.uneca.org/publications/covid-19-africa-protecting-lives-and-economies

VOAnews. Retrieved from https://www.voanews.com/science-health/coronavirus-outbreak/face-mask-culture-common-east-new-west

Watson J, Whiting PF, Brush JE (2020) Interpreting a covid-19 test result. Retrieved from https://www.bmj.com/content/369/bmj.m1808/rapid-responses. https://doi.org/10.1136/bmj.m1808

WHO (Producer) (2020) Access to COVID-19 tools (ACT) accelerator. Retrieved from https://www.who.int/publications/m/item/access-to-covid-19-tools-(act)-accelerator

World Bank. Retrieved from https://data.worldbank.org.cn/indicator/NY.GDP.MKTP.CD

Xiao Y, Torok ME (2020) Taking the right measures to control COVID-19. Lancet Infect Dis 20(5):523–524. https://doi.org/10.1016/s1473-3099(20)30152-3

Yahoo Finance. Retrieved from https://finance.yahoo.com/news/the-surprising-covid-19-stock-marketrally-could-collapse-soon-top-strategist-170111578.html

A Borderless Solution Is Needed for A Borderless Complexity, Like COVID-19, the Universal Invader

50

Kawthar Mohamed, Rangarirai Makuku, Eduardo Rodríguez-Román, Aram Pascal Abu Hejleh, Musa Joya, Mariya Ivanovska, Sara A. Makka, Md Shahidul Islam, Nesrine Radwan, Attig-Bahar Faten, Chunfeng Xiao, Leander Marquez, and Nima Rezaei

Abstract

This chapter briefly describes the universal intricacies caused by the COVID-19 pandemic, from the ineffectiveness of distance measures, the massive economic impacts, and the severe mental health challenges to the failure of finding a vaccine, a therapeutic agent or even accurately diagnosing the infection. The entire world is suffering, but every country is trying to combat this pandemic individually, and this deed is the main barrier that prevents reaching a peaceful end.

K. Mohamed
School of Medicine, Tehran University of Medical sciences, Tehran, Iran

Universal Scientific and Research Network (USERN), Manama, Bahrain

R. Makuku
School of Medicine, Tehran University of Medical sciences, Tehran, Iran

Universal Scientific and Research Network (USERN), Harare, Zimbabwe

E. Rodríguez-Román
Center for Microbiology and Cell Biology, Instituto Venezolano de Investigaciones Científicas, Caracas, Venezuela

Universal Scientific and Research Network (USERN), Caracas, Venezuela

A. P. A. Hejleh
Medical Faculty Mannheim, Heidelberg University, Heidelberg, Germany

Universal Scientific Education and Research Network (USERN), Mannheim, Germany

M. Joya
Radiology Department, Kabul University of Medical Sciences, Kabul, Afghanistan

Department of Medical Physics and Biomedical Engineering, Tehran University of Medical Sciences, Tehran, Iran
Universal Scientific Education and Research Network (USERN), Kabul, Afghanistan

N. Rezaei (ed.), *Coronavirus Disease - COVID-19*, Advances in Experimental Medicine and Biology 1318, https://doi.org/10.1007/978-3-030-63761-3_50

Keywords

Borderless · Complexity · COVID-19 · Pandemic · Solution · Universe

50.1 Introduction

50.1.1 A Borderless Solution Is Needed for a Borderless Complexity

The present world order is the result of slow historical developments of particular theories, principles, and international practices influencing each other on various relative angle values (Hershey 2017). The Westphalian Treaty of 1648 is one of the concrete historical events associated with the triumph of state sovereignty, the establishment of a community of states, and even the beginnings of collective security, which later transformed into a conceptualization of the international system (Schmidt 2011). Although this treaty is recognized by some authors, such as the League of Nations and the United Nations today (Gross 2017), it is evident that its spirit is fading away (Osiander 2003).

However, cooperation has always been a common strategy of surviving in nature at all biological levels and even at the economic level (West et al. 2007; Hirshleifer 1978). History of victory over past global pestilences, pandemics, and natural disasters ascertains that human beings are eusocial and naturally high cooperators, with the potential of resisting the forces of natural selection (Nowak 2006; Jabbari et al. 2020). Relatively, the pandemic of Coronavirus Disease 2019 (COVID-19) caused by severe acute respiratory syndrome coronavirus 2 (SARS-CoV-2), a new betacoronavirus in the family Coronaviridae, cannot escape the defeat by the intensive human effort and the extensive borderless cooperation.

This chapter seeks to discuss national border shutdowns and quarantines, international ties needed to handle the COVID-19 economic crisis, flow of experiences and experts, need for international research collaborations, unified diagnostic

M. Ivanovska
Department of Microbiology and Immunology, Faculty of Pharmacy, Research Institute, Medical University-Plovdiv, Plovdiv, Bulgaria

Laboratory of Clinical Immunology, University Hospital "St. George"- Plovdiv, Plovdiv, Bulgaria

Universal Scientific Education and Research Network (USERN), Plovdiv, Bulgaria

S. A. Makka
Neuroscience Research Center, Faculty of Medical Sciences, Lebanese University, Beirut, Lebanon

Universal Scientific Education and Research Network (USERN), Beirut, Lebanon

M. S. Islam
Department of Tissue Engineering and Applied Cell Sciences, School of Advanced Technologies in Medicine, Tehran University of Medical Sciences, Tehran, Iran

Universal Scientific Education and Research Network (USERN), Dhaka, Bangladesh

N. Radwan
Pediatric Allergy and Immunology Unit, Children's hospital, Ain Shams University, Cairo, Egypt
Universal Scientific Education and Research Network (USERN), Cairo, Egypt

A.-B. Faten
Chair of Wind Energy Technology, Faculty of Mechanical Engineering and Marine Technology, The University of Rostock, Rostock, Germany

Universal Scientific Education and Research Network (USERN), Tunis, Tunisia

C. Xiao
Universal Scientific Education and Research Network (USERN), Beijing, China

L. Marquez
College of Social Sciences and Philosophy, University of the Philippines, Diliman, Philippines

Universal Scientific Education and Research Network (USERN), Diliman, Philippines

N. Rezaei (✉)
School of Medicine, Tehran University of Medical sciences, Tehran, Iran

Universal Scientific Education and Research Network (USERN), Tehran, Iran
e-mail: rezaei_nima@tums.ac.ir

approach, therapeutic development, and strategies for vaccine development, and stress control during the COVID-19 pandemic.

50.1.2 The Borderless and the Complex Nature of COVID-19

COVID-19 is a rapidly spreading respiratory disease, which started in Wuhan (Lipsitch et al. 2020), the largest metropolitan area in China's Hubei province. On December 31, 2019, the first case was reported to the World Health Organization (WHO) Country Office (Cascella et al. 2020). On January 30, 2020, the WHO declared COVID-19 as the sixth public health emergency of international concern (pandemic), after affecting 18 countries and territories (Lai et al. 2020; Hanaei and Rezaei 2020). It took 93 days (from December 31, 2019, to April 2, 2020) for COVID-19 to reach the first one million cases; however, it only took 12 days (April 2–14, 2020) to reach the second million cases in 213 countries and territories (Organisation 2020a, b, c, d, e, f). Consequently, COVID-19 became a problem present in the world regardless of any borders.

Furthermore, the borderless complex COVID-19 has caused major socioeconomic disruptions worldwide from its healthcare burden to containment measures, like extensive testing, quarantine, and social distancing (McKibbin and Fernando 2020; Mhlanga and Ndhlovu 2020). The implementation of these measures occurred in almost every country or region around the world. As of April 1, 2020, more than 90% of countries applied restrictions of mobility, and about 40% of countries used complete border shutdowns (Connor 2020).

New strategies are needed to fight COVID-19 and its sequels. Applying cooperation and solidarity among nations would help reduce the impact of this disease, for example, supporting more vulnerable countries through donations of materials and equipment for health personnel and the general population, such as face masks, hand sanitizers, and gloves (Nacoti et al. 2020; Atkeson 2020).

Although COVID-19 has affected the whole world, its case fatality rate (CFR) and the incidence are not uniform among nations and regions (Table 50.1). Current studies estimate that CFR varies from 1% to 15% in different regions and countries (Rajgor et al. 2020). Overall, the CFR is, however, estimated to be less than 1% (Fauci et al. 2020). Therefore, such a large number of deaths related to COVID-19 indicate the high transmission rate of the viral particles even during the non-symptomatic phase of infection (He et al. 2020). Also, it might lie in virus-driven hyper inflammation as a product of a cytokine storm syndrome that also conduces to ARDS in severe cases (Mehta et al. 2020; Zhang et al. 2020b; Zhang et al. 2020a). On the other hand, older people are at a higher risk of developing acute respiratory distress syndrome (ARDS) (Koyama et al. 2016) and death, likely owing to a less vigorous immune response in this population (Wu et al. 2020). Given that the old-age population is more pronounced in wealthier countries (Western), it is expected that the impact of this pandemic might be mitigated on poorer countries (Africa) with weaker health systems but younger age structures. Current statistics support this expectation (Dowd et al. 2020).

Conclusively, the current surge of proliferation and intensification of globalization is rendering borders obsolete, and state decisions are subordinated to international institutions, like WHO. Within that scope and as regards the issue of COVID-19 and other pandemics, the globalization of diseases is becoming one reality that is so omnipresent and erosive to global public health hence calls for more collective borderless solutions. Finally, we need to learn from past experiences like pandemics of flu in 1918, which occurred in three big waves with the second wave being the most destructive, especially in the United States (source: Centers for Disease Control and Prevention 2018). It is time to focus on maximizing tools and knowledge to stop the possibility of the next pandemic wave, which is not impossible as COVID-19 seems to cause re-infection (Jabbari and Rezaei 2020).

Table 50.1 Comparison of the World Health Organization (WHO/Europe) regions

Region	Confirmed	Today	Yesterday	Case/Inhabitants	Recovered	Recovered%	Deaths	Lethality
Europe	1,020,198	22,992	28,594	1697	295,298	28.9%	98,886	9.7%
North America	776,239	27,322	34,278	1571	80,757	10.4%	40,864	5.3%
Asia	406,934	16,691	16,287	86	179,147	44.0%	14,714	3.6%
South America	77,637	4418	5579	182	27,627	35.6%	3630	4.7%
Africa	21,689	1248	1373	17	5329	24.6%	1060	4.9%
Central America and the Caribbean	12,313	326	685	132	1190	9.7%	500	4.1%
Australia Oceania	8116	5	105	199	5046	62.2%	81	1.0%

The table shows that the WHO Africa Region is the least affected, while Europe and the Americas are struggling with COVID-19 burden. At the time when this table was updated, Belgium showed a CFR of 14.7%, and other countries as Sudan, Bahamas, Burundi, and Nicaragua show CFR from 15–20%, however, with fewer cases than Belgium. Prepared with data from Coronavirus disease (COVID-19) outbreak World meters; April 18, 2020 (Worldometers 2020)

50.2 Border Shutdowns and Quarantine: Disputable Solutions

The unprecedented crisis, COVID-19, has been declared as a pandemic by the WHO on March 11, 2020. Following the emergence of this outbreak in Wuhan, the epicenter in China, in late December 2019, the virus spread rapidly to different countries, including Italy, Spain, France, Germany, the United States, Canada, the United Kingdom, and Japan (Organization 2020c, April 17). The current COVID-19 outbreak is considered as the third outbreak caused by the coronaviruses in the twenty-first century, after the severe acute respiratory syndrome (SARS) (2003) and the Middle East Respiratory Syndrome (MERS) (2013) (Wells et al. 2020). Apart from curative protocol, countries around the world have deployed sweeping measures, including lockdown to entire country or cities, shutting down airports, imposing travel restriction, emergency closure of the borders, isolation, and quarantine to block the spread of the novel coronavirus (News 2020, April). The initial strategy of airport screening was adopted in many countries in response to global spread; however, eventually, it appeared ineffective in slowing the spread of disease (Samaan et al. 2004; Wilder-Smith et al. 2003; John et al. 2005). On January 23, 2020, China implemented a strict lockdown strategy, along with the other 16 neighboring cities in Hubei province, also known as the largest Cordon sanitaire in history (Leung et al. 2020; Rachman 2020).

50.2.1 Quarantine

The word quarantine has come from the history of the black death in Europe. Repeated waves of plague swept across Europe started from the middle of the fourteenth century. It was estimated that one-third of people in Europe died by 1350 (Kilwein 1995). As a part of extreme control measures, in 1374, Viscount Bernabo of Reggio, Italy, declared that persons with plague to be shifted out of the city till they die or recover (Jewell 1857). In 1377, the Great Council of Dubrovnik (former Ragusa), Croatia, introduced a similar strategy that strangers had to stay 30 days (Trentino) in a restricted location to observe the symptoms of plague (Stuard 1992). Later on, the isolation period was extended to 40 days, called quarantine originated from the Latin word *Quaranta* meaning 40, the isolation time to prevent the spread of infection. The isolation or quarantine is considered as the greatest achievement of medieval medicine and an essential milestone of the medical heritage of Dubrovnik and Croatia (Anonymous 2002).

Since then, different countries experienced such transient border restriction and movement limitations in all pandemic threats. For example, after the discovery of HIV/AIDS in 1984, 66 of 186 countries in the world imposed some form of travel restriction (Joseph and Amon 2008, Dec 16). However, a research team led by Norbert Gilmore (1989) concluded that the adopted travel restriction on HIV transmission was "ineffective, impractical, costly, harmful, and maybe discriminatory" (Gilmore et al. 1989). International border restriction also instituted during outbreaks of SARS (2003), MERS (2012), and Ebola virus disease (EVD 2014) aimed at the prevention of the spread of infection (Errett et al. 2020).

50.2.2 The Travel Bans

Restriction, to some extent, is beneficial. According to Lainie, the Rutkow travel ban is a legal option that the government can enforce to mitigate a pandemic (Rutkow 2020, February 13). Anthony Fauci, head of the US National Institute of Allergy and Infectious Diseases, claimed, "If we had not put a travel restriction on, we would have had many, many, many more travel-related cases than we have" (Cohen and Kupferschmidt 2020). Recent research revealed that international travel bans in and out of China during December and January could minimize the viral spread. For example, China's lockdown of Wuhan city delayed the epidemic progression only for a few days within China and few weeks internationally. It was, however, helpful to flatten the curve of infection, thereby reducing the pressure on the healthcare system (Chinazzi et al. 2020).

By early February 2020, the government of several countries, including the United States, Australia, Russia, and Italy, imposed travel restrictions, and 59 airline companies limited or suspended flights to Mainland China (Sang-Hun 2020, Jan 21; Kiselyova 2020, Jan 28; Travellers from China to be denied entry into Australia 2020, Feb 2). A research team led by Matteo Chinazzi estimated the effect of travel ban implemented in Wuhan and the international travel restriction enforced by other countries in early February 2020. They utilized a metapopulational network integrated with real-world data and the airline transportation data from the Official Aviation Guide (OAG) and IATA databases in the Global Epidemic and Mobility (GLEAM) model. The model indicates that the travel ban in Wuhan was initially effective at reducing international case importation, and the growth of cases outside Mainland China delayed only for 2–3 weeks. Furthermore, the model found that a significant number of individuals exposed to SARS-CoV-2 have been traveling without being detected (Chinazzi et al. 2020).

European experts reached a consensus on February 27, 2020, that refusal of entry of people with coronavirus symptoms is not an appropriate measure to prevent the virus spread; rather, it would be counterproductive and ineffective. Experts emphasized the systematic checks for arriving people and the coordination between border guards and national authorities with real-time exchange of information (NICOLÁS 2020, Feb 28).

Paolo Bajardi and colleagues analyzed the 2009 H1N1 flu data and concluded that mobility restriction could make a delay (on an average less than 3 days) in the spread of disease by containing outbreaks of infectious diseases at their source. They, however, reached to this conclusion that "such restrictions achieved no containment, and the virus was able to reach pandemic proportions in a short time" (Bajardi et al. 2011).

50.2.3 Airport Screening

At the airport, the temperature screening during exit or entry is not a practical way to stop the international spread since infected people may be in the incubation period, asymptomatic, or have mild symptoms. Moreover, massive investment is needed for little benefits. However, disease prevention recommendation messages to travelers, collecting health declaration with contact details for the proper risk assessment, and contact tracing of incoming passengers are the more effective approaches (Organization 2020d, February 29).

A debate rages about the restricted border control of multiple countries locking down the citizens in the pandemic that have been credited, on the one hand, with delaying the disease spread and, on the other hand, assigning a very high risk to economic activities. For example, the first travel restriction enforced/enacted in January 2020 and focused on the people in Mainland China and other hotspots in Asia. The WHO observed an initial fall in the case numbers inside China accompanied by 80% reduction of case exportation from China due to this bold approach and the delay of disease spread regarded as "buying time" for 2–3 weeks for the preparedness of the rest of countries (Patel 2020).

50.2.4 Effect on the Economy

The international border restriction and quarantine due to pandemic have a significant effect on macro- and microeconomy. For example, G7 countries jointly share 60% of world supply and demand, 65% of world manufacturing, and 41% of manufacturing exports. A temporary negative supply and demand shock are attributed to regular flu, also called macroeconomic sneeze, in which quick recovery is possible. However, a pandemic like COVID-19 may produce a global, large-scale, and persistent economic breach. Furthermore, quarantine and travel restrictions will elevate the unemployment issue worldwide. According to the International Labour Organization, global unemployment could rise between 5.3 and 24.7 million based on global gross domestic product (GDP) growth estimates. The workers from the poor and developing countries will encounter the massive effect of unemployment that largely depends on the export-oriented industries. For example, Bangladesh, one of the leading apparel exporters in the world, is under the threat of order cancellation worth about $2.8 billion, and about 70% of workers are already sent home without wages in the face of nationwide lockdown for the COVID-19 pandemic (Barua 2020).

50.2.5 Disputing the Quarantine Approach

Laurent Hébert-Dufresne, working on contagion modeling in the University of Vermont's Complex Systems Center, stated that "the effectiveness of mass quarantine is disputed in general." He criticized any physical or legal barrier since tracking and testing people become harder. Quarantine, in a sense, staying home or canceling social events, might be acceptable for avoiding the transmission of infectious disease (Newman 2020, March 11).

Lessons from several pandemics (SARS, MERS, EVD, and ongoing COVID-19) indicate that border restriction and quarantine are not the ultimate solutions to mitigate an outbreak. Such attempts might provide potential time for the preparedness at the cost of massive economic disruption. Social distancing, lockdown, and quarantine should be proportional to the public health risk and economic capability. Effective collaboration in pandemic management, including equitable distribution of salvation, medical aid, and flow of information regarding development of drugs or vaccines, is essential beyond the border (Moazzami et al. 2020).

50.3 International Ties Are Needed to Handle the Economic Crisis

COVID-19 has profoundly affected the international economy (review 2020). The extent of the ultimate damage depends on how quickly the virus is contained, the steps that the authorities are taking to contain it, and how much economic support governments are willing to deploy during the outbreak and its aftermath. Generally, it seems that world responses to COVID-19 have been mostly uncoordinated and country-specific (organization).

One standard measure taken globally was closing the borders (review 2020). This measure has immediately disrupted international trade since this measure would not only deprive some countries of all of the goods and products that they cannot produce but also hurt the economies and worsen unemployment. It even gets worse when combined with full and partial lockdown, as some countries implemented.

Closing the border together with the lockdown of cities can increase the risk of disrupting access to medical supplies and destabilizing food markets in addition to the unemployment. For example, China, which is the second-largest economy of the world, strived back to life in early April after suffering a withering blow from the COVID-19 outbreak (Review).

The question arises on how to control the spread of the virus and resume the economic activities. It is challenging to resume the economic activities unless the spread of the COVID-19 is controlled. For this purpose, the public health response must focus on hastening and containing the pandemic as soon as possible. Many countries expended hospital intensive care units (ICU), prepared temporary hospitals, and overproduced respiratory protective gears and masks (Organization 2020a). Other countries, such as Germany, implemented large-scale testing of the population (Worldometer 2020). Moreover, international collaboration in communicating the evolution of the virus and the best practices is crucial to understanding how to control the spread of disease and reduce the contamination. It is also needed to provide funding for developing countries. All in all, unified leadership is the most relevant solution, as it is wiser to take a coordinated approach to boost production and meeting the needs of the most vulnerable populations and regions worldwide.

50.4 The Flow of Experiences and Experts

50.4.1 The Need for a United Diagnostic Approach

The year 2020 will be a historical year, taking a look at the COVID-19 pandemic affecting not only international healthcare systems and the scientific community but also people's everyday lives. The causative virus of COVID-19 is highly contagious, can spread quickly, and has been capable of imposing enormous health, economic, and societal impacts in any setting. Therefore, real solidarity and collaboration are essential between nations to combat against this common threat.While nations implement strict border policies as a necessary measure to control the pandemic, scientists reach out beyond their borders to initiate regional and international research collaborations. These collaborative works were launched within a few days after the outbreak started, demonstrating the fast dynamics in science. The current outbreak highlights the necessity for accurate and fast diagnostic methods (Basiri et al. 2020a). During the last couple of weeks, the laboratory diagnostic methods have evolved significantly and gradually replacing conventional gold standards (Yazdanpanah et al. 2020b). Prestigious research journals offer free access to published COVID-19 data to make it more accessible for the research community and accelerate diagnostic and therapeutic approaches. Researchers from the German Centre for Infection Research (DZIF) at Charité – Universitätsmedizin Berlin developed the first laboratory assay along with academic collaborators in Europe and Hong Kong and immediately handed over the protocol to the WHO for further publishing (Corman et al. 2020). As a result, the WHO launched the shipment and distribution of 250,000 kits to 159 laboratories. Dozens of laboratories started using the publicly available assay protocol to establish diagnostic of COVID-19, showing that data sharing in the form of diagnostic protocols is crucial for the emerging public health crisis. Laboratory testing is an integral part and constitutes the foundation of the WHO's strategy published in the Strategic Preparedness and Response Plan (Organization 2020b). Additionally, diagnostic testing remains an essential tool for tracking the viral disease and helps authorities to assess the impending burden and allocate medical resources and staff more efficiently.

Currently, testing of COVID-19 is done with reverse transcription-polymerase chain reaction (RT-PCR) based on viral genetic material from nasopharyngeal swabs. This method works by amplification of a specific sequence of the viral genome (Ohan and Heikkila 1993); thus, it is limited to the detection of an active infection rather than a resolved infection. Other possible methods for diagnosing COVID-19 are immune identification technologies like point-of-care testing (POCT) of IgM/IgG and enzyme-linked immunosorbent assay (ELISA). The urge for a fast, simple, and feasible diagnostic device led researchers to intensify the development of POCT due to its rapid turnaround time (approximately 15 min) (Kozel and Burnham-Marusich 2017). It is mainly required in resource-limited settings that lack governmental support and healthcare infrastructure.

The current pandemic has created a novel situation for researchers and laboratories. In many countries, the mass of patients exceeds the diagnostic capacities of the respective healthcare systems; consequently, the WHO has recommended two COVID-19 diagnostic kits for emergency use and facilitated their distribution worldwide. The efforts of these global institutions constitute the backbone of a unified diagnostic approach, yet it is not the only measure taken to enable the acceleration of diagnostics. Sharing prior published protocols and data, for example, on the free online archive medRxiv, would guarantee the openness and accessibility of scientific findings

(Rezaei 2020b). Addressing low- and middle-income countries, the Foundation for Innovative New Diagnostics (FIND), a Geneva-based nonprofit organization, facilitates the development and delivery of diagnostics (Sheridan 2020; Sahu et al. 2020). For that, FIND published an open call to all laboratories around the world to submit any performance data on commercially available diagnostic tests for independent evaluation. The results will be accessible to everyone via their website.

Additionally, individual laboratories contribute to the fight against COVID-19. Cancer researchers at London's Francis Crick Institute turned their lab into a COVID-19-testing facility, volunteering to help battle the pandemic (Baker 2020). Taken together, all institutions from governmental authorities, nonprofit organizations, worldwide alliances, or individual laboratories need to work hand in hand to guarantee an effective, fast, and feasible way to diagnose COVID-19 on a large scale. Although there are still countries lacking urgently needed diagnostic kits and testing infrastructure, there has been impressive progress made during the current outbreak as opposed to the 2002–2003 SARS outbreak, where it took nearly 6 months to identify and to establish assays for the detection of newly emerged coronavirus. For the future, the recently established collaborations and ties between international researchers need to be intensified to accelerate diagnostic data sharing, facilitate and unify the evaluation of novel diagnostic methods, and improve the equitable distribution of medical goods to be prepared for prospective global health problems (Kafieh et al. 2020).

50.4.2 International Collaboration: The Key to Therapeutic Development

Since December 2019, scientists have been working to find a treatment for COVID-19 (Sharifkashani et al. 2020; Rezaei 2020b; Rabiee et al. 2020; Mansourabadi et al. 2020; Pashaei and Rezaei 2020; Fathi and Rezaei 2020; Jahanshahlu and Rezaei 2020b; Basiri et al. 2020b; Saghazadeh and Rezaei 2020b; Pourahmad et al. 2020); three robust drug trials have been completed so far, but neither of them brought the acceptable results (Bin Cao 2020; Prof Mandeep R Mehra 2020; Yeming Wang 2020). At the start of each drug trial, hope springs eternal; maybe this time, the cure will be found. The fate of the COVID-19 drug discovery is ambiguous (Mohamed et al. 2020), and one of the tools that can clear up this ambiguity is an international collaboration. How could international collaboration be helpful in COVID-19 drug discovery?

To start up, ritonavir/lopinavir and remdesivir trials can precisely explain the critical need for international collaboration. The same scientists from China accomplished both of these trials; Bin Cao et al. conducted the ritonavir/lopinavir trial from late January till early February, which is precisely at the time that COVID-19 peaked in China. Although this work epitomized an organized rapid response against this pandemic, which is extremely difficult to manage in these circumstances, it was not able to come up with a cure for the COVID-19 infection (Bin Cao 2020).

After that, from early February until March, Yeming Wang et al. handled a randomized, double-blind, placebo-controlled, multicenter trial of remdesivir. This time remdesivir showed promising effects but not reached statically significance confirming the results (Yeming Wang 2020). If both of these studies were supervised by the same investigators, why one of them had great significance while the other did not have it? The answer might lie in the downswing of the COVID-19 cases in China during the remdesivir trial. What would have happened if the remdesivir study was performed in one of the countries that were in their way of hitting a peak in March like Italy, France, or the United Kingdom? An influential study with COVID-19 remedy could have been handed to the world due to conducting the remdesivir trial with plenty of cases, the main limitation that Chinese scientists face (Tsallis; India 2020; news).

By having a look at Fig. 50.1 that demonstrates the number of drug trials of two groups of countries, the countries that have passed their

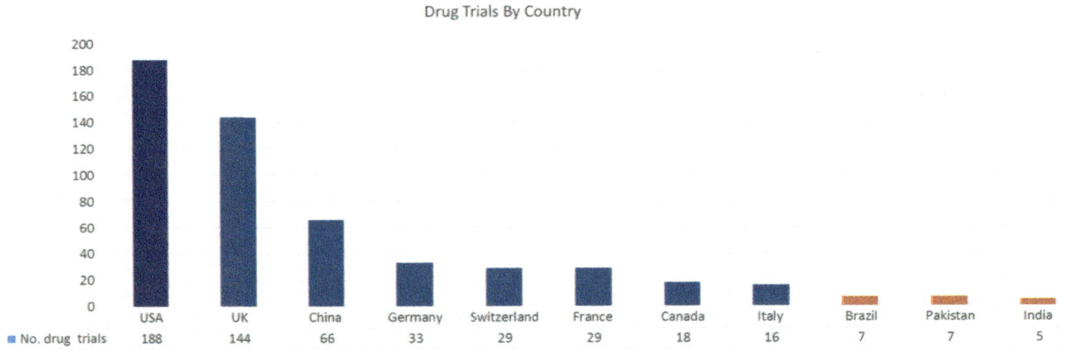

Fig. 50.1 Number of drug trials conducted by the countries that have hit their COVID-19 cases' peaks no later than May (blue columns) and the countries that are expected to hit their peaks by June (orange columns) (Tsallis 2020; India 2020; news; clinicaltrials.gov 2020)

peaks and the countries that will have reached their peaks by June; each of the United States, the United Kingdom, and China has the highest number of drug trials, whereas Brazil, India, and Pakistan have the lowest record of drug trials. The former countries represent the countries that have passed their peaks, while the former group illustrates the countries that are on their way to reach a peak (Tsallis; India 2020; news; clinicaltrials.gov 2020). Therefore, it could be beneficial to exchange the experience, the knowledge, and COVID-19 cases between the countries that hit their peaks previously like China and the countries that are nearly reaching their peak of COVID-19 cases like Brazil, in order to prevent another frustrating work (Tsallis).

Another pronounced issue that is worth mentioning is the repetition of the trials that have clear consequences like the ineffectiveness of hydroxychloroquine or chloroquine, which was approved by Mehra and his colleagues (Prof Mandeep R Mehra 2020). Although repeating the same research for getting more accurate results cannot be neglected, wasting a lot of time and effort in ascertaining a negative result is not recommended too.

COVID-19 is a newly emerged condition that is filled with uncertainty. There is no evident cure for it, and the world is in a race for the discovery of its remedy. It is better to consider this pandemic as a global crisis other than a race, to exchange data, experience, or even COVID-19 cases, in order to end this pandemic immediately. When countries individually supervise their drug trials, they might accomplish a frustrating, repetitive work that, in consequence, not only leads to misusing their sources but also reducing the pace of finding a prosperous solution.

Figure 50.2 reflects the total number of 143 drug trials that investigate chloroquine/hydroxychloroquine or both of them in treating the COVID-19 infection. There are about 70 studies in the recruiting phase, and about 59 registered trials that have not yet been recruited (clinicaltrials.gov 2020). Therefore, around 41% of the studies have not started yet, while we need to consider other therapeutics too. If international collaboration was present, the number of frustrated studies would have decreased.

50.4.3 International Collaboration: An Agile Strategy for Vaccine Production

50.4.3.1 Outbreak of COVID-19

The increase in the world population, traveling habits, and contact between people from all areas of the world have highly favored the global spreading of pathogens (Niall 2002). It will lead to an increased risk of pandemic outbreaks. Such risk further rises by the climate change that influences the distribution, abundance, and prevalence of pathogen-bearing vectors, promoting

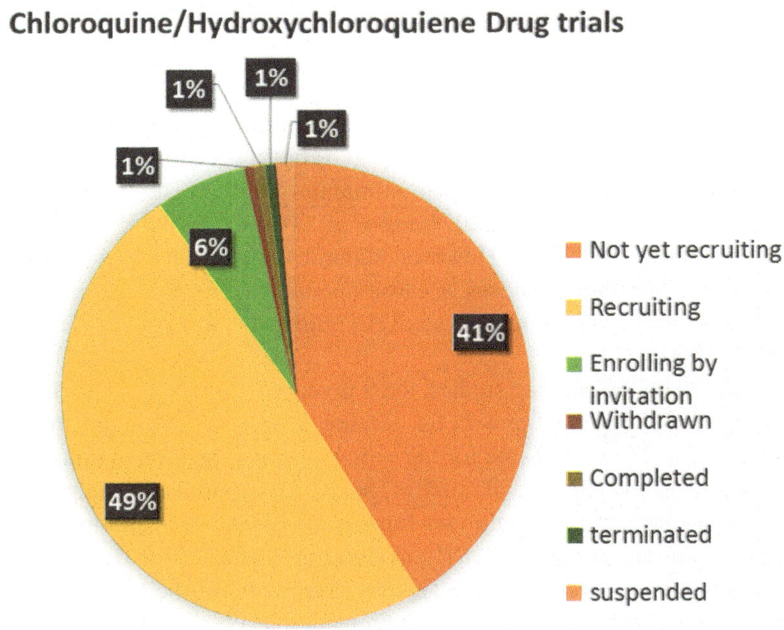

Fig. 50.2 Number of drug trials on the effects of chloroquine, hydroxychloroquine, or both of them in the treatment of COVID-19 infection by May 31, 2020

infections with a range of vector-borne diseases (Susanne 2018).

An outbreak of COVID-19 has started in China since December 2019 and has caused atypical pneumonia and has spread to 26 countries (Shang 2020). COVID-19 spreads rapidly by human-to-human transmission through respiratory droplets or direct contact (Qun 2020; Lotfi et al. 2020). The median incubation period is 3 days, and the time from symptom onset to developing pneumonia is 4 days (Qun 2020). By February 2020, a total of 75,465 cases with COVID-19 infections have been confirmed, and 2236 people have died in China (China). SARS-CoV-2 is currently infecting more than 70,000 people with about 2.7% case fatality rate (Wang N 2020). In March 2020, the WHO announced COVID-19 illness as the worldwide pandemic that is overwhelming the healthcare systems globally (Organisation 2020a, b, c, d, e, f; Fauci et al. 2020; Mahase 2020). To date, there are no available vaccine and specific treatment for COVID-19 (Shang 2020).

50.4.3.2 Structure of SARS-CoV-2

There have been seven coronaviruses (CoVs) isolated from humans (Wang et al. 2020a, b). Among them are three emerging pathogenic CoVs, including SARS-CoV, MERS-CoV, and the newly identified SARS-CoV-2 (Wang et al. 2020a, b). SARS-CoV infection in humans was first reported from China in 2002, while MERS-CoV infection in humans was first reported from Saudi Arabia in June 2012 (Ksiazek et al. 2003; Zaki et al. 2012). The fatality rate in SARS-CoV and MERS-CoV was reported to be 10% and 34.4%, respectively (Organisation 2020a, b, c, d, e, f; Du et al. 2009). Different from MERS-CoV but similar to SARS-CoV, SARS-CoV-2 can cause human-to-human transmission, but its intermediate host that leads to the current human infection and outbreak is still under investigation (Wang et al. 2020a, b).

After studying the structure of SARS-CoV-2, it was found that it shares some features with MERS-CoV and SARS-CoV. Coronaviruses have four structural proteins, namely, envelope (E), membrane (M), nucleocapsid (N), and spike (S) (Chan et al. 2020). The single-stranded RNA of SARS-CoV is packed inside the capsid formed by the N protein, while the M, E, and S proteins form the envelope surrounding the capsid (Qun 2020; Niall 2002; Wang et al. 2020a, b). The E and M proteins have essential functions in the viral assembly of a coronavirus, and the N protein is necessary for viral RNA synthesis (Shang 2020; Schoeman and Fielding 2019). The glycoprotein S of SARS-CoV-2 is responsible for virus binding and entry (Gralinski 2020). All these proteins can stimulate the immune system (Shang 2020; Jiang et al. 2005; Regla-Nava et al. 2015). However, subunit vaccines require multiple doses to cause maximum stimulation for the immune system (Eyigün and Gul 1998).

Based on the ongoing studies, the S glycoprotein can be a potential protein for vaccine development as it is responsible for the viral invasion of human cells (Jiang et al. 2020). The S glycoprotein can be cleaved into two functional subunits, an amino-terminal S1 subunit, and a carboxyl-terminal S2 subunit. The S1 subunit is responsible for virus-host cell receptor binding, which mediates the viral entry through the host angiotensin-converting enzyme two receptors initiating infection (Lu et al. 2014) (Li et al. 2005; Wan et al. 2020; McCaffery et al. 2015). In contrast, the S2 subunit is involved in virus-host membrane fusion and, consequently, delivering the viral genetic materials into the host cell through the fusion core (Lu et al. 2014; Li et al. 2005).

50.4.3.3 International Collaboration for Vaccine Development

Conventional vaccine development has successfully decreased the burden of several infections (Susanne 2018). Nevertheless, the vaccine development process is a lengthy, expensive process that typically takes multiple candidates and many years to produce a licensed vaccine (Du et al. 2009; Gouglas et al. 2018). During such a process, developers follow a linear sequence of steps, with many pauses for data analysis or manufacturing-process checks. However, these established steps are not suitable or even feasible in outbreak situations. During a pandemic, vaccine developments meet many challenges that need to be addressed differently. They need to develop efficient platforms for vaccine development and to execute many steps in parallel before confirming a successful outcome of another step, hence resulting in a significant challenge due to elevated financial risk (Gouglas et al. 2018).

The Coalition for Epidemic Preparedness Innovation (CEPI) is an international nongovernmental organization that supports the development of vaccines against epidemic pathogens on the World Health Organization (WHO/Europe) priority list. They have tried to support having ideal platforms for vaccine development that allows a timeline of 16 weeks between the identification of viral sequence to commencing clinical trials (Lurie et al. 2020). The developing platforms aim to meet different challenges for rapid vaccine development that includes identification of the best antigenic protein for the vaccine, the mode of administration, safety, the efficacy of the developed vaccine and finally the production capacity to meet global need. Over the past decade, the scientific community and the vaccine industry has been successful in the H1N1 influenza vaccine, but not in SARS, Ebola, and Zika. The success in the H1N1 vaccine was attributed to the presence of a previously well-developed platform for such viruses (Lurie et al. 2020). However, efforts are still going to meet challenges and develop a rapid response to meet emerging epidemics.

Since the announcement of the genetic sequence of SARS-CoV-2 on January 11, 2020, the global R&D activity is set to develop a vaccine against the disease, and the first COVID-19 vaccine candidate entered human clinical testing on March 16, 2020. As of April 08, 2020, the global COVID-19 vaccine R&D landscape includes 115 vaccine candidates, and it is predicted to have a vaccine by early 2021. Most of the vaccine developers are in North America, China, Australia, and Europe (Thanh Le et al. 2020). Such international collaboration is the key to the rapid vaccine development and is a step in the changes needed for vaccine development to meet new emerging epidemics and pandemics.

50.5 Mental Health Issues Derived from the COVID-19 Pandemic

50.5.1 Recognizing Challenges Posed to Global Mental Health Work by the COVID-19 Pandemic

The mental health challenges brought by the COVID-19 pandemic should not be ignored. Some believe that COVID-19-related psychological and mental problems are likely to cause a second pandemic (Choi et al. 2020). After the occurrence of such sudden public health emergencies as the COVID-19 pandemic, many psychological and mental problems, including depression, anxiety, and sleep problems, have emerged.

Populations who are, in particular, vulnerable to COVID-19-related psychological and mental problems include women (Wang et al. 2020a, b), people younger than 40 years old or older than 60 years old (Wang et al. 2020a, b; Qiu et al. 2020), medical workers (Zhang et al. 2020c; Rezaei 2020a), students (Zhou et al. 2020), patients with pre-existing physical and mental conditions (Ozamiz-Etxebarria et al. 2020; Xu et al. 2020; Hao et al. 2020), people with higher educational level (Wang et al. 2020a, b), residents living in the severely affected areas (Ni et al. 2020), pregnant women (Mirbeyk and Rezaei 2020), and other special groups (Özdin and Bayrak Özdin 2020; Saccone et al. 2020). Among these populations, the level of depression, anxiety, and stress were significantly higher than their control population(s). For example, the risk of anxiety in women during the pandemic was 3.01 (95% CI: 1.39–6.52) times higher than in men (Wang et al. 2020a, b). From this perspective, to prepare for the second pandemic in the right direction requires the identification of potentially vulnerable populations and develop practical approaches to promote the mental health of these populations.

The existing evidence suggests brief reactive psychotic disorder and panic disorder as the most common mental disorders during the COVID-19 pandemic (Bhatia et al. 2020; Zulkifli et al. 2020). Manjeet et al. reported a 28-year-old unmarried, college-educated man of upper-middle socioeconomic status with COVID-19 pandemic-induced panic disorder. Repeatedly hearing news of the COVID-19 pandemic was the precipitating factor, as demonstrated in other studies (Bhatia et al. 2020). Therefore, it is concluded that psychological and mental problems caused by the COVID-19 pandemic are not only caused by a single factor such as the pandemic as a stressor itself but other factors, including but not limited to quarantine, media behavior, and government behavior during the pandemic play role as well. Due to the diversity of precipitating factors and the complexity of pathogenesis, mental health work in the context of the COVID-19 pandemic still has a long way to go.

50.5.2 Understanding the Homogeneity and the Heterogeneity of Stress across the World: The Source of the Challenges

When an individual is exposed to prolonged, intense or unexpected, external or internal stressors, the mind and body mediate physiological, biochemical, mental, and behavioral responses, collectively referred to as stress responses. The term stress, in the most general sense, means the accumulation of emotions from situations in which one needs to adapt to specific facts, factors, and conditions of the environment in which he lives. Stress is a constant factor in today's rapidly developing world that can endanger our health if left unchecked. It affects the immune system by activating the hypothalamic-pituitary-adrenal and sympathetic-adrenal-medullary axes. Various neurotransmitters, neuropeptides, hormones, and cytokines mediate these complex interactions, which have been the subject of a field of study called psychoneuroendocrinoimmunology (PNEI). The pandemic condition and related stress would exacerbate the effect of COVID-19 on immune dysregulation and inflammation, and therefore multiple organs and systems fail to work correctly, as demonstrated in patients with a severe form of COVID-19 (Yazdanpanah et al. 2020b; Shamshirian and Rezaei 2020; Jahanshahlu and Rezaei 2020a; Sahu et al. 2020; Rokni et al. 2020; Bahrami et al. 2020; Yazdanpanah et al. 2020a; Mansourabadi et al. 2020; Saghazadeh and Rezaei 2020a; Lotfi and Rezaei 2020; Nasab et al. 2020; Saleki et al. 2020).

General adaptation syndrome (Abida et al. 2019) developed by Hans Selye involves three stages in the body's response to stress: alarm response, resistance, and recovery/exhaustion (Taylor and Sirois 2012).

Alarm response (alarm) occurs in two phases: shock is when the body is subject to changes as a result of stressors; however, the body's resistance to stressors is gradually diminishing; and anti-shock is when the threat or stressor is identified or realized and the body begins to respond and is in an alert state. During the latter stage, the locus coeruleus of the sympathetic nervous system is activated, and production of catecholamine, mainly of noradrenaline and adrenaline, begins; therefore, the fight-flight or freeze response (it is a physiological response that arises as a result of the anticipated adverse event, attack, or threat to survival) appears.

During the stage of resistance, increased secretion of glucocorticoids plays a significant role. Lipolysis and catabolic and anti-anabolic reactions result in increased concentration of fats, amino acids, and glucose in the blood. Besides, lymphocytopenia, eosinopenia, and neutrophilia might occur. Recovery occurs when the body's compensatory mechanisms have successfully prevented the effect of the stressor. Otherwise, all the resources are depleted, and, finally, the body is unable to maintain normal function (exhaustion). If the third stage is prolonged, long-term damage can lead to prolonged vasoconstriction and ischemia, which in turn leads to cellular necrosis.

Good stress (eustress) and bad stress (distress) are two different conditions. Short-term response to stress has considerable survival value. Distress is the one that leads to stress-related diseases.

It is interesting to know that people can respond differently to stress due to the difference in every person's level of discrete hormones (noradrenaline in fast response and cortisol in slow stress response) that are released. It can be because of polymorphisms in genes coding stress response and some hormones and neurotransmitters like GABA. These heterogeneities of the stress response pattern can be seen in different populations. Not all of us will become ill and develop depression and post-traumatic stress disorder (PTSD). Epigenetic modifications (methylation of DNA leading to different gene expression) because of persistently high levels of stress hormones are also of importance in stress response. Having increased gene expression in some cases can lead to visible psychological and behavioral changes in some people. The mean gene expression would be influenced by the genetic background within a population and varies from individual to individual.

During the pandemic time, some of us are going to be more resistant than others to the pathological effects of COVID-19 in terms of both physiological (Yousefzadegan and Rezaei 2020; Ahanchian et al. 2020; Babaha and Rezaei 2020) and psychological effects. As mentioned above, this might arise from our genetic background (Darbeheshti and Rezaei 2020) and its effects on the neuroimmune system. In order to protect ourselves, we must find methods to cope with it. The issue of reducing stress is not insignificant. When using cognitive-behavioral therapies to manage stress such as meditation, yoga, hypnosis, and muscle relaxation, the psychological and physiological effects of distress can be reduced. Systematic training and strategies for dealing with stressors are needed. Children must see a good model of coping stress from their parents. Creative strategies like singing, drawing, or usage of digital methods (not fake news on the internet) for social connection can be helpful too in quarantine conditions all over the world while fighting COVID-19 (WHO/Europe). With all of these actions, we will be able to prevent distress, long-term changes in our body, and the development of stress-related disorders after this pandemic.

50.6 Conclusion

It has been around 6 months since the emergence of the C0VID-19 pandemic, and we are still striving to find a light that can rescue us from the darkness of COVID-19. A tremendous effort has been made so far; the condition is relatively better than what it was at the start, but this effort is not enough to resolve the COVID-19 crisis completely. There is no individual economy, infrastructure, or even population with the ability to combat this virus; perhaps the best solution is a call for increased international cooperation and setting aside of political interests and misunderstandings, straightening of priorities in fighting the pandemic, and putting more premium on combating the adverse social effects that the COVID-19 pandemic has brought to the global community (Kawthar Mohamed 2020; Momtazmanesh et al. 2020; Moradian et al. 2020). It would increase our chance to arrive at the cross-disciplinary research of higher quality (Rzymski et al. 2020).

References

Abida W, Cheng ML, Armenia J, Middha S, Autio KA, Vargas HA, Rathkopf D, Morris MJ, Danila DC, Slovin SF (2019) Analysis of the prevalence of microsatellite instability in prostate cancer and response to immune checkpoint blockade. JAMA Oncol 5(4):471–478

Ahanchian H, Moazzen N, Faroughi MSD, Khalighi N, Khoshkhui M, Aelami MH, Haghi NSM, Rezaei N (2020) COVID-19 in a child with primary specific antibody deficiency. Clin Case Rep 00:1–4. https://doi.org/10.1002/ccr3.3643

Anonymous (2002) The old port of Dubrovnik. Emerg Infect Dis 8(1):110

Atkeson A (2020) What will be the economic impact of COVID-19 in the US? Rough estimates of disease scenarios. National Bureau of Economic Research

Babaha F, Rezaei N (2020) Primary immunodeficiency diseases in COVID-19 pandemic: a predisposing or protective factor? Am J Med Sci. https://doi.org/10.1016/j.amjms.2020.07.027

Bahrami A, Vafapour M, Moazzami B, Rezaei N (2020) Hyperinflammatory shock related to COVID-19 in a patient presenting with multisystem inflammatory syndrome in children: first case from Iran. J Paediatr Child Health. https://doi.org/10.1111/jpc.15048

Bajardi P, Poletto C, Ramasco JJ, Tizzoni M, Colizza V, Vespignani A (2011, Jan 31) Human mobility networks, travel restrictions, and the global spread of 2009 H1N1 pandemic. PLoS One 6(1):e16591

Baker N (2020) These cancer scientists turned their lab into a coronavirus-testing facility. Nature 580(7804):441

Barua S (2020) Understanding Coronanomics: the economic implications of the coronavirus (COVID-19) pandemic. SSRN Electron J. https://doi.org/10/ggq92n

Basiri A, Heidari A, Nadi MF, Fallahy MTP, Nezamabadi SS, Sedighi M, Saghazadeh A, Rezaei N (2020a) Microfluidic devices for detection of RNA viruses. Rev Med Virol n/a (n/a):e2154. https://doi.org/10.1002/rmv.2154

Basiri A, Pazhouhnia Z, Beheshtizadeh N, Hoseinpour M, Saghazadeh A, Rezaei N (2020b) Regenerative medicine in COVID-19 treatment: real opportunities and range of promises. Stem Cell Rev Rep:1–13. https://doi.org/10.1007/s12015-020-09994-5

Bhatia MS, Goyal S, Singh A, Daral A (2020) COVID-19 pandemic-induced panic disorder. Prim Care Companion CNS Disord 22(3). https://doi.org/10.4088/PCC.20l02626

Bin Cao. (2020). A trial of Loponavir-Ritonavir in Adults Hospitalized with Severe COVID-19. https://www.nejm.org/doi/full/10.1056/NEJMoa2001282. Accessed 29 May 2020

Cascella M, Rajnik M, Cuomo A, Dulebohn SC, Di Napoli R (2020) Features, evaluation and treatment coronavirus (COVID-19). In: StatPearls [Internet]. StatPearls Publishing

Chan JF, KH Kok Zhu Z, (2020) Genomic characterization of the 2019 novel human-pathogenic coronavirus isolated from a patient with atypical pneumonia after visiting Wuhan doi:https://doi.org/10.1080/22221751.2020.1719902

China NHCotPsRo Epidemic Prevention and Control of Dynamic. http://www.nhc.gov.cn/xcs/yqtb/202002/ac1e98495cb04d36b0d0a4e1e7fab545.shtml. Accessed 30 May 2020

Chinazzi M, Davis JT, Ajelli M, Gioannini C, Litvinova M, Merler S, y Piontti AP, Rossi L, Sun K, Viboud C (2020) The effect of travel restrictions on the spread of the 2019 novel coronavirus (2019-nCoV) outbreak. medRxiv

Choi KR, Heilemann MV, Fauer A, Mead M (2020) A second pandemic: mental health spillover from the novel coronavirus (COVID-19). J Am Psychiatr Nurses Assoc:1078390320919803. https://doi.org/10.1177/1078390320919803

clinicaltrials.gov (2020). https://clinicaltrials.gov/ct2/results?term=Drug&cond=COVID-19. Accessed 31 May 2020

Cohen J, Kupferschmidt K (2020) Strategies shift as coronavirus pandemic looms. American Association for the Advancement of Science

Connor P (2020) More than nine-in-ten people worldwide live in countries with travel restrictions amid COVID-19. Pew Research Center. https://www.pewresearch.org/fact-tank/2020/04/01/more-than-nine-in-ten-people-worldwide-live-in-countries-with-travel-restrictions-amid-covid-19/. Accessed 01 Apr 2020

Copyright © BloombergQuint(2020). https://www.bloombergquint.com/quicktakes/why-curbing-travel-is-no-cure-all-for-the-coronavirus-quicktake. Accessed 20 Apr 2020

Corman VM, Landt O, Kaiser M, Molenkamp R, Meijer A, Chu DK, Bleicker T, Brünink S, Schneider J, Schmidt ML (2020) Detection of 2019 novel coronavirus (2019-nCoV) by real-time RT-PCR. Eur Secur 25(3):2000045

Darbeheshti F, Rezaei N (2020) Genetic predisposition models to COVID-19 infection. Med Hypotheses 142:109818. https://doi.org/10.1016/j.mehy.2020.109818

Dowd JB, Andriano L, Brazel DM, Rotondi V, Block P, Ding X, Liu Y, Mills MC (2020) Demographic science aids in understanding the spread and fatality rates of COVID-19. Proc Natl Acad Sci:202004911. https://doi.org/10.1073/pnas.2004911117

Du L, He, Y Zhou Y, (2009) The spike protein of SARS-CoV – a target for vaccine and therapeutic development. https://doi.org/10.1038/nrmicro2090

Errett NA, Sauer LM, Rutkow L (2020) An integrative review of the limited evidence on international travel bans as an emerging infectious disease disaster control measure. J Emerg Manag (Weston, MA) 18(1):7–14

Eyigün CP YS, Gül C, (1998) A comparative trial of two surface subunit recombinant hepatitis B vaccines vs a surface and PreS subunit vaccine for immunization of healthy adults. https://doi.org/10.1046/j.1365-2893.1998.00106.x

Fathi N, Rezaei N (2020) Lymphopenia in COVID-19: therapeutic opportunities. Cell Biol Int 44(9):1792–1797. https://doi.org/10.1002/cbin.11403

Fauci AS, Lane HC, Redfield RR (2020) Covid-19 — navigating the uncharted. N Engl J Med 382(13):1268–1269. https://doi.org/10.1056/NEJMe2002387

Gilmore N, Orkin AJ, Duckett M, Grover SA (1989) International travel and AIDS. AIDS 3(1):S225–S230

Gouglas D, Thanh Le T Henderson K (2018) Estimating the cost of vaccine development against epidemic infectious diseases: a cost minimisation study. https://doi.org/10.1016/S2214-109X(18)30346-2

Gralinski LE (2020) Return of the coronavirus: 2019-nCoV. https://doi.org/10.3390/v12020135

Gross L (2017) The peace of Westphalia, 1648–1948. Am J Int Law 42(1):20–41. https://doi.org/10.2307/2193560

Hanaei S, Rezaei N (2020) COVID-19: developing from an outbreak to a pandemic. Arch Med Res. https://doi.org/10.1016/j.arcmed.2020.04.021

Hao F, Tan W, Jiang L, Zhang L, Zhao X, Zou Y, Hu Y, Luo X, Jiang X, McIntyre RS, Tran B, Sun J, Zhang Z, Ho R, Ho C, Tam W (2020) Do psychiatric patients experience more psychiatric symptoms during COVID-19 pandemic and lockdown? A case-control study with service and research implications for immunopsychiatry. Brain Behav Immun. https://doi.org/10.1016/j.bbi.2020.04.069

He X, Lau EHY, Wu P, Deng X, Wang J, Hao X, Lau YC, Wong JY, Guan Y, Tan X, Mo X, Chen Y, Liao B, Chen W, Hu F, Zhang Q, Zhong M, Wu Y, Zhao L, Zhang F, Cowling BJ, Li F, Leung GM (2020) Temporal dynamics in viral shedding and transmissibility of COVID-19. Nat Med. https://doi.org/10.1038/s41591-020-0869-5

Hershey AS (2017) History of international law since the peace of Westphalia. Am J Int Law 6(1):30–69. https://doi.org/10.2307/2187396

Hirshleifer J (1978) Competition, cooperation, and conflict in economics and biology. Am Econ Rev 68(2):238–243

India Tt (2020) COVID-19 cases to peak in India between June 21–28. https://www.google.com/amp/s/www.tribuneindia.com/news/health/covid-19-cases-to-peak-in-india-between-june-21-28-study-88895. Accessed 30 May 2020

Jabbari P, Rezaei N (2020) With risk of reinfection, is COVID-19 Here to stay? Disaster Med Public Health Prep:1. https://doi.org/10.1017/dmp.2020.274

Jabbari P, Jabbari F, Ebrahimi S, Rezaei N (2020) COVID-19: a chimera of two pandemics. Disaster Med Public Health Prep:1–2. https://doi.org/10.1017/dmp.2020.223

Jahanshahlu L, Rezaei N (2020a) Central Nervous System Involvement in COVID-19. Arch Med Res. https://doi.org/10.1016/j.arcmed.2020.05.016

Jahanshahlu L, Rezaei N (2020b) Monoclonal antibody as a potential anti-COVID-19. Biomed Pharmacother 129:110337. https://doi.org/10.1016/j.biopha.2020.110337

Jewell W (1857) Historical sketches of quarantine: Address, Delivered before the Philadelphia County Medical Society, January 28, 1857. TK and PG Collins, printers

Jiang S, He, Y Liu S (2005) SARS vaccine development. https://doi.org/10.3201/1107.050219

Jiang S, Du L Shi Z (2020) An emerging coronavirus causing pneumonia outbreak in Wuhan, China: calling for developing therapeutic and prophylactic strategies. https://doi.org/10.1080/22221751.2020.1723441

John RKS, King A, De Jong D, Bodie-Collins M, Squires SG, Tam TW (2005) Border screening for SARS. Emerg Infect Dis 11(1):6

Joseph J, Amon KWT (2008, December 16) Fear of foreigners: HIV-related restrictions on entry, stay, and residence. J Int AIDS So. https://doi.org/10.1186/1758-2652-11-8

Kafieh R, Arian R, Saeedizadeh N, Minaee S, Yadav SK, Vaezi A, Rezaei N, Javanmard SH (2020) COVID-19 in Iran: a deeper look into the future. medRxiv

Kawthar Mohamed (2020) Borderless collaboration is needed for COVID-1; a disease that knows no borders. Infect Control Hosp Epidemiol. https://doi.org/10.1017/ice.2020.162

Kilwein J (1995) Some historical comments on quarantine: part one. J Clin Pharm Ther 20(4):185–187

Kiselyova M (2020, January 28) Russia ramps up controls, shuts China border crossings over virus fears. REUTERS

Koyama S, Akbay EA, Li YY, Herter-Sprie GS, Buczkowski KA, Richards WG, Gandhi L, Redig AJ, Rodig SJ, Asahina H, Jones RE, Kulkarni MM, Kuraguchi M, Palakurthi S, Fecci PE, Johnson BE, Janne PA, Engelman JA, Gangadharan SP, Costa DB, Freeman GJ, Bueno R, Hodi FS, Dranoff G, Wong KK, Hammerman PS (2016) Adaptive resistance to therapeutic PD-1 blockade is associated with upregulation of alternative immune checkpoints. Nat Commun 7:10501. https://doi.org/10.1038/ncomms10501

Kozel TR, Burnham-Marusich AR (2017) Point-of-care testing for infectious diseases: past, present, and future. J Clin Microbiol 55(8):2313–2320

Ksiazek TG, D Erdman Goldsmith CS, et al (2003) A novel coronavirus associated with severe acute respiratory syndrome. https://doi.org/10.1056/NEJMoa030781

Lai C-C, Shih T-P, Ko W-C, Tang H-J, Hsueh P-R (2020) Severe acute respiratory syndrome coronavirus 2 (SARS-CoV-2) and coronavirus disease-2019 (COVID-19): the epidemic and the challenges. Int J Antimicrob Agents 55(3):105924–105924. https://doi.org/10.1016/j.ijantimicag.2020.105924

Leung K, Wu JT, Liu D, Leung GM (2020) First-wave COVID-19 transmissibility and severity in China outside Hubei after control measures, and second-wave scenario planning: a modelling impact assessment. Lancet 395:1382–1393

Li W, M Farzan Harrison SC (2005) Structure of SARS coronavirus spike receptor-binding domain complexed with receptor. https://doi.org/10.1126/science.1116480

Lipsitch M, Swerdlow DL, Finelli L (2020) Defining the epidemiology of Covid-19 — studies needed. N Engl J Med 382(13):1194–1196. https://doi.org/10.1056/NEJMp2002125

Lotfi M, Rezaei N (2020) SARS-CoV-2: a comprehensive review from pathogenicity of the virus to clinical consequences. J Med Virol. https://doi.org/10.1002/jmv.26123

Lotfi M, Hamblin MR, Rezaei N (2020) COVID-19: transmission, prevention, and potential therapeutic opportunities. Clin Chim Acta; Int J Clin Chem 508:254–266. https://doi.org/10.1016/j.cca.2020.05.044

Lu L, Liu Q Zhu Y, (2014) Structure-based discovery of Middle East respiratory syndrome coronavirus fusion inhibitor. https://doi.org/10.1038/ncomms4067

Lurie N, Saville M Hatchett R (2020) Developing Covid-19 vaccines at pandemic speed. https://doi.org/10.1056/NEJMp2005630

Mahase E, Kmietowicz Z (2020) Covid-19: doctors are told not to perform CPR on patients in cardiac arrest. BMJ 368:m1282

Mansourabadi AH, Sadeghalvad M, Mohammadi-Motlagh HR, Rezaei N (2020) The immune system as a target for therapy of SARS-CoV-2: a systematic review of the current immunotherapies for COVID-19. Life Sci 258:118185. https://doi.org/10.1016/j.lfs.2020.118185

McCaffery JM, Papandonatos GD, Faulconbridge LF, Erar B, Peter I, Wagenknecht LE, Pajewski NM, Anderson A, Wadden TA, Wing RR (2015) Genetic predictors of depressive symptoms in the look AHEAD trial. Psychosom Med 77(9):982–992. https://doi.org/10.1097/psy.0000000000000242

McKibbin WJ, Fernando R (2020) The global macroeconomic impacts of COVID-19: Seven scenarios

Mehta P, McAuley DF, Brown M, Sanchez E, Tattersall RS, Manson JJ (2020) COVID-19: consider cytokine storm syndromes and immunosuppression. Lancet (London, England) 395(10229):1033–1034. https://doi.org/10.1016/s0140-6736(20)30628-0

Mhlanga D, Ndhlovu E (2020) Socio-economic Implications of the COVID-19 pandemic on smallholder livelihoods in Zimbabwe

Mirbeyk M, Rezaei N (2020) The impact of COVID-19 on pregnancy and neonatal health: a systematic review

Moazzami B, Razavi-Khorasani N, Dooghaie Moghadam A, Farokhi E, Rezaei N (2020) COVID-19 and telemedicine: immediate action required for maintaining healthcare providers well-being. J Clin Virol 126:104345. https://doi.org/10.1016/j.jcv.2020.104345

Mohamed K, Yazdanpanah N, Saghazadeh A, Rezaei N (2020) Computational drug discovery and repurposing for the treatment of Covid-19: a systematic review. Available at SSRN 3583748

Momtazmanesh S, Ochs HD, Uddin LQ, Perc M, Routes JM, Vieira DN, Al-Herz W, Baris S, Prando C, Rosivall L, Abdul Latiff AH, Ulrichs T, Roudenok V, Aldave Becerra JC, Salunke DB, Goudouris E, Condino-Neto A, Stashchak A, Kryvenko O, Stashchak M, Bondarenko A, Rezaei N (2020) All together to fight COVID-19. Am J Trop Med Hyg 102(6):1181–1183. https://doi.org/10.4269/ajtmh.20-0281

Moradian N, Ochs HD, Sedikies C, Hamblin MR, Camargo CA Jr, Martinez JA, Biamonte JD, Abdollahi M, Torres PJ, Nieto JJ, Ogino S, Seymour JF, Abraham A, Cauda V, Gupta S, Ramakrishna S, Sellke FW, Sorooshian A, Wallace Hayes A, Martinez-Urbistondo M, Gupta M, Azadbakht L, Esmaillzadeh A, Kelishadi R, Esteghamati A, Emam-Djomeh Z, Majdzadeh R, Palit P, Badali H, Rao I, Saboury AA, Jagan Mohan Rao L, Ahmadieh H, Montazeri A, Fadini GP, Pauly D, Thomas S, Moosavi-Movahed AA, Aghamohammadi A, Behmanesh M, Rahimi-Movaghar V, Ghavami S, Mehran R, Uddin LQ, Von Herrath M, Mobasher B, Rezaei N (2020) The urgent need for integrated science to fight COVID-19 pandemic and beyond. J Transl Med 18(1):205. https://doi.org/10.1186/s12967-020-02364-2

Nacoti M, Ciocca A, Giupponi A, Brambillasca P, Lussana F, Pisano M, Goisis G, Bonacina D, Fazzi F, Naspro R, Longhi L (2020) At the epicenter of the Covid-19 pandemic and humanitarian crises in Italy: changing perspectives on preparation and mitigation. NEJM Catal Innov Care Deliv 1(2). https://doi.org/10.1056/CAT.20.0080

Nasab MG, Saghazadeh A, Rezaei N (2020) SARS-CoV-2-a tough opponent for the immune system. Arch Med Res. https://doi.org/10.1016/j.arcmed.2020.05.020

Newman LH (2020, March 11) No, a border wall won't stop coronavirus. Wired, news A Pakistan braced for COVID-19 peak in late June as daily death rate hits record high. https://www.arabnews.com/node/1667081/world. Accessed 30 May 2020

News B (2020, April). Coronavirus pandemic: Your Coronavirus Stories. https://www.bbc.com/news/coronavirus

Ni MY, Yang L, Leung CMC, Li N, Yao XI, Wang Y, Leung GM, Cowling BJ, Liao Q (2020) Mental health, risk factors, and social media use during the COVID-19 epidemic and cordon sanitaire among the community and health professionals in Wuhan, China: cross-sectional survey. JMIR Ment Health 7(5):e19009. https://doi.org/10.2196/19009

Niall PASJ (2002) Global mortality of the 1918-1920 "Spanish" influenza pandemic. Bull Hist Med 76(1):105–115. https://doi.org/10.1353/bhm.2002.0022

NICOLÁS ES (2020, February 28) EU experts: closing borders 'ineffective' for coronavirus. euobserver

Nowak MA (2006) Five rules for the evolution of cooperation. Science (New York, NY) 314(5805):1560–1563. https://doi.org/10.1126/science.1133755

Ohan NW, Heikkila JJ (1993) Reverse transcription-polymerase chain reaction: an overview of the technique and its applications. Biotechnol Adv 11(1):13–29

Organization WH (2020a) WHO Novel Coronavirus (COVID-19) Situation Board. World Health Organisation. https://covid19.who.int/. Accessed 14/04/2020 2020

Organization WH (2020b). https://www.who.int/emergencies/mers-cov/en/. Accessed 30 May 2020

Organization WH (2020c, April 17) Coronavirus disease (COVID-19) Pandemic. https://www.who.int/emergencies/diseases/novel-coronavirus-2019

Organization WH (2020d, February 29) Updated WHO recommendations for international traffic in relation to COVID-19 outbreak. https://www.who.int/news-room/articles-detail/updated-who-recommendations-for-international-traffic-in-relation-to-covid-19-outbreak

Organization WH (2020e) China shows COVID-19 responses must be tailored to the local context http://www.euro.who.int/en/health-topics/health-emergencies/coronavirus-covid-19/news/news/2020/4/china-shows-covid-19-responses-must-be-tailored-to-the-local-context Accessed 30 May 2020

Organization WH (2020f) Laboratory testing strategy recommendations for COVID-19: interim guidance, 22 March 2020. World Health Organization

organization Wt COVID-19 and world trade

Osiander A (2003) Sovereignty, international relations, and the Westphalian myth. Int Organ 55(2):251–287. https://doi.org/10.1162/00208180151140577

Ozamiz-Etxebarria N, Dosil-Santamaria M, Picaza-Gorrochategui M, Idoiaga-Mondragon N (2020) Stress, anxiety, and depression levels in the initial stage of the COVID-19 outbreak in a population sample in the northern Spain. Cad Saude Publica 36(4):e00054020. https://doi.org/10.1590/0102-311X00054020

Özdin S, Bayrak Özdin Ş (2020) Levels and predictors of anxiety, depression and health anxiety during COVID-19 pandemic in Turkish society: the importance of gender. Int J Soc Psychiatry:2076402092705. https://doi.org/10.1177/0020764020927051

Pashaei M, Rezaei N (2020) Immunotherapy for SARS-CoV-2: potential opportunities. Expert Opin Biol Ther:1–5. https://doi.org/10.1080/14712598.2020.1807933

Patel T., https://www.bloombergquint.com/quicktakes/why-curbing-travel-is-no-cure-all-for-the-coronavirus-quicktake Rma, BloombergQuint C (2020) The Costs and Benefits of Limiting Movement to Fight Coronavirus

Pourahmad R, Moazzami B, Rezaei N (2020) Efficacy of plasmapheresis and immunoglobulin replacement therapy (IVIG) on patients with COVID-19. SN Compr Clin Med 1–5. https://doi.org/10.1007/s42399-020-00438-2

Prof Mandeep R Mehra (2020) Hydroxychloroquine or chloroquine with or without a macrolide for treatment of COVID-19: a multinational registry analysis

Qiu J, Shen B, Zhao M, Wang Z, Xie B, Xu Y (2020) A nationwide survey of psychological distress among Chinese people in the COVID-19 epidemic: implications and policy recommendations. Gen Psychiatr 33(2):e100213. https://doi.org/10.1136/gpsych-2020-100213

Qun Li (2020) Early transmission dynamics in Wuhan, China, of novel coronavirus-infected pneumonia. https://doi.org/10.1056/NEJMoa2001316

Rabiee N, Rabiee M, Bagherzadeh M, Rezaei N (2020) COVID-19 and picotechnology: potential opportunities. Med Hypotheses 144:109917. https://doi.org/10.1016/j.mehy.2020.109917

Rachman G (2020) Coronavirus: how the outbreak is changing global politics. https://www.ft.com/content/fd8bfd8a-5a25-11ea-abe5-8e03987b7b20. Accessed 20 Apr 2020

Rajgor DD, Lee MH, Archuleta S, Bagdasarian N, Quek SC (2020) The many estimates of the COVID-19 case fatality rate. Lancet Infect Dis. https://doi.org/10.1016/s1473-3099(20)30244-9

Read more at: https://www.bloombergquint.com/quicktakes/why-curbing-travel-is-no-cure-all-for-the-coronavirus-quicktake

Regla-Nava JA, Jose L Nieto-Torres Jimenez-Guardeño JM (2015) Severe acute respiratory syndrome coronaviruses with mutations in the E protein are attenuated and promising vaccine candidates. https://doi.org/10.1128/JVI.03566-14

Review CB China's economic response to COVID-19 has helped, for now. https://review.

chicagobooth.edu/economics/2020/article/china-s-economic-response-covid-19-has-helped-now
Review Hb (2020) What Coronavirus Could Mean for the Global Economy https://hbr.org/2020/03/what--coronavirus-could-mean-for-the-global-economy Accessed 30 May 2020
Rezaei N (2020a) COVID-19 affects healthy pediatricians more than pediatric patients. Infect Control Hosp Epidemiol 1. https://doi.org/10.1017/ice.2020.139
Rezaei N (2020b) COVID-19 and medical biotechnology. Avicenna J Med Biotechnol 12(3):139
Rokni M, Hamblin MR, Rezaei N (2020) Cytokines and COVID-19: friends or foes? Hum Vaccin Immunother 16(10):2363–2365
Rutkow L (2020, February 13) Effectiveness of travel bans – readily used during infectious disease outbreaks – mostly unknown. Science Daily
Rzymski P, Nowicki M, Mullin GE, Abraham A, Rodríguez-Román E, Petzold MB, Bendau A, Sahu KK, Ather A, Naviaux A-F (2020) Quantity does not equal quality: scientific principles cannot be sacrificed. Int Immunopharmacol 86:106711
Saccone G, Florio A, Aiello F, Venturella R, De Angelis MC, Locci M, Bifulco G, Zullo F, Di Spiezio Sardo A (2020) Psychological impact of COVID-19 in pregnant women. Am J Obstet Gynecol doi:https://doi.org/10.1016/j.ajog.2020.05.003
Saghazadeh A, Rezaei N (2020a) Immune-epidemiological parameters of the novel coronavirus - a perspective. Expert Rev Clin Immunol 16(5):465–470. https://doi.org/10.1080/1744666x.2020.1750954
Saghazadeh A, Rezaei N (2020b) Towards treatment planning of COVID-19: rationale and hypothesis for the use of multiple immunosuppressive agents: anti-antibodies, immunoglobulins, and corticosteroids. Int Immunopharmacol 84:106560. https://doi.org/10.1016/j.intimp.2020.106560
Sahu KK, Siddiqui AD, Rezaei N, Cerny J (2020) Challenges for management of immune thrombocytopenia during COVID-19 pandemic. J Med Virol. https://doi.org/10.1002/jmv.26251
Saleki K, Banazadeh M, Saghazadeh A, Rezaei N (2020) The involvement of the central nervous system in patients with COVID-19. Rev Neurosci 31(4):453–456. https://doi.org/10.1515/revneuro-2020-0026
Samaan G, Spencer J, Roberts L, Patel M (2004) Border screening for SARS in Australia: what has been learnt? Med J Aust 180(5):220–223
Sang-Hun C (2020, January 21) North Korea Bans Foreign tourists over coronavirus, tour operator says … Tourists New York Times
Schmidt S (2011) To order the minds of scholars: the discourse of the peace of Westphalia in international relations Literature1. Int Stud Q 55(3):601–623. https://doi.org/10.1111/j.1468-2478.2011.00667.x
Schoeman D Fielding BC (2019) Coronavirus envelope protein: current knowledge. https://doi.org/10.1186/s12985-019-1182-0
Shamshirian D, Rezaei N (2020) Cardiovascular diseases burden in COVID-19: systematic review and meta-analysis
Shang W (2020) The outbreak of SARS-CoV-2 pneumonia calls for viral vaccines. NPJ Vaccines. https://doi.org/10.1038/s41541-020-0170-0
Sharifkashani S, Bafrani MA, Khaboushan AS, Pirzadeh M, Kheirandish A, Yavarpour Bali H, Hessami A, Saghazadeh A, Rezaei N (2020) Angiotensin-converting enzyme 2 (ACE2) receptor and SARS-CoV-2: potential therapeutic targeting. Eur J Pharmacol 884:173455. https://doi.org/10.1016/j.ejphar.2020.173455
Sheridan C (2020) Fast, portable tests come online to curb coronavirus pandemic. Nat Biotechnol. https://doi.org/10.1038/d41587-020-00010-2
source: Centers for Disease Control and Prevention NCfIaRDN (2018) 1918 Pandemic Influenza: Three Waves. source: Centers for Disease Control and Prevention, National Center for Immunization and Respiratory Diseases (NCIRD). https://www.cdc.gov/flu/pandemic-resources/1918-commemoration/three-waves.htm. Accessed April 11, 2020 2018
Stuard SM (1992) A state of deference: Ragusa (Dubrovnik) in the medieval centuries. University of Pennsylvania Press
Susanne R (2018) New vaccine technologies to combat outbreak situations. Front Immunol 9:1963. https://doi.org/10.3389/fimmu.2018.01963
Taylor SE, Sirois FM (2012) Health psychology. McGraw-Hill Ryerson, Toronto
Thanh Le T, Andreadakis Z, Kumar A (2020) The COVID-19 vaccine development landscape. https://doi.org/10.1038/d41573-020-00073-5
Travellers from China to be denied entry into Australia (2020, February 2)The Australian
Tsallis C (2020) Predicting COVID-19 peaks around the world. Front Phys. https://doi.org/10.3389/fphy.2020.00217
Wan Y, J Shang Graham R, (2020) Receptor recognition by novel coronavirus from Wuhan: an analysis based on decade-long structural studies of SARS. https://doi.org/10.1128/JVI.00127-20
Wang N, Shang J, Jiang S et al (2020a) Subunit vaccines against emerging pathogenic human coronaviruses. Front Microbiol 11:298. https://doi.org/10.3389/fmicb.2020.0029
Wang Y, Di Y, Ye J, Wei W (2020b) Study on the public psychological states and its related factors during the outbreak of coronavirus disease 2019 (COVID-19) in some regions of China. Psychol Health Med:1–10. https://doi.org/10.1080/13548506.2020.1746817
Wells CR, Sah P, Moghadas SM, Pandey A, Shoukat A, Wang Y, Wang Z, Meyers LA, Singer BH, Galvani AP (2020) Impact of international travel and border control measures on the global spread of the novel 2019 coronavirus outbreak. Proc Natl Acad Sci 117(13):7504–7509

West SA, Griffin AS, Gardner A (2007) Evolutionary explanations for cooperation. Curr Biol 17(16):R661–R672. https://doi.org/10.1016/j.cub.2007.06.004

WHO/Europe Coronavirus disease (COVID-19) outbreak – Mental health and psychological resilience during the COVID-19 pandemic. http://www.euro.who.int/en/health-topics/health-emergencies/coronavirus-covid-19/news/news/2020/3/mental-health-and-psychological-resilience-during-the-covid-19-pandemic. Accessed 30 May 2020

Wilder-Smith A, Goh KT, Paton NI (2003) Experience of severe acute respiratory syndrome in Singapore: importation of cases, and defense strategies at the airport. J Travel Med 10(5):259–262

Worldometer (2020). https://www.worldometers.info/coronavirus/country/germany/. Accessed 30 May 2020

Worldometers (2020) Coronavirus. Worldometers. https://www.worldometers.info/coronavirus/. Accessed 18 Apr 2020

Wu C, Chen X, Cai Y, Xia J, Zhou X, Xu S, Huang H, Zhang L, Zhou X, Du C, Zhang Y, Song J, Wang S, Chao Y, Yang Z, Xu J, Zhou X, Chen D, Xiong W, Xu L, Zhou F, Jiang J, Bai C, Zheng J, Song Y (2020) Risk factors associated with acute respiratory distress syndrome and death in patients with coronavirus disease 2019 pneumonia in Wuhan, China. JAMA Intern Med. https://doi.org/10.1001/jamainternmed.2020.0994

Xu H, Yang K, Yang G, Yang L, Mi Y, Cui X, Yang M, Wang D, Wang Y (2020) Explore the optimal Resolvent of medical needs and mental health for patients with lung Cancer during epidemic novel coronavirus pneumonia. Zhongguo Fei Ai Za Zhi 23(4):247–254. https://doi.org/10.3779/j.issn.1009-3419.2020.101.18

Yazdanpanah F, Hamblin MR, Rezaei N (2020a) The immune system and COVID-19: friend or foe? Life Sci 256:117900. https://doi.org/10.1016/j.lfs.2020.117900

Yazdanpanah N, Saghazadeh A, Rezaei N (2020b) Anosmia: a missing link in the neuroimmunology of coronavirus disease 2019 (COVID-19). Rev Neurosci. https://doi.org/10.1515/revneuro-2020-0039

Yeming Wang (2020) Remdismivir in adults with severe COVID-19: a randomised, double-blind, placebo-controlled, multicentre trial. https://www.thelancet.com/journals/lancet/article/PIIS0140-6736(20)31022-9/fulltext. Accessed 29 May 2020

Yousefzadegan S, Rezaei N (2020) Case report: death due to COVID-19 in three brothers. Am J Trop Med Hyg 102(6):1203–1204. https://doi.org/10.4269/ajtmh.20-0240

Zaki AM, Boheemen, S Bestebroer TM (2012) Isolation of a novel coronavirus from a man with pneumonia in Saudi Arabia. https://doi.org/10.1056/NEJMoa1211721

Zhang C, Wu Z, Li J-W, Zhao H, Wang G-Q (2020a) The cytokine release syndrome (CRS) of severe COVID-19 and Interleukin-6 receptor (IL-6R) antagonist tocilizumab may be the key to reduce the mortality. Int J Antimicrob Agents:105954. https://doi.org/10.1016/j.ijantimicag.2020.105954

Zhang C, Wu Z, Li JW, Zhao H, Wang GQ (2020b) The cytokine release syndrome (CRS) of severe COVID-19 and Interleukin-6 receptor (IL-6R) antagonist Tocilizumab may be the key to reduce the mortality. Int J Antimicrob Agents 105954. https://doi.org/10.1016/j.ijantimicag.2020.105954

Zhang WR, Wang K, Yin L, Zhao WF, Xue Q, Peng M, Min BQ, Tian Q, Leng HX, Du JL, Chang H, Yang Y, Li W, Shangguan FF, Yan TY, Dong HQ, Han Y, Wang YP, Cosci F, Wang HX (2020c) Mental health and psychosocial problems of medical health workers during the COVID-19 epidemic in China. Psychother Psychosom:1–9. https://doi.org/10.1159/000507639

Zhou SJ, Zhang LG, Wang LL, Guo ZC, Wang JQ, Chen JC, Liu M, Chen X, Chen JX (2020) Prevalence and socio-demographic correlates of psychological health problems in Chinese adolescents during the outbreak of COVID-19. Eur Child Adolesc Psychiatry. https://doi.org/10.1007/s00787-020-01541-4

Zulkifli NA, Sivapatham S, Guan NC (2020) Brief psychotic disorder in relation to coronavirus, COVID-19 outbreaks: a case report. Malays J Psychiatry 29(1):67–72

Socialization During the COVID-19 Pandemic: The Role of Social and Scientific Networks During Social Distancing

51

Sara Momtazmanesh, Noosha Samieefar, Lucina Q. Uddin, Timo Ulrichs, Roya Kelishadi, Vasili Roudenok, Elif Karakoc-Aydiner, Deepak B. Salunke, Jan L. Nouwen, Juan Carlos Aldave Becerra, Duarte Nuno Vieira, Ekaterini Goudouris, Mahnaz Jamee, Morteza Abdullatif Khafaie, Morteza Shamsizadeh, Mohammad Rasoul Golabchi, Alireza Samimiat, Donya Doostkamel, Alireza Afshar, Mohammad Amin Khazeei Tabari, Melika Lotfi, Reza Yari Boroujeni, Niloofar Rambod, Anzhela Stashchak, Alla Volokha, Dainius Pavalkis, André Pereira, Amir Hamzah Abdul Latiff, Rauf Baylarov, Bagher Amirheidari, Mojtaba Hedayati Ch, Antonio Condino-Neto, and Nima Rezaei

Abstract

In the COVID-19 era, while we are encouraged to be physically far away from each other, social and scientific networking is needed more than ever. The dire consequences of social distancing can be diminished by social networking. Social media, a quintessential component of social networking, facilitates the dissemination of reliable information and fighting against misinformation by health

S. Momtazmanesh · N. Rezaei (✉)
Universal Scientific Education and Research Network (USERN), Tehran, Iran

Tehran University of Medical Sciences, Tehran, Iran
e-mail: rezaei_nima@tums.ac.ir

N. Samieefar · R. Y. Boroujeni
Student Research Committee, School of Medicine, Shahid Beheshti University of Medical Sciences, Tehran, Iran

USERN Office, School of Advanced Technologies in Medicine, Shahid Beheshti University of Medical Sciences, Tehran, Iran

L. Q. Uddin
Department of Psychology, University of Miami, Coral Gables, FL, USA

Universal Scientific Education and Research Network (USERN), Coral Gables, FL, USA

N. Rezaei (ed.), *Coronavirus Disease - COVID-19*, Advances in Experimental Medicine and Biology 1318, https://doi.org/10.1007/978-3-030-63761-3_51

authorities. Distance learning, telemedicine, and telehealth are among the most prominent applications of networking during this pandemic. Additionally, the COVID-19 pandemic highlights the importance of collaborative scientific efforts. In this chapter, we summarize the advantages of harnessing both social and scientific networking in minimizing the harms of this pandemic. We also discuss the extra collaborative measures we can take in our fight against COVID-19, particularly in the scientific field.

Keywords

COVID-19 · Pandemic · Scientific network · Social distancing · Social network · Socialization

T. Ulrichs
Institute for Research in International Assistance, Akkon University for Human Sciences, Berlin, Germany

Universal Scientific Education and Research Network (USERN), Berlin, Germany

R. Kelishadi
Child Growth and Development Research Center, Research Institute for Primordial Prevention of Non-Communicable Disease, Isfahan University of Medical Sciences, Isfahan, Iran

USERN Office, Research Institute for Primordial Prevention of Non-Communicable Disease, Isfahan University of Medical Sciences, Isfahan, Iran

V. Roudenok
Belarusian State Medical University, Minsk, Belarus

Universal Scientific Education and Research Network (USERN), Minsk, Belarus

E. Karakoc-Aydiner
Division of Pediatric Allergy/Immunology, Marmara University Hospital, Marmara University, Istanbul, Turkey

Universal Scientific Education and Research Network (USERN), Istanbul, Turkey

D. B. Salunke
National Interdisciplinary Centre of Vaccine, Immunotherapeutics and Antimicrobials, Department of Chemistry, Panjab University, Chandigarh, India

Universal Scientific Education and Research Network (USERN), Chandigarh, India

J. L. Nouwen
Erasmus MC, University Medical Center, Rotterdam, The Netherlands

Universal Scientific Education and Research Network (USERN), Rotterdam, The Netherlands

J. C. A. Becerra
Hospital National Edgardo Rebagliati Martins, Lima, Peru

Universal Scientific Education and Research Network (USERN), Lima, Peru

D. N. Vieira
Institute of Legal Medicine and Institute of Bioethics, University of Coimbra, Faculty of Medicine, Portugal, Coimbra, Portugal

Universal Scientific Education and Research Network (USERN), Coimbra, Portugal

E. Goudouris
Pediatrics Department, Medical School, Federal University of Rio de Janeiro, Rio de Janeiro, Brazil

Universal Scientific Education and Research Network (USERN), Rio de Janeiro, Brazil

M. Jamee
Student Research Committee, Alborz University of Medical Sciences, Karaj, Iran

USERN Office, Alborz University of Medical Sciences, Karaj, Iran

M. A. Khafaie
Social Determinants of Health Research Center, Ahvaz Jundishapur University of Medical Sciences, Ahvaz, Iran

USERN Office, Ahvaz Jundishapur University of Medical Sciences, Ahvaz, Iran

M. Shamsizadeh
USERN Office, Hamadan University of Medical Sciences, Hamadan, Iran

Department of Medical Surgical Nursing, School of Nursing and Midwifery, Hamadan University of Medical Sciences, Hamadan, Iran

M. R. Golabchi · A. Samimiat
USERN Office, Isfahan University of Medical Science, Isfahan, Iran

D. Doostkamel
Faculty of pharmacy, Ardabil University of Medical Sciences, Ardabil, Iran

USERN Office, Ardabil University of Medical Sciences, Ardabil, Iran

51.1 Introduction

Social distancing is an effective measure in slowing down the spread of pandemics. The benefits of social distancing in combating COVID-19 are indisputable. Many countries implemented these measures to limit the spread of the virus, to protect vulnerable groups within the population, and to avoid the breakdown of their healthcare system by an overwhelming number of patients needing critical care (Mahase 2020). These advantages, however, come at a price. Our experience from COVID-19 and other outbreaks has shown us that social distancing and isolation can lead to several dire consequences. A tangible example of these repercussions is increased vulnerability to mental illnesses, like depression and anxiety (Hawryluck et al. 2004; Jeong et al. 2016; Holmes et al. 2020). The measures on social distancing during the COVID-19 pandemic are also contributing to psychological problems (Qiu et al. 2020; Venkatesh and Edirappuli 2020). The main components causing these repercussions included worries or concerns about one's own health and loved one's health, limiting visits from loved ones, and financial concerns (Cosic et al. 2020; Lai et al. 2020). In addition to mental illnesses, patients with other chronic or acute diseases may also be treated insufficiently or may be diagnosed late because of the effect of the COVID-19 pandemic on screening programs and referral of patients to clinicians or the emergency units (Jones et al.; Rosenbaum 2020).

The COVID-19 pandemic is most certainly not our first experience with physical and social distancing. Before the recent pandemic, a similar

A. Afshar
USERN Office, The Persian Gulf Marine Biotechnology Research Center, The Persian Gulf Biomedical Sciences Research Institute, Bushehr University of Medical Sciences, Bushehr, Iran

M. A. K. Tabari
USERN Office, Mazandaran University of Medical Sciences, Sari, Iran

Student Research Committee, Mazandaran University of Medical Sciences, Sari, Iran

M. Lotfi
USERN Office, Zanjan University of Medical Sciences, Zanjan, Iran

N. Rambod
USERN Office, Islamic Azad University Medicine Faculty, Mashhad, Iran

A. Stashchak
Kharkiv National Medical University, Kharkiv, Ukraine

Universal Scientific Education and Research Network (USERN), Kharkiv, Ukraine

A. Volokha
Pediatric Infectious Disease and Pediatric Immunology Department, Shupyk National Medical Academy for Postgraduate Education, Kiev, Ukraine

Universal Scientific Education and Research Network (USERN), Kiev, Ukraine

D. Pavalkis
Rector Medical University, Astana, Kazakhstan

Universal Scientific Education and Research Network (USERN), Astana, Kazakhstan

A. Pereira
Universal Scientific Education and Research Network (USERN), Coimbra, Portugal

University of Coimbra, Coimbra, Portugal

A. H. A. Latiff
Allergy and Immunology Centre, Pantai Hospital Kuala Lumpur, Kuala Lumpur, Malaysia

Universal Scientific Education and Research Network (USERN), Kuala Lumpur, Malaysia

R. Baylarov
Azerbaijan Medical University, Baku, Azerbaijan

Universal Scientific Education and Research Network (USERN), Baku, Azerbaijan

B. Amirheidari
Herbal and Traditional Medicines Research Center, Kerman University of Medical Sciences, Kerman, Iran

USERN Office, Kerman University of Medical Sciences, Kerman, Iran

M. H. Ch
Faculty of Medicine, Department of Microbiology, Guilan University of Medical Sciences, Rasht, Guilan, Iran

USERN Office, Guilan University of Medical Sciences, Rasht, Guilan, Iran

A. Condino-Neto
Department of Immunology, Institute of Biomedical Sciences, University of São Paulo, São Paulo, Brazil

Universal Scientific Education and Research Network (USERN), São Paulo, Brazil outbreak that prompted physical distancing on a large scale occurred more than a century ago, when the 1918 influenza pandemic (also called as the Spanish flu) struck the globe. It contaminated nearly one-third of the world's population and resulted in almost 50 million deaths. Even during the 1918 influenza pandemic, social distancing showed promising results (Canales 2020; Aimone 2010). At that time, there was no hope for any vaccine to end the pandemic, and the knowledge about immunity and strategies to fight the virus was somewhat limited. Thus, the population had to rely on distancing as the only measure and had no information on how long this measure would last. Not only was the therapeutic armamentarium limited, but back then, social interactions and exchange of information were mainly only feasible via person-to-person communications, mail, and phone.

To date, we have paved a long path from where we were in 1918. In our era, physical isolation does not and should not mean emotional and social isolation. In the twenty-first century, we have many useful tools for staying connected at any time and from any location. The internet and media have made the dissemination of information extremely easy and rapid. Millions of people use various platforms categorized as social media, which allow the rapid exchange of data. Additionally, we have gained many experiences from conducting collaborative scientific efforts and scientific networking.

51.2 Social Networking: What Is the Role of Social Media?

Social media is a crucial part of social networking, which is defined as "communication with people who share your interests using a website or other service on the internet" (Oxford 2020). With more than three billion regular users spending a considerable amount of time daily on social media, these platforms have reached a significant penetration all around the world (Statista 2019). We can potentially employ mass media to tackle many of our challenges during the COVID-19 pandemic.

Providing citizens with reliable information is a vital tool in tackling COVID-19. With its far-reaching influence, social media can be harnessed effectively for achieving this goal (Merchant and Lurie 2020). Many reliable sources, including the World Health Organization (WHO) and the Centers for Disease Control and Prevention (CDC), are using mass media to convey critical information to citizens. Social media can distribute trustworthy and critical recommendations among their users, such as the importance of handwashing and staying home. Many platforms have taken significant steps toward this aim. For instance, Facebook and Twitter have partnered with reliable sources to disseminate reliable information and have facilitated sharing WHO and CDC recommendations (Facebook 2020; Twitter 2020). Similarly, social scientific platforms such as Google Scholar and ResearchGate, by designating a specific section to the research related to COVID-19, have made it easier for clinicians and researchers to access reliable data (ResearchGate 2020; Busvine 2020).

While social media can rapidly disseminate useful information and guidelines, we cannot overlook the fact that it can also provide a tool for knowingly spreading misinformation, which can have a detrimental effect in our fight against COVID-19 (Cuan-Baltazar et al. 2020). These platforms may doubtlessly limit this flood of misinformation by imposing appropriate regulations (Limaye et al.). As WHO Director-General Tedros Adhanom Ghebreyesus has said, "We're not just fighting an epidemic; we're fighting an infodemic." WHO has launched the WHO Information Network for Epidemics (EPI-WIN) to ensure the dissemination of evidence-based information and avoid misinformation (Zarocostas 2020). Another excellent example of the taken actions in avoiding misinformation is the joint industry statement, which was announced by Facebook, Google, LinkedIn, Microsoft, Reddit, Twitter, and YouTube in March 2020 to combat misinformation (Statt 2020). Nevertheless, it is still challenging to reduce the spread of false information; therefore, further attempts and regulations are necessary.

Prominently, social networking through mass media can also play a critical role in maintaining our mental health during this challenge (Galea et al. 2020). Many countries, like Canada and the United Kingdom, or organizations have provided online counseling instead of in-person sessions to help both the public and healthcare workers in coping with the challenges the COVID-19 pandemic has created for them (Richards and Vigano 2013). Additionally, social media enables us to stay connected with our family, friends, and colleagues while we stay at home (Panahi et al. 2016). Without a doubt, maintaining social connections is an essential need of human beings.

Last but not least, while in-person classes and training had to shut down due to social distancing temporarily, online social networking played a significant role in promoting distance learning via numerous international and national resources (World_Bank 2020). In response to COVID-19, E-learning websites like Coursera, Udacity, and edX provided their users with many free courses, especially courses related to global and public health, social sciences, and courses for students, helping them to advance their knowledge and skills during this challenging time (Coursera 2020; edX 2020; Lager 2020). Not only do these classes help the public to become more skillful, but they can also help in preparing medical professionals and can be used in medical education (Walsh et al. 2018; Liang et al. 2020).

51.3 Networking With and Within the Healthcare System: How Does It Change Patient Care?

The role of telemedicine in managing the COVID-19 pandemic is undeniable when the disease can widely affect the consumers of healthcare as well as the healthcare providers (Rezaei 2020a). Telehealth and telemedicine are some of the other merits of social networking in our fight against COVID-19 (Latifi and Doarn 2020). In the COVID-19 era, the public strives to minimize unnecessary visits to hospitals and stay at home as much as possible. Using social networks can help us greatly in taking care of not only potential COVID-19 patients but also individuals who would typically visit hospitals or clinics, but because of the circumstances, preferably not to do so. Several fields, including ophthalmology, surgery, and urology, have implemented telemedicine to take care of patients during social distancing (Hakim et al. 2020; Saleem et al. 2020; Boehm et al. 2020). Online self-assessment tools are examples of implications of telehealth in managing the COVID-19 pandemic. For instance, should someone suspect that they might be infected with COVID-19, they can use online self-assessment tools instead of referring to a clinic or hospital. These tools, which are provided by governmental authorities, analyze the symptoms and possible exposure of the person to the virus and guide the individual on taking the next appropriate steps (Rabiee et al. 2020; Government-of-india 2020). Such websites and applications can lower unnecessary referrals of citizens to clinics and hospitals and can be immensely helpful for healthcare providers. Contact tracing apps are another example of using the opportunities networking provides us. These applications, however, are in their initial stages, can help authorities in detecting exposed subjects. They notify people who were in contact with COVID-19 patients and ask them to quarantine themselves (Nature 2020; Zastrow 2020). In addition to reducing unnecessary hospital and clinic visits, detecting potential COVID patients, and protecting patients, social networking can also be implemented for inpatient management. For example, an infectious disease specialist might not need to examine their patient in person to give a consult. Should we use only a tablet to establish a virtual visit, we can lower the risk of infection for both the clinician and the patients and also save on usage of personal protective equipment (PPE) (Hollander and Carr 2020; Moazzami et al. 2020). Undoubtedly, telemedicine and telehealth are still growing fields needing much further research and collaborative global efforts (Ohannessian et al. 2020).

Additionally, networking can help us immensely in overcoming the shortage of PPE, testing kits, and other necessary equipment. The endeavors which have been taken by the

WHO are examples of this. Since the beginning of the pandemic, the WHO has shared COVID-19 testing kits and PPE with more than 120 countries. The exchange platform founded by Resilinc and Premier Inc., in collaboration with Stanford Medicine, is another example of such efforts taking place at the national level. This platform matches organizations needing certain supplies with peer providers who can provide them.

51.4 Scientific Networking: Can Collaborative Scientific Efforts Help Us?

In the scientific field, networking and collaboration are considered critical elements in conducting research. Notably, it has been shown that multicenter and multinational papers commonly reach higher cite scores and impact (Chinchilla-Rodriguez et al. 2019; Breugelmans et al. 2018). Our experience with the recent epidemics or pandemics, including severe acute respiratory syndrome (SARS), H1N1 influenza, Middle East Respiratory Syndrome (MERS), and, lastly, Ebola, showed us that international collaboration is indispensable for preparedness to combat these challenges (Jabbari et al. 2020).

The COVID-19 pandemic highlights the importance of collaborative scientific efforts. A quintessential example of scientific collaborations is the rapidly advancing knowledge of COVID-19. For instance, the rapid sharing of the genomic sequence of COVID-19 paved the road for investigating this virus by many researchers all around the world at the same time (Wang et al. 2020). When it comes to the COVID-19 pandemic, we face an extremely rapid pace of publishing data, which is not comparable with our previous epidemic and pandemic experiences. As of the end of May 2020, nearly 14,000 articles were registered in PubMed sharing clinical and epidemiological features of COVID-19, in vitro and in vivo investigations aimed to find an effective treatment and potential vaccines. This immense sharing of data, which is commonly open access, eases the path for other researchers to investigate this still-evolving challenge further.

While countries have closed their borders, in the scientific field, researchers are working without borders now more than ever. Undoubtedly, conducting multiple simultaneous investigations in different laboratory settings and with widely different samples can accelerate the process of finding a solution for this challenge (Momtazmanesh et al. 2020). WHO has made remarkable endeavors to promote collaborative research. In April 2020, it publically announced the initiation of global collaboration for finding an effective vaccine against the virus (Public 2020). Another example of such actions is the launch of the SOLIDARITY trial. This trial aims to find an effective treatment for COVID-19 and is being conducted in at least ten countries (WHO 2020).

Doubtlessly, in addition to the research activities conducted to find potential vaccines and treatments, both on the national and international scale, regular and effective transfer of knowledge and supplies can be extremely helpful in fighting this virus not only at this stage but also in the future (Kafieh et al. 2020). This transfer would be particularly applicable to healthcare providers who are on the frontline in the fight against COVID-19 as well as to the educational programs of the universities for training medical students who are future physicians (Mian and Khan 2020). For this purpose, other platforms can be launched that run in parallel with the rapid publishing of open-access scientific papers.

51.5 Scientific Networking: What More Can We Do?

Despite all of our success, as a united body, we can take many extra collaborative scientific measures in our fight against COVID-19 yet. Conducting collaborative genetic studies in various ethnicities to detect genetic predispositions can be an example of such efforts. The need for such studies is justified by the role of the host immune system in the pathogenesis of COVID-19-related multisystem involvement (Saghazadeh and Rezaei 2020a; Rokni et al. 2020; Bahrami et al. 2020; Yazdanpanah et al. 2020a; Mansourabadi et al. 2020; Pashaei and Rezaei 2020; Fathi and Rezaei 2020; Babaha and Rezaei 2020; Lotfi and Rezaei 2020; Nasab et al. 2020; Saghazadeh and Rezaei 2020b; Pourahmad et al. 2020; Ahanchian et al. 2020; Yazdanpanah et al. 2020b; Jahanshahlu and Rezaei 2020a; Saleki et al. 2020; Shamshirian and Rezaei 2020), reports of severe forms of COVID-19 within some families (Yousefzadegan and Rezaei 2020), and genetic algorithm proposals for modeling COVID-19 (Darbeheshti and Rezaei 2020). Moreover, international cooperation can ease multi-site randomized clinical trials aimed to explore effective drugs, whether already available old drugs or new drugs (Mohamed et al. 2020b). These collaborations can take giant steps in developing potential treatments (Lotfi et al. 2020; Rezaei 2020b) by bringing industry, academia, and public health entities together. Not only will these investigations contribute to finding potential treatment options, but they also can help us to deal with challenges (Sahu et al. 2020) in developing and testing potential vaccines and rapid and reliable tools for diagnosis of a new infection and its differentiation from a re-infection, which especially need to happen now (Basiri et al. 2020a; Sharifkashani et al. 2020; Jahanshahlu and Rezaei 2020b; Basiri et al. 2020b; Jabbari and Rezaei 2020).

Universal Scientific Education and Research Network (USERN), founded in 2016 to promote borderless science (Rezaei 2016, 2017, 2018), is among the active organizations during the COVID-19 pandemic. As a united body, members of this organization, from different scientific fields and more than 15 countries, have announced their support for international cooperation and collaboration in combating this challenge (Mohamed et al. 2020a; Momtazmanesh et al. 2020; Rahmani 2020; Moradian et al. 2020; Rzymski et al. 2020).

To summarize, we can harness networking in three significant fields: education, research, and advocacy of the healthcare system in terms of prevention and treatment (Fig. 51.1). Social media can help us in the dissemination of reliable information and fighting against misinformation, which plays a significant role in preventing the spread of the virus. Networking can also be effectively utilized in improving healthcare and university medical education. We can use it in all the forms of education, including distance learning for the empowerment of the public and professionals, making them more prepared for the fight

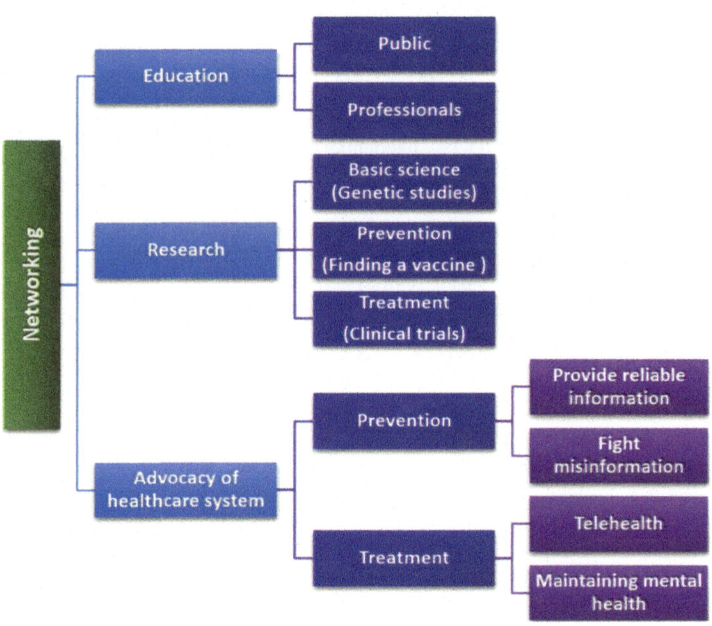

Fig. 51.1 The roles of networking in COVID-19 pandemic

against COVID-19. Its substantial role in research to find potential vaccines, treatments, and discovering genomic variants causing severe forms of the disease is undeniable.

51.6 Conclusion

To conclude, in the COVID-19 era, while we are encouraged to be physically far away from each other, we need social and scientific networking more than ever. Both social and scientific networking provide us the advantage of minimizing the harms of this pandemic and becoming more durable and more prepared for fighting it. Distance learning, telemedicine, and telehealth are among the most prominent applications of networking during this pandemic. As we more effectively harness the opportunities which are provided by networking, we become more influential in the fight against COVID-19 and similar challenges.

References

Ahanchian H, Moazzen N, Faroughi MSD, Khalighi N, Khoshkhui M, Aelami MH, Haghi NSM, Rezaei N (2020) COVID-19 in a child with primary specific antibody deficiency

Aimone F (2010) The 1918 influenza epidemic in New York City: a review of the public health response. Public Health Rep 125(Suppl 3):71–79. https://doi.org/10.1177/00333549101250S310

Babaha F, Rezaei N (2020) Primary immunodeficiency diseases in COVID-19 pandemic: a predisposing or protective factor? Am J Med Sci. https://doi.org/10.1016/j.amjms.2020.07.027

Bahrami A, Vafapour M, Moazzami B, Rezaei N (2020) Hyperinflammatory shock related to COVID-19 in a patient presenting with multisystem inflammatory syndrome in children: first case from Iran. J Paediatr Child Health. https://doi.org/10.1111/jpc.15048

Basiri A, Heidari A, Nadi MF, Fallahy MTP, Nezamabadi SS, Sedighi M, Saghazadeh A, Rezaei N (2020a) Microfluidic devices for detection of RNA viruses. Rev Med Virol n/a(n/a):e2154. https://doi.org/10.1002/rmv.2154

Basiri A, Pazhouhnia Z, Beheshtizadeh N, Hoseinpour M, Saghazadeh A, Rezaei N (2020b) Regenerative medicine in COVID-19 treatment: real opportunities and range of promises. Stem Cell Rev Rep:1–13. https://doi.org/10.1007/s12015-020-09994-5

Boehm K, Ziewers S, Brandt MP, Sparwasser P, Haack M, Willems F, Thomas A, Dotzauer R, Hofner T, Tsaur I, Haferkamp A, Borgmann H (2020) Telemedicine online visits in urology during the COVID-19 pandemic-potential, risk factors, and Patients' perspective. Eur Urol. https://doi.org/10.1016/j.eururo.2020.04.055

Breugelmans JG, Roberge G, Tippett C, Durning M, Struck DB, Makanga MM (2018) Scientific impact increases when researchers publish in open access and international collaboration: a bibliometric analysis on poverty-related disease papers. PLoS One 13(9):e0203156. https://doi.org/10.1371/journal.pone.0203156

Busvine D (2020) Science platform ResearchGate launches COVID-19 community. Reuters. https://www.reuters.com/article/us-health-coronavirus-tech-researchgate/science-platform-researchgate-launches-covid-19-community-idUSKBN21O1QD. Accessed 20 April 2020

Canales K (2020) Photos show the precautions US cities took to 'flatten the curve' during the 1918 Spanish Flu pandemic Business Insider. https://www.businessinsider.fr/us/spanish-flu-pandemic-1918-precautions-us-cities-2020-4

Chinchilla-Rodriguez Z, Sugimoto CR, Lariviere V (2019) Follow the leader: on the relationship between leadership and scholarly impact in international collaborations. PLoS One 14(6):e0218309. https://doi.org/10.1371/journal.pone.0218309

Cosic K, Popovic S, Sarlija M, Kesedzic I (2020) Impact of human disasters and COVID-19 pandemic on mental health: potential of digital psychiatry. Psychiatr Danub 32(1):25–31. https://doi.org/10.24869/psyd.2020.25

Coursera (2020) Coursera together: free online learning during COVID-19. https://blog.coursera.org/coursera-together-free-online-learning-during-covid-19/. Accessed 4 May 2020

Cuan-Baltazar JY, Munoz-Perez MJ, Robledo-Vega C, Perez-Zepeda MF, Soto-Vega E (2020) Misinformation of COVID-19 on the internet: Infodemiology study. JMIR Public Health Surveill 6(2):e18444. https://doi.org/10.2196/18444

Darbeheshti F, Rezaei N (2020) Genetic predisposition models to COVID-19 infection. Med Hypotheses 142:109818. https://doi.org/10.1016/j.mehy.2020.109818

edX (2020) edX's Response to COVID-19. https://www.edx.org/covid-19. Accessed 4 May 2020

Facebook (2020) Coronavirus (COVID-19) information hub for media. https://www.facebook.com/facebookmedia/solutions/coronavirus-resources. Accessed 4 May 2020

Fathi N, Rezaei N (2020) Lymphopenia in COVID-19: therapeutic opportunities. Cell Biol Int 44(9):1792–1797. https://doi.org/10.1002/cbin.11403

Galea S, Merchant RM, Lurie N (2020) The mental health consequences of COVID-19 and physical distancing: the need for prevention and early intervention. JAMA Intern Med. https://doi.org/10.1001/jamainternmed.2020.1562

Government-of-india (2020) Aarogya Setu mobile app. https://www.mygov.in/aarogya-setu-app/. Accessed 17 May 2020

Hakim AA, Kellish AS, Atabek U, Spitz FR, Hong YK (2020) Implications for the use of telehealth in surgical patients during the COVID-19 pandemic. Am J Surg. https://doi.org/10.1016/j.amjsurg.2020.04.026

Hawryluck L, Gold WL, Robinson S, Pogorski S, Galea S, Styra R (2004) SARS control and psychological effects of quarantine, Toronto, Canada. Emerg Infect Dis 10(7):1206–1212. https://doi.org/10.3201/eid1007.030703

Hollander JE, Carr BG (2020) Virtually perfect? Telemedicine for Covid-19. N Engl J Med 382(18):1679–1681. https://doi.org/10.1056/NEJMp2003539

Holmes EA, O'Connor RC, Perry VH, Tracey I, Wessely S, Arseneault L, Ballard C, Christensen H, Cohen Silver R, Everall I, Ford T, John A, Kabir T, King K, Madan I, Michie S, Przybylski AK, Shafran R, Sweeney A, Worthman CM, Yardley L, Cowan K, Cope C, Hotopf M, Bullmore E (2020) Multidisciplinary research priorities for the COVID-19 pandemic: a call for action for mental health science. Lancet Psychiatry. https://doi.org/10.1016/S2215-0366(20)30168-1

Jabbari P, Jabbari F, Ebrahimi S, Rezaei N (2020) COVID-19: a chimera of two pandemics. Disaster Med Public Health Prep:1–2. https://doi.org/10.1017/dmp.2020.223

Jabbari P, Rezaei N (2020) With risk of reinfection, is COVID-19 here to stay? Disaster Med Pub Heal Prep 1. https://doi.org/10.1017/dmp.2020.274

Jahanshahlu L, Rezaei N (2020a) Central nervous system involvement in COVID-19. Arch Med Res. https://doi.org/10.1016/j.arcmed.2020.05.016

Jahanshahlu L, Rezaei N (2020b) Monoclonal antibody as a potential anti-COVID-19. Biomed Pharmacoth Biomed Pharma 129:110337. https://doi.org/10.1016/j.biopha.2020.110337

Jeong H, Yim HW, Song YJ, Ki M, Min JA, Cho J, Chae JH (2016) Mental health status of people isolated due to Middle East respiratory syndrome. Epidemiol Health 38:e2016048. https://doi.org/10.4178/epih.e2016048

Jones D, Neal RD, Duffy SRG, Scott SE, Whitaker KL, Brain K Impact of the COVID-19 pandemic on the symptomatic diagnosis of cancer: the view from primary care. Lancet Oncol. https://doi.org/10.1016/S1470-2045(20)30242-4

Kafieh R, Arian R, Saeedizadeh N, Minaee S, Yadav SK, Vaezi A, Rezaei N, Javanmard SH (2020) COVID-19 in Iran: a deeper look into the future. medRxiv

Lager S (2020) Udacity's contribution to the COVID-19 crisis: one free month access to nanodegree programs udacity. https://blog.udacity.com/2020/03/one-month-free-on-nanodegrees.html. Accessed 17 May 2020

Lai J, Ma S, Wang Y, Cai Z, Hu J, Wei N, Wu J, Du H, Chen T, Li R, Tan H, Kang L, Yao L, Huang M, Wang H, Wang G, Liu Z, Hu S (2020) Factors associated with mental health outcomes among health care workers exposed to coronavirus disease 2019. JAMA Netw Open 3(3):e203976. https://doi.org/10.1001/jamanetworkopen.2020.3976

Latifi R, Doarn CR (2020) Perspective on COVID-19: finally, telemedicine at center stage. Telemed e-Heal. https://doi.org/10.1089/tmj.2020.0132

Liang ZC, Ooi SBS, Wang W (2020) Pandemics and their impact on medical training: lessons from Singapore. Acad Med. https://doi.org/10.1097/ACM.0000000000003441

Limaye RJ, Sauer M, Ali J, Bernstein J, Wahl B, Barnhill A, Labrique A Building trust while influencing online COVID-19 content in the social media world. Lancet Dig Heal. https://doi.org/10.1016/S2589-7500(20)30084-4

Lotfi M, Hamblin MR, Rezaei N (2020) COVID-19: transmission, prevention, and potential therapeutic opportunities. Clinica Chimica Acta Int J Clin Chem 508:254–266. https://doi.org/10.1016/j.cca.2020.05.044

Lotfi M, Rezaei N (2020) SARS-CoV-2: a comprehensive review from pathogenicity of the virus to clinical consequences. J Med Virol. https://doi.org/10.1002/jmv.26123

Mahase E (2020) Covid-19: UK starts social distancing after new model points to 260000 potential deaths. BMJ 368:m1089. https://doi.org/10.1136/bmj.m1089

Mansourabadi AH, Sadeghalvad M, Mohammadi-Motlagh HR, Rezaei N (2020) The immune system as a target for therapy of SARS-CoV-2: a systematic review of the current immunotherapies for COVID-19. Life Sci 258:118185. https://doi.org/10.1016/j.lfs.2020.118185

Merchant RM, Lurie N (2020) Social media and emergency preparedness in response to novel coronavirus. JAMA. https://doi.org/10.1001/jama.2020.4469

Mian A, Khan S (2020) Medical education during pandemics: a UK perspective. BMC Med 18(1):100. https://doi.org/10.1186/s12916-020-01577-y

Moazzami B, Razavi-Khorasani N, Dooghaie Moghadam A, Farokhi E, Rezaei N (2020) COVID-19 and telemedicine: immediate action required for maintaining healthcare providers Well-being. J Clin Virol Off Publ Pan Am Soc Clin Virol 126:104345. https://doi.org/10.1016/j.jcv.2020.104345

Mohamed K, Rodríguez-Román E, Rahmani F, Zhang H, Ivanovska M, Makka SA, Joya M, Makuku R, Islam MS, Radwan N, Rahmah L, Goda R, Abarikwu SO, Shaw M, Zoghi S, Irtsyan S, Ling I, Cseprekal O, Faten A-B, Hazar E, Tbalvandany CS, Grancini G, Rezaei N (2020a) Borderless collaboration is needed for COVID-19; a disease that knows no borders. Infect Control Hosp Epidemiol:1–7. https://doi.org/10.1017/ice.2020.162

Mohamed K, Yazdanpanah N, Saghazadeh A, Rezaei N (2020b) Computational drug discovery and repurposing for the treatment of Covid-19: a systematic review. Available at SSRN 3583748

Momtazmanesh S, Ochs HD, Uddin LQ, Perc M, Routes JM, Vieira DN, Al-Herz W, Baris S, Prando C, Rosivall L, Abdul Latiff AH, Ulrichs T, Roudenok V, Aldave Becerra JC, Salunke DB, Goudouris E, Condino-Neto A, Stashchak A, Kryvenko O, Stashchak M, Bondarenko A, Rezaei N (2020) All together to fight novel coronavirus disease (COVID-19). Am J Trop Med Hyg. https://doi.org/10.4269/ajtmh.20-0281

Moradian N, Ochs HD, Sedikies C, Hamblin MR, Camargo CA, Martinez JA, Biamonte JD, Abdollahi M, Torres PJ, Nieto JJ, Ogino S, Seymour JF, Abraham A, Cauda V, Gupta S, Ramakrishna S, Sellke FW, Sorooshian A, Wallace Hayes A, Martinez-Urbistondo M, Gupta M, Azadbakht L, Esmaillzadeh A, Kelishadi R, Esteghamati A, Emam-Djomeh Z, Majdzadeh R, Palit P, Badali H, Rao I, Saboury AA, Jagan Mohan Rao L, Ahmadieh H, Montazeri A, Fadini GP, Pauly D, Thomas S, Moosavi-Movahed AA, Aghamohammadi A, Behmanesh M, Rahimi-Movaghar V, Ghavami S, Mehran R, Uddin LQ, Von Herrath M, Mobasher B, Rezaei N (2020) The urgent need for integrated science to fight COVID-19 pandemic and beyond. J Transl Med 18(1):205. https://doi.org/10.1186/s12967-020-02364-2

Nasab MG, Saghazadeh A, Rezaei N (2020) SARS-CoV-2-a tough opponent for the immune system. Arch Med Res. https://doi.org/10.1016/j.arcmed.2020.05.020

Nature (2020) Show evidence that apps for COVID-19 contact-tracing are secure and effective. Nature 580(7805):563. https://doi.org/10.1038/d41586-020-01264-1

Ohannessian R, Duong TA, Odone A (2020) Global telemedicine implementation and integration within health systems to fight the COVID-19 pandemic: a call to action. JMIR Public Health Surveill 6(2):e18810. https://doi.org/10.2196/18810

Oxford (2020) Social networking. Oxford University Press. https://www.oxfordlearnersdictionaries.com/definition/english/social-networking. Accessed 4 May 2020

Panahi S, Watson J, Partridge H (2016) Social media and physicians: exploring the benefits and challenges. Health Informatics J 22(2):99–112. https://doi.org/10.1177/1460458214540907

Pashaei M, Rezaei N (2020) Immunotherapy for SARS-CoV-2: potential opportunities. Expert Opin Biol Ther:1–5. https://doi.org/10.1080/14712598.2020.1807933

Pourahmad R, Moazzami B, Rezaei N (2020) Efficacy of plasmapheresis and immunoglobulin replacement therapy (IVIG) on patients with COVID-19. SN Comprehen Clin Med:1–5

Public (2020) Public statement for collaboration on COVID-19 vaccine development. https://www.who.int/news-room/detail/13-04-2020-public-statement-for-collaboration-on-covid-19-vaccine-development. Accessed 4 May 2020

Qiu J, Shen B, Zhao M, Wang Z, Xie B, Xu Y (2020) A nationwide survey of psychological distress among Chinese people in the COVID-19 epidemic: implications and policy recommendations. Gen Psychiatr 33(2):e100213. https://doi.org/10.1136/gpsych-2020-100213

Rabiee N, Rabiee M, Bagherzadeh M, Rezaei N (2020) COVID-19 and picotechnology: potential opportunities. Med Hypotheses 144:109917. https://doi.org/10.1016/j.mehy.2020.109917

Rahmani F (2020) Advisory board, junior ambassadors, and interest groups respond to COVID19. https://usern.tums.ac.ir/News/New?title=Advisory%20Board,%20Junior%20Ambassadors,%20and%20Interest%20Groups%20Respond%20to%20COVID19. Accessed 4 May 2020

ResearchGate (2020) Combating COVID-19. ResearchGate GMBH. https://www.researchgate.net/COVID-19. Accessed 4 May 2020

Rezaei N (2016) Universal scientific education and research network (USERN):a new horizon for science. Acta Med Iran 54(1):1–3

Rezaei N (2017) Universal scientific education and research network (USERN): to make the knowledge without Borders. Acta Med Iran 55(1):1–5

Rezaei N (2018) Universal scientific education and research network (USERN): twinkling stars unite to make the world glow. Acta Med Iran 56(1):1–3

Rezaei N (2020a) COVID-19 affects healthy pediatricians more than pediatric patients. Infect Cont Hosp Epidemiol 1. https://doi.org/10.1017/ice.2020.139

Rezaei N (2020b) COVID-19 and medical biotechnology. Avicenna J Med Biotechnol 12(3):139

Richards D, Vigano N (2013) Online counseling: a narrative and critical review of the literature. J Clin Psychol 69(9):994–1011. https://doi.org/10.1002/jclp.21974

Rokni M, Hamblin MR, Rezaei N (2020) Cytokines and COVID-19: friends or foes? Hum Vaccin Immunother:1–3

Rosenbaum L (2020) The untold toll — the Pandemic's effects on patients without Covid-19. N Engl J Med. https://doi.org/10.1056/NEJMms2009984

Rzymski P, Nowicki M, Mullin GE, Abraham A, Rodríguez-Román E, Petzold MB, Bendau A, Sahu KK, Ather A, Naviaux A-F (2020) Quantity does not equal quality: scientific principles cannot be sacrificed. Int Immunopharmacol 86:106711

Saghazadeh A, Rezaei N (2020a) Immune-epidemio logical parameters of the novel coronavirus – a perspective. Expert Rev Clin Immunol

Saghazadeh A, Rezaei N (2020b) Towards treatment planning of COVID-19: rationale and hypothesis for the use of multiple immunosuppressive agents: anti-antibodies, immunoglobulins, and corticosteroids. Int Immunopharmacol 84:106560. https://doi.org/10.1016/j.intimp.2020.106560

Sahu KK, Siddiqui AD, Rezaei N, Cerny J (2020) Challenges for management of immune thrombocytopenia during COVID-19 pandemic. J Med Virol. https://doi.org/10.1002/jmv.26251

Saleem SM, Pasquale LR, Sidoti PA, Tsai JC (2020) Virtual ophthalmology: telemedicine in a Covid-19 era. Am J Ophthalmol. https://doi.org/10.1016/j.ajo.2020.04.029

Saleki K, Banazadeh M, Saghazadeh A, Rezaei N (2020) The involvement of the central nervous system in patients with COVID-19. Rev Neurosci 31(4):453–456. https://doi.org/10.1515/revneuro-2020-0026

Shamshirian D, Rezaei N (2020) Cardiovascular diseases burden in COVID-19: systematic review and meta-analysis

Sharifkashani S, Bafrani MA, Khaboushan AS, Pirzadeh M, Kheirandish A, Yavarpour Bali H, Hessami A, Saghazadeh A, Rezaei N (2020) Angiotensin-converting enzyme 2 (ACE2) receptor and SARS-CoV-2: potential therapeutic targeting. Eur J Pharmacol 884:173455. https://doi.org/10.1016/j.ejphar.2020.173455

Statista (2019) Number of social network users worldwide from 2010 to 2023 Statista website. https://www.statista.com/statistics/278414/number-of-worldwide-social-network-users/. Accessed 30 April 2020

Statt N (2020) Major tech platforms say they're 'jointly combating fraud and misinformation' about COVID-19. The Verge. https://www.theverge.com/2020/3/16/21182726/coronavirus-covid-19-facebook-google-twitter-youtube-joint-effort-misinformation-fraud. Accessed 4 May 2020

Twitter (2020) Coronavirus: staying safe and informed on Twitter. https://blog.twitter.com/en_us/topics/company/2020/covid-19.html. Accessed 4 May 2020

Venkatesh A, Edirappuli S (2020) Social distancing in covid-19: what are the mental health implications? BMJ 369:m1379. https://doi.org/10.1136/bmj.m1379

Walsh K, Sandars J, Nordquist J (2018) Technology-enhanced learning for healthcare professionals: an essential response to infectious disease pandemics. BMJ Simulat Technol Enhan Learn 4(1):1. https://doi.org/10.1136/bmjstel-2017-000236

Wang C, Horby PW, Hayden FG, Gao GF (2020) A novel coronavirus outbreak of global health concern. Lancet 395(10223):470–473. https://doi.org/10.1016/S0140-6736(20)30185-9

WHO (2020) WHO Director-General's opening remarks at the media briefing on COVID-19 https://www.who.int/dg/speeches/detail/who-director-general-s-opening-remarks-at-the-media-briefing-on-covid-19%2D%2D-18-march-2020

World_Bank (2020) Remote learning, distance education and online learning during the COVID19 pandemic: a resource list by the World Bank's Edtech Team (English). World Bank Group, Washington, DC

Yazdanpanah F, Hamblin MR, Rezaei N (2020a) The immune system and COVID-19: friend or foe? Life Sci 256:117900. https://doi.org/10.1016/j.lfs.2020.117900

Yazdanpanah N, Saghazadeh A, Rezaei N (2020b) Anosmia: a missing link in the neuroimmunology of coronavirus disease 2019 (COVID-19). Rev Neurosci. https://doi.org/10.1515/revneuro-2020-0039

Yousefzadegan S, Rezaei N (2020) Death due to COVID-19 in three brothers. Am J Trop Med Hyg

Zarocostas J (2020) How to fight an infodemic. Lancet 395(10225):676. https://doi.org/10.1016/S0140-6736(20)30461-X

Zastrow M (2020) South Korea is reporting intimate details of COVID-19 cases: has it helped? Nature. https://doi.org/10.1038/d41586-020-00740-y

Interdisciplinary Approaches to COVID-19

52

Negar Moradian, Marjan Moallemian, Farnaz Delavari, Constantine Sedikides, Carlos A. Camargo Jr, Pedro J. Torres, Armin Sorooshian, Saeid Paktinat Mehdiabadi, Juan J. Nieto, Stephane Bordas, Hamid Ahmadieh, Mohammad Abdollahi, Michael R. Hamblin, Frank W. Sellke, Jack Cuzick, Bozkurt Biykem, Michael Schreiber, Babak Eshrati, Georg Perry, Ali Montazeri, Ali Akbar Saboury, Roya Kelishadi, Amirhossein Sahebkar, Ali A. Moosavi-Movahed, Hassan Vatandoost, Mofid Gorji-Bandpy, Bahram Mobasher, and Nima Rezaei

Abstract

The coronavirus disease 2019 (COVID-19) pandemic has been a significant concern worldwide. The pandemic has demonstrated that public health issues are not merely a health concern but also affect society as a whole. In this chapter, we address the importance of bringing together the world's scientists to find appropriate solutions for controlling and managing the COVID-19 pandemic. Interdisciplinary cooperation, through modern scientific methods, could help to han-

N. Moradian · N. Rezaei (✉)
Universal Scientific Education and Research Network (USERN), The World, Tehran, Iran

Research Center for Immunodeficiencies, Children's Medical Center, Tehran University of Medical Sciences, Tehran, Iran
e-mail: rezaei_nima@tums.ac.ir

M. Moallemian
Universal Scientific Education and Research Network (USERN), The World, Tehran, Iran

Department of Clinical Nutrition and Dietetics, Faculty of Nutrition and Food Technology, National Nutrition and food technology Research Institute, Shahihd Beheshti University of Medical Sciences, Tehran, Iran

F. Delavari
Universal Scientific Education and Research Network (USERN), The World, Tehran, Iran

Department of Psychiatry, School of Medicine, University of Geneva, Geneva, Switzerland

C. Sedikides
Universal Scientific Education and Research Network (USERN), The World, Tehran, Iran

Centre for Research on Self Identity, Department of Psychology, School of Psychology, University of Southampton, Southampton, UK

C. A. CamargoJr
Universal Scientific Education and Research Network (USERN), The World, Tehran, Iran

Department of Emergency Medicine, Massachusetts General Hospital, Harvard Medical School, Boston, MA, USA

N. Rezaei (ed.), *Coronavirus Disease - COVID-19*, Advances in Experimental Medicine and Biology 1318, https://doi.org/10.1007/978-3-030-63761-3_52

dle the consequences of the pandemic and to avoid the recurrence of future pandemics.

Keywords

Cooperation · COVID-19 · Interdisciplinary · Pandemic · Problem · Solution

52.1 Introduction

Coronavirus disease 2019 (COVID-19), an infection attributed to the severe acute respiratory syndrome of coronavirus 2 (SARS-CoV-2), attained pandemic proportions in early 2020 (Jabbari et al. 2020; Hanaei and Rezaei 2020). As of June 30, 2020, there are more than 10 million people confirmed to have COVID-19 worldwide. Even this disturbing amount is supposed to be just the tip of the iceberg. The COVID-19 fatality rate is probably around 0.02% to 0.4% (Ioannidis 2020); though, due to the high infectivity of severe acute respiratory syndrome coronavirus 2 (SARS-CoV-2) causing COVID-19 (Huang et al. 2020b), the number of deaths has been high, with more than 500,000 people died from COVID-19 as of writing this. The pandemic has exposed the need for a holistic commitment to developing healthcare systems around the world; only, for example, during the pandemic, many scientists worldwide have been carrying out studies related to the pandemic (Remuzzi and Remuzzi 2020; Rzymski et al. 2020).

Public health is a practical and relevant area for the implementation of interdisciplinary research. Actions to enhance health services quality include innovative strategies like the

P. J. Torres
Universal Scientific Education and Research Network (USERN), The World, Tehran, Iran

Departamento de Matemática Aplicada, Universidad de Granada, Granada, Spain

A. Sorooshian
Universal Scientific Education and Research Network (USERN), The World, Tehran, Iran

Department of Chemical and Environmental Engineering, University of Arizona, Tucson, Arizona, USA

Department of Hydrology and Atmospheric Sciences, University of Arizona, Tucson, Arizona, USA

S. P. Mehdiabadi
Universal Scientific Education and Research Network (USERN), The World, Tehran, Iran

Physics Department, Elementary Particle, Yazd University, Yazd, Iran

Faculty of Physics, Yazd University, Yazd, Iran

J. J. Nieto
Universal Scientific Education and Research Network (USERN), The World, Tehran, Iran

Instituto de Matemáticas, Universidade de Santiago de Compostela, Santiago de Compostela, Spain

S. Bordas
Universal Scientific Education and Research Network (USERN), The World, Tehran, Iran

University of Luxembourg, Institute of Computational Engineering Sciences, Luxembourg, Cardiff University, Department of Applied and Computational Mechanics, Wales, UK

H. Ahmadieh
Universal Scientific Education and Research Network (USERN), The World, Tehran, Iran

Ophthalmic Research Center, Shahid Beheshti University of Medical Sciences, Tehran, Iran

M. Abdollahi
Universal Scientific Education and Research Network (USERN), The World, Tehran, Iran

Pharmaceutical Sciences Research Center (PSRC), The Institute of Pharmaceutical Sciences (TIPS), and School of Pharmacy, Tehran University of Medical Sciences, Tehran, Iran

M. R. Hamblin
Universal Scientific Education and Research Network (USERN), The World, Tehran, Iran

Wellman Center for Photomedicine, Massachusetts General Hospital, Harvard Medical School, Boston, USA

Laser Research Centre, Faculty of Health Science, University of Johannesburg, Doornfontein, South Africa

F. W. Sellke
Universal Scientific Education and Research Network (USERN), The World, Tehran, Iran

Division of Cardiothoracic Surgery, Department of Surgery, Rhode Island Hospital, Alpert Medical School of Brown University, Providence, RI, USA inclusion of related disciplines, such as medicine and pharmacy, molecular and cellular biology, microbiology and biochemistry, genetics, immunology, pharmacology, nutrition, psychology, epidemiology, economics, societal needs, communication, and political sciences, health, and nursing care services, physics and chemistry, geography, and statistics or computational sciences for big data management. All of these diverse fields have study viewpoints that lead to the evaluation, review, perception, interpretation of the health consequences of COVID-19 as well as coping with them (Christakos et al. 2005).

Interdisciplinary study and evidence-based treatment of COVID-19 is essential for many reasons. One reason is the need by public health policy analysts for science-based information on the efficacy and usefulness of interventions. Such analysts may assist decision-makers when addressing the needs and desires of the population. Another reason is that this new human pathogen requires interventions at multiple levels, from the individual to societal. Collaboration

J. Cuzick
Universal Scientific Education and Research Network (USERN), The World, Tehran, Iran

Cancer Research UK Centre for Epidemiology, Mathematics and Statistics, Wolfson Institute of Preventive Medicine, Queen Mary University of London, London, UK

B. Biykem
Universal Scientific Education and Research Network (USERN), The World, Tehran, Iran

Winters Center for Heart Failure Research, Cardiovascular Research Institute, Baylor College of Medicine, Houston, TX, USA

M. Schreiber
Universal Scientific Education and Research Network (USERN), The World, Tehran, Iran

Institute of Physics, Technische Universität Chemnitz, Chemnitz, Germany

B. Eshrati
Universal Scientific Education and Research Network (USERN), The World, Tehran, Iran

Centre for Communicable Diseases Control, Ministry of Health and Medical Education, Tehran, Iran

G. Perry
Universal Scientific Education and Research Network (USERN), The World, Tehran, Iran

University of Texas at San Antonio, Biology and Chemistry, One UTSA Circle, San Antonio, TX, USA

A. Montazeri
Universal Scientific Education and Research Network (USERN), The World, Tehran, Iran

Population Health Research Group, Health Metrics Research Center, Institute for Health Sciences Research, ACECR, Tehran, Iran

A. A. Saboury · A. A. Moosavi-Movahed
Universal Scientific Education and Research Network (USERN), The World, Tehran, Iran

Institute of Biochemistry and Biophysics, University of Tehran, Tehran, Iran

R. Kelishadi
Universal Scientific Education and Research Network (USERN), The World, Tehran, Iran

Child Growth and Development Research Center, Research Institute for Primordial Prevention of Non-communicable Disease, Isfahan University of Medical Sciences, Isfahan, Iran

A. Sahebkar
Universal Scientific Education and Research Network (USERN), The World, Tehran, Iran

Biotechnology Research Center, Pharmaceutical Technology Institute, Mashhad University of Medical Sciences, Mashhad, Iran

Neurogenic Inflammation Research Center, Mashhad University of Medical Sciences, Mashhad, Iran

H. Vatandoost
Universal Scientific Education and Research Network (USERN), The World, Tehran, Iran

Department of Medical Entomology and Vector Control, School of Public Health, Tehran University of Medical Sciences, Tehran, Iran

Department of Environmental Chemical Pollutants and Pesticides, Institute for Environmental Research, Tehran University of Medical Sciences, Tehran, Iran

M. Gorji-Bandpy
Universal Scientific Education and Research Network (USERN), The World, Tehran, Iran

Department of Mechanical Engineering, Babol Noshirvany University of Technology, Babol, Iran

B. Mobasher
Universal Scientific Education and Research Network (USERN), The World, Tehran, Iran

Department of Physics and Astronomy University of California, Riverside, CA, USA with an interdisciplinary focus can help achieve these goals (Kivits et al. 2019; Yu et al. 2017).

52.2 COVID-19

The twenty-first century has seen the appearance and outbreak of three previously unknown coronaviruses: severe acute coronavirus respiratory syndrome (SARS-CoV) in 2003, Middle East Coronavirus Respiratory Syndrome (MERS-CoV) in 2012, and 2019 novel coronavirus (2019-nCoV, later officially named SARS-CoV-2) in December 2019 (Tu et al. 2020). The current pandemic of COVID-19, initially seen in China's Hubei Province, has spread to almost every nation in the world (Li et al. 2020b). On January 30, 2020, the World Health Organization (WHO) announced a public health alert based on increasing cases of COVID-19 observed in China and subsequently in other areas around the world (Velavan and Meyer 2020). This global health alert came just 1 month after the first officially recorded case in Hubei, another proof of the high infectivity of this virus (WHO 2020b).

Coronaviruses are enveloped RNA viruses that mostly infect birds and mammals. Before the three coronaviruses causing outbreaks appeared, four other coronaviruses, including 229E (alphacoronavirus), NL63 (alphacoronavirus), OC43 (betacoronavirus), and HKU1 (betacoronavirus), were identified in humans. They cause mild infection in the upper respiratory tract, like the common cold, and much less frequently do they cause severe lower respiratory tract infections (Weiss and Leibowitz 2011). However the three coronaviruses causing crises of SARS, MERS, and COVID-19, in essence, invade the lower respiratory tracts, though after the entry into upper respiratory tracts (Lotfi and Rezaei 2020).

The SARS-CoV-2 virus affects not only the respiratory system but also the digestive, gastrointestinal, hepatic, and central nervous systems (Rodriguez-Morales et al. 2020; Arentz et al. 2020; Yazdanpanah et al. 2020b; Jahanshahlu and Rezaei 2020a; Saleki et al. 2020). Compared to SARS-CoV, which triggered an outbreak of SARS in 2003, SARS-CoV-2 has a higher transmission capacity (Lotfi et al. 2020). Despite SARS-CoV having a higher fatality (9.6%), the new coronavirus has caused many more deaths. The dramatic rise in reported cases makes it particularly challenging to prevent and control COVID-19.

Scientists and policymakers around the world have taken urgent steps to monitor the outbreak and carry out virologic, clinical, and epidemiological studies (Chen et al. 2020c).

52.3 The Importance of Interdisciplinary Research

The pandemic has made it clear that COVID-19 not only causes physical health damage but also poses a mental health threat and can stir major economic, social, and societal upheaval. Therefore, COVID-19 does not merely fall in the purview of medical sciences; also, it is a matter of urgency for economics, behavioral and social sciences (e.g., physical distancing rules, social isolation), and political decision-making (e.g., border control, unemployment) (Moradian et al. 2020; Dimant and Schulte 2019). The synergy of diverse fields of expertise promises to contribute to understanding, prevention, and management of the detrimental consequences of COVID-19 (Hartley and Perencevich 2020). We advocate such an interdisciplinary approach in this chapter, where researchers work jointly, but also from a disciplinary-specific base, to solve the common problem (Okumus et al. 2018; Mohamed et al. 2020a).

We provide examples below on the role of different sciences for controlling and managing the COVID-19 pandemic. Throughout, we emphasize the importance of experts' cooperation for achieving a proper solution to this pandemic and for decreasing the chances of similar natural disasters in the future.

52.3.1 Physics, Chemistry, and Engineering

COVID-19 transmission mainly happens through droplets containing the virus, which are formed in the respiratory tract of an infected person and come out of the mouth and nose when breathing, coughing, or sneezing (Huang et al. 2020c). In particular, droplet transmission or contact route of transmission occurs through physical contact with droplets on the surface or direct transmission into the mouth or eyes, while airborne transmission occurs through inhalation (Shereen et al. 2020).

The rules of physics modulate virus transmission. Virus-infected particles showing up through sneezing or coughing behave differently. After leaving the mouth or nose, the particles mainly join the other particles in the air and move with them (Bourouiba et al. 2014).

Physics can predict the particles' type of motion by examining their behavior, which depends on their size and speed. For example, small particles are suspended in the air and move upward, polluting the ventilation system in the ceiling (Bourouiba et al. 2014). Therefore, access to different models of particle motion can help to predict contamination at different levels. The chemical composition of the respirable aerosols containing the virus impacts on how much the aerosols can change in size upon inhalation owing to hygroscopic growth, which will ultimately determine if and where the aerosols will deposit in the respiratory system (Youn et al. 2016).

Either way, it is vital to disinfect all surfaces efficiently. Chemistry can change the properties of materials in nanoscale to increase the self-sanitization capacity of metal surfaces (Saccucci et al. 2018). It would be particularly useful in hospitals. Another beneficial surface decontamination solution has been recently obtained by physicists. They proposed that violet/blue (400–470 nm) light is antimicrobial against numerous bacteria and even viruses. Empirical research has demonstrated that this is a low-cost and safe opportunity to sanitize equipment, hospital facilities, emergency care vehicles, homes, and the general environment (Enwemeka et al. 2020; Kowalski et al. 2020).

Moreover, personal protective equipment (PPE) restrictions have posed health risks for medical staff. Improving PPE, such as by reusing gloves, masks, air-purifying respirators, goggles, and gowns, as well as detecting the virus and monitoring the physiological condition of the person wearing the clothes, alleviates some of those health risks (Medicine 2006; Livingston et al. 2020).

It is also advantageous to use disposable equipment to reduce the production of plastic waste. The role of chemists and engineers working together to mass produce these products is becoming more and more crucial (Huang et al. 2020a).

52.3.2 Mathematics and Computer Science

Mathematical and epidemiological modeling is essential in forecasting, detecting, and monitoring epidemics. In order to better understand and model the complexities of an outbreak like COVID-19, researchers need to consider several factors, ranging from host-pathogen interactions to host-to-host encounters and the predominant political, social, economic, and local parameters worldwide (Soheilypour and Mofrad 2018). Mathematical modelers use evidence from recent and long-past outbreaks to determine who could be affected, where vaccine efforts may be more successful, and how to minimize the spread of the virus.

For example, a recent study implicated cell phones, along with online surveys, for data collection purposes. Such self-reported clinical data can be used for preliminary screening and early identification of COVID-19 cases. Thousands of data points can be analyzed by artificial intelligence (AI) system that can assess and stratify people in terms of risk status (i.e., no risk, low risk, moderate risk, high risk). Identified high-risk cases might then be quarantined early, thereby reducing the possibility of virus transmission (Srinivasa Rao and Vazquez 2020). Another newly investigated route for data collection, and a more objective one, is the analysis of wastewater as a way to estimate the total number of infected persons in a city. Virologists can locate traces of the virus in the water and then, with an adequate network model of the wastewater system developed by mathematicians, can track the virus back to its origin to identify hot spots. On this basis, scientists can plan for appropriate action depending, for instance, on postal code (EPFL Scientists 2020; Mallapaty 2020).

In order to make reliable predictions with a mathematical model, one would need to have accurate estimates of the parameters concerned. Data fitting is the method of modifying system templates and evaluating the accuracy of fit. The introduction of an effective and consistent protocol for calculating the number of sick, deceased, and recovered cases and for unifying various protocols at the international level is essential to the calculation of accurate estimates (roland Oliphant 2020; Rafa de Miguel et al. 2020). Dynamic simulation, game theory, and spatial modeling will be of growing importance in understanding the biology of complex systems. These approaches can also illuminate the complexities of interactions between pathogens and their host (Ewald et al. 2020).

Two statistical mechanisms specific to a pandemic such as COVID-19 are stochastic and susceptible-infected-recovered (SIR). The stochastic simulation includes all statistical equations of random variables in the function assignment. In other words, the stochastic simulation uses random assignment for predictive functions, which is useful in the early stages of virus dissemination (Ming et al. 2016). The SIR model has a robust predictive capacity extended to circumstances where people infect each other explicitly (Adamu et al. 2019). However, in recent years, spatial frameworks and, in particular, network theory (as a model for road networks or flight connections) and metapopulation have proven relevant in understanding the dynamics of virus transmission (Huppert and Katriel 2013; Rihan and Anwar 2012). Network theory proposes that simple mathematical models of epidemics presume a random combination between people, which is not the case in practice. Populations have a structure, and people appear to communicate with a tiny minority of the population. Contacts or connections between entities are a network, and infectious diseases that transmit by direct interaction may propagate only through the networks and pathways. In this module, specific models of communication and transmission networks are applied to investigate the impact of the network system on the dissemination of epidemics. It is one of the most challenging modules.

Tracing apps such as STOPCOVID (in France), Corona-Warn-App (Germany), and Ariana (Luxembourg) help build the network of contacts within a population, which can have a significant impact on the design of confinement and deconfinement strategies. From a given infected person, these apps, based on low-energy Bluetooth technologies, enable the safe and privacy-aware tracing of contacts, without any global positioning system. In turn, smartphone applications can also be used to track symptoms and hence help predict the spread of the pandemic and are invaluable in providing a rational and quantitative approach to estimating the main parameters appearing in epidemiological models, be they agent-based or equation-based (Arianatech 2020).

Such systems, together with information technology (e.g., bioinformatics, big data analytics, computer intelligence, artificial intelligence, sensors, image recognition), are used worldwide by public health authorities to analyze and establish strategies for ever-emerging infectious diseases and to respond to outbreaks (Ewald et al. 2020; Seno and Dansu 2019; Siettos and Russo 2013; Dansu and Seno 2019).

52.3.3 Biological Sciences

Biological science offers a greater understanding of pathogen-host interactions via molecular biology and computational biology advances developed during the first decade of the twenty-first century (Aderem et al. 2011; Li et al. 2020a). In particular, computational methods of drug discovery can help select among the best existing drugs for COVID-19 treatment during the pandemic (Mohamed et al. 2020c).

The control of COVID-19 depends to a considerable extent on two factors. The first factor is the discovery of testing methods and kits to estimate the extent of the pandemic accurately. The second factor is the development of new vaccines. Testing kits are the product of research on biomedical sciences and engineering. Ideally, they are designed to provide high precision and fast detection for mitigating the transmission of disease in the population (Speers 2006; Chen et al. 2020b; Konrad et al. 2020).

Regarding vaccine development, scientists need to identify viral epitopes that could be used as therapeutic targets and that not only activate the immune system but also recognize the virus as a target (Ahmed et al. 2020b; Sharifkashani et al. 2020; Saghazadeh and Rezaei 2020b). More than 100 vaccines are being produced against SARS-CoV-2 by research teams in universities and companies around the world. Scientists are exploring various techniques, some of which have not historically been used in an approved vaccine. At least eight projects have now started to administer substances into participants in health trials; many others have started testing in animals. There are now more than a half-dozen candidates, including live viruses, recombinant protein subunits, and nucleic acids, that could lead to an effective vaccine against COVID-19 (Chen et al. 2020a).

A look into the future is that current molecular biology methods have the potential to offer information beyond simply identifying causative agents for recently found infectious diseases, such as COVID-19, but could also inform us about the emergence of other novel pathogens, the longevity of infectious processes in nature, the study of the origins of this and other pandemics, and the resistance mechanisms of different host classes (Rezaei 2020b; Rabiee et al. 2020). This knowledge could culminate in the creation of DNA and RNA banking for the study of pathogenic gene encoding factors (Speers 2006; Qu et al. 2020; Yousefi and Moosavi-Movahedi 2020; Wong et al. 2007; Ahmed et al. 2020a).

Various theories have also arisen from this virus by scientists looking forward to demonstrating a novel disease mechanism. According to one of the theories suggested by Jean-Laurent Casanova of Rockefeller University, immunodeficiency leads, at least in part, to severe disease and high mortality among the elderly (jean laurent casanova 2020). However, it is equally important to explain the deadly disease that is sometimes found in the younger population. For example, there might be an inherent, inborn error in immunity associated with COVID-19 infection in younger individuals (Darbeheshti and Rezaei 2020).

52.3.4 Social and Economic Sciences

The occurrence of illness and the propagation of epidemics rely on individual behavior, the social systems, and the political and societal climate that could discriminate between different sectors of the society in terms of race and socioeconomic status (Arthur et al. 2017; Cauchemez et al. 2011). Once exclusion rules and social contact limitations (e.g., spatial distances) are placed into practice (Uchenna and Ossai 2016; Parmet and Sinha 2020), social scientists must be mindful of the psychological, behavioral, and economic implications of these decisions. Social and clinical psychology, as well as psychiatry, illustrates the psychological problems (e.g., loneliness, depression) associated with social isolation (Liu et al. 2012; Hawryluck et al. 2004; Hawkley and Cacioppo 2010).

Communication of facts about COVID-19 is another critical issue. Many people doubt the veracity of information disseminated by politicians but are more receptive to information disseminated by scientists. Indeed, the presence of scientists on TV, the press, and, more generally, social media is becoming increasingly prevalent. Also, some people are susceptible to fake news (Sheera Frenkel and Zhong 2020) or conspiracy theories (Emma Grey Ellis 2020). These can be debunked with counter-persuasion techniques, such as inoculation (exposing people to a weak argument to build their defenses against a more robust argument) (Banas and Rains 2010) or nudge (prompting people to consider accuracy) (Bago et al. 2020). For more information on infodemic and misinformation, we recommend the following materials offered by the WHO (WHO 2020a, c; Organization, PAH 2020).

Moreover, COVID-19 has enormous socioeconomic consequences. With the massive and systematic rise in international travel and commerce, pandemic outbreaks can cause economic shock waves that reach way beyond the conventional health system and extend beyond the initial geographic range of the pathogen (Smith et al. 2019). During the first months of the outbreak, the economic consequences were limited to China and affected China's economy by disruption and a drop in goods production. As a result, the entire supply chain has been disrupted, given that China provides essential equipment and materials to businesses and organizations worldwide (Warwick et al. 2020). The effect of COVID-19 on the global economy has been much higher than that of SARS. In 2002, China played a minor role in the global economy, but today it has become an economic powerhouse (Bobdey and Ray 2020). Also, the virus has spread far beyond China and is having a direct impact on economies worldwide. Analysts, for example, expect that the pandemic will result in a projected annual economic loss of 2–3% of global gross domestic product (GDP) in the coming year (Oxford Economics 2020).

Economists can encourage the need for companies to become pandemic resilient by building political, organizational, and financial resilience and declare that these measures are a duty of care for society and the environment. Intergovernmental coordination is a crucial component in controlling COVID-19. Coordinated measures are required to ensure that all external causes are taken into consideration to minimize the transmission of the virus across borders and to deter future pandemics, and this is important given the risk for re-infection with COVID-19 (Jabbari and Rezaei 2020).

52.3.5 Medical Sciences

Medical sciences are expected to spearhead the effort to address the pandemic, although, as we have argued, the frontline role of medical sciences would not diminish the importance of interdisciplinary research (Moradian et al. 2020). Here, we emphasize that the borderless approach should also be imposed within the branches of medical sciences itself. For years, medical sciences have been put into the solid box of the linear model (Godin 2006), which indicates a sequence of predisposing achievements to make innovation (and more specifically, intervention) possible. However, in extraordinary circumstances, an integrated approach could yield a faster and more appropriate response. For example, the usual perspective on diseases suggests that management protocols follow the identification of the disease pathophysiology. In the case of the new pandemic, though, the race to identify the pathogenesis was launched alongside (and not before) attempts to identify treatment approaches (Cao and Li 2020).

As the number of COVID-19 cases and deaths increases globally, the healthcare policymakers have re-oriented research on prevention, identifying risk factors, and estimating their future impact (Bedford et al. 2020; Kafieh et al. 2020). As preventive and treatment approaches advance, the role of health system design is becoming critical in achieving equity and efficiency in the distribution of screening and care services (Bollyky et al. 2020). This development and fairness would be drastically anticipated at the time of reaching effective treatment or vaccines. Moreover, the frequent updates on protocols for healthcare providers and critical caregivers remain vital in providing standard care. However, the mental and physical care provided specifically for the healthcare staff has not attracted the research attention it deserves (Li et al. 2020c). Such research is especially important, because a prolonged overdrive in healthcare is anticipated (Rezaei 2020a; Moazzami et al. 2020). On another level, fast decision-making and providing palliative care could benefit from the involvement of both medical ethics and healthcare management (The Lancet 2020).

Aside from tackling strategies at a population level, an interdisciplinary response to individual care is also relevant. The virus attacks multiple systems. Patients may evince respiratory failure as well as cardiovascular complications, including myocardial injury, clot formation, vascular injury, and inflammation in different vascular beds, stroke, rhythm abnormalities, renal injury, liver failure, coagulation abnormalities, cytokine storm, vascular collapse, and shock (Rismanbaf 2020). The strain of multiple systems has challenged pharmacology (Zaim et al. 2020) and different medical disciplines, particularly cardiology (Shamshirian and Rezaei 2020), immunology (Ahanchian et al. 2020; Babaha and Rezaei 2020), oncology (Ahmadi et al. 2020), and hematology (Sahu et al. 2020). However, searching for other frequent manifestations of the disease might open new avenues for early diagnosis and treatment of the disease (Basiri et al. 2020a). Furthermore, the state of immunological hyperinflammation may play a fundamental role in the morbidity and mortality of patients, particularly younger ones (Merad and Martin 2020; Rokni et al. 2020; Bahrami et al. 2020; Yazdanpanah et al. 2020a; Mansourabadi et al. 2020; Saghazadeh and Rezaei 2020a; Pashaei and Rezaei 2020; Nasab et al. 2020). This prospect has called for the mobilization of multiple branches of medicine – including immunology (Mansourabadi et al. 2020; Pashaei and Rezaei 2020; Jahanshahlu and Rezaei 2020b; Pourahmad et al. 2020; Fathi and Rezaei 2020), personalized medicine (Basiri et al. 2020b), and genetics (Yousefzadegan and Rezaei 2020; Darbeheshti and Rezaei 2020) – in reaching a consensus over the potentially most effective therapeutic techniques. Ultimately, properly designed (and interpreted) randomized controlled trials are necessary to test the many therapies proposed for COVID-19 formally. These trials will benefit from clinical trialists, biostatisticians, and other members of effective clinical trial research teams.

As the COVID-19 infection continues to spread globally, the many branches of medical sciences are active and will keenly engage in the foreseeable future, often needing to take preemptive action. For example, as mentioned above, the pandemic will inflict mental health scars, such as anxiety and depression (Vieta et al. 2020). The anticipation of such consequences will inform a comprehensive approach concerning lockdowns, isolation strategies, treatment approaches, and follow-up care. On the other hand, the long-term somatic impact of the virus is still unknown (Morlacco et al. 2020). Medical scientists are in search of an effective plan for COVID-19 prevention, treatment, and rehabilitation. This plan rests on an interdisciplinary approach involving infection, re-infection, and immunity, epidemiology, physical distancing, recognition of the acute and asymptomatic disease, acute and chronic long-term multisystem complications, and clinical outcomes like morbidity and mortality, treatment, and vaccination.

52.4 Conclusion

Interdisciplinary research must be central to the international response to the COVID-19 pandemic, given its potential deleterious impact on individuals and society. The first half of 2020 has witnessed one of the greatest scientific collaborations, with experts crossing physical and intellectual boundaries (Mohamed et al. 2020a) and producing evidence-based guidance for responding to the pandemic (Holmes et al. 2020). COVID-19 has become a turning point for establishing the importance of interdisciplinary research. In this context, the contribution of health-related international bodies in integrating knowledge about the virus and improving the management of infections worldwide has been invaluable. These bodies include the World Health Organization (WHO), International Union of Nutritional Science (IUNS), International Union of Biochemistry and Molecular Biology (IUBMB), Food and Agriculture Organization (FAO), and Non-governmental Organizations (NGOs) as well as international consortia such as Universal Scientific Education and Research Network (USERN) (Momtazmanesh et al. 2020; Mohamed et al. 2020b). In particular, USERN has allowed the Integrated Science Association (ISA) to integrate different disciplines. ISA is built on top scientists and provides its members with a platform for moving beyond their own discipline. The Integrated Science book series is entitled to use such a platform.

Acknowledgments We would like to thank Prof. Jagan Mohan Rao for his critical comments on this chapter.

References

Adamu H, Muhammad M, Jingi A, Usman M (2019) Mathematical modelling using improved SIR model with more realistic assumptions. Int J Eng Appl Sci (IJEAS) 6. https://doi.org/10.31873/IJEAS.6.1.22

Aderem A, Adkins JN, Ansong C, Galagan J, Kaiser S, Korth MJ, Law GL, McDermott JG, Proll SC, Rosenberger C, Schoolnik G, Katze MG (2011) A systems biology approach to infectious disease research: innovating the pathogen-host research paradigm. mBio 2(1):e00325-00310. https://doi.org/10.1128/mBio.00325-10

Ahanchian H, Moazzen N, Faroughi MSD, Khalighi N, Khoshkhui M, Aelami MH, Haghi NSM, Rezaei N (2020) COVID-19 in a child with primary specific antibody deficiency

Ahmadi M, Saffarzadeh N, Habibi MA, Hajiesmaeili F, Rezaei N (2020) Colon cancer and SARS-CoV-2: impact of ACE2 expression in susceptibility to COVID-19. bioRxiv

Ahmed SF, Quadeer AA, McKay MR (2020a) Preliminary identification of potential vaccine targets for the COVID-19 coronavirus (SARS-CoV-2) based on SARS-CoV immunological studies. Viruses 12(3):254. https://doi.org/10.3390/v12030254

Ahmed SF, Quadeer AA, McKay MR (2020b) Preliminary identification of potential vaccine targets for the COVID-19 coronavirus (SARS-CoV-2) based on SARS-CoV immunological studies. Viruses 12(3). https://doi.org/10.3390/v12030254

Arentz M, Yim E, Klaff L, Lokhandwala S, Riedo FX, Chong M, Lee M (2020) Characteristics and outcomes of 21 critically ill patients with COVID-19 in Washington State. JAMA. https://doi.org/10.1001/jama.2020.4326

Ariana-tech (2020, February 1) Data science and crisis management. http://www.ariana-tech.com/

Arthur RF, Gurley ES, Salje H, Bloomfield LS, Jones JH (2017) Contact structure, mobility, environmental impact and behaviour: the importance of social forces to infectious disease dynamics and disease ecology. Philos Trans R Soc Lond Ser B Biol Sci 372(1719). https://doi.org/10.1098/rstb.2016.0454

Babaha F, Rezaei N (2020) Primary immunodeficiency diseases in COVID-19 pandemic: a predisposing or protective factor? Am J Med Sci. https://doi.org/10.1016/j.amjms.2020.07.027

Bago B, Rand DG, Pennycook G (2020) Fake news, fast and slow: deliberation reduces belief in false (but not true) news headlines. J Exp Psychol Gen. https://doi.org/10.1037/xge0000729

Bahrami A, Vafapour M, Moazzami B, Rezaei N (2020) Hyperinflammatory shock related to COVID-19 in a patient presenting with multisystem inflammatory syndrome in children: first case from Iran. J Paediatr Child Health. https://doi.org/10.1111/jpc.15048

Banas JA, Rains SA (2010) A meta-analysis of research on inoculation theory. Commun Monogr 77(3):281–311. https://doi.org/10.1080/03637751003758193

Basiri A, Heidari A, Nadi MF, Fallahy MTP, Nezamabadi SS, Sedighi M, Saghazadeh A, Rezaei N (2020a) Microfluidic devices for detection of RNA viruses. Rev Med Virol n/a(n/a):e2154. https://doi.org/10.1002/rmv.2154

Basiri A, Pazhouhnia Z, Beheshtizadeh N, Hoseinpour M, Saghazadeh A, Rezaei N (2020b) Regenerative medicine in COVID-19 treatment: real opportunities and range of promises. Stem Cell Rev Rep:1–13. https://doi.org/10.1007/s12015-020-09994-5

Bedford J, Enria D, Giesecke J, Heymann DL, Ihekweazu C, Kobinger G, Lane HC, Memish Z, Oh M-D, Sall AA, Schuchat A, Ungchusak K, Wieler LH (2020) COVID-19: towards controlling of a pandemic. Lancet (London, England) 395(10229):1015–1018

Bobdey S, Ray S (2020) Going viral – Covid-19 impact assessment: a perspective beyond clinical practice. J Mar Med Soc 22:9. https://doi.org/10.4103/jmms.jmms_12_20

Bollyky TJ, Gostin LO, Hamburg MA (2020) The equitable distribution of COVID-19 therapeutics and vaccines. JAMA. https://doi.org/10.1001/jama.2020.6641

Bourouiba L, Dehandschoewercker E, Bush John WM (2014) Violent expiratory events: on coughing and sneezing. J Fluid Mech 745:537–563. https://doi.org/10.1017/jfm.2014.88

Cao W, Li T (2020) COVID-19: towards understanding of pathogenesis. Cell Res 30(5):367–369. https://doi.org/10.1038/s41422-020-0327-4

Cauchemez S, Bhattarai A, Marchbanks TL, Fagan RP, Ostroff S, Ferguson NM, Swerdlow D (2011) Role of social networks in shaping disease transmission during a community outbreak of 2009 H1N1 pandemic influenza. Proc Natl Acad Sci 108(7):2825–2830. https://doi.org/10.1073/pnas.1008895108

Chen W-H, Strych U, Hotez PJ, Bottazzi ME (2020a) The SARS-CoV-2 vaccine pipeline: an overview. Curr Trop Med Rep:1–4. https://doi.org/10.1007/s40475-020-00201-6

Chen WH, Strych U, Hotez PJ, Bottazzi ME (2020b) The SARS-CoV-2 vaccine pipeline: an overview. Curr Trop Med Rep:1–4. https://doi.org/10.1007/s40475-020-00201-6

Chen Y, Liu Q, Guo D (2020c) Emerging coronaviruses: genome structure, replication, and pathogenesis. J Med Virol 92(4):418–423. https://doi.org/10.1002/jmv.25681

Christakos G, Olea RA, Serre ML, Yu H-L, Wang L-L (2005) Interdisciplinary public health reasoning and epidemic modelling: the case of black death. Springer, Berlin/Heidelberg

Dansu EJ, Seno H (2019) A model for epidemic dynamics in a community with visitor subpopulation. J Theor Biol 478:115–127. https://doi.org/10.1016/j.jtbi.2019.06.020

Darbeheshti F, Rezaei N (2020) Genetic predisposition models to COVID-19 infection. Med Hypotheses 142:109818. https://doi.org/10.1016/j.mehy.2020.109818

Dimant E, Schulte T (2019) The nature of corruption: an interdisciplinary perspective. German Law J 17(1):53–72. https://doi.org/10.1017/S2071832200019684

Emma Grey Ellis (2020, February 24) The coronavirus outbreak is a petri dish for conspiracy theories

Enwemeka CS, Bumah VV, Masson-Meyers DS (2020) Light as a potential treatment for pandemic coronavirus infections: a perspective. J Photochem Photobiol B Biol 207:111891. https://doi.org/10.1016/j.jphotobiol.2020.111891

EPFL Scientists (2020, April 30) COVID-19: using wastewater to track the pandemic. https://neurosciencenews.com/coronavirus-waste-water-16301/

Ewald J, Sieber P, Garde R, Lang SN, Schuster S, Ibrahim B (2020) Trends in mathematical modeling of host-pathogen interactions. Cell Mol Life Sci 77(3):467–480. https://doi.org/10.1007/s00018-019-03382-0

Fathi N, Rezaei N (2020) Lymphopenia in COVID-19: therapeutic opportunities. Cell Biol Int 44(9):1792–1797. https://doi.org/10.1002/cbin.11403

Godin B (2006) The linear model of innovation: the historical construction of an analytical framework. Sci Technol Hum Values 31(6):639–667. https://doi.org/10.1177/0162243906291865

Hanaei S, Rezaei N (2020) COVID-19: developing from an outbreak to a pandemic. Arch Med Res. https://doi.org/10.1016/j.arcmed.2020.04.021

Hartley DM, Perencevich EN (2020) Public health interventions for COVID-19: emerging evidence and implications for an evolving public health crisis. JAMA

Hawkley LC, Cacioppo JT (2010) Loneliness matters: a theoretical and empirical review of consequences and mechanisms. Ann Behav Med 40(2):218–227. https://doi.org/10.1007/s12160-010-9210-8

Hawryluck L, Gold WL, Robinson S, Pogorski S, Galea S, Styra R (2004) SARS control and psychological effects of quarantine, Toronto, Canada. Emerg Infect Dis 10(7):1206–1212. https://doi.org/10.3201/eid1007.030703

Holmes EA, O'Connor RC, Perry VH, Tracey I, Wessely S, Arseneault L, Ballard C, Christensen H, Cohen Silver R, Everall I, Ford T, John A, Kabir T, King K, Madan I, Michie S, Przybylski AK, Shafran R, Sweeney A, Worthman CM, Yardley L, Cowan K, Cope C, Hotopf M, Bullmore E (2020) Multidisciplinary research priorities for the COVID-19 pandemic: a call for action for mental health science. Lancet Psychiatry 7(6):547–560. https://doi.org/10.1016/S2215-0366(20)30168-1

Huang H, Fan C, Li M, Nie HL, Wang FB, Wang H, Wang R, Xia J, Zheng X, Zuo X, Huang J (2020a) COVID-19: a call for physical scientists and engineers. ACS Nano 14(4):3747–3754. https://doi.org/10.1021/acsnano.0c02618

Huang L, Zhang X, Zhang X, Wei Z, Zhang L, Xu J, Liang P, Xu Y, Zhang C, Xu A (2020b) Rapid asymptomatic transmission of COVID-19 during the incubation period demonstrating strong infectivity in a cluster of youngsters aged 16–23 years outside Wuhan and characteristics of young patients with COVID-19: a prospective contact-tracing study. J Infect 80(6):e1–e13. https://doi.org/10.1016/j.jinf.2020.03.006

Huang X, Wei F, Hu L, Wen L, Chen K (2020c) Epidemiology and clinical characteristics of COVID-19. Arch Iran Med 23(4):268–271. https://doi.org/10.34172/aim.2020.09

Huppert A, Katriel G (2013) Mathematical modelling and prediction in infectious disease epidemiology. Clin Microbiol Infect 19(11):999–1005. https://doi.org/10.1111/1469-0691.12308

Ioannidis J (2020) The infection fatality rate of COVID-19 inferred from seroprevalence data. medRxiv.: 2020.2005.2013.20101253. https://doi.org/10.1101/2020.05.13.20101253

Jabbari P, Rezaei N (2020) With risk of reinfection, is COVID-19 here to stay? Disas Med Pub Health Prep 1. https://doi.org/10.1017/dmp.2020.274

Jabbari P, Jabbari F, Ebrahimi S, Rezaei N (2020) COVID-19: a chimera of two pandemics. Disaster Med Public Health Prep:1–2. https://doi.org/10.1017/dmp.2020.223

Jahanshahlu L, Rezaei N (2020a) Central nervous system involvement in COVID-19. Arch Med Res. https://doi.org/10.1016/j.arcmed.2020.05.016

Jahanshahlu L, Rezaei N (2020b) Monoclonal antibody as a potential anti-COVID-19. Biomed Pharmaco Biomed Pharmaco 129:110337. https://doi.org/10.1016/j.biopha.2020.110337

jean laurent casanova (2020, April) COVID human genetic efforts. https://www.covidhge.com/

Kafieh R, Arian R, Saeedizadeh N, Minaee S, Yadav SK, Vaezi A, Rezaei N, Javanmard SH (2020) COVID-19 in Iran: a deeper look into the future. medRxiv

Kivits J, Ricci L, Minary L (2019) Interdisciplinary research in public health: the 'why' and the 'how'. J Epidemiol Community Health 73(12):1061–1062. https://doi.org/10.1136/jech-2019-212511

Konrad R, Eberle U, Dangel A, Treis B, Berger A, Bengs K, Fingerle V, Liebl B, Ackermann N, Sing A (2020) Rapid establishment of laboratory diagnostics for the novel coronavirus SARS-CoV-2 in Bavaria, Germany, February 2020. Euro Surv Bull Eur sur les maladies transmissibles Eur Comm Dis Bull 25(9). https://doi.org/10.2807/1560-7917.es.2020.25.9.2000173

Kowalski W, Walsh T, Petraitis V (2020) 2020 COVID-19 coronavirus ultraviolet susceptibility. https://doi.org/10.13140/RG.2.2.22803.22566

Li G, Fan Y, Lai Y, Han T, Li Z, Zhou P, Pan P, Wang W, Hu D, Liu X, Zhang Q, Wu J (2020a) Coronavirus infections and immune responses. J Med Virol 92(4):424–432. https://doi.org/10.1002/jmv.25685

Li Q, Guan X, Wu P, Wang X, Zhou L, Tong Y, Ren R, Leung KSM, Lau EHY, Wong JY, Xing X, Xiang N, Wu Y, Li C, Chen Q, Li D, Liu T, Zhao J, Liu M, Tu W, Chen C, Jin L, Yang R, Wang Q, Zhou S, Wang R, Liu H, Luo Y, Liu Y, Shao G, Li H, Tao Z, Yang Y, Deng Z,

Liu B, Ma Z, Zhang Y, Shi G, Lam TTY, Wu JT, Gao GF, Cowling BJ, Yang B, Leung GM, Feng Z (2020b) Early transmission dynamics in Wuhan, China, of novel coronavirus–infected pneumonia. N Engl J Med 382(13):1199–1207. https://doi.org/10.1056/NEJMoa2001316

Li Z, Ge J, Yang M, Feng J, Qiao M, Jiang R, Bi J, Zhan G, Xu X, Wang L, Zhou Q, Zhou C, Pan Y, Liu S, Zhang H, Yang J, Zhu B, Hu Y, Hashimoto K, Jia Y, Wang H, Wang R, Liu C, Yang C (2020c) Vicarious traumatization in the general public, members, and non-members of medical teams aiding in COVID-19 control. Brain Behav Immun:S0889-1591(0820)30309-30303. https://doi.org/10.1016/j.bbi.2020.03.007

Liu X, Kakade M, Fuller CJ, Fan B, Fang Y, Kong J, Guan Z, Wu P (2012) Depression after exposure to stressful events: lessons learned from the severe acute respiratory syndrome epidemic. Compr Psychiatry 53(1):15–23. https://doi.org/10.1016/j.comppsych.2011.02.003

Livingston E, Desai A, Berkwits M (2020) Sourcing personal protective equipment during the COVID-19 pandemic. JAMA. https://doi.org/10.1001/jama.2020.5317

Lotfi M, Rezaei N (2020) SARS-CoV-2: a comprehensive review from pathogenicity of the virus to clinical consequences. J Med Virol. https://doi.org/10.1002/jmv.26123

Lotfi M, Hamblin MR, Rezaei N (2020) COVID-19: transmission, prevention, and potential therapeutic opportunities. Clinica Chim Acta Int J Clin Chem 508:254–266. https://doi.org/10.1016/j.cca.2020.05.044

Mallapaty S (2020) How sewage could reveal true scale of coronavirus outbreak. Nature 580(7802):176–177. https://doi.org/10.1038/d41586-020-00973-x

Mansourabadi AH, Sadeghalvad M, Mohammadi-Motlagh HR, Rezaei N (2020) The immune system as a target for therapy of SARS-CoV-2: a systematic review of the current immunotherapies for COVID-19. Life Sci 258:118185. https://doi.org/10.1016/j.lfs.2020.118185

Medicine Io (2006) Reusability of facemasks during an influenza pandemic: facing the flu. The National Academies Press, Washington, DC. https://doi.org/10.17226/11637

Merad M, Martin JC (2020) Pathological inflammation in patients with COVID-19: a key role for monocytes and macrophages. Nat Rev Immunol. https://doi.org/10.1038/s41577-020-0331-4

Ming R-X, Liu J, Cheung WKW, Wan X (2016) Stochastic modelling of infectious diseases for heterogeneous populations. Infect Dis Poverty 5(1):107. https://doi.org/10.1186/s40249-016-0199-5

Moazzami B, Razavi-Khorasani N, Dooghaie Moghadam A, Farokhi E, Rezaei N (2020) COVID-19 and telemedicine: immediate action required for maintaining healthcare providers well-being. J Clin Virol Off Publ Pan Am Soc Clin Virol 126:104345. https://doi.org/10.1016/j.jcv.2020.104345

Mohamed K, Rodríguez-Román E, Rahmani F, Zhang H, Ivanovska M, Makka SA, Joya M, Makuku R, Islam MS, Radwan N, Rahmah L, Goda R, Abarikwu SO, Shaw M, Zoghi S, Irtsyan S, Ling I, Cseprekal O, Faten A-B, Hazar Sayar E, Soloukey C, Grancini G, Rezaei N (2020a) Borderless collaboration is needed for COVID-19 – a disease that knows no borders. Infect Cont Hosp Epidemiol:1–2

Mohamed K, Rodríguez-Román E, Rahmani F, Zhang H, Ivanovska M, Makka SA, Joya M, Makuku R, Islam MS, Radwan N, Rahmah L, Goda R, Abarikwu SO, Shaw M, Zoghi S, Irtsyan S, Ling I, Cseprekal O, Faten AB, Hazar Sayar E, Soloukey C, Grancini G, Rezaei N (2020b) Borderless collaboration is needed for COVID-19-A disease that knows no borders. Infect Cont Hosp Epidemiol:1–2. https://doi.org/10.1017/ice.2020.162

Mohamed K, Yazdanpanah N, Saghazadeh A, Rezaei N (2020c) Computational drug discovery and repurposing for the treatment of Covid-19: a systematic review. Available at SSRN 3583748

Momtazmanesh S, Ochs HD, Uddin LQ, Perc M, Routes JM, Vieira DN, Al-Herz W, Baris S, Prando C, Rosivall L, Abdul Latiff AH, Ulrichs T, Roudenok V, Aldave Becerra JC, Salunke DB, Goudouris E, Condino-Neto A, Stashchak A, Kryvenko O, Stashchak M, Bondarenko A, Rezaei N (2020) All together to fight COVID-19. Am J Trop Med Hyg 102(6):1181–1183. https://doi.org/10.4269/ajtmh.20-0281

Moradian N, Ochs HD, Sedikies C, Hamblin MR, Camargo CA, Martinez JA, Biamonte JD, Abdollahi M, Torres PJ, Nieto JJ, Ogino S, Seymour JF, Abraham A, Cauda V, Gupta S, Ramakrishna S, Sellke FW, Sorooshian A, Wallace Hayes A, Martinez-Urbistondo M, Gupta M, Azadbakht L, Esmaillzadeh A, Kelishadi R, Esteghamati A, Emam-Djomeh Z, Majdzadeh R, Palit P, Badali H, Rao I, Saboury AA, Jagan Mohan Rao L, Ahmadieh H, Montazeri A, Fadini GP, Pauly D, Thomas S, Moosavi-Movahed AA, Aghamohammadi A, Behmanesh M, Rahimi-Movaghar V, Ghavami S, Mehran R, Uddin LQ, Von Herrath M, Mobasher B, Rezaei N (2020) The urgent need for integrated science to fight COVID-19 pandemic and beyond. J Transl Med 18(1):205. https://doi.org/10.1186/s12967-020-02364-2

Morlacco A, Motterle G, Zattoni F (2020) The multifaceted long-term effects of the COVID-19 pandemic on urology. Nat Rev Urol. https://doi.org/10.1038/s41585-020-0331-y

Nasab MG, Saghazadeh A, Rezaei N (2020) SARS-CoV-2-a tough opponent for the immune system. Arch Med Res. https://doi.org/10.1016/j.arcmed.2020.05.020

Okumus F, Mv N, Koseoglu MA, Bilgihan A (2018) Interdisciplinary research in tourism. Tour Manag 69:540–549. https://doi.org/10.1016/j.tourman.2018.05.016

Organization, PAH (2020, April 30) COVID-19 factsheets: understanding the infodemic and misinformation in the fight against COVID-19

Oxford Economics (2020, March 31) Global economic prospects: coronavirus inflicts a short, sharp recession on world economy. http://resources.oxfordeconomics.com/world-economic-prospects-executive-summary

Parmet WE, Sinha MS (2020) Covid-19 – the law and limits of quarantine. N Engl J Med. https://doi.org/10.1056/NEJMp2004211

Pashaei M, Rezaei N (2020) Immunotherapy for SARS-CoV-2: potential opportunities. Expert Opin Biol Ther:1–5. https://doi.org/10.1080/14712598.2020.1807933

Pourahmad R, Moazzami B, Rezaei N (2020) Efficacy of plasmapheresis and immunoglobulin replacement therapy (IVIG) on patients with COVID-19. SN Comprehen Clin Med:1–5

Qu G, Li X, Hu L, Jiang G (2020) An imperative need for research on the role of environmental factors in transmission of novel coronavirus (COVID-19). Environ Sci Technol. https://doi.org/10.1021/acs.est.0c01102

Rabiee N, Rabiee M, Bagherzadeh M, Rezaei N (2020) COVID-19 and picotechnology: potential opportunities. Med Hypotheses 144:109917. https://doi.org/10.1016/j.mehy.2020.109917

Rafa de Miguel EM, Pacho L, Ayuso S, Ferrer I (2020, March 30) Tracking the coronavirus: why does each country count deaths differently? https://english.elpais.com/society/2020-03-30/tracking-the-coronavirus-why-does-each-country-count-deaths-differently.html

Remuzzi A, Remuzzi G (2020) COVID-19 and Italy: what next? Lancet (London, England). doi:https://doi.org/10.1016/s0140-6736(20)30627-9

Rezaei N (2020a) COVID-19 affects healthy pediatricians more than pediatric patients. Infect Cont Hosp Epidemiol:1. https://doi.org/10.1017/ice.2020.139

Rezaei N (2020b) COVID-19 and medical biotechnology. Avicenna J Med Biotechnol 12(3):139

Rihan FA, Anwar MN (2012) Qualitative analysis of delayed SIR epidemic model with a saturated incidence rate. https://doi.org/10.1155/2012/408637

Rismanbaf A (2020) Potential treatments for COVID-19; a narrative literature review. Arch Acad Emerg Med 8(1):e29–e29

Rodriguez-Morales AJ, Cardona-Ospina JA, Gutierrez-Ocampo E, Villamizar-Pena R, Holguin-Rivera Y, Escalera-Antezana JP, Alvarado-Arnez LE, Bonilla-Aldana DK, Franco-Paredes C, Henao-Martinez AF, Paniz-Mondolfi A, Lagos-Grisales GJ, Ramirez-Vallejo E, Suarez JA, Zambrano LI, Villamil-Gomez WE, Balbin-Ramon GJ, Rabaan AA, Harapan H, Dhama K, Nishiura H, Kataoka H, Ahmad T, Sah R (2020) Clinical, laboratory and imaging features of COVID-19: a systematic review and meta-analysis. Trav Med Infect Dis:101623. https://doi.org/10.1016/j.tmaid.2020.101623

Rokni M, Hamblin MR, Rezaei N (2020) Cytokines and COVID-19: friends or foes? Hum Vaccin Immunother:1–3

roland Oliphant (2020, March 31) Counting coronavirus: how are different countries calculating death tolls, and can you trust them? https://www.telegraph.co.uk/news/2020/03/31/counting-coronavirus-different-countries-calculating-death-tolls/

Rzymski P, Nowicki M, Mullin GE, Abraham A, Rodríguez-Román E, Petzold MB, Bendau A, Sahu KK, Ather A, Naviaux A-F (2020) Quantity does not equal quality: scientific principles cannot be sacrificed. Int Immunopharmacol 86:106711

Saccucci M, Bruni E, Uccelletti D, Bregnocchi A, Sarto MS, Bossù M, Di Carlo G, Polimeni A (2018) Surface disinfections: present and future. J Nanomater 2018:8950143. https://doi.org/10.1155/2018/8950143

Saghazadeh A, Rezaei N (2020a) Immune-epidemiological parameters of the novel coronavirus – a perspective. Expert Rev Clin Immunol 16(5):465–470. https://doi.org/10.1080/1744666x.2020.1750954

Saghazadeh A, Rezaei N (2020b) Towards treatment planning of COVID-19: rationale and hypothesis for the use of multiple immunosuppressive agents: anti-antibodies, immunoglobulins, and corticosteroids. Int Immunopharmacol 84:106560. https://doi.org/10.1016/j.intimp.2020.106560

Sahu KK, Siddiqui AD, Rezaei N, Cerny J (2020) Challenges for management of immune thrombocytopenia during COVID-19 pandemic. J Med Virol. https://doi.org/10.1002/jmv.26251

Saleki K, Banazadeh M, Saghazadeh A, Rezaei N (2020) The involvement of the central nervous system in patients with COVID-19. Rev Neurosci 31(4):453–456. https://doi.org/10.1515/revneuro-2020-0026

Seno H, Dansu E (2019) A model for epidemic dynamics in a community with visitor subpopulation

Shamshirian D, Rezaei N (2020) Cardiovascular diseases burden in COVID-19: systematic review and meta-analysis

Sharifkashani S, Bafrani MA, Khaboushan AS, Pirzadeh M, Kheirandish A, Yavarpour Bali H, Hessami A, Saghazadeh A, Rezaei N (2020) Angiotensin-converting enzyme 2 (ACE2) receptor and SARS-CoV-2: potential therapeutic targeting. Eur J Pharmacol 884:173455. https://doi.org/10.1016/j.ejphar.2020.173455

Sheera Frenkel DA, Zhong R (2020, March 8) Surge of virus misinformation stumps Facebook and Twitter. https://www.nytimes.com/2020/03/08/technology/coronavirus-misinformation-social-media.html

Shereen MA, Khan S, Kazmi A, Bashir N, Siddique R (2020) COVID-19 infection: origin, transmission, and characteristics of human coronaviruses. J Adv Res 24:91–98. https://doi.org/10.1016/j.jare.2020.03.005

Siettos CI, Russo L (2013) Mathematical modeling of infectious disease dynamics. Virulence 4(4):295–306. https://doi.org/10.4161/viru.24041

Smith KM, Machalaba CC, Seifman R, Feferholtz Y, Karesh WB (2019) Infectious disease and economics: the case for considering multi-sectoral impacts. One Heal (Amsterdam, Netherlands) 7:100080. https://doi.org/10.1016/j.onehlt.2018.100080

Soheilypour M, Mofrad MRK (2018) Agent-based modeling in molecular systems biology. BioEssays News

Rev Mol Cell Dev Biol 40(7):e1800020. https://doi.org/10.1002/bies.201800020

Speers DJ (2006) Clinical applications of molecular biology for infectious diseases. Clin Biochem Rev 27(1):39–51

Srinivasa Rao ASR, Vazquez JA (2020) Identification of COVID-19 can be quicker through artificial intelligence framework using a mobile phone-based survey when cities and towns are under quarantine. Infect Cont Hosp Epidemiol:1–5. https://doi.org/10.1017/ice.2020.61

The Lancet (2020) Palliative care and the COVID-19 pandemic. Lancet (London, England) 395(10231):1168–1168. https://doi.org/10.1016/S0140-6736(20)30822-9

Tu Y-F, Chien C-S, Yarmishyn AA, Lin Y-Y, Luo Y-H, Lin Y-T, Lai W-Y, Yang D-M, Chou S-J, Yang Y-P, Wang M-L, Chiou S-H (2020) A review of SARS-CoV-2 and the ongoing clinical trials. Int J Mol Sci 21(7):2657. https://doi.org/10.3390/ijms21072657

Uchenna A, Ossai O (2016) Use of contact isolation to prevent spread: ebola outbreak in a healthworkers base camp, Port Loko District, Sierra Leone, March 2015. Int J Infect Dis 45:216. https://doi.org/10.1016/j.ijid.2016.02.491

Velavan TP, Meyer CG (2020) The COVID-19 epidemic. Trop Med Int Heal TM & IH 25(3):278–280. https://doi.org/10.1111/tmi.13383

Vieta E, Pérez V, Arango C (2020) Psychiatry in the aftermath of COVID-19. Revista de Psiquiatría y Salud Mental. https://doi.org/10.1016/j.rpsm.2020.04.004

Warwick J, McKibbin, Fernando R (2020, March 2) The global macroeconomic impacts of COVID-19: seven scenarios. https://www.brookings.edu/research/the-global-macroeconomic-impacts-of-covid-19-seven-scenarios/

Weiss SR, Leibowitz JL (2011) Coronavirus pathogenesis. Adv Virus Res 81:85–164. https://doi.org/10.1016/b978-0-12-385885-6.00009-2

WHO (2020a, April 15) Coronavirus disease 2019 (COVID-19) situation report – 86. https://www.who.int/publications/i/item/coronavirus-disease-2019-(covid-19)

WHO (2020b, March 11) WHO Director-General's opening remarks at the media briefing on COVID-19

WHO (2020c, May 27) Coronavirus disease (COVID-2019) situation reports. https://www.who.int/emergencies/diseases/novel-coronavirus-2019/situation-reports

Wong DW, Oudit GY, Reich H, Kassiri Z, Zhou J, Liu QC, Backx PH, Penninger JM, Herzenberg AM, Scholey JW (2007) Loss of angiotensin-converting enzyme-2 (Ace2) accelerates diabetic kidney injury. Am J Pathol 171(2):438–451. https://doi.org/10.2353/ajpath.2007.060977

Yazdanpanah F, Hamblin MR, Rezaei N (2020a) The immune system and COVID-19: friend or foe? Life Sci 256:117900. https://doi.org/10.1016/j.lfs.2020.117900

Yazdanpanah N, Saghazadeh A, Rezaei N (2020b) Anosmia: a missing link in the neuroimmunology of coronavirus disease 2019 (COVID-19). Rev Neurosci. https://doi.org/10.1515/revneuro-2020-0039

Youn JS, Csavina J, Rine KP, Shingler T, Taylor MP, Sáez AE, Betterton EA, Sorooshian A (2016) Hygroscopic properties and respiratory system deposition behavior of particulate matter emitted by mining and smelting operations. Environ Sci Technol 50(21):11706–11713. https://doi.org/10.1021/acs.est.6b03621

Yousefi R, Moosavi-Movahedi AA (2020) Achilles' heel of the killer virus: the highly important molecular targets for hitting SARS-CoV-2 that causes COVID-19. J Iran Chem Soc:1–2. https://doi.org/10.1007/s13738-020-01939-6

Yousefzadegan S, Rezaei N (2020) Case report: death due to COVID-19 in three brothers. Am J Trop Med Hyg 102(6):1203–1204. https://doi.org/10.4269/ajtmh.20-0240

Yu D, Poon SK, Tran V, Lam MK, Hines M, Brunner M, Power E, Shaw T, Togher L (2017) Enabler for inter-disciplinary eHealthcare: a qualitative study. Stud Health Technol Inform 239:160–166

Zaim S, Chong JH, Sankaranarayanan V, Harky A (2020) COVID-19 and multi-organ response. Curr Probl Cardiol:100618. https://doi.org/10.1016/j.cpcardiol.2020.100618

Health and Art (HEART): Integrating Science and Art to Fight COVID-19

53

Nima Rezaei, Aida Vahed, Heliya Ziaei, Negin Bashari, Saina Adiban Afkham, Fatemeh Bahrami, Sara Bakhshi, Alireza Ghanadan, Atlasi Ghanadan, Nastaran Hosseini, Pariya Kafi, Reihaneh Khalilianfard, Kawthar Mohammed, Sepideh Sargoli, Kosar Tavasoli, Mahya Zare, and Amene Saghazadeh

Abstract

Netting individuals separated from each other by vast distances; the present condition of COVID-19 needs art and its extraordinary capacity to connect human beings and integrate scientific disciplines. We can predict that the COVID-19 pandemic would leave the mind lonely and vulnerable to diseases, for, on the one hand, the COVID-19 pandemic and related problems, in particular social isolation, are itself stressor. On the other hand, studies confirm the potential of COVID-19 to involve the central nervous system by affecting the immune system, either directly or indirectly. The COVID-19 condition, thus, calls for a necessary compensation of loneliness to reduce the psychological impact of the pandemic. Not only art can fulfill this purpose by meeting social affiliation needs, but also its

N. Rezaei (✉)
Research Center for Immunodeficiencies, Tehran University of Medical Sciences, Tehran, Iran

Integrated Science Association (ISA), Universal Scientific Education and Research Network (USERN), Tehran, Iran

Department of Immunology, School of Medicine, Tehran University of Medical Sciences, Tehran, Iran

Health and Art (HEART), Universal Scientific Education and Research Network (USERN), Tehran, Iran
e-mail: rezaei_nima@tums.ac.ir

A. Vahed · H. Ziaei · N. Bashari · S. A. Afkham · F. Bahrami · S. Bakhshi · A. Ghanadan · A. Ghanadan · N. Hosseini · P. Kafi · R. Khalilianfard · K. Mohammed · K. Tavasoli · M. Zare
Health and Art (HEART), Universal Scientific Education and Research Network (USERN), Tehran, Iran

S. Sargoli
Association of Science and Art (ASA), Universal Scientific Education and Research Network (USERN), Tehran, Iran

A. Saghazadeh
Research Center for Immunodeficiencies, Tehran University of Medical Sciences, Tehran, Iran

Health and Art (HEART), Universal Scientific Education and Research Network (USERN), Tehran, Iran

MetaCognition Interest Group (MCIG), Universal Scientific Education and Research Network (USERN), Tehran, Iran

N. Rezaei (ed.), *Coronavirus Disease - COVID-19*, Advances in Experimental Medicine and Biology 1318, https://doi.org/10.1007/978-3-030-63761-3_53

related creativity is a definite achievement of the performer while acting as a motivation facilitator of creation for the observer. Besides, artworks that illustrate effective hygiene behaviors and physical distancing in an easy-to-understand manner could help health information systems to control the spread of COVID-19. The integration of art with biomedical science applied for simulation of the infected population, lung imaging data, and the viral surface has been useful for prediction of the spread of disease and earlier diagnosis of COVID-19 by imaging techniques and might be a contributor to drug discovery for COVID-19. Also, arts admirably influence the immunoemotional regulatory system so that not only would it enable humanity to tolerate quarantine but also enhance antiviral immunity. More interestingly, the effects of dance have been observed in children, elderly, healthcare workers, and pregnant women, which have been of special attention during the COVID-19 pandemic. In summary, arts provide us powerful tools for tolerating the quarantine time and enhancing the immune system, educating behavioral tips for hygiene practices and physical distancing and in psychosocial care of vulnerable populations during the pandemic.

Keywords

Art · COVID-19 · Health · Metacognition · Integrated sciences · Social isolation

53.1 Introduction

Health promotion is the mission of medicine that will be well-accomplished by humanities thinking interdisciplinarity, not focusing on biomedical thinking merely. Medical humanities is an interdisciplinary field developed for this purpose. It involves different fields of science, such as literature, philosophy, ethics, history, and religion, which, when integrated, can provide opportunities for special patient-physician relationships (Stewart and Swain 2016). Patient-physician relationships can be a powerful tool for identifying and addressing patients' unmet needs (Hafizi et al. 2014). Active listening and observing the patients while they are sharing their story are the skills required for the identification of patients' problems, while empathy and self-reflection are a first step toward addressing the source of the individual problem.

Arts in practice can, in general, encourage storytelling, enhance empathy in the healthcare environment, and help patients and practitioners develop effective communication. In particular, visual and performing artists help people learn and analyze arts, by which the brain would rescue immersion in thoughts of disease, disability, and death, break the silence, and create his arts. When creating arts, people have opportunities to express their emotions, improve self-awareness and self-reflection, and build better coping mechanisms. Accordingly, the arts offer a form of psychotherapy.

While professionals steadily insist on harvesting physical and mental islands of health by deep borders, the evidence is mounting that the psychological states closely correlate to the physiological states, as such psychiatric disorders to the physical diseases (Saghazadeh and Rezaei 2019; Saghazadeh et al. 2019a). Beyond the act of art as psychotherapy, we can, therefore, expect that physical health have been improved by the treatment of mental health with the arts. As reviewed in Stuckey and Nobel (2010), current literature corroborates that art-based interventions produce benefits to both psychological and physiological health.

Uncertainty is a well-appreciated source of stress, and uncertainty conditions where a surprise is expected to cause emotional responses can affect both the brain and systemic function (Greco and Roger 2003; Monat et al. 1972; Peters et al. 2017). Under these conditions, the vertebrate brain increases the energy budgets as a means of mitigating the uncertainty. Persistent increases in cerebral energy correspond to cognitive impairment. Also, the whole body, in the attempt to meet the increased demand of the brain, faces challenges, which particularly trouble the autonomic cardiovascular function. In this manner, the selfish act of the brain can lead to cardio- and cerebrovascular events, and uncertain conditions are a trigger of this act.

In December 2019, a new coronavirus occurred to humans (hCoV). This small enveloped virus, called severe acute respiratory syndrome coronavirus 2 (SARS-CoV-2), with approximately 120 nm in diameter, could spread worldwide and has already caused a global pandemic, with more than 400,000 people died by mid-June 2020. The so-called coronavirus disease 2019 (COVID-19) has shown the ability to spread in the body and affect critical organs. It can have potentially fatal consequences, though it will either remain asymptomatic or minimally symptomatic in more than 80% of cases. In the current pandemic condition, uncertainties exist in various research areas, including the prevention, transmission, management, and treatment of COVID-19. Therefore, it has challenged not only human physical health but also his mental health. Such large numbers of incidence and death have made the 2020 pandemic of COVID-19 as a top-ranked source of stress for humans. Moreover, physical distancing is maintained as an indication of that highly contagious disease, and, consequently, social support and its power of stress-buffering under current conditions are not permitted. COVID-19 has shown a global capacity for inflicting people from patients and society to healthcare professionals (Yazdanpanah et al. 2020b). Now stress has filled the home and healthcare environments, and that the popular way of social support to relieve stress does not apply to the present time, the virtual forms of art might help to bridge the physical distances between people and then become capable of improvement of their health.

53.2 Definition of Health

The definitions of health and disease reflect the nature of human and so must dynamically relate to the degree to which humanity demands on biological and psychosocial aspects. A dynamic definition of health can be formulated as follows:

> Health is a dynamic state of wellbeing characterized by a physical, mental and social potential, which satisfies the demands of a life commensurate with age, culture, and personal responsibility. If the potential is insufficient to satisfy these demands the state is disease. (Bircher 2005)

Where the term "disease" refers to deviations of health due to different conditions people have been exposed to them. Thus, it covers the full range of states, including malady, illness, ill health, and sickness, that call for medical care. In this manner, health is identical to the potential to manage short-term, medium-term, and long-term demands that happen throughout life, and this potential would be dependent on biological and individual factors. During the COVID-19 pandemic, genetic factors seem to make some families highly susceptible to severe phenotypes of disease (Yousefzadegan and Rezaei 2020; Darbeheshti and Rezaei 2020; Ahanchian et al. 2020), and, also, more than 95% of deaths are reported over the age of 44 years. Genetic factors and age are biological factors that can protect or predispose the host to the COVID-19 (Babaha and Rezaei 2020), while personal hygiene culture is an individual factor affecting health during the disaster.

53.3 Integration of Science and Art

On the joy of understanding the way an artist has used different elements in constructing a work of art as a whole, we can understand the components by which the self is assembled (Räsänen 1998). When perceiving our self as just an element of that work of art, we can value our current position, and evaluate the individual limits of our function, and find clues to locate ourselves closer to the goals of the perfect world.

The art of art goes beyond the basic emotion of happiness by providing a privilege to think without boundaries. For this, despite its neurobiological mechanisms remained to be fully understood, art integration is a solution to bridge the gap between disciplines and figure out interdisciplinary and transdisciplinary platforms. In science and research, these platforms produce fruits that we could never find in the context of each discipline separately. For example, biomedical engineering integrates engineering science with the biological and medical sciences to solve health-related problems. Imaging and visual arts are a means of simulation and illustration that fill up the interval between engineering and biomedical sciences. Also, the integration of art with education would pave the way for the cultivation of productive styles and skills, such as empathy (Bradshaw 2016; Reif and Grant 2010), learning, and creativity (Marshall 2005). Altogether, with the integration of science and art and the possibility of scientific problems being transformed into aesthetic experiences, we can adopt a new approach to systems thinking (Marshall 2014).

Early diagnosis was and is a matter of concern since the COVID-19 pandemic emerged. Real-time reverse transcription-polymerase chain reaction (RT-PCR) testing for the causative virus, SARS-CoV-2, is challenged with accuracy, precision, rapidity, and stability issues (Li et al. 2020b; Tahamtan and Ardebili 2020; Tang et al. 2020). Analysis of X-ray and computed tomography (CT) images could help earlier and more reliable detection of patients (Tang et al. 2020). Also, visual arts applied for the education of physical distancing, traveling, mask-wearing, and hygiene behaviors, which are vital for controlling the spread of disease.

53.4 History of Health and Art (HEART)

As mentioned above, the concept of health allows us to bring the biological and psychosocial aspects in one close association flexibly. When engaged with art activities, either as an observer or as a performer, people experience improvement in short-term and also long-term emotions. Considering the reciprocal interaction between the psychological and physiological health, the result of this experience is represented by both psychological and physiological parameters. Such effects are attributed to the action of arts on strengthening creativity, imagination, self-expression, and self-reflection. This view is supported by evidence of the most noticeable effects in the case of engagement with creative and expressive art activities, including music engagement, visual arts therapy, movement-based creative expression, and expressive writing (Stuckey and Nobel 2010).

53.4.1 Universal Scientific Education and Research Network (USERN)

In 2015, the Universal Scientific Education and Research Network (USERN) was established as an independent, nongovernmental, and nonprofit organization binding humanity together with the purpose of education and research for all people who feel passionate about science, irrespective of the GPS coordinates on which they reside. USERN has provided a network to incorporate credible sources, including ideas, facilities, human, financial, and educational resources, into life and continuously strives to improve credibility in science. The USERN has connected more than 13,000 people and 100 universities across the world, though it is not 6 years old yet. The USERN board is constituted of Nobel laureates, Abel laureates, and top scientists and provides support and supervision for junior members of the USERN who actively work in the well-defined interest groups of USERN. The publication of more than 600 high-quality research papers is only one representative of the reliability of this network in the last 5 years. The USERN celebrates its congress festival each year in the week in which November 10, the World Science Day for Peace and Development, falls.

53.4.2 Health and Art (HEART)

Just after its establishment, the USERN initiated a series of Heart and Art (HEART) activities. In the HEART group, medical doctors work hand in hand with artists aimed at providing art platforms to boost the mental health of patients, especially children, and thereby helping to improve the likelihood of successful medical care and treatment. For this purpose, the HEART group uses social media to inform members of healthcare centers about the effect of art on health and introduce no-cost ways of enhancing the nonverbal communication skills to help sick people and their families to express their feelings, control their emotions, have less anxiety, and experience less pain. Also, the HEART group holds various programs making a calm and hopeful atmosphere in healthcare centers. Some programs encourage sick patients to become actively engaged, helping them to develop a healthy and happy sense of self-confidence and self-efficiency.

53.4.2.1 International Festival of Pediatric Patients' Painting (IFPPP)

The International Festival of Pediatric Patients' Painting (IFPPP) is an annual painting festival for children and adolescents aged <18 years with a history of admission to the hospital. Annually, the top three paintings in each age category (Group A, up to 6 years; Group B, 7–10 years old; Group C, 11–14 years old; and Group D, 15–18 years old) are awarded, and the top 200 paintings will be printed in a booklet. The exhibition and awarding ceremony of the IFPPP festival took place in Iran (November 2015), Hungary (October 2016), Ukraine (November 2017), Iran (August 2018), and Hungary (November 2019). However, different countries celebrate this festival annually and announce the top-ranked painting in their own country. Figure 53.1 shows the painting gallery opening of the fifth IFPPP held at more than 40 centers in Azerbaijan, Belarus, Finland, Hungary, India, Iran, Italy, Peru, Romania, Russia, Slovenia, South Africa, Turkey, Ukraine, and the USA. The sixth IFPPP will take place in Iran (November 2020).

53.4.2.2 Kick4Sick (miniWorldCup)

The Kick4Sick (miniWorldCup) is a football competition organized by the HEART group at the national level. Kick4Sick offers inpatient toddlers and kids playing football. Adults who are famous footballers (former and current), celebrities, and medical doctors perform the coaching sessions prior to the main competition. They use tips and strategies by observing which children and teens learn effective team working and know hospital as not only a place for undergoing diagnostic and therapeutic procedures but as a place for treating pain and suffering, playing, and having fun, thereby helping them to build a positive picture of the hospital in their mind. Figure 53.2 is selected from the miniWorldCup records.

53.4.2.3 HEART Band

HEART band is made up of pediatric patients who are amazing in playing music. DAL band works with pediatric musicians to direct performance (Fig. 53.3). These bands came together in two live music performances.

53.4.2.4 International Village of Games and Art

In 2018, the theme of World Mental Health Day was defined as "Young people and mental health in a changing world." The theme reflects concerns about the consequences of the changing world for mental health, especially in young people. Also, it is consistent with the research providing evidence of the recent trends of rising stress-related consequences in children, in particular, fear of failure and psychosomatic issues (Stueck and Gloeckner 2005). Immediately, the HEART group made the event of the international village of games and art aimed at enhancing children's mental health. The event team members were international medical students who could provide fun ideas for the easy introduction of national and international arts to children. The village designed spaces for gaming competitions and painting in parallel. Recycling processes of waste materials, sound recording and reproduction, live radio, and Shahnameh-Khani were performed at the place of the village in front of children and their parents. Figure 53.4

Fig. 53.1 The painting gallery opening of the fifth IFPPP held at more than 40 centers across the world

Fig. 53.2 The miniWorldCup held at the Children Hospital Medical Center, Tehran, Iran

shows some children's activities that took place in the event.

53.4.3 Association of Science and Art (ASA)

As mentioned earlier, the integration of science and art supports a wise act, making outcomes easier and more sensible to gain for both parties. Therefore, the Association of Science and Art (ASA) was established to provide a fertile ground for joined artistic and scientific activities. For this, ASA developed a series of 12-min talks prompting junior scientists to present in public coming alongside junior artists who make an artwork. The votes of the audience, including senior scientists and artists, are counted to determine the winners of these talks, known as the USERN miniature talks. These talks were first held in the year 2018, and the last minute 2018 event was a theater of science reached on December 31, 2018,

Fig. 53.3 The HEART band: the group of pediatric musicians

Fig. 53.4 The international village of games and art, Tehran, Iran. The top three images show making mud shapes, painting, and planting flowers. The image on the bottom left corner shows the introduction of space science to kids. The image on the bottom right corner shows the event notes in different languages written by the international medical students who planned the implementation of the international village of games and art, Tehran, Iran (October 2018)

Fig. 53.5 The theater of science performed by the winners and finalists of the 2018 USERN miniature talks

Fig. 53.6 The calligraphy for the poem called Bani Adam (human kind)
"Human beings are members of a whole
In creation of one essence and soul
If one member is afflicted with pain
Other members uneasy will remain
If you've no sympathy for human pain
The name of human you cannot retain"
Adapted with permission from the Association of Science and Art (ASA), USERN; Contributors: Alireza Ghanadan (calligraphy) and Saádi Shirazi (poem)

the ceremony day for the third anniversary of USERN establishment. It was performed by the winners and finalists of the 2018 USERN miniature talks (Fig. 53.5).

53.5 COVID-19 and Psychological Problems

Figure 53.6 is a calligraphy poem that states:

Human beings are members of a whole
In creation of one essence and soul
If one member is afflicted with pain
Other members uneasy will remain
If you've no sympathy for human pain
The name of human you cannot retain

The above clearly shows that merely living during the pandemic and witnessing that the world and humanity are in pain would influence people' emotions, even if not infected themselves (Fig. 53.7). The effects might be profound as much as psychological resilience matters and, therefore, present to us mental health difficulties, in particular, anxiety, depression, panic disorder, and stress.

A survey of 1210 respondents from 194 cities in China has revealed that more than 50% of the general population are affected psychologically by the initial phase of the COVID-19 outbreak and about 30% reported moderate to severe levels of anxiety (Wang et al. 2020). The female population appeared to be more vulnerable to the psychological impact of and stress and anxiety associated with the COVID-19. Also, people (i) who had symptoms of the disease, such as chills, myalgia, cough, dizziness, coryza, and sore throat; (ii) with pre-existing chronic conditions; (iii) who reported contact history in the past 2 weeks; (iv) who felt they had not received a sufficient amount of health information; (v) who were concerned about other family members, in particular, children; and (vi) who lacked confidence about the doctor's ability to detect COVID-19 were more likely to have high levels of stress and anxiety. Interestingly, engagement with precautionary behaviors, such as washing hands with soap and water, avoiding sharing of utensils during meals, and washing hands immediately after coughing, rubbing the nose, or sneezing, correlated to lower levels of stress and anxiety.

In another nationwide survey of 52,730 valid respondents from Hong Kong, Macau, and Taiwan, more than one-third of the population noted the experience of psychological distress during the initial phase of the COVID-19 out-

Fig. 53.7 The world's pain in the human eye. The figure clearly shows that merely living during the pandemic and witnessing that the world and humanity are in pain would influence people' emotions even if not infected themselves. (Adapted with permission from the Association of Science and Art (ASA), USERN; Painting by Fatemeh Bahrami)

break and that the experience was more frequent among the female population, people aged between 18 and 30 years or 60 years above, people who were highly educated and therefore highly self-aware to navigate health information, migrant workers, and people residing in regions profoundly affected by the COVID-19 (Qiu et al. 2020).

A study of college students in China during the initial phase of the outbreak confirmed that more than 20% of students experienced mild to severe forms of anxiety (Cao et al. 2020). In this population, a history of infection of closely related people, as well as the impact of infection on economic and academic activities, predicted higher levels of anxiety, while students who lived in urban areas or with parents, students whose family income was stable, and students provided with social support reported lower levels of anxiety.

The study conducted (Li et al. 2020a) machine learning analyses of the Weibo posts made by active Weibo users (n = 17,865). Based on the word frequency data, it provided evidence that daily lives, during the COVID-19 pandemic, deal with more negative emotions, including anxiety, depression, indignation, and sensitivity to social risks, while seeded with less frequent moments of positive emotions, including happiness and satisfaction, than before.

An online survey investigating the impact of COVID-19 in the healthcare environment indicated frontline medical staff being more prone to produce fear and develop depression and anxiety compared to nonclinical staff (Momtazmanesh et al. 2020a). Frontline medical staff work in the departments of respiratory diseases, emergency, infectious disease, and critical care.

In summary, mental health difficulties are common to any occupation but become different from individual to individual and from the initial phase of the pandemic to the late and post-pandemic phases. These differences are necessary to be considered for policymaking purposes.

53.6 COVID-19, Self-Regulation, and Metacognition

Metacognition refers to the higher-order cognitive skill collection developed beyond the basic, average levels of thinking. It is a means of self-regulation of thinking and action composed of metacognitive knowledge and metacognitive strategies. Metacognitive knowledge and beliefs have been shown to moderate the association between perceived stress and negative emotions such as anxiety and depression (Spada et al. 2008), and, more interestingly, this moderating effect goes much deeper than that of self-esteem (Palmier-Claus et al. 2011). As a result, metacognitive therapy could make a substantial recovery for patients with post-traumatic stress disorder (PTSD), enduring less negative affect and lower levels of anxiety and depression (Wells and Colbear 2012). Besides, metacognitive strategies enable the mind to master the cognitive processes that undertake simple to complex operations of learning (Veenman et al. 2006), including academic learning, multisensory learning, and sensory substitution or sixth sense learning (Saghazadeh et al. 2019b).

In the context of the COVID-19 outbreak, a study of 1778 health workers and public service providers classified nearly 30% of the population as the clinical or subclinical diagnosis of PTSD. Different factors predicted the diagnosis of PTSD, including a history of previous mental disorder, interpersonal problems, and sociodemographic status (Johnson et al. 2020). Of particular interest was the significant effect of negative metacognitive beliefs on emotional distress and thoughts of danger. It calls attention to metacognitive therapy as psychotherapy for psychosocial challenges associated with COVID-19.

The COVID-19 pandemic has forced educational systems to launch e-learning programs. Self-regulation learning strategies, including metacognitive capacities, time management, critical thinking, and effort regulation, would facilitate the transition from the classroom to online learning and thereby coaching academic achievement during the pandemic period (Broadbent and Poon 2015).

53.7 Social Media and COVID-19

Social media presents online platforms for interactions to happen around numerous health topics relating to patient education, health promotions, public relations, and crisis communication (Eckler et al. 2010). Social media involve several technological approaches such as blogs, microblogging (e.g., Twitter), social networking (e.g., Facebook and Patients Like Me), video- and file-sharing sites (e.g., YouTube), e-games, and wikis (Househ et al. 2014).

The public usually turns to social media for information when emerging infectious diseases (EIDs) outbreaks occur (Tang et al. 2018). The outbreak of COVID-19 is a severe global public health emergency, and dissemination of accurate and up-to-date information is a crucial part of the collective response. So, social media has become the main channel through which the public can receive information and express their beliefs and feelings (Han et al. 2020; Ahmed et al. 2020). However, it should also be noted that while social media can have positive effects in this situation, it can also have adverse effects on public health (Viviani and Pasi 2017; Waszak et al. 2018; Pulido et al. 2020).

53.7.1 Benefits of Social Media During the COVID-19

The benefits of social media in the health sector, especially in the pandemic, can be examined in terms of its impact on three groups: governments and policymakers, scientific and medical associations, and patients or population. Regarding the positive effects of social media for the governments and policymakers, it can be said that during the COVID-19 pandemic, between different governments around the world and researchers, open sharing of clinical, epidemiological, and virologic data is helping to form international public health strategies based on public responses, potential courses of action, and public opinions from social media data, and the policymakers of each country can be prepared to deal with this pandemic based on information (Ting et al. 2020; McKendry et al. 2020; Han et al. 2020; Mohamed et al. 2020a; Momtazmanesh et al. 2020a; Moradian et al. 2020b; Rzymski et al. 2020).

On the other hand, social media allows effective communication between healthcare professionals and the population to reduce the number of face-to-face visits and alleviate the clinical load of physicians in medical centers by educating patients about health issues related to the prevention and symptoms of the disease (Moazzami et al. 2020). It also has a role in facilitating communication between members of the scientific and medical communities so that information can be shared accurately and quickly and updated accordingly (Hagg et al. 2018; Ting et al. 2020; Ahmed et al. 2020).

Social networks can contribute to educating and promoting the health of the community or patients and provide them with accurate and fast information (Figs. 53.8 and 53.9). Social networks also keep people up to date with the latest news during pandemic conditions, and that can increase the level of public awareness, such as quarantine conditions and other government guidelines. Another advantage of social media is that patients can help empower and enhance their knowledge by reading the experiences of other patients. Moreover, because people need to keep a social distance during quarantine physically, digital technology allows people to communicate virtually with each other, and this will help to provide psychological support and boost people's morale (Househ et al. 2014; Ahmed et al. 2020; Depoux et al. 2020; Tang et al. 2018).

53.7.2 Threats of Social Media During the COVID-19

Because information is collected on social media by users, one of the threats in this field is the dissemination of fake information and rumors, which has dangerous consequences such as threatening global health and awareness, which is

Fig. 53.8 The benefits of social media in framing the reality of COVID-19. Social networks can contribute to educating and promoting the health of the community or patients and provide them with accurate and fast information. (Adapted with permission from the Association of Science and Art (ASA), USERN; Designed by Ehsan Keramati)

Fig. 53.9 The benefits of social media in framing the reality of COVID-19. (Adapted with permission from the Association of Science and Art (ASA), USERN; Painting by Saina Adiban Afkham)

still one of the main concerns in public health today (Tang et al. 2018; Pulido et al. 2020).

The World Health Organization (WHO) said that "We're not just fighting an epidemic; we're fighting an infodemic. Fake news spreads faster and more easily than this virus and is just as dangerous." As well, WHO warned that the outbreak of COVID-19 was accompanied by a widespread "infodemic" or an overabundance of information – some of which was correct and some were not – which made it difficult for people to find trustworthy sources and accurate information when they needed it (Lotfi et al. 2020b; Pulido et al. 2020).

Although it is helpful to give people information about epidemics, the spread of misinformation about the coronavirus, and the constant listening to media coverage of stressful events, raises fears, anxieties, and panics among the people, which in turn creates a serious problem (Pulido et al. 2020; Ahmed et al. 2020).

Headline stress disorder is the psychological disorder caused by too many news coverages, and psychologist Dr. Steven Stosny first defined it as a high emotional response to endless reports from the news media, such as feeling anxiety and stress. Although it is not a medical diagnosis, the continued anxiety or stress may cause physical functional disorders, including palpitation, chest tightness, and insomnia, and further progression may lead to physical and mental diseases, such as anxiety disorders, depression disorders, endocrine disorders, and hypertension.

In conclusion, in a situation where the only way to deal with the coronavirus is to cut off the transmission chain through quarantine, using social media to speed up shaping people's perception of the conditions created by the virus and to increase people's knowledge and awareness, both in terms of guidelines and in terms of disease prevention and treatment, will be very helpful in their accompaniment with governments (Depoux et al. 2020; Chan et al. 2020).

However, the information that is published on social media must be accurate and reliable enough to gain public trust and reduce potential psychological harm. To achieve this goal, specialist physicians have an important role to play in improving the dissemination of accurate information. The World Health Organization also has created a WHO myth busters webpage to address and correct misinformation about the COVID-19 outbreak (Dong and Zheng 2020; Mattingly II 2015; Depoux et al. 2020).

Figure 53.10 shows the brain surrounded by the flow of SARS-CoV-2. The idea behind this painting is that as the viruses, the pandemic-related news can penetrate the mind (Figs. 53.11 and 53.12), affect the thought, and cause psychological problems in people, while the concept painted in Fig. 53.13, i.e., working together, lays out the potential of global networking to fight COVID-19.

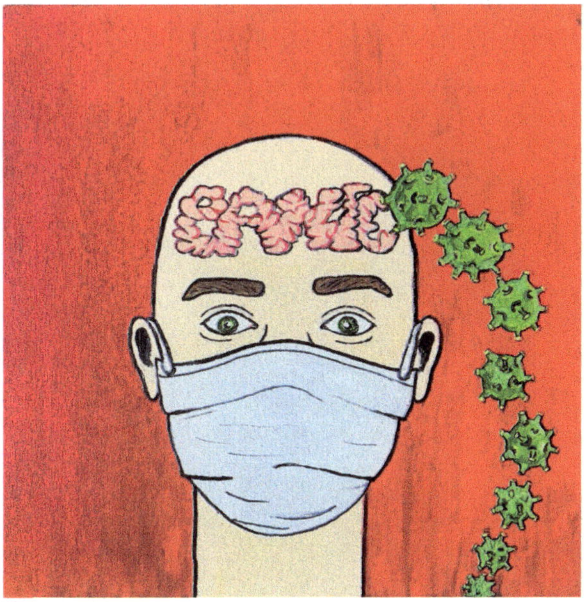

Fig. 53.10 The brain surrounded by the flow of SARS-CoV-2. The idea behind this painting is that as the viruses, the pandemic-related news can penetrate the mind. (Adapted with permission from the Association of Science and Art (ASA), USERN; Painting by Negin Bashari)

Fig. 53.11 The SARS-CoV-2 in the brain. This graphic figure clearly illustrates the capacity of the SARS-CoV-2 to enter the brain. (Adapted with permission from the Association of Science and Art (ASA), USERN; Designed by Nastaran Hosseini)

Fig. 53.12 The vulnerability of neural cells to SARS-CoV-2 infection. (Adapted with permission from the Association of Science and Art (ASA), USERN; Designed by Nastaran Hosseini)

Fig. 53.13 The potential of global networking to fight COVID-19. (Adapted with permission from the Association of Science and Art (ASA), USERN; Painting by Saina Adiban Afkham)

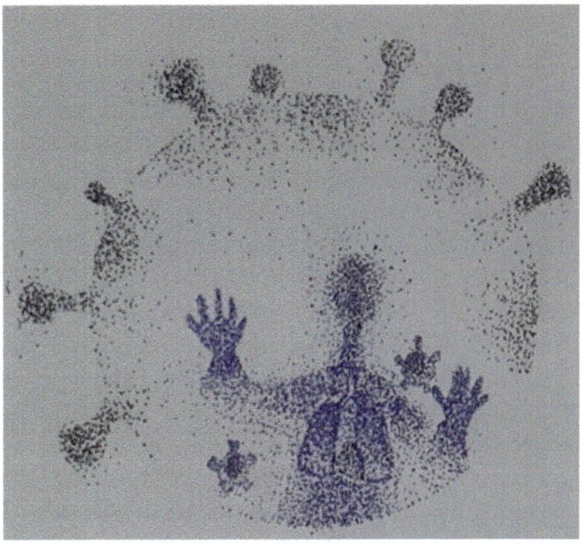

Fig. 53.14 Quarantine as the only solution currently available for the problem of COVID-19. The painting shows how the SARS-CoV-2 pandemic has trapped us in a cage, that in our eyes looks like the virus. (Adapted with permission from the Association of Science and Art (ASA), USERN; Painting by Fatemeh Bahrami)

53.8 Art to Fight COVID-19

53.8.1 Power of Art in Tolerating the Quarantine Time and Enhancing the Immune System

Months after the discovery of SARS-CoV-2, the outbreak went beyond expectations and turned into a pandemic, and the world is still waiting for a specific diagnosis, prevention, and treatment of infection (Basiri et al. 2020b; Fathi and Rezaei 2020; Jahanshahlu and Rezaei 2020b; Mohamed et al. 2020c; Saghazadeh and Rezaei 2020b; Rabiee et al. 2020; Hanaei and Rezaei 2020; Jabbari et al. 2020; Rezaei 2020b; Basiri et al. 2020a; Rezaei 2020a). Inevitably, as painted in Fig. 53.14, the SARS-CoV-2 pandemic has trapped people in a cage because quarantine was the only solution to control the first wave of the COVID-19 pandemic. There are therapeutic opportunities that look promising (Pashaei and Rezaei 2020; Sharifkashani et al. 2020). However, until the treatment is discovered, the quarantine will remain the only solution to control the disease transmission for countries undergoing the next waves of COVID-19. Separating people and restricting their movement, in particular for more than 10 days, leads to stress (Stefana et al. 2020). Quarantine-related stress causes people to overeat food, especially carbohydrate-rich food, as an attempt to relieve stress and enhance mood (Muscogiuri et al. 2020). Such eating behaviors pertain to metabolic dysfunction and inflammatory and oxidative responses (Gregersen et al. 2012), which correlate with a constellation of metabolic, autoimmune, and cardiovascular disorders (Saghazadeh et al. 2019c). Obese people are more vulnerable to adverse outcomes of COVID-19, and that the higher the inflammatory responses, the more the COVID-19 is severe (Saghazadeh and Rezaei 2020b; Bahrami et al. 2020). Therefore, on the one hand, quarantine would help reduce COVID-19 transmission among people. On the other hand, quarantine-related stress would make individuals susceptible to mental health problems in the short and long term, which, in turn, predicts the development of more severe phenotypes of COVID-19 (if the case of infection) and chronic immunometabolic disorders.

In COVID-19, different organs and systems, notably the cardiovascular system, central nervous system (CNS), and immune system, fall into a dysregulated mode (Fathi and Rezaei 2020; Lotfi and Rezaei 2020; Sahu et al. 2020; Yazdanpanah et al. 2020a; Hessami et al. 2020; Jahanshahlu and Rezaei 2020a; Lotfi et al. 2020a; Saleki et al. 2020; Mansourabadi et al. 2020; Nasab et al. 2020; Saghazadeh and Rezaei 2020a). Arts admirably influence the immuno-emotional regulatory system so that not only would it enable humanity to tolerate quarantine but also enhance antiviral immunity. Music, for example, has shown to meet social affiliation needs, lower sensitivity to stress and arousal, involve the neurocircuitry of reward, motivation, and pleasure, and regulate immune responses (Chanda and Levitin 2013). Also, a recently published systematic review demonstrated that music would cause pleasure and decrease anxiety while affecting the cardiac autonomic nervous system and enhancing heart rate variability (HRV)-related parameters (Mojtabavi et al. 2020).

Dance has also been demonstrated to regulate immune, metabolic, and hormonal responses as well (McMurray et al. 1996; Rahim et al. 2017). Interestingly, it could increase lymphocyte count, especially cytotoxic T cells, which play a role in antiviral immunity in people who underwent a regular program of dance (Rahim et al. 2017). Since there is evidence that the progression from mild to severe COVID-19 is accompanied by lymphocyte count reduction, dance might be a helpful complementary therapy for enhancing antiviral immunity. More interestingly, the effects of dance have been observed in elderly and pregnant populations (Kim et al. 2011; McMurray et al. 1996), which have been of special attention during the COVID-19 pandemic.

53.8.2 Power of Art in Educating Behavioral Tips for Hygiene Practices and Physical Distancing

The COVID-19 pandemic has brought by itself other epidemics of information usage (infodemic) and face mask usage. Governments and public health agencies have made different recommendations (Feng et al. 2020). However, most of which converge on the use of face masks in public in the cases of people who are at moderate to high risk of infection and when respiratory symptoms are present. Hand washing and physical distancing are other strategies of public health to control the spread of disease. Artworks are useful for the demonstration of how much effective hygiene practices and physical distancing can help to control the spread of disease and thereby encouraging people to do so. In particular, they can be designed in an easy-to-understand manner for children (Fig. 53.15) and those who have difficulties in learning.

Figure 53.16 is an artwork of the effect of hand hygiene on killing the SARS-CoV-2.

53.8.3 Power of Art in the Psychological Care of Healthcare Workers During the Pandemic

The recent pandemic put an extra burden on healthcare professionals who work in different specialties; this may result in higher rates of psychological problems among medical staff due to extra physical and psychological pressure during this pandemic (Walton et al. 2020). Studies revealed that the rates of anxiety, stress, PTSD, depression, insomnia, obsessive-compulsive disorders, fear, and distress are significantly higher among healthcare workers, especially in the frontline medical staff caring for critically ill patients (Walton et al. 2020; Wu et al. 2020; Pappa et al. 2020).

Fig. 53.15 Hygiene practice in children. The painting of a boy and his bear wearing facemasks shows the successful implementation of hygiene procedures. Artworks are useful for the demonstration of how much effective hygiene practices and physical distancing can help to control the spread of disease and thereby encouraging people to do so. In particular, they can be designed in an easy-to-understand manner for children. (Adapted with permission from the Association of Science and Art (ASA), USERN; Painting by Fatemeh Bahrami)

Fig. 53.16 An artwork encouraging hygiene practices to prevent the spread of SARS-CoV-2 infection. (Adapted with permission from the Association of Science and Art (ASA), USERN; Designed by Nastaran Hosseini)

Higher workload during the pandemic, caring for very sick patients, working with new strict infection protocols, less contact with their family members due to the fear of contagion, unknown duration of the pandemic, and lack of any proven treatment for the disease are some etiological factors that can predispose healthcare personnel to psychological problems (Walton et al. 2020; Wu et al. 2020). In general, they are more prone to various mental disorders because of a wide array of risk factors that are present during the pandemic (Zhang et al. 2020).

Healthcare policymakers and the government should consider protocols for reducing these problems, such as providing counseling services for healthcare workers, regular checkups for evaluating the psychological status of the personnel, and providing a supportive work climate.

The power of art and humanities in reducing stress and in enhancing the quality of healthcare activities and effectiveness has been proven; an active art program has positive effects on different aspects of working activities among healthcare workers. Regular weekly art activities had effects on the general and mental health of the nursing staff in an investigation compared to the control group; it also improved their creativity and reduced their fatigue and stress (Karpavičiūtė and Macijauskienė 2016; Sonke et al. 2015). Most of these nursing staff were encouraged to continue these art activities in the future, which helps them to have positive emotions despite their heavy work and negative environment of hospitals, which are the center of COVID-19 patients (Karpavičiūtė and Macijauskienė 2016). It has also improved communication between patients and medical staff. It provides a caring and more relaxing environment for not only medical staff but also patients (Sonke et al. 2015). The positive effect of these programs has been proven in the surgical environment in hospitals; it enhances the quality of patients' care and medical staff practices in the short-stay unit (Sonke et al. 2015). It should be considered that these programs should not have adverse effects, such as medical staff distraction in the working environment (Sonke et al. 2015).

Among different art constituents, visual art and music are the most appropriate types of art in the hospital environment. Colors are the most crucial part of the visual art that can have emotional effects. The colors with higher levels of pleasure and lower levels of excitement are suitable for conveying a sense of calm (Lankston et al. 2010). Visual art display in hospitals has positive effects on patients, visitors, and staff; it improved the clinical outcome and healing of the patients other than positive psychological effects on medical staff (Lankston et al. 2010).

On the other hand, painting or creating any artwork regarding the efforts of the healthcare workers during the outbreak by the artists for appreciating their valuable activities is very worthwhile. It causes that the medical staff perceives the sense of sympathy from the people, and it gives them the energy to keep up their activities during the crisis.

Figure 53.17 is a painting showing the pressure under which healthcare workers are during the COVID-19 pandemic. Figure 53.18 is a calligraphy that illustrates a nurse holding the SARS-CoV-2 in her hands and expresses the calligrapher's acknowledgment of nurses' efforts and sacrifices. Also, the calligrapher's sadness is behind the poem in Fig. 53.19, asking the medical doctor why you who are a physician for all people get sick.

53.8.4 Power of Art in Promoting the Development of Children During the Pandemic

In general, physical activity is an important factor in the success of making a healthy lifestyle. In particular, physical activity in childhood has been associated with a more favorable cardiovascular profile and achieving a higher peak bone mass at a later age (Boreham and Riddoch 2001). Also, research shows the positive effect of physical activity on mental health, especially reduced anxiety and enhanced self-esteem and cognitive functioning (Biddle and Asare 2011). Accordingly, guidelines recommend 60 minutes of moderate-intensity physical activity daily for children. However, children less participate in physical activity compared with previous generations. Moreover, the pandemic condition and the

Fig. 53.17 The healthcare workers: the heroes of the fight against COVID-19. (Adapted with permission from the Association of Science and Art (ASA), USERN; Painting by Sara Bakhshi)

Fig. 53.18 The calligraphy of a nurse holding the SARS-CoV-2 in her hands. The poem expresses the calligrapher's acknowledgment of nurses' efforts and sacrifices. (Adapted with permission from the Association of Science and Art (ASA), USERN; Contributors: Reihaneh Khalilianfard (calligraphy and lyrics) and Mahya Zare (painting))

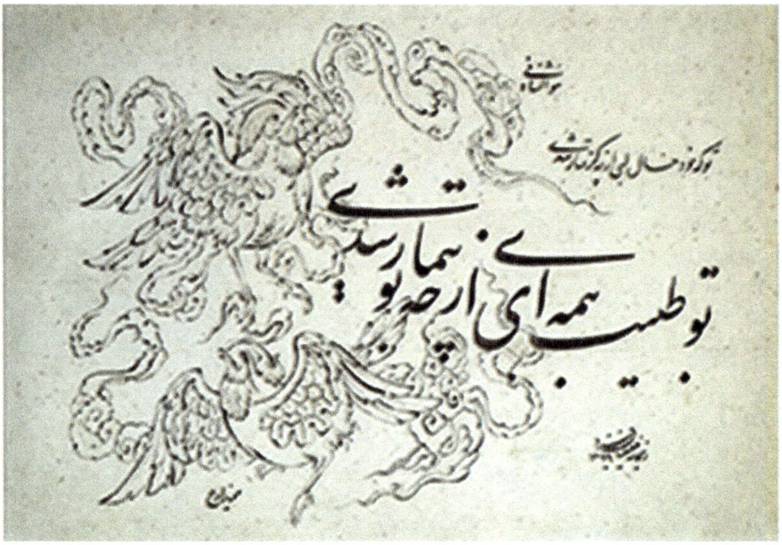

Fig. 53.19 The patients' sadness due to doctors' disease. The calligrapher's sadness is behind the poem, asking why you who are a physician for all people get sick. (Adapted with permission from the Association of Science and Art (ASA), USERN; contributors: Reihaneh Khalilianfard (calligraphy) and Mahya Zare (painting))

Fig. 53.20 The SARS-CoV-2: a sling attached to the foot of children. The painting clearly shows how the pandemic and related social distancing has restricted children' movement and affected their mode. (Adapted with permission from the Association of Science and Art (ASA), USERN Painting by Sara Bakhshi)

importance to adhere to the rules of social distancing has exacerbated this situation. As drawn in Fig. 53.20, the SARS-CoV-2 has acted like a sling attached to the foot of children.

Yoga is a mind-body movement therapy that might help to improve CNS functioning in children and adolescents (Kaley-Isley et al. 2010). Research shows the improvement of sensory-motor performance and planning and execution times in both simple and complex tasks with yoga (Galantino et al. 2008). In particular, specific yoga practices, breathing techniques, and imagination journeys apply to the pandemic condition for handling stress and developing self-regulation skills (Stueck and Gloeckner 2005; Hagen and Nayar 2014). Figure 53.21 is a painting that shows a peaceful child stretching his arms while living at home and looking out of the window.

Creative arts and plays are regarded as therapy. As reviewed in Zarobe and Bungay (2017), creative arts activities, e.g., music, dance, singing, drama, and visual arts, make children learn to be more resilient, boost confidence in their selves, build relationships, and thereby feeling less sense of loneliness and achieving a greater sense of belongingness. When occurring in the group setting, the effects are pronounced, reaching the power to promote the development of social cognition and so the personality development in children. Noteworthy mentioning is the family's role in supporting and helping children to learn through creative arts and plays (Brown 2001). And now, with the emergence of massively distributed SARS-CoV-2 (Fig. 53.22) that has limited outdoor, group activities (Fig. 53.23), the family's role in encouraging children to express their thoughts and feelings is important more than ever, and arts-based approaches greatly help them to accomplish their role. In the post-pandemic world, of course, we hope the SARS-CoV-2 does not want to stay in our home (Fig. 53.24) (Jabbari and Rezaei 2020), and its threat will be effectively neutralized (Fig. 53.25);

Fig. 53.21 The mental health-promoting effect of mindfulness-based therapy. The painting shows a peaceful child stretching his arms while living at home and looking out of the window. (Adapted with permission from the Association of Science and Art (ASA), USERN; Painting by Pariya Kafi)

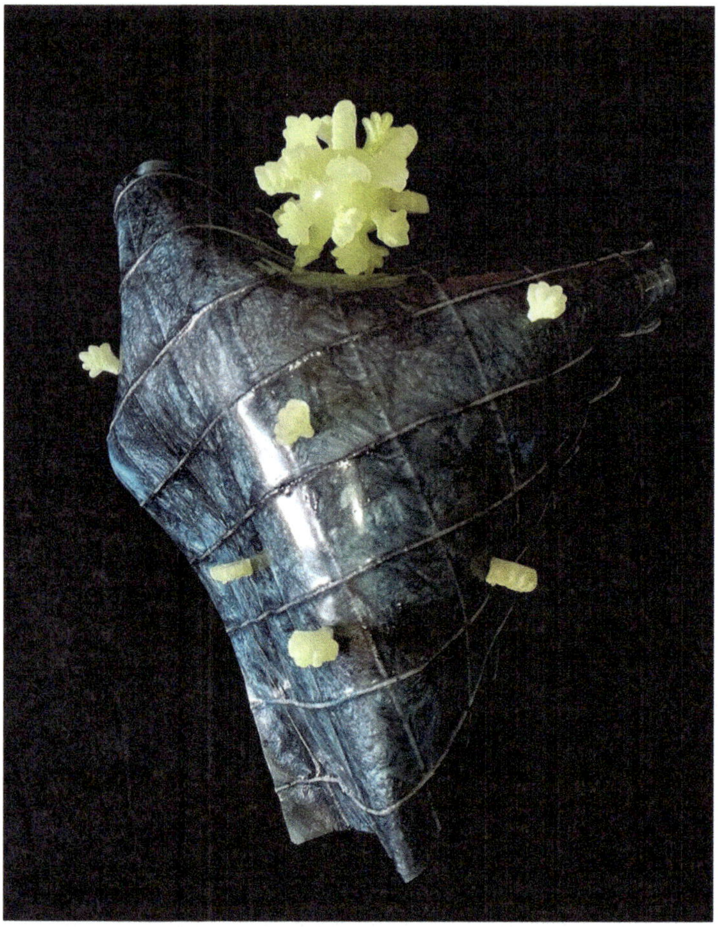

Fig. 53.22 The spacetime curvature of the SARS-CoV-2 pandemic. The theory of relativity by Albert Einstein relates the spacetime curvature to the total energy of the system or equivalently to its local mass. This paint and resin art illustrate the curvature of spacetime due to the massive distribution of SARS-CoV-2. (Adapted with permission from the Association of Science and Art (ASA), USERN; Made by Sepideh Sargoli)

Fig. 53.23 The shadow of the pandemic on the times when we were together. (Photo by Kawthar Mohammed)

Fig. 53.24 A tiny virus that happened to the world and the human's huge hope to knock out this virus. (Adapted with permission from the Association of Science and Art (ASA), USERN painting by Atlasi Ghanadan)

Fig. 53.25 A white T-shirt govash painting which illustrates a flowering tree on the earth that allowed the flowers to circulate in the air, while the threat, i.e., the SARS-CoV-2, has been neutralized. (Adapted with permission from the Association of Science and Art (ASA), USERN made by Kosar Tavasoli)

group art activities, in particular music therapy and song writings, are required as they pave the way for children to share their fear from what has happened to the world and maybe to their families. Then, it would be the music therapist's role to make children feel empowered to come up with hope-filled bright futures (Fairchild and McFerran 2019).

53.9 Conclusion

Metacognitive skills, which are a means of self-regulation of thinking and action, enable the people to master the cognitive processes that undertake simple to complex operations of learning and thereby feeling empowered to overcome negative emotions and thoughts and better manage dangerous and stressful situations. Social learning promotes the development of metacognitive skills, while the pandemic condition has limited access to social support. Therefore, the current condition calls for a necessary compensation of loneliness to reduce the psychological impact of the pandemic.

In the critical condition of the pandemic, social media has its advantages and threats. Social networks can contribute to educating and promoting the health of the community or patients and provide them with accurate and fast information. However, because the information is collected on social media by users, one of the threats in this field is the dissemination of fake information and rumors that have dangerous consequences such as threatening global health and awareness, which is still one of the main concerns in public health today.

In addition to social media, art integration into science offers a powerful tool to combat COVID-19. The art of art goes beyond the basic emotion of happiness by providing a privilege to think without boundaries. Indeed, art integration is a solution to bridge the gap between disciplines and figure out interdisciplinary and transdisciplinary platforms. In science and research, these platforms produce fruits that we could never find in the context of each discipline separately (Mohamed et al. 2020b; Momtazmanesh et al. 2020b; Moradian et al. 2020a). Also, the integration of art with education would pave the way for the cultivation of productive styles and skills, such as empathy, learning, and creativity. Altogether with the integration of science and art and the possibility of scientific problems being transformed into aesthetic experiences, we can adopt a new approach to systems thinking.

Also, arts admirably influence the immuno-emotional regulatory system so that not only would it enable humanity to tolerate quarantine but also enhance antiviral immunity. Music, for example, has shown to meet social affiliation needs, lower sensitivity to stress and arousal, involve the neurocircuitry of reward, motivation, and pleasure, and regulate immune responses. Music can be used to invite peace into our home when we live at a distance from each other. Dance has been demonstrated to regulate immune, metabolic, and hormonal responses as well and that the effects of dance have been observed in elderly and pregnant populations, which have been of special attention during the COVID-19 pandemic.

Arts are also regarded as a type of psychotherapy. In particular, the application of arts in healthcare centers is associated with reduced psychological problems, e.g., PTSD, anxiety, and depression, among healthcare workers. Creative arts and plays also make children learn to be more resilient, boost confidence in their selves, build relationships, and thereby feeling less sense of loneliness and achieving a greater sense of belongingness.

In this manner, the arts can directly enhance mental health profiles during the pandemic condition on the one hand. On the other hand, art integration into science would promote interdisciplinary and transdisciplinary thinking into the problem of COVID-19, hopefully reaching us the neutralization of the causative agent, SARS-CoV-2.

References

Ahanchian H, Moazzen N, Faroughi MSD, Khalighi N, Khoshkhui M, Aelami MH, Haghi NSM, Rezaei N (2020) COVID-19 in a child with primary specific antibody deficiency

Ahmed W, Angel N, Edson J, Bibby K, Bivins A, O'Brien JW, Choi PM, Kitajima M, Simpson SL, Li J (2020) First confirmed detection of SARS-CoV-2 in untreated wastewater in Australia: a proof of concept for the wastewater surveillance of COVID-19 in the community. Sci Total Environ:138764

Babaha F, Rezaei N (2020) Primary immunodeficiency diseases in COVID-19 pandemic: a predisposing or protective factor? Am J Med Sci. https://doi.org/10.1016/j.amjms.2020.07.027

Bahrami A, Vafapour M, Moazzami B, Rezaei N (2020) Hyperinflammatory shock related to COVID-19 in a patient presenting with multisystem inflammatory syndrome in children: first case from Iran. J Paediatr Child Health

Basiri A, Heidari A, Nadi MF, Fallahy MTP, Nezamabadi SS, Sedighi M, Saghazadeh A, Rezaei N (2020a) Microfluidic devices for detection of RNA viruses. Rev Med Virol n/a(n/a):e2154. https://doi.org/10.1002/rmv.2154

Basiri A, Pazhouhnia Z, Beheshtizadeh N, Hoseinpour M, Saghazadeh A, Rezaei N (2020b) Regenerative medicine in COVID-19 treatment: real opportunities and range of promises. Stem Cell Rev Rep:1–13

Biddle SJH, Asare M (2011) Physical activity and mental health in children and adolescents: a review of reviews. Br J Sports Med 45(11):886–895

Bircher J (2005) Towards a dynamic definition of health and disease. Med Health Care Philos 8(3):335–341

Boreham C, Riddoch C (2001) The physical activity, fitness and health of children. J Sports Sci 19(12):915–929

Bradshaw RD (2016) Art integration fosters empathy in the middle school classroom. Clear House J Educ Strat Issu Ideas 89(4–5):109–117

Broadbent J, Poon WL (2015) Self-regulated learning strategies & academic achievement in online higher education learning environments: a systematic review. Internet High Educ 27:1–13

Brown CD (2001) Therapeutic play and creative arts: helping children cope with illness, death, and grief

Cao W, Fang Z, Hou G, Han M, Xu X, Dong J, Zheng J (2020) The psychological impact of the COVID-19 epidemic on college students in China. Psych Res:112934

Chan A, Nickson C, Rudolph J, Joynt ALG (2020) Social media for rapid knowledge dissemination: early experience from the COVID-19 pandemic

Chanda ML, Levitin DJ (2013) The neurochemistry of music. Trends Cogn Sci 17(4):179–193. https://doi.org/10.1016/j.tics.2013.02.007

Darbeheshti F, Rezaei N (2020) Genetic predisposition models to COVID-19 infection. Med Hypotheses

Depoux A, Martin S, Karafillakis E, Preet R, Wilder-Smith A, Larson H (2020) The pandemic of social media panic travels faster than the COVID-19 outbreak. J Travel Med

Dong M, Zheng J (2020) Letter to the editor: headline stress disorder caused by Netnews during the outbreak of COVID-19. Heal Expect Int J Pub Particip Heal Care Heal Pol 23(2):259

Eckler P, Worsowicz G, Rayburn J (2010) Social media and healthcare: an overview. PM&R 2(11):1046–1050

Fairchild R, McFerran KS (2019) "Music is everything": using collaborative group songwriting as an arts-based method with children experiencing homelessness and family violence. Nord J Music Ther 28(2):88–107

Fathi N, Rezaei N (2020) Lymphopenia in COVID-19: therapeutic opportunities. Cell Biol Int

Feng S, Shen C, Xia N, Song W, Fan M, Cowling BJ (2020) Rational use of face masks in the COVID-19 pandemic. Lancet Respir Med 8(5):434–436

Galantino ML, Galbavy R, Quinn L (2008) Therapeutic effects of yoga for children: a systematic review of the literature. Pediatr Phys Ther 20(1):66–80

Greco V, Roger D (2003) Uncertainty, stress, and health. Personal Individ Differ 34(6):1057–1068

Gregersen S, Samocha-Bonet D, Heilbronn LK, Campbell LV (2012) Inflammatory and oxidative stress responses to high-carbohydrate and high-fat meals in healthy humans. J Nutr Metab 2012

Hafizi S, Koenig HG, Arbabi M, Pakrah M, Saghazadeh A (2014) Attitudes of Muslim physicians and nurses toward religious issues. J Relig Health 53(5):1374–1381. https://doi.org/10.1007/s10943-013-9730-1

Hagen I, Nayar US (2014) Yoga for children and young people's mental health and well-being: research review and reflections on the mental health potentials of yoga. Front Psych 5:35

Hagg E, Dahinten VS, Currie LM (2018) The emerging use of social media for health-related purposes in low and middle-income countries: a scoping review. Int J Med Inform 115:92–105

Han X, Wang J, Zhang M, Wang X (2020) Using social media to mine and analyze public opinion related to COVID-19 in China. Int J Environ Res Public Health 17(8):2788

Hanaei S, Rezaei N (2020) COVID-19: developing from an outbreak to a pandemic. Arch Med Res

Hessami A, Shamshirian A, Heydari K, Pourali F, Alizadeh-Navaei R, Moosazadeh M, Abrotan S, Shojaei L, Sedighi S, Shamshirian D (2020) Cardiovascular diseases burden in COVID-19: systematic review and meta-analysis. medRxiv

Househ M, Borycki E, Kushniruk A (2014) Empowering patients through social media: the benefits and challenges. Health Informatics J 20(1):50–58

Jabbari P, Rezaei N (2020) With risk of reinfection, is COVID-19 here to stay? Disaster Med Public Health Prep:1. https://doi.org/10.1017/dmp.2020.274

Jabbari P, Jabbari F, Ebrahimi S, Rezaei N (2020) COVID-19: a chimera of two pandemics. Disaster Med Public Health Prep:1–3

Jahanshahlu L, Rezaei N (2020a) Central nervous system involvement in COVID-19. Arch Med Res

Jahanshahlu L, Rezaei N (2020b) Monoclonal antibody as a potential anti-COVID-19. Biomed Pharmacother:110337

Johnson SU, Ebrahimi OV, Hoffart A (2020) Level and predictors of PTSD symptoms among health workers and public service providers during the COVID-19 outbreak

Kaley-Isley LC, Peterson J, Fischer C, Peterson E (2010) Yoga as a complementary therapy for children and adolescents: a guide for clinicians. Psychiatry (Edgmont) 7(8):20–32

Karpavičiūtė S, Macijauskienė J (2016) The impact of arts activity on nursing staff well-being: an intervention in the workplace. Int J Environ Res Public Health 13(4):435

Kim S-H, Kim M, Ahn Y-B, Lim H-K, Kang S-G, Cho J-H, Park S-J, Song S-W (2011) Effect of dance exercise on cognitive function in elderly patients with metabolic syndrome: a pilot study. J Sports Sci Med 10(4):671–678

Lankston L, Cusack P, Fremantle C, Isles C (2010) Visual art in hospitals: case studies and review of the evidence. J R Soc Med 103(12):490–499

Li S, Wang Y, Xue J, Zhao N, Zhu T (2020a) The impact of COVID-19 epidemic declaration on psychological consequences: a study on active Weibo users. Int J Environ Res Public Health 17(6):2032

Li Y, Yao L, Li J, Chen L, Song Y, Cai Z, Yang C (2020b) Stability issues of RT-PCR testing of SARS-CoV-2 for hospitalized patients clinically diagnosed with COVID-19. J Med Virol

Lotfi M, Rezaei N (2020) SARS-CoV-2: a comprehensive review from pathogenicity of the virus to clinical consequences. J Med Virol

Lotfi M, Hamblin MR, Rezaei N (2020a) COVID-19: transmission, prevention, and potential therapeutic opportunities. Clin Chim Acta

Lotfi M, Hamblin MR, Rezaei N (2020b) COVID-19: transmission, prevention, and potential therapeutic opportunities. Clin Chim Acta Int J Clin Chem 508:254–266. https://doi.org/10.1016/j.cca.2020.05.044

Mansourabadi AH, Sadeghalvad M, Mohammadi-Motlagh HR, Rezaei N (2020) The immune system as a target for therapy of SARS-CoV-2: a systematic review of the current immunotherapies for COVID-19. Life Sci 258:118185. https://doi.org/10.1016/j.lfs.2020.118185

Marshall J (2005) Connecting art, learning, and creativity: a case for curriculum integration. Stud Art Educ 46(3):227–241

Marshall J (2014) Transdisciplinarity and art integration: toward a new understanding of art-based learning across the curriculum. Stud Art Educ 55(2):104–127

Mattingly TJ II (2015) Innovative patient care practices using social media. J Am Pharm Assoc 55(3):288–293

McKendry RA, Rees G, Cox IJ, Johnson A, Edelstein M, Eland A, Stevens MM, Heymann D (2020) Share mobile and social-media data to curb COVID-19. Nature 580(7801):29

McMurray RG, Hackney AC, Guion WK, Katz VL (1996) Metabolic and hormonal responses to low-impact aerobic dance during pregnancy. Med Sci Sports Exerc 28(1):41–46. https://doi.org/10.1097/00005768-199601000-00012

Moazzami B, Razavi-Khorasani N, Moghadam AD, Farokhi E, Rezaei N (2020) COVID-19 and telemedicine: immediate action required for maintaining healthcare providers well-being. J Clin Virol:104345

Mohamed K, Rodríguez-Román E, Rahmani F, Zhang H, Ivanovska M, Makka SA, Joya M, Makuku R, Islam MS, Radwan N (2020a) Borderless collaboration is needed for COVID-19 – a disease that knows no borders. Infect Cont Hosp Epidemiol:1–2

Mohamed K, Rodríguez-Román E, Rahmani F, Zhang H, Ivanovska M, Makka SA, Joya M, Makuku R, Islam MS, Radwan N, Rahmah L, Goda R, Abarikwu SO, Shaw M, Zoghi S, Irtsyan S, Ling I, Cseprekal O, Faten AB, Hazar Sayar E, Soloukey C, Grancini G, Rezaei N (2020b) Borderless collaboration is needed for COVID-19-A disease that knows no borders. Infect Cont Hosp Epidem:1–2. https://doi.org/10.1017/ice.2020.162

Mohamed K, Yazdanpanah N, Saghazadeh A, Rezaei N (2020c) Computational drug discovery and repurposing for the treatment of Covid-19: a systematic review. Available at SSRN 3583748

Mojtabavi H, Saghazadeh A, Valenti VE, Rezaei N (2020) Can music influence cardiac autonomic system? A systematic review and narrative synthesis to evaluate its impact on heart rate variability. Complement Ther Clin Pract 39:101162. https://doi.org/10.1016/j.ctcp.2020.101162

Momtazmanesh S, Ochs HD, Uddin LQ, Perc M, Routes JM, Vieira DN, Al-Herz W, Baris S, Prando C, Rosivall L (2020a) All together to fight COVID-19. Am J Trop Med Hyg 102(6):1181–1183

Momtazmanesh S, Ochs HD, Uddin LQ, Perc M, Routes JM, Vieira DN, Al-Herz W, Baris S, Prando C, Rosivall L, Abdul Latiff AH, Ulrichs T, Roudenok V, Aldave Becerra JC, Salunke DB, Goudouris E, Condino-Neto A, Stashchak A, Kryvenko O, Stashchak M, Bondarenko A, Rezaei N (2020b) All together to fight COVID-19. Am J Trop Med Hyg 102(6):1181–1183. https://doi.org/10.4269/ajtmh.20-0281

Monat A, Averill JR, Lazarus RS (1972) Anticipatory stress and coping reactions under various conditions of uncertainty. J Pers Soc Psychol 24(2):237

Moradian N, Ochs HD, Sedikies C, Hamblin MR, Camargo CA Jr, Martinez JA, Biamonte JD, Abdollahi M, Torres PJ, Nieto JJ, Ogino S, Seymour JF, Abraham A, Cauda V, Gupta S, Ramakrishna S, Sellke FW, Sorooshian A, Wallace Hayes A, Martinez-Urbistondo M, Gupta M, Azadbakht L, Esmaillzadeh A, Kelishadi R, Esteghamati A, Emam-Djomeh Z, Majdzadeh R, Palit P, Badali H, Rao I, Saboury AA, Jagan Mohan Rao L, Ahmadieh H, Montazeri A, Fadini GP, Pauly D, Thomas S, Moosavi-Movahed AA, Aghamohammadi A, Behmanesh M, Rahimi-Movaghar V, Ghavami

S, Mehran R, Uddin LQ, Von Herrath M, Mobasher B, Rezaei N (2020a) The urgent need for integrated science to fight COVID-19 pandemic and beyond. J Transl Med 18(1):205. https://doi.org/10.1186/s12967-020-02364-2

Moradian N, Ochs HD, Sedikies C, Hamblin MR, Camargo CA, Martinez JA, Biamonte JD, Abdollahi M, Torres PJ, Nieto JJ (2020b) The urgent need for integrated science to fight COVID-19 pandemic and beyond. J Transl Med 18(1):1–7

Muscogiuri G, Barrea L, Savastano S, Colao A (2020) Nutritional recommendations for CoVID-19 quarantine. Eur J Clin Nutr:1–2

Nasab MG, Saghazadeh A, Rezaei N (2020) SARS-CoV-2-a tough opponent for the immune system. Arch Med Res. https://doi.org/10.1016/j.arcmed.2020.05.020

Palmier-Claus JE, Dunn G, Morrison AP, Lewis SW (2011) The role of metacognitive beliefs in stress sensitisation, self-esteem variability, and the generation of paranoia. Cogn Neuropsychiatry 16(6):530–546

Pappa S, Ntella V, Giannakas T, Giannakoulis VG, Papoutsi E, Katsaounou P (2020) Prevalence of depression, anxiety, and insomnia among healthcare workers during the COVID-19 pandemic: a systematic review and meta-analysis. Brain Behav Immun

Pashaei M, Rezaei N (2020) Immunotherapy for SARS-CoV-2: potential opportunities. Expert Opin Biol Ther:1–5. https://doi.org/10.1080/14712598.2020.1807933

Peters A, McEwen BS, Friston K (2017) Uncertainty and stress: why it causes diseases and how it is mastered by the brain. Prog Neurobiol 156:164–188

Pulido CM, Ruiz-Eugenio L, Redondo-Sama G, Villarejo-Carballido B (2020) A new application of social impact in social media for overcoming fake news in health. Int J Environ Res Public Health 17(7):2430

Qiu J, Shen B, Zhao M, Wang Z, Xie B, Xu Y (2020) A nationwide survey of psychological distress among Chinese people in the COVID-19 epidemic: implications and policy recommendations. Gen Psych 33(2)

Rabiee N, Rabiee M, Bagherzadeh M, Rezaei N (2020) COVID-19 and Picotechnology: potential opportunities. Med Hypo:109917

Rahim M, Ooi FK, Wan Abdul Hamid WZ (2017) Blood immune function parameters in response to combined aerobic dance exercise and honey supplementation in adult women. J Tradit Complement Med 7(2):165–171. https://doi.org/10.1016/j.jtcme.2016.06.001

Räsänen M (1998) Building bridges: experiential art understanding: a work of art as a means of understanding and constructing self. University of Art and Design Helsinki UIAH

Reif N, Grant L (2010) Culturally responsive classrooms through art integration. J Praxis Multicult Edu 5(1):11

Rezaei N (2020a) COVID-19 affects healthy pediatricians more than pediatric patients. Infect Cont Hosp Epidemiol:1. https://doi.org/10.1017/ice.2020.139

Rezaei N (2020b) COVID-19 and medical biotechnology. Avicenna J Med Biotechnol 12(3):139

Rzymski P, Nowicki M, Mullin GE, Abraham A, Rodríguez-Román E, Petzold MB, Bendau A, Sahu KK, Ather A, Naviaux A-F (2020) Quantity does not equal quality: scientific principles cannot be sacrificed. Int Immunopharmacol 86:106711

Saghazadeh A, Rezaei N (2019) The physical burden of immunoperception. In: Biophysics and neurophysiology of the sixth sense. Springer, pp 137–154

Saghazadeh A, Rezaei N (2020a) Immune-epidemiological parameters of the novel coronavirus – a perspective. Expert Rev Clin Immunol 16(5):465–470. https://doi.org/10.1080/1744666x.2020.1750954

Saghazadeh A, Rezaei N (2020b) Towards treatment planning of COVID-19: rationale and hypothesis for the use of multiple immunosuppressive agents: anti-antibodies, immunoglobulins, and corticosteroids. Int Immunopharmacol:106560

Saghazadeh A, Hafizi S, Rezaei N (2019a) The mental burden of Immunoperception. In: Biophysics and neurophysiology of the sixth sense. Springer, pp 111–135

Saghazadeh A, Khaksar R, Rezaei N (2019b) Learning the sixth sense. In: Rezaei N, Saghazadeh A (eds) Biophysics and neurophysiology of the sixth sense. Springer, Cham, pp 319–327. https://doi.org/10.1007/978-3-030-10620-1_29

Saghazadeh A, Mahmoudi M, Rezaei N (2019c) Nutriepigenomic immunity. In: Nutrition and immunity. Springer, pp 483–501

Sahu KK, Siddiqui AD, Rezaei N, Cerny J (2020) Challenges for management of immune thrombocytopenia during COVID-19 pandemic. J Med Virol

Saleki K, Banazadeh M, Saghazadeh A, Rezaei N (2020) The involvement of the central nervous system in patients with COVID-19. Rev Neurosci 1 (ahead-of-print)

Sharifkashani S, Bafrani MA, Khaboushan AS, Pirzadeh M, Kheirandish A, Yavarpour Bali H, Hessami A, Saghazadeh A, Rezaei N (2020) Angiotensin-converting enzyme 2 (ACE2) receptor and SARS-CoV-2: potential therapeutic targeting. Eur J Pharmacol 884:173455. https://doi.org/10.1016/j.ejphar.2020.173455

Sonke J, Pesata V, Arce L, Carytsas FP, Zemina K, Jokisch C (2015) The effects of arts-in-medicine programming on the medical-surgical work environment. Arts Heal 7(1):27–41

Spada MM, Nikčević AV, Moneta GB, Wells A (2008) Metacognition, perceived stress, and negative emotion. Personal Individ Differ 44(5):1172–1181. https://doi.org/10.1016/j.paid.2007.11.010

Stefana A, Youngstrom EA, Hopwood CJ, Dakanalis A (2020) The COVID-19 pandemic brings a second wave of social isolation and disrupted services. Eur Arch Psych Clin Neurosci:1

Stewart KA, Swain KK (2016) Global health humanities: defining an emerging field. Lancet 388(10060):2586

Stuckey HL, Nobel J (2010) The connection between art, healing, and public health: a review of current literature. Am J Public Health 100(2):254–263

Stueck M, Gloeckner N (2005) Yoga for children in the mirror of the science: working spectrum and practice fields of the training of relaxation with elements of yoga for children. Early Child Dev Care 175(4):371–377

Tahamtan A, Ardebili A (2020) Real-time RT-PCR in COVID-19 detection: issues affecting the results. Taylor & Francis

Tang L, Bie B, Park S-E, Zhi D (2018) Social media and outbreaks of emerging infectious diseases: a systematic review of literature. Am J Infect Control 46(9):962–972

Tang Y-W, Schmitz JE, Persing DH, Stratton CW (2020) Laboratory diagnosis of COVID-19: current issues and challenges. J Clin Microbiol 58(6)

Ting DSW, Carin L, Dzau V, Wong TY (2020) Digital technology and COVID-19. Nat Med 26(4):459–461

Veenman MVJ, Van Hout-Wolters BHAM, Afflerbach P (2006) Metacognition and learning: conceptual and methodological considerations. Metacogn Learn 1(1):3–14

Viviani M, Pasi G (2017) Credibility in social media: opinions, news, and health information—a survey. Wiley Interdis Rev Data Min Know Dis 7(5):e1209

Walton M, Murray E, Christian MD (2020) Mental health care for medical staff and affiliated healthcare workers during the COVID-19 pandemic. Eur Heart J Acute Cardio Care: 2048872620922795

Wang C, Pan R, Wan X, Tan Y, Xu L, Ho CS, Ho RC (2020) Immediate psychological responses and associated factors during the initial stage of the 2019 coronavirus disease (COVID-19) epidemic among the general population in China. Int J Environ Res Public Health 17(5):1729

Waszak PM, Kasprzycka-Waszak W, Kubanek A (2018) The spread of medical fake news in social media–the pilot quantitative study. Heal Pol Technol 7(2):115–118

Wells A, Colbear JS (2012) Treating posttraumatic stress disorder with metacognitive therapy: a preliminary controlled trial. J Clin Psychol 68(4):373–381

Wu PE, Styra R, Gold WL (2020) Mitigating the psychological effects of COVID-19 on health care workers. CMAJ 192(17):E459–E460

Yazdanpanah F, Hamblin MR, Rezaei N (2020a) The immune system and COVID-19: friend or foe? Life Sci:117900

Yazdanpanah N, Saghazadeh A, Rezaei N (2020b) Anosmia: a missing link in the neuroimmunology of coronavirus disease 2019 (COVID-19). Rev Neurosci. https://doi.org/10.1515/revneuro-2020-0039

Yousefzadegan S, Rezaei N (2020) Case report: death due to COVID-19 in three brothers. Am J Trop Med Hyg 102(6):1203–1204

Zarobe L, Bungay H (2017) The role of arts activities in developing resilience and mental wellbeing in children and young people a rapid review of the literature. Perspect Public Health 137(6):337–347

Zhang W-r, Wang K, Yin L, Zhao W-f, Xue Q, Peng M, Min B-q, Tian Q, Leng H-x, Du J-l (2020) Mental health and psychosocial problems of medical health workers during the COVID-19 epidemic in China. Psychother Psychosom 89(4):242–250

Printed by Printforce, the Netherlands